AMPLIFIER HANDBOOK

OTHER McGRAW-HILL HANDBOOKS OF INTEREST

AMERICAN INSTITUTE OF PHYSICS · *American Institute of Physics Handbook*
BEEMAN · *Industrial Power Systems Handbook*
BLATZ · *Radiation Hygiene Handbook*
BRADY · *Materials Handbook*
BURINGTON · *Handbook of Mathematical Tables and Formulas*
BURINGTON AND MAY · *Handbook of Probability and Statistics with Tables*
COCKRELL · *Industrial Electronics Handbook*
CONDON AND ODISHAW · *Handbook of Physics*
CROFT AND CARR · *American Electricians' Handbook*
ETHERINGTON · *Nuclear Engineering Handbook*
HENNEY · *Radio Engineering Handbook*
HENNEY AND WALSH · *Electronic Components Handbook*
HUNTER · *Handbook of Semiconductor Electronics*
HUSKEY AND KORN · *Computer Handbook*
IRESON · *Reliability Handbook*
JASIK · *Antenna Engineering Handbook*
JURAN · *Quality Control Handbook*
KNOWLTON · *Standard Handbook for Electrical Engineers*
KOELLE · *Handbook of Astronautical Engineering*
KORN AND KORN · *Mathematical Handbook for Scientists and Engineers*
KURTZ · *The Lineman's Handbook*
LANDEE, DAVIS, AND ALBRECHT · *Electronic Designers' Handbook*
MACHOL · *System Engineering Handbook*
MARKS AND BAUMEISTER · *Mechanical Engineers' Handbook*
MARKUS · *Electronics and Nucleonics Dictionary*
MARKUS · *Handbook of Electronic Control Circuits*
MARKUS AND ZELUFF · *Handbook of Industrial Electronics Circuits*
STETKA AND BRANDON · *NFPA Handbook of the National Electrical Code*
TERMAN · *Radio Engineers' Handbook*
TRUXAL · *Control Engineers' Handbook*
WALKER · *NAB Engineering Handbook*

generally differ from this ideal. They may, for example, have finite nonzero (and nonnegligible) input and output impedances. To take account of these we need an impedance in the input circuit and another impedance in the output circuit. They may have nonzero reverse transmission, that is, a voltage applied at the output port yields a nonzero voltage response at the input; this effect may be taken into account by connecting an impedance between the input and output ports. The complete equivalent circuit is discussed in Chap. 2, Sec. 4. Thus instead of the one parameter characterizing the ideal amplifier we have four parameters, which, as should be clear from the two equations describing any two-port, are sufficient to characterize the most general active two-port device.

Since controlled sources most often introduce a unilateral or unidirectional effect into a network, in general networks containing them will not satisfy the reciprocity

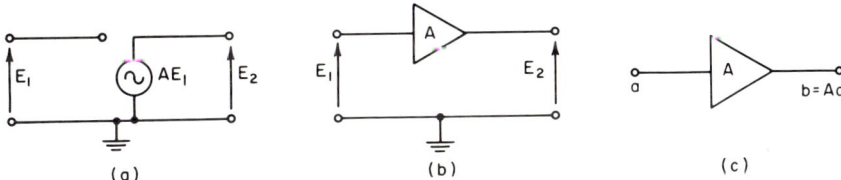

Fig. 8. Three representations of a grounded ideal amplifier.

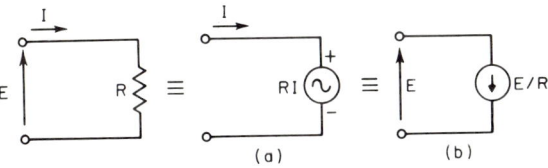

Fig. 9. Representation of a positive resistance by a controlled source. (*a*) Current-controlled voltage source. (*b*) Voltage-controlled current source.

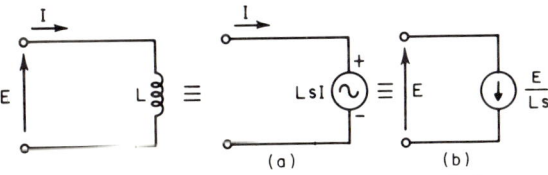

Fig. 10. Representation of a positive inductance by a controlled source. (*a*) Current-controlled voltage source. (*b*) Voltage-controlled current source.

theorem. However, controlled sources can also be used to represent one-port elements; for example, the representations of a resistance by a current-controlled source and by a voltage-controlled source are shown in Fig. 9; the corresponding representations for an inductance are shown in Fig. 10. It is thus possible to eliminate the passive elements entirely and use networks that involve only controlled and independent sources. However, this is merely of academic interest since analysis using the passive elements is convenient and simple. Instead of eliminating passive elements, we generally proceed in the other direction and attempt to replace controlled sources by passive elements wherever possible.

Though controlled sources are required for the realization of an amplifier or an oscillator, that is, a network to be used as an amplifier or oscillator must be *active*, we should not conclude that a network containing one or more controlled sources is necessarily active; it may be passive and hence incapable of giving a power gain. If it is passive, we then know that all the controlled sources may be eliminated; that is, an equivalent network can be realized using only the five passive elements.

NETWORK THEORY: DEFINITIONS AND BASIC CONCEPTS 1-13

other properties are the same as those of the independent sources; e.g., the voltage source represents a zero impedance and constrains the potential difference of the nodes to which it is connected to equal the voltage of the source. The independent sources, however, are one-port elements, whereas the controlled sources are in general two-port elements. The analysis of a network containing controlled sources is carried out in essentially the same way as that of other networks.

Since a dependent source may be controlled by a voltage or a current, there are four types of controlled sources; their representations are shown in Fig. 6. In Fig. 7 is shown a current-controlled voltage source in an actual circuit, where E_1 is an input source voltage and E_2 is to be determined.

The controlled sources may also be used for the representation of ideal amplifiers. Thus the voltage-controlled voltage source in Fig. 6 is a representation of an ideal voltage amplifier, the current-controlled current source an ideal current amplifier, the current-controlled voltage source an ideal transimpedance amplifier, and the voltage-controlled current source an ideal transadmittance amplifier. When the amplifier is a grounded quadripole, the representation is changed accordingly; for example, a grounded voltage amplifier is shown in Fig. 8a. An alternative representation for an ideal voltage amplifier with a common ground is shown in Fig. 8b, whereas in Fig. 8c the ground connection is not shown since all voltages are assumed to be measured with respect to ground. In addition the input and output may be either a current or a voltage independently; that is, this representation may be used for any of the controlled sources.

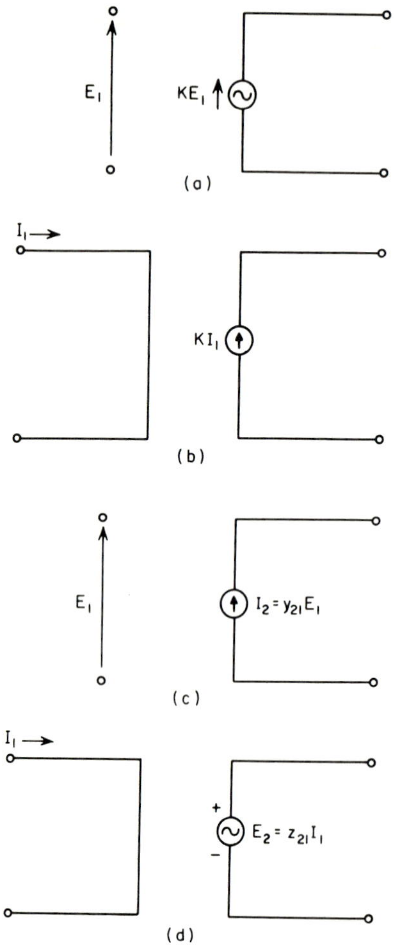

FIG. 6. Representations of the four types of controlled sources. (a) Voltage-controlled voltage source. (b) Current-controlled current source. (c) Voltage-controlled current source. (d) Current-controlled voltage source.

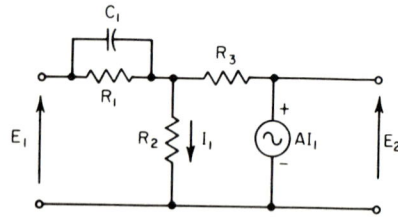

FIG. 7. Current-controlled voltage source embedded in an actual circuit.

The properties of the ideal amplifiers should be clear from the circuit models. The ideal voltage amplifier, for example, has an infinite input impedance, zero output impedance, a forward voltage ratio equal to A, and a zero reverse voltage ratio. Thus it takes no input power but can furnish an unlimited output power and therefore has an infinite power gain. In addition, the amplifier is unilateral, that is, its input port does not respond to signals applied at the output port. As we shall see when we consider equivalent circuits of active devices in Chap. 2, physical amplifiers will

1-12 AMPLIFIER FUNDAMENTALS

Now we turn to the second method for characterizing active networks, which shall be used in this book, namely, the use of sources both internal and external to the network. Two ideal sources or generators are defined, a voltage source and a current source. Their representations and properties are shown in Table 2. In order to make

Table 2. Independent Ideal Sources

Element	Representation	Internal Impedance
Voltage source	E	Zero
Current source	I	Infinite

the distinction between the representations of a voltage and a current source evident by inspection, we place an *arrow inside* the circle used to represent the current source, and use a circle with a sine wave inside it for the voltage source. The arrow indicates the assumed positive direction for the current of the current source, and for the voltage source we use plus and minus signs to indicate polarity, or the arrow is used outside the circle; the arrow direction indicates the direction of voltage *rise*, that is, it points from minus to plus.*

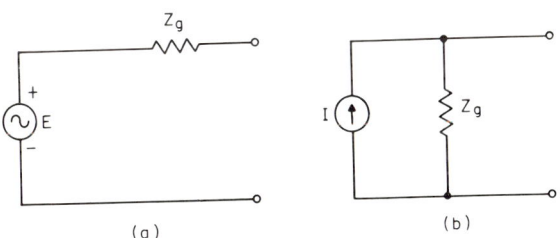

FIG. 5. Representations of independent sources used in practice.

These sources are said to be *independent* sources; this term is used because the values of the voltage of a voltage source and the current of a current source do not change irrespective of the voltages and currents in the rest of the network; thus the strength of the source is independent of the network in which the source is connected. The sources are also *ideal* in that no internal impedance elements are associated with them. Their ideal character is made evident if we connect an increasingly large resistance to the current source or a decreasingly small resistance to the voltage source; in each case an increasingly large amount of power is delivered by the source. Since practical sources cannot deliver infinite power, they must have internal impedances associated with them; representations of practical sources are shown in Fig. 5. Frequently the source impedance (also called *generator impedance*) is a pure resistance.

5.1. Controlled Sources.[4] For the representation of amplifiers and of general linear systems, in addition to the above elements another type of source is required; this is the *controlled source*. Such a source may be either a voltage source or a current source; its strength is dependent on another voltage or current which in turn is determined by the network configuration. Though the strength of these controlled sources is not independent (and thus the sources are often called *dependent*), their

---

* At the ports of a network we frequently use an arrow alone to indicate the positive direction of an applied current source or voltage source. The arrow is also used to indicate the voltage across or current through a load, but no confusion results since, when a distinction is necessary, the accompanying text makes clear whether a source or a load is connected to a port.

in Fig. 4b, where the load, that is, the impedance $Z(x_1)$ seen at point x_1, has been shown explicitly. The equations may now be written in matrix form as

$$\begin{bmatrix} E_1 \\ I_1 \end{bmatrix} = \begin{bmatrix} \cosh \gamma l & Z_0 \sinh \gamma l \\ \dfrac{\sinh \gamma l}{Z_0} & \cosh \gamma l \end{bmatrix} \begin{bmatrix} E_2 \\ I_2 \end{bmatrix} \quad (6)$$

Dividing the first of Eqs. (6) by the second and making use of the fact that $Z_L = E_2/I_2$ then gives the input impedance of the line

$$Z_{11} = \frac{E_1}{I_1} = Z_0 \frac{Z_L + Z_0 \tanh \gamma l}{Z_0 + Z_L \tanh \gamma l} \quad (7)$$

Let us now consider passive lumped-constant networks containing a finite number of elements; it is possible, even without the use of mathematics, to give some powerful characterizations of such networks. We must first distinguish between reciprocal and nonreciprocal networks, where we use the term *reciprocal* to mean that the network obeys the reciprocity theorem, which is defined in Chap. 2. For reciprocal networks the four types of elements—resistance, capacitance, inductance, and the ideal transformer—are said to form a *complete* set of elements. By this is meant that any realizable reciprocal network can be synthesized by interconnecting elements of only these types. For nonreciprocal networks a gyrator must be added to the set of permissible elements. Moreover, since two cascaded gyrators can realize an ideal transformer and since a gyrator plus an inductance (capacitance) can realize a capacitance (inductance), only three types of passive elements are required to realize any passive network. Thus resistances, gyrators, and either capacitances or inductances form a complete set.

For distributed-constant passive networks corresponding general results are not known; for example, we do not know how to characterize networks containing only transmission lines and capacitances, even if we restrict the network to contain a *single* capacitance. Results are known, however, for networks containing transmission lines and resistances.

To realize active networks energy sources are required. Before we consider such sources, it is useful to discuss briefly an alternative characterization of active networks in which they are realized with the use of only passive elements.[1] Active systems, as the reader is well aware, contain elements like transistors which do not contain an energy source but behave as if they generate energy when they are connected to batteries and small signals are applied. Physically, such systems, apart from the batteries, are passive *nonlinear* systems that may be said to be *locally* active; that is, when excited by d-c sources, they behave like active systems for small variations about an equilibrium point. Similar remarks hold for time-varying systems: the source of the variation may be considered to be not in the system proper but in some *outer* electrical or mechanical cause.

Thus all sources are external to the black box or network to be realized and the synthesis problem for the realization of active networks can be restricted to passive constant nonlinear systems. As was done for the passive linear case, one can postulate the simplest one-port or two-port systems and consider these as new basic network elements. The active-network realization problem is then solved if the new set of network elements can be shown to be a complete set, that is, if one can prove that all linear active systems can be synthesized by networks composed of these constant nonlinear network elements in combination with the five linear ones.

Some important work along this direction has been carried out. Duinker,[2,3] for example, has proposed two passive three-port elements called the traditor and conjunctor. These elements, like the linear gyrator and ideal transformer are what Duinker calls *nonenergic*, by which he means that at any time the total instantaneous power delivered to them is zero; thus they neither store nor dissipate energy. Some interesting results can be obtained using these elements, but the completeness of the new set of elements has not been proved, nor has the question of their realizability been answered. However, the approach looks promising and further work might yield significant results.

Table 1. Passive Lumped-constant Elements

Name of parameter or element	Network symbol	Units	Volt-ampere relations (in time domain)	Volt-ampere relations (in frequency domain) (initial energy is considered as zero)	Energy relations
Resistance	(R)	Ohms	$e = Ri$; $i = e/R$	$E = RI$; $I = E/R$	Dissipated power, $p = ei = Ri^2 = e^2/R$
Inductance	(L)	Henrys	$e = L\dfrac{di}{dt}$; $i = \dfrac{1}{L}\int_0^t e\, dt + i(0)$	$E = LsI$; $I = E/Ls$	Stored magnetic energy, $w = \tfrac{1}{2}Li^2$
Capacitance	(C)	Farads	$e = \dfrac{1}{C}\int_0^t i\, dt + e(0)$; $i = C\dfrac{de}{dt}$	$E = I/Cs$; $I = CsE$	Stored electric energy, $w = \tfrac{1}{2}Ce^2$
Mutual inductance	(M)	Henrys	$e_1 = L_1\dfrac{di_1}{dt} + M\dfrac{di_2}{dt}$ $e_2 = M\dfrac{di_1}{dt} + L_2\dfrac{di_2}{dt}$	$E_1 = L_1 s I_1 + M s I_2$ $E_2 = M s I_1 + L_2 s I_2$	Stored magnetic energy, $w = \tfrac{1}{2}L_1 i_1^2 + M i_1 i_2 + \tfrac{1}{2}L_2 i_2^2$
Ideal transformer	(1:n IDEAL)	....	$e_2 = n e_1$ $i_2 = -i_1/n$	$E_2 = n E_1$ $I_2 = -I_1/n$	Neither stores nor dissipates energy; total power supplied to the ports at any instant is zero: $e_1 i_1 + e_2 i_2 = 0$
Gyrator	(g)	g, the gyration conductance, is in mhos	$i_1 = g e_2$ $i_2 = -g e_1$	$I_1 = g E_2$ $I_2 = -g E_1$	Neither stores nor dissipates energy; total power supplied to the ports at any instant is zero: $e_1 i_1 + e_2 i_2 = 0$

NETWORK THEORY: DEFINITIONS AND BASIC CONCEPTS 1-9

the zeros of the denominator) *are* the natural frequencies. We postpone a detailed discussion of this point until Sec. 6.

5. ACTIVE AND PASSIVE NETWORKS

Networks may be classified as either *active* or *passive*. This is the classification we shall use. However, another means of differentiation is possible, which we shall consider briefly; namely, the class of linear passive constant networks and the class of nonlinear passive constant networks; the latter class can be shown to include linear active networks. Each class may be further subdivided into *lumped-parameter* and *distributed-parameter* networks. In the former each parameter or passive element is considered to be concentrated or *lumped* in one part of the network; space considerations and the distribution of the electric and magnetic fields are not essential in the analysis. Thus it is assumed in treating an element as lumped that at any given instant the current has the same value at every point in the element; this assumption is valid if the dimensions of the circuit element are small compared with the wavelength. In another type of system, for example, a system containing a transmission line, the distribution of electromagnetic effects along the length of the line is important. However, if voltages and currents can be defined and are of primary interest rather than the electromagnetic fields, a modified network representation is possible; this is a one-space dimension or distributed-parameter representation.

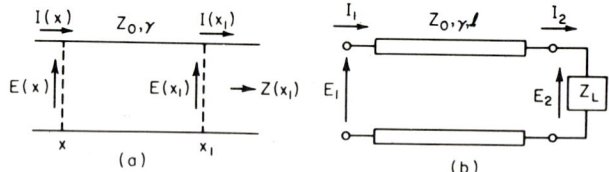

FIG. 4. Alternative representations of a length of transmission line.

In the sinusoidal steady state the instantaneous energy and the average power furnished to a passive one-port cannot be negative and hence each of the passive elements must satisfy this condition; a physical one-port element violating the condition is active. The same condition holds for an n-port. The set of passive lumped-constant network elements comprise the one-port elements—resistance, inductance, and capacitance—and the two-port elements—mutual inductance or the real transformer, the ideal transformer, and the gyrator. The schematic representations and the defining relations for the passive lumped elements are given in Table 1. A distributed-constant passive element is the lossless homogeneous transmission line; powerful synthesis procedures have been developed for realizing networks containing transmission lines (or their equivalents) and lumped-constant elements. The transmission line may be considered as a one-port element or as a two-port element. A pictorial representation is shown in Fig. 4a. Here the voltage and current anywhere along the line are represented in terms of the voltage and current at a single point x_1 by the equations

$$E(x) = E(x_1) \cosh \gamma(x - x_1) - Z_0 I(x_1) \sinh \gamma(x - x_1)$$
$$I(x) = I(x_1) \cosh \gamma(x - x_1) - [E(x_1)/Z_0] \sinh \gamma(x - x_1) \qquad (4)$$

where Z_0 is the characteristic impedance and γ the propagation function of the line. Often it is convenient to let $x = 0$ be the sending end, $x = x_1$ be the receiving end, and $l = x_1 - x$ represent the length of the line. The equations are then written as

$$E_S = E_R \cosh \gamma l + Z_0 I_R \sinh \gamma l$$
$$I_S = I_R \cosh \gamma l + (E_R/Z_0) \sinh \gamma l \qquad (5)$$

where the subscripts S and R denote the sending and receiving ends of the line, respectively. For consideration as a two-port element we may represent the line as shown procedures, however, to use the function representing the ratio of the input to the output, that is, the reciprocal of a system function. We define the reciprocal of a system function by the shortened term *reciprocal system function*. Since, as indicated above, the reciprocal of a driving-point function is also a driving-point function, the new term is needed *only for transfer functions*, and consequently the word *transfer* may be omitted in the designation of reciprocal system functions. For example, E_1/I_2, the reciprocal of a transfer admittance, has the dimension of impedance and is called a *reciprocal admittance*.

The definitions of system function and reciprocal system function are easily extended to an n-port network, such as that shown in Fig. 3. Here the Y_{rk} for $r \neq k$ are defined as the transfer admittances for an input E_k and an output I_r, while for $r = k$ the network's driving-point admittances are defined. Similar extensions hold for the impedances and the dimensionless system functions.

It is well known that the most general form of a system function for a lumped-constant network is given by a *real rational function* of s, that is, by the quotient of two *polynomials with real coefficients*. The transfer impedance Z_{21}, for example, may always be represented by

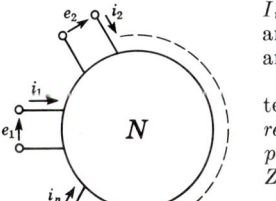

FIG. 3. General n-port network.

$$Z_{21}(s) = \frac{p(s)}{q(s)} = \frac{a_m s^m + a_{m-1} s^{m-1} + \cdots + a_1 s + a_0}{b_n s^n + b_{n-1} s^{n-1} + \cdots + b_1 s + b_0}$$

$$= H \frac{(s - s_1)(s - s_2) \cdots (s - s_m)}{(s - s_1')(s - s_2') \cdots (s - s_n')} \qquad (3)$$

where $H = a_m/b_n$ is a constant multiplier, and p and q are real polynomials; that is, the a's and b's are all real, and m and n are finite. Unless the contrary is stated, it is assumed that all common factors of the numerator and denominator have been canceled.

In terms of the derivations given in Sec. 2, system functions may be given either of two interpretations. The transfer impedance $E_2(s)/I_1(s)$, for example, represents the Laplace transform of the impulse response of the network, that is, the transform of the response $e_2(t)$ when the input $i_1(t)$ is a unit impulse; or in the second interpretation, this ratio for $s = j\omega$ is the function that gives the value of the complex amplitude of the steady-state voltage output for a unit-amplitude sinusoidal current input. Thus when we realize a system function, we accomplish the synthesis of a network that has the sinusoidal steady-state response specified by the function; or alternatively and equivalently, the realized network for a unit-impulse excitation has a time response given by the inverse Laplace transform of the system function.

The rational function, which represents the most general system function for a lumped-parameter network, is easily characterized: the *poles*, *zeros*, and *constant multiplier* of the function specify it completely. The zeros of a rational function are the values of s for which the function is zero, and the poles are the values for which the function is infinite. The *finite zeros* are therefore given by the zeros of the numerator, and the *finite* poles by the zeros of the denominator; thus Z_{21} of Eq. (3) has zeros at the points where $p(s) = 0$ and it has poles at the points where $q(s) = 0$. There is also a zero at infinity if $n > m$, the order of this zero being equal to $n - m$, and there is a pole at infinity if $m > n$, its order being equal to $m - n$. Because they completely define the function (within a constant multiplier), the zeros and poles collectively are called the *critical frequencies*, or *critical values*, of the function. The constant multiplier H is spoken of as the *level* of the function; for convenience in intermediate calculations in realizing a system function, it is often set equal to unity.

It is convenient to define the term *order*, which indicates the complexity of a rational function: the order of a rational function is equal to the degree of the numerator or denominator, whichever is greater.

At this point the question may arise about the relationship of the *natural frequencies* (or *natural modes*) of oscillation of a network and the critical frequencies of its system function. The answer is simply given: the finite poles of the system function (that is, and (loosely) an *unbalanced* network. A common grounded quadripole is the *ladder*. In other cases practical considerations, like a push-pull source, may dictate that the network to be realized be a balanced one, i.e., symmetrical with respect to a center line considered as a ground potential. The *lattice* is such a balanced network.

3.5. Functions Specifying the Network. What is desired may be a grounded quadripole for which only the transfer characteristic is specified. Or a more difficult problem may be posed by an additional requirement on the input characteristic of the network; for example, a constant-resistance input may be specified for a network that realizes a given transfer relation. (The term *constant-resistance* input is probably self-defining, but its precise definition will be given later in Sec. 7.) In some cases a network must be realized to possess the magnitude of a specified transfer characteristic; in others both the magnitude and phase may be specified.

4. NETWORK FUNCTIONS

A network to be realized may have its desired behavior specified by one or more network functions, or even by a whole matrix of network functions. We shall employ two general types of network functions, namely, *system functions* and *reciprocal system functions*. Though the first is more important for our purposes, the second is also useful.

The *system function* of a network is defined as a function of s representing the ratio of the Laplace transform of a *response, or output, variable* to the Laplace transform of an *excitation, or input, variable*. In the formation of this ratio it is assumed that the initial energy storage of the network is zero.

There are two types of system functions of interest, the *driving-point function* and the *transfer function*. Each of these may have the dimension of impedance or admittance, while the transfer function may, in addition, be dimensionless. For a driving-point function (or as it is often called, a *two-terminal function*), the input and output are measured at the *same* port; e.g., the driving-point admittance (also called the *input* admittance) at terminals 1-1' of Fig. 2 is given by $Y_{11}(s) = I_1(s)/E_1(s)$. Thus any input or output impedance or admittance is a driving-point function. For a transfer function, on the other hand, the input and output are measured at two *different* ports.

As Fig. 2 has been drawn, there is nothing to indicate which port represents the input and which the output. If we now assume that the input signal is applied to 1-1' and the output taken from 2-2', then four transfer functions may be defined:*
E_2/E_1 represents the *transfer voltage ratio*, I_2/I_1 the *transfer current ratio*, E_2/I_1 the *transfer impedance*, and I_2/E_1 the *transfer admittance*. Since the voltage and current ratios are useful only as transfer functions, being identically unity for driving-point functions, the shortened forms *voltage ratio* and *current ratio* are used for these dimensionless transfer functions. In discussions that are intended to apply to both of them, the term *transfer ratio* is often used.

Each of the four transfer functions, it is noted, represents the ratio of an output transform to an input transform; the expression E_1/E_2, since it is the ratio of an *input* to an *output*, is not a system function but represents the reciprocal of a system function. It is emphasized, however, that the reciprocal of a driving-point function is also a system function; that is, if by using a current as the input and a voltage as the output at one port, a driving-point impedance results, then the reciprocal of this function is identical with the driving-point admittance obtained by using a voltage input and calculating the current response at the same port. In short, both a driving-point function and its reciprocal are system functions, whereas the reciprocal of a transfer function is in general not a system function.

For the most part the properties of network functions will be derived in terms of system functions. It is convenient in some analysis problems and some synthesis

---
* The argument is often omitted in writing network functions; thus $E_2(s)/I_1(s)$ becomes E_2/I_1. This is done for brevity when there is no chance of confusion. Similarly this is done for functions of time t, as in Fig. 2, where we note that time functions are generally designated by lowercase letters.

The terminals may also be considered separately, and the network description given in terms of the total number of terminals rather than terminal pairs. It is important to recognize this distinction between an *n-terminal-pair* network and an *n-terminal* network. As a measure of the complexity of the two network types, it is to be noted that complete characterization of a two-port requires only four independent network functions, whereas the true four-terminal network, in which no assumption is made about pairings of terminals as ports, requires nine independent functions for its characterization.

Use of the terminal and terminal-pair designation often proves awkward, so that, as has been mentioned, *port* is being increasingly used in the literature for terminal pair. Also, to achieve brevity or avoid awkwardness, the word *pole* is often used as a synonym for *terminal*.

Thus a *dipole* is a *two-terminal* or a *one-port network*, and a *tetrapole* is a *four-terminal network*. Two other terms have been used in the literature to signify a *two-port;* they are *quadripole* and *four-pole*.

3.2. Types of Elements. The simplest network contains one type of element, e.g., a resistance attenuation pad. Next in complexity is a network with only two types of elements, called a *two-element-kind* network. Such networks are of great importance in practice, included among them being the lossless filter and many impedance-transforming networks; for brevity, a network is often described by the symbols of its elements; e.g., a lossless network containing only capacitances and inductances is called an LC network. Another two-element-kind network of importance is the RC network, which is finding ever-growing application in amplifiers used in control circuits. The RLC network is a still more general type. In all networks containing inductance an additional specification may require no mutual inductance to be present; or real transformers may be allowed but no ideal transformers; or finally, in procedures of mainly theoretical interest, ideal transformers may be permitted. When it is necessary to distinguish between RLC networks that do and do not contain real transformers, we can use the letter M; thus a general network with transformers can be designated as an $RLCM$ network.

In addition to the elements in the above passive networks, there is another passive element called the *gyrator* (see Table 1). Since passive networks cannot furnish energy in the sinusoidal steady state, networks containing active elements like voltage or current sources must be present in amplifiers. All the elements mentioned thus far are of the *lumped-parameter* type; in contrast to these lumped-parameter networks there are *distributed-parameter* networks which contain one or more distributed-parameter elements like a transmission line or a waveguide.

Finally, the preceding passive elements are all constants; thus a lumped-parameter network containing constant elements is also called a *lumped-constant* network. It is possible that such elements in a network may be time-varying so that the differential equations describing the network are still linear but have time-varying coefficients. An important class of amplifiers called parametric amplifiers depend for their operation on time-varying reactances.

3.3. Types of Terminations. A particular type of output termination may be required: the device to be controlled at the output of the network may be a pure resistance, so that a resistance termination is specified, or a shunt capacitance may be required when the network output is fed to the grid of a vacuum tube with its associated input capacitance. At the input some designs may also call for a resistance termination to provide for the finite nonzero source impedance.

The specification of a termination may further restrict the necessary and sufficient conditions for realizability of a given function; that is, not only must the general realizability conditions be satisfied, but additional conditions as well. For example, a realizable transfer function must have no poles in the right half of the complex plane; however, to be realizable as a quadripole with a *resistance termination*, it must in addition have no poles on the imaginary axis.

3.4. Network Configurations. Often a grounded terminal common to the input and output of a quadripole is a requirement; in the literature such a network has been variously called a *grounded* network or quadripole, a *three-terminal* network,

NETWORK THEORY: DEFINITIONS AND BASIC CONCEPTS

voltage, the network function representing the voltage ratio is $E_2(s)/E_1(s)$, where $E_1(s)$ and $E_2(s)$ are the Laplace transforms of $e_1(t)$ and $e_2(t)$, respectively, and it is assumed that the initial conditions are zero.

A second method starts with the sinusoidal steady state. This may be understood by considering the general quadripole of Fig. 2, for which the indicated variables are sinusoidal quantities; for example, $e_1(t)$ is given by

$$e_1 = |E_1| \cos(\omega t + \varphi) \tag{1}$$

We can then make the following obvious transformations:

$$e_1 = |E_1| \cos(\omega t + \varphi) = \mathrm{Re}\,[|E_1|e^{j(\omega t+\varphi)}] = \mathrm{Re}\,(E_1 e^{j\omega t}) = \mathrm{Re}\,(E_1 e^{st}) \tag{2}$$

where Re denotes "the real part of" the expression that follows it.* The variable s in Eq. (2) has been introduced as a representation for $j\omega$; however, by analytic continuation we can now extend the allowable values of s so that it covers the complex range: thus $s = \sigma + j\omega$. With s a "complex frequency" the voltage e_1 becomes a sinusoid with a variable amplitude, $|E_1|e^{\sigma t}\cos(\omega t + \varphi)$. Thus the complex frequency, like the ω frequency, has a direct physical interpretation; the "steady-state" current in any part of a network whose input is an exponentially decreasing sinusoid may be calculated in the same way as for an input of constant amplitude, merely by use of the complex value of the frequency of the input. For Eq. (2) it is simple, of course, to revert to the original sinusoidal input by restricting s to the imaginary axis; that is, we can again let $s = j\omega$. However, by considering the network functions as functions of a complex variable, the full power and *simplicity* of complex function theory are brought to bear on network problems.

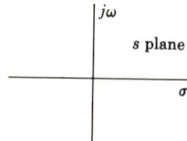

Fig. 1. The complex-frequency plane.

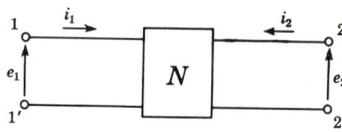

Fig. 2. A general quadripole.

The quantity E_1 in the last equation of Eqs. (2) may be written as a function of s, i.e., $E_1(s)$, and is called the complex amplitude of $e_1(t)$, so that the network functions are given in terms of specified ratios of two complex amplitudes.

3. TYPES OF NETWORKS

Design specifications may describe the characteristics of a desired network in a variety of ways. Each type of network places corresponding restrictions on the functions used to characterize it. Some of the possible network classifications follow.

3.1. Number of Accessible Terminals or Terminal Pairs. Two accessible terminals considered as a pair, at which voltage or current sources may be applied or responses measured, constitute a *terminal pair* or *port*. It is assumed that the current entering a network through one terminal of a port is always equal to the current leaving the network through the other terminal of that port; it is further assumed that two-terminal elements may be connected externally to a port but not *between* ports. A network with one terminal pair—or a *one-port network*, or more briefly, a *one-port*—may be completely described by one network function, namely, a driving-point function (about which more will be said below); a two-terminal-pair network (or a two-port) is completely described by four independent functions, two driving-point functions, and two transfer functions; and so on. Thus the number of ports of a network constitutes a measure of the complexity of the network.

* It should be noted that in our usage the letter E represents a voltage *rise*, with the arrow in the diagram showing the direction of the rise.

terminal networks coupling one or more inputs with one or more outputs. Sometimes the passive networks are an inseparable part of the amplifier so that the whole system must be designed as an entity.

We are concerned with the application of network theory to linear systems, that is, systems that may be described by linear differential equations. This does not mean that nonlinear devices do not occur in the amplifiers; in fact, as the reader is aware, the nonlinearities are basic for the operation of the amplifying devices. However, with the proper choice of an operating point and for small signals, a so-called *incremental linear* equivalent circuit may be used in the analysis and design.

The diversity of systems can be understood and designed only after a proper groundwork has been laid. It is desirable, for example, to characterize active and passive networks precisely, to know the performance capabilities of useful classes of networks, and to have techniques available for the rational and algorithmic design of these networks. To achieve these goals we first present some basic definitions, concepts, and theorems of network theory.

As is well known, network theory encompasses both *analysis* and *synthesis*. In network analysis a physical system, or more properly, its representation or network model as an interconnection of resistances, inductances, and capacitances, has one or more current or voltage sources connected to it; a mathematical function (representing, for example, a current, a voltage, or an impedance) is to be found. A solution always exists and is *unique*. In the converse problem, that of network synthesis, a function of a complex-frequency variable, or a set of these functions, is given as a specification of a desired network; a network characterized by the given function is to be found. A solution may not exist, in which case we say of the given function that it cannot be *represented by a physical network*, or alternatively, is not *physically realizable*, or more briefly still, is not *realizable*. If a solution does exist, it is *not unique*, an infinite number of solutions existing in general. Thus the designer is in the enviable position of being free to choose the most practical network from a large number of equivalent designs.

In network theory certain branches of mathematics are used with power and elegance, e.g., functions of a complex variable (briefly referred to as *function theory*), the algebra of polynomials, and matrix theory. Not only is it assumed that the reader is familiar with introductory network theory or can easily fill in any gaps in his background by use of one of the excellent texts available, but it is also desirable that the reader be acquainted with these associated fields of mathematics, for in large part network theory, and hence amplifier theory, are applied function theory and applied matrix theory. Finally, a knowledge of Laplace and Fourier transform theory, though not necessary for following most of the discussion, deepens one's understanding of it.

The literature on network theory is so vast that a fairly complete presentation is not possible or even desirable in a book of this type. However, a self-contained discussion that possesses unity and coherence is attempted; thus some fairly elementary concepts familiar to most readers are included in the definitions in order to facilitate subsequent discussions and make precise statements possible.

2. COMPLEX FREQUENCY*

We deal in network theory with functions of a complex variable $s = \sigma + j\omega$, where σ and ω are real variables. When $\sigma = 0$, the variable ranges over the imaginary axis of the s plane shown in Fig. 1. These ω values, which correspond to frequencies of the sinusoidal steady state, are, according to tradition, called *real frequencies*.

To introduce the complex variable into the network functions we may use at least two methods which lead to equivalent results. The first makes use of the Laplace transforms of the network variables; the network functions represent the specified ratios of these transforms. Thus if $e_1(t)$ is an input voltage and $e_2(t)$ the output

* This section and a number of other sections in these first chapters are adapted with the permission of the publishers from the author's book, "Network Analysis and Synthesis," McGraw-Hill Book Company, New York, 1962.

Chapter 1

NETWORK THEORY: DEFINITIONS AND BASIC CONCEPTS

LOUIS WEINBERG[*]

CONTENTS

1. Introduction.. 1-3
2. Complex Frequency... 1-4
3. Types of Networks.. 1-5
    3.1. Number of Accessible Terminals or Terminal Pairs.......... 1-5
    3.2. Types of Elements.. 1-6
    3.3. Types of Terminations.................................... 1-6
    3.4. Network Configurations................................... 1-6
    3.5. Functions Specifying the Network......................... 1-7
4. Network Functions.. 1-7
5. Active and Passive Networks.................................... 1-9
    5.1. Controlled Sources....................................... 1-12
    5.2. Negative Resistance...................................... 1-15
6. Natural Frequencies and Characteristic Polynomials.............. 1-18
    6.1. Short-circuit Stability and Open-circuit Stability........ 1-21
7. Some Definitions: Equivalent, Constant-resistance, Complementary, and Dual Networks....................................... 1-23

1. INTRODUCTION

Amplifiers are an integral part of almost all physical systems. The complete amplifier includes a source or generator that provides the input excitation and a load for receiving an enlarged and perhaps filtered or otherwise modified version of the input; most often, cascaded between the source and load, there are also active networks and passive interstage networks. The amplifiers may use any of a large number of different physical devices and appear in a wide variety of forms. In spite of their diversity, however, they share a common requirement: network theory is a useful and even indispensable tool for their analysis and design. Generally the important part of the design problem is the design of the passive networks; these networks may be two-terminal networks to serve as terminating impedances, three-terminal or four-terminal coupling networks between one input and output, or $2n$-

[*] Department of Electrical Engineering, University of Michigan, Ann Arbor, Mich.

Part I

AMPLIFIER FUNDAMENTALS

AMPLIFIER HANDBOOK

CONTENTS

9. Rare-earth Lasers	29-52
10. Semiconductor or Injection Lasers	29-58
11. The Raman Laser	29-65
12. The Organic Laser	29-67

Chapter 30. Acoustic Wave Amplifiers *Stephen W. Tehon*   **30-1**

1. Introduction	30-1
2. Principles of Operation	30-2
3. Linearized Analysis	30-3
4. Interpretation of the Linear Theory	30-5
5. Design Procedures	30-7

Chapter 31. Integrated Circuits *Richard J. Patch*   **31-1**

1. Introduction	31-2
2. Manufacturing Techniques	31-2
3. Integrated Components	31-6
4. Integrated-circuit Design Considerations	31-15
5. Multipurpose Circuits	31-20
6. Direct-coupled Amplifiers	31-32
7. Audio Amplifiers	31-39
8. Servoamplifiers	31-42
9. Precision A-C Summing Amplifiers	31-42
10. I-F Amplifiers	31-44
11. Video Amplifiers	31-45

Index follows Chapter 31.

CONTENTS

Chapter 25. Broadband Amplifiers *Jacob S. Brown* **25-1**
    1. Broadband Amplifier Characteristics 25-1
    2. Electron-tube Characterization 25-8
    3. Electron-tube Amplifiers 25-10
    4. Transistor Characterization and Amplifiers 25-43

Chapter 26. Nonlinear Amplifiers *William M. Trenholme* . . **26-1**
    1. Introduction 26-2
    2. Use of Nonlinear Device Characteristics 26-2
    3. Use of Piecewise-linear-approximation Techniques . . . 26-35
    4. Use of Servo Techniques 26-44
    5. Circuits with Amplitude Limiting 26-53
    6. Circuits with Amplitude Compression and/or Expansion . 26-58
    7. Circuits Having a Logarithmic Amplitude Response . . 26-76
    8. Circuits Having a Logarithmic Frequency or Pulse-rate Response 26-85
    9. Amplitude Squaring and Square-rooting Circuits . . . 26-87
   10. Circuits for Analog Multiplication and Division . . . 26-92

Chapter 27. Microwave Amplifiers *Carl J. Eichenauer, Jr.* . . **27-1**
    1. Introduction 27-2
    2. Functions of Microwave Amplifier Complexes 27-10
    3. Characteristics of Microwave Amplifier Complexes . . 27-19
    4. Typical Microwave Amplifier Complexes 27-27
    5. Power-amplifier Tubes and Their Assemblies 27-36
    6. Oscillators 27-45
    7. Frequency Translators 27-50
    8. Frequency Multipliers 27-54
    9. Transmission-line Elements 27-55
   10. Power Supplies 27-67
   11. Modulators and Pulsers 27-74
   12. Protective Equipment 27-87
   13. Monitoring Equipment 27-90
   14. Control Equipment 27-92

Chapter 28. Diode and Parametric Amplifiers *Chang S. Kim, Harry J. Peppiatt, and Hsiung Hsu* **28-1**
    1. Tunnel-diode Amplifiers 28-1
    2. Capacitance-diode Amplifiers 28-27
    3. Traveling-wave Parametric Amplifiers 28-41

Chapter 29. Induced-emission Amplifiers and Oscillators (Masers and Lasers) *Edmund B. Tucker* **29-1**
    1. Introduction 29-2
    2. The Beam Maser 29-2
    3. Three-level Solid-state Cavity Masers 29-11
    4. Traveling-wave Masers 29-22
    5. Masers with Intermittent Operating Characteristics . . 29-29
    6. The CW Gas Laser 29-32
    7. Metallic-vapor Lasers 29-41
    8. Solid state Lasers—Ruby 29-43

CONTENTS

    9. Types of Modulators 18-47
   10. Field-effect Transistor D-C Amplifiers 18-69

Chapter 19. Operational Amplifiers *Paul J. Bénéteau and Keats A. Pullen, Jr.* **19-1**
    1. Introduction 19-3
    2. Basic Theory of Operational Amplifiers 19-4
    3. Electron-tube-type Operational Amplifiers 19-8
    4. Differential and Common-mode Operating Considerations 19-12
    5. Transistor Operational Amplifiers 19-14
    6. Applications of Operational Amplifiers 19-24

Chapter 20. High-power Amplifiers *Stephen J. Angello, Fred J. Heath, and John D. Meng* **20-1**
    1. High-power Transistor Amplifiers 20-2
    2. Characteristics of Commercial High-power Transistors . 20-4
    3. Thermal Considerations for High-power Transistors . 20-14
    4. Applications of High-power Transistors at Audio Frequencies 20-22
    5. High-power High-frequency Transistor Amplifiers . . 20-26
    6. High-power Amplifiers Utilizing Silicon Controlled Rectifiers 20-32
    7. Amplifiers for Transmitters 20-40

Chapter 21. Magnetic Amplifiers *Harold W. Lord* **21-1**
    1. Simple Magnetic Amplifiers (Saturable Reactors) . . 21-1
    2. High-performance Magnetic Amplifiers (Amplistats) . 21-5

Chapter 22. Ionic Amplifiers *D. Raymond Fewer and Jerry D. Merryman* **22-1**
    1. Introduction 22-1
    2. Integrators 22-2
    3. Solions in D-C Amplifiers 22-6
    4. Solion Transducers 22-9
    5. Further Applications of Ionic Devices 22-12

Chapter 23. Servo Amplifiers *Fritz H. Schlereth* **23-1**
    1. Introduction 23-1
    2. Direct-coupled Servo Amplifiers 23-2
    3. Alternating-current Servo Amplifiers 23-6
    4. Equalization Networks 23-14
    5. Nonlinear Servo Amplifiers 23-16

Chapter 24. Tuned Amplifiers *John R. Boykin* **24-1**
    1. Class A Tuned Amplifiers 24-2
    2. Stability 24-2
    3. Single-tuned Circuits 24-3
    4. Double-tuned Circuits 24-8
    5. Stagger Tuning and Stagger Damping 24-10
    6. Bandwidth of Cascaded Stages 24-11
    7. Design Procedure 24-12
    8. Class B Amplifiers 24-15
    9. Class C Amplifiers 24-15
   10. Class S Amplifiers 24-18

CONTENTS

    3. Solions 15-3
    4. The Memistor 15-12
    5. Mercury Coulombmeter 15-14

Chapter 16. Ceramic Devices *Stephen W. Tehon* 16-1

    1. Introduction 16-1
    2. Poled Ferroelectric Ceramics 16-2
    3. Properties of Poled Ceramics 16-4
    4. Adiabatic and Isothermal Constants, Pyroelectricity . . 16-8
    5. Characterization of a Poled Ceramic 16-10
    6. Electromechanical Coupling Factors 16-12
    7. Energy Storage and Piezoelectric Conversion . . . 16-16
    8. Waves and Resonance 16-17
    9. Standing Waves, Equivalent Circuits and Attenuation . 16-27
   10. The Side-plated Ceramic Bar 16-32
   11. Constant-D Transducers 16-41
   12. Ceramic-disk Resonator 16-46
   13. Ceramic Transformers 16-49
   14. Delay Line Transducers 16-57
   15. Ceramic Filters 16-61

PART III. AMPLIFIER CIRCUITS

Chapter 17. Audio Amplifiers *Richard F. Shea* 17-3

    1. Introduction 17-4
    2. Single-stage Electron-tube Amplifiers 17-6
    3. Single-stage Transistor Amplifiers 17-9
    4. Transistor Bias 17-11
    5. Transformer-coupled Stages 17-15
    6. RC-coupled Stages 17-18
    7. Cascaded Stages 17-22
    8. Hearing Aids 17-24
    9. High-input-impedance Amplifiers 17-28
   10. Electron-tube Preamplifiers 17-34
   11. Transistor Preamplifiers 17-38
   12. Classes of Output Stages 17-40
   13. Class A Power Amplifiers 17-41
   14. Class A, AB, and B Push-Pull Power Amplifiers . . 17-57
   15. Typical Power Amplifiers 17-71
   16. Electron-tube Stereo Amplifiers 17-79
   17. Transistor Stereo Amplifiers 17-90

Chapter 18. D-C Amplifiers *Paul J. Bénéteau and Edward G. Nielsen* 18-1

    1. Direct-coupled Amplifiers 18-2
    2. Single-ended Electron-tube Amplifiers 18-3
    3. Direct-coupled Transistor Amplifiers 18-8
    4. Differential Electron-tube Amplifiers 18-14
    5. Differential Transistor Amplifiers 18-17
    6. Chopper and Modulated carrier Amplifiers 18-25
    7. Input and Output Switching Circuits 18-27
    8. Modulator-demodulator Performance 18-43

CONTENTS

7. Power Transistors	10-20
8. Field-effect Transistors	10-21
9. Specialized Transistors	10-23
10. Future Transistors	10-24

Chapter 11. Solid-state Switching Devices *John D. Meng* . . **11-1**

1. Four-layer Diodes	11-1
2. Silicon Controlled Rectifiers	11-12
3. Silicon Controlled Switches	11-19

Chapter 12. Tunnel Diodes and Backward Diodes *Chang S. Kim and Jerome J. Tiemann* . . **12-1**

1. Tunnel-diode Theory	12-1
2. Static Parameters of Tunnel and Backward Diodes	12-5
3. Tunnel-diode Equivalent Circuit	12-6
4. Resistive Cutoff and Self-resonance Frequencies	12-6
5. Various Tunnel Diodes	12-7
6. Packaging Considerations	12-8
7. Noise	12-9
8. Nonlinear Characteristics of Tunnel Diodes	12-12
9. Mixers	12-21

Chapter 13. Parametric Devices *Harry J. Peppiatt* . . **13-1**

1. Capacitor Diodes	13-1

Chapter 14. Induced-emission-amplifier Principles *Edmund B. Tucker* . . **14-1**

1. Introduction	14-3
2. Blackbody Radiation	14-3
3. Induced-emission Amplifiers	14-7
4. Methods of Population Inversion	14-8
5. Methods of Population Inversion in Two-level Paramagnetic Systems	14-9
6. R-F Methods of Population Inversion in Multilevel Paramagnetic Systems	14-12
7. Optical Pumping	14-15
8. Laser Inversion Methods	14-17
9. Induced-emission Amplifiers Which Do Not Require Inverted Populations	14-21
10. Noise	14-23
11. Transmission-cavity-maser Characteristics	14-25
12. Relection-cavity-maser Characteristics	14-29
13. The Traveling-wave Maser	14-30
14. Comparison of Maser Types	14-32
15. Laser Amplifier Characteristics	14-34
16. Induced-emission Oscillators	14-35
17. Cavities	14-39
18. Laser Pump Sources	14-44

Chapter 15. Ionic Devices *D. Raymond Fewer* . . **15-1**

1. Introduction	15-1
2. Electrochemical Principles of Operation	15-1

CONTENTS

Chapter 5. Filter Amplifiers *Louis Weinberg* 5-1

    1. Introduction. 5-1
    2. Some Optimum Magnitude and Phase Characteristics. . 5-4
    3. Dipole Networks Synthesized by Recognition. . . . 5-19
    4. Additional Techniques for Realization of Filter Amplifiers. 5-27

Chapter 6. Feedback *Sorab K. Ghandhi*. 6-1

    1. Introduction. 6-2
    2. Elementary Feedback Theory. 6-2
    3. Two-port Analysis of Feedback Amplifiers 6-5
    4. General Feedback Theory 6-9
    5. The Specification of Amplifier Response 6-31
    6. The Single-loop Unconditionally Stable Amplifier . . 6-34
    7. Feedback Chains 6-41

Chapter 7. Amplifier Noise *Edward G. Nielsen*. 7-1

    1. Introduction. 7-6
    2. Noise in Two-terminal Networks 7-7
    3. Noise in Two-ports 7-8
    4. Noise Factor. 7-12
    5. Noise Temperature 7-23
    6. Techniques for Measurement of Noise Parameters . . 7-29
    7. Instrumentation of Noise Measurements 7-35
    8. The Application of Noise Temperature to Amplifier and Receiver Performance Calculations 7-47
    9. Limiting Noise Conditions 7-52

PART II. AMPLIFYING DEVICES

Chapter 8. Magnetic Amplifier Devices *Harold W. Lord*. . 8-3

    1. Static Devices Which Amplify through Nonlinearity of Core Materials 8-3
    2. Magnetic Materials Properties 8-4
    3. Saturable-reactor Fundamentals 8-17

Chapter 9. Electron Tubes *Lloyd G. Mumford, Philip E. Hatfield, Theodore G. Mihran, and Daryl W. Hawkins*. . . . 9-1

    1. Introduction. 9-2
    2. Considerations in Choosing a Tube Type 9-2
    3. Electron-tube Life 9-42
    4. Receiving Tubes 9-47
    5. Microwave Tubes 9-80
    6. Transmitting Tubes 9-97

Chapter 10. Transistors *W. Crawford Dunlap* 10-1

    1. Introduction. 10-2
    2. Basic Principles of Transistors 10-3
    3. Equivalent Circuits for the Transistor 10-7
    4. Design Theory of Transistors 10-9
    5. Alloy Transistors 10-12
    6. Diffused Transistors 10-16

CONTENTS

Contributors v

Preface vii

PART I. AMPLIFIER FUNDAMENTALS

Chapter 1. Network Theory: Definitions and Basic Concepts
Louis Weinberg **1-3**

    1. Introduction 1-3
    2. Complex Frequency 1-4
    3. Types of Networks 1-5
    4. Network Functions 1-7
    5. Active and Passive Networks 1-9
    6. Natural Frequencies and Characteristic Polynomials . . 1-18
    7. Some Definitions: Equivalent, Constant-resistance, Complementary and Dual Networks 1-23

Chapter 2. Network Theorems, System Functions, and Equivalent Circuits *Louis Weinberg* **2-1**

    1. Introduction 2-1
    2. Network Theorems 2-2
    3. Parts of a System Function 2-8
    4. Equivalent Circuits 2-17

Chapter 3. Matrix Representations of Multiport Networks
Louis Weinberg **3-1**

    1. Introduction 3-1
    2. Matrices for Two-port and n-port Networks . . . 3-3
    3. Equations for Terminated Networks 3-13
    4. Power-transfer Concepts, Matching, and Potential Instability 3-22
    5. Interconnections of Quadripoles 3-26

Chapter 4. Scattering Matrices and Matrices for Multiterminal Networks *Louis Weinberg* **4-1**

    1. Introduction 4-1
    2. Scattering Matrix 4-2
    3. Transfer Scattering Matrix 4-23
    4. Scattering Matrices for Multiports 4-25
    5. Interrelationships among Network Matrices . . . 4-27
    6. Multiterminal Networks 4-32

PREFACE

tative reference work for the practicing engineer, scientist, technician, and student. Here he will find details on specific designs, together with necessary information from which he can develop desired variations. A handbook contains a profusion of reference materials, tables, specifications, and similar data, compiled from a multitude of sources to simplify the task of locating desired information.

This Handbook is divided into three logical sections, dealing respectively with fundamentals, devices, and circuits. The first section lays a foundation for the rest. Here will be found a condensation of the extensive writings by many authors on basic network theory, on feedback, and on noise, so that the reader may readily grasp these fundamentals for future reference. This will provide an exposition of these complex subjects, written in such manner as to extract the essentials and present them in readily comprehensible form.

The devices section includes descriptions of typical electron tubes, semiconductor devices, magnetic devices, and such specialized items as masers, lasers, and ionic and ceramic devices. Obviously all forms of these devices cannot be described; rather one-of-a-kind treatment is frequently employed, with adequate details given to illustrate the construction and characteristics of these forms of devices.

The bulk of the book is devoted to specific categories of circuits and, as mentioned previously, includes designs ranging from one end of the frequency and power spectra to the other, as well as incorporating practically all physical forms of devices. Again, it is physically impossible, for obvious reasons, to include all possible forms of a particular type of amplifier; thus frequently only representative circuits are included, together with specific design details. Wherever possible the design is generalized so that variations are possible, as desired, by extension of the values given.

It is sincerely felt that the contributors to this Handbook have compiled a veritable treasure chest of detailed knowledge between its pages and that the reader will find herein the answers to his needs in this all-inclusive field.

R. F. Shea

Another form of device which is commonly considered an amplifier is the magnetic amplifier, in which the output wave form is definitely not a replica of the input. Similarly, the growing use of switching devices in amplifiers, and of choppers and modulators, produces additional examples where dissimilarity between input and output exists. In some examples the input may be d-c, the output a-c; in others the reverse may be true.

Thus, to fit common usage, we must modify the rigorous definition of the term "amplifier" by inserting the word "essential" before "quality," with the understanding that this permits a considerable degradation of the input signal as long as the input power is increased. Thus an amplifier becomes essentially a device for increasing power from a low input level to a substantially higher output level, with true duplication of the form of the input signal frequently of secondary importance.

From these preliminary remarks it is obvious that the term "amplifier" can be applied to an almost infinite variety of devices and concepts, and it becomes necessary to apply certain boundary conditions to restrict this Handbook to reasonable size and scope. It is felt that the term is most commonly applied in the electrical or electronic sense rather than mechanical, pneumatic, thermal, etc., and, therefore, this is the area most widely covered in this book. Certain of the more recent developments in the field of optics, involving amplification of light by means of stimulated emission, are closely related to similar techniques employed for microwave amplication and hence are included. Likewise, at the other end of the frequency spectrum, ionic devices are finding application, and a chapter is, therefore, included on this topic.

There are also a number of circuits which are normally used in conjunction with, or incorporated within, amplifiers. Typical of these are analog circuits such as summers, multipliers and dividers, and various nonlinear circuits, e.g., compressors and expanders. It seems logical, therefore, to include these tools along with the more conventional circuits, and, consequently, chapters have been included to take care of these circuits.

Even with the above restrictions the scope is truly tremendous, encompassing direct current to gigacycles and angstroms, picowatts to megawatts; gases, solids, liquids; conductors, semiconductors, insulators, dielectrics; electron tubes, transistors and other solid-state devices, magnetic devices; electrons to magnetrons. Yet, despite this extensive scope, the treatment of each subject must be sufficiently definitive and detailed to provide adequate reference material for the seeker after broad or specialized knowledge.

A handbook differs from a textbook in that the emphasis is on practical application. Mathematical derivations must be held to a minimum and only the basic concepts given; likewise only enough theory is included to explain basic practice. Thus the handbook becomes the basic authori-

PREFACE

An amplifier has been defined as a device for increasing the power associated with a phenomenon without appreciably altering its quality, through control by the amplifier input of a larger amount of power supplied by a local source to the amplifier output.

The key words in this definition are input, output, local source, and quality. Thus, in accordance with the above, an amplifier must, in essence, deliver an enlarged replica of an input signal to the output, calling upon the source for the added power. The characteristic which is enlarged may be input power or input amplitude; however, unless there is a local source to supply power, the device is not considered to be an amplifier. This distinguishes an amplifier from a transformer, for example.

The signal may be mechanical, thermal, electrical, or optical in nature and similarly the source may take any of these forms, as long as it is capable of supplying auxiliary power. In the rigorous definition of an amplifier, the input and output signals must be similar. If one type of input signal controls a different form of output, the device is called a converter or a gate rather than an amplifier.

In actuality, few amplifiers conform completely to this rigorous interpretation of the above definition. For one thing, the presence of distortion implies that the output signal is not a linear replica of the input, and in fact many amplifiers contribute considerable distortion yet are entirely suitable for their purpose. A servo amplifier is an example of this category. True linearity is of relatively little importance here since steady-state operation is usually at a null balance and, in many cases, deliberate nonlinearity is introduced to limit peak output swings.

WILLIAM M. TRENHOLME, *Massachusetts Institute of Technology, Department of Nuclear Engineering, Cambridge, Mass. (Present affiliation: Arizona Atomic Energy Commission, Phoenix, Ariz.*

EDMUND B. TUCKER, *General Electric Co., Research and Development Center, Schenectady, N.Y.*

LOUIS WEINBERG, *Department of Electrical Engineering, University of Michigan, Ann Arbor, Mich. (Present affiliation: Department of Electrical Engineering, The City College of the City University of New York, New York, N.Y.)*

CONTRIBUTORS

STEPHEN J. ANGELLO, *Westinghouse Electric Corp., Research and Development Center, Pittsburgh, Pa.*
PAUL J. BÉNÉTEAU, *Director, Applications Engineering Department, Fairchild Research Laboratory, Palo Alto, Calif.*
JOHN R. BOYKIN, *Westinghouse Electric Corp., Defense and Space Center, Baltimore, Md.*
JACOB S. BROWN, *General Electric Co., Missile and Space Vehicle Department, King of Prussia Park, Pa.*
W. CRAWFORD DUNLAP, *NASA Electronics Research Center, Cambridge, Mass.*
CARL J. EICHENAUER, JR., *General Electric Co., Heavy Military Electronics Department, Syracuse, N.Y.*
D. RAYMOND FEWER, *Texas Instruments, Inc., Semiconductor Components Division, Dallas, Texas.*
SORAB K. GHANDHI, *Department of Electrical Engineering, Rensselaer Polytechnic Institute, Troy, N.Y.*
PHILIP E. HATFIELD, *General Electric Co., Tube Department, Owensboro, Ky.*
DARYL W. HAWKINS, *General Electric Co., Tube Department, Schenectady, N.Y.*
FRED J. HEATH, *Canadian General Electric Co., Ltd., Toronto, Ontario, Canada.*
HSIUNG HSU, *Antenna Laboratory and Department of Electrical Engineering, Ohio State University, Columbus, Ohio.*
CHANG S. KIM, *General Electric Co., Electronics Laboratory, Syracuse, N.Y.*
HAROLD W. LORD, *General Electric Co., Research and Development Center, Schenectady, N.Y.*
JOHN D. MENG, *Electronics Engineering Department, University of California, Lawrence Radiation Laboratory, Berkeley, Calif.*
JERRY D. MERRYMAN, *Texas Instruments, Inc., Semiconductor Components Division, Dallas, Texas.*
THEODORE G. MIHRAN, *General Electric Co., Research and Development Center, Schenectady, N.Y.*
LLOYD G. MUMFORD, *General Electric Co., Tube Department, Owensboro, Ky.*
EDWARD G. NIELSEN, *General Electric Co., Electronics Laboratory, Syracuse, N.Y.*
RICHARD J. PATCH, *General Electric Co., Semiconductor Products Department, Syracuse, N.Y.*
HARRY J. PEPPIATT, *General Electric Co., Communications Products Department, Lynchburg, Va.*
KEATS A. PULLEN, JR., *Ballistics Research Laboratories, Aberdeen Proving Ground, Md.*
FRITZ H. SCHLERETH, *General Electric Co., Electronics Laboratory, Syracuse, N.Y.*
RICHARD F. SHEA, *Consulting Electronics Engineer, Schenectady, N.Y.*
STEPHEN W. TEHON, *General Electric Co., Electronics Laboratory, Syracuse, N.Y.* (*Present affiliation: Tecumseh Products Co., Research Laboratory, Ann Arbor, Mich.*)
JEROME J. TIEMANN, *General Electric Co., Research and Development Center, Schenectady, N.Y.*

Various other concepts have been defined to substitute for the controlled source. Mason and Zimmerman,[5] for example, define a *unistor*, which is a three-terminal branch that is particularly adaptable for topological analysis of a network. The unistor is basically a voltage-controlled current source; as the authors point out, it represents the linear equivalent circuit for a common-grid triode in which the plate conductance is small and the grid-to-plate transconductance is the only nonnegligible parameter.

Below we shall consider another alternative for the controlled source, which, in addition, is an exceedingly useful concept that can be used in conjunction with controlled sources.

5.2. Negative Resistance. With the inclusion of controlled sources, we may represent a one-port element that has a negative value as well as an element with a positive value; this is obtained merely by reversing the polarity on the controlled source.[4] In Fig. 11 are shown current-controlled source representations of the active one-port elements, a negative resistance, a negative inductance, and a negative capacitance. Equivalent controlled-source representations can also be given which

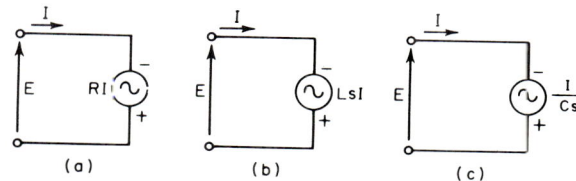

Fig. 11. Controlled-source representations of negative elements. (a) Negative resistance where $E = -RI$. (b) Negative inductance where $E = -LsI$. (c) Negative capacitance where $E = -(I/Cs)$.

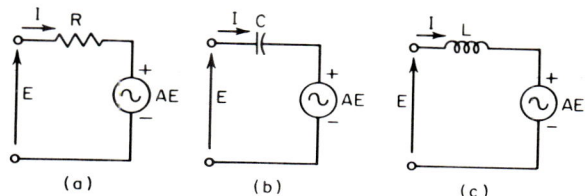

Fig. 12. Use of voltage-controlled voltage sources for alternative representations of negative elements.

are more useful in that they suggest how the negative elements can be practically realized by use of amplifiers; these are shown in Fig. 12. Considering the negative resistance, for example, we see that the input current is given by $I = (1 - A)E/R$, so that at the terminals we see an impedance $E/I = R/(1 - A)$. Since for $A > 1$ this input resistance is negative, we see that a negative resistance can be realized by a positive resistance and a voltage amplifier of gain greater than unity. Below we discuss the properties of the negative resistance; a corresponding discussion can be given for the other two types of negative elements.

Negative resistance is not only useful as a concept for analysis and synthesis but in the past decade has also acquired usefulness as a practical device. It may not be too fanciful to suggest that in the near future negative resistors will in many cases be stocked like the passive elements. It is certain that they have many practical applications where an adequate physical embodiment of negative resistance can be inserted into a network like a positive resistor. For example, to achieve filters with magnitudes that have exceedingly sharp cutoff characteristics, tuned circuits with low loss are needed, that is, high Q's are required; as a result crystals often become imperative. However, it is also possible to use an ordinary coil-capacitor tuned circuit and then connect a negative resistance across the circuit; this technique has been used successfully in field equipment.[6]

The transistor has made the active network competitive with the passive one; the tunnel diode now provides an even more direct embodiment of a negative resistance since its equivalent circuit is merely an inductance in series with the parallel combination of a capacitance and a negative resistance. The passive elements, as the reader is aware, are rugged and long-lived and can be designed to be quite stable with respect to variations in ambient conditions. Before the invention of the transistor, active networks were bulkier than passive networks, less rugged, and had characteristics that could deteriorate for any of a number of reasons, among them being insufficient cathode emission, the variation of an active parameter, or a change in the power-supply voltage. With the advent of the junction transistor, a small rugged, long-lived active package became available that could provide a fairly constant negative resistance over a desired operating range of frequencies.

The concept of a negative resistance appears to be simple, but it is only superficially so; indeed, it may be said to be deceptively simple. To understand it completely we must first understand the properties of a general impedance. The reader, with his understanding of a positive resistance, e.g., how it dissipates power, and how resistances in series or parallel may be combined, may question why multiplying a positive resistance by minus one should introduce complications. A negative resistance does combine in series and parallel in the same way as a positive resistance; however, instead of absorbing and dissipating power it supplies it. But there are complications and these are involved with the problems of infinite power and stability.

First of all, it should be clear that no physical two-terminal network can possess a negative resistance that is constant for all signal amplitudes; the negative resistance must be nonlinear since a linear voltage vs. current relationship implies an infinite power source.[7] Secondly, the negative resistance cannot be constant for all frequencies since this again implies an infinite power source. Thus the negative resistance is a constant only over a specified range of frequencies. It is stressed, however, that it is just as useful to postulate such an ideal element as it is to assume an ideal positive resistance.

The frequency dependence of a negative resistance may be clarified by an additional brief discussion. For the positive resistance it is immaterial whether a voltage generator or a current generator is connected to its terminals: Ohm's law will still hold. However, a physical impedance that is a negative resistance may be perfectly well behaved when a voltage source is connected to its terminals but may become unstable when a current source is substituted for the voltage source. This property is related to the short-circuit and open-circuit stability of a negative resistance. Its stability, as we shall see, is determined by the positions of its poles and zeros. If it is short-circuit unstable, some of its zeros are in the right half plane; if it is open-circuit unstable some of its poles are in the right half plane. But we are speaking of a *constant* resistance, which has no poles or zeros, and yet we describe it in terms of poles and zeros. We do this to emphasize that *every physical negative resistance is in reality a general active impedance.*

Thus, to summarize the frequency characteristics, the negative resistance of a physical device is the real part of a general impedance; over a certain range of frequencies, this real part is approximately constant and negative, and the imaginary part is approximately zero. Outside this range of frequencies the imaginary part becomes significant, and beyond some frequency the negative resistance becomes positive. It is thus evident that a classification of general impedances must be made in order to fit the concept of a negative resistance or impedance into a proper context. Bode gives such a classification in his classic book.[8]

Negative-resistance devices have been built; one obvious application of such devices is a bilateral amplifier or, in telephone parlance, a repeater. The negative-resistance device may be inserted in series with a transmission line or in shunt with it. If it is put in series, it must be an open-circuit stable device, whereas a shunt connection requires a short-circuit stable characteristic. Such a resistance inserted in a line would of course introduce a discontinuity and thus cause reflections. With the use of

NETWORK THEORY: DEFINITIONS AND BASIC CONCEPTS 1–17

two negative resistances, however, it is possible to introduce no discontinuity. The circuit used is the bridged-T network of Fig. 13; for this circuit it is simple to show that, irrespective of the value of ρ, the characteristic impedance of the line Z_0 is seen at port a-b and at port c-d. In addition, if the value of ρ is made negative, i.e., ρZ_0 and Z_0/ρ are negative impedances, a gain can be obtained. The multiplier ρ need not be real but may be complex; thus filters may be used to cause the gain to vary in a prescribed manner.

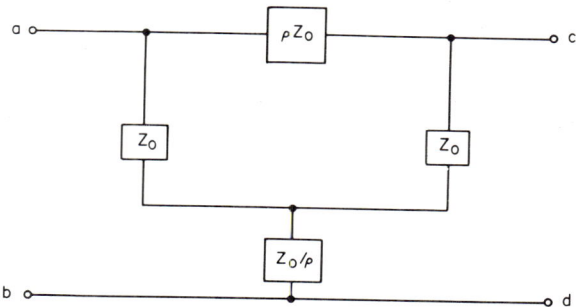

Fig. 13. A possible bilateral amplifier with both input impedances equal to Z_0.

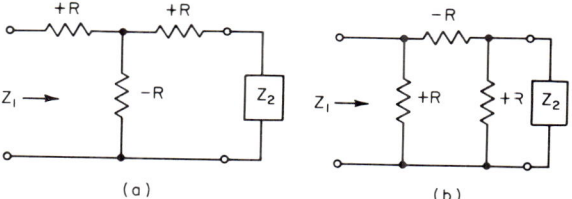

Fig. 14. Two networks each of which produces a negative impedance by use of one negative resistance.

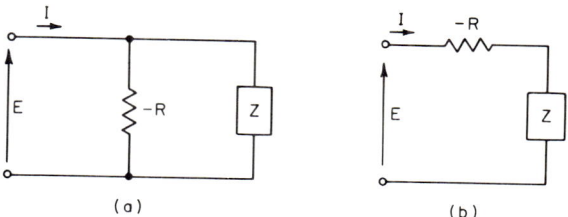

Fig. 15. Representation of any realizable active one-port by a passive impedance plus a single negative resistance.

With the negative resistance as an admissible element any negative impedance may be obtained.[8] Two simple networks for obtaining a general negative impedance are shown in Fig. 14. The input impedance of each is easily computed as

$$Z_1 = -R^2/Z_2 \qquad (8)$$

Thus any negative impedance can be produced by terminating the T or Π network by the positive inverse of the required impedance; for example, as a particular case of a general negative impedance a negative capacitance of value $C = -L/R^2$ farads is obtained if the termination is an inductance of L henrys.

The negative resistance allows us to show the simple distinction between active and passive networks, and thus guards against setting up impossible performance specifications for active networks. Every realizable driving-point function may be realized by a negative resistance in parallel with a passive impedance or in series with a passive impedance, as shown in Fig. 15.* Correspondingly, an amplifier with a flat gain, i.e., an ideal amplifier, in cascade with a passive quadripole can realize any realizable transfer function; the network is shown in Fig. 16. From this network it is clear that the only possible difference between the transfer functions of an active and passive quadripole is in the constant multiplier.

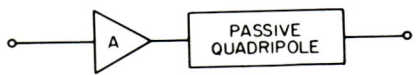

FIG. 16. Representation of the transfer function of any physical quadripole by an ideal amplifier plus a passive quadripole.

6. NATURAL FREQUENCIES AND CHARACTERISTIC POLYNOMIALS

An important property of a network (or other physical system) is its *natural behavior*. By this is meant that part of the system response whose form is dependent only on the values of the elements and their interconnection and independent of the applied force. This behavior is characterized by means of the *natural frequencies* of oscillation (or *natural modes*, or *free vibrations*) of the network.

The natural behavior is often described as the response to a *unit-impulse* excitation. This description serves to relate the natural frequencies to the critical frequencies of a system function. The inverse Laplace transform of the system function shows that the impulse response includes terms of the form $A_k e^{s_k t}$, where A_k is a constant and s_k are the zeros of the denominator polynomial. Since s_k are the natural frequencies, we conclude that the *finite poles of the system function are the natural frequencies*. For this reason the denominator is called the *characteristic polynomial* of the system, and the equation formed by setting the denominator equal to zero is called the *characteristic equation*.

Passive networks must of course be stable; that is, they have no internal energy source, and therefore an impulse response that increases without bound is impossible. Thus, since a pole in the right half of the s plane corresponds to an exponentially increasing time function, we see immediately that a system function of a passive network can have no poles in the right half plane.

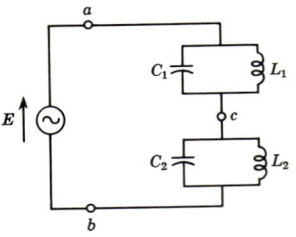

FIG. 17. Network whose natural frequencies are to be determined.

The natural frequencies have been shown to be the poles of a system function. Yet the reader is probably aware that they are often calculated as the *zeros* of a system function. The reason becomes obvious when it is pointed out that the system function whose zeros give the natural frequencies is a driving-point function. What is actually being used is the *reciprocal* of the system function that applies for the prescribed type of input. However, since the reciprocal of a driving-point function is again a driving-point function, the fact is obscured that the ratio of an input to an output is being used. For example, for the network shown in Fig. 17, where the voltage source is a unit impulse, the natural frequencies are the *zeros of the input impedance* seen looking into the network at a-b. However, the system function for a voltage input is the input *admittance*, so that its poles are the natural frequencies.

Thus *both the poles and zeros of a driving-point function* are natural frequencies of a network, *depending on the type of excitation*. Consider again the network in Fig. 17

---

* Here we do not consider degenerate cases for which no practical application has yet been found. For example, it has been shown that a driving-point function may be realized that acts simultaneously as a short and an open circuit.[9]

NETWORK THEORY: DEFINITIONS AND BASIC CONCEPTS 1-19

without any source applied. The zeros of Z, the impedance seen at a-b, in Fig. 18, are the natural frequencies for a *voltage* input; in other words, we can visualize that we cut open the network of Fig. 17 after the voltage source is short-circuited and then calculate the impedance seen at the port formed by the cut terminals. The poles of Z, on the other hand, are the natural frequencies for a current source applied at a-b. Thus, if we consider the applicable system function in each case—that is, the input admittance for a voltage input and the input impedance for a current input—then only the poles of each system function are natural frequencies.

The procedure illustrated by the above example is general. Furthermore, since the impulse response calculated in any part of the network exhibits the natural behavior, it is not necessary that the driving-point function be calculated at a point where a generator appears. In other words, this is a consequence of what was stated previously, namely, that the natural behavior is independent of the applied force. However, in order not to misinterpret this statement we must recall that all sources must be *properly* removed; that is, voltage sources are *short-circuited* and current sources are *open-circuited*. Thus, after all sources have been removed, the network may be broken into at any point: the zeros of the impedance are the natural frequencies.

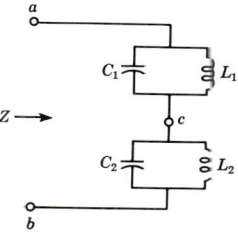

FIG. 18. Network of Fig. 17 without any source shown.

It is of value to restate the above precisely in mathematical terms. An inhomogeneous differential equation describes the network containing sources. The corresponding force-free equation is obtained by setting all forcing terms—i.e., voltage and current sources—equal to zero. This equation, which is also called the *homogeneous*, or *reduced*, equation, describes the network without sources *only if all the sources have been properly removed.*

Often it is more convenient to use a procedure dual to that of breaking into a network. One may imagine soldering leads across any node pair: now the zeros of the

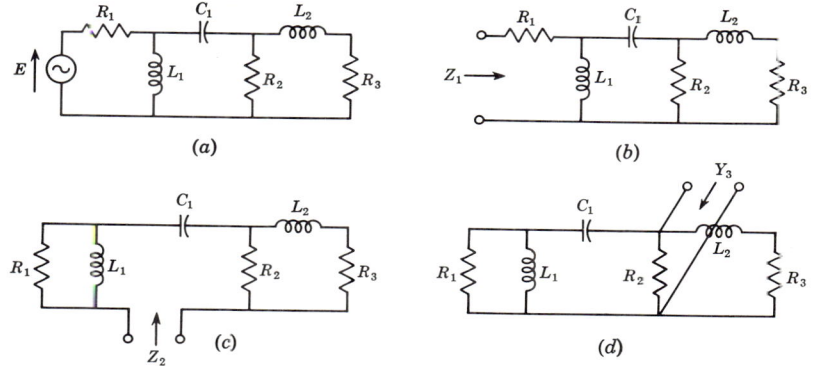

FIG 19. Methods of determining natural frequencies of a given network.

input admittance at the port formed by these soldered terminals are the natural frequencies. This is clear when one visualizes a unit-impulse current source applied across this node pair; then the pertinent system function is the *impedance* between this node pair; this confirms that the *zeros of the admittance* are the natural frequencies As an example, the natural frequencies of the network in Fig. 19a may be calculated as the zeros of Z_1 in Fig. 19b, or it may be more convenient to calculate them as the zeros of Z_2 in Fig. 19c or the zeros of Y_3 in Fig. 19d.

We have specified the denominator of a system function as a characteristic polynomial. However, as we have seen, there are other polynomials that give natural frequencies for a *different* termination of the network; thus there is more than one characteristic polynomial. For a quadripole, in fact, there are four polynomials that give the natural frequencies of the network for any combination of open- or short-circuit terminations.[10,11] These four characteristic polynomials, plus two other polynomials that give the zeros of transfer functions, completely describe the network. When the network is passive and reciprocal, the two additional polynomials are equal, so that only five polynomials completely characterize the quadripole. In fact, for a reciprocal quadripole only four of these five polynomials are independent and thus describe the network uniquely.

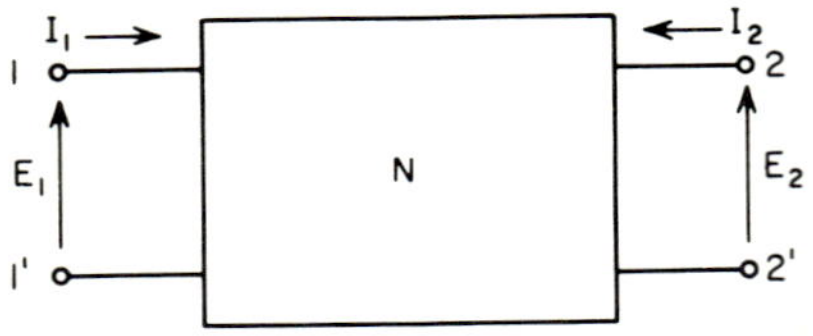

To discuss these polynomials it is convenient to introduce alternative sets of functions for characterizing the quadripole shown in Fig. 20. The reader is probably familiar with the set of *open-circuit impedances* z_{jk} defined by

$$E_1 = z_{11}I_1 + z_{12}I_2 \\ E_2 = z_{21}I_1 + z_{22}I_2 \qquad (9)$$

Fig. 20. General quadripole to be characterized by a set of two simultaneous linear equations.

or more briefly by the matrix equation

$$E] = [z] \, I] \qquad (10)$$

where $[z]$ is a square matrix and $E]$ and $I]$ represent column matrices. An alternative set of functions is the set of *short-circuit admittances* defined by

$$I_1 = y_{11}E_1 + y_{12}E_2 \\ I_2 = y_{21}E_1 + y_{22}E_2 \qquad (11)$$

or in matrix terms by

$$I] = [y] \, E] \qquad (12)$$

If $[z]$ is nonsingular, it is clear that

$$[y] = [z]^{-1} \qquad (13)$$

so that the determinants of the two matrices are each other's reciprocals,

$$|y| = 1/|z| \qquad (14)$$

Later in Chaps. 3 and 4 we shall discuss the properties of $[z]$ and $[y]$ along with those of other useful matrix characterizations of a quadripole; here we are interested only in making use of two properties in the discussion of the characteristic polynomials, namely, that the z_{jk} are calculated or measured with an open circuit at each port and the y_{jk} with a short circuit at each port.

Since the open-circuit driving-point impedance z_{11} describes the network with an open circuit at each port, the denominator of z_{11} is the characteristic polynomial for both ports open-circuited. Now consider the reciprocal of z_{11} as a system function, that is, as an output divided by an input. Thus the denominator of the admittance $y \equiv 1/z_{11}$ is the characteristic polynomial for port 2 open-circuited and port 1 short-circuited. A short circuit is required for port 1 since defining y as a system function requires a voltage source as the input.

Now let n_{jk} represent a characteristic polynomial, where the first and second subscripts refer to ports 1 and 2, respectively, and the subscripts may become s and o, representing, respectively, a short and open circuit at the designated port; for example, n_{so} is the characteristic polynomial for the quadripole with port 1 short-circuited and port 2 open-circuited. Then the previous discussion establishes that $z_{11} = n_{so}/n_{oo}$. We also let m_{12} and m_{21} represent the polynomials whose zeros are the transmission zeros of the corresponding transfer functions. Then the open-circuit impedances may be represented by

$$z_{11} = n_{so}/n_{oo} \qquad z_{12} = m_{12}/n_{oo} \\ z_{22} = n_{os}/n_{oo} \qquad z_{21} = m_{21}/n_{oo} \qquad (15)$$

An argument similar to that used on z_{11} gives $y_{11} = n_{os}/n_{ss}$. Subsequent use of the

NETWORK THEORY: DEFINITIONS AND BASIC CONCEPTS   1-21

inverse relationship of $[z]$ and $[y]$ gives the short-circuit admittances

$$y_{11} = n_{os}/n_{ss} \qquad y_{12} = -m_{12}/n_{ss}$$
$$y_{22} = n_{so}/n_{ss} \qquad y_{21} = -m_{21}/n_{ss} \tag{16}$$

Another important relationship, which follows from the determinantal relationship $|z| = 1/|y|$, is

$$n_{oo}n_{ss} = n_{so}n_{os} - m_{12}m_{21} \tag{17}$$

Using (17) now gives the determinant of the z's as

$$|z| = (n_{so}n_{os} - m_{12}m_{21})/n_{oo}^2$$
$$= n_{oo}n_{ss}/n_{oo}^2$$
$$= n_{ss}/n_{oo} \tag{18}$$

Correspondingly, we obtain

$$|y| = n_{oo}/n_{ss} \tag{19}$$

Examination of Eqs. (15) to (19) reveals several important properties. First, we note that any set of network functions such as the open-circuit impedances can be formed using only five of the polynomials; the sixth polynomial, which arises when the determinant of the set of functions is computed, may be evaluated by use of Eq. (17). Second, when the determinant is formed in the usual way, redundant canceling factors occur, which thus results in the manipulation of polynomials of unnecessarily high degree. Long division must be used to eliminate the redundant part of the numerator polynomials; this is evident in Eqs. (18). We thus observe that use of the common network functions tends to mask the polynomial manipulations and often causes unnecessary computation. Third, of the six polynomials—the four characteristic polynomials plus the two polynomials specifying the transmission zeros—only five are required to specify the quadripole uniquely. If the network is reciprocal, $m_{12} = m_{21}$ and only four of the six polynomials then completely specify the quadripole.

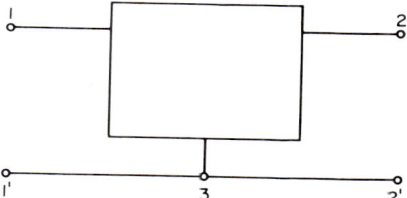

FIG. 21. Grounded quadripole formed from the three-terminal network with terminals 1, 2, and 3.

The question arises whether it is possible to describe a network in terms of characteristic polynomials alone; in other words, can we eliminate the need for direct specification of m_{12} and m_{21}. Since, as we shall show later when we consider n-terminal networks, every reciprocal network—whether it is an n-port or a $2n$-terminal network—may be completely characterized by a set of *driving-point* functions, the answer must be in the affirmative for reciprocal n-ports: the numerator and denominator of a driving-point function are, as we have seen, both characteristic polynomials. The answer is in the negative, however, for nonreciprocal n-ports.[12] We shall illustrate this for the grounded quadripole.[13]

Consider the grounded quadripole shown in Fig. 21, where in the drawing we emphasize the fact that it is a three-terminal network that is being used as a quadripole with a terminal common to the two ports. Each of its four characteristic polynomials yields the natural frequencies for the network with one of the four possible sets of terminations shown in Fig. 22.

However, when the network is to be described by the characteristic polynomials is not a quadripole but is a true three-terminal network, such as the network shown in Fig. 23a, another mode of operation is possible. It is possible to have all three terminals shorted, as is shown in Fig. 23b. Thus another characteristic polynomial must be defined for this set of conditions.[13] Except for an ambiguity about z_{12} and z_{21}, specifically, whether the function defined as z_{12} is actually z_{21} so that it is possible that the designations z_{12} and z_{21} should be interchanged, these five characteristic polynomials describe the three-terminal network and the quadripole completely.

6.1. Short-circuit Stability and Open-circuit Stability. A passive structure is stable for any passive terminations connected to its ports; however, this is not true

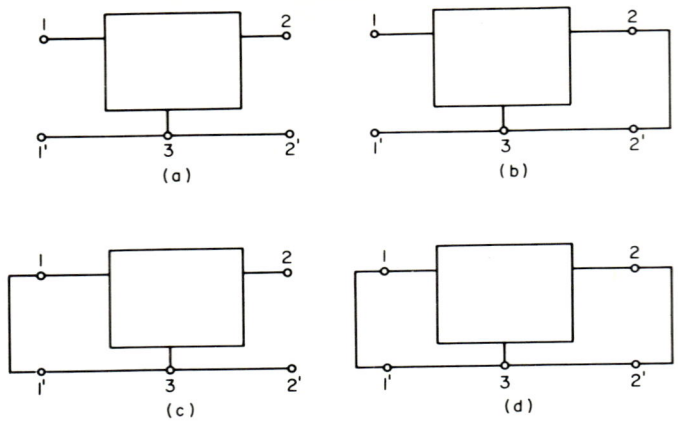

FIG. 22. Possible short-circuit and open-circuit terminations for the quadripole.

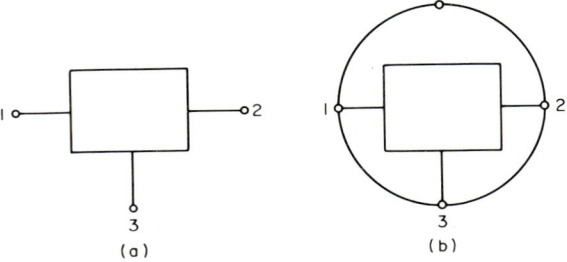

FIG. 23. Shorting of all the terminals of a three-terminal network to define a fifth characteristic polynomial.

for active networks. The stability of an active network may be specified in terms of characteristic polynomials.

Consider the active network shown in Fig. 24; suppose a passive termination is to be connected to the available port and that we wish to determine whether the complete system will be stable.

If the input impedance $Z_{in} = p(s)/q(s)$ has all its *zeros* in the left half plane, N is said to be *short-circuit* stable, whereas if some zeros are in the right half plane, it is short-circuit unstable; if a short circuit is connected to the port of a short-circuit stable network, it will not oscillate. Hence a voltage source may be connected to the port without causing instability. If the input impedance has all its poles in the left half plane, it is *open-circuit* stable; a current source may thus be connected to the port to yield a stable system. It is therefore clear that, depending on the positions of the critical frequencies of Z_{in}, the network may be open-circuit stable, or short-circuit stable, or both, or neither. Since there is no necessary relation between the zeros and poles of an active driving-point impedance, the stability upon being driven by a voltage source [for which the characteristic polynomial is $p(s)$] is independent of the stability upon being driven by a current source [for which the characteristic polynomial is $q(s)$].

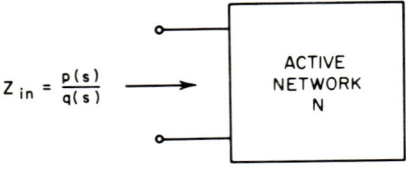

FIG. 24. Active network with input impedance $Z_{in} = p(s)/q(s)$.

It is also possible that N is neither open-circuit stable nor short-circuit stable but is short-circuit stable when $Z_1 = p_1(s)/q_1(s)$ is used as a termination. In this case,

NETWORK THEORY: DEFINITIONS AND BASIC CONCEPTS 1–23

though $p(s)$ and $q(s)$ both have zeros in the right half plane, none of the zeros of $Z_1 + Z_{in}$, that is, the numerator of this sum, namely, $pq_1 + p_1q$, lies in the right half plane.

We conclude that the stability properties can be determined in terms of the appropriate characteristic polynomial.

7. SOME DEFINITIONS: EQUIVALENT, CONSTANT-RESISTANCE, COMPLEMENTARY, AND DUAL NETWORKS

Various terms are used to describe realizable networks and functions. In some problems two or more *equivalent* networks are desired; frequently it may be necessary to realize a network *complementary* to a given network; often a *dual* network or a *constant-resistance* network is required. These terms must be precisely defined.

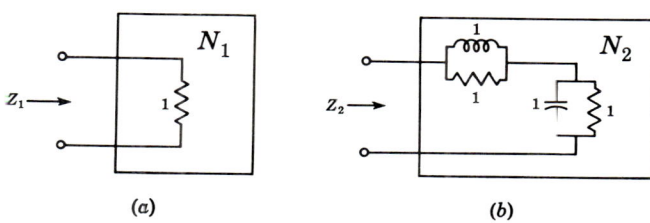

Fig. 25. Two equivalent dipoles (values in ohms, henrys, farads).

Two networks are said to be equivalent if the matrices of their corresponding system functions are equal.* Thus one dipole is equivalent to another dipole if their corresponding driving-point functions are equal, irrespective of the fact that the networks may differ greatly in their configurations and in the number of elements possessed by each. For example, the dipoles in Fig. 25a and b are equivalent; that is, $Z_1 = Z_2 = 1$. No electrical measurements at the port of one of these networks can detect whether it is N_1 or N_2.

For networks with more than one port, care must be exercised, since the term equivalent has been used in the literature in two senses. For example, the quadripoles N' and N'' are equivalent, in the sense defined above, if $[z'] = [z'']$, where the first open-circuit impedance matrix is that of N' and the second describes N''. However, in many problems with two-port networks, only a transfer function (like a transfer impedance or a voltage ratio) is specified; in this case, it is customary in the literature to say that two networks are equivalent if they have the specified transfer function. In accordance with the general definition of equivalence, however, it is clear that a symmetrical network (that is, $z_{11} = z_{22}$) and an unsymmetrical one can never be equivalent, whereas use of the restricted concept of equivalence permits such a designation.

A network whose input impedance is a constant *for all frequencies* is called a *constant-resistance network*. Such a network might be used, for example, to provide a specified filtering action and, in addition, to terminate a transmission line properly. The term *complementary* is related to constant-resistance networks. It is applied to driving-point functions or their corresponding networks. Two driving-point functions are said to be complementary if their sum is equal to a constant at all frequencies. For two impedances we have

$$Z_1(s) + Z_2(s) = R \qquad (20)$$

where R is a constant. The connection of the corresponding complementary networks is shown in Fig. 26a; here the *series* connection forms a constant-resistance network. The analogous relationship for admittances is

$$Y_1(s) + Y_2(s) = G \qquad (21)$$

* Note that it is the matrices that are *equal* and the networks that are *equivalent*. The reader should not make the error of requiring only that the matrices be equivalent.

and the networks corresponding to Y_1 and Y_2 are connected in *parallel* to form a constant-resistance network, as shown in Fig. 26b. (It is clear that if the conductance is constant, so is the resistance; the term constant-conductance network, however, is not used.)

A use for a complementary network may be indicated briefly at this point. Suppose that a quadripole is to be realized for a specified transfer characteristic; the question of how to provide for the source impedance may arise. If the source impedance is large compared with the input impedance of the network, as is usually true when a pentode tube is used as the current source, then the transfer impedance may be realized by a network that has an ideal source at the input; in other words, no specified terminating impedance need be provided at the input. Analogously, in the realization of a transfer admittance, when the source impedance is very small compared with the input impedance of the network, a situation often brought about by the use of a cathode follower as the input voltage source, then again a network may be designed that does not provide any input terminating resistance. However, practical cases arise in which the source impedance is resistive and not negligible. There are at least three ways of meeting this situation.

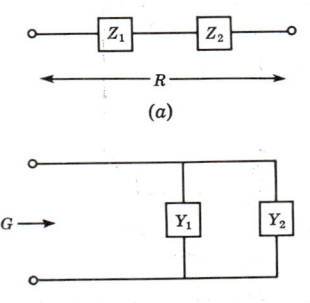

Fig. 26. Networks whose input impedance and admittance are constant.

First, for those cases in which a resistance termination is also needed at the output, the network may be designed for resistance terminations at both input and output. Secondly, though a load resistance may not be required, the network is designed to possess a resistance load. Then, by use of the reciprocity theorem, the network may be turned end for end, allowing what was the load resistance to provide for the source resistance. Finally, the network may be designed to have no terminating resistance at the input, but instead of this resistance the complement is connected to the designed network. By thus making the input impedance a constant, the resistance of the source will have no effect on the designed frequency characteristics but will introduce only a constant loss at all frequencies; in other words, the shapes of the magnitude and phase as functions of frequency are not altered. The constant loss is called a *flat loss*. To use this last method it is of course necessary that the complementary network be realizable.

Duals, or *dual networks*, are of great use in synthesis. Other terms often used to describe such networks are *reciprocal*,* or *inverse*. In the literature one finds that dual networks are defined in two different ways. The less severe definition states that an n-port network is the dual of a second n-port if the impedance matrix of the first is equal to the admittance matrix of the second, or stated in other terms, the impedance matrix of one network is the *inverse* of the impedance matrix of the other. To distinguish this definition from the second definition to be given later, it may be said to define a *restricted form of duality*, whereas the second defines *completely dual* networks.

The simplest example is that of a two-terminal network; the dual of a given dipole is one whose impedance is the *reciprocal* of that of the given network. If

$$Z_a Z_b = R^2 \tag{22}$$

then Z_b is said to be the reciprocal, or dual, of Z_a with respect to R^2.

As in the concept of equivalence, it should be observed that the internal structures of the networks are not relevant in the determination of duality by use of the restricted definition. All that is required is dual behavior at the *specified ports*. In accord with this definition, networks with mutual inductance have duals. In fact, the dual of *any* physical network is always realizable; that is, if a given matrix [b] is realizable

---

* We should distinguish this use of the term *reciprocal* from that used to specify a network obeying the reciprocity theorem; no confusion should arise in our usage since the context makes clear which meaning is intended.

as an open-circuit impedance matrix $[z]$, it is also realizable as a short-circuit admittance matrix $[y]$.

It is assumed that the reader is well acquainted with dual concepts and elements:[14,15] a voltage is the dual variable of a current, an inductance of L henrys is dual to a capacitance of L farads, a short circuit is the dual of an open circuit, a series connection is the dual of a parallel connection, and so on. Simple examples of duality for two-terminal networks are shown in the pairs of figures given by the primed and unprimed letters in Fig. 27. If the numerical equalities $L' = C$, $C' = L$, and $G' = R$ are satisfied for each pair, the networks in Fig. 27a and a', b and b', etc., are duals; that is, the impedance of one is equal to the admittance of the other.

The second definition of duality is much stricter.[15] Two networks are said to be duals if they possess dual characteristics with respect to *all possible terminal pairs that can be created*. Thus they are not only electrically dual but are also physically dual. In order for a network to be the physical dual of a given network, it must possess a *dual element* corresponding to *each* element of the given network. Thus the dual networks must have *equal numbers of elements*. Since no dual element exists for mutual inductance, a network with transformers possesses no dual. Every mesh (node) of one network becomes a node (mesh) of the dual; it is clear that an element common to two meshes becomes in the dual network the dual element common to the two corresponding nodes.

Authors restricting the term *dual* only to those networks that are *completely dual* suggest that *reciprocal* and *inverse* be applied to networks that have dual behavior only at specified terminal pairs. However, from the point of view of synthesis—that is, the realization of networks with a given behavior at specified terminal pairs—it seems preferable to use the first definition; unless the contrary is stated, we do this implicitly. Thus the terms *dual*, *reciprocal*, and *inverse* will be used interchangeably.

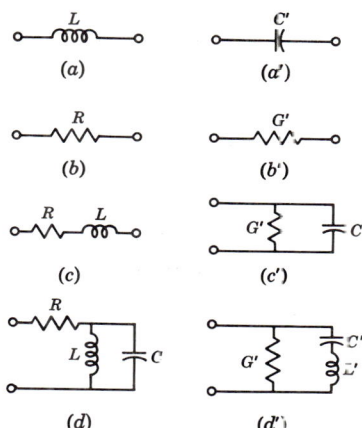

Fig. 27. Examples of two-terminal dual networks

Completely dual networks do not exist even for some networks without mutual inductance. Only *planar* networks have complete duals, where a planar network is defined as one that can be drawn on a plane without any crossed branches. Though in many texts *the* dual of a planar network is spoken of, this is actually incorrect; we should say *a* dual. Unless the planar graph has additional restrictions, it has two or more duals which are topologically different even though they are electrically equivalent.

A method for constructing a complete dual for any planar network (containing no mutual inductance) is quite important for simplifying the synthesis problem. Having realized one network, we have, in effect, also realized its dual, which may be drawn almost by inspection. To cite two practical examples, suppose that after we have realized a given Z_{21} by a planar network, we discover that it is more convenient to use a voltage source like a cathode follower rather than the original pentode current source. A new network does not have to be designed; all we need do is construct the dual network. The transfer admittance Y_{21} for this network is the same function as the original Z_{21}. Secondly, many satisfactory transistor circuits may be designed as duals of existing vacuum-tube circuits, where the three-terminal transistor is substituted as the dual of the three-terminal tube.

Facetiously, we may say that the synthesist is lucky that the technique for drawing the dual exists. Life is short, and yet there is an infinity of planar networks. However, because of the method outlined below, the synthesist need contemplate the realization of only half of the possible networks, the others being duals.

The geometric method for finding the dual of any planar network consists of the following steps:

1. Place a dot within each mesh of the network and one dot outside the network (the outer dot corresponding to the loop formed by the periphery of the network). These dots are the nodes of the dual network.
2. Between any two dots draw a dashed line through each element (or source) common to the two meshes enclosing these dots. Each of these dashed lines represents a branch of the dual network.

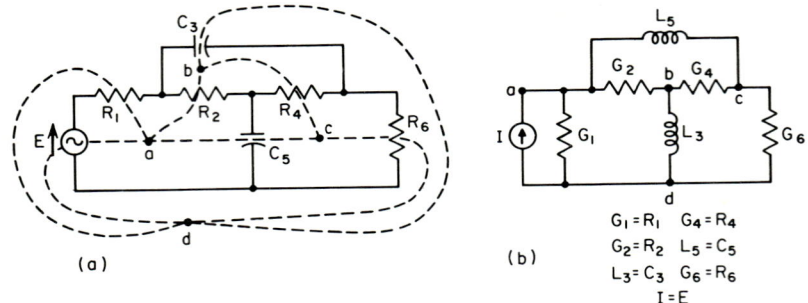

Fig. 28. Construction of the dual of a planar network.

3. The element to be placed in each branch is the dual of the corresponding element in the original network.
4. Draw the dual network with its element values, using the information obtained from the above construction.

The procedure is illustrated by the networks in Fig. 28, where the dual of the voltage-driven RC network in Fig. 28a is found to be the RL network in Fig. 28b.

REFERENCES

1. B. D. H. Tellegen, La Recherche pour une série complète d'éléments de circuit idéaux non-linéaires, *Rend. Seminario Mat. Fis. Milano*, vol. 25, pp. 134–144, 1953–1954.
2. S. Duinker, Conjunctors, Another New Class of Non-energic Non-linear Network Elements, *Philips Res. Rept.*, R 443, vol. 17, no. 1, pp. 1–19, February, 1962.
3. S. Duinker, Traditors, A New Class of Non-energic Non-linear Network Elements, *Philips Res. Rept.*, R 360, vol. 14, pp. 29–51, February, 1959.
4. W. A. Lynch and J. G. Truxal, "Principles of Electronic Instrumentation," McGraw-Hill Book Company, New York, 1962.
5. S. J. Mason and H. J. Zimmerman, "Electronic Circuits, Signals, and Systems," pp. 20–38, John Wiley & Sons, Inc., New York, 1960.
6. J. T. Bangert, The Transistor as a Network Element, *Bell System Tech. J.*, vol. 33, pp. 329–352, March, 1954.
7. H. J. Zimmerman and S. J. Mason, "Electronic Circuit Theory," John Wiley & Sons, Inc., New York, 1959.
8. H. W. Bode, "Network Analysis and Feedback Amplifier Design," D. Van Nostrand Company, Inc., Princeton, N.J., 1945.
9. H. J. Carlin and D. C. Youla, Network Synthesis with Negative Resistors, *Symposium on Active Networks and Feedback Systems*, Polytechnic Institute of Brooklyn, Microwave Research Institute, Apr. 19, 1960.
10. G. L. Matthaei, Some Simplifications for Analysis of Linear Circuits, *IRE Trans. PGCT*, vol. CT-4, no. 3, pp. 120–124, September, 1957.
11. B. D. H. Tellegen, Network Synthesis, Especially the Synthesis of Resistanceless Four-terminal Networks, *Philips Res. Rept.*, vol. 1, pp. 169–184, April, 1946.
12. L. Weinberg, Reciprocal and Non-reciprocal Systems, Characteristic Polynomials, and Driving-point Functions, *IEEE Trans. CT*, vol. CT-12, no. 2, June, 1965.
13. L. DePian, "Linear Active Network Theory," Prentice-Hall, Inc., Englewood Cliffs, N.J., 1962.
14. M. F. Gardner and J. L. Barnes, "Transients in Linear Systems," vol. I, John Wiley & Sons, Inc., New York, 1942.
15. E. A. Guillemin, "Introductory Circuit Theory," John Wiley & Sons, Inc., New York, 1953.

Chapter 2

NETWORK THEOREMS, SYSTEM FUNCTIONS, AND EQUIVALENT CIRCUITS

LOUIS WEINBERG[*]

CONTENTS

1. Introduction.. 2-1
2. Network Theorems... 2-2
   2.1. Principle of Superposition................................. 2-2
   2.2. Compensation Theorem................................... 2-3
   2.3. Thévenin's and Norton's Theorems........................ 2-3
   2.4. Reciprocity Theorem...................................... 2-5
3. Parts of a System Function.................................... 2-8
   3.1. Time Delay of a Network................................. 2-11
   3.2. Magnitude Function Specified............................. 2-13
   3.3. Real (or Imaginary) Part Specified........................ 2-14
4. Equivalent Circuits.. 2-17
   4.1. Vacuum-tube Amplifier as Two-port Network.............. 2-17
   4.2. Multielectrode Tube...................................... 2-20
   4.3. Transistor Equivalent Circuits............................ 2-22
   4.4. Common-base Configuration.............................. 2-26
   4.5. Common-emitter Configuration........................... 2-30
   4.6. Common-collector Configuration.......................... 2-30
   4.7. High-frequency Equivalent Circuits for Transistors........ 2-31

1. INTRODUCTION

In this chapter we consider some additional aspects of network theory that are useful in the analysis and design of amplifiers.

Most parts of the network problem are straightforward; the mathematical tools for treating linear systems are readily available. Grinding through the mathematical machinery in all its generality, however, may be a formidable task to carry out by hand and, for some large-scale problems, may even overload digital computers. It is fortunate that we don't often need a complete solution to a large network problem; partial analytical techniques can then be used. For this purpose network theorems are often indispensable.

[*] Department of Electrical Engineering, University of Michigan, Ann Arbor, Mich.

AMPLIFIER FUNDAMENTALS

In the design of an amplifier system we usually require the realization of only a part of a system function, or parts of many system functions; for example, a specified phase characteristic may be required. Before the design can proceed, however, it is necessary that the complete system function be determined from the specified phase. It is necessary to be familiar with the parts of a system function, and to know their interrelations and how to determine one part from another.

An amplifier that is to be designed has parameters that must be manipulated to achieve the desired properties. When the amplifier is a multistage one, the parameters of one stage generally affect those of another stage. For manipulating the parameters and determining and predicting their effects on other parameters, it is convenient to represent the active devices by models or equivalent circuits. These models are made by using the elements discussed in Chap. 1 in a configuration so that the external behavior of a multiport active device is achieved.

We discuss these three topics below, namely, network theorems, parts of a system function and their relations, and equivalent circuits for active systems.

2. NETWORK THEOREMS

For simplifying the analysis and design of amplifiers a number of theorems are available. Here we discuss briefly a few of the most important of these theorems; proofs may be found in most of the standard texts on network theory.

2.1. Principle of Superposition. The linearity property of the networks we are considering permits the very powerful principle of *superposition* to be applied. Basically, the principle restates the additive property of systems of linear equations that describe the networks; for such a system the composite solution when several driving functions are present simultaneously is obtained by adding the separate solutions for each driving function applied alone. In other words, the response of the system to each input is independent of the presence of other inputs. A major difficulty inherent in nonlinear systems arises because superposition does not hold for them. Thus if $f_o'(t)$ is the response when $f_i'(t)$ is applied, and $f_o''(t)$ is the response when $f_i''(t)$ is applied, then the response when both $f_i'(t)$ and $f_i''(t)$ are applied at the same time is given by $f_o'(t) + f_o''(t)$. Furthermore, if $bf_i'(t)$ and $cf_i''(t)$ are present, where b and c are linear operators, the response is $bf_o'(t) + cf_o''(t)$. Thus, if we wish to find a specified current when a voltage of 110 volts is applied at a port, it may be more convenient to find the current when 1 volt is applied and then multiply the current by 110. Similarly, the response of a network to an impulse function can be obtained from a known response $f_o(t)$ to a linear ramp function by differentiating $f_o(t)$ twice.

In physical systems a driving function is called a source. When removing a source, i.e., setting it equal to zero, we must take account of the impedance characteristic of the source; thus proper removal requires that a voltage source be replaced by a short circuit and a current source be replaced by an open circuit.

The convolution integral serves to state superposition in a general form; namely, the output $f_o(t)$ for any input $f_i(t)$ is obtained by convolving the impulse response with the input. Mathematically, we have

$$f_o(t) = \int_{-\infty}^{t} f_i(\tau) h(t,\tau)\, d\tau \qquad (1)$$

where $h(t,\tau)$ is the response to a unit impulse $\delta(t - \tau)$ which is applied at time τ. For any physical system a response does not occur until the impulse is applied so that $h = 0$ for $t < \tau$. If the system is a time-invariant one, $h(t,\tau) = h(t - \tau)$ and a time shift of the input causes merely a corresponding shift of the output.

We can obtain the equivalent of Eq. (1) in the frequency domain by the use of Laplace transforms. Transforming Eq. (1) yields for time-invariant systems

$$F_o(s) = F_i(s) H(s) \qquad (2)$$

where F_i and F_o are the Laplace transforms of f_i and f_o, respectively, and H, the transform of the impulse response h, is the system function.

2.2. Compensation Theorem.

In discussing dependent sources we mentioned that two-terminal elements may be replaced by such sources. This replacement of elements by sources or the replacement of sources by the corresponding elements is called the *compensation* (or *substitution*) theorem. The proof that this substitution is valid of course follows simply from the fact that the volt-ampere terminal relations have been kept invariant. Thus we can replace the current-controlled voltage source in Fig. 1a and the voltage-controlled current source in Fig. 1b by an element A whose volt-ampere relationship is $E = AI$. Correspondingly, we can replace such a two-terminal element A by either of the controlled sources in Fig. 1.

As an example of the use of the compensation theorem, consider the network containing the controlled source shown in Fig. 2a, which the reader will probably recognize as the equivalent circuit for a cathode-follower vacuum-tube amplifier. We can by making use of the relationship

$$E_g = E_1 - E_2 \qquad (3)$$

obtain the circuit with two controlled sources in Fig. 2b; now use of the compensation theorem yields the network in Fig. 2c with only one controlled source and an additional resistance of value g_m mhos.

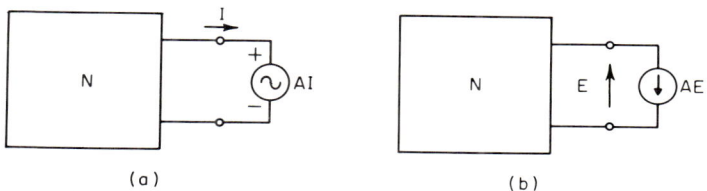

(a) (b)

Fig. 1. Networks with a controlled source for which a two-terminal element may be substituted.

2.3. Thévenin's and Norton's Theorems.

In dealing with a large and complex network we are most often concerned with determining not the currents and voltages in all parts of the network but only the current and voltage at a specified pair of terminals. For example, we may be interested in determining the maximum power that an amplifier can deliver to a variable load Z_L. It is then convenient to replace the network considered from this port, with its multiplicity of elements and sources, by a single equivalent source and a single equivalent impedance. This equivalence for an arbitrary network is shown in Fig. 3, where the node pair a-b, containing the impedance Z_L, is shown explicitly in Fig. 3a; in Fig. 3b the port a-b has been placed external to the network, that is, it has been made an accessible port. We are now interested in replacing the network to the left of a-b by an equivalent network for purposes of calculating the voltage and current at port a-b; the equivalent networks are shown in Fig. 3c and d. From the figure it is seen that one equivalent network takes the form of a voltage source in series with an impedance, and that the second one consists of a current source in parallel with an admittance. The impedance in shunt is the same as the impedance in series, that is, $Y_{eq} = 1/Z_{eq}$. It is emphasized that the equivalence holds only for determining the variables at port a-b and not those in the rest of the network; for example, the power dissipated in the network exclusive of Z_L is not in general equal to the power dissipated in Z_{eq}.

Thévenin's theorem gives the series form of network shown in Fig. 3c, and may be stated as follows. Let N be any network with a pair of terminals a-b; we assume that none of the elements of N is magnetically coupled to any of the elements of the subnetwork represented by Z_L, the impedance load at port a-b. Since we are interested only in the voltage and current at port a-b, it is the only accessible port and we can consider N a two-terminal network; N may contain sources. Let N' represent N with all its *independent* sources removed, that is, set equal to zero; the impedance of N' at a-b we designate as Z_1. Let E_{oc} be the voltage across port a-b when it is open-circuited.

AMPLIFIER FUNDAMENTALS

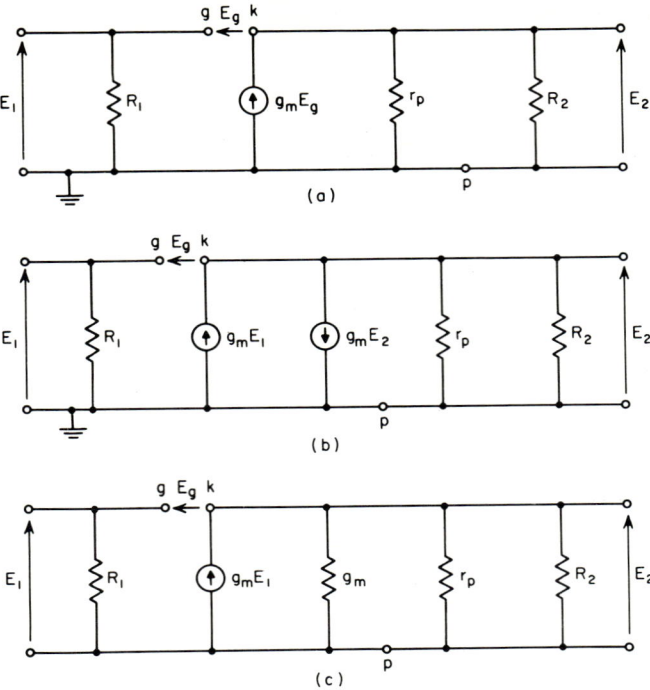

Fig. 2. Illustration of the use of the compensation theorem for removing a controlled source.

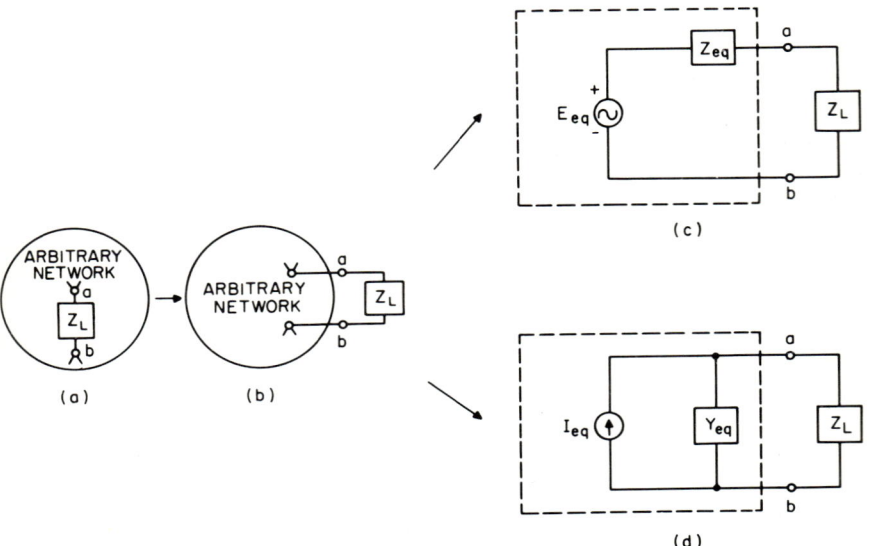

Fig. 3. Representation of a network at a pair of terminals by its Thévenin and Norton equivalents.

Then for the representation of the external behavior of N at port a-b, we can replace it by its equivalent circuit consisting of a voltage source E_{eq}, which is equal to E_{oc}, in series with an impedance Z_{eq}, which is equal to Z_1.

Norton's theorem, which yields the parallel form of network shown in Fig. 3d, may be formulated by use of the dual concepts. Again let N be any two-terminal network with internal sources, N' represent N when all independent sources have been set to zero, and $Y_1 = 1/Z_1$ be the admittance of N' at port a-b. Let I_{sc} be the current in a short circuit connected to port a-b. Then to represent the external behavior of N at port a-b, we can replace it by its equivalent circuit of a current source I_{eq}, which is equal to I_{sc}, in parallel with the admittance Y_{eq}, which is equal to Y_1.

The equivalence in each case can hold for the transient state and the steady state, provided that we take proper account of initial conditions in the network; for example, we can replace each initial condition by an appropriate voltage or current source.

Two quantities are important for defining the parameters in the Thévenin and Norton equivalent circuits: the open-circuit voltage E_{oc} at port a-b, and the short-circuit current I_{sc} that flows in a short circuit across a-b. In the Thévenin equivalent circuit the source E_{eq} is set equal to E_{oc}; from Fig. 3c we then see that

$$Z_{eq} = E_{oc}/I_{sc} \qquad (4)$$

Equation (4) provides an operational definition for determining Z_{eq}; thus we must calculate E_{oc} and I_{sc} at port a-b. An alternative method for determining Z_{eq}, as we see from the statement of the theorems, is to calculate the impedance seen looking back (i.e., to the left) into the network, when we have removed all independent sources and initial-condition sources from the network. Thus operationally we apply a voltage (or current) source at a-b and determine the current flowing into the network (or voltage across a-b) due to the applied source.

The reader is cautioned that controlled (or dependent) sources are not removed in determining Z_{eq} by the second method. In general, removal of dependent sources will lead to an incorrect Z_{eq}. To convince ourselves that errors will be introduced, we recall that the compensation theorem permits the representation of positive and negative elements by dependent sources; thus removing these sources will change the impedance characteristic of the network. In feedback networks part of the controlled source does in fact represent such two-terminal elements. The method that determines Z_{eq} from E_{oc} and I_{sc} leads to no corresponding difficulty, since no sources are removed in applying this method.

In the Norton equivalent circuit the current source I_{eq} is equal to the short-circuit current I_{sc}; the impedance in shunt is calculated in the same way as the Z_{eq} for the Thévenin circuit. Thus, having once obtained the Thévenin circuit, we can derive the Norton circuit from it by calculating

$$I_{sc} = E_{eq}/Z_{eq} \qquad (5)$$

Conversely, we can obtain E_{oc} from the Norton circuit for use in the Thévenin circuit. Conversion of a voltage source with a series impedance to the equivalent current source with a shunt impedance, or vice versa, is referred to as a *source transformation*. Source transformations when the impedance is a single element and the source is represented as a function of time are shown in Fig. 4. A single illustration when the source current is represented by I, the Laplace transform of the instantaneous source current $i(t)$, is shown in Fig. 5; here it is assumed that the initial current in the inductance L is zero.

Clearly the Thévenin and Norton circuits are not equivalent for effects in the rest of the network; for example, if port a-b is open-circuited, no power is furnished by E_{eq}, whereas I_{eq} dissipates power in Z_{eq} on open circuit. To demonstrate by a counter-example that the Thévenin circuit correctly represents only the conditions at a-b and not those in the rest of the network, the reader should calculate the power dissipated in the network of Fig. 6a and in its Thévenin equivalent in Fig. 6b and observe that they are not equal.

2.4. Reciprocity Theorem. Previously in Eqs. (10) and (12) of Chap. 1 we made use of the $[z]$ and $[y]$ matrices to define the behavior of a quadripole. There

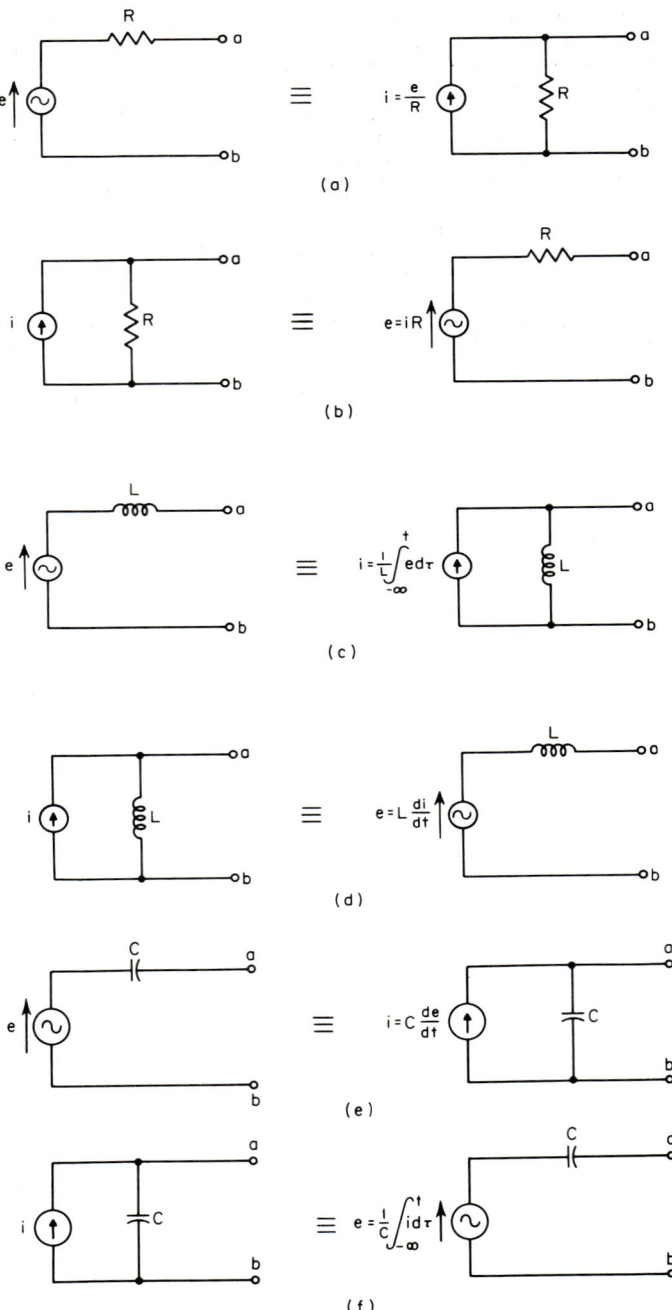

Fig. 4. Source transformations.

are four elements in each of these matrices, each of which is independent. There are networks, however, for which only three of these functions are independent; more precisely, $z_{12} = z_{21}$ and $y_{12} = y_{21}$. Such networks are said to obey the *reciprocity theorem*, and are called *reciprocal* networks. Included in such classes of networks are those containing passive elements excluding the gyrator. The proof that is generally given is made to depend on the symmetry of the loop or node equations that can be used to describe any passive network that does not contain gyrators. However, there is an elegant proof due to Tellegen that shows simply that the reciprocity theorem also applies to active networks that contain negative resistances in addition to the passive elements.[1] Thus it is not true, though it is often stated in texts, that

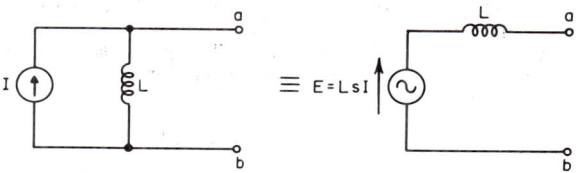

Fig. 5. Source transformation with a source represented by its Laplace transform.

z_{12} is necessarily unequal to z_{21} for active networks; active networks containing negative resistances and passive elements but no gyrators always obey the reciprocity theorem. In fact, as we shall indicate at the end of this section, the reciprocity relation applies to even more general active networks.

The reciprocity theorem can be applied not only at the ports of a network but also at any two pairs of terminals within a network. To obtain a transfer function that has the dimension of impedance, we apply a current source to a specified pair of terminals in a network and measure a voltage response (with an ideal voltmeter) at another specified pair of terminals. The dual quantity, a transfer admittance, requires that a voltage source be applied in series with a network branch and that a current flowing in another branch be the response. The reciprocity theorem states that any transfer function with the dimension of impedance or admittance remains unchanged

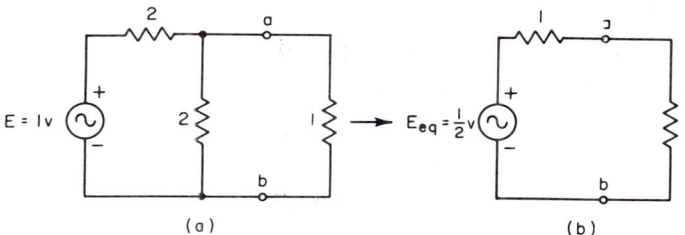

Fig. 6. Network and its Thévenin equivalent.

if the points of excitation and response are interchanged. For the transfer impedance we may consequently interchange the current source and the voltmeter and obtain a new transfer impedance which is exactly the same as the original one; for the transfer admittance the voltage source and the ammeter are interchanged to yield the same transfer admittance.

Thus, if the branches are brought out as the ports of the network, the quantities $z_{12} = z_{21}$ and $y_{12} = y_{21}$, and in general, for the matrix of an n-port network, $z_{jk} = z_{kj}$ and $y_{jk} = y_{kj}$. Of the n^2 elements in such a matrix only $n(n + 1)/2$ are independent, and these independent elements completely specify the behavior of the reciprocal network at its accessible terminal pairs.

It is important for the reader to note that the reciprocity theorem, if it is correctly interpreted, also applies to a dimensionless transfer function. When the interchange of source and measuring instrument is carried out for an admittance or impedance,

2-8   AMPLIFIER FUNDAMENTALS

the network is not changed because the impedance of a voltage source and an ammeter is zero and the impedance of a current source and a voltmeter is infinite. However, if we tried to apply the reciprocity theorem to a voltage ratio, we would be interchanging a voltmeter of infinite impedance with a voltage source of zero impedance; this leads in general to a change in the network and consequently to a different transfer function. To keep the network unchanged, we should apply a current source and measure a current response in a short circuit. Thus we have as a statement of reciprocity for a two-port with the usual definitions of positive directions of voltages and currents

$$E_2/E_1 = -I_1'/I_2' \tag{6}$$

where in the second set of measurements I_2' is the source applied to port 2 and $-I_1'$ is the short-circuit current response at port 1.

Now to conclude the section we briefly indicate Tellegen's result; in fact we explicitly extend it. Tellegen shows that the reciprocity relation for a quadripole is given in its general form by the equation

$$I_1 E_1' + I_2 E_2' = I_1' E_1 + I_2' E_2 \tag{7}$$

where the unprimed and primed variables refer to two different states of the quadripole, and the subscripts refer to ports 1 and 2 of the quadripole; here by a state we mean that the generators have one set of values in the unprimed case and another arbitrary set in the primed case. Inside the quadripole any types of linear reciprocal two-terminal elements* exist, that is, each element has a driving-point impedance that is invariant when the element is turned end for end. Thus the elements may be active—e.g., negative resistances, negative capacitances, negative inductances—and may be distributed-parameter or lumped-parameter; a branch may be nonrealizable by a physical network as long as the same impedance Z is defined for the branch in both its orientations.

Now four special cases of Eq. (7) occur:

1. With $I_2 = 0$ and $I_1' = 0$ we obtain

$$E_2/I_1 = E_1'/I_2' \tag{8}$$

2. With $E_2 = 0$ and $E_1' = 0$ we obtain

$$I_2/E_1 = I_1'/E_2' \tag{9}$$

3. With $I_2 = 0$ and $E_1' = 0$ we obtain

$$E_2/E_1 = -I_1'/I_2' \tag{10}$$

4. With $E_2 = 0$ and $I_1' = 0$ we obtain

$$-I_2/I_1 = E_1'/E_2' \tag{11}$$

Thus we have derived the four forms in which the reciprocity relation may be given, but we have also stressed that it applies to a much larger class of networks than is generally postulated.

3. PARTS OF A SYSTEM FUNCTION

Since a system function is a complex function, it has a *real part*, an *imaginary part*, a *magnitude*, and a *phase;* these may be called the four *parts* of a system function. If any one of the four parts of a system function is given over the whole range of real frequencies, the other parts can then be determined. This follows from the fact that a system function is an analytic function of the complex frequency with no poles in the right half plane. To calculate the unknown parts from a given part it is necessary

---

* However, the elements must not be degenerate or singular like the nullator (the linear time-invariant one-port with $E = I = 0$). (See H. J. Carlin, Singular Network Elements, *IEEE Trans. Circuit Theory*, vol. CT-11, no. 1, pp. 67–72, March, 1964.)

NETWORK THEOREMS AND SYSTEM FUNCTIONS

to be able to derive and manipulate the parts of a given rational function in simple ways. We give such derivations here.

In many of our calculations we examine network functions at real frequencies only, even though the functions are defined over the entire complex plane. It is cumbersome, however, to carry $j\omega$ and powers of $j\omega$ throughout a calculation; it is preferable to use the variable s until the stage in the computations at which it becomes convenient to make the substitution $s = j\omega$. For this purpose the even and odd parts of a function are often useful. Any function may be uniquely represented as the sum of its even and odd parts; thus for $h(s)$,

$$h(s) = h_e(s) + h_o(s) \tag{12}$$

where the subscripts e and o denote, respectively, the even and odd parts. These parts by definition satisfy the relations

$$\begin{aligned} h_e(s) &= h_e(-s) \\ h_o(s) &= -h_o(-s) \end{aligned} \tag{13}$$

Therefore
$$h(-s) = h_e(s) - h_o(s) \tag{14}$$

Adding the respective sides of (12) and (14) and solving yield the definition of h_e as

$$h_e(s) = \tfrac{1}{2}[h(s) + h(-s)] \tag{15}$$

whereas $h_o(s)$ is found by subtraction as

$$h_o(s) = \tfrac{1}{2}[h(s) - h(-s)] \tag{16}$$

Since a network function has real coefficients, it should be clear that its even part is purely real on the imaginary axis of the s plane and its odd part is purely imaginary. Therefore, if we now assume that $h(s)$ has real coefficients, we obtain from (15) and (16)

$$\begin{aligned} \mathrm{Re}\,[h(j\omega)] &\equiv h_e(j\omega) = \tfrac{1}{2}[h(s) + h(-s)]_{s=j\omega} \\ \mathrm{Im}\,[h(j\omega)] &\equiv (1/j)h_o(j\omega) = (1/2j)[h(s) - h(-s)]_{s=j\omega} \end{aligned} \tag{17}$$

Equations (17) of course merely express the well-known fact that twice the real part of a function is obtained if the conjugate of a given function is added to the function, and j times twice the imaginary part results when the conjugate is subtracted from the function.

Now consider any type of system function, which we designate as $A(s)$, in order to allow it to represent both driving-point and transfer functions. A representation of $A(s)$ that is frequently used is

$$\begin{aligned} A(s) &= \frac{p(s)}{q(s)} \\ &= \frac{M_1 + N_1}{M_2 + N_2} \end{aligned} \tag{18}$$

where M_1 and N_1 are, respectively, the even and odd parts of the numerator, and M_2 and N_2 play the same roles for the denominator. As shown above, at real frequencies the even parts M_1 and M_2 become the real parts of the numerator and denominator, respectively, while N_1/j and N_2/j represent the imaginary parts. Furthermore, the complex conjugate function $\bar{A}(s)$ is given by $A(\bar{s})$, since $A(s)$ has real coefficients; and for $s = j\omega$, $A(\bar{s})$ is given simply by $A(-s)$.

Rationalizing the denominator of A, we obtain the alternative forms

$$\begin{aligned} A(s)\bigg|_{s=j\omega} &= \frac{p(s)q(-s)}{q(s)q(-s)}\bigg|_{s=j\omega} \\ &= \frac{M_1 + N_1}{M_2 + N_2}\frac{M_2 - N_2}{M_2 - N_2}\bigg|_{s=j\omega} \\ &= \frac{M_1 M_2 - N_1 N_2 + N_1 M_2 - M_1 N_2}{M_2{}^2 - N_2{}^2}\bigg|_{s=j\omega} \end{aligned} \tag{19}$$

2-10   AMPLIFIER FUNDAMENTALS

The denominator, $M_2^2 - N_2^2$, since it is the square of an absolute value, is always nonnegative; it is also clear that it is an even function of ω. The real and imaginary parts of the above expression are given by

$$\text{Re}\,[A(j\omega)] = \left.\frac{M_1M_2 - N_1N_2}{M_2^2 - N_2^2}\right|_{s=j\omega}$$
$$\text{Im}\,[A(j\omega)] = \left.\frac{1}{j}\frac{N_1M_2 - M_1N_2}{M_2^2 - N_2^2}\right|_{s=j\omega} \qquad (20)$$

from which the important characteristic is again noted that the *real* part of a system function is always an *even* function of ω, whereas the *imaginary* part is an *odd* function of ω. Instead of Re $[A(j\omega)]$ and Im $[A(j\omega)]$, we can thus work with $A_e(s)$ and $A_o(s)$ whenever this is more convenient.

The system function is often represented in exponential form as

$$A(j\omega) = e^{\gamma(\omega)} \qquad (21)$$

where for transfer functions the complex function γ is referred to as the *propagation function*. The function γ may be separated into its real and imaginary parts,*

$$\gamma(\omega) = \alpha(\omega) + j\beta(\omega) \qquad (22)$$

where α is the *gain function* in nepers and β is the *phase* in radians. (Other terms used for β are *phase function, phase angle,* and *phase shift.*) Taking the natural logarithm of $A(j\omega)$, we obtain the relationships

$$\begin{aligned}\ln A(j\omega) &= \ln\,[|A(j\omega)|e^{j\beta}]\\ &= \ln|A(j\omega)| + j\beta\\ &= \alpha + j\beta\end{aligned} \qquad (23)$$

In words, the real and imaginary parts of the logarithm of a system function are given, respectively, by the gain and phase functions. Thus one may manipulate the magnitude and phase of a transfer function, or equivalently, the real and imaginary parts of the propagation function.

In order to eliminate the complication of taking a square root, it is convenient to work with the squared magnitude of a function. This is obtained by multiplying $A(s)$ by its complex conjugate; thus at real frequencies,

$$\begin{aligned}|A(j\omega)|^2 &= \left.A(s)A(-s)\right|_{s=j\omega}\\ &= \left.\frac{p(s)p(-s)}{q(s)q(-s)}\right|_{s=j\omega}\\ &= \left.\frac{M_1^2 - N_1^2}{M_2^2 - N_2^2}\right|_{s=j\omega}\\ &= \frac{C(\omega^2)}{D(\omega^2)}\end{aligned} \qquad (24)$$

the last equation of (24) being written to emphasize the fact that the magnitude function is always an even function.

The phase angle, obtained in the conventional way from the representation of the real and imaginary parts of Eq. (20), is given by

$$\beta = \tan^{-1}\frac{\text{Im}\,[A(j\omega)]}{\text{Re}\,[A(j\omega)]} \qquad (25)$$

so that
$$j\tan\beta = \left.\frac{N_1M_2 - M_1N_2}{M_1M_2 - N_1N_2}\right|_{s=j\omega} \qquad (26)$$

* Some authors adopt the convention of defining $\gamma = -(\alpha + j\beta)$, so that α is the *attenuation* or *loss* function and β is the phase *lag*.

An alternative method for finding the phase is to use a method analogous to multiplying $A(s)$ by $A(-s)$ for finding the magnitude; we merely divide $A(s)$ by $A(-s)$. This quotient, evaluated for $s = j\omega$, has a phase angle equal to 2β. Since the phases of the numerator and denominator are negatives of each other, we have

$$\beta = \measuredangle \text{ numerator of } \frac{A(s)}{A(-s)}\bigg|_{s=j\omega}$$
$$= \measuredangle(M_1M_2 - N_1N_2 + N_1M_2 - M_1N_2)\bigg|_{s=j\omega} \tag{27}$$

The reader should note from inspection of (25) that changing the level of $A(s)$ will not affect β; the constant multiplier will merely cancel from the real and imaginary parts. Thus, if a system function is to be determined from a specified phase, it will *not be unique* but will be arbitrary within a constant multiplier. It is apparent, also, that the numerator of β in (25) is an odd function of ω, whereas the denominator is even, so that β is an odd function of frequency. Thus, in summary, the real part and the magnitude are even functions, whereas the imaginary part and the phase are odd functions.

Since the phase function β is a transcendental function, it is desirable wherever possible in phase calculations to work with a related function, namely, the rational function given by the right side of Eq. (26). It is again convenient to work with a function of s rather than $j\omega$, and we define

$$\mathbf{T}(s) \equiv \frac{[p(s)q(-s)]_o}{[p(s)q(-s)]_e}$$
$$\equiv \frac{N_1M_2 - M_1N_2}{M_1M_2 - N_1N_2} \tag{28}$$

This odd function defined over the entire s plane becomes on the j axis

$$\mathbf{T}(j\omega) \equiv j\tan\beta \tag{29}$$

3.1. Time Delay of a Network. An important function that is related to the phase function is the *time delay*. There are two common methods in use for defining this characteristic of a network function: the first is the *phase delay* given by $-\beta/\omega$, and the second is the *group delay* (or *envelope delay*), defined by the derivative $-d\beta/d\omega$. We shall work almost exclusively with the time delay given by the latter definition.

Though the phase function is a transcendental function, its derivative is a rational function; for example, if

$$\beta = -\tan^{-1}\frac{\omega}{3-\omega^2} \tag{30}$$

then
$$T_d \equiv -\frac{d\beta}{d\omega} = \frac{3+\omega^2}{9-5\omega^2+\omega^4} \tag{31}$$

Not only is T_d rational, but it is always an *even* function of ω with no pole at the origin. Thus it may be expanded in a Maclaurin series. This is simply done by long division of the rational fraction in which the terms are written in ascending powers of ω, as they are in (31). For this example, the Maclaurin series is

$$T_d = \frac{1}{3} + \frac{8}{27}\omega^2 + \frac{31}{243}\omega^4 + \frac{83}{2,187}\omega^6 + \cdots$$
$$= \tfrac{1}{3}(1 + \tfrac{8}{9}\omega^2 + \tfrac{31}{81}\omega^4 + \tfrac{83}{729}\omega^6 + \cdots) \tag{32}$$

and in general a time delay is given by

$$T_d = t_0(1 + a_1\omega^2 + a_2\omega^4 + a_3\omega^6 + a_4\omega^8 + \cdots) \tag{33}$$

where t_0 represents the zero-frequency delay. As is to be expected, since T_d is an even function, the odd-order derivatives are always zero at the origin. The Maclaurin series is a convenient representation of the time delay since it makes evident the deviation from linearity of the phase function in the neighborhood of the origin. As the phase function becomes more linear, correspondingly the time delay more closely approaches its constant zero-frequency value; that is, a_1 and perhaps higher-order even derivatives will become zero.

Determination of a System Function from a Given Part. In many synthesis applications only one of the parts of a system function may be prescribed; for example, the desired magnitude characteristic of a transfer function may be given as a rational function (or this rational function may be obtained by approximating a given curve or a given transcendental function). It has already been mentioned that the parts of a system function are related and consequently cannot be specified independently.

Before the actual synthesis can proceed, it is generally necessary that the complete system function be found. That this is always possible becomes clear when it is realized that the problem is identical with one that arises in potential theory. Since the real and imaginary parts satisfy the Cauchy-Riemann differential equations at *all* points in the right half plane, they also satisfy Laplace's equation in two dimensions in this region of analyticity. Consequently they may be looked upon as potential functions and (in the sense of potential theory) are called *conjugate functions*.

One of the classical methods for determining a function that is conjugate to a given function is by use of the *Hilbert transform*.[2,3] This method is general and is useful when most of the other methods do not apply, e.g., when the given function is not specified analytically but is given graphically or, if specified analytically, is not a rational function. Consequently, we discuss the Hilbert transform later; the discussion is brief, and for a fuller treatment the reader is referred to the cited references.

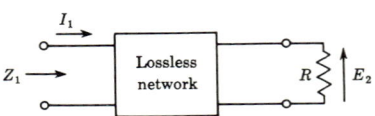

FIG. 7. Lossless network terminated in a resistance.

If one part of a *transfer* function is specified, not only are its other parts determinable, but it may also be true for certain types of networks that the specification is equivalent to specifying a part of an associated *driving-point* function. For example, a relationship between the magnitude of a transfer impedance and the real part of a driving-point impedance for a lossless network terminated in a resistance may be derived by consideration of the network shown in Fig. 7; this relationship finds much application in synthesis.

Since the coupling network is lossless, all the average power input is dissipated in the load resistance. For real frequencies this is expressed by

$$|I_1|^2 \, \text{Re} \, [Z_1(j\omega)] = |E_2|^2/R \qquad (34)$$

or

$$R \, \text{Re} \, [Z_1(j\omega)] = |E_2/I_1|^2 \qquad (35)$$

If R is normalized to 1 ohm, the above becomes

$$|Z_{21}(j\omega)|^2 = \text{Re} \, [Z_1(j\omega)] \qquad (36)$$

In words, (36) states that at real frequencies the squared magnitude of the transfer impedance of a lossless network terminated in 1 ohm is equal to the real part of the input impedance. An analogous relationship holds for admittances. Thus for such a network, even though the magnitude of the transfer function is specified, one may work with the real part of the driving-point function; and in the actual synthesis of the network, instead of working with the transfer function, we may use the associated driving-point function.

The problems that arise in determining a complete system function are the following:
(1) Given a magnitude function, determine the corresponding phase, and conversely.
(2) Given a real part, determine the corresponding imaginary part, and conversely. The real part may be that of a transfer function, a driving-point function, or a propagation function. Simple procedures exist for carrying out these tasks; they are covered in detail in the literature.[4] To illustrate the techniques we discuss below two

procedures, one for determining a system function from a given magnitude function, and another from a given real part.

3.2. Magnitude Function Specified. Suppose that a rational function of ω^2 has been given as the squared magnitude of a system function and that the complete system function $p(s)/q(s)$ is to be found. The procedures used to find the system function differ for driving-point and transfer functions; we consider the driving-point case first.

If in accordance with (24) we represent the given function as

$$|Z(j\omega)|^2 = \left.\frac{p(s)p(-s)}{q(s)q(-s)}\right|_{s=j\omega} = \frac{C(\omega^2)}{D(\omega^2)} \tag{37}$$

a procedure may then be formulated. We substitute $s^2 = -\omega^2$ in the given function; we then factor $C(-s^2)$ and $D(-s^2)$. Since, as is shown in any text on synthesis,[4] the system function must be positive real, it cannot have any zeros or poles in the right half plane. We therefore choose as $p(s)$ those factors of $C(-s^2)$ which correspond to zeros in the left half plane; similarly $q(s)$ is formed from $D(-s^2)$. In addition, the given magnitude function may have critical frequencies on the j axis; these will necessarily be of second order and thus have the form $-s^2$ or $(s^2 + a)^2$, where $a > 0$. When these occur, a first-order factor s or $s^2 + a$, respectively, is allotted to the appropriate polynomial.

For a transfer function, say Z_{21}, an additional problem arises in forming the numerator polynomial $p(s)$. Since stability requires that Z_{21} has no poles in the right half plane, its denominator is obtained in the same way as for the driving-point function. However, zeros of Z_{21} may lie anywhere in the plane. Thus, in general, the Z_{21} corresponding to the given magnitude is not unique. From each pair of zeros of $C(-s^2)$ given by $(s - s_1)(s + s_1)$, either the left half plane or the right half plane factor may be chosen; of course, when a complex zero is chosen, its conjugate must also be chosen. Consequently $p(s)$ may contain zeros only in the left half plane or only in the right half plane or zeros in both half planes so that a number of possible Z_{21} functions exist.

The system function Z_{21} may be made unique by requiring that it be *minimum phase*, that is, that it have no right half plane zeros. Then no such zeros can be chosen, and all the other possible Z_{21} functions may be considered as the minimum-phase Z_{21} multiplied by an *all-pass* function; they are obtained simply by reflection of all or some of the zeros of the minimum-phase function into the right half plane. By an all-pass function is meant one whose magnitude is constant for all frequencies. For example, if

$$|Z_{21}(j\omega)|^2 = \left.\frac{(s^4 + 2s^2 + 9)(s^2 + 3)^2}{(s^4 + 13s^2 + 49)(4 - s^2)(s^2 + 6)^2}\right|_{s=j\omega} \tag{38}$$

then, by further factoring, we obtain

$$|Z_{21}(j\omega)|^2 = \left.\frac{(s^2 + 2s + 3)(s^2 - 2s + 3)(s^2 + 3)^2}{(s^2 + s + 7)(s^2 - s + 7)(2 + s)(2 - s)(s^2 + 6)^2}\right|_{s=j\omega} \tag{39}$$

Since the degree of the denominator of (39) is 10, the *minimum* possible degree of q is 5. The minimum-phase transfer impedance is found by inspection as

$$Z_{21a} = \frac{(s^2 + 2s + 3)(s^2 + 3)}{(s^2 + s + 7)(s + 2)(s^2 + 6)} \tag{40}$$

The only other possible system function with a denominator of the fifth degree is the non-minimum-phase one obtained by reflection of one pair of zeros into the right half plane,

$$Z_{21b} = \frac{(s^2 - 2s + 3)(s^2 + 3)}{(s^2 + s + 7)(s + 2)(s^2 + 6)} \tag{41}$$

which may also be written as

$$Z_{21b} = Z_{21a} \frac{s^2 - 2s + 3}{s^2 + 2s + 3} \tag{42}$$

There are, of course, other possible system functions; these may be obtained by multiplication of the minimum-phase function Z_{21a} by an all-pass function *whose critical frequencies are not present* in the given square of the magnitude function, e.g., the all-pass function $(s-3)/(s+3)$. However, all such non-minimum-phase system functions, it will be noted, require that the degree of the numerator and denominator be greater than the minimum needed to represent a possible system function with the prescribed magnitude. This example illustrates the fact that any of the possible system functions is equal to the minimum-phase one multiplied by an all-pass function; thus for a prescribed magnitude the system function is unique within a factor given by an all-pass function.

We can now understand more fully what is meant by the term *minimum phase*. The phase that is referred to is the *phase shift* (or, as it is frequently called, the *net phase shift*) over the whole frequency range. In a minimum-phase function the *phase at zero frequency (approached from positive frequencies, if there is a discontinuity in phase at $\omega = 0$), minus the phase at infinite frequency has the minimum value that can possibly be associated with the specified magnitude function.*

3.3. Real (or Imaginary) Part Specified. Most of the discussion is given in terms of a *real* part specified as a rational function; however, the discussion also applies with slight changes to the procedures for finding the complete system function when the *imaginary* part is specified, so that a separate detailed consideration of this problem is not needed. Furthermore, to avoid unnecessary repetition, the discussion is given in terms of a realizable driving-point function of a passive network (i.e., a positive real function) with no poles on the j axis, but it applies equally well for a transfer function, except for a couple of differences that are now pointed out.

Because of the positive real character of a driving-point function it is necessary that the real part be nonnegative for all ω; the real part of the transfer function, however, need not satisfy this requirement. The other difference concerns poles on the j axis and the consequent uniqueness of the system function. A reactance function, that is, a driving-point function with all its zeros and poles on the j axis, has, as we shall see, a real part equal to a sum of impulses, with an impulse occurring at each finite j-axis pole of the reactance function; the value of the impulse is equal to π times the residue at the pole. Thus a driving-point function determined from a real part that is a rational function of frequency will have no poles on the j axis. However, the simple or multiple-order poles at infinity of a system function will not contribute to the real part. It is therefore clear that the process of determining a system function from a specified real part is *unique within an additive function that is an odd polynomial in s;* when the system function is that of a passive network, and consequently cannot have multiple-order poles at infinity, the odd polynomial consists of a single linear term. Correspondingly, since a constant has zero imaginary part, any driving-point function determined from a specified imaginary part is *unique within an additive constant.*

The question arises about what is done when the denominator of the specified real part of a driving-point function has zeros on the j axis. This may occur, for example, when the function from which the real part was originally computed is not minimum reactive but has a denominator that is not in factored form so that its non-minimum-reactive character is not evident. In this case the denominator of Re $[Z(j\omega)]$ will contain positive ω^2 zeros of the second order; these zeros will become evident when the necessary factoring of the denominator is carried out. However, the knowledge that the contribution of these poles of $Z(s)$ to the rational-function portion of the real part should be zero leads to the conclusion that the numerator must also contain these second-order zeros, and thus the singularities are removable. For each pole on the

---

* This is the negative of the definition for the phase shift usually given. We adopt it because we operate with β, the *negative* of the phase lag. By use of this definition the net phase shift of an all-pass function remains a positive quantity.

j axis we know that a corresponding impulse should be present in the real part.* It is assumed that a cancelation is carried out as a first step in the procedures to be explained. The discussion below is in terms of a driving-point function with no poles on the j axis; later in considering the Hilbert transform we shall discuss impulses in the real part of a system function.

In regard to the j-axis poles of the transfer function a distinction between the transfer impedance (or admittance) and the dimensionless function must be made. It can be shown[4] that the transfer function with the dimension of impedance or admittance may have simple poles on the j axis, including the origin and infinity. However, unlike the driving-point function, though the residues at these poles must be real, they may be positive or negative. Thus the partial-fraction terms for the finite poles will be of the same form, except for a possible sign difference, as those of a reactance function and accordingly will contribute only impulses to the real part. We therefore assume, as for the driving-point function, that if the denominator of the real part is found to have positive ω^2 zeros, then as a first step in the procedure these roots are canceled from the numerator and denominator and corresponding impulses are added to the real part.

For a dimensionless transfer function of a passive network, simple poles on the j axis are permitted, *excluding* the origin and infinity. Here, however, the residues must be pure imaginary. Consequently, for a pair of conjugate imaginary poles, a term of the form $b/(s^2 + \omega_1^2)$, in which b is real but may be positive or negative, will be present in the partial-fraction expansion. Since this term is purely real for $s = j\omega$, it would seem that a different procedure is required when a real part of a dimensionless transfer function is found to possess positive ω^2 zeros; that is, we must treat these j-axis poles of the transfer function in the same way as poles of any other type. However, the difficulty that occurred for the real part of the transfer impedance is now merely shifted to the *imaginary* part of the dimensionless transfer function; since a partial-fraction term for a finite j-axis pole contributes an impulse to the imaginary part, the discussion for the real part of the transfer impedance now applies to the imaginary part of the voltage ratio. Thus again, we exclude poles on the j axis from the discussion of the procedures and will therefore have the real and imaginary parts uniquely related within a constant.

There are several methods for finding a driving-point function from its specified real part; we discuss the one due to Bode. The Bode procedure follows from consideration of the even and odd parts of a system function;

$$\begin{aligned}Z_e &= \tfrac{1}{2}[Z(s) + Z(-s)] \\ Z_o &= \tfrac{1}{2}[Z(s) - Z(-s)]\end{aligned} \quad (43)$$

Examination of these equations reveals significant information. The poles of $Z(s)$ form a configuration in the left half plane, whereas those of $Z(-s)$ lie in the right half plane symmetrically placed with respect to the j axis, and thus Z_e {which for real frequencies is equal to Re $[Z(j\omega)]$} contains all these symmetrically distributed poles. The method for finding $Z(s)$ may now be formulated as follows:

1. Substitute $s^2 = -\omega^2$ in the given Re $[Z(j\omega)]$ and multiply this function by 2.
2. Make a partial-fraction expansion of the resulting function. This expansion will place the poles in evidence and will contain a nonzero constant term only if the given real part is an improper fraction.
3. The desired $Z(s)$ is equal to the sum of the terms with poles in the left half plane plus one-half of the constant term.

The Bode method thus yields $Z(s)$ in its partial-fraction form and is accordingly used when the subsequent synthesis procedure requires this expansion.

It is important to emphasize the reason why this simple method works, namely, that we make use of the knowledge that a system function contains no poles in the right

---

* The value of the residue at the j-axis pole may be determined by the theory of equations, as shown by Guillemin. (See E. A. Guillemin, "Synthesis of Passive Networks," John Wiley & Sons, Inc., New York, 1957.) But if we merely cancel the common j-axis factors, the system function obtained from the real part is minimum-reactive, that is, it has no poles on the j axis.

half plane. If the given real part was that of an ordinary function and not a system function, then the method would yield a *number* of possible complete functions, with no criterion for choosing one of the functions.

The method for finding the driving-point system function from a given imaginary part should be clear from the above discussion and the use of the second equation of (43).

Hilbert Transform.[5] If we represent the system function $Z(s)$ on the imaginary axis by $Z(j\omega) = R(\omega) + jX(\omega)$, its real and imaginary parts are related by the *Hilbert transforms*

$$R(\omega) = \frac{1}{\pi} \int_{-\infty}^{\infty} \frac{X(y)}{\omega - y} dy \qquad (44)$$

and

$$X(\omega) = -\frac{1}{\pi} \int_{-\infty}^{\infty} \frac{R(y)}{\omega - y} dy \qquad (45)$$

These integrals make clear that the real (imaginary) part of a system function is determined when the imaginary (real) part is specified; thus the complete system function is obtained from one of its parts. The system function may be not only an impedance but any other stable function, for example, a propagation function of a minimum-phase system.

We note that the roles played by R and X become interchanged if we change the sign of one of them. For example, changing the sign of $X(\omega)$ in (44) yields

$$R(\omega) = -\frac{1}{\pi} \int_{-\infty}^{\infty} \frac{X(y)}{\omega - y} dy \qquad (46)$$

so that $R(\omega)$ now represents the imaginary part and $X(\omega)$ the real part. This property follows from the fact that R and X satisfy the Cauchy-Riemann equations

$$\begin{aligned} \partial R/\partial \sigma &= \partial X/\partial \omega \\ \partial R/\partial \omega &= -\partial X/\partial \sigma \end{aligned} \qquad (47)$$

A second property to be noted is that the Hilbert transform is a convolution integral; this suggests an instructive method of derivation. If $h(t)$ with transform $Z(s)$ represents the response of a physical system, then it may be expressed for $t > 0$ by

$$h(t) = 2h_e(t)u_{-1}(t) \qquad (48)$$

where h_e is the even part of h and u_{-1} is the unit-step function. Since the transform of h_e for $s = j\omega$ is $R(\omega)$ and the transform of u_{-1} is $1/s$, which on the imaginary axis becomes $\pi\delta(\omega) + 1/j\omega$, where $\delta(\omega)$ is an impulse, then from (48) we have

$$Z(j\omega) = (1/\pi)R(\omega) * [\pi\delta(\omega) + 1/j\omega] \qquad (49)$$

where the asterisk is used to signify convolution.

Carrying out the indicated convolution yields

$$Z(j\omega) = R(\omega) - \frac{j}{\pi} \int_{-\infty}^{\infty} \frac{R(y) \, dy}{\omega - y} \qquad (50)$$

where in obtaining the first term on the right we have made use of the fact that the convolution of a function with an impulse gives the original function. From Eq. (50) we get the desired result

$$X(\omega) = -\frac{1}{\pi} \int_{-\infty}^{\infty} \frac{R(y)}{\omega - y} dy \qquad (51)$$

There are many alternative forms of the transforms that are useful; one of them, for example, is

$$X(\omega) = -\frac{2\omega}{\pi} \int_{0}^{\infty} \frac{R(y)}{\omega^2 - y^2} dy \qquad (52)$$

NETWORK THEOREMS AND SYSTEM FUNCTIONS   2-17

In addition, the transforms may be used to yield realizability conditions and bounds on performance like the Paley-Wiener criterion and the resistance-integral theorem; these are discussed in the reference.[5]

We now briefly apply the transforms to two cases. First we note from Eq. (51) that if $R(\omega) = \pi\delta(\omega - \omega_0)$ then we again can make use of the fact that an impulse is the perfect scanning function; that is, it yields the function scanned. Thus

$$X(\omega) = -\frac{1}{\omega - \omega_0} \qquad (53)$$

which verifies the previous assertion that a system function with a pole on the j axis has an impulse at the j-axis point as the corresponding real part.

When the real or imaginary part is a rational function, then we may use a contour integral form of the Hilbert transform

$$Z(s) = \frac{1}{j\pi} \oint \frac{sR(\lambda/j)}{\lambda^2 - s^2} d\lambda \qquad (54)$$

which may be obtained by making use of the fact that $X(\omega)$ is an odd function and thus may be represented by

$$X(\omega) = \tfrac{1}{2}[X(\omega) - X(-\omega)] \qquad (55)$$

Substitution of Eq. (49) in the above Eq. (55) yields the required integrand: the doubly infinite integral may then be converted to a contour integral. Use of the Cauchy residue theorem to evaluate the integral requires that we expand the integrand in partial fractions. Thus the computation required for this case by use of the Hilbert transform is the same as that required by the Bode procedure discussed previously.

4. EQUIVALENT CIRCUITS

For an active device like a vacuum tube or a transistor it is possible to determine experimentally the functional relations between the voltages and currents of the device, and to plot sets of static curves for these relations. Then by use of graphical analysis on these curves we can obtain output currents and voltages for specified inputs and thus determine the operation of the device as an amplifier.

However, for small-signal operation, with the device operated as a *linear* amplifier, graphical analysis is tedious and, in fact, most often unnecessary; it is more convenient to define and measure the incremental parameters of the device like its impedance or admittance matrix and then apply network theory for the calculation of the required variables. There are at least two methods for using the incremental parameters: we can define an (incremental) *equivalent circuit* whose n-port matrix is the same as that of the device, or we can specify no equivalent circuit but merely use a matrix that describes the *terminal* properties of the device; in other words, we can use an n-port or an n-terminal representation.

In this section we use the equivalent-circuit method of characterizing active devices; specifically, we consider some equivalent circuits of tubes and transistors, emphasizing the different methods of deriving and interpreting these circuits. The discussion serves to make clear how equivalent circuits of other active devices may be obtained. These equivalent circuits permit a designer to explore the capabilities of a new device without actually building an amplifier; with a knowledge of theoretical bounds on active networks they also permit him to compare the performance of a circuit incorporating the device with the best possible performance.

4.1. Vacuum-tube Amplifier as Two-port Network. If we consider the triode tube as a linear amplifier with zero grid-current flow, i.e., with an infinite input impedance, the equivalent circuit may be obtained from the circuit for the ideal voltage amplifier in Table 1, Chap. 1, merely by taking account of the finite nonzero output impedance of the plate circuit. The three parameters used in the equivalent circuits are called *incremental* or *variational* parameters; they are also called *differential coeffi-*

cients, for reasons that will become clear in the discussion. The three parameters are the *plate resistance* r_p, the (grid-to-plate) *transconductance* or *mutual conductance* g_m, and the *amplification factor* μ.

Now consider the triode amplifier shown in Fig. 8, where the total instantaneous values of the variables shown are defined as

$$e_c = E_{co} + e_g$$
$$i_b = I_{bo} + i_p$$
$$e_b = E_{bo} + e_p$$
(56)

In the above and in Fig. 8 subscripts c and b denote total grid and plate quantities, respectively, k denotes the cathode, the capital letters whose second subscript is o

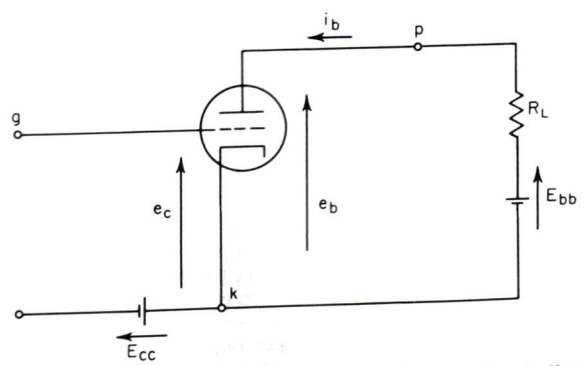

FIG. 8. Triode amplifier with total values of the variables indicated.

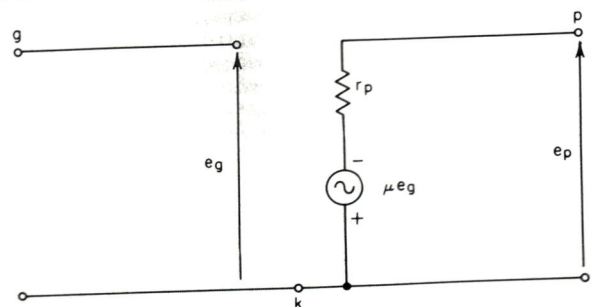

FIG. 9. Linear equivalent circuit of the triode for incremental variables.

denote the quiescent values of the variables at the operating point, and the subscripts g and p denote the *incremental* grid and plate quantities, respectively. We are concerned only with the incremental variables e_g, i_p, and e_p. The linear equivalent circuit (for low frequencies where the effects of interelectrode capacitances are negligible) for the triode (operating with a grounded cathode) is shown in Fig. 9.

This circuit and the definition of the parameters μ and r_p may be obtained graphically from the static characteristic curves or mathematically by the use of the formula for the total differential.[6] The mathematical derivation can be briefly given.

The total current i_b is a function of the two variables, the plate voltage and the grid voltage, and thus may be written as $i_b = i_b(e_b, e_c)$. If we take the total differential of i_b, we obtain

$$di_b = (\partial i_b / \partial e_b)\, de_b + (\partial i_b / \partial e_c)\, de_c$$
$$= (1/r_p)\, de_b + g_m\, de_c$$
(57)

where we have defined the plate resistance r_p and the transconductance g_m by

$$r_p = \partial e_b/\partial i_b = de_b/di_b|_{e_c=\text{const}}$$
$$g_m = \partial i_b/\partial e_c = di_b/de_c|_{e_b=\text{const}} \quad (58)$$

The differentials in Eq. (57) may be defined as the incremental components so that the equation becomes

$$i_p = e_p/r_p + g_m e_g \quad (59)$$

and therefore

$$r_p i_p = e_p + r_p g_m e_g$$
$$= e_p + \mu e_g \quad (60)$$

which describes the same equivalent circuit as that shown in Fig. 9.

In the second equation of Eq. (60) we have defined the amplification factor μ as equal to $r_p g_m$. This may be derived from Eq. (57) by setting di_b equal to zero, after which we obtain by algebraic manipulation

$$-de_b/de_c = r_p g_m \quad (61)$$

which we use as the definition of the amplification factor

$$\mu = -\partial e_b/\partial e_c$$
$$= -de_b/de_c|_{i_b=\text{const}} \quad (62)$$

The triode tube, when operating as a linear amplifier at low frequencies, therefore has only two independent linear parameters, the third being determined by the other

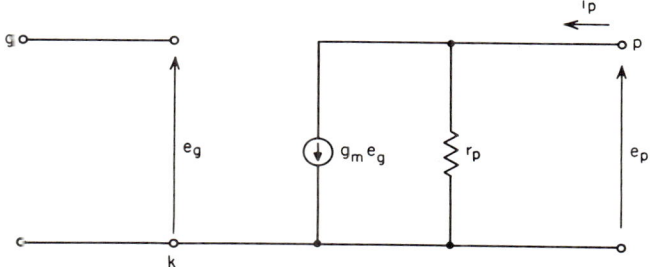

Fig. 10. Alternative triode equivalent circuit that employs a controlled current source.

two. This is so, we recall, because of our assumption that the grid current equals zero, that is, the input impedance of the tube is infinite. For the transistor, as we shall see, four independent parameters are necessary in order to define it uniquely as a two-port network.

Another equivalent circuit may be obtained by transformation of the voltage source in Fig. 9 to a current source or more directly by use of Eq. (59). This equation is the Kirchhoff current equation obtained at the plate terminal for the equivalent circuit shown in Fig. 10.

These equivalent circuits are also correct models for the pentode tube in which the suppressor grid is connected to the cathode, the screen grid is held at a fixed potential with respect to the cathode, and the tube is operating in a linear mode without grid-current flow; that is, the mode of operation is again class A_1. The screen and suppressor grids then do not affect the voltage and current increments and the pentode operates like the triode. The incremental parameters of course differ from those of the triode, and the fixed screen-grid voltage does determine the set of plate characteristics that must be used for determining the parameters.

Inspection of a set of plate characteristics of a pentode shows that along the flat portion of the i_b-e_b curves the plate resistance is very large. Thus in the equivalent circuit of Fig. 10 the resistance may often be omitted and we obtain an ideal current source in the output circuit, and the pentode is then an ideal transadmittance amplifier. In synthesis applications this ability of the pentode to serve as an ideal current source is often used to realize a given transfer impedance by a lossless network terminated in resistance at only the load end.

The vacuum tube, however, is sometimes operated with grid current flow. It then becomes necessary to determine another set of static characteristics, analogous to those for the plate circuit. The complete two-port equivalent circuit for such operation is as general as possible; that is, four independent parameters are needed to characterize it uniquely. Instead of deriving this network directly, we consider the general case of a multielectrode tube which is then specialized to the triode with grid current flow.

4.2. Multielectrode Tube. Consider the equations for the total differentials of an n-electrode tube (not counting the cathode), whose symbolic representation is shown in Fig. 11. Each current is a function of all the electrode voltages,

$$i_k = i_k(e_1, e_2, \ldots, e_n) \qquad (63)$$

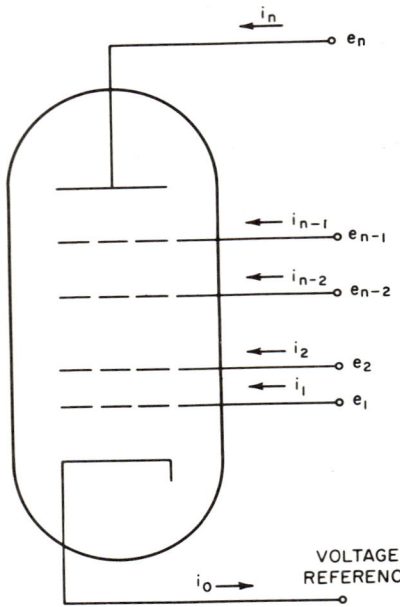

Fig. 11. Representation of an n-electrode tube.

The total differential is

$$di_k = \sum_{j=1}^{n} g_{kj}\, de_j$$

$$(k = 1, 2, \ldots, n) \qquad (64)$$

where

$$g_{kj} = \partial i_k/\partial e_j \qquad (65)$$

is defined as the transconductance from j to k, i.e., from the second subscript to the first.

When $j = k$ the quantity becomes a self-conductance and

$$r_j = 1/g_{jj} \qquad (66)$$

is defined as the self-resistance of the jth electrode. Physically g_{kj} is interpreted as the change produced in the current i_k by a unit change in e_j, all other electrode voltages being held constant.

Another quantity of interest for these multielectrode tubes is the amplification factor, which is defined as

$$\mu_{kj} = r_k g_{kj} \qquad (67)$$

From Eq. (66) it follows that

$$\mu_{jj} = 1 \qquad (68)$$

Multiplication of Eq. (64) by r_k gives

$$r_k\, di_k = \sum_{j=1}^{n} \mu_{kj}\, de_j \qquad (69)$$

In order to obtain a definition of μ_{kj} in terms of differentials let di_k equal zero and set each of the voltage differentials except de_j and de_k equal to zero; we then obtain from Eq. (69)

$$\mu_{kj} = -de_k/de_j\big|_{i_k=\text{const}}$$
$$= -\partial e_k/\partial e_j \qquad (70)$$

The above equation leads to the physical interpretation of μ_{kj} as the negative change in e_k produced by a unit change in e_j, current i_k and all voltages other than e_j being held constant.

The above can now be applied to the triode, where the subscript 2 becomes p and subscript 1 becomes g. Writing out Eq. (64) and interpreting the differentials as incremental values, we obtain

$$i_g = g_{11}e_g + g_{12}e_p$$
$$i_p = g_{21}e_g + g_{22}e_p \tag{71}$$

Multiplying the first of Eqs. (71) by $1/g_{11} = r_g$ and the second by $1/g_{22} = r_p$ yields

$$i_g r_g = e_g + \mu_{12}e_p$$
$$i_p r_p = \mu_{21}e_g + e_p \tag{72}$$

which can be interpreted simply by an equivalent circuit with two sources, as shown in Fig. 12.

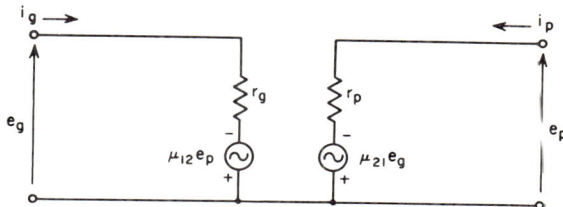

FIG. 12. Equivalent circuit representing most general type of two-port network.

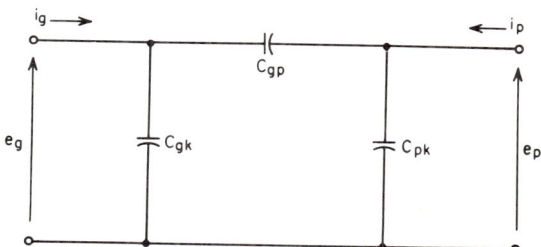

FIG. 13. Network of interelectrode capacitances.

There is an important point to be stressed in the results of the above derivation; namely, that the most general type of linear two-port circuit (tube, transistor, acoustic transducer, waveguide, parametric amplifier, etc.) can be described by two simultaneous linear equations such as those given in Eq. (71). Only four independent parameters are needed to specify completely the operation of the network at its ports.

We have used resistance in our equivalent circuits rather than general impedances; that is, we have assumed low-frequency operation where the effects of interelectrode capacitances are negligible. But before we leave the equivalent circuit of Fig. 12, it is desirable to see how it can be modified so that it holds for high-frequency operation. For such operation the total current to an electrode is not only a conduction current but also includes a displacement current; in other words, the interelectrode capacitances of the tube must be taken into account. The two-port network of these capacitances is shown in Fig. 13 and its equations are given by

$$i_g = s(C_{gk} + C_{gp})e_g - sC_{gp}e_p$$
$$i_p = -sC_{gp}e_g + s(C_{gp} + C_{pk})e_p \tag{73}$$

where s is the complex variable defined by $s = \sigma + j\omega$.

To obtain the complete equivalent circuit it is merely necessary to connect the networks of Figs. 12 and 13 in parallel at both input and output. This is done in Fig. 14, and the corresponding equations are obtained simply by adding the corresponding admittance parameters of Eqs. (71) and (73), which yields

$$i_g = [g_{11} + s(C_{gk} + C_{gp})]e_g + (g_{12} - sC_{gp})e_p \\ i_p = (g_{21} - sC_{gp})e_g + [g_{22} + s(C_{gp} + C_{pk})]e_p \quad (74)$$

The above manipulations, as the reader is well aware, may be accomplished concisely by the use of matrices, but it is instructive to show them more explicitly at this point, as we have done.

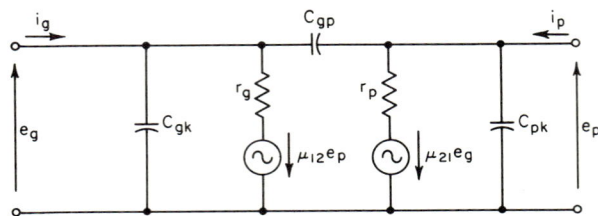

FIG. 14. High-frequency equivalent circuit of triode vacuum tube.

The complete equivalent circuit for the n-electrode tube leads to an n-port network. As a specific example[7] we consider the pentode shown in Fig. 15, in which the screen and suppressor grids are not held at constant potentials with respect to the cathode, so that is is necessary in the equivalent circuit to account for their effects on the tube currents. We shall derive an equivalent circuit that corresponds to the two-port containing two controlled sources shown in Fig. 12.

Again interpreting the differentials in Eq. (64) as incremental voltages and currents, we obtain for the plate circuit

$$i_4 = g_{41}e_1 + g_{42}e_2 + g_{43}e_3 + g_{44}e_4 \quad (75)$$

The corresponding circuit is shown in Fig. 16, where g_{44} is the incremental plate conductance, and g_{41}, g_{42}, and g_{43} are the incremental transconductances from grids 1, 2, and 3, respectively, to the plate. It should be clear that each of the controlled sources in Fig. 16 accounts for the effect of one of the grids on the plate current, whereas the conductance g_{44} accounts for the effect of the plate voltage on the plate current.

The same procedure applied to each of the three grids yields a similar equivalent circuit; then use of the fact that

$$i_0 = i_1 + i_2 + i_3 + i_4 \quad (76)$$

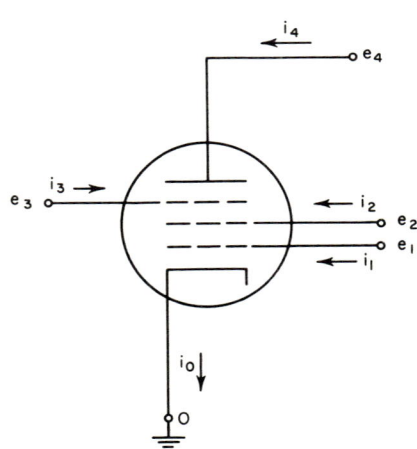

FIG. 15. Pentode tube with designated currents and voltages.

permits the connection of the four separate circuits to the cathode terminal. The complete four-port network that describes the pentode is shown in Fig. 17. By use of source transformations we can obtain an equivalent circuit in terms of controlled voltage sources; this is shown in Fig. 18.

4.3. Transistor Equivalent Circuits. Before considering the incremental circuits for a transistor used as a linear amplifier we shall explain some symbols and notation.

NETWORK THEOREMS AND SYSTEM FUNCTIONS 2–23

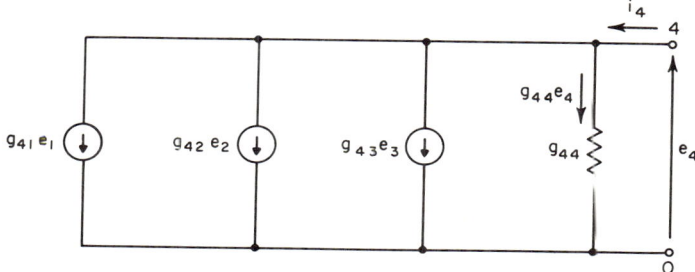

Fig. 16. Equivalent circuit for the plate electrode of a pentode.

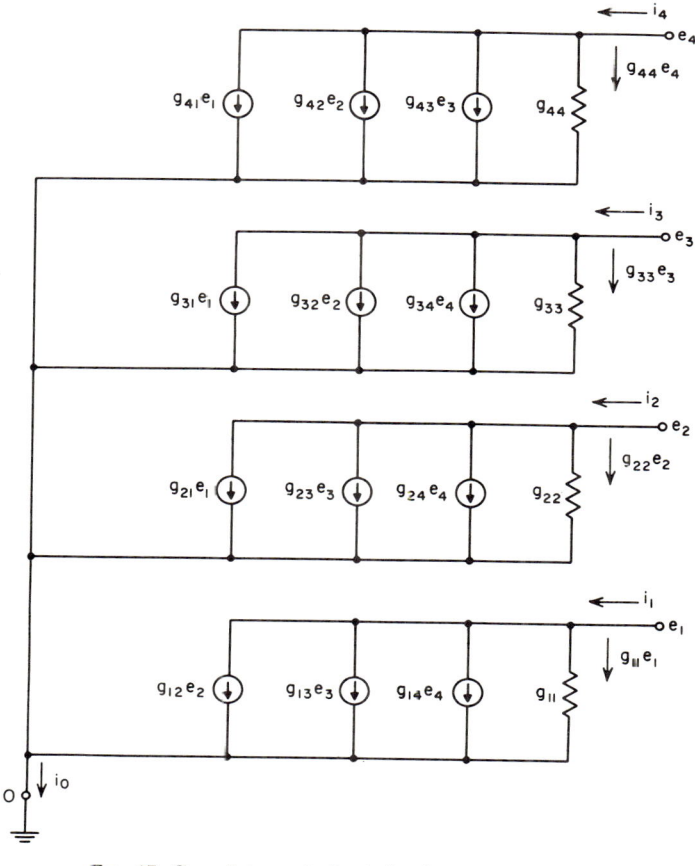

Fig. 17. Complete equivalent circuit for the pentode.

AMPLIFIER FUNDAMENTALS

A symbol for a transistor is shown in Fig. 19. The positive directions of voltage and current are indicated with the base used as a common electrode. For purposes of identification in this first use of the symbol the electrodes are marked with letters, where E, C, and B denote the emitter, collector, and base, respectively.

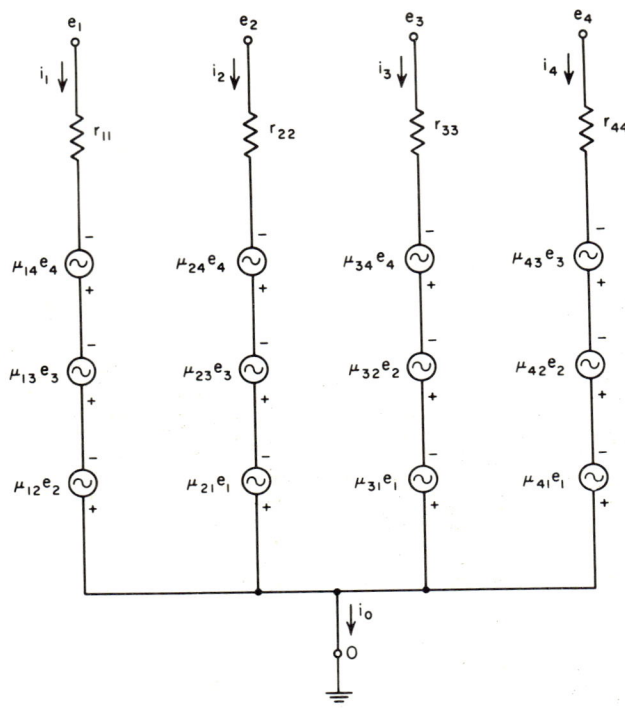

Fig. 18. Complete equivalent circuit for the pentode employing voltage sources.

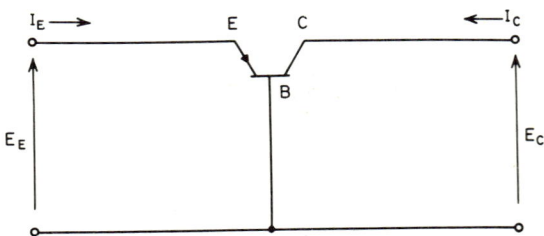

Fig. 19. Symbol for a transistor.

The arrow convention on the emitter requires explanation. If the arrow points into the base, then the base is made of n-type material; if it points out of the base, then the base is made of p-type material. Consequently, in the first case, a junction transistor is a p-n-p type; in the second, it is n-p-n; an n-p-n transistor symbol is shown in Fig. 20. The arrow on the emitter in both symbols shows the direction of positive current flow necessary to bias the transistor for amplifying action.

It is also necessary to define the symbols that are used for the currents and voltages. For the emitter current this set is:

NETWORK THEOREMS AND SYSTEM FUNCTIONS 2-25

i	Lowercase letter represents an instantaneous value of the variable		thus i_e is the instantaneous value of the incremental component of emitter current
i_E	Capital letter subscript on the lowercase letter indicates a total instantaneous value; thus i_E is the total instantaneous emitter current	I_E	Quiescent value without signal applied
		I_{ES}	d-c value with signal applied
		$I_{e(rms)}$	rms value of a-c component
i_e	Lowercase letter as subscript indicates incremental value;	$I_{e(max)}$	Maximum value of a-c component of emitter current

A similar set is used for each of the other variables and electrodes, c and C being used for the collector, and b and B for the base. Often the rms subscript is omitted so that I_e is used to represent the rms value of the incremental component of emitter current. To avoid confusion in the meanings of the symbols for the voltages, it is often desirable to use a double-subscript notation which indicates the reference or common terminal as the second subscript; thus in the common-base configuration to be discussed below the rms collector voltage—that is, the voltage rise from base to collector—is denoted by E_{cb} whereas in the common-emitter configuration the collector voltage is E_{ce}.

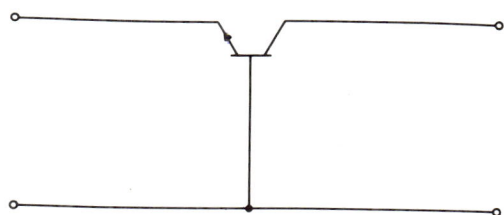

Fig. 20. Representation of an n-p-n transistor.

Since the definitions of the voltages used in the following discussion are clear from the context, double-subscript notation is not used.

Now applying the procedure that was used previously for the vacuum tube, we obtain from the total differentials of

$$e_E = e_E(i_E, i_C)$$
$$e_C = e_C(i_E, i_C) \qquad (77)$$

the equations

$$de_E = (\partial e_E/\partial i_E)\, di_E + (\partial e_E/\partial i_C)\, di_C$$
$$de_C = (\partial e_C/\partial i_E)\, di_E + (\partial e_C/\partial i_C)\, di_C \qquad (78)$$

The partial derivatives have the dimension of resistance and the transistor linear parameters may therefore be determined from the slopes of the appropriate static characteristics. Interpreting the total differentials as incremental variables we obtain

$$e_e = r_{11} i_e + r_{12} i_c$$
$$e_c = r_{21} i_e + r_{22} i_c \qquad (79)$$

where

$$r_{11} = \partial e_E/\partial i_E \qquad r_{12} = \partial e_E/\partial i_C$$
$$r_{21} = \partial e_C/\partial i_E \qquad r_{22} = \partial e_C/\partial i_C \qquad (80)$$

Thus it is clear that the transistor low-frequency parameters may be determined by measurement of the slopes of the static characteristics. Equations (80) suggest another method for measuring the linear parameters at a particular operating point, namely, use of the definitions

$$r_{11} = E_e/I_e|_{I_c=0} \qquad r_{12} = E_e/I_c|_{I_e=0}$$
$$r_{21} = E_c/I_e|_{I_c=0} \qquad r_{22} = E_c/I_c|_{I_e=0} \qquad (81)$$

where the rms values are now used for convenience. In other words, with the transistor properly biased and the collector open-circuited for a-c variables, a small current signal is applied to the emitter and the emitter voltage is measured; the ratio E_e/I_e gives r_{11}. From this measurement it should be clear why the r's are called open-circuit parameters: currents are either constrained to be zero or current sources are applied as signals.

Now we turn to equivalent circuits. It should first be emphasized that there is no single equivalent circuit which is useful for all operating conditions. On the contrary, there is a whole range of circuits whose complexity increases as we try to account for more detailed aspects of transistor operation. Below we derive some equivalent circuits for low-frequency operation of the transistor in its three possible configurations. These three are the *common-base*, *common-emitter*, and *common-collector* circuits. Often the word *grounded* is used interchangeably with the word *common*.

For the common-base connection the emitter is the input electrode, and for the other two the base is the input electrode; thus the terms are completely defined. It is of course possible to reverse the direction of signal transmission; for example, in the common-collector configuration the emitter may be used as the input and the amplifier will still yield gain. However, we shall designate this as the *reversed* common-collector connection. Of course, since the triode vacuum tube is also a three-terminal network, it too can have each of three common terminals—the common-cathode amplifier, which was previously analyzed; the common-plate amplifier, which is the well-known cathode-follower circuit, and the common-grid amplifier. The last two can be analyzed in terms of the equivalent circuit for the common-cathode amplifier, as will become clearer after the discussion of the transistor equivalent circuits; one of the differences between these configurations and those of the transistor configurations is that all the transistor configurations have internal feedback whereas the normal common-cathode amplifier has no feedback at low frequencies.

Consider again the open-circuit resistance equations

$$E_1 = r_{11}I_1 + r_{12}I_2$$
$$E_2 = r_{21}I_1 + r_{22}I_2 \tag{82}$$

where the subscript 1 designates the input to a two-port black box and 2 designates the output. If the box contains a transistor in the common-base connection, then $E_1 = E_e$, $E_2 = E_c$, and so forth. But if a common-emitter connection is used $E_1 = E_b$ and $E_2 = E_c$, where E_c is now measured with respect to the emitter, since the emitter is the common electrode. In double-subscript notation, as indicated previously, the collector voltage is E_{cb} and E_{ce} for the common-base and common-emitter connections, respectively. We shall work with the common base as the standard connection and then derive the parameters for the other connections in terms of those for the common base.

First it may be mentioned that the trivially simple (but often useful) equivalent circuit employing two active sources may be drawn by inspection of Eqs. (82); this is shown in Fig. 21a. Another equivalent circuit with two controlled sources is shown in Fig. 21b; it uses the h parameters which, since they involve mixed current and voltage variables, are called *hybrid parameters*. These parameters have attractive advantages and are therefore much used in the literature.[7,8] First of all, they are directly related to the terminal characteristics of the two-port, and can be measured easily. In addition, the reverse voltage transmission h_{12} is often negligible so that the controlled voltage source can be omitted. The hybrid circuit then becomes unilateral with all the isolation properties of a triode tube at low frequencies. We discuss the h parameters for characterizing a two-port in Chap. 3. Since our main purpose is to show how equivalent circuits are derived and interpreted, we may use any set of parameters. However, commercial specifications for transistors at present use the h parameters most frequently. These parameters are therefore used in most of the applications of equivalent circuits in this book, for example, in Chap. 10.

4.4. Common-base Configuration. We wish to obtain a low-frequency equivalent circuit for the common-base transistor configuration shown in Fig. 22a that contains only one controlled source; furthermore, it is convenient to divide the equivalent

NETWORK THEOREMS AND SYSTEM FUNCTIONS 2-27

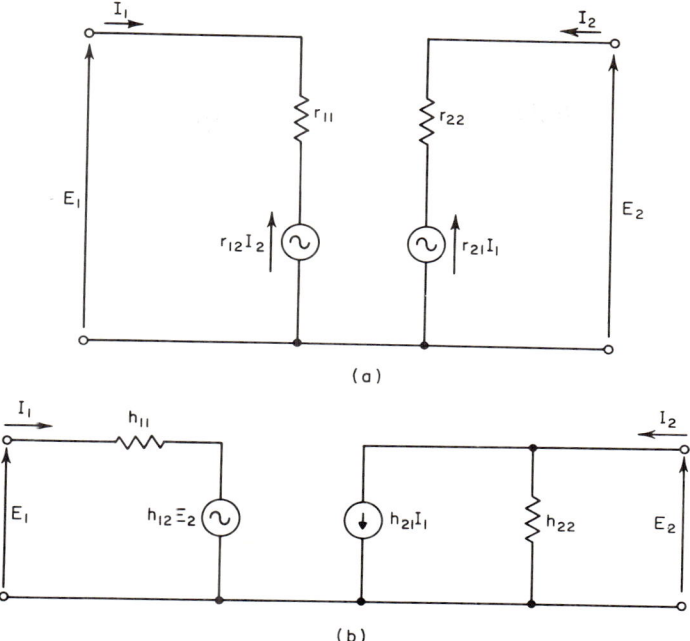

FIG. 21. Equivalent circuits using two sources.

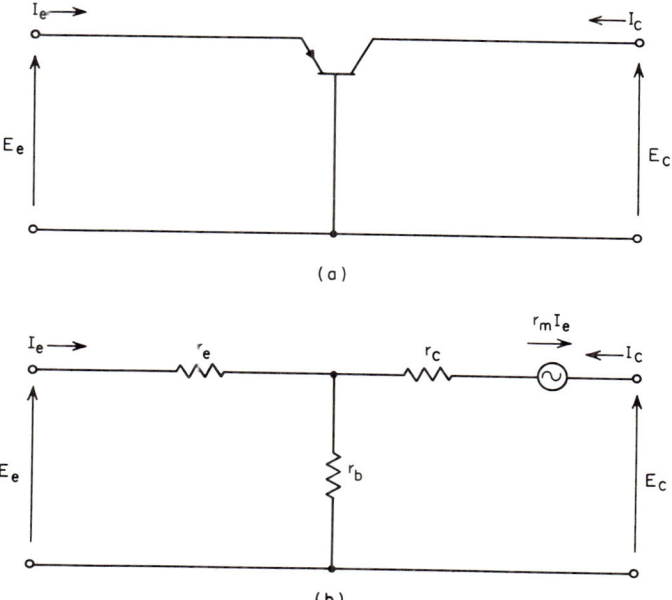

FIG. 22. An equivalent circuit for the common-base connection. (a) Transistor configuration. (b) Equivalent circuit.

circuit into a reciprocal part and a nonreciprocal part. We accomplish the latter objective by manipulating the equations characterizing the transistor

$$E_e = r_{11}I_e + r_{12}I_c$$
$$E_c = r_{21}I_e + r_{22}I_c \qquad (83)$$

so that reciprocity applies to part of it; we then find by *chance* (since a reciprocal network can still be active) that the subnetwork representing the reciprocal part of the equations is passive; that is, it can be constructed with positive resistances.

In the second of Eqs. (83) for the common-base connection we add and subtract $r_{12}I_e$:

$$E_c = r_{12}I_e + r_{22}I_c + (r_{21} - r_{12})I_e \qquad (84)$$

We now define $r_m = r_{21} - r_{12}$; this parameter has been variously called the *effective transfer resistance* and the *net mutual resistance*. Thus if we consider $E_c - r_m I_e$ as the effective voltage in one loop of a T network, we can draw the equivalent circuit shown in Fig. 22b.

It is of course also possible, and often convenient, to work with a Π network rather than a T network; this may be obtained directly from the admittance form of the equations describing the common-base transistor

$$I_e = g_{11}E_e + g_{12}E_c$$
$$I_c = g_{21}E_e + g_{22}E_c \qquad (85)$$

We merely derive the form of this circuit for completeness and then use the T network of Fig. 22b for all further discussion in this section.

Now operating on the second of Eqs. (85) as we did previously with Eqs. (83) we obtain

$$\begin{aligned} I_c &= g_{12}E_e + g_{22}E_c + (g_{21} - g_{12})E_e \\ &= g_{12}E_e + g_{22}E_c + g_m E_e \end{aligned} \qquad (86)$$

The corresponding equivalent circuit, where the g's represent conductances in mhos, is shown in Fig. 23. From this circuit we calculate directly the conductances in Eq. (85):

$$g_{11} = g_1 + g_3$$
$$g_{12} = -g_3$$
$$g_{22} = g_2 + g_3$$
$$g_{21} = g_m - g_3$$

Comparison of this circuit for a general physical amplifier in Fig. 23 with the ideal amplifier previously given in Fig. 6c, Chap. 1, makes the differences clear. However,

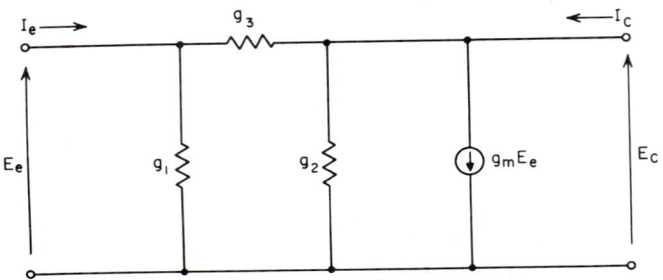

FIG. 23. Equivalent circuit for common-base transistor using a Π network of conductances.

if g_3 is zero so that there is no feedback and if g_1 or g_2 is zero, then the circuit represents what could be termed a *semi-ideal* amplifier. This term is used because for g_3 and g_1 zero the open-circuit voltage amplification is given by $-g_m/g_2$; when g_3 and g_2 are zero, the short-circuit current amplification is given by g_m/g_1.

Returning to the T network of resistances in Fig. 22 we obtain

$$r_{11} = r_e + r_b$$
$$r_{12} = r_b \; (\equiv \text{feedback resistance})$$
$$r_{22} = r_c + r_b$$
$$r_{21} = r_m + r_b$$
(87)

The resistances in the T network are also convenient in that they may be given an approximate physical significance. The resistance r_e is the forward resistance of the emitter-base diode, and consequently has a small value; r_c is the reverse resistance of the collector-base diode and is therefore high, while r_b represents an interaction effect between the emitter and collector.

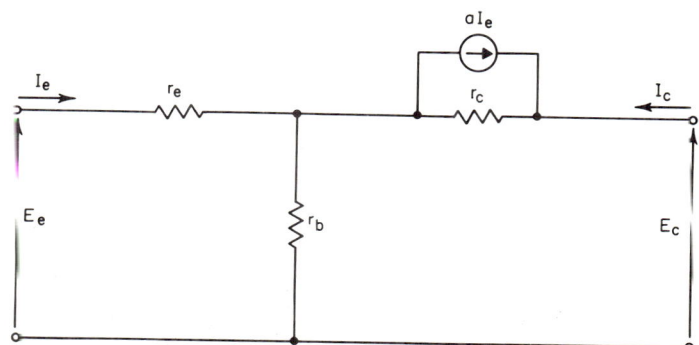

FIG. 24. Equivalent circuit using a controlled current source.

It may be more convenient in some problems to have a controlled current source rather than the voltage source $r_m I_e$. If we use a source transformation in the collector branch we obtain the equivalent circuit of Fig. 24, where

$$a = r_m/r_c \tag{88}$$

The parameter defined above is an important parameter of a transistor and is often used interchangeably with the short-circuit current amplification α. The α is defined as

$$\alpha = -\partial i_C/\partial i_E = -di_C/di_E|_{e_C=\text{const}} \tag{89}$$

or for incremental quantities as

$$\alpha = -I_c/I_e|_{E_c=0} \tag{90}$$

Setting E_c equal to zero in the second of Eqs. (83) plus some algebraic manipulation yield for the short-circuit current amplification

$$-I_c/I_e|_{E_c=0} = r_{21}/r_{22}$$
$$= \frac{r_m + r_b}{r_c + r_b} = \alpha \tag{91}$$

Now to obtain the relationship between a and α, we divide both sides of the equation for α by a to obtain

$$\frac{\alpha}{a} = \frac{r_c + r_b/a}{r_c + r_b} \tag{92}$$

For a junction transistor $r_c \gg r_b$ and $a \approx 1$, so that

$$\alpha/a \approx r_c/r_c = 1 \tag{93}$$

2-30 AMPLIFIER FUNDAMENTALS

Thus the use of α and a interchangeably is therefore justified as a very good approximation.

For vacuum tubes the equivalent circuit for the common cathode is used as the basic one, the equivalent circuits for common grid and common plate (cathode follower) being obtained from it. Similarly for transistors the two equivalent circuits for the common-emitter and common-collector connections may be derived from the one for common-base operation.

4.5. Common-emitter Configuration. For the common-emitter connection of the transistor shown in Fig. 25a it is only necessary to interchange the positions of the base and emitter arms to obtain the equivalent circuit shown in Fig. 25b.

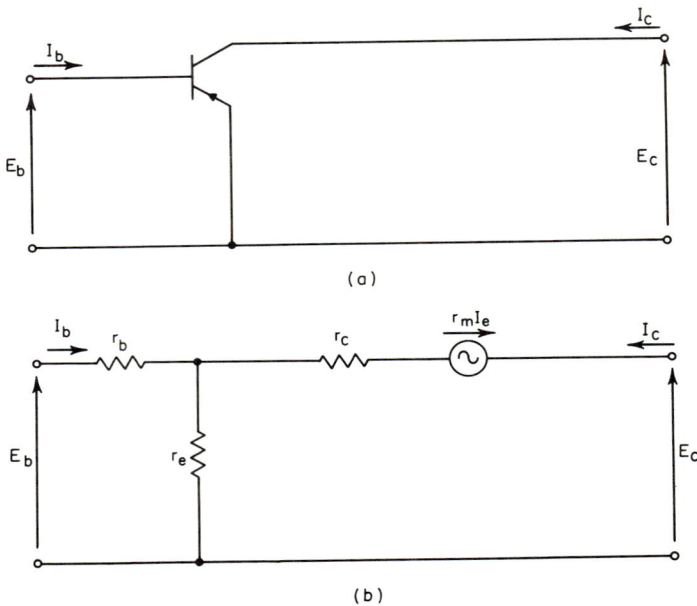

Fig. 25. Common-emitter configuration. (a) Symbolic representation. (b) Equivalent circuit.

It is convenient to represent the strength of the controlled source in terms of the input current, which is now I_b not I_e. Kirchhoff's current law applied at the internal node yields

$$I_e = -(I_c + I_b) \tag{94}$$

The equivalent circuit in Fig. 26a now has two sources, $r_m I_c$ and $r_m I_b$. However, since I_c is flowing through the sources, $-r_m$ can be substituted for $r_m I_c$; the resulting total resistance is $r_c - r_m = r_c(1 - a)$. As before, an equivalent current source may be substituted for the voltage source. The complete sequence of steps is shown in Fig. 26. The quantity $r_m/[r_c(1 - a)] = a/(1 - a)$ is often defined by a single letter in the literature, viz., $b = a/(1 - a)$. The b represents approximately the short-circuit current amplification for common-emitter operation.

4.6. Common-collector Configuration. The equivalent circuits for the common-collector transistor circuit shown in Fig. 27a may be derived simply by interchanging the emitter and collector arms in b and c of Fig. 26; these are shown in Fig. 27b and c.

It should be clear that if a positive current flow *into* the input terminals causes a positive current to flow *out of* the output terminals, there is no phase reversal. Examination of the equivalent circuits for the three connections (i.e., noting the positive direction of the internal source) shows that the common-emitter configuration is the only one that leads to a phase reversal; the other two at low frequencies give an output signal that is in phase with the input signal.

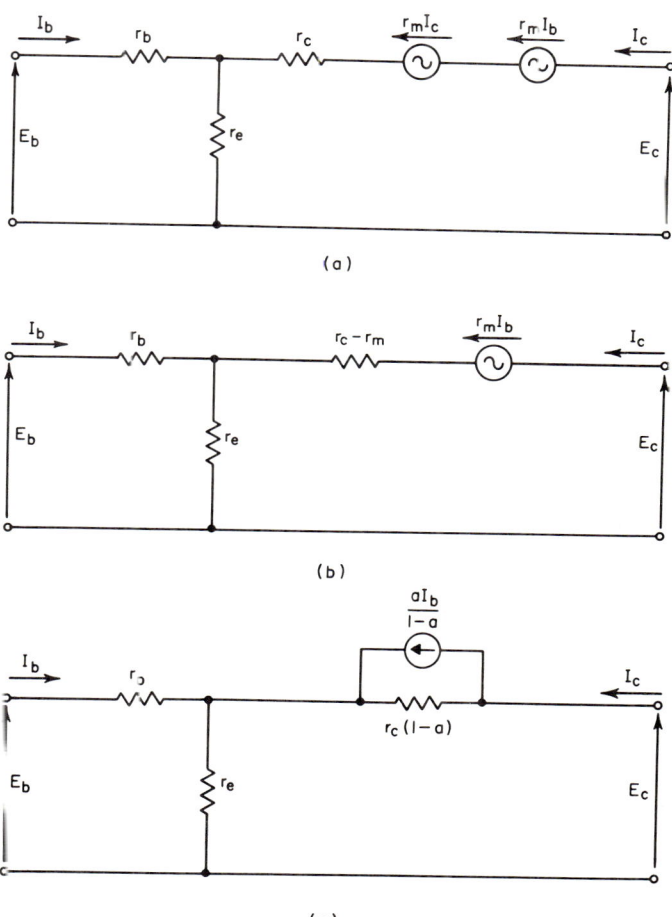

Fig. 26. Common-emitter equivalent circuits with current-controlled sources. (*a*) Circuit with two sources. (*b*) Circuit with one voltage source. (*c*) Circuit with one current source.

A mnemonic scheme (if one is desired) for remembering which transistor amplifier configurations produce phase reversals is to think of the emitter, the collector, and the base as corresponding, respectively, to the cathode, plate, and grid of the triode tube.

Inspection of the equivalent circuits also gives useful information about input and output impedances; it shows, for example, that the common-base and common-emitter connections have low input and high output impedances, whereas the reverse is true for the common-collector circuit.

4.7. High-frequency Equivalent Circuits for Transistors. Since the response at high frequencies is a very complicated phenomenon, it is difficult to formulate an

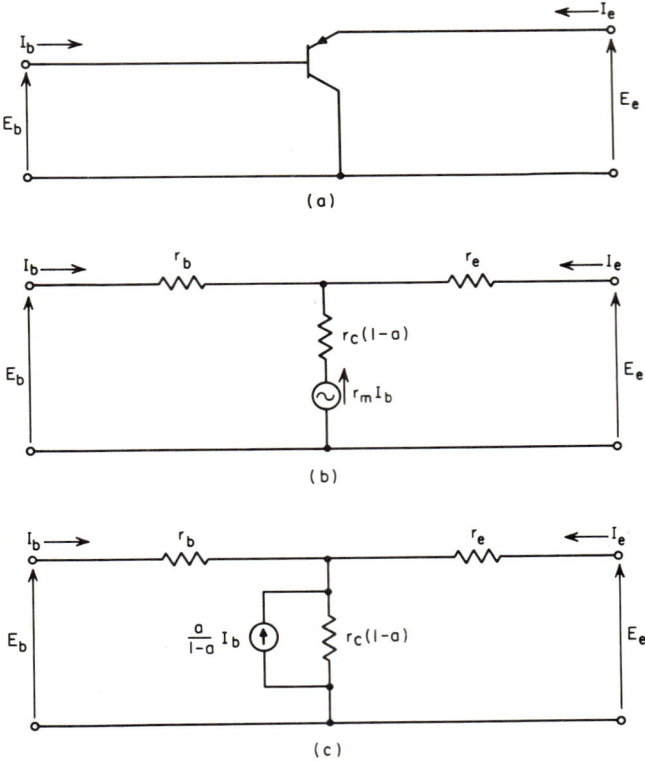

Fig. 27. Common-collector equivalent circuits. (a) Symbolic representation. (b) Circuit with voltage source. (c) Circuit with current source.

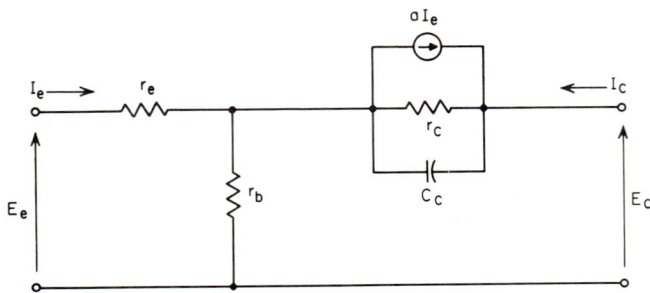

Fig. 28. Common-base high-frequency equivalent circuit.

equivalent circuit. In this section we briefly discuss some of the complexities and derive an approximate equivalent circuit that is satisfactory for many practical purposes.

Common-base Amplifier. To modify the low-frequency equivalent circuit to take account of high-frequency effects, we use the equivalent circuit for the common-base connection given in Fig. 28. It appears that it is necessary to add only a collector capacitance C_c.

We recall from our knowledge of the physics of the transistor that there is a space-charge region at the collector junction, the n-type region acting as one plate of a capacitance and the p-type as the other. The width of this region, and consequently the value of C_c, varies with the collector voltage. More fixed charge and consequently a wider spacing are necessary for a higher voltage across the junction; a wider spacing between plates causes a lower junction capacitance. There is, of course, a similar capacitance across the emitter junction, but here it is shunting a very low resistance and we assume for purposes of illustration that it is negligible in its effect.

There is another difference represented in Fig. 28, which is not evident by inspection of the circuit: the a or α is a complex quantity now. It can be shown by a mathematical analysis of the junction transistor that α is given by

$$\alpha = \text{sech}\,[(W/L_b)\sqrt{1 + j\omega\tau_m}] \tag{95}$$

where W is the width of the base region, L_b is the diffusion distance for the minority carriers in the base region, and τ_m is the lifetime of the minority carriers. A decrease of the value of α with frequency could even be predicted intuitively; the junctions are not perfectly planar and a dispersion in transit times of charge carriers is to be expected; and secondly, since the carriers diffuse across the base region, their paths and transit times are different. As the frequency increases it is possible for carriers injected by the emitter at the trough of the input signal to arrive at the collector at the same time as those injected at the signal peak, thus causing a partial cancellation.

How can the variation given in Eq. (95) be represented by lumped parameters in an equivalent circuit? Remembering that a finite network of lumped elements can yield only a rational function as the most general transfer relationship, we realize that it is impossible to represent the transcendental function of Eq. (95) by such a network.

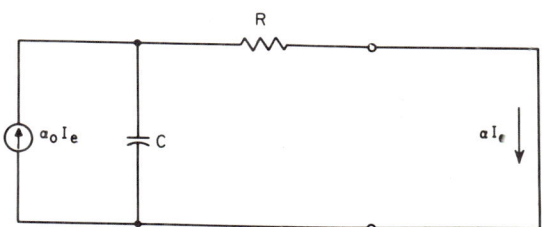

Fig. 29. Simple equivalent representation for α.

However, it can be shown that for a transmission line with negligible series inductance and shunt conductance, the ratio of the short-circuit output current to the input current is given by

$$I_o/I_i = \text{sech}\,\sqrt{j\omega RC} \tag{96}$$

where R is the total line resistance and C the total capacitance. Thus if $I_i = \alpha_o I_e$ and $I_o = \alpha I_e$, where α_o is the low-frequency value of α, the required variation can be represented in the equivalent circuit by such a transmission line. A simple approximation to this line is given by one of its RC sections; this is shown in Fig 29. This approximation is equivalent to assuming the variation

$$\alpha = \frac{\alpha_o}{1 + jf'/f_{\alpha b}} \tag{97}$$

where $f_{\alpha b}$ is the frequency at which α has fallen to $1/\sqrt{2}$ of its low-frequency value, that is, $f_{\alpha b}$ is the 3-db frequency. Therefore, to sum up, Fig. 28 can be used as the high-frequency equivalent circuit with an α variation given by Eq. (97).

The circuit may also be drawn with an internal voltage source as shown in Fig. 30. The quantity $z_m \equiv \alpha r_c$, where α is given by Eq. (97).

For some transistors and circuits the high-frequency response will be limited by C_c before the drop-off of α becomes effective; in other circuits the α characteristic determines the high-frequency response. When the circuit cutoff frequency for a common-base amplifier is not limited by C_c but is determined by α, then this frequency is approximately equal to $f_{\alpha b}$; the common-emitter amplifier, however, has a much lower cutoff frequency, being given approximately by $(1 - \alpha_o)f_{\alpha b}$.

The collector capacitance gives an output capacitance of approximately C_c for the common-base amplifier but $C_c/(1 - \alpha)$ for the common-emitter connection. Therefore, if C_c is between 5 and 50 pf, the output capacitance for the latter connection may be more than 1,000 pf. However, the frequency cutoff due to the capacitance for both the common-base and common-emitter amplifiers is approximately the same under matched conditions because the matched load impedance of the common-emitter amplifier is considerably lower. Another troublesome effect of the C_c, in addition to its loading of the output circuit, is the difficulty it causes in tuning a tuned amplifier.

Common-emitter and Common-collector Amplifiers. The high-frequency equivalent circuits of Fig. 31 for the other two configurations follow directly.

However, a little difficulty may be experienced in deriving these circuits, specifically, in showing how the collector branch is modified. The derivation of the circuit in

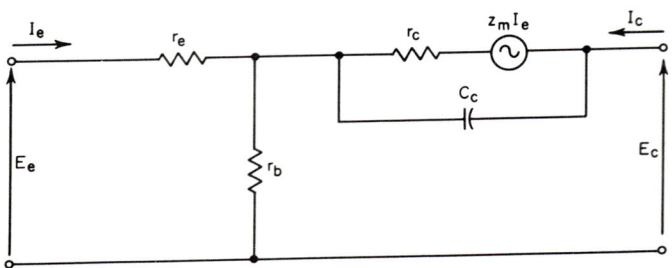

FIG. 30. Equivalent circuit for common-base amplifier.

Fig. 31b is therefore sketched. Starting with the collector branch of Fig. 28, we use Kirchhoff's law, $I_e = -(I_b + I_c)$, to obtain a branch with two current sources, from which we would like to remove the unnecessary aI_c source. We let

$$z_c \equiv \frac{r_c/j\omega C_c}{r_c + 1/j\omega C_c}$$

$$= \frac{r_c}{1 + j\omega r_c C_c}$$

$$b \equiv \frac{a}{1 - a} \qquad (98)$$

$$1 + b = \frac{1}{1 - a}$$

$$1 - a = \frac{1}{1 + b}$$

We now use the superposition theorem and disregard the aI_b source for the moment. The total current flowing through z_c, namely, $(1 - a)I_c$, causes an open-circuit voltage $z_c(1 - a)I_c = [z_c/(1 + b)]I_c$. The same voltage would of course be caused by I_c flowing through an impedance $z_c/(1 + b)$. Thus the aI_c source can be eliminated if we change the impedance level by the factor $1/(1 + b)$. However, the aI_b source must now be considered: in order to maintain the open-circuit voltage due to it (which is $aI_b z_c$) at the same value, it is necessary to multiply it by $(1 + b)$ (since z_c was divided by this factor). The source therefore becomes $a(1 + b)I_b = [a/(1 - a)]I_b = bI_b$.

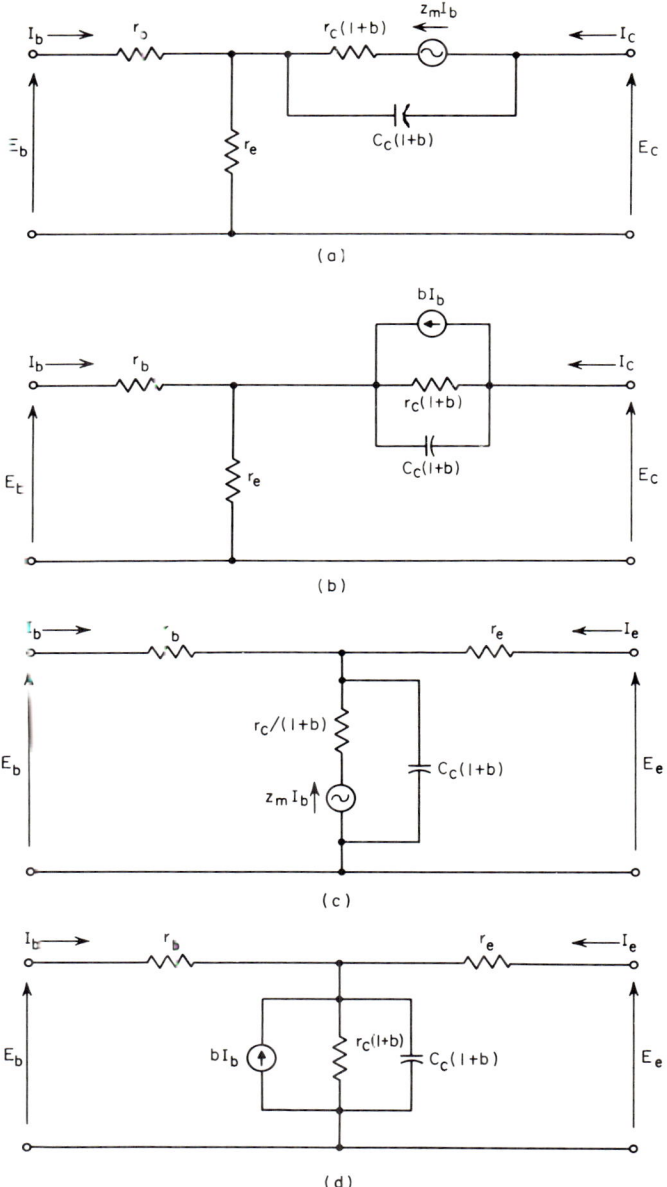

Fig. 31. Equivalent circuits for common-emitter and common-collector amplifiers. (a) Common-emitter circuit with voltage source. (b) Common-emitter circuit with current source. (c) Common-collector circuit with voltage source. (d) Common-collector circuit with current source.

2-36 AMPLIFIER FUNDAMENTALS

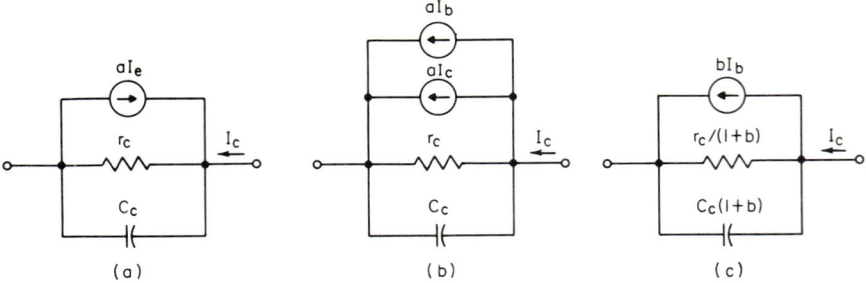

Fig. 32. Derivation of collector branch in common-emitter equivalent circuit. (a) Collector arm of common-base circuit. (b) Collector arm for common-emitter circuit (using two sources). (c) Equivalent of b with one source.

To change an impedance level by a factor it is only necessary to multiply the resistances and divide the capacitances by that factor. Thus the sequence of steps is shown in Fig. 32.

It is clear that these circuits will now present complicated input and output impedances; the method of calculation is direct, however. As an example, let us calculate the input impedance of a common-base amplifier with a load of Z_L. Substitution into the formula

$$Z_i = z_{11} - \frac{z_{12}z_{21}}{z_{22} + Z_L} \tag{99}$$

of the values

$$\begin{aligned}
z_{11} &= r_e + r_b \\
z_{12} &= r_b \\
z_{22} &= r_b + \frac{jX_c r_c}{r_c + jX_c} \quad \left(\text{where } X_c = \frac{-1}{\omega C_c}\right) \\
&= r_b + z_c \\
z_{21} &= r_b + \frac{jX_c z_m}{r_c + jX_c} \\
&= r_b + a z_c
\end{aligned} \tag{100}$$

and simplification yield

$$\begin{aligned}
Z_i &= r_e + r_b - \frac{r_b(r_b + az_c)}{r_b + z_c + Z_L} \\
&= r_e + \frac{r_b[(1 - a)z_c + Z_L]}{r_b + z_c + Z_L}
\end{aligned} \tag{101}$$

The other impedances of interest may be calculated similarly.

REFERENCES

1. B. D. H. Tellegen, A General Network Theorem, with Applications, *Philips Res. Rept.*, vol. 7, pp. 259–269, August, 1962.
2. E. A. Guillemin, "The Mathematics of Circuit Analysis," John Wiley & Sons, Inc., New York, 1949.
3. Y. W. Lee, Synthesis of Electric Networks by Means of the Fourier Transforms of Laguerre's Functions, *J. Math. Phys.*, vol. 11, no. 2, pp. 83–113, June, 1932.
4. L. Weinberg, "Network Analysis and Synthesis," McGraw-Hill Book Company, New York, 1962.
5. E. A. Guillemin, "Theory of Linear Physical Systems," John Wiley & Sons, Inc., New York, 1963.
6. T. S. Gray, "Applied Electronics," 2d ed., John Wiley & Sons, Inc., New York, 1954.
7. E. J. Angelo, Jr., "Electronic Circuits," 2d ed., McGraw-Hill Book Company, New York, 1964.
8. J. G. Linvill and J. F. Gibbons, "Transistors and Active Circuits," McGraw-Hill Book Company, New York, 1961.

Chapter 3

MATRIX REPRESENTATIONS OF MULTIPORT NETWORKS

LOUIS WEINBERG*

CONTENTS

1. Introduction.... 3-1
2. Matrices for Two-port and n-port Networks.................... 3-3
   2.1. Open-circuit Impedances and Short-circuit Admittances..... 3-3
   2.2. g Matrix and h Matrix............................. 3-7
   2.3. Chain Matrix.. 3-8
   2.4. Image Parameter Matrix............................... 3-9
3. Equations for Terminated Networks............................ 3-13
   3.1. Characterization in Terms of $[z]$, $[y]$, $[g]$, and $[h]$............. 3-14
   3.2. Chain-matrix Characterization......................... 3-20
4. Power-transfer Concepts, Matching, and Potential Instability.... 3-22
5. Interconnections of Quadripoles........................... 3-26

1. INTRODUCTION

As has been indicated previously, a network to be designed may be characterized as an n-port or n-terminal device; for different applications one or the other may be required. The number and type of equations or parameters necessary for a complete characterization depend on the type of network—e.g., whether it is a reciprocal or nonreciprocal network, whether it contains only resistances—and on the complexity of the network as measured by the number of terminals or ports it possesses. A one-port, for example, is described by one linear equation relating the port voltage and current, and thus by one parameter. When the one-port contains only resistances, the parameter is a real number; when it contains lumped elements, the parameter is a rational function of the complex frequency s; finally, when it contains distributed elements the parameter is a transcendental function of s.

The parameters used to characterize a network are conveniently presented in the form of a matrix. An nth-order matrix, as we shall see, is required for an n-port network and also for an n-terminal network; the elements of the matrix for the latter, however, are not independent.

Each type of matrix is convenient for a particular class of networks or network

* Department of Electrical Engineering, University of Michigan, Ann Arbor, Mich.

3-1

configurations, where convenience may be judged by such criteria as the elimination of unnecessary calculations in analysis, the clear indication of basic network properties and the interrelations of these properties, and the simple and accurate means available for measuring the elements of one matrix type compared with those of the other types of matrices. The open-circuit impedance matrix, for example, is most convenient as an n-port characterization when the ports are all open-circuited, the short-circuit admittance matrix when the ports are short-circuited, the chain matrix for the analysis and design of a chain of cascaded quadripoles, and the image parameter matrix for (both lumped-parameter and distributed-parameter) quadripoles cascaded on an image-match basis. When we are mainly interested in the transfer and distribution of average power throughout a network terminated in resistances or general impedances, the scattering matrix is an appropriate and powerful tool. When in addition the network consists of a cascade of component quadripoles, the transfer scattering matrix is a simpler tool for characterizing the component quadripoles than the scattering matrix.

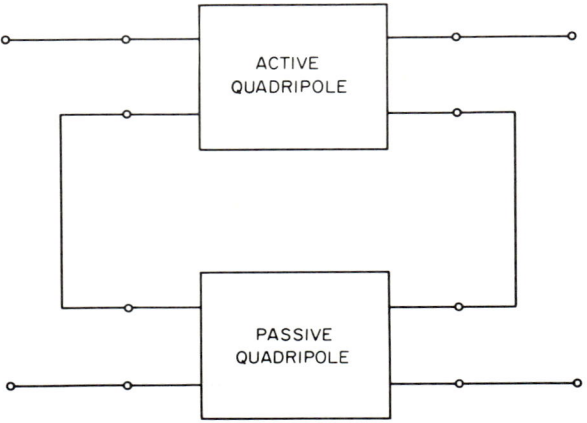

Fig. 1. Series connection of a passive quadripole with an active quadripole in order to change the properties of the latter.

These matrices are also useful for calculating in a simple manner the properties of an overall network that is formed by the interconnection of two or more n-port networks or n-terminal networks. For a number of reasons such interconnections are often used. Consider, for example, an active two-port device that is to be used in the design of a specified quadripole. The parameters of the device may change with time or with changes in ambient conditions; the device, because it is not unilateral, does not satisfy a requirement that its input impedance be independent of the load termination; or the input or output impedance may differ from the specified impedances. To correct any or all of these properties a two-port passive network may be interconnected with the active two-port; for example, a series interconnection—that is, the series connection of the input ports and the series connection of the output ports—is shown in Fig. 1. For any interconnection the parameters of the composite quadripole are obtained by a simple operation on the matrices of the individual two-ports. For each of the various interconnections a different matrix characterization of the individual quadripoles is convenient; for example, the open-circuit impedance matrix for each of the quadripoles in Fig. 1 is convenient for obtaining the matrix of the composite quadripole.

It will be shown that all the matrix types are related by linear transformations so that each may be converted to any other. Thus each type may actually be used to characterize any class of network (provided, of course, that the matrix exists); an open-circuit impedance matrix, for example, is often used in a fairly simple manner for the analysis and synthesis of resistance-terminated networks. The point to be recognized, however, is that we often sacrifice simplicity and convenience by limiting our analytical tools to one or two of these matrix types, and that network properties, almost self-evident in one type of characterization, are obscured by the linear transformation used to convert to another type of matrix. In short, we should be familiar with all these demonstrably useful types of matrices so that each seems as natural and easy to use as any other, and, in accordance with the philosophic principle known as Occam's razor (which states that entities are not to be multiplied unnecessarily), we use the simplest type for a particular class of problems.

In this chapter and the following one we present the various matrices used to characterize networks and then discuss possible interconnections of these networks and the corresponding simple operations on the network matrices of the component networks. We also show how the matrix parameters are used for obtaining additional important properties of quadripoles, for example, their input and output characteristics when terminated in a generator impedance and a simple load or the load presented by a cascaded quadripole, and formulas that specify the transfer of power by the quadripoles.

The matrices considered in this chapter are the open-circuit impedance and short-circuit admittance matrices, the g and h matrices, the chain matrix, and the image parameter matrix. We postpone to the next chapter a treatment of the scattering matrix, the transfer scattering matrix, and multiterminal networks characterized by the indefinite admittance and impedance matrices. Because a presentation of the scattering matrices, especially with complex normalization, is not yet widely available in book form, we consider them in some detail. Similarly, because the importance and simplicity of the indefinite matrices for the analysis of multistage transistor amplifiers are not stressed adequately, we give a fairly expanded treatment of their properties.

2. MATRICES FOR TWO-PORT AND n-PORT NETWORKS

Two linear equations completely characterize the behavior of a quadripole at its accessible ports; an n-port network requires n equations for its complete description. There are many different sets of equations that may be used, the type used depending on the choice of independent variables. In general, the equations of one set may be derived from those of any other set by appropriate linear transformations; the set used in a particular problem depends mainly on convenience.

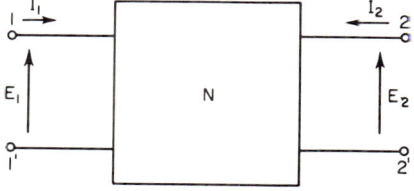

Fig. 2. General quadripole that is described by two linear equations.

In this section we describe some of the different sets of equations and corresponding matrices for n-port networks. Most of the matrix properties are illustrated by means of two-port networks, but n-port networks are also considered.

2.1. Open-circuit Impedances and Short-circuit Admittances. Two convenient sets of equations to describe the quadripole shown in Fig. 2 are given by

$$E_1 = z_{11}I_1 + z_{12}I_2$$
$$E_2 = z_{21}I_1 + z_{22}I_2 \quad (1)$$

and the inverse set of equations

$$I_1 = y_{11}E_1 + y_{12}E_2$$
$$I_2 = y_{21}E_1 + y_{22}E_2 \quad (2)$$

It is convenient to discuss these together. The z's are called the *open-circuit impedances*, whereas the y's are called the *short-circuit admittances;* either set completely characterizes the network. The definitions of these system functions are contained in the equations. From them we obtain

$$z_{11} = E_1/I_1|_{I_2=0} \qquad z_{21} = E_2/I_1|_{I_2=0}$$
$$z_{12} = E_1/I_2|_{I_1=0} \qquad z_{22} = E_2/I_2|_{I_1=0} \tag{3}$$

and

$$y_{11} = I_1/E_1|_{E_2=0} \qquad y_{21} = I_2/E_1|_{E_2=0}$$
$$y_{12} = I_1/E_2|_{E_1=0} \qquad y_{22} = I_2/E_2|_{E_1=0} \tag{4}$$

The above equations make clear why the impedances are designated as open-circuit functions, whereas the admittances are short-circuit functions: each z_{ik} represents a voltage response when all ports are open-circuited and a unit current is applied to the input port; each y_{ik} represents a current response to a unit-voltage input when all ports are short-circuited. For example, to evaluate the open-circuit forward transfer impedance z_{21}, a current source I_1 is used as the input to port 1-1' and I_2 is made to equal zero; that is, port 2-2' is open-circuited. Since the internal impedance of a current source is infinite (or an open circuit), it is clear that open-circuit conditions exist at both ports. An analogous discussion applies to the short-circuit admittances.

Extension to the n-port network is straightforward. The set of n impedance equations is given in matrix form by

$$E] = [z] I] \tag{5}$$

where

$$[z] = \begin{bmatrix} z_{11} & z_{12} & \cdots & z_{1n} \\ z_{21} & z_{22} & \cdots & z_{2n} \\ \vdots & & & \vdots \\ z_{n1} & z_{n2} & \cdots & z_{nn} \end{bmatrix} \tag{6}$$

the elements of the matrix are defined by

$$z_{sk} = E_s/I_k|_{I_1=I_2=\cdots=I_{k-1}=I_{k+1}=\cdots=I_{n-1}=I_n=0} \tag{7}$$

and $E]$ and $I]$ indicate column matrices. Again it is clear that *all* ports are open-circuited.

Since the equations

$$[z] I] = E] \tag{8}$$

may be obtained by solving for $E]$ in

$$[y] E] = I] \tag{9}$$

whenever $[y]$ is nonsingular—that is, the two sets of equations are, as was stated previously, inverse sets—the important relation

$$[z] = [y]^{-1} \tag{10}$$

is obtained

In this handbook, unless the context indicates otherwise, the lowercase z's and y's with *double-subscript* notation will be used to mean the open-circuit impedances and short-circuit admittances, respectively. These functions are very useful in synthesis: their use in characterizing a terminated quadripole will be amplified later. Here we now discuss some basic properties of $[z]$ and $[y]$, and later show how they are used for obtaining the matrix of an overall quadripole composed of an interconnection of two or more component quadripoles.

The n^2 elements in an nth-order matrix that describes a nonreciprocal n-port are all independent; for a reciprocal n-port the matrix is symmetrical, that is, for matrix A its elements $a_{ik} = a_{ki}$ for all i, k, so that only $n(n+1)/2$ elements are independent. We saw that the four parameters in an equivalent circuit for the nonreciprocal two-port contained three two-terminal bilateral elements plus one controlled source; this representation can be carried out in general so that for a nonreciprocal n-port its matrix

MATRIX REPRESENTATIONS OF MULTIPORT NETWORKS 3-5

with n^2 independent elements can be represented conveniently by $n(n+1)/2$ two-terminal elements plus $n(n-1)/2$ controlled sources.*

An important property of $[z]$ and $[y]$ concerns their principal submatrices. If $[z]$ is the open-circuit impedance matrix of an n-port network N, then the principal submatrix $[z']$ formed from $[z]$ by deleting row k and column k is the open-circuit impedance matrix of the same network with the same ports, except that port k is open-circuited and is now considered an inaccessible pair of terminals. This should be clear from physical considerations, since the same network is considered in both cases; or it may be seen mathematically, since

$$z_{11}' = z_{11} = E_1/I_1|_{I_2=I_3=\ldots=I_k=\ldots=I_n=0}$$
$$z_{21}' = z_{21} = E_2/I_1|_{I_2=I_3=\ldots=I_k=\ldots=I_n=0} \quad (11)$$

and so on for all the terms of $[z']$.

This process of deleting rows and corresponding columns can be repeated on the new matrix $[z']$. Thus a principal submatrix of order $p < n$, formed from $[z]$ by deleting the $n-p$ rows $j, k, l, \ldots$ and the same columns, is the open-circuit impedance matrix of the p-port network formed from N by open-circuiting the $n-p$ ports $j, k, l, \ldots$ and using the other p ports as the points of access to the network.

The analogous property holds for $[y]$. Here ports corresponding to the deleted rows and columns must be *short-circuited*.

It is essential for the designer to be familiar with the relationships and distinctions between the $[z]$ and $[y]$ matrices, which are used to describe the *external* properties of a network, and the loop-impedance matrix and the node-admittance matrix, which are used in analysis to describe the *internal* properties of a network. The many facets of these relationships are discussed in detail in the literature;[1] we discuss some of them more briefly here.

Consider the loop equations for a network N given by

$$\begin{bmatrix} \zeta_{11} & \zeta_{12} & \cdots & \zeta_{1l} \\ \zeta_{21} & \zeta_{22} & \cdots & \zeta_{2l} \\ \cdots & \cdots & \cdots & \cdots \\ \zeta_{l1} & \zeta_{l2} & \cdots & \zeta_{ll} \end{bmatrix} \begin{bmatrix} I_1 \\ I_2 \\ \cdot \\ I_l \end{bmatrix} = \begin{bmatrix} E_1 \\ E_2 \\ \cdot \\ E_l \end{bmatrix} \quad (12)$$

or more compactly by

$$[\zeta\, I] = [E] \quad (13)$$

It is noted that the *form* of these equations is the same as that of the open-circuit impedance equations. However, $[\zeta]$ should not be confused with an open-circuit impedance matrix even though we have formally

$$\zeta_{ik} = E_i/I_k|_{I_1=I_2=\ldots I_{k-1}=I_{k+1}=\ldots=I_l=0} \quad (14)$$

A first difference is that†

$$\zeta_{ik} = R_{ik} + L_{ik}s + 1/C_{ik}s \quad (15)$$

whereas z_{ik} is in general a more complicated rational function. Secondly, the network N described by $[\zeta]$ can have only l independent loops, whereas a network whose *external behavior* is described by $[z]$ may have any internal complexity. This fact is further obscured in the important case of a pure resistance network since the elements of both $[\zeta]$ and $[z]$ are then real numbers. Finally, it may not be possible to form an l-port network from N and thus interpret $[\zeta]$ as an open-circuit impedance matrix.

---

* This does not imply that $n(n-1)/2$ controlled sources are required in the realization of an n-port; this is merely a convenient representation derived by the same manipulation of the n equations as was performed on the two equations of the two-port. In fact, it has been shown by Sandberg (in I. W. Sandberg, Synthesis of N-port Active RC Networks, *Bell System Tech J.*, vol. 40, pp. 329–346, January, 1961) that only n controlled sources plus an RC network are required.

† For $i = k$ the constants R_{ik}, L_{ik}, and C_{ik} are nonnegative. However, when $i \neq k$ they may have any sign, depending on the relative reference directions of the currents I_i and I_k flowing through the network element.

AMPLIFIER FUNDAMENTALS

Now assume that only E_1 and E_2 are nonzero and that it is possible to make two external ports so that only I_1 flows through E_1 and only I_2 flows through E_2. We then find the elements of the $[y]$ of the resulting two-port by the equations

$$\begin{aligned} y_{11} &= A_{11}/A \\ y_{22} &= A_{22}/A \\ y_{12} &= A_{21}/A \\ y_{21} &= A_{12}/A \end{aligned} \tag{16}$$

where A is the determinant of $[\zeta]$ and A_{ik} is the cofactor of the element in the ith row and kth column. The z's may now be obtained from the y's by

$$\begin{aligned} z_{11} &= \frac{y_{22}}{|y|} = \frac{A_{22}A}{A_{11}A_{22} - A_{12}A_{21}} \\ z_{22} &= \frac{y_{11}}{|y|} = \frac{A_{11}A}{A_{11}A_{22} - A_{12}A_{21}} \\ z_{12} &= \frac{-y_{12}}{|y|} = \frac{-A_{21}A}{A_{11}A_{22} - A_{12}A_{21}} \\ z_{21} &= \frac{-y_{21}}{|y|} = \frac{-A_{12}A}{A_{11}A_{22} - A_{12}A_{21}} \end{aligned} \tag{17}$$

Since by the Jacobi theorem*

$$A A_{jk}{}^{jk} = A_{jj}A_{kk} - A_{jk}A_{kj} \tag{18}$$

where the new symbol $A_{jk}{}^{jk}$ represents the cofactor† formed by suppressing rows j and k and the same columns of $[\zeta]$, we obtain

$$\begin{aligned} z_{11} &= A_{22}/A_{12}{}^{12} \\ z_{22} &= A_{11}/A_{12}{}^{12} \\ z_{12} &= -A_{21}/A_{12}{}^{12} \\ z_{21} &= -A_{12}/A_{12}{}^{12} \end{aligned} \tag{19}$$

An analogous discussion may be given when node voltages rather than loop currents are chosen as the variables. Instead of considering the equations for a general set of voltages, we consider the node-to-datum set of equations. Here a reference, or datum, node is chosen, and a voltage variable E_k is then used to define the voltage rise from the datum to node k. The equations describing a network N with $n+1$ nodes are

$$[\eta] E] = I] \tag{20}$$

where the elements of the nth-order matrix of $[\eta]$ are of the form

$$\eta_{ik} = G_{ik} + C_{ik}s + 1/L_{ik}s \tag{21}$$

and G_{ik}, C_{ik}, and L_{ik} are nonpositive numbers when $i \neq k$ and nonnegative numbers when $i = k$. Here, however, in contrast with the loop-impedance case, $[\eta]$ may always be considered as a short-circuit admittance matrix; specifically, it describes the n-port formed from N in which all the n-ports have a common terminal. It should be clear that it is not a general form of short-circuit admittance matrix since the network N is restricted to have precisely $n+1$ nodes (each of which becomes a terminal of the n-port), and η_{ik} is at most a quadratic divided by a linear term.

---

* The reader should become familiar with the Jacobi theorem and the more general Binet-Cauchy theorem; both are exceedingly useful in network theory. Aitken[2] expresses the matter well when he says of the Binet-Cauchy theorem, "The elegance of this theorem is equalled only by the wealth of its applications."

† In accordance with the notation A_{jk}, we could have used A_{jkjk} to represent the new cofactor. However, it is convenient to introduce the notation of (18) which is used often in the literature.

Now if only I_1 and I_2 are nonzero, we can form a grounded quadripole with nodes 1 and 2 as the positive terminals of ports 1 and 2, respectively.

Solving for $[z]$ of the resulting quadripole, we have

$$\begin{aligned} z_{11} &= \Delta_{11}/\Delta \\ z_{22} &= \Delta_{22}/\Delta \\ z_{12} &= \Delta_{21}/\Delta \\ z_{21} &= \Delta_{12}/\Delta \end{aligned} \qquad (22)$$

where Δ is the determinant and Δ_{ik} a cofactor of $[\eta]$.* Correspondingly, we have

$$\begin{aligned} y_{11} &= \Delta_{22}/\Delta_{12}{}^{12} \\ y_{22} &= \Delta_{11}/\Delta_{12}{}^{12} \\ y_{12} &= -\Delta_{21}/\Delta_{12}{}^{12} \\ y_{21} &= -\Delta_{12}/\Delta_{12}{}^{12} \end{aligned} \qquad (23)$$

It should be noted that to find $[z]$, in the form given in Eqs. (19) from the loop equations, it was necessary to make intermediate use of Eqs. (16) and (17). Analogously, to determine Eqs. (23) for $[y]$ in terms of $[\eta]$, it was necessary to use Eqs. (22) and $[y] = [z]^{-1}$ in intermediate steps. It should be mentioned, however, that there are also *direct* procedures for determining any system function from a specified matrix.[1]

2.2. g Matrix and h Matrix. The open-circuit impedances are used to express the voltages of an n-port in terms of the currents, whereas the short-circuit admittances express the currents in terms of the voltages. In many situations it is convenient to express one set of *mixed* variables, that is, voltages at some ports and currents at other ports, in terms of the corresponding port currents and voltages; to do this we use the so-called *hybrid* parameters. This can be done for arbitrary sets of mixed currents and voltages for the n-port;[1] we consider here only the two-port case where the matrices are designated as $[g]$ and $[h]$.

The h matrix is defined by the set of equations

$$\begin{bmatrix} E_1 \\ I_2 \end{bmatrix} = \begin{bmatrix} h_{11} & h_{12} \\ h_{21} & h_{22} \end{bmatrix} \begin{bmatrix} I_1 \\ E_2 \end{bmatrix} \qquad (24)$$

and the g matrix by

$$\begin{bmatrix} I_1 \\ E_2 \end{bmatrix} = \begin{bmatrix} g_{11} & g_{12} \\ g_{21} & g_{22} \end{bmatrix} \begin{bmatrix} E_1 \\ I_2 \end{bmatrix} \qquad (25)$$

Again these basic equations yield the definitions of the parameters. The elements of $[h]$ are defined by

$$\begin{aligned} h_{11} &= E_1/I_1|_{E_2=0} & h_{12} &= E_1/E_2|_{I_1=0} \\ h_{21} &= I_2/I_1|_{E_2=0} & h_{22} &= I_2/E_2|_{I_1=0} \end{aligned} \qquad (26)$$

and those of $[g]$ by

$$\begin{aligned} g_{11} &= I_1/E_1|_{I_2=0} & g_{12} &= I_1/I_2|_{E_1=0} \\ g_{21} &= E_2/E_1|_{I_2=0} & g_{22} &= E_2/I_2|_{E_1=0} \end{aligned} \qquad (27)$$

The first point to note is that like the z's and y's the g's and h's are all system functions, that is, the ratio of a response to an input. For example, h_{11} is the *short-circuit* driving-point impedance at port 1, that is, the input impedance at port 1 with the terminals of port 2 shorted; thus $h_{11} = 1/y_{11}$. The open-circuit voltage ratio h_{12} is equal to the voltage at open-circuited port 1 for a unit voltage applied to port 2. Furthermore, it is clear from the definitions that satisfaction of the reciprocity theorem requires that $g_{12} = -g_{21}$ and $h_{12} = -h_{21}$.

These parameters are often useful because their use eliminates accuracy problems that are present in making measurements specified by the other sets of parameters, or

---

* Any desired transfer ratio may be obtained as the ratio of two cofactors, with the denominator a *principal* cofactor; for example, $E_k/E_i = \Delta_{ik}/\Delta_{ii}$. However, the mistake should not be made of determining E_k/E_i as Δ_{jk}/Δ_{ji} by use of the relationship $E_k/E_i = z_{kj} \times 1/z_{ij}$. As discussed in Ref. 1, since $1/z_{ij}$ is not a system function, the last equation is not valid.

when we are interconnecting two or more quadripoles in series at one port and in parallel at the other; such interconnections are discussed in Sec. 5.

2.3. Chain Matrix. Often one realizes a given system function with a quadripole formed by the cascade connection of two or more simpler quadripoles. Such a two-port network* is shown in Fig. 3 where it is specified that 1-1' is the input port. From the figure it is evident that

$$E_1 = E_1' \qquad I_1 = I_1'$$
$$E_2' = E_1'' \qquad I_2' = I_1'' \qquad (28)$$
$$E_2'' = E_2 \qquad I_2'' = I_2$$

For this form of interconnection the output voltage and the output current of the first two-port are common, respectively, with the input voltage and the input current of the second two-port network. The set of equations that is useful is given in terms of what were formerly called the *general circuit parameters*, whose matrix is now called

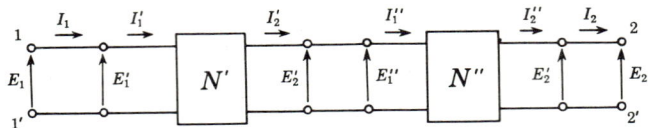

FIG. 3. A quadripole formed by the cascade connection of two component quadripoles.

the *chain matrix*. The essential property of the chain matrix is that, for a quadripole composed of two quadripoles connected in cascade, it is equal to the product of the chain matrices of the component quadripoles. The set of equations defining the chain matrix is

$$\begin{bmatrix} E_1 \\ I_1 \end{bmatrix} = \begin{bmatrix} A & B \\ C & D \end{bmatrix} \begin{bmatrix} E_2 \\ I_2 \end{bmatrix} \qquad (29)$$

whose parameters are therefore given by

$$A = E_1/E_2|_{I_2=0} \qquad C = I_1/E_2|_{I_2=0}$$
$$B = E_1/I_2|_{E_2=0} \qquad D = I_1/I_2|_{E_2=0} \qquad (30)$$

Since terminals 1-1' form the input port, it is therefore clear that each of the chain-matrix parameters is a reciprocal system function. The *reciprocal open-circuit voltage ratio* is given by A, whereas D is the *reciprocal short-circuit current ratio;* the *reciprocal short-circuit admittance* is equal to B, and C is the *reciprocal open-circuit impedance*.

Relationships of the general circuit parameters with the z's and y's may be simply derived. Using Eqs. (3) and dividing z_{11} by z_{21}, we obtain

$$z_{11}/z_{21} = E_1/E_2|_{I_2=0} \qquad (31)$$

But the right side of (31) is equivalent to A as defined in Eq. (30). Since $[y]$ is the inverse of $[z]$, the above relationship may also be written in terms of the y's as $-y_{22}/y_{21}$. Inspection of B as given in Eq. (30) shows that it is the negative of the reciprocal of y_{21} in Eq. (4). Proceeding in this way, we derive the two other relations; all four relations are collected below:

$$A = z_{11}/z_{21} = -y_{22}/y_{21} \qquad C = 1/z_{21} = -|y|/y_{21}$$
$$B = -1/y_{21} = |z|/z_{21} \qquad D = -y_{11}/y_{21} = z_{22}/z_{21} \qquad (32)$$

where $|z|$ and $|y|$ denote the determinants of the respective matrices. Algebraic manipulation of the above equations yields the z's and y's in terms of the chain-matrix

---

* It should be noted that the positive direction of the output current is defined to flow *out* of the upper terminal. This is convenient whenever the chain matrix is used, and is adopted here in all subsequent use of the chain-matrix parameters.

parameters:

$$z_{11} = A/C \qquad z_{21} = 1/C \qquad z_{22} = D/C$$
$$y_{11} = D/B \qquad y_{21} = -1/B \qquad y_{22} = A/B \qquad (33)$$

It is evident that the condition for a symmetrical quadripole—that is, for z_{11} to equal z_{22}—is $A = D$.

Now returning to the network N' in Fig. 3, we have

$$\begin{bmatrix} E_1' \\ I_1' \end{bmatrix} = \begin{bmatrix} A' & B' \\ C' & D' \end{bmatrix} \begin{bmatrix} E_2' \\ I_2' \end{bmatrix} \qquad (34)$$

which, by use of the relationships in Eq. (28), becomes

$$\begin{bmatrix} E_1 \\ I_1 \end{bmatrix} = \begin{bmatrix} A' & B' \\ C' & D' \end{bmatrix} \begin{bmatrix} E_1'' \\ I_1'' \end{bmatrix} \qquad (35)$$

Similarly for N'' we obtain

$$\begin{bmatrix} E_1'' \\ I_1'' \end{bmatrix} = \begin{bmatrix} A'' & B'' \\ C'' & D'' \end{bmatrix} \begin{bmatrix} E_2'' \\ I_2'' \end{bmatrix} \qquad (36)$$

which becomes

$$\begin{bmatrix} E_1'' \\ I_1'' \end{bmatrix} = \begin{bmatrix} A'' & B'' \\ C'' & D'' \end{bmatrix} \begin{bmatrix} E_2 \\ I_2 \end{bmatrix} \qquad (37)$$

Substitution of the matrix product given in (37) for the output voltage and current matrix of (35) yields

$$\begin{bmatrix} E_1 \\ I_1 \end{bmatrix} = \begin{bmatrix} A' & B' \\ C' & D' \end{bmatrix} \begin{bmatrix} A'' & B'' \\ C'' & D'' \end{bmatrix} \begin{bmatrix} E_2 \\ I_2 \end{bmatrix}$$
$$= \begin{bmatrix} A & B \\ C & D \end{bmatrix} \begin{bmatrix} E_2 \\ I_2 \end{bmatrix} \qquad (38)$$

The above analysis may be repeated for this overall quadripole cascaded with another quadripole. Therefore, the chain matrix of a composite quadripole formed by the cascade connection of two or more component quadripoles is given by the matrix product of the chain matrices of the component quadripoles.

2.4. Image Parameter Matrix. In the design of filters and delay networks we often make use of constant-resistance networks; this allows the cascading of quadripoles without interaction. As an example the constant-resistance lattice is shown

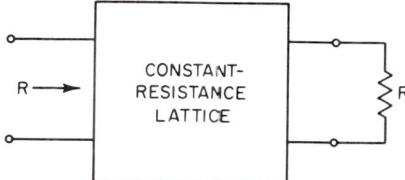

Fig. 4. Example of a constant-resistance quadripole.

in Fig. 4; such a network may be designed as a cascade of several less complex networks, as shown in Fig. 5, where because of the constant-resistance property each quadripole may be designed separately. We in effect achieve isolation, i.e., the absence of interactions between quadripoles, without the use of isolating amplifiers. Thus the design problem and also the physical-realization problem in the laboratory are simplified. The overall magnitude characteristic is simply the product of the magnitude characteristics of the separate quadripoles and the overall phase is the sum of the phases. In addition, when the constant-resistance quadripole provides an all-pass time-delay characteristic, it is possible to achieve a desired total delay and tap-off points between quadripoles at prescribed fractions of the total delay.[1]

Since at the junction of two component quadripoles or of a quadripole and a two-terminal termination, we have a *matched* condition, that is, the impedance at a junction is the same looking to the left and to the right or, stated in another way, each impedance sees its *image*, we say that we have an *image match*. This concept has been generalized so that instead of image resistances there are general impedances at each junction; the parameters defined on the basis of this concept are called *image parameters*. Though there is a certain artificiality in the concept, since the image impedances that are required as terminations are irrational functions and thus not realizable by lumped-parameter networks, the concept is still useful as a characterization of the two-port and has been applied powerfully in the classical image parameter method of filter design. In fact, the characterization is borrowed from analogy with a uniform transmission line, which, since it is a symmetrical and reciprocal structure, is completely described by two functions, a characteristic impedance and a propagation function. For unsymmetrical reciprocal networks we of course need three functions. Though it is possible to extend the image parameter characterization to nonreciprocal networks, it has found its greatest utility for the treatment of passive reciprocal networks; we therefore restrict our discussion to this case.

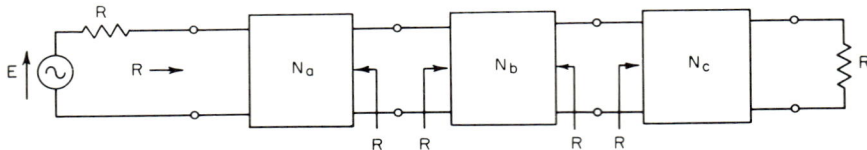

Fig. 5. Achievement of a desired quadripole as a cascade of constant-resistance component quadripoles.

It should be noted that this concept of matching, that is, terminating a network in a load equal to its image impedance, differs from matching for maximum power transfer, where, as we recall, the load must be equal to the complex conjugate of the output impedance. To distinguish the two we can refer to the image matching as *reflectionless* matching.

First we define the image impedances. From our preceding discussion we see that a two-port is image-matched if both the following requirements are satisfied:

$$Z_{11} = Z_{I1}$$
$$Z_{22} = Z_{I2}$$
(39)

where for the two-port N shown in Fig. 6a, the operational definitions of Z_{11} and Z_{22} are shown in Fig. 6b and c, respectively. Using the relationships $E_2 = -Z_{I2}I_2$ and $E_1 = Z_{I1}I_1$ in the open-circuit impedance equations, and also using the fact that reciprocal networks are assumed, we obtain by straightforward manipulation

$$Z_{11} = z_{11} - \frac{z_{12}^2}{z_{22} + Z_{I2}}$$
$$Z_{22} = z_{22} - \frac{z_{12}^2}{z_{11} + Z_{I1}}$$
(40)

Eliminating Z_{11} and Z_{22} from (40) by use of (39) and solving the resulting set of simultaneous equations yield

$$Z_{I1} = \sqrt{z_{11}/y_{11}}$$
$$Z_{I2} = \sqrt{z_{22}/y_{22}}$$
(41)

When the quadripole is symmetrical, i.e., $z_{11} = z_{22}$ and $y_{11} = y_{22}$, then $Z_{I1} = Z_{I2}$, as is required.

The image impedances may also be expressed in terms of the elements of the chain

MATRIX REPRESENTATIONS OF MULTIPORT NETWORKS   3-11

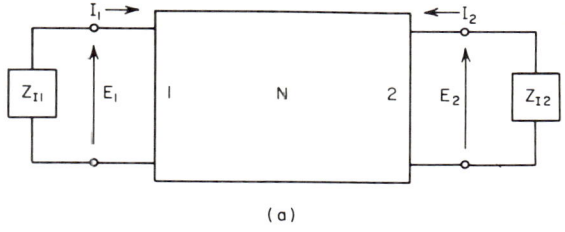

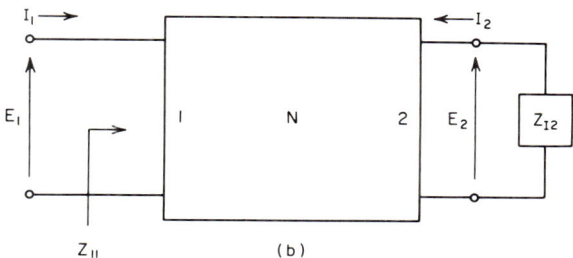

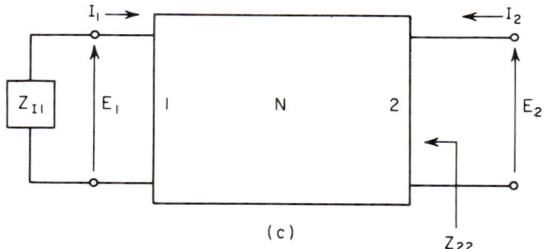

Fig. 6. Terminated two-port network with definitions of input impedances used in determining whether image-match condition exists.

matrix; by use of (33) and (41) we have

$$Z_{I1} = \sqrt{AB/CD}$$
$$Z_{I2} = \sqrt{BD/AC} \tag{42}$$

In addition, if we start with the chain-matrix equations, we can obtain for the transfer ratios of an image-matched quadripole

$$E_2/E_1 = \sqrt{D/A}\,(\sqrt{AD} - \sqrt{BC})$$
$$-I_2/I_1 = \sqrt{A/D}\,(\sqrt{AD} - \sqrt{BC}) \tag{43}$$

where $-I_2$ is the current flowing out of the upper terminal of port 2. The third function required for describing the quadripole can now be defined; if we multiply the corresponding sides of the above two equations, we obtain

$$\begin{aligned}-E_2 I_2/E_1 I_1 &= (\sqrt{AD} - \sqrt{BC})^2 \\ &\equiv e^{-2\theta} \\ &= e^{-2(\alpha_I + i\beta_I)}\end{aligned} \tag{44}$$

which serves to define the *image propagation function* θ. In accordance with usage in image parameter theory we have used a negative sign in defining the propagation function so that α_I is the *image loss function* (rather than a gain function which we used in our previous definition of the propagation function γ).

The transfer ratios in (43) can now be expressed in terms of the propagation function; using (42) to eliminate the first radical in each of Eqs. (43), we then have by the definition in (44)

$$E_2/E_1 = \sqrt{Z_{I2}/Z_{I1}}\, e^{-\theta}$$
$$-I_2/I_1 = \sqrt{Z_{I1}/Z_{I2}}\, e^{-\theta} \tag{45}$$

for an image-matched quadripole. When the quadripole is symmetrical

$$E_2/E_1 = -I_2/I_1 = e^{-\theta} \tag{46}$$

When two quadripoles are connected in cascade as shown in Fig. 7, where at the junction of the quadripoles $Z_{I1b} = Z_{I2a}$, and the terminations are image-matched, then

$$\begin{aligned} E_3/E_1 &= (E_3/E_2)(E_2/E_1) \\ &= (\sqrt{Z_{I2b}/Z_{I1b}}\, e^{-\theta_a})(\sqrt{Z_{I2a}/Z_{I1a}}\, e^{-\theta_b}) \\ &= \sqrt{Z_{I2b}/Z_{I1a}}\, e^{-(\theta_a+\theta_b)} \end{aligned} \tag{47}$$

Thus the overall quadripole has a propagation function equal to the sum of the separate propagation functions and has image impedances Z_{I1a} and Z_{I2b}. It is clear

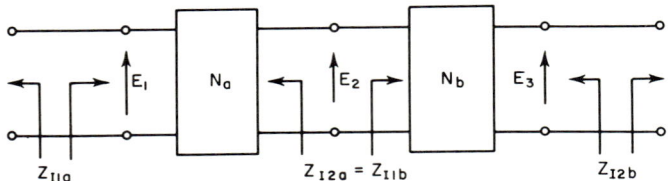

FIG. 7. Two image-matched quadripoles connected in cascade.

that this process holds in general; that is, if n quadripoles are connected in cascade so that the image impedance on the output side of one network is equal to the image impedance on the input side of the next, then for the overall network the input image impedance is equal to the input image impedance of the first network, the output image impedance is equal to the output image impedance of the last network, and the image propagation function is equal to the sum of the image propagation functions of the component networks.

It is also convenient to express the propagation function in terms of the open-circuit impedances and short-circuit admittances. Since

$$e^{-\theta} = \sqrt{AD} - \sqrt{BC} \tag{48}$$

we have

$$e^{\theta} = \sqrt{AD} + \sqrt{BC} \tag{49}$$

where use has been made of the reciprocity condition

$$AD - BC = 1 \tag{50}$$

Then we obtain simply

$$\begin{aligned} \cosh\theta &= (e^{\theta} + e^{-\theta})/2 = \sqrt{AD} \\ \sinh\theta &= (e^{\theta} - e^{-\theta})/2 = \sqrt{BC} \\ \coth\theta &= \cosh\theta/\sinh\theta = \sqrt{AD/BC} \end{aligned} \tag{51}$$

so that by use of (32)
$$\coth \theta = \sqrt{z_{11}y_{11}} = \sqrt{z_{22}y_{22}} \tag{52}$$

We can also express the chain-matrix elements in terms of the image parameters by use of Eqs. (42) and (51); we obtain

$$\begin{bmatrix} A & B \\ C & D \end{bmatrix} = \begin{bmatrix} \sqrt{Z_{I1}/Z_{I2}} \cosh \theta & \sqrt{Z_{I1}Z_{I2}} \sinh \theta \\ (1/\sqrt{Z_{I1}Z_{I2}}) \sinh \theta & \sqrt{Z_{I2}/Z_{I1}} \cosh \theta \end{bmatrix} \tag{53}$$

If we let
$$n = \sqrt{Z_{I1}/Z_{I2}} \tag{54}$$

the above matrix is like that of a transmission line with a characteristic impedance Z_{I2}, a propagation function θ, and an ideal transformer with an $n:1$ turns ratio at the input end. If we let the quadripole be symmetrical, i.e., $Z_{I1} = Z_{I2} = Z_I$, then the chain matrix is precisely that of a transmission line with characteristic impedance Z_I and propagation function θ. Thus this substantiates the statement made earlier that the image parameters are defined by analogy with a transmission line.

If two identical symmetrical quadripoles that are image-matched are connected in cascade, the overall chain matrix is the square of the matrix in (5) with $Z_{I1} = Z_{I2} = Z_I$, namely,

$$\begin{bmatrix} A & B \\ C & D \end{bmatrix}^2 = \begin{bmatrix} \cosh 2\theta & Z_I \sinh 2\theta \\ \sinh 2\theta/Z_I & \cosh 2\theta \end{bmatrix} \tag{55}$$

For n such quadripoles the chain matrix becomes

$$\begin{bmatrix} A & B \\ C & D \end{bmatrix}^n = \begin{bmatrix} \cosh n\theta & Z_I \sinh n\theta \\ \sinh n\theta/Z_I & \cosh n\theta \end{bmatrix} \tag{56}$$

In all the above we have considered that the quadripole under discussion is image-matched. If the matched condition is not satisfied, reflections will occur and the voltage ratio will of course no longer be given by (45). If the terminating impedances at ports 1 and 2 are given respectively by R_1 and R_2, it can be shown[3] that

$$\frac{E_2}{E_1} = \sqrt{\frac{Z_{I2}}{Z_{I1}}} e^{-\theta} \frac{2\sqrt{Z_{I1}R_1}}{Z_{I1} + R_1} \frac{2\sqrt{Z_{I2}R_2}}{Z_{I2} + R_2} \frac{1}{1 - \rho_1\rho_2 e^{-2\theta}} \tag{57}$$

where ρ_1 and ρ_2 are the reflection coefficients at the input and output, respectively, defined by

$$\rho_1 = \frac{Z_{I1} - R_1}{Z_{I1} + R_1}$$
$$\rho_2 = \frac{Z_{I2} - R_2}{Z_{I2} + R_2} \tag{58}$$

3. EQUATIONS FOR TERMINATED NETWORKS

Often a quadripole to be synthesized requires specified terminations; there are many different combinations that arise in practice. In some applications, for example, a resistance may be required at the output or at both the input and output, or the desired network may feed a tube whose nonnegligible input capacitance requires that the network have a shunt load capacitance. In still other applications, the load may be an antenna whose input impedance over the frequency range of interest is equivalent to an inductance and a resistance in series; here a network with the specified RL impedance as a termination is required. The source may be a transistor amplifier whose Thévenin equivalent impedance is thus required as an input termination on the quadripole; and the load may be the input impedance of another transistor amplifier. The quadripole itself may be passive or an active network with or without feedback. It becomes clear that the basic two-port with a source and complex impedance at the input and a complex impedance at the output can represent a variety of situations, and its properties are thus worthy of detailed analysis.

Two choices are open for realizing a network with specified terminations. With most synthesis procedures either can be used, and the more convenient is chosen. We can consider the complete network without the terminations shown explicitly; in this case the functions used directly in the synthesis—the open-circuit impedances, the general circuit parameters, or any others—characterize the *complete* network. We must be sure, however, to use a synthesis procedure capable of yielding the desired terminations. Use of the second method requires merely a change in point of view. Now the complete network consists of a partial network plus the terminations considered *explicitly*. The synthesis procedure requires that we first derive the functions

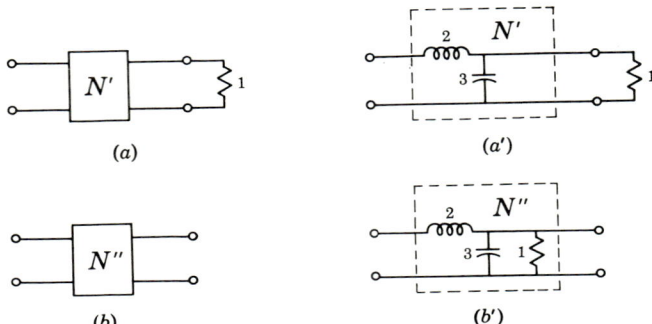

Fig. 8. Equivalent representations for a resistance-terminated network.

characterizing the *partial* network from the given system function for the *complete* network and then realize these derived functions.

A simple illustration makes the distinction clear. Suppose that a resistance-terminated network is desired to realize a given voltage ratio K. In Fig. 8a the desired network is shown as N' terminated in a 1-ohm resistance and in Fig. 8b as the open-circuited quadripole N''. The networks obtained when $K = 1/(6s^2 + 2s + 1)$ are shown in Fig. 8a' and b'. To synthesize Fig. 8a' the short-circuited admittances for N', $y_{21}' = -1/2s$, and $y_{22}' = (6s^2 + 1)/2s$ would be used; whereas for Fig. 8b' the admittances for N'' are $y_{21}'' = -1/2s$ and $y_{22}'' = (6s^2 + 2s + 1)/2s$. These admittances are obtained from K by formulas to be derived below.

Resistance-terminated quadripoles are required so often that it is useful to have equations describing them available for reference. Problems also arise in which the terminations are general impedances or admittances—for which we shall use the term *immittance*—and we shall consider these cases too. The equations are now derived, first in terms of the parameters of $[z]$, $[y]$, $[g]$, and $[h]$, and then in terms of the chain-matrix parameters.

3.1. Characterization in Terms of $[z]$, $[y]$, $[g]$, **and** $[h]$. Consider Fig. 9, which shows a quadripole terminated in a resistance. The characteristics of the complete network may be described in a number of ways; first we use the open-circuit impedances and the short-circuit admittances, and then later we also introduce the hybrid parameters. For the output terminals open-circuited, that is, for $I_2 = 0$, which occurs when R_L becomes infinite, the reciprocal voltage ratio has already been obtained in Eq. (31) by the use of the two simultaneous linear equations of Eq. (3). Therefore, the voltage ratio is

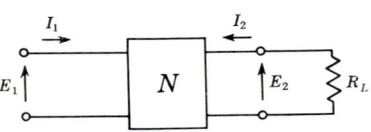

Fig. 9. Two-port network terminated in a resistance.

$$E_2/E_1 = z_{21}/z_{11} \tag{59}$$

The dual relationship, namely, the transfer current ratio for a short-circuited output, is

$$I_2/I_1 = y_{21}/y_{11} \tag{60}$$

For R_L finite and nonzero, the transfer impedance and transfer admittance may be obtained by applying Thévenin's theorem and Norton's theorem, respectively at the

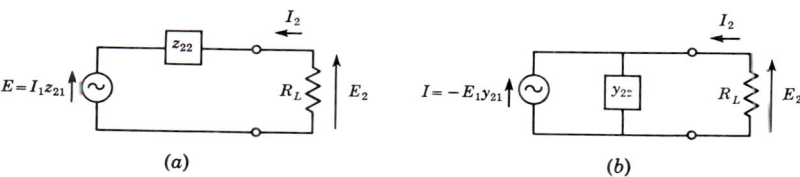

Fig. 10. Networks equivalent to that of Fig. 9, obtained by application of Thévenin's and Norton's theorems, respectively.

output terminals. The equivalent networks are shown in Fig. 10. From these a simple calculation yields the transfer impedance

$$Z_{21} \equiv E_2/I_1 = R_L z_{21}/(R_L + z_{22}) \tag{61}$$

and the dual equation for the transfer admittance

$$Y_{21} \equiv I_2/E_1 = G_L y_{21}/(G_L + y_{22}) \tag{62}$$

where the conductance $G_L \equiv 1/R_L$, and $\equiv$ means equal by definition. For R_L normalized to 1 ohm, these formulas become

$$Z_{21} = z_{21}/(1 + z_{22}) \tag{63}$$
and
$$Y_{21} = y_{21}/(1 + y_{22}) \tag{64}$$

From Eq. (62) the transfer voltage ratio for the network with a finite nonzero load is obtained as

$$K \equiv E_2/E_1 = -y_{21}/(G_L + y_{22}) \tag{65}$$

since $E_2 = -I_2/G_L$. For an open-circuited output, that is, $G_L = 0$, (65) becomes equivalent to (59).

If the termination is a general impedance $Z_L(s) = 1/Y_L(s)$, where

$$Z_L(j\omega) = R_L(\omega) + jX_L(\omega)$$

and $Y_L(j\omega) = G_L(\omega) + jB_L(\omega)$, then the above formulas apply with Z_L substituted for R_L and Y_L substituted for G_L; of course, it is no longer true when the termination is complex that G_L is equal to the reciprocal of R_L.

Another network of practical importance is a lossless passive quadripole with resistance terminations at both input and output, shown in Fig. 11. Such resistance-terminated networks are often used in matching problems where it is desired to maximize the load power at specified frequencies. It is therefore natural to measure the quality of the match in terms of the

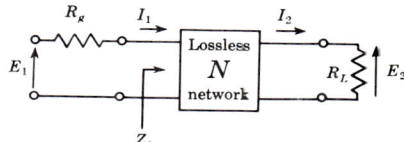

Fig. 11. Network terminated in resistance at both input and output.

load power P_L divided by the *available power* P_A, the latter being defined as the maximum power that can be delivered by a generator with an internal resistance R_g; this ratio, which will be discussed again later, is called the *transducer power gain* or the *transducer gain*. The fractional power absorbed by the load is often represented in

3-16 AMPLIFIER FUNDAMENTALS

terms of a *transmission coefficient* $t(s)$; that is,

$$|t(j\omega)|^2 = P_L/P_A \qquad (66)$$

It is important to note that for values of ω, $|t(j\omega)|$ cannot exceed unity, since the coupling network is passive and hence the load power cannot exceed the available power. The fractional power rejected by the load is expressed by a *reflection coefficient* $\rho(s)$, where for real frequencies

$$|\rho(j\omega)|^2 \equiv 1 - P_L/P_A = 1 - |t(j\omega)|^2 \qquad (67)$$

Both ρ and t may be related to impedances and admittances; the relationships are derived below. Later, in the discussion of the scattering matrix in Chap. 4, we shall use the elements of the scattering matrix to designate the identical quantities; that is, $S_{21} \equiv t$ and $S_{11} \equiv \rho$ for a quadripole terminated in its normalizing resistances. However, here it is useful to continue using ρ and t since these symbols are also used in the synthesis literature. In addition, the treatment in Chap. 4 generalizes the problem so that complex impedances may be handled; the scattering parameters also apply to an active quadripole. In Sec. 4, when we consider the different forms of power gain we also discuss the transducer gain for an active quadripole. We use still another symbol G_T in the discussion, to maintain consistency with the other symbols used. The reader should become familiar with these different symbols that are used in the literature for transducer gain and the different contexts in which they occur and the restrictions that apply; t and ρ are often used for the analysis and synthesis of passive quadripoles, S_{21} and S_{11} are used for the general quadripole when the scattering matrix is employed, and G_T is often used in the analysis and design of amplifiers.

Since N is lossless, it is clear that at real frequencies all power entering the network N is dissipated in the load; that is,

$$|I_1|^2 R_1 = |E_2|^2/R_L \qquad (68)$$

where we have used the input impedance $Z_1(j\omega) = R_1 + jX_1$. But

$$I_1 = \frac{E_1}{R_g + Z_1} \qquad (69)$$

so that

$$|I_1|^2 R_1 = \frac{|E_1|^2 R_1}{|R_g + Z_1|^2} \qquad (70)$$

Equating the right sides of Eqs. (68) and (70) yields

$$\frac{|E_2|^2}{R_L} = \frac{|E_1|^2 R_1}{|R_g + Z_1|^2} \qquad (71)$$

which may be transformed to the more convenient form

$$\frac{|E_2|^2/R_L}{|E_1|^2/4R_g} = \frac{4R_1 R_g}{|R_g + Z_1|^2} \qquad (72)$$

The left side of Eq. (72) is clearly equal to P_L/P_A and hence to $|t(j\omega)|^2$. The right side becomes

$$\frac{4R_1 R_g}{|R_g + Z_1|^2} = \frac{(R_g + R_1)^2 + X_1^2 - (R_g - R_1)^2 - X_1^2}{|R_g + Z_1|^2} = \frac{|Z_1 + R_g|^2 - |Z_1 - R_g|^2}{|Z_1 + R_g|^2}$$

$$= 1 - \left|\frac{Z_1 - R_g}{Z_1 + R_g}\right|^2 \qquad (73)$$

Thus

$$|t(j\omega)|^2 = 1 - \left|\frac{Z_1 - R_g}{Z_1 + R_g}\right|^2_{s=j\omega} \qquad (74)$$

and the function on the right side whose magnitude is squared is the *reflection coefficient at the input terminals* of the network. By use of the definition for ρ in Eq. (67) we may then write*

$$\rho = (Z_1 - R_g)/(Z_1 + R_g) \tag{75}$$

so that (74) becomes

$$|t(j\omega)|^2 = 1 - |\rho(j\omega)|^2 \tag{76}$$

The last equation is equivalent to the statement that the power transmitted through the network plus the reflected power at the input terminals is equal to the total power incident upon the input to the network.

When the impedance level of the network is normalized with respect to R_g, the reflection coefficient becomes

$$\rho = (Z_1 - 1)/(Z_1 + 1) \tag{77}$$

The inverse function is found to be

$$Z_1 = (1 + \rho)/(1 - \rho) \tag{78}$$

Below we shall derive formulas for Z_1 in terms of the two-port parameters and the output termination; then use of the preceding equations that relate ρ and t to Z_1 allows them to be expressed as functions of the two-port parameters.

Analogously, ρ may be derived in terms of the input admittance. This proceeds in a straightforward manner by use of the dual functions.

Synthesis procedures that use ρ and t start with a given transfer function which is identified with t. From t, ρ may be found, and the network design may then proceed in a number of different ways. For example, since $|t|^2$ is given by the left side of (72), then

$$t(s) = 2\sqrt{R_g/R_L}\,(E_2/E_1) \tag{79}$$

Using $E_2 = R_L I_2$, a transfer-admittance relationship is obtained as

$$t(s) = 2\sqrt{R_g R_L}\,(I_2/E_1) \tag{80}$$

If a current source with a shunt R_g is desired at the input port, a transfer impedance is required. Substitution of $I_1 = E_1/R_g$ in Eq. (79) yields for this case

$$t(s) = (2/\sqrt{R_g R_L})\,(E_2/I_1) \tag{81}$$

As a concluding remark it is pointed out that since t may be equated to a transfer impedance or admittance, the network with ports interchanged must have the same t, if the quadripole is not only passive and lossless but also reciprocal. Use of Eq. (67) then shows that $|\rho(j\omega)|^2$ for the input and the output reflection coefficients must be equal. The function $\rho(s)$ for one port, however, will in general be different from $\rho(s)$ for the other; that is, they will have the same poles but not the same zeros.

We now consider the input and output driving-point impedances and admittances, that is, the immittances, of a terminated two-port.

The driving-point impedance Z_1 at the input terminals has already been used in Eq. (99) of Chap. 2 in the discussion of equivalent circuits and also in Sec. 2.4 on the image parameter matrix, specifically, in Eq. (40). Here we wish to obtain all immittances of the two-port with general complex terminations and show that the same procedure[1] holds for any of the sets of equations of the form

$$\begin{aligned} U_{id} &= k_{11} U_{ii} + k_{12} U_{oi} \\ U_{od} &= k_{21} U_{ii} + k_{22} U_{oi} \end{aligned} \tag{82}$$

where the first subscript on the variable U_{jk} indicates whether it is an input or output variable, and the second subscript indicates whether it is a dependent or independent

---

* Since only the square of the magnitude of ρ is fixed by the preceding discussion, it is also possible to define ρ as the negative of the right side of (75), that is, $\rho = (R_g - Z_1)/(R_g + Z_1)$. Then instead of Eq. (78) we get its reciprocal, $Z_1 = (1 - \rho)/(1 + \rho)$. Thus each of two dual networks may be used to realize the prescribed transmission coefficient. Alternatively, we see that using the $-\rho$ when defining the reflection coefficient in terms of the driving-point impedance is equivalent to using $+\rho$ when defining the reflection coefficient in terms of the driving-point admittance, that is, $\rho = (Y_1 - 1)/(Y_1 + 1)$.

3–18 AMPLIFIER FUNDAMENTALS

variable; thus U_{oi} is the independent output variable. The k_{ij}, it is important to note, are system functions, that is, ratios of dependent variables to independent variables. It is clear that the equations involving $[z]$, $[y]$, $[g]$, and $[h]$ are of the form given above; however, the chain matrix, since each of its parameters is a reciprocal system function, is not covered by this discussion and is therefore treated separately.

The basic principle in using the results obtained from operations on Eqs. (82) containing generalized parameters is that the same operations give corresponding results when corresponding variables, corresponding parameters, and corresponding terminations are used. The correspondences are obtained from the defining equations, e.g., z_{11}, y_{11}, g_{11}, and h_{11} correspond to k_{11}, and from the terminated two-ports shown in

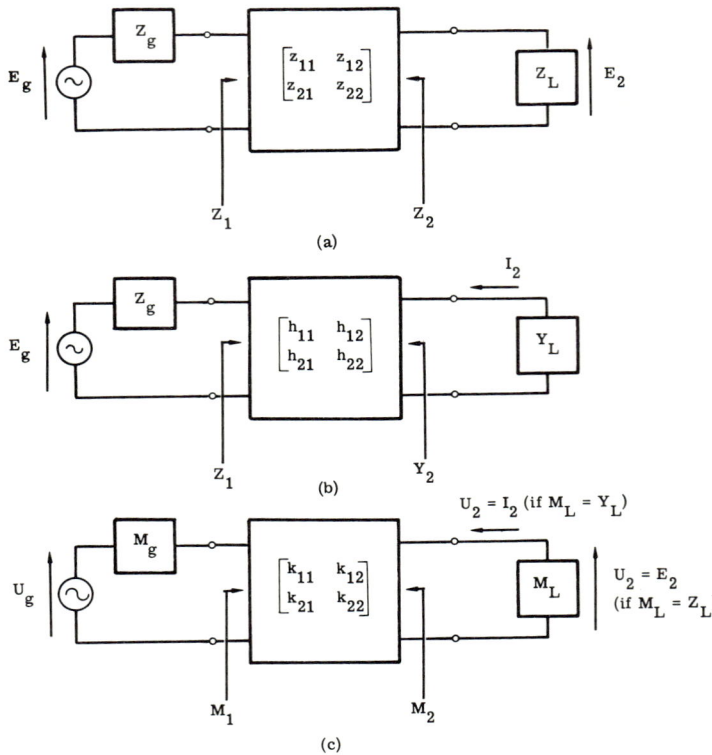

(a)

(b)

(c)

Fig. 12. Corresponding two-ports, generators, impedances, and loads for parameter sets requiring a voltage generator.

Figs. 12 and 13; from these figures we see that the input parameter M_1 for the z's is Z_1 and for the g's is Y_1.

We now solve Eqs. (82) for $M_1 \equiv U_{id}/U_{ii}$ to obtain

$$M_1 = k_{11} - k_{12}k_{21}/(k_{22} + M_L) \tag{83}$$

The corresponding input immittances for the four sets of parameters are

$$\begin{aligned} Z_1 &\equiv E_1/I_1 = z_{11} - z_{12}z_{21}/(z_{22} + Z_L) \\ Y_1 &\equiv I_1/E_1 = y_{11} - y_{12}y_{21}/(y_{22} + Y_L) \\ Z_1 &\equiv E_1/I_1 = h_{11} - h_{12}h_{21}/(h_{22} + Y_L) \\ Y_1 &\equiv I_1/E_1 = g_{11} - g_{12}g_{21}/(g_{22} + Z_L) \end{aligned} \tag{84}$$

Similarly, solving for the output immittance $M_2 \equiv U_{od}/U_{oi}$ gives

$$M_2 = k_{22} - k_{12}k_{21}/(k_{11} + M_g) \tag{85}$$

The corresponding output immittances in terms of the four parameter sets are

$$\begin{aligned} Z_2 &\equiv E_2/I_2 = z_{22} - z_{12}z_{21}/(z_{11} + Z_g) \\ Y_2 &\equiv I_2/E_2 = y_{22} - y_{12}y_{21}/(y_{11} + Y_g) \\ Y_2 &\equiv I_2/E_2 = h_{22} - h_{12}h_{21}/(h_{11} + Z_g) \\ Z_2 &\equiv E_2/I_2 = g_{22} - g_{12}g_{21}/(g_{11} + Y_g) \end{aligned} \tag{86}$$

The usefulness of the generalized parameters is thus evident; as we shall see, they are also useful in the equations for power flow.

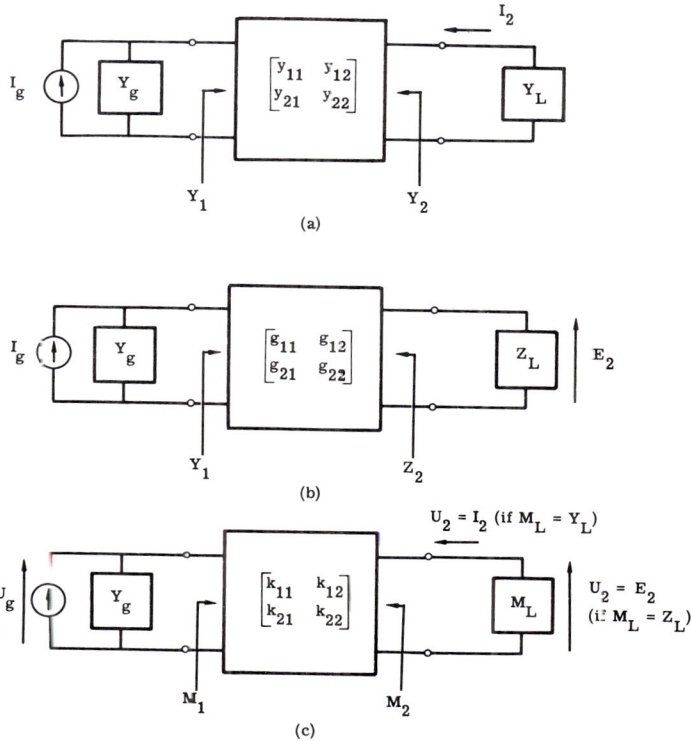

FIG. 13. Corresponding two-ports, generators, admittances, and loads for parameter sets requiring a current generator.

These equations for the input and output immittances are of great importance in amplifier design. They show quite clearly some of the difficulties in the design of a multistage amplifier, namely, the dependence of the parameters of one stage upon the variations of the parameters in other stages. For example, any variation in the load termination Z_L of a stage will affect the input impedance of that stage and hence also affect the terminations of all the preceding stages.

The equations may be manipulated into other equivalent forms. We have, for example,

$$Z_1 = z_{11} - z_{12}z_{21}/(z_{22} + Z_L) = (|z| + z_{11}Z_L)/(z_{22} + Z_L) \tag{87}$$

and if Z_L is a pure resistance of 1 ohm, the above becomes

$$Z_1 = (|z| + z_{11})/(z_{22} + 1) \tag{88}$$

3.2. Chain-matrix Characterization. Most of the discussion is given in terms of resistance terminations; however, if the terminations are general impedances, all the formulas apply if we merely substitute Z_g for R_g and Z_L for R_L. The discussion of the chain matrix in Sec. 2.3 showed that for the network in Fig. 14 under open-circuit conditions, i.e., for $R_L = \infty$, the voltage ratio is given by $1/A$ and the transfer impedance by $1/C$, whereas for $R_L = 0$ the current ratio is given by D and the transfer admittance by $-1/B$. For R_L finite and nonzero the transfer admittance may be

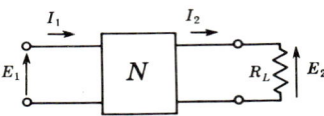
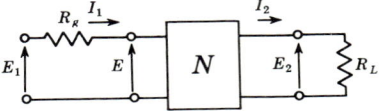

Fig. 14. Resistance-terminated network.

Fig. 15. Quadripole with resistance terminations.

obtained from the first of the set of chain-matrix equations in (29), $E_1 = AE_2 + BI_2$, by using $E_2 = R_L I_2$ to eliminate E_2. We then obtain

$$Y_{21} \equiv I_2/E_1 = 1/(AR_L + B) \tag{89}$$

The dual function is obtained similarly from the second equation of the set as

$$Z_{21} \equiv E_2/I_1 = 1/(C + D/R_L) \tag{90}$$

Substitution for I_2 in the first chain-matrix equation and subsequent manipulation yield the transfer voltage ratio

$$K \equiv E_2/E_1 = 1/(A + B/R_L) \tag{91}$$

which of course can also be obtained from (89). Analogously, the current ratio is obtained as

$$I_2/I_1 = 1/(C/R_L + D) \tag{92}$$

When $R_L = 1$ the above equations become

$$\begin{aligned} Y_{21} &= 1/(A + B) & E_2/E_1 &= 1/(A + B) \\ Z_{21} &= 1/(C + D) & I_2/I_1 &= 1/(C + D) \end{aligned} \tag{93}$$

The reader will note the dual roles played by the sums $A + B$ and $C + D$.

For networks terminated in resistance at both ends, the overall system function can be expressed in terms of the elements of the chain matrix without recourse to the reflection and transmission coefficients ρ and t. With the variables and parameters defined in Fig. 15, the description of quadripole N is given by

$$E = AE_2 + BI_2 \qquad I_1 = CE_2 + DI_2 \tag{94}$$

Since $E_1 = E + R_g I_1$, we get, by multiplying the second equation of (94) by R_g and then adding the result to the first equation,

$$E_1 = AE_2 + BI_2 + CR_g E_2 + DR_g I_2 \tag{95}$$

Making use of $E_2 = R_L I_2$ to eliminate E_2 and then solving for I_2/E_1 gives

$$Y_{21} = \frac{1}{AR_L + B + CR_g R_L + DR_g} \tag{96}$$

MATRIX REPRESENTATIONS OF MULTIPORT NETWORKS   3-21

If I_2 is eliminated from (95), the voltage ratio is obtained as

$$K = \frac{1}{A + B/R_L + CR_g + DR_g/R_L} \quad (97)$$

For a resistance-terminated network driven by a current source, as shown in Fig. 16, the system functions derived in a dual manner are

$$Z_{21} = \frac{E_2}{I_1} = \frac{1}{A/R_g + B/R_L R_g + C + D/R_L} \quad (98)$$

and

$$\frac{I_2}{I_1} = \frac{1}{AR_L/R_g + B/R_g + CR_L + D} \quad (99)$$

It is sometimes convenient to normalize with respect to R_g and at other times with respect to R_L; for each of these cases the appropriate modification of the above equations is used. When the network works into a load resistance equal to the

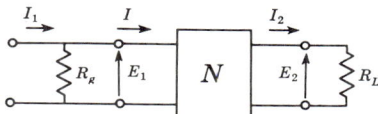

FIG. 16. Resistance-terminated quadripole driven by a current source.

generator resistance, it is possible to consider both R_L and R_g unity. For this normalized case the equations for all the transfer functions become equal to the same function, namely, $1/(A + B + C + D)$.

In chain-matrix synthesis procedures it is often simpler to use the reciprocal system function of the terminated network rather than a transfer function. This is especially true when each of the chain-matrix parameters is not a general rational function but is a polynomial. Inspection of the above equations for a network with resistances at both ends gives the equations for the reciprocal system functions

$$\begin{aligned} E_1/I_2 &= AR_L + B + CR_g R_L + DR_g \\ I_1/E_2 &= A/R_g + B/R_L R_g + C + D/R_L \\ E_1/E_2 &= A + B/R_L + CR_g + DR_g/R_L \\ I_1/I_2 &= AR_L/R_g + B/R_g + CR_L + D \end{aligned} \quad (100)$$

It is also useful to express the driving-point characteristics of a quadripole in terms of its chain-matrix parameters. We are frequently interested in the input and output impedances of a terminated network, that is, the respective impedances seen by the generator impedance and by the load impedance when the input source is removed. In these problems the terminations may not be pure resistances, and thus for generality we assume Z_g and Z_L as the terminations for N in Fig. 15. If we divide the first equation of (94) by the second, we obtain

$$\frac{E}{I_1} = \frac{AE_2 + BI_2}{CE_2 + DI_2} = \frac{AE_2/I_2 + B}{CE_2/I_2 + D} \quad (101)$$

Substitution of $Z_1 \equiv E/I_1$ and $Z_L \equiv E_2/I_2$ in the above equation yields the input impedance

$$Z_1 = \frac{AZ_L + B}{CZ_L + D} \quad (102)$$

The above form of equation, which previously arose in (77), is a type that arises repeatedly in synthesis. It is called a *linear fractional*, or *bilinear*, transformation. When the input impedance is written in terms of the load impedance in the above form, it becomes evident that the network N possesses an *impedance-transforming* property; that is, it converts a given load impedance to a desired impedance Z_1. This is a frequent use of quadripoles.

The output impedance is derived similarly. Solving Eqs. (94) for E_2 and $-I_2$ yields

$$E_2 = DE - BI_1 \qquad -I_2 = CE - AI_1 \qquad (103)$$

The same procedure as used in deriving Z_1 is now followed. Substitution of $Z_2 \equiv E_2/(-I_2)$ and $Z_g \equiv E/(-I_1)$ gives

$$Z_2 = \frac{DZ_g + B}{CZ_g + A} \qquad (104)$$

4. POWER-TRANSFER CONCEPTS, MATCHING, AND POTENTIAL INSTABILITY

In order to get maximum power output from a linear source, a conjugate match is needed; the complex load impedance, which is the variable, must be adjusted to be equal to the conjugate of the generator impedance. This means that maximum power output is obtained for resistive circuits when $R_L = R_g$. Because of considerations that are often more important than maximum power output the optimum value of load resistance is often not used. It is also well to point out that the curve of

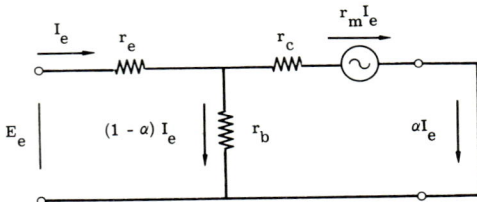

FIG. 17. Equivalent circuit for short-circuited common-base configuration.

power output as a function of load impedance is a very flat curve; for example, varying R_L from the optimum by a factor of 2 changes the output by less than 1 db.

If a device has a characteristic that is short-circuit unstable, then it will be impossible to obtain matched terminations at input and output. The point-contact transistor, because its α is greater than 1, will most often be short-circuit unstable. To see this, consider the two-port equations for the common-base circuit,

$$E_e = r_{11}I_e + r_{12}I_c \qquad E_c = r_{21}I_e + r_{22}I_c \qquad (105)$$

The second of Eqs. (105), for $E_c = 0$, yields

$$I_c = (r_{21}/r_{22})I_e = -\alpha I_e \qquad (106)$$

which, when substituted in the first equation along with the values of r_{11} and r_{12} for common-base operation, finally yields

$$E_e/I_e = r_e + (1 - \alpha)r_b \qquad (107)$$

The left-hand ratio represents the input impedance, which is given by the right-hand side as a positive quantity for $\alpha < 1$; thus the junction transistor is short-circuit stable. For the point-contact transistor, however, with an $\alpha > 1$, it is possible to have instability; this occurs when

$$(\alpha - 1)r_b > r_e \qquad (108)$$

The same short-circuit input impedance can be derived very simply in another way, that is, merely by inspection of the equivalent circuit, shown short-circuited in Fig. 17. Since αI_e flows in the output short circuit, the current in the r_b branch is

MATRIX REPRESENTATIONS OF MULTIPORT NETWORKS 3-23

given by $(1 - \alpha)I_e$. Writing the Kirchhoff voltage law around the input mesh, we can then obtain Eq. (107).

A two-port that can be made unstable by the addition of passive generator and load impedances (that are not coupled by mutual inductance) is said to be *potentially unstable*. Thus a short-circuit unstable transistor is potentially unstable, that is, for $Z_g = Z_L = 0$ the transistor has at least one natural frequency whose real part is positive or zero. For a transistor with real parameters this means that the determinant is zero or negative or that at least one of the input parameters (e.g., h_{11} and h_{22}) is negative. However, it is possible to make the complete system stable by an appropriate choice of terminations.

Consider the resistance-terminated two-port transistor shown in Fig. 18, where the transistor parameters are resistive; in accord with the discussion of natural frequencies

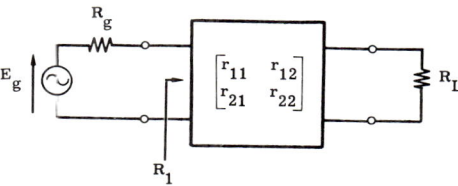

FIG. 18. Two-port network with real parameters and resistance terminations.

in Chap. 1, Sec. 6, the characteristic polynomial is given by the zeros of the impedance seen by breaking into the input circuit or equivalently by the zeros of the determinant. Thus for this resistive case, if

$$R_g + R_1 = \Delta = \begin{vmatrix} R_g + r_{11} & r_{12} \\ r_{21} & r_{22} + R_L \end{vmatrix} \leq 0$$

$$R_g r_{22} + R_L r_{11} + r_{11} r_{22} - r_{12} r_{21} \leq 0 \tag{109}$$

the system is unstable. Thus

$$\frac{r_{12} r_{21}}{r_{11} r_{22}} \geq 1 \tag{110}$$

or

$$\begin{aligned} r_{11} &< 0 \\ r_{22} &< 0 \end{aligned} \tag{111}$$

are conditions for potential instability. If (110) holds, then for $R_g = R_L = 0$ the total resistance will be nonpositive and the network can oscillate with constant amplitude or have a natural frequency that increases exponentially. For example, if Eq. (110) is satisfied by equality, then letting the generator impedance be zero and the load impedance Z_L be a parallel combination of an inductance and a capacitance resonating at ω_1 will cause the transistor to oscillate with an undamped sinusoid whose frequency is ω_1. If (110) is satisfied as an inequality, for example, the determinant is equal to -2, then letting $R_g = 0$ and Z_L be an inductance with impedance s will yield a natural frequency equal to the growing exponential e^{2t}. Similarly, if (111) is satisfied, for example, $r_{11} < 0$, then the left side of (109) can be made negative by choice of a sufficiently large positive R_L. It is therefore clear that these conditions characterize a potentially unstable transistor; that is, a passive termination exists that causes the terminated network to be unstable.

A corresponding statement[4] in terms of h parameters is that the condition for potential instability for a two-port with real parameters is

$$h_{12} h_{21} \geq h_{11} h_{22} \tag{112}$$

or h_{11} negative or h_{22} negative.

Now we consider various definitions of power transfer where matching and potential instability play a role. We make use of four different powers, which are indicated in

Fig. 19, where the designation of the terminations corresponds to the use of h parameters. The first is the maximum power available from the input generator. This is given by

$$P_{Ag} = |E_g|^2/4Z_{gr} \qquad (113)$$

for a voltage source and

$$P_{Ag} = |I_g|^2/4Y_{gr} \qquad (114)$$

for a current source, where the subscript r has been used to indicate the real part. Thus if $Z_g(j\omega) = R_g(\omega) + jX_g(\omega)$, then $Z_{gr} = R_g$.

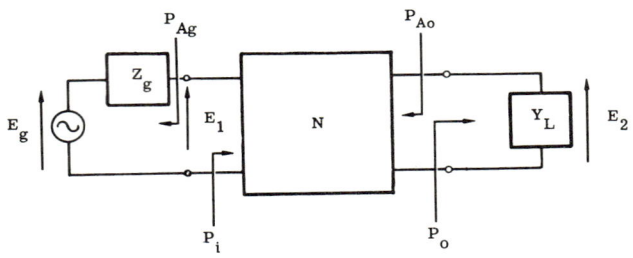

FIG. 19. Terminated two-port with four different powers indicated.

The second power is P_{Ao}, the maximum power available from the output of the two-port. If a Norton transformation is made looking back into the two-port from the output terminals, we obtain the equivalent circuit shown in Fig. 20, with values in terms of h parameters given by

$$I_N = \frac{-E_g h_{21}}{h_{11} + Z_g}$$
$$Y_N = h_{22} - \frac{h_{12}h_{21}}{h_{11} + Z_g} \qquad (115)$$

The third power is the power output P_o, that is, the power furnished to the load, and the fourth is the power P_i furnished at the input port of the quadripole. These are given by straightforward calculation as

$$P_o = |E_2|^2 Y_{Lr} = \frac{|E_g|^2 |h_{21}|^2 Y_{Lr}}{|(h_{11} + Z_g)(h_{22} + Y_L) - h_{12}h_{21}|^2}$$
$$P_i = |E_1|^2 Y_{1r} = \frac{|E_1|^2 \operatorname{Re}\,[h_{11} - h_{12}h_{21}/(h_{22} + Y_L)]}{|h_{11} - h_{12}h_{21}/(h_{22} + Y_L)|^2} \qquad (116)$$

We can now define some of the various power-gain concepts that are used. Probably the most important is the transducer gain which we introduced previously. It is designated as G_T and defined by

$$G_T = \frac{\text{power delivered to load}}{\text{power available from generator}} = \frac{P_o}{P_{Ag}} \qquad (117)$$

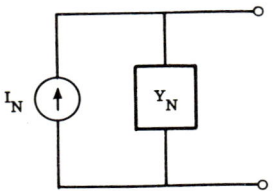

FIG. 20. Norton equivalent circuit for network looking to left at output port of two-port in Fig. 19.

The transducer gain indicates the effectiveness of the active two-port. If, for example, the generator was so badly matched at the input port that G_T was less than unity, then it is clear the active two-port could easily be dispensed with; we could synthesize a passive network that would match the generator impedance and thus deliver maximum power at some desired frequencies. From the definition we observe that G_T is a function of the source and load impedances and the two-port parameters.

MATRIX REPRESENTATIONS OF MULTIPORT NETWORKS 3-25

Useful information on power transfer is given by other measures that are easier to calculate. One is the power gain G and a second is the available gain G_A. These are defined by

$$G = \frac{\text{power delivered to load}}{\text{power input to amplifier}} = \frac{P_o}{P_i} \tag{118}$$

and

$$G_A = \frac{\text{available power at output}}{\text{available power at input}} = \frac{P_{Ao}}{P_{Ag}} \tag{119}$$

Another measure of power gain that is sometimes useful, but which we shall merely define, is the insertion power gain G_I, which indicates what the insertion of the two-port accomplishes. It is defined by

$$G_I = \frac{\text{power delivered to load with two-port inserted}}{\text{power delivered to load when it is connected directly to the source}} \tag{120}$$

The power gain G is a function only of the two-port parameters and the load; we recall that the input impedance is also a function of these parameters. Thus both are determined when these data are fixed. If we choose a source that matches the two-port, then P_i and P_{Ag} are equal and hence G_T and G are equal. However, if the generator impedance is not equal to the complex conjugate of the input impedance, G_T will be less than G.

Another upper bound is placed on G_T by G_A, which is dependent on the same parameters as the output impedance, namely, the two-port parameters and the source impedance. If the load impedance is chosen to match the output impedance of the two-port, then G_T and G_A will be equal; if not, G_T is less than G_A.

Frequently both Z_g and Z_L may be subject to our control. If the two-port is not potentially unstable, we may then make G and G_A equal, a condition which occurs for unique generator and load terminations. These terminations are chosen so that we simultaneously match the input and output. Then G_A and G achieve their maximum finite values and are equal to each other and to G_T. We thus achieve the *maximum available gain* $G_{A,max}$; this is an important parameter of a two-port. The simultaneous match cannot be achieved if the two-port is potentially unstable; in this case $G_{A,max}$ is infinite.

It should of course be clear that we are not maximizing the load power output but only the available gain ratio; with a fixed E_g and a variable Z_g and Z_L, one of the conditions for maximum power output is $Z_g = 0$.

Now we can evaluate the three power ratios using the previously evaluated powers. They are given by

$$\begin{aligned} G &= \frac{Y_{Lr}|h_{21}|^2}{\text{Re }[h_{11} - h_{12}h_{21}/(h_{22} + Y_L)]|h_{22} + Y_L|^2} \\ G_T &= \frac{4|h_{21}|^2 Y_{Lr} Z_{gr}}{|(h_{11} + Z_g)(h_{22} + Y_L) - h_{12}h_{21}|^2} \\ G_A &= \frac{|h_{21}|^2 Z_{gr}}{\text{Re }[(h_{11}h_{22} - h_{12}h_{21} + h_{22}Z_g)(h_{11} + Z_g)]} \end{aligned} \tag{121}$$

As before, it is useful to give the above formulas in terms of the generalized parameters. Using the correspondences discussed previously when we considered the input and output immittances of a two-port, we have

$$\begin{aligned} G &= \frac{M_{Lr}|k_{21}|^2}{\text{Re }[k_{11} - k_{12}k_{21}/(k_{22} + M_L)]|k_{22} + M_L|^2} \\ G_T &= \frac{4|k_{21}|^2 M_{Lr} M_{gr}}{|(k_{11} + M_g)(k_{22} + M_L) - k_{12}k_{21}|^2} \\ G_A &= \frac{|k_{21}|^2 M_{gr}}{\text{Re }[(k_{11}k_{22} - k_{12}k_{21} + k_{22}M_g)(k_{11} + M_g)]} \end{aligned} \tag{122}$$

AMPLIFIER FUNDAMENTALS

As pointed out in the reference,[4] there are a number of alternative ways for determining the optimum terminations for a two-port that has a finite maximum available gain.

1. We can determine the terminations that simultaneously conjugate-match the two ports. This solving of two simultaneous equations becomes cumbersome unless the two-port parameters are real.
2. We can find the load termination that maximizes the power gain of the two-port. To do this the generator immittance need not be considered. Having fixed the optimum load termination, we then determine the corresponding input immittance and choose its conjugate value as the source immittance.
3. We can find the source immittance that leads to the greatest available power gain of the two-port; in this case the load immittance need not be considered. Having fixed the optimum source immittance, we then determine the corresponding output immittance and select its conjugate value for the load immittance.
4. The first parts of methods 2 and 3 can be performed independently.

Examples in terms of h parameters are worked out in the reference[4] using method 1 for the resistive case and method 2 for the complex immittance case. Below we use method 1 in terms of real open-circuit impedances.

The maximum available gain occurs when we match at the output, that is, set the load impedance equal to the output impedance,

$$R_L = R_2 = r_{22} - r_{12}r_{21}/(r_{11} + R_g) \tag{123}$$

and simultaneously match at the input, that is, set the generator impedance equal to the input impedance,

$$R_g = R_1 = r_{11} - r_{12}r_{21}/(r_{22} + R_L) \tag{124}$$

Rewriting Eqs. (123) and (124) as

$$\begin{aligned} R_g R_L + R_g r_{22} &= r_{11}r_{22} + r_{11}R_L - r_{12}r_{21} \\ R_g R_L + R_L r_{11} &= r_{11}r_{22} + r_{22}R_g - r_{12}r_{21} \end{aligned} \tag{125}$$

and subtracting yields

$$R_g r_{22} - R_L r_{11} = r_{11}R_L - r_{22}R_g \tag{126}$$

Solution of Eq. (126) gives the relationship between R_g and R_L,

$$R_g = (r_{11}/r_{22})R_L \tag{127}$$

This value of R_g is substituted in the first of Eqs. (125) and R_L is now found to be

$$R_L = r_{22}\sqrt{1 - \delta} \tag{128}$$

Use of Eq. (127) with Eq. (128) gives

$$R_g = r_{11}\sqrt{1 - \delta} \tag{129}$$

The quantity $\delta = r_{12}r_{21}/r_{11}r_{22}$ may be called the stability factor: if it is equal to or greater than 1, then the transistor is short-circuit unstable, that is, potentially unstable. If these values of R_g and R_L are substituted in the formula for available gain,

$$G_A = R_g/R_L[r_{21}/(r_{11} + R_g)]^2 \tag{130}$$

we obtain the formula for maximum available gain,

$$G_{A,max} = \frac{r_{11}}{r_{22}}\left(\frac{r_{21}}{r_{11} + r_{11}\sqrt{1 - \delta}}\right)^2 = \frac{1}{r_{11}r_{22}}\left(\frac{r_{21}}{1 + \sqrt{1 - \delta}}\right)^2 \tag{131}$$

5. INTERCONNECTIONS OF QUADRIPOLES

Often in analysis problems it is necessary to determine the parameters characterizing a quadripole that is formed by an interconnection of two or more component quadripoles. It is useful to be able to express these parameters in terms of the parameters of the component quadripoles, which may be known or easily determined. In the converse problem of synthesis these same relationships may be used to break down a given system function into simpler system functions, each of which may be realized by a component quadripole. The motivation in this case is to make the realization and interconnection of the quadripoles evident almost by inspection. We now discuss the principles of the basic interconnections; we consider two component quadripoles, but it will be clear that more than two quadripoles are treated by a straightforward extension of the principles.

Quadripoles may be interconnected in any of the following basic ways:

1. Cascade
2. Series
3. Parallel
4. Series-parallel
5. Parallel-series

Since the cascade connection has already been treated in Sec. 2 in terms of the chain matrix, we say no more about it here. The other four connections are illustrated in Fig. 21.

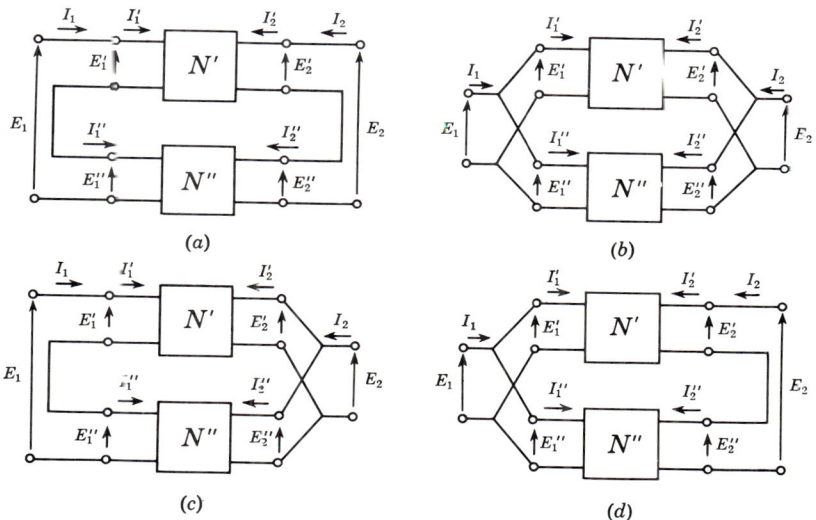

Fig. 21. Interconnections of quadripoles. (a) Series. (b) Parallel. (c) Series-parallel. (d) Parallel-series.

Suppose we wished to determine the parameters of the overall quadripole for each interconnection. The method for choosing the types of parameters proceeds from a recognition of the current or voltage variables that are common to N' and N'' in each case. We illustrate this method for the series and parallel interconnections and show that an appropriate matrix addition describes the interconnection. However, we hasten to point out that the matrix obtained by matrix addition is not a valid description of the overall quadripole unless a test is satisfied; this test is best described after the discussion of the parameter determination.

In Fig. 21a the networks N' and N'' are connected in series at both their input and output ports; consequently their port currents are equal to the respective currents of the composite network, whereas the voltages of the composite network are equal to the sums of the component voltages. In symbols, we have

$$I_1 = I_1' = I_1''$$
$$I_2 = I_2' = I_2''$$
(132)

3–28 AMPLIFIER FUNDAMENTALS

or in matrix notation, $I] = I'] = I'']$ and

$$E_1 = E_1' + E_1'' \\ E_2 = E_2' + E_2'' \tag{133}$$

or $E] = E'] + E'']$. Thus, since the currents are common, the open-circuit impedance matrices are called for. Adding the respective left and right sides of the matrix equations

$$E'] = [z'] I'] \\ = [z'] I] \tag{134}$$

and
$$E''] = [z''] I''] \\ = [z''] I] \tag{135}$$

gives
$$E] = [z] I] \tag{136}$$
where
$$[z] = [z'] + [z''] \tag{137}$$

By extension of the above argument, more than two quadripoles whose corresponding ports are connected in series form a composite quadripole whose open-circuit impedance matrix is equal to the sum of the component open-circuit impedance matrices. The short-circuit admittance matrix $[y]$ is of course given by the inverse of $[z]$.

Analogously, the networks of Fig. 21b have their corresponding ports connected in parallel so that their port voltages are common and

$$[y] = [y'] + [y''] \tag{138}$$

This argument can again be extended to more than two quadripoles. Thus the short-circuit admittance matrix of the quadripole formed by connecting in parallel the corresponding ports of two or more component quadripoles is given simply by the sum of the short-circuit admittance matrices of the component quadripoles.

The procedure for determining the matrices to be added is similar for the two other cases. For the series-parallel interconnection, the input currents and output voltages are common to N' and N'', so that the matrices to be added for each component quadripole are of the mixed form of equations, which for N' is

$$\begin{bmatrix} E_1' \\ I_2' \end{bmatrix} = \begin{bmatrix} h_{11}' & h_{12}' \\ h_{21}' & h_{22}' \end{bmatrix} \begin{bmatrix} I_1 \\ E_2 \end{bmatrix} \tag{139}$$

Similarly, the parameters for the parallel-series interconnection are given for N' by

$$\begin{bmatrix} I_1' \\ E_2' \end{bmatrix} = \begin{bmatrix} g_{11}' & g_{12}' \\ g_{21}' & g_{22}' \end{bmatrix} \begin{bmatrix} E_1 \\ I_2 \end{bmatrix} \tag{140}$$

It is now possible to discuss the test that must be satisfied. In order for the matrix additions to be valid, the current leaving one terminal of a port must be the same as that entering the other terminal. This is evidently true for a component quadripole considered by itself; it must continue to be true when the quadripoles are interconnected. In other words, a component quadripole standing alone is described by a set of parameters; when interconnected, these parameters must continue to describe this quadripole.

It is easy to find illustrations where the method of matrix addition is not valid; two are shown in Fig. 22. In each case elements are shorted by the interconnection. For example, for the series connection of Fig. 22b

$$[z'] = \begin{bmatrix} R_1' + R_3' & R_3' \\ R_3' & R_2' + R_3' \end{bmatrix} \\ [z''] = \begin{bmatrix} R_1'' + R_3'' & R_3'' \\ R_3'' & R_2'' + R_3'' \end{bmatrix} \tag{141}$$

However, $[z]$ of the composite quadripole is not given by

$$[z] = [z'] + [z''] \tag{142}$$

MATRIX REPRESENTATIONS OF MULTIPORT NETWORKS

since after the interconnection the effective $[z'']$ is

$$[z''] = \begin{bmatrix} R_3'' + \dfrac{R_1'' R_2''}{R_1'' + R_2''} & R_3'' + \dfrac{R_1'' R_2''}{R_1'' + R_2''} \\ R_3'' + \dfrac{R_1'' R_2''}{R_1'' + R_2''} & R_3'' + \dfrac{R_1'' R_2''}{R_1'' + R_2''} \end{bmatrix} \quad (143)$$

and the $[z]$ of the overall quadripole is

$$[z] = \begin{bmatrix} R_1' + R_3' + R_3'' + \dfrac{R_1'' R_2''}{R_1'' + R_2''} & R_3' + R_3'' + \dfrac{R_1'' R_2''}{R_1'' + R_2''} \\ R_3' + R_3'' + \dfrac{R_1'' R_2''}{R_1'' + R_2''} & R_2' + R_3' + R_3'' + \dfrac{R_1'' R_2''}{R_1'' + R_2''} \end{bmatrix} \quad (144)$$

These simple cases can be changed without difficulty to make the method of matrix addition valid. In the parallel connection of Fig. 22b, if we remove the series resistances from the bottom branches of the upper network and add these to the series

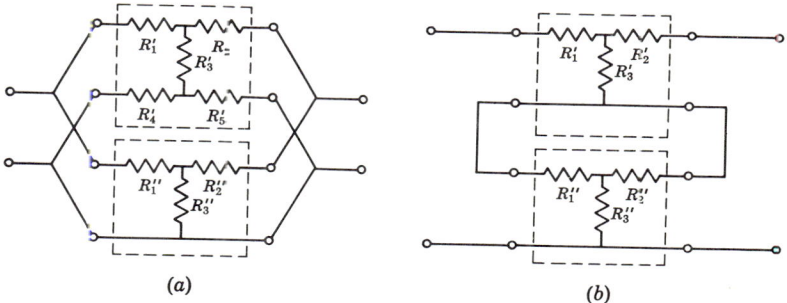

Fig. 22. Parallel and series interconnections of quadripoles for which the method of matrix addition is not valid.

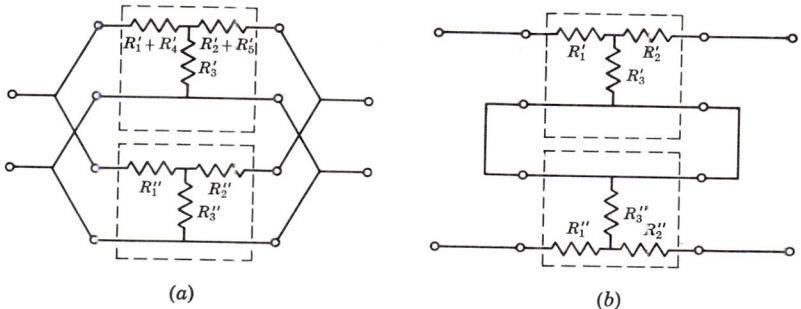

Fig. 23. Modification of component quadripoles of Fig. 22 so that matrix addition is valid.

resistances of the respective top branches, its behavior as a quadripole will be unchanged. In this form it may be placed in parallel with the lower network, and calculation shows that Eq (142) now holds, i.e., that matrix addition is valid. The networks in Fig. 22b can be modified similarly to do away with the short-circuiting action, and the interconnections for which the matrix method is now valid are shown in Fig. 23.

However, in more complicated cases the lack of validity and the remedy for it may not be evident. A general test is needed. We should recognize first that the cause of the current imbalance after the interconnection—that is, the current leaving one terminal of a port is not the same as that entering the other terminal—is a *circulating current* between the quadripoles, as shown for the series and parallel interconnections in Fig. 24. If the circulatory current is zero, the matrix method will be valid.

The test may now be formulated as an examination of the voltages around the loop in which the circulatory current flows. Consider the parallel interconnection. The additional voltage to be considered is that between terminals 1' and 2', shown for N' in

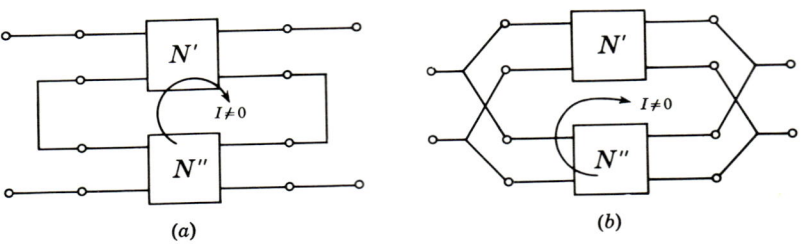

FIG. 24. Interconnections showing presence of a circulatory current that destroys validity of matrix method.

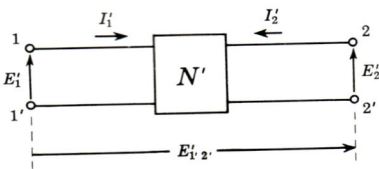

FIG. 25. Sketch showing additional voltage to be considered to determine validity of matrix method.

Fig. 25. Inspection of the parallel connection shows that if $E'_{1'2'}$ is equal to the corresponding voltage in N'', the total voltage, and hence the current around the loop, will be zero. The condition for validity is therefore

$$E'_{1'2'} = E''_{1'2'} \tag{145}$$

If we consider E_1' and E_2' as sources, we can express $E'_{1'2'}$ as

$$E'_{1'2'} = a_1'E_1' + a_2'E_2' \tag{146}$$

and

$$E''_{1'2'} = a_1''E_1'' + a_2''E_2'' \tag{147}$$

Since $E_1' = E_1''$ and $E_2' = E_2''$ for the parallel connection, the necessary and sufficient condition for $E'_{1'2'}$ to be equal to $E''_{1'2'}$ is

$$\begin{aligned} a_1' &= a_1'' \\ a_2' &= a_2'' \end{aligned} \tag{148}$$

By the use of the principle of superposition these conditions may be applied to any given problem. First we let $E_2' = E_2'' = 0$—that is, the output ports of N' and N'' are separately short-circuited—and the input ports of N' and N'' are connected in parallel. Then if Eq. (145) is satisfied, this means that

$$a_1' = a_1'' \tag{149}$$

Now we reverse the networks, i.e., let E_1' and E_2' be shorted whereas the ports 2-2' of N' and N'' are connected in parallel. Now if $E'_{1'2'} = E''_{1'2'}$, we have

$$a_2' = a_2'' \tag{150}$$

and thus the two conditions are satisfied.

MATRIX REPRESENTATIONS OF MULTIPORT NETWORKS 3-31

The two parts of the test may be interpreted physically as shown in Fig. 26, where the voltage defined by E is required to be zero. It should be evident that if $E \neq 0$, connecting the output ports in parallel will cause a circulatory current to flow. These two tests, which for many cases may be carried out by inspection, replace the necessary and sufficient conditions of Eq. (148).

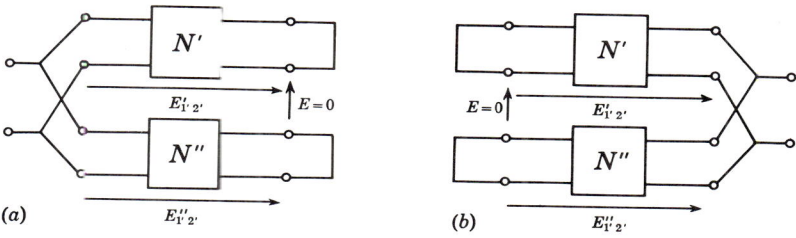

FIG. 26. Sketches showing that $E = 0$ is the condition for the validity of matrix analysis of the parallel connections.

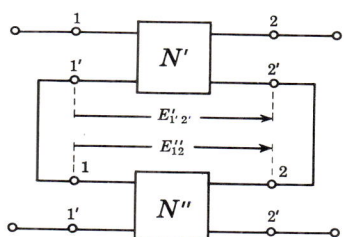

FIG. 27. Series connection defining the voltages around the center loop.

A corresponding discussion holds for the series interconnection of quadripoles. Here the currents I_1 and I_2 are considered as sources and the voltages that must be equal for the circulatory current to be zero are

$$E'_{1'2'} = b_1' I_1' + b_2' I_2' \qquad (151)$$
and
$$E''_{1'2'} = b_1'' I_1'' + b_2'' I_2'' \qquad (152)$$

These voltages are shown in Fig. 27. The necessary and sufficient conditions are given by

$$b_1' = b_1''$$
$$b_2' = b_2'' \qquad (153)$$

Again superposition is used to determine whether each of the above conditions is satisfied. We first let $I_2 = 0$ and then let $I_1 = 0$; in each case we check whether a single voltage is zero. The two tests are given a physical interpretation in Fig. 28. If $E = 0$ in both cases, the validity test is satisfied.

It is now possible to apply these tests to the connections in Fig. 22 and observe that they are not satisfied, whereas those in Fig. 23 do satisfy the tests.

The tests for the parallel-series and the series-parallel connections use one of the parallel and one of the series tests. For example, the parallel-series tests are illustrated in Fig 29; it should be clear from the figure that the voltages that must be equal to guarantee $E = 0$ are now $E'_{1'2'}$ and $E''_{1'2'}$. The reader can now formulate these two cases for himself.

It is possible that these tests are not satisfied and that the artifices used to convert Fig. 22 to Fig. 23 do not help. Then ideal transformers with a 1:1 turns ratio may be used at either the input or the output ports. An ideal transformer *forces* the current entering one terminal of a port to be equal to the current leaving the other terminal. Since the currents entering and leaving the terminals of a port of the *overall* quadripole are equal, it is necessary to use only $n - 1$ ideal transformers for an interconnection of n quadripoles. The schemes for connecting the ideal transformers for the series and parallel interconnections are shown in Fig. 30.

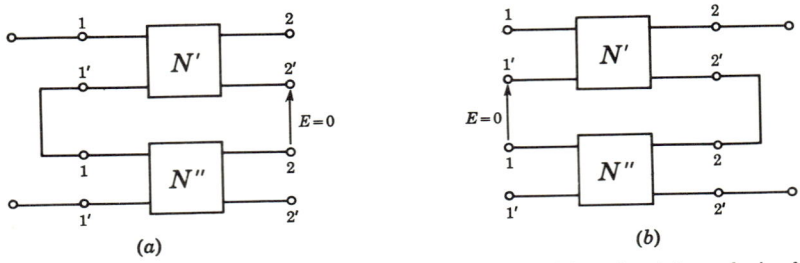

Fig. 28. Sketches showing that $E = 0$ is the condition for validity of matrix analysis of the series connection.

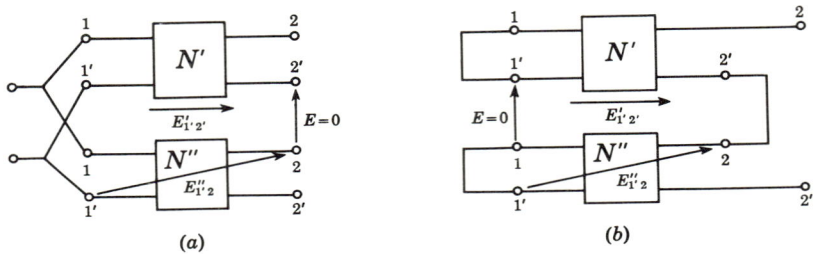

Fig. 29. Sketches illustrating the validity test for matrix analysis of the parallel-series connection.

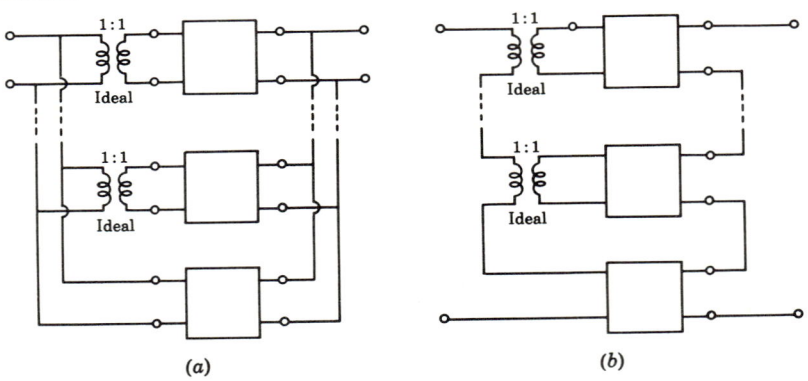

Fig. 30. Methods for using ideal transformers to make matrix analysis valid.

REFERENCES

1. L. Weinberg, "Network Analysis and Synthesis," McGraw-Hill Book Company, New York, 1962.
2. A. C. Aitken, "Determinants and Matrices," 9th ed., p. 98, Interscience Publishers, Inc., New York, 1956.
3. E. S. Kuh and D. O. Pederson, "Principles of Circuit Synthesis," McGraw-Hill Book Company, New York, 1959.
4. J. G. Linvill and J. F. Gibbons, "Transistors and Active Circuits," McGraw-Hill Book Company, New York, 1961.

Chapter 4

SCATTERING MATRICES AND MATRICES FOR MULTITERMINAL NETWORKS

LOUIS WEINBERG*

CONTENTS

1. Introduction.. 4–1
2. Scattering Matrix... 4–2
   2.1. Scattering Matrix Normalized with Respect to Complex Impedances.. 4–18
3. Transfer Scattering Matrix... 4–22
4. Scattering Matrices for Multiports.................................. 4–25
5. Interrelationships among Network Matrices.................. 4–27
6. Multiterminal Networks... 4–32
   6.1. Calculation of Transfer Impedance and Voltage Ratio...... 4–37
   6.2. Reduction of an n-terminal Network to a p-terminal Network 4–38
   6.3. Connection of Two n-poles in Parallel......................... 4–39
   6.4. Relation between Driving-point Functions and Transfer Functions of a Reciprocal n-terminal Network.................... 4–41

1. INTRODUCTION

In this chapter we continue the discussion of the characterization of n-port networks by considering the scattering matrices. We also show how to relate the n-port matrices by providing tables of relationships for all the n-port matrices considered in this chapter and in Chap. 3. We then turn to n-terminal networks for which useful characterizations are given in terms of the indefinite admittance matrix and the indefinite impedance matrix.

The scattering matrix[1,2,3] is a "natural" characterization of distributed-parameter systems, and is also a powerful tool for the analysis and synthesis of lumped-parameter n-port networks. It is natural in the sense that the concepts it uses, namely, incident and reflected waves, occur physically in waveguides and transmission lines and other wave-propagating structures. In addition, though the open-circuit impedance matrix or the short-circuit admittance matrix may not exist for certain networks, e.g., neither exists for the ideal transformer, the scattering matrix exists for every passive network.

* Department of Electrical Engineering, University of Michigan, Ann Arbor, Mich.

It is especially useful in problems involving filters, equalizers, matching networks, and various types of biconjugate networks, that is, balanced bridges and transformer hybrid networks. Finally, since the scattering parameters are direct measures of return loss and forward transmission, they are peculiarly suited for the study of feedback amplifiers.

To give a physical basis for the scattering coefficients we first describe the wave picture of scattering in a qualitative way and introduce the concepts of reflected and incident waves. Then we consider an example in which the scattering matrix is derived for a one-port and a two-port and the necessity for normalization is shown. This is followed by a general treatment of the n-port where matrix notation and normalization with respect first to real numbers and then to complex numbers are used. We show that normalization with respect to complex impedances is not only possible but highly desirable;[4] it leads, for example, to a new theory of broadband matching.[5]

We find that it is frequently useful to consider elements with more than two terminals; for example, a transistor may be considered a three-terminal active element rather than a grounded two-port. Also some distributed-parameter systems have natural representations as *n-terminal transducers*, where by such a transducer we mean a network with one set of n terminals as the input and another set of n terminals as the output. In the analysis and synthesis of passive networks it is often desirable to consider them as n-terminal networks or even as n-terminal *elements* rather than as structures whose terminals are always paired as ports. Valuable insights and ease of analysis are obtained by being able to switch from the port to the terminal characterization of a network, and vice versa.

In this chapter we show how the analysis of n-terminal networks may be carried out; we suggest some of its power but because of space limitations do not show all the advantages—e.g., we do not consider in detail the usefulness of n-terminal analysis for active circuits where cascaded transistor circuits in any configuration may be easily analyzed—but the reader may find further applications in the cited references.

In our consideration of the n-terminal network we shall use a matrix with the property that the sum of the elements of every row and of every column equals zero; because of this property such a matrix is called a *zero-sum matrix*. Though the analysis may be carried out on an n-node or an n-loop basis (where the latter designation refers to the *external* loops formed at the network terminals), we shall restrict our discussion almost entirely to the node analysis which gives rise to an nth-order admittance matrix; this matrix, in addition to its being called a zero-sum matrix, has been designated in the literature as an *indefinite admittance matrix*.[6] The analogous treatment on the loop basis, leading to an impedance matrix with the same properties as the indefinite admittance matrix, is also discussed in the literature.[7,8] An elegant treatment is given by Baranov;[9] even earlier, Campbell used many of the important formulas without giving any proofs.[10] Both nonreciprocal and reciprocal networks may be treated by this method; the only difference in the indefinite admittance matrices that arise is that the associated matrix is unsymmetric and symmetric, respectively.

2. SCATTERING MATRIX

Consider the system shown in Fig. 1, which we discuss as a distributed-parameter system like a waveguide but which can of course also represent a lumped-parameter quadripole. In this waveguide the ports $x_1 = 0$ and $x_2 = 0$ can represent different terminal planes, between which is included the two-port system N. In the wave picture of scattering depicted in Fig. 1 the total wave is considered as broken into two parts, a positive-traveling wave going in the positive direction and a negative-traveling wave going in the opposite direction. We note that the positive wave *enters* the two-port, or is *incident* upon it; thus at port 1 the positive direction of the wave entering the two-port is from left to right, whereas at port 2 it is from right to left; that is, x_2 increases to the left. The negative wave is the emergent or reflected wave from the two-port. Thus we shall call the positive wave the incident wave, designated by a subscript i, and the negative wave the reflected wave, designated by subscript r, so that E_1^+ and E_1^- become E_{1i} and E_{1r}, respectively. Frequently, too, the amplitude and phase of the transverse component of the electric field of the incident wave, measured at the port terminals, is designated by a, and b is used to designate the amplitude and phase of the reflected wave. These parameters are normalized so that $a\bar{a} = |a|^2$ represents the incident power and $|b|^2$ the reflected power, and we can define a reflection coefficient $S = b/a$. The currents and voltages are represented in terms of these field quantities, where the voltage E is a measure of the total transverse electric field

$$E = g(a + b) = ga(1 + S) \tag{1}$$

and g is some proportionality factor. Similarly, the current I is given by

$$I = (1/g)(a - b) = (1/g)a(1 - S) \tag{2}$$

where b now has a negative sign because the magnetic field is reversed in the reflected wave, and the proportionality factor is $1/g$ because of the way in which the variables are normalized with respect to power. As we shall see, the introduction of the wave

FIG 1. Two-port system with incident and reflected waves at each port.

amplitudes a and b defined above is also very convenient in low-frequency or lumped-parameter circuits where the wave nature of the solutions is not obvious.

For the two-port or, more generally, for n-port networks we have

$$\begin{aligned} E_k &= g_k(a_k + b_k) \\ I_k &= (1/g_k)(a_k - b_k) \end{aligned} \tag{3}$$

for $k = 1, 2, \ldots, n$ and g_k is a constant for each port or waveguide. The characteristic impedance of each guide is related in a simple way to g_k. To see this let $b_k = 0$, that is, assume there is no reflected wave, so that

$$\begin{aligned} E_k &= g_k a_k \\ I_k &= (1/g_k)a_k \end{aligned} \tag{4}$$

and
$$Z_{0k} = E_k/I_k = g_k^2 \tag{5}$$
or
$$g_k = \sqrt{Z_{0k}} \tag{6}$$

Thus g_k is the square root of the characteristic impedance of the kth guide.

We can always change a_k and b_k to make $Z_{0k} = 1$. If this is done, then $g_k = 1$ and

$$\begin{aligned} E_k &= a_k + b_k \\ I_k &= a_k - b_k \end{aligned} \tag{7}$$

By addition and subtraction of the above equations we get

$$\begin{aligned} a_k &= \tfrac{1}{2}(E_k + I_k) \\ b_k &= \tfrac{1}{2}(E_k - I_k) \end{aligned} \tag{8}$$

The incident and reflected waves introduced briefly above will be used later after we relate the scattering coefficients to the impedances of a one-port and a two-port in the example below. We consider reflected and incident components of voltages and currents in the example because this is a formulation frequently used in the literature.

Example 1.[11] *a.* For the one-port shown in Fig. 2 derive and explain the scattering coefficient in terms of the impedances Z_1 and Z_0. Use currents first and then voltages in the discussion. Also indicate the relation of Z_0 to the incident and reflected voltage and current components.

b. For the two-port shown in Fig. 3 derive the scattering coefficients in terms of voltages.

Solution.   a. The current I in Fig. 2 may be considered as a sum of a *matched* or *incident* current and a *mismatched* or *reflected* current. Thus for the reference directions shown we have

$$I_1 = I_{1i} - I_{1r} \tag{1}$$

The actual current I_1 is given by

$$I_1 = E_g/(Z_1 + Z_0) \tag{2}$$

However, if Z_1 is assumed for the moment to be equal to Z_0 we get only the matched component

$$I_{1i} = E_g/2Z_0 \tag{3}$$

For an arbitrary Z_1 the actual current is given in Eq. (1) as the difference between the matched and mismatched components. Using Eqs. (2) and (3) in (1) and simplifying, we get

$$\begin{aligned}I_{1r} &= I_{1i} - I_1 \\ &= \frac{E_g}{2Z_0} - \frac{E_g}{Z_1 + Z_0} \\ &= \frac{Z_1 - Z_0}{Z_1 + Z_0}\frac{E_g}{2Z_0} \\ &= \frac{Z_1 - Z_0}{Z_1 + Z_0} I_{1i}\end{aligned} \tag{4}$$

We define the scattering coefficient

$$S_{11} = I_{1r}/I_{1i} \tag{5}$$

so that

$$S_{11} = \frac{Z_1 - Z_0}{Z_1 + Z_0} \tag{6}$$

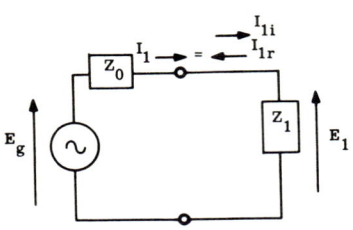

FIG. 2. One-port network for derivation of scattering coefficient.

This scattering coefficient for the one-port is called the reflection coefficient with respect to Z_0, which is called the reference impedance. We note from (6) that if an image match exists, i.e., $Z_1 = Z_0$, then $S_{11} = 0$ so that the mismatched component of current is zero.

Similarly, the reflection coefficient can be defined as the ratio of the components of the voltage E_1. If an image match exists, i.e., $Z_1 = Z_0$, the voltage across Z_1 is

$$E_{1i} = E_g/2 \tag{7}$$

For an arbitrary Z_1 the voltage across Z_1 is

$$E_1 = [Z_1/(Z_1 + Z_0)]E_g \tag{8}$$

which can be considered as the sum of an incident (or matched) component E_{1i} and a reflected (or mismatched) component E_{1r}; thus

$$E_1 = E_{1i} + E_{1r} \tag{9}$$

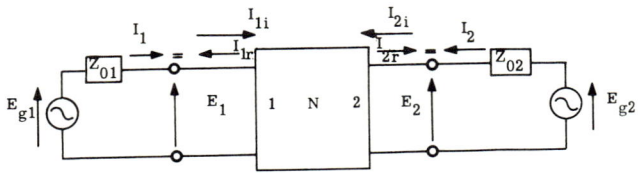

FIG. 3. Two-port network with a generator and source impedance at each port.

MATRICES FOR MULTITERMINAL NETWORKS

Solving (9) for E_{1r} and then using (7) and (8), we obtain

$$E_{1r} = E_1 - E_{1i}$$
$$= \frac{Z_1 - Z_0}{Z_1 + Z_0} \frac{E_g}{2}$$
$$= S_{11} E_{1i} \quad (10)$$

Thus for both the voltage and the current the ratio of the reflected component to the incident component is the same reflection coefficient. From Eqs. (3) and (7) the reference impedance is seen to be the ratio of the incident component of voltage to the incident component of current:

$$E_{1i}/I_{1i} = Z_0 \quad (11)$$

Similarly, from the above equation and from Eqs. (5) and (10) we have

$$E_{1r}/I_{1r} = Z_0 \quad (12)$$

b. Suppose that the impedances Z_{01} and Z_{02} are arbitrarily chosen as the reference impedances and that these are then used as the source impedances for ports 1 and 2, respectively, as shown in Fig. 3. After the voltages and currents have been separated into components as was previously done for the one-port, we are ready to evaluate these components and the scattering matrix.

We consider first the voltage at port 1. If the source at port 2 is assumed shorted, that is, $E_{g2} = 0$, then the incident voltage E_{1i} is defined as the actual voltage at port 1 when an image match exists. As previously explained in the discussion of the image parameters, an image match is said to exist if Z_1 is equal to the reference impedance Z_{01}, where Z_{11} is the input impedance for the quadripole terminated at port 2 in the reference impedance Z_{02}. For this image-matched condition

$$E_{1i} = E_{g1}/2 \quad (13)$$

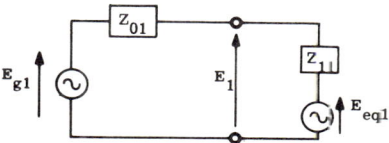

Fig. 4. Two-port of Fig. 3 where the network to the right of port 1 has been replaced by its Thévenin equivalent.

Similarly, with a source at port 2 only and for the image-matched condition $Z_{22} = Z_{02}$, where Z_{22} is the output impedance for the quadripole terminated in Z_{01} at port 1, we define the incident component of E_2 and evaluate it as

$$E_{2i} = E_{g2}/2 \quad (14)$$

Again, as in (*a*), defining each of the reflected components E_{1r} and E_{2r} as the differences between the total port voltage and the incident component,

$$E_{1r} = E_1 - E_{1i}$$
$$E_{2r} = E_2 - E_{2i} \quad (15)$$

we proceed to evaluate E_1 and E_2. We must find the contributions of both sources E_{g1} and E_{g2} to these voltages. We find E_1 by use of Fig. 4, where a Thévenin equivalent has replaced the network to the right of port 1, and the Thévenin equivalent voltage E_{eq1} is given in terms of the open-circuit impedances and the reference impedance by

$$E_{eq1} = [z_{12}/(z_{22} + Z_{02})] E_{g2} \quad (16)$$

Use of such a Thévenin transformation, it should be noted, is also necessary in the one-port case considered in (*a*) above when the one-port network contains internal sources.

From Fig. 4 we obtain simply

$$E_1 = \frac{Z_{11}}{Z_{11} + Z_{01}} E_{g1} + \frac{Z_{01}}{Z_{11} + Z_{01}} E_{eq1} \quad (17)$$

Using Eqs. (13) through (17) then gives the reflected voltage E_{1r} in terms of the incident voltages

$$E_{1r} = \frac{Z_{11} - Z_{01}}{Z_{11} + Z_{01}} E_{1i} + \frac{2Z_{01}}{Z_{11} + Z_{01}} \frac{z_{12}}{z_{22} + Z_{02}} E_{2i} \quad (18)$$

A similar procedure yields E_{2r} as

$$E_{2r} = \frac{2Z_{02}}{Z_{22} + Z_{02}} \frac{z_{21}}{z_{11} + Z_{01}} E_{1i} + \frac{Z_{22} - Z_{02}}{Z_{22} + Z_{02}} E_{2i} \quad (19)$$

The coefficients in Eqs. (18) and (19) are the scattering coefficients of the two-port with respect to the reference impedances Z_{01} and Z_{02}. It is clear that even for a reciprocal two-port, where $z_{12} = z_{21}$, the off-diagonal coefficients are not equal and thus the matrix is not symmetric. The equations can be rewritten as

$$E_{1r} = S_{11}{}^E E_{1i} + S_{12}{}^E E_{2i}$$
$$E_{2r} = S_{21}{}^E E_{1i} + S_{22}{}^E E_{2i} \qquad (20)$$

or in matrix notation

$$E_r] = [S^E] E_i] \qquad (21)$$

where $E_r]$ and $E_i]$ are column matrices and the superscript E has been used on the scattering matrix to indicate that its basis is voltage.

From Eq. (20) itself or from the preceding discussion the definition of each of the coefficients is clear. The parameter $S_{11}{}^E$ is the reflection coefficient at port 1 or the input reflection coefficient, i.e., the ratio of the reflected to the incident component of E_1 when port 2 is terminated in the reference impedance Z_{02} and $E_{2i} = 0$; that is, the source E_{g2} is zero. Under these same conditions $S_{21}{}^E$, which is the forward transmission coefficient, is equal to E_{2r}/E_{1i}. Corresponding definitions hold for the output reflection coefficient $S_{22}{}^E$ and the reverse transmission coefficient $S_{12}{}^E$.

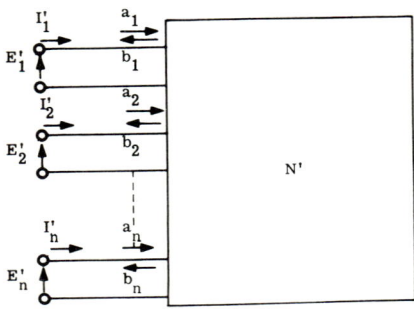

Fig. 5. n-port network used for defining the scattering matrix.

In the second part of the above example the scattering coefficients are derived in terms of voltages. As in the one-port case, the corresponding relations for the scattering matrix can be obtained with current as the basis merely by use of a dual argument. These matrices are not equal in general; they become equal only in the special case when the reference impedances are equal, i.e., $Z_{01} = Z_{02} = Z_0$. It is awkward to have unequal voltage and current scattering matrices to describe a network. It would be convenient to define a scattering matrix which is identical for the voltage and current bases. This is accomplished by *normalization* of the scattering matrix. By normalization, as we shall see below, we obtain, in addition, the desirable property that the scattering matrix for a reciprocal n-port is symmetric.

We now amplify and extend the discussion in a concise notation, but use the scattering wave amplitudes rather than the normalized voltage and current. We normalize with respect first to resistances, that is, real positive numbers, and then to complex impedances defined over the $j\omega$ axis, showing in addition that it is possible to normalize with respect to general impedances defined not only on the $j\omega$ axis but over the whole complex plane.

Consider the n-port network N' shown in Fig. 5 for which the reference resistances for ports 1, 2, ..., n have been chosen as $R_{01}, R_{02}, \ldots, R_{0n}$, respectively, and primes have been used on the port voltages and currents to designate that these variables are unnormalized. Since these numbers are used for normalization, they are also called the normalizing impedances or normalizing numbers. It should be clear that these numbers may be chosen arbitrarily but are constants of the transformations so that once chosen they are then fixed throughout the analysis. The particular problem being considered generally determines the most convenient values to choose. For example, if port k for $k = 1, 2, \ldots, n$ is driven by a generator with an internal resistance r_k, then it is usually simplest to choose $R_{0k} = r_k$ for $k = 1, 2, \ldots, n$. However, it is clearly not required that we use the scattering-matrix analysis only for networks each of whose ports is terminated in the corresponding normalizing resistance; it is only convenient to do so. Just as it is possible to use, for example, the open-circuit impedance matrix [z] for terminated networks, so it is possible to use the scattering matrix for the analysis of networks terminated in impedances that are not equal to the reference impedances.

The reference-impedance matrix is defined by the diagonal matrix

$$[R_0] = \begin{bmatrix} R_{01} & 0 & \cdots & 0 \\ 0 & R_{02} & \cdots & 0 \\ \cdots & \cdots & \cdots & \cdots \\ 0 & 0 & \cdots & R_{0n} \end{bmatrix}$$
$$= \text{diag } [R_{01}, R_{02}, \ldots, R_{0n}] \quad (9)$$

where the notation on the right of the second equation is used to save vertical space when it is necessary to show the elements of a diagonal matrix explicitly. We further define

$$[R_0]^{\frac{1}{2}} = \text{diag } [R_{01}^{\frac{1}{2}}, R_{02}^{\frac{1}{2}}, \ldots, R_{0n}^{\frac{1}{2}}]$$
$$[R_0]^{-\frac{1}{2}} = \text{diag } [R_{01}^{-\frac{1}{2}}, R_{02}^{-\frac{1}{2}}, \ldots, R_{0n}^{-\frac{1}{2}}] \quad (10)$$

The normalized* incident-wave amplitude a_k and the normalized reflected-wave amplitude b_k for the kth port are defined as the following linear combinations of the associated port voltage and current:

$$a_k = \tfrac{1}{2}(E_k'/\sqrt{R_{0k}} + \sqrt{R_{0k}}\, I_k')$$
$$b_k = \tfrac{1}{2}(E_k'/\sqrt{R_{0k}} - \sqrt{R_{0k}}\, I_k') \qquad k = 1, 2, \ldots, n \quad (11)$$

where the square roots are always chosen positive. In matrix notation we have

$$a] = \tfrac{1}{2}[R_0]^{-\frac{1}{2}}\, E'] + \tfrac{1}{2}[R_0]^{\frac{1}{2}}\, I']$$
$$b] = \tfrac{1}{2}[R_0]^{-\frac{1}{2}}\, E'] - \tfrac{1}{2}[R_0]^{\frac{1}{2}}\, I'] \quad (12)$$

We now define the normalized port voltages and currents by

$$E] = [R_0]^{-\frac{1}{2}}\, E']$$
$$I] = [R_0]^{\frac{1}{2}}\, I'] \quad (13)$$

so that Eq. (12) takes on the simple form

$$a] = \tfrac{1}{2}\{E] + I]\}$$
$$b] = \tfrac{1}{2}\{E] - I]\} \quad (14)$$

By addition and then subtraction of the above equations we also obtain the normalized voltages and currents in terms of the incident and reflected waves

$$E] = a] + b]$$
$$I] = a] - b] \quad (15)$$

These relationships, it will be recalled, were given in the introductory discussion of the scattering waves.

The $n \times n$ scattering matrix $[S]$ of the n-port N', normalized with respect to the n resistances $R_{01}, R_{02}, \ldots, R_{0n}$ is defined by the matrix equation

$$b] = [S]\, a] \quad (16)$$

Thus the definition of each scattering coefficient is given by the above equation. We have for $j, k = 1, 2, \ldots, n$

$$S_{jk} = b_j/a_k\big|_{a_1=a_2=\ldots=a_{k-1}=a_{k+1}=\ldots=a_n=0} \quad (17)$$

* In our discussion we shall not explicitly use the unnormalized wave amplitudes which are related to the normalized amplitudes by $a_k = a_k'/\sqrt{R_{0k}}$ and $b_k = b_k'/\sqrt{R_{0k}}$ for $k = 1, 2, \ldots, n$. Thus the definitions using only unnormalized variables, which sometimes occur in the literature, are

$$a_k' = \tfrac{1}{2}(E_k' + R_{0k}I_k')$$
$$b_k' = \tfrac{1}{2}(E_k' - R_{0k}I_k')$$

It is these unnormalized variables which should be used with the unnormalized scattering matrix, that is, $b'] = [S']\, a']$.

For the two-port this becomes

$$S_{11} = b_1/a_1|_{a_2=0} \qquad S_{21} = b_2/a_1|_{a_2=0}$$
$$S_{12} = b_1/a_2|_{a_1=0} \qquad S_{22} = b_2/a_2|_{a_1=0} \qquad (18)$$

As was demonstrated in Example 1 and will be shown again below, the incident wave a_j is zero when the generator voltage E_{gj} is zero and port j has a matched termination, that is, when port j of the *normalized augmented* network is terminated in a one-ohm normalizing resistance; the normalized augmented network is defined below.

It is possible and useful to give a network interpretation to the normalization and to introduce another useful network derived from the given n-port N'. If we modify N' by connecting an ideal transformer at port k of turns ratio $1:\sqrt{R_{0k}}$ for $k = 1, 2, \ldots, n$, then the modified n-port is the normalized network whose port voltages and currents are the normalized voltages and currents. This is shown in Fig. 6.

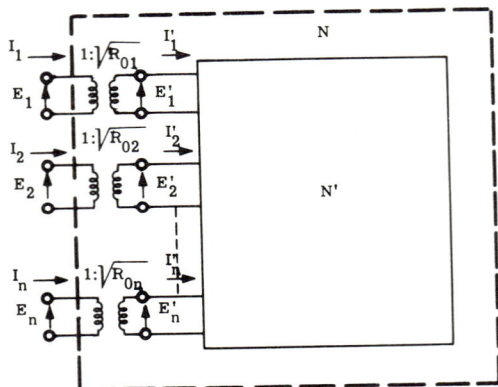

Fig. 6. Normalized network N formed from N' by addition of an ideal transformer at each port.

Now assume that we connect a voltage generator of internal voltage E_{gk}' to port k for $k = 1, 2, \ldots, n$, and choose the internal resistances of these generators to equal the respective port normalization numbers $R_{01}, R_{02}, \ldots, R_{0n}$. The network made up of the given n-port N' and the resistances R_{0k} connected in series with the respective ports is termed the *augmented* network N_a'. This network, shown in Fig. 7, has an augmented short-circuit admittance matrix $[y_a']$ defined by

$$I'] = [y_a'] E_g'] \qquad (19)$$

where $E_g']$ is the column matrix of generator voltages and $I']$ represents the currents into the ports of N_a' and N'. The matrix $[y_a']$ exists for every passive network so that it can conveniently be used, as we shall see, for calculation of $[S]$. If we now normalize the generator voltages and the currents of the augmented network, that is, if we define

$$E_g] = [R_0]^{-\frac{1}{2}} E_g']$$
$$I] = [R_0]^{\frac{1}{2}} I'] \qquad (20)$$

then Eq. (19) becomes

$$I] = [R_0]^{\frac{1}{2}}[y_a'][R_0]^{\frac{1}{2}} E_g]$$
$$= [y_a] E_g] \qquad (21)$$

where the normalized augmented admittance matrix $[y_a]$ is

$$[y_a] = [R_0]^{\frac{1}{2}}[y_a'][R_0]^{\frac{1}{2}} \qquad (22)$$

We may interpret this normalized augmented network by the addition of ideal transformers, as shown in Fig. 8; also shown is an n-port equivalent to the normalized augmented network, where each of the normalizing resistances becomes a 1-ohm resistance by being referred to the primary side of the transformer.

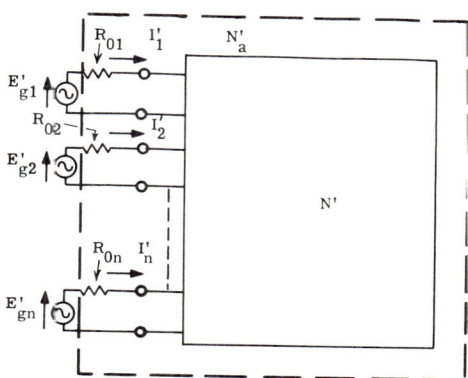

Fig. 7. Augmented n-port N_a' found from N' by the addition of R_{0k} in series with port k for $k = 1, 2, \ldots, n$.

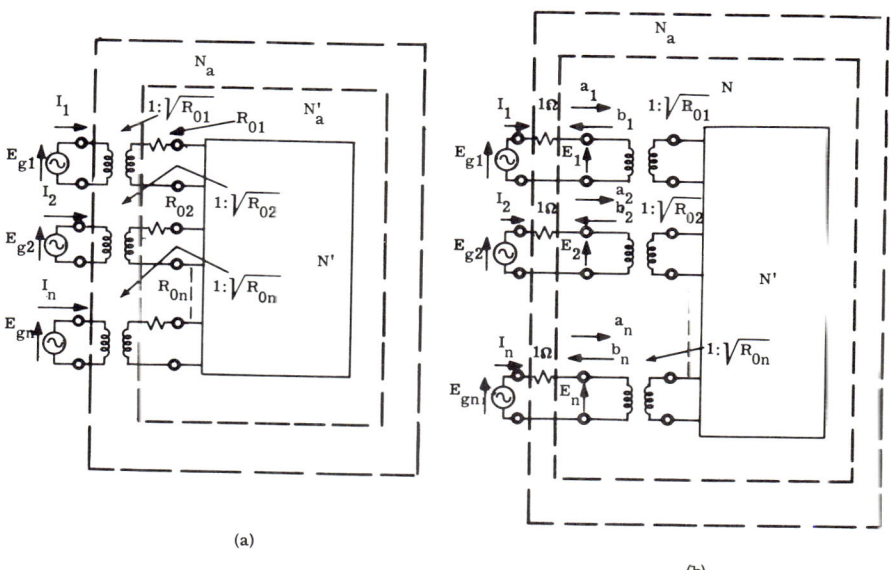

Fig. 8. Two equivalent representations of the normalized augmented network.

As we shall see, the elements of the unnormalized scattering matrix may be related directly to *short-circuit* measurements on the augmented network. A main-diagonal element S_{kk}' is given by

$$S_{kk}' = \frac{Z_{kk}' - R_{0k}}{Z_{kk}' + R_{0k}} \tag{23}$$

where Z_{kk}' is the input impedance at port k of the short-circuited augmented network, that is, $Z_{kk}' + R_{0k}$ is the reciprocal of the corresponding short-circuit driving-point admittance of the augmented network:

$$Z_{kk}' + R_{0k} = 1/y'_{akk} \qquad (24)$$

The off-diagonal element S_{kj}' is equal to twice the short-circuit voltage ratio of the augmented network; that is, we short all the voltage generators except E_{gj}' and measure the output voltage E_k' across R_{0k}; then

$$S_{kj}' = 2(E_k'/E_{gj}') \qquad (25)$$

It is clear that $S_{kj}' \neq S_{jk}'$ unless the network is reciprocal and $R_{0j} = R_{0k}$; thus the unnormalized scattering matrix is in general not symmetric for a reciprocal system; more precisely, we have the relation for a reciprocal network

$$S_{jk}'/S_{kj}' = R_{0j}/R_{0k} \qquad (26)$$

However, making the same measurements on the *normalized* augmented network, we have

$$\begin{aligned} S_{kk} &= \frac{Z_{kk} - 1}{Z_{kk} + 1} \\ &= \frac{Z_{kk}'/R_{0k} - 1}{Z_{kk}'/R_{0k} + 1} \\ &= \frac{Z_{kk}' - R_{0k}}{Z_{kk}' + R_{0k}} \end{aligned} \qquad (27)$$

so that $S_{kk}' = S_{kk}$. The element S_{kj} for $k \neq j$ is equal to twice the short-circuit transfer voltage ratio of the normalized augmented network, that is, twice the voltage ratio of the normalized network with 1-ohm terminations and fed at port j by a generator E_{gj} with a 1-ohm internal resistance. Thus for a reciprocal network

$$\begin{aligned} S_{kj} &= 2(E_k/E_{gj}) \\ &= 2(E_j/E_{gk}) \\ &= S_{jk} \end{aligned} \qquad (28)$$

so that the normalized scattering matrix is symmetric.

We therefore see how the elements of the scattering matrix may be evaluated by measurements of the short-circuited augmented network.

It should be mentioned that a completely dual formulation for the scattering matrix is possible using an augmented network formed by adding reference conductances in parallel with the ports and with the n-port driven by current generators. Then an augmented open-circuit impedance matrix is defined and the measurements of the input admittance and the transfer current ratio are made on an *open-circuited* augmented network.

A comment should be made about the two senses in which the term *matching* is used, often in a manner that may confuse the reader. When we say that the n-port is terminated in matched resistances, we mean merely that the terminations are equal to the respective normalizing numbers; in other words, we are analyzing the augmented network for which the reflection coefficient of each termination is zero. Thus if we knew the actual internal resistances of the generators to be used, we could then choose these as the normalizing numbers so that the analysis could be carried out for an n-port with matched terminations. However, this does not necessarily mean that the n-port is matched for maximum power transfer; that is, each generator is faced by an input impedance equal to the complex conjugate of its internal impedance so that it delivers its total available power as its input to the n-port. In this case we would first have to determine the set of resistances that would simultaneously match each port when used as terminations on the given n-port, then choose these as the normalizing numbers, and finally use generators with these resistances, assuming that the internal resistances are variable and under our control. Then we would have an n-port that is both matched for maximum power transfer and has matched terminations. It should thus be clear that it is possible to have an n-port that has mismatched terminations, that is, with the terminations not equal to the respective normalizing resistances, but has matched generators, that is, each generator is delivering its available power to the n-port. It is also possible for an n-port to have matched terminations but not to be matched for maximum power transfer. In fact, it may be impossible simultaneously to match all the ports of a given network for maximum power transfer; this is true, for example, for a lossless reciprocal frequency-invariant n-port with n odd. By such a frequency-invariant network is meant a network composed of ideal transformers or, equivalently in the distributed-parameter case, of multiport microwave junctions, interconnected by zero lengths of transmission line. If reactive elements are used, since they are not constant with frequency, the theorem is not true; for example, a lossless matched reciprocal five-port is realizable at a single frequency.

Now we consider some different methods for determining $[S]$ by calculation or measurement. One of these, as discussed above, is to evaluate the input impedances and the voltage ratios with each port terminated in its respective normalizing resistance; a similar approach was demonstrated in Example 1, where the reflected and incident components were evaluated directly in terms of the impedances of the network. A second method is to use the open-circuit impedance matrix $[z]$ or the short-circuit admittance matrix $[y]$ for the normalized network when these exist. The requisite formulas are derived below.

Substituting for $E]$ and $I]$ in the open-circuit impedance equations for the normalized network,

$$E] = [z] I] \tag{29}$$

the incident and reflected waves as given by Eq. (15), we obtain

$$\{a] + b]\} = [z]\{a] - b]\} \tag{30}$$

This yields by straightforward manipulation

$$b] = \{[z] + [U]\}^{-1}\{[z] - [U]\} a] \tag{31}$$

so that the scattering matrix is identified as

$$[S] = \{[z] + [U]\}^{-1}\{[z] - [U]\} \tag{32}$$

where $[U]$ is the identity (or unit) matrix.

Similarly, starting with the short-circuit admittance equations for the normalized network,

$$I] = [y] E] \tag{33}$$

we obtain

$$\{a] - b]\} = [y]\{a] + b]\} \tag{34}$$

which gives the relationship

$$b] = \{[U] + [y]\}^{-1}\{[U] - [y]\} a] \tag{35}$$

Thus the scattering matrix is given by

$$[S] = \{[U] + [y]\}^{-1}\{[U] - [y]\} \tag{36}$$

A third method is to use the admittance matrix of the normalized augmented network

$$I] = [y_a] E_a] \tag{37}$$

It is evident from inspection of Fig. 8b, since each port current flows through a 1-ohm resistance, that

$$E_a] = E] + I] \tag{38}$$

Using the first equation of Eqs. (14) then gives

$$E_a] = 2a] \tag{39}$$

Substituting in Eq. (37) the above value for E_g} and the value of I] given by the second of Eqs. (15) yields

$$\{a] - b]\} = 2[y_a] \, a] \qquad (40)$$

which becomes by rearrangement of terms

$$b] = \{[U] - 2[y_a]\} \, a] \qquad (41)$$

so that

$$[S] = [U] - 2[y_a] \qquad (42)$$

Since

$$[y_a] = \{[z] + [U]\}^{-1} \qquad (43)$$

and $[z]$ is symmetric for a reciprocal network, $[y_a]$ is also. Thus by Eq. (42) $[S]$ is symmetric if N' is a reciprocal network.

We may also derive Eq. (32) from (42). Using (43) to eliminate $[y_a]$ from (42) and then factoring $\{[z] + [U]\}^{-1}$ we obtain

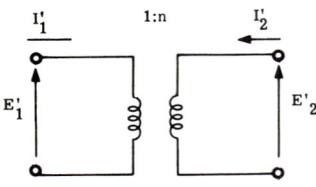

$$\begin{aligned}[S] &= [U] - 2\{[z] + [U]\}^{-1} \\ &= \{[z] + [U]\}^{-1}\{[z] + [U] - 2[U]\} \\ &= \{[z] + [U]\}^{-1}\{[z] - [U]\} \end{aligned} \qquad (44)$$

An alternative expression is obtained by postmultiplying by the matrix sum that is removed as a factor:

Fig. 9. Ideal transformer with turns ratio $1:n$.

$$\begin{aligned}[S] &= [U] - 2\{[z] + [U]\}^{-1} \\ &= \{[z] + [U] - 2[U]\}\{[z] + [U]\}^{-1} \\ &= \{[z] - [U]\}\{[z] + [U]\}^{-1} \end{aligned} \qquad (45)$$

These equivalent forms may be used to show again that $[S]$ is symmetric for a reciprocal network, since if we take the transpose of the last expression on the right in Eq. (44) we get the right-hand side of the last equation of (45).

Finally, we consider a method that, like the method using $[y_a]$, may be used when a network has no impedance or admittance matrix but may be described by an equation of the type

$$[A] \, E] = [B] \, I] \qquad (46)$$

Substituting for $E]$ and $I]$ the expressions containing the incident and reflected waves given in Eq. (15),

$$[A]\{a] + b]\} = [B]\{a] - b]\} \qquad (47)$$

Rearranging terms we obtain the scattering-matrix form

$$b] = \{[B] + [A]\}^{-1}\{[B] - [A]\} \, a] \qquad (48)$$

so that

$$[S] = \{[B] + [A]\}^{-1}\{[B] - [A]\} \qquad (49)$$

Frequently we wish to evaluate $[z]$, $[y]$, or $[y_a]$ from a given $[S]$. Solving the appropriate equations derived above, we get for $[z]$, $[y]$, and $[y_a]$

$$\begin{aligned}[z] &= \{[U] - [S]\}^{-1}\{[U] + [S]\} \\ &= \{[U] + [S]\}\{[U] - [S]\}^{-1} \\ [y] &= \{[U] + [S]\}^{-1}\{[U] - [S]\} \\ &= \{[U] - [S]\}\{[U] + [S]\}^{-1} \\ [y_a] &= \tfrac{1}{2}\{[U] - [S]\} \end{aligned} \qquad (50)$$

To illustrate the use of these methods we derive the scattering matrix normalized to arbitrary resistances R_{01} and R_{02} for the ideal transformer shown in Fig. 9. First we use the augmented admittance matrix $[y_a]$. For the augmented network of Fig. 10a, we have by simple analysis

$$[y_a'] = \begin{bmatrix} \dfrac{n^2}{n^2 R_{01} + R_{02}} & -\dfrac{n}{n^2 R_{01} + R_{02}} \\ -\dfrac{n}{n^2 R_{01} + R_{02}} & \dfrac{1}{n^2 R_{01} + R_{02}} \end{bmatrix} \qquad (51)$$

MATRICES FOR MULTITERMINAL NETWORKS 4-13

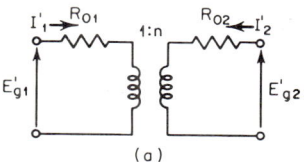

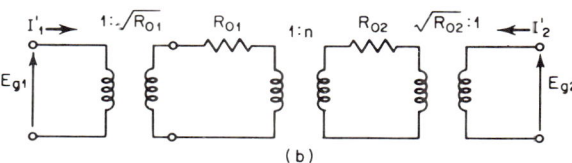

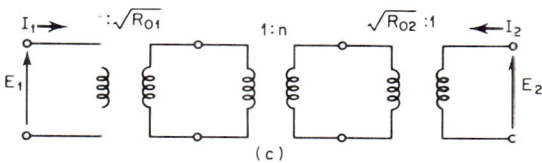

Fig. 10. Networks derived from the ideal transformer. (a) Augmented network. (b) Normalized augmented network. (c) Normalized network.

Employing Eq. (23) we obtain the normalized augmented admittance matrix:

$$[y_a] = [R_0]^{\frac{1}{2}}[y_a'][R_0]^{\frac{1}{2}}$$

$$= \begin{bmatrix} \dfrac{n^2 R_{01}}{n^2 R_{01} + R_{02}} & -\dfrac{n\sqrt{R_{01}R_{02}}}{n^2 R_{01} + R_{02}} \\ -\dfrac{n\sqrt{R_{01}R_{02}}}{n^2 R_{01} + R_{02}} & \dfrac{R_{02}}{n^2 R_{01} + R_{02}} \end{bmatrix} \quad (52)$$

Finally, using Eq. (42) we obtain

$$[S] = [U] - 2[y_a]$$

$$= \begin{bmatrix} \dfrac{R_{02} - n^2 R_{01}}{n^2 R_{01} + R_{02}} & \dfrac{2n\sqrt{R_{01}R_{02}}}{n^2 R_{01} + R_{02}} \\ \dfrac{2n\sqrt{R_{01}R_{02}}}{n^2 R_{01} + R_{02}} & \dfrac{n^2 R_{01} - R_{02}}{n^2 R_{01} + R_{02}} \end{bmatrix} \quad (53)$$

We note that if $n = \sqrt{R_{02}/R_{01}}$ the ports of the transformer are simultaneously matched by terminations equal to the normalizing numbers and

$$[S] = \begin{bmatrix} 0 & 1 \\ 1 & 0 \end{bmatrix} \quad (54)$$

If $n = 1$ so that a direct connection exists between the ports, then

$$[S] = \begin{bmatrix} \dfrac{R_{02} - R_{01}}{R_{01} + R_{02}} & \dfrac{2\sqrt{R_{01}R_{02}}}{R_{01} + R_{02}} \\ \dfrac{2\sqrt{R_{01}R_{02}}}{R_{01} + R_{02}} & \dfrac{R_{01} - R_{02}}{R_{01} + R_{02}} \end{bmatrix} \quad (55)$$

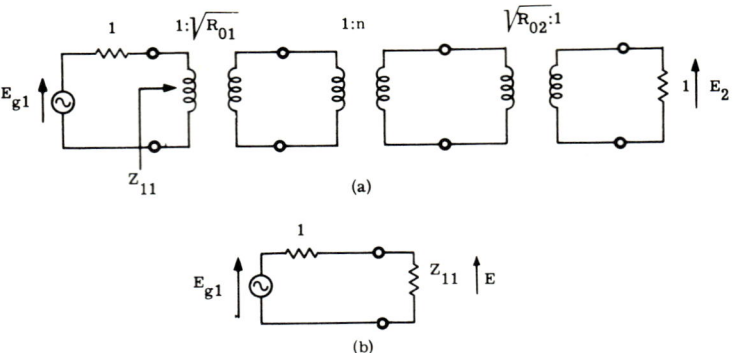

FIG. 11. Normalized augmented ideal transformer for calculating the input impedance and the voltage ratio.

We now use the last of the methods discussed above on this same problem, that is, we first put the equations in the form given in (46). By inspection of the normalized network in Fig. 10c, we obtain

$$I_1/\sqrt{R_{01}} = -nI_2/\sqrt{R_{02}}$$
$$n\sqrt{R_{01}}\,E_1 = \sqrt{R_{02}}\,E_2 \tag{56}$$

We now form

$$\begin{bmatrix} n\sqrt{R_{01}} & -\sqrt{R_{02}} \\ 0 & 0 \end{bmatrix} \begin{bmatrix} E_1 \\ E_2 \end{bmatrix} = \begin{bmatrix} 0 & 0 \\ \dfrac{1}{\sqrt{R_{01}}} & \dfrac{n}{\sqrt{R_{02}}} \end{bmatrix} \begin{bmatrix} I_1 \\ I_2 \end{bmatrix} \tag{57}$$

so that

$$[B] - [A] = \begin{bmatrix} -n\sqrt{R_{01}} & \sqrt{R_{02}} \\ \dfrac{1}{\sqrt{R_{01}}} & \dfrac{n}{\sqrt{R_{02}}} \end{bmatrix} \tag{58}$$

and

$$\{[B] + [A]\}^{-1} = \frac{\sqrt{R_{01}R_{02}}}{n^2R_{01} + R_{02}} \begin{bmatrix} \dfrac{n}{\sqrt{R_{02}}} & \sqrt{R_{02}} \\ -\dfrac{1}{\sqrt{R_{01}}} & n\sqrt{R_{01}} \end{bmatrix} \tag{59}$$

We thus obtain

$$S = \{[B] + [A]\}^{-1}\{[B] - [A]\}$$
$$= \begin{bmatrix} \dfrac{R_{02} - n^2R_{01}}{n^2R_{01} + R_{02}} & \dfrac{2n\sqrt{R_{01}R_{02}}}{n^2R_{01} + R_{02}} \\ \dfrac{2n\sqrt{R_{01}R_{02}}}{n^2R_{01} + R_{02}} & \dfrac{n^2R_{01} - R_{02}}{n^2R_{01} + R_{02}} \end{bmatrix} \tag{60}$$

To check the above elements by direct evaluation of the input impedances and the voltage ratios, we have for the normalized augmented network shown in Fig. 11a, where E_{g2} has been set to zero,

$$S_{11} = \frac{Z_{11} - 1}{Z_{11} + 1}$$
$$= \frac{R_{02}/n^2R_{01} - 1}{R_{02}/n^2R_{01} + 1}$$
$$= \frac{R_{02} - n^2R_{01}}{R_{02} + n^2R_{01}} \tag{61}$$

MATRICES FOR MULTITERMINAL NETWORKS 4-15

The other reflection coefficient is calculated similarly to give the value obtained by the previous methods. To determine $S_{21} = S_{12}$ we find by inspection of Fig. 11b that the voltage across the load referred to the primary is

$$\frac{E}{E_{g1}} = \frac{Z_{11}}{1 + Z_{11}}$$
$$= \frac{R_{02}/n^2 R_{01}}{1 + R_{02}/n^2 R_{01}} = \frac{R_{02}}{n^2 R_{01} + R_{02}} \tag{62}$$

Now reflecting the voltage E to the output, we have

$$E_2/E = n\sqrt{R_{01}/R_{02}} \tag{63}$$

so that
$$E_2/E_{g1} = n\sqrt{R_{01}R_{02}}/(n^2 R_{01} + R_{02}) \tag{64}$$
and
$$S_{21} = 2(E_2/E_{g1})$$
$$= 2n\sqrt{R_{01}R_{02}}/(n^2 R_{01} + R_{02}) \tag{65}$$

Thus all the elements of the matrix have been checked.

We now consider another example to illustrate that the performance of an n-port network is readily determined for arbitrary terminations if its scattering matrix is given and the preassigned normalizing numbers are known. Suppose that the normalized impedance $Z_k = Z_k'/R_{0k}$ terminates the kth port where $k = 1, 2, \ldots, n$. As shown in Fig. 12 for the termination Z_k the *incident* wave is b_k and the *reflected* wave is a_k; thus the reflection coefficient S_k for the load is a_k/b_k. Another way of seeing this is to note that the voltage across the load is E_k and the current into the upper terminal of the load is $-I_k$ so that the reflected wave for the termination is given by $\frac{1}{2}[E_k - (-I_k)] = \frac{1}{2}(E_k + I_k) = a_k$ and the incident wave by $\frac{1}{2}(E_k - I_k) = b_k$. Thus

$$S_k = \frac{E_k + I_k}{E_k - I_k} = \frac{Z_k - 1}{Z_k + 1} = \frac{Z_k' - R_{0k}}{Z_k' + R_{0k}} = \frac{a_k}{b_k} \tag{66}$$

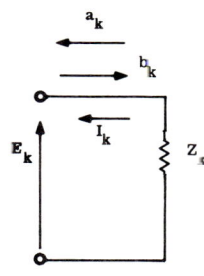

Fig. 12. Impedance Z_k used as a termination at kth port for $k = 1, 2, \ldots, n$ illustrating that for the load b_k is the *incident* wave and a_k is the *reflected* wave. E_k and $-I_k$ are the voltage and current to be used for defining its reflection coefficient.

These boundary conditions when substituted into the scattering-matrix equations

$$b] = [S] a] \tag{67}$$

will yield the reflection and transmission coefficients for the n-port terminated in these impedances.

To be more specific, suppose we wish to find the input reflection coefficient, the transmission coefficient, and the voltage ratio of a two-port terminated in a load of reflection coefficient S_2. Writing out the scattering-matrix equations and using

$$a_2 = S_2 b_2 \tag{68}$$

to eliminate a_2, we obtain

$$b_1 = S_{11}a_1 + S_{12}a_2 = S_{11}a_1 + S_{12}S_2 b_2 \tag{69}$$
and
$$b_2 = S_{21}a_1 + S_{22}a_2 = S_{21}a_1 + S_{22}S_2 b_2 \tag{70}$$

Solving the above equation for b_2 and substituting into (69) give as the input reflection coefficient of the terminated two-port

$$S_{in-1} = b_1/a_1 = S_{11} + S_{12}S_{21}S_2/(1 - S_{22}S_2) \tag{71}$$

The ratio b_2/a_1 is found from (70) as

$$b_2/a_1 = S_{21}/(1 - S_{22}S_2) \tag{72}$$

The voltage ratio E_2/E_{g1} may be found from the above equation. Since

$$E_2 = a_2 + b_2 = b_2(1 + S_2) \tag{73}$$

and for the network driven by a generator whose internal resistance equals the refer-

ence resistance
$$E_{g1} = 2a_1 \tag{74}$$

then
$$\frac{E_2}{E_{g1}} = \frac{b_2(1 + S_2)}{2a_1}$$
$$= \frac{S_{21}(1 + S_2)}{2(1 - S_{22}S_2)} \tag{75}$$

If $S_2 = 0$, i.e., the normalized load impedance $Z_2 = 1$, then Eqs. (71) and (72) reduce to
$$S_{in-1} = S_{11}$$
$$b_2/a_1 = S_{21} \tag{76}$$

For this case $a_2 = 0$ and $b_2 = E_2$ so that for the matched augmented network with $E_{g2} = 0$, that is, if a normalized unit-resistance generator excites port 1 with port 2 terminated in a 1-ohm normalized load, the ratio

$$b_2/a_1|_{S_2=0} = S_{21}$$
$$= E_2/(E_{g1}/2)$$
$$= 2(E_2/E_{g1}) \tag{77}$$

The same reasoning clearly holds for the n-port so that our previous statement is confirmed, namely, S_{kj} is equal to twice the voltage ratio of the normalized network when it is fed at port j by a unit-resistance generator and terminated at all other ports in 1-ohm loads. Using this property we see that the unnormalized network when terminated in the reference resistances yields

$$S_{kj}' = 2(E_k'/E_{gj}')$$
$$= 2\sqrt{R_{0k}/R_{0j}}\,(E_k/E_{gj})$$
$$= \sqrt{R_{0k}/R_{0j}}\,S_{kj} \tag{78}$$

so that it is clear that in general the unnormalized scattering matrix for a reciprocal network is not symmetric.

Now turning to questions of power, we consider the power transfer ratio, referred to generator available power, between ports j and k terminated in the normalizing numbers; as pointed out in Chap. 3, Sec. 3, this ratio of output power to available power is called the *transducer power gain* in the literature. If P_k is the power dissipated in R_{0k}, the load at port k, and $P_{A,j}$ is the available power at port j, we have

$$G_{kj}' = \frac{P_k}{P_{A,j}} = \frac{|E_k'|^2/R_{0k}}{|E_{gj}'|^2/4R_{0j}} = \frac{|E_k|^2}{|E_{gj}|^2/4}$$
$$= \frac{|b_k|^2}{|a_j|^2} = |S_{kj}|^2 \tag{79}$$

or
$$G_{kj}' = G_{kj} = |S_{kj}|^2 \tag{80}$$

In substituting S_{kj} for b_k/a_j in Eqs. (79) we have assumed that all ports are terminated in their respective normalizing resistances. We therefore see that $|S_{kj}|^2$ is equal to the transducer power gain from port j to port k, when each port is terminated in its reference resistance, and that, unlike the voltage ratio, this power gain has the same value for the normalized and unnormalized network. Thus when the insertion loss of a transducer, namely, $L = 20 \log 1/|S_{21}|$ in decibels, is specified for the augmented network, this immediately determines the magnitude of the scattering coefficient $|S_{21}|$.

Now for $s = j\omega$, that is, for the sinusoidal steady state, we express in terms of $[S]$ the total average power P absorbed by N'. We have from the definition of power, where we use an asterisk on a matrix to denote the complex conjugate transpose,

$$P = \text{Real} \sum_{k=1}^{n} \bar{E}_k' I_k' = \text{Real} \sum_{k=1}^{n} \bar{E}_k I_k = \text{Real } E]^* I]$$
$$= \text{Real } \{(a]^* + b]^*)(a] - b])\} = a]^* a] - b]^* b]$$
$$= a]^* \{[U] - [S]^*[S]\}\,a] \tag{81}$$

MATRICES FOR MULTITERMINAL NETWORKS 4-17

Now if N' is lossless, $P = 0$ for any a], which implies that

$$[U] - [S]^*[S] = [0] \tag{82}$$

where $[0]$ is the null matrix. Thus

$$[S]^*[S] = [U] \tag{83}$$
or
$$[S]^* = [S]^{-1} \tag{84}$$

which, in words, states that the complex conjugate transpose of $[S]$ is equal to its inverse. Since this is the definition of a unitary matrix, we have shown that for a lossless network $[S]$ is a unitary matrix. If N' is passive and dissipative, $P \geq 0$, which implies that $[U] - [S]^*[S]$ is a positive semidefinite Hermitian matrix.

The average power contributed by the kth port alone (or for a one-port the average power dissipated by the load) is

$$P_k = \text{Real } \bar{E}_k'I_k' = \text{Real } \bar{E}_k I_k = |a_k|^2 - |b_k|^2 \tag{85}$$

Thus for a one-port or for the kth port of an n-port when each port is terminated in its reference resistance

$$P_k = |a_k|^2 - |b_k|^2 = |a_k|^2(1 - |S_{kk}|^2) \tag{86}$$

We may define $|a_k|^2$ as the incident power at port k denoted by P_{ki}, and $|b_k|^2$ as the reflected power P_{kr}. Thus the power absorbed by the network at port k is given by the difference between the two powers

$$P_k = P_{ki} - P_{kr} \tag{87}$$

Since the available power at port k is given by

$$P_{A,k} = |E_{gk}'|^2/4R_{0k} = |E_{gk}/2|^2 = |a_k|^2 \tag{88}$$

we obtain

$$P_k/P_{A,k} = 1 - |S_{kk}|^2 \tag{89}$$

For passive networks $0 \leq P_k/P_{A,k} \leq 1$ for all ω so that the above gives a realizability condition on $|S_{kk}|$, namely,

$$|S_{kk}| \leq 1 \tag{90}$$

We thus see again that the reflection coefficient at port k is a measure of the deviation from maximum power transfer at this port, that is, it is a power-reflection coefficient. When it is zero, the input impedance at the port is matched to the generator impedance, and maximum power is transferred; when it is unity, the input impedance is purely reactive and the power transfer ratio is zero. Therefore, the *return loss* defined by $20 \log |1/S_{kk}|$ db is a measure of the inefficiency of a network in transferring available power.

One additional comment should now be made, preparatory for the discussion of complex normalization. The reader is probably aware from his knowledge of transmission-line and waveguide theory that a passive load, that is, one whose real part is nonnegative for all ω, may have a reflection coefficient whose magnitude is greater than 1; this appears to contradict Eq. (90). However, no contradiction exists. We have defined the scattering matrix with *real* normalizing numbers. When we have a dissipative guide, its characteristic impedance Z_0 is complex. If we use Z_0 as the normalizing impedance, then it is clear that

$$|S_{11}| = \left|\frac{b_1}{a_1}\right| = \left|\frac{Z_1 - Z_0}{Z_1 + Z_0}\right| = \left|\frac{(R_1 - R_0) + j(X_1 - X_0)}{(R_1 + R_0) + j(X_1 + X_0)}\right| \tag{91}$$

may exceed unity. Thus in this case $|S_{11}|^2$ cannot be interpreted as a power-reflection coefficient. However, it is possible to retain this useful interpretation by using a reflection coefficient defined by

$$S_{11} = \frac{Z_1 - \bar{Z}_0}{Z_1 + Z_0} \tag{92}$$

We shall see below that such a definition is obtained by normalizing with respect to complex numbers.

2.1. Scattering Matrix Normalized with Respect to Complex Impedances.*

Suppose an n-port is excited at its ports by n generators $E_{g1}, E_{g2}, \ldots, E_{gn}$ with prescribed internal impedances that are not real and positive but are complex over the frequency band of interest W. The question arises whether it is possible to describe the normalized network N in terms of a normalized scattering matrix $[S]$ so that $[S]$ still possesses the important properties discussed previously, namely, $[S]$ is a unitary matrix when N is lossless, and under matched terminations (that is, with the n-port terminated in its normalizing impedances) the transducer power gain $G_{jk}(\omega^2)$ from port k to port j is given by

$$G_{jk}(\omega^2) = |S_{jk}(j\omega)|^2 \tag{93}$$

It has been shown that this is indeed possible when each normalizing impedance has a positive real part over the frequency range W; we may thus use n complex impedances $Z_{0k}(j\omega)$ with positive real parts, for example, the complex internal impedances,

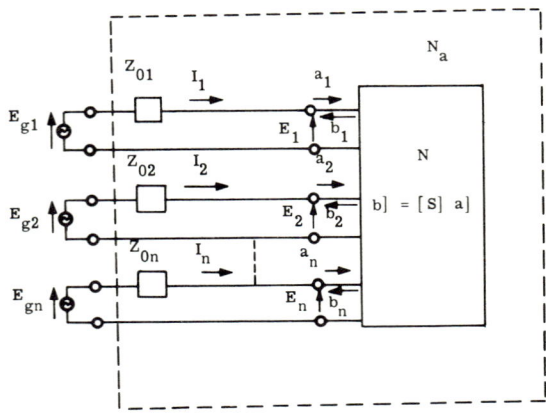

Fig. 13. Given n-port excited at its n ports by generators $E_{g1}, E_{g2}, \ldots, E_{gn}$ with respective internal impedances $Z_{01}(j\omega), Z_{02}(j\omega), \ldots, Z_{0n}(j\omega)$.

as the normalizing impedances. We consider this below. Moreover, though we do not discuss this further, it is possible to normalize the scattering matrix to a full $n \times n$ matrix $[Z_0]$ rather than use n uncoupled impedances, which leads to a diagonal $[Z_0]$. Furthermore, the scattering matrix may be normalized with respect to impedances defined not only on the imaginary axis but over the entire s plane.[12]

The network under consideration is shown in Fig. 13, where for convenience we now represent the unnormalized network by N without a prime. This network is characterized in terms of *normalized* wave amplitudes by

$$b] = [S] a] \tag{94}$$

where $[S]$ is the normalized scattering matrix. We assume that for all ω in W each of the normalizing impedances satisfies

$$\text{Re } Z_{0k}(j\omega) = R_{0k}(\omega) > 0 \qquad k = 1, 2, \ldots, n \tag{95}$$

The normalized incident- and reflected-wave amplitudes for the kth port, a_k and b_k, are defined as linear combinations of the associated port voltages and currents as follows:

$$\begin{aligned} 2\sqrt{R_{0k}}\, a_k &= E_k + Z_{0k} I_k \\ 2\sqrt{R_{0k}}\, b_k &= E_k - \bar{Z}_{0k} I_k \end{aligned} \qquad k = 1, 2, \ldots, n \tag{96}$$

* Almost the entire discussion follows that given in the Youla paper cited as Ref. 4.

with all square roots again chosen positive. Though E_k and I_k are clearly unnormalized variables, we shall for convenience omit primes throughout the following discussion. The above definition clearly reduces to our previous one when the normalization impedances Z_{0k} are real numbers, that is, when $Z_{0k} = R_{0k}$ for all k. In matrix form, we have

$$2[R_0]^{\frac{1}{2}} a] = E] + [Z_0] I]$$
$$2[R_0]^{\frac{1}{2}} b] = E] - [\bar{Z}_0] I] \tag{97}$$

where
$$[Z_0] = \text{diag } [Z_{01}, Z_{02}, \ldots, Z_{0n}]$$
$$[R_0] = \text{real } [Z_0]$$
$$= \text{diag } [R_{01}, R_{02}, \ldots, R_{0n}] \tag{98}$$

The $n \times n$ scattering matrix $[S(j\omega)]$ of N, normalized with respect to the n impedances $Z_{01}(j\omega), Z_{02}(j\omega), \ldots, Z_{0n}(j\omega)$, is defined by means of the matrix equation

$$b] = [S] a] \tag{99}$$

First we calculate the total average power P absorbed by N in terms of $[S]$. From the definition of power we have

$$P = \text{Real} \sum_{k=1}^{n} \bar{E}_k I_k$$
$$= \text{Real } \{[E]^* I]\} = \tfrac{1}{2}\{[E]^* I] + I]^* E]\} \tag{100}$$

Subtracting the second of Eqs. (97) from the first, we obtain

$$2[R_0]^{\frac{1}{2}} \{a] - b]\} = \{[Z_0] + [\bar{Z}_0]\} I] = 2[R_0] I] \tag{101}$$
so that
$$I] = [R_0]^{-\frac{1}{2}} \{a] - b]\} = \eta] - \xi] \tag{102}$$
where
$$\eta] = [R_0]^{-\frac{1}{2}} a]$$
$$\xi] = [R_0]^{-\frac{1}{2}} b] \tag{103}$$

Substituting the first of Eqs. (102) in the first of Eqs. (97) and rearranging terms yield

$$E] = \{2[R_0]^{\frac{1}{2}} - [Z_0][R_0]^{-\frac{1}{2}}\} a] + [Z_0][R_0]^{-\frac{1}{2}} b] \tag{104}$$

However, since
$$2[R_0]^{\frac{1}{2}} - [Z_0][R_0]^{-\frac{1}{2}} = [\bar{Z}_0][R_0]^{-\frac{1}{2}} \tag{105}$$

Eq. (104) becomes
$$E] = [\bar{Z}_0] \eta] + [Z_0] \xi] \tag{106}$$

Clearly, (102) and (106) imply that

$$E]^* I] = \{\eta]^*[Z_0] + \xi]^*[\bar{Z}_0]\}\{\eta] - \xi]\}$$
$$= \eta]^*[Z_0]\eta] - \xi]^*[\bar{Z}_0]\xi] - \eta]^*[Z_0]\xi] + \xi]^*[\bar{Z}_0]\eta] \tag{107}$$

Thus, using the last equation of Eqs. (100) for the real part of the complex power, we have

$$\text{Real } \{[E]^* I]\} = \eta]^*[R_0]\eta] - \xi]^*[R_0]\xi] \tag{108}$$

since $\eta]^*[Z_0]\xi]$ is the complex conjugate of $\xi]^*[\bar{Z}_0]\eta]$. Finally, (100) and (103) give

$$P = \text{Real } \{[E]^* I]\} = a]^* a] - b]^* b]$$
$$= a]^*\{[U] - [S]^*[S]\} a] \tag{109}$$

Consequently, if N is lossless over the frequency range W, $P(\omega^2) = 0$ for all ω in W for any $a]$; this in turn implies that

$$[U] - [S(j\omega)]^*[S(j\omega)] = [0] \tag{110}$$

that is, $[S(j\omega)]$ is unitary for all ω in W. Again, if N is passive and dissipative, $P(\omega^2) \geq 0$ for all $a]$ so that

$$[U] - [S(j\omega)]^*[S(j\omega)] \geq 0 \tag{111}$$

for all ω in W. By the notation in the above equation we mean that the matrix expression on the left is positive semidefinite. Thus the normalization procedure has preserved two of the most important properties possessed by a scattering matrix normalized to real positive numbers.

Now suppose that the network exists with matched terminations; more specifically, let all ports of N except the kth port be closed on their respective normalization impedances and suppose that port k is driven as shown in Fig. 14. Then

$$E_j = -Z_{0j}I_j \qquad (j \neq k) \qquad (112)$$

According to the first of Eqs. (96), the above is equivalent to

$$a_j = 0 \qquad (j \neq k) \qquad (113)$$

Hence, from (94)

$$b_j = S_{jk}(j\omega)a_k \qquad j = 1, 2, \ldots, n$$
$$= S_{jk}(j\omega)[E_{gk}/2\sqrt{R_{0k}(\omega)}] \qquad (114)$$

since, as is clear from Fig. 14,

$$E_{gk} = E_k + Z_{0k}I_k \qquad (115)$$

Equation (112), coupled with the second of Eqs. (96), yields

$$b_j = -\sqrt{R_{0j}(\omega)}\,I_j \qquad (j \neq k) \qquad (116)$$

and therefore, from (114)

$$|S_{jk}(j\omega)|^2 = \frac{R_{0j}(\omega)|I_j|^2}{|E_{gk}|^2/4R_{0k}(\omega)}$$
$$= G_{jk}(\omega^2) \qquad (j \neq k) \qquad (117)$$

the transducer power gain from port k to port j.

To determine $S_{kk}(j\omega)$ let $Z_{kk}(j\omega)$ represent the impedance seen looking into port k under matched terminations, as shown in Fig. 14. Then

$$E_k = Z_{kk}I_k \qquad (118)$$

and the division of the second of Eqs. (96) by the first yields, with the aid of (114) and (118),

$$S_{kk}(j\omega) = \frac{b_k}{a_k} = \frac{E_k - \bar{Z}_{0k}I_k}{E_k + Z_{0k}I_k} = \frac{Z_{kk} - \bar{Z}_{0k}}{Z_{kk} + Z_{0k}} \qquad (119)$$

In words, S_{kk} is the input reflection coefficient at port k with all other ports matched. Thus S_{kk} indicates the deviation at port k from a *conjugate* match, that is, the matched impedance required for generator E_{gk} to deliver its maximum power at port k of the network. It is important to note that

$$S_{kk} \neq \frac{Z_{kk} - Z_{0k}}{Z_{kk} + Z_{0k}} \qquad (120)$$

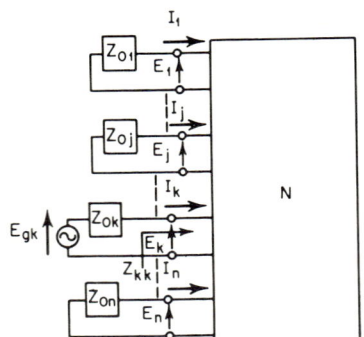

FIG. 14. Schematic of n-port under matched terminations to illustrate the meaning of the transducer power gain $G_{jk}(\omega^2)$.

which result, obtained by normalizing in the usual manner, is correct only when the normalizing impedances are real, if we wish to use the scattering coefficients in the simple power-flow formulas and to keep their useful properties, like the unitary character of the scattering matrix for lossless networks. However, as is done in image-parameter theory, we can use the same normalizing equations used for real normalization even when the normalizing impedances are complex, and thus obtain the above equation as a correct one for the reflection coefficient; but then the simple formulas for power transfer in terms of scattering coefficients no longer hold; e.g., the reflection coefficient is no longer a measure of the power reflected. It is important to understand these distinctions for real and complex normalization.

We also observe that if port j is terminated in $-\bar{Z}_{0j}(j\omega)$ instead of $Z_{0j}(j\omega)$,

$$E_j = \bar{Z}_{0j}I_j \tag{121}$$

Thus by the second of Eqs. (96), $b_j = 0$. In short, terminating a port with its respective normalization impedance obliterates the corresponding incident wave whereas termination in the negative complex conjugate of the normalization impedance obliterates the reflected wave. In the first case we have designated the port as *matched*; in the second the port is said to be *paraconjugate matched*.

According to (119), $S_{kk}(j\omega)$ is the input reflection coefficient at port k under matched conditions. We now show that for a lossless n-port N, the input reflection coefficient at port k under paraconjugate matched conditions is $(\tilde{S}_{kk})^{-1}$. By the preceding discussion paraconjugate matching of all ports except the kth means that $b_j = 0$, $j \neq k$. From (99) and the unitary character of $[S]$,

and thus
whence
$$a] = [S(j\omega)]^{-1} b] = [S(j\omega)]^* b] \tag{122}$$
$$a_j = \tilde{S}_{kj}(j\omega)b_k \qquad j = 1, 2, \ldots, n \tag{123}$$
$$b_k/a_k = (\tilde{S}_{kk})^{-1} \tag{124}$$

This result implies that the impedance seen looking into port k under paraconjugate matched conditions equals $-\bar{Z}_{kk}$, where Z_{kk} is, as defined previously, the impedance seen at port k under matched terminations.

It is possible to express $[S]$ in terms of $[y_a(j\omega)]$, the admittance matrix of N_a, the augmented n-port associated with N. The result can be shown to be

$$[S] = [U] - 2[R_0]^{\frac{1}{2}}[y_a][R_0]^{\frac{1}{2}} \tag{125}$$

The matrix $[y_a]$ exists if N is passive. If N is also reciprocal, $[y_a]$ is symmetric and therefore $[S]$ is also symmetric since $[R_0]^{\frac{1}{2}}$ is diagonal.

Similarly, if N possesses an impedance matrix $[z]$, we can show

$$[S] = [R_0]^{-\frac{1}{2}}\{[z] - [\bar{Z}_0]\}\{[z] + [Z_0]\}^{-1}[R_0]^{\frac{1}{2}} \tag{126}$$

The reader should note that the formula usually given in the literature,

$$[S] = [R_0]^{-\frac{1}{2}}\{[z] - [Z_0]\}\{[z] + [Z_0]\}^{-1}[R_0]^{\frac{1}{2}} \tag{127}$$

is incorrect if $[Z_0]$ is complex.

Now we consider an application of our formulas to negative-resistance devices such as the tunnel diode and the varactor. Since all the Z_{0k} are assumed to have positive real parts, all the $-\bar{Z}_{0k}$ have negative real parts; it is this fact that makes the method so valuable in dealing with negative-resistance devices. We shall derive a formula for the maximum transducer power gain attainable with a single negative-resistance device Z_d, embedded in a lossless environment, as shown in Fig. 15.

Let N be described by its scattering matrix $[S(j\omega)]$ normalized to $Z_{01} = Z_g$ at port 1, $Z_{02} = Z_L$ at port 2, and $Z_{03} = -\bar{Z}_d$ at port 3, where $Z_g(j\omega) = R_g(\omega) + jX_g(\omega)$, $Z_L(j\omega) = R_L(\omega) + jX_L(\omega)$, and $Z_d(j\omega) = R_d(\omega) + jX_d(\omega)$. By assumption these three normalizing impedances have positive real parts over the band of interest W. Since N is lossless,

and therefore
$$[S]^*[S] = [U] \tag{128}$$
$$a] = [S(j\omega)]^* b] \tag{129}$$

From our previous argument on matched and paraconjugate matched terminations, $a_2 = 0$, $b_3 = 0$, and $a_1 = E_{g1}/2\sqrt{R_g}$. Thus

$$E_{g1}/2\sqrt{R_g} = a_1 = \tilde{S}_{11}b_1 + \tilde{S}_{21}b_2 \tag{130}$$
$$0 = \tilde{S}_{12}b_1 + \tilde{S}_{22}b_2 \tag{131}$$
$$a_3 = \tilde{S}_{13}b_1 + \tilde{S}_{23}b_2 \tag{132}$$

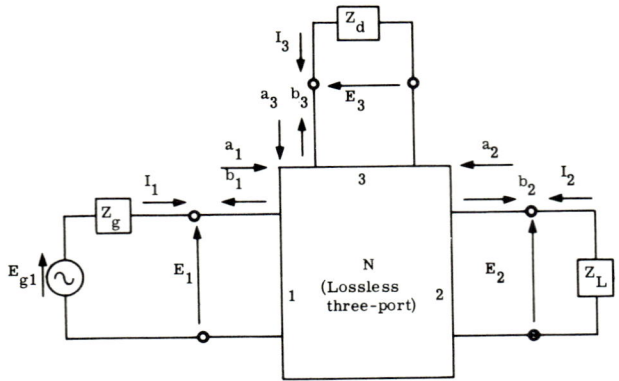

FIG. 15. Schematic of a power amplifier employing a lossless equalizer and a single negative-resistance device Z_d.

By (130) and (131)
$$b_2 = -\tilde{S}_{12}a_1/\tilde{\Delta} \tag{133}$$
where
$$\Delta = S_{11}S_{22} - S_{12}S_{21} \tag{134}$$
Hence
$$R_L|I_2|^2 = |b_2|^2 - |a_2|^2 = |S_{12}/\Delta|^2|a_1|^2 \tag{135}$$
and thus
$$G_{21}(\omega^2) = \frac{R_L|I_2|^2}{|E_{g1}|^2/4R_g} = \left|\frac{S_{12}(j\omega)}{\Delta(j\omega)}\right|^2 \tag{136}$$

However, because [S] is unitary, $|\Delta(j\omega)| = |S_{33}(j\omega)|$ and G_{21}, the transducer power gain from port 1 to port 2, reduces to

$$G_{21}(\omega^2) = |S_{12}(j\omega)/S_{33}(j\omega)|^2 \tag{137}$$

Now the unitary character of [S] implies that $|S_{12}(j\omega)| \leq 1$. Consequently,

$$G_{21}(\omega^2) \leq 1/|S_{33}(j\omega)|^2 \tag{138}$$

Since $S_{33}(j\omega)$ is the input reflection coefficient at port 3 with the other two ports matched,
$$S_{33} = \frac{Z_{33} - \bar{Z}_{03}}{Z_{33} + Z_{03}} = \frac{Z_{33} + Z_d}{Z_{33} - \bar{Z}_d} \tag{139}$$
and therefore
$$G_{21}(\omega^2) \leq \left|\frac{Z_{33}(j\omega) - \bar{Z}_d(j\omega)}{Z_{33}(j\omega) + Z_d(j\omega)}\right|^2 \tag{140}$$

where Z_{33}, the impedance faced by Z_d, is defined in the usual manner, as illustrated in Fig. 16. The above expression is quite important, for it contains the solution to the broadband problem for both the tunnel diode and the varactor with the loss and packaging inductance taken into account. The conclusion to be drawn from (140) is that the ultimate power gain is completely delimited by the equalizer impedance Z_{33} facing the negative-impedance element.

Actually to realize the maximum gain in (140) it is necessary to use a matching network N incorporating a three-port circulator since the magnitude of S_{12} must be made unity, irrespective of the magnitude of $|S_{33}|$. If N is restricted to be reciprocal, [S] must be symmetric and this solution is no longer available. In this case it can be shown that

$$|S_{12}(j\omega)| \leq \frac{1 + |S_{33}(j\omega)|}{2} \tag{141}$$

which yields

$$G_{21}(\omega^2) = \left|\frac{S_{12}(j\omega)}{S_{33}(j\omega)}\right|^2 \leq \frac{1}{4}\left(1 + \left|\frac{Z_{33}(j\omega) - \bar{Z}_d(j\omega)}{Z_{33}(j\omega) + Z_d(j\omega)}\right|\right)^2 \quad (142)$$

Consequently, for the same Z_{33}, optimum lossless, nonreciprocal equalization yields at most 6 db more gain over optimum lossless, reciprocal equalization.

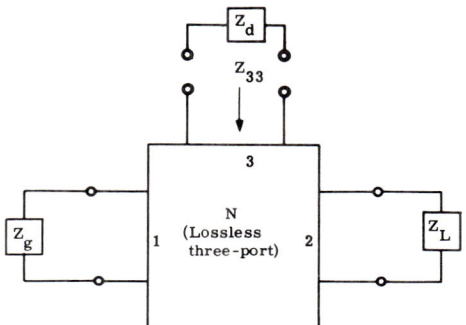

Fig. 16. Network illustrating the meaning of Z_{33}, the impedance facing the negative-resistance element Z_d.

3. TRANSFER SCATTERING MATRIX

We saw that, for problems where quadripoles are cascaded, a simple characterization of each of the quadripoles is given by the chain matrix; this is true because the chain matrix of the overall quadripole is equal to the product of the chain matrices of the component quadripoles. When the natural formulation of an analysis or synthesis problem is in terms of the scattering matrix and, in addition, the network is in the form of cascaded sections or quadripoles, it would be convenient to use a matrix that has the property of the chain matrix but still uses incident and reflected waves as its variables. Such a matrix[2] is the *transfer scattering matrix* [F].

With the chain matrix we used the output variables E_2 and $-I_2$, since the signs of the output quantities of one quadripole should be chosen so that they become the input quantities of the next quadripole. For the transfer scattering matrix we use a slightly different convention to achieve the same effect: since the reflected wave out of the first quadripole is the input wave for the second quadripole, it is clear that we merely interchange the words *reflected* and *incident without any change of sign*. Thus the matrix [F] relates the variables in the equation

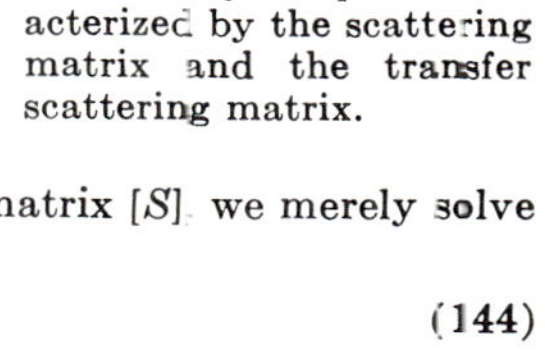

$$\begin{bmatrix} b_1 \\ a_1 \end{bmatrix} = \begin{bmatrix} F_{11} & F_{12} \\ F_{21} & F_{22} \end{bmatrix} \begin{bmatrix} a_2 \\ b_2 \end{bmatrix} \quad (143)$$

Fig. 17. Quadripole characterized by the scattering matrix and the transfer scattering matrix.

for the quadripole shown in Fig. 17.

To find the relation of [F] to the normalized scattering matrix [S] we merely solve the equations

$$\begin{aligned} b_1 &= S_{11}a_1 + S_{12}a_2 \\ b_2 &= S_{21}a_1 + S_{22}a_2 \end{aligned} \quad (144)$$

for the variables b_1 and a_1. Solving the second of the above equations for a_1, we obtain

$$a_1 = -(S_{22}/S_{21})a_2 + (1/S_{21})b_2 \quad (145)$$

4-24 AMPLIFIER FUNDAMENTALS

Substituting this in the first equation to eliminate a_1, we have

$$b_1 = -\frac{S_{11}S_{22} - S_{12}S_{21}}{S_{21}} a_2 + \frac{S_{11}}{S_{21}} b_2$$

$$= -\frac{\det [S]}{S_{21}} a_2 + \frac{S_{11}}{S_{21}} b_2 \tag{146}$$

so that

$$\begin{bmatrix} b_1 \\ a_1 \end{bmatrix} = \frac{1}{S_{21}} \begin{bmatrix} -\det [S] & S_{11} \\ -S_{22} & 1 \end{bmatrix} \begin{bmatrix} a_2 \\ b_2 \end{bmatrix} \tag{147}$$

and

$$[F] = \frac{1}{S_{21}} \begin{bmatrix} -\det [S] & S_{11} \\ -S_{22} & 1 \end{bmatrix} \tag{148}$$

A simple calculation then gives

$$\det [F] = S_{12}/S_{21} \tag{149}$$

so that $\det [F] = 1$ for a reciprocal quadripole.

Similarly, solving for $[S]$ in terms of the elements of $[F]$ gives

$$[S] = \frac{1}{F_{22}} \begin{bmatrix} F_{12} & \det [F] \\ 1 & -F_{21} \end{bmatrix} \tag{150}$$

which becomes for a reciprocal network, since $\det [F] = 1$,

$$[S] = \frac{1}{F_{22}} \begin{bmatrix} F_{12} & 1 \\ 1 & -F_{21} \end{bmatrix} \tag{151}$$

Now suppose that for the two quadripoles connected in cascade as shown in Fig. 18, we desire the relation between the input and output wave amplitudes of the complete

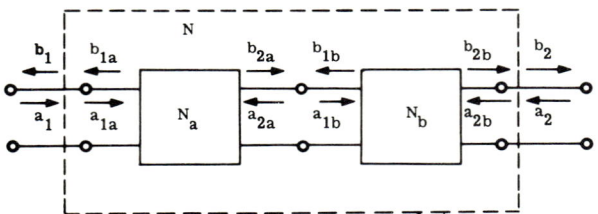

FIG. 18. Quadripoles N_a and N_b connected in cascade to form the quadripole N.

quadripole N, and we are given $[F_a]$ and $[F_b]$, the transfer scattering matrices for quadripoles N_a and N_b, respectively. We have

$$\begin{bmatrix} b_{1a} \\ a_{1a} \end{bmatrix} = [F_a] \begin{bmatrix} a_{2a} \\ b_{2a} \end{bmatrix} \tag{152}$$

which becomes by inspection of Fig. 18

$$\begin{bmatrix} b_1 \\ a_1 \end{bmatrix} = [F_a] \begin{bmatrix} b_{1b} \\ a_{1b} \end{bmatrix} \tag{153}$$

Now since

$$\begin{bmatrix} b_{1b} \\ a_{1b} \end{bmatrix} = [F_b] \begin{bmatrix} a_{2b} \\ b_{2b} \end{bmatrix} \tag{154}$$

we have by substitution of the above equation in Eq. (153)

$$\begin{bmatrix} b_1 \\ a_1 \end{bmatrix} = [F_a][F_b] \begin{bmatrix} a_{2b} \\ b_{2b} \end{bmatrix}$$

$$= [F] \begin{bmatrix} a_2 \\ b_2 \end{bmatrix} \tag{155}$$

so that

$$[F] = [F_a][F_b] \tag{156}$$

In words, the above analysis has shown that the transfer scattering matrix $[F]$ for a quadripole N that consists of N_a followed in cascade by N_b is equal to the product $[F_a][F_b]$ of the transfer scattering matrices of the component quadripoles, taken in the order in which they are cascaded. Clearly, this holds for an arbitrary finite number of quadripoles with respective matrices $[F_a]$, $[F_b]$, $[F_c]$, . . . , $[F_n]$ so that

$$[F] = [F_a][F_b][F_c] \cdots [F_n] \tag{157}$$

Now in the preceding example if we desired to determine the scattering matrix of the complete quadripole, we would use Eq. (150). For example, the input reflection coefficient is given simply by F_{12}/F_{22}, where F_{12} and F_{22} are elements of the overall transfer scattering matrix $[F]$.

One final point should be made about the analysis of cascaded quadripoles in terms of their transfer scattering matrices. An implicit assumption is that the proper normalizations are used; otherwise the product rule does not hold as formulated. It is required that the normalization number for port 2 of the first quadripole be the complex conjugate impedance of that of port 1 of the second quadripole; similarly, the normalization number of port 2 of the second quadripole is the complex conjugate impedance of that of port 1 of the third quadripole; and so on. In other words, a port common to two quadripoles has normalization numbers that are complex conjugate impedances. Then the product rule has the simple form given in Eq. (157).

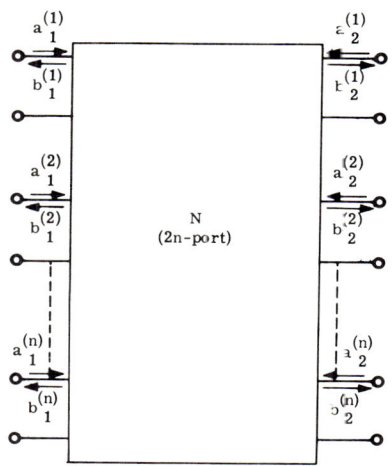

FIG. 19. Network with n input and n output ports.

4. SCATTERING MATRICES FOR MULTIPORTS[13]

The two-port may be considered a particular case of the n-port for which $n = 2$ or of the $2n$-port for which $n = 1$; it is convenient here to consider the $2n$-port, that is, a network with n input ports and n output ports, such as is shown in Fig. 19. We now show that the application of the scattering matrix and the transfer scattering matrix to the two-port may be generalized in a straightforward manner to the $2n$-port. By generalization of two matrices (which are the *Pauli spin matrix* for the z axis of a coordinate system and this matrix with its two rows interchanged) it is also shown how transfer scattering matrices may be tested to determine whether their corresponding $2n$-ports are reciprocal, passive, or lossless.

For the $2n$-port in Fig. 19 the scattering matrix is given by

$$b] = [S] a] \tag{158}$$

where $[S]$ is a $2n \times 2n$ matrix. If we define the column matrices for the input ports

$$b_1] = [b_1{}^{(1)} \; b_1{}^{(2)} \; \cdots \; b_1{}^{(n)}]'$$
$$a_1] = [a_1{}^{(1)} \; a_1{}^{(2)} \; \cdots \; a_1{}^{(n)}]' \tag{159}$$

and the corresponding matrices for the output ports

$$b_2] = [b_2{}^{(1)} \; b_2{}^{(2)} \; \cdots \; b_2{}^{(n)}]'$$
$$a_2] = [a_2{}^{(1)} \; a_2{}^{(2)} \; \cdots \; a_2{}^{(n)}]' \tag{160}$$

where the prime is used to indicate the transpose, then the scattering matrix may be written in partitioned form as

$$\begin{bmatrix} b_1] \\ b_2] \end{bmatrix} = \begin{bmatrix} [S_{11}] & [S_{12}] \\ [S_{21}] & [S_{22}] \end{bmatrix} \begin{bmatrix} a_1] \\ a_2] \end{bmatrix} \tag{161}$$

Similarly, the transfer scattering matrix may be written as

$$\begin{bmatrix} b_1] \\ a_1] \end{bmatrix} = \begin{bmatrix} [F_{11}] & [F_{12}] \\ [F_{21}] & [F_{22}] \end{bmatrix} \begin{bmatrix} a_2] \\ b_2] \end{bmatrix} \quad (162)$$

The scattering-matrix equation (161) may be rearranged into the form of the transfer scattering matrix so that the submatrices of $[F]$ are expressed in terms of those of $[S]$:

$$[F] = \begin{bmatrix} [S_{12}] - [S_{11}][S_{21}]^{-1}[S_{22}] & [S_{11}][S_{21}]^{-1} \\ -[S_{21}]^{-1}[S_{22}] & [S_{21}]^{-1} \end{bmatrix} \quad (163)$$

The above equation for the $2n$-port represents a generalization of (148) for the two-port.

We recall that for a reciprocal two-port $S_{12} = S_{21}$ and $\det [F] \equiv \Delta_F = 1$; these conditions may be generalized to the $2n$-port. A similar generalization is possible for the properties of passivity and losslessness.

The condition on $[S]$ for a reciprocal $2n$-port follows immediately from that for the two-port, namely, $[S]$ in partitioned form must be symmetric; thus

$$[S_{12}] = [S_{21}]' \quad (164)$$

To obtain the corresponding condition on $[F]$, we use one of the Pauli spin matrices defined by

$$\alpha_1 = \begin{bmatrix} 0 & -1 \\ 1 & 0 \end{bmatrix} \quad (165)$$

and evaluate the expression given by $[F]'\alpha_1[F]$, the congruence transformation of α_1, as

$$[F]'\alpha_1[F] = \begin{bmatrix} 0 & -\det [F] \\ \det [F] & 0 \end{bmatrix} \quad (166)$$

For a reciprocal two-port, since $\det [F] = 1$, we have

$$[F]'\alpha_1[F] = \begin{bmatrix} 0 & -1 \\ 1 & 0 \end{bmatrix} = \alpha_1 \quad (167)$$

which is a matrix expression for the reciprocity property for two-ports.

The generalization of α_1 for $2n$-ports is

$$A_1 = \begin{bmatrix} O_n & -U_n \\ U_n & O_n \end{bmatrix} \quad (168)$$

when O_n and U_n are the zero matrix and unit matrix of order n, respectively. Reciprocity for $2n$-ports now requires

$$[F]'A_1[F] = A_1 \quad (169)$$

The passivity of a two-port requires that the total power into the ports be nonnegative for any incident variables, and losslessness requires that the power be zero. These may be defined by use of the matrix

$$\alpha_2 = \begin{bmatrix} 1 & 0 \\ 0 & -1 \end{bmatrix} \quad (170)$$

which is obtained from α_1, defined in (165), by permuting its two rows.

For a lossless two-port we have

$$(a_1\bar{a}_1 - b_1\bar{b}_1) + (a_2\bar{a}_2 - b_2\bar{b}_2) = 0 \quad (171)$$

which may be rewritten in terms of α_2 as

$$-\mu_1]^* \alpha_2 \mu_1] + \mu_2]^* \alpha_2 \mu_2] = 0 \quad (172)$$

where we have used the definitions

$$\mu_1] = \begin{bmatrix} b_1 \\ a_1 \end{bmatrix}$$
$$\mu_2] = \begin{bmatrix} a_2 \\ b_2 \end{bmatrix} \quad (173)$$

Substituting

$$\mu_1] = [F]\mu_2] \tag{174}$$

in (171) we have

$$\mu_2]^*\{\alpha_2 - [F]^*\alpha_2[F]\}\mu_2] = 0 \tag{175}$$

which yields the matrix expression for the constraint on $[F]$ of a lossless two-port

$$[F]^*\alpha_2[F] = \alpha_2 \tag{176}$$

A similar analysis for a passive dissipative two-port is obtained by specifying that the power be nonnegative rather than zero; this yields

$$\{\alpha_2 - [F]^*\alpha_2[F]\} \geq 0 \tag{177}$$

where this is interpreted to mean that the matrix expression on the left is positive semidefinite.

The generalization to a $2n$-port follows immediately by defining the generalization of α_2 as

$$A_2 = \begin{bmatrix} U_n & O_n \\ O_n & -U_n \end{bmatrix} \tag{178}$$

We then have for the lossless $2n$-port the condition

$$[F]^*A_2[F] = A_2 \tag{179}$$

and for the general passive $2n$-port

$$\{A_2 - [F]^*A_2[F]\} \geq 0 \tag{180}$$

where again the inequality means that the matrix expression is positive semidefinite.

For the scattering matrix of a $2n$-port we obtain from our previously derived results, namely, from Eq. (83), that

$$[S]^*[S] = [U_{2n}] \tag{181}$$

that is, $[S]$ is a unitary matrix for a lossless $2n$-port, and for a passive $2n$-port

$$\{[U_{2n}] - [S]^*[S]\} \geq 0 \tag{182}$$

that is, the expression on the left is positive semidefinite.

5. INTERRELATIONSHIPS AMONG NETWORK MATRICES

The different types of characterizations considered in this chapter and the preceding one illustrate the fact that the properties of an n-port may be described by any set of n independent equations between the $2n$ variables involved, i.e., the n voltages and the n currents at the network ports. Any such system is represented by a matrix of its coefficients and is best adapted for some particular terminations of the network. In practice, many types of terminations of an n-port are possible, and it may be necessary to deal with a considerable number of different representations. Since the task of representing or evaluating the numerous coefficients may prove extremely laborious, it is desirable to find some direct method of doing this and to have tables of relationships of the most commonly used types of network matrices.

In this section we present such tables for the two-port, and we mention in passing that a method of symbolic notation for matrix analysis has been formulated.[14] This method allows one to convert from one matrix representation to another by essentially three steps; it is covered in detail in the reference.

The interrelationships for second-order matrices are given in Table 1, where it is assumed that the scattering matrix $[S]$ and the transfer scattering matrix $[F]$ are normalized to real numbers. In this table the determinant of the matrix under consideration is represented by Δ with an appropriate subscript. Thus the determinants of the open-circuit impedance matrix, the short-circuit admittance matrix, and the chain matrix are denoted by $\Delta_z \equiv z_{11}z_{22} - z_{12}z_{21}$, $\Delta_y \equiv y_{11}y_{22} - y_{12}y_{21}$, and $\Delta_{ch} \equiv AD - BC$, respectively. Since expressions for each of the determinants in terms of elements of the other matrices are also required, these are given in Table 2. By use of these

Table 1. Relations among Network Matrices

From / To	[z]	[y]	[g]
[z]		$z_{11} = \dfrac{y_{22}}{\Delta_y}$ $\quad z_{12} = \dfrac{-y_{12}}{\Delta_y}$ $z_{21} = \dfrac{-y_{21}}{\Delta_y}$ $\quad z_{22} = \dfrac{y_{11}}{\Delta_y}$	$z_{11} = \dfrac{1}{g_{11}}$ $\quad z_{12} = \dfrac{-g_{12}}{g_{11}}$ $z_{21} = \dfrac{g_{21}}{g_{11}}$ $\quad z_{22} = \dfrac{\Delta_g}{g_{11}}$
[y]	$y_{11} = \dfrac{z_{22}}{\Delta_z}$ $\quad y_{12} = \dfrac{-z_{12}}{\Delta_z}$ $y_{21} = \dfrac{-z_{21}}{\Delta_z}$ $\quad y_{22} = \dfrac{z_{11}}{\Delta_z}$		$y_{11} = \dfrac{\Delta_g}{g_{22}}$ $\quad y_{12} = \dfrac{g_{12}}{g_{22}}$ $y_{21} = \dfrac{-g_{21}}{g_{22}}$ $\quad y_{22} = \dfrac{1}{g_{22}}$
[g]	$g_{11} = \dfrac{1}{z_{11}}$ $\quad g_{12} = \dfrac{-z_{12}}{z_{11}}$ $g_{21} = \dfrac{z_{21}}{z_{11}}$ $\quad g_{22} = \dfrac{\Delta_z}{z_{11}}$	$g_{11} = \dfrac{\Delta_y}{y_{22}}$ $\quad g_{12} = \dfrac{y_{12}}{y_{22}}$ $g_{21} = \dfrac{-y_{21}}{y_{22}}$ $\quad g_{22} = \dfrac{1}{y_{22}}$	
[h]	$h_{11} = \dfrac{\Delta_z}{z_{22}}$ $\quad h_{12} = \dfrac{z_{12}}{z_{22}}$ $h_{21} = \dfrac{-z_{21}}{z_{22}}$ $\quad h_{22} = \dfrac{1}{z_{22}}$	$h_{11} = \dfrac{1}{y_{11}}$ $\quad h_{12} = \dfrac{-y_{12}}{y_{11}}$ $h_{21} = \dfrac{y_{21}}{y_{11}}$ $\quad h_{22} = \dfrac{\Delta_y}{y_{11}}$	$h_{11} = \dfrac{g_{22}}{\Delta_g}$ $\quad h_{12} = \dfrac{-g_{12}}{\Delta_g}$ $h_{21} = \dfrac{-g_{21}}{\Delta_g}$ $\quad h_{22} = \dfrac{g_{11}}{\Delta_g}$
Chain matrix	$A = \dfrac{z_{11}}{z_{21}}$ $\quad B = \dfrac{\Delta_z}{z_{21}}$ $C = \dfrac{1}{z_{21}}$ $\quad D = \dfrac{z_{22}}{z_{21}}$	$A = \dfrac{-y_{22}}{y_{21}}$ $\quad B = \dfrac{-1}{y_{21}}$ $C = \dfrac{-\Delta_y}{y_{21}}$ $\quad D = \dfrac{-y_{11}}{y_{21}}$	$A = \dfrac{1}{g_{21}}$ $\quad B = \dfrac{g_{22}}{g_{21}}$ $C = \dfrac{g_{11}}{g_{21}}$ $\quad D = \dfrac{\Delta_g}{g_{21}}$
[S]	$S_{11} = \dfrac{\Delta_z + z_{11} - (z_{22}+1)}{\Delta_z + z_{11} + z_{22} + 1}$ $\quad S_{12} = \dfrac{2z_{12}}{\Delta_z + z_{11} + z_{22} + 1}$ $S_{21} = \dfrac{2z_{21}}{\Delta_z + z_{11} + z_{22} + 1}$ $\quad S_{22} = \dfrac{\Delta_z + z_{22} - (z_{11}+1)}{\Delta_z + z_{11} + z_{22} + 1}$	$S_{11} = \dfrac{y_{22}+1-(\Delta_y+y_{11})}{y_{22}+1+\Delta_y+y_{11}}$ $\quad S_{12} = \dfrac{-2y_{12}}{y_{22}+1+\Delta_y+y_{11}}$ $S_{21} = \dfrac{-2y_{21}}{y_{22}+1+\Delta_y+y_{11}}$ $\quad S_{22} = \dfrac{y_{11}+1-(\Delta_y+y_{22})}{y_{22}+1+\Delta_y+y_{11}}$	$S_{11} = \dfrac{g_{22}+1-\Delta_g-g_{11}}{g_{22}+1+\Delta_g+g_{11}}$ $\quad S_{12} = \dfrac{-2g_{12}}{g_{22}+1+\Delta_g+g_{11}}$ $S_{21} = \dfrac{2g_{21}}{g_{22}+1+\Delta_g+g_{11}}$ $\quad S_{22} = \dfrac{g_{22}-1+\Delta_g-g_{11}}{g_{22}+1+\Delta_g+g_{11}}$
[F]	$F_{11} = \dfrac{z_{11}-\Delta_z-1+z_{22}}{2z_{21}}$ $\quad F_{12} = \dfrac{z_{11}+z_{22}-1-z_{22}}{2z_{21}}$ $F_{21} = \dfrac{z_{11}-\Delta_z+1-z_{22}}{2z_{21}}$ $\quad F_{22} = \dfrac{z_{11}+\Delta_z+1+z_{22}}{2z_{21}}$	$F_{11} = \dfrac{-y_{22}+1+\Delta_y-y_{11}}{2y_{21}}$ $\quad F_{12} = \dfrac{-y_{22}-1+\Delta_y-y_{11}}{2y_{21}}$ $F_{21} = \dfrac{-y_{22}+1-\Delta_y+y_{11}}{2y_{21}}$ $\quad F_{22} = \dfrac{-y_{22}-1-\Delta_y-y_{11}}{2y_{21}}$	$F_{11} = \dfrac{1-g_{22}-g_{11}+\Delta_g}{2g_{21}}$ $\quad F_{12} = \dfrac{1+g_{22}-g_{11}-\Delta_g}{2g_{21}}$ $F_{21} = \dfrac{1-g_{22}+g_{11}-\Delta_g}{2g_{21}}$ $\quad F_{22} = \dfrac{1+g_{22}+g_{11}+\Delta_g}{2g_{21}}$

Table 1. Relations among Network Matrices (Continued)

From → To ↓	[h]	Chain matrix
[z]	$z_{11} = \dfrac{\Delta_h}{h_{22}}$ $z_{12} = \dfrac{h_{12}}{h_{22}}$ $z_{21} = \dfrac{-h_{21}}{h_{22}}$ $z_{22} = \dfrac{1}{h_{22}}$	$z_{11} = \dfrac{A}{C}$ $z_{12} = \dfrac{\Delta_{ch}}{C}$ $z_{21} = \dfrac{1}{C}$ $z_{22} = \dfrac{D}{C}$
[y]	$y_{11} = \dfrac{1}{h_{11}}$ $y_{12} = \dfrac{-h_{12}}{h_{11}}$ $y_{21} = \dfrac{h_{21}}{h_{11}}$ $y_{22} = \dfrac{\Delta_h}{h_{11}}$	$y_{11} = \dfrac{D}{B}$ $y_{12} = \dfrac{-\Delta_{ch}}{B}$ $y_{21} = \dfrac{-1}{B}$ $y_{22} = \dfrac{A}{B}$
[g]	$g_{11} = \dfrac{h_{22}}{\Delta_h}$ $g_{12} = \dfrac{-h_{12}}{\Delta_h}$ $g_{21} = \dfrac{-h_{21}}{\Delta_h}$ $g_{22} = \dfrac{h_{11}}{\Delta_h}$	$g_{11} = \dfrac{C}{A}$ $g_{12} = \dfrac{-\Delta_{ch}}{A}$ $g_{21} = \dfrac{1}{A}$ $g_{22} = \dfrac{B}{A}$
[h]		$h_{11} = \dfrac{B}{D}$ $h_{12} = \dfrac{\Delta_{ch}}{D}$ $h_{21} = \dfrac{-1}{D}$ $h_{22} = \dfrac{C}{D}$
Chain matrix	$A = \dfrac{-\Delta_h}{h_{21}}$ $B = \dfrac{-h_{11}}{h_{21}}$ $C = \dfrac{-h_{22}}{h_{21}}$ $D = \dfrac{-1}{h_{21}}$	
[S]	$S_{11} = \dfrac{h_{11} + \Delta_h - 1 - h_{22}}{h_{11} + \Delta_h + 1 + h_{22}}$ $S_{12} = \dfrac{2h_{12}}{h_{11} + \Delta_h + 1 + h_{22}}$ $S_{21} = \dfrac{-2h_{21}}{h_{11} + \Delta_h + 1 + h_{22}}$ $S_{22} = \dfrac{h_{11} - \Delta_h + 1 - h_{22}}{h_{11} + \Delta_h + 1 + h_{22}}$	$S_{11} = \dfrac{A + B - C - D}{A + B + C + D}$ $S_{12} = \dfrac{2\Delta_{ch}}{A + B + C + D}$ $S_{21} = \dfrac{2}{A + B + C + D}$ $S_{22} = \dfrac{-A + B - C + D}{A + B + C + D}$
[F]	$F_{11} = \dfrac{-\Delta_h + h_{11} + h_{22} - 1}{2h_{21}}$ $F_{12} = \dfrac{-\Delta_h - h_{11} + h_{22} + 1}{2h_{21}}$ $F_{21} = \dfrac{-\Delta_h + h_{11} - h_{22} + 1}{2h_{21}}$ $F_{22} = \dfrac{-\Delta_h - h_{11} - h_{22} - 1}{2h_{21}}$	$F_{11} = \dfrac{A + B - C + D}{2}$ $F_{12} = \dfrac{A + B - C - D}{2}$ $F_{21} = \dfrac{A - B + C - D}{2}$ $F_{22} = \dfrac{A + B + C + D}{2}$

4–29

Table 1. Relations among Network Matrices (Continued)

From \ To	[S]	[F]
[z]	$z_{11} = \dfrac{1 - S_{22} + S_{11} - \Delta s}{1 - S_{22} - S_{11} + \Delta s}$ $z_{21} = \dfrac{2S_{21}}{1 - S_{12} - S_{11} + \Delta s}$	$z_{12} = \dfrac{2S_{12}}{1 - S_{22} - S_{11} + \Delta s}$ $z_{22} = \dfrac{1 + S_{22} - S_{11} - \Delta s}{1 - S_{22} - S_{11} + \Delta s}$	$z_{11} = \dfrac{F_{22} + F_{11} + F_{12} + F_{21}}{F_{22} - F_{11} - F_{12} + F_{21}}$ $z_{21} = \dfrac{2}{F_{22} - F_{11} - F_{12} + F_{21}}$	$z_{12} = \dfrac{2\Delta F}{F_{22} - F_{11} - F_{12} + F_{21}}$ $z_{22} = \dfrac{F_{22} + F_{11} - F_{12} - F_{21}}{F_{22} - F_{11} - F_{12} + F_{21}}$
[y]	$y_{11} = \dfrac{1 + S_{22} - S_{11} - \Delta s}{1 + S_{22} + S_{11} + \Delta s}$ $y_{21} = \dfrac{-2S_{21}}{1 + S_{22} + S_{11} + \Delta s}$	$y_{12} = \dfrac{-2S_{12}}{1 + S_{22} + S_{11} + \Delta s}$ $y_{22} = \dfrac{1 - S_{22} + S_{11} - \Delta s}{1 + S_{22} + S_{11} + \Delta s}$	$y_{11} = \dfrac{F_{22} + F_{11} - F_{12} - F_{21}}{F_{22} - F_{11} + F_{12} - F_{21}}$ $y_{21} = \dfrac{-2}{F_{22} - F_{11} + F_{12} - F_{21}}$	$y_{12} = \dfrac{-2\Delta F}{F_{22} - F_{11} + F_{12} - F_{21}}$ $y_{22} = \dfrac{F_{22} + F_{11} + F_{12} + F_{21}}{F_{22} - F_{11} + F_{12} - F_{21}}$
[g]	$g_{11} = \dfrac{\Delta s - S_{11} - S_{22} + 1}{-\Delta s + S_{11} - S_{22} + 1}$ $g_{21} = \dfrac{2S_{21}}{-\Delta s + S_{11} - S_2 + 1}$	$g_{12} = \dfrac{-2S_{12}}{-\Delta s + S_{11} - S_{22} + 1}$ $g_{22} = \dfrac{\Delta s + S_{11} + S_{22} + 1}{-\Delta s + S_{11} - S_{22} + 1}$	$g_{11} = \dfrac{F_{22} - F_{11} - F_{12} + F_{21}}{F_{22} + F_{11} + F_{12} + F_{21}}$ $g_{21} = \dfrac{2}{F_{22} + F_{11} + F_{12} + F_{21}}$	$g_{12} = \dfrac{-2\Delta F}{F_{22} + F_{11} + F_{12} + F_{21}}$ $g_{22} = \dfrac{F_{22} - F_{11} + F_{12} - F_{21}}{F_{22} + F_{11} + F_{12} + F_{21}}$
[h]	$h_{11} = \dfrac{\Delta s + S_{11} + S_{22} + 1}{-\Delta s - S_{11} + S_{22} + 1}$ $h_{21} = \dfrac{-2S_{21}}{-\Delta s - S_{11} + S_{22} + 1}$	$h_{12} = \dfrac{2S_{12}}{-\Delta s - S_{11} + S_{22} + 1}$ $h_{22} = \dfrac{\Delta s - S_{11} - S_{22} + 1}{-\Delta s - S_{11} + S_{22} + 1}$	$h_{11} = \dfrac{F_{22} + F_{11} + F_{12} + F_{21}}{F_{22} - F_{11} + F_{12} - F_{21}}$ $h_{21} = \dfrac{-2}{F_{22} - F_{11} + F_{12} - F_{21}}$	$h_{12} = \dfrac{2\Delta F}{F_{22} - F_{11} + F_{12} - F_{21}}$ $h_{22} = \dfrac{F_{22} - F_{11} - F_{12} + F_{21}}{F_{22} - F_{11} + F_{12} - F_{21}}$
Chain matrix	$A = \dfrac{-\Delta s + S_{11} - S_{22} + 1}{2S_{21}}$ $C = \dfrac{\Delta s - S_{11} - S_{22} + 1}{2S_{21}}$	$B = \dfrac{\Delta s + S_{11} + S_{22} + 1}{2S_{21}}$ $D = \dfrac{-\Delta s - S_{11} + S_{22} + 1}{2S_{21}}$	$A = \dfrac{F_{22} + F_{11} + F_{12} + F_{21}}{2}$ $C = \dfrac{F_{22} - F_{11} + F_{12} + F_{21}}{2}$	$B = \dfrac{F_{22} - F_{11} + F_{12} - F_{21}}{2}$ $D = \dfrac{F_{22} + F_{11} - F_{12} - F_{21}}{2}$
[S]		$S_{11} = \dfrac{F_{12}}{F_{22}}$ $S_{21} = \dfrac{1}{F_{22}}$	$S_{12} = \dfrac{\Delta F}{F_{22}}$ $S_{22} = \dfrac{-F_{21}}{F_{22}}$
[F]	$F_{11} = \dfrac{-\Delta s}{S_{21}}$ $F_{21} = \dfrac{-S_{22}}{S_{21}}$	$F_{12} = \dfrac{S_{11}}{S_{21}}$ $F_{22} = \dfrac{1}{S_{21}}$	

4–30

Table 2. Relationships among Determinants of Network Matrices

From \ To	Δ_z	Δ_y	Δ_g	Δ_h	Δ_{ch}	Δ_s	Δ_r
Δ_z		$\dfrac{1}{\Delta_y}$	$\dfrac{g_{22}}{g_{11}}$	$\dfrac{h_{11}}{h_{22}}$	$\dfrac{B}{C}$	$\dfrac{\Delta_s + S_{11} + S_{22} + 1}{\Delta_s - S_{11} - S_{22} + 1}$	$\dfrac{F_{22} - F_{11} + F_{12} - F_{21}}{F_{22} - F_{11} - F_{12} + F_{21}}$
Δ_y	$\dfrac{1}{\Delta_z}$		$\dfrac{g_{11}}{g_{22}}$	$\dfrac{h_{22}}{h_{11}}$	$\dfrac{C}{B}$	$\dfrac{\Delta_s - S_{11} - S_{22} + 1}{\Delta_s + S_{11} + S_{22} + 1}$	$\dfrac{F_{22} - F_{11} - F_{12} + F_{21}}{F_{22} - F_{11} + F_{12} - F_{21}}$
Δ_g	$\dfrac{z_{22}}{z_{11}}$	$\dfrac{y_{11}}{y_{22}}$		$\dfrac{1}{\Delta_h}$	$\dfrac{D}{A}$	$\dfrac{-\Delta_s - S_{11} + S_{22} + 1}{-\Delta_s + S_{11} - S_{22} + 1}$	$\dfrac{F_{22} + F_{11} - F_{12} - F_{21}}{F_{22} + F_{11} + F_{12} + F_{21}}$
Δ_h	$\dfrac{z_{11}}{z_{22}}$	$\dfrac{y_{22}}{y_{11}}$	$\dfrac{1}{\Delta_g}$		$\dfrac{A}{D}$	$\dfrac{-\Delta_s + S_{11} - S_{22} + 1}{-\Delta_s - S_{11} + S_{22} + 1}$	$\dfrac{F_{22} + F_{11} + F_{12} + F_{21}}{F_{22} + F_{11} - F_{12} - F_{21}}$
Δ_{ch}	$\dfrac{z_{12}}{z_{21}}$	$\dfrac{y_{12}}{y_{21}}$	$\dfrac{-g_{12}}{g_{21}}$	$\dfrac{-h_{12}}{h_{21}}$		$\dfrac{S_{12}}{S_{21}}$	Δ_r
Δ_s	$\dfrac{\Delta_z + 1 - z_{11} - z_{22}}{\Delta_z + 1 + z_{11} + z_{22}}$	$\dfrac{-y_{22} + 1 + \Delta_y - y_{11}}{y_{22} + 1 + \Delta_y + y_{11}}$	$\dfrac{g_{22} + g_{11} - 1 - \Delta_g}{g_{22} + g_{11} + 1 + \Delta_g}$	$\dfrac{-\Delta_h + h_{11} + h_{22} - 1}{\Delta_h + h_{11} + h_{22} + 1}$	$\dfrac{-A + B + C - D}{A + B + C + D}$		$\dfrac{-F_{11}}{F_{22}}$
Δ_r	$\dfrac{z_{12}}{z_{21}}$	$\dfrac{y_{12}}{y_{21}}$	$\dfrac{-g_{12}}{g_{21}}$	$\dfrac{-h_{12}}{h_{21}}$	Δ_{ch}	$\dfrac{S_{12}}{S_{21}}$	

4–31

4–32 AMPLIFIER FUNDAMENTALS

tables analysis is facilitated. For example, suppose we wish to determine the open-circuit impedance matrix of a quadripole made up of three component quadripoles connected in cascade. Suppose further that the open-circuit impedance matrix is known for the first two quadripoles, and the short-circuit admittance matrix for the third. By use of the tables we can calculate the chain matrix for each of the quadripoles; we then multiply these matrices in the correct order to obtain the overall chain matrix. Finally, we again make use of the relationships between the chain-matrix elements and the open-circuit impedances given in the tables to compute the desired impedance matrix.

Since the scattering matrix and the transfer scattering matrix are normalized, we must consider the impedances and admittances that are transformed to and from elements of $[S]$ and $[F]$ as normalized functions. The relations between the unnormalized and normalized functions are given by

$$z_{ij} = \frac{z_{ij}'}{\sqrt{R_{0i}R_{0j}}}$$
$$y_{ij} = y_{ij}' \sqrt{R_{0i}R_{0j}}$$
(183)

where the prime designates the unnormalized function and as indicated previously the normalization numbers are real.

The tables apply for reciprocal and nonreciprocal quadripoles. For a reciprocal system the following identities hold:

$$\begin{aligned} z_{12} &= z_{21} \\ y_{12} &= y_{21} \\ g_{12} &= -g_{21} \\ h_{12} &= -h_{21} \\ \Delta_{ch} &\equiv AD - BC = 1 \\ S_{12} &= S_{21} \\ \Delta_F &\equiv F_{11}F_{22} - F_{12}F_{21} = 1 \end{aligned}$$
(184)

Thus the formulas may be simplified for reciprocal systems; for example, using $\Delta_{ch} = 1$, we have $S_{12} = 2/(A + B + C + D)$.

6. MULTITERMINAL NETWORKS

Consider the n-terminal network shown in Fig. 20, where currents and voltages have been designated at each terminal. We assume in our discussion that the network is not degenerate; that is, two terminals are not connected to the same internal node, or what is equivalent, two terminals are not connected by a zero impedance; in addition, there are no isolated terminals and the network is connected; i.e., every pair of nodes is joined by a path formed by network branches. To analyze such a network we can, as is done in the conventional analysis, choose one terminal, say, terminal n, as a reference, and then write the node-to-datum admittance equations. This method of analysis recognizes two facts:

1. Because of the conservation of charge, the currents entering this isolated network must add up to zero; that is,

$$\sum_{k=1}^{n} I_k = 0$$
(185)

The network itself is thus like a giant node that obeys Kirchhoff's current law; hence the current I_n at the reference terminal need not be specified, since it is determined by all the other currents, being equal to the negative of their sum.

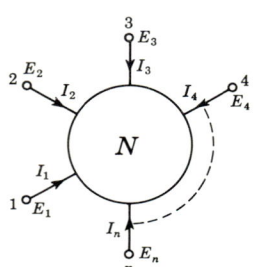

FIG. 20. Representation of a network with n accessible terminals.

2. The terminal currents are dependent on potential differences rather than on the absolute potential of the terminals. Thus the currents are invariant to the addition of an arbitrary voltage E_0 to each of the E_k. In choosing a datum terminal we have used $E_0 = -E_n$ and then in effect defined a new voltage $E_p' = E_p - E_n$; the same result is obtained by setting $E_n = 0$ and then ignoring the nth equation. Thus there has been no loss in generality in using the node-to-datum equations. As can be easily shown, for a connected, nondegenerate network the matrix of this set of equations is nonsingular.

However, let us now assume that the voltage at each terminal is measured with respect to an arbitrary but unspecified voltage, which may be external to the network; in this way all n currents and n voltages are to be used in the analysis. A number of representations are possible, but it is clear from our previous analysis—e.g., from the node-to-datum set of equations—that for a passive and reciprocal network only $n(n-1)/2$ of the parameters are independent, and for an active network there are only $(n-1)^2$ independent parameters. To make the discussion concrete and introduce the indefinite admittance matrix, we consider the analysis of a three-pole.

The three-pole to be characterized is shown in Fig. 21, where we have attached voltage sources connected to a common node external to the network. Each of the currents is due to a superposition of the effects of the voltage sources considered separately; hence we may describe the network in terms of admittances by

$$y_{11}E_1 + y_{12}E_2 + y_{13}E_3 = I_1$$
$$y_{21}E_1 + y_{22}E_2 + y_{23}E_3 = I_2 \qquad (186)$$
$$y_{31}E_1 + y_{32}E_2 + y_{33}E_3 = I_3$$

or briefly by

$$[Y_i] E] = I] \qquad (187)$$

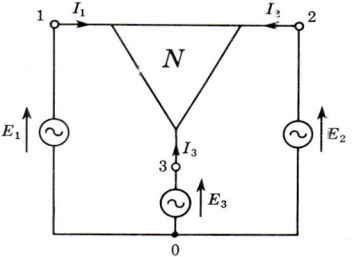

FIG. 21. Representation of a three-pole with specified terminal currents and voltages.

where $[Y_i]$ is the previously defined indefinite admittance matrix. It is evident that each element of the matrix is a short-circuit admittance, but we observe that nine admittances appear in the matrix, though we know that only $n(n-1)/2 = 3$ are independent for the reciprocal case and $(n-1)^2 = 4$ for the nonreciprocal case. To derive a set of independent admittances is straightforward.

First, assuming that the network obeys the reciprocity theorem, we have

$$y_{12} = y_{21} \qquad y_{23} = y_{32} \qquad y_{13} = y_{31} \qquad (188)$$

Now Kirchhoff's current law applied at the terminal 0 shows

$$I_1 + I_2 + I_3 = 0 \qquad (189)$$

If we add the respective left and right sides of the three equations in (186) and substitute the result for the sum of the currents in Eq. (189), we get

$$(y_{11} + y_{21} + y_{31})E_1 + (y_{12} + y_{22} + y_{32})E_2 + (y_{13} + y_{23} + y_{33})E_3 = 0 \quad (190)$$

Since the applied voltage sources are arbitrary, it follows from the above that the coefficient of each voltage must be separately zero; that is,

$$y_{11} + y_{21} + y_{31} = 0$$
$$y_{12} + y_{22} + y_{32} = 0 \qquad (191)$$
$$y_{13} + y_{23} + y_{33} = 0$$

We conclude from this analysis that only three of the admittances can be specified independently; we may choose, for example, the three driving-point admittances y_{11}, y_{22}, and y_{33} to characterize the three-pole uniquely; or equivalently, we may choose the three transfer admittances y_{12}, y_{13}, and y_{23}; this implies that these transfer and driving-point admittances are related, and later we shall give an equation expressing this general relationship. The choice of the transfer admittances leads to a representation of the three-pole by three admittances connected in a delta configuration, as shown in Fig. 22. Of course, none of these two-terminal admittances is necessarily physical, i.e., realizable as a driving-point admittance; yet this representation can be quite useful. Later we shall generalize such a representation by showing that any reciprocal n-terminal network may be represented in a similar way.

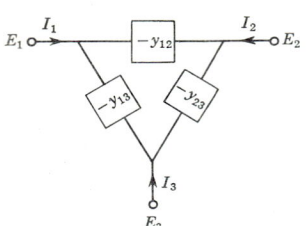

FIG. 22. Representation of a reciprocal three-pole by a Δ connection of three two-terminal admittances.

If an analogous analysis of the three-pole is carried out on an impedance basis, an equivalent representation of the general reciprocal three-pole is obtained. The two representations are given in Fig. 23.

If the three-terminal network is not reciprocal, Eq. (191) still follows from Eq. (189) as before, and thus the sum of the elements in each column is zero. We now establish that the sum of the elements in each row is also zero. Let the port voltage E_{ab} represent the voltage rise from terminal b to terminal a; we then have

$$E_{13} = E_1 - E_3$$
$$E_{21} = E_2 - E_1 \qquad (192)$$
$$E_{32} = E_3 - E_2$$

and thus
$$E_{13} + E_{21} + E_{32} \equiv 0 \qquad (193)$$

which is evident, since the three voltages form a closed loop. Now if E_0 is added to each of the terminal voltages, each port voltage, since it is the difference of two

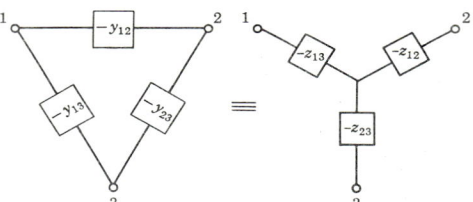

FIG. 23. Equivalent representations of a reciprocal three-pole by a connection of three two-terminal networks.

terminal voltages, remains unchanged. But the terminal currents are linear functions of the port voltages, and therefore also do not change. Let

$$E_a] = E] + E_0 1] \qquad (194)$$

where 1] is a third-order column matrix of unit elements. Then Eq. (187) becomes

$$I] = [Y_i] E_a] = [Y_i] E] + E_0[Y_i] 1] \qquad (195)$$

Therefore
$$E_0[Y_i] 1] = 0 \qquad (196)$$

for all E_0 so that
$$[Y_i] 1] = 0 \qquad (197)$$

or
$$\sum_{k=1}^{3} y_{ik} = 0 \qquad i = 1, 2, 3 \qquad (198)$$

In other words, the sum of the elements in each row is zero.

It is a simple matter to obtain the node-to-datum set of equations from those involving the indefinite admittance matrix in (186); it is important to observe that we are free to choose the datum node as any one of the three terminals. For example, if terminal 3 is chosen as the ground terminal, then $E_3 = 0$. Setting $E_3 = 0$ in Eqs. (186), we may then disregard the last equation, since it now gives redundant information. We thus get the node-to-datum equations

$$y_{11}E_1 + y_{12}E_2 = I_1$$
$$y_{21}E_1 + y_{22}E_2 = I_2 \qquad (199)$$

If Eqs. (186) are represented by the matrix equation

$$[Y_i] E] = I] \qquad (200)$$

where $[Y_i]$ is again the indefinite admittance matrix, then the operation leading to (199) consists of deleting the third row and third column of $[Y_i]$.

Similarly, if terminal 2 is chosen as the reference, row 2 and column 2 of $[Y_i]$ are deleted. It is thus a simple process to go from the representation as an n-terminal network in terms of the indefinite admittance matrix to the representation as an $(n - 1)$-port network in terms of the node-to-datum matrix, where any node may be chosen as reference. As we shall see below, it is also a simple process to proceed in the reverse direction from the node-to-datum matrix to the indefinite admittance matrix. These relationships are exceedingly useful; for example, it should be clear that for the active case they permit a simple analysis of all the possible transistor configurations.

The above analysis in terms of the indefinite admittance matrix for the three-pole may be directly generalized to the n-pole. For the n-pole we therefore have

$$[Y_i] E] = I] \qquad (201)$$

where $[Y_i]$ is an nth-order matrix of short-circuit admittances, $E]$ is the vector of terminal voltages with respect to some arbitrary reference, and $I]$ is the vector of terminal currents flowing into the network. From Eqs. (189) and (195)—that is, from the facts that the sum of the currents is equal to zero and the terminal currents are invariant with respect to the addition of an arbitrary voltage E_0 to each of the terminal voltages—it follows that

$$\sum_{i=1}^{n} y_{ik} = 0 \qquad k = 1, 2, \ldots, n \qquad (202)$$

and

$$\sum_{k=1}^{n} y_{ik} = 0 \qquad i = 1, 2, \ldots, n \qquad (203)$$

In words, the above two equations state that the sum of the elements in each column equals zero, as does the sum in each row of an indefinite admittance matrix. As has been mentioned, this property leads to the designation as a zero-sum matrix.

The zero-sum property provides the basis for converting a node-to-datum matrix to an indefinite admittance matrix: we merely add a row and a column to the node-to-datum matrix; since the sum of the elements in each row and column must be zero, each element of the added row or column is the negative of the sum of the elements in the corresponding row or column. For example, if $[Y]$ is the node-to-datum matrix of a three-port shown in Fig. 24, for which terminal 4 is the reference, then the indefinite admittance matrix of the four-terminal network is

$$[Y_i] = \begin{bmatrix} y_{11} & y_{12} & y_{13} & -y_{11} - y_{12} - y_{13} \\ y_{21} & y_{22} & y_{23} & -y_{21} - y_{22} - y_{23} \\ y_{31} & y_{32} & y_{33} & -y_{31} - y_{32} - y_{33} \\ -y_{11} - y_{21} - y_{31} & -y_{12} - y_{22} - y_{32} & -y_{13} - y_{23} - y_{33} & \sum_{i=1}^{3}\sum_{k=1}^{3} y_{ik} \end{bmatrix}$$

$$(204)$$

Hence we see that the indefinite admittance matrix, like the node-to-datum matrix, may be easily written by inspection.

A zero-sum matrix has a number of interesting properties. It is clear, first of all, that its determinant is equal to zero, since the addition of the first $(n-1)$ rows to the last row changes the latter to a row of zeros by virtue of (202). In addition, all the first-order (or more briefly, the first) cofactors of the indefinite admittance matrix are equal; that is, if we represent a first cofactor by Y_{jk}, we have for any j and k

$$Y_{jk} = Y_{nn} \qquad (205)$$

This property does not hold for the second cofactors. Equation (205) follows from the fact that any first cofactor is obtainable from any other by elementary transformations involving addition of rows and columns; this is proved in Example 2 below.

For convenience, we now summarize the properties of the indefinite admittance matrix that have been derived. These are:

1. The indefinite admittance matrix is symmetric for a passive reciprocal network and in general is unsymmetric for an active network.
2. Each of its elements is a short-circuit admittance.
3. The sum of the elements in each column is equal to zero. This is a direct consequence of Kirchhoff's current law.
4. The sum of the elements in each row is equal to zero. This follows from the fact that the node currents are dependent on potential differences between various node pairs rather than on the absolute potential of the nodes. It is not necessary to make use of this fact in deriving the property for a network that obeys reciprocity. It must be used, however, to establish the property for an active network.
5. The determinant of $[Y_i]$ is zero.
6. The first cofactors are all equal. This property is true for any zero-sum matrix. Because of it a third designation has been used for this matrix, namely, the *equicofactor matrix*.[8]
7. The matrix $[Y_i]$ may be simply transformed to a node-to-datum admittance matrix by deleting any row k and column k; this has the effect of making terminal k the reference terminal and making the network an $(n-1)$-port with all ports possessing terminal k as a common terminal. Also, because of properties 2 to 4 any node-to-datum admittance matrix can be converted to an indefinite admittance matrix by the addition of a row and a column whose elements are determined by use of properties 3 and 4.

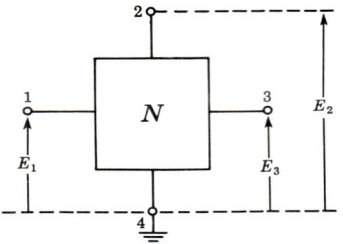

FIG. 24. Three-port network in which all ports have a common terminal.

It may be remarked at this point that the indefinite impedance matrix possesses similar properties; that is, if the external loop equations

$$E] = [Z_i] I] \qquad (206)$$

are written for an n-terminal network, then $[Z_i]$ is a zero-sum matrix. Here $I]$ represents the n loop currents, and $E]$ is the vector of the loop voltages, where each loop voltage is defined between two adjacent terminals.* These definitions are illustrated by the n-terminal network shown in Fig. 25. The sum of the elements in each row of $[Z_i]$ is zero because the voltages depend only on the *differences* between the loop currents, rather than on their absolute values; in other words, they are dependent on the *terminal* currents, which as is evident from Fig. 25, depend only on the difference between two adjacent loop currents. The sum of the elements in each column must also equal zero, since the loop voltages form a complete loop around the terminals, and hence by Kirchhoff's voltage law their sum is zero. Finally, a non-

* It should be noted that this so-called "loop analysis" is not being used in the usual sense, where it refers to the *internal* characterization of a network.

singular matrix may be obtained from $[Z_i]$ by taking the nth loop current as zero; this requires that we delete the nth row and column of $[Z_i]$.

We now discuss how various operations may be performed with the indefinite admittance matrix.

6.1. Calculation of Transfer Impedance and Voltage Ratio. One of the important advantages of the indefinite admittance matrix is that it facilitates the calculation of transfer impedances and voltage ratios from *any pair of terminals to any other pair;* this is true because the matrix is a description of the network on a terminal basis, and thus in using it we are not initially committed to a choice of ports. To show these calculations it is convenient to use a notation previously introduced for cofactors. We define $Y_{rst\ldots}{}^{klm\ldots}$ as the cofactor of the minor formed by rows $r, s, t, \ldots$ and columns $k, l, m, \ldots$, or alternatively as $(-1)^{r+s+t+\cdots+k+l+m+\cdots}$ times the determinant of the submatrix formed by suppressing rows $r, s, t, \ldots$ and columns $k, l, m, \ldots$ in $[Y_i]$. Here it is assumed that the subscripts and superscripts are in ascending order of magnitude.

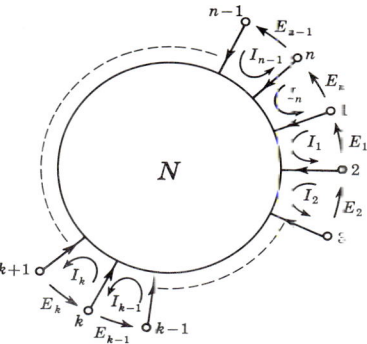

Fig. 25. Representation of an n-terminal network with its external loop currents and loop voltages.

Otherwise the number of transpositions necessary to place them in ascending order must be counted; if this is odd, the cofactor must be multiplied by -1, whereas if it is even, the cofactor is unchanged. In accordance with this notation, we have, for example, $Y_{rst}{}^{klm} = -Y_{rst}{}^{lkm} = Y_{sr}{}^{lkm}$; in addition, the first cofactor previously designated as Y_{jk} may now also be represented by $Y_j{}^k$.

Alternatively, we may express the determinants in a more explicit notation, specifically, in terms of their diagonal elements. If

$$|A| = \begin{vmatrix} a_{11} & a_{12} & \cdots & a_{1n} \\ a_{21} & a_{22} & \cdots & a_{2n} \\ \cdots & \cdots & \cdots & \cdots \\ a_{n1} & a_{n2} & \cdots & a_{nn} \end{vmatrix}$$

we represent this as $|a_{11}a_{22}a_{33}\cdots a_{nn}|$. In this notation the minors

$$\begin{vmatrix} a_{11} & a_{12} & a_{1n} \\ a_{21} & a_{22} & a_{2n} \\ a_{n1} & a_{n2} & a_{nn} \end{vmatrix} \quad \text{and} \quad \begin{vmatrix} a_{14} & a_{12} \\ a_{34} & a_{32} \end{vmatrix}$$

are represented by $|a_{11}a_{22}a_{nn}|$ and $|a_{14}a_{32}|$, respectively. Thus we have also

$$|a_{14}a_{32}| = -|a_{12}a_{34}|$$

We now assume that current I_{rq} flows into terminal r and out of terminal q and that all other currents are zero; that is, $I_r = -I_q = I_{rq}$; we also define voltage E_{km} as the potential rise from m to k; that is,

$$E_{km} = E_k - E_m \tag{207}$$

The open-circuit transfer impedance E_{km}/I_{rq} is then given by

$$z_{rq}{}^{km} = E_{km}/I_{rq} = Y_{rq}{}^{km}/Y_q{}^m \tag{208}$$

When $k = r$ and $m = q$, we obtain the driving-point impedance

$$z_{rq}{}^{rq} = E_{rq}/I_{rq} = Y_{rq}{}^{rq}/Y_q{}^q \tag{209}$$

It is clear that a four-index notation is necessary to designate the impedances. Here it is convenient to make the subscripts and superscripts correspond to those of the second cofactor; the subscripts on $z_{rq}{}^{km}$ indicate that I_r is injected into the network and $I_r = -I_q$, and the superscripts should be taken to mean that the positive direction of the output voltage is measured from terminal m to terminal k. Now by dividing the transfer impedance in (208) by the driving-point impedance in (209), we obtain the voltage ratio

$$E_{km}/E_{rq} = Y_{rq}{}^{km}/Y_{rq}{}^{rq} \qquad (210)$$

These formulas are derived in Example 3 below.

The reader should note the physical meaning of the above formulas. The indefinite admittance matrix has permitted the choice of any four terminals for the calculation of a transfer function, where the references for current and voltage are in general different terminals. By deleting column m, we have set $E_m = 0$ and chosen terminal m as the voltage reference; by deleting row q we have used terminal q as the second terminal of the port through which current is supplied to the network. Then assuming that the determinant of the remaining equations is nonzero (which it is for the nondegenerate connected networks we are considering), we obtain the desired transfer functions in terms of this determinant and a cofactor.

6.2. Reduction of an n-terminal Network to a p-terminal Network.[9] We may often desire to convert an n-terminal network into a p-terminal network where

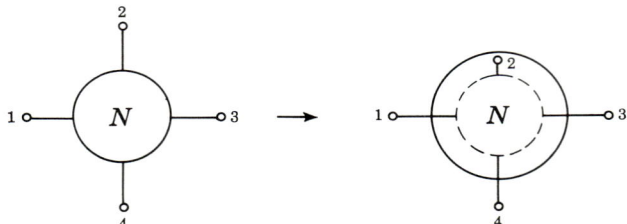

Fig. 26. Reduction of a four-terminal network to a three-terminal network.

$p < n$. This may take either of two forms: we may short-circuit two or more terminals together, or we may eliminate one or more terminals by considering them no longer accessible, i.e., by constraining their currents to be zero. In the first case, when two terminals are shorted and this double terminal is now used as a single accessible terminal, the corresponding change in the matrix is evident. Since the potentials of the two terminals are equal and their currents are added, we add the corresponding rows of the matrix together and also the corresponding columns. For example, if terminals 3 and 4 of a four-terminal network are shorted, then the indefinite admittance matrix of the new three-pole is given by

$$[Y_i] = \begin{bmatrix} y_{11} & y_{12} & y_{13} + y_{14} \\ y_{21} & y_{22} & y_{23} + y_{24} \\ y_{31} + y_{41} & y_{32} + y_{42} & y_{33} + y_{34} + y_{43} + y_{44} \end{bmatrix} \qquad (211)$$

When terminals are eliminated by setting their currents equal to zero, the situation is slightly more complicated. Suppose that terminal t is suppressed in this way; we then have an $(n-1)$-terminal network for which we wish to find the new indefinite admittance matrix $[Y_i']$ in terms of the elements of $[Y_i]$ for the n-terminal network. The formula for any element in the new matrix is given by

$$y_{jk}' = |y_{jk} y_{tt}|/y_{tt} \qquad (212)$$

where, it should be noted, the subscripts on each of the elements of $[y']$ correspond to the original designations of the terminals and not to the element position in $[y']$. For example, if a four-terminal network is converted to a three-pole by constraining I_2 to be zero and thus making terminal 2 inaccessible, as shown in Fig. 26, then the new indefinite admittance matrix is

$$[Y_i'] \equiv \begin{bmatrix} y_{11}' & y_{13}' & y_{14}' \\ y_{31}' & y_{33}' & y_{34}' \\ y_{41}' & y_{43}' & y_{44}' \end{bmatrix}$$

$$= \frac{1}{y_{22}} \begin{bmatrix} |y_{11}y_{22}| & |y_{13}y_{22}| & |y_{14}y_{22}| \\ |y_{31}y_{22}| & |y_{33}y_{22}| & |y_{34}y_{22}| \\ |y_{41}y_{22}| & |y_{43}y_{22}| & |y_{44}y_{22}| \end{bmatrix} \tag{213}$$

The formula given in (212) can be derived by the standard method of pivotal condensation;[15] this corresponds to setting $I_2 = 0$ in the system of equations, using the second equation to eliminate E_2 from the other three equations, and then neglecting the second equation.

By the same process of pivotal condensation, i.e., essentially by successive elimination of variables,[9] we can derive the general formula when more terminals are eliminated. For example, when the terminals t, q, and r are removed by making them inaccessible, the elements of the indefinite admittance matrix for the $(n-3)$-terminal network are given by

$$y_{jk}' = \frac{|y_{jk}y_{tt}y_{qq}y_{rr}|}{|y_{tt}y_{qq}y_{rr}|} \tag{214}$$

If we ignore all terminals except j and k, that is, if the n-terminal network is converted to a dipole, then its driving-point function is

$$y_{jk}' = 1/z_{jk}' = Y_j{}^k/Y_{jk}{}^{jk} \tag{215}$$

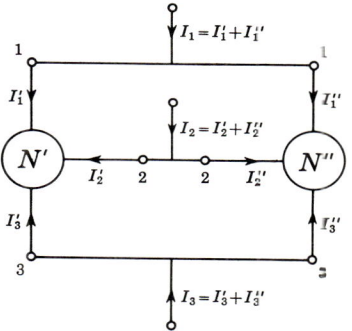

Fig. 27. Parallel connection of two three-pole networks.

It is worthwhile to point out that these formulas apply not only to n-terminal networks but also to n-port networks; the derivation is the same as that given for the n-terminal case. By use of these formulas we have extended the types of characterizations that can be automatically used for a network: the short-circuit admittance matrix describes the network under short-circuit constraints, the open-circuit impedance matrix applies for open-circuit constraints, while the admittance matrix derived above pertains to the network under mixed constraints, that is, with some of its terminals (or ports) open-circuited and the others shorted. For example, the admittance y_{jk}' in Eq. (214) is equal to the current measured in a short circuit placed between terminal j and the reference when a source of 1 volt is applied to terminal k, ports t, q, and r being open-circuited, whereas all the remaining ports are shorted.

6.3. Connection of Two n-poles in Parallel. Consider two networks, each of which has n terminals. Now suppose that these are connected in parallel; that is, we join together corresponding terminals of the network with a short circuit and then consider each of the double terminals thus formed as a terminal of the composite n-pole. This parallel connection is shown for two three-poles N' and N'' in Fig. 27. Since the currents of the joined terminals add and their potentials are equal, it is clear that

$$[Y_i] = [Y_i'] + [Y_i''] \tag{216}$$

so that a matrix element of the composite three-pole is given by

$$y_{jk} = y_{jk}' + y_{jk}'' \tag{217}$$

Now suppose that the number of terminals for two networks to be connected in parallel is different; that is, we have an m-pole and an n-pole, where $m < n$. For example, we may have a three-terminal tube to which we wish to add a load resistance R, as shown in Fig. 28.

We can always consider an m-terminal network as an n-terminal network merely by adding $n - m$ terminals that are not connected to the network. The potentials of these added terminals are arbitrary, and their currents are zero; thus this corresponds to adding $(n - m)$ rows and $(n - m)$ columns of zeros to the indefinite admittance matrix of the m-pole. For example, we consider the resistance as a three-terminal network, with terminal 1 an isolated terminal. Its indefinite admittance matrix is then

$$[Y_i] = \begin{bmatrix} 0 & 0 & 0 \\ 0 & \dfrac{1}{R} & -\dfrac{1}{R} \\ 0 & -\dfrac{1}{R} & \dfrac{1}{R} \end{bmatrix} \qquad (218)$$

Now the resistance may be connected to the tube, giving an indefinite admittance matrix for the composite network equal to the sum of the above matrix and the indefinite admittance matrix of the tube.

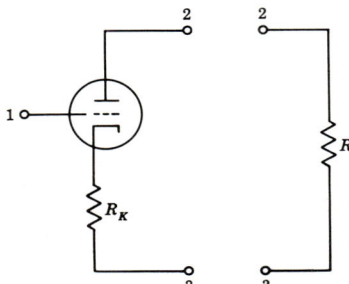

Fig. 28. Tube and load resistance whose interconnection is to be characterized by an indefinite admittance matrix.

Thus the indefinite admittance matrix or the indefinite impedance matrix of a complicated active network may be set up with little calculation and without the necessity of drawing equivalent circuits containing controlled sources. From these matrices we may obtain any other matrices or transfer functions that are desired, e.g., the node-to-datum admittance matrix is obtained by eliminating the appropriate row and corresponding column, or a desired voltage ratio is given by (210). What is required is that we know the indefinite matrices for the multiterminal active elements included in the network. It therefore is useful to have available the matrices (or tables) for frequently used elements like the tube and the transistor. These are presented below; examples in which the matrices are employed are given in the reference.[16]

The admittance-coefficient table for the triode is given by the indefinite admittance matrix

$$[Y_i] = \begin{bmatrix} 0 & 0 & 0 \\ g_m & g_p & -g_p - g_m \\ -g_m & -g_p & g_p + g_m \end{bmatrix} \qquad (219)$$

where the terminals 1, 2, and 3 correspond, respectively, to the grid, plate, and cathode, and g_m is the transconductance and $g_p = 1/r_p$ is the plate conductance. For the transistor, letting the emitter, collector, and the base correspond to terminals 1, 2, and 3, respectively, we have

$$Y_i] = \begin{bmatrix} \dfrac{r_b + r_c}{r_e(r_b + r_c) + r_b(r_c + r_m)} & \dfrac{-r_b}{r_e(r_b + r_c) + r_b(r_c + r_m)} & \dfrac{-r_c}{r_e(r_b + r_c) + r_b(r_c + r_m)} \\ \dfrac{-r_b - r_m}{r_e(r_b + r_c) + r_b(r_c + r_m)} & \dfrac{r_b + r_e}{r_e(r_b + r_c) + r_b(r_c + r_m)} & \dfrac{r_m - r_e}{r_e(r_b + r_c) + r_b(r_c + r_m)} \\ \dfrac{-r_c + r_m}{r_e(r_b + r_c) + r_b(r_c + r_m)} & \dfrac{-r_e}{r_e(r_b + r_c) + r_b(r_c + r_m)} & \dfrac{r_e + r_c - r_m}{r_e(r_b + r_c) + r_b(r_c + r_m)} \end{bmatrix}$$
$$(220)$$

where r_e, r_c, r_b, and r_m are the incremental emitter, collector, base, and mutual resistances, respectively. The impedance coefficients for the transistor with the loop currents labeled as shown in Fig. 29a are given by

$$[Z_i] = \begin{bmatrix} r_e + r_b & -r_b & -r_e \\ -r_m - r_b & r_b + r_c & r_m - r_c \\ r_m - r_e & -r_c & r_e + r_c - r_m \end{bmatrix} \quad (221)$$

For the triode, since the impedance coefficients associated with the grid terminal are infinite, it is convenient to assume that the tube has a grid-return resistance R_g connected between the grid and cathode as shown in Fig. 29b. For the loop currents numbered as shown, we have for this composite three-terminal element

$$[Z_i] = \begin{bmatrix} R_g & 0 & -R_g \\ \mu R_g & r_p & -\mu R_g - r_p \\ -(1 + \mu)R_g & -r_p & r_p + (1 + \mu)R_c \end{bmatrix} \quad (222)$$

where r_p is the plate resistance and μ is the voltage amplification factor.

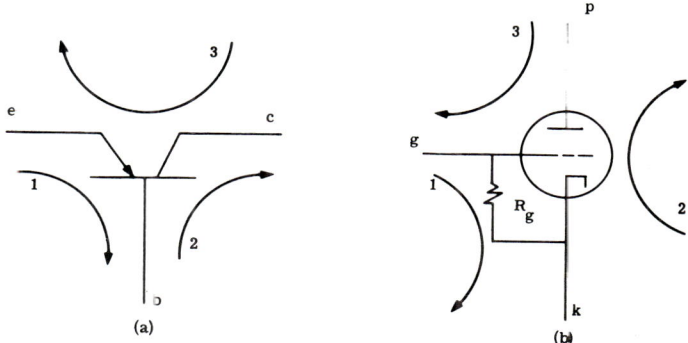

Fig. 29. Transistor and triode with loop currents designated for representation by indefinite impedance matrix.

6.4. Relation between Driving-point Functions and Transfer Functions of a Reciprocal n-terminal Network. We have shown in the three-pole that the network can be characterized by three of its independent transfer functions, or equivalently by three of its independent driving-point functions. In this subsection we wish to derive general formulas for reciprocal n-terminal networks which show that a set of $n(n-1)/2$ independent driving-point functions specify an n-terminal network uniquely. This implies that the network's transfer functions are related to these driving-point functions; we derive formulas expressing this relationship. We also show a representation of an n-terminal network due to Campbell that is a generalization of the results derived for the three-pole in terms of transfer admittances connected in a delta network; specifically, we derive a network representation of any reciprocal n-terminal network in terms of its transfer admittances. Finally, we show how the elements of the matrix describing an n-port network can be given in terms of driving-point measurements made on the same network considered as a $2n$-terminal network; thus every reciprocal n-port may be uniquely characterized by driving-point functions.

Consider an n-terminal network that is described by its indefinite admittance matrix $[Y_i]$. Now let terminal n be the reference and obtain a nonsingular matrix $[y_{jk}]$ by removing row and column n from $[Y_i]$. The inverse of this matrix is given by

$$[z_{jk}] = [y_{jk}]^{-1} \quad (223)$$

The open-circuit impedance matrix $[z_{jk}]$ characterizes the n-terminal network; each of the main-diagonal elements, z_{kk} for $k = 1, 2, \ldots, n-1$, represents the impedance seen between terminals k and n with all the other terminals open-circuited; each of the off-diagonal elements, z_{kr} for $k \neq r$, represents the transfer impedance between terminals k and r when all the terminals except r have zero currents; that is, a current source is applied between terminals r and n, the open-circuit voltage is measured between k and n, and all other terminals are left free.

We now assume that all the currents are zero except I_r and I_k; this condition is shown in Fig. 30. Then, from the set of $(n - 1)$ equations represented by

$$E] = [z_{jk}] I] \qquad (224)$$

we obtain, by setting all currents except I_r and I_k equal to zero and eliminating all the equations except the two independent ones, the rth and the kth,

$$\begin{aligned} E_r &= z_{rr}I_r + z_{rk}I_k \\ E_k &= z_{kr}I_r + z_{kk}I_k \end{aligned} \qquad (225)$$

But since $I_r = -I_k \equiv I_{rk}$, this may be substituted in the above Eqs. (225) and the second equation then subtracted from the first to give

$$(E_r - E_k)/I_{rk} = z_{rr} + z_{kk} - (z_{rk} + z_{kr}) \qquad (226)$$

The left side represents the driving-point impedance between terminals r and k, all other terminals being left free. It is convenient to define this driving-point impedance

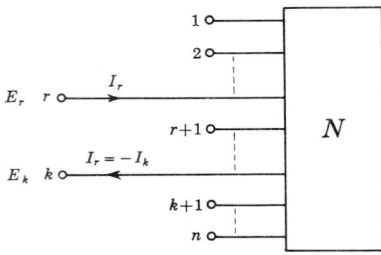

FIG. 30. Representation of an n-terminal network in which all the terminals except r and k are open-circuited.

by $S_{k,r}$. Then we define $S_{k,k} = 0$, since this measurement corresponds to both the measuring leads connected to the same terminal. It is also clear from the definition that

$$z_{kk} = S_{n,k} \qquad k = 1, 2, \ldots, n - 1 \qquad (227)$$

From (226) we then have

$$\begin{aligned} S_{k,r} &= z_{rr} + z_{kk} - (z_{rk} + z_{kr}) \\ &= S_{n,r} + S_{n,k} - (z_{rk} + z_{kr}) \end{aligned} \qquad (228)$$

which yields

$$z_{rk} + z_{kr} = S_{n,r} + S_{n,k} - S_{k,r} \qquad (229)$$

This equation holds for any active or passive network.

For a reciprocal network, using $z_{kr} = z_{rk}$, we obtain

$$z_{rk} = \tfrac{1}{2}(S_{n,r} + S_{n,k} - S_{k,r}) \qquad (230)$$

When $r = k$, we get

$$z_{kk} = S_{n,k} \qquad (231)$$

which checks (227) and shows that Eq. (230) is valid for all $r, k = 1, 2, \ldots, n - 1$.

Thus the important fact emerges that in order to characterize the reciprocal n-terminal network uniquely, it is necessary to make only driving-point measurements; specifically, we must determine all the $n(n - 1)/2$ independent $S_{r,k}$.

MATRICES FOR MULTITERMINAL NETWORKS 4-43

An analogous formula can be derived for the dual set of short-circuit admittances. We let $\hat{S}_{n,r}$ represent the admittance between terminals r and n when all the voltages except E_r equal zero; that is, all the terminals except r are shorted to terminal r. This condition is shown in Fig. 31. Thus the set of $n - 1$ equations

reduce to
so that
for this case.

$$I] = [y_{jk}] E] \qquad (232)$$
$$I_r = y_{rr} E_r \qquad (233)$$
$$y_{rr} = \hat{S}_{n,r} = I_r/E_r \qquad r = 1, 2, \ldots, n-1 \qquad (234)$$

To determine the method for measurement of $\hat{S}_{k,r}$, we use the conditions dual to those shown in Fig. 30 for the measurement of $S_{k,r}$. There all terminals except r and k

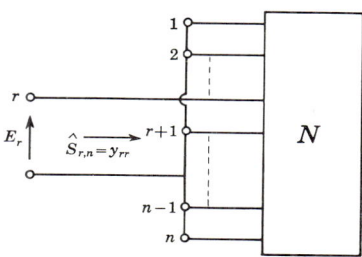

Fig. 31. Representation of short-circuit conditions for measurements of $\hat{S}_{r,n} = y_{rr}$.

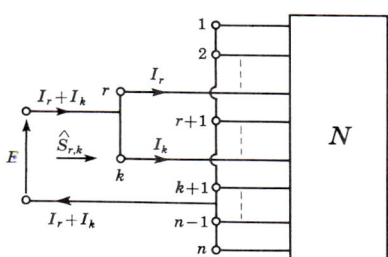

Fig. 32. Method for measuring $\hat{S}_{r,k}$ when $r \neq k$.

are open-circuited and the same current flows in r and k. Hence in this case all terminals except r and k must be shorted and the same voltage is applied to terminals r and k. Thus $E_l = 0$ where $l \neq r, k$ and $E_k = E_r = E$. These conditions are represented in Fig. 32.

For this case the set of equations in (232) reduces to

$$\begin{aligned} I_r &= y_{rr}E + y_{rk}E \\ I_k &= y_{kr}E + y_{kk}E \end{aligned} \qquad (235)$$

Using the same algebraic manipulations on the above equations as for the impedance case, we obtain

$$\hat{S}_{k,r} \equiv (I_r + I_k)/E = y_{rr} + y_{kk} + y_{kr} + y_{rk} \qquad (236)$$

which yields for a reciprocal network

$$y_{rk} = -\tfrac{1}{2}(\hat{S}_{n,r} + \hat{S}_{n,k} - \hat{S}_{k,r}) \qquad (r \neq k) \qquad (237)$$

This formula corresponds to the one given in (230).

The total number of $\hat{S}_{k,r}$ to be measured should be $n(n-1)/2$. This is easily shown to be satisfied: there are $n-1$ measurements of $\hat{S}_{n,k}$ since $k=1, 2, \ldots, n-1$; in addition, the number of $\hat{S}_{k,r}$ is given by the number of combinations of $n-1$ terminals taken two at a time. Thus the total is the required $n(n-1)/2$.

Instead of using driving-point functions for characterizing the n-terminal network, we can alternatively use all the independent transfer admittances of the indefinite admittance matrix; this leads to a useful network representation due to Campbell.[17] He showed that any reciprocal n-terminal network may be represented by a network with only n nodes, where each node is a terminal; each terminal is connected to every other terminal by precisely one branch, the driving-point admittance of this branch being given by the negative of the transfer admittance whose subscripts are determined by the terminals it joins. This representation is a generalization of the three-pole case previously represented in Fig. 22; the branches in this n-terminal network, as was pointed out for the three-pole, may be nonphysical. As an example, the general four-terminal network is shown in Fig. 33, where y_{jk} represents a short-circuit transfer admittance. The reader should check this representation; for example, by choosing terminal 4 as a reference and applying 1 volt between terminals 2 and 4, he will find that the short-circuit current flowing from 4 to 1 is given by the negative of the branch

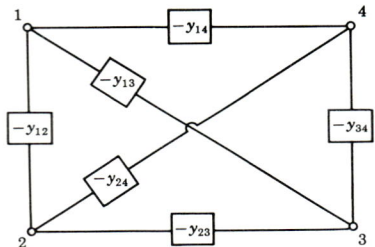

Fig. 33. Representation of any reciprocal four-terminal network by a network of four nodes and its transfer admittances.

represented as $-y_{12}$. Similarly, the short-circuit driving-point admittance between terminals 1 and 4 is correctly given by the sum of the three branches $-y_{12}$, $-y_{13}$, and $-y_{14}$, which thus satisfies the zero-sum property of the indefinite admittance matrix.

It is also useful to relate the system functions in the matrix for a general n-port network to the driving-point functions of the same network considered as a $2n$-terminal network. It is felt that in addition to the practical value of this relationship in providing an alternative method for carrying out measurements in the laboratory on an n-port device, and its theoretical value for analysis of an n-port, the formula may yield insights for understanding a presently unsolved synthesis problem, namely, the formulation of the necessary and sufficient conditions for the realization of a real symmetric matrix by an n-port network containing only resistances (and no ideal transformers). For example, since only driving-point measurements are made on the $2n$-terminal network, an obvious necessary condition on each measurement is that it is a nonnegative number.

Consider a reciprocal n-port network with an open-circuit impedance matrix $[z_{jk}]$. Since the matrix is symmetric, only $n(n+1)/2$ of its elements are independent. Now consider this network as a $2n$-terminal network, with the terminals numbered from 1 to $2n$ and with the ports so numbered that port 1 comprises terminals 1 and $n+1$, the assigned positive direction of the port voltage being from terminal 1 to terminal $n+1$. In general, port k will run from terminal k to terminal $n+k$.

As was done previously, for the representation of the $2n$-terminal network we let $S_{j,k}$ denote the measured driving-point impedance between terminals j and k, all other terminals being left free. We noted above that

$$z_{kk} = S_{k,n+k} \qquad k = 1, 2, \ldots, n \qquad (238)$$

The general formula for the elements of z_{jk} is then given by

$$z_{jk} = \tfrac{1}{2}(S_{j,n+k} + S_{k,n+j} - S_{j,k} - S_{n+j,n+k}) \tag{239}$$

which reduces to the formula in (238) when $j = k$.

As a simple check on this formula, we observe that it should reduce to that in (230) when the $2n$-terminal network is reduced to an n-terminal network. A method for converting the network to an n-terminal network is to short all the n terminals designated by $n+1, n+2, \ldots, 2n-1, 2n$, to terminal n; that is, all these terminals are made to coincide with terminal n. For this network any subscript greater than n may now be replaced by n. Equation (239) then becomes

$$\begin{aligned} z_{jk} &= \tfrac{1}{2}(S_{j,n} + S_{k,n} - S_{j,k} - S_{n,n}) \\ &= \tfrac{1}{2}(S_{n,j} + S_{n,k} - S_{j,k}) \end{aligned} \tag{240}$$

which is the same as (230). In deriving the last equation of (240) we have used the facts that $S_{r,n} = S_{n,r}$ and $S_{k,k} = 0$ for each r and k.

Finally, it should be emphasized that the above characterization in terms of driving-point functions applies only to reciprocal systems; it does not apply to nonreciprocal systems and therefore cannot be used, for example, to describe a three-terminal transistor network. It can be shown that since the driving-point functions yield expressions that are symmetric in z_{rk} and z_{kr}, they cannot serve to distinguish between the two.[18] As a consequence it is possible to have two systems characterized by identical driving-point functions but possessing different corresponding transfer functions.

Example 2. Prove that the first cofactors of an indefinite admittance matrix are equal and give an illustration using a fourth-order matrix.

Solution. To show that

$$Y_{jk} = Y_{nn} \tag{1}$$

for any zero-sum matrix of order n we follow the steps below, which transform Y_{jk} to Y_{nn}.

1. The minor $(-1)^{j+k} Y_{jk}$ is the determinant of the submatrix formed by deleting row j and column k of $[Y_i]$. Add the first $n-2$ rows of this minor to its last row and use (202) to transform the new last row to the negative of row j.

2. Add the first $n-2$ columns to the last column of the minor and use (203) to transform this new last column to the negative of column k (except that the element in the last row is positive).

3. This minor may now be converted to Y_{nn} by $n-1-j$ transpositions of rows and $n-1-k$ transpositions of columns, which multiplies its value by $(-1)^{j+k}$, so that Y_{nn} is equal to the cofactor Y_{jk}.

As an example, consider the fourth-order matrix

$$[Y_i] = \begin{bmatrix} y_{11} & y_{12} & y_{13} & y_{14} \\ y_{21} & y_{22} & y_{23} & y_{24} \\ y_{31} & y_{32} & y_{33} & y_{34} \\ y_{41} & y_{42} & y_{43} & y_{44} \end{bmatrix} \tag{2}$$

for which we show that $Y_{12} = Y_{44}$. We have

$$Y_{44} = |y_{11} y_{22} y_{33}| \tag{3}$$

and
$$-Y_{12} = |y_{21} y_{33} y_{44}| \tag{4}$$

If we add the first two rows to the last in (4) and then make use of (202), we obtain

$$-Y_{12} = \begin{vmatrix} y_{21} & y_{23} & y_{24} \\ y_{31} & y_{33} & y_{34} \\ -y_{11} & -y_{13} & -y_{14} \end{vmatrix} \tag{5}$$

4–46 AMPLIFIER FUNDAMENTALS

If we now add the first two columns to the last in (5) and use (203), we have

$$-Y_{12} = \begin{vmatrix} y_{21} & y_{23} & -y_{22} \\ y_{31} & y_{33} & -y_{32} \\ -y_{11} & -y_{13} & y_{12} \end{vmatrix}$$

$$= \begin{vmatrix} y_{21} & y_{23} & y_{22} \\ y_{31} & y_{33} & y_{32} \\ y_{11} & y_{13} & y_{12} \end{vmatrix} \tag{6}$$

Finally, we make two transpositions of rows to make the last row become the first row, and one transposition of columns to interchange the positions of the second and third columns; this yields

$$Y_{12} = \begin{vmatrix} y_{11} & y_{12} & y_{13} \\ y_{21} & y_{22} & y_{23} \\ y_{31} & y_{32} & y_{33} \end{vmatrix}$$

$$= Y_{44} \tag{7}$$

which checks the proof.

Example 3.[8] Derive the formulas given for the transfer impedance, the driving-point impedance, and the voltage ratio in Eqs. (208) to (210).

Solution. Suppose that

$$[Y_i] E] = I] \tag{1}$$

describes an n-terminal network. Since the sum of the elements in each row of $[Y_i]$ is zero, we may subtract $\sum_{i=1}^{n} y_{ti} E_m$ from the left side of each equation, where $1 \leq m \leq n$. If we then consider E_m as a reference voltage and define the new voltage variable $E_p' = E_p - E_m$, the equations become

$$\sum_{\substack{p=1 \\ p \neq m}}^{n} y_{tp} E_p' = I_t \qquad t = 0, 1, 2, \ldots, n \tag{2}$$

Finally, we constrain the network so that current I_{rq} is injected into terminal r and flows out of terminal q and all other currents are zero; that is, $I_t = 0$ except for $t = r$ and q, in which case $I_r = -I_q = I_{rq}$. The equations may now be written as

$$y_{11}E_1' + y_{12}E_2' + \cdots + y_{1,m-1}E'_{m-1} + y_{1,m+1}E'_{m+1} + \cdots + y_{1n}E_n' = 0$$
$$\cdots$$
$$y_{r1}E_1' + \cdots \cdots \cdots \cdots \cdots + y_{rn}E_n' = I_{rq}$$
$$\cdots$$
$$y_{q1}E_1' + \cdots \cdots \cdots \cdots \cdots + y_{qn}E_n' = -I_{rq} \tag{3}$$
$$\cdots$$
$$y_{n1}E_1' + \cdots \cdots \cdots \cdots \cdots + y_{nn}E_n' = 0$$

Because of the equicofactor property, the equations are linearly dependent and one equation is superfluous; let us delete the qth equation. Then, by Cramer's rule, the solution of this set of equations for any E_k' is

$$E_k' = \frac{\begin{vmatrix} y_{11} & y_{12} & \cdots & y_{1,k-1} & 0 & y_{1,k+1} & \cdots & y_{1n} \\ \cdots & \cdots & \cdots & \cdots & 0 & \cdots & \cdots & \cdots \\ y_{r1} & y_{r2} & \cdots & y_{r,k-1} & I_{rq} & y_{r,k+1} & \cdots & y_{rn} \\ \cdots & \cdots & \cdots & \cdots & 0 & \cdots & \cdots & \cdots \\ y_{n1} & y_{n2} & \cdots & y_{n,k-1} & 0 & y_{n,k+1} & \cdots & y_{nn} \end{vmatrix}}{(-1)^{q+m} Y_q^m} \tag{4}$$

where the determinant in the numerator is formed by omitting the qth row and the mth column from the determinant of $[Y_i]$ and then replacing the kth column by the vector of zero currents and I_{rq} as shown. The factor $(-1)^{q+m}$ occurs in the denominator because

Y_q^m stands for a cofactor (which of course is equal to any other first cofactor), and what we require is the minor formed by deleting row q and column m.

We now expand the determinant in the numerator of E_k' according to the cofactors of the column with $n-1$ zeros and I_{rq}. We note that I_{rq} is in the kth column if $k < m$, but in the $(k-1)$st column if $k > m$. Furthermore, the row in which I_{rq} appears is the rth row if $r < q$, but is the $(r-1)$st row if $r > q$. Finally, the determinant obtained by omitting the row and column in which I_{rq} appears is the minor corresponding to the cofactor Y_{rq}^{km} and thus equals $(-1)^{r+q+k+m} Y_{rq}^{km}$ multiplied by -1 or $+1$ according as the number of required transpositions is odd or even, that is, the number of transpositions required to bring subscript pair r and q and the superscript pair k and m to increasing order of magnitude. Hence we have

$$E_k' = \frac{\lambda \rho I_{rq}(-1)^{r+q+k+m} Y_{rq}^{km}}{(-1)^{q+m} Y_q^m} \tag{5}$$

where λ is the coefficient necessary to give the proper sign to the cofactor of I_{rq} and ρ is -1 or $+1$ as the number of transpositions is odd or even, respectively.

From the discussion we see that the following four cases occur:

$r < q, k < m$	$r > q, k < m$	$r > q, k > m$	$r < q, k > m$
$\rho = 1$	$\rho = -1$	$\rho = 1$	$\rho = -1$
$\lambda = (-1)^{r+k}$	$\lambda = (-1)^{r-1+k}$	$\lambda = (-1)^{r-1+k-1}$	$\lambda = (-1)^{r+k-1}$

(6)

It is clear that all four cases when substituted in (5) yield the single equation

$$\frac{E_{km}}{I_{rq}} = \frac{Y_{rq}^{km}}{Y_q^m} \tag{7}$$

where for convenience we have written E_{km} for E_k'. Thus we have proved (208).

When $k = r$ and $m = q$, the transfer impedance in (7) becomes the driving-point impedance

$$\frac{E_{rq}}{I_{rq}} = \frac{Y_{rq}^{rq}}{Y_q^q} \tag{8}$$

Now by dividing the transfer impedance in (7) by the driving-point impedance in (8), we get the voltage ratio

$$\frac{E_{km}}{E_{rq}} = \frac{Y_{rq}^{km}}{Y_{rq}^{rq}} \tag{9}$$

Equations (8) and (9) are identical with (209) and (210), and the proof is complete.

REFERENCES

1. H. J. Carlin, The Scattering Matrix in Network Theory, *IRE Trans. PGCT*, vol. CT-3, no. 2, pp. 88–97, June, 1956.
2. C. Montgomery, R. H. Dicke, and E. M. Purcell, "Principles of Microwave Circuits," vol. 8, chap. 5, MIT Radiation Laboratory Series, McGraw-Hill Book Company, New York, 1948.
3. R. Redheffer, Difference Equations and Functional Equations in Transmission-line Theory, in 2d ser., pp. 282–337, E. F. Beckenbach (ed.), "Modern Mathematics for the Engineer," McGraw-Hill Book Company, New York, 1961.
4. D. C. Youla, On Scattering Matrices Normalized to Complex Port Numbers, *Proc. IRE*, vol. 49, July, 1961 (correspondence section).
5. D. C. Youla, A New Theory of Broad-band Matching, *IEEE Trans. Circuit Theory*, vol. CT-11, no. 1, pp. 30–49, March, 1964.
6. J. Shekel, Matrix Representation of Transistor Circuits, *Proc. IRE*, vol. 40, no. 11, pp. 1493–1497, November, 1952.
7. T. H. Puckett, A Note on the Admittance and Impedance Matrices of an n-terminal Network, *IRE Trans. PGCT*, vol. CT-3, no. 1, pp. 70–75, March, 1956.
8. G. E. Sharpe and B. Spain, On the Solution of Networks by Means of the Equicofactor Matrix, *IRE Trans. PGCT*, vol. CT-7, no. 3, pp. 230–239, September, 1960.
9. V. Baranov, Méthode de calcule des réseaux d'impédances, *Rev. Gen. Elec.*, vol. 37, no. 11, pp. 339–351, Mar. 16, 1935.

10. G. A. Campbell, Direct Capacity Measurement, *Bell System Tech. J.*, vol. 1, no. 1, pp. 18–38, July, 1922.
11. E. S. Kuh and D. O. Pederson, "Principles of Circuit Synthesis," McGraw-Hill Book Company, New York, 1959.
12. D. C. Youla, An Extension of the Concept of Scattering Matrix, *Memo.* 52, PIBMRI, Polytechnic Institute of Brooklyn, Microwave Research Institute, Sept. 8, 1961.
13. H. J. Carlin and A. B. Giordano, "Network Theory," Prentice-Hall, Inc., Englewood Cliffs, N.J., 1964.
14. L. Weinberg, "Network Analysis and Synthesis," McGraw-Hill Book Company, New York, 1962.
15. A. C. Aitken, "Determinants and Matrices," 9th ed., Interscience Publishers, Inc., New York, 1956.
16. L. A. Zadeh, A Note on the Analysis of Vacuum Tube and Transistor Circuits, *Proc. IRE*, vol. 41, no. 8, pp. 989–992, August, 1953.
17. G. A. Campbell, Cisoidal Oscillations, *Trans. AIEE*, vol. 30, pp. 873–909, April, 1911.
18. L. Weinberg, Reciprocal and Nonreciprocal Systems, Characteristic Polynomials, and Driving-point Functions, *IEEE Trans. GCT*, vol. CT-12, no. 2, June, 1965.

Chapter 5

FILTER AMPLIFIERS

LOUIS WEINBERG*

CONTENTS

1. Introduction.. 5- 1
2. Some Optimum Magnitude and Phase Characteristics............. 5- 4
    2.1. Butterworth Polynomials................................. 5- 7
    2.2. Bessel Polynomials...................................... 5- 8
    2.3. Transitional Butterworth-Thomson Characteristic......... 5-11
    2.4. Equal-ripple Property: Chebyshev Polynomials............ 5-13
    2.5. Inverse Chebyshev Characteristic........................ 5-17
    2.6. Equal Ripple in Pass and Stopbands: Chebyshev Rational
         Functions... 5-17
3. Dipole Networks Synthesized by Recognition................... 5-19
4. Additional Techniques for Realization of Filter Amplifiers... 5-25
    4.1. Two-terminal Coupling................................... 5-25
    4.2. Two-port Coupling....................................... 5-27

1. INTRODUCTION

Amplifiers are often designed to provide a specified high gain and a specified passband and stopband characteristic. The specified characteristic may be a desired magnitude as a function of frequency (given either as a curve or in functional form) or a desired phase function or both magnitude and phase functions. Thus in addition to a gain a filtering action is required. Such amplifiers, which are appropriately called *filter amplifiers*, can be designed in many simple ways by making use of the techniques of modern network synthesis.

Since several active elements like tubes or transistors are needed to achieve the gain, their impedance properties can be made use of in providing the desired frequency specifications. The impedance character of an ideal amplifier permits it to serve as a buffer to provide isolation between stages; with such isolation a system function of high degree can be broken into a product of simple system functions of low degree, thus tremendously simplifying the network-realization problem. A pentode can often be used as an ideal transadmittance amplifier, that is, a voltage-controlled current source without parasitics, and can thus provide the desired isolation between

* Department of Electrical Engineering, University of Michigan, Ann Arbor, Mich.

stages. In fact, isolation or buffering action is satisfactorily provided as long as the active element is unilateral. If the pentode approximates a semi-ideal amplifier in that its output resistance is nonnegligible, the effect of this resistance is easily absorbed in the coupling network; that is, the coupling network is designed to be driven by a current source with a specified internal resistance. If a shunt capacitance and resistance must be taken into account at both the output of the pentode that serves as the current source for the stage and at the input of the pentode of the next stage, then the coupling network is designed to have such terminations at its input and output. As we shall see, this design is almost trivially simple when the coupling network is a one-port plate load: a simple tuned circuit consisting of the parallel connection of a capacitance, an inductance, and a resistance satisfies the termination requirements at input and output. Finally, when the active element has nonzero reverse transmission like a transistor, that is, all three parasitic impedances are nonnegligible, the desired filter and gain characteristics may still be achieved by synthesis techniques[1] but some simplicity of design is lost.

In the last-mentioned method, it is important to note, the feedback of the active element is utilized so that the properties of an RLC passive network, namely, poles

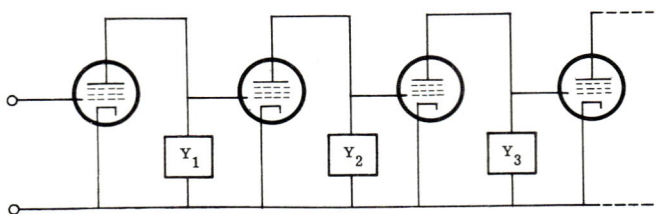

Fig. 1. Multistage amplifier in which the interstage coupling is a two-terminal shunt impedance.

that are complex and lie close to the imaginary axis, are achieved by a network requiring resistances and capacitances but no inductances. Readers familiar with microelectronics and integrated circuits will recognize how desirable this is, since resistances and capacitances may easily be fabricated as thin-film elements or parts of integrated circuits but the realization of inductances presents formidable difficulties.

This procedure of dividing a complicated transfer function into a product of simpler functions, each of which can be realized rather easily by a transistor and a simple passive network, may be especially advantageous for the realization of integrated circuits. Transistors in an integrated circuit are fairly inexpensive so that the use of a minimum number of active elements is not an important desideratum. In fact, in many integrated circuits it is possible for the transistors to be cheaper than the passive elements; then a valid criterion of design is the minimization of the passive rather than the active elements.

The problem of the realization of filter amplifiers by the methods described above is treated in detail by Wallman and others.[2] The different interstage coupling methods that are considered by Wallman are:

1. Synchronous single-tuned network
2. Stagger-tuned network
3. Double-tuned (including stagger-damped) network

Also discussed is the use of inverse feedback pairs, triples, and chains using a single-tuned circuit as plate load, but we do not consider this technique here.

As Wallman points out, each of the above has advantages and disadvantages with respect to its efficiency (as measured by its gain-bandwidth product), its gain stability, and its ease in construction, gain control, and criticalness of adjustment. For example, the synchronous single-tuned coupling scheme has minimum efficiency and maximum simplicity, whereas the others are more complicated but also more efficient. All the methods except the third use simple two-terminal loads, whereas the third uses a restricted form of quadripole as the coupling network.

The treatment in the reference is excellent, and it is therefore not necessary to repeat the results. However, it is useful to indicate briefly what is achieved there so that the additional techniques in this chapter can be fitted into a proper framework.

For the coupling scheme that uses a two-terminal shunt impedance as a plate load on each tube, the amplifier takes the form shown in Fig. 1. In this class is the *synchronously tuned* amplifier, where the load for each tube is a tuned circuit as shown in Fig. 2, and each load is tuned to the *same* frequency. As Wallman shows, an increase in the gain-bandwidth product is obtained when the resonant frequencies of each load are made different and a maximally flat magnitude characteristic is achieved; such an amplifier is called a *flat-staggered n-tuple*. A further increase in gain-bandwidth product is obtained by staggering the resonant frequencies so that an *equal-ripple* magnitude characteristic is achieved; Wallman calls this an *overstaggered* amplifier.

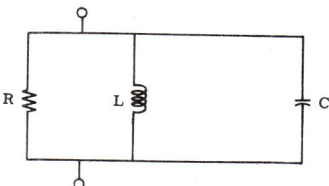

FIG. 2. Tuned circuit used as a two-terminal coupling network.

When the coupling network is generalized to a grounded quadripole, thus permitting a separation of the parasitic output capacitance of one tube from the input capacitance of the next tube, a larger gain-bandwidth product becomes possible. The amplifier

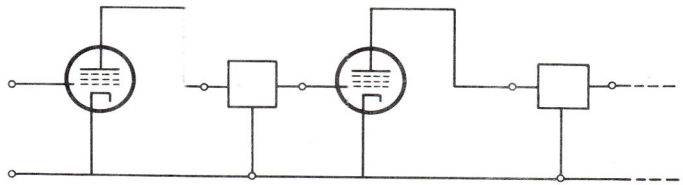

FIG. 3. Multistage amplifier in which the coupling network is a grounded quadripole.

then takes the form shown in Fig. 3. The double-tuned amplifier treated in the reference uses the quadripole shown in Fig. 4. When the resonant frequencies for all the quadripoles in the amplifier are the same, a smaller gain-bandwidth product results than for the amplifier with different resonant frequencies in each stage; the latter is designated as the *stagger-damped* amplifier.

Wallman considers in great detail the design of one magnitude characteristic, the maximally flat case, which is also called the *Butterworth* case. He also discusses but in much less detail the overstaggered case, which we call the *equal-ripple* or *Chebyshev polynomial* case. As he points out, the technique can be applied to any desired magnitude characteristic; further it can be applied to any phase characteristic; in short, to any system function. In this chapter we consider the properties of a number of optimum magnitude characteristics to which the technique can be applied, including the Butterworth and the Chebyshev polynomial characteristics, so that the amplifier designer can choose the system function that best satisfies his specifications. We also consider a desirable phase characteristic obtained from the use of Bessel polynomials.

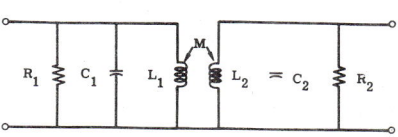

FIG. 4. Quadripole used in double-tuned amplifier.

Then we discuss some additional simple techniques for achieving a specified system function and a high gain. Some of these techniques use quadripoles as coupling networks; however, the tables that have been published reduce the design of these quadripoles almost to the simplicity of the design of the two-terminal tuned circuit.

We do not attempt to discuss all the possible synthesis procedures that can be used to design the coupling networks of the filter amplifiers; however, adequate discussions are given in any of the standard texts.[3,4] Part of what we do here is to suggest that synthesis is a general, flexible, and powerful tool for such design, and give some guide lines for its use. Furthermore, we stress that with the tables and other aids that are now available the design of complex filter amplifiers is no longer cumbersome or tedious.

2. SOME OPTIMUM MAGNITUDE AND PHASE CHARACTERISTICS

The four different magnitude characteristics, the one phase characteristic, and the combined magnitude and phase characteristic that we consider are briefly introduced with respect to their approximation of a low-pass characteristic. A more detailed discussion is available in the reference.[4]

Consider the ideal magnitude characteristic of a low-pass filter shown in Fig. 5, where the squared magnitude of the normalized transfer impedance is plotted against positive values of the normalized real frequency. In this plot all signals with frequencies in the passband $0 \leq \omega < 1$ are transmitted without loss, whereas inputs with frequencies $\omega > 1$ yield zero output. It is known (from the Paley-Wiener theorem) that such a characteristic (because it equals zero over a nonzero range of frequencies) is unrealizable by a physical network, so that it becomes necessary to approximate it.

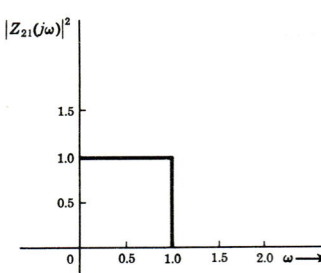

FIG. 5. Ideal low-pass filter characteristic.

The approximating function must approximate a constant in each of two ranges: unity in the range $0 \leq \omega < 1$ and zero for $\omega > 1$. Thus if the function

$$|Z_{21}(j\omega)|^2 = 1/[1 + A_n(\omega^2)] \tag{1}$$

is used, it is necessary that

$$\begin{aligned} A_n \ll 1 & \quad 0 \leq \omega < 1 \\ A_n \gg 1 & \quad \omega > 1 \end{aligned} \tag{2}$$

Since the squared magnitude must be an even function of ω, it is clear why A_n is chosen as a function of ω^2.

Butterworth suggested that $A_n = \omega^{2n}$ be used as an approximation. Consequently, the function

$$|Z_{21}|^2 = 1/(1 + \omega^{2n}) \tag{3}$$

is called a *Butterworth function;* it is reasonable to name the denominator polynomials derived from it *Butterworth polynomials* and the network realizations *Butterworth filters.* The characteristic is also called *maximally flat* for reasons that are made clear in the discussion of Butterworth polynomials. Finally, as pointed out above, in amplifier design a network with this characteristic is called a *flat-staggered n-tuple.*

For ω much smaller than unity, the value of the Butterworth function is approximately 1, but for large values of ω,

$$|Z_{21}|^2 \cong 1/\omega^{2n} \tag{4}$$

which approaches zero more and more closely as ω increases. Clearly, as the integer n is increased, the approximation is improved in both the passband and the stopband; for all values of n, however, the function is equal to $\frac{1}{2}$ at the point $\omega = 1$.

Sketches of the approximations given by the first three orders of the Butterworth function are shown in Fig. 6. It is observed that the characteristic is *monotonic* in both the passband and the stopband. It should be noted, furthermore, that as n increases, a larger part of the passband has small attenuation, and secondly, the slope of the curve in the *transition region* (or *attenuation interval*) of frequencies also increases, where the transition region is defined as the frequency range between the cutoff frequency $\omega = 1$ and the frequency at which a specified attenuation occurs. It is clear that a sharp-cutoff Butterworth filter requires a high value of n. The approximation close to $\omega = 0$ is observed to be very good, but there is an increased attenuation at the higher passband frequencies. This is true because, as shown later, the approximation is of the Taylor-series type; i.e., it is an approximation *about a point*. Since in filter applications we are interested in an approximation over a *band* of frequencies, this property is often undesirable.

The Butterworth approximation is useful for many applications. Its great advantage is mathematical simplicity, and it is shown later that in a certain sense it is a best approximation. However, for some applications it is unsuitable. Such applications may require a more uniform transmission of frequencies in the passband and a sharper rate of cutoff, or they may relate to the transient characteristic and specify a smaller overshoot in the response to a step-function input than is given by the Butterworth function.[5] Its overshoot increases with increasing n, exceeding 12 per cent for $n > 4$.

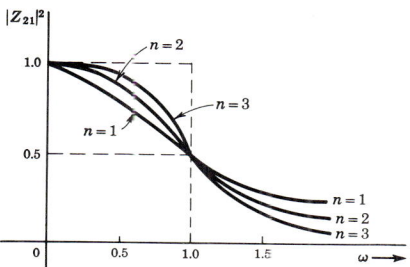

FIG. 6. Sketches of the first three orders of the Butterworth approximation to the low-pass filter.

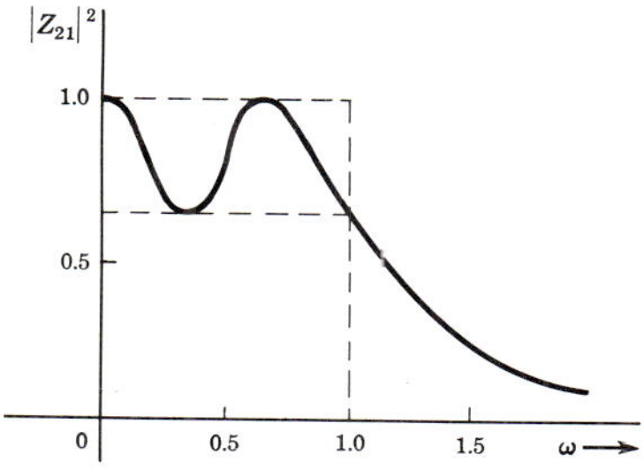

FIG. 7. An approximation with equal maxima and equal minima in the passband.

This behavior may be ascribed to the fact that the properties in the passband and the network complexity (i.e., the value of n) cannot be adjusted independently. If more uniform coverage of the passband or a sharper cutoff is desired, a large n must be used and the concomitant large overshoot must be accepted.

An approximation that gives a more uniform coverage throughout the passband (but no improvement in decreasing the overshoot of the step response) is the oscillating one shown in Fig. 7. This is an approximation over a band of frequencies which can be shown to reduce in the limit to the Butterworth approximation about a point ω as the width of the band approaches zero, i.e., as the band is reduced to the point ω_1. The magnitude response varies between equal maximum and equal minimum values in the passband and decreases monotonically outside it; the plot is thus said to have an *equal-ripple* (or *equiripple*) character in the passband. Such a characteristic with equal peaks and valleys may be obtained by the use of *Chebyshev polynomials*;[6] we define a Chebyshev polynomial by

$$T_n(\omega) = \cos(n \arccos \omega) \qquad (5)$$

in which n is a positive integer denoting the degree of the polynomial. The square of the product of this class of polynomials and a ripple factor ϵ^2 is substituted for the A_n of (1) to yield

$$|Z_{21}|^2 = 1/[1 + \epsilon^2 T_n^2(\omega)] \qquad (6)$$

The above filter function is exceedingly useful in applications where the magnitude of the transfer function is of primary concern. However, its nonlinear phase characteristic and the resulting variation of its time delay[7] preclude its use where a constant time delay is a paramount requirement. For such time-delay filters an excellent approximation is given by the use of *Bessel polynomials*, which are introduced later in this section.

Often equal ripples are desired in the stopband whereas the passband variation is permitted to be monotonic; such a characteristic may also be obtained with Chebyshev polynomials. If the function in (6) is subtracted from unity and $1/\omega$ is then substituted for ω, the new transfer impedance is

$$|Z_{21}'|^2 = \frac{\epsilon^2 T_n^2(1/\omega)}{1 + \epsilon^2 T_n^2(1/\omega)} \tag{7}$$

The subtraction from unity changes the plot of Fig. 7 to that in Fig. 8a; this curve is recognized as that of a *high-pass* filter, which is monotonic in its passband and has

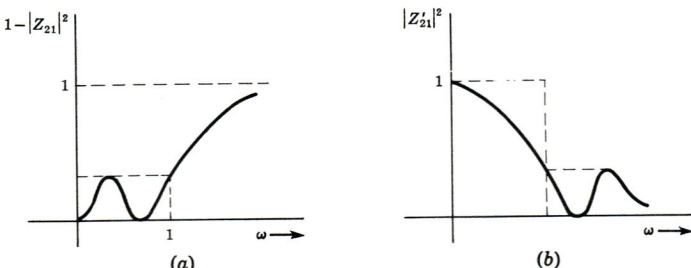

FIG. 8. Steps used in obtaining low-pass filter that is monotonic in the passband and equiripple in the stopband.

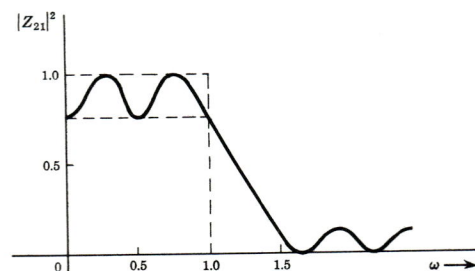

FIG. 9. Low-pass filter with an equiripple passband and an equiripple stopband.

equal ripples in its stopband. Now it is necessary to interchange the points $\omega = 0$ and $\omega = \infty$; this is accomplished by the transformation of ω to $1/\omega$, which produces the curve given in Fig. 8b. We call this an *inverse Chebyshev* characteristic, in order to distinguish it from the characteristic with an equiripple passband and a monotonic stopband. It should be noticed that the frequency normalization for this filter, as represented by (7), is not at the cutoff frequency but at the frequency where the characteristic first reaches a value equal to the peak of the stopband ripple.

The equal-ripple property may be required in both bands of a low-pass filter, as shown in Fig. 9. This may be achieved by use of the *Chebyshev rational functions* $R_n(\omega)$; Eq. (1) becomes

$$|Z_{21}|^2 = 1/[1 + \epsilon^2 R_n^2(\omega)] \tag{8}$$

Because elliptic functions are most often used in the design of such a filter, it is also called an *elliptic-function filter*.

FILTER AMPLIFIERS

If the ideal low-pass character of Fig. 5 is now relabeled as a time delay τ_d, where $\tau_d \equiv -d\beta/d\omega$, then the plot shown in Fig. 10 characterizes another class of problems. Since the time delay is constant in the passband, the phase β is linear; thus a linear phase characteristic is an alternative designation for the constant time delay. The constant time delay is ideal in the sense that, when coupled with the magnitude plot of Fig. 5, it will yield facsimile reproduction of an input signal whose frequencies are confined to the range $0 \leq \omega < 1$.

A monotonic approximation to the time delay of Fig. 10 is given by the use of *Bessel polynomials*. Just as the Butterworth function gives a maximally flat approximation to the magnitude characteristic, so the Bessel polynomials give a maximally flat time delay. A rough sketch of the first two orders of the approximation is shown in Fig. 11.

An important characteristic of all the classes of functions should be noted: each of them is unique. Stated in other terms, there is no class of polynomials other than the Butterworth that gives a maximally flat magnitude about the point $\omega = 0$; the Chebyshev polynomials are the only set of polynomials that yields a function with the equiripple property in one band and a monotonic variation in the other; the Chebyshev functions are the rational functions of lowest order that yield equal ripples in both the

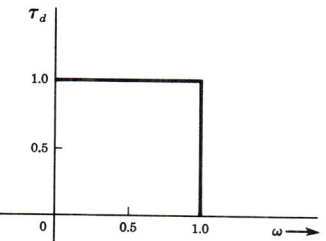

FIG. 10. Time-delay characteristic of a low-pass filter.

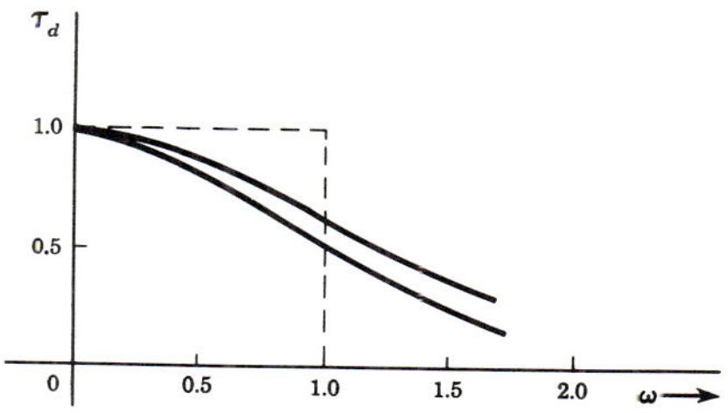

FIG. 11. Maximally flat time-delay approximations given by the first two orders of the Bessel polynomials.

passband and attenuation band, with a specified slope in the transition region; and the Bessel polynomials are the only polynomials that yield a transfer function with a maximally flat time delay. Other properties of the approximations are considered below.

2.1. Butterworth Polynomials. The property of *maximal flatness* may be understood in terms of the adjustment of the derivatives of a function, more specifically, as a process of setting the derivatives equal to zero. As mentioned previously, this concept has applicability not only to a magnitude characteristic but also to a phase characteristic. We may define the maximally flat function as one whose Maclaurin expansion has each of its first $2n - 1$ derivatives with respect to ω equal to zero.

When we require in addition to maximal flatness that all the zeros of the transfer function lie at infinity, that is, we restrict the reciprocal of the transfer function to be a simple polynomial, we then have as our maximally flat approximation

$$|Z_{21}(j\omega)|^2 = 1/(1 + d_{2n}\omega^{2n}) \tag{9}$$

If for convenience we now specify that the function equal $\tfrac{1}{2}$ at $\omega = 1$, we have the Butterworth function of order n,

$$|Z_{21}(j\omega)|^2 = 1/(1 + \omega^{2n}) \tag{10}$$

Thus we see clearly the sense in which the Butterworth function is optimum: for a transfer function whose numerator is a constant and whose magnitude is monotonic in the passband, the Butterworth function gives the flattest possible curve at the origin. It should also be clear that the higher the value of n (or what is equivalent, the larger the number of elements we are prepared to use in the network realization), the greater the degree of maximal flatness possible; but for a fixed n, the optimum flat adjustment is obtained when the maximally flat condition has been reached.

If we now substitute $s = j\omega$ in (10), we see that the poles of the Butterworth function are given by the roots of

$$s^{2n} + (-1)^n = 0 \tag{11}$$

Thus the poles are equally spaced on the unit circle; for example, for $n = 2$ the poles are shown in Fig. 12. We define $B_n(s)$, the Butterworth polynomial of order n, by

$$\frac{1}{1 + \omega^{2n}} = \frac{1}{B_n(s)B_n(-s)}\bigg|_{s=j\omega} \tag{12}$$

where $B_n(s)$ is an nth-degree polynomial that has all its zeros in the left half plane. The transfer function derived from (10) is then given as

$$Z_{21}(s) = 1/B_n(s) \tag{13}$$

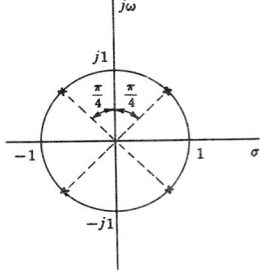

Fig. 12. Representation of poles of Butterworth function for $n = 2$, showing their uniform angular separation of $\pi/2$ radians.

This function multiplied by the constant factor H is used for the design of Butterworth filters.

Though the zeros of the Butterworth polynomials can be calculated easily with the aid of trigonometric tables, it is useful to have tables of the zeros readily available, and it is also useful to have the coefficients of $B_n(s)$ in unfactored form. The zeros and coefficients are given in Tables 1 and 2, respectively, for $n = 1, 2, \ldots, 10$.

2.2. Bessel Polynomials. A discussion similar to that for the general magnitude function applies for the time delay τ_d, which is also an even function of frequency. It can be shown that the adjustment of the coefficients of a transfer function for maximally flat delay leads to a class of polynomials called the *Bessel polynomials*, which are related to the Lommel polynomials. Thus the adjustment need not be laboriously carried out for each value of n.

The Bessel polynomials in the variable $1/s$ are defined by

$$w_n(1/s) = \sum_{k=0}^{n} (n + k)!/[(n - k)!k!(2s)^k] \tag{14}$$

or equivalently, by the recurrence relation

$$w_n(1/s) = [(2n - 1)/s]w_{n-1}(1/s) + w_{n-2}(1/s) \qquad n = 1, 2, 3, \ldots \tag{15}$$

with the initial values $w_{-1} = 1$ and $w_0 = 1$.

A polynomial in s may be defined in terms of the Bessel polynomials. This set of polynomials, which has direct applicability to the approximation problem, is defined by

$$y_n(s) = s^n w_n(1/s) \tag{16}$$

so that the equation for $y_n(s)$ obtained by substitution of (14) in (16) is

$$y_n(s) = \sum_{k=0}^{n} (2n - k)!s^k/[2^{n-k}k!(n - k)!] \tag{17}$$

The zeros of the polynomials $y_n(s)$ are all simple and lie in the left half plane, no y_n and y_{n+1} have a zero in common, and y_n has at most one real zero. Unlike the zeros

Table 1. Zeros of Butterworth Polynomials B_n

$n=1$	$n=2$	$n=3$	$n=4$	$n=5$	$n=6$	$n=7$	$n=8$	$n=9$	$n=10$
−1.0000000	−0.7071068 ± j0.7071068	−1.0000000	−0.3826834 ± j0.9238795	−1.0000000	−0.2588190 + j0.9659258	−1.0000000	−0.1950903 ± j0.9807853	−1.0000000	−0.1564345 ± j0.9876883
		−0.5000000 ± j0.8660254	−0.9238795 ± j0.3826834	−0.3090170 ± j0.9510565	−0.7071068 ± j0.7071068	−0.2225209 ± j0.9749279	−0.5555702 ± j0.8314696	−0.1736482 ± j0.9848078	−0.4539905 ± j0.8910065
				−0.8090170 ± j0.5877852	−0.9659258 ± j0.2588190	−0.6234898 ± j0.7818315	−0.8314696 ± j0.5555702	−0.5000000 ± j0.8660254	−0.7071068 ± j0.7071068
						−0.9009689 ± j0.4338837	−0.9807853 ± j0.1950903	−0.7660444 ± j0.6427876	−0.8910065 ± j0.4539905
								−0.9396926 ± j0.3420201	−0.9876883 ± j0.1564345

Table 2. Coefficients of Butterworth Polynomials $B_n = s^n + a_{n-1}s^{n-1} + \ldots + a_2 s^2 + a_1 s + 1$

n	a_1	a_2	a_3	a_4	a_5	a_6	a_7	a_8	a_9
1	1.4142136								
2	2.0000000	2.0000000							
3	2.6131259	3.4142136	2.6131759						
4	3.2360680	5.2360680	5.2360680	3.2360680					
5	3.8637033	7.4641016	9.1416202	7.4641016	3.8637033				
6	4.4939592	10.0978347	14.5917939	14.5917939	10.0978347	4.4939592			
7	5.1258309	13.1370712	21.8461510	25.6883559	21.8461510	13.1370712	5.1258309		
8	5.7587705	16.5817187	31.1634375	41.9863857	41.9863857	31.1634375	16.5817187	5.7587705	
9	6.3924532	20.4317291	42.8020611	64.8823963	74.2334292	64.8823963	42.8020611	20.4317291	6.3924532

5–9

Table 3. Exact Coefficients of the Polynomials $y_n(s) = s^n + a_{n-1}s^{n-1} + a_{n-2}s^{n-2} + \cdots + a_1s + a_0$ Used for Maximally Flat Time-delay Networks

n	a_0	a_1	a_2	a_3	a_4	a_5	a_6	a_7	a_8	a_9	a_{10}
1	1										
2	3	3									
3	15	15	6								
4	105	105	45	10							
5	945	945	420	105	15						
6	10,395	10,395	4,725	1,260	210	21					
7	135,135	135,135	62,370	17,325	3,150	378	28				
8	2,027,025	2,027,025	945,945	270,270	51,975	6,930	630	36			
9	34,459,425	34,459,425	16,216,200	4,729,752	945,945	135,135	13,860	990	45		
10	654,729,075	654,729,075	310,134,825	91,891,800	18,918,900	2,837,835	315,315	25,740	1,485	55	
11	13,749,310,575	13,749,310,575	6,547,290,750	1,964,187,225	413,513,100	64,324,260	7,567,560	675,675	45,045	2,145	66

Table 4. Zeros of Polynomials $y_n(s) = s^n w_n(1/s)$ Derived from the Bessel Polynomials for Values of n from 1 to 11

Zeros

n	
1	-1.0000000
2	$-1.5000000 \pm j0.8660254$
3	$-2.3221854;\ -1.8389073 \pm j1.7543810$
4	$-2.8962106 \pm j0.8672341;\ -2.1037894 \pm j2.6574180$
5	$-3.6467386;\ -3.3519564 \pm j1.7426614;\ -2.3246743 \pm j3.5710229$
6	$-4.2483594 \pm j0.8675097;\ -3.7357084 \pm j2.6262723;\ -2.5159322 \pm j4.4926730$
7	$-4.9717869;\ -4.7582905 \pm j1.7392861;\ -4.0701392 \pm j3.5171740;\ -2.6856769 \pm j5.4206941$
8	$-5.5878860 \pm j0.8676144;\ -2.8389840 \pm j6.3539113;\ -5.6044218 \pm j3.4981573;\ -4.3682892 \pm j4.4144425;\ -5.2048408 \pm j2.6161751$
9	$-6.2970193;\ -6.1293679 \pm j1.7378484;\ -5.6044218 \pm j3.4981573;\ -4.6384399 \pm j5.3172717;\ -2.9792608 \pm j7.2914637$
10	$-6.9220449 \pm j0.8676651;\ -3.1089162 \pm j8.2326995;\ -6.6152916 \pm j2.6115683;\ -5.9657282 \pm j4.3849471;\ -4.8862195 \pm j6.2249855$
11	$-7.6223398;\ -6.3013375 \pm j5.2761917;\ -5.1156483 \pm j7.1370208;\ -7.4842299 \pm j1.737028;\ -7.0578924 \pm j3.4890145;$
	$-3.2297221 \pm j9.1771116$

of the Butterworth polynomials $B_n(s)$, the zeros of $y_n(s)$ cannot be obtained in any simple manner but must be found by the usual process of solving a polynomial for its roots. It is useful to recognize, however, that the zeros very nearly lie on concentric circles;[5] the centers of these circles for successive values of n are equally spaced along the positive σ axis, the radii of successive circles (with respect to their centers) differing by equal amounts.

The first three polynomials evaluated by (17) are

$$\begin{aligned} y_1 &= s + 1 \\ y_2 &= s^2 + 3s + s \\ y_3 &= s^3 + 6s^2 + 15s + 15 \end{aligned} \tag{18}$$

Though the polynomials for higher n can be evaluated similarly, it is convenient to have the coefficients tabulated; in addition, the zeros are often useful in design. Tables of the coefficients and the zeros for $n = 1, 2, \ldots, 11$ are given as Tables 3 and 4, respectively.

The above polynomials, when used as the denominator of a transfer function, for example,

$$Z_{21} = H/y_n(s) \tag{19}$$

possess a time delay that is maximally flat. As two illustrations we consider y_1 and y_2, for which, respectively,

$$-\beta_1 = \arctan \omega \qquad -\beta_2 = \arctan [3\omega/(3 - \omega^2)] \tag{20}$$

Therefore,

$$\tau_{d_1} = -\frac{d\beta_1}{d\omega} = \frac{1}{1 + \omega^2} \qquad \tau_{d_2} = -\frac{d\beta_2}{d\omega} = \frac{9 + 3\omega^2}{9 + 3\omega^2 + \omega^4} \tag{21}$$

Each of the above time delays is evidently maximally flat; this is evident when one recalls that the required Maclaurin series is obtained by a simple long division.

From the above functions we see that the delay is very closely equal to the zero-frequency value up to a certain value of ω (which is an increasing function of n), and then declines smoothly for values greater than this frequency. A useful tabulation of the variation of the normalized frequency u with n is given in Table 5a, where $u = \omega t_0 = \omega/\omega_0$ and t_0 is the zero-frequency delay. Since the transfer function is the reciprocal of a polynomial whose zeros are all in the left half plane, it is clear that the magnitude characteristic associated with the maximally linear phase is a low-pass one; frequencies at which particular values of loss occur are given in Table 5b for $n = 1, 2, \ldots, 11$. These tables are most often adequate for solving practical problems.

It has been remarked that the Butterworth function is not the most efficient polynomial approximation over a *band* of frequencies; the optimum approximation given by Chebyshev polynomials is considered below. Similarly, Bessel polynomials, though yielding a good and useful approximation, are not the most efficient time-delay approximation.[8] however, no class of polynomials has been found to yield the optimum approximation.

2.3. Transitional Butterworth-Thomson Characteristic.[9] We now introduce a characteristic that is derived by merging the desirable properties of the Butterworth function and the function that uses Bessel polynomials. The discussion is brief; the interested reader should consult the reference for further details.

We have previously remarked that Butterworth filters exhibit excessive transient overshoot in their response to a unit-step input. However, the response is characterized by a good rise time. More specifically, the overshoot and rise time increase with the order of the filter; this does not lead to faithful reproduction of an input signal. Numerical values of these parameters are given for the first five orders of Butterworth filter in Table 6.

Table 5. Significant Values of u for Time Delay and Loss Characteristic of a Maximally Flat Time-delay Network

a. Time-delay Table Giving Frequencies u at Which Time Delay Deviates a Specified Value from Its Zero-frequency Value

n	u for 1% deviation	u for 10% deviation	u for 20% deviation	u for 50% deviation
1	0.10	0.34	0.50	1.00
2	0.56	1.09	1.39	2.20
3	1.21	1.94	2.29	3.40
4	1.93	2.84	3.31	4.60
5	2.71	3.76	4.20	5.78
6	3.52	4.69	5.95	6.97
7	4.36	5.64	6.30	8.15
8	5.22	6.59	7.30	9.33
9	6.08	7.55	8.31	10.50
10	6.96	8.52	9.33	11.67
11	7.85	9.49	10.34	12.84

b. Loss ($L = -20 \log |Z_{21}(ju)|$, in Decibels) Table Giving Frequencies u at Which Loss Is a Specified Number of Decibels Down from Its Zero-frequency Value

n	u for $\frac{1}{50}$ db	u for $\frac{1}{20}$ db	u for $\frac{1}{10}$ db	u for $\frac{1}{5}$ db	u for $\frac{1}{2}$ db	u for 1 db	u for 3 db
1	0.07	0.11	0.14	0.21	0.35	0.51	1.00
2	0.11	0.18	0.26	0.36	0.57	0.80	1.36
3	0.14	0.23	0.34	0.48	0.75	1.05	1.75
4	0.17	0.28	0.40	0.56	0.89	1.25	2.13
5	0.20	0.32	0.45	0.64	1.01	1.43	2.42
6	0.22	0.36	0.50	0.71	1.12	1.58	2.70
7	0.24	0.39	0.54	0.77	1.22	1.72	2.95
8	0.26	0.41	0.59	0.83	1.31	1.85	3.17
9	0.28	0.44	0.62	0.88	1.40	1.97	3.39
10	0.30	0.47	0.66	0.93	1.48	2.08	3.58
11	0.31	0.49	0.69	0.98	1.55	2.19	3.77

Table 6. Rise Time and Overshoot of Maximally Flat n-Pole Networks

n	Overshoot, %	Rise time, sec
1	0	2.20
2	4.3	2.15
3	8.15	2.29
4	10.9	2.43
5	12.8	2.56

FILTER AMPLIFIERS

The filters with maximally flat time delay, on the other hand, have an overshoot that is less than 1 per cent and decreases for order $n > 4$, tending to zero for large values of n. This is illustrated in Table 7. It is also observed from the table that the rise time is exceedingly slow.

Table 7. Rise Time and Overshoot of Maximally Flat Time-delay Networks

n	Overshoot, %	Rise time, sec
1	0	2.20
2	0.43	2.73
3	0.75	3.07
4	0.83	3.36
5	0.76	3.58

It is often desirable in practice to obtain a better transient response (with respect to overshoot and rise time) than is given by either the maximally flat magnitude or the maximally flat time-delay networks. Specifications often call for a smaller overshoot than is given by the former and a faster rise time than is exhibited by the latter. Some specifications frequently couple with these transient requirements a stipulation on the bandwidth of the steady-state magnitude characteristic. As is clear from observation of Table 5, the bandwidth of the magnitude characteristic of the maximally flat time-delay functions is much less than the bandwidth over which its phase is linear; the Butterworth function for corresponding n has a uniform magnitude response over a larger bandwidth than the Bessel polynomial function.

The question arises whether a merging of the two functions can be devised, that is, whether a set of functions can be found that have steady-state and transient characteristics that lie between those of maximally flat magnitude and maximally flat time-delay functions. This question has been answered affirmatively by the derivation of sets of polynomial filters whose characteristics vary smoothly from those of the Butterworth to those of the Bessel polynomial filters as the parameter controlling the pole position is varied.

This is done by defining the parameter m, which prescribes a suitable path passing through the poles of the Butterworth and Bessel polynomial functions. The radius vector r and the phase φ of the new poles are defined by

$$r = r_T{}^m \qquad \varphi = \varphi_B - m(\varphi_B - \varphi_T) \tag{22}$$

where the subscript B stands for Butterworth and the subscript T for maximally flat time delay, or for Thomson (who did a good deal of the original work on the maximally flat time-delay filters). For each value of m, a new set of polynomials is defined. In the reference, the new filters are designated as *transitional Butterworth-Thomson filters*.

It is clear from the definition in Eq. (22) that when $m = 1$ and $m = 0$, the maximally flat time-delay filter and the maximally flat magnitude filter are obtained, respectively. It is shown in the reference that the bandwidth of the magnitude characteristic goes up as m is decreased from unity, whereas the phase is maximally linear for $m = 1$ and becomes less linear as m decreases. In addition, the rise time becomes longer and the overshoot smaller as m is increased from zero.

Curves (for functions from the second through the fifth order) are given in the reference that summarize the data on the bandwidth, rise time, and overshoot for selected values of m. These curves illustrate that a large variety of filter specifications can be satisfied by an appropriate choice of m. Thus the transitional Butterworth-Thomson filters will probably find increasing applications.

2.4. Equal-ripple Property: Chebyshev Polynomials. An optimum function to approximate a low-pass characteristic is one of lowest order, with a magnitude characteristic that does not exceed a prescribed maximum deviation in the passband and has the fastest possible rate of cutoff outside the passband. An equivalent way of stating the latter part of the requirement is that the transition interval for reaching a prescribed attenuation is a minimum and that the attenuation in the stopband is never less than this prescribed attenuation. In terms of the symbols in Fig. 13, the gain in the passband is not less than g_{min} and in the stopband is not greater than g_a, and the transition interval $\omega_a - 1$ is a minimum; the transfer function of minimum order is required.

In some practical problems, the order n may be given and the minimum $\omega_a - 1$ is to be determined; this is an equivalent problem. In fact, for the specification of any three of the parameters—minimum gain in the passband, maximum gain in the stopband, width of transition interval, complexity of the function as denoted by n—minimizing the fourth is automatically accomplished by the use of the appropriate optimum function.

Two classes of optimum functions must be distinguished. When the transfer function is restricted to have all its zeros at infinity, that is, A_n is required to be a poly-

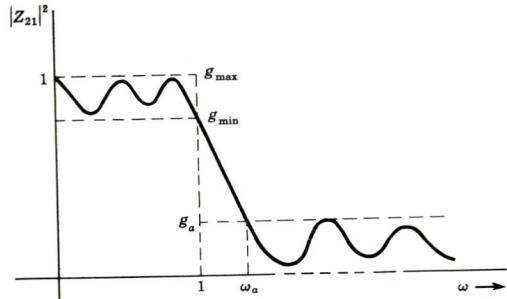

FIG. 13. Low-pass filter with prescribed limits of passband and stopband gain.

nomial, then the optimum function is given by use of Chebyshev polynomials. In this case A_n is set equal to $\epsilon^2 T_n^2(\omega)$. The characteristic then has equal ripples in the passband and is monotonic in the stopband, as previously shown in Fig. 7. Thus, of all possible transfer functions whose zeros all lie at infinity, the transfer function obtained by the use of Chebyshev polynomials is optimum in the sense that it is the function of lowest order for achieving a prescribed maximum deviation in the passband and the fastest possible rate of cutoff outside the passband.

When the transfer function is permitted to have finite zeros, that is, A_n may be a general rational function, then the optimum function becomes the Chebyshev rational function. For this case, $A_n = \epsilon^2 R_n^2(\omega)$, and the characteristic has equal ripples in both the passband and stopband, an example of which was shown in Fig. 9. As was previously indicated, these polynomials and rational functions can be shown to be unique: no other polynomials and rational functions with their respective optimum

Table 8. Chebyshev Polynomials of Order 1 to 10

n	$T_n(\omega)$
1	ω
2	$2\omega^2 - 1$
3	$4\omega^3 - 3\omega$
4	$8\omega^4 - 8\omega^2 + 1$
5	$16\omega^5 - 20\omega^3 + 5\omega$
6	$32\omega^6 - 48\omega^4 + 18\omega^2 - 1$
7	$64\omega^7 - 112\omega^5 + 56\omega^3 - 7\omega$
8	$128\omega^8 - 256\omega^6 + 160\omega^4 - 32\omega^2 + 1$
9	$256\omega^9 - 576\omega^7 + 432\omega^5 - 120\omega^3 + 9\omega$
10	$512\omega^{10} - 1,280\omega^8 + 1,120\omega^6 - 400\omega^4 + 50\omega^2 - 1$

properties exist. Knowledge of these upper bounds on performance is extremely valuable; for example, it permits the determination of how close the characteristic of a given network is to the best that can be achieved, and it shows the futility of searching for other approximating polynomials that will give better performance (as judged by its magnitude characteristic) than the Chebyshev polynomials of the same degree.

The Chebyshev polynomials of order 1 through 10 are given in Table 8. Some characteristics of the polynomials are immediately evident. They are even and odd

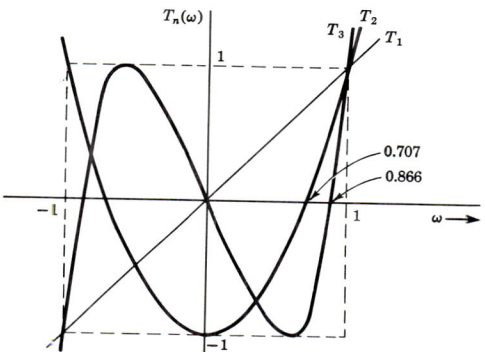

FIG. 14. Sketches of the first three Chebyshev polynomials.

functions for n even and odd, respectively. Every coefficient is an integer, the coefficient of the term of highest degree being 2^{n-1}. For n even, the absolute value of each polynomial is unity for $\omega = 0$ and $\omega = \pm 1$, and in the range $-1 \leq \omega \leq 1$ all the polynomials have the equal-ripple property, varying between a maximum of 1 and a minimum of -1. Sketches of the polynomials of the first three orders are shown in Fig. 14.

Since the polynomials take on negative as well as positive values with a maximum absolute value of unity in the passband $-1 \leq \omega \leq 1$, they are not suitable by themselves to approximate a magnitude function with a value of unity in the passband. We therefore use a ripple factor ϵ^2 as a multiplier of the square of the Chebyshev polynomial in order to limit the amplitude of its oscillation.

Because of the above properties of the Chebyshev polynomials, it is clear why the approximation is an equal-ripple one, i.e., has equal maxima and equal minima in the passband. The total number of troughs and peaks for positive ω is equal to n. Or alternatively, considering both positive and negative ω, we note that the number of peaks within the passband is equal to n, whereas the number of troughs is equal to $n - 1$; outside the band the magnitude decreases monotonically. At the edge of the passband, i.e., at the cutoff frequency $\omega = 1$, the magnitude goes through a minimum point. To illustrate these remarks about the Chebyshev approximation, a sketch of Eq. (6) for $n = 3$ and 1-db ripples is shown in Fig. 15; the 1-db ripple corresponds to $\epsilon^2 = 0.26$ or $\epsilon = 0.51$.

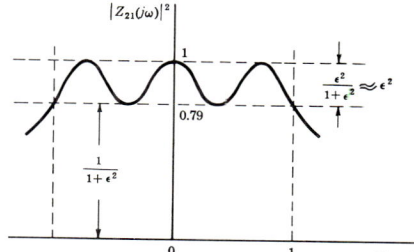

FIG. 15. Low-pass filter obtained by using the Chebyshev polynomial approximation with $n = 3$ and a 1-db ripple.

To find the complete transfer function

$$Z_{21} = H/V_n(s) \tag{23}$$

Table 9. Zeros of Polynomials V_n Derived from the Chebyshev Approximation

a. ½-db Ripple ($\epsilon = 0.3493114$, $\epsilon^2 = 0.1220184$)

$n=1$	$n=2$	$n=3$	$n=4$	$n=5$	$n=6$	$n=7$	$n=8$	$n=9$	$n=10$
−2.8627752	−0.7128122 ± j1.0040425	−0.6264565	−0.1753531 ± j1.0162529	−0.3623196	−0.0776501 ± j1.0084608	−0.2561700	−0.0436201 ± j1.0050021	−0.1984053	−0.0278994 ± j1.0032732
		−0.3132282 ± j1.0219275	−0.4233398 ± j0.4209457	−0.1119629 ± j1.0115574	−0.2121440 ± j0.7382446	−0.0570032 ± j1.0064085	−0.1242195 ± j0.8519996	−0.0344527 ± j1.0040040	−0.0809672 ± j0.9950658
				−0.2931227 ± j0.6251768	−0.2897940 ± j0.2702162	−0.1597194 ± j0.8070770	−0.1859076 ± j0.5692879	−0.0992026 ± j0.8829063	−0.1261094 ± j0.7182643
						−0.2308012 ± j0.4478939	−0.2192929 ± j0.1999073	−0.1519873 ± j0.6553170	−0.1589072 ± j0.4611541
								−0.1864400 ± j0.3486869	−0.1761499 ± j0.1589029

b. 1-db Ripple ($\epsilon = 0.5088471$, $\epsilon^2 = 0.2589254$)

$n=1$	$n=2$	$n=3$	$n=4$	$n=5$	$n=6$	$n=7$	$n=8$	$n=9$	$n=10$
−1.9652267	−0.5488672 ± j0.8951286	−0.4941706	−0.1395360 ± j0.9833792	−0.2894933	−0.0621810 ± j0.9934115	−0.2054141	−0.0350082 ± j0.9964513	−0.1593305	−0.0224144 ± j0.9977755
		−0.2470853 ± j0.9659987	−0.3368697 ± j0.4073290	−0.0894584 ± j0.9901071	−0.1698817 ± j0.7272275	−0.0457089 ± j0.9952839	−0.0996950 ± j0.8447506	−0.0276674 ± j0.9972297	−0.1013166 ± j0.7143284
				−0.2342050 ± j0.6119198	−0.2320627 ± j0.2661837	−0.1280736 ± j0.7981557	−0.1492041 ± j0.5644443	−0.0796652 ± j0.8769490	−0.0650493 ± j0.9001063
						−0.1850717 ± j0.4429430	−0.1759983 ± j0.1982065	−0.1220542 ± j0.6508954	−0.1276664 ± j0.4586271
								−0.1497217 ± j0.3463342	−0.1415193 ± j0.1580321

5-16

c. 2-db Ripple ($\epsilon = 0.7647831$, $\epsilon^2 = 0.5848932$)

−1.3075603	−0.4019082 ± j0.6893750				
	−0.3689108	−0.1048872 ± j0.9579530	−0.2183083	−0.1206298	−0.0169758 ± j0.9934868
		−0.1844554 ± j0.9230771	−0.0674610 ± j0.9734557	−0.0469732 ± j0.9817052	−0.0264924 ± j0.9897870	−0.1552958	0.0767332 ± j0.7112580
		−0.2532202 ± j0.3967971	−0.1283332 ± j0.7186581	−0.0754139 ± j0.8391009	−0.0345566 ± j0.9866139	0.0209471 ± j0.9919471	
			−0.1766151 ± j0.6016287	−0.1129098 ± j0.5606693	−0.0968253 ± j0.7912029	−0.0603149 ± j0.8723036	−0.0492657 ± j0.8962374
			−0.1753064 ± j0.2630471	−0.1331862 ± j0.1968809	−0.1399167 ± j0.4390845	−0.0924078 ± j0.6474475	−0.0966894 ± j0.4566558
					−0.1133549 ± j0.3444996	−0.1071810 ± j0.1573528

d. 3-db Ripple ($\epsilon = 0.9976283$, $\epsilon^2 = 0.9952623$)

−1.0023773	−0.3224498 ± j0.7771576				
	−0.2986202	−0.0851704 ± j0.9464844	−0.1775085	−0.0982716	−0.0138320 ± j0.9915418
		−0.1493101 ± j0.9038144	−0.2056195 ± j0.3920467	−0.0548531 ± j0.9659238	−0.0382295 ± j0.9764060	−0.0215782 ± j0.9867664	−0.1264854	−0.0401419 ± j0.8944827
			−0.1044450 ± j0.7147788	−0.0614494 ± j0.8365401	−0.0281456 ± j0.9826957	
			−0.1436074 ± j0.5969738	−0.1426745 ± j0.2616272	−0.0788623 ± j0.7880608	−0.0919655 ± j0.5589582	−0.0491358 ± j0.8701971	−0.0625225 ± j0.7098655
					−0.1139594 ± j0.4373407	−0.1084807 ± j0.1962800	−0.0752804 ± j0.6458839	−0.0787829 ± j0.4557617
						−0.0923451 ± j0.3436677	−0.0873316 ± j0.1570448

5–17

Table 10. Polynomials V_n in Expanded Form $V_n = s^n + b_{n-1}s^{n-1} + \cdots + b_1 s + b_0$

a. ½-db Ripple ($\varepsilon = 0.3493114$, $\varepsilon^2 = 0.1220184$)

n	b_0	b_1	b_2	b_3	b_4	b_5	b_6	b_7	b_8	b_9
1	2.8627752									
2	1.5162026	1.4256245								
3	0.7156938	1.5348954	1.2529130							
4	0.3790506	1.0254553	1.7168662	1.1973856						
5	0.1789234	0.7525181	1.3095747	1.9373675	1.1724909					
6	0.0947626	0.4323669	1.1718613	1.5897635	2.1718446	1.1591761				
7	0.0447309	0.2820722	0.7556511	1.6479029	1.8694079	2.4126510	1.1512176			
8	0.0236907	0.1525444	0.5735604	1.1485894	2.1840154	2.1492173	2.6567498	1.1460801		
9	0.0111827	0.0941198	0.3408193	0.9836199	1.6113880	2.7814990	2.4293297	2.9027337	1.1425705	
10	0.0059227	0.0492855	0.2372688	0.6269689	1.5274307	2.1442372	3.409268	2.7097415	3.1498757	1.1400664

b. 1-db Ripple ($\varepsilon = 0.5088471$, $\varepsilon^2 = 0.2589254$)

n	b_0	b_1	b_2	b_3	b_4	b_5	b_6	b_7	b_8	b_9
1	1.9652267									
2	1.1025103	1.0977343								
3	0.4913067	1.2384092	0.9883412							
4	0.2756276	0.7426194	1.4539248	0.9528114						
5	0.1228267	0.5805342	0.9743961	1.6888160	0.9368201					
6	0.0689069	0.3070808	0.9393461	1.2021409	1.9308256	0.9282510				
7	0.0307066	0.2136712	0.5486192	1.3575440	1.4287930	2.1760778	0.9231228			
8	0.0172267	0.1073447	0.4478257	0.8468243	1.8369024	1.6551557	2.4230264	0.9198113		
9	0.0076767	0.0706048	0.2441864	0.7863109	1.2016071	2.3781188	1.8814798	2.6709468	0.9175476	
10	0.0043067	0.0344971	0.1824512	0.4553892	1.2444914	1.6129856	2.9815094	2.1078524	2.9194657	0.9159320

5–18

c. 2-db Ripple ($\varepsilon = 0.7647831$, $\varepsilon^2 = 0.5848932$)

1	1.3075603					
2	0.6367681	0.8038164				
3	0.3268901	1.0221903	0.7378216			
4	0.2057651	0.5167981	1.2564819	0.7162150		
5	0.0817225	0.4593491	0.6934770	1.4995133	0.7064606	
6	0.0514413	0.2102706	0.7714618	0.8670149	1.7458587	0.7012257
7	0.0204228	0.1660920	0.3825056	1.1444390	1.0392203	1.9935272	0.6978929
8	0.0128603	0.0729373	0.3587043	0.5982214	1.5795807	1.2117121	2.2422529	0.6960646
9	0.0051076	0.0543756	0.1684473	0.6444677	0.8568648	2.0767479	1.3837464	2.4912897	0.6946793
10	0.0032151	0.0233347	0.1440057	0.3177560	1.0389104	1.1585287	2.6362507	1.5557424	2.7406032	0.6936904

d. 3-db Ripple ($\varepsilon = 0.9976283$, $\varepsilon^2 = 0.9952623$)

1	1.0023773					
2	0.7079478	0.6448996				
3	0.2505943	0.9283480	0.5972404			
4	0.1769869	0.4047679	1.1691176	0.5815799		
5	0.0626391	0.4079421	0.5488626	1.4149847	0.5744296	
6	0.0442467	0.1634299	0.6990977	0.6906098	1.6628481	0.5706979
7	0.0156621	0.1461530	0.3000167	1.0518448	0.8314411	1.9115507	0.5684201
8	0.0110617	0.0564813	0.3207646	0.4718990	1.4666990	0.9719473	2.1607148	0.5669476
9	0.0039154	0.0475900	0.1313851	0.5834984	0.6789075	1.9438443	1.1122863	2.4101346	0.5659234
10	0.0027654	0.0180313	0.1277560	0.2492043	0.9499208	0.9210659	2.4834205	1.2526467	2.6597378	0.5652218

from the squared magnitude of the Chebyshev characteristic requires that we determine its poles, i.e., the roots of

$$1 + \epsilon^2 T_n^2(\omega) = 0 \tag{24}$$

Though the poles may be evaluated with the aid of tables of trigonometric and hyperbolic functions, it is useful to have these poles tabulated. This tabulation has been carried out and is given in Table 9 for various ripple values and for $n = 1, 2, \ldots, 10$. Since the unfactored form of the denominator is also often needed, the coefficients of these polynomials are tabulated in Table 10.

2.5. Inverse Chebyshev Characteristic. The properties of the transfer function of the inverse Chebyshev filter given by (7), which is rewritten below (without the prime on Z_{21}),

$$|Z_{21}(\omega)|^2 = \frac{\epsilon^2 T_n^2(1/\omega)}{1 + \epsilon^2 T_n^2(1/\omega)} \tag{25}$$

complement those of the ordinary Chebyshev filter. Whereas the latter is equiripple in the passband and maximally flat in the stopband, the former has a maximally flat magnitude in the passband and equal ripples in the stopband. The maximally flat property is evident from (25). Since $T_n^2(1/\omega)$ is a polynomial of degree $2n$ in the variable $1/\omega$, upon multiplication of numerator and denominator by ω^{2n}, both the numerator and denominator become polynomials in ω, with the denominator equal to ω^{2n} plus the numerator. If n is odd, the numerator is of degree $2(n - 1)$; if n is even, the numerator and denominator are both of degree $2n$, but the function is still maximally flat; that is, $2n - 1$ derivatives of its Maclaurin expansion are equal to zero.

Using Table 8 we have, as illustrations of the inverse Chebyshev filter for $n = 2$,

$$|Z_{21}|^2 = \frac{\epsilon^2(\omega^4 - 4\omega^2 + 4)}{\omega^4 + \epsilon^2(\omega^4 - 4\omega^2 + 4)}$$

$$= \frac{\epsilon^2(\omega^4 - 4\omega^2 + 4)}{(1 + \epsilon^2)\omega^4 - 4\epsilon^2\omega^2 + 4\epsilon^2} \tag{26}$$

and for $n = 3$,

$$|Z_{21}|^2 = \frac{\epsilon^2(9\omega^4 - 24\omega^2 + 16)}{\omega^6 + 9\epsilon^2\omega^4 - 24\epsilon^2\omega^2 + 16\epsilon^2} \tag{27}$$

The tabulations in Table 9 for the poles of the Chebyshev transfer function should be useful for computing those of the inverse Chebyshev function, since it is clear that the poles of (25) are given by the reciprocals of those for the Chebyshev transfer function.

2.6. Equal Ripple in Pass and Stopbands: Chebyshev Rational Functions. The form of the Chebyshev rational function was introduced into network theory by Cauer[10] and was later used by Norton.[11] Depending on whether it is odd or even, this function is represented by one of the two forms

$$R_n(\omega) = \frac{\omega(\omega_1^2 - \omega^2)(\omega_2^2 - \omega^2) \cdots (\omega_k^2 - \omega^2)}{(1 - \omega_1^2\omega^2)(1 - \omega_2^2\omega^2) \cdots (1 - \omega_k^2\omega^2)} \tag{28}$$

and

$$R_n(\omega) = \frac{(\omega_1^2 - \omega^2)(\omega_2^2 - \omega^2) \cdots (\omega_k^2 - \omega^2)}{(1 - \omega_1^2\omega^2)(1 - \omega_2^2\omega^2) \cdots (1 - \omega_k^2\omega^2)} \tag{29}$$

and will have the equal-ripple character in both the passband and the stopband.

The integer n determines the complexity of the function; specifically, it is equal to the degree of the numerator polynomial and hence the order of the rational function R_n. The integer k gives the number of ω^2 zeros (or poles or a suitable combination of the two types of critical frequencies) that must be specified. For n odd, $n = 2k + 1$, and for n even, we have $n = 2k$.

Inspection of $R_n(\omega)$ in (28) and (29) shows that its poles are the reciprocals of its zeros. Because of this reciprocal relationship between the zeros and poles, it is clear

that
$$R_n(1/\omega) = 1/R_n(\omega) \tag{30}$$
In words, the function $R_n(\omega)$ has the important property that its value at any frequency ω_1 in the range $0 \leq \omega < 1$ is the reciprocal of its value at the reciprocal frequency $1/\omega_1$ in the range $1 < \omega \leq \infty$. Therefore, if the critical frequencies can be found such that $R_n(\omega)$ has equal ripples in the passband, it will automatically have equal ripples in the stopband.

Consideration of Fig. 9 and Eq. (8), which for convenience is repeated below,
$$|Z_{21}|^2 = 1/[1 + \epsilon^2 R_n{}^2(\omega)] \tag{31}$$
shows that $|Z_{21}|^2$ is bounded and therefore cannot have any poles for real ω, and that

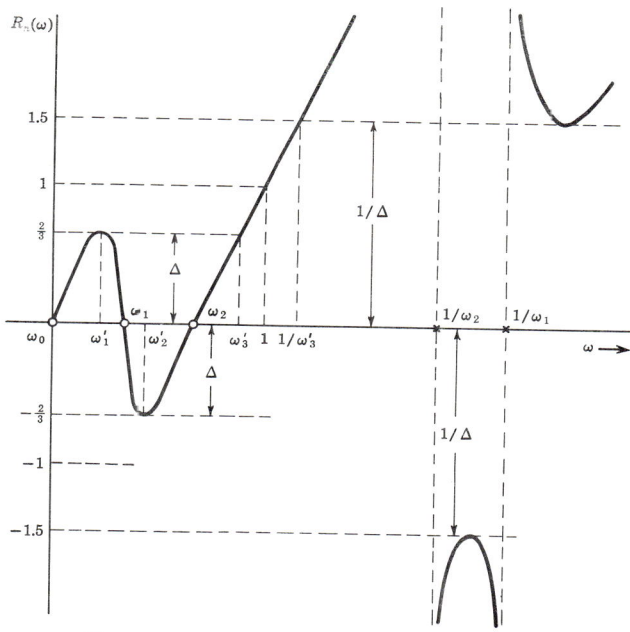

Fig. 16. Plot of odd $R_n(\omega)$ for $n = 5$ and $\Delta = \frac{2}{3}$.

$|Z_{21}|^2$ is never zero in the passband and thus its zeros for real ω must be outside the passband. But the poles of $R_n{}^2$ are the zeros of $|Z_{21}|^2$, and consequently each of the poles of R_n must be greater than unity. We thus conclude that R_n has zeros in the range of $0 \leq \omega < 1$ and poles in the complementary range $1 < \omega \leq \infty$. A sketch of an odd R_n is shown in Fig. 16 for $n = 5$ and for $\Delta = \frac{2}{3}$, where we let Δ designate the maximum amplitude about zero in the passband.

Now when the square of $R_n(\omega)$ is used in Eq. (31), it will clearly give equal ripples in the passband and equal ripples in the stopband; a plot is shown in Fig. 17, where ϵ^2 has been set equal to $\frac{1}{3}$. When the characteristic is plotted for negative and positive ω, it should be clear that the number of peaks in the passband will be equal to n, that is, equal to $2k + 1$ for odd R_n and equal to $2k$ for even R_n, whereas the number of troughs will be one less than the number of peaks.

As a concluding remark, we should observe how we get more flexibility (and consequent complexity) as we go from the Butterworth characteristic to the Chebyshev polynomial characteristic and end with the elliptic-function characteristic. This flexibility is shown by the increase in the number of parameters available to specify the filter. One parameter—the degree of the polynomial—fixes the Butterworth characteristic; two parameters—n, the degree of the Chebyshev polynomial, and ϵ^2, which determines the amplitude of the ripple—specify the Chebyshev polynomial filter; and specification of any three parameters of the four that can be specified—the order of the Chebyshev rational function, the passband ripple, the stopband ripple, and the width of the transition frequency range—is needed to determine the elliptic-function filter.

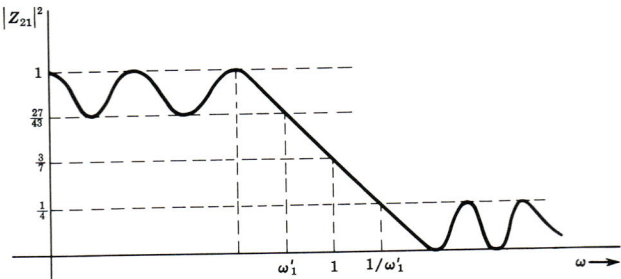

FIG. 17. Equal-ripple filter using the R_n of Fig. 16 and $\epsilon^2 = \frac{1}{3}$.

3. DIPOLE NETWORKS SYNTHESIZED BY RECOGNITION

In Sec. 1 we introduced the concept of a staggered n-tuple amplifier, that is, the design of a multistage amplifier with a simple one-port or two-port passive coupling network between the active element of one stage and the active element of the succeeding stage. In the next section we consider additional methods for the realization of a multistage amplifier with simple coupling networks. In this section we show that the design of the one-port networks can be carried out almost by inspection.

In general the design of a passive two-terminal network whose impedance or admittance is equal to a specified rational function requires knowledge of synthesis procedures. As a preliminary the given function must be checked to see that it satisfies the necessary and sufficient condition for realizability, namely, that it is a *positive real* function. A function $G(s)$ is defined to be positive real if it satisfies the following two conditions:

1. $G(s)$ is real for s real; that is, $G(\sigma)$ is purely real.
2. $\text{Re } [G(s)] \geq 0$ for $\text{Re } s \geq 0$.

If it is positive real, a realization procedure is then used to synthesize the network. However, for the realization of one-port networks that contain at most four elements, such as are useful in the design of multistage amplifiers, a simpler procedure is available; we discuss this below.

Since the basic volt-ampere relationships allow the recognition of single two-terminal elements, we are, in effect, already able to carry through simple problems in synthesis. If the reader were asked to synthesize a driving-point impedance equal to the constant 6, he would immediately draw a circuit consisting of a single resistance with a value of 6 ohms. He could, of course, also draw a circuit consisting of two resistances in parallel, each equal to 12 ohms; or further, he might, whimsically, use six resistances in a network of three parallel branches, each containing two 9-ohm resistances in series; and so on ad infinitum.

The different solutions may appear trivial, but they do illustrate the point that the solution to a synthesis problem may take an infinite number of forms; in a more complicated example the superfluous elements would not be so obvious. Similarly, the reader can already realize a driving-point admittance equal to $4s$ as a capacitance of

FILTER AMPLIFIERS

4 farads. If every driving-point function could be put in the form of an explicit sum of such elements, mere recognition could again serve for synthesis. Such synthesis could be called *synthesis by recognition*, or *direct synthesis*.

As illustrations, $Z = 4 + 3/s$ is a series connection of a resistance of 4 ohms with a capacitance of $\frac{1}{3}$ farad, while the three-element circuit $Y = 2 + s + 2/s$ is a parallel connection of a conductance of 2 mhos, a capacitance of 1 farad, and an inductance of $\frac{1}{2}$ henry. Although no synthesis procedure yields such simple sums of single elements, there are general procedures that decompose a driving-point function into sums of basic *two-element networks*, and it is well for the designer of amplifiers to be able to recognize these basic components. In fact, recognition of the rational functions corresponding to three- and four-element networks is also easy and useful.

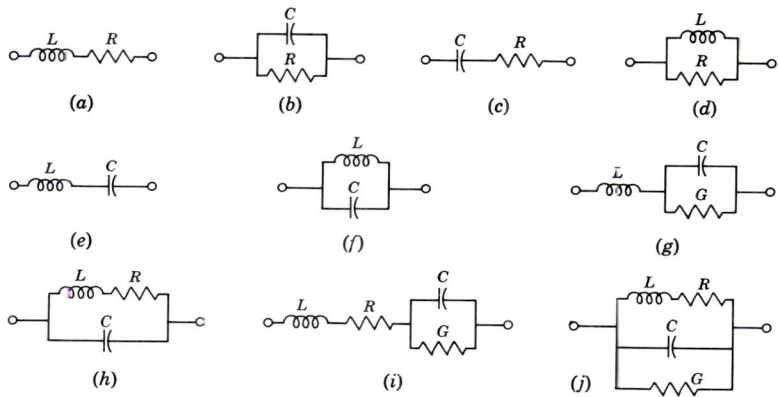

Fig. 18. Simple two-, three-, and four-element structures.

These basic substructures are shown in Fig. 18. First consider the RL two-element combination in Fig. 18a. The impedance is

$$Z = Ls + R = L(s + R/L) \qquad (32)$$

which yields the admittance

$$Y = 1/Z = 1/[L(s + R/L)] \qquad (33)$$

A function of the same mathematical form as that of (33) is given by the impedance of Fig. 18b:

$$Z = 1/[Cs + 1/R] = 1/[C(s + 1/RC)] \qquad (34)$$

It is clear that the function of Eq. (34) has a pole at $s = -1/RC$ and a zero at infinity; thus all the critical frequencies of both Z and Y lie on the negative real axis. The first basic two-element rational function of interest is therefore given by*

$$f_{RCZ} = H/(s - s_1) \qquad (35)$$

where s_1 is negative and H is the constant multiplier.

The other RC and RL two-element configurations also yield impedance and admittance functions with real critical frequencies; in these cases the origin becomes an

---

* We have used a subscript designation that refers to the *impedance* of the RC network, since the RC network is more important than the RL network in practice. In addition, the realization of any RL function can be obtained from the realization of an associated RC function; thus only the RC case need be considered in detail.

additional critical point. The admittance for Fig. 18c is

$$Y = s/[R(s + 1/RC)] \tag{36}$$

while the impedance for Fig. 18d is given by

$$Z = Rs/(s + R/L) \tag{37}$$

The next basic two-element function with critical points on the negative real axis and at the origin is

$$f_{RCY} = Hs/(s - s_1) \tag{38}$$

where, again, s_1 is negative. The subscript, it should be clear, serves as a reminder that the function represents the admittance of a basic RC network (consisting of two elements).

It will be noted that attention has been directed to the form of the rational function that has *finite nonzero* poles rather than to its reciprocal. The reason for this is that some procedures for synthesizing driving-point functions of two-element-kind networks are based on partial-fraction expansions, and this is the form in which the function will appear in such expansions. For this form of the rational function, it is important to note that the *admittance* corresponds to a *series* network (that is, the admittance of the basic series structure has finite nonzero poles) and the impedance corresponds to a *parallel* network.

The LC structures shown in Fig. 18e and f yield functions with critical frequencies on the j axis. For the series circuit in Fig. 18e the admittance is

$$Y = s/[L(s^2 + 1/LC)] \tag{39}$$

and the impedance of the parallel circuit is evaluated as

$$Z = s/[C(s^2 + 1/LC)] \tag{40}$$

These functions lead to the third basic two-element function

$$f_{LC} = Hs/(s^2 + \omega_1^2) \tag{41}$$

which has zeros at the origin and infinity and conjugate imaginary poles at $\pm j\omega_1$.

The above two-element structures have their critical frequencies restricted to a straight line in the s plane, either the nonpositive real or the imaginary axis. For a more general distribution of critical frequencies, we must go on to the three- and four-element structures shown in Fig. 18g to j. The networks shown in Fig. 18g and h may be looked upon as partial generalizations of Fig. 18e and f, respectively, in the sense that dissipation has been added to one of the reactive elements. The complete generalizations, with dissipation added to both reactances, are given in Fig. 18i and j.

As the reader can check, the structures in Fig. 18g to j lead to a fourth simple rational function of interest

$$f_{RLC} = \frac{H(s - s_1)}{(s - s_2)(s - s_3)} \tag{42}$$

in which s_1 is nonpositive, whereas s_2 is either complex or negative real. If s_2 is complex, then $s_3 = \bar{s}_2$. If s_2 is negative, then s_3 is also negative; in addition, $0 < -s_2 \leq -s_3 < -s_1$ or $0 \leq -s_1 < -s_2 \leq -s_3$, that is, the alternation of poles and zeros represented by $0 < -s_2 < -s_1 < -s_3$ is not possible. If G in Fig. 18i and R in Fig. 18j are set equal to zero, giving the three-element networks of Fig. 19, then a degenerate form of f_{RLC} results in which s_1 equals zero, namely,

$$f_{RLC} = \frac{Hs}{(s - s_2)(s - s_3)} \tag{43}$$

If f_{RLC} is written in the form

$$f_{RLC} = \frac{H(s+a)}{s^2 + bs + c} \tag{44}$$

then, as shown in the example at the end of this section, it is necessary that $a \leq b$ in order for the function to be positive real and thus realizable.

In network analysis it is customary to speak of resonance and antiresonance. For example, the frequency at which the impedance of Fig. 18e is zero is called the resonant frequency and that for which the impedance of Fig. 18f is infinite is called the antiresonant frequency. These terms can, of course, be extended to the other basic two-element configurations. It is clear, however, that the pole and zero terminology adequately characterizes all the structures.

In regard to synthesis, if a procedure has been used that separates a given driving-point expression into a sum of the above basic functions, we may find the connections and element values by working on each function separately. For f_{RCZ}, f_{RCY}, f_{LC}, and the degenerate form of f_{RLC}, we merely invert each function and carry out division to obtain a sum of single elements; then we make the appropriate series or parallel connection of the networks found for each function. In the case of a general f_{RLC} function, the simple division procedure yields a sum of simple elements plus a remainder which is an f_{RCZ} function.

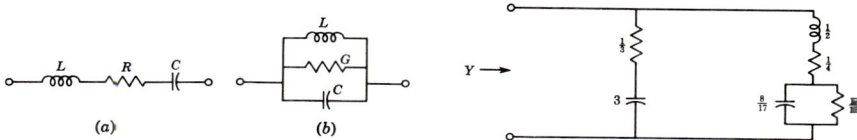

FIG. 19. Degenerate forms of the four-element configuration.

FIG. 20. Realization of admittance given in Eq. (45) (element values in ohms, henrys, farads).

As a check on the element values of f_{RCZ}, f_{RCY}, and f_{LC} functions, or as an alternative method of determining the element values by inspection, we can examine each function and its corresponding two-element network representation for very low and very high frequencies. For example, the asymptotic value of (36) as $s \to 0$ is $Y = Cs$, and the corresponding network in Fig. 18c reduces to the capacitance; thus the value of the capacitance may be determined. As $s \to \infty$, $Y = 1/R$ and the network reduces to the resistance. Similarly, (40) becomes $Z = Ls$ for $s \to 0$ and $Z = 1/Cs$ for $s \to \infty$, while the corresponding network in Fig. 18f becomes an inductance as $s \to 0$ and a capacitance as $s \to \infty$.

To illustrate the division process, we realize the admittance

$$Y = \frac{3s}{s+1} + \frac{2s+3}{s^2 + 2s + 5} = y_1 + y_2 \tag{45}$$

We then have

$$\frac{1}{y_1} = \frac{s+1}{3s} = \frac{1}{3} + \frac{1}{3s} \qquad \frac{1}{y_2} = \frac{1}{2} \frac{s^2 + 2s + 5}{s + \frac{3}{2}} \tag{46}$$

To decompose $1/y_2$, the division is carried out synthetically:

$$\begin{array}{ccc|c} 1 & 2 & 5 & -\frac{3}{2} \\ & -\frac{3}{2} & -\frac{3}{4} & \\ \hline 1 & \frac{1}{2} & \frac{17}{4} & \end{array}$$

so that

$$\frac{1}{y_2} = \frac{s}{2} + \frac{1}{4} + \frac{\frac{17}{4}}{2(s + \frac{3}{2})} = z_1 + z_2 + z_3 \tag{47}$$

Since the last term is of the f_{RCZ} type, it is also inverted to yield

$$\frac{1}{z_3} = \frac{8s}{17} + \frac{12}{17} \tag{48}$$

The element values are now evident by inspection, and the network appears in Fig. 20.

One type of function corresponding to the driving-point impedance of a four-element dipole remains to be considered, namely, the function whose form is given by the right side of (42) where s_2 is negative and, in addition, $0 < -s_2 < -s_1 < -s_3$. Since this function represents the driving-point impedance of an RC network, we designate it as a second RC function, that is,

$$f_{RCZ2} = \frac{H(s - s_1)}{(s - s_2)(s - s_3)} \tag{49}$$

In realizing this function we must also realize the three-element RC network whose admittance is given by

$$f_{RCY2} = \frac{s - s_1}{s - s_2} \tag{50}$$

where $0 < -s_1 < -s_2$.

To realize f_{RCZ2} there is a slight change in the procedure that applies for realizing f_{RLC}. We invert and divide as before but carry out only one step of division so that the quotient contains only one term instead of two. The remainder, which is the ratio of a linear term to a linear term, is now itself inverted and a division is then performed on this inverted remainder. The remainder after the division is an f_{RCZ} function, which is now treated in the usual manner. Thus the complete process requires that we first write the rational function with its polynomials arranged in descending order, and then carry out three cycles of a basic step that consists of an inversion and a division that yields one term as the quotient. It may be of incidental interest to know that this process, which is called a *continued-fraction expansion* about infinity, works on the general RC impedance of any order.

We illustrate the process by realizing the impedance

$$Z = \frac{s + 3}{(s + 2)(s + 4)} = \frac{s + 3}{s^2 + 6s + 8} \tag{51}$$

The continued-fraction expansion yields

$$\begin{array}{c}
s + 3 \overline{\smash{)}s^2 + 6s + 8} \\
\underline{s^2 + 3s} \\
3s + 8 \overline{\smash{)}s + 3} \\
\underline{s + \tfrac{8}{3}} \quad 9s \\
\tfrac{1}{3} \overline{\smash{)}3s + 8} \\
\underline{3s} \\
8 \overline{\smash{)}\tfrac{1}{3}} \\
\underline{\tfrac{1}{3}} \\
\end{array} \tag{52}$$

which allows the impedance to be written as

$$Z = \cfrac{1}{s + \cfrac{1}{\tfrac{1}{3} + \cfrac{1}{9s + \cfrac{1}{2\tfrac{1}{4}}}}} \tag{53}$$

The corresponding RC network is shown in Fig. 21.

We observe that after the first division the remainder is an f_{RCY2} function, namely,

$$Y_1 = \frac{3s+8}{s+3} \tag{54}$$

If we had continued to divide to obtain a quotient with two terms, as in the procedure for realizing f_{RLC}, we would have obtained a remainder that is negative and hence unrealizable; thus the reason for inverting before dividing is clear.

In the example the procedure for realizing the three-element f_{RCY2} is shown to be a part of the realization of the four-element f_{RCZ2}. The procedure simply consists of two cycles of the basic step of inversion and division. Thus from (52) we have immediately that the admittance Y_1 of Eq. (54) is given by

$$Y_1 = \cfrac{1}{\tfrac{1}{3} + \cfrac{1}{9s + \cfrac{1}{\tfrac{1}{24}}}} \tag{55}$$

The network is obtained by omitting the 1-farad capacitance in the network shown in Fig. 21; thus Y_1 is realized by the network in Fig. 22.

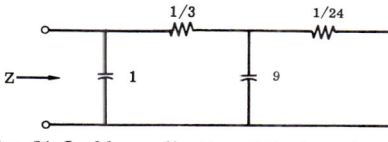
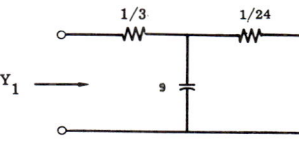

Fig. 21. Ladder realization of the impedance given in Eq. (51) (element values in ohms, henrys, and farads).

Fig. 22. Ladder realization of the admittance given in Eq. (54) (element values in ohms, henrys, and farads).

Example 1. Find networks corresponding to the following driving-point functions, interpreting each of them first as an impedance and then as an admittance:

$$2s/(s+3) \tag{1}$$
$$1/(s+4) \tag{2}$$
$$3s/(s^2+6) \tag{3}$$

Solution. Considering the functions as impedances,

$$Z_1 = 2s/(s+3) \qquad 1/Z_1 = 1/2 + 3/2s \tag{1}$$
$$Z_2 = 1/(s+4) \qquad 1/Z_2 = s+4 \tag{2}$$
$$Z_3 = 3s/(s^2+6) \qquad 1/Z_3 = s/3 + 2/s \tag{3}$$

The admittances will lead to the dual networks. All the networks are given in Fig. 23.

Example 2. Find the maximum value that the positive constant a can have in the driving-point impedance

$$Z = \frac{s+a}{s^2+3s+5}$$

in order for the impedance to be physically realizable. Make use of the fact that the most general configuration for such a rational function is given by Fig. 18j.

Solution. The admittance of the network in Fig. 18j is

$$Y = G + Cs + \frac{1}{Ls + R} = \frac{LCs^2 + (RC+GL)s + RG + 1}{Ls + R}$$
$$= \frac{C[s^2 + (R/L + G/C)s + RG/LC + 1/LC]}{s + R/L}$$

5-28   AMPLIFIER FUNDAMENTALS

which gives the impedance

$$Z = \frac{s + R/L}{C[s^2 + (R/L + G/C)s + RG/LC + 1/LC]}$$

The fraction R/L is involved as a separate term only in the numerator and as a coefficient of s in the denominator. Because all the elements must be positive, R/L cannot be greater than the coefficient of s, $R/L + G/C$, but at most can be equal to it when $G = 0$. Therefore, the maximum value of a is 3.

This property, as may be shown more rigorously, is general in that

$$f = \frac{s + a}{s^2 + bs + c}$$

is realizable as a driving-point function for positive a, b, and c only if $a \leq b$.

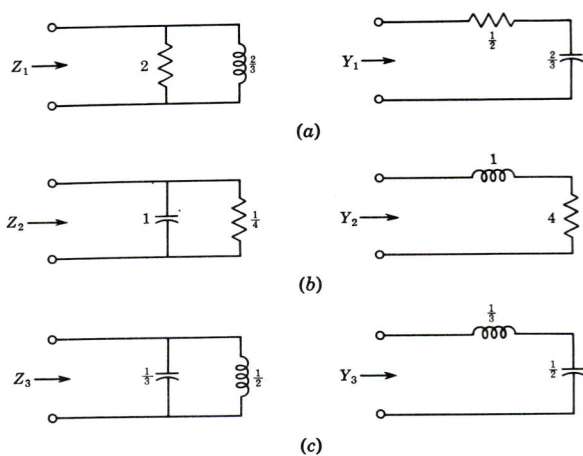

Fig. 23. Values in ohms, henrys, and farads.

4. ADDITIONAL TECHNIQUES FOR REALIZATION OF FILTER AMPLIFIERS

As mentioned previously, Wallman has shown a technique for realizing a filter amplifier with a two-terminal coupling network and Twiss has discussed filter amplifiers that use a two-port network in the form of a double-tuned transformer. These techniques may be used to realize any of the system functions with optimum characteristics discussed in Sec. 2. We now indicate some other methods of design that make use of synthesis by recognition for two-terminal coupling and the available tables for ladder networks[4,12,13,14] for two-port coupling. For generating additional required tables easily there are also explicit formulas for determining the element values of ladders with Butterworth or Chebyshev polynomial characteristics.[4]

4.1. Two-terminal Coupling.[3] Consider a minimum-phase transfer function, that is, one that has no zeros in the left half plane. By the use of *surplus polynomials* we can always write this function as a product of positive real functions, each of which is at most a linear polynomial divided by a quadratic. This restriction on the order of each rational function is made merely for simplicity so that each of the positive real functions can be realized by recognition, as explained in Sec. 3. If the designer wishes to employ the general techniques for realizing more general impedances, he can of course include cubics and quartics in a single positive real function; thus a great deal of flexibility is possible. Each of the positive real functions is realized as the two-terminal load in the plate circuit of a tube.

FILTER AMPLIFIERS

To divide a given minimum-phase transfer function into a product of realizable driving-point functions, we must decide on the value to assign to the constant in each linear surplus polynomial. To guide our choice of surplus polynomial we make use of the fact, as shown in Example 2, that the impedance

$$Z_1 = \frac{s + a}{s^2 + bs + c} \tag{56}$$

where $b, c > 0$ and $0 \leq a \leq b$, is positive real. In addition, we use any of the other functions realized in Sec. 3; for example, we know that

$$Z_2 = \frac{s + d}{s + e} \tag{57}$$

is realizable by an RC network with three elements if $d > e > 0$, and an RC four-element network can realize

$$Z_3 = \frac{s + g}{(s + f)(s + h)} \tag{58}$$

if $h > g > f > 0$.

The procedure is best illustrated by an example. Suppose that we wish to realize a low-pass maximally flat voltage ratio of the fourth order by a cascade of pentode-amplifier stages. Suppose furthermore that the required gain can be furnished by four tubes. The voltage ratio (within a constant multiplier) is determined from Table 1 as

$$K = \frac{1}{(s^2 + 0.7654s + 1)(s^2 + 1.8478s + 1)} \tag{59}$$

We now wish to use surplus polynomials to rewrite this function as a product of four positive real functions, each of which will be used as the plate load of a tube, that is,

$$K = Z_1 Z_2 Z_3 Z_4 \tag{60}$$

This is done simply by multiplying K by the surplus polynomials

$$\frac{s + 0.7654}{s + 0.7654} \frac{s + 1.8478}{s + 1.8478} \frac{s + 1}{s + 1} \tag{61}$$

which does not change K but allows a division into four positive real factors. We obtain

$$K = \frac{s + 0.7654}{s^2 + 0.7654s + 1} \frac{s + 1.8478}{s^2 + 1.8478s + 1} \frac{s + 1}{(s + 0.7654)(s + 1.8478)} \frac{1}{s + 1} \tag{62}$$

so that

$$\begin{aligned} Z_1 &= \frac{s + 0.7654}{s^2 + 0.7654s + 1} \\ Z_2 &= \frac{s + 1.8478}{s^2 + 1.8478s + 1} \\ Z_3 &= \frac{s + 1}{(s + 0.7654)(s + 1.8478)} \\ Z_4 &= \frac{1}{s + 1} \end{aligned} \tag{63}$$

Each of these impedances is realized by recognition, as discussed in Sec. 3; for example, we obtain for Z_1 by a continued-fraction expansion about infinity

$$s + 0.7654 \overline{\left| s^2 + 0.7654s + 1 \right.} \quad \frac{s}{} \tag{64}$$

(equation 64 shown as continued fraction expansion)

Some comments about the choice of the constants in the surplus polynomials are instructive. First, the choice of the constant in a surplus polynomial equal to (rather than less than) the coefficient of the linear term in a quadratic eliminates one element in the realization of Z_1; if the constant were smaller the first remainder would have a nonzero linear term and thus require one more element for its realization. Secondly, the choice of the constant 1 in the surplus polynomial $s + 1$ rather than a positive constant <0.7654 or >1.8478 makes the zeros and poles of Z_3 alternate so that it does not require an RLC network for its realization but is realizable by an RC network.

The realization of the complete amplifier is shown in Fig. 24.

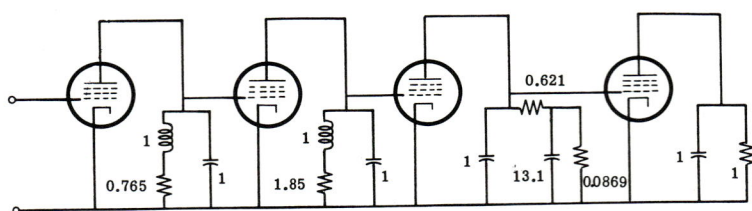

Fig. 24. Cascade of pentode stages for realization of a fourth-order Butterworth function (values in ohms, henrys, and farads).

4.2. Two-port Coupling. In contrast with the preceding procedure that uses one-port networks for coupling the amplifier stages, we now consider the use of two-port networks. In this case the given transfer function must be broken into a product of component *transfer* functions, each of which is then realized as a convenient form of network like a ladder. This procedure allows a great deal more flexibility than the preceding one at the cost of some additional complexity of design calculations. No surplus polynomials are required so that economy in number of elements is achieved. In addition, the complexity disappears if one makes use of available tables of element values for the filter characteristics considered in Sec. 2.

For example, suppose we wish to realize a Chebyshev polynomial low-pass filter with 0.2 db ripple in the passband. Suppose that the required selectivity of the magnitude characteristic is such that a cascade of two third-order filters each with 0.1 db ripple or a sixth-order filter with 0.2 db ripple is satisfactory. We therefore have a choice of using one or two tubes. If the required gain can be achieved with one tube, we then use a single stage with the sixth-order filter. However, if two tubes are required, we use a cascade of two stages, each with a third-order filter. In each case the element values of the coupling network are read off from tables[4] so that no calculations (except the usual removal of frequency and impedance normalization) are required.

If three tubes are required to give the specified gain, we then might use three second-order Chebyshev polynomial filters, each with one-third of the 0.2-db ripple, if this combination provides the specified attenuation. Again tables are used for the design. An alternative approach, which is more efficient for achieving the fastest possible filter cutoff characteristic, is to use the sixth-order Chebyshev polynomial function which we then divide into three second-order transfer functions. In this way we achieve the increased bandwidth that is derived from stagger tuning. However, we can no longer use the available tables but must realize each second-order function directly. Such a realization is a fairly simple procedure for a second-order function. In fact, one does not have to employ synthesis techniques but can adopt the simple expedient of equating coefficients.[15] This technique is used in the reference in an illustrative example that realizes a third-order maximally flat characteristic as a single-stage pentode amplifier with equal source and load resistances.

A network for realizing a second-order function—more specifically, the reciprocal of a second-degree polynomial—as a voltage ratio is shown in Fig. 25. Simple analysis gives

$$K = \frac{1}{LC[s^2 + (R/L)s + 1/LC]} \tag{65}$$

If we normalize the impedance level with respect to R, or equivalently let $R = 1$, the voltage ratio becomes

$$K = \frac{1}{LC(s^2 + s/L + 1/LC)} \tag{66}$$

The matching of coefficients with a given function of the same form then proceeds easily: from the reciprocal of the coefficient of s we obtain L, and we then use this value

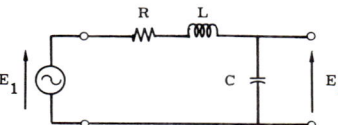   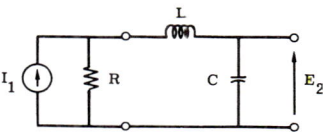

Fig. 25. Quadripole whose voltage ratio is a second-order function.

Fig. 26. Quadripole whose transfer impedance is a second-order function.

of L and the constant term to determine C; and we also have $R = 1$. Finally, we introduce the desired impedance level and cutoff frequency by removing the impedance and frequency normalizations.

Thus if we were realizing the sixth-order Chebyshev polynomial filter, we would require three networks of this form separated by two pentode tubes. To take account of the fact that the pentode is a current source, we use Norton's theorem looking back into the voltage source and the series resistance; the network in Fig. 25 then becomes the one shown in Fig. 26, where $I_1 = E_1/R$. It should be noted that if the plate resistance of the pentode is not negligible, that is, the pentode is only a semi-ideal amplifier, R may be used to provide for this internal resistance; in addition, the output capacitance of the quadripole can be used to provide for a nonnegligible input capacitance of the pentode.

It is interesting to compare this method of dividing the specified transfer function into a product of *transfer* functions with the method of division into driving-point functions, which requires the use of surplus polynomials. Even when the specified transfer function is of low order, the reduction in number of elements can be appreciable. To illustrate this, we consider the fourth-order low-pass Butterworth function of Eq. (59), which was previously realized by the network shown in Fig. 24. Though four pentodes were used in that amplifier, it is of course possible to remove the first one if the gain provided by three pentodes is sufficient.

Now let us assume that a single pentode can provide the required gain. This would not reduce the complexity of the amplifier in Fig. 24 if we wished to continue using only second-order positive real functions: three pentodes would still be required. However, one pentode will suffice if two-port coupling networks are used.

Let us assume that the voltage source driving the amplifier has an internal resistance of 600 ohms, but that no terminating resistance is required. In addition, the 3-db cutoff frequency for the low-pass filter amplifier is $\omega_0 = 10{,}000$ radians/sec.

We may write the given voltage ratio

$$K = \frac{H}{(s^2 + 0.7654s + 1)(s^2 + 1.8478s + 1)} \tag{67}$$

as
$$\begin{aligned} K &= E_2/E_1 \\ &= (E/E_1)(E_2/E) \\ &= K^{(1)}K^{(2)} \end{aligned} \tag{68}$$

where
$$K^{(1)} = \frac{H^{(1)}}{s^2 + 0.7654s + 1}$$

$$K^{(2)} = \frac{H^{(2)}}{s^2 + 1.8478s + 1} \tag{69}$$

We have included the constant multiplier in the equations, though we need realize each of the functions only within a constant multiplier, since it has been assumed that

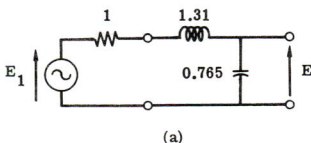

(a)

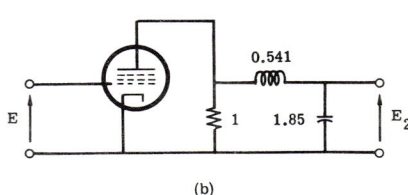

(b)

FIG. 27. Networks realizing the component voltage ratios (element values in ohms, henrys, and farads).

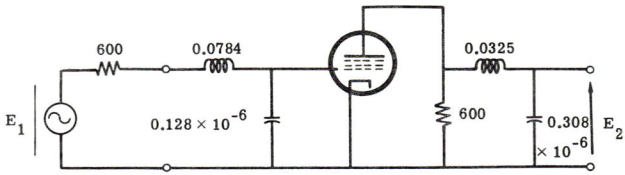

FIG. 28. Network realizing the specified voltage ratio.

other calculations have shown that the gain constant provided by the pentode is sufficient.

The networks realizing $K^{(1)}$ and $K^{(2)}$ are shown in Fig. 27a and b, respectively, where the element values have been obtained by matching of coefficients. Now we remove the normalizations to provide a 600-ohm source resistance and a 3-db bandwidth of 10,000 radians/sec; this is accomplished by multiplying the resistances by 600 and the inductances by 600/10,000, and dividing the capacitances by 600(10,000).

The complete amplifier, with element values in ohms, henrys, and farads, is shown in Fig. 28.

REFERENCES

1. I. M. Horowitz, Exact Design of Transistor RC Band-Pass Filters with Prescribed Active Parameter Insensitivity, *IRE Trans. PGCT*, vol. CT-7, no. 3, pp. 313–320, September, 1960.
2. G. E. Valley, Jr., and H. Wallman, "Vacuum Tube Amplifiers," vol. 18, MIT Radiation Laboratory Series, McGraw-Hill Book Company, New York, 1948.
3. E. A. Guillemin, "Synthesis of Passive Networks," John Wiley & Sons, Inc., New York, 1957.
4. L. Weinberg, "Network Analysis and Synthesis," McGraw-Hill Book Company, New York, 1962.
5. K. W. Henderson and W. H. Kautz, Transient Responses of Conventional Filters, *IRE Trans. PGCT*, vol. CT-5, no. 4, pp. 333–347, December, 1958.
6. P. R. Clement, The Chebyshev Approximation Method, *Quart. Appl. Math.*, vol. 11, pp. 167–183, July, 1953.
7. E. Green, "Amplitude-frequency Characteristics of Ladder Networks," pp. 15–16, 25–26, Marconi's Wireless Telegraph Co., Essex, England, 1954.
8. E. S. Kuh, Synthesis of Lumped Parameter Precision Delay Line, *Proc. IRE*, vol. 45, no. 12, pp. 1632–1642, December, 1957.
9. Y. Peless and T. Murakami, Analysis and Synthesis of Transitional Butterworth-Thomson Filters and Bandpass Amplifiers, *RCA Rev.*, vol. 18, no. 1, pp. 60–94, March, 1957.
10. W. Cauer, Die Verwirklichung von Wechselstromwiderständen vorgeschriebener Frequenzabhängigkeit, *Arch. Elektrotech.*, vol. 17, no. 4, pp. 355–388, December, 1926.
11. E. L. Norton, Constant Resistance Networks with Applications to Filter Groups, *Bell System Tech. J.*, vol. 16, pp. 178–193, April, 1937.
12. S. D. Bedrosian, E. L. Luke, and H. N. Putschi, On the Tabulation of Insertion Loss Low-pass Chain Matrix Coefficients and Network Element Values, *Proc. Natl. Electron. Conf.*, vol 11, pp. 697–717, 1955.
13. R. Saal and E. Ulbrich, On the Design of Filters by Synthesis, *IRE Trans. PGCT*, vol. CT-5, no. 4, pp. 284–327, December, 1958.
14. R. Saal, "Der Entwurf von Filtern mit Hilfe des Kataloges normierter Tiefpässe," Telefunken G.M.B.H. Backnang/Württ, Western Germany, 1961.
15. E. S. Kuh and D. O. Pederson, "Principles of Circuit Synthesis," McGraw-Hill Book Company, New York, 1959.

Chapter 6

FEEDBACK

SORAB K. GHANDHI[*]

CONTENTS

1. Introduction..... 6–2
2. Elementary Feedback Theory..... 6–2
   - 2.1. A Simple Feedback Amplifier..... 6–3
   - 2.2. Loop Gain..... 6–3
   - 2.3. Effect of Feedback on Gain..... 6–3
   - 2.4. Effect on Distortion and Noise..... 6–4
3. Two-port Analysis of Feedback Amplifiers..... 6–5
   - 3.1. The Series-Series Connection..... 6–6
   - 3.2. The Shunt-Shunt Connection..... 6–7
   - 3.3. The Series-Shunt Connection..... 6–7
   - 3.4. The Shunt-Series Connection..... 6–8
   - 3.5. Some Conclusions and Precautions..... 6–8
4. General Feedback Theory..... 6–9
   - 4.1. Return Ratio..... 6–9
   - 4.2. Return Ratio for a Vacuum Tube..... 6–11
   - 4.3. Return Ratio for a Transistor..... 6–12
   - 4.4. Return Ratio for a General Reference..... 6–14
   - 4.5. Sensitivity..... 6–15
   - 4.6. Impedance in Active Networks..... 6–17
   - 4.7. Stability of Single-loop Amplifiers..... 6–21
   - 4.8. Stability of Multiloop Amplifiers..... 6–21
   - 4.9. Unwanted Feedback Loops..... 6–23
   - 4.10. Conditional vs. Unconditional Stability..... 6–23
   - 4.11. The Bode Plot..... 6–25
   - 4.12. The Nichols Chart..... 6–27
   - 4.13. The Root-locus Method..... 6–29
5. The Specification of Amplifier Response..... 6–31
   - 5.1. The Amount of Feedback..... 6–31
   - 5.2. The Closed-loop Response..... 6–32
   - 5.3. The Gain Margin..... 6–32
   - 5.4. The Phase Margin..... 6–32
   - 5.5. The Transient Response..... 6–32
6. The Single-loop Unconditionally Stable Amplifier..... 6–34
   - 6.1. Specifications..... 6–35
   - 6.2. The Ideal Cutoff Characteristic..... 6–35

[*] Electrical Engineering Department, Rensselaer Polytechnic Institute, Troy, N.Y.

6.3. The High-frequency Asymptote.......................... 6–35
  6.4. Maximum Available Feedback........................... 6–36
  6.5. Excess Phase.. 6–37
  6.6. An Illustrative Example................................... 6–38
7. Feedback Chains... 6–41
  7.1. Vacuum-tube Feedback Chains........................... 6–42
  7.2. Transistor Feedback Chains.............................. 6–44
  7.3. Transistor Alternating Feedback Chains................. 6–49
  7.4. Transistor Feedback-pair Chains......................... 6–52
  7.5. Some Concluding Remarks................................ 6–54

1. INTRODUCTION

The previous chapters have discussed the network theory of active circuits. In them, emphasis has been placed on maximizing the gain of the active element by embedding it in an appropriately designed passive network. This follows closely the historical development of amplifiers, where initial limitations were mostly on the amplifying properties of the active elements.

Eventually, a stage was reached at which it was possible to secure any practical value of gain. As a consequence, design emphasis shifted to the achievement of other desirable characteristics. Thus, stability with variations of component parameters, stability with variations of environment and power supply, and freedom from nonlinearities began to assume increasing importance. It was at this point that the application of degenerative feedback methods, where amplifier gain was traded for one or more of these desirable properties, began to be considered seriously.

The earliest patents and publications in feedback amplifiers are those of H. S. Black[1,2] of the Bell Telephone Laboratories. Impetus in this development was derived from activity in the design of complex telephone transmission systems. During World War II, the theory of feedback amplifiers was directly applied to the design of servomechanisms and automatic control systems.[3,4] In time, the field of control systems has developed and expanded to a point where it is reasonable to consider the feedback amplifier as essentially such a system. Indeed, many of the more recent advances in feedback-amplifier theory have originated in this area of activity.[5,6]

The general theory of feedback amplifiers, applicable to all presently used devices, is derived from system theory. Here, emphasis is placed on the transfer characteristics of the individual functional blocks of the system, without regard to the precise nature of the elements that constitute these blocks. The entire system is comprised of an ensemble of such blocks.

In order that the theory may be applicable to practical devices, it becomes necessary to consider separately the situations where the devices used have either (1) substantial isolation between input and output (such as vacuum tubes) or (2) interaction between input and output (as in transistors).

Considerable simplification occurs with devices of the first type, since these may be used to synthesize noninteracting blocks at will. On the other hand, special techniques must be used in order to partition an ensemble of transistors into noninteracting blocks. Alternately, the transistor amplifier must be considered as a single unit, and the problem rapidly becomes intractable.

2. ELEMENTARY FEEDBACK THEORY

From the system point of view, the feedback amplifier may be represented by a single function block with a transfer function specifying its terminal characteristics. This block may be synthesized by the interconnection of a number of elementary noninteracting blocks. If this is done, the contribution of each individual block can be evaluated and optimized without affecting the performance of the other blocks (because of their noninteracting feature).

FEEDBACK

A preliminary understanding of many of the characteristics of feedback amplifiers (or more specifically, of *negative*-feedback amplifiers) may be obtained on the basis of system blocks with functions specified at their midband value, i.e., independent of frequency. Needless to say, such an analysis will provide no information on stability or frequency performance, both of which can be evaluated only on the basis of functional block response over the entire frequency spectrum. At a later point, therefore, a more sophisticated analysis will be necessary.

2.1. A Simple Feedback Amplifier. Figure 1 shows the block diagram of a feedback amplifier in its simplest form. Here, two noninteracting blocks are considered, an active-gain block with a transfer function of A^0 and a passive block having a transfer function β by means of which a portion of the output from the active block is returned to its input. Also included is a summing point at which the returned signal and the input are summed to give an error signal ϵ.

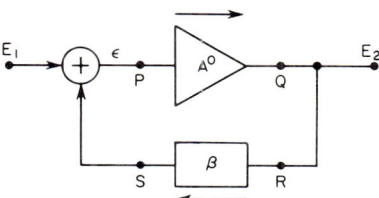

Fig. 1. A simple feedback amplifier.

Writing E_2 and E_1 as the output and input signals, respectively, gives

$$\epsilon = E_1 + \beta E_2 \tag{1}$$

whence
$$A = E_2/E_1 = A^0/(1 - A^0\beta) \tag{2}$$

This is the overall transfer function of the feedback amplifier. In a negative-feedback amplifier, the transfer function A^0 provides a phase shift of 180° in the midband. Consequently,
$$1 - A^0\beta > 1 \tag{3}$$
for all A^0 and β.

In the absence of feedback,
$$E_2/E_1 = A^0 \tag{4}$$

Thus the presence of feedback reduces the gain of the amplifier by the factor $(1 - A^0\beta)$. This term is defined as the *feedback factor*.

In the above analysis, it has been assumed that transmission through the blocks takes place only in the directions indicated by the arrows in Fig. 1. In practical feedback amplifiers, the gain block A^0 usually consists of one or more active unilateral (or almost unilateral) elements. However, the feedback block β is ordinarily composed of passive bilateral elements, and there is a possibility of bilateral transmission through this block. In most practical cases, only reverse transmission through the β loop (as shown by the arrow in Fig. 1) need be considered if the forward gain of the A^0 block is large.

2.2. Loop Gain. If the amplifier of Fig. 1 is cut at any of the points P, Q, R, or S, the transfer function between the two ends of the cut point is given by $A^0\beta$. This is defined as the *loop gain* of the amplifier.

2.3. Effect of Feedback on Gain. If $|A^0\beta| \gg 1$
$$|E_2/E_1| \simeq 1/\beta \tag{5}$$

Thus, for large values of loop gain, the overall transfer function of the amplifier is given by the reciprocal of the transmission through the feedback loop. As a result, the amplifier performance is now largely governed by the behavior of passive elements and becomes almost independent of the parameter variations of the active elements. This is perhaps the single most important advantage to be obtained by the use of negative feedback.

A precise measure of this degree of independence may be obtained by differentiating Eq. (2) and writing it in the form

$$\frac{d(E_2/E_1)}{E_2/E_1} = \frac{1}{1 - A^0\beta} \frac{dA}{A} \tag{6}$$

Thus the relative change in the gain of a feedback amplifier is $1/(1 - A^0\beta)$ times the relative change in the gain of the amplifier without feedback.

2.4. Effect on Distortion and Noise. The application of feedback also affects the distortion products generated in an amplifier, as well as extraneous noises resulting from electrostatic and electromagnetic pickup, power-supply hum, etc., generated within the feedback loop.

Figure 2 shows the block diagram of a feedback amplifier, in which the forward-gain block is subdivided into two blocks, having gains of A_1^0 and A_2^0, respectively.

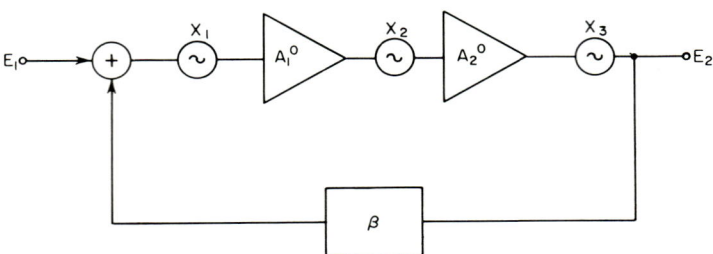

FIG. 2. Feedback amplifier with extraneous noise signals.

Signal sources X_1, X_2, X_3 are inserted at various points in the forward loop so as to represent extraneous sources at the input of the gain elements, at an interstage point, and at the end of the gain elements, respectively. For this amplifier, the output signal in the absence of extraneous noise is given by

$$E_2 = E_1 A_1^0 A_2^0 / (1 - A_1^0 A_2^0 \beta) \tag{7}$$

Furthermore, in the absence of input signal, the outputs due to the extraneous generators X_1, X_2, and X_3 are

$$\frac{A_1^0 A_2^0 X_1}{1 - A_1^0 A_2^0 \beta}, \quad \frac{A_2^0 X_2}{1 - A_1^0 A_2^0 \beta} \quad \text{and} \quad \frac{X_3}{1 - A_1^0 A_2^0 \beta}$$

respectively.

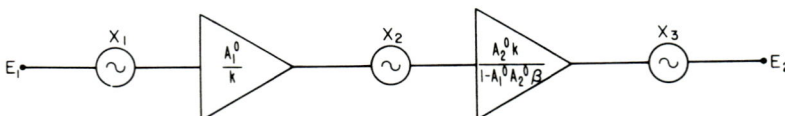

FIG. 3. Nonfeedback equivalent of Fig. 2.

By way of comparison, Fig. 3 shows an amplifier with no feedback. The amplifier has the same extraneous sources, the same overall gain, and the same input and output signal levels as that of Fig. 2. The choice of the constant k is quite arbitrary.

In the absence of input signal, the outputs due to the extraneous generators X_1, X_2, and X_3 are

$$\frac{A_1^0 A_2^0 X_1}{1 - A_1^0 A_2^0 \beta}, \quad \frac{A_2^0 X_1 k}{1 - A_1^0 A_2^0 \beta} \quad \text{and} \quad X_3$$

respectively.

If k is picked so that signal levels through both amplifiers are identical, those extraneous terms which are amplitude-dependent (e.g., distortion products) will also be identical. At this point, a comparison of the output signals shows that feedback has no effect on reducing those extraneous signals which are generated at the input, becomes more effective with signals generated within the amplifier, and is most effective in reducing those signals which are generated at the output. Thus feedback is of no value in reducing noise due to fluctuation phenomena (thermal noise, shot noise, flicker noise, etc.) which are usually troublesome in the input stages of an amplifier. On the other hand, it is very effective in reducing the effects of distortion and intermodulation, usually present in the intermediate and output stages.

3. TWO-PORT ANALYSIS OF FEEDBACK AMPLIFIERS

Consider a feedback amplifier, comprising a forward-gain two-port network and a reverse-transmission two-port network. These two networks may be interconnected in series or in parallel, on input and/or output. Hence, the following four configurations are possible:

1. Series input, series output
2. Shunt input, shunt output
3. Series input, shunt output
4. Shunt input, series output

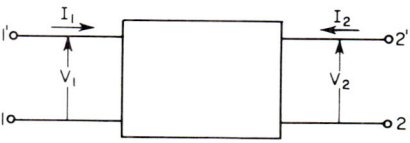

Fig. 4. Two-port network.

In each case a perfectly general analysis may be made of these interconnected networks with the aid of matrix algebra.[7] However, the slightly restrictive analysis that follows lends emphasis on the terminal properties of the amplifier, both with and without the application of negative feedback. In addition, by adequately defining terms, the effect of interconnecting the blocks is also included in the analysis.

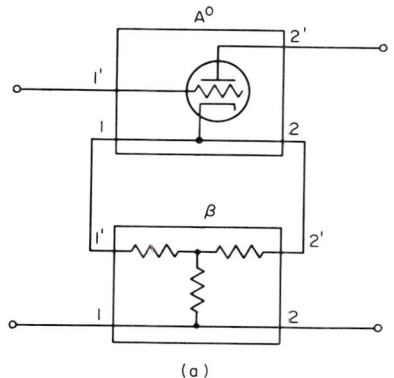 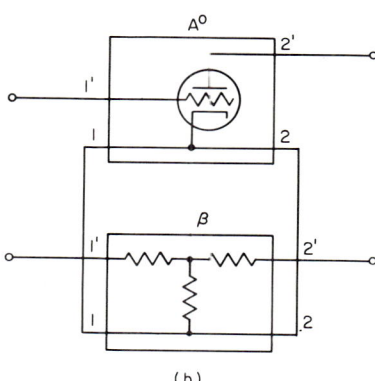

Fig. 5. Interconnected two-port networks. (a) Degenerate form. (b) Nondegenerate form.

One important qualification should be made with regard to the various interconnections that are possible with two-port networks. For any such networks, as shown in Fig. 4, the equations are set up on the basis that the current entering 1' is the same as that leaving 1, and the current entering 2' is the same as that leaving 2. While this is always true for any single two-port network, it may not be true when two such networks are interconnected. Although rules have been set up to avoid this occurrence, the situation can be noted (and corrected) by inspection when the two-port networks have a lead common to both ports. Thus, Fig. 5a shows a degenerate connection of two such networks in series-series. Clearly, the common lead of the A^0 network has shorted the β network! In Fig. 5b, the situation is corrected by altering the manner of interconnecting the A^0 and β networks.

In the following analysis, two-port networks having a common lead will be used. These are usually encountered in practice and can always be properly interconnected.

3.1. The Series-Series Connection. Figure 6 shows the series-series connection of a forward-gain block A^0 and a feedback block β. In this figure, the pertinent electrical characteristics are shown, together with the conditions under which they are obtained. Thus, K_i is a function not only of the two-port parameters of the gain block, but also of its termination (comprising the series connection of Z_l and the input impedance looking backward into the reverse transmission block). In like manner, Z_t is the reverse transfer impedance of the feedback block when terminated in the series combination of Z_g and Z_i. The precise determination of K_i and Z_t is

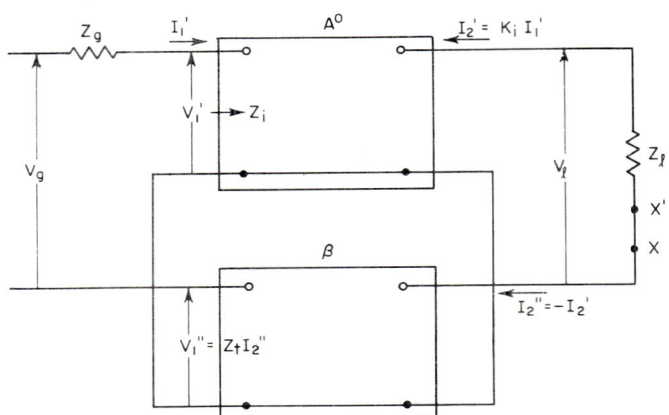

Fig. 6. The series-series connection.

thus quite complex; however, they may be readily approximated for any physical system.

For the circuit of Fig. 6, the voltage gain A_v is given by

$$A_v = \frac{-[Z_l K_i/(Z_i + Z_g)]}{1 + [Z_t K_i/(Z_i + Z_g)]} \tag{8}$$

For $Z_t = 0$, i.e., no feedback, the voltage gain is given by

$$A_v^0 = -\frac{Z_l K_i}{Z_i + Z_g} \tag{9}$$

whence
$$A_v = \frac{A_v^0}{1 - A_v^0(Z_t/Z_l)} \tag{10}$$

This is of the same form as the expression derived from transfer-function analysis. In this case, negative feedback is present if A_v^0 has a 180° phase reversal incorporated in it, i.e., if the midband value of A_v^0 is negative.

For this configuration, Z_{in}, the input impedance seen by the source, is given by

$$Z_{in} = Z_{in}^0(1 - A_v^0 Z_t/Z_l) \tag{11}$$

where Z_{in}^0 is the input impedance in the absence of feedback, i.e., when $Z_t = 0$.

If V_g is shorted, and the network is opened at XX', the impedance seen looking into XX', defined* as Z_{out}, is given by

$$Z_{out} = Z_{out}^0(1 - A_v^0 Z_t/Z_l) \tag{12}$$

As before, Z_{out}^0 is the output impedance in the absence of feedback.

* It is common practice for network theorists to define Z_{out} in this manner. Note that it is consistent with the manner of defining Z_{in}.

FEEDBACK

Thus, the voltage gain of the series-series feedback configuration is reduced by the feedback factor, while its input and output impedances are increased by this factor.

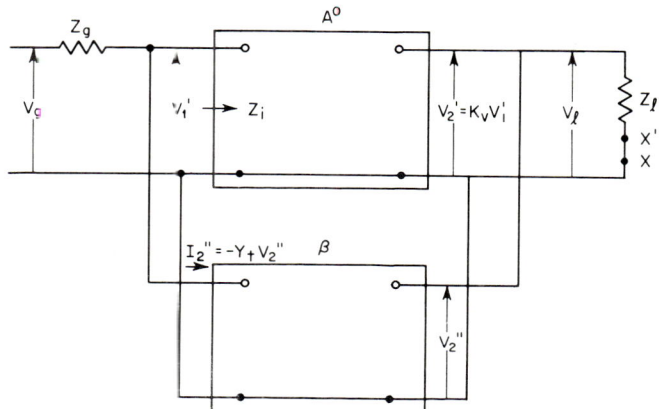

Fig. 7. The shunt-shunt connection.

3.2. The Shunt-Shunt Connection. Figure 7 shows the shunt-shunt connection for a feedback amplifier. For this configuration

$$A_v = \frac{K_v/(1 + Z_g/Z_i)}{1 - [K_v Y_t Z_g/(1 + Z_g/Z_i)]} \tag{13}$$

For no feedback, $Y_t = 0$, and

$$A_v{}^0 = \frac{K_v}{1 + Z_g/Z_i} \tag{14}$$

whence

$$A_v = \frac{A_v{}^0}{1 - A_v{}^0 Y_t Z_g} \tag{15}$$

Again, for negative feedback, the midband value of $A_v{}^0$ must be a negative quantity. The input impedance seen by the source is given by

$$Z_{in} = Z_{in}{}^0/(1 - A_v{}^0 Y_t Z_g) \tag{16}$$

In like manner, the output impedance seen at XX' is given by

$$Z_{out} = Z_{out}{}^0/(1 - A_v{}^0 Y_t Z_g) \tag{17}$$

As before, $Z_{in}{}^0$ and $Z_{out}{}^0$ are the input and output impedances with no feedback.

Thus, for the shunt-shunt connection, the gain and the input and output impedances are reduced by the feedback factor.

3.3. The Series-Shunt Connection. For this configuration (see Fig. 8)

$$A_v = \frac{K_v/(1 + Z_g/Z_i)}{1 - K_v A_t/(1 + Z_g/Z_i)} \tag{18}$$

With no feedback, $A_t = 0$, and

$$A_v{}^0 = \frac{K_v}{1 + Z_g/Z_i} \tag{19}$$

whence, in the presence of feedback,

$$A_v = \frac{A_v{}^0}{1 - A_v{}^0 A_t} \tag{20}$$

The input and output impedances are given by

$$Z_{in} = Z_{in}{}^0 (1 - A_v{}^0 A_t) \tag{21}$$

and
$$Z_{out} = Z_{out}{}^0 / (1 - A_v{}^0 A_t) \tag{22}$$

Thus, the gain and the output impedance in the series-shunt connection are reduced by the feedback factor, while the input impedance is increased by this factor.

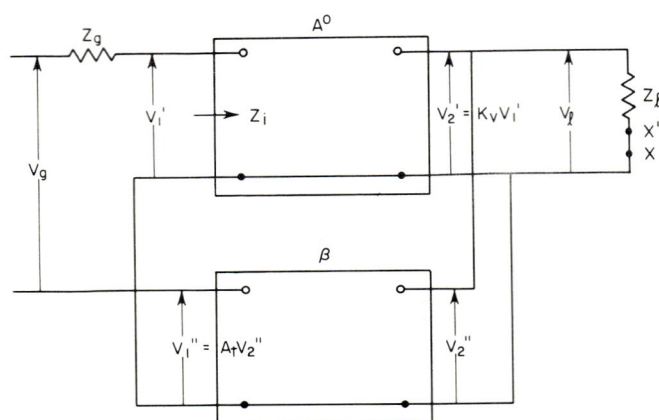

Fig. 8. The series-shunt connection.

3.4. The Shunt-Series Connection. Figure 9 shows the shunt-series feedback connection. Here,

$$A_v = \frac{-K_i Z_l / (Z_i + Z_g)}{1 + A_t K_i Z_g / (Z_i + Z_g)} \tag{23}$$

With no feedback,
$$A_v{}^0 = -K_i Z_l / (Z_i + Z_g) \tag{24}$$
whence
$$A_v = A_v{}^0 / (1 - A_v{}^0 A_t Z_g / Z_l) \tag{25}$$
Also,
$$Z_{in} = Z_{in}{}^0 / (1 - A_v{}^0 A_t Z_g / Z_l) \tag{26}$$
and
$$Z_{out} = Z_{out}{}^0 (1 - A_v{}^0 A_t Z_g / Z_l) \tag{27}$$

Thus, the gain and the input impedance of the shunt-series connection are reduced by the feedback factor, while the output impedance is increased by this factor.

3.5. Some Conclusions and Precautions. The results of the preceding sections may be summarized as follows:

1. Degenerative feedback results in reducing the gain of an amplifier. The amount of this reduction is known as the feedback factor.

2. Degenerative feedback alters the input and output impedances by the feedback factor. The impedance is reduced on the side where the feedback connection is made in shunt with the gain block, and increased on the side where this connection is made in series.

Since a shunt connection on the output side results in feeding back a signal proportional to the output voltage, this connection is often referred to as "voltage feedback." In like manner, a series connection on the output side results in feeding back a signal proportional to the output current, and is often referred to as "current feedback."

It cannot be too highly emphasized, however, that these terms *do not* completely specify the nature of the input and output impedances. As has been shown, the behavior of the feedback is determined not only by the manner in which the signal is obtained from the output, but also by the manner in which it is returned to the input.

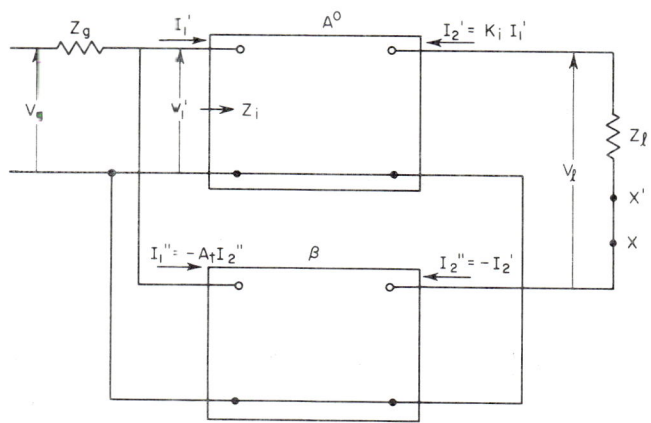

FIG. 9. The shunt-series connection.

The two-port analyses of this section have shown that negative feedback plays an important role in controlling the input and output impedances of an amplifier. As a corollary, it must be noted that a feedback amplifier's performance is completely specified only if its source and load impedances are specified.

4. GENERAL FEEDBACK THEORY

In the previous sections, the role of the A^0 block and the β block have been carefully delineated. In many practical amplifiers, such a delineation is not possible. Thus, Fig. 10 shows a vacuum-tube feedback-amplifier stage. Here, R_K is the element that provides negative feedback between input and output. However, the magnitude of this feedback, i.e., the value of β is a function not only of R_K but also of the parameters of the vacuum tube, which are normally associated with the A^0 block.

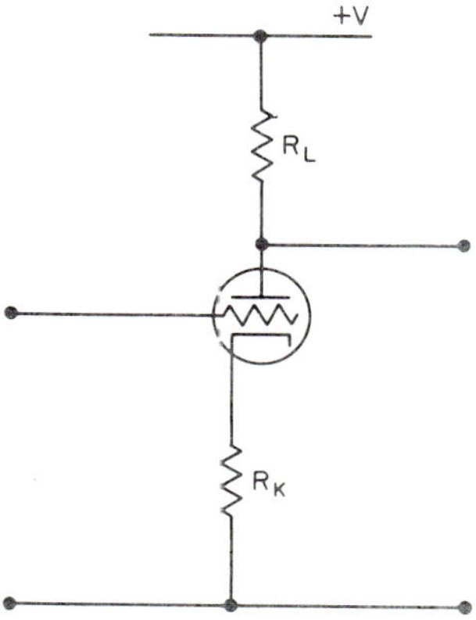

Figure 11a shows a transistor amplifier stage. While it would appear that this is only an amplifying element without feedback, its equivalent circuit, shown in Fig. 11b, indicates that the base spreading resistance r_b' provides negative feedback from output to input. In this example, therefore, the element providing feedback is completely inseparable from the gain block.

The problem is further complicated in amplifiers where multiple loops are present. At this point, it becomes necessary to utilize a more general theory of feedback, based on the network equations of the system.

FIG. 10. Vacuum-tube feedback amplifier.

4.1. Return Ratio. In order to develop a general theory, it is necessary to return to the original concept of feedback. In essence, feedback consists of taking a signal from one point in a system, and *returning it* to

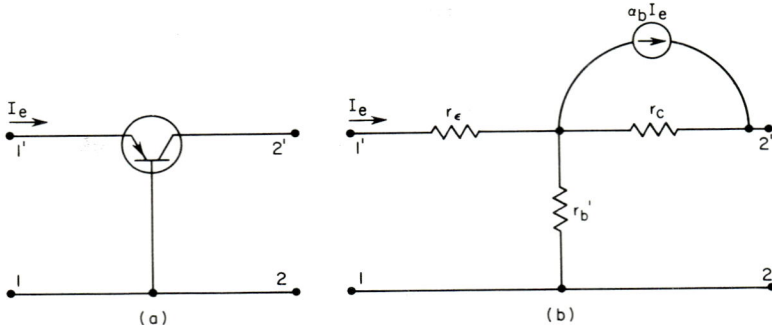

Fig. 11. The common-base transistor.

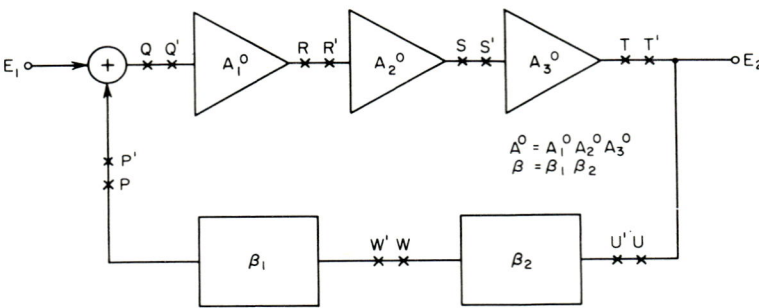

Fig. 12. Single-loop feedback amplifier.

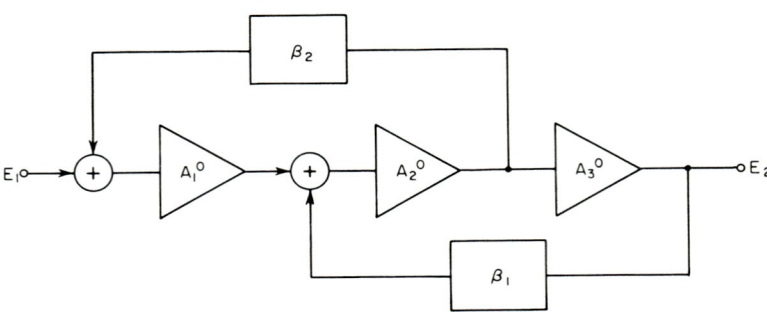

Fig. 13. Multiloop feedback amplifier.

another. The amount of the feedback involved is obtained by computing the signal that is fed around the amplifier, and back to the point of entry. Figure 12 shows a single-loop feedback-amplifier configuration in which this computation is quite straightforward. If the feedback loop is cut open at the point PP', and a unit signal is injected at P', the magnitude of the signal returned to P is given by $A^0\beta$, where A^0 and β are defined in Fig. 12.

It is important to note, however, that the same amount of signal is returned to any point at which the feedback loop is cut open. Thus, the returned signal (or rather, the return ratio, since unit signal is injected into the open loop) is the same for all points in a single-loop feedback amplifier.

The problem becomes more complex for a multiple-loop network such as the one shown in Fig. 13. Here, the feedback signal returned depends upon the point at which the cut is made.

The problem of determining the amount of feedback in a system is best handled by considering the return ratio for any *element* in the system. Since the largest parameter variations generally occur in the active elements, it is customary to consider the return ratio for one of these. It must be recognized however, that the return ratio can be obtained for both active and passive elements in a feedback system.

4.2. Return Ratio for a Vacuum Tube. Figure 14 shows a generalized feedback amplifier. In this amplifier, a single vacuum tube is shown in detail, and the rest of the circuit elements are relegated to the network N. This network may include both active and passive elements of any type (transistors, vacuum tubes, resistors, etc.). The input of the tube under consideration is opened at XY. Since this is a fictitious cut, it is entirely feasible to make it beyond the tube parasitic capacitances, i.e., at the grid of an ideal vacuum tube.

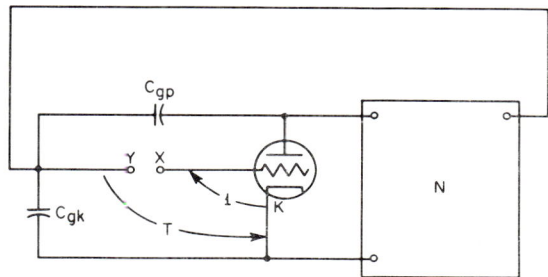

Fig. 14. Vacuum-tube feedback amplifier.

Unit voltage is impressed across the points XK, resulting in a returned voltage T across YK. Using the standard sign convention, we may define

$V_{XK} \triangleq 1$, the unit impressed voltage
$V_{KY} \triangleq T$, the return ratio
$V_{XY} = 1 + T \triangleq F$, the return difference

Using the above definitions for return ratio and return difference, Bode[8] has shown that

$$F = \Delta/\Delta^0 \qquad (28)$$

where Δ is the network determinant for the intact feedback amplifier (i.e., with the terminals XY connected) and Δ^0 is the network determinant of the amplifier when the gain of the tube in question is set to zero.* Thus Δ^0 is the network determinant for the case when the feedback loop is cut open. The return ratio T is the corresponding loop transmission and can be computed readily for any amplifier configuration. In addition, since opening the feedback loop often results in a simple cascade of A^0 and β elements, direct measurements may be made to obtain T without fear of instability.

For the feedback amplifier of Fig. 12

$$T = -A^0\beta \qquad (29)$$

* All other tube parameters are unchanged when this is done.

6-12   AMPLIFIER FUNDAMENTALS

The minus sign appears because of the convention by which T is defined. Consequently,

$$F = 1 - A^0\beta \tag{30}$$

Thus, for a simple feedback-amplifier configuration, F is readily identifiable with the feedback factor.

Figure 15a shows a feedback amplifier in which the roles of amplification and feedback are not clearly delineated. Figure 15b shows its equivalent circuit. For this circuit,

$$T = \mu R'/(R' + r_p) \tag{31}$$

and
$$F = 1 + T = [R'(1 + \mu) + r_p]/(R' + r_p) \tag{32}$$

where R' is the parallel combination of R_K and R_g.

That this is indeed the feedback factor can be shown by conventional circuit analysis. The amplifier gain without feedback is obtained by applying the input signal

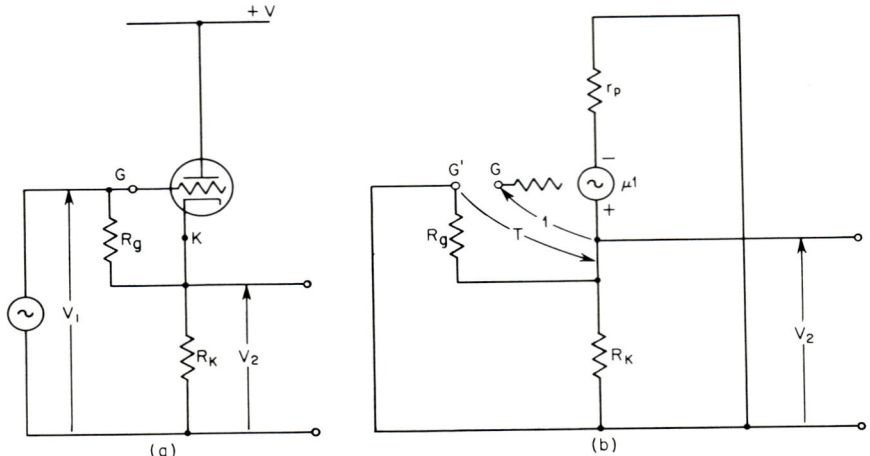

FIG. 15. Vacuum-tube amplifier configuration.

directly between the grid and cathode. Thus, the circuit is unchanged, but the feedback action of R_K is eliminated. For this case,

$$A_v{}^0 = \mu R'/(R' + r_p) \tag{33}$$

Using conventional network analysis,

$$A_v = \mu R'/[R'(1 + \mu) + r_p] \tag{34}$$

$$= \frac{A_v{}^0}{[R'(1 + \mu) + r_p]/(R' + r_p)} \tag{35}$$

whence
$$F = [R'(1 + \mu) + r_p]/(R' + r_p) \tag{36}$$

as before.

4.3. Return Ratio for a Transistor. Certain points of difference appear when it is required to obtain the return ratio for a transistor:

1. The transistor is inherently a current-controlled device. Thus, the specification and laboratory measurements of a transistor circuit are more meaningful if based on current rather than on voltage.

2. The input impedance of a transistor is low. Thus, transmission around a loop is more readily set to zero by means of a short circuit than by means of an open circuit.

FEEDBACK   6-13

In view of these differences, the analysis for transistor feedback amplifiers is carried out as follows: Fig. 16 shows a feedback amplifier in which a single transistor is delineated, and Fig. 17 shows its equivalent circuit. The rest of the circuit, comprising active and passive elements of any type (including vacuum tubes), is contained in the general network N. In order to obtain a definition for T that is consistent with the one given by Bode, it is necessary that the loop transmission be zero when the gain of the transistor is set to zero. This can occur only if the transistor input is shorted *after* incorporating Z_e and r_b' as components of the network N. (This follows the approach of Bode, who opened the grid at a fictitious point, after considering the

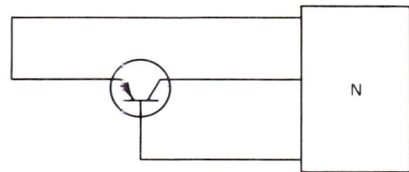

Fig. 16. Transistor feedback amplifier.

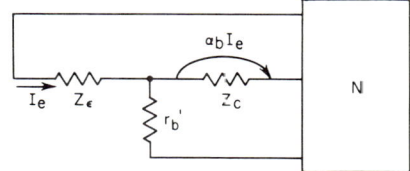

Fig. 17. Equivalent circuit of Fig. 16.

parasitic capacitances as external to the vacuum tube under consideration.) In Fig. 18, the terminals of the fictitious transistor are designated as e, b, and c, respectively. To obtain T, the terminals e and b are shorted, with unit current injected into the transistor as shown. Then, T is returned to the transistor through the feedback network.

Under these conditions, it has been shown[9] that

$$F \triangleq 1 + T = \Delta/\Delta^0 \tag{37}$$

where Δ^0 is the network determinant when the alpha of the transistor in question, and hence the transmission around the loop, is set to zero. As before, Δ is the determinant of the intact feedback amplifier.

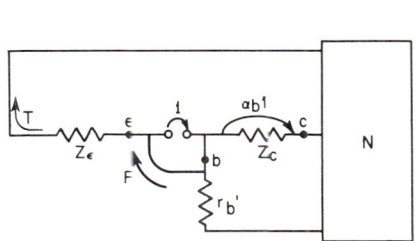

Fig. 18. F and T.

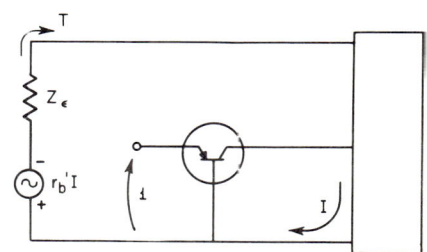

Fig. 19. Circuit for measuring T.

Laboratory measurements of T may be made for a transistor by means of the circuit of Fig. 19. Often, the voltage source $r_b'I$ may be omitted with little error.

Figure 20 shows a feedback amplifier with a transistor in the common-emitter connection. In view of the generalized nature of the network N, the equivalence with Fig. 16 is readily apparent. Consequently, the corresponding circuit is as shown in Fig. 21 together with the manner in which T is defined.

While this method lends itself to easy network analysis, it is somewhat awkward for laboratory measurements, since the feedback path is still returned to the emitter. A generalized technique for circumventing this problem is next described.

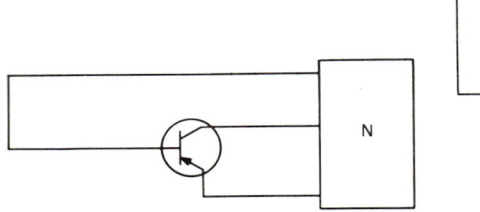
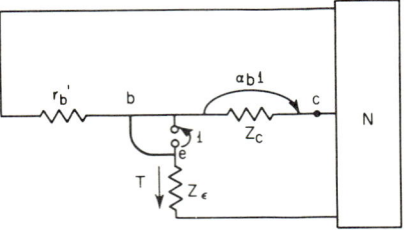

FIG. 20. Common-emitter feedback stage.   FIG. 21. Equivalent circuit of Fig. 20.

4.4. Return Ratio for a General Reference. In considering the return ratio for both tubes and transistors, the condition for zero transmission around the loop is that the gain of the active element under consideration (μ for the tube, and α for the transistor) be set to zero. Thus, the "reference condition" for this element is zero. It is possible to extend the definition of T to any general reference k. Thus, Bode has shown that, if T_k is the return ratio for a reference k, then

$$F_k = 1 + T_k = \Delta/\Delta^k \tag{38}$$

where F_k is the return difference for the reference k, and Δ^k is the value of the network determinant when the element in question is set to k.

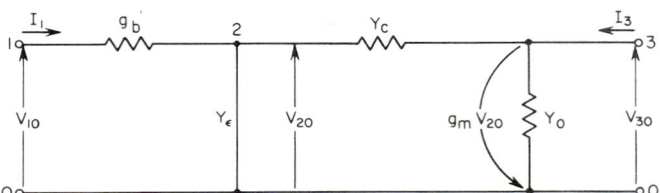

FIG. 22. The hybrid-π equivalent circuit.

This concept of return difference for a generalized reference[10] may be used in analyzing transistor circuits. Figure 22 shows the equivalent circuit of a transistor in the common-emitter connection. The hybrid π model[11,12] is used, with

$$Y_\epsilon = sC_\epsilon + (1 - \alpha_{b0})/r_\epsilon \tag{39}$$
$$Y_c = g_c + sC_c \tag{40}$$
$$g_m = \alpha_{b0}/r_\epsilon \tag{41}$$
$$Y_o = \mu_o/r_\epsilon \simeq 1/r_c(1 - \alpha_{b0}) \tag{42}$$
$$g_b' = 1/r_b' \tag{43}$$

For this circuit, the network equations are written as

$$\begin{bmatrix} I_1 \\ 0 \\ 0 \end{bmatrix} = \begin{bmatrix} g_b' & ; & -g_b' & ; & 0 \\ -g_b' ; & g_b' + Y_\epsilon + Y_c ; & -Y_c \\ 0 & ; & g_m - Y_c & ; & Y_c + Y_o \end{bmatrix} \begin{bmatrix} V_{10} \\ V_{20} \\ V_{30} \end{bmatrix} \tag{44}$$

Now,
$$A_v = V_{30}/V_{10} = \Delta_{13}/\Delta_{11} \tag{45}$$
But
$$\Delta_{13} = 0 \quad \text{if} \quad g_m = Y_c \tag{46}$$

Thus, the forward transmission through the stage is zero when g_m is set equal to Y_c. This is the reference condition for the transistor.

FEEDBACK 6–15

Figure 23 shows the equivalent circuit for the transistor in this reference condition. Return ratio and return difference for this reference are related by

$$F_{Y_c} = 1 + T_{Y_c} = \Delta/\Delta^{Y_c} \qquad (47)$$

where Δ^{Y_c} is the network determinant with the g_m of the transistor in question set equal to Y_c.

The equivalent circuit associated with the measurement of T for the common-emitter feedback amplifier of Fig. 20 is shown in Fig. 24. Here, the returned voltage

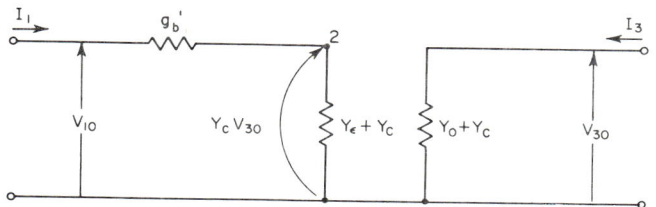

FIG. 23. The transistor in its reference condition.

is T_{Y_c}. It is possible to synthesize the current generator $Y_c V_{30}$ with the aid of an additional transistor whose properties are closely matched to those of the reference device. However, this active term is omitted in most practical cases. Furthermore, since

$$Y_\epsilon \gg Y_c \qquad (48)$$
$$\text{and} \qquad g_{b'} \gg Y_\epsilon \qquad (49)$$

the measurement simplifies to one of measuring the voltage gain around an open loop, terminated in the equivalent input impedance, as shown in Fig. 25. Here, we may

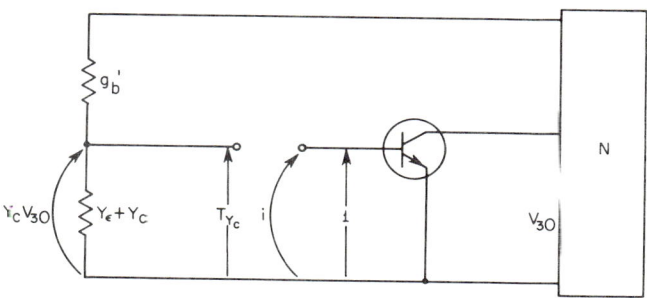

FIG. 24. The measurement technique for T_{Y_c}.

write

$$T_{Y_c} \simeq V_o/V_i \qquad (50)$$

4.5. Sensitivity. The sensitivity of an element W is a measure of the effect of a change in the element upon the overall performance of the amplifier. In general, the principal changes in the value of an element from its prescribed design value are due to

1. Parameter tolerances for a new device
2. Parameter tolerances for end of life
3. Parameter variations with temperature
4. Parameter variations with varying power-supply voltages and currents.

6-16   AMPLIFIER FUNDAMENTALS

In most amplifiers, the components that suffer the largest parameter changes are the active elements. In part, this is a reflection of the fact that these are the most complex elements of an amplifier, prescribed by a number of interrelated characteristics. Furthermore, many parameter variations are fundamental in nature and cannot be "corrected" by manufacturing processes. By way of example, the low-

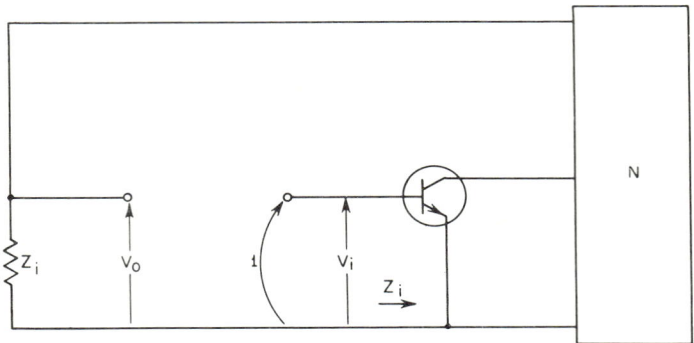

Fig. 25. Approximate measurement method for T_{Yc}.

frequency value of the input impedance of a transistor in the common-emitter connection is given approximately by

$$h_{11e} = (kT/qI_E)\alpha_{e0} \tag{51}$$

where k = Boltzmann's constant
            q = charge of an electron
            T = temperature, °K
            I_E = d-c value of the emitter current
            α_{e0} = d-c value of the common-emitter current gain

From this expression, we see that the input impedance (1) varies directly with the temperature, (2) varies inversely with power-supply current, and (3) varies directly with yet another important device parameter (current gain). This parameter is ordinarily specified by wide manufacturing tolerances.* In addition, it too varies with temperature and operating current.

The passive element, on the other hand, is generally specified by a single parameter, and various manufacturing techniques allow its control to almost any degree of precision. The evaporated thin-film resistor, for example, can be fabricated to ±0.1 per cent initial tolerance (with ±0.25 per cent tolerance at end of life) and can be processed to result in a temperature variation of less than 100 ppm per degree centigrade.

In many amplifiers, negative feedback is used to stabilize the gain characteristics and render them insensitive to changes in device parameters. The degree of this stabilization may be characterized mathematically as follows:

Let W be the amplifier parameter that is subject to change and A be the overall gain of the amplifier that is affected by this parameter change. The "sensitivity" of the element, defined as S, is given by

$$dA/A = S(dW/W) \tag{52}$$

Thus, the sensitivity of an element W with respect to A is the ratio of the percentage change in A to the percentage change in W. The effect of inverse feedback

---

* Typically, a design center value of α_{e0} with a -50 to $+300$ per cent tolerance is specified by the device manufacturer.

FEEDBACK   6-17

is to decrease this sensitivity. Thus, the ideal amplifier is one which has zero sensitivity to parameter changes.

The original definition of sensitivity, as given by Bode, used the reciprocal of this term as sensitivity. Thus, using Bode's definition, the ideal amplifier has infinite sensitivity. While the definition used here is commonly used nowadays, many of the early references use Bode's original definition.

Using our definition, it may be shown that $S = 1/F$ for any element provided that the setting of this element to zero results in zero transmission around the network. This theorem may be directly applied to obtain the sensitivity of the gain parameter of a vacuum tube or a transistor, where F may be defined for the zero reference condition.

The sensitivity relations for a general reference k are somewhat more involved. If S_k, the sensitivity for this reference, is defined by

$$dA/A = S_k[d(W-k)/k] \tag{53}$$

then the value of S_k is given by

$$1/S_k = \Delta/\Delta^k \tag{54}$$

Thus, the sensitivity for a general reference is equal to the reciprocal of the return difference for this reference. Consequently, the general theory of feedback reaffirms the results of the elementary theory, which showed that the effect of gain variations in a feedback amplifier can be reduced by the feedback factor.

The general theory for sensitivity can be readily applied to determine the effect of parameter variation in a bilateral element. For this case, we must define sensitivity for a reference k, where k is that value of the element which results in zero transmission. Such a value of k (often negative) exists for all elements and may be computed in a manner analogous to that used for computing the reference value of g_m for a transistor in Sec. 4.4.

4.6. Impedance in Active Networks. Section 3 discussed the active impedance of a feedback amplifier, based on the two-port analysis of interconnected A^0 and β blocks. In this section we shall show the results for the general case.

Figure 26 shows a multimesh active network. The impedance at any point in this network may be obtained by making a cut at that point, and "looking into" the network. Such a cut is shown at $X'X$. For this cut,

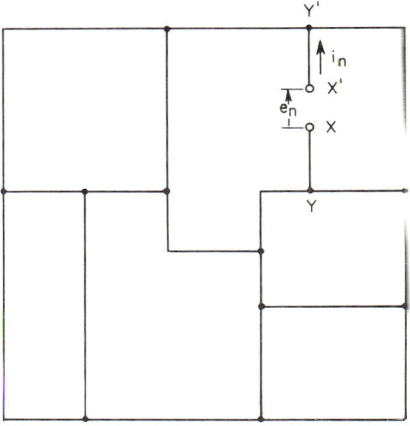

Fig. 26. General network.

$$Z_n = e_n/i_n = \Delta/\Delta_{nn} \tag{55}$$

where Δ_{nn} is the network determinant with row n, column n deleted.

Equation (55) may be rewritten as

$$Z_n = (\Delta/\Delta^0)(\Delta^0/\Delta_{nn}{}^0)(\Delta_{nn}{}^0/\Delta_{nn}) \tag{56}$$

The above terms may be interpreted as follows:

1. Δ/Δ^0 is the return difference for the network, i.e., for the circuit obtained if the terminals $X'X$ were shorted. We define this quantity as F_{sc}.

2. $\Delta^0/\Delta_{nn}{}^0$ is the value of Δ/Δ_{nn} when some element W is set to zero. If W is so picked that this results in zero transmission around the feedback loop, the term is the input impedance at $X'X$ in the absence of feedback. We define this quantity as $Z_n{}^0$.

6–18 AMPLIFIER FUNDAMENTALS

3. $\Delta_{nn}/\Delta_{nn}{}^0$ is the same as the return difference Δ/Δ^0 when row n, column n is deleted. This is the same as if the generator e_n were replaced by an open circuit. Hence, $\Delta_{nn}/\Delta_{nn}{}^0$ is the return difference when the terminals $X'X$ are left open. We define this quantity F_{oc}.

Rearranging terms gives

$$Z_n = Z_n{}^0 F_{sc}/F_{oc} \tag{57}$$

In like manner, the admittance across $Y'Y$ may be written as

$$Y_n = Y_n{}^0 F_{oc}/F_{sc} \tag{58}$$

The choice of W must be made in such a manner that there is no feedback when $W = 0$. In many feedback amplifiers this occurs when the gain of any one active element is set to zero. In some situations, where this does not happen, it is possible to carry out the analysis by setting to zero as many active elements as are needed to

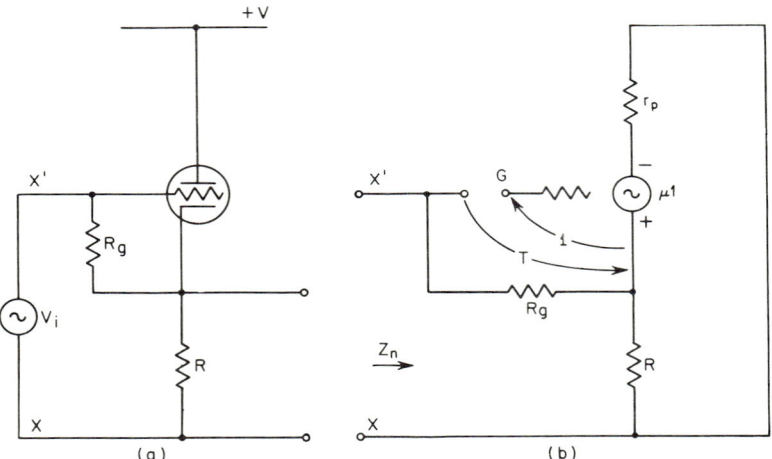

Fig. 27. Vacuum-tube feedback stage.

make the feedback inactive, and applying Eqs. (57) and (58) repeatedly as each is returned to its normal value.

Often, the value of one of the return differences reduces to unity, while the value of the other reduces to the feedback factor. For these cases, the active impedance of the network is either increased or decreased by the feedback factor. In amplifiers of the type described in Sec. 3, this is seen to be the case.

Computations for active impedance may also be made for the case where $W = k$ for zero transmission through the loop. By a similar reasoning, it may be seen that

$$Z_n = Z_n{}^k F_{sc}{}^k/F_{oc}{}^k \tag{59}$$
and
$$Y_n = Y_n{}^k F_{oc}{}^k/F_{sc}{}^k \tag{60}$$

By way of example, Fig. 27 shows a vacuum-tube feedback amplifier and its equivalent circuit. For this circuit, the input impedance without feedback is given by

$$Z_n{}^0 = R_g + Rr_p/(R + r_p) \tag{61}$$
$$T_{oc} = 0 \quad \text{and} \quad F_{oc} = 1 \tag{62}$$
$$T_{sc} = \mu R /(R' + r_p) \tag{63}$$

FEEDBACK 6-19

where R' is the parallel combination of R and R_g.

$$F_{sc} = [R'(1 + \mu) + r_p]/(R' + r_p) \qquad (64)$$

whence
$$Z_n = Z_n^0[R'(1 + \mu) + r_p]/(R' + r_p) \qquad (65)$$

where Z_n^0 is given by Eq. (61).

It is instructive to compare this result with that obtained for the voltage gain of

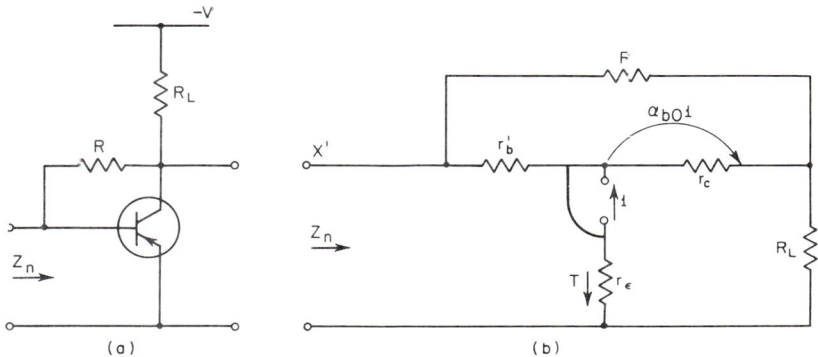

FIG. 28. Transistor feedback stage.

the same circuit at Eq. (35). We note that the voltage gain was given by the zero-feedback value divided by the feedback factor, while the input impedance is given by the zero-feedback value multiplied by this same factor. The increased input impedance results from the fact that the feedback is returned to the input in a series-connected manner.

Figure 28 considers a shunt-feedback transistor amplifier and its low-frequency equivalent circuit. For this circuit, we may assume that $r_c \gg R, \gg R_L, \gg r_\epsilon,$ and $\gg r_b'$.

Then
$$Z_n^0 \simeq \frac{(r_\epsilon + r_b')(R + R_L)}{r_\epsilon + r_b' + R + R_L} \qquad (66)$$

With terminals $X'X$ shorted,
$$T_{sc} \simeq -\alpha_{b0}r_b'/(r_\epsilon + r_b') \qquad (67)$$

With terminals $X'X$ opened,
$$T_{oc} \simeq -\frac{\alpha_{b0}(R + r_b')}{r_\epsilon + r_b' + R + R_L} \qquad (68)$$

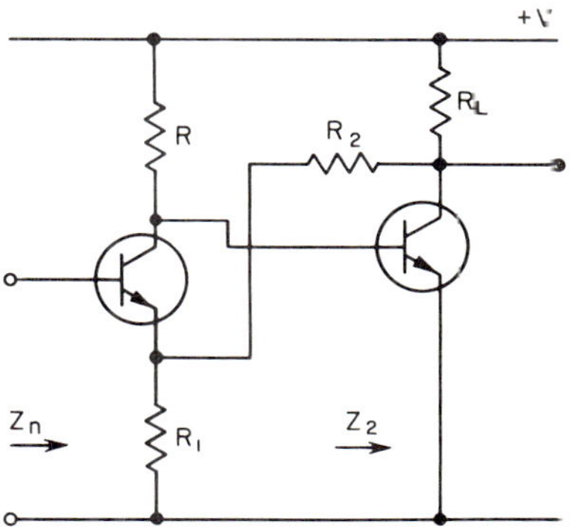

whence
$$Z_n = Z_n^0 F_{sc}/F_{oc} \qquad (69)$$
$$\simeq \frac{(R + R_L)[r_\epsilon + r_b'(1 - \alpha_{b0})]}{r_\epsilon + r_b'(1 - \alpha_{b0}) + R_L + R(1 - \alpha_{b0})} \qquad (70)$$

FIG. 29. Two-stage feedback amplifier

Figure 29 shows a more complex situation, comprising a two-stage feedback amplifier whose input impedance must be determined. For this circuit, let $R_2 \gg R_L$ and $R \gg Z_2$, where Z_2 is the input impedance of the second stage. In addition, it is assumed that the collector resistance is much larger than all other circuit elements.

While it does not matter which device is placed in its reference condition, the choice will determine the ease with which the analysis may be accomplished. Consider the case where the first stage is selected. The equivalent circuit is shown in Fig. 30.

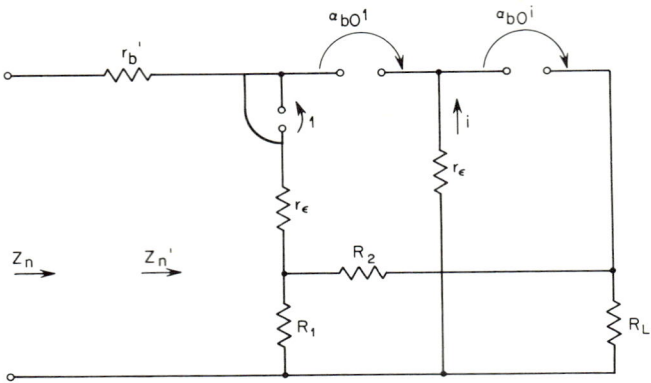

FIG. 30. Equivalent circuit of Fig. 29.

For this circuit,
$$Z_n'^0 = r_\epsilon + R_1 R_2/(R_1 + R_2) \tag{71}$$
$$T_{oc} \simeq -\alpha_{b0} \tag{72}$$

and
$$T_{sc} = \frac{\alpha_{b0}^2 R_1 R_L}{(1 - \alpha_{b0})[r_\epsilon R_1 + R_2(r_\epsilon + R_1)]} \tag{73}$$

whence
$$Z_n = r_b' + Z_n'^0 F_{sc}/F_{oc} \tag{74}$$
$$= r_b' + \frac{\alpha_{b0}^2 R_1 R_L + (1 - \alpha_{b0})[r_\epsilon R_1 + R_2(r_\epsilon + R_1)]}{(1 - \alpha_{b0})^2 (R_1 + R_2)} \tag{75}$$

It is interesting to note how this problem may be handled with reference to the

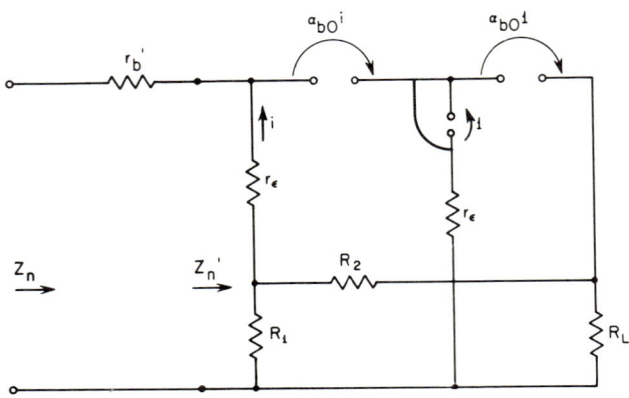

FIG. 31. Alternate version of Fig. 30.

second transistor. Figure 31 shows the equivalent circuit for this case. Here, knowledge of the input impedance of a common-emitter amplifier is used to compute $Z_n'^0$

$$Z_n'^0 = \frac{1}{1 - \alpha_{b0}} \left(r_\epsilon + \frac{R_1 R_2}{R_1 + R_2} \right) \tag{76}$$
$$T_{oc} \simeq -\alpha_{b0} \tag{77}$$

Let T_{sc1} and T_{sc2} be the return ratios for the first and second transistors, respectively. Then

$$T_{sc1} \simeq \alpha_{b0} R_1 R_L / [r_\epsilon R_1 + R_2(r_\epsilon + R_1)] \tag{78}$$

and

$$T_{sc2} \simeq -\alpha_{b0} \tag{79}$$

Writing $T_{sc} = 1 + T_{sc1} + T_{sc2}$, and collecting terms, we obtain

$$Z_n = r_b' + \frac{\alpha_{b0}^2 R_1 R_L + (1 - \alpha_{b0})[r_\epsilon R_1 + R_2(r_\epsilon + R_1)]}{(1 - \alpha_{b0})^2 (R_1 + R_2)} \tag{80}$$

as before

4.7. Stability of Single-loop Amplifiers. Stability is the single most important property that must be ensured in the design of negative-feedback amplifiers. While the term "negative feedback" connotes a reduction in gain, it is unfortunately impossible to maintain this over the entire range of frequencies. Thus, various phase shifts in an amplifier can result in the feedback's becoming positive over some range of frequencies. The study of stability must therefore be a study of amplifier properties over the frequency spectrum, both within and outside the passband.

Consider a general network into which a disturbance has been inserted. The nature of this disturbance is unimportant; it may be either external pickup or internally generated noise. In either case, the network is stable if the disturbance dies away as a function of time. Conversely, the network is unstable if the disturbance initially grows with time. (Since no disturbance can grow indefinitely in time, only the initial growth or decay of the disturbance is of interest.)

The borderline case, which rarely occurs, is when the disturbance neither grows nor decays with time. By definition, this case is considered stable.

The stability of a network is best analyzed by a consideration of the properties of the network determinant. If the zeros of Δ are plotted on the s plane (where $s = \sigma + j\omega$), the network is stable provided (1) none of the zeros can be found in the right half plane (2) zeros on the ω axis are simple.

When these zeros of Δ are not known, an alternative procedure consists of making a search of the entire right half of the s plane. If no zeros are found, the network is stable.

If Δ is a single-valued regular function of a complex variable $f(z)$, and if the point z is moved over a closed path, in the positive direction, the number of zeros minus poles of $f(z)$ encountered in this region is given by the number of times the $f(z)$ plot encircles the origin in a positive direction. If the path of integration covers the entire right half s plane, and does not encircle the origin, then Δ has no zeros in this half plane, and corresponds to a stable network.

Figure 32a shows the manner in which this contour integration is carried out in order to avoid enclosing simple zeros on the ω axis. In each case, the semicircle around the zero is of radius ϵ where $\epsilon \to 0$. Figure 32b shows a plot of the corresponding Δ in the Δ plane for a stable network.

The method for investigating the zeros of Δ for single-loop feedback amplifiers may now be described.

Since Δ^0 is the network determinant when the feedback loop is open, it represents a stable network function. Hence, Δ^0 can have no zeros in the right half s plane. Furthermore since $F = \Delta/\Delta^0$, a plot of F in the F plane will therefore indicate stability if and only if it does not enclose the origin. Finally, since $F = 1 + T$, a plot of T in the T plane can be used to indicate stability if and only if it does not enclose the point $-1 \pm j0$.

Figure 33 shows typical plots for F and T for a stable amplifier. These are shown for ω ranging from $-\infty$ to $+\infty$, corresponding to s encircling the entire right half of the s plane.

Since the transmission around an open feedback loop is readily computed for an amplifier design, or readily measured in the laboratory, it is this function T that is usually plotted to investigate stability. Such a plot is known as a Nyquist diagram for stability.

4.8. Stability of Multiloop Amplifiers. The F or T plot for a single active element provides information on the zeros of Δ, provided that $\Delta^\bullet$ is known to be stable. In a multiloop amplifier, it is possible that Δ^0 itself corresponds to an unstable situation. In this event, the plot of T does not provide information on the roots of Δ, but only on the roots of Δ/Δ^0.

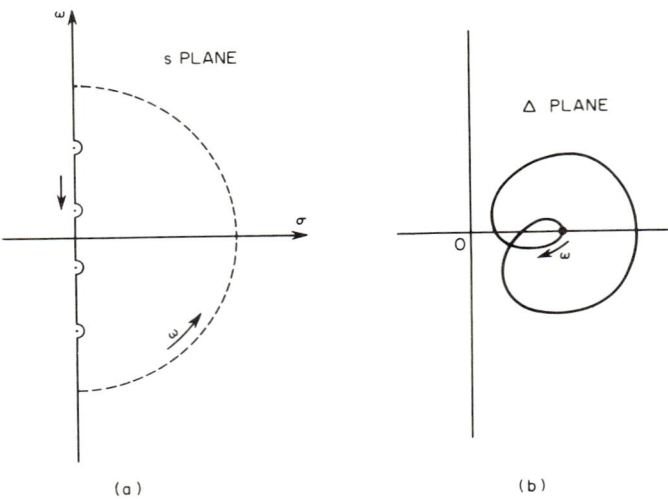

Fig. 32. The integration path and the plot of Δ.

The starting point for the stability analysis of a multiloop feedback amplifier must be a situation whose stability is ensured. This may be done by rendering as many

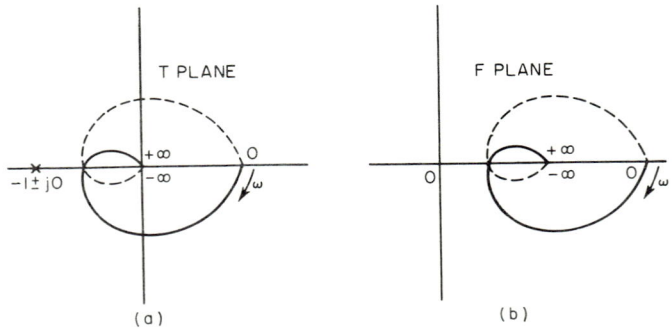

Fig. 33. The T and F plots.

devices inoperative as are necessary for this condition. Consider an n-stage amplifier, where m devices, appropriately selected, will render the feedback inoperative.

If Δ_i = network determinant when i devices are inoperative (gain = 0)
F_i = return difference for the ith device with $i - 1$ devices inoperative

Then
$$F_m = \Delta_{m-1}/\Delta_m \tag{81}$$
$$F_{m-1} = \Delta_{m-2}/\Delta_{m-1} \tag{82}$$
$$F_2 = \Delta_1/\Delta_2 \tag{83}$$
$$F_1 = \Delta/\Delta_1 \tag{84}$$

Since Δ_m is the network determinant for the stable amplifier, the location of its zeros is known. Hence, F_m provides information on the roots of Δ_{m-1}. With this information, F_{m-1} can be plotted to give information on the roots of Δ_{n-2}. The procedure is repeated successively, until the roots of Δ are known. The stability analysis may be performed as follows:

1. Render the network stable by making as many active elements inoperative as are necessary.
2. Reactivate these elements one by one, in each case plotting the T locus, until all are replaced.
3. The *algebraic* sum of all the clockwise (negative) and anticlockwise (positive) encirclements about the critical point must be zero for the amplifier to be stable.

4.9. Unwanted Feedback Loops. All active elements have parasitic coupling between their terminals. This coupling cannot be avoided, since it is a function of the geometric placement of the elements that constitute a real device. Since it is usually capacitive in nature, its effect becomes increasingly apparent at high frequencies and results in feedback over a range of such frequencies. In addition, devices such as transistors have internal resistive coupling between input and output terminals, which provides feedback over the entire frequency range.

One may conclude therefore that, in addition to the main feedback loop or loops, all amplifiers have at least as many additional feedback paths as there are active devices. In fact, if the effects of interwiring capacitances are included, the number of feedback loops becomes uncountably large!

The analysis of a feedback amplifier becomes extremely cumbersome, if not impossible, if all these feedback paths are considered. Consequently, it becomes important to determine which feedback paths can be ignored and which cannot. Since

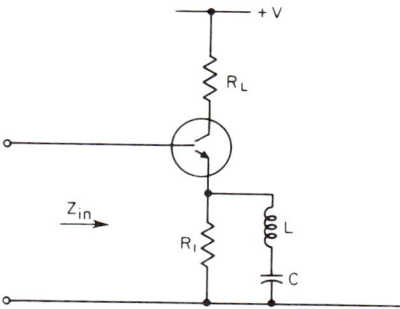

Fig. 34. Transistor stage.

many of these feedback paths are present at high frequencies alone, it is customary to reduce their effect (1) by paying careful attention to lead dress and shielding and (2) by sharply restricting the bandwidth of the amplifier to the range for which it must be designed and reducing its gain beyond this useful range.

Occasionally, a small amount of local feedback is placed around a device, in order to provide some control of its gain. The unbypassed cathode resistor is an example of such an element. Since this provides a small amount of feedback, its effect may usually be ignored in any stability analysis, and the entire stage, complete with its feedback element, may be treated as a simple gain block in an amplifier.

An important case of undesired feedback is that associated with the excess phase shift of a transistor. Thus, it may easily be shown that the input impedance of the amplifier stage of Fig. 34 shows a negative real part for frequencies where

$$1/\omega C > \omega L + (r_b' + r_e)\omega/\omega_{ab}$$

Such a stage is potentially capable of oscillation if improperly terminated on its input side.

The stability of the tuned amplifier transistor stage[13,14] has been the subject of considerable study, and various methods have been developed to render such stages unconditionally stable. In particular, both damping (by loading) and neutralization have been used to achieve this end. If a transistor amplifier uses potentially unstable stages of this type, it is general practice to render these stages unconditionally stable individually before proceeding with the design of the main feedback loop.

4.10. Conditional vs. Unconditional Stability. A stable feedback amplifier can always be made unstable by increasing the gain around the loop beyond a critical point. Thus, for any amplifier, there is a maximum amount of feedback that can be applied before the system begins to oscillate. This can be readily shown with the aid of the T plot for a single-loop amplifier. Since this is a polar plot of the loop transmission, the effect of increasing the loop gain is visualized by expanding the radial scale.

Figure 35 shows the T plots of a stable feedback amplifier, with a gain constant K_1. It is seen that if the loop gain is increased, a point is eventually reached (gain = K_3) when the amplifier becomes unstable. On the other hand, any reduction of gain results in shrinkage of the radial scale, and always results in stable operation. An amplifier having a loop characteristic of this type is defined as being unconditionally stable.

The T plot of a conditionally stable amplifier is shown in Fig. 36 for a specified loop gain. As before, increasing the gain of the amplifier can result in instability.

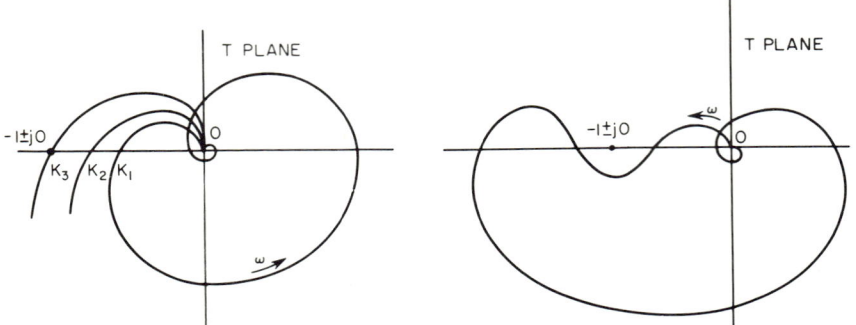

Fig. 35. Effect of increasing loop gain.

Fig. 36. T-plane plot for a conditionally stable amplifier.

However, such an amplifier can also become unstable if its loop gain is *lowered* beyond a certain critical value. Stability of this type is known as conditional stability.

It is worth remembering, at this point, that one primary reason for using negative feedback is to compensate for device gain variations over a period of time. The effect of aging usually results in a reduction in device gain and is reflected as a contraction in the T plot over a period of time. Clearly, the use of conditional feedback in such a situation is not permissible.

An additional problem presents itself in the case of conditionally stable amplifiers. If, for any reason (e.g., power-supply transient or input signal overload), the gain of the amplifier elements is temporarily reduced, the amplifier may become temporarily unstable. If this period of instability is sufficiently long, the oscillation will build up to the point where nonlinear gain limiting occurs. In this event, the amplifier will remain oscillating, even after the transient is removed.

For the above reasons, it is customary to design amplifiers which are unconditionally stable. Only this type of amplifier will be considered in succeeding sections of this chapter.

It can be shown that it is possible to put *more* feedback around a conditionally stable amplifier than around an equivalent unconditionally stable one. Thus, better operational characteristics are possible with conditionally stable systems. This technique is often used in feedback control systems, where the complexity is sufficiently high to warrant the use of safeguards against the problems described here. In such systems, it is customary to stabilize individual gain blocks separately (by unconditionally stable feedback) and use these stable blocks as elements of the loop around which conditionally stable feedback is placed.

While the analysis has considered only single-loop feedback amplifiers, these comments are applicable to multiloop systems also. However, the single-loop amplifier allows a ready visualization of its characteristics from a single T plot, while this is not the case for the multiloop system.

4.11. The Bode Plot. In preceding sections we have discussed the use of a polar Nyquist plot of T for the analysis of stability. This plot presented the magnitude and phase angle of T in a single display, with frequency as the running parameter. While such a plot gives a ready visualization of the amplifier stability, it is relatively tedious to construct. In addition, any changes in the T function are exceedingly difficult to incorporate and require a complete recomputation of the plot.

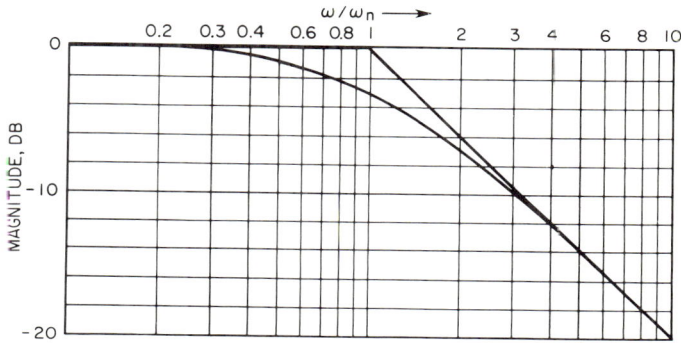

FIG. 37. Magnitude of $1/(s/\omega_n + 1)$ vs. frequency.

The Bode plot provides a convenient alternative. Here, the magnitude and phase of T are plotted independently as functions of the frequency. If frequency and amplitude are plotted on logarithmic scales, these curves may be drawn by inspection of the T function, using the highly developed straight-line-approximation methods of network theory. The method is best described by investigating the behavior of the first-order polynomial $1/(s/\omega_n + 1)$ where s is the complex frequency. Figure 37

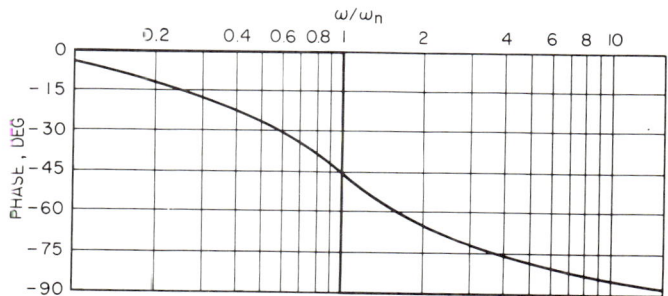

FIG. 38. Phase of $1/(s/\omega_n + 1)$ vs. frequency.

shows the plot of the magnitude of this function vs. frequency. The straight-line approximation is at a negative slope of 6 db/octave, starting at a corner frequency $\omega = \omega_n$.

Figure 38 shows the argument of this function vs. frequency. Here, the straight-line approximation consists of a step lag (negative) in phase by 90°, at $\omega = \omega_n$. In each case the actual curves are also drawn in addition to the straight-line approximation.

A network function, consisting of the ratio of a number of first-order polynomials, can be drawn by inspection, starting with the lowest corner frequency and appropriately altering the amplitude and phase as each additional corner frequency is encountered. An example will serve to illustrate this method. Figure 39 shows the attenuation and phase plots for

$$T = K/(s/10 + 1)(s/20 + 1)(s/40 + 1) \tag{85}$$

where K is the low-frequency loop gain. The appropriate corner frequencies are $\omega = 1$, 20, and 40 radians, respectively. Once the straight-line approximations are made, the actual curves are sketched in (K in decibels) with the aid of Figs. 37 and 38. From this plot, it is seen that T has a phase shift of 180° when its magnitude is $(K - 21)$ db. By adjusting K, this magnitude may be set to any desired value.

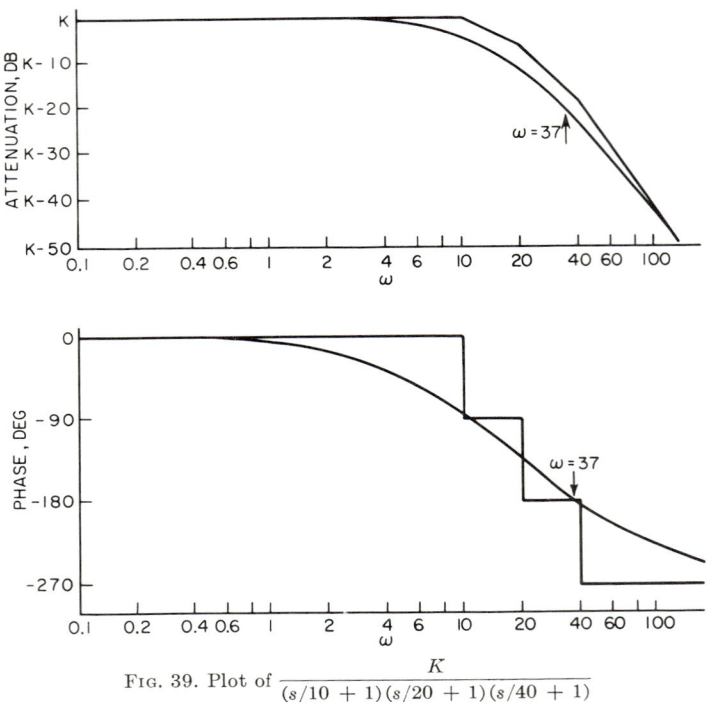

Fig. 39. Plot of $\dfrac{K}{(s/10 + 1)(s/20 + 1)(s/40 + 1)}$

Thus, it is possible to investigate the effect of varying the amount of feedback on the performance of the amplifier.

The effect of shaping the feedback function can also be readily studied. Let the T function of Eq. (85) be shaped such that

$$T = \frac{K}{(s/10 + 1)(s/20 + 1)(s/40 + 1)} \frac{s/6 + 1}{s/0.6 + 1} \tag{86}$$

The Bode plot for this function is shown in Fig. 40. It is seen here that T has a phase shift of 180° when its magnitude is $(K - 38)$ db. Thus, the gain required to cause instability in this amplifier is 17 db more than for the example of Fig. 39, where no shaping is used.

An additional function that is often encountered is the second-order polynomial $1/(s^2/\omega_n^2 + 2\zeta s/\omega_n + 1)$. Its straight-line-magnitude approximation consists of a 12 db/octave fall-off characteristic, at $\omega = \omega_n$. The phase approximation consists of a 180° phase lag at this frequency. Figures 41 and 42 show the actual plots for $T = 1/(s^2/\omega_n^2 + 2\zeta s/\omega_n + 1)$ for different values of ζ.

For all $\zeta \geq 1$, the polynomial can be written as $1/(s/\omega_1 + 1)(s/\omega_2 + 1)$. Thus, for $\zeta \geq 1$, the polynomial contributes two poles on the σ axis. For $\zeta < 1$, the polynomial contributes a complex conjugate pole pair in the s plane. For reasons that will be apparent later, ζ is often called the damping ratio.

4.12. The Nichols Chart. The Nichols chart provides a method for obtaining the closed-loop frequency response from the loop transmission. The theory underlying the use of this chart is as follows:

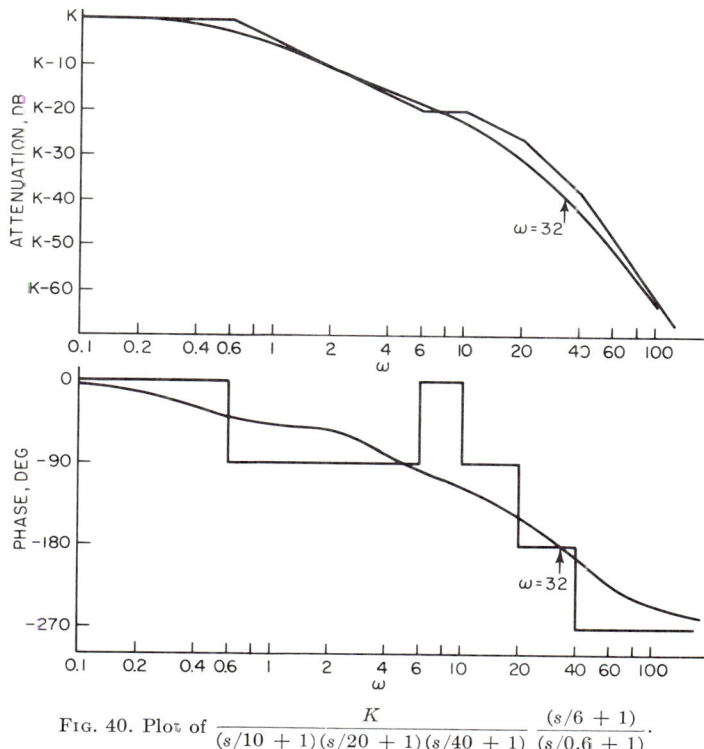

FIG. 40. Plot of $\dfrac{K}{(s/10 + 1)(s/20 + 1)(s/40 + 1)} \dfrac{(s/6 + 1)}{(s/0.6 + 1)}$.

The response of a single-loop feedback amplifier may be written as

$$\frac{E_o}{E_i} = \frac{A^0}{1 - A^0\beta} = -\frac{1}{\beta}\frac{-A^0\beta}{1 - A^0\beta} \qquad (87)$$

In the general feedback theory we have associated T with $A^0\beta$, the loop transmission. (The minus sign is introduced because of the sign convention in defining T.) Thus,

$$E_o/E_i = -(1/\beta)[T/(1 + T)] \qquad (88)$$

If we define

$$M = |T/(1 + T)| \qquad (89)$$

and

$$N = \arg [T/(1 + T)] \qquad (90)$$

the two functions M and N can be directly displayed on a Nichols chart, as shown in Fig. 43. By plotting the magnitude and phase of T on this chart, the magnitude and

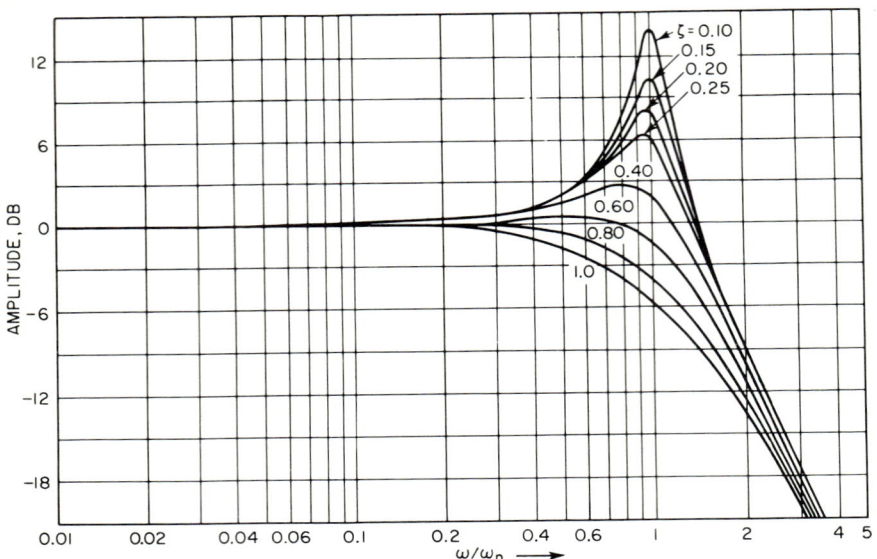

Fig. 41. Amplitude vs. frequency plot for a complex conjugate pole pair.

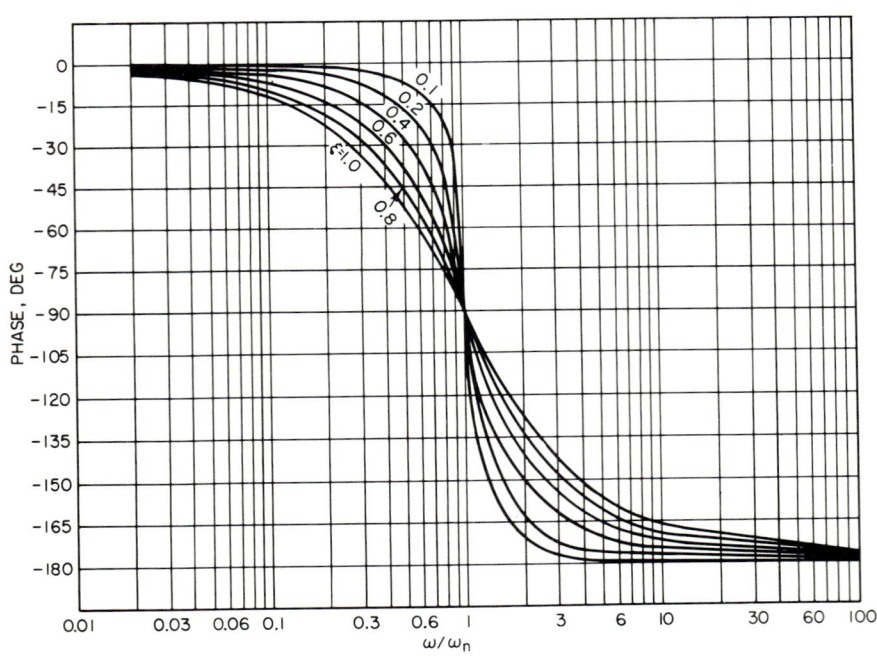

Fig. 42. Phase vs. frequency plot for a complex conjugate pole pair.

phase of $T/(1 + T)$ may be obtained directly. This may be combined with the frequency characteristic of $1/\beta$ to give the closed-loop response of the amplifier. If β is frequency-independent, as is often the case, the Nichols chart gives the closed-loop response without further adjustment.

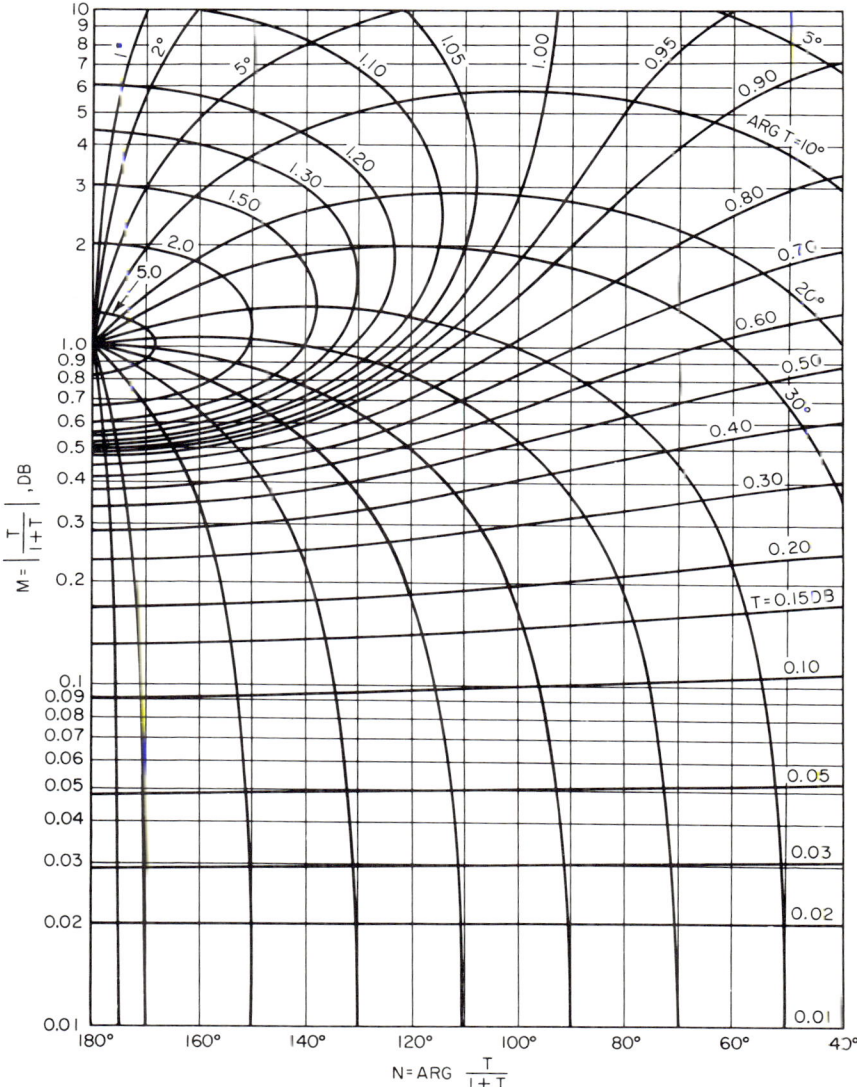

Fig. 43. The Nichols chart.

4.13. The Root-locus Method. The root-locus method may be used to study directly the roots of the closed-loop response characteristic as the gain around the loop is altered. The use of this method makes it possible to gage the relative effectiveness of the various poles of the closed-loop characteristic on the transient response of the overall system. As such, it is particularly useful in designing systems with prescribed transient responses and is specially suited[15] to the design of wideband pulse amplifiers.

A detailed description of the method, with its many ramifications and short cuts, is given in any standard text[16] on control-system synthesis. Actual design examples based on this method are more appropriately treated in Chap. 25 on Broadband Amplifiers.

The technique of the root-locus method is best illustrated by means of an example. Consider an amplifier whose loop transmission is given by

$$T = \frac{K}{(s/s_1 + 1)(s/s_2 + 1)} \tag{91}$$

where s_1 and s_2 are simple poles. For such an amplifier, the closed-loop response is given by

$$\frac{T}{1+T} = \frac{K s_1 s_2}{s^2 + s(s_1 + s_2) + s_1 s_2 (1 + K)} \tag{92}$$

The performance of this amplifier may be studied by investigating the roots of $s^2 + s(s_1 + s_2) + s_1 s_2 (1 + K)$ as the gain factor K is varied.

At $K = 0$, the roots are at $s = -s_1$ and $s = -s_2$. Thus, the gain function comprises two simple poles. As K is increased, the poles approach each other, as shown in Fig. 44, until

$$K = K_1 = (s_1 + s_2)^2/4s_1 s_2 - 1 \tag{93}$$

At this point, the poles coincide on the σ axis, at A, forming a double pole given by

$$s = -(s_1 + s_2)/2 \tag{94}$$

With further increase in K, the poles become complex and conjugate in nature. However, $\sigma = -(s_1 + s_2)/2$ for all values of $K \geq K_1$. Hence the root locus is a straight line, parallel to the $j\omega$ axis, and spaced at $(s_1 + s_2)/2$ from it.

For the case where

$$K = K_2 = (s_1 + s_2)^2/3s_1 s_2 - 1 \tag{95}$$

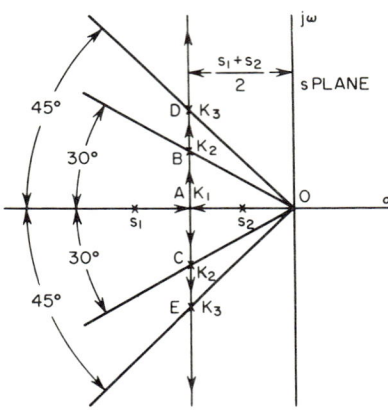

FIG. 44. Root-locus plot.

the roots are located at B and C, on the 30° radial line from the origin. An amplifier with this root location will have a maximally flat delay characteristic.

If K is increased further, until

$$K = K_3 = (s_1 + s_2)^2/2s_1 s_2 - 1 \tag{96}$$

the roots are at D and E, on the 45° radials. This corresponds to an amplifier with a maximally flat magnitude characteristic. Additional increase in K moves the poles farther along the locus, resulting in an equiripple characteristic.

The root-locus method also provides information on the stability of the amplifier. In this example, the roots always remain in the left half s plane, and hence the amplifier is stable for all values of K.

Figure 45 shows the sketch of a root locus for an unconditionally stable amplifier, where

$$T = \frac{K}{s(s+1)(s+2)} \tag{97}$$

At $K = 0$, the poles are at $s = 0$, -1, and -2. With increasing K, the pole at -2 moves farther and farther away from the $j\omega$ axis, and its effect on the amplifier performance becomes correspondingly reduced.

The behavior of the poles at 0 and -1 is quite different. With increasing K these poles approach each other, coincide on the real axis (at $K = 0.4$), and then break away to form a conjugate complex pole pair, whose effect on the system response increases with increasing K. At $K = 6$, the system becomes unstable, and it remains unstable for any further increase in K. For all values of $K \leq 6$, the amplifier is stable.

This example serves to illustrate an additional merit of the root-locus method. Here, the system performance is almost entirely dominated by the behavior of the poles at $s = 0$ and -1. Appropriately, these are known as the dominant pole pair, and improvement in the transient performance of the amplifier must be centered around their behavior.

Figure 46 shows the sketch of a root

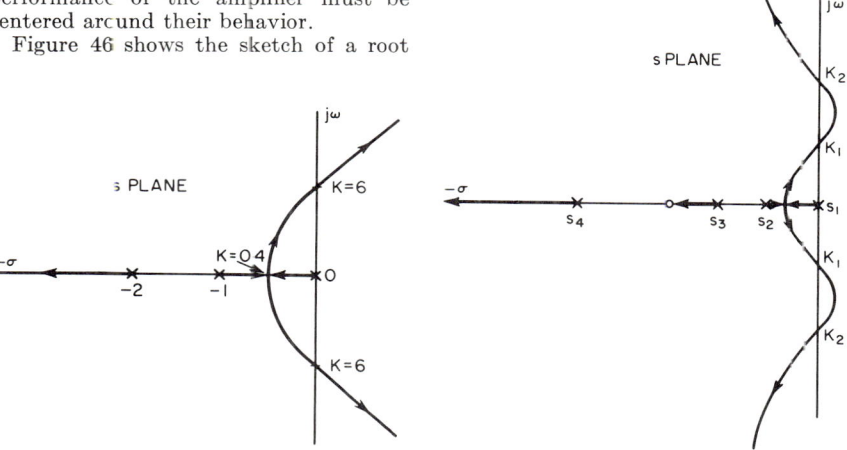

FIG. 45. Root-locus plot for an unconditionally stable amplifier.

FIG. 46. Root-locus plot for a conditionally stable amplifier

locus for a conditionally stable amplifier. Ignoring all but the dominant poles, it is seen that the amplifier is stable for values of K from zero to K_1, and from K_2 to infinity, but unstable for all values of K between K_1 and K_2. Again it should be noted that the behavior of this relatively complex system (having four poles) is almost entirely governed by that of a single pair of poles.

5. THE SPECIFICATION OF AMPLIFIER RESPONSE

In this section we shall discuss the various ways that the performance of a negative-feedback amplifier may be specified. In each case, the specification is translated into a design parameter associated with the loop transmission characteristic. The analysis is largely confined to single-loop amplifiers, and to those amplifiers which may be so treated for all practical purposes.

5.1. The Amount of Feedback. The most common use of feedback is in amplifiers which must operate with a closely prescribed characteristic, independent of variations in individual components due to aging. In general, the variation of the active components is the most serious, and negative feedback must be used to reduce the effect of this variation on the overall performance of the amplifier.

The general theory of feedback has shown that the sensitivity of an amplifier to component change is inversely related to the value of the feedback factor. Thus, given the tolerance limits on the components involved (both as a function of operating conditions and at end of life), as well as the tolerance limits on overall amplifier gain, the necessary feedback factor may be established.

The feedback factor is the amount by which the amplifier gain is reduced by the application of feedback. Thus, a 40-db amplifier with 30-db negative feedback implies an amplifier which has 70-db gain prior to the addition of feedback, and 40-db gain with feedback. This "amount of feedback" is usually defined within the amplifier passband.

5.2. The Closed-loop Response. Often, a specification is placed on the closed-loop frequency response, in addition to the sensitivity. Thus, an amplifier may be required to maintain certain tolerances on its frequency response, both in the passband and outside it. The Nichols chart provides an ideal way of handling this problem, since it allows the specification on closed-loop response to be directly translated into a specification of the loop transmission characteristic.

5.3. The Gain Margin. In some amplifier designs, a margin of safety against oscillation is specified. Even if unspecified, the designer ordinarily sets such a margin,

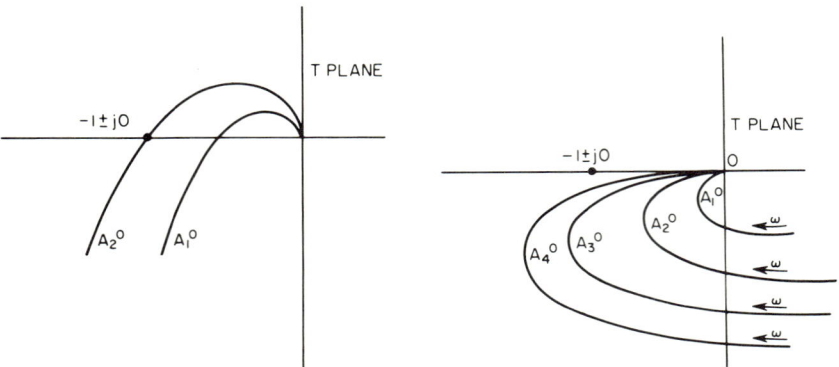

Fig. 47. Nyquist plot for an amplifier.   Fig. 48. Amplifier with infinite gain margin.

known as the *gain margin*. The gain margin is conveniently defined in terms of the Nyquist plot of T. Figure 47 shows this plot for an amplifier, in the neighborhood of the critical point $-1 \pm j0$. The same function is shown for two different values of loop gain A_1^0 and A_2^0. The feedback amplifier with the loop gain of A_1^0 is stable. In addition, the loop gain may be increased by A_2^0/A_1^0; i.e., its gain may be increased to A_2^0 before it becomes unstable. Hence the amplifier has a gain margin of A_2^0/A_1^0. This gain margin is usually expressed in decibels and is a direct measure of the gain change that may be tolerated in the loop characteristic.

5.4. The Phase Margin. Occasionally, gain margin does not provide a meaningful criterion for specifying an amplifier. Thus, an amplifier whose T plot is depicted in Fig. 48 (in the neighborhood of the critical point) has an infinite gain margin. Clearly, other criteria are necessary for such a system.

For this amplifier, increasing the open-loop gain results in the T plot coming increasingly close to the critical point. Consequently, the closed-loop characteristic will show an increasingly undamped high-frequency response, as in Fig. 49.

The "nearness" of the T plot to the critical point can be specified in terms of its angular distance from the critical point, when T has a magnitude of unity. Thus, for the T plot of Fig. 50, we can obtain a *phase margin* of ϕ_m. Thus, the amplifier corresponding to this T plot can tolerate an additional phase shift (lag) of ϕ_m before it will oscillate.

5.5. The Transient Response. It is often required to design a feedback amplifier with a specified transient response. As digital applications become more important, this type of specification becomes increasingly common. The requirements on a pulse amplifier are ordinarily given in terms of a prescribed rise time and a prescribed overshoot. Both of these are associated with the high-frequency characteristics of the amplifier.

The input and output response of a feedback amplifier may generally be written in the form of a linear differential equation with constant coefficients. Let us consider a *second-order system*, in which

$$d^2e_2/dt^2 + 2\zeta\omega_n(de_2/dt) + \omega_n^2 e_2 = \omega_n^2 e_1 \qquad (98)$$

where e_2 and e_1 are output and input signals, respectively, and ζ and ω_n are system constants.

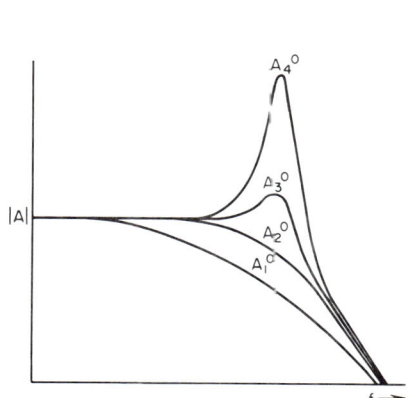

FIG. 49. Frequency response of the closed loop.

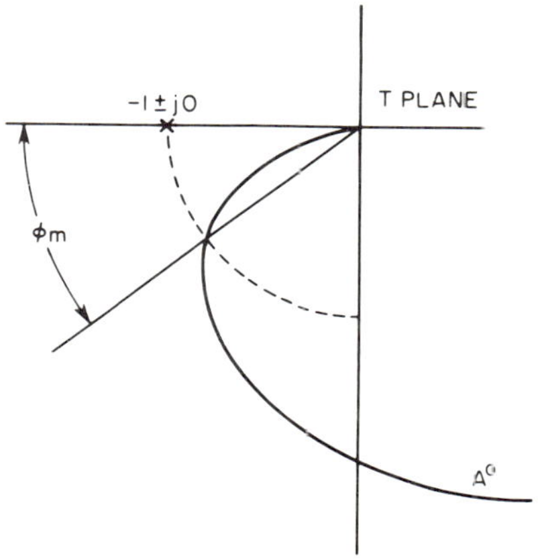

FIG. 50. Phase margin.

If a step input signal $e_1(t)$ is applied to the amplifier at time $t \geq 0$, the amplifier response is given by, for $t > 0$,

$$e_2(t) = 1 - (e^{-\zeta\omega_n t}/\sqrt{1-\zeta^2}) \sin(\omega_n \sqrt{1-\zeta^2}\, t + \cos^{-1} \zeta) \qquad (99)$$

This response is plotted in Fig. 51 and indicates overshoot for all $\zeta \leq 0.7$ (the case for which $\zeta = 0.7$ corresponds to a critically damped response). It is seen that, for all $\zeta > 0$, the response finally settles down to the steady-state value.

For an amplifier of this type, the damping ratio ζ is uniquely specified for a prescribed overshoot. This relationship is shown in Fig. 52. Since the abscissa of Fig. 51 is in terms of $\omega_n t$, the rise time is given by both ω_n and ζ.

For the second-order system, ζ is related to the phase margin ϕ_m by

$$\phi_m = \tan^{-1} 2\zeta \frac{1}{(4\zeta^4 + 1)^{\frac{1}{2}} - 2\zeta^2} \qquad (100)$$

Figure 53 is a plot of this relation. For phase margins under 40°, we may use the approximate relation

$$\phi_m \simeq 2\zeta \qquad (101)$$

where ϕ_m is in radians. Thus, the transient response of the system is uniquely determined by its bandwidth and its phase margin.

The above analysis has been developed for an amplifier whose characteristic equation has a pair of complex conjugate roots. In practical amplifiers, the input-output relation is considerably more complex, and the characteristic equation has correspondingly more roots, some of which are real, and others form complex pole pairs. In any such system, as discussed in Sec. 4.13, the high-frequency response is governed by a dominant pole pair. Thus, even though an amplifier may have a large number of poles, its transient response is very similar to that of a second-order system.

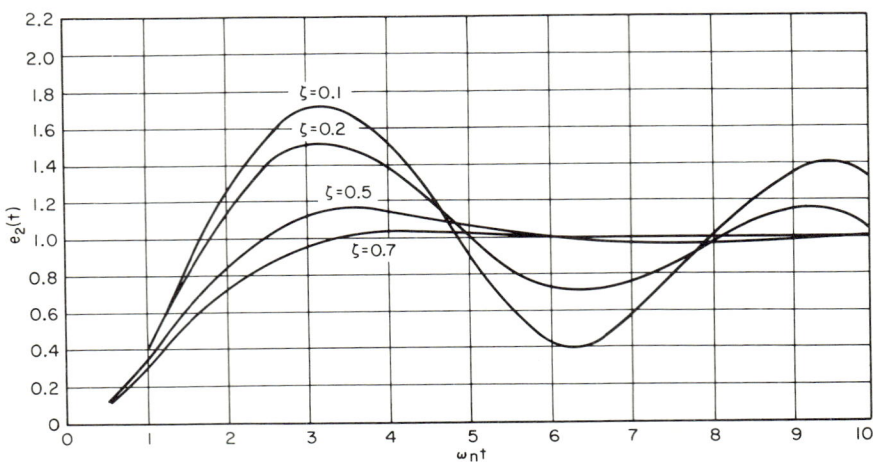

FIG. 51. Transient response of a second-order system.

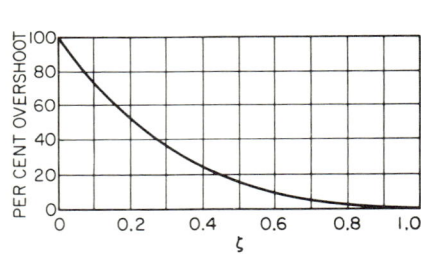

FIG. 52. Overshoot vs. damping ratio.

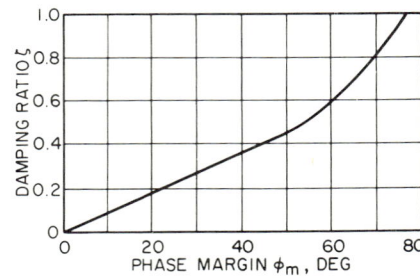

FIG. 53. Phase margin vs. damping ratio.

6. THE SINGLE-LOOP UNCONDITIONALLY STABLE AMPLIFIER

The single-loop unconditionally stable feedback amplifier is perhaps the most commonly encountered type in actual practice. Its specification and design will be detailed in this section.

A single-loop amplifier may be defined as one in which the return ratio for any active element is zero, if the gain of any *other* active element is set to its reference value. More simply stated, a single transmission path exists around the amplifier. This condition is fulfilled by means of a tandem arrangement of stages. In practice, however, the presence of small parasitic impedances between all parts of the circuit, as well as the feedback coupling between different parts of the wiring, result in a multiloop amplifier. The presence of such unwanted loops has been described in Sec. 4.9, together with techniques for minimizing or eliminating the need for their inclusion in the analysis. Once these methods are applied, the amplifier may indeed be treated as a single-loop system.

6.1. Specifications. In addition to being unconditionally stable, a common requirement is that the amplifier *remains* stable on a long-term basis. Thus, the amplifier must remain stable for a specified change of loop gain with time. A gain margin is therefore required to specify the amplifier.

A specification on phase margin can also be developed for the amplifier, based on either a prescribed steady-state or a transient response. The manner in which phase margin relates to these performance characteristics is detailed in Sec. 5.

In addition, the cutoff bandwidth is usually specified.

6.2. The Ideal Cutoff Characteristic. Our attention will focus on the high-frequency characteristics of the amplifier, for the following reasons:

1. The high-frequency characteristics of the amplifier govern its transient response.
2. High-frequency response is generally harder to control than low-frequency response. This is due to the fact that all active elements show attenuation and

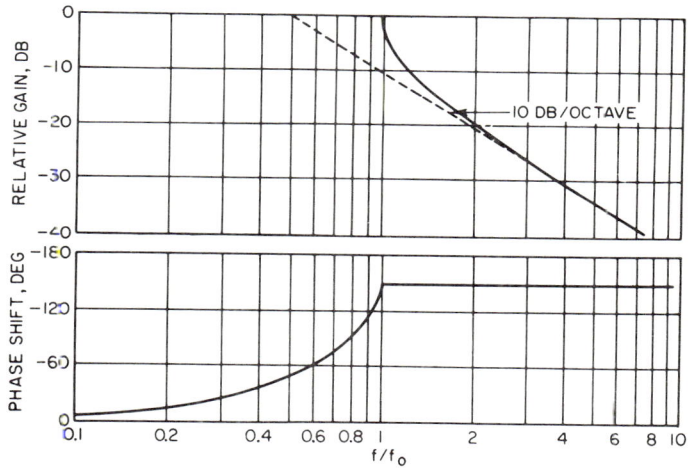

FIG. 54. The ideal cutoff characteristic. (*Courtesy of D. Van Nostrand Company, Inc.*)

phase-shift effects at the high-frequency end of their performance spectrum. On the other hand, low-frequency performance to direct current can be obtained by direct coupling.

Let the amplifier have a 3-db bandwidth of f_o, a gain margin of x db, and a phase margin of πy radians. For such an amplifier, Bode[17] has shown that the phase shift of the loop transmission characteristic will be limited to $\pi(1 - y)$ radians (i.e., the phase margin will be πy radians), if the rate at which the open-loop gain changes is limited to $12(1 - y)$ db/octave.

For a typical amplifier with a phase margin of 30°, $y = \frac{1}{6}$ and the asymptotic cutoff of the ideal Bode characteristic is 10 db/octave. Figure 54 shows the ideal loop transmission characteristic for this amplifier, for gain as well as for phase. If the rolloff is approximated by that due to a simple capacitive cutoff, it may be shown that the cutoff frequency must be set at approximately $2f_o$.

6.3. The High-frequency Asymptote. In all amplifiers, an upper frequency limit is reached beyond which it is not possible to maintain a fall-off at $12(1 - y)$ db/octave. In vacuum-tube circuits, parasitic tube capacitances determine the ultimate asymptotic behavior of T, while transit-time effects are usually the limiting factor in transistor circuits. The ultimate asymptotic behavior of any circuit takes the form of a slope of $6n$ db/octave. (In general, n is equal to or greater than the number of active elements in the circuit.)

6-36 AMPLIFIER FUNDAMENTALS

Figure 55 shows the T characteristic of an unconditionally stable amplifier with a phase margin of πy radians. Also shown is the asymptotic slope of $6n$ db/octave. At any frequency, the phase shift due to this asymptote will *add* to that due to the ideal T characteristic. In general, this will further limit the maximum amount of

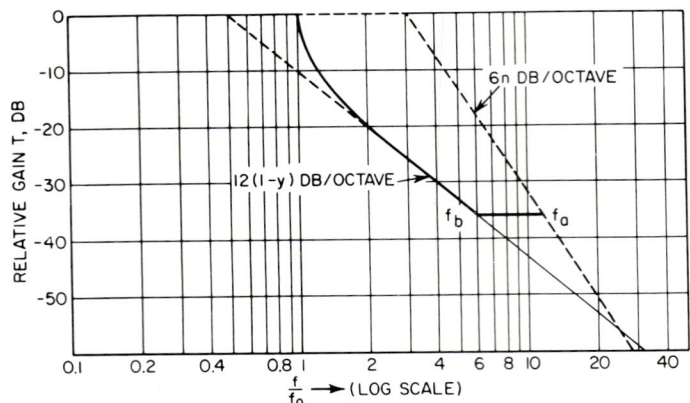

Fig. 55. Inclusion of the asymptote.

feedback that can be applied in the passband of the amplifier, in order that the phase shift at zero db can be maintained below 180°.

6.4. Maximum Available Feedback. The amount of feedback that can be used may be maximized by the inclusion of a step in the T characteristic, linking it with

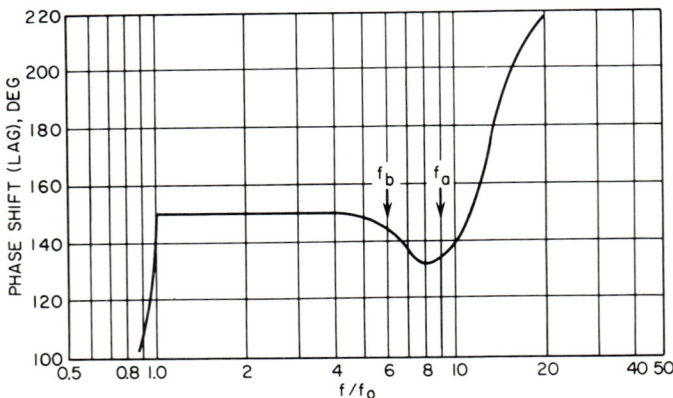

Fig. 56. Phase shift of corrected T. (*Courtesy of D. Van Nostrand Company, Inc.*)

the asymptote, as shown in Fig. 55. If f_a and f_b are selected such that

$$f_a/f_b = n/2(1 - y) \tag{102}$$

the lagging phase shift associated with the asymptote is exactly canceled by the leading phase associated with the corner frequency f_b. This cancellation is in effect to f_b and beyond, and takes the form shown in Fig. 56. Thus, at frequencies up to, and somewhat beyond, f_b, the overall phase shift is equal to or less than that of the T characteristic alone.

The gain-margin criterion is now fulfilled by setting the gain constant associated with the low-frequency value of T, so that the flat region $f_a - f_b$ occurs at the desired value. Thus, the curve of Fig. 57 is drawn for a gain margin of 9 db with a phase margin of 30° and an asymptotic characteristic of 18 db/octave. For this amplifier

$$f_a/f_b = n/2(1 - y) = 1.8 \tag{103}$$

Setting the $f_a - f_b$ line at -9 db sets the maximum value of available feedback at 29 db. If this value is above that specified by the design requirements, it may be adjusted by using devices whose asymptote occurs at a lower frequency (i.e., devices with poorer high-frequency response, and hence less expensive). Alternately, the design may be left unchanged, since it is well within specifications.

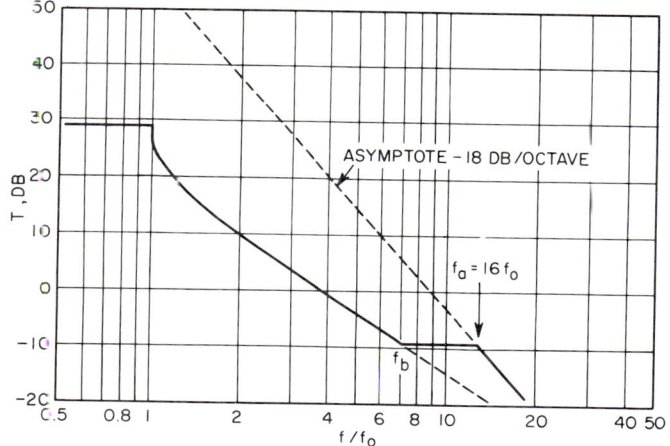

FIG. 57. The corrected cutoff characteristic. (*Courtesy of D. Van Nostrand Company, Inc.*)

If the maximum amount of available feedback is below the design requirement, it will be necessary to use devices with extended high-frequency performance, in order to move the asymptote out to a higher frequency.

It should be noted that the successful execution of this design required that control of the amplifier characteristic be maintained to approximately $16f_o$, i.e., to about 4 octaves beyond the cutoff frequency. This result has been generalized by Bode to provide a useful rule of thumb: It is necessary to control the T characteristic out to $2^{(k+1)}f_o$ in order to provide $10k$ db of feedback in the passband. This imposes a relatively stringent limitation on the design of absolutely stable single-loop amplifiers having large amounts of negative feedback.

6.5. Excess Phase. Thus far, the design has assumed a minimum phase shift consistent with loop characteristics. In most practical amplifiers, excess phase is present beyond this minimum. Perhaps the most important form of excess phase is that associated with transit-time effects in the devices themselves.

This effect is present in both vacuum tubes and transistors. However, parasitic capacitive effects in vacuum tubes are so serious that the asymptotic falloff characteristic associated with them is reached before this excess phase becomes significant. Thus, excess phase effects may be ignored in all but very high frequency multistage vacuum-tube amplifiers.

The situation is quite different for modern diffused transistors. These devices are characterized by extremely small parasitic capacitances, graded base-impurity distributions, and wide depletion layers, resulting in comparably large transit angles. It may be shown that the common-emitter current gain of a transistor can be approximated[18] by

$$\alpha_e = \alpha_{e0} \frac{e^{-jM(\omega/\omega_{ae})}}{1 + j\omega/\omega_{ae}} \quad (104)$$

Thus, the current-gain term not only provides the asymptote $1/(1 + j\omega/\omega_{ae})$ but also provides an excess linear-phase-shift term $e^{-jM(\omega/\omega_{ae})}$, in addition to that given by the single-pole expression.

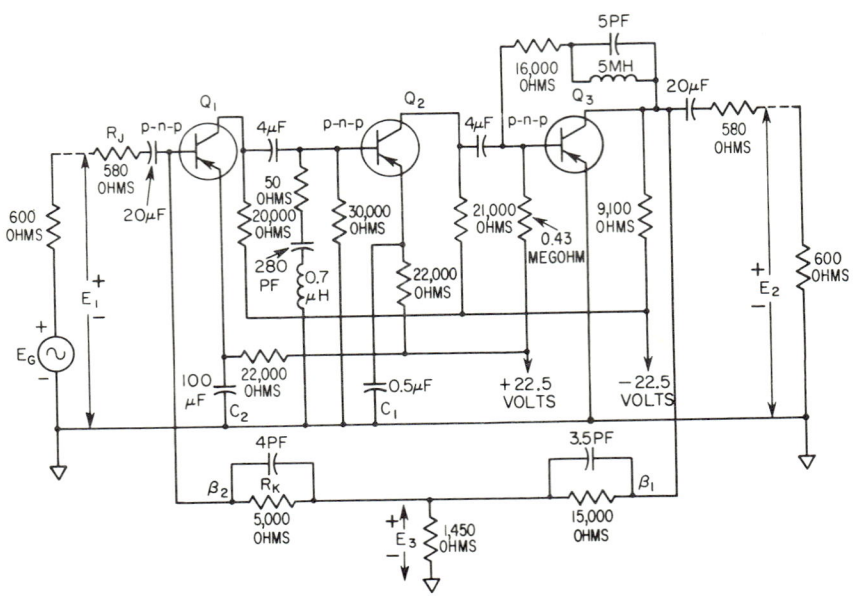

Fig 58. Three-stage feedback amplifier. (*Reproduced by permission from F. H. Blecher, "Design Principles for Single Loop Transistor Feedback Amplifiers," IRE Trans., CT-4, 1957, pp. 145–156.*)

The overall cutoff characteristic for T may be compensated for excess phase by readjusting the step at which the transition occurs from the Bode characteristic to the asymptotic characteristic. If the asymptotic characteristic falls off at $6n$ db/octave, and if f_p is the frequency at which the excess phase is $2n/\pi$ radians, then frequency compensation for excess phase is provided by setting

$$f_b = \frac{2(1 - y)}{n} \frac{f_a f_p}{f_a + f_p} \quad (105)$$

The rest of the design proceeds as before. It should be noted that this further reduces the maximum amount of feedback that can be applied around an amplifier.

6.6. An Illustrative Example. The design of a transistor amplifier incorporating these principles has been carried out by Blecher[19] and will be described in this section. Amplifier requirements were as follows:

Phase margin	30°
Gain margin	10 db
Voltage gain	36 db ± 1 db
Input and output impedance	600 ohms ± 5%
Bandwidth	2 kc–1 Mc

In order to meet the voltage-gain tolerance, about 34 db of feedback is required. A three-stage amplifier is used as shown in Fig. 58 with shunt feedback at both input and output. This reduces both input and output impedances to about 20 ohms (approximately 34 db below that obtained without feedback). Consequently, a fixed resistance of 580 ohms is used to raise the impedance to 600 ohms and render it insensitive to amplifier gain changes.

The voltage gain of the amplifier is given by

$$\frac{E_2}{E_1} = \frac{R_K}{R_j A_1} \frac{A^0\beta}{1 - A^0\beta} \tag{106}$$

where $A_1 = E_3/E_2$, and $\angle^0\beta$ is the loop transmission. For the circuit values shown, this is 36 db.

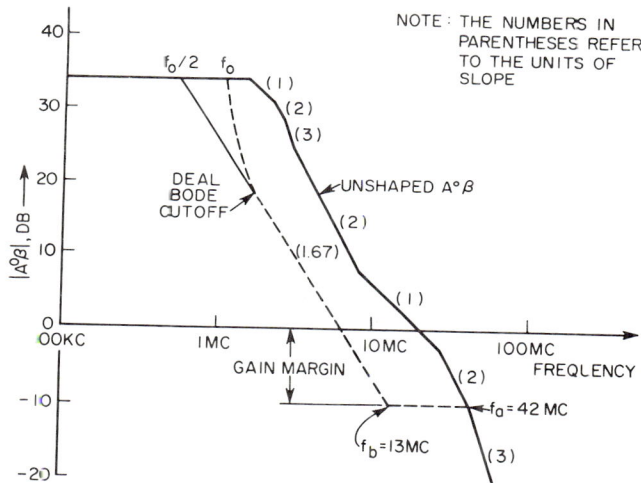

Fig. 59. Unshaped $A^0\beta$ plot. (Reproduced by permission from F. H. Blecher, "Design Principles for Single Loop Transistor Feedback Amplifiers," IRE Trans., CT-4, 1957, pp. 145-156.)

Figure 59 shows the unshaped $A^0\beta$ characteristic, obtained from a network analysis of the amplifier prior to the insertion of the various shaping networks. The ideal Bode cutoff is also shown, with the step inserted so as to compensate for an excess phase of 0.35 radian per transistor. (This value is somewhat conservative for the alloy transistors used in the design.)

Figure 60 shows the details of the main interstage shaping network, together with its frequency characteristic.

Figure 61 shows the various shaping terms used to approximate the Bode cutoff. Here, ω_{32}, ω_{42}, ω_{52}, and ω_{62} are those due to the RLC shaping network between the first and second stages, $\omega_{\beta1}$, $\omega_{\beta1}'$ are obtained by means of the 3.5-pf–15-kilohm combination in the feedback loop, while $\omega_{\beta2}$, $\omega_{\beta2}'$ refer to the 4-pf–5-kilohm combination in the feedback loop. Finally, the fillet required near the top of the useful band is approximated by means of a small amount of shunt feedback around the third stage. While this makes the amplifier a multiloop structure, this feedback is so slight that a single-loop analysis is entirely satisfactory.

Figure 62 shows the measured loop-gain characteristic for the amplifier, while Fig. 63 shows its overall response.

6-40 AMPLIFIER FUNDAMENTALS

The low-frequency response of the amplifier is shaped by means of the 0.5-μf emitter bypass capacitor C_1, which provides a 6 db/octave cutoff at 2,000 cps. No stability problems are created at the low end, since this term introduces at most a 90° phase shift. At extremely low frequencies, phase shift due to the 100-μf capacitor C_2

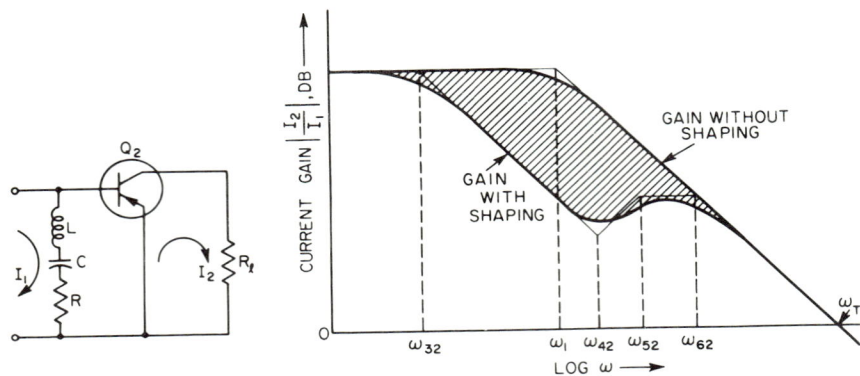

Fig. 60. Interstage shaping network. (*Reproduced by permission from F. H. Blecher, "Design Principles for Single Loop Transistor Feedback Amplifiers," IRE Trans., CT-4, 1957, pp. 145–156.*)

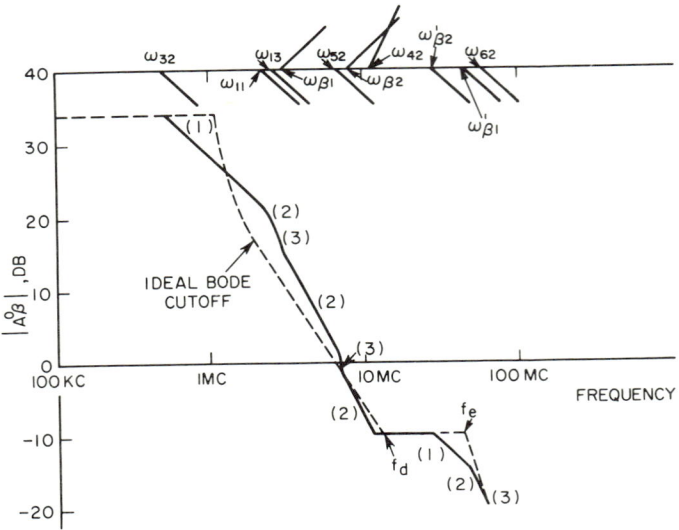

Fig. 61. Approximation to the Bode cutoff. (*Reproduced by permission from F. H. Blecher, "Design Principles for Single Loop Transistor Feedback Amplifiers," IRE Trans., CT-4, 1957, pp. 145–156.*)

becomes significant. By this time, however, $A^0\beta$ is considerably less than unity, and the amplifier stability is ensured.

The interstage shaping function (see Fig. 60) may also be provided by applying local feedback around a single stage. While this results in a multiloop structure, the analysis may still be carried out on a single-loop basis. Ghandhi[20] has used this technique to design the amplifier of Fig. 64, with local shunt feedback provided around the second stage. Overall shunt feedback is provided to result in an amplifier having a gain of 40 ± 0.3 db over a temperature range from -55 to $-125°C$. Unique features of the circuit include the use of direct coupling between stages, providing a flat gain characteristic down to direct current (not including the effect of the

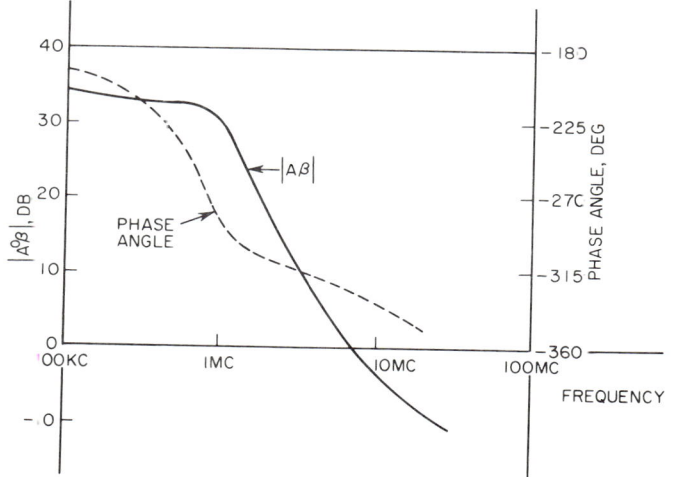

FIG. 62. Measured loop characteristics. (*Reproduced by permission from F. H. Blecher, "Design Principles for Single Loop Transistor Feedback Amplifiers," IRE Trans., CT-4, 1957, pp. 145–156.*)

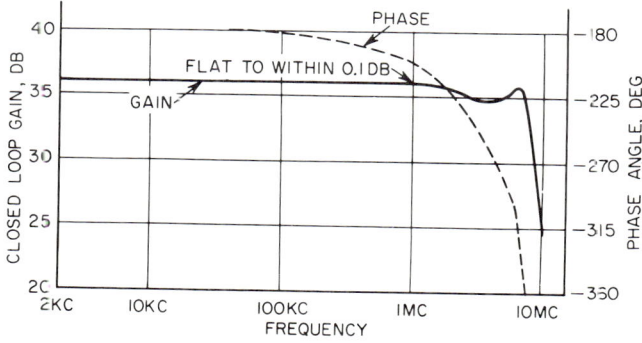

FIG. 63. Closed-loop response. (*Reproduced by permission from F. H. Blecher, "Design Principles for Single Loop Transistor Feedback Amplifiers," IRE Trans., CT-4, 1957, pp. 145–156.*)

input and output coupling elements) and the use of breakdown diodes to set the bias on the individual stages.

7. FEEDBACK CHAINS

For maximum improvement in performance characteristics, feedback should be applied between the input and output of an amplifier, i.e., over as many stages as possible. However, as has been pointed out in Sec. 6.4, the application of each 10 db

6-42 AMPLIFIER FUNDAMENTALS

of feedback requires the control of the loop characteristics for an additional octave beyond the cutoff frequency. As the number of stages in the feedback path increases, the control of the loop characteristic becomes increasingly difficult to achieve; hence, correspondingly less feedback can be applied. For this reason, it is often more practical to subdivide the amplifier into small sections, each of which has local feedback.

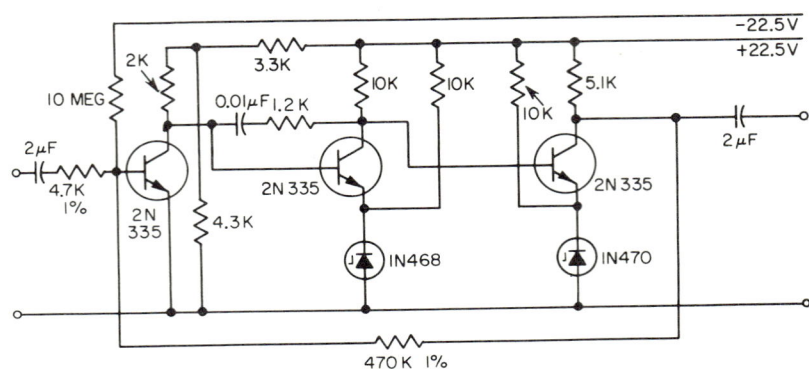

FIG. 64. Direct-coupled feedback amplifier. (*From S. Schwarz, "Selected Semiconductor Circuits Handbook," courtesy of John Wiley and Sons, Inc.*)

Such an amplifier is known as a feedback chain. Sometimes, a small amount of overall feedback may be applied around such a chain to effect an additional improvement in amplifier characteristics.

In general, the design of a transistor feedback amplifier is considerably more difficult than that of its vacuum-tube counterpart. The prime reason for this difficulty lies in the fact that a transistor amplifier is invariably operated in the frequency range where the phase shift in alpha is significant. In addition, the excess phase of a transistor may, in many cases, provide an additional 90° phase shift at high frequencies.* For this reason, the feedback chain is used more frequently with transistor amplifiers than with vacuum-tube circuits.

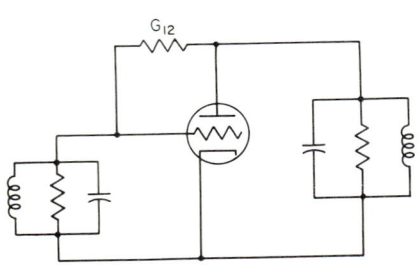

FIG. 65. The synchronous-tuned feedback stage.

7.1. Vacuum-tube Feedback Chains. Feedback chains have often been used with synchronously tuned vacuum-tube amplifiers in order to achieve the band characteristics ordinarily associated with stagger tuning. Thus, maximally flat or equiripple characteristics may be obtained by the proper application of feedback.

Figure 65 shows a single synchronously tuned stage, with feedback from output to input. Here, maximum feedback is provided at resonance since both input and output terminations are resistive and maximized at this frequency. As the frequency deviates from the band center, the amount of feedback falls off. It may be shown that a scheme of this type results in increasing the gain-bandwidth product of the stage, since the fall in midband gain is less than the effective improvement in bandwidth.

Figure 66 shows a feedback chain of this type. An important disadvantage of such a scheme lies in the loss of isolation normally existing between vacuum-tube stages.

* It is worth noting that the gain-bandwidth product of transistors is considerably higher than that of vacuum tubes.

FEEDBACK   6-43

This leads to difficulty in alignment, and also to dependence of the overall amplifier characteristic on the gain of any individual stage.

The problem can be eliminated by alternating isolation stages with feedback stages. A pair of such stages is shown in Fig. 67.

Beveridge[21] has made a detailed analysis of feedback chains and has developed

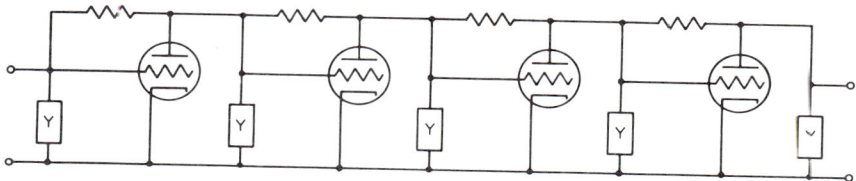

Fig. 66. The feedback chain.

procedures for designing the single feedback stage, the feedback pair, and the feedback triplet. He concludes that the feedback pair gives at once the greatest design simplicity, as well as performance comparable with the other schemes.

The equivalent circuit of the feedback pair is shown[22] in Fig. 68. For the bandpass case, let

$$Y = G + sC + 1/sL \qquad (107)$$

The center frequency ω_o is that associated with L and C.

For a maximally flat response, with bandwidth B, one obtains

$$G_{12} = (g_{m2}/2)[1 \pm \sqrt{1 - 2(BC/g_{m2})^2}] \qquad (108)$$

and

$$G = BC/\sqrt{2} - G_{12} \qquad (109)$$

Use of the negative sign in Eq. (108) usually results in higher gain.

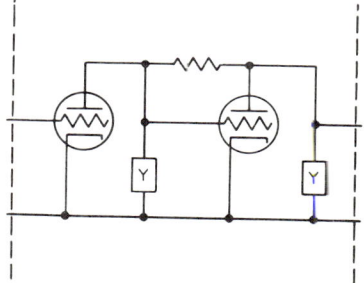

Fig. 67. The feedback pair.

The design of a complete amplifier may be carried out by partitioning its required pole locations into pairs of conjugate pole pairs, and synthesizing each such pair by

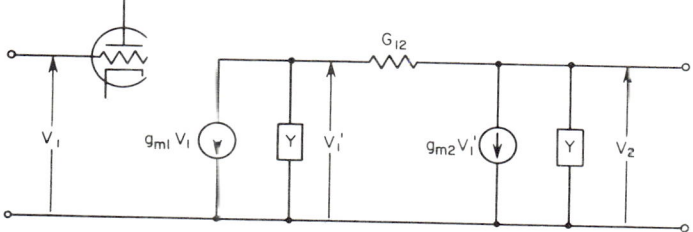

Fig. 68. Equivalent circuit for the feedback pair.

means of a feedback doublet. Figure 69 shows an example[23] of the use of this method to obtain a four-stage amplifier with a gain of 30 to 35 db, and a maximally flat bandpass characteristic centered around 21 Mc.

A similar technique may be used for synthesizing low-pass amplifiers, with the difference that a low-pass structure, given by

$$Y = G + sC \qquad (110)$$

must be employed. This technique is not commonly used, however, since alternate simpler methods are readily available.

It is worth noting that band shaping is the main reason for using feedback in this section. The amount of feedback is so slight that there is no significant improvement in amplifier characteristics. In addition, amplifier band shape is dependent on the gain of the second stage (the feedback stage) and is subject to change with aging of the vacuum tubes.

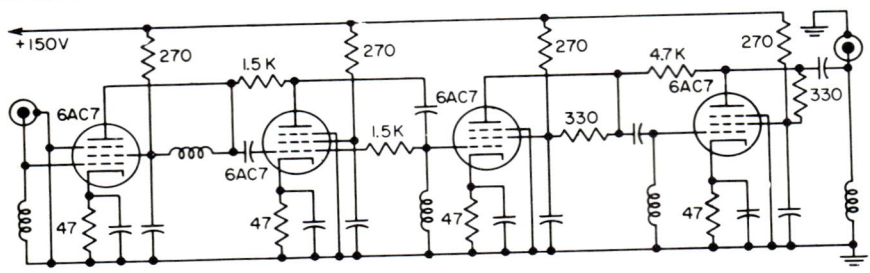

FIG. 69. Flat quadripole feedback amplifier. (*From G. E. Valley, Jr. and H. Wallman, "Vacuum Tube Amplifiers," courtesy of McGraw-Hill Book Company.*)

7.2. Transistor Feedback Chains. Howell[24] has provided an example of how a transistor feedback chain may be designed to achieve a maximum of bandwidth conservation. His design procedure is typical of that necessary to obtain the maximum performance from a device, and consists of a series of theoretical design and laboratory measurement steps.

The basic circuit used is the RL shunt feedback[25,26,27] stage of Fig. 70. It may be shown that the gain-bandwidth product of this stage is nearly equal to that obtained from the transistor alone. Thus, the use of this type of feedback allows a direct trade-off between gain and bandwidth. Figure 71 shows the performance of a single stage for varying values of L_f. Using a transistor with an f_T of 850 Mc, the single-stage response of Fig. 72 is obtained, with a bandwidth (zero-db point) of approximately 600 Mc. In theory, a cascade of such *noninteracting* stages would suffice to obtain any desired gain with a 500-Mc bandwidth.

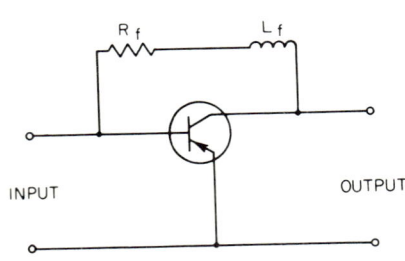

FIG. 70. Basic RL feedback stage. (*Reproduced by permission from D. W. Howell, "A Transistor Amplifier with 500-Mc Bandwidth," Stanford Electronics Laboratories, Technical Report No. 1820-1, SU-SEL-64-054.*)

The interaction between cascaded stages is shown by the measured frequency response of the amplifier using three identical stages (see Fig. 73). Here, the deviation from flat response is primarily due to the frequency dependence of source and load impedances of the individual stages.

Using the experimental data, circuit modifications of Fig. 74 may be used to eliminate the midband dip (using network X) and the high-frequency peak (using network Y). The complete stage, including bias networks, is shown in Fig. 75. Here, the inclusion of L_i and C_i permits some increase in gain by providing interstage impedance matching.

The circuit diagram of a seven-stage feedback chain, using identical* stages, is shown in Fig. 76. Its frequency-response characteristic is shown in Fig. 77.

* Final experimental adjustments, always necessary with high-frequency amplifiers, necessitated a slight departure from completely identical stages.

FEEDBACK   6-45

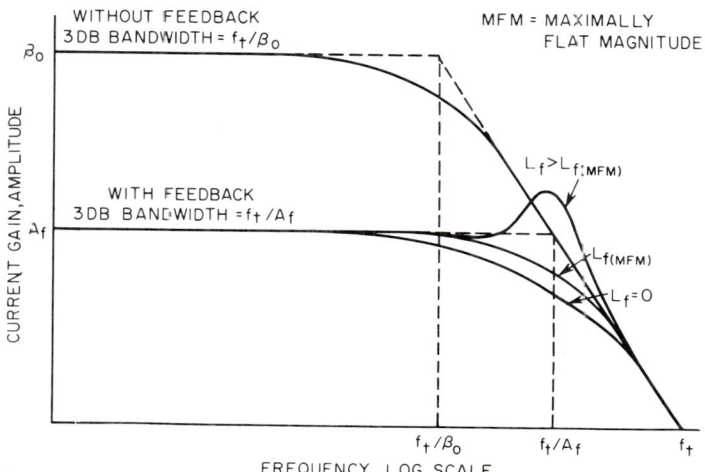

Fig. 71. Response of RL feedback stage. (*Reproduced by permission from D. W. Howell, "A Transistor Amplifier with 500-Mc Bandwidth," Stanford Electronics Laboratories, Technical Report No. 1820-1, SU-SEL-64-054.*)

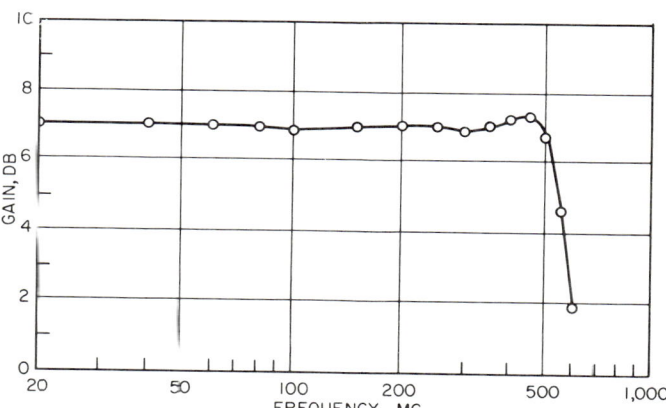

Fig. 72. Response for a single stage. (*Reproduced by permission from D. W. Howell, "A Transistor Amplifier with 500-Mc Bandwidth," Stanford Electronics Laboratories, Technical Report No. 1820-1, SU-SEL-64-054.*)

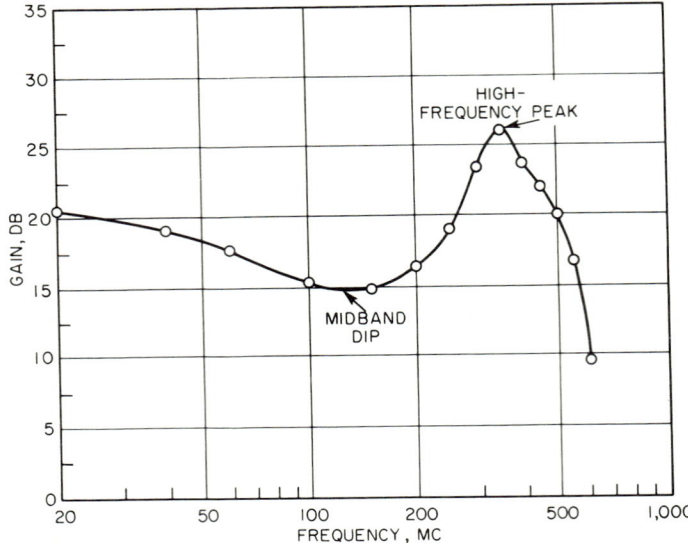

Fig. 73. Response for three identical stages. (*Reproduced by permission from D. W. Howell, "A Transistor Amplifier with 500-Mc Bandwidth," Stanford Electronics Laboratories, Technical Report No. 1820-1, SU-SEL-64-054.*)

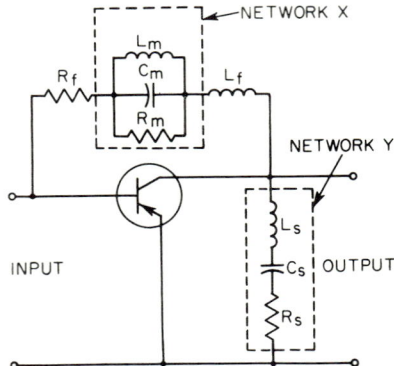

Fig. 74. Circuit modifications to eliminate the midband dip and the high-frequency peak. (*Reproduced by permission from D. W. Howell, "A Transistor Amplifier with 500-Mc Bandwidth," Stanford Electronics Laboratories, Technical Report No. 1820-1, SU-SEL-64-054.*)

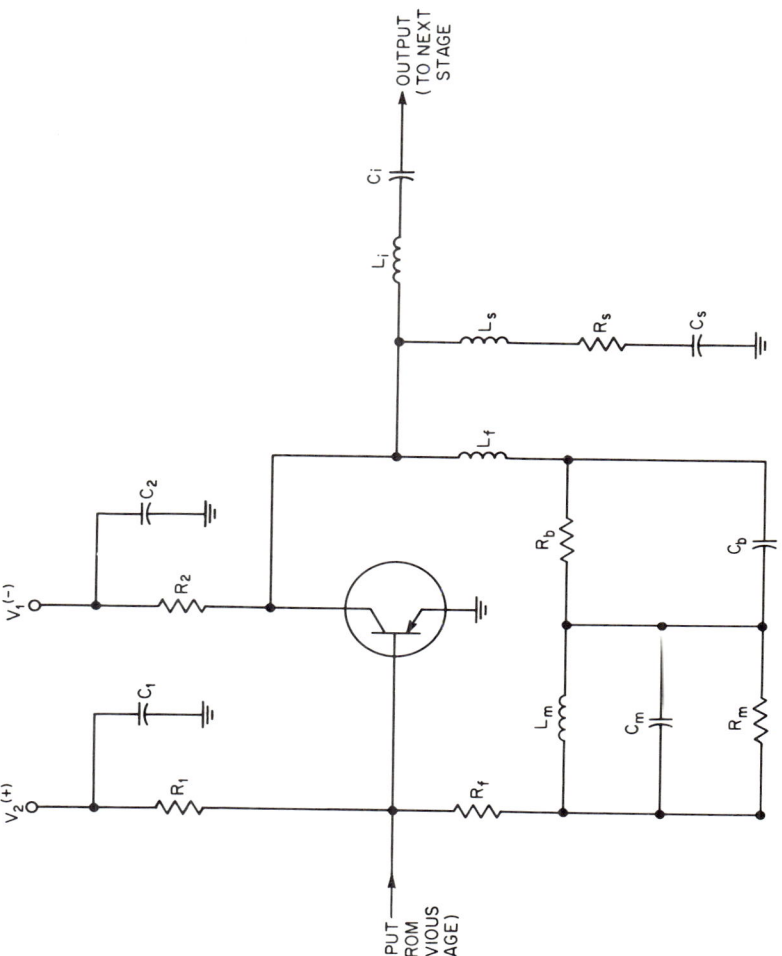

FIG. 75. Single stage with bias and matching network. (*Reproduced by permission from D. W. Howell, "A Transistor Amplifier with 500-Mc Bandwidth," Stanford Electronics Laboratories, Technical Report No. 1820-1, SU-SEL-64-054.*)

6–47

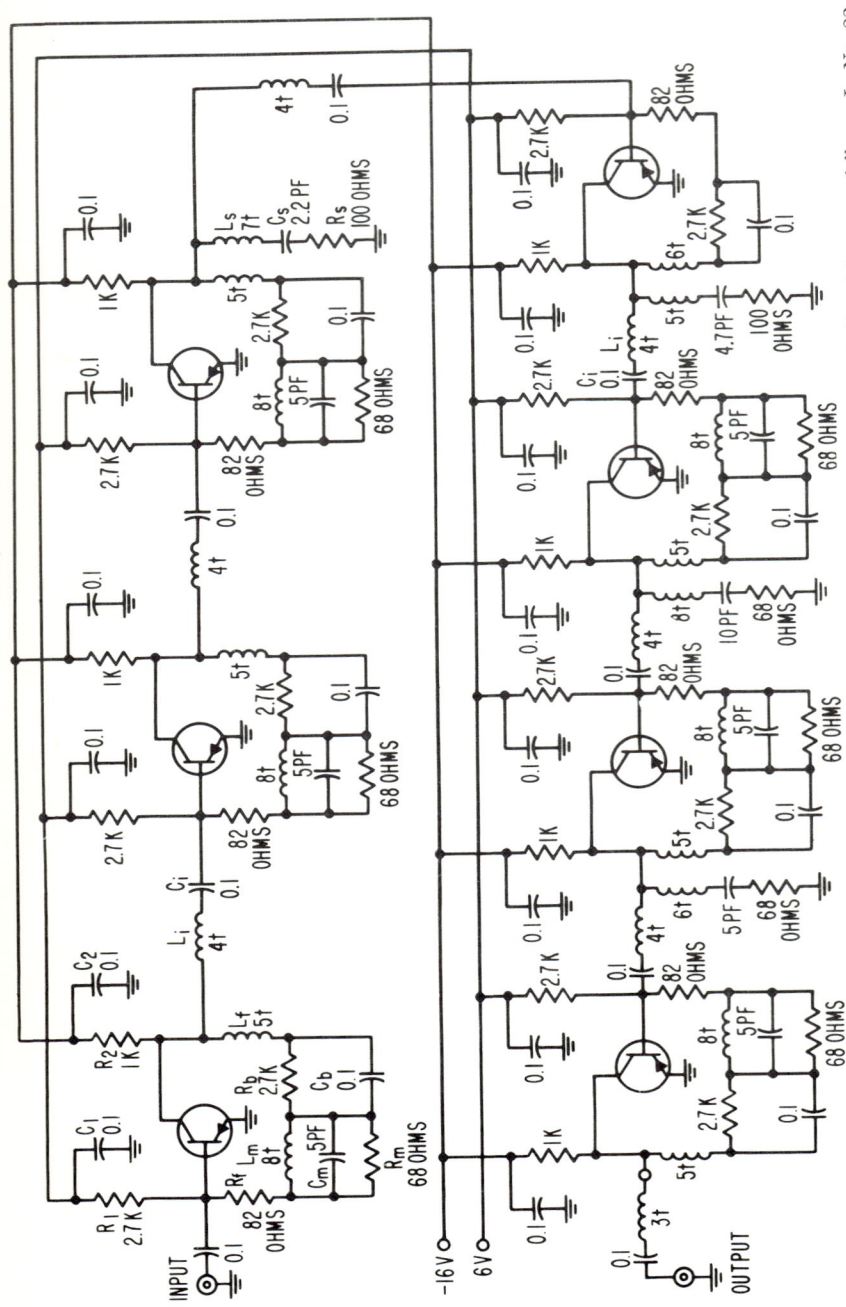

FIG. 76. The complete amplifier. Notes: 1. Inductance values are given in turns. Wire sizes and winding data are as follows: L_f No. 22 tinned wire, evenly spaced on CTC type SLST-2C4L ceramic form; stage 4: No. 22 enameled wire, close wound on form. L_m No. 28 enameled wire, close spaced and wound on R_m. L_s No. 28 enameled wire, close spaced on 22-megohm ¼-watt carbon resistor. L_i No. 22 tinned wire, air core, ⅛ in. ID, 3/16 in. long. 2. All resistors are ¼-watt carbon composition. 3. All transistors are Philco T-2872. (Reproduced by permission from D. W. Howell, "A Transistor Amplifier with 500-Mc Bandwidth," Stanford Electronics Laboratories, Technical Report No. 1820-1, SU-SEL-64-054.)

6–48

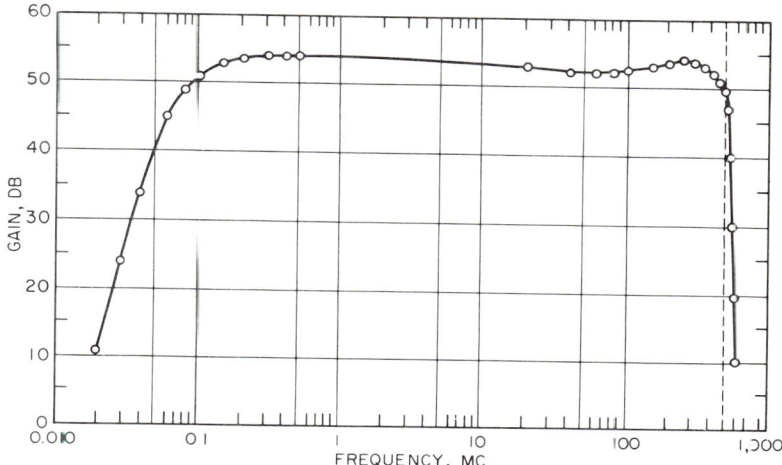

Fig. 77. Amplifier performance. (*Reproduced by permission from D. W. Howell, "A Transistor Amplifier with 500-Mc Bandwidth," Stanford Electronics Laboratories, Technical Report No. 1820-1, SU-SEL-64-054.*)

7.3. Transistor Alternating Feedback Chains. Cherry[28] has shown that the alternating feedback chain presents an exceptionally powerful engineering approach to the design of transistor feedback amplifiers. Unlike the previous techniques of Sec. 7.2, which required complete information about the characteristics of the active devices over their entire frequency range, this method presents a straightforward approach, requiring a minimum of information about the active elements. In view of its extreme simplicity, it cannot be expected to result in maximum feedback utilization; on the other hand, it presents a pragmatic design philosophy that can be readily applied. In essence, the method is based on the following principles:

1. Feedback placed around individual stages of an amplifier will eliminate the possibility of oscillation.
2. Series feedback results in a high input and output impedance but is effective only if the stage is terminated (at both input and output) in low impedances.
3. Shunt feedback results in a low input and output impedance but is effective only if the stage is terminated (at both input and output) in high impedances

Thus, the shunt- and series-feedback stages present effective terminations for each other. In addition, the gross mismatch resulting from this sequence provides isolation between stages. Hence, alternating series- and shunt-feedback stages may be cascaded with a minimum of interaction.

Consider the series-feedback stage of Fig. 78, together with its equivalent circuit in the hybrid-π form.

For this circuit, the low-frequency transconductance G_t is given by

$$G_t \triangleq \frac{I_2}{V_1} \simeq \frac{\alpha_{b0}}{R_E + r_\epsilon + r_{b'}(1 - \alpha_{b0})} \tag{111}$$

where r_b' is the base spreading resistance, typically under 100 ohms. A detailed analysis shows that, for typical values of transistor and bias elements, this result is within 5 per cent of the actual value.

Figure 79 shows a shunt-feedback stage, together with its equivalent circuit. At low frequencies, the transresistance R_t is given by

$$R_t \triangleq \frac{V_2}{I_1} \simeq \frac{R_F}{1 + (R_L + R_F)/\alpha_{e0} R_L} \tag{112}$$

6-50 AMPLIFIER FUNDAMENTALS

provided that
$$R_F \gg r_b' + r_\epsilon/(1 - \alpha_{b0}) \tag{113}$$

Again, Eq. (112) gives a value of R_t that is within 5 per cent of the exact value, for typical transistor and circuit parameters.

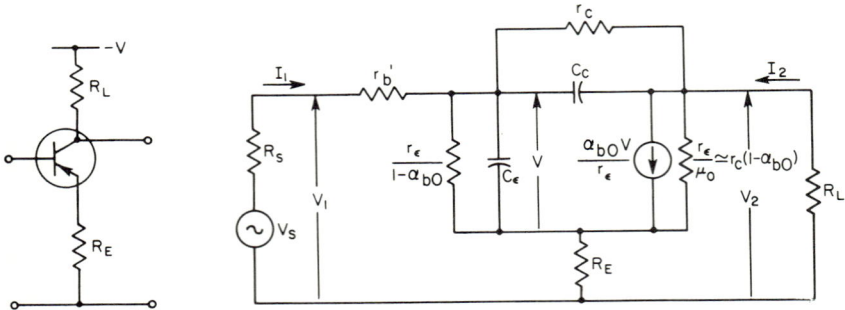

Fig. 78. The series-feedback stage.

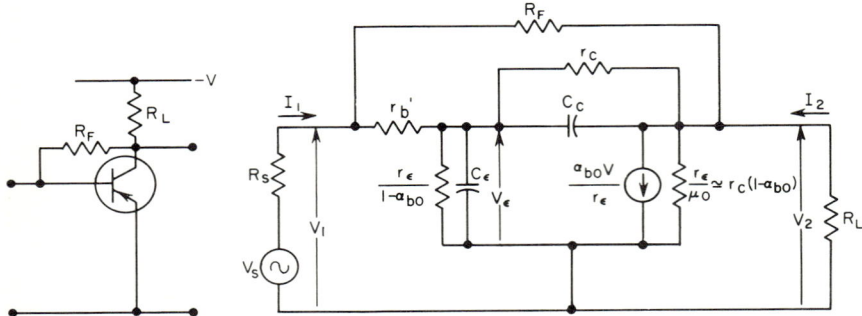

Fig. 79. The shunt-feedback stage.

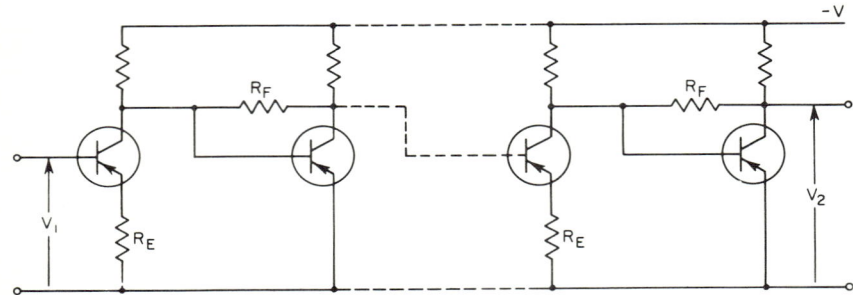

Fig. 80. Series-parallel chain.

The circuit diagram of an amplifier with alternating series- and shunt-feedback stages is shown in Fig. 80. An amplifier of this type has a high input impedance and a low output impedance, and a voltage gain given by

$$V_2/V_1 = G_{t1}R_{t2}G_{t3}R_{t4} \cdots G_{t(n-1)}R_{tn} \tag{114}$$

Alternately, an amplifier with the configuration of Fig. 81 has a low input impedance and a high output impedance. For such a circuit, current gain is more meaningful and is given by

$$I_2/I_1 = R_{t1}G_{t2}R_{t3}G_{t4} \cdots R_{t(n-1)}G_{tn} \tag{115}$$

Using these methods it is possible to build a feedback amplifier whose performance is within ±10 per cent of that obtained by an exact design procedure. Once this is done, an additional overall loop may be placed around the amplifier to stabilize its performance further. Such a loop need not involve a large amount of feedback. In

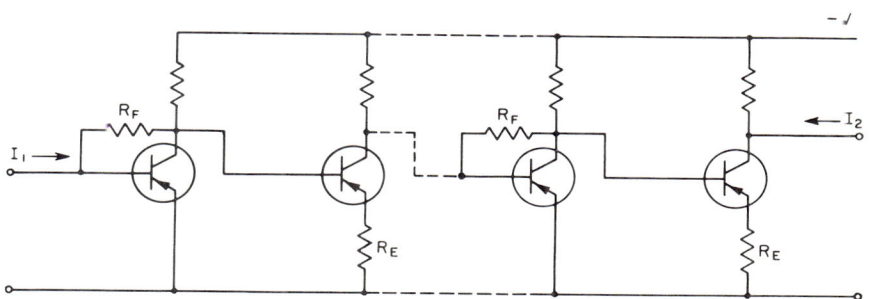

Fig. 81. Parallel-series chain.

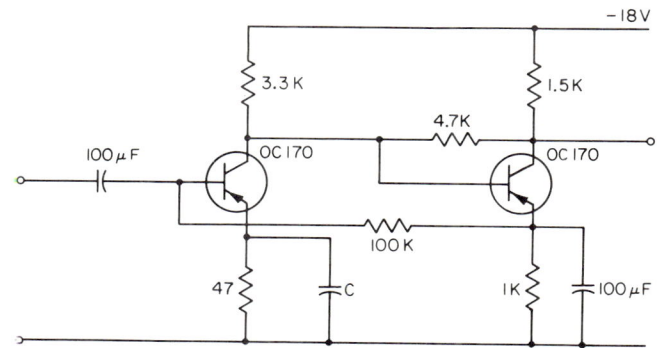

Fig. 82. An illustrative example. (*Reproduced by permission from E. M. Cherry, "An Engineering Approach to the Design of Transistor Feedback Amplifiers," Jour. Brit. IRE, Vol. 25, No. 2, Feb. 1963.*)

addition, since it is placed around elements which have already been stabilized to some extent its design is quite straightforward.

Figure 82 shows the circuit diagram of a practical two-stage video amplifier with about 40-db gain, designed as a cascade of series- and shunt-feedback stages. Overall feedback around the amplifier provides bias stabilization. A small capacitor C is experimentally selected to provide high-frequency peaking. A value of 150 pf was found to be suitable with these transistors (type OC 170, having an f_T of 70 Mc) and resulted in a frequency response of 20 Mc.

Using these principles, Reddi[29] has designed a more sophisticated amplifier, consisting of a series-feedback stage followed by a shunt-feedback stage (see Fig. 83). While the resistor R provides a small amount of shunt feedback around the first stage, its primary purpose (in conjunction with the breakdown diode Z_1) is to provide d-c bias for this stage. A similar scheme for biasing is used with the second stage. Here, however, R_f provides a significant amount of shunt feedback.

High-frequency compensation is provided by means of C_e and L_f and results in the frequency response of Fig. 84. Transistors having an f_T of 300 Mc were used for this design.

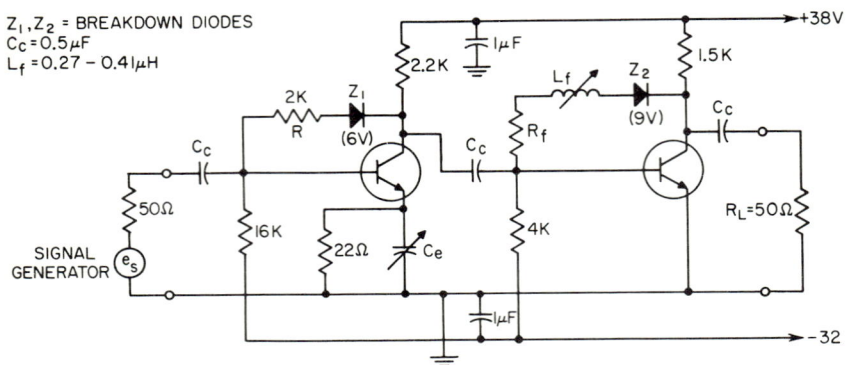

FIG. 83. A second example. (*From Application Data APP-32, "Transistor Pulse Amplifiers," by V. G. K. Reddi, courtesy of Fairchild Semiconductor.*)

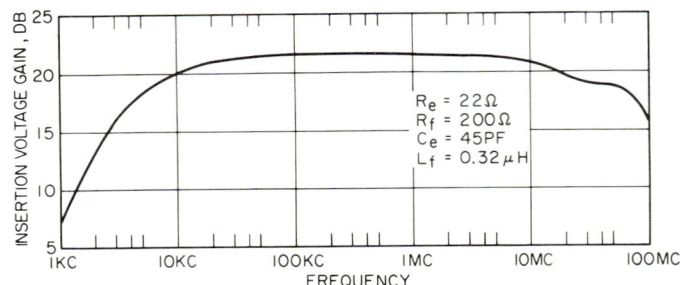

FIG. 84. Frequency response of amplifier of Fig. 82. (*From Application Data APP-32, "Transistor Pulse Amplifiers," by V. G. K. Reddi, courtesy of Fairchild Semiconductor.*)

7.4. Transistor Feedback-pair Chains. Feedback pairs[30,31] can also be used as building blocks in an alternating feedback chain. Figure 85 shows a current feedback configuration suited for this application. The circuit has a low input impedance and a high output impedance. For feedback to be fully effective, it must be fed from a high source impedance. Feedback pairs of this type may be directly cascaded, since they provide the right terminations for each other. The current gain of this type of circuit is given by

$$A_i \simeq \frac{R_F + R_E}{R_E} \frac{\alpha_{b02}}{1 + (R_F/R_E)[(R_C + \alpha_{e02}R_E)/\alpha_{e01}\alpha_{e02}R_C]} \quad (116)$$

$$\simeq R_F/R_E \quad (117)$$

for most circuit values.

Figure 86 shows a voltage feedback pair which is also suitable for direct cascading to form a feedback chain. This circuit has a high input impedance and a low output impedance and must be fed from a low-impedance source for its feedback to be fully effective. For such a feedback pair, the voltage gain is given by

$$A_v \simeq \frac{R_F + R_E}{R_E} \frac{1}{1 + (R_E + r_{e1})(R_F + R_L)/\alpha_{e02}R_E R_L} \tag{118}$$

$$\simeq R_F/R_E \tag{119}$$

For both schemes, a maximum gain of $0.1(\alpha_{e0})^2$ is possible with a ± 10 per cent uncertainty.

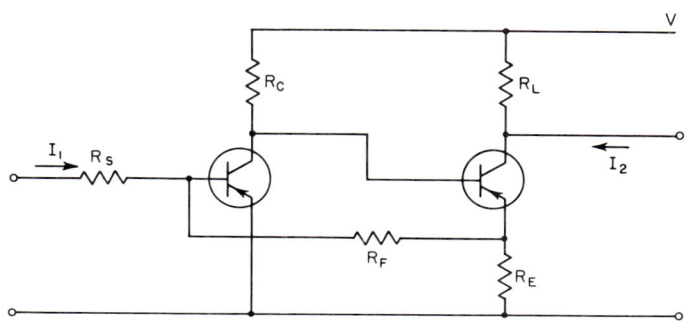

FIG. 85. The shunt-series feedback pair.

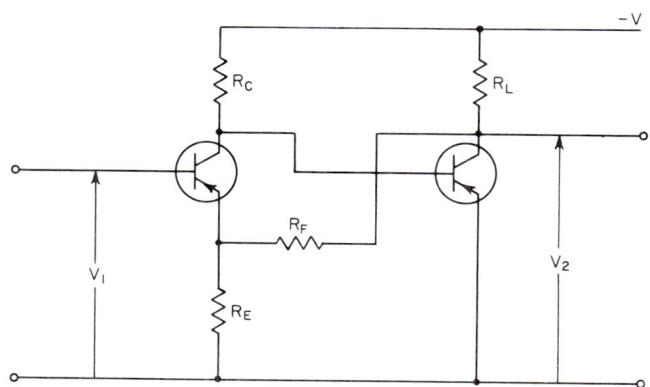

FIG. 86. The series-shunt feedback pair.

While either circuit type may be used, it is worth noting that the current feedback pair allows a convenient means of providing d-c bias to the transistors. Figure 87 shows a feedback pair that was used as part of a 90-db broadcast amplifier. For this stage,

$$A_i \simeq R_F/R_E = 70.3 \tag{120}$$

while a more accurate value,* as given by Eq. (116), is 66.3.

A unique feature of this circuit lies in the fact that the feedback resistor (33 kilohms) serves to set the d-c bias for the stages. Thus separate biasing is avoided, and with it the expense of large bypass capacitors.

* For $\alpha_{e01} = 60$ and $\alpha_{e02} = 40$.

Ghausi[32] has undertaken the design of the shunt-series pair by means of the root-locus technique. Figure 88 shows his configuration for an amplifier with 20-db gain, a closed-loop bandwidth of 35 Mc, and 35 db of feedback. The transistors used in this amplifier had an f_T of 450 Mc. Figure 89 shows the closed-loop response characteristic of the amplifier.

In theory, it is also possible to utilize feedback triples to form a chain. However, the design of these triples requires considerably more knowledge of the loop gain characteristic if the stability of the amplifier is to be ensured.

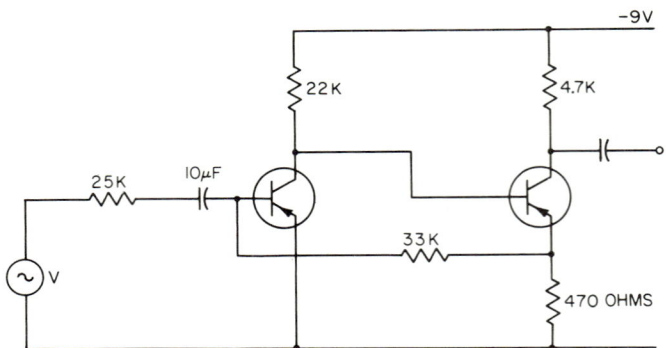

Fig. 87. A two-stage amplifier. (*Reproduced by permission from E. M. Cherry, "An Engineering Approach to the Design of Transistor Feedback Amplifiers," Jour. Brit. IRE, Vol. 25, No. 2, Feb. 1963.*)

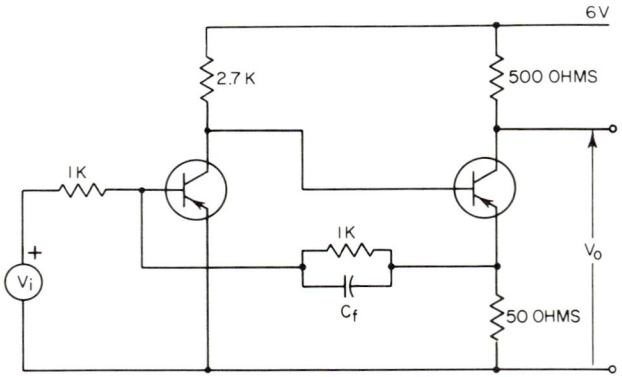

Fig. 88. A two-stage amplifier. (*Reproduced by permission from M. S. Ghausi, "Optimum Design of the Shunt-Series Feedback Pair," IRE Trans., CT-8, 1961, pp. 448–453.*)

7.5. Some Concluding Remarks. Many of the techniques described for the design of transistor feedback chains may also be used with vacuum-tube circuits. However, it should be noted that

1. The vacuum tube has relatively little interaction between stages. Consequently, it is unnecessary to use alternating feedback techniques to provide this feature.

2. The gain-bandwidth product limit of a vacuum-tube amplifier can be attained without the aid of feedback. With transistors, however, this is not the case.

3. The vacuum-tube amplifier has less excess phase than its transistor counterpart. Consequently, feedback around a number of stages is quite feasible. This overall feedback is more effective in controlling amplifier performance than is local feedback.

4. Finally, the normal manufacturing tolerances on the gain of vacuum-tube devices are much tighter than on transistor devices.* Thus, local feedback loops for partial gain stabilization are a necessity with transistor amplifiers, even if overall feedback is contemplated.

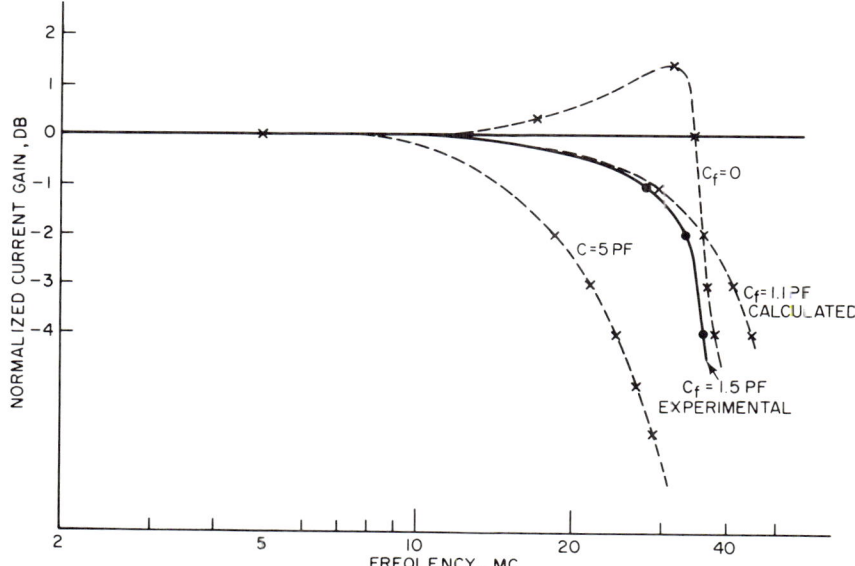

FIG. 89. Response of the amplifier of Fig. 87. (*Reproduced by permission from M. S Ghausi, "Optimum Design of the Shunt-Series Feedback Pair," IRE Trans., CT-8, 1961, pp. 448–453.*)

REFERENCES

1. H. S. Black, U.S. Patent 2,102,671.
2. H. S. Black, Stabilized Feedback Amplifiers, *Bell System Tech. J.*, vol. 13, pp. 1–18, 1934.
3. H. M. James, et al., "Theory of Servomechanisms," McGraw-Hill Book Company, New York, 1947.
4. L. A. MacColl, "Fundamental Theory of Servomechanisms," D. Van Nostrand Company, Inc. Princeton, N.J., 1945.
5. W. R. Evans, "Control-system Dynamics," McGraw-Hill Book Company, New York, 1954.
6. C. J. Savant, Jr., "Control System Design," 2d ed., McGraw-Hill Book Company, New York, 1964.
7. R. F. Shea, "Principles of Transistor Circuits," John Wiley & Sons Inc., New York, 1953.
8. H. W. Bode, "Network Analysis and Feedback Amplifier Design," D. Van Nostrand Company, Inc., Princeton, N.J., 1945.
9. S. K. Ghandhi, Transistor Feedback Amplifiers, *Proc. Natl. Electron. Conf.*, vol. 9, pp. 738–747, 1953.
10. S. S. Hakim, "Junction Transistor Circuit Analysis," John Wiley & Sons, Inc., New York, 1962.

* A common-emitter current-gain spread of 5:1 is common practice with transistor manufacturers.

11. L. J. Giacoletto, Study of p-n-p Alloy Junction Transistors from DC through Medium Frequencies, *RCA Rev.*, vol. 15, pp. 506–562, 1959.
12. R. L. Pritchard, Electric Network Representation of Transistors—a Survey, *IRE Trans. Circuit Theory*, vol. CT-3, pp. 5–21, 1956.
13. C. C. Cheng, Neutralization and Unilateralization, *IRE Trans. Circuit Theory*, vol. CT-2, pp. 138–145, 1955.
14. A. P. Stern, et al., Internal Feedback and Neutralization of Transistor Amplifiers, *Proc. IRE*, vol. 43, pp. 838–847, 1955.
15. M. S. Ghausi and D. O. Pederson, A New Design Approach for Feedback Amplifiers, *IRE Trans. Circuit Theory*, vol. CT-9, pp. 274–284, 1961.
16. J. G. Truxal, "Automatic Feedback Control System Synthesis," McGraw-Hill Book Company, New York, 1955.
17. H. W. Bode, Ref. 8, p. 455.
18. D. E. Thomas and J. L. Moll, Junction Transistor Short Circuit Current Gain and Phase Determination, *Proc. IRE*, vol. 46, pp. 1177–1184, 1958.
19. F. H. Blecher, Design Principles for Single Loop Transistor Feedback Amplifiers, *IRE Trans. Circuit Theory*, vol. CT-4, pp. 145–156, 1957.
20. S. Schwartz (ed.), "Selected Semiconductor Circuits Handbook," p. 3–39, John Wiley & Sons, Inc., New York, 1960.
21. H. N. Beveridge, Broadband Feedback Amplifiers, *IRE Conv. Record*, pt. 5, pp. 52–56, March, 1953.
22. J. M. Pettit and M. M. McWhorter, "Electronic Amplifier Circuits," McGraw-Hill Book Company, New York, 1961.
23. G. E. Valley, Jr. and H. Wallman, "Vacuum Tube Amplifiers," McGraw-Hill Book Company, New York, 1947.
24. D. W. Howell, A Transistor Amplifier with 500-Mc. Bandwidth, *Stanford Electronic Labs. Tech. Rept.* 1820-1, June, 1964.
25. D. E. Thomas, Some Design Considerations for High Frequency Transistor Amplifiers, *Bell System Tech. J.*, vol. 38, pp. 1551–1558, 1959.
26. R. S. Pepper and D. O. Pederson, Designing Shunt-peaked Transistor Amplifiers, *Electronics*, vol. 33, pp. 68–70, December, 1960.
27. E. M. Cherry and D. E. Hooper, The Design of Wideband Transistor Feedback Amplifiers, *Proc. IEE*, vol. 110, no. 2, 1963.
28. E. M. Cherry, An Engineering Approach to the Design of Transistor Feedback Amplifiers, *J. Brit. IRE*, vol. 25, no. 2, pp. 127–144, 1963.
29. V. G. K. Reddi, Transistor Pulse Amplifiers, Fairchild Application Data, APP-32/2, September 1961 (Available from Fairchild Semiconductor Corp., Mountain View, Calif.).
30. E. M. Cherry, Ref. 28, p. 134.
31. F. D. Waldhauer, Wideband Feedback Amplifiers, *IRE Trans. Circuit Theory*, vol. CT-4, pp. 178–190, 1957.
32. M. S. Ghausi, Optimum Design of the Shunt-Series Feedback Pair, *IRE Trans. Circuit Theory*, vol. CT-8, pp. 448–453, 1961.

Chapter 7

AMPLIFIER NOISE

EDWARD G. NIELSEN*

CONTENTS

List of Symbols.	7-2
1. Introduction.	7-6
1.1. Thermal Noise.	7-6
2. Noise in Two-terminal Networks.	7-7
2.1. Shot Noise.	7-7
2.2. Representation of Noise Currents and Voltages.	7-7
3. Noise in Two-ports.	7-8
4. Noise Factor.	7-12
4.1. Definition of Noise Factor.	7-12
4.2. Excess Noise Factor.	7-14
4.3. Noise Factor for Cascaded Two-ports.	7-14
4.4. Extended Noise Factor.	7-15
4.5. Behavior of Noise Factor with Source Impedance (Admittance)	7-17
4.6. Experimental Determination of Two-port Noise Parameters.	7-19
4.7. Average Noise Factor.	7-21
4.8. Noise Bandwidth.	7-22
5. Noise Temperature.	7-23
5.1. Effective Input Noise Temperature and Operating Noise Temperature.	7-23
5.2. Average Noise Temperature.	7-25
5.3. Transducers with Multiple Signal Responses.	7-25
5.4. Comparison of F, T_{op}, and T_e.	7-26
6. Techniques for Measurement of Noise Parameters.	7-29
6.1. Basic Dispersed Signal Generator Technique.	7-30
6.2. Basic Sine-wave Signal Generator Technique.	7-30
6.3. Determination of the Response-factor–Noise-bandwidth Product.	7-31
6.4. Formulas for Noise Measurements on Amplifiers.	7-32
6.5. Noise Measurements on Two-terminal Devices.	7-33
7. Instrumentation of Noise Measurements.	7-35
7.1. Sine-wave Signal Generators.	7-35
7.2. Dispersed Signal Generators.	7-36
7.3. Diode Noise Sources.	7-36

* General Electric Co., Electronics Laboratory, Syracuse, N.Y.

7.4. Thermal Noise Sources.................................... 7-39
7.5. Gas Discharge Noise Sources............................ 7-39
7.6. Automatic Noise-factor Meters......................... 7-40
7.7. Output Power Indicators—Average-reading Instruments..... 7-41
7.8. Output Power Meters—True RMS Instruments............ 7-42
7.9. Averaging Time and Measurement Bandwidth............. 7-44
7.10. Transducer and Detector Linearity...................... 7-45
8. The Application of Noise Temperature to Amplifier and Receiver Performance Calculations................................ 7-47
8.1. Output Noise Temperature of an Attenuator.............. 7-47
8.2. Operating Noise Factor................................. 7-48
8.3. Theorem for Combination of Noise Temperatures.......... 7-49
8.4. Examples of Noise-Performance Calculations............ 7-49
9. Limiting Noise Conditions..................................... 7-52
9.1. Thermal Noise at Very Low Temperatures and Very High Frequencies... 7-52
9.2. Ultimate Limits on Detector Sensitivity................. 7-53

LIST OF SYMBOLS

		First Used in Section
a	Parameter $abcd$ matrix	3
B	Bandwidth (cps)	1.1
B_o	Optimum source susceptance (mhos)	4.5
B_o	Signal bandwidth at output port (cps)	5.2
B_s	Source susceptance (mhos)	4.5
B_x	Susceptance of two-terminal device (mhos)	6.5
B_1, B_2	Susceptance of dummy loads (mhos)	6.5
$B_1, B_2, \ldots, B_n$	Noise bandwidth of responses 1, 2, ..., n (cps)	5.2
B_γ	Correlation susceptance II noisy two-port (mhos)	4.5
b	Parameter $abcd$ matrix	3
c	Parameter $abcd$ matrix	3
d	Parameter $abcd$ matrix	3
E	Thévenin voltage generator	2
E	Energy of a system	9.2
E_A	Noise voltage generator at port 1	3
E_B	Noise voltage generator at port 2	3
E_i	Input energy	9.2
E_{ni}	Source noise voltage generator	3
E_o	Output energy	9.2
e	Instantaneous voltage (volts)	2
F	Noise factor	4.1
F'	Operating noise factor	8.2
F_e	Extended noise factor	4.4
$F_{e1}, F_{e2}, \ldots, F_{en}$	Extended noise factor for transducers 1, 2, ..., n	4.4
F_o	Minimum value of noise factor	4.5
$F_1, F_2, \ldots, F_n$	Noise factor of two-ports 1, 2, ..., n	4.3
f	Frequency	1.1
f_o	Reference frequency for noise bandwidth (cps)	4.8
$f_{o1}, f_{o2}, \ldots, f_{on}$	Reference frequencies at responses 1, 2, ..., n (cps)	5.2
G	Conductance (mhos)	2
$G, \mathcal{G}(f)$	Transducer gain	4.7

AMPLIFIER NOISE

First Used in Section

G	Amplifier gain	9.2
G_A	Available gain	4.1
$G_{A1}, G_{A2}, \ldots, G_{An}$	Available gain of two-ports 1, 2, ..., n	4.3
G_b	Photoconductor bias conductance (mhos)	8.4
G_e	Exchangeable gain	4.4
$G_{e1}, G_{e2}, \ldots, G_{en}$	Exchangeable gain for transducers 1, 2, ..., n	4.4
G_n	Equivalent noise conductance (mhos)	2.2
G_n'	Noise conductance of T noisy two-port (mhos)	3
G_o	Optimum source conductance (mhos)	4.5
G_o	Transducer gain at reference frequency f_o	4.8
G_o	Transducer gain at principal response	5.4
$G_{o1}, G_{o2}, \ldots, G_{on}$	Transducer gain at reference frequencies 1, 2, ..., n	5.2
G_{pc}	Photoconductor conductance (mhos)	8.4
G_s	Source conductance (mhos)	4.5
G_s	Signal gain	5.1
G_s	Signal generator conductance (mhos)	6.5
G'_{s1B}, G'_{s2B}	Increased noise conductances, measurements 1B, and 2B (mhos)	6.5
G_u	Noise conductance of Π noisy two-port (mhos)	3
G_x	Conductance of two-terminal device (mhos)	6.5
G_x'	Noise conductance of two-terminal device (mhos)	6.5
G_γ	Correlation conductance of Π noisy two-port (mhos)	4.5
G_1, G_2	Conductance of dummy loads (mhos)	6.5
$G_1, G_2, \ldots, G_n$	Transducer gains at responses 1, 2, ..., n	5.1
g_m	Transconductance (mhos)	2.2
h	Planck's constant (6.625 × 10^{-34} joule-sec)	9.1
I	Norton current generator	2
I	Equivalent d-c shot noise current (amp)	2.1
I	Noise diode d-c anode current (amp)	7.3
I_A	Noise current generator at port 1	3
I_B	Noise current generator at port 2	3
I_b	Photoconductor bias current (amp)	8.4
I_{ni}	Source noise current generator	3
I_s	Signal current generator	6.5
I_s	Noise diode shot noise current generator	7.3
I_u	Uncorrelated noise current generator of Π noisy two-port	3
I_x	Noise current generator of two-terminal device	6.5
I_1	Two-port input current (amp)	3
I_1, I_2	Noise current generators of dummy loads 1 and 2	6.5
I_2	Two-port output current (amp)	3
i	Instantaneous current (amp)	2
i_{ni}	Instantaneous noise current of source (amp)	4.5
i_{s1A}, i_{s1B}	Instantaneous current, measurements 1A and 1B (amp)	6.5
i_u	Instantaneous uncorrelated noise current of Π noisy two-port (amp)	4.5
k	Boltzmann's constant (1.380 × 10^{-23} joule/°K)	1.1
$\mathcal{K}$	Thermal noise reduction factor	9.1
L	Loss	8.1
N_e	Equivalent transducer noise power referred to input (watts)	4.1
N_{ei}	Exchangeable thermal noise power at source (watts)	4.4

7–4   AMPLIFIER FUNDAMENTALS

		First Used in Section
N_{eo}	Exchangeable thermal noise power at output (watts)	4.4
N_{eN}	Exchangeable noise power at transducer output when source noise is zero (watts)	4.4
N_{et}	Total equivalent input noise power for cascade (watts)	4.3
$N_{e1}, N_{e2}, \ldots, N_{en}$	Equivalent transducer noise power referred to input for transducers $1, 2, \ldots, n$	4.3
N_i	Available noise power from source (watts)	4.1
N_{io}	Portion of noise power delivered to load due to noise temperatures of all sources (watts)	5.1
N_L	Output noise power resulting from load noise (watts)	5.1
N_N	Available noise power or noise power delivered to load from output when the noise of all the sources is zero (watts)	4.1
N_o	Available noise power or noise power delivered to load from output port of transducer (watts)	4.1
$N_o, N_o(f)$	Noise power delivered by output port to load (watts)	4.7
$N_o(T_o)$	Transducer output noise power for a source noise temperature of T_o at all responses (watts)	5.4
N_{TL}	Total output noise power resulting from load noise (watts)	5.4
N_{To}	Total noise delivered to output load (watts)	5.2
N_{To} (cold)	Total noise delivered to output load for a source noise temperature of T_i (cold) (watts)	6.1
N_{To} (hot)	Total noise delivered to output load for a source noise temperature of T_i (hot) (watts)	6.1
$N_{To}(T_o)$	Total noise delivered to output load for a source noise temperature of T_o at all responses (watts)	5.4
P	Mean available power (watts)	1.1
P	Available signal power (watts)	6.5
P_e	Exchangeable power (watts)	4.4
P_1	Power delivered to output load when all responses of source are at temperature T_r (watts)	6.2
P_2	Power delivered to output load when all responses of source are at temperature T_r and a signal $S_i(f_o)$ is present in the principal response (watts)	6.2
$P_{1A}, P_{1B}, P_{2A}, P_{2B}$	Output power for measurement $1A, 1B, 2A, 2B$ (watts)	6.5
p_b	Fraction of power delivered to bias resistor	8.4
p_j	Fraction of noise power delivered from jth element to output port	8.3
p_{pc}	Fraction of power delivered to photoconductor	8.4
q	Electronic charge (1.602×10^{-19} coulomb)	2.1
R	Resistance (ohms)	2
R_b	Photoconductor bias resistance (ohms)	8.4
R_n	Equivalent noise resistance (ohms)	2.2
R_o	Transducer output resistance (ohms)	4.4
R_o	Optimum source resistance (ohms)	4.5
R_{o1}, R_{o2}, R_{o3}	Output resistance for transducers 1, 2, 3 (ohms)	4.4
R_{pc}	Photoconductor resistance (ohms)	8.4
R_s	Source resistance (ohms)	4.4
R_{μ}'	Noise resistance of T noisy two-port (ohms)	3

AMPLIFIER NOISE 7–5

		First Used in Section
r	Noise diode plate resistance (ohms)	7.3
$\mathcal{R}$	Response factor	5.4
S_{ei}	Exchangeable signal power at source (watts)	4.4
S_{eo}	Exchangeable signal power at output (watts)	4.4
S_i	Available signal power from source (watts)	4.1
$S_{i1}, S_{i2}, \ldots, S_{in}$	Available signal power at responses 1, 2, ..., n (watts)	5.3
$S_i(f_o)$	Available signal power from source at frequency f_o (watts)	6.2
S_o	Available signal power at output port of transducer (watts)	4.1
S_o	Signal power delivered to output load (watts)	5.1
T	Absolute temperature (°K)	1.1
T_b	Noise temperature of bias resistor (°K)	8.4
T_e	Effective input noise temperature (°K)	5.1
T_i	Noise temperature of signal source (°K)	5.1
$T_{i1}, T_{i2}, \ldots, T_{in}$	Noise temperatures of source at responses 1, 2, ..., n (°K)	5.4
T_i (cold)	Smaller noise temperature of signal source (°K)	6.1
T_i (hot)	Larger noise temperature of signal source (°K)	6.1
T_j	Noise temperature of jth element (°K)	8.3
T_L	Temperature of passive two-port (°K)	8.2
T_{min}	Minimum input noise temperature (°K)	9.2
T_o	Standard noise temperature (290°K)	2.2
T_{op}	Operating noise temperature (°K)	5.1
T_{output}	Noise temperature of output port of transducer (°K)	8.4
T_{pc}	Noise temperature of photoconductor (°K)	8.4
T_r	Room or ambient temperature (°K)	6.2
t	Time	9.2
t_i	Time of input energy	9.2
t_o	Time of output energy	9.2
V	Voltage (volts)	8.4
V_1	Two-port input voltage (volts)	3
V_2	Two-port output voltage (volts)	3
X	Reactance (ohms)	4.4
X	Derived quantity for determination of noisy two-port parameter	4.6
X_o	Optimum source reactance (ohms)	4.5
X_s	Source reactance (ohms)	4.5
X_γ	Correlation reactance of T noisy two-port (ohms)	4.5
Y	Output power ratio for dispersed generator	6.1
Y'	Output power ratio for sine-wave generator	6.2
Y_o	Optimum source admittance (mhos)	4.5
Y_s	Source admittance (mhos)	3
Y_γ	Correlation admittance II noisy two-port (mhos)	3
$y_{11}, y_{12}, y_{21}, y_{22}$	y parameters of two-port (mhos)	3
Z_o	Optimum source impedance (ohms)	4.5
Z_s	Source impedance (ohms)	4.5
Z_γ	Correlation impedance T noisy two-port (ohms)	3
$z_{11}, z_{12}, z_{21}, z_{22}$	z parameters of two-port (ohms)	3
ϵ	Incremental loss	8.2
λ_m	Wavelength of light energy	7.5
τ	Detector time constant	7.9
τ_d	Time delay	9.2

1. INTRODUCTION

When telephone and radio were in their infancy, it was soon learned that their transmitted information was disturbed by interfering sounds. These sounds were quite naturally referred to by the conglomerate term "noise." The "noise" of course resulted from electric fluctuations introduced into the received signal. As the uses of electronic circuits expanded, the fluctuations were found to be present in circuits not directly related to the transmission of voice and acoustical sounds. By that time, the term had become so deeply entrenched that nearly all unwanted fluctuations in electric circuits were called noise.

A certain amount of noise appears in every type of electric circuit. It is heard in the output of telephones, radios, and television receivers and is seen in the television picture. It is present in computer circuits, power supplies, electric measuring instruments, and radar and sonar equipments. Noise is present in physical systems as well as electric. The noise at the output of a microphone, for example, is partially due to the random motion of the air molecules striking the diaphragm. Depending on its origin, the noise may have widely varying characteristics, from the sharp violent disturbance of a lightning stroke to the wideband uniform thermal noise spectrum of a resistor.

Noise generally impairs the functioning of electric circuits. Particular characteristics of the noise will offend particular circuits and not others. A computer may be affected principally by the peak value of the noise voltage or current whereas an electric measuring instrument might be more strongly influenced by the average noise energy. The impairment of telephone reception is dependent on the frequency, bandwidth, and other characteristics of the interference. There are many special techniques of characterizing and dealing with noise, depending upon its properties and the manner in which it influences circuit operation.

Noise is not always undesirable. Random fluctuations are often relied upon to start oscillators. Noise signals are used to test acoustic systems. The transmitter of a communications system requiring secrecy may modulate its information with noise or a noiselike signal. The signal can be demodulated only by a receiver that has the proper modulation code. These are but a few examples.

At the present time, most analytical studies involving noise have dealt with noise that has a gaussian or normal distribution or at least approximately such a distribution. This distribution is found in phenomena resulting from a large number of independent random events. Thermal and shot noise have this type of distribution. This chapter will describe means that are most useful for dealing with noise that is gaussian or reasonably like it.

1.1. Thermal Noise. The random motion of electrons in a conductor causes thermal noise. The mean-square velocity of the electrons is proportional to the absolute temperature. The moving electrons have frequent collision interactions with the molecules of the conductor. The motion of each electron from collision to collision produces a minute pulse of current. The current pulses resulting from all the moving electrons sum to produce a current in the conductor, if it forms a closed loop, or a voltage if it is open-circuited. The current or voltage is randomly fluctuating and has a d-c value of zero. The a-c component is not zero and leads to a mean available power P given by

$$P = kTB \tag{1}$$

where k is Boltzmann's constant (1.38×10^{-23} joules/°C), T is the absolute temperature in degrees Kelvin, and B is the bandwidth of the detecting system in cycles per second. Mean available power is the maximum average power that the element, normally a resistor or passive impedance, will deliver to a conjugate matched load. This power is equal to 4×10^{-21} watts/cycle bandwidth at *standard noise temperature*, 290°K or 17°C. This formula is valid at all frequencies and temperatures normally encountered in electronic circuits. However, at very high frequencies and low temperatures, such that the ratio $f/T > 1$ (f in gigacycles, T in degrees Kelvin), thermal noise decreases as described in Sec. 9.1. Because of its universality, thermal noise is frequently used as a base of reference for all noise in a system.

2. NOISE IN TWO-TERMINAL NETWORKS

Noise in a two-terminal network can be represented by a Thévenin voltage generator E in series or a Norton current generator I in parallel with the network as shown in Fig. 1. If these networks have only thermal noise, the mean-square open-circuit noise voltage is

$$\overline{e^2} = 4kTRB \qquad (2)$$

and the mean-square short-circuit noise current is

$$\overline{i^2} = 4kTGB \qquad (3)$$

The reader can easily verify these equations by connecting a conjugate matched load to each of the immittances and noting that an average power of kTB watts will be delivered to each load. The mean-square voltage is directly proportional to the resistance and the mean-square current to the conductance.

As a generalization, the networks of Fig. 1 may be considered to be the Thévenin or Norton equivalents of multielement networks. The thermal noise voltages of each resistive element add up to produce the Thévenin equivalent voltage for the network. The mean-square value of this voltage can be determined from Eq. (2) when all the elements are at the same temperature.

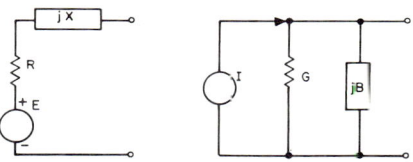

Fig. 1. Two-terminal networks with noise generators.

The reader can assure himself of this fact by taking any network, inserting a proper thermal noise generator in each resistive impedance, and determining the equivalent thermal output voltage. Power addition must be used because the noise voltages of the resistive elements are independent of each other.

2.1. Shot Noise. Another commonly encountered type of noise is shot noise. The cathode current flowing in a temperature-limited thermionic diode exhibits shot noise. Shot noise results from the random emission of electrons from the cathode. At frequencies where transit time is insignificant, the fluctuating current appearing in the anode is given by

$$\overline{i^2} = 2qIB \qquad (4)$$

where q is the electronic charge (1.6×10^{-19} coulombs), I is the direct current in amperes, and B is the bandwidth in cycles per second. Shot noise is found in virtually all types of electron tubes, semiconductor diodes, and bipolar transistors.

2.2. Representation of Noise Currents and Voltages. The mean-square values or the root-mean-square (rms) values of noise currents and voltages are of major significance in circuit and system calculations. There is no point in dealing with the average noise current or voltage since this is zero. When speaking of noise currents or voltages, the bandwidth must always be specified. It is obviously awkward to speak in terms of mean-square volts per cycle or rms volts per root cycle, especially since the quantities are usually numerically very small. Therefore, these terms are usually discarded in favor of the following terms:

Equivalent Noise Resistance. If a noise voltage has a mean-square value $\overline{e^2}$ its equivalent noise resistance R_n is

$$R_n = \overline{e^2}/4kT_oB \qquad (5)$$

T_o is, by definition, the *standard noise temperature*, 290°K. Equation (5) is a rearrangement of the equation for thermal noise voltage, Eq. (2). If a system has an equivalent noise resistance of R_n ohms, its mean-square open-circuit noise voltage is given by Eq. (2). The statement does not imply a physical resistance of R_n ohms exists anywhere in the system nor that the system is at 290°K. For example, many triode vacuum tubes have a noise resistance of approximately

$$R_n = 2.5/g_m \quad \text{ohms}$$

where g_m is the transconductance. This means that all the noise in the tube can be represented by a voltage generator, with zero internal impedance, in series with the grid. The mean-square voltage produced by this generator is equal to the mean-square thermal noise produced by a resistor of R_n ohms at *standard noise temperature*.

Equivalent Noise Conductance. If a noise current has a mean-square value $\overline{i^2}$, its equivalent noise conductance G_n is

$$G_n = \overline{i^2}/4kT_oB$$

This equation is a rearrangement of the equation for thermal noise current, Eq. (3). The noise conductance is the dual of the noise resistance described above.

Equivalent Noise Current. If a system has a mean-square noise current $\overline{i^2}$, its equivalent noise current is

$$I = \overline{i^2}/2qB$$

This is a rearrangement of the shot noise equation (4). The mean-square noise current in the system is described in terms of a direct current whether or not there is an actual direct current flowing in the system.

Noise Temperature. If the average noise power available in a system is P, the noise temperature is

$$T = P/kB \tag{6}$$

This is a rearrangement of the thermal noise equation (1). A high noise temperature corresponds to a large noise power and a low noise temperature to a small noise power. By definition, a noise temperature of 290°K yields an available power equal to standard thermal noise.

Decibels with Respect to 1 μv per Root Cycle. There is no generally accepted symbol for this quantity. It is frequently referred to simply as db with the meaning understood. If the system has an rms noise voltage $\sqrt{\overline{e^2}}$, the db noise relative to 1 μv per root cycle is given by

$$db = 20 \log_{10} \sqrt{\overline{e^2}}$$

where e is in microvolts per root cycle. This quantity has no particular physical significance as did the previous quantities. It is used when calculations must be made using the noise voltage directly.

Using Eqs. (3) and (4) with $R = 1/G$, the relationship between the equivalent noise resistance and the equivalent noise current is found to be

$$R_n = 2kT_o/qI = 1/20.1I$$

This equation is frequently used in transferring from one system to the other.

3. NOISE IN TWO-PORTS

The noise in two-ports can be represented in a variety of ways. The difficulty or ease of solving problems involving two-ports will depend strongly on the method of representing the noise. It is therefore desirable to become familiar with the common methods of representation.

One method is to break the two-port down into an equivalent circuit and assign an equivalent noise generator to each two-terminal element in the equivalent circuit.

The transmission from each two-terminal element to each of the network's ports can then be computed using standard techniques.[1] All the mean-square values of all the noise components are then summed at each port, taking into account any possible correlation that exists among them. This process is generally employed as a means to an end. When the equivalent noise at the input and output ports is determined the internal operation of the two-port is no longer considered. Rather, the noise performance of the two-port is represented by its noise factor, noise temperature, or one of a variety of noise equivalent circuits.

Noise equivalent circuits of two-ports can be closely related to the common two-port network equations. The two-port of Fig. 2 can be described by the impedance representation [Eq. (7)] if it does not have independent internal sources and if it is linear.

$$V_1 = z_{11}I_1 + z_{12}I_2$$
$$V_2 = z_{21}I_1 + z_{22}I_2 \quad (7)$$

V_1, I_1 and V_2, I_2 are the voltage and current at ports 1 and 2, respectively, as shown in the figure. Alternatively, the admittance matrix can be used.

$$I_1 = y_{11}V_1 + y_{12}V_2$$
$$I_2 = y_{21}V_1 + y_{22}V_2$$

Many other representations can of course be used also. Of special significance is the mixed *abcd* matrix,

$$V_1 = aV_2 + bI_2$$
$$I_1 = cV_2 + dI_2$$

This representation places both dependent variables at the input of the network.

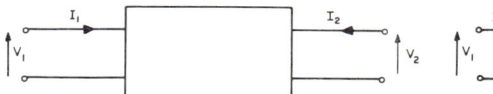 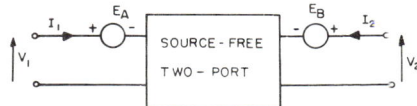

Fig. 2. Standard two-port.   Fig. 3. Two-port with internal sources represented by external voltage generators.

In each of these network representations, the coefficients are, in general, functions of frequency. Positive currents and voltages are defined in the directions shown.

When a network has internal sources, such as noise sources, they can be brought out to the terminals of the network by an extension of Thévenin's theorem.[2] The resulting equivalent circuit for the impedance representation is shown in Fig. 3. The new set of equations for the network becomes

$$V_1 = z_{11}I_1 + z_{12}I_2 + E_A$$
$$V_2 = z_{21}I_1 + z_{22}I_2 + E_B \quad (8)$$

By definition, E_A is the voltage that appears at port 1 when ports 1 and 2 are open-circuited, that is, when $I_1 = I_2 = 0$.

$$E_A = V_1|_{I_1=I_2=0}$$

The voltage at port 2, E_B, is defined similarly.

$$E_B = V_2|_{I_1=I_2=0}$$

---

[1] See, for example, H. C. Montgomery, Transistor Noise in Circuit Applications, *Proc. IRE*, vol. 40, pp. 1461–1471, November, 1942.

[2] L. C. Peterson, Bell Telephone Laboratories, unpublished memorandum, August, 1943, and 1949, reported in footnote 1.

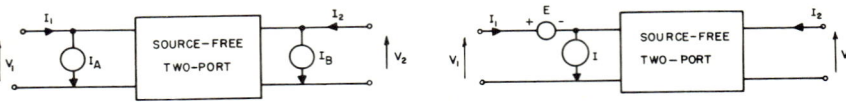

Fig. 4 Two-port with internal sources represented by external current generators.

Fig. 5 Two-port with internal sources represented by external generators at input.

It is possible for this representation to make the determination of z_{11}, z_{12}, z_{21}, and z_{22} difficult. To avoid this problem, it is practically assumed that E_A and E_B are small, such that the z parameters can be determined in the usual manner by applying signal voltages and currents that are much larger than those resulting from E_A and E_B.

The admittance two-port can be represented by the dual network (Fig. 4). Here the external generators are Norton current sources, I_A at port 1 and I_B at port 2. The network is represented by the following equations:

$$I_1 = y_{11}V_1 + y_{12}V_2 + I_A$$
$$I_2 = y_{21}V_1 + y_{22}V_2 + I_B$$

I_A is a current that flows in the input terminals when both V_1 and V_2 are equal to zero, that is, when the input and output terminals are short-circuited.

$$I_A = I_1|_{V_1=V_2=0}$$

I_B is defined similarly for the output port.

$$I_B = I_2|_{V_1=V_2=0}$$

The *abcd* matrix leads to a network with both sources brought to the input terminals (Fig. 5).[3] The equations are obtained as before.

$$V_1 = aV_2 + bI_2 + E$$
$$I_1 = cV_2 + dI_2 + I$$

Unfortunately, the generators E and I are not so easily defined for this network as for the previous ones. Writing the matrix equations for the network of Fig. 5 gives[4]

$$V_1 = z_{11}(I_1 - I) + z_{12}I_2 + E$$
$$V_2 = z_{21}(I_1 - I) + z_{22}I_2$$

By comparison with Eq. (8), the following relationships are determined:

$$E - Iz_{11} = E_A$$
$$-z_{21}I = E_B$$

Rearranging gives the desired result.

$$E = E_A - E_B z_{11}/z_{21}$$
$$I = -E_B/z_{21}$$

The generators can be determined from the admittance representation in a similar manner.

$$E = -I_B/y_{21}$$
$$I = I_A + I_B y_{11}/y_{21}$$

[3] A. G. Th. Becking, H. Groendijk, and K. S. Knol, The Noise Factor of Four-terminal Networks, *Philips Res. Rept.*, vol. 10, pp. 349–357, October, 1955.

[4] H. Rothe and W. Dahlke, Theory of Noisy Fourpoles, *Proc. IRE*, vol. 44, pp. 811–818, June, 1956.

Regardless of the representation used, four quantities are required to define the noise properties of the network completely. These quantities are two sources and the real and imaginary parts of the correlation coefficient relating these two sources. Rothe and Dahlke have suggested a means of replacing the correlation coefficient by a correlation admittance or impedance.[5] Using their technique, a noise equivalent circuit is obtained that is easily handled by analysis techniques familiar to circuit engineers. Furthermore, the parameters of the resulting noise equivalent circuit can be determined relatively easily by a series of noise-factor measurements as described in Sec. 4.6. The resulting circuit is shown in Fig. 6.

The voltage generator $\bar{E}$ is the same as the voltage generator used in the previous equivalent circuit, but the current generator I_u represents only the uncorrelated portion of the current I of the previous circuit. The effect of the correlated current is completely represented by the correlation admittance Y_γ. Y_γ is assumed to be at

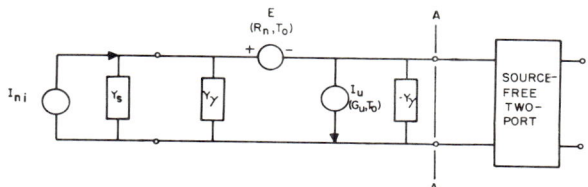

Fig. 6. Two-port with internal sources represented by uncorrelated generators and correlation admittance. (*From H. Rothe and W. Dahlke, Theory of Noisy Fourpoles, Proc. IRE, vol. 44, p. 813, June, 1956.*)

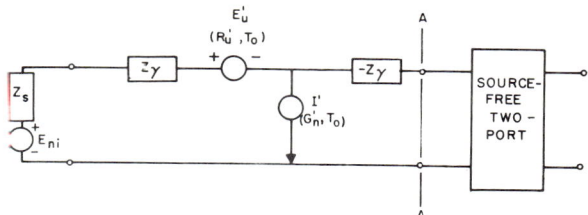

Fig. 7. Two-port with internal sources represented by uncorrelated generators and correlation impedance. (*From H. Rothe and W. Dahlke, Theory of Noisy Fourpoles, Proc. IRE, vol. 44, p. 813, June, 1956.*)

0°K; that is, it generates no noise. To avoid modifying the signal transfer properties of the network, a negative correlation admittance $-Y_\gamma$ is shunted across the output terminals. This equivalent circuit will be used in determining the behavior of the noise factor of two-ports, and a signal source (Y_s, I_{ni}) has been included for this future use. For convenience, the voltage and current generators are represented by their equivalent noise resistance and conductance, respectively, R_n and G_u.

The Π noise equivalent circuit of Fig. 6 is appropriate when the signal source is more conveniently represented by an admittance. When the signal source is more conveniently represented by an impedance, a dual equivalent circuit is available (Fig. 7). This is a T equivalent circuit and employs a correlation impedance Z_γ together with a negative correlation impedance $-Z_\gamma$ to avoid modifying the signal transfer properties of the network. As before, the generators are most conveniently represented by their equivalent noise resistance and conductance R_u' and G_n' where the prime denotes the T equivalent circuit. The relations between the parameters of the two equivalent circuits are listed in Table 1. Thus given one equivalent circuit, it is easy to translate to the other.

[5] Rothe and Dahlke, *loc. cit.*

AMPLIFIER FUNDAMENTALS

Table 1. Transformation Equations between Π and T Noise Matrices†

$$R_n = R_u' + G_n'|Z_\gamma|^2 \qquad G_n' = G_u + R_n|Y_\gamma|^2$$

$$G_u = \frac{R_u'}{|Z_\gamma|^2 + R_u'/G_n'} \qquad R_u' = \frac{G_u}{|Y_\gamma|^2 + G_u/R_n}$$

$$Y_\gamma = \frac{Z_\gamma^*}{|Z_\gamma|^2 + R_u'/G_n'} \qquad Z_\gamma = \frac{Y_\gamma^*}{|Y_\gamma|^2 + G_u/R_n}$$

Primed quantities refer to T matrix. Asterisk denotes complex conjugate.

† After H. Rothe and W. Dahlke, Theory of Noisy Fourpoles, *Proc. IRE*, vol. 44, p. 813, June, 1956.

4. NOISE FACTOR

The noise factor, or noise figure, is one of the most commonly used, and unfortunately one of the most commonly misused, terms describing a system's noise performance. The term is simple in its concept, and with a reasonable study of its properties there is little danger of its misuse. The noise factor is a single number that characterizes the noise properties of a network or system. It cannot be a complete characterization of the noise properties; it was shown in the previous section that it takes four independent quantities to specify the noise properties of circuit completely at a given operating point and temperature.[6] The source impedance or admittance must always be specified with the noise factor. It is not possible to infer the noise factor obtained using a different source impedance without having additional information. Frequently, the specified noise factor is a minimum noise factor obtained under a given set of operating conditions. The minimum noise factor indicates a more or less optimum limit. More sophisticated noise analyses show that a minimum noise factor does not always lead to the maximum signal-to-noise ratio in the system; however, the minimum noise factor is a good objective for most practical purposes.

4.1. Definition of Noise Factor. Two equivalent definitions for the noise factor of a transducer exist. The first is a measure of the amount the transducer degrades the signal-to-noise ratio of the system. The second is the ratio of the total noise in the system, referred to the input, to thermal noise. It is important to know both definitions since they represent different points of view.

The noise factor F may be defined as the quotient of (1) the available-signal-to-available-noise ratio at the input to (2) the available-signal-to-available-noise ratio at the output.[7] The signal and noise powers can be defined on a per-unit-bandwidth basis. The noise at the output includes the noise from all responses of the transducer including spurious outputs due to noise introduced at image or idler frequencies and converted to the output frequency.[8] The noise factor does not include noise generated by the load. Using S and N to represent available signal and noise powers, respectively, and subscripts i and o to represent input and output, respectively, the noise factor is

$$F = \frac{S_i/N_i}{S_o/N_o} \qquad (9)$$

The available input noise N_i is by definition thermal noise at *standard noise temperature*, 290°K, $N_i = kT_oB$. *The noise factor is always referred to thermal noise at 290°K as a base.* The noise factor is usually specified in decibels,

$$F \text{ (db)} = 10 \log S_i/N_i - 10 \log S_o/N_o$$

It specifies the decibel difference between the signal-to-noise ratio at the input of the network and the signal-to-noise ratio at the output. It thus specifies the amount by

---

[6] The characterization assumed is the determination of the mean-square noise properties of the network with arbitrary input termination. The four parameters described in the previous section will not necessarily provide other statistical information on the noise.

[7] H. T. Friis, Noise Figures of Radio Receivers, *Proc. IRE*, vol. 32, pp. 410–422, July, 1944.

[8] See Sec. 5.4.

which the network degrades the signal-to-noise ratio. This is a useful factor in determining the performance of a system, especially if the noise base of the input signal generator is actually thermal noise at standard temperature. Frequently, it is not. However, knowing the input noise base, it is not a difficult calculation to determine the degradation in signal-to-noise performance for any input noise base.

This concept of noise factor is not attractive for circuit analysis nor measurement purposes because available powers are usually not readily calculated nor measured. Fortunately, the second concept of the noise factor is readily derived. Define the available gain G_A of the transducer as the ratio of signal power available at the output of the two-port, at a specified output frequency, to the signal power available from the source at the corresponding input frequency.

$$G_A = S_o/S_i \tag{10}$$

The available gain is frequently misunderstood; so there is merit in describing a few of its important properties. First, the available gain is a function of the source impedance. Second, the source impedance does *not* have to be matched to the input impedance of the transducer and conversely. The definition is based on the available power from the signal source, but the definition does not in any way imply that the source actually delivers its maximum available power to the transducer. There is no definite relationship between the noise factor and input impedance of the transducer; however, as a rule, the minimum noise factor is achieved when the source impedance is somewhat mismatched.

Eliminating the signal powers from Eq. (9) by use of G_A,

$$F = N_o/N_i G_A \tag{11}$$

The output noise N_o is composed of two parts

$$N_o = N_i G_A + N_N \tag{12}$$

The first part, $N_i G_A$, is the amplified thermal noise from the signal source. The second term, N_N, includes all other noises in the system that are available at the output port. N_N is due to the noises generated within the transducer itself. Combining Eq. (11) with Eq. (12), the noise factor becomes

$$F = \frac{N_i G_A + N_N}{N_i G_A} \tag{13}$$

This equation corresponds to the IEEE definition of noise factor.[9] "At a specified input frequency the ratio of 1) the total noise power per unit bandwidth at a corresponding output frequency available at the output *Port* to 2) that portion of 1) engendered at the input frequency by the input termination at the *Standard Noise Temperature* (290°K)."

If a transducer has multiple responses, the numerator of Eq. (13) must contain the noise from all responses. If there is a signal in only one response, the denominator must contain only the noise from that response. If there is a signal in more than one response, the denominator may include all responses that have signals. Alternatively it may include only one response with a separate noise factor derived for each.

Suppose a receiver has a normal response and an image. If the receiver is used for communications, only the normal response will contain a signal. The denominator of Eq. (13) will contain only the noise from the normal response, but the numerator will contain the noise from both the normal and the image responses. If the receiver is used for astronomy, both responses will be employed for signal usage. The denominator should contain the noise from both responses, and this will lead to a lower noise-factor specification—for the same receiver! *If there is any possibility of confusion it should always be stated whether the noise factor is single- or multiple-response.* The handling of multiple responses is described in more detail in Sec. 5.

[9] P. A. Redhead, chairman, *et al.*, IRE Standards on Electron Tubes: Definitions of Terms, 1957, *Proc. IRE*, vol. 45, p. 1000, July, 1957.

For single-response transducers, it is convenient to interpret the factor N_N/G_A as the equivalent transducer noise referred to the input terminals, and give it the symbol N_e. The noise factor then becomes

$$F = 1 + N_e/N_i = (N_i + N_e)/N_i \qquad (14)$$

Thus, for a two-port, the noise factor may be interpreted as the total noise referred to the input divided by the thermal noise from the signal source.

4.2. Excess Noise Factor. Another measure that is gaining significance is sometimes called the excess noise factor, $F - 1$. From Eq. (14), it is seen that the excess noise factor is simply the ratio of the equivalent input noise of the amplifier to the thermal noise of the source.

$$F - 1 = N_e/N_i \qquad (15)$$

If the quantities were transferred to the output, the excess noise factor would be the ratio of that part of the output noise engendered by the amplifier divided by that part of the output noise engendered by the signal source at the *standard noise temperature*.

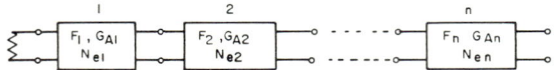

Fig. 8. Noise factor of cascaded two-ports.

4.3. Noise Factor for Cascaded Two-ports. A very convenient and useful formula can be developed for the total noise factor of a number of two-ports in cascade. Assume there are n two-ports in cascade (Fig. 8). Assume that the noise factors of the two-ports are, respectively, F_1, F_2, ..., and F_n. Similarly, the available gains are $G_{A1} \cdots G_{An}$; and the equivalent noise powers of the two-ports referred to their inputs, $N_{e1} \cdots N_{en}$. The total equivalent input noise N_{et} due to all the two-ports is then

$$N_{et} = N_{e1} + \frac{N_{e2}}{G_{A1}} + \frac{N_{e3}}{G_{A1}G_{A2}} + \cdots + \frac{N_{en}}{G_{A1}G_{A2} \cdots G_{A(n-1)}}$$

By comparison with Eq. (15), the excess noise factor is

$$F - 1 = N_{et}/N_i$$

Thus, using Eq. (15) to determine the individual N_e's, the excess noise factor of the cascade becomes

$$F - 1 = F_1 - 1 + \frac{F_2 - 1}{G_{A1}} + \frac{F_3 - 1}{G_{A1}G_{A2}} + \cdots + \frac{F_n - 1}{G_{A1}G_{A2} \cdots G_{A(n-1)}} \qquad (16)$$

For example, if an amplifier consisted of a cascade of three stages, each having a noise factor of 5 and a gain of 2, the resulting noise factor would be

$$F - 1 = 4 + \tfrac{4}{2} + \tfrac{4}{4}$$
$$F = 8$$

The application of this formula requires appropriate attention to the source impedance of each two-port. The output impedance of the first two-port is the source impedance of the second, etc. Therefore, the noise factor F_2 of the second two-port must be specified for a source impedance equal to the output impedance of the first two-port, etc. This requirement is usually fulfilled automatically in microwave designs which must be made on a constant impedance basis. However, this is not generally the case for lower-frequency circuits. If, for example, the three two-ports represent three stages in an audio amplifier, it is most likely that each stage will see a different source impedance and its noise factor must be specified accordingly.

4.4. Extended Noise Factor.[10]

The definitions of noise factor can be applied to transducers in a straightforward manner when the source resistance and the output resistance of the transducer are positive. However, in many practical situations the output resistance may be negative. This can happen, for example, with tunnel-diode amplifiers. At first glance, the source resistance would always be positive. However, when the transducer being considered is in a cascade, its source resistance is the output resistance of the previous transducer, and this may be negative.

When the source resistance or output resistance is negative, a noise factor based on available power is ambiguous because the maximum available power from a source with a negative internal resistance is theoretically infinite. This condition occurs when the real and imaginary parts of the load impedance equal, respectively, the negative of the real imaginary parts of the source impedance. This interpretation of "maximum available power" for the negative-resistance case serves no useful purpose in dealing with noise factor It is much more satisfactory to employ more mathematical concepts. The concept of maximum available power is replaced by the concept of *exchangeable power* P_e. For a Thévenin source with an internal voltage generator E and internal impedance, $Z = R + jX$ (Fig. 1), the exchangeable power is

$$P_e = \overline{e^2}/4R \qquad R \neq 0$$

R may be positive or negative but may not be equal to zero. When R is positive, the exchangeable power is equal to the maximum available power. When R is negative, the exchangeable power is negative and has the same magnitude as the case for R positive. The negative power can be interpreted as a power flowing into the source rather

Table 2. Sign Relationships for Extended Noise Factor

R_s	R_o	G_e	N_{eN}	$F_e - 1$
+	+	+	+	+
+	−	−	−	+
−	+	−	+	−
−	−	+	−	−

than out of it; however, it is often more straightforward to assume that it is a mathematical definition adopted for convenience.

An *exchangeable gain* G_e can now be defined in terms of the exchangeable signal power at the output S_{eo} and exchangeable signal power at the source S_{ei}.

$$G_e = S_{eo}/S_{ei} \qquad (17)$$

Note that G_e is negative when either S_{eo} or S_{ei} is negative. G_e is positive when S_{eo} and S_{ei} have the same sign.

An *extended noise factor* F_e can now be defined on the basis of exchangeable power and exchangeable gain. With reference to Eq. (13), the exchangeable noise factor is defined as

$$F_e = 1 + N_{eN}/N_i G_e \qquad (18)$$

where N_{eN} is the exchangeable noise power at the output of the transducer when the noise from the signal source is zero; and $N_i = kT_oB$. N_i is identical to the N_i in ordinary noise factor and *is always positive*. N_{eN} is positive when the output resistance is positive and negative when the output resistance is negative. The sign relationships for extended noise factor are summarized in Table 2 for all combinations of positive and negative source resistances R_s and output resistances R_o. Note that the quantity $F_e - 1$ is always negative when R_s is negative.

[10] H. A. Haus and R. B. Adler, An Extension of the Noise Figure Definition, *Proc. IRE*, vol. 45, pp. 690–691, May, 1957.

It can be easily shown that F_e may be interpreted in a manner directly parallel to the noise-factor interpretation [Eq. (9)] when R_s is positive.

$$F_e = \frac{S_{ei}/N_{ei}}{S_{eo}/N_{eo}} \qquad R_s > 0$$

S_{ei}/N_{ei} is the ratio of the exchangeable signal to exchangeable thermal noise at the input of the transducer, and S_{eo}/N_{eo} is the ratio of the exchangeable signal to exchangeable noise at the output.

Unfortunately this parallelism fails when R_s is negative.

$$F_e \neq \frac{S_{ei}/N_{ei}}{S_{eo}/N_{eo}} \qquad R_s < 0$$

With reference to Eqs. (12) and (17), this relationship can be written

$$\frac{S_{ei}/N_{ei}}{S_{eo}/N_{eo}} = \frac{N_{eo}}{N_{ei}G_e} = \frac{N_{ei}G_e + N_{eN}}{N_{ei}G_e}$$

Rewriting in the form of Eq. (18) gives

$$\frac{S_{ei}/N_{ei}}{S_{eo}/N_{eo}} = 1 + N_{eN}/N_{ei}G_e \tag{19}$$

The only difference between Eqs. (19) and (18) is that N_{ei} in Eq. (19) replaces N_i in Eq. (18). If R_s is positive, $N_{ei} = N_i$; but if R_s is negative, $N_{ei} = -N_i$; and the second term of Eq. (19) has the incorrect sign.

Since the extended-noise-factor and exchangeable-power concepts unavoidably incur difficulties of interpretation and are not always parallel to the concepts of noise factor, for what purpose have they been introduced? First, they have a sound mathematical basis and are used for the sophisticated treatments of linear noisy networks.[11] Second, they can be employed directly in the cascade-noise-factor equation without ambiguity whenever the source resistance of the cascade is positive. It can be shown that the cascade formula based on the extended noise factor can always be used to determine the conventional noise factor when the source impedance for the complete network has a positive real part. Thus

$$F - 1 = F_{e1} - 1 + \frac{F_{e2} - 1}{G_{e1}} + \frac{F_{e3} - 1}{G_{e1}G_{e2}} + \cdots + \frac{F_{en} - 1}{G_{e1}G_{e2} \cdots G_{e(n-1)}} \qquad R_s > 0 \tag{20}$$

One of the requirements of the cascade formula is that each added network must add noise to the system; i.e., it must increase the quantity $F - 1$. Table 3 illustrates the fact that the extended-noise-factor concept fulfills this requirement. Three transducers are shown together with their respective exchangeable noise factor, exchangeable gain, and output resistance. The source resistance for the cascade R_s is always positive, but the signs of the transducer output resistances are allowed to assume all combinations of positive and negative values. The table lists the corresponding values of the exchangeable gains and excess exchangeable noise factors $F_e - 1$. The last three columns are the first three terms of the cascade formula [Eq. (20)]. It is seen that each one of these terms is positive. This is as it should be since the addition of each network must increase the overall excess noise factor of the system regardless of its terminating impedances.

[11] H. A. Haus and R. B. Adler, "Circuit Theory of Linear Noisy Networks," The Technology Press of the Massachusetts Institute of Technology, Cambridge, Mass., and John Wiley & Sons, Inc., New York, 1959.

AMPLIFIER NOISE

Table 3. Sign Relationships for Cascade

R_s	R_{01}	R_{02}	R_{03}	G_{e1}	G_{e2}	$F_{e2}-1$	$F_{e3}-1$	$F_{e1}-1$	$\dfrac{F_{e2}-1}{G_{e1}}$	$\dfrac{F_{e3}-1}{G_{e1}G_{e2}}$
+	+	+	+	+	+	+	+	+	+	+
+	+	+	−	+	+	+	+	+	+	+
+	+	−	+	−	+	−	+	−	+	+
+	+	−	−	−	+	−	+	−	+	+
+	−	+	+	−	−	+	−	+	+	+
+	−	+	−	−	−	+	−	+	+	+
+	−	−	+	+	−	−	−	+	+	−
+	−	−	−	+	−	−	−	+	+	−

4.5. Behavior of Noise Factor with Source Impedance (Admittance). Noise factor has certain important characteristics which are true of all linear two-ports. The manner in which the noise factor varies as a function of the source impedance (admittance) is especially significant. Noise factor is one of the easiest noise parameters of a circuit or device to measure. An experimental determination of the behavior of the noise factor as a function of the source impedance (admittance) can yield all the parameters of the two-port noise equivalent circuit. This is the most straightforward method available to obtain these noise parameters. Also, since the noise-factor-source-impedance (admittance) characteristic must obey certain laws, a study of this characteristic can yield considerable information as to whether or not the noise-factor measurements are being made accurately.

It is first necessary to calculate the noise factor. The noisy two-ports of Figs. 6 and 7 lead to a set of dual noise-factor equations. The Π equivalent circuit (Fig. 6) is most satisfactory for a signal source represented by an admittance, and the derivations will be carried out using this representation. The conclusions for both circuits will be discussed in parallel, with the equations for the two representations appropriately identified.

Assume Fig. 6 is divided by the line A-A. The block to the right of A-A generates no noise. Since it is linear, its response is the same for both signal and noise; and it cannot change the signal-to-noise ratio of the system. Therefore, the noise factor computed for the elements to the left of A-A will be the same as the noise factor for the entire system. It is much simpler to deal solely with the left-hand elements, which (exclusive of the signal source) are usually referred to as the noisy two-port.

The noise available at A-A in Fig. 6 due to the source is

$$N_i G_A = \overline{i_{ni}^2}/4(G_s + G_\gamma - G_\gamma) \tag{21}$$

where G_s and G_γ are the conductances of the source and correlation admittances, respectively. In calculating the noise contributed by the transducer, it is convenient to transform the voltage e of generator E into an equivalent shunt current generator by making a Thévenin to Norton transformation of e about $Y_s + Y_\gamma$.

$$i = e(Y_s + Y_\gamma)$$

The noise power at A-A due to the transducer becomes

$$N_N = (\overline{i_u^2} + \overline{e^2}|Y_s + Y_\gamma|^2)/4(G_s + G_\gamma - G_\gamma) \tag{22}$$

By combining Eqs. (13), (21), and (22), the noise factor becomes

$$F = \frac{\overline{i_{ni}^2} + \overline{i_u^2} + \overline{e^2}|Y_s + Y_\gamma|^2}{\overline{i_{ni}^2}} \qquad (23)$$

The numerator is simply the total mean-square equivalent noise current across the input terminals. The denominator is the corresponding current from the signal source. If Fig. 7 had been used, a dual expression would have resulted with the numerator equal to the sum of voltages in series between the input terminals.

These types of equations are frequently used for calculating the noise factor of transducers when the individual noise generator mechanisms in the transducer can be represented by equivalent noise voltage or current sources. These voltage or current sources can then be transformed to the input circuit. Equation (23) is closely related to Eq. (14) since the available noise powers are directly proportional to their corresponding mean-square noise currents.

Representing the noise generators by their equivalent noise conductances or resistances, and separating the source and correlation admittances into their real and imaginary parts, gives[12]

$$F = 1 + \frac{G_u + R_n[(G_s + G_\gamma)^2 + (B_s + B_\gamma)^2]}{G_s} \qquad (24\Pi)$$

$$F = 1 + \frac{R_u' + G_n'[(R_s + R_\gamma)^2 + (X_s + X_\gamma)^2]}{R_s} \qquad (24T)$$

The Π and T notations refer to equations derived for Figs. 6 and 7, respectively. It can be seen that the noise factor has a minimum with respect to both the real and imaginary parts of the source immittance. The optimum value of the source susceptance B_o (reactance X_o) is obtained when the source susceptance (reactance) is equal to the negative of the correlation susceptance (reactance).

$$B_o = -B_\gamma \qquad (25\Pi)$$
$$X_o = -X_\gamma \qquad (25T)$$

This optimum can be easily found experimentally through the process of "noise tuning." The real part of the source immittance is held fixed at any desired value. The imaginary part is then varied until a minimum noise factor is obtained.

Noise tuning leads to

$$F = 1 + \frac{G_u + R_n(G_s + G_\gamma)^2}{G_s} \qquad (26\Pi)$$

$$F = 1 + \frac{R_u' + G_n'(R_s + R_\gamma)^2}{R_s} \qquad (26T)$$

The optimum value of the source conductance G_o (resistance R_o) can be easily obtained from Eq. (26) and is given by

$$G_o = \left(\frac{G_u + R_n G_\gamma^2}{R_n}\right)^{\frac{1}{2}} \qquad (27\Pi)$$

$$R_o = \left(\frac{R_u' + G_n' R_\gamma}{G_n'}\right)^{\frac{1}{2}} \qquad (27T)$$

The minimum value of the noise factor F_o is then

$$F_o = 1 + 2R_n(G_\gamma + G_o) \qquad (28\Pi)$$
$$F_o = 1 + 2G_n'(R_\gamma + R_o) \qquad (28T)$$

The same minimum noise factor is obtained regardless of whether the signal source is an admittance or an impedance.

[12] The source is assumed to be at *standard noise temperature*, 290°K.

Additional insight into the relationships is obtained from the magnitude of the optimum source admittance (impedance).

$$|Y_o|^2 = |Y_\gamma|^2 + G_u/R_n$$
$$|Z_o|^2 = |Z_\gamma|^2 + R_u'/G_n'$$

A possibly more useful form for the noise factor is given by

$$F = F_c + (R_n/G_s)[(G_s - G_o)^2 + (B_s - B_o)^2] \quad (29\Pi)$$
$$F = F_c + (G_n'/R_s)[(R_s - R_o)^2 + (X_s - X_o)^2] \quad (29T)$$

This can be derived from the preceding equations. When the real part of the source immittance is very much larger than the optimum, the noise factor increases in direct proportion to the real part. Conversely, when the real part is very much smaller than the optimum value, the noise factor increases inversely with the real part. The behavior is somewhat different for the imaginary part.

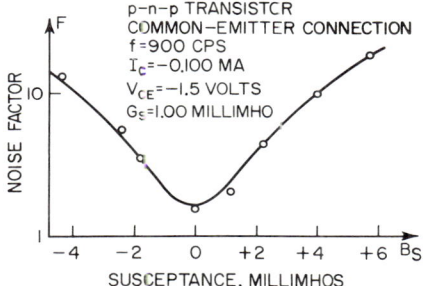

Fig. 9. Noise factor vs. source susceptance. (*From H. A. Haus, chairman, et al., IRE Standards on Methods of Measuring Noise in Linear Two Ports, 1959, Proc. IRE, vol. 48, p. 67, January, 1960.*)

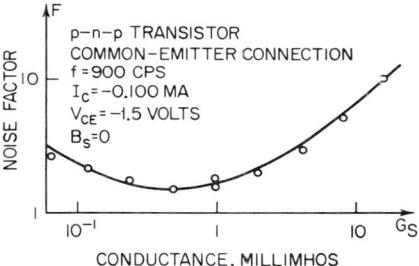

Fig. 10. Noise factor vs. source conductance. (*From H. A. Haus, chairman, et al., IRE Standards on Methods of Measuring Noise in Linear Two Ports, 1959, Proc. IRE, vol. 48, p. 67, January, 1960.*)

It is also of interest to consolidate the real and imaginary parts of the equations to yield equations in more compact form,

$$F = F_o + (R_n/G_s)|Y_s - Y_o|^2 \quad (30\Pi)$$
$$F = F_o + (G_n'/R_s)|Z_s - Z_o|^2 \quad (30T)$$

4.6. Experimental Determination of Two-port Noise Parameters. The two-port noise parameters are readily determined through a series of measurements of the noise factor. The noise-factor data are plotted on graphs and the information is derived from these graphs. The use of a graphical experimental technique is advantageous since data that do not fit the proper graphical forms signal the fact that the measurements are in error. Since the noise factor is a measurement of randomly varying quantities, the measurement result will be randomly distributed. A good fit to the random data can be obtained by drawing smooth curves through the "best average" of the experimental points. In order to illustrate the experimental procedure, graphs of data taken on a *p-n-p* transistor will be referred to.

The first step is to measure the spot noise factor* as a function of the susceptance (reactance) of the signal source. The conductance (resistance) of the signal source must be held constant for these measurements. The data are then graphed (Fig. 9). From Eq. (25), it is seen that the noise factor has a minimum when the source susceptance (reactance) is equal to the negative of the correlation susceptance (reactance). This determines the imaginary part of the correlation immittance [Eq. (25)].

The imaginary part of the source immittance is then held constant at the optimum value. The noise factor is given by Eq. (26). The noise factor is then measured as a function of the source conductance (resistance). The resulting curve (Fig. 10) has a

* See Sec. 4.7.

7-20 AMPLIFIER FUNDAMENTALS

minimum at the optimum value of the source conductance (resistance). This is the minimum noise factor obtainable since both the real and the imaginary parts of the source immittance have been optimized.

Using the data obtained, the optimum source admittance (impedance) $Y_o = G_o + jB_o$, ($Z_0 = B_o + jX_o$) is then computed. Also, the source admittance (impedance) $Y_s = G_s + jB_s$ ($Z_s = R_s + jX_s$) is computed for each of the noise-factor measurements made. Then the noise factor is plotted as a function of the quantity X (Fig. 11), where

$$X = \frac{|Y_s - Y_o|^2}{G_s}$$

$$X = \frac{|Z_s - Z_o|^2}{R_s}$$

From Eq. (30), it is seen that the locus of this equation is a straight line with a zero intercept at the minimum noise factor $F = F_o$. The line has a slope equal to R_n (G_n'). Using the data obtained from the curves, it is found that

$$B_o = -B_\gamma = 0$$
$$G_o = 0.5 \text{ millimho}$$
$$F_o = 1.55$$
$$R_n = 540 \text{ ohms}$$

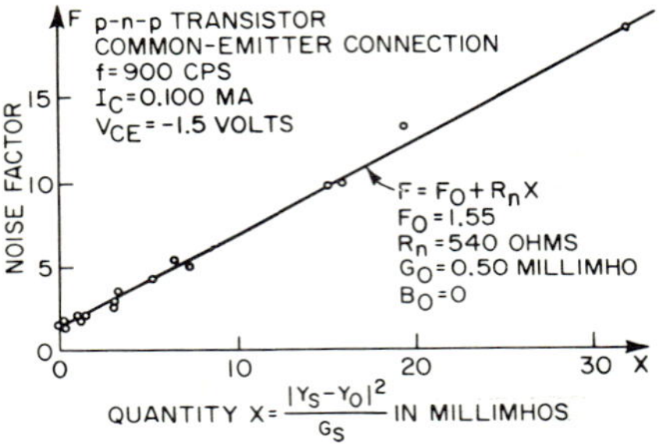

These quantities give full information for determining the noise factor for any source immittance through the use of Eqs. (29) or (30). However, two parameters of the two-port, G_γ and G_u (R_γ and R_u'), have not been determined. If needed, these quantities can be easily derived. G_γ (R_γ) is obtained from Eq. (28).

FIG. 11. Noise factor vs. quantity X. (*From H. A. Haus, chairman, et al., IRE Standards on Methods of Measuring Noise in Linear Two Ports, 1959, Proc. IRE, vol. 48, p. 67, January, 1960.*)

$$G_\gamma = \frac{F_o - 1}{2R_n} - G_o$$

$$R_\gamma = \frac{F_o - 1}{2G_n'} - R_o$$

The noise conductance G_u (resistance R_u') is obtained by combining Eqs. (27) and (28).

$$G_u = (F_o - 1)\left(G_o - \frac{F_o - 1}{4R_n}\right)$$

$$R_u' = (F_o - 1)\left(R_o - \frac{F_o - 1}{4G_n'}\right)$$

For the transistor in the example, the correlation conductance and the noise conductance are found to be, respectively,

$$G_\gamma = 0.009 \text{ millimho}$$
$$G_u = 0.13 \text{ millimho}$$

Thus all the two-port noise parameters can be derived by two sets of noise-factor measurements. One set of measurements is taken with the real part of the source immittance held constant and the imaginary part varied. The second set is taken with the imaginary part held at its optimum value and the real part varied.

If the curves do not have the specified shape, there must be an error in the measuring process. Errors can be introduced into the measurement in many ways. The transducer may become nonlinear or the output power indicator may not provide a linear response to power for all the measurements employed. Errors also result when the real and imaginary parts of the source immittance are not varied independently as required for the measurements. For example, suppose that the real part is varied by

AMPLIFIER NOISE   7-21

changing resistors in parallel with the input. If the different resistors used have different capacitances, or different lead inductances, measurement error will result unless corrections for the variations are made in the experimental process.

4.7. Average Noise Factor. The calculations thus far have described what is essentially a single-frequency noise factor. This is a noise factor determined for a sufficiently narrow fractional bandwidth that all parameters of the system, which are a function of frequency, may be considered constant. The noise factor thus computed is more specifically referred to as the *spot noise factor*.

In many systems, the signal and the noise are distributed over a fairly wide band and the system parameters will be functions of frequency. It is possible in this case to divide the spectrum into a number of bands that are sufficiently narrow to fulfill the narrowband assumptions and then determine the spot noise factor of each one of these narrow bands. The spot noise factor might be presented as a curve which is a function of frequency. Unfortunately, the usefulness of noise factor as a single-number characterization is lost.

An alternate approach is to define an average noise factor $\bar{F}$. More information is, of course, required to characterize the noise over a wide band than over an incremental band. In a wideband system, it will often be necessary to know the frequency distribution of the noise as well as the frequency distribution of the signal in order to determine adequately the interfering effect of the noise. On this basis, the average noise factor may be considered to provide less information than the spot noise factor. However, the average noise factor is used frequently.

The average noise factor is defined as[13] "the ratio of 1) the total noise power delivered by the transducer into its output termination when the *Noise Temperature* of its input termination is standard (290°K) at all frequencies, to 2) that portion of (1) engendered by the input termination." This is represented by the equation

$$\bar{F} = \frac{\int N_o(f)\, df}{\int N_i G(f)\, df} \qquad (31)$$

$N_o(f)$ is the noise power per unit bandwidth *delivered* by the output port and N_i is the noise power per unit bandwidth available from the signal source.

$$N_i = kT_o$$

$G(f)$ is the transducer gain, defined as[14] "the ratio of 1) the actual signal power transferred from the output *Port* of the transducer to its load, to 2) the available signal power from the source driving the transducer."

The transducer gain is slightly different from the available gain defined in Sec. 4.1. The numerator of the transducer gain is the power delivered to the load from the output port, whereas the numerator of the available gain is the power available at the output port. The denominators of both expressions are the same and refer to the power available from the signal source. Available gain was used in the definition of spot noise factor because it allowed the cascade-noise-factor formula to be derived easily. Power delivered to the load is used in defining the average noise factor because it is more easily measured. *The noise factor may be defined either in terms of the available noise power at the output port or in terms of the power delivered by the system into its output determination.* The two definitions are equivalent because the noise factor defines a power ratio rather than an absolute power.

The numerator of Eq. (31) is the total noise that appears at the output of the network. If there are multiple responses, the numerator includes noise contributions from all responses with their respective sources at 290°K. The denominator is the integral of the thermal noise available from the source times the transducer gain over the frequency band of interest. This includes only the principal signal responses as discussed in Sec. 4.1. N_i is a constant and can be moved outside the integral sign.

It is sometimes convenient to express the average noise factor in terms of the spot noise factor. Using an equation having the same form as Eq. (11) (but defined in

---
[13] P. A. Redhead, *loc. cit.*
[14] P. A. Redhead, *op. cit.*, p. 995.

terms of the transducer gain and the power delivered to the load), the $N_o(f)$ and N_i terms can be eliminated.

$$\bar{F} = \frac{\int FG(f)\,df}{\int G(f)\,df}$$

4.8. Noise Bandwidth. When dealing with frequency averages of noise quantities, an equivalent rectangular noise bandwidth, commonly called noise bandwidth, is usually specified. The noise bandwidth is defined by

$$B = \frac{\int G(f)\,df}{G_o}$$

G_o is the transducer gain at a reference frequency f_o, B is the noise bandwidth of the system in cycles per second, and $G(f)$ is the transducer gain at the frequency f. The

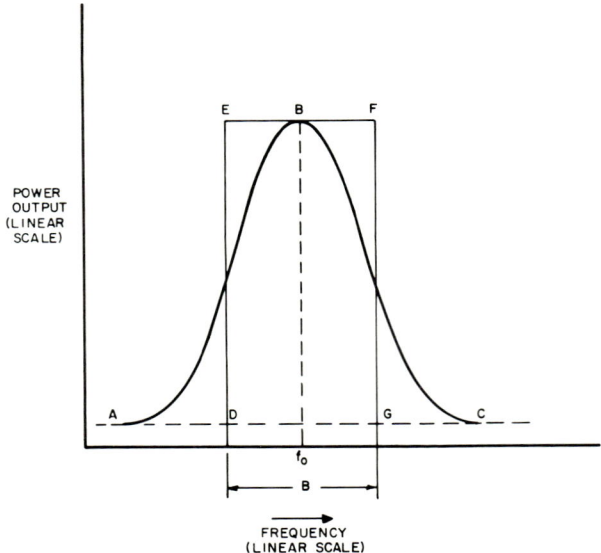

FIG. 12. Determination of noise bandwidth.

noise bandwidth is usually determined graphically (Fig. 12). The curve is obtained using a variable-frequency signal source, with a constant available output power, connected to the input of the system, and a power meter connected to the output. At frequencies lower than A and higher than C in Fig. 12, the power output meter settles to a constant residual reading. This reading is caused by the noise in the system, and in many cases may be entirely negligible. If it is not negligible, only the area of the figure above the line $ADGC$ is employed in determining the noise bandwidth.

A desired reference frequency f_o is selected. If the power-output curve is more or less symmetrical, f_o is usually taken as a frequency of maximum gain. A rectangle, $DEFG$, passing through the power-output curve at f_o is constructed. The rectangle must have the same area above $ADGC$ as the power-output curve. The width of this rectangle, B, is the noise bandwidth.

When the power-output curve is not a simple symmetrical function, the noise bandwidth B and the location of the frequency f_o become somewhat arbitrary. For example, in Fig. 13, the center frequency might be selected at f_o or f_o' or, in fact, at any other place on the graph. The reference frequencies f_o and f_c' yield, respectively, rectangles $ABCD$ and $A'B'C'D'$. The primed rectangle corresponds to a higher gain and narrower noise bandwidth than the unprimed rectangle. However, the gain bandwidth products are the same. The position of the rectangle on the abscissa has no particular significance since the factors of interest are the gain and noise bandwidth and not the specific location of the noise bandwidth. The reference frequency f_o does have significance since it defines the gain G_o.

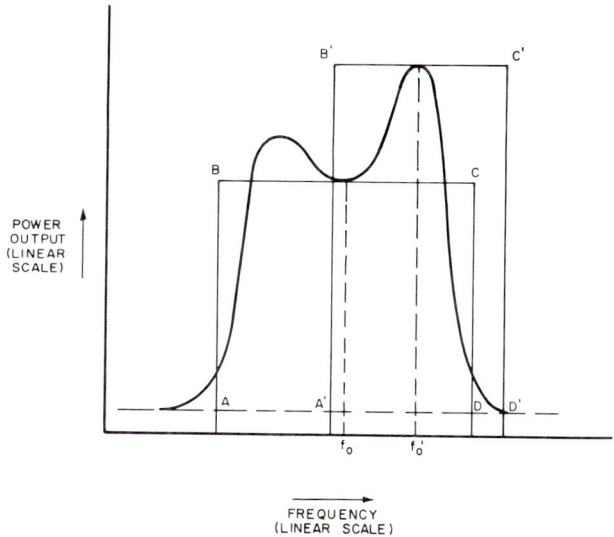

Fig. 13. Determination of noise bandwidth for nonsymmetrical curve.

5. NOISE TEMPERATURE

Noise temperature is a means of characterizing the noise properties of a one-port, a two-port, or for that matter an n-port. In its simplest form, the noise temperature of a two-port is directly related to the noise factor of the same two-port. However, as will be seen, noise temperature is defined much more broadly than noise factor. This permits the concept of noise temperature to be applied more effectively to systems of a complicated nature. For example, the noise-temperature concepts are applicable to (1) systems with signal and noise bandwidths that are different, (2) systems with multiple signal responses, and (3) systems where noise is generated at the load and reflected at the output port. Within the scope of the noise-factor definitions, these conditions are handled only with difficulty.

The noise temperature is based on thermal noise. The noise temperature at a port is "the temperature of a passive system having an available noise power per unit bandwidth equal to that of the actual *Port*, at a specified frequency.[15] Thus, if the available noise power in a 1-cycle band from a port is kT, its noise temperature is $T°K$. The temperature of a passive network at *standard noise temperature* is 290°K.

5.1. Effective Input Noise Temperature and Operating Noise Temperature. The IEEE defines two noise temperatures that are closely related to the noise factor. These are the *effective input noise temperature* T_e and the *operating noise temperature* T_{op}. In a two-port with a single input frequency and a single output frequency, T_e is related to the noise factor by

$$T_e = 290(F - 1) \qquad (32)$$

[15] P. A. Redhead, *op. cit.*, p. 1000.

Referring to Eq. (15) it is noted that the excess noise factor $(F - 1)$ is equal to the equivalent noise added by the two-port (referred to its input) divided by the thermal noise of the source at 290°K. Thus multiplying $(F - 1)$ by 290° gives an effective input noise temperature which corresponds to the noise added by the two-port.

The effective input noise temperature is defined as "the *noise temperature* in degrees Kelvin which, assigned simultaneously to the specified impedance terminations at the set of frequencies contributing to the output, at all accessible ports except the designated output port of a noise-free equivalent of the transducer, would yield the same available power per unit bandwidth at a specified output frequency at the output port as that of the actual transducer connected to noise-free equivalents of the terminations at all ports except the output port.[16]

The effective input noise temperature may be considered to define a noise power that would have to be added to each input port of a transducer to produce an output noise power from the output port equal to that developed by the transducer itself. By definition, *equal* noise powers are added simultaneously to each input port (or response) in determining the effective input noise temperature. This is to avoid ambiguity in specification and to facilitate easy experimental determination of T_e. T_e is a function of the impedance of the sources connected to its ports. However, it is not a function of the noise temperature of the sources (except for the special case of noise generated by the load at frequencies other than the output frequency; see Sec. 5.4). The use of T_e will become more apparent during the following development.

The operating noise temperature is defined for a system under actual operating conditions. The term is most simply defined for a two-port network with a single input frequency and a single output frequency and a load that produces a negligible amount of noise. If the noise temperature of the signal source is T_i, the operating noise temperature is

$$T_{op} = T_i + T_e$$

If T_i is 290°K and the above conditions hold, then from Eq. (32), T_{op} is related to the noise factor by

$$T_{op}|_{T_i=290°\text{K}} = 290° + T_e = 290F$$

For the more general case, consider a receiver under normal operating conditions. It is connected to its usual signal source and it is in its normal environment. Define the output noise power, per unit output signal bandwidth, delivered to the load as N_o. Define also the signal gain G_s as the ratio of the output signal S_o *delivered* to the output load at the specified output frequency, to the input signal S_i *available* from the input source at the specified input frequency. Using these factors T_{op} is defined for the general case by

$$T_{op} = N_o/kG_s \tag{33}$$

Note that *signal* gain is used in defining T_{op} whereas *transducer* gain or *available* gain is used in defining noise factor. The signal gain is equal to the transducer gain for a transducer with only one signal response. More broadly, a composite signal gain can be defined for transducers with multiple signal responses as shown in Sec. 5.3. This is not possible with transducer gain.

The operating noise temperature is useful because it enables the output signal-to-noise ratio S_o/N_o to be determined under actual operating conditions. The output signal-to-noise ratio is determined as a function of the operating temperature by

$$\frac{S_o}{N_o} = \frac{S_o/G_s}{N_o/G_s} = \frac{S_i}{kT_{op}}$$

Thus the signal-to-noise ratio is equal to the available signal at the input divided by the product of Boltzmann's constant and T_{op}.

It is now important to determine the makeup of the output noise power N_o. N_o includes the noise of the signal sources that is transmitted to the output, noise gener-

---

[16] H. A. Haus, chairman, *et al.*, IRE Standards on Electron Tubes: Definitions of Terms, 1962, (62 IRE 7.S2), *Proc. IRE*, vol. 51, p. 435, March, 1963.

ated within the transducer and transmitted to the output, and noise generated in the load, transmitted to the output of the transducer, and reflected back to the load. Let T_{in} be the noise temperature of the input source for the nth response. Then the portion of the output noise N_{io} due to the noise temperatures of all the sources is given by

$$N_{io} = k(T_{i1}G_1 + T_{i2}G_2 + \cdots + T_{in}G_n) \qquad (34)$$

where G_n is the transducer gain at the nth response. It is assumed that the inputs to all the responses are uncorrelated.

The output noise N_N due to the transducer itself must now be determined. This is given by

$$N_N = kT_e(G_1 + G_2 + \cdots + G_n) + N_L \qquad (35)$$

N_L is the noise generated by the load and reflected by mismatch at the output port of the transducer.[17] The load noise is usually of little consequence when the transducer has gain. Hence N_L will be neglected, except in Sec. 5.4, where it is included for completeness.

Summing Eqs. (34) and (35) gives N_o.

$$N_o = k[G_1(T_{i1} + T_e) + G_2(T_{i2} + T_e) + \cdots + G_n(T_{in} + T_e)] + N_L \qquad (36)$$

T_{op} is then determined from Eqs. (36) and (33).

5.2. Average Noise Temperature. In the characterization and measurement of the noise performance of transducers, the average noise temperature is a frequently used term. The total noise N_{To} delivered to the output load is given by

$$N_{To} = k[B_1 G_{o1}(\bar{T}_{i1} + \bar{T}_e) + B_2 G_{o2}(\bar{T}_{i2} + \bar{T}_e) + \cdots + B_n G_{on}(\bar{T}_{in} + \bar{T}_e)] \qquad (37)$$

B_n is the noise bandwidth of the nth response; G_{on} is the transducer gain at the reference frequency f_{on} of the nth response as described in Sec. 4.8; and $\bar{T}_{in}$ and $\bar{T}_e$ are average noise temperatures. The average operating noise temperature $\bar{T}_{op}$ is defined in terms of N_{To}.

$$\bar{T}_{op} = N_{To}/kB_o G_s \qquad (38)$$

Here B_o is the *signal bandwidth* at the output port. The bandwidths actually present in the transducer may not correspond to B_o, although the bandwidth of the signal response must be no smaller than B_o if there is to be no loss of information.

Combining Eqs. (37) and (38) gives the average operating noise temperature.

$$\bar{T}_{op} = \left[\frac{B_1 G_{o1}}{B_o G_s}(\bar{T}_{i1} + \bar{T}_e) + \frac{B_2 G_{o2}}{B_o G_s}(\bar{T}_{i2} + \bar{T}_e) + \cdots + \frac{B_n G_{on}}{B_o G_s}(\bar{T}_{in} + \bar{T}_e) \right]$$

Thus the average operating noise temperature is equal to the summation of the average operating noise temperatures of each response times a weighting factor. The weighting factor is the ratio of the gain–noise-bandwidth product of the particular response to the signal-gain–output-signal-bandwidth product of the signal response. In a system with a single-signal response, the terms after the first should be minimized and made negligible. This is accomplished by minimizing the noise temperatures and the gain–noise-bandwidth products of the undesirable responses. The smallest value that the $B_1 G_{o1}/B_o G_s$ factor can have is unity, and an optimum transducer must achieve this result. Thus the signal response of the transducer should ideally match the signal bandwidth and there should be no spurious responses.

5.3. Transducers with Multiple Signal Responses. The operating noise temperature for a multiple-response transducer is given by Eq. (38). The fact that signals appear in more than one response does not influence the numerator. The factor that must be modified to adapt the equation to a multiple-signal-response receiver is

---

[17] Here N_o and N_N are delivered output powers. When N_o and N_N are used to compute noise factor they may be interpreted as either available or delivered output powers so long as consistency is maintained. In Sec. 4.1, N_o and N_N were interpreted as available powers and load noise was neglected. Load noise is not included in the noise factor. In Sec 5.4, N_L is specifically subtracted from N_o in computing the noise factor.

the G_s term in the denominator. This is accomplished by normalizing G_s over all the responses. The total signal S_o delivered to the output load is

$$S_o = S_i G_s = S_{i1}G_1 + S_{i2}G_2 + \cdots + S_{in}G_n \tag{39}$$

where S_{in} is the available input signal at the nth response and the signals are not correlated. The total signal available at the input is

$$S_i = S_{i1} + S_{i2} + \cdots + S_{in} \tag{40}$$

The ratio of Eq. (39) to Eq. (40) gives the signal gain.

$$G_s = \frac{S_{i1}G_1 + S_{i2}G_2 + \cdots + S_{in}G_n}{S_{i1} + S_{i2} + \cdots + S_{in}} \tag{41}$$

Combining Eqs. (41) and (38) gives the operating noise temperature.

$$\bar{T}_{op} = \frac{N_{To}(S_{i1} + S_{i2} + \cdots + S_{in})}{kB_o(S_{i1}G_1 + S_{i2}G_2 + \cdots + S_{in}G_n)}$$

N_{To} is obtained from Eq. (37) for use with this equation. The concept of multiple responses can also be applied to T_{op} using Eqs. (33), (36), and (41).

5.4. Comparison of F, T_{op}, and T_e. F, T_{op}, and T_e, and $\bar{F}$, $\bar{T}_{op}$, and $\bar{T}_e$ treat the source, load, and bandwidth somewhat differently. Translations among these factors can be handled with little difficulty for simple systems. However, complicated systems require careful attention to the particular requirements of each factor to avoid error or ambiguity.

Salient requirements for each factor are listed below.

F and $\bar{F}$:
1. By definition, the source temperature is 290°K for all responses.
2. Load noise is not included in F or $\bar{F}$.
3. If the system has multiple responses, the numerator of F and $\bar{F}$ must contain the noise from all responses. If there is a signal in only one response, the denominator must contain only the noise from that response. If there is a signal in more than one response, the denominator may include all responses that have signals. Alternatively it may include only one response and a separate noise factor derived for each. If there is any possibility of confusion, it should always be stated whether the noise factor is single or multiple response; or T_e and $\bar{T}_e$ should be used instead of F and $\bar{F}$.
4. F and $\bar{F}$ may be defined in terms of available gain and power available at the output port or transducer gain and power delivered by the output port to the load.

T_e and $\bar{T}_e$:
1. The equivalent input noise temperature is not a function of the noise temperatures of the sources, although it is a function of the impedances of the sources.
2. The noise generated by the output load at the output frequency is not included in T_e or $\bar{T}_e$. However, noise in the load at frequencies other than the output frequency may be converted to the output frequency by the transducer and may thus contribute to T_e or $\bar{T}_e$.
3. T_e and $\bar{T}_e$ are influenced by the total number of responses of a transducer but are independent of the number of responses that are devoted to signal usage.

T_{op} and $\bar{T}_{op}$:
1. T_{op} and $\bar{T}_{op}$ specifically include noise generated by the load, i.e., noise transmitted to the transducer by the load and reflected back into the load by means of a mismatch at the output port of the transducer.
2. The source temperatures used to compute T_{op} and $\bar{T}_{op}$ are the actual noise temperatures of the transducer sources in their normal operating environment.
3. T_{op} and $\bar{T}_{op}$ are defined in terms of signal gain. A composite signal gain is determined for a multiple-signal-response transducer using the techniques of Sec. 5.3.
4. The bandwidth employed in the denominator of the $\bar{T}_{op}$ expression is the bandwidth of the output signal, rather than the bandwidth of the transducer itself. The

Table 4. Relationships of F, T_{op}, and T_e

	Transducer with one response*	Transducer with one signal response and equal image response* $T_i = T_{i1} = T_{i2}$	Transducer with one signal response and $n-1$ other responses* $T_i = T_{i1} = T_{i2} = \cdots = T_{in}$	General equations
T_{op} in terms of T_e	$T_e + T_i + \dfrac{N_L}{kG_1}$	$2(T_e + T_i) + \dfrac{N_L}{kG_1}$	$(T_e + T_i)(1 + G_2/G_1 + \cdots + G_n/G_1) + \dfrac{N_L}{kG_1}$	$T_e(1 + G_2/G_1 + \cdots + G_n/G_1)$ $+ T_{i1} + T_{i2}\dfrac{G_2}{G_1} + \cdots$ $+ T_{in}\dfrac{G_n}{G_1} + \dfrac{N_L}{kG_1}$
T_{op} in terms of F	$290(F-1) + T_i + \dfrac{N_L}{kG_1}$	$290(F-2) + 2T_i + \dfrac{N_L}{kG_1}$	$290F + (T_i - 290)(1 + G_2/G_1 + \cdots + G_n/G_1) + \dfrac{N_L}{kG_1}$	$290[F - (1 + G_2/G_1 + \cdots + G_n/G_1)] + T_{i1} + T_{i2}\dfrac{G_2}{G_1} + \cdots + T_{in}\dfrac{G_n}{G_1} + \dfrac{N_L}{kG_1}$
F in terms of T_e	$(1 + T_e/290)$	$2(1 + T_e/290)$	$(1 + T_e/290)(1 + G_2/G_1 + \cdots + G_n/G_1)$	
T_e in terms of F	$290(F-1)$	$290(F/2 - 1)$	$290[F/(1 + G_2/G_1 + \cdots + G_n/G_1) - 1]$	

*Signal is in response number 1, and $G_1 = G_s$.

Table 5. Relationships of $\bar{F}$, $\bar{T}_{op}$, and $\bar{T}_e$

	Transducer with one response* $\mathfrak{R} = 1$	Transducer with one signal response* and equal image response $\bar{T}_i = \bar{T}_{i1} = \bar{T}_{i2}$, $G_o = G_{o1} = G_{o2}$, $B = B_1 = B_2$, $\mathfrak{R} = 2$	Transducer with one signal response* and $n-1$ other responses $\bar{T}_i = \bar{T}_{i1} = \bar{T}_{i2} = \cdots = \bar{T}_{in}$	General equations
$\bar{T}_{op}$ in terms of $\bar{T}_e$	$\dfrac{B}{B_o}(\bar{T}_e + \bar{T}_i) + \dfrac{N_TL}{kB_oG_o}$	$2\dfrac{B}{B_o}(\bar{T}_e + \bar{T}_i) + \dfrac{N_TL}{kB_oG_o}$	$(\bar{T}_e + \bar{T}_i)\dfrac{B}{B_o}\mathfrak{R} + \dfrac{N_TL}{kB_oG_o}$	$\bar{T}_e\dfrac{B}{B_o}\mathfrak{R} + \bar{T}_{i1}\dfrac{B_1}{B_o} + \bar{T}_{i2}\dfrac{B_2G_{o2}}{B_oG_o} + \cdots$ $+ \bar{T}_{in}\dfrac{B_nG_{on}}{B_oG_o} + \dfrac{N_TL}{kB_oG_o}$
$\bar{T}_{op}$ in terms of $\bar{F}$	$290\dfrac{B}{B_o}(\bar{F}-1) + \bar{T}_i\dfrac{B}{B_o}$ $+ \dfrac{N_TL}{kB_oG_o}$	$290\dfrac{B}{B_o}(\bar{F}-2) + 2\bar{T}_i\dfrac{B}{B_o}$ $+ \dfrac{N_TL}{kB_oG_o}$	$290\dfrac{B}{B_o}(\bar{F}-\mathfrak{R}) + \bar{T}_i\dfrac{B}{B_o}\mathfrak{R}$ $+ \dfrac{N_TL}{kB_oG_o}$	$290\dfrac{B}{B_o}(\bar{F}-\mathfrak{R}) + \bar{T}_{i1}\dfrac{B_1}{B_o} + \bar{T}_{i2}\dfrac{B_2G_{o2}}{B_oG_o}$ $+ \cdots + \bar{T}_{in}\dfrac{B_nG_{on}}{B_oG_o} + \dfrac{N_TL}{kB_oG_o}$
$\bar{F}$ in terms of $\bar{T}_e$	$(1 + \bar{T}_e/290)$	$2(1 + \bar{T}_e/290)$	$(1 + \bar{T}_e/290)\mathfrak{R}$	
$\bar{T}_e$ in terms of $\bar{F}$	$290(\bar{F}-1)$	$290(\bar{F}/2 - 1)$	$290[\bar{F}/\mathfrak{R} - 1]$	

*Signal in response number 1, and $G_o = G_s = G_{o1}$, $B = B_1$, and $\mathfrak{R} = 1 + B_2G_{o2}/BG_o + \cdots + B_nG_{on}/BG_o$.

7–28

AMPLIFIER NOISE   7–29

output bandwidth of the transducer must be at least as large as the output bandwidth of the signal; but making the transducer output bandwidth any larger deteriorates $\bar{T}_{op}$.

The noise factor and equivalent input noise temperature do not include the load noise, so that they may be employed directly in the cascade-noise-factor relationships (Sec. 4.3). The cascade-noise-factor relationships apply only to single-response transducers. T_{op} includes the load noise because it is intended to be a practical working factor for determining the actual signal-to-noise ratio of a system. $\bar{T}_{op}$ includes the output signal bandwidth for the same reason.

Equations showing the significant relationships among F, T_e, and T_{op} are listed in Tables 4 and 5. Table 4 is applicable to spot quantities and Table 5 to average quantities. The relationships were taken from the equations of Secs. 5.1 to 5.3 and the following two additional formulas for noise factor.[18]

$$F = \frac{N_o(T_o) - N_L}{kT_o G_o}$$

$$\bar{F} = \frac{N_{To}(T_o) - N_{TL}}{kT_o G_o B} \tag{42}$$

$N_o(T_o)$ and $N_{To}(T_o)$ are specified for a uniform source temperature of 290°K at all input ports. N_{TL} is the wideband equivalent of N_L, G_o is the transducer gain corresponding to the principal (signal) response, and $G_o B$ is the average gain-bandwidth product of the principal response. It is assumed the transducer has only one *signal* response, which is in the number 1 location, and $G_o = G_1 = G_{o1} = G_s$ and $B = B_1$.

A *response factor* $\Re$ is employed in Table 5.

$$\Re = 1 + B_2 G_{o2}/BG_o + \cdots + B_n G_{on}/BG_o \tag{43}$$

Ideally, $\Re$ should be equal to unity. A response factor greater than unity indicates the system has unnecessary responses that deteriorate the noise performance. For a single-response system $\Re = 1$; for a system with equal signal and image responses, $B_2 G_{o2} = BG_o$ and $\Re = 2$.

Note that all the expressions for T_{op} included an additive term describing the noise added by the load. This term can usually be neglected except when the gain of the transducer is very small or when the output impedance of the transducer has a negative real part.

The formulations of Tables 4 and 5 are similar with one notable exception. The relationships for $\bar{T}_{op}$ include the ratios of the transducer bandwidths to the output bandwidth which are not required for the spot quantities.

6. TECHNIQUES FOR MEASUREMENT OF NOISE PARAMETERS

The quantities characterizing the noise performance of a transducer have been carefully defined to be useful, unambiguous, and capable of being readily determined by measurements. The noise quantities most readily determined by measurement are the effective noise temperature T_e, noise factor F, and gain and noise bandwidth. From a knowledge of these factors and the transducer's signal and operating environment, other quantities such as operating noise temperature T_{op}, output noise, and output signal-to-noise ratio can be calculated using the techniques of Sec. 5. The ease of making these calculations depends primarily upon the complexity of the system and its operating environment.

The measurements require a meter to measure the power delivered to the load of the transducer, a signal generator, and appropriate terminations for the inputs. The signal generators fall into two broad categories depending on whether they put out a single-frequency sine-wave signal or a dispersed signal that has a uniform energy density over a wide band. If a dispersed-signal generator is used, it must have a uniform energy density over all significant responses of the transducer, including those undesired responses which do not form a part of the signal response.

[18] See footnote 17.

AMPLIFIER FUNDAMENTALS

Noise factor and equivalent noise temperature are both functions of the input terminations; therefore, the input terminating impedance must be constant throughout the measurement. Furthermore, if the measurements are to be used to determine the operating performance, the input terminating impedances must correspond exactly to those which exist under operating conditions.

6.1. Basic Dispersed Signal Generator Technique. The dispersed signal generator must have a uniform noise temperature over all significant responses of the transducer,

$$T_i = T_{i1} = T_{i2} = \cdots = T_{in}$$

where T_{in} is the noise temperature of the input termination (output noise temperature of the dispersed signal generator) at the nth response. Substituting this into Eq. (37) gives the noise power delivered to the load N_{To}.

$$N_{To} = k(T_i + \bar{T}_e)(B_1 G_{o1} + B_2 G_{o2} + \cdots + B_n G_{on}) \tag{44}$$

Assume that the signal generator can produce two noise temperatures, T_i (hot) and T_i (cold).[19] The ratio of the two output powers is then

$$\frac{N_{To}(\text{hot})}{N_{To}(\text{cold})} = \frac{T_i(\text{hot}) + \bar{T}_e}{T_i(\text{cold}) + \bar{T}_e} = Y$$

This ratio is usually assigned the symbol Y. Solving for $\bar{T}_e$ gives

$$\bar{T}_e = \frac{T_i(\text{hot}) - Y T_i(\text{cold})}{Y - 1} \tag{45}$$

This equation is valid for all linear systems of any number of responses if the noise from the load can be neglected.

The average noise factor is given by Eq. (42). Assuming the principal response is response number 1, then $BG_o = B_1 G_{o1}$; and combining Eqs. (42), (43), and (44)

$$\bar{F} = \frac{T_o + \bar{T}_e}{T_o} \mathcal{R} \tag{46}$$

assuming the load noise is negligible. Combining Eqs. (45) and (46), the average noise factor becomes

$$\bar{F} = \frac{\left(\dfrac{T_i(\text{hot})}{T_o} - 1\right) - Y\left(\dfrac{T_i(\text{cold})}{T_o} - 1\right)}{Y - 1} \mathcal{R} \tag{47}$$

This expression is commonly used with $\mathcal{R}$ set equal to unity. However, it must be remembered that if the transducer has multiple responses the correct value of $\mathcal{R}$ must be employed in order to determine the average noise factor using a dispersed signal generator.

In common practice, the dispersed signal generator for the cold input temperature is simply a passive resistance termination at room temperature, approximately 290°K.[20] Substituting $T_i(\text{cold}) = T_o$ in Eq. (47)

$$\bar{F} = \frac{\dfrac{T_i(\text{hot})}{T_o} - 1}{Y - 1} \mathcal{R} \qquad [T_i(\text{cold}) = T_o]$$

6.2. Basic Sine-wave Signal Generator Technique. The output impedance of the sine-wave generator must match the normal source impedance of all responses of the transducer being measured. If it does not, a matching pad must be used between its output terminals and the input of the transducer being measured. The

---

[19] If the signal generator is capable of only one noise temperature, passive terminations at room temperature can be substituted to provide the second noise temperature.

[20] With very low noise transducers, it may be appropriate to make the hot source a passive resistance termination at room temperature and to obtain the cold source by refrigerating the termination. The equations can be easily modified to account for this condition.

AMPLIFIER NOISE 7-31

noise power output meter is the same as for the case of a distributed noise source. Two measurements are made. In the first measurement, the transducer is connected to a signal source having a uniform noise temperature at all frequencies corresponding to the responses of the transducer. The common way of doing this is to employ a passive network at room temperature T_r. Using Eqs. (37) and (43), the output power P_1 is

$$P_1 = kBG_o(T_r + \bar{T}_e)\Re \tag{48}$$

assuming the signal is in the first response and $BG_o = B_1G_{o1}$.

A second measurement is made by inserting a sine-wave signal with an available power $S_i(f_o)$. The output power of the transducer under this condition is P_2.

$$P_2 = kBG_o(T_r + \bar{T}_e)\Re + G_o S_i(f_o) \tag{49}$$

It is assumed that the sine-wave signal is added to the thermal noise already present in the system. Also, it is very important that the signal S_i is at the reference frequency f_o used in determining the noise bandwidth, as discussed in Sec. 4.8. Define the ratio of the two powers, $P_2/P_1 = Y'$. From Eqs. (48) and (49), T_e becomes

$$\bar{T}_e = \frac{S_i(f_o)}{kB(Y' - 1)\Re} - T_r \tag{50}$$

If the transducer has multiple responses, the gain-bandwidth products of each response must be determined in order to evaluate the denominator of the expression. The measured $\bar{T}_e$ will be optimistic for such a transducer if the response factor is neglected. It should be noted that the response factor does not need to be determined if the dispersed signal generator method is employed to measure $\bar{T}_e$, Eq. (45). Combining Eq. (46) with Eq. (50), the average noise factor becomes

$$\bar{F} = \frac{S_i(f_o)}{kBT_o(Y' - 1)} + \left(1 - \frac{T_r}{T_o}\right)\Re \tag{51}$$

Again, the gain–noise-bandwidth products of all the responses must be known. This requirement can be avoided if the room or ambient temperature is made equal to the standard temperature, $T_r = T_o$. This corresponds to a room or ambient temperature of 17°C, 63°F, which is relatively easy to achieve. Under this condition, the average noise factor becomes

$$\bar{F} = \frac{S_i(f_o)}{kBT_o(Y' - 1)} \qquad (T_r = T_o)$$

This is the equation normally employed. However, if the system has large responses outside the main signal response, then from Eq. (51), a large error in noise factor can be obtained if the room temperature is not very nearly equal to the standard noise temperature.

6.3. Determination of the Response-factor–Noise-bandwidth Product. If both a dispersed signal generator and a sine-wave signal generator are available, the product of the response factor and the noise bandwidth is easily determined. Given one of these terms, the other one can be found. If the response factor is known, the noise bandwidth can be determined by the techniques of either Sec. 4.8 or this section. If the response factor is not known, the noise bandwidth can be determined according to Sec. 4.8 and the response factor according to this section.

Combining Eqs. (45) and (50), and assuming T_i (cold) $= T_r$, the response-factor–gain-bandwidth product becomes

$$B\Re = \frac{S_i(f_o)}{k[T_i \text{ (hot)} - T_r]} \frac{Y - 1}{Y' - 1}$$

The mathematics can be simplified by making $Y = Y'$.

In a multiple-response transducer, B is the noise bandwidth that corresponds to the input frequency f_o. Thus if $\Re$ is known, the noise bandwidth of any response of the transducer, including those responses not devoted to signal usage, can be determined by the appropriate selection of f_o.

6.4. Formulas for Noise Measurements on Amplifiers. The formulas commonly used in noise measurements have been brought together in two tables. Measurements employing a dispersed signal generator are covered in Table 6, and measurements employing a single-frequency sine-wave generator are covered in Table 7.

The first column in Table 6 is for the case of arbitrary hot and cold input noise generators. The second column is for the case of the cold noise temperature equal to

Table 6. Formulas for Dispersed Signal Generator Measurements

		$Y = \dfrac{\text{noise output power for } \bar{T}_i \text{ (hot)}}{\text{noise output power for } \bar{T}_i \text{ (cold)}}$	$Y = \dfrac{\text{noise output power for } \bar{T}_i \text{ (hot)}}{\text{noise output power for } \bar{T}_i = T_o = 290°\text{K}}$
Transducer with single response (load noise negligible) $\mathcal{R} = 1$	$\bar{T}_e$	$\dfrac{\bar{T}_i \text{ (hot)} - Y\bar{T}_i \text{ (cold)}}{Y - 1}$	$\dfrac{\bar{T}_i \text{ (hot)} - 290Y}{Y - 1}$
	$\bar{F}$	$\dfrac{\left[\dfrac{\bar{T}_i \text{ (hot)}}{290} - 1\right] - Y\left[\dfrac{\bar{T}_i \text{ (cold)}}{290} - 1\right]}{Y - 1}$	$\dfrac{\dfrac{\bar{T}_i \text{ (hot)}}{290} - 1}{Y - 1}$
Transducer with one signal response and equal image response (load noise negligible) $\mathcal{R} = 2$	$\bar{T}_e$	$\dfrac{\bar{T}_i \text{ (hot)} - Y\bar{T}_i \text{ (cold)}}{Y - 1}$	$\dfrac{\bar{T}_i \text{ (hot)} - 290Y}{Y - 1}$
	$\bar{F}$	$2\dfrac{\left[\dfrac{\bar{T}_i \text{ (hot)}}{290} - 1\right] - Y\left[\dfrac{\bar{T}_i \text{ (cold)}}{290} - 1\right]}{Y - 1}$	$2\dfrac{\dfrac{\bar{T}_i \text{ (hot)}}{290} - 1}{Y - 1}$
Transducer with one signal response (BG_o) $n - 1$ other responses (load noise negligible)	$\bar{T}_e$	$\dfrac{T_i \text{ (hot)} - Y\bar{T}_i \text{ (cold)}}{Y - 1}$	$\dfrac{T_i \text{ (hot)} - 290Y}{Y - 1}$
	$\bar{F}$	$\mathcal{R}\dfrac{\left[\dfrac{\bar{T}_i \text{ (hot)}}{290} - 1\right] - Y\left[\dfrac{\bar{T}_i \text{ (cold)}}{290} - 1\right]}{Y - 1}$	$\mathcal{R}\dfrac{\dfrac{\bar{T}_i \text{ (hot)}}{290} - 1}{Y - 1}$
	$\mathcal{R}$	\multicolumn{2}{c}{$1 + \dfrac{B_2 G_{o2}}{BG_o} + \cdots + \dfrac{B_n G_{on}}{BG_o}$}

Table 7. Formulas for Sine-wave Signal Generator Measurements

		$Y' = \dfrac{\text{output power for } [kBT_r + S_i(f_o)] \text{ input}}{\text{noise output power for } kB\bar{T}_r \text{ input}}$	$Y' = \dfrac{\text{output power for } [290kB + S_i(f_o)] \text{ input}}{\text{noise output power for } 290kB \text{ input}}$
Transducer with single response (load noise negligible) $\mathcal{R} = 1$	$\bar{T}_e$	$\dfrac{S_i(f_o)}{kB(Y' - 1)} - T_r$	$\dfrac{S_i(f_o)}{kB(Y' - 1)} - 290$
	$\bar{F}$	$1 + \dfrac{S_i(f_o)}{290kB(Y' - 1)} - \dfrac{T_r}{290}$	$\dfrac{S_i(f_o)}{290kB(Y' - 1)}$
Transducer with one signal response and equal image response (load noise negligible) $\mathcal{R} = 2$	$\bar{T}_e$	$\dfrac{S_i(f_o)}{2kB(Y' - 1)} - T_r$	$\dfrac{S_i(f_o)}{2kB(Y' - 1)} - 290$
	$\bar{F}$	$2 + \dfrac{S_i(f_o)}{290kB(Y' - 1)} - \dfrac{2T_r}{290}$	$\dfrac{S_i(f_o)}{290kB(Y' - 1)}$
Transducer with one signal response (BG_o) and $(n - 1)$ other responses (load noise negligible)	$\bar{T}_e$	$\dfrac{S_i}{kB(Y' - 1)\mathcal{R}} - T_r$	$\dfrac{S_i}{kB(Y' - 1)\mathcal{R}} - 290$
	$\bar{F}$	$\dfrac{S_i}{290kB(Y' - 1)} + \left(1 - \dfrac{T_r}{T_o}\right)\mathcal{R}$	$\dfrac{S_i}{290kB(Y' - 1)}$
	$\mathcal{R}$	\multicolumn{2}{c}{$1 + \dfrac{B_2 G_{o2}}{BG_o} + \cdots + \dfrac{B_n G_{on}}{BG_o}$}

standard noise temperature, 290°K. The three double rows list $\bar{T}_e$ and $\bar{F}$ formulations for, respectively, single-response, equal-double-response, and general-multiple-response transducers. $\bar{T}_e$, $\bar{F}$, and $\Re$ are derived, respectively, from Eqs. (45), (47), and (43). For the multiple-response case, it has been assumed that the transducer has only one signal response, which is denoted as the number one response such that $G_o B = G_{o1} B_1$.

The assumptions for Table 7 are similar, except that $\bar{T}_e$ and $\bar{F}$ are derived from Eqs. (50) and (51), respectively. It is also assumed that the sine-wave generator generates an output only at the reference frequency f_o corresponding to the noise bandwidth B. The signal generator power $S_i(f_o)$ adds to the thermal noise of the termination kBT_r. The thermal noise kBT_r is available equally to all responses. In the second column of Table 7, the room temperature T_r is assumed equal to standard noise temperature, 290°K.

The tables list only average noise quantities. The corresponding noise quantities for a narrow fractional bandwidth (e.g., spot noise factor) can be determined from the

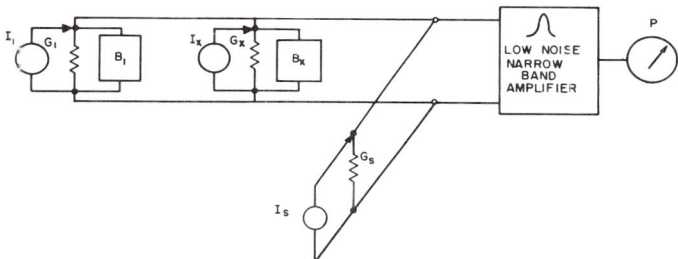

Fig. 14. Circuit for measuring noise in two-terminal devices.

same equations if a narrowband filter is included between the output of the transducer and the output power indicator, and the average bars are removed from the equations.

6.5. Noise Measurements on Two-terminal Devices. Measurements of the noise of two-terminal devices can be made using techniques similar to those employed in measuring the noise factor and noise temperature of amplifiers. The methods described in this section are designed to accommodate a wide range of practical devices including both positive- and negative-resistance devices.

The measurements require the use of a low-noise tuned amplifier and a detector that indicates average output power. The amplifier contributes its own noise to the system. In order to keep this noise contribution constant throughout the measurements, the source admittance facing the amplifier must be kept constant (unless the noise of the two-terminal device is much larger than that of the amplifier). This is achieved through the use of passive dummy loads at the input. The measurement circuit is shown in Fig. 14. The unknown is represented as an admittance $G_x + jB_x$ which has an equivalent noise current generator I_x and an equivalent noise conductance G_x'. A passive dummy load $G_1 + jB_1$ with an equivalent thermal noise generator I_1 is connected in parallel. Also connected into the circuit is a signal generator with internal admittance G_s and Norton generator I_s. A second measurement circuit (Fig. 15) is employed to evaluate the noise of the system. Figure 15 incorporates a second passive dummy load G_2, B_2, I_2 which replaces the unknown admittance and the first dummy load. In order to maintain a constant admittance facing the amplifier, it is necessary that

$$G_2 = G_x + G_1$$
$$B_2 = B_x + B_1$$
(52)

The magnitude of the dummy load admittances should be minimized to avoid loading the input circuit and masking the noise of the unknown. If G_x is a positive conductance, G_1 can be eliminated ($G_1 = 0$). If G_x is a negative conductance, as, for example, in a tunnel diode, G_1 together with the other conductances in the circuit are required to stabilize the unknown to prevent it from oscillating.[21] If G_x is negative, G_2 and B_2 may be eliminated, which requires

$$G_1 = -G_x$$
$$B_1 = -B_x$$

Both G_1 and G_2 will be carried through to the solution where either one can be eliminated by setting it equal to zero.

The signal generator may be a sine-wave generator or a dispersed signal source. Four steps, 1A, 1B, 2A, and 2B, are described in the measurement procedure. Only three steps, 1A, 1B, and 2A, or 1A, 2A, and 2B, are required. The user can determine the steps that best suit his particular need.

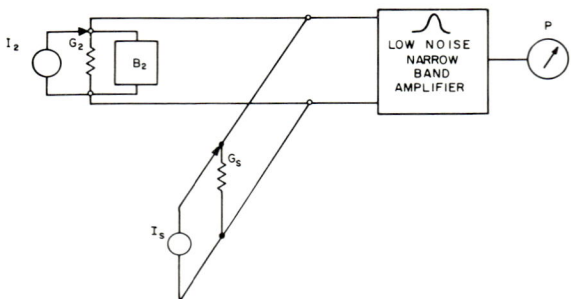

Fig. 15. Circuit for calibrating system noise in two-terminal-device measurement.

Step 1A:
Using the circuit of Fig. 14, the signal generator is turned off, and the output power P_{1A} is measured. It is assumed that the output of the signal source contains only the thermal noise in G_s at room temperature T_r.

$$\overline{i^2}_{s1A} = 4kT_r G_s B \tag{53}$$

Step 1B:
Using the circuit of Fig. 14, the output of the signal generator is increased, and an output power reading P_{1B} is obtained. It is assumed that the signal generator has increased its output conductance by an amount G'_{s1B}

$$\overline{i^2}_{s1B} = 4kT_r(G_s + G'_{s1B})B$$

Step 2A:
The circuit of Fig. 15 is employed making sure that Eq. (52) is satisfied. The generator is turned off so that its noise is thermal noise [Eq. (53)], and output power P_{2A} is measured.

Step 2B:
Using the circuit of Fig. 15, the output of the signal generator is increased as in step 1A, and output power P_{2B} is measured.

In steps 1B and 2B, the added noise conductance G_s' is determined as follows:
(a) *Sine-wave generator:*

$$G_s' = \overline{i^2}/4kT_r B$$

where $\overline{i^2}$ is the added mean-square signal current and B is the noise bandwidth of the tuned amplifier.

[21] If the unknown is an open-circuit stable device, the dual circuit to the above might be used. The unknown and dummy impedances would be connected in series.

(b) *Distributed source:*

$$G_s' = \overline{i^2}/4kT_r$$

where $\overline{i^2}$ is the mean-square output current per cycle added.

(c) *Noise diode:*

$$G_s' = qI/2kT_r$$

where I is the direct current of the noise diode.

(d) *Sine-wave generator calibrated in available output power:*

$$G_s' = PG_s/kT_rB$$

where P is the available signal output power in watts and B is the noise bandwidth of the tuned amplifier.

(e) *Distributed source calibrated in noise temperature:*

$$G_s' = G_s \frac{T - T_r}{T_r}$$

where T is the noise temperature of the source.

Using steps 1A, 1B, and 2A, the unknown noise conductance G_x' becomes

$$G_x' = \frac{1 - P_{2A}/P_{1A}}{P_{1B}/P_{1A} - 1} G'_{s1B} + G_x \qquad (54)$$

Thus the noise conductance can be determined through the use of two power ratios, the added noise conductance from the generator G'_{s1B} and the conductance G_x.

Using steps 1A, 2A, and 2B, the unknown conductance becomes

$$G_x' = \frac{P_{1A}/P_{2A} - 1}{P_{2B}/P_{2A} - 1} G'_{s2B} + G_x \qquad (55)$$

This equation is similar in form to Eq. (54). The ratio P_{2B}/P_{2A} is essentially the Y or Y' factor used in measuring the noise factor of an amplifier. If the noise factor remains relatively stable, this ratio can be determined once and need not be repeated unless the source admittance for the amplifier $G_s + G_2 + jB_2$ is changed. It can be assumed then that P_{2A}, P_{2B}, and G'_{s2B} are constant with time and independent of the unknown. A number of unknown noise conductances can then be measured by adjusting $G_1 + jB_1$ to satisfy Eq. (52), and measuring one noise power P_{1A} and the conductance G_x for each unknown.

If the output of the signal generator can be smoothly varied over a sufficient range, it may be convenient to make the power ratios in the denominators of Eqs. (54) and (55) equal to 2.

$$P_{1B}/P_{1A} = 2 \qquad P_{2B}/P_{2A} = 2$$

7. INSTRUMENTATION OF NOISE MEASUREMENTS

The instrumentation of noise measurements requires attention to factors not normally considered in other instrumentation problems. Particular attention must be given to the signal generator and the output power indicator. Also, bandwidth, averaging time, and the detector response characteristics are important factors.

7.1. Sine-wave Signal Generators. Laboratory sine-wave generators are usually satisfactory for noise measurements, although they often require the use of auxiliary loss pads. Noise factor and noise temperature are a function of source impedance. Therefore, the signal generator must properly terminate the transducer at all its significant responses. Furthermore, when the signal generator is turned off, it is assumed that the transducer's input noise temperature is room temperature. Reducing a generator's output control to zero does not usually produce this result. The proper conditions can be obtained by placing a pad, with a large attenuation, between the signal generator output and the input of the transducer. The pad is designed to terminate the transducer and isolate possible anomalies in the signal generator output impedance from the transducer. In the "off" condition, the signal generator may be disconnected from the pad to assure that there is no signal leakage.

A signal generator's available output power is normally much larger than is required for noise measurements. For example, the available thermal noise in a 10-kc band at 290°K is 4×10^{-14} mw. If the signal generator produces an output power of 1 mw, a 134-db attenuator is required to reduce the signal generator power to the level of thermal noise. It is difficult to measure accurately such a large attenuation. The slightest signal leakage via a path other than through the attenuator can cause large errors in the measurements.

7.2. Dispersed Signal Generators. A variety of dispersed signal generators are available. Signals of a nonrandom but dispersed nature are produced by repetitive-impulse generators having closely spaced harmonic outputs over a wide frequency range, and by swept-frequency sine-wave generators. Random signals are obtained from photomultipliers, gas discharge tubes, heated resistors, temperature-limited vacuum diodes, and other devices. The random noise sources are in more popular use than the nonrandom. One of their major advantages is that the characteristics of their outputs closely resemble the noise characteristics of most transducers; and this simplifies the detection problem. For noise measurements, the output of the dispersed generator must be stable and capable of being accurately calibrated. Temperature-limited vacuum diodes, gas discharge sources, and heated resistors fulfill these requirements.

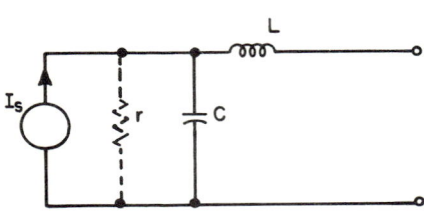

Fig. 16. Alternating-current equivalent circuit of noise diode.

7.3. Diode Noise Sources. Temperature-limited vacuum diodes, commonly called noise diodes, are employed in the frequency range from about 10 kc to several gigacycles. They are limited by $1/f$ noise at lower frequencies and by transit time at higher frequencies. An a-c equivalent circuit of such a diode is shown in Fig. 16. L is the lead inductance, C the anode-to-cathode capacitance, r the plate resistance, and I_s the shot noise current generator. L and C can be neglected at low frequencies but generally become important in the megacycle frequency range. Since the diode operates temperature-limited, its plate resistance is ideally infinite. Practically, however, the Schottky effect gives rise to a large but finite plate resistance. The noise current generator I_s develops full shot noise [Eq. (4)] where I is the d-c anode current. The noise is independent of frequency from a few kilocycles up to several hundred megacycles. At very low frequencies, $1/f$ noise sets in. Individual diodes have been used to frequencies below 1 kc; however, in general, diodes should be checked for $1/f$ noise when they are operated below about 10 kc. At high frequencies, electron transit time becomes of significance and the output noise decreases. This decrease can be evaluated and included as a correction factor. Also, the diode capacitance and lead inductance must be taken into account at high frequencies.

Diodes operating at the highest frequencies have a coaxial construction. The cathode forms the inner conductor and the anode the outer. The ratio of the inner and outer diameters is adjusted to give the proper transmission-line impedance, usually 50 ohms. A number of diodes based on this structure have been developed.[22,23,24] These diodes will operate to low gigacycle frequencies with electron-transit-time correction factors of a few decibels.

[22] R. Kompfner, J. Hatton, E. E. Schneider, and L. A. G. Dresel, The Transmission Line Diode as a Noise Source at Centimetre Wavelengths, *J. Inst. Elec. Engrs.* (London), part IIIA, vol. 93, pp. 1436–1442, 1946.

[23] H. Johnson, A Coaxial-line Diode Noise Source for U-H-F, *RCA Rev.*, vol. 8, pp. 169–185, March, 1947.

[24] H. Groendijk, A Noise Diode for Ultra-high Frequencies, *Philips Tech. Rev.*, vol. 20, pp. 108–110, Oct. 30, 1958.

AMPLIFIER NOISE

The Sylvania 5722 is in common use up to vhf. This diode has a single-ended T-5½ bulb and a maximum rated anode current of 35 ma.[25] A practical circuit suitable for a diode such as the 5722 is shown in Fig. 17. A load resistance R in parallel with the plate resistance of the diode provides the real part of the output admittance. The 5722 has been experimentally found to have a plate resistance of about 1 megohm at a plate current of 1 ma. The plate resistance varies inversely with the plate current so that at the maximum rated current, 35 ma, the plate resistance has a minimum value of $10^6/35 = 28,600$ ohms.[26] These values are subject to considerable variation and should be checked for particular diodes. Thus, for receiver measurements, where R is on the order of 50 to 1,000 ohms, the plate resistance is entirely negligible. However, it can become significant if R is higher. The inductor L_l is employed, if needed, to tune out the capacitance of the diode or to bypass the d-c anode current around R.

The d-c electron current flows from the negative lead of the power supply through the ammeter into the filament circuit. Electrons are emitted by the filament and collected by the anode. The anode current flows through the load resistance R or the inductor L_l to the positive lead of the anode supply and to ground.

The anode current is controlled by controlling the filament voltage, which varies the filament temperature. The filament is bypassed to ground at all noise frequencies. The inductors and capacitors in the filament circuit form a low-pass filter to prevent noise from being introduced by the filament supply, the anode supply, or their connecting leads. The lower output terminal of the noise diode is at both d-c and a-c ground. The upper terminal will have a d-c voltage corresponding to the IR drop in R or L_l. A d-c blocking capacitor C_c is employed if needed.

The mean-square noise current $\overline{i^2}$ at the terminals is the summation of that due to the resistor and to the noise diode.

$$\overline{i^2} = 2qIB + 4kT_rB/R \quad (56)$$

Here T_r is the temperature of R when the diode is conducting a current I.[27]

The available power P is

$$P = \overline{i^2}R/4 \quad (57)$$

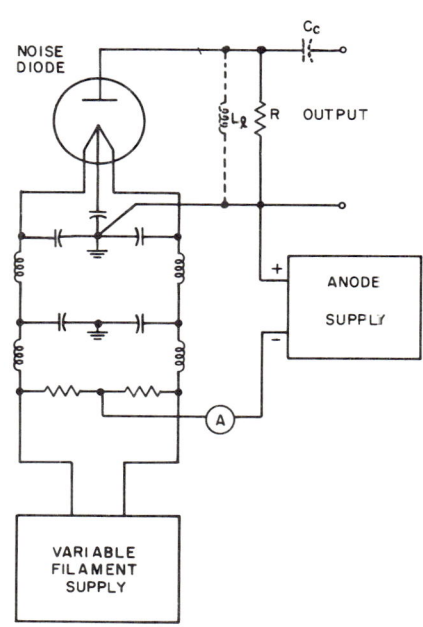

FIG. 17. Practical noise diode circuit.

neglecting the plate resistance of the diode. Combining Eqs. (6), (56), and (57), and using the notation of Sec. 6.1, the noise temperature T_i (hot) with the diode on is

$$T_i \text{ (hot)} = T_r + qIR/2k \quad (58)$$

[25] R. W. Slinkerman, Temperature-limited Noise Diode Design, *Sylvania Technologist*, vol. 2, p. 6, October, 1949.

[26] A. van der Ziel, "Fluctuation Phenomena in Semiconductors," p. 9, Butterworths Scientific Publications, London, 1959, also private communication.

[27] T_r is normally considered to be room temperature. If the d-c flow through R causes a significant temperature rise, it is advisable to use a shunt inductance L to eliminate this condition. If the condition cannot be eliminated, the temperature of the resistance must be determined for the diode currents used in the measurements and factored into the equations.

The noise temperature T_i (cold) with the diode off is T_r. The ratio Y of the transducer output power with the diode on to the transducer output power with the diode off is then obtained. Using Eqs. (45) and (58), the average effective input noise temperature $\bar{T}_e$ is

$$\bar{T}_e = \frac{qIR}{2k(Y-1)} - T_r \qquad [T_i \text{ (cold)} = T_r]$$

Using Eqs. (47) and (58), the average noise factor is

$$\bar{F} = \left[\frac{qIR}{2kT_o(Y-1)} + 1 - \frac{T_r}{T_o}\right] \mathcal{R} \qquad [T_i \text{ (cold)} = T_r] \tag{59}$$

The term $kT_o/q = 0.0250$ volt. If the room temperature is approximately equal to *standard noise temperature*, Eq. (59) becomes

$$\bar{F} = \frac{20IR}{Y-1} \mathcal{R} \qquad (T_r = T_o = 290°\text{K})$$

It is common practice to adjust the filament voltage so that the added noise power doubles the output power of the transducer being measured ($Y = 2$).

The noise factor becomes

$$\bar{F} = 20IR\mathcal{R} \qquad (T_r = T_o, Y = 2) \tag{60}$$

Similarly, the noise temperature becomes

$$\bar{T}_e = 290(20IR - 1) \qquad (T_r = T_o, Y = 2) \tag{61}$$

This is not always the most accurate method of determining the noise factor using a noise diode, the reason being that the ($Y - 1$) factor becomes a difference of two numbers that are of the same order of magnitude. A more accurate result is obtained if Y is very much larger than unity.

Commercial diode noise generators are available with built-in power supplies, a noise diode, and an anode current meter. These meters are often calibrated directly in noise factor for a specified value of R, usually 50 ohms, using Eq. (60), assuming that $\mathcal{R} = 1$, $Y = 2$, and $T_r = T_o$. Such meters can also be calibrated in terms of $\bar{T}_e$ using Eq. (61), thus eliminating the assumption regarding $\mathcal{R}$.

Unfortunately, most commercial diode noise generators must be connected to the transducer by a length of cable and can suffer errors due to cable mismatch and cable loss if it is significant. Frequently also, they use a-c filament power for the diode, or rectified direct current with a simple filter. The ripple in the filament voltage introduces an a-c component at the power-line frequencies and near harmonics into the diode current. Unless the filament circuit filtering is very good, this component is much larger than the mean-square current of the diode. This can cause severe overload in amplifiers having a significant input response at the power-line frequency or its near harmonics. Another problem is that the diode current will not remain constant at a particular setting but tends to fluctuate. This fluctuation is traceable to power-line variations and sometimes to contact problems in the high-current variable resistors used to control the filament voltage. The diode current varies exponentially with filament temperature and hence is very sensitive to small fluctuations in the filament voltage.

The above problems can be largely eliminated by assembling individual components. The cost of the assembly is relatively small. Cable losses and mismatch problems are minimized by connecting the noise diode directly to the terminals of the transducer being measured. Naturally, the transducer terminals and the noise diode should be adequately shielded. The anode voltage for a diode such as the 5722 can be supplied by a set of three 45-volt batteries, such as the Eveready W363F, connected in series.

AMPLIFIER NOISE   7–39

The d-c current is monitored by an ordinary milliammeter or microammeter. The filament voltage is conveniently supplied by a laboratory-type regulated supply.[28]

7.4. Thermal Noise Sources. Resistors and lossy terminations for waveguides or transmission lines can be heated or cooled to provide sources with noise temperatures other than room temperature. When the equivalent input noise temperature of the transducer being measured is below room temperature, it is appropriate to cool the source. Cooling with liquid nitrogen (77°K) reduces the noise by a factor of 3.8 below standard thermal noise. Liquid helium (4.2°K) reduces the noise by a factor of 69.

When the equivalent input noise temperature is above room temperature, it is appropriate to heat the source. A temperature rise of several hundred degrees centigrade is required to obtain a noise temperature significantly different from room temperature. The highest temperatures are obtained by heating the resistive element in a vacuum. Vacuum-enclosed tungsten filaments can, for example, be operated at temperatures up to about 2000 to 3000°K. Thus a noise temperature approximately ten times *standard noise temperature* can be achieved. The heating is most easily accomplished by passing a d-c or low-frequency a-c current through the resistive element. The temperature can be determined with the aid of an optical pyrometer.

In both the heated and cooled modes, the temperature change is sufficiently great to cause a large resistance change in the resistive element. In fact, the temperature of the element is perhaps most easily determined from its resistance if its temperature-resistance characteristic is known. It is also possible to devise a self-balancing bridge that will automatically keep the element at a fixed resistance (and hence a known temperature). Such a bridge would operate in a manner similar to the self-balancing bolometer bridge described in Sec. 7.8. In this type of bridge, sufficient electric power is added to the element to cause its resistance to reach a predetermined value which balances the bridge. If cooled operation is desired, the resistance can be cooled below the desired temperature, say with liquid nitrogen or with liquid helium, and raised to the desired temperature by the addition of electric energy from the bridge.

In any system of this type, care must be taken that the heating or cooling system does not introduce unwanted parasitics or noise into the resistive element.

7.5. Gas Discharge Noise Sources. A column of ionized gas can serve as an effective distributed noise source. The noise temperature of a low-pressure gas column corresponds approximately to the noise temperature of the electrons in the column. If the light energy produced by the gas is principally monochromatic with a wavelength λ_m the noise temperature T is related to the wavelength by Wien's displacement law

$$\lambda_m T = 0.289 \text{ cm } °K$$

Mercury vapor, for example, radiates primarily at a wavelength of 2.5×10^{-5} cm. The corresponding noise temperature is 11,400°K, which is a factor of 39.4 (16 db) above *standard noise temperature*, 290°K. The electrons can achieve this high temperature because they have a relatively long mean free path before collisions; and they gain more energy during their free travel than they lose during the collisions.

Mercury, argon, and neon are used in gas discharge noise lamps. Commercial fluorescent lamps which contain mercury vapor are often used. Mumford demonstrated the use of such a lamp, operated from a d-c source, in 1949.[29] Mumford tested 32 different lamps including 10 different types of fluorescent coatings and a germicidal lamp with no fluorescent coating. The noise temperatures were all within

---

[28] A low-cost commercial supply providing a maximum output of 15 volts and 1.7 amp has given excellent results when used with the 5722. The supply had a maximum ripple of 0.245 mv and a regulation the greater of 0.05 per cent or 7.5 mv. A 6-ohm resistor was used between the output of the supply and the filament of the 5722 to prevent accidental burnout. The power supply was fitted with a vernier control, having a range of 1 volt, at a small extra cost by the manufacturer. The vernier control is advisable because of the sensitivity of the anode current to the filament temperature. The power supply provided excellent results with negligible anode current drift and ripple.

[29] W. W. Mumford, A Broad-band Microwave Noise Source, *Bell System Tech. J.*, vol. 28, pp. 608–618, October, 1949.

±0.25 db of each other, which demonstrated that the source of the microwave noise energy was chiefly in the gaseous discharge rather than in the fluorescent coating. The noise spectrum of Mumford's lamps was essentially flat over the frequency range of 3.7 to 4.5 gc. Since that time, gas lamps have been used at frequencies as low as 65 Mc,[30] and commercial lamps are available to at least 40 gc.

Mumford found that the noise temperature of the fluorescent lamps was substantially independent of their operating current, while their RF conductance was directly proportional to their operating current. He found that the noise of the lamps was a weak function of the temperature of the glass tube. The excess noise temperature $(T/290-1)$ had a negative temperature coefficient of 0.055 db/°C.

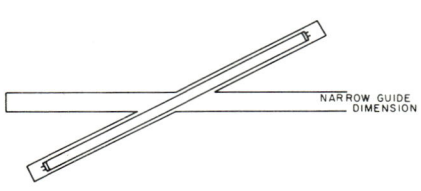

Fig. 18. Waveguide mount of gas tube.

The consistency of the noise temperature of gas discharge tubes makes them very attractive as reference microwave noise sources. Many commercial devices operating within the frequency range of approximately 0.4 to 40 gc are available. These noise sources usually have an excess noise temperature of about 15 to 16 db above standard noise. Most commonly the tubes are mounted in waveguides and their noise temperature is specified by the manufacturer. The tubes are usually mounted through the broad faces of the waveguides at a small angle with regard to the longitudinal axis (Fig. 18). This gives a good coupling of the noise energy into the waveguide. The gas in the tube also provides about a 20-db loss such that a good standing-wave ratio is achieved. Because the gas column must be matched to the waveguide, the frequency range for a particular waveguide mount is usually limited to an octave or less. Coaxial mounts may cover a wider band.

7.6. Automatic Noise-factor Meters. Automatic noise-factor meters, making use of distributed noise sources, are available commercially. A typical meter is shown

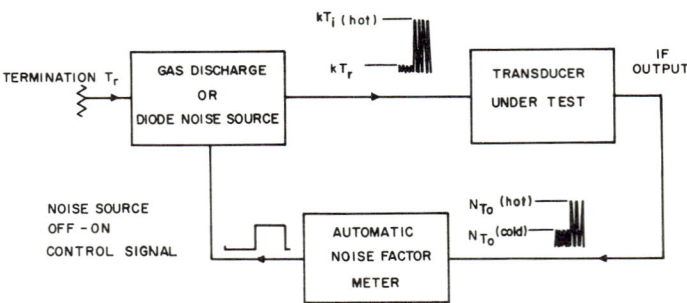

Fig. 19. Operation of automatic noise-factor meter.

in Fig. 19. A gas discharge or diode noise source is connected to the input of the transducer. An input termination at room temperature T_r is also included. The noise source is automatically pulsed off and on by a control signal from the noise-factor meter. When the noise source is turned off, it has a noise temperature $T_r = T_i$ (cold); and when it is turned on, it has a noise temperature T_i (hot).

The IF output of the transducer is connected to the noise-factor meter. The transducer delivers noise powers N_{T_o} (cold) and N_{T_o} (hot), respectively, for the off-on conditions, to the noise-factor meter. The ratio of these two powers is Y. The noise factor is determined from Eq. (47), where it is assumed that $T_r = T_i$ (cold) = T_o.

---

[30] W. R. Bennett, "Electrical Noise," p. 146, McGraw-Hill Book Company, New York, 1960.

AMPLIFIER NOISE

7-41

Expressing the noise factor in decibels gives

$$10 \log \bar{F} = 10 \log [T_i (\text{hot})/T_o - 1] + 10 \log \Re - 10 \log (Y - 1) \qquad (62)$$

The ratio $T_i (\text{hot})/T_o$ is a constant for a particular noise source. The response factor $\Re$ is a constant for a given receiver. The only variable is the ratio Y. One commercial instrument employs an AGC system to maintain $N_{T_o} (\text{hot})$ constant and provides an output meter reading corresponding to $N_{T_o} (\text{cold})$.[31] The automatic noise-factor meters are often calibrated with the assumption that $\Re = 1$, so that the second term in Eq. (62) is zero. When the response factor is different from unity, an appropriate correction factor must be added.

The transducer input frequency can be any frequency for which a distributed noise source is available for operation with the meter. The meter pulses the noise source on and off at a sufficiently rapid rate that the noise-factor reading is essentially instantaneous and continuous. The meters can even be designed to be connected continuously to operating radar receivers in such a way that they are active between the received pulses. Thus they continually measure noise factor when the receiver is not in use and can detect when troubles occur. They are also very useful when a receiver is tuned up for minimum noise factor. Their reading is independent of gain and bandwidth so that the receiver can be tuned to its optimum noise factor quickly and easily.

7.7. Output Power Indicators—Average-reading Instruments. The output power indicator is a critical item in the noise-measurement apparatus. Errors in noise measurements can frequently be traced to its improper use. The output power indicator has two basic requirements. First, it must provide a response proportional to the power delivered to a load; or it must provide a response proportional to something that is proportional to power; for example, the square of the voltage across the known resistance. Second, it must average this response over a suitable time period.

The first requirement may seem to be rather trivial yet it is the one most commonly neglected. Most a-c voltmeters and wave analyzers are average-responding rectifier instruments. They employ stable a-c amplifiers followed by diode rectifiers and a d-c meter. Their rectifier output is normally the average of the a-c signal. These meters incorporate a correction factor in the scale reading that permits them to read rms volts for a sine wave. However, if the incoming signal is not a sine wave, they will read a value 0.707/0.636 times the average value of the incoming signal. Such meters may be suitable for noise measurements if the proper precautions are taken; but the integrity of measurements made using them is always subject to question and should be tested.

It is generally safer to apply such an instrument to the dispersed noise generator method of measuring noise than to the sine-wave generator method.[32] In the noise generator method, the output meter always detects noise, the noise of the transducer plus the noise of the source in one case, and the noise of the transducer plus the noise of the generator in the other case. If the character of the noises in the two instances is similar, the ratio of the meter indications will usually be directly proportional to the ratio of the rms noise voltages.

If a sine-wave generator is used, the output in one case is noise and in the other case predominately a sine wave. However, the average-to-rms ratio of a gaussian noise voltage is 1.05 db less than the average-to-rms ratio of a sine wave.[33] Therefore, approximately 1 db must be added to the noise voltage measurements to avoid obtaining an optimistic measure of noise performance.

---

[31] H. C. Poulter, An Automatic Noise Figure Meter for Improving Microwave Device Performance, *Hewlett-Packard J.*, vol. 9, p. 2, January, 1958.

[32] It is assumed that the dispersed generator develops gaussian noise. An impulse generator, for example, which generates periodic sharp spikes would not be satisfactory. The impulse generator provides a uniform output over a very wide band, but its output does not have the proper amplitude and phase distributions. Furthermore, the signal from an impulse generator is difficult to detect properly because of its very large peak-to-rms ratio (see Sec. 7.10).

[33] H. C. Montgomery, *loc. cit.*

The time constant of the rectifier meters can be extended to several seconds by placing large electrolytic or tantalum capacitors across the meter movement. Several thousands of microfarads are usually required. Care must be taken that the presence of the capacitors does not interfere with the calibration of the meter or that the leakage currents of the capacitors do not influence the readings. This can usually be checked by noting whether or not the presence of the capacitors changes the meter indication for a sine-wave input. In some instances, the meter will be connected directly at the output of the rectifiers such that the presence of the capacitors will change the meter to a peak-reading instrument. This should generally be avoided since the meter will respond to random peaks of the noise. The capacitors may also cause overload of the output amplifiers of the meter, which would be required to supply larger peak currents than usual.

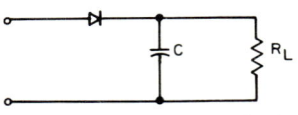

Fig. 20. Diode quadratic detector.

In summary, a-c meters are very commonly used for low-frequency noise measurements. However, their accuracy for noise measurements should always be checked against a true power or square-law instrument or by other means.

7.8. Output Power Meters–True RMS Instruments. The common types of instruments that read true power delivered to a load are those which respond to magnetic forces such as electrodynamometers, those which employ quadratic devices, such as diode detectors, and those which respond to heat such as thermistor, barretter, and thermocouple instruments. Dynamometers are seldom used because they have a restricted bandwidth and require a large driving power.

Diode detectors become approximately quadratic when operated at low signal levels—less than about 0.1 volt. The basic difference between the average-responding rectifiers described in Sec. 7.7 and quadratic diode detectors is the magnitude of the a-c signal.

The diode detector of Fig. 20 will provide an output current proportional to the square of the input voltage over about two decades of input voltage. The range of quadratic operation depends upon the load R_L. Capacitor C together with R_L provide the circuit time constant. The output signal is usually very small. A galvanometer is commonly used in place of R_L as the indicating instrument. Alternatively, a stable d-c amplifier and meter may be employed.

Thermistors are thermally sensitive resistors with a very large negative temperature coefficient of resistance. They can be had in all sizes from small vacuum-enclosed beads with a time constant on the order of milliseconds to large rod and disk structures designed for compensation of ambient temperature changes. Barretters employ a fine wire or film that is heated by the a-c signal energy. They have a positive coefficient of resistance. Milliampere-range instrument fuses and small incandescent light bulbs are often used as barretters. Instruments that employ thermistors or barretters to measure power are frequently called bolometers.

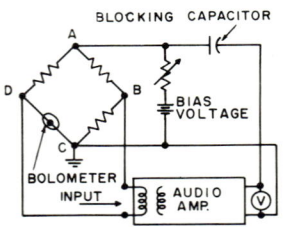

Fig. 21. Self-balancing bolometer bridge. (*From "Electronic Measurements" by Terman and Pettit. Copyright 1952. McGraw-Hill Book Company. Used by permission.*)

Both thermistors and barretters are difficult to use effectively. Their variable-resistance characteristic makes them hard to couple into and out of; and their temperature sensitivity makes them very susceptible to drift. Fortunately, these difficulties have been largely overcome in modern bolometer and calorimetric power meters.

In bolometers, the impedance variation is overcome through the use of a self-balancing bridge such as that shown in Fig. 21. The thermally sensitive element or bolometer is incorporated as one leg of the bridge. When the thermally sensitive element is at ambient temperature, the bridge is unbalanced in a direction to cause positive feedback to the amplifier system. The system then oscillates. The oscillatory signal heats the thermally sensitive element, and the bridge is driven to near balance. A signal, at another frequency, from an external source introduced into the thermally sensitive element heats it further, driving the bridge closer to balance and reducing the oscillation. The change in oscillation level is directly proportional to the power introduced and is registered on an a-c voltmeter V calibrated to read input power. Since the external power replaces the internal oscillatory power, and the bridge is always maintained at near balance, the resistance of the thermally sensitive element is held nearly constant.

Temperature sensitivity is reduced by using two identical thermally sensitive elements in thermal contact, but not electric contact. The external power is introduced into only one element such that the resulting unbalance is an indication of the power. The two thermal elements can be employed in self-balancing bridges similar to the type described previously.[34] Commercial instruments of this type are available.

A somewhat different technique is employed in the calorimetric power meter (Fig. 22). Here the input signal energy raises the temperature of a load resistance which is in thermal contact with a circulating fluid. The fluid transfers a portion of the energy to an "input temperature gauge," which is a thermally sensitive resistive element. The circuit of Fig. 22 employs two identical thermally sensitive elements operating in a bridge configuration. The circulating fluid maintains the two elements in identical thermal environments. The bridge is driven by an internal oscillator. Any bridge unbalance causes a signal to be delivered to an internal amplifier. The amplifier drives a "comparison load" and raises its temperature. The fluid transfers a portion of this energy to the second thermally sensitive element (comparison temperature gauge), thereby changing its resistance and balancing the bridge. The power delivered by the amplifier to the comparison load is very nearly equal to the input signal power dissipated in the input load. The power delivered to the comparison load is appropriately metered. The bridge configuration minimizes drift and simplifies calibration of the input power.

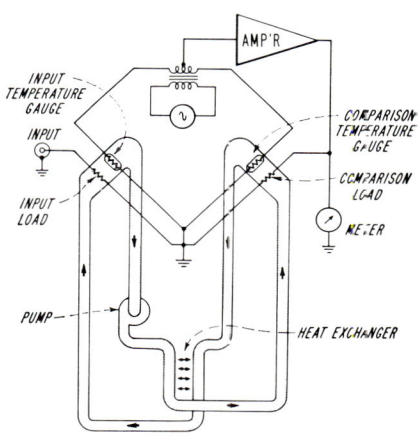

Fig. 22. Calorimetric power meter. (*From B. P. Hand, An Automatic D-C to X-Band Power Meter for the Medium Power Range, Hewlett-Packard J., vol. 9, p. 1, August, 1958.*)

The indirect heating provides three advantages for noise measurements. The internal bridge signal cannot be confused by an external signal of approximately the same frequency; the meter can therefore operate at all frequencies down to direct current. The load resistances do not have to be of thermally sensitive material, and they can be designed to provide a good impedance termination over a very wide band of frequencies. Calorimetric power meters will generally handle larger signals than bolometers and are less subject to overload damage.

Thermocouple instruments have long been used in the precise measurement of a-c power. The instruments employ a junction of two dissimilar metals in thermal contact (but not necessarily electric contact) with a heater element. The junction is connected in series with a d-c meter and another similar junction that is not in thermal contact with the heater. The input signal raises the temperature of the heater and

[34] C. C. Bath and H. Goldberg, Self-balancing Thermistor Bridge, *Proc. Natl. Electron. Conf.*, vol. 3, pp. 47–57, November, 1947.

its junction. This causes the thermal emf generated at the hot junction to be different from that generated at the cold junction; and the meter measures a current flow proportional to the difference between the two emfs.

For greatest sensitivity, the heater and thermocouple junctions are constructed of very fine wire and are enclosed in a vacuum. The vacuum allows the heater to be operated at a much higher temperature than is possible in air. Furthermore, the vacuum provides good thermal insulation of the elements and thus increases their thermal sensitivity. The d-c output of the thermocouple is metered by a galvanometer. For high sensitivity, the heater is operated near burnout. A power overload of as little as 50 per cent may cause evaporation of the heater material and a consequent permanent change in the sensitivity of the instrument. Significantly greater overloads will result in complete burnout.

The sensitivity of thermocouples is independent of frequency at all frequencies where skin effect, stray capacitances, standing waves, and the like are insignificant. Also, for d-c inputs, there may be some leakage between the heater and the junction which causes an apparent change in sensitivity. This is corrected for by taking the average of the readings obtained with normal and reverse input signal polarities. The

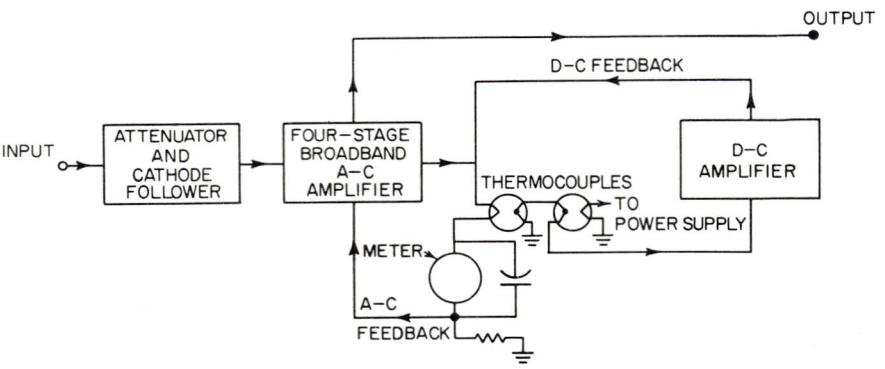

Fig. 23. Thermocouple voltmeter. (*Courtesy of the John Fluke Manufacturing Co., Inc., Seattle, Wash.*)

sensitivity of a thermocouple will change somewhat with ambient temperature because of the change in temperature of the cold junction.

With proper instrumentation, the disadvantages of thermocouples can largely be circumvented. Figure 23 shows an instrument that operates from 10 cps to 7 Mc with an input voltage range of 0.1 mv to 300 volts. The thermocouple is preceded by a precision video feedback amplifier, a precision attenuator, and a high-impedance cathode-follower input circuit. The video amplifier limits overload signals and thus protects the thermocouple. The output of the thermocouple is connected to a sensitive d-c feedback amplifier and meter. The d-c feedback permits the thermocouple to operate at an approximately constant level.

7.9. Averaging Time and Measurement Bandwidth. A suitable time averaging for the detected signal is extremely important. The relative error for a single measurement of the noise power in a narrow band B is $(2B\tau)^{-\frac{1}{2}}$, where τ is the time constant of the indicating instrument.[35] The error is thus inversely proportional to the one-half power of the product of the bandwidth and time constant. Narrowband measurements require a proportionally longer averaging time for a given accuracy than wideband measurements. For bandwidths of several megacycles, the time constant inherent in most detection and metering circuits is usually sufficient. However, for bandwidths of, say, 10 cycles, averaging times of several tens of seconds must be employed to get fractional decibel measurement accuracy.

[35] A. van der Ziel, "Noise," p. 53, Prentice-Hall, Inc., Englewood Cliffs, N.J., 1954.

AMPLIFIER NOISE

The point is seldom recognized that it is usually desirable to have as wide a bandwidth as possible, consistent with other requirements, since this leads to a shorter averaging time and greater accuracy. In narrowband systems, where many noise measurements have to be made, the time required to make the measurements can be an important factor. Furthermore, it is difficult to obtain a conventional meter that has a time constant of more than a fraction of a second. While a meter can be modified by placing large capacitors across the movement, it is usually difficult to obtain a time constant of more than a few seconds by this method. Thus wide bandwidths increase the speed of measurement and simplify the instrumentation.

When the above method of obtaining a long averaging time is inadequate, considerably longer averaging times can be obtained by using analog computer techniques. These will not be discussed here.

Generally, noise-measuring instruments will have a bandwidth that is either large compared with the noise bandwidth of the transducer being measured or small compared with its noise bandwidth. If the bandwidth of the measuring instrument is large, it will measure the average noise parameters of the transducer. If the bandwidth of the measuring instrument is small, it will measure spot noise parameters. For transducers having a narrow fractional bandwidth, such as i-f amplifiers, the average noise measured for the center of the band will usually correspond quite closely to the spot noise measured for the center of the band.

7.10. Transducer and Detector Linearity. For meaningful noise measurements, the transducer being tested should respond linearly to the input signal. It may change the phase, the frequency spectrum, or the amplitude, but it should not distort the signal.

The most common source of error is the AM detector. Its detected a-c output should never be used for noise-factor or noise-temperature measurements. Its detected output represents the a-c envelope of the noise rather than the noise itself. If the AM detector is a diode detector, it may be tempting to use its d-c output as an indication. Generally, such a detector would have to meet the requirements for average or quadratic operation as described in Secs. 7.7 and 7.8. Unless it is designed for this purpose, a diode detector is apt to suffer from drift and d-c offset; and its response will probably lie somewhere between linear and quadratic. Thus it is unlikely that the average AM detector could be used in other than relatively crude noise measurements or perhaps in comparison measurements of similar instruments.

The most satisfactory place to detect the noise in a superheterodyne receiver is usually somewhere in the center of the final i-f amplifier. AGC circuits in the receiver must be deactivated. The noise-measuring instrument should meet the requirements described in Secs. 7.7 or 7.8. Also, the instrument should not appreciably load the i-f amplifier or distort its bandwidth. The noise-measuring instrument should be located sufficiently far in advance of envelope detectors, limiters, or other nonlinear elements in the receiver to avoid the possibility of detecting a distorted wave. Yet if average noise quantities are being measured, the noise-measuring instrument should be located sufficiently close to the end of the i-f strip that the i-f bandwidth is properly established. The connection of the noise-measuring instrument to the receiver thus usually involves compromises of a number of conflicting factors.

Amplifiers and systems handling noise are more apt to overload than comparable systems handling CW signals. One of the main problems of handling noise is that its peak-to-rms ratio is theoretically infinite. Thus no matter how large are the signal-handling capabilities of an instrument, it will on the average be overloaded for some fraction of the time. Table 8 shows the percentage of the time that a given peak-to-rms ratio is exceeded for gaussian noise.

If, for example, a system is not to be overloaded more than 1 per cent of the time, it must handle a peak signal that is at least 2.576 times the rms. As the peak-to-rms capabilities are increased, the fraction of the time the system is overloaded decreases very rapidly. For example, for a peak-to-rms capability of 3.89, the system is overloaded 0.01 per cent of the time. This corresponds to 0.36 sec/hr on the average! As a rule of thumb, a capability of handling a peak-to-rms ratio of 4 is usually considered adequate for gaussian noise.

AMPLIFIER FUNDAMENTALS

Table 8. Peak Factors for Gaussian Noise*

% of time peak is exceeded	$\dfrac{\text{Peak}}{\text{rms}}$	Peak factor in db $= 20 \log_{10} \dfrac{\text{peak}}{\text{rms}}$
10.0	1.645	4.32
1.0	2.576	8.22
0.1	3.291	10.35
0.01	3.890	11.80
0.001	4.417	12.90
0.0001	4.892	13.79

* From W. R. Bennett, "Electrical Noise," p. 44, McGraw-Hill Book Company, New York, 1960.

The peak-factor table will change with different types of noise, although the results may not be too different from gaussian noise. For example, the envelope of narrowband gaussian noise has a Rayleigh distribution. This distribution is obtained when the narrowband gaussian noise is applied to an ideal diode detector with an RC filter. The peak factors for the Rayleigh distribution are shown in Table 9. This table is based on the rms value of the a-c component of the detected wave, which is equal to 0.655 times the rms value of the a-c input wave. Table 9 shows that Rayleigh noise has a smaller peak-to-rms ratio than gaussian noise and hence is less apt to cause overload.

Table 9. Peak Factors for Rayleigh Noise*

% of time peak is exceeded	$\dfrac{\text{Peak}}{\text{rms}}$	Peak factor in db $= 20 \log_{10} \dfrac{\text{peak}}{\text{rms}}$
10.0	1.517	3.62
1.0	2.146	6.63
0.1	2.558	8.39
0.01	3.034	9.64
0.001	3.392	10.61
0.0001	3.675	11.31

* From W. R. Bennett, "Electrical Noise," p. 48, McGraw-Hill Book Company, New York, 1960.

Another source of possible overload occurs when the output of a wideband amplifier is connected to a narrowband filter driving, for example, a power meter (Fig. 24). For white noise, the signal at the output of the amplifier is greater than the signal at the output of the filter by the ratio of the noise bandwidths of the amplifier and filter. The amplifier must be able to provide an output power equal to the indication on the power meter, times the ratio of the noise bandwidths of the amplifier and filter, times

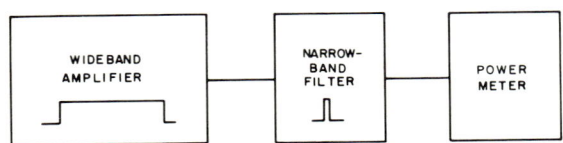

Fig. 24. Potential overload situation of a wideband amplifier driving a narrowband filter.

the loss in the filter. In addition, it must have an adequate peak-to-rms capability as described in the preceding paragraphs.

8. THE APPLICATION OF NOISE TEMPERATURE TO AMPLIFIER AND RECEIVER PERFORMANCE CALCULATIONS

The concept of noise temperature is being used increasingly to describe the noise performance of complete receiving systems including the antenna, transmission lines, coupling networks, etc. The techniques have been developed primarily for use with low-noise microwave receiving systems operating on a matched-impedance basis. However, many of the concepts can be employed equally well with radio-frequency receiving systems and even audio amplifiers where matched impedances are the exception rather than the rule. The extension of the concepts to the lower-frequency ranges can significantly simplify the calculations and often permit a better physical understanding.

8.1. Output Noise Temperature of an Attenuator. Consider the output noise temperature of an attenuator (Fig. 25). The attenuator has a loss L, is at a temperature T_L, and is connected to a source with a noise temperature T_i. L is defined as the ratio of the signal power available from the signal source to the signal

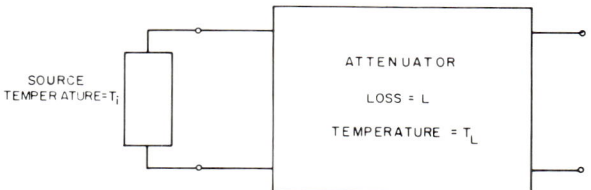

Fig. 25. Attenuator and source.

power available from the output of the attenuator. If the source and attenuator are at the same temperature T, such that $T_i = T_L = T$, the available thermal noise at the output of the attenuator is kTB. A portion of this noise is due to the signal source and a portion due to the attenuator. The portion due to the signal source is obviously kTB/L and the portion due to the attenuator is currently an unknown quantity X. The summation of these two quantities is the output power.

$$kTB = kTB/L + X$$

Solving for X gives the contribution of the attenuator

$$X = kTB(1 - 1/L)$$

Combining the above equations and dropping the kB factor gives the output noise temperature in terms of the contribution from the signal source and the contribution from the attenuator.

$$T = T/L|_{\text{source}} + T(1 - 1/L)|_{\text{attenuator}}$$

The source and the attenuator may now be given different temperatures T_i and T_L, respectively.

$$T = T_i/L + T_L(1 - 1/L) \tag{63}$$

The derivation of this equation is not intended to be rigorous but rather to illustrate the relative contributions to the output temperature.

Equation (63) becomes most useful when the concept of an attenuator is broadened. For example, the attenuator may actually be a lossy section of a transmission line or waveguide at a uniform temperature T_L. It may be a region of uniform lossy space through which a signal must be transmitted before it reaches an antenna.[36] It may also be an arbitrary passive two-port preceding an amplifier. The fact that L is defined in terms of available powers does not require that the input or output impedance be matched. Examples of the use of this formula will appear later.

8.2. Operating Noise Factor. It is important to know the effect of a network on the signal-to-noise ratio of a system. An *operating noise factor* F' may be defined that provides this information. F' is defined as the quotient of (1) the available signal to available noise ratio at the input to (2) the available signal to available noise ratio at the output.

$$F' = \frac{S_i/N_i}{S_o/N_o} \tag{64}$$

This definition is the same as the definition of conventional noise factor except that the source noise temperature T_i is allowed to be a variable. With conventional noise factor, the noise temperature of the source must be 290°K. The relationship of the input and output signals is

$$S_o = S_i/L \tag{65}$$

Using Eqs. (63), (64), and (65) and assuming N_i is proportional to T_i and N_o is proportional to T, the operating noise factor of a passive network at temperature T_L is

$$F' = 1 + (L - 1)T_L/T_i \tag{66}$$

If the network is lossless, $L = 1$; and the network can add no noise to the system. The signal-to-noise ratio at the output will then be equal to the signal-to-noise ratio at the input. Also, if the temperature of the network is 0°K, the operating noise factor will be unity. The output signal will, of course, have been attenuated by the factor L, but the noise from the source will have been attenuated by the same factor and no noise will have been added. Thus the signal-to-noise ratio at the output will remain unchanged. Also, since L, T_L, and T_i are all positive, real numbers, the minimum value of F' is unity; the signal-to-noise ratio at the output cannot be improved over that of the input.

A case of special importance occurs when the signal source is at a relatively low temperature, and the passive network has a small loss but is at relatively high temperature. The loss is conveniently represented by

$$L = 1 + \epsilon \tag{67}$$

Substituting Eq. (67) into Eq. (63), and assuming $\epsilon \ll 1$

$$T \approx T_i(1 - \epsilon) + \epsilon T_L = T_i + \epsilon(T_L - T_i)$$

If T_L is very much larger than T_i, the output noise temperature is increased over the input by an amount ϵT_L.

Suppose, for example, that an antenna with a noise temperature of 10°K is connected to a low noise receiver by a coaxial cable. The coaxial cable is at 300°K and has a loss of 0.2 db (1.047). Thus $\epsilon = 0.047$ and

$$T \approx 10 + 0.047(300 - 10) = 23.6$$

Thus the addition of 0.2 db of loss at a practical operating temperature has increased the noise temperature of the system by more than a factor of 2. From Eq. (66) the operating noise factor is

$$F' = 1 + \epsilon T_L/T_i \tag{68}$$

Inserting the above numbers, it is seen that the signal-to-noise ratio is deteriorated by a factor of 2.41 (3.8 db).

$$F' = 1 + 0.047 \times \tfrac{300}{10} = 2.41$$

[36] For systems with a nonuniform temperature see J. L. Pawsey and R. N. Bracewell, "Radio Astronomy," pp. 94–101, Oxford University Press, Fair Lawn, N.J., 1955.

AMPLIFIER NOISE

8.3. Theorem for Combination of Noise Temperatures. An extremely useful relationship can be derived for a system having any number n of lossy elements where the noise temperature of the jth element is T_j.[37]

The network must be reciprocal and therefore cannot contain isolators, circulators, or other nonreciprocal active devices.[38] The system has an output port, and each of the lossy elements transmits a fraction p_j of its noise power to the output port. The fraction p_j may have any value from zero to unity. The summation of all the contributions produces an available power $p = kTB$ at the output port, where T is the noise temperature of the output port.

The contribution of each lossy element to T must be determined. Assume that the system is operated in the reverse direction, and a matched source is connected to the output port. This source delivers unit power, say 1 watt, into the system. Since the source is matched, there is no reflection and the entire power is delivered to the network and dissipated in it. A power of p_j watts is delivered to each lossy element. The sum of all the powers must equal unity

$$\sum_{j=1}^{j=n} p_j = 1$$

When the network is returned to its normal condition, with each lossy element contributing a fraction of its available power to the output, Eq. (69) can be written.

$$T = \frac{\sum_{j=1}^{j=n} p_j T_j}{\sum_{j=1}^{j=n} p_j} = \sum_{j=1}^{j=n} p_j T_j \qquad (69)$$

where T_j is the noise temperature of the jth element. This equation states that the output noise temperature of a multielement system can be determined by summing the products of the noise temperature of each lossy element with a weighting factor. The weighting factor has a value of p_j where $0 < p_j < 1$. The weighting factor is determined by operating the network in reverse, inserting a power into the output port, and determining the fraction of the power delivered to the lossy element in question.

8.4. Examples of Noise-Performance Calculations. An example of the use of the previous equations is given in Fig. 26.[39] Figure 26 shows a receiving system with an antenna whose main beam is aimed at a portion of the sky having a 200°K average noise temperature, block 6. This is coupled through blocks 5 and 4 to the antenna main beam, block 3. Block 5 is a volume of atmosphere, assumed to be at a uniform temperature of 100°K. This atmosphere has a loss of 1.12. Block 4 is a similar volume of uniform-temperature atmosphere with a loss of 1.2 at a temperature of 250°K. The antenna coupling to the main beam is assumed to be 0.5. This means that if the antenna is operated in reverse, as a transmitting antenna, one-half the transmitted energy will be transmitted into the main beam. The other one-half of the energy will appear in side lobe couplings to space and the earth, block 2. It is assumed that these couplings have an average noise temperature of 250°K. The output of the antenna is connected to a receiver through a transmission line with a loss of 1.08 at a temperature of 290°K, block 1. The resultant noise temperature, due to all the elements facing the receiver, must now be determined. Using the technique of Sec. 8.3, a unit power, 1 watt, is fed into the output terminals of the system as shown at the bottom of Fig. 26. The transmission line has a loss of 1.08 such that 0.926 watt

---

[37] A. E. Siegman, Thermal Noise in Microwave Systems, Part 1, *Microwave J.*, vol. 4, pp. 81–90, March, 1961.

[38] A related formulation can be derived for systems containing nonreciprocal elements. See footnote 37, pp. 87–90.

[39] Example taken from H. H. Grimm, Computing Noise Levels in Microwave Receiver Systems, *Electronics*, a McGraw-Hill Publication, vol. 34, p. 53, Aug. 4, 1961.

7–50 AMPLIFIER FUNDAMENTALS

is delivered to the antenna and 0.074 watt is dissipated in the transmission line. The antenna delivers one-half the power, 0.463 watt, to the space and earth side lobes and an equal amount to the main beam. Continuing the calculation, the uniform atmosphere, block 4, dissipates 0.077 watt, the uniform atmosphere, block 5, 0.041 watt, and 0.345 watt is delivered to the main beam background. The temperature of each lossy element is then weighted according to the fractional power dissipated in it; and the weighted temperatures are summed.

$$\begin{aligned}
T_6 p_6 &= 200 \times 0.345 = 69.0 \\
T_5 p_5 &= 100 \times 0.041 = 4.1 \\
T_4 p_4 &= 250 \times 0.077 = 19.3 \\
T_2 p_2 &= 250 \times 0.463 = 115.8 \\
T_1 p_1 &= 290 \times 0.074 = 21.5 \\
& \overline{229.7}
\end{aligned}$$

The noise temperature of all the elements referred to the input of the receiver is thus 229.7°K. The earth and space side lobes are the major contributors to the noise temperature and thus seriously degrade the performance of the receiving system.

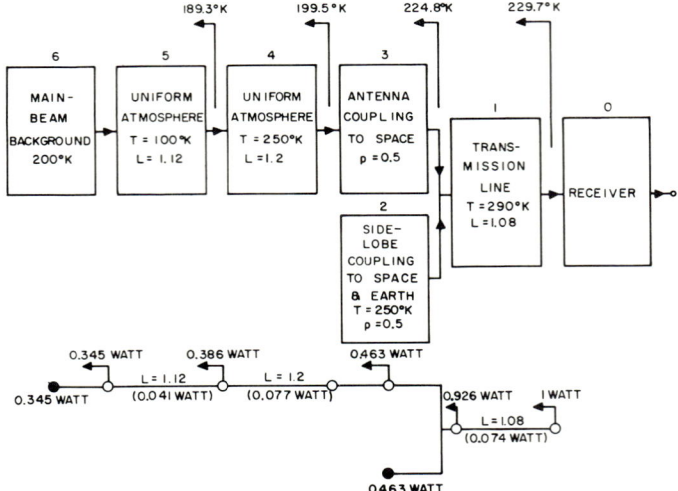

FIG. 26. Block diagram of receiving system.

As an alternate method, the performance may be calculated using the technique of Sec. 8.1. Using Eq. (63), the temperature at the output of block 5 is calculated to be 189.3°K (refer to top of Fig. 26).

$$T = 200/1.12 + 100(1 - 1/1.12) = 189.3°K$$

Similarly, the noise temperature at the output of block 4 is 199.5°K. The temperature at the output of blocks 2 and 3 is determined from Eq. (69).

$$199.5 \times 0.5 + 250 \times 0.5 = 224.8°K$$

Finally, using Eq. (63) again, the output temperature of the transmission line is found to be 229.7°K, which agrees with the previous calculation.

What effect does the system have on the signal-to-noise ratio? A signal appearing at the output of block 6 will be attenuated in blocks 1 to 5 and noise will be added. The deterioration in signal to noise caused by each block, with the exception of blocks

AMPLIFIER NOISE 7-51

2 and 3, can be determined from Eq. (68). However, it is easier to determine the overall performance using Eqs. (64) and (65). Combining Eqs. (64) and (65), the degradation in signal-to-noise ratio is found to be

$$F' = LN_o/N_i = LT_{\text{output}}/T_i$$

The overall loss, which is the product of the losses of each block in the main beam path, is easily determined. The input and output noise temperatures have been determined previously. The resultant signal-to-noise degradation is 3.33 (5.2 db)

$$F' = \frac{1.12 \times 1.2 \times (1/0.5) \times 1.08 \times 229.7}{200} = 3.33$$

The concepts can be applied equally well to low-frequency lumped-constant systems. Consider the example of Fig. 27. The signal source is a sensitive IR photo-

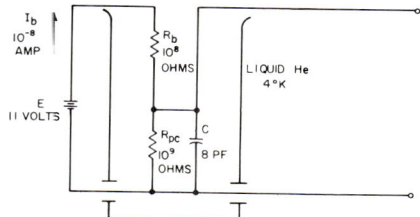

Fig. 27. Photoconductor circuit.

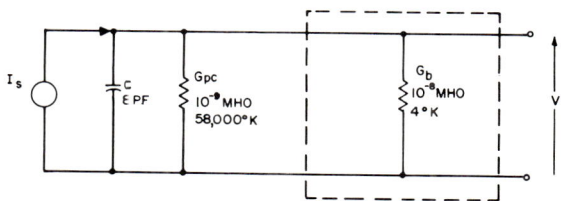

Fig. 28. Alternating-current equivalent circuit.

conductor R_{pc} with a resistance of 10^9 ohms. Bias is provided by a resistor R_b of 10^8 ohms and a bias battery E of 11 volts. A bias current I_b of 10^{-8} amp flows. The output circuit is shunted with a net stray capacitance C of 8 pf. The photoconductor and bias resistor are immersed in a liquid helium dewar which maintains the temperature at 4°K. Not shown in the figure are the necessary windows in the dewar to admit the IR radiation to the photoconductor.

Assume the photoconductor has shot noise proportional to its direct current [Eq. (4)]. Combining Eqs. (3) and (4), the noise temperature is found to be 58,000°K.

$$T = \frac{qI_b}{2kG_{pc}} = \frac{1.6 \times 10^{-19} \times 10^{-8}}{2 \times 1.38 \times 10^{-23} \times 10^{-9}} = 58,000°K$$

where G_{pc} is the conductance of the photoconductor, 10^{-9} mhos. The bias resistance R_b develops thermal noise at 4°K. An a-c equivalent circuit can now be drawn (Fig. 28) assuming that the battery is an a-c short circuit. The photoconductor and bias resistor appear in parallel and are shown as conductances. A signal generator I_s is shown for purposes of computing gain.[40]

[40] An incoming IR signal causes a change in the conductance of the photoconductor. It can be shown that this produces an equivalent signal current generator I_s which is proportional to the bias current.

At first glance, it would appear that the bias conductance virtually short-circuits the photoconductor and therefore drastically degrades the noise performance. However, it will be seen that the bias conductance reduces the signal and the noise of the signal source in proportion and adds very little noise of its own. Therefore, it deteriorates the signal-to-noise ratio very little.

Using the technique of Sec. 8.3, assume that a voltage V is applied to the output terminals of the network. This will cause a power $V^2 G_{pc}$ to be dissipated in G_{pc} and a power $V^2 G_b$ to be dissipated in G_b. The total power dissipated is $V^2(G_{pc} + G_b)$, such that the fraction of the power dissipated in G_{pc} and G_b is given, respectively, by

$$p_{pc} = \frac{G_{pc}}{G_{pc} + G_b} = \frac{10^{-9}}{10^{-9} + 10^{-8}} = 0.0909$$

$$p_b = \frac{G_b}{G_{pc} + G_b} = \frac{10^{-8}}{10^{-9} + 10^{-8}} = 0.909$$

From Eq. (69), the output noise temperature of the system is 5276°K.

$$\begin{aligned} p_{pc} T_{pc} &= 0.0909 \times 58{,}000 = 5272 \\ p_b T_b &= 0.909 \times 4 = \underline{4} \\ & 5276 \end{aligned}$$

Note that the bias resistor increases the output noise temperature by only 4°K, less than 0.1 per cent.

The solution may also be found by the technique of Sec. 8.1. The bias resistor is considered to be an attenuator, shown dotted in Fig. 28. The loss of the attenuator is equal to the ratio of the power available at the signal source to the power available at the attenuator's output terminals.

$$L = \frac{\overline{i_s^2}/4 G_{pc}}{\overline{i_s^2}/4(G_{pc} + G_b)}$$

$$L = \frac{G_{pc} + G_b}{G_{pc}} = 11$$

From Eq. (63), the noise temperature at the output is 5276°K.

$$T = 58{,}000/11 + 4(1 - \tfrac{1}{11}) = 5276°K$$

Using Eq. (66), the operating noise factor is found to be 1.0007 (0.003 db).

$$F' = 1 + (11 - 1)4/58{,}000 = 1.0007$$

Thus the bias resistor has little effect on the signal-to-noise performance.

9. LIMITING NOISE CONDITIONS

Under most conditions of operation, the equations and techniques developed thus far can be used without reservation. However, at extremely high frequencies and low temperatures, the concepts have to be modified. First, the available thermal noise power is not constant with frequency but decreases as the frequency increases. The decrease is most prevalent at low absolute temperatures and is discussed in Sec. 9.1. Second, there has been nothing to indicate that an ideal "noiseless" amplifier could not be built, in the material presented thus far. Actually, while this ideal can be approached very closely at ordinary radio and microwave frequencies, it is physically impossible to build an amplifier that contributes no noise to a system. The minimum effective input noise temperature that can be achieved increases with increasing frequency such that at optical frequencies it may be several thousands of degrees Kelvin. The noise temperature limit is determined through the application of Heisenberg's uncertainty principle, as discussed in Sec. 9.2.

9.1. Thermal Noise at Very Low Temperatures and Very High Frequencies. The available thermal noise power from a passive element is normally considered to be

AMPLIFIER NOISE

kTB. This is true at normal radio and microwave frequencies for temperatures that are at least a few degrees above absolute zero. At higher frequencies and lower temperatures, the available thermal noise becomes significantly less than kTB. The complete expression for thermal noise, obtained from quantum thermodynamics, is shown below.

$$P = \frac{hfB}{\exp(hf/kT) - 1} \qquad (70)$$

where h is Planck's constant (6.63×10^{-34} joule-sec).[41,42]

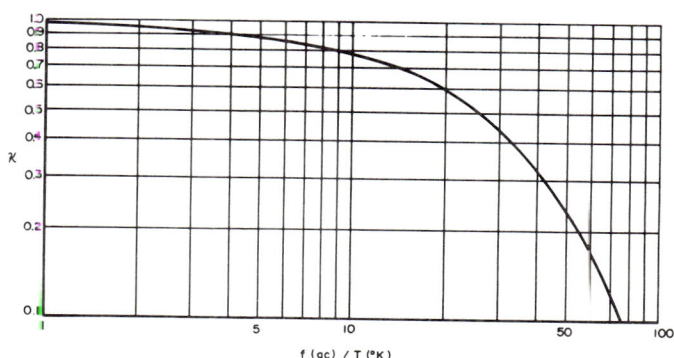

Fig. 29. Thermal noise reduction factor versus f/T.

For a waveguide system with more than one mode, this power is available from each mode that the system propagates. For $hf/kT \ll 1$ the equation reduces to the familiar kTB. It is convenient to write Eq. (70) in the form of

$$P = \mathcal{K} \, kTB$$

where $\mathcal{K}$ is the *thermal noise reduction factor*.

$$\mathcal{K} = \frac{hf/kT}{\exp(hf/kT) - 1}$$

The factor $\mathcal{K}$ is plotted as a function of $f(gc)/T(°K)$ in Fig. 29. Note that the reduction is less than 10 per cent for f/T ratios smaller than 4. For example, if the thermal source is cooled to liquid helium (4.2°K) the reduction factor is 0.994 (-0.025 db, not shown on graph) for $f = 1$ gc, 0.94 (-0.26 db) for $f = 10$ gc, and 0.54 (-2.7 db) for $f = 100$ gc. As the state of the art moves toward higher frequencies and lower temperatures, increasing attention will have to be given to the thermal noise reduction factor.

9.2. Ultimate Limits on Detector Sensitivity.[43,44,45,46] Amplifiers such as the maser and parametric amplifier are approaching the fundamental noise limit. This

---

[41] A. E. Seigman, *loc. cit.*

[42] H. Nyquist, Thermal Agitation of Electric Charge in Conductors, *Phys. Rev.*, 2d ser., vol. 32, pp. 110–113, July, 1928.

[43] H. Heffner, The Fundamental Noise Limit of Linear Amplifiers, *Proc. IRE*, vol. 50, pp. 1604–1608.

[44] K. Shimoda, H. Takehashi, and C. H. Townes, Fluctuations in Amplification of Quanta with Application to Maser Amplifiers, *J. Phys. Soc. Japan*, vol. 12, pp. 686–700, June, 1957.

[45] M. W. Muller, Noise in a Molecular Amplifier, *Phys. Rev.*, vol. 106, pp. 8–12, April, 1957.

[46] W. H. Louisell, A. Yariv, and A. E. Siegman, Quantum Fluctuations and Noise in Parametric Processes, *Phys. Rev.*, vol. 124, pp. 1646–1654, December, 1961.

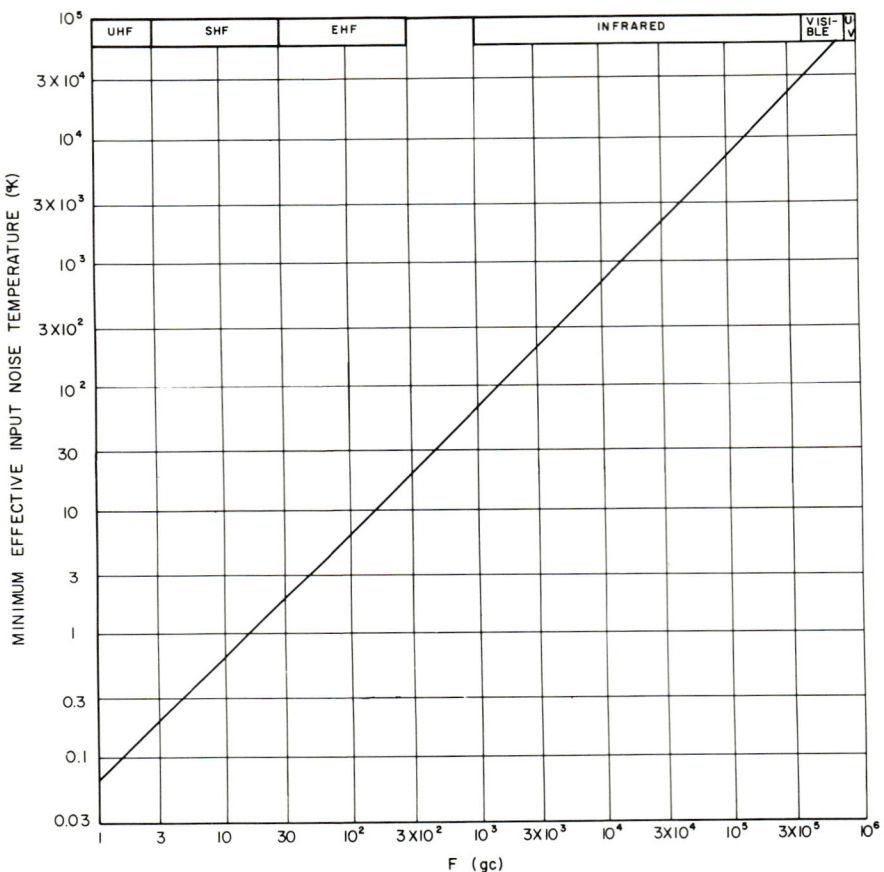

Fig. 30. Input noise temperature versus frequency.

limit is not zero. The fundamental limit is determined by the application of Heisenberg's uncertainty principle. The principle states that two canonically conjugate quantities such as the energy of a system E and the precise time t at which it possesses this energy cannot be measured with absolute certainty. The minimum uncertainty is given by

$$\Delta E \Delta t = h/4\pi \qquad (71)$$

It can be shown that a noiseless amplifier cannot exist by showing that such an amplifier would violate the uncertainty principle. Assume that this hypothetical noiseless amplifier is connected to the output of an energy source, and an ideal detector is connected at the output of the amplifier. The ideal detector measures the output energy and time with a minimum uncertainty $\Delta E_o \Delta t_o = hf/4\pi$. The amplifier has a gain G and a time delay τ_d such that the corresponding input energy E_i and time t_i are

$$E_i = E_o/G$$
$$t_i = t_o - \tau_d$$

The uncertainty of the energy and time from the source then become

$$\Delta E_i \Delta t_i = \Delta E_o \Delta t_o / G \qquad (72)$$

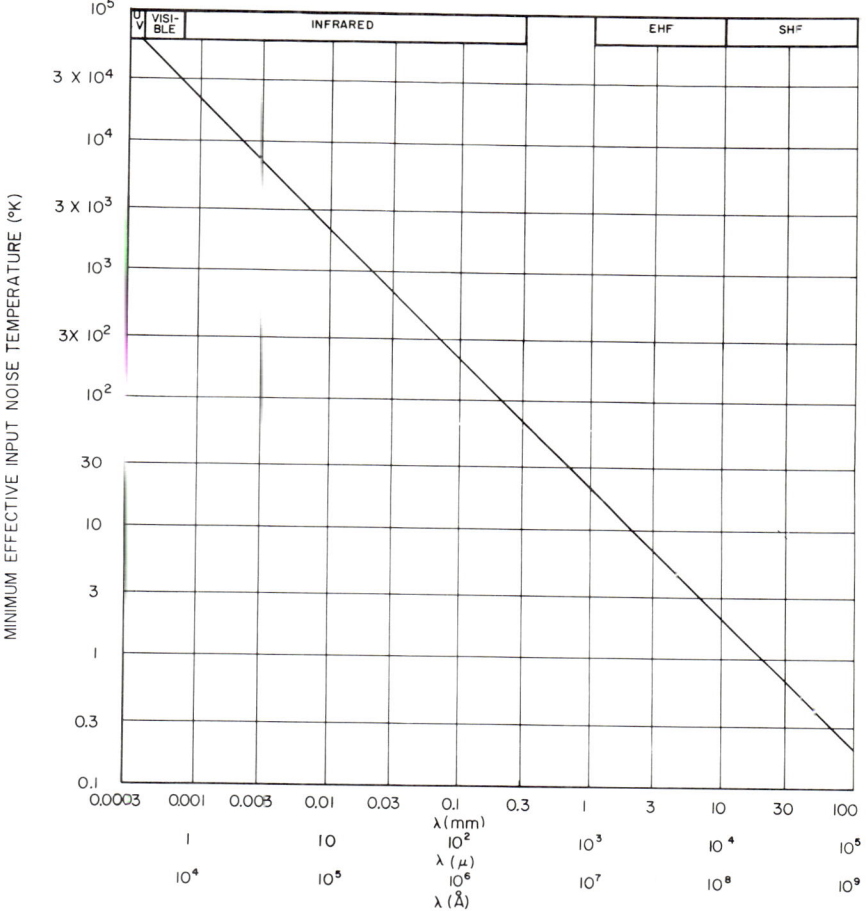

Fig. 31. Input noise temperature versus wavelength.

Combining Eqs. (71) and (72) gives

$$\Delta E_i \Delta t_i = h/4\pi G$$

which clearly violates the uncertainty principle if $G > 1$. Therefore, the amplifier must add noise to the system in order that the uncertainty of the time and energy from the source may be no less than Eq. (71). If it is assumed that the amplifier adds white gaussian noise to the system, it can be shown that its minimum effective input noise temperature is [47]

$$T_{min} = \left(\ln \frac{2 - 1/G}{1 - 1/G} \right)^{-1} \frac{hf}{k}$$

The corresponding noise power is determined by inserting T_{min} into Eq. (70).
The minimum input noise temperature for high-gain amplifiers becomes

$$T_{min} = hf/k \, (\ln 2) \quad (G \to \infty)$$
$$T_{min} = 0.0693f \quad (f \text{ in gc})$$

[47] H. Heffner, loc. cit.

This effective minimum input noise temperature is plotted as a function of frequency in gigacycles in Fig. 30. For frequencies up to 15 gc, the minimum noise temperature is less than 1°K. Thus at ordinary microwave frequencies, the minimum achievable input noise temperature can be assumed equal to 0°K for practical purposes. However, at optical frequencies, it becomes very large. For example, red light, which corresponds to a frequency of approximately 4×10^5 gc, is detected with a minimum noise temperature of 28,000°K. For convenience, the minimum noise temperature is also plotted in terms of wavelength in Fig. 31.

The results of this section raise the question, of what value is a low-noise amplifier if it cannot reduce the detection uncertainty? The answer is simply that the inclusion of the amplifier permits the use of a less-than-ideal detector, i.e., a detector with an uncertainty greater than the minimum set by Eq. (71).

BIBLIOGRAPHY

Bennett, William R.: "Electrical Noise," McGraw-Hill Book Company, New York, 1960.
Davenport, Wilbur B., Jr., and William L. Root: "An Introduction to the Theory of Random Signals and Noise," McGraw-Hill Book Company, New York, 1958.
Freeman, J. J.: "Principles of Noise," John Wiley & Sons, Inc., New York, 1958.
Haus, Hermann A., and Richard B. Adler: "Circuit Theory of Linear Noisy Networks," The Technology Press of Massachusetts Institute of Technology, Cambridge, Mass., and John Wiley & Sons, Inc., New York, 1959.
Sucher, Max, and Jerome Fox (ed.): "Handbook of Microwave Measurements," 3d ed., vol. III, pp. 865–898, John Wiley & Sons, Inc., New York, 1963.
Smullin, Louis D., and Hermann A. Haus (ed.): "Noise in Electron Devices," The Technology Press of Massachusetts Institute of Technology, Cambridge, Mass., and John Wiley & Sons, Inc., New York, 1959.
van der Ziel, Aldert: "Noise," Prentice-Hall, Inc., Englewood Cliffs, N.J., 1954.

Part II

AMPLIFYING DEVICES

Chapter 8

MAGNETIC AMPLIFIER DEVICES

HAROLD W. LORD*

CONTENTS

1. Static Devices Which Amplify through Nonlinearity of Core Materials.. 8-3
   1.1. Introduction... 8-3
   1.2. Saturable Reactor...................................... 8-4
   1.3. Amplistat... 8-4
2. Magnetic Materials Properties............................... 8-4
   2.1. General Discussion..................................... 8-4
   2.2. D-C Magnetization Curve and Static Hysteresis Loop...... 8-4
   2.3. Dynamic (A-C) Hysteresis Loops and Their Measurement... 8-8
   2.4. Typical Hysteresis Loops of Selected Materials.......... 8-12
3. Saturable-reactor Fundamentals.............................. 8-17
   3.1. General Discussion..................................... 8-17
   3.2. Series-connected Saturable Reactor with Free Even-harmonic Currents... 8-17
   3.3. Parallel-connected Saturable Reactor with Free Even-harmonic Currents... 8-19
   3.4. Generalized Characteristics of Saturable-reactor Circuits with Free Even-harmonic Currents................................ 8-22
   3.5. Series-connected Saturable Reactor with Suppressed Even-harmonic Currents....................................... 8-23
   3.6. Core and Coil Configurations........................... 8-25
   3.7. Summary... 8-27

1. STATIC DEVICES WHICH AMPLIFY THROUGH NONLINEARITY OF CORE MATERIALS

1.1. Introduction. If the term *magnetic amplifier device* were considered in its broad sense, it would include magneto-electro-mechanical devices such as electromagnetic relays, magneto-electro-dynamic devices such as Ward-Leonard drives and amplidynes, and static magnetic amplifiers such as saturable reactors and magamps or amplistats. In keeping with the general context of this handbook, only *static magnetic amplifiers* will be treated here.

* General Electric Co., Research and Development Center, Schenectady, N.Y.

AMPLIFYING DEVICES

The basic principles of operation of the three general types of magnetic amplifiers are so widely different that about the only thing they have in common is a magnetic circuit which includes a ferromagnetic core material. Static magnetic amplifier devices depend upon the nonlinearity of the magnetization characteristic of the core material to produce amplification. The amplifying properties of the other types of magnetic amplifiers are produced by physical principles which are so different that each type is in a separate and distinct specialty field. A bibliography pertaining to a few of the nonstatic magnetic amplifier devices is listed at the end of this chapter.

1.2. Saturable Reactor. A *saturable reactor* has been defined[1] as ". . . an adjustable inductor in which the current versus voltage relationship is adjusted by control magnetomotive forces applied to the core." A *magnetic amplifier* has been defined[1] as "a device using *saturable reactors* either alone or in combination with other circuit elements to secure amplification or control." A *simple magnetic amplifier* has been defined[1] as "a *magnetic amplifier* consisting only of *saturable reactors*." Saturable reactors are therefore the heart of *magnetic amplifiers*, and the term *magnetic amplifier* usually refers to a *static* magnetic amplifier. In the rest of this chapter and Chap. 21, the term *magnetic amplifier* will be used to denote a *static magnetic amplifier*.

1.3. Amplistat. The term Amplistat is one of several trade names used to denote a so-called *self-saturating magnetic amplifier*. Other trade names include Magnestat and MagAmp.

By definition,[1] "*Self-saturation* in a magnetic amplifier refers to the connection of half-wave rectifying circuit elements in series with the output windings of the saturable reactors." It has been shown[2] that saturable reactors can be considered to have negative feedback. Half-wave rectifiers in series with the load windings will block this intrinsic feedback. An Amplistat is therefore *a parallel-connected saturable reactor with blocked intrinsic feedback*. This definition avoids the term *self-saturation* which, although extensively used, does not have a very sound physical basis.

Amplistats inherently provide a much higher gain than do saturable reactors (simple magnetic amplifiers). Transfer characteristics of Amplistats are markedly affected by characteristics of the circuit elements (including the load), the magnetic characteristics of the saturable-reactor cores, and the circuit configuration. An Amplistat is not a simple magnetic device; so the basic theories of operation of Amplistats will be treated in Chap. 21.

2. MAGNETIC MATERIALS PROPERTIES

2.1. General Discussion. A study of the theory of magnetic amplifiers requires some knowledge of ferromagnetism and in particular the relations between the magnetic flux density in the core and the current which must flow in the windings on the core to establish and/or change the flux density from one value to another. Except in rare instances, it is only necessary to know the rms values of exciting current and core loss vs. peak flux density at a given frequency. Transformer theory involves no more than this until an explanation of the shape of the exciting current is sought. Then an understanding of magnetic hysteresis is required. Magnetic hysteresis plays such an important part in magnetic amplifier theory that an understanding of the theory requires a considerable amount of detailed knowledge concerning magnetic hysteresis and the effects of rates of change of flux upon the hysteresis loop.

2.2. D-C Magnetization Curve and Static Hysteresis Loop. Figure 1 shows a d-c magnetization curve and a static hysteresis loop with significant points or portions of the curves identified by words or letter symbols having the following meanings:

B Flux density, gauss
B_m Maximum operating flux density, gauss
B_s Residual flux density, gauss
$B\mu_m$ Flux density at $\mu = \mu_m$
H Magnetizing force, oersteds
H_c Coercive force, oersteds
H_s Magnetizing force required to produce B_s

μ	Normal permeability (B/H), gauss/oersted
μ_d	Differential permeability, gauss/oersted
μ_m	Maximum of normal permeability, gauss/oersted

Meanings of additional letter symbols:

B_i	Intrinsic flux density, gauss
μ_v	Permeability of vacuum, gauss/oersted

In Fig. 1, the distances from the origin to H_c and to H_s do not show the proper ratio of these values for a good core material for magnetic amplifiers. The true ratio of H_s/H_c is in the range of 100 to 500.

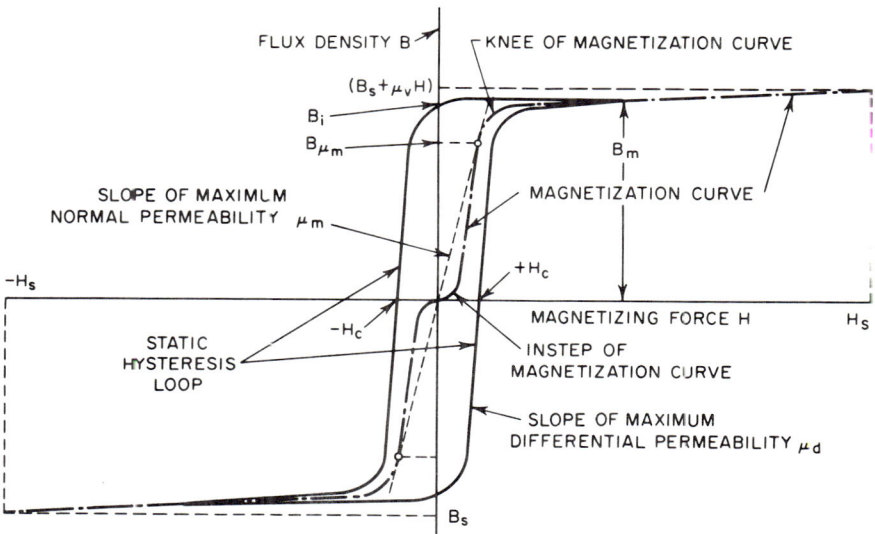

Fig. 1. Magnetization curve and static hysteresis loop with identification of significant points and portions.

If a core material is demagnetized and the magnetizing force is gradually increased, the flux density will increase along the *magnetization curve*. The initial increase of B with respect to H is gradual, indicating a low initial permeability. The curve portion between the origin and the "instep" is often called the "toe" of the *magnetization curve*. After the "instep," small increases of H produce large increases in B until the "knee" is reached and there is an abrupt decrease in the slope of the magnetization curve. The part of the magnetization curve above the "knee" is called the *saturation region*. The saturation flux density $(B_s + \mu_v H)$ is that density at which the slope of the magnetization curve has decreased to unity, i.e., $dB/dH = 1$.

The magnetic flux density B of Fig. 1 is the flux density in the core material and consists of two components:

1. A flux density $\mu_v H$ which would be present in the material if it possessed no ferromagnetic properties
2. A flux density B_i which is due to the intrinsic magnetization of the material

Mathematically expressed, this is

$$B = \mu_v H + B_i \qquad \text{gauss} \tag{1}$$

Table 1. Magnetic Amplifier Core Materials, Typical D-C Magnetic Properties

No.	Nominal composition	Gauge, in.	Saturation flux density B_s, gauss	Magnetizing force for B_s, H_s, oersteds	Max normal permeability μ_m, gauss/oersted	Flux density for μ_m, $B_{\mu m}$, gauss	Coercive force H_c, oersteds	Residual flux density B_r, gauss	From peak flux density B_m, gauss	$\dfrac{B_{\mu m}}{B_s}$, unity	$\dfrac{B_r}{B_s}$, unity
1	97 Fe, 3 Si (oriented crystal)	0.014	19,700	800	60,000	7,500	0.10	13,500	15,500	0.38	0.69
2	50 Ni, 50 Fe (oriented crystal)	0.002	15,500	15	125,000	13,000	0.1	14,700	15,500	0.85	0.95
3	79 Ni, 17 Fe, 4 Mo	0.002	8,800	25	80,000	5,300	0.035	6,500	8,000	0.60	0.74
4	79 Ni, 16 Fe, 5 Mo	0.002	7,900	10	250,000	3,700	0.013	4,300	6,700	0.47	0.54
5	75 Ni, 18 Fe, 5 Cu, 2 Cr	0.014	6,650	50	105,000	1,800	0.015	2,400	5,000	0.27	0.36
6	65 Ni, 35 Fe (domain oriented)	0.002	13,650	15	400,000	11,500	0.02	12,200	12,500	0.89	0.91
7	50 Fe, 48 Co, 2 Va (domain oriented)	0.004	22,200	15	100,000	19,000	0.18	21,000	22,000	0.86	0.95

The maximum value of B_i is the saturation flux B_s. If the intrinsic saturation curve of B_i versus H is plotted, since $\mu_v = 1$, B_s will occur when the slope (dB_i/dH) of the intrinsic saturation curve becomes zero.

If the flux is cycled several times between flux-density values of $+B_m$ and $-B_m$, illustrated in Fig. 1, and then the magnetizing force H is varied *slowly* so as to cause the flux density to vary slowly from $+B_m$ to $-B_m$ and then back to $+B_m$, the B versus H curve will follow a B-H loop similar to the solid-line loop depicted in Fig. 1. It is evident that in this loop the flux always lags behind the magnetizing force. This

Table 2. Some Trade Names of Alloys Listed in Table 1
1. Silectron, Hypersil, Trancor, Magnesil, Microsil
2. Deltamax, Hypernik V, Orthonol, Carpenter 49, Square Mu
3. 4-79 Mo-Permalloy, Molypermalloy, Hymu 80
4. Supermalloy
5. Mumetal
6. 65-Permalloy
7. Vanadium Permendur, Supermendur

phenomenon is known as hysteresis, a word meaning "a lagging behind." Consequently, the loop shown in Fig. 1 is known as the hysteresis loop for the magnetic material. If the loop is obtained by changing the magnetizing force very slowly, eddy-current effects are considered to be negligible and the hysteresis loop so obtained is called a static or d-c hysteresis loop.

Table 1 shows some of the d-c magnetic properties of several magnetic alloys of interest to the designers of magnetic amplifiers, and Table 2 lists the various trade names given to the correspondingly numbered alloys. Magnetic characteristics of alloys are markedly affected by the processing by which the core material is produced; hence the values given in Table 1, for μ_m, $B_{\mu m}$, H_c, and B_r and in the last two columns, can only be considered as approximate values. They are sufficiently representative to provide a means for an initial selection of one or more core materials for a proposed application.

In some magnetic materials useful in magnetic amplifiers the static hysteresis loops may have extremely steep sides. The differential permeability μ_d over these portions of the hysteresis loop can be extremely high (exceeding 10^6). Measuring the static hysteresis loops of such materials must be done with great care, and they are most readily obtained by means of a hysteresigraph.[3] This instrument automatically plots, on rectangular coordinates, the B-H loop of a test core.

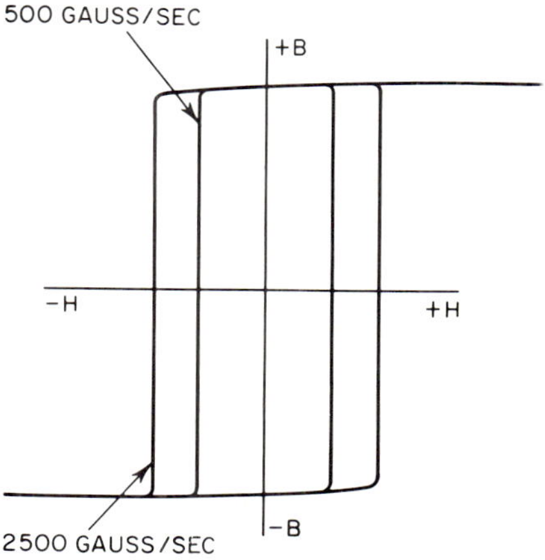

Fig. 2. Dynamic hysteresis-loop-widening effect.

Hysteresigraphs have been built which have sufficient long-time stability to permit elapsed times of several minutes for completing a loop and also sufficient speed to permit one to be completed within a few seconds. An instrument with this capability has been used to demonstrate a magnetic effect which is important to the theory of operation of certain Amplistats.

The curves of Fig. 2 were made on a very stable hysteresigraph. The vertical portions were made by setting the magnetizing force H to the indicated values and timing with a stop watch the change in flux density B. The test core was a spirally wound core of 0.005-in. 65-Permalloy. Saturation density B_s of 65-Permalloy is about 13,500 gauss. The time to traverse each vertical portion of the inner loop was 50 sec.

This loop is the d-c or static hysteresis loop. Corresponding portions of the outer loop were traversed in 10 sec. Of interest here is the fact that a small increase of H in excess of that required by the static loop does not cause an instant or even a very rapid change in flux.

Classical concepts of skin effects cannot account for the observed widening of the hysteresis loops in 0.005-in. core material at such low rates of change of flux density. The eddy currents which flow and thereby inhibit the rapid flux change are generally believed to be associated with magnetic domain[4] boundary movements. A given H in excess of that required by the static hysteresis loop is balanced out by eddy currents associated with the domain boundary velocities. The degree of this widening effect is

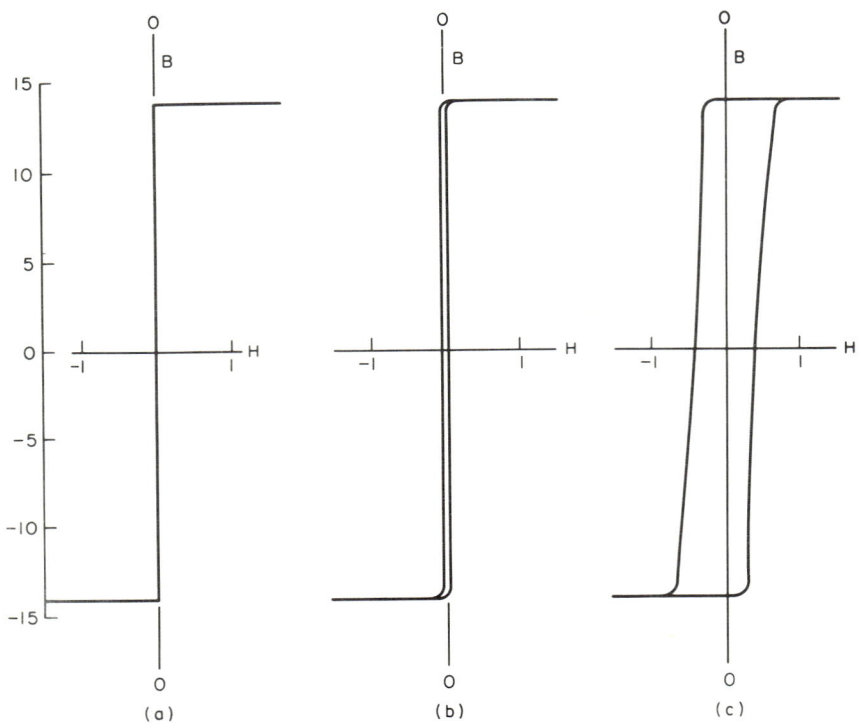

Fig. 3. Comparison of hysteresis loops.   (a) Ideal.   (b) Static.   (c) Dynamic.

determined by such factors as the ease with which domains can be nucleated, the crystal structure, the presence or absence of small impurities, the volume resistivity, and the lamination thickness. Details of the interactions which determine this widening are beyond the scope of this book, but the manifestations of the interactions, i.e., the shape of the static hysteresis loop, and the alteration of its shape when the flux density is required to change rapidly, are important to the operation of magnetic amplifiers.

2.3. Dynamic (A-C) Hysteresis Loops and Their Measurement. Figure 2 shows a hysteresis loop which was recorded in about 20 sec (the outer loop). This can be considered the dynamic hysteresis loop of 0.005-in.-thick 65-Permalloy operating at a frequency of $\frac{1}{20}$ cps. Figure 3a shows an ideal hysteresis loop often used in analyzing magnetic amplifiers. Figure 3b is a plot of the inner loop of Fig. 2 to a different scale. Figure 3c shows the dynamic hysteresis loop of the same 0.005-in.-

thick 65-Permalloy core excited so as to produce a 400-cycle sinusoidal flux in the core. The 400-cycle loop is much wider than the static hysteresis loop and departs from rectangularity. Its coercive force H_c is about twenty times the coercive force of the static loop. Chapter 21 gives the reason a core material such as this would be a poor choice for use in high-performance Amplistats and shows the roles that both static and dynamic hysteresis loops play in determining their transfer characteristics.

Dynamic hysteresis loops can be measured with reasonably good accuracy by an oscilloscope provided that the amplifiers for both the vertical and the horizontal axes have adequately low phase shift over a band of frequencies extending two to three decades each side of the test frequency.[5] Figure 4 is a reproduction of Fig. 1 of that reference, showing a block diagram of an excitation and test circuit which is often used to display the dynamic hysteresis loop of a core material upon the screen of a cathode-ray oscilloscope. This circuit includes means for impressing upon the vertical, or B, axis of the oscilloscope a signal which is proportional to the flux density in the core and impressing upon the horizontal, or H, axis of the oscilloscope a signal which is proportional to the magnetizing force acting on the core specimen under test. The B-axis

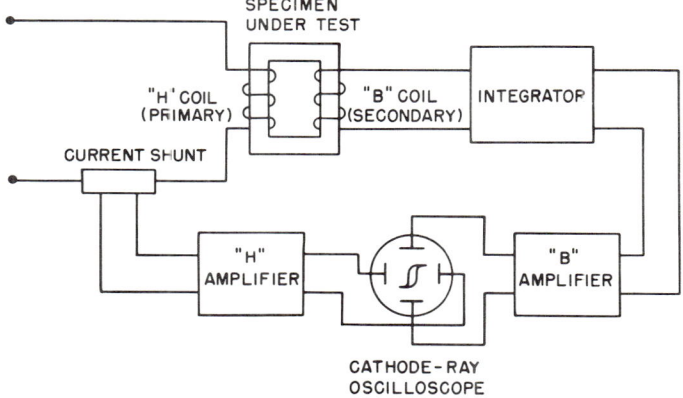

FIG. 4. Circuit for displaying dynamic hysteresis loops.

channel includes a secondary (B) coil wound on the specimen under test, an integrating circuit, and a linear amplifier, the output of which supplies voltage to the vertical deflection plates. The H-axis channel consists of an amplifier to amplify the relatively low voltage drop across a resistor (current shunt) in series with the primary or exciting (H) coil, to provide an adequate deflection voltage to the horizontal deflecting plates of the oscilloscope. The Lissajous figure shown by this system on the cathode-ray-tube screen then will be indicative of the shape of the dynamic hysteresis loop of the specimen under test.

The display of highly rectangular hysteresis loops with an error less than 5 per cent or one trace width, whichever is greater, requires amplifiers of the following minimum performance.[5]

Each amplifier is assumed to be a three-stage amplifier and is specified upon the basis that a good wideband amplifier is used for the other axis. Somewhat poorer phase-response characteristics could be tolerated if both amplifiers were identical and had a constant time delay. However, this possibility has been ignored in arriving at the requirements given here.

In order to keep the phase shift to a satisfactorily low value, each stage of the H-axis amplifier should have less than 45° of phase shift at a low frequency $\frac{1}{100}$ of the test frequency and at a high frequency 250 times the test frequency. With conventional types of resistance-capacitive coupling networks and with screen and cathode circuits that do not contribute to the frequency-response characteristics, the amplitude response per stage will be down 3 db at these indicated frequencies.

The requirements for the B-amplifier phase shift are very stringent on low-frequency response and a little less stringent than that of the H amplifier on the high-frequency end. If a single resistance-capacitance integrating circuit having a phase angle at operating frequency of 89.75 to 90° is used, the low-frequency phase response per amplifier stage must have a phase shift of less than 45° at $\frac{1}{300}$ of the operating frequency. However, if suitable compensation is provided by an additional corrective integrating network per amplifier stage, the low-frequency phase response need only be less than 45° at $\frac{1}{30}$ of the operating frequency.

The high-frequency phase shift of this B amplifier should be less than 45° in each stage at 100 times operating frequency.

With a suitable broadband B amplifier, a simple single resistance-capacitance network of proper values will provide satisfactory integration. It should provide, across the capacitor, a voltage which lags the input voltage by 89.75 to 90° at the operating frequency.

When additional integrating types of resistance-capacitive networks are added to equalize for excessive low-frequency phase advance or lead due to the coupling networks in the amplifier, the equalizing networks should provide, at operating frequency, a total overall lagging phase of 90° ± 0.25° for the combined integrator and entire B axis amplifier circuit.

The Tektronix type 536 is an oscilloscope which meets the above criteria for measurements in the range of 60 cycles to several kilocycles. The cathode-ray tube of this instrument has identical vertical and horizontal deflection sensitivities and the main deflection amplifiers are identical. It will also accept identical plug-in preamplifiers in the two deflection channels. With type 53/54G preamplifiers, the bandwidth is direct current to 10 Mc.

When hysteresis loops of large-area cores are measured, a suitable RC integrator circuit supplied by a B coil consisting of a few turns wound directly over the core or corebox will provide sufficient output voltage to the vertical preamplifier of the Tektronix 536 oscilloscope. When measuring small cores, especially those having low values of H_c, care must be taken to assure that the integrator circuit does not place so much load on the B coil that it affects the accuracy of the H measurement. One way to accomplish this is to use a small number of turns on the B coil and feed the voltage from it into a modified Miller-type integrator-amplifier. Figure 5 is the circuit diagram of an integrator-amplifier which has been used satisfactorily in the range of 60 to 1,000 cps.

Values of B in gauss for a given core under test can be obtained by calculation and an overall calibration of the oscilloscope and integrating circuit or amplifier. Small d-c biases and memory effects make the measurement of the true instantaneous flux density in cores rather uncertain. But the total flux-density change ΔB can be measured quite accurately; and if the core test brings the flux density well into the saturation region for at least one portion of the loop, the rest of the dynamic hysteresis loop can be referenced to the known values of the static hysteresis loop in the saturation region. This is possible because the tips of the static and dynamic hysteresis loops coincide for a given value of H in the saturation region.

Calibration of the Y axis (flux density) is accomplished by determining the *rectified average* value of an a-c *calibrating* voltage $E_{av,c}$ of any convenient frequency f_c which produces a certain peak-to-peak vertical deflection. (If the calibration voltage is a pure sine wave, any a-c voltmeter may be used to measure the $E_{av,c}$ by dividing the indicated voltage by 1.11.) The flux-density change ΔB in a test core which will produce the same peak-to-peak vertical deflection when tested in a circuit similar to Fig. 4 is calculated by

$$\Delta B = \frac{E_{av,c}}{f_c} \times \frac{10^8}{2AN_B} \qquad \text{gauss} \qquad (2)$$

where A = net core area of test core, cm^2
N_B = number of turns in B coil (Fig. 4)

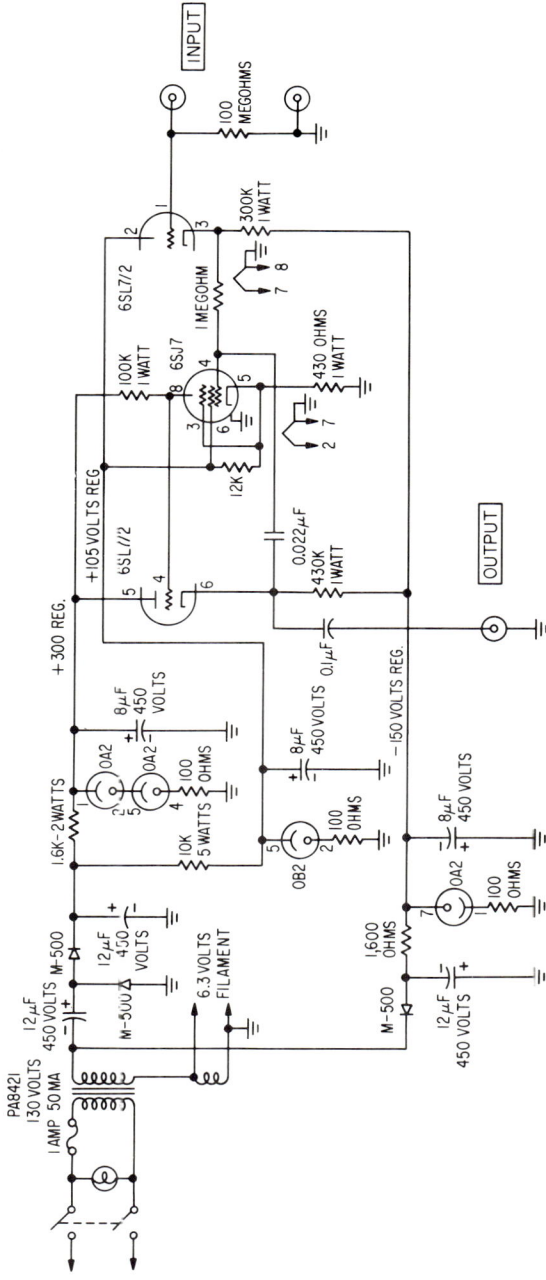

Fig. 5. Integrator-amplifier for *B* portion of hysteresis-loop display circuit.

In an integrating circuit such as an RC network, it can be shown that the output voltage of the network will have a constant value for all inputs having the same value of E_{av}/f. Also, the average voltage induced in the B test coil is

$$E_{av} = 2\Delta B A N_B f \times 10^{-8} \qquad (3)$$

and
$$\Delta B = \frac{E_{av}}{f} \times \frac{10^8}{2AN_B} \qquad (4)$$

which shows that, for any given value of ΔB, the value of E_{av}/f is a constant.

Since E_{av}/f is a constant for a given value of ΔB and since the output voltage of the integrator is a constant for a given value of E_{av}/f, after calibration at a convenient frequency f_c, the circuit of Fig. 4 will indicate the ΔB of the core under test for all frequencies within the linear frequency-response range of the integrator-amplifier and/or the oscilloscope amplifier without further calibration. Moreover, it will indicate correctly the ΔB of the core under test even though the waveshape of the voltage induced in the B coil is highly distorted. This insensitivity to distortion is important if the shape of the dynamic hysteresis loop is measured under conditions of *sine-current* excitation.

Values of H in oersteds for a given core under test are determined by the following equation:

$$H = 0.4\pi N_H i_H/l_m \qquad \text{oersteds} \qquad (5)$$

where l_m = mean length of magnetic path of the core, cm
N_H = turns on test H coil (Fig. 4)
i_H = instantaneous current in H coil

The voltage e_H applied to the oscilloscope X axis in Fig. 4 is the voltage drop across R_H, and $i_H = e_H/R_H$. Therefore, the magnetizing force acting on the core at any instant is measured by observing the value of the horizontal deflection voltage indicated by the oscilloscope, using the following equation:

$$H = 0.4\pi N_H e_H/l_m R_H \qquad (6)$$

Oscillographic measurements of dynamic hysteresis loops can be sufficiently accurate to provide data for designing magnetic amplifiers. The equipment required to make the measurements in this manner can usually be had in an engineering laboratory. It is an excellent tool to study the influence of the magnetic amplifier circuitry upon the actual operating hysteresis loops as discussed in Chap. 21. In those cases where greater accuracy may be required, equipment is available, designed specifically for tracing dynamic hysteresis loops directly on an XY recorder.[6,7]

2.4. Typical Hysteresis Loops of Selected Materials. The hysteresis loops shown here are primarily for use as guides in *selecting* a type of material for an application and making preliminary design calculations. Some of them illustrate the effects of materials processing and laminar thickness upon the shape of the hysteresis loops. Manufacturers of core materials *for magnetic amplifiers* have been encouraged to present their magnetic materials data according to certain AIEE (now IEEE) Standards and Recommended Practices.[8]

In designing magnetic amplifiers, the pertinent magnetic characteristics of the selected type of core material should be obtained from the manufacturer. Spirally wound tape cores with toroidally wound coils were used to secure the data for all the following hysteresis loops.[9] The loops are for sinusoidal flux waveshapes.

Figure 6 shows the hysteresis loops for 0.002-in. Supermalloy. This is a high-nickel-content unoriented alloy characterized by high values of maximum permeability, a rounded knee, and relatively low values of saturation flux densities. The loops are extremely narrow. This alloy is particularly suited for use in magnetic amplifiers for amplifying very low levels of signal power. (Note: The portions of the dynamic curves above the knee are probably in error since the d-c loop normally should fall inside or coincide with the a-c loop in this region. Also, observe that both scales of this figure are different from those of the following figures.)

Figure 7 shows hysteresis loops for 0.002-in. oriented silicon steel. This material is

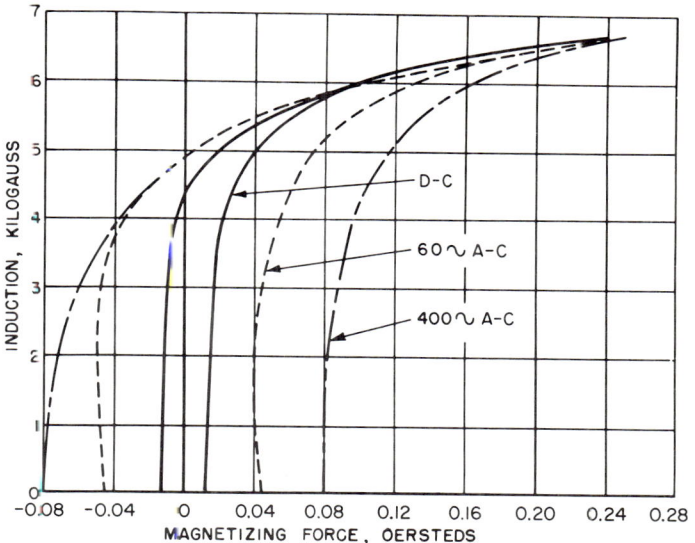

Fig. 6. Hysteresis loops for 0.002-in. Supermalloy.

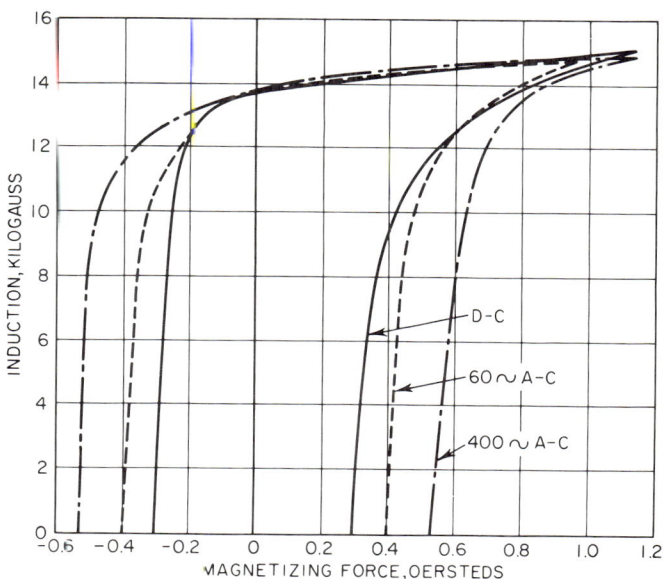

Fig. 7. Hysteresis loops for 0.002-in. oriented silicon steel.

suitable for medium and large power-handling magnetic amplifiers. In comparable thicknesses, it costs less than any of the other alloys shown herein, is more available, and does not use nickel or other critical materials. Since the hysteresis loops of this material are wider (for comparable thicknesses) than those of the other alloys shown, magnetic amplifiers constructed of this material will have poor performance in terms of gain and dynamic response relative to that of amplifiers constructed of the other higher-permeability narrower-hysteresis loop materials. It is, however, an excellent material for use in simple magnetic amplifiers.

Figure 8 shows hysteresis loops for 0.002-in. 65-Permalloy. Note particularly the very narrow d-c hysteresis loop; yet the dynamic loops are comparable in width with those for Deltamax, as shown in Fig. 9. The so-called hourglass or wasp-waisted shape shown by the a-c loops of Fig. 8 has been observed to occur with highly oriented

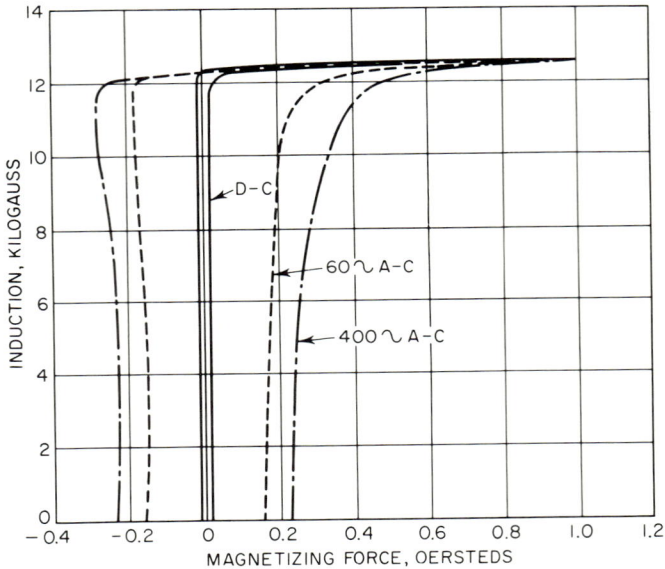

FIG. 8. Hysteresis loops for 0.002-in. 65-Permalloy.

alloys. This tendency seems to be particularly pronounced when they are domain-oriented alloys. Orientation in 65-Permalloy is obtained by a magnetic anneal; so it is a domain-oriented material. At present, 65-Permalloy is not available commercially. In a magnetic amplifier it yields no better performance than that of Deltamax and its saturation density is a little lower than that of Deltamax.

Figure 9 shows hysteresis loops for a specimen of Deltamax which has been annealed at a temperature slightly above that which produces optimum rectangularity of the d-c loop. However, the dynamic loops shown are more rectangular; so this specimen represents about the best overall characteristic for this material for use in high-performance magnetic amplifiers.

Figure 10 shows the results obtained when the original specimen used for the Fig. 9 data was given a magnetic anneal. This treatment produced a material whose d-c hysteresis loop is not only narrower but more rectangular. However, the comparable dynamic hysteresis loops are actually wider than they were before the magnetic annealing treatment, and tests in a magnetic amplifier showed a corresponding deterioration of performance.

The effect of annealing temperature upon the shape and width of the d-c hysteresis

MAGNETIC AMPLIFIER DEVICES

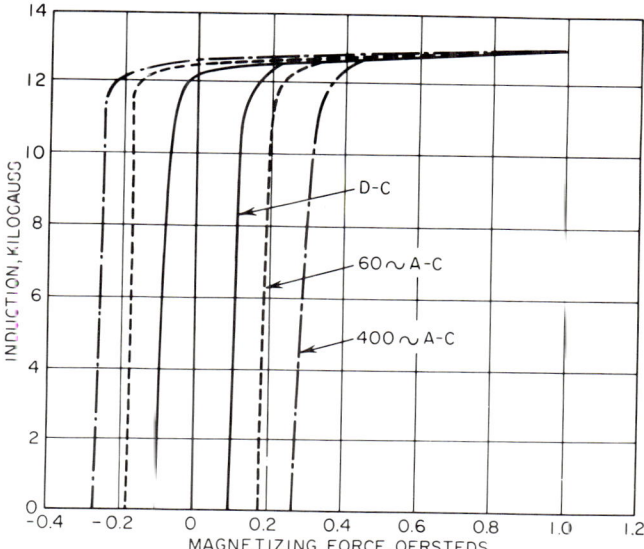

Fig. 9. Hysteresis loops for 0.002-in. Deltamax.

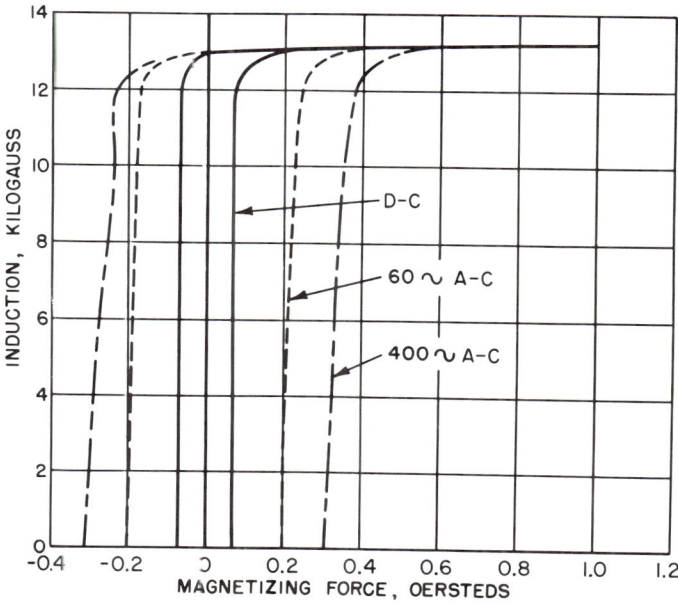

Fig. 10. Hysteresis loops for 0.002-in. Deltamax with magnetic anneal.

8-16 AMPLIFYING DEVICES

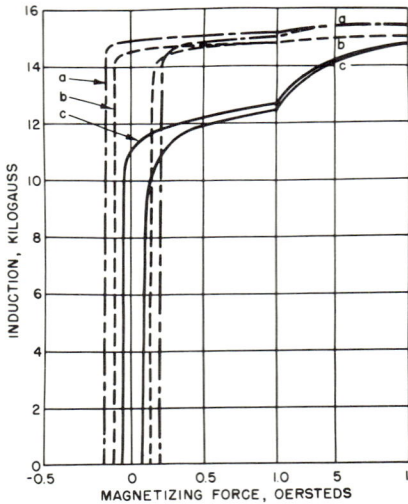

FIG. 11. Effect of annealing temperature upon the shape and width of static hysteresis loops of Deltamax. (a) Annealed at 1000°C. (b) Annealed at 1075°C. (c) Annealed at 1200°C.

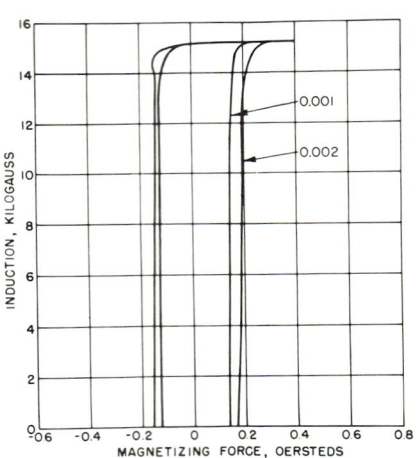

FIG. 12. Effect of gauge upon the 60-cycle hysteresis loop for Deltamax.

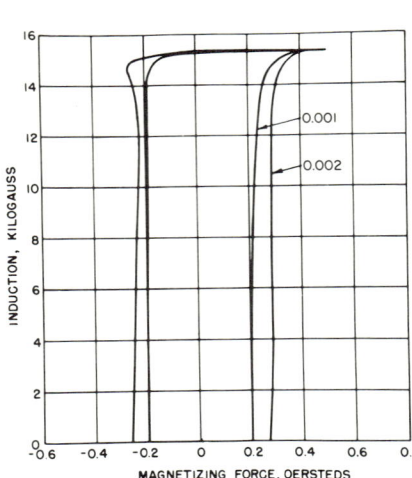

FIG. 13. Effect of gauge upon the 400-cycle hysteresis loop for Deltamax.

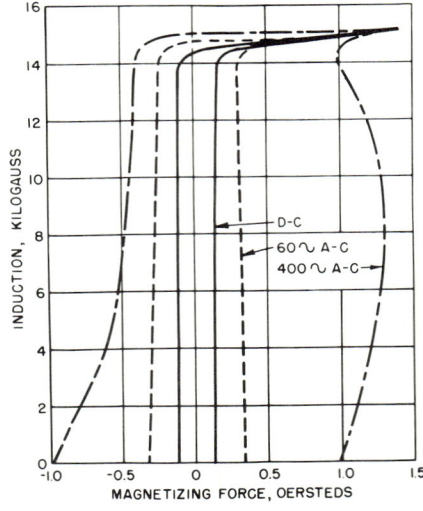

FIG. 14. Hysteresis loops for 0.010-in. Deltamax.

loops is shown by Fig. 11. These three specimens were all from the same laboratory-produced ingot.

The material used for curve a was annealed at 1000°C, for curve b at 1075°C, and for curve c at 1200°C. Dynamic hysteresis loops have been omitted in this comparison. The increase of H_c for the dynamic loops was roughly the same for the two specimens annealed at the lower temperatures, and was about 40 per cent less for the one annealed at 1200°C.

Figures 12 and 13 show the effect of gauge upon the 60- and 400-cycle dynamic hysteresis loops, respectively, of 0.002- and 0.001-in. Deltamax. The two specimens used for this comparison were selected on the basis that their d-c hysteresis loops were substantially identical. Thus any difference in their dynamic characteristics can be attributed to their difference in thickness. Note that even at 60 cycles there is some advantage in favor of the thinner material.

Figure 14 further illustrates the effect of gauge and is convincing evidence of the poor performance of 0.010-in.-thick materials of high permeability at 60 cycles when compared with thin-gauge specimens of such materials. These loops illustrate why it is poor practice, even for 60-cycle magnetic amplifier cores, to use high-permeability materials such as Deltamax, Monimax, Supermalloy, Mumetal, and Molypermalloy in thicknesses exceeding 0.005 in. For frequencies above about 300 cycles, such thick-gauge alloy cores are little or no better than comparable gauges of ordinary silicon steel materials.

3. SATURABLE-REACTOR FUNDAMENTALS

3.1. General Discussion. To save time and space, the term *saturable reactor* will be abbreviated to SR; this abbreviation corresponds to the abbreviations PT for *potential transformer* and CT for *current transformer*. This abbreviation is also used in Ref. 2.

In many analyses, the SR is viewed as having an impedance whose magnitude is gradually changed by direct current. In the analyses used herein, the SR is considered to have a very high impedance throughout one part of the half-cycle of supply voltage and a very low impedance throughout the rest of the half-cycle. The change from one impedance level to the other is assumed to be very abrupt. The phase angle at which the impedance changes is controlled by a direct current.

Two types of operation, representing limiting cases, will be discussed. They are identified by the terms *free even-harmonic currents* and *suppressed even-harmonic currents*. (The meanings of these terms will become apparent later.) The study of intermediate cases is quite complex and will not be included. Reference 10 covers such a study, but use of one or the other of the two extreme cases is sufficiently accurate for most practical applications.

The treatment will be limited to resistive loads as that type of load is the most usual for SR applications. Reference 11 discusses inductive d-c loads and Ref. 12 discusses inductive a-c loads.

3.2. Series-connected Saturable Reactor with Free Even-harmonic Currents. Basically, a SR circuit consists of the equivalent of two identical single-phase transformers. Figure 15 shows two transformers, SR_A and SR_B, interconnected in a manner to form a rudimentary series-connected SR circuit.

The two series-connected windings in series with the load are designated by the term *gate windings*,[1] and the two series-connected windings to which the d-c control power is applied are referred to as *control windings*.[1] Note from the dots which indicate polarity that the gate windings are connected in series *additive* and the control windings are connected in series *subtractive*. By reason of this connection, the *fundamental* power frequency voltage and all *odd* harmonics will not appear across the total of the two control windings; but any *even* harmonic induced in one control winding will be *additive* with respect to the corresponding *even* harmonic induced in the other control winding.

For this connection, SR_A and SR_B are normally so designed that each gate winding will accommodate one-half of the supply voltage without producing a peak flux density in the core which exceeds the knee of the magnetization curve, provided that no direct current flows in the control windings. With zero direct current in the control windings and assuming that SR_A is identical to SR_B, one-half of the supply voltage will appear across each gate winding and the net voltage induced in the control circuit will be zero. Under this condition, the two SR's operate as transformers over the entire portion of each half-cycle.

If a direct current is supplied to the control circuit, the flux in the core of each SR will reach the saturation region during a part of each cycle, SR_A during half-cycles of one polarity and SR_B during half-cycles of the opposite polarity. The interval during which a SR core is operating in the saturation region is designated as the *saturation interval*.[1] (The arrows associated with each SR indicate the conventional direction of current flow when that SR saturates.) The portion of each cycle during which a

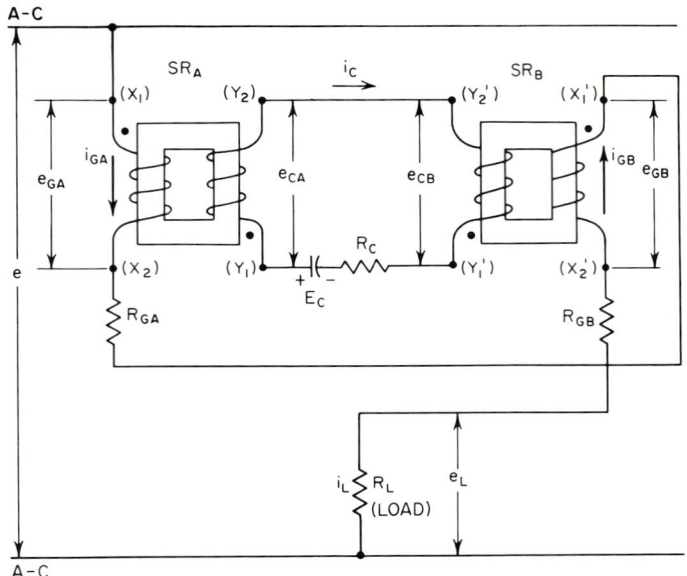

Fig. 15. Series-connected saturable-reactor circuit.

SR core is operating in the unsaturated interval and is absorbing all, or a part of, the a-c supply voltage is designated as the *exciting interval*.[1]

During the saturation interval, a SR presents a low impedance across the terminals of the gate winding and also across the terminals of the control winding on the same core. During the exciting interval, an SR operates precisely like an ordinary transformer. An exciting voltage applied to the terminals of any winding will then "see" an impedance consisting of the core-excitation requirements in parallel with any loads reflected from all other windings.

When a SR core operates in the saturation region during part of only one half-cycle, and there is a load or other impedance in series with the gate winding, even-harmonic voltages are induced in all windings on that core. In the series connection, one SR gate winding can be the series impedance for the other SR and large even-harmonic voltages will appear across the individual gate windings and control windings when direct current flows in the control windings. The amount of even-harmonic *current* which will flow in the control circuit of the series-connected SR circuit of Fig. 15 will depend upon the impedance of the control circuit to the harmonic voltages. If the control-circuit impedance between terminals (Y_1) and Y_1') is low with respect to the harmonic voltages induced in the two control coils (usually referred to as a *relatively low* control-circuit impedance), the harmonic currents can flow freely in this circuit and the SR circuit is identified by the term *free even-harmonic currents*. If the control-circuit impedance is high with respect to the induced harmonic voltages (a *relatively high* control-circuit impedance), the harmonic current flow is suppressed and the SR circuit is identified by the term *suppressed even-harmonic currents*. (The parallel-connected SR circuit will later be shown to be of the *free even-harmonic currents* type even when the control-circuit impedance is relatively high.)

Referring to Fig. 15, if R_C is relatively low and the source impedance of the d-c control voltage is low, then the circuit will be of the *free even-harmonic currents* type and the two SR's are coupled together very tightly. If one SR core is operating in the saturation region (saturation interval) it will place a low-impedance load across the control winding of the second SR, which is operating in the exciting interval, and this low impedance will be reflected by transformer action to the second SR. By this transformer action and the flow of even-harmonic current through the control circuit the saturation interval of one SR causes the impedance of both the series-connected gate windings to fall to low values and allow current to flow to the load. During those portions of the exciting intervals that both SR cores are operating in the high-permeability region of the dynamic hysteresis loop, the only current flowing through the gate circuit to the load is the small current required by the excitation of the SR cores. The current flowing in the control circuit during exciting intervals is substantially zero.

With the SR's saturating on alternate half-cycles, the same amount of current will be admitted to the load each half-cycle and there will be no even-harmonic current flow in the load circuit.

If the exciting current is so low as to be negligible, and if N_G are the turns in each gate winding and N_C the turns in each control winding, the application of the law of equal ampere-turns for transformers provides the following expressions for Fig. 15.

For the saturation interval of SR_A (first saturation interval)

$$i_{GB}N_G + i_C N_C = 0 \tag{7}$$

For the saturation interval of SR_B (second saturation interval)

$$i_{GA}N_G + i_C N_C = 0 \tag{8}$$

Inspection of Fig. 15 shows that during the first saturation interval $i_{GB} = -i_L$; hence

$$i_C N_C = i_L N_G \tag{9}$$

and during the second saturation interval, when i_L is negative with respect to the first saturation interval

$$i_C N_C = i_L N_G \tag{10}$$

This shows that the instantaneous ampere-turns in the control windings are equal to the instantaneous ampere-turns in the gate windings for both saturation intervals. Figure 16 has been drawn to illustrate these relationships.

If I_L denotes the average value of the load current as measured by a d'Arsonval-type instrument after the load current has been rectified, and I_C is the average value of the control current, it can be shown by integration of Eq. (10),

$$I_C N_C = I_L N_G \tag{11}$$

This is the law of equal ampere-turns for the series-connected SR with resistive load. It is also valid, as will be shown, for the parallel-connected SR, after I_L has been replaced by $I_L/2$ to account for the division of the load current between the two parallel-connected gate windings.

3.3. Parallel-connected Saturable Reactor with Free Even-harmonic Currents. Figure 17 shows the circuit diagram for the parallel-connected SR. Each

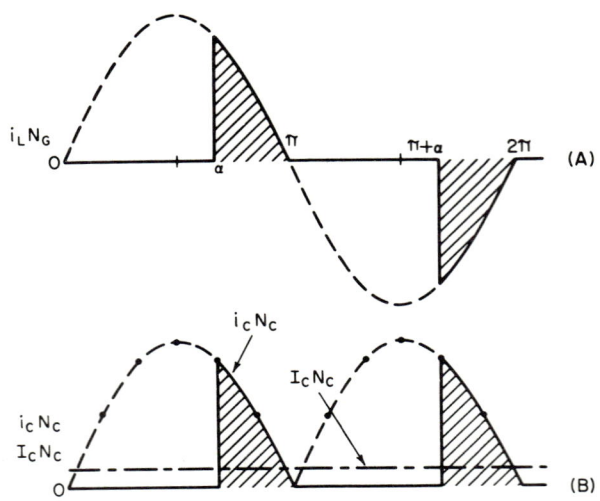

FIG. 16. Currents in series-connected saturable-reactor windings for free even-harmonic current condition. (A) Gate current. (B) Control-circuit current.

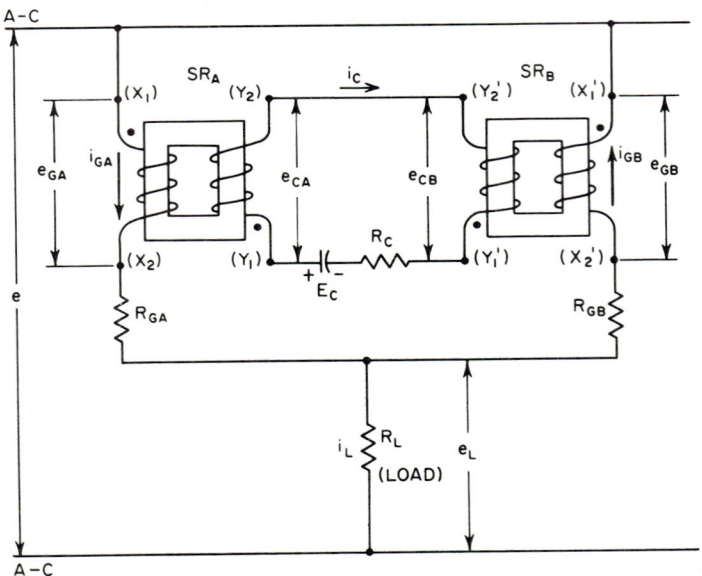

FIG. 17. Parallel-connected saturable-reactor circuit.

gate winding is connected directly between the a-c supply and the load resistance, thus providing two *parallel* paths through the SR circuit.

With the two gate windings connected in parallel, there is a low-impedance path for the free flow of even-harmonic currents even though the impedance of the control circuit is relatively high. Thus the parallel-connected SR is always of the *free flow of even-harmonic currents* type. The cores will operate in the same manner as described in Sec. 3.2 and the waveshapes of the current to the *load* will therefore be the same as is shown by Fig. 16; the currents in the gate windings and in the control circuit will depend upon the relation of the total control-circuit resistance R_C' to the gate winding resistance R_G, where

$$R_C' = R_C(N_G/N_C)^2 \qquad (12)$$

If $R_C' \ll R_G$, when one SR is in the exciting interval it can act as a transformer to connect its gate-winding resistance effectively in parallel with the gate-winding resistance of the saturated core SR. The load current will then divide into two practically equal components so that

$$i_{GA} = i_{GB} = i_L/2 \qquad (13)$$

This illustrates the fact that, whereas in the series-connected SR, the full load current must of necessity flow through each gate winding, in the parallel-connected SR with $R_C' \ll R_G$, only one-half of the load current flows through each gate winding.

The expression corresponding to Eq. (10) is then

$$i_C N_C = i_L N_G/2 \qquad (14)$$

and it follows that over the proportional operating range, where the saturation-interval duration is less than 180°,

$$I_C N_C = I_L N_G/2 \qquad (15)$$

This is the relation which was anticipated at the close of Sec. 3.2.

If $R_C' \gg R_G$, during the saturation intervals the impedance of the parallel path provided by the transformer action of the unsaturated SR is high compared with the resistance of the saturated gate winding. Most of the load current may then be expected to flow alternately through each gate winding as shown by Fig. 18B and C. If we consider i_{GA}' and i_{GB}' only as flowing through the gate windings, each gate winding would have an average d-c component of current flowing in it and would require that there be a direct-voltage source to supply the resistance voltage drops in the windings. The alternative is to have a d-c component flowing around the gate circuit in the opposite direction, as shown by I_L', whose average value equals the average value of i_{GA}' and i_{GB}'. Neglecting the small core-excitation requirements, the net ampere-turns acting on a core must be zero during an exciting interval; then

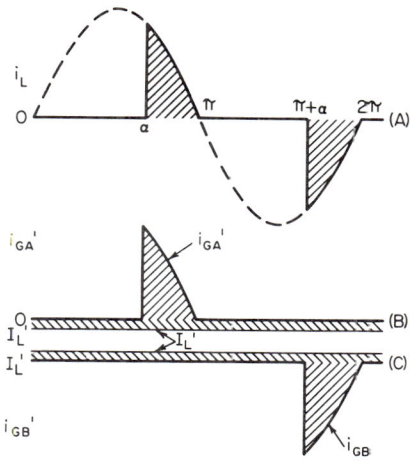

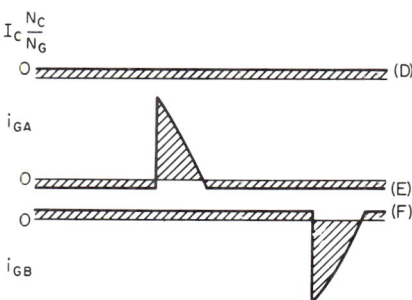

Fig. 18. Currents in parallel-connected saturable-reactor windings for free even-harmonic current condition. (A) Load current. (B) Components of load current in SR_A gate winding. (C) Components of load current in SR_B gate winding. (D) Control-circuit current. (E) Gate current of SR_A. (F) Gate current of SR_B.

$$I_L' N_G + I_C N_C = 0 \qquad (16)$$

It is obvious from Fig. 18 that $I_L' = -I_L/2$. From Eq. (16)

$$I_C N_C = I_L N_G/2 \qquad (17)$$

This equation is identical to (15); so the two extreme cases, i.e., $R_C' \ll R_G$ and $R_C' \gg R_G$, lead to the same relation between load ampere-turns and control ampere-turns.

Figure 19 shows oscillograms indicating the load current and the gate current of a parallel-connected SR circuit, constructed of two toroidally wound cores of Deltamax. The control-circuit resistance was such that $R_C' = 400 R_G$. Note the similarity to the waveshapes of Fig. 18A and E. An oscilloscope amplifier which would have preserved d-c components, had they been present, was used to secure the oscillograms. This clearly demonstrates the validity of the concept of a d-c component, flowing in the gate

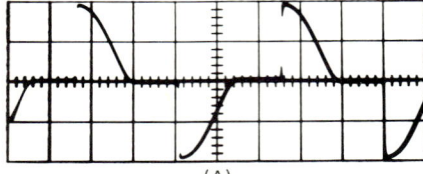
(A)

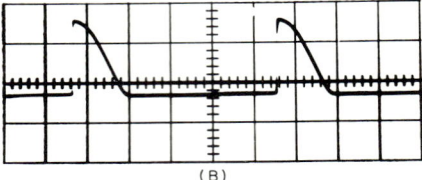
(B)

Fig. 19. Oscillograms of parallel-connected saturable-reactor circuit. (A) Load current. (B) Gate current of one SR.

windings, which is equal and opposite to the rectified-average value of half the load current.

3.4. Generalized Characteristics of Saturable-reactor Circuits with Free Even-harmonic Currents. The law of equal ampere-turns has been shown to hold for all SR's, operating in the proportional region, when the even-harmonic currents are free to flow in either the control circuit or paralleled gate windings. The proportional region for this type of operation extends from the current minimum point ($I_C = 0$) to the region where the load current is limited by resistance. With a given supply voltage E, if R_O is the total output resistance, i.e., the sum of the effective gate circuit resistance and the load resistance, the maximum possible load current (I_{Lm}) is then

$$I_{Lm} = E/R_O \qquad (18)$$

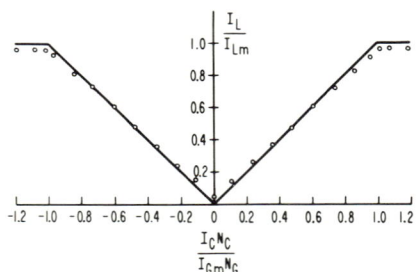

Fig. 20. Generalized control characteristic of saturable-reactor circuits with free even-harmonic currents.

If Eq. (11) is rewritten $I_L = I_C N_C/N_G$ and both sides are divided by I_{Lm}, the result is the dimensionless equation

$$I_L/I_{Lm} = I_C N_C/I_{Lm} N_G \qquad (19)$$

Equation (19) can be generalized to cover both series-connected and parallel-connected SR circuits, by replacing I_{Lm} in the right-hand side by I_{Gm} where, for the series-connected SR, $I_{Gm} = I_{Lm}$ and, for parallel-connected SR circuits, $I_{Gm} = I_{Lm}/2$. The result is

$$I_L/I_{Lm} = I_C N_C/I_{Gm} N_G \qquad (20)$$

Figure 20 shows the results of the foregoing theory plotted according to Eq. (20). The small circles indicate experimental results from a SR circuit constructed with toroidally wound cores of 0.002-in. Deltamax. The sections OA and OA' are called proportional regions [Eqs. (11), (15), and (17)], and sections AB and $A'B'$ are called resistance-limited regions [Eq. (18)]. In the region of the origin O, the departure of the experimental points from the theoretical characteristic is due to the exciting current. In the resistance-limited region the measured load current does not quite reach

I_{Lm} because of leakage inductance and the saturation inductance of the gate windings. The departures from the theoretical characteristics increase when the width of the dynamic hysteresis loop increases, when the knee of the saturation curve is more rounded, and when the slope in the saturation region increases. A decrease in the maximum effective permeability of the core, as by introduction of air gaps in the magnetic circuit, will increase the departure from the theoretical characteristics in the region of the origin O.

3.5. Series-connected Saturable Reactor with Suppressed Even-harmonic Currents. Most applications of this connection are for operation into loads having a low impedance relative to the effective impedance of the SR gate circuit. An initial analysis of this circuit is clearer if it is assumed that the load resistance is zero, that the gate windings have zero resistance and inductance, and that the hysteresis loop of the core is the ideal one shown by Fig. 3A. The SR's will be assumed to be so designed that, with zero control current, each gate winding will accommodate one-half of an alternating supply voltage E, without producing a peak flux density in the core which exceeds the knee of the magnetization curve.

Suppression of even-harmonic currents in the circuit shown in Fig. 15 can be accomplished either by making R_C very large, by introducing an inductance in series with R_C, or by substituting for E_C a current source I_C. If a constant current I_C is flowing in the control windings of the two SR's and a sinusoidal alternating voltage E is applied to the serially connected gate windings at the peak of the voltage wave, it can be shown[13] that there will be no transient. Thus, under steady-state operation, the current through the gate windings will be an alternating square wave which lags the applied voltage by 90°.

Figure 21 shows the portions of the ideal B-H loop over which each SR core operates when an unvarying direct current I_C flows in the control circuit. Figure 22 shows the waveshapes of (A) control current, (B) input voltage, (C) load current, and (D) voltage between control-circuit terminals (Y_1) and (Y_1'). Note that the voltage is applied to the circuit at time $\pi/2$. The letters

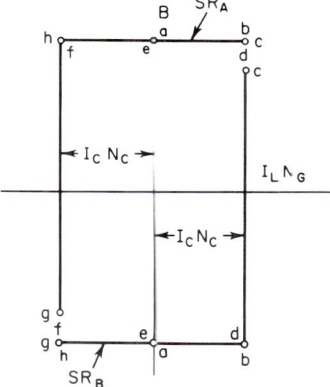

Fig. 21. Idealized operating hysteresis loops of series-connected saturable reactors for suppressed even-harmonic current condition.

(a) through (h) in Figs. 21 and 22 help to relate the operating points on the hysteresis loops to the correspondingly lettered portions of the current and voltage waves.

When the control current I_C is applied, but before the application of an a-c supply voltage, the core of SR_A is in the positive saturation region at point a and, by reason of the connections, the core of SR_B is in the negative saturation region at point a. When the alternating voltage E is applied at $\pi/2$ radians, the current i_L through the circuit instantly increases until the ampere-turns in the gate windings of SR_B equal the ampere-turns in its control winding (point b). SR_B then begins to operate in the excitation region toward point c and limits the gate winding current i_G to

$$i_G = i_C(N_C/N_G) = i_L \tag{21}$$

This rise in gate current also flows through N_{GA}, driving its core farther into saturation to point b. SR_A has zero impedance in this part of the idealized hysteresis loop; so all the applied alternating voltage appears across the gate winding of SR_B. The flux density in the core of SR_B then becomes less negative, reverses polarity, and reaches a positive value corresponding to point c at π radians. As shown by Fig. 22D, during this period a negative quarter-cycle of voltage e_{CB} is induced in N_C such that

$$e_{CB}(N_G/N_C) = e \tag{22}$$

8–24 AMPLIFYING DEVICES

SR_B is functioning as a transformer during the period $\pi/2$ to π and energy is delivered *to* the control circuit *from* the a-c power source.

At π radians the voltage of the a-c power source becomes negative. The core of SR_B is still in the excitation period at point c and can continue to act as a coupling between the a-c supply and the constant-current control circuit. The flux-density in SR_B during this period changes from point c to point d, reaching the latter at $3\pi/2$ radians. The control circuit delivers power to the a-c source during this period and maintains the gate current positive and of the same value as given by Eq. (21).

When SR_B again reaches the saturation region, it ceases to transfer energy from the control circuit to the gate circuit. With SR_B having low impedance, the negative

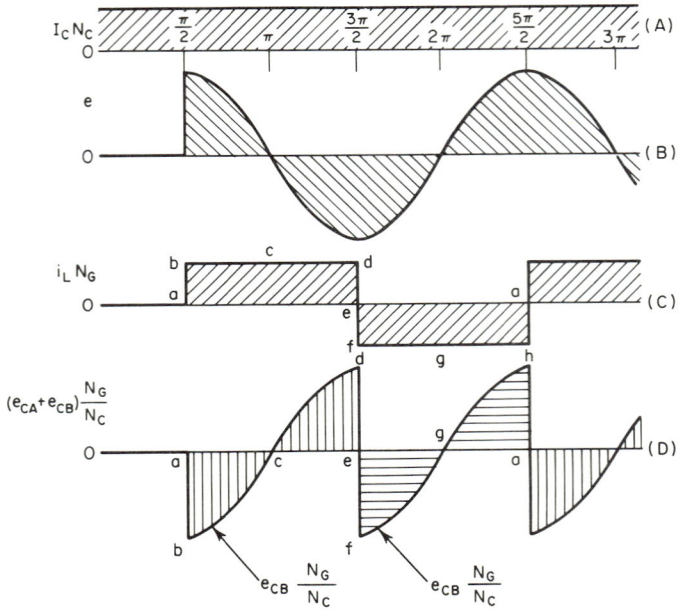

Fig. 22. Currents and voltages in series-connected saturable reactors for suppressed even-harmonic current condition. (A) Control-circuit current. (B) Applied input voltage. (C) Gate winding and load current. (D) Voltage across series-connected control coils.

polarity of supply voltage can take over and force i_L to collapse and to go negative until it becomes limited, by SR_A going out of saturation, to a current

$$-i_L = i_C(N_C/N_G) \tag{23}$$

From $3\pi/2$ to $5\pi/2$, the roles of the two SR's are reversed. As a result, SR_A transfers energy to the control circuit from $3\pi/2$ to 2π. From 2π to $5\pi/2$ it transfers the energy from the control circuit back to the a-c supply because the polarity of the a-c supply voltage is positive but the current through the circuit still flows in the negative direction. At $5\pi/2$, SR_A again reaches saturation and the next succeeding cycle of operation begins. Since the starting conditions for the second cycle are identical to those of the first cycle, under steady-state conditions the second and all following cycles will be the same as the first cycle.

If load resistance R_L is increased from the initially assumed value of zero, the current through the circuit remains rectangular, over a range of resistances, but the phase angle at which it reverses polarity advances. The voltage drop across the load reduces the voltage across the gate windings, thereby reducing the flux-density swing in the SR cores.

If R_L is a constant resistance and the control current is varied, there is a range over which the load current is rectangular, but the phase angle as well as the amplitude of current will vary.

When the voltage drop in the load plus gate winding resistance is low, relative to the a-c supply voltage, the current through the load circuit is rectangular and the law of equal ampere-turns applies to the suppressed even-harmonic currents operation just as it did for the free even-harmonic currents cases. The circuit is then operating in the proportional mode. The limit of this mode occurs when the load voltage amplitude and its phase angle is such that it equals the instantaneous supply voltage at the instant the load voltage reverses polarity. The load current ceases to be proportional to control current beyond this point and begins to be resistance-limited.[13]

Figure 23 shows oscillograms which verify the analysis when operating into a very-low-resistance load.

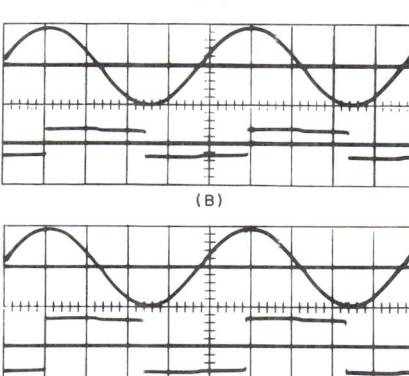

FIG. 23. Oscillograms of load current into zero load resistance for series-connected saturable reactors with suppressed even-harmonic currents. (A) Low output current. (B) Intermediate output current. (C) High output current.

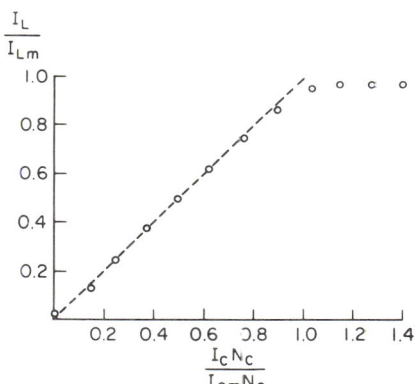

FIG. 24. Generalized control characteristic of saturable-reactor circuits for suppressed even-harmonic condition.

Figure 24 is a normalized plot of the control characteristic when operating into a fixed load resistance. Figure 25 is an oscillogram of load voltage waveshapes, for the same circuit parameters, at three levels of control current.

3.6. Core and Coil Configurations. There are a number of different core configurations suitable for saturable reactors. Some core configurations dictate the coil configuration. Others permit a variety of arrangements of the gate windings in relation to the control windings. Depending upon the application, some core and coil configurations are superior to others.

Perhaps the simplest form of SR, other than the use of two single-phase transformers, is illustrated by Fig. 26. Usually, the two coils on the outer legs are made the gate windings with the single center coil the control winding. The gate windings are connected in series or in parallel as desired, with their interconnections always made in a manner which will cause the path of the flux in the core arising from the a-c excitation to be only through the outside legs and the yokes, as shown by the arrows. As a result, no fundamental and odd-harmonic voltages will be induced in the control winding wound around the center leg. An equivalent of the *series-connected* SR can be obtained by using the coil on the center leg as the equivalent of two gate windings connected in series and using the serially connected outer coils as the control winding.

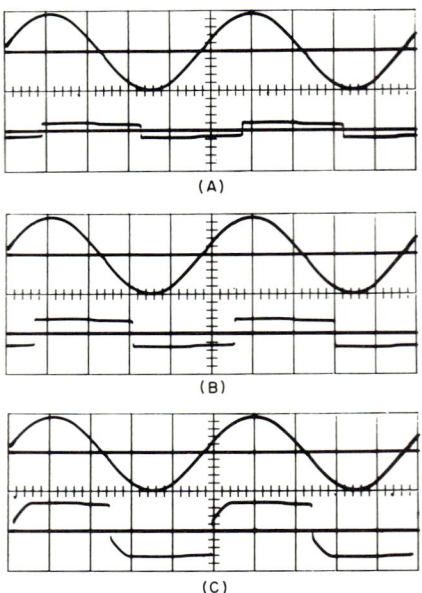

Fig. 25. Oscillograms of load currents into finite load resistance for series-connected saturable reactors with suppressed even-harmonic currents. (A) Low output current. (B) Intermediate output current. (C) High output current.

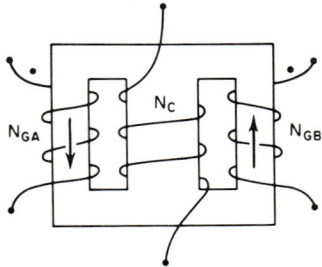

Fig. 26. Saturable-reactor core and coil configuration with three-legged core.

The core-and-coil configuration of Fig. 26 has two characteristics which may be undesirable. First, the leakage reactance between the gate windings and the control windings is inherently high and will cause a substantial departure from the ideal control characteristic when operating at or near the upper end of the proportional range; and second, a hysteresis effect in the core of the control-circuit leg will cause the control characteristic to be double-valued.[14]

A four-legged construction which is widely used in SR simple magnetic amplifiers is shown in Fig. 27. The gate windings are interconnected so that the fundamental and odd-harmonic components of a-c flux flow only in the two center legs and across the center yokes. Only the d-c components of flux, due to direct current in the control windings, flow through the outer legs. The gate windings are wound over the full length of the two center legs and the control winding is wound over them for their entire length. This coil configuration provides a low leakage reactance between the gate windings and the control winding. The a-c core loss and exciting current are low relative to other core configurations because of the decreased core volume and mean length of magnetic circuit which is subject to a-c magnetization.

When a SR core consists of two ring-shaped cores, the core-and-coil configuration often consists of two gate windings, one wound around each core in toroidal fashion, and, after they are placed side by side coaxially, a control winding toroidally wound over both gate windings simultaneously. This configuration is widely used in

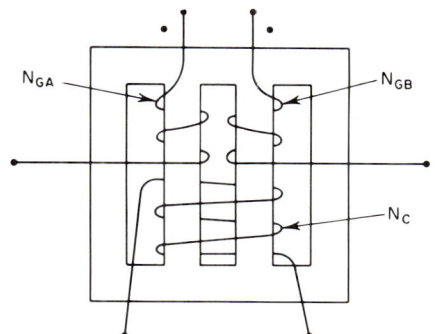

Fig. 27 Saturable-reactor core and coil configuration with four-legged core.

Amplistat circuits. The magnetic properties of many core materials, when used in the form of uncut strip-wound cores, are usually superior to those attained by the same material when used in other core configurations.

3.7. Summary. The SR has been shown to be a device which under certain conditions functions as a transformer and under other conditions functions as a low impedance, switching from one condition to the other for portions of each cycle. For the current drawn from the a-c supply to be free of even-harmonic components, two magnetic circuits are required. The *law of equal ampere-turns* has been shown to hold for various connections and for conditions of *free even-harmonic currents* and *suppressed even-harmonic currents*. The *law of equal ampere-turns* can be shown to hold also for all intermediate cases.

REFERENCES

1. "The International Dictionary of Physics and Electronics," D. Van Nostrand Company, Inc., Princeton, N.J., 1956.
2. H. F. Storm, "Magnetic Amplifiers," chap. 5, John Wiley & Sons, Inc., New York, 1955.
3. P. P. Cioffi, A Recording Fluxmeter of High Accuracy and Sensitivity, *Rev. Sci. Instr.*, vol. 21, no. 7, pp. 624–628, 1950.
4. Ref. 2, p. 25.
5. H. W. Lord, Dynamic Hysteresis Loop Measuring Equipment, *Trans. AIEE*, vol. 71, pt. I, pp. 269–271, 1952.
6. F. Koppelman, The New AEG Vectometer, *AEG Progr.*, no. 3, pp. 77–81, July, 1953.
7. E. Kittl, An Accurate Electronic Tracer for Dynamic Characteristics of Magnetic Materials, *Trans. AIEE*, vol. 74, pt. I, pp. 407–418, 1955.
8. Standards for the Presentation of Data on Magnetic Amplifier Core Materials, AIEE Committee Report, *Trans. AIEE*, vol. 74, pt. I, pp. 598–599, 1955.

9. Harold W. Lord, Dynamic Hysteresis Loops of Several Core Materials Employed in Magnetic Amplifiers, *Trans. AIEE*, vol. 72, pt. I, pp. 85–88, 1953.
10. H. M. Gale and P. D. Atkinson, A Theoretical and Experimental Study of Series-connected Magnetic Amplifier, *Proc. Inst. Elec. Engrs. (London)*, vol. 96, pt. I, pp. 99–124, May, 1949.
11. Ref. 2, chaps. 11 and 12.
12. T. G. Wilson, Series-connected Magnetic Amplifier with Inductive Loading, *Trans. AIEE*, vol. 71, pt. I, pp. 101–110, 1952.
13. Ref. 2, chap. 8.
14. Ref. 2, p. 110.

BIBLIOGRAPHY ON AMPLIDYNES

Alexanderson, E. F. W., *et al.*: The Amplidyne Generator, A Dynamo Electric Amplifier for Power Control, *Trans. AIEE*, vol. 59, pp. 937–939, 1940.

Bower, J. L.: Fundamentals of the Amplidyne Generator, *Trans. AIEE*, vol. 69, pp. 139–145, 1950.

Pestarini, J. M.: "Metadyne Statistics," John Wiley & Sons, Inc., New York, 1952.

Chapter 9

ELECTRON TUBES

LLOYD G. MUMFORD,[*] **PHILIP E. HATFIELD,**[*]
THEODORE G. MIHRAN,[†] **and DARYL W. HAWKINS**[‡§]

CONTENTS

1. Introduction	9–2
2. Considerations in Choosing a Tube Type	9–2
2.1. General	9–2
2.2. Electron Currents in Tubes	9–2
2.3. Performance Limitations	9–6
2.4. Manufacturer's Data	9–13
3. Electron-tube Life	9–42
3.1. General	9–42
3.2. Heater Voltage	9–43
3.3. Power Dissipation	9–44
3.4. Envelope Temperature	9–46
3.5. Cathode Interface	9–46
3.6. Heater-Cathode Voltage	9–47
3.7. Plate Voltage	9–47
4. Receiving Tubes	9–47
4.1. Scope and Definition	9–47
4.2. The Electron Tube as a Network Element	9–48
4.3. Receiving-tube Construction	9–49
4.4. D-C and VLF Amplifier Tubes	9–58
4.5. Audio-frequency Amplifier Tubes	9–58
4.6. R-F and I-F Amplifier Tubes	9–61
4.7. UHF and Gridded Microwave Tubes	9–80
5. Microwave Tubes	9–80
5.1. Introduction to Transit-time Tubes	9–80
5.2. Klystron Amplifiers	9–81
5.3. Traveling-wave Tubes	9–88
5.4. Crossed-field Amplifiers (M-type Amplifiers)	9–91
5.5. The Adler Tube (Quadrupole Amplifier)	9–96

[*] General Electric Co., Tube Dept., Owensboro, Ky.
[†] General Electric Co., Research and Development Center, Schenectady, N.Y.
[‡] General Electric Co., Tube Dept., Schenectady, N.Y.
[§] Sections 2, 3, and 4 were written by L. G. Mumford and P. E. Hatfield, Section 5 by T. G. Mihran, and Section 6 by D. W. Hawkins.

9-2 AMPLIFYING DEVICES

  6. Transmitting Tubes................................. 9-97
    6.1. Introduction................................... 9-97
    6.2. Construction................................... 9-98
    6.3. Ratings.. 9-100
    6.4. Pulse Ratings.................................. 9-101
    6.5. Life... 9-101
    6.6. Cooling.. 9-102

1. INTRODUCTION

The electron tube can undoubtedly be given credit, more than any one other device, for the tremendous advances which have been made in amplifiers in this century. In spite of intense competition from transistors and other solid-state devices, the electron tube is still very much in the picture and, in fact, represents the only means for accomplishing some forms of amplification or operating under certain extremes of environment, e.g., extreme heat or nuclear irradiation.

Electron tubes are used over the complete gamut of powers and frequencies encompassed by this book, with the possible exception of light wavelengths. So-called receiving tubes operate with inputs in picoamperes and down to direct current. At the other extreme, power tubes operate at many kilowatts of power, and various forms of special configurations operate up into the gigacycle range of frequencies. Many diverse phenomena find application in electron tubes, including electron-beam shaping and deflection, secondary-emission effects, etc.

It would obviously be impossible to describe all possible forms electron tubes can and have taken. An attempt has been made in the following sections, however, to describe the basic types and to catalog their properties, so that the user can use them intelligently.

2. CONSIDERATIONS IN CHOOSING A TUBE TYPE

2.1. General. Over five thousand electron-tube types have been developed and marketed since the first triode "audion" was produced. There are currently over two thousand tube types available, the majority of which are suitable for amplifier service of one kind or another. Thus the amplifier designer must choose the types for his application from a vast list of tube types of a number of manufacturers, taking into account suitability, cost, and present and future availability. Assistance may be obtained from charts of tubes classified by potential applications and lists of tubes approaching obsolescence, but the final choice must be made by the designer.

The designer's choice is made more difficult by a large number of hidden factors which affect the success of tube applications. For example, one of the most insidious problems in tube application is the tolerance of short-term abuse by tubes, which often traps the user into misapplication and seriously shortened tube life.

To apply tubes intelligently, it is necessary to know why the tube manufacturer imposes various limitations on the tubes, the risks involved in approaching these limiting values, and the precautions to be taken to ensure satisfactory performance and tube life.

The purpose of this section is to describe the factors that affect tube application, give some of the fundamental performance limitations, and assist in the interpretation of manufacturer's data.

2.2. Electron Currents in Tubes. *Electron emission* by a hot cathode is basic to all the tube types covered in this chapter. The cathode may be directly heated by passing current through it (filamentary cathode) or it may be indirectly heated by heat conduction or radiation from an insulated heater (indirectly heated cathode) or by operating in an environment with a high ambient temperature (TIMM[1]).

[1] Thermionic integrated micro modules; see Sec. 4.

ELECTRON TUBES 9-3

Filamentary cathodes may be pure tungsten, tungsten with thorium dispersed through it (thoriated tungsten), or nickel coated with alkaline-earth oxides. Indirectly heated cathodes are usually nickel, coated with alkaline-earth oxides, although some oxide cathodes have the alkaline-earth oxides contained within a heated chamber with a porous tungsten plug which allows active material to diffuse to its surface for a continuously replenished emitting layer (dispenser cathode), while others use metal mesh embedded in the oxide coating or metallic particles dispersed through the coating to increase the conductivity of the cathode (matrix cathode).

The tube designer selects the cathode to be used in a particular tube type, but the user often has the choice of tubes with several types of cathodes for a given application and should be aware of the advantages and disadvantages of each.

Filamentary cathodes of tungsten and thoriated tungsten are used in relatively large power-amplifier tubes, where high cathode currents and high anode voltages are required and cathode bombardment would be likely to destroy a coated cathode. Both types of tungsten cathodes operate at high temperatures: the pure tungsten at 2300°K and the thoriated tungsten from 1800 to 2200°K.

Oxide-coated filamentary cathodes, which operate in the temperature range of approximately 1000 to 1100°K, are used in some low-power transmitting tubes, power rectifiers, and some receiving tubes, especially battery-operated ones. These cathodes reach operating temperature rapidly and consume less power than equivalent indirectly heated cathodes. Where standby power consumption is important, as in some mobile equipment, filamentary-cathode tubes offer almost instantaneous warmup. A disadvantage is the lack of isolation of the emitter from the heater supply. This restricts circuit applications to those where the cathode may operate at heater potential and makes this type of tube unsuitable for low-level amplification if alternating current is to be used as the heater supply.

Indirectly heated cathodes, which operate in the temperature range of approximately 975 to 1025°K, are used in the majority of receiving tubes and in some low- to medium-power transmitting tubes. The disadvantage of a relatively long warmup time, compared with filamentary cathodes, is offset by increased flexibility in application because of the electrically isolated emitter. Tubes employing this type of cathode are suited for both low-level and power amplification.

The cathode current that results from cathode emission may be continuous, it may vary about a steady-state level because of an input signal, or it may occur in bursts with quiescent periods between the bursts (pulsed emission). The total potential cathode current is a function of the type of cathode, the area of the active cathode surface, and the cathode temperature. The allowable steady-state cathode current may be determined from the foregoing criteria by the tube manufacturer and is often given in the manufacturer's data (see Sec. 2.4). When cathode current is pulsed, much larger current amplitudes are allowable. However, there are factors that make the determination of maximum allowable values difficult. Since higher than usual plate voltages and positive grid voltages are used to obtain high pulse currents from the cathode, two factors which can affect the life of the tube must be considered. High plate voltages cause the plate and the tube envelope walls to be bombarded with higher-energy electrons than when lower plate voltages are used. This tends to dislodge absorbed gases, which in turn may "poison" the cathode and reduce its emission capabilities. High positive grid voltages may also cause gas release and, in addition, may cause particles of cathode coating to be torn from a coated cathode by electrostatic attraction, because of the proximity of the grid and cathode. Therefore, whenever possible, tubes for pulsed amplifier service should be selected from among those specifically rated for pulse operation by the tube manufacturer.

Although the cathode is the primary source of electrons in a tube, other tube elements may emit electrons as either primary or secondary emission. Primary emission occurs when a tube element becomes hot enough to emit electrons, while secondary emission results from the collision of high-velocity electrons with a tube element, or even with the tube envelope. Primary emission is most likely to occur if cathode-coating material is deposited on elements other than the cathode and these elements are heated by excessive power dissipation. The control grid may receive enough heat directly from the cathode to emit, even though there is no control-grid dissipation. Emission by elements other than the cathode is usually detrimental to the operation of a tube, and methods for decreasing it are incorporated in many tubes. However, the user must observe certain design precautions (see Electrode Currents, below) to minimize detrimental emission. Secondary emission is used to advantage in secondary-emission pentodes.

Electrode Currents. Plate current flow is fundamental to all electron-tube operation; the magnitude of the current must be controlled to operate the tube at a favorable point on its characteristic curve without exceeding the cathode-current capability or the plate dissipation rating. This is done by selecting values of plate voltage, screen voltage (for a tube having this element), and control-grid voltage, within the maximum ratings of the tube, that allow the desired performance, plus restricting the signal to the control grid in large-signal amplifiers where excessive drive may result in too high values of plate current.

Screen current must be carefully controlled as most screens have relatively low dissipation ratings, and failure to check screen current in all modes of operation of a circuit may result in excessive screen dissipation and drastically shortened tube life. This is especially true in audio power amplifiers where high input-signal voltages may drive the screen current up to values far in excess of the zero-signal value, and in r-f power amplifiers where removal of the load from the plate circuit of the tube may cause the screen current to soar.

Grid current is probably the least understood of the internal phenomena in electron tubes. In some classes of service, such as class B and class C amplifiers, grid current is essential to the operation of the circuit, and the principal precaution that the circuit designer must take is to limit the grid current to values within the grid-current rating of the tube. However, in the majority of small-signal amplifiers grid current is not essential and is usually detrimental to circuit operation and tube life; one exception to this is the use of so-called "contact potential" bias on some low-level amplifiers.

Grid currents may be classified as "positive" or "negative," with the positive being those which are used to produce negative bias and the negative being those which are detrimental. The negative currents include gas current, grid emission, and leakage current. Gas current flows because positive gas ions are attracted to the negative-biased grid and are neutralized by electrons flowing from the negative pole of the external circuit to the grid. If there is a resistive component in the external grid circuit, a voltage will be developed across it which is of a polarity to cause the grid to be made more positive with respect to the cathode and thus cause a reduction in negative bias. In the worst case the condition will become cumulative; that is, the reduction of bias will cause more current to flow, more heat dissipation, and more evolution of gas, and the tube will "run away" to ultimate destruction. It is for this reason that manufacturers recommend maximum values of grid-circuit resistance (see Sec. 2.4). Although it is commonly recognized that vacuum tubes do not have perfect vacuums, it is not generally realized that gas currents in tubes usually result from gas being released by the metal parts or the envelope because of heat or electrolysis. Thus a given tube may show gas current under high dissipations, envelope temperatures, or anode voltages but none at lower temperatures and voltages.

Grid emission current results from emissive material having migrated from the cathode to the grid because of excessive cathode temperature, although some deposition during initial tube processing and migration during life are inevitable. Because of the proximity of the control grid to the cathode (less than 0.001 in. in some tubes and commonly 0.002), conditions are favorable for the grid to become hot enough to emit thermal electrons. The resulting current flows in the same direction in the external grid circuit as gas current and thus acts to make the grid more positive. Emission from the screen grid can likewise occur if the screen is allowed to overheat; this emission will cause the screen to become more positive if the screen is being supplied through a screen-dropping resistor. Emission from either the control grid or the screen can cause runaway and ultimate destruction of the tube.

Grid leakage currents flow across the structure supporting the elements of the tube. Mica and ceramics are used in the supporting structures of tubes, and both these materials are excellent insulators; however, sublimation of metallic elements and their subsequent condensation on the insulators can produce leakage paths. Leakage can occur between any of the elements that contact the insulators, but the most serious conditions result when there is leakage to the grid from other elements. Cathode-to-grid leakage reduces the input resistance of the tube to values far lower than the normal value, with accompanying detrimental effects on circuit operation. By far the worst detriments are leakage paths between the control grid and the plate or screen. These paths cause current flow in the external grid circuit in such a direction as to cause the control grid to become more positive and thus produce the same effects as gas current and emission current. Leakage current is also more erratic and is therefore a source of noise in amplifiers operating at low signal levels.

Contact-potential grid current is the most complex and least understood. Although it often exists in combination with the other grid currents, it is treated here as it would occur in an ideal case in the absence of the other currents. The term "contact potential" has three meanings, depending on the group using the term. To the physicist it is the difference in potential established between two metals with different work functions in contact with each other; to the tube manufacturer it is the negative potential required to bias the grid of a tube to the point where 0.1 μa or less of grid current flows; and to the tube user it is a biasing mechanism, inherent in the tube when a large resistance is used in the grid circuit. The last meaning is the one of most interest here; it will be termed "contact-potential bias" and defined as that potential which is established across the external resistive component of the grid circuit of an electron tube because of the flow of thermally emitted electrons from the cathode being collected by the grid.

Thermal electrons from the cathode are emitted with an initial velocity related to the temperature of the cathode and its work function. Some of these electrons have enough energy to reach the grid, even though the grid is negative, and thus develop a retarding field. This retarding field is dependent upon the work function of any emissive material that may be deposited on the grid and the temperature of the grid. Thus the potential that exists as a bias between the grid and cathode resembles the physicist's contact potential in that it does depend on the difference in work functions of the cathode and grid. Actually, the tube manufacturer attempts to deposit a judicious amount of emissive material on the grid during processing to establish a work function which will reduce the contact-potential bias to a reasonable value. In fact, continued deposition of emissive material and consequent change in contact-potential bias accounts for some changes in characteristics during the life of a tube that cannot be accounted for by changes in cathode emission capability. The effect of temperature on contact-potential bias may be seen in the plot of contact-potential current vs. grid voltage at several values of heater voltage, shown in Fig. 1.

In using contact-potential bias, several disadvantages should be noted. When a tube is operated with contact-potential bias, its input resistance is no longer in the order of thousands of megohms but is determined by the slope of the tangent to the grid-current curve at the operating bias. (Values of 100,000 to 240,000 ohms are shown in Fig. 1.) This will result in the tube loading its input circuit more than might be anticipated. If the tube is biased, by other means, outside the contact-potential bias region but is driven into it on positive peaks, the signal source will also be unduly loaded, resulting in distortion in audio amplifiers and lowered Q when tuned circuits are involved. It is particularly important to remember that when high-mu triodes and very sharp-cutoff pentodes with limited signal-handling capabilities are used, contact potential restricts that capability even more. A common mistake is to operate high-mu triodes at quite low plate voltages; when this is done, the contact-potential bias may be sufficient to cut the plate current almost to zero.

When contact-potential bias is used, and the most common use is in the low-level stages of audio amplifiers, a high value of grid circuit resistance (commonly 10 megohms) is used for the bias to be developed across. It is important to remember that the input resistance may be considerably lower than the value of the grid resistor and that signal-handling capability is limited to input signals no larger than approximately 300 mv. Finally, although tube manufacturers attempt to control contact potential in those types commonly used in this mode, there may be considerable variation during life, and this should be taken into account in initial circuit design. The use of contact-potential bias with tubes not recommended for this type of service by the tube manufacturer is inadvisable.

Heater-cathode leakage is a flow of current between the heater and cathode. This current flow is not the result alone of a simple resistive path with linear characteristics but is rather a semiconductor phenomenon occurring between the heater and the cathode through the aluminum oxide coating on the heater. Because the heater and its coating commonly operate at 1200°C or higher, any small amount of impurity in the aluminum oxide may lower the Fermi level and thus allow conduction. Furthermore, ion migration of impurity elements in the nickel cathode sleeve, or the heater wire, during the life of the tube can supply the impurity elements needed for semiconductor action. Since ion migration is enhanced by high temperature and high d-c voltage between heater and cathode, tubes should be operated at the value of heater voltage (or current) recommended by the tube manufacturer and with the lowest possible d-c voltage between heater and cathode, in order to ensure minimum heater-cathode leakage during the life of the tube.

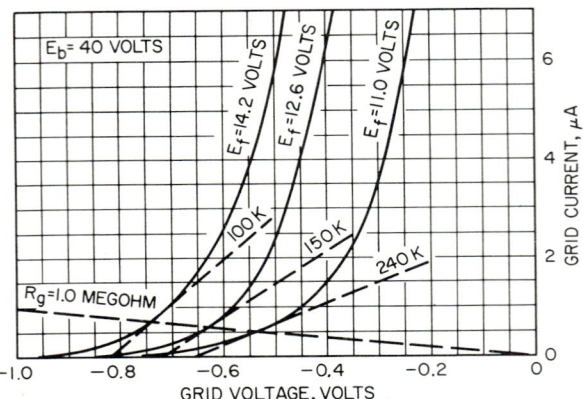

Fig. 1. Effect of heater voltage on contact-potential grid current in a high-mu triode.

Heater-cathode leakage is often a source of hum in a low-level (10 mv or lower input signal) amplifier stage operated with alternating current on the tube heater, since the leakage current through any cathode-circuit resistance appears as a signal on the stage. Because heater-cathode leakage saturates at a relatively low voltage, a "bias" voltage is sometimes applied between the heater and cathode to "swamp out" the effect. This is not recommended because its effectiveness varies, and it can only be regarded as a last resort where d-c operation of the heater would be preferable.

Even though hum is not a problem, heater-cathode leakage effectively limits the value of impedance between these elements to an extent not readily apparent from the schematic of the tube. Many common receiving tubes may have heater-cathode leakage as high as 20 μa or more with 10 volts applied. For circuits that are especially critical of leakage, tube types having special controls and having leakage specifications published by the manufacturer should be used.

2.3. Performance Limitations. Electron tubes have performance limitations that the user should be aware of. Some of these are probably fundamental in the operation of hot-cathode electron devices and show little promise of being amenable to design improvement at present. With these, there is little the circuit designer can do, other than to avoid misapplication or to be aware that results obtained may be less than optimum. Other limitations may be the result of improper circuit design, poor mechanical layout, or improper tube choice, and these limitations can definitely be overcome.

Noise is one of the common limiters of performance and is here defined as an undesired output from an amplifier, excluding distortion products and oscillation, which may be generated within the amplifier or inadvertently coupled into its input. This includes hum and microphonics, which are not normally classified as "noise."

Hum caused by insufficient filtering of plate, screen, and control grid supplies is not treated here, as the remedies for this are obvious. At least three other sources of hum should be recognized. They are modulation of the electron stream in the tube by an external magnetic field, heater-to-grid leakage, and heater-cathode leakage (see above for heater-cathode leakage). These three sources are of importance when the input signal to the tube in question is 10 mv or less; higher-level stages are seldom influenced by them.

The magnetic field around an average power transformer has an intensity of 5 to 10 gauss at a distance of 2 in. from the core and increases inversely as the square of the distance from the transformer. The effect of this alternating field varies with the type of tube and the associated circuit. As might be expected, metal tubes show the least effect and unshielded glass tubes the most. Pentodes are more susceptible than triodes and also show increased hum-to-signal output with increased values of plate load resistance. Rotation of the tube socket for minimum hum may be helpful if the transformer is so oriented that the flux vector is not parallel to the tube axis. (This is more effective with vertical-mounting transformers than with the half-shell type.) The safest method is to mount the low-level stages as far as possible from the power transformer, certainly not less than 2 in.

Heater-to-grid leakage will result in a signal on the grid when alternating current is used on the heater. This leakage may occur internally through the tube structure, or externally through the tube socket. The effect of leakage through the socket is dependent on both the impedance of the leakage path and the grid-circuit impedance since these two form a voltage divider. Thus low grid-circuit impedance and high leakage-path impedance produce minimum hum at the grid. The designer should use high-quality tube sockets and keep the grid-circuit impedance low. Double-ended tubes, such as the 6J7, are excellent in this respect, although somewhat out of fashion; the grid terminal is well isolated from the heater terminals, and in addition, its metal envelope provides magnetic as well as electrostatic shielding.

Microphonics occur when mechanical vibration of the internal elements of the tube cause a corresponding modulation of the electron current. Usually, movement of the grid, the grid laterals, or the cathode cause the most trouble, although low-mu triodes may show some effect because of movement of the plate. All the metal parts in a tube will exhibit resonances at various audio frequencies with the grid assembly, cathode, and plate resonating in the range of approximately 100 to 1,000 cycles and the individual grid wires resonating in the range from a few thousand up to 20,000 cycles, when very short grid wires under tension are employed. These resonances can be particularly troublesome when low-level amplifier tubes are mounted in such a manner that acoustic coupling can occur from a source of high-intensity sound such as a loudspeaker.

To avoid trouble with microphonics, tubes recommended for low-level amplifier service with low microphonic output should be used, and in addition, shock mounting of the tube, and its location remote from a loudspeaker may be helpful.

Noise (the word is used here in its more restrictive meaning) generated in an amplifier may be divided into two general classes: thermal noise and tube noise. Tube noise, in turn, may be divided into shot noise, induced grid noise, and partition noise. Since tubes are being considered, nothing is given here on thermal noise, although it may determine ultimate amplifier sensitivity.

Shot noise arises from the random nature of electron emission by hot cathodes. Each electron that leaves the cathode and arrives at the plate of a tube produces a pulse of current when it strikes the plate. The total of these pulses constitutes the plate current. The random arrival of electrons causes variations in plate current, akin to those resulting from a signal modulating the electron stream, and thus produces a modulated output from the tube which, when detected, appears as noise.

Induced grid noise is produced by the passage of electrons through the grid of a tube in their journey from cathode to plate. As an electron approaches the grid, the induced pulse of grid current has one polarity, and as the electron recedes from the grid and approaches the plate, the pulse polarity is opposite. Since the flow of electrons from cathode to plate is random, and these electrons must pass the grid on their way to the plate, random variations in grid current occur. These variations in grid current appear, in amplified form, as noise in the plate circuit of the tube. Random variation in the number of electrons passing the grid is not the only source of induced grid noise: random variation in electron velocities, caused by the variation in the path followed by the individual electrons, also produces noise.

In multigrid tubes an electron leaving the cathode and moving through the control grid and toward the plate may reach the plate or may strike one of the additional grids and not arrive at the plate at all. Therefore, a random distribution of electrons between plate and other positive-potential elements will occur and produce random variations in plate current. This partition noise is in addition to shot noise and induced grid noise, both of which are also present in multigrid tubes.

The most common measure of noise is the noise figure or noise factor, which assesses effective sensitivity of a vacuum-tube amplifier to weak signals in terms of the signal-to-noise power ratio. This noise figure may be measured for a practical amplifier, or it may be calculated from certain measurements that can be made on the tubes used in the amplifier.

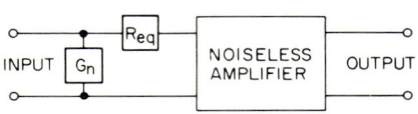

FIG. 2. Equivalent circuit for noise parameters.

Measured noise figures are obtained by determining the behavior of an amplifier when a known or controllable source of noise power is connected to the amplifier input. This noise source may be a temperature-limited diode or a gas-discharge diode. If a temperature-limited diode is used, the diode temperature is set to a value, by adjusting the filament voltage, which causes a noise output from the amplifier double that obtained without the noise source. The diode plate current is read and the noise figure calculated with the following relationship:

$$NF = 20 I_d R_s \tag{1}$$

where NF = noise figure (power ratio)
I_d = diode plate current, amp
R_s = source resistance, ohms

If a gas-discharge diode is used, the noise output is constant and cannot be varied; therefore, the noise output of the amplifier is read with and without the noise source turned on, and the noise figure calculated with the following relationship:

$$NF = T/(Y - 1) \tag{2}$$

where NF = noise figure (power ratio)
T = excess noise temperature of gas-discharge diode expressed as a power ratio
Y = ratio of noise power output with lamp on to that with lamp off

These noise figures are normally converted to decibels to render them more useful.

Calculation of the potential noise performance of tubes involves the concept of equivalent noise parameters.[2] Briefly, this concept considers the circuit of a vacuum-tube amplifier as a four-pole network having a noiseless amplifier with parameters describing the total noise in the input circuit (see Fig. 2). These parameters are the equivalent noise resistance R_{eq} and the equivalent noise conductance G_n. The equivalent noise resistance describes the effect on plate-circuit noise of having a resistance in the grid circuit whose random electron currents constitute a source of noise; the equivalent noise conductance is a measure of induced grid noise of the tube. Both

[2] H. Rothe and W. Dahlke, Theory of Noisy Fourpoles, *Proc. IRE*, vol. 44, pp. 811–818, June, 1956.

these parameters may be measured and average numerical values for any triode or triode-connected pentode determined. These values may be used to predict the optimum noise figure obtainable with a given tube type and the optimum source resistance at which this noise figure exists.[3]

The noise parameters vary in the following manner: R_{eq} is independent of operating frequency but varies with the operating parameters of the tube, while G_n is essentially independent of these operating parameters but varies with the square of the frequency. Thus both R_{eq} and G_n may be measured at one frequency under a particular set of operating conditions for a tube, and the values obtained may be used to calculate these noise parameters at other frequencies under the same operating conditions.

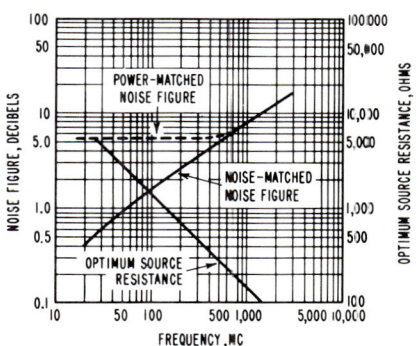

FIG. 3. Calculated noise performance of a metal-ceramic high-mu triode.

The noise figure obtained from an amplifier depends to some degree upon the impedance transformation between the signal source and the input of the amplifier stage. The impedance transformation that produces maximum transfer of power is not necessarily the one that produces the lowest noise figure. This means that a deliberate mismatch may be made at the input of an amplifier for best noise figure, with the resulting power loss made up in later stages. Therefore, there is an optimum source resistance at a given frequency which results in a minimum noise figure. This optimum resistance may be calculated when the values of R_{eq} and G_n are known.

$$R_s \text{ (optimum)} = (f_o/f)(R_{eq}/G_n) \quad (3)$$

where R_s (optimum) = optimum source resistance, ohms
f_o = frequency, Mc, at which G_n was measured
f = desired frequency of operation, Mc
R_{eq} = equivalent noise resistance, ohms
G_n = equivalent noise conductance, mhos

In addition, the minimum noise figure attainable with the optimum source resistance may be calculated from the same noise parameters:

$$NF \text{ (min)} = 1 + 2(f/f_o)R_{eq}G_n \quad (4)$$

where NF (min) = minimum attainable noise figure
f_o = frequency, Mc, at which G_n was measured
f = desired frequency of operation, Mc
R_{eq} = equivalent noise resistance, ohms
G_n = equivalent noise conductance, mhos

Figure 3 shows the relationship between power-matched noise figure, "noise-matched" noise figure, and optimum source resistance for a high-performance metal-ceramic triode.

If the optimum value of source resistance is not used, the noise parameters may still be used in calculating a theoretical noise figure:

$$NF = 1 + \frac{5G_t}{G_s} + \frac{R_{eq}(G_s + G_t)^2}{G_s} \quad (5)$$

$$G_t \cong G_n/5 \quad (6)$$

and $$G_s = 1/R_s \quad (7)$$

[3] C. Metelmann, Noise Parameters in VHF-UHF Circuit Design, *Electron. Ind.*, vol. 18, pp. 90–93, July, 1959.

where NF = noise figure
G_t = transit-time conductance, mhos
G_s = source conductance, mhos
R_s = source resistance, ohms
R_{eq} = equivalent noise resistance, ohms
G_n = equivalent noise conductance, mhos

Table 1 lists the equivalent noise parameters of a number of modern tubes, together with the operating conditions under which they were measured.

Table 1. Equivalent Noise Parameters
(Measured at 90 Mc)

Tube type	R_{eq}, ohms	G_n, μmhos	E_b, volts	R_k, ohms	G_m, μmhos	I_b, ma
6AM4	260	600	200	100	9,800	10
6BC8	600	320	150	220	6,200	10
6BN4	420	390	150	220	6,930	9.0
6BS8	390	330	150	220	7,300	10
6CE5*	650	1,200	200	180	5,700	11
6CY5*	525	640	125	150	6,640	10
6688*	120	1,160	150	82	19,000	15
7077	350	140	150	82	10,000	7.5

* Triode connected.

Tubes most suitable for low-noise operation are usually so designated by the tube manufacturer, and typical noise figures are often included in his data sheets. The noise figure obtained is governed by the basic structure of the tube, the circuit in which the tube is used, and the operating conditions. In general, minimum noise is exhibited at the maximum transconductance obtainable with a grid bias sufficient to prevent grid current flow. However, a range of operating conditions, over which noise figures remain uniform, usually exists. Information on this range may be plotted in the form of noise contours on the characteristic curves of a tube. Noise contours for a high-mu ceramic-metal planar tube are presented in Fig. 4.

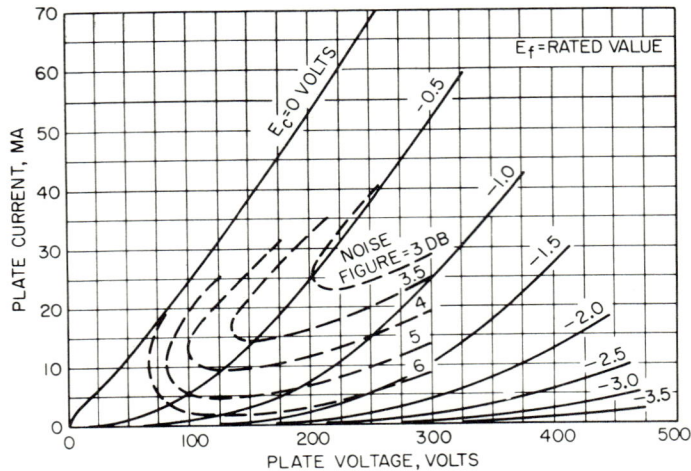

Fig. 4. Noise contours for a metal-ceramic high-mu triode.

Shot noise and partition noise, because of their completely random nature, are of uniform amplitude over the useful frequency spectrum, while induced grid noise increases with frequency. It might be deduced from this that, although the noise problem is present at all frequencies, it becomes more severe as the operating frequency is increased. Unfortunately, the existence of another type of noise, which is inversely proportional to frequency, extends the problem to the other end of the spectrum. This noise, which is called "flicker noise," is troublesome in the range from audio frequencies to direct current.

The exact cause of flicker noise is not known, but it has been attributed to a number of phenomena occurring in the grid-cathode region of tubes: these include ionization and deionization of molecules, movement of interstitials in the cathode, dying and forming of emission centers on the cathode, changes in spacing due to temperature changes, crystal growth in the cathode, changes in availability of donor centers, formation of cathode interface layers, and photoelectric effects.

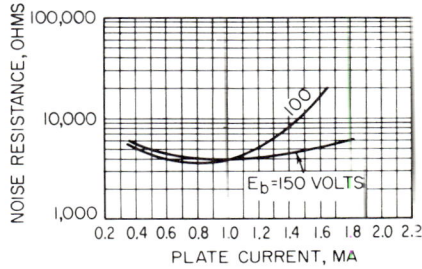

Fig. 5. Variation in low-frequency noise with plate current—high-mu triode.

Flicker noise appears to be minimized by close grid-cathode spacing in tubes, plus careful processing; therefore, tube choice is best based on manufacturers' recommendations. However, plate-current levels also play a part, as inspection of Fig. 5 will show. In this example, noise reached a minimum at a plate current of 1 ma and then increased as the plate current was decreased. The plate voltage at which the tube is operated is also significant, as may be seen in Fig. 6.

Leakage noise usually results from currents flowing from the plate or screen grid to the control grid, through leakage paths built up across the mica spacers by condensation of conductive material from the hot cathode. The resistance of these paths is

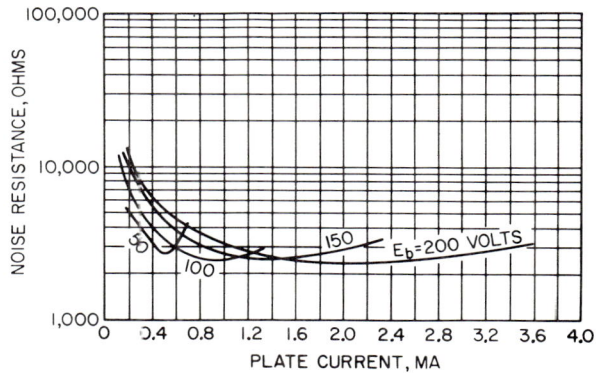

Fig. 6. Variation in low-frequency noise with plate current at various plate voltages—high-mu triode.

noisy, and contact of the resistance paths with the control elements of the tube is erratic, giving rise to very severe noise. Since these leakage paths build up during the life of the tube, it is important to choose tubes designed to minimize this problem if signal levels on the order of 10 mv in the audio range are expected.

Grid Film. In some tubes a layer of insulating material builds up on the grid from the sublimation products of the cathode during processing. This produces a phenomenon known as "blackout" or "poor grid recovery." It was first noticed during the early days of radar, when transmitted pulses would cause the radar receiver to be "blacked out" and thus fail to display the return pulses. The phenomenon appears to be the result of the buildup of an electric charge on the insulating film, which causes the grid to appear more negative than the amount of externally applied bias will account for. The effect may sometimes be noticed when a time-dependent signal, such as a television raster, is viewed: the charge on the grid will spontaneously discharge and cause an unexplained discontinuity in the observed signal.

Grid films cause the most difficulty in amplifiers that have their grids driven positive and thus draw large positive grid currents—for example, pulsed amplifiers. When a tube having poor grid recovery receives a positive grid pulse, it acts as though there were a parallel RC circuit in series with the grid, and a "backswing" in plate current occurs, whose duration depends on the "time constant" of the apparent RC circuit. The backswing may be observed with an oscilloscope, and its effect may also be noted by the average plate current of the tube being lowered during pulsing. Figure 7 shows the plate-current waveform of a tube having poor grid recovery. The delay time resulting from grid films varies widely, and the circuit designer can do little to overcome the effect; he must depend on the tube manufacturer to control grid films.

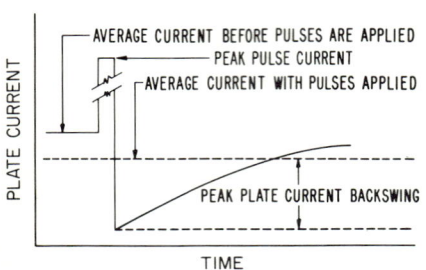

FIG. 7. Plate current of a pulsed tube with "blackout."

Frequency and Bandwidth Limitations. As the frequency to be amplified is increased, the gain obtainable with conventional negative-grid vacuum tubes decreases until a frequency is reached where it becomes unity. The frequency at which this occurs has been extended over the years by improvements in tube design, but it still constitutes a real barrier to relatively low-cost amplifier design.

However, with proper circuitry some present-day negative-grid tubes will produce power gains of as much as 8 db at frequencies as high as 7,000 Mc.

Degradation of performance with frequency is brought about by dielectric losses, finite values of lead inductance and interelectrode capacitance, and transit-time effects.

Dielectric losses have been decreased by tube designs which confine the dielectric material to portions of the tube where dielectric stresses are minimized and by the use of dielectric materials with the lowest possible losses. Initially, losses were decreased by the use of glass bases (miniatures, subminiatures, and acorn tubes) instead of plastic ones. In recent years, with the trend away from general-purpose tubes and toward tubes for specific applications, ceramics have been used for the bases, insulators, and envelopes of tubes designed for uhf use (metal-ceramic planars, nuvistors, and pencil tubes).

As the operating frequency is increased, the inductances and capacitances inherent in the tube structure become an increasing portion of the tuned circuit of an amplifier stage; this continues until the tuned circuit "disappears" into the tube, no further tuning is possible, and thus a limiting frequency is reached. Furthermore, as the frequency increases even the reactance of the relatively short leads in the tube becomes great enough to decrease the magnitude of the driving signal appearing across the tube elements. In addition, cathode lead inductance can act as an impedance common to both grid and plate circuits and thus produce undesired feedback.

In narrow-bandwidth applications with lumped-constant circuits the input and output capacitances of a tube are limiting factors in performance only if the frequency of operation is in the uhf range. In wideband circuits the bandwidth is a function of the combined circuit and tube capacitances, with the tube capacitances often the major portion, and input and output capacitances may be quite important, even though the center frequency of the amplifier is not in the vhf or uhf range. In many amplifier circuits the grid-to-plate capacitance of the tube sets a limit on the gain obtainable in the amplifier without excessive feedback across the tube. As a means of assessing the usefulness of tubes as amplifiers in both narrowband and wideband applications, a number of "merit figure" formulas have been devised which attempt to take into account the effect of capacitances. These formulas are applied to specific tube types in Sec. 4.

The effects of lead inductance are minimized in tubes designed for uhf operation by the use of large-diameter leads, multiple leads, and planar element construction which allows connections to external circuitry to be made around the periphery of contact disks or over the entire surface of cylinders. Tubes with planar element construction are especially advantageous since they are readily adapted to operation in resonant cavities where much higher frequency operation may be achieved than with lumped-constant circuits.

With increased operating frequency, the time required for electrons to travel from cathode to plate becomes a greater percentage of the time required for one cycle of the input signal; this gives rise to transit-time effects. The principal effects are an increase in the input conductance as the square of the frequency and a decrease in the transconductance. The first effect requires increased driving power for a given degree of excitation, and the second effect reduces the gain obtained from the tube.

Transit-time effects are lessened in tube design by decreasing the interelectrode spacing and in tube operation by increasing the plate voltage. Since decreasing the interelectrode spacing increases the interelectrode capacitances, electrode dimensions must be decreased to maintain low capacitances. Unfortunately, decreasing the size of the electrodes reduces the dissipation capabilities of the tube and the cathode emitting area. However, the use of heat sinks, especially in cavity circuitry where they can be large and quite effective, will allow the operation of relatively small-electrode tubes at dissipation levels which allow considerable useful power output. In tube operation the plate voltage level is limited by the possibility of arcing occurring, and in close-spaced tubes this level may be relatively low. Figure 8 shows the variation in input resistance of a subminiature pentode as the operating frequency is varied. Figure 9 shows variations in resistance and capacitance for a close-spaced metal-ceramic triode, both hot and cold.

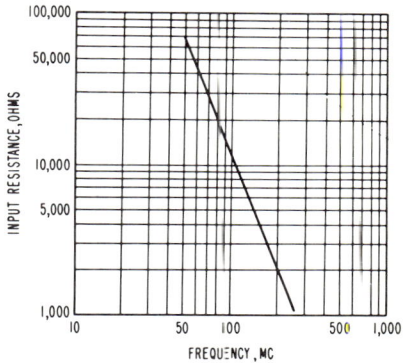

FIG. 8. Input resistance as a function of frequency for a subminiature pentode.

2.4. Manufacturer's Data. *General.* This handbook contains condensed data on the more widely used tube types. While this is useful in making a preliminary selection of possible tube types for a specific application, serious design work is aided by having full ratings, characteristic curves, characteristics limits, and life end points available. The best sources of this information are the data sheets and tube handbooks made available by the tube manufacturers. However, to obtain maximum utility from these sources, knowledge of how to interpret the manufacturer's data is essential.

Information Presented. Most tube data sheets, excluding the "condensed data" variety, give the following information, not necessarily in this order:

1. Tube type number
2. Tube class
3. Features or principal use
4. Physical dimensions
5. Terminal connections

6. Interelectrode capacitances
7. Filament or heater characteristics
8. Ratings
9. Characteristics
10. Characteristic curves

In addition, data for some tube types include:

11. Characteristics limits
12. Special tests
13. Life-test end points

These classes of information are treated in detail below.

The tube type number may be one selected by the tube manufacturer according to some system of his own, or it may have been assigned by the Joint Electron Device

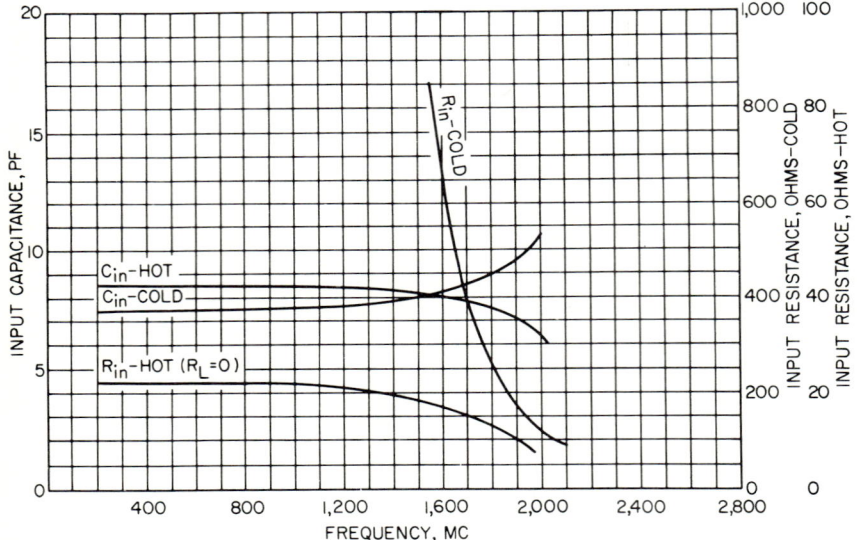

Fig. 9. Input resistance and input capacitance as functions of frequency for the 7768 planar triode.

Engineering Council of the Electronic Industries Association, a cooperative association of manufacturers. The terms "JEDEC designation" or "EIA designation" are often used for the latter. JEDEC designations assigned to tubes normally intended for use in home-entertainment devices, such as radio and television receivers, consist of one or more digits giving the approximate filament or heater voltage, one or more letters having no special significance with respect to the nature of the tube, one or two digits giving the number of active elements ("active" elements are here defined to include the heater and internal shields) available for external connection, and on occasion suffix letters. The suffix letters have the following meanings:

G: This indicates a tube with a glass bulb and an octal base. It is used on some of the older tubes that were developed as substitutes for metal tubes having the same basic designations and characteristics.

GT: This indicates a glass tube with a straight-sided bulb and an octal base, which is electrically equivalent to the tube having the same basic designation but a G suffix.

A, B, C, D, etc.: These letters, assigned in this order, represent various modifications of the basic tube type. The modifications must be such that an A may be substituted for the basic type, a B for an A or the basic type, a C for a B or an A or the basic type, etc. However, the reverse of this order must not be assumed; often the later suffix types have higher dissipation ratings, and a circuit designed to take advantage of these ratings might cause the allowable dissipation of an early suffix type to be exceeded if it were substituted for the proper one.

W: This suffix letter is used for certain military types.
X: This indicates a tube having a low-loss base.
Y: This indicates a tube having an intermediate-loss base.

Most tubes intended for industrial or military use are assigned numerical designations, although suffix letters A, B, C, etc., may be used with the same meaning that they have in the home-entertainment series of designations. Various series of numbers have been used, the latest having started in 1949 with 5500. These numbers are assigned in numerical order and have no relation to the class of tube or its characteristics.

The tube class is designated by a name which may indicate the number of active elements (here "active" elements do not include heaters and internal shields), the function of the tube, or some construction feature. Thus we have diodes, triodes, tetrodes, pentodes, and heptodes (Greek derivatives have been favored here); gated-beam discriminators, used for FM detection; and sheet-beam tubes, named for the manner in which electron flow occurs. Multiple-section tubes are described by combinations of the foregoing names: for example, triode-pentode, twin triode, triode-heptode, and pentode-gated-beam discriminator.

A paragraph is often included which describes features of the tube and indicates whether or not the tube was designed for a specific application. Obviously, a tube can often be used for purposes other than those spelled out, but it is certainly well to heed warnings as to applications that are not recommended. For example, a tube described as being unsuitable for use in applications critical as to microphonics should not be designed into a low-level audio-amplifier stage.

The physical dimensions given are normally maximum values, since these are usually of most interest to the designer. However, where minimum values are important, as with tubes having top caps which require flexible leads, these are given. Bulb size is often stated as T-5 1/2, T-6 1/2, T-9, etc. These are standardized values and indicate the nominal diameter in eighths of an inch; thus a T-9 bulb has a nominal diameter of $\frac{9}{8}$ or $1\frac{1}{8}$ in. Occasionally, only the JEDEC (or EIA) standardized outline number is listed for the tube type (JEDEC 5-1, JEDEC 9-1, JEDEC 12-52, etc.): dimensions may then be obtained from the standardized outlines which are usually grouped in a special section of the tube manual.

Terminal connections are given in tabular form, often accompanied by a basing diagram which carries a JEDEC designation (JEDEC 7AC, JEDEC 9DE, JEDEC 12BF, etc.). Decisions as to connections to be made to tube socket clips should be based on the tabulated data rather than the diagram, since the diagram does not show internal shields or carry warnings concerning the use of socket clips that match "internal connection" terminals.

The notation "internal connection—do not use" is a source of some confusion in tube application. It is used on terminals that the tube manufacturer uses as "anchor points" for the tube structure or reserves for later use if he feels that it may be necessary to modify the construction of the tube. Thus, even though inspection shows which element is connected to one of these terminals, the terminal should not be used, since the tube manufacturer may remove this internal connection in future constructions, or another tube manufacturer may use the terminal as an anchor point for a different electrode. If inspection shows no internal connection to a terminal, the socket clip should not be used as a circuit tie point, since the manufacturer of the tube may use the terminal later, or another manufacturer may use it at present.

Some tube terminals are designated "no connection," and the socket clips corresponding to these terminals may be used as circuit tie points unless notes are appended forbidding it. These notes are usually added when the "no connection" terminals are employed to increase the distance between terminals with a large potential difference.

If a tube has two or more similar sections (twin triode, triple triode, dissimilar double pentode, etc.) the sections are numbered in the following manner: the highest section number is assigned to the section having an electrode connected to the lowest-numbered base pin or terminal, and successively lower section numbers are assigned according to the sequence in which the connections of the corresponding electrode in all other sections are made to successively higher-numbered base pins. The section numbers bear no relationship to the physical position of the sections in the tube or their position on the basing diagram. For example, the section shown at the center of the basing diagram for a three-section triode is not necessarily the center section in the actual tube.

Interelectrode capacitances are given on tube data sheets for nearly all tube types. The capacitances listed and the tolerances within which these capacitances are maintained depend upon the primary application of the tube. Minimum capacitance listings for all tubes suitable for amplifier use are grid-to-plate, input, and output. Where capacitances are not considered critical, they may be classed as "approximate" on the data sheet. An example of this is the capacitance listing for an audio output tube, where variations in the capacitances will have little effect on circuit performance. Grid-to-plate capacitances are often listed as maximum values if they are smaller than 0.02 pf. This allows the circuit designer to evaluate the grid-to-plate capacitance as a potential undesired feedback path with the assurance that the published value is the worst possible case. For larger values of grid-to-plate capacitance, especially where neutralization may be required, center values are usually given. Input and output capacitances are nearly always center values. Multiple-section tubes intended for applications where capacitive coupling between sections may be a source of undesired feedback usually have cross capacitances between sections stated. These are usually given as maximum values for the same reason that the grid-to-plate capacitances are so given. The terms "input capacitance," "output capacitance," etc., are usually supplemented by a listing of the tube elements between which the capacitance values listed have been measured. Any elements not listed were connected to ground. This is especially important for multiple-element tubes where elements, such as internal shields, of one section may provide part of the input or output capacitances of the other section. Capacitance values may be published as having been measured with or without an external tube shield, although for most of the recently developed tube types capacitances measured without a shield are given unless capacitances measured with a shield are absolutely necessary for the intended application. Unless it is specifically stated in the data, all capacitances shown have been measured with no voltages applied to the tube ("cold"). For some closely spaced, high-performance tubes both "cold" and "hot" capacitance values may be published, as there is a measurable difference between these values (see Fig. 9).

Two terms which appear on data sheets and have specific meanings to the tube manufacturer are often misinterpreted and sometimes used interchangeably by those outside the tube industry. These terms, "rating" and "characteristic," are defined here before their use on data sheets is covered.

A "rating" is a limiting value of voltage, current, dissipation, environmental stress, etc., beyond which it is not prudent to go if reasonable tube life is expected. A "characteristic" is a property of a tube, inherent in its design, such as its ability to exhibit a specific transconductance, draw a specific plate current, or deliver a specific power output when stated electrode voltages are applied. Thus a tube has "rated" plate and screen voltages that define, according to the rating system used, the maximum permissible values; and "characteristic" values of plate current, screen current, and transconductance which are exhibited when the tube is operated under stated conditions, within the limits established by the ratings.

The heater or filament voltage given on a data sheet is the stated value at which the tube should be operated to exhibit its characteristic properties. The tolerance in heater voltage listed is a rating, intended to ensure optimum life of the tube; the high limit protects against heater burnout, use-up of emissive material, and excessive sublimation, while the low limit maintains emission at a satisfactory level. The heater current listed is the current that will flow when the stated filament voltage is applied to a "bogey" tube and is often called the "bogey" heater current. A true bogey tube is one that exhibits all the characteristic values listed on the data sheet;

however, the heater characteristics are based on tubes that may be bogey for this characteristic only.

Ratings. Ratings may be assigned to tubes according to three systems: design-center, design-maximum, and absolute-maximum. In addition, the absolute-maximum ratings may vary according to three classes of service: continuous commercial service (CCS), intermittent commercial and amateur service (ICAS), and intermittent mobile service (IMS).

The rating systems are usually defined in notes on the tube data sheets similar to those below:

Design-center ratings are limiting values of operating and environmental conditions applicable to a bogey electron tube of a specified type as defined by its published data and should not be exceeded under normal conditions.

The tube manufacturer chooses these values to provide acceptable serviceability of the tube in average applications, making allowance for normal changes in operating conditions due to rated supply-voltage variation, equipment component variation, equipment control adjustment, load variation, signal variation, environmental conditions, and variations in the characteristics of the tube under consideration and of all other electron devices in the equipment.

The equipment manufacturer should design so that initially no design-center value for the intended service is exceeded with a bogey tube under normal operating conditions at the stated normal supply voltage.

Design-maximum ratings are limiting values of operating and environmental conditions applicable to a bogey electron tube of a specified type as defined by its published data and should not be exceeded under the worst probable conditions.

The tube manufacturer chooses these values to provide acceptable serviceability of the tube, making allowance for the effects of changes in operating conditions due to variations in the characteristics of the tube under consideration.

The equipment manufacturer should design so that initially and throughout life no design-maximum value for the intended service is exceeded with a bogey tube under the worst probable operating conditions with respect to supply-voltage variation, equipment component variation, equipment control adjustment, load variation, signal variation, environmental conditions, and variations in the characteristics of all other electron devices in the equipment.

Absolute-maximum ratings are limiting values of operating and environmental conditions applicable to any electron tube of a specified type as defined by its published data and should not be exceeded under the worst probable conditions.

The tube manufacturer chooses these values to provide acceptable serviceability of the tube, making no allowance for equipment variations, environmental variations, and the effects of changes in operating conditions due to variations in the characteristics of the tube under consideration and of all other electron devices in the equipment.

The equipment manufacturer should design so that initially and throughout life no absolute-maximum value for the intended service is exceeded with any tube under the worst probable operating conditions with respect to supply-voltage variation, equipment component variation, equipment control adjustment, load variation signal variation, environmental conditions, and variations in the characteristics of the tube under consideration and of all other electron devices in the equipment.

The first of these ratings systems, design-center, is seldom used today for new tube types, although many of the older types are still rated according to this system. At present, most tubes intended for use in home-entertainment service (radio, television, phonographs, etc.) are rated under the design-maximum system, and those intended for industrial or military service are rated under the absolute-maximum system. The latter systems are in most common use because the design-center system, which was adequate when "radio" was the principal application of tubes, tends to give the circuit designer a false sense of security and may cause him to neglect evaluation of all operating conditions with regard to exceeding the ratings, since a condition considered "normal" by the designer may not have been anticipated by the tube manufacturer in establishing the ratings.

In both the design-maximum and absolute-maximum systems the responsibility for determining the worst probable conditions and the determination of conformance to the ratings under these conditions are the sole responsibility of the equipment designer. The two systems differ in the kind of tube to be used in evaluating conformance to ratings; under the design-maximum system a bogey tube of the type being evaluated must be used, and when conformance is obtained with this tube, any tube within the spread of characteristics controlled by the tube manufacturer will satisfy the requirements of the rating system; while under the absolute-maximum rating system "limit" tubes that display the characteristics spread controlled by the tube manufacturer must be used to ensure that any tube will meet the requirements of the rating system.

The three subsystems, CCS, ICAS, and IMS, of the absolute-maximum rating system are applied to tubes used in transmitting equipment and are intended to take into account the wide variety of duty cycles and tube life expected in this service. For example, a broadcast transmitter may be operated 18 hours each day, 365 days of the year; an amateur transmitter may be operated only an hour a day and probably not every day, while a mobile communications transmitter may be operated for only a few seconds out of each hour.

Continuous commercial service (CCS) is defined as that type of service in which normal life and reliability of performance under continuous conditions are the prime considerations.

Intermittent commercial and amateur service (ICAS) is defined to include the many applications where the transmitter design factors of minimum size, light weight, and considerably increased power output are more important than long tube life. In this service life expectancy may be one-half that obtained in continuous commercial service.

Under the ICAS classification are such applications as the use of tubes in amateur transmitters, and the use of tubes in equipment where transmissions are of an intermittent nature. The term "intermittent" is used to identify operating conditions in all applications other than amateur in which no operating or "on" period exceeds 5 min and every "on" period is followed by an "off" or standby period of at least the same or greater duration.

Intermittent Mobile Service (IMS) is defined to include those applications, such as aircraft, where the transmitter design factors of minimum size, light weight, and exceedingly high power output for short intervals are the primary requirements, even though the average life expectancy of the tubes used in such transmitters is reduced to about 100 hr. (This 100-hr figure represents actual transmission time.)

Tube ratings for IMS service are established on the basis that the transmissions have maximum "on" periods of 5 min followed by "off" periods of at least 5 min provided the total "on" time of such periods does not exceed 10 hr during the life of any tube.

Ratings listed for a typical tube suitable for amplifier service usually include the following:

Maximum plate voltage
Maximum screen voltage or maximum screen-supply voltage
Maximum grid (or grid-1) voltage
Maximum plate dissipation
Maximum screen dissipation
Maximum heater-cathode voltage
Maximum grid-circuit resistance

If the tube is intended for service other than small-signal, class A, the following may also be included:

Maximum grid (or grid-1) current
Maximum direct cathode current
Maximum peak cathode current
Maximum envelope temperature

If the principal application of the tube involves severe environmental conditions, limiting values of these may be included, although they are usually grouped in a separate place in the data sheet rather than being placed with the other ratings.

ELECTRON TUBES   9-19

The purpose of ratings, as defined in all three of the rating systems, is to ensure acceptable serviceability of the tube. Acceptable serviceability must be defined according to the intended application; expected life may be only a few minutes or it may be many years. In the first case, there may be a tendency to design up to the limits where catastrophic failure is risked, while in the second case the cumulative effect of repeated excursions outside ratings is often ignored. Before discussing methods of determining conformance to ratings, it is worthwhile to discuss the significance of the individual ratings.

The maximum plate voltage rating listed, unless otherwise indicated, may be taken to be the maximum permissible value of d-c voltage that may be applied to the plate of the tube. This in turn may be interpreted as allowing a peak plate voltage during operation of twice the d-c value. If the tube is rated for pulse service, a peak pulse plate voltage will also be listed. The rated plate voltage must, in most instances, be grossly exceeded for catastrophic breakdown to occur. Exceptions to this are the types of tubes intended for plate-pulsed service or for high-voltage-regulator service where flashover might occur at plate voltages only slightly in excess of the rated value. However, operation of medium-voltage tubes at plate voltages in excess of the rated value will increase glass electrolysis, especially if envelope temperatures run high, and lead to premature tube failure.

Screen voltage may be given either as the maximum screen voltage or as the maximum screen supply voltage. In some applications the screen may be fed from the plate supply through a dropping resistor. If AGC is applied to the tube and the screen-dropping resistor has been selected to apply rated screen voltage in the absence of AGC, the presence of high levels of AGC will cause the screen voltage to approach the plate supply voltage, and thus presumably exceed the rated screen voltage. This contingency is taken into account when a screen supply voltage rating is given. However, this rating is also linked to the screen dissipation rating, and this relationship is described below in the discussion of screen dissipation ratings.

The grid (or grid-1) voltage rating may be expressed as positive d-c, negative d-c, positive peak, or negative peak. Nearly all class A amplifier tubes have a positive d-c grid voltage rating of zero volts, since there is little reason to operate with positive grid bias, and such operation would cause excessive grid current flow. Negative d-c grid voltage ratings are seldom greater than 50 volts for tubes used in small-signal class A amplifier applications. Those tubes intended for class C service have higher ratings as required by the application. Tubes intended for pulsed-amplifier service may have both positive and negative peak grid voltage ratings. Since grid current flows when positive pulses are applied to the grid, the maximum duty cycle of the pulse is often appended, as a footnote, to the peak positive grid voltage rating to limit allowable grid dissipation. Excessive grid voltage may cause arcing between grid and cathode or may pull coating from the cathode by electrostatic attraction. The danger of these occurrences is greatest in the very close spaced, extremely high transconductance types.

The plate dissipation rating is intended to control temperature rise of the plate during operation. The heat-dissipating capability of the plate depends upon the plate area, the plate material, the plate surface, and the possibility of connecting the plate to an external heat sink. Excessive plate temperatures cause gas release, secondary emission, and if carried far enough may result in plate deformation or even melting. In small-signal tubes the first and second occurrences are most likely to be encountered, while in power tubes the latter occurrences are not unknown when the plate dissipation rating is grossly exceeded.

The screen dissipation rating has the same general purpose as the plate dissipation rating, since excessive temperature rise of the screen is also detrimental to tube life. In small-signal tubes where a maximum screen voltage rating is given, the screen may be operated up to the maximum rated screen dissipation value. If a maximum screen supply voltage rating is given, the allowable screen dissipation is a function of the supply voltage. With this type of rating the full screen dissipation is allowable if the screen voltage does not exceed one-half the maximum rated screen supply voltage. At voltages from this point up to the full screen supply voltage, the allowable dissipation decreases along a parabolic curve, reaching zero at full voltage. This relationship is usually shown on a screen rating chart similar to Fig. 10. (Some tube manufacturers construct the charts with specific values of screen voltage and screen dissipation, rather than in terms of percentages of the maximums.)

To ensure that the requirements of the screen rating chart will be met, the minimum value of screen-dropping resistor may be calculated with the following formula:

$$R_{g2} = (E_{cc2})^2/4P_{g2} \tag{8}$$

where R_{g2} = screen-dropping resistor
E_{cc2} = screen supply voltage
P_{g2} = maximum screen dissipation rating

Note that E_{cc2} is the actual screen supply voltage to be used rather than the maximum screen supply voltage rating.

In audio power amplifiers screen current, and consequently screen dissipation, rises with the input signal. Because of this, screen dissipation rather than plate dissipation usually limits the power-handling capability of an audio output tube. If an amplifier is to be designed for operation with a continuous sine-wave signal, care must be taken that the screen dissipation rating is not exceeded. If the amplifier is to be used for speech and music, maximum driving voltage is seldom applied to the audio output tube for sustained periods, and more liberal screen dissipation ratings are permissible. In recognition of this, some audio output tubes have two maximum screen dissipation ratings, one for continuous operation and the other for peaks of speech and music; the peak rating is usually twice the continuous rating. Since no averaging time is specified, conformance to the peak rating is subject to considerable individual interpretation. One method of assessment is to apply a sine-wave input signal and increase it until the peak screen dissipation rating is reached, taking care to remain at this point as short a time as possible. The allowable peak input signal may then be taken as the peak value of the sine-wave signal that produced maximum rated screen dissipation.

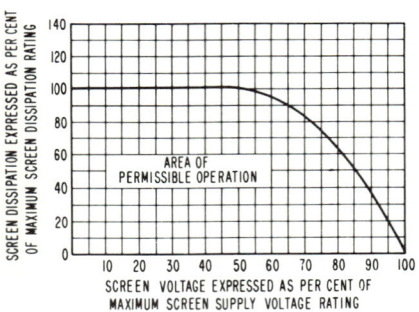

Fig. 10. Screen rating chart.

Removal of plate voltage from a tube with screen voltage applied will cause the screen dissipation rating to be grossly exceeded and may result in melting of screen wires. This is seldom a problem with small-signal tubes operating with the screen and plate supplied by a common power supply, but where separate power supplies are used, as is often done with power tubes, precautions should be taken to ensure removal of screen voltage if the plate voltage supply fails.

Heater-cathode voltage ratings are given for operation with the heater both positive and negative with respect to the cathode. Most present-day home-entertainment types are rated to withstand a d-c voltage of 100 volts and a total of d-c and peak of 200 volts when the heater is positive with respect to the cathode, and a total of d-c and peak of 200 volts when the heater is negative with respect to the cathode. These ratings are the result of the extensive use of tubes in line-operated series-string circuits in radio, television, and phonograph applications where heater-cathode voltage stresses at these levels may be encountered. Since this type of circuit is not common in industrial and military applications, tubes for these services often have heater-cathode voltage ratings lower than those of the home-entertainment types. However, where typical applications demand high heater-cathode ratings, the tubes will have them, although high heater-cathode voltages are never conducive to long tube life. Exceeding the heater-cathode ratings can result in arcing and heater burnout in the worst cases and in increased heater-cathode leakage in any event.

ELECTRON TUBES

As mentioned earlier (see Sec. 2.2), negative grid currents in a tube will cause the grid to become more positive with respect to the cathode if there is resistance in the grid circuit. The magnitude of the effect is in part dependent upon the amount of resistance in the grid circuit. Because of this, maximum allowable grid-circuit resistances are listed as ratings. These ratings are usually based on life tests, carried out with resistance in the grid circuit. Separate ratings are usually given for fixed-bias and cathode-bias operation. The allowable maximum resistance is usually higher for cathode-bias operation than for fixed-bias operation, since cathode-biased tubes are to a degree self-protected from thermal runaway when excessive grid current flows. If only one maximum grid-circuit resistance rating is given it may be assumed that the rating applies to both fixed-bias and cathode-bias operation.

If no resistance value is given, it is possible to calculate maximum allowable values with empirical formulas[4] based on the assumption that a 10 per cent change in plate current due to bias shift caused by grid current flow is acceptable. A further assumption is that in a usable tube the gas content will be at a low enough level that a maximum grid current flow of $\frac{1}{3}$ μa/watt of plate input will not be exceeded. The calculated values are intended to prevent thermal runaway and are usually more conservative than those determined by life test. Formulas are given below for both triodes and multigrid tubes.

For triodes:

$$R_g = (300{,}000/E_b)(1/g_m + R_k + R_L/\mu) \qquad (9)$$

where R_g = maximum grid-circuit resistance, ohms
E_b = plate voltage, volts
g_m = transconductance, mhos
R_k = cathode-bias resistor, ohms
R_L = load resistance, ohms
μ = amplification factor

For tetrodes or pentodes:

$$R_g = (30{,}000/E_b)(1/g_m + R_k + I_{c2}R_{c2}/I_k\mu_t) \qquad (10)$$

where R_g = maximum grid-1 circuit resistance, ohms
E_b = plate voltage, volts
g_m = transconductance, mhos
R_k = cathode-bias resistor, ohms
I_{c2} = screen current, amp
R_{c2} = screen-circuit resistance, ohms
I_k = cathode current, amp
μ_t = triode amplification factor (grid 1 to grid 2)

When tubes are used with contact-potential bias, grid-circuit resistances much larger than those normally recommended for fixed- or cathode-bias operation are often used, a value of 10 megohms being quite common. Maximum grid-circuit resistance values have seldom been published for this class of service, since apparently very little difficulty has been encountered in present-day tubes with negative grid current shortening tube life. This may be due to counteraction of the negative grid current by the increased positive grid current produced as the bias shifts in the positive direction. In addition, operation in this mode is usually with a large value of load resistor, and this limits the plate current to a safe value. However, there are a number of disadvantages in the use of contact-potential bias, and these are detailed under Sec. 2.2.

Maximum grid current ratings are listed for tubes intended for class C service. These ratings are intended to limit grid dissipation to safe values, since excessive grid heating will cause grid emission and under the worst conditions may cause grid deformation and even melting.

In some applications, notably pulse amplifiers, cathode emission capability in addition to element dissipation limitations may govern the power-handling ability of a tube. Tubes intended for pulse service normally have peak cathode-current ratings.

---

[4] J. H. Robb, Some Methods of Estimating Maximum Grid Circuit Resistance, E.I. 35, Tube Department, General Electric Co., Owensboro, Ky., November, 1960.

It is best to use only tubes recommended by the tube manufacturer for pulse service; however, if a direct cathode current rating is given, an approximation of the allowable peak current may be obtained from the pulse rating chart shown in Fig. 11. An additional limitation is that the peak current should not exceed $\frac{1}{2}$ amp/watt of heater power.

The maximum envelope temperature rating for glass tubes is intended to restrict the temperature below the level where excessive cathode contaminants are released from the glass and where, in the presence of high voltages on the electrodes, glass electrolysis is enhanced. In metal-ceramic tubes the rating is intended primarily to restrict sublimation of cathode material which may deposit on insulating surfaces and decrease the insulation resistance of the tube. However, exceeding the rating grossly can result in ceramic cracking.

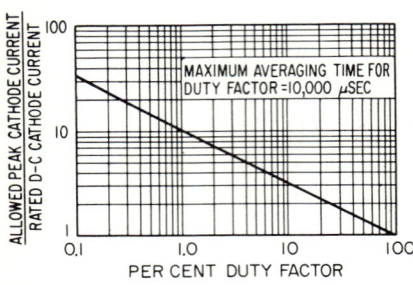

Fig. 11. Pulse rating chart.

Maximum envelope temperature is usually specified as being measured at the hottest point on the envelope. Thus the tube user is faced with the problems of both finding the hottest point and measuring the temperature at this point. However, the zone in which the hottest spot will occur can often be predicted with some degree of accuracy. The hottest spot on a glass tube is usually located opposite that portion of the plate where the electron stream impinges. In single-section tubes this is ordinarily opposite the broad side of the plate at a height corresponding to the center of the plate (see Fig. 12). The presence of radiators on the plate and the getter flash on the glass may modify this. Multisection tubes make prediction more difficult, since they are not symmetrical with respect to the radius of the envelope, and the relative power dissipations in the sections will determine the temperature distribution. The hottest spot on a metal-ceramic tube is usually located at the plate stud, unless the plate is connected to a heat sink, and

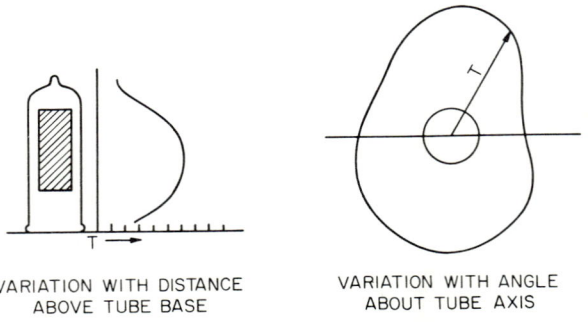

VARIATION WITH DISTANCE ABOVE TUBE BASE

VARIATION WITH ANGLE ABOUT TUBE AXIS

Fig. 12. Envelope temperature variation for a typical glass tube.

if the latter is true the hot spot may be located on the ceramic tube body near the plate stud.

The location of the hottest spot on any tube could be determined by cementing a number of thermocouples to the tube and taking readings close together in time as the tube warmed up. However, this would be a rather complicated process, and the common method is to use temperature-sensitive lacquer[5] for the determination. The procedure is to coat the general area where the hottest spot is believed to be with a lacquer having a melting point in the vicinity of the rated maximum bulb temperature, operate

[5] Tempilaq, manufactured by Tempil Corp., New York, N.Y.

ELECTRON TUBES   9-23

the tube at rated plate dissipation in an oven, and slowly raise the oven temperature, watching for the spot where the temperature-sensitive lacquer first melts. This spot may be considered the hottest spot.

Temperature measurements to determine conformance to maximum envelope temperature ratings may be made with a thermocouple attached to the tube envelope by means of cement,[6] with a thermocouple attached to a thin wire loop that fits around the envelope, or with a conduction-band thermocouple. The latter method is the one most commonly used in the tube industry. A phosphor-bronze band with thermocouple wires welded to opposite sides (see Fig. 13) is slipped over the tube and positioned so as to cover the previously determined hot spot. The temperature is then measured with element dissipations equal to those expected in the contemplated application of the tube. Consideration should be given to the ambient temperature in which the tube will operate, and if necessary, the tube should be enclosed in an oven duplicating this temperature when measurements are made. If the hot-spot location has not been determined in advance, the conduction-band thermocouple may be moved about on the tube envelope, with readings taken at each position, until the hot spot has been found. With metal-ceramic tubes the conduction-band method is not always suitable, especially if the tube is quite small, and measurements are usually made with a thermocouple cemented to the hot spot.

In preliminary circuit design the operating parameters can usually be chosen so that operation within the tube ratings will be maintained under quiescent signal conditions with normal voltages applied. However, consideration of possible voltage variations, parts tolerances, and input signal variations will show the designer that conformance to ratings can best be checked

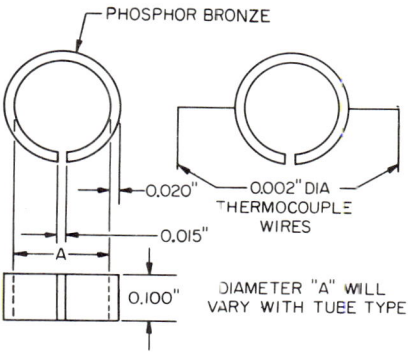

FIG. 13. Conduction-band thermocouple.

by making measurements on a working circuit under a variety of conditions and, if necessary, readjusting the operating conditions.

The first step is to list all probable variations in operating conditions and circuit constants and determine whether these variations will be in a direction to cause voltages and dissipations to approach or recede from the maximum rated values. The second step is to determine whether the degree to which various ratings are approached is significant. The third step is to measure the effects which appear to be significant, especially those whose magnitude cannot be readily predicted.

As an example, consider a pentode tube, rated under the design-maximum system, operated as an intermediate-frequency amplifier, and supplied from the power line through an unregulated heater and plate power supply (see Fig. 14). An increase in power-line voltage will increase the heater voltage, the plate voltage, the screen voltage, and the heater-cathode voltage. The increased plate and screen voltages will, in turn, increase the plate and screen dissipations, respectively. A value of cathode-bias resistor R_k lower than the design value will lower the grid bias and thus increase the plate and screen currents and dissipations, while a value of cathode-bias resistor higher than the design value may increase the heater-cathode voltage. A value of decoupling resistor R_b lower than the design value will increase the plate and screen voltages and dissipations. A value of screen resistor R_{sg} lower than the design value will increase the screen voltage and dissipation. This increase in screen voltage will, in turn, increase the plate current and dissipation. Increased AGC voltage will increase the grid-1 voltage, the screen voltage, and the plate voltage.

To assess the significance of the foregoing changes, it is necessary to know the probable limits of line voltage excursion and parts tolerances. It is customary to assume

[6] Sauereisen No. 1 Cement, Manufactured by Sauereisen Cements Co., Pittsburgh, Pa.

that domestic power lines are maintained within plus or minus 10 per cent of a center value of 117 volts.[7] This means that under the worst probable conditions the heater, plate, and screen supply voltages may rise 10 per cent. In practice, since an unregulated power supply will have appreciable internal resistance, the increased plate and screen currents resulting from the increased output voltage of the power supply will increase the internal voltage drop and tend to limit d-c output voltage increases to less than 10 per cent. Obviously, if it is known that line voltage will vary more than 10 per cent, the larger variation must be used in the evaluation. In addition, if the power supply is being used to supply other circuits which may be switched off and thus allow the unregulated output voltage to increase, this must also be taken into account.

Parts tolerances are usually determined by considerations of circuit performance and economics and are usually known by the time conformance to ratings is being determined. For this example, resistor tolerances of plus or minus 10 per cent may be assumed.

If the heater of the tube is operated with one side grounded, the heater-cathode voltage in the example is relatively low under normal conditions, and it is highly unlikely that any combination of conditions will cause the heater-cathode voltage rating to be approached; therefore, measurement of heater-cathode voltage is not necessary. If the heater is operated in a series-string circuit directly from the power

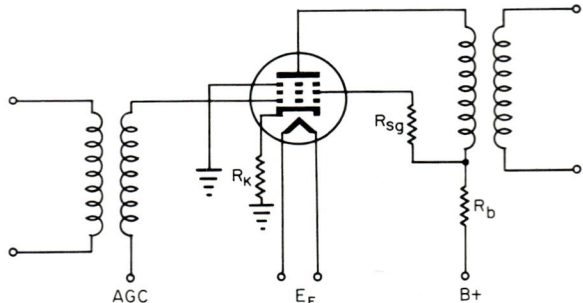

Fig. 14. Intermediate-frequency amplifier schematic.

line, as is done in transformerless radio and television receivers, more attention must be paid to heater-cathode voltages.

This leaves the heater voltage, grid-1 voltage, screen voltage, plate voltage, screen dissipation, and plate dissipation to be measured or calculated. The procedure is to set up the circuit with a bogey tube. Although a true bogey tube should be bogey for all characteristics, this is difficult to achieve, and a tube which exhibits bogey plate and screen currents and cutoff may be considered satisfactory for this particular circuit. The resistors R_k, R_b, and R_{sg} should have the minimum expected resistances. The line voltage should be set at the maximum probable value, normally 129 volts. The AGC voltage should be the minimum expected value. Under the foregoing conditions measure the heater voltage, plate voltage, screen voltage, plate current, and screen current, and calculate the plate and screen dissipations. Now adjust the AGC voltage to the maximum probable value and measure the grid-1, screen, and plate voltages. If all the voltages and dissipations measured and calculated are below the maximum rated values, the tube is operating in a manner to satisfy the rating system.

As a second example, consider a push-pull, class AB1, audio power amplifier (see Fig. 15), operated from the 117-volt line through an unregulated power supply. The principal difference between the rating evaluation of these tubes and the i-f amplifier tube is the necessity for taking into account the effect of signal input.

[7] Over the years this center value has been rising, and recent surveys show the center to be near 120 volts. If the designer elects to use a value of 120 volts, or higher, he should adjust the high and low limits accordingly.

ELECTRON TUBES 9–25

The bogey tubes used should be bogey for plate current, screen current, and single-tube class A power output.

Filament voltage, plate voltage, screen voltage, plate current, and screen current should be measured with the line voltage at the maximum probable value, the cathode-bias resistor at its minimum probable value, and with no signal input to the amplifier. The plate and screen dissipations should be calculated and conformance of these plus the plate and screen voltages should be determined. The plate voltage, screen volt-

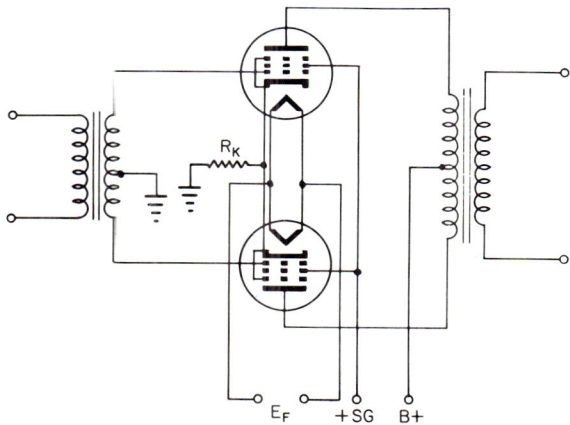

FIG. 15. Push-pull audio power amplifier schematic.

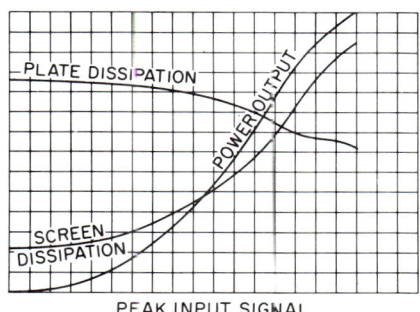

FIG. 16. Power output, plate dissipation, and screen dissipation vs. input signal for 7581-A's in push-pull. (*Courtesy of General Electric Co.*)

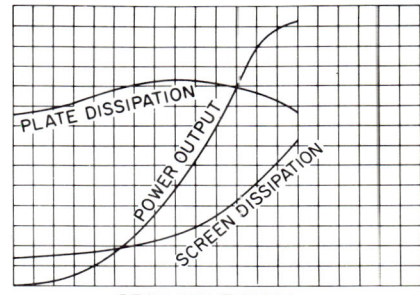

FIG. 17. Power output, plate dissipation, and screen dissipation vs. input signal for 7189-A's in push-pull. (*Courtesy of General Electric Co.*)

age, plate current, screen current, and power output should now be measured with the line voltage at the maximum probable value, the cathode-bias resistor at its minimum probable value, and the input signal at its maximum probable value. The latter figure may be taken as the maximum value at which the distortion level is acceptable. The screen dissipation should be calculated and the plate dissipation determined by subtracting the power output from the calculated plate input. Most power output pentodes show variations in plate dissipation, screen dissipation, and power output with variations in signal input similar to the curves in Fig. 16. However, some pentodes exhibit the behavior shown in the curves of Fig. 17. Note that in this figure the plate dissipation rises as the input signal is increased, reaches a maximum at a

9-26 AMPLIFYING DEVICES

point below the maximum power output point, and then falls. This means that measurement of plate dissipation at zero signal and maximum signal may not be sufficient to determine conformance to ratings, and that measurements should also be made at intermediate values of input signal.

Although many tubes do not have envelope temperature ratings, it is well to check envelope temperature in the final piece of equipment if the equipment is housed in an enclosure and/or the tubes are shielded, especially if the equipment is to be operated in high ambient temperatures. Tube shields can cause the envelope temperature to be considerably increased unless the shields are those specifically designed to remove heat from the tube. As a rule of thumb, the envelope temperature of a conventional lime-glass-bulb tube should not exceed 175°C unless otherwise rated.

Those tubes which have envelope temperature ratings are normally ones that have a fairly high plate dissipation rating, and it is especially important to determine the conformance to bulb temperature ratings of these.

It should be borne in mind that the foregoing procedures are for the purpose of determining conformance to ratings and not for determining satisfactory circuit performance. Before rating conformance is determined, the tube should be operating properly in the circuit. If ratings are being exceeded, adjustments in operating parameters to bring the tube within the ratings must also result in satisfactory performance under the new conditions or the circuit must be redesigned or abandoned.

Characteristics. The complete characteristics of a tube are best presented in graphical form, and this is done for many tube types. However, to allow a quick assessment of a tube's capabilities and its comparison with similar tubes, a specific operating point is usually selected by the tube manufacturer, and the characteristics are presented at this point. If the point is chosen for convenience and is not necessarily a recommended operating point, the data presented may be labeled "average characteristics." These characteristics are basic to the tube and not to the tube combined with a circuit. This method of stating characteristics is used for tubes, such as those intended for horizontal-deflection amplifier service, which are difficult to compare under the conditions of their intended service unless all circuit details are known. For some tubes the characteristics shown are classified as "typical operation" and the class of operation such as class A, class AB1, class B, or class C is also given. These characteristics describe the tube in relation to its circuit performance. "Typical operation" parameters may be taken as values that may be applied directly in circuit design to produce predictable results. Obviously, these are not the only permissible operating conditions, since the only limitations imposed are the ratings of the tube and the performance desired; within these limits there may be many combinations of parameters that are satisfactory.

Average characteristics are usually given under conditions that fall within the maximum ratings of the tube, but this is not always true, and a check should be made before a tube is operated for an extended period at the published points. Typical operation points given should be within the ratings in all cases, but a careful designer will want to check these too against the maximum ratings.

Both average characteristics and typical operation sections of a data sheet list the plate and screen voltages, bias, and plate and screen currents drawn by the tube with these voltages applied. In addition to these static characteristics, the amplification factor, transconductance, and plate resistance are usually given. For convenience, these coefficients are usually measured with an a-c signal in the audio-frequency range; however, the results correlate quite well with results obtained at radio frequencies unless limiting factors such as decreased input resistance or transit time enter in. Those tubes which show typical operation, such as audio output tubes, also show values of power output and distortion. If the characteristics shown are for a class of service in which changes in input signal cause changes in plate and screen currents, the plate and screen currents are usually listed for both zero-signal and maximum-signal conditions.

The characteristics published for a tube type are based on a bogey tube. The values are selected by determining the median values for each of the characteristics on a number of lots of the tubes. All the bogey values are rarely found in one tube. The

ELECTRON TUBES   9-27

circuit designer should be aware of this and always keep in mind that the characteristics published are subject to variations that are controlled within certain limits by the tube manufacturer. As an example of the variations that may be encountered, consider the characteristics of the 6C4, a medium-mu triode designed for general-purpose use. The characteristics listed for a bogey tube are:

Heater current (at E_f = 6.3 volts), amp	0.15
Plate voltage, volts	250
Grid voltage, volts	−8.5
Amplification factor	17
Plate resistance, ohms	7,700
Transconductance, μmhos	2,200
Plate current, ma	10.5

A typical manufacturer's specification for the 6C4 lists the limits to which the foregoing characteristics are held:

	Min	Max
Heater current (at E_f = 6.3 volts), amp	0.138	0.162
Amplification factor	15.5	18.5
Plate resistance, ohms	6,200	9,200
Transconductance μmhos	1,750	2,650
Plate current, ma	6.5	14.5

These figures show that for the 6C4 there may be a variation in plate current, under the same operating conditions, of more than 2 to 1, and a variation in transconductance of $1\frac{1}{2}$ to 1, between high- and low-limit tubes. Figure 18 shows the variation in transconductance for 100 type 6C4 tubes; some extremely close-spaced, high-performance tubes show even greater variation in characteristics.

Obviously, designing around a bogey tube with no thought given to the effect of tube characteristic variation is unwise. Thus, in any circuit which appears to be critical of tube characteristics, the circuit designer should try to obtain limit tubes for the important characteristics and use these in evaluating circuit performance. Even though tube selection may be used initially in completing a piece of equipment, it is always possible that tubes near specification limits may find their way into the equipment as replacements.

The effect of characteristic spread on circuit performance may be lessened by selection of the proper biasing scheme. Figures 19 through 24 show characteristic spreads with different biasing methods for a high-mu triode having characteristics similar to one section of a 12AX7. The spreads represent a sample of 27 tubes, all of which were within specification limits. The biasing methods are: fixed bias, Figs. 19 and 22; contact-potential bias, Figs. 20 and 23; and cathode bias, Figs. 21 and

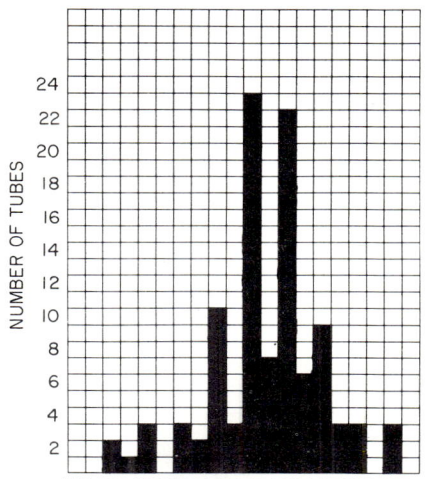

FIG. 18. Variation in transconductance, within specification limits, of a lot of 100 6C4's. (*Courtesy of General Electric Co.*)

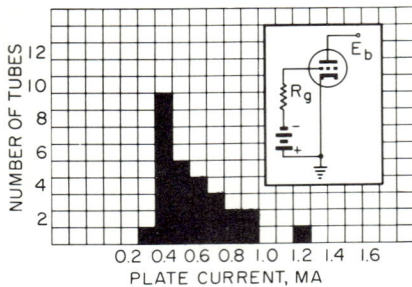

Fig. 19. Plate current spread with fixed bias—high-mu triode. (*Courtesy of General Electric Co.*)

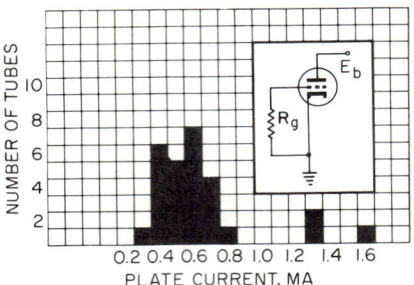

Fig. 20. Plate current spread with contact-potential bias—high-mu triode. (*Courtesy of General Electric Co.*)

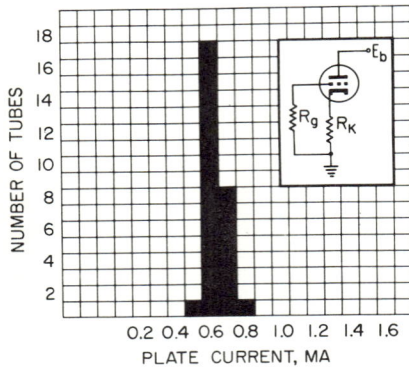

Fig. 21. Plate current spread with cathode bias—high-mu triode. (*Courtesy of General Electric Co.*)

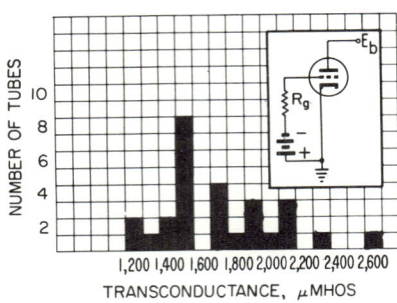

Fig. 22. Transconductance spread with fixed bias—high-mu triode. (*Courtesy of General Electric Co.*)

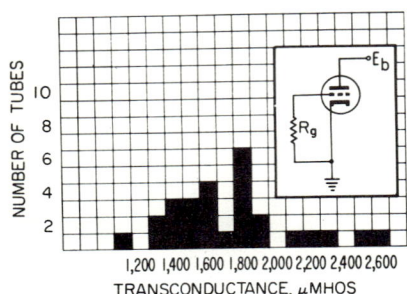

Fig. 23. Transconductance spread with contact-potential bias—high-mu triode. (*Courtesy of General Electric Co.*)

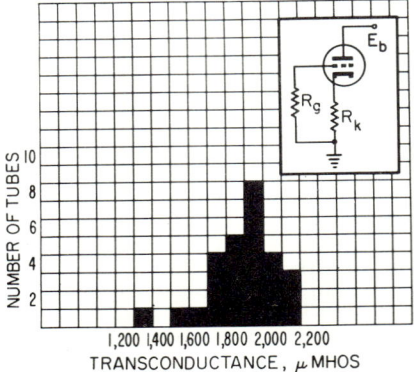

Fig. 24. Transconductance spread with cathode bias—high-mu triode. (*Courtesy of General Electric Co.*)

ELECTRON TUBES 9–29

24. Inspection of these figures shows that cathode bias produces a narrower spread of both plate current and transconductance than the other biasing methods.

The problem of obtaining uniform characteristics in uhf amplifier tubes with very close spacing between grid and cathode is such that fixed-bias operation is not recommended at all for some types by the tube manufacturer. Cathode bias is usually used

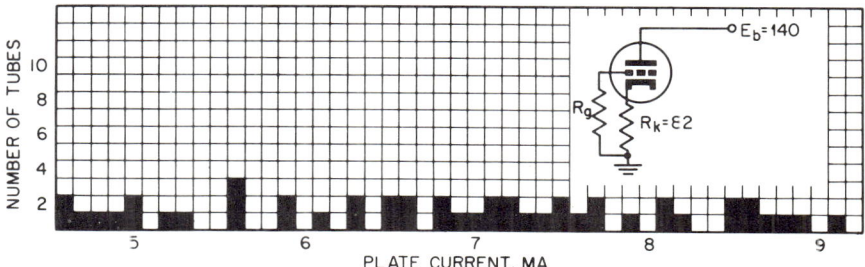

FIG. 25. Plate current spread with cathode bias—type 7462. (*Courtesy of General Electric Co.*)

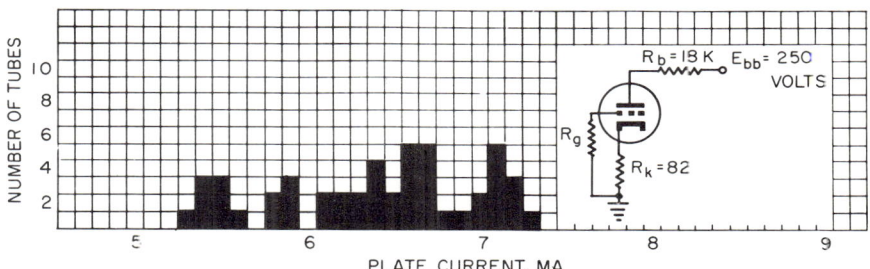

FIG. 26. Plate current spread with cathode bias and plate resistor—type 7462. (*Courtesy of General Electric Co.*)

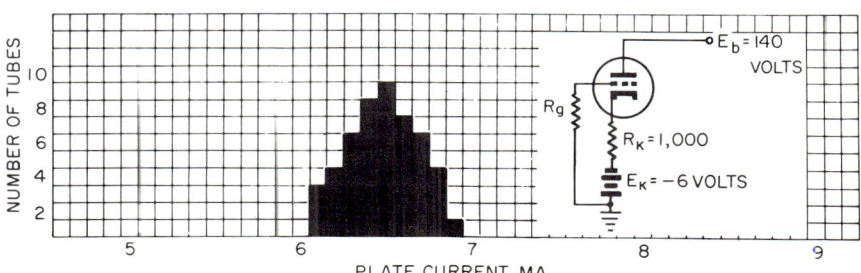

FIG. 27. Plate current spread with cathode bias and a bucking cathode voltage—type 7462 (*Courtesy of General Electric Co.*)

and several variations in the basic cathode-bias scheme are used to reduce characteristic spread. Figures 25 through 27 show the spread of plate current with three cathode-biasing methods for a close-spaced, metal-ceramic, planar triode, the 7462. Figure 25 shows the results with simple cathode bias, Fig. 26 with cathode bias plus a plate resistor and the plate voltage elevated to compensate for the drop in the plate resistor, and Fig. 27 with a cathode resistor plus a bucking fixed cathode voltage.

Inspection of these figures shows that minimum spread is obtained with the method that combines a cathode resistor with a bucking voltage. The same results as those shown in Fig. 27 may be obtained by using a cathode resistor combined with fixed positive bias on the grid, as shown in Fig. 28.

In some applications of close-spaced, high-performance tubes the most important characteristic, aside from the ability to deliver gain in the circuit, is the noise figure. Noise figure is usually optimum at a specific value of bias, and the combination of cathode bias and a resistor in the plate circuit will give the narrowest spread of this

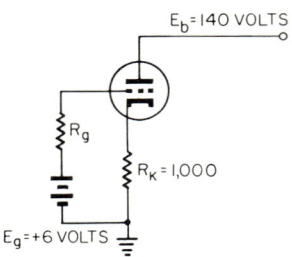

FIG. 28. Alternate method for applying bucking voltage with cathode bias.

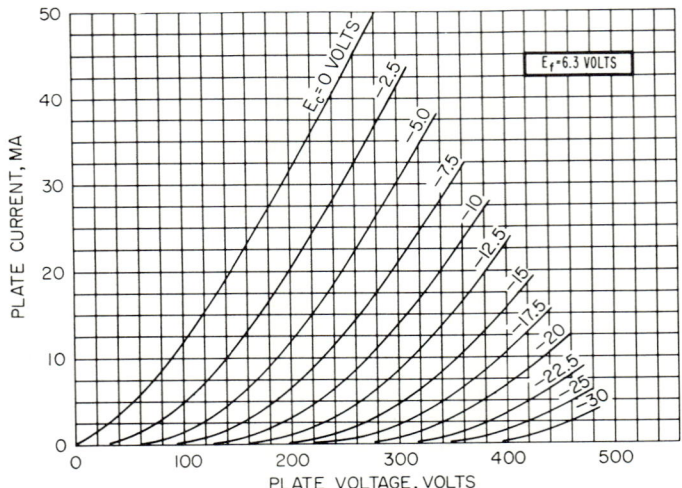

FIG. 29. Typical triode plate family—type 6C4. (*Courtesy of General Electric Co.*)

characteristic, since this combination tends to keep the tube current relatively constant and thus hold the bias constant.

When the characteristics of a tube suitable for amplifier service are presented in graphical form, the most common plot is the plate current as a function of the plate voltage at various bias levels. If the tube is a tetrode or pentode, a fixed value of screen voltage is specified and, in addition, the screen current is presented in the same manner, often on the same graph sheet with the plate current. These curves are usually referred to as "plate families" and "screen families" (see Figs. 29 and 30). As a supplement to these curves, the "transfer" curves are often published. For a triode, these show the plate current as a function of the grid voltage at various plate voltage levels and are called "plate transfer" curves (see Fig. 31). For a tetrode or pentode, plate and screen currents are plotted as functions of grid-1 voltage at various screen voltage levels and a fixed value of plate voltage (see Figs. 32 and 33). These

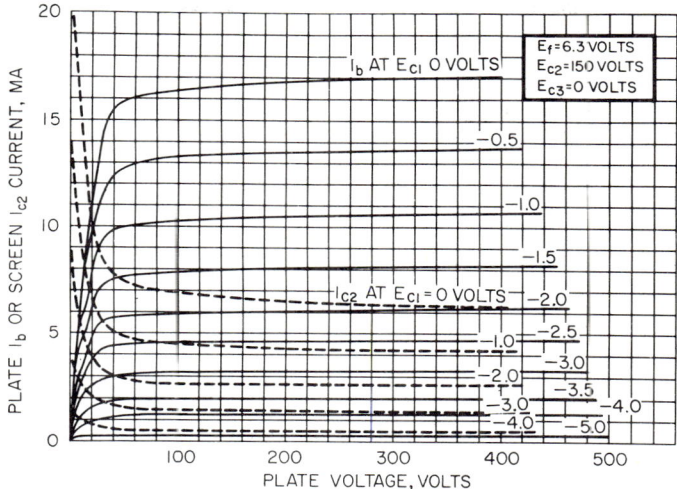

Fig. 30. Typical pentode plate and screen families—type 6AU6. (*Courtesy of General Electric Co.*)

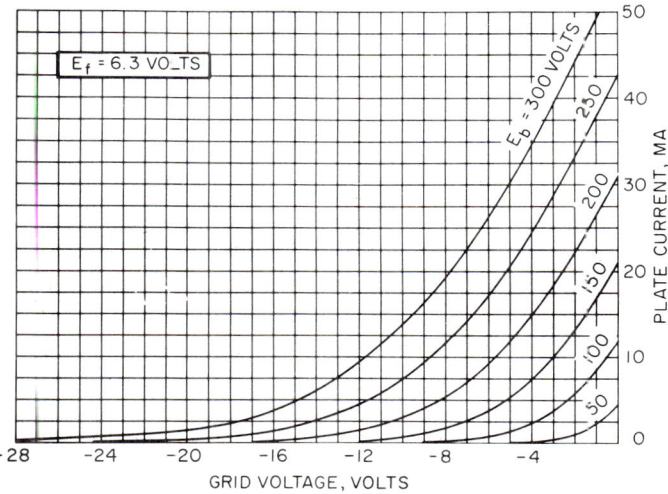

Fig. 31. Typical triode plate transfer curves—type 6C4. (*Courtesy of General Electric Co.*)

are usually plotted on separate graph sheets, and the screen curves are similarly referred to as "screen transfer" curves.

Since there are more operating parameters that may be varied with tetrodes and pentodes than with triodes, "zero bias" plate and screen families are sometimes presented. On these curves the plate and screen currents vs. plate voltage are plotted at zero grid-1 voltage and various screen voltage levels (see Fig. 34).

If the tube for which characteristic curves are published is one whose grid may be driven into the positive region in its recommended mode of operation, plate current, screen current, and grid current vs. plate voltage may be plotted at various levels of positive grid voltage (see Figs. 35 to 37). In addition, for this type of tube the data are often rearranged in the form of "constant current" curves. In these curves, grid voltage vs. plate voltage is plotted at constant values of plate current and grid current.

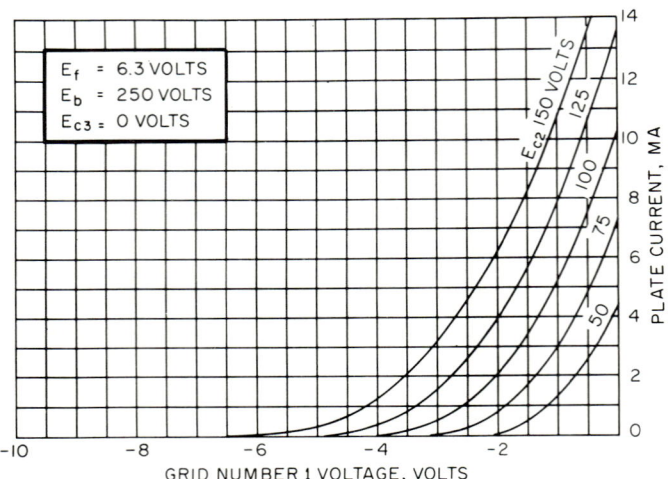

FIG. 32. Typical pentode plate transfer curves—type 6AU6. (*Courtesy of General Electric Co.*)

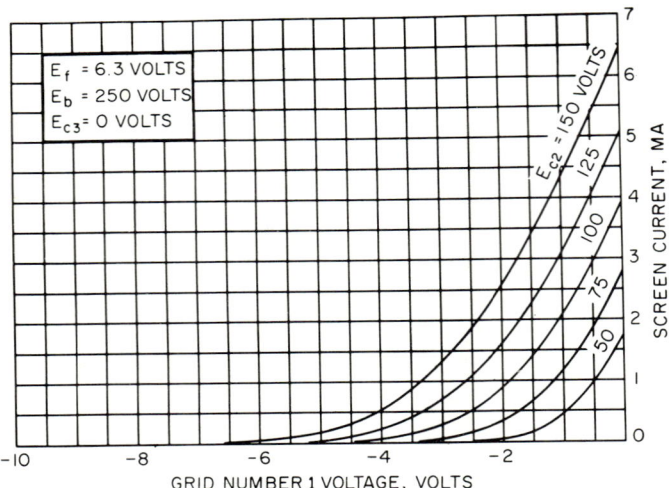

FIG. 33. Typical pentode screen transfer curves—type 6AU6. (*Courtesy of General Electric Co.*)

ELECTRON TUBES 9-33

If the tube is a tetrode or pentode, a fixed value of screen voltage is used, and the screen constant-current curves are added (see Figs. 38 to 40).

Other curves less frequently published are those which show the dynamic characteristics of tubes, including power output and harmonic distortion. Examples of those are shown in Figs. 41 to 43.

All the foregoing curves are useful for determining various tube characteristics under operating conditions other than those listed on the tube data sheet as average characteristics. In addition, they may be used to predict circuit operation by graphical

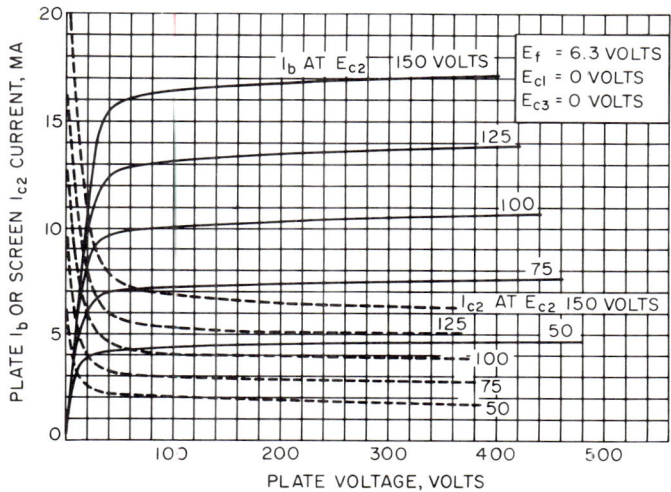

FIG. 34. Typical pentode zero-bias plate and screen families—type 6AU6. (*Courtesy of General Electric Co.*)

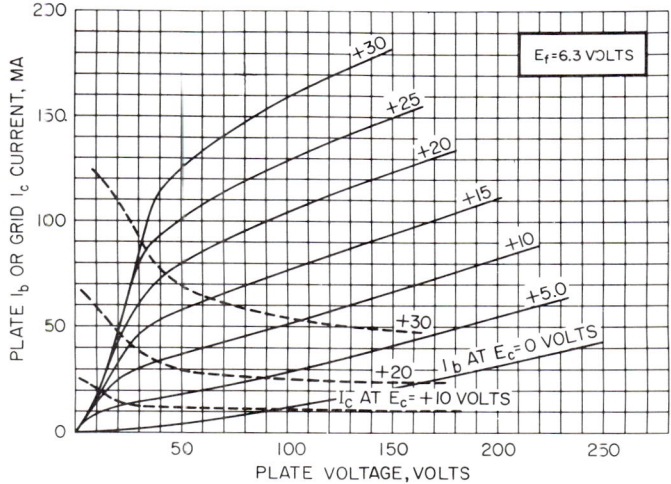

FIG. 35. Typical triode positive-bias plate and grid families—type 6C4. (*Courtesy of General Electric Co.*)

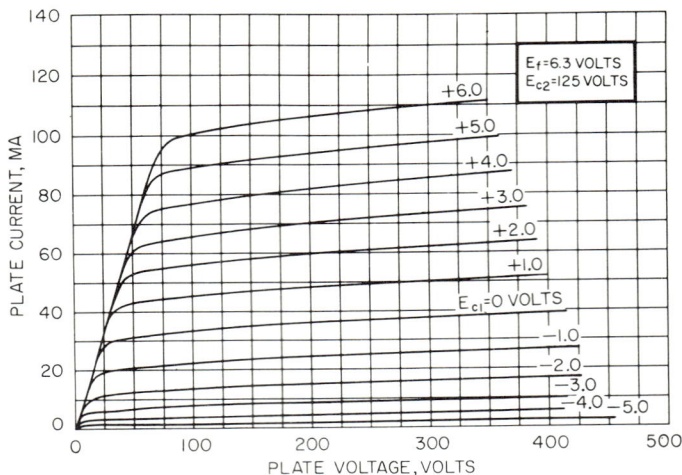

Fig. 36. Typical pentode positive-bias plate family—type 6AU8-A. (*Courtesy of General Electric Co.*)

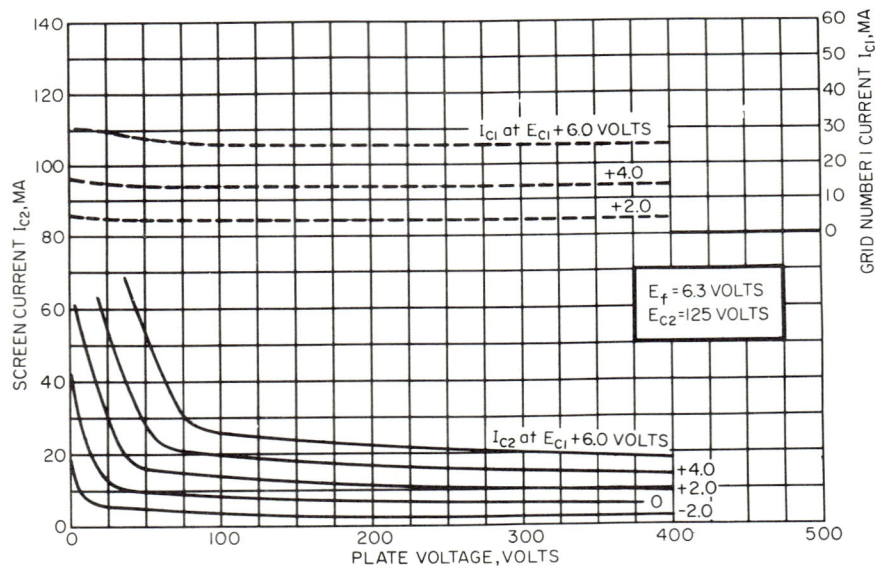

Fig. 37. Typical pentode positive-bias grid-1 and screen families—type 6AU8-A. (*Courtesy of General Electric Co.*)

constructions.[8] The constant-current curves are especially useful for determining circuit operation of radio-frequency power amplifiers.[9,10]

[8] F. Langford-Smith, "Radiotron Designer's Handbook," 4th ed., Wireless Press, Sydney, Australia, 1957.
[9] "Reference Data for Radio Engineers," 4th ed., International Telephone and Telegraph Corp., New York, 1956.
[10] Tube Performance Computer, *Application Bulletin* 5, Eitel-McCullough, Inc., San Carlos, Calif., 1952.

ELECTRON TUBES

Because of the number of parameters that may be varied in preparing curves, it is inevitable that a circuit designer may wish to use a tube under conditions for which characteristics cannot be determined from available curves. Fortunately, approximations of the characteristics at points other than those published may be obtained by the use of conversion factors. These factors are based on the relationship between plate voltage and plate current in an ideal diode, where the plate current varies as the three-halves power of the plate voltage. Thus, if the plate current of a diode at a certain plate voltage is known, the plate current at a voltage 0.8 of the first voltage may be determined by multiplying the known plate current by 0.72, which is 0.8

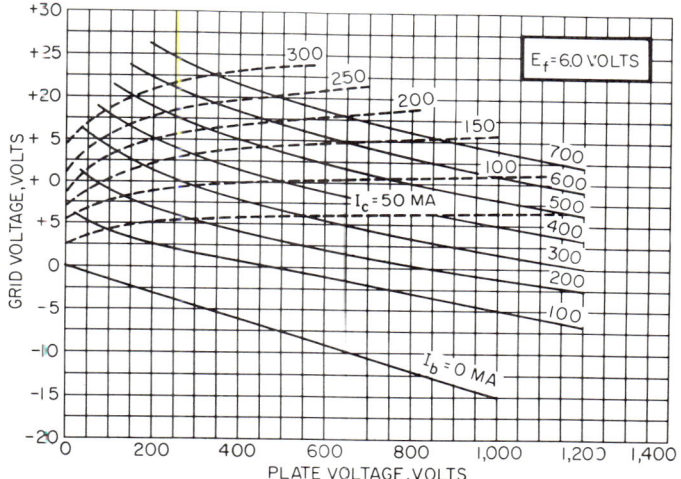

FIG. 38. Typical triode constant-current plate and grid curves—type 3C39-A. (*Courtesy of General Electric Co.*)

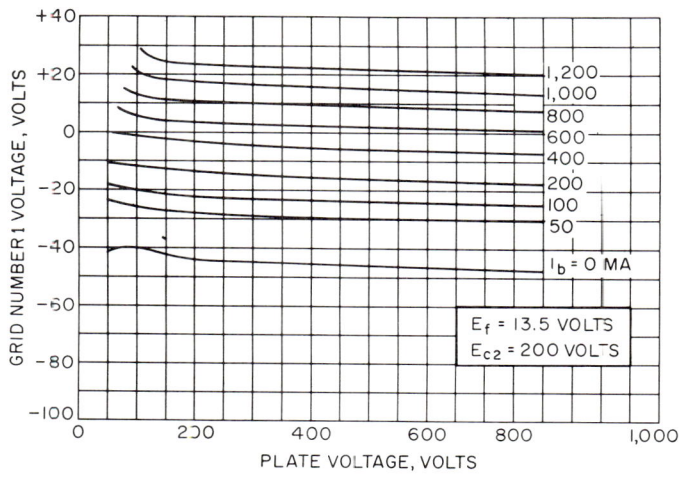

FIG. 39. Typical pentode constant-current plate curves—type 7984. (*Courtesy of General Electric Co.*)

9-36 AMPLIFYING DEVICES

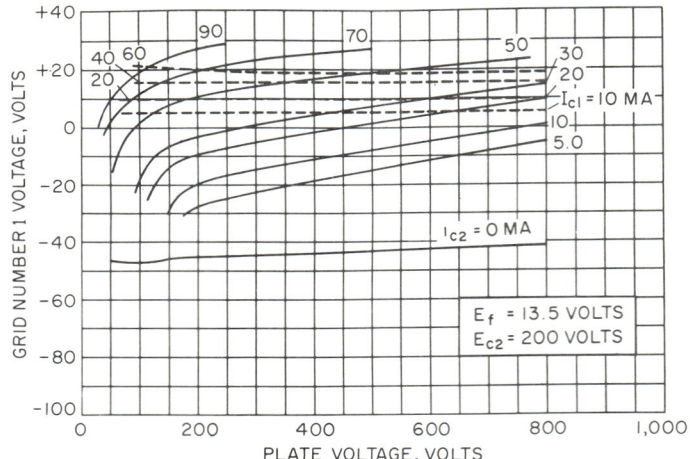

Fig. 40. Typical pentode constant-current grid-1 and screen curves—type 7984. (*Courtesy of General Electric Co.*)

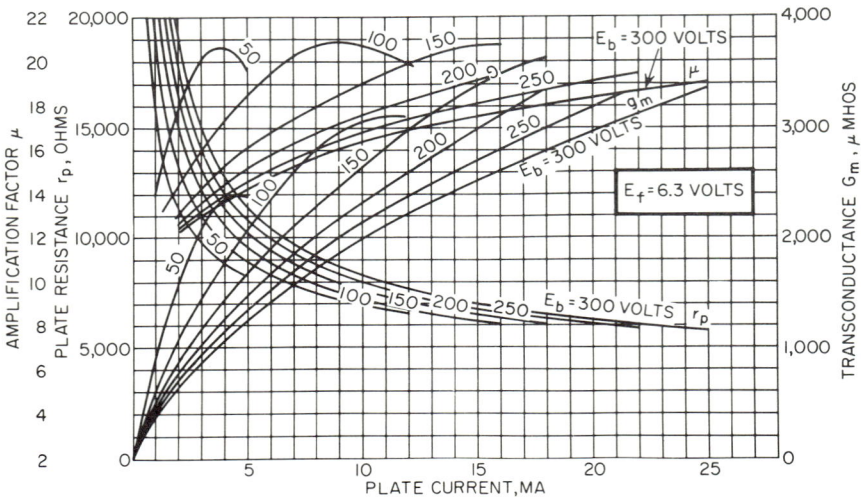

Fig. 41. Typical triode amplification factor, transconductance, and plate resistance transfer curves—type 6C4. (*Courtesy of General Electric Co.*)

raised to the three-halves power. This method may be extended to triodes, tetrodes, and pentodes with acceptable accuracy by simultaneously varying all the electrode voltages by the same percentage. It may also be used to determine other tube characteristics, since they are all related to the electrode voltages and currents and their interdependent changes. Thus the power output may be determined by multiplying the known power output by a factor which represents the voltage-multiplying factor raised to the five-halves power. In similar fashion, the transconductance may be

ELECTRON TUBES

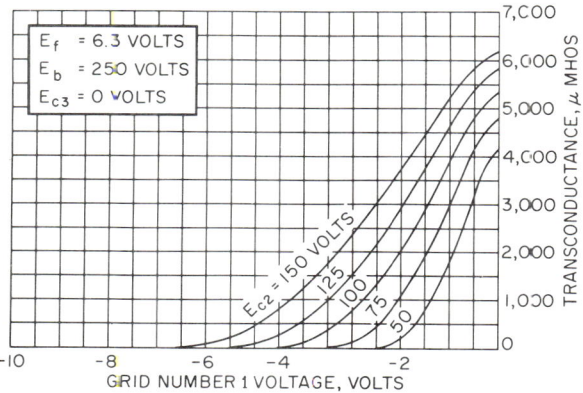

Fig. 42. Typical pentode transconductance transfer curves—type 6AU6. (*Courtesy of General Electric Co.*)

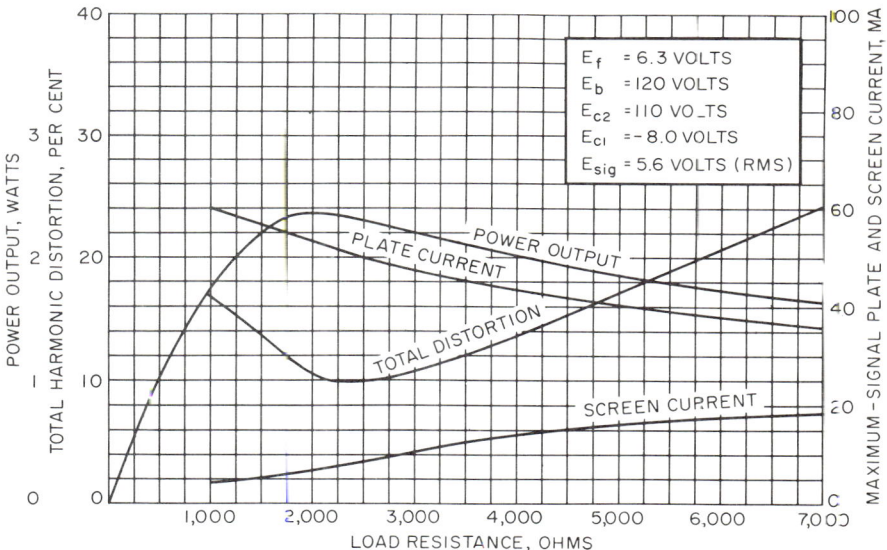

Fig. 43. Typical pentode power output and distortion curves—type 6CU5. (*Courtesy of General Electric Co.*)

converted by a factor equal to the voltage-multiplying factor raised to the one-half power, and the plate resistance by a factor equal to the voltage-multiplying factor raised to the negative one-half power. Figure 44 is a conversion-factor chart which allows ready determination of the various factors from the basic voltage-multiplying factor. The example below illustrates the use of the chart.

The 50EH5 is a power output pentode with the following characteristics and typical operation listed:

AMPLIFYING DEVICES

Plate voltage, volts	110
Screen voltage, volts	115
Cathode-bias resistor, ohms	62
Peak a-f grid-1 voltage, volts	3.0
Plate resistance, ohms	11,000
Transconductance, μmhos	14,600
Zero-signal plate current, ma	42
Zero-signal screen current, ma	11.5
Load resistance, ohms	3,000
Maximum-signal power output, watts	1.4

Assume that characteristics are desired with the screen voltage reduced to 100 volts. The new value (100 volts) divided by the published value (115 volts) gives a voltage conversion factor F_e of 0.87. Enter the chart of Fig. 44 at 0.87 and determine the values of the current conversion factor F_i, power conversion factor F_p, resistance conversion factor F_r, and transconductance conversion factor F_{gm}. For a value of F_e of 0.87 these are $F_i = 0.81, F_p = 0.71, F_r = 1.07,$ and $F_{gm} = 0.93$. Each of these factors

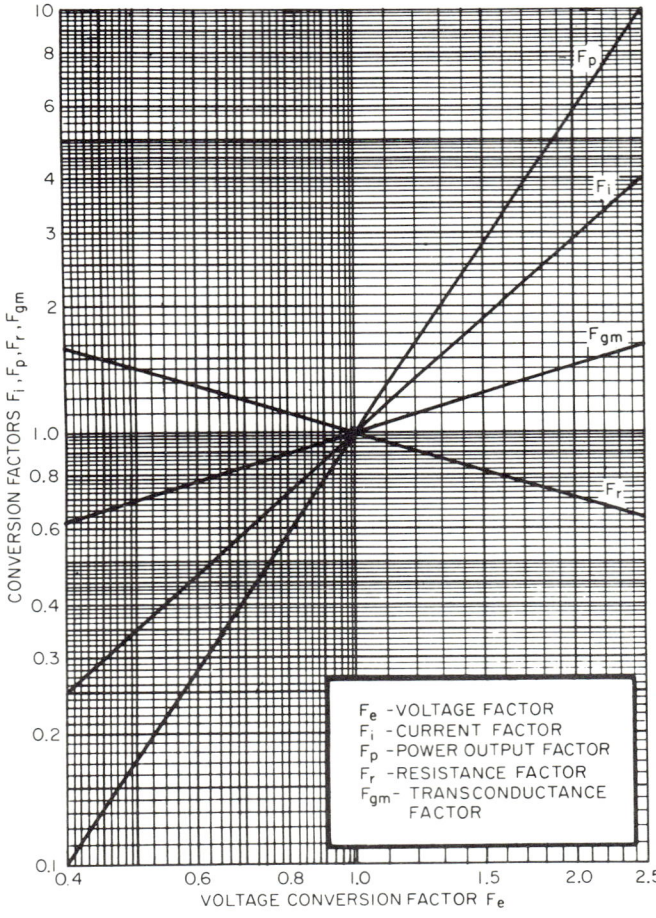

FIG. 44. Characteristics conversion-factor chart. (*Courtesy of RCA.*)

is used to convert the corresponding parameter: F_e for the plate and a-f grid voltages, F_i for the plate and screen currents, F_r for the plate and plate load resistances, F_{tm} for the transconductance, and F_p for the power output. The new value of cathode-bias resistor is calculated by noting that the original bias resistor produced a bias of approximately 3.3 volts at a cathode current of 53.5 ma. Thus the new bias must be 0.87 times 3.3 or approximately 2.9 volts. This voltage divided by the new cathode current of 43.3 ma gives a cathode-bias resistor value of 67 ohms. The new characteristics for the 50EH5 are:

Plate voltage, volts	96
Screen voltage, volts	100
Cathode-bias resistor, ohms	67
Peak a-f grid-1 voltage, volts	2.9
Plate resistance, ohms	11,800
Transconductance, μmhos	13,600
Zero-signal plate current, ma	34
Zero-signal screen current, ma	9.3
Load resistance, ohms	3,200
Maximum-signal power output, watts	1.0

If, in addition to the plate family of a tetrode or pentode, zero-bias curves are available, it is possible to plot new curves for values of screen voltage other than the one at which the plate family was plotted. This can be done because the ratio of plate current at any bias value to that at zero bias remains the same when the screen voltage is changed, if the bias is multiplied by the ratio of the two screen voltages. In practice, this will produce curves at odd values of bias, but interpolation may be used to draw curves at other bias values.

As an example, consider the 6AU6. A plate family, plotted with a screen voltage of 150 volts (Fig. 45), and a zero-bias family (Fig. 46) are available; it is desired to convert one or more of the plate-family curves to a screen voltage of 100 volts.

The voltage conversion factor is obtained by dividing the screen voltage at which curves are desired (100 volts) by the voltage (150 volts) at which curves are available. This gives a factor of 0.67, which is used to convert the grid-bias voltages to the new conditions. For the first new point, any plate voltage on the existing curves may be chosen, and for this example 250 volts is used. Plate current at 250 volts and zero

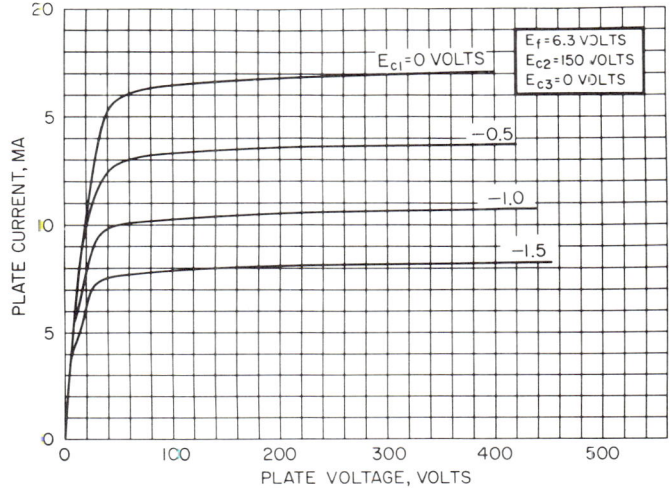

FIG. 45. Partial plate family for type 6AU6. (*Courtesy of General Electric Co.*)

bias (16.9 ma) and 250 volts at −1.0 volt bias (10.6 ma) are determined from the plate-family curves (Fig. 45). The plate current at −1.0 volt bias divided by the plate current at zero bias gives the ratio (0.63) which is the current ratio.

The zero-bias curve at a screen voltage of 100 volts (Fig. 46) is replotted on a new sheet (Fig. 47, top curve). The voltage (250 volts) at which the current readings were

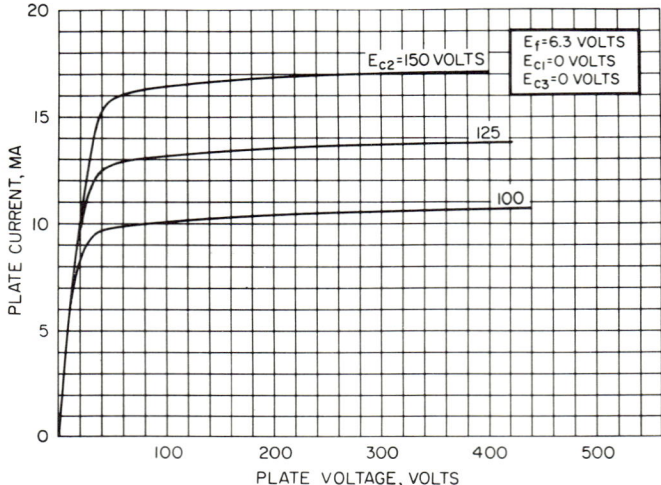

Fig. 46. Partial zero-bias plate family for type 6AU6. (*Courtesy of General Electric Co.*)

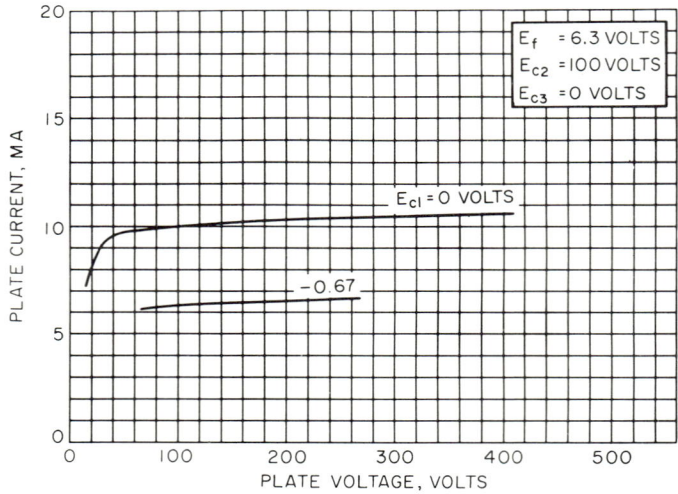

Fig. 47. Derived curves for type 6AU6. (*Courtesy of General Electric Co.*)

obtained from Fig. 45 is multiplied by the voltage conversion factor (0.67) to give the voltage (167 volts) at which a current reading should be obtained from the zero-bias curve in Fig. 47. This current (10.3 ma) multiplied by the current ratio (0.63) gives the current for one point on a new curve at a screen voltage of 100 volts, a plate voltage of 167 volts, and a bias of −0.67 volt (−1.0 times 0.67). Additional points may be obtained and the curve at −0.67 volt bias plotted as in Fig. 47.

When using characteristic curves, the circuit designer should be aware of a number of limitations. First, the curves are based on the characteristics of a bogey tube, and deviations from these curves may be expected in any lot of tubes. Second, the curves sometimes present characteristics that are not closely controlled in manufacture, and even greater deviation from the values displayed on the curves may be expected for some of these. For example, grid current is one such characteristic which may vary between wide limits. Third, the use of conversion factors produces only approximate values, and the further from the original curves these values are extended, the more approximate they become. The limits of usable voltage conversion factors are from about 0.7 to 1.5. In using these factors, caution must also be observed in staying well to the right of the knee of the curves when converting points for tetrodes and pentodes. This is especially true of tetrodes and beam-type pentodes, where secondary-emission effects make prediction of characteristic changes hazardous. Fourth, curves are often extended outside the steady-state dissipation limits of the tube in order to show its potentialities with pulsed or other low-duty-factor signals.

Special Tests and Ratings. Nearly all data sheets published for tubes contain, as a minimum, the information discussed thus far. When the tubes are those intended for critical industrial and military applications, additional data are usually provided to give the characteristics limits to be expected, to cover the behavior of the tubes under severe environmental stress, and often to include information on life tests.

The characteristics limits may be given as initial characteristics limits or as both initial limits and limits after operation for a stated number of hours. The characteristics listed usually include most of those given in the average characteristics section of the data sheet, but include in addition those characteristics which must be controlled for proper operation and life in critical circuits. For example, the data sheet for the 6201 of one manufacturer lists the bogey values of the following characteristics: heater current, interelectrode capacitances, amplification factor, plate resistance, transconductance, plate current, and grid voltage for plate current cutoff. The characteristics limits, however, are listed for the following: heater current, interelectrode capacitances, plate current, plate current difference between sections, transconductance, transconductance at reduced heater voltage, amplification factor, grid voltage for plate current cutoff, pulse cathode current, pulse cathode current at reduced heater voltage, negative grid current, heater-cathode leakage current, interelectrode leakage resistance, grid emission current, and low-frequency vibrational output. Of the foregoing, the heater current, transconductance, transconductance at reduced heater voltage, pulse cathode current, negative grid current, heater-cathode leakage current, and interelectrode leakage resistance limits are given as initial values and as values at the end of 1,000 hr of life-test operation; initial limits only are given for the other characteristics.

Interpretation of these data is difficult without some knowledge of statistical quality control and a familiarity with armed services specifications for electron tubes. The latter is especially true, since the limit data published are usually those called for in the applicable armed services specification, and manufacturers often pattern their data presentation after that used in these specifications. In addition, the sampling plans used and the quality levels prescribed are usually in line with the same specifications.[11] Because of widespread familiarity with armed services specifications by designers working in the military field, material necessary to the interpretation of limits is often omitted from manufacturer's data sheets, and reference is made to MIL-E-1 for further information; obviously, this publication[12] is essential to a circuit designer applying tubes to military equipment.

The application of statistical quality control to electron-tube manufacture is outside the scope of this chapter, and reference should be made to a suitable text.[13]

Severe environmental stresses include shock, fatigue, vibration, high altitude, and nuclear radiation. The data presented on the effect of these stresses on tubes are also patterned after the armed services specifications.

[11] Sampling Procedures and Tables for Inspection by Attributes, MIL-STD-105 Military Standard.

[12] General Specification for Electron Tubes, Military Specification MIL-E-1.

[13] Eugene L. Grant, "Statistical Quality Control," 3d ed., McGraw-Hill Book Company, New York, 1964.

3. ELECTRON-TUBE LIFE

3.1. General. The circuit designer's problems with tube characteristics are not solved when he determines that his circuit will work with the characteristics spread inherent in the tube type he selects. He must further determine what effect the normal degradation of characteristics with life will have on operation of the circuit. In addition, he must not lose sight of the possibility that some of the detrimental electrode currents not normally considered tube characteristics (see Sec. 2.2) may increase, and some of the fundamental performance limitations (see Sec. 2.3) may become more restrictive during life. To make these evaluations the circuit designer needs both life-test information and some idea of what he considers satisfactory life for the tubes. The latter must be based on performance limits of the equipment in which the tubes are to be used, the permissible intervals between tube replacements, or the desired life of the equipment if the tubes cannot be readily replaced.

Life-test end points are published on the data sheets for many of the four-digit types, and may be obtained for nearly all other tube types from the tube manufacturer. However, these life tests are usually run under conditions that place the maximum stress, within the ratings, on the tubes. Thus, they do not always constitute a measure of tube life under less severe conditions. Unfortunately, there are no empirical formulas for converting these life-test results to projected life under conditions well within the ratings that a prudent designer might choose. In addition, economic considerations preclude checking all the possible degradations which might affect all possible uses of a tube type. However, when a tube type is specifically designed for a particular use, it is reasonable to assume that the life tests will reflect the requirements of that use.

Some of the factors that shorten tube life are obvious, and all life tests confirm their effects. Others, while not so obvious, have been agreed upon as to their effects. As an example of the latter, at one time it might have seemed obvious that if operation at high levels of plate current tended to exhaust the emission capabilities of an oxide-coated cathode early in life, then operation at lower current levels should extend cathode life, and the lower the current the longer the life. Only with the extensive use of large numbers of tubes in circuits where the tube remained cut off for long periods was the fallacy of extending this concept to very low cathode currents discovered: operation at cutoff tends to increase cathode interface impedance which, in its own way, is just as detrimental to long life as cathode emission failure. To make matters more confusing, the exact effects on life of some operating conditions are not fully agreed upon, and clear-cut proof one way or the other is seldom obtained from life tests. Finally, generalizations concerning life can seldom be applied to *all* tube types because there is considerable variation in materials and processing methods used in tubes among manufacturers. Even tubes of a specific type from different manufacturers are not necessarily comparable. Perhaps the only generalization possible is that best life is usually obtained when a tube is operated within its ratings. However, no rule can be laid down as to how far within ratings optimum life is obtained, especially since some ratings are unfortunately not realistic.

Life-test data may be presented in such a manner as to indicate the percentage of tubes that fail or the percentage of tubes that survive during the test, or the degradation of a specific characteristic during the test. Tubes that fail are termed "inoperatives" and are those which show any of the following defects:

1. Air leaks
2. Shorted elements
3. Open elements
4. Degradation of a major characteristic, such as transconductance, power output, or plate current, to an arbitrary low percentage (usually 10 per cent) of the initial bogey value of the characteristic
5. Excessive grid current—usually a value greater than 200 μa
6. Excessive heater-cathode leakage—usually a value greater than 200 μa

Presentation of the degradation of characteristics on data sheets is usually in terms of the minimum value (or the maximum value if degradation involves an increase) of the characteristic that may be expected after a specified time of operation. This time is often 1,000 hr, although for very long life tubes it may be 10,000 hr. The value of the characteristic shown at the end of the stated time does not necessarily hold for all individual tubes but rather for a sufficient number in a given lot of tubes to make that lot acceptable under a statistical sampling plan. In some instances average degradation for a lot may be presented graphically as shown in the succeeding sections.

In the following sections examples of plotted data are shown for several tube types; information on a number of additional types may be obtained from two books[14,15] devoted exclusively to life factors.

3.2. Heater Voltage. The effects of three levels of heater voltage during life test of a miniature twin triode are illustrated here to show graphical presentation of life-test data. Figure 48 shows survival with respect to inoperatives and Fig. 49 the degradation of transconductance during life tests. For the lots represented for this

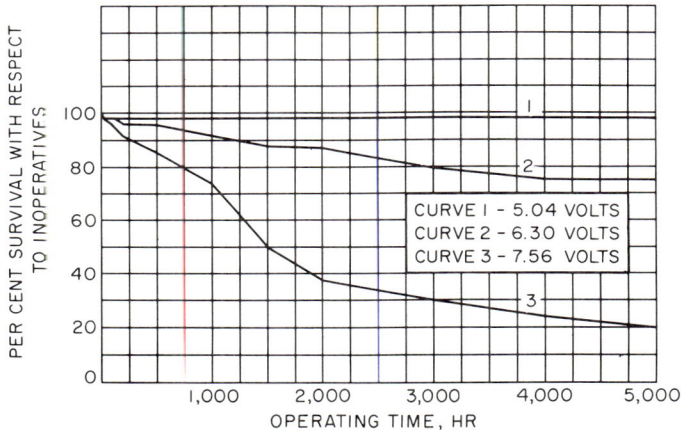

FIG. 48. Survival of the 5670 at three levels of heater voltage. (*Courtesy of General Electric Co.*).

particular tube type, Fig. 48 shows that the smallest number of inoperatives occurred with the minimum heater voltage and the largest number with the maximum heater voltage. Figure 49 shows that transconductance underwent the least change with the lowest heater voltage, although it started at a lower-than-bogey level, while at higher heater voltages it started at a higher level but slumped more. Thus operation at low heater voltage appears attractive to preserve a low failure rate and to achieve stability of transconductance, while operation at elevated heater voltage allows achievement of high initial values of transconductance and a reasonably high level of transconductance at the end of life. However, there are certain disadvantages in operation both below and above bogey heater voltage which are considered here.

Operation below bogey heater voltage involves the following considerations. First, the emission capability of a tube plotted against heater voltage shows a "knee" in the curve, below which emission is seriously limited. This knee is usually below 5.7 volts for 6.3-volt heater types, and those tubes which have data published on special tests and ratings often list the performance at 5.7 volts, most often in the form of limits for

---

[14] Walsh and Tsao, "Electron Tube Life Factors," Engineering Publishers, Elizabeth, N.J., 1959.
[15] Edwards, Lammers, and Zoellner, "Subminiature Electron Tube Life Factors," Engineering Publishers, Elizabeth, N.J., 1961.

the major characteristic such as transconductance. Second, operation at lowered heater voltage requires regulation of the heater voltage supply to ensure that the voltage does not drop below the selected value of lowered heater voltage, as the knee may be approached or even passed under low line conditions without regulation, with consequent degradation of characteristics below acceptable values. Third, greater variations in electrical characteristics usually occur from tube to tube at reduced heater voltage than at bogey heater voltage.

Operation above bogey heater voltage involves the following considerations. First, high heater voltage will shorten the life of the heater. Catastrophic failure of the heater may be due to poor welds or to fracture of the heater wire because of crystal growth within the wire. The latter type of failure is usually due to the promotion of excessive crystal growth by elevated heater temperature. Second, high heater voltage raises the cathode temperature, causing excessive sublimation of the cathode coating material, which may be deposited on various parts of the tube with the following results: (1) Loss of material from the cathode, especially activating agents, will reduce the emission capabilities of the cathode. (2) Reduced emission, if it falls to low

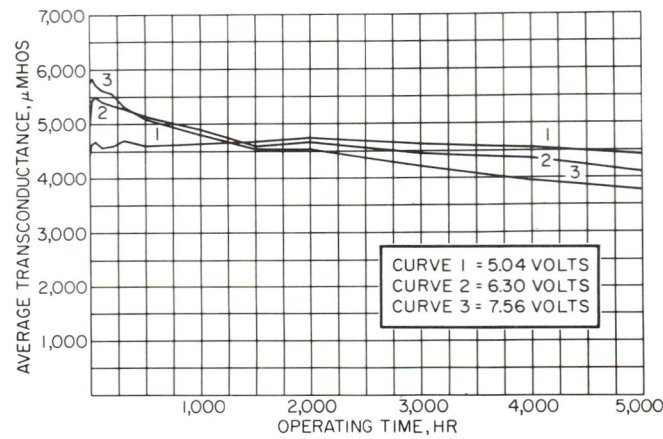

Fig. 49. Transconductance variations of the 5670 at three levels of heater voltage. (*Courtesy of General Electric Co.*)

enough levels, will degrade nearly all characteristics. Even before this level is reached, reduced emission will affect pulse operation. Reduced emission may also increase the noise figure, since low noise figure is favored by a dense space-charge cloud which tends to "smooth" electron flow to the plate. (3) Emission from elements upon which the emissive material is deposited may occur, provided the elements are heated sufficiently. (4) Leakage paths will be established on insulators, which may result in unwanted interelement currents. Third, excessive heat from the cathode will be transmitted to adjacent elements, especially the No. 1 grid. This heat will increase emission from the grid, which will be aggravated by the emissive material deposited on the grid as a result of the excessive cathode temperature. Fourth, high heater voltage encourages migration of impurities from the heater or the cathode sleeve to the heater coating, and this may increase heater-cathode leakage. Fifth, high heater voltage shifts the contact potential bias produced (see Fig. 1).

In summary, operation at bogey heater voltage is recommended unless extremely long life at a relatively low performance level is desired.

3.3. Power Dissipation. The effects of three levels of combined plate and screen dissipation on survival and power output during life test of an audio power pentode are shown in Figs. 50 and 51. Power output was measured at the conditions required to

ELECTRON TUBES   9-45

obtain the dissipations indicated for each curve; since these conditions were selected with the dissipations as the prime requirements, power output is not optimum, and varies with the level of dissipation achieved. Figure 50 shows that for the lots tested of this particular tube type, excessive dissipation increased the number of failures However, an interesting result is that the failure rate for the lowest dissipation was

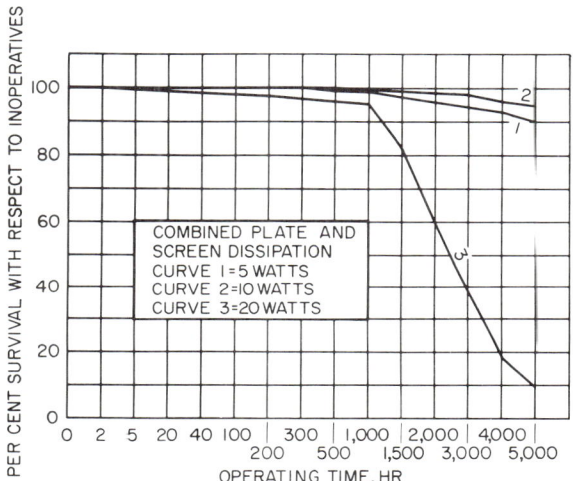

FIG. 50. Survival of the 6005 at three levels of combined plate and screen dissipation. (*Courtesy of General Electric Co.*)

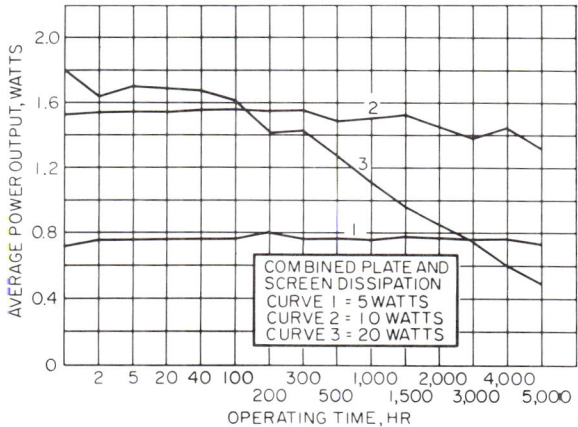

FIG. 51. Power output variations of the 6005 at three levels of combined plate and screen dissipation. (*Courtesy of General Electric Co.*)

greater than for the intermediate dissipation. Although this anomaly is not typical for all tubes, it does illustrate some of the difficulties involved in predicting tube life. Figure 51 shows that power output slumped more at the highest dissipation level; however, in some applications not requiring long life, the increased initial power output might make high-dissipation operation preferable.

9-46　AMPLIFYING DEVICES

Maximum dissipation, neglecting the dissipation in the heater, usually occurs at the plate. The chief damaging effects of excessive plate dissipation are release of gas, plate emission, plate deformation, and heating of the tube envelope. Screen dissipation is second to plate dissipation in magnitude, and excessive values may also result in gas release and emission. Grossly excessive dissipation at the screen may melt individual screen wires. Grid dissipation is a problem only with tubes that are driven to grid current, such as oscillators and class C amplifiers, and the principal effect is grid emission.

In general, high dissipation tends to shorten tube life. When high dissipation is required in an application, tube life may be lengthened by removing as much of the heat as possible from the tube. If the tube is one having an external anode, circuitry can sometimes be arranged to allow direct connection of the anode to a heat sink or to a radiator which in turn may be cooled by forced air. Heat may be removed from glass tubes by using tight-fitting heat-radiating shields.

3.4. Envelope Temperature. There is almost universal agreement that high envelope temperatures shorten tube life. The degree to which it is shortened in one tube type may be gauged from Figs. 52 and 53, which show failure rate and degradation

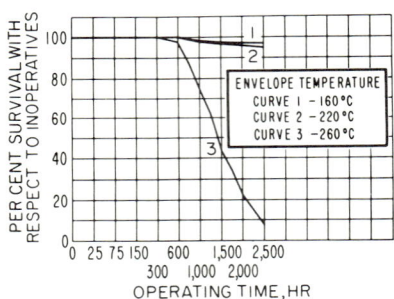

Fig. 52. Survival of the 5840 at three levels of envelope temperature. (*Courtesy of General Electric Co.*)

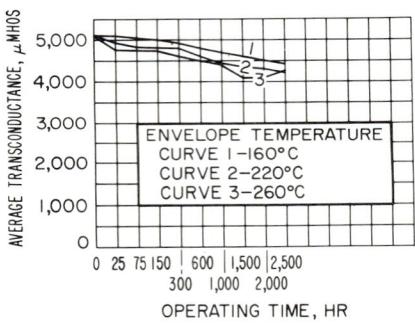

Fig. 53. Transconductance variations of the 5840 at three levels of envelope temperature. (*Courtesy of General Electric Co.*)

of transconductance of a subminiature pentode at several levels of envelope temperature. It is probable that two factors are involved in the shortening of life by high temperatures: (1) Higher temperatures tend to dislodge occluded gas from the envelope, which serves to increase gas current in the tube and to contaminate the cathode. (2) Electrolysis of the glass, in glass-envelope tubes, due to electron bombardment, occurs more readily at elevated envelope temperatures.

Some of the same measures recommended for reducing plate temperature are effective for reducing envelope temperature. Heat removed from the plate will not be transmitted to the envelope and air-blast cooling of the plate radiator may be arranged so that some of the air is diverted to the purpose of cooling the envelope. Heat-radiating shields are especially effective in cooling glass envelopes, provided they are properly designed.

3.5. Cathode Interface. Early in the application of tubes in computer circuits it was found that operation of many tubes for long periods with the plate current cut off produced a condition where, when an attempt was made to pulse the tube, emission appeared to be quite low. This phenomenon, known as "cathode interface" or "sleeping sickness," was found to be due to a resistive layer which built up between the nickel sleeve of the cathode and the cathode coating. The resistive layer, which is the result of chemical combination of the barium oxide in the cathode coating and the silicon in the cathode sleeve, acts as an unwanted resistor in the cathode circuit which produces cathode bias. Because there is capacitance associated with the resistance, interface shows frequency-dependent characteristics, and transconductance and gain vary with frequency.

The barium orthosilicate layer responsible for interface builds up most rapidly when cathode temperatures are high. Therefore, high heater voltage operation favors interface formation. In addition, operation for long periods under conditions of plate current cutoff accentuates the effect.

Aside from proper control of heater voltages, the user's only recourse is to choose tube types designed for low interface which include interface controls in their specifications. This is particularly important where application requires the tube to operate with little or no plate current for long periods of time (hours or days) with sudden demands for greater plate currents, as in computers or other intermittent-pulse applications.

3.6. Heater-Cathode Voltage. Although oxide-coated cathode-type tubes have a wide variation in heater-cathode voltage ratings, they seldom vary in their actual capability. Universally, they use a fired aluminum oxide coating on the heater wire as insulation. And, although impurity levels are controlled to a high degree, this insulation suffers varying degrees of semiconductor activity depending upon heater temperature, heater-cathode voltage, and past history of these factors in the manufacturer's processing.

Unfortunately, the oxide coating appears to act as an electrolyte, allowing migration of metallic ions from the heater wire or the inner surface of the cathode, depending upon heater-cathode voltage polarity. When the cathode is positive with respect to the heater, the tendency is for the insulation to become a semiconductor and heater-cathode leakage increases with life. When the cathode is negative with respect to the heater, the metallic-ion migration to the heater wire causes islands of crystal formation, making the wire brittle. This phenomenon accounts for some of the effects noted in some large computers in which gradual application of heater voltage was found beneficial in reducing the incidence of open heaters; the real culprit was heater-cathode voltage.

The rate at which the foregoing effects progress is accelerated exponentially as heater voltage (temperature) is increased—another reason for operating heater voltage as closely as possible to the rated center.

Oddly enough, the effects go forward to some extent even at zero voltage between heater and cathode but are accelerated as voltage is raised.

3.7. Plate Voltage. This rating is difficult to generalize upon because the effect varies markedly between tube types. Aside from the obvious effect on plate dissipation and temperature, plate voltage also affects the rate of liberation of gas molecules and cathode contaminants. It appears that gas molecules may be tightly enough bound to the anode surface to not be dislodged by temperatures achieved within the anode ratings. However, at higher anode voltages, electron velocities may be great enough to dislodge these gas molecules even though the tube is operated within dissipation ratings.

Secondly, electrons escape the anode potential field and strike the walls of the envelope, causing gas evolution by electrolytic dissociation of the calcium oxide. This effect is also greater at high anode voltages, depending upon the geometry of the structure, and is also enhanced by high bulb temperatures.

Thus the life of a tube *may* be enhanced by operating at anode potentials below the rated maximum and will probably be degraded by operating above the rating.

4. RECEIVING TUBES

4.1. Scope and Definition. In this section various types of amplifier service are given and receiving tubes suitable for application are listed, together with their major ratings and characteristics. Obviously, all types which may be used cannot be listed; therefore, an attempt was made to select them on the following basis:

1. Recommendation of the tube manufacturer. The tube manufacturer should be aware of the advantages of a tube in a specific application and should be eager to avoid possible misapplications which reflect unfairly on the quality of his product.

2. Wide use in a particular service. If a tube has been used by many designers in many similar circuit designs, the tube would appear to be a satisfactory choice for that service.

3. Availability. Scanning of comprehensive tube lists will turn up types which appear to have excellent characteristics for a particular application. Caution must be exercised in designing around some of these without investigating availability; some may no longer be available, and some may never have been available except in sample quantities. An effort has been made to list only tubes available at the time of compilation of this handbook (with the exceptions noted below), preferably tubes made by more than one manufacturer.

4. Superior performance. Some tube types may be difficult to obtain because of limited applications, approaching obsolescence, or because they are available from a single source. If these tubes have distinct advantages, they are listed, even though they may be hard to find. In the tube tables that follow, these types are marked with an asterisk as a warning to the circuit designer to determine present availability before proceeding too far with the design.

5. Cost. No attempt was made to take cost into account; the circuit designer should investigate this before making a final decision to incorporate a particular tube type.

The term "receiving tube" is a misnomer, since receiver service is a small part of the field covered by the tubes discussed in this section. "Low-power vacuum tubes" is a better term, but the traditional term is used here since it is well established. However, it is seldom precisely defined, and the definition used here is arbitrary. Receiving tubes are defined as those having no greater than 50 watts continuous plate dissipation rating, since this limit encompasses most tubes manufactured in conventional receiving-tube form.

4.2. The Electron Tube as a Network Element. The commonly published characteristics of electron tubes and their definitions are:

$$\mu = \partial E_p / \partial E_g | I_p \tag{11}$$

Amplification factor μ is a measure of the relative control of electron current to the plate I_p, exercised by the plate as compared with that of the control grid. Thus a triode with a μ of 100 takes 100 times as great a change in plate voltage E_p to effect a given change in plate current I_p as a change in grid voltage E_g which will cause the same change in plate current. The internal amplification factor of a pentode expresses this same relationship between the screen grid acting as a plate and the control grid.

$$g_m = \partial I_p / \partial E_g | E_p \tag{12}$$

Transconductance is the rate of change of plate current I_p with a change in grid voltage E_g with the plate voltage E_p constant. In a pentode the transconductance at the cathode is greater than at the plate by the ratio of cathode current to plate current.

$$r_p = \partial E_p / \partial I_p | E_g \tag{13}$$

Dynamic plate resistance r_p is the effective resistance calculated by noting the change in plate current I_p caused by a change in plate voltage E_p with the grid voltage E_g held constant. It is not to be confused with the d-c resistance found by dividing the d-c plate voltage by the direct plate current.

$$\mu = g_m r_p \tag{14}$$

Amplification factor, transconductance, and dynamic plate resistance are related to one another as shown by Eq. (14).

In the three equations above,

E_p = plate voltage, volts
E_g = grid voltage, volts
I_p = plate current, amp
r_p = dynamic plate resistance, ohms
g_m = transconductance, mhos

Comparison of the foregoing "standardized" method of expressing tube characteristics with four-terminal network theory shows that the Y-parameter expressions of a four-pole are most easily adapted to tube characteristics.

Tubes are basically three-pole networks with the inactive elements grounded. The three basic poles are:
1. Cathode
2. Control grid
3. Plate or anode

Since there are three poles, there are six possible connection schemes by which any two poles may be chosen as the input pair and the remaining pole and either of the first two poles as the output pair. Of these six, only three have particular application:

1. Input signal between control grid and cathode, output signal between plate and cathode. This is called the common-cathode or grounded-cathode configuration and is the most widely used arrangement. It exhibits practically infinite input resistance (up to several megacycles), a wide range of output resistance, a high voltage gain, and a nearly infinite power gain. In this configuration, the output signal is 180° out of phase with the input signal.

2. Input signal between control grid and cathode, output signal between control grid and plate. This is called the grounded-grid or common-grid configuration and is used at very high frequencies so that the grid acts as an electrostatic shield between input (cathode) and output (plate). It exhibits low input resistance, high voltage gain, and high power gain. This configuration can be used as a step-up impedance transformer.

3. Input signal between control grid and plate, output signal between cathode and plate. This is called the common-plate or cathode-follower configuration. It exhibits practically infinite input resistance, low output resistance, less than unity voltage gain, but nearly infinite power gain. This configuration is used as a step-down impedance transformer.

The three foregoing configurations are shown in terms of their conventional characteristics and in Y-parameter notation for four-pole networks below.

Configuration	y_{11}	y_{12}	y_{21}	y_{22}
Common-cathode............	0	0	g_m	$\dfrac{1}{r_p}$
Grounded-grid...............	$\dfrac{\mu + 1}{r_p}$	$\dfrac{-1}{r_p}$	$-\left(\dfrac{\mu + 1}{r_p}\right)$	$\dfrac{1}{r_p}$
Cathode-follower............	0	0	$-g_m$	$-\left(\dfrac{\mu - 1}{r_p}\right)$

4.3. Receiving-tube Construction. It is assumed that many users of this handbook are familiar with the construction features of conventional tubes. However, in recent years a number of unconventional tube designs have been developed, all of them with certain application advantages inherent in their construction. These tubes may differ in internal features from conventional tubes, or they may be unconventional in their external appearance. To allow comparison of the new with the old, mechanical and electrical features of both are briefly described here.

In a conventional glass tube the cathode sleeve is the center of the tube structure. The sleeve contains the heater, which is covered by an insulating coating and folded or coiled so that it may be inserted in the sleeve, and is surrounded concentrically by the other elements, starting with the No. 1 grid and ending with the plate. The elements are supported at one end by being welded to stem leads which provide electrical connections through the glass base of the tube and at the other end by a mica disk which centers the elements in the glass bulb and also maintains the elements concentric to one another (Fig. 54). Although all the elements in tubes have undergone improvements in recent years, the control grid has been the one whose changes have brought about the most noticeable results.

The grids of tubes are the least satisfactory elements from the standpoints of ruggedness and reproducibility. Electrically, the ideal grid would be nonexistent until called upon to modify electron flow, and even then it should have so little surface as to be incapable of intercepting electrons. The grid consists, most generally, of a spiral of wire, with two side rods to provide support for the spiral, whose wires are termed grid laterals. Since the grid laterals are not under tension, and in fact are usually shaped to a circular or oval form, they must be heavy enough to hold their shape. Consequently, to minimize obstruction of the electron stream the number of turns permissible is limited. High transconductance per milliampere of plate current, with its accompanying high performance, is achieved by having very close spacing between the No. 1 grid and the cathode. However, as the grid is moved closer to the cathode, the grid wires must be moved closer together to maintain good control of the electron stream. If the wires are moved closer together, their thickness compared with their spacing increases and shadowing of the electron stream occurs. Thus the wires must be made thinner as they are moved together. These requirements would be difficult to meet with conventional grid structures, and the *frame grid* was designed to solve this problem (Fig. 55).

The frame grid consists of a rigid frame on which very small diameter grid wire is wound under tension. The frame supplies all the support, and by choice of materials with the proper temperature coefficients, the tension on the grid wires may be maintained over a considerable tempera-

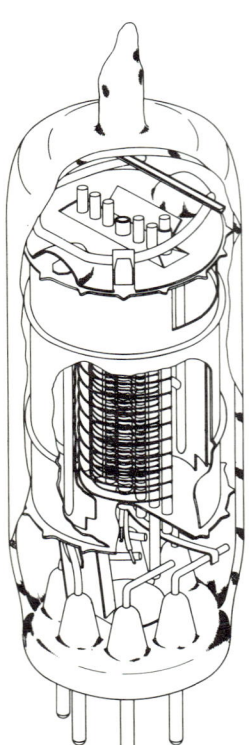

Fig. 54. Cutaway view of a conventional miniature tube. (*Courtesy of General Electric Co.*)

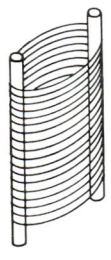
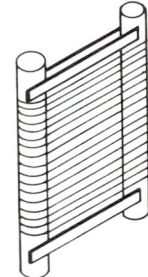

Fig. 55. Comparison of a conventional wound grid (left) and a frame grid (right). (*Courtesy of General Electric Co.*)

ture range. The frame grid allows high transconductance to be achieved and helps maintain uniform spacing throughout life. One disadvantage of frame-grid construction is that the close grid-cathode spacing that produces high transconductance also results in high input capacitance unless the tube is scaled down in size to reduce all capacitances.

Some typical frame-grid tubes are the 6HA5, a triode for grounded-cathode vhf r-f amplifier use; the 6EH7 and 6EJ7, pentodes for i-f amplifier use; the 12GN7 for video amplifier use; the 6DJ8, a twin triode for cascode r-f amplifier use; and the 6688, a pentode for wideband amplifier use.

Although the frame grid is a radical innovation in grid design, other modifications in internal structure have been used to improve performance, some of which are not necessarily new. For example, the beam pentode, the first of which was the 6L6, incorporates beam plates to direct the flow of electrons to a desired portion of the plate (see Fig. 56). This is done in the beam pentode to reduce secondary-emission effects. However, the same type of beam plates have been used to concentrate the electron stream in high-performance triodes for other reasons. To secure satisfactory operation in the vhf and uhf regions, the electron transit time must be made as short as possible

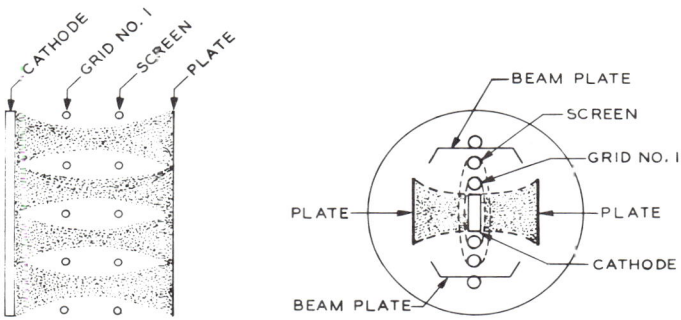

FIG. 56. Electron paths in a beam pentode. (*Courtesy of General Electric Co.*)

and uniform over the entire active surface. To do this, the element spacing must be very close, which increases the interelectrode capacitances unless the elements are made quite small. If the plate is made very small, its heat-radiating surface is reduced to a point where the allowable plate dissipation is below an acceptable value for the intended service unless some provision is made for an external heat sink. To alleviate these problems, the structure shown in Fig. 57 has been used. Here the plate is formed so that a portion of it is quite close to the grid. The rest of the plate is far enough away from the grid to reduce interelectrode capacitance. At the same time, the plate has sufficient area to allow a reasonable dissipation capability. The electron stream is guided by the beam plates to follow the shortest path to the plate to reduce transit time. In addition, the beam plates provide some shielding to decrease the interelectrode capacitance further. Two tubes that employ this construction are the 6GK5 and 6HA5, both of which are also frame-grid tubes.

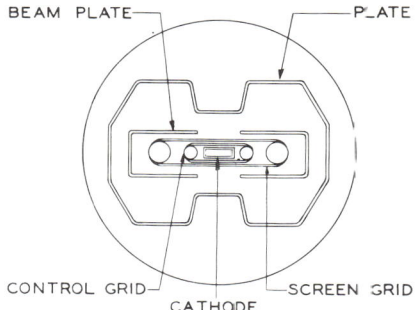

FIG. 57. Element arrangement of the 6HA5. (*Courtesy of General Electric Co.*)

In most tubes employing a screen grid, it is desirable to hold the screen current as low as possible. In power output tubes the screen current tends to rise rapidly as the plate voltage is swung below the screen voltage, and excessive dissipation occurs in the screen, which has limited power-handling capabilities. This problem has been attacked by using "aligned" grids. That is, the No. 1 grid and the screen have the same number of turns and are mechanically aligned so that each turn of the screen grid is shadowed by a turn of the No. 1 grid. This reduces the number of electrons striking the screen, especially since beam plates are also incorporated in these types to direct the electron stream, and hence the screen current. Examples of aligned-grid tubes are the 6L6-GC, the 6AQ5-A, the 6005, and the 50C5, all of which are audio power output tubes.

In pentodes used for vhf amplification, reduction of screen current is worthwhile not only to reduce the screen dissipation but also to reduce partition noise (see Sec. 2.2). It is not practical to attempt to align the No. 1 and screen grids in small, high-transconductance tubes with high turns per inch because of mechanical problems. A recent solution is embodied in the *shadow-grid* beam pentode. In this construction, a closely wound grid is used for the No. 1 grid for high transconductance, and a shadow grid, aligned with the screen grid, is placed between the No. 1 grid and the screen grid. This shadow grid effectively reduces the screen current, and also allows operation of the screen at the same voltage as the plate, thus eliminating the necessity for screen-dropping resistors. Figure 58 shows the electron flow in a shadow-grid tube. Examples of shadow-grid tubes are the 6FG5, 6FS5, and 6GU5. The 6GU5 also has a frame grid for the No. 1 grid, which gives it very high transconductance and low noise figure.

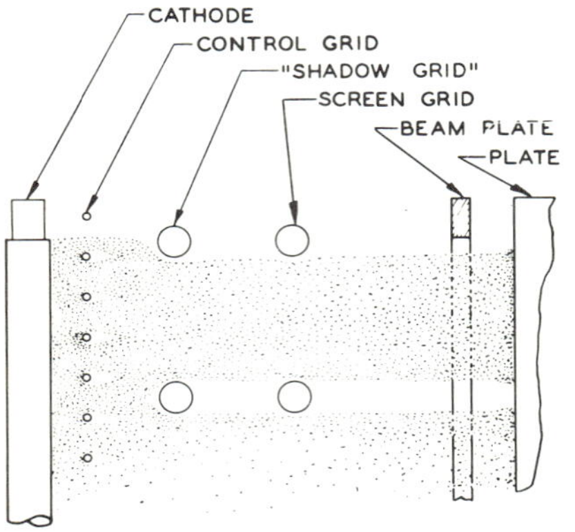

In order to allow the proper operation of AGC systems, the No. 1 grid of certain pentodes is wound with a variable pitch or, sometimes, with a "window" in the center of the grid. This gives the tube a semi-remote-cutoff characteristic but also has some disadvantages. As the tube approaches cutoff, most of the cathode current flows through the widely spaced turns. This results in inefficient use of the cathode area, overheating of screen-grid wires adjacent to the widely spaced wires of the No. 1 grid, and a number of other operating disadvantages that are detailed below under Sec. 4.6. To overcome these difficulties,

FIG. 58. Electron paths in a shadow-grid beam pentode. (*Courtesy of General Electric Co.*)

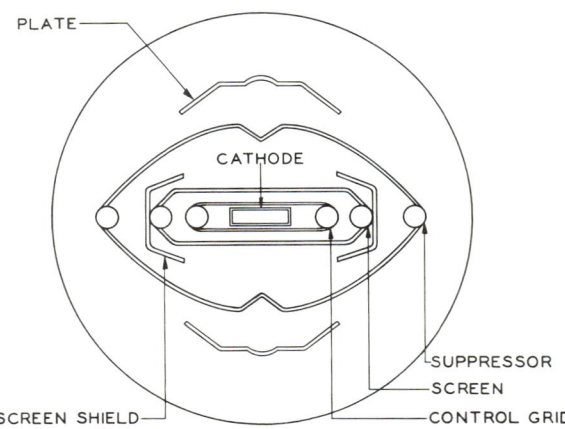

FIG. 59. Element arrangement of the Ulug tube. (*Courtesy of General Electric Co.*)

a Canadian, M. E. Ulug, has invented a novel pentode[16] (Fig. 59) in which the major AGC voltage is applied to the No. 3 grid. When AGC voltage is applied to the No. 3 grid, fewer electrons penetrate to the plate and the transconductance of the tube

[16] M. E. Ulug, Novel Tube and Circuit Designs for Automatic Gain Controlled Broad Band Intermediate Frequency Amplifiers, *IRE Trans. Broadcast Television Receivers*, November, 1962.

decreases. These electrons are deflected toward the screen so as to keep the cathode current relatively constant. This would result in excessive heating of the screen lateral wires with conventional screen construction, but in the 9BJ11 (one production type incorporating this principle), screen-grid shields are incorporated to intercept most of these electrons on a surface capable of better radiating the developed heat.

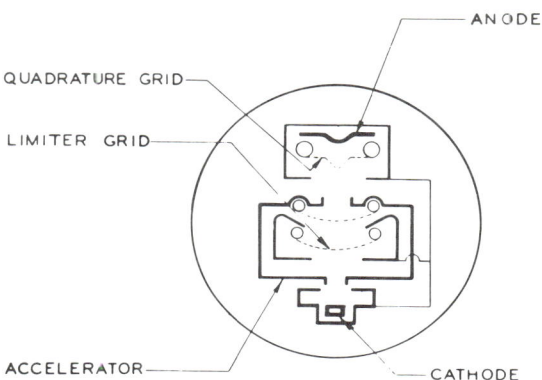

Fig. 60. Element arrangement of the 6BN6. (*Courtesy of General Electric Co.*)

An almost complete departure from conventional cathode-grid-plate construction is found in the tubes classified as gated-beam, sheet-beam, and beam-deflection tubes. These tubes use techniques similar to those employed in electrostatic-deflection cathode-ray tubes for shaping and directing the electron stream.

A typical gated-beam tube is the 6BN6,[17] whose element arrangement is shown in Fig. 60. In operation the plate and accelerator are operated at positive potential, and the limiter and quadrature grids act as control grids. When electrons leave the cathode, they are accelerated toward the boxlike structure of the accelerator. Whether or not the electrons pass through this structure is dependent on the voltage at the limiter grid. If this grid is sufficiently negative, plate current flow is cut off. If the voltage on the grid is such that the electron stream passes through the accelerator, control may still be obtained through the quadrature grid, which will deflect the electron flow to the outside surface of the accelerator if the quadrature grid is sufficiently negative. If it is not, the electrons will flow to the plate. Although the 6BN6 has been used mostly for limiter-discriminator service in FM detection, it has other possible applications.

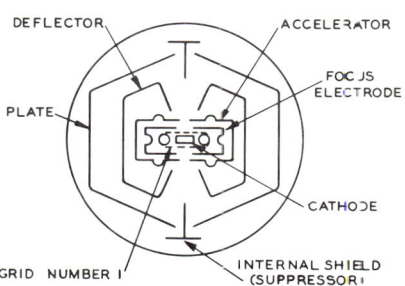

Fig. 61. Element arrangement of the 6AR8. (*Courtesy of General Electric Co.*)

The 6AR8[18] is a typical sheet-beam tube (see Fig. 61). In this tube the electrons pass from the cathode to either of the two plates in the form of a "sheet beam." The magnitude of the electron stream is controlled by a No. 1 grid. The stream is shaped by focus electrodes and accelerated by the accelerator. The deflectors serve to deflect

[17] A. P. Haase, A New One-tube Limiter-discriminator for FM, *Tele-Tech*, February, 1950.
[18] R. Adler and C. Heuer, Color Decoder Simplifications Based on a Beam-deflection Tube, *IRE Trans. Broadcast Television Receivers*, January, 1954.

the accelerated beam entirely or in part to either of the two plates, depending on the polarity and magnitude of the voltage placed on the deflectors. The suppressor tends to reduce coupling between the plates caused by interchange of secondary-emission electrons.

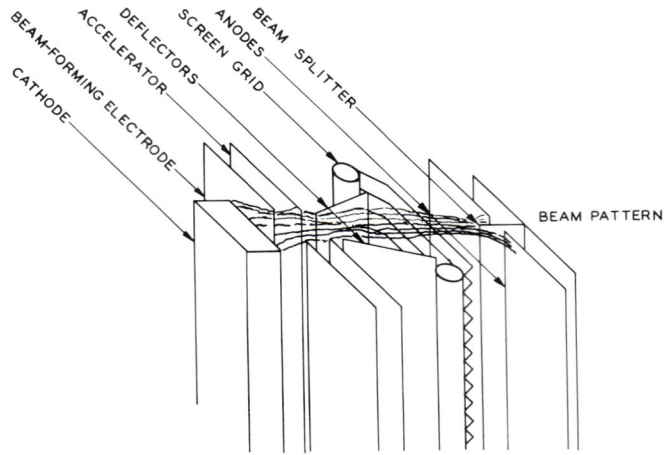

Fig. 62. Element arrangement of the 7763. (*Courtesy of General Electric Co.*)

Another sheet-beam tube, the 7763, differs somewhat in construction from the 6AR8, and is intended for high phase-fidelity i-f limiter service.[19] The electron source is again a cathode together with a beam-forming electrode and an accelerator. The latter two electrodes have slot apertures which form the "sheet beam" of electrons. Element spacing and geometry are such that accelerator current is held to a minimum. Beyond the accelerator are a pair of deflectors. After these are a half-grid screen and a pair of plates. Between the plates is a beam splitter. Figure 62 shows the element arrangement of the 7763. The a-c signal to be amplified and limited is fed to the deflectors, and output is taken from the plates. Limiting occurs as soon as the input voltage reaches a point where the beam is being deflected, on each cycle, completely from one plate to the other. The beam splitter, which is operated at cathode potential, assists in switching the beam from one plate to the other and provides shielding between the plates. The screen serves to shield the deflectors from the plate.

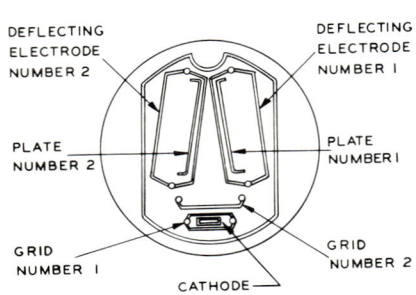

Fig. 63. Element arrangement of the 7360.

Another tube using an electron beam is the "beam-deflection" 7360,[20] which differs in its internal structure from both the 6AR8 and the 7763. The 7360 element arrangement is shown in Fig. 63.

Construction features least capable of detection by visual inspection are the use of better materials. One area where improvement has been made is in heater construction. Heater failures are usually due to crystal growth in the heater wire and conse-

[19] E. R. Wingrove, Phase Stable Limiting I-F Amplifiers Using Beam Deflection Tubes, *Proc. 4th MIL-E-CON*, 1960.
[20] M. B. Knight, A New Miniature Beam-deflection Tube, *RCA Rev.*, June, 1960.

quent embrittlement. The use of rhenium alloyed with the tungsten normally used for this wire reduces the tendency of the tungsten to crystallize and extends heater life. High temperature is also detrimental to heater life, and heater temperatures have been lowered by coating the heater wire with a dark coating to produce a "dark heater." This dark coating allows better heat transfer to the cathode so that for a given cathode temperature a lower heater temperature will be required.

Plate materials have also been improved to allow higher plate dissipations. Normally the electron stream strikes only certain portions of the plate, and at these points the maximum heat is developed with consequent formation of "hot spots." This may be cured by using a material with good heat conductivity to make the temperature of all parts of the plate relatively uniform. However, the heat produced not only must be conducted over the entire plate but must also be radiated. To do both these jobs, some high-dissipation-rated tubes use plates made from a material having a copper core sandwiched between two layers of aluminum-clad steel; the copper furnishes the proper heat conduction and the aluminum-clad steel the radiation. As an example of the difference that modern construction makes on dissipation ratings, the 6L6 originally had a dissipation rating of 19 watts; the present 6L6-GC has a rating of 30 watts.

In the four-digit series of tubes there are many types that have special construction features which give them advantages over the home-entertainment types from which they are often derived. For example, doubled micas are frequently used to improve vibration characteristics, and getter flash shields between the top mica and the getter ring reduce deposition of getter material on the top mica with consequent improvement in leakage characteristics. Short cage assemblies are often used to increase rigidity of the tube structure, and gold or silver plating are sometimes used on grids to reduce grid emission.

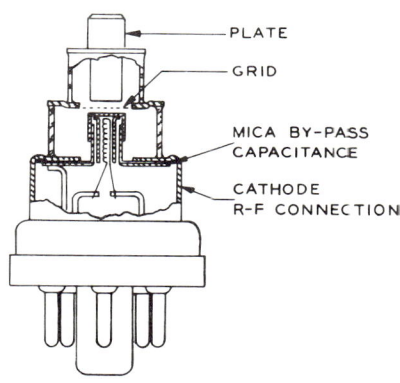

Fig. 64. Lighthouse tube. (*Courtesy of General Electric Co.*)

Tubes having an unconventional appearance include lighthouse tubes, metal-ceramic tubes, rocket tubes, pencil tubes, nuvistors, compactrons, novars, neonovals, and TIMM's.

The lighthouse tubes (see Fig. 64) represent one of the early attempts to enhance high-frequency performance of tubes by decreasing interelectrode spacing and reducing the electrode lead inductance. The planar electrode structure used in these tubes has been carried over in more modern tubes such as the metal-ceramics and rocket tubes. All three of these tube styles have the cathode, grid, and plate arranged in parallel planes; they differ in their external appearance, size, and materials of construction. The lighthouse tubes were designed in the octal-tube era and have this style of base. However, only heater leads and d-c cathode connections are brought to this base. All r-f connections are made to the sleeve, disk, and cap which connect to or support the cathode, grid, and plate, respectively. This allows low-inductance leads which, when combined with the close spacing and low interelectrode capacitances, make the tubes useful at vhf and uhf. The general shape of the tubes also makes it possible to fit them directly into coaxial circuits. The plate is relatively small, and connection to a heat sink, usually inherent in coaxial cavities, is required to take advantage of the rated plate dissipation.

Typical lighthouse tubes are the 2C40, a triode for r-f amplifier service up to 3,300 Mc; and the 2C43, a triode for pulsed service up to 3,300 Mc.

The rocket tubes also employ the planar structure but are smaller and do not have conventional bases (Fig. 65). Like the lighthouse tubes they have envelopes which are glass. Typical rocket tubes are 2C37, a triode for amplifier and frequency-multiplier service up to 3,300 Mc; and the 5768, a triode for amplifier service up to 3,000 Mc.

The metal-ceramics differ from the lighthouse and rocket tubes in having envelopes which in part are ceramic. (Strictly speaking, none of these tubes has an envelope in the conventional sense.) One of the early metal-ceramics, the 2C39B, was descended directly from the 2C39, which was a glass tube, sometimes called an "oilcan tube" because of its shape. The 2C39B differs principally in having ceramic substituted for glass. The advantages of ceramic are its lower r-f losses and greater ability to stand high temperatures.

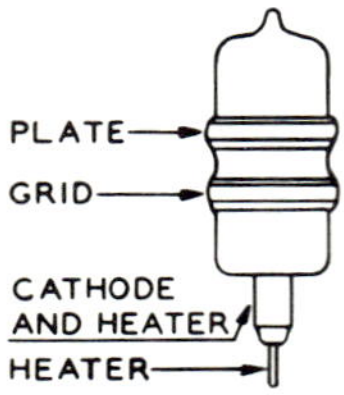

Ceramics have been used as bases and as internal insulators for many years, but the ceramic-metal tubes described here differ from previous tubes in that ceramics are used to form part of the vacuum chamber, and ceramic-to-metal sealing is used in their construction. This style is used in tubes over nearly the entire power dissipation range of electron tubes. However, we are here concerned with receiving tubes. A typical metal-ceramic is the 7077, which is a small triode for uhf amplifier service. A cross-section view of the 7077 is shown in Fig. 66. The very close spacings give high transconductance and short transit time, resulting in high gain and low noise figure. The plate of the tube, as well as the contact rings for the grid and cathode, are fabricated from titanium, which acts as a continuous getter during the life of the tube. The grid is a form of frame grid, with the wires stretched on a circular frame instead of a rectangular one. The cathode is planar, and useful emission is obtained from the whole top surface. The ceramic insulators, which hold the plate, grid ring, and cathode ring in rigid alignment, have excellent insulation resistance and very low losses at high frequencies. The terminals of the 7077 are designed for low-inductance connection to coaxial circuits, although the tube can be placed in sockets and used for other purposes than coaxial uhf amplifier service. In addition, the tube is small, tolerant of higher temperatures than glass tubes, and is shock- and radiation-resistant.

FIG. 65. Rocket tube.

For lower-frequency operation ceramic-metal tubes are available with lug rather than coaxial connections. These include the 7296, the 7588, and the 7625.

The nuvistors are also small tubes which, like the metal-ceramics, use metal and ceramic as materials of construction. However, the nuvistors use a cantilevered coaxial construction rather than a planar one. The nuvistor elements are attached to cones which in turn are supported on short rigid rods which extend through the ceramic base as connection pins. A metal shell encloses the elements (see Fig. 67). High transconductance, low noise figure, and resistance to shock and radiation are also features of these tubes. Lack of any insulating material near the active portions of the elements and the use of ceramic at the base make the insulation resistance high and the dielectric losses very low. Typical nuvistors are the 6CW4, a triode intended for vhf amplifier service; the 7586, a triode intended for general industrial applications; and the 7587, a tetrode intended for industrial applications.

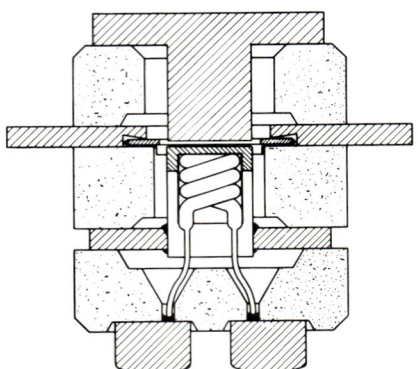

FIG. 66. Cross section of the 7077. (Courtesy of General Electric Co.)

The pencil tube is a coaxial-element tube in which plate- and cathode-contact cylinders extend out in opposite directions from a central grid-contact disk (see Fig. 68). Insulation between the cylinders and the disk may be either glass or ceramic, depending upon the tube type. The elements are close-spaced, and low-inductance

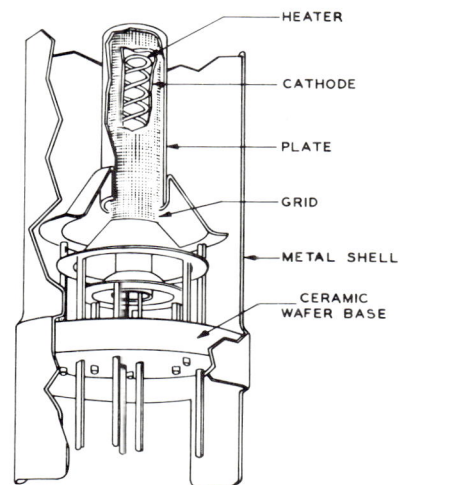
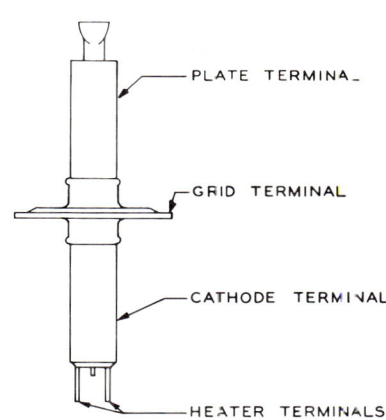

Fig. 67. Cutaway view of a nuvistor. (*Courtesy of RCA.*)

Fig. 68. Pencil tube.

connections may be made to them. The relatively large grid-contact disk provides additional isolation by shielding between the cathode and plate in grounded-grid circuits. The pencil tubes may be used in either coaxial or lumped-constant circuits. Typical tubes are the 5675, a triode for grounded-grid amplifier operation up to 3,000 Mc; and the 7552, a triode for amplifier use up to 1,500 Mc.

Many of the recently introduced receiving tubes have unconventional outward appearances but use conventional internal construction. Among these are the compactrons, the novars, and the neonovals.

The compactrons differ from conventional tubes in having a 12-pin glass-button base, using the same size pins as the 7- and 9-pin miniatures. The envelope may be either T-9 or T-12 (see Fig. 69). The number of pins available on the base allows from one to four functions per envelope. When the compactron style is used for a single-function tube, the extra base pins available may be used to assist in conducting away heat or in providing multiple connections for low impedance. The use of a glass-button base reduces base losses and allows a lower seated height than permitted by equivalent octal types. Typical compactrons are the 6AV11, a triple triode; the 6AF11, a dissimilar-double-triode pentode; the 6AR11, a twin pentode; and the 6AL11, a dissimilar double pentode.

Novars and neonovals both use glass-button bases and differ in pin-circle diameters. The novars are either T-9 or T-12 with a 9-pin base having a pin-circle diameter of 0.687 in. and

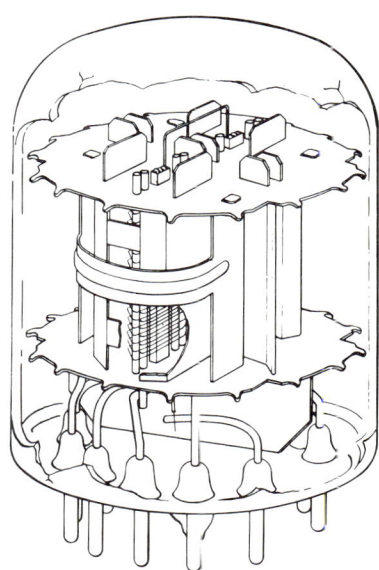

Fig. 69. Cutaway view of a compactron twin pentode. (*Courtesy of General Electric Co.*)

pins with the same diameter as those used on 7- and 9-pin miniatures. Examples of novars are the 6JE6, a beam pentode; and the 7868, a beam pentode.

All the foregoing glass types in T-9 envelopes have an advantage over comparable tubes in T-6 1/2 bulbs of having larger bulbs for better heat radiation.

A very unusual configuration for tubes is one called TIMM,[21] an acronym for "thermionic integrated micro modules." As the name implies, the device is actually a complete electronics circuit, made by stacking small ceramic cylinders, each containing a resistor, capacitor, diode, or triode. The unique feature of the resulting circuit is that it is designed to be operated in a thermally insulated package at 580°C. The temperature is that needed for cathode emission, and the "tubes" contained in the modules do not have individual heaters for their cathodes. The entire package is brought up to operating temperature by a source external to the tubes, such as a heating element contained within the package or by heat transfer from a hot body such as a rocket engine. After operation is started, heat dissipated in the tubes and components may be used to assist in maintaining the operating temperature.

The advantages of the TIMM concept lie in the small size of the modules, the high thermal efficiency, and the ability to withstand severe nuclear radiation for prolonged periods.

4.4. D-C and VLF Amplifier Tubes. In amplifiers with response down to zero frequency or to subaudio frequencies, such as operational amplifiers or geophysical equipment, the critical stages are the input ones, and the principal requirements for suitable tubes are low grid current, low flicker noise, low microphonics, stability of characteristics, and for dual-section tubes, balanced plate current. Unfortunately, the demand for tubes specifically intended for these services has been limited, and few types are available designed to meet all the requirements. Those which appear most suitable are listed in Table 2.

While many tube types are stabilized by operation, by the manufacturer, for various time intervals under specified conditions, the resulting characteristics stabilization seldom extends to other operating conditions. Therefore, stabilization of tubes for d-c and subaudio service at the actual operating conditions that the circuit designer selects will probably be required for satisfactory operation.

A special form of d-c amplifier tube is the electrometer tube, whose principal function is to convert the extremely high resistance input circuit of an electrometer to lower values where more common tubes may be used for further amplification. Here the requirements of extremely low grid current and extremely high insulation resistance make the use of tubes specifically designed or selected for the purpose almost mandatory. The required characteristics are obtained by special constructions and by operation with very low filament power to minimize sublimation and positive ion emission, and with very low plate voltage to minimize ionization of the small amount of residual gas that cannot be removed from any vacuum tube. Special external bulb coatings are often used to decrease leakage across the glass under conditions of high humidity, and thus maintain very high input resistance. To preserve the high resistance inherent in these tubes, the bulb and base of the tubes should never be touched with the bare hands. Electrometer tubes are listed in Table 3.

Although not specifically recommended for this service, some of the metal-ceramics have been suggested for electrometer use because of their high insulation resistance and their freedom from gas through the gettering action of the titanium parts. As a result of these qualities, they may be operated at higher plate voltages, and consideration of this by the designer might be worthwhile.

4.5. Audio-frequency Amplifier Tubes. Tubes for audio-frequency amplifier service may be divided into voltage-amplifier tubes and power-amplifier tubes. Desirable characteristics for voltage-amplifier tubes include low hum, low microphonics, high gain, and characteristics stability at low plate current levels. The high- and medium-mu tubes are used for the input stages and lower-mu tubes are used for higher voltage level stages and as phase inverters. The cathode follower, strictly speaking, is not a voltage amplifier, but the circuit finds use in low-level audio cir-

[21] J. E. Beggs, W. Grattidge, P. J. Molenda, A. P. Haase, and A. F. Dickerson, Thermionic Integrated Micromodules, *Electronics*, May 15, 1959.

Table 2. Tubes for D-C and Very-low-frequency Amplifier Service

Tube type	Tube style	Classification	Heater voltage	Heater current, amp	Max plate voltage and dissipation (each section)	E_h	E_o	I_p, ma	g_m, μmhos	μ	r_p, ohms	Features
5755	9-pin miniature	Twin triode	6.3/12.6	0.36/0.18	250/1.0	110	−0.95	0.15	500	70	140,000	Characteristics controlled for plate current balance, drift, and grid current. Type specifically intended for this service
5751	9-pin miniature	Twin triode	6.3/12.6	0.35/0.175	330/0.8	250	−3.0	1.0	1,200	70	58,000	High gain. Characteristics controlled for plate current balance
5965	9-pin miniature	Twin triode	6.3/12.6	0.45/0.225	330/2.2	150	−2.0	8.5	7,000	47	6,700	Characteristics controlled for plate current balance
6832*	8-lead subminiature	Twin triode	6.3	0.4	165/0.1	100	0	0.8	1,050	26	. . . . . .	Characteristics controlled for plate current balance, drift, and grid current. Type specifically intended for this service
7625	Metal-ceramic	Triode	6.3	0.215	275/0.85	150	$R_k = 1,000$	0.95	1,400	80	57,000	High gain. Low microphonics

* Check availability before proceeding with circuit design.

Table 3. Tubes for Electrometer Service

Tube type	Tube style	Classification	Filament voltage	Filament current, ma	Max plate and screen voltage	Max cathode current, μa	Avg characteristics							
							E_b	E_{c2}	E_{c1}	I_p, μa	I_{g2}, μa	g_m, μmhos	μ	I_{g1}, μa
VX-55	4-lead sub-miniature	Triode	1.25	10	50	...	7.5	...	−2.2	95	...	110	2.1	5×10^{-8}
CK587*	5-lead sub-miniature	Pentode	0.625	10	45/45	100	8	5.5	−2.0	6	2.5	14	...	2×10^{-9}
5800*	5-lead sub-miniature	Tetrode	1.25	10	50	500	4.5	−3	3.4	12	10^{-9}	15	1	300
5803*	4-lead sub-miniature	Triode	1.25	10	...	500	7.5	...	−1.7	100	...	150	2	2×10^{-8}
5886	5-lead sub-miniature	Pentode	1.25	10	22.5/22.5	300	8.5	4.5	−2.0	6	3.6	14	...	3×10^{-9}
5889	5-lead sub-miniature	Pentode	1.25	7.5	45/45	300	12	4.5	−2.0	4	4	10	250	3×10^{-9}

* Check availability before proceeding with circuit design.

cuits, and some of the voltage-amplifier tubes are quite suitable. Specifically a combination of high amplification factor and transconductance, found in the 12AT7, provides a tube suitable for this service.

The most popular low-level tube is the 12AX7 and its various derivatives. These include the 6EU7, the 5751, and the 7025. The 6C4 and its derivatives, the 12AU7 and the 5814-A, are most popular for phase-inverter service. Low-level amplification and phase-inverter use are combined in a dissimilar double triode, the 7247, which contains a 12AX7 triode and a 12AU7 triode. Pentodes are also used as low-level voltage amplifiers, and they are represented in Table 4 by the 5879 and 7543. Low-level amplification and phase-inverter service are also combined in a triode-pentode, the 7199.

Table 4 lists tubes recommended for audio-frequency voltage amplification. Since resistance-coupled amplifier circuits are the most common ones in this service, a number of tables showing potential operating conditions and the available gain are included as Tables 5 through 10.

Tubes for audio power-amplifier service are usually chosen on the basis of available power output and allowable power dissipation. Although distortion is important, the distortion figure given on nearly all data sheets represents the distortion without inverse feedback, and ultimate distortion depends upon the amount of feedback used. In low-cost equipment where a minimum of stages preceding the power output stage is desirable, power sensitivity of the output tube is also important. There are several ways to state the sensitivity, and in the data here the power output in milliwatts is divided by the peak grid-signal voltage squared to obtain a sensitivity figure. The power output tubes in Table 11 are arranged in order of increasing power output.

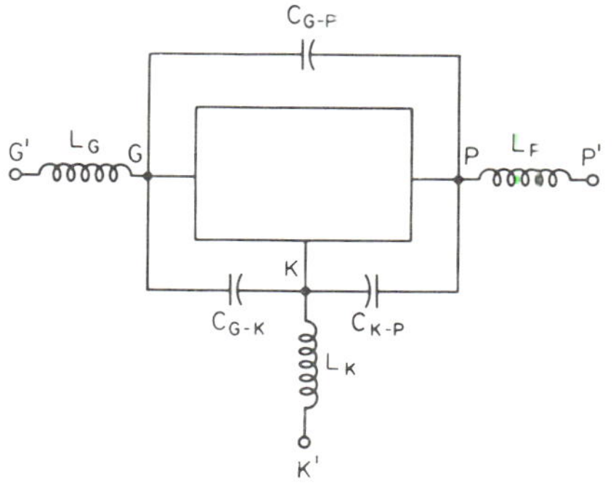

4.6. R-F and I-F Amplifier Tubes. This classification covers tubes that are used from frequencies starting in the audio range and running up to approximately 300 Mc. It includes tubes that may be used for either narrowband or wideband service.

Fig. 70. Schematic of three-pole representing a triode.

Heretofore, only the conductive portions of tube characteristics have been dealt with. In many applications, however, the internal physical dimensions of tubes become quite important, and interelectrode capacitance and lead inductance effects must be taken into consideration.

A schematic representation of the three-pole constituting a triode is shown in Fig. 70. In this drawing, C_{g-p} is the grid-to-plate capacitance, C_{g-k} is the grid-to-cathode capacitance, and C_{k-p} is the cathode-to-plate capacitance. G, K, and P represent the grid, cathode, and plate, respectively, but G', K', and P' represent the actual terminals of the tube to which the external circuitry must be connected, and the internal leads in the tube must be represented by inductances, designated as L_g, L_k, and L_p, respectively.

The effects of these characteristics are frequency-dependent, with those involving capacitances becoming evident first as frequency is increased. The input and output capacitances, which are the grid-to-cathode and plate-to-cathode capacitances in common-cathode operation, plus circuit capacitances, determine the high-frequency rolloff of audio and video amplifiers. The input-to-output capacitance, grid-to-plate, gives rise to the Miller effect which, because of the gain of the tube and the inherent 180° phase difference in input and output signals, causes a relatively great increase in input capacitances in the common-cathode configuration. The Miller-effect relationship is

$$C_{in} = C_{g-k} + (A - 1)C_{g-p} \qquad (15)$$

Table 4. Tubes for Audio-frequency Voltage-amplifier Service

Tube type	Tube style	Classification	Heater voltage	Heater current, amp	Max plate voltage and dissipation (each section)	Max screen voltage and dissipation	E_b	E_{c2}	E_{c1}	I_p, I_{g2}, ma ma	μ	g_m, μmhos	r_p, ohms	Features
6AV6	7-pin miniature	Duplex-diode triode	6.3	0.3	300/0.5	......	250 100	......	−2.0 −1.0	1.2 0.5	100 100	1,600 1,250	62,500 80,000	Single high-mu triode—similar to one section of 12AX7
6C4	7-pin miniature	Triode	6.3	0.15	300/3.5	......	250 100	......	−8.5 0	10.5 11.8	17 19.5	2,200 3,100	7,700 6,250	Single medium-mu triode—similar to one section of 12AU7
6EU7	9-pin miniature	Twin triode	6.3	0.3	300/1.2	......	250 100	......	−2.0 −1.0	1.2 0.5	100 100	1,600 1,250	62,500 80,000	High-mu twin triode—similar to 12AX7 except for 6.3-volt-only heater and basing which is symmetrical for stereo use—controlled for hum
12AT7	9-pin miniature	Twin triode	6.3/12.6	0.3/0.15	300/2.5	......	250 100	......	R_k = 200 ohms R_k = 270 ohms	10 3.7	60 60	5,500 4,000	10,900 15,000	Twin triode—combines high mu and high transconductance
12AU7-A	9-pin miniature	Twin triode	6.3/12.6	0.3/0.15	300/2.75	......	250 100	......	−8.5 0	10.5 11.8	17 20	2,200 3,100	7,700 6,500	Medium-mu twin triode
12AX7	9-pin miniature	Twin triode	6.3/12.6	0.3/0.15	300/1.0	......	250 100	......	−2.0 −1.0	1.2 0.5	100 100	1,600 1,250	62,500 80,000	High-mu twin triode
5751	9-pin miniature	Twin triode	6.3/12.6	0.35/0.175	330/0.8	......	250	......	−3.0	1.0	70	1,200	58,000	High-mu twin triode—for use under severe environmental conditions ("premium" or "reliable" tube)
5814-A	9-pin miniature	Twin triode	6.3/12.6	0.35/0.175	330/3.0	......	250 100	......	−8.5 0	10.5 11.8	17 19.5	2,200 3,100	7,700 6,250	Medium-mu twin triode similar to 12AU7-A—for use under severe environmental conditions ("premium" or "reliable" tube)

5879	9-pin miniature	Sharp-cutoff pentode	6.3	0.15	330/1.25	165/0.25	250	100	−3.0	1.8 0.4	1,000	......	2,000,000	Pentode—high gain
6072-A	9-pin miniature	Twin triode	6.3/12.6	0.35/0.175	330/1.6	....	250	...	−4.0	3.0 ...	1,750	44	25,000	Medium-mu twin triode—for use under severe environmental conditions ("premium" or "reliable" tube)
6201	9-pin miniature	Twin triode	6.3/12.6	0.3/0.15	330/2.7	....	250	...	$R_k = 200$ ohms	 10	5,500	60	10,900	Twin triode—combines high mu and high transconductance—similar to 12AT7 ("premium" or "reliable" tube)
7025	9-pin miniature	Twin triode	6.3/12.6	0.3/0.15	300/1.0	....	250 100	... ...	−2.0 −1.0	1.2 ... 0.5 ...	1,600 1,250	100 100	62,500 80,000	High-mu twin triode similar to 12AX7—controlled for hum
7199	9-pin miniature	Triode-pentode	6.3	0.45	Pentode 330/3.0 Triode 330/2.4	165/0.6	220 215	130 ...	$R_k = 62$ ohms −8.5	12.5 3.5 9.0 ...	7,000 2,100	 17	400,000 8,100	Triode similar to 12AU7—controlled for hum
7247	9-pin miniature	Dissimilar double triode	6.3/12.6	0.3/0.15	Section 1 330/1.2 Section 2 330/3.0	....	250 250	... ...	−2.0 −8.5	1.2 ... 10.5 ...	1,600 2,200	100 17	62,500 7,700	Section 1 similar to 12AX7—section 2 similar to 12AU7—controlled for hum
7543	7-pin miniature	Sharp-cutoff pentode	6.3	0.3	300/3.0	150/0.65	250	150	$R_k = 68$ ohms	10.6 4.3	5,200	......	1,000,000	Pentode—high gain—similar to 6AU6—controlled for hum
6C10	Compactron	Triple triode	6.3	0.6	330/1.0	....	250	...	−2.0	1.2 ...	1,600	100	62,500	Three 12AX7 sections
6AV11	Compactron	Triple triode	6.3	0.6	330/2.75	....	250	...	−8.5	10.5 ...	2,200	17	7,700	Three 12AU7 sections

9-63

Table 5. Resistance-coupled Amplifier Chart for the 6AV6, 6C10, 6EU7, 12AX7, 7025, and 7247 (Section 1)*

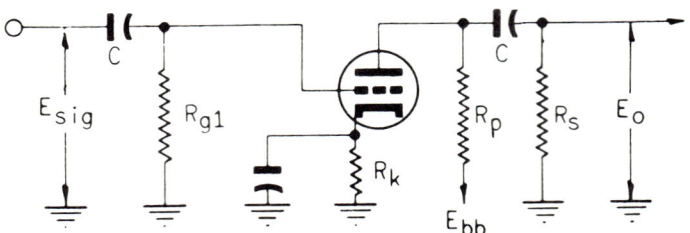

Note: Coupling capacitors C should be selected to give desired frequency response. R_k should be adequately bypassed.

Notes:
1. E_o is maximum RMS voltage output for 5 per cent total harmonic distortion.
2. Gain measured at 2.0 volts RMS output.
3. For zero-bias data, generator impedance is negligible.

R_p Meg-ohms	R_s Meg-ohms	R_{g1} Meg-ohms	E_{bb} = 90 Volts			E_{bb} = 180 Volts			E_{bb} = 300 Volts		
			R_k	Gain	E_o	R_k	Gain	E_o	R_k	Gain	E_o
0.10	0.10	0.1	1,700	31	5.0	1,000	40	15	760	43	30
0.10	0.24	0.1	2,000	38	6.9	1,100	46	20	900	50	40
0.24	0.24	0.1	3,500	43	6.5	2,000	54	18	1,600	58	37
0.24	0.51	0.1	3,900	49	8.6	2,300	59	24	1,800	64	47
0.51	0.51	0.1	7,100	50	7.4	4,300	62	19	3,100	66	39
0.51	1.0	0.1	7,800	53	9.1	4,800	64	24	3,600	69	46
0.24	0.24	10	0	37	3.9	0	53	15	0	62	32
0.24	0.51	10	0	44	5.4	0	60	19	0	67	41
0.51	0.51	10	0	44	5.0	0	61	17	0	69	35
0.51	1.0	10	0	49	6.4	0	66	21	0	71	41

* Courtesy of General Electric Co.

Table 6. Resistance-coupled Amplifier Chart for the 12AT7 and 6201*

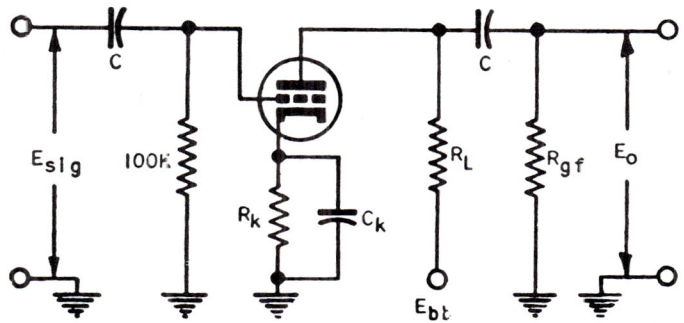

Notes:
1. E_o is maximum RMS voltage output for approximately 5 per cent total harmonic distortion.
2. Gain is measured for an output voltage of 2 volts RMS.
3. R_k is in ohms; R_L and R_{gf} are in megohms.
4. Coupling capacitors C should be selected to give desired frequency response. R_k should be adequately bypassed.

		LOW-IMPEDANCE DRIVE (APPROXIMATELY 200 OHMS)							
R_L	R_{gf}	E_{bb} = 90 Volts			E_{bb} = 180 Volts			E_{bb} = 300 Volts		
		R_k	E_o	Gain	R_k	E_o	Gain	R_k	E_o	Gain
0.10	0.10	1,600	5.3	26	1,100	12	31	1,000	22	32
0.10	0.24	1,800	7.8	29	1,400	17	33	1,200	30	33
0.24	0.24	3,800	7.2	28	2,800	16	32	2,300	28	34
0.24	0.51	4,200	9.4	30	3,300	20	33	2,800	35	33
0.51	0.51	8,000	8.3	28	5,600	18	31	4,900	31	33
0.51	1.0	9,600	10	29	6,700	23	32	6,000	38	33
		HIGH-IMPEDANCE DRIVE (APPROXIMATELY 100K OHMS)							
R_L	R_{gf}	E_{bb} = 90 Volts			E_{bb} = 180 Volts			E_{bb} = 300 Volts		
		R_k	E_o	Gain	R_k	E_o	Gain	R_k	E_o	Gain
0.10	0.10	2,000	9.9	25	1,200	17	31	900	35	33
0.10	0.24	2,400	13	27	1,400	28	33	1,200	47	33
0.24	0.24	4,700	12	27	2,900	25	32	2,300	42	34
0.24	0.51	5,300	15	28	3,600	31	33	2,900	52	34
0.51	0.51	9,300	13	27	6,000	27	31	5,000	45	33
0.51	1.0	11,000	16	28	7,100	33	32	6,400	55	34

*Courtesy of General Electric Co.

Table 7. Resistance-coupled Amplifier Chart for the 6AV11, 6C4, 12AU7A, 5814A, 7199 (Triode Section), and 7247 (Section 2)*

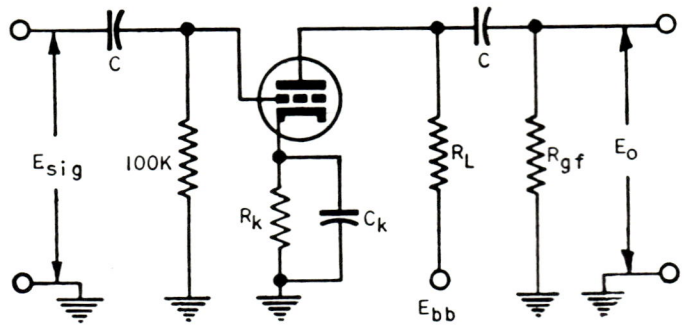

Notes:
1. E_o is maximum RMS voltage output for approximately 5 per cent total harmonic distortion.
2. Gain is measured for an output voltage of 2 volts RMS.
3. R_k is in ohms; R_L and R_{gf} are in megohms.
4. Coupling capacitors C should be selected to give desired frequency response. R_k should be adequately bypassed.

		\multicolumn{9}{c}{LOW-IMPEDANCE DRIVE (APPROXIMATELY 200 OHMS)}
R_L	R_{gf}	\multicolumn{3}{c}{E_{bb} = 90 Volts}	\multicolumn{3}{c}{E_{bb} = 180 Volts}	\multicolumn{3}{c}{E_{bb} = 300 Volts}
		R_k	E_o	Gain	R_k	E_o	Gain	R_k	E_o	Gain
0.10	0.10	3,900	10	10	3,600	20	11	3,500	30	11
0.10	0.24	5,000	14	11	4,700	27	12	4,400	41	12
0.24	0.24	9,400	13	11	8,700	25	11	8,700	38	12
0.24	0.51	11,000	17	11	11,000	32	12	11,000	48	12
0.51	0.51	19,000	15	11	18,000	29	12	18,000	43	12
0.51	1.0	24,000	19	11	23,000	37	12	23,000	54	12

		\multicolumn{9}{c}{HIGH-IMPEDANCE DRIVE (APPROXIMATELY 100K OHMS)}
R_L	R_{gf}	\multicolumn{3}{c}{E_{bb} = 90 Volts}	\multicolumn{3}{c}{E_{bb} = 180 Volts}	\multicolumn{3}{c}{E_{bb} = 300 Volts}
		R_k	E_o	Gain	R_k	E_o	Gain	R_k	E_o	Gain
0.10	0.10	2,600	11	12	2,000	22	13	1,800	31	13
0.10	0.24	3,400	16	12	2,800	32	13	2,600	44	14
0.24	0.24	7,200	15	12	5,800	29	13	5,000	41	13
0.24	0.51	9,400	19	12	8,400	37	13	7,000	52	13
0.51	0.51	17,000	16	12	15,000	33	13	13,000	46	13
0.51	1.0	22,000	20	12	20,000	42	13	18,000	58	13

* Courtesy of General Electric Co.

Table 8. Resistance-coupled Amplifier Chart for the 5751*

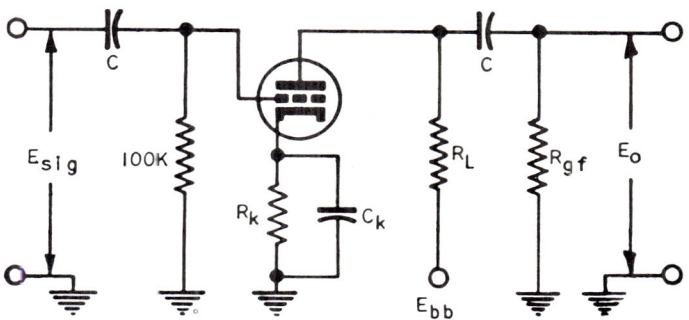

Notes:
1. E_o is maximum RMS voltage output for approximately 5 per cent total harmonic distortion.
2. Gain is measured for an output voltage of 2 volts RMS.
3. R_k is in ohms; R_L and R_{gf} are in megohms.
4. Coupling capacitors C should be selected to give desired frequency response. R_k should be adequately bypassed.

R_L	R_{gf}	\multicolumn{3}{c	}{E_{bb} = 90 Volts}	\multicolumn{3}{c	}{E_{bb} = 180 Volts}	\multicolumn{3}{c	}{E_{bb} = 300 Volts}
		R_k	E_o	Gain	R_k	E_o	Gain	R_k	E_o	Gain

LOW-IMPEDANCE DRIVE (APPROXIMATELY 200 OHMS)

R_L	R_{gf}	R_k	E_o	Gain	R_k	E_o	Gain	R_k	E_o	Gain
0.10	0.10	1,800	4.7	28	1,100	15	35	960	31	36
0.10	0.24	2,000	5.9	34	1,500	21	40	1,200	43	43
0.24	0.24	3,800	6.5	36	2,300	18	45	1,900	38	47
0.24	0.51	4,300	8.5	40	2,700	24	47	2,300	47	50
0.51	0.51	7,400	7.2	40	4,800	20	48	3,700	38	51
0.51	1.0	8,400	9.3	44	5,600	25	50	4,500	49	53

HIGH-IMPEDANCE DRIVE (APPROXIMATELY 100K OHMS)

R_L	R_{gf}	R_k	E_o	Gain	R_k	E_o	Gain	R_k	E_o	Gain
0.10	0.10	2,700	7.4	26	1,600	18	33	1,200	34	36
0.10	0.24	3,100	9.7	31	1,900	25	39	1,500	47	42
0.24	0.24	5,300	9.7	35	3,200	24	43	2,500	43	46
0.24	0.51	6,100	12	38	3,700	30	46	3,000	54	49
0.51	0.51	10,000	11	38	6,000	26	46	4,600	47	50
0.51	1.0	11,000	13	42	6,900	31	49	5,400	59	53

* Courtesy of General Electric Co.

Table 9. Resistance-coupled Amplifier Chart for the 6072A*

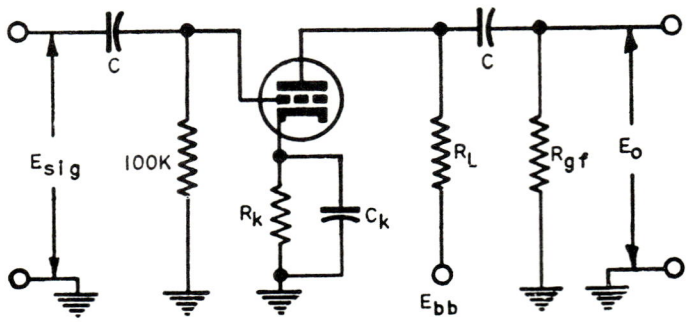

Notes:
1. E_o is maximum RMS voltage output for approximately 5 per cent total harmonic distortion.
2. Gain is measured for an output voltage of 2 volts RMS.
3. R_k is in ohms; R_L and R_{gf} are in megohms.
4. Coupling capacitors C should be selected to give desired frequency response. R_k should be adequately bypassed.

| | | LOW-IMPEDANCE DRIVE (APPROXIMATELY 200 OHMS) ||||||||
| R_L | R_{gf} | E_{bb} = 90 Volts ||| E_{bb} = 180 Volts ||| E_{bb} = 300 Volts |||
		R_k	E_o	Gain	R_k	E_o	Gain	R_k	E_o	Gain
0.10	0.10	1,900	6.9	22	1,300	18	25	1,000	34	27
0.10	0.24	2,100	9.6	25	1,500	24	28	1,300	45	29
0.24	0.24	4,200	8.2	26	2,700	20	28	2,200	36	30
0.24	0.51	4,800	11	27	3,100	25	28	2,700	45	31
0.51	0.51	8,800	8.6	26	6,000	20	29	4,700	36	30
0.51	1.0	10,000	11	27	7,200	25	29	6,000	45	31
		HIGH-IMPEDANCE DRIVE (APPROXIMATELY 100K OHMS)							
R_L	R_{gf}	E_{bb} = 90 Volts			E_{bb} = 180 Volts			E_{bb} = 300 Volts		
		R_k	E_o	Gain	R_k	E_o	Gain	R_k	E_o	Gain
0.10	0.10	2,600	8.8	21	1,600	20	24	1,300	36	26
0.10	0.24	3,000	12	23	1,900	27	27	1,600	48	28
0.24	0.24	5,500	11	24	3,500	24	27	2,800	41	29
0.24	0.51	6,200	13	25	4,100	29	28	3,400	51	30
0.51	0.51	11,000	11	25	6,800	25	28	5,500	49	30
0.51	1.0	12,000	14	26	8,100	31	29	6,700	54	30

*Courtesy of General Electric Co.

Table 10. Resistance-coupled Amplifier Chart for the 7543*

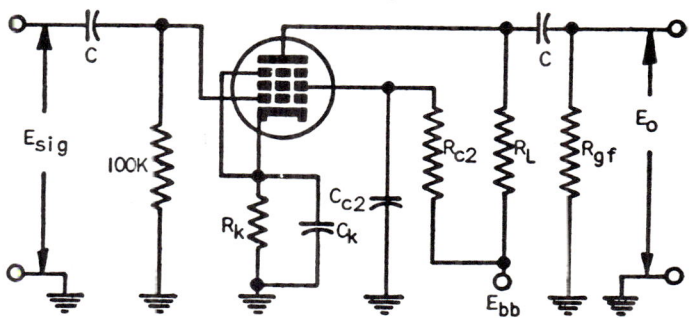

Notes:
1. E_o is maximum RMS voltage output for approximately 5 per cent total harmonic distortion.
2. Gain is measured for an output voltage of 2 volts RMS.
3. R_k is in ohms; R_{c2}, R_L, and R_{gf} are in megohms.
4. Coupling capacitors C should be selected to give desired frequency response. R_k and R_{c2} should be adequately bypassed.

		\multicolumn{12}{c}{LOW-IMPEDANCE DRIVE (APPROXIMATELY 200 OHMS)}
R_L	R_{gf}	\multicolumn{4}{c}{E_{bb} = 90 Volts}	\multicolumn{4}{c}{E_{bb} = 180 Volts}	\multicolumn{4}{c}{E_{bb} = 300 Volts}
		R_k	R_{c2}	E_o	Gain	R_k	R_{c2}	E_o	Gain	R_k	R_{c2}	E_o	Gain
0.10	0.10	960	0.1	13	68	610	0.2	27	96	480	0.2	47	120
0.10	0.24	1,000	0.2	16	93	630	0.2	35	130	480	0.2	60	160
0.24	0.24	2,900	0.3	12	88	1,700	0.4	25	120	820	0.6	44	200
0.24	0.51	3,600	0.4	14	110	1,800	0.5	31	170	960	0.7	53	240
0.51	0.51	5,300	0.9	10	110	4,000	0.9	23	160	2,100	1.1	38	230
0.51	1.0	4,600	1.1	12	125	3,800	1.1	25	200	1,800	1.3	44	300
		\multicolumn{12}{c}{HIGH-IMPEDANCE DRIVE (APPROXIMATELY 100K OHMS)}
R_L	R_{gf}	\multicolumn{4}{c}{E_{bb} = 90 Volts}	\multicolumn{4}{c}{E_{bb} = 180 Volts}	\multicolumn{4}{c}{E_{bb} = 300 Volts}
		R_k	R_{c2}	E_o	Gain	R_k	R_{c2}	E_o	Gain	R_k	R_{c2}	E_o	Gain
0.10	0.10	1,000	0.2	13	70	560	0.2	26	100	380	0.2	47	130
0.10	0.24	1,100	0.2	17	100	630	0.2	34	140	470	0.2	59	180
0.24	0.24	1,900	0.6	15	100	1,100	0.7	29	170	890	0.7	46	210
0.24	0.51	2,200	0.7	17	140	1,200	0.8	36	210	990	0.7	57	260
0.51	0.51	3,000	1.6	17	120	1,700	1.8	33	200	1,200	1.8	54	290
0.51	1.0	3,200	1.8	21	140	1,800	2.0	41	240	1,300	1.9	68	350

* Courtesy of General Electric Co.

Table 11. Tubes for Audio-frequency Power-amplifier Service

Tube type	Tube style	Classification	Heater voltage	Heater current, amp	Max plate voltage and dissipation	Max screen voltage and dissipation	Amplifier class	E_b	E_{c2}	E_{c1}	Peak signal voltage at grid 1	Zero signal I_p, ma	Max signal I_p, ma	Zero signal I_{g2}, ma	Max signal I_{g2}, ma	Load resistance, ohms	Power output, watts	Distortion, %	Sensitivity
6EH5	7-pin miniature	Power pentode	6.3	1.2	135/5.0	117/1.75	A1	110	115	$R_k = 62$	3.0	42	42	11.5	14.5	3,000	1.4	7	155
					150/5.0	130/1.75	AB1 2 tubes p.p.	140	120	$R_k = 68$	9.4 g-g	47 total	51 total	11 total	17.7 total	6,000 p-p	3.8	5	
50C5	7-pin miniature	Beam pentode	50	0.15	135/5.5	117/1.25	A1	110	110	−7.5	7.5	49	50	4.0	8.5	2,500	1.9	9	34
6BF11	Compactron	Dissimilar double pentode (section 2 similar to 6DT6)	6.3	1.2	Section 1 165/6.5	Section 1 150/1.8	Section 1 A1	145	110	−6.0	6.0	36	40	3.9	9.0	3,000	2.4	10	67
6J10	Compactron	Pentode— gated-beam discriminator (discriminator section similar to 6BN6)	6.3	0.95	Pentode section 275/10	Pentode section 275/2.0	Pentode section A1	250	250	−8.0	8.0	35	39	2.5	7.0	5,000	4.2	10	66
6T9	Compactron	Triode-pentode (triode similar to 12AX7)	6.3	0.93	Pentode section 275/12	Pentode section 275/2.0	Pentode section A1	250	250	−8.0	8.0	35	39	2.5	7.0	5,000	4.2	10	66

6AQ5-A	7-pin minia-ture	Beam pentode	6.3	0.45	275/12	275/2.0	A1 A1 AB1 2 tubes p.p.	250 250 180 180 250 250	−12.5 −8.5 −15	12.5 8.5 30 g-g	45 29 70 total	47 30 79 total	4.5 3.0 5.0 total	7.0 4.0 13 total	5,000 5,500 10,000 p-p	4.5 2.0 10	8 8 5	29
6005	7-pin minia-ture	Beam pentode	6.3	0.45	275/11	275/2.2	A1 A1 AB1 2 tubes p.p.	250 250 180 180 250 250	−12.5 −8.5 −15	12.5 8.5 30 g-g	45 29 70 total	47 30 79 total	4.5 3.0 5.0 total	7.0 4.0 13 total	5,000 5,500 10,000 p-p	4.5 2.0 10	8 8 5	29
6BQ5	9-pin minia-ture	Power pentode	6.3	0.76	300/12	300/2.0	A1 AB1 2 tubes p.p.	250 250 300 300	−7.3 $R_k = 130$	6.1 28.2 g-g	48 72 total	49.5 92 total	5.5 8.0 total	10.8 22 total	5,200 8,000 p-p	5.7 17	10 4	153
6L6-GC	Glass octal	Beam pentode	6.3	0.9	500/30	450/5.0	A1 A1 A1 2 tubes p.p. AB1 2 tubes p.p.	350 250 300 200 250 250 270 270 450 400	−18 −12.5 −14 −17.5 −37	18 12.5 14 35 g-g 70 g-g	54 48 72 134 total 116 total	66 55 79 155 total 210 total	2.5 2.5 5.0 11 total 5.6 total	7.0 4.7 7.3 17 total 22 total	4,200 4,500 2,500 5,000 p-p 5,600 p-p	10.8 6.5 6.5 17.5 55	15 11 10 2 1.8	33
7868	Novar	Power pentode	6.3	0.8	550/19	440/3.3	A1 AB1 2 tubes p.p.	300 300 450 400	−10 −21	10 42 g-g	60 40 total	75 145 total	8.0 5.0 total	15 30 total	3,000 6,600 p-p	11 44	13 5	30

where A is the ratio of output voltage to input voltage (voltage gain) in the amplifier circuit.

The Miller-effect relationship shows the advantage of the pentode, with its much lower C_{g-p}, over the triode. However, as higher frequencies are reached, even the low C_{g-p} of the pentode becomes intolerable. When this range is reached, external feedback circuits may be used to neutralize the effect over a usable frequency range.

There is another capacitance effect which is sometimes confused with the Miller effect but is totally unrelated. This effect is the higher input capacitance a tube exhibits when conducting heavily as opposed to when it is not conducting. This change in capacitance is explained by the concept of the electrons filling the grid-cathode space during conduction and causing the relative permittivity to be greater than during nonconduction. A rule of thumb generally followed is to consider the hot, conducting grid-to-cathode capacitance to be at least 50 per cent greater than the cold, nonconducting value usually listed on tube data sheets. This effect varies only with conduction and is totally independent of voltage gain. However, changes in conduction change the voltage gain so that variation of conduction varies the magnitude of the Miller effect as well as that of the space-charge effect. The variation of capacitance due to space charge shows up in the detuning effect on the input tuned circuits of i-f amplifiers when AGC voltage is applied to the No. 1 grid. To compensate for this effect, negative feedback is used in the form of specific values of unbypassed resistance. The value of resistance needed depends upon the specific tube type. The variation in input capacitance with changes in grid voltage for a typical i-f tube with various values of cathode resistor is shown in Fig. 71. Although the unbypassed cathode is effective in stabilizing this capacitance effect, it also reduces the available gain because of degenerative feedback. The effective transconductance of a tube with an unbypassed cathode resistor is given by

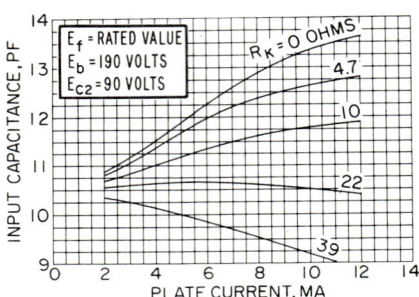

FIG. 71. Variation of input capacitance of an i-f pentode with grid bias.

$$g_m \text{ eff.} = \frac{g_m}{1 + g_m R_k (I_k/I_p)} \qquad (16)$$

To avoid the loss of gain occasioned by use of an unbypassed cathode resistor, a tube design[22] which allows AGC voltage to be applied to the No. 3 grid has been described. The type 9BJ11 double pentode contains one such pentode, specifically designed for this type of service. Although any pentode with a third grid brought out to a separate base pin (some contain only a space-charge-forming window or beam plates, or the No. 3 grid is internally connected to the cathode) might be used in this way, the 9BJ11 has design features that uniquely suit it to this use. First, the No. 3 grid has more lateral wires per inch than most third grids, giving a high transconductance and, therefore, sensitivity to change in AGC voltage. Second, the No. 3 grid is shaped so that as it is biased negatively by the AGC voltage, the electrons are diverted outward to be collected by metal plates attached to the screen grid which help to dissipate the heat that this extra screen current produces. Third, this diversion of electrons, rather than reflection back to the lateral-wire portion of the screen, prevents penetration of deflected electrons into the grid-1 cathode region. The effect of such penetration is to produce feedback which varies with the AGC voltage. A unique feature of this tube is its ability to develop an apparent positive bias on the No. 3 grid when a tuned circuit is connected to this grid.[23] Although

[22] Ulug, *op. cit.*
[23] Ulug, *loc. cit.*

the gain of a pentode is increased by applying positive voltage to the No. 3 grid, such bias voltage is generally not readily available, particularly in AGC circuits. Ulug has shown that if a circuit resonant at a frequency above the passband of the amplifier is inserted in the grid-3 circuit, the effect on the gain is the same as applying a positive bias. Although Ulug's experimental tube showed no appreciable narrowing of the passband or other regenerative effects, the 9BJ11 is not entirely free of these; their magnitude must be determined in a specific circuit by the circuit designer.

In the high-frequency region, commonly above 10 Mc, a tube which is supposed to exhibit infinite input resistance does in fact exhibit a very real positive input resistance. In many cases its magnitude is so low as to cause excessive loading on the source impedance of the driving stage. This effect in common-cathode configurations is one of feedback due to the voltage developed across the cathode-lead inductance. Similar undesirable effects occur in the other two generally used configurations. The voltage produced across the cathode-lead inductance is in part transferred, with altered phase, into the grid circuit through the grid-cathode capacitance. The resultant effect is the presentation of a real, positive conductance between the external grid terminal and ground. The expression for the magnitude of this effect is

$$R_{in} = \frac{1}{\omega^2 L_k C_{g-k} g_m'} \qquad (17)$$

where g_m' is the transconductance of the cathode current. In pentodes where the g_m published in data sheets is that of the plate current,

$$g_m' = g_m \left(\frac{I_k}{I_p}\right) \qquad (18)$$

where I_k and I_p are the d-c cathode and plate currents, respectively.

The power absorbed by this real, positive input resistance, since it cannot be dissipated in the reactive elements of interelectrode capacitance and lead inductance, is passed on into the output circuit, where it is dissipated at the plate (and screen in multigrid tubes) and partially in the external plate load.

A form of Miller effect is sometimes used to compensate for the cathode-lead inductance loading in pentodes. Examination of the Miller effect shows that an inductive "plate load" will result in a real but *negative* input resistance at the grid. The relationship for this effect is

$$R_{in} = \frac{1}{\omega^2 L_p C_{g-p} g_m} \qquad (19)$$

If a small inductance is inserted in the screen lead of a pentode, this inductance, the grid-1 to screen capacitance, and the screen transconductance form a similar negative resistance which may be used to compensate for the cathode-lead inductance loading; this method is termed screen-grid regeneration. Types 6EH7 and 6EJ7 have inductance built in to produce this compensation at the i-f frequencies used in television receivers.

Another method of decreasing cathode-lead inductance loading is through tube construction which makes the tube a true four-pole by providing two separate cathode leads. With this type of tube, the input signal is applied between the grid and one cathode lead, and the output signal is taken from between the plate and the other cathode lead. By this means, the feedback path is decreased and input loading is decreased, but this is done at the expense of eliminating common grounds for the input and output circuits. Tubes which employ this cathode-lead construction include the 6EH7, 6EJ7, and 6HA5. A unique feature of the 6EH7 and 6EJ7 is the fact that input loading may be low enough to require resistive loading of tuned circuits in order to maintain stability and wide enough bandwidth. The advantage in this lies in the fact that resistive loading is more stable and predictable from tube to tube.

Another component of input resistance becomes significant in the vhf range above 100 Mc. This effect is that of transit-time loading, which occurs when the frequency

Table 12. Tubes for R-F and I-F Amplifier Service

Tube type	Tube style	Classification	Heater voltage	Heater current, amp	Max. plate voltage and dissipation	Max. screen voltage and dissipation	E_b	E_{c2}	E_{c1}	I_p, ma	I_{c2}, ma	g_m, μmhos	μ	r_p, megohms	Figure of merit (GBW), Mc	Features and application
6AU6-A	7-pin miniature	Sharp-cutoff pentode	6.3	0.3	330/3.5	330 (supply)/0.75	250	150	$R_k = 68$	10.6	4.3	5,200	...	1.0	75	Widely used—readily available—used for i-f service
6BA6	7-pin miniature	Remote-cutoff pentode	6.3	0.3	300/3.0	300 (supply)/0.6	250	100	$R_k = 68$	11	4.2	4,400	...	1.0	60	Widely used—readily available—used for r-f and i-f service with AGC
6BZ6	7-pin miniature	Semiremote-cutoff pentode	6.3	0.3	330/2.3	330 (supply)/0.55	125	125	$R_k = 56$	14	3.6	8,000	...	0.26	125	Used for i-f service with AGC
6CY5	7-pin miniature	Sharp-cutoff tetrode	6.3	0.2	180/2.0	180 (supply)/0.5	125	80	-1.0	10	1.5	8,000	...	0.1	145	Used for vhf amplifier service
6DJ8	9-pin miniature	Sharp-cutoff twin triode	6.3	0.365	130/1.8 (each section)	...	90	...	-1.3	15	...	12,500	33	...	250	Frame grid—used in cascode amplifier service
6DK6	7-pin miniature	Sharp-cutoff pentode	6.3	0.3	300/2.0	150/0.5	125	125	$R_k = 56$	12	3.8	9,800	...	...	175	Used in i-f amplifier service
6DS4	Nuvistor	Triode	6.3	0.135	135/1.5	...	70	...	0	7.0	...	12,500	68	0.005	260*	Physically small—used in vhf amplifier service
6EH7	9-pin miniature	Remote-cutoff pentode	6.3	0.3	250/2.5	250/0.65	200	90	-2.0	12	4.5	12,500	...	0.5	180	Frame grid—used in i-f amplifier service with AGC
6EJ7	9-pin miniature	Sharp-cutoff pentode	6.3	0.3	250/2.5	250/0.9	200	200	-2.5	10	4.1	15,000	...	0.35	200	Frame grid—used in i-f amplifier service
6ES8	9-pin miniature	Remote-cutoff twin triode	6.3	0.365	130/1.8 (each section)	...	90	...	-1.2	15	...	12,500	...	...	...	Frame grid—used in cascode amplifier service with AGC
6EW6	7-pin miniature	Sharp-cutoff pentode	6.3	0.4	330/3.1	330 (supply)/0.65	125	125	$R_k = 56$	11	3.2	14,000	...	0.2	180	Used in i-f amplifier service
6GK5	7-pin miniature	Gain-controlled triode	6.3	0.18	200/2.5	...	135	...	-1.0	11.5	...	15,000	78	0.005	250	Frame grid—used in vhf r-f amplifier service with AGC
6GU5	7-pin miniature	Shadow-grid pentode	6.3	0.22	300/3.0	150/0.15	275	135	-0.4	10	0.17	15,500	...	0.165	235	Frame grid—low screen current—low partition noise—used in vhf r-f amplifier service

9–74

Type	Base	Class	E_f	I_f	E_b/I_b	E_{c2}/I_{c2}	E_{c1}	g_m	r_p	μ	C_{in}	C_{out}	C_{gp}	Merit	Remarks
6HA5	7-pin minia-ture	Triode	6.3	0.18	220/2.6		135	-1.0	11.5		14,500	72		270	Frame grid—used in vhf r-f amplifier service
8BM11	Compactron	Dissimilar double pentode	8.4	0.45	160/2.2 (section 1) 160/2.2 (section 2)	160/0.55 (section 1) 160/0.55 (section 2)	125 125	Section 1 $R_k = 56$ Section 2 $R_k = 56$	14 12	3.6 3.8	8,000 9,800		0.26	135 150	Section 1 similar to 6BZ6—section 2 similar to 6DK6. Used in 2-stage i-f amplifier in TV receivers
9BJ11	Compactron	Dissimilar double pentode	9.6	0.45	160/2.8 (section 1) 160/2.2 (section 2)	160/1.1 (section 1) 160/0.55 (section 2)	110 125	Section 1 $E_{c3} = 0$ $E_{cc1} = 0$ $R_{g1} = 100$ kilohms Section 2 $R_k = 120$	6.0 7.0 8.4 2.5		7,500 9,200			100 135	Section 1 remote-cutoff—designed for AGC to be applied to No. 3 grid (U lug section). Section 2 sharp-cutoff. Tube intended for i-f amplifier in TV receivers
5654	7-pin minia-ture	Sharp-cutoff pentode	6.3	0.175	200/1.65	155/0.55	120	-2.0	7.5 2.5		5,000		0.34	95	"Reliable" 6AK5—used in broadband amplifier service
6386	9-pin minia-ture	Remote-cutoff twin triode	6.3	0.35	300/1.5 (each section)		100	Each section $R_k = 200$	9.6		4,000	17	0.004	110	"Reliable"—designed for low cross modulation—used in cascode amplifier service
6688	9-pin minia-ture	Sharp-cutoff pentode	6.3	0.3	210/2.7	175/0.6	190 160	$+9.0$ and $R_k = 630$	13 3.3		16,500		0.09	250	Frame grid—"reliable"—used in broadband amplifier service
7462	Metal-ceramic planar	Triode	6.3	0.24	250/1.1		150	$+6.0$ and $R_k = 910$	7.2		10,500	94	0.009	350*	Tube has lugs for direct mounting in printed-circuit boards
7586	Nuvistor	Triode	6.3	0.135	110/1.0		75	$R_k = 100$	10.5		11,500			250*	Physically small—used in vhf r-f amplifier service
7587	Nuvistor	Sharp-cutoff tetrode	6.3	0.15	250/2.2		125 50	$R_k = 68$	10 2.7		10,000		0.2	200	Physically small—used in r-f and i-f amplifier service
7588	Metal-ceramic planar	Triode	6.3	0.4	300/5.5		200	$+6.0$ and $R_k = 270$	24		45,000	175	0.004	1,100*	Tube has lugs for direct mounting in printed-circuit boards
8113	7-pin minia-ture	Sharp-cutoff tetrode	6.3	0.2	180/2.0	180 (supply) 0.5	120	-2.0	10 2.3		7,000		0.02	130	"Reliable" 0CY3

* Merit figure based on operation in a neutralized-triode circuit.

of the input signal at the grid is sufficiently high that an appreciable change in signal phase takes place during the transit time of an electron past the grid. At low frequencies the displacement current that flows out of the grid circuit as an electron approaches the grid and then flows back after the electron passes results in a net power flow of zero because the average grid voltage has not changed. However, at high frequencies the displacement current flows out at one instantaneous grid voltage and back in at a different instantaneous grid voltage. The net power is not zero under these conditions and the difference appears as a power-consuming element of input resistance which varies inversely as the square of the frequency.

A feedback effect not commonly recognized is that caused by mutual inductive coupling between adjacent leads within the tube. An example of this occurs in the 6AU6, a type which is frequently used as an i-f amplifier at frequencies up to 30 Mc. At higher frequencies, the mutual coupling from the No. 3 grid lead to the No. 1 grid lead feeds back output voltage due to capacitive currents resulting from the relatively high capacitance from the plate to the No. 3 grid and to the outer shield which is also tied to the No. 3 grid lead.

Special construction features are incorporated in tubes to be used in the frequency range above 300 Mc. Very short leads and/or multiple leads are used to minimize lead inductance, while small structures are used to minimize capacitances. Close spacings are used to decrease transit time, and these spacings are kept as near uniform as possible to minimize transit-time variation, since such variation gives rise to noise as well as reducing the gain.

Table 12 lists tubes suitable for r-f and i-f service. Since both gain and bandwidth are of importance in many applications, a figure of merit has been included in the table. However, it is almost impossible to formulate a merit figure that describes all conditions of operation, for several reasons detailed here, and caution should be observed in picking tubes purely on the basis of a merit figure.

Because of the variations in performance brought about by the use of different coupling circuitry, it is difficult to generalize on the gain of various tubes intended for r-f and i-f service. However, the most commonly used figure of merit is the gain-bandwidth factor (GBW). Although not rigorously applicable to all coupling systems and specifically intended for transitionally coupled, double-tuned stages with equal Q's,[24] the following formula for GBW is used in Table 12:

$$\text{GBW} = \frac{g_m}{2\pi \sqrt{2C_p C_s}} \qquad (20)$$

where GBW = gain bandwidth, cps
g_m = transconductance of the tube, mhos
C_p = output capacitance of the tube, farads
C_s = input capacitance of the tube, farads

To account for stray wiring capacitances, an additional 2.5 pf has been added to both the input and output capacitances in calculating the values of GBW used in Table 12. Obviously, the values of capacitance do not take into account any capacitances generated at the input because of feedback paths; it is assumed that these will be compensated for by circuitry. However, if no compensation is used, the GBW must be recalculated with the effective value of capacitance. Furthermore, if unbypassed cathode resistance is used for stabilizing space-charge capacitance effects in AGC stages, the GBW must be recalculated using the degenerated effective value of g_m.

A number of tubes suitable for use as r-f power amplifiers are included in the classification "receiving tubes." Tubes for this application are selected by considering the power output required, the allowable plate dissipation, the operating frequency, the driving power required, and the heater voltage. Power-amplifier tubes are usually rated differently for each type of service for which the tube is intended (see Sec. 2.4) and typical operation given within each rating system. The data shown in Table 13 include the CCS ratings and typical operation under this system. However, most of

[24] R. W. Landee, D. C. Davis, and A. P. Albrecht, "Electronic Designers' Handbook," p. 7-45, McGraw-Hill Book Company, New York, 1957.

Table 13. Tubes for R-F Power-amplifier Service

Tube type	Tube style	Classification	Heater voltage	Heater current, amp	Class of service	Max plate voltage and dissipation	M_{ax} screen voltage and dissipation	Max frequency for full input	E_b	E_{c2}	E_{c1}	Driving power, watts	I_p, ma	I_{g2}, ma	I_{g1}, ma	Power output, watts
2E26	Octal	Beam pentode	6.3	0.8	Class C telegraphy	500/10	200/2.5	125 Mc	400	190	−30	0.12	75	11	3.0	20
807	Octal	Beam pentode	6.3	0.9	Class C telegraphy	600/25	300/3.5	60 Mc	600	250	−45	0.3	100	8.0	4.0	40
5686	9-pin miniature	Beam pentode	6.3	0.35	Class C telegraphy	250/7.5	250/3.0	160 Mc	250	250	−50	0.15	40	10.5	2.0	5.0
5763	9-pin miniature	Beam pentode	6.3	0.75	Class C telegraphy	300/12	250/2.0	50 Mc	300	250	−60	0.35	50	5.0	3.0	7.0
6146-B	Octal	Beam pentode	6.3	1.125	Class C telegraphy	600/27	250/3.0	60 Mc	600	200	−70	0.3	150	10	2.8	63
7984	Compactron	Beam pentode	13.5	0.58	Class C telegraphy	600/20	250/3.0	175 Mc	315	165	−74	2.0	150	8.5	3.7	26.5
8156	Compactron	Beam pentode	13.5	0.3	Class C telegraphy	500/11	250/2.5	175 Mc	250	185	−70	1.0	100	11	3.5	14

Table 14. Gridded Tubes for UHF and Microwave Service

Tube type	Tube style	Classification	Heater voltage	Heater current, amp	Max plate voltage and dissipation	Class of service	E_b	E_c	I_p, ma	g_m, μmhos	μ	R_p, ohms	Typical operation	Application and remarks
6AM4	9-pin miniature	Triode	6.3	0.225	200/2.0	Class A, uhf grounded grid	200	$R_k = 100$	10	9,800	85	8,700	....	Designed for uhf TV r-f amplifier use
5768	Rocket tube	Triode	7.0	0.45	250/4.0	CW amplifier grounded grid	150	$R_k = 100$	9.0	9,600	88	9,150	Grounded-grid amplifier at 500 Mc. $E_b = 150$, $R_k = 100$, $I_p = 7.0$, $g_m = 9,000$, μ = 90, noise figure = 9 db, power gain = 16 db	Used as a CW amplifier in the 1,000- to 3,300 Mc range
6299	Metal-ceramic planar	Triode	6.3	0.3	200/2.0	Grounded-grid amplifier	175	Adj. for $I_p = 10$	10	15,000	110	7,300	Grounded-grid amplifier at 1,200 Mc. $E_b = 175$, E_c adj. for $I_p = 10$, noise figure = 8.5 db, power gain = 17 db, bandwidth = 10 Mc	Used as an r-f amplifier to 3,000 Mc
6442	Metal-ceramic planar	Triode	6.3	0.9	350/8.0	CW power amplifier	350	−4.25	35	16,500	50	......	Class C amplifier at 1,000 Mc. $E_f = 5.7$, $E_b = 250$, $I_g = 6.0$, $I_p = 23$, driving power = 0.35 watt, power output = 2.8 watts	Used as an r-f power amplifier to 2,500 Mc
7077	Metal-ceramic planar	Triode	6.3	0.24	250/1.1	Grounded-grid amplifier	250 through 18 kilohms	$R_k = 82$	6.5	10,000	90	9,000	Grounded-grid amplifier at 450 Mc. $E_b = 250$ through 18 kilohms, $R_k = 82$, $I_p = 6.5$, noise figure = 5.5 db, power gain = 14.5 db, bandwidth = 7.5 Mc	Low noise—small—adapted to coaxial circuits
7552	Pencil tube	Triode	6.3	0.225	250/2.5	Grounded-grid amplifier	125	$R_k = 50$	13	13,500	80	......	Grounded-grid amplifier at 1,100 Mc. $E_b = 150$,	Used as an r-f amplifier to 1,500 Mc

7768	Metal-ceramic planar	Triode	6.3	0.4	330/5.5	Broadband r-f amplifier	200	+6.0 and $R_k = 270$	24	50,000 225 4,500	$R_k = 50$, $I_b = 13.5$, noise figure = 12.5 db, power gain = 16 db, bandwidth = 10 Mc	High gain band width
7815	Metal-ceramic planar	Triode	6.0	1.0	3,500/10 (pulse rating)	Plate-pulsed r-f amplifier					Plate-pulsed amplifier at 1,100 Mc. $E_f = 6.0$, $E_b = 1,700$ volts, PD = 3.5 μsec, duty factor = 0.001, peak $I_p = 1.9$ amp, peak power output = 1,500 watts	Used as a plate-pulsed r-f amplifier to 3,000 Mc
7910	Metal-ceramic planar	Triode	6.3	0.275	1,200/1.5 (pulse rating)	Plate-pulsed r-f amplifier	125	$R_k = 82$	11.5	16,000 75		Used as a plate-pulsed r-f amplifier to 7,500 Mc
7911	Metal-ceramic planar	Triode	6.3	0.55	3,000/6.5 (pulse rating)	Plate-pulsed r-f amplifier	200	$R_k = 100$	23	25,000 58 2,300		Used as a plate-pulsed r-f amplifier to 6,000 Mc
7913	Metal-ceramic planar	Triode	6.3	0.4	330/5.5	R-f power amplifier	200	$R_k = 47$	25	40,000 100 2,500		Used as an r-f power amplifier
8058	Nuvistor	Triode	6.3	0.135	150/1.5	Grounded-grid r-f amplifier	110	$R_k = 47$	10	12,400 70 5,600	Grounded-grid amplifier at 1,200 Mc. $E_b = 110$, $R_k = 47$, $I_p = 10$, noise figure = 12.2 db, power gain = 10.5 db bandwidth = 12 Mc	Small-tube outline specifically designed for grounded-grid circuitry

AMPLIFYING DEVICES

the tubes also have ICAS and some have IMS ratings. These generally higher ratings were not included because of space limitations, but they may be obtained from the manufacturers' data sheets.

4.7. UHF and Gridded Microwave Tubes. In the frequency range above 300 Mc the ability of conventional glass miniature tubes to amplify declines rapidly as the internal active surfaces become electrically farther removed by lead inductance from the circuit and dielectric losses and capacitances in the glass surrounding the leads further complicate matters. Even tubes with multiple leads such as the 6AM4 and EC88 are only marginally successful as small-signal amplifiers up to 900 Mc. The most successful tubes for this use are the planar constructed tubes, especially those with ceramic insulation instead of glass. In these tubes all the important factors are optimized—low lead inductance; low-loss, low-capacitance insulators; small active areas for low interelectrode capacitances; close electrode spacings for small transit angles; and high transconductance to extend the useful frequency range against mounting circuit losses.

Up to 1,000 Mc the planar tubes hold their own against solid-state devices as small-signal amplifiers, but above this frequency they begin to suffer by comparison with the latter from the standpoint of noise figure. However, tubes do maintain advantages of temperature tolerance, stability, and wide dynamic signal-handling capability. The 7077 is an example of a planar tube for this frequency range. While not planar in construction, a double-ended Nuvistor, the 8058, is also suitable for this range.

For CW power-amplifier applications the useful frequency range of planar tubes is considerably greater, extending up to approximately 4,000 Mc. There are two reasons for this: (1) Tubes used as small-signal amplifiers lose to competing devices primarily because of noise rather than because of lack of gain, and this noise is not usually a problem in CW power-amplifier service. (2) Higher plate voltages reduce transit time, thus extending the range of useful gain. Recent innovations in cathode design in some types allow an increase in cathode current density, which increases the available power gain for a given tube structure size. An example of a planar tube for this service is the 7486.

Tubes may be operated as pulsed amplifiers at even higher frequencies. Because of the even higher plate voltages permissible in pulsed operation and the high cathode current densities permitted by the short duty cycle of pulsed operation the useful frequency extends well above 4,000 Mc. The 7486 seems to be suited for this use, also.

Tubes for uhf and microwave service are listed in Table 14.

5. MICROWAVE TUBES

5.1. Introduction to Transit-time Tubes. The fundamental difference between the microwave amplifying tubes discussed in the previous sections and those discussed in the following four sections is that the present devices *make use of transit time to achieve density modulation* rather than obtaining it by the valving action of a control grid. It is noteworthy that a major disorder of space-charge control tubes at high frequencies, namely, the effect of transit time, forms the very basis of operation of this generically new class of high-frequency amplifying tubes. This class is characterized by interaction times on the order of *cycles of the r-f frequency* rather than fractions of a cycle, as is typical in space-charge control tubes.

In order to prevent space-charge forces from dispersing the electrons in the physically long region over which such an interaction takes place, it is nearly always necessary to utilize a d-c magnetic field in addition to the d-c electric field. Transit-time tubes can be divided into two distinct types, depending on whether the magnetic field is parallel to the direction of d-c electron drift (*O type*) or perpendicular to the direction of drift (*M type*). The major representatives of O-type tubes are the *klystron* and the *traveling-wave tube*. M-type tubes are represented by crossed-field amplifiers such as the *amplitron*, the *dematron*, the *bimatron*, and the *bitermitron*.

There is a fundamental electrical difference between O-type tubes and M-type tubes. In O-type devices the *d-c kinetic energy* of the electron stream is converted into r-f output power, whereas in M-type tubes the r-f output power is derived from the

ELECTRON TUBES 9-81

d-c potential energy of the electrons in the interaction space. In this respect M-type devices resemble the conventional class C amplifier, for in both types of tubes electrons are transferred from the cathode, a position of high d-c potential energy, to the anode circuit, a position of low d-c potential energy, without gaining d-c kinetic energy. This loss of potential energy appears instead as r-f output power. Like triode amplifiers, M-type devices are characterized by medium gain (10 to 20 db) and high efficiency (50 to 80 per cent) In contrast, O-type devices provide high gain (30 to 50 db) and medium efficiency (30 to 50 per cent).

The electron-beam parametric amplifier, also known as the Adler tube or the quadrupole amplifier, works on an entirely different principle from O- and M-type tubes. In this device the d-c drift energy of the electron beam is untapped. Rather, the r-f output power stems from the *pump*, which is an r-f generator supplying r-f power at a frequency different from the signal frequency. By avoiding the d-c power of the beam as the source of amplified r-f energy, the inherent noise associated with the temperature of the d-c beam is also avoided. The result is a voltage amplifier with an extremely low noise figure.

Table 15 summarizes the properties of O-type, M-type, and parametric amplifiers. The third column lists the source from which the amplified output signal derives its

Table 15. Properties of O-Type, M-Type, and Parametric Amplifiers

Type	Representative tubes	Source of energy	Direction and type of source energy
O....................	Klystron Traveling-wave tube	D-c power	Longitudinal Kinetic energy
M....................	Amplitron Dematron Bitermitron Bimatron (TPOM)	D-c power	Transverse Potential energy
Parametric...........	Adler tube	R-f pump power	Transverse Kinetic energy
Space-charge control.....	Triode Tetrode	D-c power	Longitudinal Potential energy

energy. The type of source energy is listed in the fourth column together with the direction in which the electrons move in giving up their energy, compared with the direction of their drift. Space-charge control tubes have been added to Table 15 for completeness.

The operation of klystron amplifiers will be described in Sec. 5.2, traveling-wave tubes in Sec. 5.3, crossed-field devices in Sec. 5.4, and the electron-beam parametric amplifier in Sec. 5.5.

5.2. Klystron Amplifiers. The klystron tube is a very stable microwave power amplifier which provides high gain at medium efficiency. Although low-noise klystron amplifiers have been built, helix-type traveling-wave tubes (Sec. 5.3) are more commonly used for purposes of voltage amplification.

Physical Configuration. A sketch of a typical broadband multicavity klystron is shown in Fig. 72. The main portion of the tube consists of a number of resonant cavities linked together with sections of metallic pipe called *drift tubes*. The drift tubes are designed to be nonpropagating at the frequency of operation; thus the output cavity of a klystron is very effectively isolated from the input cavity without the need of internal attenuation, a very important consideration for high-power amplifiers.

An electron beam originates from an electron gun and passes through the succession of cavities and drift tubes, ultimately striking the collector. The main body of the tube, including the collector, is usually operated at ground potential; the cathode and its surrounding focus electrode are operated at a large negative potential. An anode with an aperture faces the cathode, and with the aid of the focus electrode, the electron beam is directed into the first drift tube. An axial magnetic field is used to keep the beam from spreading as it traverses the tube. In one commonly used beam collimation scheme, called *Brillouin focusing*, this is accomplished by passing the beam through an aperture in a magnetic plate which shields the bulk of the magnetic field from the gun region. The transverse component of the magnetic field in the aperture gives the beam a rotating motion as it travels. The transverse component of beam velocity interacts with the longitudinal magnetic field in the drift space to produce an inward force which by proper choice of the magnetic field strength can be adjusted to cancel exactly the outward space-charge repulsion force of the electron beam. By

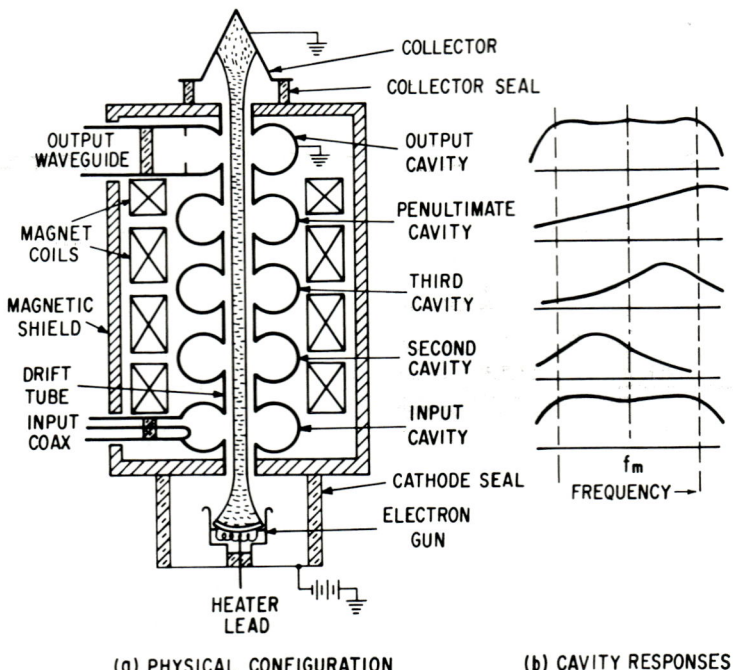

FIG. 72. Broadband multicavity klystron.

this means the electron beam can be made to travel long distances, e.g., many feet, with less than 1 per cent current interception on the drift-tube walls. The collector is usually tapered to increase the area available for heat dissipation and to discourage secondary electrons from traveling backward through the tube.

The input and output lines of a klystron usually take the form of coaxial lines or rectangular waveguide, although circular waveguide is occasionally used to increase the power-handling capacity. Three major ceramic seals are used in klystrons: (1) the main cathode seal, which stands off the high voltage; (2) the r-f input seal, often a coaxial seal; and (3) the output window, usually a slab of ceramic placed at right angles to the waveguide. A fourth seal is sometimes used to insulate the collector from the main body of the tube, enabling collector current to be monitored or the collector potential to be depressed in order to enhance efficiency.

Cavity Tuning. The tuning pattern of the cavities of a multicavity klystron is dependent upon whether gain or efficiency is the prime objective. For maximum gain all cavities are tuned to the midband resonant frequency f_m; this is called *synchronous tuning*. It is more common for efficiency and bandwidth considerations to dictate cavity tuning and Q. Typical cavity tuning for optimum efficiency over a broad band of frequencies is shown in Fig. 72b. In this case only the input and output cavities are tuned synchronously. Quite often these cavities are double- or triple-tuned by the use of additional resonant cavities placed in series with the input and output lines. The cavity immediately preceding the output cavity is called the *penultimate* cavity, and for reasons discussed below, it is adjusted to give optimum power amplification by tuning it to a frequency *above* the band of frequencies to be amplified. The remaining cavities are usually stagger-tuned across the operating band, comprising stages of voltage amplification.

Bunch Formation. Although the electrons which leave the electron gun of a klystron are uniform in density, the cooperative action of the cavities and drift tubes

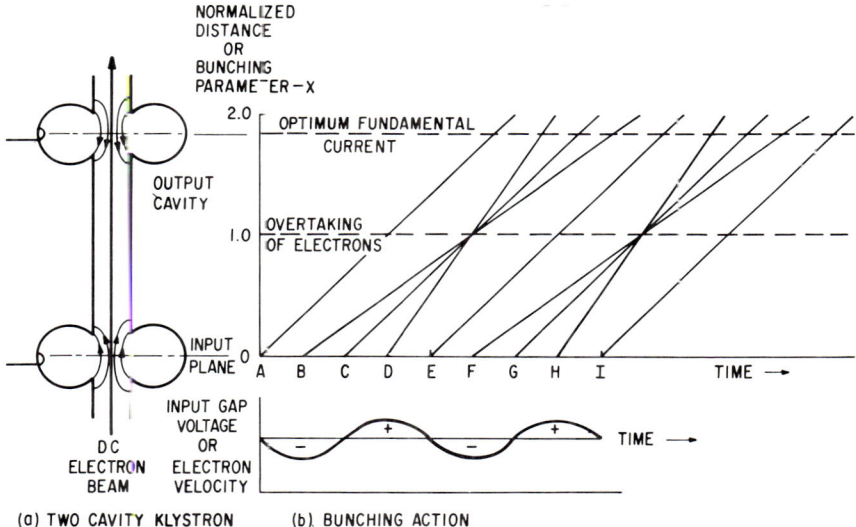

Fig. 73. Electron bunching in a two-cavity klystron.

causes a large degree of density modulation to appear in the beam at the output gap. This action, called *bunching*, may be understood with reference to Fig. 73. At the input plane the d-c electron beam is exposed to the sinusoidally varying electric field which appears across the input gap. This electric field causes the velocities of the electrons to vary sinusoidally in the fashion shown by the sine wave sketched below the graph. Electrons which pass through the gap when the r-f voltage is zero travel in the drift region at a velocity corresponding to the d-c beam voltage; such trajectories are indicated by the straight lines whose starting times are marked by the letters A, C, E, G, and I. Electrons leaving at times B and F travel at a reduced velocity; hence their trajectories have a smaller slope. Electrons leaving at times D and H on the other hand, have a slope greater than the d-c slope. The result of this r-f velocity modulation is that electrons tend to bunch progressively around the d-c electrons with initial times C and G. Conversely, electrons with starting times A, E, and I find themselves at the center of regions whose charge density is less than the d-c charge density. Thus the velocity modulation of a uniform electron beam leads, upon subsequent drift in a field-free region, to density modulation.

It is interesting that if the drive voltage is made large enough to overcome space-charge forces, electrons starting at B, C, and D (or F, G, and H) can actually overtake one another at a critical distance down the drift tube. If a snapshot were taken along the drift tube of a typical klystron, it would look like that sketched in Fig. 74. Note the progressive buildup in current density, culminating in a sharp spike of current at crossover, $X = 1$. Continued drifting beyond crossover results in a double-spiked current waveform which is found to possess optimum fundamental component of current at a distance 1.84 times the crossover distance.

A simplified non-space-charge analysis of the bunching process in klystrons leads to the following equation for the normalized magnitudes of the fundamental and harmonic components of current:

$$I_n/I_0 = 2J_n[n\pi N(V_1/V_0)] = 2J_n(nX) \tag{21}$$

where I_n = peak nth harmonic current
$J_n(X)$ = nth-order Bessel function of the first kind
I_0 = d-c current
$N = xf/v_0$ = drift length measured in cycles transit time for a d-c electron
x = physical drift length
f = operating frequency
v_0 = velocity of d-c electron
V_1 = a-c voltage across input gap
V_0 = d-c beam voltage
$X = \pi N V_1/V_0$ = ballistic bunching parameter, or normalized distance.

The variation of the fundamental component of current ($n = 1$), the second harmonic ($n = 2$), and the ninth harmonic ($n = 9$) as given by Eq. (21) is plotted in

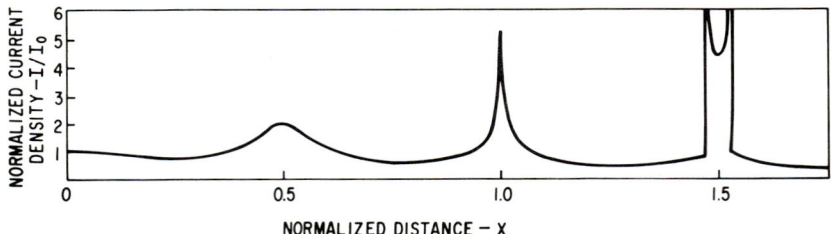

Fig. 74. Instantaneous current density along the drift tube of a klystron.

Fig. 75 as a function of the ballistic bunching parameter X. The fundamental current normalized to I_0 increases linearly for small X, reaches the value 0.88 at crossover, and saturates at a value of 1.16, corresponding to a theoretical current conversion efficiency of 58 per cent. The harmonic currents are surprisingly large, and it is interesting to note that they reach their maximums at somewhat smaller values of the bunching parameter, corresponding to somewhat shorter drift lengths or lower drive levels than for optimum fundamental current.

Energy Extraction. The energy-extraction process which takes place in the output gap of a klystron can be understood with reference to Fig. 76. The output cavity can be represented by an LC resonant circuit as shown in the upper left-hand sketch. The simplest form of resonant cavity is a hollow cylindrical can operating in the TE_{10} mode as shown in the upper right-hand sketch. The electric field lines are at their strongest on the axis and are the counterpart of the electric field lines that exist across the capacitor of the simple LC circuit. In order to allow for the passage of an electron beam through the cavity, holes are placed in the top and bottom surfaces of the cylinder as shown in the lower left-hand sketch. The E lines terminate on the charges at the edges of the holes and fringe into the central region where they provide a predominantly axial electric field. This field is relatively uniform across the radius, provided the hole is not too large. In order to improve the coupling between this electric field and the electron beam, opposing reentrant cylinders customarily are used to make the gap shorter, as shown in the lower right-hand sketch.

In a klystron the axial electric field of the output cavity is exposed to a series of electron bunches which are timed to arrive with a frequency which is exactly equal to the resonant frequency of the output cavity. Furthermore, the phase of the output voltage swing adjusts itself to oppose the motion of the electrons through the gap,

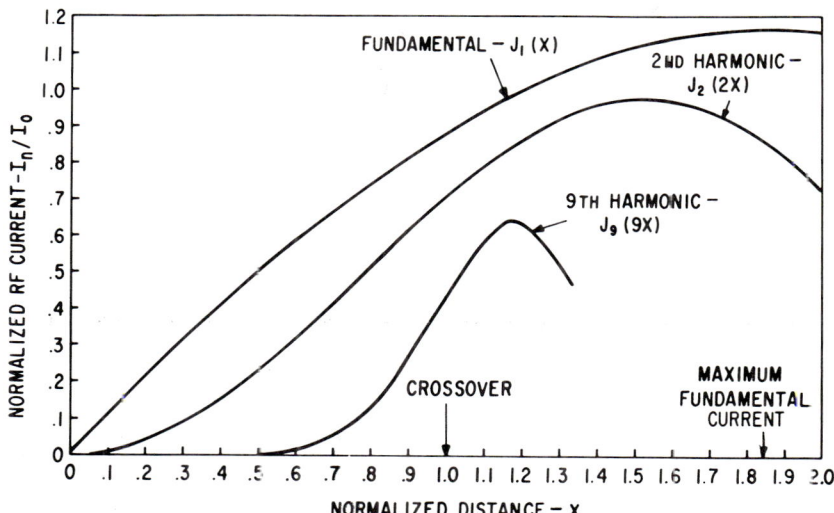

Fig. 75. Fundamental and harmonic currents vs. bunching parameter.

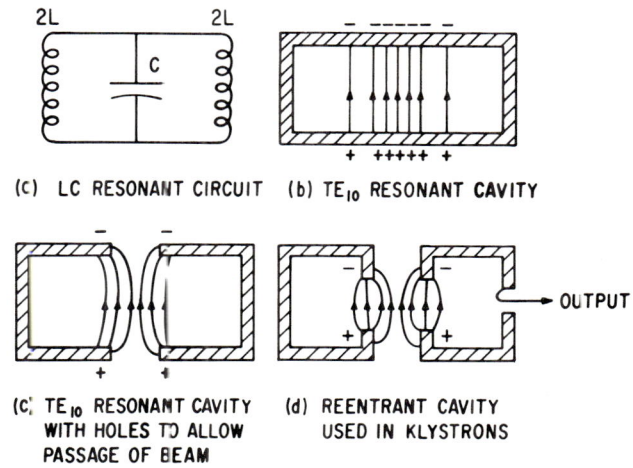

Fig. 76. Evolution from LC circuit to klystron cavity.

which means that the field has a maximum decelerating value when an electron bunch is passing through the gap. One half-cycle later the field gives the electrons maximum acceleration, but since the beam is bunched periodically, fewer electrons are accelerated than are decelerated. Hence there is a net power flow from the beam to the r-f cavity fields. This energy may be removed with a coupling loop, or an iris followed by a waveguide, and constitutes the useful output power of the tube. As a result of this energy-extraction process, the effective current induced in the output cavity is very nearly equal to the fundamental component of current in the bunched electron beam as given by Eq. (21).

Output Efficiency. The efficiency corresponding to the maximum value of fundamental current may be determined as follows: The maximum current that is induced in the output cavity is $2I_0 J_1(1.84)$. The peak value of the r-f voltage across the output gap should not exceed the d-c beam voltage or else electrons would be reflected in the output gap and would travel backward through the tube, disturbing the focusing of the beam and possibly causing oscillation. Thus the expression for the maximum theoretical efficiency of a two-cavity (single-drift-tube) klystron becomes

$$\eta_{max} = \frac{I_{max} V_{max}}{2 I_0 V_0} = \frac{2 I_0 J_1(1.84) V_0}{2 I_0 V_0} = 0.58 \quad (22)$$

where I_{max} and V_{max} are the maximum attainable values of output circuit current and voltage, respectively. The factor of 2 is needed in the denominator because the circuit quantities are always expressed as peak values.

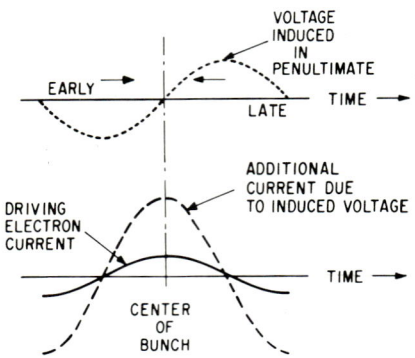

Fig. 77. Current enhancement by penultimate cavity tuned to high side.

Penultimate-cavity Action. In practice a two-cavity klystron is found to have an efficiency only about one-half that derived above. The reasons for this discrepancy are complex and are related to the fact that space charge has been neglected in the above discussion. In order to increase the operating efficiency of practical klystrons to the 35 to 45 per cent range, it is necessary to add a third cavity between the input and output cavities. Furthermore, it is necessary to tune the middle or *penultimate* cavity to a frequency higher than the operating frequency, i.e., to *tune it to the high side*. The reason for the enhancement of efficiency under these conditions is illustrated in Fig. 77. The lower solid sinusoid depicts the electron current which is present at the middle cavity of a klystron as a result of the initial velocity modulation and subsequent drift in the first drift tube. This electron current induces a lagging voltage in the penultimate cavity which, because it is being operated below its resonant frequency, looks like an inductive reactance. This induced voltage is shown as a dotted line in the upper sketch. As has already been demonstrated with the aid of Fig. 73, the d-c electron which immediately precedes the bunching-voltage maximum constitutes the center about which the other electrons will bunch; therefore, the new electron current developed in the final drift tube as a result of the penultimate voltage will be exactly in phase with the electron current which arrived at the penultimate cavity. This enhancement of the electron bunching leads to an efficiency improvement of 10 to 20 percentage points.

Because of the falling impedance characteristic of a detuned resonant circuit, the gain of a three-cavity klystron tuned for maximum efficiency is less than that when tuned for maximum gain. This behavior is illustrated in Fig. 78. With slight penultimate detuning output power rises rapidly with drive but saturates at a level considerably below that attainable with large penultimate detuning. In the latter case, however, the slope of the curve is less, indicating that the gain has been reduced. Addition of more intermediate cavities does not result in further improvement of efficiency but simply in increased gain.

ELECTRON TUBES   9-87

Multiple-beam Klystron. A tube has recently been devised in which n klystron beams are effectively coupled together by means of periodically loaded multiwavelength resonators. Such a tube, called the *multiple-beam klystron*, is capable of giving an *n-fold increase in power output* over single-beam capability with no degradation of other operating characteristics such as gain, bandwidth, and heat dissipation. A

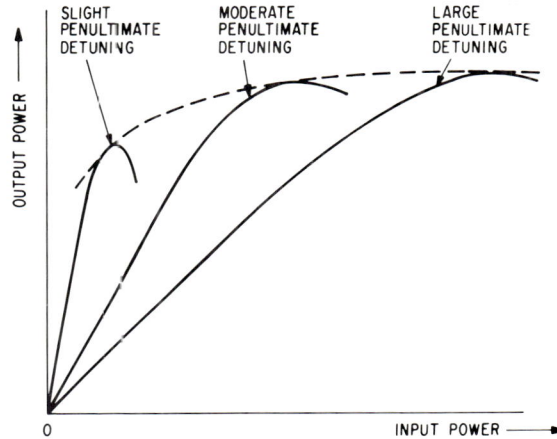

Fig. 78. Effect of penultimate detuning on efficiency and gain of a klystron.

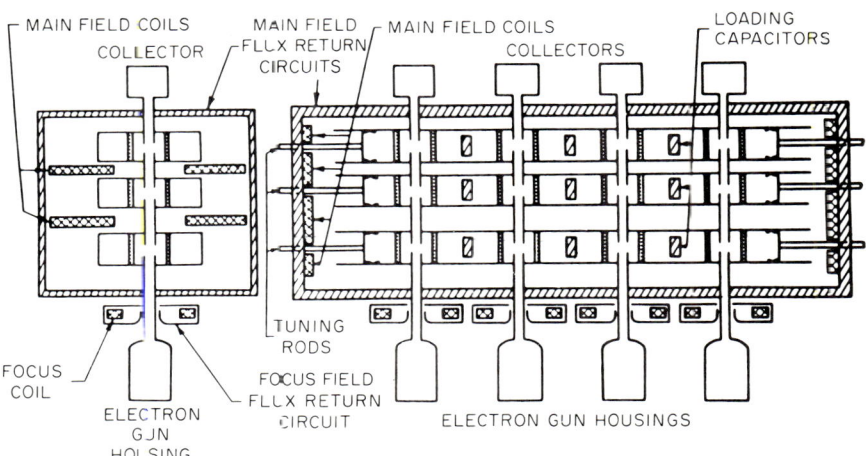

Fig. 79. Multiple-beam klystron.

four-beam multiple-beam klystron is illustrated in Fig. 79. Viewed from the end its cross section is identical with that of a conventional single-beam klystron. However, the side view discloses that four electron beams are spanning three waveguides. These waveguides are periodically loaded by the interaction gaps and by additional dummy capacitors placed midway between gaps. The waveguides are terminated by tunable shorting plungers at either end, and are operated with $\pi/2$ radians phase shift from beam gap to loading capacitor. This mode of operation leads to maximum mode separation, minimum attenuation, and excellent end-to-end coupling.

5.3. Traveling-wave Tubes.
Whereas the electron beam in a klystron travels for the most part in regions free from r-f electric fields, in the traveling-wave tube the beam is continually interacting with an r-f electric field which propagates along an external circuit surrounding the beam. To obtain amplification this circuit must propagate a wave whose phase velocity is nearly synchronous with the d-c velocity of the electron beam. D-c velocity in the nonrelativistic region below 10 kv is given by the expression

$$v_0 = 5.93 \times 10^7 \sqrt{V_0} \quad \text{cm/sec} \tag{23}$$

where V_0 is the d-c beam voltage. For instance, a 10,000-volt beam travels at approximately one-fifth the velocity of light.

Physical Configuration. The phase velocity in a waveguide which is uniform in the direction of propagation is always greater than the velocity of light. However, this velocity can be reduced below the velocity of light by introducing a periodic vari-

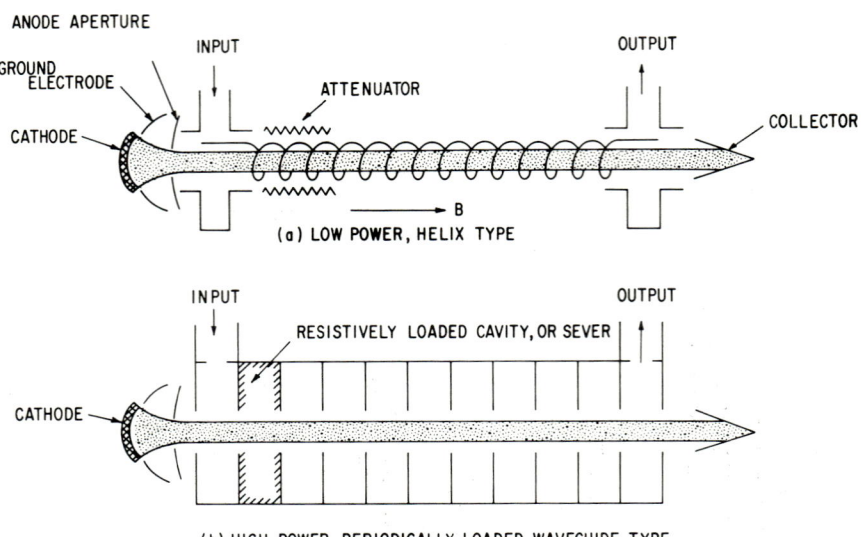

FIG. 80. TWT amplifiers.

ation of the circuit in the propagating direction. The simplest form of periodic variation is obtained by wrapping the circuit in the form of a helix whose pitch is equal to the desired slowing factor, as shown in Fig. 80. The electron beam is focused and constrained to flow along the axis of the helix by means of a three-element electron gun and an axial magnetic field, as in the klystron. The magnetic field may be provided either by a solenoid or by permanent magnets. Periodic permanent magnet (PPM) focusing has been developed extensively because of the great saving in weight.

Although the helix is an excellent slow wave circuit for tubes which operate at low and medium power levels, periodically loaded waveguide circuits are preferred at high power levels for two reasons. First, it is difficult to remove heat from a helix, and second, to operate at high beam voltages the pitch of the helix must be reduced to such a degree that radiation from the circuit becomes a limiting factor. In contrast, the periodically loaded waveguide circuit shown in Fig. 80b is completely enclosed, and the periodic loading elements, in this case apertured disks, provide a direct path for rapid heat transfer to the outer housing which may readily be cooled. The slowing of the wave in the periodically loaded guide occurs because a small amount of power is reflected at each loading element. This leads to a resonant type of phase-velocity reduction, contrasted to the nonresonant slowing action of a helix.

In both types of slow wave circuits energy may flow in either direction, leading to the possibility of oscillation. This tendency is minimized by placing resistive material near the input end of the slow wave circuit, as illustrated in Fig. 80. In the helix tube, the resistance takes the form of an attenuating coating placed on insulators adjacent to the helix. In the periodically loaded waveguide, one or more low-Q cavities are used which essentially sever the slow wave circuit. Such lossy sections completely absorb any backward-traveling wave. The forward-traveling wave on the circuit is also absorbed to a great extent, but the signal is carried past the attenuator by the modulated electron stream and is reinstituted on the slow wave circuit to the right of the attenuator. Attenuators which are capable of absorbing large backward-traveling wave powers, such as may develop in case of mismatch at the load, are an important problem for the high power TWT designer.

Mechanism of Amplification. The longitudinal component of electric field along the axis of a slow wave structure continually interacts with the electron beam to

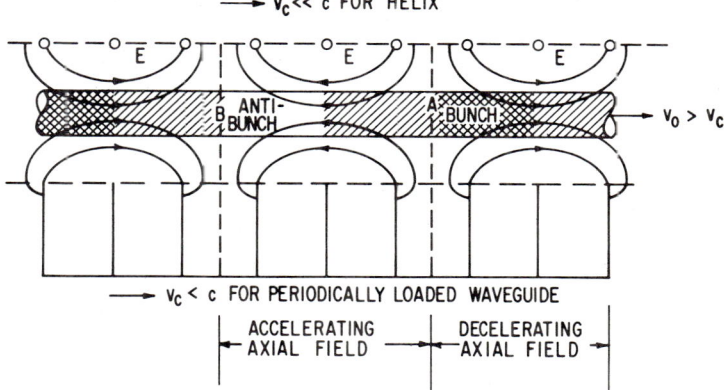

Fig. 81. Amplification mechanism in TWT amplifiers.

provide the gain mechanism of TWT's. This interaction mechanism may be understood with reference to Fig. 81. In the upper part of this composite figure a helix is shown providing the r-f electric fields which fringe into the region occupied by the electron beam. In the lower portion of this figure a periodically loaded waveguide circuit is shown providing very similar fields. Consider first the case where the electron velocity v_b is exactly synchronous with the circuit phase velocity v_c. In this case the electrons experience a steady d-c electric force which tends to bunch them around position A and debunch them around position B. In this case as many electrons are accelerated as are decelerated; hence there is no net energy transfer between the beam and the circuit. To get amplification in the TWT the electron beam is adjusted to travel slightly *faster* than the circuit wave. The bunching and debunching mechanisms just discussed are still at work, but the bunch and antibunch now move slightly ahead of the circuit wave, as shown in Fig. 81. Under these conditions more electrons are in the decelerating field to the right of A than are in the accelerating field to the right of B. The energy balance is no longer maintained, for now more electrons are decelerated than are accelerated. Thus energy flows from the beam to the circuit, and the circuit wave grows.

Current Bunching and Energy Extraction. The bunching process in TWT's results in current waveforms which resemble those developed in a klystron. In Fig. 82 the electron density and efficiency in a typical TWT are plotted as a function of distance. In this figure the shaded regions represent the positions of r-f decelerating fields. For the first five cycles the electron bunch grows and advances into the r-f decelerating region. In the middle portion of the tube however, the r-f circuit field has grown so strong that the bunch is forced backward in phase. This is the main region of energy extraction. In the latter part of the tube the r-f circuit power saturates, accompanied by steep density fronts and sharp density spikes which are reminiscent of the klystron current patterns of Fig. 74.

There is no elementary theory which can predict the saturation efficiency of a TWT. The upper plot in Fig. 82 shows efficiency vs. distance as calculated on a high-speed digital computer. The fundamental component of r-f power rises to a value representing 36 per cent power conversion efficiency before the electron bunch becomes so dispersed that the power output decreases. Efficiencies obtained in practical TWT's range from 20 to 40 per cent.

Low-noise TWT's. Traveling-wave-tube applications range from high-power tubes capable of the performance discussed above to voltage amplifiers with extremely low noise figures. The lowest TWT noise figures achieved to date range from less than

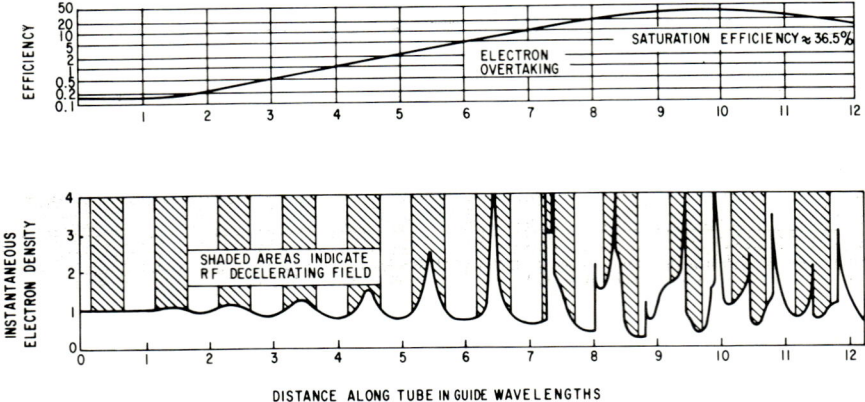

Fig. 82. Variation of electron density and efficiency with distance in a TWT.

3 db at S band to less than 4 db at X band. These figures have been obtained in tubes which utilize a special electron gun, called the Currie gun. Two unusual features of this gun are: (1) The "ground" electrode of Fig. 80 is run at a slightly positive potential, which results in a beam of electrons which is hollow rather than solid. (2) A number of low-voltage accelerating electrodes are placed between the ground electrode and the anode aperture to allow the beam velocity to be varied in the gun region. The mechanism of the reduction of noise figure to the vicinity of 6 db can be explained by linear theory. The further reduction of noise figure to below 3 db appears to be the result of nonlinear mixing effects in the multivelocity region of the electron beam immediately adjacent to the cathode. This process is not clearly understood; hence it is not possible to predict an ultimate lower limit for noise figure in TWT's.

The ω-β Diagram. Low-power TWT's with helical circuits operate over bandwidths of an octave or more, whereas high-power periodically loaded waveguide circuits can rarely provide more than 20 per cent bandwidth. The reason for this is illustrated in Fig. 83, where ω, the operating radian frequency, is plotted as a function of βd, the phase shift per circuit period. On the left is the ω-β diagram of a helix. This type of diagram is very useful because at a given ω the slope of a line drawn from the origin to the curve is proportional to the *phase velocity* of the circuit at that frequency, i.e.,

$$v_p = \omega/\beta \tag{24}$$

Furthermore, the slope of a tangent to the curve at any point is proportional to the *group velocity* of the circuit at that point, i.e.,

$$v_g = d\omega/d\beta \tag{25}$$

At very low frequencies the phase velocity along a helix is equal to the velocity of light c. Over a broad range of intermediate frequencies, however, the phase velocity approaches the velocity of light divided by the pitch of the helix. An electron beam traveling at this latter velocity will be synchronous with the phase velocity of the wave over an extremely large range of frequencies, as shown by the shaded region of Fig. 83a.

In contrast, the ω-β diagram for a typical forward wave periodically loaded waveguide circuit is shown to the right in Fig. 83. It is evident that the phase velocity shows a fairly rapid change with frequency; hence for a given beam velocity (a line with fixed slope from the origin) near-synchronous conditions will prevail over a relatively small range of frequencies, as shown by the shaded region. The width of this region can be optimized by using negative mutual inductive coupling between the

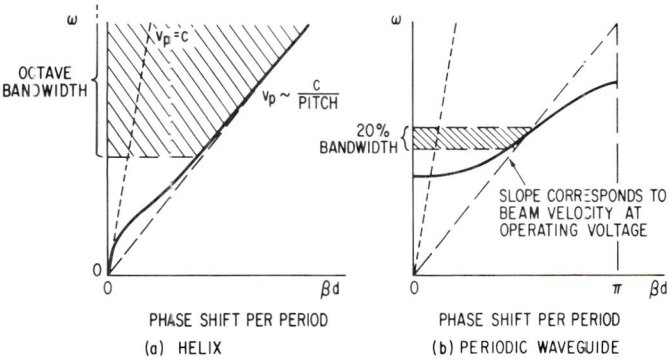

FIG. 83. ω-β diagrams of low-power and high-power TWT circuits.

cavities of the slow wave circuit, as in, for example, the *cloverleaf* circuit, but the bandwidths achieved thus far with such circuits still fall considerably short of the octave bandwidth of the helix.

5.4. Crossed-field Amplifiers (M-type Amplifiers). *Physical Configurations.* Amplifiers which make use of perpendicular or "crossed" d-c electric and magnetic fields are called M-type amplifiers. M-type amplifiers have been built in a great variety of physical configurations, all of which involve the use of a nonreentrant slow wave circuit. The most important of these are shown in Fig. 84. The first column of sketches depicts devices which utilize the electrons from a continuous, or distributed, cathode in a reentrant or recirculating manner. The slow wave circuits, not shown in detail in these sketches, are represented by the outer dashed circular lines. The advantage of recirculating the electrons is that they can be utilized in an extremely efficient manner. The efficiency of reentrant-beam M-type devices is high, 70 to 80 per cent. A disadvantage of reentrant-beam devices is that the gain must be kept low, in the vicinity of 10 db, to prevent oscillation. The most common reentrant-beam M-type amplifier is the amplitron, which utilizes a reentrant electron beam in conjunction with a backward-wave circuit (see below). It is also possible to construct a forward-wave crossed-field amplifier (FW-CFA). There are no injected-beam devices utilizing beam reentrancy because the gun electrodes would interfere with the circulation of the electrons.

The nonreentrant-beam circular devices shown in the second column and the linear-beam devices shown in the third column are very closely related, as can be seen. The linear devices have the advantage of minimizing r-f feedback, which in the circular

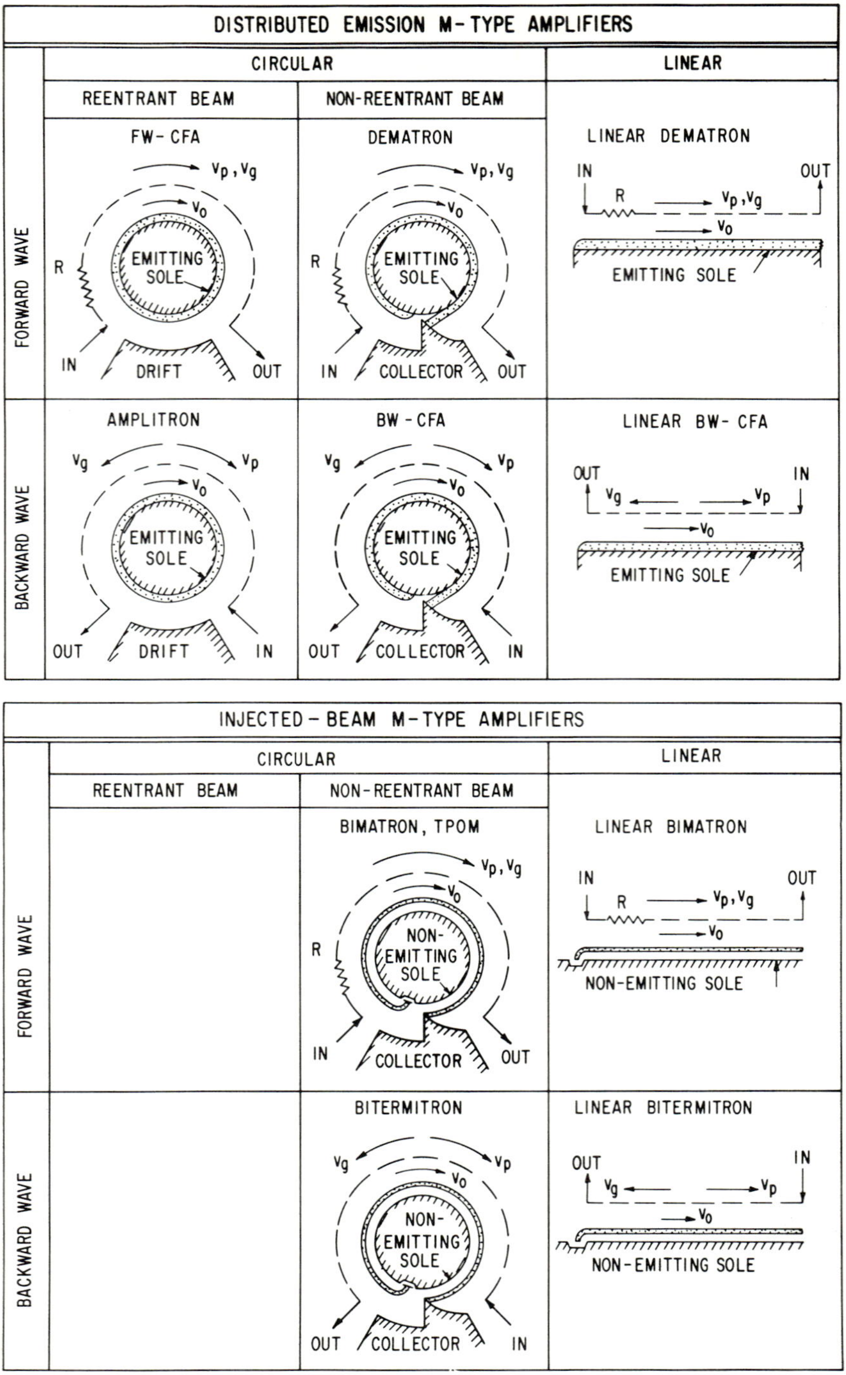

Fig. 84. M-type amplifiers.

devices places a limitation on gain and bandwidth. It is possible to use either forward- or backward-wave interaction in nonreentrant-beam devices. In addition, either distributed emission or beam injection may be employed to supply the electrons. The *dematron* (distributed emission magnetron amplifier) utilizes a forward wave space harmonic circuit (see below) and a distributed cathode. The *bimatron* (beam injection magnetron amplifier) utilizes a forward-wave circuit, but with an injected beam of electrons. This arrangement is also called a *TPOM*. It is possible to use a backward-wave circuit with a distributed nonreentrant beam. A final arrangement, the *bitermitron*, utilizes a backward-wave circuit with an injected beam. Forward-wave devices require an attenuator near the input end of the tube, as in the TWT, and gains in the vicinity of 20 db may be obtained in these tubes. Backward-wave devices are essentially locked oscillators and do not require an attenuator. However, their gain must be kept low (10 db) to prevent self-oscillation.

In injected-beam devices the electron beam travels at a velocity equal to the phase velocity of the wave on the circuit, i.e., the synchronous velocity. In distributed-emission devices only the uppermost layer of electrons is synchronous with the circuit field because the lower layers travel at reduced velocities. The lower layers can be considered to be a source of electrons which move upward as the outer synchronous layer is depleted. Distributed-emission devices are useful for high-power pulsed

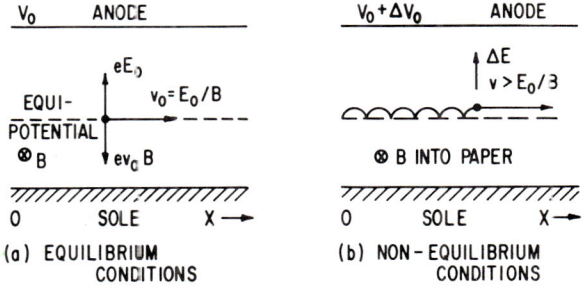

FIG. 85. D-C electron motion in crossed electric and magnetic fields.

applications since beam perveances can be obtained which are typically one to two orders of magnitude greater than those of injected-beam M-type or O-type amplifiers. Furthermore, these devices allow cold-cathode operation, provided some means of starting the secondary-emission process is included. Such class B operation is highly desirable because no pulse modulator is required.

Energy-exchange Mechanism. The energy-exchange mechanism in M-type devices may be understood with reference to Fig. 85. Consider first the d-c beam conditions in a linear magnetron, as shown in the sketch to the left. An electron beam is injected from an electron gun to the left (not shown) with velocity v_0 parallel to the ground plane or "sole" electrode. An electron in the beam is acted upon by the d-c electric field E_0 with an upward force whose magnitude is given by eE_0, where e is the charge on an electron. Also present on the particle is a magnetic force of magnitude ev_0B acting to move the beam downward. The magnetic field strength B is chosen such that with no r-f signal the two opposing forces exactly cancel one another. When these forces are equal the forward drift velocity v_0 is equal to E_0/B. Notice this unusual property of equilibrium electron motion in crossed fields: the electrons do not move in accelerated motion in the direction of the electric field but rather move with constant velocity perpendicular to it. In general, *in crossed-field devices electrons tend to move along equipotential lines rather than along electric field lines.* This statement is still valid if the electric field is increased slightly above its equilibrium value, as in Fig. 85b. In this case, when the additional E field is applied, the electron begins to rise, but the magnetic force acts to return the electron to its original height above the sole. The forward motion carries the particle to the right during this time, and the net result of the additional E field is a small additional drift velocity along the X direction upon which is superimposed a small cycloidal motion, which may be ignored in the present discussion. Thus even under nonequilibrium conditions the electrons tend to move along equipotential lines.

Consider now the case where the additional electric field is produced by a slow wave circuit above the sole and parallel to it. Let the phase velocity of the circuit be chosen equal to E_0/B. In a frame of reference traveling at this velocity the d-c electric field disappears and the magnetic field causes no force because the velocity is zero. The r-f electric field leads to an electrostatic force field acting in a variety of directions, as shown in Fig. 86a. These fields, when they are weak compared with the d-c electric field, cause the electrons to move in tight cycloids along the equipotentials of the circuit field, as indicated by the arrows tangent to the dashed lines. For instance, electron e is in a decelerating electric field and starts to move backward, but the magnetic field interacts with the backward velocity and the result is that at a later time this particle has moved upward, as shown in Fig. 86b. Similarly, electrons a and i move downward, while electrons c and g move toward one another. Thus the initially flat sheet of electrons is warped upward under the maximum decelerating field position. Furthermore, electrons are pulled into this half-cycle, that is, they are *phase-focused* or

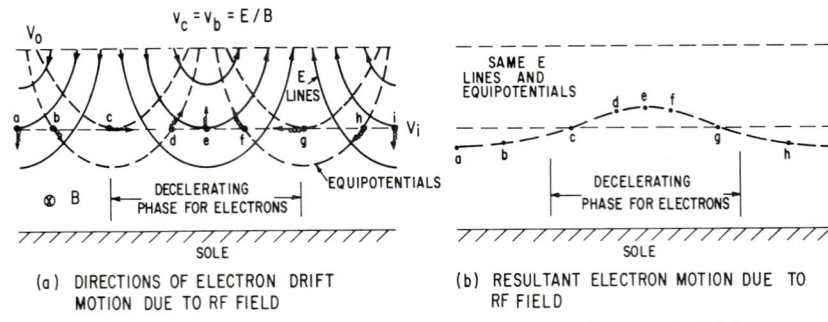

(a) DIRECTIONS OF ELECTRON DRIFT MOTION DUE TO RF FIELD

(b) RESULTANT ELECTRON MOTION DUE TO RF FIELD

Fig. 86. R-F electron motion in crossed electric and magnetic fields.

bunched, whereas electrons exposed to the accelerating half-cycle of electric field move downward and are debunched. Notice that in Fig. 86b electrons c and g are well within the decelerating phase whereas they originally were at its edges. Ultimately, a spoke of electrons forms under the portions of the circuit carrying the decelerating field, and the electrons in this spoke drift slowly toward the anode. The most important feature of M-type interaction is that, although many electrons move upward and finally reach the positive electrode, in doing so *they gain only a small amount of kinetic energy*. The electrons, however, have lost potential energy of the amount $e(V_0 - V_i)$, where V_i is the voltage of the d-c equipotential line on which they were moving initially. This amount of energy is delivered to the longitudinal component of the electric field and represents useful output power. A highly desirable feature of this interaction is that as the electrons give up energy, they do not fall out of phase, as in the O-types amplifiers; on the contrary, as they lose energy they are concentrated more and more in the energy-extracting region. It is this property that makes the efficiency of M-type devices high, namely, 50 to 80 per cent.

In the above discussion space charge has been ignored. The effect of space charge is to cause the electrons in the spoke to rotate. This unfortunately can lead to an instability which under some conditions can dominate the interaction. In order for the circuit field to predominate over the space-charge field, the circuit field must be relatively strong at the input of the amplifier, particularly in distributed-emission devices. In high-perveance crossed-field amplifiers gains of 5 to 15 db are found to be the maximum that can be obtained before the instability due to space-charge rotation intrudes causing a large noise content and leading to loss of circuit interaction. In

ELECTRON TUBES 9-95

M-type devices currently under development, the characteristics of the interaction region are tapered to overcome this problem, and stable gains in excess of 20 db are anticipated.

Backward-wave Circuits. The physical configuration of a typical backward-wave circuit is shown in Fig. 87a. This is an *interdigital* circuit, so named because the

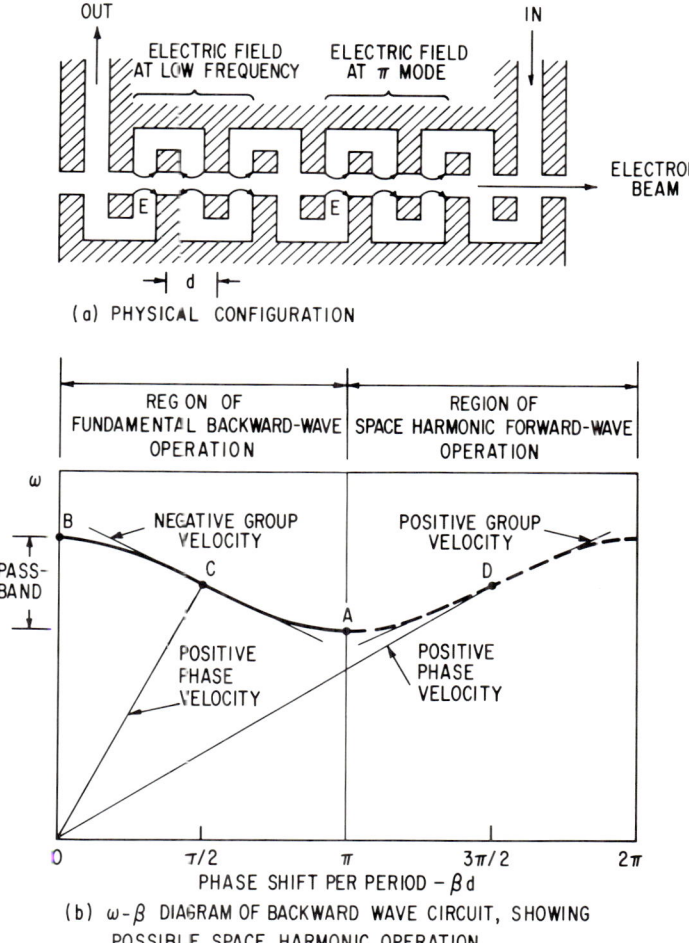

Fig. 87. ω-β diagram of backward-wave circuit, showing possible space harmonic operation.

vanes of the slow wave circuit are interleaved. The electrons may either interact with the longitudinal component of electric field which fringes upward from such a circuit or, alternately, pass down a hole cut through the tips of the vanes, interacting directly with the electric field between the vanes. In such a circuit there will be little phase shift with distance at low frequencies as a wave travels the serpentine path between the vanes. However, because of the *geometrical configuration* of the circuit, the electric field reverses its direction once each circuit period. As a result, at low frequencies the electron beam is exposed to an electric field which effectively has π radians phase shift per circuit period. On the ω-β diagram for this circuit shown in Fig. 87b, operation at the low-frequency cut-on of the waveguide is indicated by point A. Note the contrast with the forward-wave circuit of Fig. 83, discussed in connection with a forward-wave TWT amplifier for which there is zero phase shift per section at the lowest propagating frequency.

As the frequency is raised, there is less phase shift per section until at a frequency for which the path length from one interaction gap to the next is one-half wavelength, the E field instantaneously has the same direction at each gap, i.e., zero phase shift per section. This frequency is the upper cutoff frequency of the circuit and is indicated by point B in Fig. 87b. The interesting characteristic shown by this diagram is that although the phase velocity of the circuit is positive, as indicated by the positive slope of a line drawn from the origin to the curve, the group velocity is negative, as indicated by the slope of a tangent to the curve. Since this circuit has a negative group velocity in the region from 0 to π phase shift per period, it is called a fundamental backward-wave circuit.

Space Harmonics. It is not necessary to interact with a slow wave circuit in the fundamental region from 0 to π phase shift per section. If the vanes of the circuit have a width that is larger than the distance between the vanes, as is the case in Fig. 87a, the circuit is rich in what are called *space harmonics*. The first space harmonic of the interdigital circuit is shown as a dashed line in Fig. 87, corresponding to phase shifts per period of from π to 2π. Note that the first space harmonic of a fundamental backward circuit is a forward wave. A unique feature of the space-harmonic operation indicated by point D is that relatively large bandwidths can be obtained because the phase velocity of the space harmonic is uniform over a fairly wide band of frequencies. Unfortunately, the impedance of space harmonics is considerably lower than that of the fundamental wave, but in high-power tubes where d-c beam impedance is low, this is not a serious limitation.

5.5. The Adler Tube (Quadrupole Amplifier). In the Adler tube extremely low noise operation is obtained by utilizing the principle of parametric amplification acting on a transversely deflected electron beam. Effective noise temperatures less than 100°K are readily achieved using only milliwatts of r-f pump power acting on beams traveling at very low voltage (less than 10 volts) carrying only tens of microamperes of current.

A sketch of an Adler tube, or quadrupole amplifier, is shown in Fig. 88. A pencil beam of electrons is projected parallel to a magnetic field toward a collector electrode. The beam first passes through an input coupler, typically a pair of deflection plates connected to a parallel-resonant circuit. The beam passes next through a quadrupole pumping structure, and finally through an output coupler, which is similar to the input coupler. The magnetic field strength is chosen so that the cyclotron frequency of the electrons $f_c = eB/2\pi m = 2.80 \times B$ Mc/sec is equal to the signal frequency. Under this synchronous condition, an electron entering the input coupler is deflected transversely in a spiral of linearly increasing radius, as shown. This type of motion is called a *fast cyclotron wave*. If no pump signal were applied to the quadrupole structure, the radius of the spiraling electron would remain constant in the quadrupole region. At the output coupler the spiraling electron induces a voltage which tends to oppose its motion; thus the transverse motion is damped out and the transverse kinetic energy is converted to r-f output power.

To make such a device amplify, the quadrupole pump section between the two couplers is driven from an external r-f source at twice the cyclotron frequency. Alternate elements of the quadrupole structure are strapped together (not shown) to provide a tangential electric field which rotates at the cyclotron frequency, as shown to the right in Fig. 88a. An electron spiraling in this structure in the optimum phase position finds itself in a continually accelerating electric field and moves outward, continuing always to rotate at f_c. In the quadrupole structure the field intensity increases linearly with radius; thus the force exerted on an electron is proportional to the radius of the circle on which it moves. Under synchronous conditions, this causes the radius of the optimum-phase electron to increase exponentially with time. Since the radius of the trajectory is a measure of the r-f signal carried by the electron beam, the effect of pumping at twice the signal frequency is to establish signal gain.

The unique feature of the Adler tube is that as the electron beam passes through the input coupler the transverse beam noise is removed at the same time that the signal power is introduced, as depicted in Fig. 88b. The longitudinal noise of the beam is of no consequence since the longitudinal energy of the beam is not tapped. Stable gains of 20 db/cm of quadrupole length can be achieved without danger of oscillation because there is no link between the output and input circuits other than the forward-moving electron beam. There is no burnout problem because an excessive input signal simply causes the beam to be intercepted by the quadrupole structure where its extremely low kinetic energy is readily dissipated.

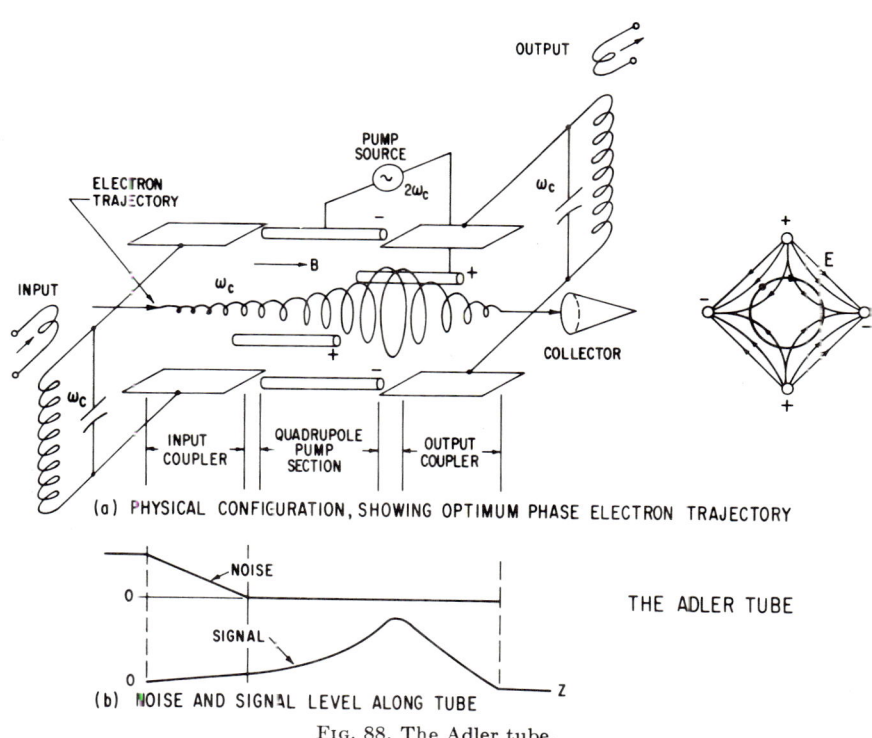

Fig. 88. The Adler tube.

6. TRANSMITTING TUBES

6.1. Introduction. Transmitting tubes, like other negative-grid power-amplifier tubes, have been widely applied for several decades. Consequently they are available in many hundreds of types for power levels up to several hundreds of kilowatts, and frequencies to a few gigacycles. For r-f power generation between audio frequencies and about 200 Mc, they are used almost exclusively where required power exceeds that available from solid-state devices. As frequencies increase, the maximum power that can be made available from these devices decreases because of well-known limitations, such as electron transit time and practical circuitry that can be fitted to the tube. At 1,200 Mc, available CW power is currently on the order of 1 kw.

Transmitting tubes can be effectively applied for CW, FM or AM CW, and for pulsed service. They are generally characterized by good efficiencies (up to 70 per cent or so under favorable conditions), and good though not phenomenal gain and reasonably high bandwidths. Up to 10 per cent or more bandwidth is readily obtainable at a sacrifice in efficiency. They offer potentially excellent phase stability.

6.2. Construction.

Most transmitting tubes for use below 100 Mc are constructed with glass envelopes. The use of ceramic materials in tube construction is becoming increasingly widespread at all frequencies because of performance advantages. Ceramic construction generally allows better tube processing during manufacture and consequently higher allowable operating temperatures. Low tube-envelope dielectric losses at high temperature become a particular advantage at uhf frequencies. While there are available types of glass with very low dielectric losses they can be sealed to a limited number of usable tube metals. Consequently high-power uhf transmitting tubes are typically of metal and ceramic construction.

Cathodes in transmitting tubes have many variants but are of two basic types: thoriated tungsten and oxide-coated. The choice depends on many factors. Thoriated tungsten provides maximum resistance to voltage breakdown and electrical damage due to use and abuse. Oxide cathodes provide maximum efficiency, peak emission capability, and generally greater resistance to mechanical abuse.

Thoriated tungsten is no longer limited to filamentary cathodes. Many uhf transmitting tubes now employ thoriated tungsten cathodes in cylindrical and strip forms. This material provides a long-life emitter capable of about 100 ma emission per watt of heating power at a typical operating temperature of 1675°C.

Oxide-coated cathodes are being applied at increasingly higher power and frequency levels as material and processing technology improves. Tubes capable of several kilowatts at uhf are available. The oxide cathode is usually constructed with a nickel base which can be made in any desired shape. Typically operating at about 800 to 900°C, this cathode provides amperes of potentially usable emission per watt of heating power. Usable d-c emission is limited to tens of milliamperes per watt of heating power.

Grids in transmitting tubes vary from fine wire mesh to massive bar construction. Typically a cylindrical cage of helical or axial elements is used. Control grids may be spaced a few thousandths of an inch from the cathode in uhf tubes or over a tenth of an inch in powerful low-frequency tubes. The choice of grid material varies. For tubes with high-temperature cathodes, such as thoriated tungsten, a high-strength low-vapor-pressure refractory metal such as molybdenum or tungsten is used. To prevent primary grid emission a covering of platinum is commonly employed in high-powered or closely spaced uhf types. Tetrodes typically use a screen grid of the same material. Platinum is among the better materials used for minimizing primary grid emission in the presence of thoriated tungsten. It does have a high secondary-emission ratio which accounts for negative screen currents in some tubes.

Grids in tubes with low-temperature oxide cathodes are usually gold-covered molybdenum or even copper alloys where the grid temperature can be kept sufficiently low. Gold is an excellent suppressor of primary grid emission in the presence of an oxide cathode. Like platinum it also has a high secondary-emission ratio, frequently resulting in negative screen currents.

The internal anode or plate element in radiation-cooled glass transmitting tubes generally is made of molybdenum, tantalum, or graphite except in the very low-powered tubes. These anodes are frequently designed to operate at visible temperature at maximum ratings. A grayish coating frequently seen on this type anode usually contains zirconium. This material serves a dual purpose by increasing anode heat-radiation capability and serving as a high-temperature getter of gas. Tubes with external anodes forming a portion of the envelope structure usually employ oxygen-free high-conductivity copper anodes.

Figure 89 shows a section of a small uhf transmitting tube about 4 in. long. It is capable of delivering several hundred watts of power and is usable to 1,250 Mc or so. A few ratings and operating parameters are as follows:

Plate voltage, volts d-c................	2,000 max
Plate current, ma d-c.................	300 max
Screen voltage, volts d-c..............	320 max
Plate dissipation, watts...............	500
Cathode-heating power, watts..........	23

A cylindrical unipotential oxide-coated cathode is employed with gold-plated molybdenum grids forming concentric cages of aligned axial wires around the cathode. The envelope is entirely of metal and ceramic construction. Cathode power is provided by a tungsten coil heater. The one-piece finned anode is designed for forced-air cooling.

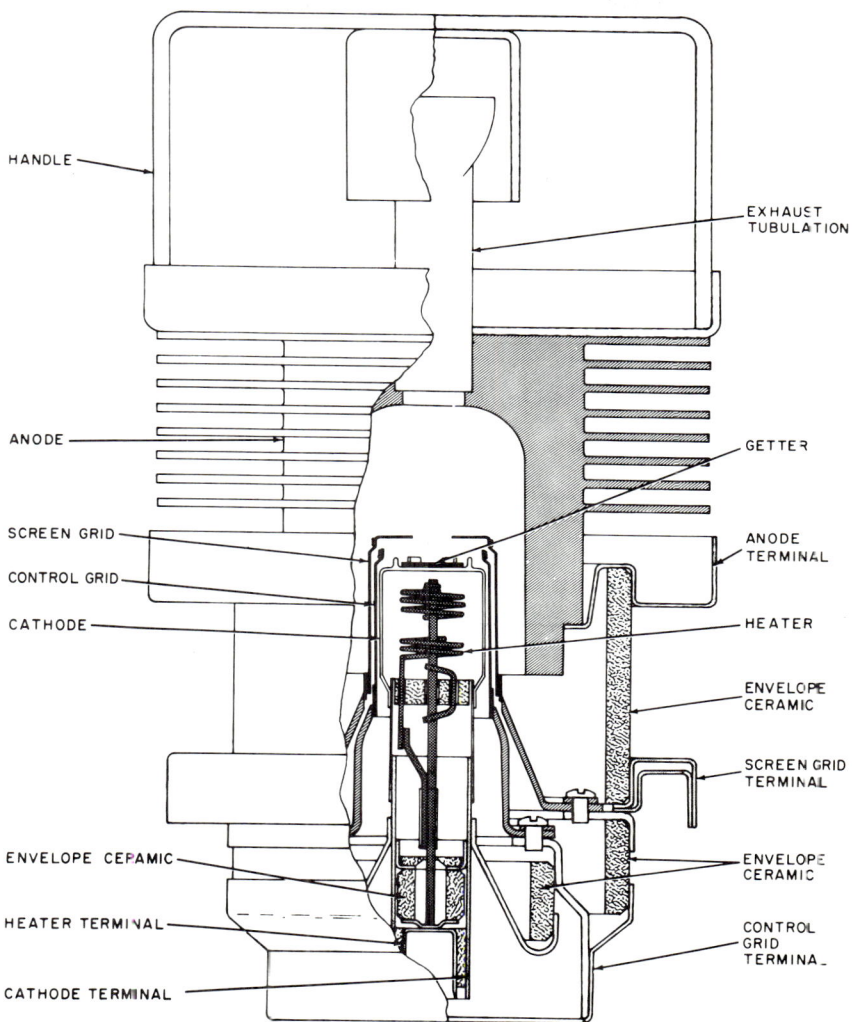

Fig. 39. Transmitting tube, GL-6283. (*Courtesy of General Electric Co.*)

The getter is a titanium disk mounted on top of the cathode. During manufacture gases are removed through the exhaust tubulation. The end of the severed tubulation is protected from damage by an inverted cup attached to the tube handle. Designed for uhf, it has minimized capacitance and lead inductance through use of concentric terminals. Tubes of this general design are ideally suited for grounded-grid operation since screen-grid and control-grid terminals are adjacent and readily bypassed to ground.

6.3. Ratings.

Almost all transmitting tubes are rated by the absolute-maximum system. This system[25] places considerable responsibility on the equipment designer and user for selection of operating conditions and environment that are at all times consistent with ratings. In addition, individual ratings are independent of each other. For instance, the maximum allowable d-c power input may be less than the product of the maximum allowable direct plate current and d-c plate voltage. The maximum usable value of a current, voltage, temperature, or other parameter is frequently thus limited to a lower than rated value by another rated item. The primary advantage of this rating system is that it allows a tube to be utilized nearer its ultimate performance. Where the tube manufacturer is required to second-guess both equipment manufacturer and user ranges of operating parameters, ratings necessarily include safety factors that cover the worst expected conditions.

Probably the most important single factor in rating determination for transmitting tubes is the limiting temperature of the tube elements. This is usually apparent to the user in the plate dissipation ratings where dissipated power can usually be measured, calculated, or estimated to a fair degree of accuracy. The effects of temperature in determining ratings for grid dissipation and cathode power are usually somewhat more subtle. These elements are frequently thermally isolated from external seals or connections so that most heat is removed from them by radiation. Further, in the case of grids in uhf tubes, it is extremely difficult for the user to measure accurately the amount of power being dissipated during operation. With tube characteristic curves it is possible to obtain grid dissipation values by calculation if the operating conditions are accurately known. The tube manufacturer frequently makes such determinations, usually verified by actual performance. Knowing the limiting wattage capability of the control grid, for example, is not of much value to the average user; so the ratings are defined in terms of measurable parameters such as maximum d-c grid bias and d-c grid current. These are not always reliable indicators of grid dissipation; so the bias and grid current ratings specified may be necessarily conservative for some operating conditions. In the case of screen grids, the d-c input is readily measured but is frequently an unreliable measure of actual power dissipated by the screen grid. Secondary emission is very common in screen-grid tubes. This seldom affects performance,[26] but it does mask the real screen-grid current interception and dissipation by subtracting from the d-c screen current meter reading. Also in grounded-grid circuits, the screen-grid-to-cathode voltage varies with the signal voltage; this adds an a-c component that increases screen dissipation. A screen-grid dissipation rating can be safely applied only when these effects are known to be negligible. As in the case of the control grid, screen ratings are usually based on measurable parameters. In this case, d-c screen input is usually specified, which also requires considerable measure of conservatism to be applied in a tube rating. Usually a careful analysis using tube characteristics is required or a check with the manufacturer may be in order. Fortunately, most applications require only a fraction of the allowable d-c screen input.

---

[25] The absolute-maximum rating system as defined by the Electronic Industries Association is as follows:

Absolute-maximum ratings are limiting values of operating and environmental conditions applicable to any electron tube of a specified type as defined by its published data, and should not be exceeded under the worst probable conditions.

The tube manufacturer chooses these values to provide acceptable serviceability of the tube, making no allowance for equipment variations, environmental variations, and the effects of changes in operating conditions due to variations in the characteristics of the tube under consideration and of all other electron devices in the equipment.

The equipment manufacturer should design so that initially and throughout life no absolute-maximum value for the intended service is exceeded with any tube under the worst probable operating conditions with respect to supply-voltage variation, equipment component variation, equipment control adjustment, load variation, signal variation, environmental conditions, and variations in the characteristics of the tube under consideration and of all other electron devices in the equipment.

[26] One exception is the case of screen-grid modulation. Secondary screen-grid screen emission can contribute to extreme screen voltage-current nonlinearity, introducing modulation distortion unless screen modulator impedance is kept at a low value.

Cathode ratings are usually straightforward on low-frequency tubes. Adherence to the manufacturer's filament voltage and current recommendations will generally provide optimum tube life and performance. Whenever a range of heater or filament voltages is specified, there is a life advantage to be obtained with operation at lower values of voltage, provided the voltage regulation in application is adequate to keep the voltage within the specified range. A general rule of thumb that can be used is that a 5 per cent reduction in heater voltage will increase cathode life 100 per cent. This is reasonably accurate with thoriated tungsten tubes but is quite approximate with oxide-coated tubes whose complex chemistry makes accurate predictions difficult.

Cathode, or plate current ratings may be determined by either plate dissipation or emitter limitations. In the case of the thoriated tungsten cathode with a typical emission of 100 ma/watt at rated heater voltage, usable emission may be 25 to 50 per cent of this value because of linearity limitations. Since most r-f amplifiers require a peak-to-average r-f cathode current ratio between 3:1 and 5:1, the allowable d-c cathode current can be as little as 5 per cent and is seldom more than 15 per cent of the available peak cathode emission.

Cathodes of transmitting tubes that operate at uhf frequencies are subject to electron bombardment or backheating. Heater voltage ratings on uhf power tubes generally make provision for backheating but usually are phrased in general terms because of the numerous factors that can affect the amount of backheating. For any given tube operating at a specified power and frequency the amount of backheating, if present, is aggravated by the use of excessive d-c grid-bias voltages and circuit impedance mismatches.

The external envelopes of transmitting tubes are subject to temperature limitations for rather obvious reasons such as mechanical failure or excessive r-f dielectric losses resulting in puncture or fracture and release of gas within the tube. The measurement of envelope temperatures of tubes operating at high voltage or in r-f fields is best accomplished by applying very small dots of temperature-sensitive paint or liquid to the envelope surface. A number of very satisfactory materials are on the market. When properly used, they are little affected by the presence of r-f fields and provide very useful temperature determinations for the equipment designer.

Another important factor in tube ratings, particularly in high-power or pulse types, is voltage breakdown strength, both internal and external. Breakdown strength is usually adversely affected by a low-impedance plate power supply. Damage caused by breakdown ranges from mere user inconvenience to outright tube failure.

Outline dimensions are a part of transmitting-tube ratings which are frequently misunderstood or misinterpreted. Tubes with complex outline drawings usually have a single major reference plane for dimensioning. A minimum tolerance buildup is usually obtained if tube contact locations are dimensioned from this plane. Problems can also be avoided by making connections and contacts only on indicated surfaces that have dimensions with tolerances or a limit on the outline. Eccentricities between circular tube contacts and other external elements are generally expressed as TIR (total indicator reading) when the tube is revolved about a specified center line. This value is equal to twice the eccentricity, or displacement of the center of the tube contact from the arbitrary center line.

6.4. Pulse Ratings. Many tubes have pulse and CW versions that are very similar in appearance but generally receive different testing and processing during manufacture. Pulse widths, particularly in tubes with oxide-type cathodes, have presented a very complex tube rating problem. The allowable pulse width for most pulse-rated tubes can have any value, provided other operating conditions are appropriately selected. For example, a typical tube construction will carry well-established ratings for CW operation and for specific short pulse conditions up to perhaps 5 or 20 μsec. Between these extremes are an infinite number of pulse width, duty factor, and peak cathode current combinations that might be applied. Attempts are being made to apply some logical interpolation method but no generally accepted system has been found.

6.5. Life. Even rough life approximations require reasonable definition of many operating parameters.

The most significant factor affecting tube life in normal applications is cathode heating power. A normal application is one in which end of life is determined by depletion of the cathode and not by sudden-death causes such as mechanical or severe overload failures.

Thoriated tungsten cathodes have very predictable life under controlled conditions of environment. Normal end of life in this cathode occurs when the surface layer of tungsten carbide is depleted. In well-processed tubes operated conservatively, this decarburation rate is determined by the operating temperature of the cathode rather than the cathode emission current used. During life, the filament current of a directly heated thoriated tungsten cathode rises steadily. It is possible to use this characteristic to make rough life determinations. The initial stabilized current must be accurately known and very precise metering utilized since the total current change is usually only a few per cent. In filamentary types the cathode current loading can have a secondary effect on emitter life. This current adds to the normal heating of the filament, raising its temperature. The extra heating is not uniform since this current varies from zero to maximum along each filament strand. Some life reduction occurs but can be minimized by avoiding excessive filament voltages.

Oxide-type cathode life is affected by innumerable internal and external environmental factors. Life is greatly affected by manufacturing processing, internal tube materials, and tube construction. From the user standpoint, oxide cathode life, unlike that of thoriated tungsten, is detrimentally affected by increased cathode emission current density requirements. No precise relationship exists but the total d-c cathode current in CW operation, or rms current in pulsed service, provide some relative indication of expected life. Root-mean-square cathode current is given by the following equation:

$$I = \sqrt{I_k^2 D} \tag{26}$$

where I_k is the total d-c cathode current during the pulse and D is the duty factor or product of pulse length and repetition rate.

Since so many life factors are at work on the average oxide cathode, good life-expectancy determinations require operation under expected conditions or conditions that are quite similar.

6.6. Cooling. Transmitting tubes run the gamut of cooling methods. For frequencies below 100 Mc and power levels up to 1 kw or so, radiation-cooled tubes with glass envelopes are extensively used and are generally most economical where size or other environmental requirements are not critical. Frequently some auxiliary forced-air cooling is specified to maintain the tube envelope temperature within ratings.

For tubes operating in the vhf-uhf region, forced-air cooling is most common. Tube manufacturers typically specify air-cooling requirements to limit the maximum anode or radiator hub temperature. Where such a temperature is specified it is wise to determine the actual radiator temperature at the location specified as well as specified air flow and back pressure. Air ducting and baffling configurations sometimes make accurate air-volume measurements difficult. Altitude and air temperature should not be overlooked when air-cooled tube ratings are considered. Most manufacturers give sea-level air-volume requirements, but specified incoming-air temperatures are not uniform. Altitude corrections can be made with reasonable accuracy by maintaining a constant mass flow of air. For instance, at 10,000 ft standard air density is only 0.051 lb/cu ft or 69 per cent of the sea-level value, and an air volume 145 per cent that at sea level is required. The back pressure required should be taken from the tube manufacturers' flow-pressure curve. Values beyond the curve can be estimated by assuming that back pressure varies as the square of air volume required, and is independent of altitude.

The effect of incoming-air temperatures above normal ambient can also be considered on a mass-flow basis similar to altitude correction. The volume of air required for a given mass flow varies directly with the absolute temperature, i.e., degrees centigrade plus 273. Calculations for higher than normal air ambients are useful only below the maximum rated ambient-air temperature. If this temperature is not specified, a curve relating air flow, plate dissipation, and radiator temperature rise is required for usable calculations in high ambients.

Chapter 10

TRANSISTORS

W. CRAWFORD DUNLAP*

CONTENTS

1. Introduction.. 10–2
2. Basic Principles of Transistors............................... 10–3
    2.1. Principles of Semiconductors........................... 10–3
    2.2. Transistor Action..................................... 10–4
    2.3. *p-n* Junctions... 10–5
    2.4. Transistor Parameters.................................. 10–6
3. Equivalent Circuits for the Transistor....................... 10–7
4. Design Theory of Transistors................................. 10–9
    4.1. General... 10–9
    4.2. Emitter Efficiency.................................... 10–10
    4.3. Transport Factor...................................... 10–10
    4.4. Surface-recombination Effects......................... 10–11
    4.5. High-frequency Transistor Effects..................... 10–11
    4.6. Temperature Response.................................. 10–11
    4.7. Noise in Transistors.................................. 10–12
5. Alloy Transistors.. 10–12
    5.1. General... 10–12
    5.2. Impurity Distribution................................. 10–12
    5.3. Structure of the Alloy Transistor..................... 10–13
    5.4. Performance of Alloy Transistors...................... 10–14
    5.5. Surface-barrier and Related Transistors............... 10–14
    5.6. Applications of Alloy Transistors..................... 10–16
6. Diffused Transistors... 10–16
    6.1. Mixed Alloy-diffused Transistors...................... 10–16
    6.2. The Mesa Transistor................................... 10–17
    6.3. Planar Transistors.................................... 10–18
7. Power Transistors.. 10–20
8. Field-effect Transistors..................................... 10–21
9. Specialized Transistors...................................... 10–23
    9.1. Tetrodes.. 10–23
    9.2. Avalanche Transistors................................. 10–23
    9.3. Grown-junction Transistors............................ 10–24
10. Future Transistors.. 10–24
    10.1. Ultimate Capabilities of Transistors................. 10–24
    10.2. New Materials.. 10–25
    10.3. Multiple Transistors and Integrated Circuits......... 10–25
    10.4. Thin-film Transistors................................ 10–26

*NASA Electronics Research Center, Cambridge, Mass.

1. INTRODUCTION

It is the purpose of the present chapter to discuss basic theory and design principles of transistors, and to describe the present state of the art in fabrication technique and performance of transistors. No exhaustive discussion of theory will be attempted, and the bibliography indicates a number of easily available references on transistor theory. The purpose shall not be to educate the device designer, but to educate the device user as to the properties and limitations of transistors. The bibliography also indicates a number of extensive treatises on transistor circuits and applications.

The transistor, invented in 1948, is now well established as a standard component for electronic circuitry. Estimates indicate that for the year 1962 about 200 million transistors were sold for use in electronic circuitry. Most of these were in one of the

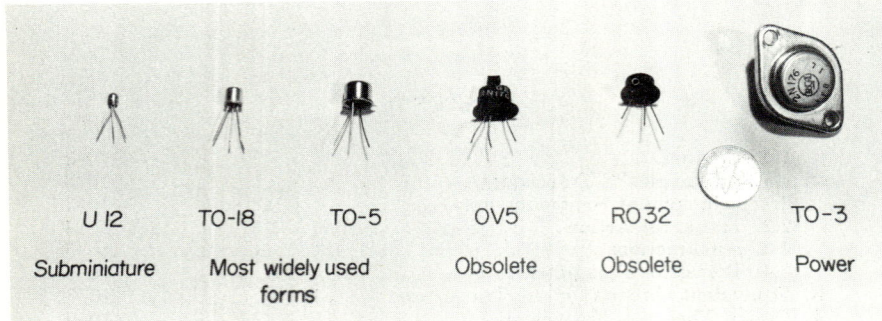

Fig. 1. Picture of transistor packages.

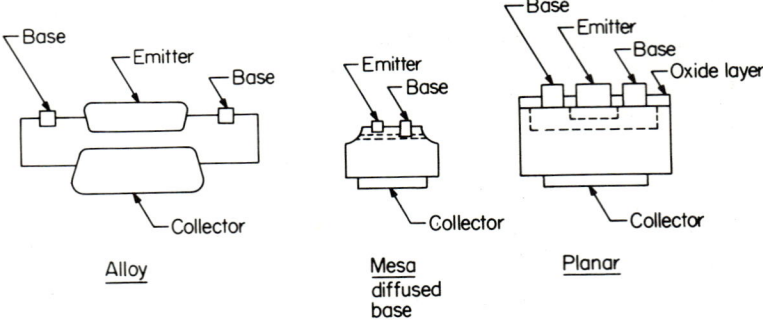

Fig. 2. Cross section of structures.

five forms shown in Fig. 1. The different packages refer mostly to frequency range and power-handling capability.

Figure 2 shows the cross section of the structure of several of the most widely used transistor types, namely, the alloy-fused, the diffused mesa, and the planar. The transistor began as the so-called "point-contact" variety, which is now obsolete and of historical interest only. The first truly practical transistors were of the "grown-junction" type, which is also essentially obsolete, although many are still sold.

Figure 3 shows such a grown-junction transistor in a basic measuring circuit. Because of its simple geometry, the grown-junction type is the easiest to discuss. Reference to Fig. 2 shows that all the other types contain the same basic structures, albeit in somewhat different arrangement.

The transistor is defined in IRE Standards as "an active three-terminal solid-state device." Its operation is somewhat analogous to that of a vacuum tube. The thin center section, called the *base* region, can be considered a control element which obstructs the flow of current between the *emitter* and the *collector*, which are analogous in some ways to the cathode and the plate of a vacuum tube.

2. BASIC PRINCIPLES OF TRANSISTORS

2.1. Principles of Semiconductors.

Following is a very brief review of some of the semiconductor terms used in describing transistor operation. The reader is referred to one of the many standard texts on semiconductors for further details. The *junction transistor* is a form of *bipolar* transistor, in contrast to the *field-effect transistor*, which is a form of *unipolar* transistor. As the terms imply, the bipolar transistor employs *two* types of current carrier, while the unipolar uses only one. These carriers are the *electron* and the *hole*, which have, respectively, negative and positive current-carrying character. Under some circumstances both carriers can exist in the same material, specifically in *semiconductors*, which form an intermediate class of conductors about halfway in conductivity between metals and dielectrics. However, electrons and holes tend to annihilate one another; hence their coexistence is often unstable, and a specific time, called *recombination* time, or *lifetime*, is often used to describe the time they will coexist in a given material.

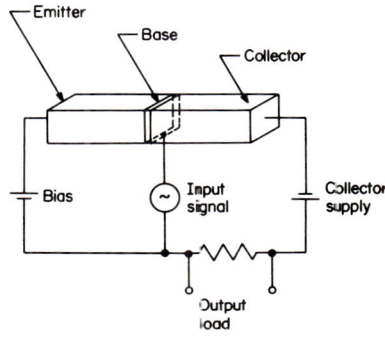

FIG. 3. Basic circuit for transistors and grown-junction structure (common-emitter).

Electrons exist permanently in what are called *n-type semiconductors*, which are materials having *impurity contents* such that the impurity atoms tend to give off or donate electrons. Holes can exist permanently in *p-type semiconductors*, in which the impurity atoms take up, or *accept* electrons. The boundary region between two n- and p-type semiconductor regions is called a *p-n junction*, and a junction transistor as seen in Fig. 3 consists of two p-n junctions in close proximity (1 mil or less).

Germanium and silicon are the important transistor materials. They are the only materials known, even though hundreds have been investigated, which have the necessary purity and structural perfection to permit of transistor action meeting present commercial standards. Gallium arsenide is the only semiconducting compound which seems to have much chance of finding any commercial application.

Germanium as it occurs in the completely pure refined state is not suitable for transistors but must be *doped* with impurities, namely, with *donors* to form the suitably formed *n region*, and with *acceptors* to form a suitably formed *p region*. The studies leading to the most effective means for these processes have formed the main bulk of transistor development during the last ten years, and we can now say that practically all the problems of producing transistors meeting very strict design specifications have been solved, at least in principle.

To give an idea of the subtlety of the transistor field, we might state that the amount of impurity to be added to a cubic centimeter of pure germanium to form a typical collector region is on the order of only one-billionth of a gram! In addition to this extreme sensitivity to impurities, germanium suitable for transistors must possess crystal perfection greatly exceeding that required for any other commercial device. In fact, germanium and silicon, through the impetus given by transistor requirements, have become probably the most perfectly grown crystal materials in science or technology today. New crystal-growing techniques have had to be developed during the past few years to meet these needs, and these developments in turn have enriched the whole field of materials technology.

2.2. Transistor Action.

To take a probably oversimplified point of view toward transistor action, we consider the common-emitter connection, in which signals are applied to the base region, as in Fig. 3. Now the *p-n* junction is characterized by an *electric field* built into the material because of the differences in structure of the *n*- and *p*-type impurities, and this *space-charge layer* permits a highly asymmetrical current flow depending on the sign of applied voltage.

The intermediate *p* region in Fig. 3 acts somewhat like the grid of a triode vacuum amplifier, with which we assume the reader has more familiarity than with transistors. The transfer of current from the low-voltage low-impedance emitter circuit to the high-voltage high-impedance collector circuit gives a power gain. A simple hydraulic analogy to transistor action can be made, in that the base control action is likened to

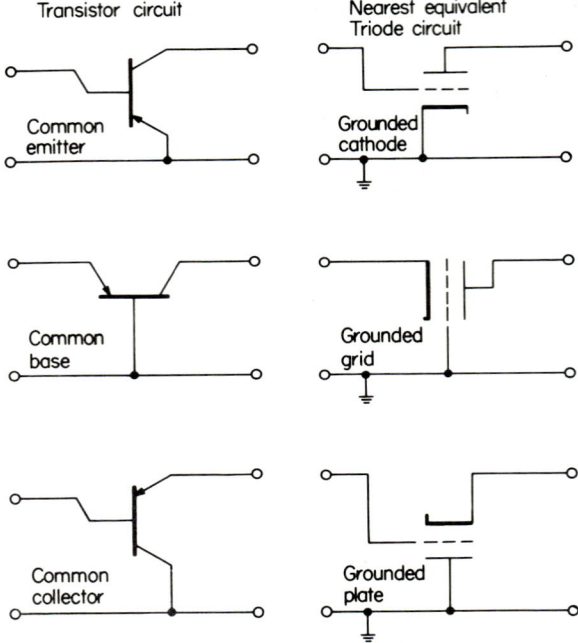

Fig. 4. Transistor–vacuum tube circuit analogs.

the action of a gate at the top of a dam, controlling water which can then, in falling over the dam, deliver many times the power involved in operating the gate. The various connections for a transistor are shown in Fig. 4, with the vacuum-tube analog.

One difference between transistor and vacuum-triode action is that the vacuum triode deals with one type of charge carrier only. In the semiconductor triode, however, electrons from the emitter do not normally exist in the *p*-type base region. However, they can be *injected as minority carriers* into the *p* region, where they can coexist for a time equal to the lifetime before recombining. This is one of the reasons for a very thin base, since the carriers must *diffuse* to the collector for collection before this time elapses; otherwise an efficiency-reducing loss of carriers by the direct recombination will occur. With base thickness in modern transistors of a few microns, it is possible to have efficiency of traversal of the emitter-injected carriers as high as 99 per cent or more, and for the collector current to follow changes of signal at frequencies of many hundreds of megacycles.

2.3. p–n Junctions. As can be seen from Fig. 3, the transistor comprises two p-n junctions in close proximity. Thus the emitter-base circuit is one p-n junction diode, and the collector-base circuit is another p-n junction diode. To understand transistor action it is helpful to review the properties of the junction diode first.

Figure 5 shows a p-n junction diode characteristic, here taken as the collector diode. The *reverse current* through the diode is called, in transistor nomenclature, I_{CBO}, meaning the current between base and collector with emitter open-circuited. It consists of a low voltage range, a long *saturation-current* range, in which the current is essentially independent of voltage, and the *breakdown* range, indicated by the symbol BV_{CBO} (the term BV, meaning breakdown voltage, is actually not mathematically acceptable as a symbol, but it is widely used). The value BV is of course a limiting value over which damage to the transistor may result. In the voltage range just below BV_{CBO}, current increases rapidly with voltage through *avalanche multiplication*.

As indicated, the diode characteristic is very sensitive to temperature. This results in one of the major problems of transistor circuit design. Eventually transistor materials may be found having negligible temperature dependence, but this is not in sight at the present time.

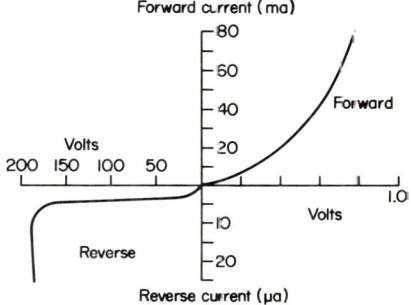

FIG. 5. p-n junction characteristics.

The emitter diode in transistor action is biased in the opposite direction, or *forward current direction*. It is seen that the current increases rapidly with very small applied voltages, and is also dependent on temperature.

Proceeding from two isolated diodes to the characteristics obtained when we apply forward voltages to the emitter and reverse biases to the collector, we get the transfer

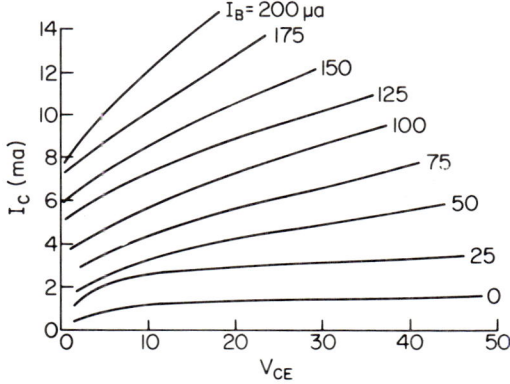

FIG. 6. Output transistor characteristics—common-emitter connection.

characteristics indicated in Fig. 6. There we see a number of curves for the collector current, similar to those of Fig. 5, except that the emitter current has been added, through transfer of carriers, from the emitter to the collector circuit. The only difference, a small *base* current, represents the loss of carriers from the emitter which do not get into the collector. This base current, like the grid current of a vacuum triode, is kept to a minimum for good transistors and should preferably be on the order of 1 per cent or less of the emitter current. Figure 7 shows other characteristics of importance for analysis of transistor performance.

2.4. Transistor Parameters. It will be impossible within the scope of the present section to describe in detail all the transistor characteristics which may be important in applications. Some of the more important transistor parameters, and their symbols, are discussed below. For more details, the reader is referred to one of the books available on transistor circuits.

1. I_{CBO} is the collector current when the collector is biased in the back or high-resistance direction, relative to the base, and the emitter is left open. The quantity I_{CBO} represents, usually, an unwanted leakage of current in the collector circuit and must be minimized.

2. The corresponding quantities, I_{CEO}, I_{CES}, I_{CER}, have the corresponding meanings, with S referring to "shorted" rather than "open," and R referring to emitter connected to the base through a resistance R.

3. Breakdown voltage is usually referred to, rather improperly, by the symbol BV. Thus, we have BV_{CBO}, BV_{CEO}, BV_{CES}, and other corresponding terms, all of which are used under various conditions.

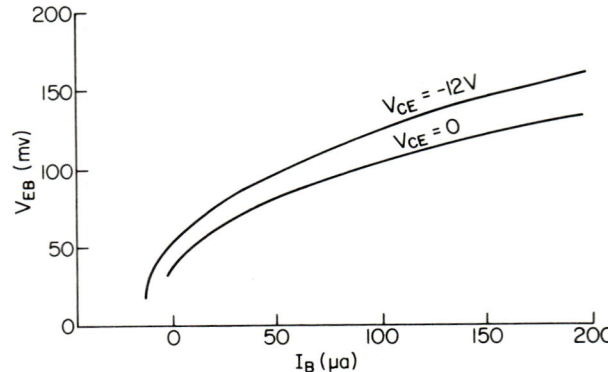

FIG. 7. Input transistor characteristics.

A related quantity is "punch-through" voltage V_{PC}. At this value of voltage the space-charge layer of the collector, which widens with increased voltage, expands into the base and rather large current flow ensues.

Breakdown voltage has generally, but not always, a well-defined value. In some transistors, having a "soft" breakdown, the approach is so gradual that the value is not well defined. Therefore, it is usually defined in an arbitrary way as the voltage at which the current reaches a certain value. The value of breakdown is also dependent upon the thermal conditions of the transistor, the heat sink available, and the ambient temperature.

4. h_{fe} is the common-emitter short-circuit amplification factor (lowercase subscripts refer to small-signal, capitals to static values). A related term is h_{fb}, the common base current amplification: $\Delta I_C = \Delta I_E h_{fb}$, and $h_{fe} = (-h_{fb})/(1 + h_{fb})$. A further discussion of the h parameters will be given below after a discussion of the equivalent circuit of a transistor.

The current amplification alpha may be expressed by the relation $\alpha = \gamma \beta M$, where γ is the *emitter efficiency*, namely, the fraction of the emitter current which is carried by the minority carrier; β is the *transport factor*, namely, the fraction of minority carriers in the base which manage to arrive through diffusion at the collector region; and M the *collector multiplication factor*. M is not of significance for most applications of the transistor, since it is very close to unity unless the voltage is close to the breakdown value. The process of avalanche multiplication causes the carriers to multiply by collisions, each carrier perhaps producing many additional ones as the electric field causes it to accelerate to the point where it can excite other carriers to a free state. Transistors operated so as to utilize the avalanche process for high speed or high gain are called *avalanche-mode* transistors.

5. C_e is the capacitance in the emitter circuit, which is important for amplifiers operated at very high frequency. It may be considered a consequence of the diffusion process, leading to storage or depletion of carriers in the vicinity of the emitter-base junction. There is also a capacitance in the emitter related to the existence at the junction of the space-charge, or dielectric layer.

6. C_c is the corresponding capacitance in the collector circuit. It is most closely related to the thickness of the dielectric layer forming the collector junction, and the diffusion term is usually negligible because of high reverse biases usually used.

7. r_b is the base resistance, generally determined by the fact that a small current must flow through the rather narrow base region (often only a few microns in width) to reach the base contact. Thus high current amplifications, which require narrow base widths, also tend to raise the base resistance, which is undesirable. r_t' is a related base spreading resistance which is more important for circuit applications.

8. The frequency response is generally designated by f_T, the frequency at which the current amplification h_{fe} becomes equal to 1. The time constant $r_b'C_c$ is another parameter which limits the frequency response, and the corresponding frequency will be the limiting frequency if smaller than f_T. The symbol $f_{\alpha b}$ is sometimes used to denote the frequency at which the current amplification in the common-base connection drops to 0.707 of its low-frequency value. This is the "alpha cutoff frequency."

3. EQUIVALENT CIRCUITS FOR THE TRANSISTOR

The engineer using transistors in amplifiers is primarily interested in the behavior of the device as it interacts with other circuit elements. For this purpose he would

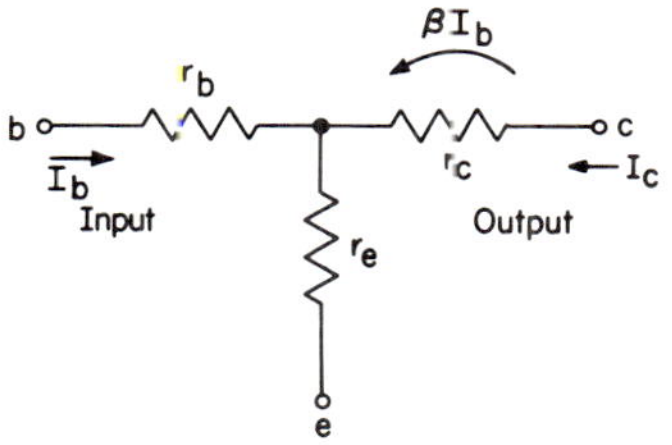

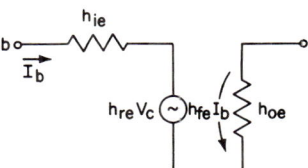

FIG. 8. Transistor equivalent circuit—common emitter.

FIG. 9. h-parameter equivalent circuit—common-emitter connection.

like to be able to represent it by a simple collection of standard elements such as resistors, capacitors, and a voltage or current source to represent the active or amplifying element. It is usual to think of the transistor as a four-terminal black box, having the simple equivalent circuit shown in Fig. 8. This circuit is valid only for very low frequency conditions, and is called the "T-equivalent" circuit. Many other representations are possible, depending upon the precision with which the operating characteristics are to be presented for varying conditions. We shall deal here only with the common-emitter connection, by far the most widely used.

The active source is a current source having the value $I = \beta I_b$, and is represented by the curved arrow around the collector resistance.

The most widely used parameters for the presentation of the properties of transistors are called the "h parameters," of which we have mentioned several examples without explanation. These parameters are widely discussed in detail in various references, and we give here only a brief outline.

The equivalent circuit most suitable for discussion of the h parameters is that given in Fig. 9. This circuit is in two parts, the one containing a voltage source, the other a current. The basic equations from which the h parameters are derived are:

$$V_1 = h_{11}I_1 + h_{12}V_2 \qquad I_2 = h_{21}I_1 + h_{22}V_2 \tag{1}$$

Because these equations are not symmetrical in V or I as are the systems using z or y parameters, the h parameters are called "hybrid" parameters.

The notation of the hybrid parameters for the common-emitter circuit is as follows:
Input resistance:

$$h_{11} = h_{ie} = (\partial V_{BE}/\partial I_B)V_{CE=\text{const}} = r_b' + (1 + h_{fe})r_e \tag{2}$$

Feedback factor:

$$h_{12} = h_{re} = (\partial V_{BE}/\partial V_{CE})I_{B=\text{const}} = (r_e/r_c) \tag{3}$$

Current amplification factor:

$$h_{21} = h_{fe} = (\partial I_C/\partial I_B)V_{CE=\text{const}} = \alpha_{fe} \tag{4}$$

Output admittance:

$$h_{22} = h_{oe} = (\partial I_C/\partial V_{CE})I_{B=\text{const}} = (1/r_c) \tag{5}$$

In the subscript notation, now standardized, the first letter such as i, r, f, and o refers to input, reverse transfer, forward transfer, and output characteristics, respec-

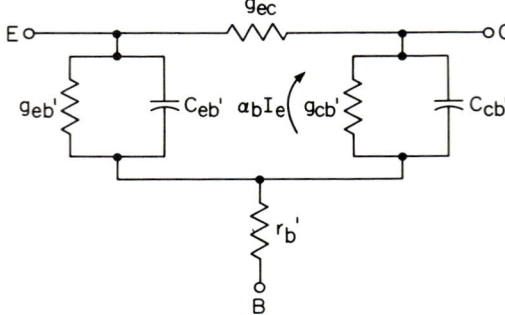

Fig. 10. High-frequency equivalent circuit.

tively. The second subscript refers to common-emitter (e) connection, although it might also be b for common-base or c for common-collector.

For those not familiar with partial derivatives, the expression for h_{ie}, for example, simply means the *slope* of the characteristic of V_{BE} as a function of I_B, taken with the collector voltage constant. In other words, the h parameters can be obtained easily from the two sets of characteristics for a transistor shown in Figs. 6 and 7.

Thus h_{ie} and h_{oe} represent the slopes of the two characteristics at desired values of V_{CE} or I_B. On the other hand, h_{re} is obtained by calculating numerically or graphically the change in V_{BE} produced by the change of V_{CE} (30 volts in this case) for a chosen constant value of I_B. A similar calculation gives h_{fe} as the ratio of change in I_C to change in I_B.

From the curves of Figs. 6 and 7, as an example, the following values of the h parameters are obtained:

$$h_{ie} = 210 \text{ ohms}$$
$$h_{re} = 2 \times 10^{-3}$$
$$h_{fe} = 60$$
$$h_{oe} = 1{,}800 \text{ } \mu\text{mhos}$$

These characteristics are adequate to describe the response of the transistor for almost all low-frequency applications.

On the other hand, for high-frequency applications more complex equivalent circuits are required. One of the most commonly used is shown in Fig. 10. It is useful for common-base connections, whereas the circuit of Fig. 11, which is called a "hybrid-pi" connection, is most useful for common-emitter connection at higher frequencies.

The following are the relations between the indicated circuit elements and the low-frequency parameters of the transistor for the two cases:

a.
$$g_{eb'} \cong 1/r_e \tag{6}$$
$$C_{eb'} \cong C_e \cong 1/r_e \omega_{ab} \tag{7}$$
$$g_{cb'} \cong g_c \tag{8}$$
$$g_{ec} \cong \mu_0/r_e \tag{9}$$
$$\alpha_b \cong \alpha_{bo}/(1 + j\omega/\omega_{ab}) \tag{10}$$
$$C_{cb'} \cong C_C \tag{11}$$

b.
$$g_{b'e} \cong (1 - \alpha_{bo})/r_e \tag{12}$$
$$C_{b'e} \cong C_e \cong 1/r_e \omega_{ab} \tag{13}$$
$$g_{ce} \cong \mu_0/r_e \tag{14}$$
$$g_{cb'} \cong g_c - g_{cd}/2 \cong g_c \tag{15}$$
$$C_{cb'} \cong C_c - C_{cd}/2 \cong C_c \tag{16}$$
$$g_m \cong \alpha_{eo}/r_e \tag{17}$$

In these equations r_e is the emitter diffusion resistance, μ_o the low-frequency reverse-current transfer ratio, α_b the low-frequency short-circuit forward current transfer

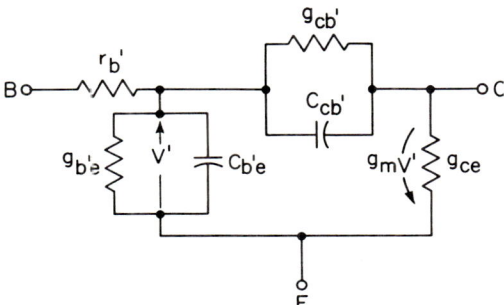

Fig. 11. Hybrid π equivalent circuit.

ratio, g_c the collector conductance, C_c the collector capacitance, g_{cd} the collector diffusion conductance, C_{cd} the collector diffusion capacitance, ω_{ab} the cutoff frequency, that is, the angular frequency at which α_b has dropped to 0.707 of its low-frequency value α_{bo}. For further discussion of these and other equivalent circuits, the reader is referred to Chap. 2, Sec. 4.3.

4. DESIGN THEORY OF TRANSISTORS

4.1. General. We have in previous sections given a general account of the physical principles of transistors. In this section we go into a little more depth in indication of the relation between the operating characteristics of a transistor and the physical parameters of the junctions involved.

The basic diagram to be considered is that of Fig. 12. This is the energy diagram showing the way an electron energy varies as it crosses the region between emitter and collector. For simplicity we use the n-p-n case, so that we have electrons in the emitter section moving across the p-type base region, being collected at the collector.

Electrons must cross over an energy hill to enter the n-type material; without help they cannot do it. This help comes from two sources—temperature-derived energy and applied voltage. Because of the temperature-induced variations in energy among the electrons, a few electrons will always be able to enter the p region by thermal excitation. We thus see the important fact that, while we do not like the temperature sensitivity of transistors, temperature is essential for transistor operation. Most transistors will not work when cooled so low that there is no thermal energy available.

Another important fact about the particle distributions is that *diffusion* of electrons tends to force electrons into the p region. This is because there is a large gradient of electrons between p and n regions, and gradients cause them to try to spread out uniformly. Once in the base, carriers must diffuse all the way in order to reach the collector.

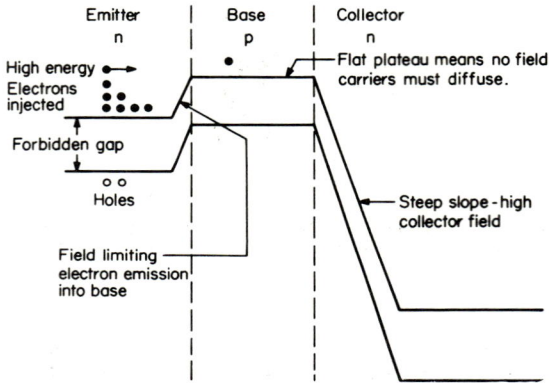

Fig. 12. Energy diagram, theory of transistors.

Calculation of the properties of a transistor in terms of the properties of the semiconductor and of the specific geometrical and other parameters of the specific transistor is essentially one of translating the above ideas into quantitative terms.

4.2. Emitter Efficiency. Among the basic factors to be understood are the emitter efficiency and the transport factor. Emitter efficiency is determined by the ratio of minority (desired) emission current into the base to the majority (unwanted) carrier emission in the base.

For a transistor with resistivity ρ_e in the emitter, ρ_b in the base, width W for the base region, and recombination rate in the emitter region determined by the "diffusion length" L_{pe},

$$\gamma = \frac{1}{1 + \rho_e W / \rho_b L_{pe}} \tag{18}$$

If the resistivity in the emitter is 1 ohm-cm, in the base $\rho_b = 5$ ohm-cm, $W = 0.002$ cm, and $L = 0.005$ cm, then $\gamma = 0.92$—that is, 92 per cent of the current emitted into the base region is suitable for transfer into the collector. If on the other hand, the emitter resistivity is reduced to 0.01 ohm-cm, then $\gamma = 0.992$ and a correspondingly more efficient transistor can be made.

4.3. Transport Factor. For the calculation of the transport factor, we must consider the relative effects of diffusion as a means of transporting carriers across the base, and of recombination as a source of removal before they reach the collector. For a simple geometry such as the grown-junction transistor, with base width W and diffusion length (which determines recombination rate) for minority carriers in the base of L_{nb}, then the transport factor is found to be, if an approximation is made assuming recombination relatively small,

$$\beta = 1 - W^2/2L_{nb}^2 \qquad (19)$$

For a more general expression, not involving a small recombination factor,

$$\beta = \text{sech}\,(W/L_{nb}) \qquad (20)$$

which reduces to the former expression when $W \ll L_{nb}$.

4.4. Surface-recombination Effects. Besides loss of carriers due to recombination in the bulk, surface effects are quite important. In fact, surface effects are probably responsible for more transistor failures than any other single mechanism. The surface recombination has the characteristics of a velocity s such that the current density due to this recombination is ens, where e is the electronic charge and n is the minority excess (over the equilibrium value) carrier density just underneath the surface. For some simplified conditions the β due to surface recombination is given by

$$\beta = 1 - sA_sW/A_eD_n \qquad (21)$$

where A_s is the surface area of the base region immediately around the emitter junction and A_e is the emitter-junction area.

4.5. High-frequency Transistor Effects. Since diffusion through the base region is an important part of transistor action, and since diffusion takes appreciable time, the transport factor is frequency-dependent. Solution of the simplified time-dependent diffusion equations gives

$$\beta = \text{sech}\,(W/L)(1 + j\omega\tau)^{\frac{1}{2}} \qquad (22)$$

where ω is the angular frequency involved ($2\pi f$) and τ is the lifetime in the base.

Likewise the emitter efficiency varies with frequency in a somewhat similar way

$$\gamma = \left[1 + \frac{\rho_e W(1 + j\omega\tau_p)^{\frac{1}{2}}}{\rho_b L_{pe}(1 + j\omega\tau_n)^{\frac{1}{2}}}\right]^{-1} \qquad (23)$$

The emitter junction is nearly equivalent to a conductance and capacitance in parallel, for many conditions. This capacitance is called the diffusion capacitance as distinguished from the junction capacitance related to the capacitance in the space-charge layer itself.

The diffusion capacitance can be calculated for simplified conditions to be

$$C_D = W^2 I_E/2D \qquad (24)$$

where D is the diffusion length of minority carriers in the base, and from which it is seen that for very thin base regions the value can be very small.

4.6. Temperature Response. Temperature sensitivity of transistors is much greater than that of vacuum tubes, and this is one of the things which has held back transistor development. For example, the collector resistance drops by as much as 5 per cent per °C in the range 20 to 60°C, and for many applications this is difficult to tolerate unless compensation schemes are used.

It is difficult to make general statements about the temperature response of the transistor parameters. For example, the h_{fe} of various types may increase or decrease with temperature.

Another fact about transistors that must be kept in mind is that they have definite restrictions on the low temperatures they may be held at before interaction between emitter and collector disappears. This temperature, of course, is far below 0°C.

The temperature range of transistors can be changed by using materials of different band gap. Thus silicon, with a band gap of 1.12 ev, has a higher temperature response than germanium, with a gap of 0.72 ev; and gallium arsenide with a gap of 1.50 ev operates at even higher temperatures. Silicon carbide has been proposed for transistors at temperatures up to 500°C, but successful development has not been made. Indium antimonide, with a gap of 0.18 ev, has been proposed for transistors to operate at 100°K and below.

4.7. Noise in Transistors. All amplifiers produce unwanted and usually random output variations which tend to obscure the wanted signal output. Transistors are no exception, although truly phenomenal improvement in the noise properties has been achieved since the early days of transistors. Noise in transistors is of several types: The first, $1/f$ noise, is a type of noise which decreases inversely as the frequency at which the noise is measured, and which probably is related to surface conditions on the semiconductor material. As the frequency goes up, the noise becomes less $1/f$ in type

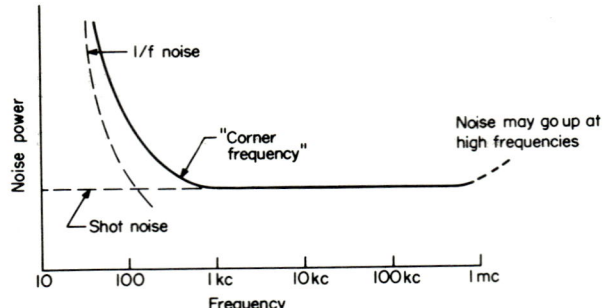

Fig. 13. Noise vs. frequency.

and more white noise, associated with generation and recombination in the space-charge region. Since carriers are being continually excited into electrical activity from the valence band and in turn recombining, a fluctuation in carrier density appears which shows up as a noise voltage. The noise also includes a component called shot noise, which results from the corpuscular nature of electricity, and the discrete passage of individual electrons or holes across junctions. Figure 13 shows a typical noise spectrum for a junction transistor.

5. ALLOY TRANSISTORS

5.1. General. Figure 2 may be referred to for a general view of the cross section of an alloy transistor.

Germanium-alloy transistors contain metallic dots of indium, generally alloyed with aluminum or gallium, or both. The size of the dots will depend on the power level. The alloying may be carried on at any temperature from about 300 to about 800°C or even above.

Alloy transistors began with germanium, p-n-p types being the most common, n-p-n types becoming available only several years later. Silicon-alloy types began with n-p-n varieties.

With increasing attention being given to diffusion and planar techniques, alloy transistors are tending to become obsolete. However, at present writing they are still the cheapest and most widely used type of transistor, ranging from the subminiature types of Fig. 1 to high-power units having a rating of 50 to 60 amp.

5.2. Impurity Distribution. p-n-p germanium-alloy transistors employ a wafer of germanium whose resistivity is chosen so as to provide the desired breakdown voltage. The impurity distribution is indicated in Fig. 14.

TRANSISTORS   10–13

It is seen that the emitter contact is more highly doped than the collector—this is generally achieved by mixing gallium and aluminum with indium since these metals are much more soluble in Ge than is In. The high doping level gives a high emitter efficiency and makes possible a more uniform current amplification vs. emitter current characteristic, as indicated in Fig. 15.

Alloys for *n-p-n* germanium transistors are generally based on lead or tin—85 per cent lead with 15 per cent antimony is a widely used alloy. In general the problems of *n-p-n* germanium transistors due to alloy problems, cracking, and instability are somewhat greater than those for *p-n-p* units using the soft metal indium.

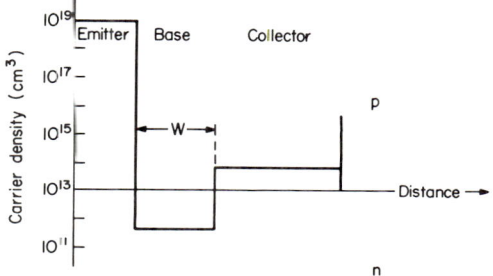

Fig. 14. Impurity distribution, alloy transistor (impurity content and carrier density are equivalent).

5.3. Structure of the Alloy Transistor. When indium is heated upon a germanium wafer, cooled, and then sectioned so as to reveal the structure of the junction, it is found that the *p-n* junction is single-crystalline like the basic structure of the germanium itself. The phenomenon of *regrowth* makes possible the good characteristics observed. Indium dissolves in the germanium to form an indium-germanium solution. When the wafer is cooled, germanium heavily doped with indium recrystallizes on the unattacked germanium base and forms a crystal which is practically indistinguishable from the original material as far as crystalline perfection is concerned.

Besides the phenomena of solution and regrowth, another effect can take place under some conditions. This is the *diffusion* of indium out of the regrown region into the germanium base.

If there were to be a completely sharp drop in indium content between the regrown region and the base, we would have what is called a completely *abrupt* or *sharp* junction, which is desirable for many purposes. Such junctions can be detected by measurement of the *capacitance* of the junction as a function of voltage. Abrupt junctions

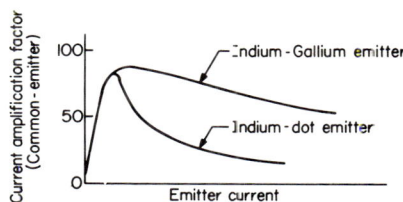

Fig. 15. β vs. emitter current.

have a capacitance which varies as $1/V^{\frac{1}{2}}$. Diffusion of the impurity atoms, on the other hand, produces a gradation of indium content which, if pronounced enough, gives a *graded* junction. One example of a graded junction is one in which there is a linear crossover in the content of the impurities responsible for the *n* and *p* characteristics, respectively. Such linearly graded junctions have a capacitance which varies as $1/V^{\frac{1}{3}}$.

The question of whether alloying or diffusion is of the most importance in a particular transistor is usually settled by comparing the *diffusion depth* of the indium or other impurity with the depth of the *space-charge layer*. Even with an absolutely abrupt junction, the voltage drop in the potential diagram can take place only over a distance which is finite, and indeed usually on the order 10^{-4} cm in germanium. If the diffusion depth is less than this value, the junction is considered to be an alloy junction, whereas if the diffusion depth is several or more times the width of the space-charge layer we have a diffused junction.

5.4. Performance of Alloy Transistors. Alloy transistors do not, in general, yield as desirable characteristics as diffused types. This is due, among other things, to the fact that the junctions for emitter and collector must be formed independently. Although it has been found possible, by using the 111 plane of crystal orientation for the basic wafer, to produce quite flat alloy-germanium interfaces, it is nevertheless not possible to make them flat enough and parallel enough to get close separation on a production basis. Thus alloy transistors are confined to base widths that are relatively large, and the frequency response is correspondingly limited, usually to 1 Mc or less.

In our discussion of the alloy transistor we now take up the basic questions for the designer and user of transistors. These include: Which of the various device parameters we have discussed are essential for the device performance in the particular application? To what extent can the particular technique, in this case the alloy, provide simultaneously in a given device the required set of operating parameters? To what extent can this be done at a price consistent with the nature of the application?

Among the parameters that can be controlled are wafer resistivity, wafer lifetime or diffusion length, surface conditions and surface recombination, doping level and resistivity in both the emitter and collector contacts, distance to base contact, base width (between emitter and collector), emitter and collector areas, nature of the base contact, and a few others. It is generally assumed that wafer resistivity is constant throughout the chip, although in the alloy-diffused transistor we shall see some of the advantages of varying resistivity continuously between emitter and collector.

Among the operating characteristics required, most of which have been defined already, are (1) collector voltage before breakdown to be above a certain value, say 50 volts, (2) collector capacitance to be below a certain value, say 5 pf, (3) frequency cutoff to be above a certain value, say 10 Mc, (4) punch-through to be above a certain value, (5) leakage I_{CBO} not to exceed a certain value.

For p-n-p germanium transistors, the various parameters are connected with device properties by the following simple equations.[1] The fact that real transistors have properties reasonably close to the predictions is what makes a design theory practical and useful.

$$f_{\alpha b} = 2.67/W^2 \quad \text{Mc, } W \text{ in mils} \tag{25}$$

$$BV_{CBO} = 87\rho_b{}^{0.72} \tag{26}$$

$$PV_C = 630W^2/\rho_b \tag{27}$$

$$C_c = 0.071d^2/\sqrt{V_{CB}\rho_b} \quad \text{pf, } d = \text{diameter of collector} \tag{28}$$

For example, if $f_{\alpha b} = 10$ Mc is needed, a value of $W = \sqrt{0.267} = 0.5$ mil is required. Such a value is difficult to attain reproducibly in germanium, although it can be done. Because of the lower mobility and diffusion coefficients for carriers in silicon, the base width to achieve 10-Mc operation has to be even smaller.

For high reverse voltage, a high resistivity is required. However, the punch-through voltage varies *inversely* as the wafer resistivity, and *directly* as W, so that decreasing W and increasing ρ_b to improve frequency response and increase collector breakdown have a drastic effect in reducing punch-through voltage.

The above information will give the reader, it is hoped, a feeling for the kinds of design compromises which determine what is attainable in a specific device. Because of the many combinations possible, it is also easy to see why there are available such a bewildering variety of specific types in each general device family. Table 1 gives operating parameters of a few typical germanium-alloy transistors.

5.5. Surface-barrier and Related Transistors. The "surface-barrier" transistor, while technically not an alloy type, is related to it, and has evolved into alloy types such as the "MADT" (MicroAlloy Diffused). The surface-barrier transistor was originally designed to provide high-frequency performance through special electro-

---

[1] For further details, see L. P. Hunter (ed.), "Handbook of Semiconductor Electronics," 2d ed., McGraw-Hill Book Company, New York, 1962.

Table 1. Germanium *p-n-p* Junction Transistors, Average Characteristics Measured at 27°C

Type	Typical application	Collector voltage, volts	Emitter current, ma	Avg collector resistance, megohms	Avg base resistance, ohms	Avg emitter resistance, ohms	Avg base current amplification factor	Avg cutoff current, ~ μa	Max noise factor, db	Alpha cutoff frequency, Mc
2N64	a-f r-f amplifier	−6	1.0	2.0	700	25	45	6	22	0.8
2N106	Low-noise a-f amplifier	−2.5	0.5	1.0	700	50	25	6	6	0.8
2N111A	Medium-gain, i-f amplifier	−6	1.0	1.7	50	25	25	1	...	3
2N130	a-f i-f amplifier	−6	1.0	2.0	350	25	22	6	25	0.6
2N138	High-gain a-f output	−6	1.0	2.0	1,800	25	140	6	...	1.4

10–15

lytic processing techniques which gave an extremely thin base region over a small portion of the area. By the same electrolytic procedures, so that the equipment did not need to be changed, the metallic contacts which provided the emitter and collector electrodes were then deposited. Metals such as cadmium and indium were found to work well, and no heat was needed. Since alloying did not take place, the junctions are "natural" junctions created by surface states on the germanium, to which the metallic layers merely make contact. Such transistors have been widely used, although their high costs have made them less and less competitive. Also, the frequencies attainable are no longer competitive with values obtained with mesa and planar techniques.

The microalloy process was found to be a valuable modification to the surface-barrier technique as a means of improving emitter efficiency and reliability of the surface-barrier transistor. Evaporated layers of indium when heated produced an extremely thin layer for emitter and collector regions, combined with the thin base regions carried over from the surface-barrier technology.

This line of development has also continued into diffusion technology through the "ECDC" process (ElectroChemical-Diffused Collector), by which diffusion techniques are used to form the collector-base junction, while the electrolytic process, combined with microalloying, is used to form the emitter-base junction as previously described.

5.6. Applications of Alloy Transistors. Germanium-alloy transistors at the present writing are the most widely used type of transistor in the electronics field. Originally first applied in the hearing-aid industry, their use has spread throughout electronics, particularly in those areas where low cost is a primary factor and temperature and other operating restrictions are not too severe. Probably the largest single area of application is in the computer field. Germanium power transistors are also sold in large numbers, particularly for automobile radios. Portable personal radios consume a considerable fraction of the germanium-alloy transistors sold.

Silicon-alloy transistors are not sold in anywhere nearly so large numbers as germanium units, partly because they have historically been more expensive. They are more useful where temperature limitations are severe, since germanium-alloy units must be operated so that the junction temperature is not over 85°C for long life, whereas silicon can be used with the junction at 125 to 150°C or even somewhat higher. Silicon-alloy transistors have been subjected to increasing competition from silicon diffused and planar transistors, which are expected to dominate the market.

6. DIFFUSED TRANSISTORS

6.1. Mixed Alloy-diffused Transistors. The drift transistor, the post-alloy diffused transistor, and the p-n-i-p transistor are transitional types combining alloy and diffusion techniques.

The alloy transistor transfers carriers from emitter to collector using the diffusion of carriers—there is little or no electric field in the base region, so that little help can be obtained from the field for aiding motion of the minority carriers to the collector. The reason for this is in the constant resistivity of the base region. If, on the other hand, there is a variation of resistivity in the base, there may be an assisting electric field. The drift transistor, employing a diffused region in the base, is probably the first example of the *diffused* transistor. The profile of resistivity is shown in Fig. 16. Thus the drift transistor is a mixed alloy-diffused transistor.

The sweeping field which is effective in aiding the movement of carriers is given by the following equation:

$$E = -\frac{dV}{dx} = -\frac{kT}{q}\frac{1}{n_b}\frac{dn_b}{dx} = \frac{kT}{q}\frac{1}{\rho_b}\frac{d\rho_b}{dx} \tag{29}$$

and is essentially determined by the slope of the resistivity vs. distance curve. The resulting transit time is given by integrating $dx/\mu E$ over the base region, where μ is carrier mobility. For a diffusion front the functional dependence of impurity concentration will be an error function, and the integration is made graphically or numerically.

The transit time so determined varies as W/D_p, where D_p is the diffusion coefficient for the minority carriers. The transit time for diffusion of carriers in a constant-resistivity base varies as W^2/D_p. This means that as base width is reduced the advantage of the drift becomes less and less. Increased control over diffusion techniques makes possible such narrow W's that transit time in itself is no longer the limiting factor in frequency response, which is more likely to be limited by $r_b'C_c$, the time constant in the base-collector circuit.

Another type of mixed diffused transistor is the post-alloy diffused transistor, in which the *base* region is *formed*, rather than being merely modified, by diffusion. The impurity profiles are shown in Fig. 17. It is seen that as a result of a subsequent diffusion out of the emitter region, a very thin base region can be formed as a result of the excess of donor impurity centers over acceptors in the narrow region.

Because of the inflexibility of the diffusion process, it is difficult to combine the drift principle with the post-alloy diffusion principle, and the result is that the fields occurring in the base region of the post-alloy diffused transistor may actually hinder rather than help the motion of minority carriers through the base.

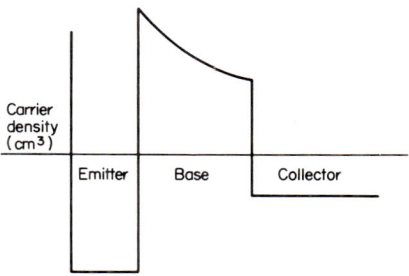

Fig. 16. Profile of carrier density—drift transistor.

Still another type of transistor may be formed by the use of diffusion from an alloy region. This is the *p-n-i-p structure*, which is designed to improve the high-frequency performance by the reduction of the collector capacitance. If the base wafer of the semiconductor is close to intrinsic in resistivity value, the *p-n-i-p* structure can be formed as in the post-alloy diffused device by alloying the emitter, post-alloy diffusing the base, and alloying the collector. With the recent trend toward completely diffused devices, most of the mixed alloy-diffused devices discussed above are becoming

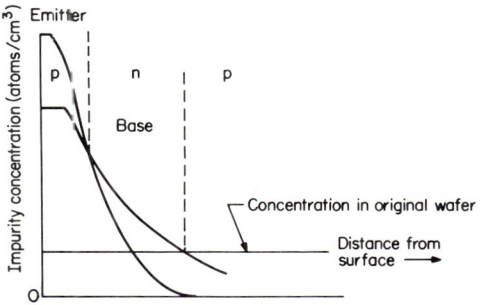

Fig. 17. Concentration profiles of the *p*- and *n*-type impurities in a *p-n-p* diffused-base germanium double-diffused transistor.

obsolete, or at least are being used in more and more restricted and specialized applications.

6.2. The Mesa Transistor. The drift transistor is an alloy type into which diffusion regions were incorporated so as to reduce transit time in the base. The diffused transistor, on the other hand, is a device in which alloy regions may or may not be used at all. Collector, base, and emitter regions can all be formed by the diffusion process, sometimes simultaneously, sometimes sequentially, according to the individual manufacturer.

10–18 AMPLIFYING DEVICES

Figure 2 showed the construction of a typical high-frequency "mesa" structure, the first commercial type of diffused-base transistor. The mesa is formed by etching techniques so as to form a very small raised portion containing the essential junctions. In the original mesa types, only a single diffused junction was formed, generally a few microns from the surface. The emitter region was made by alloying a contact such as aluminum into the n-type base region which in turn was formed by diffusing phosphorus into the germanium or silicon.

Later transistors were made by a double-diffusion process, the emitter junction being a boron diffusion onto a phosphorus diffused layer. The base and emitter leads were then "thermal-compression bonds" designed to make fine ohmic contacts with as little damage to the crystal as possible.

Because of the small size of the emitter and base stripes, the mesa transistor made possible the first invasion of the kilomegacycle region. Experimental units capable of oscillating at 1,500 Mc were developed shortly after the design was developed in the laboratory.

In the following we give a review of some of the operating characteristics of diffused-base transistors, of the amplifier type.

Table 2 gives a summary of the operating characteristics of typical germanium mesa transistors. These devices are particularly useful for amplifiers in the r-f region to 20 Mc.

Table 2

Parameter	Symbol	Conditions	Min	Typical	Max
Collector-base breakdown, volts	BV_{CBO}	$I_C = 0.10$ ma	-15		
Collector-emitter breakdown, volts	BV_{CES}	$I_C = 0.10$ ma	-15		
Emitter-base breakdown, volts	BC_{EBO}	$I_E = 0.10$ ma	2.5		
Collector cutoff current, μa	I_{CBO}	$V_{CB} = -6.0$ volts	......	...	3.0
Current transfer ratio	h_{FE}	$V_{CE} = -0.30$ volt $I_C = 10$ ma	25	40	
Base emitter voltage, volts	V_{BE}	$I_B = 1.0$ ma $I_C = 10$ ma	-0.34	...	-0.44
Collector saturation voltage, volts	V_{CE}	$I_B = 1.0$ ma $I_C = 10$ ma	......	...	-0.20

6.3. Planar Transistors. The planar transistor is a variety of diffused transistor in which oxide layers play a fundamental role. Because the oxides are used for insulating, masking, and passivating functions, and since germanium oxides are unsuitable for one or all of these, all planar devices now in use are silicon devices. The planar process gives indications of being the most significant advance in the technology of transistors since the development of the junction transistor. It will also probably result in the obsolescence sooner or later of all germanium transistors and their replacement by silicon. This is already being seen in the threat being posed by small silicon high-frequency transistors to germanium mesas, and in other areas.

As seen in Fig. 2, the planar transistor involves no structure fundamentally different from the diffused base or the mesa. Emitter-base and collector-base junctions are made by double diffusion, and bonded or fused contacts are made on the emitter and base so formed.

The primary difference between planar and previous techniques is that a heavy silicon dioxide layer (usually 1 micron in thickness) is formed on the base wafer before any diffusions are done. Then masking techniques (which we shall not go into here) are used, usually by photolithographic-etching processes, so as to create windows in the oxide through which diffusion takes place. The oxide effectively inhibits the diffusion in those areas which have the oxide covering. The base region is diffused in one step; then, usually, an oxide layer is grown over the new silicon surface, a new mask is used to form another window in the oxide for the emitter, and the emitter is diffused into the base region, the base width being controlled by the emitter diffusion time.

Another photoetch step exposes the base and emitter areas for evaporation of metal contacts, to which lead wires are generally bonded by thermal-compression bonding techniques.

These steps are generally carried out en masse on up to several hundred units on a single slice before mounting. The dice are usually scribed to initiate a crack, then separated by rolling over a rubber cylinder, after which they are mounted on headers for lead assembly, etc.

The standard configuration for the emitter-base combination is a pair of fine stripes usually a few mils wide by 6 to 10 mils long, separated by a mil or so. For extremely high frequency units the separation may be down to less than a few tenths of a mil, approaching the limits of control with present masking and etching techniques.

Other types of planar devices designed to deliver power at high frequencies may have other characteristic shapes for the emitter-base profile. Since at higher current levels the edge of the emitter is the current-carrying region, emitter periphery is more important than emitter area, and variations on the comb structure are likely to be used. Among these are the "star" configuration, "shark's tooth," "mapleleaf," and various other shapes used by different manufacturers.

Using such techniques, production transistors are now available with an upper frequency limit of 1.5 to 2 gc and a bandwidth of 1 gc. The planar technique has not been remarkably successful in providing high yields, however, and as a result the devices are expensive and are used almost exclusively for military applications.

One of the most important implications of the use of the oxide technique is in the "passivated" surfaces that result. All the junctions that come to the surface are protected from any direct contact with external ambients, both during the production of the device and during operation, by the heavy oxide layer. This has led to remarkably better characteristics, such as higher gain and lower I_{CBO} than with many of the older techniques. On the other hand, early hopes that the oxide layer would make possible a "packageless" transistor protected indefinitely from even extreme environments by the oxide layer have not been realized. The coating, while strong and relatively impervious, is not absolutely so, and external influences such as water vapor will

Table 3

Characteristic	Conditions	Symbol	Min	Max
Collector cutoff current, μa d-c........	V_{CB} = 6 volts d-c I_E = 0	I_{CBO}	...	0.5
Collector cutoff current, μa d-c........	V_{CE} = 30 volts d-c V_{BE} = 0	I_{CES}	...	100
Emitter cutoff current, μa d-c.........	V_{EB} = 1 volt d-c I_C = 0	I_{EBO}	...	100
Static forward-current transfer ratio....	V_{CE} = 6 volts d-c I_C = 2 ma d-c	h_{FE}	10	
Base spreading resistance, ohms........	V_{CE} = 6 volts d-c	r_b'	...	15
Collector capacitance, pf...	V_{CB} = 6 volts d-c	C_c	...	0.7
Small-signal short-circuit forward-current transfer ratio, db.............	V_{CE} = 6 volts d-c I_C = 2 ma d-c f = 200 Mc	h_{fe}	15	
Power gain, small signal, db..........	V_{CE} = 6 volts d-c I_C = 2 ma d-c f = 200 Mc	P_g	20	
Noise figure, db.....................	V_{CE} = 6 volts d-c I_C = 2 ma d-c f = 200 Mc	NF	...	8

eventually contact and injure the junctions if exposure at high concentrations is continued long enough.

Another modification of the planar transistor which appears extremely promising is the use of the epitaxial technique. The epitaxial method is one of growing from the vapor phase a thin layer of single-crystal material of specified properties, on a substrate of different properties. For transistors, in order to obtain high frequency response with a low saturation voltage it is desirable to make the transistor on a very thin (a few microns) layer of silicon of high resistivity on top of a very low resistivity substrate (a few hundredths of an ohm-cm resistivity). In order that excess leakage will not occur, and low noise may be achieved, there must be almost perfect continuation of the lattice from the old to the new material, and this provides many of the problems of producing epitaxial material of high quality for transistors. More and more of the high-frequency low-noise amplifier transistors are now being made by the diffused-base, planar, epitaxial technique.

The TO-18 package is probably most widely used for high-frequency planar transistors.

Details of the design objectives of a typical development program for a very-high-frequency planar epitaxial transistor are summarized in Table 3. The basic characteristic of the unit chosen is that it have 20-db gain at 200 Mc.

7. POWER TRANSISTORS

Power transistors may be defined as those with collector currents of 1 amp or more. Heat dissipation for such devices requires that they have larger cases than the standard JEDEC 30 or similar packages; the TO-3 package of Fig. 1 is typical of the medium-power transistor.

A number of excellent references on power transistors are available, and we shall give only an elementary summary of the present status of theory, design, and application of power transistors as amplifiers.

The first power transistor to be described in the literature was probably that of R. N. Hall, who used an emitter comb interleaved with a collector-comb structure, on the same side of a germanium wafer. The base formed the alloy contact to the entire opposite face. The more modern structure is to have the emitter interleaved with the base, or to have emitter and base parallel bars in a long linear structure, depending on the power required.

The elementary theory of transistors assumes not too high a concentration of minority carriers in the base region, whereas for power applications conditions are such that both the minority and majority carrier densities may be many times that of the low-power case. This requires new design theory, which has not been completely well worked out as yet. Also, base currents may be so high that one-dimensional calculations are generally of little value.

Because of the large areas required for power dissipation, power transistors have historically been restricted to the audio range of frequencies, but with the advent of diffused-base techniques and more sophisticated thermal designs so that power can be combined with relatively small area, the power-frequency combinations available to designers have been moving up steadily. In a recent paper, Hilibrand points out the progress since 1958. Figure 18 shows the new development in this direction.

The chart shows that while we have not progressed much in strictly high power as such, 200 watts remaining about the highest available at low frequencies, at high frequencies much progress has been made, with recently developed transistors being capable of delivering the order of 10 watts at 100 Mc. It is probable that units for 1 watt at 500 Mc will be available soon.

Until recently all power transistors available commercially have been germanium-alloy units. However, diffused types, and especially silicon diffused types are becoming more available. The latter probably will make the germanium units obsolete because of the higher temperature capabilities of silicon.

For power transistors it appears that the n-p-i-n structure is the most satisfactory compromise for power work, where frequency response, breakdown voltage, linearity,

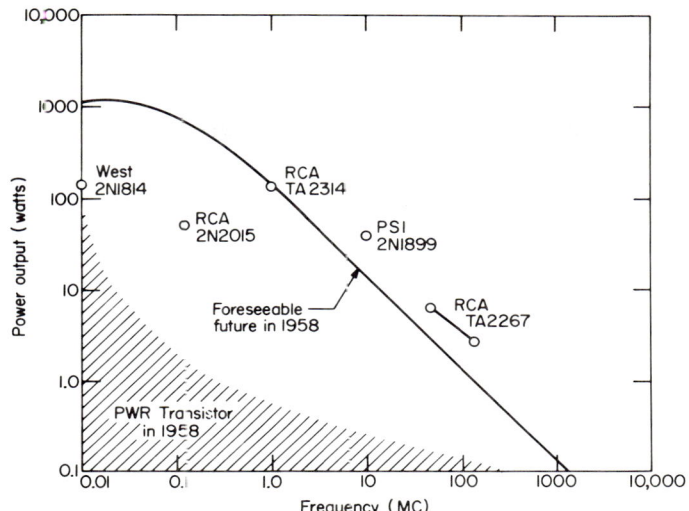

FIG. 18. Power vs. frequency chart. (*After Hilibrand; courtesy of NEREM Record and IEEE.*)

and reliability must all be considered. Triple-diffusion techniques seem to be gaining ground as the preferred technique. However, planar techniques as such do not seem to have taken hold as a preferred power-transistor fabrication method. A recently announced group of 5-amp n-p-n silicon planar power transistors for 30 watts at 100°C has the properties given in Table 4.

Table 4

n-p-n silicon	BV_{CBO}, volts	BV_{CEO}	BV_{EBO}	Gain
2N2877	80	60	8	20–60
2N2878	80	60	8	40–120
2N2879	100	80	8	20–60
2N2880	100	80	8	40–120

Among the most troublesome problems of power-transistor design today is the phenomenon of *second breakdown*, so called because it appears as a second break in the IV curve, beyond the normal breakdown voltage of the collector. Its cause is not clearly known, but its manifestation is an irreversible deterioration in the characteristics, accompanied usually by melting of the semiconductor and contact inside the transistor. Its appearance is highly sporadic and apparently not easily referrable to design characteristics of the device. Because the second breakdown is a slow thermal process, having buildup times on the order 10 msec, it is possible to use power transistors in pulse applications at which the device would be inoperable if used continuously.

8. FIELD-EFFECT TRANSISTORS

The field-effect, or unipolar, transistor has become a commercially interesting device only within the last several years. Although described in the literature as early as 1952, relatively little development was done on this area.

The field-effect transistor operates on a different principle from that of the bipolar type, although it uses silicon and germanium and embodies p-n junctions. The sketch of Fig. 19 illustrates its operation. The power circuit includes the body of a semiconductor wafer, through the two contacts called the source and the drain, respectively. Laterally there are at least one, but usually two, p-n junctions which collectively form the control grid or gate. The basic principle involves constriction of the current flow paths resulting from space-charge widening of the p-n junctions resulting from application of control voltage or signal to the gate. At a certain voltage called the pinch-off voltage, the depletion layers meet and current flow becomes close to zero. Proper design calls for the pinch-off voltage to be not too great.

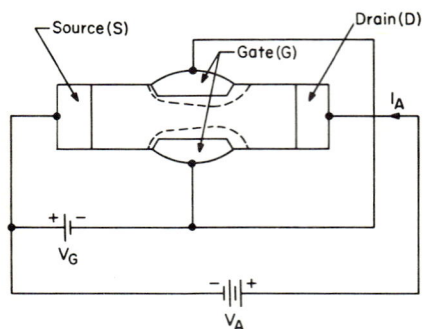

FIG. 19. Field-effect transistor.

Various geometries have been used for field-effect transistors. The early models used the planar construction shown in both grown-junction and alloyed types. The French development called the Tecnetron and various types also developed in the United States used a cylindrical body with a circular girdle forming the gate.

With the advent of diffusion and planar techniques, silicon field-effect transistors having characteristics superior to others for certain applications are on the market. These devices have very thin current-control paths, and low pinch-off voltages. They have extremely high input resistances, and in fact are more nearly akin to vacuum tubes in input resistance than other transistors, values of $\sim 10^{10}$ ohms being fairly easy

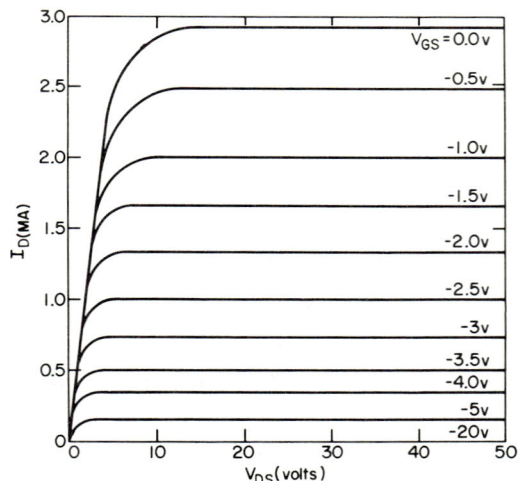

FIG. 20. Characteristics of field-effect transistor.

to realize. Some of the characteristics are shown in Fig. 20, and some of the typical operating parameters in Table 5.

Another type of field-effect transistor which is obtaining increased attention is the MOS (Metal-Oxide-Semiconductor) structure. This device is related to the field-effect transistor previously described, except that there is no deliberately formed junction region on, say, the silicon. Instead, the gate electrode is on an insulating layer, usually silicon dioxide, over a natural silicon surface. The application of bias to the gate produces either a conducting layer underneath the oxide, which corresponds to the conducting region in the usual field device, or an insulating region which leads to a decreased conductance between the source and drain. The first mode of operation is called *enhancement*, the second the *depletion* mode. The effect represents the device embodiment of the first form of transistor proposed by Shockley and his group just before the invention of the point-contact type. At that time, surface states prevented obtaining satisfactory characteristics, at least for germanium; but now, with modern methods of surface preparation, it appears that the high impedance, low power, satisfactory sensitivity, and suitability for mass production may

Table 5

Parameter at 25°C	Conditions	Limits	Typical silicon field-effect transistor
Transconductance g_m, μmhos......	$V_{GS} = 0$ $V_{DS} = 3$ volts	Min Typical Max	50 75 300
Drain current I_A, μa.............	$V_{GS} = 0$ $V_{DS} = 3$ volts	Min Typical Max	50 100 250
Gate leakage current I_{CGA}, μa.....	$V_{DSG} = -10$ volts	Typical Max	-0.003 -0.1
Gate leakage current I_{GA}, μa...... Noise figure, db..................	$V_{GD} = -3$ volts $f = 1$ kc $R_s = 3$ megohms $V_{GS} = 0$ $V_{DS} = 3$ volts	Typical Typical Max	-0.001 3.5 5.0
Amplification factor	$V_{GS} = 0$ $V_{DS} = 3$ volts	Min Typical	2 10

make the MOS a valuable supplement to the transistor designer's repertory. Instability due to ionic processes in the oxide layer appears to have been one of the factors limiting commercial application.

9. SPECIALIZED TRANSISTORS

In this category we include relatively little used transistors which may be minor modifications of standard types, as in the case of the avalanche-mode transistors, or simply obsolete types such as the tetrode and the grown-junction types whose capabilities relative to cost do not make them competitive in today's market.

9.1. Tetrodes. The tetrode represented an early attempt in junction technology to reduce the base resistance of the transistor by attaching a second lead to the base. This made possible the attainment of higher frequencies, and greater efficiency in the use of the transistor, since lateral base currents were not so likely to bias off part of the emitter and thus render it unused. Tetrodes have been made using grown-junction transistors, as well as the little used "rate-grown" and "melt-back" varieties. Transistors oscillating in the hundreds of megacycles during the early days of the transistor were made by these techniques.

9.2. Avalanche Transistors. The trend in transistor technology has been for high voltages to move toward lower ones. In computer switching, for example, 50- to 100-volt operation of transistors has given way to operation at 1 to 6 volts. In addition most transistors are not used at voltages which are a considerable fraction of the breakdown value, since they tend to become noisy and unstable. An exception is the avalanche-mode transistor, which is deliberately operated at voltages close to breakdown to obtain faster speed, even if noise is greater. While such transistors are mainly used as switches, with which we are not particularly concerned here, they can also be used as pulse amplifiers.

Rise times of 1 to 2 nsec are easily attainable with avalanche-mode transistors, but since such values have also recently been obtained by mesa and planar techniques, the avalanche transistor is not being used much commercially.

9.3. Grown-junction Transistors. The first junction transistors produced in the laboratory were of the grown-junction type. Having a structure essentially that shown in Fig. 3, the basic element consisted of a small bar of germanium about $\frac{1}{4}$ in. long and 10 to 20 mil square in cross section. In the middle was the base region, containing the region of opposite conductivity type. To make the transistor, all that needed to be done was to make three contacts, one to each end piece, the third to the base region.

Bars for grown-junction transistors are cut out of Czochralski-grown single-crystal material. During the growth process, the crystal, which originally contained, say, some p-type impurity, was counterdoped with arsenic or antimony so that it began to grow n-type. After growth had continued for a mil or two, the growth was again reversed by addition of more p-type impurity such as gallium or indium, and the crystal was finished this way. Since the emitter needed to be of low resistivity, the collector was grown first, the emitter last. The high concentrations in the emitter side made it impossible to grow more than one junction per crystal. This difficulty was partially obviated with the "rate-growing" technique of making grown junctions, since change of growth rate by a temperature change or a change in pulling rate also gave junctions if the original crystal had both n- and p-type materials in the proper amounts.

Because of difficulties of control, the grown junctions had to have base widths of 1 to 2 mils or more, and the cost of growing the crystal made them lose ground competitively when alloying and diffusion were shown to be effective. Silicon grown-junction transistors, however, held a good share of the market for some years, and many are still sold.

No particular discussion of the details of the operating characteristics of these devices will be given here, and the interested reader is advised to consult the literature.

10. FUTURE TRANSISTORS

10.1. Ultimate Capabilities of Transistors. With continued development of transistors to have higher frequency response, lower noise, higher power, and greater reliability, the question comes up as to the ultimate capabilities of transistors as amplifying devices. We have already mentioned that structures for transistors amplifying to 200 to 1,000 Mc are already so small that they are approaching the limits of what can be done by present masking and etching techniques.

An analysis of these questions has been made by Goldey. He considers that the performance of transistors is defined by the following criteria, and makes the following judgments as to the possibilities of future development.

1. Frequency response. The frequency cutoff f_T calculable on basic design theory is on the order 10 gc. Present transistors are not within this limit by a factor of 5, although germanium units with f_T of 4 gc have been built in the laboratory. Thus although there remains some improvement possible, germanium units in particular appear to be approaching the maximum attainable. Silicon units because of greater difficulties with technology have not reached the values of germanium, and hence more room for improvement is available.

2. Power capability. There appears to be considerable room for improving transistors above the 200- to 300-watt range at various frequencies. Thermal design as well as electrical and mechanical design will be important for this area.

3. Low power. Computers having more and more switching elements may require

lower and lower power operation. Laboratory transistors have been made with gains on the order of 100 at current levels of 1 μa, with f_T's on the order of 1 gc at 0.3 ma and 500 Mc at 100 μa. There appears to be no reason why comparable performance may not be attainable at even lower current levels. It seems safe to conclude that low-power operation, particularly for switching, presents many opportunities for future development of transistors.

4. Noise. As indicated previously, noise in transistors consists primarily of shot noise, thermal noise, and surface or $1/f$ noise. The first two are well known and based upon fundamental limitations such as the discrete nature of electric charge. The $1/f$ noise, while it has been reduced, still remains as a poorly understood aspect of transistors. Since present transistors have been improved to the extent that $1/f$ noise is no longer a problem above 100 cycles, at least in some types, these do not present much of a problem. Many high-frequency transistors, however, exhibit noise well above these levels, and for them further improvements are possible and needed.

5. Reliability. Reliability is a much more difficult topic to assess than the others. Ultimate capabilities based upon such considerations of failure as diffusion of impurities or field drift of impurities which might eventually destroy the junction yield such long lifetimes for transistors as to be unrealistic. Surface problems and problems of packaging so as to prevent leakage and contamination by water and other materials are probably the most critical ones for the reliability of devices. Improvement in reliability is likely to be considerable in future years as improved processing techniques are developed and as encapsulation and passivation techniques are improved.

10.2. New Materials. At the present time only silicon and germanium are significant transistor materials. The only competitors on the horizon are gallium arsenide, and to a lesser extent indium antimonide. Gallium arsenide has high temperature capabilities, but transistors made from it have not been shown to be capable of performing as well as silicon or germanium. The conductivity of gallium arsenide is not well understood, and appears to be very difficult to control, possibly because atomic defects in the crystal as well as foreign impurities may contribute to the conductivity. Indium antimonide transistors can be used only at very low temperatures (78°K). It appears to be a safe prediction that silicon and germanium will remain the only significant transistor materials for the foreseeable future. This does not mean that new amplifier devices based on other principles than known transistors may not come along which will supplant present devices.

10.3. Multiple Transistors and Integrated Circuits. Within the last several years the packaging of a number of components in a single transistor case and based upon elements constructed within a single wafer of silicon has become commonplace, and such devices, called *integrated circuits*, *micrologic* circuits, electronic functional blocks, etc., are now commercially available from a number of suppliers. In their most common form they are digital circuits, such as flip-flops and various types of gates. Since analog circuits such as linear amplifiers have been found to be very much more difficult to fabricate on the monolithic-block approach, compromises have had to be made. Such circuits are generally called "hybrid" circuits, and are made by the juxtaposition of a number of *separate* wafers within a small package, each containing one, but often several, of the elements of the linear amplifier circuit.

This area probably will be the most significant area of transistor development within the next few years. Although the devices are small, the pressures tending to promote the development most strongly are actually the need for lower cost and greater reliability for the increasingly complex electronic systems which are now in design and production. In addition, a related requirement for easier trouble shooting and simpler maintenance of complex systems is also forcing development in this direction.

Integrated circuits open up a new area of engineering and development problems. They will require a much more integrated study of new circuitry that is compatible with the new devices. Components on a silicon substrate, for example, cannot be manufactured with the close tolerances of separate components, and circuits must be designed to allow for wider variations of resistor values, transistor leakage currents, and gains. Capacitor values that can be created with the integrated circuit techniques may be limited in range and difficult to control.

Concomitant with the development of semiconductor integrated circuits is the development of amplifier circuits using thin-film components with transistor chips soldered or bonded in separately. To date, in spite of much study, it has not been possible to develop a thin-film amplifier device which is competitive with the transistor made on a small silicon chip. To summarize what has been done on thin-film transistors, we append the following concluding section.

10.4. Thin-film Transistors. A number of thin-film devices have been made which show amplifier properties, although it is dubious whether they should be called transistors. None of the presently considered thin-film devices, for example, operates on the bipolar diffusion mechanism, because of the poor crystalline structure of films and the apparent impossibility of achieving a lifetime long enough to allow penetration by diffusion through a layer thick enough to support any blocking voltage. On the other hand, *field-effect* thin-film transistors have been made having good amplifying properties. In particular, cadmium sulfide thin-film transistors have been made with gains of 20 db and frequency cutoffs in the region of 10 Mc.

The thin-film devices which have been proposed include "hot-electron" devices. These provide for the penetration of the thin film, by the "tunneling" through the thin films, of electrons which have energy in the lattice of several volts in excess of their thermal energy. Such electrons, it has been shown both experimentally and theoretically, can penetrate gold films as thick as 600 angstroms without undue losses due to collision with the lattice. Although the effect has been demonstrated, practical devices, such as the "metal-base" transistor, have yet to be developed. Further progress in this field, however, is likely to be rapid.

BIBLIOGRAPHY

Biondi, F. J. (ed.): "Transistor Technology," vols II and III, D. Van Nostrand Company, Inc., Princeton, N.J., 1958.

Bridgers, H. E., J. H. Scaff, and J. N. Shive (eds.): "Transistor Technology," vol. I, D. Van Nostrand Company, Inc., Princeton, N.J., 1958.

Dunlap, W. C.: "An Introduction to Semiconductors," John Wiley & Sons, Inc., New York, 1957.

Evans, J.: "Fundamental Principles of Transistors," D. Van Nostrand Company, Inc., Princeton, N.J., 1962.

Goldey, J. M.: Are Transistors Approaching Their Maximum Capabilities? *NEREM Record*, Boston, November, 1962, p. 125.

Hilibrand, J.: Transistors for High Power at High Frequency, *NEREM Record*, Boston, November, 1962, p. 128.

Hunter, L. P. (ed.): "Handbook of Semiconductor Electronics," 2d ed., McGraw-Hill Book Company, New York, 1962.

Lo, A. W., *et al.*: "Transistor Electronics," Prentice-Hall, Inc., Englewood Cliffs, N.J., 1955.

Middlebrook, R. D.: "Theory of Junction Transistors," John Wiley & Sons, Inc., New York, 1957.

Nussbaum, Allen: "Semiconductor Device Physics," Prentice-Hall, Inc., Englewood Cliffs, N.J., 1962.

Shea, R. F. (ed.): "Principles of Transistor Circuits," John Wiley & Sons, Inc., New York, 1953.

Shea, R. F. (ed.): "Transistor Circuit Engineering," John Wiley & Sons, Inc., New York, 1957.

Shive, J. N.: "Semiconductor Devices," D. Van Nostrand Company, Inc., Princeton, N.J., 1959.

Shockley, W.: "Electrons and Holes in Semiconductors," D. Van Nostrand Company, Inc., Princeton, N.J., 1950.

Texas Instruments, Inc.: "Transistor Circuit Design," McGraw-Hill Book Company, New York, 1963.

Valdes, L. B.: "The Physical Theory of Transistors," McGraw-Hill Book Company, New York, 1961.

Warschauer, D. M.: "Semiconductors and Transistors," John Wiley & Sons, Inc., New York, 1957.

Chapter 11

SOLID-STATE SWITCHING DEVICES

JOHN D. MENG*

CONTENTS

1. Four-layer Diodes.. 11–1
   1.1. Introduction... 11–1
   1.2. Construction of the Four-layer Diode........................... 11–3
   1.3. Device Ratings and Characteristics............................. 11–3
   1.4. Triggering Methods... 11–7
   1.5. Turn-off Techniques.. 11–7
   1.6. Typical Circuits... 11–9
2. Silicon Controlled Rectifiers....................................... 11–12
   2.1. Introduction... 11–12
   2.2. Construction of the SCR.. 11–12
   2.3. Device Characteristics... 11–13
   2.4. Triggering Requirements.. 11–14
   2.5. SCR Turn-off Techniques.. 11–17
   2.6. Typical SCR Circuits... 11–17
3. Silicon Controlled Switches... 11–19
   3.1. Introduction... 11–19
   3.2. Construction of the SCS.. 11–19
   3.3. Ratings and Characteristics of the SCS......................... 11–19
   3.4. Triggering Requirements.. 11–20
   3.5. Turn-off Techniques.. 11–20
   3.6. Typical Circuits... 11–20

1. FOUR-LAYER DIODES (TWO-LEAD FOUR-LAYER DEVICES)

1.1. Introduction. Four-layer diodes, characterized by four layers of semiconductor material, three junctions, and two leads or terminals, are built to have two stable states in the forward-biased direction (anode positive). One is a very high resistance state, typically on the order of 10 to 1,000 megohms, and the other is a very low resistance state, typically less than 20 ohms. The high-forward-resistance state is typified, within the rated voltage capability of the device, by leakage currents ranging from small fractions of a microampere for low-current devices at cold temperatures

* Electronics Engineering Dept., University of California, Lawrence Radiation Laboratory, Berkeley, Calif.

to a few milliamperes for very high current devices at elevated temperatures. These currents are substantially constant for a wide range of anode-to-cathode voltage.

The low-forward-resistance state is characterized by a low forward voltage drop, on the order of a volt for silicon devices. This forward voltage drop is very nearly constant for a wide range of forward anode current. When reverse-biased, these devices act like typical rectifiers, having a low value of reverse leakage current flow over a

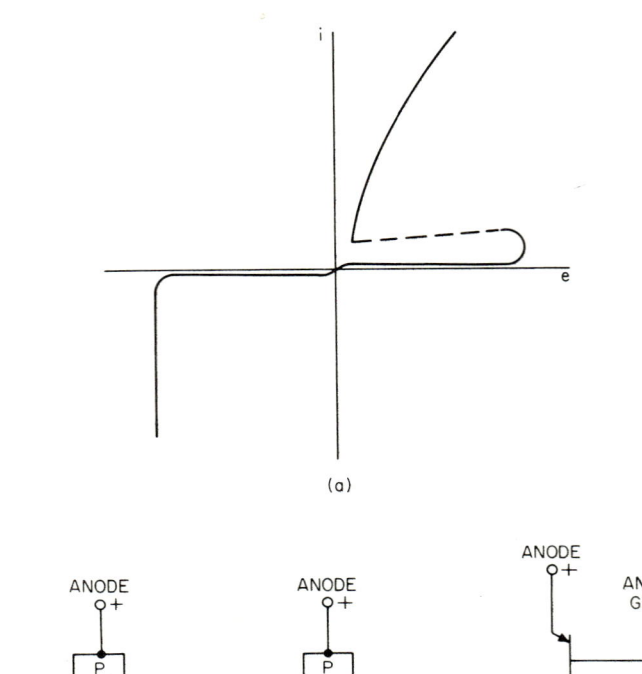

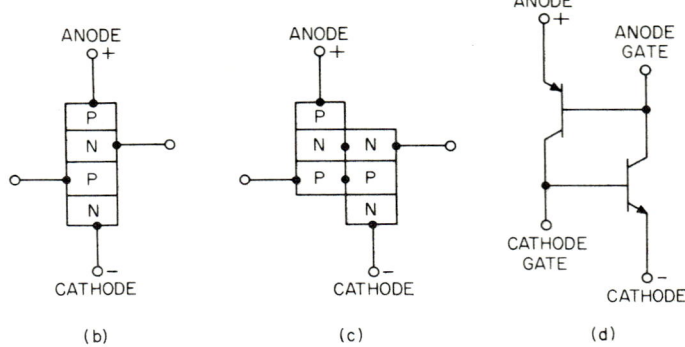

Fig. 1. The e-i curve and the development of the two-transistor analogy for four-layer devices.

wide voltage range. A typical V-I curve for a four-layer device is shown as Fig. 1a. Figure 1b through 1d demonstrates the derivation of the two-transistor analogy of this device.

Leads may be connected to each of the four layers of the device, as shown, each additional lead adding its share of versatility and flexibility to the unit, but each also adding some complications. The simplest device, that having only two leads (an anode and a cathode), will be considered first. Everything which is true of this two-lead device will also, with a very few exceptions, be true of three- and four-lead devices.

From the two-transistor analogy for the four-layer diode (Fig. 2) it may be easily seen how the two stable states exist. Referring to Fig. 2,

and
so
$$i_1 = i_2\beta_2$$
$$i_2 = i_1\beta_1$$
$$i_1 = i_1\beta_1\beta_2 \tag{1}$$

If we now say that the two i_1's of Eq. (1) are not the same, but that one occurs at time t and the other at time $t + \Delta t$, and if we call these values of $i_1 I$ and I' respectively, then Eq. (1) may be rewritten as

$$I' = I\beta_1\beta_2 \tag{2}$$

It is evident that $\beta_1\beta_2$ is the loop gain of the system. If $\beta_1\beta_2$ is greater than 1, the system becomes regenerative. In other words, I', the value of i_1 at time $t + \Delta t$, will be greater than I, the value of i_1 at time t. I, and hence i_1, will increase indefinitely until saturation of both transistors occurs.

If $\beta_1\beta_2$ is less than 1, the system becomes degenerative, and the currents through the transistors decrease to zero. The device is constructed such that $\beta_1\beta_2$ is less than 1 with little or no anode current flowing. The β's are made very current-dependent and increase with increasing anode current such that at some relatively small current level, called the holding current of the device, the $\beta_1\beta_2$ product becomes greater than 1 and the unit is said to have turned on.

1.2. Construction of the Four-layer Diode. Four-layer devices are generally silicon devices, employing either an all-diffused or an alloy-diffusion manufacturing process. The current tendency is toward all-diffused devices because of the uniformity of characteristics realized by better process controls and the potentially inexpensive manufacturing process involved. Diffusion techniques allow hundreds or possibly thousands of devices to be constructed simultaneously on a single silicon crystal

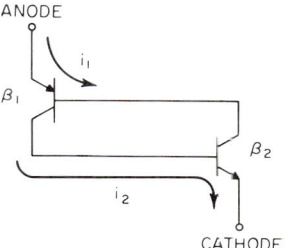

Fig. 2. Currents in the four-layer diode structure.

wafer. Alloy processes generally require extensive individual handling and processing of units. After the silicon is diffused, the unit is completely formed electrically, and the only remaining task is the mechanical mounting of the pellet and the attachment of leads. This is generally accomplished by means of soldering or some other thermal bonding technique. In high-current high-reliability devices it is generally found necessary to match the pellet thermally to its mounting to prevent long-term destruction from thermal fatigue. For low-current devices (under 1 amp), or in cases where initial economy is particularly important, less attention may be paid to thermal cycling problems.

Low-current devices are available in lead-mounted packages ranging from subminiature diode packages to signal transistor cases. High-current devices are generally packaged in cases suitable for attachment to heat sinks, via either stud, screwing or press-fitting means.

1.3. Device Ratings and Characteristics. The addition of a fifth semiconductor layer to a four-layer diode, giving a p-n-p-n-p structure, makes possible a symmetrical (a-c) switch, having e-i characteristics as shown in Fig. 3.[1]

As can be seen, the device reverse characteristics are similar to its forward characteristics, both of which are typical of a four-layer diode's forward characteristics. Since the operation of this device is identical to that of two four-layer diodes connected back to back, there is no need to discuss it as a separate entity.

A distinction must be made between device ratings and device characteristics.

[1] R. W. Aldrich and N. Holonyak, Jr., Two-terminal Asymmetrical and Symmetrical Silicon Negative Resistance Switches, *J. Appl. Phys.*, vol. 30, no. 11, pp. 1819–1824, November, 1959.

11-4 AMPLIFYING DEVICES

Ratings are assigned by the manufacturer, and operation of the device beyond the published ratings may result in permanent damage to it.

Characteristics are measurable properties of a device and are inherent in its design and construction.

Ratings and characteristics may be divided into four general categories. They are either currents, voltages, times, or thermal properties. The first two mentioned may be explained thoroughly by examining a typical characteristic curve for the device, such as that in Fig. 4. Times associated with the devices will require consideration of an additional dimension or axis, and thermal properties involve the mechanical nature of the device. Additionally, currents may be classified as either leakage currents, transient currents, or operating currents. These will all be considered in detail.

Leakage Currents. i_S, *Forward Saturation Characteristic (Forward Leakage Current)*. This is the instantaneous forward anode current at a given positive anode voltage and at a given junction temperature with the device in the off state. Forward leakage current increases with increasing junction temperature.

i_R, *Reverse-leakage-current Characteristic*. This is the instantaneous anode current with a given negative anode potential applied and at a given junction temperature. The magnitude of the reverse leakage current will increase with increasing temperature.

In addition to instantaneous values for leakage currents, some manufacturers specify average values for a full cycle of applied alternating current.

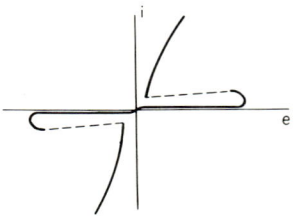

Fig. 3. The e-i characteristic curve for a p-n-p-n-p device.

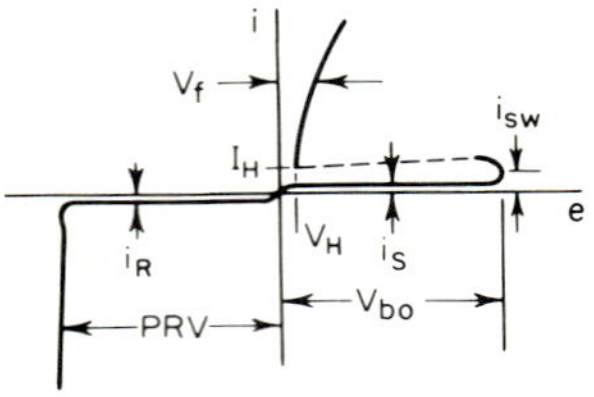

Fig. 4. Currents and voltages as they appear on the four-layer diode specification sheet.

Operating Currents. i_H, *Holding-current Characteristic*. When the forward current through a four-layer diode in the on state falls below a certain value, the voltage across the device will increase abruptly, and the device will turn off. This current value is known as the holding current. For the device to remain in the on state, its anode current must be maintained above its holding current. Holding current decreases with increasing temperature.

Forward Surge Current Rating. This rating specifies the maximum amount of forward current the unit will handle nondestructively for a stated short time interval. This time interval is generally not longer than a few milliseconds.

Reverse Current Rating. This rating specifies the maximum amount of current which may be handled nondestructively by the device in the reverse direction (anode negative). In devices not specifically designed to handle significant amounts of reverse current nondestructively, such a rating is generally not specified by the manufacturer, but it is instead necessary to guarantee that significant amounts of reverse current (amounts in excess of normal leakage currents) will never be allowed to flow.

I_f, *Forward Current Rating.* This rating is the maximum nondestructive current the device will handle in the forward direction while in the on state. It is a steady-state rating. The forward current-handling capability of a four-layer diode is determined by the temperature rise of the internal parts of the device. Since its forward drop is nonlinear, heat generated within it is not a simple function of current passing through it. Consequently, steady-state current ratings will in general be dependent on the current waveform.

Because the forward voltage drop of the four-layer diode may be approximated by a constant voltage E over a reasonably large range of currents, the heating caused by junction voltage drops is approximately proportional to *average* current. In the leads and other simple conductors, heating is proportional to rms current. Consequently, forward current ratings are generally given a dual representation. A device will have both a maximum permissible *average* forward current rating (corrected for different current waveshapes) and a maximum permissible rms forward current rating. The former is determined by junction characteristics, and the latter is a function of the device leads and mechanical contacts.

Still another factor enters into these ratings. This is the capability of the device to transfer the heat it generates to its environment. This will be considered under the heading of thermal characteristics.

Voltage Breakdown Characteristics. V_{bo}. The forward breakover voltage is the positive anode potential above which the anode current will tend to increase very abruptly. If the V_{bo} is exceeded momentarily and the anode current is allowed to exceed holding current, the four-layer diode will be turned on. The forward current through the device at the V_{bo} point is termed the switching current i_{sw}. V_{bo} may either rise or fall in any particular device over a moderate temperature range. When the temperature increase forces the forward leakage current to increase beyond the switch-on current of the device, however, the V_{bo} will decrease very rapidly with a continued temperature increase.

PRV. The peak reverse voltage is the magnitude of negative anode voltage above which the magnitude of negative anode current rises very sharply. In devices not specifically rated to handle appreciable amounts of reverse current, this voltage should never be exceeded. Allowing it to be exceeded will in general result in the destruction of the device. The cause of this is the concentration of the reverse current in a very small area. Since there is no appreciable reduction in the voltage across the device with large increases in reverse current (as opposed to what happens when the holding current is exceeded in the forward direction), small amounts of reverse current may lead to large amounts of power dissipation in the diode. If this power is generated in a small volume of material in the vicinity of the junction, very high temperatures will be generated in a matter of microseconds, and the material in this spot may melt, destroying the device. PRV will increase with increases in temperature in devices designed to handle appreciable amounts of reverse power. Its temperature dependence in other devices is indeterminate.

Operating Voltage Characteristics. V_f, *Forward Voltage Drop.* This characteristic is the voltage developed across the device when it is in the on state. It may be specified either as a curve of forward voltage vs. anode current or as a dynamic resistance. The forward voltage drop of a four-layer diode will decrease with increases in junction temperature.

V_H, holding voltage, is the voltage above which the anode must be held in order for the diode to remain on. This characteristic is not very often specified, as it may easily be derived from the forward-voltage-drop curve and the holding-current value for the device. This voltage, in a manner similar to that of the forward voltage drop, will decrease with increasing temperature.

Thermal Characteristics. R_T, *Steady-state Thermal-resistance Characteristic.* This characteristic, with units of degrees centigrade per watt, is the temperature difference between any two physical positions for each watt of power being generated at one of the positions. In the case of semiconductors, this may be specified from the junction to the device stud, or possibly (in the case of lead-mounted devices) from the junction to ambient. With stud-mounted devices, the heat sink will be the largest determining factor in the thermal resistance from the stud to ambient.

r_t, *Transient Thermal Resistance.* For short pulses of power, the physical mass of the device and its heat sink will make the thermal resistance appear to be less than it actually is. It will take a finite time for the device to reach thermal equilibrium, in other words, and before it reaches equilibrium, the physical mass of the device will prevent the attainment of excessive temperatures. For this reason, the surge capability of a device is generally much greater than its steady-state capability. To specify the device surge capability in a general manner, an apparent thermal resistance vs. time curve may be generated. This value defines junction temperature rise above a reference value vs. time for a step function of power applied.

Maximum Storage-temperature Rating. This is the maximum temperature at which the device may be stored without any eventual deterioration of its characteristics. It is typically 125 to 200°C for silicon devices.

Maximum Operating Junction-temperature Rating. This is the maximum temperature a spot on the junction could withstand nondestructively for a short period of time (less than 50 μsec). This temperature may be as high as 600°C for some devices.

Times. *Delay-time Characteristic.* This is the time from the application of the trigger signal to the drop of the anode-to-cathode voltage to 90 per cent of its full value.

Fall-time Characteristic. This is the time from when the anode voltage reaches 90 per cent of its full value to when it reaches 10 per cent of its full value.

Turn-on time is the sum of delay time and fall time.

Turn-off time is the time which the device requires to return to its forward blocking characteristic following conduction.

The above times are, in many cases, a function of the remainder of the circuit. In the cases where the circuit does not limit them, however, all but the turn-off times are typically fractions of a microsecond for four-layer diodes. Turn-off time is generally many times longer than turn-on time and, in fact, is in some instances more than two orders of magnitude longer.

Miscellaneous. In this category is junction capacitance, a significant quantity in circuits developing or using fast rise times. Junction capacitance will decrease with increasing junction reverse bias.

The above ratings and characteristics will not all be found on any single four-layer diode specification sheet. This list should adequately cover the ratings of all available four-layer diode specification sheets, however.

To illustrate better the form four-layer diode specifications may take, a few examples are listed below from the "Catalog of Shockley 4-Layer Diodes."[2] Each type of diode comes in two series, one specified at temperature extremes and known as the "MIL-LINE Series," the other being a "Commercial" unit.

The available switching (breakover) voltages range from 20 to 200 volts, and the tolerance on the 20-volt units is ±4 volts whereas that on 200-volt units is ±20 volts. All the units are available with holding currents ranging from 1 to 45 ma.

For the type E, 10-amp pulse current unit, the remaining specifications are:

Switching Current (I_s): <125 μa
Holding Voltage (V_H): 0.5 to 1.2 volts
Current Carrying Capacity: 150 ma steady dc. Maximum peak current rating 10 amperes —dependent on duty factor, repetition rate, pulse duration, and ambient temperature
Turn-on Time: 0.1 μs (circuit will determine exact switching time)
Turn-off Time: 0.2 μs (circuit will determine exact switching time)
Leakage Current (I_{lk}): <15 μa (measured at 0.75 V_s)
Dynamic Resistance On (R_{on}): <2 ohms at I_h + 25 ma and <0.3 ohm at 5 amperes
Capacitance: <100 pf (exact value depends on V_s and applied voltage)
Ambient Temperature Operating Range: −40° to 65°C
Reverse Breakover (Avalanche) Voltage (V_{rb}): >60% of nominal switching voltage (V_s). Voltages applied in the reverse direction in excess of V_{rb} may cause damage to the device

RATINGS ABOVE APPLY AT 25°C

This unit is available in the subminiature diode glass package.

At the other end of the scale is the type G unit designed for 75-amp pulse currents. This unit comes in a stud-mount type of package and has the following characteristics:

Switching Current (I_s): <250 μa
Holding Voltage (V_H): 0.5 to 1.2 volts
Current Carrying Capacity: 5 amperes steady dc. Maximum peak current rating 75 amperes—dependent on duty factor, repetition rate, pulse duration and ambient temperature
Power Rating: 5 watts
Derating: Current carrying capacity and power rating derated from 100% at 25°C to 25% at 105°C.

[2] "Catalog of Shockley 4-layer Diodes," Shockley Transistor Unit of Clevite Transistor, Stanford Industrial Park, Palo Alto, Calif.

SOLID-STATE SWITCHING DEVICES 11-7

Turn-on Time: 0.1 μs (circuit will determine exact switching time)
Turn-off Time: 0.2 μs (circuit will determine exact switching time)
Leakage Current (I_{lk}): <35 μa (measured at $0.75 V_s$)
Dynamic Resistance On (R_{on}): <0.3 ohm at 5 amperes
Capacitance: <200 pf (exact value depends on V_s and applied voltage)
Ambient Temperature Operating Range: −60° to +105°C
Reverse Breakover (Avalanche) Voltage (V_{rb}): >60% of nominal switching voltage (V_s). (>50% of nominal switching voltage at 105°C.) Voltages applied in the reverse direction in excess of V_{rb} may cause damage to the device

1.4. Triggering Methods. A conventional four-layer diode will turn on only if its anode-to-cathode voltage is made to exceed its V_{bo}. This task is simplified somewhat by the fact that as the rate of rise of positive voltage from anode to cathode is made larger, the V_{bo} of the four-layer diode will decrease. The reason is that as the rate of rise of voltage is increased, the capacitive current through the reverse-biased junction increases. This current acts to increase the betas of the two equivalent transistors, increasing leakage and effectively lowering the V_{bo} of the unit. The basic techniques used for triggering four-layer diodes are illustrated in Fig. 5. Figure 5a demonstrates triggering from a d-c level. As the voltage E is increased until it exceeds the V_{bo} of the device, the unit will switch into its conducting state.

Pulse triggering is illustrated in Fig. 5b, c, d, e, and f. As is shown in Fig. 5b, a pulse transformer may be connected in series with the unit in such a way that the pulse increases the positive potential applied to the unit. The pulse must be of a magnitude such that the device's V_{bo} will be exceeded for a long enough time to turn the diode on

Figure 5c demonstrates how a pulse may be directly coupled across the unit from either a pulse transformer or some other source. The small inductor in series with the device decouples the pulse from the remainder of the circuit, preventing the remainder of the circuit from loading the pulse.

Figure 5d and e demonstrates how a positive or negative pulse may be applied across a diode in series with the unit. The pulse voltage adds to the supply voltage to increase the anode-to-cathode voltage to a value in excess of its V_{bo}.

In Fig. 5f, the pulse exceeds the V_{bo} of one unit. The voltage across this one unit will then drop, increasing the voltage across the other units in the series string. By properly selecting units, they will all switch on when the first one does. This is particularly useful in very high voltage applications.

In all the pulse firing circuits, particular care must be exercised when the load is inductive. The rate of current buildup through the inductor is a limitation on how short the turn-on pulse can be. The turn-on pulse must be present for a long enough period of time to allow the current through the diode to reach its holding current. Otherwise, the diode will turn back off after the trigger pulse is removed.

Four-layer diodes have been produced which will trigger from such things as radioactive sources or light sources. These devices rightfully fall into a different category from that of four-layer diodes and will be considered in Sec. 2 under the heading of Silicon Controlled Rectifiers (Pylistors).

1.5. Turn-off Techniques. To turn off the four-layer diode, the current flowing through it must be reduced below its holding current for a long enough time to allow the device to recover (turn-off time). The basic techniques in common use are illustrated in Fig. 6. When the supply voltage is a-c, as illustrated in Fig. 6a, the reversal of the supply voltage will turn the unit off. This is perhaps the simplest technique.

If the supply is full-wave-rectified and of a sufficiently low frequency, the time during which the supply voltage is zero will be long enough to allow the unit to commutate. The same is true of a half-wave supply. If the load is inductive in these cases, it will often be necessary to have a "free-wheeling" diode across it as is shown in Fig. 6b. The diode provides a path for the inductive current decay, allowing the four-layer diode to commutate when the supply voltage goes through zero.

When the supply voltage does not reverse or go through zero, the turn-off problem gets more difficult. A very direct turn-off method is shown in Fig. 6c. By increasing the value of the resistor R, the current flowing through the device can be reduced below its holding current. A logical extension of this circuit is the replacement of the resistor with a switch.

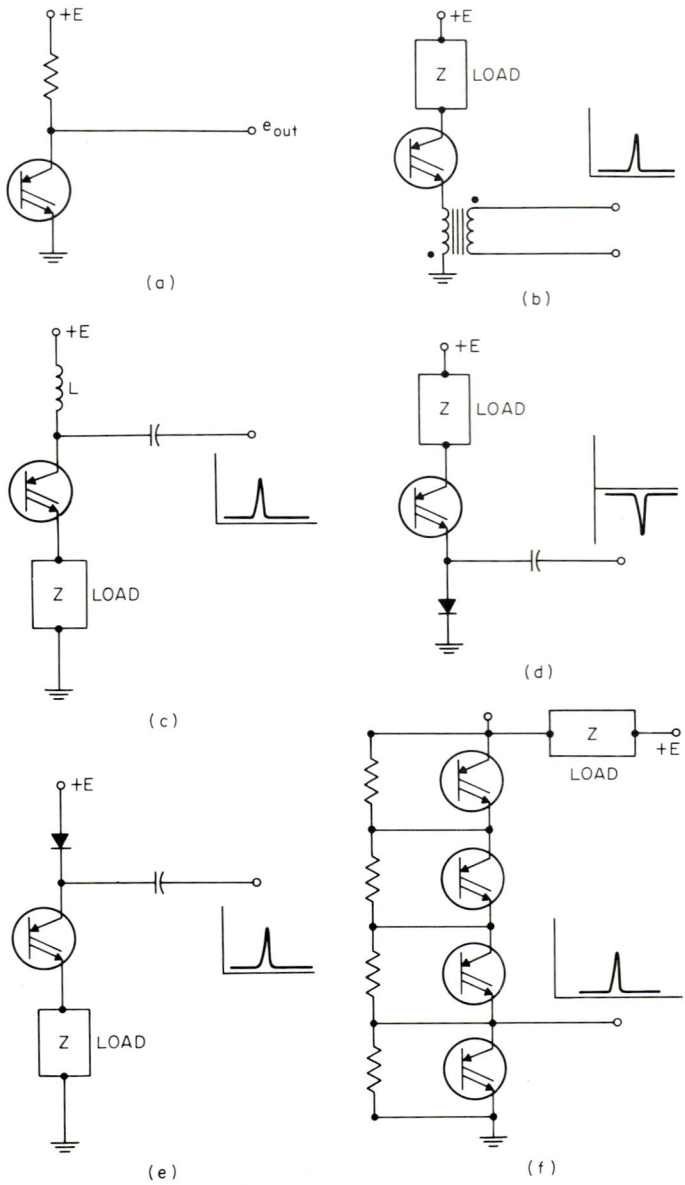

Fig. 5. Ways of triggering four-layer diodes.

SOLID-STATE SWITCHING DEVICES   11-9

The circuit of Fig. 6d uses a second four-layer diode to turn off the first one. If D_1 is on originally, capacitor C will charge to supply voltage as shown. Firing D_2 connects this capacitor directly across D_1, back-biasing D_1 until the capacitor is recharged in the reverse direction.

In the circuit of Fig. 6e, when the diode is switched on, capacitor C is charged resonantly through inductor L. The capacitor voltage will rise above the supply voltage, and the current in the circuit will try to reverse, switching off the diode. The power is supplied to the load in pulses in this circuit.

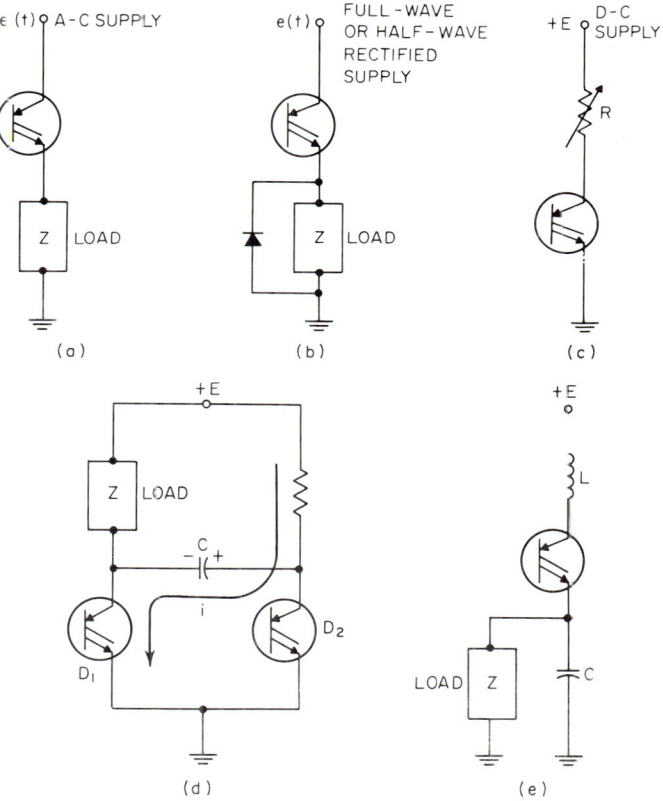

Fig. 6. Methods of turning off the four-layer diode.

1.6. Typical Circuits. A few circuit examples should serve to illustrate the use of the four-layer diode. It is, operationally, similar to a gas discharge tube. The forward voltage drop is much smaller for the diode than in the on state, however, and the speed of operation for the solid-state device is much faster than that of the gas device. A few basic circuit examples are illustrated in Fig. 7.[3] Figure 7a is an illustration of a sawtooth oscillator. E is above the V_{bo} of the four-layer diode, and the value of R_1 is made large enough to limit the current through the four-layer diode while it is in the on state to a value less than its holding current. With the diode initially off, capacitor C will charge toward the supply voltage E. R_2 is used as a current-limiting resistor when the diode has just switched on and therefore will generally be a great deal smaller in value than resistor R_1. The voltage drop across R_2 will be very small during the

[3] Circuits courtesy of Shockley Transistor Unit of Clevite Transistor, Palo Alto, Calif.

charging period and can be ignored in most applications. When the voltage across the four-layer diode, which is very nearly equal to the voltage across the capacitor, reaches the V_{bo} of the diode, the diode will switch to the on state, dumping the capacitor charge through the resistor R_2. The discharge period can be made very short and

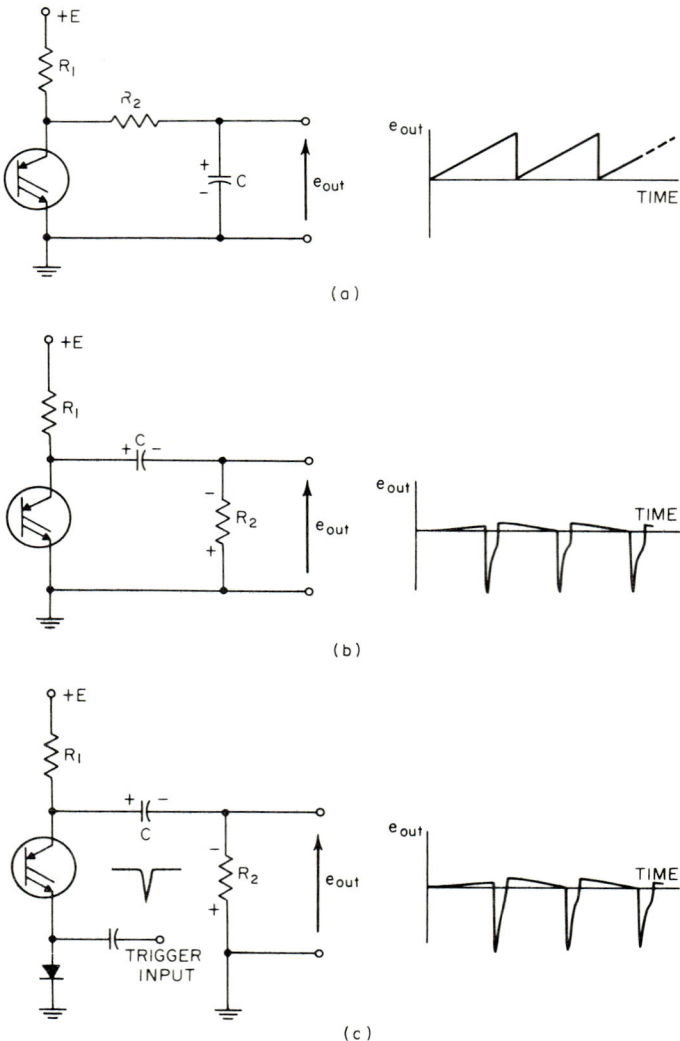

Fig. 7. Some basic applications of the four-layer diode.

will in fact be dependent on the value of R_2 for any given capacitor value. As the capacitor discharges, the current flowing from it, and therefore through the diode, decreases. When the sum of the capacitor discharge current and the current flowing through R_1 from the supply drops below the holding current of the diode, the diode switches off and the cycle repeats itself.

SOLID-STATE SWITCHING DEVICES   11-11

This circuit, then, is a simple example of an application using the d-c turn-on technique and the current-starvation turn-off technique.

By simply reversing the relative positions of R_2 and the capacitor, the sawtooth oscillator becomes a pulse generator as shown in Fig. 7b. The output in this case is generated from the capacitor discharge through R_2. The peak pulse amplitude will very nearly be equal to the V_{bo} of the diode. The circuit operation is identical in every respect to that of the circuit in Fig. 7a.

By reducing the supply voltage to a value less than the V_{bo} of the device being used, it may be used as a pulse amplifier. This is illustrated in Fig. 7c. The circuit is identical to that of Fig. 7a and b, but a diode has been added to make it practical to trigger the diode from an external pulse generator. This is an example of a circuit using pulse triggering and current starvation as a means of commutation.

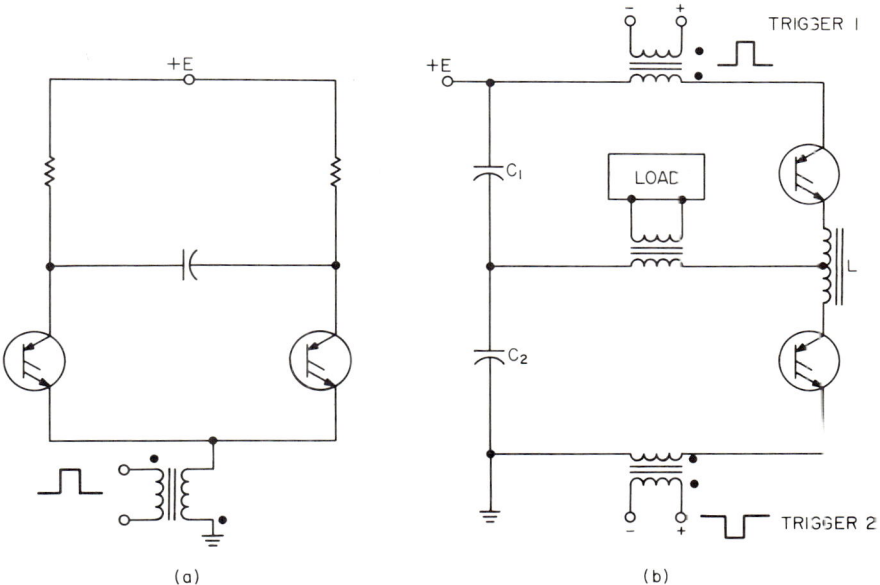

FIG. 8. Two basic types of inverters.

Figure 8a is a simple flip-flop using four-layer diodes. One of the diodes must be turned on initially for the circuit to operate, and the trigger pulse must not last beyond the turn-off interval supplied by the capacitor. This circuit is identical to the one discussed in Fig. 6d.

The sine-wave inverter of Fig. 8b is an example of the use of pulse triggering and resonant turn-off. When the circuit is initially at rest (both diodes off), about half of the supply voltage will appear across each capacitor. If a trigger pulse is then supplied to input 1, the top four-layer diode will fire, resonantly charging C_2 and discharging C_1 to a voltage in excess of supply voltage, and supplying a half sine wave of current to the load. When the current through the diode tries to reverse, the diode will commutate. A second trigger pulse is then supplied, this time to input 2. This results in the resonant discharge of C_2 and the charging of C_1. If the triggering frequency is the same as the resonant charging frequency, the output will be a sine wave. For the circuit to operate, the trigger pulses must be of a considerable amplitude. The reason for this is that initially the voltage across either four-layer diode is, at best, supply voltage. This may ring up to twice supply voltage after the circuit starts operating. Consequently, the V_{bo} of the device must be in excess of twice supply voltage, and the trigger pulse must be capable of triggering the diodes under both circumstances. This may require a trigger pulse amplitude in excess of the supply voltage amplitude.

2. SILICON CONTROLLED RECTIFIERS (FOUR-LAYER THREE-LEAD DEVICES)

2.1. Introduction. The silicon controlled rectifier (SCR), referred to as the "pylistor" in some international circles, is essentially a four-layer diode with an additional lead attached to it. Referring back to Fig. 1, the third lead may be attached either to the n region in the center of the sandwich or to the p region. As it turns out, it is easier to build a device using the sandwiched p region as the gate. This is known as a p-n-p-n structure as opposed to a possible n-p-n-p structure. The p-n-p-n structure is the one in common use today.

With the gate open-circuited, the SCR displays an e-i characteristic identical to that of the four-layer diode, shown in Fig. 1a. If we apply a current through the gate-cathode junction, the V_{bo} of the SCR decreases rapidly, as shown in Fig. 9.

For an explanation of this phenomenon, return to Fig. 1d. The SCR gate current is base current to the n-p-n transistor in the two-transistor analogy. This current will increase the beta of the n-p-n transistor sufficiently to make the loop gain greater than 1, turning the SCR on. Once the SCR is on, it behaves just as a four-layer diode

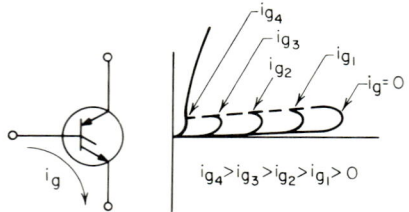

Fig. 9. Changes in SCR forward breakover voltage with increasing gate current.

does. The SCR has the advantage that it can be turned on from a low-voltage low-current pulse supply. The only voltage that is needed is what is required to break over the bottom p-n junction in the structure. Trigger current requirements may range from microamperes for low-current devices to over 100 ma for devices capable of handling in excess of 100 amp average.

A number of special SCR types exist. Notable among them are the device which is capable of being turned off with a negative gate current and the device which can be triggered from light. Gate-turn-off devices (GTO's) are still in their infancy. Devices are available which will control several amperes with turn-off gains over 20.

Light-activated switches (LAS's) have been around for a longer time than have GTO's, and they currently come in two forms. One type features a gate lead in addition to the ability to be triggered from a light source. This lead can be used to adjust the sensitivity of the device to light. The simpler type of LAS comes in a glass subminiature-diode package and has only two leads, an anode and a cathode. When light strikes the surface of its silicon pellet, it turns on from the current generated.

2.2. Construction of the SCR. SCR construction is very similar to four-layer diode construction in many respects, as might be expected. Both have four layers of semiconductor material and three junctions. At least two of the junctions are generally formed by diffusion. The gate junction is sometimes formed by means of the alloy process. In the SCR, this has been a very common procedure. The trend in the industry has rapidly moved toward an all-diffused structure for reasons mentioned earlier (in the discussion of the construction of four-layer diodes).

In stud-mounted SCR's, the *anode* is, by virtue of the mechanical difficulties involved in attaching the gate lead to the device, connected to the stud. Whereas reverse diodes are commonly available to ease heat-sinking problems for the mechanical designer, the reverse SCR is not. The added manufacturing expense makes it impractical currently to sell a reverse SCR at a price comparable with that of its mate.

Physically, the SCR looks just like the four-layer diode. The difference, of course, is the fact that the SCR has a gate lead or terminal protruding from its body. The gate connection is often physically close to the cathode lead or terminal. In very high current SCR's, not only is the gate lead brought out, but a second internal connection to the cathode is often brought out, making it physically simpler to connect the gate signal source to the SCR gate. Rather than making it necessary to connect a signal lead to the SCR high-current lead, this is done internally by the SCR manufacturer, in other words. The result is a simplified overall mechanical connection to the device.

2.3. Device Characteristics. SCR's are specified in a manner very similar to that in which four-layer diodes are specified. With the following additions, the same terms apply to both devices.

Voltage Rating. PFV, Peak Forward Voltage Rating. This rating is the maximum forward voltage with which the SCR may be turned on nondestructively by exceeding its V_{bo}. If the SCR is anode-fired (fired by applying a voltage in excess of its V_{bo} between its anode and cathode), the voltage at which the firing occurs must not be in excess of the PFV rating of the device. Otherwise, the device may be permanently damaged. This rating has no meaning for a four-layer diode because the only way to fire a four-layer diode is to exceed its V_{bo}. The V_{bo} of a four-layer diode is therefore necessarily below its PFV. SCR manufacturing technology has recently advanced to the extent that it is possible to manufacture devices with an infinite PFV rating and V_{bo}'s of at least 800 volts. It is to be expected that even higher V_{bo} SCR's will appear in the marketplace in the near future. This rating is unaffected by temperature variations.

Gate Characteristics and Ratings. V_{gt}, *Gate Trigger Voltage Characteristic, D-C.* This is defined as the SCR gate-cathode voltage required to produce the SCR gate trigger current. This voltage will decrease with increasing temperature.

I_{gt}, *Gate Trigger Current Characteristic, D-C.* This is the minimum direct gate current required to cause the SCR to switch from the off to the on state at a given anode-to-cathode voltage. At higher temperatures, this current decreases.

V_{grm}, *Peak Reverse Gate Voltage Rating.* This is the peak allowable reverse gate voltage (gate negative).

P_{gm}, *Peak Gate Power-dissipation Rating.* This is the maximum permitted peak gate power dissipation between the gate and the cathode.

P_g (av), *Average Gate Power-dissipation Rating.* This is the maximum gate-to-cathode power dissipation permitted, the power being averaged over a complete cycle.

Minimum Light to Trigger Characteristic. This is the minimum light level which is required for triggering a light-sensitive unit at the stated light color temperature, LAS junction temperature, and anode voltage. Light is to an LAS what gate current is to a regular SCR.

A_{go}, *Turn-off Gain Characteristic.* This characteristic is the ratio of SCR anode current to gate current required to turn the SCR off. This characteristic is rarely specified for anything but units specifically designed to display a substantial A_{go} (GTO's). It is generally specified over a specific range of anode currents.

In addition to the above additions, the definition of holding current must be expanded.

I_{ho}, *SCR Holding-current Characteristic, Gate Open.* This is the minimum anode current required to maintain the SCR in the on state, the gate being open-circuited. In the case of the LAS, this characteristic is measured in total darkness.

In the case of low-current devices, all characteristics are often specified with a fixed value of resistance connected between the gate and the cathode. This is used to stabilize the devices by draining off some of the normal leakage current as negative gate current. This is necessary because the leakage currents may be of the same order of magnitude as the current required to fire the SCR.

dv/dt Rating. This rating is the maximum rate of application of forward voltage for which the device will remain in the off state under maximum rated conditions of forward anode voltage and at the maximum rated junction temperature. The turn-on mechanism resulting from too rapid rate of application of forward voltage is the current through the reverse-biased (center) junction capacitance.

di/dt Rating. This rating specifies the maximum permissible rate of anode current rise at turn-on. This must be specified because the gate terminal is generally connected to the silicon in a small-area spot. When gate signal is applied, the entire silicon junction will not immediately begin conducting. The small section around the gate connection will begin conducting first, and then the conducting area will spread to the remainder of the junction over a finite time interval, on the order of 20 μsec for a typical 100-amp SCR. The problem may be remedied by connecting a saturating inductor in series with the SCR anode, and arranging the number of turns and core size so that very little current will flow for a period of about 20 μsec after the SCR is turned on. Then the core will switch the high current after the SCR has turned on completely, and no destructive aftereffects will be felt by the SCR as long as it is operated within its other ratings. Low-current devices (1 amp and under) are relatively insensitive to *di/dt* because of their small junction area and lower current capabilities.

The above characteristics and ratings specify the differences between SCR's and four-layer diodes. SCR's possess the very real advantage of extreme ease in triggering but will not in general withstand the *dv/dt* or *di/dt* that a four-layer diode will withstand. The tendency, of late, has been for SCR manufacturers to improve substantially the *dv/dt* and *di/dt* capabilities of their devices.

The abbreviated characteristics of two typical SCR's are listed below.

GE Type C5 Series
Maximum allowable ratings:
    PFV............................ 500 volts
    rms forward current............ 1.6 amp
    d-c forward current............ 1.6 amp (85°C case temperature)
    Peak one-cycle surge current... 18 amp
    Peak gate power................ 0.1 watt
    Avg gate power................. 0.01 watt
    Peak gate current.............. 0.1 amp
    Peak gate voltage.............. 6 volts (forward and reverse)
    Storage temperature............ −65 to +150°C
    Operating junction temperature. −65 to +125°C
Characteristics:
    Max holding current............ 2.0 ma ($T_j = 25$°C) (With 1,000 ohms or less connected from gate to cathode)
    Firing characteristics......... 1 volt at 350 μa will fire all units
    Forward voltage drop........... Under 1.5 volts at 1.0 amp of anode current
    Leakage currents............... Under 100 μa

GE C50 Series
Maximum allowable ratings:
    Transient peak reverse voltage. 35–600 volts
    Repetitive PRV................. 25–400 volts
    Repetitive PFV................. 500–600 volts
    rms forward current............ 110 amp
    Avg forward current............ Depends on conduction angle (70 amp at 180°)
    Peak one-cycle surge current... 1,000 amp
    Peak gate power................ 5 watts
    Peak gate current.............. 2 amp
Characteristics at maximum ratings:
    Max forward voltage drop (full cycle avg) 0.80 volt (70 amp average, 180° conduction angle)
    Max thermal resistance......... 0.4°C/watt (junction to stud)
    Typical holding current........ 20 ma

The above specifications are abbreviated but are complete enough to indicate the type of information which will generally be found on specification sheets.

2.4. Triggering Requirements. The SCR, as has been mentioned earlier, is very simple to trigger. The light-activated switch (LAS), although it has been placed in the same category as an SCR, must be considered as a separate entity where triggering is concerned. The normal SCR, one having a gate lead, will be considered first.

Figure 10 shows a typical curve of gate characteristics as it might appear on the manufacturer's specification sheet. This particular curve was taken from a 2N681 series specification sheet. The curve contains (shown as a shaded area) all the possible firing points of the SCR. Notice that there is a value of gate current and gate voltage below which the SCR is guaranteed not to fire. The curve covers the entire temperature range of the device and provides a rough indication of the behavior of the device gate characteristics with variations in temperature. When designing a firing circuit for the SCR represented in Fig. 10, one which will supply 80 ma at 3 volts will fire the device under all temperature conditions which it is rated to withstand.

For the LAS, a minimum light to fire vs. color temperature curve could be constructed which would serve the same purpose.

As pulse firing is often desirable, it is worthwhile to consider the effects of pulse shape and duration on the firing characteristics of the SCR. It will suffice to say that the

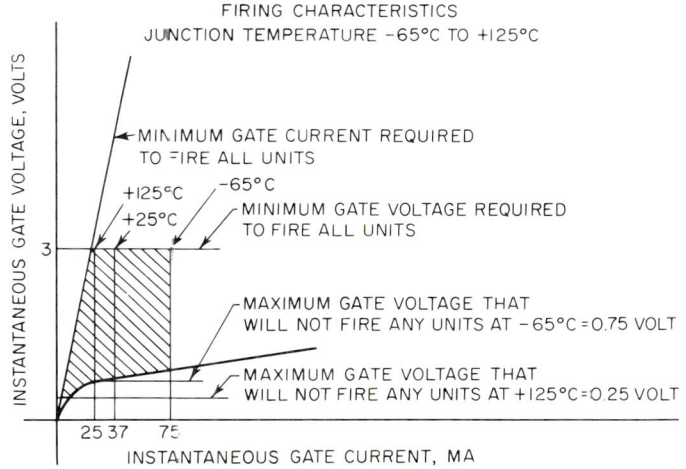

Fig. 10. A typical SCR gate firing specification; 2N681 series SCR.

characteristics as shown on the device specifications will not change substantially for any pulse duration in excess of 10 μsec. For narrower pulses than this, the gate-current requirements for triggering increase quite rapidly. To obtain the fastest possible turn-on of the device, it is desirable for the gate signal to have a rise time which is as fast as is practical to obtain, and to arrange to have the gate overdriven substantially for the first few microseconds. These techniques will tend to shorten the delay time.

A similar result is true of the LAS. A fast-rise-time high-intensity pulse of light will tend to shorten the LAS delay time.

The application of a negative potential between the SCR gate and cathode will often improve the SCR dv/dt capability significantly.

In the LAS, the physical area of the pellet which is exposed to the activating light is very small, and the rating of the unit is based on a light source giving off parallel rays of light. Consequently, a simple lens or concentrating reflector system can improve the triggering threshold of the device by several orders of magnitude. The LAS is most sensitive to light in the infrared region and demonstrates a response similar to that of a silicon photocell.

The number and types of trigger circuits for conventional SCR's are nearly innumerable. Figure 11 illustrates just a few. The circuit of Fig. 11a is a unijunction firing circuit. While a detailed description of the operation of the unijunction transistor is beyond the scope of this chapter, it will suffice here to say that the control voltage charges the capacitor C through the resistor R_2. When the voltage across C reaches the firing voltage of the unijunction, the capacitor is suddenly discharged through the primary of the pulse transformer. The resulting pulse at the secondary terminals fires the SCR.

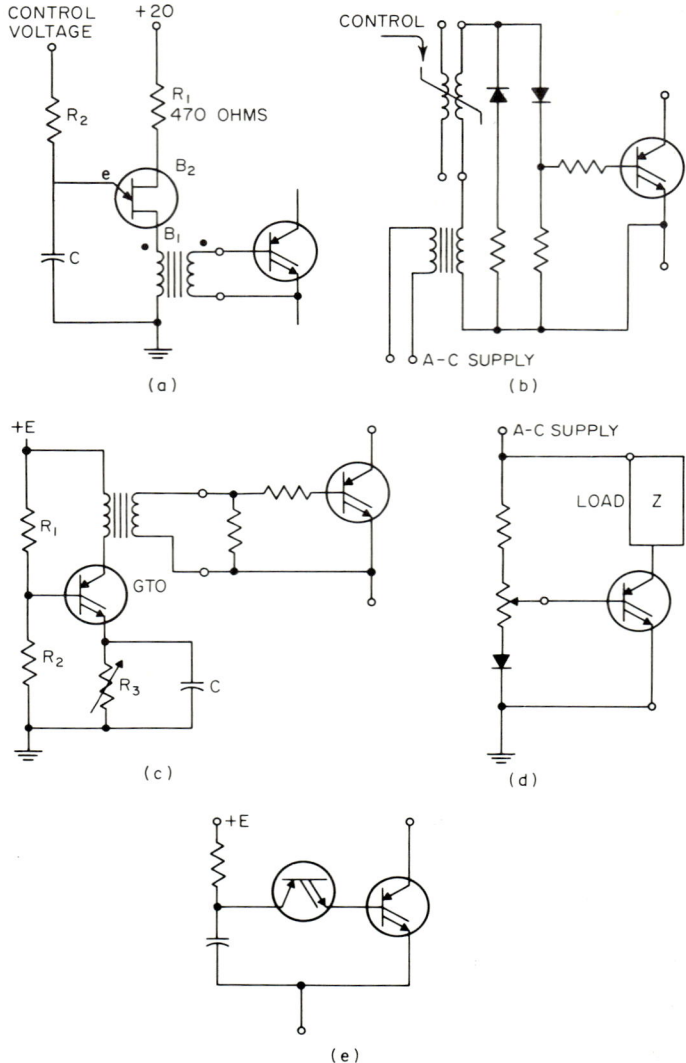

FIG. 11. SCR firing circuits.

Figure 11b illustrates a magnetic firing circuit.

Figure 11c illustrates how firing pulses may be generated with the help of a gate-turn-off (GTO) device. The R_1, R_2 resistor combination produces a voltage at the gate of the GTO which fires it. When the GTO fires, capacitor C charges to supply voltage, providing the output pulse. This results in the gate-cathode junction of the GTO being reverse-biased, turning the GTO off. This type of circuit is particularly adaptable to inverter applications, as the output pulse has a very short rise time and typically will produce several amperes at 10 to 20 volts output.

Figure 11d illustrates the use of the a-c supply signal directly to fire the SCR, and Fig. 11e illustrates the use of a four-layer diode as the trigger element. These circuits are not meant to be all-inclusive. As was mentioned earlier, nearly infinite variations of firing circuits exist, and this small selection barely scratches the surface.

The important facts to remember concerning firing circuits are:

1. Where fast SCR turn-on is desired, use a gating signal having a fast rise time and an initially high amplitude.

2. Where the SCR load is inductive, use a triggering signal which is long enough to allow holding current to build up through the inductive load.

3. A slight negative bias on the gate of the SCR will generally improve its dv/dt capability significantly.

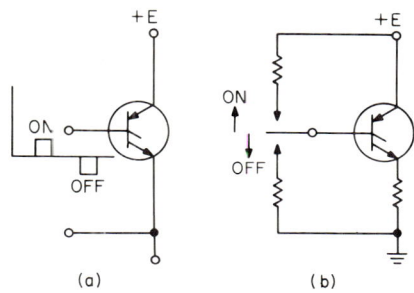

Fig. 12. Turning a GTO off and on.

2.5. SCR Turn-off Techniques. In the case of conventional SCR's (not GTC's), turning the device off is accomplished exactly as it was done with the four-layer diode. The current flowing through the SCR must be dropped below its holding current for a period at least equal to the SCR's turn-off time. In the case of the GTO, the device's holding current is increased by the application of negative gate current, allowing the device to be turned off by increasing its holding current to a value in excess of its anode current. In addition to using the same turn-off techniques applicable to the four-layer diode and the regular SCR, others are available to the GTO. The basic technique is illustrated in Fig. 12a where a positive pulse is used to trigger the device on and a negative pulse is used to turn it off. In the case of Fig. 12b, the load for the GTO is in the cathode lead, making it possible to turn the device off merely by grounding the gate lead.

It is of interest to note what seem to be inherent differences between conventional SCR's and GTO's having comparable steady-state current ratings. The GTO tends to have less gate sensitivity for turn-on, higher holding current, and higher forward drop than does the SCR. On the other side of the picture, the GTO generally exhibits better dv/dt capability and faster switching speeds. The GTO is generally rated for operation in d-c circuits, while the SCR is rated for use in a-c circuits where commutation is not a problem.

2.6. Typical SCR Circuits. The PFV rating on a SCR and the fact that SCR's do not generally have their V_{bo}'s specified to any close tolerance are the barriers preventing the clipping off of the gate lead or the attachment of the gate lead to the SCR cathode and the subsequent use of the SCR as a four-layer diode. Recent technological developments have allowed the manufacture of SCR's with infinite PFV ratings, and devices can be obtained from manufacturers with V_{bo}'s specified to within reasonable tolerances. These facts make it practical to use some types of SCR's as four-

11–18 AMPLIFYING DEVICES

layer diodes. Consequently, all the circuits which were presented as examples of the use of the four-layer diode are equally applicable to SCR's. The use of a SCR as a four-layer diode has the advantage of a much-improved di/dt rating and much faster turn-on than that possessed by the SCR used conventionally. Additionally, the SCR is available with very high breakdown voltages.

Figure 13a demonstrates the use of a SCR as a rectifier. When used this way the SCR has the advantage of a much faster recovery time than a rectifier having the same current rating.

Figure 13b illustrates the use of a SCR as a phase-control device. This is probably the most popular use of the conventional SCR. The a-c input voltage takes care of

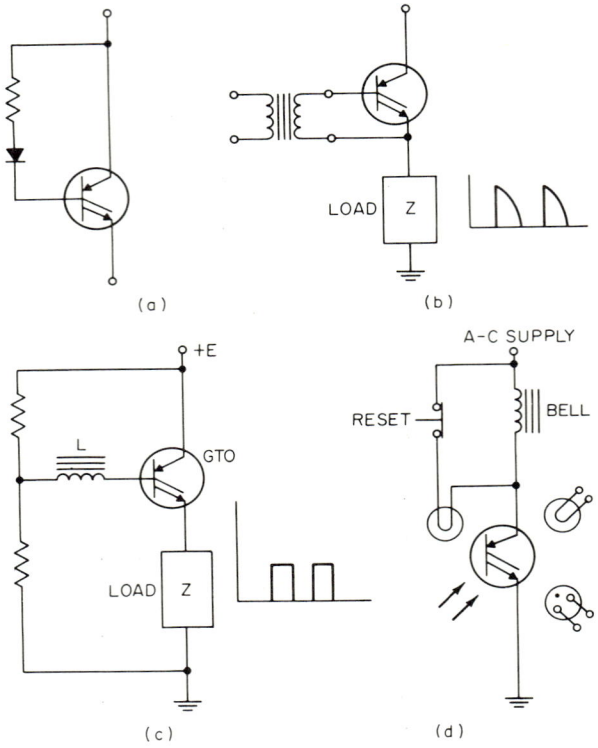

FIG. 13. Some typical SCS circuits.

commutation. The trigger input comes at the same point in each cycle of the a-c input voltage. By varying the phase of the trigger signal, the output power delivered to the load may be varied. The power gain in a single stage of this nature, when one considers that the trigger pulse need be no more than about 10 μsec wide and must supply only 3 volts at $\frac{1}{2}$ amp for the largest available SCR, may be in the millions. This circuit is in widespread use as the power-amplifier stage in such things as motor controls, heater supplies, and regulated power supplies.

Figure 13c illustrates the use of a GTO as a d-c–to–a-c converter (inverter). The inverter load is placed in the cathode lead of the device. The inductor connected in the gate lead produces the necessary delays for obtaining the proper pulse width at the output. When the power supply is initially connected, current builds up through the inductor and fires the GTO. Full supply voltage then appears across the load, back-biasing the gate-cathode junction. The current through the inductor decays to zero and begins to reverse. When the magnitude of the negative gate current (through the inductor) reaches a sufficient value, the GTO is turned off and the cycle repeats itself.

Figure 13d illustrates a use of a light-sensitive SCR. Any of the light inputs will fire the LAS, activating the bell and lighting the lamp connected across the bell. The lamp which is connected across the bell is placed physically close to the LAS to hold it in the on state until the reset button is pushed.

These examples are illustrative of an enormous number of uses for these devices. For detailed information on these and other uses, the reader is referred to the applications material published by the device manufacturers.

3. SILICON CONTROLLED SWITCHES (FOUR-LAYER FOUR-LEAD DEVICES)

3.1. Introduction. The silicon controlled switch (SCS) is a four-layer device with a lead attached to each of the four semiconductor layers. The units which have been produced to date are low-voltage low-current devices. They are useful primarily in signal applications as opposed to power and control applications, and display a staggering amount of versatility. According to one manufacturer, it can perform as an "extremely sensitive NPNP controlled switch (complement of SCR), extremely sensitive PNPN controlled switch (SCR configuration), NPN Silicon transistor, PNP Silicon transistor, N-type negative resistance (Trigistor, Transwitch), S-type negative resistance (Binistor), four layer diode (Shockley Diode), [and] zener diode."[4] High-power SCS's have not yet been manufactured, and the production of such devices would present a number of unique manufacturing problems.

3.2. Construction of the SCS. Silicon controlled switches currently being manufactured are the result of the alloying of a fourth lead to a silicon n-p-n diffused-base transistor to form the third n-p junction. The transistor diffused base region becomes the SCS cathode gate. The p region of the additional alloyed junction becomes the device anode. What would normally be the transistor emitter becomes the SCS cathode, and what would normally be the transistor collector becomes the SCS anode gate. For an elaboration on the subject of SCS construction, the reader is referred to the transistor manual listed in footnote 4.

3.3. Ratings and Characteristics of the SCS. Since the SCS is a SCR with an added lead, all the ratings and characteristics which were defined for the SCR apply to the SCS. The additional gate lead on the SCS requires some additional specification. The three characteristics which are specified for the anode gate are reverse breakdown voltage, dynamic impedance at several current levels (three for currently available devices), and turn-off gain.

The device characteristics are particularly sensitive to the gate connections, as may be seen from the following chart.

	Both open	G_c shorted	G_a shorted	Both shorted
V_{bo}, volts	0–20	70	50	80
PRV, volts	80	80	4	0.5
I_h, ma	0	0.5–1	2–3	2–3
Anode-gate ma to fire	0–0.008	0.2–1		
Anode-gate volts to fire	0.4–0.6	0.6–0.9		
Cathode-gate ma to fire	0–0.001	....	0.01–0.06	
Cathode-gate volts to fire	0.4–0.6	....	0.5–0.8	

[4] "General Electric Transistor Manual," 6th ed., p. 335.

3.4. Triggering Requirements. The anode gate of the SCS features extreme sensitivity. Otherwise, it is analogous to the gate on a SCR and may be treated as such. To fire the SCS by means of the anode gate requires a negative signal with respect to the anode. Although the anode gate does not feature quite the sensitivity that the cathode gate does, the operation of the two as far as firing circuitry is concerned is analogous.

3.5. Turn-off Techniques. In addition to having the capability of being turned off in a manner similar to that used to commutate SCR's or GTO's, the SCS may be turned off by applying a positive current to the anode gate equal in magnitude to the anode current.

3.6. Typical Circuits. Two circuits which illustrate the use of the anode gate of the SCS are contained in Fig. 14. Figure 14a illustrates how a SCS may be connected to operate as a unijunction transistor (UJT) would. The resistance divider

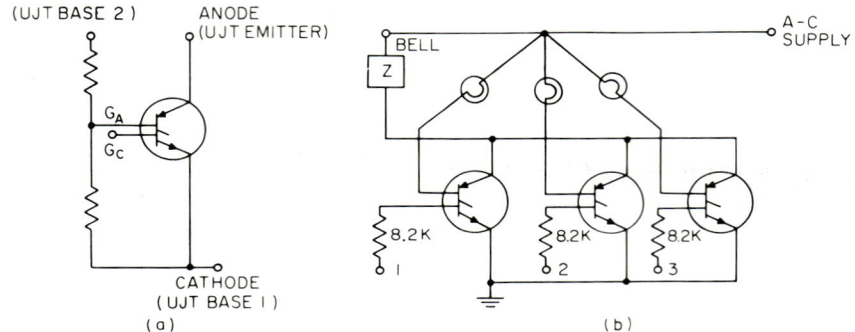

Fig. 14. Circuits using the anode gate of the SCS.

reverse-biases the anode–anode gate junction of the device until the voltage at the anode (acting as the UJT emitter) exceeds the voltage at the anode gate. At this point, the SCS fires. The anode voltage at which the SCS fires is determined by the voltage supply for the resistor divider.

Figure 14b illustrates another manner in which the anode gate connection may be used. The unit which is on will have anode gate current flowing, providing an individual indicator for each circuit.

BIBLIOGRAPHY

"Application Notes on the Shockley Four-layer Diode," Shockley Transistor Division of Clevite Transistor, 1801 Page Mill Road, Palo Alto, Calif.

"Application Notes on Silicon Controlled Rectifiers," General Electric Co., Application Engineering Center, Auburn, N.Y.

"Application Notes on the SCS and Unijunction Transistor," General Electric Co., Semiconductor Products Department, Electronics Park, Syracuse 1, N.Y.

"General Electric SCR Manual," 2d ed., General Electric Company, Rectifier Components Department, Auburn, N.Y., 1961.

"General Electric Transistor Manual," 6th ed., General Electric Co., Semiconductor Products Department, Electronics Park, Syracuse 1, N.Y., 1962.

Chapter 12

TUNNEL DIODES AND BACKWARD DIODES

CHANG S. KIM* and JEROME J. TIEMANN†‡

CONTENTS

1. Tunnel-diode Theory.. 12–1
   1.1. Introduction... 12–1
   1.2. Tunneling Phenomenon................................. 12–2
   1.3. Backward Diodes....................................... 12–5
2. Static Parameters of Tunnel and Backward Diodes............ 12–5
3. Tunnel-diode Equivalent Circuit............................. 12–6
4. Resistive Cutoff and Self-resonance Frequencies.............. 12–6
5. Various Tunnel Diodes....................................... 12–7
6. Packaging Considerations.................................... 12–8
7. Noise... 12–9
8. Nonlinear Characteristics of Tunnel Diodes.................. 12–12
   8.1. Introduction... 12–12
   8.2. The I-V Characteristic Equation................... 12–13
   8.3. Operation at the Inflection Point...................... 12–15
   8.4. Experimental Results—Operation at Inflection Point.... 12–18
   8.5. Operating at Other Than Inflection Point of I-V Characteristic... 12–20
9. Mixers.. 12–21

1. TUNNEL-DIODE THEORY

1.1. Introduction. A tunnel diode is a semiconductor device in which quantum-mechanical tunneling leads to a region of a negative slope in the forward direction of the current-voltage characteristics. A tunnel diode (tunneling phenomenon) was first reported by Esaki in 1958.[1]

Tunnel diodes are made from highly doped thin semiconductor junctions whose thicknesses are on the order of 100 angstroms using materials such as Ge, GaSb, and GaAs. Since the tunneling mechanism which dominates its characteristic is a majority-carrier phenomenon, minority-carrier current is of minor importance over a large part of the characteristic. Therefore, parameters such as diffusion length and

---

* General Electric Co., Electronics Laboratory, Syracuse, N.Y.
† General Electric Co., Research and Development Center, Schenectady, N.Y.
‡ Sections 1 to 7 were written by C. S. Kim and Sections 8 and 9 by J. J. Tiemann.

lifetime, which limit very high frequency operation in minority-carrier devices, have little effect on a tunnel diode. The frequency limitations in tunnel-diode circuits come rather from other associated terminal effects such as junction capacitance, the series resistance, and the series inductance of the diode. Tunnel diodes are capable of operating from direct current to the millimeter wave range.

Unlike other negative-resistance devices such as dynatrons and p-n-p-n transistors, the I-V characteristic of a tunnel diode does not depend on any device parameters that are subject to aging, and can be made relatively independent of temperature. Thus the device stability is excellent. Highly doped tunnel diodes are one or two orders of magnitude better than ordinary transistors in resistance to nuclear radiation.

Tunnel diodes are relatively low noise devices if operation is in the negative-conductance region. The noise associated with a tunnel diode in this region is shot noise of the d-c diode current and will be discussed later.

A backward diode is a semiconductor device in which quantum-mechanical tunneling leads to a current-voltage characteristic with a reverse current greater than the forward current for equal and opposite applied voltages in some voltage range centered about the origin. As will be seen later, a backward diode is essentially a tunnel diode. The distinction between tunnel diodes and backward diodes is based on circuit-application emphasis of the negative-resistance property (tunnel diode) or the low-level rectification property (backward diode).

Tunnel diodes can be operated as amplifiers, oscillators, mixers, converters, and switches in logic circuits. Tunnel diodes have simple and rugged structure and the size is very small.

1.2. Tunneling Phenomenon. Esaki[1] explained the theoretical model of a tunnel diode in the following way. There are two tunneling currents across the junction in opposite directions, namely, the forward and backward currents. As seen in Fig. 1 these currents are due to the fact that an electron occupying a state on one side of the junction can cross the barrier by means of the quantum-mechanical tunnel effect if there is an appropriate unoccupied state available on the other side. The forward current I_f consists of the current flowing from the conduction band to the valence band while the backward current I_b is the current flowing from the valence band to the conduction band.

The electron potential plot with zero bias vs. junction thickness is shown in Fig. 1. We define $\rho_c(E)$ as the density of available states at energy E in the conduction band. The number of electrons of allowed state in an energy range between E and $E + dE$ in the conduction band becomes $\rho_c(E)\, dE$. The Fermi function $f_c(E)$, which obeys the Fermi-Dirac statistical distribution, must be used to obtain the probability of finding electrons at energy state E in the conduction band. Then the number of electrons occupying states between E and $E + dE$ becomes $f_c(E)\rho_c\, dE$. In a similar manner, the Fermi function $f_v(E)$ and the density-of-states function $\rho_v(E)$ for the valence band are also defined. The probabilities of a particular energy state being empty in the valence and conduction bands are given, respectively, by $[1 - f_v(E)]$ and $[1 - f_c(E)]$.

The number of holes occupying states between E and $E + dE$ now becomes $[1 - f_v(E)]\rho_v(E)$. The number of electrons tunneling from a state E in the conduction band to the corresponding state in the valence band is proportional to the product of $f_c(E)\rho_c(E)$ and $[1 - f_v(E)]\rho_v(E)$. Similarly, the number of electrons at energy level E which tunnel from the valence band to the conduction band is proportional to the product of $f_v(E)\rho_v(E)$ and $[1 - f_c(E)]\rho_c(E)$.

The functions $P_{c \to v}$ and $P_{v \to c}$ are now defined as the tunneling probabilities of an electron moving, respectively, from an energy state E in the conduction band through the potential barrier to the corresponding state in the valence band and from an energy state E in the valence band to the corresponding state in the conduction band. The forward and backward tunnel currents I_f and I_b then become

$$I_f = A \int_{E_c}^{E_v} f_c(E)\rho_c(E) P_{c \to v}[1 - f_v(E)]\rho_v(E)\, dE \tag{1}$$

$$I_b = A \int_{E_c}^{E_v} f_v(E)\rho_v(E) P_{v \to c}[1 - f_c(E)]\rho_c(E)\, dE \tag{2}$$

where A is an appropriate arbitrary constant.

TUNNEL DIODES AND BACKWARD DIODES

It is assumed that*

$$P_{c \to v} = P_{v \to c} = \text{const}$$

The total current I flowing through the junction is given by

$$I = I_f - I_b = B \int_{E_c}^{E_v} [f_c(E) - f_v(E)] \rho_c(E) \rho_v(E) \, dE \quad (3)$$

where $B = P_{c \to v} A = P_{v \to c} A$.

It is possible to evaluate Eq. (3) by assuming ρ's and fixing the location of the band edges with respect to the Fermi level. Figure 2 shows the various energy plots, similar to that in Fig. 1, under various bias conditions.

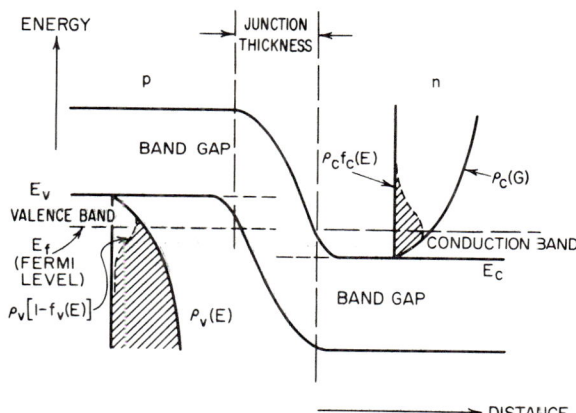

FIG. 1. Energy plots of a tunnel diode at zero bias.

At zero junction bias, I_f and I_b are equal in magnitude but opposite in sign and thus the total junction current I becomes zero. As the forward junction bias is increased, the difference in energy levels of the conduction bands and of the valence bands of the P and N regions decreases. This results in a decrease of the backward tunnel current since

$$\rho_v(E) f_v(E) \rho_c(E) [1 - f_c(E)] \, dE \dagger$$

becomes smaller. However, the forward tunnel current, given by

$$\rho_c(E) f_c(E) \rho_v(E) [1 - f_v(E)] \, dE$$

increases until the centers of gravity of the $\rho_c(E) f_c(E)$ and $\rho_v(E)[1 - f_v(E)]$ curves are just crossed. The energy plots and I-V characteristic for the tunnel diode in this condition are shown in regions 1 and 2 of Figs. 2 and 3, respectively.

Increasing the forward junction bias further, the forward junction current, which is now almost equal to the total junction current because of a small backward current, decreases, and gives rise to a negative conductance, as shown in region 3 of Fig. 3. As the forward bias is increased still further, the forward current will become almost zero, as shown in region 4 of Fig. 3. Beyond this forward bias level, the total junction current will be made up of electrons and holes that are thermally excited from the conduction band of the n region to the conduction band of the p region and from the valence band of the n region to the valence band of the p region, respectively; this

---

* $P_{c \to v}$ and $P_{v \to c}$ are functions of E and the bias voltage across the junction. See Refs. 2 and 3 for further information.

† The relative position of the two curves $\rho_v(E) f_v(E)$ and $\rho_c(E) [1 - f_c(E)]$ is a function of bias and strongly influences the values of tunneling currents.

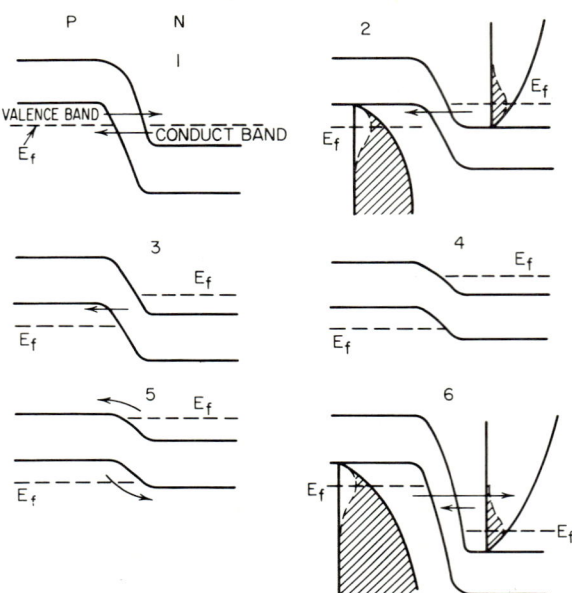

FIG. 2. Energy plots under various bias conditions. The number diagrams correspond to the numbered points on the V-I curves of Figs. 3 and 4.

current corresponds to the forward diode current in an ordinary semiconductor diode. This operation is represented by 5 of Figs. 2 and 3.

If an increasing reverse bias is applied to the tunnel-diode junction, the difference in energy levels between corresponding conduction bands and valence bands of the p and n regions will increase and will result in a monotonic increase of the backward tunnel current as a function of this back-bias voltage. The forward tunnel current under these conditions will decrease rapidly. Region 6 of Figs. 2 and 3 represents the reverse bias operation of a tunnel diode. I, I_f, and I_b are plotted as a function of bias voltage V in Fig. 3. I_{eq} in Fig. 3 is the sum of I_f and $|I_b|$ and is related to the overall shot noise in a tunnel diode, as will be discussed in detail later. However, as seen in Fig. 4, a plot of the measured I-V characteristic of a typical diode and the corresponding plot of the calculated I-V curve show some discrepancy.

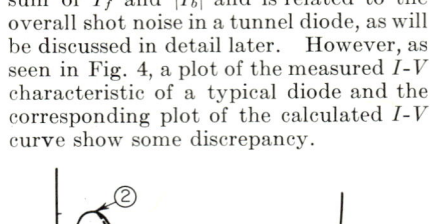

FIG. 3. Theoretical tunnel-diode characteristics.

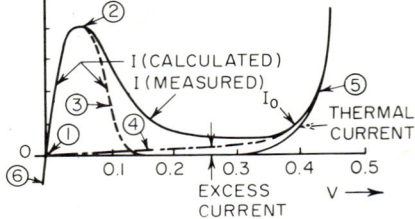

FIG. 4. Theoretical and measured tunnel-diode characteristics.

TUNNEL DIODES AND BACKWARD DIODES 12-5

Ideally, if there were no tunneling, as in the case of the ordinary diode, the measured I-V curve would become I_o of Fig. 4. In the negative-conductance region, the measured current is larger than I_o and the difference of these currents is called the *excess current*. The excess current may be caused by unknown impurity concentrations and by loss of electron energy through interactions with other entities (phonons and photons, etc.).

1.3. Backward Diodes. In a tunnel diode, the voltages corresponding to points 2, 3, 4, and 5 in Fig. 2 are essentially determined by the semiconductor materials and their doping levels. The corresponding current values are determined by the cross-sectional area and doping level of the junction. In practice, the tunnel junction is started with a large cross-sectional area and then etched until the peak current (point 2) has the desired value. Tunnel diodes can be made with peak currents anywhere from a few microamperes to several amperes.

As shown in Figs. 2 and 3, in the reverse bias direction, a large current flow increases continuously as a function of reverse-bias voltage. The magnitude of the reverse tunneling current is much larger than the corresponding tunneling currents in the forward region. A tunnel diode used to take advantage of this large change in current with polarity of applied voltage is called a "backward diode." The term "backward" means that the diode conducts heavily with negative rather than positive applied voltage. Figure 5 shows a comparison of germanium "backward" and conventional diodes.

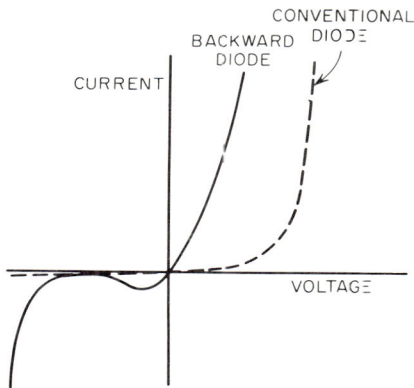

Fig. 5. Comparison between backward and conventional diode characteristics.

It can be seen that a backward diode has a lower voltage drop at a given current than a conventional diode. This low forward drop is advantageous in some tunnel-diode circuits, in transistor circuits, and as a mixer. In a backward diode the backward I-V characteristic is of primary interest while the relatively small variation in the negative-resistance region is of secondary interest.

2. STATIC PARAMETERS OF TUNNEL AND BACKWARD DIODES*

Static parameters[4] of tunnel- and backward-diode characteristics are defined with reference to the static characteristic of Fig. 6.

Peak Current I_P. The current at which the slope of the current-voltage characteristic changes from positive to negative as the forward voltage is increased.

Peak Voltage V_P. The voltage which corresponds to I_P on the I-V characteristic.

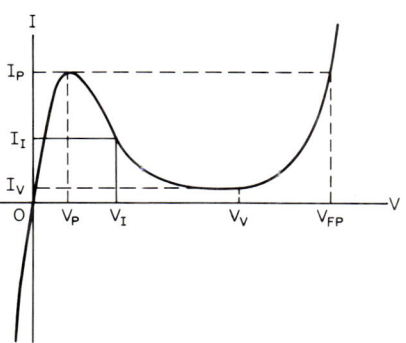

Fig. 6. Static parameters and I-V curve of a tunnel diode.

Valley Current I_V. The current at which the slope of the I-V characteristics changes from negative to positive as the forward voltage is increased.

Valley Voltage V_V. The voltage which corresponds to I_V.

Forward-voltage Point V_{FP}. The voltage, greater than V_V, at which the current is equal to I_P.

* Definitions and symbols of the proposed IEEE Standards, IRE 28.4.14 PS1, are used.

AMPLIFYING DEVICES

Inflection-point Voltage V_I. The voltage at which the slope of I-V characteristic reaches its most negative value.

Inflection-point Current I_I. The current corresponding to the inflection-point voltage.

3. TUNNEL-DIODE EQUIVALENT CIRCUIT

As mentioned in Sec. 1 the tunneling effect is extremely fast. The theoretical limit on the frequency response is on the order of 10^{12} to 10^{13} cps. Since the I-V characteristic is nonlinear, it is convenient to use the incremental conductance $g = dI/dV$ when discussing small-signal operation. However, when the device is characterized by its terminal admittance this quantity must be considered a function of frequency, even though the junction parameters may be independent of frequency.

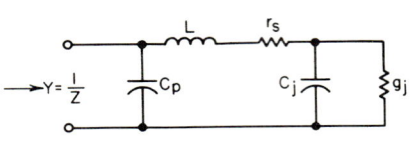

Fig. 7. Tunnel- and backward-diode equivalent circuit.

Characterization of tunnel and backward diodes is simplified by the use of the equivalent circuit of Fig. 7, the elements of which are independent of frequency over a broad range. In addition, the elements C_p, L, and often r_s are independent of bias.

The terms necessary to discuss these elements are as follows:[4]*

Case Capacitance C_P. The residual capacitance between the diode terminals when the p-n junction is opened (the junction is removed).

Junction Capacitance C_j. The small-signal capacitance associated with the p-n junction alone, at a specified bias and frequency.

Incremental Resistance. The reciprocal of the slope of the I-V characteristic.

Series Resistance r_s. The portion of the incremental resistance (under specified conditions of bias and frequency) which is external to the junction.

Junction Conductance g_j. The reciprocal of the incremental resistance less the series resistance.

Series Inductance L. The self-inductance of a diode in a specified circuit configuration.

4. RESISTIVE CUTOFF AND SELF-RESONANCE FREQUENCIES

As the figures of merit of a tunnel diode, it is convenient to use quantities defined as "resistive cutoff frequency" and "self-resonance frequency." The resistive cutoff frequency f_r is defined as the frequency at which the real part of the terminal impedance Z of the tunnel diode (see Fig. 7) goes to zero, at the bias where the magnitude of junction conductance $|g_j|$ becomes maximum in the negative-conductance region. Mathematically it becomes

$$f_r = (|g_j|/2\pi C_j)\sqrt{1/|g_j|r_s - 1} \qquad (4)$$

where $|g_j|$ is maximum in the negative-conductance region.

The real part of Z is negative below f_r and becomes positive above f_r. Thus f_r provides the highest possible oscillating frequency for a tunnel diode. At other bias points in the negative-conductance region, since $|g_j|$ becomes smaller, f_r is not so high as that of Eq. (4).

The self-resonance frequency is defined as the frequency at which the imaginary part of the terminal admittance Y becomes infinite (excluding $f = \infty$). The series inductance of a diode can be specified only if the way in which a diode is mounted in a particular circuit is specified.[5]

Infinite Y implies that the terminal is ideally shorted. However, in reality, a diode has a finite size, and at high frequency cannot be considered a point device. That is,

---

* Again, those definitions and symbols proposed by IEEE Standards, IRE 28.4.14 PS1, are used.

the admittance is not an ideal short circuit at high frequency. The admittance of the smallest-volume closed metallic cavity (called the minimum-volume closed cavity) which is built around the encapsulated diode is the best one can achieve. Therefore, in reality, the self-resonance frequency can be obtained when the diode is embedded in a minimum-volume closed cavity. A typical minimum-volume closed cavity is shown in Fig. 8. Actually, the evaluation of series inductance comes from the measurement of the resonant frequency of a diode in a particular cavity.[5]

The simple mathematical expression for the self-resonance frequency f_x can be obtained from Fig. 7, with $\text{Im } |Y| = \infty$, by

$$f_x = (1/2\pi) \sqrt{1/LC_j - g_o^2/C_j} \tag{5}$$

where g_o is the average conductance* given by

$$|g_o| = C_j r_s / L \tag{6}$$

It is interesting to relate f_r and f_x in terms of the junction area A. Since the voltage characteristic of a tunnel diode is determined by the energy level of the junction material, for a given material the power-handling capability is proportional to the diode current, i.e., the junction area. g_j, C_j, and $1/r_s$ can be considered to be proportional to A. However, as will be discussed later, $1/r_s$ is approximately proportional to $\sqrt{A}$ for a junction having a square or round cross section at microwave frequencies where the skin effect is pronounced. Therefore, f_r will be reduced slightly as a function of A.

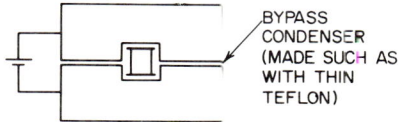

FIG. 8. Minimum-volume closed cavity for a pill-type diode.

Considering L as independent of A, $1/LC_j$ will be inversely proportional to A. Of course, g_o is proportional to A. Consequently, f_x will decrease much faster than f_r as a function of A. As A increases, $1/LC_j$ can be smaller than $(g_j/C_j)^2$, and consequently f_x will become imaginary, leading to an undesirable instability.

Even if f_x is real, but is smaller than the desired operating frequency, rather complicated circuit techniques must be resorted to in order to overcome undesirable instability.† Therefore, for small-signal high-frequency operation, it is desirable to make f_r and f_x as high as possible. In order to make f_r high, it is necessary to make $|g_j|/C_j$ high and $r_s|g_j|$ as small as possible. This requires that the semiconductor material be very heavily doped and that it have a reasonably high mobility on both sides of the junction. The most suitable materials are Ge, GaAs, and GaSb. In order to make f_x high, it is necessary to make L as small as possible and at the same time keep the junction area A very small (on the order of 1 mil or less for a microwave diode). To increase stability, manufacturers very often produce tunnel diodes having f_x larger than f_r.

As will be discussed later the noise characteristic of a tunnel diode is related to its resistive cutoff frequency f_r. In order to have low-noise operation, it is desirable to use a diode with a very high resistive cutoff frequency (at least three times as high as the operating frequency).

5. VARIOUS TUNNEL DIODES

Three kinds of tunnel diodes are commercially available. These are Ge, GaAs, and GaSb diodes, each having a different band gap. Typical V-I characteristics of diodes made with three different materials are quite different, as shown in Fig. 9. Junction conductances $g_j = dI/dV$ of these diodes are given in Fig. 10.

---

* g_o is the average conductance[6] for a large-signal oscillation and is equivalent to g_j for a small oscillation.

† More detailed discussion on stability and instability will be given in Chap. 28.

12-8 AMPLIFYING DEVICES

As can be seen in Fig. 9, the difference between valley and peak voltages of different materials increases in the order GaSb, Ge, and GaAs.

For a given peak current, GaAs diodes are preferred for higher-power oscillators (because of the larger negative-conductance region). On the other hand, from a low-noise standpoint, GaSb is preferred since the smaller $I/|g_j|$ ratio near the inflection point provides less shot-noise contribution in an amplifier, as will be shown later. Curves of $I/|g_j|$ vs. bias voltage for typical Ge and GaSb diodes are shown in Fig. 11. The flat portion of g_j, which corresponds to the dynamic range in a negative-conductance amplifier, is the smallest for GaSb. Therefore, if a larger dynamic range is needed, diodes made from other materials are preferred.

Although a diode with large voltage swing would normally be preferable from logic applications, a word of caution is in order on the use of GaAs tunnel diodes, which are

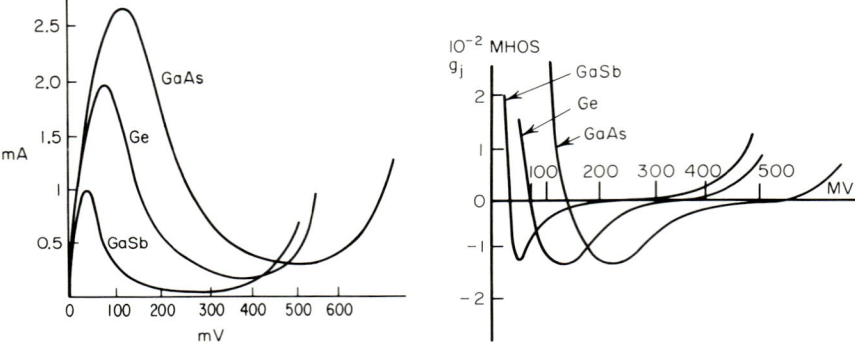

FIG. 9. Characteristics of Ge, GaSb, and GaAs tunnel diodes.

FIG. 10. g_j-V characteristics of tunnel diodes of Fig. 9.

subject to a gradual degradation of both I_p and I_v whenever they are operated in the thermal-injection region of their I-V characteristic.

Since GaAs has poor noise characteristic, Ge and GaSb diodes have been mostly used in low-noise applications of tunnel diodes.

Comparing Ge and GaSb diodes, Ge tunnel diodes are mechanically stronger than GaSb diodes, and because of their higher current density, Ge diodes have an inherently greater frequency capability. For example, Ge tunnel diodes have achieved $|g_j|/C_j$ values of 10^{12} while the best reported $|g_j|/C_j$ value for GaSb diodes has been in the neighborhood of 10^{10}.

Typical Ge tunnel-diode parameters pertinent to S-band and C-band amplifiers are as follows:

$\qquad$ S band $\qquad\qquad\qquad\qquad$ C band

$|r_j| = \dfrac{1}{|g_j|} = 70$ ohms $\qquad$ $|r_j| = \dfrac{1}{|g_j|} = 70$ ohms
$C_j < 1$ pf $\qquad\qquad\qquad\qquad$ $C_j < 0.5$ pf
$r_s < 0.1|r_j|$ $\qquad\qquad\qquad\quad$ $r_s < 0.1|r_j|$
$L < 0.1$ nh $\qquad\qquad\qquad\quad$ $L < 0.1$ nh

6. PACKAGING CONSIDERATIONS

The effects of the junction area on the resistive cutoff frequency f_r and the self-resonance frequency f_x have been discussed briefly. More detailed discussion of the effect of geometrical dimensions of a device on device performance will be given in this section.

It is well understood that the high-frequency resistance is inversely proportional to the square root of the area for a round cross section, because of the skin effect.[7,8] Thus r_s is a function of frequency. On the other hand, g_j and C_j are proportional to A and independent of frequency.

If a different diode geometry is used, both r_s and L can be reduced for a given junction area without changing the value of g_j and C_j.* This can be achieved by increasing the ratio of the junction's perimeter to its area. For example, as shown in Fig. 12, instead of one large-area diode, several junction diodes with the same net junction area can be used in the case where a diode strip-line circuit is used. Here, the contributions to magnetic flux produced by adjacent diodes cancel between adjacent diodes. Thus the series inductance of a multijunction diode can be made smaller than that of the single-junction diode. r_s of a multiple-junction diode also becomes smaller than that of the single-junction diode with no variation of overall g_j and C_j. The height of a diode should be held to a minimum to reduce L and r_s as much as possible.

An ideal geometry for a lumped (nondistributed) diode is to form a coaxial ring (bound) diode which is used in a coaxial transmission line. As shown in Fig. 13, a ring junction is formed along the circumference, with current flowing in the radial direction. If the ring diode is placed between the inner and outer conductors of a coaxial transmission line, the operational mode will be a TEM wave, and the diode junction current must flow in the radial direction. Thus, since the junction is enclosed in a ring, the total magnetic flux surrounding the junction will be zero and the series inductance L of the diode will also be zero. At the same time, if the axial and radial dimensions of the junction are made small as compared with its circumference, it is possible to reduce skin effect to a minimum.

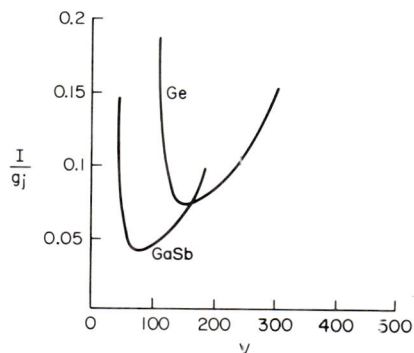

Fig. 11. $I/|g_j|$ vs. voltage for Ge and GaSb tunnel diodes.

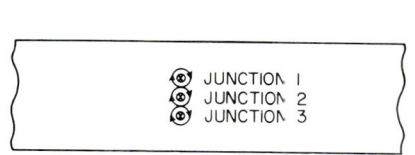

Fig. 12. Multiple-junction tunnel diode.

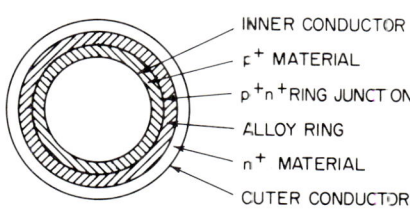

Fig. 13. Coaxial-ring tunnel diode.

Thus, using special junction geometry, lumped tunnel-diode performance, such as higher-frequency and higher-power operation, can be improved.

If device technology improves, a distributed-junction tunnel diode can be considered, and associated circuits present interesting problems. However, since no such device is available, discussions on this subject will not be included.

7. NOISE

As mentioned previously, there are two tunneling currents, namely, the forward current I_f and the backward current I_b. Tiemann[9] has pointed out that noise contri-

---

* The junction dimension is considered to be much smaller than the wavelength of the operating signal.

butions from these two currents are shot noise. The equivalent shot-noise current I_{eq} is given by the sum of I_f and I_b. Referring to Eqs. (1) and (2), I_{eq} becomes

$$I_{eq} = I_f + I_b = B \int_{E_c}^{E_v} [f_c(E) + f_v(E) - 2f_c(E)f_v(E)]\rho_c(E)\rho_v(E)\, dE \quad (7)$$

The plot of I_{eq} is shown in Fig. 3.

If the contributions from two currents I_f and I_b are uncorrelated, the total shot-noise current $\overline{i_j^2}$ due to junction currents can be expressed by

$$\overline{i_j^2} = 2eI_{eq}\,\Delta f \quad (8)$$

where e = electron charge = 1.6×10^{-19} coulombs
Δf = incremental bandwidth

It should be pointed out that the total equivalent shot noise generated by the two tunnel currents at the origin (where thermal equilibrium is maintained) is thermal noise $\overline{i_t^2}$.

$$\overline{i_t^2} = 4kTG\,\Delta f$$
$$= \overline{i_j^2} = 2eI_{eq}\,\Delta f = 4eI_f\,\Delta f = 4e|I_b|\,\Delta f$$

where
$$G = \partial I/\partial V]_{V=0} \quad (9)$$

k = Boltzmann's constant = 1.38×10^{-23} joule/degree
T = absolute temperature, °K

In the negative-conductance region, I_b is negligibly small and $\overline{i_j^2}$ can be expressed by

$$\overline{i_j^2} = 2eI_f\,\Delta f = 2eI_{DC}\,\Delta f \quad (10)$$

In an actual case, the direct current in the negative-conductance region contains the excess current. If the noise due to the excess current is shot noise and uncorrelated with I_f, as is the case with most good diodes, then the total shot noise $\overline{i_j^2}$ in the negative-conductance region again becomes $2eI_{DC}\,\Delta f$. However, for a poor diode, actual noise is much higher than $2eI_{DC}\,\Delta f$ in the negative-conductance region. Noise measurements were performed to a few megacycles, much above the $1/f$ noise region. It was reported that the actual shot noise could be smaller than that of Eq. (10) because of the smoothing effect. In this case, the right-hand term of Eq. (10) would be multiplied by the smoothing constant γ,[10] which is smaller than 1.

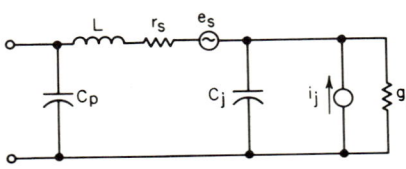

Fig. 14. Noise equivalent circuit of a tunnel diode.

In the early stages of tunnel-diode development, it was hoped that tunnel diodes were $1/f$-noise-free devices, since diodes were so highly doped that no surface effect was noticeable. However, there was $1/f$ noise in tunnel diodes. Furthermore, placing trap states by contaminating impurities such as Au, Mn, and Cu in GaAs tunnel diodes not only changed the excess current but also greatly increased shot noise and especially $1/f$ noise.[11]

A noise equivalent circuit for a tunnel diode is shown in Fig. 14. It is the same as the small-signal equivalent circuit shown in Fig. 7, except that noise generators have been added. In addition to the shot-noise current generator $\overline{i_j^2}$ contributed by the junction, a thermal-noise voltage generator $\overline{e_s^2}$ contributed by r_s is added, where

$$\overline{e_s^2} = 4kTr_s\,\Delta f \quad (11)$$

TUNNEL DIODES AND BACKWARD DIODES

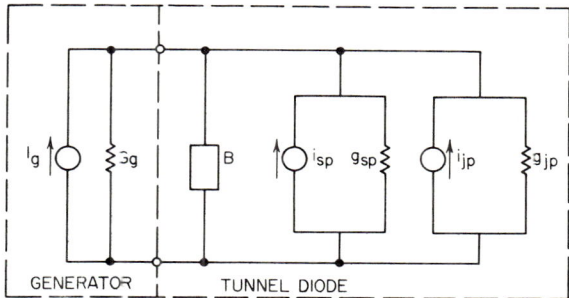

FIG. 15. Parallel noise equivalent circuit.

Rearranging Fig. 14 into a parallel equivalent circuit, Fig. 15 can be obtained. The following are identities between the parameters of Figs. 14 and 15:

$$g_{jp} = \frac{1}{|Z_s|^2 g_j x}$$

$$g_{sp} = \frac{r_s}{|Z_s|^2}$$

$$B = \omega C_p - \frac{\omega L - (1/\omega C_j)(1 - 1/x)}{|Z_s|^2} \qquad (12)$$

where
$$x = 1 + \omega^2 C_j^2/g_j^2$$
$$|Z_s|^2 = (r_s - 1/|g_j|x)^2 + [\omega L - (1/\omega C_j)(1 - 1/x)]^2 \qquad (13)$$

and
$$\overline{i_{jp}^2} = 4kT(G_{eq}/|g_j|)|g_{jp}|\,\Delta f \qquad (14)$$

where
$$\overline{i_{sp}^2} = 4kT g_{sp}\,\Delta f$$
$$G_{eq} = eI_{eq}/2kT$$
$$= 20 I_{eq} \quad \text{at room temperature} \qquad (15)*$$

As seen in Chap. 7, a noise factor, as well as an extended noise factor of a network including an active network with one or more negative-conductance device, is independent of the load and a second stage (if cascaded). Then the noise factor of a tunnel-diode equivalent circuit can be defined,†

$$F = 1 + \frac{\overline{i_{sp}^2} + \overline{i_{jp}^2}}{\overline{i_g^2}}$$
$$= 1 + \frac{(g_{sp} + G_{eq}|g_{jp}|/|g_j|)T}{G_g T_o} \qquad (16)$$

where $\overline{i_g^2} = 4kT_o G_g\,\Delta f$: thermal noise current from the generator
T_o = generator temperature
T = tunnel-diode temperature

If G_g becomes $|g_{jp}| - g_{sp}$, which is the criterion for infinite gain of an ideal tunnel-diode amplifier, the noise factor of Eq. (16) can be expressed by [refer to Eqs. (12) to (16)]

$$F = 1 + \frac{(g_{sp} + G_{eq}|g_{jp}|/|g_j|)T}{(|g_{jp}| - g_{sp})T_o} \qquad (17)$$

* The equivalent noise conductance is defined by Eq. (15). This is derived by equating the shot-noise current of Eq. (18) to its equivalent thermal noise $4kTG_{eq}\,\Delta f$.
† If there is frequency conversion such that the noise from a load or a second stage at the other spectrum band is converted into the spectrum band which is under consideration, this noise-factor definition does not apply.

and for $T = T_o$

$$F_o = \frac{1 + G_{eq}/|g_j|}{(1 - r_s|g_j|)[1 - (f/f_r)^2]} \qquad (18)$$

Equation (18) shows that F_o is related to $G_{eq}/|g_j|$, $1 - r_s|g_j|$, and f/f_r. For low F_o, $r_s|g_j|$, f/f_r, and $G_{eq}/|g_j|$ should be made small.

If I_{eq} could be approximated by I_{DC} in the negative-conductance region as mentioned before, $I_{DC}/|g_j|$ would be determined by the material used, with some variation due to doping level. Typical $I_{DC}/|g_j|$ values for Ge and GaSb are shown in Fig. 11. A typical $r_s|g_j|$ value is about 0.1. Therefore, if $f/f_r = \frac{1}{3}$, F_o of Eq. (18) becomes, for Ge ($I_{eq}/|g_j| = 0.06$) and GaSb ($I_{eq}/|g_j| = 0.04$),

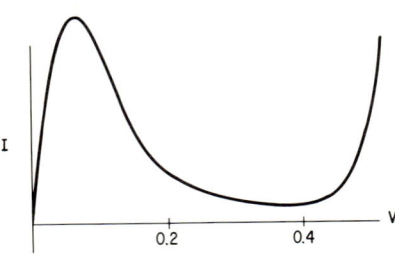

$$F_o = \frac{1 + 20 \times 0.06}{(1 - 0.1)(1 - \frac{1}{3}^2)}$$
$$= 2.75 = 4.4 \text{ db for Ge}$$

and $$F_o = \frac{1 + 20 \times 0.04}{(1 - 0.1)(1 - \frac{1}{3}^2)}$$
$$= 2.25 = 3.5 \text{ db for GaSb}$$

Tunnel diodes are capable of operating at microwave frequencies in amplifiers or mixers with relatively low noise. The noise factor of a tunnel diode will not be so good as that of parametric circuits. However, if simplicity (operation with a d-c source), compactness, and economy are of major concern, tunnel diodes are useful devices for microwave applications.

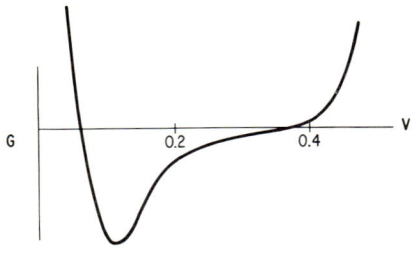

Fig. 16. *I-V* and conductance characteristics of Ge tunnel diode at room temperature.

8. NONLINEAR CHARACTERISTICS OF TUNNEL DIODES

8.1. Introduction. In the previous sections it has been either implicitly or explicitly assumed that the negative conductance is essentially constant in the vicinity of the operating point. This assumption does not cause any serious errors in the qualitative features of amplifiers that are considered in those sections provided that the quiescent operating point is chosen at the inflection point of the *I-V* characteristic. At this point, the negative conductance is independent of the operating point to first order in $(V - V_I)$. For this reason the region around the inflection point is often referred to as the "linear region" of the *I-V* characteristic. Strictly speaking, of course, no portion of the tunnel-diode characteristic is truly linear, and in this section we shall examine some of the consequences of the nonlinearities of the tunnel-diode characteristic. We shall consider operating points both in the linear region and also outside the linear region. Plots of the conductance vs. bias for tunnel diodes of the more common semiconductor materials are shown in Figs. 16, 17, and 18. As we shall see, many of the properties of small-signal tunnel-diode amplifiers are determined by nonlinear effects—even in the limit of zero signal amplitude!

The reason that nonlinear effects are so important for small-signal amplifiers has to do with the fact that the tunnel-diode amplifier operates by canceling the major portion of the total positive conductance of the source and load. Since the amplification depends only on the algebraic sum of the positive conductances of the source plus the load and the negative conductance of the tunnel diode, the gain of the amplifier can change by a large amount if any of the conductances of the circuit changes ever by a relatively small amount.

If the signal itself displaces the bias point by a small amount, the cancelation between the positive conductances and the negative diode conductance can also be grossly upset even though the negative conductance has been changed only slightly. If the amplifier is operating at the inflection point (where the negative conductance attains its largest value), this effect manifests itself as a clipping of the peaks of the amplified wave, and as a fall-off of the gain as the amplitude is increased. If some other operating point is chosen, the portions of the signal waveform which move the operating point away from V_I are amplified less while those which move the operating point closer to V_I are enhanced. The bias dependence of the gain also complicates the noise picture. If one takes the viewpoint that noise fluctuations are equivalent to fluctuations in the bias point and notes that the gain of the amplifier depends on

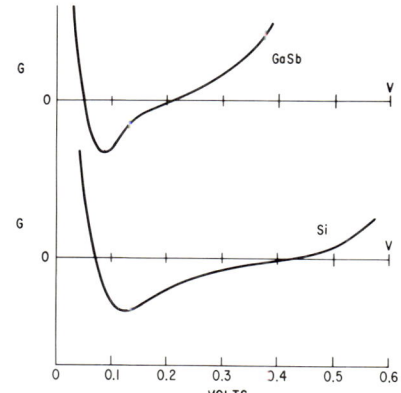

Fig. 17. Conductance curves of GaSb and Si tunnel diodes at room temperature.

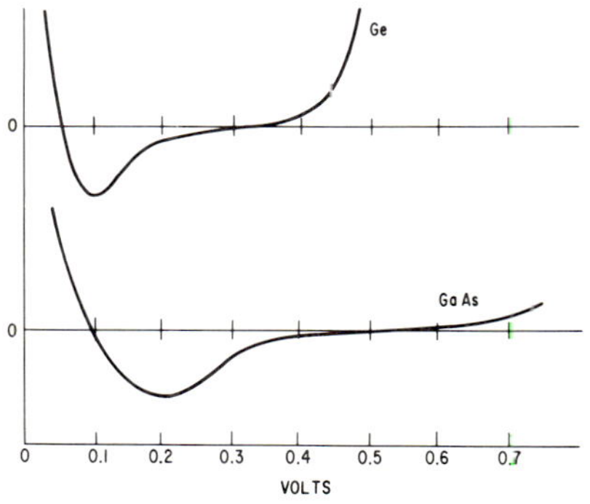

Fig. 18. Conductance curves of Ge and GaAs tunnel diodes at room temperature.

the bias point, it is clear that the gain of the amplifier will fluctuate because of the noise, and that the signal will consequently be amplitude-modulated by the noise.

In this section we shall analyze several specific effects of nonlinearity of tunnel-diode amplifiers. In particular, we shall calculate the manner in which the gain of the amplifier changes with increasing signal amplitude, both for inflection-point operation and elsewhere. Expressions for distortion and maximum gain will be given, and such questions as how to optimize dynamic range, linearity, and gain will be considered. We shall then consider the operation of a tunnel-diode amplifier as a mixer, and shall calculate the conversion efficiency for various modes. Finally, the cross modulation of the signal with the noise at other frequencies will be discussed.

8.2. The I-V Characteristic Equation. We shall start by expanding the I-V characteristic in a power series about the operating point. If V_o is the operating-point bias voltage, let

$$I(V) = I_o(V_o) - G_o(V - V_o) + \alpha(V - V_o)^2 + \beta(V - V_o)^3 + \cdots \quad (19)$$

where I_o, V_o, and G_o refer to the current, the voltage, and the negative conductance at the operating point, respectively. Note that we have not assumed that $V_o = V_I$. The constants α and β are those which produce the best fit of the form assumed in Eq. (19) at V_o. If we keep terms up to the third order, we are in effect fitting a parabola to the conductance plots shown in Figs. 16, 17, and 18.

The circuit we shall analyze is shown in Fig. 19. Actually, this circuit is more generally applicable than it looks. If we had to consider the lead inductance of the diode, or if we had a complicated-looking source or load, there is nothing to prevent us from focusing our attention on the diode junction itself and replacing all the remaining circuitry by the appropriate Norton generator. The parameters of this generator would depend on the transformation properties of the diode package and the circuit, and would in general be a function of frequency, but once these properties are calculated, the performance of the amplifier can be determined from the circuit of Fig. 19.

With reference to Fig. 19, Kirchhoff's law can be applied to the current:

$$i_s = e(Y_s + Y_l + S_j - G_o) + \alpha e^2 + \beta e^3 \tag{20}$$

where $e = (V - V_o)$ is the displacement of the diode voltage from the operating point V_o produced by the signal current i_s; Y_s and Y_l are the admittances of the source and load, respectively, and S_j is the susceptance of the diode junction. For this analysis, we shall assume S_j is independent of V. (This is not necessarily a good assumption.)

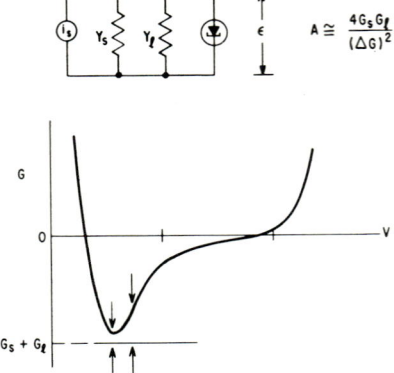

Fig. 19. Small signal amplifier.

At this point some simplification would result if we assumed that, over the frequency band of interest, the total susceptance across the generator (including that due to the tunnel-diode junction capacitance) was equal to zero. This assumption is equivalent to taking the real parts of the transformed admittances of the source and load to be independent of frequency. In cases of practical interest, both the real parts and the imaginary parts of the admittances depend on frequency, and we shall keep our expression sufficiently general to apply to these cases.

Let us define

$$Y_s + Y_l + S_j - G_o = \Delta Y = \Delta G + jS \tag{21}$$

where
$$\Delta G = G_s + G_l - G_o$$
and
$$S = S_s + S_l + S_j$$
with
$$Y_s = G_s + jS_s;\ Y_l = G_l + jS_l$$

With these definitions, Eq. (20) can be rewritten as

$$i_s = e\,\Delta Y + \alpha e^2 + \beta e^3 \tag{22}$$

We shall set

$$i_s = i_1 \sin \omega t \tag{23}$$

and look for $e(t)$ in the form

$$e(t) = e_1 \sin(\omega t) + e_2 \cos(2\omega t) + e_3 (\sin 3\omega t) + \cdots \tag{24}$$

As we shall see, these are all the terms necessary to understand the behavior of an amplifier using an ordinary tunnel diode.

Inserting Eqs. (23) and (24) into Eq. (20) and expressing products of sines and cosines in terms of sums of sines and cosines of the sums and differences of the arguments, we get after equating coefficients at each frequency

$$i_1 = e_1 \Delta Y_1 + \alpha(-e_1 e_2 + e_2 e_3) + \tfrac{3}{4}\beta[e_1^3 + 2e_1(e_2^2 + e_3^2) - e_3(e_1^2 + e_2^2)] \quad (25)$$
$$0 = e_2 \Delta Y_2 + \alpha(-\tfrac{1}{2}e_1^2 + e_1 e_3) + \tfrac{3}{2}\beta(e_1^2 - \tfrac{1}{6}e_1 e_3 + \tfrac{1}{2}e_2^2 + e_3^2) \quad (26)$$
$$0 = e_3 \Delta Y_3 + \alpha e_1 e_2 - \tfrac{1}{4}\beta[e_1^3 - 6e_3(e_1^2 + e_2^2) + 3(e_1 e_2^2 - e_3^2)] \quad (27)$$

where $\Delta Y_n = \Delta Y$ at the nth harmonic. Here, only the terms involving time dependences up to $\sin 3\omega t$ have been kept.

In these expressions $\Delta Y_1 \neq \Delta Y_2 \neq \Delta Y_3$ because S is assumed to be a function of frequency.

Furthermore, because of the transformation properties of diode lead inductance, etc., the values G_s and G_l, as seen at the diode junction, will also depend on frequency.

8.3. Operation at the Inflection Point. We shall first apply Eqs. (25) through (27) to the case where the tunnel diode is biased at the inflection point. At this bias point, $\alpha = 0$. Hence e_2 is equal to zero. Thus, to first order in β, Eqs. (7) and (9) become*

$$i_1 = e_1[\Delta Y_1 + \tfrac{3}{4}\beta(e_1^2 - e_1 e_3 + 2e_3^2)] \quad (28)$$
$$0 = e_3 \Delta Y_1 + \beta(-\tfrac{1}{4}e_1^3 + \tfrac{3}{2}e_1^2 e_3 + \tfrac{3}{4}e_3^3) \quad (29)$$

Fortunately the solution of Eq. (29) is such that $e_3 \ll e_1$ for all values of e_1; so the term in e_3^3 can be ignored to obtain a first approximation. The result is

$$e_3 = \tfrac{1}{4}\beta \frac{e_1^3}{\Delta Y_1 + \tfrac{3}{2}\beta e_1^2} \quad (30)$$

In the small-signal limit ($e_1 \to 0$), e_3 is negligible with respect to e_1. In the large-signal limit, ΔG is negligible with respect to $\tfrac{3}{4}\beta e_1^2$, and $e_3 \approx \tfrac{1}{6}e_1$. This implies that Eq. (30) is correct to within 1 per cent for all values of e_1. If this result is inserted in Eq. (28), we have

$$i_1 = e_1\left[\Delta Y_1 + \tfrac{3}{4}\beta\left(e_1^2 - \frac{e_1^4}{4}\frac{\beta}{\Delta Y_1 + \tfrac{3}{2}\beta e_1^2}\right)\right] \quad (31)$$

In the small-signal limit, the power gain is

$$\lim_{e \to 0} [A] \equiv A_o = \frac{4 G_s G_l}{(\Delta G)^2 + S^2}$$

* We should like to point out that by starting from Kirchhoff's law we avoided a mistake which is easy to make. If we had used the relation for the small-signal quantities i and e

where
$$i = e[G_s + G_l - G(e)]$$
$$-G(e) = G_o + 3\beta e^2 \quad (N1)$$

(see Fig. 16), we would have obtained the wrong answer. The trouble can be seen by starting from the relation

$$di = de[G_s + G_l - G(e)] \quad (N2)$$

If we integrate this expression between zero and the small-signal quantities i and e, we get

$$i = e(G_s + G_l) - \int_0^e G(e')\,de' \quad (N3)$$

Since we are considering nonlinear effects, it is not permissible to set

$$\int_0^e G(e')\,de' = eG(e) \quad (N4)$$

as is implied in Eq. (N1).

where $A = W_l/W_s$ is the ratio of the power delivered to the load conductance divided by the maximum power available from the source. For the case of finite output voltage, ΔY is replaced by the expression in square brackets in Eq. (31), and the gain is reduced by the amount

$$\frac{A}{A_o} = \frac{(\Delta G)^2 + S^2}{\Delta G + \frac{3}{2}(\beta W_l/G_l) - \frac{3}{4}(\beta W_l/G_l)^2[\Delta Y_3 + 3(\beta W_l/G_l)]^{-1}} \tag{32}$$

where the relation $W_l = \frac{1}{2}e_l^2 G_l$ has been used.

Equation (32) tells us several things. First of all, as long as the output voltage is small enough so that $(3/4)\beta e_1^2 \ll \Delta G$, the gain is constant and the output vs. input would be a straight line on a loglog plot with unit slope. The gain is down by 3 db when

$$\frac{A}{A_o} = \frac{1}{2} = \frac{(\Delta G)^2 - S^2}{(\Delta G + \frac{3}{2}\beta W_l/G_l - 3\beta^2 W_l^2/4\Delta Y_3 G_l^2)^2 + S^2} \tag{33}$$

We note that the 3-db point occurs when the *output* power has a certain relation to ΔG. Thus, if we had a constant input power and raised the gain by decreasing ΔG, the output power would increase, and eventually we would have sufficient output to cause saturation *no matter how small the input power happens to be*. This fact therefore places a limit on the amount of gain at input that can be achieved without exceeding a 3-db fall-off. If one is willing to accept a larger distortion of the signal waveform, one can achieve a correspondingly larger gain, but for any given level of acceptable signal distortion there will be a corresponding maximum achievable gain.

There is not much more one can learn from this expression without making some simplifying assumptions. Unfortunately, the assumptions which produce the most simplifications, and which consequently provide the greatest insight, are not likely to be more than qualitatively correct for many of the situations encountered in practice. We shall discuss this point later on, when we compare experimental results to the approximate expressions.

Let us first confine our attention to the case where $S \ll \Delta G$. Let us also assume that $\Delta Y_3 = \Delta Y_1$ (i.e., G_s, G_l, G_o are all independent of frequency). Then Eq. (33) implies

$$\Delta G = 3.03(\beta W_l/G_l) \tag{34}$$

so that
$$i_1 = 4.53(\beta W_l/G_l)e_l \approx 6.4\beta(W_l/G_l)^{\frac{3}{2}} \tag{35}$$

Making use of the relation
$$A = W_l/W_s = 4W_l G_s/i_1^2$$

(where $W_s = i_1^2/4G_s$ is the maximum power available from the source), we find

$$A_{sat} = 0.10(G_s G_l^3/\beta^2 W_l^2) = (0.10 G_s G_l^3/\beta^2 W_s^2)^{\frac{1}{3}} \tag{36}$$

If we maximize this expression with respect to G_s subject to the constraints that W_s remains constant and $G_s + G_l$ remains constant at a value greater than G_o, we find

$$G_s \gtrsim \tfrac{1}{4}G_o \qquad G_l \gtrsim \tfrac{3}{4}G_o$$

Inserting these quantities into Eq. (36) we get finally

$$(A_{max})_{3\,\text{db}} = 0.27(G_o^2/\beta W_s)^{\frac{2}{3}} \tag{37}$$

There are several things to be learned from Eq. (37). First of all, we notice that the gain does not increase with decreasing signal power as fast as W^{-1}. This means that, even though the permissible gain becomes infinite in the limit of zero signal, the maximum output power goes to zero in this limit.

In fact, the maximum permissible output power goes like

$$W_l \sim W_s^{\frac{1}{3}} \tag{38}$$

(see Fig. 20).

Also, we see that, since β and G_o both scale in proportion to the peak current for diodes of a given material, one should operate at the minimum impedance level consistent with circuit-design techniques in order to minimize saturation effects. Furthermore, if we operate under this condition, then in comparing diodes of different materials, we should choose the material with the best ratio of G/β. Thus a materials figure of merit in this respect is

$$F = (G/\beta)|_{VI} \tag{39}$$

In the limit of large output power the relation between input and output again becomes simple. Rather than use Eq. (32) to obtain this limit, it is simpler to go back

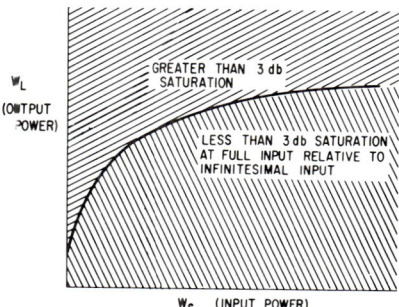

Fig. 20. Curve of 3-db saturation level.

to Eqs. (25) and (27) and neglect ΔY from the outset. We have, then,

$$i_1 = \tfrac{3}{4}\beta(e_1{}^3 - e_1{}^2 e_3 + 2e_1 e_3{}^2) \tag{40}$$
$$0 = (\tfrac{1}{4}e_1{}^3 + \tfrac{3}{2}e_1{}^2 e_3 + \tfrac{1}{2}e_3{}^3) \tag{41}$$

The second of these equations tells us that e_3/e_1 is small, and upon neglecting the term in $e_3{}^3$, we get

$$e_3/e_1 \cong \tfrac{1}{6}$$

Inserting this into Eq. (40) we get

$$i_1 = \tfrac{2}{3}\beta e_1{}^3$$

which leads to

$$W_s = i_1{}^2/4G_s = \tfrac{8}{9}(\beta^2 W_L{}^3/G_s G_L{}^3) \tag{42}$$

Therefore, the curve of output vs. input should again approach a straight line on a loglog plot with a slope of 1/3.

The intersection of this straight line with that which applies in the small-signal limit occurs when

$$W_s = \tfrac{8}{9}(\beta^2 W_L{}^3/G_s G_L{}^3) = W_L/A_o = W_L\,(\Delta G)^2/4G_s G_L \tag{43}$$

or
$$\Delta G = (4\sqrt{2}/3)(\beta W_L/G)$$

at this point, the actual gain would have fallen by the amount

$$\frac{\tfrac{3}{9}2}{(4\sqrt{2}/3 + \tfrac{3}{2} - \tfrac{1}{4})^2} = 0.36 = -4.5 \text{ db} \tag{44}$$

There are several reasons why one would want to operate at the inflection point—not the least of which is that the gain is independent of fluctuations in the bias voltage to first order. This means that the gain is likely to be more stable, and it also means that low-frequency excess noise (which can be looked at as a bias fluctuation) does not modulate the signal as severely.

8.4. Experimental Results—Operation at Inflection Point. Before we turn our attention to operating points at other than the inflection point, let us look at some experimental results which are intended to check the analysis presented here and give some idea of what can be done in practice. Several amplifiers designed to operate with a variable transformation ratio between both the input terminals and the diode and between the diode and the output terminals were built. The object was to vary both G_s and G_l without having to modify the impedances of the primary source or the detector which served as the ultimate load.

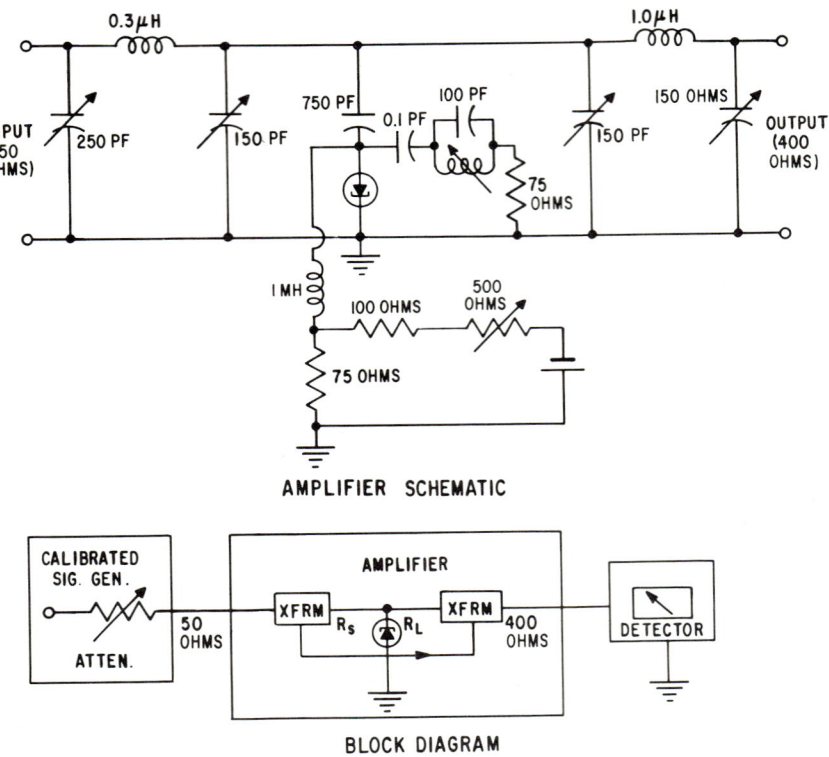

FIG. 21. Amplifier block diagram and schematic.

The circuit diagram is shown in Fig. 21 together with a block diagram of the test setup. The amplifier was tested at gains as high as 70 db, but in order to minimize adjustment, while providing a reasonably high gain, the data presented here were taken at a small-signal gain of under 60 db.

The first experiment was designed to check Eq. (38). The source impedance seen by the diode was varied in order to produce different values of gain.

The signal power was increased until the gain was 3 db down from its value for very small signals. The values of input power and output power for this condition are recorded in Fig. 22. The data fall on a line whose slope equals 1/3 to within experimental error even though no attempt was made to adjust G_l and G_s to their optimum values. In fact, with an eye toward low-noise performance, G_l was chosen to be less

TUNNEL DIODES AND BACKWARD DIODES 12-19

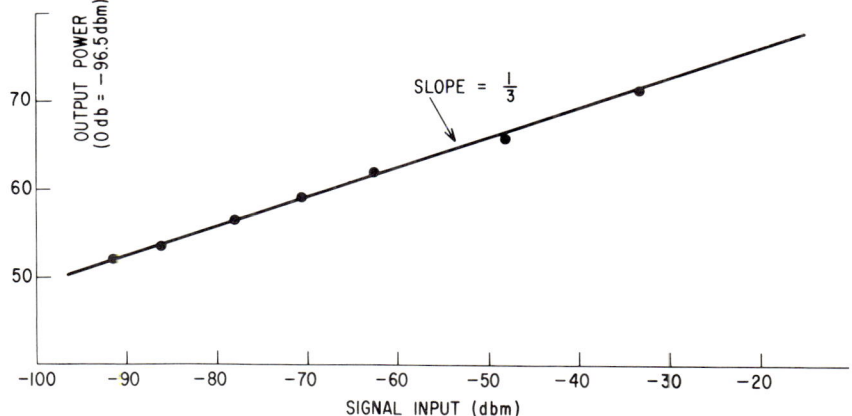

FIG. 22. Gain down 3 db from small-signal value.

than G_s.[9] Under these conditions one can see from Eq. (36) that by varying the source impedance we are producing a variation proportional to $G_s^{\frac{1}{4}}$. Since G_s is varied only by a small percentage of itself to produce all the gains, for which data are presented, the effect of variations in G_s is negligible.

The next experiment was designed to check Eq. (32). At very small signals, the gain of the amplifier was set at several values ranging from 20 to 50 db gain. The output vs. input was plotted to the point where the detector was overloaded (see Fig. 23). As expected, there were two linear regions having slopes of unity and 1/3,

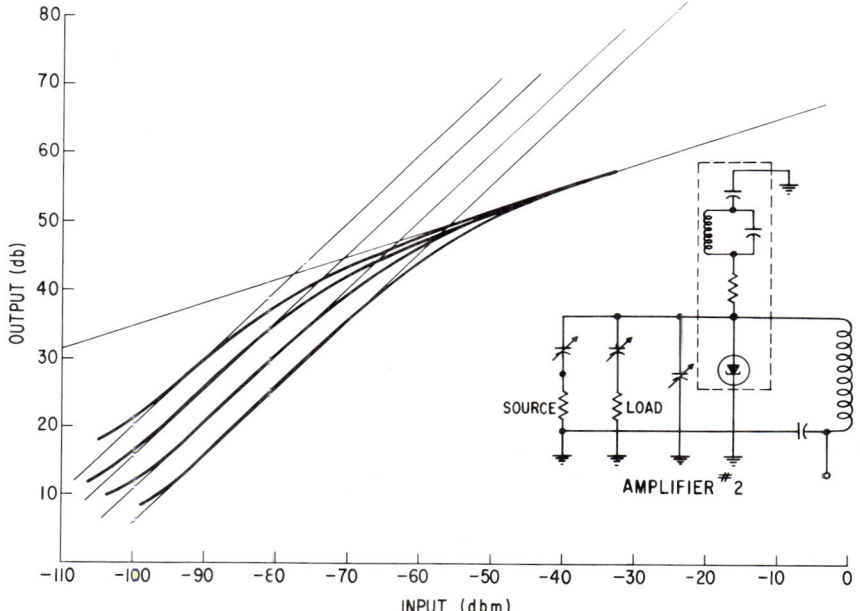

FIG 23. Amplifier response: diode B-12.

12-20 AMPLIFYING DEVICES

respectively. The only discrepancy was that the experimental amplifier was down only 3 db instead of 4.6 db at the point where the linear regions intersected. This would be expected if the susceptance was larger than ΔG. It is not unreasonable that this is the case since the tuning controls were adjusted always for maximum response instead of for zero susceptance. The two situations do not coincide because the real part of the admittance is a function of the capacitor settings. Thus the maximum response always occurs slightly off resonance on the side where the conductance is smaller.

Distortion. By combining Eqs. (34) and (30), we find that at the saturation limit

$$e_3 = 0.083 e_o \tag{45}$$

which corresponds to a third harmonic power content of 0.7 per cent. This value drops off very rapidly as the output voltage drops below that of the saturation limit.

8.5. Operating at Other Than Inflection Point of I-V Characteristic. For this case, the second harmonic distortion is going to dominate the small-signal behavior (except when the gain is extremely high).

It is reasonable to approximate this situation by ignoring the existence of β. If we set $\beta = 0$ in Eqs. (25), (26), and (27), they become

$$i_1 = e_1 \Delta G + \alpha(-e_1 e_2 + e_2 e_3) \tag{46}$$
$$0 = e_2 \Delta G + \alpha(-\tfrac{1}{2} e_1^2 + e_1 e_3) \tag{47}$$
$$0 = e_3 \Delta G + \alpha e_1 e_2 \tag{48}$$

The solution of these equations can be expressed as a power series in α, and the solutions up to α^2 are

$$e_1 = (i_1/\Delta G)\{1 - \tfrac{1}{2}[\alpha i_1/(\Delta G)^2]^2 + \cdots\}^{-1} \tag{49}$$
$$e_2 = (i_1/\Delta G)\{\tfrac{1}{2}[\alpha i_1/(\Delta G^2)] + \cdots\} \tag{50}$$
$$e_3 = (i_1/\Delta G)\{\tfrac{1}{2}[\alpha i_1/(\Delta G)^2]^2 + \cdots\} \tag{51}$$

In this case there is an increase in the gain at the fundamental frequency, but it is only of second order in α. In fact, the increase is exactly equal in amplitude to the third harmonic amplitude. Thus, distortion is possibly even more important than the change in gain in determining the maximum permissible gain.

If we apply the same criterion as before for determining the permissible gain, we write

$$\sqrt{A_o/A} = 1/\sqrt{2} = \{1 - \tfrac{1}{2}[\alpha i_1/(\Delta G)^2]^2\} \tag{52}$$

and the solution to this to first order in α is

$$\Delta G \approx 1.31(\alpha e_1) \tag{53}$$

The gain at this point is

$$A_{max} = 4 G_s G_e/(\Delta G)^2 = 1.08(G_l/\alpha)(G_s/W_s)^{\tfrac{1}{2}} \tag{54}$$

and the output power is

$$W_l = W_s A_{max} = 1.08(G_l/\alpha)(G_s W_s)^{\tfrac{1}{2}} \tag{55}$$

By comparing with Eq. (36) it can be seen that the operating point is poorer than the case studied previously if the main consideration is to get the most output power for a given input power. The best possible choices of G_l and G_s this time are

$$G_s = \tfrac{1}{3} G \qquad G_l = \tfrac{2}{3} G$$

which gives

$$A_{max} = 0.415(G^{\tfrac{3}{2}}/\alpha W_l^{\tfrac{1}{2}}) \tag{56}$$

As before, we can calculate the fraction of the power appearing in the second harmonic.

Equations (50) and (53) yield

$$e_2 = 0.381 e_1 \qquad (57)$$

which corresponds to a second harmonic power content of 14 per cent.

The higher-order terms in the expansion of the conductance characteristic have an interesting effect which is shown in Fig. 24. Note that as the bias point is changed (and the source conductance is modified to keep the small-signal gain constant) the saturation changes from sublinear to superlinear and finally to a situation where oscillation occurs. What is happening here is that the relative magnitude of the fourth-order term is being increased with respect to the third-order term. At the bias point where the two terms are comparable in magnitude, the dynamic range for linearity is increased by about 17 db.

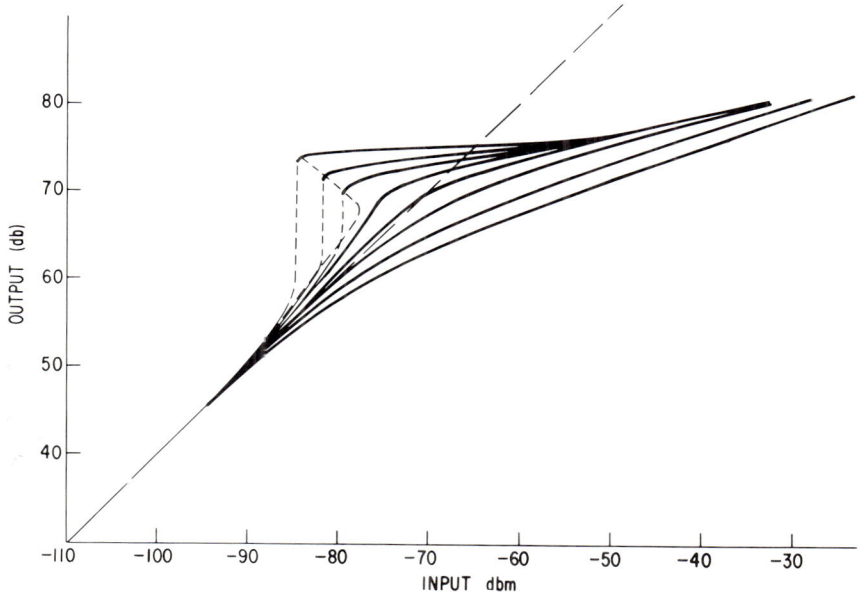

Fig. 24. Response of amplifier away from inflection point.

9. MIXERS

Both the third-order nonlinearity at the inflection-point amplifier of case 1 and the second-order nonlinearity of case 2 can be used to obtain mixing. The situation in case 1 is likely to be found for autodyne converters, since it is easiest to obtain self-oscillation at the bias point of maximum negative conductance. We shall treat this case first.

Case 1. Mixing at an Inflection Point. We shall be dealing with an equivalent circuit something like the one shown in Fig. 25. We shall simplify this considerably in order to make the analysis more transparent. We first note that at all the frequencies of interest the admittance seen by the tunnel diode is real. Furthermore, if all the tuned circuits have passbands substantially smaller than the separation of the frequencies f_s, f_o, f_q, then the admittance seen by the diode at any one of the frequencies is determined by only one of the tuned circuits. We can then separate the behavior of the circuit as regards the three frequencies.

We shall write the equivalent circuit of the converter at the three frequencies as in

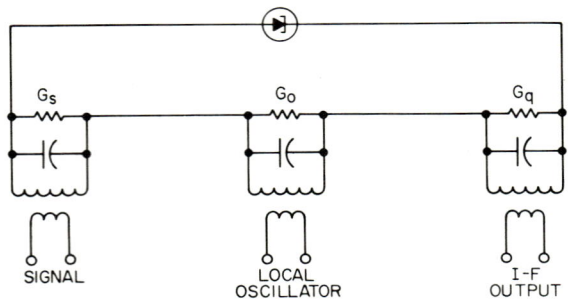

FIG. 25. Equivalent circuit of mixer.

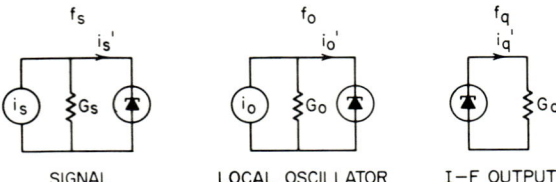

FIG. 26. Equivalent circuits at signal, local oscillator, and i-f frequencies.

Fig. 26, where the diode characteristic will be assumed independent of frequency. We have

$$i_s + G_s e_s + i_s'(e_s) = 0 \tag{58}$$
$$i_o + G_o e_o + i_o'(e_o) = 0 \tag{59}$$
$$e_q G_q - i_q'(e_q) = 0 \tag{60}$$
$$i_s' = i_s - G_s e_s \tag{61}$$
$$i_o' = i_o + G_o e_o \tag{62}$$
$$i_q' = G_q e_q \tag{63}$$

where the total diode current is

$$i = (i_s' + i_o' - i_q') \tag{64}$$

the diode voltage is

$$e = (e_s + e_o + e_q) \tag{65}$$

and

$$i = -Ge + \beta e^3 + \cdots$$
$$= -Ge + \beta e_o^3 + 3\beta e_o^2 (e_s + e_q) + \cdots \tag{66}$$

We assume voltages of the form

$$e_o = V_o \sin \omega_o t \tag{67}$$
$$e_s = V_s \sin \omega_s t \tag{68}$$
$$e_q = V_q \sin \omega_q t \tag{69}$$

so that

$$i = -Ge + (\beta V_o^3/4)(3 \sin \omega_o t - \sin 3\omega_o t)$$
$$\quad + (3\beta V_o^2/2)(1 - \cos 2\omega_o t)(V_s \sin \omega_s t + V_q \sin \omega_q t) + \cdots$$
$$= -Ge + (3\beta V_o^2/2)(e_s + e_q) + (3\beta V_o^2/4)e_o$$
$$\quad - (3\beta B_o^2/4)(V_s \sin \omega_q t + V_q \sin \omega_s t) + \cdots \tag{70}$$

where ω_q has been chosen:* $\omega_q = \omega_s - 2\omega_o$ and all time dependencies other than those

* Note that this corresponds to a local oscillator frequency one-half as high as the usual choice. If the usual choice ($\omega_o = \omega_s \pm \omega_q$) had been used, there would have been no output unless the bias were moved away from the inflection point.

for which there are no tuned circuits present have been dropped. Thus we have

$$I_s - G_s V_s = -GV_s + (3\beta V_o^2/2)V_s - (3\beta V_o^2/4)V_q \tag{71}$$
$$I_o - G_o V_o = -GV_o + 3\beta V_o^3/4 \tag{72}$$
$$I_q - G_q V_q = GV_q - (3\beta V_o^2/2)V_q + (3\beta V_o^2/4)V_s \tag{73}$$

where $i_s = I_s \sin \omega_s t$, etc. Equation (73) can be solved for the output voltage

$$V_q = (G_q - G_o + 3\beta V_o^2/2)I_q \tag{74}$$

In the case of an autodyne, $I_o = 0$ and we have

$$\Delta G_o = 3\beta V_o^2/4 \tag{75}$$

where $\Delta G_o = G - G_o$ is now the amount by which the diode negative conductance exceeds the oscillator circuit conductance. Hence

$$V_q = + \frac{\Delta G_o V_s}{\Delta G_q + 2\Delta G_o} \tag{76}$$

which approaches $1/2V_s$ for large ΔG_o. Also, $I_s = [\Delta G_s + (2 + V_q/V_s)\Delta G_o]V_s$.

That is, the finite amplitude of the local oscillator and i-f voltages present an effective conductance at the signal frequency equivalent to

$$G_{eff} = (2 + V_q/V_s)\Delta G_o \tag{77}$$

The maximum conversion gain permissible is again limited by saturation considerations, and is about 6 db less than the maximum gain as a straight-through amplifier.

An important property of a mixer, especially when conditions of ultimate gain are sought, is whether or not there is intermodulation between the desired frequency band and others which contain unwanted noise power. Because all resistances and batteries as well as the tunnel diode itself generate a noise spectrum which has an excess low-frequency component (i.e., flicker or $1/f$ noise), it is of interest to calculate the intermodulation between the signal and frequencies near zero.

We shall proceed by regarding the low-frequency noise as a perturbation on the local oscillator. Thus Eqs. (65) and (66) become

$$e = (e_s + e_q + e_o + e_n) \tag{78}$$
$$i = -G_o e + \beta e_o^3 + 3\beta(e_o + e_n)^2(e_s + e_q) + \cdots \tag{79}$$

If we now look at the terms having the time dependence of the output frequency we find

$$I_q = -G_o V_q + (3\beta/2)V_o^2 V_s + (3\beta/2)[e_n(t)]^2 V_q \tag{80}$$

The last term in Eq. (80) is the intermodulation, and a single low-frequency component contributes to the output of one octave higher, at an amplitude proportional to the square of the amplitude of the low-frequency component.

Case 2. Mixing Elsewhere Than at Inflection Point. We now examine the case where $i = -G_o e + \alpha e^2 + \cdots$

where

$$\begin{aligned} e &= e_s + e_o + e_q + e_n \\ &= V_s \sin \omega_s t + V_o \cos \omega_o t + V_q \sin \omega_q t + e_n(t) \end{aligned} \tag{81}$$

combining, we have

$$i = -G_o e + \alpha e_o^2 + V_o V_s \sin \omega_q t + \alpha V_o V_q \sin \omega_s t$$
$$+ 2\alpha e_n(t)V_q \sin \omega_q t + 2\alpha e_n(t)V_s \sin \omega_s t \tag{82}$$
$$I_q = G_q V_q = +G_o V_q + \alpha V_o V_s + 2\alpha e_n(t)V_q \tag{83}$$
$$I_s = G_s V_s = -G_o V_s + \beta V_q V_o + 2\alpha e_n(t)V_s \tag{84}$$

Here we see that the noise term is linear, rather than quadratic, in $e_n(t)$. It is therefore less attractive to operate away from the inflection point from the point of view of low-frequency excess noise, even though it is better to operate at a slightly higher bias from the point of view of minimizing the effect of shot noise.*

REFERENCES

1. L. Esaki, New Phenomenon in Narrow Ge P-N Junctions, *Phys. Rev.*, vol. 109, p. 603, January, 1958.
2. E. O. Kane, Observation of Direct Tunneling in Germanium, *Phys. Rev. Letters*, vol. 3, pp. 466–468, November, 1959.
3. E. O. Kane, Electron-electron Scattering in N-P Tunneling, *Bull. Am. Phys. Soc.*, vol. 5, p. 160, March, 1960.
4. "General Electric Tunnel Diode Manual," General Electric Co, Syracuse, N.Y.
5. W. B. Hauer, Definition and Determination of the Series Inductance of Tunnel Diodes, *IRE Trans. Electron Devices*, ED-8, pp. 470–475, November, 1961.
6. C. S. Kim and A. Brandli, High-frequency High-power Operation of Tunnel Diodes, *IRE Trans. Circuit Theory*, CT-8, pp. 416–425, December, 1961.
7. S. Ramo and J. R. Whinnery, "Fields and Waves in Modern Radio," p. 210, John Wiley & Sons, Inc., New York, 1953.
8. M. E. Hines, "High Frequency Limitation of Solid State Devices and Circuits," Digest of Technical Papers of 1963 International Solid State Circuit Conference.
9. J. J. Tiemann, Shot Noise in Tunnel Diode Amplifiers, *Proc. IRE*, vol. 48, no. 8, p. 1418, August, 1960.
10. R. La Rosa and C. R. Wilhelmsen, Theoretical Justification for Shot-noise Smoothing in the Esaki Diode, *Proc. IRE* (correspondence), vol. 48, p. 1903, November, 1960.
11. N. Holonyak and C. S. Kim, Application of Tunneling to Active Diodes, Scientific Rept. 66, AF 19(604)–6623, Oct. 30, 1961.
12. E. G. Nielsen, Noise Performance of Tunnel Diodes, *Proc. IRE*, vol. 48, no. 11, p. 1903, November, 1960.
13. R. S. Pucel, Theory of the Esaki Diode Frequency Converter, *Solid-State Electron.*, vol. 3, p. 167, 1961.

* For a further discussion of the noise properties of tunnel-diode mixers see, for example, Ref. 13.

Chapter 13

PARAMETRIC DEVICES

HARRY J. PEPPIATT*

CONTENTS

1. Capacitor Diodes.. 13-1
   1.1. Junction Elastance vs. Charge Current and Voltage......... 13-2
   1.2. Small-signal Equivalent Circuit.......................... 13-4
   1.3. Large-signal Analysis.................................... 13-10

1. CAPACITOR DIODES

Almost any semiconductor diode exhibits nonlinear capacitance effects to a certain extent. However, at the present time there is a large variety of diodes (frequently referred to as varactors) manufactured specifically for use in parametric amplifiers. The characterization of these devices for use in parametric amplifiers is somewhat complicated and time-consuming. For this reason very few diode manufacturers supply sufficient information for the device designers. Hence, in the next few sections a method of characterizing capacitor diodes will be treated in some detail. This method leads logically into a method[1] for adjusting and/or testing a completed parametric-amplifier design.

Anyone who undertakes a thorough study of the "simple" p-n junction is soon confronted with a variety of phenomena which are quite complicated and in some cases have not been fully explained. Many of these phenomena are of some importance to the amplifier designer, but in the main they have only second-order influence on the operation of a well-designed parametric amplifier and for this reason will be neglected in the synthesis and analysis of parametric amplifiers to be treated in a later chapter. Some of these phenomena are listed below along with some comments and references.

Diffusion Capacitance. This capacitance (sometimes called storage capacitance) is significant when the diode is operated in or near the forward-bias region. It is usually frequency-dependent, although in the case of the graded junction the dependence is slight.[2] In some cases it has been suggested[3] that it can improve amplifier performance, especially the gain-bandwidth product.

Resonance-response Anomalies. It can be shown[4] that when a nonlinear reactance is near resonance and is driven hard with a voltage source, the frequency-response curve becomes extremely asymmetric, as illustrated in Fig. 1.

* General Electric Co., Communications Products Dept., Lynchburg, Va.

It should be understood that this phenomenon results because of the nonlinear nature of the reactance (which of course is the underlying principle of parametric amplification) and hence cannot be entirely eliminated from parametric amplifiers. Intuitively, one can see that the effect is reduced by lowering the Q of the resonant circuit. Fortunately, the Q of the pump circuit of a parametric amplifier is low if the amplifier is pumped at the optimum pump frequency. Conversely, it is evident that this effect is prevalent when an amplifier is pumped at much less than its optimum pump frequency. The effect manifests itself in anomalous tuning effects and bias stability problems.[5]

High-frequency Avalanche Phenomena. When a high-frequency signal is applied to a *p-n* diode and minority carriers are injected during a forward portion of the cycle, it is believed[6] that substantial avalanche current can flow in the reverse direction during the reverse portion of the cycle. This occurs if the lifetime of minority carriers is large compared with the r-f period and if conditions[7] for avalanche multiplication prevail. If the r-f level or bias is such that the forward peaks do not produce the normal forward current, then this anomalous avalanche effect can be

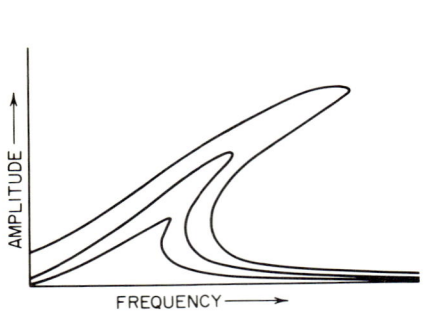 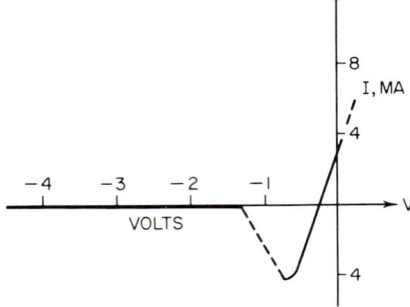

Fig. 1. Asymmetric response of a nonlinear tuned circuit at three different levels.

Fig. 2. The d-c characteristic of a germanium diode with high-level r-f applied (1.5 watts).

observed by noting the d-c (or average) reverse current of the device while the r-f is being applied. This reverse current can be an order of magnitude larger than the normal d-c reverse current. The writer has observed this effect in a high-level switching application. In Fig. 2, the measured current vs. voltage d-c characteristic is shown for an incident r-f power of 1.5 watts at a frequency of 6 gc. Note that the anomalous reverse current reaches a value slightly greater than -4 ma as the current drops back to its normal value (approximately at -1.2 volts). The dotted region of course represents a negative resistance in the bias circuit.

This effect can occur in a parametric amplifier if the diode is being pumped at too high a level. In such case, in addition to possible bias-circuit instabilities, the noise may be excessively high. There is little direct evidence that the pump level can be lowered to avoid this avalanche effect without also reducing the dynamic quality factor produced by the diffusion and depletion capacitances. However, the fact that many amplifiers have been built which give very closely their predicted performance would lead to believe that this is so.

Nonlinear Resistance. The bulk resistance of the diode varies with the depletion-layer capacitance. This variation can be quite small if the overall length of the diode is large with respect to the width of the depletion layer at breakdown.

1.1. Junction Elastance vs. Charge Current and Voltage. The small-signal junction capacitance at any bias point of a capacitance diode has been analyzed in great detail in various articles (see references at the end of chapter). A simple approximate treatment is given here which is more from the point of view of elastance S.[8] Obviously, the use of S rather than capacitance C ($= 1/S$) should have no deep significance in the mathematical sense, but it does provide added insight into practical aspects, as will be evident in what follows.

First, the abrupt-junction diode will be discussed. As is well known, in this case the net impurity concentration changes abruptly from type n to type p, as illustrated in Fig. 3.

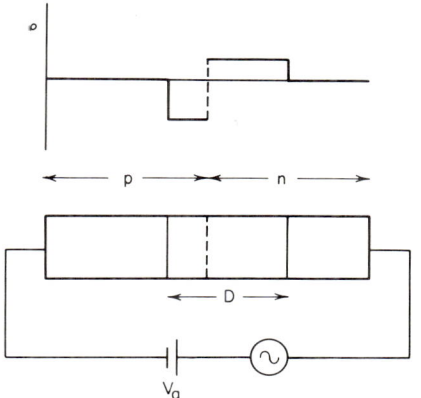

FIG. 3. Abrupt p-n junction and its d-c charge distribution in the reverse-bias condition.

For a sufficiently small a-c voltage applied to the junction, the junction acts as an elastance given by

$$S = D/\epsilon A \qquad (1)$$

where D is the depletion-layer thickness, A is the area of the junction, and ϵ is the permittivity of the dielectric. The total uncompensated d-c charge on either the p or n side of the diode is

$$\tfrac{1}{2}\int |\rho| A \, dl = Q' \qquad (2)$$

where ρ is the charge density in the depletion layer and the integration is carried out through the length D of the depletion layer. If the concentration is constant within both the p and the n regions, it is obvious that Q' is proportional to D, i.e.,

$$Q' = kD \qquad (3)$$

where k is a constant for a given abrupt-junction diode. Substituting this value of D into Eq. (1) gives for the elastance

$$S = Q'/\epsilon A k \qquad (4)$$

That is, the small-signal elastance is proportional to the absolute value of the d-c charge in either the p or the n side of the depletion layer. This total d-c charge can be altered by the d-c bias voltage V_a. The maximum value of Q' ($= Q_B'$) occurs at breakdown, and of course, this gives the maximum a-c elastance S_{max} where

$$S_{max} = Q_B'/\epsilon A k \qquad (5)$$

and by combining Eqs. (4) and (5)

$$S = (Q'/Q_B')S_{max} \qquad (6)$$

At zero applied d-c voltage there is a finite depletion charge Q_φ, say, that is set up to balance the diffusion forces at the junction. Hence, the total charge Q' can be split

13-4 AMPLIFYING DEVICES

into two parts Q and Q_φ, and the expression for the elastance can be written

$$S = [(Q + Q_\varphi)/(Q_B + Q_\varphi)]S_{max} \tag{7}$$

where Q is the d-c charge set up on either p or n side by the applied d-c voltage V_a only.

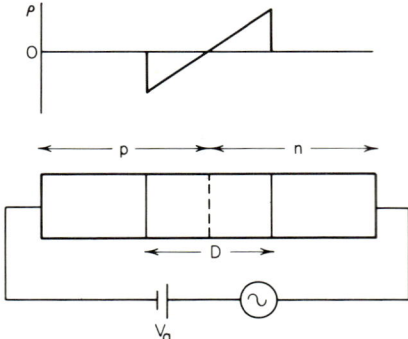

FIG. 4. Graded p-n junction and its d-c charge distribution in the reverse-bias condition.

In the case of a linear graded junction as illustrated in Fig. 4, the relation between S and Q is derived in a similar way, i.e.,

$$Q' = \tfrac{1}{2}\int|\rho|A\,dl \tag{8}$$

where the integration is carried out through the depletion layer. In this case, since ρ varies linearly with distance

$$Q' \propto D^2 \tag{9}$$

and combining Eqs. (1) and (9) we obtain

$$S = \sqrt{Q'/Q_B'}\,S_{max} \tag{10}$$

or

$$S = \sqrt{(Q + Q_\varphi)/(Q_B + Q_\varphi)}\,S_{max} \tag{11}$$

The relations between S and Q are illustrated in Fig. 5.

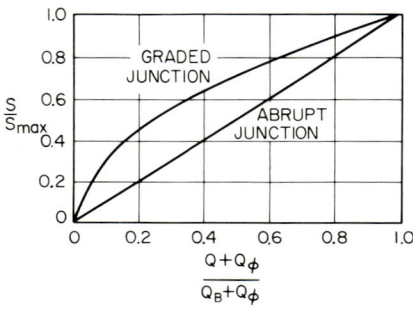

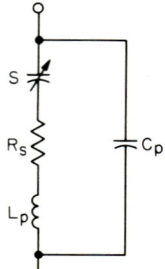

FIG. 5. Small-signal elastance vs. depleted charge.

FIG. 6. Approximate lumped equivalent of capacitance diode.

1.2. Small-signal Equivalent Circuit. Although most capacitance diodes are used at microwave frequencies, the diode packages in the main are sufficiently small that the diode junction plus package can be represented by the small-signal lumped equivalent circuit shown in Fig. 6.

L_p represents the package inductance and C_p represents the shunt package capacitance. Here, the junction conductance and leakage conductance are neglected. Rather complicated measurement procedures are required to determine the element values in this circuit. It will be evident from a later discussion on parametric amplifiers that each of these quantities plays an important role in the design of a practical amplifier.

The relation between S and the d-c voltage at the diode junction can be obtained by solving Poisson's equation in the depletion layer. In the case of the abrupt junction, $V_a + \varphi$ is proportional to D^2; hence

$$S = D/\epsilon A = \sqrt{(V_a + \varphi)/(V_B + \varphi)}\, S_{max} \qquad (12)$$

where φ is the contact potential, V_a is the applied voltage, and V_B is the applied breakdown voltage. In the linear-graded junction $V_a + \varphi$ is proportional to D^3 and hence

$$S = [(V_a + \varphi)/(V_B + \varphi)]^{\frac{1}{3}} S_{max} \qquad (13)$$

These relations between S and the d-c voltage are illustrated in Fig. 7.

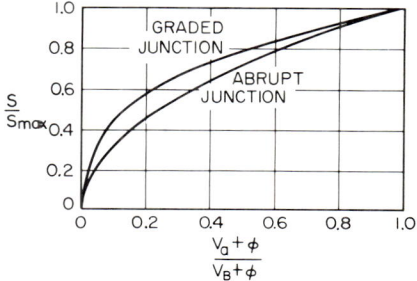

Fig. 7. Small-signal elastance vs. the d-c voltage at the junction.

The diode equivalent circuit is more complicated than it may appear at first sight since in a parametric-amplifier application the junction elastance is varied by the pump source. As an aid in displaying this nonlinear aspect of the equivalent circuit several Smith chart plots of the normalized impedance of the circuit as a function of bias voltage are plotted in Fig. 8, with bias ranging from 10 μa reverse current to 10 μa forward current. These plots were computed for a diode with an R_S of 2 ohms, a capacitance swing of 7.2 to 0.85 pf, and an L_p of 4.34×10^{-9} henrys. The reactances are normalized to 50 ohms. The plots show the effect of shunt capacitances of 0, 0.4, and 1 pf. Note that near 2 gc the d-c bias is capable of varying the reactance equally on each side of the zero reactance point. This we shall define as the "series-resonant" frequency f_{sr}.

At all frequencies, for $C_p = 0$, the plot remains on the 0.04 constant-resistance circle. Note that at 1 gc (below f_{sr}) the effect of C_p is to shift the plot to higher-resistance points with corresponding reduction in reactance swing. At frequencies higher than f_{sr}, the points move in to lower-resistance points with a correspondingly greater reactance swing. Also at even higher frequencies C_p causes the diode to swing from inductive to capacitive reactance near the high-impedance end of the chart, indicating of course a parallel-resonance phenomenon (see $C_p = 1$ pf at 3 gc). The plots of impedance represent those which would be obtained from a slotted-line impedance measurement with the diode placed on the end of the line if the proper reference point is chosen. In practice it is difficult to choose this reference point, especially at the higher frequencies. In Fig. 9, measurements are presented for an actual diode with characteristics similar to those of the "diode" computed above. Here note the general similarity between the computed plots and the measured. However, in the forward-bias region, the impedance turns inward, and although a portion of this can be attributed to the shunt capacitance C_p, it appears that there is also a slight increase in the series resistance, which might be expected.

This measured diode junction has excellent high-frequency characteristics and is capable of good performance in a 5- to 10-gc amplifier. However, the series induct-

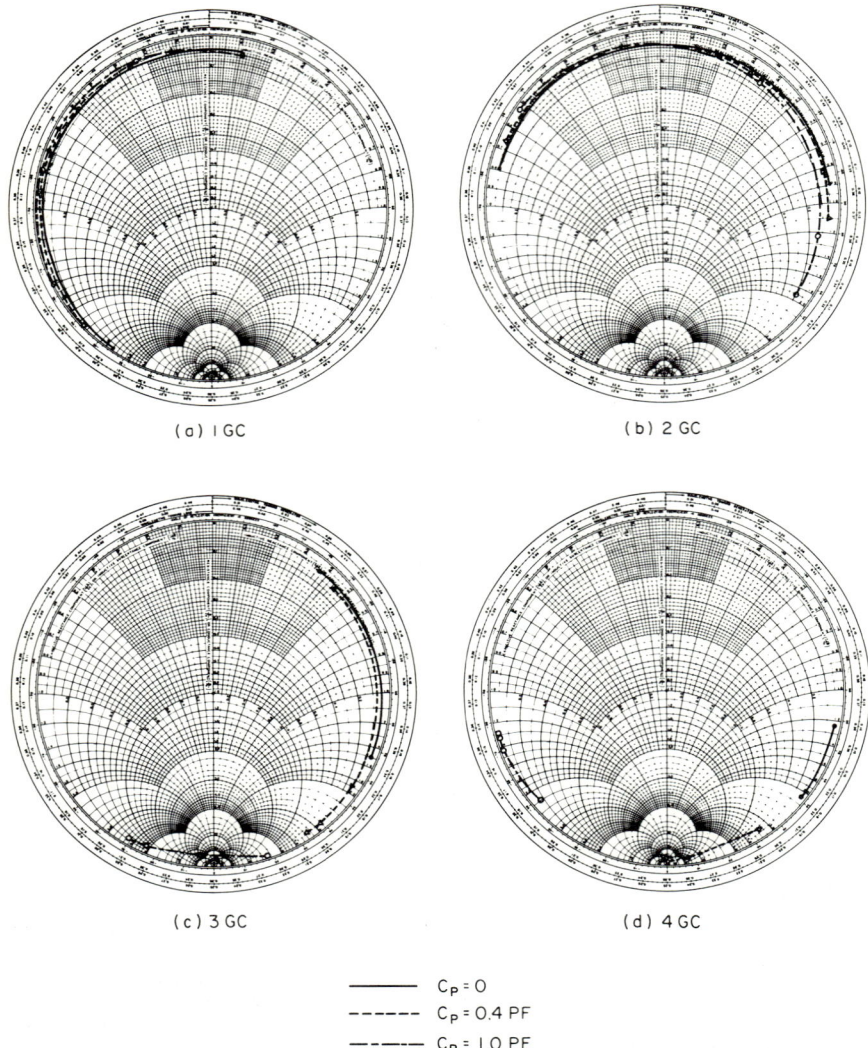

FIG. 8. Computed small-signal impedance of capacitance diode.

ance and shunt capacitance are too large for best operation in this frequency range. That is, somewhat complicated circuits would be necessary to achieve the proper impedance levels and tuning conditions at the signal and idler frequencies. In addition, this diode is not too suitable in the vhf range for a different reason. Here, the

PARAMETRIC DEVICES   13-7

impedance levels would be rather high and circuit losses would probably degrade the noise figure from that which the diode is inherently capable of giving.

In Fig. 10, the computed impedance of a diode with the following parameters is shown:

$$C_p = 0.4 \text{ pf}$$
$$L_p = 2 \text{ nh}$$
$$R_s = 4 \text{ ohms}$$
$$C_{max} = 1/S_{min} = 15 \text{ pf}$$
$$C_{min} = 1/S_{max} = 1.7 \text{ pf}$$

This represents the characteristics of a typical uhf unit. The solid curve shows the impedance variation with $C_p = 0$ at a frequency of 2.036 gc, which is also the series-resonant frequency for $C_p = 0$. With $C_p = 0.4$ pf, the series resonance is reduced to

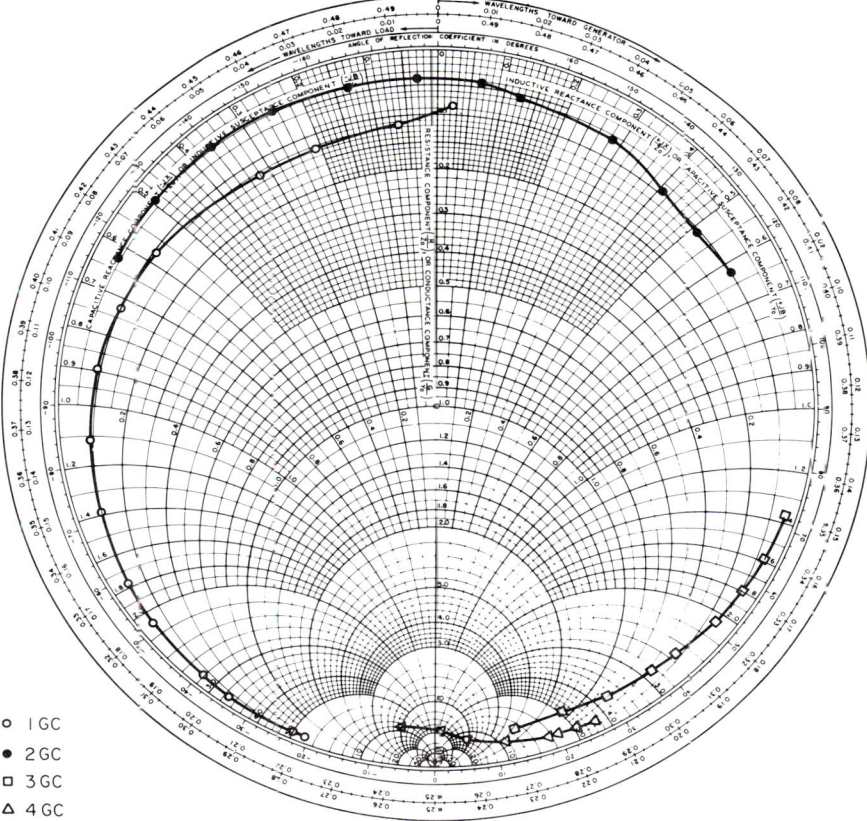

○ 1 GC
● 2 GC
□ 3 GC
△ 4 GC

FIG. 9. Measured small-signal impedance of capacitance diode.

1.954 gc. The circular points are obtained by rotating the computed points until a best fit on a constant-resistance circle with equal positive and negative reactance swing is obtained. The series resonance obtained in this way is about 2.4 gc. This indicates that if impedance measurements are taken on a slotted line with the diode at the end of the line and the points rotated for best fit on a constant-resistance circle considerable error will result in estimating the series-resonant frequency. A better procedure would be first to remove the admittance due to C_p from the measured points (assuming the measurement is made near a multiple of a half wavelength from the diode). The resultant points should lie near the constant-resistance circle, and the frequency corresponding to equal reactance swing is approximately the true series-resonant frequency.

One of the important questions which arises in connection with the use of capacitance diodes in parametric amplifiers is "is there a single quantity or figure of merit which accurately describes the quality of a diode for parametric-amplifier operation?" From

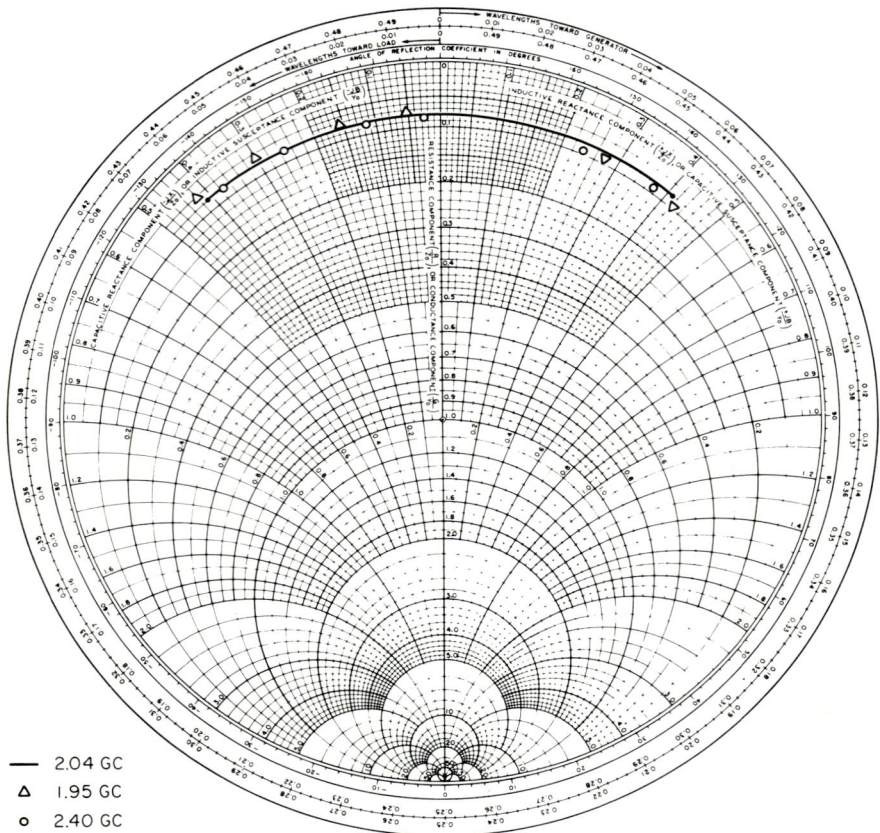

— 2.04 GC
△ 1.95 GC
○ 2.40 GC

Fig. 10. Computed impedance of a typical uhf unit.

the number of different figures of merit which have been proposed one must conclude that there is not one all-encompassing definition. It is clear that both the elastance variation of the diode and the series resistance must be included in the definition. That is, a diode with a very low R_s and negligible elastance change is of little use, as is a diode with a large elastance change and a large R_s. From the Smith charts presented above, it is possible to define a figure of merit which is adequate from the amplifier designer's point of view and which is easily measured. This quantity is defined as

$$\omega_Q = (1/4R_s)(S_{max} - S_{min}) \tag{14}$$

or
$$f_Q = (1/8\pi R_s)(S_{max} - S_{min}) \tag{15}$$

PARAMETRIC DEVICES

In Chap. 28 it is shown that ω_Q plays a central role in the synthesis of parametric amplifiers.

If the shunt capacitive admittance is removed from the measured impedance or if it is negligible f_Q can be expressed in terms of quantities obtained from the Smith charts, i.e.,

$$f_Q = \frac{f}{4R_s/R_g} \frac{\Delta X}{R_g} \qquad (16)$$

$$f_Q = \frac{f}{4r_s} \Delta x \qquad (17)$$

where Δx is the total normalized reactance change measured on the constant-resistance circle, r_s is the normalized resistance read from the circle, f is the frequency at which the measurement is made, and R_g is the impedance of the line. From the Smith chart in Fig. 10, the f_Q of that particular diode is

$$f_Q = \frac{2.036 \times 0.82}{4 \times 0.08}$$
$$= 5.2 \text{ gc}$$

To illustrate the accuracy with which the measurement of f_Q can be made a high-frequency pill diode was measured. The diode was placed at the very end of a high-quality high-frequency 50-ohm coaxial slotted line. The dimensions of the diode are about twenty times smaller than the wavelength at the highest frequency used. Measurements were made at 7, 8, 8.5, 10, and 11 gc. A shunt capacitance of 0.15 pf was arrived at and the corresponding admittance was removed. The results of the series-resistance measurement at 7 and 11 gc are shown in Table 1. The average resistance is 2.8 ohms with an average deviation of about ±0.26 ohms, or approximately ±10 per cent. The bias was adjusted in the range −5.8 to +0.64 volts. The direct current at the extremes in bias voltage was 10 μa. The series resistance increased to a maximum of about 7.8 ohms in the forward-bias region. This increase in R_s in the forward direction is illustrated in the Smith charts of Fig. 11. The definition of f_Q given in Eq. (17) assumes that R_s is a constant independent of the bias voltage. The analysis of the combination of a nonlinear resistance and a nonlinear capacitance is extremely complicated. For this reason we are tempted to define f_Q' as follows:

$$f_Q' = (f/4r_s') \Delta x$$

where Δx is the total normalized reactance swing and r_s' is the reverse-bias normalized series resistance.

Table 1. Series Resistance of Pill-type Capacitance Diode, Ohms

Bias voltage, volts	Frequency, gc				
	7	8	8.5	10	11
−5.8	2.53	2.66	3.00	2.80	2.20
−4.0	2.45	2.67	2.60	3.00	2.15
−2.0	2.50	2.66	2.85	3.00	2.40
−1.0	2.50	2.76	3.00	3.10	2.45

It can be seen that there is excellent agreement in the measurements. From the analysis in Chap. 28, it can be shown that a constant-resistance capacitance diode with an f_Q of 44.1 gc could be used in a room-temperature parametric amplifier to produce a noise figure of 1 db at 5 gc or 3 db at 15 gc. Of course, the pill diode discussed here would probably fall short of this performance because of the increase in its resistance in the forward-bias region. It should be emphasized that f_Q depends only on the S_{max}, S_{min}, and R_s, and it can be measured by simple slotted-line techniques using small signals. The discussion does not say anything about the manner in which the elastance varies with the bias voltage, and this is important in the application of the device to parametric amplifiers. However, the writer is of the opinion that the law of elastance variation should be discussed in connection with the pump circuitry used in the amplifier design and should not be included in a fundamental figure-of-merit quantity. This point will be discussed in further detail in the next section where high-level operation of the capacitance diode is considered.

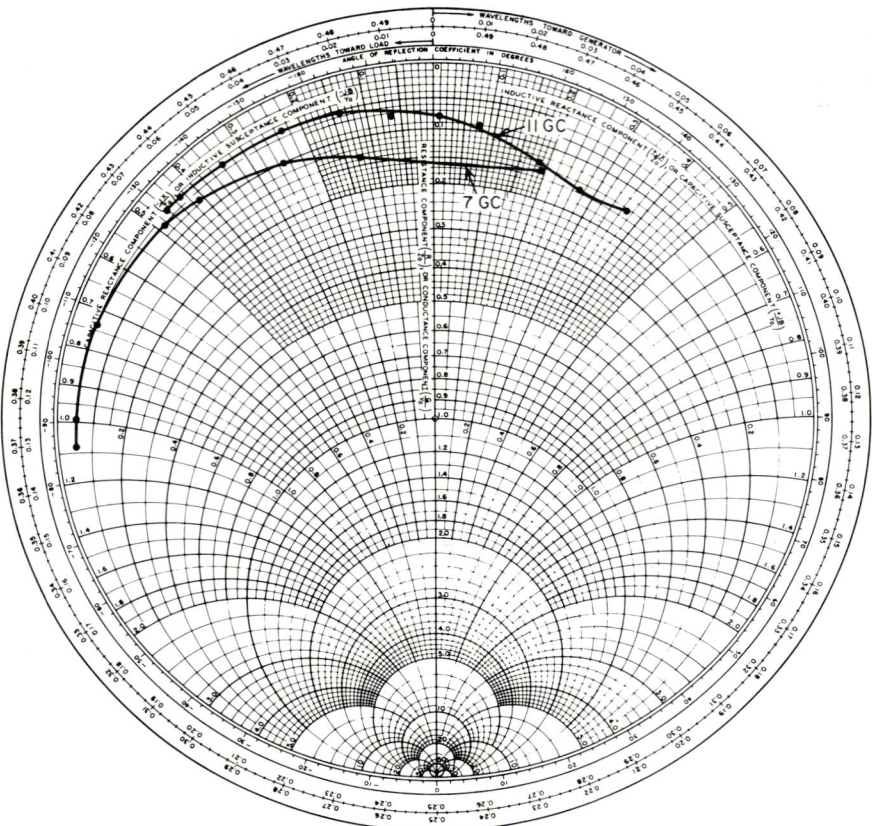

Fig. 11. Measured small-signal impedance of pill-type capacitance diode.

1.3. Large-signal Analysis. To determine the necessary characteristics of the pump circuit in a parametric amplifier a large-signal analysis must be carried out. The Smith chart plots presented previously apply only to the case where the bias voltage is altered at a slow rate and a small r-f signal is applied to the diode terminals. However, there is no known practical limit to the rate at which the junction elastance S can be varied. Hence we can write down the following relations between junction elastance and the r-f charge and voltage:

$$S(t) = \frac{q(t) S_{max}}{Q_B + Q_\varphi} = \frac{q(t)}{Q_B{}'} S_{max}$$
$$= \sqrt{v(t)/(V_B + \varphi)}\, S_{max} = \sqrt{v(t)/V_B{}'}\, S_{max} \qquad (18)$$

for an abrupt junction and

$$S(t) = \sqrt{q(t)/Q_B'}\, S_{max}$$
$$= [v(t)/V_B']^{\frac{1}{2}} S_{max} \qquad (19)$$

for a graded junction.

In a parametric-amplifier application, the d-c bias is fixed and the junction elastance is varied by a high-frequency signal at a level sufficient to drive the junction elastance over its useful range. Because of the nonlinear nature of S, it is obvious that any circuit representation of the diode must contain parameters which are level-sensitive. Hence the concept of complex impedance, a linear concept, is of little significance. One must return to the basic formulation of circuit laws to initiate the analysis. First, consider a very simple case of an abrupt junction driven by a current generator (see Fig. 12). Of course, we assume the generator has infinite impedance at all frequencies.

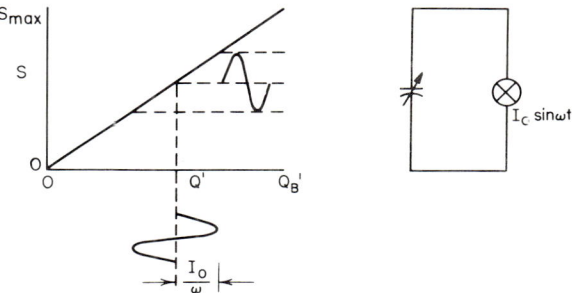

Fig. 12. Abrupt-junction diode driven by a current source.

Hence we can say that the current through the diode is $I_o \sin \omega t$ and that the instantaneous charge is

$$q = \int I_o \sin \omega t \, dt$$
$$= (-I_o/\omega) \cos \omega t + \text{const} \qquad (20)$$

and assume the d-c bias is such that the constant is $Q_B'/2$ and hence

$$q = -(I_o/\omega) \cos \omega t + Q_B'/2 \qquad (21)$$

where $I_o/\omega \leq Q_B'/2$. The instantaneous voltage across the junction is given by the relation in Eq. (18), and therefore,

$$v(t) = (S^2/S^2_{max}) V_B'$$
$$S(t) = (S_{max}/Q_B') q(t)$$

and since $v(t)$ becomes

$$v(t) = (1/Q_B'^2) q^2 V_B'$$
$$= (V_B'/Q_B'^2)[-(I_o/\omega) \cos \omega t + Q_B'/2]^2$$
$$= (V_B'/Q_B'^2)[(I_o^2/\omega^2)(\tfrac{1}{2} + \cos \omega t/2) - (I_o Q_B'/\omega) \cos \omega t + Q_B'^2/4]$$
$$= (V_B'/Q_B'^2)[(I_o^2/2\omega^2) \cos 2\omega t - (I_o Q_B'/\omega) \cos \omega t + I_o^2/2\omega^2 + Q_B'^2/4] \qquad (22)$$

If the diode is driven fully, then

$$v(t) = V_B'(\tfrac{1}{8} \cos 2\omega t - \tfrac{1}{4} \cos \omega t + \tfrac{3}{8}) \qquad (23)$$

The bias voltage in this case is $\tfrac{3}{8} V_B'$; for example, for a unit with a 6-volt breakdown voltage, and a 0.5-volt contact potential, the bias would be 2.48 volts or 1.98 volts negative.

The addition of R_s in series with the junction capacitance alters the above analysis very little if one assumes that the source is large at all harmonic frequencies other than the fundamental. Since the peak current for full elastance swing is $Q_B' \omega/2$ the power absorbed by R_s is $(Q_B'^2\omega/8)R_s$ and since $V_B' = \frac{1}{2}S_{max}Q_B'$ this power is $\frac{1}{2}V_B'^2(\omega^2 R_s/S^2_{max})$. Expressed in terms of $\omega_Q \simeq (1/4R_s)S_{max}$ the power is $\frac{1}{32}(V_B'^2/R_s)(\omega^2/\omega_Q^2)$.

In a later chapter, it is shown that the optimum pump frequency of a parametric amplifier is near ω_Q and the pump power in this case would be approximately $\frac{1}{32}(V_B'^2/R_s)$.

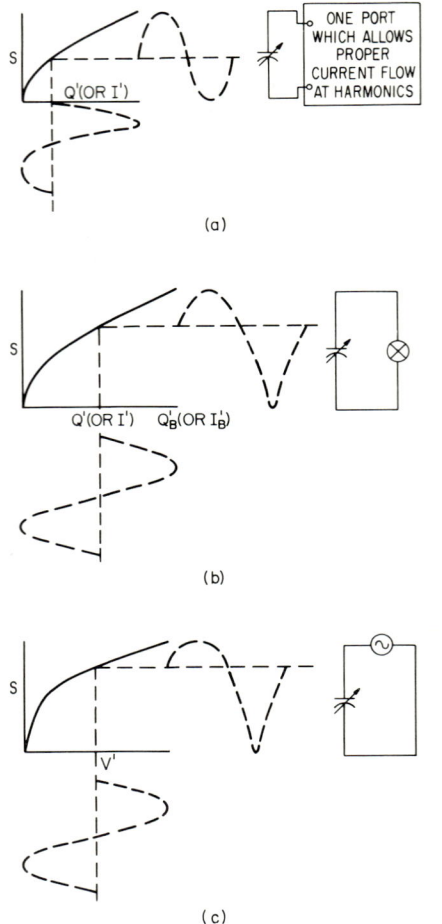

FIG. 13. (a) Sinusoidal elastance. (b) Sinusoidal current. (c) Sinusoidal voltage.

The ratio of the voltage to the current at the fundamental frequency is

$$\frac{I_o(V_B'/\omega Q_B')\cos\omega t}{I_o \sin\omega t}$$

which is analogous to a capacitive reactance of

$$V_B'/\omega Q_B' = (-1/2)(S_{max}Q_B'/\omega Q_B') = -S_{max}/2\omega$$

Hence to achieve full elastance swing with the minimum pump power the internal impedance of the pump source must be an inductive reactance of $S_{max}/2\omega_o$ in series with a resistance equal to R_s. Also, of course the source voltage must be $2V_B'\omega R_s/S_{max}$ for proper match and proper elastance swing.

In summary, in the above paragraphs, we have obtained the necessary data to design the pump circuit for an abrupt-junction-diode parametric amplifier, assuming the diode junction is open-circuited at all harmonics of the pump frequency. The addition of the series inductance alters the above analysis only slightly. It should be noted that the analysis does not predict any unusual resonance response which would occur if the same circuit were driven with a voltage source.[4]

Of course, a diode can be pumped in an infinite number of ways. The analysis of the abrupt-junction current pumping is the simplest because of the linear relation between S and Q (or i). The solution of other cases involves complex Fourier analysis. In Fig. 13, a few cases are illustrated graphically.

In the simplest parametric amplifiers, the elastance change at the fundamental (or pump) frequency is of particular importance. For a given total elastance change, the pumping circuit which gives the largest peak-to-peak elastance variation at the fundamental is desirable. However, the requirements on pump power and the complexity of the pumping circuit may be prohibitive. For example, square-wave pumping of the junction gives a peak-to-peak elastance swing at the fundamental which is about 27 per cent greater than the total swing, but it requires an infinite amount of

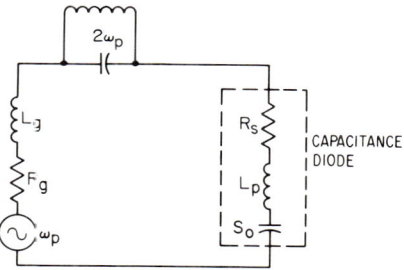

FIG. 14. A circuit for achieving sinusoidal current pumping of an abrupt-junction capacitance diode.

pump power. The peak-to-peak swing in elastance can be made equal to the total elastance variation if a sinusoidal source is used and the source impedances at the harmonics are adjusted to give the proper current or voltage waveform at the junction. The case of the abrupt junction has been discussed. It is the simplest in detail and is obtained by open-circuiting the harmonic terminations; i.e., a sinusoidal current waveform is forced. If a voltage source (zero impedance at all harmonics) is used, the fundamental peak-to-peak variation is less than the available swing (see Fig. 13).

In addition, since the resistance R_s is physically in series with the junction it does not allow for true voltage-source operation. In the case of sinusoidal capacitance,[8] the peak-to-peak elastance change is again less than the total change available and a complicated voltage or current waveform is required.

Hence, from both practical and theoretical considerations, sinusoidal current pumping seems desirable. In the case of the abrupt junction this gives a peak-to-peak at the fundamental which is equal to the total elastance change available, and in the case of the graded junction the peak-to-peak[8] is as much as 85 per cent of the total. A circuit for achieving sinusoidal current pumping with an abrupt-junction diode is sketched in Fig. 14.

REFERENCES

1. K. Kurokawa, On the Use of Passive Circuit Measurements for the Adjustment of Variable Capacitance Amplifiers, *Bell System Tech. J.*, vol. 41, pp. 361–381, January, 1962.

2. A. van der Ziel, "Solid State Physical Electronics," pp. 291–295, Prentice-Hall, Inc., Englewood Cliffs, N.J., 1957.
3. K. Siegel, Comparative Figures of Merit for Available Varactor Diodes, *Proc. IRE*, vol. 49, pp. 809–810, April, 1961.
4. W. J. Cunningham, "Introduction to Nonlinear Analysis," pp. 173–186, McGraw-Hill Book Company, New York, 1958.
5. J. C. McDade, Explanation of One Type of RF Induced Negative Resistance in Junction Diodes, *Proc. IRE*, vol. 50, p. 91, January, 1962.
6. I. Hefini, Effect of Minority Carriers on the Dynamic Characteristic of Parametric Diodes, *Electron. Eng.*, vol. 32, pp. 226–227, April, 1960.
7. K. G. McKay, Avalanche Breakdown in Silicon, *Phys. Rev.*, vol. 94, pp. 877–884, May, 1954.
8. P. Penfield and R. P. Rafuse, "Varactor Applications," The M.I.T. Press, Cambridge, Mass., 1962.

Chapter 14

INDUCED-EMISSION-AMPLIFIER PRINCIPLES

EDMUND B. TUCKER[*]

CONTENTS

1. Introduction.. 14-3
2. Blackbody Radiation... 14-3
   2.1. Mode Density and Energy per Mode..................... 14-3
   2.2. Absorption, Induced and Spontaneous Emission.......... 14-4
   2.3. Effect of Line Shape................................... 14-6
   2.4. Interaction between a Field and a System of Energy Levels—Attenuation and Amplification.......................... 14-6
3. Induced-emission Amplifiers................................. 14-7
   3.1. General Requirements for Induced-emission Amplifiers.... 14-7
   3.2. Classification of Induced-emission Amplifiers............ 14-7
   3.3. Energy Levels... 14-8
4. Methods of Population Inversion............................. 14-8
   4.1. Two-level Gas Maser................................... 14-8
5. Methods of Population Inversion in Two-level Paramagnetic Systems... 14-9
   5.1. The Rotating Coordinate System....................... 14-10
   5.2. Sudden Field Reversal................................ 14-10
   5.3. The 180° Pulse....................................... 14-10
   5.4. Adiabatic Fast Passage................................ 14-11
6. R-F Methods of Population Inversion in Multilevel Paramagnetic Systems... 14-12
   6.1. Paramagnetic Three-level Systems..................... 14-12
   6.2. Push-Push Pumping................................... 14-14
   6.3. Push-Pull Pumping.................................... 14-14
   6.4. Staircase Inversion................................... 14-14
   6.5. Relaxation-time Modification by Spin-Spin Coupling...... 14-15
7. Optical Pumping... 14-15
   7.1. Ground-state Population Inversion by Polarization-dependent Transitions....................................... 14-16
   7.2. Ground-state Inversion Using Laser Pumping........... 14-17
8. Laser Inversion Methods.................................... 14-17
   8.1. Optical Pumping...................................... 14-17
   8.2. Gas Discharge Inversion—Electron Impact.............. 14-18
   8.3. Resonant Transfer or Inelastic Atom-Atom Collisions 14-19

[*] General Electric Co., Research and Development Center, Schenectady, N.Y.

AMPLIFYING DEVICES

    8.4. Molecular Dissociation (Dissociative Excitation Transfer). 14–20
    8.5. Semiconducting Junction Inversion. 14–20
9. Induced-emission Amplifiers Which Do Not Require Inverted Populations. 14–21
    9.1. Double-quantum Maser. 14–21
    9.2. The Raman Laser. 14–21
10. Noise. 14–23
11. Transmission-cavity-maser Characteristics. 14–25
    11.1. Gain. 14–25
    11.2. Bandwidth. 14–26
    11.3. Noise. 14–28
    11.4. Gain Modulation. 14–28
12. Reflection-cavity-maser Characteristics. 14–29
    12.1. Gain. 14–29
    12.2. Bandwidth. 14–29
    12.3. Noise. 14–29
    12.4. Gain Modulation. 14–30
13. The Traveling-wave Maser. 14–30
    13.1. Gain. 14–30
    13.2. Bandwidth. 14–32
    13.3. Noise. 14–32
    13.4. Gain Modulation. 14–32
14. Comparison of Maser Types. 14–32
15. Laser Amplifier Characteristics. 14–34
16. Induced-emission Oscillators. 14–35
    16.1. Tuning. 14–35
    16.2. Frequency Stability. 14–37
    16.3. Oscillation Conditions and Pump Power Requirements. 14–37
    16.4. Laser-oscillator Power Output. 14–39
17. Cavities. 14–39
    17.1. Maser Cavities. 14–39
    17.2. Laser Cavity—Normal Modes. 14–39
    17.3. Mirrors—Metal Films. 14–42
    17.4. Mirrors—Dielectric Coatings. 14–42
18. Laser Pump Sources. 14–44

LIST OF SYMBOLS

B	bandwidth
β	Bohr magneton = 0.92732×10^{-20} erg/gauss
c	velocity of light
E	electric field amplitude
$\mathcal{E}$	energy
$f(T)$	Planck expression for noise power per unit bandwidth
F_N	noise figure of amplifier
g	spectroscopic splitting factor or g factor
$g(\nu)$	line-shape function
G	gain
h	Planck's constant = 6.6252×10^{-27} erg-sec
$\hbar$	$h/2\pi$
H	magnetic field amplitude *and* in paramagnetic formulas, where custom demands it, for the magnetic flux density (usually denoted by B)
$\mathbf{k}$	wave vector of absolute value $2\pi/\lambda$
k	Boltzmann constant = 1.38042×10^{-16} erg/deg
M	magnetization
N	phonon or photon population
$\bar{N}_\nu$	number of photons or phonons per mode
n	level populations (number of ions occupying a given level)

P	power
p_ν	number of modes per unit volume per unit frequency interval
r	power-reflection coefficient
T	temperature
$T_{N\,eff}$	effective noise temperature of amplifier
T_1	spin-lattice relaxation time (longitudinal relaxation time)
T_2	dephasing time or transverse relaxation time
t	time
τ	relaxation time
v	phase velocity
v_g	group velocity
w	transition probability due to thermal effects and accounting for relaxation time T_1
W	transition probability due to applied fields
γ	gyromagnetic ratio $ge/2mc$
ν	frequency
λ	wavelength
ω	angular frequency $2\pi\nu$
ρ_ν	energy density of radiation field
ξ	filling factor
$\chi = \chi' + i\chi''$	magnetic susceptibility

1. INTRODUCTION

The basic principles of induced-emission amplifiers are to be found in Einstein's treatment of the equilibrium between a system of energy levels and thermal radiation. The now famous A and B coefficients, accounting, respectively, for spontaneous and stimulated emission, are certainly better understood by many people today thanks to the work of Weber, Townes, and Schawlow, who were responsible in one way or another for getting the field of induced-emission amplifiers started. This is another example of the development of an instrument, from principles well understood by a number of people, primarily for the purpose of more precise scientific study. The practical implications with all the attendant urgency caused the field to expand very rapidly as measured by the effort expended and the results obtained. The uses are specialized, and it is difficult to generalize except for some of the basic principles.

After a brief discussion of the basic concepts, this chapter covers the fundamentals of population inversion. System properties which are peculiar to or are important only in the case of induced-emission amplifiers such as the modes of a Fabry-Perot interferometer are dealt with briefly. In many cases, the circuit illustrations of Chap. 29 may be helpful in more clearly defining the practical problems and their solutions.

2. BLACKBODY RADIATION

Many of the concepts and formulas required in the detailed discussion of induced-emission amplifiers are conveniently introduced by considering thermal radiation in a black box.

2.1. Mode Density and Energy per Mode. The equilibrium electromagnetic (heat) radiation in a box depends only on the temperature. The number of normal modes is readily determined if one assumes that a wave equation such as that for the vector potential **A**

$$\nabla^2 \mathbf{A}(\mathbf{r},t) - (1/v^2)(\partial^2/\partial t^2)\mathbf{A}(\mathbf{r},t) = 0 \tag{1}$$

(where v = velocity) governs the field and furthermore that each component of the field can be expanded in a Fourier series as a function of $\mathbf{k}\cdot\mathbf{r}$ where $\mathbf{k}$ is the wave vector $(2\pi/\lambda)$ and $\mathbf{r}$ the distance.

$$\mathbf{A}_i(\mathbf{r},t) = \sum_k \mathbf{A}_{ik}(t)e^{i\mathbf{k}\cdot\mathbf{r}} \tag{2}$$

$\mathbf{A}_{ik}(t)$ is a constant which varies with $\mathbf{k}$ and the summation is over the allowed values of $\mathbf{k}$. Assuming cyclic boundary conditions (or using some similar restriction) the values of $\mathbf{k}$ are limited to

$$k_x = 2\pi p/L \qquad k_y = 2\pi q/L \qquad k_z = 2\pi r/L \tag{3}$$

where p, q, r are integers and L^3 is the volume of the box. A sinusoidal dependence on time for the components of the field, and these three equations lead to the result that

$$\omega^2/v^2 = k^2 \equiv k_x^2 + k_y^2 + k_z^2 \tag{4}$$

If the modes are considered in k space a volume $dk_x\, dk_y\, dk_z = (2\pi/L)^3$ will correspond to a single mode. The number of modes with k values of from zero up to k_{max} will be given by the volume of a sphere in k space divided by $(2\pi/L)^3$ or

$$\frac{\frac{4}{3}\pi k^3}{(2\pi/L)^3} = \frac{k^3 L^3}{6\pi^2} = \frac{4}{3}\frac{\pi \nu^3 L^3}{v^3} \tag{5}$$

(where ν = frequency) for each polarization direction. The last equality is the result of the fact that $k = 2\pi/\lambda = 2\pi(\nu/v)$. If different polarizations travel with different velocities v_i the number of modes per unit volume becomes

$$\tfrac{4}{3}\pi\nu^3(1/v_1^3 + 1/v_2^3 + 1/v_3^3) \tag{6}$$

In the case of light only two transverse modes traveling at the same velocity c are possible and the familiar $\tfrac{8}{3}\pi\nu^3/c^3$ is the number of modes per unit volume.

Taking the derivative of Eq. (6) with respect to frequency gives the number of modes per unit volume per unit frequency interval p_ν,

$$p_\nu = 4\pi\nu^2(1/v_1^3 + 1/v_2^3 + 1/v_3^3) \tag{7}$$

which is equal to $8\pi\nu^2/c^3$ for electromagnetic radiation with two polarizations. Consideration of the mode density is very important when considering the amplification, oscillation, and noise properties of the amplifiers with which this chapter is concerned.

In thermal equilibrium the average energy per mode is given by Planck's distribution law as

$$\bar{\mathcal{E}}_\nu = \frac{h\nu}{\exp(h\nu/kT) - 1} \tag{8}$$

For the case of $h\nu \ll kT$ this reduces to the classical value of kT. h is Planck's constant and the product $h\nu$ is the energy of a single quantum of energy of frequency ν. The number of quanta per mode $\bar{\mathcal{E}}_\nu/h\nu$ is simply

$$\bar{N}_\nu = \frac{1}{\exp(h\nu/kT) - 1} \tag{9}$$

The energy density of the radiation field ρ_ν is given by $p_\nu \bar{\mathcal{E}}_\nu$ with the value

$$\rho_\nu = \frac{8\pi\nu^2}{v^3}\left(\frac{h\nu}{\exp(h\nu/kT) - 1}\right) \tag{10}$$

If a second system is placed in this box filled with blackbody radiation, after sufficient time elapses for equilibrium to be established, the energy absorbed per unit of time by this new system from the radiation field exactly balances the energy emitted by this same system to the field. In the next section the implications of such a balance for a simple system of energy levels will be discussed.

2.2. Absorption, Induced and Spontaneous Emission. If a system (Fig. 1) consisting of two levels with energies $\mathcal{E}_1$ and $\mathcal{E}_2$ and a frequency difference $(\mathcal{E}_2 - \mathcal{E}_1)/h = \nu$ is allowed to interact with the thermal-radiation field, the balance of energy absorption and energy emission suggests that the details of the absorption and emission process be investigated. Absorption in this two-level system requires the elevation of a particle (perhaps an ion) from $\mathcal{E}_1$ to $\mathcal{E}_2$ utilizing energy $h\nu$ from the radiation field.

Emission of energy is the result of the reverse process, and in thermal equilibrium the energy loss must equal the energy gain. Assuming a total of n particles, n_1 in the lower level and n_2 in the upper level, and labeling the transition probabilities as W_{12}, W_{21} for the probability of a particle going from level $\mathcal{E}_1$ to $\mathcal{E}_2$ and vice versa we have

$$n_1 W_{12} h\nu = n_2 W_{21} h\nu \tag{11}$$

for the energy balance. The transition probability for absorption W_{12} is proportional to the energy density $\rho(\nu)$ of the radiation field at frequency ν, but the emission probability W_{21} must include a term representing the tendency of an $\mathcal{E}_2 \to \mathcal{E}_1$ event occurring even in the absence of a radiation field—spontaneous emission. That is,

$$n_1 B_{12} \rho(\nu) h\nu = n_2 h\nu [B_{21}\rho(\nu) + A] \tag{12}$$

where A is the spontaneous-emission probability and $B\rho(\nu)$ replaces W as the induced probability.

In thermal equilibrium the ratios of the level populations must be governed by the Boltzmann ratio

$$n_2/n_1 = e^{-h\nu/kT} = e^{-(\mathcal{E}_2 - \mathcal{E}_1)/kT} \tag{13}$$

Combining the results of Eq. (13) with (12):

$$\rho(\nu) = \frac{A}{e^{h\nu/kT} B_{12} - B_{21}} \tag{14}$$

and equating this to Eq. (10):

$$B_{21} = B_{12} = B$$
and $$A/B = 8\pi h\nu^3/v^3 \tag{15}$$

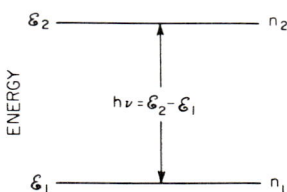

Fig. 1. System of two energy levels. The n's refer to populations.

The equality of B_{21} and B_{12} indicates that the probability of absorption is equal to the probability of induced emission. The physical significance of this relationship is important and will be mentioned later a number of times. It is immediately evident that the spontaneous emission is responsible for maintaining the balance-of-energy exchange in thermal equilibrium with $n_1 > n_2$. Referring to the definition of $W_{12} = B\rho(\nu)$, the second part of the above equation together with Eqs. (9) and (10) results in

$$W_{12}/A = B\rho(\nu)/A = \bar{N} \tag{16}$$

Thus the induced-emission rate into a given mode is $\bar{N}$ times the rate of spontaneous emission into the same mode. It can be shown by a quantum-mechanical derivation that

$$B\rho(\nu) = (e^2 \omega^3 / 2\pi \hbar v^3) |X_{ab}|^2 \bar{N}_\nu = A \bar{N}_\nu \tag{17}$$

where X_{ab} is the matrix element involved in the transition.

The fact that the induced absorption and emission must be proportional to the incident energy, i.e., $\bar{N}_\nu$, is not unexpected and throws more light on the interpretation of Eq. (16). If the total emission from the system is calculated quantum-mechanically, the result is equivalent to Eq. (12) using (16) with total emission proportional to $\bar{N}_\nu + 1$. The spontaneous emission into a mode is equivalent to the induced emission produced by one quantum of radiation in that mode while the induced transitions (absorption and emission) are proportional to the actual number of quanta per mode.

The question of spontaneous emission is an important one for induced-emission amplifiers since it is the major source of noise in them. Recognizing that the rate of spontaneous emission is given by Eq. (17) with $\bar{N}_\nu$ replaced by 1 it is evident that the spontaneous-emission rate is proportional to ω^3. The magnitude of this factor changes drastically from the maser range to the laser range [$\omega^3 = (2\pi)^3 \times 10^{27}$ at 1,000 Mc/sec and $\omega^3 \simeq (2\pi)^3 \times 10^{45}$ at $\lambda = 3,000$ Å]. This will obviously be important in the consideration of noise.

Even though Eq. (6) indicates less than a single mode in a volume corresponding to that of a microwave maser cavity, the cavity itself ensures there being at least one mode. Because of this enhanced mode density in microwave cavities, spontaneous emission is increased and the relaxation times are more rapid than one calculates for a similar unenclosed volume. For the frequencies involved in lasers, the mode density from Eq. (6) is very nearly the cavity mode density, and a large number of modes are responsible for spontaneous emission.

2.3. Effect of Line Shape. The above discussion is adequate if $\rho(\nu)$ is constant over the frequency range considered. There are cases when this assumption is not justified, and it is necessary to consider the frequency characteristics of the emission or absorption line. The line-shape function is generally represented by $g(\nu)$, and the two most common forms are the

Lorentz line
$$g(\nu) = \frac{2/T_2}{(\omega - \omega_0)^2 + 1/T_2{}^2} \tag{18a}$$

with full width at half maximum amplitude $\Delta\omega = 2/T_2$, when T_2 = transverse lifetime or the lifetime for dephasing, and the

Gaussian line
$$g(\nu) = 2T_2 e^{-\left[T_2{}^2 \frac{(\omega_0 - \omega)^2}{\pi}\right]} \tag{18b}$$

of full width at half maximum amplitude $\Delta\omega = 1.476 \times (2/T_2)$. These are both normalized to satisfy the usual criterion that

$$\int_0^\infty g(\nu)\, d\nu = 1 \tag{19}$$

The Lorentz line shape is generally associated with a line whose width is determined by its lifetime, while the Gaussian line results from Doppler shifts, local field inhomogeneities, nonidentical spins, etc.

All the previous equations, e.g., (15), (16), (17), involve the radiation to or from modes at the frequency ν. Rewriting them multiplied by $g(\nu)$ will convert the meaning to radiation to or from the modes in width $d\nu$ weighted according to the line-shape function. Equation (16) then becomes

$$\frac{W_{12} g(\nu)\, d\nu}{A g(\nu)\, d\nu} = \bar{N} = \frac{\rho(\nu) v^3}{8\pi h \nu^3} \tag{20}$$

In many induced-emission systems the spectral range of radiation is small compared with the line width. Because in the case of a traveling interaction we are dealing with an energy flux $I(\nu)\, d\nu = v\rho(\nu)\, d\nu$ which, for a monochromatic frequency, becomes I_ν (watts/sq cm) $= v\rho(\nu)$ it will be convenient to rewrite (20) in the form

$$W_{12}' = (v^2/8\pi h\nu^3) A g(\nu) I_\nu \tag{21}$$

where W_{12}' is the total transition rate using a monochromatic source.

2.4. Interaction between a Field and a System of Energy Levels—Attenuation and Amplification. For simplicity, consider the same two energy levels of Fig. 1 interacting with a field. The field need not be specified as long as it has an interaction with the system of energy levels; the same arguments hold for electromagnetic radiation interacting with electron energy levels through electric or magnetic interaction and for lattice vibrations interacting with the energy levels of a paramagnetic spin system. The frequency of the field is determined in this case by $h\nu = \varepsilon_2 - \varepsilon_1$, the condition for resonance. From the results of Sec. 2.2, it is evident that the field produces two effects. It gives up energy to the system causing transfer of ions (let us refer to the levels as ionic) from ε_1 to the higher energy ε_2. At the same time ions are induced to emit energy $h\nu$ and drop from level ε_2 to level ε_1. This induced emission is coherent with and indistinguishable from the incident energy. Since the probability for induced emission is the same as the probability for absorption, the net effect of radiation on the system depends on the populations n_1 and n_2 of the lower and upper levels, respectively. The total absorption is proportional to n_1, the induced emission to n_2, and the net effect for the thermal-equilibrium situation will be an absorption proportional to $\Delta n = n_1 - n_2 = n_1(1 - e^{-h\nu/kT})$. This is the well-known absorption of incident energy by any system of energy levels in thermodynamic equilibrium and accounts for the absorption of light passing through a gas containing atoms whose levels are resonant with the light and for the absorption of energy by magnetic spin levels in paramagnetic resonance. The lower the temperature the larger Δn and the larger the absorption or attenuation.

If in some fashion the populations of the energy levels can be inverted, i.e., n_2 made larger than n_1, the induced emission will exceed the absorption, the radiation field will gain energy, and we have an amplifier. This is the fundamental principle of induced-emission amplifiers. The basis of understanding lies in Einstein's A and B coefficients. It should be emphasized that physically the same processes occur in attenuation, for there is in that case a cancellation of some absorption by induced emission. The difference lies in the populations of the energy levels.

3. INDUCED-EMISSION AMPLIFIERS

3.1. General Requirements for Induced-emission Amplifiers. The requirements for any induced-emission device are:

1. Appropriate energy levels
2. A system for inverting level populations (note exception in Sec. 9)
3. A cavity with high enough Q that the gain from the inverted level system is sufficient to overcome cavity losses
4. Provision for coupling the signal frequency in and out

3.2. Classification of Induced-emission Amplifiers. Induced-emission amplifiers fall into two broad categories separated by time of development and frequency range. The maser, an acronym for "microwave amplification by stimulated emission of radiation," was first proposed by Townes et al.[1] as the name for the induced-emission amplifiers which utilize electric or magnetic dipole transitions with associated resonant frequencies of from 10^8 to 10^{11} cps. A number of different types of masers have been developed, each with some characteristic of circuit or inversion scheme which differentiates it from others. We shall consider the following categories: active materials—gases, solids—two and three-level; circuit properties—cavity, traveling wave. Obviously these are not exclusive categories, and a traveling-wave maser may utilize a three-level solid-state material.

The same principle of stimulated emission applied to amplification of optical and infrared frequencies was proposed by Townes and Schawlow.[2] Since that time, many acronyms have been coined for these devices, but the name laser (light amplification by stimulated emission of radiation) has become the accepted term covering infrared and optical frequencies. The proposed names iraser (infrared amplification by stimulated emission of radiation), loser (light oscillation . . .), etc., have not found favor. In the laser category we shall include gas lasers pumped by using a gas discharge or optical pumping and solid-state lasers with optical pumping or semiconductor junction inversion. These will be specified more precisely in the section on population inversion.

It will be observed that very few of the laser devices are used as amplifiers. They do oscillate and therefore have the gain capabilities of amplifiers if the feedback is eliminated. This is a difficult problem, but it is solvable. However, the main attraction of the laser is its coherent, relatively monochromatic output. There is no other light source available with this property, and until this is exploited the mere amplifying ability takes second place.

There are a few stimulated-emission devices not covered by the above categories. For example, the phonon maser has been demonstrated and the X-ray maser has been proposed,* but rather than using new acronyms the term maser is now interpreted as

---

* See articles by E. B. Tucker, L. Gold, and D. Marcuse in "Quantum Electronics III," listed in the bibliography.

emphasizing stimulated emission rather than the frequency range and is used with a clarifying word. In some circles the term *optical maser* is preferred to laser.

3.3. Energy Levels. The energy levels appropriate for maser work either are of the general type described in Sec. 4 for the ammonia-gas-beam maser or are paramagnetic levels. The latter are energy levels, due to the magnetic moments of unpaired electrons, in the outer shells of impurity ions of a crystalline lattice under the influence of a d-c magnetic field $\mathbf{H}_0$ either by itself or combined with the internal crystalline field. For a discussion of these energy levels reference should be made to any of the standard references given in the bibliography at the end of the chapter. Two-level masers may utilize the two energy levels due to a single unpaired electron or two levels of a multilevel system. The three- (or more) level inversion schemes require an ion with spin ≥ 1. (The number of levels is given by $2S + 1$.) In addition, the multilevel pumping schemes require transitions to nonadjacent levels. The probability for such transitions must be determined and will normally exist if the magnetic field is not aligned with the axis of the crystalline field.

Lasers make use of the normal optical energy levels, and reference to the standard works on optics and spectroscopy is in order for information on this subject. It is sufficient to state that, although these levels are affected by a magnetic field, laser action does not require a magnetic field but makes use of the energy levels resulting from ionic configuration (and crystalline electric fields when applicable).

4. METHODS OF POPULATION INVERSION

If one considers the inverted population of two energy levels in the light of the Boltzmann ratio of Eq. (13), the temperature characterizing the fact that $n_2 > n_1$ must be negative. For this reason, spin systems with inverted populations are often said to be at a negative temperature. This is a concept which must be used with considerable care and which is discussed in Sec. 10.

The methods of population inversion may be divided quite naturally into two types: The first consists of inversion methods making use of the two levels between which the stimulated emission is to take place; the second requires the presence of more than two levels and utilizes the extra levels in the inversion process. The following sections, devoted to methods of inversion, are divided in this fashion.

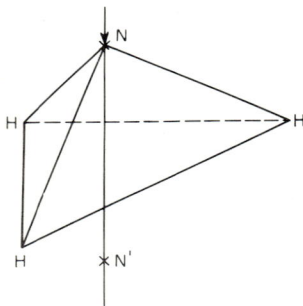

Fig. 2. The ammonia molecule with alternative positions, N and N' of the ammonia molecule.

4.1. Two-level Gas Maser. The first maser[1] utilized a population-inversion system which is conceptually very simple—the molecules in the lowest energy state are discarded. The gas used is ammonia, which possesses two energy levels distinguished from one another by the inversion of the nitrogen position (Fig. 2). The most probable state of NH_3 is the 3-3 state; the hydrogens have $3\hbar$ units of angular momentum and the nitrogen oscillates from one position to the other along the line of symmetry. The result of this inversion motion is that two levels, $\mathcal{E}_{1S}$ and $\mathcal{E}_{1A}$, separated by $23,870.14 \times 10^6$ cps make up the ground state. If NH_3 molecules in these two states are passed through an inhomogeneous electric field, the two levels are affected differently. The energies for the two levels may be expressed as

$$\begin{aligned}\mathcal{E}_{1S} &= \mathcal{E}_{1S}{}^0 - P_0\epsilon - P_0{}^2 E^2/\Delta W_1 \\ \mathcal{E}_{1A} &= \mathcal{E}_{1A}{}^0 - P_0\epsilon + P_0{}^2 E^2/\Delta W_1\end{aligned} \tag{22}$$

The subscripts $1S$ and $1A$ distinguish the two doublets. $\mathcal{E}_{1S}{}^0$ and $\mathcal{E}_{1A}{}^0$ are the unperturbed ground doublet levels, lower and upper, respectively. $P_0\epsilon$ is the change in energy due to a change in size induced by the electric field inhomogeneity. The last term represents the electric field acting on an electric dipole moment induced in the molecule. The sign of this last term is different for the two states, with the result that in order to minimize the energy those molecules in $\mathcal{E}_{1S}$ seek the strong-field region of an inhomogeneous electric field while those in $\mathcal{E}_{1A}$ gravitate toward the weak-field region.

Suppose a beam of NH_3 molecules is passed through a quadrupole electric field such as may be generated by the four-bar geometry of Fig. 3. The field near the axis

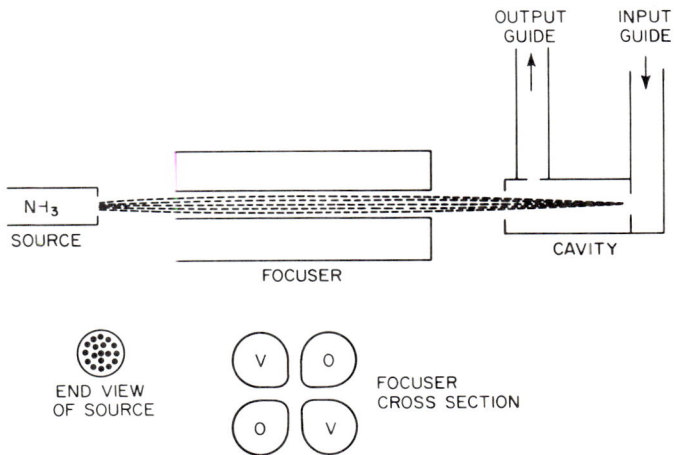

FIG. 3. Essentials of beam maser including quadrupole focuser cross section. (*After Gordon, Zieger, and Townes,*[1] *published by permission of the editor, Physical Review.*)

approximates hyperbolic cylinders and the electric field gradient is given by

$$\nabla|E| = G\mathbf{r}_1 \qquad (23)$$

where $\mathbf{r}_1$ = unit radius vector
G = const

The field gradient is thus directed radially. (Similar geometries are utilized in the so-called quadrupole or alternate-gradient focusing in accelerators and for particle-focusing nuclear-physics experiments.) The $\mathcal{E}_{1S}$ molecules experience a force outward and the molecules in the upper $\mathcal{E}_{1A}$ level are accelerated toward the axis. Thus the field arrangement of Fig. 3 acts as a focuser for the upper-level while defocusing the lower-level molecules. Collection of the upper-level ions in some fashion then can provide the necessary inverted population. The system is compatible with CW operation since the separation is continuous and a supply of molecules in the upper level is available as long as the source supplies molecules to the quadrupole focuser.

5. METHODS OF POPULATION INVERSION IN TWO-LEVEL PARAMAGNETIC SYSTEMS

Two-level paramagnetic systems differ from the gas system above in that the population of one level cannot be disposed of. The energy levels are properties of ions dispersed throughout the lattice of the host crystal. The populations of the levels in thermal equilibrium are governed by the Boltzmann ratio so that for the two-level system of Fig. 1 the population ratio is

$$n_2/n_1 = e^{-h\nu/kT}$$

The best population inversion which can be produced here is to exchange level populations, putting the number n_2 in level 1 and the number n_1 in level 2. Since the initial population difference is greater at low temperatures better inverted populations are achieved the lower the temperature.

5.1. The Rotating Coordinate System. The rotating frame of reference was first introduced by Rabi, Ramsey, and Schwinger.[3] The action of the spins under the influence of the d-c and r-f magnetic fields is considered in the frame rotating at the same rate as the spin about the d-c magnetic field. Actually, the spin is acted upon by both the d-c and r-f fields (H_0 and H_1, respectively). The resultant motion is a fast precession about the d-c field and a slow precession about H_1 as well. The motion is very complicated, and considerable simplification is obtained by observing it from the

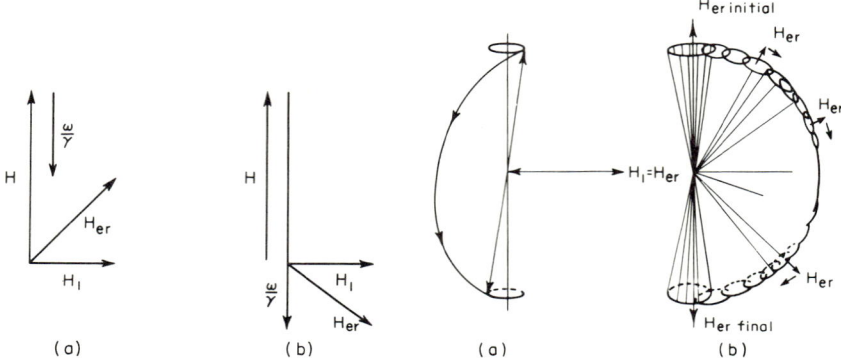

FIG. 4. Vectors in the rotating frame of reference. (a) Below resonance. (b) Above resonance.

FIG. 5. Spin moment in rotating system of reference. (a) 180° pulse inversion. (b) Adiabatic rapid passage.

system rotating with the effective component of H_1 at an angular frequency ω. The effective field in the rotating system is

$$H_{er} = H + H_1$$

but the d-c field H is reduced by the angular rotation from the value H_0 to $(H_0 - \omega/\gamma)$. The result in the rotating system is

$$|H_{er}| = [(|H_0| - \omega/\gamma)^2 + |H_1|^2]^{\frac{1}{2}} \qquad (24)$$

where γ = gyromagnetic ratio. Figure 4 illustrates these quantities. H_1 is most effective at right angles to H_0 as shown. In this system the spin precesses about H_{er} and at resonance, when $H_{er} = H_1$, about H_1. The motion of the spin is therefore much simplified.

5.2. Sudden Field Reversal. The principle of this method is that of reversing the magnetic field fast enough that the spins cannot follow and are left in an inverted-population situation. To be successful the field must be reversed in a time short compared with the precession time of the spins. Even for fields of a few gauss this requires inversions in times on the order of 10^{-8} sec. This inversion scheme has not been successful in electron paramagnetic systems but was utilized by Purcell and Pound[4] to invert the nuclear spins of LiF in the first demonstration of a negative temperature, and one in which excess of induced emission over absorption was demonstrated.

5.3. The 180° Pulse. From quantum-mechanical considerations the probability that a transition from a level $\mathcal{E}_1$ to a level $\mathcal{E}_2$, separated by a frequency ν_{12}, has taken place because of the action of a sinusoidal field of frequency ν applied for a time t is

$$W \frac{\sin^2\{[(\nu_{12} - \nu)^2 + W]^{\frac{1}{2}} \pi t\}}{(\nu_{12} - \nu)^2 + W} \qquad (25)$$

where $W = |\langle 1|M|2\rangle|^2 E^2/\hbar^2$ for an electric-dipole transition or $W = |\langle 1|M|2\rangle|^2 H^2/\hbar^2$ for a magnetic-dipole transition and $|\langle 1|M|2\rangle|$ represents the appropriate matrix element in each case, electric dipole and magnetic dipole, respectively. If $\nu = \nu_{12}$ and the field is applied for a time t given by

$$W^{\frac{1}{2}}t = \tfrac{1}{2} \tag{26}$$

the populations of the two levels will have been interchanged and an inverted population been achieved. It is assumed that no relaxation takes place during the process. This is explained in more physical form in the diagrams of Fig. 5a and by consideration of the rotating frame of reference. The magnetization **M** is directed along the axis of rotation. **H**₁, the resonant r-f field, is suddenly applied. The spin precesses about the applied r-f magnetic field **H**₁ (only one of the circulating polarized components is effective in producing transitions) at a rate

$$\omega_1 = g\beta|\mathbf{H}_1|/\hbar \tag{27}$$

After a period of time corresponding to a precession of 180° the magnetic moment of the system is directed oppositely to the d-c magnetic field **H**₀. A transfer back to the nonrotating system makes this evident.

The restrictions on such an inversion are:
1. The inversion must be fast compared with T_1.
2. **H**₁ must be greater than the spin-spin interaction field, which accounts for the homogeneous-line width or spin-packet width of an inhomogeneous line, or else the inversion will not successfully be achieved because the magnetization does not rotate uniformly.
3. The amplitude of **H**₁ and the length of pulse are extremely critical. The combination of the above requirements has limited the use of the 180° pulse inversion to experimental use in laboratories.

5.4. Adiabatic Fast Passage. The inversion system used for most two-level work is known as adiabatic fast passage, a name which is somewhat contradictory. As is implied by the word adiabatic, changes in the conditions which cause the inversion are slow compared with the precession of the electron moment about the d-c magnetic field. Again, however, the inversion must take place in a time short compared with T_1, the spin-lattice relaxation time.

Once more it is most convenient to consider the rotating coordinate system. As discussed above, if the rotating coordinate system is made to rotate at an angular frequency ω, the effective magnetic field in the rotating system is given by Eq. (24). The angle between the axis of rotation (direction of d-c field) and the effective field is given by (see Fig. 4)

$$\theta = \arctan |\mathbf{H}_1|/(|\mathbf{H}_0| - \omega/\gamma)$$

This effective field is nearly aligned with the axis of rotation for frequencies far from resonance while at resonance when $\mathbf{H}_0 = \omega/\gamma$ they are perpendicular to one another. As the frequency is varied from well below to well above resonance, the effective field in the rotating system changes direction by 180° from being aligned with the field **H**₀ to being antialigned. The spins at the same time precess about the effective field at a frequency $\omega_r = \gamma \mathbf{H}_{er}$ and will therefore tend to follow the effective field. It is immaterial whether the frequency or magnetic field is varied to cause the rotation of the effective field direction. In practice it is easier to vary the magnetic field. Adiabatic conditions are assured by keeping

$$\begin{aligned} d\mathbf{H}/dt &< \omega_p \mathbf{H}_1 \\ \text{or} \quad d\mathbf{H}/dt &< \gamma \mathbf{H}_1^2 \end{aligned} \tag{28}$$

since the precession frequency $\omega_p \simeq \gamma \mathbf{H}_1$. Once the rotation is complete, transformation out of the rotating coordinate system leaves the spin moment inverted with respect to **H**₀. As in the case of the pulse inversion, **H**₁ must be kept larger than spin-spin and local fields in order that the spin moment may precess about it during the inversion. Obviously, the whole process must be carried out in a time short compared with the spin-lattice relaxation time T_1 or temperature equilibrium with the lattice will be maintained and the inversion will not succeed. The path of the magnetization, in the rotating frame of reference, during the inversion is sketched in Fig. 5b.

It should be noted that all the inversion schemes for two-level systems are intermittent. This is true of the gas-beam system as well as for paramagnetic systems, although in that case the transport of the upper-level molecules into the cavity on a continuous basis leads to continuous operation. The inversion will last until the relaxation time returns it to an infinite spin temperature (where $n_1 = n_2$) on the way to thermal equilibrium with the lattice. Furthermore, the inversion obtained depends on the population difference existing before the inversion process begins. Unless the system has returned to thermal equilibrium with the lattice or other surroundings less than optimum inversion is bound to result.

Ways of surmounting the intermittency of these systems have been suggested. Most have involved the motion of material from an inverting to an operating section of the apparatus on a continuous basis. None of these systems appears to have been used successfully except for the beam maser as noted above.

6. R-F METHODS OF POPULATION INVERSION IN MULTILEVEL PARAMAGNETIC SYSTEMS

The methods used for inversion of the population in a pair of levels in multilevel systems are remarkably alike. It will be seen below that appropriate relaxation times are the real secret of population inversion.

6.1. Paramagnetic Three-level Systems.

An inversion method proposed by Bloembergen[5] was the impetus for much of the early solid-state maser work. This scheme is the one used in most paramagnetic masers, and the physical principles are identical to those of the inversion methods for other multilevel systems.

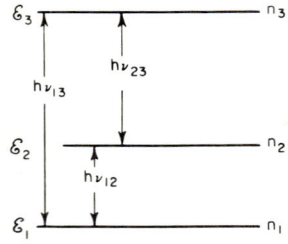

Fig. 6. System of three energy levels.

Consider the three-level paramagnetic-spin energy-level system represented by Fig. 6. Qualitatively, the inversion procedure is extremely simple. Transition $\varepsilon_1 \to \varepsilon_3$ is saturated; i.e., the populations n_1, n_3 are made equal by use of power at a frequency ν_{13}. If the spin-lattice relaxation time T_1 is shorter for the transition $\varepsilon_3 \to \varepsilon_2$ than for $\varepsilon_2 \to \varepsilon_1$ the population of level 2 will take the value

$$n_2/n_3 = e^{h\nu_{23}/kT}$$

But since

$$n_1 = n_3 \qquad n_2/n_1 = e^{h\nu_{23}/kT} \qquad \text{or} \qquad n_2 > n_1 \qquad (29)$$

Thus the populations of level 2 and 1 are favorable for amplification at frequency ν_{12}. It must be emphasized that in order to have electromagnetic transition probabilities between all these levels the levels cannot be pure spin states but must be mixed, as is the case in an axial crystalline field with the magnetic field at an angle to the crystal axis.

A quantitative evaluation of the situation can be obtained by considering the transition probabilities of the system. Those due to thermal vibration of the lattice are represented by w's

$$w_{12} = w_{21}e^{-h\nu_{12}/kT} \qquad w_{23} = w_{32}e^{-h\nu_{23}/kT} \qquad w_{13} = w_{31}e^{-h\nu_{13}/kT} \qquad (30)$$

These follow from the thermal-equilibrium conditions of the spin system. The direct spin-lattice relaxation time for two levels of the spin system is equal to $1/w$.

The effect of the pump energy at frequency ν_{13} is to produce a saturation through the transition probability W_{13}. The transition probability for the signal energy will be denoted by W_{12} but the power level is lower, so that $W_{12} \ll W_{13}$. Note that the

INDUCED-EMISSION-AMPLIFIER PRINCIPLES

capital W's refer to the transition probabilities due to impressed pump and signal fields while the lowercase w's refer to the probabilities responsible for the spin-lattice relaxation times. The populations of the levels n_1, n_2, n_3 must satisfy the relation $n_1 + n_2 + n_3 = n$, the total number of spins. For the condition $h\nu/kT \ll 1$ the dn/dt or rate equations are easily formulated. (Even though this condition is not always met, no added physical meaning is derived by using more exact methods.)

$$dn_1/dt = n_3 w_{31} + n_2 w_{21} - n_1(w_{12} + w_{13}) + W_{13}(n_3 - n_1) + W_{12}(n_2 - n_1)$$
$$dn_2/dt = n_1 w_{12} + n_3 w_{32} - n_2(w_{23} + w_{21}) + W_{12}(n_1 - n_2) \quad (31)$$
$$dn_3/dt = n_1 w_{13} + n_2 w_{23} - n_3(w_{32} + w_{31}) + W_{13}(n_1 - n_3)$$

Utilizing the simplification $\exp(h\nu/kT) \simeq 1 + h\nu/kT$, assuming W_{13} greater than W_{12} and the w's, and taking in second order $n_1 = n_2 = n_3 = n/3$

$$dn_1/dt = w_{31}[n_3 - n_1 + (n/3)(h\nu_{31}/kT)] + w_{21}[n_2 - n_1 + (n/3)(h\nu_{21}/kT)]$$
$$+ W_{13}(n_3 - n_1) + W_{12}(n_2 - n_1)$$
$$dn_2/dt = w_{21}[n_1 - n_2 - (n/3)(h\nu_{12}/kT)] + w_{32}[n_3 - n_2 + (n/3)(h\nu_{23}/kT)]$$
$$+ W_{12}(n_1 - n_2) \quad (32)$$
$$dn_3/dt = w_{31}[n_1 - n_3 - (n/3)(h\nu_{13}/kT)] + w_{32}[n_2 - n_3 - (n/3)(h\nu_{32}/kT)]$$
$$+ W_{13}(n_1 - n_3)$$

The solution of these equations for equilibrium $(dn/dt = 0)$ is

$$n_1 - n_2 = \frac{1}{3}\frac{hn}{kT}\frac{w_{21}\nu_{21} - w_{32}\nu_{32}}{w_{21} + w_{32} + W_{12}} = -(n_2 - n_3) \quad (33)$$

The population is inverted if $n_2 > n_1$ or for

$$w_{32}\nu_{32} > w_{21}\nu_{21} \quad (34a)$$

and in terms of relaxation times

$$\nu_{32}/(T_1)_{32} > \nu_{21}/(T_1)_{21} \quad (34b)$$

For equal frequencies this requires $(T_1)_{21} > (T_1)_{32}$, as expected from the qualitative argument above. The frequency factors can help considerably, and it is evident that it is easier to invert level systems in which $\nu_{32} > \nu_{21}$, which requires the pump frequency ν_{13} to be more than double the signal frequency ν_{21}. Aside from these considerations it does not matter whether the signal transition is the lower or the upper one.

The power emitted by the inverted spin system is

$$P_M = (n_2 - n_1)h\nu_{21}W_{12} = \frac{1}{3}\frac{h^2 n\nu_{21}}{kT}\left(\frac{w_{32}\nu_{32} - w_{21}\nu_{21}}{w_{21} + w_{32} + W_{12}}\right)W_{12} \quad (35)$$

The transition probability W_{12}, at resonance, is given by

$$W_{12} = \hbar^{-2}|\langle 1|M|2\rangle|^2 H^2(\nu_{12}) g(\nu)_{max} \quad (36)$$

in which $H(\nu_{12})$ represents the r-f magnetic field strength of frequency ν_{12} at the crystal' $\langle 1|M|2\rangle$ is the matrix element for the transition depending on the orientation of the field $H(\nu_{12})$, and $g(\nu)_{max}$ is the normalized line-shape amplitude [see Eq. (18a)].

The magnetic quality factor Q_M, which is a direct analog of the circuit Q, is defined as the ratio of the energy stored in the cavity to the energy loss or gain per radian in the magnetic material. It is given by

$$Q_M = \frac{\omega \langle H^2(\nu_{12})\rangle_{av} V_c}{8\pi P_M}$$
$$= \omega \frac{\langle H^2(\nu_{12})\rangle_{av} V_c}{8\pi \Delta n h\nu_{12} W_{12}}$$
$$= \frac{\hbar}{8\pi}\frac{\langle H^2(\nu_{12})\rangle_{av} V_c}{\Delta n|\langle 1|M|2\rangle|^2 H^2(\nu_{12})_{crystal} g(\nu)_{max}} \quad (37)$$

where $\langle H^2(\nu_{12})\rangle_{av}$ is the average of $H^2(\nu_{12})$, the magnetic field at frequency ν_{12}, over the cavity, $H(\nu_{12})_{crystal}$ is the r-f magnetic field at the crystal (assumed uniform), and V_c is the crystal volume. Q_M becomes negative for an inverted spin system where instead of loss there is gain due to the spins. Note that for maximum gain small negative values of Q_M are required.

In some cases it is possible to take advantage of a fourth level to improve the pumping efficiency. The effect is such that inversion ratios may be about twice those for the three-level pumping system discussed above.

6.2. Push-Push Pumping. In an energy-level scheme such as that illustrated in Fig. 7 the pump frequency may be made to correspond to ν_{12} and ν_{24} at the same

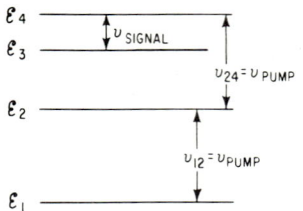
FIG. 7. Four-energy-level system suitable for push-push pumping.

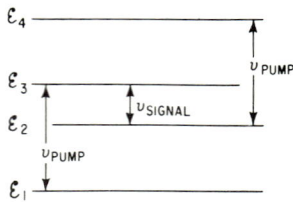
FIG. 8. Four-energy-level system suitable for push-pull pumping.

time. The effect in the top three levels is similar to that in the conventional pumping. The gain is an increased relaxation rate from level 3 due to the nonequilibrium between levels 1 and 3. The population of level 3 is set by the relaxations to both levels 1 and 2, and the effect is similar to having a higher-frequency pump.

6.3. Push-Pull Pumping. A level system such as that shown in Fig. 8 is susceptible to pumping as indicated with $\nu_{13} = \nu_{24} = \nu_{pump}$. Inversion efficiency is improved by the additional relaxation of level 4 to level 1 in parallel with the relaxation to level 3. Again the result is similar to that achieved by the use of a higher-frequency pump.

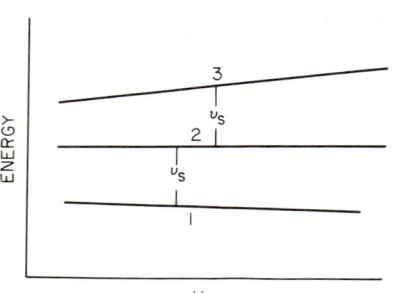
FIG. 9. Energy levels for staircase inversion.

This push-pull pumping is particularly useful in paramagnetic systems of spin $S \geq \tfrac{3}{2}$ with axial crystalline fields. In that case at an angle $\theta = \cos^{-1} 1/\sqrt{3}$ the required symmetry for push-pull pumping exists for all values of H.

6.4. Staircase Inversion. In a multilevel system it is possible to carry out successive adiabatic rapid passages on adjacent transitions. Such a process has been termed *staircase inversion*.[6] It can be illustrated with respect to Fig. 9. Suppose, using a constant-frequency radio frequency of amplitude H_1, sufficiently large to satisfy adiabatic-fast-passage conditions, the populations of the bottom two levels are inverted by sweeping the magnetic field up through the resonant transition ν_S. The magnetic field sweep, if continued, will then proceed to invert the populations of levels 2 and 3 at the same frequency. Thus to a first approximation level populations which were n_3, n_2, n_1 become n_1, n_3, n_2, inverting both the 2-3 and the 1-3 transition if the time between the two successive inversions is less than the spin-lattice relaxation time appropriate to the population excess in level 2.

Actually it is not quite so simple, because as Wagner, Castle, and Chester[7] have pointed out, the inversion efficiency is not 100 per cent. If one defines an inversion efficiency

$$\alpha_{ij} = \frac{n_i - n_j}{n_j - n_i} = \frac{\text{population difference after passage}}{\text{population difference before passage}}$$

the level populations as assumed in Table 1 on the left, at thermal equilibrium after the two inversions are converted to those in the middle, starting with the basic assumption that the two inversion efficiencies are the same. The final population difference

$$n_3 - n_1 = \Delta \left(\frac{\alpha^2 + 6\alpha - 3}{4} \right) \tag{38}$$

is positive (inverted) only for $\alpha \geq 0.46$. Extension to triple inversion of a four-level system requires $\alpha \geq 0.69$ for inversion of the 4-1 transition. Cross relaxation taking

Table 1

Initial population	Population after successive passages	Population after simultaneous passage
$1 - \Delta$	$1 + \Delta \left(\frac{\alpha^2 + 4\alpha - 1}{4} \right)$	$1 + \alpha\Delta$
1	$1 - \Delta \left(\frac{\alpha^2 + 2\alpha + 1}{4} \right)$	1
$1 + \Delta$	$1 + \Delta \left(\frac{1 - \alpha}{2} \right)$	$1 - \alpha\Delta$

place between the inversions can be most troublesome—decreasing the overall efficiency.

On the other hand, coincident levels inverted at the same time, i.e., a simultaneous rapid passage on both, result in inversion of the level populations except for the factor α. The populations are those shown in the right column of Table 1 and

$$n_3 - n_1 = 2\alpha\Delta \tag{39}$$

6.5. Relaxation-time Modification by Spin-Spin Coupling. The selection of a suitable maser material will normally involve consideration of the observed relaxation times as well as the signal and the pump frequencies. Equations (34a) and (34b) are the criteria by means of which the relaxation times may be judged suitable. Of course, other factors associated with the use of a crystal in a microwave cavity, such as dielectric loss, conductivity, and frequency shift due to dielectric constant, must be considered as well.

In cases where the relaxation time at the third frequency, the first and second being those of the signal and pump transitions, is not sufficiently fast it is sometimes possible to improve the situation. If a second resonant system, possessing a fast relaxation time in a transition resonant at the desired field and at a frequency corresponding to this third frequency, is introduced into the crystal in sufficient concentration, spin-spin coupling[8] between the two species may result in a much faster relaxation time for the third frequency.

Scovil, Feher, and Seidel[9] have successfully used a 0.2 per cent concentration of Ce^{3+} ions to reduce the relaxation time of the appropriate level of Gd^{3+} in lanthanum ethyl sulfate. The reduction of relaxation time for a 0.5 per cent Gd^{3+} concentration was as high as a factor of 7.[10] The resulting maser was the first three-level solid-state maser reported, and the cavity configuration is given in Sec. 3.4 of Chap. 29.

7. OPTICAL PUMPING

There are two aspects of optical pumping of interest for induced-emission amplifiers. The first, applicable to masers, involves the population inversions created by the optical energy within the ground state. The second, utilized in lasers, is the consideration of the population inversions induced in the optical levels themselves.

7.1. Ground-state Population Inversion by Polarization-dependent Transitions. The production of population differences in the ground state depends on the differences in transition probability, between the various levels involved, due to

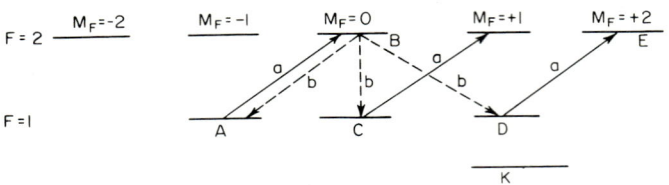

FIG. 10. Optical pumping transitions.

the polarization of the light. For purpose of illustration, consider the hypothetical energy-level system of Fig. 10. Assume that this level system is irradiated with light of the appropriate frequency ($\simeq 10^{15}$ cps) with the light directed along the magnetic field. Right-hand circularly polarized light possesses one unit $h/2\pi = \hbar$ of angular momentum and can cause transitions in which the M_F quantum number increases by one unit as illustrated by transitions a. Left-hand circularly polarized light produces only the opposite transitions, i.e., $\Delta M = -1$, such as $F = 1$, $M_F = 0$ to $F = 2$, $M_F = -1$ (F is the total angular momentum $= J + I$ the sum of angular momentum of the electrons and the nucleus). Linearly polarized or unpolarized light in this same direction consisting of mixtures of the two circular polarizations can cause both $\Delta M_F = +1$ and $\Delta M_F = -1$ transitions. (Light polarized linearly parallel to the field H_0 and propagating at right angles to it produces $\Delta M_F = 0$ transitions.)

To consider one possibility, the effect of the right-hand circularly polarized light is to cause transitions from A to B. The population in B tends to return to thermal equilibrium via spontaneous radiation for which $\Delta M_F = \pm 1$ or 0 transitions are possible. The possible transitions for this spontaneous return are labeled b. Thus population pumped from A returns about $\frac{1}{3}$ to A, $\frac{1}{3}$ to C, and $\frac{1}{3}$ to D. The right-hand circularly polarized light tends to move the population in the ground state to the high values of M_F. The population pumped from $D(F = 1$ $M_F = +1)$ to $E(F = 2$ $M_F = +2)$ must return to D via the spontaneous-radiation route. After a period of time it is expected that the populations of the

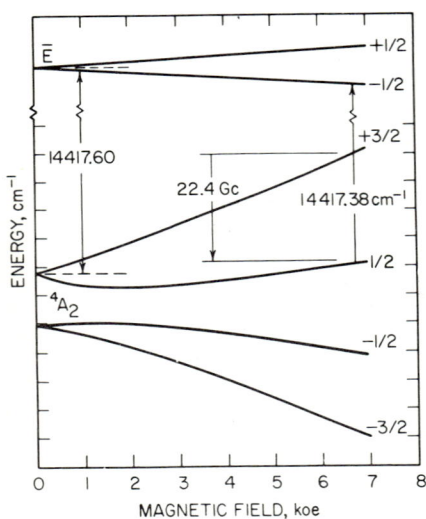

FIG. 11. Laser pumping of ruby ground-state levels. (*After Devor, D'Haenens, and Asawa,*[11] *published with the permission of the editor, Physical Review Letters.*)

$F = 1$ level will be $D > C > A$. If a microwave transition DK exists such that D is the upper level and if the relaxation time of DK is slow enough an inverted population will result.

Such optical pumping has not found great use because of the difficulty of finding suitable energy-level systems. The above illustration is obviously a simplification because one must in practice consider the optical pumping effect on level K and any others in the ground state. Furthermore, optical lines are generally broad, so that more than one excited state may be involved and in that case the spontaneous-emission selection rules can be altered.

Until the availability of lasers, optical lines were so broad that only optical pumping systems such as this depending on differing transition probabilities were possible. In the next section the use of laser pumping of paramagnetic levels will be discussed.

7.2. Ground-state Inversion Using Laser Pumping. The frequency width of laser beams is such that it is possible to pump one of the paramagnetic levels of the ground state selectively and in essence make a three-level maser with an optical pump. Figure 11 illustrates this method as used in ruby.[11] The ruby laser output of 14,417.6 cm^{-1} corresponds very closely to the indicated pump transition at a field of about 7,000 gauss. The laser then acts as the pump of the three-level system consisting of the $-\frac{1}{2}$ level of $\bar{E}$ and the $+\frac{3}{2}$, $+\frac{1}{2}$ levels of $^4A_2$, and the inverted population is obtained between the top two levels of the ground state at 22.4 gc. This, then, is three-level operation with an optical pump.

8. LASER INVERSION METHODS

8.1. Optical Pumping. This inversion method, first suggested by Townes and Schawlow,[2] has been used extensively for solid-state lasers but has been less popular in gaseous systems. The principles of optical pumping methods are very similar to those of inversion in the three-level maser system. Figure 12 illustrates the energy levels of cesium in which the populations of the 6D levels may be inverted with respect to the $6S_\frac{1}{2}$ ground state by saturating the 3,888-angstrom transition $6S_\frac{1}{2}$ to $8P_\frac{1}{2}$. The pumping transition in this case is a sharp one, requiring a strong emission line coinciding in energy with the absorbing transition for effective pumping. In this particular case one of the helium lines (3^3P-2^3S) corresponds to the cesium 3,888-angstrom line within half its Doppler width. Such coincidences with emission lines strong enough for effective pumping, though rare, are necessary for this type of optical pumping of gases. In solids, however, pumping can be accomplished more easily because many of the pump transitions are absorption bands.

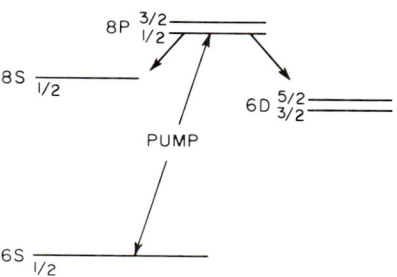

Fig. 12. Cesium energy levels.

A schematic representation of two types of energy-level systems with absorption-band pumping transition are illustrated in Fig. 13. The three-level system of Fig. 13a is identical to the maser three-level schemes except for the broad transition into which

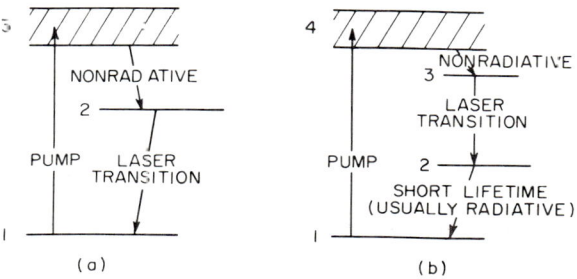

Fig. 13. Laser energy-level schemes using pumping into an absorption band. (a) Three-level system. (b) Four-level system with transition terminating on a level appreciably above the ground state.

pump energy is fed. The transition 3-2 is nonradiative with a fast transition rate w_{32}. The conditions for inversion are given by Eqs. (34), assuming saturation of the 1-3 transition.

The four-level system of Fig. 13b is much more readily inverted because, if the spacing of levels 1 and 2 is greater than kT with a fast relaxation time, level 2 will be sparsely populated and the 2-3 transition readily inverted. Again the 3-4 spontaneous decay must be fast and nonradiative.

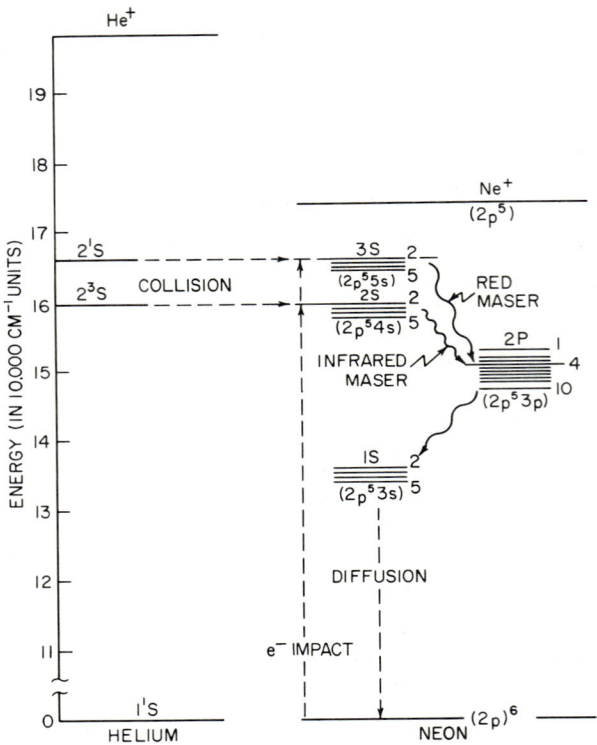

FIG. 14. Helium and neon energy levels. Dominant collision excitation and laser transitions are shown. (*After Bennett,*[12] *published with permission of the editor, Applied Optics.*)

8.2. Gas Discharge Inversion—Electron Impact. The probability for an electron to raise an ion to an excited state m in the process of a collision is given by

$$\left| \int \exp i\mathbf{K} \cdot \mathbf{r} \psi_m^* \psi_o \, d\mathbf{r} \right|^2$$

where $\mathbf{K}$ is the change in electron-propagation vector during the collision. On expanding the exponential, the second term (the first one vanishes because of orthogonality of wave functions) is proportional to the electric-dipole transition probability between the levels o and m. The transition probability is then closely related to the optical pumping transition probability between the same levels, and simultaneously, the spontaneous transition probability will be high (given by the Einstein A coefficient).

Inverted populations in pure helium discharges have been known for some time, and if in a discharge some of the higher levels are excited (they are effectively excited for electron energies well above threshold) in the process of decay to the ground state through a number of intermediate levels, it is to be anticipated that some of the pairs of levels will have relaxation times which are conducive to inversion. Because of the proportionality of excitation to radiation in the gas discharge system, sets of levels where the lowest must depopulate rapidly by radiation to the ground state cannot be inverted. For pure neon, however, direct electron impact preferentially populates the $2S$ and $3S$ levels (Fig. 14), and the maser transition is as indicated ($2S_2$-$2P_4$). The lower level has a very short radiative lifetime to the $1S$ level which decays by diffusion and to some extent by collisions. Maser action has been observed in this and similar cases.

8.3. Resonant Transfer or Inelastic Atom-Atom Collisions. Again referring to Fig. 14 we see that there is a near coincidence between the helium 2^3S and the neon $2S$ level. It has been known for some time[13] that resonant collisions, in this case,

$$He^* + Ne \rightarrow Ne^* + He$$

have a high probability, a probability which increases with decrease of excitation energy difference between the levels. In the He-Ne case this type of transition

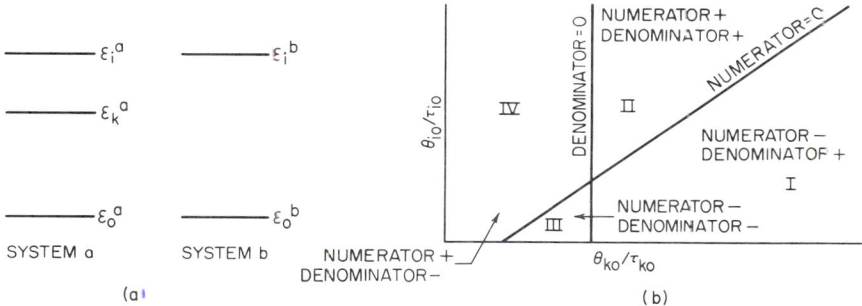

FIG. 15. (a) Nomenclature and energy levels used in analyzing inversion of levels $\epsilon_i{}^a - \epsilon_k{}^a$ by gas discharge inversion as a function of properties of gas b, assuming resonant transfer of energy. (b) Graphical presentation of Eq. (42a) showing the demarcation between major regions of inversion and noninversion.

accounts for about 50 per cent of the pumping action—the remainder is direct excitation of the Ne$2S$ by electron impact. There can be considerable advantage to pumping in this way because the undesired electron-collision effects in the lower Ne levels are bypassed by the resonant transfer process.

Basov and Krokhin,[14] and others, have considered theoretically the addition of a second gas to the laser material and have derived criteria for the effectiveness of the addition as measured by inversion. The Maxwellian velocity distribution assumed for the electrons has been confirmed for Xe plasmas[15] up to electron energies of 21 volts at a temperature of 9 volts ($\simeq 10{,}000°K$).

Consider (Fig. 15a) an energy-level system a with levels $\epsilon_i{}^a$ and $\epsilon_o{}^a$ together with a second system b with ground state $\epsilon_o{}^b$ and excited level $\epsilon_i{}^b$ coinciding in energy with $\epsilon_i{}^a$. System a also has a level $\epsilon_k{}^a$ below $\epsilon_i{}^a$ which serves as the lower of the inverted levels ($\epsilon_i{}^a \rightarrow \epsilon_k{}^a$). The rate of change of population $n_i{}^a$ in level $\epsilon_i{}^a$ is

$$dn_i{}^a/dt = -n_i{}^a(1/\theta_{i0} + 1/\tau_{i0} + 1/t_{ab}) + n_o{}^a(1/\theta_{i0} + 1/t_{ba}) \qquad (40)$$

where $1/\theta_{0i} = \overline{\rho_e \sigma_{0i} v}$ is the probability of electronic excitation by electron collisions of the first kind

$1/\theta_{i0} = \overline{\rho_e \sigma_{i0} v}$ is the probability of de-excitation by electron collisions of the second kind

$1/\tau_{i0}$ = total probability for transitions from level i except by collisions of the second kind

$1/t_{ab} = n_o{}^b \overline{\sigma_{ab}v}$ is the probability for resonant excitation of a b atom by an a atom

$1/t_{ba} = n_i{}^b \overline{\sigma_{ba}v}$ is the reverse of the above

σ's = cross sections

θ, t, and τ = lifetimes for the respective processes

ρ_e, n_i, n_o = densities, electron and ion, respectively

v = relative velocity prior to collision

$\overline{\sigma v}$ = average product over the kinetic-energy distribution

On the assumption of a Maxwellian distribution of electron energies

$$\theta_{i0}/\theta_{0i} = \exp - (\mathcal{E}_i/kT_e) \qquad (41)$$

It is worth pointing out that if one uses a temperature to characterize the level populations of system a its temperature will be lower than the electron temperature T_e because of interatomic collisions and the radiative processes (included in $1/\tau$).

Using an equation similar to (40) for $n_k{}^a (1/t_{ba} = 1/t_{ab} = 0)$ the equilibrium condition of $dn/dt = 0$ requires

$$\frac{(1 + \theta_{k0}/\tau_{k0})[1 + (\theta_{i0}/t_{ba}) \exp(\mathcal{E}_i{}^a/kT_e)]}{1 + \theta_{i0}/\tau_{i0} + (\theta_{i0}/t_{ba}) n_o{}^b/n_i{}^b} > \exp\left(\frac{\mathcal{E}_i{}^a - \mathcal{E}_k{}^a}{kT_e}\right) \qquad (42)$$

as the condition for an inverted population. Deviations from the assumed Maxwellian distribution will alter this inequality in detail. However, as might be expected, the result depends on the densities of the pumping gas b

$$n_o{}^b/n_i{}^b = \exp(\mathcal{E}_i{}^b/kT_b)$$

For graphical purposes, Eq. (42) can be rearranged

$$\frac{\theta_{i0}}{t_{ba}} \gtreqless \left[\exp\left(-\frac{\mathcal{E}_i{}^a}{kT_e}\right)\right] \frac{\{\exp[(\mathcal{E}_i{}^a - \mathcal{E}_k{}^a)/kT_e]\}(1 + \theta_{i0}/\tau_{i0}) - (1 + \theta_{k0}/\tau_{k0})}{1 + \theta_{k0}/\tau_{k0} - (n_o{}^b/n_i{}^b) \exp(-\mathcal{E}_k{}^a/kT_e)} \qquad (42a)$$

(Use > for denominator positive, < if negative.) One may plot θ_{i0}/τ_{i0} against θ_{k0}/τ_{k0}, and various areas are demarcated by the lines representing the denominator and the numerator equaling zero as shown in Fig. 15b. In region I the inequality is satisfied even for $t_{ba} \to \infty$, or without gas b, because of fast relaxation from level $\mathcal{E}_k{}^a$. Gas b is required to satisfy the inequality in region II; i.e., $1/t_{ba}$ must be larger. On the left-hand side of the vertical line determined by setting the numerator equal to zero the gas b is cold

$$\exp\left(\frac{\mathcal{E}_i{}^b}{kT_b} - \frac{\mathcal{E}_k{}^a}{kT_e}\right) > \left(1 + \frac{\theta_{k0}}{\tau_{k0}}\right) > \left(1 + \frac{\theta_{i0}}{\tau_{i0}}\right) \exp \frac{\mathcal{E}_i{}^a - \mathcal{E}_k{}^a}{kT_e}$$

and in region III it inhibits the population inversion. Population inversion is impossible in region IV because of the ineffective relaxation from level $\mathcal{E}_k{}^a$.

8.4. Molecular Dissociation (Dissociative Excitation Transfer). In some cases it is conceivable for an excited atom to transfer its energy to a molecule, leaving it in an excited state which dissociates into component atoms at least one of which is excited. This excitation method has been used in neon-oxygen lasers with emission from the excited oxygen resulting from the mechanism.

$$\text{Ne}(^3P_1, {}^3P_0) + O_2 \to O(3^3P, 3^5P) + O(2^3P) + \text{Ne}$$

Emission from 3^3P_2 to 3^3S_1 at 8,446 angstroms is observed.[16] The resonance requirements are not so stringent as for the resonant transfer of energy process, and hence the mechanism may have considerable application.

8.5. Semiconducting Junction Inversion. The use of a semiconducting junction to achieve inverted level populations[17] is the best example of conversion from direct current to coherent optical output. (Some gas discharge lasers accomplish this too.) If a p-n junction is forward-biased to cause electron current to flow from degenerate n to degenerate p type, the electrons in the conduction band of the n type on crossing into the p type find themselves in a region with empty states in the valence band, as shown in Fig. 16. This is precisely the inverted population required for laser action. On returning to the conduction band the electrons emit energy in the optical region. It is not clear whether this is an interband transition or one of slightly lower energy involving an impurity level. The current densities required for oscillation are high (10 to 1,000 amp/sq cm), and the resulting input powers of 10^8 to 10^9 watts/cu cm in the extremely thin junction region are high. Operation is generally on a pulsed basis because of power losses.

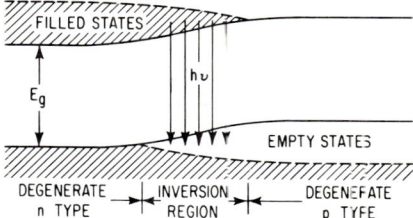

Fig. 16. Energy levels at p-n junction illustrating semiconductor inversion. (*After Hall*,[17] *published by permission of the editor, Solid-State Electronics.*)

The efficiencies are high, with some estimates of optical power outputs being almost 100 per cent of the d-c input.

9. INDUCED-EMISSION AMPLIFIERS WHICH DO NOT REQUIRE INVERTED POPULATIONS

9.1. Double-quantum Maser. Javan,[18] in 1957, published a proposal for a three-level maser operating via the mechanism of a double-quantum transition without the requirement for population inversion between levels. The essentials of such a system may be seen in Fig. 17. Pumping is carried out on the 1-3 transition just as in the case of the usual three-level system, but the emission of ν_s is connected with the probability of two quanta being involved, simultaneous absorption of ν_p and emission of ν_s. The probability for such a transition has been discussed by Hughes and Grabner[19] and can be written as

$$W \propto \frac{\langle \phi_1 | H' | \phi_3 \rangle \langle \phi_3 | H' | \phi_2 \rangle}{\varepsilon_0 - \varepsilon_{13}}$$

where $\varepsilon_0 - \varepsilon_{13}$ is the energy difference between the actual position of level 3 and the utilized energy of level 3. Coincidence is not needed, and if one of the probabilities in the numerator is very large the overall double-quantum probability can be of reasonable size. (This generally will occur only for large values of r-f fields.) The reverse double-quantum process is also allowed, but thermal Boltzmann populations favor the absorption of ν_p and emission of ν_s; i.e., inversion is not required and thermal-equilibrium populations are satisfactory. Spontaneous-emission terms have been omitted, and therefore this corresponds to the stimulated Raman effect considered below.

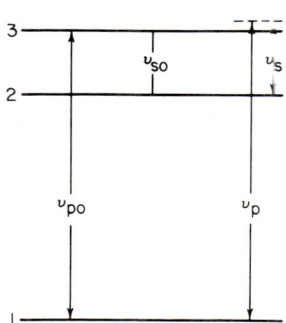

Fig. 17. Energy levels for discussion of double-quantum maser. ν_{po} and ν_{so} indicate pump and signal frequencies when level 3 is used; ν_p and ν_s are pump and signal frequencies for more general case where effective pump level is near to level 3.

Experimental results verifying this prediction have been obtained by Yajima[20] in a gas system (HCOOH). Calculated line shapes are given in Fig. 18, and the experimental results are in qualitative agreement showing the double peak. It is interesting to note that the signal line may, under the same pumping conditions, show both absorption and emission. The absorption is near the normal absorption frequency but the emission is pulled, by the deviation of the pump frequency ν_p from ν_{po}, away from the normal absorption. Although an emissive condition was observed experimentally net gain was not achieved.

9.2. The Raman Laser. Coherent emission from organic liquids by virtue of a Raman-effect action was first observed[21] when a nitrobenzene Kerr cell was being used in a giant pulse laser. In addition to the expected laser frequency, coherent emission is obtained at a lower frequency, the difference in frequency corresponding to a Raman transition. The operating mechanism is similar to Javan's double-quantum amplifier and requires no inversion of the population.

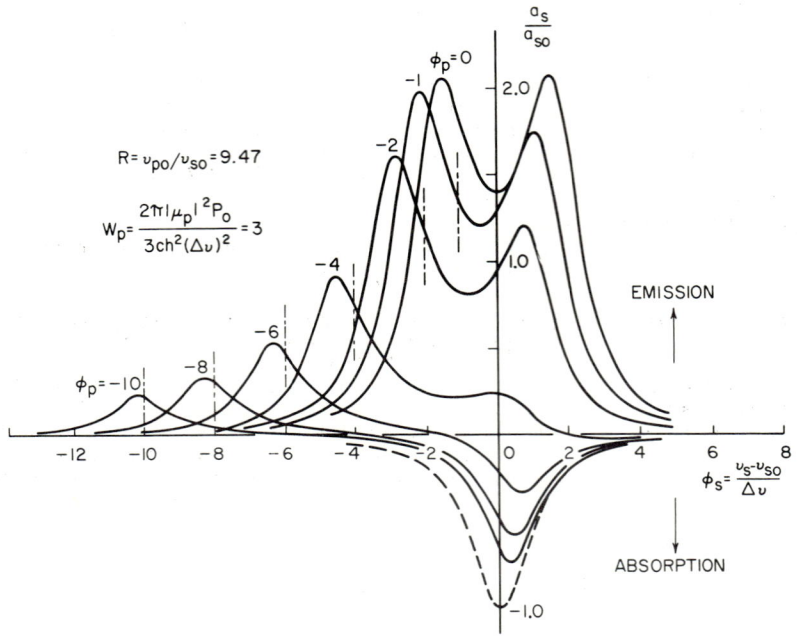

FIG. 18. Calculated line shapes for double-quantum maser under various pump conditions. $\phi_p = (\nu_p - \nu_{po})/\Delta\nu$ is a measure of the off-resonant pump frequency, $W_p = 2\pi|\mu_p|^2 P_o/3ch^2(\Delta\nu)^2 S$ is a pump parameter, and $\phi_s = (\nu_s - \nu_{so})/\Delta\nu$ is a measure of the signal deviation from the "resonant" value. (After Yajima,[20] published by permission of the editor, The Physical Society of Japan.)

Consideration of the stimulated Raman effect is illuminating. Suppose light of angular frequency ω_i is incident on a material with a Raman transition ω_R. Normally two frequencies $\omega_i - \omega_R$, the Stokes line, and $\omega_i + \omega_R$, the anti-Stokes line, are observed in the output. The intermediate state is a virtual one; hence there is no limitation on the excitation frequency. Considering the levels of Fig. 19, the net change in N_β, the number of photons at frequency $\omega_i - \omega_R$ due to N_α incoming photons may be written as[22]

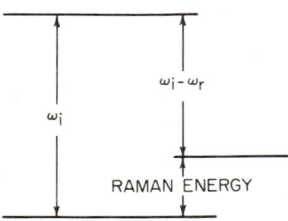

FIG. 19. Levels involved in Raman transitions.

$$\dot{N}_\beta = N_\alpha(1 + N_\beta)S_{ij}(\alpha,\beta)n_i - N_\beta(1 + N_\alpha)S_{ji}(\alpha\beta)n_j \tag{43a}$$

The matrix elements S_{ij}, S_{ji} are equal to one another. In the normal Raman effect only the spontaneous-emission term is considered and therefore the N_β is omitted on the right-hand side of (43). In that case only the first term of (43a) is obtained and the stimulated Raman effect is absent. Equation (43a) can be rearranged to give

$$\dot{N}_\beta = N_\alpha n_i - N_\beta n_j + (n_i - n_j)N_\alpha N_\beta S_{ij} \tag{43b}$$

INDUCED-EMISSION-AMPLIFIER PRINCIPLES   **14**-23

The third term is the one of interest since it implies, if $n_i - n_j$ is positive, a positive value of $\dot{N}_\beta$ proportional to N_β. This gain is moreover proportional to the incident intensity, i.e., to N_α. In normal Raman experiments the intensity within the Raman line width is too small to make this gain appreciable, but for the laser input, concentrated in essentially a pure mode at a single frequency, this third term can produce sufficient gain to overcome the losses at the frequency ω_R, and hence maser or laser action results in the typical coherence and line narrowing. It should be emphasized that coherence of the incident radiation is not required—only a high intensity in a narrow bandwidth.

10. NOISE

The noise output of an induced-emission amplifier will contain components due to the well-recognized sources of noise, the thermal radiation from input and output waveguides and the cavities. In addition, one must consider the contribution of the amplifying material, which is frequently spoken of as being at a negative temperature

The term negative temperature must be used cautiously[23] and only under the condition that thermal equilibrium in each level is established in a time which is fast compared with the radiation lifetime. Another statement of the same stipulation is that the transverse magnetization must be small. The spin temperature is then defined in terms of the populations of the levels as

$$T_M = (h\nu/k) \log_e (n_1/n_2) \qquad (44)$$

For the inverted population required for most amplification applications T_M is negative.

The noise contributed by the maser material is, physically, spontaneous emission from the upper level. Nyquist's consideration of the thermal equilibrium between a resistance and radiation at the same temperature showed that the mean-square noise voltage across a resistance must be given by

$$\overline{e^2} = R \frac{h\nu \, d\nu}{\exp(h\nu/kT_M) - 1} = R \frac{h\nu \, d\nu}{n_1/n_2 - 1} \qquad (45)$$

The resistance may be considered to be that of the spin-system absorption.[24-29] One may then qualitatively argue that this formula must hold also for negative temperature by substituting $R \propto n_1 - n_2$. Then (45) becomes

$$\overline{e^2} \propto (n_1 - n_2)(n_1/n_2 - 1)^{-1} h\nu \, d\nu = n_2 h\nu \, d\nu$$

The important feature to note in Eq. (45) is that the sign of $\overline{e^2}$ is unchanged by a change in the sign of R (since the denominator has the same sign as R). The conclusion is that the value of $\overline{e^2}$, or the noise, is always positive and proportional to n_2, as it must be if it is due to spontaneous emission. The noise from the spin system increases steadily from $0°$ to $-0°$, from zero to being proportional to $n/2$ at $\infty°$ and then proportional to n at $-0°$. This is the origin of the statement that $-0°$ is hotter than an infinite temperature. At the infinite temperature the noise from the spin system is small compared with the noise from circuit components which, of course, is infinite at that temperature.

The term

$$\frac{h\nu}{\exp(h\nu/kT) - 1} = f(T) \qquad (46)$$

of Eq. (45) is the Planck expression for the noise power per unit bandwidth from a matched source at $T°K$. Its value for both positive and negative temperatures is given in Fig. 20 along with that of the classical value kT for the same quantity [to which (46) reduces for $h\nu/kT \ll 1$]. At $+0°K$ the quantum limit is identical to the classical limit, i.e., zero. At $-0°K$, however, the quantum limit is $h\nu$ as compared with the classical value of 0. This represents the physical situation of having the whole population in the upper state from which the random spontaneous emission will contribute noise. The limiting temperature on the classical plot equivalent to the $-0°K$ quantum limit is $T = +(h\nu/k)°K$, and quantum amplifiers are often said to have this limit. The large value of $f(T)$ for high temperatures is not at odds with the limited noise output from the spin system at those same high temperatures because R disappears at $T = \infty°$ according to the previous assumptions.

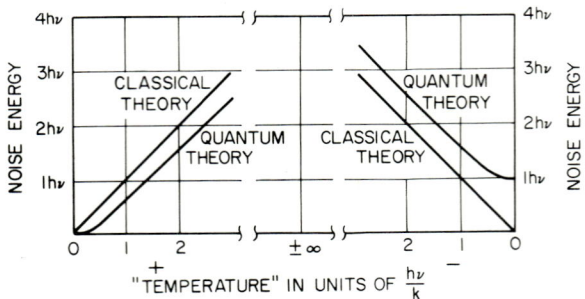

Fig. 20. Noise energy [Eq. (46)] at positive and negative temperatures. (*After Ditchfield,*[30] *published by permission of the editor, Solid-State Electronics.*)

The limiting-temperature sensitivity, the positive temperature needed to give an input equivalent to a single quantum, is $1.45(h\nu/k) \simeq 7 \times 10^{-11}\nu°K$. Assuming that doubling the output is detectable, the limiting-temperature (T_{eff}) sensitivities are as shown in Table 2 for several frequencies. At low frequencies the noise limit imposed by spontaneous emission is extremely small, making possible low-noise amplifiers up to about 10^{12} cps. At the higher frequencies, specifically in the optical range, the limiting temperatures are so high that induced-emission devices are not important as low-noise components. In fact, prior to the development of induced-emission amplifiers, devices (such as photomultipliers and photoconductors) capable of detecting single

Table 2. Limiting-temperature Sensitivities

ν, cps	λ, cm	T_{eff}
10^6	3×10^4	7×10^{-5}
10^9	30	7×10^{-2}
10^{12}	3×10^{-2}	70
10^{15}	3×10^{-5} = 3,000 angstroms	7×10^4

quanta in the optical region existed and were widely used. Lasers are not low-noise amplifiers and their noise properties will not be discussed further. Masers, on the other hand, are useful in a limited number of situations when very low noise amplifiers are justified. Their noise properties are important and will be considered in detail as the various types of masers are discussed.

The noise of an amplifier may be characterized in two ways—by the noise figure or by the effective noise temperature. The ratio of input signal-to-noise/output signal-to-noise is the noise figure. According to IEEE standards the input T_o must be at 290°K. Mathematically this definition is (for $h\nu \ll kT$)

$$F_n = (N_p)_{output}/GkT_oB \qquad (47)$$

where G is the amplifier gain, N_p the noise power, and B the bandwidth. Note that $F_n > 1$ unless the amplifier contributes no noise to the output. Because of the low-noise properties of the masers F_n will differ only slightly from unity, and it is usually more convenient to refer to the noise temperature of the amplifier. The noise temperature is related to the noise figure by the statement

$$T_{N\,eff} = (F_n - 1)T_o \qquad (48)$$

where T_o is the input temperature. $T_{N\,eff}$ is the effective temperature of the amplifier measured by the noise it contributes to the system. The noise temperature is always given as a positive temperature; and for negative temperatures, because of the difference between positive and negative temperatures (Fig. 20), the equivalent positive value will be slightly larger than the absolute value of the negative temperature of the system.

The noise power output may be written as

$$(N_p)_{output} = GB[f(T_{input}) + f(T_N)] \qquad (49)$$

where $f(T)$ is the Planck expression of Eq. (46). For the usually assumed condition that $h\nu \ll kT$, Eq. (49) becomes

$$(N_p)_{output} = GBk(T_{input} + T_N) \qquad (50)$$

11. TRANSMISSION-CAVITY-MASER CHARACTERISTICS

The maser types, of which there are broadly speaking three—transmission cavity, reflection cavity, and traveling wave—have somewhat different characteristics independent of the means of population inversion used. A comparison of the gain, bandwidth, and noise properties of the three maser types will be made in the following sections, where they are considered in the order given above.

11.1. Gain.
This configuration was discussed by Gordon, Zeiger, and Townes[1] in the first paper giving maser characteristics in detail. A subsequent treatment by Stitch[31] is most illuminating, and we shall generally follow it.

The properties of the transmission-cavity maser may be determined by reference to the equivalent circuit or representation of Fig. 21. The cavity as marked is considered to be lossless with actual cavity losses of $1/Q_o$ and the negative loss or gain from the maser material of $-4\pi\xi\chi''(\nu)$ represented by the side arm C with coupling $1/Q_o - 4\pi\xi\chi''(\nu)$. The Q_M of Eq. (37) can be expressed in terms of χ'' by the relation $Q_M^{-1} = 4\pi\xi\chi''(\nu)$. The ξ is a filling factor and the $\chi''(\nu)$ is

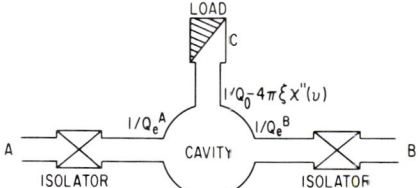

FIG. 21. Equivalent circuit or representation of a transmission-cavity maser. Input and output couplings are represented by $1/Q_eA$ and $1/Q_eB$, respectively. The losses in the cavity and the effect of the maser material are taken care of by the third arm (C) with its fictitious coupling of $1/Q_o - 4\pi\xi\chi''(\nu)$ to a matched load. The term $1/Q_o$ accounts for the cavity losses while the $-4\pi\xi\chi''(\nu)$ represents the negative resistance due to the inverted population of the maser material.

the absolute value of the imaginary part of the susceptibility $\chi = \chi' - i\chi''$, which gives rise to loss or gain. The negative sign is explicitly written. The power emitted from the beam at frequency ν is given by $P = 8\pi^2\nu\mathcal{E}\xi\chi''$ where $\mathcal{E}$ represents the stored energy in the cavity. The filling factor ξ is a measure of the fraction of the stored energy which is in contact with the maser material ($0 > \xi \geq 1$; if the cavity is filled $\xi = 1$)

$$\xi = \frac{\int_{sample} H^2\, d\tau}{\int_{cavity} H^2\, d\tau} \qquad (51)$$

where $d\tau$ is the volume element. The power transmitted through a cavity with $\chi'' = 0$ is

$$P_B(\nu) = \frac{P_o}{Q_e{}^A Q_e{}^B}\left[\left(\frac{1}{2Q_L}\right)^2 + \left(\frac{\nu - \nu_c}{\nu_c}\right)^2\right]^{-1} \tag{52}$$

$$1/Q_L = 1/Q_e{}^A + 1/Q_e{}^B + 1/Q_o \tag{53}$$

The presence of the maser material affects both terms in the denominator; hence both χ'' and χ' are needed. χ', the dispersive part of the susceptibility, is related to χ'' by the Kramers-Kronig relation

$$\chi'(\nu) = \frac{1}{\pi}\int_0^\infty \frac{\chi''(\nu')\,d\nu'}{\nu' - \nu} \tag{54}$$

and is equivalent to the reactive part of a resonant impedance. The power transmitted through the cavity including the emissive effect of χ'' and the frequency shift due to χ' is

$$P_{B'}(\nu) = \frac{P_o}{Q_e{}^A Q_e{}^B}\left\{\left(\frac{1}{2Q_L} - 2\pi\xi\chi''(\nu)\right)^2 + \left[\frac{\nu - \nu_c + 2\pi\nu_c\xi\chi'(\nu)}{\nu_c}\right]^2\right\}^{-1} \tag{55}$$

The gain at midband $\nu = \nu_c$ is then

$$G = P_{B'}(\nu)_o/P_o = (4/Q_e{}^A Q_e{}^B)[1/Q_L - 4\pi\xi\chi''(\nu_o)]^{-2} \tag{56}$$

Large gain corresponds to $2\pi\xi\chi''(\nu_o) \simeq 1/2Q_L$. The properties of the transmission-cavity maser can best be ascertained from this equation by expanding the loaded Q defined in Eq. (53). Dividing Eq. (56), numerator and denominator, by $-Q_o{}^{-1} + 4\pi\xi\chi''(\nu_o)$ and defining

$$\begin{aligned}\gamma_A &= \{Q_e{}^A[4\pi\xi\chi''(\nu_o) - Q_o{}^{-1}]\}^{-1} \\ \gamma_B &= \{Q_e{}^B[4\pi\xi\chi''(\nu_o) - Q_o{}^{-1}]\}^{-1} \\ \gamma_o &= \{Q_o[4\pi\xi\chi''(\nu_o) - Q_L{}^{-1}]\}^{-1}\end{aligned} \tag{57}$$

The gain G may be written as

$$G = 4\gamma_A\gamma_B(\gamma_A + \gamma_B - 1)^{-2} \tag{58}$$

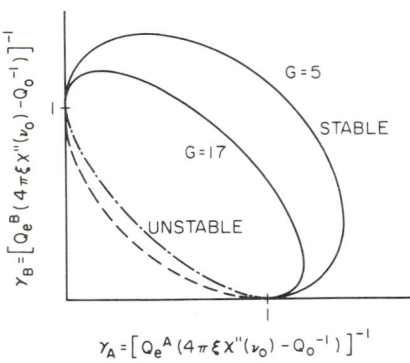

FIG. 22. Curves of constant gain for transmission-cavity maser. (*After Stitch,*[31] *published with permission of the editor,* Journal of Applied Physics.)

γ_A and γ_B are now measures of input and output coupling. Obviously the same gain G may be obtained for a number of combinations of γ_A and γ_B. Two of the ellipses characteristic of this relation are shown in Fig. 22. The dotted portions of the curves correspond to oscillation conditions where the loading is insufficient to keep $Q_L{}^{-1} > 4\pi\xi\chi''(\nu_o)$, i.e., $\gamma_A + \gamma_B < 1$. For a given value of $4\pi\xi\chi''(\nu_o) - 1/Q_o$ the external Q's may be adjusted to give a wide variety of results. The discussion of the following sections will develop criteria for picking various operating points.

11.2. Bandwidth. Because of the characteristics of $\chi''(\nu)$ and $\chi'(\nu)$, being those of absorption and dispersion, they can be approximated by

$$\chi''(\delta) = \chi_o(1 - 2Q_b{}^2\delta^2 + \cdots)$$
$$\chi'(\delta) = \chi_o(2Q_b\delta + \cdots)$$
$$\delta = (\nu - \nu_o)/\nu_o \qquad Q_b = \nu_o/2\Delta\nu_b \qquad \chi_o = \chi''(\nu_o)$$

where $2\Delta\nu_b$ = full width at half amplitude of the resonance line or susceptibility.

In this approximation Eq. (55) becomes

$$P_B'(\nu) = \frac{P_o}{Q_e{}^A Q_e{}^B} \left[\left(\frac{1}{2Q_L} - 2\pi\xi\chi_o + 4\pi\xi\chi_o Q_b{}^2 \delta^2 \right)^2 + \left(\frac{\nu - \nu_c + 4\pi\nu_c\xi\chi_o Q_b \delta}{\nu_c} \right)^2 \right]^{-1} \quad (59)$$

The variation of χ'' with frequency is very small near resonance and the cavity Q's and gains are such that this variation of χ'' may be neglected. Setting $\delta = B/2\nu_o$, i.e., equal to the half fractional bandwidth of the maser, and using B as the bandwidth of the maser amplifier, the gain may be rewritten as

$$G = \frac{4}{Q_e{}^A Q_e{}^B} \left[\left(\frac{1}{Q_L} - 4\pi\xi\chi_o \right)^2 + (1 + 4\pi\xi Q_b\chi_o)^2 \frac{B^2}{\nu_o{}^2} \right]^{-1} \quad (60)$$

The bandwidth criterion is that the reactive component (the second term in the denominator) is equal to the resistive component (the first term in the denominator) since then G is half its maximum value. At this point

$$B = \frac{\nu_o(1/Q_L - 4\pi\xi\chi_o)}{1 + 4\pi\xi Q_b\chi_o} \quad (61)$$

From Eq. (60) the numerator may be related to the midband gain of the maser so that

$$B = \frac{2\nu_o}{1 + 4\pi\xi Q_b\chi_o} \left(\frac{1}{GQ_e{}^A Q_e{}^B} \right)^{\frac{1}{2}} \quad (62)$$

$$= 2\nu_o \frac{4\pi\xi\chi''(\nu_o) - Q_o{}^{-1}}{1 + 4\pi\xi Q_b\chi_o} \left(\frac{\gamma_A\gamma_B}{G} \right)^{\frac{1}{2}}$$

$$B = S \left(\frac{\gamma_A\gamma_B}{G} \right)^{\frac{1}{2}} \quad (63)$$

For future use S is defined as used here to be

$$S = 2\nu_o \frac{4\pi\xi\chi''(\nu_o) - Q_o{}^{-1}}{1 + 4\pi\xi Q_b\chi_o} \quad (63a)$$

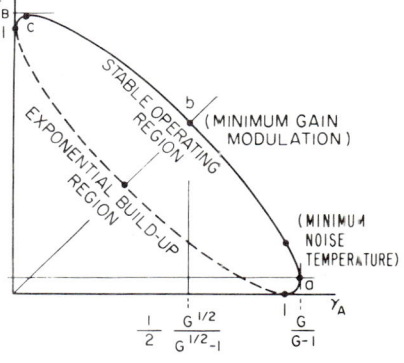

Fig. 23. Optimum operation points for transmission-cavity maser. The characteristics at a, b, and c are given in Table 3. (After Stitch,[31] published with permission of the editor, Journal of Applied Physics.)

The product $\gamma_A\gamma_B$ can be evaluated for a given ellipse and an investigation made of the properties. Results for the extreme values and for the $\gamma_A = \gamma_B$ point are illustrated in Fig. 23 with the values in Table 3. They will be discussed later after noise and gain modulation can be included.

Table 3

Point	$G^{\frac{1}{2}}B$	Noise $f(T_N)$	Gain modulation
a	$\dfrac{SG^{\frac{1}{2}}}{G-1}$		$\dfrac{2(1+\gamma_o)}{\gamma_A + \gamma_B - 1}$
b	$\dfrac{S}{2}\dfrac{G^{\frac{1}{2}}}{G^{\frac{1}{2}}-1}$	$\begin{cases} \dfrac{Q_e{}^A}{Q_o} f(T) + Q_e{}^A(-4\pi\xi\chi_o)f(T_{eff}) \\ \dfrac{Q_e{}^A}{Q_o} f(T) + Q_e{}^A(-4\pi\xi\chi_o)f(T_{eff}) \\ + \left[\dfrac{1}{G} - 1 + (4\pi\xi\chi_o - Q_o{}^{-1})Q_e{}^A\right] f(T_B) \end{cases}$	$\dfrac{2(1+\gamma_o)}{2\gamma_A - 1}$
c	$S\dfrac{G^{\frac{1}{2}}}{G-1}$		$\dfrac{2(1+\gamma_o)}{\gamma_A + \gamma_B - 1}$

11.3. Noise. Considering the schematic circuit of Fig. 24, the available output noise power per unit bandwidth [Eq. (49) with $B = 1$] including contribution of spontaneous emission is[31]

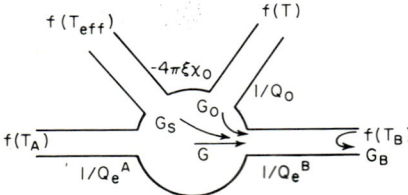

FIG. 24. Schematic for noise considerations in a transmission maser. The couplings are the $1/Q$ values indicated while the gains for the various processes are

$$G = \frac{4(1/Q_e^A)(1/Q_e^B)}{(1/Q_L - 4\pi\xi\chi_o)^2}$$

$$G_s = \frac{4(-4\pi\xi\chi_o)(1/Q_e^B)}{(1/Q_L - 4\pi\xi\chi_o)^2}$$

$$G_o = \frac{(1/Q_e^B)(1/Q_o)}{(1/Q_L - 4\pi\xi\chi_o)^2}$$

$$G_B = \frac{(1/Q_e^B - 1/Q_e^A + 4\pi\xi\chi_o - 1/Q_o)^2}{(1/Q_L - 4\pi\xi\chi_o)^2}$$

$(N_p)_{output} = G[f(T_A) + (G_o/G)f(T)$
$\qquad + (G_s/G)f(T_{eff}) + (G_B/G)f(T_B)]$

From the definition of the noise temperature, on noting that $f(T)$ replaces kT, we have

$$f(T_N) = (G_o/G)f(T) + (G_s/G)f(T_{eff}) + (G_B/G)f(T_B) \quad (64)$$

Using the values of G given with relation to Fig. 24 this becomes

$$f(T_N) = (Q_e^A/Q_o)f(T) + (-4\pi\xi\chi_o)Q_e^A f(T_{eff}) + [1/G - 1 + (4\pi\xi\chi_o - 1/Q_o)Q_e^A]f(T_B) \quad (65)$$

Just as in the previous qualitative discussion of noise, the negative sign of the $-4\pi\xi\chi_o$ is compensated for by the negative value of $f(T_{eff})$ for negative temperatures so that the noise contributed by the maser material is positive. It is apparent from this expression for the noise output that it will be a minimum for a minimum Q_e^A or maximum γ_A if it is assumed that all quantities but the couplings are fixed. This point is marked on Fig. 23 and corresponds to the value of $\gamma_A = G/(G - 1)$ and $\gamma_A/\gamma_B = G$. It is important to note that at this point the coefficient of $f(T_B)$ is zero, indicating no noise contribution from the output regardless of match or temperature. At this point the value of Eq. (65) becomes

$$f(T_N) = [(G - 1)/G][\gamma_o f(T) + (1 + \gamma_o)(-)f(T_{eff})] \quad (66)$$

For constant gain and with γ_A maximized the noise is controlled by the value of γ_o, and in fact γ_o^{-1} might be called a figure of merit. If γ_o^{-1} is large enough $f(T_N)$ is equal to $(-)f(T_{eff})$ and is unaffected by the cavity temperature itself. That is,

$$\exp(h\nu/kT_N) - 1 = -[\exp(h\nu/kT_{eff}) - 1]$$

and for

$$|h\nu/kT_{eff}| \ll 1 \qquad T_N \equiv |T_{eff}| \quad (67)$$

11.4. Gain Modulation. Any change of gain in a random fashion is equivalent to noise, and being regenerative in nature, masers are very susceptible to gain-fluctuation problems. Variation of the inverted population because of pump power fluctuation or noise is the most commonly expected source of difficulty.

If the gain modulation is defined as

$$M = \frac{\Delta G/G}{\Delta[\xi\chi''(\nu_o)]/\xi\chi''(\nu_o)} \quad (68)$$

in terms of the change in $G, (\Delta G)$, produced by a change in $\xi\chi''(\nu_o)$, the value for the transmission maser from Eq. (56) is

$$M = \frac{2(1 + \gamma_o)}{(\gamma_A + \gamma_B - 1)} \quad (69)$$

which has a minimum value for $\gamma_A = \gamma_B$.

The points of interest in connection with the transmission-cavity maser are illustrated in Fig. 23 and Table 3. It is evident that one thing can be traded for another up to a point. Maximum bandwidth and minimum gain modulation occur at point b, but optimum noise temperature requires accepting the narrower bandwidth of point a. Point c has the same bandwidth and gain modulation as a but much higher noise.

12. REFLECTION-CAVITY-MASER CHARACTERISTICS

There are fewer variables to be considered in the reflection-cavity maser than were necessary in the transmission-cavity maser. Figure 25 illustrates schematically the system used. The circulator serves the same purpose as the output isolator of Fig. 21, that is, to prevent the noise from the output termination from reaching the amplifier cavity where it would be amplified and appear as an increase in effective noise temperature.

12.1. Gain. The gain of a reflection cavity is obtained by determining the reflection coefficient with the active maser material in the cavity. The gain is equal to the reflection coefficient and in this case is

$$G = \left[\frac{1/Q_A - 1/Q_o + 4\pi\xi\chi''(\nu_o)}{1/Q_A + 1/Q_o - 4\pi\xi\chi''(\nu_o)}\right]^2$$
$$= \left(\frac{\gamma_A + 1}{\gamma_A - 1}\right)^2 \qquad (70)$$

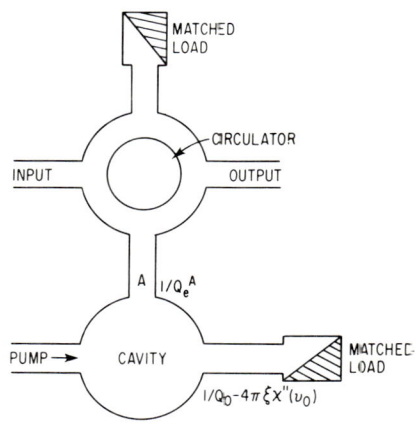

FIG. 25. Schematic of reflection-cavity maser.

The gain is directly dependent on the coupling, and maximum gain occurs as expected at the point where the cavity and external losses are just equal to the output of the maser material.

12.2. Bandwidth. Substituting from Eq. (70) for the numerator of Eq. (61), the bandwidth is given as

$$B = 2\nu_o \frac{4\pi\xi\chi''(\nu_o) - Q_o^{-1}}{1 + 4\pi\xi Q_b \chi_o} \frac{1}{G^{\frac{1}{2}} - 1} = S \frac{1}{G^{\frac{1}{2}} - 1} \qquad (71)$$

S is defined by Eq. (63a). For typical values of $Q_o \simeq 10^4$, $4\pi\xi\chi''(\nu_o) = 10^{-3}$ (i.e. $Q_M = 10^3$), $G^{\frac{1}{2}} > 1$, and we can neglect the last term in both numerator and denominator of S, arriving at an approximate bandwidth

$$B = 2\nu_o 4\pi\xi\chi''(\nu_o)G^{-\frac{1}{2}}$$

which numerically becomes

$$2\nu_o \times 10^{-4}$$

for a gain of 100 and the assumed value of $4\pi\xi\chi''(\nu_o)$.

12.3. Noise. The equivalent representation for noise purposes is Fig. 26 with the input connection A. Here the available output noise power is

$$(N_p)_{output} = Gf(T_A) + G_o f(T) + G_s(-)f(T_{eff})$$

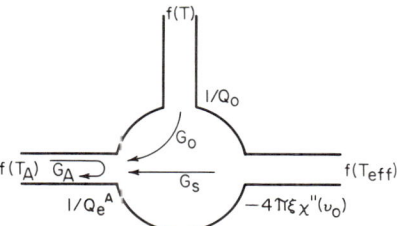

FIG. 26. Schematic of reflection-cavity maser for noise analysis.

and as for the transmission case this leads to a relation between $f(T_N)$ and the actual temperatures

$$f(T_N) = (G_o/G)(f(T)) + (G_s/G)(-)f(T_{eff}) \qquad (72)$$

This is the same available output power as for the transmission-cavity maser except that it lacks the $f(T_B)$ term of Eq. (64). Since the minimum noise-temperature condition in that case occurred when this source of noise made zero contribution it is not surprising that the result of evaluating Eq. (72) is identical to the first two terms of Eq. (65) or to Eq. (66) itself. Again Eq. (67) applies if γ_o^{-1} and G are large, while for

$$h\nu/kT_{eff} > 1 \quad T_N = (h\nu/k)\ln 2 = 1.4(h\nu/k)$$

12.4. Gain Modulation. Using the appropriate substitution in Eq. (68) the gain modulation becomes

$$M_R = (1 + \gamma_o)[(G - 1)/G^{\frac{1}{2}}] \tag{73}$$

13. THE TRAVELING-WAVE MASER

The most effective method of increasing the bandwidth is to use a nonresonant structure. In order, however, to make use of the low-noise properties the gain must be high enough that the noise of succeeding stages is unimportant. Both the gain and bandwidth requirements can be met by utilizing a traveling-wave structure. As the circuit bandwidth is broadened, the resonant line width of the maser crystal will become a limiting factor. Even then, there are ways, such as varying the direction of the crystal axis, by means of which the bandwidth may be improved.

The electronic traveling-wave tube derives its gain by requiring the electron beam to have a velocity slightly greater than the phase velocity of the electromagnetic wave on the slow-wave structure. Its close relation, the linear accelerator, requires the reverse; that is, the beam travels at the slower of the two velocities. In both cases, the desired result, gain or acceleration, is proportional to the length or the total time over which the interaction takes place. The traveling-wave maser has no synchronization requirement with another energy component (assuming that the spins are uniformly inverted); however, the gain depends on induced emission and hence energy density. It will be seen that the requirement of increasing the interaction with the maser crystal compels one to choose a structure with a slow *group* velocity.

The treatment which follows and most of the data on TW masers are due to DeGrasse, Schulz du Bois, and Scovil.[33]

13.1. Gain. Let us assume some microwave structure in the z direction at least partly filled with maser material. The change of power with distance is then given by

$$dP/dz = -\tfrac{1}{2}\omega \int_{A_m} H \cdot \chi'' \cdot H^* \, dA \tag{74}$$

The integration is carried out over the maser material cross section A_m. The power in the waveguide is

$$P = (v_g/8\pi) \int_{A_s} H^2 \, dA$$

where v_g is the group velocity in the waveguide of cross section A_s. The gain is then, inserting the negative sign of χ'' for an inverted population,

$$\begin{aligned} dP/P &= +(4\pi\omega/v_g)\chi''\xi dz \\ G &= \exp[+(4\pi\omega\chi''\xi l/v_g)] \\ G_{db} &= +27.3(4\pi\nu\chi''\xi l/v_g) \quad \text{db} \end{aligned} \tag{75}$$

ξ is the filling factor, here defined as

$$\xi = \frac{\int_{A_m} H \cdot \chi'' \cdot H^* \, dA}{\chi'' \int_S H^2 \, dA} \tag{76}$$

or, for a circularly polarized wave,

$$\xi_+ = \frac{\int_{A_m} H_+ \cdot \chi'' \cdot H_+^* \, dA}{\chi'' \int_{A_s} (H_+^2 + H_-^2) \, dA}$$

It is convenient to measure velocity as a function of the velocity of light and to measure length in free-space wavelengths. A factor $Z = c/v_g$ and a length $L = l\nu/c$ may be used in the gain formula to give

$$G_{db} = -27.3(LZ/Q_M) \quad \text{db} \tag{77}$$

using the definition of $1/Q_M = -4\pi\xi\chi''$.

The circular polarization is of considerable value in obtaining unidirectional effects. In order to prevent a traveling-wave device from oscillating, loss for the backward direction sufficient to overcome the round-trip gain (including reflection losses) is needed. In this case, since regions of both right- and left-hand circular polarization exist in most guide structures, the difference between the effects of the two circular polarizations can be used to get both unidirectional gain and loss, the gain being in the forward direction and the loss in the backward direction. This loss is in addition to the dielectric and structure losses which contribute to $1/Q_o$ and which are not sensitive to the direction of power flow. The nonreciprocity is due to the fact that, at the same point in the structure, power which in one direction gives rise to right-handed circular polarization will be responsible for left-handed polarization when traveling the opposite direction. The nonreciprocity of gain is then governed by filling-factor ratios or

$$R_g = \int_{A_m} H_+^2 \, dA \Big/ \int_{A_m} H_-^2 \, dA \tag{78}$$

where A_m = area of amplifying material

It is assumed that for the forward direction H_+ reacts with the maser material to produce gain and in the reverse direction H_+ interacts with the absorbing material. Referred to the forward direction, losses will be placed in regions of H_- and the isolator figure of merit will be

$$R_i = \int_{A_i} H_-^2 \, dA \Big/ \int_{A_i} H_+^2 \, dA \tag{79}$$

where A_i = isolator area

Both paramagnetic and ferrimagnetic materials possess this property of responding differently to right- and left-handed polarizations (with respect to the applied d-c magnetic field direction). For example, the same material used as the active maser material may be used in higher-concentration form (the pump power can then not invert the levels because of the relaxation times) to provide the backward attenuation simply by locating the two types of crystal at the appropriate points in the slow-wave structure. The matching of energy levels is automatically taken care of by using the same orientation for all crystals. In this case, $R_i \simeq R_g$. Ferrimagnetic materials, which have some advantage in the loss available per unit volume, may be used in smaller quantity and a slightly better value of R_i obtained. The value of the resonance magnetic field is made to coincide with that required for the paramagnetic maser material by shaping the sample.[34] One big advantage of the ferrimagnetic isolator is that it does not interact with the pump whereas the higher-concentration paramagnetic material, with unfavorable relaxation times for spin inversion, does absorb pump power rather well in its unsaturated spin levels.

The forward-gain and reverse-loss equations are then

$$\begin{aligned}\text{Gain} &= 27.3LZ[+4\pi\chi_+''\xi_+ - 4\pi\chi_i''(\xi_i^-/R_i) - \alpha_o] \quad \text{db} \\ \text{Loss} &= 27.3LZ[-4\pi\chi_+''(\xi_+/R_g) + 4\pi\chi_i''\xi_i^- + \alpha_o] \quad \text{db}\end{aligned} \tag{80}$$

with α_o being the structure attenuation. The first term refers to the maser material and the second to the isolation crystal.

13.2. Bandwidth. The width of the resonance line itself is the main limiting factor to be considered in the bandwidth if the slow-wave structure itself is properly designed with a broadband frequency characteristic. The paramagnetic resonance line may vary in shape from Lorentzian to Gaussian depending on crystal perfection, crystal type, concentration, etc. If the line shape is Lorentzian, the line shape is given by Eq. (18a), and for that case, setting $B_b = 2(\nu - \nu_o) = (\pi T_2)^{-1}$, the full width at half maximum for χ'', the 3-db bandwidth is

$$B = B_b \sqrt{3/(G_{db} - 3)} \qquad (81)$$

Since a number of crystals are normally used in a traveling-wave maser structure it is possible to broaden the response by varying the orientation of the crystals, giving a stagger-tuned effect similar to that used in broadband i-f amplifiers. In this case the maximum gain will be sacrificed for bandwidth, Eq. (81) will apply to each section, and the response curves for each section must be added to get the overall characteristic.

It is worthy of note that, with the broadband slow-wave structures possible for obtaining the group velocity needed, for a reasonable length of device the center frequency of operation may be shifted by changing the magnetic field and the pump frequency. In practice, such a shift may be as much as ten times the paramagnetic resonance width and the order of a hundred times the resonance width of a cavity suitable for a cavity maser.

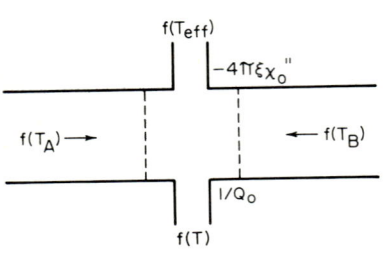

Fig. 27. Schematic of traveling-wave maser section for noise analysis.

13.3. Noise. The noise properties of an increment of length dz may be treated (Fig. 27) in a fashion similar to that illustrated for the transmission cavity maser in Fig. 24. The noise inputs are reinterpreted and in place of a cavity a section of the maser length is considered. By using methods similar to those above the effective noise temperature is found to be[30]

$$f(T_N) = \frac{G-1}{G} \frac{\alpha_c f(T) + \alpha_M(-)f(T_{eff})}{\alpha_M - \alpha_c} \qquad (82)$$

where $\alpha_c = 4\pi\chi_i''(\xi_i^-/R_i) + \alpha_o$
$\alpha_M = +4\pi\chi_+''\xi_+$

α_c then is the loss in the forward direction and α_M the gain in the same direction. The traveling-wave maser has been idealized to the extent that, as in previous sections, no account is taken of losses in the input waveguide, and the structure is taken to be completely unidirectional in that output noise is not considered.

13.4. Gain Modulation. Little attention has been paid to the necessity of using a pump of one kind or another to invert the populations. The problems associated with this in practice are considered in Chap. 29. The same gain-modulation problem can exist here which was considered in the case of cavity masers. Consideration of the gain equation (75) immediately leads to a gain-modulation value of

$$M_{TW} = \log_e G \qquad (83)$$

14. COMPARISON OF MASER TYPES

A summary of the results obtained in the preceding sections is given in Table 4. It is immediately obvious that the transmission-cavity maser falls short of the reflection-cavity type by a factor of 2 at its optimum points. For this as well as for practical reasons of dewar and cryostat construction, the reflection cavity maser has been used

Table 4

	Reflection cavity	Transmission cavity	Transmission cavity (optimum)	Traveling wave
Gain	$\left(\dfrac{\gamma_A + 1}{\gamma_A - 1}\right)^2$	$4\gamma_A\gamma_B(\gamma_A + \gamma_B - 1)^{-2}$		$\exp LZ\left(\chi_+'' \xi_+ - \chi_{i+}'' \dfrac{\xi_{i-}}{R_i} - \alpha_0\right)$
Bandwidth	$\dfrac{S}{G^{\frac{1}{2}} - 1}$	$S\left(\dfrac{\gamma_A\gamma_B}{G}\right)^{\frac{1}{2}}$	$\dfrac{1}{2}\left(\dfrac{S}{G^{\frac{1}{2}} - 1}\right)$ b^*	$B_b\sqrt{\dfrac{3}{G_{db} - 3}}$
Noise temperature	$\dfrac{G - 1}{G}[\gamma_0 f(T)$ $+ (1 + \gamma_0)(-)f(T_{eff})]$	$\dfrac{1}{\gamma_A}[\gamma_0 f(T)$ $+ (1 + \gamma_0)(-)f(T_{eff})$ $+ \left[\dfrac{1}{G} - 1 + \dfrac{1}{\gamma_A}\right]f(T_B)$	$\dfrac{G - 1}{G}[\gamma_0 f(T)$ $+ (1 + \gamma_0)(-)f(T_{eff})]$ a^*	$\dfrac{G - 1}{G}\left[\dfrac{\alpha_c f(T) + \alpha_M(-)f(T_{eff})}{\alpha_M - \alpha_c}\right]$
Gain modulation	$(1 + \gamma_0)\left(\dfrac{G - 1}{G^{\frac{1}{2}}}\right)$	$\dfrac{2(1 + \gamma_0)}{\gamma_A + \gamma_B - 1}$	$\dfrac{2(1 + \gamma_0)}{G^{\frac{1}{2}} - 1}$ b^*	$\log_e G$
$G^{\frac{1}{2}}B$	$S\left(\dfrac{G^{\frac{1}{2}}}{G^{\frac{1}{2}} - 1}\right)$	$S(\gamma_A\gamma_B)^{\frac{1}{2}}$	$\dfrac{S}{2}\left(\dfrac{G^{\frac{1}{2}}}{G^{\frac{1}{2}} - 1}\right)$ b^*	$B_b G_b^{\frac{1}{2}}\sqrt{\dfrac{3}{G_{db} - 3}}$

$\gamma_A = \{Q_e{}^A[4\pi\xi\chi''(\nu_0) - Q_0^{-1}]\}^{-1}$
$\gamma_B = \{Q_e{}^B[4\pi\xi\chi''(\nu_0) - Q_0^{-1}]\}^{-1}$
$\gamma_0 = \{Q_0[4\pi\xi\chi''(\nu_0) - Q_0^{-1}]\}^{-1}$
$S = 2\nu_0\left(\dfrac{4\pi\xi\chi''(\nu_0) - Q_0^{-1}}{1 + 4\pi\xi Q_0\chi_0}\right)$

a^* and b^* refer to the points in Fig. 23.

$B_b = 2\Delta\nu =$ full width at half amplitude for paramagnetic line
$Q_b = \nu_0/B_b$
$Z = c/v_g$, $L = l\nu/c$
$\alpha_c = 4\pi\chi_i''\xi_i/R_i + \alpha_0$
$\alpha_M = 4\pi\chi''\xi$

14-33

almost exclusively in practice. In cases where good isolators are available but good circulators unavailable there may be reason to consider the transmission-type circuit.

In the following, a comparison between the traveling-wave and reflective-cavity masers will be sketched. In gain bandwidth, and gain modulation, the TW maser is superior. The gain is dependent on the length of structure and is really limited only by need and necessity of providing sufficient reciprocal loss to prevent oscillation. The bandwidth is on the order of the paramagnetic line width; in fact for a gain of 30 db it is one-third the line width. The result can be a bandwidth of several tens of megacycles for a system employing ruby. The gain-bandwidth product $G^{\frac{1}{2}}B$ increases as the gain increases rather than approaching a constant value characteristic of the cavity-type masers. The gain-modulation factor is also lowest for the traveling-wave maser, signifying less trouble from fluctuations of pump power than in the cavity types. More problems are associated with the pump in the TW maser than in the others where a dual-mode cavity is quite straightforward.

15. LASER AMPLIFIER CHARACTERISTICS

The physical dimensions of laser crystals are generally many times the wavelength of the radiation involved. It is therefore more useful to consider the amplification to be traveling-wave in nature. Addition of reflection coefficients at the ends will provide the feedback necessary for conversion to an oscillator. The considerations are similar to the considerations of Sec. 13.

Suppose radiation of an intensity $I(\nu\theta\varphi P)$ of frequency ν, polarization P, and direction θ,φ passes through a medium with two energy states (Fig. 1) with level populations n_1, n_2 and degeneracies g_1, g_2. The change in intensity due to interaction between the radiation field and the energy-level system is[35]

$$\frac{dI(\nu\theta\varphi P)}{dt} = W(\theta\varphi P)\left[\frac{I(\nu\theta\varphi P)\nu^3}{v^2}\left(n_2 - \frac{g_2}{g_1}n_1\right) + h\nu n_2\right]g(\nu) \quad (84)$$

where $W(\theta\varphi P)$ is the spontaneous-emission probability per unit solid angle, for the direction $\theta\varphi$ with polarization P, and v is the velocity of propagation. For an isotropic W, the relation to lifetime and hence to A is

$$8\pi W = 1/\tau = A$$

The last term is the pure spontaneous emission from level 2 and may be dropped since we are interested in gain or loss. Equation (84) is the equivalent of (74) except for this term. The gain over the length $l(\theta\varphi)$ is then

$$G_{db}(\nu\theta\varphi P) = 10 \log e\{A(\theta\varphi P)[\lambda_o^2 l(\theta\varphi)/\epsilon][n_2 - (g_2/g_1)n_1]\}g(\nu) \quad (85)$$

λ_o is the free-space wavelength and ϵ is the dielectric constant at frequency ν. For a gaussian line the gain then becomes

$$G_{db}(\nu\theta\varphi P) = 10 \log e \left[A(\theta\varphi P)\frac{\lambda_o^2 l(\theta\varphi)}{\epsilon}\left(n_2 - \frac{g_2}{g_1}n_1\right)\right] \frac{2}{\Delta\nu}\sqrt{\frac{\ln 2}{\pi}}\exp\left[-\frac{(\nu-\nu_o)^2}{\Delta\nu^2}4\ln 2\right] \quad (86)$$

and defining the bandwidth B as the frequency width to the points when the gain is down 3 db

$$B = B_b\sqrt{\frac{\log[G_{db\ max}/(G_{db\ max} - 3)]}{\log 2}} \quad (87)$$

where B_b is the full width at half maximum for the resonance line.

INDUCED-EMISSION-AMPLIFIER PRINCIPLES

It is worth noting that for the normal laser crystal there is appreciable reflection at the crystal-air interface. If this is represented by a power-reflection coefficient r at each end, the apparent one-pass gain observed for a system with gain in both directions is

$$G = \frac{(1-r)^2 G_o}{1 + r^2 G_o^2 - 2rG_o \cos \varphi} \quad (88)$$

where φ is the relative phase shift for a complete (double) traversal of the amplifier. Because of random phasing due to the tolerances involved it is usually safe to average φ around zero obtaining

$$G = \frac{(1-r^2) G_o}{1 - r^2 G_o^2} \quad (89)$$

This will be the observed single-pass gain with G_o given by (85) as modified by the line shape $g(\nu)$, e.g., (86) for the gaussian line.

16. INDUCED-EMISSION OSCILLATORS

The gas maser and lasers in general are of interest as oscillators. Their great attribute is the frequency stability and in the case of the laser the additional property of coherence (possessed by no other source in this frequency range). Fundamentally, the frequency is set, certainly within the line width, by the transitions, or combinations of transitions in the Raman laser, utilized for the gain process. In contrast to the solid-state maser (which is not of general interest as an oscillator) the energy levels used for the oscillators are fixed by the atomic or molecular-energy levels, or the band-gap and impurity levels of the semiconductor, and are tuned only with considerable difficulty. The frequency characteristic achieved depends on mechanical and dimensional stability, cavity Q's, pumping properties, and the success in spurious-mode elimination.

16.1. Tuning. The frequency of oscillation depends on both Q_b, the quality factor of the transition, and Q_c, that of the resonator, in the following manner:[36]

$$\nu = \frac{\nu_b Q_b + \nu_c Q_c}{Q_b + Q_c} \quad (90)$$

for homogeneously broadened Lorentz line shapes when ν is the oscillation frequency and the subscript b refers to the active material and c to the cavity. Thus ν depends on the relative widths ($Q = \nu/\Delta\nu$) of the molecular transition and the cavity. If $Q_b > Q_c$ the cavity has a relatively minor influence and the oscillation frequency is

$$\nu \simeq \nu_b + (\nu_c - \nu_b)(Q_c/Q_b) \quad (91)$$

In the ammonia maser Q_c/Q_b is typically 10^{-3}. The frequency can be pulled only within the width of the molecular transition, in fact a few thousand cycles.

The situation in the gas laser is the reverse of the above, and in addition, the cavity width and mode separations are such that several modes may oscillate at the same time within the Doppler broadened line width. This is illustrated in Fig. 28. The frequency of oscillation is then predominantly determined by ν_c by approximating (90)

$$\nu \simeq \nu_c + (\nu_b - \nu_c)(Q_b/Q_c) \quad (92)$$

The interval between modes ($\delta\nu$) is then slightly shifted from the value calculated by considering the cavity modes and becomes

$$\delta\nu = \delta\nu_c(1 - Q_b/Q_c) \quad (93)$$

Experimental verification of this relation has been obtained by Bennett,[38] who also observed some power-dependent effects. These originate in hole burning and dispersion and may be of importance in frequency-stabilizing schemes on page 14-36.

14-36 AMPLIFYING DEVICES

The inverted population of the laser transition results in variation of refractive index and hence velocity dispersion for frequencies within $\Delta\nu_b$. The variation of refractive index n for both positive and negative temperatures is illustrated in Fig. 29. Within the line width $\Delta\nu_b$ the modes will be moved closer together for the inverted populations. The resonance frequencies are given approximately by $p(c/2b)(1/n)$, and hence modes above resonance are shifted down and those below up. The beats between three modes situated at frequencies marked by the a's and b's in Fig. 29 will

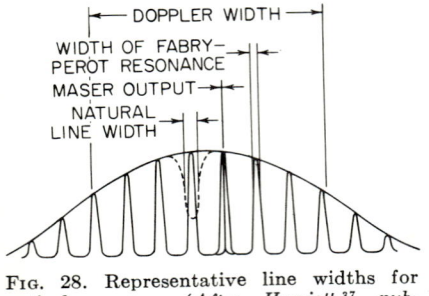

FIG. 28. Representative line widths for optical masers. (*After Herriott*,[37] *published with permission of the editor, Journal of Optical Society of America.*)

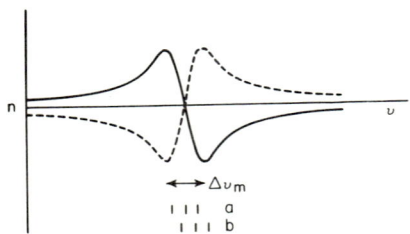

FIG. 29. Variation of refractive index near resonance. Solid curve represents the behavior at positive temperatures and the dotted curve at negative ones. The frequency markers a and b relate to Fig. 30.

be as shown in Fig. 30. For the symmetric case a the beats between adjacent modes coincide but for the asymmetric case b two separate beat notes are observed. This shift due to the nonlinearity of the refractive index may be useful in feedback systems to stabilize the laser frequency with respect to the line center.

For the gas laser, where the Doppler broadened line is inhomogeneously broadened, hole-burning effects add complications. These give rise to power-dependent mode splittings and "hole-repulsion effects," which have been discussed in the literature.[38] The inhomogeneity of the line is also given as the cause of the simultaneous oscillation

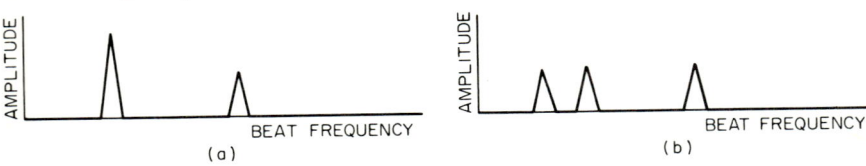

FIG. 30. Schematic of observed beat frequencies. (*a*) For symmetrical frequency modes as in *a* of Fig. 29. (*b*) For unsymmetrical situation of *b* in Fig. 29. The high frequency corresponds to beats between the outside modes while the lower-frequency ones correspond to beats between adjacent modes. The latter split in case *b* because of dispersive effects.

in a number of modes in some cases, although the differing spatial distribution of fields associated with simultaneously oscillating modes will also allow oscillation in a number of modes simultaneously even in the case of an homogeneous line if lateral diffusion is slow. Normally, a number of modes at frequencies differing from those attributed to the length of the cavity, because of refractive-index changes and power-sensitive hole-burning effects, oscillate at the same time. The number can be reduced to one by simply reducing the gain to the point where only one mode is above the threshold or by selectively inserting loss into unwanted modes.[39] Kogelnik[40] found that use of an iris centered on the optic axis limits the oscillation to the *highest*-order mode which diffraction and other losses will allow to oscillate in a gas laser.

The frequency of an individual mode may be varied by changing the cavity length without changing the angle of tilt. In practice this is difficult to accomplish in a controlled fashion, but it does often occur as the result of thermal effects.

Electric and magnetic fields both produce tuning changes in many solid-state and gas lasers. The electric field effect has been used as a type of Q switch in ruby.[41]

Magnetic fields of 6.8 kilogauss have been used to shift the frequency of a HeXe maser by 38 Doppler widths (38 × 210 Mc).[42] This presumably is accomplished by the oscillation jumping modes so as to remain within the oscillation bandwidth of the gas. Both the frequency and threshold in an InAs laser shift with magnetic field.[3]

16.2. Frequency Stability. The fundamental limit in the width of the oscillating output of either the maser or laser is the spontaneous emission into the single oscillating mode. The width attributable to this source was first derived for the ammonia maser,[1] and is given by

$$\Delta \nu = \frac{8\pi h \nu}{P} \left(\frac{\Delta \nu_c \, \Delta \nu_b}{\Delta \nu_c + \Delta \nu_b} \right)^2 \tag{94}$$

For the ammonia maser, $\Delta \nu_b$, the width of the active transition, is about 3,000 cps, much less than the cavity width $\Delta \nu_c$, and (94) can be approximated by

$$\Delta \nu = (8\pi h \nu / P)(\Delta \nu_b)^2 \tag{95}$$

For a power output of $P = 10^{-10}$ watts at the transition frequency of 24 gc, $\Delta \nu \simeq 10^{-2}$ cps or $\nu/\Delta \nu \simeq 5 \times 10^{12}$. The frequency stability usually attainable is an order of magnitude or more less than this, with the pulling of the frequency by cavity detuning a major source of trouble.

For the laser with its high-Q cavity the appropriate relation for $\Delta \nu$ is

$$\Delta \nu = (8\pi h \nu / P)(\Delta \nu_c)^2 \tag{96}$$

For a cavity with $\Delta \nu_c = (c\alpha/2\pi b) = 1 \times 10^6$ ($\alpha = 0.02$ per pass), and a power output of 10^{-3} watts = 10^4 erg/sec, $\Delta \nu \simeq 6 \times 10^{-3}$ cps or $\nu/\Delta \nu = 1.6 \times 10^{16}$. The thermal fluctuations of the laser length in most cases set a higher limit on the frequency width.[44] The fractional frequency width due to this cause is

$$\Delta \nu / \nu = (kT/YV)^{\frac{1}{2}} \tag{97}$$

where k is the Boltzmann constant, T the temperature, Y the Young's modulus of the support material, and V the volume of the support. A Young's modulus of 20×10^{11} dynes/sq cm and a support volume of 500 cu cm result in a fractional frequency width of about 6.5×10^{-15} because of this thermal motion. This then sets the width limitation, one which Javan[44] has come within a factor of 10 of achieving. A short-term stability of 8 parts in 10^{14} was obtained under relatively ideal conditions (in an underground vault on solid rock), and the main cause of this width was microphonics. Since the frequency depends directly on the length b, the fractional length stability must equal that of the frequency.

A situation which comes much closer to checking on Eq. (96) is the experiment of beating together two modes of the same laser. Length variations tend to affect each mode in the same way. Javan et al.[45] have under these conditions measured drifts of a few cycles (actually 4) over a period of several seconds, implying a constancy in each mode to within 2 cycles.

Results for the solid-state lasers and semiconducting-junction lasers are not so good. This is to be expected from the smaller cavities used, but the results are made worse by the spiking and thermal heating which occur during the pulse in intermittent operation. These factors are discussed in Chap. 29.

16.3. Oscillation Conditions and Pump Power Requirements. The conditions under which oscillations will exist in a given maser or laser mode are that the gain from the amplifying medium must equal the losses of energy from that mode. The losses include absorption in the medium, absorption in and transmission out of the cavity, mode conversion, etc. In the case of the maser oscillator, cavity losses and coupled output power are the main losses. For the laser one can enumerate mirror absorption and transmission, diffraction, mode conversion due to imperfect mirrors, absorption in the laser material, and scattering from inhomogeneities as possible loss mechanisms.

Spontaneous emission (not to be confused with relaxation mechanisms) is a negligible sink for pump power in the maser. The oscillation condition is simply an equality between power emitted by the beam or inverted spin system and the losses. This may be expressed as

$$\Delta n h\nu W_{12} = 2\pi\nu Q(E^2V/8\pi) \qquad (98)$$

where Δn = population difference $n_2 - n_1$
W_{12} = induced transition probability
$E^2V/8\pi$ = energy stored in the cavity
V = effective cavity volume

Since $W_{12} \simeq |\pi\mu Et/h|^2$ where μ is the electric-dipole moment for the transition and t is the time the molecule spends in the field E, (98) for the gas maser becomes

$$\Delta n = hV/4\pi^2\mu^2 t^2 Q \qquad (99)$$

This is the number of excess upper-state molecules required per second to maintain minimum oscillation conditions. If an output coupling coefficient is inserted, so that the power coupled out is included, the loaded Q must be used.

In the frequency range characteristic of the laser because of the high spontaneous-emission probability [Eq. (17) and Table 2] and the high mode density, spontaneous emission is the sink for a considerable part of the energy lost. However, the inversion conditions for a single-mode oscillation are similar to those given above for the maser.

It is convenient to assign a Q to the system such that all the losses including scattering, diffraction, and transmission are given by $(\omega/Q) \times$ (stored energy). This is equivalent to defining a photon lifetime of $Q/\omega = l/\alpha c$, when α is the attenuation ($\ll 1$) per traversal of the length l. The equivalent of (98) for the start of oscillation condition is

$$h\nu[n_2 - n_1(g_2/g_1)]W_{12} = -(I/ct_{photon}) \qquad (99a)$$

where the weight factors g_2 and g_1 are introduced to account for degeneracies of levels 2 and 1, respectively (with total populations n_2 and n_1), W_{12} is the induced transition probability, and the stored energy is written as intensity/velocity, I/c. Substituting (21) for W_{12} the threshold condition is given by

$$n_2 - n_1(g_2/g_1) = 8\pi t_{spont}\nu^2/c^3 g(\nu_c)t_{photon} \qquad (100a)$$

after replacing A by $1/t_{spont}$. The "critical inversion density" defined by (100a) depends on $1/t_{photon} = \omega/Q$ and except for expressing a number density is equivalent to (99) for the beam maser. Occasionally, this threshold condition is expressed in terms of the oscillator strength f related to t_{spont} by the relation

$$ft_{spont} = (mc/8\pi^2 e^2)(g_2/g_1)\lambda_o^2 = 1.51(g_2/g_1)\lambda_o^2$$

In terms of the oscillator strength (100a) becomes

$$n_2 - n_1(g_2/g_1) = [12.08\pi\nu^2/c^3 g(\nu_c)t_{photon}][(g_2/g_1)(\lambda_o^2/f)] \qquad (100b)$$

Different shape factors $g(\nu)$ will result in the following forms:

Lorentzian [Eq. (18)]

$$n_2 - n_1(g_2/g_1) = (4\pi^2\nu^2\Delta\nu/c^3)(t_{spont}/t_{photon}) \qquad (100c)$$

Gaussian [Eq. (18)]

$$n_2 - n_1(g_2/g_1) = (4\pi^2\nu^2/c^3)[\Delta\nu/(\pi \ln 2)^{\frac{1}{2}}](t_{spont}/t_{photon}) \qquad (100d)$$

The maintenance of these critical inversion densities is conditioned to a large extent on the spontaneous emission emitted within the total line width. For the four-level scheme where n_1 is effectively zero the power required to maintain the threshold condition is given for a Lorentzian line by

$$P = n_2(h\nu/\tau_{spont}) = 4\pi^2 h\nu^3 \Delta\nu/c^3 t_{photon} \qquad (101a)$$

and for a gaussian by

$$P = 4\pi^2 h\nu^3 \Delta\nu/c^3 t_{photon}(\pi \ln 2)^{\frac{1}{2}} \qquad (101b)$$

in units of power per unit volume.

For three-level systems where the population of the lower level is not negligible, the spontaneous emission, being proportional to the total population of the upper level, will be many times that represented by (101). In fact, for 1 per cent inversion, the ratio will be $\simeq 100$.

In any case, the above are minimum estimates. Since the pumping is to a higher level in the typical three- or four-level system, the above equations must use the quantum $h\nu$ appropriate to the pump energy. Moreover, the possibility of a number of spontaneous transitions would require a sum of terms of the type (101a) or (101b).

16.4. Laser-oscillator Power Output. The output power of the laser oscillator consists of the induced emission from the inverted upper level. Three- and four-level masers fall into different categories because the required upper-state populations are so different, as indicated in the above section.

For the three-level system the population of the upper level is approximately $n/2$ where n is the total population. The total number of useful transitions may be written as $P_{pump}\eta/h\nu_{pump} - n/2\tau_{spont}$. The first term including the quantum efficiency and the pump power absorbed P_{pump} gives the number of upper-level ions produced, from which must be subtracted the nonuseful spontaneous emission represented by the second term. The output power then is[46]

$$P_{out} = (P_{pump}\eta/h\nu_{pump} - n/2\tau_{spont})h\nu_{sig} \qquad (102a)$$
$$P_{out} = (P_{pump} - P_{min})\eta(\nu_{sig}/\nu_{pump}) \qquad (102b)$$

where P_{min} is defined as $nh\nu_{pump}/2\tau_{spont}\eta$, representing the minimum pump power required to maintain an inverted population of $n/2$ at threshold. For the three-level case, Eq. (102a) may be rewritten as

$$P_{out(3)} = (P_{pump}/P_{min} - 1)(Nh\nu_{sig}/2\tau_{spont}) \qquad (102c)$$

This same formulation is easily extended to the four-level case by observing that the inversion required is not represented by $n/2$ in the upper level but rather by a much smaller number which will be represented by n_c. In direct analogy to the above the equation for the output power may be written as

$$P_{out(4)} = (P_{in}/P_{min} - 1)(n_c h\nu_{sig}/\tau_{spont}) \qquad (103)$$

Thus a knowledge of the amount by which the pump power exceeds the threshold quite simply determines the output power.

17. CAVITIES

17.1. Maser Cavities. The maser cavities can be designed in a straightforward fashion by reference to any of the standard microwave engineering references on the subject. In a few cases lumped-constant circuits have been employed. The only precaution suggested is that both the dielectric constant of the crystal and that of liquid helium, if it is to fill the cavity, must be taken into account.

17.2. Laser Cavity—Normal Modes. At first sight the laser cavities are simply adaptations of the optical Fabry-Perot interferometer with its well-known properties. Calculations of the losses from configurations such as those illustrated in Fig. 31 have shown that the plane-wave assumption of the Fabry-Perot instrument is not justified here.[47,48,49]

14-40 AMPLIFYING DEVICES

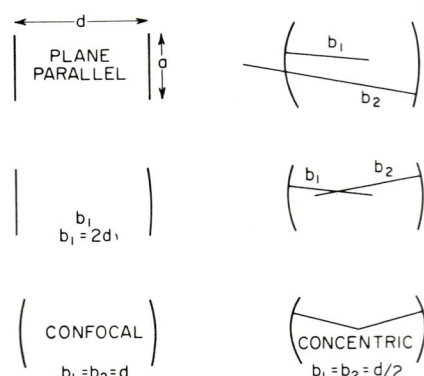

Fig. 31. Some possible laser mirror configurations. a and d refer to transverse mirror dimensions and separations even for curved geometries.

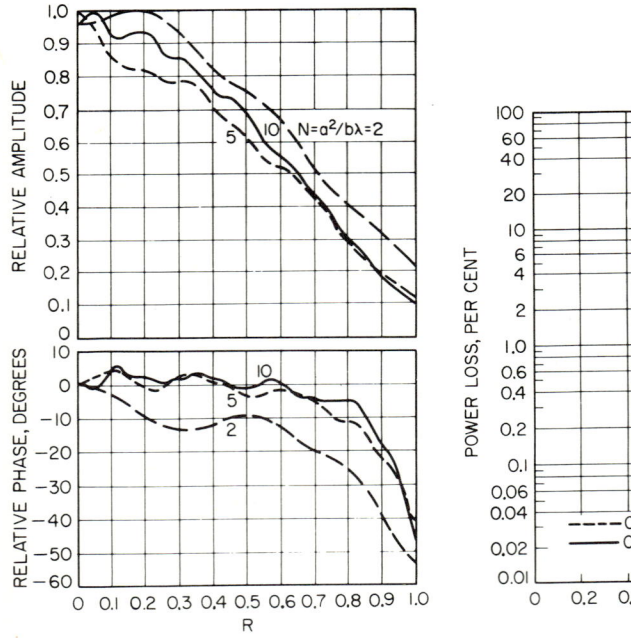

Fig. 32. Relative field-intensity amplitude and phase distributions for the dominant or low-loss TEM_{00} using circular plane mirrors. $N = a^2/b\lambda$ is the Fresnel number. (*After Fox and Li,*[47] *reprinted by permission of the copyright owner, American Telephone and Telegraph Company and the authors. This article originally appeared in the Bell System Technical Journal, vol. 40, March, 1961.*)

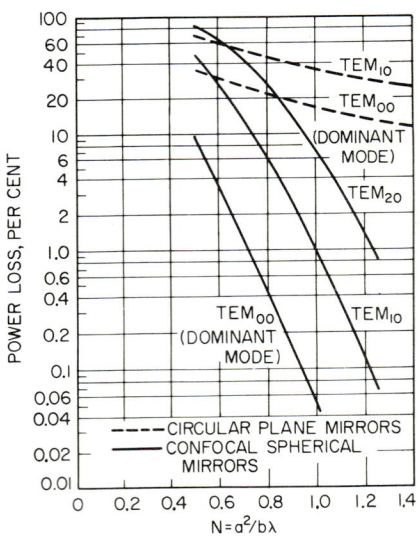

Fig. 33. Power loss as a function of Fresnel number N for confocal spherical and plane circular mirrors. (*Reprinted by permission of the copyright owner, American Telephone and Telegraph Company, and the authors. This article originally appeared in the Bell System Technical Journal, vol. 40, March, 1961.*)

The normal modes of such a resonator are defined as amplitude and phase distributions across a mirror face which except for an attenuation factor repeat themselves after a round trip consisting of reflection by the other mirror. Starting with a plane wave and allowing for diffraction losses, it is found that the amplitude distribution across the mirror face tends to drop considerably at the edges, leaving in the lowest mode (an even one) maximum amplitude at the center. The diffraction losses are lower than the plane-wave approximation would indicate because of the low amplitude near the outside edges. The amplitude and phase distributions as calculated for a circular plane mirror are shown in Fig. 32. A positive relative phase shift (a lead relative to the free-space electrical length of 360 b/λ degrees) implies a phase velocity slightly larger than the speed of light, as is found in metal waveguides. A negative phase shift is the result then of a slower phase velocity. As an example, the phase shift per transit indicates a lead of 1.59° for infinite-strip mirrors. In analogy with cavity modes the terminology is such that for rectangular surfaces the TEM_{mn} modes are characterized by m and n nulls along the rectangular axis while for circular symmetry the m and n refer to the order of radial and angular variation, respectively. (Fox and Li[47] have used the reverse order.) In both plane-parallel mirrors and the confocal configurations (Fig. 31) the TEM_{00} mode is the one with the lowest loss.

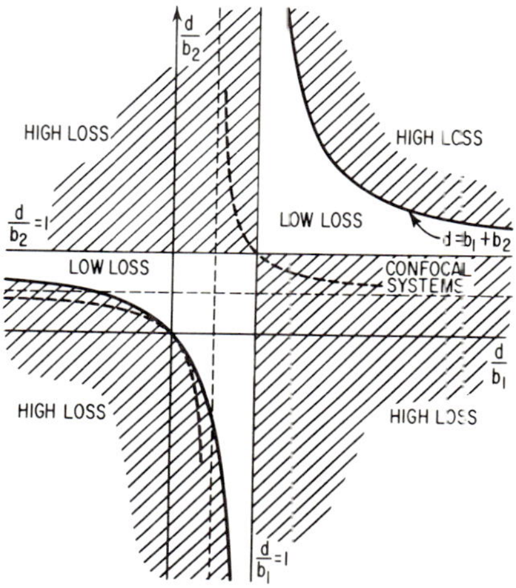

For the plane-parallel mirrors the TEM_{00} modes will be separated by $c/2d$ cps (or 75 Mc for $d = 50$ cm). Odd radial modes, e.g., TEM_{10}, are slightly higher in frequency and have higher diffraction losses as do all the higher-order modes. The confocal mirror arrangement has its lowest-loss TEM_{00} modes separated by $c/4d$, but there is a high degree

Fig. 34. Stability diagram for spherical mirrors of radius b_1 and b_2 with spacing d. (Reprinted by permission of the copyright owner, American Telephone and Telegraph Company, and the authors. This article originally appeared in the Bell System Technical Journal, vol. 41, July, 1962.)

of degeneracy which is removed by slight variations from the confocal arrangement. The existence of the various modes has been beautifully demonstrated by Rigrod.[50]

The confocal system has by far the lowest loss, as may be seen in Fig. 33, where power loss per transit is plotted against the Fresnel number $a^2/d\lambda$. This also illustrates the advantage of using system dimensions resulting in a large value of N.

The confocal arrangement, however, is an anomalously low-loss adjustment.[48] If the mirror spacing d is intermediate between the two radii of curvature, the losses become high. These losses have been compared with the instabilities or blow-ups which have become well known in the focusing of beams using periodic arrays of lenses. The stable and unstable regions of the numerous possible combinations of radii and spacings are illustrated by Fig. 34. It is obvious that the confocal system is critical. The two low-loss regions are equivalent—giving patterns which are the complex conjugates of one another—in one case the b's $>$ d's while in the other the d's $>$ b's; spot size, losses, and frequency responses are identical if the condition

$$1/b_1 + 1/b_2 = 2/d$$

is satisfied.

The operation of the Fabry-Perot as an optical interferometer is normally not affected by these considerations because the Q's obtained in these instruments result in mode widths greater than the mode separations. A plane wave which must be a composite of a number of these modes is not then decomposed into its component modes.

17.3. Mirrors—Metal Films. In addition to the diffraction losses considered above and the absorptions in the component parts of the system, the effectiveness of the reflection from the mirror surfaces is an important factor in the resonator Q. The typical mirror (first surface) uses either a thin metallic layer, a multilayer dielectric coating, or a combination of metal and dielectric. The transparency required in one

Table 5. Reflectance of Freshly Evaporated Films of Aluminum,[52] Silver,[51] Gold,[51] Copper,[51] and Rhodium[51] from 0.22 to 10 Microns

λ, micron	Al	Ag	Au	Cu	Rh
0.220	.....	28.0	27.5	40.4	58.5
0.240	.....	29.5	31.6	39.0	61.3
0.250	.....	30.4	33.2	37.0	63.0
0.260	.....	29.2	35.6	35.5	65.0
0.280	.....	25.2	37.8	33.0	68.5
0.300	92.08	17.6	37.7	33.6	71.2
0.315	.....	5.5	37.3	35.5	73.0
0.320	.....	8.9	37.1	36.3	73.6
0.340	.....	72.9	36.1	38.5	75.5
0.360	.....	88.2	36.3	41.5	77.0
0.380	.....	92.8	37.8	44.5	77.4
0.400	91.94	94.8	38.7	47.5	77.6
0.450	91.75	96.6	38.7	55.2	77.2
0.500	91.62	97.7	47.7	60.0	77.4
0.550	91.57	97.9	81.7	66.9	78.0
0.600	91.17	98.1	91.9	93.3	79.1
0.650	90.57	98.3	95.5	96.6	79.9
0.700	89.77	98.5	97.0	97.5	80.4
0.750	88.62	98.6	97.4	97.9	81.2
0.800	86.76	98.6	97.7	98.1	82.0
0.850	86.77	98.7	97.8	98.3	82.8
0.900	89.08	98.7	98.0	98.4	83.5
0.950	92.43	98.8	98.1	98.4	84.2
1.0	94.02	98.9	98.2	98.5	85.0
1.5	97.42	98.9	98.2	98.5	88.2
2.0	97.79	98.9	98.3	98.6	90.5
3.0	98.05	98.9	98.3	98.6	92.5
4.0	98.26	98.9	98.3	98.7	94.0
5.0	98.43	98.9	98.3	98.7	94.5
6.0	98.56	98.9	98.3	98.7	94.8
7.0	98.66	98.9	98.4	98.7	95.2
8.0	98.72	98.9	98.4	98.7	95.5
9.0	98.74	98.9	98.4	98.8	95.8
10.0	98.76	98.9	98.4	98.8	96.0

The reflectance of a good evaporated coating is always higher than that of a polished or electrolytically produced surface of the same material. One of the main conditions for preparing a high-quality reflection coating by evaporation in high vacuum is a high rate of deposition or fast evaporation of the metal.

mirror to allow some of the radiation to escape as useful output is either the film transparency or an uncoated area of the surface.

Metallic films are simpler than multilayer dielectric films but usually not so good. The properties of metallic films are illustrated by the contents of Table 5. A comparison of the absorption coefficient and reflectance will illustrate the generality that for metals a high reflectance and a high absorption coefficient go hand in hand.

17.4. Mirrors—Dielectric Coatings. The use of a dielectric film to cut down the amount of power reflected from glass surfaces is well known. A similar technique

INDUCED-EMISSION-AMPLIFIER PRINCIPLES 14-43

can be used to enhance the reflection if the proper choice of refractive index is made. The typical reflecting surface consists of an odd number of λ/4 thick dielectric layers, with alternately high-low-high-low-high, etc., refractive indices directly on an optical flat. The calculated[53] result for such a filter is shown in Fig. 35 for dielectric layers of

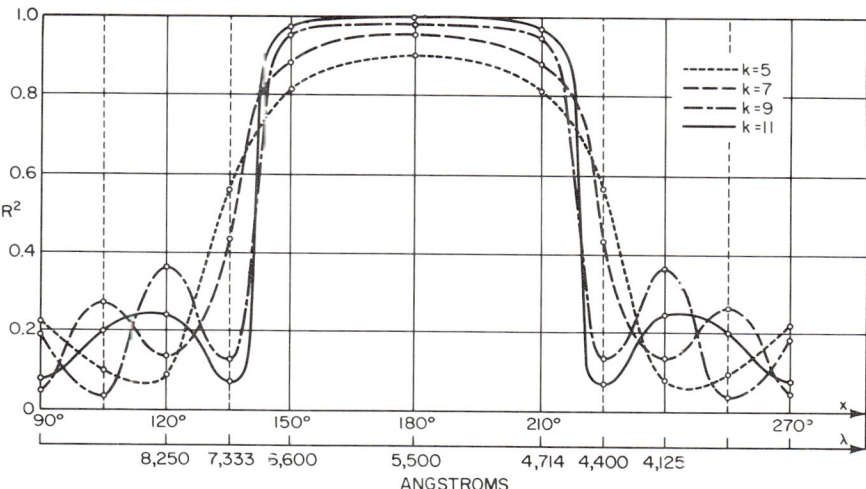

FIG. 35. Theoretical reflectivity as a function of wavelength or phase for λ/4 layers of alternating refractive index 2.30, 1.36, 2.30, etc., on glass, for 5, 7, 9, 11 layers. (*After Vasicek,*[53] *published by permission of Interscience Publishers Inc., New York.*)

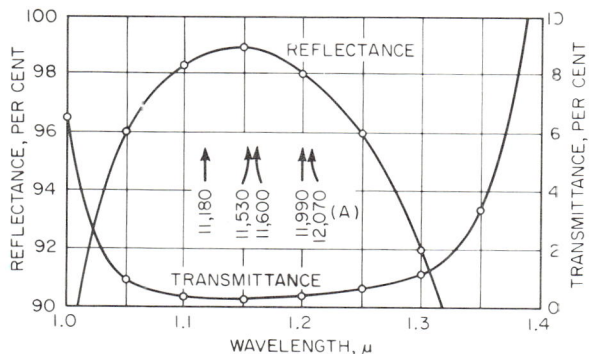

FIG. 36. Experimentally obtained reflection and transmittance curves for a 13-layer composite dielectric film alternating ZnS ($n = 2.30$) and MgF$_2$ ($n = 1.38$). (*After Herriott,*[37] *published by permission of the editor, Optical Society of America.*)

refractive index 2.30 (ZnS or TiO$_2$) and 1.36 (LiF). The maximum reflectances are given in the accompanying Table 6. Excellent reflectance over a wide bandwidth is obtainable for this type of filter. The actual behavior of a 13-layer dielectric film coating (as used by Herriott[37]) is illustrated in Fig. 36. The bandwidth is not so square-shouldered, but a power reflectance of 98.9 per cent ±0.2 per cent and transmittance of 0.3 per cent were obtained, leaving a loss due to absorption and scattering of 0.8 per cent.

Table 6

No. of layers	Reflectance, %
1	33.97
3	71.23
5	89.70
7	96.57
9	98.88
11	99.65
13	99.89
15	99.96

18. LASER PUMP SOURCES

The light sources for pumping the pulsed lasers have generally been pulsed rare-gas-filled flash tubes. The efficiencies increase with the molecular weight, and xenon tubes are in the vicinity of 50 lumen-sec/watt-sec. The spectral output of a GE FT214 is shown in Fig. 37. This is typical of the pulsed xenon lamps; it exhibits a

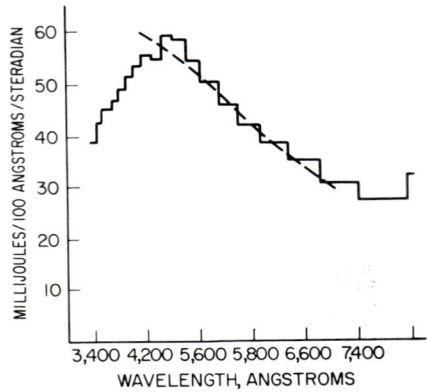

Fig. 37. Spectral radiation of a GE FT-214 operating at 2,000 volts, 50 μf. (*Courtesy of the General Electric Photo Lamp Department.*)

color temperature of about 6000°K and can be compared with the dotted curve which represents the spectral distribution of a 7450°K blackbody. When operated in pulsed fashion (10^{-4} sec duration) the gas comes to thermal equilibrium at a high temperature, and the duty cycle is limited by the power dissipation in the tube. Quartz tubulations are normally used because of their high temperature characteristics.

For continuous operation mercury-vapor (GE AH6) and mercury-xenon (Hanovia 941B) lamps operated at high pressure have been successfully used. The choice of lamp will depend on the required pump transition. The most critical examples are the gaseous systems where a close match between emission line of the pumping source and absorption line is required for efficient operation. Figure 38 depicts the outputs of two high-pressure continuously operating lamps and illustrates the difference between high-pressure xenon and high-pressure mercury.

The output of double-pulsed flash lamps[54,55] has been investigated. The xenon flash tube is pulsed to form a stable discharge at low current densities and then pulsed as shown in Fig. 39. Figure 40 shows the relative intensities for the two current densities. There obviously is good reason for using the higher value, and without the prepulsing such a current pulse normally shatters the tube.

The use of exploding wires as pump sources has been explored by Stevenson *et al.*[56] Wires exploded in small-diameter tubes evacuated to 1 to 20 microns of Hg radiate as blackbodies while line spectra are characteristic of larger enclosures. In the visible, air-exploded wires have spectral radiance about that of a xenon (GE FT524) lamp, but

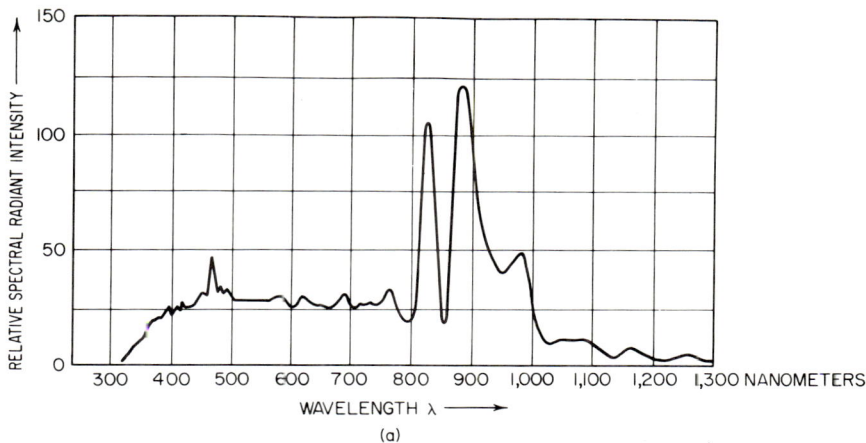

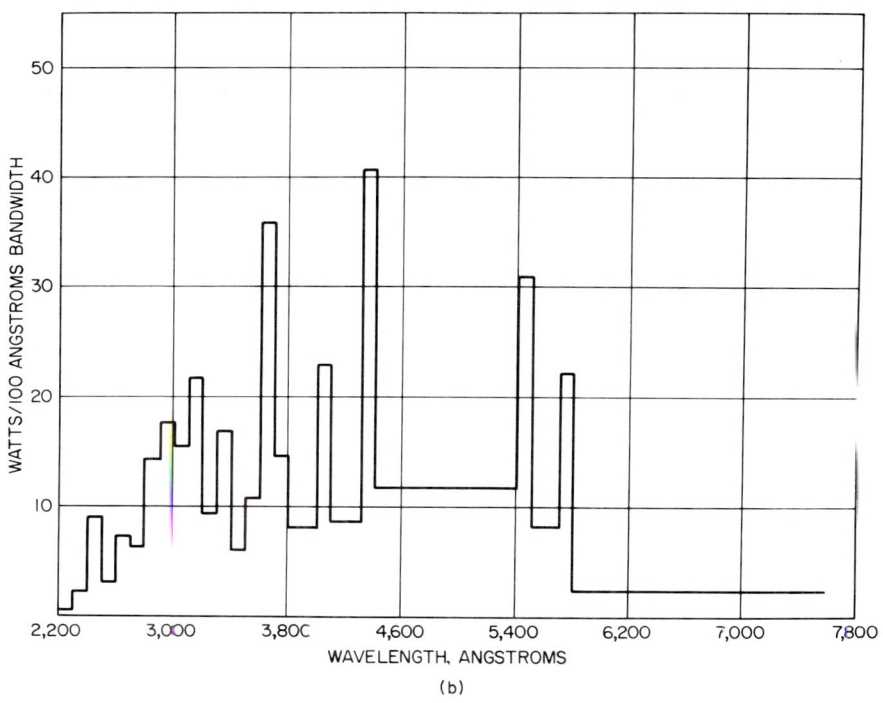

Fig. 38. (a) Relative spectral radiation intensity of a high-pressure xenon lamp (Osram XBF 6000W/1). (*Published with the permission of the Osram Co., Berlin and Munich.*) (b) Output of a high-pressure mercury lamp (GE BH6). 1 nm (nanometer) = 10^{-9} m = 10 angstroms.

they exceed it by an order of magnitude in the ultraviolet. Vacuum-exploded wires are four to eight times better than the air-exploded wires. Tungsten, tantalum, and molybdenum wires of 0.005 to 0.010 in. diameter have been tested at energy inputs

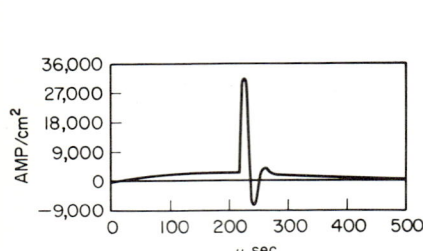

FIG. 39. Current as a function of time for double-pulse operation of a flash lamp. (After Emmett and Schawlow,[55] published with permission of the editor, Applied Physics Letters.)

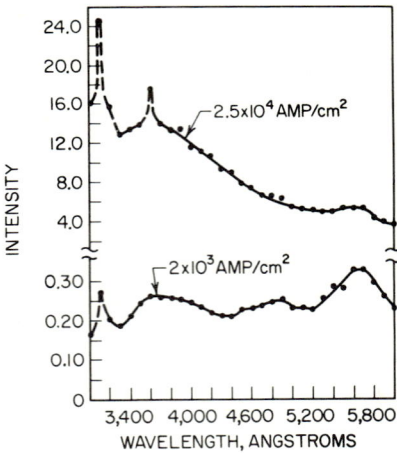

FIG. 40. Output vs. wavelength for standard single-pulse and for double-pulse operation. (After Emmett and Schawlow,[55] published with permission of the editor, Applied Physics Letters.)

of more than 3,000 joules. This pumping scheme appears to be well suited to ultraviolet pumping of a pulsed nature.

REFERENCES

1. J. P. Gordon, H. J. Zeiger, and C. H. Townes, *Phys. Rev.*, vol. 99, p. 1264, 1955.
2. A. L. Schawlow and C. H. Townes, *Phys. Rev.*, vol. 112, p. 1940, 1958.
3. I. I. Rabi, N. F. Ramsey, and J. Schwinger, *Rev. Mod. Phys.*, vol. 26, p. 167, 1954.
4. E. M. Purcell and R. V. Pound, *Phys. Rev.*, vol. 81, p. 279, 1951.
5. N. Bloembergen, *Phys. Rev.*, vol. 104, p. 324, 1956.
6. A. E. Siegman and R. J. Morris, *Phys. Rev. Letters*, vol. 2, p. 302, 1959.
7. P. E. Wagner, J. G. Castle, Jr., and P. F. Chester (C. H. Townes, ed.), "Quantum Electronics," p. 509, Columbia University Press, New York, 1960.
8. N. Bloembergen, S. Shapiro, P. S. Pershan, and J. O. Artman, *Phys. Rev.*, vol. 114, p. 445, 1959.
9. H. E. D. Scovil, G. Feher, and H. Seidel, *Phys. Rev.*, vol. 105, p. 760, 1957.
10. G. Feher and H. E. D. Scovil, *Phys. Rev.*, vol. 105, p. 760, 1957.
11. D. P. Devor, I. J. D'Haenens, and C. K. Asawa, *Phys. Rev. Letters*, vol. 8, p. 432, 1962.
12. W. R. Bennett, Jr., *Appl. Opt.*, suppl. 1, p. 24, 1962.
13. H. S. W. Massey and E. H. S. Burhop, "Electronic and Ionic Impact Phenomena," Oxford University Press, Fair Lawn, N.J., 1952.
14. N. G. Basov and O. N. Krokhin, *J. Exptl. Theoret. Phys. (U.S.S.R.)*, vol. 39, p. 1777, 1960; *Soviet Phys. JETP (English Transl.)*, vol. 12, p. 1240, 1961.
15. Sol Aisenberg, *Appl. Phys. Letters*, vol. 2, p. 87, 1963.
16. W. R. Bennett, Jr., W. L. Faust, R. A. McFarland, and C. K. N. Patel, *Phys. Rev. Letters*, vol. 8, p. 470, 1962.
17. R. N. Hall, *Solid-State Electron.*, vol. 6, p. 405, 1963.
18. A. Javan, *Phys. Rev.*, vol. 107, p. 1579, 1957.
19. W. Hughes and L. Grabner, *Phys. Rev.*, vol. 79, p. 829, 1950.
20. Tatsuo Yajima, *J. Phys. Soc. Japan*, vol. 16, p. 1594, 1961; vol. 16, p. 1709, 1961.
21. Gisela Eckhardt, R. W. Hellwarth, F. J. McLung, S. E. Schwartz, D. Weiner, and E. J. Woodbury, *Phys. Rev. Letters*, vol. 9, p. 455, 1962.

22. R. W. Hellwarth, *Phys. Rev.*, vol. 30, p. 1850, 1963.
23. Norman F. Ramsey, *Phys. Rev.*, vol. 103, p. 20, 1956.
24. N. Bloembergen (C. J. Gorter, ed.), "Progress in Low Temperature Physics," vol. III, North Holland Publishing Company, Amsterdam, 1961.
25. R. V. Pound, *Annals of Physics*, vol. 1, p. 24, 1957.
26. M. W. P Strandberg, *Phys. Rev.*, vol. 106, p. 617, 1957.
27. M. W. Muller, *Phys. Rev.*, vol. 106, p. 8, 1957.
28. J. Weber. *Rev. Mod. Phys.*, vol. 31, p. 681, 1959.
29. K. Shimoda, H. Takahasi, and C. H. Townes, *J. Phys. Soc. Japan*, vol. 12, p. 686, 1957.
30. C. R. Ditchfield, *Solid-State Electron.*, vol. 4, p. 171, 1962.
31. M. L. Stitch, *J. Appl. Phys.*, vol. 29, p. 782, 1958.
32. John C. Slater, "Microwave Electronics," D. Van Nostrand Company, Inc., Princeton, N.J., 1950.
33. R. W. DeGrasse, E. O. Schulz du Bois, and H. E. D. Scovil, *Bell System Tech. J.*, vol. 38, p 305, 1959.
34. C. Kittel, *Phys. Rev.*, vol. 71, p. 270, 1947.
35. J. E. Geusic and H. E. D. Scovil, *Bell System Tech. J.*, vol. 41, p. 1371, 1962.
36. C. H. Townes, "Advances in Quantum Electronics," Columbia University Press, New York, 1961.
37. Donald R. Herriott, *J. Opt. Soc. Am.*, vol. 52, p. 31, 1962.
38. W. R. Bennett, Jr., *Phys. Rev.*, vol. 126, p. 580, 1962.
39. H. Kogelnik and W. W. Rigrod, *Proc. IRE*, vol. 50, p. 220, 1962.
40. H. Kogelnik, *Appl. Phys. Letters*, vol. 2, p. 51, 1962.
41. W. Kaiser and H. Lessing, *Appl. Phys. Letters*, vol. 2, p. 206, 1963.
42. Richard L. Fork and C. K. N. Patel, *Appl. Phys. Letters*, vol. 2, p. 180, 1963.
43. I. Melngailis and R. H. Rediker, *Appl. Phys. Letters*, vol. 2, p. 202, 1963.
44. T. S. Jaseja, A. Javan, J. Murray, and C. H. Townes, *Phys. Rev.*, vol. 133A, p. 1221, 1964.
45. A. Javan, E. A. Ballik, and W. L. Bond, *J. Am. Opt. Soc.*, vol. 52, p. 96, 1962.
46. A. Yariv, *Proc. IEEE*, vol. 51, p. 1723, 1963.
47. A. G. Fox and T. Li, *Bell System Tech. J.*, vol. 40, p. 453, 1961.
48. G. D. Boyd and J. P. Gordon, *Bell System Tech. J.*, vol. 40, p. 489, 1961.
49. G. D. Boyd and H. Kogelnik, *Bell System Tech. J.*, vol. 41, p. 1347 1962.
50. W. W. Rigrod, *Appl. Phys. Letters*, vol. 2, p. 51, 1963.
51. G. Hass, *J. Opt. Soc. Am.*, vol. 45, p. 945, 1955; "APS Handbook."
52. H. E. Bennett, M. Silver, and E. J. Ashley, *J. Opt. Soc. Am.*, vol. 53, p. 1089, 1963.
53. A. Vasicek "Optics of Thin Films," North Holland Publishing Company, Amsterdam, 1960; Interscience Publishers, Inc., New York.
54. J. L. Emmett and R. W. Hellworth, *Bull. Am. Phys. Soc.*, ser. II, vol. 7, p. 615, 1962.
55. J. L. Emmett and A. L. Schawlow, *Appl. Phys. Letters*, vol. 2, p. 204, 1963.
56. M. J. Stevenson, W. Reuter, N. Braslau, P. P. Sorokin, and A. J. Landon, *J. Appl. Phys.*, vol. 34, p. 500, 1963.

BIBLIOGRAPHY

Chang, W. S. C. (ed.): "Lasers and Applications," Ohio State University Press, Columbus, Ohio, 1963.
Fox, Jerome (ed.): "Optical Masers," Polytechnic Press, New York, 1963.
Grivet, Pierre, and Nicolas Bloembergen (eds.): "Quantum Electronics III," Dunod, Paris, Columbia University Press, New York, 1964.
Lengyel, Bela A.: "Lasers," John Wiley & Sons, Inc., New York, 1962.
Mitchell, Allan C. G., and Mark W. Zemansky: "Resonance Radiation and Excited Atoms," Cambridge University Press, Cambridge, England, 1934.
Optical Society of America: "Optical Masers," Applied Optics Supplement no. 1, 1962.
Quantum Electronics, *Proc. IEEE*, vol. 51, pp. 1–294, 1963.
Siegman, Anthony E.: "Microwave Solid-State Masers," McGraw-Hill Book Company, New York, 1964.
Singer, J. R. (ed.): "Advances in Quantum Electronics," Columbia University Press, New York, 1961.
Singer, J. R.: "Masers," John Wiley & Sons, Inc., New York, 1959.
Townes, C. H. (ed.): "Quantum Electronics," Columbia University Press, New York, 1960.
Troup, Gordon: "Masers and Lasers," Interscience Publishers, Inc., New York, 1963.
Vuylsteke, Arthur A.: "Elements of Maser Theory," D. Van Nostrand Company, Inc., Princeton, N.J., 1960.

Chapter 15

IONIC DEVICES

D. RAYMOND FEWER*

CONTENTS

1. Introduction... 15-1
2. Electrochemical Principles of Operation......................... 15-1
   2.1. Conduction in Electrolytic Solutions....................... 15-2
   2.2. Electrodes... 15-2
   2.3. Concentration Potential.................................... 15-2
   2.4. Polarization... 15-2
3. Solions.. 15-3
   3.1. Basic Operation.. 15-3
   3.2. Solion Tetrode... 15-5
   3.3. Two-terminal Integrators................................... 15-10
   3.4. Solion Transducers... 15-11
4. The Memistor.. 15-12
   4.1. Operation.. 15-14
5. Mercury Coulombmeter.. 15-14
   5.1. Basic Operation.. 15-15
Appendix. Excerpts from Some Commercial Solion Data Sheets....... 15-17
   A. Solion Tetrodes Types SE 100 and SE 110..................... 15-17
   B. Solion Two-terminal Integrator Type SV 150.................. 15-19

1. INTRODUCTION

The types of devices discussed in this chapter are those which function by virtue of the control of the flow of ions in an electrolytic solution. This control is usually obtained by electric fields, electrode geometry, selection of solution paths, and sometimes by controlling the flow of the electrolytic solution. Some of the many varieties which represent an illustrative cross section of these devices will be discussed.

Strictly speaking, many of these devices are not considered amplifiers. They do process signals in various ways, and many of them possess gain.

2. ELECTROCHEMICAL PRINCIPLES OF OPERATION

The material contained in this section is a simplified discussion of the basic electrochemical principles which are involved in the operation of ionic devices. More detailed treatments may be found in the standard electrochemical texts.[1]

---
* Texas Instruments Incorporated, Semiconductor-Components Division, Dallas, Tex.
[1] G. Kortum and J. O'M. Bockris, "Electrochemistry," vols. I and II, Elsevier Publishing Company, New York, 1951.

2.1. Conduction in Electrolytic Solutions. The current carriers in an electrolytic solution are ions. These ions are usually formed by dissolving certain chemicals in a suitable solvent. The chemical dissociates into atoms or molecules of positive and negative charge. For example, potassium iodide (KI) in water is an electrolytic solution in which the potassium iodide dissociates to positively charged potassium ions (K^+) and negatively charged iodide ions (I^-).

The solution conductance is a function of many variables, such as length and cross section of the solution path, number and type of ions present, solvent viscosity, and temperature. The solutions behave ohmically in many devices.

All ions possess a large mass compared with that of the electron. Therefore, the transient behavior of devices relying on ionic motion for their operation is much slower than that of electron devices.

2.2. Electrodes. Electric circuits are coupled to the electrolytic solution by electrodes which may be inert (do not take part in the chemical reaction) or active (take part in the chemical reaction). At the electrodes the mechanism of conduction changes from ionic to electronic. Thus there is an electrochemical process at the solution-electrode interface where the valences of the ions are altered. At the anode, ions lose electrons and thus are oxidized; at the cathode, ions gain electrons and thus are reduced. The amount of oxidation and reduction which takes place at the electrodes is proportional to the amount of charge which is transferred in the circuit. This is a general statement of Faraday's law.

The devices discussed in this chapter are considered to have active electrodes when metal is plated out on the cathode or goes into solution at the anode. When oxidized and reduced ions are both present in the solution, the electrode is assumed to have no chemical effect or is an inert electrode. The inert electrode simply serves as a conductive surface at which electrons can be interchanged between the solution and the external circuit. An electrochemical system of this sort is called a *redox system*.

2.3. Concentration Potential. An electrode will assume a potential in a solution which is dependent on the activities of the various ionic species in its vicinity. When two similar electrodes are placed in regions of a solution differing in their ionic activities, a potential difference will exist between them. If these two electrodes are joined external to the solution by a conducting path, a current will flow. This current will be in a direction which will tend to equalize the activities. Two cases are considered in our discussion: (1) Ionic activities are the same at all electrodes and thus the potential between any two electrodes is zero. (2) The concentration of only one species of ion will vary from electrode to electrode. Since ionic activity varies directly with concentration, the potential of one electrode relative to another is a function of the concentration of the ionic species in the vicinity of each electrode. This concentration potential between two electrodes can be expressed as

$$E_{c_{12}} = (RT/nF) \ln (c_1/c_2) \tag{1}$$

where R = gas constant = 8.3 joules/°C/mole
T = absolute temperature, °C
n = number of electrons involved in the reaction
F = the Faraday = 96,500 coulombs per equivalent
c_1 = concentration of ionic species at electrode 1
c_2 = concentration of ionic species at electrode 2
$RT/F \cong 26$ mv at 25°C

The polarity of $E_{c_{12}}$ depends on the type of ions involved.

2.4. Polarization. A polarized electrode is one that hinders the flow of electric current between the electrode and the solution in which it is immersed. Chemical or electrochemical barriers exist which must be overcome by the current carriers in order to allow current to flow across the electrode-solution interface. Forcing a net current flow across this interface requires an added voltage drop above and beyond that needed to overcome the ohmic losses in the solution. This additional voltage is called the *overvoltage*.

The overvoltage is related to the current density at the electrode. For a given electrode and solution, there is a critical current density at which the overvoltage tends toward infinity. In many of the devices considered in this chapter, aqueous solutions are used. If the overvoltage becomes too large in these devices, hydrogen is evolved at the cathode. This is an irreversible process, and the devices discussed here are not operated in this region.

3. SOLIONS

The name solion is coined from the words "solution" and "ion." It is a name used to designate a family of electrochemical devices which function by physically and electrically controlling a reversible electrochemical, or redox, reaction. Solion devices have been made in a variety of ways to perform as integrators, amplifiers, and transducers. They were originally conceived and studied at the U.S. Naval Ordnance Laboratory[2,3] and have recently become commercially available as a design tool for the circuit designer.[4,5]

3.1. Basic Operation. Solions can use a variety of redox systems. The most commonly used electrolyte consists of an aqueous solution of iodine and potassium iodide. The amount of potassium iodide in the system is large compared with the amount of iodine. Thus the iodine is the controlling species since a small variation in the amount of iodine will produce a greater effect than the same variation in the amount of potassium iodide. Platinum is the most commonly used material for the inert electrodes. The chemistry of the system is such that the potassium iodide dissociates in water to yield potassium ions and iodide ions:

$$KI \to K^+ + I^- \qquad (2)$$

In the presence of iodide ions, the iodine exists predominantly as a triiodide ion:

$$I_2 + I^- \rightleftharpoons I_3^- \qquad (3)$$

The triiodide ion will be referred to as iodine in this discussion.

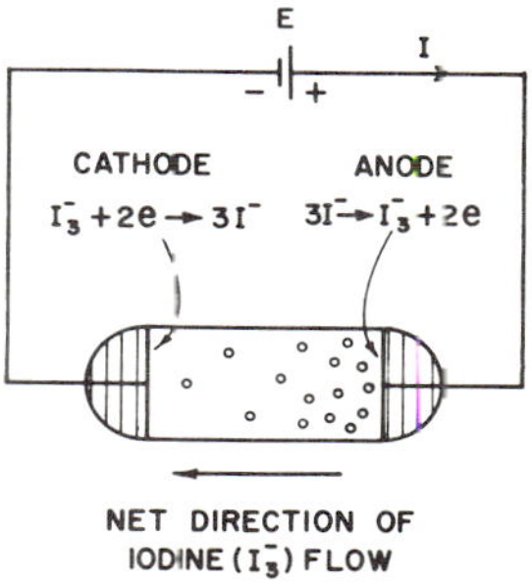

Fig. 1. The iodine-iodide redox system.

Consider the case of a simple solion two-electrode device connected as shown in Fig. 1. Voltage-current characteristics of such a device are shown in Fig. 2, where iodine concentration is assumed to be variable in some manner. If the voltage were held constant at, say, 0.5 volt, a linear relationship between current and iodine concentration would be observed as shown. The behavior of the iodine ions, during current flow, is shown in Fig. 1. The other ions in the solution are not shown since their concentration changes relatively little over the volume of the solution. The following reactions take place at the electrodes:

Cathode reaction (reduction)

$$I_3^- + 2e \to 3I^- \qquad (4)$$

Anode reaction (oxidation)

$$3I^- \to I_3^- + 2e \qquad (5)$$

The electrode reactions are the reverse of each other and proceed at the same rate so that there is no net depletion of iodine. The system is completely reversible and

---

[2] "Solion Principles of Electrochemistry and Low-Power Electrochemical Devices," U.S. Department of Commerce, Office of Technical Services, PB 131931 (U.S. Naval Ordnance Laboratory, Silver Springs, Md.), August, 1958.
[3] Nelson N. Estes, Solions, Their Characteristics and Commercial Applications, *IEEE Trans. Ind. Electron.* vol. IE-10, no. 1, pp. 91–100, May, 1963.
[4] Solions are presently being manufactured by Self-Organizing Systems, Inc., Box 9918, Dallas, Tex., 75214.
[5] D. R. Fewer, Solions, A Family of Electrochemical Devices, *Electron. Weekly*, no. 96, p. 11, July 4, 1962.

AMPLIFYING DEVICES

thus can be a closed one. The rate of these reactions is usually controlled by the amount of iodine reaching the cathode. Since iodine concentration is the controlling factor, it is sometimes convenient to disregard the iodide (I^-) ions and consider the anode as an iodine (I_3^-) source and the cathode as an iodine sink.

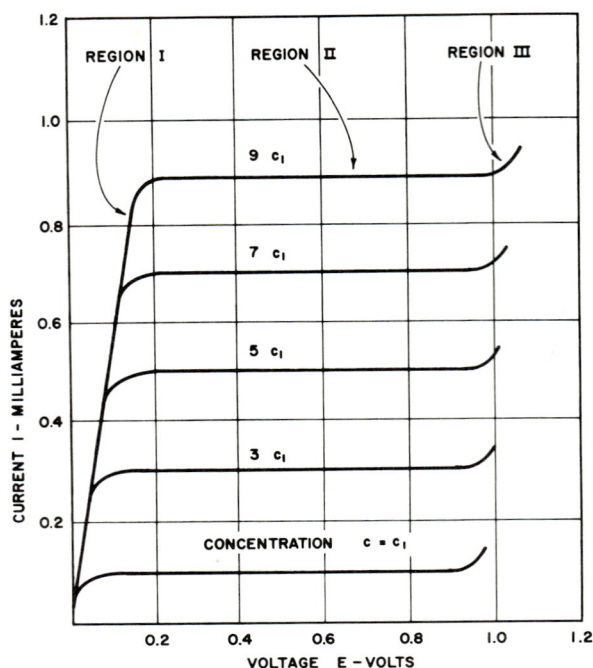

Fig. 2. Polarization curves for the solion redox system.

Three distinct regions exist in the voltage-current characteristics, as shown in Fig. 2. In region I, the device behaves ohmically; the amount of iodine reduced at the cathode, and hence the current, is directly proportional to the voltage. The resistance is the resistance of the solution, which is essentially independent of the iodine concentration; it depends almost exclusively on the potassium iodide concentration. In region II, the current depends on the iodine concentration. The voltage is sufficiently high to deplete the iodine in the immediate vicinity of the cathode, and the current is determined by the ability of the iodine to diffuse from the bulk of the solution to the cathode. This diffusion process is relatively independent of the voltage. Thus the current is essentially independent of the voltage and is diffusion-limited. The device is not allowed to operate in region III since the voltage is sufficiently high to cause hydrogen evolution and the chemical reaction is irreversible.

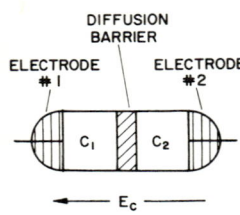

Fig. 3. Solion system which allows measurement of concentration potential.

A concentration potential will exist between two electrodes when the concentration of iodine at these electrodes differs. This potential can be measured at the electrodes of a device such as that shown in Fig. 3. The diffusion barrier prevents the diffusion of iodine between the anode and cathode compartments but permits a conductive electrolytic path between the electrodes. The potential between the two electrodes is, from Eq. (1),

$$E_c = E_a \log_{10} (c_1/c_2) \qquad (6)$$

where E_c = concentration potential
E_a = constant, approximately 30 mv at 25°C[6]
c_1, c_2 = iodine concentration at the electrodes

If the electrodes are shorted, a current will flow which will tend to equalize the concentration of iodine in the compartments.

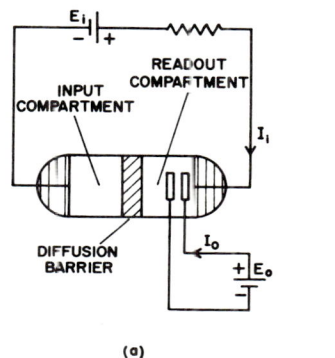 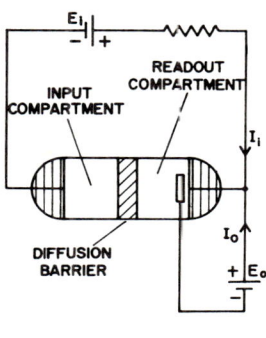

FIG. 4. Basic solion integrators. (a) Four-terminal. (b) Three-terminal.

3.2. Solion Tetrode. A symbolic four-terminal device which can function as an integrator or an amplifier is represented in Fig. 4a. The device has an input compartment, a readout compartment, a permeable barrier between these two compartments, and two readout electrodes. The input circuit is operated at sufficiently low current density so that neither input electrode is polarized and the input impedance is ohmic, as shown in region I, Fig. 2. The readout electrodes sense the iodine concentration in the readout compartment. They are biased to operate in the polarized condition, as shown in region II, Fig. 2. A refinement of this device is shown in Fig. 4b, where one of the input electrodes serves as a common electrode for input and output circuits. This is possible since the output circuit operates with only the cathode polarized.

A solion tetrode connected as a simple integrator is shown in Fig. 5. The symbol for the device is that used in recent technical publications. The commercial tetrode has a reservoir (input) compartment, an integral (output) compartment and four electrodes: input, shield, readout, and common. The input, readout, and common electrodes function as described above. The shield electrode is introduced to prevent iodine diffusion between the reservoir and integral compartments. It is situated near the readout electrode and biased so as to keep the iodine concentration low in its vicinity.

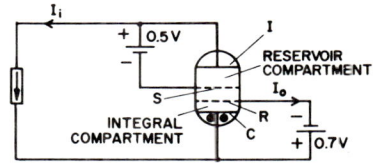

FIG. 5. Solion tetrode operating as an integrator. Electrodes: I, input; C, common; S, shield; R, readout.

Typical commercial tetrodes use variations of the iodine, potassium iodide, aqueous electrolyte system previously discussed. Other additional chemicals are sometimes used to enhance such performance factors as temperature range and reliability. The basic redox system, however remains unchanged. Variations in chemical concen-

---

[6] The constant E_a is derived from Eq. (1), where the base of the logarithm is changed and the value of n for iodine is 2.

15-6   AMPLIFYING DEVICES

trations and device geometry allow the production of a family of tetrodes. The electrodes are flat platinum disks. The shield and readout electrodes are perforated to provide a conducting path between input and common electrodes. The readout (and shield) perforations are sufficiently small to inhibit diffusion of iodine into and out of the integral compartment. A drawing of a typical tetrode is shown in Fig. 6.

The tetrode operates as an integrator as shown in Fig. 5. Iodine is transferred from the reservoir to the integral compartment when input current flows in the direction shown. The current flows internally from the common to the input electrode. The resistance between these electrodes is the ohmic resistance of the solution, the effective cross section of which is determined by the size of the perforations in the shield and readout electrodes. The readout–common electrode circuit produces an

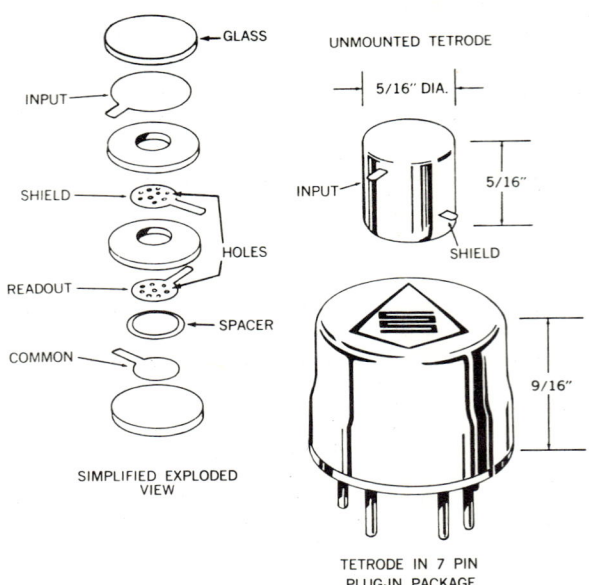

Fig. 6. Drawing of solion tetrode.

output current which is proportional to the amount of iodine in the integral compartment. A high impedance exists between these two electrodes since the cathode (readout electrode) operates in the polarized condition. The shield does not take part in the basic operation. By Faraday's law,

$$c = AQ_i \tag{7}$$

or
$$c = A \int^t I_i(t)dt \tag{8}$$

where c = concentration of iodine in the integral compartment
 A = const
 Q_i = charge transferred in the input circuit
 $I_i(t)$ = input current as a function of time
 t = time

The output current is proportional to the concentration:

$$I_o = Bc \tag{9}$$

where B = const

IONIC DEVICES

By combining Eqs. (8) and (9), we can write

$$I_o = K \int^t I_i(t)dt \tag{10}$$

where $K = AB$ = integrator sensitivity

The concentration of iodine in the integral compartment can be decreased, and hence the output current decreased, by reversing the direction of the input current. The concentration of iodine in the reservoir remains essentially constant for all conditions of operation since the volume of this compartment is much larger than that of the readout compartment.

The performance of the tetrode as an integrator can be examined more closely by referring to the static characteristics of the device.[7] The output characteristics are shown in Fig. 7, where input charge is the parameter. The zero value of the input

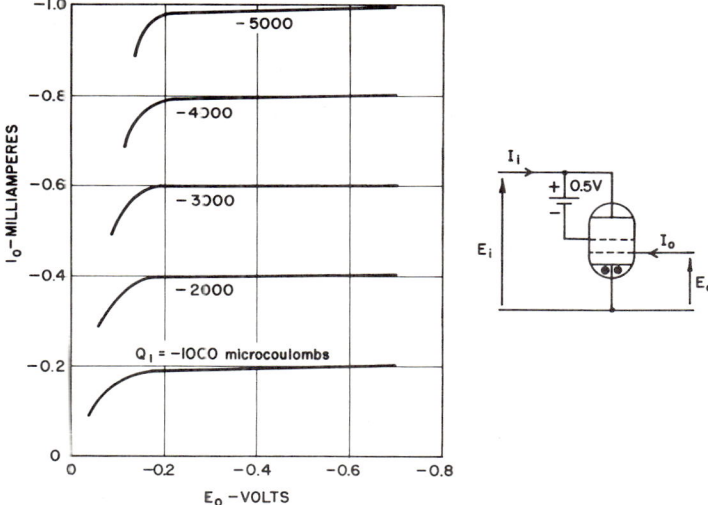

FIG. 7. Solion tetrode output characteristics.

charge is that value corresponding to zero output current. Linear operation is obtained to output currents of 1 ma or more, permitting a load resistance up to several hundred ohms. A useful method of evaluating the performance of the tetrode as an integrator is by examining the curve shown in Fig. 8. To obtain this response, the device is driven by a rectangular current pulse. Transient effects are apparent in the output current at the beginning and at the end of the input pulse. An effect, indicative of charge storage, at the beginning is due to capacitive feedthrough. This is discussed in more detail below. Overshoot can occur, at the end, for a number of reasons. Deviation from ideal-device geometry is a major one. This effect is generally reduced as the input drive is reduced.

Input characteristics are shown in Fig. 9, where input voltage is plotted against input charge with input current as the parameter. Two effects, previously discussed, combine to produce the input voltage: one is the IR drop across the electrolyte; the other is the concentration potential. From Fig. 9, the input resistance is

$$R_i = \left.\frac{\Delta E_i}{\Delta I_i}\right|_{Q_i = \text{const}} \cong 750 \text{ ohms} \tag{11}$$

---

[7] Excerpts from some typical data sheets of commercially available solion devices are shown in the Appendix.

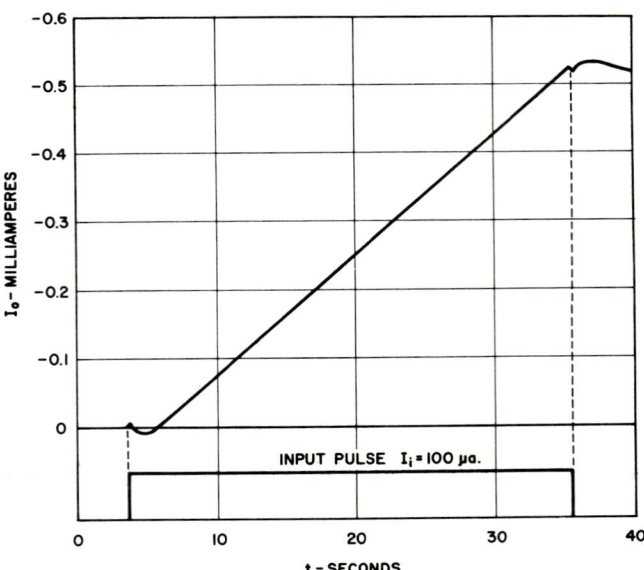

Fig. 8. Integral of an input current pulse.

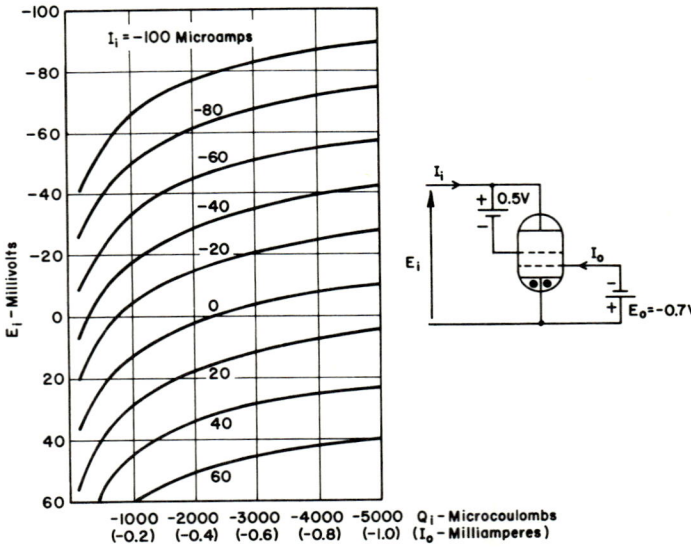

Fig. 9. Solion tetrode input characteristics.

The concentration of iodine is essentially a constant at the input electrode, and the concentration at the common electrode is directly proportional to the input charge. An equilibrium charge Q_x, which is a constant associated with a given device, is defined using Eq. (6):

$$E_c = -E_a \log_{10} (Q_i/Q_x) \tag{12}$$

where $Q_x = Q_i$ (when $E_c = 0$) = equilibrium charge

IONIC DEVICES

This equation can be written in terms of the output current:

$$E_c = -E_a \log_{10}(I_o/I_x) \tag{13}$$

where $I_x = I_o$ (when $E_c = 0$) = equilibrium output current

A d-c equivalent circuit of the solion tetrode is shown in Fig. 10. The resistance and the voltage generator in the input circuit are those described above. The output circuit is represented by a current generator of value given by Eq. (10), shunted by a resistance R_o of value determined from the slope of the output characteristics. The sensitivity is a function of temperature and varies approximately 2.4 per cent per °C over the operating range.

The solion tetrode can be used as an amplifier. Small-signal a-c equivalent circuits can be derived which are useful for design purposes. At low frequencies such that $\omega \ll K$ the equivalent circuit shown in Fig. 10 is useful. Variations of this circuit are shown in Fig. 11a and b, which are a-c equivalent circuits. The various parameters of these circuits can be derived from the above equations. Using Eq. (10) for the a-c signals, we may write

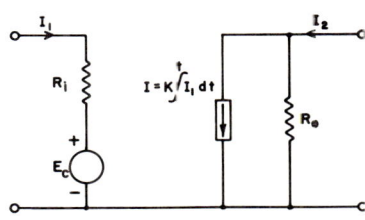

Fig. 10. Solion tetrode d-c equivalent circuit.

$$i_2 = K \int^t i_1 \, dt$$

By substituting $i_1 = I_{1m} \exp(j\omega t)$ and integrating

$$i_2/i_1 = \beta = K/j\omega \tag{14}$$

The output circuit is biased at a current I_o so that the total output current is

$$I_2 = I_o + i_2$$

Similarly, the input voltage generator is biased at a voltage E_c so that the total voltage is

$$E = E_c + e_c$$

The effect of small variations in the voltage generator can be expressed as a capacitance as follows: From Eq. (13),

$$E = E_c + e_c = -(RT/2F) \ln(I_2/I_x) = -(RT/2F) \ln[(I_o + i_2)/I_x] \tag{15}$$

Since $|i_2| \ll |I_o|$ we may write

$$e_c = -(RT/2F)(i_2/I_o) \tag{16}$$

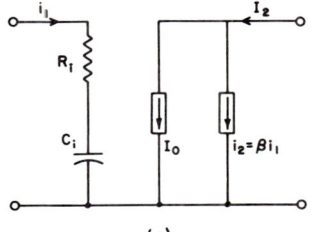
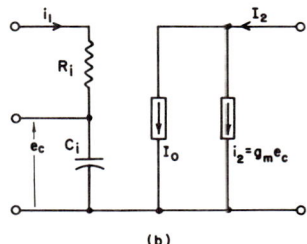

Fig. 11. Solion tetrode low-frequency equivalent circuits.

15–10   AMPLIFYING DEVICES

and the impedance in series with the input resistance R_i can be written using Eq. (14)

$$Z = e_c/i_1 = -(RT\beta/2FI_o) = -(RTK/2FI_o j\omega)$$
or
$$Z = 1/j\omega C_i \tag{17}$$

where $C_i = -(2F/RT)(I_o/K)$
or $\quad C_i = -(77 I_o/K) \quad$ at 25°C

In Fig. 11b the output current generator is expressed as $i_2 = g_m e_c$ where, from Eq. (16),

$$g_m = -(2F/RT)I_o = -77 I_o \quad \text{at 25°C} \tag{18}$$

The equivalent circuits discussed above are useful when considering the solion tetrode as an integrator. The device can be considered as a combination low-pass filter and integrator in series. Since the current gain β approaches $K/j\omega$ as ω approaches zero the device functions as a nearly perfect integrator for input signals having high-frequency components up to several kilocycles.

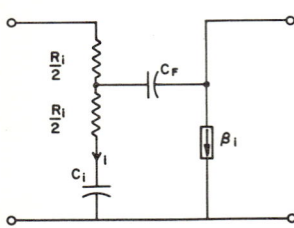

Fig. 12. Solion tetrode medium-frequency equivalent circuit.

Studies of solion equivalent circuits by J. D. Merryman[8] lead to the equivalent circuit shown in Fig. 12. The circuit is the same as those described above with the addition of the capacitance C_F.[9] This capacitance represents the "double-layer" capacitance of the cathodically polarized readout electrode. The position of C_F relative to R_i has yet to be determined. However, if it is assumed that it joins the middle of R_i the equivalent circuit shown in Fig. 12 is useful up to frequencies where $\omega \cong K$.

Several effects, previously unexplained, can be explained by referring to this circuit. Output resistance varies inversely with solion sensitivity. From Fig. 12,

$$R_o = 1/KC_F \tag{19}$$

Charge storage, which appears at the onset of integration, is simply that charge necessary to charge C_F when the output voltage changes by ΔV. Thus

$$\text{Charge storage} = Q_S = \Delta V C_F \tag{20}$$

Solion tetrodes are used in circuits where the frequencies are much greater than that for which $\omega = K$. The equivalent circuit shown in Fig. 12 is useful, qualitatively, for frequencies up to several times K. Higher-frequency equivalent circuits have not yet been determined. Four-pole parameters have been measured to 1 kc. The h parameters for the SE 110 are shown in Fig. 13a, b, c, and d.

3.3. Two-terminal Integrators. Two-terminal solions have been designed in a variety of configurations which function as integrators. The basic operation is essentially the same as that described above. Readout, however, is accomplished by measuring the concentration potential between the two electrodes or by observing the color of the solution in the vicinity of one or both electrodes.

The device can be represented as shown in Fig. 14. Although the iodine-iodide redox reaction is used, the chemical composition of the electrolyte differs from that of the solion tetrode. Two immiscible solutions are used; one acts as a barrier between the two electrode compartments, which are filled with the other solution. Usually the solution in the electrode compartments is organic while that of the barrier is aqueous. The solubility of iodine in the organic solution is much greater than that in the aqueous solution. Segregation coefficients up to several thousands are obtained so that the aqueous solution acts as an extremely efficient barrier to diffusion. As a result integrals can be stored in this device with little degradation for periods up to several months.

[8] To be published.
[9] Approximately 10 to 15 µf in present commercial devices.

IONIC DEVICES

The data sheet of a commercial two-terminal integrator (SV 150) is shown in the Appendix. This device can be "read out" electrically or visually. The electrode compartments are symmetrical.

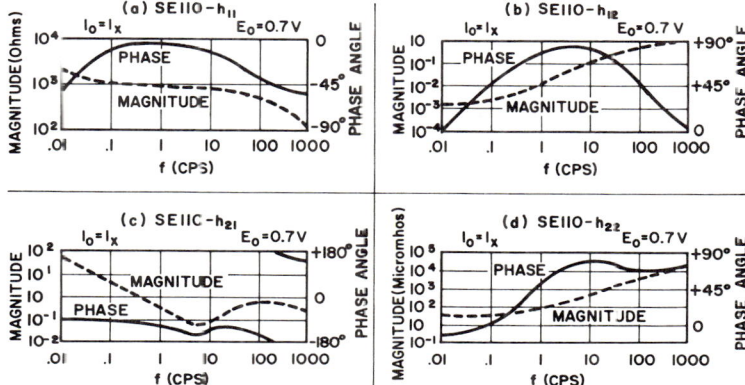

FIG. 13. Solion tetrode four-pole parameters.

3.4. Solion Transducers.[10,11] Solion devices can be constructed to convert mechanical to electrical energy. This is accomplished by measuring the flow of electrolytic solution past a cathodically polarized electrode. The sensitivity and response of such devices can be controlled by the cathode configuration.

A simple transducer is shown in Fig. 15. Two compartments, each containing anodes immersed in the electrolyte, are separated by a small orifice so that there is a continuous electrolytic path throughout the system. The cathode is placed in the orifice. The electrolyte can be similar to that used in the tetrode above.

Under static conditions an equilibrium current will flow which is diffusion-limited at the cathode. Thus a region surrounding the cathode is depleted in iodine. If the anode cham-

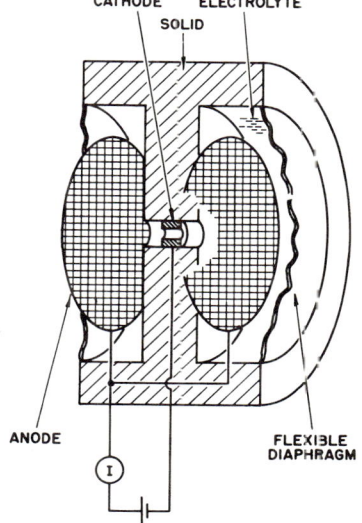

FIG. 15. Solion transducer.

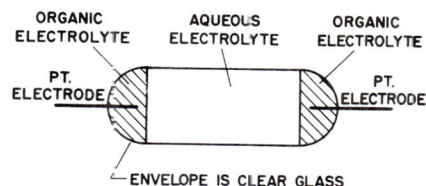

FIG. 14. Solion two-terminal integrator.

bers are deformed in a way to cause a flow of solution through the orifice, then "iodine-rich" solution appears at the cathode which reacts with the solution to cause the current increase in the external circuit.

---

[10] A. F. Wittenborn, Analysis of a Logarithmic Solion Acoustic Pressure Detector, *J. Acoust. Soc. Am.*, vol. 31, no 4, pp. 475–478, April, 1959.

[11] G. T. Kemp, "Low Frequency Solion Linear Detector," Preprint No. 119-LA-1, Instrument Society of America.

15-12 AMPLIFYING DEVICES

The current output as a function of fluid flow can be made a function of cathode configuration. If the cathode is designed to reduce all the iodine which flows through the orifice then the response is linear and is given by

$$I = FN(dv/dt) \times 10^{-3} \tag{21}$$

where I = current, amp
F = the Faraday = 96,500 coulombs per equivalent
N = normality of iodine
dv/dt = flow rate, cu cm/sec

When all the iodine is not reduced currents smaller than that given above are obtained. The "reduction efficiency" of the cathode can be made a function of the flow rate to give various responses.

Detectors have been made in a variety of configurations. They are usually similar to that shown in Fig. 15, where flexible diaphragms determine the low-frequency performance and the solution and cathode configuration determine the high-frequency performance. Linear, logarithmic, and other types of output can be obtained over

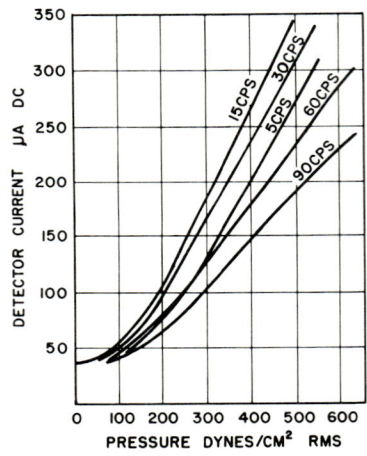

FIG. 16. Pressure response of a solion transducer.

several decades of pressure. Figure 16 shows the output current as a function of fluctuating pressure applied to one of the diaphragms.

In addition to relatively high output currents the solion transducers, in general, produce a power gain. This gain can be defined as the ratio of electric power out to mechanical (acoustical) power in and is given by

$$P_o/P_i = 10 R_L F^2 N^2 / R_o \tag{22}$$

where R_L = load resistance, ohms
R_o = orifice resistance, acoustic ohms

Power gains from 10^3 to 10^6 can be obtained for a reasonable range of parameters.

The simple solion transducer has an output current which is unidirectional. For a-c output when pressure fluctuates above and below an equilibrium value, two cathodes can be used. The orifice is designed so that flow in one direction produces a current in one cathode and flow in the opposite direction produces a current in the other.

4. THE MEMISTOR

The memistor is a variable resistance with memory. It is a three-terminal device in which the resistance between two terminals is controlled by the current flowing in the third terminal. It is a sealed plating cell containing two electrodes and an elec-

IONIC DEVICES 15–13

trolytic solution. One electrode can be considered a substrate upon which metal is deposited. The second electrode is a metal source. Two connections to the substrate electrode allow its resistance to be measured. This resistance can be reversibly controlled by the current passing between the metal-source electrode and the substrate electrode.

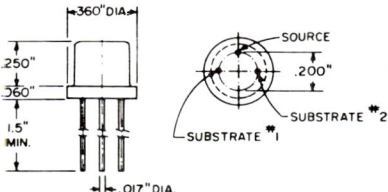

WEIGHT: 1½ gm.

RESISTANCE RANGE: From less than 1 ohm to greater than 25 ohms.

SPEED: Typical—25 ohms to 2 ohms, 15 seconds; range—5 seconds to 25 seconds.

TYPICAL FREQUENCY RESPONSE: Essentially flat from dc to 10 megacycles.

TYPICAL PULSE RISE TIME: Less than 20 nanoseconds.

TEMPERATURE COEFFICIENT:
Maximum 0.4% of resistance value per °C.
Typical 0.25% of resistance value per °C.
Minimum 0.075% of resistance value per °C.

OPERATING LIMITS:

TEMPERATURE RANGE: For either storage or operation, −10°C. to 60°C.

MAXIMUM PLATING CURRENT: 0.25 ma.

MAXIMUM PLATING (DE-PLATING) VOLTAGE: 0.75 volts.

MAXIMUM READOUT VOLTAGE:
(a) without impairing stability, 50 millivolts ac RMS.
(b) without impairing life, 200 millivolts RMS.

MAXIMUM POWER DISSIPATION: 50 milliwatts.

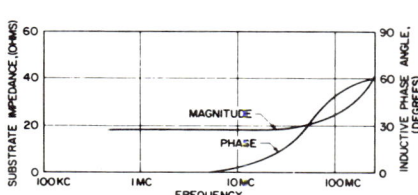

FREQUENCY RESPONSE CURVES
Magnitude and phase curves shown above were measured on a Boonton Radio Model 250-A, R X Meter.

STABILITY AT 25°C:

	1 hour	48 hrs	1 week
high	0.06 mho	0.08 mho	0.08 mho
typical	0.05 mho	0.07 mho	0.05 mho
low	0.00 mho	0.00 mho	−0.02 mho

Drift after plating from 20 ohms to 2 ohms at 0.2 ma.

	1 hour	48 hrs	1 week
high	0.005 mho	0.02 mho	0.02 mho
typical	0.00 mho	±0.005 mho	±0.01 mho
low	−0.005 mho	−0.02 mho	−0.02 mho

Drift after de-plating from 1 ohm to 2 ohms at 0.2 ma.

0.01 mho drift is less than 1% of total range.

The drift after plating is approximately proportional to the conductance change. This drift is smaller for smaller conductance changes, and can be reduced by connecting a resistor of 10K or less between the source and one of the substrate leads, and by plating with a reduced plating current

FIG. 17. Specifications of Memistor M-2CR.

Two versions of this device, models M-2CR and M-3CR, are available commercially. They are similar in characteristics. The M-3CR is the more recent model and represents some improvements electrically. It is also somewhat smaller, being encased in a conventional TO-9 package. Some specifications of the M-3CR are shown in Fig. 17. The device is represented by the symbol shown in Fig. 18.

AMPLIFYING DEVICES

4.1. Operation. According to Faraday's law, the amount of metal deposited on the substrate is proportional to the integral of the current flowing between electrodes. When the memistor is connected as shown in Fig. 19 the conductance will vary as a

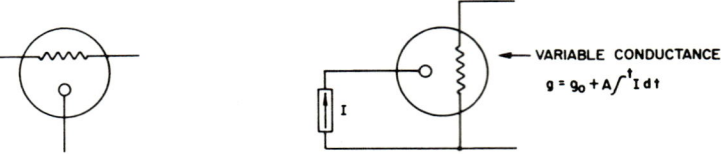

Fig. 18. Memistor circuit symbol.   Fig. 19. Memistor operation.

function of the driving current. This conductance is the sum of the unplated substrate conductance and the conductance of the deposited metal film. This can be expressed as

$$g = g_o + A \int^t I(t)dt \qquad (23)$$

where g_o = unplated substrate conductance
A = const
$I(t)$ = current
t = time

Commercial devices are available[12] where g_o is 0.02 mho and A is 25 mhos/coulomb. The conductance can be nondestructively sensed by low-voltage alternating current which can range in frequency from 60 cycles to several megacycles but should not exceed 0.1 rms volt.

Constant-current plating curves for the M2CR memistor are shown in Fig. 20. In the linear range, integration accuracy is about 5 per cent. Long-term drift is less

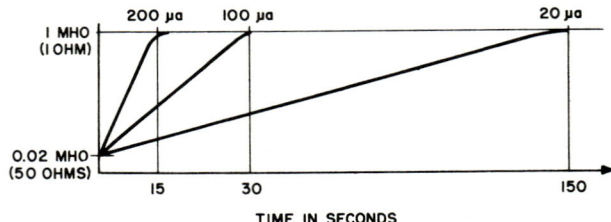

Fig. 20. Constant-current plating curves for the M-2CR.

than 1 per cent per week. The device operates over a temperature range from -15 to $100°C$. The variation of the plated resistance with temperature is approximately 0.25 per cent per °C.

Saturation takes place when the substrate is heavily plated. This effect determines the upper limit of integration. At the lower limit of integration, when the substrate is completely stripped and shows maximum resistance, further current flow causes the input voltage to rise. When this voltage exceeds 0.75 volt, gas evolution occurs which can alter or destroy the device.

The average input current should not exceed 0.25 ma d-c. The input resistance is approximately 1,000 ohms over the operating range.

5. MERCURY COULOMBMETER

The mercury coulombmeter is another type of integrating device. They are fabricated from precision-bore capillary tubing which is filled with two columns of mercury

---

[12] Memistor Corp., 270 Polaris Ave., Mountain View, Calif.

IONIC DEVICES   15-15

separated by a small gap which contains an aqueous electrolytic solution. Nickel electrodes make contact with each mercury column.

The device is shown in Fig. 21. When direct current is applied to the electrodes current flowing through the device causes mercury at the anode to be electrochemically transferred across the gap to the cathode at a rate proportional to the current. Mercury transfer effectively moves the gap along the tube length. The total gap movement, with uniform glass bore, is directly proportional to the amount of charge transferred, or gap displacement is proportional to the time integral of the current.

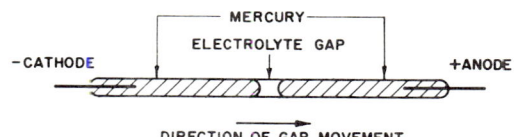

Fig. 21. Mercury coulombmeter.

5.1. Basic Operation. The electrolyte solution which forms the indicating gap is an aqueous solution of a soluble iodide salt and mercuric iodide. The mercuric ion is in the form of the tetraiodide complex which is extremely stable in the presence of excess iodide ions:

$$Hg^{++} + 4I^- \rightarrow Hg\ I_4^{--} \tag{24}$$

The soluble iodide salt serves as the supporting electrolyte, thereby reducing the ohmic resistance of the solution and providing excess iodide ions to maintain complete complexing of mercuric ions. The electrode reactions are reversible and symmetrical:

Cathode reaction (reduction)

$$Hg\ I_4^{--} + 2e \rightarrow Hg + 4I^- \tag{25}$$

Anode reaction (oxidation)

$$4I^- + Hg \rightarrow Hg\ I_4^{--} + 2e \tag{26}$$

resulting in a net transfer of mercury across the electrolyte.

The mass of mercury transferred can be determined as a function of the current through the device:

$$m = (W/nF) \int^t I(t)dt \tag{27}$$

where m = mass, grams
W = molecular weight
n = valence
F = the Faraday
$I(t)$ = current, amp
t = time, sec

The mass can also be written

$$m = \rho L \pi d^2 / 4 \tag{28}$$

where ρ = density
L = gap displacement
d = capillary bore diameter

Thus the gap displacement can be written

$$L = B \int^t I(t)dt \tag{29}$$

where $B = 4W/n\rho F \pi d^2$

The constant B is called the "meter constant" and, for example, equals 0.096 in. of gap travel/ma-hr for a bore diameter of 0.015 in.

15–16 AMPLIFYING DEVICES

An equivalent circuit for a mercury coulombmeter is shown in Fig. 22. The ohmic resistance of the electrolyte is represented by r_e. The parallel combination of C_{mt} and r_{mt} represent the effects of concentration polarization since the Hg I_4^{--} is more highly concentrated at the anode than at the cathode. This effect becomes increasingly predominant as the current density decreases where r_{mt} can be an order of magnitude

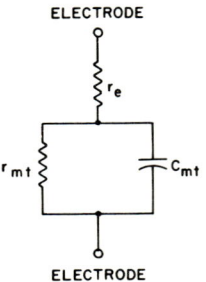

Fig. 22. Equivalent circuit of the mercury coulombmeter.

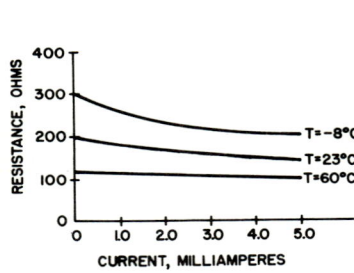

Fig. 23. Cell resistance vs. a-c current (2,000 cps).

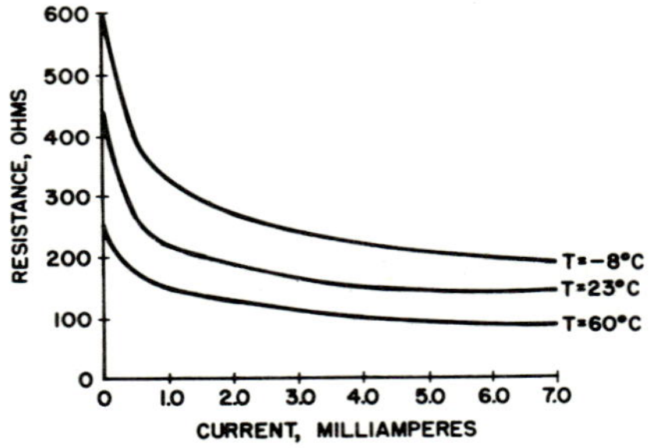

Fig. 24. Cell resistance vs. d-c current.

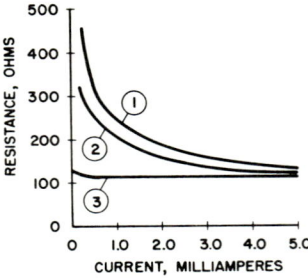

Fig. 25. Cell resistance as a function of position at room temperature. (1) D-C resistance—cathode up—meter vertical. (2) D-C resistance—cathode down—meter vertical. (3) A-C resistance—meter vertical (2,000 cps).

larger than r_e and C_{mt} can be several thousand microfarads. As current density increases both C_{mt} and r_{mt} approach zero.

Some characteristics of a commercially available mercury coulombmeter known as the "Curtis miniature ampere-hour meter" are shown in Figs. 23 to 25.[13] The device operates over a temperature range from −20 to 125°C for currents up to 5 ma.

[13] Curtis Instruments, Inc., 351 Lexington Ave., Mount Kisco, N.Y.

APPENDIX
EXCERPTS FROM SOME COMMERCIAL SOLION DATA SHEETS*
A. Solion Tetrodes Types SE 100 and SE 110
MECHANICAL DATA
Standard 7 pin miniature tube socket.

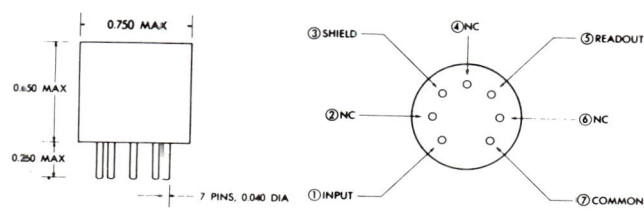

TYPICAL CHARACTERISTICS

SE 100

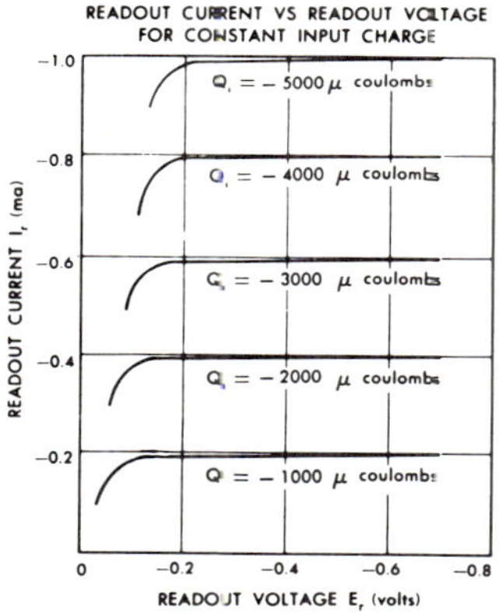

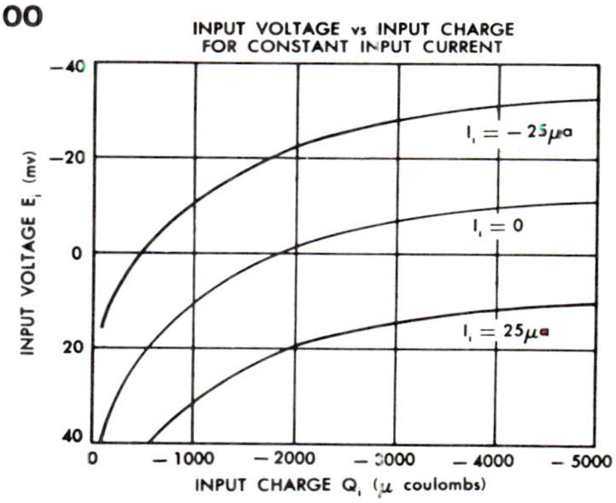

SE 110

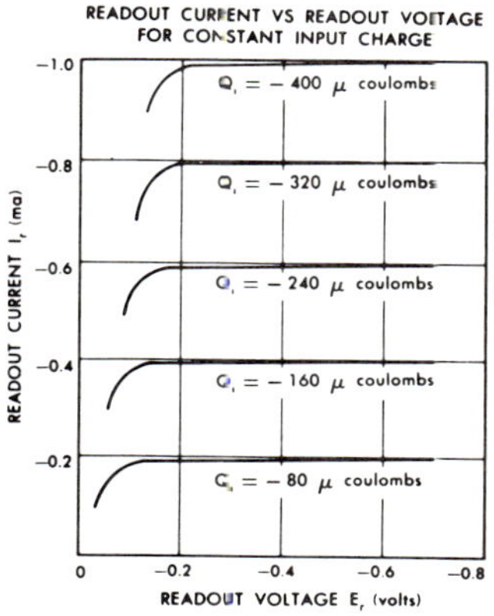

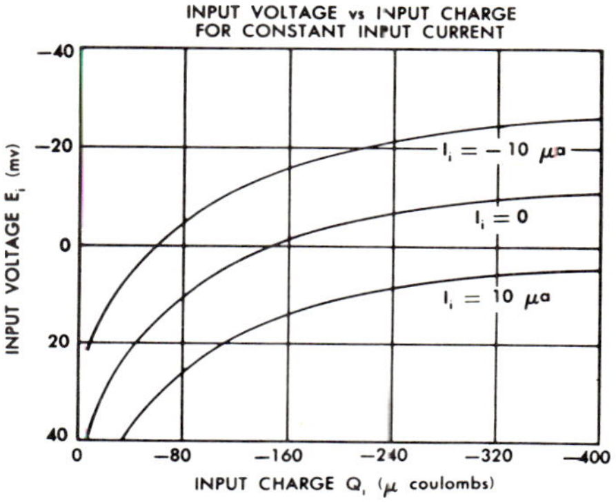

* By permission of Self-Organizing Systems Inc., Box 9918, Dallas 35, Tex. 75214.

AMPLIFYING DEVICES

Solion Tetrodes Types SE 100 and SE 110 (Continued)

ELECTRICAL CHARACTERISTICS at $T_A = +25°C$ and $E_s = -0.5v$

PARAMETER			TEST CONDITIONS	MIN.	TYP	MAX.	UNITS
K	Integrator Sensitivity	SE 100	$E_r = -0.7v$	0.16	0.22	0.28	μ amperes per μ coulomb
		SE 110	$E_r = -0.7v$	1.8	2.6	3.4	
ΔI_{rv}	Voltage Sensitivity	SE 100	$E_r = -0.7v$ to $-0.5v$ $I_i = 0$ $I_r = -0.8$ ma		1.3	4	μ amperes
		SE 110			8	25	
I_s	Shield Current	SE 100	$E_r = -0.7v$		6	20	μ amperes
		SE 110	$E_r = -0.7v$		2	10	
I_x	Equilibrium Output Current		$E_r = -0.7v$ $I_i = 0$ $E_i = 0$		350		μ amperes
R_i	Equivalent Input Resistance	SE 100	$I_r = I_x$ $I_i = -25$ μa		900		ohms
		SE 110	$I_r = I_x$ $I_i = -10$ μa		1500		
ΔI_{rt}	Output Transient Response	SE 100	$E_r = -0.7v$ $I_i = -25$ μa to 0		0.5		% of correct current value
		SE 110	$E_r = -0.7v$ $I_i = -10$ μa to 0		1.0		

Definitions:

E_s Potential of shield electrode with respect to the input electrode
K   See equivalent circuit
ΔI_{rv} Change in readout current for a 0.2 volt change in readout voltage (input circuit open)
I_s Self explanatory
I_x Value to which readout current will settle if pin 1 and pin 7 are shorted together
R_i See equivalent circuit
ΔI_{rt} Maximum deviation of readout current from true value during the first minute after abrupt decrease of input current signal.

MAXIMUM RATINGS

	PARAMETER			UNITS
E	Maximum Voltage between any two electrodes		0.75	v.
I_i	Input Current (NOTE 1)	SE 100	50	μa
		SE 110	25	
I_r	Readout Current (NOTE 1)		1	ma
T_A	Ambient Operating Temperature Range		0 to +60	°C
T	Storage Temperature Range (Note 2)		−20 to +60	°C

NOTE 1: These currents are based on 1% or better linearity and reproducability. Most devices are linear up to 2 ma readout current. Higher currents may be used, but note that maximum voltage between any two electrodes should not exceed 0.75 volts.

NOTE 2: Tetrodes available on special order for storage to −40°C.

RUGGEDNESS: The solion tetrodes are very rugged devices. Tests have been performed at shocks up to 150 g each axis and vibration of 20 g each axis at frequencies of 30 - 2000 cps without failure. The shock and vibration required to cause damage are unknown, but should be well beyond the test values.

TYPICAL DRIFT CHARACTERISTICS

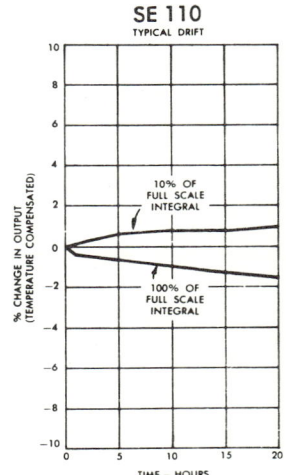

SE 110 TYPICAL DRIFT

IONIC DEVICES 15-19

B. Solion Two-terminal Integrator Type SV 150

MECHANICAL DATA

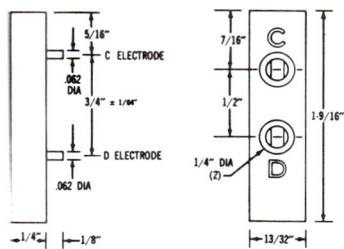

ELECTRICAL CHARACTERISTICS @ $T_A = +25°C$

PARAMETER		MIN.	TYP.	MAX.	UNITS
Q_t	Total Charge (Note 1) (Note 2)	0.036	0.040	0.044	coulomb
R_s	Ohmic Resistance		3	5	k ohm
I_i	Maximum Input Current (Note 3)			± 5	μamp
T_r	Reset Time (Note 4)		15		minutes
T_s	Settling Time (Note 5)		1		hour
$\frac{dQ}{dt}$	Drift Rate (-40 mv $\leq E_c \leq +40$ mv)		3		$\frac{\mu coulomb}{day}$

NOTE 1: Total charge is the charge required to move all of the iodine from one compartment to the other.
NOTE 2: Tolerance of ± 0.002 coulomb also available. The type number is SV 150A.
NOTE 3: Larger currents may be integrated under intermittent conditions as long as the maximum voltage is not exceeded.
NOTE 4: Time required to move at least 99% of the charge to one compartment with 0.4 volts applied.
NOTE 5: Time required for E_c (see curve) to reach steady-state condition after removal of input. Low input currents allow E_c to approximate steady state while charging.

MAXIMUM RATINGS

	PARAMETER	SV 150	UNITS
E	Maximum Voltage	± 0.75	V.
T_A	Ambient Operating Temperature Range	-25 to $+65$	°C
T_s	Storage Temperature Range	-65 to $+65$	°C

RUGGEDNESS: The type SV 150 integrator is an extremely rugged device. Tests have been performed at shocks up to 150 g each axis and vibration of 20 g each direction at frequencies of 30-2000 cps without failure. The shock and vibration required to cause a failure are unknown, but should be well beyond the test values.

Solion Two-terminal Integrator Type SV 150 (Continued)

TYPICAL CHARACTERISTIC CURVE

Curves on individual devices will vary slightly, due to differences in total charge, and may be displaced up or down somewhat if the size of the two compartments is not identical. The logarithmic charge scale is convenient when electrical readout is used.

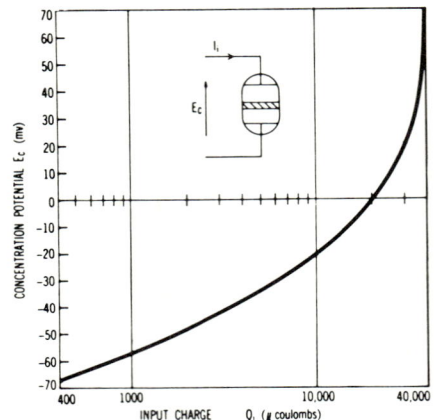

TYPICAL VOLTAGE AND COLOR VARIATION

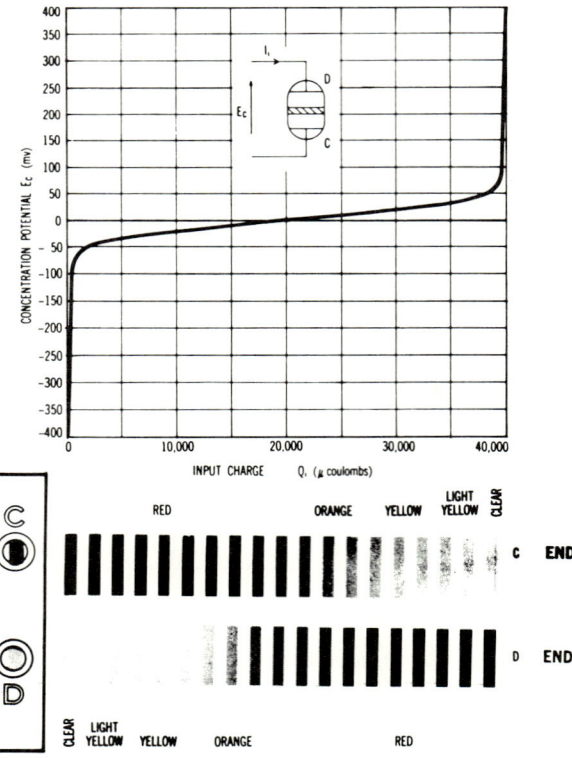

Chapter 16

CERAMIC DEVICES

STEPHEN W. TEHON*

CONTENTS

1. Introduction.. 16- 1
2. Poled Ferroelectric Ceramics................................ 16- 2
3. Properties of Poled Ceramics................................ 16- 4
4. Adiabatic and Isothermal Constants: Pyroelectricity........... 16- 8
5. Characterization of a Poled Ceramic.......................... 16-10
6. Electromechanical Coupling Factors........................... 16-12
7. Energy Storage and Piezoelectric Conversion.................. 16-16
8. Waves and Resonance... 16-17
   8.1. Plane Longitudinal Wave: Propagation Perpendicular to the Poling Axis.. 16-19
   8.2. Plane Longitudinal Wave: Propagation Parallel to the Poling Axis.. 16-21
   8.3. Plane-polarized, Plane Shear Wave: Propagation Parallel to Poling Axis, Particle Motion along the X Axis............. 16-22
   8.4. Plane-polarized, Plane Shear Wave: Propagation Perpendicular to the Poling Axis, Particle Motion along the Z Axis.. 16-22
   8.5. Plane-polarized, Plane Shear Wave: Propagation and Particle Motion Perpendicular to the Poling Axis.................. 16-23
   8.6. Extensional Wave in a Length-polarized Thin Bar......... 16-23
   8.7. Extensional Wave in a Width-polarized Thin Bar.......... 16-25
   8.8. Torsional Wave along a Circumferentially Poled, Round Cylinder.. 16-25
9. Standing Waves, Equivalent Circuits, and Attenuation........ 16-27
10. The Side-plated Ceramic Bar................................. 16-32
11. Constant-D Transducers...................................... 16-41
12. Ceramic-disk Resonator....................................... 16-46
13. Ceramic Transformers... 16-49
14. Delay Line Transducers....................................... 16-57
15. Ceramic Filters.. 16-61

1. INTRODUCTION

In multistage amplifiers, the means for coupling between active elements may include impedance transformation for maximum power gain and may also include

* General Electric Co., Electronics Laboratory, Syracuse, N.Y.

means for providing selectivity. Ceramic filters,[1] which utilize piezoelectric coupling between electrical and mechanical energy, can provide both selectivity and impedance transformation.

The ceramic materials used for transformers and filters are polycrystalline aggregates of ferroelectric grains.[2-8] Chemical compositions in common use include barium titanate ($BaTiO_3$), lead metaniobate ($PbNb_2O_6$), and lead zirconate titanate (PZT*). A number of other elements frequently are added, to provide specific electrical or electromechanical properties. The ceramic materials, formed from a slurry of the powdered base material and a binder material in water, are shaped and dried in desired form. The shaped forms are then fired at temperatures on the order of 1200°C, which drive off the binder materials, and the basic materials form a hard, dense sintered mass. On microscopic examination,[9] the material may be seen to consist of irregularly shaped grains interlocked with occasional voids. Under polarized light, individual grains may be seen to be subdivided into numbers of domains. The domains are influenced by temperature and electric field and are primarily involved in the unusual electrical and electromechanical properties of ferroelectric ceramics which make them suitable for use in filters and transformers.

A crystal, selected from a ceramic body or grown in single-crystal form, has a periodic crystalline structure.[10] Barium titanate, one of the simplest and most thoroughly studied crystals, has a tetragonal structure. Within a single domain, the titanium ions are all shifted along the c axis, away from the center of the unit cube formed by barium ions by about one twenty-fifth of a lattice unit. Each Ti^{4+} ion forms the positive charge in a dipole, with negative charge formed by the associated Ba^{++} and face-centered O^{--} ions as a group. A single domain can be regarded as a chain structure of elementary dipoles joined side by side and end to end, and in such a representation the domain can then in turn be regarded as one resultant dipole. The dipole moment at room temperature is about 0.16 coulomb-meter/cu m. This is also the polarization charge density, 0.16 coulomb/sq m, at a surface normal to the c axis.

2. POLED FERROELECTRIC CERAMICS

After the sintering operation at high temperature, the ceramic consists of a large number of grains, and each grain consists of many domains. The freshly fired material could conceivably have a large polarization dipole moment and bound surface charge density, but these are not present. Instead, during the firing the domains which form within grains when cooling are randomly oriented, as are the c-axes of the many grains. Therefore, the average polarization is zero.

If electrodes are attached to opposing faces of a ferroelectric sample, an internal electric field can be applied by means of a voltage between the electrodes. Large amounts of charge can be observed to flow when the field is applied, and the effective dielectric constant for such large charge storage is phenomenally large. If the material is barium titanate at room temperature, the dielectric constant is on the order of 1,000. In mks units, the permittivity of free space is

$$\epsilon_0 = 8.854 \times 10^{-12} \text{ farad/m} \tag{1}$$

Field intensity E is measured in volts per meter, and displacement D is in coulombs per square meter:

$$D = K_R \epsilon_0 E \tag{2}$$

where K_R is the relative dielectric constant.

The total charge density on an electrode is equal to D in the immediately adjacent space.† The electrode charge terminates the lines of force which would be present if the adjacent space were a vacuum and also neutralizes the bound charge due to

---

* PZT is a designation trademarked by Clevite Corporation.
† Rationalized units.

polarization in the material occupying that space:

$$D = \epsilon_0 E + P \tag{3}$$

The relative dielectric constant can be written from Eqs. (2) and (3) as

$$K_R = 1 + P/\epsilon_0 E \tag{4}$$

The large dielectric constant of a ferroelectric material results from the large density of charge involved in the dipole moment. The dielectric behavior is characterized by a Curie temperature, above which the permittivity follows the Curie-Weiss law

$$\epsilon = c/(\theta - \theta_C) \tag{5}$$

where c is a constant for the material, θ is temperature, and θ_C is the Curie temperature. From Eq. (5), it can be seen that as temperature approaches the Curie point from above, the permittivity approaches infinity. This is an indication of the polarization "catastrophe" which occurs spontaneously at the Curie point.

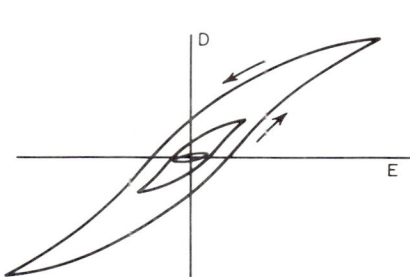

Fig. 1. Hysteresis in polycrystalline barium titanate.

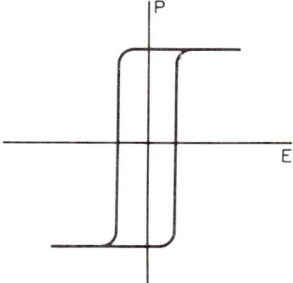

Fig. 2. Dielectric hysteresis along the c axis of single-domain barium titanate.

When spontaneous polarization occurs, domains are formed. The domain wall patterns remain relatively constant, as long as temperature stays below the Curie point and no strong fields are applied. However, if electric fields are applied, the domain walls tend to move, and new domains nucleate, all in directions such that the polarization vector increases along the applied field direction. For moderate or small field intensities, the domain wall motion and polarization follow reversible trends, but for large fields, hysteresis is present in the domain formation as well as in the locus of polarization with field intensity.

For large intensities, a plot of polarization vs. electric field intensity is very similar to the plot of magnetization vs. magnetic field intensity for a ferromagnetic material. Figure 1 is taken from experimental data for a barium titanate ceramic sample which was fitted with plated silver electrodes. This represents the sum of dipole moments taken over a large number of domains, in many individual grains within the sample.

Figure 2 is a hysteresis loop for a single crystal of barium titanate oriented with its faces perpendicular to a c axis. In comparison with the loop for a ceramic, it can be seen that the domain motion in a single crystal is abrupt and that the hysteresis curve follows a remarkably square-cornered loop. Single crystals have been proposed as memory elements, since they can be very tiny, can switch in very short time intervals, and have relatively small coercive fields which are required for switching. They are not in general use for several reasons, which include the circuit problems in setting up large arrays in switching matrices, and the heating due to the large area of the hysteresis loop traversed in each cycle.

In ferroelectric ceramic material, domains form within the grains along the c axes closest to applied field intensity. The field along any individual c axis is less than the applied field intensity, except in the rare case where they are parallel. Over the large assemblage of domains, switching therefore occurs gradually throughout the ceramic, and the total polarization vector follows a smooth hysteresis loop as in Fig. 1.

The appearance of spontaneous polarization at the Curie temperature can be regarded as evidence of nearly infinite permittivity at the Curie point. With infinite permittivity, any finite field intensity would produce infinite dipole moment, or finite polarization would require only vanishingly small field intensities.

While the simple Curie-Weiss law is not quite able to describe the exact situation at the Curie point, nevertheless domains are quite easily controlled with applied fields at this temperature. Barium titanate ceramics can be "poled" by maintaining field intensities of about 20,000 volts/in. in the ceramic bodies during cooling through the Curie point. For adequate insulation, heating may be applied by a silicone oil bath. PZT and lead metaniobate ceramics, with Curie points considerably higher than the 120°C for barium titanate, develop considerable conductivity at their Curie temperatures, and so are usually poled with fields of about 100 kv/in. at temperatures in the range from 100 to 200°C.

After poling, a ceramic still contains a large number of domains with many different orientations. However, the net polarization vector produced by these domains is not zero but is quite large and is parallel to the direction in which the poling field was applied. After the poling process has been completed, a small amount of domain relaxation takes place at a logarithmically slowing rate of aging. The aging rate can be determined for any required property, such as dielectric constant, as the fractional change in that property per decade of time. Accelerated aging at elevated temperatures is possible for many ceramic materials. In some forms of PZT and lead metaniobate, extremely small amounts of aging have been reported.

3. PROPERTIES OF POLED CERAMICS

Before poling, a ferroelectric ceramic is isotropic: its macroscopic physical properties are independent of direction. Poling rearranges the structure in transverse isotropy, establishing a polar axis with unique properties. All properties measured within a plane perpendicular to the polar axis are independent of the azimuthal angle.

In a right-hand rectangular coordinate system, it is customary[11] to direct the z axis parallel to the remanent polarization, and to place the x axis and y axis at any convenient orientation in an azimuthal plane. With this convention, the space orientation of a vector, with respect to the transversely isotropic symmetry of the poled ceramic, is specified by the XYZ components of the vector. Again by established custom, the x, y, and z subscripts denoting components along the coordinate axes are usually replaced with the numerical subscripts 1, 2, and 3, respectively.

The electrical quantities of interest in and around the ceramic are the field intensity E, the displacement D, and the polarization P. These are not independent but are interrelated through Eq. (3). Fundamental studies of domain behavior are most readily described in terms of E and P; application studies linking geometry to external circuits are more readily described by E and D. E, D, and P are vector quantities, each having three components and each describing a condition at a point.

The mechanical quantities of interest are stress T and strain S. Stress has the dimensions of force per unit area, in mks units, newtons per square meter. Stress may be applied by force normal to a surface, tangent to the surface, or in combination. Strain is dimensionless; longitudinal strain is a fractional elongation, which may be measured as inches per inch, or meters per meter, and shear strain is an angular distortion which is measured in radians. Both stress and strain are properties describing conditions in a small neighborhood of a point, depending for their definition on surfaces which bound that neighborhood. There are three independent components of longitudinal stress, corresponding to tensile forces, and three independent components of shearing stress. By custom,[12] subscripts 1, 2, 3 are assigned to the longitudinal stresses and 4, 5, 6 to the shearing stresses:

T_1: tensile stress producing elongation along X
T_2: tensile stress producing elongation along Y
T_3: tensile stress producing elongation along Z
T_4: transverse stress producing shearing closure of lines along Y and Z, in an X plane
T_5: transverse stress producing shearing closure of lines along Z and Z, in a Y plane
T_6: transverse stress producing shearing closure of lines along X and Y, in a Z plane

The corresponding fractional elongations are the longitudinal strain components S_1, S_2, S_3, and the corresponding shearing closures are the shear strain components S_4, S_5, S_6. Stress and strain are tensor quantities describing conditions in a neighborhood of a point.

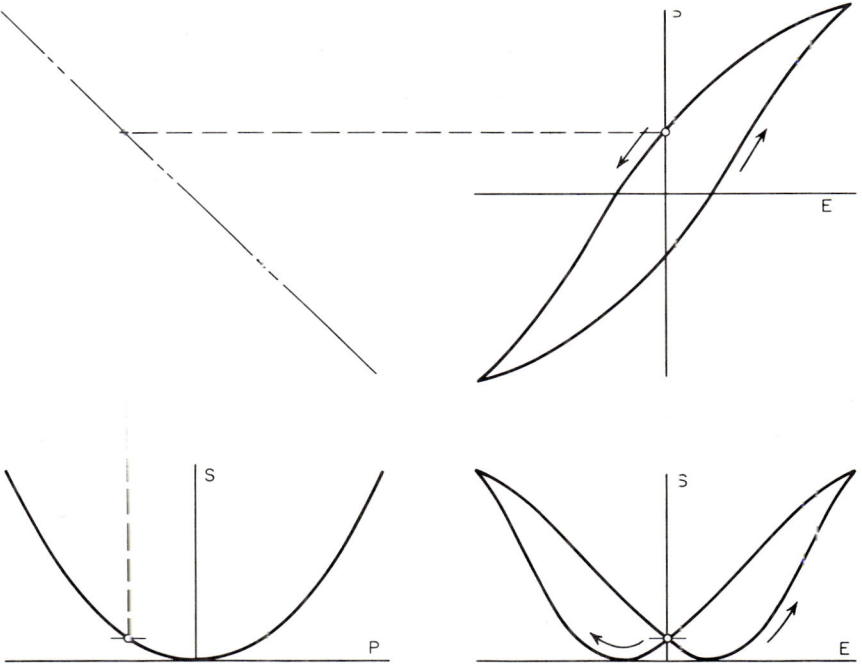

FIG. 3. Large-signal hysteresis and electrostriction in unpoled electrostrictive ceramic.

Before a ceramic is poled, application of electric fields produces polarization through domain wall motion, and the distortion of the molecular structure with polarization produces strain in the ceramic. The strain is proportional to the square of polarization.[13,14] On the other hand, in an unpolarized ceramic there is no ordering of the domains, and while applying a strain will produce domain wall motions and changes of polarization within the microscopic domains, the electrical effects taken over many randomly oriented domains and grains are zero. Therefore, the unpoled ceramic has the property of *electrostriction*, characterized by elongation parallel to an applied field and by the absence of voltage generation under mechanical pressure. The electrostrictive strain is longitudinal along an axis parallel to the macroscopic polarization vector, and follows the square of polarization quite closely, in magnitude.

Figure 3 represents, in its three projections, the interrelations of electric field, polarization, and strain. For simplicity, only one hysteresis loop is shown. The double-valued hysteresis function results in the double-valued "butterfly" curve of strain as a function of electric field. These curves are typical of large field variations.

A poled ceramic, with its domains directed as completely as possible along what then becomes the z axis, has been given an ordered sense of polarization which results in production of a net electric field under applied pressure. For small signals, hysteresis is negligible, and the electromechanical generation of field under pressure, or of strain under electric field, is linear. The linear effect, which is *piezoelectric* coupling, is truly reversible, and there is a form of the reciprocity theorem which accompanies linearity as a direct consequence of conservation of energy.[15,16] With linearity for small signals in the piezoelectric, dielectric, and mechanical properties, the state of the ceramic can be described by equations of linear proportionality, such as

$$\begin{aligned} S &= sT + dE \\ D &= dT + \epsilon E \end{aligned} \quad (6)$$

In Eqs. (6), the coefficient d appears twice, because of reciprocity.

In general, six components of the strain tensor must be described, as well as three components of the displacement vector, six of the stress tensor, and three of the field intensity vector. These are linearly interrelated by means of nine equations, which can be written in the form[17]

$$\begin{aligned}
S_1 &= s_{11}^E T_1 + s_{12}^E T_2 + s_{13}^E T_3 + s_{14}^E T_4 + s_{15}^E T_5 + s_{16}^E T_6 + d_{11} E_1 + d_{21} E_2 + d_{31} E_3 \\
S_2 &= s_{12}^E T_1 + s_{22}^E T_2 + s_{23}^E T_3 + s_{24}^E T_4 + s_{25}^E T_5 + s_{26}^E T_6 + d_{12} E_1 + d_{22} E_2 + d_{32} E_3 \\
S_3 &= s_{13}^E T_1 + s_{23}^E T_2 + s_{33}^E T_3 + s_{34}^E T_4 + s_{35}^E T_5 + s_{36}^E T_6 + d_{13} E_1 + d_{23} E_2 + d_{33} E_3 \\
S_4 &= s_{14}^E T_1 + s_{24}^E T_2 + s_{34}^E T_3 + s_{44}^E T_4 + s_{45}^E T_5 + s_{46}^E T_6 + d_{14} E_1 + d_{24} E_2 + d_{34} E_3 \\
S_5 &= s_{15}^E T_1 + s_{25}^E T_2 + s_{35}^E T_3 + s_{45}^E T_4 + s_{55}^E T_5 + s_{56}^E T_6 + d_{15} E_1 + d_{25} E_2 + d_{35} E_3 \\
S_6 &= s_{16}^E T_1 + s_{26}^E T_2 + s_{36}^E T_3 + s_{46}^E T_4 + s_{56}^E T_5 + s_{66}^E T_6 + d_{16} E_1 + d_{26} E_2 + d_{36} E_3 \\
D_1 &= d_{11} T_1 + d_{12} T_2 + d_{13} T_3 + d_{14} T_4 + d_{15} T_5 + d_{16} T_6 + \epsilon_{11}^T E_1 + \epsilon_{12}^T E_2 + \epsilon_{13}^T E_3 \\
D_2 &= d_{21} T_1 + d_{22} T_2 + d_{23} T_3 + d_{24} T_4 + d_{25} T_5 + d_{26} T_6 + \epsilon_{12}^T E_1 + \epsilon_{22}^T E_2 + \epsilon_{23}^T E_3 \\
D_3 &= d_{31} T_1 + d_{32} T_2 + d_{33} T_3 + d_{34} T_4 + d_{35} T_5 + d_{36} T_6 + \epsilon_{13}^T E_1 + \epsilon_{23}^T E_2 + \epsilon_{33}^T E_3
\end{aligned} \quad (7)$$

The array of coefficients can be written as a 9×9 square matrix. Of the 81 terms, only those along the principal diagonal from s_{11}^E to ϵ_{33}^T and half of those remaining can be independent. Reciprocity requires that the matrix is always symmetrical; so terms mirrored about the principal diagonal are equal. Therefore, only 45 terms can be independent.

The dashed lines in Eqs. (7) indicate a means of dividing the nine equations of state into groups, each containing coefficients of the same type. The partitioned matrix equation then takes the form

$$\begin{bmatrix} S \\ \hline D \end{bmatrix} = \begin{bmatrix} s^E & d_t \\ \hline d & \epsilon^T \end{bmatrix} \begin{bmatrix} T \\ \hline E \end{bmatrix} \quad (8)$$

The submatrix s^E is the 6×6 array of elastic-compliance coefficients relating strain coefficients to stress coefficients under the condition that the electric field intensity vector remains constant. A typical coefficient is defined by the equation

$$s_{ij}^E = \frac{\partial S_i}{\partial T_j}\bigg|_{E=\text{const}} \quad (9)$$

The reciprocity theorem also requires s^E to be a symmetrical matrix, and so there are, at most, 21 independent elastic-compliance coefficients.

The submatrix ϵ^T is the 3×3 array of permittivity coefficients for constant stress. A typical coefficient is

$$\epsilon_{ij}^T = \frac{\partial D_i}{\partial E_j}\bigg|_{T=\text{const}} \quad (10)$$

The permittivity matrix is symmetrical to satisfy reciprocity, and so there are, at most, 6 independent permittivity coefficients.

The submatrix d is 3 rows by 6 columns; its transpose is d_t, 3 columns by 6 rows. Both d and d_t appear in the array of coefficients in the equation of state (7), and

CERAMIC DEVICES   16–7

reciprocity does not limit the possible independence of terms. Therefore, there may be, in the most general case, 18 independent piezoelectric coefficients.

The partitioned equation of state (8) is most conveniently written as a pair of linear matrix equations

$$S = s^E T + d_t E$$
$$D = dT + \epsilon^T E \qquad (11)$$

This is not the most convenient pair of equations for some types of calculations. Other forms can be written with different choices of dependent and independent variables, and standard notations for the submatrices of coefficients have been adopted,[15] as summarized in Table 1. Any pair of equations can be derived from any other pair,

Table 1
Piezoelectric Equations of State

$T = c^D S - h_t D$ $T = c^E S - e_t E$
$E = -hS + \beta^S D$ $D = eS + \epsilon^S E$

$S = s^D T + g_t D$ $S = s^E T + d_t E$
$E = -gT + \beta^T D$ $D = dT + \epsilon^T E$

Piezoelectric Matrix Relationships

$c^D = (s^D)^{-1} = c_t{}^D$ $c^E = (s^E)^{-1} = c_t{}^E$
$h = \beta^S e = gc^D$ $e = \epsilon^S h = dc^E$
$\beta^T = (\epsilon^T)^{-1} = \beta_t{}^T$ $\epsilon^S = (\beta^S)^{-1} = \epsilon_t{}^S$
$g = hs^D = \beta^T d$ $d = es^E = \epsilon^T g$
$c^D - c^E = h_t e = e_t h$ $s^E - s^D = g_t d = d_t g$
$\beta^S - \beta^T = gh_t = hg_t$ $\epsilon^T - \epsilon^S = ed_t = de_t$

and so the submatrices are interrelated. Connecting matrix equations are also shown in Table 1. These are easily derived. For example, from Eqs. (11), a solution for the vector matrix E is

$$E = (\epsilon^T)^{-1}(D - dT) \qquad (12)$$

where the -1 superscript indicates an inverse matrix. Equations (11) and (12) then provide the forms

$$S = [s^E - d_t(\epsilon^T)^{-1}d]T + d_t(\epsilon^T)^{-1}D$$
$$E = -(\epsilon^T)^{-1}dT + (\epsilon^T)^{-1}D \qquad (13)$$

With the formal definition that $(\epsilon^T)^{-1} = \beta^T$, Eqs. (13) can be identified with the defined form

$$S = s^D T + g_t D$$
$$E = -gT + \beta^T D \qquad (14)$$

Similar derivations serve to derive all the interrelations given in Table 1.

Usually, in working with poled ceramics, it is possible to simplify the coefficient matrices to a considerable degree, by utilizing the transverse symmetry in an orientation where the z axis is the axis of polarization. This is the normal orientation, unless otherwise specifically stated. With this symmetry, many of the cross-coupling coefficients are zero, and some nonzero coefficients are equal within a matrix. All the elastic matrices s^E, s^D, c^E, and c^D have coefficients following the pattern of

$$s^D = \begin{bmatrix} s_{11}{}^D & s_{12}{}^D & s_{13}{}^D & 0 & 0 & 0 \\ s_{12}{}^D & s_{11}{}^D & s_{13}{}^D & 0 & 0 & 0 \\ s_{13}{}^D & s_{13}{}^D & s_{33}{}^D & 0 & 0 & 0 \\ 0 & 0 & 0 & s_{44}{}^D & 0 & 0 \\ 0 & 0 & 0 & 0 & s_{44}{}^D & 0 \\ 0 & 0 & 0 & 0 & 0 & s_{66}{}^D \end{bmatrix} \qquad (15)$$

with the additional relationships that

$$s_{66} = 2(s_{11} - s_{12})$$
$$c_{66} = \tfrac{1}{2}(c_{11} - c_{12}) \tag{16}$$

In any system of axes rotated with respect to these above, about the z axis, the same coefficients apply because of the transverse isotropy. In any other system of rotated axes involving a modified z axis, new coefficients must apply because the z-axis symmetry does not apply. Neither the form of (15) nor the relations in Eqs. (16) will then hold.[18]

With the normal z axis, parallel to the permanent polarization, the permittivity matrices ϵ^T and ϵ^S and the inverse permittivity matrices β^T and β^S have coefficients in the pattern of

$$\epsilon^T = \begin{bmatrix} \epsilon_{11}{}^T & 0 & 0 \\ 0 & \epsilon_{11}{}^T & 0 \\ 0 & 0 & \epsilon_{33}{}^T \end{bmatrix} \tag{17}$$

The piezoelectric matrix d, as well as the matrices g, h, and e, have the form

$$d = \begin{bmatrix} 0 & 0 & 0 & 0 & d_{15} & 0 \\ 0 & 0 & 0 & d_{15} & 0 & 0 \\ d_{31} & d_{31} & d_{33} & 0 & 0 & 0 \end{bmatrix} \tag{18}$$

These patterns of coefficients, as was the case for the elastic-moduli matrix, hold only for orientation of the z axis along the direction of poling. In the simplified form, as compared with the general equations (7), the full equations for z-axis poling can be written as relatively abbreviated expressions, by following the matrix forms above for substitution into the matrix equations given in Table 1. For example,

$$\begin{aligned}
S_1 &= s_{11}{}^D T_1 + s_{12}{}^D T_2 + s_{13}{}^D T_3 + g_{31} D_3 \\
S_2 &= s_{12}{}^D T_1 + s_{11}{}^D T_2 + s_{13}{}^D T_3 + g_{31} D_3 \\
S_3 &= s_{13}{}^D T_1 + s_{13}{}^D T_2 + s_{33}{}^D T_3 + g_{33} D_3 \\
S_4 &= s_{44}{}^D T_4 + g_{15} D_2 \\
S_5 &= s_{44}{}^D T_5 + g_{15} D_1 \\
S_6 &= 2(s_{11}{}^D - s_{12}{}^D) T_6 \\
E_1 &= -g_{15} T_5 + \beta_{11}{}^T D_1 \\
E_2 &= -g_{15} T_4 + \beta_{11}{}^T D_2 \\
E_3 &= -g_{31} T_1 - g_{31} T_2 - g_{33} T_3 + \beta_{33}{}^T D_3
\end{aligned} \tag{19}$$

4. ADIABATIC AND ISOTHERMAL CONSTANTS: PYROELECTRICITY

In very slowly changing conditions, any heat generated by distorting a piezoelectric material is transferred with little temperature change; the material therefore remains under essentially *isothermal* circumstances.

Under rapidly changing distortion, very little heat is transferred, and the material is essentially *adiabatic*.

An extension of piezoelectric theory[15,18,19] takes into account the effects of entropy change $\Delta\sigma$ and temperature change $\Delta\theta$ in reversible processes within a body at mean temperature θ_0. From thermodynamic considerations, it can be shown that the usual partitioned matrix equations

$$\begin{aligned} T &= c^D S - h_t D \\ E &= -h S + \beta^S D \end{aligned} \tag{20}$$

then become the extended set

$$\begin{aligned}
\Delta\theta &= \frac{\theta_0}{C^{SD}} \Delta\sigma + q_t{}^D S + p_t{}^S D \\
T &= q^D \Delta\sigma + c^{D\sigma} S - h_t{}^\sigma D \\
E &= p^S \Delta\sigma - h^\sigma S + \beta^{S\sigma} D
\end{aligned} \tag{21}$$

CERAMIC DEVICES 16-9

Under high-frequency adiabatic conditions, $\Delta\sigma = 0$. The last two of Eqs. (21) then reduce to the usual form (20), if the superscript σ, denoting constant entropy, is deleted from $c^{D\sigma}$, h^σ and $\beta^{S\sigma}$.

$\Delta\theta$, θ_0, C^{SD}, and $\Delta\sigma$ are scalar quantities. $\Delta\sigma$ and the specific heat C^{SD} are measured in joules per cubic meter per degree K. In general,

$$q_t{}^D = [q_1{}^D \quad q_2{}^D \quad q_3{}^D \quad q_4{}^D \quad q_5{}^D \quad q_6{}^D] \tag{22}$$

is a second-rank tensor quantity, describing reciprocal constant-D coupling between strain and temperature (*thermal expansion*) or between entropy change and stress (*thermal pressure*). The strong degree of symmetry in a poled ceramic, with z the axis of polarization, requires the form

$$q_t{}^D = [q_1{}^D \quad q_2{}^D \quad q_3{}^D \quad 0 \quad 0 \quad 0] \tag{23}$$

The matrix
$$p_t{}^S = [p_1{}^S \quad p_2{}^S \quad p_3{}^S] \tag{24}$$

describes reciprocal clamped coupling between dielectric displacement and temperature change, or between entropy change and electric field intensity; these are aspects of the *pyroelectric effect*. In a poled ceramic,

$$p_t{}^S = [0 \quad 0 \quad p_3{}^S] \tag{25}$$

Pyroelectric effects are observed along only the polar axis. Only piezoelectric materials with polar axes can exhibit pyroelectricity,[20] although there are many nonpolar piezoelectric crystals, such as quartz.

Under quasistatic conditions, $\Delta\theta = 0$. Solution of the first of Eqs. (21) for $\Delta\sigma$ leads to an alternate form, which is convenient for description of isothermal relations:

$$\begin{aligned}\Delta\sigma &= (C^{SD}/\theta_0)(\Delta\theta - q_t{}^D S - p_t{}^S D) \\ T &= (C^{SD}/\theta_0) q^D \Delta\theta + c^{D\theta} S - h_t{}^\theta D \\ E &= (C^{SD}/\theta_0) p^S \Delta\theta - h^\theta S + \beta^{S\theta} D\end{aligned} \tag{26}$$

where
$$\begin{aligned} c^{D\sigma} - c^{D\theta} &= q^D q_t{}^D (C^{SD}/\theta_0) \\ h^\theta - h^\sigma &= p^S q_t{}^D (C^{SD}/\theta_0) \\ \beta^{S\sigma} - \beta^{S\theta} &= p^S p_t{}^S (C^{SD}/\theta_0) \end{aligned} \tag{27}$$

In a poled ceramic, the relations between adiabatic and isothermal coefficients, calculated from Eqs. (23), (25), and (27), are relatively simple:

$$\begin{aligned} c_{11}{}^{D\sigma} - c_{11}{}^{D\theta} &= c_{12}{}^{D\sigma} - c_{12}{}^{D\theta} = c_{21}{}^{D\sigma} - c_{21}{}^{D\theta} = c_{22}{}^{D\sigma} - c_{22}{}^{D\theta} = (C^{SD}/\theta_0)(q_1{}^D)^2 \\ c_{13}{}^{D\sigma} - c_{13}{}^{D\theta} &= c_{31}{}^{D\sigma} - c_{31}{}^{D\theta} = c_{23}{}^{D\sigma} - c_{23}{}^{D\theta} = c_{32}{}^{D\sigma} - c_{32}{}^{D\theta} = (C^{SD}/\theta_0) q_1{}^D q_3{}^D \\ c_{33}{}^{D\sigma} - c_{33}{}^{D\theta} &= (C^{SD}/\theta_0)(q_3{}^D)^2 \end{aligned} \tag{28}$$

$$h_{31}{}^\theta - h_{31}{}^\sigma = p_3{}^S q_1{}^D (C^{SD}/\theta_0)$$
$$h_{33}{}^\theta - h_{33}{}^\sigma = p_3{}^S q_3{}^D (C^{SL}/\theta_0) \tag{29}$$
$$\beta_{33}{}^{S\sigma} - \beta_{33}{}^{S\theta} = (p_3{}^S)^2 (C^{SD}/\theta_0) \tag{30}$$

Only the longitudinal coefficients, with subscripts other than 4, 5, and 6, show any difference in value between adiabatic and isothermal conditions; permittivity differs only in the longitudinal 33-term, along the axis of polarization.

Strictly speaking, truly isothermal conditions are not maintained during quasistatic changes, nor do high-frequency vibrations represent an adiabatic process. Within the ordinary limits of accuracy, however, they are excellent approximations; and since the differences between adiabatic and isothermal values of the elastic, dielectric, and piezoelectric constants are quite small, it is customary to neglect them. Numerical values of the constants are commonly derived from mixed sets of static and dynamic measurements, and the superscripts σ and θ are not ordinarily used.

It should not be assumed, however, that the thermal effects are always negligible. Large electric fields can be generated through pyroelectric coupling, as a result of large changes in temperature. Nye[20] has estimated that the difference between pyroelectric coefficients p^T at constant stress and p^S at constant strain is on the order of 100 per cent in nonferroelectric crystals; in ferroelectric materials, all forms of energy coupling tend to be enhanced, and p^T is relatively large. The difference here is so pronounced that p^S has been associated with the so-called *primary* (or "true") *pyroelectric effect*, which is small, and p^T with the *secondary pyroelectric effect*, which is pronounced.

5. CHARACTERIZATION OF A POLED CERAMIC

If we assume that the ceramic is to be utilized under adiabatic conditions, then rapidly changing mechanical and electrical coordinates are interrelated by the constraints imposed by the properties of mass and the linear proportionalities contained in any of the general 9 × 9 elasto-piezo-dielectric matrices. Because of the transverse isotropy introduced in poling, there are only five nonzero independent elastic constants, three piezoelectric constants, and two dielectric constants. These 10 values, plus the one scalar value of density, completely determine the characteristics of the material under adiabatic vibrations.

Table 2 provides the necessary formulas for conversion from one matrix system to another. While it is convenient, for numerical calculations, to have all 10 of the constants lying within one matrix as a consistent set, it is not generally practicable to

Table 2. Poled Ceramic Coefficients

$s_{11}{}^E - s_{11}{}^D = s_{12}{}^E - s_{12}{}^D = d_{31}g_{31}$
$c_{11}{}^D - c_{11}{}^E = c_{12}{}^D - c_{12}{}^E = e_{31}h_{31}$
$s_{33}{}^E - s_{33}{}^D = d_{33}g_{33} \qquad c_{33}{}^D - c_{33}{}^E = e_{33}h_{33}$
$s_{13}{}^E - s_{13}{}^D = d_{31}g_{33} \qquad c_{13}{}^D - c_{13}{}^E = e_{31}h_{33}$
$s_{44}{}^E - s_{44}{}^D = d_{15}g_{15} \qquad c_{44}{}^D - c_{44}{}^E = e_{15}h_{15}$
$s_{66}{}^E = s_{66}{}^D \qquad c_{66}{}^D = c_{66}{}^E$
$\epsilon_{11}{}^T = 1/\beta_{11}{}^T \qquad \epsilon_{11}{}^S = 1/\beta_{11}{}^S \qquad \epsilon_{33}{}^T = 1/\beta_{33}{}^T \qquad \epsilon_{33}{}^S = 1/\beta_{33}{}^S.$
$\epsilon_{11}{}^T - \epsilon_{11}{}^S = d_{15}e_{15} \qquad \beta_{11}{}^S - \beta_{11}{}^T = h_{15}g_{15}$
$\epsilon_{33}{}^T - \epsilon_{33}{}^S = d_{33}e_{33} + 2d_{31}e_{31}$
$\beta_{33}{}^S - \beta_{33}{}^T = h_{33}g_{33} + 2h_{31}g_{31}$
$d_{31} = \epsilon_{33}{}^T g_{31} = e_{31}(s_{11}{}^E + s_{12}{}^E) + e_{33}s_{13}{}^E$
$d_{33} = \epsilon_{33}{}^T g_{33} = e_{33}s_{33}{}^E + 2e_{31}s_{13}{}^E$
$d_{15} = \epsilon_{11}{}^T g_{15} = e_{15}s_{44}{}^E$
$e_{31} = d_{31}(c_{11}{}^E + c_{12}{}^E) + d_{33}c_{13}{}^E = \epsilon_{33}{}^S h_{31}$
$e_{33} = d_{33}c_{33}{}^E + 2d_{31}c_{13}{}^E = \epsilon_{33}{}^S h_{33}$
$e_{15} = d_{15}c_{44}{}^E = \epsilon_{11}{}^S h_{15}$
$g_{31} = h_{31}(s_{11}{}^D + s_{12}{}^D) + h_{33}s_{13}{}^D = \beta_{33}{}^T d_{31}$
$g_{33} = h_{33}s_{33}{}^D + 2h_{31}s_{13}{}^D = \beta_{33}{}^T d_{33}$
$g_{15} = h_{15}s_{44}{}^D = \beta_{11}{}^T d_{15}$
$h_{31} = g_{31}(c_{11}{}^D + c_{12}{}^D) + g_{33}c_{13}{}^D = \beta_{33}{}^S e_{31}$
$h_{33} = g_{33}c_{33}{}^D + 2g_{31}c_{13}{}^D = \beta_{33}{}^S e_{33}$
$h_{15} = g_{15}c_{44}{}^D = \beta_{11}{}^S e_{15}$

For $F = D$ and for $F = E$:
$c_{44}{}^F = 1/s_{44}{}^F$
$c_{66}{}^F = 1/s_{66}{}^F$
$\dfrac{1}{c_{11}{}^F} = (s_{11}{}^F - s_{12}{}^F)\left[1 + \dfrac{(s_{13}{}^F)^2 - s_{12}{}^F s_{33}{}^F}{(s_{13}{}^F)^2 - s_{11}{}^F s_{33}{}^F}\right]$
$\dfrac{1}{s_{11}{}^F} = (c_{11}{}^F - c_{12}{}^F)\left[1 + \dfrac{(c_{13}{}^F)^2 - c_{12}{}^F c_{33}{}^F}{(c_{13}{}^F)^2 - c_{11}{}^F c_{33}{}^F}\right]$
$\dfrac{1}{c_{12}{}^F} = -(s_{11}{}^F - s_{12}{}^F)\left[1 + \dfrac{(s_{13}{}^F)^2 - s_{11}{}^F s_{33}{}^F}{(s_{13}{}^F)^2 - s_{12}{}^F s_{33}{}^F}\right]$
$\dfrac{1}{s_{12}{}^F} = -(c_{11}{}^F - c_{12}{}^F)\left[1 + \dfrac{(c_{13}{}^F)^2 - c_{11}{}^F c_{33}{}^F}{(c_{13}{}^F)^2 - c_{12}{}^F c_{33}{}^F}\right]$
$\dfrac{1}{c_{13}{}^F} = -\dfrac{s_{33}{}^F}{s_{13}{}^F}(s_{11}{}^F + s_{12}{}^F) + 2s_{13}{}^F$
$\dfrac{1}{s_{13}{}^F} = -\dfrac{c_{33}{}^F}{c_{13}{}^F}(c_{11}{}^F + c_{12}{}^F) + 2c_{13}{}^F$
$\dfrac{1}{c_{33}{}^F} = s_{33}{}^F - \dfrac{2(s_{13}{}^F)^2}{s_{11}{}^F + s_{12}{}^F} \qquad \dfrac{1}{s_{33}{}^F} = c_{33}{}^F - \dfrac{2(c_{13}{}^F)^2}{c_{11}{}^F + c_{12}{}^F}$

CERAMIC DEVICES 16-11

make the experimental measurements for all coefficients in a set. It is sufficient to measure 10 independent constants selected from *any* of the matrices—the test for independence being simply that none of the measured values can be derived from the other measurements by use of the conversion formulas, or equivalently that all matrix constants can be computed from the measured values. A consistent set, for example, would be

$$s_{11}^E \quad s_{12}^E \quad s_{13}^E \quad s_{33}^E \quad s_{55}^E \quad d_{31} \quad d_{33} \quad d_{15} \quad \epsilon_{11}^T \quad \epsilon_{33}^T$$

A mixed set, much more readily found by direct measurements, is the independent collection

$$s_{11}^E \quad s_{12}^E \quad s_{33}^D \quad s_{55}^D \quad s_{55}^E \quad c_{33}^D \quad d_{31} \quad d_{33} \quad \epsilon_{11}^T \quad \epsilon_{33}^T$$

Under ordinary circumstances, the most accurate and convenient methods of measurement are electrical. Permittivity is measured in terms of electrical capacitance at appropriate frequencies and with electrodes placed for generation of the desired field directions. Elastic constants and piezoelectric constants are computed from precisely measured values of frequency at which the samples display mechanical resonance under electric excitation. Voltage and current meters, impedance and admittance bridges, and digital frequency counters are excellent measuring equipment. The combinations of independent constants which can be measured by resonance techniques determine the data normally available. Recommended measurement procedures have been summarized in several IRE/IEEE Standards on Piezoelectric Crystals.[11,12,21,22,23]

In general, the resonance experiments require sample shapes and electrode configurations which exhibit vibrations controlled by single coefficients rather than by many. These are also the forms most easily adapted to practical applications. The common vibrations can be grouped into three classes: longitudinal, shear, and planar. Within each class, there are a number of variations, determined by sample shape and electrode configuration. By class, the elastic and piezoelectric constants *primarily* involved are as follows:

Longitudinal: $\quad s_{11}^E \quad s_{33}^D \quad c_{33}^D$
$\quad\quad\quad\quad\quad\quad\quad d_{31} \quad d_{33} \quad h_{33}$
Shear: $\quad\quad\quad s_{55}^E \quad c_{55}^D$
$\quad\quad\quad\quad\quad\quad\quad d_{15} \quad h_{15}$
Planar: $\quad\quad\quad s_{11}^E \quad s_{12}^E$
$\quad\quad\quad\quad\quad\quad\quad d_{31}$

A complete set of measurements, which provides 10 independent constants, can be carried out with four samples; one is fabricated in shape, polarization, and electroding to exhibit planar resonance; one exhibits shear resonance; and two different longitudinal forms are required. Since elements all falling within one class require only a portion of the 10 values, it is possible to design a wide variety of useful devices without knowledge of the complete set.

Some coefficients appear so frequently in work with these classes that convenient alternate terminology is commonly used. The reciprocals of s_{11}^E and s_{33}^D appear in expressions for the frequency of mechanical resonance in thin ceramic bars and are analogous to *Young's modulus* for isotropic materials. Since the coefficients have appreciably different numerical values, it is convenient to use two values of Young's modulus:

$$Y_1^E = 1/s_{11}^E \tag{31}$$
$$Y_3^D = 1/s_{33}^D \tag{32}$$

A ratio which appears regularly is s_{12}^E/s_{11}^E; if a thin rod is stretched along its axis, lying in the x direction, then the ratio of cross contraction to axial elongation is s_{12}^E/s_{11}^E. Since s_{12} is always a negative number and s_{11} is always positive, it is convenient to follow the definition for *Poisson's ratio*, which is used for isotropic materials

$$\sigma^E = -s_{12}^E/s_{11}^E \tag{33}$$

Poisson's ratio is always a positive number and for most poled ceramics has a value of about 0.3.

6. ELECTROMECHANICAL COUPLING FACTORS

Perhaps the most convenient and descriptive means for characterizing a piezoelectric ceramic is the *electromechanical coupling factor* (or coupling coefficient). It bears exact correspondence to the coupling coefficient in electric-circuit theory.

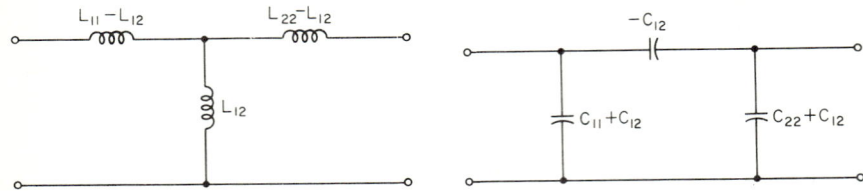

FIG. 4. Reactive coupling networks.

In a two-port electrical network containing only elements of the same type (i.e., L, or C, or R), with four-pole equations

$$E_1 = I_1 z_{11} + I_2 z_{12}$$
$$E_2 = I_1 z_{12} + I_2 z_{22} \tag{34}$$

and

$$I_1 = E_1 y_{11} + E_2 y_{12}$$
$$I_2 = E_1 y_{12} + E_2 y_{22} \tag{35}$$

the coupling coefficient is defined as

$$k = z_{12}/\sqrt{z_{11} z_{22}} = y_{12}/\sqrt{y_{11} y_{22}} \tag{36}$$

If the network is a magnetic transformer, represented in Fig. 4, the energy stored is

$$U = \tfrac{1}{2}(L_{11} - L_{12})I_1^2 + \tfrac{1}{2}L_{12}(I_1 + I_2)^2 + \tfrac{1}{2}(L_{22} - L_{12})I_2^2$$
$$= \tfrac{1}{2}L_{11}I_1^2 + L_{12}I_1 I_2 + \tfrac{1}{2}L_{22}I_2^2$$
$$= U_1 + 2U_{12} + U_2 \tag{37}$$

U_1 is energy-dependent only on I_1, U_2 is dependent only on I_2, and U_{12} is mutually dependent on I_1 and I_2. The coupling coefficient is

$$k = L_{12}/\sqrt{L_{11} L_{22}} = \sqrt{U_{12}^2/U_1 U_2} \tag{38}$$

In a two-port capacitive network,

$$U = \tfrac{1}{2}(C_{11} + C_{12})E^2 - \tfrac{1}{2}C_{12}(E_1 - E_2)^2 + \tfrac{1}{2}(C_{22} + C_{12})E_2^2$$
$$= \tfrac{1}{2}C_{11}E_1^2 + C_{12}E_1 E_2 + \tfrac{1}{2}C_{22}E_2^2$$
$$= U_1 + 2U_{12} + U_2 \tag{39a}$$

The coupling coefficient is

$$k = y_{12}/\sqrt{y_{11} y_{22}} = C_{12}/\sqrt{C_{11} C_{22}} = \sqrt{U_{12}^2/U_1 U_2} \tag{39b}$$

In a piezoelectric material, coupling coefficient is defined in terms of energy, in precisely the same way. The energy term U_1 involves only elastic parameters, U_2 involves only electric fields and permittivities, and U_{12} is a mutual function of stresses, fields, and piezoelectric constants. The electromechanical coupling factor is defined as

$$k = \sqrt{U_{12}^2/U_1 U_2} \tag{40}$$

CERAMIC DEVICES 16-13

Coupling factor is a function of the relations between the elastic and dielectric variables, and depends on the boundary conditions. The situation is comparable with that of an n-port reactive network, which can be a two-port coupling network when the remaining $n - 2$ ports are terminated; combinations of terminations correspond to combinations of boundary conditions. In a piezoelectric material, a port involves one stress T_i and the corresponding strain S_i, or one field intensity E_i and the corresponding dielectric displacement D_i. There are thus a total of nine ports—six mechanical and three electrical—or 10 if temperature and entropy change are considered. Normally, for adiabatic conditions, the thermal port is "open-circuited," corresponding to zero entropy change.

At each port, there is an *extensive* variable (S, D, or σ) and an *intensive* variable (T, E, or θ). An increment of energy to the system can be supplied at every port: if all variables change,

$$\delta U = \sum_{i=1}^{6} S_i \, \delta T_i + \sum_{j=1}^{3} E_j \, \delta D_j + \theta \, \delta\sigma \tag{41}$$

The total internal energy of the system is the integral of δU.

There are an unlimited number of coupling factors, if the terminal constraints are unlimited. If only short-circuit (constant intensive variable) and open-circuit (constant extensive variable) boundary conditions are considered, there are 36×2^7 possible adiabatic coupling factors and an equal number of isothermal coupling factors. Many are zero. For example, in a poled ceramic, there is no elastic or dielectric coupling between longitudinal and shear strains.

Coupling factors are usually applied to simple combinations of boundary conditions. The IRE Standards on Piezoelectric Crystals—1961[11] specifically name five, in connection with poled ceramics:

k_{31}: transverse coupling factor
k_{33}: longitudinal coupling factor
k_t: thickness coupling factor
k_p: planar coupling factor
k_{15}: shear coupling factor

These are defined in Table 3, with appropriate boundary conditions and order-of-magnitude indications of frequency ranges over which they have found use. Representative values for ceramic PZT-4 given here are from values given by Mason,* where eight additional coupling factors k_{31}', k_{31}'', k_{31}''', k_{33}', k_p', k_h, k_{i3}, and $k_{i1} (= k_{15})$ are defined.

Derivation of coupling-factor formulas from Eqs. (40) and (41) is straightforward for single-stress or single-strain elements. For a k_{31} ceramic element, the corresponding nonzero elastic variables are T_1 and S_1, and nonzero electrical variables are E_3 and D_3. The algebraic equations, following matrix forms in Table 1, are

$$\begin{aligned} S_1 &= s_{11}^E T_1 + d_{31} E_3 \\ D_3 &= d_{31} T_1 + \epsilon_{33}^T E \end{aligned} \tag{42}$$

Internal energy is

$$\begin{aligned} U &= \int (T_1 \, \delta S_1 + E_3 \, \delta D_3) \\ &= \int [T_1(s_{11}^E \, \delta T_1 + d_{31} \, \delta E_3) + E_3(d_{31} \, \delta T_1 + \epsilon_{33}^T \, \delta E_3)] \\ &= s_{11}^E \int T_1 \, \delta T_1 + d_{31} \int \delta(E_3 T_1) + \epsilon_{33}^T \int E_3 \, \delta E_3 \\ &= \tfrac{1}{2} s_{11}^E T_1^2 + d_{31} E_3 T_1 + \tfrac{1}{2} \epsilon_{33}^T E_3^2 \\ &= U_1 + 2U_{12} + U_2 \end{aligned} \tag{43}$$

The coupling factor is given by the equation

$$k_{31}^2 = \frac{(\tfrac{1}{2} d_{31} E_3 T_1)^2}{(\tfrac{1}{2} s_{11}^E T_1^2)(\tfrac{1}{2} \epsilon_{33}^T E_3^2)} = \frac{d_{31}^2}{s_{11}^E \epsilon_{33}^T} \tag{44}$$

Table 3. Coupling Factors

Element	Application and frequency range, cps	Equations and boundary conditions	Coupling factor: PZT-4 value†
TRANSVERSE LONGITUDINAL BAR	Low-frequency transducer: $1-10^6$ Portion of bender element: $1-10^4$ Low-frequency resonator: 10^4-10^6	$\dfrac{\partial E_3}{\partial x} = 0$ $T_2 = T_3 = 0$ $\begin{cases} S_1 = s_{11}{}^E T_1 + d_{31} E_3 \\ D_3 = d_{31} T_1 + \epsilon_{33}{}^T E_3 \end{cases}$	$k_{31} = \dfrac{d_{31}}{\sqrt{s_{11}{}^E \epsilon_{33}{}^T}}$ 0.334 Transverse
LONGITUDINAL BAR	Transducer: 10^3-10^7 Resonator: 10^4-10^6	$\dfrac{\partial D_3}{\partial z} = 0$ $T_1 = T_2 = 0$ $\begin{cases} S_3 = s_{33}{}^E T_3 + d_{33} E_3 \\ D_3 = d_{33} T_3 + \epsilon_{33}{}^T E_3 \end{cases}$	$k_{33} = \dfrac{d_{33}}{\sqrt{s_{33}{}^E \epsilon_{33}{}^T}}$ 0.70 Longitudinal
LONGITUDINAL PLATE	Transducer: $1-10^8$ Resonator: 10^6-10^7	$\dfrac{\partial D_3}{\partial z} = 0$ $S_1 = S_2 = 0$ $\begin{cases} T_3 = c_{33}{}^D S_3 - h_{33} D_3 \\ E_3 = -h_{33} S_3 + \beta_{33}{}^S D_3 \end{cases}$	$k_t = \dfrac{h_{33}}{\sqrt{c_{33}{}^D \beta_{33}{}^S}}$ 0.513 Thickness longitudinal
TORSION BAR	Transducer: 10^5-10^7 Resonator: 10^4-10^6	$\dfrac{\partial D_1}{\partial x} = 0;\ E_2 = E_3 = 0$ $\begin{cases} S_5 = s_{44}{}^E T_5 + d_{15} E_1 \\ D_1 = d_{15} T_5 + \epsilon_{11}{}^T E_1 \end{cases}$	$k_{15} = \dfrac{d_{15}}{\sqrt{s_{44}{}^E \epsilon_{11}{}^T}}$ 0.71 Shear

Table 3. Coupling Factors (Continued)

Element	Application and frequency range, cps	Equations and boundary conditions	Coupling factor: PZT-4 value†
SHEAR PLATE	Transducer: 1–10^8 Resonator: 10^6–10^7	$\dfrac{\partial D_1}{\partial x} = \dfrac{\partial}{\partial y} = E_3 = 0$ $\begin{cases} T_5 = c_{44}{}^D S_5 - h_{15} D_1 \\ E_1 = -h_{15} S_5 + \beta_{11}{}^S D_1 \end{cases}$	$k_{15} = \dfrac{h_{15}}{\sqrt{c_{44}{}^D \beta_{11}{}^S}}$ Shear 0.71
PLANAR PLATE, DISK (TWO-DIMENSIONAL)	Resonator: 10^4–10^6	$\dfrac{\partial E_3}{\partial x} = \dfrac{\partial E_3}{\partial y} = 0$ $S_1 = s_{11}{}^E T_1 + s_{12}{}^E T_2 + d_{31} E_3$ $S_2 = s_{12}{}^E T_1 + s_{11}{}^E T_2 + d_{31} E_3$ $\left(D_3 = d_{31}(T_1 + T_2) + \epsilon_{33}{}^T E_3 \atop \dfrac{S_1 + S_2}{2} = \dfrac{s_{11}{}^E + s_{12}{}^E}{2}(T_1 + T_2) + S_3 E_3 \right)$	$k_p = d_{31} \sqrt{\dfrac{2}{(s_{11}{}^E + s_{12}{}^E)\epsilon_{33}{}^T}}$ $= k_{31} \sqrt{\dfrac{2}{1 - \sigma}}$ Planar 0.58
THIN RING	Transducer: 10^4–10^6 Resonator: 10^4–10^6	$\dfrac{\partial E_3}{\partial x} = \dfrac{\partial E_3}{\partial y} = 0$ $S_1 = s_{11}{}^E T_1 + d_{31} E_3$ $D_3 = d_{31} T_1 + \epsilon_{33}{}^T E_3$	$k_{31} = \dfrac{d_{31}}{\sqrt{s_{11}{}^E \epsilon_{33}{}^T}}$ Transverse 0.334

† See also Table 7.

For multiple stress elements, the derivation is slightly more complicated. For example, there are two nonzero stresses and corresponding strains involved with the planar coupling coefficient:

$$\begin{aligned} S_1 &= s_{11}{}^E T_1 + s_{12}{}^E T_2 + d_{31} E_3 \\ S_2 &= s_{12}{}^E T_1 + s_{11}{}^E T_2 + d_3 E_3 \\ D_3 &= d_{31}(T_1 + T_2) + \epsilon_{33}{}^T E \end{aligned} \tag{45}$$

It is convenient to combine the two strain equations, so that only a symmetrical pair remains. Let $S_p = S_1 + S_2$ represent the *planar strain*, and $T_p = T_1 = T_2$ represent the *planar stress*:

$$\begin{aligned} S_p &= 2(s_{11}{}^E + s_{12}{}^E) T_p + 2 d_{31} E_3 \\ D_3 &= 2 d_{31} T_p + \epsilon_{33}{}^T E_3 \end{aligned} \tag{46}$$

From Eqs. (46), the direct calculation of internal energy, following the steps taken in Eqs. (43), leads to the formula for the coupling factor

$$k_p{}^2 = \frac{2 d_{31}{}^2}{(s_{11}{}^E + s_{12}{}^E)\epsilon_{33}{}^T} = \frac{2 k_{31}{}^2}{1 - \sigma^E} \tag{47}$$

It may be noted that energy calculation is not necessary to obtain the formula for a coupling factor. Once the equations have been reduced to a symmetrical pair, k^2 is given by the ratio of the product of the mutual constants to the product of elastic constant by dielectric constant.

7. ENERGY STORAGE AND PIEZOELECTRIC CONVERSION

Under appropriately chosen conditions, the piezoelectric equations can be reduced to a symmetrical pair, with the form

$$\begin{aligned} S &= s^E T + dE \\ D &= dT + \epsilon^T E \end{aligned} \tag{48}$$

The effective short-circuited elastic constant is

$$s^E = \frac{S}{T}\bigg|_{E=0} \tag{49}$$

The open-circuited elastic constant, calculated from Eqs. (48), is

$$\begin{aligned} \frac{S}{T}\bigg|_{D=0} &= \frac{s^E T + d(D - dT)/\epsilon^T}{T}\bigg|_{D=0} \\ &= s^E(1 - k^2) \end{aligned} \tag{50}$$

It can be seen that short-circuiting the electrodes reduces the elastic stiffness of a piezoelectric element.

Energy per unit volume delivered to an elastic body through application of stress T is $\frac{1}{2} S T^2$. If the body is piezoelectric, then the energy supplied by mechanical input is $\frac{1}{2} s^E T^2$ if the electrodes are short-circuited and is $\frac{1}{2} s^E (1 - k^2) T^2$ when the electrodes are open-circuited. The additional energy supplied to the short-circuited body can be regarded as a piezoelectrically stored portion.

Figure 5 indicates a method for transferring energy from a mechanical source to an electrical load. The elastic variables S and T are shown in one plane and the dielectric variables D and E in another plane. Along the path OA, energy $\frac{1}{2} s^E T_A{}^2$ is supplied mechanically; E is held at zero by short-circuited electrodes. At A,

$$\begin{aligned} E &= 0 \\ T &= T_A \\ S &= s^E T_A \\ D &= d T_A \end{aligned} \tag{51}$$

From A to B, the applied stress is removed with the electrodes open-circuited. Energy returns to the mechanical source of stress, with strain S following the path corresponding to the open-circuited elastic constant $s^E(1 - k^2)$. The energy returned is $\frac{1}{2}s^E(1 - k^2)T_A^2$; the energy still stored is $\frac{1}{2}s^E k^2 T_A^2$. At B,

$$\begin{aligned} T &= 0 \\ D &= dT_A \\ E &= (d/\epsilon^T)T_A \\ S &= (d^2/\epsilon^T)T_A \end{aligned} \qquad (52)$$

From B to O, the body is left unconstrained to maintain zero stress, while the electrodes are discharged into an electrical load. The effective permittivity is ϵ^T and so the electrical energy delivered is

$$\begin{aligned} -\int E \, \delta D &= -\int \epsilon^T E \, \delta E \\ &= \tfrac{1}{2}(d^2/\epsilon^T)T_A^2 \\ &= \tfrac{1}{2}k^2 s^E T_A^2 \end{aligned} \qquad (53)$$

The cycle is completed, with S, T, D, and E returned to zero. Over one cycle of the closed path $OABO$, a fraction k^2 of input energy has been transformed to electrical

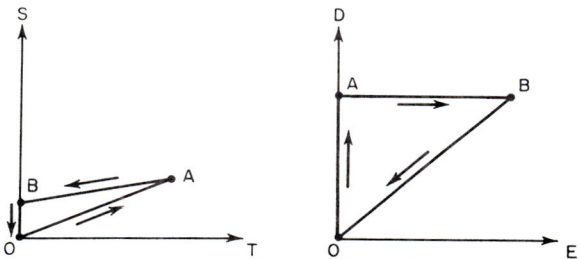

OA : APPLY STRESS T. ELECTRODES SHORTED, E = 0.
AB : REMOVE STRESS. ELECTRODES OPEN, D = CONSTANT.
BO : WITH ZERO STRESS, DISCHARGE ELECTRODES.

FIG. 5. A step-wise cycle which transfers the fraction k^2 of mechanically supplied energy to an electrical load.

form. A fraction $(1 - k^2)$ of the input energy has been stored and then returned to the mechanical source.

By a reverse process, electrical energy supplied to the body along a locus of charging under zero stress, then discharging under constant strain, and finally relieving the accumulated stress into a mechanical load, can be divided into a transformed fraction k^2 and a circulated fraction $(1 - k^2)$.

Both these processes take place along idealized loci, which are only approximated in practical piezoelectric devices but which represent the nature of piezoelectric power transformation. In all cases, the circulated fraction of energy represents potential insertion loss. It is the relatively large value of coupling factor k in a ferroelectric ceramic which makes the material especially useful for energy transducers.

8. WAVES AND RESONANCE

Ferroelectric ceramics have been used in nonresonant devices such as phonograph pickups, accelerometers, and spark sources for internal-combustion engines. Many other applications, which include ceramic filters, transformers, and ultrasonic delay-line transducers, depend on mechanical resonance for their operation. Even the non-resonant devices are limited in high-frequency response by resonances.

Mechanical resonance is produced by standing waves of vibrations, which may be highly complicated except in elements of a few simple shapes. Most practical devices are built in these shapes to limit the effects of resonances to a controllable degree of complexity. Thin plates, rods, disks, and cylinders have been favored, in shapes such that just one dimension of a device controls the standing-wave pattern.

In an unbounded isotropic* solid, bulk stress waves propagate at speeds determined by density and elasticity of the medium. There are two distinct velocities,[24] a *dilatational* velocity

$$c_L = \sqrt{c_{11}/\rho} \tag{54}$$

for waves with particle motion parallel to the direction of propagation, and a *distortional* wave velocity

$$c_S = \sqrt{c_{44}/\rho} \tag{55}$$

for waves with particle motion perpendicular to the direction of propagation.

Stress waves can propagate along parallel, converging, or diverging paths. A plane dilatational wave involves pure longitudinal strains; a surface of equal phase forms a plane wavefront perpendicular to the direction of propagation. A plane distortional wave involves pure shear strains, with wavefronts which are perpendicular to the direction of propagation; the transverse particle motion is parallel to the wavefront planes and can involve one or more planes of polarization.

Gases and most liquids have no measurable shear stiffness and so will propagate only dilatational pressure waves. Electromagnetic wave propagation involves only transverse vectors and is analogous to distortional-stress wave propagation. Simultaneous propagation of dilatational and distortional waves is a unique property of solids and a few highly viscous liquids, because of the existence of both longitudinal and shear elastic moduli.

In common solids, c_L is on the order of 5,000 m/sec, and c_S is always smaller, usually by a factor of about 2. Table 4 summarizes measured values for a number of isotropic materials.

Crystals, cold-worked metals, magnetized ferromagnetic materials, and poled ferroelectric ceramics are appreciably anisotropic. Stress-wave propagation is complicated by the variation, with orientation, of elastic constants and wave velocities. Spherical and cylindrical waves are not propagated except in certain useful cases, and plane waves generally propagate in three distinct modes, none of which is purely longitudinal or shear. Waves of the three modes sharing a common plane for wavefront involve mutually perpendicular particle motions, but the directions of propagation are not, in general, perpendicular to the wavefront. However, in an anisotropic medium there can be particular directions for which two of the wave velocities are equal and motion is perpendicular to the direction of propagation, indicating a pure shear wave. There can also be directions for which motion for one mode is parallel to the direction of propagation, indicating a pure longitudinal wave.

In quartz crystals, for example, pure longitudinal waves will propagate along the Z axis, although the other two transverse modes diverge conically rather than following the axis. The X-axis direction provides propagation for a pure longitudinal mode, and two transverse modes, a slow shear and a fast shear. The Y, AC, and BC axes are all pure shear mode directions. Plates cut with major surfaces perpendicular to all these but the Z axis, which is not piezoelectrically coupled, are useful quartz resonators.

The transverse isotropy of poled ceramics is relatively simple. Along the Z axis of poling, a longitudinal plane wave with velocity $\sqrt{c_{33}/\rho}$ and a shear wave with

---

* An isotropic solid has only three different nonzero elastic constants: $\mu = c_{44} = c_{55} = c_{66}$, $\lambda = c_{12} = c_{13} = c_{23}$, and $\lambda + 2\mu = c_{11} = c_{22} = c_{33}$. λ and μ are known as Lamé's constants.

CERAMIC DEVICES

Table 4. Longitudinal and Shear Stress Wave Velocities in Isotropic Solids*

Material	c_L, m/sec	c_S, m/sec
Aluminum, rolled	6,420	3,040
Beryllium	12,890	8,880
Brass, yellow, 70 Cu, 30 Zn	4,700	2,110
Copper, rolled	5,010	2,270
Gold, hard-drawn	3,240	1,200
Lead, rolled	1,960	690
Magnesium, drawn, annealed	5,770	3,050
Nickel	6,040	3,000
Platinum	3,260	1,730
Silver	3,650	1,610
Steel, K9	5,941	3,251
347 stainless steel	5,790	3,100
Tin, rolled	3,320	1,670
Titanium	6,070	3,125
Tungsten, drawn	5,410	2,640
Zinc, rolled	4,210	2,440
Fused silica	5,968	3,764
Lucite	2,680	1,100
Nylon 6-6	2,620	1,070
Polyethylene	1,950	540
Polystyrene	2,350	1,120

* From "American Institute of Physics Handbook," 2d ed., McGraw-Hill Book Company, New York, 1963.

velocity $\sqrt{c_{44}/\rho}$ can propagate. Along the X axis (i.e., in any direction normal to Z) a longitudinal wave can propagate with velocity $\sqrt{c_{11}/\rho}$, as well as two shear waves, one with velocity $\sqrt{c_{44}/\rho}$ and particle motion parallel to Z and one with velocity $\sqrt{c_{66}/\rho}$ and particle motion perpendicular to Z.

A quantitative description of wave propagation and mechanical resonance stems from equations of motion, written from Newton's second law of motion. An element of volume $\delta x\, \delta y\, \delta z$ subjected to a force δF is accelerated along a direction parallel to δF with particle motion u:

$$\delta F = \rho(\partial^2 u/\partial t^2)\, \delta x\, \delta y\, \delta z \tag{56}$$

The force is applied by the stresses acting on the faces of the volume element. Figure 6 shows the longitudinal and shear stresses acting in the x direction. Similar sets act in the y and z directions. If u, v, w are the particle displacements in the x, y, z directions, respectively, then the equations of motion are

$$\begin{aligned}
\rho(\partial^2 u/\partial t^2) &= \partial T_1/\partial x + \partial T_6/\partial y + \partial T_5/\partial z \\
\rho(\partial^2 v/\partial t^2) &= \partial T_6/\partial x + \partial T_2/\partial y + \partial T_4/\partial z \\
\rho(\partial^2 w/\partial t^2) &= \partial T_5/\partial x + \partial T_4/\partial y + \partial T_3/\partial z
\end{aligned} \tag{57}$$

Wave equations are derived from Eqs. (57) for particular boundary conditions, through substituting for the stress components appropriate strains obtained from the piezoelectric equations. A number of useful cases for poled ceramics follow.

8.1. Plane Longitudinal Wave: Propagation Perpendicular to the Poling Axis. Motion is perpendicular to the z axis; for convenience, it is assumed along the x axis. For a plane wave, all variables must be uniform across a wavefront. It follows that

$$S_2 = S_3 = S_4 = S_5 = S_6 = 0 \tag{58}$$
$$\partial T_i/\partial y = \partial T_i/\partial z = 0 \quad (i = 1, 2, \ldots, 6) \tag{59}$$
$$\partial D_j/\partial y = \partial D_j/\partial z = \partial E_j/\partial y = \partial E_j/\partial z = 0 \quad (j = 1, 2, 3) \tag{60}$$

Since only longitudinal strain exists, there are no elastically coupled shear stresses. However, T_4 and T_5 can be piezoelectrically induced by field components along x and y.

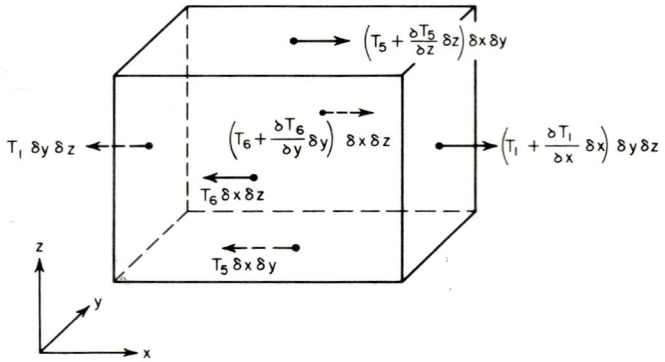

Fig. 6. Stresses accelerating a small volume element in the x direction.

These do not enter into the equations of motion (57) for particle motion u. Therefore, the equations of motion reduce to

$$\partial^2 u/\partial t^2 = \partial T_1/\partial x \tag{61}$$

From the matrix equations in Table 1, the most appropriate set for single-strain conditions is

$$\begin{aligned} T &= c^D S - hD \\ E &= -hS + \beta^S D \end{aligned} \tag{62}$$

Symmetry conditions for poled ceramics, with Eqs. (58), (59), and (60), reduce the matrix equations to the set

$$\begin{aligned}
T_1 &= c_{11}{}^D S_1 - h_{31} D_3 \\
T_2 &= c_{12}{}^D S_1 - h_{31} D_3 \\
T_3 &= c_{13}{}^D S_1 - h_{33} D_3 \\
T_4 &= -h_{15} D_2 \\
T_5 &= -h_{15} D_1 \\
E_1 &= \beta_{11}{}^S D_1 \\
E_2 &= \beta_{11}{}^S D_2 \\
E_3 &= -h_{31} S_1 - \beta_{33}{}^S D_3
\end{aligned} \tag{63}$$

The first of Eqs. (63) provides the relation

$$\partial T_1/\partial x = c_{11}{}^D (\partial^2 u/\partial x^2) - h_{31}(\partial D_3/\partial x) \tag{64}$$

In an insulating medium, without free charge,*

$$\partial D_1/\partial x + \partial D_2/\partial y + \partial D_3/\partial z = \operatorname{div} D = 0 \tag{65}$$

and so, with Eq. (60), it follows that

$$\partial D_3/\partial x = 0 \tag{66}$$

The equation of motion (61) therefore becomes

$$\rho(\partial^2 u/\partial t^2) = c_{11}{}^D(\partial^2 u/\partial x^2) \tag{67}$$

* In a piezoelectric semiconductor, div $D \neq 0$. Piezoelectric interaction between stress and moving charge can produce acoustic amplification, which is discussed in Chap. 30.

This has the form of the wave equation for one dimension,

$$\partial^2 u/\partial t^2 = c^2(\partial^2 u/\partial x^2)$$

where
$$c = \sqrt{c_{11}{}^D/\rho} \tag{68}$$

is the velocity of propagation.

The wave equation is satisfied by a displacement

$$u = f_1(x - ct) + f_2(x + ct) \tag{69}$$

in which f_1 and f_2 are arbitrary (differentiable) functions; f_1 represents a wave propagating in the $+x$ direction and f_2 a wave in the $-x$ direction. If motion varies sinusoidally with time, then Eq. (69) can also be written in the form

$$u = A_1 e^{j\omega(t-x/c)} + A_2 e^{j\omega(t+x/c)} \tag{70}$$

where A_1 and A_2 are complex numbers denoting amplitude and phase of the two traveling waves.

The strain is calculated from Eq. (69) or (70):

$$\begin{aligned} S_1 &= \partial u/\partial x \\ &= f_1' + f_2' \\ &= -(j\omega/c)A_1 e^{j\omega(t-x/c)} + (j\omega/c)A_2 e^{j\omega(t+x/c)} \end{aligned} \tag{71}$$

Stresses and field intensity, from Eqs. (63), are

$$\begin{aligned} T_1 &= c_{11}{}^D f_1' + c_{11}{}^D f_2' - h_{31} D_3 \\ T_2 &= c_{12}{}^D f_1' + c_{12}{}^D f_2' - h_{31} D_3 \\ T_3 &= c_{13}{}^D f_1' + c_{13}{}^D f_2' - h_{33} D_3 \\ E_3 &= -h_{31} f_1' - h_{31} f_2' + \beta_{33}{}^S D_3 \end{aligned} \tag{72}$$

Strain consists of just the two traveling-wave components. Stress and field intensity consist of not only traveling waves but also a stationary component proportional to D_3. D_3 is not a traveling wave, even though it is piezoelectrically coupled to the wave and may have any value, including zero, as determined by charge distributions outside the dielectric medium.

8.2. Plane Longitudinal Wave: Propagation Parallel to the Poling Axis. Particle motion is along the z axis. Only a single strain is present:

$$S_1 = S_2 = S_4 = S_5 = S_6 = 0 \tag{73}$$

Uniformity across a wavefront requires that

$$\begin{aligned} \partial T_i/\partial x &= \partial T_i/\partial y = 0 & (i = 1, 2, \ldots, 6) \\ \partial D_j/\partial x &= \partial D_j/\partial y = \partial E_j/\partial x = \partial E_j/\partial y = 0 & (j = 1, 2, 3) \end{aligned} \tag{74}$$

The equations of motion (57) reduce to

$$\rho(\partial^2 w/\partial t^2) = \partial T_3/\partial z \tag{75}$$

From the piezoelectric relations,

$$\begin{aligned} T_1 = T_2 &= c_{13}{}^D S_3 - h_{31} D_3 \\ T_3 &= c_{33}{}^D S_3 - h_{33} D_3 \\ T_4 &= -h_{15} D_2 \\ T_5 &= -h_{15} D_1 \\ T_6 &= 0 \\ E_1 &= \beta_{11}{}^S D_1 \\ E_2 &= \beta_{11}{}^S D_2 \\ E_3 &= -h_{33} S_3 + \beta_{33}{}^S D_3 \end{aligned} \tag{76}$$

In an insulating ceramic, div $D = 0$:

$$\partial D_3/\partial z = 0 \tag{77}$$
$$\rho(\partial^2 w/\partial t^2) = c_{33}{}^D(\partial^2 w/\partial z^2) \tag{78}$$

This is the wave equation, for velocity

$$c = \sqrt{c_{33}{}^D/\rho} \tag{79}$$

and motion of functional form

$$w = f_1(z - ct) + f_2(z + ct) \tag{80}$$

Strain, stresses, and field intensity are

$$\begin{aligned} S_3 &= \partial w/\partial z = f_1' + f_2' \\ T_1 = T_2 &= c_{13}{}^D(f_1' + f_2') - h_{31}D_3 \\ T_3 &= c_{33}{}^D(f_1' + f_2') - h_{33}D_3 \\ E_3 &= -h_{33}(f_1' + f_2') + \beta_{33}{}^S D_3 \end{aligned} \tag{81}$$

T_4, T_5, T_6, E_1, and E_2 are not coupled with the wave.

8.3. Plane-polarized, Plane Shear Wave: Propagation Parallel to the Poling Axis, Particle Motion along the X Axis. For convenience, let the x-axis lie along the direction of particle motion. From (57), the equation of motion is

$$\rho(\partial^2 u/\partial t^2) = \partial T_1/\partial x + \partial T_6/\partial y + \partial T_5/\partial z \tag{82}$$

In terms of displacements u, v, w along the x, y, z axes, the shear strains are

$$\begin{aligned} S_4 &= \partial w/\partial y + \partial v/\partial z \\ S_5 &= \partial w/\partial x + \partial u/\partial z \\ S_6 &= \partial v/\partial x + \partial u/\partial y \end{aligned} \tag{83}$$

Displacements v and w are zero, and u is uniform over wavefronts parallel to the xy plane. The single strain is therefore

$$S_5 = \partial u/\partial z \tag{84}$$

The piezoelectric equations involving shear stresses, following Table 1, are

$$\begin{aligned} T_4 &= -h_{15}D_2 \\ T_5 &= c_{44}{}^D S_5 - h_{15}D_1 \\ T_6 &= 0 \\ E_1 &= -h_{15}S_5 + \beta_{11}{}^S D_1 \\ E_2 &= -h_{15}S_4 + \beta_{11}{}^S D_2 \\ E_3 &= \beta_{33}{}^S D_3 \end{aligned} \tag{85}$$

T_1, T_2, T_3, T_4, T_6, E_2, and E_3 are not coupled with the wave.
The wave equation, for an insulating medium, is thus

$$\rho(\partial^2 u/\partial t^2) = \partial T_5/\partial z = c_{44}{}^D(\partial S_5/\partial z) = c_{44}{}^D(\partial^2 u/\partial z^2) \tag{86}$$

Phase velocity is

$$c = \sqrt{c_{44}{}^D/\rho} \tag{87}$$

8.4. Plane-polarized, Plane Shear Wave: Propagation Perpendicular to the Poling Axis, Particle Motion along the Z Axis. For convenience, take the x axis in the direction of propagation. The equation of motion, from (57), is

$$\rho(\partial^2 w/\partial t^2) = \partial T_5/\partial x + \partial T_4/\partial y + \partial T_3/\partial z = \partial T_5/\partial x \tag{88}$$

The shear strain is

$$S_4 = \partial w/\partial y \tag{89}$$

CERAMIC DEVICES 16–23

and the pertinent piezoelectric equations are

$$T_4 = c_{44}{}^D S_4 - h_{15} D_2$$
$$E_2 = -h_{15} S_4 + \beta_{33}{}^S D_2 \tag{90}$$

The equation of motion, for an insulating medium, is

$$\rho(\partial^2 w/\partial t^2) = c_{44}{}^D(\partial^2 w/\partial z^2) \tag{91}$$

Phase velocity is

$$c = \sqrt{c_{44}{}^D/\rho} \tag{92}$$

8.5. Plane-polarized, Plane Shear Wave: Propagation and Particle Motion Perpendicular to the Poling Axis. Take the x axis parallel to propagation, with particle motion parallel to the y axis. The equation of motion is

$$\rho(\partial^2 v/\partial t^2) = \partial T_6/\partial x + \partial T_2/\partial y + \partial T_4/\partial z = \partial T_6/\partial x \tag{93}$$

The single shear strain is

$$S_6 = \partial v/\partial x = s_{66} T_6 = (1/c_{66}) T_6 \tag{94}$$

and the wave equation is

$$\partial^2 v/\partial t^2 = c_{66}(\partial^2 v/\partial z^2) \tag{95}$$

Phase velocity is

$$c = \sqrt{c_{66}/\rho} \tag{96}$$

This wave is not piezoelectrically coupled to the electric field, and so occurs only under mechanical excitation.

8.6. Extensional Wave in a Length-polarized Thin Bar. In a thin bar, free from applied forces on the lateral surfaces, all stresses are customarily assumed equal to zero except those transmitted along the bar. With the bar oriented along the z axis, and vibrating longitudinally, only stress T_3 is not zero. The equation of motion is

$$\rho(\partial^2 w/\partial t^2) = \partial T_3/\partial z \tag{97}$$

Appropriate piezoelectric equations are

$$S_1 = S_2 = s_{13}{}^E T_3 + d_{31} E_3$$
$$S_3 = s_{33}{}^E T_3 + d_{33} E_3$$
$$S_4 = S_5 = S_6 = 0 \tag{98}$$
$$E_1 = E_2 = 0$$
$$D_3 = d_{33} T_3 + \epsilon_{33}{}^T E_3$$

In an insulating medium,

$$\partial D_3/\partial z = 0 = d_{33}(\partial T_3/\partial z) + \epsilon_{33}{}^T(\partial E_3/\partial z) \tag{99}$$

E_3, which is not independent of z, must be eliminated from the expression for T_3:

$$S_3 = s_{33}{}^E T_3 + d_{33}[(D_3 - d_{33} T_3)/\epsilon_{33}{}^T]$$
$$= s_{33}{}^E (1 - k_{33}{}^2) T_3 + (d_{33}/\epsilon_{33}{}^T) D_3 \tag{100}$$

It follows that the wave equation is

$$\rho \frac{\partial^2 w}{\partial t^2} = \frac{1}{s_{33}{}^E(1 - k_{33}{}^2)} \frac{\partial^2 w}{\partial z^2} \tag{101}$$

with phase velocity

$$c = 1/\sqrt{s_{33}{}^E(1 - k_{33}{}^2)\rho} \tag{102}$$

A general solution for stress T_3 consists of two traveling waves. Strain S_3 consists of two traveling-wave components, plus a stationary component $d_{33} D_3/\epsilon_{33}{}^T$. The field intensity is

$$E_3 = (1/\epsilon_{33}{}^T)(D_3 - d_{33}T_3) \tag{103}$$

and so consists of two traveling-wave components plus a stationary component.

The thin-bar assumption leads to a solution for particle motion and velocity which is independent of bar thickness. The solution is rigorously true only for bars with vanishing cross section but is quite accurate for all cross sections much less than a stress wavelength in extent. In thicker bars, the variation in stress over the cross section, and the kinetic energy of mass in lateral motion, cannot be neglected. A

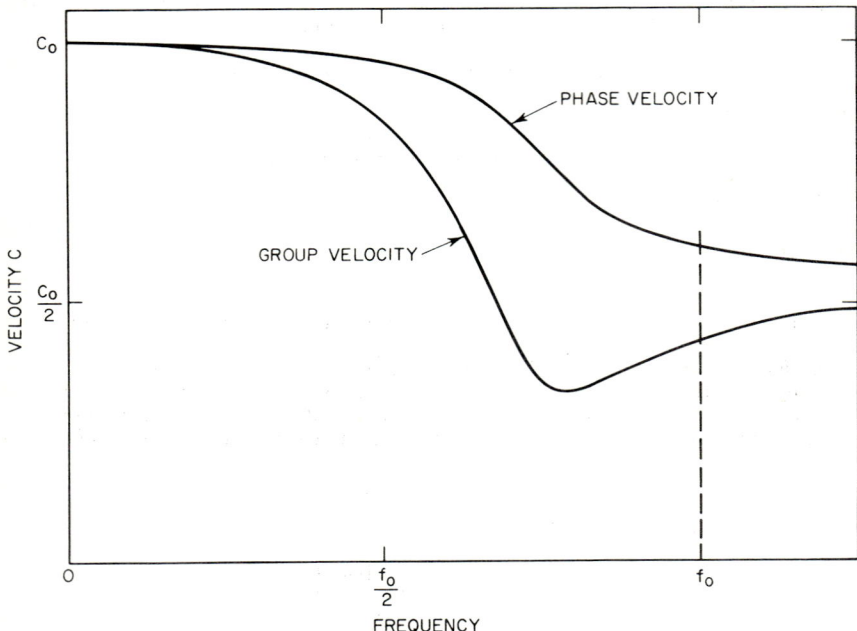

FIG. 7. Dispersion in an elastic strip with finite thickness $\omega = c_0/f_0$.

rigorous solution for bars with finite cross section is a complicated boundary-value problem, which has been solved only for infinite bar lengths, in media which are isotropic and which have only circular, rectangular, or finite thickness—infinite-width cross sections. It is found that phase velocity is not independent of cross section but dependent on the ratio of wavelength to the transverse dimensions. An infinite number of distinct propagation modes exist for longitudinal waves. All except the lowest mode are associated with cutoff frequencies; below the cutoff frequency, energy in a mode is not propagated as a wave but is attenuated with distance.

Figure 7 shows phase velocity and group velocity for the lowest longitudinal mode in infinitely wide strips. When velocities vary with frequency and wavelength, the wave is said to be dispersive, and distinction must be made between the phase velocity

$$c_p = \omega/\beta \tag{104}$$

and the group velocity

$$c_g = d\omega/d\beta = c_p - \lambda(dc_p/d\lambda) \tag{105}$$

CERAMIC DEVICES

where β is phase shift per unit length. Wavelength is

$$\lambda = c_p/f = 2\pi/\beta \tag{106}$$

For the lowest mode, with limiting velocity c_0 as frequency approaches zero [cf. Eq. (102)], dispersion is most pronounced in the frequency range for which bar diameter d is approximately one-half wavelength. A first-order approximation valid for lower frequencies was first derived by Rayleigh:

$$c_p \doteq c_0[1 - (\pi^2/4)\sigma^2(d/\lambda)^2] \tag{107}$$

From Eq. (105), it follows directly that

$$c_g \doteq c_0[1 - (3\pi^2/4)\sigma^2(d/\lambda)^2] \tag{108}$$

The Rayleigh approximation is reasonably accurate for velocity corrections as large as about 5 per cent.

8.7. Extensional Wave in a Width-polarized Thin Bar. If a thin bar is electroded along two opposing sides, the electrodes form equipotential surfaces and the field intensity along the rod is essentially zero. The transverse field intensity, directed between the electrodes, is equal to the quotient of interelectrode voltage divided by the bar width between electrodes.

For transverse longitudinal coupling, the bar is poled with voltage applied between electrodes. With the z axis parallel to the polarization and the x axis directed along the rod, the components of field intensity are subject to the boundary conditions

$$E_1 = E_2 = 0 \tag{109}$$
$$\partial E_3/\partial x = \partial E_3/\partial y = \partial E_3/\partial z = 0 \tag{110}$$

These *constant-E* conditions are a distinctive feature of a side-plated bar, as opposed to the constant-D conditions which hold in the absence of short-circuiting electrode lengths.

With extensional wave propagation along the bar, the single stress is T_1, and the equation of motion is

$$\rho(\partial^2 u/\partial t^2) = \partial T_1/\partial x \tag{111}$$

The appropriate piezoelectric equations are

$$\begin{aligned} S_1 &= s_{11}{}^E T_1 + d_{31}E_3 \\ S_2 &= s_{12}{}^E T_1 + d_{31}E_3 \\ S_3 &= s_{13}{}^E T_1 + d_{33}E_3 \\ D_3 &= d_{31}T_1 + \epsilon_{33}{}^T E_3 \end{aligned} \tag{112}$$

The wave equation is therefore

$$\rho(\partial^2 u/\partial t^2) = (1/s_{11}{}^E)(\partial^2 u/\partial x^2) \tag{113}$$

for phase velocity

$$c = \sqrt{1/s_{11}{}^E \rho} \tag{114}$$

In general, dispersion of extensional wave velocity follows the pattern of decreasing velocity with a decreasing ratio of wavelength to the transverse dimensions of the bar.

8.8. Torsional Wave along a Circumferentially Poled, Round Cylinder. Figure 8 illustrates the selection of axes at a point on the circularly poled tube. It is convenient to use cylindrical coordinates x, r, θ to specify the location of any element of volume, and rectangular coordinates x, y, z at each point for evaluation of the material properties from the piezoelectric equations. A cylinder or a solid rod which can be regarded as a special case with zero inner radius can be fabricated by assembling prepolarized segments, by sequentially poling sectors with removable electrodes, or by slowly rotating the cylinder about its axis with its surface in contact with charged rollers.

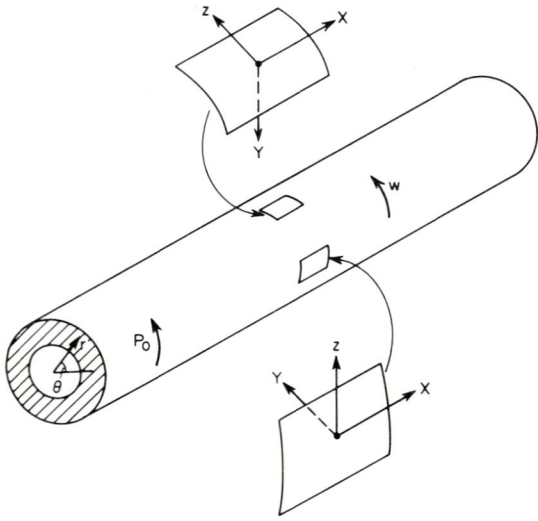

Fig. 8. Circumferentially polarized hollow cylinder.

Torsion is a pure shear with circular symmetry about the x axis of Fig. 8. In *simple* torsion, each circular cross section rotates uniformly, as a rigid surface. If rotation in the θ direction is equal to ζ, then motion of a particle at radius r is

$$w = r\zeta \tag{115}$$

where w is in the z direction.

The shear strain

$$S_5 = \partial u/\partial z + \partial w/\partial x = \partial w/\partial x = r(\partial \zeta/\partial x) \tag{116}$$

varies with r and x and is independent of θ. Corresponding piezoelectric equations are

$$\begin{aligned} T_5 &= c_{44}{}^D S_5 - h_{15} D_1 \\ E_1 &= -h_{15} S_5 + \epsilon_{11}{}^S D \end{aligned} \tag{117}$$

In an insulating ceramic

$$\begin{aligned} D_2 &= D_3 = 0 \\ \partial D_1/\partial x &= 0 \end{aligned} \tag{118}$$

The equation of motion is therefore

$$\rho \partial^2 w/\partial t^2 = \partial T_5/\partial x = c_{44}{}^D (\partial^2 w/\partial x^2) \tag{119}$$

The velocity of propagation is the shear velocity

$$c = \sqrt{c_{44}{}^D/\rho} \tag{120}$$

This mode is an exact solution, with velocity independent of cylinder diameter, and so phase and group velocity are equal and independent of frequency. Particle motion is purely circular, with maximum motion at the surface, where $w = a\zeta$. Field intensity is E_1, directed along the axis.

In a solid, isotropic rod, motion of this type represents the lowest mode of torsional wave propagation; higher-order modes involve variation of ζ with r. The higher-order modes are dispersive, with cutoff frequencies determined by wavelength-to-diameter ratios.

9. STANDING WAVES, EQUIVALENT CIRCUITS, AND ATTENUATION

In a bar or plate, with effective elastic constant c_{mn}, and without piezoelectric coupling, the wave equation indicates traveling-wave solutions for particle motion, of the form

$$u = (A_1 e^{-j\omega x/c} + A_2 e^{j\omega x/c})e^{j\omega t} \tag{121}$$

Stress and particle velocity in these waves are

$$\begin{aligned} T &= c_{mn}(\partial u/\partial x) = j\omega\rho c(-A_1 e^{-j\omega x/c} + A_2 e^{j\omega x/c})e^{j\omega t} \\ \dot{u} &= j\omega u = j\omega(A_1 e^{-j\omega x/c} + A_2 e^{j\omega x/c})e^{j\omega t} \end{aligned} \tag{122}$$

When the end surfaces, at $x = 0$ and $x = l$ as indicated in Fig. 9, are subjected to arbitrary constraints of externally applied forces and velocities, these constraints determine the values of the constants A_1 and A_2. Across any surface, including an end surface, force and velocity are continuous. Across a plane wave with area A, force is AT, in the direction of *tensile* stress.

Surface force may be expressed as a *phasor F*, with positive direction along *compressive* stress $-T$, and particle velocity may be expressed as a corresponding phasor V, with positive direction parallel to that of F. F and V are, in general, complex numbers.

At the surface $x = 0$, the applied force and velocity phasors are

$$\begin{aligned} F_1 &= -(TA/e^{j\omega t}) = j\omega A\rho c(A_1 - A_2) \\ V_1 &= \dot{u}/e^{j\omega t} = j\omega(A_1 + A_2) \end{aligned} \tag{123}$$

At $x = l$,

$$\begin{aligned} F_2 &= -(TA/e^{j\omega t}) = j\omega A\rho c(A_1 e^{-j\omega l/c} - A_2 e^{j\omega l/c}) \\ V_2 &= -(\dot{u}/e^{j\omega t}) = -j\omega(A_1 e^{-j\omega l/c} + A_2 e^{j\omega l/c}) \end{aligned} \tag{124}$$

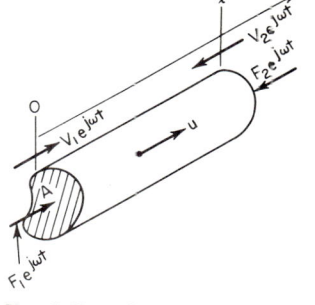

FIG. 9. Boundary constraints on a solid vibrating element.

The traveling-wave amplitudes A_1 and A_2 can be expressed, from Eqs. (123) and (124), in terms of end constraints, such as

$$\begin{aligned} A_1 &= \frac{V_1 + F_1/A\rho c}{j2\omega} = \frac{-V_2 + F_2/A\rho c}{j2\omega} e^{j\omega l/c} \\ -A_2 &= \frac{-V_1 + F_1/A\rho c}{j2\omega} = \frac{V_2 + F_2/A\rho c}{j2\omega} e^{-j\omega l/c} \end{aligned} \tag{125}$$

Since there are only two arbitrary wave amplitudes, only two of the end constraints can be arbitrarily specified, and we can write

$$\begin{aligned} F_1 + A\rho c V_1 &= (F_2 - A\rho c V_2)e^{j\omega l/c} \\ F_1 - A\rho c V_1 &= (F_2 + A\rho c V_2)e^{-j\omega l/c} \end{aligned} \tag{126}$$

which, rearranged, become the lossless transmission-line equations

$$\begin{aligned} F_1 &= F_2 \cos(\omega l/c) + (-V_2)A\rho c j \sin \omega l/c \\ V_1 &= (F_2/A\rho c)j \sin(\omega l/c) + (-V_2) \cos \omega l/c \end{aligned} \tag{127}$$

The constant expression $A\rho c$ appears as the analogous transmission-line characteristic impedance, force F corresponds to voltage, and particle velocity corresponds to current. Table 5 summarizes a few of the correspondences of the *electromechanical analogy*, which provides a systematic method for relating electrical and mechanical systems with mathematically identical characteristics.

16-28 AMPLIFYING DEVICES

Table 5. Equivalent Quantities in the Electromechanical Analogy

BASIC QUANTITIES

FORCE (NEWTONS) — VOLTAGE (VOLTS)
DISPLACEMENT (METERS) — CHARGE (COULOMBS)
VELOCITY (M/SEC) — CURRENT (AMPERES)
MECHANICAL IMPEDANCE — IMPEDANCE (OHMS)
(NEWTONS-SEC/M)
MASS (KILOGRAMS) — INDUCTANCE (HENRYS)
COMPLIANCE (M/NEWTONS) — CAPACITANCE (FARADS)
MECHANICAL RESISTANCE — RESISTANCE

EQUIVALENT STRUCTURES

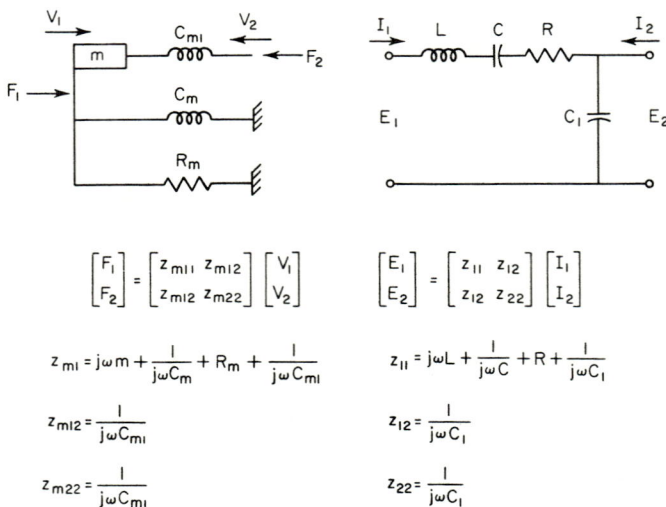

$$\begin{bmatrix} F_1 \\ F_2 \end{bmatrix} = \begin{bmatrix} z_{m11} & z_{m12} \\ z_{m12} & z_{m22} \end{bmatrix} \begin{bmatrix} V_1 \\ V_2 \end{bmatrix} \qquad \begin{bmatrix} E_1 \\ E_2 \end{bmatrix} = \begin{bmatrix} z_{11} & z_{12} \\ z_{12} & z_{22} \end{bmatrix} \begin{bmatrix} I_1 \\ I_2 \end{bmatrix}$$

$$z_{m1} = j\omega m + \frac{1}{j\omega C_m} + R_m + \frac{1}{j\omega C_{m1}} \qquad z_{11} = j\omega L + \frac{1}{j\omega C} + R + \frac{1}{j\omega C_1}$$

$$z_{m12} = \frac{1}{j\omega C_{m1}} \qquad z_{12} = \frac{1}{j\omega C_1}$$

$$z_{m22} = \frac{1}{j\omega C_{m1}} \qquad z_{22} = \frac{1}{j\omega C_1}$$

Following the analogy, an equivalent circuit for the elastic bar or plate can be drawn, as in Fig. 10. For simplicity, the mechanical values of force, velocity, and impedance are used directly in the equivalent circuit. Linear circuit analysis may be used with the equivalent circuit, to determine relations between the terminal quantities. As a derivation of the circuit from Eqs. (127), it is sufficient to note that*

$$z_{11} = \frac{F_1}{V_1}\bigg|_{V_2=0} = \frac{A\rho c}{j \tan \omega l/c} = z_{22}$$

$$z_{12} = \frac{F_2}{V_1}\bigg|_{V_2=0} = \frac{A\rho c}{j \sin \omega l/c} = z_{21} \qquad (128)$$

$$z_{11} - z_{12} = \frac{A\rho c}{j}\left(\frac{\cos \omega l/c - 1}{\sin \omega l/c}\right) = A\rho c j \tan \frac{\omega l}{2c}$$

* Two useful trigonometric identities for equivalent circuit analysis are

$$\tan \frac{\theta}{2} = \frac{1 - \cos \theta}{\sin \theta} = \frac{\sin \theta}{1 + \cos \theta}$$

$$\tanh \frac{\Gamma}{2} = \frac{\cosh \Gamma - 1}{\sinh \Gamma} = \frac{\sinh \Gamma}{\cosh \Gamma + 1}$$

CERAMIC DEVICES

Comparison with electrical transmission lines indicates that a mismatched acoustical transmission path will display multiple resonances, at fundamental and harmonic resonant frequencies. An open-circuited line, for which the output current is zero, corresponds to an acoustical path with a clamped surface at $x = l$, where $V = 0$. An elastic body cannot be truly clamped, since there are no materials with infinite stiffness, and so this condition can only be approached. The mechanical input impedance of the clamped path, which can be calculated from the equivalent circuit, or

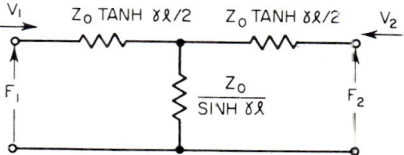

FIG. 10. Electrical analog circuit for an elastic element.

calculated from Eqs. (127), or simply written from knowledge of electrical transmission-line theory, is

$$Z_{in} = \frac{F_1}{V_1}\bigg|_{V_2=0} = \frac{Z_0}{j \tan \omega l/c} \tag{129}$$

An acoustical path with a free end surface ($F = 0$) corresponds to a short-circuited line, with input impedance

$$Z_{in} = jZ_0 \tan \omega l/c \tag{130}$$

There are resonant frequencies for which Z_{in} is zero, and antiresonant frequencies for which Z_{in} is infinite. At a resonant frequency $f_r = \omega_r/2\pi$, the acoustical path is an integral number of wavelengths

$$\omega_r l/c = n\pi \tag{131}$$
and
$$f_r = nf_1 = nc/2l \tag{132}$$

where $n = 1, 2, 3, \ldots$

For the resonant free bar, forces F_1 and F_2 at the ends are both zero, and so from Eqs. (125) the end velocities are equal in magnitude:

$$V_1 = -V_2 e^{jn\pi} = j2\omega A_1 = j2\omega A_2 \tag{133}$$

Equation (121), with this value of A_1, gives the particle velocity for any point x along the path

$$\dot{u} = (V_1/2)(e^{-j\omega x/c} + e^{j\omega x/c})e^{j\omega t} = V_1 \cos (\omega x/c)e^{j\omega t} \tag{134}$$

Vibration has maxima at the end surfaces, is symmetrical about the midpoint of the path, and is zero at nodal points for which $\cos \omega x/c$ is zero. The nodes occur at half-wave spacings and are commonly used in resonant structures as points of attachment for mounting supports and electrical connections.

Antiresonances occur at frequencies f_a for which the path length is an odd number of quarter wavelengths:

$$f_a = (2m - 1)c/2l \qquad m = 1, 2, 3, \ldots \tag{135}$$

From Eq. (125), at any of the antiresonant frequencies

$$j2\omega A_1 = -j2\omega A_2 = F_1/A\rho c = -V_2 e^{jm\pi/2} \tag{136}$$

and particle velocity is

$$\dot{u} = (F_1/2A\rho c)(e^{-j\omega x/c} - e^{j\omega x/c})e^{j\omega t} = \frac{jF_1}{A\rho c} \sin (\omega x/c)e^{j\omega t} \tag{137}$$

The antiresonant nodes include the driven end and surfaces separated from that end by half wavelengths. Motion has odd symmetry about the midpoint. The driving-point impedance is very large, in comparison with $A\rho c$, at an exact frequency of antiresonance.

In any physical solid, vibration is accompanied by power loss, produced by conversion of strain energy into heat, by radiation of energy into supports, and (for longitudinal waves) by radiation into surrounding air. The effects of end-face radiation can be estimated[25] from equivalent circuit calculations, taking the end-face loads as resistances equal to $A(\rho c)'$ for air. Since the acoustic impedance of air is about 30,000 times lower than that of a typical piezoelectric solid, the loading is very light and produces appreciable damping only in very low loss materials such as quartz.

Internal friction causes mechanical hysteresis and loss, associated with deformations of the solid. This can be described quantitatively, for time variations of the form $e^{j\omega t}$, by an elastic constant whose value is a complex number, e.g., $s_{kl} = s_{kl}' + js_{kl}''$. The associated mechanical quality factor is defined as

$$Q_m = s_{kl}'/s_{kl}'' \tag{138}$$

Q_m is rarely less than 75 in piezoelectric ceramics and may be as large as 3,000. Therefore, in order of increasing approximation,

$$\begin{aligned} s_{kl} &= s_{kl}'(1 + j/Q_m) \\ &= s_{kl}' \sqrt{1 + 1/Q_m{}^2}\, e^{j\tan^{-1} 1/Q_m} \\ &\doteq s_{kl}' e^{j/Q_m} \\ &\doteq s_{kl}' \end{aligned} \tag{139}$$

For most calculations involving appreciable damping from external mechanical loads, Q_m can be assumed infinite, as in the last approximation above. For calculation of unloaded resonant behavior or of efficiency, the preceding approximation provides better than 1 per cent accuracy.

A solid can have as many different values of Q_m as it has different elastic constants. However, Q_m tends to vary considerably, from sample to sample of otherwise identical material, and so is not completely cataloged for all s_{ij} and c_{ij} but rather is categorized for types of resonant structure.

In an ideal lossless solid, a wave travels as $e^{-j\beta z}$, where, for example,

$$j\beta = j\omega/c = j\omega/\sqrt{s_{kl}\rho} \tag{140}$$

In a lossy solid, the wave propagates as $e^{-\gamma z}$, where

$$\gamma = \alpha + j\beta = j\omega/\sqrt{(s_k' + js_k'')\rho} \tag{141}$$

Therefore,

$$\alpha \doteq \beta/2Q_m \tag{142}$$

Attenuation is

$$\begin{aligned} &1/2Q_m \text{ nepers/radian} \\ &\pi/Q_m \text{ nepers/wavelength} \\ &27.29/Q_m \text{ db/wavelength} \\ &27.29/Q_m\lambda = 27.29f/Q_m c \text{ db/m} \\ &27.29f/Q_m \text{ db/sec of propagation time} \\ &27.29/Q_m \text{ db/cps/sec} = \text{db/kc/msec} = \text{db/Mc/}\mu\text{sec} \end{aligned} \tag{143}$$

Mechanical impedances, taking into account loss, become hyperbolic functions of $(\alpha + j\beta)l$:

$$\begin{aligned} z_{11} &= z_{22} = Z_0/\tanh \gamma l \\ z_{12} &= z_{21} = Z_0/\sinh \gamma l \\ z_{11} &- z_{12} = Z_0 \tanh \gamma l/2 \\ Z_0 &= A\rho c \doteq A\rho(c')(1 + j/2Q_m) \doteq A\rho c' \end{aligned} \tag{144}$$

The free-ended input impedance $jZ_0 \tan \beta l$ becomes

$$\begin{aligned} Z_{in} &= Z_0 \tanh (\alpha + j\beta)l \\ &= Z_0 \frac{\sinh \alpha l \cos \beta l + j \cosh \alpha l \sin \beta l}{\cosh \alpha l \cos \beta l + j \sinh \alpha l \sin \beta l} \end{aligned} \tag{145}$$

which, rationalized, can be written

$$Z_{in} = Z_0 \frac{\sinh \alpha l \cosh \alpha l + j \sin \beta l \cos \beta l}{\sinh^2 \alpha l + \cos^2 \beta l}$$

$$\doteq Z_0 \frac{\alpha l + j \sin \beta l \cos \beta l}{(\alpha l)^2 + \cos^2 \beta l} \tag{146}$$

At frequencies well away from a resonance or antiresonance, the loss terms in αl are much smaller than the terms in βl and can be neglected. At resonance, for the nth harmonic, βl is $n\pi$, and

$$Z_{in} \doteq Z_0 \alpha l = Z_0 n\pi / 2 Q_m \tag{147}$$

At the mth antiresonance, βl is $(2m - 1)\pi/2$, and

$$Z_{in} \doteq Z_0/\alpha l = Z_0[4Q_m/(2m - 1)\pi] \tag{148}$$

Impedance at frequencies near a resonance is

$$Z_{in} \doteq Z_0 \alpha d + j \tan \beta l \doteq Z_0 \left[\frac{n\pi}{2Q_m} + jn\pi \left(\frac{\omega - \omega_r}{\omega_r} \right) \right] \tag{149}$$

The imaginary and real parts of Z_{in} are equal at ω_1 and ω_2, where

$$\omega_r/(\omega_2 - \omega_1) \doteq Q_m \tag{150}$$

Near an antiresonance,

$$\frac{1}{Z_{in}} = \frac{1}{Z_0} \frac{\sinh \alpha l \cosh \alpha l - j \sin \beta l \cos \beta l}{\sinh^2 \alpha l + \sin^2 \beta l}$$

$$\doteq \frac{1}{Z_0} (\alpha l - j \cot \beta l)$$

$$\doteq \frac{1}{Z_0} \left[\frac{(2m-1)\pi}{4Q_m} + j \frac{(2m-1)\pi}{2} \left(\frac{\omega - \omega_a}{\omega_a} \right) \right] \tag{151}$$

The fractional bandwidth, between frequencies for which the magnitude of impedance is reduced by $\sqrt{2}$, is $1/Q_m$.

The impedance at frequencies near a resonance can be represented by a series RLC equivalent circuit, in which

$$\begin{aligned}
R &= n(Z_0 \pi / 2 Q_m) \\
L &= Q_m R/\omega_r = n(Z_0 \pi / 2 \omega_r) = n(A\rho c/4 f_r) = n(A\rho \lambda/4) = A\rho l/2 \\
C &= 1/\omega_r^2 L = 2/\omega_r^2 A \rho l = 2l/(n\pi)^2 A \rho c^2 = [2/(n\pi)^2](l/A)s_{ij}
\end{aligned} \tag{152}$$

where s_{ij} is the effective elastic constant. R, ωL, and $1/\omega C$ are mechanical impedances, with units of mechanical ohms. L is half the total mass of the resonator; C is the fraction $2/n^2\pi^2$ of its static elastic compliance.

At frequencies near an antiresonance, the input impedance is approximately that of a parallel RLC circuit, in which

$$\begin{aligned}
R &= \frac{1}{(2m-1)} \frac{4Q_m Z_0}{\pi} \\
C &= \frac{Q_m}{\omega_a R} = \frac{(2m-1)\pi}{4\omega_a A\rho c} = \frac{(2m-1)\pi l}{4(\omega_a l/c)A\rho c^2} = \frac{ls_{ij}}{2A} \\
L &= \frac{1}{\omega_a^2 C} = \frac{1}{(2m-1)^2 \pi^2} \frac{8}{} A\rho l
\end{aligned} \tag{153}$$

C is half the static compliance, and L is the fraction $8/\pi^2(2m-1)^2$ of the mass.

10. THE SIDE-PLATED CERAMIC BAR

The transversely poled ceramic bar, shown in Fig. 11, is used as a ceramic filter resonator and as an ultrasonic transducer. The side electrodes maintain a constant electric field E_3 throughout the bar.

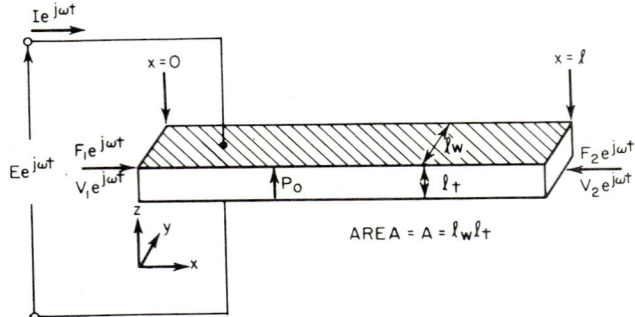

Fig. 11. Constant-E bar transducer.

Motion along x can be written, in general, as the sum of two traveling waves. Taking possible attenuation into account makes the propagation constant

$$\gamma = \alpha + j\beta = \beta/2Q_m + j\beta = j\omega\sqrt{s_{11}^E \rho} \tag{154}$$

and the characteristic impedance

$$Z_0 = A\sqrt{\rho/s_{11}^E} = A(\gamma/j\omega s_{11}^E) \tag{155}$$

Then the particle displacement is given by the equation

$$u = (A_1 e^{-\gamma x} + A_2 e^{\gamma x})e^{j\omega t} \tag{156}$$

From Eqs. (112), the longitudinal stress is

$$T_1 = \frac{1}{s_{11}^E}\frac{\partial u}{\partial x} - \frac{d_{31}}{s_{11}^E}E_3$$
$$= (\gamma/s_{11}^E)(A_1 e^{-\gamma x} + A_2 e^{\gamma x})e^{j\omega t} + (d_{31}/s_{11}^E l_t)Ee^{j\omega t} \tag{157}$$

The end forces F_1, F_2 and end velocities V_1, V_2 are boundary values

$$V_1 = \left.\frac{\dot{u}}{e^{j\omega t}}\right|_{x=0} = j\omega(A_1 + A_2)$$
$$V_2 = \left.\frac{\dot{u}}{e^{j\omega t}}\right|_{x=l} = j\omega(A_1 e^{-\gamma l} + A_2 e^{\gamma l}) \tag{158}$$

Simultaneous solution of Eqs. (158) gives

$$\begin{aligned}A_1 &= (V_1 e^{\gamma l} - V_2)/j2\omega \sinh \gamma l \\ A_2 &= (V_2 - V_1 e^{-\gamma l})/j2\omega \sinh \gamma l\end{aligned} \tag{159}$$

With these values, the expression for stress becomes

$$\begin{aligned}T_1 &= \frac{\gamma}{j\omega s_{11}^E}\frac{(V_2 - V_1 e^{\gamma l})e^{-\gamma x} + (V_2 - V_1 e^{-\gamma l})e^{\gamma x}}{2\sinh \gamma l}e^{j\omega t} + \frac{d_{31}}{s_{11}^E l_t}Ee^{j\omega t} \\ &= \frac{Z_0}{A}\frac{-V_1 \cosh \gamma(l-x) + V_2 \cosh \gamma x}{\sinh \gamma l}e^{j\omega t} + \frac{d_{31}}{s_{11}^E l_t}Ee^{j\omega t}\end{aligned} \tag{160}$$

and particle velocity is

$$\dot{u} = j\omega u = \frac{(V_1 e^{\gamma l} - V_2)e^{-\gamma x} + (V_2 - V_1 e^{-\gamma l})e^{\gamma x}}{2 \sinh \gamma l} e^{j\omega t}$$
$$= \frac{V_1 \sinh \gamma(l-x) + V_2 \sinh \gamma x}{\sinh \gamma l} e^{j\omega t} \quad (161)$$

The end forces are

$$F_1 = -\frac{AT_t}{e^{j\omega t}}\bigg|_{x=0} = Z_0 \left(\frac{V_1}{\tanh \gamma l} - \frac{V_2}{\sinh \gamma l}\right) + \frac{d_{31}l_w}{s_{11}^E} E$$
$$F_2 = -\frac{AT_t}{e^{j\omega t}}\bigg|_{x=l} = Z_0 \left(-\frac{V_1}{\sinh \gamma l} + \frac{V_2}{\tanh \gamma l}\right) + \frac{d_{31}l_w}{s_{11}^E} E \quad (162)$$

Current into the electrode is

$$Ie^{j\omega t} = -\iint\limits_{\substack{\text{electrode}\\ \text{surface}}} (\partial D_3/\partial t)\, dx\, dy = -j\omega l_w \int_0^l D_3\, dx \quad (163)$$

From Eqs. (112) and (160),

$$D_3 = d_{31}T_1 + \epsilon_{33}^T E_3$$
$$= \frac{d_{31}Z_0}{A} \frac{-V_1 \cosh \gamma(l-x) + V_2 \cosh \gamma x}{\sinh \gamma l} e^{j\omega t} + \left(\frac{d_{31}^2}{s_{11}^E} + \epsilon_{33}^T\right)\frac{E}{l_t} \quad (164)$$

The electromechanical coupling coefficient appropriate for this geometry is

$$k_{31} = d_{31}/\sqrt{s_{11}^E \epsilon_{33}^T} \quad (165)$$

With this definition, and Eq. (164), evaluation of the integral in Eq. (163) gives

$$I = (d_{31}/s_{11}^E l_t)(V_1 + V_2) + \epsilon_{33}^T(1 - k_{31}^2)(l_w l/l_t)E \quad (166)$$

Equations (162) and (166) are a solution of the boundary-value problem; they give all relations between F_1, V_1, F_2, V_2, E, and I which are imposed by the piezoelectric bar. If any three of these are specified (as independent variables), the equations determine the other three, uniquely.

Equivalent circuits are often used to represent the restraints imposed by a piezoelectric element. Three ports are needed. For convenience and physical significance, let

$$C_0 = \epsilon_{33}^T(1 - k_{31}^2)(l_w l/l_t) \quad \text{units of farads}$$
$$\Phi = d_{31}/s_{11}^E l_t \quad \text{units of newtons/volt} = \text{amp/m/sec}$$
$$= (\text{mechanical ohms/electrical ohms})^{1/2} \quad (167)$$

C_0 is the interelectrode capacitance when both ends of the bar are clamped to prevent longitudinal motion. Φ is a numerical and dimensional electromechanical conversion constant. By direct calculation, it can be shown that, if $\omega_0/2\pi$ is the frequency at which the bar is half wavelength resonant and $\omega_s = (2m - 1)\omega_0$ is any odd-integer multiple of ω_0, then

$$\frac{(2m-1)\Phi^2}{\omega_s C_0 Z_0} = \frac{\Phi^2}{\omega_0 C_0 Z_0} = \frac{1}{\pi}\frac{k_{31}^2}{1-k_{31}^2} \quad \text{(dimensionless number)} \quad (168)$$

Figure 12 gives a three-port circuit, equivalent to the bar in the sense that it imposes the constraints given by Eqs. (162) and (166).

If the bar is used as a filter element, then no end surface loading is applied. In many transducer applications, the two ends are loaded with equal mechanical impedances. Both situations can be analyzed from the equivalent circuit, terminated at

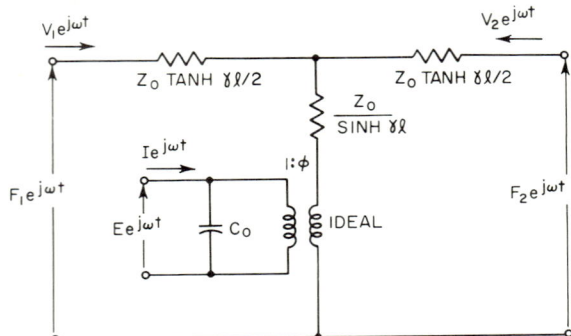

FIG. 12. Equivalent circuit for a constant-E transducer.

each mechanical port by an impedance Z_L. The mechanical impedance seen from the ideal transformer is then

$$Z_M = \frac{Z_0}{\sinh \gamma l} + \frac{1}{2} Z_0 \tanh \gamma l + \frac{1}{2} Z_L$$
$$= \frac{Z_0}{2 \tanh \gamma l/2} + \frac{1}{2} Z_L \qquad (169)$$

The electrical input admittance is therefore

$$Y_{in} = j\omega C_0 + \Phi^2/Z_M$$
$$= \omega_0 C_0 \left[j \frac{\omega}{\omega_0} + \left(\frac{2}{\pi} \frac{k_{31}^2}{1 - k_{31}^2} \right) \frac{1}{Z_L/Z_0 + (1/\tanh \gamma l/2)} \right] \qquad (170)$$

The end velocities and end forces are equal:

$$V_1 = V_2 = \frac{V_1 + V_2}{2} = \frac{E\Phi}{2Z_M} = \frac{Ed_{31}}{2s_{11}^E l_t Z_M}$$
$$F_1 = F_2 = V_1 Z_L = \frac{Ed_{31}}{2s_{11}^E l_t} \frac{Z_L}{Z_M} \qquad (171)$$

The hyperbolic tangent appears regularly in expressions describing the performance of transducers with internal loss. Since the equivalent circuit consists of the clamped capacitance in parallel with the motional impedance, the shunting capacitive current is the smallest part of total electrode current when the mechanical impedance is a minimum. This normally occurs at resonant frequency, for which $\tanh \gamma l$, $\tanh \gamma l/2$, or one of their reciprocals, represents a frequency-varying factor with minimum value (as a mechanical impedance), or a maximum value (as a mechanical admittance). It is therefore desirable to have a simplified but accurate approximation at frequencies near a pole or a zero.

Without loss, $\tanh \gamma l$ becomes $j \tan \beta l$, with zeros at $\beta l = n\pi$ and poles at $(2m - 1)\pi/2$. With loss, the zeros and poles move slightly to the left of the real frequency axis, in the complex frequency plane. Near a zero, where $\tanh \gamma l$ is small, Eq. (149) gives the approximation

$$\tanh \gamma l \doteq \alpha l + j \tan \beta l \doteq (n\pi/2Q_m)(1 + j 2Q_m \xi)$$

where
$$\xi = \frac{\beta l - n\pi}{n\pi} = \frac{\omega - \omega_s}{\omega_s} \qquad (172)$$

CERAMIC DEVICES

From this equation, near a zero of tanh γl, the large-valued function coth γl can be expressed as

$$\frac{1}{\tanh \gamma l} \doteq \frac{1}{\alpha l + j \tan \beta l} \doteq \frac{2Q_m}{n\pi} \frac{1}{1 + j2Q_m \xi}$$
$$\doteq \frac{Q_m}{n\pi}\left(1 + \frac{1 - j2Q_m\xi}{1 + j2Q_m\xi}\right) = \frac{Q_m}{n\pi}(1 + e^{-j2\theta})$$

where
$$\theta = \tan^{-1}(2Q_m\xi) \tag{173}$$

Near a pole of tanh γl, from Eq. (151),

$$\frac{1}{\tanh \gamma l} \doteq \alpha l - j \cot \beta l \doteq \frac{(2m-1)\pi}{4Q_m}(1 + j2Q_m\xi)$$

where
$$\xi = \frac{\beta l - (2m-1)\pi/2}{(2m-1)\pi/2} = \frac{\omega - \omega_a}{\omega_a} \tag{174}$$

and the large-valued function tanh γl appears as

$$\tanh \gamma l \doteq \frac{1}{\alpha l - j \cot \beta l} \doteq \frac{2Q_m}{(2m-1)\pi}(1 + e^{-j2\theta}) \tag{175}$$

In Eqs. (172) and (174), the function is the sum of a real constant, plus an imaginary term directly proportional to frequency separation from the critical value ω_r or ω_a. This is plotted as a locus on the complex plane in Fig. 13, using ω_c as a general notation for the critical frequency. The angle θ is $\pm 45°$ at the "half-power" points ω_1 and ω_2 determined by Q_m.

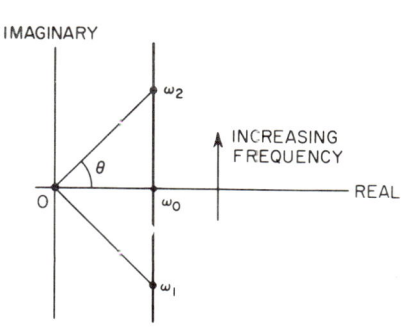

FIG. 13. Complex-plane locus for small values of tanh γl and coth γl.

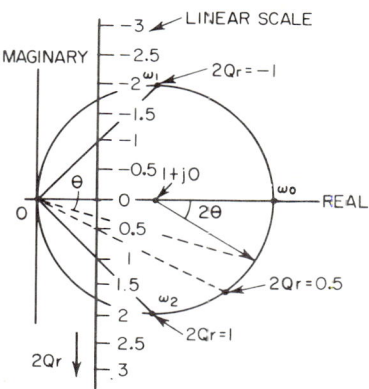

FIG. 14. Complex-plane locus for large values of tanh γl and coth γl.

Equations (173) and (175) present the function as a circle on the complex plane, as plotted in Fig. 14 for the normalized expression $1 + e^{-j2\theta}$. The complex value of the function is given by the radius vector from the origin, at angle $-\theta$, and with magnitude $1/\sqrt{1 + (2Q_m\xi)^2}$. The frequencies ω_1 and ω_2, lying at $\theta = \mp 45°$, are at the upper and lower extremes of the circle. A frequency scale around the circle can be established by projection from the origin onto any convenient vertical line; at projections through ω_1 and ω_2, ξ is correspondingly $-1/2Q_m$ and $1/2Q_m$, and the vertical scale is linear in $2Q_m\xi$. Also, two points on the circle with equal imaginary parts correspond to reciprocal values of $2Q_m\xi$, and two points with equal real parts correspond to opposite algebraic signs of $2Q_m\xi$.

With these expansions and loci, it is relatively easy to describe the admittance of a transducer with symmetrical resistive loading, as a function of frequency. Near a resonance, from Eq. (170), the input admittance is large. Equation (174) gives

$$Y_{in} \doteq \omega_0 C_0 \left\{ j\frac{\omega}{\omega_0} + \frac{2}{\pi} \frac{k_{31}^2}{1 - k_{31}^2} \frac{1}{R_L/Z_0 + [(2m-1)\pi/4Q_m](1 + j2Q_m\xi)} \right\}$$

$$\doteq \omega_0 C_0 \left\{ j\frac{\omega}{\omega_0} + \frac{1}{\pi} \frac{k_{31}^2}{1 - k_{31}^2} \frac{1}{R_L/Z_0 + [(2m-1)\pi/4Q_m]} (1 + e^{-j2\theta}) \right\}$$

$$\doteq \omega_s C_0 \left\{ j\frac{\omega}{\omega_s} + \frac{\omega_0}{\pi\omega_s} \frac{k_{31}^2}{1 - k_{31}^2} \frac{1}{R_L/Z_0 + [(2m-1)\pi/4Q_m]} (1 + e^{-j2\theta}) \right\}$$

where
$\theta = \tan^{-1} 2Q\xi$
$\omega_s/\omega_0 = 2m - 1$

$$Q = \frac{(2m-1)\pi/4}{R_L/Z_0 + [(2m-1)\pi/Q_m]} = \frac{1}{1/Q_m + [4/(2m-1)\pi](R_L/Z_0)} \quad (176)$$

Equation (176) is sketched in the loci of Fig. 15 for comparatively large and small values of Q. For $Q \gg 1$, it may be assumed that $\omega/\omega_s \doteq 1$ over the entire circle; the diameter is shown large enough to make the admittance inductive in a small range of frequencies about ω_2. For smaller Q, the loop represents a larger frequency range, and changing values of ω/ω_s distort the circle, introducing a cycloidal cusp; the diameter is smaller, and the admittance never becomes inductive. Since

$$Q = \omega_s/(\omega_2 - \omega_1) \quad (177)$$

the diameter is given by the expression

$$\frac{D}{\omega_0 C_0} = \frac{2}{\pi} \frac{k_{31}^2}{1 - k_{31}^2} \frac{4Q}{(2m-1)\pi}$$

$$= \frac{8}{(2m-1)\pi^2} \frac{k_{31}^2}{1 - k_{31}^2} \frac{\omega_s}{\omega_2 - \omega_1} \quad (178)$$

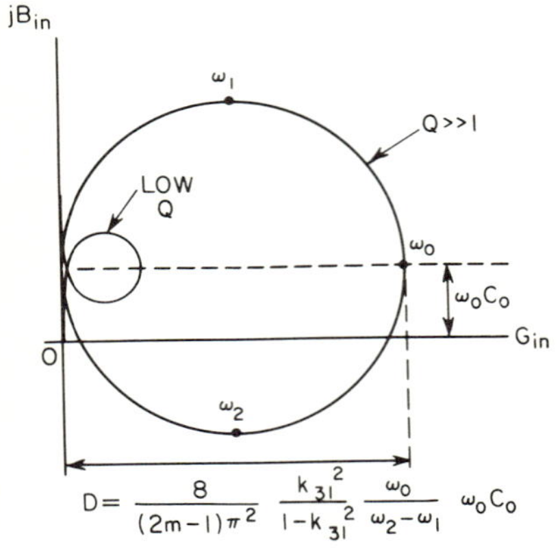

FIG. 15. Admittance loci for high-Q and low-Q transducers.

If the admittance is inductive at any point on the circle, then $D/\omega_s C_0$ must be greater than 2.

The curves of Fig. 16 were graphically constructed from a circle diagram, drawn with $D/\omega_s C_0 = 2.8$. For fundamental mode resonance, this corresponds to a resonator with the value

$$3.46 = k_{31}^2 Q/(1 - k_{31}^2) \doteq k_{31}^2 Q \quad (179)$$

Three rays drawn from the origin locate six frequencies,* which have physical significance for a resonator and can be measured with considerable accuracy:

At f_m, $|Y|$ is maximum
At f_n, $|Y|$ is minimum
At f_s, G is maximum (motional resonance)
At f_p, series resistance $G/(G^2 + B^2)$ is maximum
At f_r, $jB = 0$ and G is large (electrical resonance)
At f_a, $jB = 0$ and G is small (electrical antiresonance)

The ray $Of_n f_m$ is a diameter, passing through the center of the circle.

* The symbols for frequency used here are specified in the 1957 IRE, and proposed 1965 IEEE, Standards on the Piezoelectric Vibrator: Definitions and Methods of Measurement.

CERAMIC DEVICES 16-37

Since end-face loading reduces Q, the bandwidth of energy transfer increases with the value of mechanical load resistance, and correspondingly the diameter of the admittance circle decreases. Without end loading, Q_m can be measured by determining f_1 and f_2 with an admittance bridge; an approximate value can be found with a signal

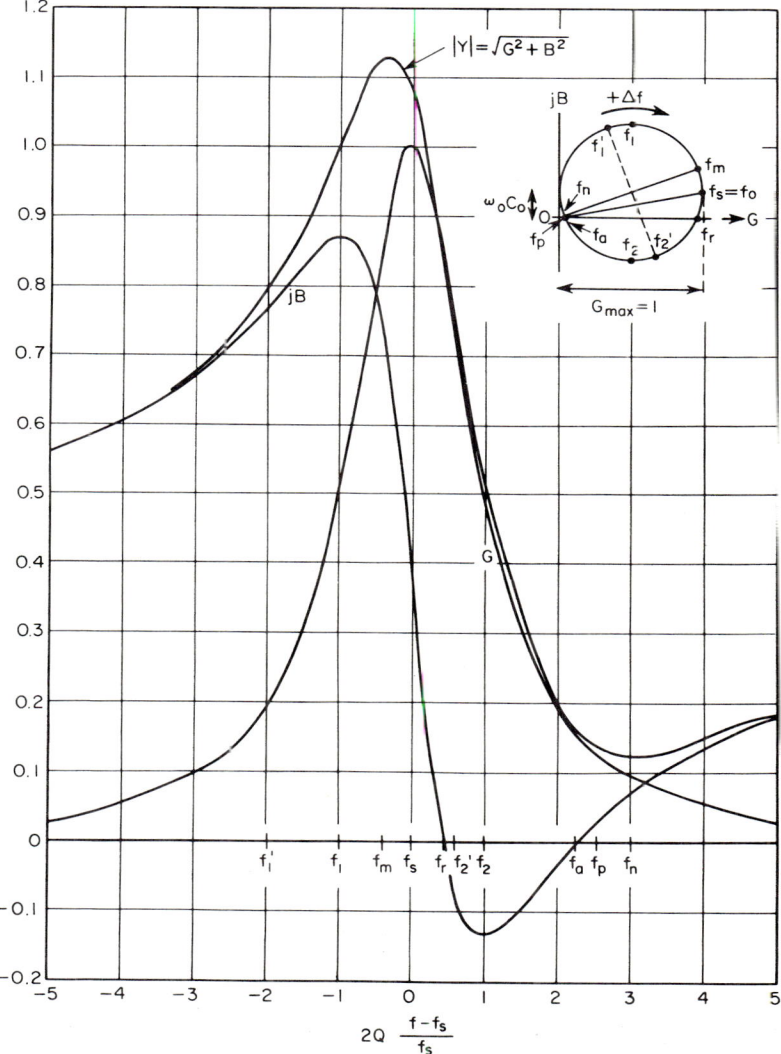

Fig. 16. Resonator admittance at frequencies near resonance.

generator and vacuum-tube voltmeter, in the transmission circuit of Fig. 17. R_L and R_g should be much smaller than the resonator impedance, so that current flow is proportional to resonator admittance. Maximum current flows at f_m, and 0.707 of maximum current flows at f_1' and f_2'. Q_m is somewhat greater than $f_m/(f_2' - f_1')$, which is almost equal to $f_s/(f_2 - f_1)$ if $k_{31}^2 Q_m$ is considerably greater than 2.

Coupling coefficient can also be measured in terms of points on the admittance circle. From Eq. (178),

$$k_{31}^2 = \frac{1}{1 + [8Q_m/\pi^2(2m-1)](\omega_0 C_0/G_{max})} \tag{180}$$

The constants Q_m, $\omega_0 C_0$, and G_{max} can be measured with the admittance bridge and a frequency counter or accurately calibrated signal generator.

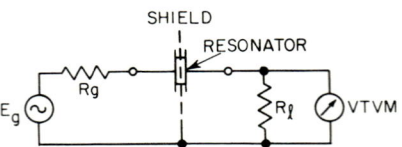

FIG. 17. Transmission circuit for resonator measurements.

Alternatively, k_{31} can be measured approximately, with the transmission circuit. Equation (170), for $Z_L 0$, gives

$$\frac{Y_{in}}{\omega_0 C_0} \doteq j\frac{\omega}{\omega_0} + \frac{2}{\pi}\frac{k_{31}^2}{1-k_{31}^2}\tanh\frac{\gamma l}{2} \tag{181}$$

Near resonance, $\tanh \gamma l/2$ is large-valued, and when $\cot^2 \beta l/2$ is large compared with $(\alpha l/2)^2$, Eq. (175) gives

$$\frac{Y_{in}}{\omega_0 C_0} \doteq \frac{1}{\pi}\frac{k_{31}^2}{1-k_{31}^2}\frac{\alpha l}{(\alpha l/2)^2 + \cot^2 \beta l/2} + j\left(\frac{\omega}{\omega_0} + \frac{2}{\pi}\frac{k_{31}^2}{1-k_{31}^2}\frac{1}{\cot \beta l/2}\right) \tag{182}$$

Equation (182) is accurate at f_a if the locus is a circle, so that $k_{31}^2 Q_m$ is on the order of 1, or larger. At f_a, the input admittance is real, and

$$\beta l/2 = \frac{2m-1}{2}\pi\frac{f_a}{f_0}$$
$$\cot\frac{\beta l}{2} = -\tan\left(\frac{2m-1}{2}\pi\frac{f_a - f_0}{f_0}\right) \tag{183}$$

Therefore

$$\frac{k_{31}^2}{1-k_{31}^2} \doteq \frac{\pi}{2}\frac{f_a}{f_0}\tan\left(\frac{2m-1}{2}\pi\frac{f_a - f_0}{f_0}\right)$$
$$\doteq \frac{\pi}{2}\frac{f_n}{f_m}\tan\left(\frac{2m-1}{2}\pi\frac{f_n - f_m}{f_m}\right) \tag{184}$$

Equation (184) is commonly used as the basis for measurement of coupling coefficient k_{31},[11,23] at fundamental resonance for which $m = 1$. A plot of the equation, such as Fig. 18, is convenient for quick evaluation of data. For small coupling factors, $f_n - f_m$ is small and

$$k_{31}^2 \doteq \frac{\pi^2}{4}\frac{f_n - f_m}{f_m} \tag{185}$$

Figure 19 shows the derivation of an equivalent circuit for the symmetrically loaded element, from the general circuit of Fig. 12. R_1, equal to $1/G_{max}$ in the admittance

CERAMIC DEVICES

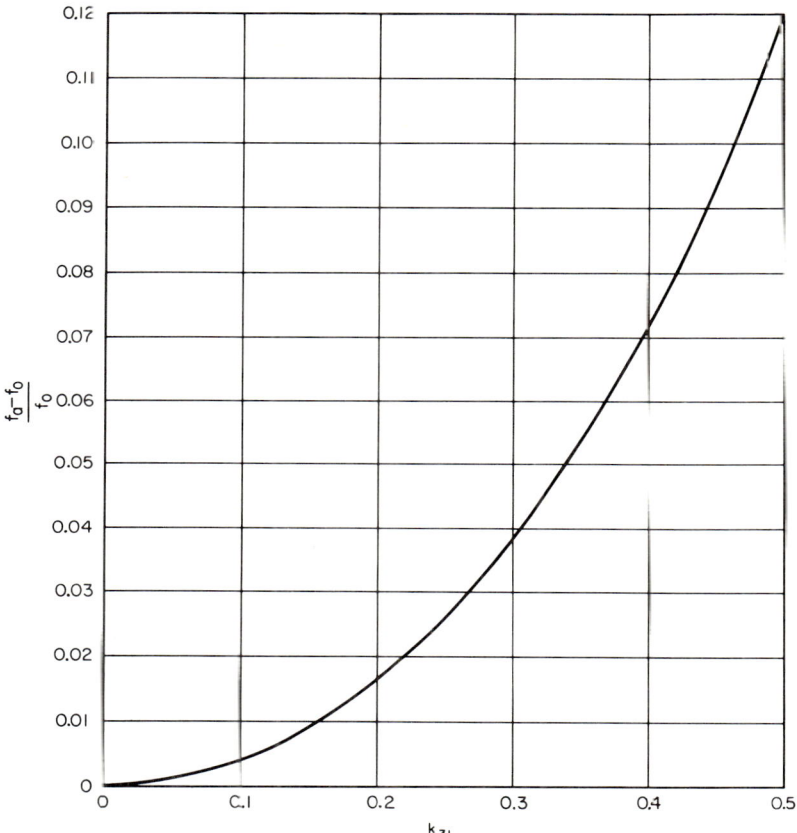

Fig. 18. Fundamental mode resonance-antiresonance bandwidth for constant-E k_{31} resonators.

locus, determines Q in the $R_1 L_1 C_1$ branch. From Eqs. (178) and (168):

$$R_1 = \frac{1}{D} = \frac{1}{\omega_c C_0} \frac{\pi^2(2m-1)}{8Q_m} \left(\frac{1}{k_{31}^2} - 1\right) = \frac{\pi(2m-1)}{8Q_m} Z_0$$

$$\frac{1}{\omega_c C_1} = Q_m R_1 = \frac{1}{\omega_0 C_0} \frac{\pi^2(2m-1)}{8} \left(\frac{1}{k_{31}^2} - 1\right) = \pi(2m-1)Z_0 \qquad (186)$$

$$\omega_0 L_1 = \frac{1}{\omega_0 C_1}$$

The capacitance ratio, characteristic of a resonator or transducer, is independent of loading:

$$r = \frac{C_1}{C_0} = \frac{8}{\pi^2(2m-1)} \frac{k_{31}^2}{1 - k_{31}^2} \qquad (187)$$

For a lossless element, with resonance at ω_0 and antiresonance at ω_a,

$$\omega_a^2 = (1/L_1)(1/C_1 + 1/C_0) = \omega_0^2(1 + r) \qquad (188)$$

16-40 AMPLIFYING DEVICES

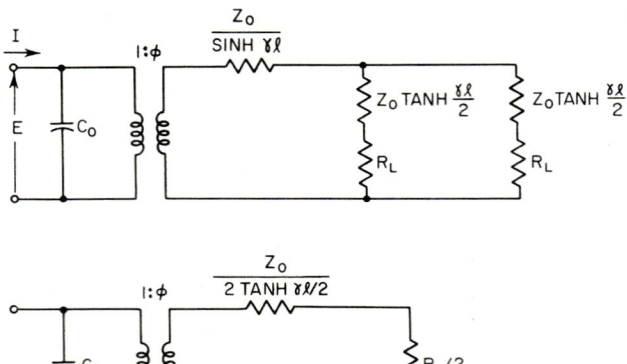

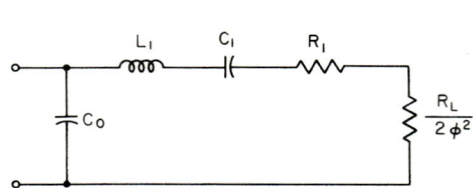

FIG. 19. Application of equivalent circuit analysis to the symmetrically loaded constant-E transducer.

Bandwidth from f_0 to f_a is thus given by the expression

$$\frac{f_a - f_0}{f_0} = \frac{f_a^2 - f_0^2}{f_0(f_a + f_0)}$$
$$\doteq \frac{\omega_a^2 - \omega_0^2}{2\omega_0^2} = \frac{r}{2} \qquad (189)$$

The locus of impedance on the complex plane, as the reciprocal of an admittance circle, is a circle with diameter

$$Z_{max} - Z_{min} = 1/Y_{min} - 1/Y_{max} = 1/(\omega_s C_0)^2 R_1 \qquad (190)$$

and with center at

$$\tfrac{1}{2}(1/Y_{min} + 1/Y_{max}) = 1/2(\omega_s C_0)^2 R_1 + 1/j\omega_s C_0 \qquad (191)$$

The *figure of merit*, defined for a resonator as

$$M = 1/\omega_s R_1 C_0 = D/\omega_s C_0 = Q_m/r \qquad (192)$$

can be determined by transmission measurements of Y_{max} and Y_{min}, when the locus is a circle:

$$Y_{max} = \sqrt{(\omega_s C_0)^2 + (1/2R_1)^2} + 1/2R_1$$
$$Y_{min} = \sqrt{(\omega_s C_0)^2 + (1/2R_1)^2} - 1/2R_1 \qquad (193)$$
$$\frac{Y_{max} - Y_{min}}{\sqrt{Y_{max} Y_{min}}} = \frac{Z_{max} - Z_{min}}{\sqrt{Z_{max} Z_{min}}} = \frac{1}{\omega_s C_0 R_1} = M$$

11. CONSTANT-D TRANSDUCERS

The characteristics of high capacitance and low impedance which make the side-plated transducer suitable for low-frequency applications such as sonar transducers bandpass filters, ultrasonic cleaning and machining equipment, microphones, phonograph pickups, and accelerometers become disadvantages at frequencies much above 1Mc. The majority of higher-frequency transducers, for ultrasonic delay lines, high-frequency filters, nondestructive test equipment, and instruments for medical diagnosis and treatment, are built in configurations with end-face electrodes.

There are distinct differences produced by these electrode positions, aside from impedance level. Transverse electrodes are equipotential surfaces extending lengthwise in the direction of stress-wave propagation. They create constant field intensity throughout the transducer, while surface charge density in the electrodes, and dielectric displacement in the piezoelectric material, vary according to patterns of stress. End-face electrodes, however, place no restriction on the lengthwise distribution of E. The absence of conduction charge in the ceramic assures zero divergence of D, and uniformity of stress over the cross section therefore assures that D has the same magnitude and direction everywhere in the transducer. The constant-D condition is accompanied by lengthwise variation of E, in accordance with stress patterns, and by impedance changes which, in high-coupling-coefficient materials, shift the apparent resonance frequency by significant amounts.

Table 6 summarizes the analytical description of the four basic transducers which comprise the constant-D class. These are idealized, to the extent that single strain or single stress conditions are assumed, corresponding to plates with infinite width, or rods with vanishingly small cross section. Practical transducers are usually plates with width considerably more than one wavelength, or rods with thickness considerably less than a wavelength.

For each of the four constant-D transducers, solution of the equation of motion follows a pattern, which is illustrated here with the analysis of the thickness-shear plate.

The wave equation is

$$\rho(\partial^2 w/\partial t^2) = c_{44}(\partial^2 w/\partial x^2) \tag{194}$$

With cross-sectional area A, mechanical end variables $F_1 e^{j\omega t}$, $V_1 e^{j\omega t}$, $F_2 e^{j\omega t}$, and $V_2 e^{j\omega t}$, and electrical terminal variables $E e^{j\omega t}$ and $I e^{j\omega t}$, the solution begins as in Eqs. (154) through (162), with the results

$$F_1 = Z_0(V_1/\tanh \gamma l + V_2/\sinh \gamma l) + (h_{15}/j\omega)I$$
$$F_2 = Z_0(V_1/\sinh \gamma l + V_2/\tanh \gamma l) + (h_{15}/j\omega)I$$

where

$$Z_0 = A \sqrt{\rho c_{44}{}^D} = A\rho c$$
$$\gamma = j(\omega/c)$$
$$I e^{j\omega t} = j\omega A D_1 \tag{195}$$

and l is transducer length.

The terminal voltage is evaluated by integrating field along the transducer:

$$E = \frac{1}{e^{j\omega t}} \int_0^l E_1 \, dx = \frac{1}{e^{j\omega t}} \int_0^l [-h_{15}(\partial w/\partial x) + \beta_{11}{}^S D_1] \, dx$$
$$= (h_{15}/j\omega)(V_1 + V_2) + (l\beta_{11}{}^S/j\omega A)I$$
$$= (1/j\omega C_0)[I + \phi(V_1 + V_2)]$$

where
$$C_0 = A/\beta_{11}{}^S = \epsilon_{11}{}^S(A/l) = \epsilon_{11}{}^T(1 - k_{15}{}^2)(A/l)$$
$$\phi^2/\omega_0 C_0 Z_0 = k_{15}{}^2/\pi \tag{196}$$

Table 6 and Fig. 20 give the equivalent circuit parameters for the constant-D transducers. The circuit differs from the constant-E circuit by the presence of a negative capacitance $-C_0$. For small values of coupling coefficient, the reactance due to $-C_0$ is much smaller than Z_0/ϕ^2 and so can be neglected if the transducer faces are damped by mechanical impedance comparable with Z_0.

For the larger values of coupling factor encountered with ceramic transducers, $-C_0$ is not negligible. If the reflected mechanical impedance at the electrical side of the

Table 6. Constant-D Transducers

	Thickness expander	Thickness shear	Extensional	Torsional
Mode				
Shape	Plate	Plate	Rod, bar, tube	Circular rod, tube
Wave propagation	Along Z	Along X	Along Z	Along X
Polarization	Thickness	Width	Length	Circumference
Boundary conditions	Single strain: S_3 $\frac{\partial D_3}{\partial w} = 0$	Single strain: S_5 Single stress: T_5 $\frac{\partial D_3}{\partial x} = 0$	Single stress: T_3 $\frac{\partial D_3}{\partial w} = 0$	Single stress: T_5 Single strain: S_5 $\frac{\partial D_1}{\partial x} = 0$
Primary particle motion	Along Z	$\perp$ thickness, along Z	Along Z	Circumferential, along Z
Equation of motion	$\rho \frac{\partial^2 w}{\partial t^2} = \frac{\partial T_3}{\partial z}$	$\rho \frac{\partial^2 w}{\partial t^2} = \frac{\partial T_5}{\partial x}$	$\rho \frac{\partial^2 w}{\partial t^2} = \frac{\partial T_3}{\partial z}$	$\rho \frac{\partial^2 w}{\partial t^2} = \frac{\partial T_5}{\partial x}$
Piezoelectric equations	$T_3 = c_{33}{}^D S_3 - h_{33} D_3$ $E_3 = -h_{33} S_3 + \beta_{33}{}^S D_3$	$T_5 = c_{44}{}^D S_5 - h_{15} D_1$ $E_1 = -h_{15} S_5 + \beta_{11}{}^S D_1$	$T_3 = \frac{1}{s_{33}{}^D} S_3 - \frac{g_{33}}{s_{33}{}^D} D_3$ $E_3 = -\frac{g_{33}}{s_{33}{}^D} S_3 + \beta_{33}{}^T \left(1 + \frac{g_{33}{}^2}{s_{33}{}^D \beta_{33}{}^T}\right) D_3$	$T_5 = c_{44}{}^D S_5 - h_{15} D_1$ $E_1 = -h_{15} S_5 + \beta_{11}{}^S D_1$
Coupling factor	$k_{33}{}^2 = \dfrac{h_{33}{}^2}{c_{33}{}^D \beta_{33}{}^S}$	$k_{15}{}^2 = \dfrac{h_{15}{}^2}{c_{44}{}^D \beta_{11}{}^S}$	$k_{33}{}^2 = \dfrac{g_{33}{}^2}{\beta_{33}{}^T s_{33}{}^D \left(1 + \dfrac{g_{33}{}^2}{\beta_{33}{}^T s_{33}{}^D}\right)} = \dfrac{g_{33}{}^2}{\beta_{33}{}^T s_{33}{}^E}$	$k_{15}{}^2 = \dfrac{h_{15}{}^2}{c_{44}{}^D \beta_{11}{}^S}$
Wave velocity	$c_l = \sqrt{\dfrac{c_{33}{}^D}{\rho}}$	$c_s = \sqrt{\dfrac{c_{44}{}^D}{\rho}}$	$c = \sqrt{\dfrac{1}{\rho s_{33}{}^D}}$	$c_s = \sqrt{\dfrac{c_{44}{}^D}{\rho}}$
C_0	$\epsilon_{33}{}^S \dfrac{A}{l}$	$\epsilon_{11}{}^S \dfrac{A}{l} = \epsilon_{11}{}^T (1 - k_{15}{}^2) \dfrac{A}{l}$	$\epsilon_{33}{}^T (1 - k_{33}{}^2)$	$\epsilon_{11}{}^S \dfrac{A}{l}$
$\dfrac{\phi^2}{\omega_0 C_0 Z_0}$	$\dfrac{k_t{}^2}{\pi}$	$\dfrac{k_{15}{}^2}{\pi}$	$\dfrac{k_{33}{}^2}{\pi}$	$\dfrac{k_{15}{}^2}{\pi}$
$R_{100}C_0$	$\dfrac{\pi^2(2m-1)}{8 Q_m k_t{}^2}$	$\dfrac{\pi^2(2m-1)}{8 Q_m k_{15}{}^2}$	$\dfrac{\pi^2(2m-1)}{8 Q_m k_{33}{}^2}$	$\dfrac{\pi^2(2m-1)}{8 Q_m k_{15}{}^2}$
$r = \dfrac{C_0}{C}$	$\dfrac{\pi^2(2m-1)}{8 k_t{}^2}$	$\dfrac{\pi^2(2m-1)}{8 k_{15}{}^2}$	$\dfrac{\pi^2(2m-1)}{8 k_{33}{}^2}$	$\dfrac{\pi^2(2m-1)}{8 k_{15}{}^2}$

CERAMIC DEVICES

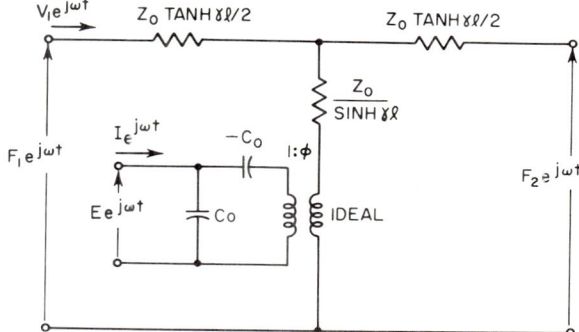

Fig. 20. Equivalent circuit for a constant-D transducer.

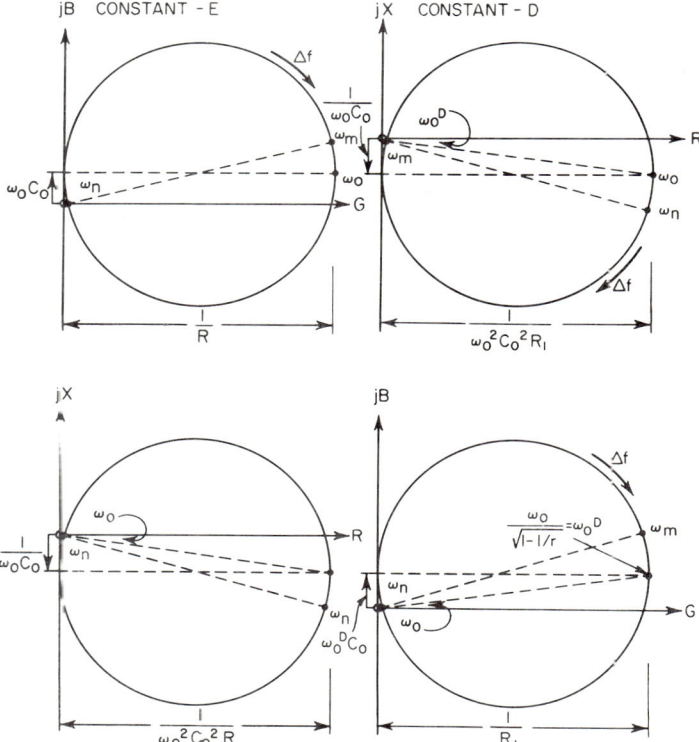

Fig. 21. Resonance loci for constant-E and constant-D resonators.

ideal transformer is represented, in terms of a frequency function $G(\omega)$, as $G(\omega)Z_0/\phi^2$, then the input impedance is

$$Z_{in} = \frac{(1/j\omega C_0)[-1/j\omega C_0 + G(\omega)Z_0/\phi^2]}{1/j\omega C_0 - 1/j\omega C_0 + G(\omega)Z_0/\phi^2}$$
$$= 1/j\omega C_0 + [1/(\omega C_0)^2][\phi^2/G(\omega)Z_0] \qquad (197)$$

This indicates that $G(\omega)$ is effectively inverted, as seen electrically through piezoelectric coupling. Mechanical resonance, corresponding to small $G(\omega)$, appears at the electrical terminals as a large resistance in series with C_0. If the mechanical admittance circle is measured electrically, it is observed as a motional-impedance circle, as shown in Fig. 21.

Mechanical resonances, which appear at the electrical terminals as fundamental and odd harmonics, are measured as high input impedances. The apparent resonances, measured as low impedances at frequencies below mechanical resonances, are not exactly harmonically related.

Alternative equivalent circuits are shown in Fig. 22, for $m = 1$ at fundamental resonance.

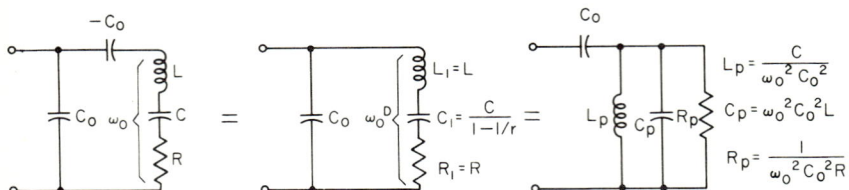

FIG. 22. Equivalent circuits for constant-D resonators.

The input impedance of a free resonator is

$$Z_{in} = \frac{1}{j\omega C_0} + \frac{1}{(\omega C_0)^2} \frac{2\Phi^2 \tanh \gamma l/2}{Z_0}$$
$$= (1/\omega C_0)[-j + (2/\pi)(\omega_0/\omega)k^2 \tanh \gamma l/2] \qquad (198)$$

where $\omega_0/2\pi$ is the fundamental mechanical resonance frequency. Electrical resonances occur at the frequencies in odd-harmonic sequence, for which $\beta l = (2m - 1)\pi$ and m is an integer; $\tanh \gamma l/2$ is large, and from Eq. (175)

$$Z_{in} \doteq \frac{1}{\omega C_0}\left[-j + \frac{2}{\pi}\frac{\omega_0}{\omega}k^2 \frac{1}{\alpha l/2 - j\cot(\beta l/2)}\right]$$
$$\doteq \frac{1}{\omega C_0}\left[-j + \frac{4Q_m}{(2m-1)\pi^2}\frac{\omega_0}{\omega}k^2(1 + e^{-j2\theta})\right] \qquad (199)$$

The maximum resistance at a mechanical resonance, where $\omega = (2m - 1)\omega_0$, is

$$R_p = \frac{1}{(2m-1)\omega_0 C_0} \frac{8Q_m k^2}{(2m-1)^2 \pi^2} \qquad (200)$$

The equivalent circuits of Fig. 22, for any one of the odd-harmonic resonances at $\omega = (2m - 1)\omega_0$, have values

$$\frac{1}{(2m-1)\omega_0 C_p} = (2m-1)\omega_0 L_p = \frac{R_p}{Q_m}$$
$$= \frac{1}{(2m-1)\omega_0 C_0} \frac{8k^2}{(2m-1)^2 \pi^2}$$
$$R_1 = \frac{1}{(2m-1)^2 \omega_0^2 C_0^2 R_p} = \frac{2m-1}{\omega_0 C_0} \frac{\pi^2}{8Q_m k^2} \qquad (201)$$
$$\frac{1}{(2m-1)\omega_0 C} = (2m-1)\omega_0 L_1 = Q_m R_1 = \frac{2m-1}{\omega_0 C_0} \frac{\pi^2}{8k^2}$$
$$r = \frac{C_0}{C} = \frac{C_p}{C_0} = \frac{(2m-1)^2 \pi^2}{8k^2}$$
$$\frac{C_0}{C_1} = r - 1$$

CERAMIC DEVICES

If the resonator is symmetrically loaded, with a mechanical resistance R_L at each end face, then only R_1 and R_p change. A resistance

$$R_L/2\Phi^2 = (1/\omega_0 C_0)(R_L/2Z_0)(k^2/\pi) \tag{202}$$

is added in series with R_1 and a resistance

$$\frac{2\Phi^2}{(2m-1)^2 \omega_0^2 C_0^2 R_L} = \frac{1}{(2m-1)^2 \omega_0 C_0} \frac{2Z_0}{R_L} \frac{\pi}{k^2} \tag{203}$$

is added in parallel with R_p. A given value of R_L therefore has successively less effect in lowering Q, at higher harmonic resonances.

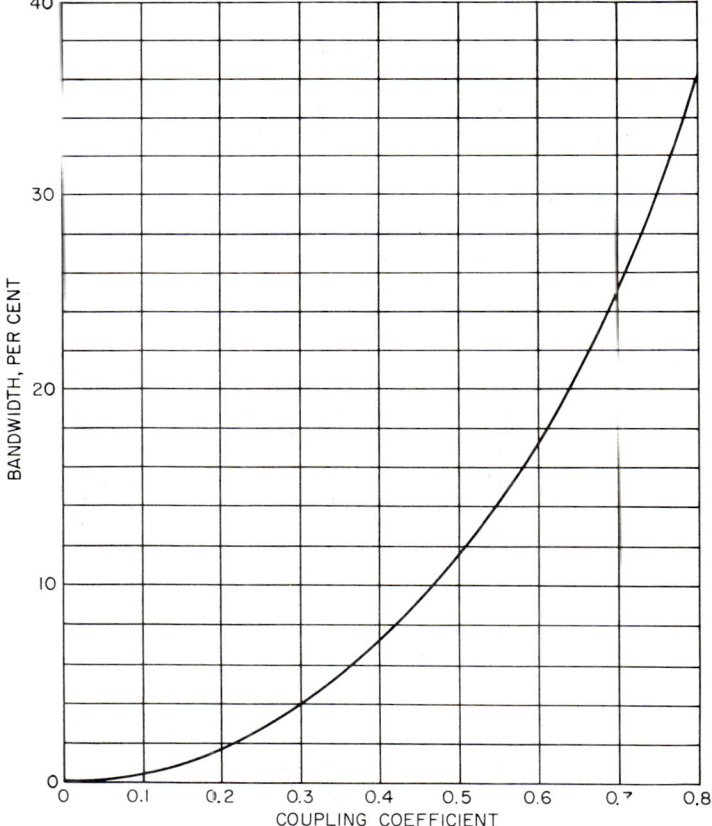

FIG. 23. Bandwidth for constant-D resonators.

The value of coupling coefficient, using a free resonator in the transmission circuit, is generally determined from the measured values of ω_m at lowest impedance and ω_n at highest impedance near fundamental resonance ω_0. Input reactance, from Eq. (199), is

$$X_{in} = \frac{1}{\omega C_0}\left[-1 + \frac{2}{\pi}\frac{\omega_0}{\omega}k^2\frac{\cot\beta l/2}{(\alpha l/2)^2 + (\cot\beta l/2)^2}\right] \tag{204}$$

At series input resonance ω_r, $X_{in} = 0$, and if the locus is a circle

$$(\alpha l/2)^2 \ll (\cot \beta l/2)^2$$

$$\tan \frac{\beta l}{2} = \cot \frac{\pi}{2} \frac{\omega_0 - \omega_r}{\omega_0}$$

$$X_{in} = 0 = \frac{1}{\omega_r C_0}\left(-1 + \frac{2}{\pi}\frac{\omega_0}{\omega_r} k^2 \cot \frac{\pi}{2}\frac{\omega_0 - \omega_r}{\omega_0}\right) \tag{205}$$

Since

$$\frac{\omega_0 - \omega_r}{\omega_0} \doteq \frac{\omega_n - \omega_m}{\omega_n} = \frac{\Delta f}{f_n} \tag{206}$$

the coupling coefficient is usually calculated[11,23] from the equation

$$k^2 = \frac{\pi}{2}\frac{f_m}{f_n} \tan \frac{\pi}{2}\frac{\Delta f}{f_n} \tag{207}$$

Figure 23, plotted from this equation, can be used to determine k from the measured value $\Delta f/f_n$; note that f_n, greater than f_m, is the frequency for large input impedance. For an end-plated bar, C_0 is generally just a few picofarads; so care must be taken to minimize stray capacitance between the electrodes. A grounded metallic shield, placed around the bar, is generally required.

12. CERAMIC-DISK RESONATOR

A thin ceramic disk, polarized in thickness and electroded on its faces, exhibits resonances with radial particle motion at frequencies which are not in harmonic order. The fundamental and first two overtone resonance frequencies are in the approximate ratio 1:2.63:4.18. In these vibrations, transverse piezoelectric coupling involves both radial and circumferential strains, and so the planar coupling factor is larger than k_{31} by a factor involving Poisson's ratio of transverse strain.

The analysis of a disk with full face electrodes has been given in detail by Mason,[13] and so is given here only in brief summary.

Motion with radial symmetry is most readily described in cylindrical coordinates. The z axis is taken through the disk thickness along the direction of bias polarization, r is measured radially from the center, and θ is the azimuthal angle. The corresponding displacements are u_z, u_r, and u_θ. The strains expressed in cylindrical coordinates are

$$\begin{aligned}
S_{rr} &= \partial u_r/\partial r \\
S_{\theta\theta} &= (1/r)(\partial u_\theta/\partial \theta) + u_r/r \\
S_{zz} &= \partial u_z/\partial z \\
S_{\theta z} &= (1/r)(\partial u_z/\partial \theta) + \partial u_\theta/\partial z \\
S_{rz} &= \partial u_r/\partial z + \partial u_z/\partial r \\
S_{r\theta} &= \partial u_\theta/\partial r - u_\theta/r + (1/r)(\partial u_r/\partial \theta)
\end{aligned} \tag{208}$$

The corresponding stresses T_{rr}, $T_{\theta\theta}$, T_{zz}, $T_{\theta z}$, T_{rz}, $T_{r\theta}$ have the customary significance: T_{ij} is force per unit area in the i direction, exerted on a face with outward normal in the j direction.

Newton's equations of motion[26] are

$$\begin{aligned}
\rho \ddot{u}_r &= \frac{\partial T_{rr}}{\partial r} + \frac{1}{r}\frac{\partial T_{r\theta}}{\partial \theta} + \frac{\partial T_{rz}}{\partial z} + \frac{T_{rr} - T_{\theta\theta}}{r} \\
\rho \ddot{u}_\theta &= \partial T_{r\theta}/\partial r + (1/r)(\partial T_{\theta\theta}/\partial \theta) + \partial T_{\theta z}/\partial z + T_{r\theta}/r \\
\rho \ddot{u}_z &= \partial T_{rz}/\partial r + (1/r)(\partial T_{\theta z}/\partial \theta) + \partial T_{zz}/\partial z + T_{rz}/r
\end{aligned} \tag{209}$$

The cylindrical-coordinate piezoelectric equations of state for the ceramic are given by matrices identical to those for rectangular coordinates:

$$\begin{bmatrix} S_{rr} \\ S_{\theta\theta} \\ S_{zz} \\ S_{\theta z} \\ S_{rz} \\ S_{r\theta} \\ \hline D_r \\ D_\theta \\ D_z \end{bmatrix} = \begin{bmatrix} s_{11}^E & s_{12}^E & s_{13}^E & 0 & 0 & 0 & 0 & 0 & d_{31} \\ s_{12}^E & s_{11}^E & s_{13}^E & 0 & 0 & 0 & 0 & 0 & d_{31} \\ s_{13}^E & s_{13}^E & s_{33}^E & 0 & 0 & 0 & 0 & 0 & d_{33} \\ 0 & 0 & 0 & s_{44}^E & 0 & 0 & 0 & d_{15} & 0 \\ 0 & 0 & 0 & 0 & s_{44}^E & 0 & d_{15} & 0 & 0 \\ 0 & 0 & 0 & 0 & 0 & s_{66} & 0 & 0 & 0 \\ \hline 0 & 0 & 0 & 0 & d_{15} & 0 & \epsilon_{11}^T & 0 & 0 \\ 0 & 0 & 0 & d_{15} & 0 & 0 & 0 & \epsilon_{11}^T & 0 \\ d_{31} & d_{31} & d_{33} & 0 & 0 & 0 & 0 & 0 & \epsilon_{33}^T \end{bmatrix} \begin{bmatrix} T_{rr} \\ T_{\theta\theta} \\ T_{zz} \\ T_{\theta z} \\ T_{rz} \\ T_{r\theta} \\ \hline E_r \\ E_\theta \\ E_z \end{bmatrix} \quad (210)$$

Boundary conditions for the thin disk in radial vibration are

$$u_\theta = 0$$
$$T_{zz} = T_{\theta z} = T_{rz} = T_{r\theta} = 0 \quad (211)$$
$$E_r = E_\theta = 0 \quad E_z = Ee^{j\omega t}/l_t$$

where $Ee^{j\omega t}$ is the interelectrode voltage and l_t is disk thickness.

With these conditions, the radial equation of motion reduces to

$$\rho\ddot{u}_r = -\omega^2 \rho u_r = \frac{\partial T_{rr}}{\partial r} + \frac{T_{rr} - T_{\theta\theta}}{r}$$
$$= \frac{s_{11}^E}{(s_{11}^E)^2 - (s_{12}^E)^2}\left(\frac{\partial^2 u_r}{\partial r^2} + \frac{1}{r}\frac{\partial u_r}{\partial r} - \frac{u_r}{r^2}\right) \quad (212)$$

This is Bessel's equation of the first order, with a general solution of the form

$$u_r = [AJ_1(\omega r/v) + BN_1(\omega r/v)]e^{j\omega t}$$
where $\quad v = \sqrt{\rho s_{11}^E[1 - (\sigma^E)^2]}$
and $\quad \sigma^E = -s_{12}^E/s_{11}^E \quad (213)$

σ^E is Poisson's ratio. J_1 and N_1 are Bessel's functions of order one, of the first and second kinds, respectively. Radial motion at the center of the disk is zero. $J_1(0)$ is zero, but $N_1(0)$ is infinite, and so $B = 0$ in Eq. (213):

$$u_r = AJ_1(\omega r/v)e^{j\omega t} \quad (214)$$

The two nonzero strains are thus

$$S_{rr} = \partial u_r/\partial r = A(\omega/v)J_0(\omega r/v)e^{j\omega t}$$
$$S_{\theta\theta} = u_r/r = (1/r)AJ_1(\omega r/v)e^{j\omega t} \quad (215)$$

Radial stress, calculated from the equations of state with these values of strain, must vanish at the rim of the disk where $r = a$. This condition determines the value of the constant

$$A = \frac{(1 + \sigma^E)d_{31}}{\dfrac{\omega}{v}J_0\left(\dfrac{\omega a}{v}\right) - \dfrac{(1 - \sigma^E)J_1(\omega a/v)}{a}} \quad (216)$$

and the stresses by solution from the equations of state are

$$T_{rr} = -\frac{d_{31}E_z}{1-\sigma^E}\left[1 - \frac{\dfrac{\omega}{v}J_0\left(\dfrac{\omega r}{v}\right) - \dfrac{(1-\sigma^E)J_1(\omega r/v)}{r}}{\dfrac{\omega}{v}J_0\left(\dfrac{\omega a}{v}\right) - \dfrac{(1-\sigma^E)J_1(\omega a/v)}{a}}\right]$$

$$T_{\theta\theta} = -\frac{d_{31}E_z}{1-\sigma^E}\left[1 - \frac{\dfrac{\sigma^E\omega}{v}J_0\left(\dfrac{\omega r}{v}\right) + \dfrac{(1-\sigma^E)J_1(\omega r/v)}{r}}{\dfrac{\omega}{v}J_0\left(\dfrac{\omega a}{v}\right) - \dfrac{(1-\sigma^E)J_1(\omega a/v)}{a}}\right] \quad (217)$$

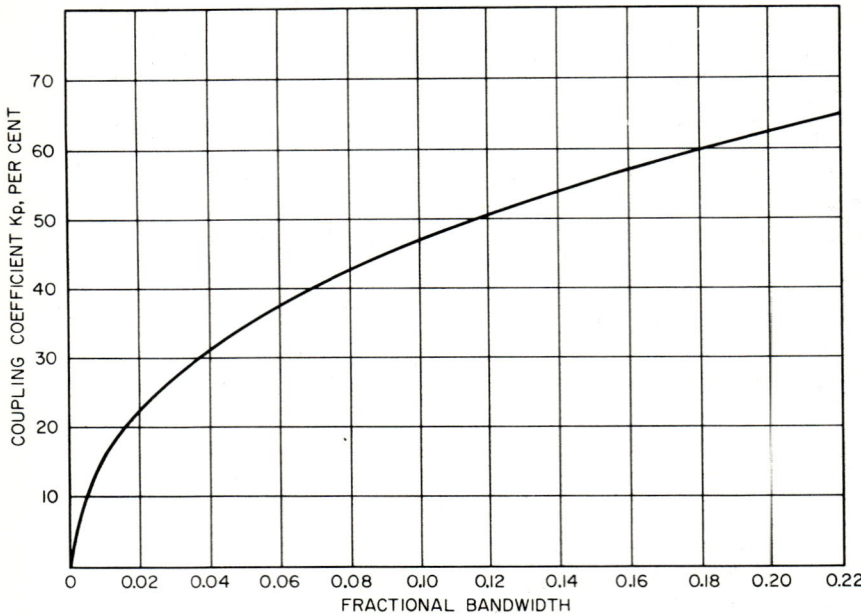

Fig. 24. Bandwidth for radial mode disk resonators.

The surface charge, evaluated by integrating D_z in the equation of state, with these stress values, over the electroded face area πa^2, is

$$Q = E_z C_0 \left[1 + \frac{k_p^2}{1 - k_p^2} \frac{(1 + \sigma^E) J_1(\omega a/v)}{(\omega a/v) J_0(\omega a/v) - (1 - \sigma^E) J_1(\omega a/v)} \right]$$

where

$$k_p^2 = \frac{2 d_{31}^2}{(1 - \sigma^E) s_{11}^E \epsilon_{33}^T} = \frac{2 k_{31}^2}{1 - \sigma^E}$$

and the "radially clamped capacitance" is

$$C_0 = \pi a^2 \epsilon_{33}^T (1 - k_p^2) \qquad (218)$$

The input admittance is therefore $j\omega Q/E_z$, which is infinite when

$$(\omega a/v) J_0(\omega a/v) - (1 - \sigma^E) J_1(\omega a/v) = 0 \qquad (219)$$

For $\sigma = 0.27$, Mason gives the lowest root

$$R_1 = \omega a/v = 2.03 \qquad (220)$$

so that the resonant frequency is

$$f_r = \frac{2.03}{2\pi a} \sqrt{\frac{1}{s_{11}^E \rho [1 - (\sigma^E)^2]}} \qquad (221)$$

The antiresonant frequency, evaluated from the first terms in a Taylor's series expansion for the Bessel's functions in the equation for input admittance, is given by the equations

$$f_a = f_r (1 + \Delta f/f_r)$$
$$\frac{\Delta f}{f_r} \doteq \frac{k_p^2}{1 - k_p^2} \frac{1 + \sigma^E}{R_1^2 - [1 - (\sigma^E)^2]} \qquad (222)$$

CERAMIC DEVICES   16–49

Solution of this equation for coupling factor,

$$k_p^2 \doteq \frac{\Delta f}{f_r} \frac{R_1^2 - [1 - (\sigma^E)^2]}{1 + \sigma^E} \qquad (223)$$

can be used to determine k_p from measured Δf and f_r, since the value of R_1 varies only negligibly for values of σ^E found in ceramics. Use of the curve in Fig. 24 plotted from the full solution is more accurate for large coupling factors. In turn, therefore, measurement of k_p and k_3 in equally poled samples of the same ceramic material can determine σ^E.

Some types of disk resonators with divided ring-and-dot electrodes are used as ceramic transformers. This geometry can be analyzed as a disk and an outer ring in firm mechanical contact. The solution of the radial equation in the ring then requires an extra boundary condition, which appears since in Eq. (213) $BN_1\,(\omega r/v)$ then no longer vanishes. Details of this analysis have been given recently by Munk.[27]

The equivalent circuit for a disk resonator should describe input admittance as the susceptance $j\omega C_0$ in parallel with a motional admittance. Near a resonance, the motional admittance is a series $R_1L_1C_1$ circuit, resonant at f_r, so that

$$\begin{aligned}
\omega_r^2 &= 1/L_1C_1 \\
Q_m &= \omega_r L_1/R_1 \\
\omega_a^2 &\doteq (1/L_1)(1/C_1 + 1/C_0) = \omega_r^2(1 + C_1/C_0)
\end{aligned} \qquad (224)$$

13. CERAMIC TRANSFORMERS*

If the electrodes of a piezoelectric element are subdivided, then vibrations can be produced with an applied alternating voltage across an input set of electrodes, and

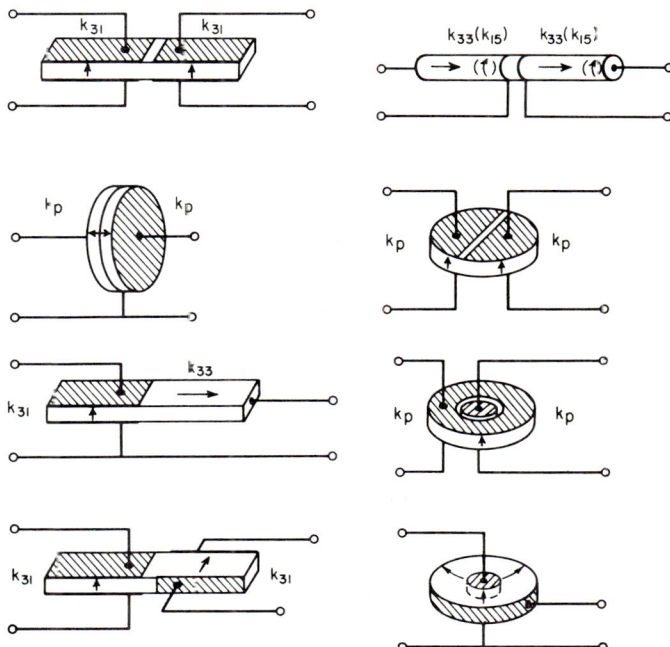

FIG. 25. Ceramic transformer geometries.

* Much of the material given here on ceramic transformers, previously unpublished, was generated from work supported by the U.S. Army Signal Corps under contracts DA 36-039-sc-75014 and DA 36-039-sc-78346 and by the General Electric Company's Electronics Laboratory.

16–50 AMPLIFYING DEVICES

voltage piezoelectrically induced by the vibrations appears across the output electrode set. The element performs as a tuned transformer, providing maximum transmission at frequencies of mechanical resonance. Depending on the electrode configuration, the input and output impedances can be alike or widely different, with corresponding ratios of voltage and current. Materials for this purpose have been ferroelectric ceramics, since desirable characteristics include high coupling coefficient, high mechanical Q, high dielectric constant, and controlled direction of polarization.

Figure 25 shows ceramic transformers of several basic types. Theoretical analysis of the performance can be carried out from equivalent circuits for the individual component transducers, joined together at the terminals representing force and velocity at each interface. Figure 26 illustrates this technique for a symmetrical transformer such as the $k_{31} - k_{31}$ bar element. The equivalent circuit of Fig. 12, for a transformer with total length l and transducer length $l/2$, is used with $F_1 = 0$ for each of the half lengths. The interconnection divides the circuit into symmetrical halves.

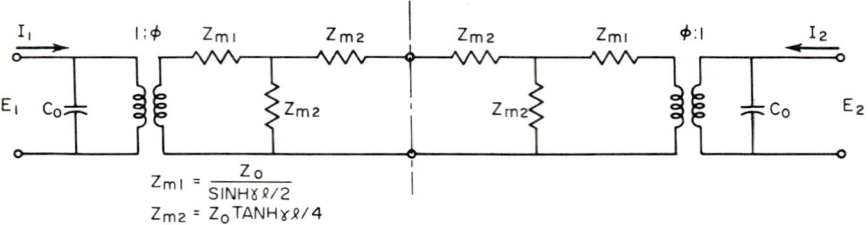

Fig. 26. Equivalent circuit for a symmetrical ceramic transformer.

Transmission from a generator to a load can be calculated for any ceramic transformer by means of straightforward circuit calculations. This can be a complicated process, however, since the complete circuit, with terminations, forms five meshes. If the two component transducers are reasonably alike in mechanical properties, then the frequency-response characteristics are essentially those of the symmetrical transformer.

Bartlett's bisection theorem[28] provides a direct solution for the transformer represented by a circuit with symmetrical mechanical parts, i.e., with equal values of Z_0 and γl. In Fig. 27, Z_m' is the impedance of *one-half* of the mechanical circuit, ending at the interface junction. From the bisection theorem, the impedances Z_x and Z_y forming the equivalent lattice are $(Z_m')_{sc}$ and $(Z_m')_{oc}$, respectively:

$$Z_x = \frac{Z_0}{\sinh \gamma l/2} + Z_0 \tanh \gamma l/4$$

$$= \frac{Z_0}{\sinh \gamma l/2} + Z_0 \frac{\cosh \gamma l/2 - 1}{2 \sinh \gamma l/2}$$

$$= \frac{Z_0}{2 \tanh \gamma l/4}$$

$$Z_y = \frac{Z_0}{\sinh \gamma l/2} + Z_0 \tanh \gamma l/4$$

$$= \frac{Z_0}{\tanh \gamma l/2} \qquad (225)$$

From the lattice, an equivalent half lattice with phase-inverting transformer can also be drawn. These circuits illustrate the basic characteristics of transformer

CERAMIC DEVICES

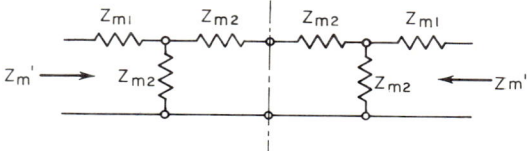

BISECTION THEOREM

$$(Z_m')^{OC} = Z_{m1} + Z_{m2} = \frac{Z_0}{\text{TANH } \gamma\ell/2}$$

$$(Z_m')^{SC} = Z_{m1} + \tfrac{1}{2} Z_{m2} = \frac{Z_0}{2 \text{ TANH } \gamma\ell/4}$$

LATTICE EQUIVALENT

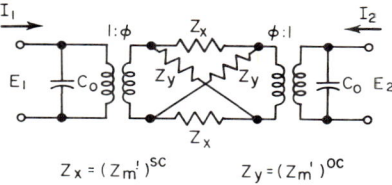

$Z_x = (Z_m')^{SC} \qquad Z_y = (Z_m')^{OC}$

LADDER EQUIVALENT

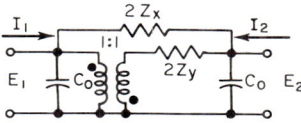

Fig. 27. Equivalent circuits based on mechanical symmetry.

response as sketched in Fig. 28. Z_y is series-resonant at the fundamental resonance f_0 for the entire transformer, and at odd harmonics, providing resonant transmission with phase reversal. Z_x is series-resonant at $2f_0$ and odd harmonics of $2f_0$, providing in-phase transmission. Transmission nulls are found at $4f_0$ and harmonics of $4f_0$. On a decibel scale, the response plotted vs. frequency over a wide spectrum thus appears as periodically repeating three-tined fork patterns. In practice, the trident is usually found to be well defined only over the first pattern, because of multiple-coupled resonances which arise at higher harmonics from interactions with other dimensions of the filter. Within the trident the dips are influenced by choice of terminating impedances, and by transmission through stray capacitance, but a dip represents roughly 20 db in practical elements.

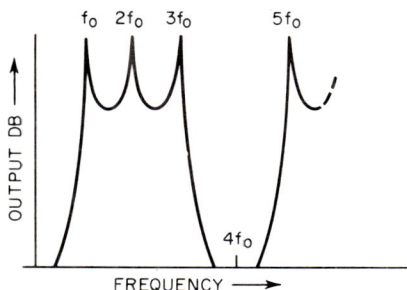

Fig. 28. Basic frequency response of a longitudinal ceramic transformer.

The admittance between input and output terminals is $1/2Z_x - 1/2Z_y$. A zero of transmission can be produced by bridging an admittance $-(1/2Z_x - 1/2Z_y)$ between the input and output terminals. Since Z_x and Z_y contain resistance components due to mechanical loss in the ceramic and its mounting supports, a complete null can be provided only if $1/2Z_x - 1/2Z_y$ includes negative resistance, which occurs in the vicinity of a resonance with phase reversal. An RC bridging network is required for a null at frequencies below this resonance, and an RL network can produce a null

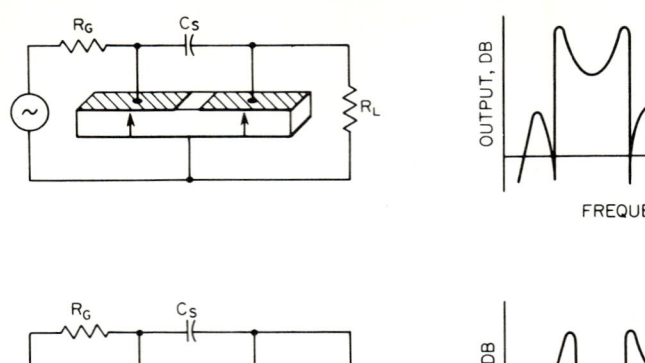
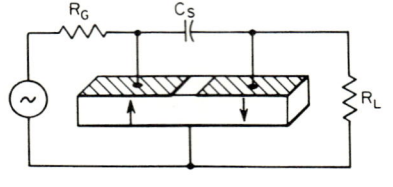
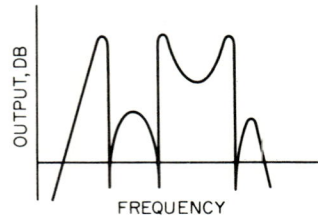

FIG. 29. Effects of bridging capacitance on the frequency response of a ceramic transformer.

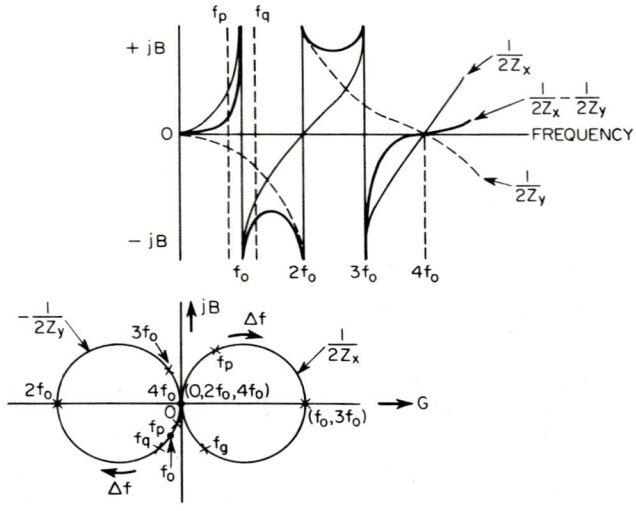

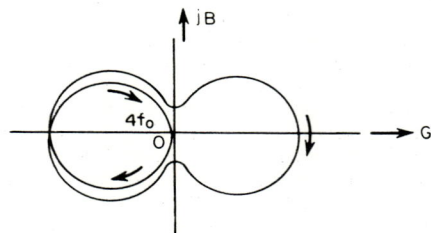

FIG. 30. Transmission loci for ceramic transformers.

16–52

CERAMIC DEVICES 16-53

above resonance. Since losses are small, C or L can produce sharp transmission dips, although not complete nulls. In addition, C produces a sharp dip above an in-phase resonance, and L below an in-phase resonance. Figure 29 illustrates typical distortion of the trident pattern for relative senses of ceramic polarization.

The locus of $1/2Z_x$ on the admittance plane sketched in Fig. 30 is a resonance circle in the right half plane, rotating at half the rate of the left half plane circle locus of $-1/2Z_y$. The sum, representing total mutual admittance, forms a figure 8, closing at the origin for frequencies 0 and $4f_0$. Addition of a bridging admittance, such as $j\omega C_s$ due to stray capacitance, shifts that locus and produces a transmission zero if the

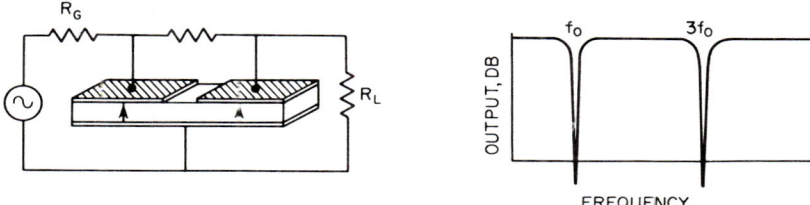

Fig. 31. Ceramic notch filter.

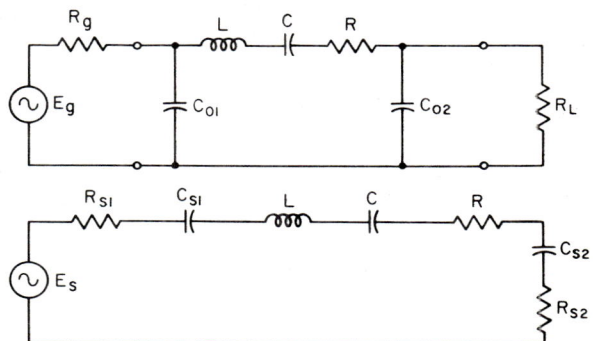

Fig. 32. Simplified transformer representations.

locus then passes through the origin. A bridging resistance, chosen to null near f_0 or $3f_0$, produces a notch filter, as in Fig. 31. Q for this characteristic is measured at the frequencies for which the output is down by 3 db from the uniform response level. If the sense of poling is reversed in one half length, then notches fall at $2f_0$ and its multiples.

For frequencies near a resonance, the nonresonant lattice arm can be neglected, and the resonant arm can be represented by a series LCR branch, as in Fig. 32. Analysis of the effects of RC terminations is then simplified considerably by replacing R_fC_{01}, and R_LC_{02}, with series equivalents evaluated at the resonant frequency f_r. If R_p and C_p are the shunt values, then the series impedance is

$$R_s + \frac{1}{j\omega_r C_s} = \frac{1/R_p + j\omega_r C_p}{(1/R_p)^2 + (\omega_r C_p)^2} \tag{226}$$

so the equivalent element values are

$$R_s = \frac{R_p}{1 + (\omega_r C_p R_p)^2} = \frac{R_p}{1 + Q_T^2}$$

$$C_s = \frac{C_p}{1 + 1/Q_T^2} \tag{227}$$

where Q_T is the termination circuit Q.

Using these values, the resonant frequency of the loop is $\omega_m/2\pi$, where

$$\omega_m^2 = (1/L)(1/C + 1/C_{s1} + 1/C_{s2})$$
$$= \omega_r^2(1 + C/C_{s1} + C/C_{s2}) \qquad (228)$$

The 3-db bandwidth of output voltage corresponds to the effective circuit Q,

$$Q_{eff} = \frac{\omega_m L}{R + R_{s1} + R_{s2}} \equiv \frac{1}{1/Q_m + 1/Q_1 + 1/Q_2} \qquad (229)$$

where $Q_1 = \omega_m L/R_{s1}$ and $Q_2 = \omega_m L/R_{s2}$ are associated with coupled resistive loading at input and output, respectively.

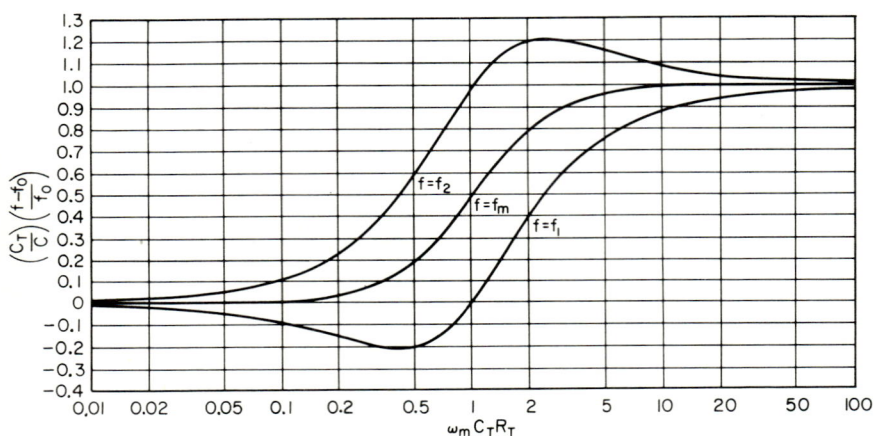

FIG. 33. Effects of terminating resistance on transformer response.

Figures 33 and 34 show the effects on frequency f_m and 3-db frequencies f_1 and f_2 produced by changing either R_G or R_L and either C_{01} or C_{02}. Increasing values of C_0 always reduces bandwidth and center frequency. Increasing R_L or R_G always increases frequency, but the bandwidth is maximum when the shunt resistance equals the reactance of the corresponding C_{01} or C_{02}. The minimum center frequency produced when both input and output are shorted by the terminations is the mechanical resonance frequency f_r. The maximum center frequency $f_r \sqrt{1 + C/C_{01} + C/C_{02}}$ is produced when R_L and R_G are infinite. If both halves of the ceramic transformer are driven effectively in parallel, with correctly phased voltages to excite resonances, then the parallel load is a ceramic resonator, with series resonance at f_r and parallel resonance at $f_r \sqrt{1 + C/C_{01} + C/C_{02}}$. C_{01} and C_{02} may include external circuit capacitances, but when they are just the clamped capacitances of the ceramic, then parallel resonance is determined by the ceramic coupling factor.

Power available from the generator with resistance R_G, into a matched load resistance, is

$$P_{avail} = |E_G|^2/4R_G \qquad (230)$$

Power delivered to a load resistance R_L, not necessarily equal to R_G, through a symmetrical ceramic transformer is

$$P_L = \frac{|E'|^2 R_{s2}}{(R_{s1} + R + R_{s2})^2} = \frac{\dfrac{|E_G|^2}{1 + Q_{T1}^2}\dfrac{R_L}{1 + Q_{T2}^2}}{(\omega_m L/Q_{eff})^2} \qquad (231)$$

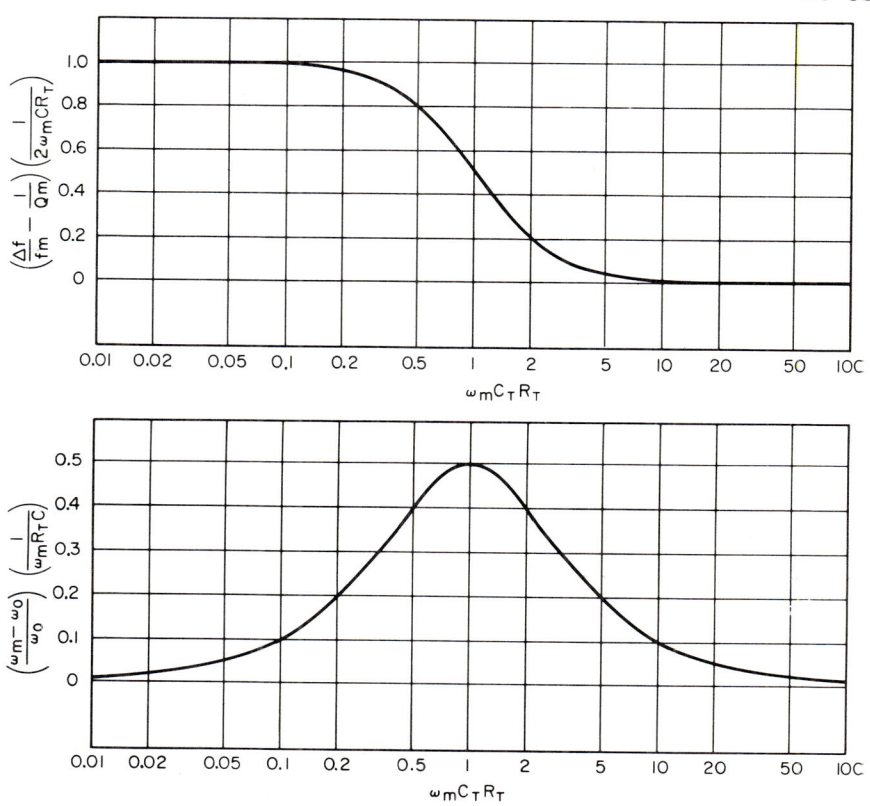

FIG. 34. Effects of terminating capacitance on transformer response.

Then the *transducer loss*, in decibels, is

$$\text{T.L.} = 10 \log \frac{P_{avail}}{P_L}$$
$$= 10 \log \frac{(\omega_m L)^2 (1 + Q_{T1}^2)(1 + Q_{T2}^2)}{4 R_L R_G Q^2_{eff}}$$
$$= 10 \log \frac{(\omega_m L)^2}{4 R_{s1} R_{s2} Q^2_{eff}}$$
$$= 10 \log \frac{Q_1 Q_2}{4 Q^2_{eff}} \quad (232)$$

When $R_L = R_G$, then for the symmetrical transformer $Q_1 = Q_2$, and Eq. (232) reduces to

$$\text{T.L.} = 20 \log \frac{1}{1 - Q_{eff}/Q_m} \quad (233)$$

This is plotted in Fig. 35. Since, for example, a value $Q_{eff} = 10$ is possible, using high-coupling-coefficient ceramic with Q_m of about 500, very small values of insertion loss are possible.

Unsymmetrical transformers can be built to provide impedance matching between unequal terminations. Since efficiency is highest when the terminating resistance and

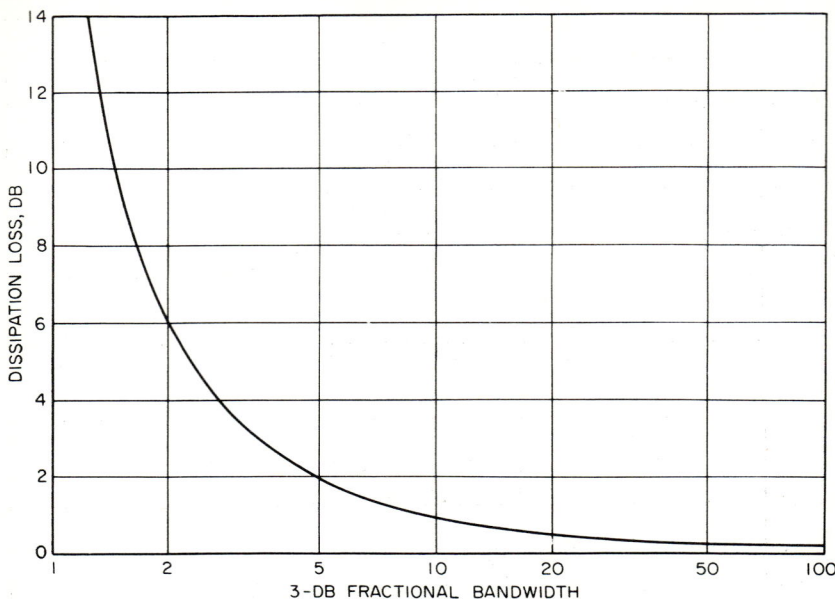

Fig. 35. Transducer loss for a ceramic transformer.

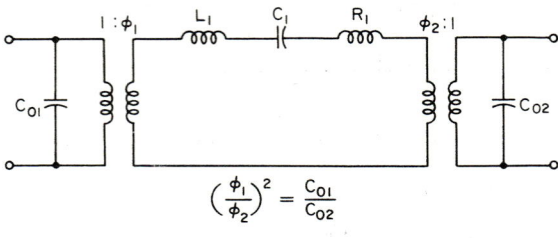

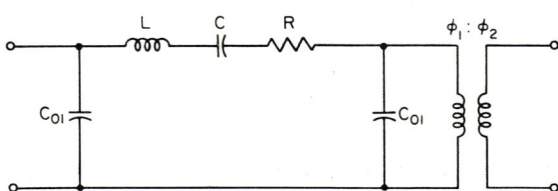

Fig. 36. Equivalent circuits for a ceramic transformer with mechanical symmetry, and equal coupling factors, but unequal clamped capacitances.

corresponding clamped capacitance are equal, a transverse transformer with large input capacitance and small output capacitance can match a low-impedance generator to a high-impedance load. If the two component transducers have equal coupling factors and are mechanically equivalent, then the transformer equivalent circuit contains an ideal transformer with turns ratio (squared) equal to the capacitance ratio,

CERAMIC DEVICES 16–57

as in Fig. 36. The unsymmetrical transformer is then equivalent to a symmetrical transformer in cascade with an ideal matching transformer.

If the component transducers have unequal coupling factors, as in the $k_{31} - k_{33}$ bar, but equal mechanical properties, then the turns ratio N is also dependent on the ratio of coupling factors. From the input terminals, with output short-circuited or open-circuited as convenient, the admittance is that of a resonator, and the capacitance ratio C_{01}/C can be computed in terms of coupling factor or determined from measurement of the series and parallel resonance frequencies. From the output end, the effective capacitance ratio calculated or measured is $N^2 C_{02}/C$. Since C_{01} and C_{02} can be separately computed or measured, this determines N. Rosen[29] has discussed the $k_{31} - k_{33}$ transformer for voltage step-up applications in detail, Lungo and Sauer and[30] have described disk transformers, and Munk has given a detailed analysis of constant-E disk transformers. A number of configurations are disclosed in patents by Rosen, Fish, and Rothenberg.[31]

14. DELAY LINE TRANSDUCERS

An ultrasonic delay line consists of an input transducer, an acoustic delay medium, and an output transducer. The delay medium, a solid or liquid propagation path, provides essentially all the signal delay, and the transducers are necessary appendages which launch and detect stress waves, introducing phase shift and—in most cases—a considerable amount of insertion loss. Except for special-purpose devices such as dispersive delay lines, which are designed to modify an input signal in a prescribed way, ultrasonic delay lines are designed to store information with a minimum of distortion. However, since power in an acoustic wave is $V^2 Z_0$ in terms of particle velocity, or $\omega^2 u^2 Z_0$ in terms of frequency and particle displacement, transmission of a uniform power spectrum requires increasingly large displacements at decreasing frequencies and becomes impractical in the d-c limit. Transducers and coupling networks are limited to moderately narrow passbands of efficient operation, and so information-storage delay lines employ modulated signals with sidebands lying within their passbands or digitally coded video pulse chains to minimize distortion of the signal information.

Reciprocity ensures that an input transducer and an output transducer have identical response characteristics, as measured by a corresponding function, provided that the electrical and mechanical terminations are correspondingly identical for the transducers. For example, if the transducer is represented by a four-terminal equivalent network, in which one terminal pair displays input voltage and current and the other pair displays force and particle velocity at the interface between transducer and delay medium, then the mutual impedances representing, respectively, voltage divided by velocity and force divided by current are equal. Since these impedances describe the functions of transmitting and receiving stress-wave energy, it is sufficient to compute the response of a transducer in only one mode of operation.

Figure 37 illustrates a few commonly employed configurations. Figure 38 is a general cross-sectional view of a transducer, loaded at its front face with the delay medium and at its back face with an absorptive backing which can be used to increase bandwidth through the added damping it imposes on the transducer. Figure 39 is the equivalent circuit for the transducer, terminated electrically in a circuit represented by the current I_g in parallel with the termination admittance Y_g. Characteristic impedances Z_B, Z_t, and Z_C are computed as the products of transducer area times ρc for, respectively, the backing, transducer, and delay media.* The impedance Z_e represents the capacitive impedance $1/j\omega C_0$ or, when the transducer is dielectrically lossy, the impedance $1/(G_a + j\omega C_0)$; the element $(-Z_e)$ is included for a constant-D transducer, short-circuited for a constant-E transducer.

From this equivalent circuit, performance of the transducer can be computed by straightforward application of Kirchhoff's laws. In this process, it is convenient to

* Figure 39 is an idealization in that the interfaces are presented as ideal bonds. Considerable development has been devoted to fabrication methods for attaching the transducer without deleterious effects.

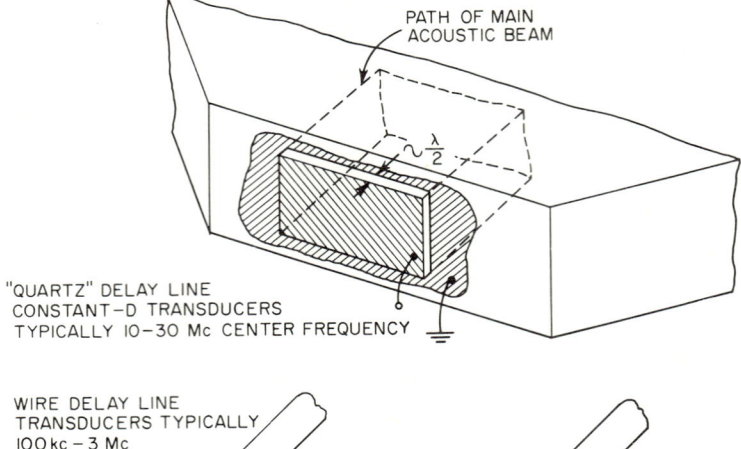

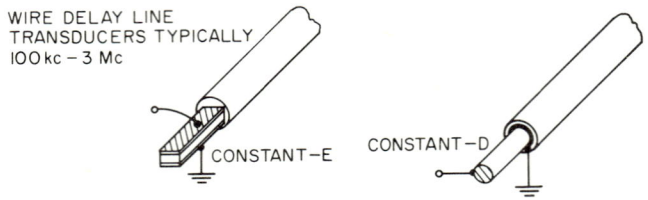

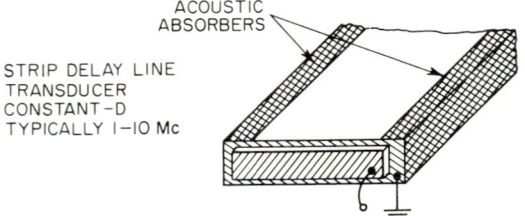

Fig. 37. Delay-line transducer configurations.

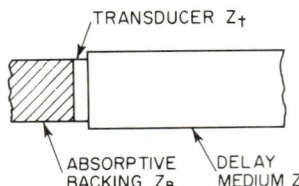

Fig. 38. Transducer cross section.

represent combinations of terms by these symbols:

$$K_0 = Z_0/Z_t$$
$$K_B = Z_B/Z_t$$
$$\xi = \frac{Z_0 + Z_t \tanh \gamma l/2}{Z_B + Z_t \tanh \gamma l/2} = \frac{K_0 + \tanh \gamma l/2}{K_B + \tanh \gamma l/2}$$
$$\nu = K_0 + \coth \gamma l \tag{234}$$

CERAMIC DEVICES   16-59

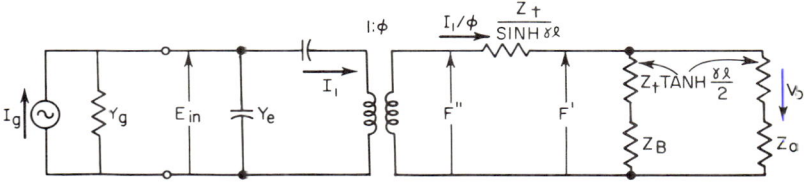

Fig. 39. Equivalent circuit for a delay-line transducer with absorptive backing.

At half-wave transducer resonance, $\beta l = \pi$. Since the transducer is heavily damped by Z_B and Z_0, the contribution from internal transducer friction to bandwidth is small, and so γl may generally be replaced with $j\beta l$. The variable ξ then approaches a simple limit at resonance:

$$\xi \doteq \frac{K_0 + j \tan \beta l/2}{K_B + j \tan \beta l/2} \xrightarrow[\substack{\omega = \omega_0 \\ \beta = \pi}]{} 1 \tag{235}$$

The variable ν appears in the following analysis in the expression $\nu + \xi/\sinh \gamma l$, which becomes

$$\nu + \frac{\xi}{\sinh \gamma l} = K_0 + \frac{1}{\tanh \gamma l} + \frac{1}{\sinh \gamma l} + \frac{\xi - 1}{\sinh \gamma l}$$

$$= K_0 + \frac{1}{\tanh \gamma l/2} + \frac{K_0 - K_B}{K_B \sinh \gamma l + \coth \gamma l - 1} \to \frac{K_0 + K_B}{2} \tag{236}$$

With these preliminaries, the circuit analysis can be written concisely:

Constant-E Transducer

$$\begin{aligned}
F' &= V_0(Z_0 + Z_t \tanh \gamma l/2) \\
I_1/\phi &= V_0 + \frac{F'}{Z_B + Z_t \tanh \gamma l/2} = V_0(1 + \xi) \\
F'' &= F' + \frac{I_1}{\phi} \frac{Z_t}{\sinh \gamma l} = V_0 Z_t \left(\nu + \frac{\xi}{\sinh \gamma l}\right) \\
E_{in} &= \frac{V_0}{\phi} Z_t \left(\nu + \frac{\xi}{\sinh \gamma l}\right) \\
I_{in} &= I_1 + E_{in} Y_e = \phi V_0 \left[1 + \xi + \frac{Y_e Z_t}{\phi^2}\left(\nu + \frac{\xi}{\sinh \gamma l}\right)\right] \\
Y_{in} &= \frac{I_{in}}{E_{in}} = Y_e \left(1 + \frac{\phi^2}{Y_e Z_t} \frac{1 + \xi}{\nu + \xi/\sinh \gamma l}\right) \\
I_{in} &= \phi V_0 (1 + \xi) \frac{Y_{in}}{Y_{in} - Y_e} \\
I_g &= \phi V_0 \left[1 + \xi + \frac{(Y_e + Y_g)Z_t}{\phi^2}\left(\nu + \frac{\xi}{\sinh \gamma l}\right)\right] \\
E_{in} &= \frac{I_g}{Y_{in} + Y_g} = \frac{I_g}{Y_g + Y_e\left(1 + \frac{\phi^2}{Y_e Z_t} \frac{1 + \xi}{\nu + \xi/\sinh \gamma l}\right)}
\end{aligned} \tag{237}$$

From Eq. (168) for the constant-E transducer,

$$\frac{\phi^2}{Y_e Z_t} = \frac{1}{j\pi} \frac{\omega_0}{\omega} \frac{k^2}{1 - k^2} \tag{238}$$

At fundamental resonance $\omega = \omega_0$, the limits for a lossless transducer then become

$$E_{in} \to \frac{V_0}{\phi} Z_t \frac{K_0 + K_B}{2}$$
$$I_{in} \to \phi V_0 \left[2 + \frac{1}{j\pi} \frac{k^2}{1 - k^2} \frac{K_0 + K_B}{2} \right]$$
$$Y_{in} \to j\omega_0 C_0 + \frac{\omega_0 C_0}{\pi} \frac{k^2}{1 - k^2} \frac{4}{K_0 + K_B} \quad (239)$$
$$I_g \to \phi V_0 \left[2 + j\pi \left(1 + \frac{Y_g}{j\omega_0 C_0}\right) \left(\frac{1}{k^2} - 1\right) \frac{K_0 + K_B}{2} \right]$$

Frequency response can be calculated from Eqs. (237) and insertion loss can be calculated at resonance, which approximates band center, from Eqs. (239). Figure 40, computed as response in decibels with a constant driving voltage E_{in} and no backing ($Z_B = 0$) illustrates the effects of acoustic impedance ratios Z_0/Z_t on bandwidth; if $Z_0 \gg Z_t$, quarter-wave resonances become apparent. A considerable variation in

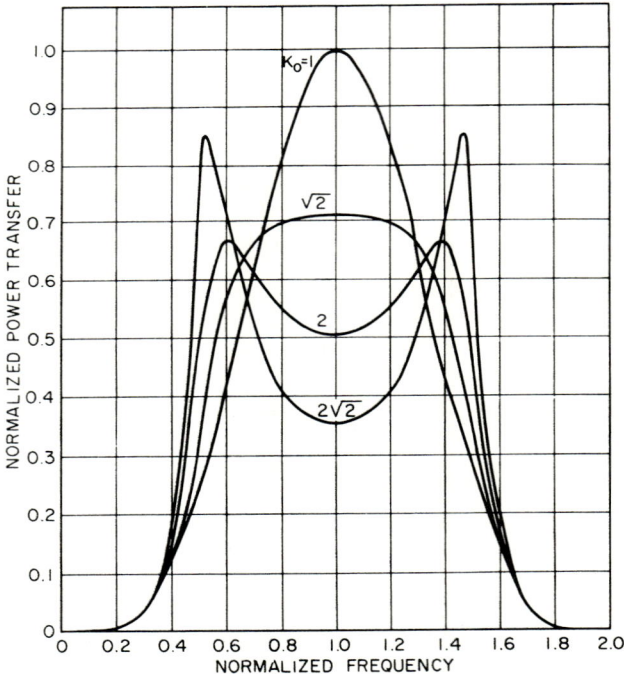

FIG. 40. Response of a free-backed transducer, from a voltage source, or into a short-circuit load.

frequency response can be produced by varying values of K_0, K_B, and Y_g. Computed curves for several combinations are given by Roth,[32] and by Brockelsby et al.[33]

Constant-D Transducer. Definitions of ξ and ν are unchanged:

$$\begin{aligned}
I_1/\phi &= V_0(1 + \xi) \\
F'' &= V_0 Z_t(\nu + \xi/\sinh \gamma l) \\
E_{in} &= (V_0/\phi) Z_t[\nu + \xi/\sinh \gamma l - (\phi^2/Y_e Z_t)(1 + \xi)] \\
I_{in} &= (V_0/\phi) Z_t Y_e(\nu + \xi/\sinh \gamma l) \\
Z_{in} &= \frac{1}{Y_e}\left(1 - \frac{\phi^2}{Y_e Z_t} \frac{1 + \xi}{\nu + \xi/\sinh \gamma l}\right) \\
I_g &= (V_0/\phi) Z_t[(Y_g + Y_e)(\nu + \xi/\sinh \gamma l) - Y_g(\phi^2/Y_e Z_t)(1 + \xi)]
\end{aligned} \quad (240)$$

CERAMIC DEVICES   16-61

From Table 6,
$$\phi^2/Y_e Z_t = (1/j\pi)(\omega_0/\omega)k^2 \qquad (241)$$

At resonance,

$$\begin{aligned}
E_{in} &\to \frac{V_0}{\phi} Z_t \left(\frac{K_0 + K_B}{2} + j\frac{2k^2}{\pi} \right) \\
I_{in} &\to \frac{V_0}{\phi} Z_t (j\omega_0 C_0) \frac{K_0 + K_B}{2} \\
Z_{in} &\to \frac{1}{j\omega_c C_0} + \frac{1}{\omega_0 C_0} \frac{k^2}{\pi} \frac{4}{K_0 + K_B} \\
I_g &\to \frac{V_0}{\phi} Z_t \left[\left(Y_g + \frac{1}{j\omega_0 C_0} \right) \frac{K_0 + K_B}{2} + jY_g \frac{2k^2}{\pi} \right]
\end{aligned} \qquad (242)$$

The impedance-inverting characteristics introduced by $(-Y_e)$, noted in the earlier analysis of constant-D transducers, are apparent from a comparison between Eqs. (237) and (240) for E_{in} and I_{in}.

15. CERAMIC FILTERS

A ceramic bandpass filter utilizes mechanical resonance and electromechanical coupling in piezoelectric ceramic resonators, or transformers, to provide a sharply

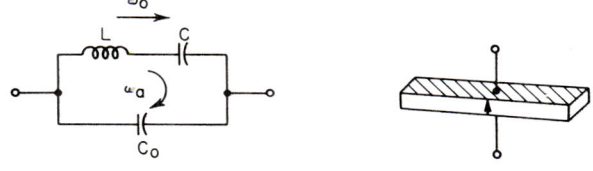

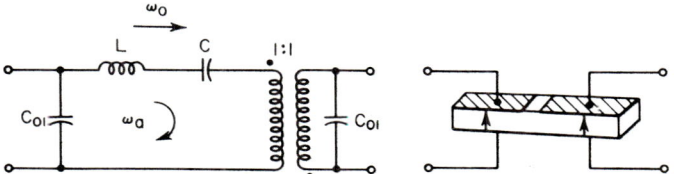

FIG. 41. Simplified equivalent circuits for a ceramic filter resonator, and a symmetrical filter transformer.

tuned frequency characteristic with uniform response over the passband, and rapid cutoff at the band edges. It is usually designed to operate between resistive input- and output-circuit terminations. Each piezoelectric element in the filter provides the equivalent of a single-tuned resonant circuit, decoupled by the shunt electrical capacitance C_0. At frequencies near the passband, the simple equivalent circuits in Fig. 41 are adequate for representation of performance for the resonator or the transformer. Here, the resistance due to mechanical losses is neglected, and each element is treated as a purely reactive network; when desirable, a resistance can be inserted in each resonant branch, with value $\omega_0 L/Q_m$. The equivalent circuit for the transformer, provided that its component transducers have exact symmetry, has equal capacitances C_0 and an ideal 1:1 transformer which may provide 180° phase shift. In the circuit of Fig. 41, for example, the phase inversion is needed to represent behavior at the fundamental mode of resonance, since the element will resonate with electrical drive through the parallel combination of the input terminals; since the element with this parallel drive will not resonate at the second-harmonic resonance, the circuit should be used without a phase-reversing transformer in this frequency range.

Image-impedance filter analysis is commonly used in the design of ceramic filters. The image impedance Z_I and its reciprocal Y_I are real numbers over the filter passband and imaginary numbers over the stop bands. These values are readily determined

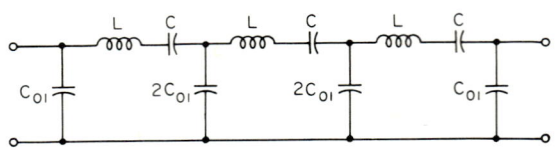

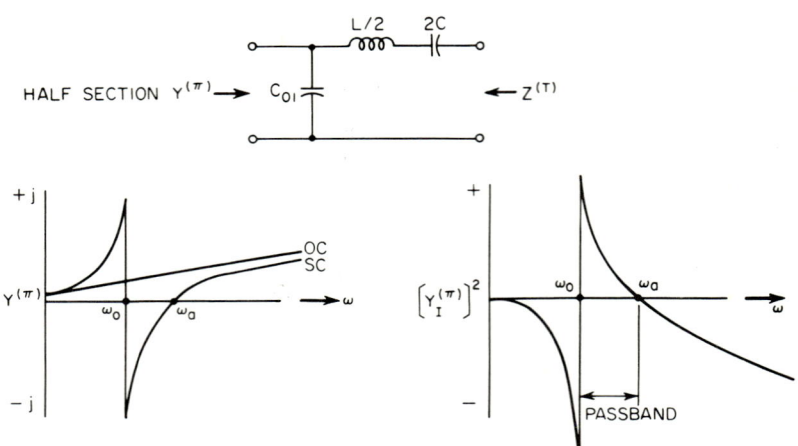

Fig. 42. Equivalent circuit analysis of a ceramic filter formed with three cascaded ceramic transformers.

from open-circuited and short-circuited impedances of sections, or half sections, making up the filter network:

$$Z_I^2 = Z_{oc}Z_{sc}$$
$$Y_I^2 = 1/Z_I^2 = Y_{oc}Y_{sc} \tag{243}$$

Figure 42 gives the equivalent circuit for a three-transformer filter and for one half section. For the half section,

$$Y_{oc}^{(\pi)} = j\omega C_{01}$$

$$Y_{sc}^{(\pi)} = j\omega C_{01} + \frac{1}{j\omega L/2 + 1/j\omega 2C}$$

$$= j\omega C_{01} + \frac{j\omega 2C}{1 - \omega^2 LC}$$

$$= j\omega C_{01} \frac{1 - \omega^2 LC + 2C/C_{01}}{1 - \omega^2 LC}$$

$$= j\omega C_{01} \frac{\omega^2 - \omega_a^2}{\omega^2 - \omega_0^2}$$

where $\qquad \omega_0^2 = 1/LC$

$$[Y_I^{(\pi)}]^2 = \omega^2 C_{01}^2 \frac{\omega_a^2 - \omega^2}{\omega^2 - \omega_0^2} \tag{244}$$

CERAMIC DEVICES

$[Y_I^{(\pi)}]^2$ is positive, and so $Y_I^{(\pi)}$ is real, in the passband

$$\omega_0 \leq \omega \leq \omega_a \tag{245}$$

Figure 42 indicates the same result, by the use of reactance sketches. The image admittance ranges from a short circuit at ω_0 to an open circuit at ω_a. At the midband frequency chosen as

$$\omega_c = \sqrt{\frac{\omega_0^2 + \omega_a^2}{2}} \tag{246}$$

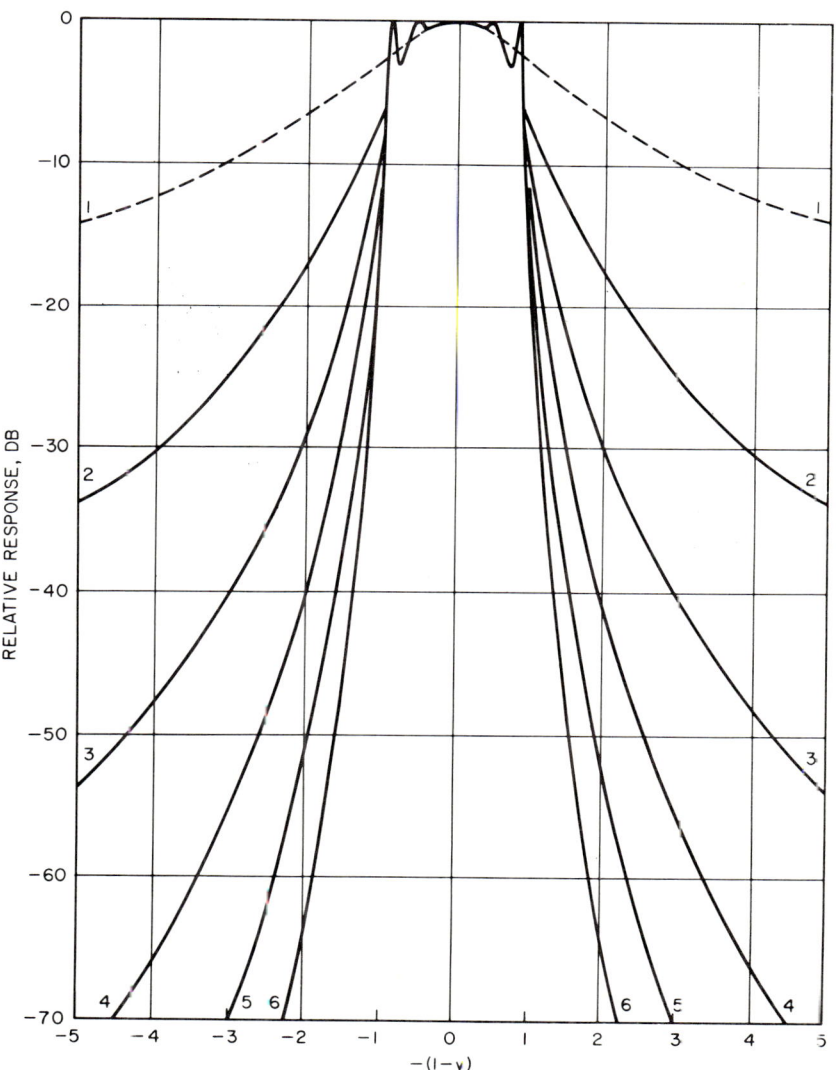

FIG. 43. Theoretical response for cascaded transformers, terminated in nominal resistance.

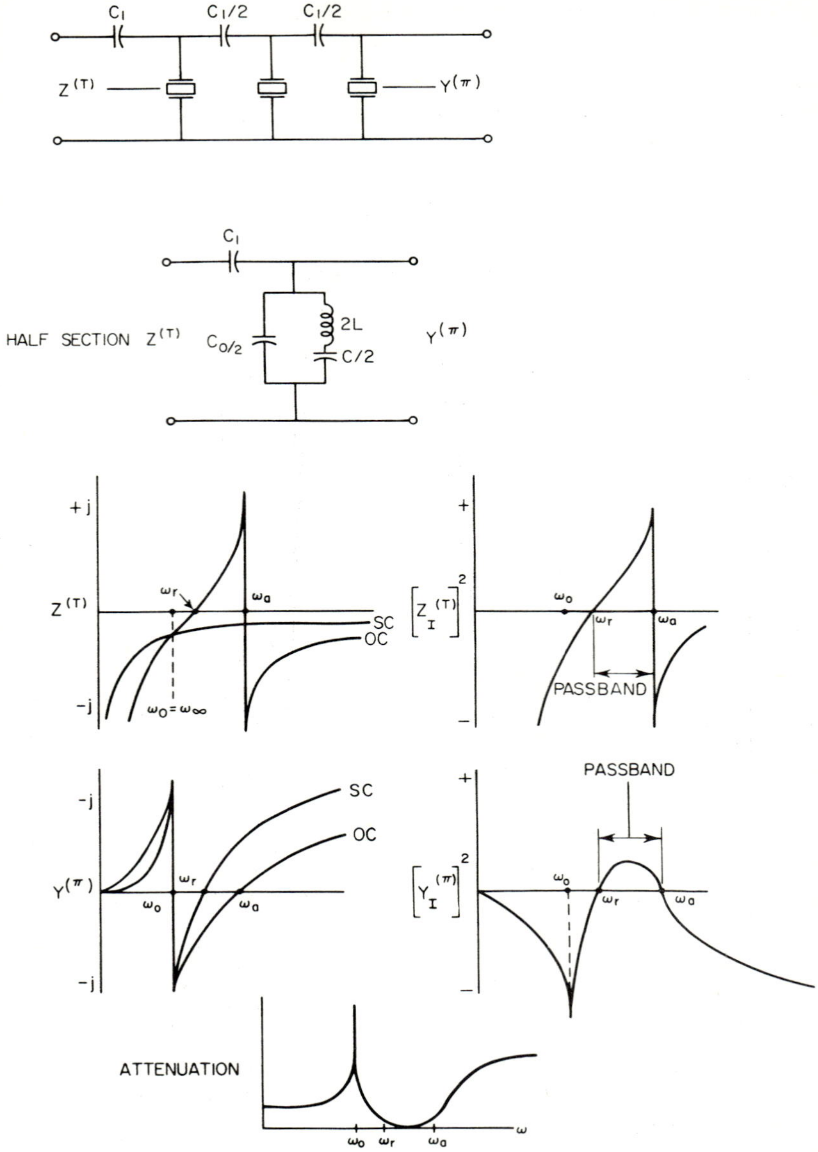

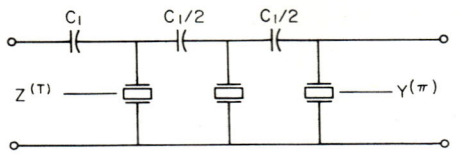

FIG. 44. Series-C, shunt-resonator ladder filter, shown with both mid-series and mid-shunt terminations.

the "nominal" terminating resistance is

$$R_{I0}^{(\pi)} = 1/\omega_c C_{01} \qquad (247)$$

Since a fixed resistance cannot match the filter image impedance closely over the passband, the transmission between terminating resistances exhibits ripple. A single transformer terminated in $R_{I0}^{(\pi)}$ provides peak output at ω_c. As additional trans-

CERAMIC DEVICES 16-65

formers are added to the filter, more ripples appear. Figure 43 shows calculated response curves for several cascaded filters.

Figure 44 shows a filter formed by series capacitors and shunt resonators. The passband extends from ω_r to ω_a, where ω_r is slightly greater than the resonator resonance frequency ω_0:

$$\omega_r^2 = \omega_0^2 \left(1 + \frac{C}{C_0 + 2C_1}\right)$$
$$\omega_a^2 = \omega_0^2(1 + C/C_0) \tag{248}$$
$$\omega_a^2 - \omega_r^2 = (\omega_a^2 - \omega_0^2)\left(1 - \frac{1}{1 + 2C_1/C_0}\right)$$

As C_1 increases, ω_r approaches ω_0, widening the passband. At the midband frequency

$$\omega_c^2 = \sqrt{\frac{\omega_r^2 + \omega_a^2}{2}} \tag{249}$$

The mid-series image impedance is

$$Z_{I0}{}^{(T)} = \frac{1}{\omega_c C_1} \sqrt{1 + 2C_1/C_0}$$

and the mid-shunt image admittance is

$$Y_{I0}{}^{(\pi)} = \omega_c \frac{C_1 \sqrt{1 + 2C_1/C_0}}{2(1 + C_1/C_0)} \tag{250}$$

The mid-shunt terminals can be matched closely over the passband with fixed resistances. At ω_c, the insertion loss is infinite, since the resonator becomes a short circuit. The attenuation outside the passband is small when $(\omega_r - \omega_0)$ is small; so a fairly large number of sections must be used to provide high stop-band attenuation. Cutoff is then extremely sharp at the low-frequency end of the passband.

Figure 45 shows the resonator-capacitor ladder structure with resonators forming the series elements. Here, the passband extends from ω_0 to ω_r, which is slightly below ω_a:

$$\omega_r^2 = \omega_0^2(1 + C/C_0 + 2C/C_2) \tag{251}$$

As C_2 increases, ω_r approaches ω_a. At the midband frequency

$$\omega_c = \sqrt{\frac{\omega_0^2 + \omega_r^2}{2}} \tag{252}$$

the mid-shunt image admittance is

$$Y_{I0}{}^{(\pi)} = \omega_c C_2 \sqrt{1 + 2C_0/C_2}$$

and the mid-series image impedance is

$$Z_{I0}{}^{(T)} = \frac{1}{\omega_c C_2} \frac{\sqrt{\frac{1}{2} + C_0/C_2}}{1 + C_2/C_0} \tag{253}$$

Figure 46 shows a ladder structure formed entirely with ceramic resonators. Points of infinite attenuation lie below the passband at $\omega_{\infty 1}$ because of series resonance in Y_B, and above the passband at $\omega_{\infty 2}$ because of parallel resonance in Y_A. Both the mid-shunt image admittance $Y_I{}^{(\pi)}$ and the mid-series impedance $Z_I{}^{(T)}$ can be matched fairly well with a terminating resistance, and so ripple is small in multiple-section filters.

From the reactance sketches, it can be seen that

$$\begin{aligned}\omega_{0A} &= \omega_{aB} \\ \omega_{0B} &= \omega_{\infty 1} \\ \omega_{aA} &= \omega_{\infty 2}\end{aligned} \tag{254}$$

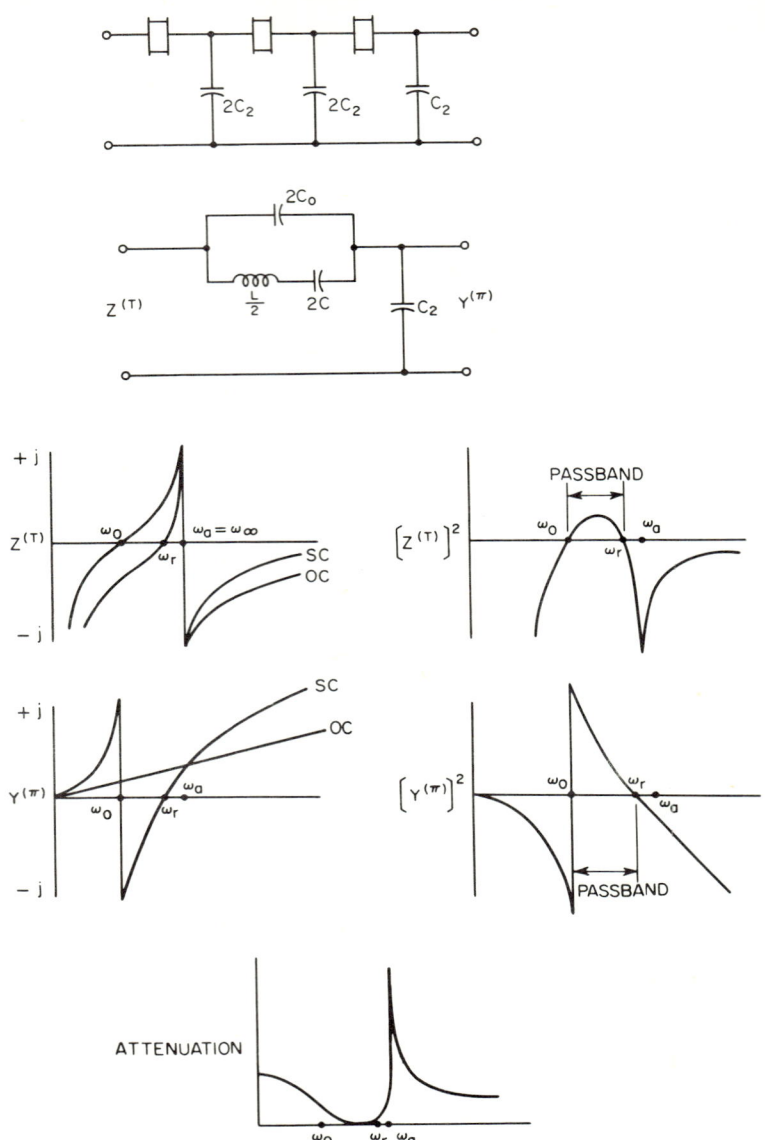

Fig. 45. Series-resonator, shunt-C ladder filter.

From the L section, defining $p = C_{0A}/C_{0B}$,

$$Z_{sc}{}^{(T)} = Z_A = \frac{1}{j\omega C_{0A}} \frac{\omega^2 - \omega_{0A}{}^2}{\omega^2 - \omega_{\infty 2}{}^2}$$

$$Z_{oc}{}^{(T)} = Z_A + Z_B = \frac{1+p}{j\omega C_{0A}} \frac{(\omega^2 - \omega_1{}^2)(\omega^2 - \omega_2{}^2)}{(\omega^2 - \omega_{0A}{}^2)(\omega^2 - \omega_{\infty 2}{}^2)}$$

(255)

CERAMIC DEVICES

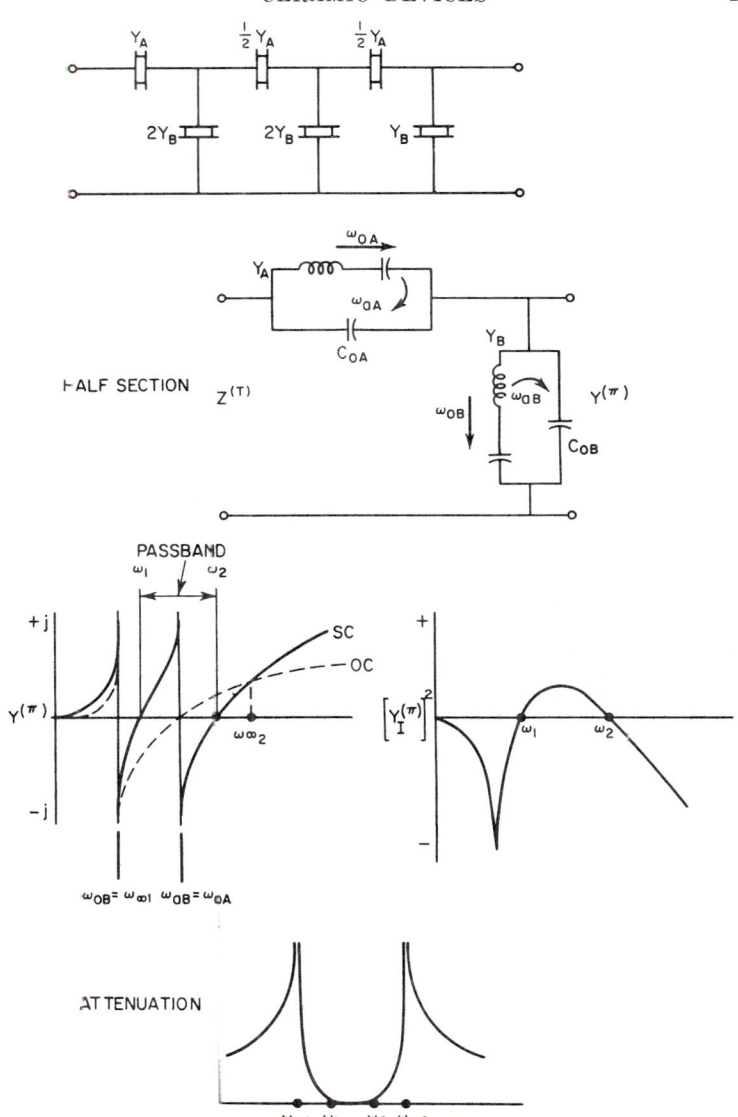

FIG. 46. Resonator ladder filter.

Therefore,

$$[Z_I^{(\tau)}]^2 = \frac{1+p}{\omega^2 C_{0A}^2} \frac{(\omega^2 - \omega_1^2)(\omega_2^2 - \omega^2)}{(\omega_{\infty 2}^2 - \omega^2)^2} \quad (256)$$

In the same fashion,

$$[Y_I^{(\pi)}]^2 = \omega^2 C_{0B}^2 (1 + \dot{p}) \frac{(\omega^2 - \omega_1^2)(\omega_2^2 - \omega^2)}{(\omega^2 - \omega_{\infty 1}^2)^2} \quad (257)$$

Curran and Gerber[34] have described the performance and design procedures for ceramic ladder filters. The attenuation, remaining small over the passband and sharply cut

Table 7. Properties of Commonly Used Piezoelectric Ceramics. Low-signal Parameters at 25°C Unless Otherwise Specified

	k_p	k_{31}	k_{33}	k_{15}	k_t	$\epsilon_{33}^T/\epsilon_0$	$\epsilon_{33}^S/\epsilon_0$	$\epsilon_{11}^T/\epsilon_0$	$\epsilon_{11}^S/\epsilon_0$	d_{33}	d_{31}	d_{15}	e_{33}	e_{31}	e_{15}	s_{33}^E	s_{11}^E	s_{44}^E	s_{66}	s_{33}^D	s_{11}^D	s_{44}^D
										\multicolumn{3}{c}{$10^{-12} C/N$}	\multicolumn{3}{c}{m^2/C}	\multicolumn{7}{c}{$10^{-12} m^2/N$}
PZT-4*	0.58	0.33	0.70	0.71	0.51	1,300	635	1,475	730	289	−123	496	15.1	−5.2	12.7	15.5	12.3	39.0	32.7	7.90	10.9	19.3
PZT-5*	0.60	0.34	0.705	0.685	0.49	1,700	830	1,730	916	374	−171	584	13.8	−5.4	12.3	18.8	16.4	47.5	44.3	9.46	14.4	25.2
PZT-5H*	0.65	0.39	0.75	0.675	0.505	3,400	1,470	3,130	1,700	593	−274	741	23.3	−6.5	17.0	20.7	16.5	43.5	42.6	8.99	14.1	23.7
PZT-6A*	0.42	0.25	0.54	...	0.39	1,050	730	...	...	189	−80	...	12.5	...	...	13.0	10.7	27.8	...	10.1	...	...
PZT-6B*	0.25	0.145	0.375	0.377	0.30	460	386	475	407	71	−27	130	7.1	−0.9	4.6	9.35	9.28	24.0	...	8.05	8.8	24.2
PZT-7A*	0.51	0.30	0.66	0.67	0.50	425	235	840	460	150	−60	362	9.5	−1.9	9.2	13.9	10.7	39.5	27.8	7.85	9.7	21.8
PZT-8*	0.50	0.295	0.62	0.55	0.44	1,000	600	...	...	218	−93	...	...	...	...	13.9	11.6	29.6	...	8.5	10.1	...
PZT-2*	0.47	0.28	0.63	0.51	0.51	450	260	990	504	152	−60	440	9.0	−1.9	9.8	13.9	11.6	45.0	29.9	9.0	10.7	22.9
BaTiO₃	0.36	0.21	0.50	0.48	0.38	1,700	1,260	1,450	1,115	190	−78	260	17.5	−4.3	11.4	9.5	9.1	22.8	23.6	9.0	8.7	17.5
95 w % BaTiO₃, 5 w % CaTiO₃	0.33	0.19	0.48	0.48	0.38	1,200	910	1,300	1,000	149	−58	242	13.5	−3.1	10.9	9.1	8.6	22.2	22.4	7.0	8.3	17.1
95 w % BaTiO₃, 5 w % CaTiO₃, 0.75 w % CoCO₃–NRE-4†																	8.1					
PbNb₂O₆‡	0.31	0.18	0.46	0.46	0.36	1,400	1,110	...	...	150	−59	...	...	...	...	25.4	11.5	...	...	21.8	10.9	...
Pb₀.₆Ba₀.₄Nb₂O₆	0.07	0.45	0.38	...	0.37	225	85	...	...	85	~9	...	...	...	...	...	...	...	...	...	...	...
Na₀.₅K₀.₅NbO₃ (hot pressed)¶	0.38 0.46	0.22 0.27	0.55 0.605	0.645	0.38 0.46	1,500 496	306	1,300 938	545	220 127	−90 −51	306	11.3	...	11.3	10.1	8.1 8.27	27.1	...	6.4	7.6	15.5

	\multicolumn{4}{c}{$10^{10} N/m^2$}	Q_M	Q_E	\multicolumn{4}{c}{Cycle m/sec}	\multicolumn{2}{c}{m/sec}	Curie point, °C	Density, 10^3 kg/m³	Static tensile strength, psi	Rated dynamic tensile strength, psi	Change in N_1 per time decade	Change in k_p per time decade	Change in ϵ_{33}^T per time decade	Change in N_1 −60 to +85°C
	c_{33}^E	c_{11}^E	c_{33}^D	c_{11}^D			N_1	N_{31}	v_3^D	v_4^D									
PZT-4	11.5	13.9	15.9	14.5	500	250	1,650	2,000	4,600	2,630	328	7.5	11,000	3,500	+1.5%	−2.3%	−5.8%	4.8%
PZT-5A	11.1	12.1	14.7	12.6	75	50	1,400	1,890	4,350	2,260	365	7.75	11,000	4,000	+0.2%	−1%	−1%	2.6%
PZT-5H	11.7	12.6	15.7	13.0	65	50	1,420	1,800	4,560	2,375	193	7.5	11,000	4,000	+0.25%	−0.35%	−1.5%	9%
PZT-6A	13.1	15.5	15.5	...	450	50	1,770	2,140	4,570	...	335	7.45	11,000	3,500	<0.1%	−0.2%	−0.6%	<0.2%
PZT-6B	16.3	16.8	17.7	16.9	1,300	110	1,920	2,225	4,820	2,340	~350	7.55	11,000	3,500	<0.1%	<0.1%	−0.6%	<0.2%
PZT-7A	13.1	14.8	17.5	15.7	600	60	1,750	2,100	4,800	2,490	350	7.6	11,000	3,500	−0.08%	0.0%	+2.0%	2.9%
PZT-8	...	...	...	...	1,000	250	1,700	...	...	...	300	7.6	11,000	3,500	+1.0%	−2.0%	−2.8%	~2%
PZT-2	11.3	13.5	14.8	13.6	680	200	1,680	2,090	4,410	2,400	370	7.6	11,000	3,000	+0.6%	−1.8%	−4.1%	−1.5%
BaTiO₃	14.6	15.5	17.1	15.0	300	300	2,200	2,520	5,470	3,160	115	5.7	7,500	3,000	+1.1%	−2.5%	−0.8%	19%
95 w % BaTiO₃, 5 w % CaTiO₃	15.0	15.8	17.7	15.9	400	170	2,290	2,740	5,640	3,240	115	5.55	7,500	3,000	+0.5%	−1.8%	...	~18%
95 w % BaTiO₃, 5 w % CaTiO₃, 0.75 w % CoCO₃	...	...	...	...	...	...	2,310	2,760	...	...	105	5.7	7,500	3,000	+0.4%	−1.9%	−1.3%	~15%
PbNb₂O₆	...	...	...	...	11	100	...	...	...	...	570	6.0	...	...	...	...	...	...
Pb₀.₆Ba₀.₄Nb₂O₆	...	...	...	...	250	100	1,915	...	...	...	260	5.9	...	...	...	...	...	2.5%
Na₀.₅K₀.₅NbO₃ (hot pressed)	...	...	...	...	240	70	2,570	...	...	...	420	4.46	...	...	...	...	...	3.3%

* Trademark, Clevite Corporation.
† General Electric Company.
‡ D. Schofield and R. F. Brown, An Investigation of Some Barium Titanate Compositions for Transducer Applications, *Can. J. Phys*, vol. 35, no. 5, pp. 594–607, May, 1957.
¶ R. E. Jaeger and L. Egerton, Hot Pressing of Potassium-Sodium Niobates, *J. Am. Ceram. Soc.*, vol. 45, pp. 209–213, May, 1962.

off at $\omega_{\infty 1}$ and $\omega_{\infty 2}$, is relatively small beyond $\omega_{\infty 1}$ and $\omega_{\infty 2}$. The filter usually employs large numbers of cascaded sections to achieve sufficient stop-band attenuation, since this is of the order of 10 db per section.

For filters, ceramic materials have been developed with low aging and temperature stability of about 0.5 kc, at 450 kc, over wide temperature ranges.

APPENDIX

Table 7, reproduced by permission from data furnished by H. Jaffe and D. Berlincourt of Clevite Corporation, summarizes the small-signal characteristics of a representative group of ceramic materials. Since exact values depend on temperature, degree of polarization, and aging after the poling process, it is often necessary to carry out specific measurements on individual samples when highly accurate values are required. This tabulation presents data for variants of the predominant ferroelectric ceramics, barium titanate, lead zirconate titanate (Clevite ceramics carry the trademarked designation PZT), lead metaniobate, lead barium niobate, and sodium potassium niobate. Materials in these classes are available from a number of commercial sources, with specific performance data.

In Table 7, N is a constant used to specify the product of sample size times resonance frequency. N_1 is the product of bar length times fundamental resonance frequency for a half-wave k_{31} constant-E thin bar, and thus equals one half the corresponding acoustic velocity. N_{3t} applies to a half-wave plate under constant-D conditions. The velocities $v_3{}^D$ and $v_4{}^D$ refer, respectively, to acoustic waves along a thin k_{33} end-plated bar and to shear waves in a plate or along a torsional rod. Aging data are presented in terms of per cent change per decade of time, since aging tends to take place logarithmically after the poling process.

REFERENCES

1. D. R. Curran and D. J. Koneval, "Miniature Ceramic Band Pass Filters," *Proc. National Electronics Conf.*, Chicago, 1961, vol. 17, pp. 514–520.
2. E. C. Herry, "Ceramics, Key to Electronic Progress," *Ceram. Ind.*, April, 1958.
3. W. P. Mason, "Piezoelectric Crystals and Their Application to Ultrasonics," D. Van Nostrand Company, Inc., Princeton, N.J., 1950, chap. 12.
4. W. P. Mason, "Physical Acoustics," Academic Press, New York, 1964, vol. I, pt. A.
5. R. Bechmann, "Elastic, Piezoelectric and Dielectric Constants of Polarized Barium Titanate Ceramics," *Proc. IRE*, vol. 46, April, 1958.
6. D. Berlincourt and H. H. A. Krueger, "Domain Processes in Lead Titanate Zirconate and Barium Titanate Ceramics," *J. Appl. Phys.*, vol. 30, pp. 1804–1810, November, 1959.
7. G. Goodman, "Ferroelectric Properties of Lead Metaniobate," *J. Am. Ceram. Soc.* vol. 38, no. 11, pp. 368–372, 1953.
8. H. Jaffe, "Piezoelectricity," Encyclopaedia Brittanica, 1961.
9. J. A. Hooton and W. J. Merz, "Etch Patterns and Ferroelectric Domains in Barium Titanate Single Crystals," *Phys. Rev.*, vol. 98, no. 2, pp. 409–413, 1955.
10. G. Shirane, F. Jona, and R. Pepinsky, "Some Aspects of Ferroelectricity," *Proc. IRE*, vol. 43, pp. 1738–1793, December, 1955.
11. "IRE Standards on Piezoelectric Crystals: Measurements of Piezoelectric Ceramics, 1961," *Proc. IRE*, vol. 49, pp. 1162–1169, July, 1961.
12. "Standards on Piezoelectric Crystals, 1949," *Proc. IRE*, vol. 37, pp. 1378–1395, December, 1949.
13. W. P. Mason, "Electrostrictive Effect in Barium Titanate Ceramics," *Phys. Rev.*, vol. 74, pp. 1134–1147, 1948.
14. H. G. Baerwald, "Thermodynamic Theory of Ferroelectric Ceramics," *Phys. Rev.*, vol. 105, pp. 480–486, 1957.
15. J. F. Haskins and J. S. Hickman, "A Derivation and Tabulation of the Piezo-electric Equations of State," *J. Acoust. Soc. Am.*, vol. 22, pp. 584–588, September 1950.
16. H. W. Katz, "Solid State Magnetic and Dielectric Devices," John Wiley & Sons, Inc., New York, 1959, Appendix 1.
17. W. L. Bond, "The Mathematics of the Physical Properties of Crystals," *Bell System Tech. J.*, vol. 22, pp. 1–72, 1943.

18. W. G. Cady, "Piezoelectricity," McGraw-Hill Book Company, Inc., 1946.
19. W. Voigt, "Lehrbuch der Kristallphysik," B. G. Teubner, Leipzig, 1st ed. 1910, 2nd ed. 1928.
20. J. F. Nye, "Physical Properties of Crystals," Oxford University Press, London, 1957, chap. 10.
21. "IRE Standards on Piezoelectric Crystals—The Piezoelectric Vibrator: Definitions and Methods of Measurement, 1957," *Proc. IRE*, vol. 45, pp. 353–358, March, 1957.
22. "*IRE* Standards on Piezoelectric Crystals: Determination of the Elastic, Piezoelectric, and Dielectric Constants—The Electromechanical Coupling Factor, 1958," *Proc. IRE*, vol. 46, pp. 764–778, April, 1958.
23. (Proposed) "IEEE Standards on Piezoelectric Crystals—The Piezoelectric Vibrator: Definitions and Methods of Measurement, 1965," 65 IEEE 14.P.
24. H. Kolsky, "Stress Waves in Solids," Oxford University Press, London, 1953; Dover Publications, Inc., 1963.
25. W. P. Mason, "Electromechanical Transducers and Wave Filters," 2d ed., D. Van Nostrand Company, Inc., Princeton, N.J., 1948, p. 251.
26. A. E. H. Love, "A Treatise on the Mathematical Theory of Elasticity," 4th ed., Cambridge University Press, London, 1934, p. 90.
27. E. C. Monk, "The Equivalent Circuit for Radial Modes of a Piezoelectric Ceramic Disc with Concentric Electrodes," *Philips Res. Repts.*, vol. 20, pp. 170–189, 1965.
28. E. A. Guillemin, "Communication Networks," John Wiley & Sons, Inc., 1935, vol. II, p. 439.
29. C. A. Rosen, "Ceramic Transformers and Filters," Ph.D. thesis, Syracuse University, 1956.
30. A. Lungo and F. Sauerland, *IRE 1961 Nat. Conv. Record*, vol. 9, pt. 6, pp. 189–203.
31. C. A. Rosen, U.S. Patent 2,974,296. K. A. Fish, C. A. Rosen, H. C. Rothenberg, U.S. Patent 2,830,274.
32. W. Roth, "Piezoelectric Transducers," *Proc. IRE*, vol. 37, pp. 750–758, July, 1949.
33. C. F. Brockelsby, J. S. Palfreeman, and R. W. Gibson, "Ultrasonic Delay Lines," Iliffe Books, London, 1963.
34. D. R. Curran and W. J. Gerber, "Piezoelectric Ceramic I. F. Filters," *Electronic Components Conf. Proc.*, 1959, pp. 160–165.

Part III

AMPLIFIER CIRCUITS

Chapter 17

AUDIO AMPLIFIERS

RICHARD F. SHEA[*]

CONTENTS

1. Introduction... 17–4
   1.1. Network Equations for Impedances and Amplification..... 17–5
   1.2. Gain... 17–5
2. Single-stage Electron-tube Amplifiers........................ 17–6
   2.1. Matrix Representation of the Tube...................... 17–6
   2.2. Frequency Characteristic of the Tube Amplifier......... 17–7
   2.3. Effect of Cathode Resistance........................... 17–7
   2.4. The Cathode Follower................................... 17–7
   2.5. Mutual Conductance..................................... 17–8
   2.6. Typical Tube Amplifiers................................ 17–8
3. Single-stage Transistor Amplifiers........................... 17–9
   3.1. The Transistor as a Network Element................... 17–10
4. Transistor Bias... 17–11
   4.1. Bias Equations for the Two-battery Arrangement........ 17–11
   4.2. Bias Equations for the Single-battery Arrangement..... 17–12
   4.3. Graphical Determination of Bias....................... 17–12
   4.4. Significance of the Stability Factors................. 17–14
5. Transformer-coupled Stages.................................. 17–15
   5.1. Transformer-coupled Tube Stage........................ 17–15
   5.2. Transformer-coupled Common-base Stage................. 17–15
   5.3. Transformer-coupled Common-emitter Stage.............. 17–16
   5.4. Transformer-coupled Common-collector Stage............ 17–17
6. RC-coupled Stages... 17–18
   6.1. RC-coupled Tube Amplifiers............................ 17–18
   6.2. RC-coupled Transistor Amplifiers...................... 17–18
   6.3. Example of RC-coupled Common-emitter Amplifier........ 17–20
   6.4. Effect of Impedance in the Emitter Circuit............ 17–21
7. Cascaded Stages... 17–22
   7.1. Example of Multistage Amplifier....................... 17–23
8. Hearing Aids.. 17–24
   8.1. Otarion X-100 Listener................................ 17–24
   8.2. Otarion OL-3 Hearing Aid.............................. 17–25
   8.3. Otarion Super-9 Hearing Aid........................... 17–28

[*] Consulting Electronics Engineer, Schenectady, N.Y.

AMPLIFIER CIRCUITS

- 9. High-input-impedance Amplifiers..................... 17-28
  - 9.1. Components of Input Impedance.................. 17-28
  - 9.2. Effect of Bias Resistors......................... 17-29
  - 9.3. Effect of Emitter Resistance..................... 17-29
  - 9.4. Effect of the Collector Resistance of the First Transistor... 17-30
  - 9.5. Example of Application of Bootstrapping.............. 17-30
  - 9.6. Complete Circuit of High-impedance Amplifier.......... 17-33
  - 9.7. High-impedance Amplifier Using Field-effect Transistors... 17-34
- 10. Electron-tube Preamplifiers............................. 17-34
  - 10.1. Pickup Characteristics............................. 17-34
  - 10.2. Other Special Preamplifier Characteristics........... 17-35
  - 10.3. Example of Tube Preamplifier....................... 17-35
- 11. Transistor Preamplifiers................................ 17-38
  - 11.1. Example of Transistorized Preamplifier.............. 17-38
- 12. Classes of Output Stages............................... 17-40
- 13. Class A Power Amplifiers............................... 17-41
  - 13.1. Class A Tube Power Amplifiers..................... 17-41
  - 13.2. Output Transformer Design Considerations........... 17-43
  - 13.3. Class A Transistor Power Amplifiers............... 17-43
  - 13.4. Graphical Determination of Class A Operating Point... 17-43
  - 13.5. Distortion in Class A Amplifiers................... 17-45
  - 13.6. Peak Clipping................................... 17-46
  - 13.7. Biasing Transistor Power Amplifiers............... 17-46
  - 13.8. Temperature Stabilization with Thermistors and Diodes.. 17-49
  - 13.9. Thermal Runaway................................ 17-51
  - 13.10. Transistor Cooling; Thermal Resistance........... 17-53
- 14. Class A, AB, and B Push-Pull Power Amplifiers........... 17-57
  - 14.1. Class AB Tube Amplifiers.......................... 17-57
  - 14.2. Class AB and B Transistor Amplifiers.............. 17-60
  - 14.3. Static Characteristics, Class B Push-Pull Amplifiers.. 17-60
  - 14.4. Bias in Class B Amplifiers........................ 17-64
  - 14.5. Output Power, Gain, Efficiency.................... 17-65
  - 14.6. Comparison between Different Configurations........ 17-67
  - 14.7. Design Charts for Power Amplifiers................ 17-67
- 15. Typical Power Amplifiers............................... 17-71
  - 15.1. Tube Power Amplifiers............................ 17-71
  - 15.2. Tube Amplifiers without Output Transformers....... 17-72
  - 15.3. Transistor Power Amplifiers...................... 17-73
- 16. Electron-tube Stereo Amplifiers......................... 17-79
  - 16.1. Sherwood Stereo Amplifier........................ 17-79
  - 16.2. General Electric Stereo Amplifier................. 17-81
- 17. Transistor Stereo Amplifiers............................ 17-90

1. INTRODUCTION

As a preliminary to the consideration of audio-amplifier design it is desirable to review the basic equations for the terminated two-port network, the fundamental component of any amplifier. The more usual multistage amplifiers are formed from combinations of single stages; thus their performance may readily be calculated from the combination of their component stages.

Consider the elementary circuit of Fig. 1, which shows the network as the familiar "black box," with the usual input and output terminations. The input is supplied from a generator V_g through a source impedance Z_g. The output is terminated in the load impedance Z_l. V_1 and V_2 are the input and output voltages, respectively, and I_1 and I_2 are the corresponding input and output currents, with the positive convention signifying current into the network.

AUDIO AMPLIFIERS

Table 1. Properties of Terminated Network

	Using z parameters	Using y parameters	Using h parameters
Z_i	$z_{11} - \dfrac{z_{12}z_{21}}{z_{22} + Z_l}$	....	$h_{11} - \dfrac{h_{12}h_{21}Z_l}{1 + h_{22}Z_l}$
Y_i	....	$y_{11} - \dfrac{y_{12}y_{21}Z_l}{1 + y_{22}Z_l}$	....
A_v	$\dfrac{z_{21}Z_l}{z_{11}(z_{22} + Z_l) - z_{12}z_{21}}$	$\dfrac{-y_{21}Z_l}{1 + y_{22}Z_l}$	$\dfrac{-h_{21}Z_l}{h_{11}(1 + h_{22}Z_l) - h_{12}h_{21}Z_l}$
A_i	$\dfrac{-z_{21}}{z_{22} + Z_l}$	$\dfrac{y_{21}}{y_{11}(1 + y_{22}Z_l) - y_{12}y_{21}Z_l}$	$\dfrac{h_{21}}{1 + h_{22}Z_l}$
Z_o	$z_{22} - \dfrac{z_{12}z_{21}}{z_{11} + Z_g}$		
Y_o	....	$y_{22} - \dfrac{y_{12}y_{21}Z_g}{1 + y_{11}Z_g}$	$h_{22} - \dfrac{h_{12}h_{21}}{h_{11} + Z_g}$

1.1. Network Equations for Impedances and Amplification. Using the most familiar z, y, and h parameters we may express the properties of this network as shown in Table 1. This table gives the equations for input and output impedances, current and voltage amplification, and also the corresponding input and output admittances for certain sets of parameters. In Fig. 1, A_v, the voltage amplification, is the ratio V_2/V_1. The current amplification A_i is the ratio I_2/I_1. The overall voltage amplification V_2/V_g may be obtained from the amplification A_v by multiplying by the input attenuation $Z_i/(Z_i + Z_g)$.

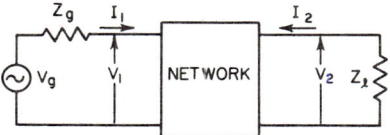

Fig. 1. Elementary network circuit.

1.2. Gain. The term *gain* is used to indicate ratios of two powers. Its use for voltage and current ratios is generally deprecated. The most useful measure of an amplifier's overall performance is the *transducer gain*, defined as the ratio of output power to the power available from the generator. If a generator has an internal voltage V_g and resistance R_g the maximum power it can deliver to a load is $V_g^2/4R_g$. This is the available power and is delivered into a load equal to the internal resistance R_g. An alternative method of expressing the available power, when the generator is considered as a current source I_g applied across an internal conductance G_g, is given by $I_g^2/4G_g$.

The transducer gain may be obtained from the following equations:

$$G_t = \frac{4|z_{21}|^2 R_g R_l}{|G_t = |(z_{11} + Z_g)(z_{22} + Z_l) - z_{12}z_{21}|^2} \quad (1)$$

$$G_t = \frac{4|y_{21}|^2 G_g G_l}{|(y_{11} + Y_g)(y_{22} + Y_l) - y_{12}y_{21}|^2} \quad (2)$$

$$G_t = \frac{4|h_{21}|^2 R_g G_l}{|(h_{11} + Z_g)(h_{22} + Y_l) - h_{12}h_{21}|^2} \quad (3)$$

In the above equations R_g is the real part of Z_g, the source impedance, and R_l is the real part of Z_l, the load impedance. Similarly G_g is the real part of Y_g, the source admittance, and G_l is the real part of Y_l, the load admittance. The above equations are completely general and are applicable to any network. In the following sections the specific application to tube and transistor circuits will be covered.

2. SINGLE-STAGE ELECTRON-TUBE AMPLIFIERS

Figure 2 shows the familiar electron tube and its equivalent circuit. The input voltage V_1 is applied to the grid and an amplified voltage μV_1 appears in the output circuit through the output impedance r_p. The negative sign on the output voltage implies phase reversal. While the tube will operate without a resistance applied to the grid, a leak is normally required, and is shown as R_1 in Fig. 3.

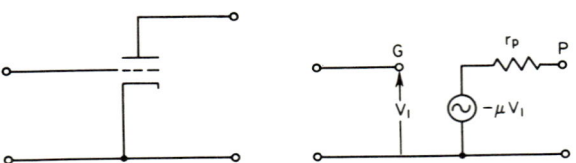

Fig. 2. Electron-tube symbol and equivalent circuit.

2.1. Matrix Representation of the Tube. The same symbols previously used can be applied to the tube by the following equivalences:

$$z_{11} = R_1 \quad z_{12} = 0 \quad z_{21} = -\mu R_1 \quad z_{22} = r_p$$
$$y_{11} = 1/R_1 \quad y_{12} = 0 \quad y_{21} = \mu/r_p \quad y_{22} = 1/r_p$$
$$h_{11} = R_1 \quad h_{12} = 0 \quad h_{21} = \mu R_1/r_p \quad h_{22} = 1/r_p$$

Figure 3 shows an a-c representation of the tube with its associated input connections and load. The input consists of the source voltage V_g, the source resistance R_g, and the coupling capacitor C. R_l represents the load resistance. Also shown on this figure is the equivalent circuit, with the tube network shown enclosed within

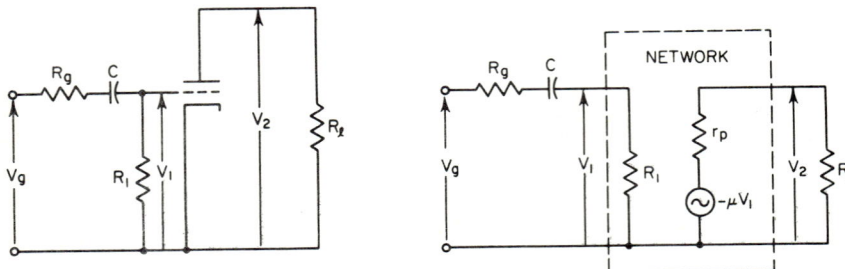

Fig. 3. Terminated electron-tube circuit.

the dotted lines. Any of the network equations may be used to obtain the overall performance of this circuit. For example, using the Z matrix representations we obtain, for the voltage amplification,

$$A_v = \frac{z_{21} Z_l}{z_{11}(z_{22} + Z_l) - z_{12} z_{21}} \tag{4}$$

Substituting R_1, 0, $-\mu R_1$, and r_p for the corresponding z parameters we obtain

$$A_v = \frac{-\mu R_1 R_l}{R_1(r_p + R_l) - 0}$$
$$= \frac{-\mu R_l}{r_p + R_l} \tag{5}$$

which is the familiar equation for voltage amplification. The same result could have been obtained using any of the other matrix forms.

The above equation gives the voltage amplification V_2/V_1. To obtain the overall voltage amplification V_2/V_g requires further multiplication by the attenuation in the input circuit, $R_1/(R_1 + R_g + 1/j\omega C)$. Thus the equation for the overall voltage amplification becomes

$$\frac{V_2}{V_g} = \frac{-\mu R_1 R_l}{(r_p + R_l)(R_1 + R_g + 1/j\omega C)} \quad (6)$$

It will be noted that the load resistance R_l was substituted for the impedance Z_l in the above equations. This use of a resistive load is normally justified in audio amplifiers, except at high frequencies, where the load resistance is shunted by various capacitances, the impedance of which must now be taken into consideration. At very low frequencies the impedance of the coupling capacitor following the plate resistor also becomes appreciable. However, the grid resistor of the succeeding tube is normally on the order of several megohms, and thus the effect of the following network on the output impedance of the tube is relatively minor, although it must be considered in obtaining the transfer ratio from the plate to the following grid.

2.2. Frequency Characteristic of the Tube Amplifier. In audio amplifiers the only components which appreciably affect the frequency characteristic are the

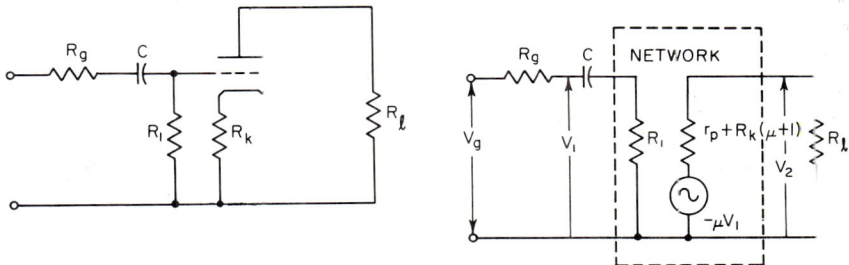

Fig. 4. Electron tube with degenerative cathode resistor.

coupling capacitor C and, at high frequencies, such items as transformer leakage inductance and stray capacitances. These latter items are usually under the control of the transformer designer; thus the major item to be considered is the capacitor C. The 3-db drop-off frequency will be that at which the impedance of C equals the sum of $R_g + R_1$, or

$$\omega_{3\,\mathrm{db}} = 1/C(R_g + R_1) \quad (7)$$

2.3. Effect of Cathode Resistance. Figure 4 shows a tube amplifier with the addition of a resistor R_k in the cathode lead, along with the equivalent circuit representation of this arrangement. It will be noted that the only effect of inserting this resistor is that the plate resistance has been increased from r_p to $r_p + R_k(\mu + 1)$. All the equations presented in the preceding section are applicable by the substitution of $r_p + R_k(\mu + 1)$ wherever r_p occurs.

2.4. The Cathode Follower. Figure 5 shows the circuit of the familiar cathode follower. It differs from the previous circuit in that the output is now taken from the cathode, across the resistor R_k. The equations for the cathode follower in the matrix form are

$$\begin{aligned}
z_{11} &= R_1 & z_{12} &= 0 \\
z_{21} &= \mu R_1 R_k/[r_p + R_k(\mu + 1)] \\
z_{22} &= r_p R_k/[r_p + R_k(\mu + 1)] \\
y_{11} &= 1/R_1 & y_{12} &= 0 & y_{21} &= -\mu/r_p \\
y_{22} &= [r_p + R_k(\mu + 1)]/r_p R_k \\
h_{11} &= R_1 & h_{12} &= 0 & h_{21} &= -\mu R_1/r_p \\
h_{22} &= [r_p + R_k(\mu + 1)]/r_p R_k
\end{aligned} \quad (8)$$

Again, these equivalences may be inserted in the equations relating the matrix parameters to amplifier performance, as before. For example, the voltage amplification may be obtained from any of the sets of parameters and gives

$$A_v = \frac{\mu R_k R_l}{r_p(R_k + R_l) + R_k R_l(\mu + 1)} \tag{9}$$

The output impedance is of particular importance in the use of the cathode follower, since it is frequently used to supply a low-impedance driving source. It may be readily calculated from the network equations for Z_o, as follows:

$$\begin{aligned}Z_o &= z_{22} - z_{12}z_{21}(z_{11} + Z_g) \\ &= r_p R_k / [r_p + R_k(\mu + 1)]\end{aligned} \tag{10}$$

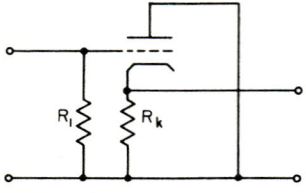

Fig. 5. The cathode follower.

Figure 6 shows the equivalent circuits for the cathode follower in two forms. In (a) the input is shown as a conductance $1/R_1$ and the output is also a conductance $[r_p + R_k(\mu + 1)]/r_p R_k$. The amplified input is shown as a current generator $\mu V_1/r_p$ across the output conductance. This representation corresponds to the admittance set of equations, using the y parameters. In (b) the circuit is shown in the more familiar voltage-generator form, with a generator having a voltage $\mu R_k V_1/[r_p + R_k(\mu + 1)]$ in series with a source resistance $r_p R_k/[r_p + R_k(\mu + 1)]$. The latter value gives the clue to the main utility of the cathode follower, since it indicates that the output impedance may be quite low, generally on the order of $r_p/(\mu + 1)$. Thus the output impedance may be reduced as much as two orders of magnitude below that of the usual amplifier.

2.5. Mutual Conductance. The mutual conductance of an electron tube is the rate of change of plate current with grid voltage and is indicated by the symbol g_m.

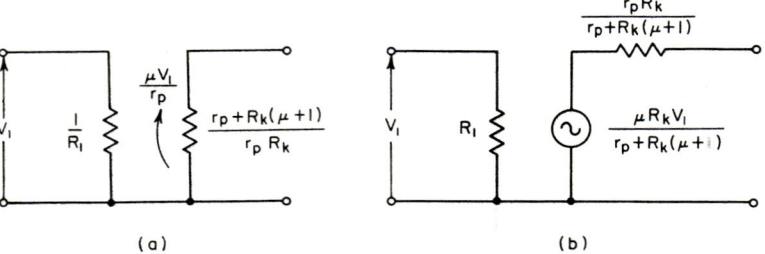

Fig. 6. Equivalent circuits for the cathode follower.

It is usually given in units of micromhos on tube specification sheets. It is numerically equal to the ratio μ/r_p and may be substituted for this ratio in the equations above. For example, Eq. (5) for the voltage amplification may also be written as follows:

$$A_v = \frac{g_m R_l}{1 + R_l/r_p} \tag{11}$$

2.6. Typical Tube Amplifiers. Figure 7 shows the circuit of a typical single-stage tube amplifier. A bias network consisting of the resistor R_k and capacitor C_k is inserted in the cathode lead. The plate is supplied from the battery V_{BB} through the resistor R_p. The output is coupled to the plate through the coupling capacitor C, R_o representing the load. A signal V_i is applied to the input grid and an output V_o is taken across the load resistor R_o. The values of the capacitors C and C_k are determined by the desired low-frequency cutoff.

AUDIO AMPLIFIERS

Typical values of these resistors and capacitors and resultant performance are given below for three representative tubes, with the capacitors chosen for 100-cycle cutoff.

	6SN7	12AU7	6CN7
V_{BB}, volts	180	180	180
R_p, megohms	0.25	0.10	0.22
R_o, megohms	1.0	0.47	0.47
R_k	5,300	3,600	4,100
C_k, µf	1.5	1.1	1.7
V_o, volts	33	40	34
A_v	53	12	42

As a check, the values of μ, r_p, g_m obtained from the 6CN7 tube specifications are 70, 5,600 ohms, and 1,200 micromhos, respectively, at a plate current of 1 ma. Inserting these values into the equation for voltage amplification and using a value of 150,000 ohms (220,000 in parallel with 470,000) for R_l gives an amplification of $70 \times 150 \div 206 = 51$. This compares reasonably well with the above value of 42.

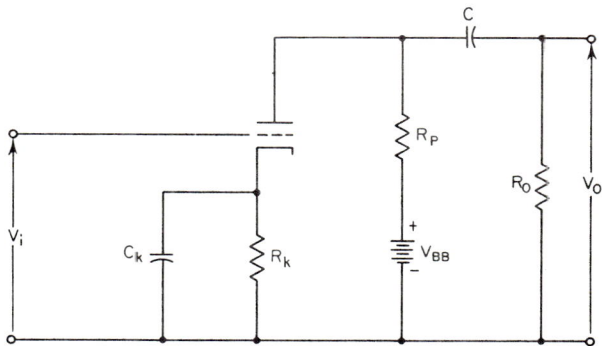

FIG. 7. Typical single-stage tube amplifier.

As stated above, the capacitors were chosen for 100-cycle cutoff. If a lower-frequency cutoff is desired the above values should be increased proportionally.

3. SINGLE-STAGE TRANSISTOR AMPLIFIERS

Figure 8 shows the basic connection of a transistor in the common-base configuration. In Fig. 8a the two-battery bias arrangement is shown, with the emitter supplied through the resistor R_1 from the battery V_{EE} and the collector supplied from the

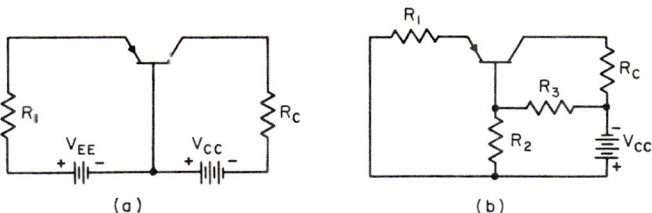

FIG. 8. Basic connections of common-base transistor stage.

17–10 AMPLIFIER CIRCUITS

battery V_{CC} through the collector resistor R_C. R_1 may be a resistor inserted to provide bias or it may be the d-c resistance of a transformer winding, for example. Likewise R_C may be the d-c resistance of an output transformer winding. Figure 8b shows the single-battery arrangement, with R_1 representing whatever d-c resistance is in the emitter circuit and R_C the d-c resistance in the collector circuit. A base voltage is set up by the resistors R_2 and R_3. A later section will describe the means by which the operating biases are set to the desired values by choice of these resistors.

Figure 9a and b shows similar arrangements for the common-emitter and common-collector configurations, with the resistors performing the same functions as in the previous circuits. In the common-collector configuration the resistor R_C may be omitted.

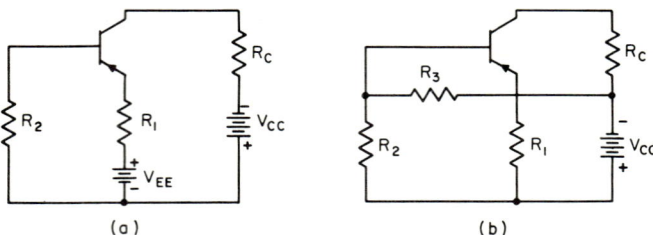

FIG. 9. Connections for common-emitter stage.

The above circuits are general in nature and are not intended to represent actual amplifier connections other than with respect to bias arrangement. Subsequent sections will deal in detail with actual circuits. Also these figures show the transistors as *p-n-p* versions. If *n-p-n* transistors are used the only changes are the polarities of the batteries.

3.1. The Transistor as a Network Element. The three most common network representations have been described previously. In transistor specifications a modification of the *h* parameters is most widely used. Figure 10 shows the conventional representation of the *h* parameter network. The input mesh includes an input resistance h_{11} and a feedback voltage $h_{12}V_2$. The polarity signs indicate that the generator polarity will be as shown if the output voltage V_2 has a polarity such that the arrow end is more positive than the lower end. In the output circuit a conductance h_{22} is shunted by a current generator $h_{21}I_1$ in the direction shown when the input current I_1 is in the direction shown.

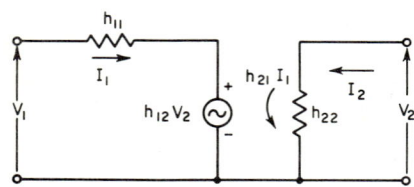

FIG. 10. The *H* equivalent circuit.

As stated above, the *h* parameters used in transistor specifications are modifications of the network representation employing numerical subscripts. The following system is used:

$$h_i \equiv h_{11} \qquad h_r \equiv h_{12} \qquad h_f \equiv h_{21} \qquad h_o \equiv h_{22}$$

In the majority of transistor specifications only the common-base parameters are given, with the exception of the common-emitter value of h_{21} or h_{fe}. Thus, to obtain the parameters for the other configurations, one must usually convert from the specification values, using network conversion tables. Table 2 gives the equations for the three sets of *h* parameters in terms of the common-base values, as commonly found in specifications.

AUDIO AMPLIFIERS 17-11

Table 2. h Parameters of the Three Configurations

	Common-base	Common-emitter	Common-collector
h_{11}	h_{ib}	$h_{ib}(h_{fe} + 1)$	$h_{ib}(h_{fe} + 1)$
h_{12}	h_{rb}	$h_{ib}h_{ob}(h_{fe} + 1) - h_{rb}$	1
h_{21}	h_{fb}	h_{fe}	$-(h_{fe} + 1)$
h_{22}	h_{ob}	$h_{ob}(h_{fe} + 1)$	$h_{ob}(h_{fe} + 1)$

Note: If h_{fe} is not given it may be calculated from the relationship $h_{fe} = -h_{fe}/(1 + h_{fe})$.

In the equivalent circuits to be shown in conjunction with the various transistor circuits in subsequent sections the above relationships will be used to give the circuit components in terms of these specification parameters.

4. TRANSISTOR BIAS

Figures 8 and 9 illustrated two common methods of biasing transistor amplifiers. In the arrangements where two batteries are used, as in Fig. 9a, there are resistances inserted in the emitter and collector leads and usually there is resistance in the base lead, as from the transformer or the base bias resistor in RC-coupled circuits. In the single-battery arrangements there is an additional resistor connected from the battery or the collector to the base. It is desirable to be able to calculate the various currents from a knowledge of the transistor parameters and the bias resistors. An additional factor of considerable importance in the determination of the bias is the current I_{CBO}, the current in the collector with the emitter open. This current is very temperature-sensitive, and transistor specification sheets usually give its value at room temperature, together with its variation with temperature, so that its value at some other temperature may readily be obtained. The following equations will give reasonable approximations for the desired operation point, based on the assumption that the base-emitter voltage V_{BE} may be neglected, as well as all transistor parameters other than h_{fb}. These approximations are normally adequate for design purposes; however, a more accurate graphical method is also presented in a subsequent section.

4.1. Bias Equations for the Two-battery Arrangement. Using the symbols shown in Fig. 9a the currents and voltages may be calculated as follows:

$$I_E = \frac{V_{EE} - R_2 I_{CBO}}{R_1 + R_2(1 - h_{fb})} \tag{12}$$

$$I_B = -\frac{R_1 I_{CBO} + V_{EE}(1 + h_{fb})}{R_1 + R_2(1 + h_{fb})} \tag{13}$$

$$I_C = \frac{(R_1 + R_2)I_{CBO} + h_{fb}V_{EE}}{R_1 + R_2(1 + h_{fb})} \tag{14}$$

$$V_{BG} = V_{EG} = R_2 \frac{R_1 I_{CBO} + V_{EE}(1 + h_{fb})}{R_1 + R_2(1 + h_{fb})} \tag{15}$$

$$V_{CG} = V_{CC} - \frac{R_C[(R_1 + R_2)I_{CBO} + h_{fb}V_{EE}]}{R_1 + R_2(1 + h_{fb})} \tag{16}$$

$$V_{CB} = V_{CC} - V_{EE}\frac{R_2(1 + h_{fb}) + R_C h_{fb}}{R_1 + R_2(1 + h_{fb})} - I_{CBO}\frac{R_1 R_2 + R_C(R_1 + R_2)}{R_1 + R_2(1 + h_{fb})} \tag{17}$$

In the above equations G refers to a virtual ground, at the connection between the low side of the resistor R_2 and the common side of the two batteries.

In addition, there are several stability factors which provide a measure of the variability of the operating point with temperature and with supply voltage. Since

17-12 AMPLIFIER CIRCUITS

I_{CBO} is the most temperature-sensitive transistor parameter it is used as the control variable, rather than temperature.

$$S_{I_E} = \frac{dI_E}{dI_{CBO}} = \frac{-R_2}{R_1 + R_2(1 + h_{fb})} \tag{18}$$

$$S_{I_C} = \frac{dI_C}{dI_{CBO}} = \frac{R_1 + R_2}{R_1 + R_2(1 + h_{fb})} \tag{19}$$

$$S_V = \frac{dV_{CB}}{dI_{CBO}} = -\frac{R_1 R_2 + R_C(R_1 + R_2)}{R_1 + R_2(1 + h_{fb})} \tag{20}$$

$$S_{V_1} = \frac{dV_{CB}}{dV_{CC}} = 1 \tag{21}$$

$$S_{V_2} = \frac{dV_{CB}}{dV_{EE}} = -\frac{R_C h_{fb} - R_2(1 + h_{fb})}{R_1 + R_2(1 + h_{fb})} \tag{22}$$

Note: V_{CC} will be negative, V_{EE} positive, and I_{CBO} negative in p-n-p transistors. The reverse will be true in n-p-n transistors.

4.2. Bias Equations for the Single-battery Arrangement. Referring to the *b* versions of Figs. 8 and 9 the currents and voltages may be obtained from the following equations:

$$I_E = \frac{-R_2 V_{CC}}{R_1 R_2 + R_1 R_3 + R_2 R_3(1 + h_{fb})} - \frac{R_2 R_3 I_{CBO}}{R_1 R_2 + R_1 R_3 + R_2 R_3(1 + h_{fb})} \tag{23}$$

$$I_B = \frac{R_2 V_{CC}(1 + h_{fb})}{R_1 R_2 + R_1 R_3 + R_2 R_3(1 + h_{fb})} - \frac{R_1(R_2 + R_3) I_{CBO}}{R_1 R_2 + R_1 R_3 + R_2 R_3(1 + h_{fb})} \tag{24}$$

$$I_C = \frac{-h_{fb} R_2 V_{CC}}{R_1 R_2 + R_1 R_3 + R_2 R_3(1 + h_{fb})} + \frac{(R_1 R_2 + R_1 R_3 + R_2 R_3) I_{CBO}}{R_1 R_2 + R_1 R_3 + R_2 R_3(1 + h_{fb})} \tag{25}$$

$$V_{BG} = V_{EG} = \frac{R_1 R_2 V_{CC}}{R_1 R_2 + R_1 R_3 + R_2 R_3(1 + h_{fb})}$$
$$+ \frac{R_1 R_2 R_3 I_{CBO}}{R_1 R_2 + R_1 R_3 + R_2 R_3(1 + h_{fb})} \tag{26}$$

$$V_{CG} = V_{CC}\left[1 + \frac{R_2 R_C h_{fb}}{R_1 R_2 + R_1 R_3 + R_2 R_3(1 + h_{fb})}\right]$$
$$- \frac{R_C(R_1 R_2 + R_1 R_3 + R_2 R_3) I_{CBO}}{R_1 R_2 + R_1 R_3 + R_2 R_3(1 + h_{fb})} \tag{27}$$

$$V_{CB} = V_{CC} \frac{R_1 R_3 + R_2 R_3(1 + h_{fb}) + R_2 R_C h_{fb}}{R_1 R_2 + R_1 R_3 + R_2 R_3(1 + h_{fb})}$$
$$- \frac{R_1 R_2 R_3 + R_C(R_1 R_2 + R_1 R_3 + R_2 R_3)}{R_1 R_2 + R_1 R_3 + R_2 R_3(1 + h_{fb})} I_{CBO} \tag{28}$$

$$S_{I_E} = \frac{dI_E}{dI_{CBO}} = \frac{-R_2 R_3}{R_1 R_2 + R_1 R_3 + R_2 R_3(1 + h_{fb})} \tag{29}$$

$$S_{I_C} = \frac{dI_C}{dI_{CBO}} = \frac{R_1 R_2 + R_1 R_3 + R_2 R_3}{R_1 R_2 + R_1 R_3 + R_2 R_3(1 + h_{fb})} \tag{30}$$

$$S_V = \frac{dV_{CB}}{dI_{CBO}} = -\frac{R_1 R_2 R_3 + R_C(R_1 R_2 + R_1 R_3 + R_2 R_3)}{R_1 R_2 + R_1 R_3 + R_2 R_3(1 + h_{fb})} \tag{31}$$

$$S_{V_1} = \frac{dV_{CB}}{dV_{CC}} = \frac{R_1 R_3 + R_2 R_3(1 + h_{fb}) + R_2 R_C h_{fb}}{R_1 R_2 + R_1 R_3 + R_2 R_3(1 + h_{fb})} \tag{32}$$

Note: The same polarity conventions apply as for the two-battery arrangement.

4.3. Graphical Determination of Bias. A more accurate determination of the operating point, including the effect of the actual value of the emitter-base voltage V_{EB}, may be obtained by using the static characteristics of the transistor. Figure 11

AUDIO AMPLIFIERS   17-13

shows the single-battery circuit once more, with the various currents and voltages identified. The steps to follow in obtaining the operating point are as follows:

1. Construct a family of I_C versus V_{CE} curves for the region of interest from the manufacturer's specification sheet.
2. Draw the effective load line from the point V_{CC}, having a slope equal to $R_C + R_1(h_{FE}+1)/h_{FE}$. (*Note:* h_{FE} is the d-c value, equal to the ratio I_C/I_B.)

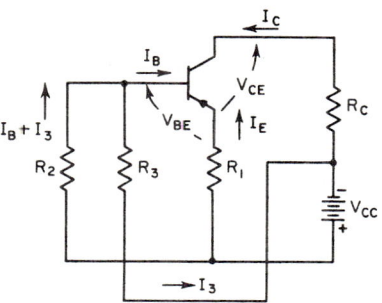

Fig. 11. Single-battery bias circuit.

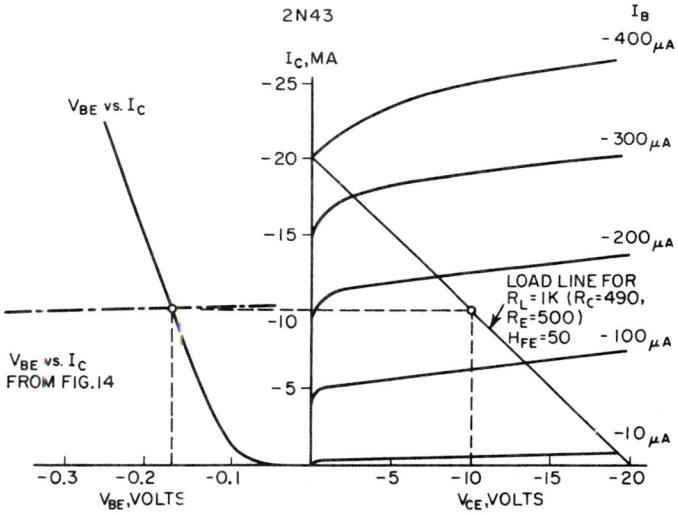

Fig. 12. Collector and emitter families for 2N43 in common-emitter configuration. Quiescent operating point: $I_B = -170$ μa, $I_E = 10.2$ ma, $I_C = -10.0$ ma, $V_{BE} = 0.177$ volts, $V_{CE} = 10$ volts.

3. Construct a plot of I_C and V_{CE} versus I_B from points along this load line.
4. From these curves and the specification-sheet curve of V_{BE} versus I_B plot V_{BE} versus I_C for points on the load line.
5. Also plot V_{BE} versus I_3 as obtained from the following equation:

$$V_{BE} = \frac{R_2 V_{CC}}{R_2 + R_3} - \frac{I_C}{h_{FE}} \left[R_1(h_{FE}+1) + \frac{R_2 R_3}{R_2 + R_3} \right] \tag{33}$$

The intersection of the last two curves will determine the operating point.
The above procedure is illustrated in Figs. 12 to 14. Figure 12 shows the collector

17–14   AMPLIFIER CIRCUITS

family for the 2N43 transistor, together with a load line corresponding to the following conditions:

$$V_{CC} = -20 \text{ volts} \qquad h_{FE} = 50$$
$$R_1 = 500 \text{ ohms} \qquad R_2 = 10{,}000 \text{ ohms}$$
$$R_3 = 20{,}000 \text{ ohms} \qquad R_C = 490 \text{ ohms}$$

These values give an effective load-line resistance of 1,000 ohms, according to step 2 above.

Figure 13 shows the variation of I_C and V_{CE} versus I_B, as obtained by picking various points off the load line. Knowing the values of V_{CE} corresponding to each value of I_B and also the corresponding values of V_{BE} from the emitter family the curve of V_{BE} versus I_C was plotted in the left-hand quadrant of Fig. 12.

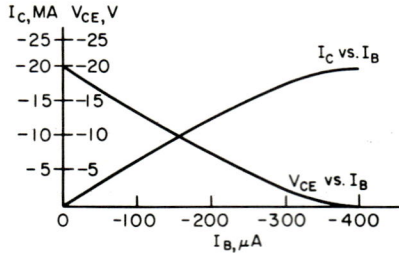

Fig. 13. Collector current and collector-to-emitter voltage vs. base current.

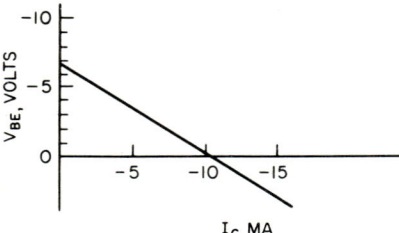

Fig. 14. Base-to-emitter voltage vs. collector current.

Figure 14 is a plot of Eq. (33), and a small segment of this curve is also plotted in Fig. 12 in the left-hand quadrant. The intersection of the two curves gives the operating point, which turns out to be

$$I_C = -10 \text{ ma} \qquad I_E = 10.2 \text{ ma} \qquad I_B = 0.17 \text{ ma}$$
$$V_{CE} = -10 \text{ volts} \qquad V_{BE} = -0.177 \text{ volt}$$

Usually the approximate method described in Eqs. (12) to (32) will supply reasonably accurate values for the great majority of cases; however, if greater accuracy is required the graphical method will usually supply it.

4.4. Significance of the Stability Factors. The various stability factors given above provide a means of determining the shift of operating point with temperature, and thus the variation of gain or impedance. While the various transistor parameters are not in themselves excessively responsive to temperature, they do vary considerably with emitter current and collector-emitter voltage. If these vary excessively with temperature they will, in turn, produce a variation in the transistor parameters. It is this secondary effect of temperature which usually produces whatever major effect there is as temperature varies. Thus, if we have criteria by which we can anticipate the effects of variation of I_{CBO} with temperature, we can then determine how much shift in this quantity can be tolerated. The stability factors give us the relationships between variation of I_{CBO} and, for example, I_C. If we know how much variation we can tolerate in I_C we can then determine the maximum possible variation of I_{CBO}, and knowing the relationship between I_{CBO} and temperature, we can then set limits for temperature. The lower the stability factors the less will be the variation of transistor operating point with temperature. However, we do not obtain anything for nothing; thus lower stability factors require more power dissipation and hence lower the efficiency. Thus there is always a compromise between stability and power consumption. In low-level stages this usually is not a serious consideration; however, in power stages it becomes an important factor.

AUDIO AMPLIFIERS

5. TRANSFORMER-COUPLED STAGES

Up to this point we have dealt with the amplifier stage in general terms without regard to the method employed to couple signals into and out of the tube or transistor. Two forms of coupling are commonly used: transformer and resistance-capacitor coupling, abbreviated to RC coupling.

5.1. Transformer-coupled Tube Stage. In the case of electron tubes little complication is introduced by the method of coupling. The input impedance of the tube can be taken as infinite at audio frequencies and our major requirement is to employ the proper transformer design to ensure obtaining the desired frequency response, efficiency, and distortion. Thus the design of transformer-coupled tube amplifiers becomes more a matter of transformer design than anything else.

With transistors there are a number of other considerations, as well as the transformer design itself. The d-c resistance of the windings becomes a factor, as it is

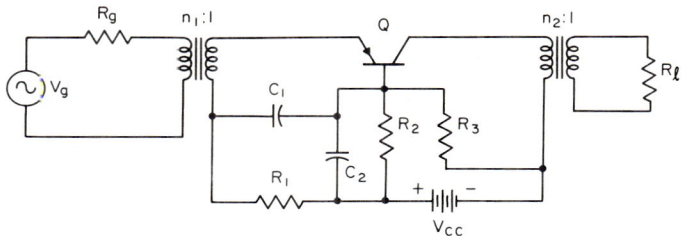

FIG. 15. Transformer-coupled common-base stage.

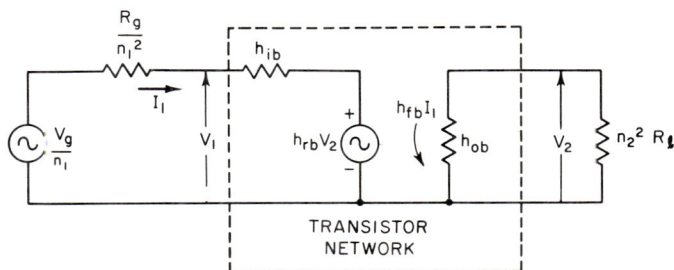

FIG. 16. Equivalent circuit of transformer-coupled stage.

inserted in one of the leads of the transistor and thus contributes to the stability factors, as indicated previously. The transistor has a finite and relatively low input resistance; hence this must be considered in the design of the transformer. The transistor is also a bilateral device, the load on which affects the input impedance, and the source impedance influences the output impedance. Thus amplifier design with transistors is somewhat more complicated than with tubes.

5.2. Transformer-coupled Common-base Stage. Figure 15 shows the circuit of a single-stage transistor amplifier employing transformer coupling for both input and output. The more common single-battery bias arrangement is used, with the same numbering system for the bias resistors as was used previously. Two bypass capacitors are shown, C_1 and C_2, to avoid the degenerative effects of resistance in the emitter or base leads. Occasionally it will be found adequate to dispense with capacitor C_2. This inserts a resistance equal to R_1 and R_2 in parallel in series with the collector load. However, if the load is much higher than this impedance the effect will be minor.

Figure 16 shows the equivalent circuit of this arrangement. The input signal voltage is reduced by the input transformer ratio, the source impedance by the ratio squared. The load resistance is multiplied by the output transformer ratio squared. Since this circuit utilizes the common-base configuration the appropriate set of parameters, as given in Table 2, are used. The transistor network is indicated by the dashed-line enclosure. Thus the circuit becomes a two-port network, with a source of V_g/n_1 operating through a source impedance of R_g/n_1^2, and with a load impedance of $n_2^2 R_l$. The network equations presented earlier in the chapter may be used to obtain

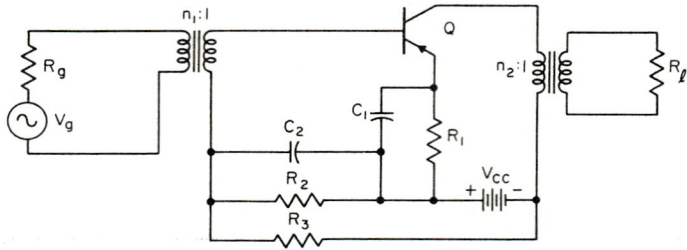

Fig. 17. Transformer-coupled common-emitter stage.

the impedances, amplifications, and gain. For example, the input impedance of the transistor network is given by

$$Z_i = h_{11} - h_{12}h_{12}Z_l/(1 + h_{22}Z_l)$$
$$= h_{ib} - h_{rb}h_{fb}n_2^2 R_l/(1 + h_{ob}n_2^2 R_l) \tag{34}$$

Similarly, the voltage amplification is given by

$$A_v = -h_{21}Z_l/[h_{11}(1 + h_{22}Z_l) - h_{12}h_{21}Z_l]$$
$$= -h_{fb}n_2^2 R_l/[h_{ib}(1 + h_{ob}n_2^2 R_l) - h_{rb}h_{fb}n_2^2 R_l] \tag{35}$$

The other network equations may be used in similar manner to obtain the output impedance, current amplification, and gain.

5.3. Transformer-coupled Common-emitter Stage. Figure 17 shows the circuit of the common-emitter transistor amplifier with transformer coupling, and

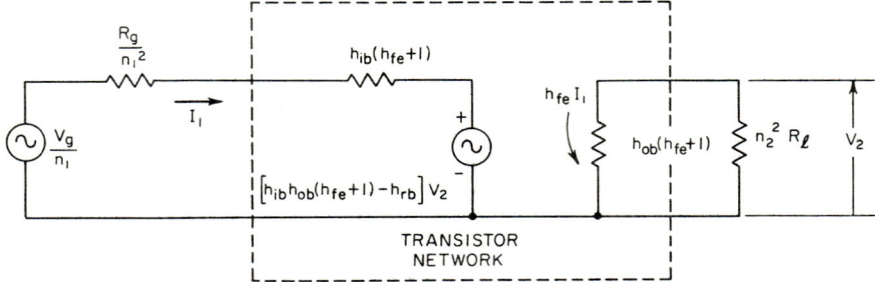

Fig. 18. Equivalent circuit of transformer-coupled common-emitter stage.

Fig. 18 shows the corresponding equivalent circuit. Again, the single-battery bias arrangement is shown, with two bypass capacitors across the bias resistors. As before, capacitor C_2 may often be omitted and C_1 connected from the low side of the transformer to the emitter.

Referring to the equivalent circuit, we note that it is essentially identical to that for the common-base stage, with the exception of the values for the transistor network inside the dotted lines. The values for the transistor parameters are now those corre-

AUDIO AMPLIFIERS

sponding to the common-emitter stage, as per Table 2. Again, substituting the appropriate terms in the network equations of Table 1 we may obtain the impedances, amplifications, and gain.

$$Z_i = h_{11} - h_{12}h_{21}Z_l/(1 + h_{22}Z_l)$$
$$= h_{ib}(h_{fe} + 1) - \frac{[h_{ib}h_{ob}(h_{fe} + 1) - h_{rb}]h_{fe}n_2^2 R_l}{1 + h_{ob}(h_{fe} + 1)n_2^2 R_l} \quad (36)$$

$$A_v = -h_{21}Z_l/[h_{11}(1 + h_{22}Z_l) - h_{12}h_{21}Z_l]$$
$$= \frac{-h_{fe}n_2^2 R_l}{h_{ib}(h_{fe} + 1)[1 + h_{ob}(h_{fe} + 1)n_2^2 R_l] - [h_{ib}h_{ob}(h_{fe} + 1) - h_{rb}]h_{fe}n_2^2 R_l} \quad (37)$$

Similarly, the other equations may be solved by inserting the appropriate transistor parameters into the network equations. If the solution of more than one equation is desired it is simpler to solve for the transistor parameters and then insert the

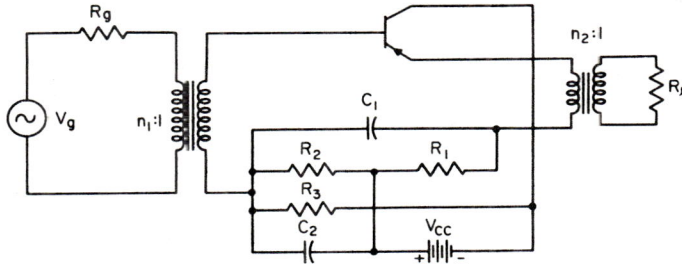

FIG. 19. Transformer-coupled common-collector stage.

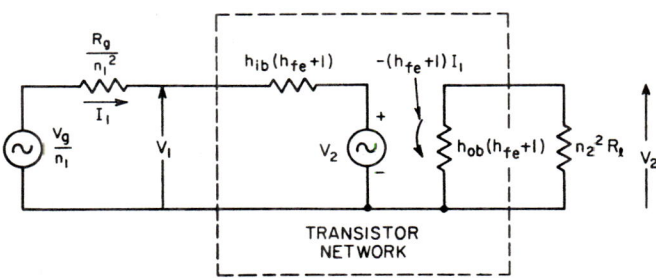

FIG. 20. Equivalent circuit of transformer-coupled common-collector stage.

resultant values into the network equations than to work out complete solutions similar to Eqs. (36) and (37).

5.4. Transformer-coupled Common-collector Stage. Figure 19 illustrates the transformer-coupled common-collector stage. The bias arrangement is the same essentially as before, and the same remarks apply concerning the capacitors. Figure 20 gives the corresponding equivalent circuit, with the transistor parameters for the common-collector stage. One very noticeable difference in this network is that the value of h_{12} for the common-collector stage is unity, thus simplifying the equations considerably.

For this circuit,

$$Z_i = h_{ib}(h_{fe} + 1) + (h_{fe} + 1)n_2^2 R_l/[1 + h_{ob}(h_{fe} + 1)n_2^2 R_l] \quad (38)$$

$$A_v = \frac{(h_{fe} + 1)n_2^2 R_l}{h_{ib}(h_{fe} + 1)[1 + h_{ob}(h_{fe} + 1)n_2^2 R_l] + (h_{fe} + 1)n_2^2 R_l} \quad (39)$$

17–18 AMPLIFIER CIRCUITS

While the above equations are somewhat simpler than those for the common-emitter stage, it is mainly a matter of degree and they are still quite formidable. Again, the easiest procedure would be to solve first for the values of the common-collector parameters, using the equations of Table 2, then insert these values into the proper equations of Table 1 to obtain the desired solutions.

Since one of the major advantages of transistors is their small size, much of this advantage would be lost by using transformers. In particular, for high-fidelity systems the bulk of the transformers required to obtain the desired frequency response would far exceed that of all the rest of the equipment. For this reason transformer-coupled transistor amplifiers are rather infrequently used, and practically all transistor audio amplifiers use some form of RC coupling. The design and performance of the RC-coupled stage will be discussed in the next section.

6. RC-COUPLED STAGES

Resistance-capacitance coupling is considerably more common than transformer coupling in audio amplifiers and will be analyzed in considerable detail in the following sections.

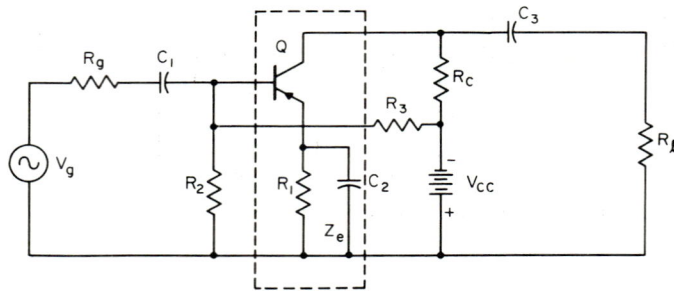

FIG. 21. RC-coupled common-emitter stage.

6.1. RC-coupled Tube Amplifiers. These were discussed in Sec. 2.6, and the reader is referred to that section for performance data on typical tube amplifiers. From a network standpoint several important characteristics stand out. The input impedance is that of the grid leak and is not affected by the tube or its terminations. The power gain is potentially infinite, although reduced to finite values by the grid leak. Likewise, the current amplification is potentially infinite, although practically it is determined by the grid leak also. The voltage amplification is usually in the range of 10 to 100, as indicated by the values shown for the three tubes illustrated. The output voltage may be quite high, as the plate battery is normally several hundred volts. Thus, in the examples shown, output voltage up to 40 volts is obtainable. Because of the relatively large values of tube impedances, the values of the coupling and bypass capacitors necessary in tube amplifiers are comparatively low, as illustrated by the examples in that section. This contrasts to transistor amplifiers, where these capacitors usually run into the microfarads.

6.2. RC-coupled Transistor Amplifiers. The common-base configuration is not commonly used in this type of circuit, because it has less than unity current amplification and hence would produce no gain unless operating into an impractically high output impedance. Our analysis in the following section will therefore deal with the common-emitter configuration mainly, and with the common-collector version, the so-called emitter follower (conforming somewhat to the tube cathode follower).

Figure 21 shows a typical RC-coupled common-emitter stage. The source V_g is coupled through the source resistance R_g and coupling capacitor C_1 to the transistor. R_l the load resistor, is coupled to the transistor through the capacitor C_3. The usual three resistors control the bias, and the collector battery V_{CC} is coupled to the collector through the resistor R_C. The emitter bias network, consisting of the resistor R_e shunted by the capacitor C_2, is shown connected within the transistor network for reasons which will be explained later. The other resistors are shown outside the network as thus they may be combined with the external impedances.

Figure 22 shows the equivalent circuit for this arrangement. The two bias resistors R_2 and R_3 have been combined into the one resistor R', representing their parallel value $R_2R_3/(R_2 + R_3)$. The collector resistor R_C is shown outside the network, and V_2 is the output voltage appearing across it. V_o is the portion of this voltage which appears across the load resistance R_l.

The parameters shown for the transistor inside the dotted enclosure are those corresponding to the common-emitter configuration from Table 2. This therefore assumes that the emitter external impedance Z_e (the series equivalent of R_1 and C_2 in parallel) is essentially zero. The effects of its not being zero will be considered subsequently.

This circuit may be analyzed by obtaining the effective load of the transistor, formed by the elements R_C, C_3, and R_l, inserting this into the network equations for

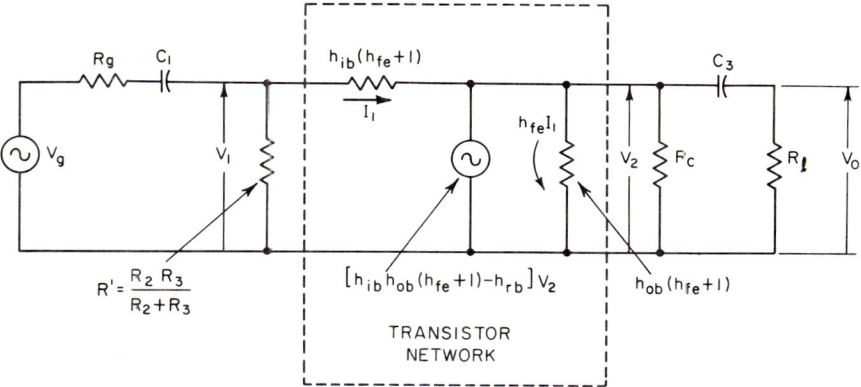

Fig. 22. Equivalent circuit for RC-coupled amplifier stage. Note: See text for modifications of transistor parameters due to nonzero value of Z_e.

impedance, amplification, or gain, then calculating the ratio of input impedance to total input circuit impedance, to get the ratio of input voltage to the source voltage V_g.

The general solution for the input impedance is

$$Z_i = \frac{h_{11} + \Delta^h R_C + j\omega C_3[h_{11}(R_C + R_l) + \Delta^h R_C R_l]}{1 + h_{22}R_C + j\omega C_3(R_C + R_l + h_{22}R_C R_l)} \tag{40}$$

(*Note:* The determinant $\Delta^h = h_{11}h_{22} - h_{12}h_{21}$.)
and for the overall amplification

$$\frac{V_o}{V_g} = \frac{-j\omega C_3 R_l h_{21}}{(1 + j\omega C_3 R_l)(\Delta^h + \Delta' Z_g/R' + h_{22}Z_g) + \{[1 + j\omega C_3(R_C + R_l)]/R_C\}(h_{11} + h_{11}Z_g/R' + Z_g)} \tag{41}$$

The first portion of the denominator is relatively minor, and the above equation can be shortened to the following with little loss of accuracy:

$$\frac{V_o}{V_g} \simeq \frac{-j\omega C_3 R' R_C R_l h_{21}}{[1 + j\omega C_3(R_C + R_l)](h_{11}R' + h_{11}Z_g + R'Z_g)} \tag{42}$$

For the common-emitter stage with zero impedance in the external emitter circuit we substitute in the above equations the following h parameters:

$$h_{11} = h_{ib}(h_{fe} + 1) \qquad h_{12} = h_{ib}h_{ob}(h_{fe} + 1) - h_{rb} \qquad h_{21} = h_{fe} \qquad h_{22} = h_{ob}(h_{fe} + 1)$$

and for the total source impedance Z_g we substitute $R_g + 1/j\omega C_1$.

6.3. Example of RC-coupled Common-emitter Amplifier. Consider the type 2N335 transistor, operating at the specification operating point of $V_{CB} = 5$ volts, $I_E = 1$ ma in the circuit of Fig. 21. From the specification sheet we obtain the following values for the transistor parameters (*Note:* If the desired operating point were at some other operating point than the standard values or at some temperature other than room temperature it would be necessary to obtain revised values of the

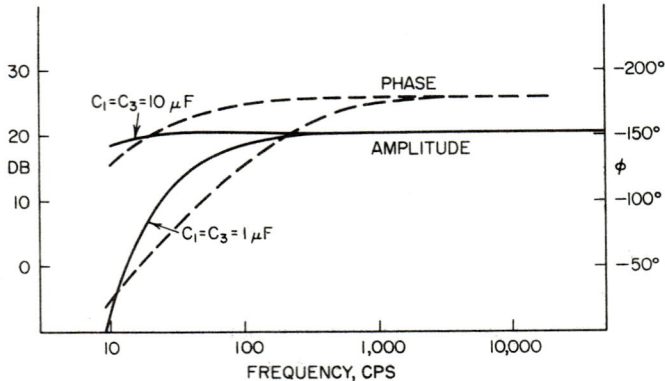

Fig. 23. Effect of coupling capacitor on amplitude and phase.

parameters by applying appropriate correction factors from the curves supplied in the specification sheets before inserting in the following equations):

$$h_{ib} = 43 \text{ ohms} \qquad h_{ob} = 0.15 \times 10^{-6} \text{ mhos}$$
$$h_{rb} = 3.0 \times 10^{-4} \qquad h_{fe} = 60$$

Substituting these in the equations above we obtain the following values for the common-emitter parameters:

$$h_{11} = 43 \times 61 = 2{,}620 \text{ ohms}$$
$$h_{12} = 43 \times 0.15 \times 10^{-6} \times 61 - 3.0 \times 10^{-4} = 0.93 \times 10^{-4}$$
$$h_{21} = 60$$
$$h_{22} = 0.15 \times 10^{-6} \times 61 = 9.15 \times 10^{-6} \text{ mhos}$$
$$\Delta^h = 2{,}620 \times 9.15 \times 10^{-6} - 60 \times 0.93 \times 10^{-4} = 0.0184$$

Let us also take the following values for the bias resistors and collector resistor: $R_1 = 1{,}000$ ohms, $R_2 = 4{,}700$ ohms, $R_3 = 27{,}000$ ohms, $R_C = 3{,}300$ ohms. These represent reasonably common values. Finally, let us use identical values of 1,000 ohms for the load resistance R_l and the generator resistance R_g and determine the effect of the capacitors C_1 and C_3 on the frequency response. It is assumed that C_2 is large enough for the present to bypass the emitter resistor R_1 effectively so that the external emitter impedance is zero. The net value of the bias resistors R_2 and R_3 in parallel is $R' = 4.7 \times 27 \div 31.7 = 4.0$ kilohms.

Figure 23 shows the variation of amplitude and phase of the amplifier stage of Fig. 21, calculated using Eq. (41). The value of input impedance, as calculated from

AUDIO AMPLIFIERS   17-21

Eq. (40), remained constant at 2,620 ohms, indicating that the output load was too low to make the input impedance appreciably different from the short-circuit value h_{ie}. Curves are shown for two values of coupling capacitors, 1 and 10 μf, and it is obvious that capacitors of the latter size are necessary if the low-frequency response is to be preserved. The very considerable phase shift introduced by inadequate capacitance is also indicated.

6.4. Effect of Impedance in the Emitter Circuit. In many practical circuits it is not possible to avoid the insertion of impedance in the emitter lead. For example, in the circuit of Fig. 21 the bias resistor R_1 and its bypass capacitor C_2 form an impedance Z_e, which will be reflected into the input circuit, multiplied by the h_{fe} of the transistor approximately. We must now use equations which reflect the changes produced by this added factor. The expressions for the revised h parameters with Z_e added are

$$h_{11} = \frac{h_{ib} + Z_e}{h_{ob}Z_e + 1/(h_{fe} + 1)} \tag{43}$$

$$h_{12} = \frac{h_{ob}(h_{ib} + Z_e)}{h_{ob}Z_e + 1/(h_{fe} + 1)} - \frac{h_{rb}}{1 + h_{ob}Z_e(h_{fe} + 1)} \tag{44}$$

$$h_{21} = \frac{h_{fb} - h_{ob}Z_e(h_{fe} + 1)}{1 + h_{ob}Z_e(h_{fe} + 1)} \tag{45}$$

$$h_{22} = \frac{h_{ob}}{h_{ob}Z_e + 1/(h_{fe} + 1)} \tag{46}$$

As an example of the effect of Z_e the h parameters were calculated for the same transistor as before, using values of $R_1 = 1,000$ ohms and $C_2 = 10$ and 100 μf. These values produced no effect on h_{21} or h_{22}; however, the changes in the other two parameters and in the determinant were considerable, as the following tabulation will indicate.

h_{11}

Frequency	$C_2 = 10$	$C_2 = 100$
0	63,000∠0	63,000∠0
10	53,700∠−30.3°	10,340∠−66.3°
100	10,340∠−66.3°	2,810∠−20.1°
1,000	2,810∠−20.1°	2,620∠0
10,000	2,620∠0	2,620∠0

h_{12}

Frequency	$C_2 = 10$	$C_2 = 100$
0	$91.7 \times 10^{-4}∠0$	Not calculated
10	$77.7 \times 10^{-4}∠-31.6°$	
100	$14.5 \times 10^{-4}∠-77.3°$	
1,000	$1.7 \times 10^{-4}∠-56.7°$	
10,000	$0.93 \times 10^{-4}∠0$	

The overall amplification was calculated using the approximation equation (42), and the results are plotted in Fig. 24. The effect of inadequate emitter bypass is very evident, and it is obvious that capacitors on the order of 100 μf or larger are required to obtain good low-frequency response. Another expedient frequently used is to substitute breakdown diodes for the bypass capacitors. This provides a reasonably low dynamic impedance at the lowest frequencies, without the requirement of bulk.

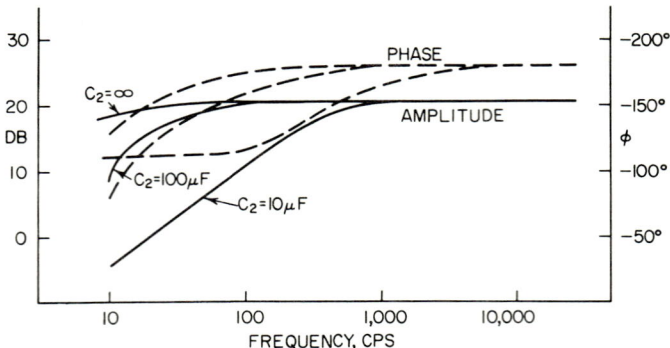

FIG. 24. Effect of emitter bypass capacitor. $C_1 = C_3 = 10$ μf.

7. CASCADED STAGES

Since the great majority of transistor (and tube) amplifiers consist of more than one stage it becomes necessary to consider the problems involved in cascading two or more stages. This may be attacked in a number of ways. For example, the theory of network matrix manipulations may be applied. Consider, for example, the simple block diagram of Fig. 25, which shows two networks connected in cascade. These networks may be both active, e.g., amplifier stages, or one or more may be passive. Thus one may be the coupling network preceding or following an active network.

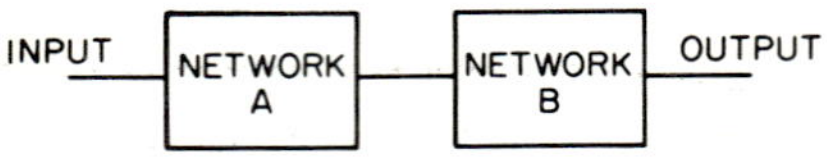

The method of attack is to convert the network parameters into the a matrix form, then to multiply the two matrices to obtain an overall a matrix expression, which may now be reconverted to any of the other matrix representations, if desired, although the impedances, amplifications, and gain may be found from the a matrix parameters directly. For example, if the h parameters are available the a parameters may be calculated from the following equations:

$$a_{11} = -\Delta^h/h_{21} \qquad a_{12} = -h_{11}/h_{21}$$
$$a_{21} = -h_{22}/h_{21} \qquad a_{22} = -1/h_{21}$$

FIG. 25. Block diagram of two cascaded stages.

This process of parameter conversion, multiplication, and reconversion is extremely tedious, and to simplify the process Table 3 has been prepared. This table permits

Table 3. Summary of Matrix Terms for Two Tandem Networks

$$[z] = \begin{bmatrix} \dfrac{z_{11a}z_{11b} + \Delta^z_a}{z_{22a} + z_{11b}} & \dfrac{(z_{11a}z_{11b} + \Delta^z_a)(z_{22a}z_{22b} + \Delta^z_b)}{z_{21a}z_{21b}(z_{22a} + z_{11b})} - \dfrac{z_{11a}\Delta^z_b + \Delta^z_a z_{22b}}{z_{21a}z_{21b}} \\ \dfrac{z_{21a}z_{21b}}{z_{22a} + z_{11b}} & \dfrac{z_{22a}z_{22b} + \Delta^z_b}{z_{22a} + z_{11b}} \end{bmatrix}$$

$$[y] = \begin{bmatrix} \dfrac{y_{11a}y_{11b} + \Delta^y_a}{y_{22a} + y_{11b}} & \dfrac{y_{11a}\Delta^y_b + \Delta^y_a y_{22b}}{y_{21a}y_{21b}} - \dfrac{(y_{11a}y_{11b} + \Delta^y_a)(y_{22a}y_{22b} + \Delta^y_b)}{y_{21a}y_{21b}(y_{22a} + y_{11b})} \\ \dfrac{-y_{21a}y_{21b}}{y_{22a} + y_{11b}} & \dfrac{y_{22a}y_{22b} + \Delta^y_b}{y_{22a} + y_{11b}} \end{bmatrix}$$

$$[h] = \begin{bmatrix} \dfrac{h_{11a} + \Delta^h_a h_{11b}}{1 + h_{22a}h_{11b}} & \dfrac{h_{11a}h_{22b} + \Delta^h_a \Delta^h_b}{h_{21a}h_{21b}} - \dfrac{(h_{11a} + \Delta^h_a h_{11b})(h_{22a}\Delta^h_b + h_{22b})}{h_{21a}h_{21b}(1 + h_{22a}h_{11b})} \\ \dfrac{-h_{21a}h_{21b}}{1 + h_{22a}h_{11b}} & \dfrac{h_{22a}\Delta^h_b + h_{22b}}{1 + h_{22a}h_{11b}} \end{bmatrix}$$

AUDIO AMPLIFIERS

calculation directly of the z, y, or h parameters of a two-stage amplifier, knowing the corresponding parameters of the component stages. For example, suppose we wish to calculate the performance of a two-stage common-emitter amplifier for which we have obtained the h parameters of the individual stages. (It is assumed that the effects of bias resistors, etc., have been incorporated into the networks.) We insert the values into the proper places in the table and obtain the overall h parameters directly. Thus the parameter h_{11} of the two-stage amplifier is given as $(h_{11a} + \Delta^h_a h_{11b})/(1 + h_{22a}h_{11b})$ and the other terms are obtained similarly. In these equations the subscript a denotes the first network, b the second.

The other common method of solving two-stage amplifiers is to work them out stage by stage, starting at the end. The output load is applied to the last stage, and the appropriate equations are used to obtain the amplification and input impedance for this stage. This input impedance is then combined with the bias resistors to provide the output load for the next to the last stage. The process is repeated stage by stage until the input is reached. While the process is tedious, it is relatively straightforward.

One factor must be borne in mind in either of the above methods, which is that the

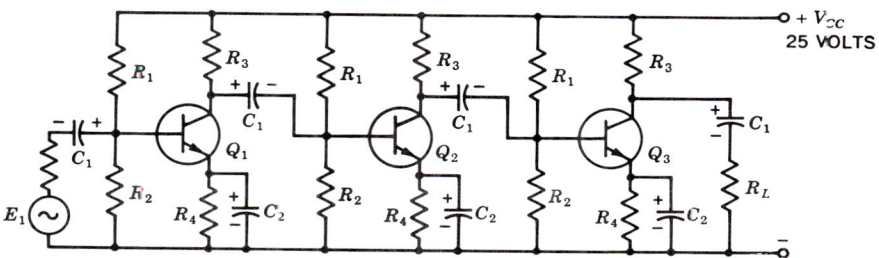

FIG. 26. Schematic diagram of three-stage amplifier. (*Reproduced with permission of Texas Instruments Incorporated, from "Transistor Circuit Design," McGraw-Hill Book Company, New York, 1963.*)

correct parameters must be used corresponding to the actual bias conditions of each stage. Usually the current and voltage will be reduced in the early stages where the signal levels are low; hence these stages will usually require considerable modification of the parameter values given for the specification operating point of 5 volts, 1 ma.

7.1. Example of Multistage Amplifier. Figure 26 shows the circuit of a three-stage amplifier, incorporating the principles described in the preceding section. The component values are as follows:

R_1 = 16,000 ohms R_2 = 6,200 ohms R_3 = 1,600 ohms
R_4 = 1,000 ohms R_L = 560 ohms Q_1, Q_2, Q_3 2N1565
C_1 = 10 μf C_2 = 100 μf

Note: The above symbols do not correspond to those used previously in the text, and the reader should make the proper exchanges if he wishes to use the bias equations, for example, to compute currents, voltages, or stability factors.

This amplifier was designed to operate over a range of −55 to +125°C. The output voltage swing was 2 volts peak-to-peak, and the frequency response was down 3 db at approximately 200 cps and 2 Mc.

The overall gain at 1,000 cycles was as follows:

Temperature, °C	Gain, db
−55	83
+25	88
+125	91

(The above circuit and information are given by courtesy of Texas Instruments, Inc.)

8. HEARING AIDS

Many special requirements are involved in the design of hearing aids. They must be of minimum size and weight, and this becomes especially true if the aid is to be incorporated in some of the extremely close quarters involved, for example, in eyeglasses, where the hearing aid is built into the frame and side bars. Battery economy is another factor of utmost importance, plus the ability to operate on small cells. This leads also to such circuit refinements as variation of the battery drain with listening level, so that the user with moderate hearing loss does not have to change batteries as often as the person with greater loss. While hearing aids can hardly be classed as high-fidelity devices, a good degree of true reproduction is still required, particularly on speech. Also means is usually incorporated to modify the frequency response to accommodate the different hearing-loss curves of the wearer.

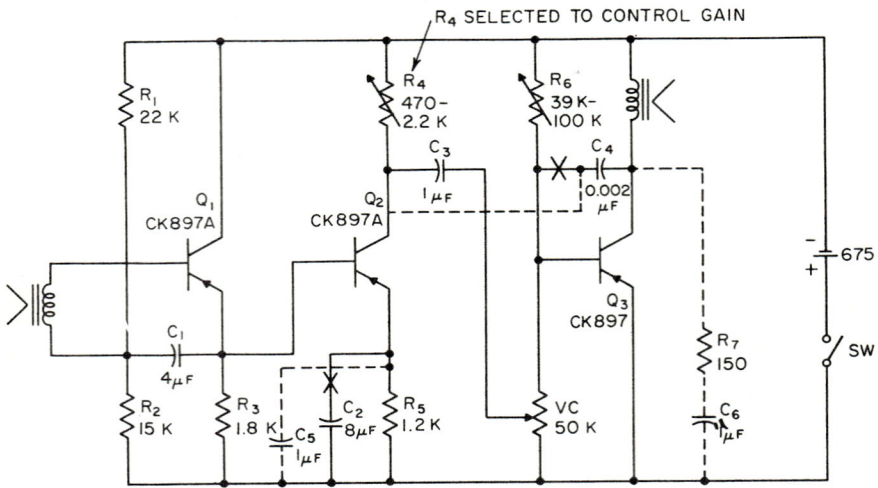

FIG. 27. Schematic diagram of Otarion X-100 Listener. (*Reproduced by courtesy of Otarion Electronics, Inc., Ossining, N.Y.*)

The emphasis on light weight, small size, and low battery consumption has been aided considerably with the advent of the transistor, so that this device has now completely taken over the hearing-aid field. In addition much ingenuity has gone into miniaturizing the devices, and the hearing-aid industry has been in the forefront in the application of the latest techniques of microminiaturization, including integrated circuits. As a result it is now possible to construct the complete transistor circuit on an assembly no larger than the transistors themselves were originally.

In the following section a number of typical hearing aids will be described.

8.1. Otarion X-100 Listener. This hearing aid is representative of the very small units designed for mounting in eyeglass frames, and is a very popular model. Figure 27 shows the schematic of this hearing aid. Three transistors are employed, giving sufficient gain for the person with a moderate amount of hearing loss. The first stage is an emitter follower, with the lower bias resistor R_2 connected effectively in parallel with the emitter resistor R_3 by the capacitor C_1, thus avoiding degeneration in the bias resistor. The second stage uses the common-emitter configuration, as also does the third stage. The volume control is used in the manner shown to avoid passing direct current through the contact, thus avoiding noise. The resistor R_4, shown as variable, is selected to control gain for the actual requirements of the user, thus holding battery drain to the actual minimum required. The resistor R_6 is preselected to give

AUDIO AMPLIFIERS   17-25

a collector current for transistor Q_3 of 1.5 ma. Three variations of this model are available, with the frequency response tailored to best fit the user's requirements by connecting certain components as shown on the schematic. The combinations are as follows:

Red-dot model. Disconnect lead of C_4 from base of Q_3 and connect to collector of Q_2. This effectively connects the high-frequency feedback through C_4 to the full output of Q_2; thus the volume control enters the feedback loop.

Green-dot model. Remove C_4 and add C_6 and R_7 as shown. This provides a means of decreasing high-frequency peaks.

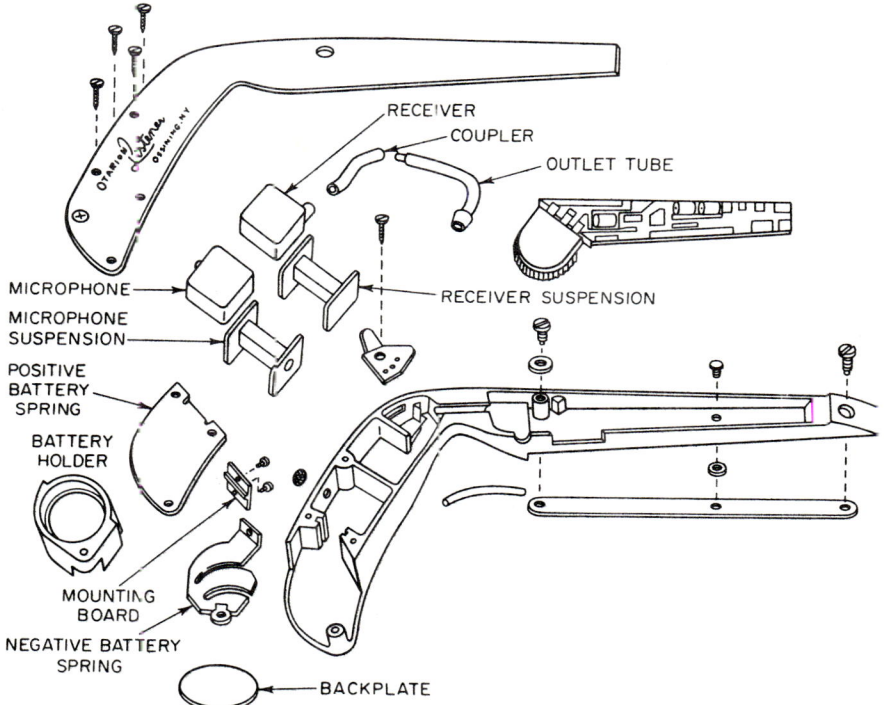

Fig. 28. Cutaway of X-100 hearing aid. (*Reproduced by courtesy of Otarion Electronics, Inc., Ossining, N.Y.*)

Silver-dot model. Replace C_2 with C_5 and connect C_4 as in the red-dot model. Since this provides a smaller bypass capacitor across the emitter resistor, the low-frequency degeneration becomes greater and the low-frequency response is lowered. At the same time, the high frequencies are reduced by feedback through C_4. Thus this arrangement accentuates the middle range, giving the impression of maximum loudness.

It can be seen that by the above changes in the basic model a number of different frequency responses can be obtained.

Figure 28 shows a cutaway of the X-100 hearing aid, illustrating the considerable ingenuity required to fit a complete hearing aid into the frame of a pair of eyeglasses.

8.2. Otarion OL-3 Hearing Aid. This is a more elaborate model, utilizing five transistors and push-pull class B output. Figure 29 shows the circuit of this model. The first stage is again an emitter follower, and it is now connected to either the micro-

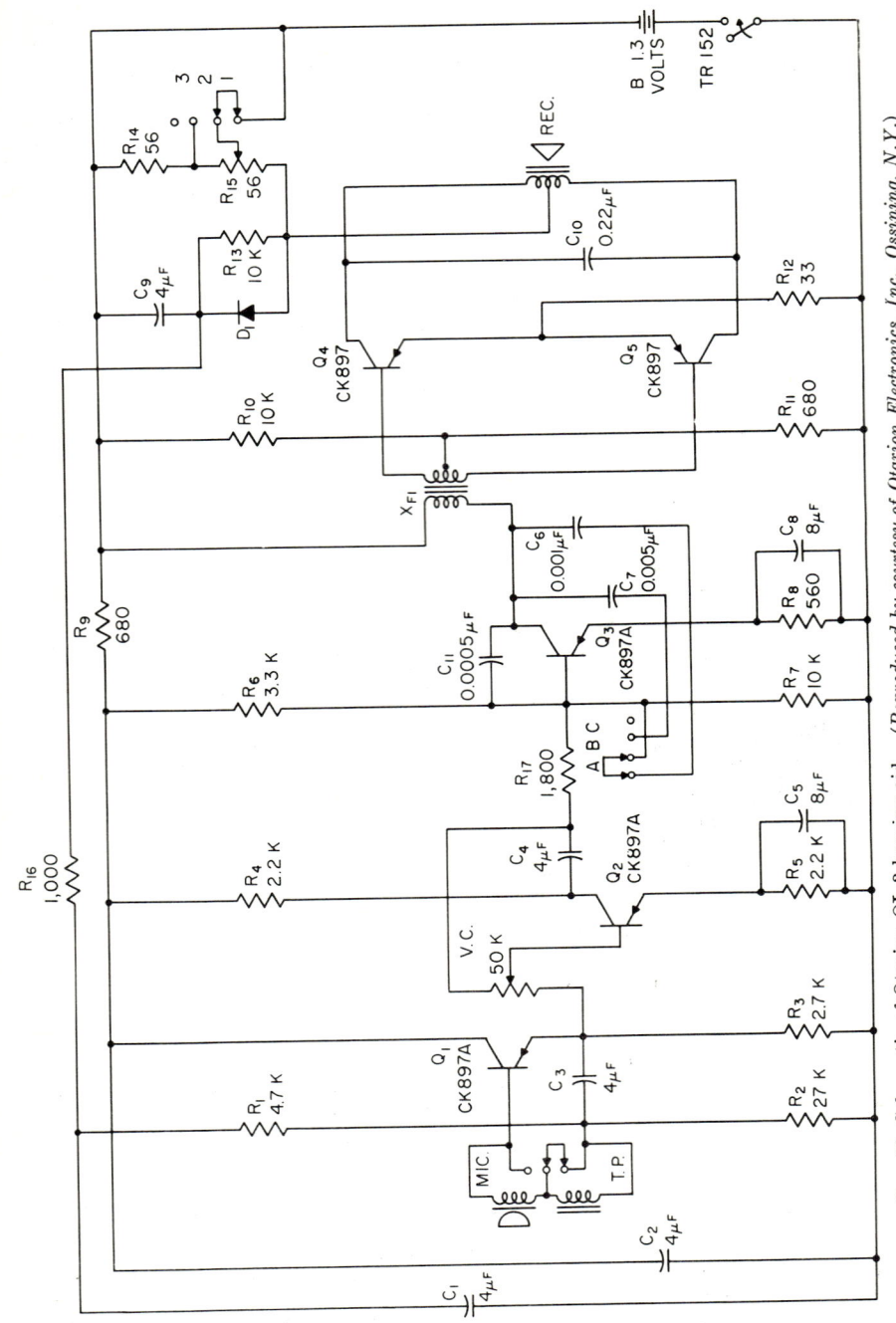

FIG. 29. Schematic of Otarion OL-3 hearing aid. (*Reproduced by courtesy of Otarion Electronics, Inc., Ossining, N.Y.*)

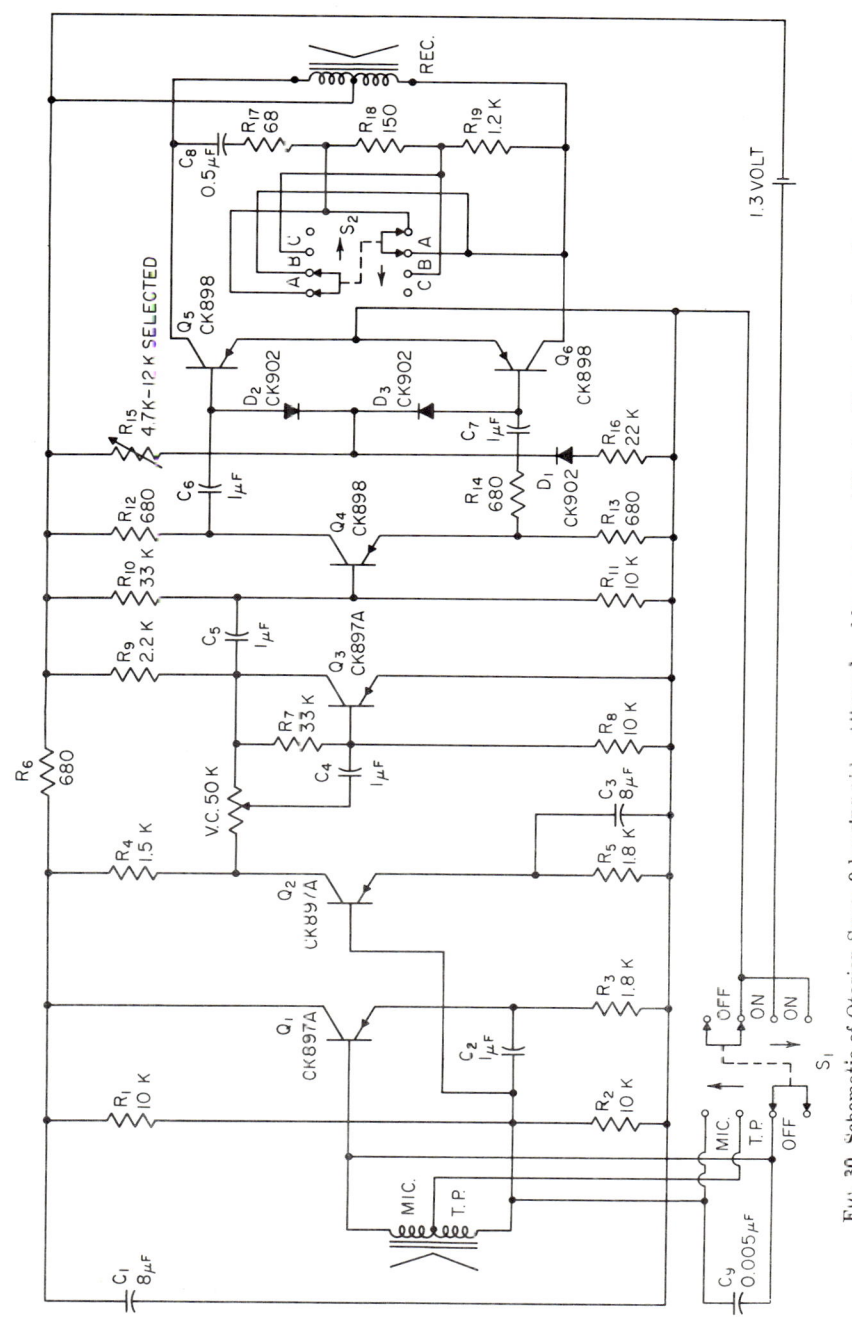

FIG. 30. Schematic of Otarion Super-9 hearing aid. (*Reproduced by courtesy of Otarion Electronics, Inc., Ossining, N.Y.*)

phone or to the telephone pickup by the selector switch. This enables the user to pick up telephone conversation by magnetic induction, to the exclusion of background noise. The volume-control circuit is quite novel, combining volume adjustment with variation of feedback. At one extreme, the base of Q_2 is connected directly to the output of the emitter follower and the feedback is a minimum. As the volume is reduced the feedback increases; thus improvement in quality is possible for those who do not need the full range of amplification. Three settings of the frequency response are available to the user by varying the amount of capacitive feedback around Q_3 by paralleling C_{11} with either C_6 or C_7 or both together.

Another desirable feature in this model is the incorporation of a form of AVC to provide overload control, combined with battery saving. Resistors R_{14} and R_{15} may be inserted in series with the supply to the output stages, thus reducing the battery drain where full power capability is not needed. In addition, when these resistors are inserted, a portion of the audio output appears across them. This is rectified by the diode D_1 and provides a d-c voltage which is applied to the base of Q_1 through R_{16} and R_1. Thus the gain is reduced on peaks, minimizing unpleasant blasting effects.

8.3. Otarion Super-9 Hearing Aid. This model employs six transistors, also push-pull output, eliminating the coupling transformer by using a transistor phase inverter. Figure 30 shows the schematic of this model.

The front part of the circuit is similar to the others previously described. The principal novelty resides in the manner employed to obtain the push-pull output. One output is taken from the emitter of Q_4, the other from its collector, this providing the requisite two phases to drive the output stages. The diodes D_2 and D_3 act to discharge the coupling capacitors C_6 and C_7 on alternate half-cycles when the output stages are not conducting. Bias for the output stages is provided by the diode D_1 and the resistors R_{15} and R_{16}. This provides compensation for temperature variation of the emitter-base voltage of the output transistors, thus avoiding crossover distortion. This is covered in considerable detail in Sec. 14. The resistor R_{14} provides the same source impedance for the lower output stage as for the upper, making up for the lower output impedance of the emitter load. Frequency adjustment is provided by the proper combination of the three resistors and the capacitor shunting the output.

One last remark concerning these devices is in order. When one looks at the parts lists of these hearing aids, in particular the total of microfarads of capacitance involved, he cannot but obtain a great respect for the ingenuity of the designers of these devices, who have long been in the forefront of the trend toward microminiaturization. The reader is also referred to the chapter on integrated circuits, to obtain an insight into the potentialities of these techniques.

9. HIGH-INPUT-IMPEDANCE AMPLIFIERS

Frequently a need arises for an amplifier having input impedance much higher than may be obtained by more conventional means. For example, some phonograph pickups require loads approaching a megohm, and occasionally it is desirable to have input impedance on the order of 100 megohms or more. While this order of input impedance may be obtained without too much difficulty using electron tubes, the normal value of input impedance of transistor amplifiers is several orders of magnitude below this, unless special devices, e.g., field-effect transistors, are used. Thus, if transistors are to be used for this type of application, special circuit modifications become necessary. Mostly this takes the form of so-called "bootstrapping," whereby feedback is employed to introduce a current opposing that which the various loss elements would otherwise consume, thus effectively raising the input impedance. Techniques for accomplishing this will be described in the following section.

9.1. Components of Input Impedance. There are three components of importance in determining the input impedance. These are the bias resistors, the effect of the emitter load impedance, and the collector circuit impedance. In effect these three effects appear as parallel input impedances, and it is necessary to reduce the effect of all three before we can achieve truly high input impedance. These effects will now be analyzed, with the steps necessary to reduce them.

9.2. Effect of Bias Resistors. With the usual method of applying bias to transistor amplifiers, such as the bias circuit of Fig. 21, the two resistors of the bias divider R_1 and R_2 are in parallel as far as alternating current is concerned. It was pointed out in the section on bias and operating-point stability that these resistors must not be too large or stability will suffer. The use of silicon transistors, with their substantially lower leakage current, enables us to employ higher values of bias resistors than would be possible otherwise. However, we are still limited to values much less than the megohm range we desire to achieve. By employing the bootstrapping technique we can effectively raise this component of input impedance several orders of magnitude.

Figure 31 illustrates the technique. In this circuit R_b and R_2 represent the usual base bias resistors, which now return to ground through resistor R_3 instead of directly, as usually. An additional transistor Q_2 has been added, with bias resistor R_1 and emitter resistor R_e. The capacitor C_1 couples the output from Q_2 back into the low end of the input bias network, thus producing a feedback signal essentially equal and opposite

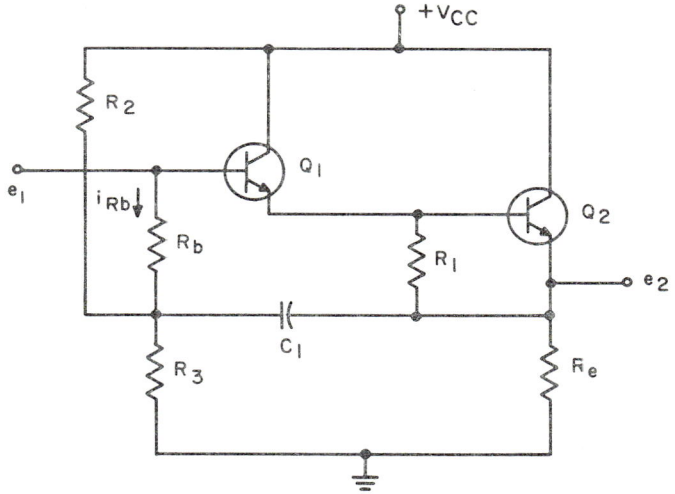

FIG. 31. Use of bootstrapping to raise input impedance. (*Reproduced through courtesy of the General Electric Co., Advanced Electronics Center, Ithaca, N.Y.*)

to the input current through the bias resistors and effectively raising their value as a result. The expression for the effective value of R_b when bootstrapped in this manner is

$$R_b' = R_b/(1 - A_{v1,2}) \quad (47)$$

where $A_{v1,2}$ is the voltage amplification from the input base to the output emitter. This amplification will be very nearly unity; thus the input bias resistance value may readily be increased by two orders of magnitude by this technique.

The increase in input impedance will fall off as the coupling capacitor C_1 becomes less than an effective short circuit. Its value must be such that its reactance is less than 1 per cent of the effective parallel resistance of R_2 and R_3 at the lowest frequency of interest.

9.3. Effect of Emitter Resistance. From material presented previously it will be recalled that the external resistance in the emitter circuit is effectively magnified by the factor $h_{fe} + 1$ (see Sec. 6.4). Thus one method of increasing the input impedance is to insert an impedance in the emitter lead. An unbypassed emitter bias resistor produces this effect; however, we are again limited by practical considerations to a few thousand ohms, which will produce an input resistance only on the order of hundreds of kilohms at best. By using the two-transistor Darlington configuration we can increase the effective value of h_{fe} and thereby obtain a considerable further improvement to an input impedance of about $R_e(h_{fe1} + 1)(h_{fe2} + 1)$. In the usual Darlington arrangement one transistor is working at extremely low collector current (since its emitter current is the base current of the second unit); thus its operating point will not be such as to optimize the value of h_{fe}. This can be improved considerably by adding a shunting resistor between the base and emitter of the second transistor, as shown in Fig. 32. This increases the emitter current of the first transistor to a more nearly optimum value. It can be seen that if transistors are used with h_{fe} on the order of 100 the net effective value of the emitter resistance can be raised to something over 100 megohms with quite reasonable values of R_e.

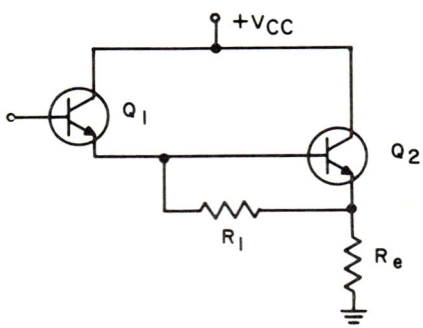

Fig. 32. Addition of shunt resistor to increase gain. (*Reproduced through courtesy of the General Electric Co., Advanced Electronics Center, Ithaca, N.Y.*)

9.4. Effect of the Collector Resistance of the First Transistor. This resistance, the reciprocal of the conductance h_{ob}, is also effectively across the input circuit and therefore becomes another limitation. In effect this resistance is active from the collector to the base, acts in somewhat the same manner as the bias resistors, and therefore should be amenable to the same treatment, i.e., bootstrapping. This is accomplished in the same manner, by

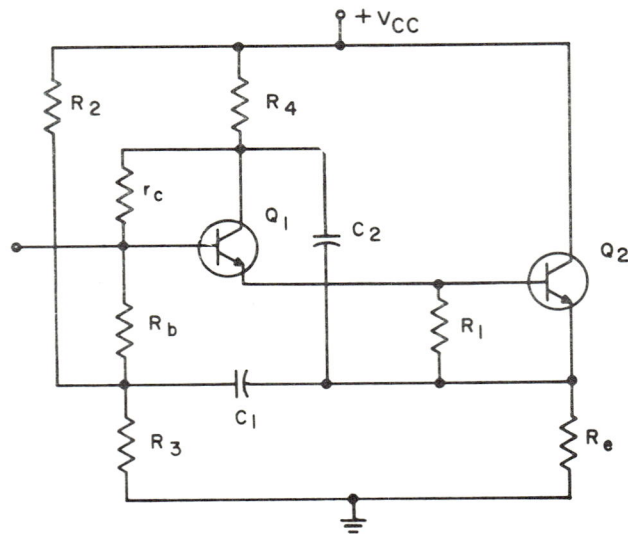

Fig. 33. Use of bootstrapping to reduce collector loading. (*Reproduced through courtesy of the General Electric Co., Advanced Electronics Center, Ithaca, N.Y.*)

adding a capacitor from the second transistor output back to the first collector, as shown in Fig. 33.

9.5. Example of Application of Bootstrapping. The following illustrates the application of the previously described steps to a Darlington configuration employing the 2N1613 transistors and a supply voltage of 28 volts. Figure 34 shows the variation of h_{fe} with collector current and temperature for this transistor. It is arbitrarily assumed that the maximum value of voltage available across the emitter resistor R_e is 20 volts, leaving 8 volts for the transistor. By determining the value of R_e for any desired collector current for the above drop and then obtaining the value of h_{fe} corresponding to this current, the effective multiplied value of bias resistance may be obtained. This is plotted in Fig. 35, where R_e'' represents the effective resistance at the base terminal of Q_2 due to R_e. A collector current of 100 μa is used for the second stage, which gives 200,000 ohms for R_e. A value of 62 kilohms is used for R_1 between the base and emitter of Q_2, thus giving about 10 μa for the collector current of Q_1. The betas of the two transistors for the above operating points are 17 and 25, which produce an effective input resistance of

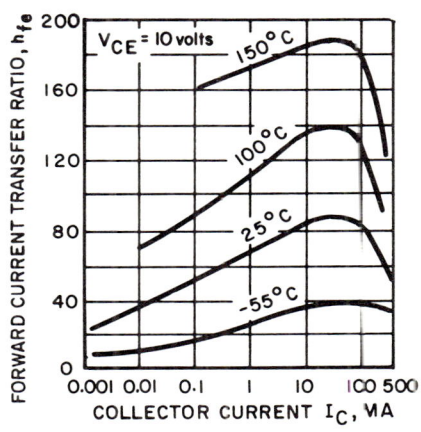

$$R_e' = (17 + 1)(25 + 1) \, 2 \times 10^5 \cong 93 \text{ megohms}$$

FIG. 34. Variation of h_{fe} with collector current. (*Reproduced through courtesy of the General Electric Co., Advanced Electronics Center, Ithaca, N.Y.*)

The assumed collector current of 10 μa and h_{fe} of 17 for Q_1 indicate a base current of 0.06 μa. For such low base current the bias resistor in the base can be quite high without producing excessive drop; hence a value of 1 megohm is used for this resistor. The completed circuit is now shown in Fig. 36.

In the above calculations the value of 200,000 was used for R_e. This is not the true value, however, since this resistor is effectively shunted by the bias resistors R_2, R_3,

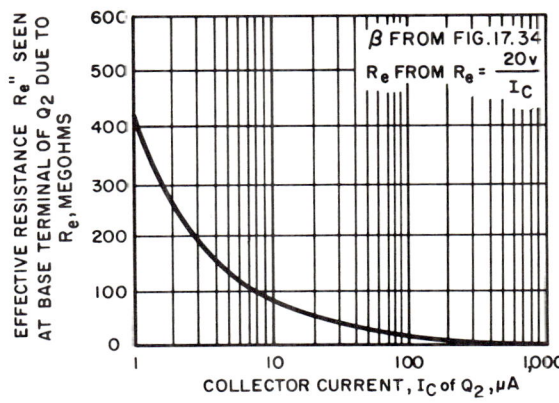

FIG. 35. Effective resistance at base of Q_2 vs. collector current. (*Reproduced through courtesy of the General Electric Co., Advanced Electronics Center, Ithaca, N.Y.*)

and R_4, to provide the load R_e'. Using the values of bias resistors indicated we can now refigure R_e, which now comes out to be 55,000 ohms instead of 200,000.

Using the formulas for the resistance multiplication we can now figure the resultant components of the input resistance. These turn out to be

$$R_b' = 100 \text{ megohms} \qquad r_c' = 500 \text{ megohms} \qquad R_e' = 25 \text{ megohms}$$

17-32 AMPLIFIER CIRCUITS

and the parallel combination of the three is

$$(25 \times 100 \times 500)10^6 \cong 19 \text{ megohms}$$

A further increase can be obtained by the addition of a third transistor Q_3, as shown in Fig. 37. This transistor supplies negative feedback to the emitter of Q_2 and raises the effective value of the total emitter resistance by approximately the current ampli-

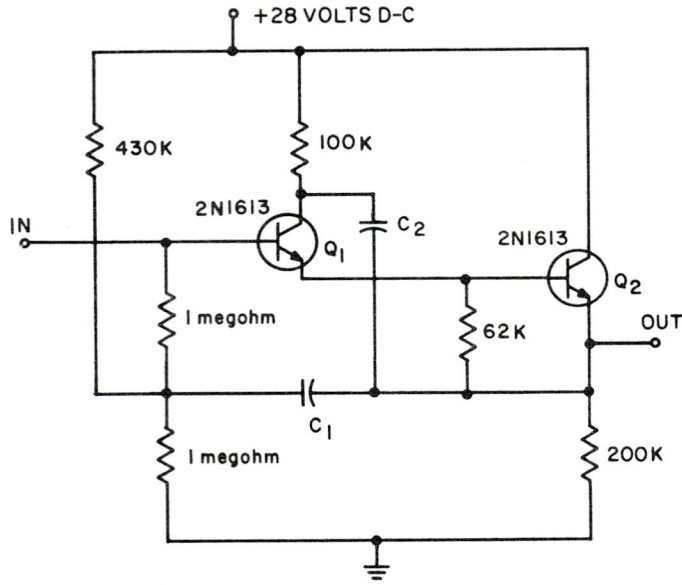

Fig. 36. Combination bootstrapped amplifier. (*Reproduced through courtesy of the General Electric Co., Advanced Electronics Center, Ithaca, N.Y.*)

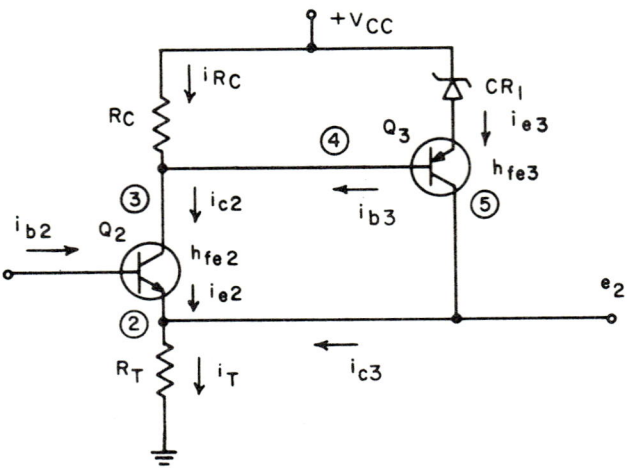

Fig. 37. Use of feedback to raise emitter load. (*Reproduced through courtesy of the General Electric Co., Advanced Electronics Center, Ithaca, N.Y.*)

AUDIO AMPLIFIERS 17-33

fication of Q_3. This provides a means of increasing the factor R_e' considerably, and this was the major limiting component of the net input resistance. The addition of this stage also raises the amount of bootstrapping of the other components by about

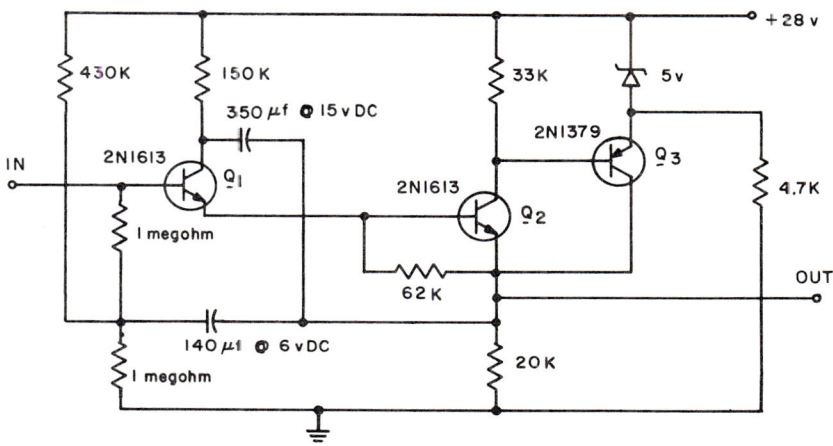

FIG. 38. Complete circuit of three-transistor bootstrapped amplifier. (*Reproduced through courtesy of the General Electric Co., Advanced Electronics Center, Ithaca, N.Y.*)

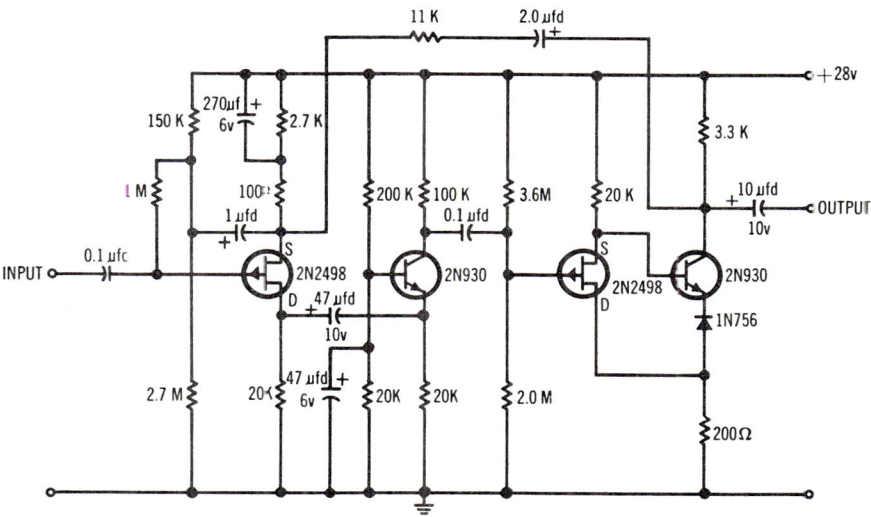

FIG. 39. Circuit of high-impedance amplifier using field-effect transistors. (*Reproduced through the courtesy of Texas Instruments Incorporated, Dallas, Tex.*)

the same amount and hence permits attaining input impedances well in excess of 100 megohms.

9.6. Complete Circuit of High-impedance Amplifier. Figure 38 shows the complete circuit of a three-transistor bootstrapped amplifier incorporating all the principles described in the previous section. It has a calculated value of input resistance of 375 megohms. Its measured input impedance was 275 megohms between the limits of 0.5 and 2,000 cycles.

9.7. High-impedance Amplifier Using Field-effect Transistors.[1] The newly introduced field-effect transistor provides another method of obtaining high input impedance, together with wide bandwidth. Figure 39 shows the circuit of an amplifier which combines two field-effect transistors with two n-p-n high-gain low-noise transistors. Input impedance greater than 30 megohms is obtained with a noise figure under 3 db over a wide range of generator resistance. The input stage is bootstrapped by feedback from the output, in manner similar to that described previously in this section. The second stage employs the common-base configuration, which then drives another field-effect transistor. Another 2N930, this time using the common-emitter configuration, completes the circuit. Figure 40 shows the noise

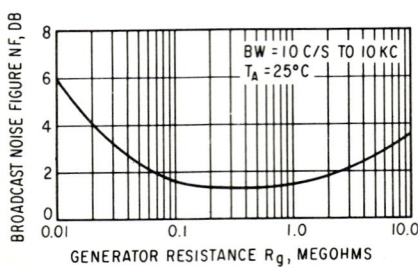

Fig. 40. Noise figure of field-effect amplifier. (*Courtesy of Texas Instruments Incorporated, Dallas, Tex.*)

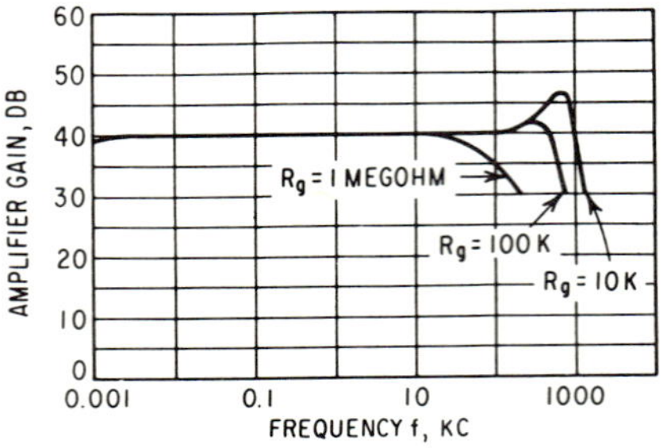

Fig. 41. Frequency response of field-effect amplifier. (*Courtesy of Texas Instruments Incorporated, Dallas, Tex.*)

figure as a function of generator resistance, and Fig. 41 shows the gain vs. frequency for three values of generator resistance.

10. ELECTRON-TUBE PREAMPLIFIERS

A preamplifier is utilized to amplify an input signal to a level sufficient for further amplification, either at some point remote from the original source or at such level that extraneous effects, such as noise, become insignificant. Probably the most common use of preamplifiers is in relation to phonograph or tape reproduction, and in this connection several special conditions are encountered, dictated by such factors as recording level and frequency characteristics, hum and other spurious signals, and pickup characteristics.

10.1. Pickup Characteristics. Phonograph pickups generally fall in one of two categories, magnetic and ceramic. In magnetic pickups the reluctance is varied as the stylus moves and its output is proportional to velocity. Thus for constant-amplitude recording the output will increase with frequency at a rate of 6 db/octave. The internal impedance of this type of pickup is inductive; thus a considerable amount of frequency shaping is produced as the resistive loading is varied.

The ceramic pickup is capacitive in nature and is therefore a high-impedance source and requires a high value of load resistance for proper response. This type of pickup is responsive to the amplitude of the stylus movement.

The difference between these two types of pickup must be considered when designing a preamplifier to operate from either or both types, and from records or tape. The standard industry recording characteristic is illustrated in Fig. 42, also the NAB tape characteristic. This standard RIAA recording characteristic indicates the proper system frequency response when using an inductive pickup. In other words, to com-

---

[1] Courtesy of Texas Instruments Incorporated.

pensate for the recording characteristic necessary to avoid excessive groove cutting at low frequency it becomes necessary to increase the low-frequency response of the preamplifier by 18 db at the low-frequency end as compared with 1,000 cycles, and to drop off another 20 db at 20,000 cycles to compensate for the pickup characteristic. This frequency compensation is usually achieved by either the pickup loading or the frequency characteristic of the amplifier or both.

10.2. Other Special Preamplifier Characteristics. Besides the correction for the recording characteristic compensation is required for other factors. One of these is the variation of the ear's frequency response with loudness, as noted long ago by Fletcher. As the volume is reduced it becomes necessary to raise the relative levels of the low and high frequencies. This is usually accomplished by combining a frequency-selective network with the volume control, such that the relative response of the lows and highs changes as the volume is changed. Tone controls, for both bass and treble, are necessary to compensate for variations in individual recordings as well as for different listener tastes. Scratch filters are usually also incorporated

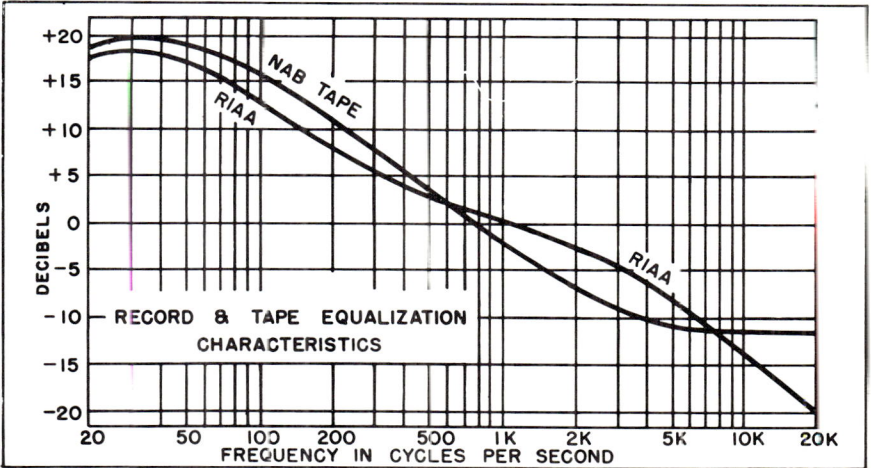

Fig. 42. NAB and RIAA recording characteristics. (*Reproduced by courtesy of Sherwood Electronic Laboratories, Inc., Chicago, Ill.*)

to reduce this effect, particularly when playing the older records, and rumble filters are provided to reduce the effect of this defect of the turntable.

10.3. Example of Tube Preamplifier. Figure 43 shows the circuit of one of the two channels of a typical tube stereo preamplifier, incorporating the features described above, as well as several others.[2] The input is designed to accommodate phonograph pickup, tape, AM-FM tuner, and auxiliary inputs. The input is applied to the input of the first tube and then amplified, and the frequency characteristic is modified to suit the desired recording characteristic by the feedback network between the plate and grid of V_2. The gain control has an RC filter connected to its slider which varies the frequency characteristic with volume-control setting as shown in Fig. 44. It will be noted that the bass is raised almost 24 db relative to mid-frequency at the lowest volume, and that the high-frequency end is raised about 6 db. A scratch filter is incorporated after the volume control to reduce the gain above about 5,000 cycles. The bass and treble controls consist of RC networks which alter the impedance between the two halves of the second stage with frequency, producing as much as $+17, -20$ db bass variation and ± 20 db treble variation. Figure 45 shows the

[2] Sherwood Model S-5000 II stereo preamplifier-amplifier.

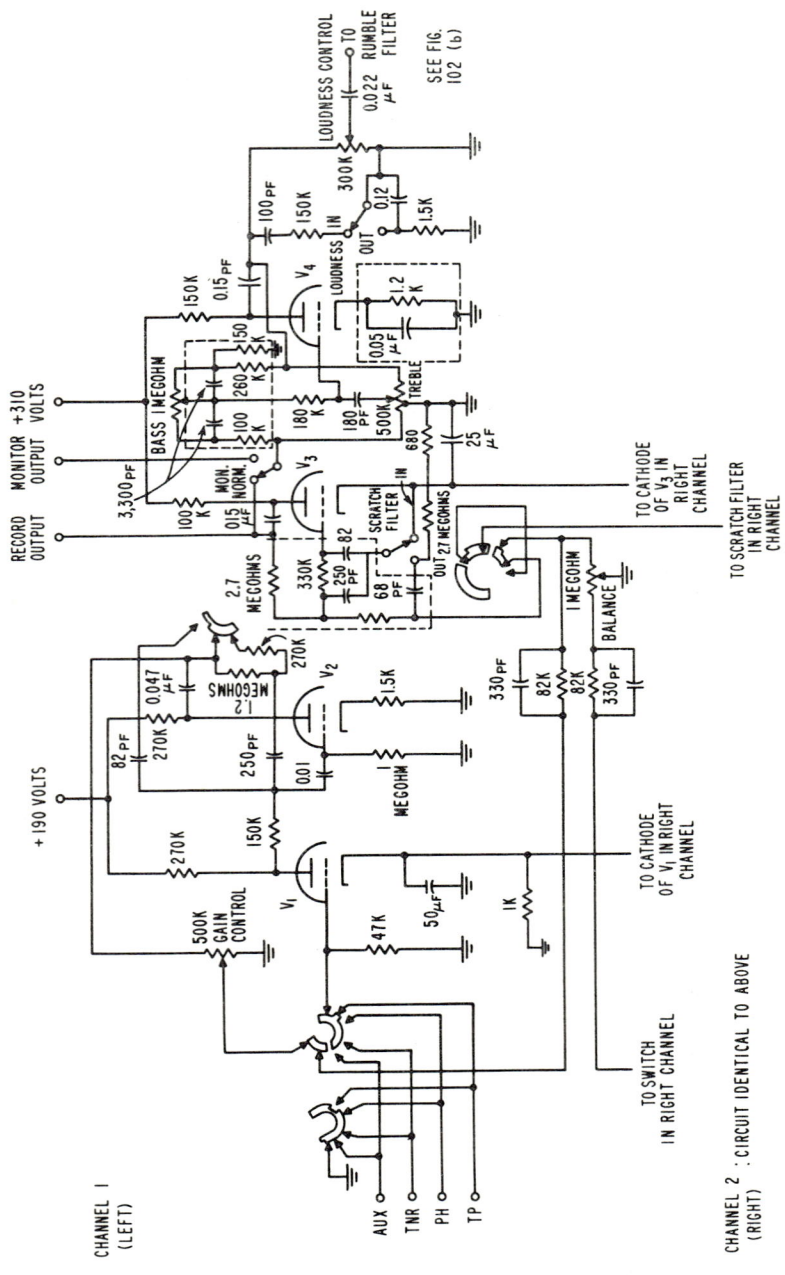

FIG. 43. Schematic of two-channel stereo preamplifier. All tubes 12AX7, ECC 83 shared between two channels. (*Courtesy of Sherwood Electronic Laboratories, Inc., Chicago, Ill.*)

AUDIO AMPLIFIERS 17-37

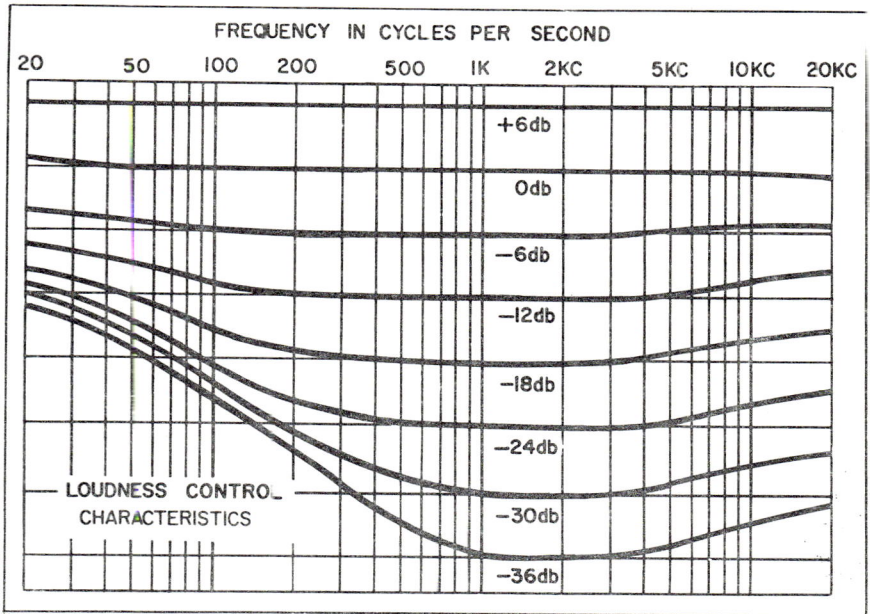

FIG. 44. Preamplifier loudness-control characteristics. (*Courtesy of Sherwood Electronic Laboratories, Inc., Chicago, Ill.*)

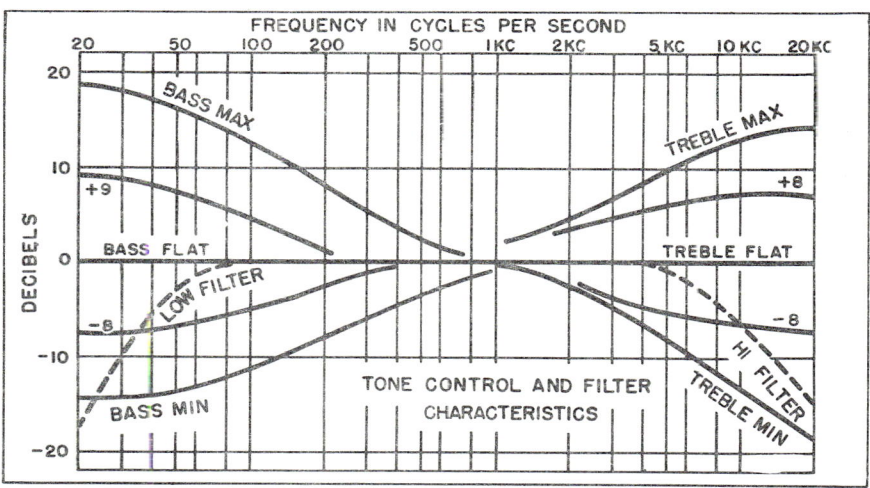

FIG. 45. Effects of preamplifier tone controls and filters. (*Courtesy of Sherwood Electronic Laboratories, Inc., Chicago, Ill.*)

variation of frequency response with the position of the tone controls, also the effects of the scratch and rumble filters. Another feature included in this amplifier is a so-called "presence" switch, which raises the response from about 500 cycles up, with maximum boost at around 3,000 cycles, as shown in the figure. The purpose of this control is to dramatize vocalists and certain instruments so they stand out from the other music. The characteristics of this preamplifier are as follows:

Frequency response (complete stereo amplifier): (36 watts) 20 cps to 20 kc ± ½ db.
Tone-control response: Flat setting, 20 cps to 20 kc ± ½ db.
Tone-control range: 15 kc, 17-db boost or cut. 40 cps, 16-db boost, 19-db cut.
Rumble filter: 27 cps, 17-db rejection; 70 cps less than 1 db down.
Preamplifier equalizer curves: AES/RIAA phono and NAB tape.
Sensitivity: Radio 0.25 volt, tape 1.4 mv, phono 1.2 mv; all inputs are adjustable with level control.
Maximum input capability: Phono, 200 mv for less than 1 per cent distortion. Radio, adjustable with level control.
Maximum hum and noise: Volume control minimum, 100 db (weighted) below rated output. Radio input (controls maximum), 90 db (weighted) below rated output. Phono input (controls flat), 60 db below rated output, 72 db below 10 mv (equivalent to ½ μv referred to input grid).
Interchannel crosstalk: Less than −50 db at 1 kc.

The complete stereo amplifier of which this preamplifier is a part, is described in further detail in Sec. 16.

11. TRANSISTOR PREAMPLIFIERS

All the previous remarks regarding frequency compensation, tone controls, and other features apply equally well to transistor preamplifiers. In addition, there is one other feature that is peculiar to this type of device. Since the transistorized preamplifier lends itself so well to miniaturization, e.g., by such means as integrated circuits, it becomes possible to make the whole preamplifier so small that it can be mounted within the tone arm. This has the advantage of reducing the lead length from pickup to amplifier to a minimum and thereby also minimizing hum pickup and other effects of extraneous capacitance.

11.1. Example of Transistorized Preamplifier.[3] Figure 46 shows the circuit of a typical transistorized preamplifier. In this example the preamplifier contains a total of six transistors and could conceivably have been separated at some intermediate point, with the remainder going over to the output portion of the complete amplifier. Just how much of the amplifier preceding the output stages comprises the so-called preamplifier is largely a matter of personal preference, or possibly chassis layout or other considerations, such as described above. In this case the designer chose to include everything ahead of the driver as part of the preamplifier.

The input may take any of three forms, auxiliary tuner, ceramic phono pickup, or magnetic pickup. The five positions of the input switch are

1. Auxiliary tuner
2. Ceramic phono
3. Ceramic phono + rumble filter
4. Magnetic phono
5. Magnetic phono + rumble filter

The rumble filter consists of an *RC* feedback arrangement for ceramic pickups, and a modification of the interstage coupling network for magnetic pickups. The first pair consists of two common-emitter stages, employing complementary transistors (one *n-p-n* and one *p-n-p*) and with feedback between the second collector and first emitter to shape the frequency response to the required curve (see the previous section for the

---

[3] From An Ultra-fidelity Transistorized Stereo Amplifier, Motorola Semiconductor Products, Inc., *Bull.* 1-12-61-1M.

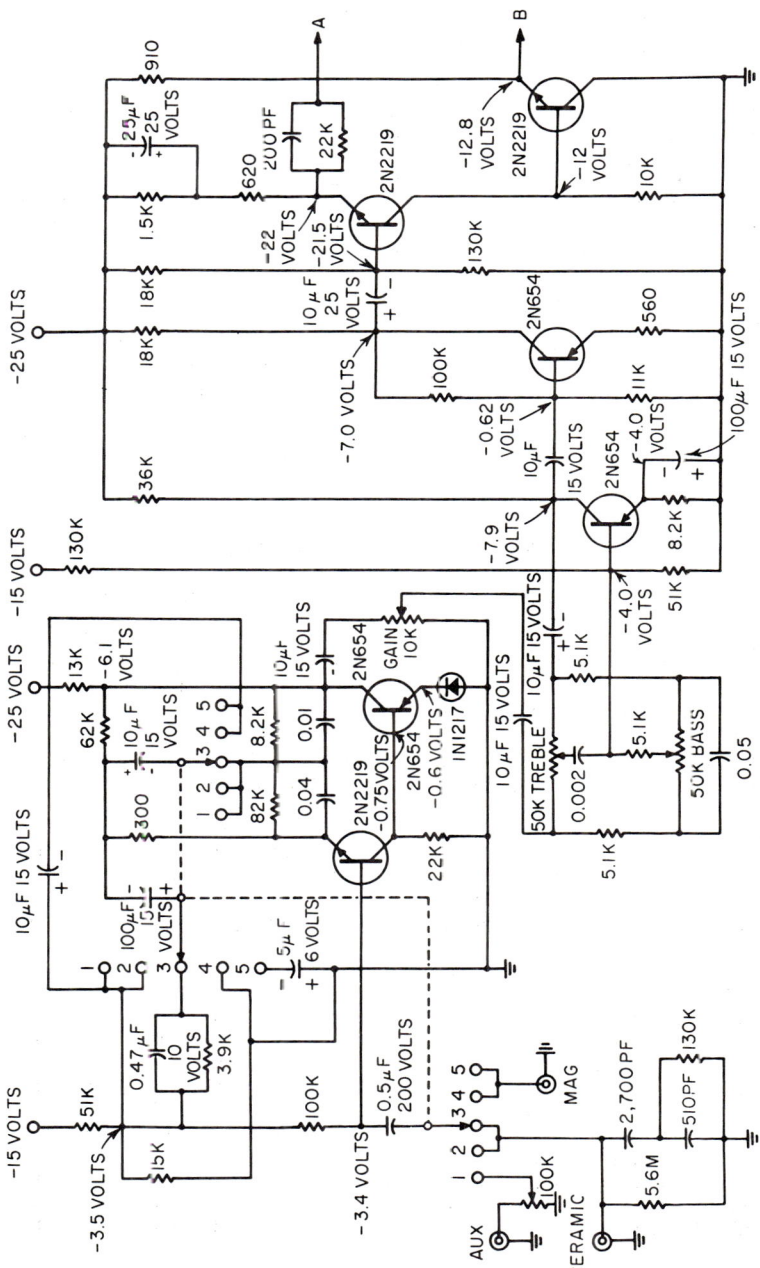

Fig. 46. Schematic of transistor preamplifier. Switch positions: (1) Auxiliary tuner. (2) Ceramic phono. (3) Ceramic phono + rumble filter. (4) Magnetic phono. (5) Magnetic phono + rumble filter. (Reproduced from Bull. 1-12-61-1M by courtesy of Motorola Inc., Semiconductor Products Division, Phoenix, Ariz.)

RIAA curve). This doublet is followed by the volume control and treble and bass tone controls, which operate in much the same manner as those previously described under the tube preamplifier.

The next portion of the amplifier consists of three common-emitter stages, followed by an emitter follower, which supplies the signal to the output driver at terminal B. As before, both p-n-p and n-p-n transistors are used in a complementary arrangement. This permits use of only one polarity of battery without the necessity of using abnormally large bias resistors. The 560-ohm resistor in the emitter lead of the second transistor and the 620-ohm resistor in the emitter lead of the third are not bypassed, in the latter case to provide a point to which feedback may be applied at the point A from the output of the power stage.

The characteristics of this preamplifier are:

Frequency response (complete amplifier, 1-watt level): $+0$, -1 db from 17 cps to 24,000 cycles.

Sensitivity (for 30 watts output, complete amplifier): Magnetic input, 2.25 mv. Ceramic input, 45 mv. High auxiliary, 450 mv. Low auxiliary, 45 mv.

Tone-control range, referenced to 1 kc: Bass control, ± 18 db (at 20 cps). Treble control ± 18 db (at 20 kc).

Equalization: Magnetic, per RIAA curve for 500-mh cartridge with recommended load impedance of 50 kilohms. Ceramic, per RIAA curve for 400-pf cartridge (input impedance approximately 2 megohms).

12. CLASSES OF OUTPUT STAGES

Power stages are commonly classed as A, AB, B, and C according to the following definitions.

Class A implies linear operation with 360° operation of a sine wave. The operating point is normally set at about the center of the operating range, and in operation the output signal produces approximately equal excursions toward the two axes.

Class B operation implies that the tubes or transistors are biased to cutoff, such that only 180° conduction can occur. Push-pull operation is necessary, with each half of the output amplifier alternately supplying the two halves of the output signal. This form of output stage is used quite extensively in transistor amplifiers because of the saving in power consumption.

Class AB operation is intermediate between class A and class B and implies that there is a small forward bias and the output stages are not completely cut off. This arrangement is also often used with transistor amplifiers to minimize crossover distortion, as will be shown in a later section.

Class C operation implies that the stage is biased beyond cutoff, allowing less than 180° conduction.

The above classes of operation are illustrated in Fig. 47.

Only class A operation is possible with single-ended amplifiers; push-pull amplifiers may be operated in any of the three methods. Further, there is often no sharp distinction between the various modes of operation, and we notice such symbols as A_1, AB_1 and AB_2, denoting intermediate categories of operation.

Class A operation is generally lowest in distortion, although this is usually more a matter of the amount of feedback employed than the type of operation. Class AB and B are more economical of power, as noted above, particularly during periods of no signal; hence these modes of operation are more popular in those cases where the duty cycle is low, as in the case where most of the signal is in the form of speech. This latter advantage is of less importance with tubes than with transistors, because of the considerable amount of filament power consumed by the tube filament at all times.

In class AB and B operation power-supply regulation assumes considerable importance, since the currents taken by the tubes or transistors vary with input signal level and hence the effective supply voltages will drop unless the regulation is good. This factor frequently becomes a limiting item from an economic viewpoint and often

AUDIO AMPLIFIERS   17–41

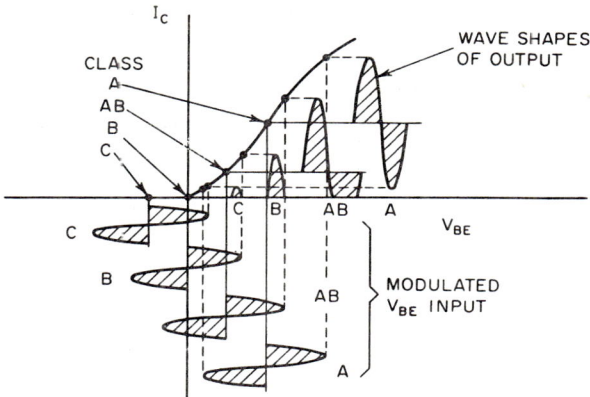

FIG. 47. Different classes of amplifier operation. (*Reproduced from "Power Transistor Handbook" through courtesy of Motorola Inc., Semiconductor Products Division, Phoenix, Ariz.*)

dictates a compromise in operational characteristics. This point will be covered in more detail in a later section.

13. CLASS A POWER AMPLIFIERS

In this section we shall describe single-ended class A amplifiers; push-pull class A amplifiers will be considered in the next section, along with the other forms of push-pull amplifiers.

13.1. Class A Tube Power Amplifiers. Figures 48 and 49, respectively, illustrate typical single-stage tube amplifiers employing transformer coupling and resistance coupling. The tubes illustrated are beam-power pentodes; however, the same

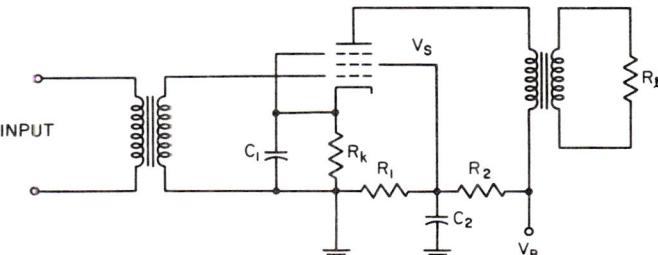

FIG. 48. Transformer-coupled tube power amplifier.

circuit applies to triode operation, with the omission of the screen supply. Transformer coupling is shown for the output, as resistance coupling is relatively infrequently used in power stages.

While the screen has been shown as supplied from a comparatively stiff bleeder, if the operation is strictly class A, so that the plate and screen currents do not vary with signal, the resistor R_1 in Fig. 48 and the resistor R_2 in Fig. 49 may be omitted, with some saving in power consumption. If class A_1 is used there will be some current shift and improved regulation will be necessary. Under this condition it may also be necessary to bleed some current through the cathode resistor R_k to prevent bias shift; this is often accomplished by returning the lower screen bleeder resistors to the cathode instead of to ground.

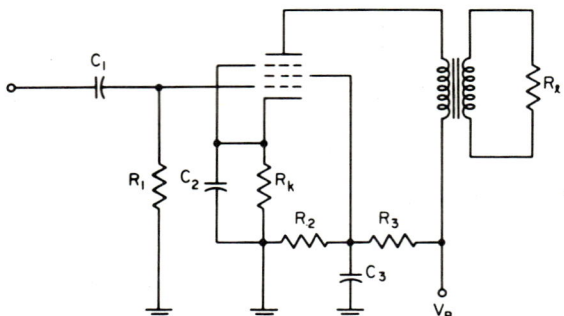

FIG. 49. Resistance-coupled class A tube amplifier.

As examples of typical power tubes the specification data of two beam-power pentodes are reproduced below.

6GK6

Characteristics, class A_1 amplifier:

Plate supply voltage, volts	250
Grid-2 supply voltage, volts	250
Cathode resistor, ohms	135
Mu factor, grid 2 to grid 1	19
Plate resistance (approx), ohms	38,000
Transconductance, micromhos	11,300
Plate current, ma (zero signal)	48
Grid-2 current, ma (zero signal)	5.5

Typical operation:

Peak a-f grid-1 voltage, volts	7.3
Max signal plate current, ma	50.6
Max signal grid-2 current, ma	10
Effective load resistance, ohms	5,200
Total harmonic distortion, %	10
Max signal power output, watts	5.7

Maximum circuit values:

Grid-1 circuit resistance, megohms:	
For fixed-bias operation, max	0.3
For cathode-bias operation, max	1

5881

Typical operation and characteristics, class A_1:

Plate voltage, volts	250	300	350
Grid-2 voltage, volts	250	200	250
Grid-1 voltage, volts	−14	−12.5	−18
Peak grid-1 a-f voltage, volts	14	12.5	18
Zero-signal plate current, ma	75	48	53
Max signal plate current, ma	80	55	65
Zero-signal grid-2 current, ma	4.3	2.5	2.5
Max signal grid-2 current, ma	7.6	4.7	8.5
Plate resistance (approx), ohms	30,000	35,000	48,000
Transconductance, micromhos	6,100	5,300	5,200
Load resistance, ohms	2,500	4,500	4,200
Total harmonic distortion, %	10	11	13
Max signal power output, watts	6.7	6.5	11.3

Maximum circuit values:

Grid-1 circuit resistance, megohms:	
For fixed-bias operation, max	0.1
For cathode-bias operation, max	0.5

AUDIO AMPLIFIERS 17–43

The 5881 may also be operated triode-connected with grid 2 connected to the plate, with the following characteristics:

5881, Triode-connected, Class A₁

Typical operations and characteristics:

Plate voltage, volts...............................	250	300
Grid-1 (control-grid) voltage, volts...............	−18	−20
Peak a-f grid-1 voltage, volts.....................	18	20
Zero-signal plate current, ma......................	52	78
Max signal plate current, ma.......................	58	85
Amplification factor...............................	8	
Transconductance, micromhos........................	5,250	
Load resistance, ohms..............................	4,000	4,000
Total harmonic distortion, %.......................	6	5.5
Max signal power output, watts.....................	1.4	1.8

Maximum circuit values:

Grid-1 circuit resistance, megohms:
  For fixed-bias operation, max............................... 0.1
  For cathode-bias operation, max............................. 0.5

It will be noted that in all the above examples the plate and screen currents do not remain constant, since these examples employ class A₁ operation. The variation in plate current is not very great, but the screen current varies by as much as 3:1 or more. Unless the screen supply regulation is stiff enough to take care of this variation the above values of power output will not be realized.

Another characteristic of power amplifiers to be noted is the wide difference between plate resistance and load resistance, often as much as 10:1. This is typical of power amplifiers, indicating that the output load is chosen for power output rather than power gain. It will also be noted that the peak power output occurs for the condition that the input signal just drives the control grid into zero bias.

13.2. Output Transformer Design Considerations. The turns ratio of the output transformer is determined by the desired load resistance to be presented to the tube and by the value of the secondary load resistance, and will be equal to the square root of the ratio of these resistances. The primary inductance is the limiting factor determining the low-frequency cutoff point, and should have an impedance equal to the reflected primary load at the desired 3-db cutoff frequency. The leakage inductance performs a similar function at the high-frequency cutoff point, and its impedance should be equal to or less than the primary impedance at that frequency. Good construction usually calls for interleaving the secondary windings and using high-grade transformer steel.

13.3. Class A Transistor Power Amplifiers. Power amplifiers utilizing transistors employ many of the same techniques previously described for low-level stages, although many of the values differ considerably in magnitude. For example, power stages frequently employ resistors in series with the emitter leads; however, the value is usually on the order of a few ohms, rather than the large values commonly employed at low levels. Again, the concept of stability factor is important here, as it dictates the permissible dissipation in the transistor for the desired operating temperature range. Other major differences are: The collector current is now usually in the ampere range. The base-to-emitter voltage is on the order of a volt occasionally. Base current is in milliamperes rather than microamperes. Input impedance is on the order of an ohm to a few ohms, even in the common-emitter configuration. Output impedances are very low; in fact this is one of the major advantages of transistor power amplifiers, in that they permit operation directly into speaker voice coils without the need for an output transformer, increasing both gain and frequency response thereby.

13.4. Graphical Determination of Class A Operating Point. A technique similar to that described in Sec. 4.3 may be used to determine the operating point

17-44 AMPLIFIER CIRCUITS

of a class A stage from the input and output characteristics. Figures 50 and 51 illustrate this technique. Both d-c and a-c conditions are obtainable from the curves of Fig. 51.

The curves in the third quadrant pertain to the input and relate the base current to the base-to-emitter voltage. The curve labeled V_E versus I_B gives the drop across the resistor R_E in the emitter lead as a function of the base current I_B. This line is

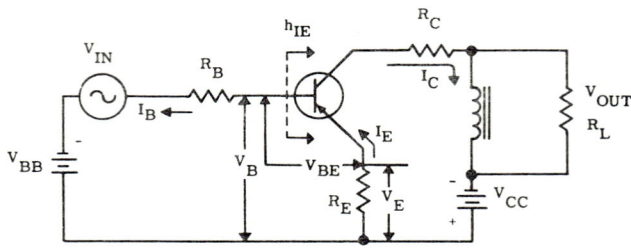

FIG. 50. Typical class A transistor amplifier with bias. (*Courtesy of Motorola Inc., Semiconductor Products Division, Phoenix, Ariz.*)

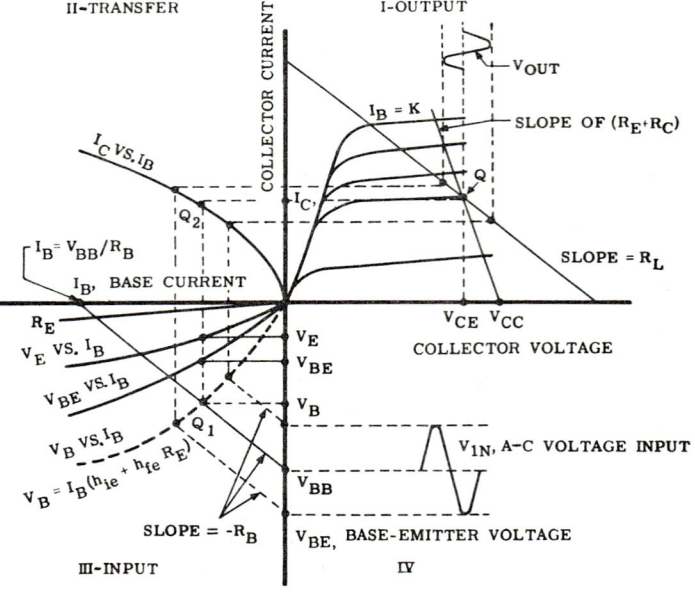

FIG. 51. Graphical analysis of typical class A operating conditions. (*Courtesy of Motorola Inc., Semiconductor Products Division, Phoenix, Ariz.*)

curved because of the drop-off of h_{FE} with current. The V_{BE} versus I_B curve is the usual input curve obtained from the specification sheet. The dotted curve V_B versus I_B is the sum of these two and is the total base voltage as a function of base current. It will be noted that the resistor R_E is effectively multiplied by the amplification h_{FE} in much the same manner as in the small-signal case.

The I_C versus I_B curve in the second quadrant would be a straight line if it were not for the fall-off of h_{FE} with increasing collector current, and its departure from linearity is a measure of this characteristic.

The usual collector family is shown in the first quadrant and the crowding of curves with increasing current is evident, another sign of fall-off of h_{FE}.

There are three load lines, two d-c and one a-c. In the third quadrant there is a line from the base supply voltage V_{BB} to a point corresponding to a current V_{BB}/R_E. Since the input operating point must fall on both this line and the curved dotted line, it must be at the intersection of these two curves, at the point Q_1. This then establishes the operating base voltage and base current under quiescent conditions. Projecting upward to the second quadrant we obtain the intersection with the I_C versus I_B curve, which gives us the quiescent collector current.

There are two load lines in the first quadrant, the d-c load line having a slope equal to approximately $R_E + R_C$ and an a-c load line of slope R_L. The former is drawn through the supply voltage V_{CC}. Projecting over from the second quadrant we obtain the intersection with the d-c load line; this determines the voltage between collector and emitter V_{CE}. The a-c load line is drawn through this quiescent operating point Q.

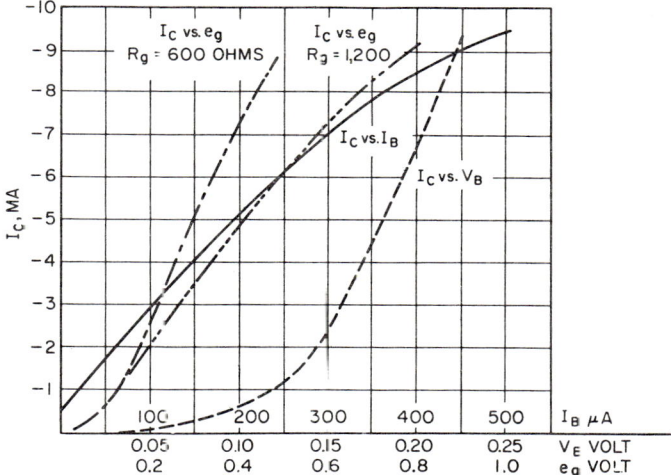

FIG. 52. Output current vs. input current and voltage, common-emitter configuration. (Reprinted with permission from R. F. Shea, "Transistor Audio Amplifiers," John Wiley & Sons, Inc., New York, 1955.)

The a-c operating conditions may also be obtained from these curves. A sinusoidal input signal is plotted in the fourth quadrant and its swings are projected to the corresponding intersections in the third quadrant, thence upward to the second quadrant, giving the collector current swing. Projecting this over to the a-c load line in the first quadrant now gives the collector voltage excursion, which is the output voltage. The effects of the cumulative nonlinearities are evident from these figures, the output waveshape being definitely nonsinusoidal.

13.5. Distortion in Class A Amplifiers. As noted above, nonlinearity in the various transistor characteristics can produce considerable distortion, as evidenced by departure of the input-output relationship from linearity. This is illustrated in Fig. 52, which shows curves of collector current vs. base current, base voltage and generator voltage, through various values of generator resistance. The solid curve I_C versus I_B shows the typical curvature of the common-emitter stage, due to fall-off of h_{FE}. The curve of I_C versus V_B shows a reverse form of distortion, in this case due to the input nonlinearity. It would be the sort of performance which would be obtained if the amplifier were driven from a low-impedance source. (Note: These curves are for lower-power operation than we normally think of for power amplifiers but are included to illustrate the effects of generator impedance and are qualitatively true for high-power stages also.) The curves for 600- and 1,200-ohm generator impedances show regions of reasonably good linearity, although the full swing could not be utilized without distortion. This indicates that there is usually an optimum value of source resistance from the standpoint of distortion. Usually this will be somewhat less than the input impedance, which is approximately equal to h_{ie}.

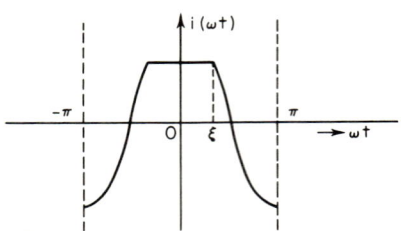

FIG. 53. Peak clipping of a sine wave. (*Reprinted with permission from R. F. Shea, "Transistor Audio Amplifiers," John Wiley & Sons, Inc., New York, 1955.*)

13.6. Peak Clipping. The ultimate limitation in the operation of a power stage is when the signal excursion extends to either the transistor saturation region or cutoff. In either case a phenomenon called peak clipping results, evidenced by an output waveform of the shape shown in Fig. 53, where the sinusoid has been clipped on the top. The distortion due to this form of clipping can be obtained by means of a Fourier analysis and the various terms obtained from the following equations:

$$\text{Fundamental} = (1/\pi)(\pi - \xi - \sin 2\xi/2) \tag{48}$$

$$n\text{th harmonic} = \frac{1}{n\pi}\left[\frac{\sin (n+1)\xi}{n+1} - \frac{\sin (n-1)\xi}{n-1}\right] \tag{49}$$

In a typical case, where $\xi = 20°$ we have

Second harmonic = 0.86 per cent
Third harmonic = 0.80 per cent
Fourth harmonic = 0.74 per cent

This indicates that for hard clipping, as illustrated in the figure, the harmonics are of about the same intensity. In the usual case, however, the clipping is not this sharp, the corners being rounded. Typical experimental results for an amplifier operating at a collector efficiency of about 58 per cent are:

Second harmonic 2.1 per cent   Fifth harmonic 0.6 per cent
Third harmonic 2.0 per cent   Sixth harmonic 0.1 per cent
Fourth harmonic 0.4 per cent   Seventh harmonic 0.1 per cent

Obviously, clipping at only one end indicated that potential power output is being lost, and the operating point should be readjusted to equalize the distortion at both extremities. For low distortion it is advisable to avoid approaching either extremity of swing. Further, an adequate safety margin must be left to accommodate the shift in operating point which will occur with temperature. This point will be covered in more detail in a subsequent section.

13.7. Biasing Transistor Power Amplifiers. A practicable method of obtaining the voltage V_{BB} of Fig. 50 is by means of the familiar single-battery arrangement, with the base voltage obtained by means of a voltage divider across the collector battery. This is illustrated in Fig. 54, where R_1 and R_2 form the base voltage divider, R_3 is the d-c resistance of the transformer winding, and R_E is the emitter resistor, shunted by the capacitor C_E to avoid degeneration.

This circuit can be reduced to the simple form of Fig. 55 by assuming the base current is negligible compared with the bleeder current and substituting the Thévenin equivalent of V_{BB} and R_B', where

$$R_B = R_1 R_2/(R_1 + R_2) \tag{50}$$
$$R_B' = R_B + R_3 \tag{51}$$
$$V_{BB} = R_B V_{CC}/R_2 \tag{52}$$

AUDIO AMPLIFIERS

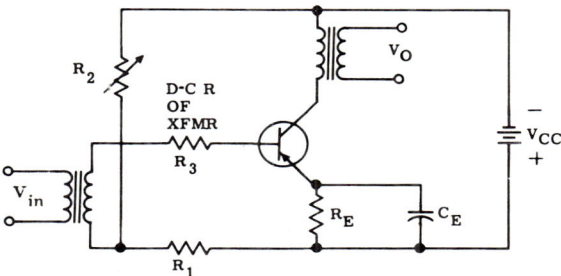

FIG. 54. Basic single-ended one-battery power amplifier. (*Reprinted from the "Power Transistor Handbook" through the courtesy of Motorola Inc., Semiconductor Products Division, Phoenix, Ariz.*)

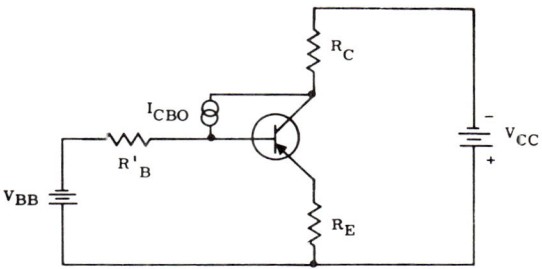

FIG. 55. Equivalent d-c circuit of Fig. 54. (*Courtesy of Motorola Inc., Semiconductor Products Division, Phoenix, Ariz.*)

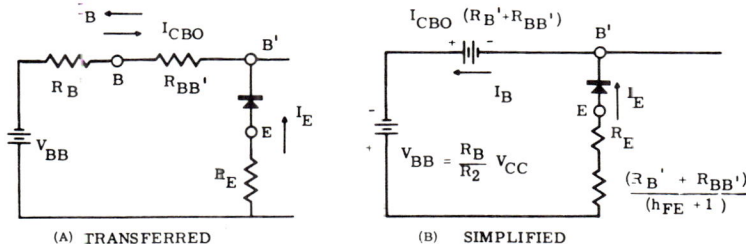

FIG. 56. Equivalent circuits. (*Courtesy of Motorola Inc., Semiconductor Products Division, Phoenix, Ariz.*)

Figure 56 shows the circuit with the transistor replaced by the equivalent base resistance R_{BB}' and a diode, also with the temperature-sensitive current I_{CBO}. A simplified version of this circuit is also shown, where the resistors have all been transferred to the emitter and the effect of I_{CBO} is simulated by a battery in the base lead. This circuit may be solved to obtain the emitter current:

$$I_E = \frac{V_{BB} - V_{B'E} + I_{CBO}(R_B' + R_{BB}')}{R_E + (R_B' + R_{BB}')/(h_{FE} + 1)} \qquad (53)$$

The stability factor may be obtained by differentiating Eq. (53) with respect to I_{CBO}, giving

$$S_I \cong \frac{1}{R_E/(R_{B'} + R_{BB'}) + 1/(h_{FE} + 1)} \qquad (54)$$

The voltage stability factor is

$$S_V \cong -S_I R_E \tag{55}$$

It is difficult to hold S_I below 10 in power-amplifier stages, which indicates that a shift on the order of 100 ma is possible for the emitter and collector currents with germanium power transistors. Thus the stability factor provides a measure of the permissible temperature excursion without excessive shift of operating point.

The above method of determining bias is good enough usually to provide a starting point in designing power stages. However, it is frequently desirable to be able to choose the bias resistors more accurately. The following method is taken from the "Motorola Power Transistor Handbook." Another extensive treatment may be found in the Texas Instruments, Inc., book ("Transistor Circuit Design," pp. 206–219, 220–237, McGraw-Hill Book Company, New York, 1963).

From data sheets and supplementary information the following must be known: $h_{FE\,min}$, $h_{FE\,max}$, $V_{BE\,min}$, $V_{BE\,max}$, $I_{CBO\,min}$, $I_{CBO\,max}$, $I_{E\,max}$, $I_{E\,min}$.

The emitter resistor R_E is chosen to produce a voltage between 0.5 and 1.0 volt, in order to reduce the effect of changes in the emitter diode voltage. The above voltages and currents take into account all variations due to transistor characteristic spreads and temperature change from $T_{J\,min}$ to $T_{J\,max}$. The relation between the transistor dissipation and junction temperature is given by the following equation:

$$T_{J\,max} - T_{A\,max} = \theta_{JA} V_{CE} I_{E\,max} \tag{56}$$

where θ_{JA} is the total junction-to-ambient thermal resistance and T_A is the ambient temperature. This equation says, in effect, that the temperature rise within the transistor is equal to the product of the internal dissipation and the thermal resistance. Knowing the maximum permissible junction $T_{J\,max}$ and the maximum ambient temperature $T_{A\,max}$, we can calculate the maximum allowable thermal resistance for any value of dissipation:

$$\theta_{JA} = (T_{J\,max} - T_{A\,max})/V_{CE} I_{E\,max} \tag{57}$$

This will allow determination of the size of the heat sink and any other factors affecting thermal resistance. This will be covered in more detail in a later section. The minimum junction temperature may be obtained similarly:

$$T_{J\,min} = V_{CE} I_{E\,min} \theta_{JA} + T_{A\,min} \tag{57a}$$

The maximum and minimum values of emitter current may be obtained from the following equations:

$$I_{E\,max} = \frac{(R_B V_{CC})/R_2 - V_{BE\,min} + I_{CBO\,max} R_B'}{R_E + R_B'/(h_{FE\,min} + 1)} \tag{58}$$

$$I_{E\,min} = \frac{(R_B V_{CC})/R_2 - V_{BE\,max} + I_{CBO\,min} R_B'}{R_E + R_B'/(h_{FE\,min} + 1)} \tag{59}$$

Usually the I_{CBO} term is insignificant in Eq. (58) since this condition is for the low-temperature extreme.

The equivalent base resistance R_B' may be obtained from the following equation:

$$R_B' = \frac{(I_{E\,max} - I_{E\,min})R_E + V_{BE\,min} - V_{BE\,max}}{I_{E\,min}/(h_{FE\,min} + 1) - I_{E\,max}/(h_{FE\,max} + 1) + I_{CBO\,max} - I_{CBO\,min}} \tag{60}$$

The value of the resistor R_2 may now be determined from the following equation:

$$R_2 = \frac{(R_B' - R_3)V_{CC}}{I_{E\,max}[R_E + R_B'/(h_{FE\,max} + 1)] + V_{BE\,min} - I_{CBO\,max} R_B'} \tag{61}$$

R_3 in the above equation is the d-c resistance of the transformer winding. Finally, R_1 can be obtained from the following equation:

$$R_1 = (R_B' - R_3)R_2/(R_2 + R_3 - R_B') \tag{62}$$

AUDIO AMPLIFIERS 17-49

Example. Given the following values:

$V_{CE} = 12$ volts
$I_{E\ min} = 500$ ma
$I_{E\ max} = 600$ ma
$T_{A\ min} = 25°C$
$T_{A\ max} = 40°C$
$T_{J\ max} = 100°C$
$R_E = 2$ ohms, based upon dropping 1 volt across R_E
$R_3 = 1.4$ ohms (assumed value of d-c resistance of driver transformer secondary)

The maximum value of thermal resistance is

$$\theta_{JA} = \frac{T_{J\ max} - T_{A\ max}}{V_{CE} \times I_{E\ max}} = \frac{100 - 40}{12 \times 0.6} = 8.3°C/\text{watt}$$

The minimum junction temperature is

$$T_{J\ min} = V_{CE} \times I_{E\ min} \times \theta_{JA} + T_{A\ min} = 12 \times 0.5 \times 8.3 + 25 = 75°C$$

Thus the junction will vary from 75 to 100°C.

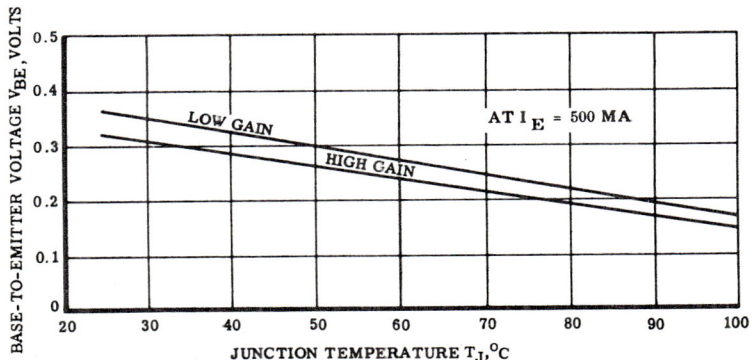

FIG. 57. Typical power transistor V_{BE} versus T_j. (*Courtesy of Motorola Inc., Semiconductor Products Division, Phoenix, Ariz.*)

Figures 57 and 58 show typical variation of V_{BE} and h_{FE} with temperature. From these it is found that our extreme values are

$V_{BE\ max} = 0.23$ volt (at 75°C)
$V_{BE\ min} = 0.15$ volt (at 100°C)
$h_{FE\ max} = 90$ (at 75°C)
$h_{FE\ min} = 50$ (at 100°C)

From the specification sheet we also find that the minimum I_{CBO} at 75°C is 1 ma and at 100°C it is 20 ma. Now, substituting,

$$R_B' = \frac{(0.6 - 0.5)2 + 0.15 - 0.23}{0.5/51 - 0.6/91 + 0.02 - 0.001} = 5.4 \text{ ohms}$$

$$R_2 = \frac{4 \times 12}{0.5[2 + 5.4/(50 + 1)] + 0.23 - 0.001 \times 5.4} = 37.8 \text{ ohms}$$

$$R_1 = 4 \times 37.8/(37.8 - 4) = 4.5 \text{ ohms}$$

Thus our bias system is specified.

13.8. Temperature Stabilization with Thermistors and Diodes. Occasionally it will be found that adequate temperature stabilization is not possible without excessive power dissipation, under which condition it would be desirable if R_1, for

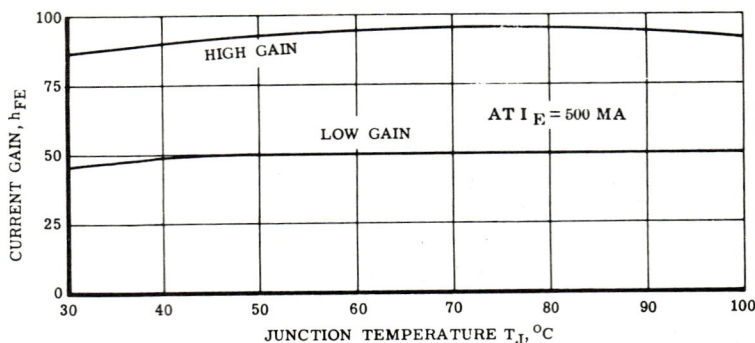

FIG. 58. Typical power transistor h_{FE} versus T_j. (*Courtesy of Motorola Inc., Semiconductor Products Division, Phoenix, Ariz.*)

example, could be made variable with temperature. This may be accomplished by means of a thermistor, a device whose resistance has a negative temperature coefficient. Figure 59 shows a thermistor connected in parallel with R_1, so that adequate temperature stability is achieved. Two methods may be utilized to obtain the curve of resistance vs. temperature. In one the value of R_1 can be calculated to maintain the emitter current within specified limits as temperature varies. In the other the values of resistance can be calculated at one temperature, using the method previously presented; then the variation of R_1 may be experimentally determined for constant current. Once the desired curve of resistance vs. temperature is obtained the corresponding curve of the thermistor may be calculated. Then, knowing the typical variation of thermistor resistance with temperature, the values of padding resistors can be calculated to give the desired overall curve.

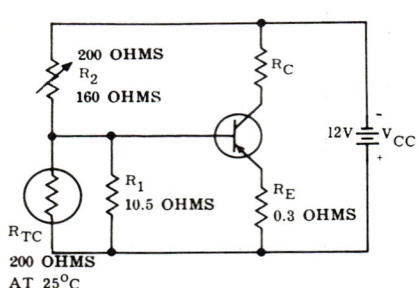

FIG. 59. Thermistor control. (*Courtesy of Motorola Inc., Semiconductor Products Division, Phoenix, Ariz.*)

The use of one or more diodes to provide stabilization is also a very efficient method of temperature-compensating power transistors. Two such circuits are illustrated in Figs. 60 and 61. The single-diode method is based on the assumption that the forward drop across the diode (which has a negative temperature coefficient of about 1.8 mv/°C) will compensate for the change in V_{BE}.

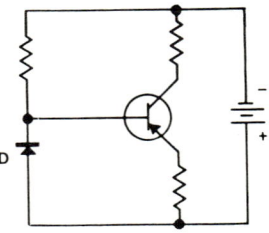

FIG. 60. Single-diode compensation. (*Courtesy of Motorola Inc., Semiconductor Products Division, Phoenix, Ariz.*)

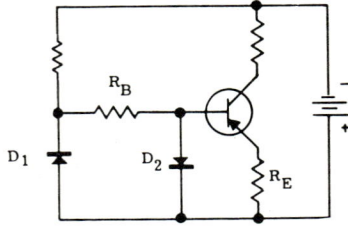

FIG. 61. Two-diode compensation. (*Courtesy of Motorola Inc., Semiconductor Products Division, Phoenix, Ariz.*)

The two-diode method uses one diode D_1 in the same manner as the single diode in the method discussed above. The second diode D_2 is reverse-biased and should have a leakage current somewhat larger than the transistor I_{CBO}. Diode D_2 allows an I_{CBO} path which is in series with the transistor collector diode. This provides a low voltage drop for I_{CBO} without affecting normal bias settings. When D_1, D_2, R_B, and R_E are at optimum values, excellent stability will usually result. Normally these values are determined by experimental methods, although graphical methods such as have been described above may be utilized.

13.9. Thermal Runaway. Thermal runaway occurs when the junction temperature builds up because the heat sink is not capable of dissipating heat fast enough. If all other transistor characteristics are temperature-independent, the I_{CBO} increase with temperature causes a regenerative effect on junction temperature.

Consider that the term $S_I \times I_{CBO} \times V_{CE}$ is the d-c power loss in excess of the normal loss $I_C V_{CE}$. This additional power dissipation causes a rise in junction temperature which increases I_{CBO}, in turn creating more power dissipation, and so on. The final junction temperature can be expressed as

$$T_J = T_A + \theta_{JA} I_E V_{CE} + \theta_{JA} S_I V_{CE} (I_{CBO})_{T_J} \qquad (63)$$

I_{CBO} is an exponential function of temperature. For example, in germanium transistors the equation for I_{CBO} is

$$(I_{CBO})_{T_J} = (I_{CBO})_{T_0} \exp\left[(T_J - T_0) 8{,}350 / T_0 T_J\right] \qquad (64)$$

where T_0 is some reference temperature and $(I_{CBO})_{T_0}$ and $(I_{CBO})_{T_J}$ are the values of I_{CBO} at temperatures T_0 and T_J, respectively. The value 8,350 is a constant associated with the germanium junction. Silicon transistors will have a different constant.

The junction temperature, as a function of operating conditions, is obtained from the following equation:

$$T_J = T_A + \theta_{JA} I_E V_{CE} + \theta_{JA} S_I V_{CE} (I_{CBO})_{T_0} \exp\left[(T_J - T_0) 8{,}350 / T_0 T_J\right] \qquad (65)$$

T_0 and T_J in the exponential term must be expressed in degrees Kelvin. All other temperatures are in degrees centigrade. Figure 62 shows a plot of T_J versus T_A and dramatically demonstrates the effect of thermal runaway.

Thermal runaway will occur when the rate of increase of junction temperature is greater than the rate of increase of ambient temperature. The condition for thermal runaway may be found by differentiating Eq. (65). This gives

$$T_R = \frac{dT_J}{dT_A} = \frac{1}{1 - [8{,}350/(T_J)^2] \times \theta_{JA} S_I V_{CE} (I_{CBO})_{T_J}} \qquad (66)$$

Thermal runaway occurs when the denominator of Eq. (66) approaches zero, or when $8{,}350/(T_J)^2 \times \theta_{JA} \times S_I \times V_{CE} \times (I_{CBO})_{T_J}$ approaches unity. Therefore, to prevent thermal runaway,

$$\theta_{JA} S_I V_{CE} (I_{CBO})_{T_J} < (T_J)^2 / 8{,}350 \qquad (67)$$

For Motorola power transistors $T_{J\,max}$ is 100°C or 373°K, and the criterion for thermal runaway is

$$\theta_{JA} S_I V_{CE} (I_{CBO})_{T_J} < (373 \times 373)/8{,}350 = 16.7$$

This means that to avoid thermal runaway for a transistor having a maximum $T_J = 100$°C, the product of thermal resistance, current stability factor, collector-to-emitter voltage, and I_{CBO} at 100°C must be less than 16.7.

The allowable ambient temperature is then

$$T_A = T_J - \theta_{JA} I_E V_{CE} - \theta_{JA} S_I V_{CE} (I_{CBO})_{T_J} \qquad (68)$$

and the maximum allowable thermal resistance is

$$\theta_{JA\,max} = \frac{T_{J\,max} - T_{A\,max}}{I_E V_{CE} + S_I V_{CE}(I_{CBO})_{T_{J\,max}}} \qquad (69)$$

Example. Given a Motorola 2N1531 in a class A circuit where

$$\begin{aligned} I_E &= 0.5 \text{ amp} \\ V_{CE} &= 10 \text{ volts} \\ T_{A\,max} &= 50°C \\ T_{J\,max} &= 100°C \end{aligned}$$

Permissible change of I_E due to distortion reasons is $\Delta I_E = 50$ ma from 25°C ambient Find permissible S_I and θ_{JA} to avoid thermal runaway.

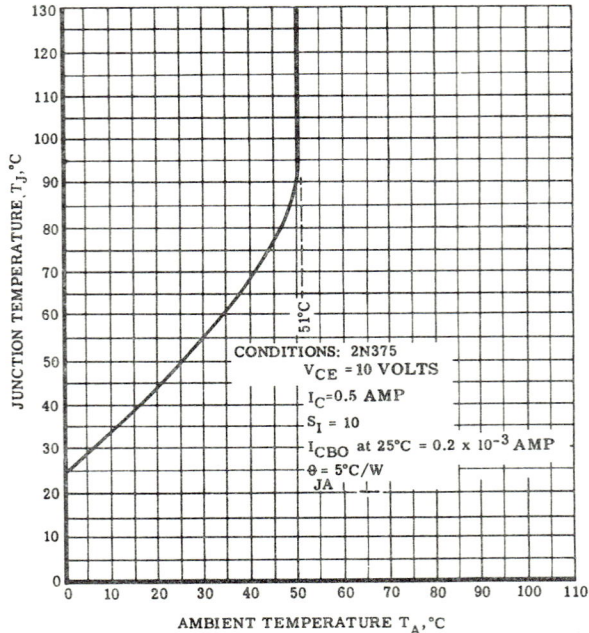

FIG. 62. Thermal runaway for a typical condition. (*Courtesy of Motorola Inc., Semiconductor Products Division, Phoenix, Ariz.*)

Solution

$$S_I = \Delta I_E / \Delta I_{CBO}$$
$$I_{CBO} \text{ at } 25°C = 0.2 \times 10^{-3} \text{ amp}$$
$$\text{At } 100°C = 12 \times 10^{-3} \text{ amp}$$
$$\Delta I_{CBO} = 11.8 \times 10^{-3} \text{ amp}$$
$$S_I = 50 \div 11.8 = 4.24$$

Thermal-runaway criterion:

$$S_I V_{CE} \theta_{JA} I_{CBO} < 16.7$$
$$4.24 \times 10 \times 6.66 \times 12 \times 10^{-3} = 3.38$$

(6.66 is the thermal resistance of this transistor without additional heat-dissipative means.) The above indicates that there is some additional thermal resistance possible without causing runaway. This may be obtained from Eq. (69), which gives

$$\theta_{JA\,max} = \frac{100 - 50}{0.5 \times 10 + 4.24 \times 10 \times 12 \times 10^{-3}}$$
$$= 9.10°C/watt$$

Thus about 2.5°C/watt may be tolerated in external thermal resistance before thermal runaway will occur. The subject of heat sinks will be covered in the next section.

13.10. Transistor Cooling; Thermal Resistance. Adequate cooling is of prime importance in assuring reliable operation of transistorized power amplifiers. The three methods of heat transfer which are significant in power-transistor cooling are conduction, convection, and radiation. Forced-circulation cooling is relatively infrequently used in transistorized equipment.

Conduction is defined as the transfer of heat through and by means of matter without the obvious motion of the matter. It is a function of the thermal conductivity of the material, the cross-sectional area normal to the direction of flow, and the thermal differential.

Convection is defined as the transfer of heat by moving matter. The fluid used for convection absorbs heat by conduction and then physically moves away, carrying the heat with it. Free or natural convection occurs when the fluid, on being heated, becomes less dense and is caused to rise by the cooler fluid below it, thus producing a flow of heat upward. In forced convection the fluid is made to move by external means.

Radiation is defined as the process by which energy is transmitted through space without the necessary presence of matter. The rate at which a body emits radiation is a function of the temperature and the surface condition.

Thermal resistance can be defined as the opposition of a substance to the flow of heat energy. It can be expressed mathematically as

$$\theta_T = \Delta T/P/A/l \qquad (70)$$

where θ_T is the thermal resistance of the substance in question at a particular temperature, °C/sq in./watt-in.; ΔT is the temperature gradient existing between the extremes of the substance, °C; P is the amount of heat generated, watts; A is the cross-sectional area perpendicular to the direction of heat flow, sq in.; and l is the length of the path of heat flow, in.

The thermal resistance of a transistor is usually specified as a maximum value, within which production transistors will fall. In some cases the thermal resistance is given for the collector-base junction to the case; in other cases it is specified to the ambient. The user must exercise care in noting which of the references are specified. Also the interface to which the transistor transfers heat is important in determining the thermal resistance. In designs where the transistor case must be insulated from the mounting plate an insulating washer, e.g., mica, is inserted between the stud and the plate. The thermal resistance will then be a function of the thickness of the mica (or other insulator) washer, as well as of the interface pressure.

The similarity between thermal and electrical resistance may be visualized from Fig. 63, which shows the transistor junction as the heat source and air as the ultimate heat sink. Two intermediate points are shown, the transistor case and the heat sink to which the transistor is mounted and which transfers the heat to the ambient air. The thermal resistances between the case and heat sink and between the heat sink and air may be measured or obtained from the vendors, and added to the transistor thermal resistance to obtain the total. If this total comes within the safe limits specified by the design (see Sec. 13.9) the design will be adequate. Any additional margin will add to the reliability of the design.

Figure 64 shows the variation of thermal resistance as a function of mica washer thickness, while Fig. 65 shows the effects of interface pressure, also of using Dow Corning 200 silicone oil to fill the interface region and avoid the presence of an air space. The effects are obvious and indicate the importance of using a minimum thickness of insulation, where one is necessary, and of ensuring proper tightening of the mounting nut.

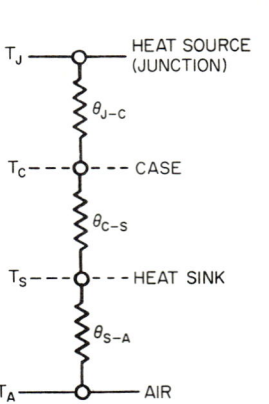

Fig. 63. Thermal circuit. (*Reproduced with permission from Westinghouse Electric Corp., Tech Tip no. 14, May 1, 1962.*)

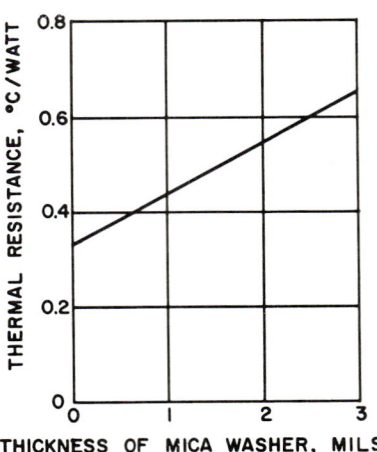

Fig. 64. Thermal resistance as a function of mica washer thickness. (*Reproduced with permission from A. D. Abel, Power Transistor Cooling, Solid State J., vol. 2, no. 10, pp. 21–33, October 1961.*)

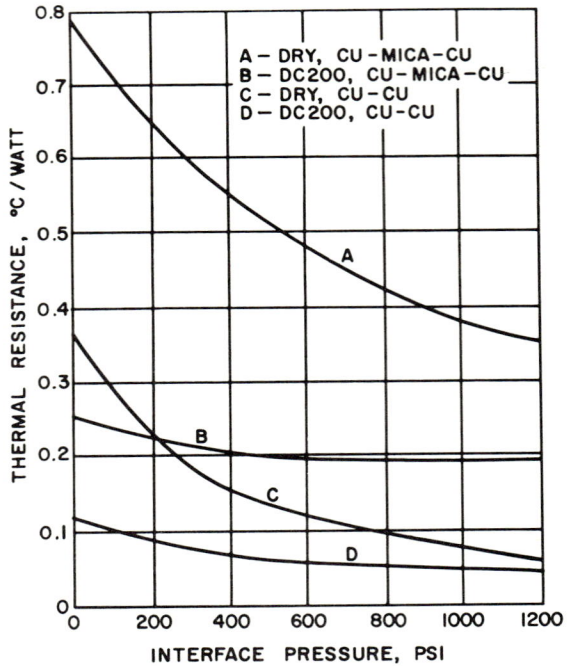

Fig. 65. Thermal resistance as function of pressure. (*Reproduced with permission from A. D. Abel, Power Transistor Cooling, Solid State J., vol. 2, no. 10, pp. 21–33, October 1961.*)

AUDIO AMPLIFIERS

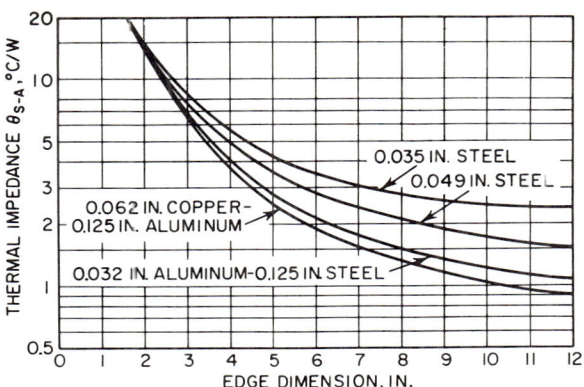

Fig. 66. Thermal impedance to ambient air of flat square plates. Natural convection—vertical mounting, painted or anodized surfaces. (*Reproduced with permission from Westinghouse Electric Corp., Tech Tip no. 14, May 1, 1962.*)

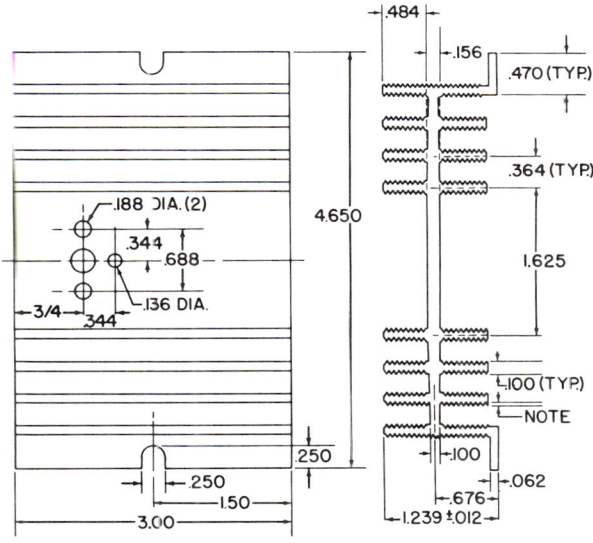

NOTE: 90° GROOVES ARE .020(MAX.) DEEP

Fig. 67. Physical dimensions of Deibert Blinn No. X-123-R-3A heat dissipator. (*Reproduced with permission from A. D. Abel, Power Transistor Cooling, Solid State J., vol. 2, no. 10, pp. 21–33, October, 1961.*)

The necessity of using an adequate heat sink in transistor power stages cannot be overemphasized. In its simplest form such a heat sink may take the form of a flat metal plate, preferably mounted vertically, to obtain maximum convection transfer. Figure 66 shows the thermal resistance of flat square plates of copper, aluminum, or steel of various thicknesses, as a function of plate dimensions. While copper is best from the standpoint of thermal resistance, it is very heavy, and a thicker, but lighter, plate of aluminum is equally effective. This figure also indicates that such simple heat sinks are inadequate where thermal resistance below 1°C/watt is required. For resistance below this value more complex shapes are needed. Figures 67 and 68 show

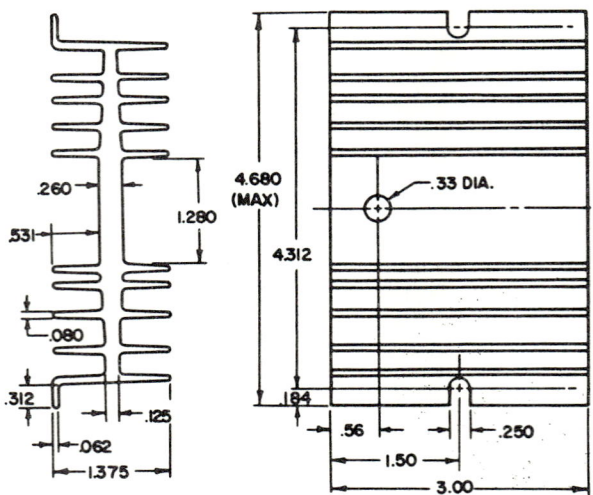

Fig. 68. Physical dimensions of Anderson heat dissipator No. 20107-1-G. (*Reproduced with permission from A. D. Abel, Power Transistor Cooling, Solid State J., vol. 2, no. 10, pp. 21–33, October, 1961.*)

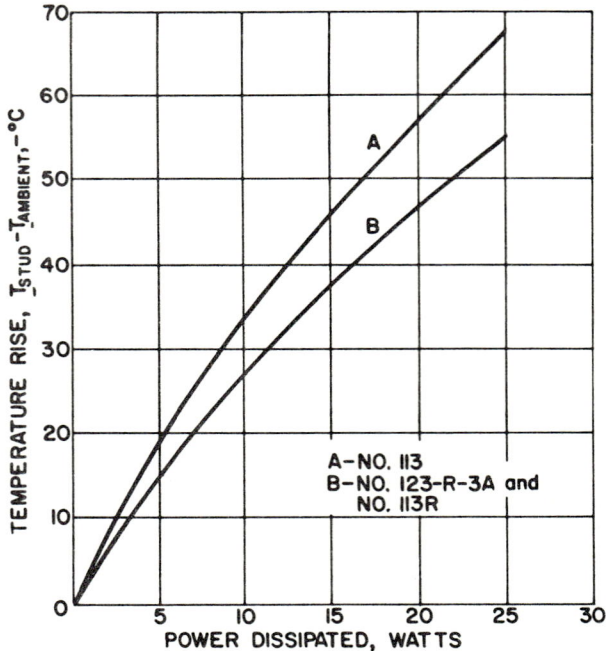

Fig. 69. Free-convection characteristics of Delbert Blinn heat sinks. (*Reproduced with permission from A. D. Abel, Power Transistor Cooling, Solid State J., vol. 2, no. 10, pp. 21–33, October, 1961.*)

two typical heat sinks capable of dissipating high amounts of heat, and Figs. 69 and 70 show curves of temperature rise vs. power dissipated from which their thermal resistance can be calculated, or these curves may be used directly to give the stud temperature of the transistor, for any given value of heat dissipation.

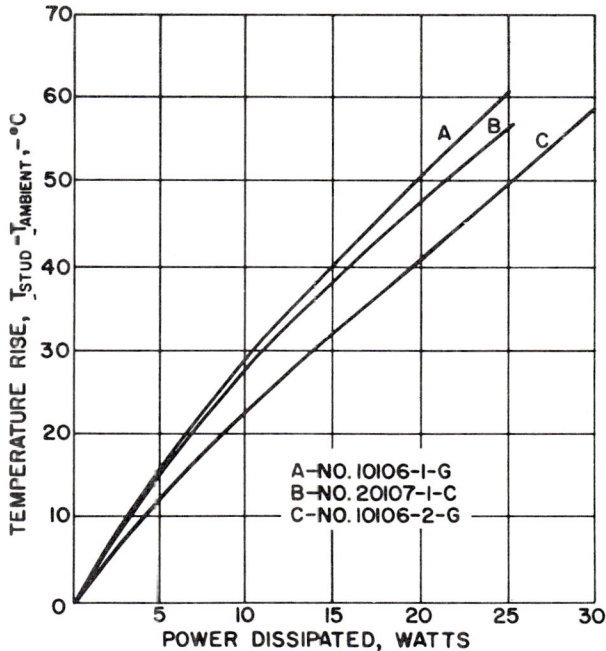

FIG. 70. Free-convection characteristics of Anderson heat sinks. (*Reproduced with permission from A. D. Abel, Power Transistor Cooling, Solid State J., vol. 2, no. 10, pp. 21–33, October, 1961.*)

14. CLASS A, AB, AND B PUSH-PULL POWER AMPLIFIERS

Class A stages connected in push-pull differ very slightly in design theory from single-ended stages, and the remarks on such designs under Sec. 13 apply equally well to push-pull stages. In tube amplifiers the combined current of both stages will usually flow through one common cathode or bias resistor; hence its resistance will be half that of the single-ended stage. Effective load resistance from plate to plate or collector to collector will be four times the single-ended value, as will be the effective input resistance offered by the stage to the driver. Since these variations are relatively minor and the design of such push-pull stages is therefore relatively straightforward, no additional discussion will be given for this arrangement. The following sections will deal with the design aspects of push-pull tube and transistor class AB and B stages. Class C stages are seldom used in audio-amplifier design and will not be discussed here.

14.1. Class AB Tube Amplifiers. In the section on single-ended class A amplifiers we gave the characteristics of the type 6GK6 and 5881 beam tubes, together with their performance in a typical class A stage. The specification sheets for these tubes also provide typical operating data for push-pull class AB_1 operation.

AMPLIFIER CIRCUITS

6GK6

Maximum ratings, design-maximum values:

Plate supply voltage, volts max................................	600
Plate voltage, volts max..	330
Grid-2 supply voltage, volts max................................	600
Grid-2 (screen-grid) voltage, volts max..........................	330
Grid-1 (control-grid) voltage:	
Negative-bias value, volts max.............................	100
Cathode current, ma max..	65
Grid-2 input, watts max	
Peak..	4
Average...	2
Plate dissipation, watts max....................................	13.2
Peak heater-cathode voltage, volts max:	
Heater negative with respect to cathode.....................	100
Heater positive with respect to cathode.....................	100

Typical operation (values are for two tubes):

Plate supply voltage, volts........................	250	300
Grid-2 supply voltage, volts.......................	250	300
Cathode resistor, ohms............................	130	130
Peak a-f grid-1 to grid-1 voltage, volts............	22.4	28
Zero-signal plate current, ma......................	62	72
Max signal plate current, ma.......................	75	92
Zero-signal grid-2 current, ma	7	8
Max signal grid-2 current, ma......................	15	22
Effective load resistance (plate to plate), ohms.....	8,000	8,000
Total harmonic distortion, %......................	3	4
Max signal power output, watts....................	11	17

Maximum circuit values:

Grid-1 circuit resistance, megohms max:	
For fixed-bias operation.....................................	0.3
For cathode-bias operation...................................	1

5881

Push-Pull Power Amplifier, Class A$_1$

Maximum ratings, design-center values:

Plate voltage, volts max..	400
Grid-2 (screen-grid) voltage, volts max..........................	400
Grid-2 input, watts max..	3
Plate dissipation, watts max....................................	23
Peak heater-cathode voltage, volts max..........................	200

Typical operation (two tubes):

Plate voltage, volts...............................	250	270
Grid-2 voltage, volts..............................	250	270
Grid-1 (control-grid) voltage, volts................	−16	−17.5
Peak a-f grid-1 to grid-1 voltage, volts............	32	35
Zero-signal plate current, ma......................	120	134
Max signal plate current, ma.......................	140	155
Zero-signal grid-2 current, ma.....................	10	11
Max signal grid-2 current, ma......................	16	17
Plate resistance (approx per tube), ohms...........	24,500	23,500
Transconductance (per tube), micromhos...........	5,500	5,700
Effective load resistance (plate to plate), ohms.....	5,000	5,000
Total harmonic distortion, %......................	2	2
Max signal power output, watts....................	14.5	17.5

AUDIO AMPLIFIERS   **17-59**

Maximum circuit values:
    Grid-1 circuit resistance:
        For fixed-bias operation, megohms max........................... 0.1
        For cathode-bias operation, megohms max....................... 0.5

Push-Pull A-F Power Amplifier, Class AB₁

Maximum ratings, design-center values:

    Plate voltage, volts max... 400
    Grid-2 (screen-grid) voltage, volts max.............................. 400
    Grid-2 input, watts max.. 3
    Plate dissipation, watts max... 23
    Peak heater-cathode voltage, volts max............................. 200

Typical operation (two tubes):

Plate voltage, volts.	360	360
Grid-2 voltage, volts.	270	270
Grid-1 (control-grid) voltage, volts.	−22.5	−22.5
Peak a-f grid-1 to grid-1 voltage, volts.	45	45
Zero-signal plate current, ma.	88	88
Max signal plate current, ma.	132	140
Zero-signal grid-2 current, ma.	5	5
Max signal grid-2 current, ma.	15	11
Effective load resistance (plate to plate), ohms.	6,600	3,800
Total harmonic distortion, %.	2	2
Max signal power output, watts.	26.5	18

Maximum circuit values:
    Grid-1 circuit resistance:
        For fixed-bias operation, megohms max........................... 0.1
        For cathode-bias operation, megohms max....................... 0.5

Note: The type of input coupling used should not introduce too much resistance in the grid-1 circuit. Transformer- or impedance-coupling devices are recommended.

Push-Pull A-F Power Amplifier, Class AB₂

Maximum ratings, design-center values:

    Plate voltage, volts max... 400
    Grid-2 (screen-grid) voltage, volts max.............................. 400
    Grid-2 input, watts max.. 3
    Plate dissipation, watts max... 23
    Peak heater-cathode voltage, volts max............................. 200

Typical operation (two tubes):

Plate voltage, volts.	360	360
Grid-2 voltage, volts.	225	270
Grid-1 (control-grid) voltage, volts.	−18	−22.5
Peak a-f grid-1 to grid-1 voltage, volts.	52	72
Zero-signal plate current, ma.	78	88
Max signal plate current, ma.	142	205
Zero-signal grid-2 current, ma.	3.5	5
Max signal grid-2 current, ma.	11	16
Effective load resistance (plate to plate), ohms.	6,000	3,800
Total harmonic distortion, %.	2	2
Max signal power output, watts.	31	47

Maximum circuit values:
    Grid-1 circuit resistance:
        For fixed-bias operation, megohms max.............. 0.1
        For cathode-bias operation........................ Not recommended

Note: Driver stage should be capable of supplying the specified driving power at low distortion to the No. 1 grids of the AB₂ stage. To minimize distortion, the effective resistance per grid-1 circuit of the AB₂ stage should be held at a low value. For this purpose, the use of transformer coupling is recommended.

It is interesting to compare the operation of these power tubes under these different modes of operation. For example, the data provide a comparison between all modes, triode-connected class A to push-pull class AB$_2$, for the 5881. The power output ranges from about 4 watts for a pair of tubes connected in the first arrangement to 47 watts for the last. The three categories of class AB give from 14.5 to 47 watts. It will also be noted that, as the mode progresses more toward the cutoff state, the change in plate and screen-grid currents becomes more pronounced; for example, in the class AB$_2$ operation the plate current changes by more than 2:1 and the screen current by more than 3:1. This implies that stiffer supplies must be used; otherwise the power output will be limited below the above figures because of reduced supply voltages at maximum power. Also the use of cathode biasing becomes impractical at this type of operation, and a stiff bias becomes necessary for this supply, too. Thus, while we reduce the power required of the amplifier as we go to AB$_2$, the increase in bleeder requirements counterbalances some of this potential reduction. It is interesting to note that the effective load resistance did not change very markedly as class of operation varied, being more dependent upon particular operating conditions than upon class of operation. It should also be noted that one does not get anything for nothing in these modes of operation, and that the cost of getting greater output power in class AB$_2$ operation is greater driving-power requirement; thus the power requirements and distortion of the driver become of increased importance.

Some examples of typical push-pull tube amplifiers will be given in the section on typical power amplifiers, and also in the section dealing with high-fidelity stereo amplifiers.

14.2. Class AB and B Transistor Amplifiers. In general, transistor amplifiers are neither class AB nor class B in the sense that is normally applied in tube amplifiers. Class AB generally denotes a condition nearer to class A than B, and such operation would not secure the large economy possible with transistors. Because of the absence of filament power, transistors become very attractive for operation near cutoff; however, crossover distortion makes operation as fully class B unsatisfactory, as will be shown later.

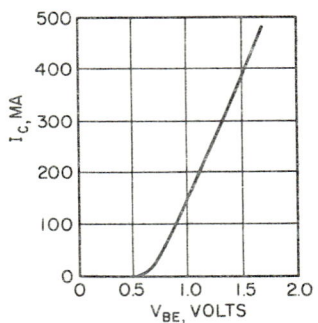

Fig. 71. Input characteristic of 2N2107 transistor. $V_{CE} = 15$ volts, $T_A = 25°C$.

A class B transistor amplifier is capable of around 75 per cent efficiency. This factor, together with the elimination of the filament supply, makes such an amplifier particularly attractive for portable equipment and other uses where battery economy is desirable. In general, it is more difficult to achieve the same degree of low distortion obtainable with class A operation; however, there are a great many applications in which distortion on the order of 5 per cent is entirely tolerable, and this can readily be achieved using properly designed class B amplifiers. Lower distortion than this can be obtained by means of feedback, in the same manner as for tubes. In the following description the designation class B will be used, although it will be shown that proper operation requires a small amount of forward bias.

14.3. Static Characteristics, Class B Push-Pull Amplifiers. Figures 71 and 72 show input and collector characteristics for the type 2N2107 medium-power transistor, chosen to illustrate typical class B operation. It will be noted that the range of currents and voltages is quite a bit higher than has been shown previously. The input curve has a considerable offset, about 0.5 volt, characteristic of silicon transistors, and the collector current has a range up to almost ½ amp. The collector family shows several features quite markedly: the saturation voltage of about 2 volts, the typical crowding of curves as the collector current increases, a contour of 5-watt dissipation. The 2N2107 is rated at only 1 watt dissipation in free air at 25°C, with a thermal resistance of 8 mw/°C. The maximum permissible junction temperature is 150°C. The above thermal resistance is for the relatively low-power operation where a heat sink is not employed. To achieve a 5-watt dissipation it becomes necessary to utilize a relatively large heat sink and it is also desirable to restrict operation to such intermittent operation as class B, where the average dissipation will usually be low.

A load line, corresponding to a load of 100 ohms per transistor, is also shown on the collector family, tangent to the 5-watt dissipation contour. In class B amplifiers the quiescent operating point is located near cutoff; however, it is necessary to apply a small forward bias to avoid crossover distortion, as will be shown later. Thus the initial setting will be near the low end of the load line, say with a base current of about 0.1 to 0.2 ma, collector current about 10 ma, V_{CE} about 44 volts (this implies a supply voltage V_{CC} of 45 volts in this example). As the input signal is applied the collector current will increase, up to the point where the load line intersects the saturation voltage limit, at about $V_{CE} = 5$ volts. Before this, however, the output will have become excessively nonlinear, because of the crowding of the characteristics.

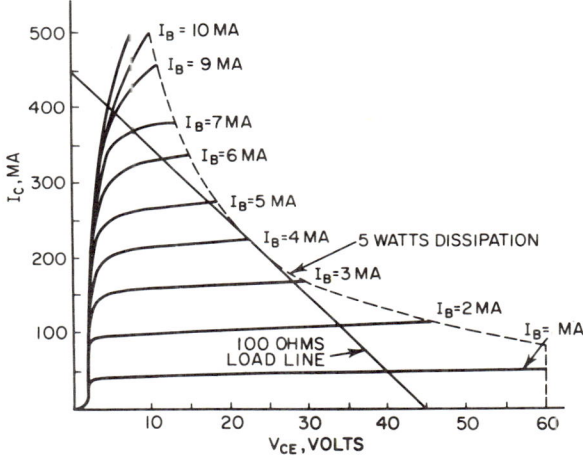

Fig. 72. Collector characteristics of 2N2107 transistor. $T_A = 25°C$.

The dynamic behavior of the class B stage can be determined from these curves by the following method: (1) Values of the collector current I_C are obtained for various values of base current I_B by taking points along the load line of Fig. 72. These are plotted in Fig. 73 as the solid I_C versus I_B curves. Mirror images are constructed to obtain the push-pull characteristic. (2) The corresponding values of the emitter-base voltage V_{BE} are obtained from Fig. 71. To this is added a voltage corresponding to the drop produced by the base current passing through the source resistance (in this example taken as 100 ohms). This gives the dotted curves, labeled I_C versus V_G. (3) If there is resistance in series with the emitter (and there usually is) an additional drop is added equal to the product of emitter current I_E ($I_E = -I_C - I_B$) and the resistance R_E. This is shown as the dot-dash curve.

Several interesting effects are evident from this figure. Most obvious is the nonalignment of the upper and lower halves of the composite curve, this misalignment being moderate for the I_C versus I_B curve but severe for the other two. This is the basis for the oft-mentioned crossover distortion. The cure for this distortion is to provide an amount of forward bias such that the opposite halves now become extensions of each other. This is illustrated in Fig. 74, where the two curves for the condition of $R_G = 100$ ohms and $R_E = 1$ ohm have been displaced 0.73 volt, so that now the linear portions of the two halves are continuations of each other. Under this condition the crossover distortion will be removed.

17-62 AMPLIFIER CIRCUITS

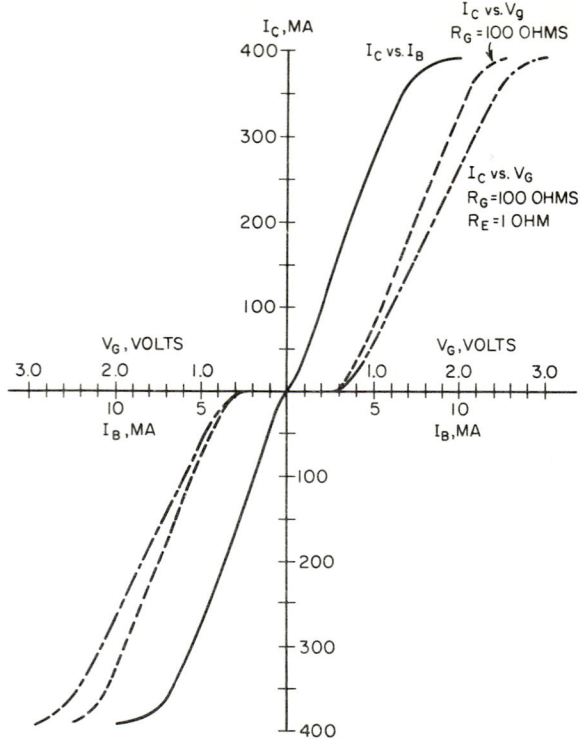

Fig. 73. Class B output vs. input curves, showing effect of crossover distortion.

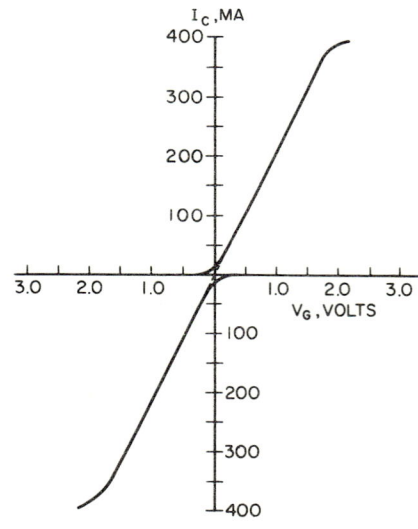

Fig. 74. Elimination of crossover distortion by forward bias. I_C versus V_G. $R_G = 100$ ohms, $R_E = 1.0$ ohms, $V_B = 0.73$ volts.

AUDIO AMPLIFIERS   17-63

Figure 75 shows how the output waveform would look if crossover distortion were present, and how it is avoided by the above method. Crossover distortion is particularly undesirable since it occurs at low levels, as contrasted to most forms of distortion, which usually occur at high levels.

There is also high-level distortion in class B amplifiers, as shown by the nonlinearity at the extremes of the preceding curves. This distortion is caused by the crowding of the collector family curves at high currents, and limits the usable swing to something less than the full range of the characteristics. Thus from inspection of these

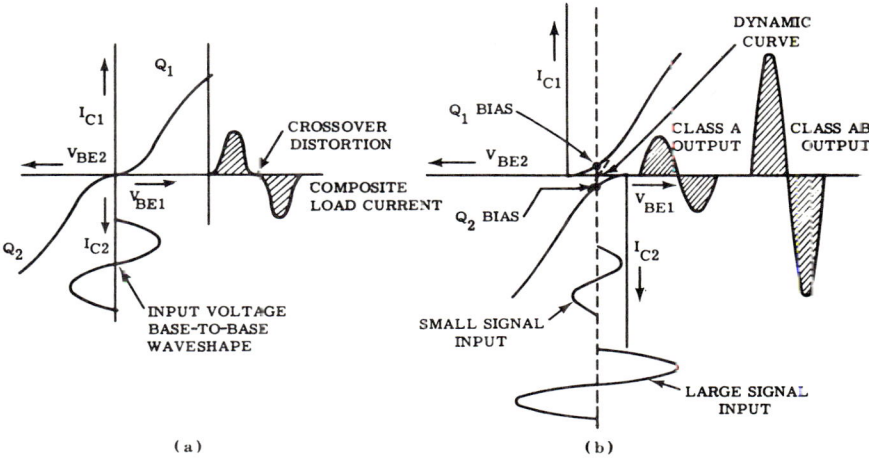

FIG. 75. Output waveforms, illustrating crossover distortion. (a) Class B. (b) Class A and AB. (Reproduced with permission from Motorola Inc., Semiconductor Products Division, "Power Transistor Handbook.")

figures we can determine the maximum current and voltage excursions possible without excessive distortion. We obtain the following values:

Peak collector current $I_C' = 0.37$ amp
Peak base current $I_B' = 7$ ma
Peak emitter current $I_E' = -0.377$ amp
Minimum $V_{CE} = 8$ volts
V_{CE} excursion $= 36$ volts
Peak generator voltage $V_G' = 1.77$ volts
Load resistance $= 100$ ohms per transistor

From the above we calculate the power output

$$P_o = (36 \times 0.37)/2 = 6.65 \text{ watts per transistor}$$

and the power gain

$$G = (36 \times 0.37)/(1.77 \times 0.007) = 1{,}070 = 30 \text{ db}$$

The input resistance is not a constant; thus only a peak value (corresponding to the peak excursion) is meaningful. This is obtained as the ratio of peak input voltage to peak base current. The input voltage to the transistor is not the above value of 1.77 volts, however, since this includes the drop in the 100-ohm source resistance. The transistor input peak resistance is therefore equal to

$$1.77/0.007 - 100 = 153 \text{ ohms per transistor}$$

17-64 AMPLIFIER CIRCUITS

The above will serve to give an insight into class B behavior. In the following section practical methods of operating in class B will be described, and equations will be given for the characteristics derived graphically in the above section.

14.4. Bias in Class B Amplifiers. As illustrated above graphically, it is necessary to provide some forward bias for class B amplifiers to avoid crossover distortion. This can be done most easily by the arrangement shown in Fig. 76a. This shows the collector supply voltage V_{CC} also supplying the bias for the base through the divider made up of resistors R_1 and R_2. It should be noted that no bypass capacitors are used in this circuit, as compared with class A operation. This is because the current is not constant; thus a capacitor would become charged and would shift the bias. The effective collector-to-collector load is here shown as R_{CC}, which is related to the secondary load R_L by the transformer turns ratio squared.

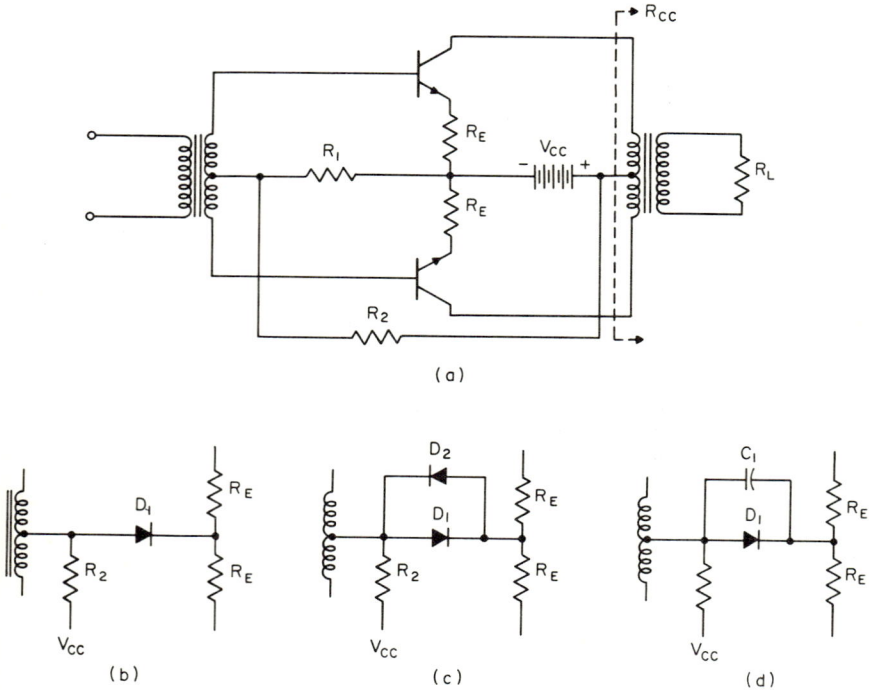

FIG. 76. (a) Basic class B schematic. (b–d) Alternative biasing arrangements.

Figure 73 showed that there is considerable effect of the emitter-base nonlinearity, as evidenced by the displacement of the two halves of the push-pull curve. This necessitates the use of a small forward bias to align these curves. Remembering that the emitter-to-base voltage V_{EB} varies with temperature it becomes obvious that this compensation can be effective at only one temperature, unless some means is employed to provide temperature-dependent compensation.

Figure 76a showed the bias obtained by means of a fixed divider; therefore, the compensation will be unbalanced as temperature varies. The modifications shown in (b), (c), and (d) illustrate techniques commonly employed to achieve wide-range temperature compensation.

In Fig. 76b a diode is used to replace resistor R_1. The current through it is adjusted by the bleeder resistor to be sufficient to avoid reverse current on signal peaks. This requires extra power from the supply voltage. The additional diode shown in Fig.

AUDIO AMPLIFIERS

76c provides a path for any reverse current during input peak swings. Alternatively, a capacitor, as shown in Fig. 76d, will absorb these peaks. While it was stated previously that bypass capacitors are not used in class B amplifiers because of rectification, in this case the use is permissible since the diode is essentially a constant-voltage device.

14.5. Output Power. Gain, Efficiency. The maximum class B power output, assuming that operation is possible from cutoff to zero voltage, would be given by the equation

$$P_{o\,max} = 2(V_{CC})^2/R_{CC} \tag{71}$$

Figure 77 is a nomograph of Eq. (71), for supply voltages of 6 to 28 volts. It should be noted that V_{CE} is used here instead of V_{CC}. This implies that something less than the full voltage swing is permitted. In our previous example, for instance, we used only 36 volts out of an available 45 because of distortion limitations. Further, this

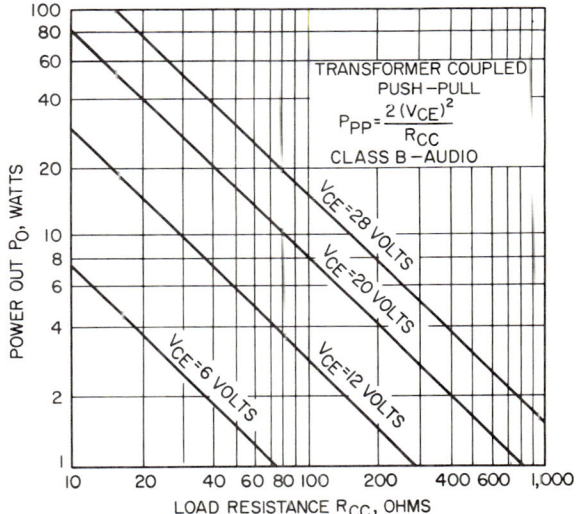

Fig. 77. Push-pull power output–load resistance nomograph. (*Reproduced from Motorola Inc., Semiconductor Products Division, "Power Transistor Handbook."*)

nomograph does not include the effect of transformer losses, and the output power delivered to the load will be less by the transformer efficiency. (Note: In some circuits it is possible to eliminate the output transformer. Such circuits will be described in a later section.)

The output power may be expressed in terms of the transformer turns ratio and secondary load resistance:

$$P_{o\,max} = (N_2/N_1)^2 \times (V_{CC})^2/2R_L \tag{72}$$

Again the useful value of voltage will normally be something less than the supply voltage, as described above.

The input resistance per transistor may be approximated by the short-circuited input resistance h_i. Thus in the common-emitter amplifier previously described we can get a good approximation for h_{ie} from the input curve and the relation between I_C and I_B. The common-base h_{ib} is obtained approximately by the slope of the linear portion of the input curve. This turns out to be 2 ohms in Fig. 71. To obtain the common-emitter value now we multiply this by h_{FE}. This can be obtained from Fig. 72 as the ratio of the peak collector-current swing, 370 ma, to the corresponding base-current swing, 7 ma, which gives $h_{FE} = 53$. Thus $h_{ie} = 2 \times 53 = 106$ ohms. This compares reasonably well with our previous figure of 153 ohms, considering the nonlinear nature of this resistance. The specification sheet gives 175 ohms as the typical value at 8 ma, 10 volts, and h_{FE} as having a range of 30 to 90.

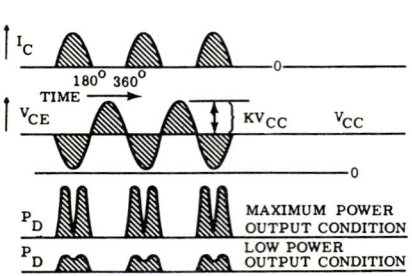

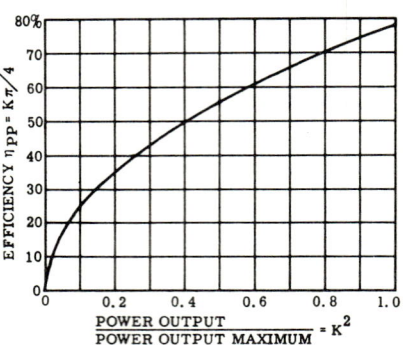

Fig. 78. Class B push-pull collector current, voltage and power dissipation waveshapes. (*Reproduced from Motorola Inc., Semiconductor Products Division, "Power Transistor Handbook."*)

Fig. 79. Class B push-pull efficiency as a function of output power. (*Reproduced from Motorola Inc., Semiconductor Products Division, "Power Transistor Handbook."*)

The power gain may be obtained from the following equation:

$$G = (h_{fe})^2 R_{CC}/4h_{ie} \tag{73}$$

In our graphical example $h_{fe} \cong h_{FE} = 53$, $h_{ie} \cong 106$ ohms, $R_{CC} = 400$ ohms (100 per transistor); thus

$$G = (53 \times 53 \times 400) \div 4 \times 106 = 2{,}650$$

This is about $2\frac{1}{2}$ times the gain calculated by the previous graphical method. The latter was based on so-called peak input resistance, which is considerably different from the dynamic resistance used above.

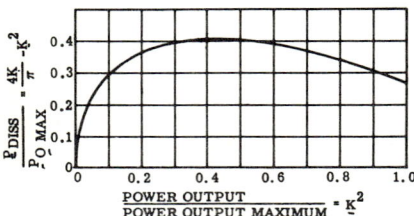

Fig. 80. Class B push-pull power dissipated as a function of power output. (*Reproduced from Motorola Inc., Semiconductor Products Division, "Power Transistor Handbook."*)

The efficiency of a class B amplifier is not a linear function of power output as is the case with class A operation. This can be visualized by reference to Figs. 78, 79, and 80. These show typical waveforms of current, voltage, and power; efficiency vs. power output; and power dissipated vs. power output, respectively. Each half of a class B push-pull stage conducts during alternate half-cycles; thus the current in each collector is in the form of half sinusoids. The voltage, however, goes through a full sinusoid on each transistor, even though it is cut off, because of the transformer coupling to the other half of the stage. Thus the instantaneous power dissipation is the product of these two curves and will show two peaks, occurring approximately at the points where the product of the sine and $1 - \text{sine}$ are a maximum. At this point the dissipation will be approximately one-half the power output, $3\frac{3}{4}$ times the average dissipation. If K represents the portion of V_{CC} corresponding to the collector excursion, the power output is

$$P_o = (K V_{CC})^2/R_{CC} \tag{74}$$

and the power dissipated per transistor is

$$P_D = [4K(V_{CC})^2/\pi R_{CC}](1 - K\pi/4) \tag{75}$$

and efficiency is

$$\eta = 1/(1 + P_D/P_o) \tag{76}$$

The maximum efficiency occurs when $K = 1$, or for full swing, under which condition efficiency becomes equal to $\pi/4$ or 78.5 per cent. This theoretical maximum is seldom realized, although efficiency of around 75 per cent can be achieved.

14.6. Comparison between Different Configurations. The previous section has dealt mainly with the common-emitter configuration, which is most widely used. However, on some occasions either the common-base or the common-collector configuration may prove advantageous. In many respects all three configurations are similar; they deliver about the same power output, have the same maximum efficiency, and require about the same load resistance. They differ mainly, as might be expected, in their input resistances and power gain. The input resistance of the common-base

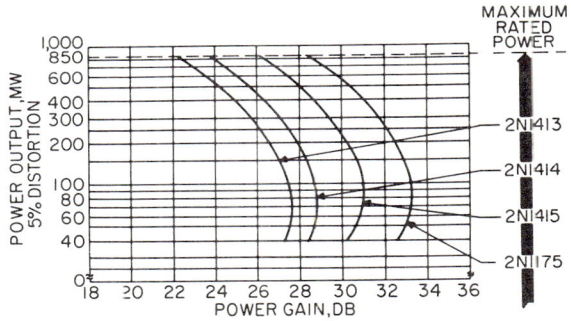

FIG. 81. Typical power gain for class B push-pull amplifiers, 12-volt supply. (*Reproduced with permission from General Electric Semiconductor Products Dept., "Transistor Manual," 6th ed., Syracuse, N.Y.*)

configuration is approximately equal to h_{ib}, as compared with h_{ie} for the common-emitter stage. The common-collector stage has a high input resistance, approximately equal to the peak collector voltage swing divided by the peak base current. The common-base gain will be less than that of the common-emitter stage by about h_{fe}. The gain of the common-collector stage is approximately equal to h_{fe}. Thus, under some conditions, the common-collector stage may have greater gain than the common-base stage, although less than the common-emitter. Another feature of the common-collector stage is its low output impedance; thus it may become attractive in some cases where very low impedance is desirable, as for speaker damping. Certain combinations of complementary pairs, using p-n-p and n-p-n units, have been described, having definite advantages in simplicity.

14.7. Design Charts for Power Amplifiers.[4] Figures 81 through 84 have compiled much of the preceding information into design charts from which amplifiers can be designed with a minimum of computation. Figure 81 shows power output vs. power gain for four typical medium-power transistors, working from a 12-volt supply, and Fig. 82 shows a similar set of curves for single-ended class A amplifiers, for use as drivers of the class B stage. The use of these charts can be illustrated by an example. Assume a 300-mw output is desired from a 12-volt amplifier consisting of a driver and push-pull output pair. Also the signal source has an available power output of 30×10^{-9} watts. Overall power gain required is $300 \times 10^{-3} \div 30 \times 10^{-9}$ or 10×10^6 or 70 db. Assume the transformer has an efficiency of 75 per cent. The required collector-to-collector power is then $300/0.75 = 400$ mw. From Fig. 81 a pair of

[4] "General Electric Transistor Manual," chap. 8.

2N1415's in class B push-pull has a power gain of approximately 28 db at 400 mw. This is a numerical gain of 650; so the input power required by the stage is $\frac{400}{650} = 0.82$ mw. The remaining power gain to be obtained from the driver is 70 db − 28 db = 42 db. From Fig. 82 the 2N322 has a power gain of 42.5 at a power output of 0.82 mw. The output transformer primary impedance is obtained

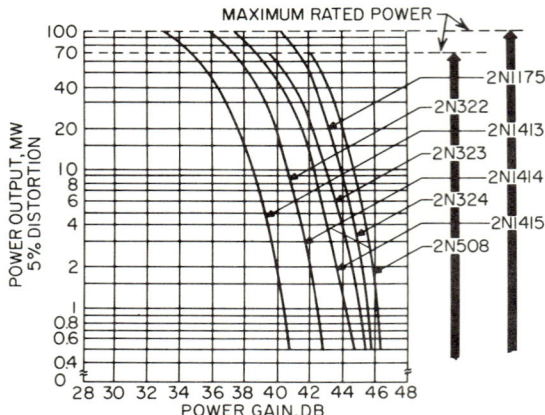

Fig. 82. Typical power gain for class A single-ended amplifiers, 12-volt supply. (*Reproduced with permission from General Electric Semiconductor Products Dept., "Transistor Manual," 6th ed., Syracuse, N.Y.*)

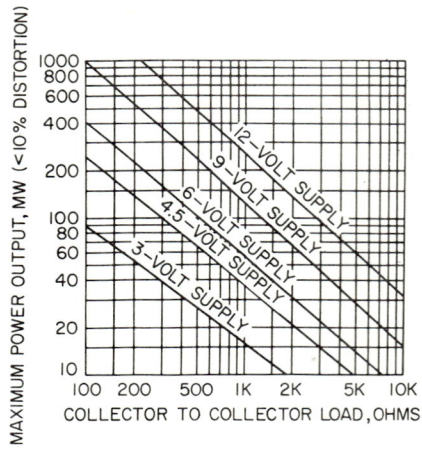

Fig. 83. Design chart for output transformer in class B push-pull audio amplifiers. (*Reproduced with permission from General Electric Semiconductor Products Dept., "Transistor Manual," 6th ed., Syracuse, N.Y.*)

from Fig. 83 and gives a value of 600 ohms maximum collector-to-collector load resistance. A standard 500-ohm, CT output transformer may be used with the secondary impedance adjusted to provide the proper load impedance. Reference is now made to Fig. 84, for the driver transformer. This gives 40,000 ohms as the driver-transformer primary impedance. However, as low as 20,000 ohms can still be used and still have 42 db gain. The secondary must be center-tapped with a total impedance of 800 to 5,000 ohms. When this procedure is used for commercial

AUDIO AMPLIFIERS   17-69

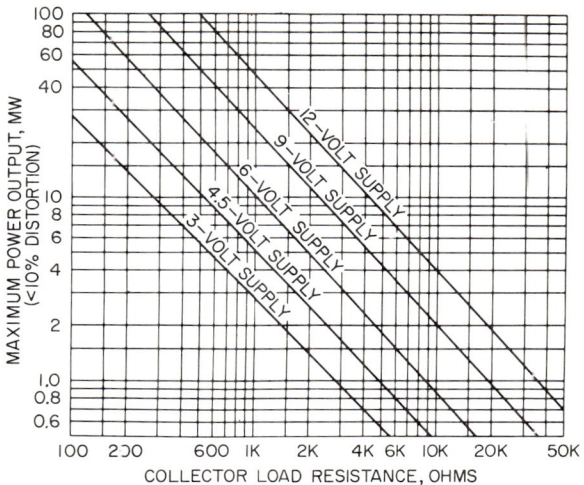

FIG. 84. Design chart for output transformer in class A single-ended amplifier. (*Reproduced with permission from General Electric Semiconductor Products Dept., "Transistor Manual," 6th ed., Syracuse, N.Y.*)

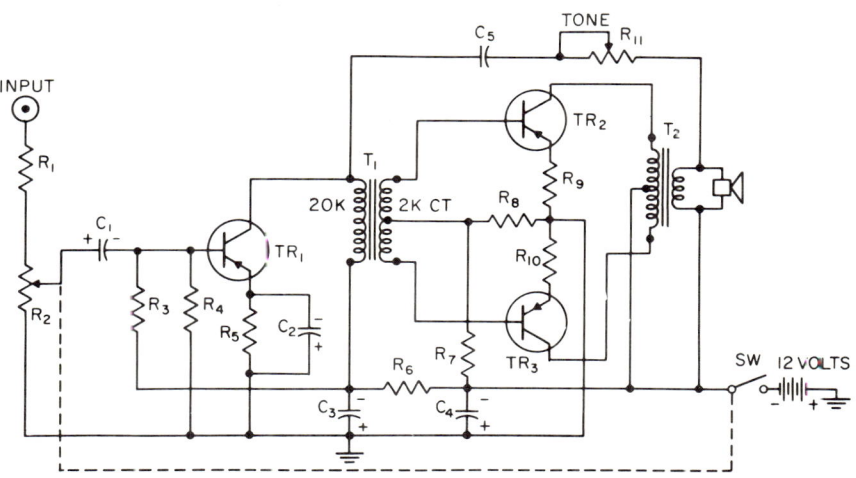

FIG. 85. Schematic of 12-volt phono amplifier.

R_1 = 220,000 ohms
R_2 = volume control 10,000-ohm ½-watt audio taper
R_3 = 190,000 ohms
R_4 = 10,000 ohms
R_5 = 470 ohms
R_6 = 220 ohms
R_7 = 2,700 ohms
R_8 = 33 ohms
R_9, R_{10} = 10 ohms

R_{11} = 25 kilohms linear
C_1 = 6 μf, 12 volts
C_2 = 100 μf, 3 volts
C_3, C_4 = 50 μf, 12 volts
C_5 = 0.01 μf
TR_1 = GE 2N322
TR_2, TR_3 = GE 2N1415
T_1 = 20 kilohms/2 kilohms CT
T_2 = 500 ohms CT/VC

Maximum power output at 10 per cent harmonic distortion, 300 mw. For use with magnetic cartridge omit R_1. All resistors ½ watt. (*Reproduced with permission from General Electric Semiconductor Products Dept., "Transistor Manual," 6th ed., Syracuse, N.Y.*)

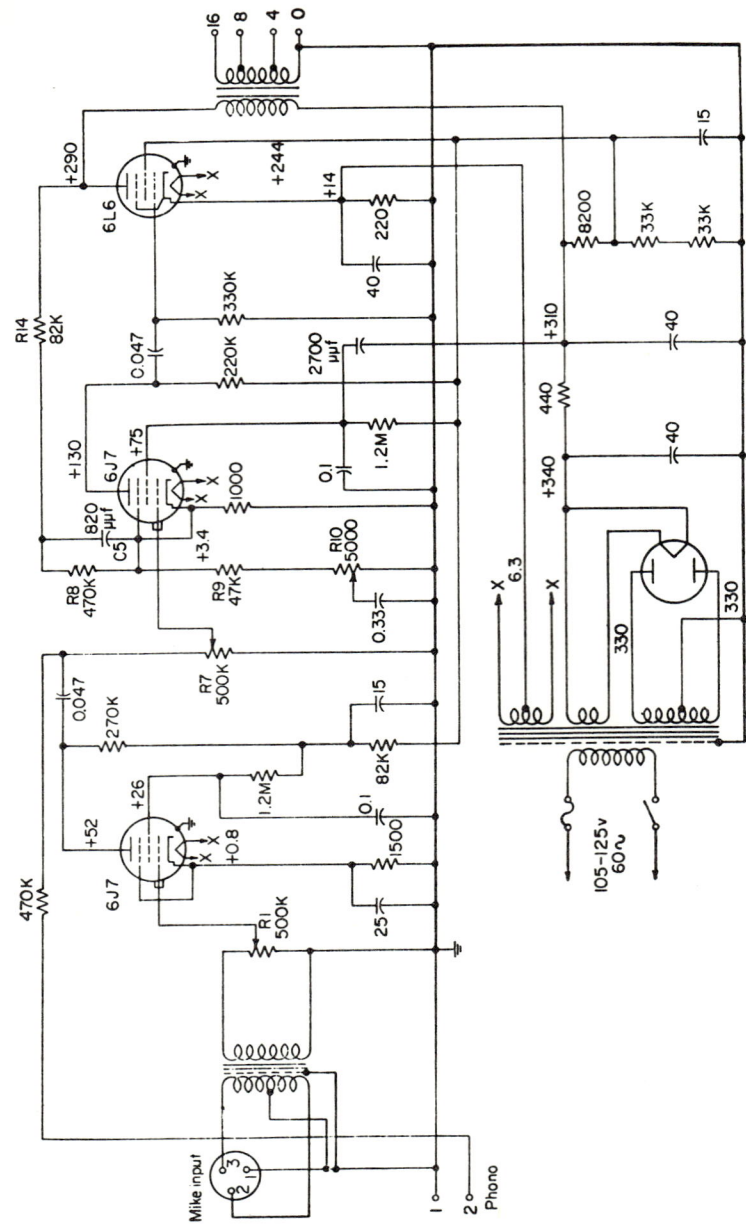

FIG. 86. Small public-address amplifier, single-ended 6L6 output, 6 watts (RCA MI-12238B type SA-6A). [From Keith Henney (ed.), "Radio Engineering Handbook," 5th ed., McGraw-Hill Book Company, New York, 1959.]

17–70

designs, it must be remembered that it represents full battery voltage, typical power gain and input impedance, and therefore does not account for end-limit points.

Figure 85 shows a circuit designed according to the above procedure, to meet the stated specifications.

15. TYPICAL POWER AMPLIFIERS

In this section a number of tube and transistor amplifiers will be described briefly which employ the principles described in the foregoing sections. Other complete systems will be described in the sections on high-fidelity stereo systems.

15.1. Tube Power Amplifiers. Figure 86 shows the schematic of a small public-address amplifier, employing single-ended class A operation with a 6L6 output tube. This small amplifier has less than 5 per cent distortion at 6 watts output from 400 to 10,000 cycles. At 100 cycles the distortion is 8 per cent. The input transformer is balanced to ground for a 125- to 600-ohm microphone. Volume controls R_1 and R_7 are for microphone and crystal phono inputs, respectively. Negative feedback to the

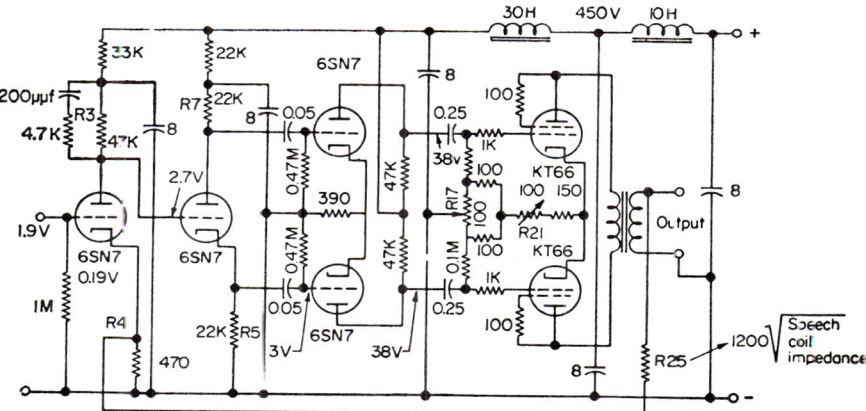

FIG. 87. Williamson high-fidelity amplifier schematic. [*From Keith Henney (ed.), "Radio Engineering Handbook," 5th ed., McGraw-Hill Book Company, New York, 1959.*]

cathode of the second 6J7 includes network R_{14}, R_8, C_5, R_9, and R_{10}. The tone control R_{10} decreases the high-frequency response as desired.

Figure 87 shows the circuit of an early version of the well-known Williamson amplifier. This was one of the earliest of the truly high-fidelity amplifiers and still exemplifies good design principles. Some of the features of this design are negligible nonlinear distortion up to maximum rated output, linear response from 10 cps to 70 kc, negligible phase shift in the audible range, good transient response, and low output resistance. It is an all-triode power amplifier requiring 2 volts across 1 megohm for 15 watts output. The input 6SN7 is direct-coupled to the phase-splitter 6SN7 grid, whose plate and cathode, in turn, feed the output stage. From the low-impedance loudspeaker terminals 20 db is obtained and applied to the unbypassed cathode resistor R_4 of the input triode. Feedback resistor R_{25} depends on the output transformer connection used. For 1.7 ohms output R_{25} is 1,500 ohms, for 9 ohms it is 3,600 ohms, and for 16 ohms it is 4,800 ohms.

Figure 88 shows the frequency response, with and without feedback, also the phase shift from 1 to 1,000,000 cycles. This amplifier takes negative-feedback voltage around four stages, a severe test of design, and one of the outstanding features of this amplifier.

Preliminary adjustments are made by R_{21} to bring the total plate current of the KT66's to 125 ma, and with no signal input, by connecting a 10-volt d-c meter across

17-72 AMPLIFIER CIRCUITS

the plates of these tubes and by adjusting R_{17}, equal plate currents are indicated when the voltmeter reads zero.

A combination triode-pentode circuit can be used by changing the standard Williamson circuit as shown in Fig. 89. The important change is the output transformer, which may be used with 6L6, 807, 5881, and KT66 tubes. The primary impedance

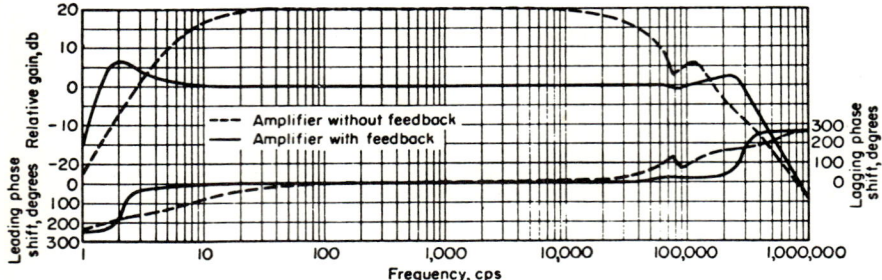

FIG. 88. Williamson high-fidelity amplifier; loop-gain and phase-shift characteristics. [*From Keith Henney (ed.), "Radio Engineering Handbook," 5th ed., McGraw-Hill Book Company, New York, 1959.*]

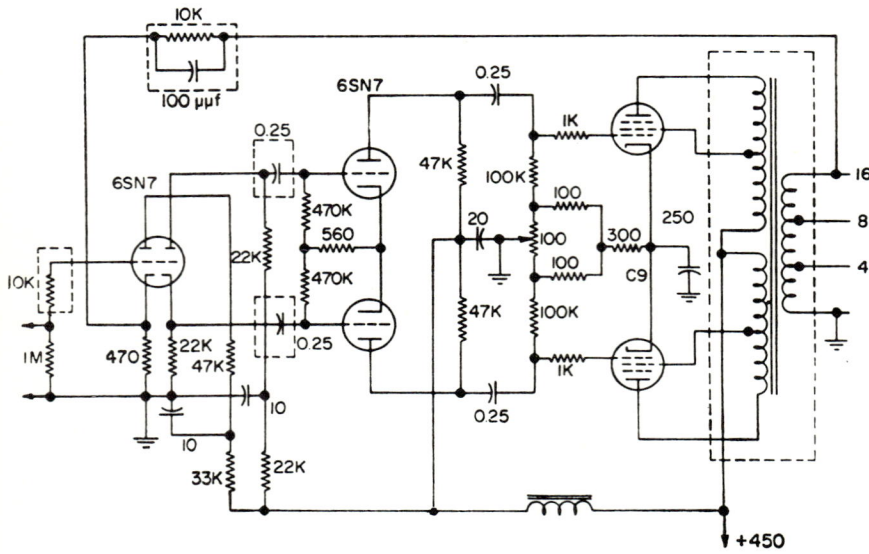

FIG. 89. "Ultralinear" Williamson amplifier. [*From Keith Henney (ed.), "Radio Engineering Handbook," 5th ed., McGraw-Hill Book Company, New York, 1959.*]

is 6,600 ohms plate-to-plate, and 1,220 ohms screen-to-screen, to provide the optimum combination of triode and pentode characteristics; the dotted boxes in Fig. 89 show changes from the original Williamson circuit.

The designers recommend C_9, a 250-μf bypass across the output-tube cathode resistors. The output power is about 30 watts with 1-volt drive. This circuit may also be adapted to use push-pull 6550 tubes to provide 60 watts output.

15.2. Tube Amplifiers without Output Transformers. One very desirable objective in power-amplifier design is to eliminate the output transformer. If this can be accomplished a number of advantages will result; the output power and

AUDIO AMPLIFIERS

efficiency will be increased by eliminating the loss which would normally occur in the transformer, and the frequency response will be improved at both extremes. The low-frequency response will improve because of absence of the loading effect of the transformer primary inductance, while the high-frequency response will likewise be improved by eliminating the effect of transformer leakage inductance.

Figure 90 illustrates a circuit for accomplishing this objective. A bridge-type output stage is used, employing two type 6AS7G dual triodes. A 400-ohm loudspeaker can be driven directly from this bridge, as shown, producing 8 watts at 7 per cent intermodulation distortion without feedback and 0.8 per cent with feedback. The output stage has four driven points, and a special plate circuit driver 12AU7 is employed.

15.3. Transistor Power Amplifiers. A number of typical transistorized power amplifiers will be described in this section, ranging from the relatively modest output of 5 watts up to 125 watts. The latter is by no means indicative of the power limit

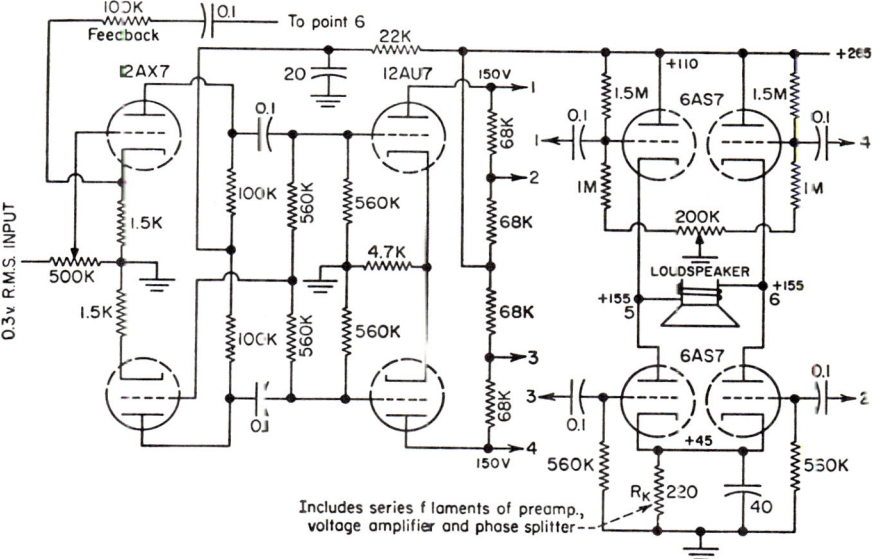

FIG. 90. Transformerless amplifier. Note bridge-type circuit to match voice-coil output: 8 watts into 400-ohm voice coil. [*From Keith Henney, (ed.), "Radio Engineering Handbook," 5th ed., McGraw-Hill Book Company, New York, 1959.*]

of transistor power amplifiers, and amplifiers employing power transistors and silicon controlled rectifiers and having power output considerably in excess of these figures are described in Chap. 20.

Figure 91 gives the circuit of a two-stage amplifier which utilizes a 3N47 or 3N51 tetrode power transistor in the output stage. This amplifier is capable of better than 5 watts output with total harmonic distortion of about 1.5 per cent. Most typical common-emitter audio power stages suffer from high nonlinear distortion and inadequate frequency response. The use of a tetrode power transistor makes possible considerable improvement in audio power stages without sacrificing the high gain and favorable driving conditions inherent in the common-base configuration. The tetrode geometry, with independent connections to the base region of the transistor, permits the current-transfer characteristic to be linearized to a considerable degree.

The advantages of direct coupling in eliminating interstage components and improving frequency response are also very significant. This circuit does not rely on heavy use of negative-loop feedback to achieve low-distortion performance.

17-74   AMPLIFIER CIRCUITS

Base 2 of the tetrode is connected to a reverse-biasing source consisting of a 6-volt supply and a 220-ohm series resistor. Base 1 is forward-biased in the conventional manner with a 1,000-ohm potentiometer R_6 connected to a 12-volt collector supply. The combination of these forward- and reverse-biasing networks will establish a quiescent no-signal collector current of 1.2 to 1.4 amp d-c. The input signal is divided between the two base connections by means of a fixed 47-ohm and a 150-ohm resistor network. The performance is tabulated below for a load of 10 ohms.

Total harmonic distortion at 5 watts output = 1.5 per cent at 100 cps
Frequency response (3 db down points) = 12 cps and 18 kc
Current amplification = 800
Output impedance = 10 ohms
Input impedance = 300 ohms

A 50 per cent reduction in power results if the load is 5 or 20 ohms instead of the above value of 10 ohms. If it is necessary to use such values a tapped choke can be used.

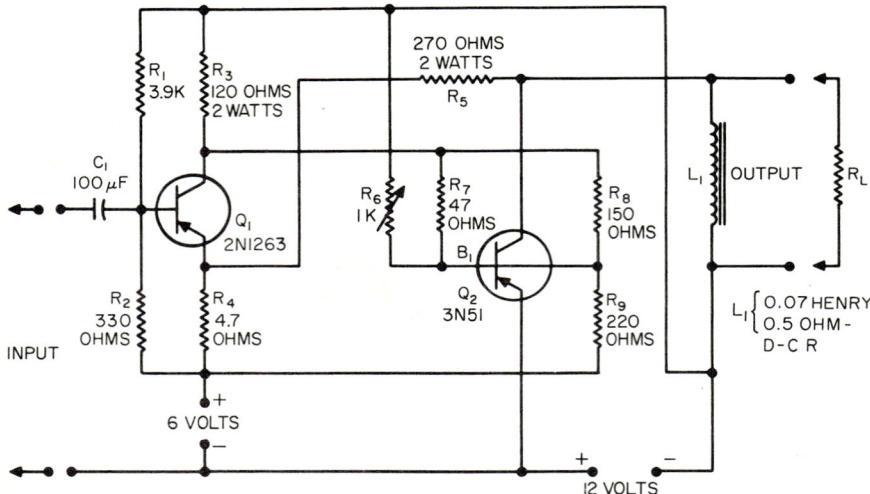

FIG. 91. Two-stage direct-coupled 5-watt class A power amplifier. (*Reproduced with permission from J. A. Lostetter, Two-stage Direct-coupled 5-watt Class-A Power Amplifier, Honeywell Application Note AN2A.*)

About 6 db of negative feedback is used to match the output impedance of the amplifier to the load and further reduce the distortion. The feedback path consists of a 270-ohm resistor connected between the collector of the output stage and the emitter of the driver stage. This path includes the emitter current of the driver stage. Because there is practically no phase shift between the emitter and collector currents of this stage in the useful passband of the amplifier, the phase shift around the feedback path is essentially that of the output stage only.

Figure 92 shows the circuit of a 10-watt amplifier utilizing class B output. Also shown is the circuit of the power supply which can be used in place of the 12-volt battery. This supply exhibits excellent regulation and low ripple, which is an important requirement when class B amplifiers are used.

Since the common-collector output stage requires a low-impedance driving source, a common-collector driver stage is used. Transformer coupling is used since it provides excellent d-c stability and permits reserve driving voltage to be obtained. The d-c winding resistance of the output transformer is used as an emitter resistor to sta-

AUDIO AMPLIFIERS 17-75

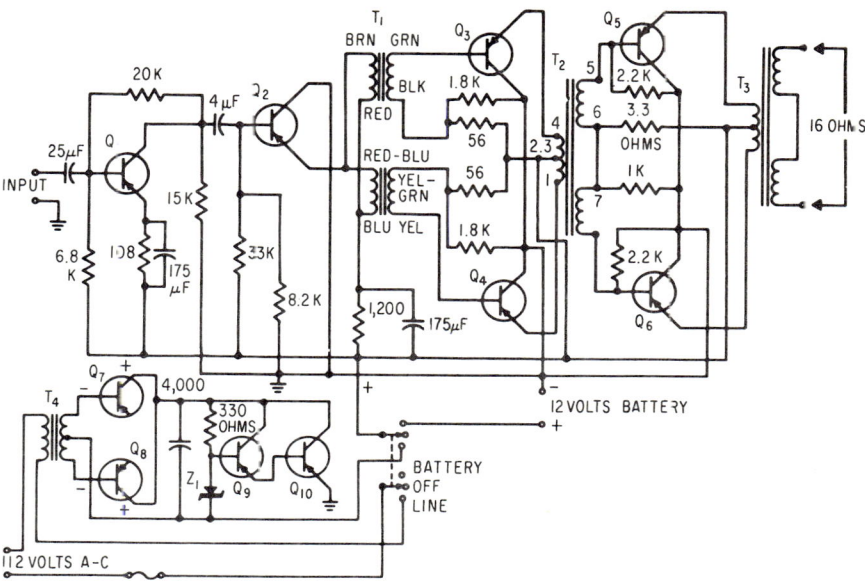

Fig. 92. Schematic of 10-watt high-fidelity transistor amplifier. Transistors Q_1Q_2 2N652, $Q_{3,4,5,6}$, 2N376. Transformers, T_1 Chicago Std. TA-4, T_2 Chicago Std. TA-13. T_3 Chicago Std. TA-14. Power supply, $Q_{7,8,9,10}$ 2N176, Z_1 1.5M14Z5, T_4 Chicago Std. RT-204. (*Reproduced with permission from Motorola Inc., Semiconductor Products Division, "Power Transistor Handbook."*)

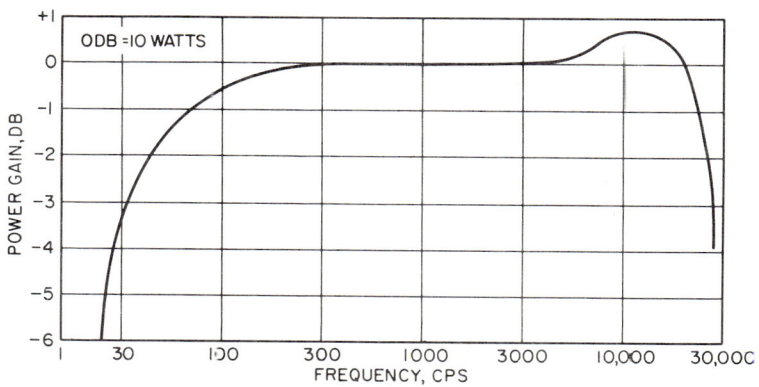

Fig. 93. Frequency response of 10-watt amplifier. (*Courtesy of Motorola Inc., Semiconductor Products Division, Phoenix, Ariz.*)

bilize the output stage. The driver operates push-pull class A for low distortion and to eliminate d-c saturation effects in the transformer, thus reducing the size and cost of this component. Since the driver must also be driven from a low-impedance source, another common-collector stage is used. It is also transformer-coupled, to provide voltage step-up and phase inversion for the push-pull drivers. The first stage is a voltage amplifier.

Frequency response is shown in Fig. 93 and harmonic distortion in Fig. 94. Inter-

17-76　　　　　　　　AMPLIFIER CIRCUITS

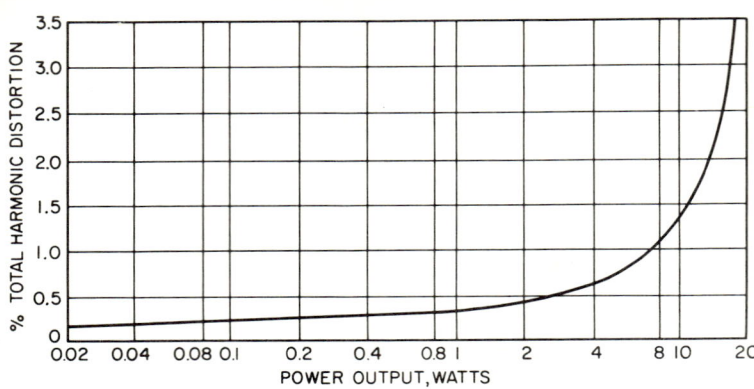

Fig. 94. Distortion vs. power output for 10-watt amplifier. (*Courtesy of Motorola Inc., Semiconductor Products Division, Phoenix, Ariz.*)

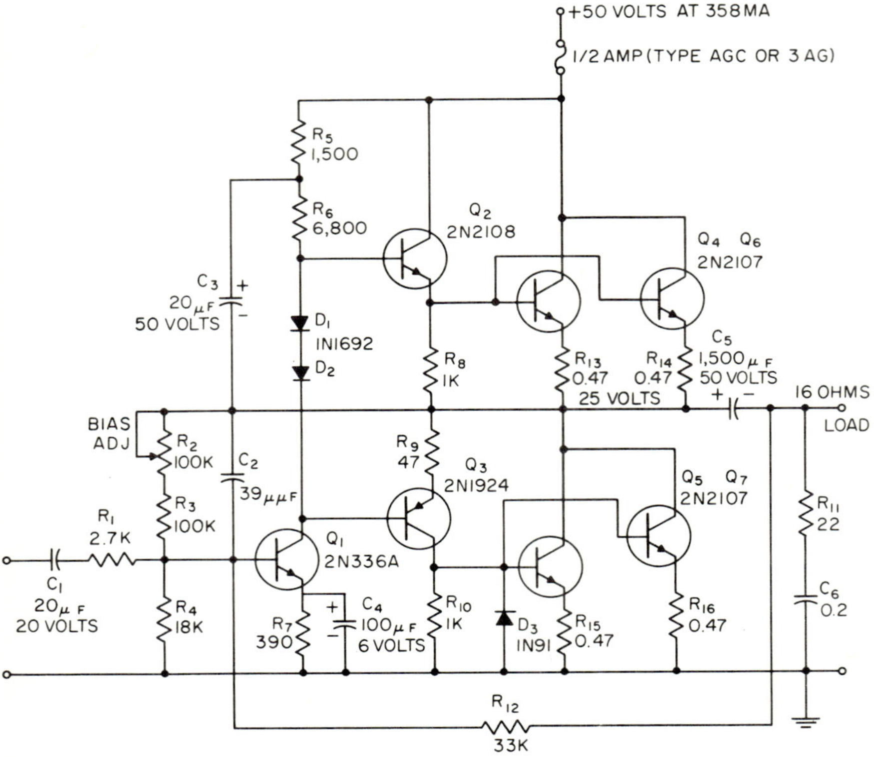

Fig. 95. Schematic of 12-watt power amplifier. Note: All resistors $\frac{1}{2}$ watt. (*Reproduced from "General Electric Transistor Manual," 6th ed., courtesy of General Electric Co., Semiconductor Products Department.*)

AUDIO AMPLIFIERS   17-77

modulation distortion at 10 watts output is 2.9 per cent, measured by the SMPTE method, using frequencies of 60 and 6,000 cps mixed 4:1.

The d-c drain is 180 ma, with no signal applied. Most of this is drawn by the push-pull driver. Total drain increases to 1.3 amp under full power output.

Figure 95 illustrates the use of paralleled output transistors to reduce the saturation resistance and thus increase available power output, without making the V_{CE} rating of the input transistor marginal. This amplifier is capable of 12 watts output into a 16-ohm load.

There are a number of unusual features about this circuit. It employs direct coupling, thus provides excellent low-frequency response; the driver transistors are

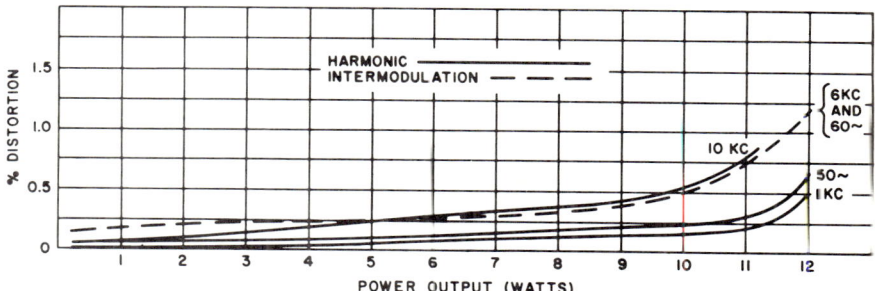

FIG. 96. Distortion vs. power output for 12-watt amplifier. (*Courtesy of the General Electric Co., Semiconductor Products Dept.*)

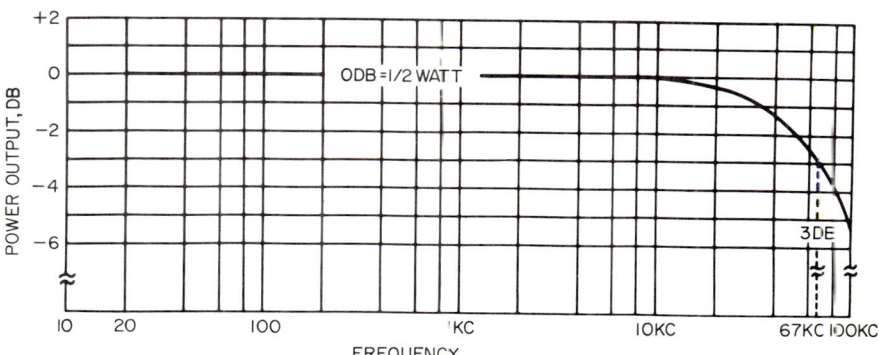

FIG. 97. Frequency response of 12-watt amplifier. (*Courtesy of the General Electric Co., Semiconductor Products Dept.*)

connected in a Darlington arrangement with the output transistors (collectors connected together, output bases fed from the driver emitter) to increase current amplification; Q_2 and Q_3 are also operated class B, as well as the output transistors, and the two halves of the amplifier are effectively connected in parallel across the load, with phase inversion supplied by driving Q_4 from the emitter of Q_2, Q_5 from the collector of Q_3.

Figure 96 shows the variation of harmonic and intermodulation distortion with power output and frequency. The load can vary between 10 and 30 ohms without excessive increase in distortion or loss of power, although the optimum load is around 20 ohms. Figure 97 shows the frequency response of this amplifier.

AMPLIFIER CIRCUITS

The final circuit to be described in this section again employs the tetrode power transistors, this time in a high-power arrangement capable of delivering 125 watts into the load. The circuit is shown in Fig. 98. The unusual features are the use of tetrodes for both stages and the common-base arrangement of the final stage. Thus, in the preceding pages we have seen examples of power amplifiers using common-emitter stages, common-collector stages, and now common-base stages. The principal object in using the common-base configuration for the final stage is that this type of stage requires only about 1 volt peak drive, thus permitting operation of the common-emitter driver at a low voltage level. It will be noted that the supply

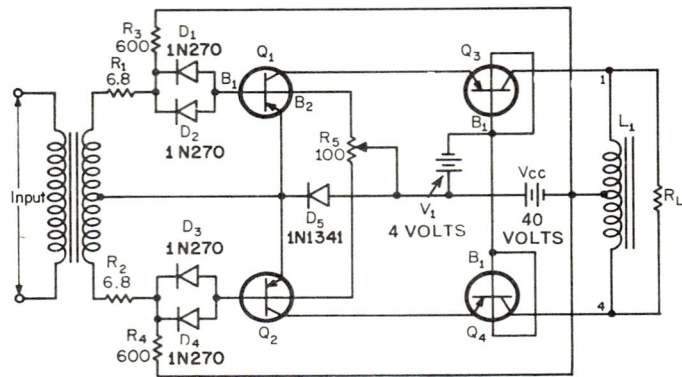

FIG. 98. Schematic of 125-watt class B push-pull transistor amplifier. Q_1 and Q_2 = 3N47 or 3N51 Honeywell transistors. Q_3 and Q_4 = 3N48 or 3N52 Honeywell transistors (triode connected). T_1 = input transformer—driving-source impedance should be 10 to 20 ohms —d-c resistance in range of 2 to 4 ohms. L_1 = output choke or transformer as required. Amplifier is designed to deliver 125 watts into 20.72 ohms collector to collector of Q_3 and Q_4 (5.18 ohms collector to center tap). (*Reproduced with permission from J. A. Lostetter, A Solid State 125-watt Linear Power Amplifier, Semiconductor Products, vol. 4, no. 10, pp. 35–39, October, 1961.*)

voltage required is only 40 volts, in spite of the high power rating of this amplifier. The following results were obtained with this circuit at 400 cycles (unless otherwise stated) and working into a load of 20.7 ohms:

Power output, watts.................	125	50
Power gain, db.....................	24.3	25.5
Cutoff frequency, −3 db, kc.........	14	13
Total harmonic distortion, %:		
400 cycles......................	3.2	3.6
7.5 kc..........................	1.8	2.5
Efficiency, %......................	70	42

I_{CQ} = 180 ma, total for all transistors.

Figure 99 shows the total harmonic distortion as a function of frequency, at 50 and 125 watts. Figure 100 shows the distortion vs. output power at two frequencies. The third harmonic is dominant, with the second harmonic about half the third.

It is possible to obtain 250 watts with this basic arrangement, using pairs of transistors. The input transistors are operated in parallel, although with separate base and emitter resistors to help equalize the current through the transistors, and the collectors of the driver transistors connect directly to the emitters of the output stages, the collectors of which are connected together and to one side of the load.

AUDIO AMPLIFIERS

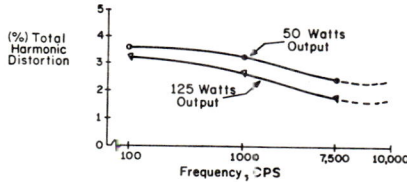

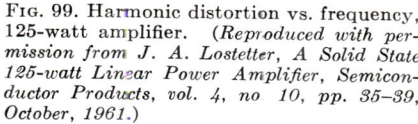

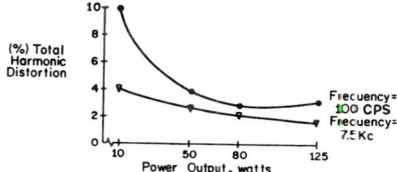

FIG. 99. Harmonic distortion vs. frequency, 125-watt amplifier. (*Reproduced with permission from J. A. Lostetter, A Solid State 125-watt Linear Power Amplifier, Semiconductor Products, vol. 4, no 10, pp. 35–39, October, 1961.*)

FIG. 100. Harmonic distortion vs. power output, 125-watt amplifier. (*Reproduced with permission from J. A. Lostetter, A Solid State 125-watt Linear Power Amplifier, Semiconductor Products, vol. 4, no. 10, pp. 35–39, October, 1961.*)

16. ELECTRON-TUBE STEREO AMPLIFIERS

In this section and the next a number of representative stereo systems, both tube and transistor, will be described. These incorporate many of the features which were described in the preceding sections of this chapter, and illustrate modern design techniques.

Stereo amplifiers are an extension of the earlier high-fidelity designs, with the utilization now of two complete channels to provide the stereo effect. By this means, the illusion of sound location is created; greater effectiveness is given to orchestral music by differentiation of the various sections, and an impression of motion can be created. This is accomplished by using two completely separate channels of communication, from microphone to loudspeaker, whether the transmission medium be radio waves or a phonograph recording. In the former either two separate transmissions may be used, e.g., TV and FM simultaneously, or the two channels may be used to provide two forms of modulation on the same carrier, e.g., amplitude and frequency or phase. In the recordings again two modes of reproduction are used, with two distinct groove modulations. The pickup must be specially designed for such reproduction, so that one output will be responsive only to one channel, the other to the second channel. Of course, mono phono operation may be obtained by simply using one channel, or both in parallel.

16.1. Sherwood Stereo Amplifier. The preamplifier portion of this system has already been described in considerable detail (see Sec. 10). This equipment has a rating of 40 watts music from each channel, 36 watts continuous, 72 watts peak on stereo; or on mono 80 watts music power, 72 watts continuous, 144 watts peak, at $1\frac{1}{2}$ per cent intermodulation distortion (60 cps, 7 kc, 4:1). Figure 101 provides an overall block diagram of the system. Each channel is identical with the other, and consists of input switching to accommodate phono, tape, tuner, or auxiliary inputs; preamplifiers and frequency equalization circuits for the phono and tape inputs; a function selector which interchanges the speakers, allows reproduction of mono programs on both speakers, and permits use of the stereo pickup on monaural recordings; further amplification, with associated filters and tone controls; loudness and balancing controls, driver stages, and push-pull output stages.

Figure 102 shows the schematic of this amplifier, showing both channels. Figure 102a is the same previously shown as Fig. 43 and is repeated here for convenience. Figure 102b shows the output portion. The input switch on the left selects any of the five inputs and connects them to the appropriate amplifier. The phono and tape inputs go to the preamplifier V_1, whereas the auxiliary and tuner inputs are routed to amplifier V_3. The RIAA or NAB frequency compensations are accomplished by means of feedback around tube V_2 through capacitors C_3, C_4, C_5 and resistors R_9 and R_{10}. The gain control varies the signal applied to amplifier V_3, through the scratch filter, which may be cut in or out as desired. The tone controls, inserted between tubes V_3 and V_4, vary the treble or bass response independently. The treble control

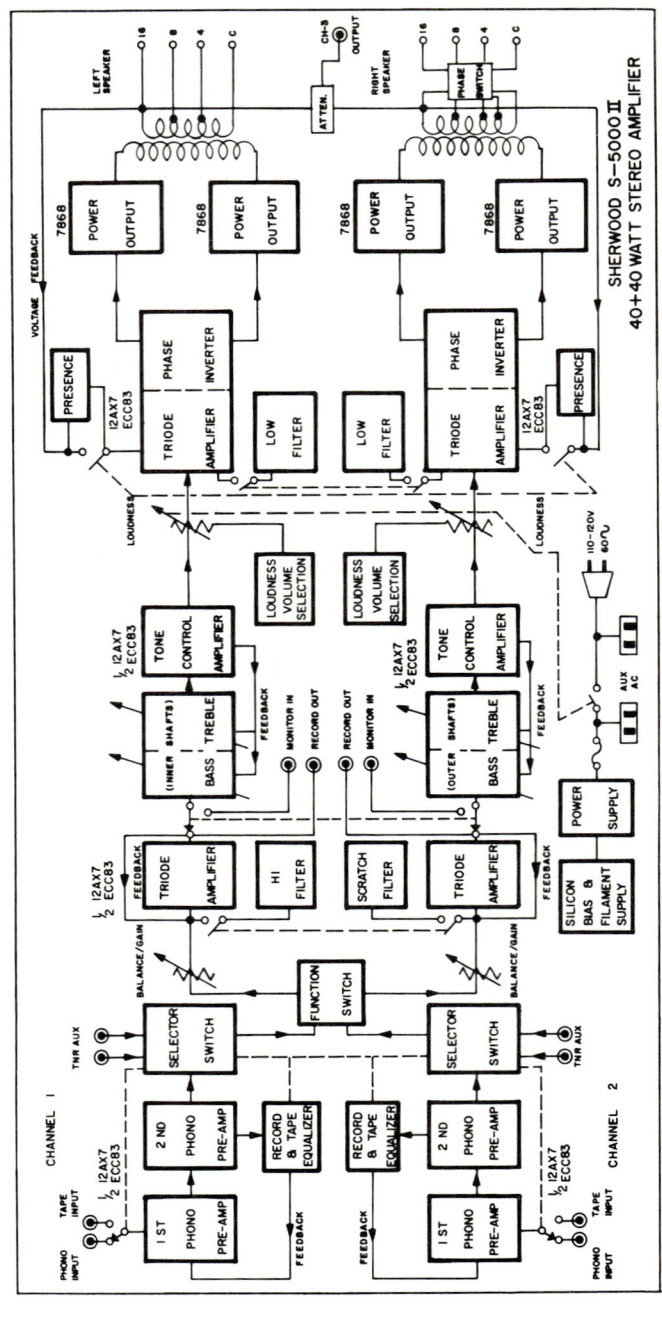

FIG. 101. Block diagram of Sherwood stereo amplifier. *(Reproduced through the courtesy of the Sherwood Electronic Laboratories, Inc., Chicago, Ill.)*

AUDIO AMPLIFIERS 17-81

functions either by adding to the high-frequency response when in its extreme left-hand position (on the schematic) by connecting the 180-pf capacitor between the plate of output of V_3 and the grid of V_4, or by reducing the response by using this capacitor to produce degenerative coupling from the plate of V_4, in the other extreme position. The bass control similarly varies the coupling to the preceding tube or the feedback from the plate, by varying the input from one 3,300-pf capacitor to the other.

It will be noted that there are separate loudness and gain controls. The loudness control has associated with it an RC network which varies the frequency response as the loudness is varied, thus compensating for the variation in frequency response of the human ear as a function of intensity. In operation the gain control is set at that position which produces relatively loud intensity with the loudness control on full; then the volume is thereafter adjusted with the latter control.

A rumble filter is available, following the loudness control, by means of which rumble due to turntable eccentricity or other causes may be reduced. The driver stages act as phase inverters, to provide the push-pull output for the output tubes, and the balance controls are provided to enable setting the two halves of the push-pull pair to equal outputs. Feedback is provided from the 16-ohm output terminals to the cathode of V_5.

Figure 103 indicates the harmonic and intermodulation distortion vs. power output, and Fig. 104 shows the frequency response and distortion as a function of frequency at 1 watt and 36 watts output. The effects of loudness and tone controls on frequency response have been shown in the section on preamplifiers (see Figs. 44 and 45).

Figure 105 shows the overall acoustic response using this amplifier and a three-speaker arrangement. The speakers are a 12-in. high-compliance woofer, an 8-in. cone midrange speaker with sealed fiberglass fill backplate, and one 3½-in. specially designed ring-radiator supertweeter, also with sealed fiberglass-fill backplate. Crossover points are 600 and 3,500 cps with 12 db/octave attenuation.

The performance specifications of this system are:

Power output: Stereo, each channel 40 watts music power (36 watts continuous, 72 watts peak). Mono, 80 watts music power (72 watts continuous, 144 watts peak), at 1½ per cent IM distortion (60 cycles, 7 kc, 4:1).

Outputs: 16-, 8-, and 4-ohm left and right speakers; 2 recording, third channel.

Inverse feedback: 16 db.

Damping factor: 5:1.

Frequency response: (36 watts) 20 cps to 20 kc ± ½ db.

Tone-control response: Flat setting, 20 cps to 20 kc, ± ½ db.

Tone-control range: 15 kc 17-db boost or cut. 40 cps, 16-db boost, 19-db cut.

Rumble filter: 27 cps, 17-db rejection; 70 cps less than 1 db down.

Sensitivity: Radio 0.25 volt, tape 1.4 mv, phono 1.2 mv; all inputs are adjustable with level control.

Maximum input capability: Phono, 200 mv for less than 1 per cent distortion. Radio, adjustable with level control.

Maximum hum and noise: Volume control minimum 100 db (weighted) below rated output. Radio input (controls maximum), 90 db (weighted) below rated output. Phono input (controls flat), 60 db below rated output, 72 db below 10 mv (equivalent to ½ μv referred to input grid).

Interchannel crosstalk: Less than −50 db at 1 kc.

Power consumption: 110 to 120 volts, 60 cps, 150 watts, 1.3 amp.

Tube complement: four 7868, six 12AX7/ECC83, four silicon rectifiers.

16.2. General Electric Stereo Amplifier. This amplifier is designed to deliver 50 watts music power per channel, and illustrates some of the problems faced by designers in their attempts to balance performance with cost.

The output tubes are the GE type 7355's, used in class AB_1. With 400 volts on the plates, 320 volts on the screens, and a −36-volt bias, two of these tubes can be driven to 50 watts of continuous output power with less than 5 per cent harmonic distortion. The plate-to-plate reflected load is 4,000 ohms. Peak-to-peak drive voltage of 72 volts is required.

Since the output stage operates in class AB_1, there is considerable variation of plate

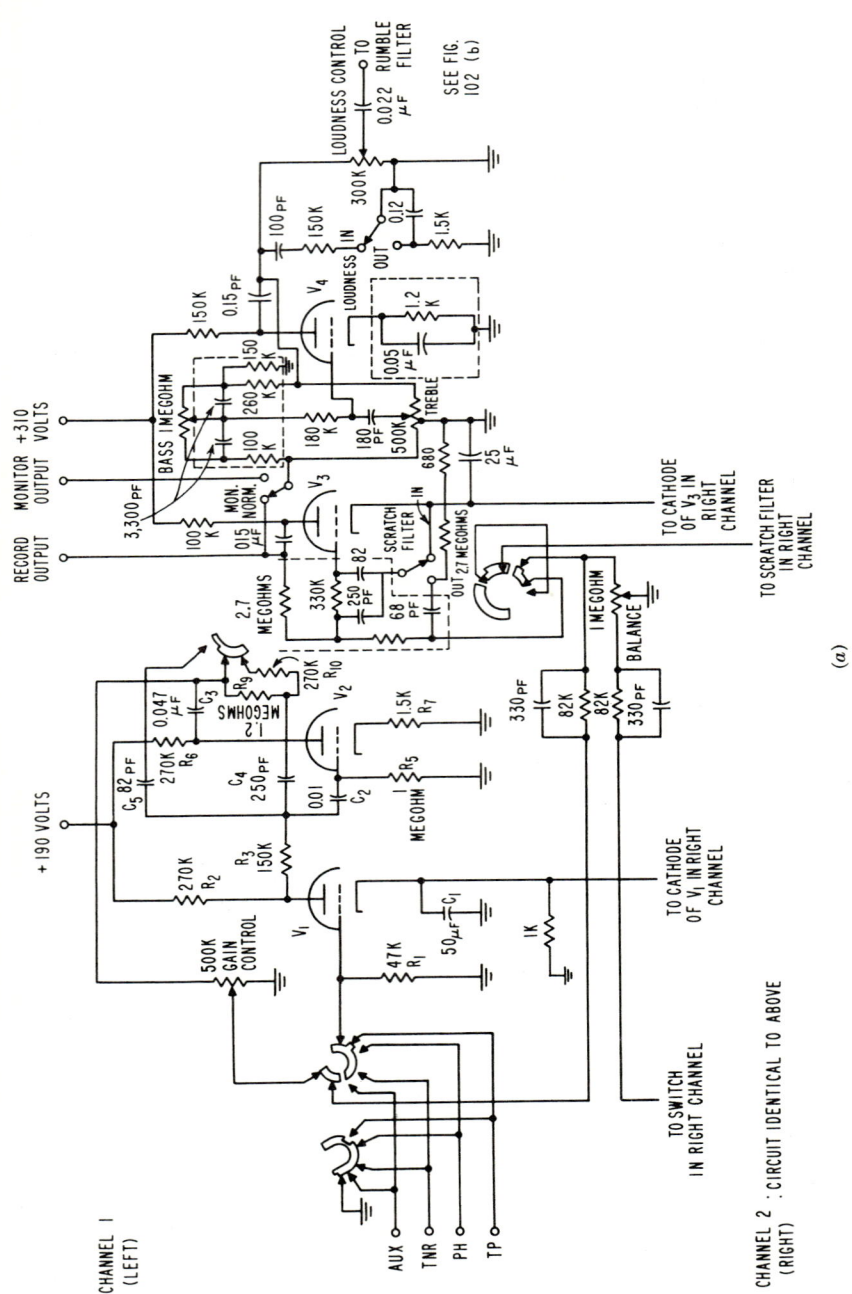

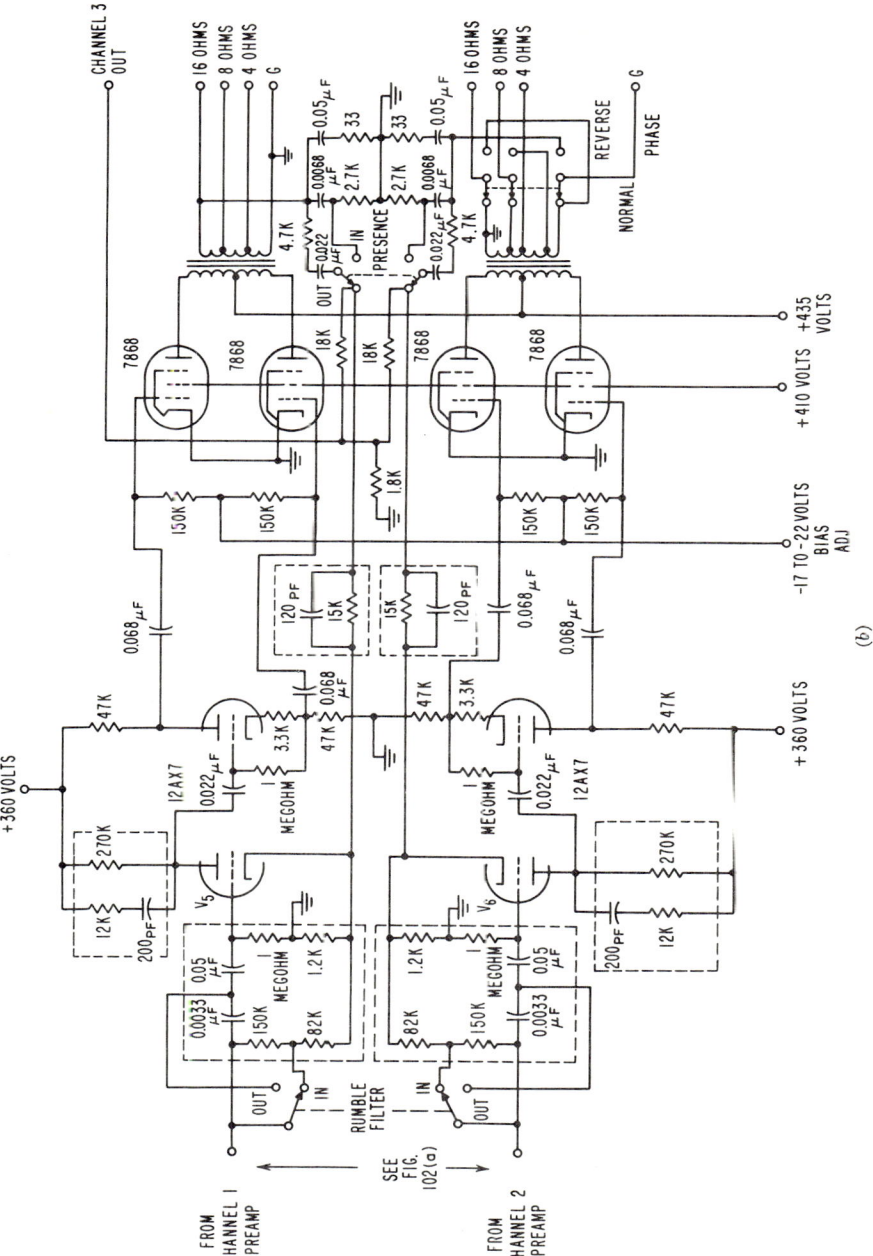

Fig. 102. Schematic diagram of Sherwood stereo amplifier. (*Courtesy of Sherwood Electronic Laboratories, Inc., Chicago, Ill.*)

17-83

17-84 AMPLIFIER CIRCUITS

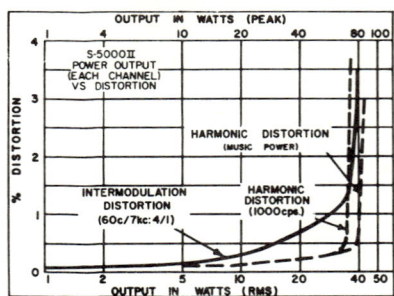

FIG. 103. Distortion vs. power output. (*Courtesy of Sherwood Electronic Laboratories, Inc., Chicago, Ill.*)

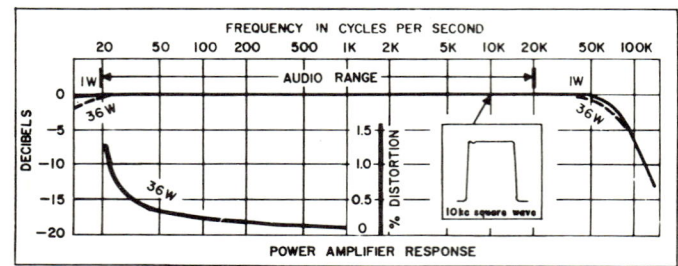

FIG. 104. Frequency response. (*Courtesy of Sherwood Electronic Laboratories, Inc., Chicago, Ill.*)

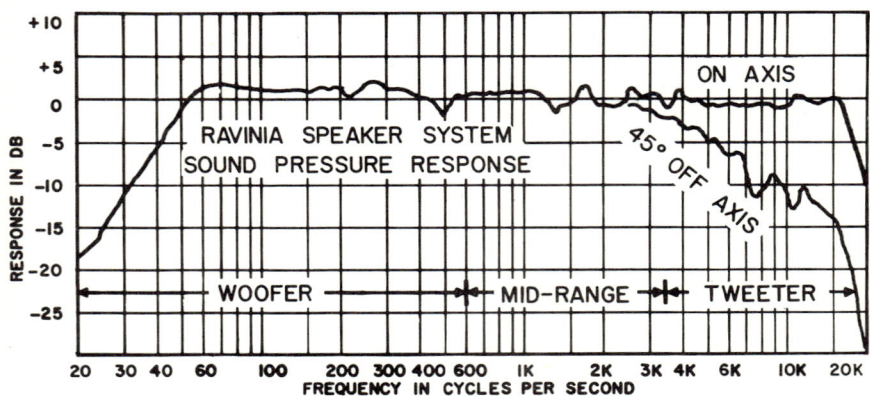

FIG. 105. Sound pressure response, Sherwood stereo system. (*Courtesy of Sherwood Electronic Laboratories, Inc., Chicago, Ill.*)

AUDIO AMPLIFIERS

and screen current with input signal and hence output power. This poses severe design problems in supplying adequately regulated voltages for the plates and screens. At full continuous output from both channels, the power supply would have to supply 400 ma of plate current and 60 ma of screen current for the output stages. This is illustrated in Fig. 106, which shows this variation for two tubes. These values must be doubled if both channels are working under the same conditions.

If this rating of continuous power output were to be met it would require a very well-regulated power supply for the plate and screen voltages, with the screen requiring a separate supply. The transformer windings would have to be low in resistance and the rectifier diodes would have to have low forward voltage drops. These requirements would add considerably to the overall cost of the system. A study was therefore made of the actual requirements of the system, and it was concluded that continuous ratings of the above values were not actually needed for satisfactory reproduction of music, as long as some means of energy storage was incorporated, such that full power could be maintained for the actual duration of music peaks. This was accomplished by means of heavy storage capacitance on the screens (100 μf), which produces discharge time constants ranging from 4 sec at zero signal to 0.5 sec at 50 watts output.

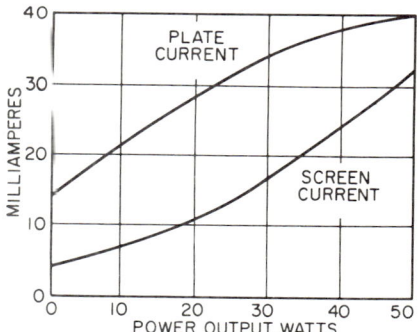

FIG. 106. Plate and screen currents (two tubes) vs. music power output. (*Reproduced by permission of General Electric Co. Audio Products Dept., Decatur, Ill.*)

This technique has been found entirely satisfactory, as evidenced by listening tests, where it was compared with arrangements with constant voltage supplies.

Figure 107 shows the circuit of the preamplifier portion of this system. The input selector switches from the dual stereo phono cartridge to the stereo FM tuner input. The loudness controls are compensated so that the frequency response is varied with loudness setting, as described previously in the section on preamplifiers. The treble and bass tone controls operate in similar manner to those of the Sherwood amplifier, although negative feedback is not employed here. A scratch filter can also be inserted to improve the reproduction from older records.

Figure 108 shows the schematic of the output portion of the amplifier, including the driver-phase inverters and push-pull output stages. The power supply for this portion of the amplifier is also shown, and the 100-μf capacitor on the screen supply may be noted. An RC network is connected between the two plates. This was used to avoid an oscillation due to the output transformer primary leakage inductance. Reducing this leakage reactance permits removal of this network and also improves the frequency response. Feedback of 14 db is employed in this design, from the output transformer secondary to the cathode of the driver tube.

Figure 109 shows the output of this amplifier at 5 per cent distortion, as a function of frequency, with the oscillation-suppression network connected, and Fig. 110 without it but with a transformer having a lower leakage inductance. Figure 111 shows the frequency response of the system, including the effects of the tone controls. Figure 112 shows distortion vs. power output.

17-86 AMPLIFIER CIRCUITS

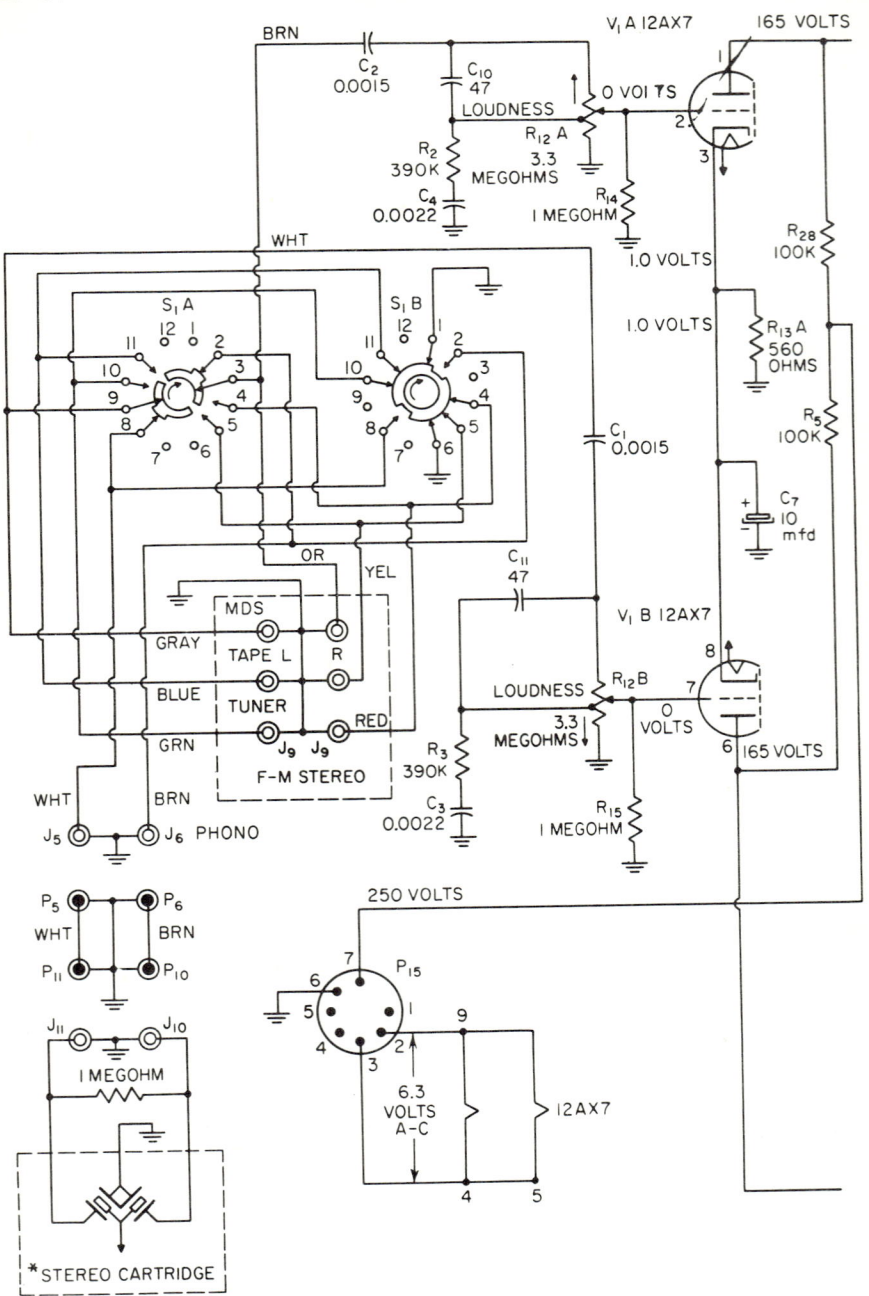

FIG. 107. Schematic of preamplifier for 100-watt stereo amplifier. *Ceramic-type output Audio Products Dept., Decatur, Ill.)

AUDIO AMPLIFIERS 17-87

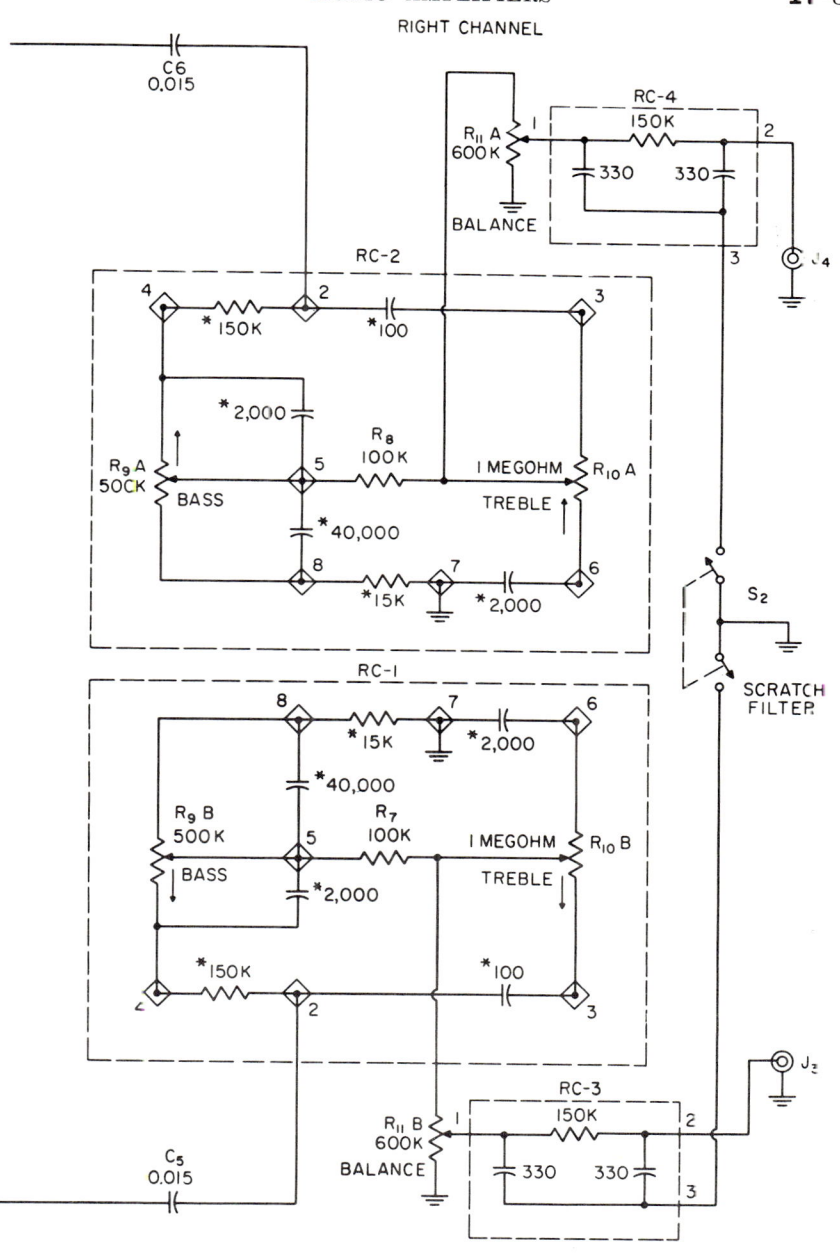

0.4 volt from average recording level. (*Reproduced by permission of General Electric Co.*)

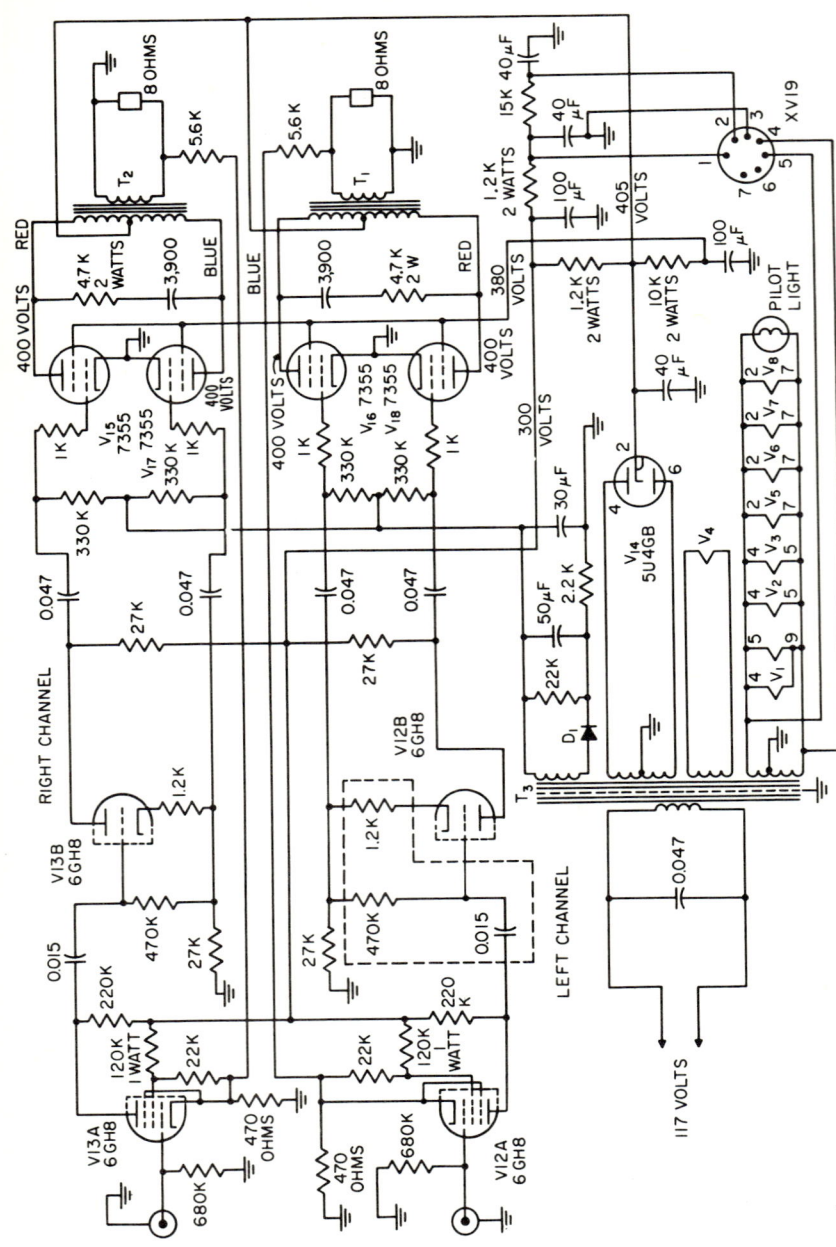

Fig. 108. Amplifier schematic. (*Reproduced by permission of General Electric Co. Audio Products Dept., Decatur, Ill.*)

17-88

AUDIO AMPLIFIERS 17-89

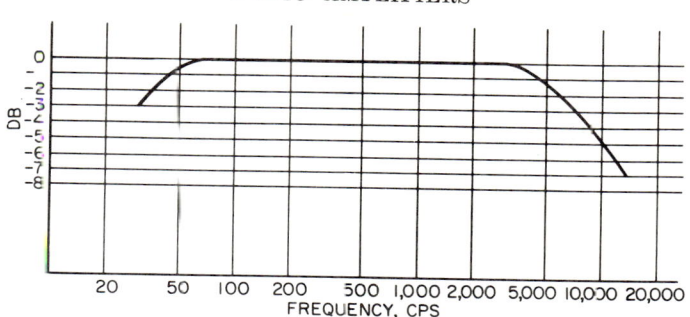

FIG. 109. 5 per cent distortion vs. frequency, 100-watt stereo amplifier. 0 db = 50 watts music power. Output transformer: 4,000 ohms plate-to-plate to 8 ohms, open-circuit primary impedance at 60 cps 50 volts rms = 20,000 ohms, short-circuit inductance referred to each primary half = 20 mh, 400-cps efficiency = 95 per cent. (*Reproduced by permission of General Electric Co. Audio Products Dept., Decatur, Ill.*)

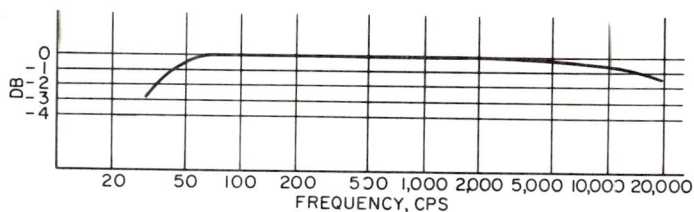

FIG. 110. Frequency response, improved output transformer. 5 per cent distortion output vs. frequency. 0 db = 50 watts music power. Output transformer: 4,000 ohms plate-to-plate to 8 ohms, open-circuit primary impedance at 60 cps 50 volts rms = 20,000 ohms, short-circuit inductance referred to each primary half = 5 mh, 400 cps efficiency = 95 per cent. (*Reproduced by permission of General Electric Co. Audio Products Dept., Decatur, Ill.*)

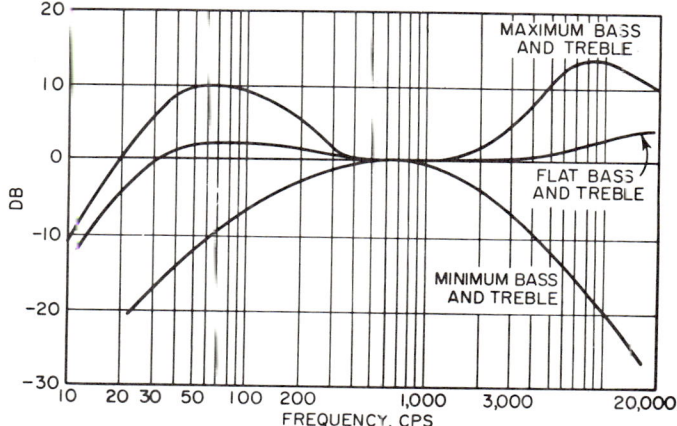

FIG. 111. Effects of tone controls, 100-watt stereo amplifier. (*Reproduced by permission of General Electric Co. Audio Products Dept., Decatur, Ill.*)

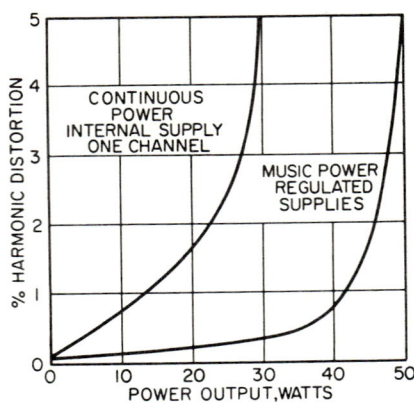

Fig. 112. Distortion vs. power output, 100-watt stereo amplifier. (*Reproduced by permission of General Electric Co. Audio Products Dept., Decatur, Ill.*)

17. TRANSISTOR STEREO AMPLIFIERS[5]

The following will describe a transistorized stereo system, incorporating several unusual features. In particular, this circuit permits direct excitation of the speaker without the necessity of an output transformer, thus saving considerably in cost, weight, and efficiency, and improving frequency response.

The unique arrangement of the output transistors is shown in Fig. 113. In effect, the two transistors are connected in series across the supply voltage E_S, with the voltage at the junction between the two set at the value E_{MP} by the voltage divider, R_1, R_2, R_3, and R_4. The input signals are imposed between points on this voltage divider and the respective bases, as shown, with opposite phase, in order to drive the transistors alternately. The collector-emitter voltages of the two transistors vary between a near-zero voltage and the supply voltage (at full swing), with opposite phase, as shown, and the transistors share the load current, each supplying alternate half-cycles.

One feature of this arrangement is that the peak collector-emitter voltage is equal to the supply voltage, rather than twice it, as in most class B amplifiers. Collector-emitter voltages of more than one-half the power supply voltage E_S are applied only when the device is reverse-biased, so that BV_{CEX}, which is normally much higher than BV_{CES}, is the pertinent parameter. The bias dividers are referenced to the collector of each transistor and are both a-c and d-c degenerative. The a-c degeneration lowers the input impedance and gain but also improves the d-c stability of the circuit.

The maximum continuous power output from the circuit is a function of the supply voltage E_S, the collector-emitter saturation voltage of the transistors $V_{CE\,sat}$, the series emitter resistors R_E, and the reactance of the coupling capacitor at the frequency in question and the load resistor. This can be expressed in the following form:

$$P_{out} = R_L \left(0.707 \frac{0.5 E_S - V_{CE\,sat}}{R_E + \sqrt{R_L^2 - X_{C_1}^2}} \right)^2 \qquad (77)$$

The idling current of each transistor is adjusted to approximately 35 ma. The forward bias applied to achieve this idling current overcomes the offset voltage of the base-emitter junction and minimizes the crossover distortion at low levels.

The importance of providing a low d-c resistance in the base circuit of the output stages cannot be overemphasized. This, and adequate heat-sinking of the output devices, is absolutely essential to realize the full reliability potential of a transistor-

---

[5] An Ultra-fidelity Transistorized Stereo Amplifier, *Motorola Bull.* 1-12-61-1M.

type amplifier. The requirement for low d-c resistance in the base circuit makes it practically mandatory to use a coupling transformer. The importance of the a-c source impedance has been emphasized in the section on class B amplifiers, where it was shown that there is an optimum source resistance from the standpoint of distortion. In this design it was found that a source impedance of about 20 ohms was correct.

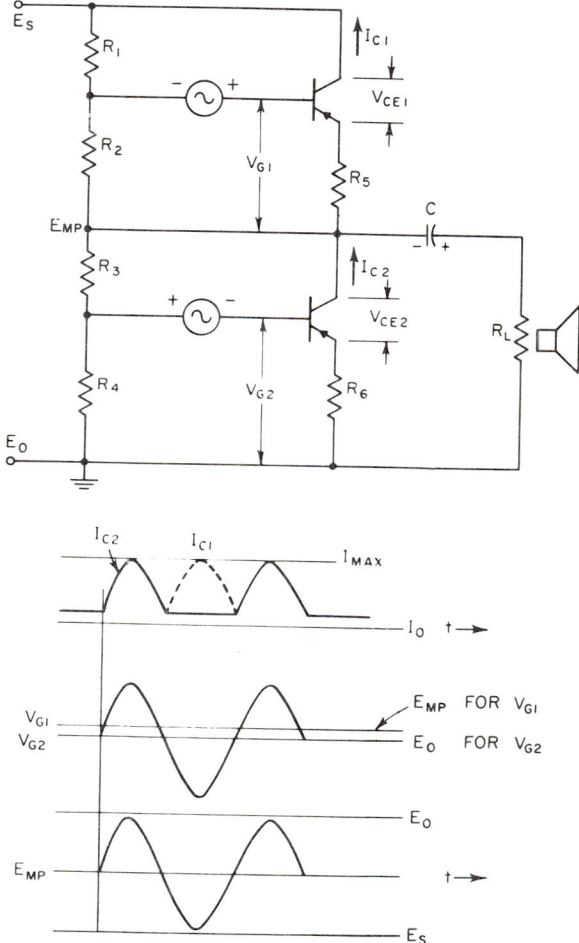

FIG. 113. Totem-pole operation. (*Reproduced with permission from D. W. Taylor, An All-transistor Ultra-fidelity Stereophonic Amplifier, Motorola Semiconductor Products Division.*)

Figure 114 shows the complete schematic of the output portion of this stereo amplifier. The preamplifier has been described elsewhere (see Sec. 11) and its schematic is shown in Fig. 46. The output stage of this preamplifier is an emitter follower. This supplies a signal to the input terminal B, the base of the driver transistor. This, too, is an emitter follower, which feeds the coupling transformer primary. Thus the source feeding this transformer is a cascaded emitter follower, which thus presents a very low impedance to the transformer, on the order of the desired 20-ohm input

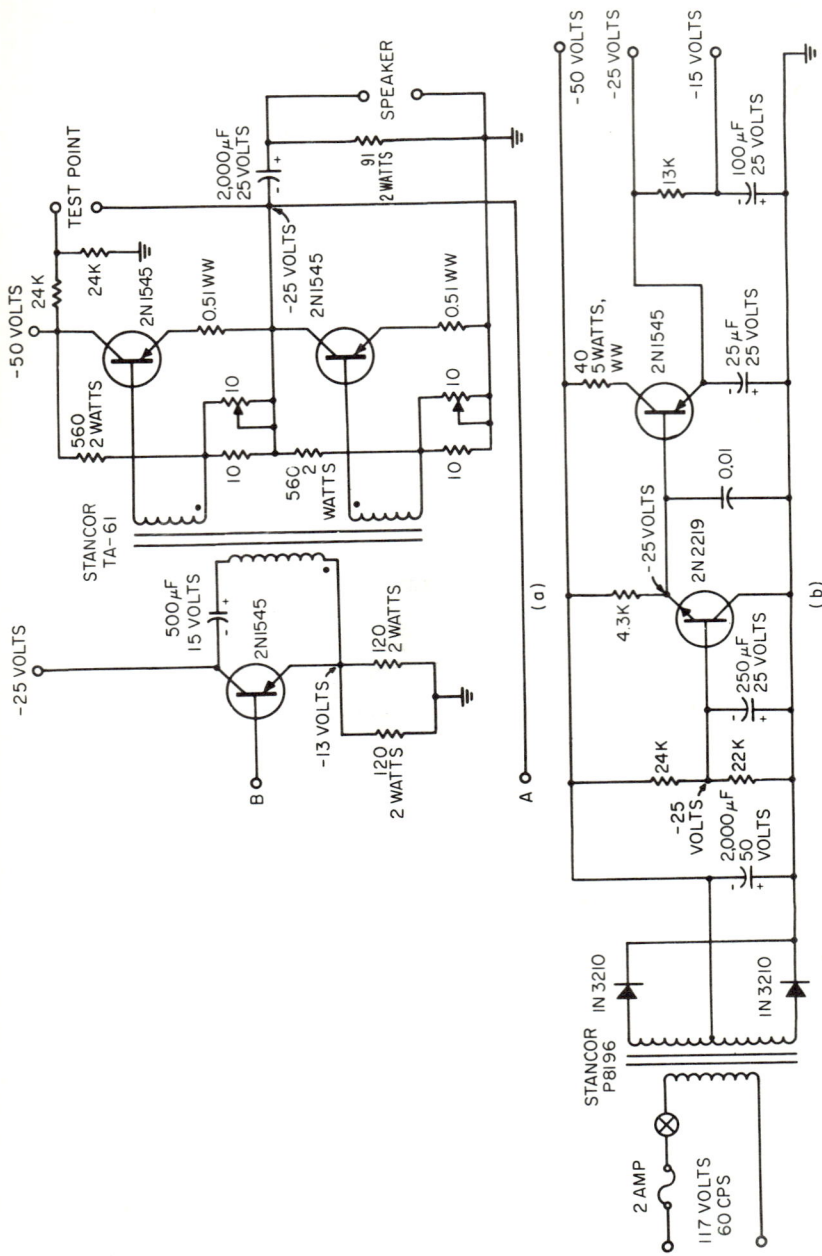

FIG. 114. Schematic of transistor stereo amplifier. (a) Power amplifier (two required). (b) Power supply. (*Reproduced with permission from Motorola Bull. 1-12-61-1M.*)

17-92

impedance for the output stages. The driver transformer can therefore be of a simple design, with a 1:1:1 ratio. Trifilar winding of a small number of turns of heavy wire produces a transformer with a vanishingly small amount of leakage inductance and a winding resistance under 2 ohms per winding. Capacitor coupling from the driver emitter to the transformer primary serves the dual purpose of blocking direct magnetizing current and assuring a very conservative d-c stability factor for the driver

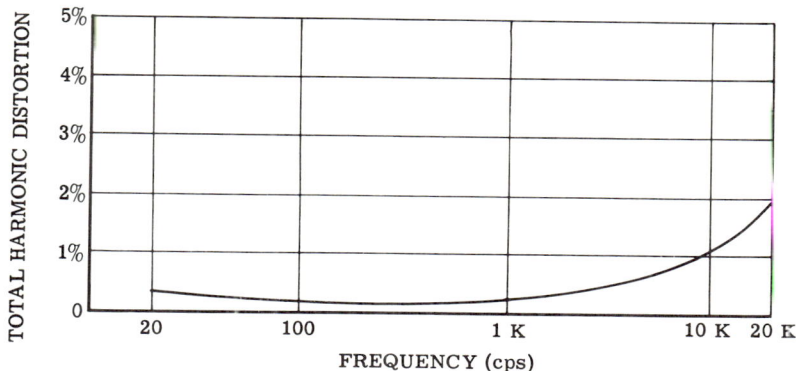

Fig. 115. Total harmonic distortion vs. frequency. Continuous 20-watt rms output, 8-ohm load, single channel. (*Reproduced with permission from Motorola Bull. 1-12-61-1M.*)

transistor. A series-resonant circuit is formed by the coupling capacitor and the transformer inductance at approximately 40 cycles. While this is well damped by the emitter resistors it produces a slight rise in the open-loop response curve. There is no other a-c coupling within the feedback loop; therefore, a wide a-c stability margin exists at low frequencies and the response rise is removed by feedback.

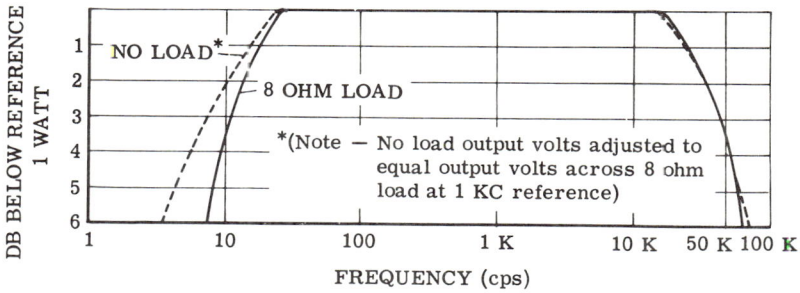

Fig. 116. Amplifier frequency response 1-watt level. (*Reproduced with permission from Motorola Bull. 1-12-61-1M.*)

Figure 115 shows the total harmonic distortion of this amplifier as a function of frequency. Figure 116 shows the frequency response with no load and with an 8-ohm load. Figure 117 shows the available power output as a function of frequency, and indicates that the full power output is available over the greater part of the useful range.

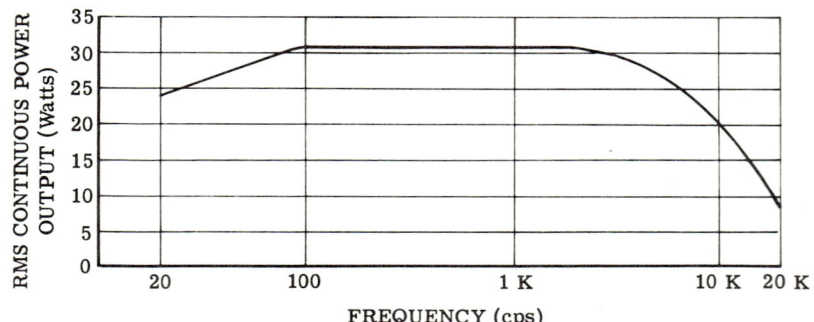

Fig. 117. Power bandwidth for 1 per cent total harmonic distortion, single channel—8-ohm load. (*Reproduced with permission from Motorola Bull. 1-12-61-1M.*)

The specifications of this amplifier are as follows:
Rated power output: 60 watts (125 watts peak), 30 watts per channel.
Maximum (rms) continuous power output for two channels and 1 per cent total harmonic distortion at 1 kc:

4 ohms	72 watts
8 ohms	53 watts
16 ohms	33 watts

Maximum (rms) continuous power out, for one channel and 1 per cent total harmonic distortion at 1 kc:

4 ohms	45 watts
8 ohms	31 watts
16 ohms	18½ watts

Peak music power out, for two channels and 1 per cent total harmonic distortion:

4 ohms	180 watts
8 ohms	128 watts
16 ohms	72 watts

Frequency response: +0, −1 db from 17 cps to 24,000 cps (1-watt level).
IM distortion: <0.6 per cent (60 cps and 7 kc, 4:1 ratio, 20 watts rms into 8-ohm load).
Hum and noise (gain adjusted for 30 watts output into 8 ohms): Magnetic, 76 db below 30 watts. Ceramic 74 db below 30 watts.
Sensitivity [(rms) for 30 watts output, 8-ohm load, 1 kc]: Magnetic input, 2.25 mv. Ceramic input, 45 mv. High auxiliary, 450 mv. Low auxiliary, 45 mv.
Feedback: 12-db gain reduction at 1 kc.
A-C stability margin: >20 db (8-ohm resistive load).
Output impedance:

20 cps	2.4 ohms
1 kc	2.2 ohms
20 kc	2.2 ohms

Tone-control range, maximum boost and attenuation (referenced to 1 kc): Bass control, ±18 db (at 20 cps). Treble control, ±18 db (at 20 kc).
Equalization: Magnetic, per RIAA curve for 500-mh cartridge with recommended load impedance of 50 kilohms. Ceramic, per RIAA curve for 400-pf cartridge (input impedance approximately 2 megohms).

While the preceding description has illustrated one design of transistorized stereo amplifier many other excellent combinations are possible; in fact, the combinations are almost unlimited. For example, the preamplifier described above, or any of the others described in Sec. 11, could be used to drive the high-power transistor amplifier described in Sec. 15.3, with resultant power output of up to 250 watts.

BIBLIOGRAPHY

Abel, A. D.: Power Transistor Cooling, *Honeywell Application Lab. Rept.* ALR-3, Jan. 4, 1962.
Bendix Aviation Corp.: Transistor Circuits I, *Application Note* 1-03159.
Ekiss, J. A.: Transistor AC Amplifiers with High Input Impedance, *Philco Application Lab. Rept.* 622, October, 1961.
Ekiss, J. A.: A High Input Impedance AC Transistor Amplifier Utilizing the 2N1428 and 2N1429, *Philco Application Lab. Rept.* 641.
"General Electric Transistor Manual," 6th ed., pp. 113–140, 1962.
Greiner, R. A.: "Semiconductor Devices and Applications," McGraw-Hill Book Company, New York, 1961.
Henney, K.: "Radio Engineering Handbook," 5th ed., chap. 15, McGraw-Hill Book Company, New York, 1959.
Hunter, L. P.: "Handbook of Semiconductor Electronics," 2d ed., chap. 11, McGraw-Hill Book Company, New York, 1962.
Jones, D. V.: Efficient High-quality Program Amplifier Circuits Using the Industrial Silicon Series 2N2107, 2N2108 and 2N2196, *General Electric Co. Application Note* 90.3, April, 1962.
Linvill, J. G., and J. F. Gibbons: "Transistors and Active Circuits," McGraw-Hill Book Company, New York, 1961.
Lostetter, J. A.: Two-stage Direct-coupled 5-watt Class-A Power Amplifier, *Honeywell Application Note* AN2A, July 25, 1960.
Lostetter, J. A.: A Solid State 125-watt Linear Power Amplifier, *Honeywell Application Lab. Rept.* ALR-1, October, 1961.
"Motorola Power Transistor Handbook," 1st ed. pp. 45–100, 1961.
Shea, R. F.: "Principles of Transistor Circuits," John Wiley & Sons, Inc., New York, 1953.
Shea, R. F.: "Transistor Audio Amplifiers," John Wiley & Sons, Inc., New York, 1955.
Shea, R. F.: "Transistor Circuit Engineering," John Wiley & Sons, Inc., New York, 1957.
Sherwood Electronics Laboratories, Inc., Chicago, Ill., Operating Instructions for Model S-5000 80 Watt Stereo Dual Amplifier-Preamplifier.
Starke, H. F.: Transistorized Preamplifiers, *Raytheon Tech. Inform. Bull.* TIS-117-T, May, 1959.
Taylor, D. W.: An All-transistor Ultra-fidelity Stereophonic Amplifier, Motorola Semiconductor Products, Inc., Phoenix, Ariz.; also *Bull.* 1-12-61-1M.
Texas Instruments, Inc.: "Transistor Circuit Design," pp. 197–237, McGraw-Hill Book Company, New York, 1963.
Westinghouse Electric Corporation: Silicon Power Semiconductors, Heat Dissipation Considerations, *Tech Tip* no. 14, May 1, 1962.

Chapter 18

D-C AMPLIFIERS

PAUL J. BÉNÉTEAU* and EDWARD G. NIELSEN†

CONTENTS

1. Direct-coupled Amplifiers.................................... 18-2
2. Single-ended Electron-tube Amplifiers....................... 18-3
   2.1. Interstage-coupling Methods........................... 18-3
   2.2. Level Controls.. 18-5
   2.3. Gain Controls... 18-5
   2.4. Single-stage Amplifier................................ 18-6
   2.5. Drift... 18-6
   2.6. Typical Design Example................................ 18-7
3. Direct-coupled Transistor Amplifiers........................ 18-8
   3.1. General... 18-8
   3.2. Various Circuit Configurations........................ 18-8
   3.3. Study of the Drift of a Single Stage.................. 18-10
   3.4. Study of the Gain Stability of a Single Stage......... 18-12
   3.5. Typical Design Examples............................... 18-12
4. Differential Electron-tube Amplifiers....................... 18-14
   4.1. General... 18-14
   4.2. Various Circuit Configurations........................ 18-14
   4.3. Drift Considerations.................................. 18-15
   4.4. Common-mode Rejection................................. 18-16
   4.5. Typical Design Example................................ 18-16
5. Differential Transistor Amplifiers.......................... 18-17
   5.1. General... 18-17
   5.2. Various Circuit Configurations........................ 18-17
   5.3. Analysis of Various Stages............................ 18-22
   5.4. Parameters Which Affect Drift......................... 18-22
   5.5. Gain Stability.. 18-23
   5.6. Common-mode Rejection................................. 18-24
   5.7. Typical Designs....................................... 18-24
6. Chopper and Modulated-carrier Amplifiers.................... 18-25
7. Input and Output Switching Circuits......................... 18-27
   7.1. Capacitor-coupled Input Choppers...................... 18-27

* Società Generale Semiconduttori, s.p.a., Agrate (Milano), Italy.
† General Electric Co., Electronics Laboratory, Syracuse, N.Y.
Sections 1 to 5 and 10 were written by P. J. Bénéteau, and Sections 6 to 9 were written by E. G. Nielsen.

18-2   AMPLIFIER CIRCUITS

>   7.2. Transformer-coupled Input Choppers.................... 18–33
>   7.3. Output Switching Circuits............................. 18–39
> 8. Modulator-demodulator Performance........................ 18–43
> 9. Types of Modulators....................................... 18–47
>   9.1. Electromechanical Choppers............................ 18–47
>   9.2. Transistor Choppers................................... 18–49
>   9.3. Diode Choppers.. 18–58
>   9.4. Variable-capacitance Diode Modulators................. 18–61
>   9.5. Vibrating-reed Capacitance Choppers................... 18–64
>   9.6. Photoconductor Choppers............................... 18–65
>   9.7. Hall-effect Modulators................................ 18–68
> 10. Field-effect Transistor D-C Amplifiers.................... 18–69
>   10.1. General.. 18–69
>   10.2. Important Points for D-C Amplifier Design Using FET's. 18–69
>   10.3. Typical Design Example............................... 18–70

1. DIRECT-COUPLED AMPLIFIERS*

Direct-coupled amplifiers are those whose frequency response extends from zero frequency to some upper limit. They are widely used in many types of instrumentation, analog computers, power supplies, and control equipment where amplification of very slowly varying signals may be required. Vacuum tubes have previously been used almost exclusively in these types of amplifiers with or without mechanical choppers, but the rapid advance of transistor technology has tended to emphasize the use of semiconductor components in new designs. However, there are still some areas of application, notably in high-impedance amplifiers, where the use of vacuum tubes is warranted, although further development of high-input-impedance semiconductors may remove many of these difficulties.

Vacuum tubes have the important relative advantage that, with high source impedances, the drift at the input of an amplifier caused by the flow of current through the source impedances into the input terminals is very much reduced; i.e., grid current is less than base current plus I_{CBO}. Secondly, the characteristics of a vacuum tube are much less temperature-sensitive than those of a transistor.

These advantages are counterbalanced by the dependence of drift on heater voltage, tube aging, lower reliability, interstage coupling difficulties, and, of course, all the usual disadvantages exhibited in conventional electronic circuits employing tubes as compared with transistors. In the discussion that follows, both tubes and transistors will be considered.

The most difficult problems encountered in d-c amplifiers are the drift and gain variations which occur as functions of temperature and other parameters. Gain variations can, of course, be practically eliminated by feedback arrangements, but operating-point instability poses a more difficult problem. Since a displacement of the operating point cannot be distinguished from the application of a signal at the input, it is thus essential to keep both the gain and the operating point as constant as possible. In cases where this cannot otherwise be done with sufficient precision, a circuit arrangement with a chopper is used. The chopper can either be used to modulate the signal, which is then amplified in an a-c and therefore low-drift amplifier,[1] or it can be used in a stabilizing arrangement with a separate amplifier as first proposed by Goldberg.[2] These techniques are discussed in Secs. 6 to 9.

The purpose of this chapter is, first, to discuss straight d-c amplifiers using vacuum tubes or transistors. The drift-determining parameters will be considered in turn as will gain stability. Balanced arrangements which reduce the drift to the difference

---

* The suggestions and experimental work of Messrs. Alderisio, Riva, Murari, and Quinzio of Società Generale Semiconduttori helped greatly in the preparation of Secs. 1 to 4 and 10.

of two parameters will then be studied, and practical circuit-design information and examples will be given.

2. SINGLE-ENDED ELECTRON-TUBE AMPLIFIERS

The design of single-ended d-c amplifiers employing electron tubes poses several difficult problems: the choice of the interstage coupling method, level and gain controls, and the controlling of the different factors influencing drift, namely, supply and heater voltage variation, grid current, and component aging. These will now be considered.

2.1. Interstage-coupling Methods. There are many methods used for interstage coupling. Among these can be listed:

1. Normal plate-grid coupling using a progressively higher supply voltage. This can consist of separate supplies or, more conveniently, can be a resistive divider as shown in Fig. 1. A certain amount of feedback is caused by this arrangement, and care should be taken that the amplifier is not unstable. The separate power supply method is preferable in this regard although it too can cause feedback difficulties,

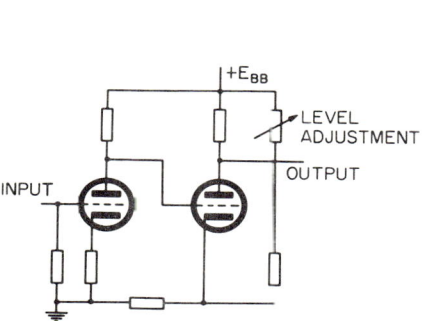

Fig. 1. Normal interstage coupling using progressively higher power supplies.

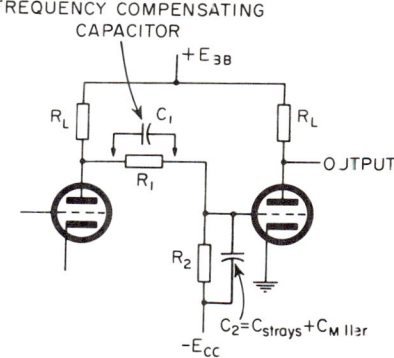

Fig. 2. The use of a negative supply for interstage coupling.

particularly in the frequency region in which the power supply output impedance begins to rise; this can occur at relatively low frequency where the amplifier gain may still be very high. Generally method 1 is not recommended.

2. The use of a negative supply. This can be applied in several ways, one of which is shown in Fig. 2. This arrangement has the important advantage that a phase-advance network is readily available by inserting a capacitor C_1 across the resistor R_1 greater than that required to compensate the phase retard caused by the R_2C_2 combination. The disadvantage is, of course, the loss of voltage gain due to the $R_2/(R_1 + R_2)$ attenuation.

3. The use of coupling devices having high d-c resistance and low a-c resistance, e.g., thyrites, gas tubes, or zener diodes. Such a circuit is shown in Fig. 3. The system with a gas tube has the advantage of having little or no gain loss since the gas tube maintains an essentially constant voltage drop and transmits to the cathode, unattenuated, any level change appearing at the anode.

Some serious disadvantages with this scheme are the noise levels generated, which are generally very high, the dependence of tube characteristics on external parameters such as the light level, and the fact that the tube must carry a minimum current to maintain a reasonable voltage stability. Either the load and grid resistors can be made sufficiently small to accommodate the current required, and thus reduce the gain, or a system can be used such as shown in Fig. 4 in which a cathode follower is placed between the gain stage and the gas tube. This eliminates the difficulty of gain loss but at the expense of an extra stage.

4. The use of a pentode with the input on the screen grid rather than on the control grid as shown in Fig. 5. This is an attractive method with the obvious advantage of

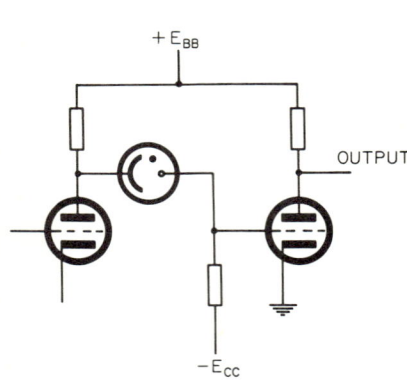

Fig. 3. The use of a gas tube for interstage coupling.

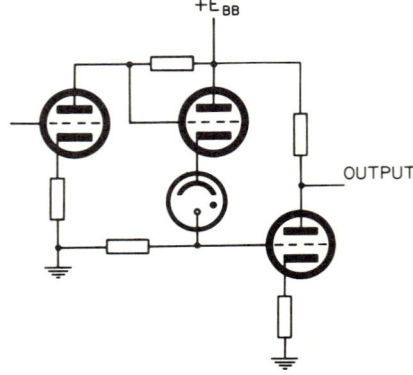

Fig. 4. The use of a cathode follower to improve the performance of the gas-tube circuit.

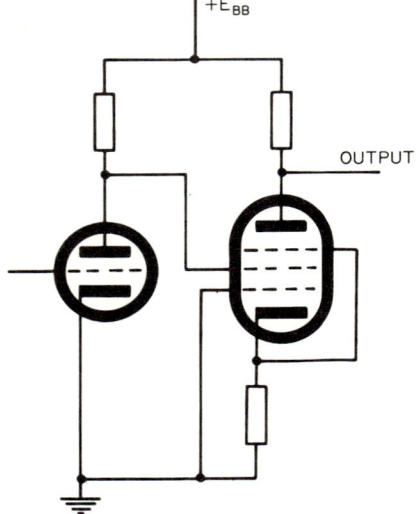

Fig. 5. Screen-grid coupling.

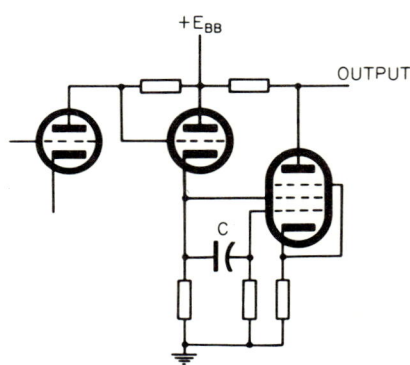

Fig. 6. The use of a cathode follower to drive the screen grid.

simplicity; the disadvantages are those of gain loss due to both the very low screen g_m of a pentode and the lower input impedance which tends to load down the preceding stage. A cathode follower is frequently inserted between the driver and the pentode in order to remove this latter difficulty. With this circuit, shown in Fig. 6, a frequency-compensating capacitor C can be introduced between the screen and control grids, to divert part of the signal at high frequency to the control grid and thus

D-C AMPLIFIERS

increase gain. Overshoot in the pulse response may of course occur, depending on the corner frequency chosen.

2.2. Level Controls. In a single-ended amplifier, it is frequently required to vary the d-c level of the amplifier to compensate for the parameter variations without varying the gain. This can be done in several ways, depending upon the type of coupling network used.

1. In the coupling method utilizing a resistive divider to boost the successive power supplies, the cathode voltage of the second stage can be varied by means of the variable series resistance, as shown in Fig. 1.

2. In the method utilizing a negative supply, the level can be varied by adjusting $-E_{CC}$ or alternatively by replacing R_2 of Fig. 2 by a triode whose current is varied by means of a potentiometer in the grid circuit such as shown in Fig. 7. A further advantage of this method is that the signal is attenuated less than in the resistive case.

3. In the gas-tube case, a level control can be made by a potentiometer connected across the tube in the manner shown in Fig. 8. Since a level shift on the anode of

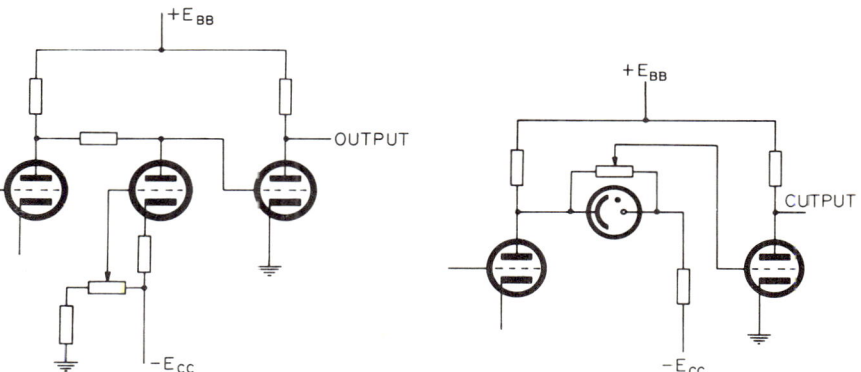

FIG. 7. A level control.　　FIG. 8. A level control for use with a gas tube.

the tube is automatically transmitted to the cathode without attenuation, the wiper of the potentiometer always receives the level change appearing on the anode, and therefore no gain change results at the different level settings.

4. In the screen-grid case, a level setting can be obtained by inserting a potentiometer in the control grid and thus controlling the plate current. The g_m will change to some extent using this system and the gain will not be completely constant but may be sufficiently so if a linear portion of the tube characteristic is utilized.

2.3. Gain Controls. Gain controls should be located in the amplifier such as not to change the d-c level with different gain settings. In general this means that no voltage difference should exist between the ends of any gain-control potentiometer in the absence of any input signal. In cases where the d-c level changes with gain setting, the level can be reset to the desired value with the level control; this, however, is an undesirable feature, particularly in a tube amplifier where frequent adjustments may be required because of aging. Some possible circuit arrangements are:

1. In the case where any point of the amplifier is at zero volts without an input, a potentiometer can be inserted between that point and ground and the wiper taken to the next stage. This can be done conveniently on the input grid, although for low-level signals noise and drift considerations may require that the gain be controlled at a higher level farther in the amplifier.

2. Most other methods use two potentiometers in which one sets zero voltage difference between the ends of the gain-control potentiometer, such as shown in Fig. 9a and b, while the other serves as a normal gain control.

18-6 AMPLIFIER CIRCUITS

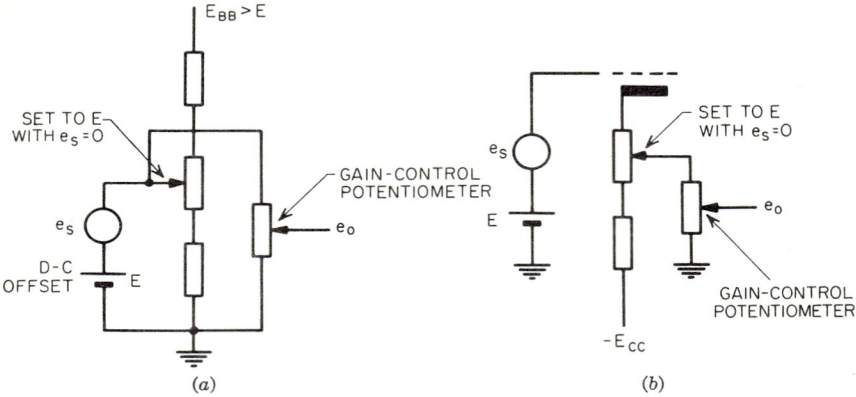

FIG. 9. (a) Gain control where input steady-state level is not zero. (b) Alternate arrangement.

2.4. Single-stage Amplifier. A single-stage triode amplifier is shown in Fig. 10. The equation for the output voltage e_o can be written as

$$e_o = E_{BB} - (e_s - i_g R_g)[\mu R_L/(R_L + r_p + \mu R_k)] \qquad (1)$$

where r_p is the tube plate resistance, μ is the amplification factor, i_g is the (negative) grid current, and R_L, R_k, and R_g are the plate, cathode, and grid resistors, respectively. The voltage gain, excluding the $i_g R_g$ term, is

$$A_v = \mu/(1 + r_p/R_L + \mu R_k/R_L) \qquad (2)$$

Thus the r_p/R_L and R_k/R_L ratios should be small, as should I_g and R_g, while the μ should be high. The limit on R_L is that as it becomes large to fulfill the conditions of Eq. (2), the plate current falls, r_p rises, and μ falls a little. There is thus an optimum R_L/r_p ratio depending on the tube used. Increasing R_L by increasing E_{BB} while maintaining the same plate current also increases the gain. It should be noted that the drift due to $i_g R_g$ cannot be distinguished from the signal, very much the same as in a transistor amplifier. This is discussed below in Sec. 2.5.

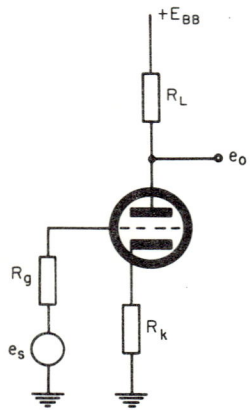

FIG. 10. A single stage.

2.5. Drift. Drift is caused in a vacuum-tube amplifier by a variety of causes including tube and component aging, grid current, and variation of heater and B+ power supplies. Each of these causes can be quite serious; and in such amplifiers, where good drift characteristics are required, differential chopper-stabilized types[2] are usually used.

Filament-voltage Variations. The rough figure that has been widely used for the equivalent input voltage variation due to a heater supply change of +10 per cent is +100 mv; i.e., changing the filament voltage from 12.0 to 13.2 volts has the same effect as increasing the signal on the grid by 100 mv. This, of course, is a serious cause of drift, and either a differential amplifier or a compensation scheme is normally used; the obvious solution of a closely regulated heater supply is not normally very attractive. Various compensation methods are possible;[3,4] a common one is shown in Fig. 11 in which the compensating tube's plate-current change due to a filament-voltage change results in a different grid-cathode potential for the amplifier tube which can improve the drift by a factor of about 25; the plate resistor of the compensating tube is inserted to maintain equal plate potentials and thus to improve the tracking of the pair. The potentiometer tap should be at about one-third from the top.[3] The use of two tubes in the same envelope reduces the differential equivalent voltage drift due to the likelihood of the tubes being similar. Gray[3] has measured for individual tubes an equivalent input drift of about 200 mv, 14 mv (15 times improvement) for pairs picked randomly and 8 mv (25 times improvement) for pairs in one envelope, all for 20 per cent change in heater voltage. Modern tubes probably can increase this match still further.

Power-supply Variation. The variation of d-c level with power-supply variation is such that closely regulated supplies are usually required. A change of ΔE_{BB} appears on the plate of the input tube attenuated by a factor $r_p/(R_L + r_p)$. For example, a 1 per cent change in a 250-volt supply might appear as a 1-volt change on the first plate, which may be serious depending on the gain following the input stage. In a differential amplifier, both plates go up or down together and much less effect is observed.

Grid Current. The drift due to grid-current flow is precisely analogous to that found in a transistor; although the grid current of a tube is, of course, much smaller than the base current of a transistor, the effect on drift is frequently as serious because of the higher grid impedances.

The grid current of a vacuum tube is a function of the tube and of the grid-cathode bias. Generally, with a positive bias, the grid current is positive and increases very rapidly; and as the bias becomes negative, the grid current decreases below zero until the grid-cathode voltage is approximately -2 volts; at lower biases, the current is negative and its magnitude increases slowly. Therefore, in a low-level amplifier, the input tube should be biased at about -2 volts to reduce this grid-current variation; the initial difference can be compensated by the level control. The effect, of course, is directly proportional to the size of the grid-leak resistor, and if it can be made smaller without undue loss of gain, this should be done. Electrometer tubes have extremely low leakage currents on the order of 10^{-4} μμa, and should be used if drift due to grid leakage is very serious, i.e., with very high impedance sources.

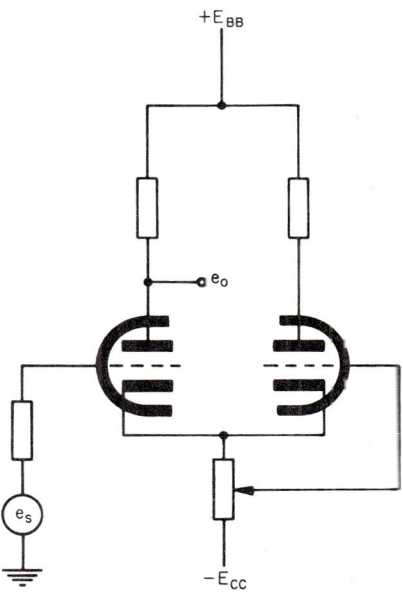

FIG. 11. A compensation method for filament supply variation.

2.6. Typical Design Example

Design Example. A d-c amplifier input stage is to be built with a voltage gain of about 25 to handle an input signal of about 100 mv. The source resistance is 10 kilohms.

Solution. Use one-half of a 5751 tube, a reliable high-gain double triode. For low grid current use a V_{gk} of -1.5 volts; select a 250-volt supply for reasonably high gain and a load of 240 kilohms. According to the tube characteristics, the plate current is 0.55 ma and therefore $R_k = 1.5/0.55 = 2.7$ kilohms. The r_p has started to rise at that current but the curve shows it is still sufficiently low to preserve high gain. The grid current is specified as 0.4 μa maximum at $V_{gk} = -3$ volts. At -1.5 volts it will be less than this, but to make the drift due to this current no more than 100 mv at the input, make $I_g R_g = 50$ mv maximum. Thus $R_g = 50$ mv/0.4 μa = 125 kilohms. Use $R_g = 100$ kilohms. If this possible drift is excessive a tube should be selected with a lower maximum grid current or further gain loss should be taken by making the grid resistor still smaller. The gain can be

18-8   AMPLIFIER CIRCUITS

calculated from Eq. (2), and this gives, for $\mu = 66$ and for $r_p = 70$ kilohms,

$$A_v = -66/[1 + \tfrac{70}{240} + (66)(2.7)/240] = -32.5$$

Considering the loading of the input, this is reduced a further 10 per cent.

The other important point is the drift due to the power supply and to the heater voltage change. Regarding the power supply, the output voltage is about (30) (100 mv) = 3 volts, and since r_p = 70 kilohms and R_L = 240 kilohms, this is equivalent to a supply variation of $3(R_L + r_p)/r_p$ or 13.3 volts. Thus the regulation required from the power supply will have to be calculated depending on the permissible drift; a 1 per cent regulated supply will give a possible drift of about 18.8 per cent. As regards the heater supply problem, either it will have to be held to better than 5 per cent, which would give an input drift of less than about 50 mv, or the second half of the tube can be used to compensate the first half as explained in Sec. 2.5, method 2.

Many other examples are given elsewhere;[3-4] Valley and Wallman[3] is particularly useful and includes much specific data on tube characteristics.

3. DIRECT-COUPLED TRANSISTOR AMPLIFIERS

3.1. General.   Direct-coupled transistor amplifiers present many advantages and disadvantages relative to vacuum-tube types. The main advantages are the relative ease of interstage coupling due to the availability of *n-p-n* and *p-n-p* transistors and zener diodes and, secondly, the much better performance that can be obtained, at least for balanced types with low source impedance, using the later types of silicon planar transistors. The main disadvantage is the much greater temperature dependence of the transistor parameters, notably current gain h_{FE}, leakage current I_{CBO}, and base-emitter voltage V_{BE}. In the early types of transistors, mostly germanium alloy, these characteristics were unstable and unpredictable. Silicon grown-junction devices were generally much better in most characteristics. Subsequently, many types of planar transistors presented vast improvements in better characteristics, such as extremely low I_{CBO}, very well matched parameters such as V_{BE} tracking between two transistors, and reliability. Furthermore, the availability from several manufacturers of two or more separate transistors on one header has greatly simplified the problem of keeping the transistors at the same temperature; thus, in high-performance amplifiers, differential input stages are generally used in order to reduce the drift to the difference of the parameters of two devices which can be kept very small using modern components.

Various configurations will be studied in this section which can be used either in input or in interstage circuits. The equation for a single stage will be worked out and the factors that influence drift and gain stability will be discussed. Typical circuits will be designed. Discussions of balanced stages will be deferred until Sec. 5.

3.2. Various Circuit Configurations.   A very serious problem in direct-coupled vacuum-tube amplifiers is the progressively higher voltage that appears on the plate of each succeeding tube. Thus the output voltage level is higher than the input, making feedback difficult.

An obvious configuration that can be used in transistor circuits to overcome this difficulty is the cascading of *n-p-n–p-n-p* transistors, as shown in Fig. 12, usually with feedback applied from output to input to stabilize the operating point and the gain.

In this manner the output can be designed for a zero volt level and thus feedback can be applied. If only two power supplies are available, it is difficult to design this circuit to get a reasonable gain, because, since the input resistance of each stage is high owing to the emitter resistance, the voltage gain of each stage is given approximately by the ratio of collector to emitter resistors and is thus very small. A typical design is shown in Sec. 3.6. Furthermore, germanium *n-p-n* and silicon *p-n-p* transistors have always been more difficult to make than their counterparts and are not generally of so high performance.

Another configuration that is very practical is that shown in Fig. 13 in which the required voltage drop is effected by a zener diode, similar to the gas-tube circuit discussed in Sec. 2. This has the advantage that good voltage gain can be obtained

D-C AMPLIFIERS

and the use of *p-n-p* silicon or *n-p-n* germanium transistors can be avoided. Zener diodes are somewhat noisy, particularly at low currents, and this may be a disadvantage; this situation can be improved by placing a resistor of about 1 kilohm between the base and emitter of the transistor following the zener which increases the current through it and thus removes it from the noisy knee region. In this

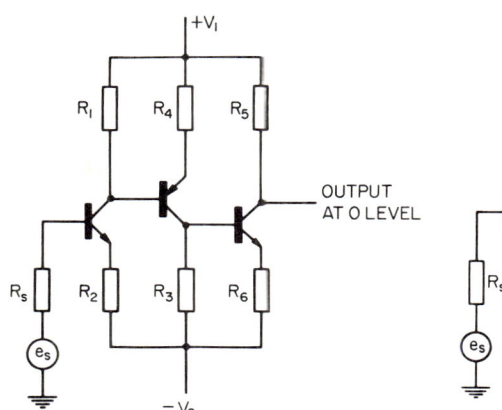

FIG. 12. A circuit configuration capable of overall feedback.

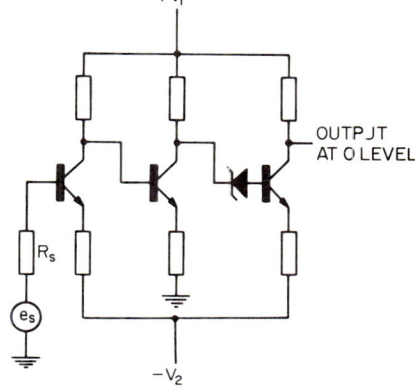

FIG. 13. Another possible configuration using a zener diode.

manner, the gain of the circuit is only slightly diminished because of the low dynamic impedance of the base-emitter diode. Another disadvantage if feedback is not to be used is the tolerance of the zener voltage; it may be that in some cases a diode would have to be used with, say, a 5 per cent tolerance in order to keep the output level near the desired value. Finally the temperature coefficient of the zener voltage may cause some difficulties, but with very careful design a zener with a positive coefficient can be used to compensate the negative V_{BE} coefficient of the transistor. A typical design is given in Sec. 3.6.

A very simple configuration is shown in Fig. 14 where the transistors, all *n-p-n*, are directly coupled. The V_{BE}'s of the second and third transistors serve as the V_{CE}'s of the first and second ones, respectively. The operation of the transistors not too far from the saturation region makes their gain smaller and their distortion higher, but this is partially offset by the simplicity of the circuit and the fact that practically all the signal current is inserted into the next stage instead of being lost in the load resistor, as was the case for the other circuit shown. Finally

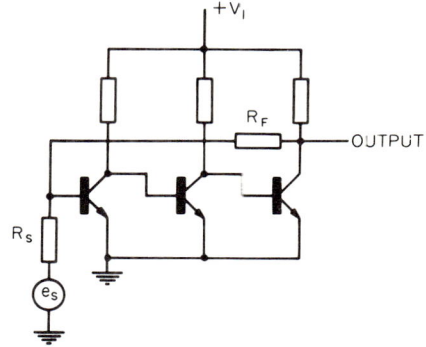

FIG. 14. A simple configuration.

this circuit depends on the $V_{BE\ sat}$ being higher than the $V_{CE\ sat}$, which of course depends on the type of transistor used and the temperature range involved since they have opposite temperature coefficients; this is most practical for silicon transistors, particularly epitaxial types. The voltage level of the output collector is approximately $V_{3E\ sat}$ times R_F/R_s because of the voltage-divider action of those two resistors and is thus temperature-sensitive. However, the voltage gain is very high and the circuit can be useful in an operational amplifier. A typical design is described in Sec. 3.6.

Another possibility which is considerably more complicated but practical in many cases is to replace the load resistors of the transistors of Fig. 12 with constant-current sources as shown in Fig. 15. In this manner, as in the previous case, almost all the signal current is diverted into the base of the next stage and thus is amplified, rather than being lost in the load resistor; furthermore the transistor can be operated in a high-gain low-distortion region. The impedance viewed from a collector can be several megohms, which is obviously negligibly large compared with the following transistor's input impedance, and thus the voltage gain becomes very high. It is to be noted that the voltage gain of the stages can be so high that the constant-current sources can be easily saturated because their V_{CE} is reduced to a very small value. The

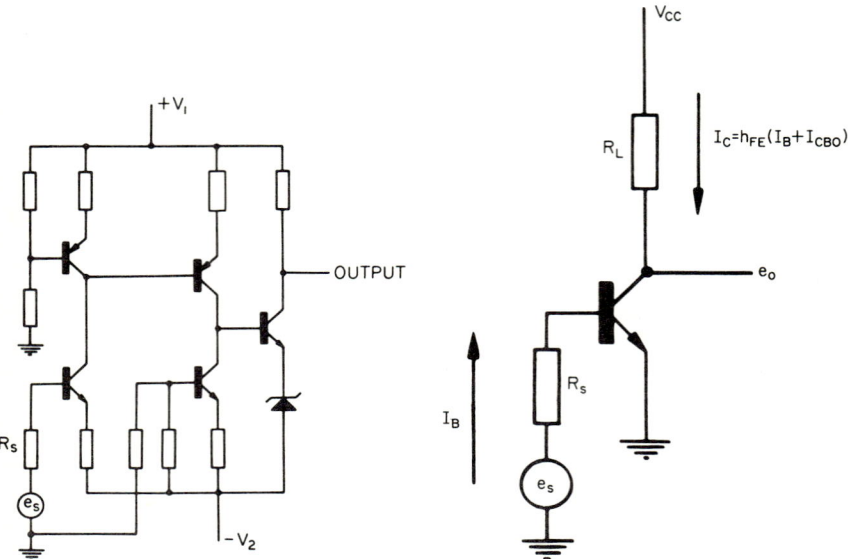

FIG. 15. A more complicated circuit using constant-current sources instead of collector resistors.

FIG. 16. A single stage.

effective impedance on the collectors should be carefully calculated to preclude this possibility.

Many other circuits are possible, although probably the most important ones have been described. Others will be shown in Sec. 5, and several have been discussed by Chaplin and Owens.[5,6]

3.3. Study of the Drift of a Single Stage. A single n-p-n stage is shown in Fig. 16 where the emitter resistance is assumed zero. Normally, large emitter resistors cannot be used in d-c amplifiers because of their degeneration of voltage gain; thus the reduction of the effect of varying I_{CBO} and h_{FE} normally effected in an a-c amplifier by a bypassed emitter resistor and small base-biasing resistors is not possible in a d-c amplifier.

The equation for the output voltage as a function of the input voltage can be written as follows:

$$I_B = (e_s - V_{BE})/R_s \tag{3}$$
$$I_C = h_{FE}(I_B + I_{CBO}) \tag{4}$$
$$e_o = V_{CC} - I_C R_L \tag{5}$$

Solving for e_o gives

$$e_o = (V_{BE} - e_s)(R_L h_{FE}/R_s) - I_{CBO} h_{FE} R_L + V_{CC} \tag{6}$$

D-C AMPLIFIERS   13-11

It is thus seen that
1. A change in V_{BE} is indistinguishable from a change in the signal voltage e_s.
2. A term $I_{CBO}h_{FE}R_L$ appears which is also indistinguishable from e_s.
3. The three parameters which affect the output voltage e_o other than, of course, e_s and V_{CC} are V_{BE}, h_{FE}, and I_{CBO}. We shall now examine each parameter in detail.

V_{BE}. This parameter depends on emitter current[7] and, more important, falls with temperature at the approximate rate of 2 mv/°C. In a normal d-c amplifier built with silicon planar transistors, this usually becomes the most important parameter and balanced arrangements are used to reduce its effect by a factor of about 2 orders of magnitude (see Sec. 5). In an amplifier built with germanium transistors, this will not always be the most important parameter depending on the temperature range and the source resistance involved, as the I_{CBO} term may become very important.

In an unbalanced amplifier, where parameter matching is unimportant, there is no advantage in using any particular type of transistor for V_{BE} reasons because all types exhibit about the same behavior.

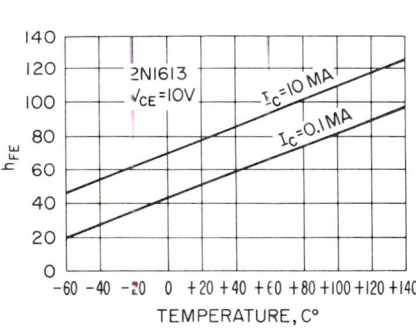

FIG. 17. Typical h_{FE} variation as a function of temperature.

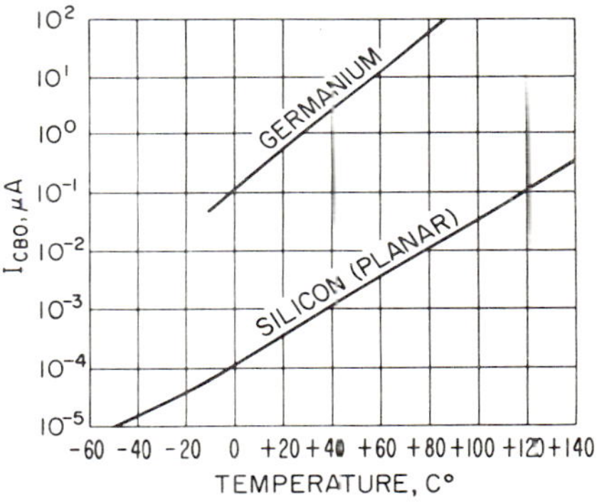

FIG. 18. Typical I_{CBO} behavior for germanium and silicon transistors.

h_{FE}. This parameter also depends on emitter current in a calculable manner[8] and varies widely with temperature. The temperature coefficient of the variation depends on the current level involved, the temperature range, and the type of transistor used. The variation at very low currents (1 to 100 μa) over a temperature range (say −55 to +125°C) is usually more unpredictable than that at a higher current level and over a narrower temperature range. A curve of h_{FE} variation is shown in Fig. 17. Here again, as in the V_{BE} case, no transistor type offers any particular advantage over other types, as behavior is very similar from type to type, except that at low currents, planar transistors generally have a higher h_{FE} than other types.

I_{CBO}. This can be the most important parameter in a germanium-transistor amplifier and is usually, though not always, negligible in a silicon planar amplifier; the variation with temperature is usually greater with a silicon transistor, but the initial value is very much lower. The leakage current doubles approximately every 8 to 10°C, and thus in an amplifier designed to operate from 25 to 85°C, the current can increase by more than 2^6 or 64 times. Typical curves for germanium and silicon transistors are shown in Fig. 18. In this case, certain transistor types are grossly worse than others; for example, a germanium transistor might exhibit an I_{CBO} of 1 μa while a silicon planar might be 0.5 na.

To conclude, the drift that can be obtained in a single-ended amplifier built with even the best transistors is usually quite poor and, without compensation, is no better than about 2 mv/°C referred to the input; this compares with figures of 10 μv/°C or better that can readily be obtained with differential pairs, which are therefore usually used. There are, however, cases where this performance may be adequate and where the factors mentioned may be of interest.

3.4. Study of the Gain Stability of a Single Stage. Equation (6), which is very useful for the calculation of drift due to the known behavior of V_{BE}, is less so for the calculation of gain stability. A more useful equation would be

$$i_b = e_s/R_s + r_{bb'} + (h_{fe} + 1)r_e \qquad (7)$$

Combining with Eqs. (4) and (5) and considering only the signal term yields

$$e_o/e_s = h_{fe}R_L/[R_s + r_{bb'} + (h_{fe} + 1)r_e] \qquad (8)$$

The gain stability depends, therefore, on three parameters assuming, of course, stable source and load resistors: the current gain h_{fe}, the base-spreading resistance $r_{bb'}$ and the intrinsic emitter resistance r_e which is equal to kT/qI_E or 26.6 ohms at $T = 25°C$ and at $I_E = 1$ ma.

The base-spreading resistance is on the order of a few tens of ohms and is therefore usually unimportant in this application; for high source resistances, the expression reduces to approximately $-h_{fe}R_L/R_s$ and therefore varies as h_{fe}.

It can be seen that for maximum gain stability an external emitter resistor R_E should be used, and thus the expression becomes

$$\begin{aligned} e_o/e_s &\cong -h_{fe}R_L/(h_{fe} + 1)(R_E + r_e) \\ &\cong -R_L/R_E \end{aligned} \qquad (9)$$

for $R_E \gg r_e$, $(h_{fe} + 1)R_E \gg R_s$ and $h_{fe} \gg 1$. Unfortunately, if these conditions are met, the voltage gain will be small and probably overall feedback would be a better solution. An alternative would be to use one n-p-n-p-n-p pair as described in Sec. 5; in this case the gain stability is excellent because of the much higher open-loop gain but the -2 mv/°C drift due to V_{BE} remains.

3.5. Typical Design Examples

Design Example 1. A very simple amplifier is required with unimportant drift characteristics, a source impedance of 1 kilohm, a fairly stable voltage gain of about 10, and with both input and output levels near zero volts. The noise figure should be less than 10 db.

Solution 1. See Fig. 19. Use n-p-n and p-n-p silicon transistors with normal characteristics, e.g., 2N1613 and 2N1131. The current level of the first stage, chosen for low noise, is 0.3 ma, while the second and third stages are at 1.0 and 3.0 ma. $e_0 = 0$; therefore, $R_5 = \frac{12}{3} = 4$ or 3.9 kilohms, Choose the emitter voltage as -8 volts for reasonable voltage gain (i.e., small emitter resistors); thus $R_6 = (12 - 8)/3 = 1.3$ kilohms. The collector voltage of Q_2 is about -7.3 volts; thus $R_3 = 4.7$ kilohms. Choosing collector voltage of Q_1 as 9 volts selects

$$R_1 = (12 - 9)/0.3 = 10 \text{ kilohms and } R_4 = (12 - 9.7)/1 = 2.2 \text{ kilohms}$$

R_2 is chosen for good voltage gain in Q_1 and $R_2 = 1.6$ kilohms. R_7 and R_8 are chosen to have a small parallel value compared with R_2 for high voltage gain in Q_1 and

$$R_8R_7/(R_7 + R_8) = 0.1$$

Thus $R_8 = 100$ ohms and $R_7 = 910$ ohms. Note that the input resistance of each stage, being $h_{fe}R_E$, is somewhat higher than the load resistors and thus the open-loop voltage gain is very roughly the product of the three collector-emitter resistor ratios reduced by about 10 per cent per stage, or

$$A_V \cong \frac{(10/1.6)(4.7/2.2)(3.9/1.3)}{1.1 \times 1.1 \times 1.1} = 30$$

Thus using $10 = 30/[1 + 30(R_s/R_F)]$,

$$R_F = 15 \text{ kilohms}$$

For gain setting use $R_F = 14$ kilohms plus a 2.5 kilohms potentiometer. A somewhat more efficient circuit could be devised by increasing or eliminating the collector resistor of Q_1 and diverting most or all the collector current into the base of Q_2. This technique can be readily used in an a-c amplifier where the emitter resistor can be bypassed, thus making

D-C AMPLIFIERS

a very low a-c input impedance while keeping a high d-c impedance, but in a d-c amplifier, the standing currents become excessively large very fast. For example, in this case, the standing current of Q_3 would have been $(0.1)(h_{FE_2})(h_{FE_3})$ or about 250 ma or more, which is obviously impractical. This explains the frequent choice of "voltage-type" operation which, though inefficient, is more practical than normal "current-type" operation in d-c amplifiers, where complementary circuitry or zener diodes cannot be used for one reason or another.

The circuit of Fig. 15 overcomes this problem by the addition of constant-current sources in lieu of the load resistances.

Design Example 2. The same is desired using only n-p-n transistors, but a zener diode is permissible.

Solution 2. See Fig. 20. The design proceeds exactly as before with the attempt to maintain large collector-emitter resistor ratios. With the collector of Q_2 at 5 volts and

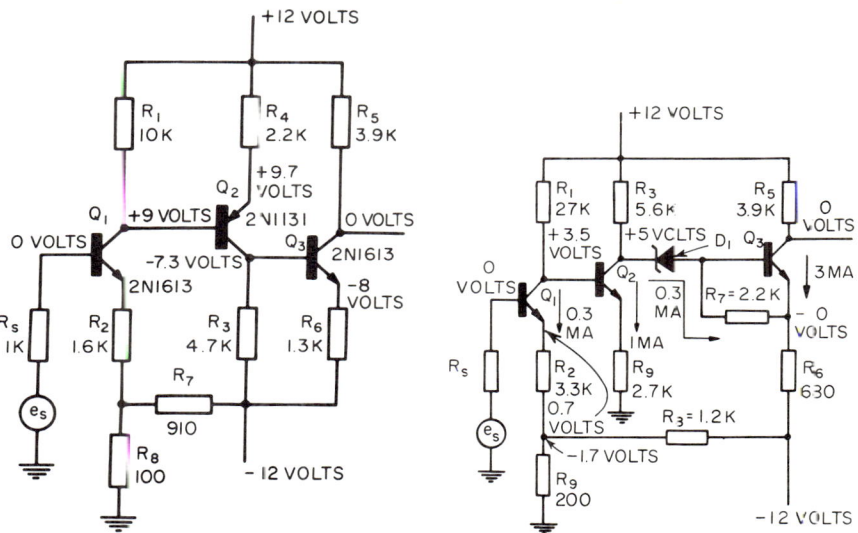

FIG. 19. Design example 1.   FIG. 20. Design example 2.

emitter of Q_3 at -10 volts, the zener diode D_1 will therefore have to be about 15 volts. The current through it without R_7 would be 3 ma/h_{FE_3} or about 60 μa, which is not sufficient for noise and voltage stability reasons. Thus, for a zener current of 0.3 ma,

$$R_7 = 0.7/0.3 \text{ or about } 2.2 \text{ kilohms}$$

This does not materially affect the gain. The voltage gain, open-loop, is about equal to the product of the collector-emitter resistor ratios less about 25 per cent per stage to take account of the loading, or about

$$(27/3.3)(5.6/2.7)(3.9/0.68)/(1.25 \times 1.25) \simeq 62$$

Feedback can be applied as in Example 1.

Design Example 3. A very simple amplifier is required with a voltage gain of 10 ± 5 per cent, but with unimportant drift characteristics.

Solution 3. See Fig. 21. The open-loop gain should be greater than 2,000 for a ± 5 per cent gain stability. Three stages give a gain of $h_{FE_1} h_{FE_2} h_{FE_3} R_L/R_{IN}$. If $R_s = 1$ kilohm and $R_1 = 10$ kilohms so that I_{C_1} is about 1 ma, then

$$A_V \cong (40)(40)(40)(1 \text{ kilohm})/(1 \text{ kilohm})$$
$$= 64,000$$

For a gain of 10, $R_F = 10 R_s$. The current gain of the amplifier being very high, the feed-

18-14 AMPLIFIER CIRCUITS

back current is much higher than the base current of the input transistor and therefore the output voltage is about 10 times V_{BE} and shifts with temperature.

Because of very high gain, oscillation may occur and a capacitor could be inserted between the base and collector of Q_2 or Q_3 to prevent it. Its required size could be calculated by using the methods of Sec. 3 of Chap. 19.

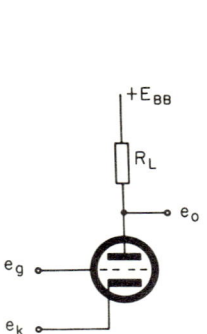

FIG. 21. Design example 3.

4. DIFFERENTIAL ELECTRON-TUBE AMPLIFIERS

4.1. General. Differential amplifiers offer the advantage of reducing the performance of an amplifier to the difference of the parameters of two devices. It is usually more difficult to maintain as much control over the absolute characteristics of a device as over their relative differences. For example, two tubes manufactured on the same day in the same plant are much more likely to have characteristics matched to within ±10 per cent than to have values of μ, say, of 60 ± 10 per cent. Further, even if the characteristics are accurately known, it is frequently difficult to compensate for certain of them without another tube; for example, the equivalent input signal due to a change in a heater voltage is most easily compensated by having another tube in the first stage, even though the drift caused by the change may be accurately known.

In this section, various circuit configurations will be discussed and analyzed, showing their relative advantages. Drift-determining factors will be discussed as in the single-ended case, and a typical design problem will be shown.

4.2. Various Circuit Configurations. *Two-input Tube* (see Fig. 22). This is the simplest possible differential amplifier and has the advantages of simplicity and no

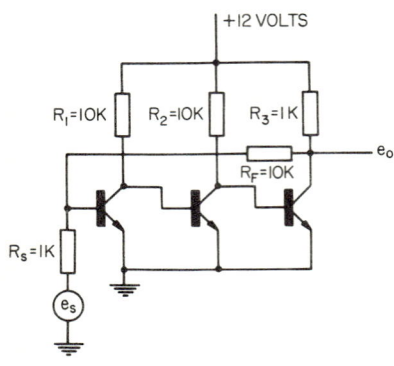

FIG. 22. A simple differential amplifier.

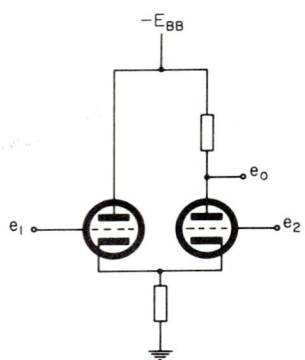

FIG. 23. Cathode-follower drive for cathode input.

phase inversion for the cathode signal. The equation for the output voltage e_o as a function of e_g and e_k can be written as[3]

$$e_o = E_{BB} - R_L \{[E_{BB} + \mu e_g - (\mu + 1) e_k]/[r_p + R_L + (\mu + 1)R_K]\} \quad (10)$$

where e_g, e_k are the input voltages at the terminals of the tube and R_K is the cathode signal source resistance. Note that, since the input impedance at the cathode is very low, the voltage e_k at the cathode terminal can be different from the actual signal voltage, depending on the source impedance R_K. Secondly, the common-mode rejection cannot be very good with this circuit being equal only to the μ of the tube.

Cathode-follower Input. In order to eliminate the problem of low-impedance drive for the cathode input of Fig. 22, a cathode follower can be inserted between the signal and the cathode of the amplifying tube such as shown in Fig. 23. This circuit has other advantages, including a big reduction in the drift due to the heater-voltage variation and a greater linearity due to the cancellation of the plate-resistance variation. The equation for this circuit will be very similar to that calculated previously with the exception that e_c and R_K will have to be modified according to the gain and output impedance of the cathode follower.

Long-tailed Pair. The circuit shown in Fig. 24 is the most commonly used differential amplifier and, under ideal conditions, offers much less sensitivity to component and power-supply variations than other circuits; for this reason it is usually used even though the input may be single-ended. The exact equation for the output e_o as a function of the input $(e_1 - e_2)$ is rather complicated; it can be written as

$$e_o = \frac{2R_L(e_1 - e_2)}{\left(\frac{r_{p1} + R_L}{\mu_1 + 1} + \frac{r_{p2} + R_L}{\mu_2 + 1}\right) + \frac{(r_{p1} + R_L)(r_{p2} + R_L)}{R_K(\mu_1 + 1)(\mu_2 + 1)}}$$

$$+ \frac{R_L\left(e_1 \frac{r_{p1} + R_L}{\mu_1 + 1} - e_2 \frac{r_{p2} + R_L}{\mu_2 + 1}\right)}{\frac{(r_{p1} + R_L)(r_{p2} + R_L)}{(\mu_1 + 1)(\mu_2 + 1)} + R_K\left(\frac{r_{p1} + R_L}{\mu_1 + 1} + \frac{r_{p2} + R_L}{\mu_2 + 1}\right)} \quad (11)$$

For the simplified case of identical tubes and a large R_K such that

$$R_K \gg (R_L + r_p)/(\mu + 1)$$

Eq. (11) can be simplified and

$$A_V \cong (\mu + 1)R_L/(R_L + r_p) \quad (12)$$

or about the same as a simple triode.

Closer investigation of Eq. (11) yields several interesting results. The second term of the equation indicates that for identical tubes, the numerator may go to zero, and there is no voltage offset or drift whatsoever if the input voltage $(e_1 - e_2)$ is zero. Secondly, the importance of R_K can be seen from both the first and second term. If R_K is large compared with $(R_L + r_p)/(\mu + 1)$, the second term will tend to be small and thus tube mismatch will not be too important except, of course, for common-mode rejection (see Sec. 4.4). Also in this case the first term will reduce to Eq. (12) and maximum gain will occur.

4.3. Drift Considerations. Similar comments apply to the drift factors for differential amplifiers as were made for single-ended amplifiers (see Sec. 2.5) with the important difference that the tight requirements on most parameters required in the latter are very much reduced. For example, the output is essentially independent of at least small changes in the

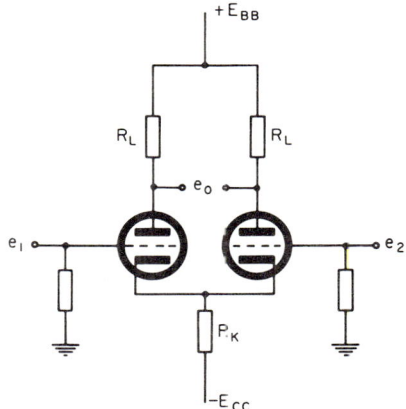

FIG. 24. Conventional long-tailed pair.

B+ supply while in a single-ended amplifier, any change is transmitted to the output reduced only by a factor $r_p/(r_p + R_L)$ or typically by perhaps 75 per cent. Thus if the B+ changes by 4 volts, 1 volt might appear at the output of a typical single-ended amplifier while very much less than that would appear in a balanced amplifier depending on the match between the tubes. Secondly, the drift due to changes in the heater voltage is very much reduced, particularly if two tubes in one envelope are used. The differences in all these parameters are so marked that designers should nearly always select a differential type when drift problems are encountered.

4.4. Common-mode Rejection. An important feature of many types of differential amplifiers, e.g., comparators, is their ability to discriminate against d-c levels common to both inputs while amplifying their differences. This ability to reject common levels is called common-mode rejection, or CMR. For the circuit of Fig. 24, the CMR for similar tubes can be expressed approximately as

$$\text{CMR} = [(\mu)(\mu + 1)/(\mu_1 - \mu_2)]\{1/[1 + (r_p + R_p)/2R_K(\mu + 1) + (r_{p_1} - r_{p_2})/2R_K(\mu + 1)]\} \quad (13)$$

where the parameters without subscripts refer to the approximate value of the parameters of the two tubes. Thus two tubes having identical μ's and r_p's have infinite CMR, and therefore, dual tubes should be selected if CMR is important. Secondly, both terms include R_K, and therefore R_K should be as large as possible. Unfortunately, as R_K increases the cathode current decreases for equal supply voltages and r_p rises, and therefore the second term of (13) does not necessarily go to zero.

A commonly used expedient is to use a constant-current source in the cathode as shown in Fig. 25. The tube can be a triode or alternatively can be a pentode; the latter, of course, has an exceedingly high output impedance. A third point of interest is that, for a given percentage match in vacuum tubes, those having high μ are superior since the first term is directly proportional to the product of the μ's of both tubes. Of course, it may be that in a given instance a match is difficult to obtain with high-gain tubes, and the determining factor will then be the $\mu^2/(\mu_1 - \mu_2)$ term of Eq. (13), which can be obtained by assuming a large R_K.

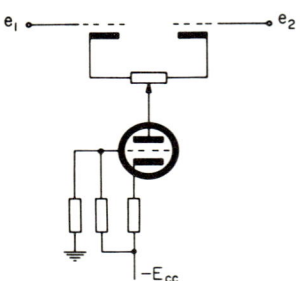

FIG. 25. Zero setting and constant-current source.

Typically, if the μ's are about 60 with a difference of 20 per cent, the CMR will be about 50 db. In this respect, transistors offer a big advantage; CMR's of 100 db are commonplace with later transistor types.

4.5. Typical Design Example

Design Example. An amplifier is required with a differential gain of about 50, a common-mode rejection of about 50 db, and as much insensitivity as possible against power-supply and heater variations.

Solution. See Fig. 26. Select the long-tailed pair of Fig. 24 because of the common-mode and power-supply insensitivity problems.

A gain of about 50 is required; using a double triode with $\mu = 60$ and $r_p = 70$ kilohms at $i_p = 0.55$ ma, $V_{gk} = -1.5$ volts and using Eq. (12)

$$50 \simeq 61/[1 + (70/R_L)]$$

giving $R_L = 320$ kilohms. Using a V_{pk} of 120 volts gives

$$E_{BB} = 120 + (0.55)(320) = 296 \text{ volts}$$

Thus select $R_L = 330$ kilohms and $E_{BB} = +300$ volts.

The grid-current argument used in Sec. 2.6 still applies, although now the problem is reduced to the difference between both tubes. Use $R_g = 100$ kilohms.

The difficult problem is that of CMR. As a first try we can use a constant-current source of 2(0.55) or 1.1 ma in the cathodes. Using a 10 kilohm resistor in the cathode of the current source gives a cathode voltage of -139 volts if a -150 volt supply is used. From the tube characteristics, $V_{gk} = -1$ volt is required, giving grid resistors of 140 and 10 kilohms. The only problem is the selection of tubes to give the required CMR. Using a μ of 60, according to Eq. (13), $CMR = \mu/(\Delta\mu/\mu)$ and for CMR of 50 db or 300,

$$\Delta\mu/\mu = \tfrac{60}{300} \text{ or } 20 \text{ per cent}$$

This is of course a difficult requirement. No zero adjustment is inserted in the amplifier, as this tends to spoil the already difficult common-mode problem; the zeroing should be done further in the amplifier.

The diagram of the complete amplifier is shown in Fig. 26. Many other design examples are given in the references.[3,4]

5. DIFFERENTIAL TRANSISTOR AMPLIFIERS

5.1. General. The possibilities of differential transistor amplifiers have been enhanced by the availability of silicon planar transistors from several manufacturers. This is due to the fact that in these amplifiers, the performance depends on the *difference* of the device parameters, and planar transistors have a much greater lot uniformity than have other devices, particularly as regards the base-emitter voltage V_{BE}. Generally the performance that is presently available from unselected devices is about as good as that which had previously been available using highly selected ones; furthermore, with some manual compensation on each amplifier at two temperatures, performance can equal or exceed chopper or chopper-stabilized types with, of course, a less complex circuit. Further improvement in lot uniformity and compensation techniques promise to improve this performance further.

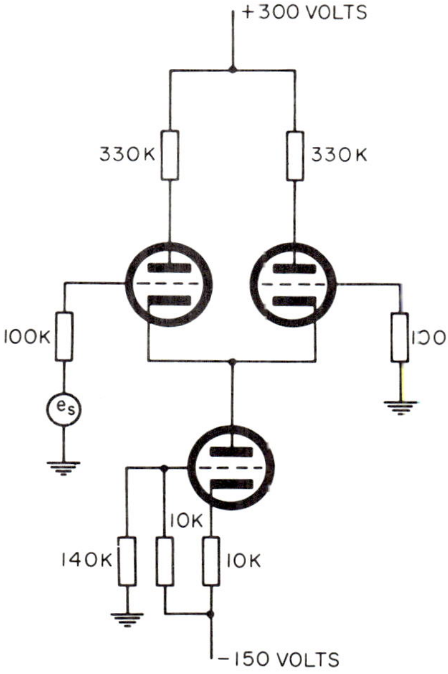

Some difficulty still exists with high-impedance stages,[9] particularly in chopper stabilized cases, and here vacuum-tube amplifiers may still maintain an advantage; whether field-effect transistors will change this picture remains unclear at the time of writing (early 1963).

FIG. 26. Design example.

In this section as in that on single-ended amplifiers, different possible circuit configurations will be described; they will then be analyzed, and the drift and gain-determining factors will be listed. Typical designs will be indicated and the performance that can be expected will be stated.

5.2. Various Circuit Configurations. Among the various circuit configurations that have been widely used, the following can be listed along with their advantages and disadvantages:

Ordinary Long-tailed Pair (see Fig. 27). This is the transistor equivalent of the vacuum-tube circuit discussed in Sec. 4 and was first described by Slaughter.[10] The advantages are that the performance of the amplifier is now reduced to the difference of the parameters of the two transistors; in other words, two very poor transistors would furnish excellent performance if they were identical. This is particularly important in the case of the base-emitter voltage, which contributes a drift of about -2 mv/°C in the unbalanced case; this figure can be improved by a factor of more than 200. The main disadvantage of this arrangement is its insufficient gain; thus, for a given emitter current, the differential base currents that must flow through the

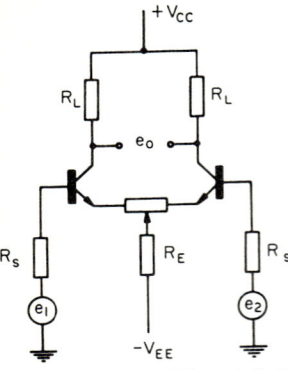

FIG. 27. A normal long-tailed pair.

source resistances frequently contribute seriously to the drift, and secondly, the gain stability is not so good as would be desirable because of the low open-loop gain.

The common-mode rejection of this circuit can be considerably improved by replacing the emitter resistor with a constant-current source as in a vacuum-tube amplifier.

p-n-p-n-p-n **pairs** (see Fig. 28a). This is a circuit first suggested for d-c amplifiers by Hilbiber[7] and subsequently described by Bénéteau.[11] Essentially it can be considered as a normal differential pair but with transistors having very high current gain $h_{FE_1} h_{FE_2}$; in this manner, the current in the source resistor is diminished by h_{FE} and the open-loop current gain is increased by h_{FE}, thus making a much more stable amplifier for a given closed-loop gain. Finally the drift in this circuit is quite low because of the V_{BE} of the *n-p-n* transistors becoming the main parameter, the *p-n-p* transistor characteristics being relatively unimportant except that the drift can be considerably reduced by having similar h_{FE}'s. This is due to the fact that different h_{FE}'s in the *p-n-p*'s cause the *n-p-n* currents to be unsymmetrical and thus the V_{BE} tracking between the *n-p-n*'s may not be so good as would be desirable. The performance can be greatly improved by furnishing either resistors or constant-current sources from the *n-p-n* collectors to the V_{CC} supply such as to make the *n-p-n* collector currents less dependent on the *p-n-p* h_{FE}'s.

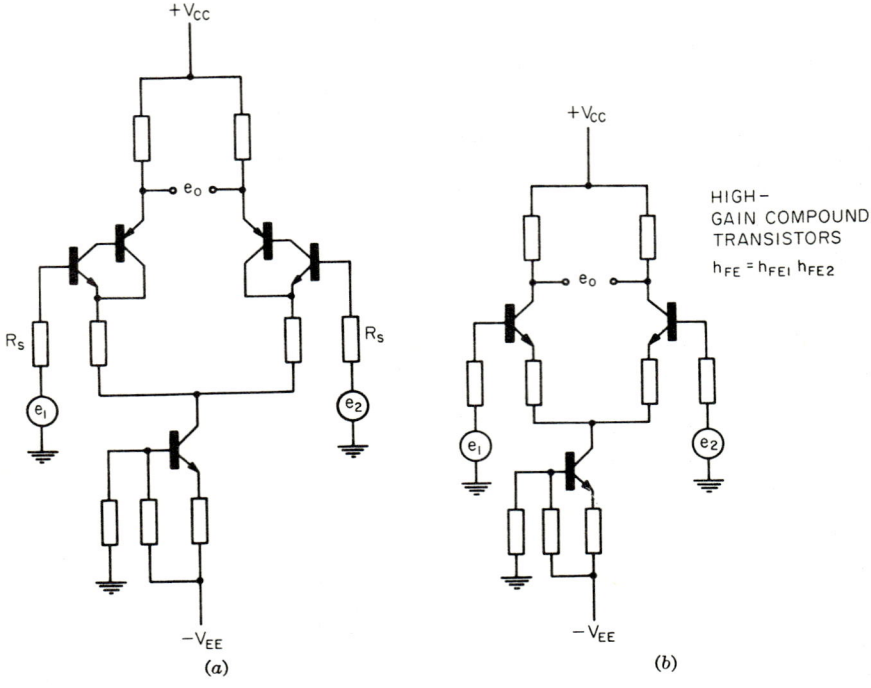

FIG. 28. (a) An *n-p-n-p-n-p* pair. (b) Equivalent circuit.

D-C AMPLIFIERS

The disadvantages of this circuit consist of the use of p-n-p transistors which are not so freely available with good characteristics as n-p-n's in silicon, and a relatively high collector-emitter voltage on the n-p-n tending to cause higher internal heating and thus drift than if a lower voltage were used. Some compensation exists because of the falling V_{CE} with increasing I_C in the n-p-n, but unless both n-p-n transistors are on the same header difficulty may occur. Drifts of the order of 10 μv/°C have been obtained with this circuit using unselected transistors.[11]

Darlington Pairs (see Fig. 29). This circuit uses the well-known Darlington combination to effect reduction of the source current by a factor of h_{FE} in the same manner as was done with the n-p-n-p-n-p pair quoted above. The advantage, therefore, is a lower drift due to reduced input current; unfortunately, the base-emitter diodes being additive, the possible drift due to V_{BE} variation is doubled. This circuit, therefore, may prove useful with high-impedance sources but is not generally to be recommended for source impedances less than a few kilohms. Further, the currents in the input transistors depend on the h_{FE}'s of the output transistors and are thus unequal; hence, as in the n-p-n-p-n-p pair, transistors not selected for h_{FE} will not track so well as those which are so selected.

More Involved n-p-n-p-n-p Pairs (see Fig. 30). Hilbiber[7] has used more complicated circuitry with the idea of increasing loop gain and of stabilizing the V_{CE} and the I_C of the input pair. He notes that to maintain matched characteristics the operating point of each transistor in the pair should be as stable as possible, and with his circuit, drifts of the order of 3 to 5 μv/°C have been obtained over the very wide temperature range of -70 to $+125$°C. The zener diodes serve to compensate the negative V_{BE} coefficient of the p-n-p's and thus maintain a constant V_{CE} on the drift-determining n-p-n's. The disadvantages of this scheme are rather complicated circuitry, matching requirements, and having more

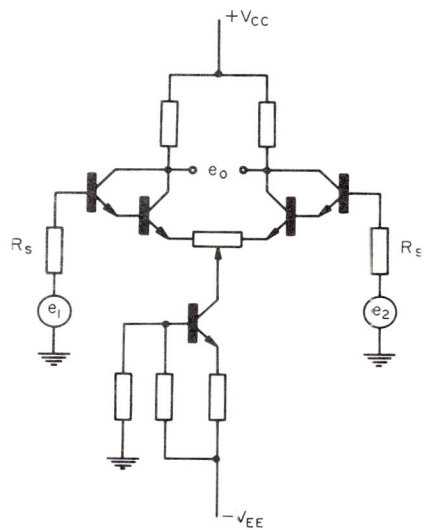

Fig. 29. Darlington input circuit used to reduce current through source resistance.

than one element on one header; the exceptional gain and drift performance may be worth the price, however, in critical circuits, although much better performance can be obtained in compensated amplifiers at the expense of an initial high-temperature adjustment.

All-n-p-n Circuit (see Fig. 31). An n-p-n circuit[12] that gives the advantage of high gain without resorting to p-n-p's and without obtaining additive effects due to V_{BE} is shown in Fig. 31. Essentially, this is again equivalent to a transistor with very high current gain and thus has a reduced source current. Drifts of less than 10 μv/°C have been obtained with this amplifier for unmatched transistors and less than 1.6 μv/°C with matched transistors for zero source impedance with both input transistors immersed in an oil bath. It has the big advantage that the dissipation in the input stage is kept very small by having a very low V_{CE} which is, in fact, the V_{BE} of the succeeding stage; unfortunately this varies with temperature, changing the V_{CE} of the input pair and thus contributing somewhat to drift as explained by Hilbiber.[7]

A Compensated Circuit (see Fig. 32a). A great deal of interest has been shown in compensated stages[13-16] in which an adjustment is made on an amplifier at a high and at room temperature; this results in more complicated production calibrations, but an order-of-magnitude improvement in drift figures can be obtained. For exam-

18–20 AMPLIFIER CIRCUITS

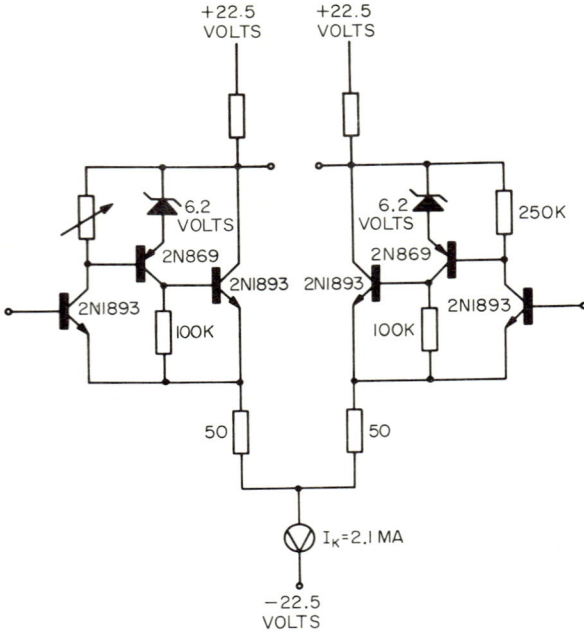

Fig. 30. High-performance complementary circuit.

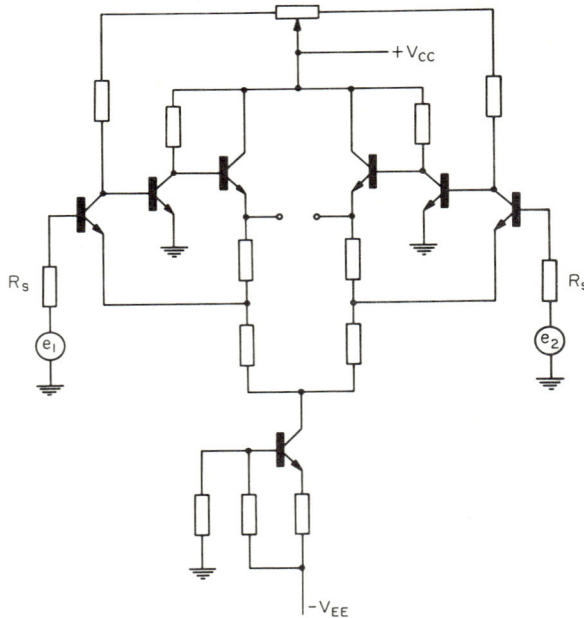

Fig. 31. An all–*n-p-n* high-performance circuit.

ple, a production amplifier[14] is currently available with a guaranteed total drift of less than 25 μv over a temperature range of 0 to 60°C using straight d-c techniques with no choppers of any kind. One possible method of drift compensation (see Fig. 32a) involves the use of a temperature-dependent current source which feeds into a potentiometer inserted between the emitters. The current source is designed for zero output at room temperature, and when the amplifier is brought up to its maximum temperature, the potentiometer is adjusted to insert this temperature—varying current more into one transistor than into the other such as to bring the output back to zero. Since the drift is, in general, fairly linear over at least some temperature range, very good compensation can thus result. Many temperature-varying current sources are possible; a practical one is shown in Fig. 32b.

The p-n-p transistor is a current source essentially independent of temperature while the n-p-n, having constant base current, varies its collector current depending

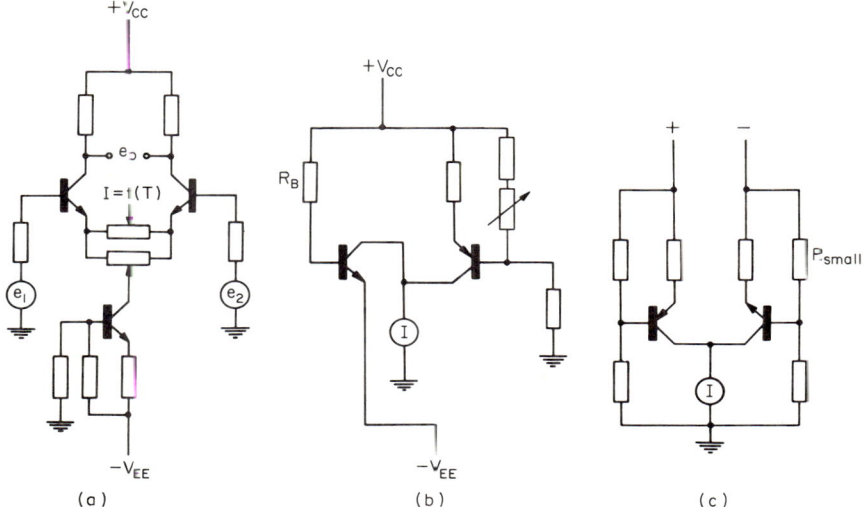

Fig. 32. (a) One of several compensation schemes. (b) A possible temperature-dependent current source. (c) Another possible circuit.

on its h_{FE}, which is a fairly linear function over the temperature ranges in which these amplifiers are required to operate. The difference is thus temperature-dependent and can go above and below zero. Another scheme which is somewhat superior is shown in Fig. 32c; it uses the fact that V_{BE} in the n-p-n transistor is quite linear with temperature and avoids the I_{CBO} problem of the previous circuit.

Several other compensation schemes are possible; a practical one* is to change the current of one transistor in the pair more than that of the other one and thus change its $\Delta V_{BE}/\Delta T$ since this is a current-dependent coefficient. Most experienced designers will probably devise their own system which will have some particular advantage over the types shown. Essentially, a very linear voltage can be obtained from either a diode or a base-emitter junction having fixed current; the problem is then to transform this temperature-dependent voltage into a current. A similar transistor is better than a diode as the voltage or current-generating element because its thermal time constant is similar to that of the drift-determining transistor; thus, if the ambient temperature changes suddenly, the compensating transistor will be more likely to instantaneously follow the variation of the other one.

* Suggested by B. Murari of S.G.S.

5.3. Analysis of Various Stages. *Normal Long-tailed Pair.* It can be shown readily that the output e_o as a function of the differential input of the differential circuit can be written approximately as

$$e_o \cong A[(e_1 - e_2) + (V_{BE_2} - V_{BE_1}) + (I_{CBO_2} - I_{CBO_1})R_s \\ + (h_{FE_1} - h_{FE_2})(V_{EE}R_s/2R_E h_{FE_1} h_{FE_2}) + (1 - 2K)(V_{EE}R_2/2R_E)] \quad (14)$$

where
$$A = 2R_L/[R_2' + R_s(1/h_{FE_1} + 1/h_{FE_2})]$$
$$R_2' = R_2 + r_e$$

and KR_2, $(1-K)R_2$ is the division of the zeroing potentiometer. The terms can be considered as follows: A is the voltage gain, $(e_1 - e_2)$ is the signal voltage, $(V_{BE_1} - V_{BE_2})$ is the base-emitter differential voltage which cannot be separated from the signal, $(I_{CBO_1} - I_{CBO_2})R_s$ is the drift due to the differential I_{CBO}'s, and $(1 - 2K)V_{EE}R_2/2R_E$ is the differential voltage caused by the constant current not flowing across two equal resistors. This latter term is of course used to balance the initial V_{BE} difference. The $(h_{FE_1} - h_{FE_2})$ term can best be understood by rewriting it as

$$(h_{FE_1} - h_{FE_2})(V_{EE}R_s)/2R_E h_{FE_1} h_{FE_2} = [(h_{FE_1} - h_{FE_2})/h_{FE_1} h_{FE_2}](I_E R_s/2) \\ = I_E R_s/2h_{FE_1} - I_E R_s/2h_{FE_2} = I_{B_1} R_s - I_{B_2} R_s$$

Thus, if the emitter current $I_E = V_{EE}/R_E$ splits up into two parts, the equivalent voltage on the base of Q_1 will be represented by the first term while that on the base of Q_2 will be represented by the second term, the differential voltage being the difference between the two. The importance of equal h_{FE}'s or at least of the very high equivalent h_{FE}'s obtained in compound connections is thus seen; alternatively, very low base currents should be used which mean low collector currents and high h_{FE}'s. Finally, the dependence of the gain term on h_{FE} can be seen, and thus the gain will always be a strong function of temperature unless R_2 is made excessively large. This, of course, can be compensated by feedback.

p-n-p–n-p-n Pairs. It can again be readily shown[11] that the output voltage e_o can be written as (see Fig. 28a)

$$e_o \cong -(R_1/R_2)[(e_1 - e_2) + (V_{BE_2} - V_{BE_1}) + (I_{CBO_2} - I_{CBO_1})R_s] \quad (15)$$

with the only assumptions made being that

$$h_{FE} \gg 1 \quad \text{and} \quad h_{FE_2} h_{FE_1} R_1 \gg R_s$$

Thus, as in the previous case, the V_{BE} and I_{CBO} effects are indistinguishable from the signal and the same criteria therefore apply as in the long-tailed pair circuit. In this case, of course, the h_{FE}'s do not enter into the gain equation, and thus much more gain stability can be obtained, although, as explained previously, matched h_{FE}'s for the *p-n-p*'s are important for low drift.

Darlington Pairs (see Fig. 29). This is exactly the same as that done for the normal long-tailed pair, with the difference that h_{FE}'s should be the compound h_{FE}'s of two cascaded stages, the V_{BE}'s should be the cascaded V_{BE}'s of the same stages, and the I_{CBO}'s should be primarily those of the input stages.

5.4. Parameters Which Affect Drift. From Eqs. (14) and (15) it can be seen that matched characteristics are essential if low drift is to be obtained. Completely matched characteristics are obviously not possible with any transistor process, even on a pilot-line basis; the thing to do, therefore, is to see which characteristics are closely matched in production and to attempt to design circuits which will minimize the effects of the others.

No process exists which will match I_{CBO}'s on a production basis; fortunately, certain types of silicon planar transistors, specifically those which have not been treated for low storage time (i.e., types that are not made for saturated switching circuits), have very low leakage currents that are frequently negligible in normal circuits. At room temperature, for example, a leakage current of 0.5 na is typical for a 2N1613.

Nevertheless, it may happen that high-temperature high-source-impedance amplifiers are required where the leakage current is important, and in those cases, some attempt may be made to select equal I_{CBO}'s.

Similarly, it is difficult to match h_{FE}'s on a production basis, although planar technology permits selection before mounting. Thus dual transistors are available commercially from several manufacturers with current gain matching to 10 per cent. Furthermore, for planar transistors that have similar V_{BE}'s and current gains, it has been found[18] that the behavior of gain as a function of temperature is very reproducible between devices. Thus the fourth term of Eq. (14) can be reduced by using these mounted duals or, alternatively, by selecting two transistors of the same type with similar V_{BE}'s and h_{FE}'s.

The feature of planar transistors that is of particular interest to the designer of d-c amplifiers is the very close match, in production, of the base-emitter voltage temperature coefficient. Hilbiber[7] gives an example of a distribution where 90 per cent of a given lot has a temperature-coefficient difference between any two transistors of less than 4 per cent. These are, of course, unselected transistors, and the match can therefore be improved still further by utilizing some degree of selection. Thus some of the dual transistors that are available have a maximum temperature coefficient of 10 μv/°C, which means that they track to better than 1 part in 200.

An important further point which is not always understood is the extreme importance of keeping both transistors of a differential pair at the same temperature. This can best be seen by considering a temperature difference of 0.01°C between two transistors which are absolutely identical otherwise. Because of the V_{BE} temperature coefficient of about -2mv/°C, the V_{BE} difference will thus be $(2)(0.01)$ mv or 20 μv, compared with the 10 μv/°C guaranteed on some dual transistors. Extremely good heat sinking is thus seen to be essential. In this respect the dual transistors offer the considerable advantage of having both transistors on the same header, thus minimizing their temperature difference. To accomplish the same heat sinking with conventional methods is obviously much more difficult. However, even in dual transistors, differences of 10 to 20 μv are quite possible due to temperature differences on the order of 0.01°C.

Another factor contributing to drift is the change of collector current or voltage on a pair of transistors. For example, two transistors might be perfectly matched at a given current level, but if this current changes, the possibility exists that they will not be matched at this new level. For that reason, when designing very high performance d-c amplifiers, it is wise to attempt to keep the emitter currents and the collector-emitter voltages as nearly constant as possible as functions of temperature.

To summarize, the drift-determining factors are V_{BE}, h_{FE}, and I_{CBO}. New transistor technology improves the match possible for the first two parameters, while the leakage current has been reduced essentially to the bulk leakage of the device and thus is very low. The seriousness of the temperature-difference problem can be alleviated by the use of dual transistors on one header.

5.5. Gain Stability. In an ordinary long-tailed pair, the gain stability is a function of the ratio of the resistor R_2 connected between the emitters and the source resistance R_s. If a large emitter resistance is used, the gain drops but the stability rises. Specifically, from Eq. (14) we can write approximately

$$A_F \cong (2R_L/R_s)/(R_2/R_s + 2/h_{FE}) \tag{16}$$

Thus, for values of R_2/R_s large with respect to $2/h_{FE}$ the gain is small but stable. This can be seen in Fig. 33 for a typical case of an amplifier utilizing transistors with a current gain of 50 and a temperature range of 25 to 100°C.

For R_2/R_s greater than about 1, the gain variation over the temperature range is about 2 per cent, while it is about 15 per cent for R_2/R_s of 0.1 and 80 per cent for R_2/R_s of 0, i.e., no feedback at all. This is the main disadvantage of the normal differential pair, and in most practical amplifiers the emitter resistor is made large enough to compensate for V_{BE} differences and then overall feedback is used to stabilize the gain.

18-24 AMPLIFIER CIRCUITS

Compound combinations largely eliminate this difficulty, however. Equation (15) gives the result that at least to a first approximation the gain is independent of transistor parameters. This is, of course, at the expense of two extra transistors, and clearly a designer would have to balance the advantages of a compound pair with those which could have been obtained had he used two normal differential amplifiers with feedback to stabilize the gain or, alternatively, had used some of the more complicated circuits that have been proposed. Theoretically, overall feedback is always more efficient than local feedback, but many practical considerations probably dictate that a compound pair would be used where gain stability is important. In any case, there is no clear-cut answer as to which is preferable, and much would depend on the circuit application and the designer's individual preference.

5.6. Common-mode Rejection. An important parameter of differential amplifiers is the degree of their ability to reject signals common to both inputs. For example, one might have a low-frequency a-c signal riding on a d-c level and apply this compound signal to one base while the same fixed bias was applied to the other base. Care would have to be taken that the response of the amplifier to the common d-c level would be arbitrarily low. It can be shown[17] that the common-mode rejection (CMR) of an externally balanced differential-amplifier pair with equal source resistance can be written approximately as

$$\text{CMR} \cong V_{EE}/\Delta V_{BE} \qquad (17)$$

where V_{EE} is the equivalent voltage $I_E R_E$ and ΔV_{BE} is the difference between base-emitter voltages of the transistors. For a dual transistor, ΔV_{BE} is less than 5 mv on some types, e.g., 2N2060, while for high-voltage current-source transistors, e.g., 2N1893, the output impedance can be well over 1 megohm. Thus, for a current

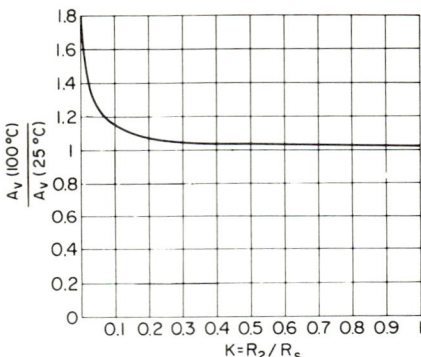

Fig. 33. Typical case of gain stability as a function of R_2/R_S ratio.

of 500 μa the equivalent V_{EE} is over 500 volts, and the CMR can be over 100,000 or 100 db. Values of over 80 db are common with carefully designed circuits.

5.7. Typical Designs

Design Example 1. Assume a 3-cps signal of about 20 μv comes from a 1-kilohm source but riding on a variable d-c level that might be as high as 20 mv. The gain stability is desirable but not too important, as overall feedback will be used, but the common-mode rejection is very important. Design a simple amplifier to do the job.

Solution 1. See Fig. 34. The normal long-tailed pair will be used, along with a matched dual for the amplifier portion. If we want the ratio of the signal to the d-c level shift at the output to be at least 20 db, then the CMR = 20 db + 20 log (20 × 10⁻³/20 × 10⁻⁶) or 80 db. The emitter current, chosen for noise reasons, is to be 50 μa total or 25 μa in each stage. Thus $R_4 = \frac{6}{50} = 120$ kilohms. Taking a bleeder current of 50 μa, $R_2 = 5.4/50 = 110$ kilohms and $R_3 = 6.6/50 = 130$ kilohms. The h_{ob} of the 2N1893 transistor used as the current source is given as 0.5 μmho maximum or 2 megohms minimum. Thus the equivalent V_{EE} is (2 × 10⁶)(50 × 10⁻⁶) = 100 volts. With a ΔV_{BE} max of 5 mv (2N2060), the CMR = 100/5 × 10⁻³ = 20,000 or 86 db, which is very close to the minimum acceptable of 80 db. However, this is the figure calculated from the worst-case transistor characteristics and therefore is probably safe; if a greater safety margin is desired, the current will have to be increased at the expense of noise.

Using a 6-volt V_{CE}, $R_1 = (12 - 6)/25$ or 240 kilohms. The required size of R_E can be calculated by

$$R_E = \Delta V_{BE}/I_E = 5 \times 10^{-3}/25 \times 10^{-6} = 200 \text{ ohms}$$

where I_E is the current in a single stage.

Use 2.5 kilohms to give some gain stability, although 200 ohms could be used with higher

D-C AMPLIFIERS 18-25

gain. The voltage gain will be, approximately,

$$Av \cong 2R_L/(2r_e + R_E) = (2 \times 240 \text{ kilohms})/[2(26.6/0.025) + 2.5 \text{ kilohms}] = 100$$

The drift may be satisfactorily small depending on the temperature range involved. If it is not, then some compound scheme should be used such as shown in Example 2 below or a compensation scheme can be used as explained in Section 5.2 under A Compensated Circuit.

Design Example 2. A differential amplifier is required with a very stable gain of 20, drift on the order of 10 $\mu v/°C$ referred to the input, and CMR of more than 80 db. Supply voltages of ± 28 volts.

Solution 2. See Fig. 35. Use the n-p-n-p-n-p pair arrangement of Fig. 28a. Selecting p-n-p emitter currents of about 1 ma and d-c level on the p-n-p emitters at about 14 volts for large dynamic swing, $R_1 = 14$ kilohms. The base current of the p-n-p's is about 1 ma/40 = 25 μa. For very low drift, select p-n-p's with equal h_{FE}'s as explained in Sec. 5.2. The current in the emitter of the current source is about 2 ma. Selecting $R_2 = 10$

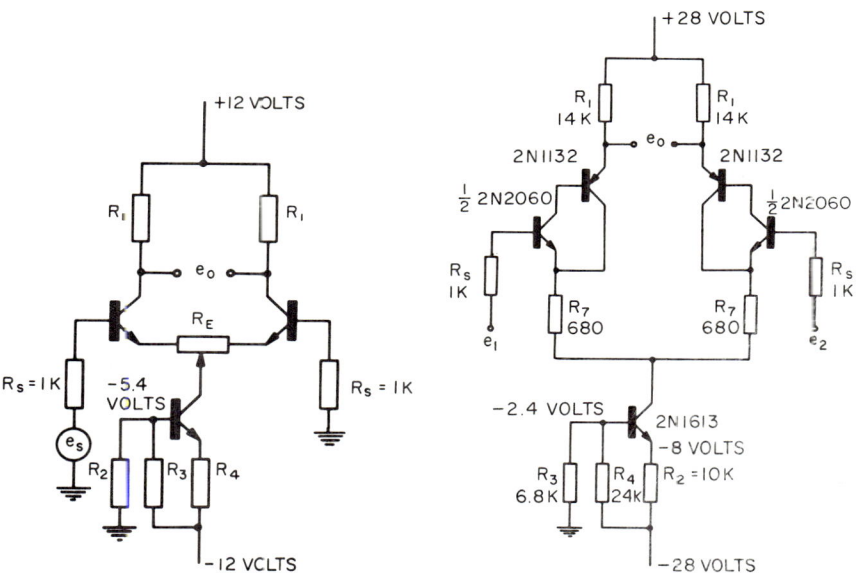

FIG. 34. Design example 1.　　　　FIG. 35. Design example 2.

kilohms makes $R_3 = 6.8$ kilohms and $R_4 = 24$ kilohms. The gain is equal to the ratio of R_1/R_7; thus $R_7 = 14$ kilohms/20 = 700 ohms. Select $R_7 = 680$ ohms. Using a value of h_{ob} of about 1 μmho, the equivalent $V_{EE} = (2 \text{ ma})(1 \text{ megohm}) = 2,000$ volts. Thus CMR $= V_{EE}/\Delta V_{BE} = 2,000/5 \times 10^{-3} = 400,000$ or over 100 db. The zeroing can either be done farther down the amplifier or between the gain-setting resistors, but in the latter case the CMR might deteriorate depending on the n-p-n match.

6. CHOPPER AND MODULATED-CARRIER AMPLIFIERS

Electronic amplifying devices generally operate more effectively with a-c signals than with d-c signals. Amplifiers operating at direct current encounter problems of drift and d-c offset. Unpredictable drift in d-c amplifiers is generally a much more severe problem than the corresponding random noise at the input of an a-c amplifier. If it is assumed that drift and noise in a d-c amplifier are synonymous and that the drift is limited in frequency by the bandwidth of the d-c amplifier, it is almost always possible to build an a-c amplifier that has a lower equivalent input noise than an amplifier having the same bandwidth but providing gain down to direct current.

18–26   AMPLIFIER CIRCUITS

The d-c offset can be another problem in d-c amplifiers that are not of the modulated-carrier type. In some cases the input is at a different reference voltage, with respect to the ground reference, than the output. In other cases, the input and the output of the d-c amplifier will both be referenced to ground and there will be zero voltage offset between them. In either case, special attention must be given to the biasing circuits of the d-c amplifier to ensure that these conditions are met. This is generally not too difficult with transistor amplifiers if the offset potentials are not larger than the collector breakdown voltage of the transistors. However, it causes considerable difficulty with electron-tube amplifiers and often results in very inefficient coupling circuits between the tubes.

Closely related to the requirement for a voltage offset between the input and output terminals is the requirement for inputs and outputs with completely independent references or ground terminals. The use of well-designed long-tail pairs and other common-mode-rejection techniques, as described in Sec. 4, may provide a satisfactory solution.

Greatly improved flexibility and performance with regard to drift, d-c offset, and input-output isolation can be achieved through the use of a chopper or modulated-carrier amplifier (Fig. 36). The input signal is fed into a modulator which converts it

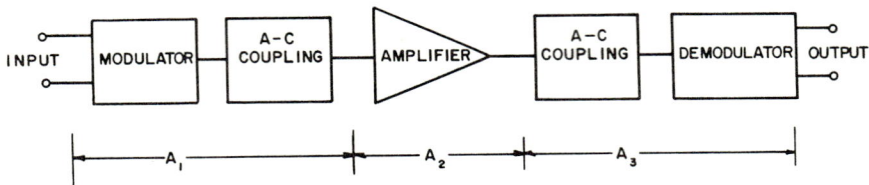

Fig. 36. Modulated-carrier amplifier.

to alternating current. The modulators considered in this chapter are AM modulators. Their output contains a carrier, its harmonics, and pairs of sidebands corresponding to the frequencies of the input signal. The carrier and certain of its harmonics will generally be suppressed. Many such modulators employ switch action to obtain more or less rectangular waves whose amplitude is proportional to the signal amplitude. These are called choppers.

The output of the modulator in Fig. 36 is a-c-coupled to an amplifier, and the output of the amplifier is a-c-coupled to a demodulator which reconstructs the input signal. In some cases, the demodulator is not included and the a-c output signal is used directly.

The amplifier is conventional and the modulators and demodulators impose only a few special requirements on it. The chief requirements are that the amplifier:

1. Have sufficiently low noise
2. Have adequate gain and stability at the carrier frequency, sideband frequencies, and sideband harmonics
3. Provide adequate output power
4. Provide a carrier delay that is at least stable and preferably negligible

The first two requirements are usually not difficult to meet. As a rule, modulators are sufficiently noisy that their noise dominates the noise of a well-designed amplifier. Also, the carrier and important sideband frequencies are usually limited by the modulator characteristics to frequencies that can be amplified with little difficulty. The third requirement may be somewhat more difficult to achieve because the demodulator may be inefficient and will generally impose a highly nonlinear load upon the amplifier. The fourth requirement is necessary when the demodulator at the output must operate synchronously with the modulator at the input.

The a-c coupling is achieved with capacitors or transformers. Such factors as d-c

D-C AMPLIFIERS   18-27

isolation, sensitivity to ground currents, and loading of the output of the amplifier are strongly influenced by the type of coupling and the type of modulator and demodulator.

With reference to Fig. 36, the amplification A of a chopper or modulated-carrier amplifier is equal to the product of the voltage transfer ratio of the input modulator A_1, the amplification of the amplifier A_2, and the voltage transfer ratio of the output demodulator A_3:

$$A = A_1 A_2 A_3 \tag{18}$$

The voltage transfer ratio of the input modulator is defined as the ratio of the peak-to-peak voltage delivered to the input impedance of the amplifier divided by the d-c input voltage. The peak-to-peak voltage is employed because the waveform is generally of a rectangular nature, which may be not symmetrical, and all direct current is removed by the a-c coupling network. The amplification A_2 is expressed as the ratio of the peak-to-peak Thévenin voltage at the output divided by the peak-to-peak voltage delivered to the input. The voltage ratio of the output demodulator A_3 is defined as the ratio of the d-c output voltage to the peak-to-peak Thévenin voltage delivered by the amplifier. The use of these expressions in determining the overall voltage amplification of the chopper amplifier is described in Sec. 8.

The upper frequency of amplification for the overall system is generally significantly lower than the carrier frequency. However, the upper frequency of operation can be significantly higher than the carrier frequency when the d-c amplifier is used in conjunction with an a-c amplifier using the Goldberg scheme (see Chap. 19).

7. INPUT AND OUTPUT SWITCHING CIRCUITS

Probably the most important type of d-c-to-a-c modulator is the chopper. The design of the chopper switching circuit has a major effect upon the performance of the chopper amplifier. While there are a large number of special-purpose chopper circuits, most designs can be derived from a limited number of basic circuits. A chopper circuit may employ single-pole single-throw (SPST), single-pole double-throw (SPDT), double-pole double-throw (DPDT), or other switching arrangements. The switching circuits may be capacitance-coupled or transformer-coupled to the amplifier. Usually, capacitance-coupled circuits employ SPST or SPDT switches. Transformer-coupled circuits generally employ SPDT or more complex switching arrangements, which can be afforded when the weight and expense of a transformer are permitted. This section deals with input switching configurations and methods of describing their performance.

7.1. Capacitor-coupled Input Choppers. Figure 37 shows eight basic input-chopper-circuit variations for capacitive coupling to the input of the amplifier. R_1 is the internal resistance of the signal source, and R_3 is the input resistance of the amplifier. A second source is shown in each circuit. When a second source is employed, the output of the chopper is proportional to the voltage difference of the two sources. The second source is shown to indicate how each of these circuits may be used to perform the function of d-c voltage differencing. When only a single input is required, the second source is simply short-circuited, as shown by the dotted line in each of the figures. R_2 is the internal resistance of the second source or a resistance built into the chopper circuit. As will be seen, R_2 has an optimum value that provides maximum gain of the chopper circuit when R_1 and R_3 are set. In a number of cases this optimum value is zero.

The circuits of Fig. 37A to D all employ SPST switches. These are the simplest chopper circuits available but when properly designed provide remarkably good performance.

Circuit A employs a shunt switch. When a voltage is connected to input 1 and the switch is open, the voltage charges the coupling capacitor through R_1 and R_3. When the switch is closed, the coupling capacitor discharges through R_1 and R_2 in parallel, and R_3. The capacitor is usually sufficiently large that the voltage across it stabilizes to a nearly constant value. The result is an alternating rectangular wave

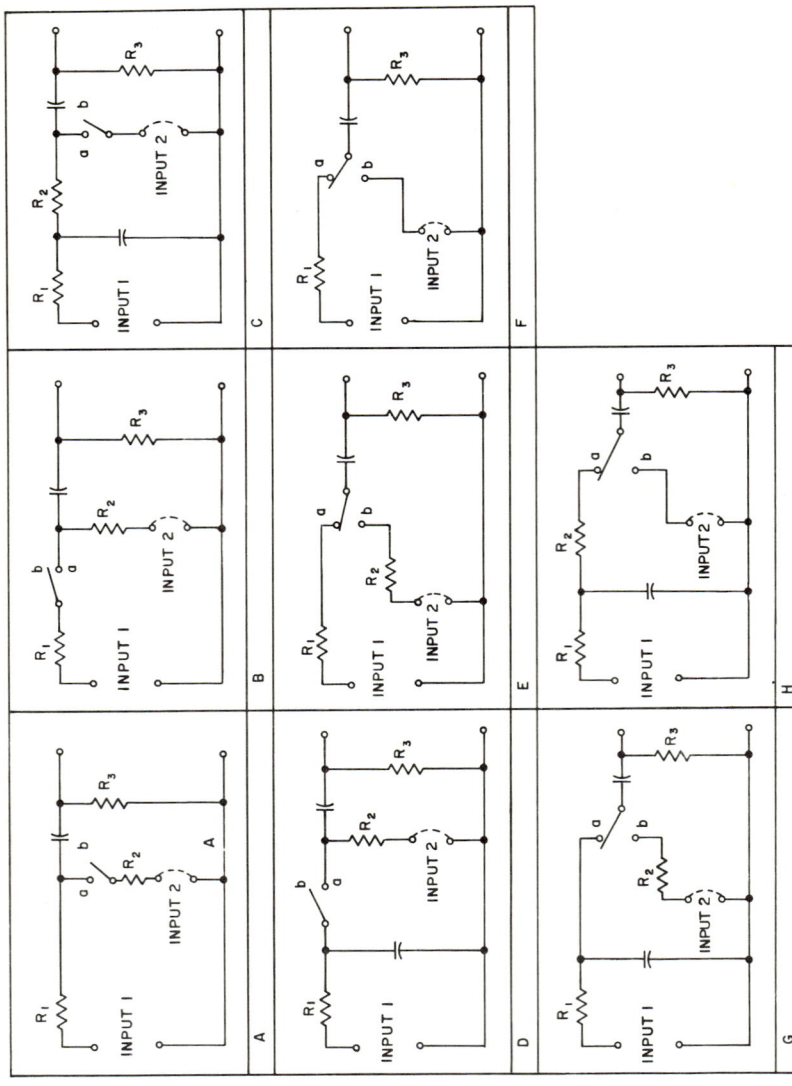

FIG. 37. Capacitor-coupled input-chopper circuits. (*A*) SPST shunt. (*B*) SPST series. (*C*) SPST shunt with filter. (*D*) SPST series with filter. (*E*) SPDT underlap. (*F*) SPDT overlap. (*G*) SPDT underlap with filter. (*H*) SPDT overlap with filter. (*Adapted from "The Contact Modulator," Airpax Electronics Incorporated, Cambridge, Md., 1963.*)

D-C AMPLIFIERS

of voltage developed across R_3. Inputs 1 and 2 are in series for d-c purposes. The gain of the circuit from input 1 to the output is the same as the gain from input 2. This is necessary if the output is to be zero when the voltage inputs to 1 and 2 are equal.

Circuit B employs a series switch. From a configuration point of view circuits A and B are identical. For d-c purposes, the two inputs can be considered short circuits; and the coupling capacitor looking to the left sees a resistor (R_1 or R_2) to ground and in parallel a switch in series with a resistor (R_2 or R_1) to ground.

Circuits C and D employ a filter in input 1. The filter capacitor is assumed to be sufficiently large that the voltage across it is constant and does not vary with the on-off positions of the switch. The filter serves two purposes. First, it isolates the chopping action from the source and thereby permits the current drain from the signal source to be steady direct current rather than a current that is fluctuating in synchronism with the chopping action. Second, it prevents unwanted high-frequency noise signals from being transferred from the signal source to the chopper.

The most serious interfering signal is generally one at the chopping frequency. Such a signal can easily arrive from the signal used to drive the chopper switch on and off. For example, if the chopper is an electromechanical switch driven by a magnetic coil, an unwanted signal at the chopping frequency may be introduced in the input through magnetic coupling, electrostatic coupling, or a common impedance to ground. Also, choppers are frequently operated in synchronism with power-line frequencies such that power-line sources of unwanted signals at the chopping frequencies may be present. In these simple types of chopper circuits, a synchronized interference signal is in essence rectified by the chopping switch and appears as a d-c offset. The pickup may be so large as to completely overload a sensitive a-c amplifier following the chopper.

Circuits E to H employ SPDT switches. If the switch is of an electronic type, say a transistor, it will often change state sufficiently fast that the contact may be assumed to transfer from a to b instantaneously. However, if the switch is mechanical, a finite time is required to transfer from one position to the other. This transfer is described as *underlap* or *break-before-make* (BBM) when the arm disconnects from a before it connects to b. In this case the switch is completely open twice during the cycle. The operation is termed *overlap* or *make-before-break* (MBB) when the switch connects to b before it disconnects from a.

Circuits E and F, which represent, respectively, underlap and overlap modes of operation, are in essence the same except that R_2 has been eliminated from circuit F. When R_2 is included in circuit F, the equations describing circuit operations become rather unwieldy and beyond the scope of this text. Circuits G and H are corresponding circuits with a filter added to input 1.

The equations for the input chopper circuits are listed in Table 1. The waveforms for the circuits are shown in the second column. The SPST switch is on for a fraction a and off for a fraction $b = 1 - a$ of the cycle. The SPDT underlap switches are in position a for a fraction a of the cycle and in position b for a fraction b of the cycle such that

$$a + b < 1$$

The SPDT overlap switches are overlapped for a total fraction c of the cycle such that

$$a + b - c = 1$$

Note that the periods a and b are defined somewhat differently for underlap and overlap operation.

The voltage transfer ratio is given in the third column and the input resistance in the fourth. The voltage transfer ratio is the same for both inputs 1 and 2 when both are used. The input resistance is simply the d-c input voltage divided by the average input current. For a maximum voltage transfer ratio, R_2 should be equal to R_2 (opt), column 5. The maximum voltage transfer ratio is given in the last column.

It is of interest to determine the circuits that provide the larger voltage transfer ratios. The maximum voltage ratios for all circuits are functions of a, b, c, and the

Table 1. Equations for Capacitor-coupled Input-chopper Circuits*

Circuit	Waveform (positive d-c input)	Voltage transfer ratio	Input resistance	R_2 (opt) for max voltage transfer ratio	Max voltage transfer ratio
A. SPST shunt		$\dfrac{1}{1 + a\dfrac{R_1}{R_3} + \dfrac{R_2}{R_1} + \dfrac{R_2}{R_3}}$	$R_1 + \dfrac{R_2}{1-a} + \dfrac{1-a}{a}\dfrac{R_1 R_3}{R_1 + R_3}$	0	$\dfrac{1}{1 + aR_1/R_3}$
B. SPST series		$\dfrac{1}{1 + \dfrac{R_1}{R_2} + a\dfrac{R_2}{R_3} + \dfrac{R_1}{R_3}}$	$\dfrac{R_1}{a} + R_2 + \dfrac{1-a}{a}\dfrac{R_2 R_3}{R_2 + R_3}$	$(R_1 R_3/a)^{\frac{1}{2}}$	$\dfrac{1}{1 + 2(aR_1/R_3)^{\frac{1}{2}} + R_1/R_3}$
C. SPST shunt with filter		$\dfrac{1}{1 + a\left(\dfrac{R_1}{R_2} + \dfrac{R_2}{R_3} + \dfrac{R_1}{R_3}\right)}$	$R_1 + R_2 + \dfrac{1-a}{a}\dfrac{R_2 R_3}{R_2 + R_3}$	$(R_1 R_3)^{\frac{1}{2}}$	$\dfrac{1}{1 + 2a(R_1/R_3)^{\frac{1}{2}} + aR_1/R_3}$
D. SPST series with filter					

18–30

Table 1. Equations for Capacitor-coupled Input-chopper Circuits* (*Continued*)

Circuit	Waveform (positive d-c input)	Voltage transfer ratio	Input resistance	R_2 (opt) for max voltage transfer ratio	Max voltage transfer ratio
E. SPDT underlap		$\dfrac{1}{1 + \dfrac{ab}{a+b}\left(\dfrac{1}{a}\dfrac{R_1}{R_3} + \dfrac{1}{b}\dfrac{R_2}{R_3}\right)}$	$\dfrac{R_1}{a} + \dfrac{R_2}{b} + \dfrac{a+b}{ab}R_3$	0	$\dfrac{1}{1 + \dfrac{b}{a+b}\dfrac{R_1}{R_3}}$
F. SPDT overlap		$\dfrac{1}{1 + b\dfrac{R_1}{R_3}}$	$\dfrac{R_3 + bR_1}{c\dfrac{R_3}{R_1} + ab}$		
G. SPDT underlap with filter		$\dfrac{1}{1 + \dfrac{ab}{a+b}\left(\dfrac{R_1}{R_3} + \dfrac{1}{b}\dfrac{R_2}{R_3}\right)}$	$R_1 + \dfrac{R_2}{b} + \left(\dfrac{a+b}{ab}\right)R_3$	0	$\dfrac{1}{1 + \dfrac{ab}{a+b}\dfrac{R_1}{R_3}}$
H. SPDT overlap with filter		$\dfrac{1}{1 + c\dfrac{R_1}{R_2} + b\dfrac{R_2}{R_3} + ab\dfrac{R_1}{R_2}}$	$R_1 + \dfrac{R_3 + bR_2}{c\dfrac{R_3}{R_2} + ab}$	$(R_1 R_3 c/b)^{\frac{1}{2}}$	$\dfrac{1}{1 + 2\left(bc\dfrac{R_1}{R_3}\right)^{\frac{1}{2}} + ab\dfrac{R_1}{R_3}}$

* Adapted from "The Contact Modulator," Airpax Electronics Incorporated, Cambridge, Md., 1963.

ratio R_1/R_3. The SPST circuits, in order of decreasing voltage transfer ratio, are A, C, D, and B.

If c is not too large, the SPDT circuits having larger voltage transfer ratios are, in descending order, G, F, and E. The form of the voltage transfer ratio equation for circuit H is not the same as the others; however, for small overlap or underlap, circuit H is similar to circuit G. Also, for small overlap or underlap, circuits E and F are

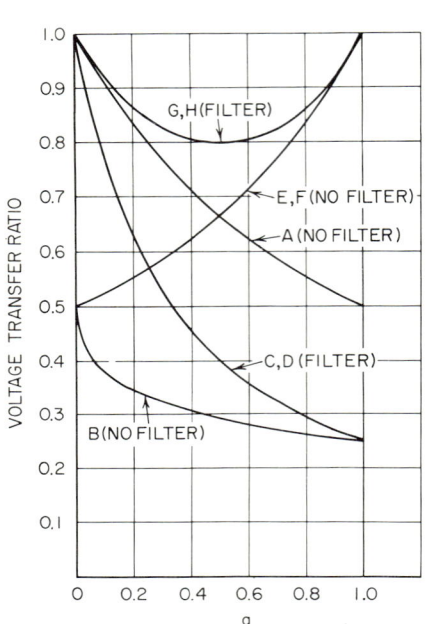
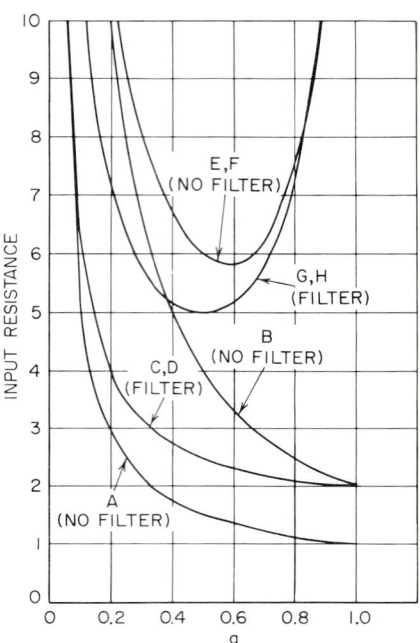

Fig. 38. Voltage transfer ratio of input-chopper circuits.

Fig. 39. Input resistance of input-chopper circuits.

comparable with A, which represents the best of the SPST circuits. The approximate relative standing is

```
           DPDT            SPST
        G, H (for c small)
          F—similar ↘
                         A
          E—similar ↗
                        C, D
                         B
```

An indication of the performance of the input-chopper circuits can be obtained by assuming values for the components and calculating the performance using the equations of Table 1. Assume that
$R_1 = R_3 = 1$.
R_2 is optimum.
Overlap and underlap are negligible in double-throw operation.
The voltage transfer ratio of the input choppers as a function of a is shown in Fig. 38. Double-pole circuits G and H have the largest voltage transfer ratio; it is equal to unity for a equal to unity and zero, and has a minimum value of 0.8 at $a = 0.5$. The input resistance of these circuits (Fig. 39) is second to the highest, which implies they draw relatively little current from the signal source. Both these chopper circuits

D-C AMPLIFIERS

employ an input filter. When the filter is not used, as in circuits E and F, the voltage ratio is lower, as shown in Fig. 38. However, the input resistance is somewhat higher.

Circuit A, the shunt SPST chopper without filter, provides the highest voltage transfer ratio of the SPST circuits. Its voltage transfer ratio is the mirror image of the voltage transfer ratio of the SPDT circuits E and F. However, as shown in Fig. 39, its input resistance is much lower and hence it draws considerably more current from the signal source. The series chopper, circuit B, provides the minimum voltage transfer ratio. For this reason, and because of the difficulties of floating a switch off ground, the series chopper is seldom used. The shunt and series SPST switches with filter, C and D, provide a voltage ratio and input resistance intermediate between the series and shunt SPST switches without filter, respectively, B and A. In general, the SPDT circuits provide a higher voltage transfer ratio than the SPST, but the advantage may not be too great if the shunt switch without filter is employed.

All the chopper circuits in Fig. 37 have a common disadvantage in that their ground reference is the same as that of the amplifier. It is possible to place the ground reference of the modulators at a different d-c potential than that of the amplifier by breaking the ground loop with a capacitor at A in Fig. 37A, for example. Alternatively the two-input voltage-differencing system could be used. However, in both cases the bottom lead of the chopper circuit would still have to be at the same a-c potential as the ground return of the amplifier.

Complete isolation can be obtained by employing the flying capacitor circuit (Fig. 40a). When both switch arms are to the left, capacitor C is charged to a voltage corresponding to the voltage at input 1. When both arms are to the right, capacitor C transfers this voltage to the input of the amplifier. It can be shown that this circuit is equivalent to the simplified SPST circuit shown in Fig. 40b for voltage transfer ratio and impedance calculations.

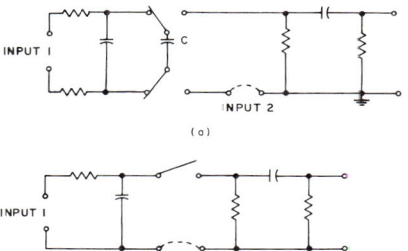

FIG. 40. Flying-capacitor chopper. (*a*) Circuit diagram. (*b*) Equivalent circuit. (*From "The Contact Modulator," Airpax Electronics Incorporated, Cambridge, Md., 1963.*)

7.2. Transformer-coupled Input Choppers. Transformer-coupled choppers are used when the advantages gained outweigh the simplicity, low cost, and light weight of capacitor-coupled choppers. Transformer-coupled choppers can have the following advantages:

1. Balanced or push-pull windings can be employed to minimize interference effects.
2. The a-c and d-c ground reference of the amplifier can be isolated from that of the signal source.
3. The amplification of the a-c amplifier can be stabilized by applying a-c feedback to the transformer windings.
4. The turns ratio can be adjusted to maximize the gain and efficiency of the chopper.

A transformer-coupled chopper may employ overlap or underlap switching. Underlap switching interrupts the current flow through the transformer and permits the possibility of large transient spikes. These can be eliminated through the use of a proper buffer capacitor. Overlap switching eliminates this problem but usually makes the chopper design more difficult, especially when an electromechanical type of chopper is employed.

When the cost and weight of a transformer are allowed, the added complexity of SPDT or DPDT switching is generally also permitted. Typical circuits for use with an overlap switch are shown in Fig. 41.

Figure 41a employs a DPDT switch and Fig. 41b a SPDT switch with a center-tapped winding. The SPDT circuit is more common because the switch simplicity

18-34 AMPLIFIER CIRCUITS

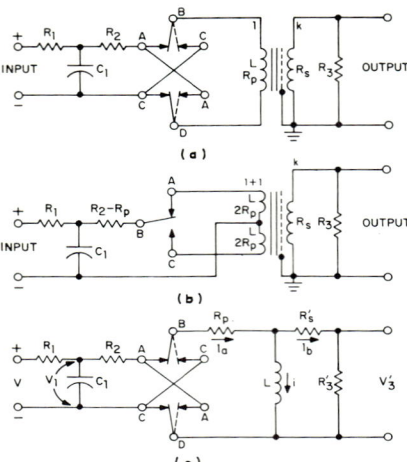

FIG. 41. Transformer-coupled chopper with overlap switch. (a) DPDT. (b) SPDT. (c) Equivalent circuit. (*From "The Contact Modulator," Airpax Electronics Incorporated, Cambridge, Md., 1963.*)

generally outweighs the added transformer complexity. The input filter is desirable in the circuit because the capacitor supplies the output current and magnetizing current of the transformer and thus minimizes waveform droop. The filter capacitor C_1 is assumed to be large enough such that the voltage across it is essentially d-c. Both circuits can be analyzed using the equivalent circuit of Fig. 41c with the nomenclature of Fig. 41a and b. A common analysis is permissible except when the two switches of the DPDT circuit are so far out of synchronism that the source becomes open-circuited, e.g., BA and DA are connected and BC and DC are open-circuited.

The switching sequence and time nomenclature are shown in Fig. 42. The time for a complete cycle is t_1. The time that each switch arm is contacting just one contact is t_2. The time that each switch is short-circuited and contacting both contacts is $2t_3$. In situations where the switches are not in exact synchronism,

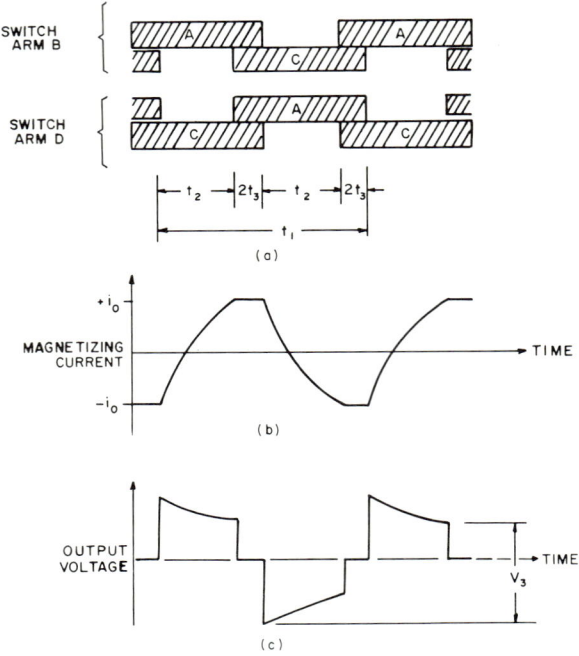

FIG. 42. Switching sequence and waveforms for overlap switch. (a) Switching sequence. (b) Magnetizing current. (c) Output voltage. (*b and c from "The Contact Modulator," Airpax Electronics Incorporated, Cambridge, Md., 1963.*)

D-C AMPLIFIERS

the total time per half-cycle that either or both are short-circuited is called $2t_3$. The waveform of the magnetizing current is shown in Fig. 42b. The magnetizing current is driven from a value $-i_o$ to $+i_o$ by the voltage V_1, across C_1, through stopper resistance R_2 and primary winding resistance R_p. During time t_3 the magnetizing current flows through the winding resistance and the short-circuited contacts of the switch, and remains essentially constant at $\pm i_o$.

Table 2. Equations for Transformer-coupled Input-chopper Circuits with Overlap Switches*

Property	Expression	Additional assumptions
Voltage transfer ratio	$\dfrac{2}{k\left(\dfrac{R_2}{R_3}+\dfrac{R_1}{R_3}\right)+\dfrac{1}{k}\left(1+4\dfrac{t_3}{t_1}\dfrac{R_1}{R_2}\right)+\dfrac{1}{6k}\dfrac{t_2}{t_1}\dfrac{R_1}{R_2}B^2}$	None
Voltage transfer ratio	$\dfrac{2}{k\left(\dfrac{R_2}{R_3}+\dfrac{R_1}{R_3}\right)+\dfrac{1}{k}\left(1+4\dfrac{t_3}{t_1}\dfrac{R_1}{R_2}\right)}$	L infinite
R_2 (opt) for max voltage transfer ratio	$\sqrt{4\dfrac{t_3}{t_1}\dfrac{R_1 R_3}{k^2}}$	L infinite
k (opt) for max voltage transfer ratio	$\sqrt{\dfrac{R_3}{R_1+R_2}\left(1+4\dfrac{t_3}{t_1}\dfrac{R_1}{R_2}\right)}$	L infinite
R_2 (opt opt)	$R_1\sqrt{4\dfrac{t_3}{t_1}}$	R_2 and k both optimum L infinite
k (opt opt)	$\sqrt{\dfrac{R_3}{R_1}}$	
Voltage transfer ratio (opt opt)	$\dfrac{1}{1+\sqrt{4\dfrac{t_3}{t_1}}}\sqrt{\dfrac{R_3}{R_1}}$	R_2 and k both optimum $\sqrt{4\dfrac{t_3}{t_1}} > \dfrac{1}{24}$ no additional restriction on L
Term describing output waveform	e^{-Bt/t_2}	None

Assumed in all expressions:
Winding resistances negligible.
Source never open-circuited.

$B = \dfrac{R_2 R_3/k^2}{R_2+R_3/k^2}\dfrac{t_2}{L} \lesssim 1$ (for droop $\lesssim 63$ per cent).

* Adapted from "The Contact Modulator," Airpax Electronics Incorporated, Cambridge, Md., 1963.

The output voltage is shown in Fig. 42c. It has a noticeable droop due to the stopper resistance R_2 and the finite value of the transformer inductance L. The a-c output voltage V_3 is assumed to correspond to the average of the output voltage at the beginning and end of t_2. The voltage transfer ratio of the circuit and the optimum values for R_2 and the turns ratio k are listed in Table 2. The complete expression for the voltage transfer ratio is shown in the first row. If L is sufficiently large, the last

18-36 AMPLIFIER CIRCUITS

term in the denominator can be eliminated, giving the equation in the second row. The value of L that permits this simplification is a rather long expression but can be easily derived from the voltage transfer ratio expression and the equation for B at the bottom of the table. The terms R_2 and k may be optimized individually; but for a maximum voltage transfer ratio, they should be optimized together as shown in the fifth and sixth rows of the table.

In a practical design one generally starts with the source resistance R_1, the load resistance R_3, and the switching times t_1, t_2, and t_3. With this information and the

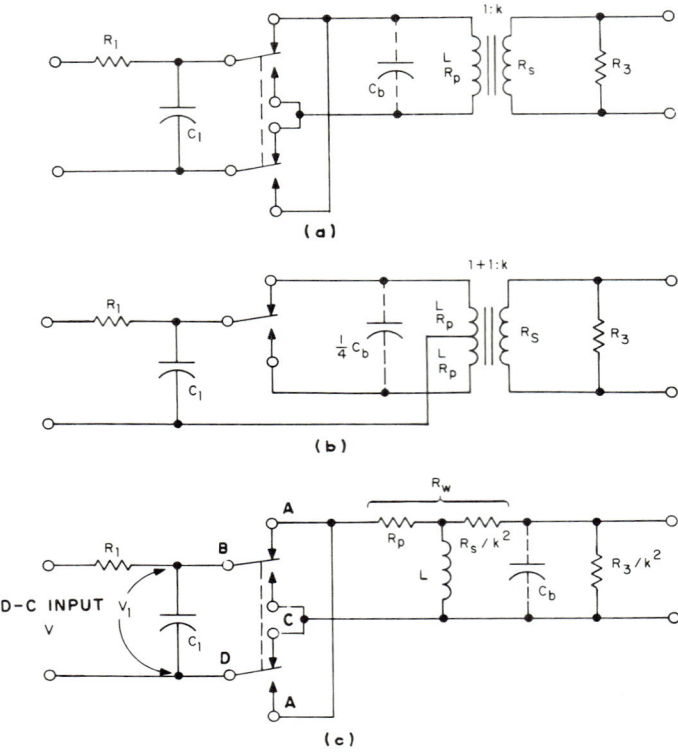

FIG. 43. Transformer-coupled choppers with underlap switch. (a) DPDT. (b) SPDT. (c) Equivalent circuit. (*From "The Contact Modulator," Airpax Electronics Incorporated, Cambridge, Md., 1963.*)

equations of Table 2, the values of R_2, k, and the voltage transfer ratio can be determined assuming L is infinite. The minimum value of L that can be employed is determined by setting a limit on the droop of the output voltage waveform. Using the equation at the bottom of Table 2, the voltage during t_2 is proportional to e^{-Bt/t_2}. At the beginning of t_2, $t = 0$ and

$$V_o \propto e^{B(0/t_2)} = 1 \qquad (19)$$

At the end of t_2, $t = t_2$ and

$$V_o \propto e^{-Bt_2/t_2} = \epsilon^{-B} \qquad (20)$$

The droop is

$$1 - e^{-B}$$

For a given droop this equation can be used to determine the maximum value of B;

and using the expression for B at the bottom of Table 2, an approximate minimum allowable value for L can be determined.

When an underlap switch is employed the transformer circuits of Fig. 43 are generally used. Figure 43a employs DPDT switching and Fig. 43b SPDT switching. These circuits frequently require a buffer capacitor C_b. An input filter R_1, C_1 is also generally employed. R_2 (Fig. 41) is unnecessary with an underlap switch. R_2 was employed with the overlap switch to prevent the switch from short-circuiting the filter.

The equivalent circuit for the two modulators is shown in Fig. 43c. L is the primary inductance, R_p the primary winding resistance, and R_s/k^2 the secondary winding

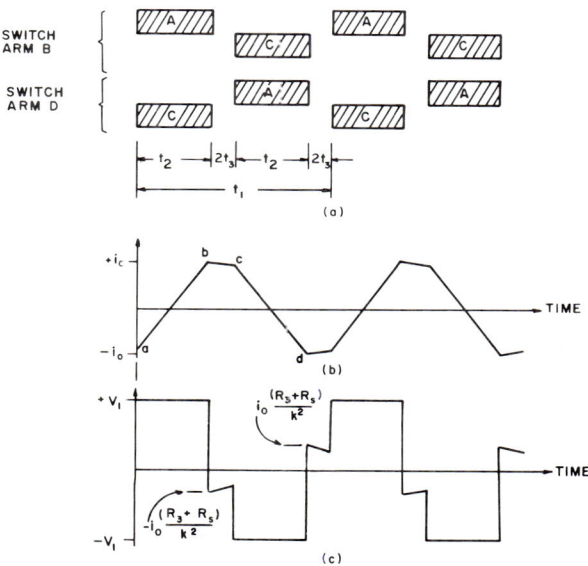

FIG. 44. Switching sequence and waveforms for underlap switch. (a) Switching sequence. (b) Magnetizing current. (c) Output voltage referred to primary. (*Adapted from* 'The Contact Modulator,'' Airpax Electronics Incorporated, Cambridge, Md., 1963.)

resistance referred to the primary. The total equivalent series resistance R_w is therefore

$$R_w = R_p + R_s/k^2 \tag{21}$$

R_3/k^2 is the load resistance referred to the primary. Equivalent parallel transformer losses can be lumped with this resistance.

The period that switch arms B and D are connected to each of the contacts A and C is illustrated in Fig. 44a. Also shown are the definitions of the cycle time t_1, the contact time t_2, and the open time $2t_3$.

Figure 44b shows the magnetizing current in the transformer. This waveform applies to a transformer which has a large primary inductance, and in which the leakage inductance can be neglected. At the start of the cycle a, the current starts at a negative value and rises linearly to a positive value $+i_o$, at b. This rise is due to the voltage V_1, across C_1, operating through a small primary winding resistance R_p. When the switch is open, during $2t_3$, the magnetizing current decays through load resistance R_3/k^2 in series with the secondary winding resistance R_s/k^2. Because the load resistance R_3/k^2 is usually a high resistance, decay occurs rather quickly. At the time of the switch closure c, the inductor is connected to a voltage V_1 of the opposite polarity, and the current decays linearly to its lower value $-i_o$, at d.

Assuming the voltage drop across R_p is small, the voltage is a flat-top wave, equal to $\pm V_1$, during the contact periods (Fig. 44c). During the open periods, the magnetizing current flows through the load resistance, giving an initial voltage $i_o(R_3 + R_s)/k^2$. The magnetizing current decays with a time constant $Lk^2/(R_3 + R_s)$. It is assumed that the decay during t_3 is not large, which requires that $Lk^2/(R_3 + R_s) > 2t_3$.

The circuit performance is largely dependent upon the inductance of the transformer. Significant circuit equations are tabulated in Table 3. Whether or not the

Table 3. Equations for Transformer-coupled Input Chopper with Underlap Switching*

	Large L	Medium L	Small L (with buffer capacitor)
Assumptions	$L \gg (R_3 + R_s)t_2/2k^2$	$L \approx \frac{(R_3 + R_s)t_2}{2k^2}$ $t_3 < t_1/10$	$L < \frac{(R_3 + R_s)t_2}{2k^2}$ $t_3 < t_1/10$
Input resistance (looking into C_1)	$(R_3/k^2 + R_w)t_1/2t_2$	$(R_3/k^2 + R_w)$	
Voltage transfer ratio	$\dfrac{2R_3}{k(2R_1t_2/t_1 + R_w) + R_3/k}$	$\dfrac{2R_3}{k(R_1 + R_w) + R_3/k}$	
k (opt)	$\left(\dfrac{R_3}{2R_1t_2/t_1 + R_w}\right)^{\frac{1}{2}}$	$\left(\dfrac{R}{R_1 + R_w}\right)^{\frac{1}{2}}$	
Voltage transfer ratio	$\left(\dfrac{R_3}{2R_1t_2/t_1 + R_w}\right)^{\frac{1}{2}} = k$ (opt)	$\left(\dfrac{R_3}{R_1 + R_w}\right)^{\frac{1}{2}} = k$ (opt)	

* From "The Contact Modulator," Airpax Electronics Incorporated, Cambridge, Md., 1963.

inductance is considered to be large is governed principally by the transformer voltage during the open-switch period t_3. As noted above, this voltage is directly proportional to the magnetizing current i_o. At the end of the dwell period this current is

$$i_o = t_2 V_1/2L \tag{22}$$

When a buffer capacitor is not employed, the step during the open-switch period should not be allowed to become larger than the voltage during the closed period. This means that

$$i_o (R_3 + R_s)/k^2 \leq V_1 \tag{23}$$

combining the two above equations gives

$$L \geq (R_3 + R_s)t_2/2k^2 \tag{24}$$

This equation provides the boundary conditions for large, medium, and small L.
Large L implies

$$L \gg (R_3 + R_s)t_2/2k^2 \tag{25}$$

This implies that the intermediate step during $2t_3$ is negligible compared with V_1. It results in the equations shown in the first column of Table 3. The transformer turns ratio k has an optimum value given in the fourth row; and this results in the optimum voltage transfer ratio in the fifth row. The optimum voltage transfer ratio is equal to the optimum turns ratio.

Under the assumption of medium L,

$$L \approx (R_3 + R_s)t_2/2k^2 \qquad (26)$$

This implies that the voltage during t_3 is approximately equal to V_1. The circuit equations for this condition are shown in the second column of Table 3. The optimum transformer ratio is given in the fourth row and simply represents matching the source and load impedances in the usual sine-wave sense. The optimum voltage transfer ratio is given in the fifth row and is equal to the optimum turns ratio.

The condition for small L is

$$L < (R_3 + R_s)t_2/2k^2 \qquad (27)$$

The intermediate step during t_3 will then exceed V_1. However, this situation is prevented by employing a buffer capacitor C_b. C_b in parallel with L and in parallel with the circuit shunt resistance forms a parallel-resonance circuit. During the open period t_3, the circuit has a relatively high Q. The magnetizing current serves to shock the circuit into a resonance oscillation. The oscillation is allowed to continue for only a fraction of a cycle, whereupon the switch closes and the normal conditions return. The capacitor is adjusted so that the instantaneous value of the oscillation voltage is just V_1 when the switch closes. The resulting voltage waveform is shown in Fig. 45b. The magnetizing current for this condition is shown in Fig. 45a.

If the capacitor is too small, the voltage waveform of Fig. 45c results. If it is too large the waveform of Fig. 45d results. The ideal buffering capacitance can be determined from Fig. 46. This figure gives a good estimate of the required buffering capacitance provided that

$$\frac{L}{R_3/k^2 + R_u} \leq \frac{t_1}{5} \text{ to } \frac{t_1}{10} \qquad (28)$$

The curve of Fig. 46 describes the equation

$$\tfrac{1}{2}t_2(LC_b)^{-\tfrac{1}{2}} = \cot t_3(LC_b)^{-\tfrac{1}{2}} \qquad (29)$$

If, for example, $4t_3/t_1 = 0.1$, then from Fig. 46, $C_b \approx 5.7 \times 10^{-3} t_1/L$ farads. Theoretically there is no lower limit on L. However, as L decreases, the circulating current increases and the I^2R losses in the transformer increase. Also, the buffering action becomes more critical; and it becomes impossible to obtain a value of buffering capacitance that is sufficiently close to the ideal for all conditions of parameter drift in the circuit.

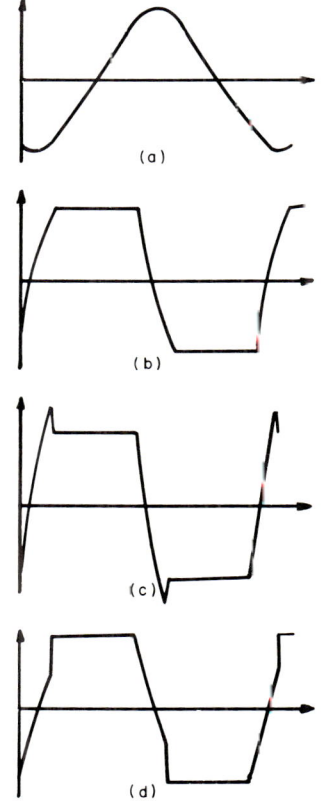

FIG. 45. Waveforms with buffering capacitor. (a) Magnetizing current. (b) Output voltage referred to primary, C_b ideal. (c) Output voltage referred to primary, C_b too small. (d) Output voltage referred to primary, C_b too large. (*From "The Contact Modulator," Airpax Electronics Incorporated, Cambridge, Md., 1963.*)

7.3. Output Switching Circuits. The output demodulator may be either transformer- or capacitor-coupled. Transformer coupling is generally employed when the requirements call for:

1. High output power, voltage, or current
2. High efficiency
3. Ground return of d-c output independent of ground return of a-c amplifier

AMPLIFIER CIRCUITS

The transformer permits the impedance of the amplifier output to be matched to obtain maximum power output. Furthermore, it provides a low-resistance d-c path to provide bias current for a tube or transistor output stage. In conventional output coupling circuits, it permits the collector (plate) of the transistor (tube) to swing both above and below the supply voltage and thus permits a greater output power for a given supply voltage than RC coupling. Transformers are generally employed with some type of double-throw switching to provide greater efficiency and eliminate the

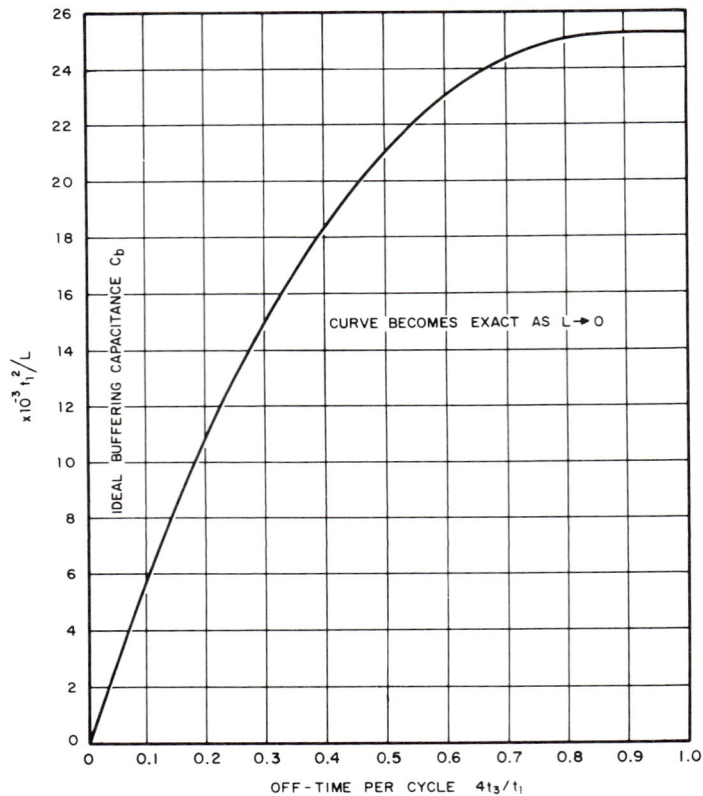

FIG. 46. Ideal buffering capacitance. (*From "The Contact Modulator," Airpax Electronics Incorporated, Cambridge, Md., 1963.*)

direct magnetizing current from the output winding. When only ground independence (factor 3 above) is required, and the other factors are not controlling, the transformer may be eliminated by employing a flying capacitor output circuit similar to Fig. 40a.

A capacitor-coupled output, on the other hand, offers simplicity, low cost, and light weight. Furthermore, it avoids the possibility of troublesome inductive voltage surges. For this reason, capacitive output coupling is common, especially in low-level amplifiers.

Two types of transformer-coupled output demodulators are shown in Fig. 47. Figure 47a employs a center-tapped transformer and a SPDT switch. Figure 47b employs an untapped transformer and a DPDT switch. A filter is generally required. The filter may or may not employ a series resistor. The series resistor tends to reduce

D-C AMPLIFIERS 18-41

the efficiency and voltage regulation. However, its elimination can result in high peak capacitor charging currents. Such currents may overload the output stage of electronic amplifiers.

A capacitor-coupled output demodulator is shown at the top of Table 4. The output of the amplifier that drives the demodulator is represented by a Thévenin voltage generator that develops a rectangular wave V and has a series internal resistance R_1. The coupling capacitor C_c is assumed to be sufficiently large that it has a constant d-c voltage across it. A filter R_2, C is employed. The switch is of the shunt type. A demodulator employing a series switch could also be used; but, as was shown previously, the series switch circuit is usually less desirable than the shunt.

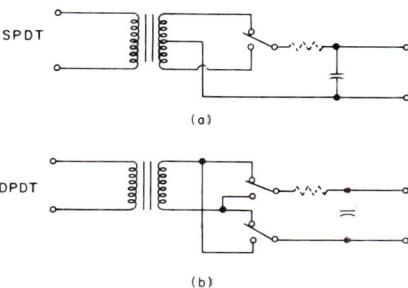

Fig. 47. Transformer-coupled output demodulators.

The performance of the SPST shunt demodulator is given in Table 4. Because of the coupling capacitor, only the a-c component of the drive voltage V is of significance. This voltage has two states d and e. The switch S also has two states, close, y, and open, z. Ideally, the switch and the drive voltage should be in synchronism; practically they will be somewhat out of synchronism. This permits four circuit states ($V = e$ and $S = z$ or y, and $V = d$ and $S = z$ or y). The duration of each of these states is given by fractions p, q, r,

Table 4. Capacitor-coupled Output Demodulator with SPST Switch*

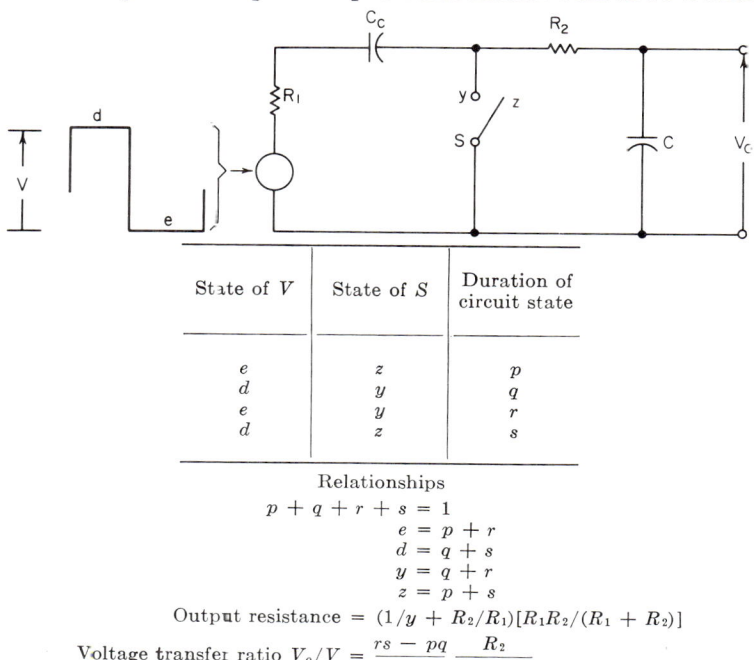

State of V	State of S	Duration of circuit state
e	z	p
d	y	q
e	y	r
d	z	s

Relationships
$$p + q + r + s = 1$$
$$e = p + r$$
$$d = q + s$$
$$y = q + r$$
$$z = p + s$$

Output resistance $= (1/y + R_2/R_1)[R_1 R_2/(R_1 + R_2)]$

Voltage transfer ratio $V_o/V = \dfrac{rs - pq}{y} \dfrac{R_2}{R_1 + R_2}$

* Adapted from I. C. Hutcheon, Performance Calculations for D.C. Chopper Amplifiers, *Electron. Eng.*, vol. 30, pp. 477–478, August, 1958.

18-42 AMPLIFIER CIRCUITS

and s, as shown in the middle table. These letters represent fractions of a cycle such that their sum is equal to unity, as shown in the lower table. Also, if e is identified as a fraction of a cycle, then $e = p + r$ as shown; and similar equations can be written for d, y, and z. The output resistance and voltage ratio, in terms of these symbols are given in the lower table.

The output resistance is not a function of e and d. It is determined only by the switch closing y and the circuit resistances. The voltage transfer ratio, however, is a function of all terms. The output resistance decreases with decreasing R_2, decreasing

Table 5. Capacitor-coupled Output Demodulator with SPDT Switch*

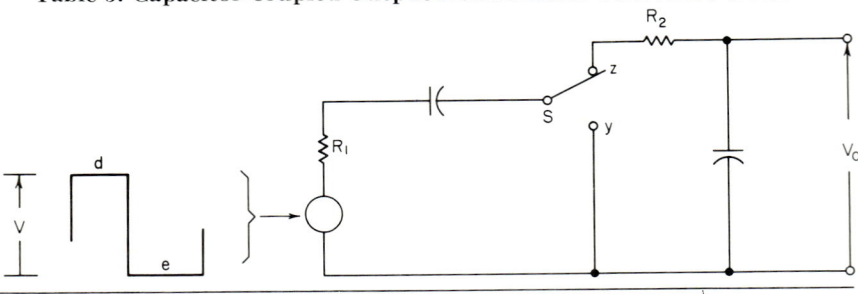

	State of S			Duration of circuit state
State of V	Overlap or underlap negligible	Underlap y and z	Overlap y and z	
e	z	z	z	p
d	y	y	y	q
e	y	y	y	r
d	z	z	z	s
e	...	...	y and z	t
d	...	...	y and z	u
e	...	Arm between y and z	...	v
d	...	Arm between y and z	...	w

Relationships

	Overlap or underlap negligible	Underlap y and z	Overlap y and z
	$p + q + r + s = 1$	$p + q + r + s + v + w = 1$	$p + q + r + s + t + u = 1$
$e =$	$p + r$	$p + r + v$	$p + r + t$
$d =$	$q + s$	$q + s + w$	$q + s + u$
$y =$	$q + r$	$q + r$	$q + r + t + u$
$z =$	$p + s$	$p + s$	$p + s + t + u$
Output resistance $=$	$\dfrac{R_1}{z(1-z)} + \dfrac{R_2}{z}$	$\dfrac{y+z}{yz} R_1 + \dfrac{R_2}{z}$	$\dfrac{R_1 + yR_2}{(y+z-1)(R_2/R_1) + yz}$
Voltage transfer ratio $V_o/V =$	$\dfrac{rs - pq}{yz}$	$\dfrac{rs - pq}{yz}$	$\dfrac{s(r+t) - p(q+u)}{(y+z-1)(R_2/R_1) + yz}$

* Adapted from I. C. Hutcheon, Performance Calculation for D.C. Chopper Amplifiers, *Electron. Eng.*, vol. 30, pp. 477–478, August, 1958.

D-C AMPLIFIERS

R_1, and increasing y. The voltage transfer ratio decreases with decreasing R_2, and with increasing y. The effect of R_2 can be understood by noting that capacitor C charges through R_1 and R_2; and hence the charge placed on it, when the switch is open, is inversely proportional to the sum of these resistances. When the switch is closed, the capacitor discharges through R_2, such that the charge removed is inversely proportional to R_2; hence the average voltage is proportional to the ratio of R_2 to $R_1 + R_2$.

When the switch operates in near synchronism with the drive voltage, either r and s will be simultaneously large and p and q simultaneously small or conversely, depending upon the phasing. These conditions give rise to the maximum value for the magnitude of the numerator in the voltage transfer ratio expression.

A capacitor-coupled demodulator with SPDT switch is shown at the top of Table 5. The remainder of the circuit is similar to the circuit of Table 4. If the switch has no underlap or overlap four circuit states exist as before. If it has significant underlap or overlap, two additional circuit states are added. The relationships illustrating circuit performance are outlined at the bottom of the table. The best performance is obtained when the switch operates in synchronism with the drive voltage, and the underlap and overlap are negligible. The voltage transfer ratio of the SPDT circuit with negligible overlap or underlap is greater than the SPST circuit by the value of $1/z$ ($z < 1$) and the elimination of the term $R_2/(R_1 + R_2)$ from the expression. The underlap operation is similar to the overlap or underlap negligible operation when the underlap is small. The overlap operation is somewhat more complicated because the short-circuit condition of the switch drains the filter capacitor through R_2.

8. MODULATOR-DEMODULATOR PERFORMANCE

The input modulator and output demodulator are strongly interdependent. Economies of switching contacts frequently can be achieved by proper integration of the two units. The demodulator is also particularly dependent upon the waveform presented to it; and this of course is basically determined by the modulator.

A number of input-output switching circuit combinations are shown in Fig. 48. Circuit 1 A-ST (input circuit A, Fig. 37, output switch single-throw) connects the input and output terminals to ground in synchronism. If it is assumed that the phase shift in the amplifier is zero, and the switches operate exactly in synchronism, the following relationships can be written:

$$a = e = y = r$$
$$b = d = z = s \qquad (30)$$
$$z = 1 - y$$

Using these relationships, the voltage transfer ratio and output resistance of the output demodulator can be determined from the equations of Table 4. These are given in the first and third rows of Table 6, respectively. The equations in the first column are derived in terms of the output switch closure y. The equations in the second column are derived in terms of the input switch closure a.

In circuit 3, Fig. 48, the input and output switches operate in reverse synchronism; the output is open-circuited when the input is short-circuited and conversely. With reference to Table 4 this gives the following relationships:

$$a = e = z = p$$
$$b = d = y = q \qquad (31)$$
$$z = 1 - y$$

The voltage transfer ratio and output impedances are determined from Table 4 as before and listed in the first and third rows of Table 6. The negative sign before the circuit number denotes a phase reversal. The equations for the other circuits of Fig. 48 have been determined similarly and listed in Table 6. Note that the voltage transfer ratio for the double-throw output switching modes is unity and is not a function of the source and filter resistances of the switch.

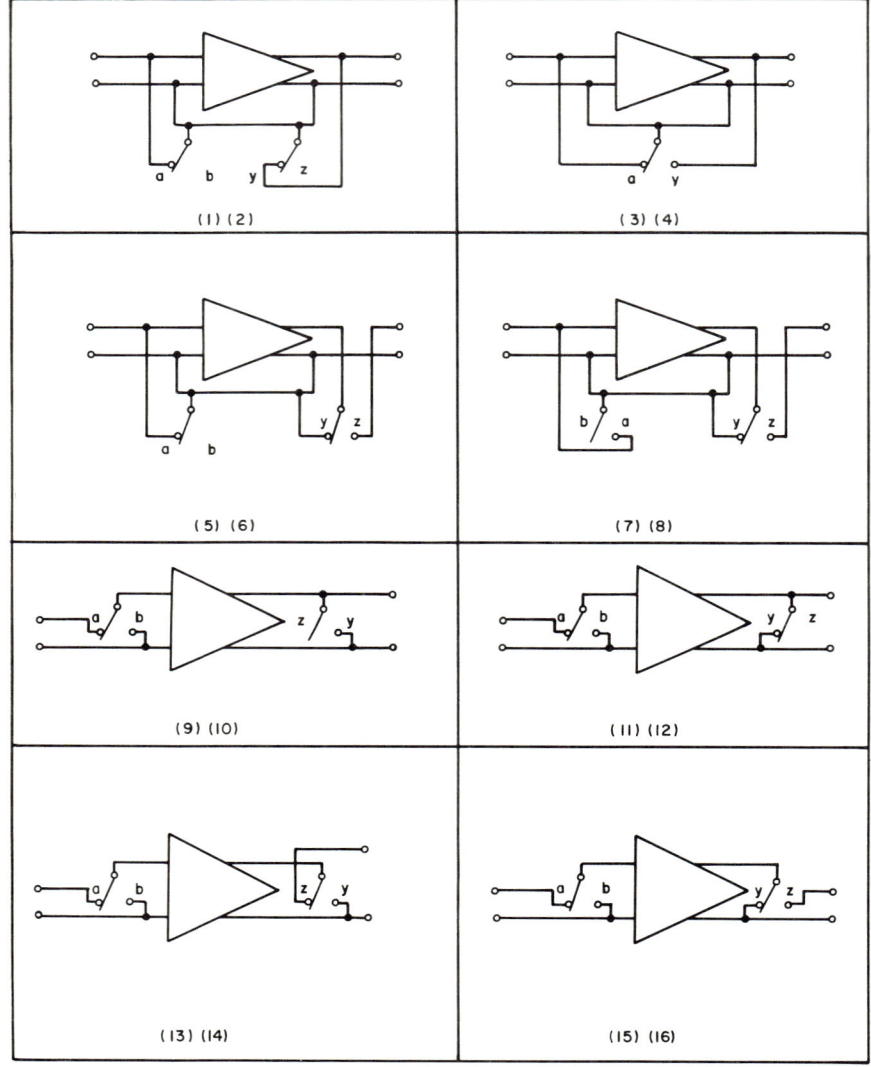

Fig. 48. Modulator—demodulator combinations. (1) A-ST. (2) C-ST (filter). (3) A-ST (reverse). (4) C-ST (filter reverse). (5) A-DT. (6) C-DT (filter). (7) A-DT (reverse). (8) C-DT (filter reverse). (9) E,F-ST. (10) G,H-ST (filter). (11) E,F-ST (reverse). (12) G,H-ST (filter reverse). (13) E,F-DT. (14) G,H-DT (filter). (15) E,F-DT (reverse). (16) G,H-DT (filter reverse).

Assuming that $R_1 = R_2$, the voltage transfer ratios of the output switches as a function of y are shown in Fig. 49. The voltage transfer ratio of the SPDT switch configuration is unity (positive or negative) independent of y. The voltage transfer ratio of the SPST switch is directly proportional to $(1 - y)$; that is, it is directly proportional to the fraction of the time the output is open-circuited. For equal source and filter resistances, the maximum voltage transfer ratio of the SPST switch is 0.5; and the voltage transfer ratio for a symmetrical wave, $y = 0.5$, is 0.25.

D-C AMPLIFIERS   18-45

Assuming that $R_1 = R_2 = 1$, the output resistances of the output switch configurations are shown in Fig. 50 as a function of y. The SPST switch configurations have the lower output resistance, which is expected since the output is short-circuited for a fraction of the time. The SPST output resistance approaches unity as the fraction of closure time for the switch approaches unity. As the fraction of closure time for the switch approaches zero, the output resistance approaches infinity. The SPDT output switch configuration has a minimum output resistance of 5.8 at $y = 0.414$; and it approaches infinity as y approaches zero or unity.

Table 6. Output Demodulator Equations*

	In terms of output switch closure	In terms of input switch closure	
Voltage transfer ratio SPST	$(1 - y) \dfrac{R_2}{R_1 + R_2}$	$a \dfrac{R_2}{R_1 + R_2}$	$(1 - a) \dfrac{R_2}{R_1 + R_2}$
Circuits	1, 2, −3, −4, 9, 10, −11, −12	1, 2, −11, −12	−3, −4, 9, 10
Voltage transfer ratio SPDT	\multicolumn{3}{c}{1 (independent of switch closures)}			
Circuits	\multicolumn{3}{c}{5, 6, −7, −8, 13, 14, −15, −16}			
Output resistance SPST	$\left(\dfrac{1}{y} + \dfrac{R_2}{R_1}\right) \dfrac{R_1 R_2}{R_1 + R_2}$	$\left(\dfrac{1}{a} + \dfrac{R_2}{R_1}\right) \dfrac{R_1 R_2}{R_1 + R_2}$	$\left(\dfrac{1}{1 - a} + \dfrac{R_2}{R_1}\right) \dfrac{R_1 R_2}{R_1 + R_2}$
Circuits	1, 2, −3, −4, 9, 10, −11, −12	1, 2, −11, −12	−3, −4, 9, 10
Output resistance SPST	$\dfrac{R_1}{y(1 - y)} + \dfrac{R_2}{1 - y}$	$\dfrac{R_1}{a(1 - a)} + \dfrac{R_2}{1 - a}$	$\dfrac{R_1}{a(1 - a)} + \dfrac{R_2}{a}$
Circuits	5, 6, −7, −8, 13, 14, −15, −16	5, 6, −15, −16	−7, −8, 13, 14

* Input and output switches in perfect synchronism, instantaneous switching with SPDT.

Using the techniques of the preceding sections, the overall voltage amplification of a chopper amplifier can now be determined. The voltage amplifications of the 16 circuit configurations of Fig. 48 are illustrated in Fig. 51. The closure time a of the input switch is the independent variable. Series SPST switches have not been included because of their poor voltage transfer ratio as compared with shunt SPST switches. The assumptions of Figs. 38 and 49 are made here also. They are:

Input: $R_1 = R_3$.
R_2 is optimum.
Overlap and underlap are negligible in double-throw operation.
Output: $R_1 = R_2$.
Input and output are in perfect synchronism.
Overlap and underlap are negligible in double-throw operation.

18-46 AMPLIFIER CIRCUITS

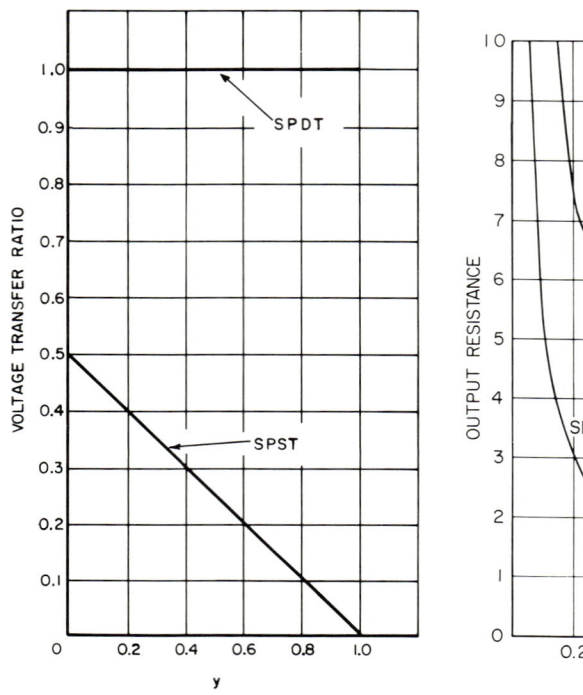

Fig. 49. Demodulator voltage transfer ratio.   Fig. 50. Demodulator output resistance.

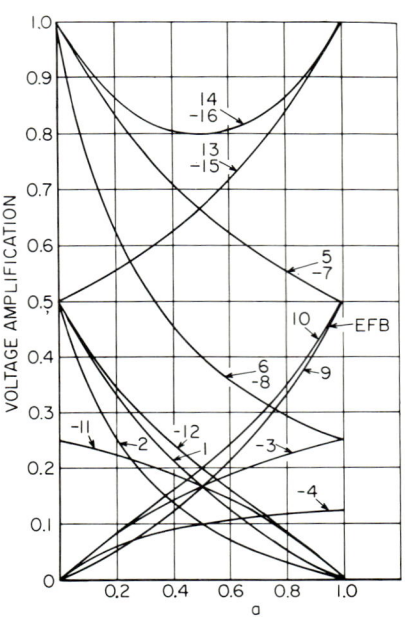

Fig. 51. Voltage amplification of chopper amplifiers.

D-C AMPLIFIERS   18-47

The voltage amplification of the amplifier is unity with zero phase shift. The circuits having the maximum voltage amplification are 14 and −16. The minus sign implies reverse synchronization and a phase reversal between input and output. Circuits 14 and −16 employ double-throw input circuits G or H with filter and a double-throw output circuit. The same switch combinations without filter, circuits 13 and −15, have less amplification. Circuits 5 and −7 have one less contact than 13 and −15 but produce the mirror image of the voltage amplification. The simplest circuits, 3 and −4, have about the smallest amplification. However, they are frequently used because of their simplicity

9. TYPES OF MODULATORS

9.1. Electromechanical Choppers. Electromechanical choppers employ mechanical switches that are driven by electrical means. While motor-driven commutators fit into this category, the most common electromechanical choppers employ a vibrating reed with a magnetic drive. The contacts on the reed make and break

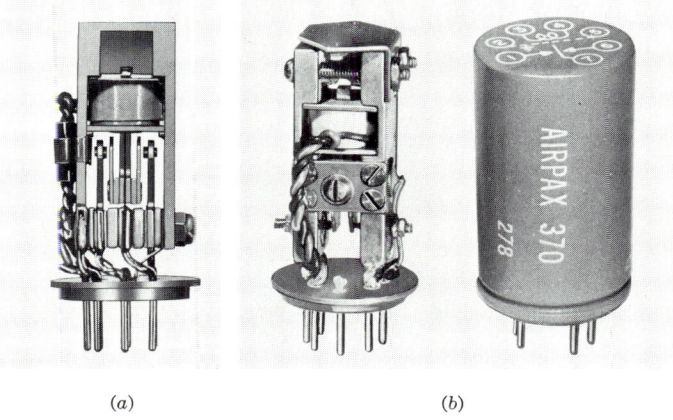

(a)   (b)

FIG. 52. Electromechanical chopper. (a) Typical electromechanical chopper. (b) Balanced armature chopper. (*From "The Contact Modulator," Airpax Electronics Incorporated, Cambridge, Md., 1963.*)

with stationary contacts in synchronism with the driving frequency. A typical electromechanical chopper is illustrated in Fig. 52a.

An equivalent circuit is shown in Fig. 53a. Single-pole double-throw operation is probably the commonest. A greater number of contacts can be employed, but this is generally not desirable. The larger number of contacts require careful mechanical matching to achieve completely synchronous operation. When a greater number of contacts are needed it is often simpler to employ a number of similar SPDT choppers. Equal or better contact synchronization can often be obtained by employing a number of similar SPDT choppers with a common drive voltage than can be obtained by employing multiple contacts on a single chopper.

Important electromechanical-chopper characteristics are outlined in Table 7. The closed resistance of the contacts can be made less than 0.1 ohm and the open resistance greater than 10^{10} ohms. The offset or null voltage of the choppers can be made as low as 1 μv with extreme precautions; however, more common values are tenths of a millivolt to a few millivolts. Common drive frequencies are 50, 60, and 400 cycles. Special choppers can be obtained that operate down to a few cycles per second or up to about 1,000 cps. The drive frequency is usually limited to about a ±10 per cent range because of the electrical and mechanical phase shifts of the chopper with frequency. By its very nature, the chopper is almost perfectly linear and can

18-48 AMPLIFIER CIRCUITS

Table 7. Electromechanical-chopper Characteristics

	Typical	Special design
Contacts....................	SPDT	Wide range
Contact resistance, closed.........	<0.1 ohm	
Contact resistance, open.........	>10^{10} ohms	
Offset or null.................	0.1–1 mv	1 μv
Drive frequency...............	50, 60, 400 cps	A few cps to 1,000 cps
Noise.......................	1 mv p-to-p	1 μv p-to-p and up
Temperature..................	−65 to +100°C	Up to 200°C

handle voltages from noise level up to a few volts, normally, although some choppers can handle up to 100 volts. The noise level of high-quality choppers varies from a few microvolts peak-to-peak for very low noise choppers to several millivolts peak-to-peak. Low-level choppers are generally rated at a contact current of 1 or 2 ma. An operating

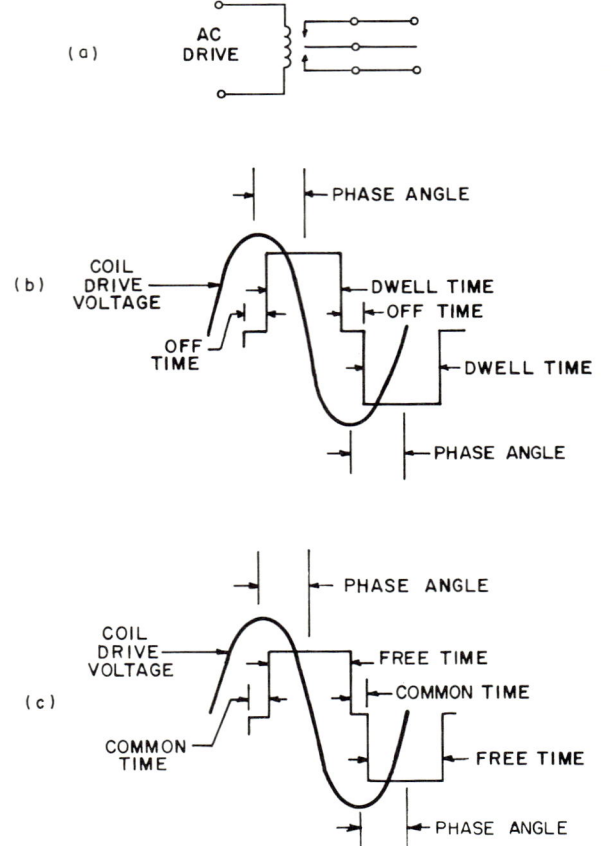

FIG. 53. Electromechanical chopper nomenclature. (*a*) Schematic. (*b*) Underlap nomenclature. (*c*) Overlap nomenclature. (*After "The Contact Modulator," Airpax Electronics Incorporated, Cambridge, Md., 1963.*)

D-C AMPLIFIERS

temperature range of −65 to +100°C is usual, although some choppers can go up to 200°C. Temperature changes cause changes primarily in the phase characteristics. Choppers can be sensitive to vibration, which can modulate their on-off characteristics. For situations of high vibration, special balanced-armature choppers can be obtained. A balanced-armature chopper is illustrated in Fig. 52b.

The phase characteristics of a chopper are highly important when it must operate synchronously. Figure 53 illustrates the phase characteristics of the SPDT underlap chopper. The sine-wave drive voltage and contact closure are shown. The contact closure lags the drive voltage by an angle of about 30 to 60°. The lag is due partially to the lagging current in the drive coil and partially to the lag of the mechanical system. The *dwell* time is the closure time of the contacts. The *off* time is the time when both contacts are open. The *dwell* times for the two halves of the cycle may be unequal; the *off* times may also be unequal.

A nomenclature for overlap operation is shown in Fig. 53c. The *free* time is the time when one contact is closed and the other is open. The *common* time is the time when both contacts are closed. The corresponding times for each half of the cycle may not be equal as with the underlap switch.

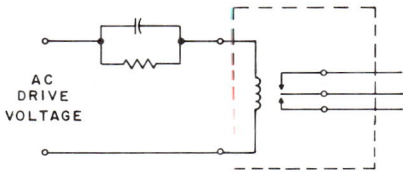

FIG. 54. Phase-correction circuit.

The phase angle between the coil drive voltage and the switch operation must be corrected in many applications. If this angle is less than 90°, the correction can be made by employing an *RC* circuit in series with the drive coil (Fig. 54). In designing this circuit, attention should be given to the change in drive-coil impedance when the switch goes from a nonoperating to an operating condition. The drive in the nonoperating condition must be sufficient to start the contacts operating.

9.2. Transistor Choppers. Transistor choppers have the advantages of high efficiency, high speed, long life, compactness, and light weight when compared with

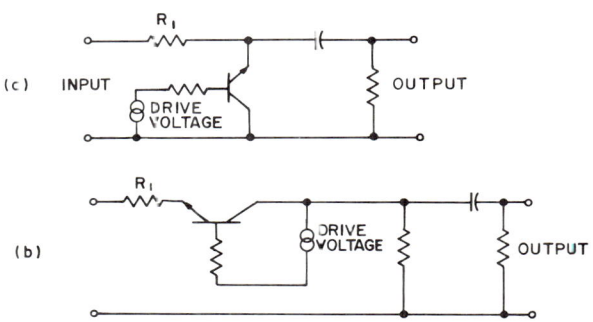

FIG. 55. Transistor choppers. (a) Shunt. (b) Series.

electromechanical choppers. However, when compared on the basis of series resistance, open-circuit shunt resistance, offset voltage, offset current, isolation between contacts and drive source, linearity, and the presence of switching spikes, transistors are inferior to high-quality electromechanical choppers. In spite of this, transistors are being used in increasing numbers for the generally less critical applications where their performance characteristics are adequate.

Both bipolar and unipolar transistors can be used as choppers. Their properties are quite different.

Bipolar-transistor Choppers. A shunt-connected bipolar-transistor chopper is shown in Fig. 55a. The drive voltage, connected between the base and collector, alternately turns the transistor off and on. When the transistor is off it is virtually an open circuit; when it is on it is virtually a short circuit. The transistor is in the inverted connection; that is, the emitter and collector are reversed such that the drive signal is injected between base and collector rather than between base and emitter as is usual. The reasons for employing this connection will become apparent.

A series-connected bipolar-transistor chopper is shown in Fig. 55b. The drive voltage, connected between base and collector, generally must be provided by a transformer.

An equivalent circuit for the "on" transistor is shown in Fig. 56a. Between the emitter and collector terminals, the device appears as a resistance r_S in series with two emfs. The resistance r_S is the sum of three resistances.

$$r_S = r_E' + r_C' + r_{EC} \qquad (32)$$

Here, r_E' and r_C' are the parasitic ohmic resistances in, respectively, the emitter and

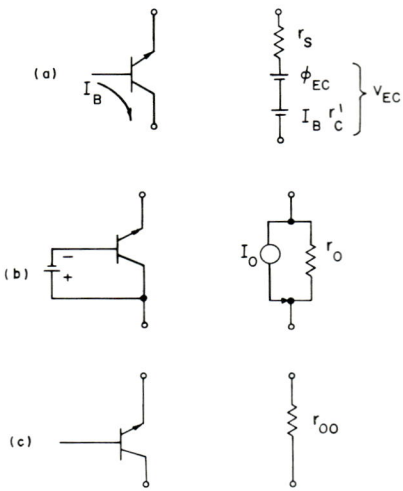

FIG. 56. Equivalent circuits of a transistor switch. (a) Transistor on. (b) Transistor off, base reverse-biased. (c) Transistor off, base open.

collector; and r_{EC} is the resistance of the active transistor. Usually r_E' and r_C' are less than 1 ohm, but they may be considerably higher in some cases. The resistance r_{EC} is given for all conditions of fixed base and emitter current by

$$r_{EC} = \pm \frac{kT}{eI_B} \left[\frac{1 - \alpha_I}{\alpha_I \left(1 - \dfrac{I_E}{I_B}\dfrac{1-\alpha_I}{\alpha_I} + \dfrac{I_{CBO}}{\alpha_N I_B}\right)} + \frac{1 - \alpha_N}{1 + \dfrac{I_E(1-\alpha_N)}{I_B} + \dfrac{I_{CBO}}{I_B}} \right] \qquad (33)$$

The positive sign is for an n-p-n transistor and the negative for a p-n-p.
The symbols are defined as follows:

I_B base current $\qquad T$ absolute temperature
I_E emitter current $\qquad \alpha_I$ inverse α
I_C collector current $\qquad \alpha_N$ normal α
e electronic charge $\qquad I_{CBO}$ collector reverse
k Boltzmann's constant $\qquad\qquad$ saturation current

The sign conventions are very important. The currents I_B, I_E, and I_C are defined as positive when they flow into the transistor and negative when they flow out. The reverse saturation currents I_{CBO} and I_{CEO} are positive for an n-p-n transistor and negative for a p-n-p transistor. The α's in this case are the d-c α's (h_{FB}) and are always positive. The *normal* α is equal to the ratio of the short-circuit collector current (less I_{CBO}) to the emitter drive current. The *inverse* α is determined similarly, with the emitter functioning as a collector and the collector as an emitter.

For the "on" condition, the transistor should be driven sufficiently hard to result in the following inequalities:

$$\begin{aligned} |I_B| &\gg |I_E|(1 - \alpha_I)/\alpha_I \\ |I_B|\alpha_N &\gg |I_{CBO}| \\ |I_B| &\gg |I_E|(1 - \alpha_N) \\ |I_B| &\gg |I_{CBO}| \end{aligned} \qquad (34)$$

Combining these inequalities with Eq. (33) gives an approximate value for r_{EC}.

$$r_{EC} \approx (kT/e|I_B|\alpha_I)(1 - \alpha_N\alpha_I)$$
$$r_{EC} \approx (26/I_B\alpha_I)(1 - \alpha_N\alpha_I) \qquad \text{(room temperature, } I_B \text{ in milli-amperes)} \qquad (35)$$

Increasing I_B decreases the series resistance.

The internal offset voltage ϕ_{EC} is

$$\phi_{EC} = \mp \frac{kT}{e} \ln \left[\frac{\alpha_N \left(1 - \dfrac{I_E}{I_B} \dfrac{1 - \alpha_I}{\alpha_I}\right) + \dfrac{I_{CBO}}{I_B}}{1 + \dfrac{I_E}{I_B}(1 - \alpha_N) + \dfrac{I_{CBO}}{I_B}} \right] \qquad (36)$$

With the assumptions of Eq. (34) this becomes

$$\phi_{EC} \approx \mp (kT/e) \ln \alpha_N \qquad \text{(negative n-p-n, positive p-n-p)}$$
$$\phi_{EC} \approx \pm 26(1 - \alpha_N) \quad \text{mv} \qquad \text{(room temperature } \alpha_N > 0.9, \qquad (37)$$
$$\text{positive n-p-n, negative p-n-p)}$$

The $I_B r_C'$ voltage of Fig. 56a is the voltage drop of the base current through the series collector resistance.

As α_N and α_I approach unity, the series resistance and the internal voltage both decrease. If the transistor were in the normal connection, α_N in would be replaced by α_I. This would give a larger offset voltage since normally α_N is greater than α_I. For this reason, the inverted connection is almost always used in transistor switches.

Typical "on" characteristics are as follows: $I_B = 0.5$ ma, $\alpha_N = 0.97$, $\alpha_I = 0.6$, $r_E' = 0.3$ ohms, and $r_C' = 0.3$ ohms. From Eq. (35), $r_{EC} = 9$ ohms; and from Eq. (32), $r_S = 9.6$ ohms. From Eq. (37) $\phi_{EC} = 0.8$ mv; and $I_B r_C' = 0.15$ mv, such that the "on" voltage of the transistor is 0.95 mv.

The "off" condition is shown in Fig. 56b. The transistor is reverse-biased and has an equivalent circuit consisting of a resistance r_O in parallel with an offset current generator I_O. The resistances r_E' and r_C' are still present but are negligible. For silicon transistors r_O is on the order of megohms or tens of megohms. For germanium transistors it is usually one or two orders of magnitude less.

The offset current is given by

$$I_O = I_{CBO} \frac{\alpha_I}{\alpha_N} \frac{1 - \alpha_N}{1 - \alpha_N\alpha_I} \qquad (38)$$

For $\alpha_I < \alpha_N$, I_O is always less than the collector saturation current I_{CBO}. It is important to note that α_N and α_I are measured at the operating current of the transistor I_O. In good passivated silicon transistors, I_{CBO} and I_O are on the order of a few nanoamperes at room temperature; α_N and α_I are then very small, and I_O is not very much less than I_{CBO}. However, with germanium transistors, or silicon transistors at high temperature, where the saturation current is larger, I_O may be significantly less than I_{CBO}.

If the offset current is a problem, the mode of operation of Fig. 56c can be employed. The base is open-circuited in this case. The current generator must go to zero, since there is no energy applied to the transistor. The output resistance r_{OO} is significantly less than for the case of Fig. 56b. It is given by

$$r_{OO} = (kT/e|I_{CBO}|)[1 + (\alpha_N/\alpha_I) - 2\alpha_N] \tag{39}$$

Note that r_{OO} is inversely proportional to $|I_{CBO}|$. Therefore, it will decrease by a factor of 2 for every 8 to 11°C rise in temperature if $|I_{CBO}|$ follows the usual pattern and α_N and α_I are more or less constant.

Switching Transients. Switching transients of the transistor are a source of offset and can cause serious overload conditions in the a-c amplifier following the chopper. Figure 57 illustrates the transient performance of the shunt chopper with an n-p-n transistor. The circuit is shown in Fig. 57a and the base drive voltage in Fig. 57b. The waveform at the collector will have the general form shown in Fig. 57c. Of special significance are the transient spikes when the transistor is turned off and on. The exact shape of the waveform, especially during the transient periods, can be expected to vary considerably depending upon transistor type, drive waveform, voltage and current levels, circuit impedances, etc. The transients are caused by (1) the base drive which is coupled to the emitter by the emitter-base capacitance, (2) the charging and discharging of the emitter and collector junction capacitances as the transistor is turned off and on, and (3) the charge that must be injected into the base and collector or emitter regions during the "on" time and removed during the "off" time (the charge-storage effects). The charge-storage effects, furthermore, give rise to delay in the output signal. Unfortunately, the positive and negative transients are not the same and they drift with temperature. If the charge difference between the positive and negative transients is ΔQ, the transients can be considered to result in an equivalent offset current ΔI_O, where $\Delta I_O = \Delta Q/T$ and T is the period of the "off" time. The transients and the delay set a maximum on the chopping frequency.

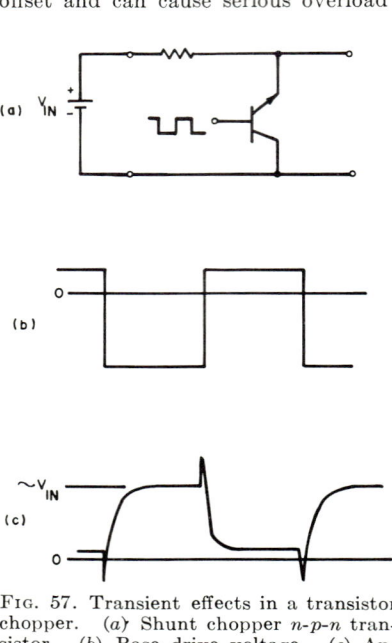

FIG. 57. Transient effects in a transistor chopper. (a) Shunt chopper n-p-n transistor. (b) Base drive voltage. (c) Approximate waveform at emitter.

The transient effects can be limited in several ways. The spikes are proportional to the voltage swing on the base, and they can be minimized by making this voltage swing as small as possible, consistent with other requirements. The charge-storage effects are proportional to the base current during the "on" condition. They can be minimized by making the "on" base current no larger than is necessary to ensure transistor saturation and sufficiently low values of V_{EC} and r_{EC}. Finally, selecting a transistor with minimum rise and fall times is desirable. These requirements imply the transistor should have small capacitances and a large cutoff frequency.

With a reasonably optimum design and high-speed transistors, the magnitude of the spikes can be kept to a few millivolts. However, under very bad conditions, the magnitude of the spikes may be almost as large as the base drive voltage. When the output of the chopper is fed into a sensitive amplifier, the spikes may overload the amplifier. This must not be allowed to occur. It is possible in some cases to remove the spikes by clipping. When an output demodulator is employed, the spikes cause large transients in the output circuit. In some cases, the transients may be rendered ineffective by keying the demodulator to operate only during the period of the cycle when they are negligible.

Since to minimize transients, the base current in the "on" condition and the base voltage in the "off" condition must be minimized, what are the constraints on the base drive? Consider first the "on" condition. The key is given by Eqs. (33) to (36). The resistance r_{EC} is inversely proportional to $|I_B|$. Therefore, $|I_B|$ should be sufficiently large to reduce r_{EC} to a value that is acceptably small. This is the first constraint. The second constraint comes from the inequalities of Eq. (34). These inequalities should be preserved for minimum r_{EC} and ϕ_{EC}. If I_{CBO} is negligible and $\alpha_I < \alpha_N$ the first inequality $|I_B| \gg |I_E|(1 - \alpha_I)/\alpha_I$ is controlling. For example, if $\alpha_I = 0.5$, the inequality becomes $|I_B| \gg |I_E|$. Here, I_E is the algebraic sum of the signal current and any other possible emitter currents that flow because of the charge or discharge of coupling or other capacitors in the circuit. In the simple shunt chopper of Fig. 58, $I_E = V_{IN}/R_1$ when the transistor is on. The inequalities of Eq. (34) are essentially independent of the polarity of I_E such that the signal current may flow through the transistor in either direction with the same restrictions on $|I_B|$.

For the "off" condition assume that a square wave $\pm V_{IN}$ is applied to the input of the shunt chopper (Fig. 58) and a voltage $-V_{BE}$ is applied between the base and ground. When V_{IN} is positive (of opposite polarity to V_{BE}) the emitter-base junction bias and the magnitude of V_{BE} need be only large enough to reverse-bias the collector (about zero volts). In this condition care must be given to the possibility of emitter-base breakdown. The sum of the voltages $V_{IN} + V_{BE}$ appears across the emitter-base junction. To prevent breakdown, $V_{IN} + V_{BE} < BV_{EBO}$. When V_{IN} is negative, V_{BE} must be sufficiently greater than $|V_{IN}|$ to ensure that the emitter-base junction will remain reverse-biased. Thus, if the input signal is only positive, the base does not have to be driven very far in the reverse direction. If, however, the input voltage can be negative, the base must be driven at least as negative, and preferably slightly more negative, than the signal. If the input signal can have either polarity, the maximum signal-handling ability is achieved when

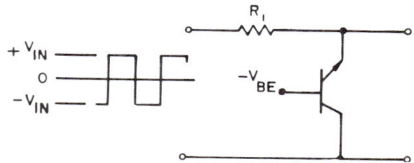

Fig. 58. Voltage limits for a transistor chopper.

$$V_{IN} = V_{BE} = BV_{EBO}/2$$

Offset-compensation Techniques. For a particular transistor at a particular temperature and operating point, the effects of the offset voltage and current can be compensated by injecting into the circuit, respectively, a voltage and current of the opposite polarity. If the offset voltage and current of the transistor are nulled out when the transistor is first put into operation, the resulting offset after a period of time will generally be less than if they had not been nulled out. Furthermore, a detailed study of the transistor characteristics can lead to an operating point where, at least, the offset voltage is relatively stable with temperature.

Figure 59a shows a shunt transistor chopper with a voltage injected into the collector that is exactly equal and opposite to the "on" voltage of the transistor. In the same circuit, a diode is employed in the base drive lead. The diode becomes reverse-biased when the transistor is switched to the "off" condition and thus removes the base drive during this period. Also, during this period, the injected collector offset voltage is removed. Thus this circuit eliminates the offset current by removing all drive potentials and it nulls out the offset voltage by inserting an equal and opposite emf. Practically, an offset current does flow because of the leakage current of the diode. The effects of this current are reduced by making $R_1 > R_2$. Furthermore, the transistor offset current may be substantially less than the remaining base current that leaks into it.

A more practical circuit is shown in Fig. 59b. CR_1 serves the same purpose as the diode in Fig. 59a. CR_2 is a breakdown diode that stabilizes the "on" drive voltage for the transistor. The potentiometer sets the level of null voltage developed across the resistor in the collector return. Both the base drive and the collector drive, which must be out of phase, are supplied by one transformer.

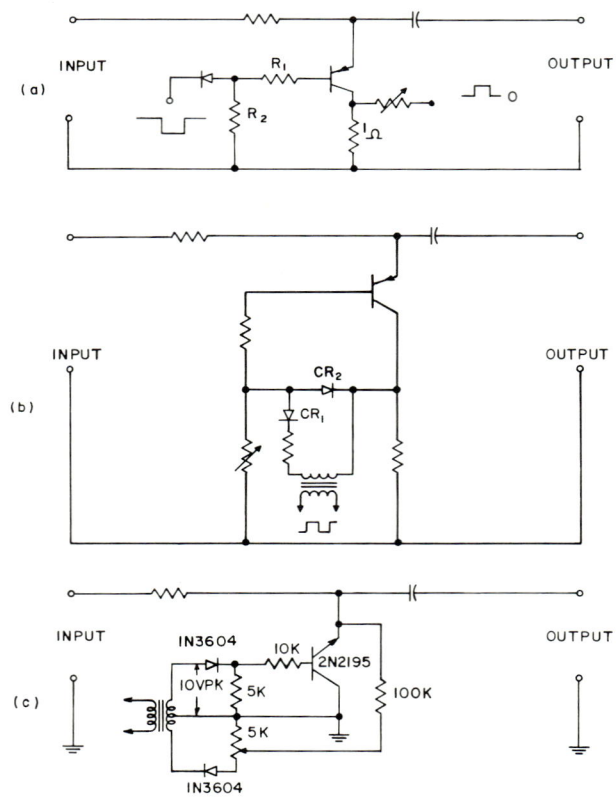

Fig. 59. Offset-compensated choppers. (a) Offset-voltage-compensation circuit. [From G. B. Chaplin and A. R. Owens, A Transistor High-gain Chopper-type D. C. Amplifier, Proc. Inst. Elec. Engrs. (London), pt. B, vol. 105, p. 259, January, 1958.] (b) Variation of offset-voltage-compensation circuit. [From I. C. Hutcheon and D. Summers, A Low-drift Chopper-type D. C. Amplifier with High-gain and Large Dynamic Range, Proc. Inst. Elec. Engrs. (London), pt. B, vol. 107, p. 455, September, 1960.] (c) Offset-current-compensation circuit. (From J. Giorgis, A Transistor Chopper Circuit for High-Z Sources, Electron. Equipment Eng., vol. 11, p. 51, January, 1963.)

A means of balancing out I_O is shown in Fig. 59c. The upper half of the center-tapped transformer drives the base in the same manner as in Fig. 59a. The lower half of the transformer, connected through the 100-kilohm resistor, provides an adjustable current for nulling the residual I_O. The maximum open-circuit resistance of the switch is limited by the 100-kilohm resistor. If both current and voltage offset compensation are required, elements of the circuits of Fig. 59a, b, and c can be combined.

Balanced-transistor Choppers. Improved performance, in the presence of varying environmental conditions, can be obtained by using pairs of "identical" transistors. The degree of improvement that can be obtained over a single uncompensated transistor is of course dependent upon how closely the balanced pairs can be matched. With selected pairs of transistors, improvements of an order of magnitude certainly can be obtained. If the transistors are fabricated by microelectronic techniques on a single chip immediately adjacent to each other, considerable further improvement is possible.

Some balanced-transistor configurations can also reduce the effect of the output transients by bucking the transient of one transistor against that of the other. Also, balanced-transistor choppers may have the ability to handle larger input signal voltages or may offer the possibility of handling signal voltages of either polarity, when this would not be possible with the same base drive to an "equivalent" single transistor. The possible advantages gained with a balanced configuration can be determined by a careful study of the circuit, taking into account the transient and signal-limiting conditions as well as the balance conditions.

Figure 60 shows four balanced-pair configurations in the first column. In Fig. 60a, two transistors Q_1 and Q_2 are connected in series opposition. The balance conditions are illustrated by the equivalent circuits for the two switching states in the second column.

A series-shunt switch configuration is shown in Fig. 60b. The balance condition for this switch combination is somewhat less obvious; however, it can be seen from the equivalent circuits in the second column that the two circuit states are equivalent if the source resistance at the input terminals is very much smaller than r_S. This circuit corresponds to the basic input switching circuit of Fig. 37e and is a very efficient input-chopper circuit.

Figure 60c is the shunt version of Fig. 60a. The potentiometer in the base of Q_2 is a fine-balance adjustment.

Figure 60d shows a configuration that also incorporates a balance-adjustment potentiometer. From the equivalent circuits in the second column it can be seen that the summation of R_1 and the input source resistance must equal R_1' for a completely symmetrical balance condition. If the a-c amplifier has an input resistance R_4 it can be shown that maximum power transfer from the signal source occurs when

$$R_1 = R_1' = R_2 = R_3 = R_4 \qquad (40)$$

It is assumed here that R_1 incorporates the internal resistance of the signal source.

Unipolar Field-effect Transistor Choppers. The switching "contacts" of the unipolar field-effect transistor (FET) are the drain and the source; and the drive is applied between the gate and the source. The "on" condition is illustrated by Fig. 61a. The values of "on" resistance r_S vary from a few hundred ohms to 10,000 ohms or more, depending upon the transistor type. The "on" resistances of the FET's are generally higher than those of the bipolar transistors. The "on" resistance for a bipolar transistor can be as low as a few ohms if the transistor is driven sufficiently hard. Bipolar transistors therefore tend to be more suitable for low-impedance circuits. Bipolar transistors, however, have unavoidable series emfs in the "on" condition (Fig. 56a). The unipolar FET has no series emfs (except for thermal emfs at the junctions of dissimilar materials within the device and its connections). The reason for this difference is that the bipolar transistor must be driven to the "on" state; whereas the FET is 'on" when the driving potential V_{GS} is reduced to zero. This makes the FET especially attractive for situations where a low offset voltage is required.

The "off" condition is illustrated by Fig. 61b. The "off" resistance is very large, on the order of 100 megohms or greater with silicon FET's. It can frequently be neglected. The "off" condition is achieved by making the gate-to-source voltage greater than the pinch-off voltage of the transistor. The transistor then has pentode-like characteristics even down to a drain-source voltage of zero volts. When the magnitude of the gate-source voltage is less than the pinch-off voltage, the output resistance r_O is roughly proportional to the gate-source voltage and ranges from thousands of ohms to hundreds of thousands of ohms. The offset current I_O is somewhat less than the gate reverse current but is approximately proportional to it.

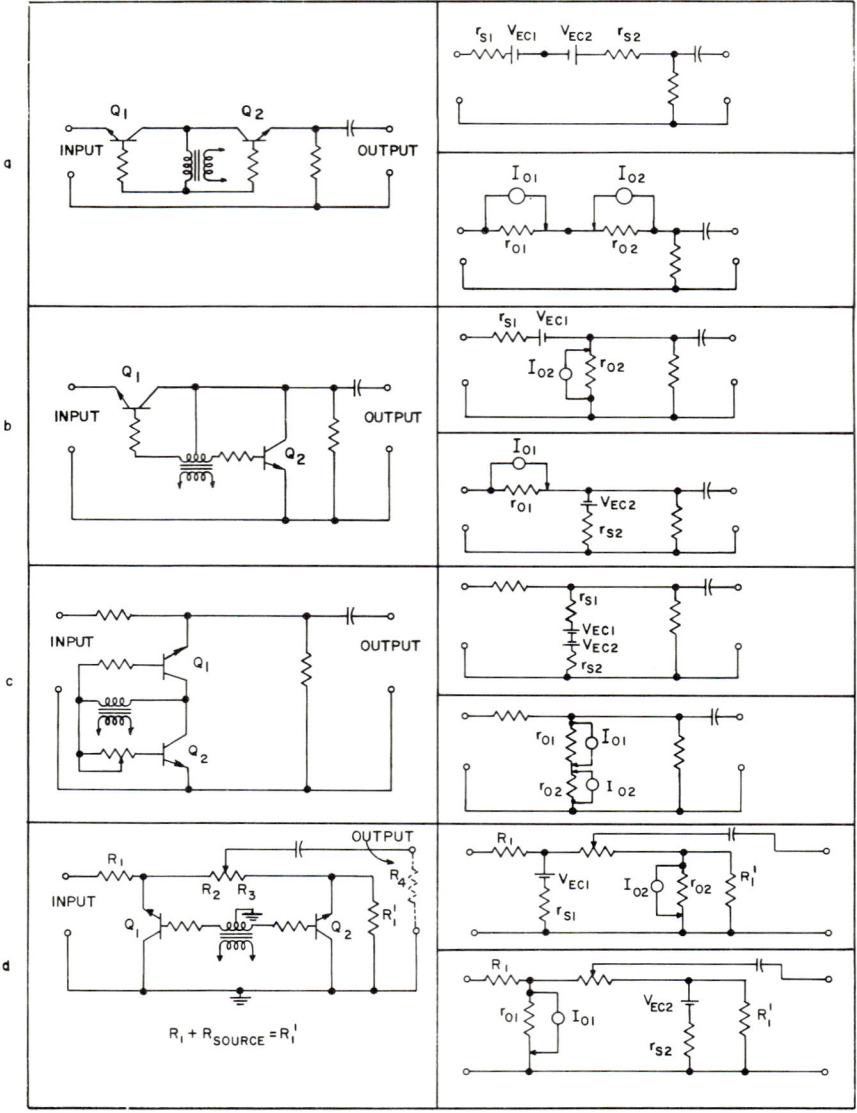

Fig. 60. Balanced-pair transistor chopper. (*a and b after R. L. Bright, Junction Transistors Used as Switches, Commun. Electron., no. 17, p. 119, March, 1955. d after D. C. Amplifier with Balanced Chopper, Mullard Tech. Commun., vol. 5, p. 99, April, 1960.*)

I_O will have about the same temperature characteristics as the gate reverse current. (In the absence of leakage, the reverse current of silicon-junction devices doubles for every 10°C rise in temperature.) The offset current is on the order of tens of picoamperes in good FET's at room temperature. For an offset current of 50 picoamperes and a signal-source resistance of 100,000 ohms, the corresponding offset voltage would be 5 μv, exclusive of possible contributions from thermal emfs and output transient effects.

The circuit of a shunt chopper with a P-channel FET is shown in Fig. 62a. The input signal to the chopper is V_{IN} and the output V_O. The gate is driven by a positive-going square wave with a magnitude equal to V_D. This wave is illustrated in Fig. 62b. The "off" time occurs when $V_{GS} = V_D$ and the "on" time when $V_{GS} = 0$. The approximate output voltage V_O is shown in Fig. 62c. During the "off" period, $V_O = V_{IN} r_O/(R_1 + r_O)$. During the "on" period, $V_O = V_{IN} r_S/(R_1 + r_S)$.

The transient behavior of the output wave results from the internal capacitances of the FET and the stray wiring capacitances. The FET has a distributed structure,

Fig. 61. Field-effect transistor chopper. (a) Drive and equivalent circuit "on" condition. (b) Drive and equivalent circuit "off" condition.

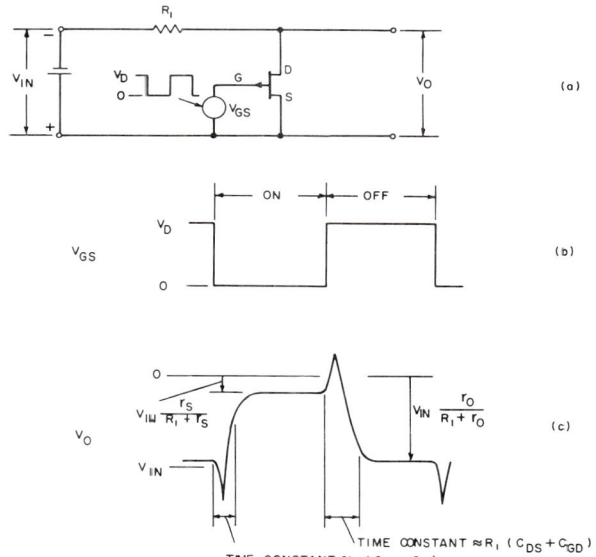

Fig. 62. FET shunt chopper operation. (a) Circuit. (b) Gate-source drive voltage. (c) Output voltage.

but it is approximately represented by the lumped-parameter equivalent circuit of Fig. 63. The voltage across the gate junction v_i controls the resistance $r(v_i)$. Parasitic series resistances r_G', r_D', and r_S' are in the gate, drain, and source, respectively.

The time for the output wave to switch from on to off is determined roughly by the time constant $R_1(C_{DS} + C_{GD})$, where it is assumed that

$$r_O = r_D' + r(v_i) + r_S' \qquad r_G', r_D' \ll R_1 \ll r_O \qquad (41)$$

The time for the output wave to switch from off to on is determined roughly by the time constant $r_S(C_{DS} + C_{GD})$, where it is assumed that

$$r_S = r_D' + r(v_i) + r_S' \qquad r_G', r_D' < r_S \ll R_1 \qquad (42)$$

The capacitances C_{GD} and C_{GS} couple spikes to the output when the gate drive voltage switches. The spikes can be minimized by selecting an FET with minimum gate-to-drain and gate-to-source capacitances and a minimum pinch-off voltage (which permits V_D to be minimized). It is possible to obtain some reduction of the spikes by neutralizing the gate-to-drain capacitance. However, the amount of reduction that can be achieved is limited because the gate-to-drain capacitance varies with the voltages applied to the transistor as well as with the temperature.

The transients result in an equivalent offset voltage. The effect can be reduced somewhat by clipping the spikes or by deactivating the demodulator (at the output of the amplifier) when the transients are present. The transients can also be reduced by employing a trapezoidal drive voltage to the gate. By using low switching frequencies and a wave of this nature, the transients can often be made negligible.

A sine-wave drive may be also employed, but it should be restricted so that it does not pass through zero. If the peak-to-peak amplitude of this sine wave is less than the pinch-off voltage of the transistor, the output of the circuit will be a modulated

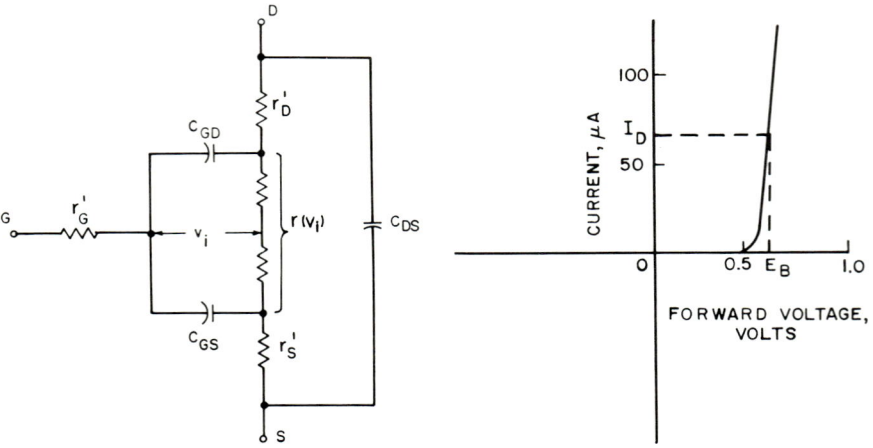

FIG. 63. Approximate equivalent circuit for an FET.

FIG. 64. Silicon-diode characteristic.

sine wave. This results because the drain-to-source resistance varies approximately linearly with the gate-to-source voltage in the region from zero volts to pinch-off. If the peak-to-peak drive voltage is greater than the pinch-off voltage, the output will appear as a clipped sine wave.

9.3. Diode Choppers. Diodes are commonly employed in suppressed-carrier modulators for multiple-channel systems. They also serve effectively in choppers to convert direct current to alternating current. Figure 64 shows the VI curve of a typical silicon diode at room temperature. The forward conduction starts at about 0.5 volt of forward bias whereupon the incremental series resistance becomes very small (as indicated by the steep slope of the forward characteristic). Reverse currents in the nano- and even picoampere range are available such that they cannot be seen on the scale of the graph. Because of the relatively high "on" voltage of silicon and other types of diodes, they are almost invariably used in pairs or in quads when employed as low-level choppers.

Figure 65a shows a typical arrangement of four diodes as a low-level balanced chopper. When the driving signal is positive at the top, diodes $B1$ and $B2$ are turned on and $A1$ and $A2$ turned off. The input signal is then connected between the center tap of the transformer and the right-hand terminal B. The "on" diodes are represented by emfs E_{B1} and E_{B2}. This is illustrated in Fig. 65b. When the driving signal

D-C AMPLIFIERS   13-59

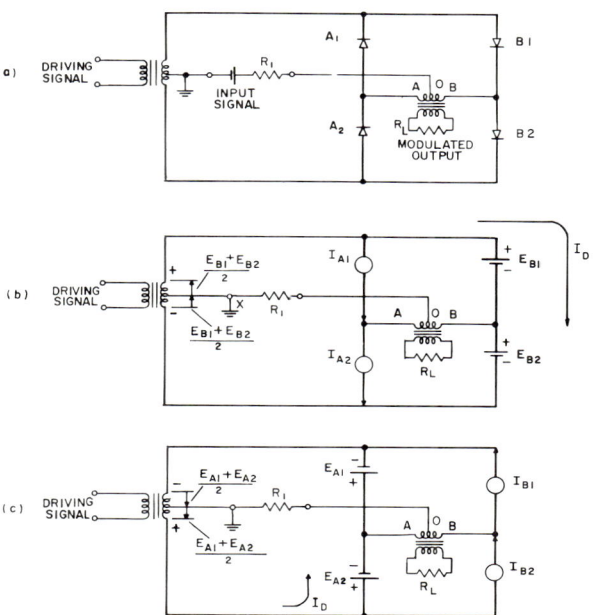

FIG. 65. Ring modulator. (a) Circuit diagram. (*After N. F. Moody, A Silicon Junction Diode Modulator, Electron. Eng., vol. 28, p. 95, March, 1956.*) (b) B diodes on. (c) A diodes on.

reverses polarity, the A diodes turn on and the B diodes off as illustrated in Fig. 65c. The input signal is then connected between the center tap of the transformer and the left-hand terminal A, thus reversing the polarity of the signal at the modulator output.

In practice the diodes will not be identical such that their forward voltages will not be equal and their reverse currents will also not be equal. The net offset voltage V_{BO} driving the output transformer is

$$V_{BG} = \frac{E_{B1} + E_{B2}}{2} - E_{B1} = \frac{E_{B2} - E_{B1}}{2} \qquad (43)$$

During the other half-cycle the net offset voltage V_{AO} is

$$V_{AO} = \frac{E_{A1} + E_{A2}}{2} - E_{A2} = \frac{E_{A1} - E_{A2}}{2} \qquad (44)$$

V_{BO} and V_{AO} are brought into phase by the demodulation process such that the net offset V_O is the average of the two (assuming a symmetrical driving wave such that V_{BO} and V_{AO} each account for one-half of the cycle time)

$$V_O = \tfrac{1}{2}(V_{BO} + V_{AO}) = \tfrac{1}{4}(E_{B2} - E_{B1} + E_{A1} - E_{A2}) \qquad (45)$$

For a given set of diodes, rearranging them in the quad will change the resulting offset voltage.

The equivalent offset current is determined in a similar manner. The reverse diodes in Fig 65b have reverse currents I_{A1} and I_{A2}, respectively, flowing. These subtract to give an offset current I_{AO}

$$I_{AO} = I_{A1} - I_{A2} \qquad (46)$$

for one half-cycle. The offset current I_{BO} for the other half-cycle is determined similarly,

$$I_{BO} = I_{B2} - I_{B1} \tag{47}$$

The resultant offset current I_O is the average of the two.

$$I_O = \tfrac{1}{2}(I_{AO} + I_{BO}) = \tfrac{1}{2}(I_{A1} - I_{A2} + I_{B2} - I_{B1}) \tag{48}$$

The multiplier is one-half for the offset currents and one-fourth for the offset voltages. Offset voltages on the order of 1 mv are achievable. The offset currents that can be achieved are somewhat less than the reverse currents. Improved uniformity can be gained by employing diodes fabricated immediately adjacent to each other on a single chip such that they will have more uniform characteristics. Naturally, as temperature increases, the reverse current increases and the achievable offset current increases.

The error equations (45) and (48) are written in terms of a single point on the operating characteristic of the diodes. This is appropriate if the driving source is a square wave and the signal is small compared with the driving source, such that each

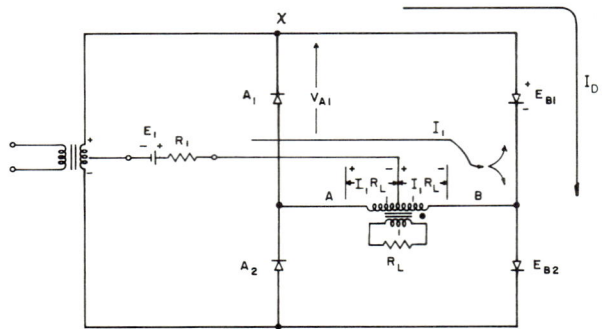

Fig. 66. Limits on maximum signal of ring modulator.

diode is effectively operated at a single point. If the driving source is a sine wave, the unbalance should be considered over the full forward and reverse operating region of each diode. In this case, the representation of the diode in the forward direction by an emf in series with a resistance would generally be quite accurate, except near the knee of the IV characteristic. Some consideration of this is given by Moody.[18] The maximum signal level will be limited by either the drive current or the forward voltage of the diodes. This can be seen from Fig. 66. Assume a signal emf E_1 drives a current I_1 into the modulator. Assume that the B diodes are turned on by a drive current I_D. Since the A diodes are turned off, no significant amount of signal current flows out through the A terminal of the transformer. Virtually all flows out through the B terminal. At B it divides, part going to the upper branch of the modulator and part to the lower branch. If it is assumed that the split is approximately equal, a current of $I_1/2$ goes upward through diode B_1 and downward through diode B_2. The signal current through the upper diode is in opposition to the drive current, and the diode will turn off unless

$$I_1/2 < I_D$$

When a sine-wave drive current is used, the restriction becomes even greater since the drive current is less than its peak value except for one instance during each half-cycle. In this case the peak drive current is usually made a factor of 5 to 10 greater than the peak signal current in order to provide assurance that the diodes are turned on for a major fraction of the half-cycle.

D-C AMPLIFIERS

The diode forward voltage limit can be determined from Fig. 66. If a 1:1:1 transformer is used, the signal current I_1 develops a voltage I_1R_L, with the polarity indicated, across the right half of the transformer primary. An equal voltage is developed across the left half by transformer action. The voltage that turns the A diodes off is determined by summing the voltages around the loop. For example, the voltage that turns $A1$ off, V_{A1}, is, summing from X to A to the right around the loop, $V_{A1} = E_{B1} - 2I_1R_L$. Diode $A1$ turns on approximately when $V_{A1} = E_{B1} = -E_B$. Thus $I_1R_L < E_B$ is a limit for the signal current–load resistance product, where the load resistance is the value transformed across one-half of the output transformer primary.

The source resistance for the drive current should be much greater than the forward resistance of the diodes. This can be seen from Fig. 64. The diode voltage E_B is relatively insensitive to drive current in the forward region. However, the drive current is extremely sensitive to a small change in the drive voltage E_B. This is especially significant since changes in temperature tend to translate the diode curve to the right or to the left, and similar diodes will often have similar VI curves with some translation to the right or to the left.

The modulator described may be used with all types of diodes; however, silicon diodes are the most appropriate for low-level operation. Germanium diodes have higher leakage currents. The older multicrystalline rectifiers such as copper oxide and selenium have been used in large quantities in the telephone industry. However, these diodes tend to have greater leakage, less uniform characteristics, and generally poorer characteristics than silicon diodes. Vacuum diodes may also be employed. They are usually less satisfactory because of the need for supplying heater power, possible leakage in heater-cathode insulation, possible hum pickup from the heater, and the fact that

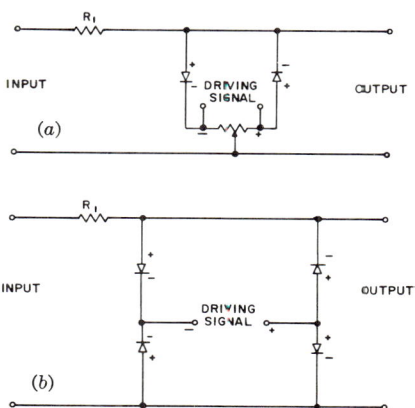

Fig. 67. Diode modulator circuits. (a) Two-diode shunt modulator. (b) Four-diode shunt modulator.

the thermal-emission velocities from the cathode provide an offset current flow. Furthermore, this current varies considerably with age.

Two diode modulator circuits are shown in Fig. 67. Figure 67a and b are half-wave shunt choppers. The polarity of the driving signal and the voltage drops across the diodes are indicated for the "on" condition. These drops are in opposition to each other so that there will be ideally zero volts between the top and bottom lines of the chopper when the diodes are on. The offset conditions are similar to those of Fig. 65. Figure 67b is similar to Fig. 67a except that it employs a quad of diodes instead of a pair and the driving signal does not require a center tap.

9.4. Variable-capacitance Diode Modulators. Variable-capacitance diode modulators may be the most satisfactory solution for applications requiring the following characteristics:

1. Low noise
2. High impedance
3. Maximum current sensitivity
4. Minimum current drift
5. Elimination of electromechanical devices

Threshold sensitivities of 10^{-10} amp from 0 to 65°C have been obtained.[19] Improved current sensitivities will be obtained as diodes with greater uniformity and lower leakage are developed.

A typical variable-capacitance modulator is shown in Fig. 68a. The modulator portion of the circuit is shown within the dashed lines. A pair of variable-capacitance semiconductor diodes CR_1 and CR_2 are driven by a pump or driving signal E_P at a frequency f_P. The output capacitor C and inductor L are series-tuned to resonate with the average capacitance of the diodes at the pump frequency. The output inductor may be tapped, as shown, to obtain an optimum impedance match. When there is no d-c input, the pump voltage across the two diodes is perfectly balanced such that no carrier voltage appears between point A and ground. This balance is obtained by means of the potentiometer in the modulator circuit. When a d-c signal is applied to the input, the capacitance of one diode increases and that of the other diode decreases. This unbalances the bridge and causes an a-c signal at the pump frequency to appear at point A. This signal is passed through the series LC circuit and appears at the output terminals.

The circuit is actually more than a simple modulator because it can provide power gain through parametric action. In parametric amplifier terms it is a double-sideband up-converter. The frequency spectrum is shown in Fig. 68b. An input frequency f_1 is converted to a double-sideband suppressed-carrier signal centered at the

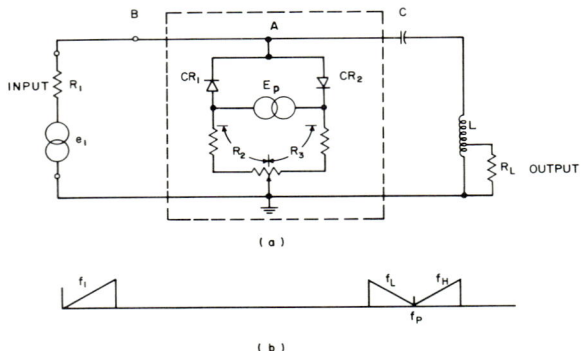

Fig. 68. Variable-capacitance diode modulator. (a) Modulator. (b) Frequency spectrum.

pump frequency f_P. Lower and upper sideband frequencies f_L and f_H are produced. By way of comparison, a microwave parametric amplifier might employ the lower sideband as the output frequency (inverting converter), or the upper sideband as the output frequency (noninverting converter), but it would normally not employ both sideband frequencies. The d-c modulator or low-frequency parametric up-converter must employ both sidebands because they are so close that they cannot be separated by filters. Theory shows that the upper sideband reflects a positive admittance to the input and the lower sideband a negative admittance to the input. When both sidebands are employed, the admittance reflected to the input corresponds to the sum of that due to the upper and the lower. The negative admittance can cause oscillation if the input and output load impedances are too high or if the pump drive voltage E_P is too large.

An approximate equivalent circuit for the modulator is shown in Fig. 69a. This equivalent circuit neglects the positive and negative admittances reflected to the input through parametric action.* The diodes are combined in parallel in the equivalent circuit such that they have an average capacitance $2C_{CR}$. The resistors R_2 and R_3 have also been combined in parallel. The left-hand portion of the circuitry is at the input frequency, which is approximately d-c. The input voltage e_1 develops a voltage e_{CR} across the diodes. In general e_{CR} is smaller than e_1. This is true at direct current because of the leakage of the diodes (not shown in the equivalent circuit) and the consequent voltage drops across R_2 and R_3, in parallel, and R_1. When e_1 is

* For a discussion of parametric amplifier theory, see Chap. 28.

D-C AMPLIFIERS

an alternating voltage, e_{CR} is further reduced relative to e_1, because of the capacitances $2C_{CR}$ and C (L is assumed to be essentially a short circuit at the input frequency). The voltage e_{CR}, at the input frequency, is converted to sidebands centered at the pump frequency. These are represented by the emf $e(e_{CR})$ at the right-hand side of the equivalent circuit. This emf is given by

$$e(e_{CR}) = Ke_{CR} \cos \omega_P t \tag{49}$$

The constant K is a function of the diode characteristics and the amplitude of the pump signal. Hoge* has shown K is given by

$$K = \ddot{E}_P/2mV_o \tag{50}$$

for a zero-biased diode with a capacitance function

$$C_{CR} = F(V + V_o)^{-1/m} \tag{51}$$

Here F is a constant, V is the voltage across the diode, V_o is the contact potential

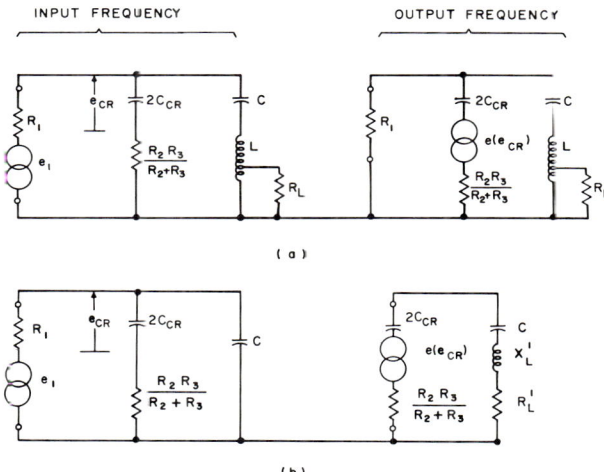

FIG. 69. Equivalent circuits. (a) Approximate equivalent circuit. (b) Simplified version of a.

of the diode, $1/m$ is a constant which is equal to one-half for an abrupt junction and may have a different value for other junction geometries, and E_P is the magnitude of the pump voltage. If e_{CR} is d-c, the output frequency is equal to f_P, and the phase of the output signal changes 180° when e_{CR} reverses sign. If e_{CR} is a sine wave $E_{CR} \cos \omega_1 t$, the output voltage is a double-sideband suppressed-carrier signal,

$$e(e_{CR}) = \tfrac{1}{2}KE_{CR}[\cos (\omega_P + \omega_1)t + \cos (\omega_P - \omega_1)t] \tag{52}$$

The modulator then operates as a parametric up-converter. It provides gain and low noise.[20]

The output circuit, principally C, loads the modulator at the input frequency and reduces the value of $e_{CR} \cos \omega_1 t$ relative to $e_1 \cos \omega_1 t$. One of the objectives of the modulator design is to reduce this parasitic loading. For this reason C should be made small compared with $2C_{CR}$. Both C and $2C_{CR}$ capacitively shunt the circuit such that if a low input capacitance is to be obtained at the input frequency, the capacitances of C and $2C_{CR}$ must be minimized.

* Ref. 19, pp. 40–42.

The source resistance R_1 loads the output circuit. This causes little difficulty if R_1 is very much larger than the real part of the resistance reflected from R_L through L and C. This is often the case. If this relationship cannot be maintained, a low-pass filter should be inserted in the circuit at B in Fig. 68a.

If the output frequency is fairly high compared with the input frequency, the simplified equivalent circuit of Fig. 69b can usually be employed. Here L and R_L have been eliminated at the input frequencies, and R_1 at the output frequencies. L and R_L have been reduced to their series equivalent X_L' and R_L' at the output frequencies. For maximum power transfer to the load, X_L' should tune out the series reactance of C and $2C_{CR}$.

The output frequencies should be much larger than the input in order to obtain a high gain and to minimize the interaction of the input and output frequencies. Output frequencies ranging from about 10 kc to tens of megacycles are not uncommon for input frequencies from direct current to low radio frequency.

9.5. Vibrating-reed Capacitance Modulators. Vibrating-reed capacitance modulators employ an electromagnet to drive a reed or diaphragm that modulates a capacitance. A vibrating-reed capacitance modulator provides a higher input resistance and a greater sensitivity to charge or current than any other type of d-c–to–a-c modulator. Commercially available modulators provide insulation resistances greater than 10^{15} ohms and contact-potential drifts of less than 0.1 mv per 24 hr.

A typical modulator is shown in Fig. 70. The signal source is represented by a current generator I in parallel with a resistor R_1 and a capacitance C_1. C is the variable-capacitance modulator. A coupling capacitor C_c couples the varying voltage across C to the input of a low-noise amplifier. The input impedance is represented by R_3 and C_3. The operation is similar to that of the variable-capacitance diode modulator described in the previous section. In the usual case,

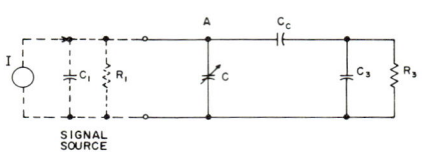

FIG. 70. Vibrating-reed capacitance modulator.

$$R_1 \gg R_3 \qquad C \geq C_3 \qquad C_c \geq C_3 \qquad (53)$$

This results in maximum sensitivity.[21] Because of the high source resistance, this type of modulator usually has a long input time constant, and it is suitable only for d-c measurements. For example, assume the nodal capacitance C_n at point A in Fig. 70 is 40 pf. C_n is given by

$$C_n = C_1 + C + C_c C_3/(C_c + C_3) \qquad (54)$$

where C is the average capacitance of the capacitance modulator. If R_1 is 10^{15} ohms (and the leakage of the modulator and coupling capacitors can be neglected), the input time constant is 11 hours! An input current I_1 might then be determined by the rate at which it charges the input capacitance. To obtain reasonable response times, it is necessary to make the capacitances of Fig. 70 as small as possible. The response time can be significantly improved by using series feedback to increase the input impedance effectively (and correspondingly decrease the input capacitance).*

An important figure of merit for a vibrating-reed capacitance modulator is its *conversion efficiency*. The conversion efficiency is the ratio of the a-c voltage output to the d-c voltage input.

$$\text{Conversion efficiency} = \frac{\text{a-c output voltage}}{\text{d-c input voltage}}$$

Conversion efficiencies, based on *rms* output voltage, of 5 to 20 per cent are obtainable. A high conversion efficiency is desirable. The vibrating frequencies range typically from 60 to 2,000 cps.

* Ref. 21, pp. 299–301.

D-C AMPLIFIERS

A major source of drift and offset in vibrating-reed capacitance modulators is their contact potential. This potential appears as an emf in series with the contacts. The contact potential results from a number of sources. One is the work function of the plates. Capacitor plates will have a contact-potential difference equal to the difference of their work functions. Such contact potentials will exist between all capacitances in the system, including the source capacitance, the vibrating-reed capacitance, the circuit stray capacitance, and the coupling capacitance. In addition, ions in the atmosphere surrounding the input elements will be collected by the contact potentials. These ions are caused by various sources of incident radiation. Dielectric currents, developed by mechanical or electrical stresses applied to the high-impedance insulating material, are another source of offset. To minimize these effects, vibrating-reed capacitors are housed in sealed enclosures of a small volume containing an inert gas such as argon. The critical elements of the capacitor are especially cleaned, polished, and plated to minimize surface oxidation and contamination, such as surface gas absorption and dust particles. In addition, mechanical stresses on the insulators are minimized.

When these precautions are not taken, vibrating-reed capacitors will have excessive offset voltages and currents and they will drift rapidly. When such precautions are taken, the offset voltage due to contact potential can be reduced to below 20 mv and drift in offset voltage to less than 0.1 mv

The sources of noise are similar to those of the variable-capacitance diode modulator described in Sec. 9.4. When the modulator is followed by an amplifier, noise will be developed at the amplifier output through four processes:

1. Very low frequency noise generated by the signal source and the input circuit to the left of C_c will be modulated by the varying capacitance and appear as sidebands very close to the driving frequency.

2. Very low frequency noise developed across R_3 and C_3 of the amplifier will be coupled to the modulating capacitor by C_c and will appear as sidebands similar to those described under 1.

3. Noise at the driving frequency generated by the signal source and input circuit to the left of C_c will be coupled by C_c to the amplifier and amplified directly. This noise should be greatly attenuated as a result of the inequalities of Eq. (53).

4. Noise generated at the input of the amplifier at the driving frequency will be amplified directly and appear at the output of the amplifier.

The noise described under 1 should be principally that of the signal source and should represent the basic limitation of any amplification or modulation technique. The noise described under 2 is minimized by employing an amplifier with minimum equivalent input noise and by means of the inequalities of Eq. (53). The noise described under 3 is minimized through the inequalities of Eq. (53). The noise described under 4 is minimized by using a low-noise amplifier. Further improvement might be achieved by using a driving-frequency bandpass filter at the point between A and the input of the amplifier, as was done in the variable-capacitance diode modulator. However, such a filter is very difficult to realize because of the low frequencies and high impedances involved, and the need to minimize capacitances to ground.

9.6. Photoconductor Choppers. Photoconductors are light-sensitive conductances. When not exposed to light, they have a very large resistance. Their resistance under no light can range from hundreds of thousands of ohms to hundreds of megohms. In the presence of light, their resistance drops several orders of magnitude. Resistances as low as 15 ohms can be achieved. The characteristics that are achieved are dependent upon the materials, temperature, device construction, light intensity, and light wavelength.

Because the switching action is performed by light, almost perfect electrical isolation can be achieved between the switching drive source and "the contacts." Thus photoconductors need introduce no emfs into the switching circuit other than usual thermal emfs. They do not need to be used in balanced configurations. Their main limitations are their fairly high "on" resistance and their relatively slow switching speed, which is on the order of milliseconds. Also, their resistance is temperature-

18-66 AMPLIFIER CIRCUITS

sensitive, which means that the input-output voltage transfer ratio of the photoconductor choppers can be a function of temperature.

Optimum performance of a photoconductor chopper requires careful design and construction. This is illustrated by the assembly shown in Fig. 71. Here two pairs of input modulator and output demodulator photoconductors are mounted in a die-cast metal block. Two neon lamps, especially manufactured to obtain uniform light output, drive the photoconductors. One lamp drives each pair. One lamp is on when the other is off. The demodulator photoconductors are designed for high-level operation, and their resistance varies from thousands of ohms when illuminated to hundreds of thousands of ohms when dark. The modulator photoconductors are designed for low-level operation, and their resistance varies from tens of thousands of ohms when illuminated to several megohms when dark. The input modulators are electrostatically shielded from the neon lamp by a stainless-steel mesh. This isolation is extremely important because of the large difference in signal levels present at the neon bulb and at the input modulator. For example, if the voltage drive to the neon bulb is 60 volts, and it is required that the stray drive signal introduced

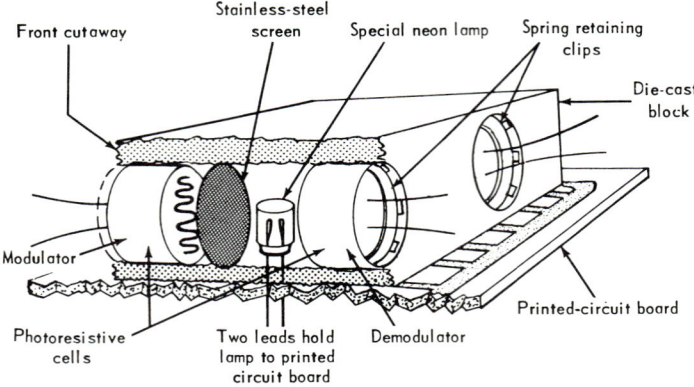

FIG. 71. Photoconductor modulator. Used in Dymec model DY-2460A operational amplifier. (*From J. A. Rose, Photochopper Reduces Amplifier Drift, Elec. Design News, vol. 8, p. 34, August, 1963.*)

into the photochopper be less than 0.2 mv, the isolation between the bulb and photochopper must be 170 db. Fortunately, such isolation can be achieved by physical separation and electrostatic shielding. The output demodulator does not require such a great isolation because of its high signal level.

Figure 72 shows an operational amplifier that employs this photochopper. This amplifier is of the Goldberg type, as described in Ref. 2. The chopper amplifier, at the upper left, employs series and shunt photochoppers $R1$ and $R2$ that operate out of phase. The demodulator employs series and shunt photoconductors $R3$ and $R4$ that operate out of phase with each other and in synchronism, respectively, with $R2$ and $R1$. Synchronism is no problem since the input and output photoconductors are driven from common light sources. The configuration of this chopper amplifier corresponds to Fig. 48, circuit 15. The operation of the amplifier is as follows:[22]

Neon lamps $I1$ and $I2$ are excited by relaxation oscillator, which operates at nominal 225 cps. With ±10-percent variation in supply voltage, chopping frequency stays within limits of 190 cps to 260 cps. This frequency is not synchronous with power line or other external frequencies. Amplifier input is applied to modulating photoresistors, $R1$, $R2$ and to base of $Q5$ of wideband amplifier. In chopper amplifier, $R1$ and $R4$ operate in synchronism with neon lamp $I1$ while $R2$ and $R3$ operate from neon lamp $I2$. $R1$ and $R2$ alternately switch amplifier input from summing point to common. High input impedance

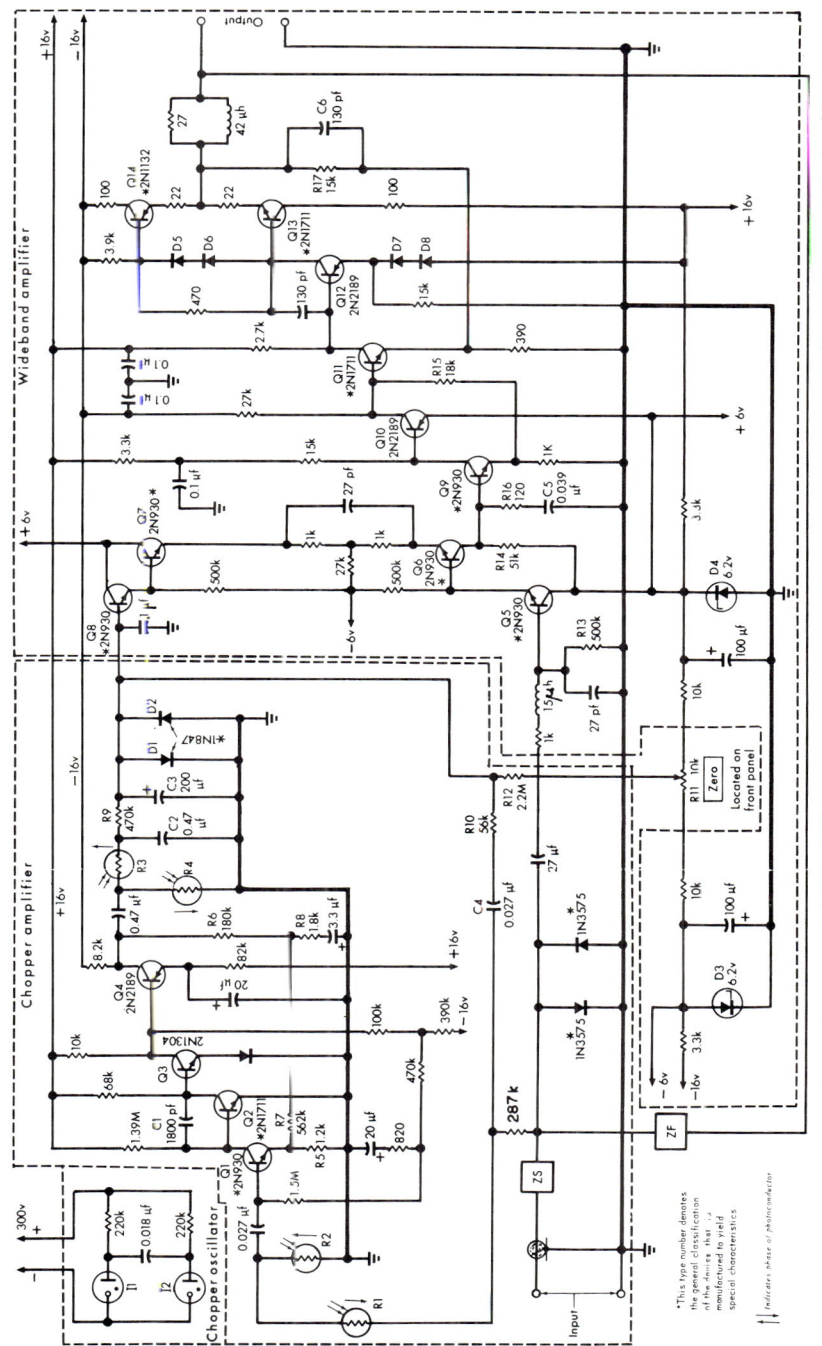

FIG. 72. Operational amplifier. Dymec model DY-2460A (From J. A. Rooc, *Photochopper Reduces Amplifier Drift*, Elec. Design News, vol. 8, p. 3b, August, 1963.)

AMPLIFIER CIRCUITS

in chopper amplifier is achieved by negative current feedback in Q1 emitter resistor R5. Additional voltage feedback is obtained from Q4 collector through R6, R7 to R5, providing gain and bias-point stability. C1 is an integrating feedback element around Q2 which reduces high frequency noise in Q2 and Q3. Signal is amplified through Q2, Q3 and Q4 and supplied to demodulating photoresistors R3 and R4. R3 and R4 operate synchronously with input photoresistors to demodulate amplified square wave. Output is filtered to a smooth d-c output by C2, R9 and C3. Zero-potentiometer R11 supplies either negative or positive voltage through R12 to output of chopper amplifier. This zero voltage cancels fixed d-c offset voltages at input. Parallel back-to-back diodes D1 and D2 clamp chopper amplifier output during extreme overload conditions. This clamping protects wideband-amplifier differential input against excessive voltage.

Q5 and Q6 are input amplifiers to wideband differential amplifier with controlled response from d-c to beyond 1 Mc. High input impedance is obtained by using emitter-follower cascaded connections: Q8 to Q7, and Q5 to Q6. Stability against temperature change is provided by differential connection. Q5, Q6, Q7 and Q8 are selected low-noise silicon transistors operating at low collector currents and voltages, providing minimum noise output. Breakdown diodes D3 and D4 produce stable and decoupled supply voltages for differential input and zero control. R13 is bias return for Q5 through Q8, which are biased through chopper-stabilizer output circuitry. Voltages supplied to differential input produce output signals across R14 and Q6 collector. Additional amplification is provided by Q9 and Q10 connected as complementary common-emitter pair using negative feedback through R15. Combination R14 and C5 gives rolloff of 6 db/octave above 100 cps. Common-emitter amplifiers Q11, Q12 and complementary emitter followers Q13 and Q14 form feedback triplet. This feedback improves output impedance and triplet gain stability. Complementary emitter followers provide push-pull output with equal output impedance for either polarity output. Silicon diodes D5 and D6 have forward voltages approximately equal to base-emitter junction voltages of silicon transistors Q13 and Q14. Diodes supply correct difference in base potential so that as one emitter follower moves toward cutoff, its complement begins to conduct. In this manner, crossover distortion is completely eliminated. Diodes D7 and D8 comprise low-impedance source, supplying stable bias for Q12 emitter. Triplet feedback C6 and R17 give rolloff above 50 kc. This rolloff counteracts R16 in series with C5 in frequency-cutoff network, R14, C5. Thus, compensation network results in a 6 db/octave slope from 100 cps to beyond 1 Mc.

The performance that can be achieved is illustrated by the specifications of this amplifier.*

Zero drift (values referred to summing point):
Constant temperature, 1 μv/week max
Temperature coefficient: 0.5 μv/°C max
Input noise (values referred to summing point, resistance to ground < 100 kilohms):
4 μv peak-to-peak max, 0 to 1 cps
Chopping frequency: 190 to 300 cps

9.7. Hall-effect Modulators. A Hall-effect modulator consists of a Hall-effect generator in the field of an electromagnet as shown in Fig. 73. The generator is a wafer of semiconductor material, such as indium antimonide, with leads on two pairs of opposite faces. A current I is applied to the leads on one pair of faces (x-axis leads, Fig. 73). The electromagnet develops a field with a flux density B along the y axis, perpendicular to the current flow. The magnetic field deflects the current flow within the semiconductor in the $\pm z$ direction, depending upon the polarities. This causes a Hall-effect voltage $\pm V_H$ to appear across the z-axis leads. When there is no magnetic field or no current flow, V_H is ideally zero. V_H is directly proportional to the product of the flux density and the x-axis current and is inversely proportional to the thickness t of the wafer

$$V_H = RBI/t \qquad (55)$$

The constant of proportionality R is the Hall coefficient.

The Hall generator can be employed as a chopper by applying the d-c input to the x-axis leads and chopping the magnetic field. The output voltage is proportional to the product of the input current and flux density. Maximum output current is achieved when the flux-density excursions are made as large as possible. The flux-

* Courtesy of Dymec Division of Hewlett Packard Co., Palo Alto, Calif.

density excursions are ultimately limited by saturation of the magnetic material in the core.

An alternate mode of operation is achieved by applying a chopped drive current I to the semiconductor and applying the signal current to the winding of the magnetic core. Maximum amplification is achieved with a maximum chopped drive current I, and this is limited by heating in the semiconductor. This mode of operation has an advantage in that the number of turns on the core can be adjusted to provide a match to the internal impedance of the signal source. This mode of operation is troubled by the remanence of the magnetic core. The remanence can be overcome by applying a small alternating current to a core winding.

Using an indium-antimonide Hall generator, an effective air gap of 0.3 mm, a ferrite core with 70,000 turns, and a chopped drive current of 200 ma, an output sensitivity of 70 $\mu v/\mu a$ has been achieved.[23] The offset and drift of the Hall generator are limited by the nonuniformities of the semiconductor wafer and the difficulty of placing the Hall-effect electrodes at exactly opposite points on the wafer. The wafer has a substantial resistance such that the current I produces a voltage gradient along the x axis. A voltage will appear between Hall electrodes even when no magnetic field is applied, if they are not placed opposite each other. This voltage will be temperature-sensitive and will cause drift, even if it is nulled out at a particular temperature.

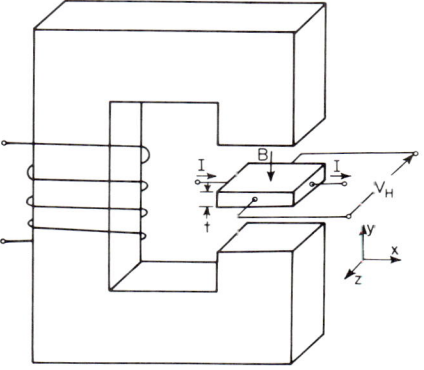

Fig. 73. Hall-effect modulator.

10. FIELD-EFFECT TRANSISTOR D-C AMPLIFIERS

10.1. General. Field-effect transistors (see Chap. 10) can be used in certain classes of d-c amplifiers that formerly required vacuum tubes. These devices have a number of advantages including high impedance at normal temperature, a very high theoretical power gain, a frequency response determined essentially by the Miller capacitance, low noise, and very low offset voltage when used as a chopper. Their disadvantages include difficulty of manufacture, very serious temperature dependence of input impedance, drain and leakage currents, and to a much lesser extent transconductance g_m. The performance that can be obtained in straight d-c amplifiers is generally very much worse than that which can be obtained with normal transistor differential amplifiers. In some cases, however, notably in very high impedance amplifiers used at room temperature, they can be used to advantage.

10.2. Important Points for D-C Amplifier Design Using FET's. 1. The design equations for an FET are in many ways similar to those of a pentode. For example, the low-frequency voltage gain A_V can be written as

$$A_V = -g_m R_L/(1 + g_m R_s) \qquad (56)$$

where g_m is the transconductance, R_L is the load resistance, and R_s is the source (cathode) resistance. There are, however, many differences: the g_m is much lower by a factor of about 10, the capacitances are much bigger, the leakage currents are much more serious, and the drain-current variation with temperature must be taken into account.

2. In many designs where a variation of ambient temperature is to be taken into account, both the leakage-current and the drain-current variations are very important. The former depends on the transistor type, with a silicon planar device conserving its advantage as in the case of a conventional transistor. However, if a

field-effect device is to be used at all, the source impedance will likely be quite high and the drift due to the leakage will therefore normally be high.

The drain-current variation of roughly 1 per cent/°C is usually the most serious cause of drift. This means that, in an unbalanced amplifier, the drift over a temperature range of 10°C is about 10 per cent of the operating current times the load resistance; this results in usually unacceptably high drift figures worse than those which can be obtained with conventional transistor amplifiers.

3. It is theoretically possible to match field-effect transistors in a differential pair to balance the effects of the leakage- and drain-current variations; however, at the time of writing, this is not a practical possibility. It is difficult to predict how soon the new devices in development and improved circuit techniques will overcome these difficulties.

10.3. Typical Design Example

Design Example. A very high impedance amplifier is required to operate at room temperature from a source impedance of 5 megohms with a gain of 10. The CMR should be as good as possible. Supplies of ±28 volts are available.

Solution. See Fig. 74. Assume FET's are available with $g_m = 0.5$ ma/volt; $I_o = 1$ ma, $V_p = 5$ volts. Because of the drift caused by the leakage current and the source resistance,

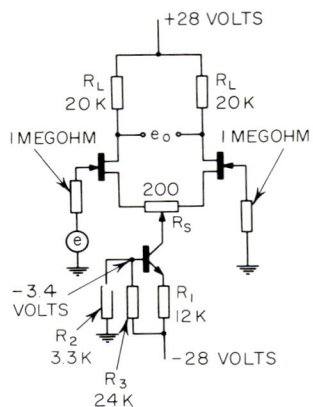

FIG. 74. Design example.

use a differential pair. For convenience, use zero gate-source bias. Because of the CMR requirement use a conventional transistor current source. Using Eq. (56),

$$A_V = (0.5)(R_L)/[1 + (0.5)R_s] = 10$$

Assuming $R_s = 200$ ohms, $R_L = 20$ kilohms, giving $V_{DS} = 8$ volts. For the current source, selecting the voltage of the emitter of the current source as -4 volts gives

$$R_1 = 12 \text{ kilohms}$$

Then, $R_2 = 3.3$ kilohms and $R_3 = 24$ kilohms.

References 24 to 26 are useful for information regarding FET characteristics.

REFERENCES

1. Williams et al., DC Amplifier Stabilized for Zero and Gain, *Trans. AIEE*, vol. 67, pp. 47–57, 1948.
2. E. A. Goldberg, Stabilization of Wideband Direct-Current Amplifiers for Zero and Gain, *RCA Rev.*, pp. 296–300, June, 1950.
3. G. E. Valley, Jr. and H. Wallman, "Vacuum Tube Amplifiers," MIT Radiation Laboratory Series, McGraw-Hill Book Company, New York, 1948.

4. R. W. Landee, et al., "Electronic Designers' Handbook," McGraw-Hill Book Company New York, 1957.
5. G. B. B. Chaplin and A. R. Owens, Some Transistor Input Stages for High-gain D.C. Amplifiers, *Proc. Inst. Elec. Engrs. (London)*, vol. 105, pt. B, no. 21, pp. 249–257, May, 1958.
6. G. B. B. Chaplin and A. R. Owens, A Transistor High-gain Chopper-type D.C. Amplifier, *Proc. Inst. Elec. Engrs. (London)*, vol. 105, pt. B, no. 21, pp. 258–266, May, 1958.
7. D. F. Hilbiber, A New D.C. Transistor Differential Amplifier, Solid State Circuits Conference, Philadelphia, Pa., Feb. 15, 1961. Also available as TP-16 from Fairchild Semiconductor, Mountain View, Calif.
8. G. Riva et al., Distortion in Transistor Amplifiers, *Proc. Inst. Elec. Engrs. (London)*, vol. 3, March, 1964; also available as Nota Tecnica 24, Società Generale Semiconduttori, Agrate/Milano, Italy.
9. P. J. Bénéteau and B. Murari, D.C. Amplifiers Using Transistors, *Electron. Eng.*, April 1963, pp. 257–259.
10. D. W. Slaughter, The Emitter-coupled Differential Amplifier, *IRE Trans. Circuit Theory*, vol. CT-3, no. 1, pp. 51–53, March, 1956.
11. P. J. Bénéteau, The Design of High Stability D.C. Amplifiers, *Semicond. Prod.*, vol. 4, no. 2, pp. 27–30, February, 1961.
12. Standard Telephone and Cables, Ltd., Some Applications of Planar Epitaxial Transistors, Application Report available from STC, Transistor Division, Footscray, Kent, England.
13. R. D. Middlebrook and A. D. Taylor, Differential Amplifier with Regulator Achieves High Stability, Low Cost, *Electronics*, July 28, 1961.
14. Rochar-Electronique, Montrouge (Seine), France.
15. R. Charbonnier, private communication.
16. P. J. Bénéteau et al., Drift Compensation in D.C. Amplifiers, Solid State Design, May, 1964.
17. W. M. De Matteis and J. W. Halligan, Designing Transistorized Differential Amplifiers, *Electron. Design*, Aug. 16, 1962, pp. 52–55.
18. N. F. Moody, A Silicon Junction Diode Modulator, *Electron. Eng.*, vol. 28, pp. 94–100, March, 1956.
19. R. R. Hoge, A Sensitive Parametric Modulator for DC Measurements, 1960 *IRE Natl. Conv. Record*, pt. 9, pp. 34–42.
20. J. R. Biard, "Low Frequency Reactance Amplifier," Digest of Technical Papers, 1960 International Solid State Circuits Conference, pp. 88–89.
21. H. Palevsky, R. D. Swank, and R. Grenchick, Design of Dynamic Condenser Electrometers, *Rev. Sci. Inst.*, vol. 18, pp. 298–314, May, 1947.
22. J. A. Rose, Photochopper Reduces Amplifier Drift, *Elec. Design News*, vol. 8, p 35, August, 1963.
23. E. Wolfendale, D.C. Amplifiers, *Proc. Inst. Elec. Engrs. (London)*, vol. 106B, suppl. 18, pp. 1339–1340, 1959.
24. T. B. Martin, Circuit Applications of the Field-effect Transistor *Semicond. Prod.*, March, 1962, p. 30.
25. Field-effect Transistors, *Application Note* no. 1, Amelco, Inc., Mountain View, Calif. (Published in Solid State Design, vol. 4, no. 1, p. 12, January, 1963.)
26. N. G. Bechtel, "A Circuit and Noise Model of the Field-effect Transistor," Digest of Technical Papers, 1963 Solid State Circuits Conference, p. 92.

Chapter 19

OPERATIONAL AMPLIFIERS

PAUL J. BÉNÉTEAU* and KEATS A. PULLEN, JR.[†],

CONTENTS

1. Introduction	19-3
2. Basic Theory of Operational Amplifiers	19-4
2.1. The Use of Laplace Transform Techniques	19-4
2.2. Transition Equalizers for Operational Amplifiers	19-7
3. Electron-tube-type Operational Amplifiers	19-8
4. Differential and Common-mode Operating Considerations	19-12
5. Transistor Operational Amplifiers	19-14
5.1. General	19-14
5.2. Analysis of Operational Amplifiers	19-14
5.3. Amplifier-design Considerations	19-19
5.4. Typical Amplifier	19-20
5.5. Operational Amplifiers Using FET's	19-23
6. Applications of Operational Amplifiers	19-24
6.1. Use of D-C and Narrowband Amplifiers in Regulated Power Supplies and Analog Recorders	19-24
6.2. Use of Stabilized Amplifiers in Frequency-selective Filters	19-26
6.3. Variable-selectivity Circuits and Q Multiplication	19-28
6.4. Operational Amplifiers in Oscillators	19-31
6.5. Integrators and Differentiators	19-31
6.6. Other Special Operational Circuits	19-34
6.7. Data Requirements on Operational Amplifiers	19-36

LIST OF SYMBOLS

A	Operational gain; amplifier	A_c	Stabilized, or closed-loop, amplification
$A(s)$	Laplace transform of amplifier gain	A_f	Closed-loop transfer immittance
A'	Amplification scaling ratio	$A_i(\omega)$	Amplifier current amplification at frequency ω
A_c	Common-mode amplification	A_m	Minimum value of amplification

* Società Generale Semiconduttori, s.p.a., Agrate (Milano), Italy.
† Ballistic Research Laboratories, Aberdeen Proving Ground, Md.,
Sections 1 to 4, inclusive, and Section 6 were written by K. A. Pullen, and Section 5 was written by P. J. Bénéteau.

19-1

19-2 AMPLIFIER CIRCUITS

Symbol	Definition
A_o	Open-loop amplification
$A_v(0)$	Wideband amplifier d-c voltage amplification
$A_v(\omega)$	Wideband amplifier voltage amplification at frequency ω
$A_v'(0)$	Chopper amplifier d-c voltage amplification, including the factor $\frac{1}{2}$ for half-wave demodulation
$A_v'(\omega)$	Chopper amplification and filter voltage amplification at frequency ω, including the factor $\frac{1}{2}$ for half-wave demodulation
A_Δ	Differential gain
A_1, A_2, A_3	Arbitrary constants
A_1, A_2	Amplifier designations
a_j, a_k	Arbitrary constants
a_1, a_2, a_3, a_4	Arbitrary constants
B_i	Fraction of output current fed back to input
B_v	Fraction of output voltage appearing at input with input open-circuited
b_j, b_k	Arbitrary constants
b_1, b_2, b_3, b_4	Arbitrary constants
C	Capacitor; capacitance
C_{eff}	Effective or equivalent capacitance
C_{ob}	Collector-base diffusion capacitance
$D(s)$	Laplace transform of denominator expression
d	Discrimination ratio
d_0, d_1, etc.	Arbitrary constants in power-series expansion
$E_e(s)$	Laplace transform of error voltage
e_d	Drift-correction voltage from chopper amplifier output filter
e_e	Error voltage
e_i	Input voltage
e_L	Voltage at the output of the operational amplifier across the load
e_o	Output voltage
e_o'	Output voltage at alternate terminals
e_1	Voltage at input of operational amplifier
$F(s)$	Laplace transform of $f(t)$
f_M	Cutoff frequency due to Miller (C_{ob}) effect
$f(t)$	General function of time
f_t	Frequency at which the common-emitter current amplification is 1
f_β	Frequency at which the common-emitter current amplification is -3 db from its low-frequency value
g_m	Transconductance
g_m'	Transconductance
g_p	Plate conductance
g_p'	Plate conductance
K	Gain factor
K	Overshoot factor
k	Drift voltage referred to input of wideband amplifier
L	Inductance
L_x	Equivalent inductance of quartz crystal
$N(s)$	Laplace transform of numerator of transfer function
Q	Ratio of stored energy to energy dissipated per cycle
Q_o	Reference value of Q
Q_{eff}	Effective value of Q
R	Resistor; resistance
R_k	Cathode resistance
R_k'	Cathode resistance
R_L	Load resistance
R_s	Standard of resistance
R_x	Crystal resonant resistance
r	Active resistance in circuit
s	Laplace (complex-frequency) variable; the operator $j\omega$
t	Time
$u(t)$	Unit, or step, function
v_c	Common-mode voltage
v_Δ	Differential voltage
v_1, v_2	Input signal voltage
X_t	Transfer function
Y_o	Output admittance
Y_t	Transfer admittance
y_f	Forward admittance
y_i	Input admittance
y_o	Output admittance
$Z(s)$	Laplace transform of impedance
Z_f	Wideband amplifier feedback impedance
Z_f'	Chopper amplifier feedback impedance

Z_i	Impedance from operational amplifier input to wideband amplifier summing junction		plifier summing junction
		ϵ'	Voltage at chopper amplifier summing junction
Z_i'	Resistance from operational amplifier input to chopper amplifier summing junction	ζ	Damping factor
		ω	Radian frequency
		ω_o	Nominal radian frequency
Z_L	Load impedance		
Z_o	Output impedance of wideband amplifier	ω_1, ω_2	First and second corners respectively, of wideband amplifier
Z_o'	Output impedance of chopper amplifier	ω_1', ω_2'	First and second corners, respectively, of chopper amplifier and filters
α	Division ratio of metering circuit		
$\beta(s)$	Laplace transform of feedback gain	$\omega_{3\,db}$	Cutoff frequency
		ω_β	Radian frequency at which common-emitter short-circuit current amplification is 0.707 of low-frequency value
β_o	Low-frequency common-emitter current amplification		
$\Delta A, \Delta A_o, \Delta A_c$	Change in A, A_o, A_c		
ΔQ	Change in Q factor	ρ	Percentage accuracy
ϵ	Voltage at wideband am-		

1. INTRODUCTION

Operational amplifiers are those which have been built in assemblies or building blocks; technically, all the subassemblies of an amplifying system might be classed as operational amplifiers. Actually, the title operational amplifier is commonly reserved for a two- or more stage assembly which has been designed for insertion into other equipment. Operational amplifiers are frequently used in digital and analog computers, servosystems, various types of feedback networks, impedance converters, and a wide range of similar applications.

There are two important areas for consideration in the discussion of operational amplifiers, the first being the design and application of the assemblies per se, and the second the consideration of the optimum way of utilizing the assemblies in order to maximize the efficiency and the precision with which the overall system may be used. For example, in the application of a nonlinear amplifier, one mode of use may require extremely precise matching of the nonlinear characteristics of two associated paths through the amplifier, whereas a slightly modified mode of operation may require only matching in a relatively small operating area. Since the latter method of operation clearly will relax the tolerance requirements which must be placed on the nonlinear system, it is of importance to determine the best mode of operation.

Operational amplifiers can be designed for d-c low-frequency operation on a narrowband basis, they can be designed for d-c wideband operation, or they can be designed for either narrow- or wideband a-c operation. In addition, some applications require that the basic operational amplifier be highly stabilized internally, and others require only relatively high gain, the stabilization being developed in the completed assembly. Also, the application may require that the amplifier be embedded in a feedback system providing a combination of forward and inverse feedback in order to achieve optimum results.

The range of amplification required in an operational amplifier may be as small as unity or less, and it may be as large as several hundred thousands or millions. Normally, the amplification of the basic amplifier without feedback is in excess of 10,000, and the stabilized amplification is significantly less than 100. Stabilization of a-c amplifiers is rather easy to achieve, as degenerative action can be made effective and an a-c drift problem is seldom encountered. Stabilization of d-c amplifiers is more complicated, however, largely because of the drift due to leakage current, contact potential, or other causes, which can be introduced in the first, or input, circuit of the amplifier. As a result, more complex amplifiers, frequently including vibrating or solid-state choppers, must be used where d-c stability is of prime importance.

2. BASIC THEORY OF OPERATIONAL AMPLIFIERS

2.1. The Use of Laplace Transform Techniques.
As is the case with ordinary servomechanisms, consideration of many networks in which operational amplifiers are used is easier to handle in terms of poles and zeros and the associated Laplace transform techniques rather than in the time domain. For this reason, a brief, specially oriented review of the most useful features of the Laplace transform method, as it applies to operational amplifiers, follows.

Laplace transform techniques are used in the solution of network and servo problems because they make possible an estimate of the behavior of a system for all time after the application of a signal, and they frequently express the estimate in polynomial terms which can be readily manipulated. The behavior just after application of the signal is "measured" or weighted more heavily than later effects by the application of a negative-exponential weighting function. The transform converts complex integrals into simple polynomials which can then be restored to their time-domain form after simplification.

Table 1. Transform Responses

Function	$f(t)$	$F(s)$	Functional order (negative of exponent of s)
Impulse	$du(t)/dt$	1	0
Step	$u(t)$	$1/s$	1
Ramp	$tu(t)$	$1/s^2$	2
Parabolic	$t^2 u(t)$	$1/s^3$	3

$u(t)$ is the step function, having a value zero prior to the time $t = 0$, and unity thereafter.

In order to take account of an application of a signal, or a signal epoch, it is necessary to have a function which has a zero value up to the epoch and a unity value thereafter. This function may then be combined with typical signals to create excitation functions. The excitation functions, in Laplace form, may be multiplied by the transfer function of the network, again in Laplace form, to obtain the Laplace transform of the output signal. This makes it possible to separate, at least in part, the study of the network properties from the response to a given excitation function.

It is important to study the error responses of a network in terms of both the excitation function and the network response, as the final error in the response depends on the characteristics of the applied signal. Typical signals applied to operational amplifiers may include impulse functions, step functions, various order ramp functions, exponential waveforms, sinusoidal waveforms, etc. Systems in which the response to step and ramp functions is important are commonly classified as servomechanisms, and those in which response to exponential and sinusoidal functions is important are called filters and/or oscillators.

A list of typical excitation functions which are useful with servomechanisms and often appear in the consideration of operational amplifiers is included in Table 1. This list is far from complete, but it contains the principal relations which will be useful in the discussion to follow. Some others which are commonly used are the sinusoidal function, the cosinusoidal function, and damped-wave functions (with both positive and negative damping).

The Laplace transforms for the input-output relation and the input-error relation

OPERATIONAL AMPLIFIERS

of the stabilized operational amplifier of Fig. 1 may be written in the form

$$e_o/e_i = A(s)/[1 + A(s)\beta(s)] \quad (1)$$
$$e_e/e_i = 1 - A(s)/[1 + A(s)\beta(s)]$$
$$= 1 - 1/[\beta(s) + 1/A(s)] \quad (2)$$

where $A(s)$ is the amplification and equalization acting in the forward part of the loop and $\beta(s)$ that in the feedback, or return, part. If the output terminal for the system and the output terminal for the feedback path differ, then the useful amplification may be A' times the value given in Eq. (1), where A' is defined by the equation

$$e_o'/e_o = A' \quad (3)$$

In this equation, e_o' is the output voltage at the actual output terminal and e_o the output at the input to the feedback network.

Now, from Eq. (2), if the d-c value of β is unity, then the error e_e is zero at direct current if and only if either $A(s)$ has a constant term approaching infinity, or if its expansion in a Laurent series (series in s^{-k}, where $k > 0$) has no term in s^0.*

The use of an infinite value of the constant term in $A(s)$ requires an infinite amplification (steady-state) and makes the operation of the amplifier extremely subject to drift. Consequently, the form taken by $A(s)$ should be

$$A(s) = (1/s)(A_1 + A_2/s + A_3/s^2 + \cdots) \quad (4)$$

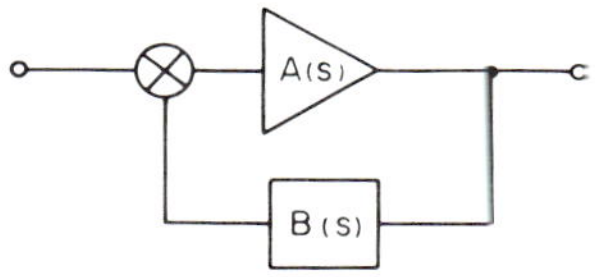

Fig. 1. Block diagram of stabilized operational amplifier.

and thus the use of an integrator along with some kind of filter is indicated.

The expansion in the parentheses on the right of Eq. (4) can sometimes be converted into the form of a continued fraction. In this form, it represents a ladder network. Then the transfer admittance may be expressed in the form

$$Y_t = 1/Z(s) \quad (5)$$

where Y_t is a transfer admittance and $Z(s)$ is a polynomial in s and is linear in the various impedances of the network. As long as $Z(s)$ is a positive-real function, a passive network can be designed which at least approximates these specifications.

Practically, an electromechanical integrator appears to have nearly ideal properties for use in these networks, since it does not require development of "infinite" amplification in the network to assure zeroing of position error when such an integrator is properly used, as do fully electrical "integrators" using pole-zero networks. Typically, such an integrator consists of a motor-driven potentiometer, with a regulated voltage supply connected across the potentiometer to permit a controlled but variable output voltage as a consequence of change of shaft position. Such a combination provides almost perfect electrical integration as long as the voltage applied to the potentiometer is very precisely stabilized.

In order to learn a little more about the potential ways that effective use of the operational amplifier may be obtained, it is instructive to solve Eqs. (1) and (2) for the theoretical limitations on error subject to the following assumptions:†

$$A(s) = K(a_1 s + b_1)/(a_2 s + b_2) \quad (6)$$
$$\beta = 1$$

Under these conditions, the values of a_1, a_2, and b_1 are optional, and only the value of

---

* In a transfer function the limit value of the transfer ratio as time approaches infinity is given by letting $s \to 0$. This then gives the steady-state response.
† This $A(s)$ is the simplest form in which a properly bounded transfer function satisfying the Wiener-Paley criterion on realizability can be represented. Strictly, only the functions satisfying this criterion are realizable.

b_2 is specifically established; its value must be zero in order to eliminate steady-state error. Simplifying gives

$$A(s) = K(a_1/a_2) + K(b_1/a_2 s) \tag{7}$$

This equation shows that in addition to the amplifier, a linear one, the forward part of the loop may include both a linear network and an integrator in parallel. Clearly, a transmission zero may be introduced in this way into the forward path of the network without degrading the steady-state zeroing of error, and it will turn out that the network can also be used in the compensation of other errors.

The polynomial-quotient form of the error function can be converted into the input-output relation of Eq. (1) in the form

$$[e_e(s)/e_i(s)] + 1 = e_o/e_i \tag{8}$$

If the denominator of the transfer function for both the input-output relation and the input-error relation takes the form

$$D(s) = d_0 + d_1 s + d_2 s^2 + \cdots \tag{9}$$

then the output function (before expansion in a power or a Laurent series) takes the form

$$N(s)/D(s) = 1 + E_e(s)/D(s)$$
$$= [d_0 + d_1 s + d_2 s^2 + \cdots + E_e(s)]/(d_0 + d_1 s + d_2 s^2 + \cdots) \tag{10}$$

where $N(s)$ is the numerator for (e_o/e_i); and $E_e(s)$ is the corresponding numerator for the error polynomial. Consequently, in order to make an operational amplifier show zero error for any specified type of input function, it is necessary to design the circuit so that the numerator function $E_e(s)$ has zero coefficients in terms of the corresponding excitation function of Table 1. For a servomechanism having position, velocity, and acceleration components of excitation, therefore, $E_e(s)$ must have zero coefficients in the first three terms in its polynomial expansion. If, for example, a zero steady-state position error is required, the network should be designed to make e_0 identically zero. Similarly, for a zero steady-state velocity error, then both e_0 and e_1 should be identically zero in the equation

$$E_e(s) = e_0 + e_1 s + e_2 s^2 + \cdots \tag{11}$$

where e_0 is the position error, e_1 is the velocity error, etc. In similar fashion, if the excitation function is sinusoidal and of frequency ω_0, the corresponding representation of $E_e(s)$ would have a zero term in that frequency.

Equation (10) shows that under ramp-type excitation the condition required of the numerator $N(s)$ for zero error is the existence of identical coefficients in the numerator and denominator for all the excitation orders for which zeroing is required.* Analysis has already shown that the condition for zeroing the position error is to make the coefficients for the zero-order terms equal in the numerator and the denominator; clearly, relations can be derived for zeroing any number of other coefficients as well. The next few paragraphs consider the requirements for these conditions.

The condition established in Eq. (7) for b_2 identically zero does not provide for reduction of other orders of error. Under this condition, the transfer function takes the form

$$e_0/e_i = (a_1 s + b_1)/[a_1 s + (a_2 s/K) + b_1] \tag{12}$$

In this equation, if K is very large so that $a_1 \gg a_2/K$, then rather good (brute-force) compensation of both the position and the velocity errors can be expected to result. But significantly better results can be obtained with a simple pole-zero filter network in the feedback path. Inserting in Eq. (1) the feedback transfer function

$$\beta = (a_3 s + b_3)/(a_4 s + b_4) \tag{13}$$

---

* The usefulness of this condition apparently was first shown by L. H. King.[11]

OPERATIONAL AMPLIFIERS   19-7

for the selected feedback filter network $\beta(s)$ immediately shows that, for $\beta(0) = 1$, it is required that $b_3 = b_4$, and secondly, that exact zeroing of the velocity error is possible for the following relation among the remaining network coefficients:

$$b_3 = (a_4 - a_3)b_1/a_2 \qquad (14)$$

Since with passive networks the values of all these coefficients must be positive, this equation shows that the value of a_4 must be greater than that of a_3, and as a result, the filter in the feedback path must be a low-pass filter. The additional equations which must be solved to obtain a complete elimination of error, in terms of an arbitrary bandwidth ω_o, are

$$a_4 = a_1 b_3 K / b_1 \qquad (15)$$

and the frequency and damping equations, which enter only coincidentally, since the transfer equation reduces to unity identically, are

$$\omega_o^2 = b_1 b_3 / (a_1 a_3 + a_2 a_4 / K) \qquad (16)$$
$$2\zeta\omega_o = (a_1 b_3 + a_3 b_1 + a_2 b_3 / K)/(a_1 a_3 + a_2 a_4 / K) \qquad (17)$$

where ω_o is the nominal cutoff frequency, and zeta the damping for the denominator polynomial. Initial values for ζ and for ω_o may be selected at convenient values since

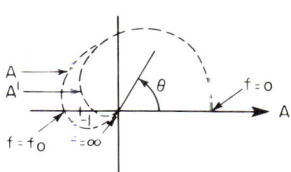

FIG. 2. Typical locus of gain as a function of frequency.

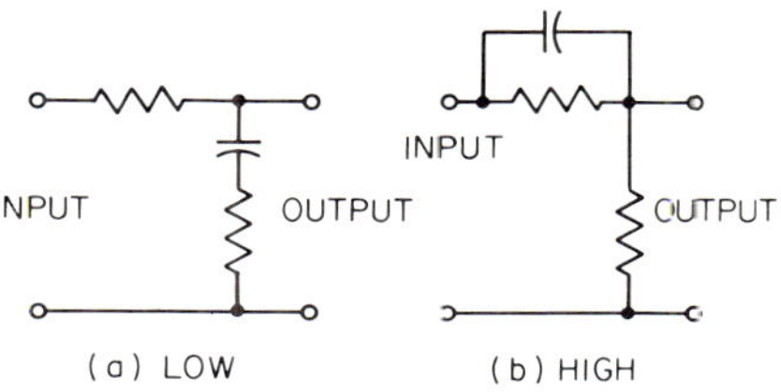

FIG. 3. Typical pole-zero networks.

the final values obtained for the coefficients superpose zeros for the numerator on the denominator poles, $-\zeta \pm j\omega_o$.

Further analysis of the kinds of compensating networks required with operational amplifiers shows that probably the best way to obtain position, velocity, and acceleration correction on a system is the use of a configuration having a total of two ideal integrators, with the forward signal being a combination of directly amplified, amplified and integrated once, and amplified and integrated twice. The amplification adjustments control the stability with this arrangement. Also, further analysis may be made using pole-zero networks in the forward and the feedback sections of the loop, and it can be quickly verified that high-pass pole-zero networks are usually required in the forward part of the loop, and low-pass in the feedback part.

The values of amplification required in the forward part of the network often are quite critical. For this reason, it is normally necessary to apply stabilization to the amplifiers used in order to assure that the stability of the overall system is in accordance with requirements. This internal feedback should be designed to provide as constant a phase characteristic as possible for the combined circuit.

2.2. Transition Equalizers for Operational Amplifiers. One of the critical problems in the design of any amplifier to which feedback is applied is the control of "poles and zeros" in order to minimize the possibility of instability, as a result of the accumulation of at least two shifts of 90° (in addition to the normal 180° in an active device), giving a response curve which may enclose the $A\beta = -1$ point (Fig. 2). A network which has only poles or zeros naturally develops the 90° of phase shift in the attenuation region, whereas a "pole-zero" network such as those shown in Fig. 3, by providing both a pole and a zero, gives a phase shift which is less than 90° and will have significantly less shift both above and below the frequency region enclosed by the pole and the zero.* Typical phase and amplitude curves for these networks are shown in Fig. 4. These networks may be cascaded, using isolation amplifiers if necessary, and the pole of one may be "canceled" by the zero of the next to give significantly increased attenuation.

As is evident from Fig. 3, there are two basic forms of the pole-zero network, one behaving like a shunted low-pass network and the other like a shunted high-pass network. Either of these networks will have the same form of transfer function:

$$Y_t = (a_j s + b_j)/(a_k s + b_k) \qquad (18)$$

where the a's and the b's are functions of the network components. For a low-pass network, $b_j = b_k \ (=1)$ identically, and $a_k > a_j$. The d-c transmission for this network is unity.

For the high-pass pole-zero network, $b_k > b_j$, and usually $a_j = a_k$. This means that there is a gain less than unity at direct current with the passive configurations (since $s = 0$ at direct current). If the gain function is extracted, this equation is similar to Eq. (18), with the restrictions that $(a_j/b_j) > (a_k/b_k)$, or the coefficient of the variable s in the numerator is greater than that in the denominator.

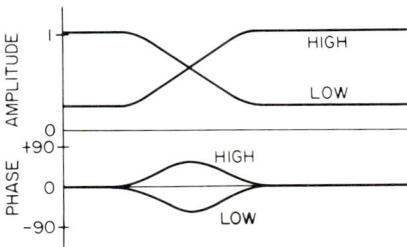

Fig. 4. Phase and amplitude curves for pole-zero networks.

3. ELECTRON-TUBE-TYPE OPERATIONAL AMPLIFIERS

Typically, the simplest operational amplifiers consist of two amplifier stages, and more complex configurations are often assembled from pairs of stages. This has become common practice since it has been shown mathematically that the most satisfactory stabilized building block consists of an amplifier with two active (amplifying) stages. The available gain can be controlled most effectively and parasitic defects rendered least important by such an arrangement. The basic amplifiers in this form normally have differentially connected input stages, particularly when d-c response is required. A typical schematic wiring diagram and a photograph of an operational-amplifier building block, for a Philbrick 4-K2W, are shown in Fig. 5.† There are two input terminals on these amplifiers, one positive (noninverting) and one negative (inverting). The output is considered to have a positive polarity. The input differential amplifier is designed to be insensitive to common-mode input voltages, and the output amplifier consists of a simple amplifier and a cathode follower in cascade.

The differential amplifier normally may have a stage gain as high as 100 or possibly a little more when used with a 12AX7A tube. It can develop this amount of amplification only when an antisymmetric, or push-pull, input signal is used. If the signal is inserted on only one input, the gain of the stage is at most half the nominal value as a result of the degenerative action which results from the common-mode suppression caused by the common-cathode resistor. The supply voltage required with these amplifiers, because of their d-c coupling, must include both a positive and a negative component of roughly equal value, typically plus and minus 250 volts. The maximum amplification available from the amplifier as a whole is smaller by a factor of between 2 and 4 than that available from the corresponding a-c amplifier. Some gain typically is lost through the use of the common-mode rejection circuit also.

The simple dual-triode operational amplifier, of which the 4-K2W is typical, will have an open-loop amplification of between 5,000 and 20,000 at direct current, the

---

* This network represents the simplest low- or high-pass filter satisfying the Wiener-Paley criterion.

† Circuit and photograph courtesy of George A. Philbrick Researches, Inc.

OPERATIONAL AMPLIFIERS   19-9

(a)

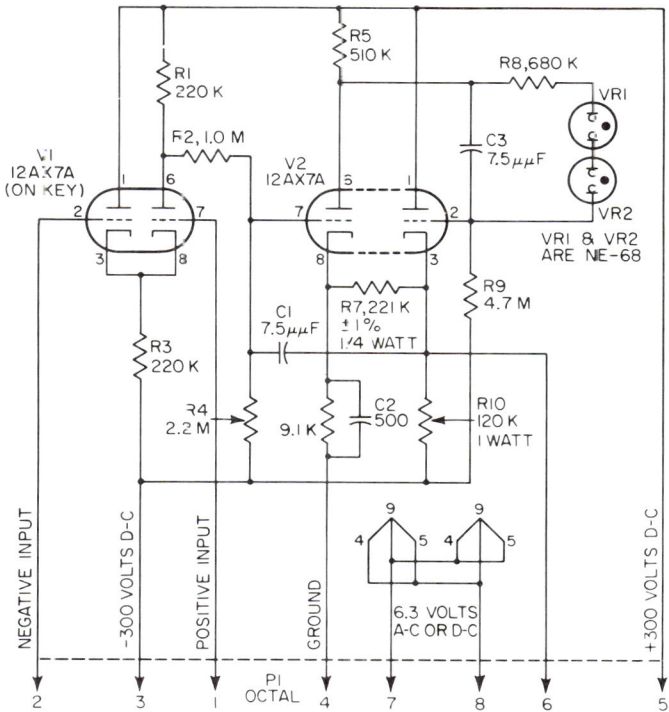

NOTE: UNLESS OTHERWISE STATED RES. ARE ±5%, 1/2 WATT

(b)

FIG. 5. Typical operational amplifier. (a) Drawing of Philbrick model 4-K2W amplifier. (b) Schematic of Philbrick 4-K2W amplifier. (*Courtesy of Philbrick Researches, Inc.*)

value tapering off to approximately unity at a frequency in excess of 100 kc. The frequency round-off encountered depends partly on the kind of signal source from which the amplifier derives its input signal and partly on the precise network configuration. When the differential amplifier is balanced and neutralized, and when it functions as a cathode-coupled amplifier, the rolloff will be 6 db/octave. If the signal source has a very small source impedance, then also the rolloff can be 6 db/octave. It may be as high as 12 db/octave, however. When the input amplifier is fully symmetrical, the rolloff can usually be limited to 6 db/octave, as the critical limitation then comes from Miller effect on the second amplifier.

In some applications, the degree of differential operation which can be achieved using simple common-mode rejection circuits in conventional operational amplifiers is not sufficient to meet the demands of the application. In such a case, a preliminary isolation circuit may be required. The procedure which must be followed then depends on the exact nature of the voltage applied to the operational amplifier. If the signal of concern consists of a range of frequencies which can be handled by a transformer, a circuit such as that shown in Fig. 6 may be used. Typically, the problem here is to measure the loss of alternating voltage across the standard resistor R_s. The transformer is capacitively coupled to prevent saturation effects due to direct current and to permit the development of a direct voltage across the standard resistor. Typically, the transformer selected should have an open-circuit impedance between 1,000 and 10,000 times that represented by R_s, and it should be loaded by a resistor of such size that it will reflect a loading resistance to the primary side which is between 100 and 500 times R_s. The minimum operating frequency of the capacitor-transformer combination should be less than a twentieth of the minimum frequency encountered in normal operation, and the maximum more than twenty times the maximum encountered.

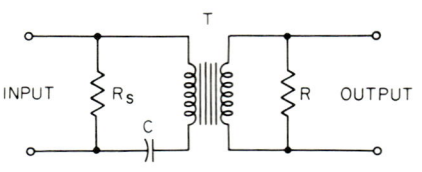

FIG. 6. Signal separation circuit using transformer.

The loading resistance on the secondary side of this transformer is very important, since it assures a constant transformation ratio for the transformer and also minimizes both the distortion and the phase-shift introduced by the transformer. Failure to include the loading resistor on a coupling transformer has frequently resulted in unsatisfactory operation of transformer-coupled amplifiers using tubes.

In applications in which the balance is required to extend to direct current, the problem is more severe, particularly when the required discrimination ratio is high. The problem in this instance is to increase the differential voltage without permitting the common-mode voltage to interfere with the operation.

The discrimination ratio d of an operational amplifier measures the ratio of the sensitivity of the circuit to the differential signal voltage to that for the common-mode voltage, and may be defined in terms of the percentage accuracy required ρ and the ratio of the differential voltage v_Δ to the common-mode voltage v_c in the form

$$d = \rho v_\Delta / v_c \tag{19}$$

If the amplifier is 1,000 times less sensitive to common-mode signal than to differential signal, and the precision required is 1 per cent, then the maximum permitted common-mode voltage is ten times the differential voltage.

In applications in which the common-mode component of voltage is large, a configuration somewhat like that shown in Fig. 7 may be used to increase the ratio of the differential voltage to the apparent common-mode voltage. Two power supplies, one special, and at least two operational amplifiers are required, since the amplifier A_1 is used as a repeater for the common-mode voltage (it has unity gain), and the second amplifier A_2 amplifies the differential voltage, using the special power supply for its power. The amplifier A_1 controls the neutral potential on the special power supply, maintaining it at the signal voltage of one of the metering terminals for v_A. Under these conditions, the maximum value of the discrimination ratio applied to the amplifier A_2 is 2, and the use of an operational amplifier for A_2 which has been stabilized to provide an overall amplification of 100 can usually solve the common-mode problem.

Operational amplifiers are often used in conjunction with chopper-stabilized preamplifiers, which may be used to simulate integrators. The output of these amplifiers is heavily bypassed, and they are effective only at very low frequencies or direct current as a consequence. They actually function as a-c amplifiers, since the input

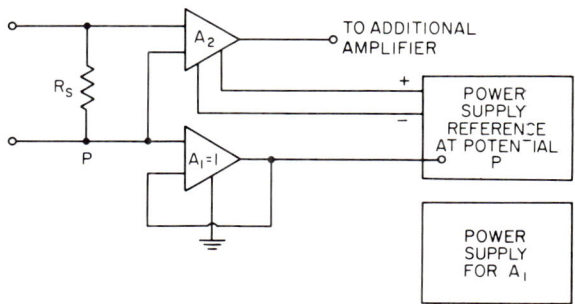

Fig. 7. Wideband circuit for common-mode suppression.

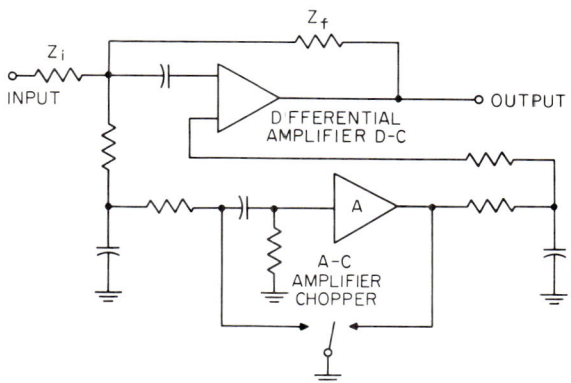

Fig. 8. The Goldberg d-c amplifier circuit.

signal is periodically grounded by the chopper, and the output is correspondingly grounded alternately with the input. Since internally these preamplifiers are a-c rather than d-c, they can be stabilized by standard feedback techniques. There is normally no local d-c feedback around these amplifiers, and they have relatively small inherent drift, but they may develop drift because of the high overall feedback factor and the large time delay in the chopper output.* These units are commonly used to reduce the offset in standard operational amplifiers. Guaranteed open-loop gains of a minimum of 100 million with design centers as high as 1 billion can be produced with these preamplifiers and associated high-performance operational amplifiers.

A modification of the d-c stabilizing circuit just described, often called the Goldberg circuit (Fig. 8), may be used for stabilization of d-c wideband amplifiers.[18]

* For a discussion of this drift in RC oscillators, see Ref. 31.

AMPLIFIER CIRCUITS

This circuit in effect uses two amplifiers, one being a standard operational amplifier and the second an a-c–coupled version of the standard amplifier. The latter is used with a full-wave chopper which is connected to maintain the d-c output at the final output terminals at the proper level through adjustment at the auxiliary input terminal of the d-c operational amplifier.

The real advantage of the Goldberg-type circuit is that it makes possible the use of an a-c–coupled input circuit and uses an auxiliary chopper-stabilizing circuit to maintain the d-c level at the appropriate value. Good d-c and a-c response are both available with such an arrangement.

It should be noted that all the usual chopper-stabilized circuits such as those described above achieve their stabilization through establishing a condition of "almost balance." As a result they are in effect differencing two large numbers, and small drift components are of tremendous importance.

In addition to the variety of assemblies strictly classified as operational amplifiers, there is an extensive family of related circuits which are often used with these amplifiers and which are packaged similarly for use in construction of more elaborate systems. The complete assemblies might be called operational assemblies, and they include in addition to operational amplifiers such configurations as stabilized oscillators, controllable oscillators, counting or countdown circuits, shaping circuits including differentiators and integrators, product circuits or mixers, and balanced-product circuits, or phase detectors. Some of these configurations are considered in later paragraphs of this chapter, and others in other chapters.

4. DIFFERENTIAL AND COMMON-MODE OPERATING CONSIDERATIONS

The operational amplifier normally is supplied with a differentially connected input stage in order that it will respond to the voltage difference between two measuring points, and it also is designed to provide a high degree of common-mode suppression, so that it will have a minimum of response to a common voltage on the two terminals. The common-mode rejection of an operational amplifier is dependent to a large extent on two factors, the first being the accuracy of symmetry of the input amplifier, and the second being the degree of degeneration included for common-mode signals.

It is important to divide the signal voltage applied to the amplifier input into a differential component and a common-mode component. These two components are defined by the equations

$$v_\Delta = (v_1 - v_2)/2 \qquad (20)$$
$$v_c = (v_1 + v_2)/2 \qquad (21)$$

where v_1 and v_2 are the input voltages on inputs 1 and 2. The common-mode suppression resistor, which normally is either a cathode or an emitter degeneration resistor, is inactive with respect to the differential signal, as the current changes in the two active devices theoretically balance. This resistor is fully effective on the common-mode component, however, as the two active devices behave as if they were in parallel, and they apply in-phase components of current across the resistor. The amplification equations, for a triode tube circuit, for the two components of signal are therefore[35]

$$A_\Delta = -g_m R_L/(1 + g_p R_L) \qquad (22)$$
$$A_c = -g_m R_L/[1 + (g_m + g_p)R_k + g_p R_L] \qquad (23)$$

where A_Δ is the differential amplification and A_c the common-mode amplification. The discrimination ratio of an amplifier may be determined from these equations.

Ideally, the desired value of cathode resistor is one approaching infinity, so that the degenerative gain will approach zero. This condition may be approximated by using a cathode-degenerated triode in place of the normal cathode resistor as shown in Fig. 9. Under these conditions, although the static impedance of the circuit is finite, the dynamic impedance may well approach an infinite value. In other words,

OPERATIONAL AMPLIFIERS

its cathode resistance appears to be

$$R_k = [1 + (g_m' + g_p')R_k']/g_p' \qquad (24)$$

where g_m' and g_p' are the values for the triode serving as the common-cathode resistance, and R_k' is the actual resistance in the cathode circuit for the degenerated triode. Values of R_k approaching 1 megohm or more are comparatively easy to obtain.

In terms of the amplification equations, the discrimination ratio for the amplifier is

$$d = A_\Delta/A_c \qquad (25)$$

Typical values of discrimination ratio as high as 100 to 1,000, or 40 to 60 db, or more may be obtained in commercially available amplifiers.

Superficially, it would appear that if one input of the operational amplifier were grounded and a signal applied to the second, equal and opposite differential currents would flow. This is not exactly true, however, as the approximate value of the common-mode voltage must be developed across the suppression resistor,

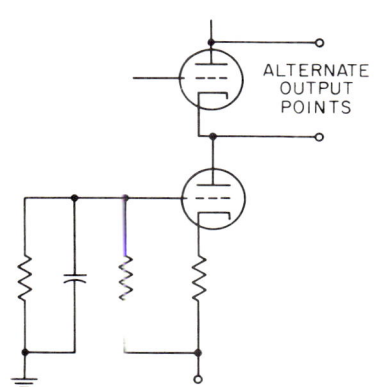

Fig. 9. Dynamic common-mode suppression circuit.

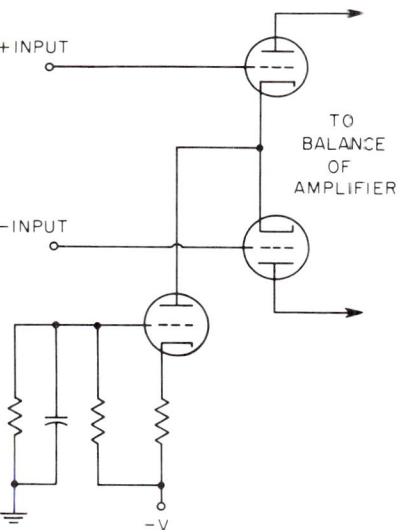

Fig. 10. Typical tube common-mode suppression circuit.

and enough unbalance in current must flow to provide this voltage. The higher the magnitude of the dynamic suppression resistance, the smaller the required current differential and, at the same time, the greater the rejection ratio. For this reason, the suppression resistance in a circuit which is required to have a very high discrimination ratio frequently uses either a tube or a transistor connected as a degenerated load (Fig. 10). These configurations yield exceedingly high magnitudes of dynamic impedance and at the same time permit the flow of significant amounts of static current. A common-mode loss is commonly encountered when these circuits are used.

A high value of rejection ratio in an amplifier would appear to be all that is required in the isolation of the differential signal from the common-mode signal. Unfortunately, however, common-mode rejection circuits are commonly used when reasonably precise measurements are required, and a fairly small common-mode voltage sometimes is sufficient to cloud a differential reading more than can be permitted. Under these conditions, the circuit of Fig. 7 already described is used to increase the effective value of v_Δ/v_c.

5. TRANSISTOR OPERATIONAL AMPLIFIERS*

5.1. General. Operational amplifiers using transistors are much more difficult to design than those using electron tubes. This is due to many factors, including the drift caused by the base current flowing through the source impedance, the much higher capacitances generally encountered in transistors, the frequency corners introduced by the current amplification fall-off and the temperature sensitivity of transistor parameters. They do, however, also offer many advantages including usually much lower open-loop drift, at least for low source impedances, and generally all the advantages offered by transistors as compared with electron tubes. In this section the alternatives available to the circuit designer will be discussed as regards the type of input stage to be utilized; the factors affecting circuit performance will be discussed, two main types of drift-reduction schemes will be considered, and a typical design will be indicated. Field-effect transistor operational amplifiers will be briefly considered.

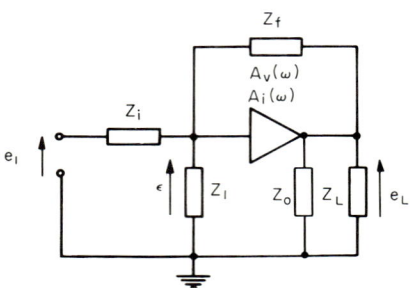

Fig. 11. A general operational amplifier.

5.2. Analysis of Operational Amplifiers. *Operational-amplifier Methods.* A perfectly general operational amplifier is shown in Fig. 11; Z_1 is the input impedance of the amplifying device (normally but not necessarily high for vacuum tubes and low for transistors), Z_o is the output impedance of the device, $A_v(\omega)$ and $A_i(\omega)$ are the voltage and current amplifications, respectively, of the device, Z_L is the load, and Z_i and Z_f are the gain-determining operational resistors. The equations representing the amplifier are

$$e_L = \epsilon A_v(\omega) \tag{26}$$
$$(e_1 - \epsilon)/Z_i = (\epsilon - e_L)/Z_f + \epsilon/Z_1 \tag{27}$$

Defining B_v as ϵ/e_L with $e_1 = 0$ gives

$$B_v = \frac{(Z_i Z_1)/(Z_i + Z_1)}{(Z_i Z_1)/(Z_i + Z_1) + Z_f} = \frac{Z_i Z_1}{Z_f Z_1 + Z_1 Z_i + Z_i Z_f} \tag{28}$$

Solving Eqs. (26), (27), and (28) yields

$$A = \frac{e_L}{e_1} = \frac{-Z_f}{Z_i} \frac{1}{1 - \dfrac{Z_f Z_1 + Z_1 Z_i + Z_i Z_f}{A_v(\omega) Z_1 Z_i}} \tag{29}$$

or, rewritten using Eq. (28),

$$A = \frac{e_L}{e_1} = \frac{-Z_f}{Z_i} \frac{1}{1 - 1/A_v(\omega) B_v} \tag{30}$$

For large $A_v B_v$ products, Eq. (30) reduces to

$$A = e_L/e_1 \cong -Z_f/Z_i \qquad |A_v(\omega) B_v| \gg 1 \tag{31}$$

For example, for an accuracy of 0.1 per cent, $A_v B_v$ should be larger than 1,000.

* Section 5 has been taken mostly from Ref. 4; acknowledgments are therefore due to P. J. Bénéteau's coauthors R. Q. Lane and Larry Blaser. Acknowledgment is also due to Fairchild Semiconductor for permission to reprint the salient features of the article, and to Società Generale Semiconduttori for assistance in the preparation of the manuscript.

OPERATIONAL AMPLIFIERS

For transistor amplifiers, $A_i(\omega)$ is usually of more interest than $A_v(\omega)$.

$$A_v(\omega) = A_i(\omega) \frac{(Z_o Z_L)/(Z_o + Z_L)}{Z_1}$$
$$= A_i(\omega) Z_L'/Z_1 \tag{32}$$

From Eq. (28), for $Z_1 \ll Z_i, Z_f$,

$$B_v|_{z_1 \text{ small}} \cong Z_1/Z_f$$

and defining $B_i = I_f/I_L \approx -Z_L/Z_f$, the expression for the gain becomes, using Eq. (30),

$$\frac{e_L}{e_1} = \frac{-Z_f}{Z_i} \frac{1}{1 - 1/A_i B_i} \tag{33}$$

For large $A_i B_i$ products, Eq. (33) reduces to

$$e_L/e_1 \cong -Z_f/Z_i \qquad |A_i B_i| \gg 1 \tag{34}$$

as in the high-impedance case of Eq. (31). A low-input-impedance amplifier having a fixed $A_i B_i$ product will thus be indistinguishable from a high-input-impedance amplifier having the same value of $A_v B_v$.

Choice of Input Impedance. The designer of operational amplifiers has the choice of input impedance of the amplifier. The maximum power gain in transistor amplifiers usually occurs in the common-emitter configuration, and for this type of amplifier, the voltage amplification times the input impedance is approximately constant.[19] Therefore, the term $A_v(\omega) Z_1$ in the denominator of Eq. (29) is approximately constant, from which it follows that the most efficient operational amplifier, using transistors, will be the one with the lowest input impedance possible. However, even though this case is the optimum one, it will not necessarily be used in practice, for the following reasons.

A comparison of Eqs. (31) and (33) shows that the quantity to consider for the high-impedance amplifier is $A_v B_v$ while that for the low-impedance amplifier is $A_i B_i$. Remembering that the operational gain $A = -Z_f/Z_i$, these feedback factors B_v and B_i can be expressed approximately as

$$B_v \cong 1/(1 - A)$$
$$B_i \cong (-1/A)(Z_L/Z_i) \tag{35}$$

where the assumption is made that the amplifier input impedance is negligibly high or low, respectively, compared with Z_i. For stability of the amplifier, both $A_v B_v$ and $A_i B_i$ will be required to have less than 360° phase shift at the frequency where they become less than 1 or 0 db. B_v and B_i are normally frequency-independent in the worst case; when the usual 180° phase shift of the inverting amplifier is taken into account, the added phase shift is usually kept less than 135° for an adequate phase margin of 45°. If the operational gain is to be changed, Z_f is usually the controlling impedance and both B_v and B_i change, as is shown in Eq. (35). Unfortunately, the phase margin changes as the operational gain changes; this holds to a different extent for high-impedance and for low-impedance amplifiers, as is shown in Fig. 12. It is seen that the variation of B_i, and thus of phase margin, with operational gain is far more serious than that of B_v over the gain range of 0.1 to 10. If such a large gain range is not required, the current amplifier may be suitable, but in general the A_i will have to be programmed along with the variable Z_f to keep the $A_i B_i$ product more or less constant. The same problem arises, of course, with the high-impedance amplifier but the permissible operational-gain range is very much wider. For example, a gain range of 0.33 to 3 changes B_v by 3:1, which is perhaps tolerable, compared with a change in B_i of 9:1, which is probably not. This problem does not, of course, enter in the case of a fixed-gain amplifier, and Z_1 should be made small in this case.

To summarize, the designer of transistor operational amplifiers has an infinite choice of input impedance. The one with lowest possible input impedance is the most efficient, while the one with highest possible input impedance can have its gain varied most safely. There are, obviously, intermediate values which form the best compromise between these two factors. By considering Eq. (29) it can be shown that there is little advantage in making Z_1 less than $(0.1Z_f)/(1-A)$, where A is the operational gain. If a large range of operational gains is contemplated, it may well be that a value of Z_1 greater than the above would be chosen to ensure an adequate phase margin.

Frequency Response of Operational Amplifiers. For the purpose of this analysis, it will be assumed that the amplifier will have a forced rolloff at a frequency ω_1 because of an RC insertion (usually at a frequency below 100 cps) plus an undesirable rolloff at a frequency ω_2 because of the open-loop frequency response. It is, of course, desirable for phase-margin (stability) reasons to have ω_2 as high as possible and thus control the rolloff solely by the forced ω_1. See Fig. 13.

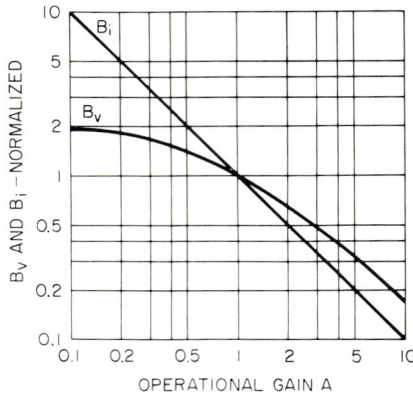

FIG. 12. Normalized feedback fraction variations.
$$Z_1 = Z_f/10(1-A)$$

FIG. 13. Assumed open-loop asymptotic response of amplifier.

The operational gain $A(\omega)$ can be written as

$$A(\omega) = \frac{A_v(\omega)}{1 - A_v(\omega)B_v} \qquad (36)$$

Letting the d-c open-loop gain be $A_v(0)$, letting $s = j\omega$, and using

$$A_v(s) = \frac{A_v(0)}{(1+s/\omega_1)(1+s/\omega_2)}$$

it can easily be shown that for $\omega_1 \ll \omega_2$ and $|A_vB_v| \gg 1$, Eq. (36) reduces to

$$A(s) = \frac{A_v(0)\omega_1\omega_2}{s^2 + \omega_2 s - A_v(0)B_v\omega_1\omega_2} \qquad (37)$$

The closed-loop 3-db frequency ω_{2c} can be calculated by noting that the numerator is not a function of s. ω_{2c} is then very nearly given by the lower of the two poles and

$$\omega_{2c} \cong \frac{\omega_2}{2}\left[1 - \sqrt{1 + \frac{4A_v(0)B_v\omega_1}{\omega_2}}\right] \qquad (38)$$

For the case where ω_2 is very large, the root can be approximated by $1 + 2A_v(0)B_v\omega_1/\omega_2$, and ω_{2c} is then given by

$$\omega_{2c} \cong |A_v(0)|B_v\omega_1 \qquad (39)$$

This means, of course, that only for a 6 db/octave slope of open-loop response is there direct gain-bandwidth trading. If the influence of ω_2 is not negligible, then Eq. (39) is not valid, and the more exact expression of Eq. (38) will have to be used.

Pulse Response of Operational Amplifiers. A sufficient but not necessary condition for no pulse overshoot can be very easily obtained from Eq. (37) by setting the discriminant of the denominator to a quantity greater than or equal to zero. This ensures that the poles of the gain function do not leave the real axis,[43] i.e., $b^2 - 4ac \geq 0$.

$$\omega_2{}^2 + 4|A_v(0)|B_v\omega_1\omega_2 \geq 0 \tag{40}$$

or $\omega_2 \geq 4|A_v(0)|B_v\omega_1$. This is usually a rather stringent requirement, since if good pulse response is required, either ω_1 or $A_v(0)B_v$ will be high. Experimental data given in Sec. 5.3 indicate the overshoot which is obtained in a practical case when ω_2 is less than the critical value.

Drift Stabilization. One of the principal requirements of an operational amplifier is that output drift be small. The early stages of the amplifier are the main sources of drift since they are followed by the remaining gain of the amplifier. In this respect, drift is analogous to noise, and if it is possible to precede the amplifier by a high-gain low-drift amplifier, the drift at the output for a given operational gain will be greatly reduced. This is analogous in a receiver to using a high-gain low-noise preamplifier before a stage of noisy mixing in order to increase the signal-to-noise ratio at the receiver output.

The causes of drift in transistor amplifiers are discussed in Secs. 4 and 5 of Chap. 18. Briefly they are in order of importance for planar transistors:

1. The negative temperature coefficient of the base-emitter voltage V_{BE} of about -2 mv/°C
2. The positive temperature coefficient of the common-emitter current amplification h_{FE}
3. The positive temperature coefficient of the collector-base inverse leakage current I_{CBO}

With careful design, drifts on the order of 10 μv/°C can be obtained with uncompensated circuits and $0.5 - 1$ μv/°C with compensated circuits over a broad temperature range. It may be, however, that the drifts which can be obtained are still too high or that no high-temperature adjustment is desired on a production amplifier; a drift-reduction scheme must then be used. Two such schemes have been described[5,18,30] and are shown below for particular application to transistor circuits.

1. *Goldberg Method.* Figure 14 shows a drift-reduction scheme suggested by Goldberg[18] in which the high-gain low-drift amplifier is formed by an a-c amplifier with half-wave modulation and demodulation. This scheme is analyzed in Ref. 4, and only the result will be quoted here.

e_L = output voltage at load
e_1 = input signal
ϵ = voltage at summing junction
k = drift potential
e_d = correction voltage from chopper amplifier

C_1R_4 forms a filter to reduce the transmission of the chopped drift voltage into the summing junction.

R_3 isolates C_1 from Z_1 in order to avoid loading the amplifier input at frequencies where C_1 has little reactance.

$$e_L \cong -\frac{e_1 Z_f}{Z_i} + \frac{kZ^3}{Z_1 Z_i Z_3[1 - A_v'(\omega)]} \tag{41}$$

where
$Z_3 = R_3 + R_4 + Z_1'$
$Z^3 = Z_1 Z_3 Z_i + Z_f Z_3 Z_i + Z_f Z_1 Z_3 + Z_f Z_1 Z_i$

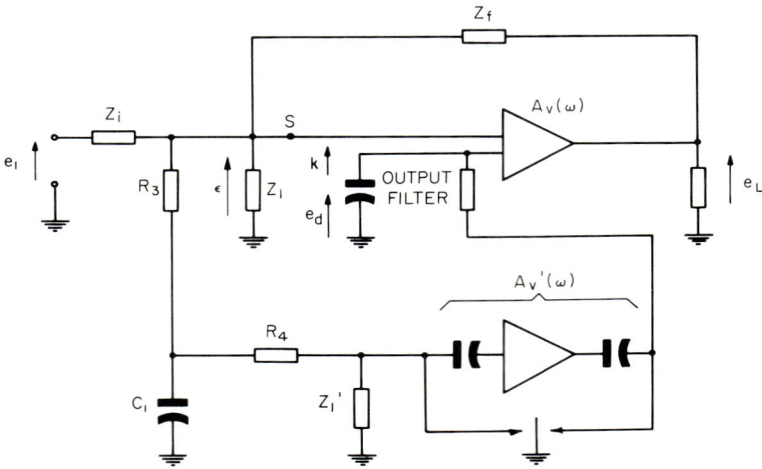

Fig. 14. Goldberg drift-reduction scheme. *Note:* Primed values always refer to chopper circuit.

$$e_L = -e_1 Z_f/Z_i + kZ^3/Z_1 Z_i Z_3 [1 - A_v'(\omega)]$$

where $Z_3 = R_3 + R_4 + Z_1'$ and $Z^3 = Z_1 Z_3 Z_i + Z_f Z_3 Z_i + Z_f Z_1 Z_i + Z_f Z_1 Z_3$.

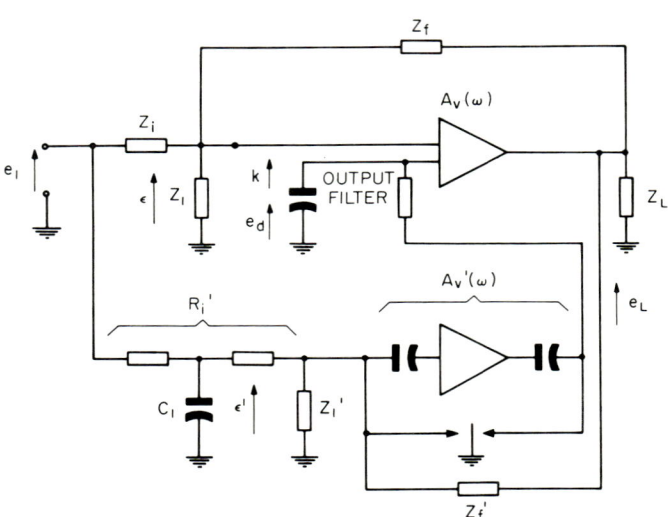

Fig. 15. Alternative drift-reduction scheme.

$$e_L \simeq -e_1 Z_f'/Z_i' + k/A_v'(\omega) Z_i' Y'$$

where $Y' = Z_1'/(Z_f' Z_i' + Z_1' Z_i' + Z_1' Z_f')$.

OPERATIONAL AMPLIFIERS

Equation (41) shows that the drift is reduced directly as $A_v'(\omega)$, the gain of the chopper amplifier, provided $A_v'(\omega) \gg 1$.

The scheme shown in Fig. 14 suffers one main disadvantage, that is, the inability of the chopper amplifier to correct for errors due to base current flowing in Z_i and Z_f. A common remedy is to insert a low-leakage capacitor at point S. This eliminates the problem, but if the amplifier is driven to saturation the capacitor can acquire a charge, thus blocking A_1 and requiring that a shorting or reset switch be placed across the capacitor.

2. An alternative method suggested by Blecher[5] and Okada[30] is shown in Fig. 15 in which separate summing resistors are used for the low-drift amplifier in order to avoid the input capacitor. These impedances Z_i' and Z_f' should have the same ratio as Z_i and Z_f, i.e., $Z_f/Z_i = Z_f'/Z_i'$.

e_1 = input signal
e_L = output voltage at load
ϵ = voltage at direct-coupled amplifier summing junction
ϵ' = voltage at chopper amplifier summing junction
k = drift potential
e_d = correction voltage from chopper amplifier

$$e_L = -\frac{e_1 Z_f'}{Z_i'} + \frac{k}{A_v'(\omega) Z_i' Y'} \quad (42)$$

where
$$Y' = \frac{Z_1'}{Z_f' Z_i' + Z_1' Z_i' + Z_1' Z_f'}$$

Equation (42) shows that the drift is again reduced directly as $A_v'(\omega)$.

The scheme of Fig. 15 suffers the disadvantage of more complicated switching if a range of operational functions is to be performed but is, however, free of the blocking problem.

Stability of Drift-reduction Schemes. The stability problem for these drift-reduction schemes has been analyzed;[4] the results are:
Goldberg scheme:*

$$\omega_2' \leq 4\omega_1'|A_v'(0)|$$
$$\omega_1 \geq \frac{4\omega_2' Z^3}{|A_v(0)| Z_1 Z_3 Z_i} \quad (43)$$

Alternative scheme:*

$$\omega_2 \geq 4|A_v(0)| B_v \omega_1 \quad (44)$$

where the meanings of the symbols are given at the beginning of this chapter. In this manner, the stability of the system can be readily calculated. Note that Eqs. (40) and (44) are identical.

5.3. Amplifier-design Considerations. Wideband Amplifier. The main design requirements of the wideband amplifier are generally as stated below in order of relative importance:

1. The amplifier should have one dominant rolloff only; any other corner frequencies must be higher than $4A_v(0) B_v \omega_1$. If this is not practicable then the unwanted negative corners will have to be canceled with positive corners by using lead networks or by placing a positive corner in the dominant rolloff network.

There are two principal causes of negative corner frequencies. These are (1) the frequency dependence of common-emitter current amplification, which produces a corner at $f_\beta = f_t/\beta_0$; and (2) the collector-to-base capacitance of the transistor, which

---

* It is to be noted that the terms ω_2' and ω_1' have been interchanged from those of Ref. 4; although these symbols are used correctly in that text, there is an error in the List of Symbols. In the equations of Ref. 4, ω_2' and ω_1' refer to the first and the second corner, respectively, of the chopper amplifier plus filter response; in order to preserve consistency with the terminology used for the wideband amplifier, this has been inverted in this section. Secondly, the stability criteria given in Ref. 4 for the alternative scheme are incorrect.

when operating in the common-emitter configuration, is multiplied by $(1 - A_v)$ where A_v is the stage voltage amplification.* This Miller capacitance appears in shunt with the input impedance of the stage, causing a corner at approximately

$$f_M = \frac{1}{2\pi(1 - A_v)C_{ob}h_{ie}} \tag{45}$$

When using "fast" transistors (i.e., $f_\beta > 1$ Mc/sec) the corners due to Miller capacitance are usually more troublesome than those due to f_β. However, by ensuring that each common-emitter stage drives into a low load impedance the voltage amplification of each stage may be restricted, thereby reducing the factor by which C_{ob} is multiplied. This occurs naturally in a cascade of common-emitter stages if each following stage has a higher collector current than the preceding stage. A large voltage amplification may then be obtained in a later or in the final stage by using the common-base configuration which is free of the Miller-capacitance problem because it has no phase inversion.

2. This amplifier should have minimum drift before stabilization since this reduces the gain of the chopper amplifier for a given overall drift. A further advantage of the reduced chopper-amplifier gain is the reduced separation required between ω_1' and ω_2' as discussed in Ref. 4. Differential input stages minimize drift and also have the advantage of much better common-mode rejection.

3. The amplifier should have low noise at the output. This can be readily achieved by placing the dominant rolloff at the output in order to reduce the high-frequency noise components developed in the whole wideband amplifier. Unfortunately, placing the rolloff at the output reduces the available output at high frequencies unless a very large quiescent current is maintained. The positioning of the rolloff is therefore a matter of compromise. When it is moved toward the input the noise at the output increases, but so does the swing available at high frequencies.

Chopper Amplifier. The chopper amplifier, being a-c–coupled, has negligible drift and therefore may have a single-ended input. If the assumption is again made that $A_v'(\omega)Z_1'$ is a constant in the common-emitter configuration, inspection of Eqs. (41) and (42) shows that the maximum drift reduction of the operational amplifier occurs as Z_1' tends to zero. However, precisely the same considerations as in Sec. 5.2 under Choice of Input Impedance apply in this case, namely, that the input impedance will generally be chosen to suit the compromise between low-impedance high-efficiency amplifiers and high-impedance very stable amplifiers. Specifically, there is little advantage in making Z_1' less than $(0.1Z_f')/(1 - A')$; as previously stated, it may well be that a value of Z_1' greater than the above would be chosen to ensure sufficient phase margin over the contemplated range of operational gains.

The bandpass of the a-c amplifier should be sufficiently wide to ensure negligible phase shift of the modulation; i.e., the upper half-power point should be much greater than that of the output filter ω_1'. The output impedance should also be low.

Generally, the chopper amplifier is of straightforward design, and few precautions need be taken.

5.4. Typical Amplifier. Circuit Description. A typical amplifier[4] is shown in Fig. 16 where Q_1 and Q_2 serve as a wideband input stage loaded by the input impedance of Q_3. The input impedance of the amplifier itself is about 1 kilohm. The use of input p-n-p's is in order to simplify the design of the output stage which consists of one n-p-n and two p-n-p's; since high-voltage silicon planar p-n-p's are not so readily available as n-p-n's, this is necessary to prevent breakdown on large positive output swings. The forced rolloff ω_1 is caused by the 820-pf capacitor between the collector and the base of Q_3. The a-c amplifier is conventional; the summing resistors are used in pairs to avoid the problem of the base current of the input stage.

The amplifier performance is shown in Figs. 17 to 20.

---

* Generally, the low-frequency value of A_v is used. However, under certain conditions of cascaded transistor amplifiers, the fact that A_v is generally complex will lead to a more complicated input network consisting of a capacitor in series with a parallel RC.

OPERATIONAL AMPLIFIERS

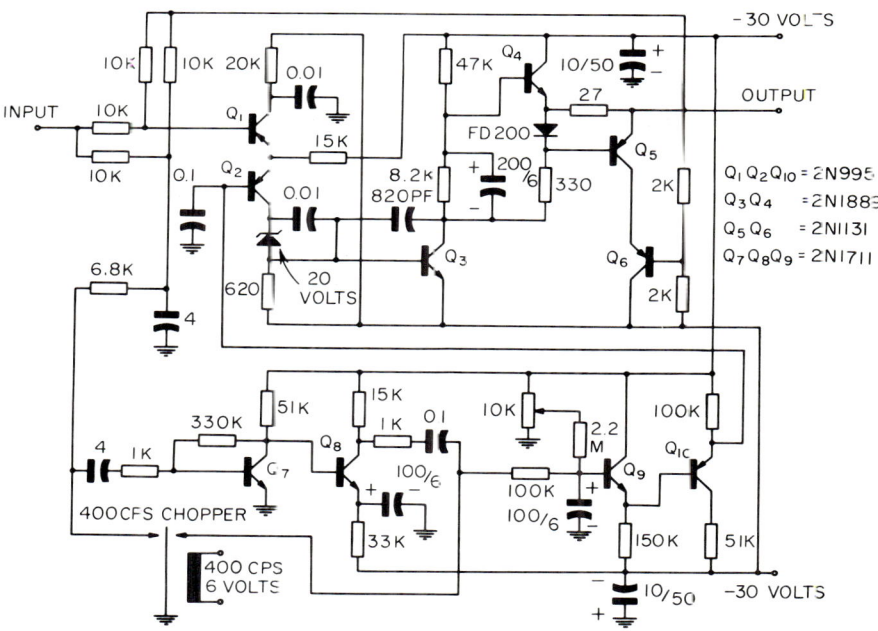

FIG. 16. A typical amplifier. Q_1, Q_2, Q_{10} = 2N995; Q_3, Q_4 = 2N1889; Q_5, Q_6 = 2N1131; Q_7, Q_8, Q_9 = 2N1711.

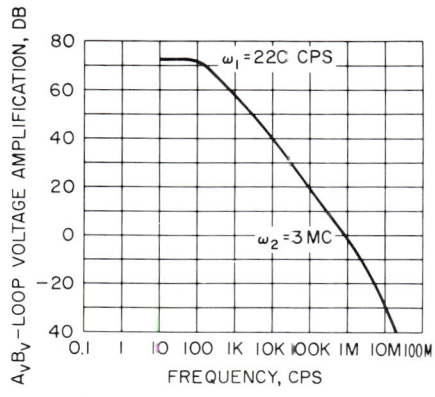

FIG. 17. Frequency response.

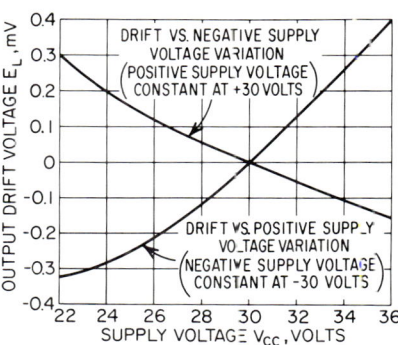

FIG. 18. Drift vs. power supply variations.

Stability Calculations. It is interesting to calculate the stability of this amplifier as a demonstration of Eq. (40).

From Fig. 17, $\omega_1 = 220$ cps, $\omega_2 = 3$ Mc/sec, and $A_v B_v = 70$ db or 3,200. Thus the condition

$$\omega_2 \geq 4|A_v'(0)|B_v \omega_1$$

reduces to

$$3 \times 10^6 \geq 4(3,200)(1K/10K)(220)$$

or

$$3 \times 10^6 > 2.82 \times 10^5$$

19–22 AMPLIFIER CIRCUITS

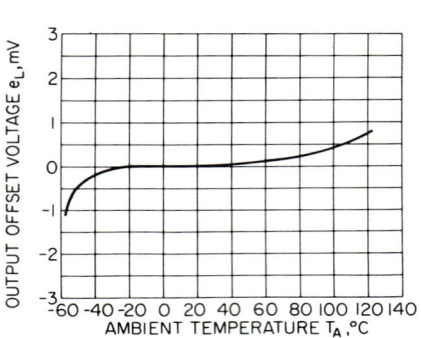

FIG. 19. Drift vs. temperature variation.

FIG. 20. Pulse response, showing effect of second corner ω_2.

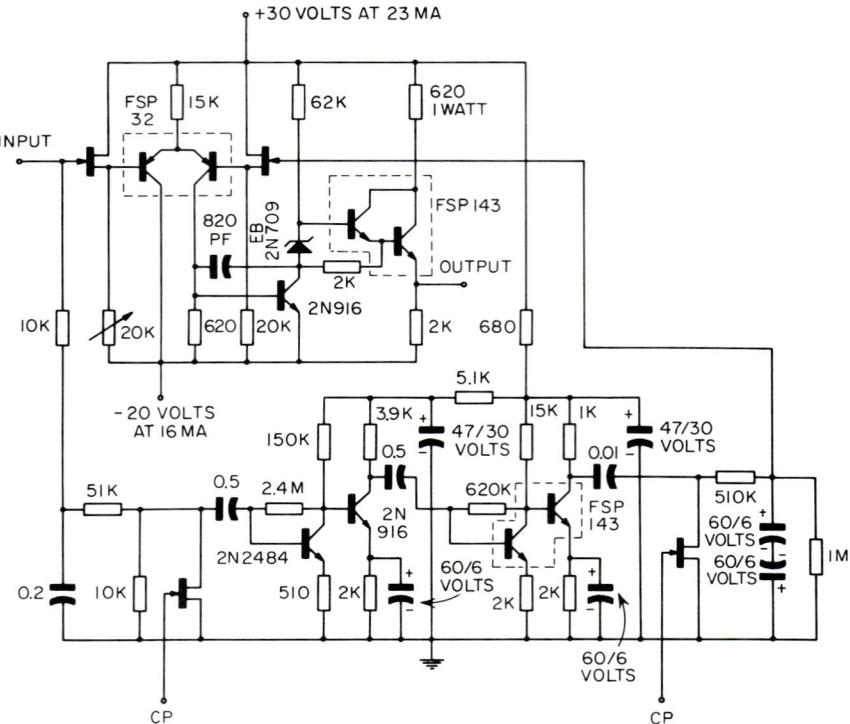

FIG. 21. FET operational amplifier. FSP is two p-n-p's similar to the 2N869. FSP 143 is two 2N916's in one can. FET's are FSP 400.

Therefore, no overshoot in the pulse response should be expected, as is shown in Fig. 20 for $K = 4$ or for $\omega_2 \geq 4A_v B_v \omega_1$. When ω_2 becomes smaller than this value, overshoot can be seen to exist. Such good correlation may not always exist because of the difficulty of measuring ω_2 accurately in a high-gain wideband amplifier, but Eq. (40) is nevertheless a very useful guide.

5.5. Operational Amplifiers Using FET's. *General.* There are many applications which require operational amplifiers with very high input impedance, on the order of 1 to 10 megohms. Transistors can be used in these applications, but because of the drift caused by the base-current flow through the large source resistance, it may be difficult to use these circuits in some cases. Although compensation methods discussed in Chap. 18, Sec. 5.2 can be used, it is difficult to reduce the drift much below 15 μv/°C with compensation for a 1-megohm source while a 10-megohm source will, of course, be much worse. If this figure is still too high, then another scheme must be used or a higher chopper-amplifier gain must be obtained.

Field-effect transistors may be suitable in such applications if the temperature range is restricted; as discussed in Chap. 18, Sec. 10, they have serious limitations because of their leakage currents and drain-current change with temperature for constant gate-drain control voltage. They

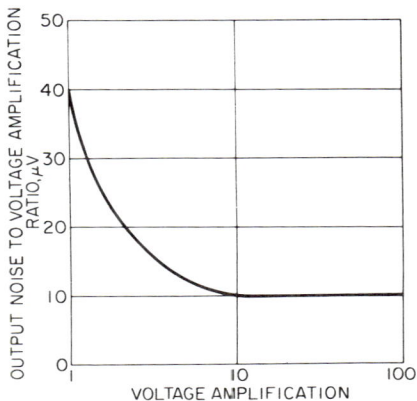

Fig. 22. FET amplifier noise.

are, of course, well suited for chopper applications because of their very low offset voltage.

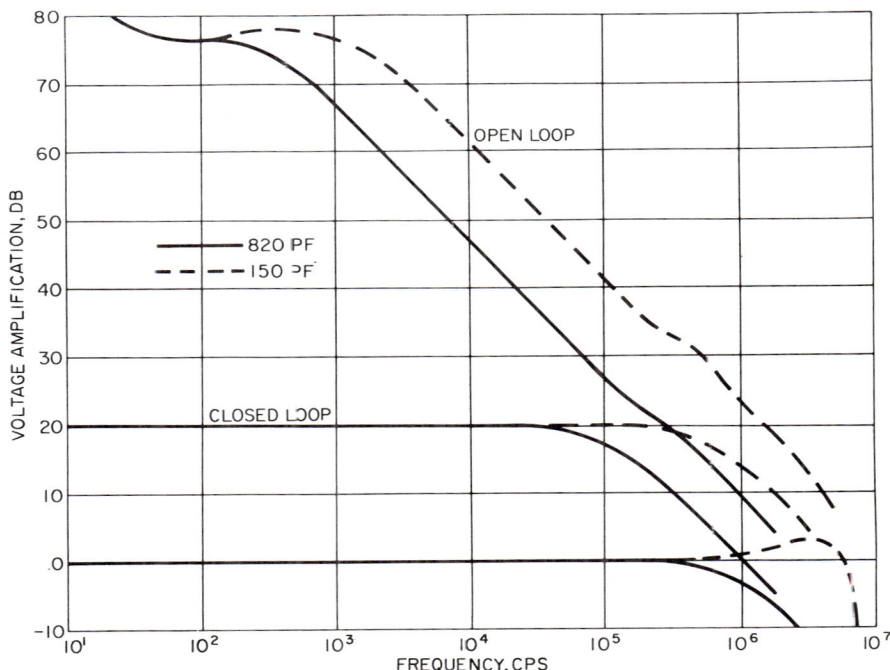

Fig. 23. FET amplifier frequency response.

Example of FET Operational Amplifier.* An example of an FET design is shown in Fig. 21 for a source impedance of 10 kilohms. The top amplifier is the wideband amplifier while the bottom one is the a-c amplifier chopped by the input and output FET's. An ingenious bootstrapping technique is used to obtain high and stable voltage amplification; the 2N709 base-emitter diode is used as a zener since it was found to have low noise in avalanche. The chopper scheme is a conventional application of the Goldberg technique with an input and an output filter.

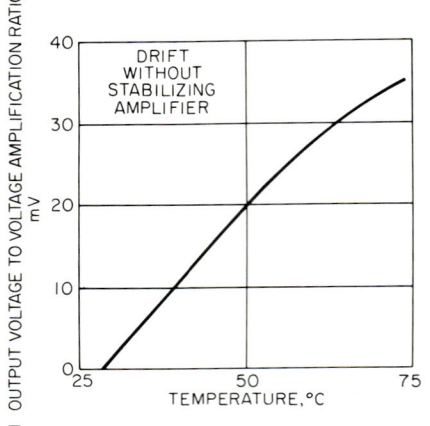

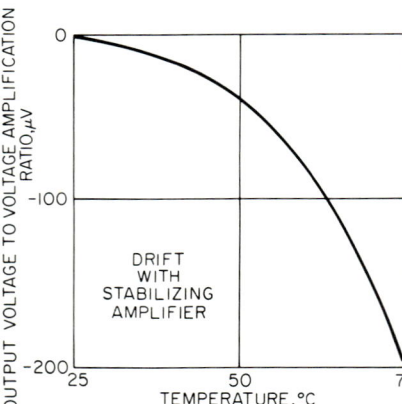

FIG. 24. FET amplifier drift.

The limitations of FET's can best be seen from a study of the amplifier's performance shown in Figs. 22 to 24. The drift without stabilization is on the order of 0.8 mv/°C, or about 80 times worse than that possible with good transistor pairs. Even with the stabilization loop the drift is 4 μv/°C, only about twice as good as uncompensated transistors or about five to ten times *worse* than that which can be obtained with transistors using compensation techniques but *without* a chopper amplifier.

This picture is nevertheless changing rapidly with some of the new devices which are becoming available.

6. APPLICATIONS OF OPERATIONAL AMPLIFIERS

6.1. Use of D-C and Narrowband Amplifiers in Regulated Power Supplies and Analog Recorders. Probably two of the most important uses of operational amplifiers in d-c and very-low-frequency (narrowband) applications are in regulated power supplies and in direct-indicating analog recorders. The words "direct indicating" here are meant to include both the indicating milliammeter and the servo-controlled bridge-type recorders, in which operational amplifiers are used to drive servomotors to establish a bridge balance condition. (Operational amplifiers are used with direct-recording milliammeters primarily to permit significant increases of sensitivity, lower full-scale current range, or smaller voltage lost in the metering circuit.) Both these applications are based on the use of one or more types of operational amplifiers.

The basic configuration on which most regulated power supplies are constructed is shown in Fig. 25. The major components include the reference amplifier, which may be a typical operational amplifier, the control circuit, the metering circuit, and the reference standard. The metering circuit usually consists of either a current-metering resistor in series with one of the load connections or a voltage-dividing network (or both). The metering circuit is designed to provide the control signal which the reference amplifier uses in providing the regulating action.

The control circuit adjusts the amount of voltage, current, or power available from

* Submitted by courtesy of R. L. Foglesong of Fairchild Semiconductor, who designed the circuit.

the supply either by controlling a variable resistor (for d-c-type control) or by controlling a saturable reactor (for a-c-type control). Typical control circuits are shown in Fig. 26, where part *a* shows a d-c type, and part *b* an a-c type. These circuits are normally transistorized (with zener diodes for reference standards) for d-c control at low voltages, and use tubes, thyratrons, silicon controlled rectifiers, or magnetic amplifiers in phase-control circuits for high voltages.

The regulating circuit used with regulated power supplies is normally intended only to correct the slow variations in the output voltage, and a large output capacitance is used across the terminals of the supply to provide for the short-time variations in current. This greatly reduces the demand which must be made on the regulator and makes possible the use of narrowband d-c operational amplifiers and

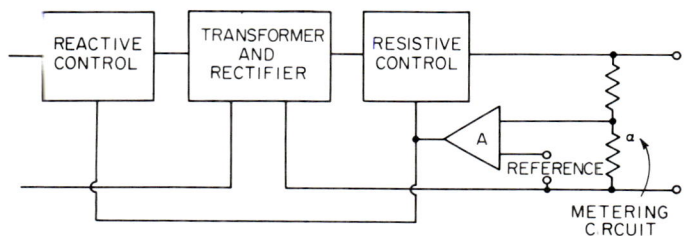

Fig. 25. Basic power-supply regulator circuit.

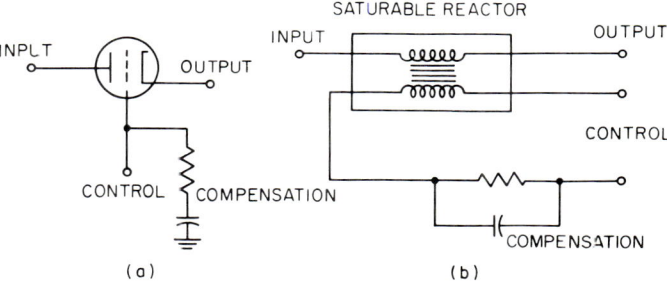

Fig. 26. Typical control circuits.

control circuits. The equation for the output admittance of the power supply, based on Fig. 26a, is

$$Y_o = i_L/v_L = (\alpha A - 1)y_f + y_o \qquad (46)$$

where y_f and y_o are the small-signal forward and output admittances, respectively, for the active device used as a control element, A is the gain of the operational amplifier, and α is the division ratio of the voltage divider. For this derivation, y_i was neglected compared with y_f; if it is not negligible, then y_f may be replaced by $(y_i + y_f)$.

Equation (46) for the output admittance shows that the y_o of the active device can make a significant contribution to the overall value of Y_o. Since, however, this y_o measures the variation of device current with the voltage applied to its plate or collector, this term also causes a cross coupling, and the introduction of an incremental change of the supply voltage, through y_o, causes a fluctuation in the output. For this reason, it is desirable to use devices having small values for the dynamic output admittance y_o. For a similar reason, it is undesirable to use devices having a relatively small ratio of y_f/y_o (a low-mu device) as very large values of αA must then be used to keep the effect of y_o negligible in comparison.

Several important differences must be kept in mind in the selection of control devices for regulated power supplies. Tube and transistor regulators of the kind just considered dissipate considerable amounts of power, but they are relatively fast-acting, in that they can start to correct for changes in output conditions within microseconds or milliseconds at the very most. This is not the case with phase-controlled, or reactive, regulators like saturable reactors, magnetic amplifiers, and switching circuits like thyratrons and silicon controlled rectifiers. With them several cycles of supply voltage may be required to generate sufficient change to adjust for changes of load. These supplies regulate excellently against slow variations, but not for fast.

The operational amplifiers required with the reactive class of regulators perform largely the same function as those used with the variable-resistance class. They probably differ in one important respect, however, namely, in the kind of frequency compensation required. The stabilization of the typical operational amplifier in a variable-resistance regulator requires the use of a pole-zero low-pass network in the feedback path in order to reduce the loop amplification at high frequencies. Otherwise, these circuits tend to oscillate. With the reactance-type controller, however, there is sufficient lag in the system that some emphasis of the high-frequency components of current in the load might be helpful. A network configuration something like that shown in Fig. 26b may provide the required compensation in the forward part of the control circuit.

The nature of the feedback loop used with regulated power supplies is such that fully stable operation is required. As a result, internal reduction of the gain of the operational amplifier itself is undesirable (unless needed for stabilization) since it will reduce the regulating capabilities of the complete network. For this reason, internal feedback loops and chopper stabilization are seldom used in this application.

The use of operational amplifiers in analog recorders, on the other hand, requires both compensation and stabilization. In this application, both a_1 and b_2 normally are zero, since it is impossible to include a linear component in a system which gives an analog recording through the use of a potentiometer-balance indicating system. This change forces some modifications in the manner of compensation and operation. Solving for the condition for zero position and velocity error gives the result

$$b_2 = b_1(a_4 - a_3)/a_2 \qquad (47)$$

This relation is derived from the transfer equation

$$A_f = (b_1/a_2 s)/[1 + b_1(a_3 s + b_3)/a_2 s(a_4 s + b_3)] \qquad (48)$$

where A_f is the closed-loop forward transfer immittance. Subject to Eq. (47), the roots of the denominator are

$$s = -(b_1/2a_2) \pm [(b_1/2a_2)^2 - (b_1 b_3/a_2 a_4)]^{\frac{1}{2}} \qquad (49)$$

As is evident, the damping is dependent on the ratio of $b_1/2a_2$ (and is positive for passive elements) and the frequency is dependent on the radical term. For over-damped conditions, the reciprocal of the time constant is approximately linear with the gain, whereas for underdamped conditions, it is approximately linear with the square root of the gain.

Acceleration compensation can be introduced by the use of a high-pass pole-zero network in the forward branch of the network. This increases the order of the denominator to third, and the numerator to second.

6.2. Use of Stabilized Amplifiers in Frequency-selective Filters. Operational amplifiers are commonly used as "negative immittance converters" in connection with frequency-selective circuits and with oscillators. In this kind of application, the amplifier is used in its own internal feedback circuit in order to obtain precisely controlled gain and phase characteristics, and it is then coupled to an associated frequency-selective circuit to provide the required operating characteristics. In this kind of application, a high degree of linearity is usually required of the amplifier.*

Operational amplifiers can be used with RC networks to produce low-pass or high-pass filters or bandpass or band-rejection filters. In these applications, the responses available are tabulated in Table 2. In this table, the stabilized amplification is identified by A_c, and the resistor and capacitor values by R and C, respectively.

Table 2. Characteristics of RC Operational Filters

Configuration	Diagram	Gain characteristic Positive	Gain characteristic Negative	Stabilization range
RC low	Fig. 27	$C_{eff} = -C/(A_c - 2)$	$C/(A_c + 2)$	$0 < A_c < \infty$
RC high	Fig. 28	$(2 - A_c)C$	$(2 + A_c)C$	$0 < A_c < \infty$
RC type I	Fig. 29	Pass $\omega CR = 1$	Reject	$0 < A_c < 3$
RC type II	Fig. 30	Reject $\omega CR = 1$	Pass	$0 < A_c < 1.5$

C_{eff} is the equivalent capacitance after taking account of Miller multiplication. The word pass means bandpass and reject means band reject. It is of interest to note that both the RC low- and the RC high-pass filters theoretically can be constructed with the operational amplifier having either positive or negative values of amplification and that the corner frequency is a function of the magnitude of the amplification. For this reason, it is only necessary to vary the gain to shift the corner frequency of either the low-pass or the high-pass version of this circuit. With the low-pass version, an increase of gain increases the corner frequency, and the converse with the high-pass. In fact, the corner frequency is approximately that frequency

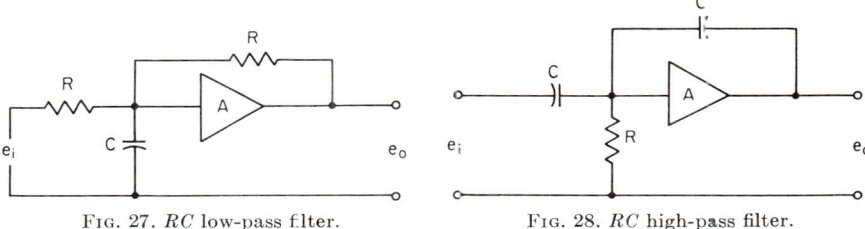

FIG. 27. RC low-pass filter.   FIG. 28. RC high-pass filter.

at which the open-loop amplification of the complete network (split apart to give an input and an output point, as in Fig. 31) decreases to unity.

It is interesting to note that the RC low- and high-pass filters which have positive amplification in excess of 2 appear to have negative effective capacitance. In other words, these networks convert a positive reactance into a related negative reactance. In order to stabilize the value of the reactance, the value of the open-circuit amplification may be stabilized by applying feedback around the operational amplifier.

The two filter networks RC type I and RC type II are related, but they have very

---

* As long as the overall feedback level is less than is required for oscillation, high linearity is helpful. When a limitation of oscillation amplitude is required in addition to stabilized gain, however, then some nonlinearity must be retained. (In the absence of nonlinearity, there is no way of achieving exact balance in the "differencing of two large numbers," and stabilization is not possible.) A limiter circuit is an ideal way of introducing the required nonlinearity. For further details, see Ref. 31.

interesting differences which can be extremely important to the user. Clearly, the reversal of the polarity of the amplification with these filters changes the network from a band-rejection to a bandpass network, or vice versa. The difference is more than this, however. With the type I filter, the feedback is limited to a narrow band of frequencies in the neighborhood of that given by $\omega CR = 1$. If the feedback is negative, band rejection results, and frequencies outside the specified neighborhood are little affected. This means that there is little if any suppression of either harmonics or subharmonics in the network as a whole, and it cannot be used very effec-

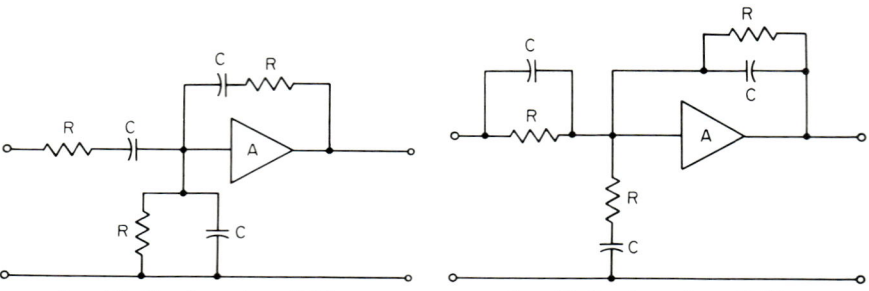

FIG. 29. Bandpass type I filter.   FIG. 30. Bandpass type II filter.

tively as a wave-analyzer circuit unless it is used with an amplifier having an extremely high degree of linearity.

With the type II filter, however, the feedback exists throughout the operating range of the circuit, and it is maximum outside the nominal operating band. If, then, this network is used to provide negative feedback, and a separate positive-feedback path is provided which is controllable, good control of distortion may be obtained, and at the same time, either an increase in the sharpness may be obtained by positive feed-back, or an oscillatory condition may be generated.

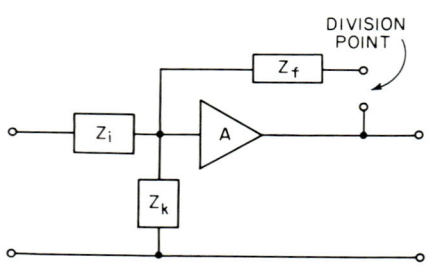

FIG. 31. Loop disruption of feedback circuit.

6.3. Variable-selectivity Circuits and Q Multiplication. The RC types I and II circuits show that it is possible to generate variable-selectivity circuits by means of operational amplifiers. There is a great variety of circuits which can be used to generate this basic characteristic, and at the same time, there are very precise criteria which must be fulfilled in order to develop this characteristic in a reliable manner. Consequently, it is important to examine the requirements for a given amount of "Q multiplication," as it is often called, in order that the design limitations are clearly defined.

This operation of "Q multiplication" is really that of moving a pair of conjugate complex poles *toward* the imaginary axis. The closer the original poles are to the axis, and as a consequence the smaller the distance through which the poles must be moved, the more stably these poles can be moved. This situation is a consequence of the fact that it is comparatively easy to make small changes in the net level of energy returned by feedback, but if the level of energy returned by feedback becomes almost equal to the original level the net input energy becomes a small difference between two relatively large values, and all the problems of differencing approximately equal numbers will exist in the network. The result is high sensitivity and poor stability.

Figure 32 shows a typical parallel-tuned circuit operating with an operational amplifier having an amplification of A_o and a feedback factor of β to give a typical closed-loop amplification of A_c. This amplifier may be so arranged that the product of its forward admittance Y_f and the parallel-resonant impedance of the tuned circuit is approximately unity; then a condition approaching oscillation will occur if positive feedback exists. Based on this, it is possible to establish the amount of multiplication of Q factor which may be developed in a given configuration, subject to specified limitations. Typically, for example, it is convenient to know that the Q factor will be within 10 per cent of a given value for a specified range of amplification of the operational amplifier. Assuming that the oscillation condition corresponds to a value of A_c of unity, the equations for the multiplication of the Q factor are

$$Q/Q_o = 1/(1 - A_c) \quad (50)$$

where the value of A_c is defined in terms of the equation

$$A_c = A_o/(1 + A_o\beta) \cong A_o/(1 + A_o) \quad (51)$$

where the condition for the second part is that the value of β is unity. Under these conditions, Table 3 may be prepared in terms of the minimum value of the open-loop amplification A_m to limit the required Q-factor stability to 10 per cent. The amount of stabilized amplification required from the amplifier to produce a given multiplication of Q factor apparently is more than that required to comply with the stability requirement, with the result that a larger range of variation of $\Delta A/A$ than 10 per cent can be permitted within the 10 per cent variation of Q factor.

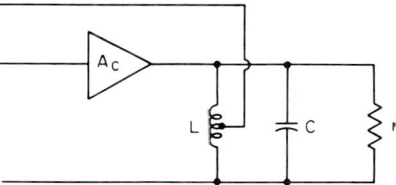

Fig. 32. Typical parallel-tuned oscillator circuit.

Table 3. Multiplication of Q Factor in Terms of Amplification

Q/Q_o	2.5	3.33	5.0	10	20	50	100
A_c	0.6	0.7	0.8	0.9	0.95	0.98	0.99
A_m	0.9	1.6	3.2	8.1	19	48	98

The exact form of the equations required in any application of an operational amplifier to a Q-multiplier circuit controls the exact amount of stabilized amplification which is required for the circuit concerned. Again, based on Fig. 32, the Laplace transform of the transfer function may be written as

$$[sC + (1/r) + (1/sL)]V = i_o = Vy_f \quad (52)$$

or

$$s^2LC + (1/r - y_f)sL + 1 = 0 \quad (53)$$

Since y_fr represents the stabilized gain, it may be replaced by A_c, and the form of the equation may be changed to express the multiplication factor in terms of $[r/(1 - ry_f)]$. In the present instance, the resulting equation is

$$(Q/Q_o)(1 - A_c) = 1 \quad (54)$$

Differentiating, and taking $\Delta Q/Q = 0.1$, gives the equation for ΔA_c in the form

$$\Delta A_c = 0.1(1 - A_c)^2 \quad (55)$$

In addition, when the value of β is unity, the substitution may be made:

$$(1 - A_c) = 1/(1 + A_o) \quad (56)$$

and Eq. (55) may be written in the form

$$\Delta A_c = (\Delta A_o/A_o)A_c^2/A_o \quad (57)$$

or, assuming that $(\Delta A_o/A_o)$ has the value 0.1,

$$A_o = 0.1 A_c^2/\Delta A_c \tag{58}$$

These equations may be solved to give the value of $(\Delta A_o/A_o)$, giving the results in Table 4.

In practical applications of this basic principle of use of operational amplifiers, it is convenient to design the circuit to require a value of A_c which is a little greater than unity in order to make small adjustments in the magnitude of the gain conveniently.

Table 4. Limit Values on Amplification for Q Multiplier

A_e	A_m	ΔA_c	$\Delta A_o/A_o$
0.6	1.5	0.04	0.167
0.7	2.33	0.03	0.147
0.8	4.0	0.02	0.125
0.9	9.0	0.01	0.111
0.95	19.0	0.005	0.105
0.98	49.0	0.002	0.102
0.99	99.0	0.001	0.100

A typical configuration which makes this possible is shown in Fig. 33; circuits with which it may be used are shown in Fig. 34. Part a of this figure shows the configuration commonly known as the Q-multiplier circuit, and b shows a possible way of using this circuit with a quartz crystal as the stabilizing element. For most effective use, the internal-resistance components of the inductor and the capacitor should be less than or equal at most to the resonant resistance of the crystal itself. Then the effective Q factor for the circuit may be as much as

$$Q_{eff} = \omega L_x/(r + R_x)(2 - A_c) \tag{59}$$

where r is the coil resistance, R_x is the crystal resistance, ωL_x is the effective inductive reactance of the crystal, and A_c is the stabilized amplification of the operational

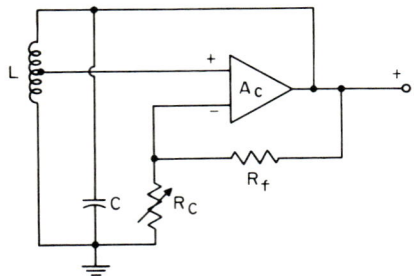

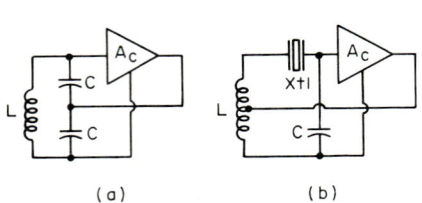

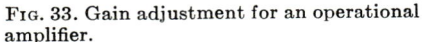

FIG. 33. Gain adjustment for an operational amplifier.

FIG. 34. Controlled-gain oscillators.

amplifier. The term $(2 - A_c)$ is based on the minimum amplification for oscillation of 2.

The coupling of signals into circuits such as that shown in Fig. 34 is best accomplished magnetically, and the output may be taken directly from the operational amplifier. The input may also be coupled into the input of the amplifier if the decoupling can be made sufficiently effective. These circuits are very susceptible to loading, since any form of loading will absorb power and decrease the effective Q factor for the circuit as a whole.

6.4. Operational Amplifiers in Oscillators. Only comparatively minor changes are required to convert a frequency-selective filter into an oscillator. The amplification required of the associated operational amplifier must of necessity be somewhat larger than that required with the equivalent filter circuit, just large enough to make the damping term either slightly negative or zero.

The behavior of this damping term is of extreme importance in the development of an oscillator based on an operational amplifier, inasmuch as the stabilization of any oscillator is a function of the variation of damping with instantaneous signal amplitude. Three noncompatible conditions exist in an oscillator circuit, the first that the frequency stability of the oscillator is a function of the constancy of the damping term for the complete oscillator, the second that the stability is a function of keeping the open-loop damping to a minimum, and the third that the amplitude stability of the oscillator is increased by increasing the amplitudes of the odd-harmonic components of the signal, and the odd-harmonic components in the expansion of the transfer function for the operational amplifier.

This problem of stabilization of both the frequency and the amplitude for an oscillator, whether based on an operational amplifier as the active network or on an amplifier specially designed for the purpose, is a very difficult one. It is necessary that the average gain in the amplifier (typically the degenerated gain, or A_c) must in some manner decrease as the amplitude of oscillation builds up; otherwise, stabilization of amplitude cannot result. This decrease may be achieved either by a general reduction of amplification with the increase of amplitude (an effect commonly obtained in tube amplifiers by making the bias on a grid more negative as the amplitude increases), or it may be accomplished by using an amplifier having a transfer function whose second derivative is negative. In the latter condition, the increase of amplification as the instantaneous signal voltage changes in one direction is significantly less rigid than the decrease when the instantaneous signal voltage changes in the other direction. This in effect introduces a negative third-harmonic component, which decreases the effective amplification at both the positive and negative limits compared with a linear variation. (The first derivative component does not affect the limiting conditions and as a result may be ignored.)

The stabilization of the signal amplitude of an oscillator using an operational amplifier is rendered difficult because of the fact that the stabilization of the amplifier to bring its amplification to the required value also almost completely eliminates the nonlinearity. The introduction of a thermal regulator such as a thermistor or a small lamp bulb can result in the required control of amplification, but such a regulator can introduce unstable poles under conditions of extremely high linearity.[31] As a result, best operation can be obtained by a combination of a symmetrical limiter and a thermistor regulator in conjunction with the operational amplifier.

6.5. Integrators and Differentiators. *Integrators.* The establishment of a true electrical analog of the mathematical operations of integration and differentiation is essential in the development of many kinds of automatic computing systems for providing close approximations to the correct solution to a problem in a minimum length of time. Both these analogs can be approximated with the help of operational amplifiers used in conjunction with either simple RC integration or differentiation circuits, for a crude approximation, or by the use of compensated networks for a more precise representation.*

Development of networks showing ideal integration and ideal differentiation is physically impossible in theory (at least in ordinary electronic networks) in a true lumped-constant network as a consequence of the Wiener-Paley criterion. This criterion states in effect that, for a network to be realizable, it must not have either a zero response or an infinite response between the frequency limits of zero and infinity, and certain restrictions are placed on the behavior of the network as the frequency approaches the limit values. In a practical way, the approximation of an integrator

---

* A concise discussion of the compensation of these networks is given in Ref. 45.

characteristic is normally easier to achieve than that of a differentiator as a consequence of the manner in which the energy is normally distributed within a signal spectrum, and the effect of integration or differentiation on the distribution.

Typically, a simple integrator circuit consists of an operational amplifier bridged by a feedback network in which the "integrating" capacitor is in the feedback path (Fig. 35). The use of this kind of arrangement makes it possible (in effect) to increase the size of the capacitor to $(1 - A)$ times its actual size, where A is the phasor amplification of the amplifier and is significantly larger than unity. In other words, if the negative terminal of the amplifier is used as the input and feedback point, then the magnitude of the capacitance multiplication is $(1 + |A|)$, whereas with the positive terminal it is $(1 - |A|)$.

The amplification available in the operational amplifier is both a help and a hindrance in the development of integration. It is a help in that it offsets much of the loss which must of necessity be accepted with passive integrators. Since the effective capacitance is roughly directly proportional to the magnitude of the amplification of the operational amplifier, however, it is clearly essential that internal stabilization be applied to the operational amplifier.

The internal stabilization required on the operational amplifier can hardly be applied across the same terminals as are used for the introduction of the capacitive feedback. The development of a positive capacitance for the integrator requires the use of the feedback capacitance either from a positive output to a negative input, or vice versa, and the gain stabilization requires a similar connection, only using a resistive network. For this reason, an operational amplifier used for this function should have both a positive and a negative output as well as a positive and a negative input. Under these conditions, stable integration is potentially available. As long as the inputs for the operational amplifier draw no current which will change the charge in the feedback capacitor, the integrator will have reasonably good static stability and will perform as an integrator. Static drift is the principal problem encountered in electronic integrators, and it is largely unavoidable because of the leakage current encountered in the input amplifier and the excess noise experienced at low frequencies.

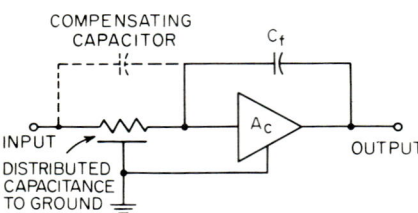

FIG. 35. Integrating circuit.

The low-frequency noise is at least partially outside the feedback loop in the operational amplifier and as a result can introduce a small amount of noise modulation onto the amplification of the amplifier and thereby affect the multiplicative factor on the integrator. In addition, the leakage current may experience either degenerative or regenerative amplification, and introduces problems of unknown magnitude thereby. The best approach to control of these effects is through reduction of impedance level, something which just cannot be done when an RC circuit is used for integration. Two alternatives suggest themselves, the first being the use of a storage battery in a partially charged condition, and the second the use of an electric-motor-driven potentiometer. The former is very subject to temperature and aging effects and is rather nonlinear, whereas the latter has promise, particularly since it can have a long integration time and has an almost infinite memory. Motors used in this way have an equivalent effective capacitance which may be considerably in excess of 1,000 μf, and besides, they can be built with an integral tachometer generator for use as a feedback element.

Several methods of use of motor-driven potentiometers suggest themselves: the first makes direct use of the a-c output of a chopper-stabilized amplifier or preamplifier to drive a two-phase motor at a speed proportional to the amplifier output. The motor then drives the potentiometer. This method is simple and, although largely free of drift, may be inaccurate, as no stabilization of the amplifier-motor combination is available (Fig. 36). A second, more elaborate, method uses the tachometer for

OPERATIONAL AMPLIFIERS

feedback as noted above, and causes the chopper to compare the input voltage against the tachometer output voltage. The rate of turning of the motor is then very nearly proportional to the unbalance voltage at the amplifier input, and the output will provide a voltage which is quite accurately the integral of the input voltage, and it is almost completely free of drift (Fig. 37).

The design of any integrating amplifier must be so compensated that integration is available at all frequencies, and the response of the system as a whole must vary uniformly at 6 db/octave. This means that a combination response must be developed, one in which motor-potentiometer or chopper-rectification integration is effective to a certain frequency, at which point it becomes ineffective and an RC circuit takes over. The transfer frequency must of necessity be small compared with the chopping rate in the circuit, and as a result, a relatively high chopping rate will be required.

It is necessary to determine the number of volt-seconds per radian per volt response for both the chopper-driven portion of the integrator and the RC portion, and to

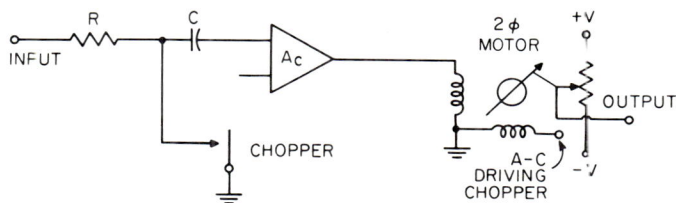

Fig. 36. Basic motor-type integrating circuit.

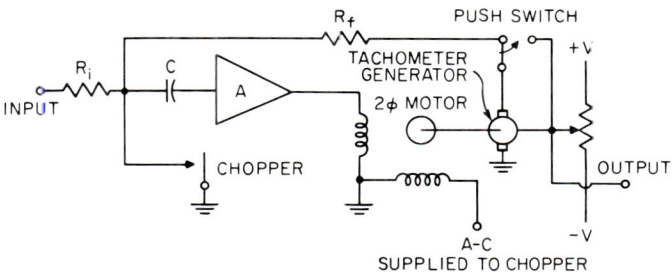

Fig. 37. Improved motor-type integrating circuit.

adjust the constants to achieve a match of slope in the transition region. A parallel combination of the two circuits is the most effective method of arranging the transition, with the control being achieved on the integrating potentiometer in one of the ways indicated in Fig. 38.

The amplification vs. frequency characteristic for a typical universal stabilized amplifier, such as the Philbrick USA-4J unit, closely approaches that to be expected of an integrating amplifier, with the result that they may be used in applications where only moderate precision is required. This characteristic has been designed into the units, and they typically have as much as a billion gain at direct current.

Differentiators. The use of operational amplifiers as differentiators follows a similar pattern to that for their use as integrators, but the achievement of satisfactory differentiation is significantly more difficult than the achievement of satisfactory integration. The active differentiator does not experience a drift problem, but the range of frequencies over which effective differentiation can be obtained is restricted by the fact that the available gain in an amplifier by nature decreases with increase of frequency. As a consequence, careful consideration is required in order to assure that a circuit will behave as a differentiator to all frequency components of the impressed signal. A good discussion of compensation of amplifiers to optimize differentiation and integration characteristics is given in Ref. 45.

The maximum operating frequency of a differentiating amplifier is a function of the gain-frequency curve for the amplifier. In an amplifier which has integration characteristics (Fig. 39) the selection of the turnover frequency, from integration to differentiation, is accomplished by taking the square root of its gain-bandwidth product, and with the amplifier which shows uniform gain below a specified frequency and integration rolloff above, the turnover frequency is usually placed at the corner frequency. A differentiation-bandwidth times amplification product, similar to the gain-bandwidth product, can be established, in that the wider the differentiation bandwidth, the smaller the available gain.

The two typical curves of frequency response show that for an operational amplifier which is intended for use as a differentiator, it is important that the design maximize the amount of available gain at any one frequency, or that the amplifier used be essentially an integrating amplifier. Some degeneration may be used on individual stages, or it may be used in a way which does not cause a decrease of the amplification at a

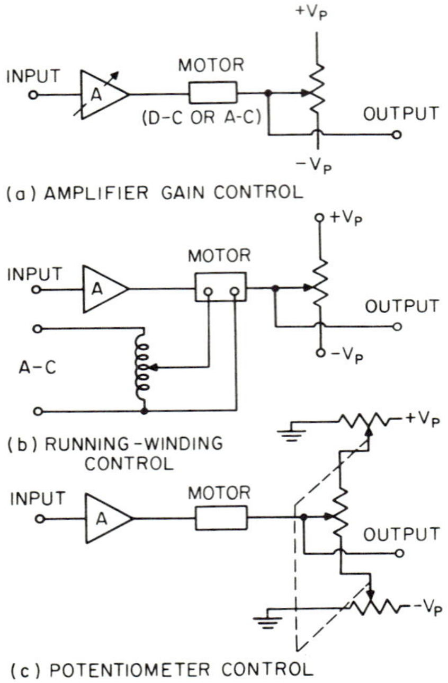

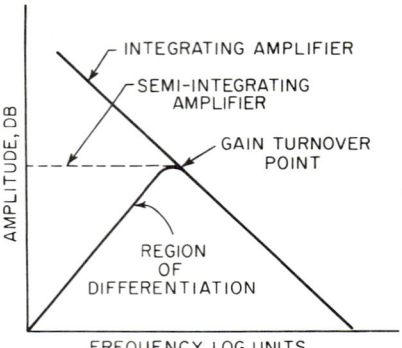

FIG. 38. Integration control circuits.

FIG. 39. Frequency response of operational amplifier with integration characteristics.

given frequency below that nominally available in accordance with the gain-bandwidth product. Otherwise, significantly reduced differentiation efficiency can result. (The phrase integrating amplifier here is used to indicate that the amplifier by its own inherent characteristics has a gain rolloff of 6 db/octave of frequency increase over most of its operating range. Since a maximum gain-bandwidth product is required of this amplifier, the rolloff should be developed only through internal capacitances of the active devices rather than through the action of an integrating capacitor.)

6.6. Other Special Operational Circuits. Precision stabilized oscillators and voltage-controllable oscillators use special amplifier circuits designed for their particular applications. A conventional operational amplifier is seldom used for this function as each configuration requires somewhat special tailoring to achieve its design objectives. As a result, mass production is not generally practical. The stabilized oscillator is designed to be coupled to a specified kind of frequency-selection circuit, and it provides a high degree of linearity and a very closely controlled value of amplification. It is usually used with a quartz crystal as the frequency-selecting element,

OPERATIONAL AMPLIFIERS

and its design is such that at the specified frequency of operation, its loop gain is unity, and its rate of change of phase shift with frequency is as large as possible. In addition, it provides either zero degrees or a multiple of 180° phase shift at the operating frequency. The multiplier is odd if the coupling network introduces an inversion of phase, and even if it does not.

Special shaping circuits such as are used in counters and both analog and digital computers are really specialized and simplified operational amplifiers. They are all special-purpose assemblies rather than units, purposely designed to have considerable flexibility. Generally these circuits are highly nonlinear, and they accomplish their intended functions largely because of the coordinated way in which they use linear and nonlinear properties. They may be passive in nature, or use amplification only to overcome loss, or they may be active, using amplification to achieve rapid changes of state of operation. In either case, a simple operational amplifier may be an important component of the circuit. Shaping circuits are often associated with countdown circuits, in that the wave to be counted must first be limited and differentiated and clamped in order to convert it into a series of unipolarity pulses before the pulses can be counted in an electronic counter.

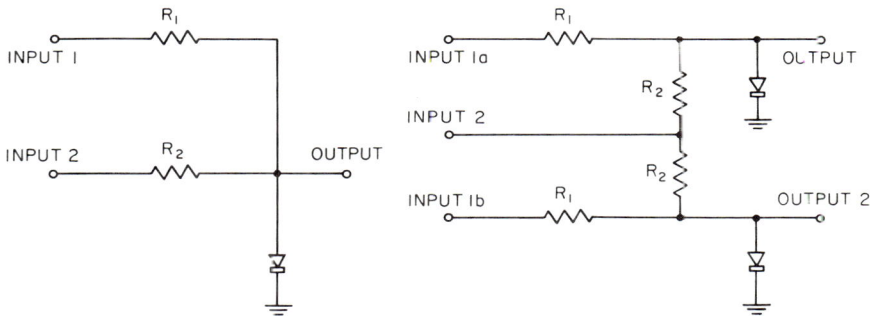

FIG. 40. Mixer circuit.   FIG. 41. Balanced mixer circuit.

Countdown circuits, of which the best-known are flip-flops, are the most widely used single operational entity available. These circuits are multivibrators of a very special kind, in which response to direct current is available in order that the state of the circuit may be maintained until the next counting pulse is received. The circuit regeneratively changes its state on the receipt of a properly formed counting pulse. These circuits are useful for the generation of balanced square waves from sinusoidal signals in addition to their normal counting function, and they can be used for controlling switching matrices. Both tube and transistor versions of these circuits are available, and they can be made in magnetic-core versions as well.

Product, or mixer, circuits (Fig. 40) and balanced-product, or phase-detector, circuits (Fig. 41) are highly nonlinear amplifiers using diodes or other nonlinear devices in a way which takes advantage of their nonlinear nature. The simple product, or mixer, circuit is normally used for frequency conversion under conditions which do not require optimum processing techniques.

The balanced-product detector is used where it is desired to measure the relative phase between two signals or where it is desirable to minimize the effects of noise on the overall system. This detector is the heart of the tracking filter which has become so important in weak-signal reception in satellite communication systems.

The balanced-product detector utilizes two input signals, one the signal being processed and the other a reference signal. Normally, the reference signal has an amplitude which is large compared with the signal being processed, in order that strictly linear time-variant (LTV) detection may be achieved. It can be shown mathematically that the reference signal may not have any (low-frequency) amplitude or phase modulation without causing serious deterioration of the ability of the detector to extract signal from noise, but the reference signal may include harmonics of the fundamental frequency without creating difficulty.

With frequency-changing, or mixing, operations, the reference signal may be modulated under strictly controlled conditions, but it should be free of harmonics in order to minimize the number of spurious responses which can be generated. Since frequency difference, or rate of change of phase, is the important parameter in this application, a controlled pattern of progression of phase in the reference voltage is important in order to provide for smooth variation of phase at the output, and a concentrated signal to appear within the required pass band.

In the phase detector, it is important to have good linearity of phase change about some specified null point. Also, the null point must be independent of the amplitudes of both the received signal and the reference signal. These conditions can be obtained most effectively by interposing a wave-shaping circuit and a frequency divider in the reference-voltage generator, as in that way a precise square-wave reference voltage can be developed, and a minimum of parasitic sensitivity results. When a phase detector is provided with such a square-wave reference voltage, the stability and the balance of components are much less critical than when a sinusoidal reference voltage is used, as in the latter condition, the matching of diodes in particular must be very precise for satisfactory results.

6.7. Data Requirements on Operational Amplifiers. Although it is clearly evident from the examination of a typical specification sheet for an operational amplifier that a large amount of information is provided, and in fact a majority of the data needed are available, in some applications some additional data would be helpful. For example, a listing of the corner frequency or frequencies for the amplifier could be quite useful in setting up differentiators. Likewise, the nominal positions of the positive and negative corner frequencies in universal amplifiers might be helpful, along with tolerances to be expected where a pole and a zero have been superposed to maintain integration in a transition region.

In addition, for Q multipliers and similar applications, it frequently might be convenient for the user to have a plot of nominal amplification of an operational amplifier as a function of positive and negative differential bias at the input of the amplifier. This plot should be made for balanced input, and for unbalanced input for each input with the other grounded. This information could be quite useful in the establishment of magnitudes of feedback required. Some kind of idea of the approximate upper and lower limits for these values would be particularly helpful, since then the data would be directly useful in the establishment of design limits for specific circuits.

REFERENCES

1. Airpax Electronics Staff, "The Contact Modulator," Airpax Electronics, Inc., Cambridge, Md., 1962.
2. Beckman/Berkeley Division, Digital and Analog Instrumentation Catalog 705, 1957.
3. H. W. Bode, "Network Analysis and Feedback Amplifier Design," D. Van Nostrand Company, Inc., Princeton, N.J., 1945.
4. P. J. Bénéteau, R. Q. Lane, and Larry Blaser, *IRE Conv. Record*, Mar. 29, 1962, pt. 9, pp. 173–185.
5. F. H. Blecher, Transistor Circuits for Analog and Digital Systems, *Bell System Tech. J.*, March, 1956, pp. 295–332.
6. G. S. Brown and D. P. Campbell, "Principles of Servomechanisms," John Wiley & Sons, Inc., New York, 1948.
7. Burr-Brown Research Corp. Release no. 259, The 1500 Series Operational Amplifiers, Burr-Brown Research Corp., Tucson, Ariz.
8. W. F. Caldwell, G. A. Korn, et al., A Precision Amplitude Distribution Amplifier, *IRE Trans. PGEC*, June, 1960, p. 252.
9. D. K. Cheng, "Analysis of Linear Systems," Addison-Wesley Publishing Company, Inc., Reading, Mass., 1959.
10. R. V. Churchill, "Operational Mathematics," 2d ed., McGraw-Hill Book Company, New York, 1958.
11. D. Deford, Stabilized Follower Amplifier, *Applications Bulletin* 12-19-57, G. A. Philbrick Researches, Inc., 1957.

12. G. Doetsch, "Laplace Transforms," D. Van Nostrand Company, Inc., Princeton, N.J., 1963.
13. "Theorie und Anwendung der Laplace Transformation," Dover Publications, Inc., New York, 1943.
14. K. Eklund, Use of Operational Amplifiers in Accelerator Beam Control Systems, *Rev. Sci. Instr.*, vol. 30, no. 5, 1959.
15. K. Eklund, Use of Operational Amplifiers in Precision Current Regulators, *Rev. Sci. Instr.*, vol. 30, no. 5, 1959.
16. C. E. Foiles, J. P. Hartmann, et al., Analog Computer Reference Supply, reprinted from *Electronic Design* by G. A. Philbrick Researches, Inc., 21, May 10, 1959.
17. M. F. Gardner and J. L. Barnes, "Transients in Linear Systems," John Wiley & Sons, Inc., New York, 1942.
18. E. A. Goldberg, Stabilization of Wide-band Direct-current Amplifiers for Zero and Gain, *RCA Rev.*, June, 1950, p. 296.
19. R. O. Gregory, "Design Considerations in a Chopper-stabilized Transistor Operational Amplifier," presented at IRE Maecon, Kansas City, Mo., Nov. 15–16, 1960.
20. IRE Standards on Letter Symbols, New York, 1956.
21. IRE Standards on Letter Symbols and Mathematical Signs, New York, 1948 (reprinted 1957).
22. J. G. Jaeger, "An Introduction to the Laplace Transformation," Methuen's Monographs on Physical Subjects, London, 1959.
23. W. J. Karplus, A New Active-passive Network Simulator for Transient Field Problems, *Proc. IRE*, January, 1961, p. 268.
24. L. H. King, Reduction of Forced Error in Closed-loop Systems, *Proc. IRE*, vol. 41, no. 8, p. 1037, 1953.
25. R. Legros and A. V. J. Martin, "Transform Calculus for Electrical Engineers," Prentice-Hall, Inc, Englewood Cliffs, N.J., 1961.
26. J. R. MacDonald, A New Integrating Circuit and Electronic Analog for Transient Diffusion and Flow, G. A. Philbrick Researches, Inc., Application Bulletin, Boston, Mass., Jan. 15, 1958.
27. N. W. McLachlan, "Complex Variable and Operational Calculus with Technical Applications," Cambridge University Press, New York, 1939.
28. F. D. Murnaghan, "The Laplace Transformation," Spartan Books, Washington, D.C., 1963.
29. D. D. Nye, Jr., The Design of Chopper-stabilized Amplifiers, *Elec. Design News*, March, 1963, p. 102.
30. R. H. Okada, Stable Transistor Wideband DC Amplifiers, *Trans. AIEE*, March, 1960, p. 26.
31. B. M. Olver, The Effect of μ-Circuit Non-linearity on Amplitude Stability of RC Oscillators, *Hewlett-Packard J.*, vol. 11, no. 8–10, April–June, 1960.
32. Ordnance Corps Publication ORDP 20-293, Surface-to-Air Missile Series, Pt. 3, Computers, part of "Ordnance Engineering Design Handbook," Washington, D.C., 1962.
33. Philbrick Staff, Notes on Operational Amplifiers (part of Analog Computor Techniques Applied to Industrial Instrumentation and Control), IRE-PGEC talk to PGEC chapter, New York, 1958.
34. K. A. Pullen, Patent Disclosure, 1963.
35. K. A. Pullen, "Conductance Design of Active Circuits," John F. Rider, Publisher, Inc., New York, 1959.
36. K. A. Pullen, The Dynamic Characteristics of Phase-lock Receivers, *Ballistic Research Labs. Rept* 1093, Aberdeen Proving Ground, Md., 1959.
37. K. A. Pullen, "Theory and Applications of Topological and Matrix Methods," John F. Rider, Publisher, Inc., New York, 1962.
38. H. J. Reich, "Functional Circuits and Oscillators," D. Van Nostrand Company, Inc., Princeton, N.J., 1961.
39. B. Seddon, Analog Methods, Reprint no. 1627 by G. A. Philbrick Researches, Inc., from 12th Annual Symposium Computors in the Process Industry, 1960.
40. B. Seddon, "Capabilities of Some Non-linear Instrument Circuits for Low-level Transients," IEEE paper CPA-63-552, 1963.
41. D. H. Sheingold, "An Extensive Set of Unpublished Notes," G. A. Philbrick Researches, Inc., Boston, Mass.
42. E. Stephens, "The Elementary Theory of Operational Mathematics," McGraw-Hill Book Company, New York, 1937.
43. John L. Stewart, "Circuit Theory and Design," chap. 10, John Wiley & Sons, Inc., New York, 1958.
44. R. J. Strehlow, A Solid-state Operational Amplifier of High Stability, *Hewlett-Packard J.*, vol. 14, no. 3–4, p. 1, 1963.

45. Tektronix Field Information Department, Introduction to Operational Amplifiers, *Tektronix Service Scope*, no. 19, February–April, 1963.
46. Tektronix Field Information Department, Sampling Oscilloscopes and the Slide-back Balanced Bridge Technique, *Tektronic Service Scope*, August, 1962.
47. W. J. Thompson, "Laplace Transformation," Prentice-Hall, Inc., Englewood Cliffs, N.J., 1960.
48. R. W. Thorpe, Transistorized Second-order Filter Provides Low Damping, *Elec. Design News*, March, 1963.
49. C. J. Tranter, "Integral Transforms in Mathematical Physics," Methuen Monograph, London, 1959.
50. J. G. Truxal, "Automatic Feedback Control System Synthesis," McGraw-Hill Book Company, New York, 1955.
51. M. E. Van Valkenburg, "Network Analysis," Prentice-Hall, Inc., Englewood Cliffs, N.J., 1955.
52. M. E. Van Valkenburg, "Introduction to Modern Network Synthesis," John Wiley & Sons, Inc., New York, 1961.

Chapter 20

HIGH-POWER AMPLIFIERS

STEPHEN J. ANGELLO, * **FRED J. HEATH,** † **and JOHN D. MENG** ‡,§

CONTENTS

1. High-power Transistor Amplifiers.............................	20-2
1.1. Introduction...	20-2
1.2. Circuit Configurations for Transistor High-power Amplifiers.	20-2
1.3. Circuit Efficiency.......................................	20-3
1.4. Amplifier Design..	20-4
2 Characteristics of Commercial High-power Transistors..........	20-4
2.1. The JEDEC Registration-data Format....................	20-5
2.2. Mechanical Aspects of High-power Transistors............	20-5
2.3. Discussion of Power-transistor Ratings and Characteristics..	20-8
2.4. Measurement of Power-transistor Characteristics..........	20-9
2.5. High-power Transistor Rating Methods...................	20-12
2.6. Typical Power Transistors...............................	20-14
3. Thermal Considerations for High-power Transistors..............	20-14
3.1. Temperature–Thermal-resistance Relations................	20-14
3.2. Determination of Junction Temperature...................	20-15
3.3. Selection of a Heat Sink.................................	20-16
3.4. Temperature-Power Relations in Dynamic Loading.........	20-18
4. Applications of High-power Transistors at Audio Frequencies....	20-22
4.1. Parallel Operation of Transistors.........................	20-23
4.2. A 10-kw Sonar Amplifier.................................	20-24
4.3. A 200-watt Audio Power Amplifier.......................	20-26
5. High-power High-frequency Transistor Amplifiers...............	20-26
5.1. A Class C 100-watt 10-Mc Power Amplifier................	20-26
5.2. A Class C 100-watt 3-Mc Power Amplifier.................	20-30
5.3. A Switching-type Linear High-power Amplifier............	20-31
6. High-power Amplifiers Utilizing Silicon Controlled Rectifiers....	20-32
6.1. Introduction..	20-32
6.2. Typical Transfer Characteristics.........................	20-33
6.3. Inherent Delays and Phase Shifts........................	20-35
6.4. Improving Overall Linearity............................	20-37

* Westinghouse Electric Corp., Research and Development Center, Pittsburgh, Pa.
† Canadian General Electric Co., Ltd., Toronto, Ont., Canada.
‡ University of California, Electronics Engineering Dept., Lawrence Radiation Laboratory, Berkeley, Calif.
§ Sections 1 to 5, inclusive, were written by S. J. Angello, Section 6 by J. D. Meng, and Section 7 by F J. Heath.

20-2　AMPLIFIER CIRCUITS

    6.5. Open-loop Stability.. 20–37
    6.6. Closed-loop Stability...................................... 20–38
    6.7. Some Practical Circuits and Their Characteristics.......... 20–38
  7. Amplifiers for Transmitters................................... 20–40
    7.1. Tube Selection.. 20–40
    7.2. Power-supply Considerations............................... 20–44
    7.3. Standard Broadcast Transmitters........................... 20–47
    7.4. Superpower Transmitters................................... 20–54
    7.5. FM and TV Transmitters.................................... 20–58
    7.6. Short-wave Transmitters................................... 20–61

1. HIGH-POWER TRANSISTOR AMPLIFIERS

Sections 1 to 5 will supplement the sections in Chap. 17 dealing with transistor amplifiers. The discussion is restricted to power amplifiers with collector dissipation of 100 watts and above.

1.1. Introduction. Segregation of "high-power" amplifiers from "power" amplifiers at 100 watts is, of course, arbitrary. Special problems concerned with impedance levels, temperature stability, temperature cycling, and device ratings arise in the very high power levels over 100 watts.

The major emphasis of these sections will be upon circuit applications and associated mechanical and thermal problems. At the outset, however, it will be instructive to consider device aspects which bear upon power and frequency limits to be expected.

The power-handling capability of a transistor is dependent upon (1) the efficiency of the circuit in which it operates, (2) the provisions for heat removal to maintain a stable junction temperature, and (3) the signal frequency.

J. M. Early[1] has proposed an equation which describes the limits of a transistor as follows:

$$PZf^2_{max} = K^2 \qquad (1)$$

where P is output power, or any closely related quantity, Z is the load impedance, and f_{max} is a characteristic frequency. K is an approximate constant which is equal to

$$K = \frac{Q_B V_{Sl}}{\epsilon} \qquad \text{volts/sec}$$

Q_B is the charge required to terminate a breakdown field, V_{Sl} the scattering-limited velocity of current carriers, and ϵ the dielectric constant. K has a value on the order of 10^{12}.

This expression can be seen to be reasonable, because the power output of transistors increases with addition of units in parallel, while at the same time impedance decreases; thus, PZ is constant. Further, f_{max} varies inversely as the transistor diameter, or f^2_{max} varies inversely as the area, and at the same time, power increases as the area; hence Pf^2_{max} is constant.

There is a theoretical maximum power density which has been pointed out, again, by J. M. Early.[2] For p-n-p germanium transistors this is about 2×10^5 watts/sq cm. Silicon and n-p-n germanium are slightly higher.

Equation (1) is degraded by a factor of 10 because of current crowding at the emitter edge. In spite of this, it is evident that quite high power is possible at very high frequencies. The equation implies that there is no frequency limit for a given power if one can utilize a small load impedance. High power, then, is obtained by a combination of circuit and device design ingenuity.

Power levels above 200 watts are now possible, and much more can be expected in the future.

---

[1] Private communication. Presented orally as "Speed and Semiconductor Devices" at the IRE National Convention, March, 1962.
[2] J. M. Early, Maximum Rapidly-switchable Power Density in Junction Triodes, *IRE Trans. Electron Devices*, vol. ED-6, pp. 322–325, July, 1959.

1.2. Circuit Configurations for Transistor High-power Amplifiers.

Contemporary circuit designs favor the common-emitter connection with base input and collector output. This connection affords the highest power gain. In addition, the current amplification is higher than for the common-base connection, and the voltage amplification is higher than for the common-collector connection.

The two-port network parameters are discussed in detail in Chap. 17, Sec. 1.1. For convenience, we shall summarize the relations which are used in this chapter.

The two-port under discussion is shown in Fig. 1. There are six possible arrays[3] of voltages and currents which characterize this two-port network. Because of con-

FIG. 1. Two-port network with small-signal voltage and current applied.

venience in parameter measurements and suitability for applications, we chose the set involving the h parameters. These are

$$v_1 = h_{11}i_1 + h_{12}v_2 \\ i_2 = h_{21}i_1 + h_{22}v_2 \tag{2}$$

The condition $v_2 = 0$, that is, with the output a-c short-circuited gives:

Small-signal value of the short-circuit input impedance:

$$h_{11} = h_i = \frac{\partial v_1}{\partial i_1}\Big|_{v_2=0} \tag{3}$$

Small-signal short-circuit forward current transfer ratio:

$$h_{21} = h_f = \frac{\partial i_2}{\partial i_1}\Big|_{v_2=0} \tag{4}$$

The condition $i_1 = 0$, that is, with the input a-c open-circuited gives:

Small-signal value of the open-circuit reverse voltage transfer ratio:

$$h_{12} = h_r = \frac{\partial v_1}{\partial v_2}\Big|_{i_1=0} \tag{5}$$

Small-signal value of the open-circuit output admittance:

$$h_{22} = h_o = \partial i_2/\partial v_2|_{i_1=0} \tag{6}$$

If the transistor is connected in the network as common emitter, base input the symbols become h_{ie}, h_{fe}, h_{re} and h_{oe}. Table 2 in Chap. 17 gives the transformation among the h parameters for common emitter, common base, and common collector.

1.3. Circuit Efficiency.

One factor governing the amount of power which a given transistor can control is the efficiency of the circuit in which it is operating. Modes of circuit operation for amplifiers are divided into classes, viz., A, A₁, AB, AB₁, B, C. These are discussed in detail in Chap. 17, Secs. 12 to 14. Maximum theoretical efficiencies vary from 50 per cent for class A to 78.5 per cent for class B push-pull. Class C amplifiers can operate in the region of 80 per cent, because they are usually connected to supply the losses in a loaded tuned circuit.

---

[3] L. E. Getgen, Application of Matrix Algebra to Circuit Design, *Electro-Technol.*, February, 1963, pp. 70–79.

20-4　AMPLIFIER CIRCUITS

1.4. Amplifier Design. The basic design principles which establish circuit-element values and bias conditions for audio output stages are to be found in the following sections in Chap. 17:

- 4. Transistor bias
- 13.3. Class A transistor power amplifiers
- 13.4. Graphical determination of class A operating point
- 13.7. Biasing transistor power amplifiers
- 13.8. Temperature stabilization with thermistors and diodes
- 14.4. Bias in class B amplifiers
- 14.7. Design charts for power amplifiers
- 15. Typical power amplifiers

Special designs appropriate for very high power transistors will be discussed in Secs. 4 and 5.

2. CHARACTERISTICS OF COMMERCIAL HIGH-POWER TRANSISTORS

The design for any transistor circuit depends strongly upon the terminal characteristics of the transistor or transistors involved. The practical design problem is a kind of inductive logic. Given the desired function, create a circuit configuration which will fulfill the specified requirements. As is usual with inductive problems, there is more than one possible solution. The skill and intuition of the designer are then brought to bear to obtain the best solution for the application in hand. To do

Table 1. Joint Electron Device Engineering Council Registration Data
(Silicon power transistor, manufacturer, Westinghouse Electric Corporation
JEDEC type 2N1833)

I. General description
   This transistor is an n-p-n silicon power transistor designed primarily for high-power switching applications and inverters. It is intended for industrial and military class of service.
II. Mechanical data
   A. The transistor has an outline as per Fig. 2
   B. Terminal location is shown in Fig. 2
   C. Maximum stud torque 100 in.-lb. Nonlubricated
III. Absolute-maximum ratings
   A. Temperature
      1. Storage T_A − 65 to +175°C
      2. Operating junction T_J − 65 to +175°C
   B. Maximum voltage
      1. V_{CEX}. See Table 2
      2. V_{CBO}. See Table 2
      3. V_{EBO} 15 volts d-c
   C. Maximum current
      Peak collector current I_C max. 30 amp d-c max
      Recommended continuous operating current I_C. See Table 2. I_B 10 amp d-c max
   D. 1. Maximum power dissipation 250 watts at a T_C of 60°C max
      2. Derating factor 2.17 watts/°C
IV. Electrical characteristics
   A. Static
      1. Collector current
         I_{CEX} at V_{CEX}. See Table 2 and $V_{BE} = -1.5$ volts, $T_C = 175$°C. 30 ma max
      2. Emitter current
         I_{EBO} at $V_{BE} = -15$ volts. 25 ma max $T_C = 175$°C.
      3. D-C forward current transfer ratio h_{FE}. See Table 2
      4. D-C collector-to-emitter saturation voltage V_{CE}(sat). See Table 2
      5. D-C base-to-emitter voltage V_{BE}. See Table 2
      6. Thermal impedance θ_{JC}. 0.45°C/watt max
   B. Dynamic. See Table 2

an effective job, the designer must have at his command definitive and pertinent information concerning the terminal properties of devices. The wider the selection, the more likely he is to reach a satisfactory design.

This section will describe in detail the kinds of terminal data pertinent to high power transistors and will give enough background information to make intelligent design possible.

2.1. The JEDEC Registration Data Format. To foster a trend toward device standardization the Joint Electron Device Engineering Council[4] (JEDEC) registers transistors under a series of numbers beginning with the prefix "2N." Encapsulation outlines are well standardized (see Sec. 2.2), and a user can develop more than one source of supply for a given transistor.

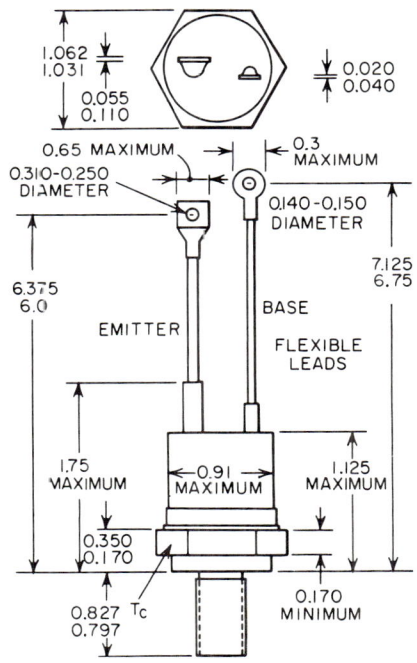

Fig. 2. Outline of power-transistor enclosure with connections.

Table 1 gives the registration data for the Westinghouse type number 2N1833.[5] This chart shows the format used in all registrations. Sections 2.2 to 2.6 discuss items of the format in detail.

2.2. Mechanical Aspects of High-power Transistors. *Enclosures.* Standard enclosures for semiconductor devices are listed by JEDEC under a series of "TO" (transistor outline) numbers. Many power transistors are packaged in the TO-3 and TO-36 types. Other types being used are TO-10, TO-13, TO-37, and TO-41. Figure 3 shows outlines of these encapsulations. We shall not attempt to list all power-transistor encapsulations in use, because the field is developing too fast. Some very high power transistors are new enough that standards have not been agreed upon. Figure 4 shows a cross-sectional view of a metal-ceramic type of high-power encapsu-

---
[4] Electronic Industries Association, 32 Green St., Newark 2, N.J.
[5] Original data sheet gave 18 type numbers. Only one is given for the sake of brevity.

Table 2

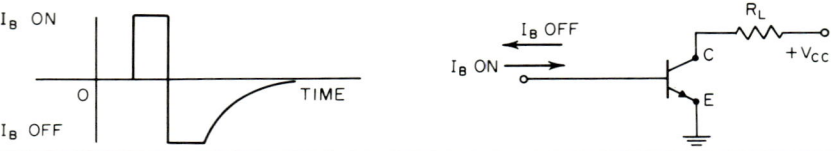

			Static Characteristics							
JEDEC no.	V_{CEX} $V_{BE} = -1.5$ volts $T_C = 175°C$	V_{CBO} $T_C = 175°C$	h_{FE}(min) $V_{CE} = 4$ volts d-c $T_C = 25°C$				Max r_{CE}(sat) $T_C = 25°C$		Max V_{BE} $T_C = 25°C$	
			I_C	Min h_{FE}	I_C	I_B	r_{CE} (sat)	I_C	I_B	Max V_{BE}
2N1833	200 volts d-c	200 volts d-c	25 amp d-c	10	25 amp d-c	5 amp d-c	0.06 ohm	25 amp d-c	5 amp d-c	2.5 volts d-c

		Dynamic Characteristics						
JEDEC no.	Recommended operating current I_C	Test conditions for "turn-on" and "turn-off" time values					$t_d + t_r$ turn-on time (Fig. 9) max	$t_s + t_f$ turn-off time (Fig. 9) max
		I_C	T_C	V_{CC}	I_B off	I_B on		
2N1833	25 amp d-c	25 amp d-c	25°C	12 volts d-c	5.0 amp	5.0 amp	20 μsec	25 μsec

lation. Figure 5 is a cutaway. Many others are in use. Manufacturer's data sheets always give outline drawings of enclosures.

Power-transistor enclosures are usually attached to a form of heat sink. These heat sinks are discussed in detail in Sec. 3.1.

Mounting of the Semiconductor to the Heat Sink. A very important mechanical aspect of power transistors is the mounting of the semiconductor wafer on the heat sink. Silicon and germanium are both hard and brittle crystalline materials. The coefficients of expansion of copper and aluminum, the preferred heat-sink materials, are substantially larger than the coefficients of silicon and germanium. For this reason, silicon large-area junction wafers are mounted on molybdenum or tungsten or powder compacts which match the expansion as closely as possible. In early silicon power rectifiers (circa 1956) the moly support was soft-soldered to the copper heat sink. A severe cyclic application, such as an electric welder, resulted in failure after a few thousand cycles. In a careful study of these failures, W. B. Green[6] found that the failure was caused by fatigue of the soft-soldered joint. Data on junctions of several sizes over a range of junction temperature are summarized by

$$ND^2(\Delta T)^2 = C^2 \qquad (7)$$

where N is the number of cycles to failure, D is the diameter, ΔT is the difference between the junction temperature and the heat-sink temperature, and C is a constant

---

[6] W. B. Green, A Fatigue-free Silicon Device Structure, *Trans. AIEE Commun. Electron.*, vol. 80, no. 54.

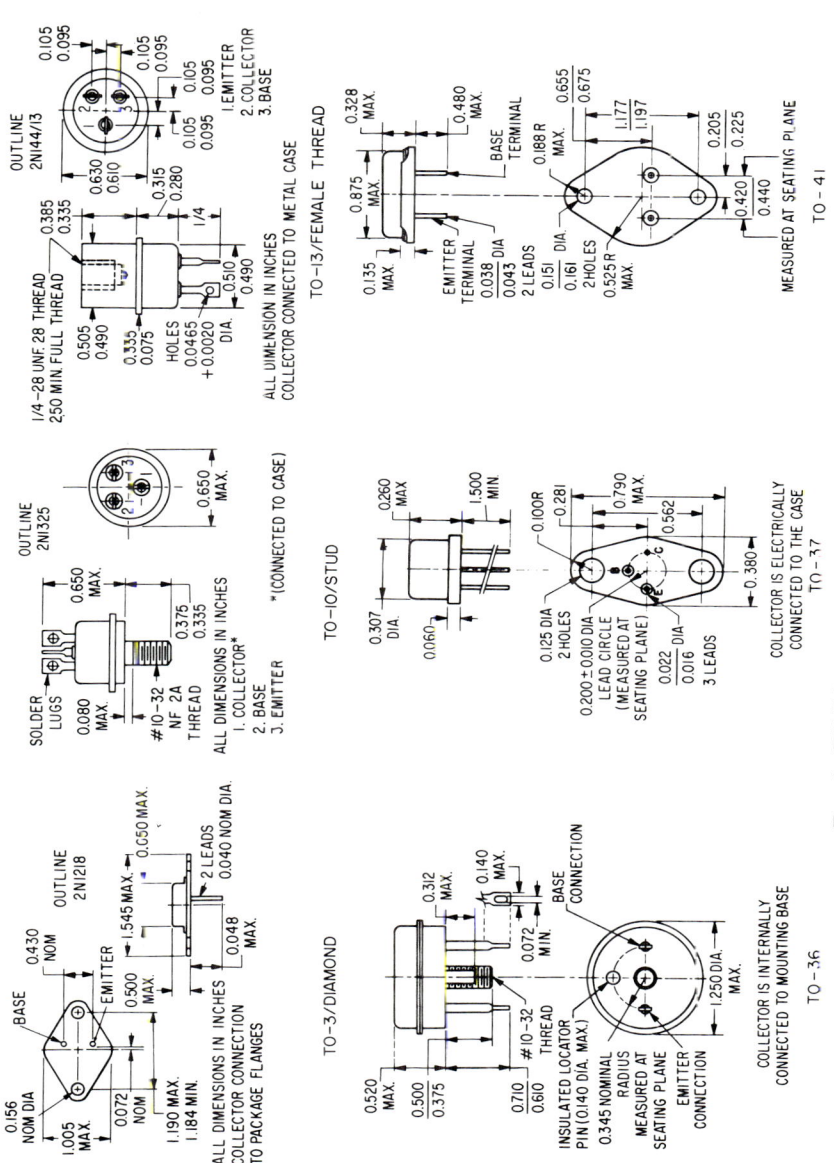

Fig. 3. JEDEC power-transistor encapsulations.

20-7

20-8 AMPLIFIER CIRCUITS

about 3,500. Similar thermal-stress failures in metals reported in the literature show

$$N^{\frac{1}{2}} \Delta \epsilon_P = C_1 \tag{8}$$

where $\Delta \epsilon_P$ is change of plastic strain per cycle. Plastic strain depends upon differential coefficient of expansion, ΔT, and diameter. The key to solve the fatigue failure, then, is to provide a structure which is never plastically deformed. The difference of expansion is taken up in the heat sink elastically. This is accomplished by supporting the silicon on a very strong (thick) tungsten base and hard-soldering this to copper. The copper is not strained plastically in this system and can therefore survive an indefinite and very large number of thermal cycles. Transistor wafers follow the same considerations as power-rectifier wafers.

2.3. Discussion of Power-transistor Ratings and Characteristics. The "rating" of a power transistor implies a limiting value for some parameter which, if exceeded, will result in degradation or failure of the transistor. Table 1 gives four ratings:

1. Collector junction temperature limits
2. Maximum collector and emitter voltages
3. Maximum collector current
4. Maximum power dissipation

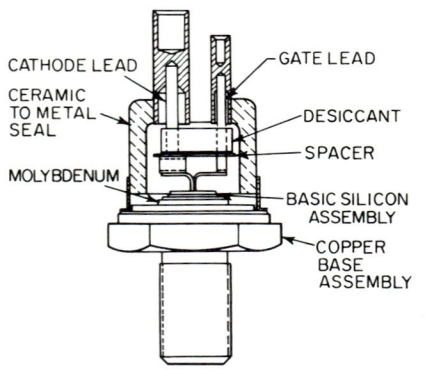

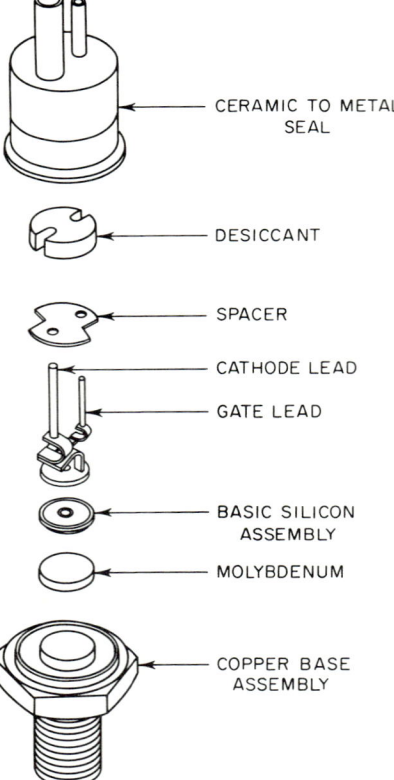

Fig. 4. Cross section of power transistor.   Fig. 5. Assembly of power transistor.

Collector junction temperature is a basic limiting parameter for a transistor. It is determined by life-testing groups of transistors at a series of temperature levels. The upper limit has to do with operating stability, both dynamic and aging. The lower limit bears upon mechanical stability under differential contraction stresses. The hermetic seal can be a factor here, as well as the device structure.

This parameter is of sufficient importance to warrant special, detailed discussion in Sec. 3.

Every junction stressed with voltage in the high-resistance direction has a turnover point or a breakdown voltage. Since there are three possible connections of the three-terminal transistor, there will be three possible breakdown voltages with the third terminal bias condition specified. For example,

V_{CEX} is the d-c voltage between the collector and the common emitter with a bias of X volts between base and emitter. The breakdown point for this parameter is written BV_{CEX}.

V_{CBO} is the d-c voltage between the collector and the common base with the emitter open-circuited.

V_{EBO} is the d-c voltage between the emitter and the common base with the collector open-circuited.

By writing the remaining combinations, it can be seen that these three voltages suffice to give all the breakdown conditions.

The maximum rating of a given transistor structure may be less than the breakdown voltage. The reason for this is the occurrence in most large-area transistor structures of a phenomenon called "second breakdown." The second breakdown is characterized by an abrupt reduction in V_{CE} at some collector current when the transistor is swept through the V_{CE} versus I_C characteristics.[7] A hot spot is formed which results in an internal short circuit. Good data sheets should give the voltage-current boundaries for this phenomenon. More is said about determining ratings in Sec. 2.5.

Maximum collector current is determined either by junction heating to the maximum allowable temperature or by the drop in current amplification[8] h_{FE} (in the usual grounded-emitter connection) below a usable value.

Maximum power dissipation is directly related to the junction temperature rise and is approximately

$$P = V_{CE}I_C \quad \text{or} \quad P = V_{CE}I_E$$

in the common-emitter connection. This neglects dissipation due to base-emitter bias current. Power dissipation and the derating factor are discussed in Sec. 3.

2.4. Measurement of Power-transistor Characteristics. As the JEDEC chart shows, power-transistor measurements are both static and dynamic. Static characteristics are:

1. *Output:* Collector current I_C vs. collector-to-emitter voltage V_{CE} for values of base current I_B.

Note: Capital-letter symbols are used because these values are d-c and also large-signal. All these characteristics are for the common-emitter connection. This output family of characteristics is very important for power-transistor circuit design. This set is usually given in two forms; the overall large-signal set, and the expanded-abscissa set to show the saturation (high-current low-voltage) region.

2. *Input:* Base current I_B vs. base-emitter voltage V_{BE} for values of case temperature T_C at a fixed value of V_{CE}.

3. *Current Transfer:* Collector current I_C vs. base current I_B for values of T_C at a fixed V_{CE}. Values of h_{FE} are also plotted vs. I_C for values of T_C.

4. *Transconductance:* Collector current I_C vs. base-to-emitter voltage V_{BE} for values of T_C at a fixed V_{CE}.

5. Junction temperature and thermal impedance are discussed in Sec. 3.

The preceding four characteristics are illustrated in Fig. 6 for various types of power transistors. A survey of manufacturer's data sheets shows that no standard format of curve plotting is followed, and data sheets vary in the amount of data given. For example, in addition to the basic four characteristics, the saturation resistance in ohms vs. base current I_B may be given for various T_C at a given I_C. This is relevant for switch applications. Another useful set of curves for stability considerations is

---

[7] H. A. Schaft and J. C. French, Characteristics of Second Breakdown and Transistor Failure, *Symposium on the Physics of Failure in Electronics*, Sept. 26, 1963, RADC and Armour Research Foundation.

[8] This is often wrongly called "current gain"; a "gain" according to IEEE standards involves power ratios only.

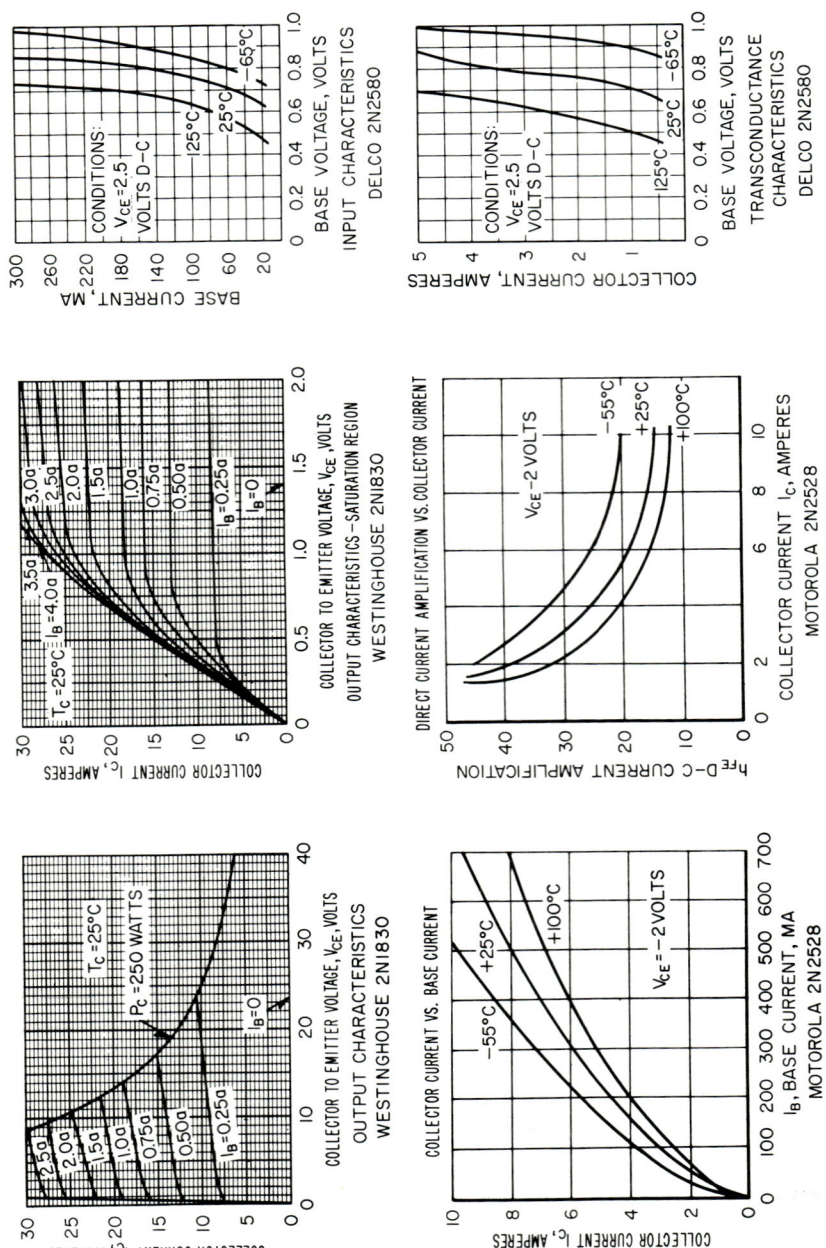

Fig. 6. Power-transistor characteristics, common-emitter.

I_{CBO} versus T_C for various V_{CB} (common-base connection). This characteristic is highly temperature-dependent. The corresponding common-emitter characteristic is I_{CEX} versus T_C for fixed V_{CEX}.

Figure 7 illustrates the essential features of a circuit for measuring the output characteristics. The power transistor under test is connected to an adequate heat sink maintained at 25°C. Silicone grease is commonly used between the case and the fixture to reduce thermal drop. A pulse type of measurement is made using half sine waves of a 60-cps supply with a repetition rate of 10 per second. The rate of change is slow enough to be equivalent to a static test, and heating is reduced.[9]

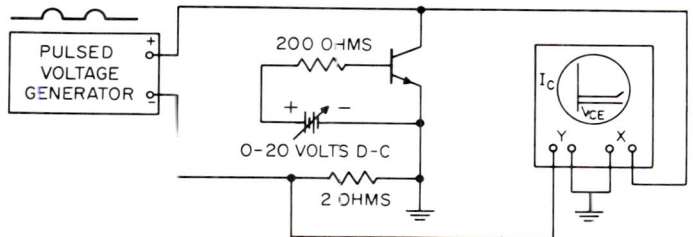

FIG. 7. Test circuit for output characteristics.

Input characteristics may be measured by a d-c measurement arrangement as shown in Fig. 8. The proper bias polarities must be observed. This same circuit is used to measure the static large-signal forward-current transfer ratio h_{FE}. The desired voltage V_{CE} is applied and the base current is varied to obtain the desired I_C. The ratio I_C/I_B is h_{FE} at I_C. Alternatively, I_C can be plotted vs. I_B. For the transconductance, collector current is plotted vs. base voltage V_{BE}.

For all the above measurements, the heat-sink temperature can be regulated at the desired case temperatures. Measurements at the higher temperatures should be made

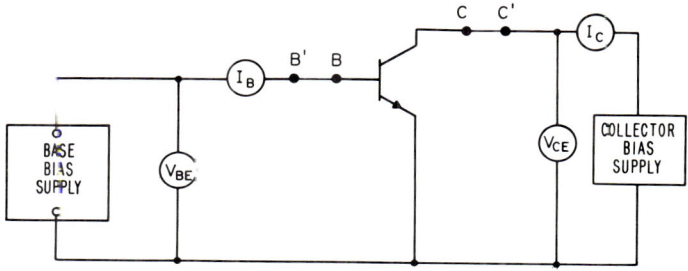

FIG. 8. General d-c measurement arrangement.

using a pulsed bias source to prevent excessive junction temperature, and to maintain as near as possible a constant difference between junction and case temperature.

The dynamic characteristics most often given are the switching times, and the frequency at which h_{fe} is 0.707 times the low-frequency value.

The switching speed of a transistor may be characterized by the delay, rise, storage, and fall times t_d, t_r, t_s, and t_f, respectively. Input and output pulse characteristics are shown in Fig. 9. The common measuring arrangement is shown in Fig. 10. The device is driven by constant current in the common-emitter connection with a specific load resistance and collector bias supply voltage. The resistance in the base lead

[9] Texas Instruments, Industrial Products Group, Pulse-testing Range Extended for Transistors, Diodes, *Electron. Design News*, April, 1963, p. 17.

(including the pulse generator) should be much larger than the forward input resistance and much smaller than the reverse input resistance of the transistor. The output waveform is monitored by an oscilloscope to determine the various switching times.

The common-emitter current-amplification cutoff frequency can be measured with the circuit shown in Fig. 11. With a transistor mounted on a heat sink at 25°C, the

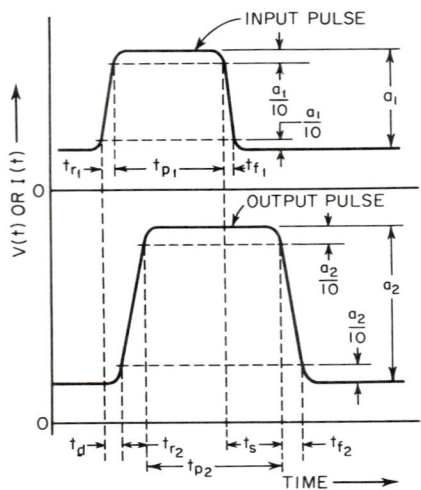

FIG. 9. Pulse characteristics

a_1 = input pulse amplitude t_d = pulse delay time
a_2 = output pulse amplitude t_s = pulse storage time
t_r = pulse rise time t_f = pulse fall time
t_p = pulse time

collector voltage and current are adjusted for the desired values. In order to prevent heating, the transistor must be mounted on an adequate heat sink. Set the generator[10] to 400 cps and adjust the drive to give undistorted output across the load. The output voltage shall be 30 db or more above any noise voltages. Increase the frequency, keeping the generator output voltage constant, until the output voltage

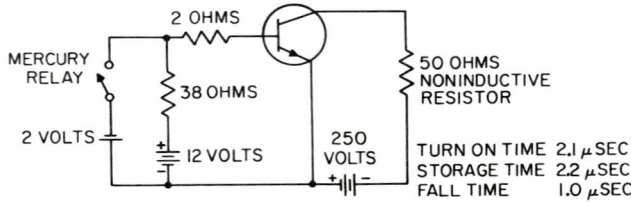

FIG. 10. Test circuit for switching times of the Delco 2N2580 power transistor.

drops to 0.707 of the original value. The cutoff frequency of h_{fe} can be read directly from the signal generator.

2.5. High-power Transistor Rating Methods. This discussion of rating methods emphasizes the safe application of high-power transistors, and does not attempt to relate characteristics to device design.

Pulse testing of power transistors in the common-emitter connection to reveal the

[10] These instructions are given by Delco Radio Transistor Test Procedures, Mar. 1, 1960.

V_{CE} versus I_C relation (Sec. 2.4) shows various breakover characteristics depending upon the base bias. Figure 12 shows typical curves. In general, a high-power transistor will not be safely operated in a region beyond a breakover voltage. The reason is that, in general, the breakover point is followed by a negative-resistance portion of the characteristic. At some point on the curve beyond the breakover (first break-

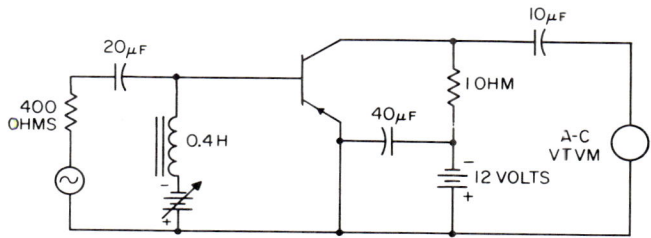

FIG. 11. Cutoff frequency test circuit for h_{fe}.

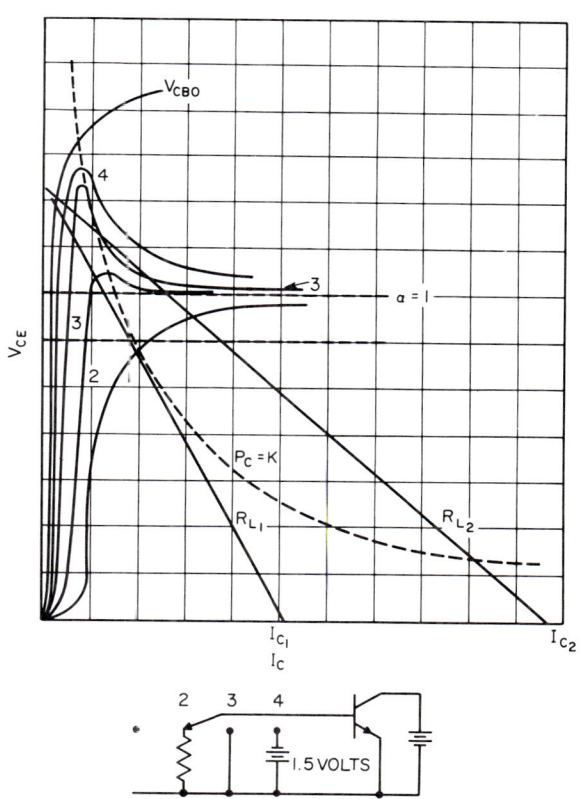

FIG. 12. Typical V_{CE} versus I_C curves with breakover for various base biases.
    1   V_{CEO} base open. $I_B = 0$. $R_{BE} = \infty$
    2   V_{CER} resistance between B and E. R_{BE} = finite
    3   V_{CES} base shorted to emitter $V_{BE} = 0$. $R_{BE} = 0$
    4   V_{CEX} reverse bias between B and E

down) a "second breakdown" occurs which is destructive (see Sec. 2.3). There will be a locus of second-breakdown trigger points which must be avoided for any duration.

In the steady state it is obvious that operation must be within the maximum-power-dissipation hyperbola. Switching transients may go out of this boundary so long as the maximum junction temperature is not exceeded. The excursion of the load line is important, because stable points may occur which lock the transistor at a point outside the dissipation hyperbola. Two resistance load lines are shown in Fig. 12. One is safely below the hyperbola; the other crosses the V_{CES} curve at point A. This stable point is in the unallowed dissipation region. Nonlinear load lines are not usually amenable to simple calculation. It is recommended in this case that the actual load line for the circuit under development be displayed along with the V_{CE} versus I_C family to be sure that no unsafe regions are traversed. This should be carried out for the temperature range specified for the application.

2.6. Typical Power Transistors. Table 3 shows data on several power transistors supplied by various vendors.

Table 3. Some Contemporary Power Transistors

Type no. Si or Ge	Mfr. power, watts	I_C max amp	V_{CEO}, volts	V_{CE} sat, volts at I_C amp	h_{FE} at I_C amp	V_{BE}, volts	f_{ae}, kc	θ_{JC}, °C/watt
2N2580 Si	Delco 150	5	400	1.2 at 5	10 min at 5	1.5	75	0.5
2N1833 Si	Westinghouse 250	30	200	1.5 at 25	10 min at 25	2.5	19	0.45
MP507A Ge	Motorola 170	60	75	0.2 at 50	20 typical at 50	1.5	3.6	0.35
2N1653 Ge	Bendix 100	25	120	1.0 max at 25	35 min at 10	1.5	1.5	0.2
3TX002 Si	Clevite 60	5	(V_{CBO}) 100	1.0 at 5	30 min at 5	2.0	(f_T) 150 Mc	2.5
2N2338 Si	RCA 150	10			10 typical at 10			
2N1899 Si	PSI 125	10	(V_{CBO}) 140	0.5 at 10	21 typical at 10	1.0	10 Mc (30 watts)	1
2N2884 Si	Fairchild 1		(V_{CBO}) 40		20 min at 0.1		(f_{ab}) 500 Mc	

3. THERMAL CONSIDERATIONS FOR HIGH-POWER TRANSISTORS

Maximum limits shown on data sheets are for constant-heat-dissipation conditions. The parameter under discussion in this section is junction temperature. For the purposes of this discussion the junction temperature T_J°C has a maximum value given in the data sheet which may not be exceeded—even for a very short time.

3.1. Temperature–Thermal-resistance Relations. Heat is generated at the junction and is conducted to the heat sink through a number of resistances. The usual resistances are (1) The material between the junction and the transistor case. (2) The material between the case and the heat radiator. This is usually a silicone grease film unless the case is to be electrically insulated from the radiator. In the latter case, it is a mica wafer or other insulating layer. (3) The material from the insulator to the ambient medium. This includes the radiator itself and the film boundary layer between the radiator and the medium.

Heat flow through a thermal resistance causes a temperature drop. The usual physical tables give a thermal conductivity defined as

$$q \text{ (watts/sq cm)} = k \frac{(T_J - T_C)(°C)}{l \text{ (cm)}} \qquad (9)$$

where q is heat flux, $T_J - T_C = \Delta T$ is the temperature difference between junction and case, l is the flow path between the high and low temperature ($\Delta T/l$ is a gradient), and k is thermal conductivity in watts/cm/°C. From the junction to the case is a total heat flow qA, where A is the area. From the equation above,

$$qA_{JC} = Q = (A_{JC}k_{JC}/l_{JC}) \Delta T_{JC} \tag{10}$$

The quantity Ak/l is a thermal conductance in watts/°C. Since heat is conserved, we have from the junction to the ambient

$$Q = \varkappa_{JC} \Delta T_{JC} \quad \text{junction to case}$$
$$Q = \varkappa_{CS} \Delta T_{CS} \quad \text{case to sink}$$
$$Q = \varkappa_{SA} \Delta T_{SA} \quad \text{sink to ambient}$$

The total temperature drop ΔT_{JA} is

$$\Delta T_{JA} = \Delta T_{JC} + \Delta T_{CS} + \Delta T_{SA} \tag{11}$$
now
$$\Delta T_{JA}/Q = \Delta T_{JC}/Q + \Delta T_{CS}/Q + \Delta T_{SA}/Q \tag{12}$$
or
$$1/\varkappa_{JA} = 1/\varkappa_{JC} + 1/\varkappa_{CS} + 1/\varkappa_{SA} \tag{13}$$

It is seen to be more convenient to use the reciprocal of conductance, or resistance, viz.,

$$\theta_{JA} = \theta_{JC} + \theta_{CS} + \theta_{SA} \quad \text{°C/watt} \tag{14}$$

As an example, consider that a certain transistor has a θ_{JC} of 0.45°C/watt and the maximum junction temperature is 175°C. If the case temperature is 60°C, the maximum power dissipation is

$$P_{max} = Q_{max} = \frac{(175 - 60)°C}{0.45°C/\text{watt}} = 255 \text{ watts} \tag{15}$$

Figure 13 is a derating curve for the example. Such a curve is usually provided in data sheets and makes possible the calculation of safe power dissipation for any case temperature. The 255-watt point is plotted at 60°C for the top curve. The slope is then 0.45°C/watt. For 110°C case temperature, the permissible dissipation is 140 watts.

It is clear that the constant-dissipation condition requires the solution of two problems: (1) the determination of T_J for a given case temperature T_C to find θ_{JC} and (2) the choice of a heat radiator which can maintain the required case temperature under the given ambient conditions. The first problem is taken up in Sec. 3.2 and the second in Sec. 3.3.

3.2. Determination of Junction Temperature. Since the junction of a transistor is within the device structure, it is obvious that some indirect measurement is necessary.

The simplest way to infer junction temperature is to use the forward voltage drop of the base-collector junction carrying a small constant current as a thermometer. The relation between voltage drop and junction temperature can be calibrated by a series of equilibrium oven-temperature measurements. The voltage drop will vary by approximately 1.8 mv/°C. To determine junction temperature under a calculated collector dissipation, the collector current is switched off, and the forward measurement is made as quickly as possible.

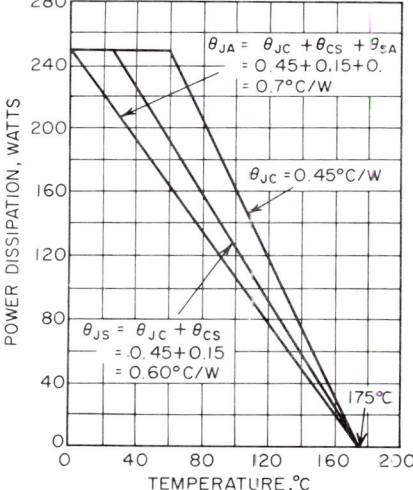

Fig. 13. Derating curve for the example of Sec. 3.1.

A much more sophisticated method has been devised by Nelson and Iwersen.[11] The temperature-rise measurement is accomplished by determining the common-emitter static-collector characteristics at known base-layer (i.e., junction) temperatures by means of a pulse technique which causes negligible junction heating above the controlled oven temperature. These curves are compared with the same V_{CE} versus I_C characteristics taken at equilibrium with a known header temperature and steady currents for which the power dissipation is known. These two sets for equal base currents intersect when h_{FE} (hence, the base-region temperature) is the same under pulsed and steady biasing. At the intersection, the base-layer temperature,

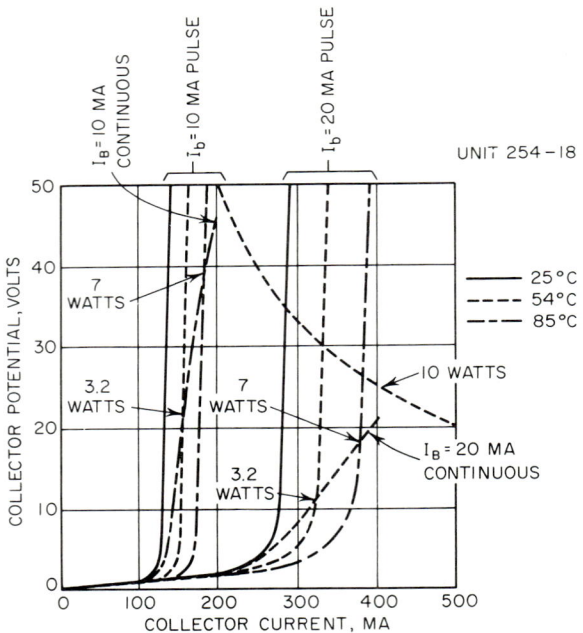

FIG. 14. V_{CE} versus I_C for pulsed and steady values.

header temperature, and power dissipation are known; hence, temperature rise of the junction above the case vs. dissipation can be determined. Figure 14 illustrates the method for a power transistor in which the intersection occurs at 85°C for 7 watts dissipation. The header was 33°C; hence the thermal resistance θ_{JC} is about 7.5°C/watt.

3.3. Selection of a Heat Sink. The central problem in the selection of a heat sink is to calculate the maximum permissible sink-to-ambient resistance θ_{SA} for the heat-flow condition.

For example, consider a type 2N1015 silicon transistor dissipating 40 watts and operating in an ambient temperature of 55°C. The data sheet for the 2N1015 shows 40 watts dissipation at a case temperature of 120°C. We have the case-to-ambient drop

$$(\Delta T)_{CA} = 120°C - 55°C = 65°C$$

---

[11] J. T. Nelson and J. E. Iwerson, Measurement of Internal Temperature Rise of Transistors, *Proc. IRE*, vol. 46, no. 6, pp. 1207–1208, June, 1958.

and the maximum permissible case-to-ambient resistance

$$\theta_{CA} = 65°C/40 \text{ watts} = 1.62°C/\text{watt}$$

The $\frac{5}{16}$-in.-24 stud on this unit will have $\theta_{CS} = 0.4°C/\text{watt}$, hence

$$\max \theta_{SA} = \max \theta_{CA} - \theta_{CS} = 1.62 - 0.4 = 1.22°C/\text{watt}$$

Various manufacturers of heat sinks give values of θ_{SA}. It is important to note the ambient conditions, that is, whether convected air or forced air. In the latter case the air velocity is needed. Other important factors are size, shape, material, surface finish and color, and orientation. It is not practical to cover all cases of heat transfer to fluids in this handbook, because of the complexity of the subject. Instead, some representative curves and tables will be given and a reference to an extensive treatise on the subject.[12]

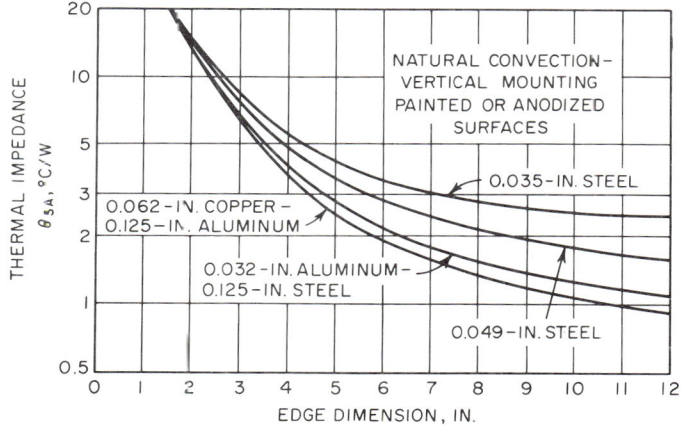

FIG. 15. Thermal resistance to ambient air of flat, square plates.

A common type of heat sink is a flat, square plate of aluminum, colored black. Figure 15 gives θ_{SA} vs. edge dimension for several types of square plate: steel, copper, and aluminum of various thicknesses. Natural convection and vertical mounting are specified. For the preceding example θ_{SA} of 1.2°C/watt is provided by a 9- by 9-in. aluminum plate $\frac{1}{8}$ in. thick.

Ralph Greenburg[13] has published a comprehensive study of the parameters of transistor heat sinks. We shall not reproduce his results here, because most of the θ_{SA} values are useful for units under 100 watts.

A simple calculation will show that θ_{SA} must have a low value (less than 1) for power devices over 100 watts. Before working an example, we shall discuss the contact between the transistor case and the heat sink. Usually, this is bolted with a silicone grease film in between. If electrical insulation is required, a mica washer about 3 mils thick is inserted. Table 4 shows some representative assemblies and the corresponding θ_{CS}.

[12] E. R. G. Eckert and R. M. Drake, Jr., "Heat and Mass Transfer," 2d ed., chap. 3, McGraw-Hill Book Company, New York, 1959.
[13] Greenburg, R., Factors Influencing Selection of Commercial Power Transistor Heat Sinks, *Solid State Design*, vol. 3, no. 7, pp. 25–31, July, 1962.

Table 4. Thermal Resistance Between Case and Sink θ_{CS} for Various Assemblies

Stud size	Recommended stud torque, in.-lb	Heat-sink thickness copper or aluminum, in.	θ_{CS} Applied directly to heat sink °C/watt	θ_{CS} With 0.003-in. mica washer, °C/watt
No. 10—32	15	0.125	1.0	5.5
$\frac{1}{4}$ in.—28	30	0.125	0.7	2.4
$\frac{5}{16}$ in.—24	50	0.125	0.4	2.0
$\frac{1}{2}$ in.—20	125	0.25	0.2	
$\frac{3}{4}$ in.—16	600	0.25	0.085	

The JEDEC data sheet of Sec. 2 lists for the 2N1833 the following:

Assume power dissipation 200 watts, max $T_C = 60°C$.
Derating factor 2.17 watts/°C or $\theta_{JC} = 1/2.17 = 0.46°C/watt$.
Assume T_A (ambient) $= 10°C$ $\theta_{CS} = 0.2°C/watt$ from Table 4.
The total thermal resistance from junction to ambient

$$\theta_{JA} = (T_J - T_A)/Q$$
$$= (175°C - 10°C)/200 \text{ watts}$$
$$= 0.87°C/watt$$

Now
$$\theta_{JA} = \theta_{JC} + \theta_{CS} + \theta_{SA}$$
$$0.87 = 0.46 + 0.2 + \theta_{SA}$$

hence
$$\theta_{SA} = 0.21°C/watt$$

This low value would require a good water-cooled heat sink, and obviously, no insulating washer could be used. It may be concluded that dissipation of powers over 200 watts becomes a difficult problem in heat-transfer design through the package and the heat sink.

A useful chart has been provided by J. R. Welling[14] to determine contact resistance between case and sink. It is given in Fig. 16. His definition of thermal resistance is slightly different from the conventional definition. His thermal resistance must be divided by the area (in square inches) to obtain the conventional θ. The area is the transistor-mounting base area neglecting nut and washer area.

3.4. Temperature–Power Relations in Dynamic Loading. Single Pulse. If step pulses of power are applied to a transistor junction, the peak power in a short interval of time may exceed the steady-state rating if the junction temperature stays below T_J max. Because the material between the junction and the case has a heat capacity, the step of power will cause T_J to rise in a exponential fashion. The relation is

$$T_J - T_C = \Delta T = P\theta_{JC}[1 - \exp(-t_1/\tau_{JC})] \qquad (16)$$

where P is the peak power of the applied pulse, t_1 is the duration of the pulse, and τ_{JC} is a thermal time constant. Problems of this and other types are solved in texts dealing with transient thermal effects.[15] The determination of τ_{JC} can be seen from consideration of dimensions. The heat capacity is

$$C_{JC} = \text{watt-sec}/°C \qquad (17)$$

[14] J. R. Welling, Determination of Transistor Case-to-Mounting Plate Thermal Resistance, *Elec. Design News*, June, 1962, pp. 146–150.

[15] H. Carslaw and J. C. Jaeger, "Conduction of Heat in Solids," 2d ed., Oxford University Press, Fair Lawn, N.J., 1959.

HIGH-POWER AMPLIFIERS   20-19

hence, τ_{JC} sec $= C_{JC}$ watt-sec/°C times θ_{JC} °C/watt. Typical numbers for a high-power transistor would be

$$\theta_{JC} = 0.45°C/\text{watt}$$
$$C_{JC} = 0.5 \text{ watt-sec}/°C$$
hence
$$\tau_{JC} = 225 \text{ msec}$$

In the case of a very long time constant compared with the pulse duration, i.e. $t_1 \ll \tau_{JC}$, we have

$$T \cong P\theta_{JC}(t_1/\tau_{JC}) \tag{18}$$

where the exponential expression of Eq. (16) has been approximated.

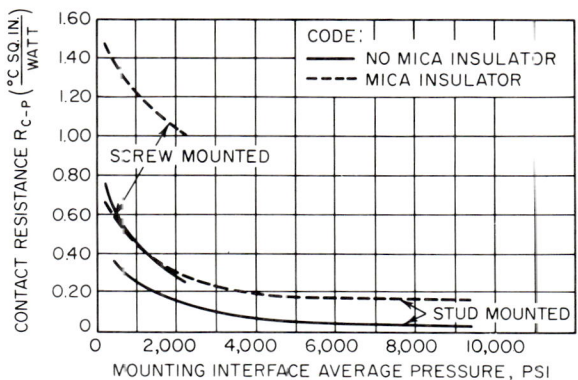

FIG. 16. Design curves for transistor mounting to a metal conduction sink. Mounting interface pressure calculated by the following empirical formula:

$$S = 5NT/A_b D$$

where S = mounting interface pressure, psi
N = number of screws
A_b = mounting interface area, sq in.
D = nominal screw diameter, in.
T = working torque per screw, in. lb

The constant 5 is based on dry screw threads. Nominal reduction factor when DC-4 grease is used is 40 per cent.

In the steady state ($t_1 \gg \tau_{JC}$), Eq. (16) becomes

$$\Delta T = P_0 \theta_{JC} \tag{19}$$

where P_0 is the steady-state power. Since $(\Delta T)_{max} = T_{J\,max} - T_{C\,max}$ in both the steady-state and pulsed cases, we may equate Eqs. (18) and (19) to obtain

$$P_{max} = (\tau_{JC}/t_1) P_{0\,max} \tag{20}$$

Thus the permissible peak power (of square-pulse form) which can be tolerated is inversely proportional to the pulse duration. The quantity

$$\tau_{JC}/t_1 \tag{21}$$

20-20 AMPLIFIER CIRCUITS

is a multiplying factor for the steady-state power dissipation to give permissible peak power.

Repetitive Pulses. If the junction is pulsed with the waveform shown in Fig. 17, the temperature rise is

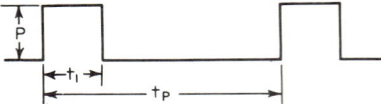

FIG. 17. Pulse width and period.

$$\Delta T = P\theta_{JC} \frac{1 - \exp(-t_1/\tau_{JC})}{1 - \exp(-t_p/\tau_{JC})} \quad (22)$$

The pulse repetition has a rate f, and the duty cycle is $t_1/t_p = t_1 f$.

Equation (22) may be rewritten

$$\frac{P}{\Delta T/\theta_{JC}} = \frac{1 - \exp(-t_p/\tau_{JC})}{1 - \exp(-t_1/\tau_{JC})} = \frac{P}{P_0} \quad (23)$$

Here the power multiplier n is indicated, viz.,

$$P = \frac{1 - \exp(-t_p/\tau_{JC})}{1 - \exp(-t_1/\tau_{JC})} P_0 = nP_0 \quad (24)$$

Figure 18 shows a plot of n vs. pulse width t_1 in milliseconds. This applies only to transistors with $\tau_{JC} = 225$ msec. Several interesting points are evident. For any

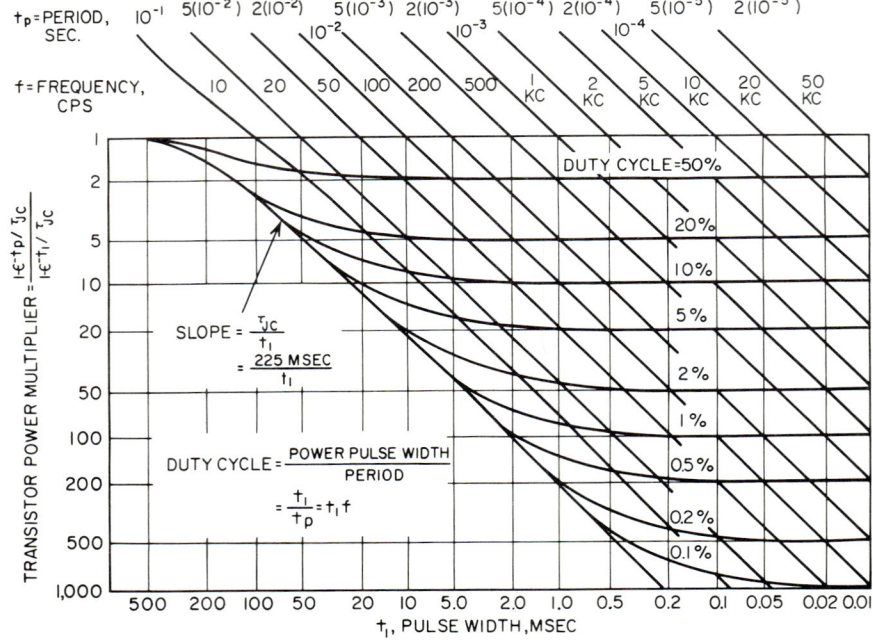

FIG. 18. Safe operating duty cycles for 30-amp transistors. Safe voltage and current limits must be observed. Square-wave input assumed.

finite duty cycle, the pulse width can be increased until the period t_p is three or more times larger than 225 msec. At this point, the pulse is relatively steady-state and the factor n is the term (21). Areas to the left of this line are therefore forbidden. Another fact is that for any given duty cycle the multiplying factor n becomes constant.

An example of the use of the plot will be given: Is a pulse width of 100 msec permissible with a power multiplication of 4? Find the intersection of the vertical line at 100 msec, and the horizontal line from $n = 4$. The point lies below the limit line and is not permissible. An n of 2 would be allowed.

Other frequencies of collector pulse are shown in addition to the zero frequency or steady state. For each line, t_p = duty cycle times t_1. Also, along each line nt_1 = const, or the energy content of the pulse is constant.

Finite Heat Sink. So far, only the ΔT has been calculated for various pulse conditions. The central problem as discussed in Sec. 3.3 is to determine a heat sink for the heat-flow condition which will keep the junction below the maximum allowable temperature. According to the foregoing discussion, the permissible peak power P is

$$P = \frac{T_{J\,max} - T_C}{\theta_{JC}} n \qquad (25)$$

where n is found from Fig. 17 for the desired pulse-width and duty-cycle conditions. The unspecified quantity here is the case temperature T_C. If there were no thermal resistance between the case and heat sink, then T_C would be the ambient temperature. This is the "infinite sink." In Sec. 3.3, the case temperature was determined for steady heat flow, finite θ_{CS} and θ_{SA}, and a given ambient. This discussion will provide formulas for the pulse heat-flow case. The basic problem is to find ΔT from case to ambient. This can be done by using

$$(\Delta T)_{CA} P \theta_{CA} \frac{1 - \exp(-t_1/\tau_S)}{1 - \exp(-t_p/\tau_S)} \qquad (26)$$

where τ_S is the time constant for the heat sink and we have combined $\theta_{CS} + \theta_{SA}$ to θ_{CA}. The peak power P is the same as for Eq. (22). The temperature difference is from case to ambient. Hence

$$T_C - T_A = (\Delta T)_{CA} \qquad (27)$$
Now
$$(\Delta T)_{CA} = P\theta_{CA}/n_S \qquad (28)$$

where n_S is analogous to n for the transistor [Eq. (24)]. The new formula for peak power is

$$P = \frac{T_J - T_A}{\theta_{JC}/n + \theta_{CA}/n_S} \qquad (29)$$

A set of curves similar to Fig. 18 for the heat sink are needed to find n_S.

If a steady power dissipation is superposed on the pulses, the effect is to increase the ambient temperature by $(\Delta T)_{SS} = \theta_{JA} P_{SS}$, where P_{SS} is the steady-state power. Thus we have

$$P = \frac{T_J - T_A - \theta_{JA} P_{SS}}{\theta_{JC}/n + \theta_{CA}/n_S} \qquad (30)$$

Actual Pulse Conditions. It has been emphasized that the preceding discussion has been for square-wave pulses on the junction. The actual pulse wave shape depends upon the load. For example, Fig. 19 shows the wave shapes for the case of a resistance load, and the turn-on and turn-off times are small compared with the on and off periods. The wave shape is triangular rather than square. In this case, calculation assuming a square wave is conservative because of the higher energy content of the pulse. If for some reason, a more accurate determination must be made, the exact solution of the thermal differential equation for the system must be obtained using the actual wave shape.

The figure also shows that the quantity t_p is one-half the period of the input waveform on the base-emitter terminals.

At higher frequencies (pulse-repetition rates), the rise and fall times tend to increase the duty cycle, lowering the permissible power. At the limit, near h_{fe} cutoff, the wave shape is altered until the unit will no longer switch.

20-22 AMPLIFIER CIRCUITS

The discussion has been carried out in terms of a switch, but the principles apply as well to class C power amplifiers if the square wave is an acceptable approximation to the real collector pulse waveform. In most cases it will yield a conservative system design. Other classes of audio amplifiers are designed not to exceed the steady-state power hyperbola.

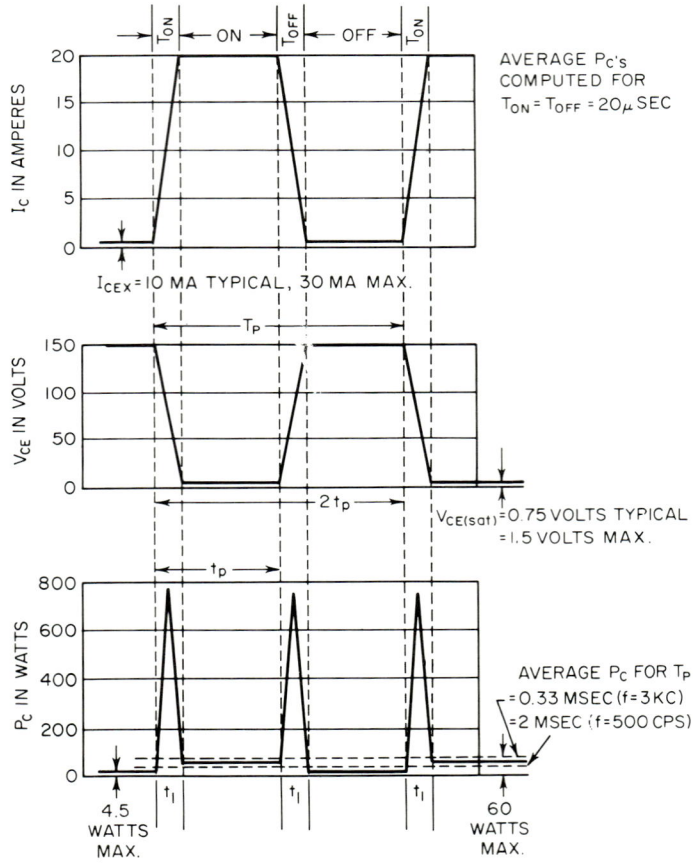

FIG. 19. Typical wave shapes for resistive load line.

Special Caution. The preceding considerations are for temperature rise *only*. If V_{CE} or I_C are exceeded before $T_{J\ max}$ is reached, the other ratings take precedence. In other words, *no rating may be exceeded*.

4. APPLICATIONS OF HIGH-POWER TRANSISTORS AT AUDIO FREQUENCIES

Power-transistor applications in the audio range have been reported up to 10 kw. Higher powers are being developed, but the work is classified at the time of this writing. Since the highest power dissipation available at this time is 250 watts it is obvious that parallel operation is necessary to meet such a requirement. We shall begin by discussing parallel and series operation.

4.1. Parallel Operation of Transistors. This discussion will be confined to frequencies below about 20 kc, but it will apply to any power transistor. Figure 20 shows the parallel connection with appropriate power supplies and resistors. The problem is to choose the resistors so that the collector currents of the various transistors equalize as much as possible.

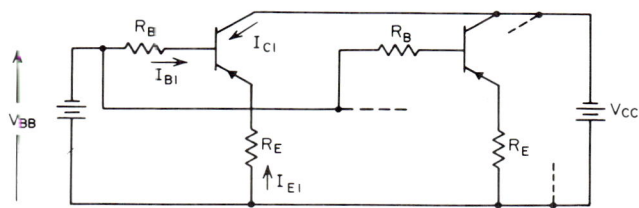

FIG. 20. Parallel connection of power transistors.

The basic equations are

$$V_{BB} = V_{BE} - R_E I_{E1} + R_B I_{B1} \tag{31}$$

Now, since $I_{C1} = h_{FE} I_{B1}$ and $-I_{E1} = I_{B1} + I_{C1}$, we have

$$V_{BB} = V_{BE} + I_{C1}[R_E(1 + 1/h_{FE}) + R_B/h_{FE}] \tag{32}$$

Since $1/h_{FE}$ is usually less than 5 per cent it is neglected. Equation (32) shows that an R_B can be found with the same effect as R_E through the value of h_{FE}. We shall consider $R_B = 0$ and proceed with

$$\begin{aligned} V_{BB} &= V_{BE_1} + I_{C_1} R_E \\ V_{BB} &= V_{BE_2} + I_{C_2} R_E \quad \text{etc.} \end{aligned} \tag{33}$$

The selection of R_E is illustrated in Fig. 21. This is a transconductance curve

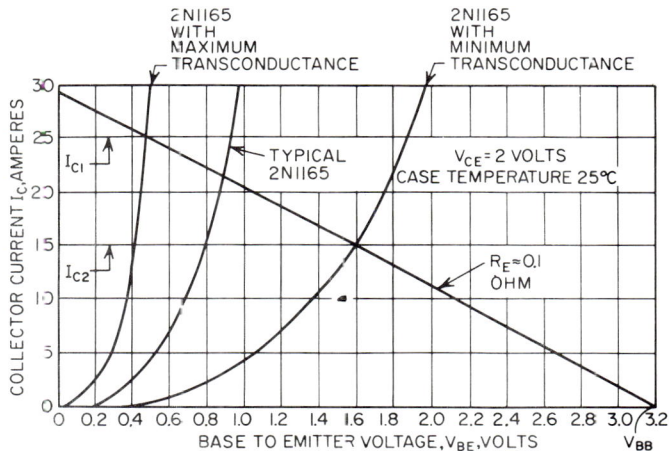

FIG. 21. Current division by emitter resistor.

(cf. Sec. 2.4) showing the maximum spread of I_C versus V_{BE} due to the allowable spread for the 2N1165 and the temperature-excursion requirement of the application.

The procedure would be to choose an acceptable current for the high-gain transistor. Either an acceptable V_{BB} (drive voltage) can be selected or a desired current for the low gain limit can be chosen. In the first case, a straight line is drawn between V_{BB} and the preassigned high-limit transistor current. The resistance of R_E is then determined by the slope of the line. In the latter case, the two selected current points determine a straight line. The slope is the required R_E and the intersection with $I_C = 0$ gives the necessary V_{BB}. In the example, $I_{C_1} = 25$ amp, $I_{C_2} = 15$ amp or a total of 40 amp. The $R_E = 0.1$ ohm and V_{BB} is 3.2 volts.

Current equalization is paid for by the losses in R_E. In this case the loss is

$$P_{loss} = I_{C_1}^2 R_E = 62.5 \text{ watts} \tag{34}$$

Stabilization with temperature change is, however, aided by R_E. This is because an increase in I_B (due to collector-base junction leakage current) is multiplied by h_{FE} and appears as an increasing voltage drop across R_E. This lowers V_{CE}, tending to

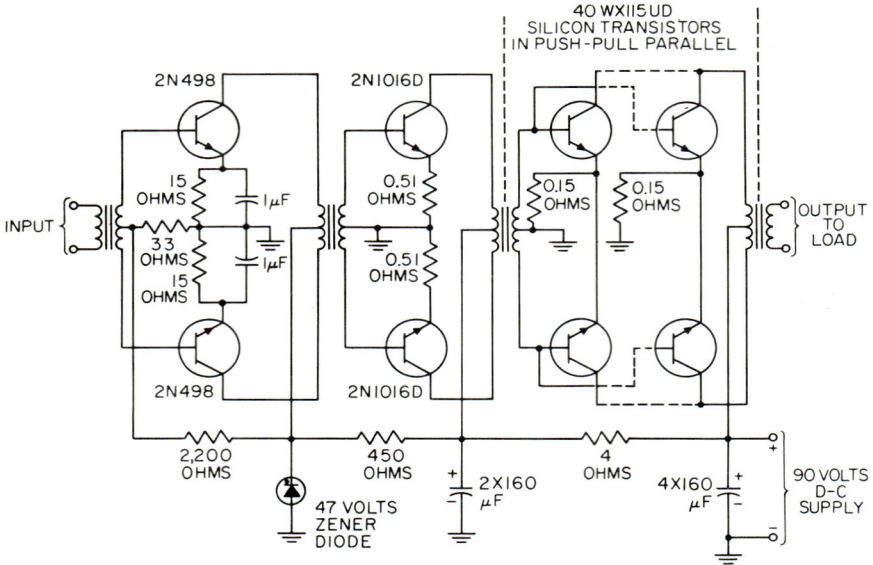

Fig. 22. Schematic of 10-kw sonar amplifier.

lower the leakage. Hence the resistor R_E has a negative-feedback effect with amplification h_{FE}. Using a base resistor instead of R_E would reduce the losses, but the negative stabilizing feedback would be absent.

4.2. A 10-kw Sonar Amplifier.[16] Various applications of sonic and ultrasonic waves are natural applications for high-power transistors. There is a need in military sonar equipments for high-powered transmitters that are reliable, efficient, and of minimum physical size. These requirements are fulfilled by the use of transistors in the design of high-power audio driver amplifiers. The present state of the art in power-transistor manufacture has made feasible the design of pulsed audio amplifiers with powers ranging up to 10 kw.

Pulsed sonar is a high-efficiency type of application for transistors because of the duty cycle of 10 per cent or less. Electron-tube amplifiers have substantial heater power on all the time regardless of the duty cycle. It turns out that transistor amplifiers work at about 50 per cent compared with less than 1 per cent for vacuum tubes.

[16] V. Vartanian, Ten KW Sonar Amplifier—Use of Transistors Reduces Size, Weight, *Undersea Technol.*, May/June, 1961.

HIGH-POWER AMPLIFIERS 20-25

Preceding the amplifier to be described was a 1-kw amplifier[16] weighing under 10 lb. This unit is in production and can be used in multiple owing to the modular construction.

The design criteria for the 10-kw amplifier were:

1. Pulsed power operation at 10,000 watts with pulse lengths of 250 msec and a 10 per cent duty cycle
2. Operating ambient temperature range of 0 to 55°C
3. Widest practical frequency range
4. Less than 1 cu ft volume and less than 40 lb

The circuit chosen was a class B push-pull amplifier, as shown in Fig. 22. This selection gives higher efficiency and power capability than a single-ended amplifier. It is obvious that transistors must be operated in parallel, and the selection was 40 WX115UD units. Load sharing was accomplished by using 0.15-ohm emitter resistors in the manner described in Sec. 4.1. Cooling was provided by a blower with a capacity of 150 cu ft/min. The maximum temperature rise on the transistor studs was 20°C above ambient. The upper frequency response was limited by the transistors to 12 kc. The lower limit was set by the transformer size and was designed to be 2 kc.

The power output and efficiency as a function of frequency are shown in Fig. 23. The curves are for 25°C, but little change is noted at the temperature extremes of 0 to 55°C. No temperature compensation over that afforded by the emitter resistors was necessary.

The load, being a piezoelectric crystal, is highly reactive. To minimize heating due to a capacitive load line, the load was made effectively resistive by tuning with toroidal inductors wound on molybdenum Permalloy powder cores. For pulse-shaping and power-supply considerations, refer to the original paper.

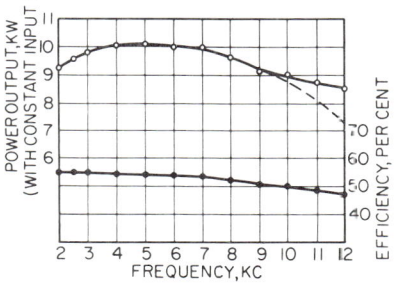

Fig. 23. Performance characteristics of 10-kw sonar amplifier.

Design Details.[17] The output stage was designed around the WX115UD (now designated 2N1812), which has the following data:

$I_{C\ max}$ = 30 amp d-c
V_{CEX} = 200 volts d-c, V_{BE} = −1.5 volts, T_J = 175°C
$I_{B\ max}$ = 10 amp d-c
V_{EB} = 15 volts maximum
θ_{JC} = 0.45°C/watt
τ_{JC} = 120 msec
Thermal capacity = 0.6 watt-sec/°C
$h_{FE\ min}$ = 10 at I_C = 10 amp
h_{fe} frequency cutoff $\cong$ 17 kc

The conservative supply voltage was chosen to be 90 volts d-c, which under load would be near 85 volts. The peak a-c swing on the collector is

$$V_{CE\ max} = 2V_{CC} - V_{min}$$

where V_{CC} is the collector power supply, and V_{min} is due to V_{CE} sat, the voltage drop across R_E (=0.15 ohm), and the voltage drop across transformer winding and wiring resistance.

[17] Personal communication from V. Vartanian.

20-26 AMPLIFIER CIRCUITS

The power output is

$$P = \frac{I_{max}(V_{CE\ max}/2)}{2}$$

Here $P = 10,000$ watts + losses, or $I_{max} \cong 300$ amp. If the peak current per transistor is selected at 15 amp, then 20 in parallel are required for 300 amp. Since the circuit is push-pull, there are 40 transistors.

The cooling requirements were calculated from the measured efficiency of 50 to 55 per cent. With 10 kw output the total losses are 9 to 10 kw. Approximately 1,200 watts are dissipated in emitter resistors, transformer, wiring, and driver stage; hence, the losses in the output transistors are 8,800 watts maximum. Average power dissipation per transistor is $8,800/40 = 220$ watts. In the steady state the junction temperature rise is 0.45°C/watt times 220 watts or 100°C.

For an "on" pulse of 0.25 sec

$$\Delta T|_{0.25sec} = 100°C\ (1 - \exp 0.25/0.12)$$
$$= 85°C$$

Hence the case temperature should not exceed $T_C = 150°C - 85 = 65°C$. This is a conservative figure. T_J can be 175°C for silicon transistors. A heat sink can be calculated from the discussion of Sec. 3.3.

The driver stage can be designed similarly by noting that the h_{FE} is about 10; hence an I_b peak of about $\frac{300}{10} = 30$ amp is required. The transistor requires a V_{BE} of about 2.5 volts d-c for the 15-amp peak I_C. Another 1.5 volts d-c drop across the emitter resistor gives a total of 4 volts d-c. Allow another 0.7 volt for transformer resistance, and the driver power for one parallel set is

$$P_{drive} = 30 \times 0.707\ \text{amp d-c times } 4.7\ \text{volts d-c}$$
$$\cong 100\ \text{watts}$$

The total drive required is approximately 200 watts.

4.3. A 200-watt Audio Power Amplifier.[18] In addition to sonar applications, high-power transistor audio amplifiers are also used for a wide-range low-distortion source of a-c power for laboratory applications. Electrodynamic vibration exciters can be powered this way, and stadium public-address systems are a good application.

Figure 24 is an oscillator–driver–power-amplifier combination for 200 watts with a design center at 25 kc. A single-ended driver is used in this case with a tapped input transformer. The parallel connection is similar to that described in Sec. 4.2. This designer elected to add resistance in the base leads also. Current equalization at low power loss is achieved, but the emitter resistors are necessary for temperature stabilization.

5. HIGH-POWER HIGH-FREQUENCY TRANSISTOR AMPLIFIERS

High-power high-frequency transistor amplifiers are designed according to the principles of Sec. 4 and Chap. 17, Sec. 15.3, with some additional considerations stemming from the high frequency. These are described in the two examples which follow.

5.1. A Class C 100-watt 10-Mc Power Amplifier.[19] Radio-frequency power amplifiers are normally operated in class B or class C to obtain high circuit efficiency. Figure 25 shows typical class C operating conditions. This type of operation requires a tank circuit with a loaded Q of about 5.

The circuit shown in Fig. 26 illustrates the kinds of considerations which are necessary to generate 100 watts of r-f power output at 10 Mc. The 2N1899 transistor is employed in a common-emitter circuit with series-tuned input and output.

[18] *Application Note*, Delco Radio, January, 1963.
[19] R. W. Lewinski, Class C 100-watt 10-megacycle Power Amplifier, *PSI Application Notes* 6A.

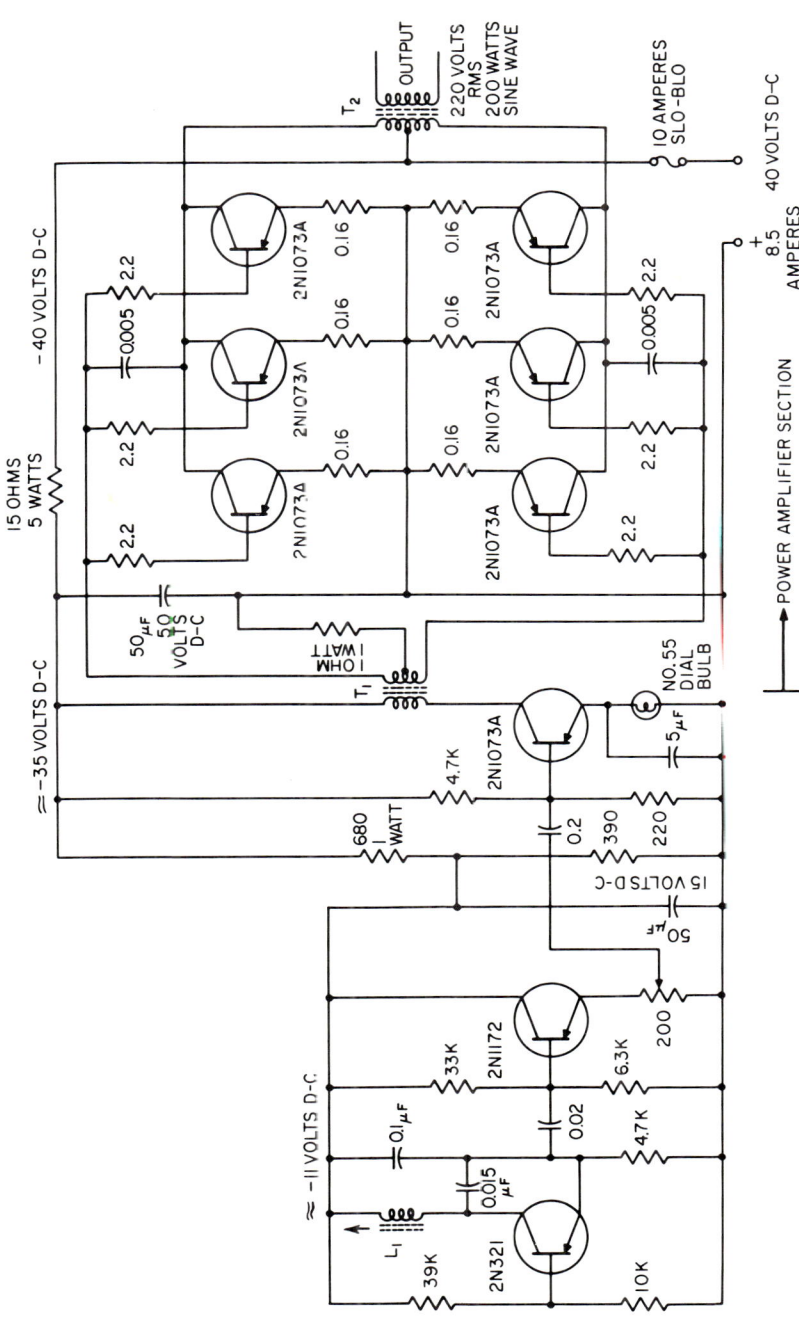

FIG. 24. Oscillator and power amplifier for 200 watts (resistive load) and 25 kc. L_1 4 mh, 600 t no. 32, ⅞ in. long on ½-in. slug-tuned coil form. T_1 primary, 51t no. 28; secondary 12t no. 20; tap at 6t. Core Allen Bradley E1102 C142A, W03 material. T_2 primary, 22t no. 14, tap at 11t; secondary 95t no. 20. Core Allen Bradley E2750 E179A, W07 material. Mount 2N1073A transistors on 7270725 heat sinks. Resistors are ½ watt unless specified otherwise. Capacitance specified in microfarads. Do not allow *heat-sink* temperature to exceed 65°C.

20-28   AMPLIFIER CIRCUITS

The collector of the transistor is grounded to achieve the lowest possible thermal resistance to the chassis. Output circuits are placed on one side of the chassis and input circuits on the other to avoid feedback coupling. Inductive elements are wound on ferrite cores to reduce the size and to minimize magnetic coupling. Critical cores are oriented 90° with respect to each other.

The need to form a low-dissipation a-c path around the 0.5-ohm emitter resistor presents an interesting components problem. To provide 0.05 ohm at 10 Mc requires a capacitance of 0.32 μf. A ceramic-disk capacitor of this size resonates at about 2 Mc, showing that there is too much parasitic inductance for this capacitor to be useful. A 0.01-μf disk capacitor resonates at 16 Mc. This design compromised between low impedance and numbers of capacitors in parallel by choosing 20 to provide 0.08 ohm.

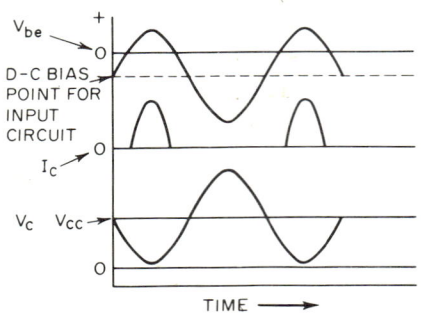

Fig. 25. Waveforms in class C operation.

Capacitors are mounted on each side of parallel copper strips to reduce lead length; further, flat strips are used to reduce skin-effect inductance. A noninductive resistor

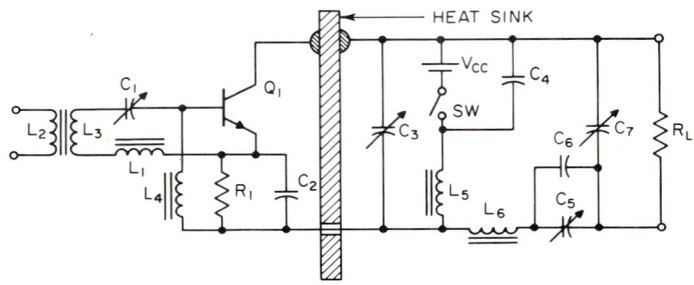

Fig. 26. 10-Mc class C 100-watt amplifier.

List of parts:

Q_1	2N1899
C_1, C_5, C_7	170–780 pf, El Menco trimmers, type 469
C_3	25–280 pf, El Menco trimmer, type 464
C_2	20 capacitors, 0.01 μf, 25 volts, Erie, disk ceramic in parallel
C_4	6 capacitors, 0.01 μf, 25 volts, Erie, disk ceramic in parallel
C_6	300 pf, El Menco mica capacitor
L_1*	0.8 μh, Q = 140, 5 turns no. 18, CF114-Q2, General Ceramics
$T_1: L_2$	20 turns no. 20, CF111-Q2, General Ceramics
L_3	5 turns no. 18, CF111-Q2, General Ceramics
L_4	3.5 μh, 20 turns no. 20, CF111-Q3, General Ceramics
L_5	2 μh, Q = 120, 8.5 turns no. 16, CF111-Q2, General Ceramics
L_6	0.7 μh, Q = 240, 3 turns no. 14, 3 cores, CF111-Q2, General Ceramics
N	$N_1/N_2 = 4$
R_1	0.3 ohms (eight × 3.9 ohm, 5%, 1 watt, carbon resistors in parallel)
R_L	Coaxial resistor, 51.5 ohms, 500 watts

* Preferred core: CF112-Q2.

of 0.5 ohm is required to be attached so that the capacitors are not overheated. This is provided by eight carbon resistors of 3.6 ohms each in parallel. Resistors may be placed at the end of the string, because little r-f current flows through them.

HIGH-POWER AMPLIFIERS   20-29

The design procedure we shall use is as follows:

1. Given the collector supply voltage V_{CC} and the load power required, determine the series equivalent load resistance.
2. Given the actual load resistance find the value of C_7.
3. Values of L_6 and $C_5 + C_6$ are calculated for the largest feasible Q of the resonant circuit.
4. The parallel combination L_5 and C_3 are calculated to give a resonant resistance 10 times the load.

Determination of Load Resistance. The peak-to-peak voltage across the tank circuit is approximately $2V_{CC}$, where V_{CC} is the collector supply voltage. To calculate the load resistance R_L', the peak-to-peak voltage is expressed in terms of the rms voltage across R_L', that is,

$$P_L = V_L^2/R_L' \tag{35}$$

Now the rms voltage is

$$2V_{CC}/2\sqrt{2} = 0.707 V_{CC}$$
hence
$$P_L = V_{CC}^2/2R_L' \tag{36}$$

For the amplifier of Fig. 27,

$$R_L' = (50 \text{ volts})^2/2 \times 100 \text{ watts}$$
$$= 12.5 \text{ ohms}$$

Determination of Capacitor C_7. The actual load R_L in Fig. 27 is a coaxial resistor of 50 ohms. Capacitor C_7 transfers this to the equivalent series resistance R_L' required to give the desired output power.

A simple RC circuit transformation gives

$$C_7 = (R_L/R_L' - 1)^{1/2}/\omega R_L \tag{37}$$

The calculation is

$$C_7 = \frac{(50 \text{ ohms}/12.5 \text{ ohms} - 1)^{1/2}}{2 \times 3.14 \times 10 \times 10^6 \times 50 \text{ ohms}}$$

$$C_7 = 552 \text{ pf}$$

Determination of L_6 and $C_5 + C_6$. Since the Q value of the resonant circuit is $\omega L_6/R_L'$, we determine L_6 from

$$L_6 = QR_L'/\omega \tag{38}$$

This inductance resonates with $C_5 + C_6$ in series with the equivalent capacitance C_7'. This composite capacitance C_X is computed from

$$C_X = C_7'(C_5 + C_6)/(C_5 + C_6 + C_7') \tag{39}$$

This can be rearranged to give

$$C_5 + C_6 = C_X C_7'/(C_7' - C_X) \tag{40}$$

The value of C_X is determined from

$$\omega C_X = 1/\omega L_6 \tag{41}$$

The key to the solution is the choice of the Q value. Experimental work showed that the highest practical Q was 3.2. High values subjected the available components to excessive voltages. Thus,

and
$$L_6 = 3.2 \times 12.5 \text{ ohms} = 40 \text{ ohms}$$
$$L_6 = 40/6.28 \times 10^7 = 0.64 \text{ }\mu\text{h}$$

Having the value of L_6, we find
$$C_X = 1/\omega(\omega L_6) = 1/(6.28 \times 10^7) \times 40 \text{ ohms}$$
$$C_X = 398 \text{ pf}$$

The value of C_7' is computed from the expression
$$C_7' = \frac{1 + (\omega C_7 R_1)^2}{R_L \omega(\omega C_7 R_L)} \qquad (42)$$
$$C_7' = 738 \text{ pf}$$

Finally
$$C_5 + C_6 = \frac{(398 \text{ pf}) \times (738 \text{ pf})}{738 - 398}$$
$$C_5 + C_6 = 865 \text{ pf}$$

Parallel Combination L_5 and C_3. We assume
$$\omega L_5 = 10 R_L' = 125 \text{ ohms}$$
Then
$$L_5 = 125/6.28 \times 10^7 \cong 2 \text{ }\mu\text{h}$$
Also
$$\omega C_3 = \tfrac{1}{125} \text{ ohms}^{-1}$$
$$C_3 \cong 128 \text{ pf}$$

5.2. A Class C 100-watt 3-Mc Power Amplifier.[20] This discussion of Fig. 27 supplements the discussion of Sec. 5.1 by showing a different form of output coupling, and an illustration of operation of transistors in parallel.

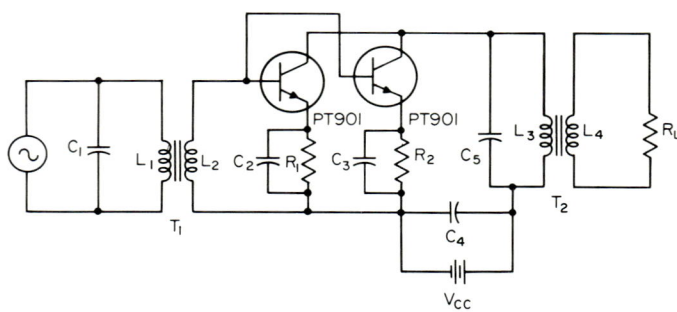

FIG. 27. Class C 100-watt 3-Mc amplifier.

Components list and operating specifications:

T_1 wound on ferrite core (OD = $\tfrac{13}{32}$ in.; ID = $\tfrac{15}{64}$ in.)
$L_1 = 1.7$ μh; 10 turns no. 14 wire $\qquad L_2 = 0.20$ μh; 3 turns no. 14 wire
T_2 wound on CTC PL.S5 coil form with red tuning slug
$L_3 = 0.15$ μh; 4 turns no. 14 wire $\qquad L_4 = 0.95$ μh; 11 turns no. 18 wire
$C_1 = 0.001$ μf $\qquad\qquad\qquad\qquad\qquad C_2 = C_3 = 0.1$ μf
$C_4 = 0.3$ μf $\qquad\qquad\qquad\qquad\qquad C_5 = 0.023$ μf
$R_1 = R_2 = 1.3$ ohms (two 2.7 ohm 1-watt resistors in parallel)
$R_L = 50$ ohms
$V_{CC} = 47$ volts $\qquad\qquad\qquad\qquad f = 3.3$ Mc
$G_P = 16$ db $\qquad\qquad\qquad\qquad\quad P_{in} = 160$ watts d-c
$I_C = 3.4$ amp d-c $\qquad\qquad\qquad\quad P_{out} = 100$ watts a-c
Collector circuit efficiency = 63 per cent

According to the components list transformer T_2 has a turns ratio of 2.75. The 50-ohm load resistance is then reflected into the collector circuit as 6.6 ohms. The impedance of $1/C_5$ is 21 ohms at 3.3 Mc. This implies a Q value of 3.2, which is the value used in the previous example.

[20] K. V. Ramanathan, Class C 100-watt 3-megacycle Power Amplifier, *PSI Application Note* 9A.

HIGH-POWER AMPLIFIERS 20–31

The values of R_1 and R_2 are selected to divide the load as evenly as possible following the discussion of Sec. 4.1.

5.3. A Switching-type Linear High-power Amplifier.[21] If a power transistor is operated in the switching mode, the power dissipation is low. For example, the PSI PT-901 has a saturation resistance of about 0.2 ohm. For an average of about 4 amp the loss is 8 watts. It is possible to control about 180 watts; hence the high efficiency of approximately 95 per cent.

The circuit of Fig. 28 shows an arrangement of two PT-901 transistors with a sine-wave base-drive yielding a sine-wave output at point Y of 50 kc with a square wave (switching mode) at point X. This mode of operation restricts the output to the frequency of the LC circuit.

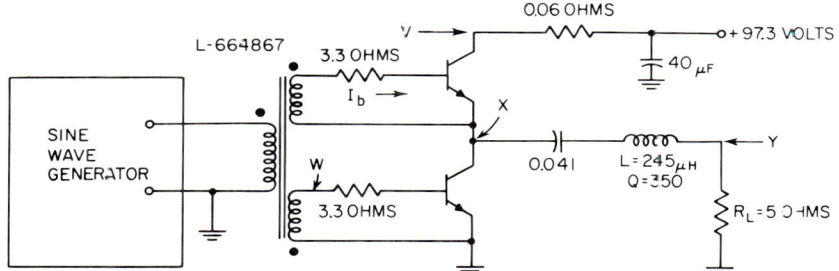

FIG. 28. Schematic diagram of final amplifier using sine-wave driving source.

Sample Design Calculation. If the collector supply voltage is 97.3 volts, and I_C is 11 amp peak, the load resistance R_L is calculated as follows:

Average square-wave voltage = $9.73/2$ = 48.6 volts
Peak voltage of fundamental component = 48.6×1.27 = 61.8 volts

Since I_C peak is 11 amp we have 11 amp = 61.8 volts/$(R_L + 0.26$ ohm$)$, where the saturation resistance is 0.2 ohms and a filter resistor of 0.06 ohms has been added. Hence $R_L = 5.2$ ohms.

Experiment shows that the current through the collector-emitter circuit is a sine wave while the voltage at X is a square wave. The average current through one transistor is then 11 amp/π = 3.5 amp.

The total power input to the circuit is 48.6 volts times 3.5 amp times 2 = 340 watts. The output power is $(I_C \text{ rms})^2 R_L$, or $(0.707 \times 11 \text{ amp})^2 \times 5.2$ = 314 watts. The overall efficiency is 314 watts/340 watts = 92.3 per cent.

The input drive power is calculated from a measured value of I_b = 1 amp peak, and a value of V_{be} of 1.1 volts from the data sheet. Peak power is then 1.1 volts times 1 amp = 1.1 watts. The peak power dissipated in the resistors is $(1 \text{ amp})^2$ times 3.3 ohms = 3.3 watts. The total peak power is 4.4 watts. For sine-wave drive the average power is 2.2 watts.

ACKNOWLEDGMENTS

The following organizations contributed data sheets and helpful information:

1. The Bendix Corporation, Holmdel, N.J.
2. Delco Radio Division, General Motors Corp., Kokomo, Ind.
3. The General Electric Company, Syracuse, N.Y.
4. Minneapolis-Honeywell Regulator Co., Riviera Beach, Fla.
5. Motorola Incorporated, Phoenix, Ariz.

[21] Personal communication from M. I. Jacob, Electronics Division, Westinghouse Electric Corp., Baltimore, Md.

6. Pacific Semiconductors Inc., Lawndale, Calif.
7. Radio Corporation of America, Somerville, N.J.
8. Raytheon Company, Portsmouth, R.I.
9. Silicon Transistor Corporation, Long Island, N.Y.
10. Sylvania Electric Products Inc., Woburn, Mass.
11. Westinghouse Electric Corp., Youngwood, Pa.

The author wishes to express thanks to Andrew Kruper, Robert Murray, Jr., and Al Mulica of Westinghouse, Youngwood, for contributions of data.

BIBLIOGRAPHY

Clark, M. A.: Power Transistors, *Proc. IRE*, vol. 46, no. 6, pp. 1185–1204, June, 1958.
Hunter, L. P.: "Handbook of Semiconductor Electronics," McGraw-Hill Book Company, New York, 1962.
"Motorola Power Transistor Handbook," 1961.
New, T. C.: Advances in Alloyed Silicon Power Transistors, *Commun. Electron.*, September, 1962, pp. 279–284.

6. HIGH-POWER AMPLIFIERS UTILIZING SILICON CONTROLLED RECTIFIERS[22]

6.1. Introduction. The SCR is a switching device, and conventional units are presently available which will switch powers ranging from milliwatts to kilowatts in

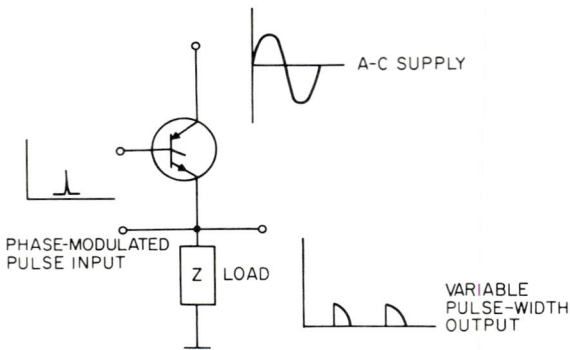

Fig. 29. Phase-controlled (constant-frequency) SCR amplifier.

microseconds at voltages in excess of 800 volts and at currents in hundreds of amperes. The device may be switched on with short low-voltage low-current pulses, leading to the practical realization of astronomical power gains for a single unit.

Being a switching device, with a low value of forward voltage drop while in the "on" state, the SCR is inherently efficient. Efficiencies in excess of 85 per cent are very common in circuits utilizing SCR's as amplifiers. Efficiencies in excess of 95 per cent are not uncommon.

The limitations of SCR's in amplifier circuits are the maximum switching speed, and the effects of voltage and current waveforms generated as the result of the use of the device. The maximum practical operating frequency for a single SCR is on the order of 5 kc. Above this frequency, the circuit complexity increases in order to provide protection for the device from voltage and current wavefronts. In addition, the power lost in switching rapidly becomes very appreciable at frequencies above this value. By using a number of SCR's operating sequentially, however, it is practical to operate efficiently in the hundred-kilocycle range.

[22] See "Silicon Controlled Rectifier Manual," 2d ed., General Electric Company, Rectifier Components Dept., Auburn, N.Y., 1961.

HIGH-POWER AMPLIFIERS   20-33

Rapidly changing currents and voltages, common at high frequencies, not only reduce the time available to the device to turn off but may cause overheating and under some conditions may generate spurious firing. Two basic types of SCR amplifiers exist. One is the constant-frequency variable-pulse-width amplifier, and the other is generally the chopper or inverter amplifier which produces variable frequency pulses of constant width. The phase-controlled amplifier, which is of the first type, is illustrated in Fig. 29. The input to the SCR gate is a phase-modulated pulse referenced to the frequency of the anode supply. The output of this particular system is a chopped half-sine wave, as shown. By varying the phase angle of the triggering signal, varying fractions of the half-sine wave will be applied to the load.

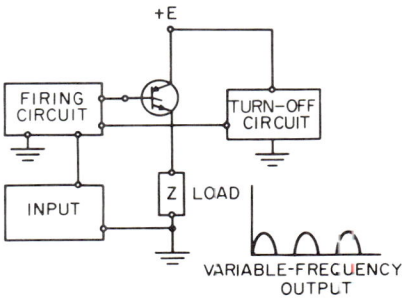

Figure 30 illustrates the latter type of SCR amplifier. The trigger circuit turns on the SCR. At a constant time later, the SCR is turned off by the turn-off circuitry. By varying the frequency of the firing signal, the power output to the load is varied.

FIG. 30. Inverter-type (constant-pulse-width) SCR amplifier.

These, then, are the two basic types of SCR amplifier. One is a form of pulse-width modulation accomplished by phase modulation of the gating signal, and the other is frequency modulation.

In summary, it becomes practical to use SCR's as amplifiers in cases where high voltages and high powers are being switched at relatively low frequencies. The two

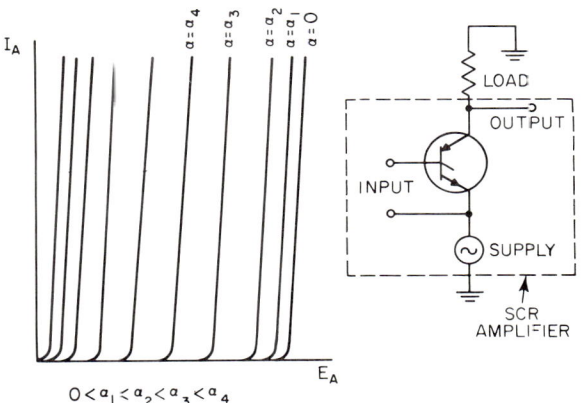

FIG. 31. Typical SCR amplifier characteristics.

basic types of SCR amplifier are variable-pulse-width constant-frequency and variable-frequency constant-pulse-width systems.

6.2. Typical Transfer Characteristics. If an SCR amplifier of either type were to have its characteristics plotted in a manner similar to that in which vacuum-tube characteristics are plotted, the result would be a set of curves very similar to those obtained when vacuum-tube characteristics are plotted. That is, the amplifier output would be very nearly a constant voltage for any particular input signal, as is shown in Fig. 31. The plate resistance of the tube would correspond to the power-supply impedance and the effective SCR impedance in series.

The major differences between the two are that the SCR input parameter is not a voltage but some characteristic of a series of pulses; that the "plate resistance" of any such SCR system can be made extremely small by using a very stiff a-c supply; and that the SCR is very much frequency-limited when compared with the vacuum tube.

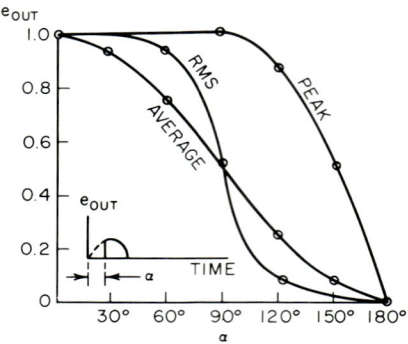

FIG. 32. Transfer characteristics of a sine-wave phase-control system.

Peak output = 1 for $\alpha \leq 90°$
= $\sin \alpha$ for $90° \leq \alpha \leq 180°$
Average output = $1 + \cos \alpha$ $0 \leq \alpha \leq 180°$
rms output = $\sqrt{1 - \alpha/\pi + \sin 2\alpha/2\pi}$
$0° \leq \alpha \leq 180°$

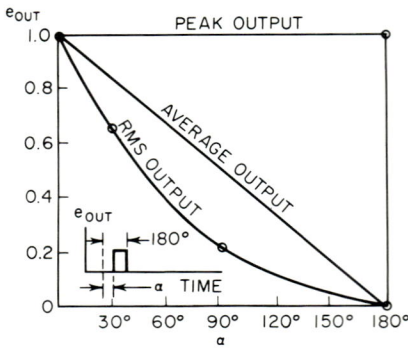

FIG. 33. Transfer characteristics of a rectangular-wave phase control system.

Peak output = 1 $0° < \alpha < 180°$
Average output = $1 - \alpha/180$
$0° \leq \alpha \leq 180°$
rms output = $\sqrt{1 - \alpha/180}$
$0° \leq \alpha \leq 180°$

In Fig. 31 the parameter α would be the delay angle in a phase-control system. In a frequency-variable system, α would be related to the triggering frequency. In either case, α may be generated as a function of some current or voltage; so a gain may be

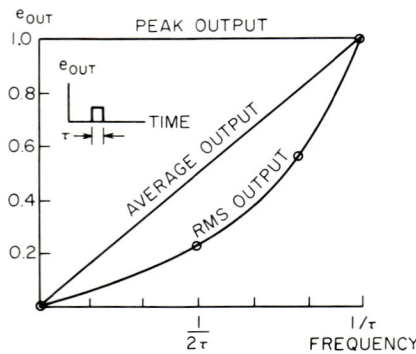

FIG. 34. Transfer characteristics of a variable-frequency SCR system.

Peak output = 1 $0 < f \leq 1/\tau$
Average output = $f\tau$ $0 \leq f \leq 1/\tau$
rms output = $\sqrt{f\tau}$ $0 \leq f \leq 1/\tau$

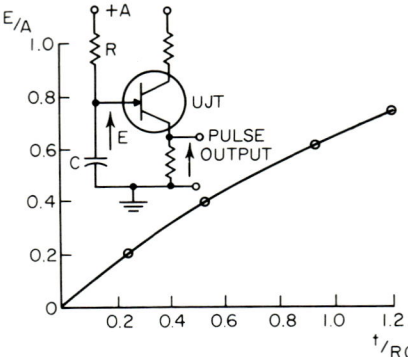

FIG. 35. Transfer characteristics of a unijunction firing circuit. E is unijunction firing potential. t is time beween pulse outputs.

established for most trigger circuit–SCR combinations in a manner parallel to that used in the case of the vacuum tube.

It is more general, however, to present the curves with α as either a phase angle or a frequency. Then, the various trigger circuits may be presented with phases or fre-

HIGH-POWER AMPLIFIERS 20-35

quencies as functions of voltages or currents, and it will be a relatively simple matter to match a trigger circuit to an output stage.

If the SCR amplifier were plotted exactly as a tube device is plotted, one fact would immediately be evident. This is that the vertical portion of the curves, those indicating current flow, would be very nearly perfectly vertical for any particular SCR. The current axis is quite useless, therefore, and we may as well simplify the circuit design by simply plotting voltage output at the knee of the curves vs. the parameter α. A consideration of the waveform of the output of the amplifier, which is a pulse train of some form, makes it evident that plots not only of average output voltage but also of rms and peak outputs will be useful in covering all situations.

Figure 32 illustrates the curves for a phase-control type of variable-pulse-width system where the supply voltage is a sine wave or a series of half-sine-wave pulses (full-wave-rectified sine wave).

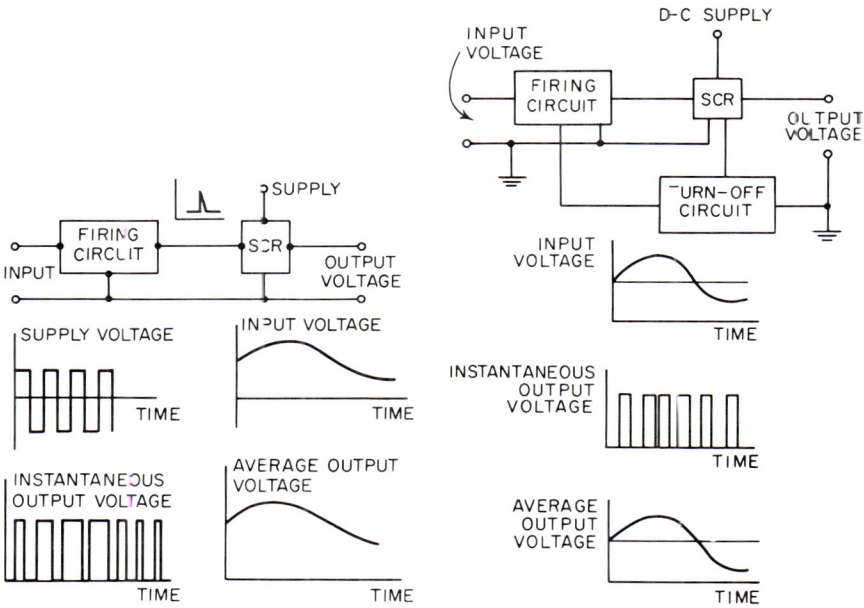

FIG. 36. The phase-control system.   FIG. 37. A frequency-modulated amplifier.

Figure 33 illustrates the curves for a variable-pulse-width system in which the output is a series of rectangular pulses of constant frequency.

Figure 34 is the set of curves for a variable-frequency system in which the pulses are assumed to be rectangular.

The unijunction transistor is an example of a firing circuit operating from a voltage input. The input voltage charges a capacitor through a resistor, and the capacitor suddenly discharges through the unijunction when its voltage reaches a predetermined value. The transfer characteristic of this type of firing circuit is shown in Fig. 35.

6.3. Inherent Delays and Phase Shifts. In phase-controlled (variable-pulse-width constant-frequency) systems, the output of the amplifier will be a voltage whose value will be dependent on the input voltage, as is illustrated in Fig. 36.

Exactly the same thing can be accomplished in a variable-frequency constant-pulse-width system, as is illustrated in Fig. 37.

20-36 AMPLIFIER CIRCUITS

Being a switching device inherently, it is generally not possible for the SCR itself to introduce significant delays in a system at the low frequencies at which it must operate. Instead, any significant delays or phase shifts will be generated in the firing circuit. Take, for example, the firing circuit illustrated in Fig. 38. The RC network will be one of the main phase shifts present in the entire amplifying system.

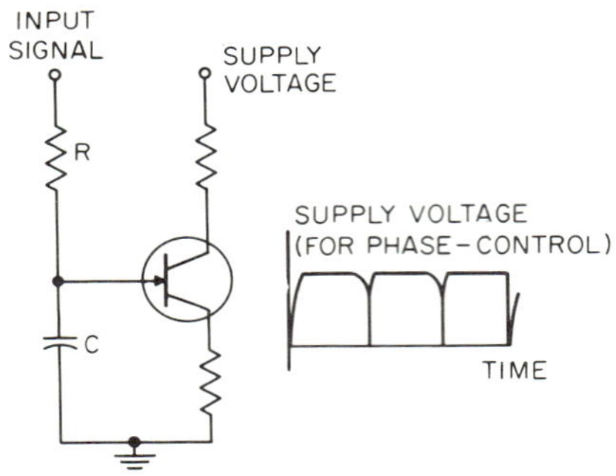

This trigger circuit may be used for either basic amplifier type. By synchronizing it with the power-supply frequency as shown, it becomes a phase-control trigger source. If the supply is made d-c, the trigger pulses will be frequency-variable with the input signal level.

It is interesting to note that phase control (pulse-width modulation) in the case of the SCR amplifier is very similar to amplitude modulation. If the input signal is close in frequency to the constant-frequency signal being fed to the SCR, it can easily be seen that the output of the system will be the power-supply frequency, amplitude-modulated with the difference frequency of the power supply and the input.

Fig. 38. Typical firing circuit.

For very low frequencies of input signal (relative to the power-supply frequency) the output will be a reasonable reproduction (with proper filtering). For very high

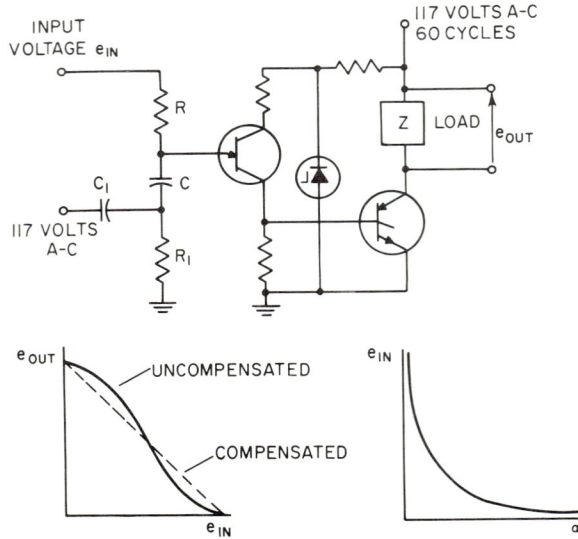

Fig. 39. Complete SCR amplifier system.

frequencies of input signal relative to the supply frequency, the output signal will be proportional to the average input voltage level.

For example, if the power-supply frequency were 400 cps, the range of reasonably linear amplification (after filtering) would be from direct current to about 40 cps.

In summary, inherent delays and phase shifts in SCR amplifier systems are generally introduced by the SCR firing circuit. The firing-circuit time constants in phase-

HIGH-POWER AMPLIFIERS

controlled systems are, for the most part, dictated by the power-supply frequency. In frequency-modulated systems, firing-circuit characteristics are dictated by the nominal operating frequency of the system.

6.4. Improving Overall Linearity. Nonlinearities in an SCR amplifier system may be inherently present in either the power section, the trigger section, or both. The commonest example of a nonlinear power system is that of the very popular phase control of a half-sine-wave pulse train. The output voltage vs. phase angle (transfer function of the SCR) is a cosine curve (Fig. 32). It is quite linear over a limited range of operation. However, if the circuit is intended to be used to its utmost power-handling capability, and linearity is a necessity, the linear range of the SCR may be extended with relative simplicity. It would be necessary simply to alter the response of the trigger circuit to compensate for the nonlinearity of the power circuit.

As an example of this, consider the amplifier of Fig. 39. If we assume, for the moment, that the trigger delivers a pulse to the SCR, and that a plot of the phase angle α of this pulse vs. the input voltage is a straight line, then what would need to be done to extend the linear range of the amplifier would be to adjust the sensitivity with a cosine function. This can be done by adding an RC network, such as R_1C_1 to the firing circuit. This adds a cosine function to the emitter signal in the proper sense, achieving the desired result.

We assumed, in the above example, that the trigger circuit was initially linear. This is generally not the case, however. The trigger circuit illustrated in Fig. 39, for instance, will have a hyperbolic response to input voltage, as shown. Compensation for this can take the form of making the value of R in Fig. 39 small enough that the voltage across the capacitor becomes an increasing exponential rather than a ramp.

6.5. Open-loop Stability. It is possible for unstable conditions to exist in an open-loop SCR amplifier. One very common unstable condition arises when the amplifier is driving a load which presents a back emf to the SCR. Two such situations are depicted in Fig. 40. The first is a SCR driving a capacitive type of load.

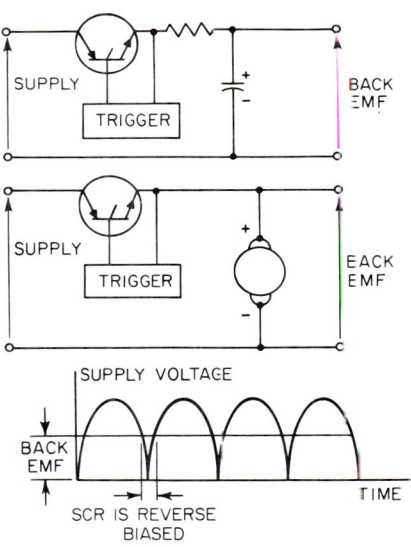

Fig. 40. Potentially unstable load conditions.

The second is a shunt-wound motor load. Both these loads present a voltage to the SCR which will back-bias it while the a-c supply is still positive. It is possible, therefore, for a trigger pulse to be fed to the SCR when it is back-biased. The SCR, of course, will not fire until a later pulse is supplied. The output voltage will drop because the SCR was not fired early enough in the cycle, making it possible for the next early pulse to fire the SCR. The voltage at the output will be forced to vary at a relatively low frequency, therefore, simply because of the type of load. A simple cure for this problem is shown in Fig. 41. The additional rectifier and resistor will allow the SCR to be controlled over nearly the entire 180°, which is possible with a simple resistive load, and this control angle will be independent of any back emf at the load.

An alternate solution is to make the trigger signal wide enough to be certain that it will be present coincidently with the SCR being forward-biased. A magnetic firing circuit will accomplish this very nicely.

Another possibly unstable condition can occur when the output is a-c feeding an inductive load, as is shown in Fig. 42. If the trigger circuit supply voltage is taken from across the SCR's rather than directly from the a-c supply, the result may be severe unbalance in the output signal. The reason is that the current flowing in the inductance of the transformer primary will not stop flowing when the voltage goes through zero at the end of a half-cycle but will continue flowing until some time after the voltage across the inductance has reversed. The reference time interval for the firing circuit begins where the voltage for its supply rises above zero. The SCR's in an inductive circuit will have no voltage across them until the current through them becomes very small, and this time will not correspond to a supply voltage zero point. The effect is a time-reference change for the trigger circuit in such a way that instability and the resulting severe unbalance can occur.

When phase-controlling SCR's with loads which may be inductive, three general rules should be followed.

1. Effective transient protection should be in place across the SCR's.
2. The trigger signal to the SCR should be long enough to allow holding current to build up in the SCR when it is intended that it be triggered.

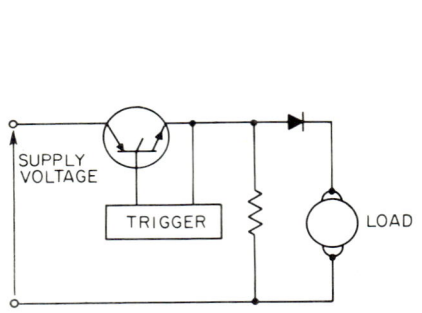

FIG. 41. Prevention of instability.

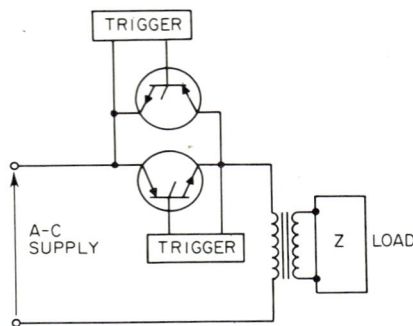

FIG. 42. Potential instability with inductive load.

3. The reference for the triggering circuit should be obtained directly from across the a-c supply.

6.6. Closed-loop Stability. Closed-loop stability in a SCR amplifier system is easier to predict than the open-loop stability in most instances. Before considering what will happen when the feedback loop is closed on itself, one must make certain that there are no potential open-loop stability problems. Then the total phase shift and gain of the loop can be calculated as a function of frequency, and necessary adjustments can be made to ensure stable operation.

The phase shifts in the system will be those generated in the trigger circuit and those generated in the load. The frequency dependence of the gain of the amplifier, as was discussed earlier, is a function of the time constants in the trigger circuit. Stability is guaranteed by a gain of less than 1 at frequencies which are phase-shifted 180°.

6.7. Some Practical Circuits and Their Characteristics. SCR amplifiers have found widespread use in equipment ranging from transmitters to motor controls. Some typical workable circuits are illustrated in Figs. 43 through 47.

Figure 43 illustrates the use of a gate-turn-off device (GTO) in a constant-pulse-width variable-frequency system. The supply voltage is d-c. The input signal is fed to the gate of the GTO through the secondary of transformer T. This fires the GTO, and the GTO anode current increases. The anode-current rise is limited by the primary impedance of transformer T. The voltage developed across the transformer primary is fed via the secondary winding to the gate to speed turn-on of the GTO.

HIGH-POWER AMPLIFIERS 20-39

When the capacitor voltage rises above e_{in}, the gate-cathode junction of the GTC is reverse-biased and the GTO begins to turn off. The reduction in anode current is fed back to the gate via the transformer and assists in a rapid turn-off. Consequently, the GTO turns on when e_{in} is greater than e_{out} and turns off when e_{out} is greater than e_{in}. The voltage across the output capacitor will swing about 10 volts typically under any safe load or supply-voltage condition.

Figure 44 illustrates an incandescent-light control. It could just as well be used to control the voltage across a heater element or some other load of a resistive nature. The firing circuit, utilizing a unijunction transistor, converts the input voltage to phase-variable pulses. These are used to fire the SCR.

The motor-control circuit of Fig. 45 illustrates the use of an extremely simple feed-back technique. The motor is placed in the cathode lead of the SCR, and the gate signal is developed between the SCR gate and the negative load terminal (ground). This gate signal is a half sine wave divided down directly from the a-c supply voltage. While the SCR is off, a voltage will appear across the motor armature which will oppose the positive flow of gate current. This armature back emf is generated from the residual magnetism in the motor, and generally will be less than 10 volts. When the gate voltage exceeds the motor back emf, the SCR will be turned on. The result is that if the back emf rises, the SCR is off for a longer period

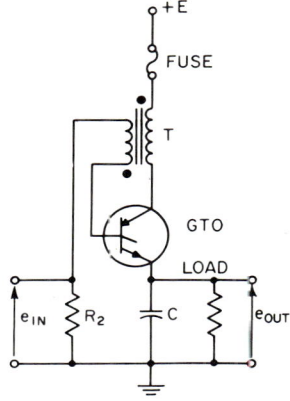

FIG. 43. Regulated d-c power supply using gate-turn-off device.

of time, reducing the voltage fed to the motor. The reverse is also true, and the result is a tendency for the armature speed of rotation to be regulated.

In Fig. 46, the output voltage of the alternator is divided by the resistor R_1 and the potentiometer R_2, and filtered by capacitor C. The voltage across C is compared with the voltage across reference diode CR_1 via transistor Q. If the voltage across C

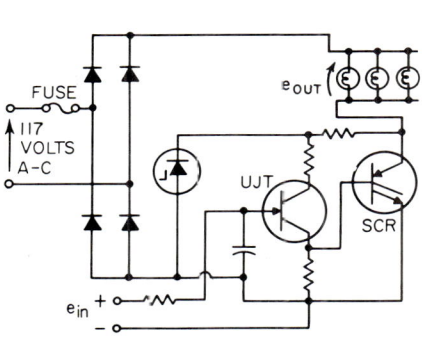

FIG. 44. Circuit for incandescent-light control.

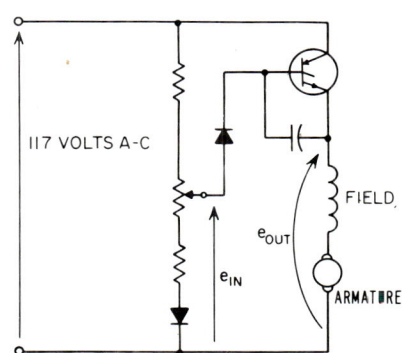

FIG. 45. Universal motor control.

is high, Q will turn off, turning off the SCR. The reverse is also true, effecting regulation of the alternator output. The voltage to the field is a burst of pulses at intervals. This circuit is very useful where both economy and reasonably good performance are desired.

Figure 47 illustrates the use of a phase-control amplifier in regulating the speed of a shunt-wound d-c motor. Current i_c is the control current flowing into the unijunction

20-40 AMPLIFIER CIRCUITS

firing circuit. Current i_f is the feedback current, the level of which is determined by the back emf of the motor.

The circuits illustrated in this section were chosen because of their very typical nature, their wide acceptance, and their general usefulness. For literature pertaining

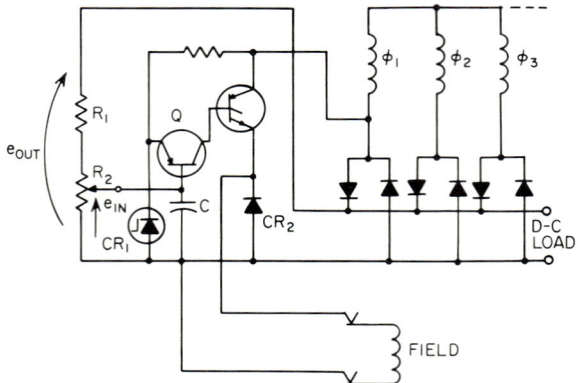

FIG. 46. Alternator regulator.

to a wide expanse of specialized uses for SCR's, the reader is referred to the applications groups maintained by the various SCR manufacturers.

When designing amplifiers with SCR's, nearly any transfer characteristic is practical from some combination of trigger circuit and power configuration. Consequently, design procedure is largely an art for instances where the desired results are not very

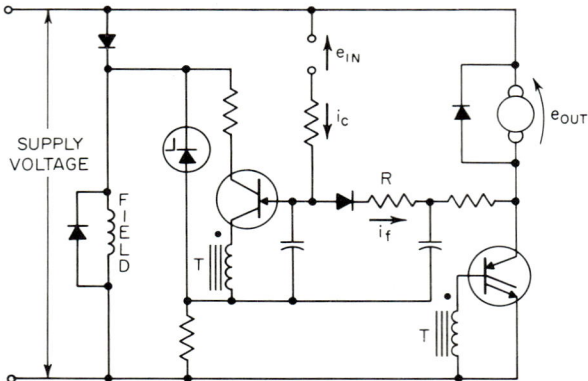

FIG. 47. Shunt-wound d-c motor control.

straightforward. For this reason, SCR manufacturers generally maintain staffs of applications people to assist prospective users of the devices.

7. AMPLIFIERS FOR TRANSMITTERS

7.1. Tube Selection. *Power Output.* The primary consideration in selecting the tubes for a high-power amplifier is their ability to produce the required output. If the amplifier is to operate continuously at full output for extended periods, and into

HIGH-POWER AMPLIFIERS

a constant and predictable load, the selection is, from this standpoint, relatively simple, as this condition is the basis of the manufacturer's ratings.

In the case of modulated-signal amplifiers, whether for music- or voice-modulated amplitude modulation, single sideband, or independent sideband operation, the peak power as well as the average power has to be taken into account in the calculations of amplifier performance.

Pulse ratings of amplifier tubes[23,24] are somewhat more difficult to deal with, since both pulse duration and duty cycle are significant parameters, in addition to hold-off voltage, average dissipation, etc. Generally speaking, for pulse durations greater than 1 or 2 msec, or for duty factors greater than 0.05, tubes in pulse service should be operated within their continuous ratings. On request, manufacturers will generally supply pulse-rating data on their tubes. Frequently the designer is willing to accept relatively short (1,000 hr) tube life in order to achieve the desired performance in a particular application. Under these circumstances, the filament voltage may be raised to 5 or 10 per cent above "normal" and the peak cathode current doubled.

Tube Dissipation. The relationship between output power and tube dissipation for class A, class B, and class C amplifiers is well known for normal loads, as encountered in tuned amplifiers, modulators, etc., and is discussed in other sections of the handbook. In many industrial applications, such as in electronic power supplies, the load may have a relatively small power factor and draw a large leading or lagging reactive component of current. This will result in considerably higher than normal tube dissipation, and where the possibility of this type of load is anticipated, the amplifier tubes must be chosen accordingly.

Operating Voltages. In the interests of greater stage efficiency, long tube life, and minimum drive power, it is desirable to operate tubes at close to their maximum rated plate voltages. On the other hand, power-supply costs and the costs of other components rise rapidly with increase in voltage, because of the need for increased clearances and better insulating materials. The production of X rays in tubes operating at plate voltages of 15 kv and higher must be recognized as an additional hazard to the operator.

Cathode Emission Capability. The emission from the cathode of a tube is the source of the electrons supplied to all its electrodes. When the emission has become insufficient to supply the current required for normal operation of the amplifier, the tube has reached the end of its life.

In the case of tubes with pure tungsten filaments, it was quite common practice to operate the filaments at reduced voltage in order to obtain maximum tube life. The operating level was determined by reducing the filament voltage to the value at which the distortion of the output signal at 100 per cent modulation had increased noticeably. The voltage was then increased by 1 per cent, or 2 per cent above the value thus determined, for normal operation. Life of these tubes was frequently reported to be well over 35,000 hr. It is generally considered that the life of a pure tungsten filament will be doubled if the filament voltage can be dropped to 5 per cent below normal. Conversely, an increase of 5 per cent will halve the life. The emission will be reduced 30 per cent by the 5 per cent reduction in filament voltage and will be increased 40 per cent by the 5 per cent increase in voltage.

The thoriated tungsten filament is not quite so straightforward in its behavior. It is operated at a lower temperature and is not so likely to fail by evaporation of the tungsten. The operating temperature should be quite closely controlled, however, to maintain the required balance between the required emission temperature and the rate of evaporation of thorium from the surface of the filament. Peak currents drawn should not exceed half of the maximum that the filament is capable of emitting. Where it is important to have extremely long life, a tube with an excess filament emission capability should be used, and it should be operated with its filament voltage reduced to 95 per cent of normal.

[23] *Pulse Service Notes*, Eimac Application Bulletin 3, Eitel-McCullough, Inc., San Carlos, Calif.

[24] "Machlett Pulse Tubes for High Voltage, High Power, Video and RF Pulsing," Machlett Laboratories, Springdale, Conn.

High-frequency Capabilities. Most of the larger power tubes are rated for service at maximum ratings for frequencies up to 30 Mc and at reduced ratings for higher frequencies. The ratings for higher frequencies are usually indicated in the data sheets issued by the manufacturer. Where unusual operating conditions are expected, it is often worth while consulting with the tube manufacturer before coming to a decision on a particular tube type. High-frequency limitations on tubes are affected by the following factors:

1. Dielectric losses in the tube insulation
2. Electrical resonances in the filament, or grid structure
3. Lead inductance
4. Transit-time losses
5. Back heating of the cathode

Tubes of special design for vhf and uhf operation are available and are used where the application warrants their higher cost.

Drive Requirements. In selecting tubes for high-power amplifiers, the requirement for drive power from the intermediate-power amplifier is of considerable importance. A final amplifier of high power gain, although in itself more expensive, may result in lower overall equipment cost. In any case, it is desirable that the intermediate-power amplifier output capability be sufficient to provide the required drive power for the final amplifier with a good margin. For high-frequency operation, input grid losses and circuit losses can become quite large. In some instances, circuits may be loaded to improve the stability of the system under fault conditions, such as loss of drive.

Most power tubes display a negative grid current in a part of the constant-current characteristic chart. This area is an advantage in a class C amplifier, where it may reduce the driving power somewhat, and at the same time, it is an area to be avoided in audio amplifiers as it is a potential source of distortion. When it is necessary to operate an audio amplifier in this region, this effect may be minimized with dynatron rectifiers and suitable resistance-capacitance bias networks.

Cooling. Four basic methods of cooling are used in high-power amplifier systems, viz., radiation-convection cooling, air cooling, water cooling, and evaporation cooling. The cooling method used depends on the quantity of heat involved and a number of other factors, which will be discussed in the following paragraphs.

Radiation-Convection Cooling. Radiation-convection cooling is used for most of the smaller tubes, with plate dissipation ratings up to 100 watts. Radiation-convection–cooled tubes are available with plate dissipation ratings up to 2 kw. These tubes are of internal anode construction, and in the higher ratings, at least, the plates are operated at a dull red heat when dissipating rated power. Usually, a small blower is required to provide additional cooling for the bases and plate seals.

Air Cooling. Air cooling is used in most equipment in the 5- to 50-kw output power range. It has the advantage of simplicity and comparative ease of maintenance. It is not seriously affected by extremes of temperature, although precautions should be taken to avoid blowing extremely cold (below −30°F) air on tubes and other components. When the incoming air is above 100°F, it may be necessary to increase the air flow in order to maintain normal operating temperatures.

For a given cooling-system design, the maximum usable air-intake temperature is reduced by 3.5°F for every 1,000 ft increase in elevation.

Air filters and the air passages in air-cooled tubes must be kept clean in order to provide adequate cooling over extended periods. This can entail considerable effort by the maintenance staff for equipment located in dusty environments.

Air filters must be large enough to operate with a reasonably low pressure drop. As a general rule the filter area should be 2 sq ft for every 1,000 cu ft/min of air flow.

Air which has been heated at the anodes of the output tubes of an amplifier should be removed from the equipment by ducts or other means to prevent excessive heating of other components. In some equipments, the cooling air is circulated through the cabinet before reaching the anodes of the output tubes and is ducted directly from the output tubes to the outside.

HIGH-POWER AMPLIFIERS 20–43

While a minimum of inflammable material is used in modern electronic equipment, it is advisable to employ thermal detectors or similar devices to shut down the blower system in case of fire. This is particularly important for remotely controlled equipment. In colder weather, the exhaust from an air-cooled equipment may be used to supplement other sources of heat for the building. Conversely, if a building is air-conditioned, it will generally be desirable to keep the air system for the equipment separate from that of the rest of the building, in order to avoid overloading the air-conditioning system.

Air cooling has the disadvantage of creating considerable acoustic noise. This can be minimized by acoustic treatment of the equipment and associated air ducts. Usually, it is desirable to keep air-flow velocities below 1,000 ft/min to keep the acoustic noise within reasonable limits.

Water Cooling. Water cooling has been extensively used for many years—particularly for equipment in which the power dissipated is 50 kw and more. It has been replaced in smaller equipment with air-cooling systems. It does have the advantage of being relatively quiet in operation and fairly economical of space.

The connections to the water supply are through two pieces of ceramic, plastic, or glass tubing, each 1.5 to 2 in. in diameter and 8 to 10 ft in length. The water is pumped under pressure (usually about 80 psi) through the system, and the rate of flow is usually between 0.5 and 0.2 gal/min for each kilowatt of plate dissipation. Should a leak occur in such a system, a great deal of water would be sprayed over everything before it could be brought under control.

In order to protect the plumbing fittings on the high-voltage side of the insulating tubes from serious electrolytic corrosion, "targets" of platinum wire are used. The target is mounted on a brass rod, extends into the insulating tubing from the fittings, and is the point where most of the electrolytic corrosion will occur. It is made of platinum wire 0.040 in. in diameter and 4 in. long. Its length should be checked semiannually, and it should be replaced when its remaining length is $\frac{3}{8}$ in. Associated metal parts should also be checked, and corroded parts replaced, as required.

The leakage current through an insulating column of this sort will be approximately 8 ma initially, and the distilled water should be replaced when it has increased to 20 ma. Algae growths may be the cause of increased conductivity of the water and in any case should be removed to avoid blocking of water passages, screens, etc. A very thin deposit of scale on the surface of the anode will seriously reduce its heat-transfer capabilities. In an emergency, a 10 per cent solution of hydrochloric acid will dissolve the scale from the anode. The anode should be rinsed thoroughly in water following the use of the acid, and care should be taken to prevent the acid solution from coming into contact with the metal near the copper-to-glass seal. In climates where there is danger of freezing, the water system can be protected by adding 15 per cent pure ethylene glycol to the water. (Inhibited antifreeze solutions such as are used in automobiles cannot be used because of the increase in electrical conductivity caused by the inhibitors.)

Evaporation Cooling.[25] Evaporation cooling has been used in Europe for many years, and suitable tubes are now available from a number of American tube manufacturers. Its initial application in North America has been in short-wave broadcast transmitters, operating at a carrier level of 250 kw. European applications include industrial high-frequency generators (30 to 120 kw), 50- to 200-kw short-wave broadcast transmitters, and FM and TV transmitters of 10 to 12 kw output.

The evaporation-cooling systems are very simple and highly efficient. They are very quiet in operation and can be made without moving parts other than a water-level regulator. Since the system is self-pumping, no water pump is required in most installations. The latent heat of vaporization of water is used, so that the flow of water required is approximately one-twentieth that required for a water-cooled system.

As the system is at atmospheric pressure and the total volume of water is very much reduced, the water damage resulting from a leak in the system would be very small.

[25] The Care and Feeding of Vapor Phase Cooling, *Application Bulletin* 11, Eitel McCullough, Inc. C. Beurtheret, The Vapotron Technique, *Rev. Tech. C.F.T.H.*, no. 24, December, 1956.

Insulation requirements for the water system are provided by a 15-in. length of $\frac{1}{2}$-in. glass tubing for the incoming water and a 15-in. length of 2-in. glass tubing for the vapor. A target is provided with the $\frac{1}{2}$-in. tubing but is not required with the 2-in. tubing.

Protection against frost can be obtained by the use of 15 per cent pure ethylene glycol in the water.

Since the condenser (air-heat exchanger) operates at 100°C, it can be one-half the size of the heat exchangers used with water-cooled tubes.

In industrial applications, the availability of heat at 100°C makes the recovery of energy from this source quite attractive and may result in considerable improvement in the cost of a complete operation.

In the evaporation-cooling system, the water is continually being redistilled; so there is a minimum of contamination, and only pure distilled water is returned to the boiler. Because of this action, and the higher temperatures encountered by the water in its operating cycle, the periods between shutdowns for draining and cleaning will be at least twice as long as for the conventional water-cooling system.

7.2. Power-supply Considerations. Power supplies are discussed in some detail in Chap. 27, Sec. 10. The following is of particular interest to the designer of high-power electron-tube amplifiers.

Three-phase Power Supplies. The power supply for a high-power electron-tube amplifier may be subjected to considerable variation in load in normal operation, and while it is usually required to have good regulation characteristics, it must also withstand the effects of near-short-circuits, caused by flashovers in the antenna system or tubes due to lightning, etc. In some instances, "crowbar" circuits are used to dissipate the stored energy in the power-supply filter, to protect the tubes and other components.

Protection of the rectifiers against the effects of such short circuits may be obtained in three ways:

1. By the use of rectifiers of high short-circuit current ratings
2. By the addition of current-limiting reactors in series with the low-voltage winding of the rectifier transformers (or by the use of transformers with additional reactance built into the design)
3. By the use of high-speed circuit breakers or vacuum interrupters which will interrupt the primary supply within 8 msec or less.

The best economic balance must be arrived at for each application, and this balance will change from time to time as the costs of each element of the system vary.

Rectifier Cells. Costs of semiconductor rectifiers have continued to drop rapidly over the past few years, while the available peak reverse voltage (PRV) ratings per cell have continued to rise. Cells with 1,000 PRV ratings are now quite common and are economical to apply. Series strings of up to 200 or more cells may be used to obtain the desired PRV ratings. Capacitors are used across each cell to ensure equal sharing of reverse voltages.

Short-circuit current ratings have also increased so that single-cycle ratings of twenty or thirty times normal rated current are readily available. Since a rectifier in a three-phase circuit will conduct for only one-third of the time, the surge-current rating may be only seven to ten times the power-supply output rating if the rectifier cell is used at its maximum rated current. In many cases it will be economical to operate cells much below their normal current rating in order to obtain the desired short-circuit rating. The additional cost of the rectifier cells may be less than the cost of current-limiting reactors and/or high-speed interrupters.

Current-limiting Reactance.[26,27] As mentioned above, the short-circuit d-c output of a power supply may be limited by means of reactance built into the trans-

---

[26] I. K. Dortort, Extended Regulation Curves for 6-phase Double Way, and Double Wye Rectifiers, *Trans. AIEE*, vol. 72, pt. 1, pp. 192–202, 1953.

[27] Witzke, Kresser, and Dillard, Influence of A.C. Reactance on Voltage Regulation of Phase Rectifiers, *Trans. AIEE*, vol. 72, pt. 1, pp. 244–253, 1953.

HIGH-POWER AMPLIFIERS

former(s) used to supply the rectifier or by means of external current-limiting reactors, which add to the reactance in the transformers.

A reactance of 0.1 per unit will limit the short-circuit direct current of a power supply to approximately eleven times full-load current (see curves 2 and 3, Fig. 48). Similarly, a reactance of 0.2 per unit will limit the short-circuit direct current to approximately 5.5 times full-load current.

Definitions. Per unit reactance is the supply-frequency reactance of an inductance expressed as a multiple of the resistance which would draw full-load current when connected to the supply.

Per unit kvar is the supply-frequency susceptance of a capacitor connected in shunt with the load, expressed as a multiple of the conductance of the rated load.

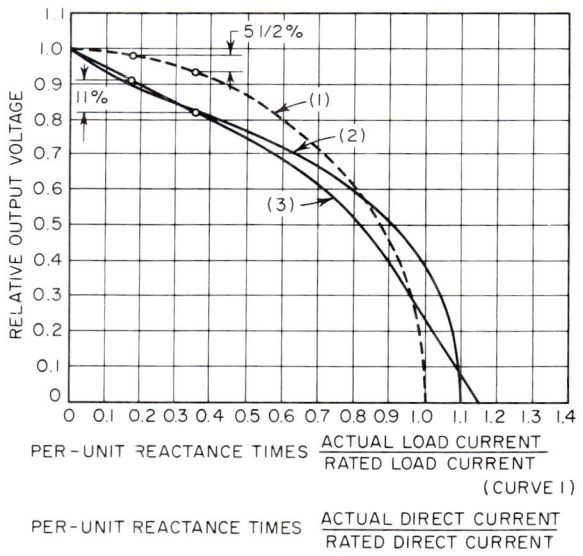

FIG. 48. Rectifier regulation curves. (1) Resistance + inductance (no rectifier) (2) Rectifier with noninductive load; (3) Rectifier with inductive load. (*Curves 2 and 3 from Dortort, Trans. AIEE, 1953.*)

Distribution-type transformers have about 0.01 per unit resistance and 0.015 to 0.035 per unit reactance inherent to the design of the transformer. Custom-made transformers may be obtained with about 0.055 per unit reactance at little or no extra cost. Where more reactance than this is required, there will be an additional charge to have it built into the transformer. In such cases, an external reactor may be used. The external reactor has the added advantage of being adjustable, and it can be removed completely should the primary source have an unusually high reactance.

Regulation. The regulation of a three-phase rectifier power supply (from zero to full load) is expressed by the following formula:

$$\text{Per unit regulation} = \text{per unit } R + \text{per unit } X/2$$

where per unit R is the per unit resistance of the transformer, including both primary and secondary resistance, and per unit X is the per unit reactance of the transformer including external limiting reactances and the reactance of the supply line.

20-46 AMPLIFIER CIRCUITS

This formula is accurate to within 5 per cent of the calculated regulation for reactances of up to 0.50 per unit (see Fig. 48). This figure compares the regulation curves of a three-phase rectifier, (2) and (3), with the regulation obtained with a transformer connected to a resistive load. The effect of winding resistance is neglected in each case for simplicity. It is readily seen that with a given transformer reactance, the regulation of the rectifier at normal loads could be greatly improved if it could be made to behave more like a resistance. This improvement can be achieved by means of capacitors, as described in the text.

It should be recognized that (with high reactances) the ripple at the rectifier output will be higher than that shown in the usual "rectifier circuit constant" charts,[28] which are based on transformers of negligible reactance.

Regulation Improvement.[29] The regulation of a power supply with current-limiting reactance can be improved by the addition of capacitors connected to the a-c side of the rectifiers (see Fig. 49). The capacitors preserve the voltage waveform at the input to the rectifiers under normal load conditions yet have negligible effect on the short-circuit currents. The kvar of the capacitors may be varied over a wide range, provided that resonances with the reactance at the harmonics of the supply frequency are avoided. In the case of a three-phase rectifier, resonance is most likely

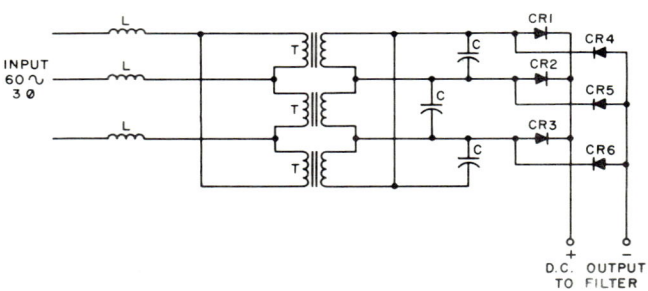

Fig. 49. Rectifier with added capacitance.

to be a problem at the fifth and seventh harmonics of the supply frequency. The presence of a harmonic resonance can readily be observed by measurement of the harmonic currents in the supply line when the rectifier is operated with and without the capacitors.

In a particular case, 0.14 per unit kvar was used with a reactance of 0.20 per unit. The regulation improved from 0.10 to 0.035 per unit.

In selecting a capacitor for this application, it must be recognized that, in addition to the supply-frequency component, the current through the capacitor also includes harmonics. Having selected the *capacity* required, on the basis of "per unit kvar," it is necessary to calculate its rating, taking into account the "harmonic kvar." This is done by adding the kvar for each harmonic (remembering that the voltage associated with a harmonic current varies inversely as the frequency). In the example mentioned, it was found that the kvar required was 1.5 times the supply-frequency kvar.

The above calculation is a good guide to capacitor rating, but it is worthwhile to make a check after the unit has been placed in operation. The temperature rise of the case should be measured after several hours of continuous operation. If the rise in temperature is 5°C or less, the capacitor is adequate.

There will be an increase in the no-load voltage due to the use of the capacitor.

[28] "Semi-conductor Rectifier Components Guide," 2d ed., General Electric Co., 1962.
[29] A. Schmidt, Jr., Capacitors in Power Systems with Rectifier Loads, *Trans. AIEE*, vol. 82, pt. 1, pp. 14–17, 1953.

Per unit rise in E_o = per unit reactance

$$\times \text{ per unit kvar } \times \frac{\text{rms capacitor current (measured)}}{\text{60-cycle component of capacitor current}}$$

In the case mentioned, this amounts to

$$0.20 \times 0.14 \times 1.7 = 0.0475 \text{ per unit}$$

For moderate kvar, no increase in transformer ratings is required. The capacitor current is in quadrature with the supply-frequency load current, and in addition, the capacitor bypasses most of the harmonic currents formerly drawn from the transformer.

High-speed Circuit Breakers. The regularly used magnetic/hydraulic circuit-breaker switches will clear a fault in 40 msec, and this is quite adequate for smaller power supplies where there is a good margin of safety provided by the short-circuit rating of the rectifier, and the current-limiting impedances.

For high-powered equipment, it is often desirable to use vacuum interrupters, with special trip solenoids operated by silicon controlled rectifiers. They will, when used with suitable sensing devices, open the main supply circuit within 8 msec.

Current-limiting fuses are usually used as back-up devices in the supply lines feeding high-powered equipments.

Power-supply Filters.[30] In designing power-supply filters, it is necessary to consider starting transients, keying transients, resonant frequency, and stored energy, in addition to the more obvious requirement for attenuation of the rectifier ripple frequencies. Starting transients in power-supply filters are generally controlled by means of resistances in series with the filter capacitors. These resistances are shorted out, after a time delay. Keying transients are minimized by using relatively large capacitances and small inductances. Where low audio frequencies are involved, as in AM and TV transmitters, it is necessary to choose the inductance L and capacitance C so that their resonant frequency is well below 30 cycles.

In some applications the stored energy in the output capacitor $0.5CV^2$ may be a serious consideration in case of a flashover in the tube. The rate of discharge may be limited by a resistance in series with the power supply.

7.3. Standard Broadcast Transmitters. A standard broadcast transmitter is an equipment designed to produce a signal at a specified frequency in the range 540 to 1,600 kc, and amplitude-modulated in accordance with an a-f voltage applied to its input terminals. Standard broadcast transmitters are rated according to their unmodulated carrier output power. The standard nominal ratings are 250 watts, 1, 5, 10, and 50 kw. The corresponding minimum power output capabilities are 275 watts, 1.1, 5.5, 10.6, and 53 kw. The additional output is required by government regulations to compensate for anticipated losses in the antenna array and the feed system.

Most standard broadcast transmitters on the market today are designed for remotely controlled operation. The use of semiconductor rectifiers, air-cooled tubes, and simpler, more reliable components and circuits has made this possible. Close control of the transmitter building temperature is no longer necessary, but the building layout and construction should be such as to minimize the risk of unauthorized entry or the spread of fire in the event of a major failure in the transmitter.

With few exceptions, all transmitters use single-ended r-f amplifier circuitry, and modulation occurs in the final class C r-f amplifier. The audio-frequency power is supplied by a push-pull class B or class AB modulator. The exceptions are the 50-kw RCA BTA50G "Ampliphase" transmitter, which uses a phase-to-amplitude system of modulation; the Continental type 315C/316C, which uses screen-grid modulation of the final amplifier; and the Continental type 317C, which uses the 316C as a driver for its high-efficiency linear amplifier.

[30] Reuben Lee, "Electronic Transformers and Circuits," 2d ed., John Wiley & Sons, Inc., New York, 1955.

AMPLIFIER CIRCUITS

Basic Elements of a Transmitter. The block diagram in Fig. 50 shows the basic elements of a transmitter. The performance characteristics of typical standard broadcast transmitters are shown in Table 5. In discussing the basic elements of transmitters, the contribution of each to the achievement of these characteristics will be noted.

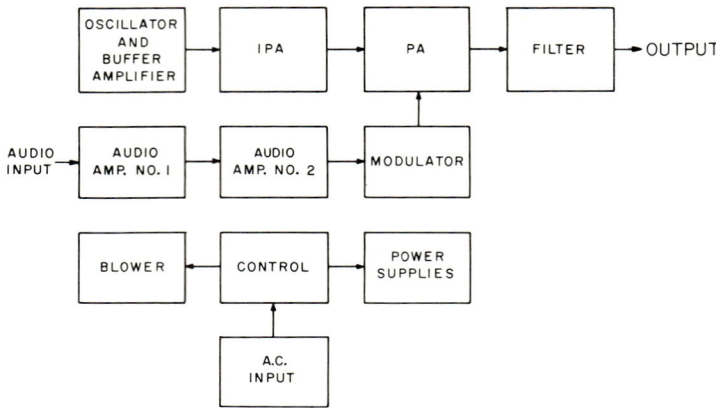

FIG. 50. Transmitter block diagram.

Oscillator. The oscillator in a standard broadcast transmitter is always crystal-controlled and is usually of the electron-coupled Colpitts type, as shown in Fig. 51. Duplicate crystals and trimmer capacitors (C3) are provided so that, in the event of

Table 5. Typical Transmitter Characteristics

	50 kw	5/10 kw	1 kw
A-F input level, dbm, for 100% modulation.	+10	+10	+10
A-F input impedance, ohms.	150/600	150/600	150/600
Frequency response, 30–10,000 cps, db.	±1.5	±1.5	±1.5
Distortion (max) 50–7,500 cps, %.	3	3	3
Carrier shift (max), %.	5	5	5
Load impedance, ohms.	50–230	50–230	50–230
Carrier-frequency stability, cps.	±5	±5	±5
Hum and noise (db below 100% modulation).	60	60	60
R-F harmonic output (db below carrier).	83	80	80
Power input:			
Frequency, cps.	50/60	50/60	50/60
Phase.	3	3	1
Voltage.	460	208/230	208/230
Power input at 100% modulation, kw.	145–155	15–40	4–4.5
Size (range):			
Length.	12–16 ft	5.5–11 ft	22–38 in.
Depth.	4.5–6 ft	22–40 in.	22–33 in.
Height.	78–84 in.	72–84 in.	76–84 in.
Weight, lb.	12,000–15,000	2,200–5,500	700–1,700

HIGH-POWER AMPLIFIERS

failure of the crystal, the transmission may be continued with a minimum of delay. Most crystals are maintained in temperature-controlled ovens, although a number of designers now specify unheated crystals. The latter are satisfactory over a limited ambient temperature range but are more costly in the first instance.

The choice of oscillator tube is quite varied, ranging from the 807 and 6146 to the 6AG7, the 6AU6/6136, and the 6AK5. In each case the circuit is designed to operate with a minimum crystal dissipation, in order that the carrier-frequency stability of ±5 cycles may be assured. Careful shielding, in the high-powered transmitters especially, is required to minimize the effects of stray feedback from the output circuit and antenna system.

Buffer. The buffer, as its name implies, is used to provide additional isolation between the oscillating crystal and the higher-powered intermediate-power amplifier (IPA) and output stages. It may be operated as a class A, B, or C amplifier, depending on the gain required and the tube used. The 6146 and the 807 tubes are most commonly used for this application, although in some cases, a first buffer using a 12BY7A, a 6SJ7, or a 5763 is used. The extra stage is useful when Conelrad operation is required, as it can be operated with broadband interstage coupling and a

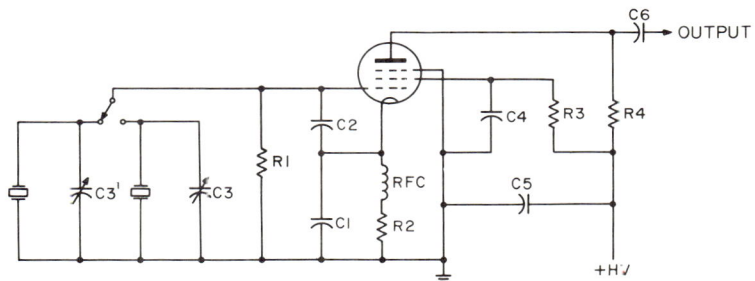

FIG. 51. Schematic of Colpitts oscillator.

minimum readjustment is required when the frequency is changed. The above tubes are all tetrodes or pentodes and are operated without neutralization.

Intermediate-power Amplifier. The intermediate-power amplifier is sometimes omitted in transmitters with outputs of 1 kw and less but is always used in transmitters with outputs of 5 kw and more. The most commonly used tubes for this application are tetrodes of the 4-65, 4-125, and 4-250 family. In some cases two tubes in parallel are used to achieve the required drive for the final amplifier.

The 50-kw transmitters generally require more drive for the final amplifier than can be provided with one intermediate-power amplifier, and an additional stage is used with an output capability of 3 to 5 kw. This may be a neutralized triode stage using the ML6693 or several 833A's in parallel. Alternatively, a tetrode such as the 4CX5000A or the 6076 may be used.

The Power Amplifier (Modulated Amplifier). The power amplifier is in many respects the most significant part of the transmitter, since its drive requirements determine the characteristics of the preceding r-f amplifier chain. It is common practice—especially in 50-kw transmitters—to use the same tube type for the power amplifier as for the modulator. This has the advantage of reducing the cost of spares and enables the operator to get the maximum tube life by moving the tubes from the r-f sockets to the audio sockets as their peak emission begins to fall off.

Typical 50-kw transmitters use a pair of 6427's in parallel in the power amplifier and a second pair in the modulator. The amplifier is operated with a plate supply of 8,500 to 9,000 volts, and the plate current for the amplifier is between 8.0 and 8.5 amp, with a carrier output of 53 kw. The methods used for calculation of amplifier performance and operating conditions are discussed in detail in Chap. 24.

Calculations should be made for the positive peak of modulation condition as well as for carrier conditions. A few trials may be necessary in order to determine the best drive and bias conditions to give the desired plate efficiency and modulation linearity. In some instances, modulation of the IPA as well as the power amplifier is employed. The IPA is usually modulated about 20 or 30 per cent when the power amplifier is modulated 100 per cent. When a modulated IPA is used, the average output of the IPA is usually lower than it would be if IPA modulation were not used.

Transmitters of 5/10-kw rating generally use two triodes in parallel for 10-kw output and a single triode when supplied for 5-kw rating. Tube types most commonly used are the 3X2500 and the 5762. Other tubes used at the 10-kw level are the 3CX10, 000A3, and the 6425. At the 5-kw level, the 4CX5000A and the 6421 have recently come into use. These transmitters generally operate with a plate supply of between 5,000 and 6,000 volts, and the tube efficiency is between 75 and 85 per cent.

Most 1-kw transmitters use a pair of 4-400's, or a pair of 6156's in the PA. These tubes require very little drive power, virtually no neutralization, and tubes of the same type may be used as modulators. They are usually operated with a plate supply of 3,000 volts. The PA screen current is supplied through a high-resistance dropping resistor, and the screen voltage is self-modulated, because of the variation in screen current with plate modulation.

Triode power amplifiers in broadcast transmitters are generally coil-neutralized; i.e., a coil is connected from the output circuit at the tube plate to the grid of the tube. This coil is adjusted to resonate with the grid-plate capacitance of the tube, or tubes, at the carrier frequency. This form of neutralization generally works quite well.

Filter. The final amplifier is connected to the output terminals of the transmitter by means of a filter designed to provide the necessary attenuation of r-f harmonic output (approximately 75 to 80 db of harmonic attenuation is required). In addition, this filter must provide the desired load impedance for the plates of the final amplifier tubes when operating into the specified transmitter load impedance.

In designing a filter for the final amplifier of a transmitter, it must be recognized that the tube output capacitance, the blocking capacitor, the neutralizing circuit, and plate r-f choke all affect the impedance at the input to the filter. Voltage ratings must be sufficient to allow a safety margin, after taking into account the positive peaks of modulation and the effects of some mismatch of the antenna system.

The size of the blocking capacitor from the tube plate to the input of the filter is dictated in part by the amount of capacitance that can be placed across the secondary of the modulation transformer without upsetting its performance at high audio frequencies. The bypass capacitor for the opposite end of the plate r-f choke is in parallel with the blocking capacitor from an audio point of view and adds to the capacitance load on the modulation transformer.

From an r-f point of view, the blocking capacitor is in series with the input capacitor of the filter. At the low-frequency end of the broadcast band (540 kc), these two capacitors may be of the same capacitance. While it is true that the filter can be adjusted to compensate for the series reactance of the blocking capacitor at the carrier frequency, this compensation has little effect at harmonic frequencies. This means that the r-f voltage waveform at the tube plate will be quite distorted.

Transmitters of higher power rating are generally equipped with some form of monitoring device or reflectometer which will trip off the carrier in the event of a disturbance to the impedance of the load. For example, a lightning stroke or even the buildup of static electricity on a tower may cause an arc to form across one of the tuning components at the tower base. The reflectometer will usually act quickly enough to prevent damage to the components involved and will restore normal operation without a noticeable interruption to the program.

The bandpass characteristics of the antenna array, combined with the length of transmission line between the transmitter and the power divider, will have an effect on the distortion of the modulated signal. In the event that this is suspected, a measurement of load impedance, as seen from the plates of the PA, should be made at carrier frequency and at ± 10 kc from the carrier frequency. For minimum distortion, the conductance of the load should be constant over the frequency range

HIGH-POWER AMPLIFIERS

measured. The susceptance may vary from zero at the carrier frequency to ±35 per cent of the conductance at the ±10-kc sideband frequencies. Alternatively, the resistance of the load should be constant and the reactance may be allowed to vary ±35 per cent of the resistance. If there is a serious problem, the condition may be corrected by inserting a T network in series with the transmission line. Its phase shift can be adjusted for optimum performance and will be between 0 and 45° at the carrier frequency.

Audio-frequency Circuits. The input signal level required for 100 per cent modulation is specified to be +10 dbm ±2 db at an impedance level of 600 ohms. This signal is usually supplied by a program amplifier or a limiting amplifier, with a gain of 40 db. Resistive pads are used to adjust the levels as required and to obtain isolation between the line and the amplifier and between the amplifier and transmitter.

The majority of broadcast transmitters employ two voltage-gain stages ahead of the modulators. Where class AB_1 modulator operation is used, the modulator is driven directly by the second stage. Where class AB_2 or class B_2 modulators are used, an extra stage is usually required to provide the necessary low-impedance modulator drive.

Balanced (push-pull) circuits are used throughout, and feedback of from 6 to 20 db is used from the plates of the modulator tubes to the grids of the first stage. Unbypassed cathode resistors may be used in the first two stages to provide some degenerative feedback in each stage. A common cathode resistor provides a self-balancing effect and reduces the gain in the parallel mode of operation. This parallel-mode gain reduction is useful in reducing the effect of the modulation transformer primary leakage reactance on the system distortion, when feedback is used.[31]

The first stage is commonly a pair of pentodes such as the 6136/6AU6 or the 6SJ7 although double triodes such as the 6SN7 and the 5814 have also been used.

The second stage must be capable of handling relatively large voltage swings (2,400 volts peak-to-peak for the grids of 3X3000's), and tetrodes such as the 4-125A 4-250A, 4CX250B, and 4CX300A are commonly used in this stage.

Where modulators requiring grid current at the peak of the grid swing are used, it is necessary to drive them from a low-impedance source such as a cathode follower. Suitable tubes for this application are the 833, the 304TL, or the 813.

3X3000's operated in class AB_1 are most commonly used as modulators in 5- and 10-kw transmitters. The 3X2500 or the 5762 is used where class AB_2 operation is desired. In the higher-powered (50-kw) transmitters, the 6427 is most common.

Blower. The blower is the obvious evidence of the problems of cooling and ventilation in a transmitter. The design of the cooling and ventilating system is of considerable significance to the life expectancy of individual components. The use of a single large blower rather than a number of smaller blowers has the advantage of simplifying the control and interlock system. Parts and tubes that dissipate a relatively small amount of power will operate at much lower temperatures if there is even a small flow of cool air past them.

The relatively hot air which has been used to cool final amplifiers and modulators should be removed from the transmitter by short air ducts. A little flow of cool air over r-f filter components may permit the use of smaller components in some applications.

The air system must be so designed that it will operate satisfactorily under all foreseeable conditions of operation, or if not, warning notices should be provided to caution the operator against trying to operate under incorrect conditions.

Control. The control circuits of modern transmitters are designed for remote operation and are so arranged that the closing of a single contact will cause the blower and then the filaments and bias supplies to be energized in sequence, followed by the low-, intermediate-, and high-voltage supplies. In the event of an overload, the control circuit will shut down, then restore operation automatically. Most control circuits will recycle in this way for two or three times if an overload does not clear during the first shutdown. If it persists to the extent of causing more than three overload trips in less than a minute the circuits will lock out and must be reset by the operator.

[31] Private communication from Dr. R. G. de Buda.

20–52 AMPLIFIER CIRCUITS

In designing the control circuit, the sequence of operations during both startup and shutdown must be given careful consideration to avoid the generation of harmful surge voltages and currents in the associated power supplies and amplifiers in the transmitter. The basic control circuit is usually a type of circuit which employs electrical interlocks to prevent incorrect sequencing.

A Typical 50-kw Standard Broadcast Transmitter. Figure 52 shows in simplified form the schematic diagram of a typical 50-kw standard broadcast transmitter. It is the General Electric type BT50A. This was the first large transmitter to employ germanium rectifiers for all its power supplies. It is interesting to note that the germanium rectifiers, which were so new in 1957, are now being replaced, in new production, with silicon rectifiers, because of their lower cost and ready availability on today's market.

As may be seen from the schematic diagram, the r-f drive circuits consist of a 6146 crystal oscillator, 6146 buffer, 6156 first intermediate-power amplifier, and 6623 second intermediate-power amplifier. Approximately 3 kw of power is available to drive the modulated amplifier, which consists of two 6427's operated in parallel.

The second intermediate-power amplifier is modulated about 30 per cent when the modulated amplifier is modulated 100 per cent. Both the second intermediate-power amplifier and the modulated amplifier are coil-neutralized. A capacitor is connected in parallel with the coil in the intermediate-power amplifier, for fine adjustment.

The audio circuits consist of a pair of 6136's operated as a push-pull voltage amplifier, followed by a pair of 6156's similarly connected. Four 304TL's are connected as cathode followers (two in parallel for each grid of the type 6427 modulator tubes). In order to minimize hum from the filaments of the modulated amplifier, they are supplied from Scott-connected transformers, which provide filament voltages that are in quadrature with each other. As a further step in reducing filament-induced hum, the audio feedback for the low (hum) frequencies is obtained from the cathode of the modulated stage. Feedback at the higher frequencies is obtained from the primary of the modulation transformer.

The mechanical design is simple, but effective. The main cabinet consists of three cubicles, of sheet-steel construction, each 54 in. square by 84 in. high, bolted together to form an assembly 13 ft 6 in. in length.

The external equipment includes a blower and the heavy power supply and modulation components. Specifically, they include the modulation transformer, the modulation reactor, the high-voltage supply transformers, and filter reactor, plus the current-limiting reactors, delta-wye switch, and plate contactor.

The total weight of the transmitter, including external components, is approximately 13,000 lb. It is available for operation from 460, 575, 2,400, or 4,160 volts three-phase, 50- or 60-cycle power.

The equipment within the cabinets is arranged for good accessibility, good circuit isolation, and a readily traced sequence of stages. The left-hand cubicle contains the air inlet and filter system, the control circuits, and the low- and intermediate-voltage power supplies, plus the rectifier for the 9,000-volt supply.

The first three stages of the audio system are in the upper front part of the center cubicle, and the modulators are in the rear. The r-f drive circuit is in the lower front part of this cubicle.

The right-hand cubicle contains the modulated amplifier and its associated output filter, including its tuning and loading controls.

The air-cooling system of this transmitter is a negative-pressure system. The air is drawn in through intake filters in the top of the rectifier cubicle and is then distributed over the rectifier cooling fins to the interior of the transmitter. The exhaust is a large duct in the lower rear of the center and right-hand cubicles.

The modulators and PA tube supports are mounted directly on this duct, so that the air, after absorbing the heat from these stages, is removed by the shortest possible path from the transmitter, minimizing the heating of other components from this source.

Two smaller ducts are used: one to ventilate the front half of the center cubicle,

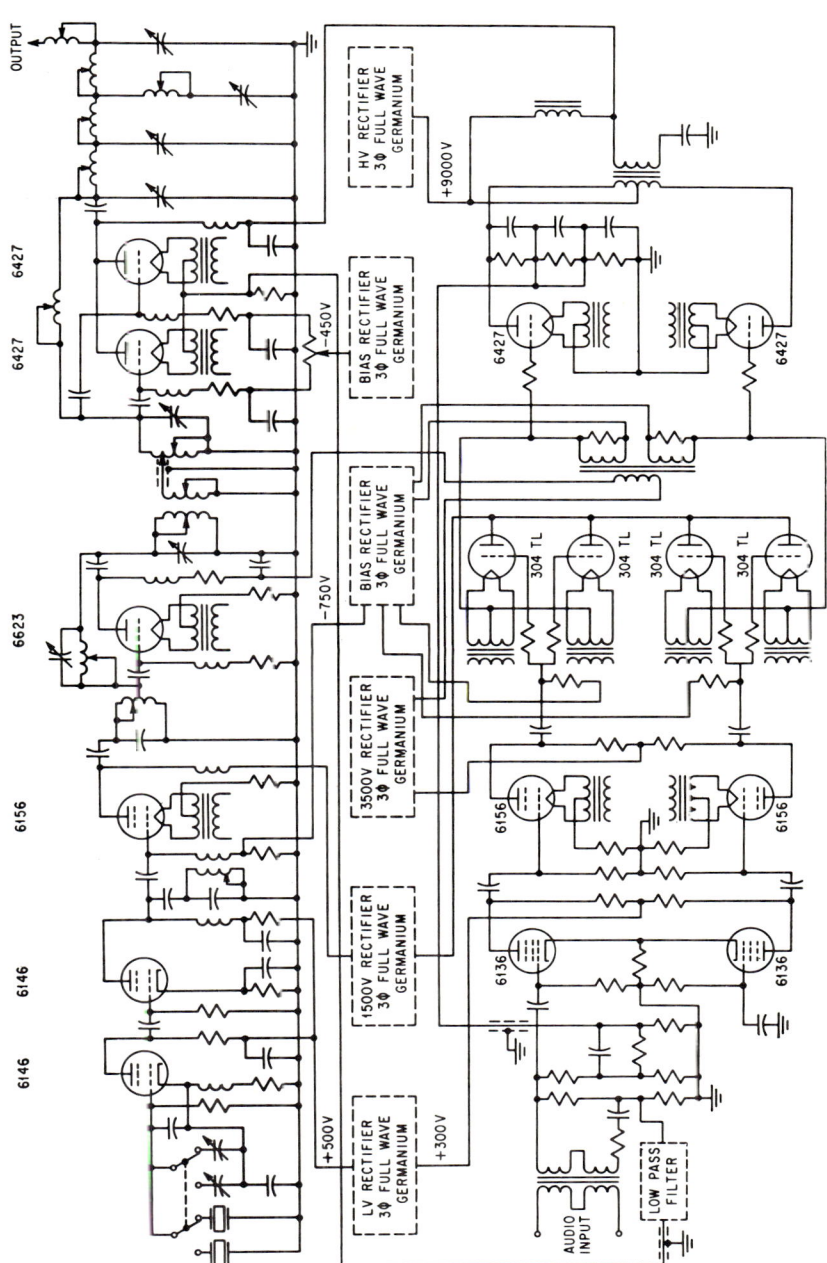

FIG. 52. Simplified schematic of General Electric type BT50A transmitter. (*Courtesy of General Electric Co.*)

cooling the low- and intermediate-power tubes and components; the other to ventilate the PA output filter boxes in the right-hand cubicle.

The blower is driven by a 7.5-hp motor and produces an air flow of 4,000 cu ft/min against a static pressure (in the external duct) of 4 in. of water.

7.4. Superpower Transmitters.[32] The superpower transmitter is, by our standards, one with an average output power greater than 100 kw. Special problems arise in equipment in this category because of the high voltages employed and the greater stored energy—both electrostatic and electromagnetic—which must be dissipated in the event that a fault occurs in a major component. Use of large blocks of power means that high-voltage power lines (4,160 or 13,200 volts) are connected directly to the equipment. The design of input transformers and associated switching facilities is a power-substation problem, with the added complications imposed by the electronic equipment.

It must be recognized that the high-voltage surges due to lightning strokes, etc., are more likely to cause trouble in the equipment, which is no longer protected by the attenuation of additional transformers normally used to step the line voltage down to 460 or 575 volts. Equipment used on 4,160-volt lines is all built to withstand voltage surges of 60 kv. This is referred to as the BIL (basic impulse level). This rating refers to the ability of the equipment to withstand, without damage, a voltage impulse which rises to 60 kv in 1.5 μsec and falls to 10 per cent of this value in 40 μsec.

Stored energy in capacitors is necessary in any equipment employing power-supply filters and particularly so in radars and similar equipment employing pulsed output circuits, where large capacitor banks are made up of many capacitors connected in parallel, with each capacitor storing (say) 1,000 watt-sec ($CV^2/2 = 1,000$ watt-sec, if $C = 0.67$ μf and $V = 55,000$ volts). A large amount of energy may be dissipated in one capacitor, if it breaks down, and the energy from the others is dissipated in the defective unit.

It is indicated by G. E. Tallmadge[33] that 200-watt wire-wound resistors (of 100,000 ohms resistance) will withstand 55,000 volts momentarily and will stand a discharge of 2,900 watt-sec per resistor without damage. Such resistors will limit the current through shorting bars to a satisfactory level and at the same time will give a visual indication of a shorted capacitor without permitting fault currents high enough to explode the defective capacitor.

For very short pulses, the inductance of the leads in a capacitor bank may be significant even if noninductive resistors are used. Special 50-kv capacitors are available, which are built with three sections in series. If one "blows," the remaining two will hold off normal voltages, while a pressure switch on the case, activated by gas pressure due to the arc, will shut down the transmitter. In some cases, the capacitor bank is built as a distributed line, or pulse-forming network.

If an arc occurs in the final amplifier, a fault-diverting device (commonly called a crowbar) may be used. For moderate-sized power supplies, a hydrogen thyratron or an ignitron[34] may be used. For larger supplies, a three-ball gap in air is used. One ball is attached to each terminal of the supply. The spacing is sufficient so that there will not be an arc with maximum d-c voltage on the bank. A third ball is placed near the high-voltage ball. When a fault occurs, a pulse of 200 to 800 kv is applied to the third ball. This starts an arc, which in turn causes an arc between the two balls, shorting the bank and protecting the final amplifier within a few microseconds.

Protection on large transformers and iron-cored reactors is obtained by means of individual ball gaps, or horn gaps, although the designer of these elements must be advised if gaps are going to be used, as the windings must be designed to withstand the surges generated. Such transformers and reactors are usually oil-filled and must be located outdoors or in suitably designed fireproof vaults.

As a maintenance procedure, the oil in the large units should be tested periodically

[32] J. O. Weldon, A 600 Kilowatt High Frequency Amplifier, *IRE* Trans. Communication Systems, CS5, no. 1, March, 1957.

[33] G. E. Tallmadge, High Power Transmitter Design, *Electron. Ind.*, July, 1962.

[34] P. H. G. Van Vlodrop, Protecting Transmitter Tubes by Short-circuiting Destructive Currents via an Ignitron, *Philips Lab. Rept.* EIV 6101, December, 1960.

HIGH-POWER AMPLIFIERS

to ensure that its dielectric strength has not been reduced by moisture getting into it or by the presence of contaminants due to an arc. The oil may be reprocessed with a filter press and restored to its original condition when necessary.

In very large equipment, the size of the components alone makes it necessary to use large separations between them. Ground connections should be run radially from the circuit with the most severe surge-current problems, which should be made the central ground point of the whole system. Inductances can be kept to a minimum by the use of thin wide strips of copper (40 in. wide by 0.10 in. thick). Closed ground loops are to be avoided if possible. Coaxial shields should generally be grounded at one end of the cable only.

In a large installation, 60-cycle hum is likely to be a problem in monitoring circuits unless the desired signals are kept quite large (20 or 30 volts). Isolating transformers or chokes wound with coaxial cable on iron or ferrite cores may be necessary to obtain the desired separation of signals from hum components.

Meters often require special attention, with isolation of the whole movement and case or by shunting the movement with a pair of silicon diodes to limit surge voltages which may appear across the movement. Small capacitors mounted on the meter studs may serve to keep low-level radio frequency out of the movement. In other instances, a shield may be required over the movement, grounded to one terminal of the meter.

Amplifier Characteristics. Individual amplfiers are often built with power output capabilities of up to 500 kw, and their outputs are combined, where higher levels of power are required (see Table 6). In one amplifier as many as eight tubes have been used in push-pull parallel to achieve the required output power. The recent introduction of the Vapotron* technique has made it possible to achieve output power of this order with fewer tubes and correspondingly simpler equipment.

Table 6. Output-tube Complements of Superpower Amplifiers

Equipment	Tube type	Quantity
2-megawatt vlf (Continental) (consists of 4 550-kw linear amplifiers), 550-kw vlf linear amplifier	ML6697	8
1-megawatt AM (Continental) (consists of 2 500-kw linear amplifiers), 500-kw linear amplifier	ML5682	8
1-megawatt vlf (RCA) (consists of 2 500-kw linear amplifiers), 500-kw linear amplifier	RCA5831	2
250-kw short-wave transmitter (GE):		
Modulated amplifier	ML7432	2
Modulator	ML7432	2
250-kw short-wave transmitter (Continental):		
Modulated amplifier	F8388	2
Modulator	F8388	2
600-kw P.E.P. short-wave linear amplifier (Continental)	RCA6949	1

The General Electric 250-kw Short-wave Broadcast Transmitter. The General Electric 250-kw short-wave broadcast transmitter occupies an area approximately 22 ft long by 12 ft deep and 11 ft high. External electrical components occupy an area 24 ft long by 13 ft deep. The blower and condenser (heat exchanger) occupy an area 12 ft by 9 ft.

The block diagram in Fig. 53 indicates the tube lineup and the interconnections of the major electrical components. Note that the r-f circuits, including the harmonic filters and the TVI filter, are all single-ended, with a balun at the output to the antenna system, which converts from a 75-ohm coaxial system to a 600-ohm bal-

* Registered at U.S. Patent Office by C.F.T.H. (La Compagnie Française Thompson-Houston, Paris, France).

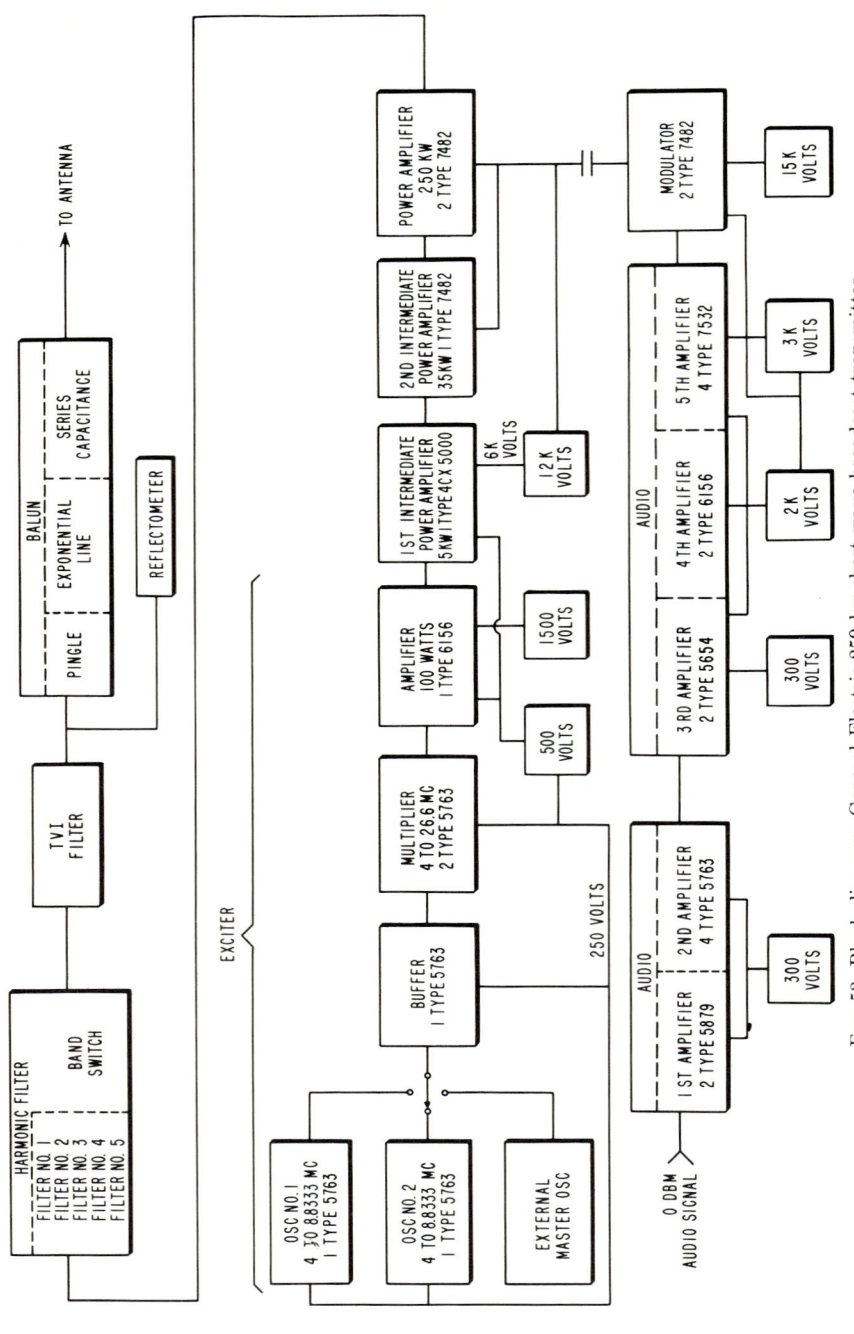

FIG. 53. Block diagram, General Electric 250-kw short-wave broadcast transmitter.

anced output. This arrangement is economical of materials in the filters and provides the required harmonic attenuation at minimum cost.

The modulators, power amplifiers, and second intermediate amplifier are all the same type of tube (ML7482). This tube is a vapor-cooled triode, with a rated dissipation of 200 kw. The fifth audio-amplifier tubes (type F7532) are also vapor-cooled and have a plate dissipation rating of 10 kw per tube.

It is interesting to note that a vapor-cooled tube dissipating 200 kw continuously requires a water flow of only 1.4 gal/min, as compared with a flow of 40 to 50 gal/min for a conventional water-cooled tube dissipating the same power.

The power amplifier employs two ML7482's operated in parallel in a grounded-grid circuit. The second intermediate-power amplifier employs a single ML7482 in a grounded-grid circuit. Both the second intermediate amplifier and the power amplifier are plate-modulated.

The plate tuning circuit used in the power amplifier is similar to that used in the second intermediate-power amplifier and is shown in simplified form in Fig. 54. It is a highly stable, self-shielding cavity resonator type of circuit. The inductive tuning element is a coil made from 2-in. copper tubing, mounted on a rigid supporting frame. The inductance is varied by means of a shorting cylinder, which is equipped with a traveling contact for shorting out the unused portion of the coil.

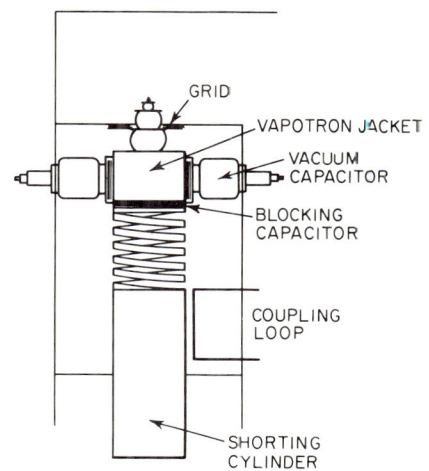

FIG. 54. Plate cavity.

The Vapotron® water jacket provides some of the required tuning capacity, and this is supplemented by means of variable-vacuum capacitors, which permit operation at the optimum capacitance at each frequency from 4 to 26.5 Mc. The modulator circuit is quite unique, in that the modulator plate supply voltage is operated with the positive terminal at ground potential. This permits the use of an autotransformer, with a grounded center tap as the modulation transformer. This greatly reduces the cost of the transformer and permits much better operating characteristics because of the reduced leakage reactance, lower insulation requirements, and simpler construction. The modulation transformer weighs only 5,000 lb. This compares with a weight of 1,500 lb for the modulation transformer used in a 50-kw transmitter.

The Continental Type 621 H-F Amplifier. This is a class B linear amplifier, rated at an output of 300 kw average power, 600 kw peak envelope power, single sideband, reduced carrier. It is capable of continuously tuning over the frequency range of 4 to 30 Mc.

It is a single-tube amplifier and uses the RCA 6949 shielded-grid beam triode. This tube is of unusual design. It has very low plate-to-grid capacitance (20 pf) because of the shielded-grid construction. The input capacitance is unusually large (1,300 pf) for the same reason. The large input capacitance requires a very high r-f grid current at 30 Mc (approximately 250 amp rms) and at the same time requires a very low inductance tuning coil to resonate with it. The desired inductance and current-carrying capability are obtained by means of two 1.5-in. copper-tubing conductors, which are wound in interlaced fashion about a central metal cylinder. A sliding contact on each conductor is fixed to a rotating contact ring, which in turn contacts an outer cylinder. The contact ring slides on insulated rollers, which support it on the double coil. The input circuit has a π configuration, with the input capacitance of the tube acting as the output capacitor of the π. The input capaci-

AMPLIFIER CIRCUITS

tance is provided by an outer grounded cylinder, which is concentric with the cylinder to which the contact ring is connected.

Short additional concentric cylinders are mounted on the base of the assembly and on the contact ring itself so that, as the ring is rotated to increase the inductance, for lower-frequency operation, the additional cylinders are meshed, adding to the capacitance at the input to the π. Variable-vacuum capacitors are also connected to the grid of the tube and are connected to the ring drive, so that at lower frequencies they add to the grid input capacitance, maintaining the proper ratio of capacitance to inductance in the circuit.

The output circuit is electrically similar to the input circuit but makes use of an unusual toroidally wound inductor. The size of the inductor is 40 in. outside diameter, 30 in. long, and 14 in. inside diameter. It is wound with flat copper strap. It has two inside terminals and two outside terminals. This coil is self-shielding and the external field from it is very small. The four segments of the coil operate in parallel, each segment carrying 104 amp of the total tank current of 416 amp.

The inductance is varied by means of a shorting disk, made up of 12 wedge-shaped segments. The segments are equipped with multiple contact fingers, which carry approximately 10 amp per contact. The 12 segments are supported by an insulating ring, which is moved up and down by means of three brass lead screws.

7.5. FM and TV Transmitters. The continuing trend to remotely controlled and operated transmitters is apparent in modern FM and TV transmitters, which are usually factory-equipped with the required telemetering facilities and have their control circuits designed for remote-control operation.

The use of "building blocks," or "modules," for power increases is common in both FM and television transmitters. This makes for great flexibility in the design of FM and TV broadcasting facilities. As antennas of power gains from 0.5 to 20 or 25 are in common use, and gains as high as 40 to 60 are available in the uhf TV frequency range, many combinations of antenna gain and transmitter power are available to the system designer.

It is common practice for a broadcaster to start with a 250-watt FM transmitter or a 500-watt television transmitter and an antenna of moderate gain (say 3 to 6). As his market builds up and competition begins to make itself felt, he will add one or two amplifiers to his transmitter to improve the coverage in his fringe areas. In some instances, he will obtain the improved service by installing an antenna of higher gain (usually on a higher supporting tower). The control circuits are designed in "ladder" fashion, so that the proper sequence of operations is obtained as each amplifier, with its control circuits and power supplies, is added to the existing equipment.

FM Transmitters. The FM exciter usually employs a "serrasoid" modulator. This is a type of phase modulator in which a sawtooth is generated at a frequency which is controlled by a conventional crystal oscillator. The sawtooth voltage is applied to the grid of a tube which is biased so that conduction begins at about halfway up the sawtooth. Modulation is achieved by the application of the audio signal to the tube bias, varying the point on the sawtooth at which conduction starts. The plate voltage of this tube is differentiated, and the resulting output pulses are amplified, then applied to a series of frequency multipliers. The output pulses are modulated approximately ± 75 cycles at 100 per cent modulation. For FM broadcast use, this must be multiplied by 1,000 to achieve the required ± 75 kc. Most modulators of this type operate at frequencies ranging from 90 to 200 kc.

Input RC combinations are used to provide the required conversion from phase modulation to frequency modulation, plus the desired preemphasis of the higher audio frequencies.

The multiplier chain usually consists of doublers and triplers. The interstage circuits are double-tuned to obtain the required frequency response and phase linearity. The output power from the exciter is usually 10 watts. This is sometimes used (as in educational broadcasting stations) without further amplification.

Power amplifiers for FM are rated at 250 watts, 1, 5, 10, 15, and 20 kw. Other output ratings, such as 100 watts, 3, 7.5, 25, 35, and 50 kw, are available but are not commonly used.

HIGH-POWER AMPLIFIERS   20-59

Tetrodes of the 4CX250B, 4CX1000A, 4CX5000A family are most frequently used for these amplifiers. Grounded-grid circuits are frequently used to minimize the possibility of instabilities which may occur in grounded-cathode stages. A well-designed grounded-cathode stage has the advantage of greater power gain, and with modern tetrodes, there is little risk of instability even at 100 Mc.

TV Transmitters. A television transmitter is usually two separate transmitters operating into a common antenna system. The visual transmitter is rated in terms of the output power during the peak of a synchronizing pulse. (This is sixteen-ninths of the "black picture" power.) The visual signal is transmitted as a "vestigial sideband" signal; i.e., the carrier and both sidebands up to a frequency of 0.75 Mc are transmitted, but for higher-frequency components the lower sideband is attenuated, and the lower sideband frequencies removed from the carrier by 1.25 Mc or more are required to be down by 20 db. Upper sideband signals up to a frequency of 4.2 Mc are transmitted at full amplitude.

A black picture element is transmitted as a positive peak of modulation (75 per cent of the synchronizing signal peak voltage), and a white picture element is transmitted as a negative peak of modulation (10 per cent of synchronizing signal peak voltage). The most common transmitter ratings are 500 watts, 2, 5, 10, 25, 35, and 50 kw. Other ratings frequently seen are 1, 6, 11, and 20 kw.

The rating of the aural (FM) transmitter has been 50 to 70 per cent of the visual transmitter rating for many years. Recent reviews of system performance, including extensive field trials, have resulted in a decision to reduce the aural transmitter power to 10 to 20 per cent of the visual transmitter rating.

The Canadian General Electric Co. produces a "modular" television transmitter, which can be operated with both visual and aural signals traversing a common chain of amplifiers. When operated in this manner, the transmitter ratings are 15 and 150 watts, and 2 kw. An identical set of amplifiers may be added to the transmitters, and when operated as separate aural and visual transmitters, they are rated at 30 and 400 watts, and 5 kw, respectively.

This type of transmitter has an added degree of flexibility in that either or both chains of amplifiers can be used. The output rating of a single chain is roughly 40 per cent of the rating for two—quite a respectable emergency condition.

All stages are class B linear amplifiers and are coupled with double-tuned circuits. Generally speaking, they are adjusted for slightly more than critical coupling (3 per cent dip in the center of the passband). A lower sideband trap is coupled to the plate circuit of the final stage, to achieve the required lower sideband attenuation. Harmonic filters are used in the transmission line connected to the diplexer (if used) or to the antenna.

Automatic gain control for both the aural signal and the visual signal (clamped to the "back porch") is used to minimize the effect of power-supply-voltage variation, as well as variation in the stage gains due to other causes.

Amplifiers for FM and TV Transmitters. The distinguishing characteristic of amplifiers for FM and TV applications is the integration of the tubes themselves into the tuned circuits. It is not quite so apparent on the low television frequencies (54 to 88 Mc), although even here it is done in the amplifiers of higher power ratings.

The single-tube amplifier has the simplest configuration (see Fig. 55). The tube socket is mounted in a metal plate, which is at r-f ground potential and serves as a separating shield between the plate circuit and the grid-cathode circuit. Usually, the control grid and the screen grid are grounded, with the drive power being supplied at the cathode of the tube. The plate of the tube is connected to a large-diameter metal tube, which serves in the dual roll of inner conductor for a resonant coaxial line, and as an air duct, exhausting the cooling air from the plate of the tube.

The plate-to-grid or plate-to-screen-grid capacitance of the tube provides most, if not all of the required capacitance of the plate tuning circuit. Adjustment for resonance is obtained by adjustment of a shorting bar or movable plate which is connected to the inner conductor by means of a clamping ring and to the outer rectangular conductor by means of spring contact fingers. Fine tuning is sometimes obtained by an adjustable plate or a variable capacitor connected between the high-voltage (plate) end of the inner conductor and the outer conductor.

Output from the amplifier is obtained by means of a coupling loop and a 50-ohm coaxial line.

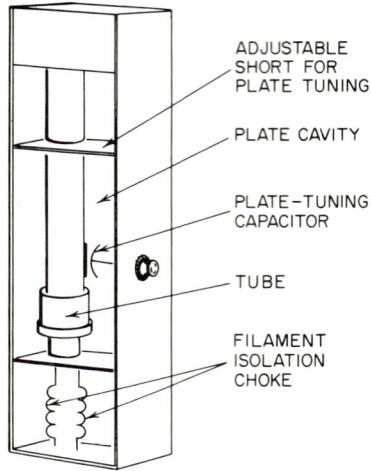

Fig. 55. Single-tube television amplifier.

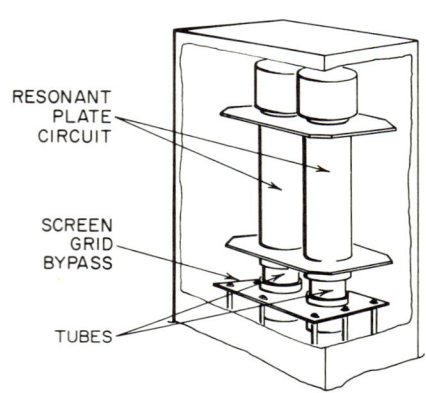

Fig. 56. Two-push-pull-tube television amplifier.

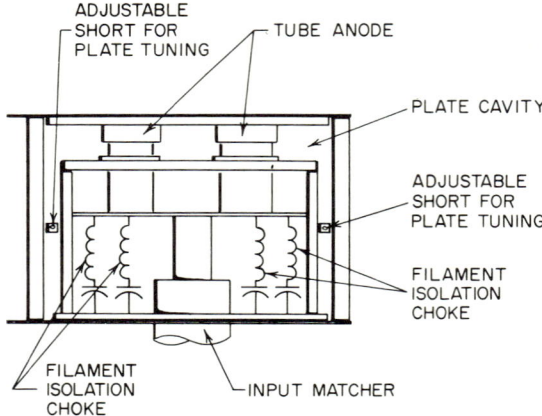

Fig. 57. Two-parallel-tube television amplifier.

The input circuit of a grounded-grid amplifier is heavily loaded by the cathode-grid impedance of the tube and almost "untuned."

Radio-frequency chokes or filament leads that are bypassed to ground approximately one-fourth wavelength from the feed point are used. The low impedance of this circuit arises from the fact that the current flowing in it is the same current as that flowing in the plate (output circuit), but the voltage is relatively small and equal to the grid-cathode voltage swing required for the proper operation of the tube. The power required to drive this circuit appears in the output as feed-through power.

Two-tube amplifiers are frequently operated in push-pull, with the plate circuit operated as a shielded two-conductor resonant line (see Fig. 56). In other cases, the tubes are operated in parallel (see Fig. 57), the plate circuit appears to be a very low impedance resonant line, and it becomes more appropriate to refer to it as a cavity resonator. An excellent example of this configuration appears in the General Electric type TF4A 20-kw television amplifier, which employs two type 7007 tubes in parallel, and the General Electric type TF5A 50-kw amplifier, which employs two type 6251 tubes in parallel.

The use of three tubes in parallel is not quite so common but has been done quite successfully in the General Electric TF14A amplifier, which employs three type 7007 tubes in a Y-shaped cavity.

A design employing five tubes in parallel is the RCA TT50AH 50-kw amplifier, which requires five type 6166 tubes. The RCA TT25CH employs seven type 5762 tubes in parallel to produce an output of 25 kw. For higher output power, the output from two amplifiers can be combined in a bridge-type network.

7.6. Short-wave Transmitters. The short-wave frequencies, in common with the rest of the electromagnetic spectrum, are very heavily loaded with the signals of many services. These include point-to-point communications, aircraft and ship services, etc. In order to obtain some relief from the resulting interference, operators in this frequency range (2 to 30 Mc) are shifting to single-sideband operation. This shift in modulation system involves the receivers as well as the transmitters, and we are in the midst of a transition period, which will continue into the 1970s.

All new short-wave transmitters (excepting those used for broadcast service) are capable of single-sideband operation. Much of the improvement in interference conditions which can be achieved by complete suppression of the carrier is being lost at the present time, as the carrier is often transmitted at or near full amplitude in order to operate satisfactorily with existing receivers.

The suppression of the undesired sideband is considered adequate by CCIR standards if it is 35 db below the peak signal transmitted in the desired sideband.

Present-day exciters are capable of producing a signal in which the unwanted sideband and the carrier are suppressed to a level of -60 db or better. This means that signals transmitted outside the band of the desired sideband frequencies do not originate in the exciter but are generated in the linear amplifier chain and are spurious signals due to odd-order distortion in the amplifiers. Tests for distortion of this type are made by the "two-tone" method, in which the transmitter is modulated by two tones of equal amplitude, adjusted to give the rated peak envelope power of the transmitter under test. A narrowband test receiver is used to measure the intermodulation products, which are spaced by multiples of the frequency difference between the two tones, above and below the desired signals. The level achieved in older designs is -25 to -30 db, referred to one of the test tones.

Modern transmitters are generally able to meet the -35-db figure specified by CCIR.

Some very ingenious methods have been devised to provide the wide range of tuning required, while retaining the desired LC ratios in the tuned circuits.

Because of the daily changes in propagation conditions, it is usually necessary to shift from one frequency to another several times each day. In some designs, this requirement is met by using a number of pretuned amplifier systems, with a single power supply and exciter. This arrangement has the advantages of very rapid change-over and ease of maintenance of the r-f circuits, which can be individually serviced during their "standby" periods.

Alternatively, multiple tap switches may be used to change circuit inductances, while mechanical motor-operated tuning devices such as the Philips 'Instantuner'® may be used to adjust the tuning capacitors to the required settings. These devices require about 15 sec, on the average, to reach any of 10 or 12 preset positions.

Where it can be justified, a servosystem can be used to adjust one or more stages for proper operation. Usually, as with the Collins 'Autotune'® system, the servo is

Table 7. Typical Television Transmitter Characteristics

1. *Characteristics Common to All Power Ratings:*

	Visual	Aural
Type of emission	A 5	F 3
R-F load impedance	50 ohms	50 ohms
Maximum VSWR of load	1.1	1.5
Input impedance	75 ohms	600/150 ohms
Input level	0.7 volt peak-to-peak	+10 ± 2 dbm
Amplitude vs. frequency response:		
Reference	0.2 Mc	1 kc
50 to 15,000 cycles		±1 db
Upper sideband:	+1 db −1.5 db (+0.5Mc to +3.58 Mc) +1 db −3 db at +4.18 Mc −20 db max at +4.75 Mc	
Lower sideband:	+1 db −1.5 db at −0.5 Mc −20 db max at −1.25 Mc −42 db max at −3.58 Mc	
Variation of frequency response with brightness	±2 db	
Carrier-frequency stability	±1 kc	±1 kc (ref. visual)
Modulation capability	12.5 ±2.5% (Reference White)	±50 kc
Audio-frequency distortion (±25 kc deviation)		1.5% max, 50 to 15,000 cycles
FM noise, below ±25 kc deviation		60 db
AM noise rms	40 db below 100% modulation	50 db below carrier
Amplitude variation over one picture frame	Less than 5% of the peak of sync level	
Regulation of output	7% max	
Carrier harmonic attenuation	At least 60 db	At least 60 db

Color-transmission Characteristics

Burst vs. subcarrier phase	±7° max
Subcarrier phase vs. brightness	±7° max
Subcarrier amplitude	15% max
Envelope delay vs. frequency (referred to standard curve)	±0.08 μsec (0.2 to 4.18 Mc) ±0.04 μsec at 3.58 Mc

2. *Characteristics Affected by Power Ratings:*

	Output power rating (sync peak)				
	50 kw	25 kw	10 kw	5 kw	2 kw
Power input:					
Frequency, cycles	50/60	50/60	50/60	50/60	50/60
Phase	3	3	3	3	3
Voltage	460	240/208	240/208	240/208	240/208
Power input, (black picture), kw	165	105	42	26	11
Size					
Length, in.	*	....	140	72	72
Depth, in.	....	....	79	79	79
Height, in.	84	84	84	84	84
Weight, lb	19,000 to 25,000	12,000 to 15,000	6,000 to 7,000	4,000 to 5,200	3,000 to 3,600

* Transmitters of 25 kw and higher ratings are usually made of so many individual cubicles and external components of large size that their dimensions cannot be defined in a short tabulation such as the above table.

HIGH-POWER AMPLIFIERS

Table 8. Typical FM Transmitter Characteristics

1. *Characteristics Common to All Power Ratings:*

Frequency range	88 to 108 Mc
Frequency stability	±1,000 cycles
Audio-frequency response*	±1 db from 50 to 15,000 cycles
Distortion (50 to 15,000 cycles)	Less than 1%
FM noise level (ref. ±75 kc deviation)	−65 db
AM noise level	−55 db rms
Carrier harmonic attenuation	80 db
Modulation capability	±100 kc
Audio input level	+10 ± 2 dbm
Output load impedance	50 ohms

2. *Characteristics Affected by Power Ratings:*

	Power output				
	15 kw	10 kw	5 kw	1 kw	250 watts
Power input:					
Frequency, cycles	50/60	50/60	50/60	50/60	50/60
Phase	3	3	3	1	1
Voltage	240/208	240/208	240/208	240/208	240/208
Power input, kw	25 to 30	18 to 22	9 to 12	2 to 3	0.7 to 1.2
Size (range):					
Length, in.	120	48 to 78	48 to 76	24 to 46	24
Depth, in.	28	22 to 37	22 to 27	22 to 28	22 to 25
Height, in.	84	72 to 84	69 to 84	69 to 84	69 to 84
Weight, lb	3,000	2,000 to 2,600	1,500	500	450

* Audio-frequency response may be measured with or without the standard preemphasis network in the input to the modulator. FM noise and distortion measurements are always made at the output of a standard 75 µsec deemphasis network following the FM detector.

positioned to roughly the correct position by means of a preset potentiometer. It is then controlled by a discriminator, which measures the phase relationship between the amplifier input and output voltages and causes the servo to bring them into phase. This system is particularly useful in adjusting the output tuning and loading controls, to compensate for variation in antenna impedance, etc. In the past, operation in the upper half of this range of frequencies (15 to 30 Mc) gave rise to problems with the tubes of medium- to high-power ratings. Grid, or filament, structures tended to resonate at the operating frequency, or its second or third harmonic. In addition, the high interelectrode capacitances represented quite low impedances, making it difficult to obtain the desired operating condition for the tubes.

In this respect, modern triodes and tetrodes operate very well at these and higher frequencies. Where more than one tube is needed for the final amplifier, it is more economical of components to operate with the tubes in parallel. It also reduces to a minimum the number of possible spurious modes of operation.

The multiple-tube designs usually have the advantage of using tubes that are in quantity production and operate at lower plate voltages for a given output capability. This reduces problems associated with corona and arc-overs. On the other hand, the individual metering and adjustments for balancing the characteristics of the tubes make for complexity, and in addition, it is generally necessary to use parasitic suppressors extensively to avoid spurious oscillations. The use of grounded-grid circuits is almost universal for the high-power stages of short-wave transmitters. These circuits are very stable, and little or no neutralization is required. For linear amplifiers, this circuit has the advantage of providing inherent negative r-f feedback, which helps to minimize distortion.

Tetrodes, while they may be operated with both grids and screens held at ground potential from an r-f viewpoint, should have the normal d-c potentials on these elements, in order to maintain their operating characteristics and avoid damage to the tubes.

In short-wave broadcast operation, where the grounded-grid amplifier may be operated as a high-level plate-modulated amplifier, the driver as well as the final amplifier must be modulated, in order to obtain low distortion and to avoid overdriving the grid on negative peaks of modulation.

BIBLIOGRAPHY

FM Transmitters

Armstrong, E. H.: A Method of Reducing Disturbances in Radio Signalling by a System of Frequency Modulation, *Proc. IRE*, May, 1936, p. 689.
Carson, J. R.: Notes on the Theory of Modulation, *Proc. IRE*, February, 1922. Reprinted in *Proc. IRE*, June, 1963, p. 893.
Day, J. R.: Serrasoid F. M. Modulator, *Electronics*, October, 1948, pp. 72–76.
Hollis, W. C.: Design of Transmission Line Tank Circuits, *Electronics*, May, 1947, pp. 130–134.

Short-wave Transmitters

Bruene, W. B.: Linear Power Amplifier for SSB Transmitters, *Electronics*, August, 1955.
Dettman, M. C.: Automatic-tuning Communication Transmitter, *Elec. Commun.*, December, 1953.
Eitel-McCullough, Inc.: Single Sideband, *Application Bulletin* 9.
van Iterson, P. W. L., C. A. Snijders, and H. A. Teunissen: Multipurpose Continuously Tunable 5 and 10 KW HF Transmitters, *Philips Telecom. Rev.*, vol. 23, no. 1, October, 1961.
Morcom, W. J.: A High Power Communication Transmitter, *Electron. Eng.*, June, 1954.
Single Sideband Issue, *Proc. IRE*, vol. 44, no. 12, December, 1956.
Sturgess, H. E., and F. W. Newson: Single-sideband Multichannel Operation of Short-wave Point-to-point Radio Links, *(British) Post Office Elec. Engrs. J.*, vol. 46, pt. 3, October, 1953; pt. 4, January, 1954.

Chapter 21

MAGNETIC AMPLIFIERS

HAROLD W. LORD*

CONTENTS

1. Simple Magnetic Amplifiers (Saturable Reactors)............... 21–1
    1.1. General Discussion...................................... 21–1
    1.2. Medium- and High-power Saturable-reactor Circuits....... 21–2
    1.3. Low-power Saturable-reactor Circuits................... 21–4
2. High-performance Magnetic Amplifiers (Amplistats)............ 21–5
    2.1. General Discussion..................................... 21–5
    2.2. Current (Ampere-turn)-controlled Half-wave Circuits..... 21–6
    2.3. Full-wave A-C Output (Doubler) Circuit with High-impedance Control Circuit................................... 21–8
    2.4. Full-wave A-C Output (Doubler) Circuit with Low-impedance Control Circuit................................... 21–10
    2.5. Validity Checks of Theory.............................. 21–12
    2.6. Control Characteristics of Low-impedance Control-circuit Amplistats.. 21–15
    2.7. Control Characteristics of High-impedance Control-circuit Amplistats.. 21–17
    2.8. Effects of Nonideal Circuit Elements on Current-controlled Amplistats.. 21–17
    2.9. Voltage-controlled (Controlled Flux Reset) Circuits...... 21–17
    2.10. Operating Amplistats into Inductive Loads.............. 21–25
    2.11. Push-Pull Amplistat Circuits.......................... 21–26
    2.12. Feedback-circuit Considerations....................... 21–29
    2.13. Amplistat-design Considerations....................... 21–30

1. SIMPLE MAGNETIC AMPLIFIERS (SATURABLE REACTORS)

1.1. General Discussion. A *saturable reactor* (SR) is a rugged and reliable means for securing amplification. It is actually a modulator, the supply frequency being the "carrier" frequency, which is modulated by a signal of lower frequency and power. One of the early (1915) applications of SR's was for voice modulation of a carrier frequency[1] for radio transmitter.

* General Electric Co., Research and Development Center, Schenectady, N.Y.

21–2 AMPLIFIER CIRCUITS

High-power applications of SR's include the control of incandescent lamps in theater lighting, electric-furnace temperature controls, and speed control of wound-rotor induction motors.

Applications of SR's for amplification of low signal levels include null detectors and d-c transformers. (The latter are also used at high levels for measuring d-c currents in the kiloampere range, as will be described in connection with Fig. 4.) The *second-harmonic modulator* connection is often used for extremely low signal level applications.

Diagrams of circuits including SR's are, in many instances, simplified by using a modification of the conventional symbol for a transformer as illustrated in Fig. 1. The modification consists of the two short lines at the ends of the two parallel lines usually used to represent the iron core. The result is a suggestion of an idealized hysteresis loop to indicate that the flux in the core reaches into the saturation region under operating conditions.

Another symbol often used for a SR is shown in Fig. 2. The coils labeled A and B are the gate coils, and F_1 and F_2* indicate control windings. F_1 and F_2 are control

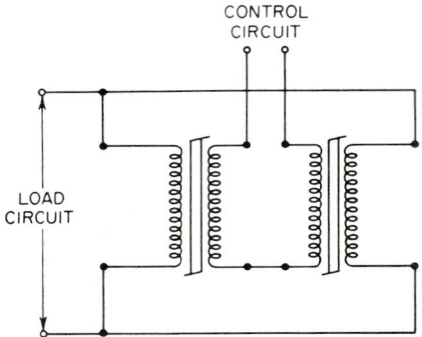
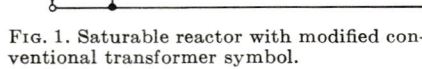
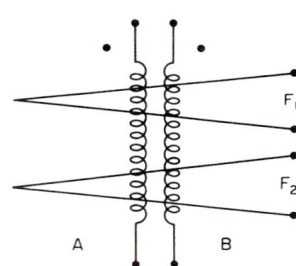

Fig. 1. Saturable reactor with modified conventional transformer symbol.

Fig. 2. Simplified symbol for saturable reactor.

windings which link both gate coils in such a manner that no fundamental supply frequency or odd-harmonic voltages appear across their terminals. If desired, SR's can be controlled by the resultant ampere-turns of a multiplicity of control windings. The symbol shown by Fig. 2 is readily adapted to the indication of two or more control windings. The dots on the gate coils indicate relative a-c polarities which do not introduce fundamental supply frequency and odd-harmonic voltages into the control circuit.

1.2. Medium- and High-power Saturable-reactor Circuits. For most applications of SR's to simple magnetic amplifiers, the effects of the a-c exciting current required by the core are not critical. Any good grade of transformer core material is usually suitable, but the oriented-silicon transformer grades will be superior to the nonoriented grades.

Design procedures for SR's for simple magnetic amplifier circuits are very similar to transformer design procedures. For controlling a given load, the SR gate windings have the same ampere-turns as the secondary winding of a transformer of equal volt-ampere rating; and the control winding has approximately the same ampere-turns as the primary winding. Thus the physical size of a SR to control a given load is about the same as a transformer of the same rating. This fact provides a useful basis for selecting the core size for an initial design of a SR.

The number of turns N_G in each gate winding is related to the core area by the

---

* F stands for "field" winding, in analogy to the control field of an amplidyne or other d-c exciter.

MAGNETIC AMPLIFIERS

usual equation for a transformer

$$E_G = 4NBAf10^{-8} \tag{1}$$

where E_G = half-cycle average voltage rating of the gate windings
$N = N_G$ for a parallel-connected SR
$ = 2N_G$ for a series-connected SR
$B \simeq B_{\mu m}$, the flux density at which μ is a maximum, gauss
A = net area of the core leg associated with N_G, sq cm
f = frequency of a-c supply voltage, cps

The cross-section area of the conductors for the gate windings is determined by the maximum effective load current I_{Lme}, keeping in mind that $I_G = I_L$ for the series-connected SR and $I_G = I_L/2$ for the parallel-connected SR. Normal transformer-design procedures are used to relate the conductor areas to conductor currents. For a 100-va SR, the allowable current density may be as high as 2,200 amp/sq in. This current-density figure will drop to about 1,700 amp/sq in. for a 1-kva SR, to about 1,200 amp/sq in. for a 10-kva SR, and to about 800 amp/sq in. for a 100-kva SR.

From the current gain desired, application of Chap. 8, Eq. (11) or (15), as appropriate, will determine the number of turns required in the control winding. The minimum conductor cross-section area is that determined by the heating effects of the control current. If the control-coil resistance for the minimum-area conductor is higher than desired, as determined by the control current and available control voltage, then a larger conductor can be used and the power gain will be increased at the expense of speed of response.

It has been shown[2] that the control time constant τ_C of the control winding of a *series-connected* SR is

$$\tau_C = (1/4f)(R_O/R_C') \tag{2}$$

where f is cps, R_O is output resistance, and R_C' is resistance of control circuit referred to gate winding; and it has also been shown that the rise and fall of load current follows the control current. This time constant is not valid when a SR is operating largely in the exciting region (minimum gate-current region).

Similarly, it has been shown[3] that the control time constant of a *parallel-connected* SR is

$$\tau_C = (1/f)(R_O/R_C') \tag{3}$$

Equation (3) is identical to Eq. (2) if the factor of 2 in the ratio of gate turns to control turns is taken into account, along with Chap. 8, Eq. (12). In the parallel-connected SR the closed loop formed by the parallel-connected gate windings couples to the control winding so that the total time constant τ is

$$\tau = \tau_C + \tau_G \tag{4}$$

where τ_G is the time constant of the gate loop. An expression similar to Eq. (3) gives

$$\tau_G = (1/f)(R_O/2R_G) \tag{5}$$

where R_G is the resistance of one gate winding. From Eqs. (4), (3), and (5), one obtains for the time constant of the parallel-connected SR

$$\tau = (R_O/f)(1/R_C' + 1/2R_G) \tag{6}$$

It is obvious from Eq. (2) that the series-connected SR can be "forced" to respond in shorter times by including an external series resistance in the control circuit. This is obtained at the expense of power gain. The parallel-connected SR is inherently slower in response compared with the series-connected SR and cannot be "forced" to be much faster than an equivalent series-connected SR without forcing.[4]

SR's are often applied to the control of three-phase power circuits. Details of their operation in polyphase circuits are quite complex and beyond the scope of this handbook. When the load is resistive, it is often connected in Y. If the neutral point of the load is returned to the neutral of a Y-connected secondary of the power-supply transformer, each SR will operate as a single-phase SR. (Note: The primary of the power-supply transformer should be Δ-connected unless supplied by a three-phase four-wire system.)

When SR's are used to control three-phase power to the primaries of three-phase transformers which have rectifier loads on the secondaries, unstable regions of operation are often encountered over a portion of the control range. A controlled-rectifier circuit which is least subject to instability is shown schematically in Fig. 3. The secondaries of transformers T_1, T_2, and T_3 are Δ-connected. Each transformer primary is connected in series with a SR, and the three sets of series circuits of SR's and primaries are Δ-connected.

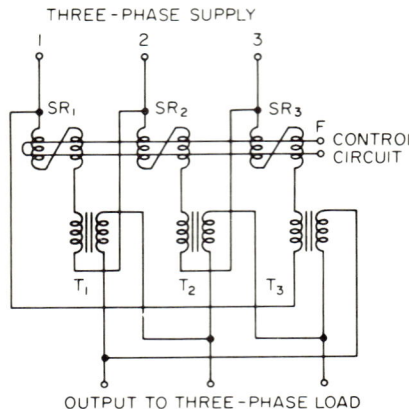

FIG. 3. Three-phase saturable-reactor-controlled rectifier.

When three single-phase SR's are to be operated on three-phase power, it is often advantageous to connect the three individual control windings in series. This connection cancels out the second- and fourth-harmonic voltages and their multiples in the control circuit. With only the sixth-harmonic voltage and multiples thereof present, a relatively small inductor can suppress the flow of harmonic currents in the control circuit. This provides a good means for securing rectangular current waves into three independent but equal single-phase loads.

Figure 3 shows a single control winding linking all three SR's. This is readily accomplished in most SR core configurations by assembling the gate coils in a manner which permits a single control coil to be wound around all the gate coils.

1.3. Low-power Saturable-reactor Circuits. The equality of ampere-turns in the gate windings and control windings of SR's provides the basis for several low-power circuit applications of SR's.

It was shown in Chap. 8, Eq. (21), that when operating in the proportional region, the instantaneous ampere-turns in the gate winding are equal to the instantaneous ampere-turns in the control circuit. The rectified gate-winding current, which is also the load current, will therefore be a replica of the control current. This circuit will function as a current transformer for measuring large direct currents, or direct currents of higher potential, by operating the rectified output into a d-c ammeter. Figure 4 shows the circuit arrangement for measuring large direct currents flowing in a busbar or cable. The even-harmonic currents introduced into the d-c loop are usually so small relative to the direct current that they can be considered as "suppressed."

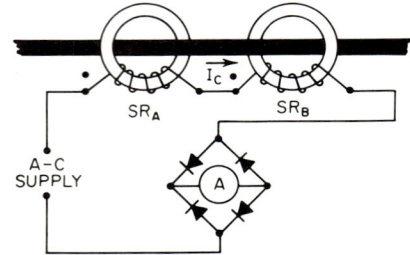

FIG. 4. Direct-current instrument transformer.

Direct-current transformers are particularly suitable for measuring direct currents in the range of 1,000 to 50,000 amp. They can be made to provide high accuracy by adhering to the following specifications: the cores are rings of very high permeability magnetic material; the gate coils are uniformly wound over the entire ring; the spacings between the ring cores and the d-c bus periphery are maintained uniform by mounting them concentric with a round busbar; and the cores are mounted near the center of a relatively long straight section of busbar.

The theory of the series-connected SR presented in Chap. 8 showed that, when the control current is zero and the two SR's are identical, the even-harmonic voltage induced in the control circuit is zero. If the cores are constructed of one of the low-coercive-force high-permeability alloys, the presence of even harmonics in the control circuit is a very sensitive indicator of the existence of a direct current in the control windings of a series-connected SR. Furthermore, the phases of the even harmonics are indicative of the direction of the direct current in the control winding.[5] A series-connected SR can therefore be designed to function as a very sensitive null detector.

The very low power d-c output of devices such as thermocouples and photovoltaic cells can be amplified with a high degree of stability by the combination of a SR null detector, a second-harmonic detector/discriminator and a high-gain d-c amplifier as shown by Fig. 5.[6] The low-level d-c signal to be measured is fed into a first control winding which is designed for high power gain. An inductance may have to be included in the input circuit to suppress the flow of even-harmonic currents. The gate windings are excited from an a-c supply at a frequency of about 1,000 cycles. The harmonic voltage from a second control winding is fed to the second-harmonic detector/discriminator, the d-c output of which is amplified by a d-c amplifier. The output current of the d-c amplifier is fed to the load in series with a third control winding connected so that the ampere-turns due to the load current oppose the ampere-turns of the d-c signal current in the first control winding. With a high-gain d-c amplifier, the ampere-turns in the third control winding are practically equal to those in the first control winding. The turns ratio

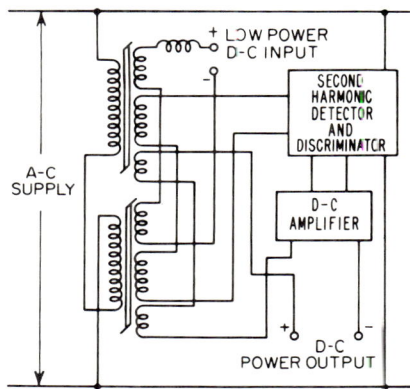

Fig. 5. Saturable-reactor null detector in low-power input d-c amplifier.

of these two control windings determine the overall current amplification of the system. The d-c amplifier can be designed for almost any amount of power, as needed to supply the load to be controlled by the low-power d-c signal.

2. HIGH-PERFORMANCE MAGNETIC AMPLIFIERS (AMPLISTATS)

2.1. General Discussion. If the a-c output of a simple SR circuit is rectified and fed back to a second control winding, the gain of the SR circuit is increased for one polarity of signal and decreased for the other polarity. If the ampere-turns in the feedback winding are equal to the ampere-turns of the gate windings, the gain in one direction becomes very high and the extrinsic positive feedback is equal to the negative feedback intrinsic to the SR.

A much simpler and more efficient way to accomplish the same high gain characteristic is to block the intrinsic negative feedback in the SR by introducing a rectifier in series with each gate winding of a parallel-connected SR. However, when this is done, the law of equal ampere-turns no longer applies, and the transfer characteristic of the magnetic amplifier is influenced greatly by the magnetic characteristics of the SR cores, the reverse leakage resistance of the rectifiers, and the configuration of the magnetic amplifier circuit. When silicon diodes are used, the influence of the magnetic characteristic of the SR cores is predominant. For the circuit analysis presented here, the effects of rectifier reverse currents are neglected.

Many of the methods of analysis applied to the Amplistat include the concept that it is a magnetomotive-force (ampere-turn)-controlled device. Other methods consider it a voltage-controlled device during the period that the core flux is being "reset." Either or both of these concepts will be used here to describe the operation of Amplistats, depending upon the circuit under discussion.

The qualitative, graphic method which will be used to describe the operating fundamentals of Amplistats was first published in Refs. 7 and 8. It provides a clear description of the mechanics of Amplistat operation, the influence of magnetic amplifier circuitry upon the operating dynamic hysteresis loops, and the effects of differences in the dynamic loop widths upon the transfer characteristics of Amplistats.

Rectangular-hysteresis-loop core materials are nearly ideal materials for the cores of SR's in Amplistat circuits. The hysteresis loops included in Chap. 8 show that several of the rectangular-loop core materials maintain the characteristic rectangular shape over a range of frequencies from direct current to 400 cycles. The principal effect of operating the cores at higher frequencies (greater rates of change of flux density) is an overall widening of the rectangular loop. This widening effect and the fact that the flux does not instantly change when the magnetizing force H acting upon a core of rectangular-loop material is slightly in excess of the d-c coercive force H_c (Chap. 8, Sec. 2.2 and Fig. 2) are the two key effects involved in the operating mechanics of Amplistats.

Descriptions of the operation of Amplistats have developed several terms which are listed and defined here for ready reference.[9]

Firing. In a magnetic amplifier, the transition from the unsaturated to the saturated state of the saturable reactor during the conducting or gating alternation. Firing is also used as an adjective modifying phase or time to designate when the firing occurs.

Gate Angle (Firing Angle). The angle at which the gate impedance changes from a high to a low value. This angle is also called the firing angle.

Gating. The function or operation of a saturable reactor or magnetic amplifier which causes it, during the first portion of the conducting alternation of the a-c supply voltage, to block substantially all the supply voltage from the load; and during a later portion allows substantially all the supply voltage to appear across the load. The "gate" is said to be virtually closed before firing and substantially open after firing.

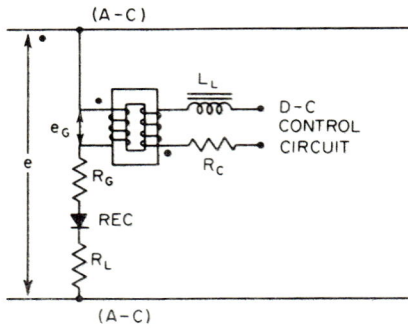

Fig. 6. Half-wave Amplistat.

Reset, Degree of. The reset flux level expressed as a percentage or fraction of the reset flux level required to just prevent firing of the reactor in the subsequent gating alternation under given conditions.

Reset Flux Level. The difference in saturable-reactor core flux level between the saturation level and the level attained at the end of the resetting alternation.

Resetting (Presetting). The action of changing saturable-reactor core flux level to a controlled ultimate reset level which determines the gating action of the reactor during the subsequent gating alternation. The terms resetting and presetting are synonymous in common usage.

Resetting Half-cycle. That half-cycle of the magnetic amplifier a-c supply voltage during which resetting of the saturable reactor may take place.

2.2. Current (Ampere-turn)-controlled Half-wave Circuits. Half-wave Amplistats per se are useful for many circuit applications. The circuit shown in Fig. 6 is often analyzed as a basic element of the so-called doubler and full-wave circuits. These circuits combine two similar cores which are *assumed* (sometimes erroneously) to operate into a common load resistance without any effect on their individual control characteristics as obtained for the simple half-wave circuit. Since a single core provides no means for balancing out the fundamental components, the half-wave circuit must always have a relatively high-impedance control circuit. Therefore, when current-controlled, this is one circuit in which the current supplied by the control circuit is relatively constant at all instants throughout the cycle and so requires the consideration of eddy-current effects in order to explain its operation when rectangular-hysteresis-loop core materials are employed.

Figure 7A shows idealized static and dynamic major hysteresis loops and minor operating loops for a core material *when used in this circuit;* Figure 7B and C shows one cycle of operation of the circuit for minimum output and for "firing" angles of 120 and 0° (full output). The shaded areas emphasize the 120° "firing" conditions. Curve B indicates the induced voltage in the gate winding of the inductor, while curve C indicates the current through the gate winding. In this case, the rectifiers are assumed to be ideal in the inverse direction, and the voltage drop across the load and rectifier during the exciting interval is assumed to be almost negligible. Also, the applied voltage is made just sufficient to cause the peak-to-peak flux-density swing at minimum output to be a fraction less than two times the saturation density.

Consider first the condition of zero control current. After one cycle of operation following application of the alternating voltage, the core will remain in the region of $+B_r$ to point 3, and the voltage across the coil will be substantially zero at all times,

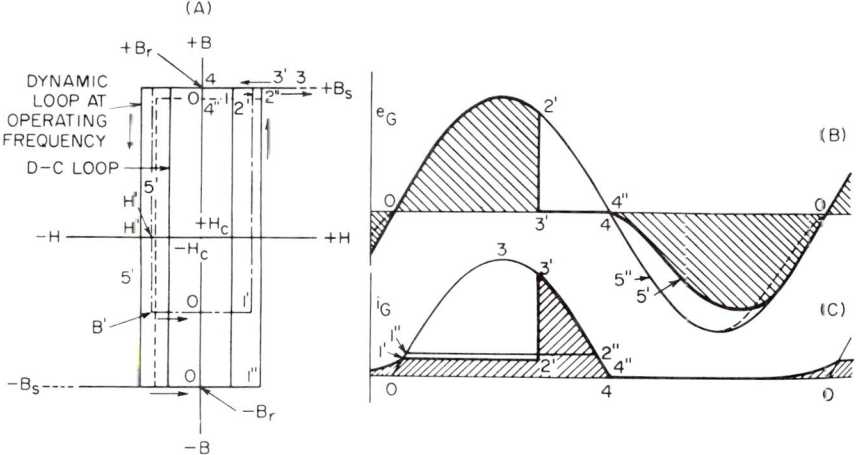

Fig. 7. For half-wave Amplistat. (A) Idealized static and dynamic hysteresis loops. (B) Gate-winding voltage. (C) Gate-winding current.

just as though it were a closed switch. The current through the gate winding will be half-cycles of sine waves, as shown by the light solid curve of Fig. 7C, and limited only by the total circuit and load resistance.

Now suppose the control current is adjusted to such a value as to provide a magnetizing force corresponding to H' of Fig. 7A. During the negative half-cycle, in which the rectifier is blocking, this value of H' will produce in the SR core a negative volt-time area corresponding to the shaded portion in the negative half of Fig. 7B. The flux density in the core will therefore be reset to B' during this period. When the next positive half-cycle occurs, the gate coil can absorb the positive volt-time area as shaded. This positive area is equal to the negative area provided during the preceding negative (resetting) half-cycle; hence "firing" is delayed until 2'. The gate current during the blocking or exciting interval has a value corresponding to the total width of the minor hysteresis loop. This is true since the current in the gate winding not only must supply the ampere-turns required by core excitation but, in addition, must balance out the magnetizing force (ampere-turns) due to the control coil current which continues to flow during the positive half-cycle. After the core saturates, or fires, the current through the gate winding rises to 3', being limited for the rest of the half-cycle, the saturation interval, only by the total circuit and load resistance.

If the control current is further increased to correspond to H'', the core will be completely reset to $-B_s$ during the negative half-cycle. Then, during the subsequent positive half-cycle, the core flux will just fall short of reaching saturation at $+B_s$; so the SR will not "fire." This is the condition of minimum output. For this condition, the current through the gate coil during the positive half-cycles has a value which is determined by the total width of the dynamic hysteresis loop at the operating frequency, and the control current required to establish this condition is determined by approximately one-half the width of this dynamic loop.

Consider now the effect of further increasing the control current so that it exceeds the value corresponding to H''. There will then be a *tendency* for the flux to reset to the $-B_s$ value more rapidly than is required by the supply voltage. However, a greater dB/dt will tend to induce a voltage in the gate winding which exceeds the supply voltage during the resetting half-cycle. Under such circumstances the rectifier REC will conduct during the reset period. The dB/dt in the core is thereby

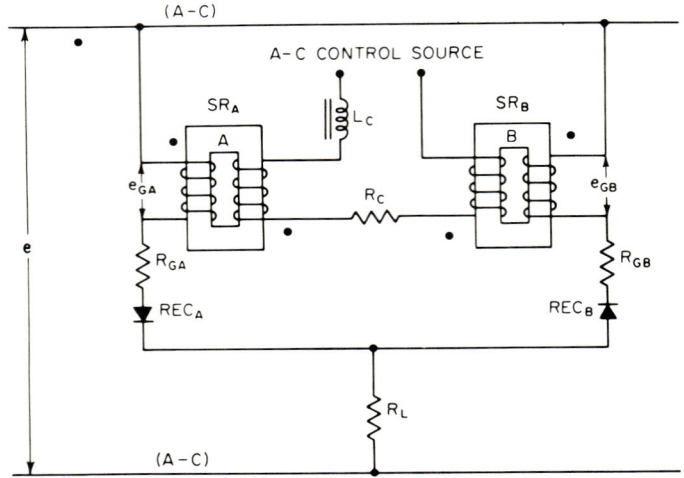

Fig. 8. Full-wave a-c output Amplistat circuit with high-impedance control circuit.

limited to an amount which exceeds the supply voltage by the relatively small resistance drop in the rectifier and load, owing to the *excess* of current supplied to the control winding beyond that required to reset the core in one half-cycle. This shows that in this half-wave circuit *the net magnetizing force acting upon the core* can exceed the value obtained from the major dynamic hysteresis loop by only a relatively small amount. Nearly all the magnetizing force supplied by the control circuit in *excess of this value* is opposed by an equal amount in the gate winding, through conduction of the rectifier during the normally nonconducting half-cycle. In this half-wave circuit the available volt-time area during the reset period consists of one half-cycle of the applied a-c voltage. This amount of volt-time is available *regardless* of the firing angle during the forward or conducting half-cycle.

2.3. Full-wave A-C Output (Doubler) Circuit with High-impedance Control Circuit. When two of the simple half-wave circuits are connected so as to supply alternating current to a common load resistor, as shown in Fig. 8, the resulting circuit is called the "doubler circuit." This circuit is one which at one time was erroneously assumed to have control characteristics similar to the simple half-wave circuit when the control impedance is relatively high. A description of the operating conditions throughout a period of one cycle will serve to show why such an assumption is in error. Figure 9 illustrates the static major and operating dynamic minor loops and the gate-winding voltage and current waveforms for minimum output and for "firing" angles of 120 and 0° (full output). The shaded areas emphasize the 120° "firing" angle conditions. In this case, the rectifiers are assumed to be ideal in the inverse direction only, and the voltage across the load due to the exciting interval current is assumed to be almost negligible. The applied voltage is made just sufficient to cause the peak-to-peak flux-density swing at minimum output to be a fraction less than two times the saturation density.

When the control current is zero, both cores will remain in the region of $+B_s$ (swinging from $+B_r$ to 3) after one cycle of operation following application of alternating voltage. The induced voltage in each coil then will be substantially zero at all times. The current through each gate winding will be half-cycles of sine waves, as shown by the light solid curve of Fig. 9C, one carrying current for half-cycles of one polarity and the other for half-cycles of the opposite polarity. The load current, therefore, will be full cycles of alternating current, the combined circuit in series with the load resistance acting as though it were a closed swtich. The current will be

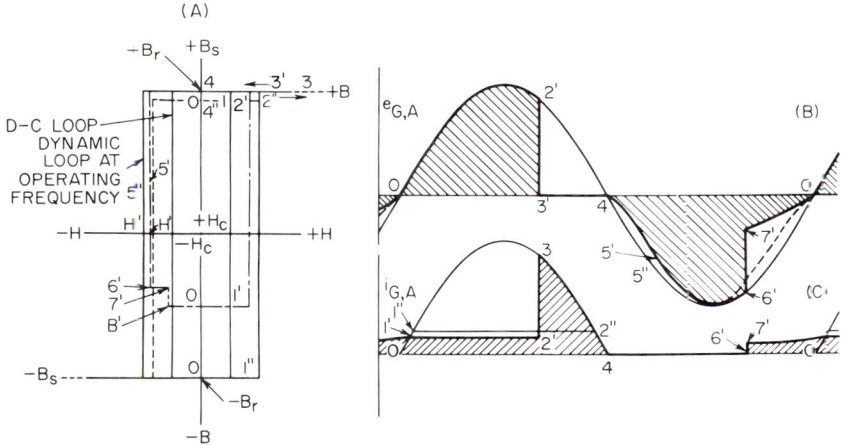

Fig. 9. For full-wave a-c Amplistat with high-impedance control circuit. (A) Idealized static and dynamic hysteresis loops. (B) Gate-winding voltage. (C) Gate-winding current.

limited only by the sum of the load resistance and the total resistance of one gate-winding circuit. This latter resistance includes the forward resistance of one rectifier and the resistance of one gate winding.

Now suppose the control current is adjusted to such a value as to provide a magnetizing force corresponding to H' of Fig. 9A. If we consider only the final steady-state conditions for the time being, then, during the half-cycle in which a rectifier is blocking, this value of H' will produce in the SR core associated with this rectifier a negative volt-time area corresponding to the shaded portion in the negative half of Fig. 9B. The flux density in the core will therefore be reset to B' during this period. This negative volt-time area curve has a large "nick" taken out of it, starting at a point in this half-cycle at which the other core "fires" and applies a large part of the line voltage across the common load resistance. For the remainder of this negative half-cycle, the negative volt-time area available for reset is no longer the full supply voltage but is the relatively small amount determined by the voltage drop across the rectifier and R_G of the branch circuit delivering current to the load. The net magnetizing force required by the lower dB/dt for the remainder of the reset period is shown as 7' in Fig. 9A. The excess of control current, above this value, forces conduction of the rectifier associated with the core being reset for the remainder of this reset period. This rectifier current has been termed preconduction current. (Some authors call it "backfiring.")

Suppose that a given core is to be reset to a certain value of average flux density in either the half-wave or the doubler circuit. The reduced volt-time area available for reset in the doubler circuit then makes it necessary that the average dB/dt for the early portion of the reset period be higher for the doubler circuit than for the simple half-wave circuit. A higher value of dB/dt for a given core will require a higher magnetizing force. Therefore, this concept indicates that for a given firing angle, other than one close to cutoff, a given core in a doubler circuit with high control-circuit impedance should require higher values of control current than it does in a half-wave circuit. This is in agreement with the observed results obtained when a core is first tested in a half-wave circuit and then combined with a similar core in a doubler circuit having a high-impedance control circuit. Figure 10 shows the results of such a test.

2.4. Full-wave A-C Output (Doubler) Circuit with Low-impedance Control Circuit. Another author[10] has presented an excellent and extensive mathematical analysis, of which those portions describing the steady-state conditions are applicable to this circuit. The analysis there presented is based upon a rectangular dynamic hysteresis loop but assumes that the control current is zero during the firing period. The description of the circuit operation which follows will include the effect of the static loop upon the control-characteristic curve and the waveform of the current flowing in the control circuit. All postsaturation inductive effects will be neglected.

Figure 11 shows schematically a doubler circuit with a control circuit made to have a low impedance to the flow of alternating currents by means of a capacitor C_C across the control source. If the control source has an inherently low impedance relative to the control-winding resistance R_C capacitor C_C is not required.

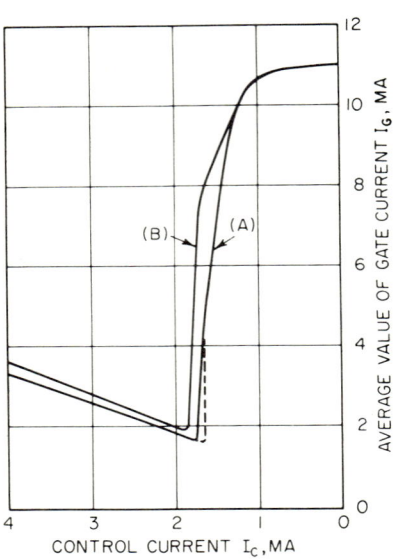

Fig. 10. Amplistat control characteristics. (A) Half-wave circuit. (B) Full-wave a-c output circuit.

When the control-circuit impedance is such as to allow the free flow of even-harmonic current components in the control windings, the mode of operation is quite different from that described for the cases in which the control-circuit impedance is so high as to suppress the flow of a-c components. With the control windings of the SR's connected together, as shown by Fig. 11, the control circuit couples the two cores together electromagnetically. When both cores are operating within their unsaturated regions, a voltage impressed upon the gate winding of one SR will induce a voltage in its control winding. Since the control windings are virtually connected in parallel with respect to such induced voltages, this induced voltage will be impressed upon the control winding of the other SR. The core areas and control-winding turns normally are made respectively equal for the two SR's of a particular magnetic amplifier; so this coupling effect forces the rate of change of flux density (dB/dt) in the two cores to be substantially the same at any given instant. The current which flows in the control circuit at any instant must therefore be such as to maintain this equality of dB/dt for the two cores. If the hysteresis loop is assumed to have zero slopes in the saturation regions, and if all air-core inductive effects are neglected, then, when either one of the cores saturates, the dB/dt in the saturated core is reduced substantially to zero and no dB/dt is permitted in the unsaturated core.

MAGNETIC AMPLIFIERS 21-11

An explanation of the manner in which this circuit operates is simplified if resistance effects are assumed negligible and if a particular steady-state firing angle is assumed. The control-current amplitude and waveshape required to maintain the assumed steady-state condition then may be determined from the magnetic characteristics of the core.

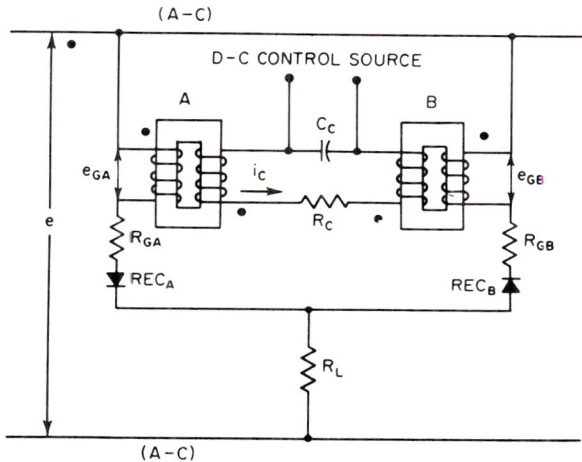

Fig. 11. Full-wave a-c output Amplistat circuit with low-impedance control circuit.

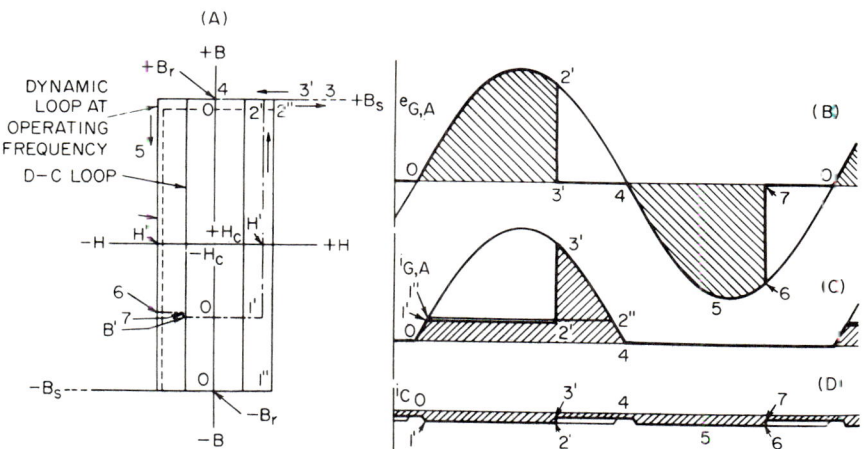

Fig. 12. For full-wave a-c output Amplistat with low-impedance control circuit. (*A*) Idealized static and dynamic hysteresis loops. (*B*) Gate-winding voltage. (*C*) Gate-winding current. (*D*) Control-winding current.

If a firing angle of 120° is assumed, then at the start of a half-cycle, when SR_A is to hold off by absorbing the applied voltage (the exciting interval) flux density in core *A* will be at *B'* and point zero on the operating minor hysteresis loop shown by Fig. 12*A*. (Note that the wider loop with solid lines is the major dynamic hysteresis loop for the core at the operating frequency. The d-c hysteresis loop is the narrower loop shown with solid lines.) During the exciting interval of SR_A, the flux density in core A changes from B' to $+B_s$ at point 2' along the dot-dashed line. Figure 12B, C, and D shows the corresponding winding voltages and currents for SR_A. However, during this same period, SR_B, which has just completed its saturation interval, must have a similar flux-density change but in the reset direction; so it swings down from $+B_r$ to point 6 (slightly less than B', owing to the effect of winding resistances) over the outer knee of the major hysteresis loop for core B. Since rectifier REC_B is blocking during this period, all the magnetizing force H'' required to reset core B must be supplied by current in its control winding. The instantaneous current flowing in the control-windings circuit during this period must therefore be proportional to H'' as shown in Fig. 12A. The control winding of SR_A then acts as the primary. If these two windings have unity turns ratio, then, since the current in the primary or gate winding during this hold-off period will have to supply its own exciting current as well as that for core B, the current in this gate winding will be proportional to $H' + H''$ or proportional to the total width of the operating minor hysteresis loop.

When the flux in core A reaches saturation density at 2' of Fig. 12, this core "fires." The dB/dt in core A then becomes zero; so no voltage will be induced in its control winding. Since the average voltage across capacitor C_C is only that required to overcome the relatively low resistance of the control circuit for the average value of control-circuit current, the voltage now impressed upon the control winding of SR_B is relatively small; hence the flux in core B, which has been reset to a value corresponding to point 6, continues to reset at a very low rate to B' during the conduction period of SR_A. Since the dB/dt in core B is very low, the current in the control circuit drops to substantially that corresponding to the static loop during this period, as shown at 3' in Fig. 12D.

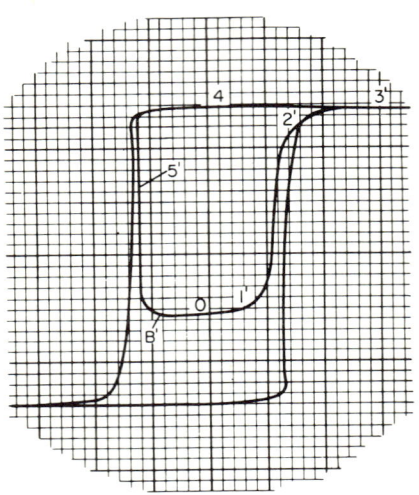

FIG. 13. Major and operating minor dynamic hysteresis loops for circuit of Fig. 6.

The effect of including a static hysteresis loop of finite width in the analysis of Ref. 10 is to shear the top of the control-characteristic curve to the left by an amount which is proportional to the coercive force of the static loop. The values of load current and control current for minimum output (cutoff) would be unaffected. The gain would be a function of the difference between the dynamic coercive force and static coercive force.

2.5. Validity Checks of Theory. The operating minor hysteresis loops and current and voltage waveforms of experimental models of each of the magnetic amplifier circuits described above were viewed on an oscilloscope screen. Vacuum-tube rectifiers were used for the experimental checks. The same specimen core of 0.002-in. Deltamax was used for all these tests. Good correspondence was obtained between the idealized loops and experimental loops for cases employing a high-impedance control circuit.

Figure 13 shows the major dynamic hysteresis loop and the operating minor hysteresis loop obtained for the simple half-wave circuit of Fig. 6 when firing at approximately 120°. Similarly shaped loops were obtained when this SR was combined with a similar SR to form the conventional full-wave d-c output circuits. The control-characteristic curves, when normalized to an equivalent single-core SR basis, were practically identical with the control characteristic for the half-wave circuit. An analysis of the steady-state operation of the half-wave circuit is therefore applicable

MAGNETIC AMPLIFIERS   21-13

to single-phase magnetic-amplifier circuits of the high-impedance control-circuit type when the normal output of the amplifier is a direct current. The normal output of the doubler circuit is an alternating current, and the half-wave circuit analysis does not apply to it even though the a-c output is subsequently rectified.

Figure 14 shows the major and operating minor hysteresis loops obtained for the doubler circuit of Fig. 8 when "firing" occurs at approximately 120°. These should be compared with Fig. 9. The static coercive force of this specimen core was approximately 0.09 oersted and the dynamic coercive force at 60 cycles was approximately 0.17 oersted, a ratio of 1/1.9. The ratio of magnetizing forces indicated between points 7' and B' to that at 5' as shown by the operating minor loop of Fig. 14 is 1/1.5. The lower ratio of the observed value is probably due to the slight widening, over the static loop, required in order to provide the small dB/dt called for during the period from 7' to B'.

When the doubler circuit of Fig. 11 was tested, the operating hysteresis loops were as shown by Fig. 15 for a 120° firing angle. Note that the bottom of the operating

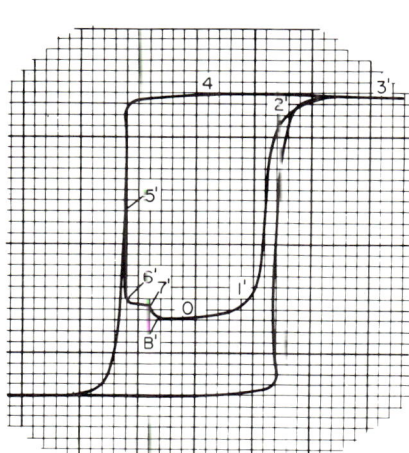 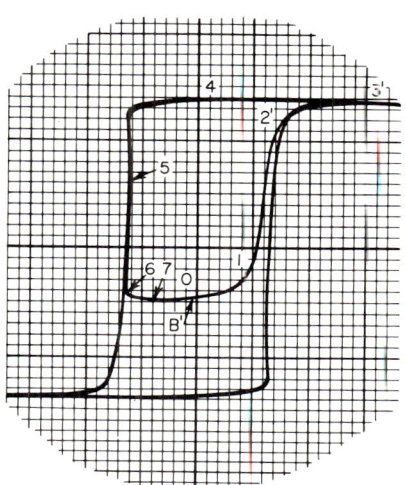

FIG. 14. Major and operating minor dynamic hysteresis loops for circuit of Fig. 8.

FIG. 15. Major and operating minor dynamic hysteresis loops for circuit of Fig. 11.

minor loop does not have the nick, as was predicted by the assumed idealized conditions. Also, Fig. 16 shows that the control current during the conduction period of the gate winding does not remain constant, as was predicted by the assumed idealized conditions. Both these instances of poor correspondence between the predicted and actual results are due to the fact that postsaturation inductive effects were neglected. By neglecting this inductance, the voltage induced in the control coil of the saturated core was assumed to be zero during the saturation period. The control voltage was then assumed to be the only voltage acting upon the control circuit during the saturation period of each half-cycle. Figure 16 shows that the current tends to decrease and may even reverse polarity during the saturation period. This is good evidence that, during the latter half of each half-cycle, the decreasing load current flowing in the gate winding induces a voltage in the control circuit which opposes the signal voltage and may be greater than the signal voltage. This actually causes the core being reset to lose a small amount of its maximum reset flux charge during the saturation period, instead of gaining some as was predicted by the idealized assumptions. The amplitude of the instantaneous voltage induced in the control coil on the saturated core is a function of the slope of the BH curve above the knee of the curve and of the instantaneous rate of change of load current. The amount of current that flows in the control circuit during this period is therefore dependent not only upon the control voltage and width of the static hysteresis loop, but also upon the saturation-region inductance, load current, and inductance of the unsaturated control coil during this period. This latter inductance may be relatively low. The differential permeability of the core across the bottom of the minor hysteresis loop of

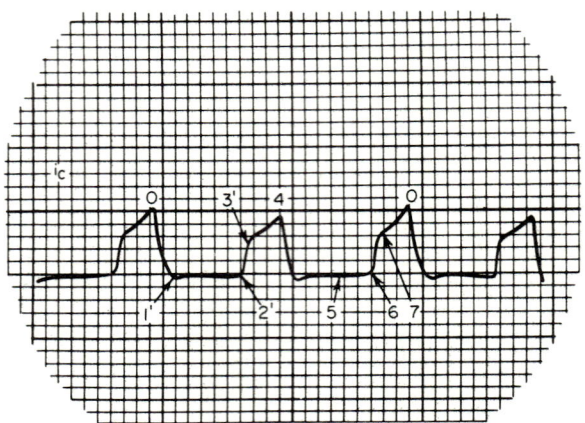

FIG. 16. Control-winding current for circuit of Fig. 11.

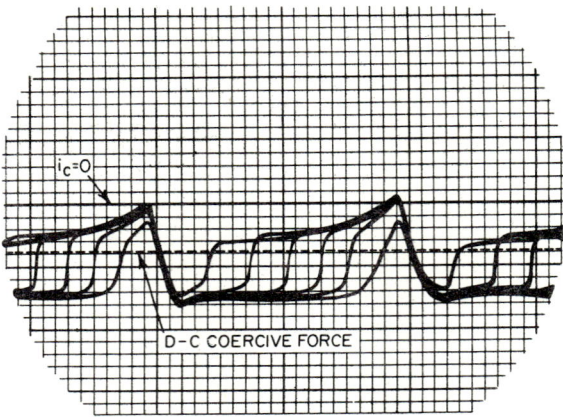

FIG. 17. Composite oscillogram of control-winding current for circuit of Fig. 11 with Deltamax cores.

rectangular-loop materials is shown by Fig. 15 to be rather low. The unsaturated core is operating in this region when the saturated core is operating above the knee of its hysteresis loop.

It therefore appears that a *rigorous* analysis of Amplistat types of circuits having low control-circuit impedances should include the effects of saturation-region inductance. This would include both the equivalent air-core inductance and the slope of the BH curve above the knee of the curve. The differential permeability across the bottom of the minor loop also becomes an important factor to include in such an analysis. However, in spite of these limitations, the analysis of Ref. 10 has been shown to yield reasonably accurate results for materials such as grain-oriented silicon and overannealed Deltamax.

When the static hysteresis loop width of a very rectangular-loop core material exceeds approximately 10 per cent of the dynamic loop width, a more rigorous analysis is required in order to predict with reasonable accuracy the control-characteristic curve of magnetic amplifiers using such materials. Figure 17 is a composite oscillogram of the control-circuit current waveform for representative Deltamax cores in the circuit of Fig. 11 when firing occurs at several different angles of delay. This indicates that a mathematical analysis that will predict the average value of current for each firing angle probably will be very difficult to achieve. Figure 16 also demonstrates that both the dynamic and static hysteresis loops are important factors in determining the control characteristics of Amplistats.

Observation of the steady-state operating minor loops for other single-phase circuits of the low-impedance control-circuit type, such as the full-wave d-c output circuits, show only very minor deviations from those of Fig. 15 for the doubler circuit. An analysis or study which yields useful information for *one* low-impedance control-circuit type should be applicable generally to *all* low-impedance control-circuit types.

Figure 18 shows an example of the arrangement of the instrumentation for displaying the dynamic operating hysteresis loops of SR cores when operating in a magnetic amplifier circuit. The two current-viewing resistors R_1 and R_2 are kept low relative to R_G and R_C, respectively. R_1' is included to preserve symmetry of the gate circuits. Resistors R_1 and R_2 are chosen to satisfy the following equation:

$$R_1/R_2 = N_G/N_C \qquad (7)$$

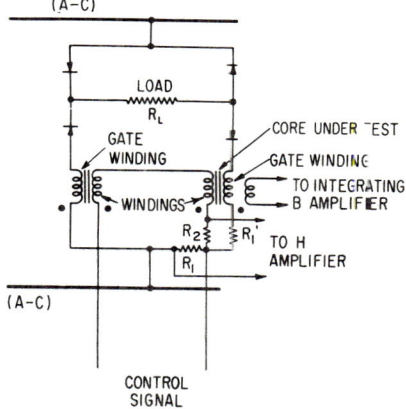

Fig. 18. Circuit for displaying operating dynamic hysteresis loops of full-wave bridge Amplistat.

The circuit interconnections and the coil polarities are arranged so that currents which are produced by transformer action in one SR winding because of currents flowing in the other winding produce voltage drops which are of the same polarity with respect to their common junction. When R_1 and R_2 satisfy Eq. (7), currents produced by transformer action will produce equal voltage drops across R_1 and R_2. Thus the instantaneous difference between the voltages across R_1 and R_2 is proportional to the instantaneous value of the magnetizing force acting on the core. The oscilloscope amplifiers and the integrating amplifier should comply with the specifications enumerated in Chap. 8, Sec. 2.3.

2.6. Control Characteristics of Low-impedance Control-circuit Amplistats. In Sec. 2.4 it was pointed out that Ref. 10 presented an analysis which assumed that the control current is zero during the firing period. The width of the rectangular d-c hysteresis loop was thereby assumed to have no part in determining the Amplistat control characteristic. Curve A of Fig. 19 is a control characteristic as predicted by the referenced analysis. To a first approximation, it is a cosine curve which has its maximum at zero control current and its minimum sheared upward to a value of gate current determined by the width of the dynamic hysteresis loop.

The graphical analysis presented in Sec. 2.4 would predict the control characteristic to be as shown by curve B of Fig. 19. Curve B is equivalent to shearing curve A to the left, the top being moved by an amount determined by the width of the d-c hysteresis loop. Neither curve A nor curve B corresponds well with the small circles which are plotted from a measured characteristic curve. Curve C was constructed by shearing the top of curve A to the left by an amount determined by one-half the width of the d-c hysteresis loop. It provides a reasonably good approximation of the measured control characteristic. Tests using rectangular-loop core materials covering a range of ratios of d-c to a-c hysteresis loop widths have shown equally good correlation of the measured control characteristics with a calculated control characteristic determined by using half of the d-c coercive force ($H_c/2$) and the sine-flux a-c coercive force H_{csf}.

Using the preceding data, the control characteristic of a full-wave bridge or doubler Amplistat with low control-circuit impedance is related to the magnetic characteristics of a rectangular-loop core material by the following equations:

Maximum a-c supply voltage, rms

$$E_e = 4.44 N_G B_{\mu m} A f 10^{-8} \qquad (8)$$

Maximum load current (rectified average)

$$I_{Lm} = E_e/[1.11(R_L + R_G)] \qquad (9)$$

Note: In the case of Amplistats, R_G includes the equivalent forward resistance of the rectifier.

Minimum load current

$$I_{LX} = 2I_X = 2H_{csf}l/0.4\pi N_G \qquad (10)$$

where I_X is the exciting current, amp, of one core, H_{csf} is in oersteds, and l is mean length of magnetic circuit, cm.

Control current for $H_c/2$ (upper end of control curve)

$$I_C = H_c l/0.4\pi N_C \qquad (11)$$

Control current for H_{csf} (minimum load-current point)

$$I_C = H_{csf}l/0.4\pi N_C \qquad (12)$$

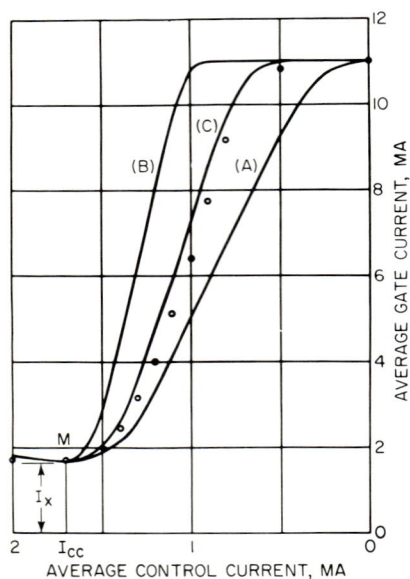

FIG. 19. Predicted control characteristics of Amplistats with low-impedance control circuits. (A) Neglecting effect of static hysteresis loop. (B) Assuming static hysteresis loop fully effective. (C) Empirically assuming half width of static hysteresis loop is effective.

The need for including the d-c hysteresis loop width as well as the *sine-flux* dynamic hysteresis loop width in the above equations is demonstrated by Fig. 20A and B. These figures shows composite oscillograms of the control-current waveshapes for low-impedance control-circuit Amplistats which were identical except for the type of core material in the SR's. The curves in Fig. 20A were obtained with SR cores of an alloy similar to 65-Permalloy (Chap. 8, Table 1, alloy 6) and Fig. 20B curves were obtained with SR cores of Deltamax (alloy 2). Comparisons of the d-c and dynamic sine-flux hysteresis loops shown in Chap. 8, Figs. 8 and 9 show that both materials have approximately the same dynamic loop widths; but the widths of the d-c loops are very different, the 65-Permalloy loop being much narrower. The approximately equal amplitude of control currents during the exciting intervals indicates that the dynamic hysteresis loops of all four SR cores were of approximately the same width. Figure 20A shows that during the saturation intervals the Amplistat control currents for the 65-Permalloy cores drop to lower values than Fig. 20B shows for the Deltamax cores. This is in agreement with the narrower d-c loop width of 65-Permalloy. Also, observe in Fig. 20A the upward bowing of the control current during the exciting interval. The bowing corresponds to the narrowing, at low flux densities, of the sine-flux dynamic hysteresis loops shown in Chap. 8, Fig. 8. The bowing also indicates that the sine-flux dynamic hysteresis loop should be used in preference to sine-current dynamic hysteresis loops for calculating Amplistat performance characteristics. The latter loops are wider than sine-flux loops and never show negative slopes.

2.7. Control Characteristics of High-impedance Control-circuit Amplistats. Assuming ideal rectifiers and rectangular-loop core materials, Amplistats operating from high-impedance control circuits deliver maximum load current until the magnetizing force of the control current during the reset half-cycle exceeds the d-c coercive force H_c of the core. Minimum load current is reached when the control current is slightly in excess of that which produces a magnetizing force equal to the dynamic coercive force of the core. The shape of the control curve between the two extremes of the normal control characteristic has been shown in Sec. 2.3 to be affected by the Amplistat circuit configuration. Its shape is also affected to some degree by the differences in shape of the dynamic hysteresis loops of the cores which are shown in Chap. 8, Figs. 8 and 9. For these reasons, no precise means have been devised to predetermine the shape of the control characteristics of high-impedance control-circuit Amplistats. The shapes of the curves shown by Fig. 10 are fairly typical of these types of Amplistat circuits. Curve A also applies to full-wave bridge and full-wave center-tap circuits.

2.8. Effects of Nonideal Circuit Elements on Current-controlled Amplistats. A rounding of the knee of the hysteresis loops will cause a rounding of the knee of the control characteristic in the region of maximum output. If the rounding is extensive, as shown by Figs. 6 and 7 of Chap. 8, the maximum output will require a positive control current.

If the magnetization curve approaches saturation gradually, as shown by Chap. 8, Fig. 11C, the saturation region inductance will be undesirably high and will cause an increased gate-winding impedance. The increased gate impedance voltage drop at maximum output current then will cause a reduction of the ratio of maximum load voltage to supply voltage.

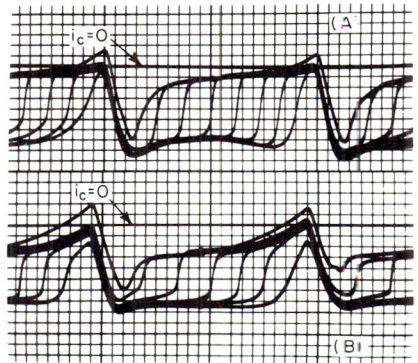

FIG. 20. Composite oscillograms of control-winding current for circuit of Fig. 11. (A) With 65-Permalloy cores. (B) With Deltamax cores.

A decrease in rectifier reverse resistance will shear the top of the control characteristic to the right, thereby reducing the gain. Selenium rectifiers are satisfactory for Amplistats built of SR's having oriented silicon and Supermendur cores (alloys 1 and 7). Germanium rectifiers work well with SR's having Deltamax and 65-Permalloy cores (alloy 2). Amplistats with SR's having cores of any of the other alloys listed in Chap. 8, Table 2 will generally show a substantial increase in gain when silicon rectifiers are substituted for germanium rectifiers. If the rectifier reverse current is greater than the exciting current a positive control current will be required to produce maximum output voltage.

Reference 11 gives more complete details regarding the effects of rectifier reverse currents and hysteresis loop shapes upon Amplistat control characteristics.

2.9. Voltage-controlled (Controlled Flux Reset) Circuits. Although the preceding Amplistat circuits were identified as current-controlled, it was shown in Sec. 2.4 that when the impedance of the control circuit is low, the two SR's are coupled together by the control circuit. Under these conditions when one SR is operating in its *exciting* or *prefiring* interval, the induced voltage in its control winding impresses a voltage upon the control winding of the other core and thereby resets that core. In the next half-cycle the roles of the two SR's are reversed. Each SR is therefore reset under the control of a voltage, and the average control current must be an amount

21-18　AMPLIFIER CIRCUITS

which is determined by the exciting requirements of the SR cores during the resetting periods. Under steady-state conditions the flux reset in one SR core is therefore equal to the flux change in the other SR core during its exciting interval.

In Sec. 2.2 the half-wave circuit of Fig. 6 is shown with a direct current as the control and the circuit operation is described with the assumption that the control signal is provided by a constant-current source. In that type of operation, the control current is the independent variable and the induced voltages in the SR windings during the reset period are determined by the magnetic properties of the core, except for the limitations due to load-circuit interactions.

Figure 21 is similar to Fig. 6 except that the control signal is a voltage source.[12,13] The independent variable is the aforementioned voltage, and the current flowing in the control circuit is determined by the core excitation requirements, except for certain limitations due to load-circuit interactions. The control circuit includes an a-c bias voltage e_2 which is of the same phase as e_1 and polarized as shown by the dots, where

$$e_2 = e_1(N_C/N_G) \quad (13)$$

Like polarities of windings also are indicated by dots.

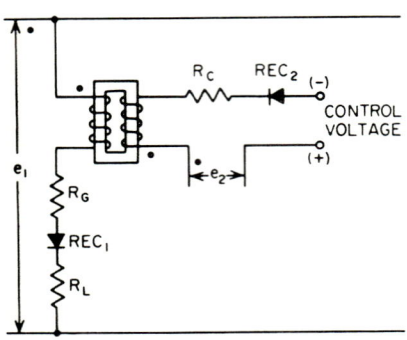

Fig. 21. Voltage-controlled Amplistat circuit.

If the control voltage, a direct voltage of the polarity shown in the parentheses, is zero, the SR core will reset completely during each reset half-cycle and the output current to the load R_L will be a minimum value. For a given SR, the minimum current will be one-half the minimum current for the half-wave current-controlled Amplistat since REC_2 prevents the flow of current in the control winding during the exciting interval. When the control voltage is increased to a value greater than the peak value of e_2, no resetting occurs and the output current to the load is a maximum.

Note that the current required to reset the SR core flows through the voltage-control source in a direction which is opposite to a normal load current from a source of direct voltage.

This type of circuit also can be controlled by a variable impedance instead of a control voltage. The variable impedance can be any one of a number of devices such as a photoresistive device, a transistor, or another half-wave Amplistat. When a series impedance is used as a controlling element, the exciting-current requirements of the core are important factors in determining the overall control characteristic.

The response time of a single-stage voltage-controlled Amplistat is a maximum of one cycle and a minimum of a half-cycle. If a step change in control signal occurs during the normal reset half-cycle, the full corresponding change in output appears in the next half-cycle. When amplifier stages are cascaded, there is an additional delay in response of ½ cycle per stage.

Voltage-controlled half-wave Amplistats can be combined to form full-wave center-tap, full-wave bridge, and push-pull circuits. The conventional doubler circuit for full-wave a-c output does not provide 1 cycle or less response time in the turn-off direction because of the same restrictions on resetting which affect the current-controlled circuit.

Owing to the presence of rectifiers in the input, the inherent drift level of voltage-controlled Amplistats may become high. For this reason, this type of circuit is usually restricted to applications where relatively large signal powers are available.

Two or three stages of voltage-controlled half-wave Amplistats provide high power gains and a minimum of time delay. With half-wave operation, the need for matching pairs of cores is eliminated. Figure 22 depicts a two-stage voltage-controlled Amplistat with the first stage controlled by a transistor Q. The transistor current

MAGNETIC AMPLIFIERS 21-19

amplification and maximum output for normal ranges of input signals must be sufficient to provide the current required to reset SR_2. When SR_2 is on full (Q cut off) the gate current flowing in N_{G2} is that required to reset SR_1. The voltage ratios of e_1/e_2 and e_2/e_3 must all satisfy Eq. (13) and $e_2' = e_2$.

Rectifier REC_4, resistor R_B, and voltage source e_2' function as a synchronous switch[14] to prevent the voltage induced in N_{C1} during its exciting interval from affecting the proper resetting of SR_2 during the same half-cycle. By this means, a voltage of $2e_2$ appears in the gate circuit of SR_2 during its resetting interval and thus prevents harmful "backfiring" of rectifier REC_2.

Although the half-wave circuits are convenient for cascade amplifiers, it is often desirable to provide full-wave output from the final stage. The fundamentals of voltage-controlled circuits can be employed to provide circuits which convert a half-wave control signal to a full-wave output.[8] The means for accomplishing this conversion will be discussed in some detail as they provide good illustrations of the fundamentals of voltage-controlled circuits.

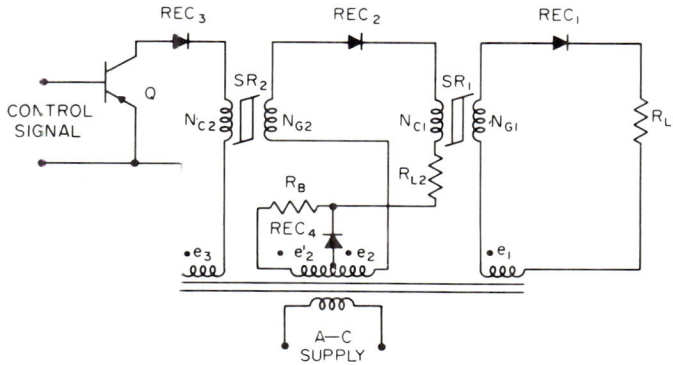

Fig. 22 Two-stage voltage-controlled Amplistat with transistor preamplifier.

In order to provide a full-wave output, the output magnetic amplifier must have two or more SR's which are approximately identical both magnetically and electrically. Assuming that both cores are identical and that the gate windings have equal turns, a properly controlled steady-state single-phase full-wave output will be obtained under the following condition: The flux density in the one core at the start of the half-cycle (gating period) during which it exercises control must be the same as that in the other core at the start of its gating period. Now, in the usual controlled-flux-reset type of circuit, the flux level in a given core can be set only during the half-cycle period that the rectifier associated with its gate winding is normally blocking (reset half-cycle). A half-wave signal of a phase relationship which is suitable for controlling one of the SR's is therefore unsatisfactory for directly controlling the alternate SR. The problem, therefore, may be solved by arranging the circuit so that one SR, which may be called the "master" SR, is controlled directly by the half-wave signal. An interconnecting circuit is then provided which will cause the flux in the core of the second, or "slave" SR to be reset *during its normal reset half-cycle* by the same amount as that set by the signal in the master SR core. Interconnecting circuits which provide this slave action on the part of the second SR are preferably unilateral, since the action of the slave SR should not influence or override the control exercised upon the master SR by the signal.

One of the commonly used magnetic-amplifier circuits for providing an a-c output is the so-called "doubler circuit." Figure 23 shows the half-wave-controlled type of doubler circuit. It is essentially the current-controlled doubler circuit with the control winding omitted on one SR and the rectifier eliminated from the gate-winding

21–20   AMPLIFIER CIRCUITS

circuit of this same SR (see Fig. 6). In spite of these omissions, over most of the control range the half-wave-controlled circuit of Fig. 6 operates in much the same way as the conventional doubler circuit.

In Fig. 23, the two-winding SR_A and rectifier REC_A form a half-wave self-saturated type of circuit. The control circuit, as shown, has resistor R_C to provide a relatively high impedance to the flow of current in the control circuit caused by voltages induced in the control coil through transformer action from the gate winding. An inductor or a combination of elements such as rectifiers and alternating and/or unidirectional half-wave voltages of proper time phase and polarity could be substituted for R_C and the control source.

SR_B of Fig. 23 is the "slave" SR. Its core should be very similar to the core of the "master" SR_A, both in physical dimensions and in magnetic characteristics. However, it need not match the magnetic characteristics of the core of SR_A as closely

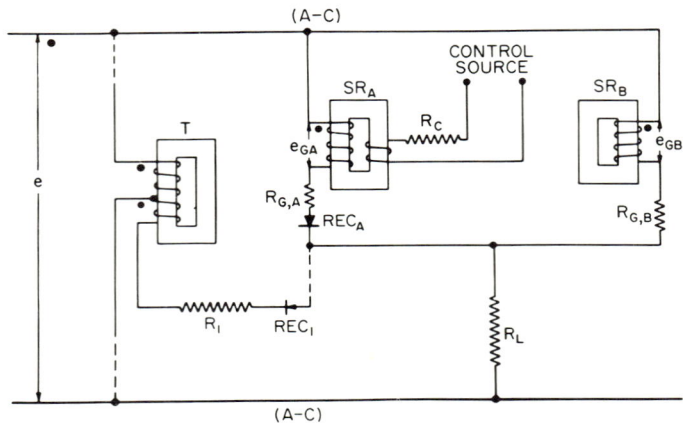

FIG. 23. Half-wave-controlled doubler circuit.

as is required for current-controlled conventional full-wave self-saturated magnetic-amplifier circuits. SR_B is wound with a gate winding only. This should be wound to the same specifications as the gate winding of SR_A.

The purpose of the circuit consisting of transformer T, resistor R_1, and rectifier REC_1 will be explained later. For the moment, we shall assume this circuit to be removed by disconnecting at the dashed lines.

The general manner in which SR_B acts to provide the desired slave action is quite simple. Note that its winding connects the load resistor R_L directly across the supply voltage. An inductor cannot have a steady-state *direct-voltage* component across it which is greater than the resistive drop of a direct current flowing through it. This slave SR, therefore, must pass such currents as are required to maintain a voltage across the load which is predominantly alternating. Any attempts of the controlled SR_A and its associated rectifier REC_A to supply a large direct-voltage component to the load during one half-cycle will be nearly completely offset by a current through SR_B during the next half-cycle.

Figure 24 shows current and voltage waveforms which depict the operation of the circuit of Fig. 23. These waveshapes are derived from idealized rectangular hysteresis loops by a method similar to that used in Sec. 2.2. The shaded portions indicate the waveforms when SR_A is controlled so as to have a firing angle of 120°. During the prefiring (exciting) period 0 to 1 that SR_A is holding off, the flux in core B is being reset by the same amount that was reset in core A during the previous half-cycle by the control signal. During this period, the gate winding of SR_B is for all practical purposes connected in parallel with the gate winding of SR_A; hence the volt-time area will be practically the same for both windings during this period. The only difference during this period is that the magnetic flux in core A is changing from the value of reset flux toward saturation, and the flux in core B is resetting from the residual flux-density value toward a reset value. After SR_A fires, the flux in the core of SR_B continues to reset at a much lower rate, as shown by Fig. 24C between points 1 and 2. This rate is determined by the voltage drop across rectifier REC_A and the resistance of the gate winding of SR_A. During the next half-cycle, from points 2 to 3, SR_B will therefore delay firing for a little more than 120° of this half-cycle. During this same time interval, the flux in core A can be reset by the signal, as shown by Fig. 24A. When SR_B fires at point 3, the resetting of the flux in core A is terminated, except for that provided by the resistance voltage drop across SR_B, even though the signal current may continue to try to reset this flux.

If the firing angle of SR_A is advanced or retarded, the firing angle of SR_B will likewise advance or retard. However, SR_B is not a perfect slave to inductor A. The voltage drop across rectifier REC_A and resistance R_{GA} causes some resetting of core B even when SR_A fires at 0°; so the firing angle of inductor SR_B will not advance completely to the 180° point. Also, when SR_A holds off for a complete half-cycle (cutoff), the exciting current of two cores is flowing through load R_2 during the half-cycle that SR_A holds off. Rectifier REC_A, however, allows only the exciting current of SR_B to flow during the half-cycle that SR_B holds off. This requires the firing angle of SR_B, when SR_A is at cutoff, to lag that of SR_A by a little less than 180°, in order that the average direct voltage across the load may be zero.

Neither of these errors in slave action is large enough to be serious in practical circuits. The delay of firing of SR_B, when SR_A is firing at zero degrees, is small if the forward drop of rectifier REC_A is low. If desired, improved slave action in the cutoff region, and a reduction of approximately 2:1 in the load current at cutoff, can be obtained by using the circuit consisting of transformer T, resistor R_1, and rectifier REC_1 shown in Fig. 23. The transformer should have an approximate 1:1 turns ratio

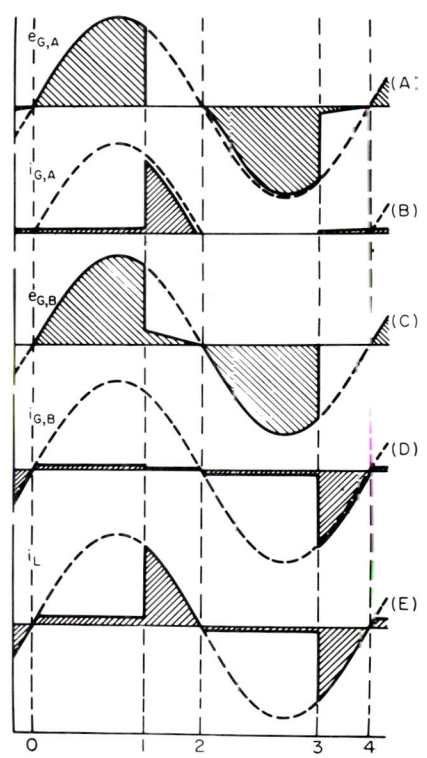

Fig. 24. Idealized waveforms for circuit of Fig. 23. (A) Gate voltage of SR_A. (B) Gate current of SR_A. (C) Gate voltage of SR_B. (D) Gate current of SR_B. (E) Load current.

between primary and secondary. Resistor R_1 should be of such value as to pass an average value of current through rectifier REC_1 which approximately equals the average current through rectifier REC_A at cutoff. With the a-c polarities indicated by the large dots on the diagram of Fig. 23, these two rectifiers will be conducting during the same half-cycle of supply voltage. Rectifier REC_1, therefore, will *reduce* the average current in the load during the half-cycle REC_A conducts, thereby reducing the average current that inductor B must pass during the next half-cycle of supply voltage. If resistor R_1 is made too low, the firing of the slave SR will lag the firing of SR_A more than 180° even in the region approaching cutoff.

Although the slave action of SR_B responds instantaneously to SR_A except for the required half-cycle of delay time, this circuit has speed-of-response characteristics similar to those exhibited by any a-c output or doubler circuit. SR_A will respond to a properly phased control signal in the turn-on sense in as short a time as $\frac{1}{2}$ cycle. In the turn-off direction, however, the firing of SR_B exerts a loading effect upon the resetting of SR_A. This reduces the speed of response to such an extent that several cycles of power-supply frequency may be required to reach cutoff from the full "on" condition.

When the output of an Amplistat is direct current, the speed of response is not limited by the loading effects inherent in the a-c output circuits. It is therefore of interest to apply this same slave action to a circuit whose output is normally direct current, rather than to rectify the a-c output of a circuit such as that of Fig. 23.

Figure 25 shows a full-wave center-tap Amplistat circuit with the flux-reset type of slave action applied to it. In this figure, SR_A is controlled by any desired means suitable for a half-wave type of circuit, including half-wave signals which occur only

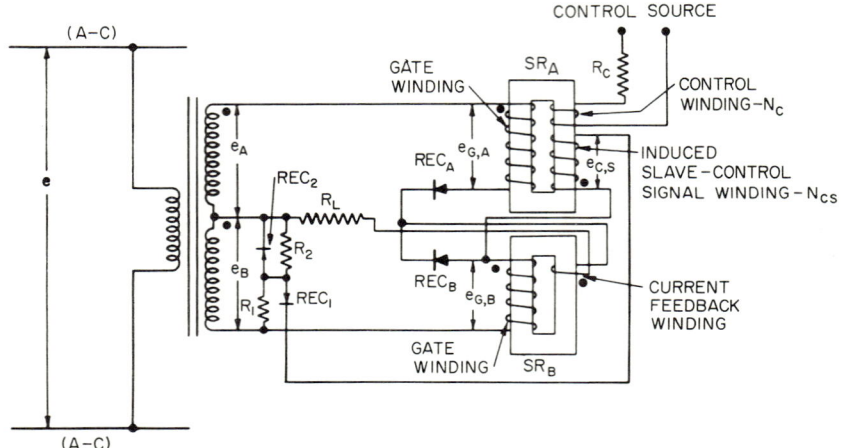

FIG. 25. Full-wave center-tap Amplistat with slave action provided by induced voltage from master inductor.

during the reset half-cycle for this core. In addition to the customary gate winding and control winding, SR_A includes a winding labeled "induced slave-control winding—N_{CS}." The turns of the latter winding equal the turns of the gate winding, but the cross-sectional area of conductor need be only a few per cent of that of the gate winding. SR_B is similar to SR_A minus the control windings and may include a current feedback winding of some sort, the purpose of which will be explained later. Rectifiers REC_A and REC_B are the main power-handling rectifiers. Rectifiers REC_1 and REC_2 are units of comparatively low current rating, since they are in the control circuit. Their functions, along with those of resistors R_1 and R_2, will be explained later.

Figure 26 indicates the voltage and current waveforms of certain parts of the circuit when SR_A is controlled during its reset half-cycle so as to fire at a phase angle of approximately 120°.

Figure 26A shows that during the period 0 to 1, when SR_A is holding off, there will be a voltage induced in winding N_{CS} whose volt-time area equals that of the gate winding during the hold-off period. This, in turn, is shown equal to the volt-time area of the voltage induced in these windings during the reset half-cycle of this SR. Now, one end of winding N_{CS} is connected directly to one end of the gate winding of SR_B, and the other end of winding N_{CS} connects through rectifier REC_1

and resistor R_1 to the other end of the gate winding of SR_B. Rectifier REC_1 is so polarized as to permit the voltage of winding N_{CS}, during the hold-off period 0 to 1 of SR_A, to impress a resetting voltage across the gate winding of SR_B. With both the gate windings and winding N_{CS} having equal turns, and neglecting for the moment the effect of resistance drops, the volt-time area of reset supplied to SR_B during its reset period just equals the hold-off volt-time area of SR_A as shown by Fig. 26C, points 0 to 1. When it becomes SR_B's turn to control the current to the load during the next half-cycle of supply voltage (rectifier REC_B conducting), SR_B will delay its firing so as to fire 180° after the firing angle of SR_A, thus providing the desired slave action.

Rectifier REC_1 prevents loading of winding N_{CS} during the reset period of SR_A (points 2 to 3), but it will not prevent the hold-off voltage of the SR_B gate winding from causing an unwanted reset of SR_A. Such a "kick-back" action is successfully prevented by conduction of rectifier REC_2 during this half-cycle (points 2 to 4). This applies an inverse voltage to rectifier REC_1, approximately equal to e_B, in addition to that from winding N_{CS}. Then, since voltage e_{GB} cannot exceed voltage e_B, rectifier REC_1 will not conduct during this half-cycle even though the voltage of winding N_{CS} during this period is zero. The purpose of resistor R_1 is to limit the current through rectifier REC_2 so that the voltage across rectifier REC_2 in the conducting direction will be small compared with that across resistor R_1.

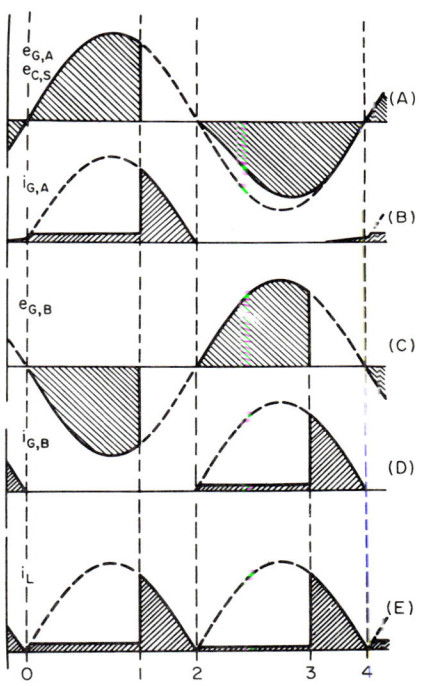

FIG. 26. Idealized waveforms for circuit of Fig. 25. (A) Gate voltage of SR_A. (B) Gate current of SR_A. (C) Gate voltage of SR_B. (D) Gate current of SR_B. (E) Load current.

Although voltage drops during the resetting of SR_B were neglected in the above explanation, their effects cannot be ignored. The flow of SR_B reset exciting current through the forward resistance of rectifier REC_1, resistor R_1, and the winding resistances causes a voltage drop which reduces the amount that SR_B will be reset by the resetting induced voltage in winding N_{CS} when SR_A is holding off. The effect of this voltage drop is to cause the firing angle of SR_B to lag that of SR_A by less than 180°. Resistor R_2 provides a simple means for compensating for such effects. If it is of the proper value a small sinusoidal component of current will be fed to resistor R_1, of the correct polarity to compensate for the resistive voltage drops caused by SR_B reset exciting current. A proper value for resistor R_2 is indicated when the SR_B firing angle lags that of SR_A by 180° at cutoff.

It is well known to those familiar with Amplistats that rectifier leakage currents tend to reset the cores of their associated SR's and thereby to act as a negative feedback. In this circuit, their effect is to cause the firing angle of SR_B to lag that of SR_A by more than 180°. Thus, when SR_A reaches maximum output (firing angle of 0°), SR_B will not have reached full output. This effect may be minimized by passing the load current through a series-connected feedback winding of sufficient turns to provide a positive feedback which equals the negative feedback caused by rectifier leakage. Similar compensation may be provided by a shunt feedback circuit when the application involves a variable load impedance. A feedback winding, hav-

21-24 AMPLIFIER CIRCUITS

ing turns that fall within the range of 2 to 5 per cent of the turns of the gate winding, when connected in series with a suitable resistor and with this series circuit connected across the load, will provide satisfactory performance without excessive loading effects because of induced voltages in the feedback winding.

Slave action of the second inductor may be obtained by a somewhat different mode of operation, as shown by Fig. 27. An auxiliary secondary winding on the power transformer T, identified as "reset voltage-supply winding N_R," supplies a resetting voltage to "slave-control winding N_{CS}" on inductor SR_B during the normal reset half-cycle. With the a-c polarities as indicated by large dots, the current required to reset SR_B flows in the series circuit composed of the forward or conduction direction through rectifier REC_1, the load as represented by R_L, and winding N_{CS}.

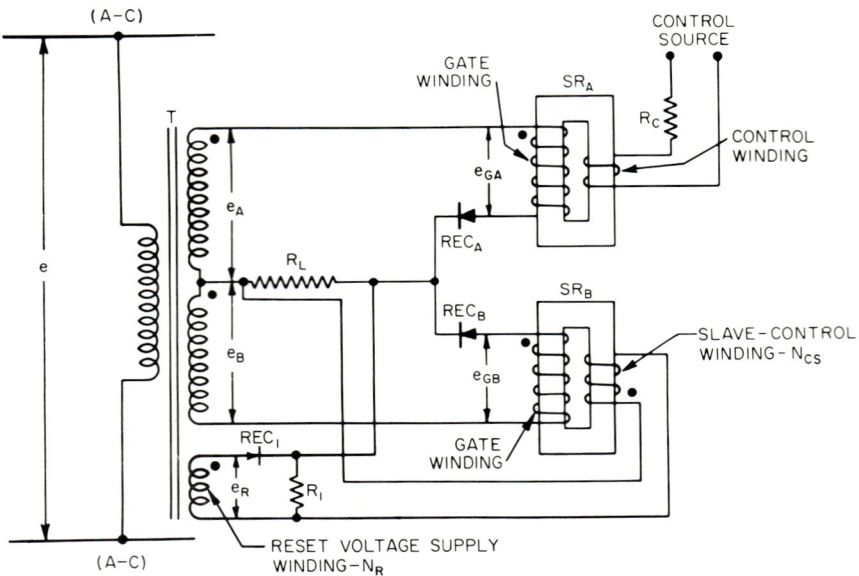

FIG. 27. Full-wave center-tap Amplistat with slave action controlled by load voltage.

During the prefiring (exciting) interval of the conduction period for REC_A (period 0 to 1 in Fig. 26) the voltage across load R_L is relatively low. A large portion of voltage e_R will then be impressed across winding N_{CS} to reset SR_B.

After SR_A fires (period 1 to 2 in Fig. 26), the voltage across load R_L acts to reduce the net voltage across winding N_{CS}. If the voltage e_R is such that, after SR_A fires, the instantaneous voltage across load R_L always exceeds e_R, an inverse voltage will appear across rectifier REC_1 during the period 1 to 2. As a result, no further resetting of SR_B will occur for the remainder of the half-cycle.

Now, by a proper choice of the turns of winding N_{CS}, the flux in core B can be made to reset during the period 0 to 1 by an amount that will delay the firing angle of SR_B so as to lag the firing angle of SR_A by 180°. The turns of winding N_{CS} will be approximately correct if the turns ratio of the gate winding to winding N_{CS} is 110 per cent of the ratio of e_B to e_R. It is better to have this ratio a bit too high rather than too low. Any tendency of SR_B to reset too rapidly will cause conduction of rectifier REC_B during the reset period 0 to 1 of Fig. 26. This will cause the current in the reset circuit to increase; and the voltage supplied by winding N_R, which is in excess of that required to reset SR_B properly, will appear as additional voltage drops across series-resistive elements of the slave control circuit.

Leakage currents of rectifier REC_B will impair the operation of this circuit in the same way as that described for the circuit of Fig. 25. Resistor R_1 in the control circuit of Fig. 27, in conjunction with winding N_{CS}, provides a shunt-type positive feedback circuit for SR_B to oppose the negative feedback effect of the leakage current of rectifier REC_B.

Both the modes of slave operation that have been described for the full-wave center-tap type of circuit can be used equally well for the full-wave bridge type of circuit.

The result of a speed-of-response test of a half-wave-controlled full-wave d-c output Amplistat circuit is shown by Fig. 28. This oscillogram was obtained by a dual-beam cathode-ray oscilloscope with synchronized sweeps. The upper trace shows the half-wave resetting signal applied to voltage-controlled SR_A. A half-wave positive-bias circuit assured full conduction of the master inductor in the absence of a resetting signal. The lower trace shows the full-wave output voltage with a resistance load. Using the signal voltage as a time reference, with the time periods as shown by letters, the sequence of operation may be followed by starting at a. From a to b, no reset signal voltage (upper trace) is applied to the control winding of SR_A; so it fires (lower trace) for a complete half-cycle from b to c, and slave SR_B likewise fires for the complete half-cycle from c to d. During the reset period c to d for SR_A, again no reset signal voltage is applied; so both SR's again fire for the intervals d to e and e to f, respectively. But while SR_B is firing during the period e to f, a half-cycle of reset signal voltage is applied to the control winding as shown by the upper trace. Consequently, master SR_A holds off for practically the whole half-cycle for period f to g (lower trace), and slave SR_B duplicates this performance from g to h. The upper trace shows that SR_A again is reset during period g to h; so again both inductors hold off for the complete half-cycles of h to i and i to j (lower trace). However, during period i to j, the upper trace shows that master SR_A received a resetting control voltage for approximately one-fourth cycle; so it is only partially reset. As a result, the lower trace shows that master SR_A fires for the last half of the half-cycle period from j to a, and slave SR_B does nearly the same during period a to b. This same sequence of events was repeated in this test for every five cycles of supply frequency, corresponding to a repetition rate of 12 per second.

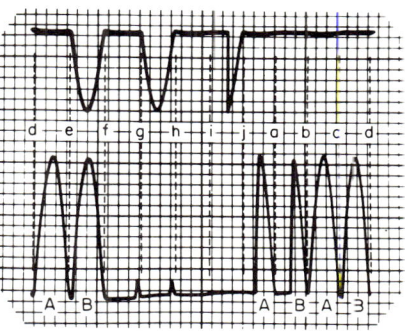

FIG. 28. Oscillogram showing speed of response provided by circuit of Fig. 23.

The preceding explanation of the sequence of events shown by Fig. 28 clearly demonstrates that, in this full-wave output circuit, the "master" SR responds to a signal with the half-cycle-per-stage time-delay characteristic of the voltage-controlled type of magnetic amplifier. The "slave" SR is shown to require only an additional half-cycle of time delay for completing its slave action. The oscillogram also shows the response time to be the same for both increasing and decreasing outputs.

2.10. Operating Amplistats into Inductive Loads. Operation into inductive loads, such as motor or generator fields, relay coils, and control windings of SR's, may seriously modify the performance characteristics of certain Amplistat circuits. An adverse effect upon the half-wave circuit can be expected as even a half-wave rectifier does not operate well into an inductive load. The full-wave center-tap circuit is adversely affected by the continued flow of current through the SR's immediately after the a-c supply voltage reverses at the end of the normal forward conducting half-cycle.

Just as adding a second rectifier, reverse-connected across the inductive load, provides good operation for half-wave rectifiers, it also provides good operation of half-wave Amplistats into inductive loads. Figure 29 is a simplified circuit diagram of this

21–26   AMPLIFIER CIRCUITS

circuit. REC_2 is the added rectifier, often designated as a *free-wheel, bypass,* or *shunt rectifier*. The control characteristic of this Amplistat will be substantially the same as though it had a resistive load, but the response time will be longer because of the L/R time constant of the load.

The effect of an inductive load upon a full-wave center-tap Amplistat is to shift the control characteristic to the left and to be double-valued over a portion of the characteristic. The latter produces an unstable or snap-action type of characteristic.[15] A shunt rectifier of low forward voltage drop is effective in eliminating the snap action, and the control characteristic will be the same as for resistive loads.

A full-wave bridge type of Amplistat will be free of snap action if the rectifiers have low forward drop. The reason for this is that two rectifiers of the bridge can act as a shunt rectifier. Figure 30 shows a full-wave bridge Amplistat with inductive load. Rectifiers REC_3 and REC_4 serve as a shunt rectifier. If their forward drop is not sufficiently low to eliminate snap action, an additional low-drop shunt rectifier is connected in parallel with the load.

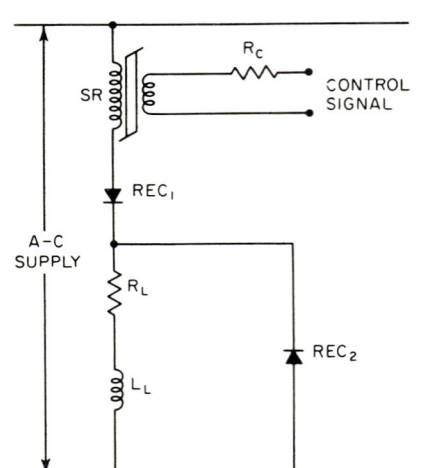
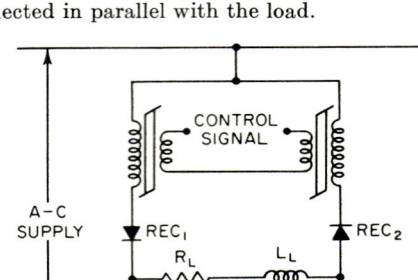

Fig. 29. Circuit for operating a half-wave Amplistat into an inductive load.

Fig. 30. Full-wave bridge Amplistat with inductive load.

2.11. Push-Pull Amplistat Circuits. The several Amplistat circuits described in preceding sections have been single-ended circuits. Without exception, any two of the circuits discussed can be combined to provide a push-pull output. Two half-wave circuits can be combined to provide a reversible-phase a-c or reversible-polarity d-c output. Two a-c output circuits (*doubler circuits*) can be combined to provide a reversible-phase a-c output; and two d-c output *center-tap* circuits or two *bridge* circuits can be combined to provide a reversible-polarity d-c output.

True push-pull operation of magnetic amplifiers into a common electric load requires some series resistance in each single-ended magnetic amplifier to limit the circulating current in the gate windings during those intervals when both magnetic amplifiers are conducting simultaneously. Such current-limiting resistors are referred to as *ballast resistors* or *dummy-load resistors*. In low-power Amplistat circuits, the resistance of the gate windings may be sufficient to provide the current-limiting action. High-power applications usually employ ballast resistors external to the Amplistat as the power dissipated in them is seldom less than equal to the maximum power delivered to the load (50 per cent efficiency) and often exceeds five times the maximum load power (efficiency of 16 per cent).

Figure 31 shows two half-wave Amplistats connected in push-pull. When no control signal is applied, the bias circuit delays the firing angle of each SR to some predetermined angle, such as 90°. With zero signal input, even though each SR delivers current to ballast resistors R_{D1} and R_{D2}, the voltage across R_L is zero since both SR's are *simultaneously* in either their excitation interval or their saturation interval.

MAGNETIC AMPLIFIERS 21-27

When a control signal of a given polarity is applied, the firing angle of one SR advances and of the other SR retards, thus producing a pulsating d-c voltage of a particular polarity across R_L. A reversal of the polarity of the control signal causes a corresponding reversal in polarity of the d-c voltage across R_L. Maximum output of either polarity occurs when the firing angle of one SR is at 0° and the other SR is at cutoff.

The control characteristic of the push-pull circuit is the composite of the individual control characteristics of the two single-ended Amplistats. Symmetry of push-pull Amplistat control characteristics requires good matching of the magnetic characteristics of the SR cores. Good linearity of control characteristic is had when each SR fires at an angle of 90° with zero control signal. A lower standby loss can be obtained by increasing the bias to retard the zero signal firing angle. This method of reducing the standby loss, if carried too far, will cause the control characteristic to be nonlinear in the zero signal region.

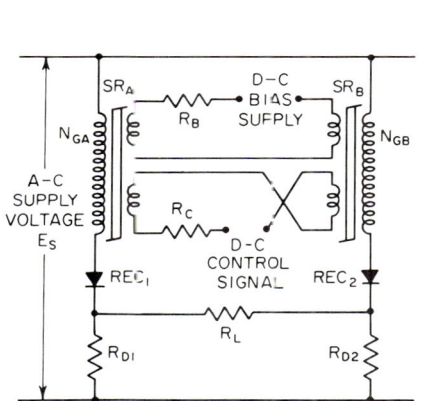

Fig. 31. Half-wave push-pull Amplistat.

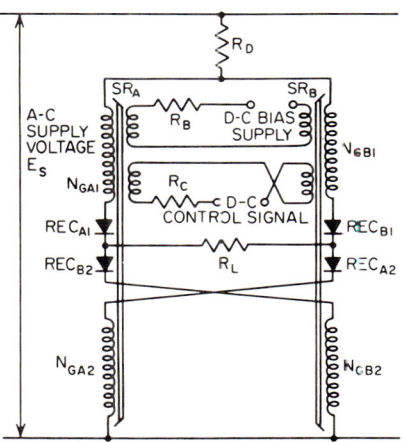

Fig. 32. Half-wave push-pull Amplistat of improved efficiency.

Assuming negligible gate-winding resistances, it can be shown that for maximum overall efficiency the ballast resistors R_{D1} and R_{D2} should be equal to the load resistance R_L. The maximum efficiency is approximately 16 per cent.

A more efficient half-wave push-pull circuit is shown in Fig. 32. Each SR has two gate windings, and the four rectifiers are connected in a bridge arrangement with a gate winding in series with each rectifier. Only one ballast resistor R_D is required. The bias and control circuits are the same as for the half-wave push-pull circuit of Fig. 31.

Operation of this circuit is essentially the same as that described in connection with Fig. 31. The higher overall maximum efficiency of 50 per cent occurs when $R_D = R_L$ (neglecting gate-winding resistances). The reason for this is that this circuit has only one ballast resistor, which is in series with the load, whereas the circuit of Fig. 31 has a ballast resistor in series with the load and another one of equal value in parallel with the serially connected load and ballast resistors. For a given maximum output load voltage and current, both circuits require the same supply voltage E_s, but the circuit of Fig. 31 requires three times the current from the supply.

Considerations of the volt-ampere ratings of the SR's for the two circuits show that, although the current ratings must be 3:1, the volt-ampere ratings are only 3 to 2. When one SR is delivering maximum voltage to R_L, each gate winding of the biased-off SR must be capable of holding off the full half-cycle of voltage across R_L.

The SR's required for the circuit of Fig. 32 have two gate coils, each of which must withstand the voltage across R_L and carry the current through R_L. The SR's required for Fig. 31 have one gate winding which must withstand the voltage across R_L and carry three times the current through R_L. For a given power output, the volt-ampere rating of SR's for the push-pull circuit of Fig. 31 is only 50 per cent more than for the more complex circuit of Fig. 32.

The addition of one more core and gate winding to each SR and a reversed rectifier to the circuit of Fig. 31 converts it to the push-pull a-c output circuit. Figure 33 shows this circuit in simplified form with the control windings omitted. This circuit provides a variable-amplitude reversible-phase a-c output in response to reversing-polarity and varying-amplitude d-c signals. To secure the maximum theoretical efficiency of 16 per cent, the resistances of R_{D1} and R_{D2} should equal R_L.

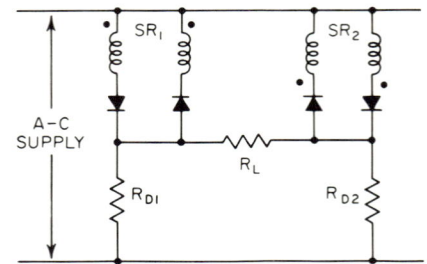

Fig. 33. Full-wave push-pull a-c output Amplistat.

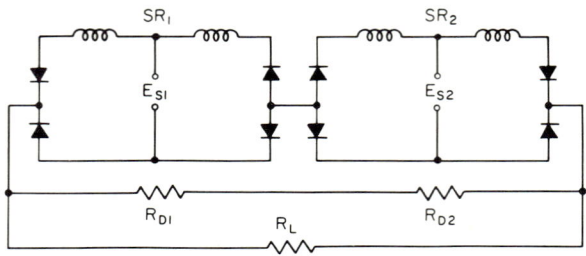

Fig. 34. Full-wave push-pull bridge Amplistat for reversible d-c output.

Figure 34 shows in simplified form a full-wave push-pull bridge circuit for reversible d-c output. The two a-c supply voltages E_{S1} and E_{S2} are of equal voltages and of the same phase. One may be secured directly from a main a-c supply and the other through a 1:1 isolation transformer. Again, for maximum theoretical efficiency the ballast resistances should be equal in value to the load resistance and the maximum efficiency is about 16 per cent.

As for the half-wave push-pull circuit, a more complex circuit will provide a higher maximum theoretical efficiency of 50 per cent. The circuit is shown in simplified form by Fig. 35. The circuit can be constructed of four identical bridge Amplistats, requiring eight matched cores. Some simplification can be achieved if each of four matched cores are wound with twin gate windings as was indicated in Fig. 32 for the half-wave SR's. One full-wave SR can then serve as SR_{A1} and SR_{A2} and the other full-wave SR can serve as SR_{B1} and SR_{B2}. The four isolated a-c voltage sources can be four equal-voltage secondaries of one transformer.

Each Amplistat of Fig. 35 directs its current upward. To produce full-wave rectified load current of one polarity, Amplistat SR_A is turned on and SR_B is turned off by the control signal. Load current will then flow through the series circuit consisting of the two halves of SR_A, the load resistance, and the dummy load resistance. Signal windings and bias windings on the two Amplistats are arranged to provide the required selective operation of the Amplistats for both half-cycles of the a-c supply. The same criterion for biasing applies as was described for the half-wave push-pull circuits. Load current of the opposite polarity is obtained when SR_B is turned on and SR_A is turned off by a reversal of signal polarity.

2.12. Feedback-circuit Considerations. Section 2.1 described the Amplistat as a SR circuit in which the intrinsic feedback is blocked by rectifiers in the gate-winding circuit. Starting with an Amplistat circuit, either positive or negative feedback can be applied to modify its control characteristic. Amplistats can also be used as part of a feedback control system.

If a small amount of positive feedback is applied to an Amplistat, for example, by feeding back a portion of the output voltage or current so as to be additive with respect to the control signal, a snap-action type of characteristic is produced. Negative feedback reduces the gain but can reduce the response times and provide an output which is a linear function of the control signal over the normal operating range of power output.

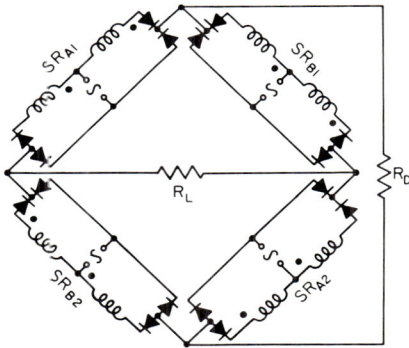

Fig. 35. Full-wave push-pull bridge Amplistat of improved efficiency.

Amplistats provide a means for isolating the feedback circuit from the signal circuit. In an ampere-turn-controlled Amplistat, the feedback current or voltage signal can be fed to one set of control windings and the control signal impressed upon another set of control windings. The net signal acting to control the output of the Amplistat is the algebraic sum of the control ampere-turns due to the feedback signal and the control ampere-turns due to the control signal. This is called *magnetic feedback* and is shown in Fig. 36.

Although electrically isolated, there is some inductive coupling between the feedback circuit and the control circuit. The effect of this coupling between control windings can increase the response time of the Amplistat when the reflected impedance of the circuits connected to the control windings is low in relation to the exciting impedance of the core.

Except for some transient effects due to magnetic coupling, magnetic feedback has no effect upon the impedance which the control circuit presents to the signal source. For some Amplistat applications, it may be desirable to have the Amplistat present an impedance to the signal source which is much higher or much lower than is normally realized in a practical Amplistat. This can be accomplished by the proper use of electric feedback. Amplistat circuits of this type have been designated by the term *magnetic amplifier of the self-balancing potentiometer type.*[16] If a high input resistance is desired, a portion of the output voltage is resistance-coupled to a resistor in series with the control circuit so as to produce a large negative (inverse) feedback. The negative feedback reduces the voltage gain but raises the input resistance.

If a low input resistance is desired, negative electric feedback voltage proportional to load current is introduced in series with the control circuit.[16] A critical amount of external positive feedback is used to increase the gain of the Amplistat to the point of instability before the negative electric feedback loop is closed. The negative electric feedback stabilizes the overall Amplistat circuit, and the positive magnetic feedback greatly reduces the power required to change the operating point of the Amplistat. The overall result is a high-gain single-stage magnetic amplifier which has excellent linearity over the normal range of power output.

2.13. Amplistat-design Considerations. Amplistat circuits are chosen in preference to simple saturable reactors for applications where the performance of the latter in regard to power gain and speed of response is not adequate. Since the magnetic characteristics of the cores of SR's used in Amplistats are the dominant factors in determining the Amplistat performance, SR's for Amplistats are usually designed around core materials of good magnetic characteristics.

The selection of core materials for large Amplistats must often be a compromise between material cost and performance. The material numbered 1 in Chap. 8,

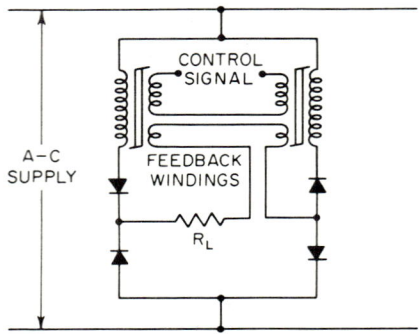

Fig. 36. Full-wave bridge Amplistat with magnetic (current) feedback.

Table 1 is generally used for large Amplistats. Number 2 material is superior to No. 1 but is much more costly. It is widely used in medium-power and 400-cycle Amplistats. Material numbered 7 is useful in the medium size range, but its high cost is seldom justified. Core materials numbered 3, 4, and 5 are good materials for low-power high-performance Amplistats. When operation is at frequencies above about 10 kc, rectangular-hysteresis-loop ferrites may be competitive with fractional-mil-thick tape cores of Supermalloy.

Selection of the core configuration also depends upon the size of the Amplistat. Toroidal shapes are superior but, to avoid hand winding, they require special toroid winders. When the operating frequency will tolerate lamination thicknesses of 0.003 in. and above, special overlapping types of U punchings are nearly equal magnetically to toroidal cores. The core configurations shown by Chap. 8, Figs. 25 and 26 are not suitable for SR's in Amplistat circuits.

The cross section of conductors for the gate windings of SR's for Amplistats is determined in the same manner as for a rectifier transformer winding in which the load currents are half waves. Thus, for a given rms current I_{Le} in the load, the rms current in each gate winding is $0.707 I_{Le}$. Suggested current densities are as listed in Sec. 1.2.

The number of turns N_G in the gate winding is related to the core area by Eq. (1) with $N = N_G$. The voltage E_G in Eq. (1) may be less than the average supply voltage E_S in some Amplistat circuits. For example, push-pull circuits are generally operated with a supply voltage which is two times the voltage at which each Amplistat would be operated single-ended. This much overexcitation can be tolerated in

MAGNETIC AMPLIFIERS

push-pull circuits since the output voltage to the load is zero when both Amplistats are operating in their saturation interval.

A single control winding on an Amplistat SR needs a net conductor area and turns product which is only 2 to 5 per cent of the product for the gate winding. If multiple control windings are required, and especially when the net control acting on the core is the differential of two large opposing current signals, the ratio of cross-sectional area of total control winding to total gate winding is considerably increased. Many designs are based upon a core window allowance of 20 per cent for the control windings.

In Amplistat circuits, a relatively high leakage reactance between the gate windings and the control windings can be tolerated. This permits core and coil configurations which have the gate windings on one leg of an overlapping U type of core and the control windings on the opposite leg. The two cores of a full-wave Amplistat having gate windings on one leg of each core can have a common control winding around the other legs of each core. Similarly, the four cores of a push-pull circuit can be assembled with all four of the control-circuit legs within a common control coil. In both cases, proper polarity relationships between the gate windings and the control windings must be observed. For push-pull operation, this method of assembly does not provide for bias coils for the individual Amplistats. Bias currents can be introduced into each gate winding by means of resistors shunting the rectifiers.[17] Alternatively, two bias windings, each common to the control legs of a pair of cores, can be included within a single control coil for both Amplistats.

When the Amplistat is designed around toroid cores, the preferred core and coil configuration is as described at the end of Sec. 3.5. For push-pull Amplistat circuits, the gate windings and bias windings can be arranged in that manner and then the two Amplistats assembled side by side coaxially and a single control coil wound to include all four cores. Limitations of toroid winders may not permit this latter type of coil and core configuration.

Once the size of the cores for the Amplistat and the turns in the gate winding N_G have been established from the maximum load current I_{Lm} and nominal voltage rating E_e of the gate windings, the control-windings specifications can be determined. To simplify the calculations, the gains and time constant of the Amplistat can be determined at a firing angle of $90°$ ($\alpha = \pi/2$). For Amplistats with d-c output, this is in the approximate center of the linear portion of the control characteristic when controlled by a d-c signal. If the contribution of the exciting current to the load at $\alpha = \pi/2$ is neglected, the effects of d-c hysteresis loop width are negligible, and the dynamic hysteresis loop is rectangular, then the following relationships are valid:

Maximum possible average load current I_{Lm} ($\alpha = 0$)

$$I_{Lm} = E/R_O \tag{14}$$

where $R_O = R_L + R_G$ and where R_L is the load resistance and R_G is the gate-winding resistance plus the forward resistance of the rectifiers. E is the average value of the supply voltage ($E = 0.9E_e$).

The average load voltage

$$E_L = E(R_L/R_O)[(1 + \cos \alpha)/2] \tag{15}$$

For $\alpha = 90°$,

$$E_L = ER_L/2R_O \tag{16}$$

Minimum average load current I_{LX} is the total exciting current of two cores (except in the case of voltage-controlled Amplistats)

$$I_{LX} = 2I_X = 2H_{csf}l/0.4\pi N_G \tag{1C}$$

where H_{csf} is the coercive force of the sine-flux dynamic hysteresis loop at the operating frequency and l is the mean length of the magnetic circuit, in centimeters. The sine-flux hysteresis loops of some core materials are published by materials manufacturers. Others give data based upon the AIEE (now IEEE) standard[18] constant-current-flux-reset core test method. The value of H_2 for ΔB_2 secured by this test method is a good approximation of H_{csf}.

The control current for minimum output

$$I_C = H_{csf}l/0.4\pi N_C \tag{12}$$

Power gain G_P and time constant τ_C are important parameters of Amplistats. These are related to each other for a given Amplistat by the figure of merit or dynamic power gain G_D where

$$G_D = G_P/\tau_C \quad \text{per sec} \tag{17}$$

The power gain is related to G_{AT}, the ampere-turn gain[19]

$$G_P = G_{AT}^2(N_C/N_G)^2(R_L/R_C)k_f^2 \tag{18}$$

If only average d-c quantities are considered k_f, the form factor (ratio of rms to average values) is unity.

With the simplifying assumptions indicated, at $\alpha = 90°$ it can be shown that

$$G_{AT} = (I_{Lm}N_G/2I_C N_C)\pi \tag{19}$$

Since it has been shown that

$$I_C = I_X N_G/N_C \tag{20}$$
$$G_{AT} = (I_{Lm}/2I_X)\pi \tag{21}$$

The time constant in the normal operating region, with no circuits magnetically coupled to the control windings, is[20]

$$\tau_C = \frac{1}{2f}\frac{1}{2I_X}\frac{E}{R_C}\left(\frac{N_C}{N_G}\right)^2 \pi \sin \alpha \tag{22}$$

The time constant is not strictly a "constant" as it varies with the firing angle α and is a maximum when $\alpha = 90°$. This does not prevent the application of τ_C to Bode or Nyquist diagrams because they can be based upon incremental quantities.

Equation (22) enables the designer to determine τ_C when the exciting current I_X of one-half of the Amplistat is obtainable by test or calculated from basic dynamic properties of the SR core material. If the calculated time constant is longer than desired, it can be decreased by increasing R_C by an external series resistor. This will reduce the power gain G_P since R_C is in the denominator of Eq. (18). Equation (22) does not take into account transportation lag effects or time delays due to circuits, such as bias-winding supplies coupled to the control circuit. Transportation lag effects may cause large increases in the time delay if the Amplistat is operated underexcited [$E < E_G$ of Eq. (1)].

When $\alpha = 90°$, Eq. (22) becomes

$$\tau_C = \frac{1}{2f}\frac{1}{2I_X}\frac{E}{R_C}\left(\frac{N_C}{N_G}\right)^2 \pi \tag{23}$$

From Eqs. (17), (18) (with $k_f = 1$), (24), (21), and (14),

$$G_D = \pi f I_{Lm} R_L/I_X R_0 \tag{24}$$

Note that the dynamic power gain is directly proportional to frequency f and inversely proportional to the exciting current of the core under sine-flux conditions at the frequency f. This shows why it is advantageous to operate Amplistats at high frequency, providing core materials are available which have narrow dynamic hysteresis loops.

The design steps suggested for Amplistats with d-c control are in summary:

1. Select a core material using Chap. 8, Tables 1 and 2 and the guides given at the beginning of Sec. 2.13.

2. Select a core size which, when wound with the required gate coil determined by Eq. (1) and the wire size required by the effective value of gate current ($I_{Ge} = 0.707 I_{Le}$), leaves approximately 20 per cent of the useful core window area for control windings.

3. Determine the dynamic power gain G_D from Eq. (24).

4. If G_D is greater than the application requires, one stage will be sufficient and the control-winding turns can be determined using Eq. (12).

5. If G_D is less than the application requires, two stages will be required. In this case, the best results are secured if the power gains of the two are nearly equal. It may be advantageous to use different core materials for the two stages.

The control characteristic of the final design can be estimated from the calculated ampere-turn gain G_{AT}, the minimum and maximum currents $2I_x$ and I_{Lm}, and the control-characteristic curve shapes indicated in Secs. 2.2 through 2.7.

Voltage-controlled Amplistat-design procedure differs from the above only in respect to the control circuits. With the requirement that the same volts per turn be applied to the control windings at cutoff as is applied to the gate windings, the d-c power gain is

$$G_P = G_{AT} R_L / R_O \qquad (25)$$

The ampere-turn gain G_{AT} is the same as for current-controlled Amplistats and is a maximum for $\alpha = 90°$ as given by Eq. (21).

With sufficient control voltage available to reset the core flux in one half-cycle, the time constant of the control circuit of voltage-controlled Amplistats has little or no meaning. A single-stage amplifier will respond to a randomly applied signal with a maximum delay of 1 cycle. Cascades of voltage-controlled amplifiers add delays of $\frac{1}{2}$ cycle for each stage added after the first stage.

REFERENCES

1. H. F. Storm, "Magnetic Amplifiers," chap. 23, John Wiley & Sons, Inc., New York, 1955.
2. Ref. 1, pp. 148–150.
3. Ref. 1, p. 164.
4. Ref. 1, p. 165.
5. Ref. 1, p. 82.
6. Ref. 1, p. 402.
7. Harold W. Lord, The Influence of Magnetic Amplifier Circuitry upon the Operating Hysteresis Loops, *Trans. AIEE*, vol. 72, pt. I, pp. 721–728, 1953.
8. Harold W. Lord, Magnetic-amplifier Circuits with Full-wave Output and Half-wave Control Signals, *Trans. AIEE*, vol. 73, pt. I, pp. 265–270, 1954.
9. Terms and Definitions for Magnetic Amplifiers, AIEE Committee Report, *Trans. AIEE*, vol. 77, pt. I, pp. 429–437, 1958.
10. H. F. Storm, Theory of Magnetic Amplifiers with Square-loop Core Materials, *Trans. AIEE*, vol. 72, pt. I, pp. 629–640, November, 1953.
11. Ref. 1, pp. 286–297.
12. R. A. Ramey, On the Mechanics of Magnetic Amplifier Operation, *Trans. AIEE*, vol. 70, pt. II, pp. 1214–1223, 1951.
13. R. A. Ramey, On the Control of Magnetic Amplifiers, *Trans. AIEE*, vol. 70, pt. II, pp. 2124–2128, 1951.
14. David L. LaFuze, "Magnetic Amplifier Analysis," p. 142, John Wiley & Sons, Inc., New York, 1962.
15. Ref. 1, chap. 20.
16. William A. Geyger, "Magnetic Amplifier Circuits," 2d ed., McGraw-Hill Book Company, New York, 1957.
17. Ref. 14, pp. 96, 102, 106.
18. Test Procedure for Toroidal Magnetic Amplifier Cores, AIEE Standard 432, January, 1959.
19. Ref. 1, p. 108.
20. Ref. 1, p. 268.

Chapter 22

IONIC AMPLIFIERS

D. RAYMOND FEWER* and JERRY D. MERRYMAN*

CONTENTS

1. Introduction.. 22-1
2. Integrators... 22-2
    2.1. Solion Integrators...................................... 22-2
3. Solions in D-C Amplifiers..................................... 22-6
    3.1. Basic Principles.. 22-6
    3.2. A Practical Solion Instrument Amplifier................. 22-7
    3.3. Solion-stabilized Wideband Amplifiers................... 22-9
4. Solion Transducers 22-9
    4.1. Solion Microbarometer................................... 22-10
    4.2. Negative Feedback with Solion Transducers............... 22-12
5. Further Applications of Ionic Devices......................... 22-12
    5.1. Solion Timer.. 22-13
    5.2. Capacitive Potentiometer................................ 22-13
    5.3. Solions in Adaptive Circuits............................ 22-14

1. INTRODUCTION

The ionic devices discussed in Chap. 15 can be used in a variety of configurations to perform many functions useful to the circuit designer. The solion tetrode, for example, can perform amplification in the commonly accepted sense but many ionic devices, including solion tetrodes, perform basically as integrators. In a sense they may be considered as potential substitutes for operational amplifiers using shunt capacitor feedback.

For many low-frequency applications, ionic devices can provide substantial improvement over the more commonly used devices. For instance, a circuit to stabilize the d-c potential at the summing junction of an operational amplifier might require a matched pair of silicon transistors to chop the d-c error signal, stages of a-c amplification followed by a demodulating stage. A solion circuit can perform this function with a solion tetrode, a transistor, and a small bias cell. Since no chopping is used no filtering is required. Circuits of this type will be discussed.

The frequency characteristics of ionic devices are functions of ion mobility (or diffusivity) and show high-frequency cutoff points ranging from a few cycles down to

* Texas Instruments Incorporated, Semiconductor Components Division, Dallas, **Tex.**

a fraction of a cycle. Some solion tetrodes, for example, show a frequency for zero db current gain, $f = K/2\pi$, where K can be less than 0.2 sec^{-1}. On the other hand, ionic devices functioning as integrators can accurately process input signals with a frequency content up to megacycles. The output signal simply lags the integral of the input signal.

When ionic devices are used in broadband circuits it is necessary to know their frequency performance over the complete range of interest since they may act as parasitics, in some cases, at frequencies beyond a few cycles. Thus it is necessary to characterize them over many decades of frequency. In the low-frequency region high-persistence oscilloscopes or X-Y recorders are useful for this purpose.

2. INTEGRATORS

The "natural" function of most ionic devices is that of integration. When the function to be integrated can be represented as a current the principles discovered by Faraday permit very accurate integration. In Chap. 15, devices which allow various forms of readout were discussed. Solion readout signals can be current, voltage, or visual; coulombmeters read out by gap position; memistors read out as resistance values. This section discusses practical integrators which can be used for a wide variety of purposes.

2.1. Solion Integrators. The solion tetrode connected in a basic integrator circuit was discussed in Chap. 15. Figure 1 shows a somewhat more elaborate circuit which incorporates features basic to solion integrator circuit design. When the switches are in the INTEGRATE position the output is a voltage which is proportional to the time integral of the input current. When the input loop resistance is high relative to the input resistance of the solion the input current is

$$I_{in} = \frac{E_g \pm E_c}{R_g + R_1} \tag{1}$$

where E_g = source voltage
R_g = source resistance
R_1 = input series resistance
E_c = concentration potential appearing at the solion input terminals

The concentration potential introduces an error current I_e in the input signal, the magnitude of which is

$$I_e = E_c/(R_g + R_1) \tag{2}$$

An increase in source voltage, together with an increase in R_1 or R_g, allows the same input-current level to be maintained while decreasing the error current. Input errors can also be reduced by operating the circuit over a region where concentration voltage is small. In some cases, advantage can be taken of the fact that the concentration voltage varies from positive to negative over the solion operating range. Thus positive and negative errors can be used to reduce the net effect.

The shield supply voltage is obtained by means of a voltage divider R_2, R_3 across the voltage supply E_s. In most applications the shield bias supply must be floating so that it does not provide a leakage path between input and common electrodes. The shield voltage should be between -0.3 and -0.5 volt relative to the input electrode. The voltage and the impedance of the shield supply are not critical in most applications. The shield current is usually only a few microamperes. Thus it is usually practicable to bias the shield continuously, which greatly reduces the time required for the device to reach electrochemical equilibrium.

Voltages around 0.75 volt and above impressed between any two solion electrodes can alter or damage the device. The danger of exceeding safe operating voltages is always present at the input where constant-current sources are usually used. A pair of high-quality silicon diodes D_1 and D_2 connected as shown in Fig. 1 will protect the device from accidental damage. The primary requirement on the diodes is that their leakage currents, which represent an error signal similar to that discussed above, should be small.

Another method of biasing solion electrodes is shown by the arrangement of the readout bias supply in Fig. 1. The supply voltage E_o is divided across the resistor P_5 and the stabistor D_3.

The sensitivity of the solion tetrode varies with temperature as shown in Fig. 2. The variation with temperature is consistent and reproducible from device to device. Thermistors can be used for temperature compensation as shown in Fig. 1, where a single thermistor in combination with the resistor R_6 and the output load can compensate for sensitivity changes to better than ±1 per cent over a temperature range from +20 to +40°C.

Switches are placed in the HOLD position to remove the source from the input circuit. This reduces the error current I_e as given in Eq. (2) to zero. The holding then becomes a function of the device properties and input circuit parasitic leakage paths.

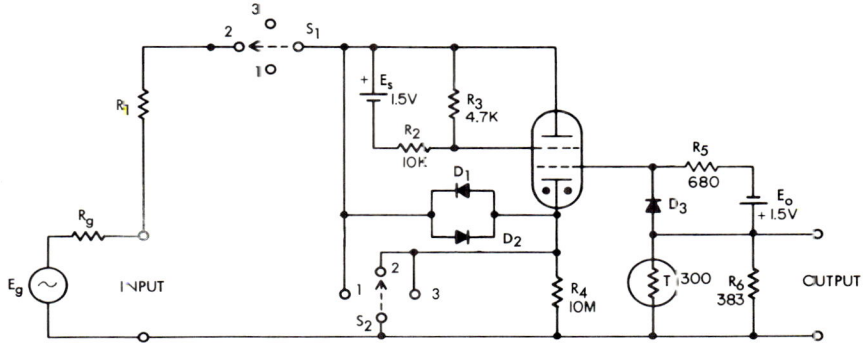

FIG. 1. Simple solion integrator circuit. S_1 and S_2 are ganged. The positions are (1) reset (2) integrate, (3) hold.

The switches are placed in the RESET position to return the integrator to some reference condition. The source is removed from the circuit and the input is shorted The reference condition is determined by the value of R_4. Since the voltage across the input terminals is zero the voltage across R_4 must be equal and opposite to the concentration voltage E_c at equilibrium. This determines an equilibrium or reference value of readout current i_o such that

$$I_o = -E_c/R_4 \tag{3}$$

The reversible properties of the solion as an integrating device can be used to make a wide variety of integrating instruments. The ability to integrate in opposite directions, depending on the direction of the input current, allows the design of circuits which "fold" the integral i.e., integrate up and down between two values of output current while the signal current remains unchanged in direction. When this function is combined with some method of permanently registering the number of folds, signals existing for an indefinitely long time can be integrated. Integrals obtained in this manner are retained indefinitely. The accuracy approaches that of the indicator used to register the number of folds.

A block diagram of a commercial solion integrator is shown in Fig. 3.[1] A somewhat simplified circuit diagram is shown in Fig. 4. The instrument measures chromatograph or other signals being recorded on a strip-chart recorder. The input to the integrator is taken from a retransmitting potentiometer on the recorder. The integral can be continuously displayed on the second channel of a two-channel recorder. Provision is also made to display the signal and the integral on a single-channel recorder. For this operation integral "folds" are indicated by a marker pen

[1] Self-Organizing Systems, Inc., P.O. Box 9918, Dallas, Tex. 75214.

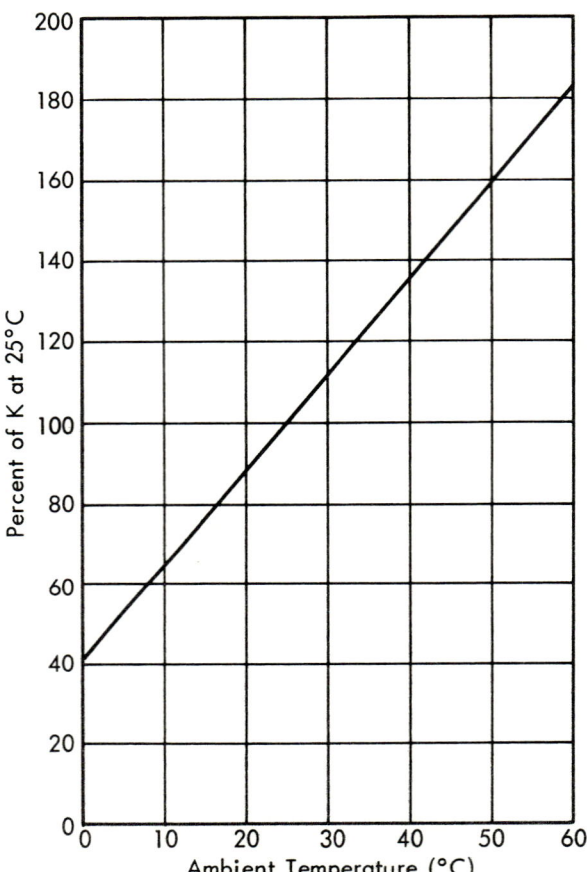

Fig. 2. Dependence of solion tetrode sensitivity on temperature.

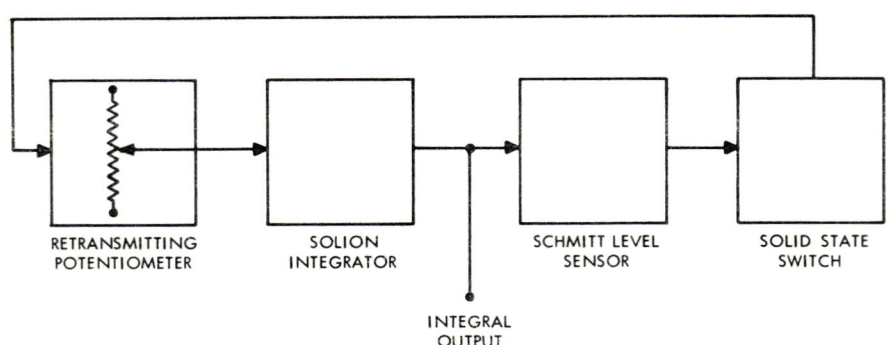

Fig. 3. Block diagram of a commercial solion integrator.

IONIC AMPLIFIERS

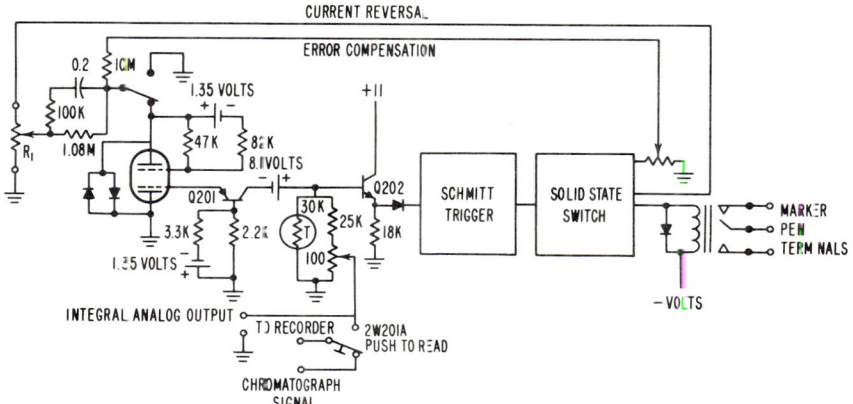

Fig. 4. Circuit diagram of a commercial solion integrator.

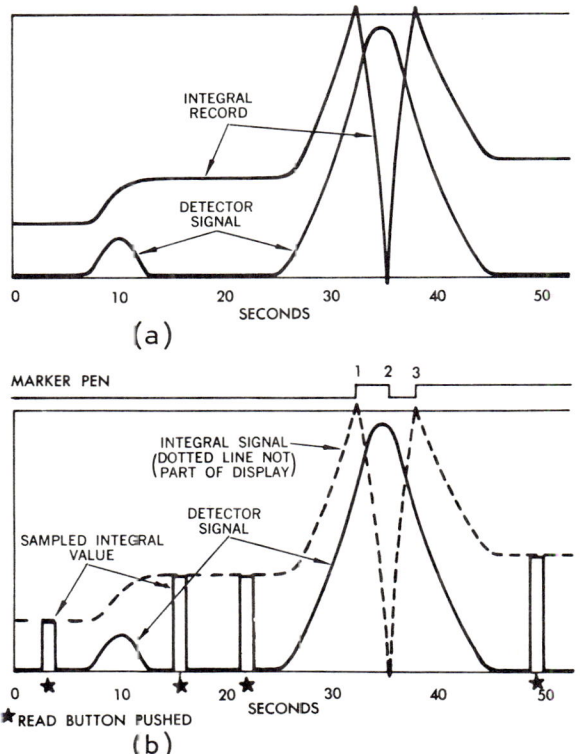

Fig. 5. Recorder signal and integral displays. (a) Two-pen recorder display. (b) Single-pen recorder display.

and a button is pushed to record the interpolated value between "folds" of the integral on the same channel that records the signal. Two-channel and single-channel displays are illustrated in Fig. 5a and b.

In Fig. 4 the retransmitting potentiometer R_1 is positioned by the recorder amplifier and provides an input to the integrator. The solion integrator drives a common-base transistor stage $Q201$, which provides a low-impedance load for the solion as well as gain. The emitter follower $Q202$ provides matching to the input of the Schmitt trigger circuit. The trigger circuit controls the solid-state switch to reverse the polarity of the voltage applied to the retransmitting potentiometer. The Schmitt trigger circuit operates within its hysteresis region to maintain the solion output current within its upper and lower design limits.

Two types of error, peculiar to circuits of this type, are corrected by special circuits. The current flowing through the end resistance which sometimes occurs in retransmitting potentiometers causes a signal which, when integrated, shows as output drift. The direction of this drift changes each time the Schmitt trigger operates. A complementary solid-state switch injects a compensating current into a summing junction at the solion input. This current is adjusted to be equal and opposite to that causing the error signal. The second type of error occurs at the moment of operation of the Schmitt trigger circuit. After current reversal the output current indicates an apparent loss of input charge during the switching operation. The magnitude of the loss was found to be proportional to the magnitude of the input current at the time of reversal. This effect is due, partly, to the feedback capacitor C_F as shown in Fig. 12 of Chap. 15 and to the rather long ion transit times in the readout compartment of the solion. The addition of a compensating capacitor across the input resistor modifies the charge at each current reversal. These compensating circuits reduce integration errors by an order of magnitude.

3. SOLIONS IN D-C AMPLIFIERS

The solion tetrode is well suited to use in low-drift d-c amplifiers. Single-stage gains of several hundred are obtainable, and drift compares favorably with chopper-stabilized amplifiers. Low power consumption of solion amplifier circuits allows their use with batteries, resulting in portability and freedom from 60-cycle noise.

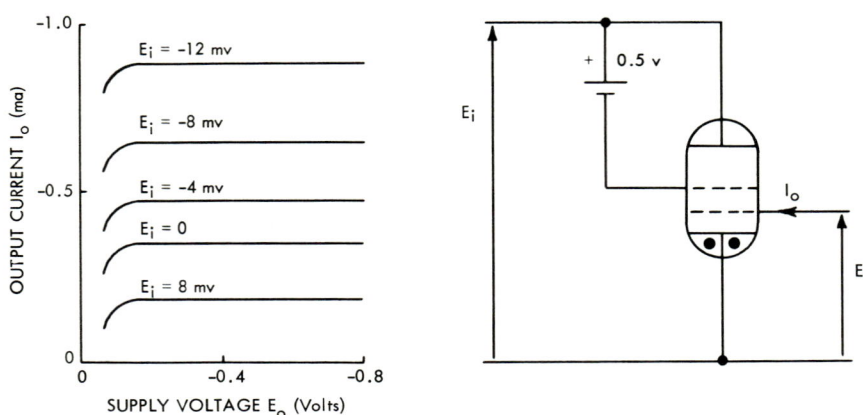

FIG. 6. Solion tetrode output characteristics.

3.1. Basic Principles. The solion has previously been described as an integrator because the output current is proportional to the charge (time integral of current) moved in the input circuit. When the characteristics are plotted as a function of input voltage, rather than charge, the pentode-like curves of Fig. 6 are obtained.[2] These curves show high output impedance and a high transconductance that increases with output current.

The usefulness of the solion as a low-level d-c amplifier is based on the excellent stability of these static characteristics with time and temperature. Solion tetrodes show, typically, drifts in operating point of 3 μv/day, referred to the input voltage. The temperature coefficient of the solion input voltage is about one-seventh that of silicon transistors; pairs match to within 1 $\mu v/°C$ over a wide temperature range. Spurious input current (current drawn through the input source) required to maintain a fixed operating point is 10^{-10} to 10^{-9} amp.

The basic solion amplifier circuit (Fig. 7) is similar to typical differential arrangements used with vacuum tubes and transistors. R_1 sets the operating point.

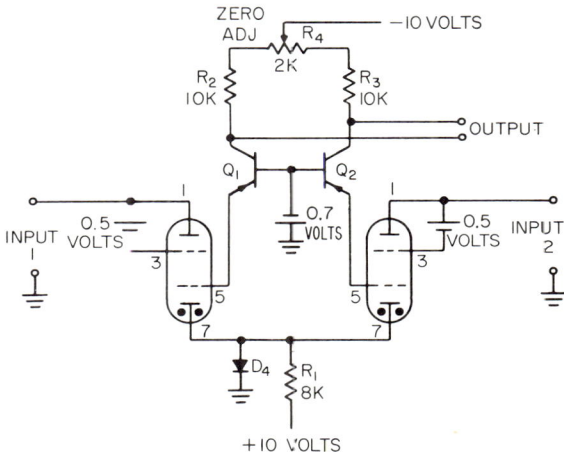

Fig. 7. Basic solion amplifier circuit.

Common-base transistors Q_1 and Q_2 have unity current gain but serve to limit the solion voltage to 0.5 volt while allowing use of large load resistances for high gain. The tap of R_4 is adjusted so that, at the zero point, the two solions will have equal input to common bias voltages E_i rather than equal operating currents I_o. This is a criterion for low temperature coefficient.[3] Voltage gain of this circuit is 500. As expected from the equivalent circuit frequency response is flat from direct current to about 0.01 cycle. Input impedance is low at other than very low frequencies.

3.2. A Practical Solion Instrument Amplifier. The frequency response of the amplifier circuit just described is too restricted for most applications. It can be improved considerably by the use of negative feedback. Figure 8, the complex

[2] The relation between input voltage and output current under static conditions is given by Eq. (13) in Chap. 15.

[3] The concentration-voltage equation

$$E_i = -(RT/nF) \ln (I_o/I_x) \tag{4}$$

may be differentiated to yield

$$\left.\frac{\partial E_i}{\partial T}\right|_{I_o \text{ const}} = -\frac{RT}{nF} \frac{\partial \ln I_x}{\partial T} - \frac{R}{nF} \ln \frac{I_o}{I_x} \tag{5}$$

The value of $\partial \ln I_x / \partial T$ is essentially the same for all solions of the same type. It follows that, for zero differential temperature coefficient, two solions must have operation currents I_o in the same respective ratio as equilibrium currents I_x. This implies equal bias voltages E_i.

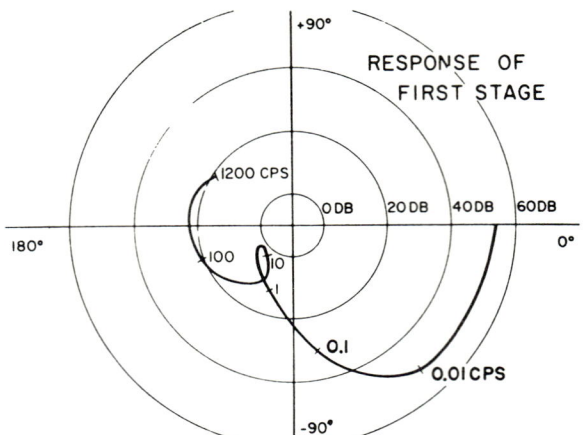

Fig. 8. Basic solion amplifier frequency response.

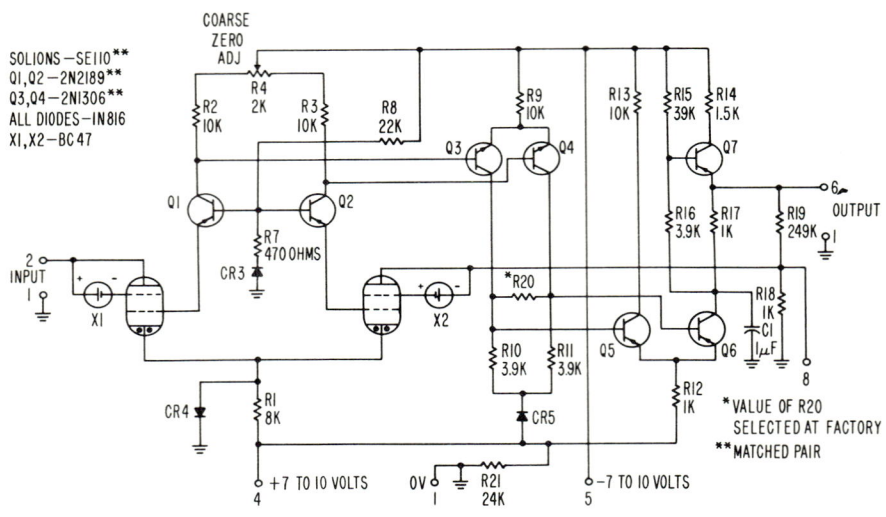

Fig. 9. Practical solion d-c amplifier.

response[4] of the basic stage, shows that feedback factors of several hundred are possible without stability problems.

[4] Evidently, the response shown in Fig. 8, lying in three quadrants, is somewhat at variance with the simple equivalent-circuit representations shown in Chap. 15. This is because of two effects. First, there is the capacitance C_F between input and output terminals (Chap. 15, Fig. 12). Second, because of diffusion processes, the current-gain operator is actually given by

$$\beta = [\cosh (1 + j) \sqrt{\omega/K} - 1]^{-1} \qquad (6)$$

This function is largely responsible for the "excess phase." For frequencies $\omega < K$, this function reduces to $K/j\omega$, which is the ideal integrator function usually given.

IONIC AMPLIFIERS

Figure 9 shows how stages of transistor amplification can be employed, with feedback, to improve the frequency response and the input impedance. The added transistor stages raise the open-loop gain to about 10^6 and provide high-level single-ended output. Closed-loop gain is set at 250 by R_{18} and R_{19}. The frequency response is essentially flat to 3 cps. Equivalent input noise is 3 μv rms, and operating temperature range is -5 to $+40°C$.

3.3. Solion-stabilized Wideband Amplifiers. A solion amplifier arrangement can be used to stabilize the d-c operating point of an operational amplifier in much the same manner that chopper stabilizers are used. In this case, the low-pass filter characteristics of the solion are an asset. The solion stabilizer does not statically load the summing junction, and there is no chopper noise that can be conducted or radiated into the amplifier or backward into the source.

A particular circuit,[5] useful when several amplifiers are to be stabilized, is shown in Fig. 10. The stabilizer arrangement is electrically equivalent to a differential solion

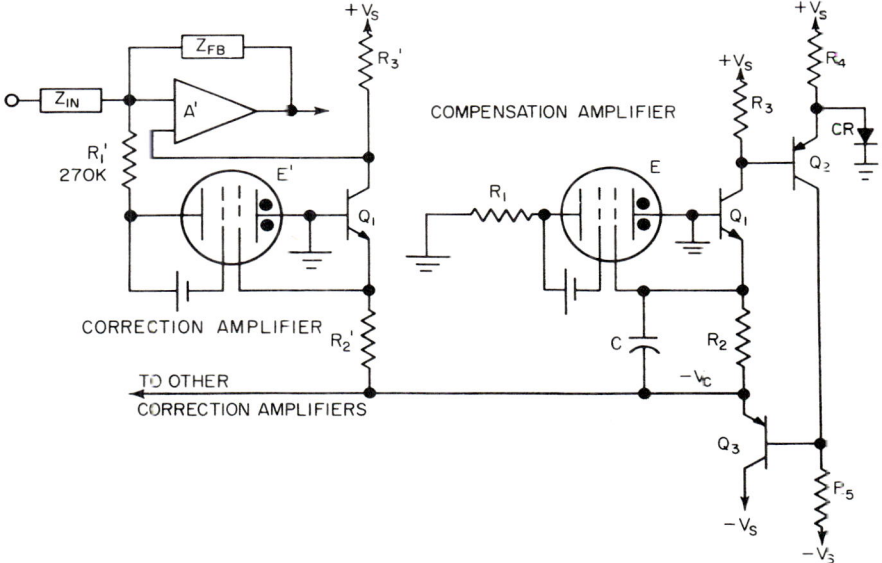

Fig. 10. Solion stabilization amplifier.

amplifier. Only one compensation amplifier is required for a number of correction amplifiers, however, and the parts savings over individual differential stabilizers is obvious. A suitable value for the summing junction isolation resistor R_1 is 270 kilohms, and the other resistors are chosen with regard to the supply voltages available. The capacitor C is required for loop stability in the compensation amplifier circuit.

4. SOLION TRANSDUCERS

In Chap. 15 it was mentioned that solions can be made to function as sensitive transducers by the electrochemical detection of very small rates of fluid flow through an orifice. The detection characteristic of the device may be linear, logarithmic, or some other function, depending largely upon the electrochemical collection efficiency of the orifice structure. In one important class of instruments, used from very low

---
[5] This circuit is due to D. W. Perkins of the General Electric Co.

frequencies up to a few cycles per second, the cathodes are made to have very high collection efficiencies in order to give linear characteristics. Two cathodes are used to produce balanced push-pull output. This class includes the solion angular accelerometer and the solion microbarometer.

4.1. Solion Microbarometer. This device is used to measure low-frequency fluctuations in atmospheric pressure. The diagram in Fig. 11 illustrates the basic properties of this device. Since the two-diaphragm solion transducer is a differential-pressure device, one diaphragm is referenced to an air chamber in order to measure atmosphere-pressure variations outside the chamber. The chamber is provided with a slow leak R_c which controls the low-frequency response of the device. The high-frequency response is determined by the properties of the solion.

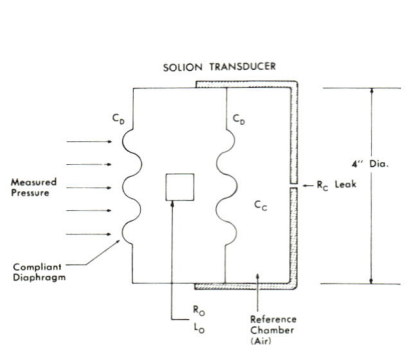
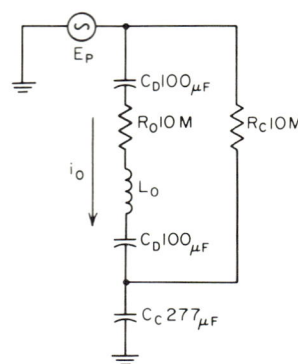

Fig. 11. Solion transducer used as a microbarometer.

Fig. 12. Microbarometer electrical analog.

In analyzing the acoustic performance of this device use is made of the electrical analogies as shown in Fig. 12. Let pressure be represented by voltage (volt = dyne/sq cm) and fluid flow be represented by current (ampere = cu cm/sec). E_p is the external applied pressure and i_o is the fluid flow through the measuring orifice [which can be converted to a real electric current by Chap. 15, Eq. (21)].

Similarly, we can speak of acoustic "circuit elements" having resistances or reactances. (The acoustic ohm has dimensions of dyne-sec/cm^5.) The acoustic resistance of a capillary is given by

$$F = 8l\eta/\pi r^4 \tag{7}$$

where η = viscosity of the fluid medium
  l = length of capillary
  r = radius

Such a relationship is used for determining the effective resistances of the leak R_c, and orifice R_o. The orifice resistance might range from a few thousand ohms to many megohms depending on the intended use of a particular solion. Developmental devices used in microbarometers use two cathodes to detect the direction of fluid flow. The combined resistance can be as high as several megohms.

Stored potential or kinetic energy, due to deflection of diaphragms or energy of high-velocity motion in the small orifice, gives rise to capacitive or inductive reactance, respectively. For example, the inductance of a simple capillary is given approximately as

$$L_o = \rho l/\pi r^2 \tag{8}$$

where ρ = density of the fluid
Capacitance of the reference cavity may be taken as

$$C_c = V/\gamma P \tag{9}$$

IONIC AMPLIFIERS 22-11

where V = total volume
P = pressure in the cavity
γ = the ratio of the specific heats of air

Compliance of diaphragms is a complex function of form, but it is given empirically for a particular type as

$$C_D = K d^m t^{-n} \qquad (10)$$

where K = constant
d = diameter
t = thickness

For example, $m = 4$ and $n = 1.6$ for experimental devices, and K is approximately 2.56×10^{-10} dyne-cm/in.$^{2\,4}$. C_D will be in cgs units when d and t are in inches.

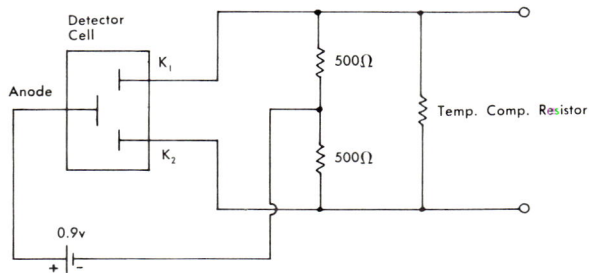

Fig. 13. Solion transducer direct-drive circuit. Output to recorder (0.1 volt/mb). Frequency response 0.0005 to 0.5 cps.

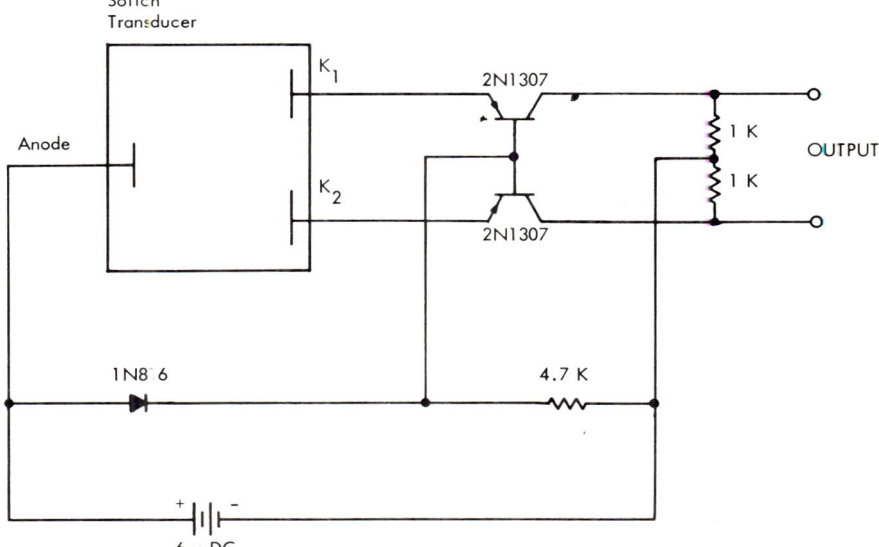

Fig. 14. Solion transducer circuit with high output voltage.

The electrical analog of Fig. 12 has the appearance of a series-resonant circuit. The parameter values, as defined above, provide relatively flat response from 0.0005 to 0.5 cps. A circuit suitable for direct drive of many recorders is shown in Fig. 13. A circuit yielding higher output voltages is shown in Fig. 14.

AMPLIFIER CIRCUITS

4.2. Negative Feedback with Solion Transducers. Negative feedback may be applied to a solion transducer by some other type of electromechanical transducer (coil and magnet, for example) and will result in extension of frequency response and dynamic range; at the same time, distortion and nonlinearity are reduced.

If the loop gain A of the system is made very large, essentially a mechanical balance is obtained where the solion is used as a null detector. The system calibration then approaches that of the electrical-to-mechanical transducer, which may be more permanent or easier to obtain. The high sensitivity and inherent power gain of the solion are then obtained at the same time that undesirable solion characteristics (temperature coefficient of sensitivity, for example) are suppressed.

In many cases, large $A\beta$ can be obtained even without an amplifier ($A = 1$), since β may be considerably in excess of unity, and is given by

$$\beta = \sqrt{\varepsilon G} \qquad (11)$$

where ε = efficiency (power) of the electrical-to-mechanical transducer
G = power gain of the solion as given in Eq. (22), Chap. 15

5. FURTHER APPLICATIONS OF IONIC DEVICES

Electrochemical devices, because they function basically as integrators, can be used to perform a variety of functions. Mercury coulombmeters, solions, and memistors are used in a variety of circuits to perform as timers, elapsed-time indicators, charge indicators for batteries, dosimeters, pulse counters, signal averagers, function gener-

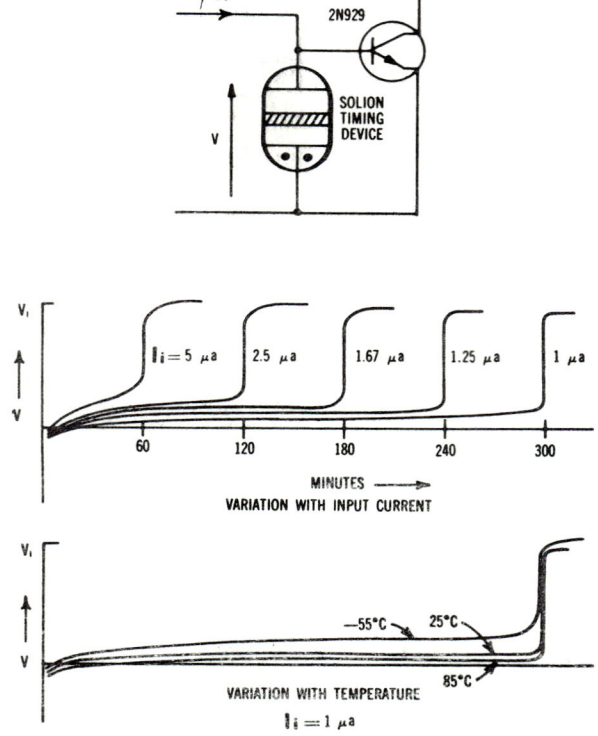

FIG. 15. Solion timer characteristics.

ators, flowmeters, servo testers, exposure meters, signal analyzers, planimeters, low-frequency noise meters, adaptive circuit elements, and many other functions.

Memistors and mercury coulombmeters exhibit no back emf or concentration potential. In many circuit applications this is an advantage. On the other hand, the concentration potential exhibited by the solion can be useful in such applications as the timer discussed below.

5.1. Solion Timer. The solion two-terminal integrator can be used in a timing device to provide a simple low-power method of obtaining time delays from a few minutes to several days. Accuracy of about 5 per cent is obtainable over a wide temperature range. The integrator contains a fixed amount of charge which is moved from one compartment to the other by a current. When all the charge has been moved there is an abrupt increase in voltage which can be used, for example, to fire a controlled rectifier. The characteristics of a simple timing circuit are shown in Fig. 15. A practical circuit requiring approximately 30 min of reset time is shown in Fig. 16.

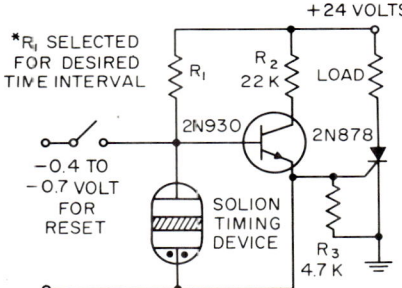

Fig. 16. Practical solion timer circuit.

During timing the controlled rectifier is off and the 2N930 transistor is not conducting because the voltage from base to emitter, across the two-terminal solion used as a timing device, is lower than the threshold of the transistor. The 2N930 has low leakage and high gain at low current. The charge in the solion is moved by the current through R_1 until it is nearly all moved. Then, the impedance of the solion increases rapidly, causing all the current from R_1 to flow into the base of the transistor. This signal is amplified by the transistor to provide a reliable trigger for the controlled rectifier.

5.2. Capacitive Potentiometer. The mercury coulombmeter described in Chap. 15 can be used as a capacitive potentiometer to provide an a-c readout of charge transfer. The glass capillary which encloses the gapped mercury column is coated with a conducting film which produces a combination of cylindrical capacitors as shown in Fig. 17a. The a-c equivalent circuit is shown in Fig. 17b. The value of

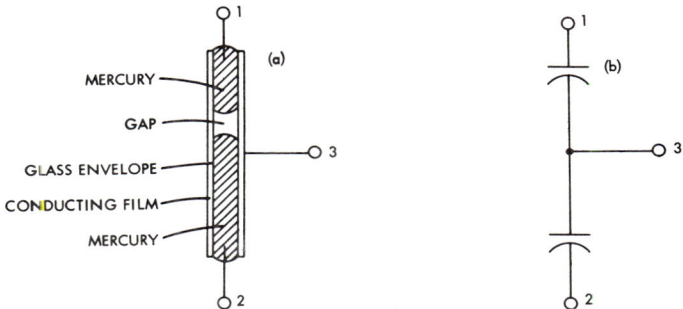

Fig. 17. The mercury coulombmeter as a capacitive potentiometer. (a) Physical arrangement. (b) Equivalent circuit.

each capacitor varies linearly with gap position. The sum of the two capacitances remains constant so that when the device is used as a voltage divider the output voltage is a linear function of gap position. The time integral of a d-c input can be read out as an a-c signal by a circuit represented by that shown in Fig. 18. If the a-c input voltage e_{in} is held constant the a-c output voltage e_{out} is proportional to time integral of the direct input current I_{in},

$$e_{out} = [C_1/(C_1 + C_2)]e_{in} = KC_1 = K \int^t I_{in}\, dt \qquad (12)$$

where $C_1 + C_2 = \text{const}$
$e_{in} = \text{const}$
$K = e_{in}/(C_1 + C_2)$

For practical circuits the frequency of the a-c signal and the frequency content of the d-c signal must be taken into consideration. The simple arrangement such as the blocking capacitor and choke shown in Fig. 18 may be replaced by filters.

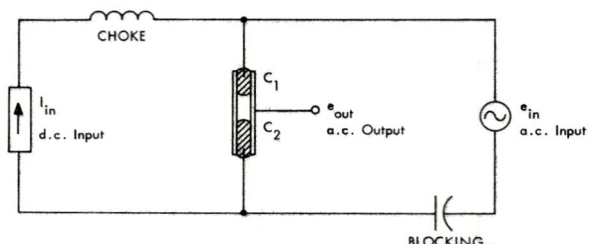

Fig. 18. Capacitive potentiometer circuit.

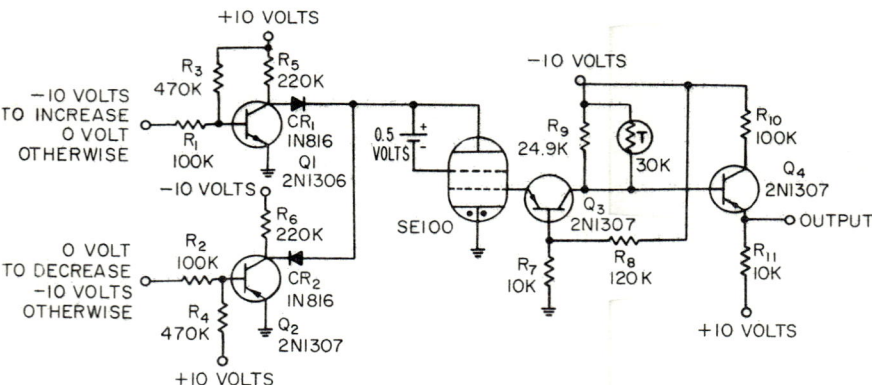

Fig. 19. Solion analog memory circuit.

5.3. Solions in Adaptive Circuits. Because the solion tetrode is an analog memory device that provides continuous d-c nondestructive readout, it is usable as the adaptive element in self-organizing computers, or "learning machines." An electronically adjustable memory element is the key component in any such system. Important properties of an adaptive element are first, the time required for adaptation from one level to another, and second, the stability of the adapted level after the learning process. Typical solion analog memory circuits can be made to traverse full scale in less than 1 min and to exhibit stabilities of 1 per cent per week afterward.

An example of the adaptive use of solions is the CHILD[6] (Cognitive Hybrid Intelli-

[6] Lt. John P. Choisser and Lt. John W. Sammon, Jr., A New Concept in Artificial Intelligence, Proceedings, National Aerospace Electronics Conference, Dayton, Ohio, May, 1963.

gent Learning Device) concept. One machine consists of 360 CHILD cells, each of which uses three type SE-100 solion tetrodes as the memory elements. The solions are used, respectively, to establish an upper acceptance threshold, a lower acceptance threshold, and a weight, to be assigned to each input variable. The learning process consists of iteratively changing the acceptance regions and weights (according to predetermined logic) until the desired function is performed. Since the solion-determined thresholds may be established by comparator circuits, and the weight-determining solion can operate a simple clamp circuit, the speed of the machine, after learning, depends only on the nonsolion circuitry.

Figure 19 shows a typical solion analog memory circuit. When the increase or decrease control terminals are actuated, current through either R_5 or R_6 causes the solion current to be adjusted up or down at a fixed rate. A swing of nearly 10 volts is available at the output terminal.

Chapter 23

SERVO AMPLIFIERS

FRITZ H. SCHLERETH[*]

CONTENTS

1. Introduction 23-1
2. Direct-coupled Servo Amplifier 23-2
   2.1. General 23-2
   2.2. Application of D-C Amplifiers to Servo Amplifiers 23-4
3. Alternating-current Servo Amplifiers 23-6
   3.1. General 23-6
   3.2. Applications of A-C Amplifiers to Servo Amplifiers 23-12
4. Equalization Networks 23-14
5. Nonlinear Servo Amplifiers 23-16
   5.1. Servo Amplifiers Utilizing Thyratrons 23-18
   5.2. Servo Amplifiers Utilizing Silicon Controlled Rectifiers .. 23-20
   5.3. Servo Amplifiers Utilizing Transistors in the Switching Mode . 23-24

1. INTRODUCTION

The main purpose of a servo amplifier[†] is to supply rated power to the load, the servo actuator. Servo amplifiers fall into two major categories, d-c and a-c, each of which can be linear or nonlinear. Nonsaturating d-c amplifiers and carrier amplifiers are examples of the former while magnetic, thyratron, transistor, and SCR amplifiers are examples of the latter.

It should be noted that servo amplifiers are not very different from other amplifiers considered in this handbook. For example, the principles involved in the design of d-c amplifiers are entirely applicable to the design of d-c servo amplifiers, and the same applies to a-c audio amplifiers and a-c servo amplifiers. However, there are design problems which are peculiar to servo amplifiers, and it will be the purpose of this chapter to discuss these particular points in detail. These problems are mainly associated with the need to drive high power into inductive loads and the need to

---

[*] General Electric Co., Electronics Laboratory, Syracuse, N.Y.

[†] The definition for the term *servo amplifier* is taken from Grabbe, Ramo, and Wooldridge, "Handbook of Automation, Computation and Control," vol. 3, John Wiley & Sons, Inc., New York, 1961, p. 623, which reads, "The role of a servo amplifier is to receive the error output signal from the comparator, at a fairly low power level, and to perform amplification so that sufficient wattage is available to power the servo actuator."

23-1

23-2 AMPLIFIER CIRCUITS

provide equalization networks at carrier frequencies. Other characteristics peculiar to servo amplifiers are that the bandwidth can be narrow so that tuned circuits can be used, and that linearity, distortion, and fidelity are of relatively minor importance.

2. DIRECT-COUPLED SERVO AMPLIFIER

A direct-coupled or d-c amplifier has the capability of amplifying zero-frequency signals. However, the passband of a d-c amplifier need not be limited to zero-frequency signals, and in some applications it will include the entire video band. In servo applications the d-c amplifier is used primarily to drive d-c actuators and to shape servo transfer functions.

The design of a d-c amplifier for servo applications is not significantly different from the design of d-c amplifiers for other applications, and thus for a complete discussion of the subject the reader is referred to the chapter on d-c amplifiers. The discussion in this section will be limited to those problems in d-c amplifier design which have a particular bearing on servo amplifiers.

2.1. General. The basic problem associated with d-c amplifiers is to design the bias circuits such that the output voltage is set at some reference level (usually zero volts) and such that this reference-level setting is maintained at zero volts, i.e., so that the amplifier is free from drift. Drift variations are indistinguishable from the signal and hence cannot be tolerated. Typical causes of drift are variations in power-supply voltages; changes in characteristics of the elements making up the circuit, due to either temperature or aging, variation in plate, screen, or grid voltages in the case of electron tubes, and variations in emitter-base voltages, current amplification, and leakage currents in the case of transistor circuits.

Variations in the input stages of the amplifier will be most serious in terms of output drift because of the gain of the amplifier. Also, the magnitude of the output drift will be a function of the gain of the amplifier, so that it is common practice to refer to the drift in terms of equivalent input drift. Typical values of equivalent input drift vary greatly with the type of circuits used, the quality of components, etc., so that it is not practical to give a value for the expected drift without qualification.

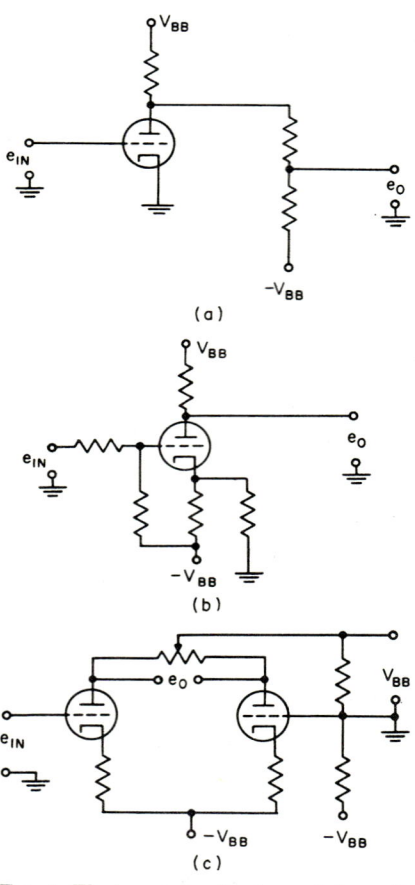

FIG. 1. Electron-tube direct-coupled amplifiers showing zero-level adjustment. [(a) from G. E. Valley, Jr., and H. Wallman, "Vacuum Tube Amplifiers," McGraw-Hill Book Company, New York, 1948. Used by permission. (b) from G. A. Korn and T. M. Korn, "Electronic Analog and Hybrid Computers," McGraw-Hill Book Company, New York, 1964. Used by permission.]

It should be noted that drift at the output of an amplifier is due to "noise" within the amplifier and thus the application of negative feedback can significantly reduce the effects of the various causes of drift.

The only case in which this will not be true is when the source of noise is not within the feedback loop. In general it is not easy to determine whether a source of noise at the input is within or without the feedback loop, and each case should therefore be separately analyzed.

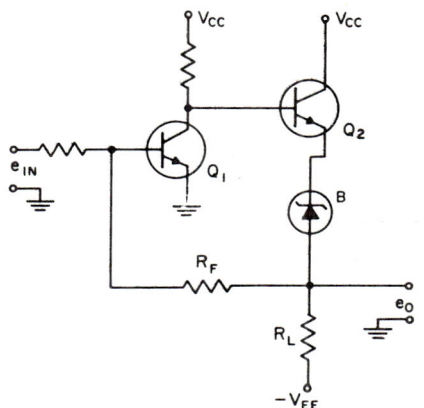

FIG. 2. Circuit illustrating the use of a zener diode for reference-level adjustment.

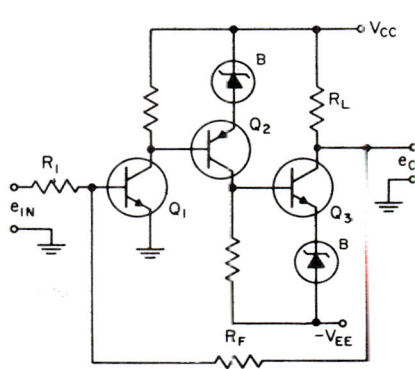

FIG. 3. Circuit illustrating use of n-p-n and p-n-p transistors for reference-level adjustment.

The particular methods utilized in adjusting the input and output reference voltages are quite varied, and the circuits shown in Figs. 1 to 3 illustrate several methods by which this can be accomplished. These are, of course, by no means inclusive. Note that it is necessary to employ both positive and negative supplies. A comparison of the tube and transistor circuits shows that the design is somewhat eased by the existence of both p-n-p and n-p-n transistor types (Fig. 3).

In the design of the circuits of Fig. 1 it is necessary to ensure that the resistor values are chosen so as to set the output voltage near zero volts. In the circuit of Fig. 2 the function of the reference diode is to raise the emitter of Q_2 and hence the collector of Q_1 to a level above zero volts. Feedback current is supplied through R_F in order to stabilize the bias. In the circuit of Fig. 3, p-n-p and n-p-n transistors are used in order to set the output voltage near zero volts. Current feedback is again used to stabilize the bias point.

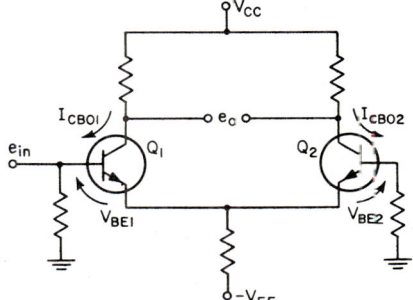

FIG. 4. Long-tailed pair.

There is one basic method for the minimization of drift in d-c amplifiers, to design the input stages so as to balance the various factors causing the drift against each other. A common circuit for accomplishing this is the so-called "long-tailed pair." A transistor version of this circuit is shown in Fig. 4. In this circuit Q_1 is "balanced against" Q_2 such that if the variations in the emitter-base voltages, current amplification, and leakage currents tend to track one another, the drift at the output will be minimized.

The use of modulators is another convenient manner in which the input to the amplifier can be balanced.* Modulators can be balanced to a much higher degree of accuracy than circuits of the type shown in Fig. 4, so that a higher d-c stability results.

Figure 5 shows the block diagram of a typical d-c amplifier which utilizes a modulator and demodulator to obtain d-c amplification. However, the use of modulators is by no means a panacea, because a whole new set of problems is introduced by the use of these devices. Care must be taken that the switching noise is kept to a minimum, lest the amplifier be overloaded by transients; that the phase shift within the amplifier is kept to a minimum so the demodulator drive is in phase with the signal to be demodulated; that the proper output filters are designed so as to eliminate

Fig. 5. D-c amplifier utilizing a modulator and demodulator.

ripple at the modulating frequency; etc. Nevertheless, modulated systems play a very important part in the design of d-c amplifiers.

2.2. Application of D-C Amplifiers to Servo Amplifiers. One important example of the use of a d-c servo amplifier is for the amplification of a d-c signal for the purpose of driving an a-c servomotor. A natural solution is to modulate the d-c signal and then use the amplified a-c signal to drive the motor directly. Figure 6 shows a typical system of this type. The amplifier consists of a conventional servo amplifier, and the discussion of the design of circuits of this type will be deferred to the section on a-c servo amplifiers.

The modulator could consist of a mechanical chopper, a diode modulator, or a transistor modulator. Typical diode and transistor-modulator circuits are shown in Figs. 7 and 8. In these circuits the devices are either in a high-impedance state or in a low-impedance state, so

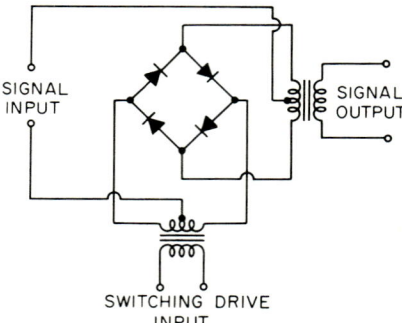

Fig. 6. Servo amplifier with chopper input.   Fig. 7. Diode modulator.

that the d-c input voltage is alternately applied to or disconnected from the load.

The offset and drift errors in these circuits are due to the residual voltage across the devices when conducting, and to the leakage currents through the devices when nonconducting. In most present-day devices, the offset voltage is a greater source of error. Typical values for the offset voltage over a temperature range of 0 to 50°C in the case of the diode-modulator are 5 mv for a matched quadruplet of diodes. For the transistor circuits the offset voltage is much lower, being on the order of 0.25 mv for a matched pair over the same temperature range.

In designing these circuits it is necessary to ensure that the diode or transistor remain in the saturated state for the largest value of signal current to be expected. This determines the level of drive signal required. A serious design problem is the elimination of components of the drive signal at the output due to leakage paths.

* It is perhaps a bit pedantic to refer to modulators in this manner, but the fact that they shift the signal spectrum is really incidental to the reason for obtaining higher d-c stability.

SERVO AMPLIFIERS

These will limit the minimum signal that can be handled. If the modulating signal is a sine wave, one component of leakage will be due to the modulator's not being switched instantaneously. One method of eliminating this is to modulate with square waves, but then droop in the transformer response becomes significant, and coupling of the fast signal transitions through transformer and transistor stray capacitances can cause significant amounts of unwanted signal at the output.

Figure 7 shows a typical diode modulator. The drive signal is supplied by the

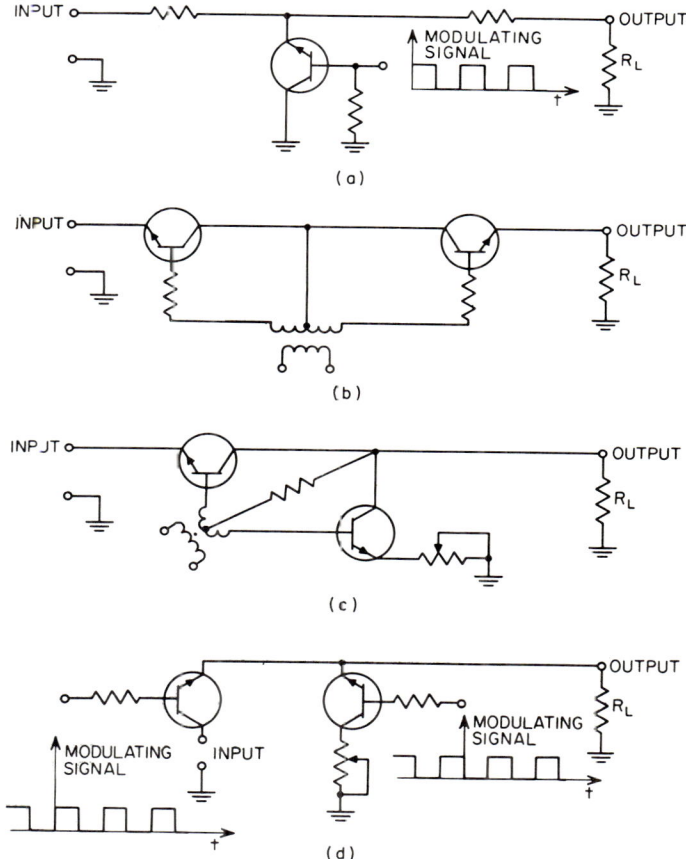

FIG. 8. Transistor modulators.

transformer and the d-c signal path is as shown. The resistors in series with the diodes swamp any differences that might exist in the diode impedances.

Figure 8 shows various methods of using a transistor chopper. Figure 8a shows the transistor used as a shunt switch. In this case it is necessary that the ratio of the "off" resistance be large with respect to the resistance in the "on" condition. Typically the "on" resistance is 10 ohms, and the "off" resistance is 10 megohms, so that the condition is easily met.

Figure 8b, c, and d shows the transistor-modulator circuits having the same operating principle as the first, but using two transistors, to balance the offset voltage. Circuit d is useful because it does not require transformer drive. However, the drive

23-6 AMPLIFIER CIRCUITS

current is flowing through the source so that a low-impedance d-c source is required. Note that in each case the switching signal is applied between collector and base in order to obtain lower offset voltage.

Figure 9 shows the "off" and "on" equivalent circuit of the transistor switch indicating the expression for the magnitude of the offset voltage. Note that if a single transistor switch is used, the offset voltage will be a linear function of the temperature, except insofar as the current amplification of the transistor varies with temperature. If two transistors are used in a balanced configuration it is necessary to ensure that the current amplification vs. temperature variation of each of the transistors is matched. The degree of matching that can be obtained is such that it is possible to obtain an error voltage of less than ± 0.25 mv over a temperature range from -50 to $+100°C$. The potentiometer shown in Figure 8c and d compensates for the component of modulating signal which flows through the signal source impedance.

For small d-c motors with lower power requirements it is possible to use a conventional d-c amplifier. The difficulty that arises when higher-power applications

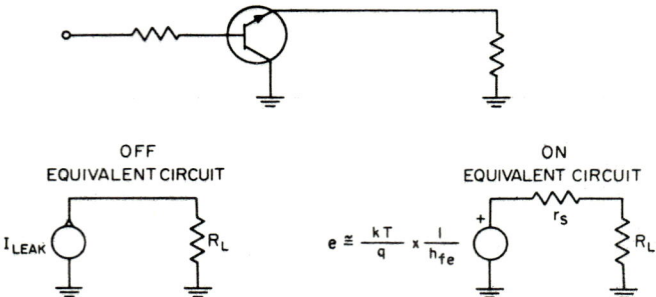

Fig. 9. Transistor chopper. r_s = bulk resistance. k = Boltzmann's constant. T = temperature, °K. q = electron charge. h_{fe} = common-emitter current-amplification factor.

are encountered is that the d-c amplifier is necessarily biased class A so that the amplifier becomes very inefficient if not impossible to design because of power dissipation in the output transistor.

One other method that can be used to drive high-power d-c motors is to use the pulse-width-modulation techniques. In systems of this type the motor is used as a low-pass filter and the d-c power to the load is changed by varying the width of the pulses applied to the motor. The section on nonlinear amplifiers will discuss circuits of this type in more detail.

3. ALTERNATING-CURRENT SERVO AMPLIFIERS

The major advantage of a-c servo amplifiers is the freedom from drift that is achieved by the use of a carrier system. Since typical carrier frequencies are 60 and 400 cps, the design of an a-c amplifier for servo applications is very similar to the design of audio amplifiers and many of the same principles can be applied. This section will be limited to a general discussion of the a-c amplifiers and a detailed discussion of those problems which have a particular bearing on a-c servo-amplifier design.

3.1. General. The following is a list of the important characteristics that a servo amplifier must possess:
1. Capability of driving rated power into load
2. Low output impedance for good motor performance
3. High input impedance for versatility in the signal sources from which the amplifier can operate
4. Relatively stable and adjustable voltage amplification
5. Small phase shift

6. The capability of supplying rated power into the load with small phase shift when saturated

On the other hand, linearity and low distortion are of relatively minor importance.

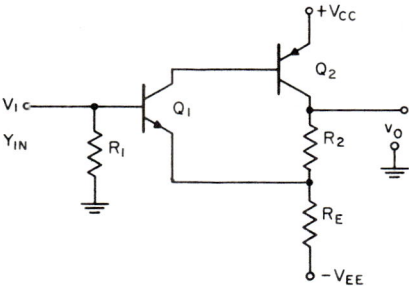

Fig. 10. Two-stage complementary-pair series feedback.

$$Y_{in} = \frac{1}{R_1} + \frac{1}{R_E h_{fe1}(1 + h_{fe2}) + h_{ie1}}$$

$$\frac{v_o}{v_1} = \frac{R_2 + R_E}{R_E + h_{ie1}/h_{fe1}h_{fe2}}$$

$$Z_o = \frac{R_1 + h_{ie1}(1 + h_{fe1})}{h_{fe1}h_{fe2}} \left(1 + \frac{R_2}{R_E}\right)$$

$$(h_{ie1} + R_1) < h_{fe1}h_{fe2}R_E$$

Figures 10 through 23 show typical a-c amplifier and servo-amplifier circuits. The equations for input and output immittances and voltage amplification are given on some of these figures.

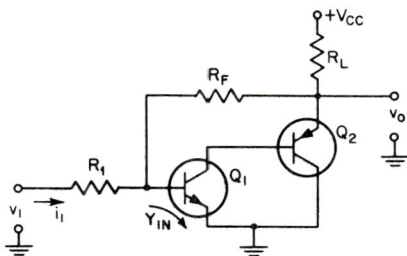

Fig. 11. Two-stage complementary-pair shunt feedback.

$$Y_{in} \cong \frac{1}{R_F} + \frac{1}{h_{ie1}} + \frac{h_{fe1}(1 + h_{fe2})R_L/R_F}{h_{ie1}}$$

$$\text{Let } A_0 = \frac{h_{fe1}(1 + h_{fe2})}{(h_{ie1} + R_F)/R_L}$$

$$\frac{v_o}{v_1} \cong \frac{-R_F/R_1}{1 + 1/A_0}$$

$$Z_0 = \frac{[1/h_{fe1}(1 + h_{fe2})][h_{ie1}(1 + R_F/R_1) + R_F]}{1 - 1/A_0}$$

Feedback is commonly used to achieve many of the desirable servo-amplifier characteristics, such as increased input impedance, low noise, stability of gain, a method of changing gain, and reduction of output impedance. While a complete discussion of feedback amplifiers is not the subject to this chapter, it is hoped that Figs. 10

through 13 will provide some insight into the types of feedback that can be used, and the limitations, in terms of the maximum open-loop gain, that can be expected. Circuits of this type are found in almost all servo amplifiers.

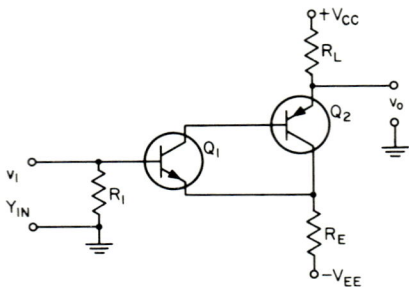

FIG. 12. Two-stage complementary-pair series feedback.

$$Y_{in} = 1/R_1 + 1/[R_E h_{fe1}(1 + h_{fe2}) + h_{ie1}]$$

$$\frac{v_o}{v_1} = \frac{R_L}{R_E + h_{ie1}/h_{fe1}(1 + h_{fe2})}$$

$$Z_0 \cong R_L$$

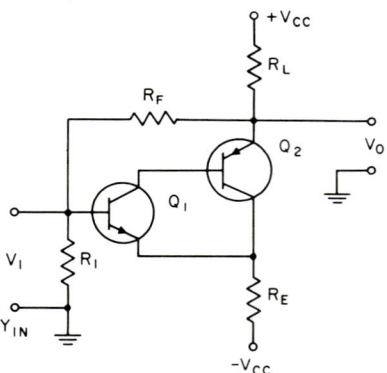

FIG. 13. Two-stage complementary-pair series shunt feedback.

$$Y_{in} = \frac{1}{R_1} + \frac{1}{R_L + R_F}\left[1 + \frac{R_F + h_{fe1}(1 + h_{fe2})R_L}{h_{ie1} + h_{fe1}(1 + h_{fe2})R_E}\right]$$

$$\frac{v_o}{v_1} = -\frac{\dfrac{h_{fe1}(1 + h_{fe2})R_L}{h_{ie1} + h_{fe1}(1 + h_{fe2})R_E} + \dfrac{R_L}{R_F}}{1 + \dfrac{R_L}{R_F}}$$

$$Z_o = \frac{R_L(1 + R_F/R_1)[h_{ie1} + h_{fe1}(1 + h_{fe2})R_E] + R_F R_L}{h_{ie1} + R_F + h_{fe1}(1 + h_{fe2})(R_E + R_L)}$$

The design of a transistor feedback amplifier is an exacting task, and for a complete discussion see the chapter on feedback. However, in the design of servo amplifiers it is usually possible to make many simplifying assumptions (primarily in the form of the transistor equivalent circuit) since the requirements of servo ampli-

SERVO AMPLIFIERS

fiers do not approach the ultimate device limits, except in the power-dissipation requirements.

In most cases it is valid to assume that the transistor current amplification h_{fe} is independent of frequency and that there is no internal coupling between input and

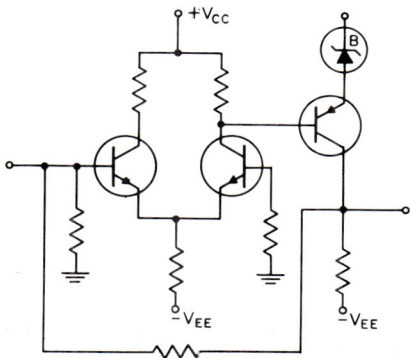

Fig. 14. Complementary amplifier with high d-c stability.

output; i.e., h_{re} is zero. With these assumptions it is possible to write the overall amplification of the device in product form by inspection so that a Bode diagram can be drawn to determine the stability. (The Bode diagram is a plot of the magnitude and phase of the open-loop transfer function vs. log frequency.)

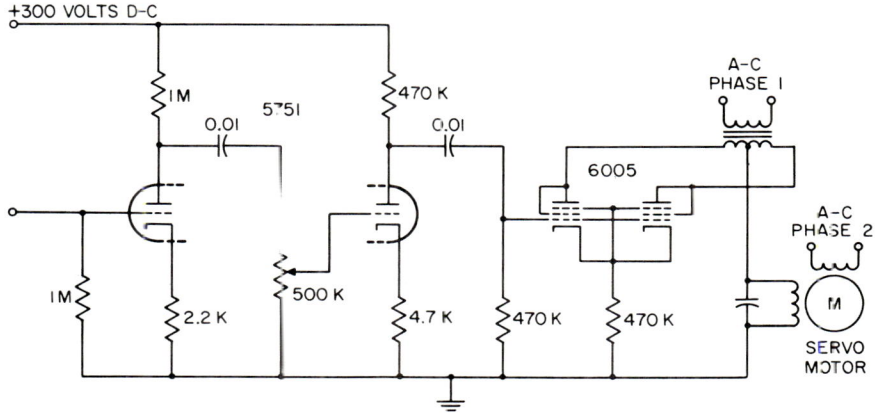

Fig. 15. Servo-amplifier output stage with a-c plate excitation. 6 watts output. (*Macson Corp., used by permission.*)

For example, it would be important to draw a Bode diagram of A_o defined in Fig. 11. In this case it is usually possible to make the approximation that $h_{ie} \ll R_F$ and that $h_{fe} \gg 1$, although in the latter case care should be taken that this approximation is valid over the entire frequency range of interest. Note also that this expression contains R_L, which in the case of a servo amplifier would be inductive, or perhaps tuned. The phase shift introduced by this element usually cannot be neglected.

23–10 AMPLIFIER CIRCUITS

The circuits shown in Figs. 14 to 23 are examples of electron-tube and transistor servo amplifiers. There are several general points to note about these circuits. The servo amplifier consists of a voltage amplifier followed by a power output stage. The

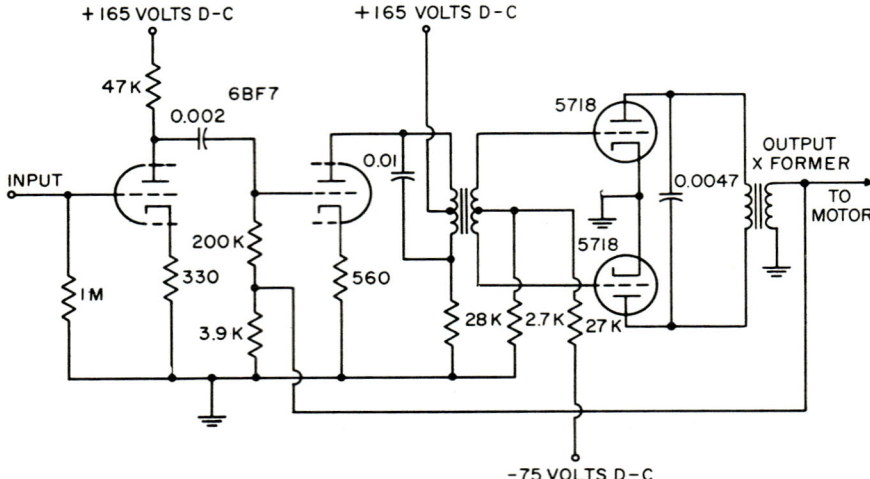

FIG. 16. Servomotor amplifier. 6 watts output. (*Maxson Corp., used by permission.*)

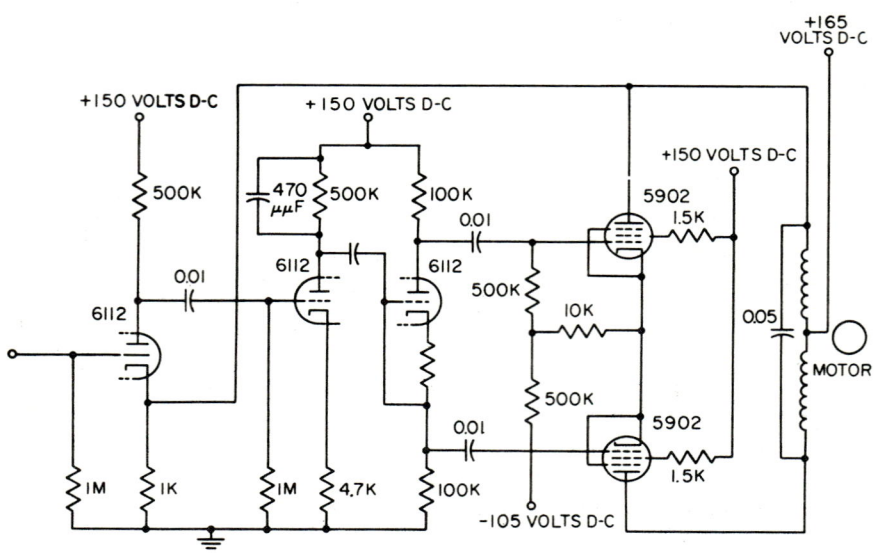

FIG. 17. Servomotor amplifier. Output stage d-c-coupled to motor; cathode-follower phase inverter. 3 watts output. (*Maxson Corp., used by permission.*)

load is the servomotor, which is inductive and can be driven either directly or through a transformer. In either case it is desirable to tune the output stage in order to improve the overall efficiency. If a push-pull output stage is used, and the motor is driven directly, it is necessary to have a center tap on the motor winding.

SERVO AMPLIFIERS   23-11

A comparison of the electron-tube circuits with the transistor circuits will show that the tube circuits tend to be a-c–coupled while the transistor circuits tend to be d-c–coupled. The reasons for this are the difficulty with which the tube circuits are

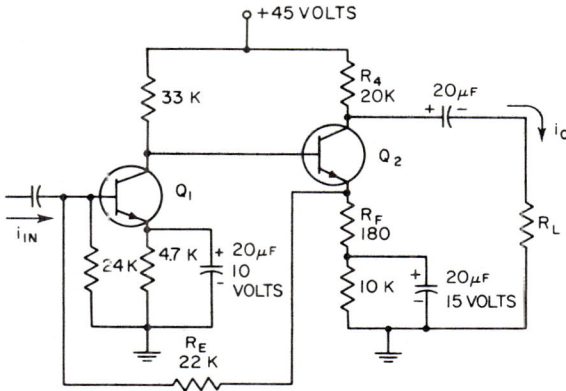

Fig. 18. 400-cycle preamplifier for operation in ambients of -55 to $+125°C$. ("*General Electric Transistor Manual*," 6th ed., used by permission.)

$$Q_1 = Q_2 = \text{GE 2N335, 4C30, 4C31, or 2N336}$$
$$i_0/i_{in} \cong R_E/R_F \quad \text{for} \quad R_L/R_4 \ll 1$$

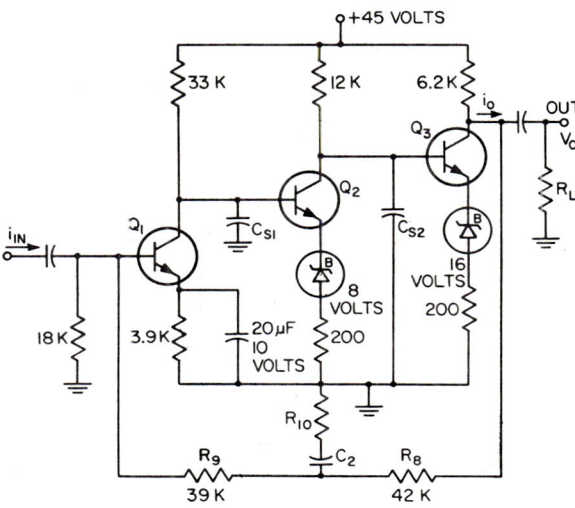

Fig. 19. Three-stage direct-coupled 400-cps preamplifier. ("*General Electric Transistor Manual*," 6th ed., used by permission.)

$$v_0/i_{in} \cong R_8 R_9/R_{10} \quad Q_1 = Q_2 = Q_3 = \text{2N335}$$

d-c–coupled and the fact that since the tube circuits are high-impedance, the coupling capacitors are of reasonable size, even at 60 cps. On the other hand, the transistor circuits are relatively easily d-c–coupled, and since the transistors are low-impedance devices, the coupling capacitors are large in magnitude and therefore undesirable.

23-12 AMPLIFIER CIRCUITS

Additional advantages of d-c coupling are reduction in the total number of components, minimum phase shift, and the capability of fast return to normal operation after the long periods of saturation to which servo amplifiers are normally subjected.

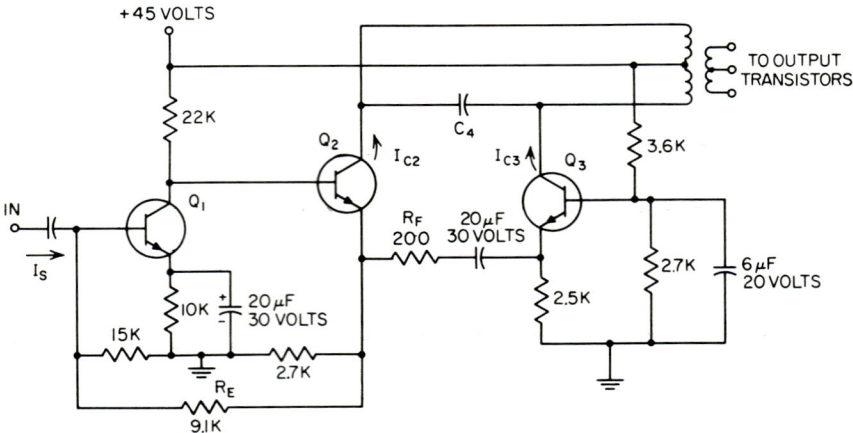

Fig. 20. Stable 400-cps driver. ("*General Electric Transistor Manual,*" 6th ed., used by permission.)

Q_1 = 2N335A
Q_2 = Q_3 = 2N656A, 2N2017, or 2N2018
C_4 adjust to tune transformer
I_{C2} = I_{C3} = 10 ma
$I_{C2}/I_S \cong R_E/R_F$

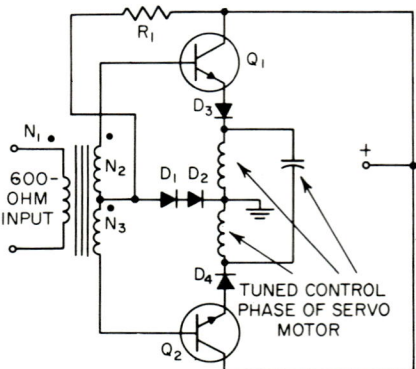

Fig. 21. Servomotor drive circuit (1 to 4 watts). ("*General Electric Transistor Manual,*" 6th ed., used by permission.)

Transformer — $N_1:N_2:N_3$ = 1:1:1
D_1 = D_2 = D_3 = D_4 = 1N1692
Q_1 = Q_2 = GE 2N2202, 2N2203, 2N2204, 2N2196, or 2N2197

Note: Use filtered or unfiltered *positive supply.*

3.2. Applications of A-C Amplifiers to Servo Amplifiers. In many cases the load on a servo amplifier is a servomotor, which presents to the amplifier a reactive load impedance. The motor impedance is a function of speed, but only the resistive component changes markedly, the parallel inductance remaining essentially constant.

Thus it is common practice to operate the output stage as a tuned stage by placing a capacitor in parallel with the motor. This makes the load look resistive and increases the efficiency of the output stage.

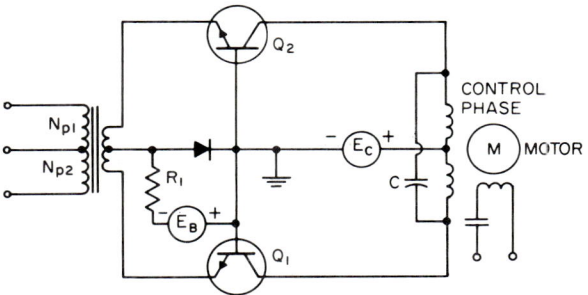

FIG. 22. Common-base servo output stage. ("*General Electric Transistor Manual,*" 6th ed., used by permission.)

E_C and E_B are full-wave rectified 400 cps.
$Q_1 = Q_2 = $ GE 2N2203
$D_1 = $ GE 1N676
$N_p/N_s \gg 1$
C is adjusted for maximum stalled torque

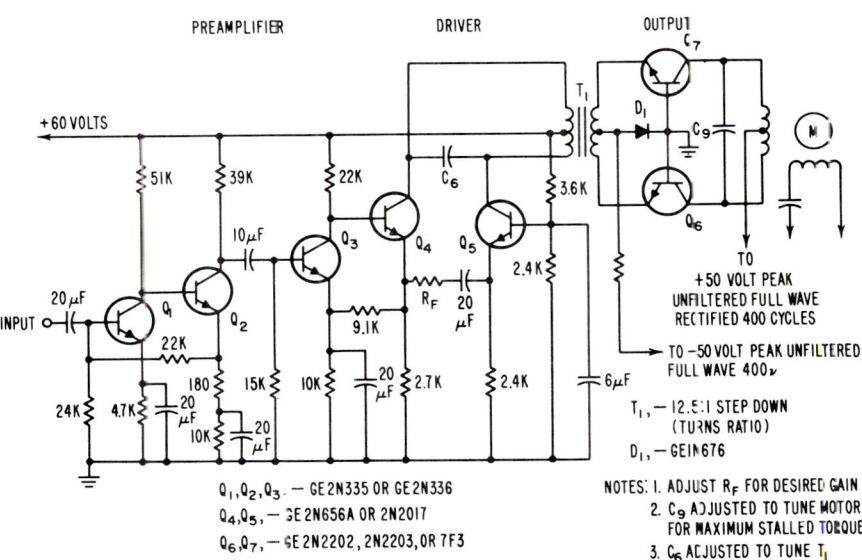

FIG. 23. 3-watt 400-cycle servo amplifier −55 to 125°C operation. ("*General Electric Transistor Manual,*" 6th ed., used by permission.)

The output stages for transistor servo amplifiers can employ the common-base, common-emitter, or common-collector configuration. Feedback of the motor control voltage to the driver or preamplifier is usually very difficult if high open-loop gain is desired because phase shifts of the motor and transformer (if one is used) make stabilization very difficult. One method of deriving a feedback signal is to operate the output stages in the common-emitter configuration and utilize the voltage across small resistors in the emitter circuit. In this way the feedback signal is independent of the motor characteristics.

A second technique which results in stable output-stage gain is illustrated in Fig. 21. The circuit shows an emitter-follower push-pull amplifier and offers the advantages of a low-impedance drive for the servomotor. The efficiency of this circuit exceeds 60 per cent with a filtered d-c supply and will be higher if an unfiltered a-c supply is used.

Another technique which results in a stable output amplifier over a wide temperature range and which is compatible with low-gain transistors is illustrated in Fig. 22. In this case a common-base stage is used and the driver is coupled to the output stage by means of a step-down transformer, so that the current amplification is obtained in the transformer rather than in the transistor.

In servo-amplifier output stages it is common to use an unfiltered d-c collector supply in order to reduce the collector dissipation. Under this condition the transistor dissipation is given by

$$P_D = (E^2_{cm}/4R_L)[a - a^2(1 + R_S/R_L)] + P_L \tag{1}$$

where P_L = dissipation due to leakage current during the half-cycle when the transistor is turned off
a = fraction of the maximum signal present, which varies from 0 to 1
R_S = saturation resistance
R_L = load resistance
E_{cm} = peak value of the unfiltered collector supply voltage

Note that according to this expression, if P_L is negligible and $R_S/R_L \ll 1$, the maximum dissipation occurs when $a = \frac{1}{2}$ or when the signal is 50 per cent of the maximum. For amplifiers that are used in position servos, the signal under steady-state conditions is either zero or maximum, which are the points of least dissipation.

The output stages of servo amplifiers are usually operated in the class B mode so that the driver circuit is required to provide phase inversion. This is usually accomplished with a transformer. A simple means for performing this is to place the primary of the output transformer in the collector of a common-emitter transistor. However, in some cases the maximum d-c ratings of the transformer and the quiescent-current requirement of the amplifier are not compatible. The circuit of Fig. 20 is useful in this respect since by the use of a modified "long-tailed pair" configuration it is possible to cancel the quiescent currents of Q_2 and Q_3 magnetically in the transformer.

The allowable power dissipation in a transistor stage is limited by the temperature to which the junction of the transistor rises during steady operation. In order to compute the allowable power dissipation it is necessary to determine the maximum allowable junction temperature and the thermal resistance between collector and ambient air (usually given in °C/watt). The power dissipation is then

$$P_d = (T_j - T_A)/\theta_{ja} \tag{2}$$

where P_d = power dissipation
T_j = junction temperature
T_A = ambient temperature
θ_{ja} = thermal resistance between junction and ambient air

The thermal resistance between junction and ambient air includes the thermal resistance between junction and case in addition to all resistances external to the case due to heat sinks, washers, etc.

The circuit of Fig. 23 is included to show a complete servo amplifier capable of driving a 3-watt servomotor in an ambient of −55 to 125°C. The amplification can be adjusted from 20,000 to 80,000 amp/amp by adjusting the value of R_F in the driver circuit.

4. EQUALIZATION NETWORKS

Although not directly related to the problem of servo-amplifier design, all servosystems require equalization networks in order to improve the dynamic character-

SERVO AMPLIFIERS

istics of the servosystem. The most common method of providing this equalization is at signal baseband (i.e., before modulation), and in this case the networks take the form of the familiar RC and RL networks. The use of these networks is usually coupled with the design of the d-c amplifier since the filter networks can be placed in the feedback loop of these amplifiers as well as at the input in order to obtain the desired response characteristics. This subject is covered in detail in the chapter on d-c amplifiers and will not be repeated here.

The design of equalization networks for a-c servos is considerably more difficult than for the d-c case. The signal spectrum consists of a double-sideband suppressed-carrier wave, and thus equalization networks must be designed which operate directly on this signal. However, in order to operate on the signal without distortion it is necessary to design a network with arithmetic symmetry about the carrier frequency.

LOW-PASS PROTOTYPE		CARRIER NETWORK	
ELEMENT	IMPEDANCE	ELEMENT	IMPEDANCE
(inductor)	sL	$L/2$, $2/\omega_c^2 L$	$L\left[\dfrac{s^2+\omega_c^2}{2s}\right]$
(capacitor)	$\dfrac{1}{sC}$	$C/2$, $\dfrac{2}{\omega_c^2 C}$	$\dfrac{1}{C}\left[\dfrac{2s}{s^2+\omega_c^2}\right]$
(resistor)	R	R	R
$G(s)$		$G\left(\dfrac{s^2+\omega_c^2}{2s}\right)$	
DATA TRANSFER FUNCTION		NETWORK TRANSFER FUNCTION	

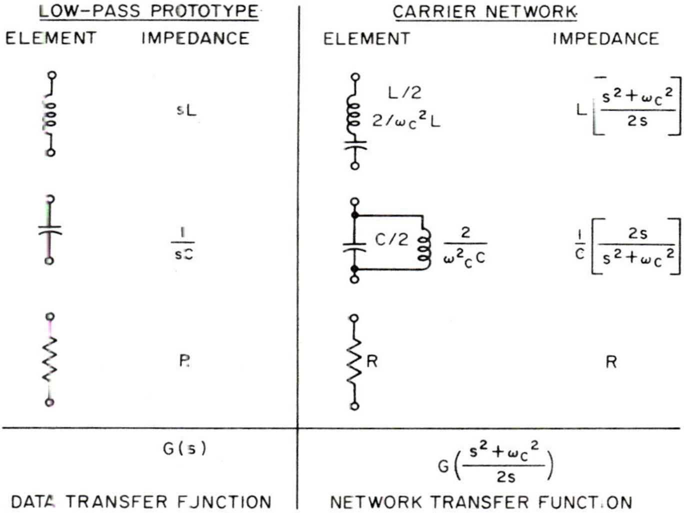

FIG. 24. Transfer characteristics of notch networks. (*Ref. 5, used by permission.*)

It can be shown that this is not possible for linear passive networks, except in the case of zero frequency. Thus an approximation must be made, and the simplest of these approximations results in geometric symmetry about the carrier frequency. This is valid if the bandwidth of the information is small with respect to the carrier frequency. For a 400-cps servo, typical data bandwidths would be on the order of 20 cps, so that the approximation is valid in most practical cases. There are two standard methods of obtaining networks with geometric symmetry:

1. *The low-pass–bandpass transformation.* This can be used regardless of the form of the data transfer function and results in RLC networks. Figure 24a shows the transfer characteristic of such networks, and Fig. 25 shows examples of some typical low-pass-to-bandpass transformations.

2. *The approximation by a function with poles on the real axis.* This results in RC networks such as bridged-T and parallel-T. The advantage of these networks is that only resistors and capacitors are used, which results in a saving in size. A disadvantage is that they are difficult to tune and to keep in tune with temperature variations.

The design of an a-c compensating network by the low-pass–bandpass transformation method consists of the following steps:

1. Determine the required data transfer function.
2. Synthesize the corresponding d-c network, the so-called "low-pass" prototype.

23-16　　　　　　　AMPLIFIER CIRCUITS

The synthesis should take the form of an *RL* network if possible, to permit the association of series R and L into real inductors.

3. Using the transformation of Fig. 24a draw the corresponding bandpass network. The transformation changes all the circuit elements as follows:

a. Each inductance is replaced by a series-resonant circuit, the new inductance being $L/2$ and the resonant frequency being the carrier frequency ω_c.

b. Each capacitance is replaced by a parallel-resonant network, the new capacitance being $C/2$ and the resonant frequency being ω_c.

c. All resistances are invariant.

(For a discussion of *RC* lead networks, see Ref. 5, pp. 407 and 656.)

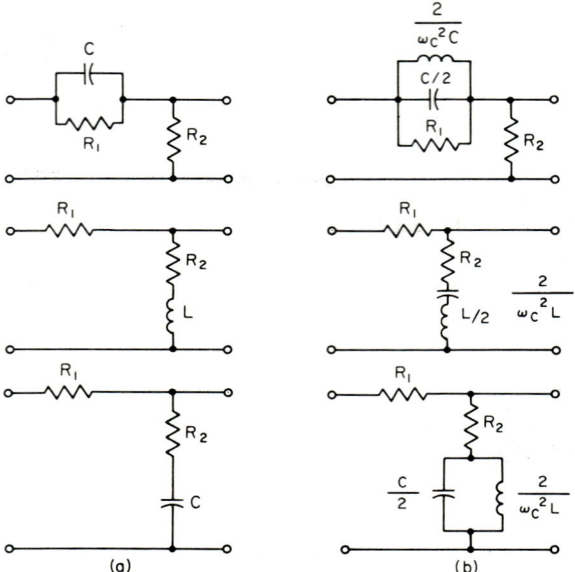

Fig. 25a. Transformations used in a-c compensating network design. (*Ref. 5, used by permission.*)

Fig. 25b. Typical low-pass–bandpass transformations. (*Ref. 5, used by permission.*)

Another approach to the design of compensation networks is the use of the loaded demodulator as a circuit element. Figure 26 shows the circuit of a network of this type. The advantage of these networks is independence of the carrier frequency, but accentuation of the harmonics is still present and in addition there is the problem of noise generation by the demodulator. The latter effect can be minimized by selective filtering of the higher harmonics.

5. NONLINEAR SERVO AMPLIFIERS

In many servo applications the average value rather than the instantaneous value of the load current is the quantity to be controlled. Since it is very difficult to design high-power linear d-c amplifiers because of the excessive power dissipation in the active devices, switching circuits have been designed that are capable of changing the duty cycle of the signal applied to power the actuator and hence obtain effective d-c amplification in this manner. This technique is convenient because the loads that are to be driven are highly reactive and hence are capable of filtering the a-c components of the driving waveform.

SERVO AMPLIFIERS 23-17

The advantage of operation in this mode is that the active devices are operating either in an "off" condition or in a highly conducting condition, and in both these states the power dissipation in the device is at a minimum. The device passes through

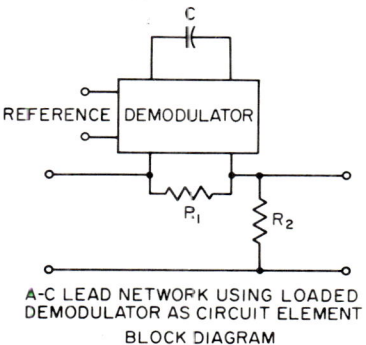

A-C LEAD NETWORK USING LOADED
DEMODULATOR AS CIRCUIT ELEMENT
BLOCK DIAGRAM

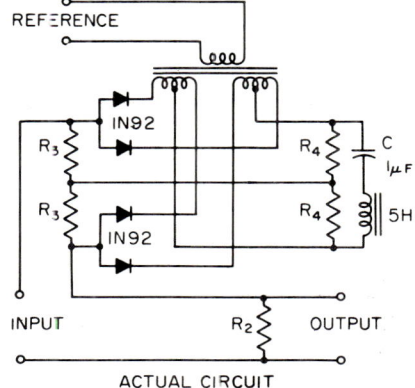

ACTUAL CIRCUIT

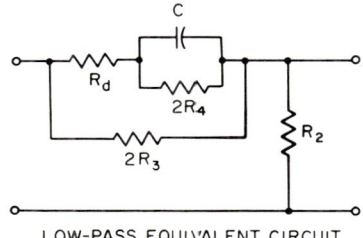

LOW-PASS EQUIVALENT CIRCUIT
R_d = DIODE FORWARD RESISTANCE PLUS
HALF OF REFERENCE WINDING

FIG. 26. A-C lead network using loaded demodulator as circuit element. (*Ref. 5, used by permission.*)

the point of highest power dissipation in a very short time, so that on the average the power dissipation in the device is low.

Some devices normally used for operation in this mode are thyratrons, silicon controlled rectifiers (SCR), and transistors.

Generally, transistors and SCR's can be used in similar applications, but thyratrons are used in the higher-power applications. The major advantage of SCR's over transistors at the present time is that the SCR's are capable of higher-voltage operation.

5.1. Servo Amplifiers Utilizing Thyratrons. The thyratron is a gas tube containing a grid, cathode, and anode. The operation of the device from a phenomenological point of view is that the tube remains in a nonconducting state unless the grid potential exceeds some critical value, whereupon the tube becomes highly con-

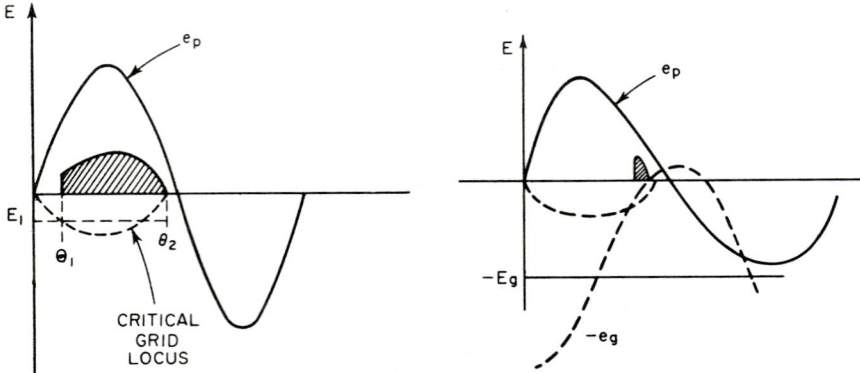

FIG. 27. Critical grid locus of thyratron, with example of firing angle θ_1.

FIG. 28. Bias-phase control.

ductive. Once the tube has fired, the grid loses control over the operation, and the only way the tube can revert to a nonconducting state is to remove the anode voltage. The turn-on (ionization) time of the tube is on the order of 0.1 μsec while the turn-off (deionization) time is on the order of 5 μsec.

One method of operation of the device is to apply an alternating voltage to the anode of the thyratron so that the tube is extinguished during each negative half-cycle and the grid regains control during the positive half-cycle. This method of operation is illustrated by the curve of Fig. 27. In this figure the anode voltage is

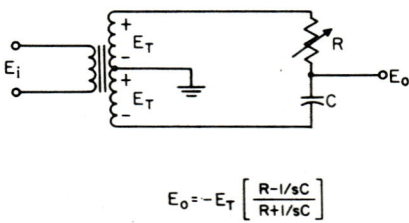

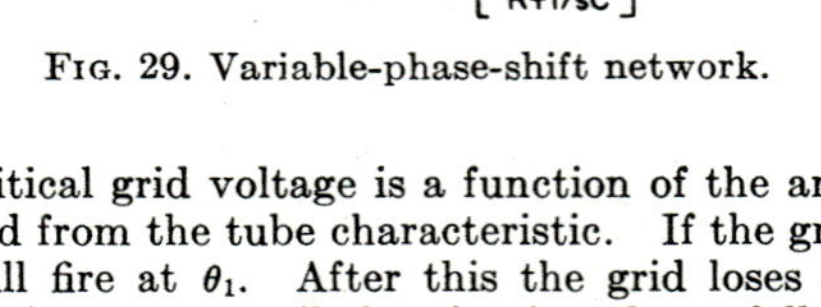

$$E_o = -E_T \left[\frac{R - 1/sC}{R + 1/sC} \right]$$

FIG. 29. Variable-phase-shift network.

plotted as e_p. The critical grid voltage is a function of the anode voltage as shown and may be determined from the tube characteristic. If the grid bias voltage is then set at E_1, the grid will fire at θ_1. After this the grid loses control, and the tube remains in the conducting state until the circuit voltage falls below the tube drop at θ_2. Thus it is seen that by modulation of the firing point it is possible to vary the average, or d-c, value of the load current.

From the curve of Fig. 27, it can be seen that the range over which the firing angle can be controlled is rather limited. A technique that is used to improve the control is to use a combination of bias and phase control. This is shown in Fig. 28. The

SERVO AMPLIFIERS

23-19

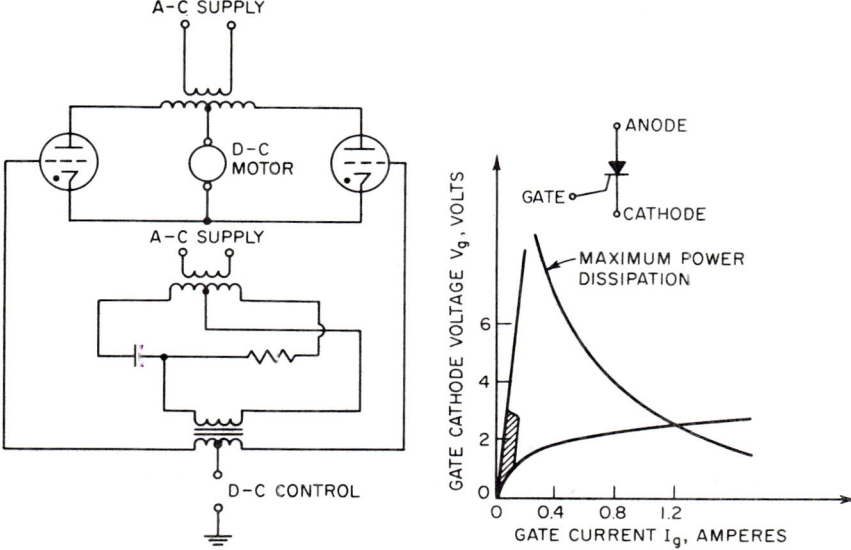

FIG. 30. Thyratron full-wave amplifier for unidirectional control of d-c motor.

FIG. 31. Graphical symbol and gate and firing characteristic. ("*General Electric SCR Manual*," 2d ed., used by permission.)

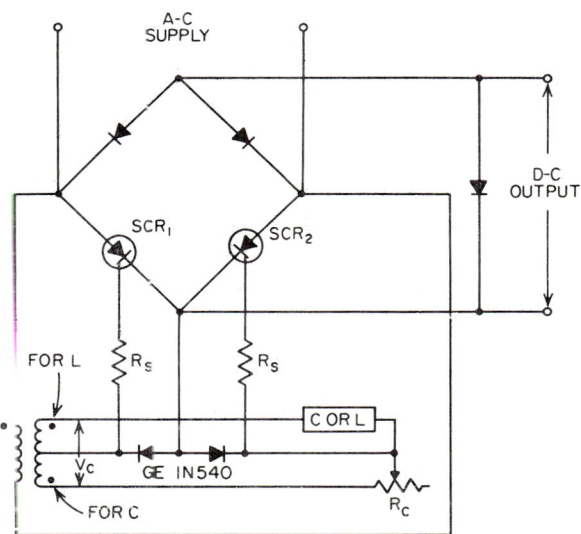

FIG. 32. *RC* or *RL* phase-shift control network. ("*General Electric SCR Manual*," 2d ed., used by permission.)

grid voltage consists of a fixed d-c voltage E_g in addition to an a-c component e_g which is shifted in phase with respect to the anode signal. In this manner it is possible to obtain much narrower firing angles than would be possible with d-c bias control alone. Also since the a-c waveform is quite steep at the point of intersection with the locus, the firing angle is reasonably independent of the tube characteristics.

23-20   AMPLIFIER CIRCUITS

Figure 29 shows a convenient network that can be used to control the phase of the output voltage with respect to the input voltage. An important feature of this network is that the voltage stays constant as the phase is changed.

Figure 30 shows a typical thyratron control circuit. The firing angle of the tube is controlled by both the phase and the d-c control voltage as shown.

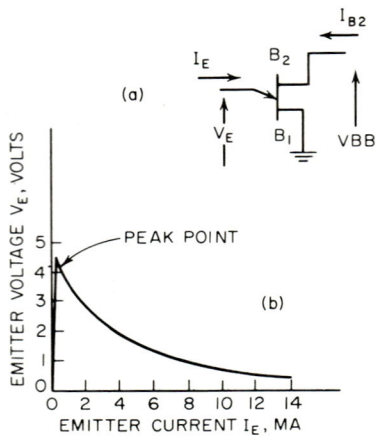

Fig. 33. Unijunction transistor. (a) Symbol. (b) Typical characteristic curve. (*"General Electric SCR Manual,"* 2d ed., used by permission.)

5.2. Servo Amplifiers Utilizing Silicon Controlled Rectifiers. In terms of the terminal characteristics, the operation of the SCR is very similar to the thyratron. There exists a certain gate voltage and current that must be exceeded before the device will fire, and once fired the gate loses control over the device. The device can then be turned off by interrupting the main path of current flow. There are

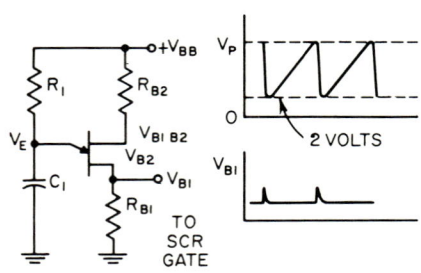
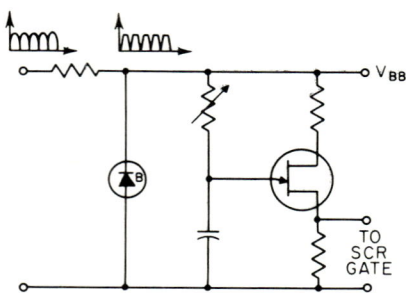

Fig. 34. Basic unijunction relaxation oscillator circuit (astable multivibrator). (*"General Electric SCR Manual,"* 2d ed., used by permission.)

Fig. 35. Circuit for synchronizing to a-c line. (*"General Electric SCR Manual,"* 2d ed., used by permission.)

many subtleties to proper operation of the SCR, and for a complete discussion of these refer to the chapter on solid-state switching devices. The emphasis of the discussion in this section will be on the use of the SCR as a servo actuator.

The gate and firing characteristics of a SCR are normally presented in the form of a curve such as shown in Fig. 31. For reliable firing under all temperature conditions, the gate signal source should be designed to produce voltages and currents in excess of

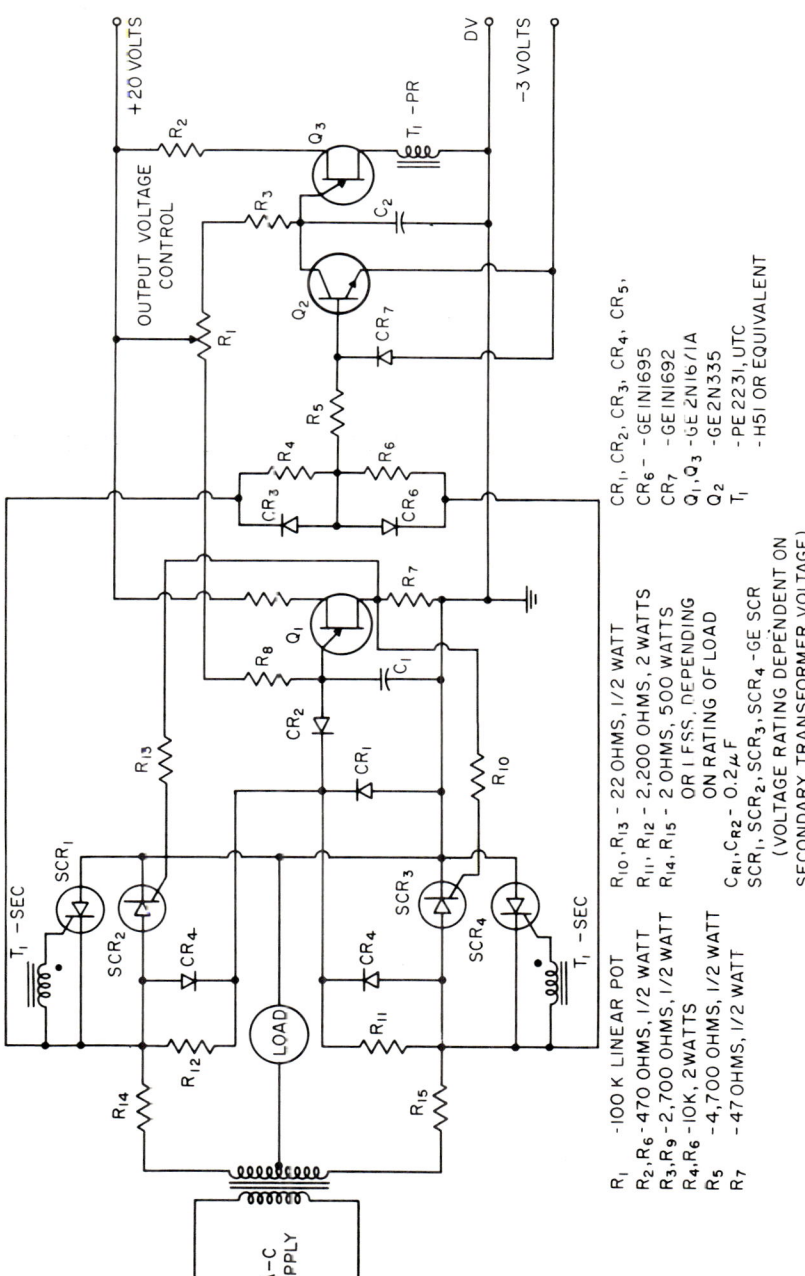

FIG. 36. Full-wave reversing drive. ("*General Electric SCR Manual,*" 2d ed., used by permission.)

23-22 AMPLIFIER CIRCUITS

the values indicated by the shaded region in Fig. 31, without exceeding the maximum power-dissipation curve.

There are several firing circuits which are commonly used in the operation of SCR's. One is the a-c phase-shift firing circuit depicted in Fig. 32. Note that this circuit includes the phase-shift control of Fig. 29 in order to achieve control of the firing point. In addition to ensuring that the gate power, voltage, and current limitations are not exceeded, the following criteria can be used to achieve a positive firing characteristic and positive phase control:

1. $V_c > 25$ volts
2. $1/2\pi fC$ or $2\pi fL \leq R_s$
3. $R_c \geq 10/2\pi fC$ or $10 \times 2\pi fL$

Because of the frequency dependence of this type of phase-shift circuit, the selection of adequate L or C components becomes easier at higher operating frequencies.

A silicon unijunction transistor (UJT) is another device which is very useful in firing circuits for SCR's. It has the advantages of stable operations over a wide temperature range and a peak current rating of 2 amp. For further discussion of the unijunction transistor refer to the section on solid-state switching devices. In simple

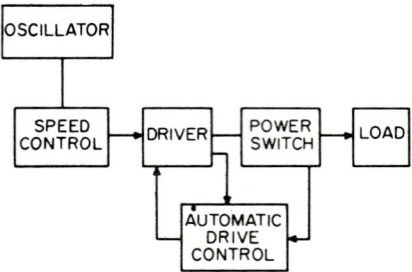

FIG. 37. Switching-mode amplifier with automatic drive control. (*Ref. 4, used by permission.*)

terms the operation of the device can be described in terms of the input characteristic shown in Fig. 33. This is a typical negative-resistance characteristic; and thus, depending on the load line, the device can display bistable, monostable, or astable operation.

A typical trigger circuit for a SCR consists of an astable multivibrator as shown in Fig. 34. In this circuit the capacitor C_1 is charged through R_1 until the emitter voltage reaches V_p at which time the UJT turns on and discharges C_1 through R_{B1}. When the emitter voltage reaches a value of about 2 volts, the emitter ceases to conduct, the UJT turns off, and the cycle is repeated. The period of oscillation is determined in part by the characteristics of the device, and the product R_1C_1. The pulse signal at V_{B1} is suitable for driving a SCR.

In utilizing the UJT as a firing device when the SCR is used as an a-c switch, it is necessary to synchronize the firing pulses with the primary a-c power source. The basic trigger circuit of Fig. 34 can be fired at any intermediate part of the cycle by reducing the interbase voltage V_{B1B2} or the supply voltage V_{BB}. A method of achieving this is shown in Fig. 35. A full-wave rectified signal obtained from a rectifier bridge is used to supply both the power and the synchronizing signal to the firing circuit. The zener diode is used to clip and regulate the peak of the alternating current, as indicated in the figure. At the end of each half-cycle, the voltage at the base 2 of the UJT is reduced to zero, causing the UJT to fire. The capacitor is thus discharged at the beginning of each half-cycle and the firing circuits are synchronized with the line.

The circuits described to this point can be used for applying both armature power and field excitation to d-c machines.

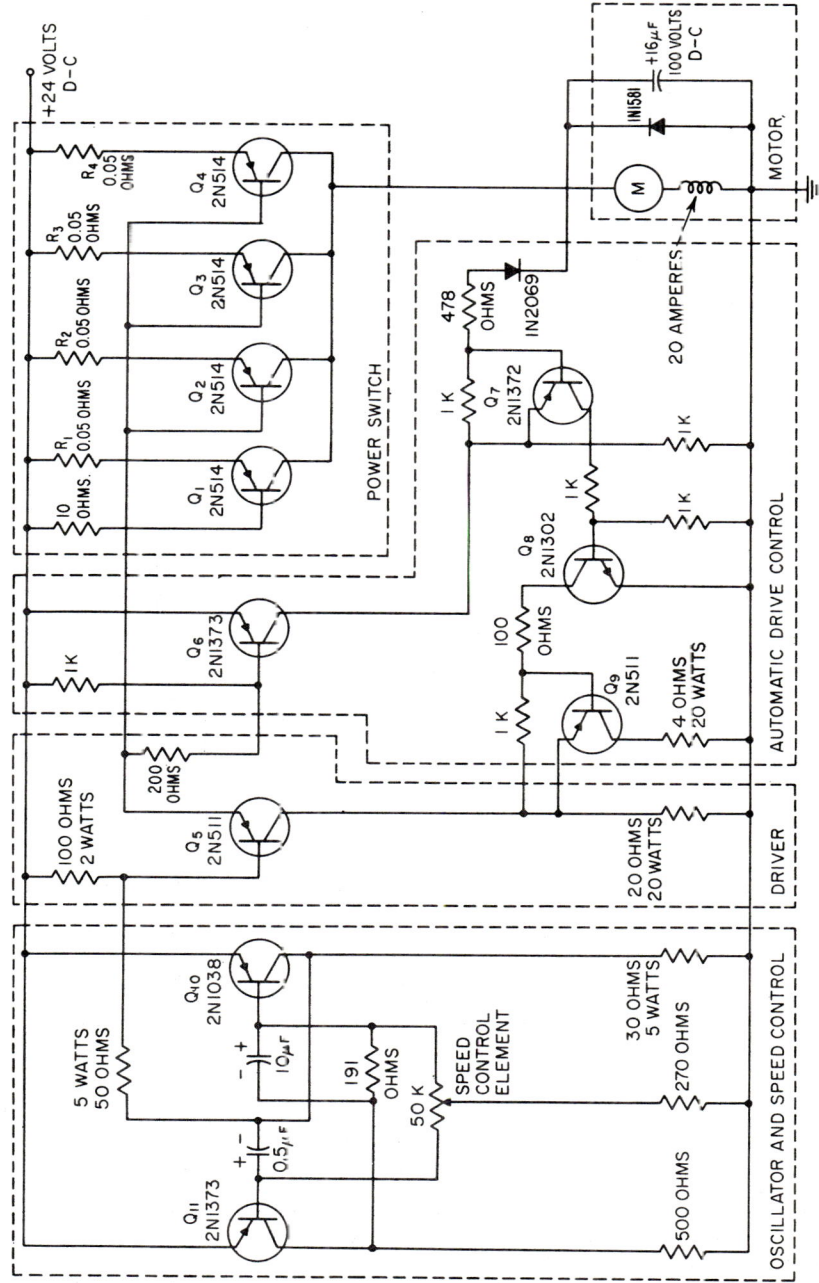

Fig. 38. Complete schematic diagram of speed-control system. (*Ref. 4, used by permission.*)

23-23

A full-wave reversing control or servo as shown in Fig. 36 can be designed around two SCR's with common cathode (SCR_2, SCR_3) and two SCR's with common anodes (SCR_1, SCR_4). In this circuit, SCR_2 and SCR_3 are fired by UJT Q_1. Since SCR_1 and SCR_4 have electrically isolated cathodes, the gate signal pulse generated by UJT Q_3 is coupled to the SCR gates by isolated secondary windings on transformer T_1. Transistor clamp Q_2 synchronizes the firing of Q_3 to the anode voltages across SCR_1 and SCR_4.

Potentiometer R_1 can be used to regulate the polarity and the magnitude of output voltage across the load. With R_1 at its center position, neither UJT fires and no output voltage appears across the load. As the arm of R_1 is moved to the left Q_1 and its associated SCR's begin to fire. At the extreme left-hand position of R_1, full output voltage appears across the load. As the arm of R_1 is moved to the right of center, similar action occurs except that the polarity across the load is reversed.

5.3. Servo Amplifiers Utilizing Transistors in the Switching Mode. Transistors can be very conveniently used in the design of nonlinear servo amplifiers.

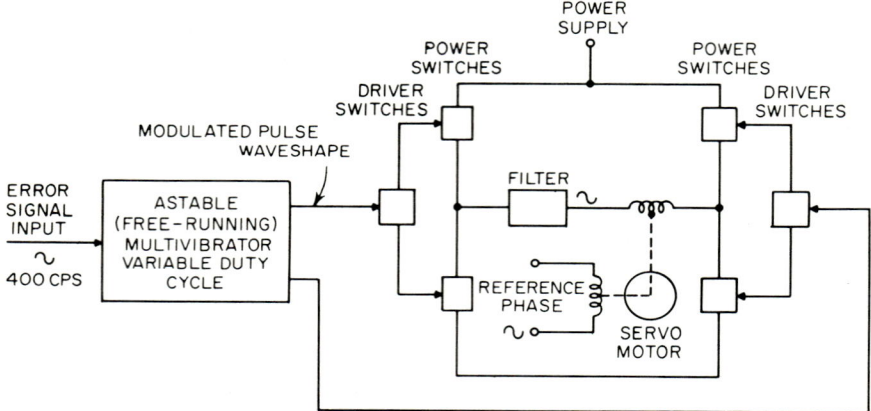

Fig. 39. Switching transistor servo-amplifier block diagram. (*Ref. 4, used by permission.*)

The basic principle is to drive the transistor into the saturated condition or the cutoff condition with a variable-duty-cycle signal. When the transistor is in the saturated condition, current flows into the load, the average value of the current being equal to the ratio of the time in saturation per cycle to the total time per cycle. It is important to keep the power dissipation in the transistors as low as possible, so that it is necessary to ensure that sufficient drive current is available at the base of the power-switching transistors, so that for the largest value of load current, and smallest value of h_{fe} expected, the transistor remains in saturation. However, since this value of base drive current can be quite large, it is desirable for greater efficiency in the output-stage driver circuits to provide for automatic variation in base current as the collector current is varied. The block diagram of such a system is shown in Fig. 37.

A circuit diagram utilizing a system of this type is shown in Fig. 38. The power-switch stage of this circuit consists of four parallel 2N514 transistors. Four transistors are necessary because the current required by the motor in a starting or stalled condition approaches 100 amp, and the 2N514 transistors can carry an absolute maximum of 25 amp. Special precautions must be taken to ensure that there is an equal division of current between the four transistors, and this is accomplished by the emitter resistors. Emitter resistors provide the compensation for variable h_{FE}, $V_{CE(\text{sat})}$ and forward transfer admittance Y_{FE}. However, the size of these resistors is a compromise between circuit efficiency and equalization of collector currents.

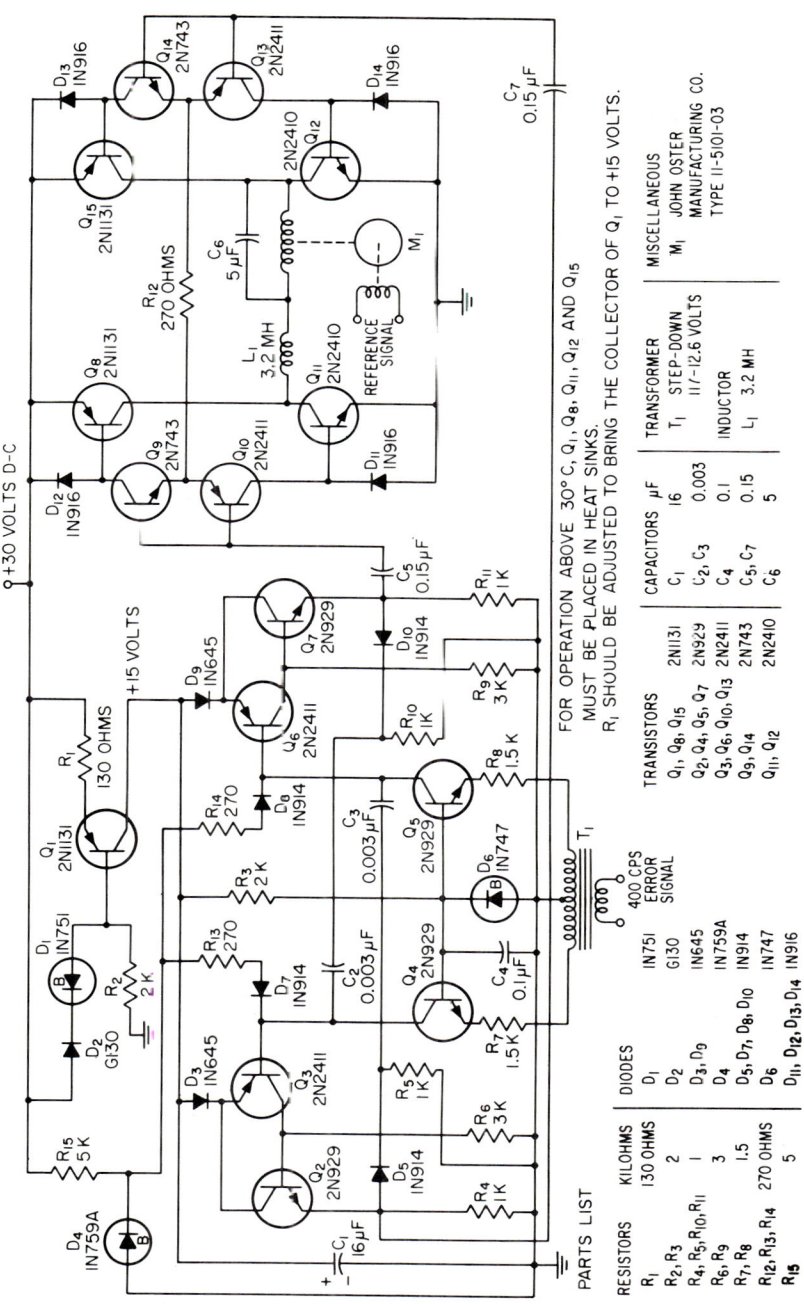

FIG. 40. Complete circuit diagram, switching-mode servo amplifier. (*Ref. 4, used by permission.*)

23-26   AMPLIFIER CIRCUITS

The rectifier and capacitor across the motor are required to minimize the possibility of damage to the power transistor when they switch off the inductive motor load.

Figure 39 shows the block diagram of a servo amplifier employing switching transistors. An actual circuit diagram is shown in Fig. 40. An alternate configuration for the final stage is shown in Fig. 41. In this circuit the base current is supplied through the square-loop core transformer. The diodes across the transistors are for the purpose of protecting against high-voltage transients.

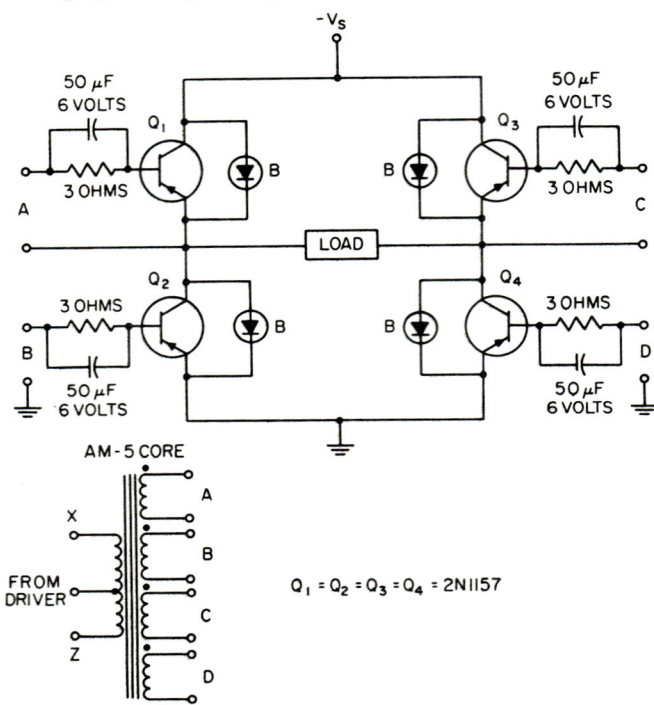

Fig. 41. Switching bridge amplifier.

REFERENCES

1. "General Electric Transistor Manual," 6th ed., Semiconductor Products Department, General Electric Company, Syracuse, N.Y.
2. "General Electric SCR Manual," 2d ed., Semiconductor Products Department, General Electric Company, Syracuse, N.Y.
3. R. F. Shea (ed.), "Transistor Circuit Engineering," John Wiley & Sons, Inc., New York, 1957.
4. Texas Instruments Incorporated, "Transistor Circuit Design," McGraw-Hill Book Company, New York, 1963.
5. John G. Truxal (ed.), "Control Engineers' Handbook," McGraw-Hill Book Company, New York, 1958.
6. R. R. Benedict, "Introduction to Industrial Electronics," Prentice-Hall, Inc., Englewood Cliffs, N.J., 1951.
7. E. Taylor and J. Burnett, Use Thyratrons to Control Higher Power AC Servomotors, *Control Eng.*, April, 1959, pp. 118–122.
8. J. A. Walston and J. E. Setliff, Designing Servo Amplifiers, *Electronics*, February, 1963, pp. 62–63.
9. M. Bodnar, Versatile Servo Amplifier, *Electronics*, January, 1963, pp. 44–45.
10. J. Burnett, Applying Thyratrons to Control, *Control Eng.*, January, February, 1957, pp. 73–77, 89–96.

Chapter 24

TUNED AMPLIFIERS

JOHN R. BOYKIN*

CONTENTS

1. Class A Tuned Amplifiers.................................... 24-2
2. Stability... 24-2
3. Single-tuned Circuits.. 24-3
4. Double-tuned Stages... 24-8
5. Stagger Tuning and Stagger Damping....................... 24-10
6. Bandwidth of Cascaded Stages.............................. 24-11
7. Design Procedure... 24-12
8. Class B Amplifiers.. 24-15
9. Class C Amplifiers.. 24-15
10. Class S Amplifiers... 24-18

Introduction. Tuned amplifiers as discussed in this chapter are amplifiers designed to amplify a band of frequencies that is small compared with the center frequency. The equations presented are all based on arithmetic symmetry rather than geometric symmetry and should give sufficient accuracy with fractional bandwidths up to about 0.1. Where the bandwidth is much greater than 0.1 the calculation becomes more complicated because the more exact methods are required. Such amplifiers are discussed in Chap. 25.

Objectives of Tuning. Tuned circuits are used in the design of both vacuum-tube and transistor amplifiers to obtain higher gain, to obtain controlled phase response, and to reject harmonics and other unwanted signals. Both types of devices have, as an inherent part of their input and output impedances, a certain amount of capacitance. This capacitance, along with the stray capacitance of the connecting circuit, can seriously limit the gain as the frequency is increased. The addition of a shunt inductance can resonate these capacitances and increase the gain attainable at these high frequencies. Such a circuit becomes the simplest form of tuned amplifier and will be used as the reference to which all other tuned amplifiers are compared. With this circuit the bandwidth, as measured between the half-power points, will be equal to twice the bandwidth amplified by the same circuit operating into a load composed of resistance only.

In some cases it is desirable not only to amplify a fairly wide band of frequencies but also intentionally to reject all other frequencies. Where this rejection is desira-

* Westinghouse Electric Corp., Defense and Space Center, Baltimore, Md.

ble, additional capacitance may be added, or more stages used, or more complicated coupling circuits incorporated, or all of these.

In amplifiers where the phase shift at a certain frequency or frequencies is important, tuning of the stages can be used to exert control over the phase.

Low-level amplifiers are usually quite inefficient, but as the power is increased, it becomes more and more important to resort to methods to reduce the losses and the resulting heating in the amplifying devices. To achieve this increased efficiency, most high-powered amplifiers are operated in a mode where the plate or collector current does not flow over the entire r-f cycle. This results in large components of harmonics in the plate or collector current. In the case of class B audio amplifiers, these harmonics are essentially all even-order harmonics and are canceled by operating two tubes or transistors in push-pull. Tuned circuits used in the load circuit, however, can provide the harmonic rejection required so that single-ended amplifier circuits can be used. Likewise, where nonlinear loads must be fed, such as the grid of a class B_2 or class C amplifier, a tuned circuit may be used to supply the peaks of current while maintaining the voltage as essentially sinusoidal.

Another way of achieving high efficiency by the use of tuning is by employing a series-tuned circuit in the load impedance. In this case, the load current will be maintained as a sinusoidal wave while the voltage waveform will be square because of the high impedance to the harmonics presented by the series-tuned circuit. Such a circuit is highly successful as a high-efficiency transistor amplifier because of the low saturation resistance attainable with switching-type transistors.

1. CLASS A TUNED AMPLIFIERS

Efficiency is not usually associated with the design of class A tuned amplifiers; therefore, the requirement for tuning is normally for the purpose of obtaining the highest gain for a given bandwidth in each stage, or for the purpose of shaping the amplitude or phase response. The maximum gain that can be used in each stage may be limited either by the gain-bandwidth product in wideband amplifiers, or by stability considerations, as is usually the case in narrowband amplifiers. Both tubes and transistors have internal feedback paths, and even though all external feedback is eliminated, the maximum gain with good stability may be limited by the characteristics of the device and the impedances facing the input and output, rather than by the gain-bandwidth product. Neutralization can be used where the feedback path remains very nearly constant. With modern pentodes, however, the feedback capacitances within the tube are so small that it is usually useless to neutralize the stage because the stray capacitances, which may be subject to change, can become a significant factor that cannot be readily accounted for in the basic circuit design. Neutralization of transistor stages is usually even less successful. Changes in ambient and internally generated temperature can cause large changes in feedback resulting in distortion of the passband or even in oscillations.

2. STABILITY

The stability factor of an amplifier as shown in Fig. 1 can be calculated by

$$\rho = \frac{2(g_{11} + G_G)(g_{22} + G_L)}{M(1 + \cos \theta)} \qquad (1)$$

where ρ is the stability factor, and the amplifying device has the following parameters:

$$y_{11} = g_{11} + jb_{11}$$
$$y_{22} = g_{22} + jb_{22}$$
$$y_{12}y_{21} = M(\cos \theta + j \sin \theta)$$

where $M = |y_{12}y_{21}|$
$\theta = \angle y_{12}y_{21}$

TUNED AMPLIFIERS

In a transistor y_{12} is y_r and y_{21} is y_f and the load and source admittances are

$$Y_G = G_G + jB_G$$
$$Y_L = G_L + jB_L$$
$$Y_i = G_i + jB_i$$
$$Y_o = G_o + jB_o$$

When $\rho = 1$, the circuit may be on the verge of oscillation. With $\rho > 1$ the circuit will be stable.[1] It is suggested that transistor amplifiers be designed with $\rho > 8$ because of variation of the parameters with temperature. It is obvious from Eq. (1) that it is not always possible to design a multistage amplifier with lossless coupling that will have the desired stability. A design approach that results in close to optimum gain for a given stability factor is to make the coupling transformation turns ratio

$$N = \sqrt{g_{11}/g_{22}} \qquad (2)$$

A shunt conductance is then added, usually in the form of loss in the coupling network. This shunt conductance provides equal mismatch to both the output of the

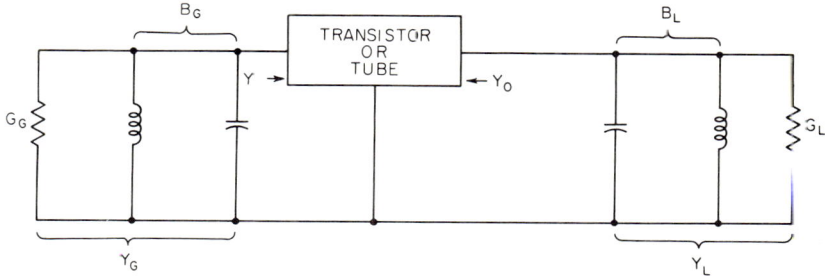

Fig. 1. Amplifier with drive and load admittances.

previous stage and the input of the following stage to provide high gain with a given stability.[2] Another advantage of this approach is that it provides a high degree of isolation between stages and makes the response much more a function of the coupling circuits than of the terminal conditions of the transistors.

The above approach also applies to vacuum tubes. At the lower frequencies g_{11} may not be clearly defined in a vacuum tube since it is the grid-cathode conductance and may be very low. Where the practical transformation ratio is limited to a ratio less than is desirable from Eq. (2) the following approximate relation may be used for pentode tubes:

$$\rho = 2/R_G R_L g_m \omega C_{gp} \qquad (3)$$

where g_m is the transconductance of the tube and C_{gp} is the grid-plate capacitance.

Since the characteristics of vacuum tubes are less subject to change with temperature than those of transistors, a somewhat lower stability factor can be tolerated. A stability factor of 4 or more is suggested.

3. SINGLE-TUNED CIRCUITS

If the plate or collector contains a single resonant circuit as shown in Fig. 2a the response of the amplifier as a function of frequency will be that of the impedance-vs.-frequency curve of a simple parallel-resonant circuit. The equivalent circuit is shown

[1] G. S. Bahrs, Stable Amplifiers Employing Potentially Unstable Transistors, 1957 IRE Conv. Record, pt. 2, pp. 185–189.
[2] R. M. Frazier, Jr., "Methods of Designing and Cascading Unneutralized Tuned Transistor Amplifiers," presented at the Solid State Circuits Conference, February, 1958.

in Fig. 2b. The capacitor C_B in Fig. 2a is for d-c blocking and its effect is neglected in Fig. 2b. The gain at the center frequency f_o will be as though the only impedance is the parallel value of the two resistances R_g and R_l. The center frequency f_o is defined by

$$f_o = 1/[2\pi \sqrt{L(C + C_g)}] \tag{4}$$

The bandwidth of the circuit is

$$\mathcal{B} = f_o/Q \tag{5}$$
$$Q = \omega(C + C_g)/(1/R_g + 1/R_l)$$

Throughout this chapter the bandwidth is denoted by $\mathcal{B}$ and is the bandwidth included between the two points having 0.707 of the voltage at the center frequency, which are also referred to as the half-power points.

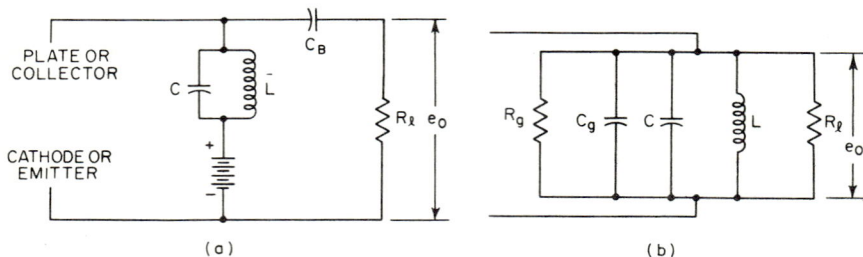

FIG. 2. Single-tuned interstage coupling with equivalent circuit.

When a single-tuned circuit is used as the coupling between two stages some transformation ratio other than 1 is usually desired. When vacuum tubes are used the ratio is usually step-up, but with transistors the ratio is usually step-down.

The gain-bandwidth product for a single-tuned single-stage amplifier is given by

$$\mathcal{G}\mathcal{B} = g_m/2\pi C$$

where C is the minimum shunt capacitance that can be realized and g_m is the forward

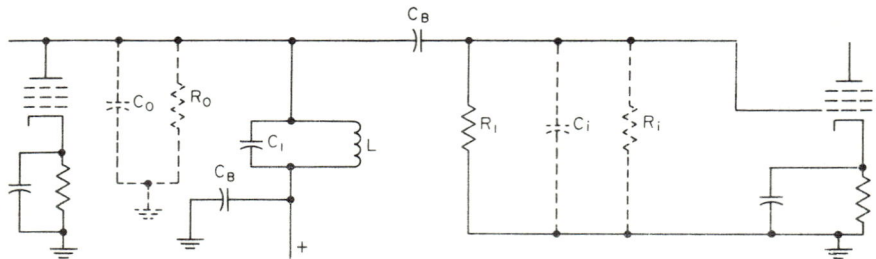

FIG. 3. Single-tuned amplifier stage (series-fed).

$$f_o = \frac{1}{2\pi \sqrt{L(C_1 + C_o + C_i)}}$$

$$\mathcal{G} = \frac{g_m}{1/R_1 + 1/R_o + 1/R_i + 1/2\pi f_o L Q_u}$$

$$\mathcal{B} = 2f_o^2 L(1/R_o + 1/R_1 + 1/R_i) + f_o/Q_u$$

transconductance of the amplifying device. $\mathcal{G}$ is the voltage gain from the input element of one stage to the input element of the following stage. In this equation the assumption is made that the load impedance is low compared with the plate or col-

TUNED AMPLIFIERS

lector impedance. This is usually the case in the wider-band amplifiers where the gain-bandwidth product is a prime consideration. The gain-bandwidth product for a single-tuned single-stage amplifier will be considered the standard with which all others will be compared. The ratio of the bandwidth of any other single stage to that of the single-tuned stage is called the gain-bandwidth factor. By definition the gain-bandwidth factor of a single-tuned stage is 1.

Figures 3 to 6 show several single-tuned coupling networks, with their equations,

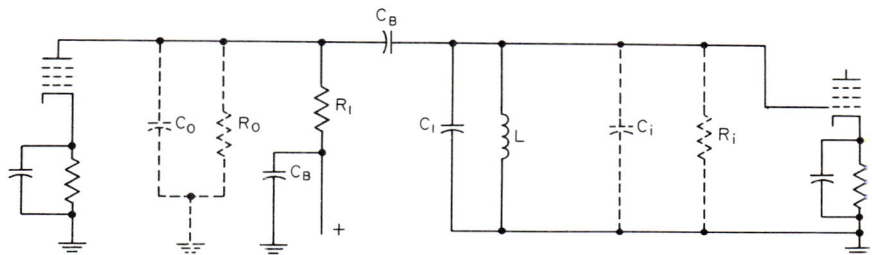

FIG. 4. Single-tuned amplifier stage (shunt-fed).

$$f_o = \frac{1}{2\pi \sqrt{L(C_1 + C_o + C_i)}}$$

$$\mathcal{G} = \frac{g_m}{1/R_1 + 1/R_o + 1/R_i + 1/2\pi f_o L Q_u}$$

$$\mathcal{B} = 2\pi f_o^2 L(1/R_o + 1/R_1 + 1/R_i) + f_o/Q_u$$

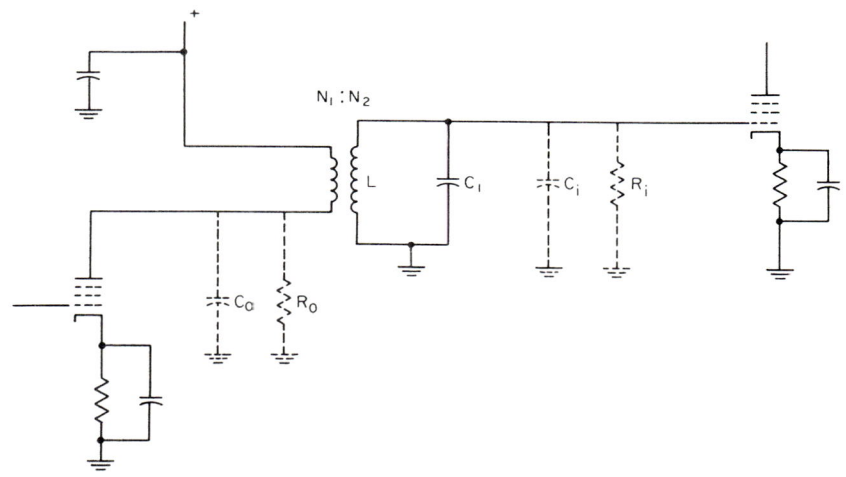

FIG. 5. Single-tuned transformer-coupled stage (tightly coupled transformer).

Optimum turns ratio = $N_1/N_2 = R_o/R_i$

$$f_o = \frac{1}{2\pi \sqrt{L[C_1 + C_i + (N_1/N_2)^2 C_o]}}$$

$$\mathcal{G} = \frac{g_m}{N_1/N_2 R_o + (N_2/N_1)(1/R_i + 1/2\pi f_o L Q_u)}$$

$$\mathcal{B} = 2\pi f_o^2 L(1/R_i + N_1^2/N_2^2 R_o) + f_o/Q_u$$

AMPLIFIER CIRCUITS

usable with vacuum tubes. The networks shown in Figs. 3 and 4 have a 1:1 transformation ratio while those shown in Figs. 5 and 6 generally have a step-up ratio to obtain higher gain for a given stability factor. The capacitors C_o and C_i and the resistors R_o and R_i shown dotted in these and succeeding figures are the output and input capacitances and resistances of the preceding and following stages, respectively, and Q_u is the unloaded Q of the inductance. Figures 7 to 9 show coupling networks

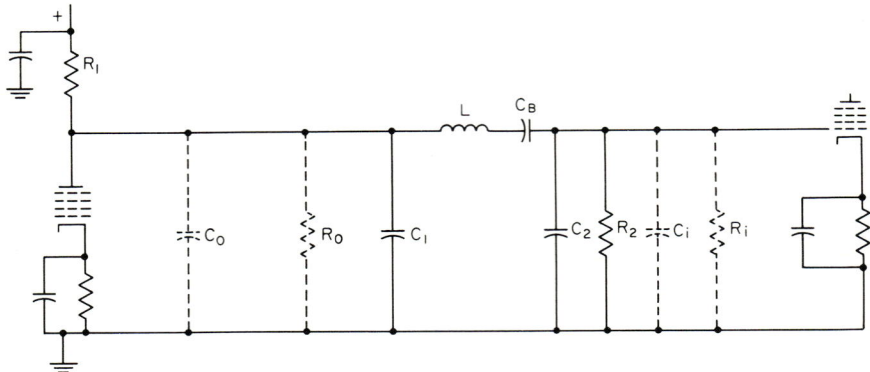

FIG. 6. Single-tuned pi-coupled stage.

Optimum ratio $= (C_2 + C_i)/(C_1 + C_o) = \sqrt{R_o/R_i}$

$$f_o = \frac{1}{2\pi \sqrt{\frac{(C_1 + C_o)(C_2 + C_i)L}{C_1 + C_o + C_2 + C_i}}}$$

$$\mathcal{G} = \frac{g_m}{\frac{(R_1 + R_o)(C_2 + C_i)}{R_1 R_o (C_1 + C_o)} + \frac{(R_2 + R_i)(C_1 + C_o)}{R_2 R_i (C_2 + C_i)} + \frac{(C_2 + C_i)}{(C_1 + C_2 + C_o + C_i)(2\pi f_o L Q_u)}}$$

$$\mathcal{B} = 2 f_o^2 L \left[\frac{(C_1 + C_2)^2}{C_1^2 R_o} + \frac{(C_1 + C_2)^2}{C_1^2 R_i} + \frac{(C_1 + C_2)^2}{C_2^2 R_2} + \frac{(C_1 + C_2)^2}{C_2^2 R_i} \right] + \frac{f_o}{Q_u}$$

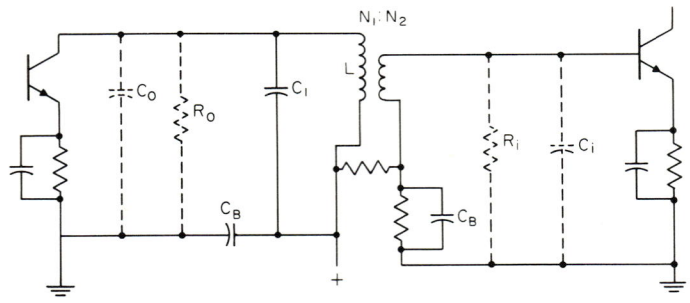

FIG. 7. Single-tuned transformer-coupled transistor amplifier stage.

Optimum turns ratio $= N_1/N_2 = \sqrt{R_o/R_i}$

$$f_o = \frac{1}{2\pi \sqrt{L(C_1 + C_o + C_i N_2^2/N_1^2)}}$$

$$\mathcal{G} = \frac{y_{21}}{N_1/N_2 R_o + N_2/N_1 R_i + N_1/2\pi f_o L Q_u N_2}$$

$$\mathcal{B} = 2\pi f_o^2 L (1/R_o + N_2^2/N_1^2 R_i) + f_o/Q_u$$

TUNED AMPLIFIERS

that are applicable to transistor amplifiers, with their equations. For both tubes and transistors the equations are approximate but should give sufficient accuracy where the bandwidth is less than one-tenth the center frequency. Where transformer coupling is used, the assumption is made that the primary and secondary windings are closely coupled so that the leakage reactance is held to a minimum. If the primary and secondary windings are not closely coupled, the leakage reactance may resonate with the tube or transistor capacitances to cause a high impedance to be presented

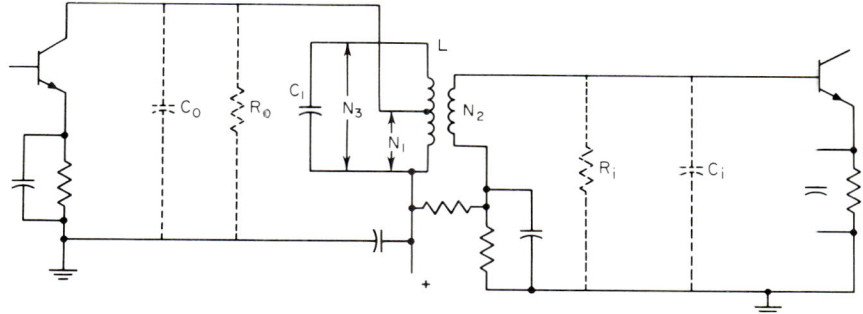

FIG. 8. Single-tuned transformer with tapped primary.

$$\text{Optimum turns ratio} = N_1/N_2 = \sqrt{R_o/R_i}$$

$$f_o = \frac{1}{2\pi \sqrt{LC_1 + N_1{}^2 C_o/N_3{}^2 + N_2{}^2 C_i/N_3{}^2}}$$

$$\mathcal{G} = \frac{y_{21}}{N_1/N_2 R_o + N_2/N_1 R_i + N_3{}^2/2\pi f_o Q_u N_1 N_2}$$

$$\mathcal{B} = 2\pi f_o{}^2 L (N_1{}^2/N_3{}^2 R_o + N_2{}^2/N_3{}^2 R_i) + f_o/Q_u$$

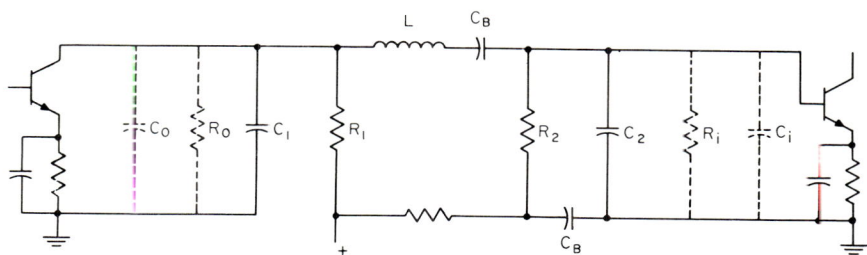

FIG. 9. Single-tuned pi-network-coupled transistor stage.

$$\text{Optimum ratio} = \frac{C_2 + C_i}{C_1 + C_o} = \sqrt{\frac{R_o}{R_i}}$$

$$f_o = \frac{1}{2\pi \sqrt{\frac{L(C_1 + C_o)(C_2 + C_i)}{C_1 + C_o + C_2 + C_i}}}$$

$$\mathcal{G} = \frac{y_{21}}{\frac{(R_1 + R_o)(C_2 + C_i)}{R_1 R_o (C_1 + C_o)} + \frac{(R_2 + R_i)(C_1 + C_o)}{R_2 R_i (C_2 + C_i)} + \frac{C_2 + C_i}{(C_1 + C_2 + C_o + C_i)(2\pi f_o Q_u L)}}$$

$$\mathcal{B} = 2\pi f_o{}^2 L \left[\frac{(C_1 + C_2)^2}{C_1{}^2 R_o} + \frac{(C_1 + C_2)^2}{C_1{}^2 R_i} + \frac{(C_1 + C_2)^2}{C_2{}^2 R_2} + \frac{(C_1 + C_2)^2}{C_2{}^2 R_i} \right] + \frac{f_o}{Q_u}$$

24-8 AMPLIFIER CIRCUITS

to the tube or transistor at some frequency outside of the intended passband. This is a frequent cause of spurious oscillations.

4. DOUBLE-TUNED STAGES

In some cases it will be found that the use of single-tuned circuits will not give sufficient bandwidth with the required gain either because the existing circuit capacitances are too high or because the required rejection of off-channel signals dictates a

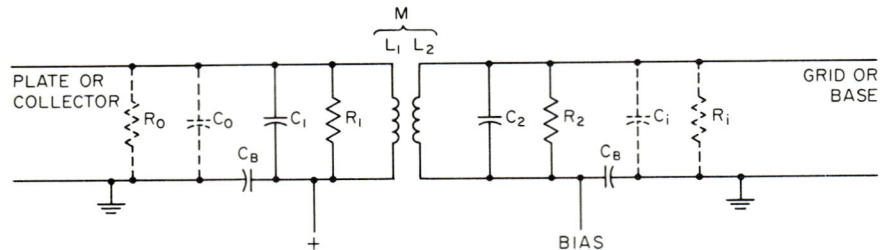

FIG. 10. Double-tuned inductively coupled circuit.

high Q. The double-tuned circuits shown in Figs. 10 and 11 can be used to increase the bandwidth in either case. In Fig. 10, M is the mutual inductance between the primary and secondary windings of the transformer. Where the existing circuit capacitances are already too high, the capacitors C_1 and C_2 would be omitted. At frequencies far removed from the center frequency, the slope of the gain-vs.-frequency curve will be 12 db/octave, thus giving better rejection of off-channel signals. The response near the center frequency is dictated by the amount of mutual coupling

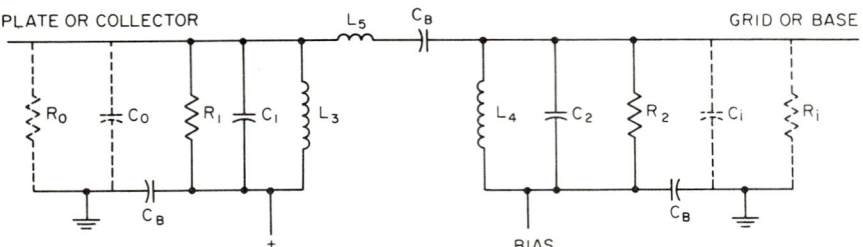

FIG. 11. Double-tuned pi-network coupling.

between the two resonant circuits. As the coupling between the primary and secondary windings is increased, the response at the center frequency first rises and then falls off as shown in Fig. 12. The point at which the amplitude response reaches a maximum is called the *point of critical coupling*. With a small amount of coupling the response curve is peaked. As the coupling is increased the response becomes flat at the center, and as the coupling is further increased the response becomes double-humped. The point of maximum flatness is called the *point of transitional coupling*. If the Q's of the primary and secondary windings are equal, the point of transitional coupling coincides with the point of critical coupling.

TUNED AMPLIFIERS   24-9

As a matter of interest it should be noted that the circuit in Fig. 11 is equivalent to that in Fig. 10 when the following transformations are made:

$$L_1 = \frac{L_3L_4 + L_3L_5}{L_3 + L_4 + L_5}$$

$$L_2 = \frac{L_3L_4 + L_4L_5}{L_3 + L_4 + L_5}$$

$$M = \frac{L_3L_4}{L_3 + L_4 + L_5}$$

While transformers designed for single-tuned-amplifier use require very high mutual inductance which results in very close coupling, those designed for use in double-tuned circuits having a coupling coefficient k require a mutual inductance as given by the following equation:

$$M = k\sqrt{L_1L_2}$$

where M is the mutual inductance of the two transformer windings.

To simplify the equations, the equivalent circuit shown in Fig. 13 will be used, combining all impedances of like type from Fig. 10, with the result that

$$R_a = \frac{1}{1/R_o + 1/R_1 + 1/2\pi f_o Q_p L_1}$$

$$R_b = \frac{1}{1/R_i + 1/R_2 + 1/2\pi f_o Q_s L_2}$$

$$C_a = C_o + C_1$$

$$C_b = C_i + C_2$$

where Q_p and Q_s are the unloaded Q's of the primary and secondary windings, respectively. Critical coupling k_c occurs

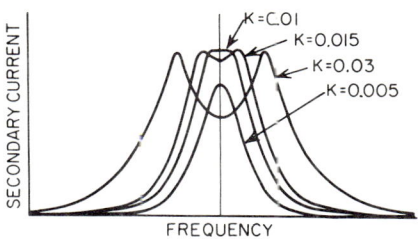

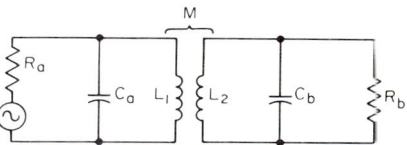

FIG. 12. Curves showing typical variation of primary and secondary currents with frequency for different coefficients of coupling when the primary and secondary windings are separately tuned to the same frequency. (From F. E. Terman, "Electronic and Radio Engineering," McGraw-Hill Book Company, New York, 1955.)

FIG. 13. Equivalent double-tuned circuit.

when

$$k_c = 1/\sqrt{Q_1Q_2} = 2\pi f_o \sqrt{L_1L_2/R_aR_b} \qquad (6)$$

where Q_1 and Q_2 are the loaded Q's of the primary and secondary windings, respectively.

The maximum attainable grid-to-grid, or base-to-base, midband gain occurs at critical coupling and is

$$G = g_{21}\sqrt{R_aR_b}/2 \qquad (7)$$

Transitional coupling k_t occurs when

$$k_t = \sqrt{\tfrac{1}{2}(1/Q_1^2 + 1/Q_2^2)} = \pi f_o \sqrt{2(L_1/R_a)^2 + 2(L_2/R_b)^2} \tag{8}$$

When the loaded Q's of the primary and secondary windings are equal, critical and transitional coupling are equal, as mentioned earlier.

With equal Q's in the primary and secondary windings and with transitional coupling, the bandwidth between the half-power points is

$$\mathcal{B} = 1/(\pi \sqrt{2R_1R_2C_1C_2}) \tag{9}$$

The gain-bandwidth product is

$$\mathcal{GB} = \sqrt{2}\, g_{21}/4\pi\, (\sqrt{C_1C_2}) \tag{10}$$

The gain-bandwidth factor of a single stage is $\sqrt{2}$.

If the primary and secondary Q's are not equal, transitional coupling occurs at greater coupling than critical and will exhibit lower midband gain than with critical

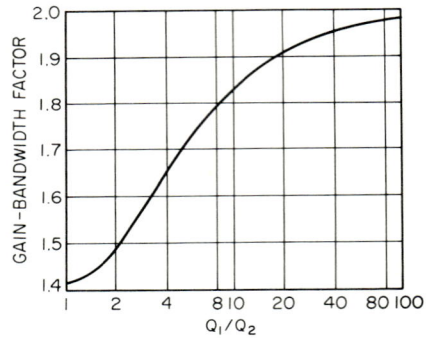

Fig. 14. Gain-bandwidth factor of transitionally coupled double-tuned circuit as a function of Q ratio. [From G. E. Valley, Jr. and H. Wallman (eds.), "*Vacuum Tube Amplifiers*," McGraw-Hill Book Company, New York, 1948.]

coupling. A case of interest is when the Q of the one circuit is much greater than the Q of the other and can be considered to be infinite. If R_1 is allowed to become infinity and if the coupling is made transitional, the midband gain is

$$\mathcal{G} = \sqrt{2}\, R_2 g_{21} \sqrt{C_2/C_1} \tag{11}$$

The gain-bandwidth product is

$$\mathcal{GB} = g_{21}/(2\pi \sqrt{C_1C_2}) \tag{12}$$

The gain-bandwidth factor is 2.

Figure 14 can be used to determine the gain-bandwidth factor for transitionally coupled stages with Q ratios other than 1 or infinity.

5. STAGGER TUNING AND STAGGER DAMPING

The response curve of one double-tuned stage can be duplicated by using two single-tuned stages tuned to slightly different frequencies. Similarly, the response of one triple-tuned stage, one quadruple-tuned stage, etc., can be obtained by respec-

TUNED AMPLIFIERS

tively using three, four, or more single-tuned stages tuned to different frequencies. This technique is called *stagger tuning*. The case of most interest is where the response curve is maximally flat. Figure 15 shows the center frequencies and relative bandwidths required of the component stages to give flat response for a group of two three, or four stages. The frequencies are given as the fraction of the desired bandwidth ⓑ away from the center frequency f_o. The bandwidths are given as the fraction of the desired bandwidth ⓑ. A group of stages tuned in this way will have the same response curve, transient response, and skirt selectivity as one stage having double, triple, or quadruple tuning. The gain-bandwidth factor of a stagger-tuned stage is 1.

Another coupling system that is often used is called *stagger damping*, where the individual stages all have the same center frequency but not the same shape of the response curve. A method that is most often used is an even number of stages with alternate overcoupling and undercoupling so arranged as to result in an overall flat response. With this method all the transformers have the same coupling coefficient, and each is loaded on one side only. For a stagger-damped pair, where ⓑ is the bandwidth required for the pair of stages,

$$k = ⓑ/f_o$$

One transformer is loaded on one winding so that

$$Q = 1.31 f_o / ⓑ$$

and the other is loaded on one winding so that

$$Q = 0.541 f_o / ⓑ$$

With both transformers the unloaded winding must have a Q that is high compared with the loaded winding. If this is not possible, it is usually not desirable to use stagger-damped tuning.

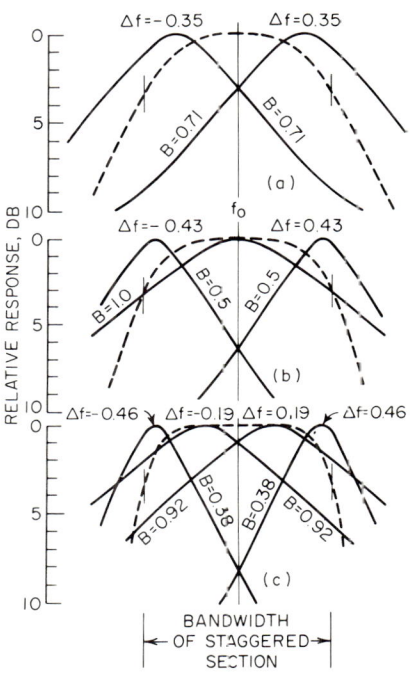

FIG. 15. Response curves of staggered (a) pair, (b) triple, and (c) quadruple with responses of component stages. f and B of component stages are expressed as fractions of bandwidth of group. Dotted line is response curve of group

6. BANDWIDTH OF CASCADED STAGES

When two or more stages of an amplifier are operated in cascade, the resulting bandwidth is narrower than that of any one stage alone. The degree of this shrinkage of bandwidth is dependent on the steepness of the response curve of the individual stages or groups of stages, and the number of stages or groups. If the amplifier consists of n sections, each section being a single synchronous-tuned or transitionally coupled double-tuned stage, or a stagger-tuned or stagger-damped pair or triple, and each section contains m tuned circuits, then the ratio of the bandwidth of the whole amplifier to the bandwidth of a single section is

$$\frac{\text{Bandwidth of amplifier}}{\text{Bandwidth of section}} = (2^{1/n} - 1)^{1/2m} \qquad (13)$$

where m is determined as follows:

Stages in each section	m
One synchronous-tuned	1
One double-tuned or a staggered pair	2
One triple-tuned or a staggered triple	3
One quadruple-tuned, a staggered quadruple, or a stagger-damped pair	4

A very important consideration in designing tuned amplifiers is the rejection of unwanted signals. A graphical method is presented in this section which can be used in determining the shape of the frequency-response curve and in assuring adequate rejection of the unwanted signals. In the method presented herein, the response curve of each tuned circuit in the amplifier is 6 db/octave. Also, to assure a straight-line plot, semilogarithmic graph paper must be used with the abscissa being in decibels and the ordinate in Δf, which is the deviation from the center frequency f_o. The plot to be made is the frequency response of one-half of the passband. The other half may be determined by arithmetic symmetry. A point P_1 is placed at -3 db at Δf equal to one-half of the bandwidth of the amplifier. Another point P_2 is placed at $-3n$ db at Δf equal to one-half of the bandwidth of each section of the amplifier. A line is then drawn through P_2 with a slope of $-6mn$ db/octave, where m and n are as previously defined. The curve can then be sketched in, joining points P_1, P_2, and 0 db.

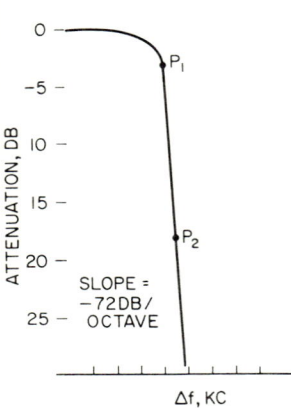

Fig. 16. Approximate frequency response of amplifier with six double-tuned stages or six staggered pairs.

A sketch showing the results of using the above procedure is presented as Fig. 16. It is a rough approximation of the selectivity curve of an amplifier designed to be maximally flat.

7. DESIGN PROCEDURE

The actual design of an amplifier can proceed as follows: First select the center frequency and type of tube or transistor and a desired stability factor, and using Eqs. (1) or (3), determine the maximum load impedance in each stage at that frequency consistent with the desired stability. Then select a circuit from Figs. 3 to 11 and with its gain equation determine the stage gain. From the required gain of the amplifier determine the minimum number of stages to give this total gain. From the gain-bandwidth product find the maximum possible bandwidth of a single-tuned stage and, applying the bandwidth shrinkage for cascaded synchronous-tuned stages, determine the maximum possible bandwidth of the overall amplifier. If this bandwidth is sufficient, determine the skirt selectivity. If this, too, is sufficient the amplifier should be designed on this basis. Generally, where it can be used, the synchronous single-tuned amplifier is highly satisfactory because of its excellent transient response, phase characteristics, and ease of tuning. It is necessary only to peak each stage. If this design results in too poor a skirt selectivity, it will be necessary to add capacitance to increase the Q of the tuned circuits while keeping the resistance constant. When the desired bandwidth is reached, if there is insufficient selectivity, it will be necessary to use multiple tuning as either multiple-tuned coupling circuits, staggered pairs or triples, or stagger-damped stages. Stagger tuning is often used in the lower-priced amplifiers where skirt selectivity is of lesser importance than gain or bandwidth. This type of amplifier not only has the advantage of low cost but also provides a degree of flexibility that other amplifiers do not have. It has the disadvantage of having poor overload characteristics; that is, certain portions of the passband overload first and cause severe ringing. It is most readily tuned by using a sweep-signal generator and an oscilloscope to observe the shape of the passband.

TUNED AMPLIFIERS 24-13

Where selectivity is a problem, it often is desirable to use double-tuned stages. This design results in twice as many tuned circuits per stage, with consequent steep sides of the selectivity curve. If the coupling is transitional, a sweep-signal generator and an oscilloscope are used to align it; however, in one popular method of design, the coupling is made slightly less than transitional so that alignment may be made by simply peaking each tuning control. This design also results in slightly better transient response, which is a result of the more gentle rolloff in amplitude response.

In many cases it will be found that gain per stage is not limited by the stability but rather by the capacitance existing in the circuit. In this event it is necessary to increase the number of stages and reduce the gain per stage by lowering the load impedances. Usually, synchronous single-tuned stages are not recommended under these conditions, but rather one should turn to double-tuned circuits with only enough added capacitance to cause the proper impedance match.

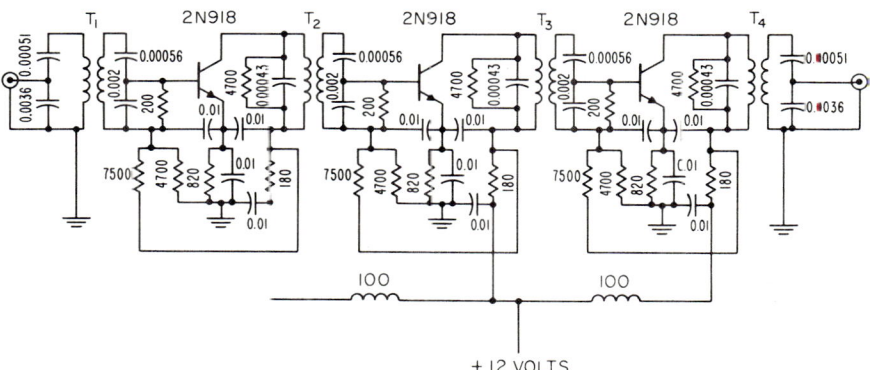

Fig. 17. Three-stage bandpass amplifier. Center frequency 10.7 Mc. Input and output impedances 50 ohm coaxial. Power supply +12 volts d-c. All capacitances in microfarads. All resistances in ohms. All inductances in microhenrys. Transformers T_1, T_2, T_3, and T_4 are all identical and have primaries and secondaries that are identical, and each is adjustable over the range of 0.475 to 0.525 μh, with each winding having an unloaded Q of 100 and having a mutual inductance between primary and secondary of 0.01 μh.

Example. The circuit as shown in Fig. 17 can be analyzed in the following manner: Each of the transistors has its base returned to the junction between a 4,700-ohm resistor and a 7,500-ohm resistor. Each of these pairs of resistors acts as a voltage divider for the 12-volt supply. Neglecting the current drawn by the base of the transistor, the voltage E_B applied to the base will be

$$E_B = \frac{4,700 \times 12}{4,700 + 7,500} = 4.6 \text{ volts}$$

The type 2N918 transistor is a silicon type; therefore, the drop to be expected in the emitter diode will be on the order of 0.6 volt, leaving approximately 4 volts between the emitter and ground. With an emitter-resistor value of 820 ohms this results in a collector current of about 5 ma for each of the transistors, plus approximately 1 ma for each pair of divider resistors, giving a total current drain for the amplifier of about 18 ma. There will be slightly less than 1 volt drop across the 180-ohm collector circuit decoupling resistor, leaving a total of about 7 volts between the collector and emitter of the transistor.

From the published curves and data for the type 2N918 transistor at approximately the above operating conditions, the following characteristics are determined:

$$y_{11} = y_{ie} = 1.8 \times 10^{-3} + j0.6 \times 10^{-3} \text{ mhos}$$
$$y_{22} = y_{oe} = 0.08 \times 10^{-3} + j0.15 \times 10^{-3} \text{ mhos}$$
$$y_{12} = y_{re} = 0.01 \times 10^{-3} + j0.08 \times 10^{-3} \text{ mhos}$$
$$y_{21} = y_{fe} = 100 \times 10^{-3} - j20 \times 10^{-3} \text{ mhos}$$

Calculation of the coupling networks is accomplished using the equivalent circuit shown in Fig. 13. In the circuit shown in Fig. 17, the loading on the secondaries of all the transformers, as well as on the primary of T_1, is tapped across one of a pair of series capacitors. These resistor and capacitor values must be transformed into the equivalent value appearing across the coil and considered to be in parallel with the shunt resistance because of the unloaded Q of the coil.

Table 1 shows the results of the intermediate steps in the calculations of the characteristics of the amplifier. Lines 1, 2, 5, and 6 are determined by calculating the equivalent shunt impedances facing the coils. The inductances of the coils, lines 2 and 3, are the inductances required to resonate these capacitors at a frequency of 10.7 Mc, since the amplifier is specified as synchronous-tuned. Lines 7 and 8 are calculated from the inductance and the value of the loaded Q. Lines 9 and 10 are determined by the value of the inductance of the coil and the parallel value of the shunt resistance facing the coil and the resistance due to Q_u for that coil.

Line 11 shows that the coupling coefficient for each stage is slightly less than transitional. This allows synchronous tuning to be accomplished by simply peaking each stage. The coefficient of coupling is, however, close enough to transitional to use the equations for transitionally coupled stages for the purpose of calculating the bandwidth. Furthermore, it will be seen from Table 1 that the characteristics of each of the four transformers are so close to those of the others that sufficient accuracy will be obtained by considering them as four identical transformers of 1:1 turns ratio and with each winding faced by a capacitance equal to the average of all the actual capacitances. The average of these capacitances is 437 pf. Likewise, the average effective resistance shunting each of these windings is 1,661 ohms. From Eq. (9) the bandwidth of each transformer is found to be 310 kc. The bandwidth of the amplifier is determined from Eq. (13) (where $n = 4$ and $m = 2$) to be 204 kc. Since there are eight tuned circuits, each with 6 db/octave slope, the slope of the skirts of the selectivity curve will be 48 db/octave from the center frequency. The response curve can be approximated as 0 db at the center frequency, minus 3 db at 102 kc each side of center, minus 12 db 155 kc each side of center (3 db for each coupled circuit) and minus 60 db (-12 db $-$ 48 db) at frequencies 310 kc each side of center frequency.

Since the input and output impedances of the amplifier are equal, it is easier to make the calculations of gain considering voltage only. The values of R_a and R_b in Fig. 13 will be the total values of resistance transformed to the value as it would appear at the input terminals of the transformer rather than as it would appear across the entire coil. Lines 12 and 13 of Table 1 show these transformed values of R_a and R_b for each of the four transformers.

The voltage gain from the input terminals to the base of the first transistor is R_b/R_a using the values of R_a and R_b associated with T_1, and is equal to 1.77 or 4.9 db.

Table 1

	T_1	T_2	T_3	T_4
Primary capacitance, pf	446	430	430	430
Secondary capacitance, pf	437	437	437	446
Primary inductance, μh	0.495	0.513	0.513	0.513
Secondary inductance, μh	0.505	0.505	0.505	0.495
Shunt resistance facing primary, ohms	3,250	3,410	3,410	3,410
Shunt resistance facing secondary, ohms	3,080	3,080	3,080	3,250
Primary resistance due to Q_u, ohms	3,330	3,450	3,450	3,450
Secondary resistance due to Q_u, ohms	3,400	3,400	3,400	3,330
Primary loaded Q	49.3	49.8	49.8	49.8
Secondary loaded Q	47.9	47.9	47.9	49.3
k/k_t	0.95	0.94	0.94	0.97
Fig. 13 R_a, ohms	25.2	1,715	1,715	1,715
Fig. 13 R_b, ohms	78.1	78.1	78.1	25.2

The voltage gain from the base of the first transistor to the base of the second transistor is found using Eq. (7):

$$\mathcal{G} = \frac{g_{21}\sqrt{R_a R_b}}{2} = \frac{102 \times 10^{-3}\sqrt{1{,}715 \times 78.1}}{2} = 18.7 \text{ or } 24.4 \text{ db}$$

TUNED AMPLIFIERS   24–15

The gain of the remaining stages is found in a similar manner to be 25.4 and 20.5 db. The total insertion gain for the amplifier is then

$$4.9 \text{ db} + 24.4 \text{ db} + 25.4 \text{ db} + 20.5 \text{ db} = 75.2 \text{ db}$$

8. CLASS B AMPLIFIERS

In transmitters, the amount of power involved is large enough that the efficiency of a stage becomes a matter of importance, and class B amplifiers as well as class C and class S amplifiers discussed in the following sections, can be used to reduce the losses. Class B amplifiers are used where the power is high enough for the efficiency to be a consideration, and the amplifier is required to be linear. Linear in this sense means that the envelope of the output wave is proportional to the envelope of the input wave. Where the amplifier is tuned, it is not necessary for the stage to be push-pull to cancel the even harmonics because the Q of the tank circuit provides the required suppression of harmonics. As in other class B amplifiers the grid or base bias is set to a value that is at the point of plate-current or collector-current cutoff, so that with no incoming signal, the power input to the stage is very low. As the signal level is increased, the input power increases linearly with the drive voltage while the output power increases as the square of the drive voltage. The point of maximum efficiency, therefore, is at full power output. The idealized efficiency = $\pi E_{cm}/4E_b$, where E_{cm} is the peak plate or collector swing and E_b is the d-c plate or collector supply voltage. The maximum theoretical efficiency of a class B amplifier at full power output with a sine-wave signal is therefore 78.5 per cent. Transistors operated in this mode can give efficiencies very close to theoretical. Tubes usually operate at a lower efficiency of about 65 per cent. Calculation of the operation of a class B amplifier is the same as that for a class C stage and will be discussed later.

When class B linear stages are used to amplify signals that vary in amplitude such as single-sideband suppressed-carrier signals, the plate or collector current and the grid or base current vary over a wide range with time. For this reason the supply voltages must have good regulation, and self-biasing cannot be used. Since the drive voltage is operating into a very nonlinear load, it too must have good regulation. This is usually attained by using a driver with excess power and loading the grid or base of the class B stage with a swamping resistor. If the stage is used to amplify an amplitude-modulated signal with carrier, the carrier is set to the 50 per cent amplitude point, which corresponds to the 25 per cent power point. As the signal is modulated the average power output increases because of the power in the sidebands, but the input power, as averaged over one modulation cycle, remains constant. The sideband power is therefore subtracted from the dissipation. Since the input power remains constant, the plate or collector current remains constant, so that with this type of signal cathode bias may be used.

With amplifiers of this type, inverse feedback is often added by providing a demodulator for the output signal and feeding back the demodulated signal, out of phase, to the modulator stage. This will materially reduce the amount of harmonic distortion present in the envelope so that less power need be dissipated in the swamping resistor, with the result that less total driving power is required.

9. CLASS C AMPLIFIERS

Class C amplifiers are usually used to amplify constant-amplitude signals, which may or may not be keyed. The fact that they are operated near plate- or collector-current saturation provides very high operating efficiency because the narrow current pulse occurs at a part of the r-f cycle where the voltage across the tube or transistor is low, and consequently, most of the power drawn from the plate or collector supply results in useful energy rather than in dissipated energy. Unlike class A and class B amplifiers, class C stages have a theoretical upper limit of 100 per cent, for as the current pulse is made narrower the efficiency will correspondingly increase. In actual practice, however, if too high an efficiency is demanded, it is found that the available power output from a given tube or transistor type becomes too low to be practical. Usually it is found that a plate or collector current pulse of 90 to 150° represents an average compromise. The shorter pulse of 90° would be used where efficiency is of prime importance, resulting in low gain. The longer pulses of 150° would be used where high gain is more important than high efficiency. Narrow current pulses also have higher peak values for a given power output, which can be a particular disadvantage in a transistor amplifier where the limiting factor may be the peak current-carrying capability of the transistors used. The operating efficiency of a class C stage usually falls in the range of 70 to 85 per cent.

In order to preserve the sinusoidal shape of the plate or collector waveform, the loaded Q of the plate or collector tank circuit should be at least 6 for a push-pull stage

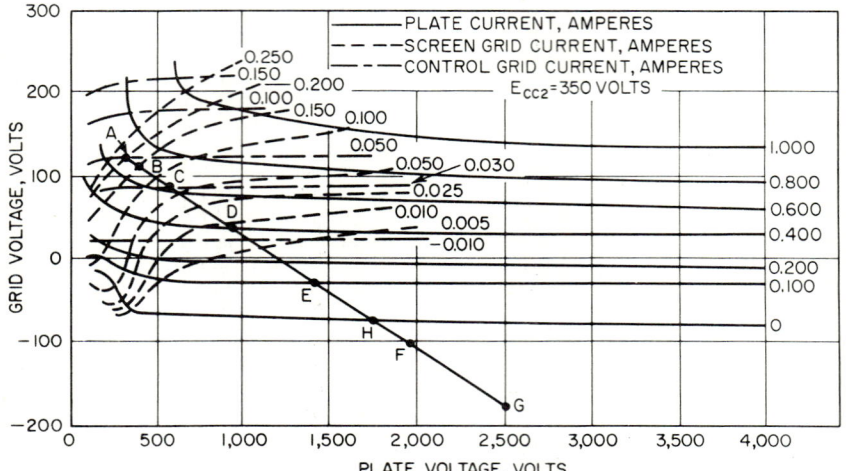

Fig. 18. Load line plotted on constant-current curves for use in Chaffee's harmonic analysis. (*From R. W. Landee, D. C. Davis, and A. P. Albrecht, "Electronic Designers' Handbook," McGraw-Hill Book Company, New York, 1957.*)

and 12 or more for a single-ended stage. The Q should not be made too high or the bandwidth and efficiency will be adversely affected. The tank-circuit efficiency is

$$\eta = (1 - Q_u)/Q_l \tag{14}$$

where Q_u is the unloaded Q and Q_l is the loaded Q. The bandwidth between the half-power points is

$$\mathcal{B} = f_o/Q_l$$

Because the input and output waveforms of a class C or class B amplifier are both forced to be sinusoidal by the tuned circuits, the best characteristic curves of the devices to use are the so-called "constant-current curves." On these curves the output (plate or collector) voltage is plotted as the abscissa and the input (grid or base) voltage is plotted as the ordinate. Each curve is the locus of points having equal plate or collector current. Superimposed is a series of curves, each made up of points having equal grid or base currents. (In the case of tubes that have a screen grid, there may be a third set of curves, each being the locus of points having equal screen current.) The load line, as plotted on a curve of this type, is a straight line passing through the quiescent point, which is the point defined by the grid or base bias as the ordinate and the plate or collector supply voltage as the abscissa. In addition, the load line has a slope equal to the ratio of the peak grid or base tank voltage swing to the peak plate or collector tank voltage swing.

A common method of analyzing the operation of a stage is known as "Chaffee's harmonic analysis."[3] Figure 18 shows a constant-current curve for a vacuum tube.

[3] E. L. Chaffee, A Simplified Harmonic Analysis, *Rev. Sci. Instr.*, vol. 7, p. 389, October, 1936.

TUNED AMPLIFIERS

The quiescent point is point G, with the plate voltage as the abscissa and the grid bias as the ordinate. The point A is the peak condition and is found by subtracting the peak plate tank-circuit voltage from the plate supply voltage for the abscissa, and by adding the peak grid tank-circuit voltage to the grid bias as the ordinate. A straight line drawn between these two points defines the load line. Points B, C, D, E, and F represent the instantaneous voltages at uniformly spaced times, in this case each 15°. These points are located by the following relations:

$$GB = 0.966GA \qquad GE = 0.500GA$$
$$GC = 0.866GA \qquad GF = 0.259GA$$
$$GD = 0.707GA$$

For each point on the load line, the corresponding plate current, grid current and screen current can be determined and tabulated. From this tabulation, the d-c and peak values of the fundamental, second, and third harmonics for the current in each element can be determined from the following relations:

$$I_{dc} = 0.0833(0.5A + B + C + D + E + F)$$
$$I_{h_1} = 0.0833(A + 1.93B + 1.73C + 1.41D + E + 0.52F)$$
$$I_{h_2} = 0.0833(A + 1.73B + C - E - 1.73F)$$
$$I_{h_3} = 0.0833(A + 1.41B - 1.41D - 2E - 1.41F)$$

For either tube or transistor stages, the plate or collector power input can be calculated as the product of the plate or collector supply voltage multiplied by the I_{dc} for the plate or collector. The screen input is the product of the screen voltage and the screen I_{dc}. In the case of the control element, the product of the bias voltage and the I_{dc} is negative and is the power dissipated in the biasing resistor or bias supply. The power output of the stage is

$$P_o = e_{p\ max}I_{h_1}/2 \qquad (15)$$

where $e_{p\ max}$ is the peak of the r-f voltage across the output tank circuit and I_{h_1} is the peak of the fundamental current in the plate or collector. The total power input to the grid or base is

$$P_{in} = e_{g\ max}I_{h_1}/2 \qquad (16)$$

where $e_{g\ max}$ is the peak of the r-f voltage across the grid or base tank and I_{h_1} is the peak of the fundamental current in the grid or base. The plate or collector dissipation is then the difference between the plate or collector power input and P_o. The grid or base dissipation is the difference between P_{in} and the bias power. The screen power is all dissipation.

A variation of class C operation using a circuit similar to that in Fig. 19 can be used to obtain higher efficiency.[4] In this circuit, the grid-drive waveform is made to be essentially a 120° wide pulse. The voltage waveform that appears on the plate, or collector in the case of transistors, is a mixture of fundamental and third harmonic, with the amplitude and phase relationships so arranged as to result in an almost flat-topped wave as shown in Fig. 20. The fact that the voltage across the tube or transistor is low during the entire current pulse accounts for the high efficiency. During the early and late parts of the current pulse, energy is stored in the third-harmonic tank circuit. During the middle of the pulse this energy is transferred to the fundamental tank circuit.

The efficiency attainable with this type of operation can be estimated by the following procedure: First, on the characteristic curves for the tube or transistor, locate the point of peak plate or collector swing. At this point the instantaneous power dissipated by the plate or collector is calculated as the product of the instantaneous plate or collector voltage and the instantaneous plate or collector current. The average dissipation will be approximately one-third of this value. The power input to the stage will be approximately one-third of the product of the instantaneous plate or

[4] V. J. Tyler, A New High Efficiency High Power Amplifier, *Marconi Rev.*, no. 130, vol. 21, third quarter, 1958.

collector voltage and the supply voltage. The useful power output is the difference between the two.

In the operation of such a stage it will be found that the peak voltage across the fundamental tank circuit will be approximately 1.2 times the peak plate or collector voltage swing, while the peak voltage across the third-harmonic tank circuit will be about 0.2 times the plate or collector voltage swing. The circuit should be designed

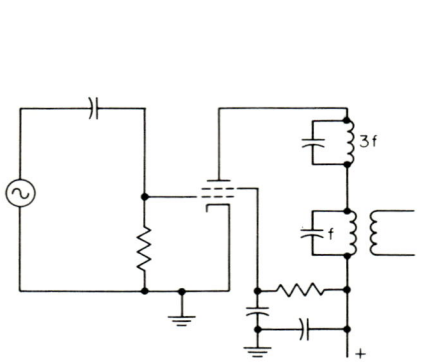

Fig. 19. High-efficiency class C amplifier.

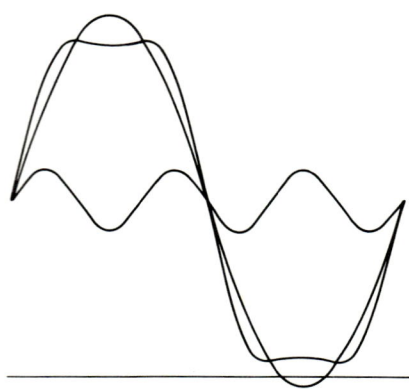

Fig. 20. Mixture of fundamental and third-harmonic voltages as measured on the plate of the high-efficiency amplifier.

so that the circulating volt-amperes in the harmonic tank circuit is about one-fifteenth of the value in the fundamental tank circuit.

In practice it is found that the grid or base can be driven with a waveform composed of a mixture of fundamental and third harmonic, and biased well beyond cutoff so that the current flows for approximately 120°. This results in a circuit diagram as shown in Fig. 21.

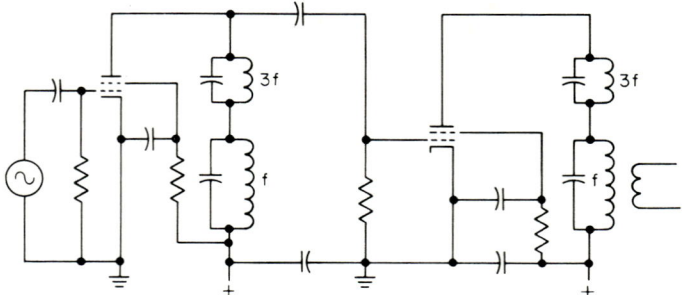

Fig. 21. High-efficiency amplifier with driver.

10. CLASS S AMPLIFIERS

Transistors have a very low saturation resistance; that is, the rated peak collector current can be drawn through a transistor where the instantaneous value of the collector-to-emitter voltage is on the order of approximately 1 volt. This fact has led to a new class of amplifiers known as class S. The term class S has been applied to transistors in the switching mode where the load line is purely resistive and therefore is a straight line. In this chapter it will be extended to cover load lines that are not pure resistances because the load circuit contains a series-tuned-circuit in series with the load resistor. Thus, while the load circuit presents a resistive impedance to the fundamental component of the collector voltage, the impedance presented to the harmonics is primarily reactive and is so high that the current which flows is essentially a sine wave and, therefore, the load line is curved.

Figure 22 shows a typical class S amplifier of the tuned-load type. In operation, the drive voltage causes the two transistors to conduct alternately, each over a period of 180°. The load impedance, being a series-tuned circuit, presents such a high impedance to the harmonic currents that the resultant voltage at the junction between the two transistors is a square wave of voltage. Since the load current, as determined by the resonant load circuit, must be a sine wave, the current waveform present in each transistor is one-half of a sine wave. The two diodes, $CR1$ and $CR2$, do not normally conduct during any portion of the cycle, but they serve to protect the transistors from surges due to transients. Figure 23 shows the voltage and current waveforms present in the circuit of Fig. 22.

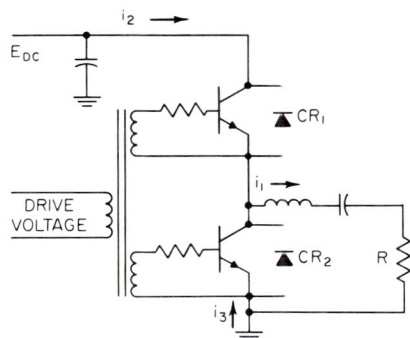

Fig. 22. Class S amplifier.

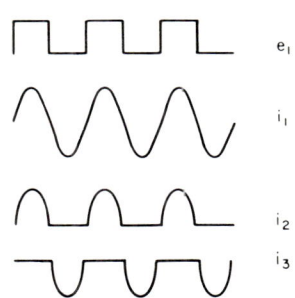

Fig. 23. Voltage and current waveforms in class S amplifier with series-tuned load circuit.

Neglecting the voltage drop in the transistors, which is usually no more than 1 to 2 per cent, the fundamental current through the load is

$$I_1 = 2E_{dc}/\pi R \quad \text{(peak)}$$

where R is the total resistance in the load circuit including the effective resistance due to losses in the tuned-circuit coil and capacitor.

Figure 24 shows a plot of the load line of a class S amplifier of this type compared with the plot of the load line using a resistive load. Note that the use of the tuned circuit allows the use of peak voltage and peak current combinations that could not be used with a purely resistive load circuit. The combination of peak collector voltage and peak collector current, as shown in Fig. 24, when used with a resistive load, passes through the latch-up region and can be destructive to the transistor. The same two operating points when used with a series-resonant circuit stay well clear of the danger area. Furthermore, since it is impossible to realize instantaneous switching, the transistors operate with lower average losses when the tuned circuit is used. Also, since there are no harmonic currents in the load circuit, all the power generated is at one frequency, the fundamental. In circuits that are used in the output stages of radio transmitters, this results in higher overall efficiency of the transmitter since any harmonic power generated is generally not useful.

The class S amplifier is almost a constant-voltage source, and consequently, it is unique in that the impedance it presents to a load that draws full power is low compared with the impedance the load presents. In the case of class C amplifiers, the source impedance may be either equal to or higher or lower than the load impedance. For example, when tetrode or pentode tubes are operated as class C amplifiers, the source impedance they present is usually high compared with the load impedance; however, when triode tubes are operated as class C amplifiers, the source impedance they present is likely to be lower than the load impedance. In any event, the source impedance is apt to vary widely with small changes in drive voltage. Because amplifiers of conventional design act as sources with somewhat unpredictable source impedance, it has been necessary to design the loads so that they will present an almost constant load impedance over the entire bandwidth to be used. In cases where electrically short antennas are to be fed at low frequencies, the inherent Q of the antenna is sometimes so high, when using conventional design practice, that swamping resistors have to be added to reduce the overall Q of the load circuit so that the bandwidth can be realized. This can result in a considerable amount of wasted power. With class S amplifiers, however, one can take advantage of the fact that the source

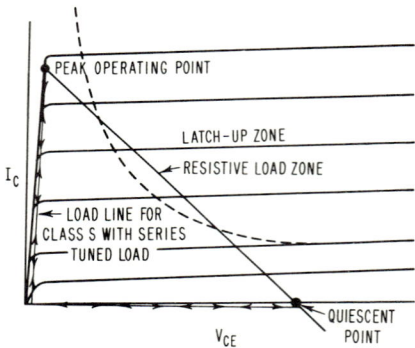

Fig. 24. Operation of class S amplifier with series-tuned load circuit.

impedance is very low, and is fixed in value, to improve the operating efficiency when the amplifier is used to amplify certain types of waveforms. The use of the following system is possible only where the antenna is to be continuously driven, such as in a phase-modulated, a phase-reversal-keyed, or a phase-shift-keyed signal, and is not applicable to on-off keying or amplitude modulation. The reason that it is not applicable to on-off keying is that during the time that the driving signal is removed from the transmitter, the output impedance is no longer the low fixed value that existed during the on time. Also, the improvement in efficiency is only for waveforms that have a nonuniform distribution about the center frequency with most of the energy close to the center frequency. With a spectrum having most of the energy concentrated at the edges of the passband, the efficiency could be worse than with the conventional output network. A coupling network of this design involves adding some swamping resistance, but not so much as with the conventional practice.

Figure 25 shows an output coupling of this type. R_a and C_a represent the impedance of the antenna at the center frequency f_o. $\mathcal{B}_a$ is the natural bandwidth of the antenna when resonated to the center frequency with a simple inductance. $\mathcal{B}_s$ is the bandwidth required for the signal. L_1, L_2, C_1, and C_2 are determined by the equations below. A swamping resistor r_s is added in series with the output of the transmitter, the value of which is the difference between the transmitter output impedance r_t and the value of r, calculated below. In this circuit, the transmitter is designed as though it is to be used with the output feeding through a simple series-resonant circuit into a load resistance equal to r.

$$r = r_t + r_s = \frac{[\sqrt{2}\,(\mathcal{B}_s/\mathcal{B}_a) - 1]\,R_a}{(\mathcal{B}_s/\mathcal{B}_a)^2 - \sqrt{2}\,(\mathcal{B}_s/\mathcal{B}_a) + 1}$$

$$L_1 = R_a + 1/2\pi f_o C_a$$
$$C_2 = 1/2\pi f_o R_a$$

$$C_1 = \frac{C_2}{\{r/R_a{}^2 2\pi f_o C_a[\sqrt{2}\,(\mathcal{B}_s/\mathcal{B}_a) - 1]\} - 1}$$

$$L_2 = (1/2\pi f_o)(R_a + 1/2\pi f_o C_1)$$

The above network has a Butterworth response when used in conjunction with a class S amplifier. The value of the resistance r as calculated in the above equation

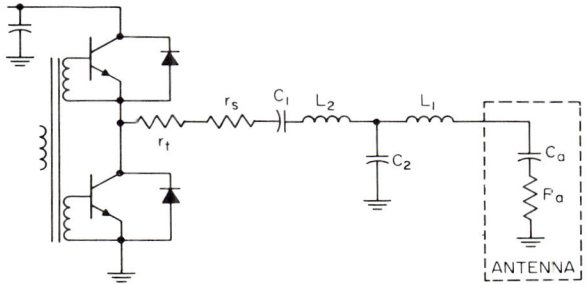

FIG. 25. Class S amplifier with broadband coupling.

causes only about half the losses that would be suffered if the antenna were broadbanded by the addition of a simple series resistor and used to transmit a signal with a $\sin x/x$ spectral distribution that is operated with the first nodes at the half-power points of the combined antenna and coupling system. It should be noted that the peak current drawn from the transmitter will be approximately the same for either system. The transmitter must be designed to handle the same peak values of current whether the broadbanding is done by the addition of a simple series resistor or by the addition of the special coupling network. The only saving is in the average power consumed and the heat dissipated. When using this circuit with phase-reversal keying and other modulations of this general type, it is wise to design the coupling network with a bandwidth no more than is needed because excess bandwidth increases the peak current required from the transmitter, and in cases where the limit on the output of the transmitter is determined by the peak current rating of the transistors, extra transistors would be required.

Chapter 25

BROADBAND AMPLIFIERS

JACOB S. BROWN*

CONTENTS

1. Broadband Amplifier Characteristics	25–1
1.1. High-frequency Response	25–2
1.2. Low-frequency Response	25–7
2. Electron-tube Characterization	25–8
3. Electron-tube Amplifiers	25–10
3.1. Low-frequency Response	25–10
3.2. High-frequency Response	25–13
3.3. Stagger Peaking	25–30
3.4. Phase Compensation	25–32
3.5. Electron-tube Distributed Amplifiers	25–34
3.6. Electron-tube Terminating Techniques	25–38
3.7. Electron-tube Phase Inversion	25–44
3.8. Electron-tube Power Amplifiers	25–45
4. Transistor Characterization and Amplifiers	25–45
4.1. Low-frequency Response for Transistors	25–48
4.2. High-frequency Response for Transistors	25–50
4.3. Stagger Peaking with Transistor Circuits	25–55
4.4. Wideband Feedback Amplifiers Using Transistors	25–57
4.5. Transistor Distributed Amplifiers	25–78
4.6. Transistor Driving Stages	25–84
4.7. Transistor Power Amplifiers	25–89

1. BROADBAND AMPLIFIER CHARACTERISTICS

Broadband amplifiers amplify a large range of frequencies about equally well. They are also used to amplify a narrow spectrum that may move about in the frequency domain, or whose location in the frequency domain is unknown. On the other hand, they are more often used to amplify broad-spectrum signals. The demands on the amplifier are very much dependent not only on the breadth of the spectrum to be amplified but on the type of signal, for example, pulse or frequency diversity system.

* General Electric Co., Missile and Space Vehicle Dept., King of Prussia, Pa.

One can therefore ask whether to describe an amplifier response in terms of amplitude and phase vs. frequency or time response and time delay. The point is well taken, in particular when it is the response of a nonperiodic or a pulse signal that is under consideration. Since there is a direct relationship between amplitude and phase shift vs. frequency and time response and time delay, the approach one should use depends on many factors including the experience, training, and past viewpoints of the designer, the instruments available, and the job to be done. In most cases, either technique of design is adequate, but when the application is demanding, the design and testing should be carried out in the domain (frequency or time) in which the objective is most demanding. Historically, by education and where instrumentation is simplest, we have been prepared principally for the amplitude vs. frequency

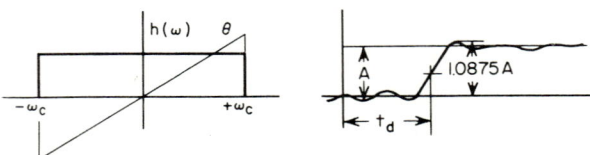

Fig. 1. Idealized constant-amplitude low-pass response and its effect on a unit-step function.

viewpoint. This is presently in the process of change. In the material to follow, an effort will be made to carry on both viewpoints side by side, but stressing particular viewpoints where considered helpful.

1.1. High-frequency Response. Early theoretical studies of low-pass responses considered two ideal cases.[1] They are the constant-amplitude and gaussian-amplitude response, the phase shift changing linearly with frequency (that is, the time delay is constant with changing frequency). The response of the two idealized low-pass characteristics to unit-step excitation is shown in Figs. 1 and 2. Notice that the constant-amplitude low-pass response has a rather severe overshoot (approximately 9 per cent) while the gaussian amplitude response has none whatever. The gaussian response is,

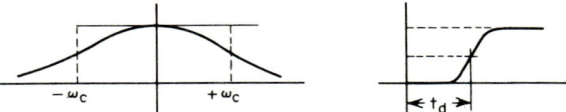

Fig. 2. Idealized gaussian low-pass response and its effect on a unit-step function.

however, slightly slower. Extensive theoretical studies have been made of the separate effects of nonlinearities of amplitude response and phase response.[2,3,4]

These separate effects can be demonstrated by starting with the normalized low-pass response function

$$A(j\omega)/A(0) = \exp(a^m\omega^m - jb^n\omega^n) \tag{1}$$

In this expression $A(0)$ is the function at zero frequency; a and m (where m is an even integer) are constants that describe the amplitude-frequency function; while b and n (where n is an odd integer) are constants that describe the phase-frequency

---

[1] Glen M. Glasford, "Fundamentals of Television Engineering," pp. 204–206, McGraw-Hill Book Company, 1955.

[2] H. A. Wheeler, The Interpretation of Amplitude and Phase Distortion in Terms of Paired Echoes, *Proc. IRE*, vol. 27, pp. 359–385, June, 1939.

[3] M. J. DiToro, Phase and Amplitude Distortion in Linear Networks, *Proc. IRE*, vol. 36, pp. 24–36, January, 1948.

[4] J. T. Bangert, Practical Applications of Time Domain Theory, *IRE Wescon Record*, 1959, pt. 3, pp. 29–38.

function. Figures 3 and 4 illustrate the amplitude-frequency response, the response to an impulse function, and the response to a unit step for various values of m, n being zero (that is the phase shift is zero with frequency). These responses show that the impulse response of such a function is symmetrical about its maximum value while the response to the unit step is symmetrical about its 50 per cent value in an odd-function sense. Further, with more rapid change of amplitude with frequency ($m = 4$) an overshoot will be obtained. As m increases without limit, the amplitude-frequency characteristic becomes rectangular, for which a symmetrical overshoot is obtained for the unit-step function of almost 9 per cent, as previously shown in Fig. 1. Nonzero but linear phase-frequency characteristic would result in only delay of these responses.

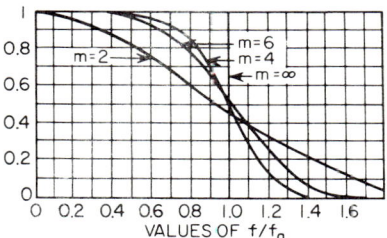

Fig. 3. Amplitude-frequency response for values of m of 2, 4, 6, and ∞ phase-shift constant with frequency. (*From M. J. DiToro, Phase and Amplitude Distortion in Linear Networks, Proc. IRE, January, 1948.*)

If now $m = 0$ and $n = 1$, the case of constant amplitude-frequency characteristic and phase shift linearly proportional to frequency, there will be no distortion, only a time delay determined by the rate at which phase shift is changing. The case of $n = 3$ shown in Fig. 5 shows severe ringing in the impulse response and the unit-step response. Further, the responses lose all symmetry shown for the case of only amplitude-frequency nonuniformity. As n becomes larger, the ringing and asymmetry become progressively more violent.

Summarizing briefly: A nonconstant amplitude response results in rise-time degradation of the response to the unit-step and impulse function; while phase shift that

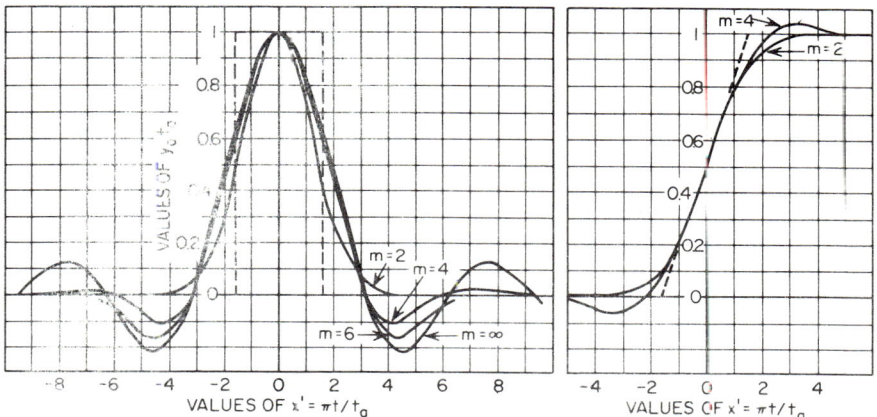

Fig. 4. The response to impulse and unit-step functions of the amplitude responses of Fig. 3. (*From M. J. DiToro, Phase and Amplitude Distortion in Linear Networks, Proc. IRE, January, 1948.*)

does not change linearly with frequency leads to asymmetry in the impulse response and overshoot as well as oscillatory response, both of which become greater as the nonlinearity of the phase shift with frequency becomes greater. It follows then that for nonoscillatory response to a unit step or an impulse signal, more attention should be given to shaping of the frequency-phase response.

An empirical relationship between bandwidth and rise time can be described by the formula

$$t_r B = 0.35 \text{ to } 0.45$$

where t_r = rise time between 10 and 90 per cent
B = bandwidth from 0 to the upper 3-db frequency

The lowest value describes the performance of circuits with little or no overshoot, while the 0.45 value holds for circuits with about 5 per cent overshoot. It is of interest to note that for the ideal rectangular low-pass response $t_r B$ is 0.51 and the overshoot is 9 per cent while for the gaussian amplitude response, a $t_r B$ of 0.41 is obtained with no overshoot.

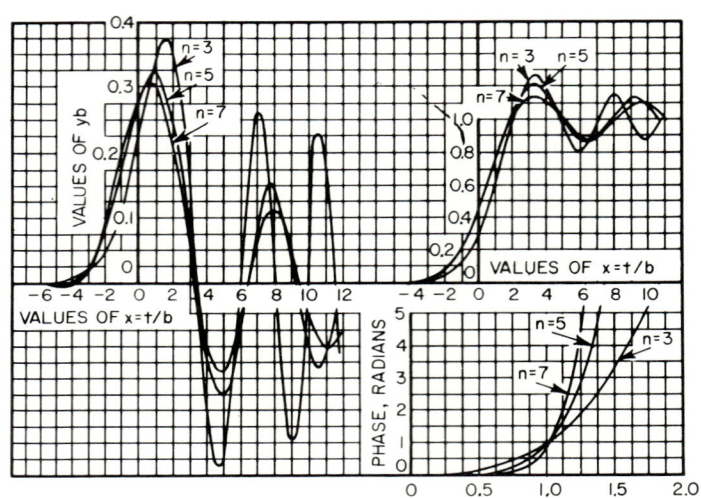

Fig. 5. Constant-amplitude curved frequency-phase response, m of 3, 5, 7, the impulse and unit-step responses. (*From M. J. DiToro, Phase and Amplitude Distortion in Linear Networks, Proc. IEE, January, 1948.*)

The ideal responses considered in the studies referred to above are not realizable.[5] They do, however, give an insight into the factors that should be considered in designing amplifiers.

The practical frequency responses to which circuitry has been designed are approximations of the above ideal frequency response. The principal approximations that have been studied are the maximally flat amplitude response,[6,7] the Chebyshev amplitude response, the maximally flat phase-delay response,[8,9] and the equal-ripple phase-delay response.[10] The simplest of these is maximally flat amplitude response, and

[5] G. E. Valley, Jr., and H. Wallman, "Vacuum Tube Amplifiers," pp. 721–727, McGraw-Hill Book Company, New York, 1948.
[6] Butterworth, On the Theory of Filter Amplifiers, *Wireless Eng.*, vol. 7, p. 8536, October, 1930.
[7] V. D. Landon, Cascade Amplifiers with Maximal Flatness, *RCA REV.*, vol. 5, p. 247, January, 1941.
[8] W. E. Thompson, Delay Networks Having Maximally Flat Frequency Characteristics, *Proc. IEE*, pt. 3, vol. 96, p. 487, November, 1949.
[9] L. Storch, An Application of Modern Network Synthesis to the Design of Constant Time Delay Networks with Low Q Elements, *IRE Conv. Record*, pt. 2, Circuit Theory, pp. 105–117, 1954.
[10] E. A. Guillemin, "Synthesis of Passive Networks," p. 639, John Wiley & Sons, Inc., New York, 1957.

the techniques become progressively more complicated in the order described. For this reason, the maximally flat amplitude response has often been used even though another response would better meet requirements. An idea of the complexity involved can be obtained by considering the pole locations required for the maximally flat amplitude and the Chebyshev responses as shown in Fig. 6. Notice that the poles of the maximally flat amplitude response lie on a circle with equal angles between radii from the origin to the poles except for the poles at the highest frequency and lowest frequency. The angle between the radii to these poles and the $j\omega$ axis is half the angle between the other radii. The poles of the Chebyshev amplitude response have the same imaginary values $j\omega$, but the real component σ changes so it lies on a semiellipse of the same span of $j\omega$ as the maximally flat amplitude response circle at the $j\omega$ axis. The greater complexity of the maximally flat phase-delay response is illustrated in Fig. 7, where no set pattern exists as compared with the previous two examples for the location of poles of different-order networks. The last of the approximations to be considered, equal ripple in time delay, is obtained with a number of poles to the left and parallel to the $j\omega$ axis. The end poles must be positioned in such a way as to give the desired transfer characteristic.

It is of interest to consider a simple comparative example of several three-pole functions and some responses. Figure 8 shows (1) a triple pole on the negative real axis, (2) a maximally flat configuration, (3) a Thompson approximation for most constant-phase response in a Taylor-series sense, and (4) a three-pole nonoscillatory configuration. The transfer characteristics for each circuit are listed as are certain transient responses, namely, the time necessary for the response to a unit-step function to reach prescribed amounts of error from the final value. The maximum overshoot is also listed. The illustration indicates that, as the real part of two of the three poles becomes smaller, the rise time decreases, but the tendency toward overshoot increases

Comparisons of the above response characteristics have shown that, as the frequency response more closely takes on the rectangular low-pass characteristics, the overshoot becomes more severe. This is because of the relationship between the amplitude and the phase responses, which is defined by the Cauchy-Riemann conditions. The restrictions described by the above relationship result in increased curvature of the high-frequency phase-frequency characteristic as the amplitude characteristic becomes flatter with the consequence of increased overshoot. The realizable response which has a minimum of overshoot for a most rapid rise time also has a rounded rather than a rectangular high-frequency amplitude response, as one could judge by comparing the pole locations of Figs. 6 and 7. This has also been demonstrated by techniques that designed the networks directly to give the desired responses.[11,12]

There is no general expression to describe the effect on rise time of cascading a

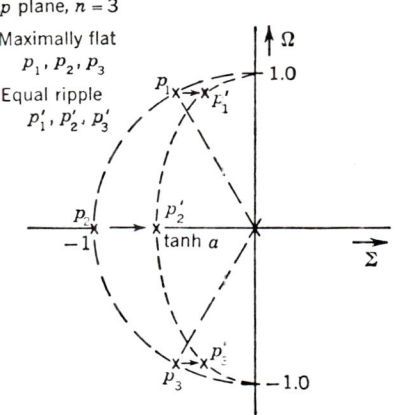

Fig. 6. Pole locations for a maximally flat amplitude response (the semicircle) and an equal-ripple response (lying on the ellipse). (From J. M. Pettit and M. M. McWhorter, "Electronic Amplifier Circuits," McGraw-Hill Book Company, New York, 1961. Used with permission.)

[11] W. H. Huggins, Network Approximation in the Time Domain, *AF Cambridge Res. Labs. Rept.* E 5048A, October, 1939.
[12] F. A. Muller, High Frequency Compensation of RC Amplifiers, *Proc. IRE*, vol. 42, no. 8, pp. 1271–1276, August, 1954.

number of networks. Several rules have, however, been stated.[13] They follow, partly reformulated.[14]

1. In circuits having little or no overshoot, the overall rise time t_{r_t} is approximately

$$t_{r_t} = \sqrt{t_{r_1}^2 + t_{r_2}^2 + t_{r_3}^2 + \cdots + t_{r_n}^2} \qquad (2)$$

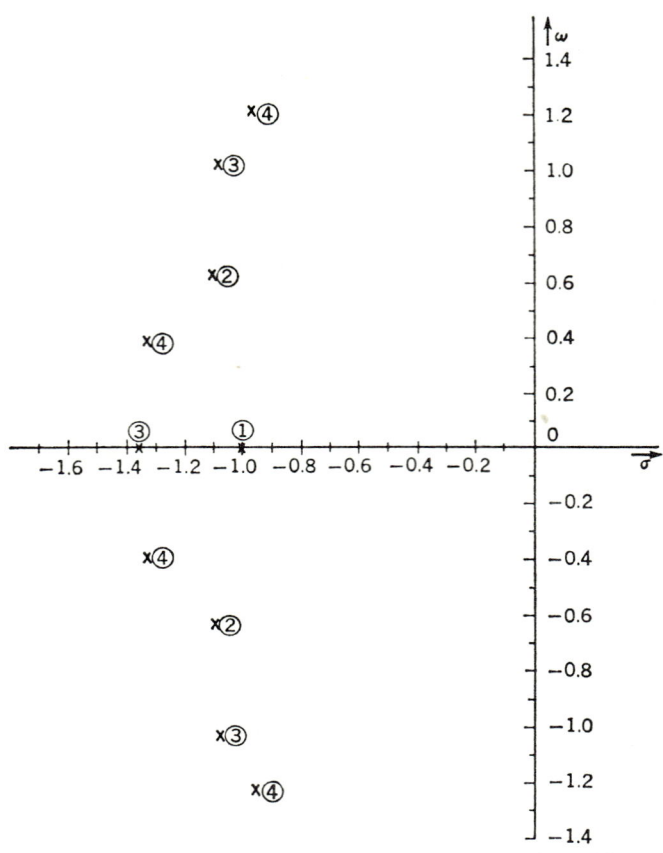

FIG. 7. Normalized pole locations for a one-pole, two-pole, three-pole, and four-pole maximally linear phase response. (*From J. M. Pettit and M. M. McWhorter, "Electronic Amplifier Circuits," McGraw-Hill Book Company, New York, 1961. Used with permission.*)

For the special case of $t_{r_1} = t_{r_2} = t_{r_3} = \cdots t_{r_n}$ a better approximation is

$$t_{r_t} = 1.1 t_{r_1} \sqrt{n}$$

2. In circuits having little or no overshoot (about 1 per cent), the total overshoot for n stages is essentially that of a single stage.

[13] Ref. 5, pp. 77, 78; also W. C. Elmore, The Transient Response of Damped Linear Networks with Regard to Wide Band Amplifiers, *J. Appl. Phys.*, vol. 19, pp. 55–63, January, 1948.
[14] J. M. Pettit and M. M. McWhorter, "Electronic Amplifier Circuits," p. 109, McGraw-Hill Book Company, New York, 1961.

3. If the overshoot of a single stage is on the order of 5 to 10 per cent then the total overshoot goes up as the square root of the number of stages.

3a. When the stage overshoot is 5 to 10 per cent, total rise time is somewhat less than that given by rule 1. An appendix to rule 1 is that the minimum overall rise time for a given gain is obtained if all stages are the same. Another is that, to obtain a given overall gain with minimum overall rise time, the gain per stage is $A_v = 1.65$. One study dealing with the case of electron-tube amplifiers[12] (it has not yet been

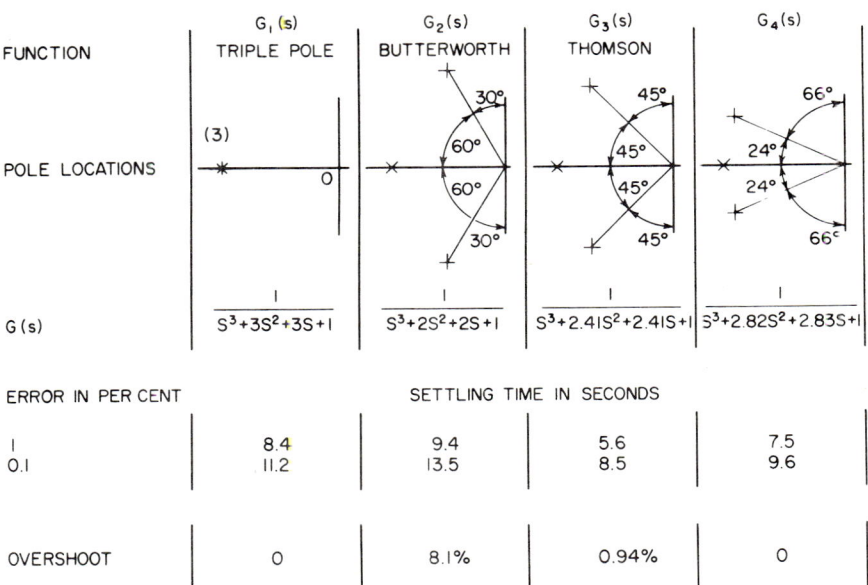

Fig. 8. Four third-order functions—their transfer characteristics, settling time, and overshoots. (*Reprinted from Electronics, May 31, 1963.*)

verified for transistor amplifiers) used inductance shunt stagger peaking and resulted in the rise time increasing as $n^{\frac{1}{4}}$.

4. The following three characteristics of an amplitude go together.[13]

   a. Small overshoot (no more than 1 to 2 per cent)
   b. Amplitude-frequency characteristic approximately gaussian
   c. Phase linear over the passband

1.2. Low-frequency Response. The low-frequency response of an amplifier can be affected by transformers as well as several RC networks. In an RC-coupled amplifier, this distorted response could be affected by RC coupling networks and RC biasing networks. Considerable distortion, in the form of sag of a square wave, can take place even though the sine-wave response for a frequency of the same pulse rate shows little deterioration. This is due to different time delays of the square-wave harmonic terms or, in other words, the nonlinearity of the low-frequency phase-frequency characteristic. If the square-wave repetition rate or the width of a narrow pulse is such that the sag produced by each of the RC networks is small, say of a few per cent, then it can be shown that the total sag is approximately equal to the sum of the sag produced by each of the RC networks. As the total sag increases, the approximation becomes poorer.

Transformer coupling produces similar results as the transformer behaves like an *RC* network (in its low-frequency and square-wave response), the biasing *RC* networks behaving the same as when the amplifier stages are *RC*-coupled.

The response of these networks can be described in a useful rule:

For an amplifier made up of stages having imperfect transmission[15] of direct current, the initial downward slope of the step-function response is the sum of the initial slopes of the component stages; that is, the slopes add arithmetically. This is illustrated in Fig. 9. This figure shows further that, for the case of similar time constants, there is an additional base-line crossing for each additional capacitance-coupled

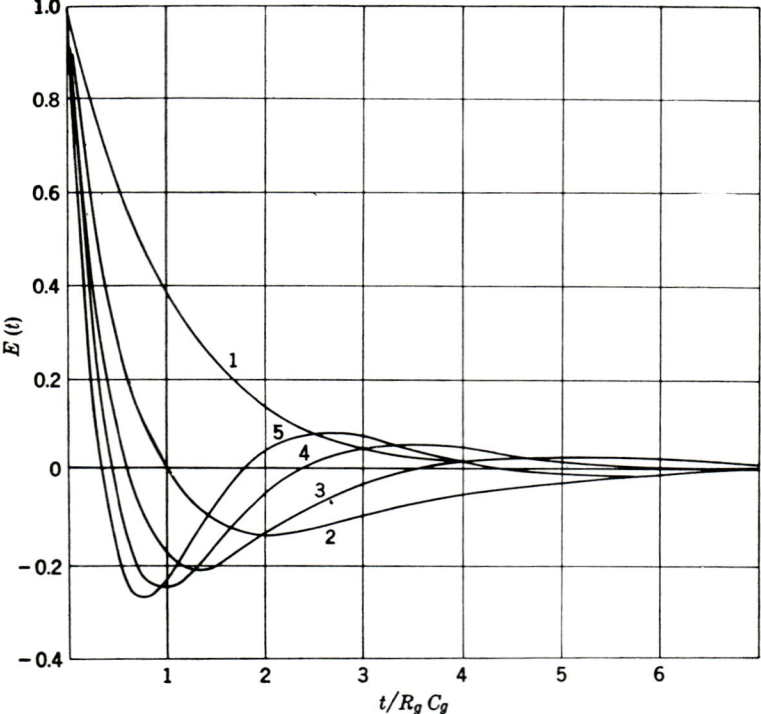

Fig. 9. Response to a step function of *n* capacitor-coupled stages of the same time constant. (*From G. E. Valley, Jr., and H. Wallman, "Vacuum Tube Amplifiers," McGraw-Hill Book Company, New York, 1948. Used with permission.*)

stages. In the case where an amplifier would have such a large signal input that limiting would take place on the first portion of the input signal, the above-mentioned secondary effects could quickly snowball to overshoots as large as the signal pulse. They could, in some cases, be confused with subsequent pulses.

2. ELECTRON-TUBE CHARACTERIZATION

The equivalent circuits of the electron tubes and transistors that are to be used are described below. They are approximations that have been found useful. For more detailed description of equivalent circuits as well as the physics of operation, refer to Chaps. 9 and 10.

[15] Ref. 5, p. 86.

A low-frequency equivalent circuit for the electron tube is shown in Fig. 10. This equivalent circuit applies for the triode as well as the tetrode and pentode provided that the screen grid is connected to the cathode through a very low impedance at the frequency of interest and the pentode grid is connected to the cathode by a frequency-independent low-impedance circuit. As the frequency of operation increases, the conductance shown in the input circuit as a dotted line begins to become important (at approximately 10 Mc for useful wideband electron tubes). This term, which increases as the square of the frequency, has two components, one due to inductance of the cathode lead and the other due to the finite transit time of the electrons between grid and plate.[16] The effect of inductance (L_k) in the cathode lead of an electron tube can be described if one assumes $1/\omega L_k \gg C_{gk}$, by the equation

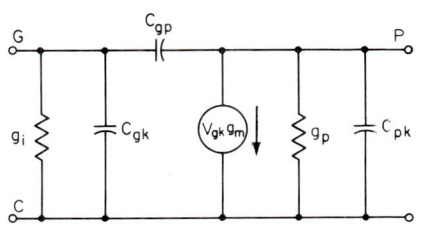

FIG. 10. Equivalent circuit of the common-cathode electron tube. C_{gk} = grid-to-cathode capacitance. C_{gp} = grid-to-plate capacitance. C_{pk} = plate-to-cathode capacitance. g_p = incremental plate-to-cathode conductance. g_m = incremental transconductance between grid voltage and plate current.

$$Y_{11} \approx \frac{j\omega C_{gc}}{1 + jg_m\omega L_k} \approx j\omega C_{gk}[1 + jg_m\omega L_k - (g_m\omega L_k)^2 - j(g_m\omega L_k)^3 \cdots] \quad (3)$$

Higher-order terms become negligible since $g_m\omega L_k \ll 1$; therefore, the significant conductive-loading term is $g_m\omega^2 C_{gk}L_k$.

As one considers even higher frequencies of operation, the equivalent circuits become more complicated as the inductances in the other leads become more significant. In most cases, however, these inductances cause no difficulties if they are neglected.

The equivalent circuits of the electron tube in two other configurations, the common-grid connection and the common-plate connection, respectively, are shown in Fig. 11. Note that in Fig. 11b, the configuration known as the cathode follower, the current generator is $V_i g_m$, not $-V_{gk}g_m$ as in the other two cases. Because of this, the output

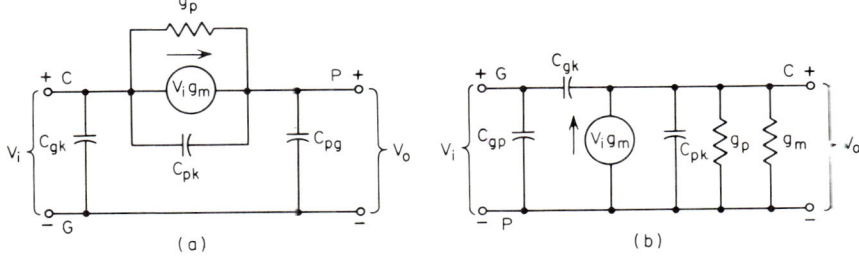

FIG. 11. (a) Equivalent circuit of common-grid electron tube. (b) Equivalent circuit of common-plate electron tube.

impedance contains a second conductive term of value g_m. In both the above circuits, there is a conductance between grid and cathode due to transit time. The effects of inductance in the cathode terminal are no longer of the same importance. In the case of the common-grid amplifier, the inductance in the grid becomes the important term while the plate lead inductance becomes the significant term for the common-plate configuration.

[16] M. J. O. Stritt and A. van der Zeil, The Cause for the Increase of the Admittance of Modern High Frequency Amplifier Tubes on Short Waves, *Proc. IRE*, vol. 26, p. 1011, August, 1938.

25-10 AMPLIFIER CIRCUITS

Before going on it should be noted that, for triode electron tubes, g_m will range over values of approximately 1,000 to 50,000 micromhos. The corresponding values of $1/g_p$ will range over values of 10,000 ohms (assuming a μ of 10) to 1,000 ohms (assuming a μ of 50). Expected values of C_{gk} and C_{gp} will both range over values of 3 to 15 pf. For tetrodes and pentodes, the expected g_m would have same range as the triodes. The μ and $1/g_p$ would have such values that they would not be important in the calculations. The input capacitance C_{gk} would have about the same range as the triode, but C_{gp}, if the screen is bypassed with a low impedance to the cathode, would be considerably smaller, ranging over values from 10^{-3} to 4×10^{-2} pf which in many cases are negligible.

3. ELECTRON-TUBE AMPLIFIERS

3.1. Low-frequency Response. The simplest broadband amplifier is shown in Fig. 12. The voltage amplification of such an amplifier under the conditions that the

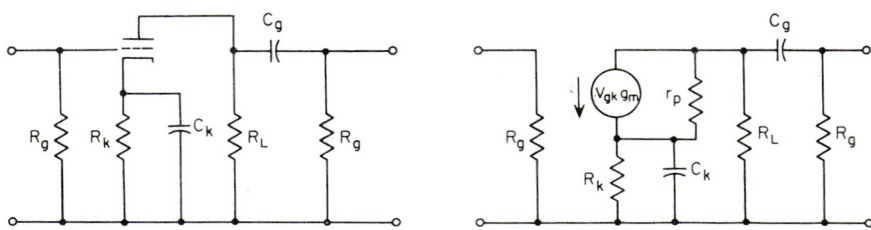

FIG. 12. Schematic diagram of electron-tube amplifier and equivalent circuit showing parameters affecting low-frequency response.

reactances are negligible can be shown to be

$$A_v = \frac{-g_m}{1/r_p + 1/R_L + 1/R_g} \tag{4}$$

Except in the narrowband case $R_g \gg R_L$. Further, in most cases the tube is a tetrode or pentode so that $R_L \ll r_p$. Therefore, the midband voltage amplification is approximately

$$A_v \approx -g_m R_L. \tag{5}$$

If the low-frequency effects are now considered the above becomes

$$A_v = -g_m R_L \frac{1}{1 + 1/j\omega C_g R_g} \frac{1 + 1/j\omega C_k R_k}{1 + (1 + g_m R_k)/j\omega C_k R_k} \tag{6}$$

Figure 13 shows the effects of the above frequency-describing terms separately and in combined effects for two examples of relative time constants.

In this figure, the time constant $C_g R_g$ determines the low-frequency half-power frequency of the amplifier, neglecting the effects of R_k and C_k. The effect of a finite value of C_k is shown by the doublet with the low-frequency break frequency of $1/C_k R_k$ and the higher break frequency of $R_k C_k/(1 + g_m R_k)$. The two figures illustrate the effect on the overall response of different $C_g R_g$-$C_k R_k$ relationships.

The above conditions have assumed that, if the tube was a pentode, the screen was bypassed with an infinite capacitance. It can be shown that, for the practical case of a finite screen capacitor, an effect is obtained that is similar in quality to the effect of the cathode resistor-capacitor combination. It would appear as an additional product term in Eq. (6) and would be written

$$\frac{1 + 1/j\omega C_s R_s}{1 + (r_s + R_s)/j\omega r_s R_s C_s} \tag{7}$$

where C_s is the capacitance of the screen bypass capacitor, R_s is the resistance in series with the screen supply voltage, and r_s is the incremental impedance of the screen grid with all other electrodes grounded. If the plate resistance of the tube connected as a triode is known (plate, screen, and suppressor connected in parallel) then an approximate value of r_s can be determined. It is $r_s = r_p I_t / I_s$ where I_t is the plate current of the triode and I_s is the d-c screen current. In practice, the effect of degeneration in the screen circuit causes no significant problem as it can easily be made to take place at a frequency much lower than the effects due to C_k and R_k. The combined effect of the cathode-circuit time constant and the screen-circuit time constant could, however, cause low-frequency oscillation in feedback amplifiers unless carefully controlled.

When designing these amplifiers for a maximum bandwidth, it is desirable to reduce shunt capacitance from plate to ground to a minimum. It might also for some other

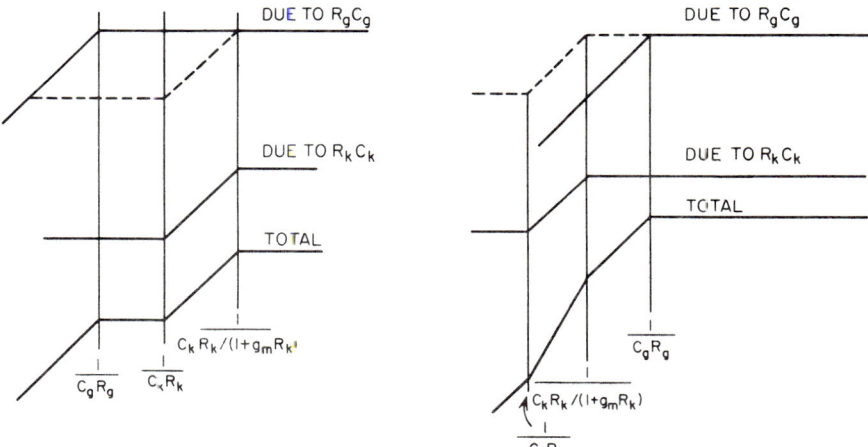

Fig. 13. Bode plots of low-frequency response of the amplifier shown in Fig. 12 illustrating effect of different time constants.

reason, such as leakage of the coupling capacitor, be desirable to use a capacitor of minimum capacitance for C_g and yet maintain the frequency-amplitude response uniform to as low a frequency as possible.

The above treatment of low-frequency performance can be presented by describing the response to a square wave. This is more desirable when the time response rather than the frequency response is of interest. It can be shown that a small amount of sag (0.1 or less) can be described by the formula

$$E_s/E = T/2.75 R_g C_g \qquad (8)$$

where the terms are described in Fig. 14 and it is assumed that only the coupling capacitor C_g and grid resistor R_g are affecting the response. Further, it can be shown that this sag in the response to a square wave takes place even though the amplitude characteristic has dropped by only a very small amount. For example, a tilt of 0.1 will take place with only a drop-off of 0.38 per cent in the sine-wave amplitude characteristic. Careful examination shows that this effect is caused by the phase shift of the lower frequencies not being proportional to frequency.

A circuit that compensates for the above effect is shown in Fig. 15. If the condition $C_g R_g = C_d R_d R_L/(R_d + R_L)$ is fulfilled the gain A_v relative to the midband

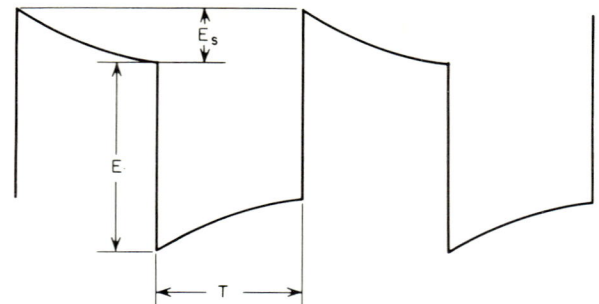

Fig. 14. Effect of coupling capacitor and resistor on response to a square wave when time constant is too short.

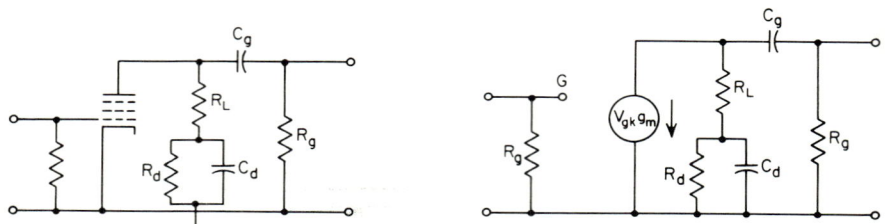

Fig. 15. Schematic and equivalent circuit showing low-frequency compensating circuit.

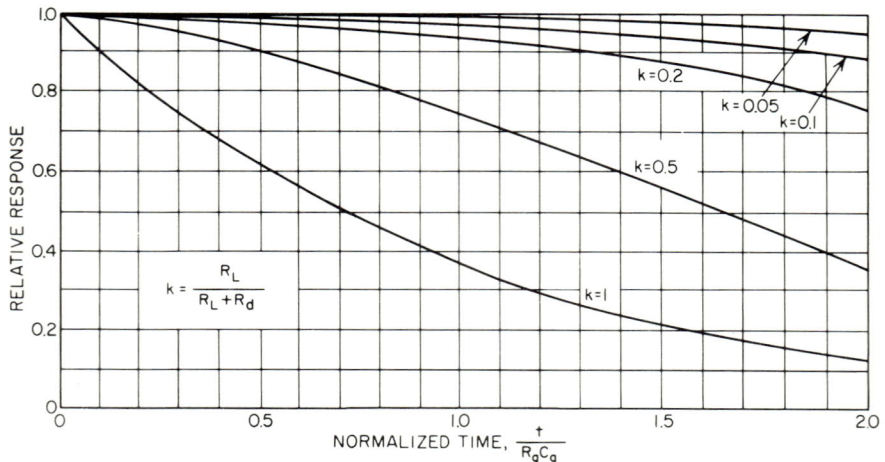

Fig. 16. Response to a step function of low-frequency compensated amplifier. (*From G. M. Glasford, "Fundamentals of Television Engineering," McGraw-Hill Book Company, New York, 1955. Used with permission.*)

gain A_{vo} is written

$$\frac{A_v}{A_{vo}} = \frac{1}{1 - j(1/\omega R_g C_g)[R_L/(R_L + R_d)]} \tag{9}$$

This equation has the identical form as before except that the half-power frequency is now $(1/2\pi R_g C_g)[R_L/(R_L + R_d)]$ compared with $1/2\pi R_g C_g$ before.

From this formula, it follows that improved compensation is obtained with increased values of R_d. The value of R_d is related to R_L and restricted by considerations of the operating plate current of the tube. R_L, on the other hand, is dependent on the

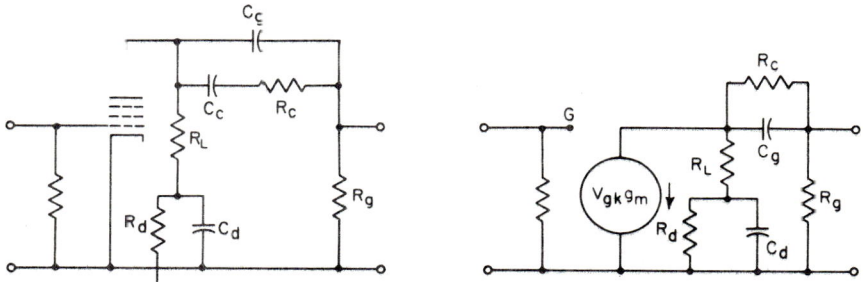

Fig. 17. Schematic and equivalent circuit of a circuit giving additional low-frequency compensation.

desired high-frequency response. Figure 16 illustrates the effect of R_d on the response to a unit-step function.

Another method of low-frequency compensation is shown in Fig. 17. In the equivalent circuit, C_c is neglected, assuming it is large enough so its reactances can be neglected over the frequency range of interest, or assuming that the amplifiers are direct-coupled. If $R_L/R_d = R_g/R_c$ and $C_d/C_g = R_c/R_d$ the response will be flat down to the frequency at which the reactance of C_c becomes a factor. The relationships above

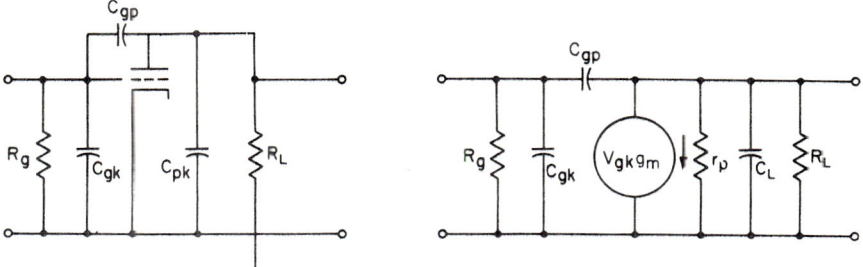

Fig. 18. Schematic and equivalent circuit illustrating the high-frequency response factors.

are somewhat critical if proper compensation is to be obtained. This compensation circuit should be used with either precision components or provisions to adjust the circuit relationships.

Other factors that are significant in the use of this compensation circuit are the choice of R_L and the parasitic capacitance due to the value of C_c. Both these factors are related to the high-frequency behavior of the amplifier.

3.2. High-frequency Response. If the high-frequency response of the circuit of Fig. 12 is considered the effect of C_k and C_g can be neglected as their reactances would be negligible in wideband amplifiers. The schematic diagram of Fig. 18 shows the circuit with the factors that influence the high-frequency response only.

The input admittance for the above amplifier is described to a good approximation by

$$Y_{in} = 1/R_g + j\omega[C_{gk} + (1 + A_v)C_{gp}] \tag{10}$$

where
$$A_v = \frac{-g_m}{1/r_p + 1/R_L + j\omega C_L} \tag{11}$$

If the amplifiers of this kind are cascaded, the C_L of each stage contains a term dependent on A_v and C_{gp} of the following stage. If the bandwidth is calculated

$$\text{BW}(\omega) = \frac{1/r_p + 1/R_L}{C_{gk} + C_{pk} + (1 + A_v)C_{gp}} \tag{12}$$

is obtained. Using only the real terms of A_v, which gives a conservative result, the gain bandwidth can be shown to be

$$A_v\text{BW} = \frac{g_m}{C_{gk} + C_{pk} + C_{gp} + g_m C_{gp}/(1/r_p + 1/R_L)} \tag{13}$$

Notice that the gain-bandwidth product depends on R_L so long as C_{gp} is a factor. Further, for the same g_m, C_{pk}, and C_{gk}, the gain-bandwidth product is greatest if $A_v C_{gp}$ is negligible. For the tetrode and pentode when C_{gp} is negligible, one obtains

$$A_v\text{BW} = g_m/(C_{gk} + C_{pk}) = g_m/C \tag{14}$$

which is a characteristic of these electron tubes only. Returning for the moment to the formula for A_v but considering the use of a tetrode or pentode noting that, for $R_L \ll r_p$, r_p and C_{gp} are negligible,

$$A_v = g_m/(1/R_L + j\omega C_L) \tag{15}$$
and
$$\text{BW}(\omega) = 1/R_L C_L \tag{16}$$

It follows then that, if the gain is increased by increasing R_L, the bandwidth decreases and vice versa.

A number of techniques can be used to overcome the above limitation. The first measure is to add other circuit elements to the resistor to compensate for shunt capacitance. These networks can become very complicated, but the major improvement takes place with the first two elements added. These two circuit elements generally are a series inductance and another shunt capacitance connected across the resistor or the inductor. Examples of these circuits are shown later in this section (see Figs. 19, 22, and 36). A limiting increase in speed or bandwidth for a two-terminal network is indicated to be 2.12 for an infinite number of circuit elements.[13]

The next improvement is obtained by separating the output capacitance of one tube from the input capacitance of the next with an inductor. This technique evolved into a large number of networks based upon filter theory.[17] The improvement takes place slowly with increasing complexity, and while the maximum improvement in rise time or bandwidth has not been clearly established, it is probably about 4.

The circuits referenced above[18,19] are shown in ascending complexity, starting with simple inductive compensation and finally combining modified inductive compensation with series inductive compensation. Their normalized performance is illustrated in Figs. 19–34, which give the schematics, amplitude and time-delay characteristics and responses to step functions of these combinations of elements.

[17] Harold A. Wheeler, Wide-band Amplifiers for Television, *Proc. IRE*, vol. 27, p. 429, July, 1939.
[18] Ref. 1, chap. 5.
[19] Ref. 5, chaps. 2 and 3.

BROADBAND AMPLIFIERS

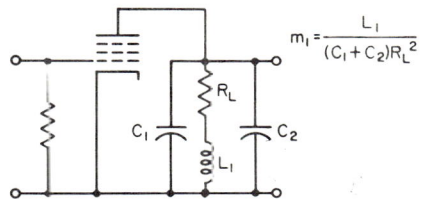

Fig. 19. Schematic of shunt-peaked amplifier.

$$m_1 = \frac{L_1}{(C_1+C_2)R_L^2}$$

(a) Relative gain

(b) Time delay characteristic

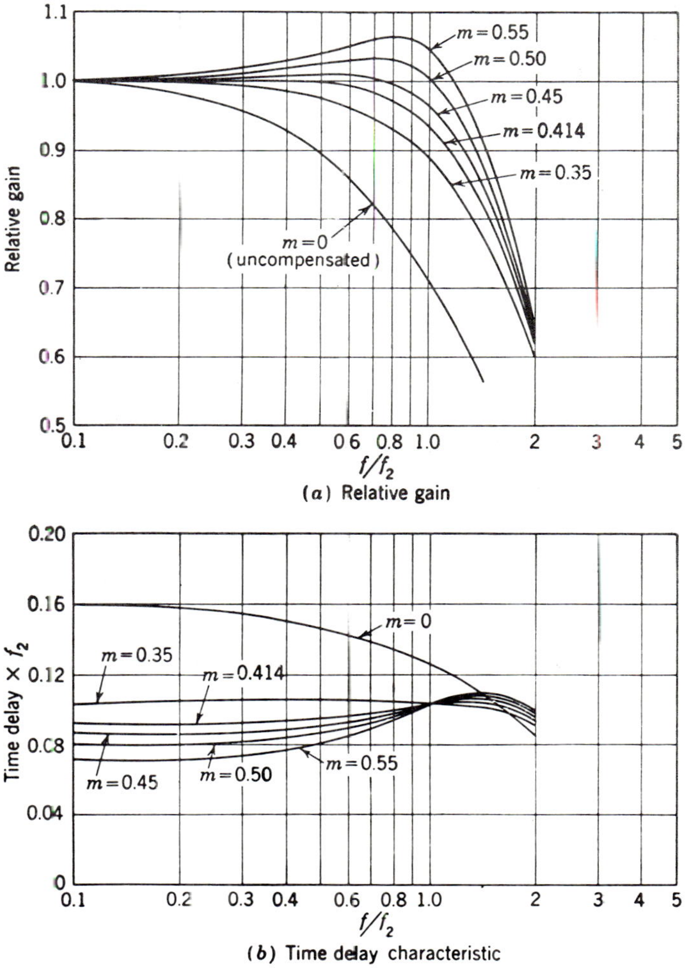

Fig. 20. Relative amplitude and time-delay characteristics of shunt-peaked amplifier for several values of compensation. (*From G. M. Glasford, "Fundamentals of Television Engineering," McGraw-Hill Book Company, New York, 1955. Used with permission.*)

25-16 AMPLIFIER CIRCUITS

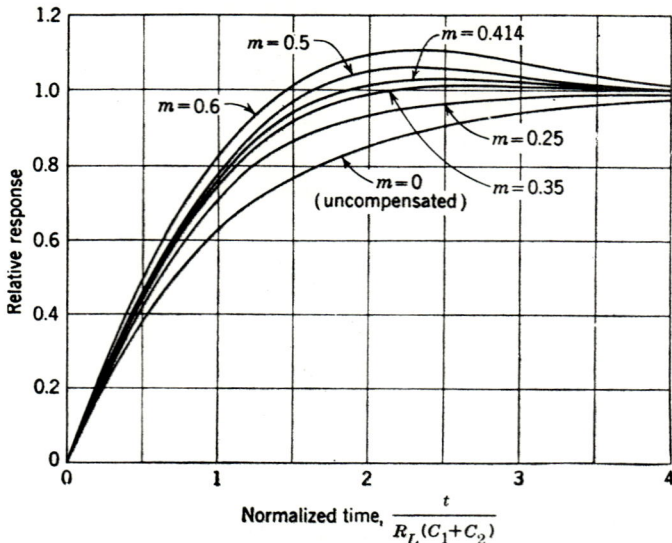

Fig. 21. Response to step function of shunt-compensated amplifier. (*From G. M. Glasford, "Fundamentals of Television Engineering," McGraw-Hill Book Company, New York, 1955. Used with permission.*)

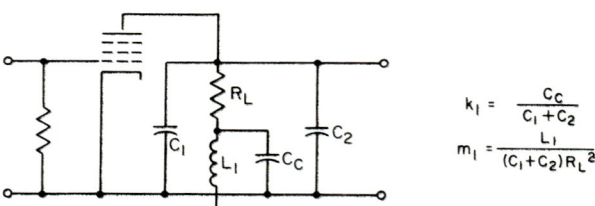

$$k_1 = \frac{C_C}{C_1+C_2}$$

$$m_1 = \frac{L_1}{(C_1+C_2)R_L^2}$$

Fig. 22. Schematic of modified shunt-peaked amplifier.

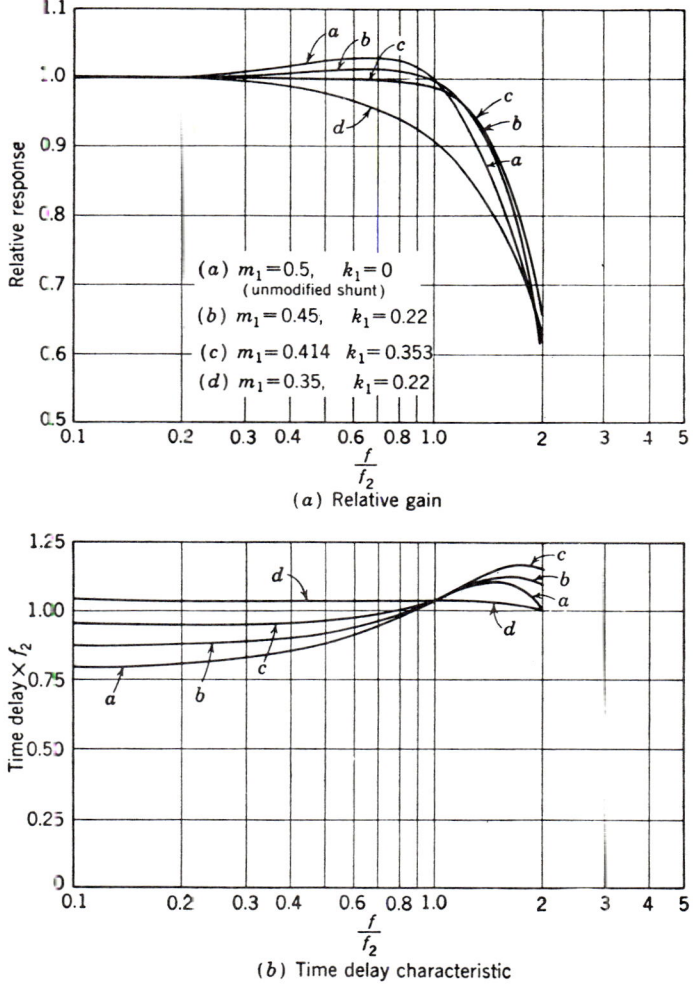

FIG. 23. Relative-amplitude and time-delay characteristics of modified shunt-peaked amplifier. (*From G. M. Glasford, "Fundamentals of Television Engineering," McGraw-Hill Book Company, New York, 1955. Used with permission.*)

AMPLIFIER CIRCUITS

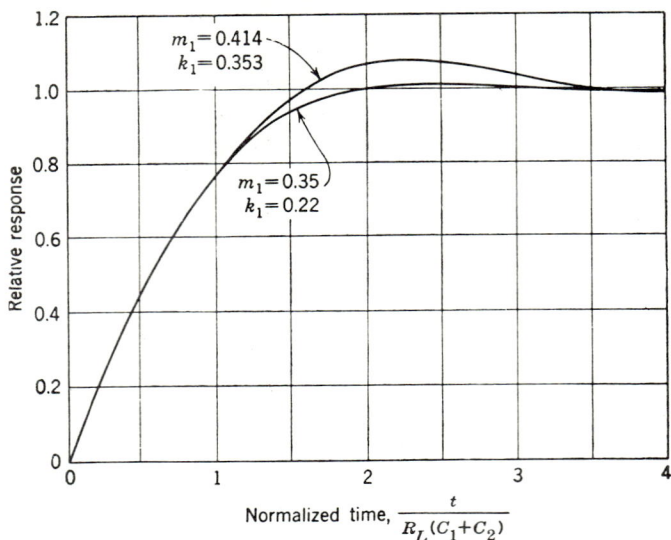

Fig. 24. Response to step function of modified shunt-peaked amplifier. (*From G. M. Glasford, "Fundamentals of Television Engineering," McGraw-Hill Book Company, New York, 1955. Used with permission.*)

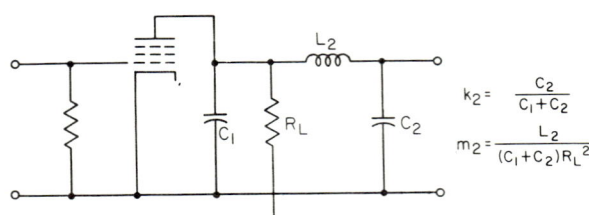

Fig. 25. Series-peaked amplifier circuit.

BROADBAND AMPLIFIERS

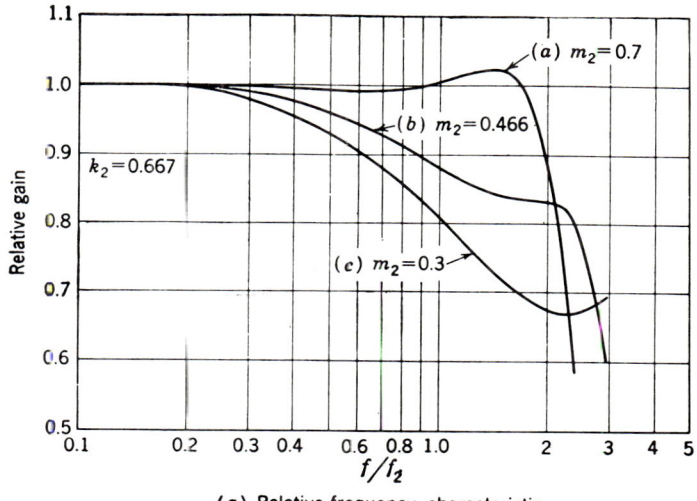

(a) Relative frequency characteristic

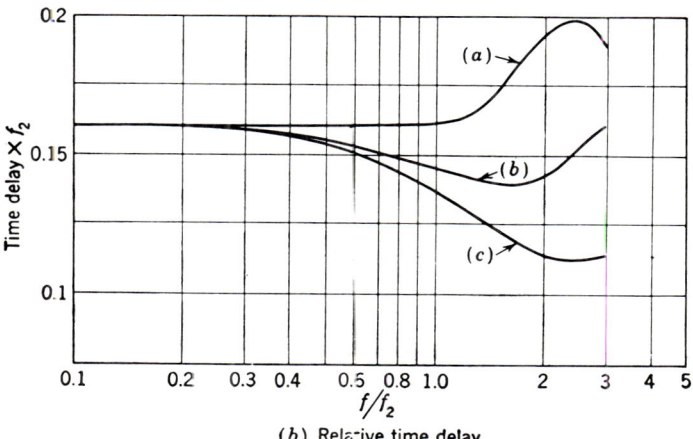

(b) Relative time delay

FIG. 26. Relative frequency vs. amplitude characteristics and relative time-delay frequency characteristics of series-peaked amplifier. (*From G. M. Glasford, "Fundamentals of Television Engineering," McGraw-Hill Book Company, New York, 1955. Used with permission.*)

25-20 AMPLIFIER CIRCUITS

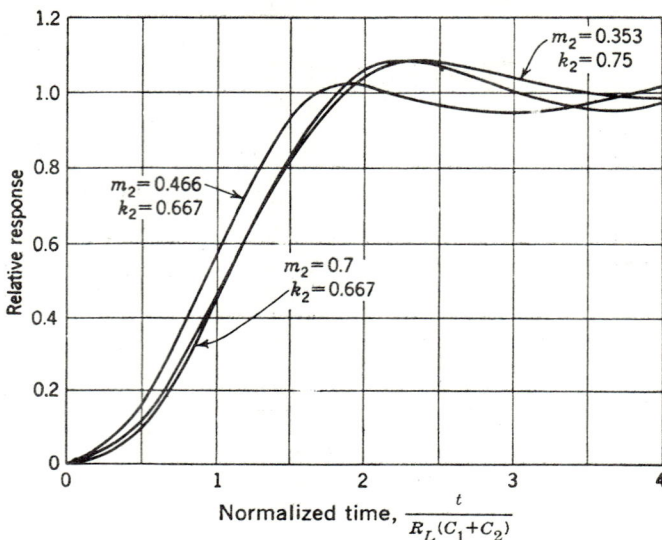

Fig. 27. Response to step function of series-peaked amplifier. (*From G. M. Glasford, "Fundamentals of Television Engineering," McGraw-Hill Book Company, New York, 1955. Used with permission.*)

BROADBAND AMPLIFIERS 25-21

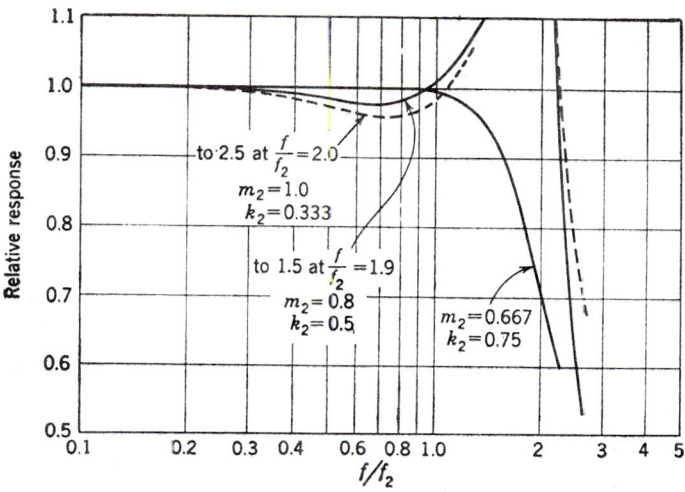

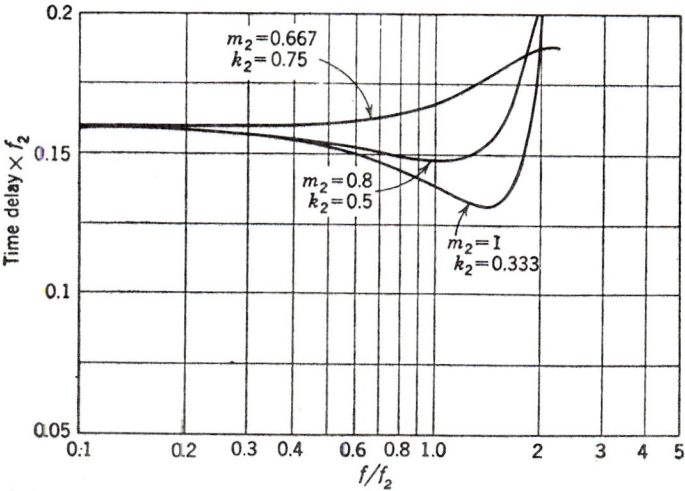

Fig. 28. Relative-amplitude frequency characteristics and time-delay frequency characteristics for series-peaked amplifier for different capacitance ratios. (*From G. M. Glasford, "Fundamentals of Television Engineering," McGraw-Hill Book Company, New York, 1955. Used with permission.*)

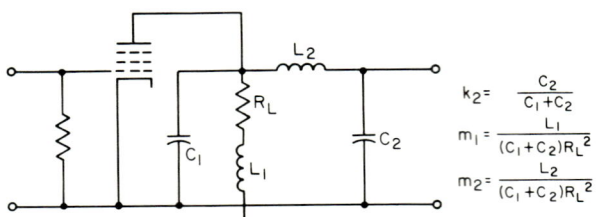

Fig. 29. Schematic for series-shunt-peaked circuit.

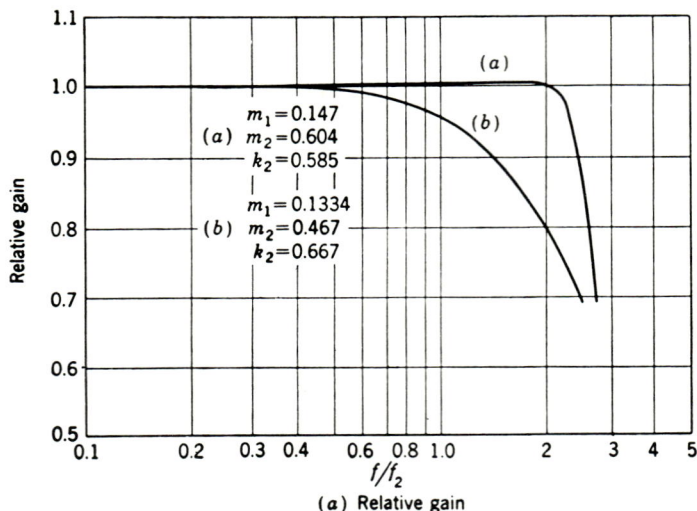

(a) Relative gain

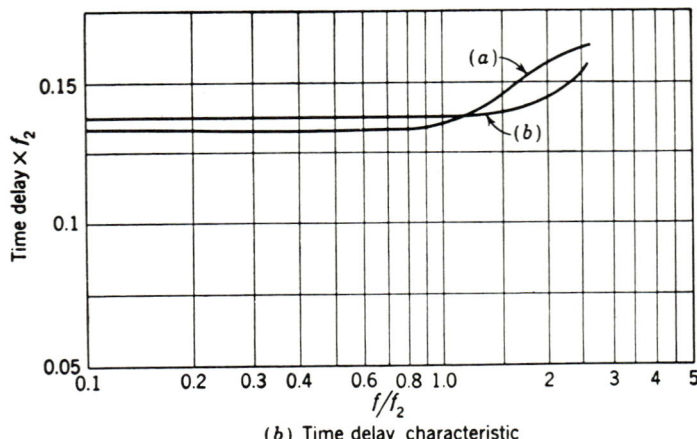

(b) Time delay characteristic

Fig. 30. Relative-amplitude and time-delay characteristic vs. frequency for series-shunt-peaked amplifier. (*From G. M. Glasford, "Fundamentals of Television Engineering," McGraw-Hill Book Company, New York, 1955. Used with permission.*)

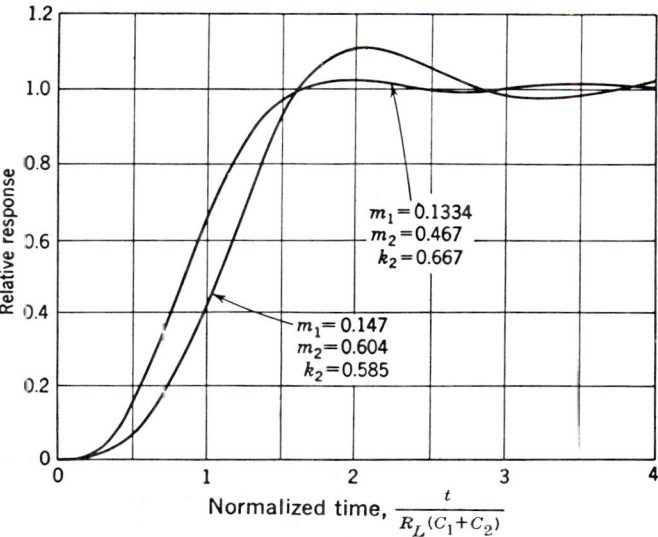

Fig. 31. Response to step function of series-shunt-peaked amplifier. (*From G. M. Glasford, "Fundamentals of Television Engineering," McGraw-Hill Book Company, New York, 1955. Used with permission.*)

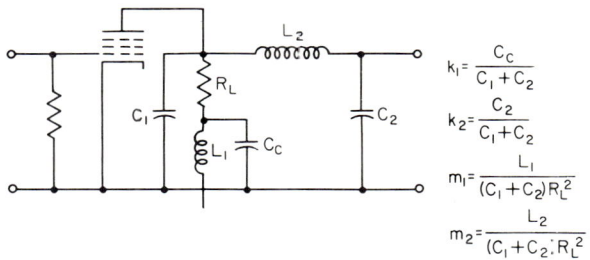

Fig. 32. Schematic of modified series-shunt-peaked amplifier.

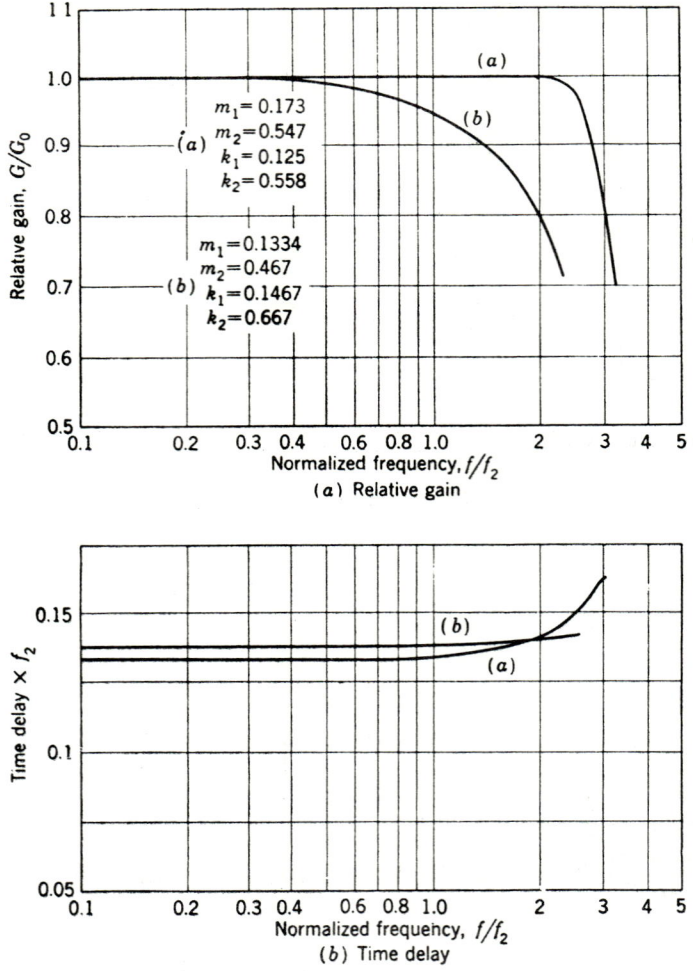

Fig. 33. Relative-amplitude and time-delay frequency characteristics for series-shunt-peaked amplifier. (*From G. M. Glasford, "Fundamentals of Television Engineering," McGraw-Hill Book Company, New York, 1955. Used with permission.*)

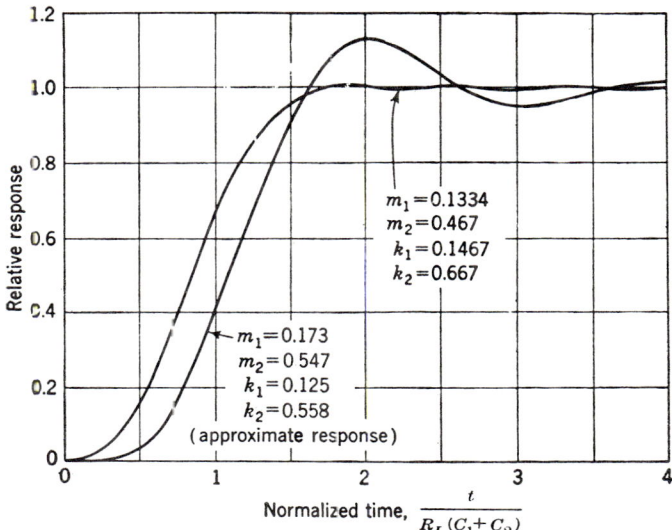

FIG. 34. Response to a unit-step function of a modified series-shunt-peaked amplifier. (From G. M. Glasford, "Fundamentals of Television Engineering," McGraw-Hill Book Company, New York, 1955. Used with permission.)

Among these circuits the performance of any of the series-peaked circuits is very dependent on the ratio of the input to output capacitance. In many practical situations, it may be found that the desired capacitance ratio of these circuits is the reciprocal of that available. By the reciprocity theorem, the input and output of the networks may then be simply reversed as illustrated in Fig. 35.

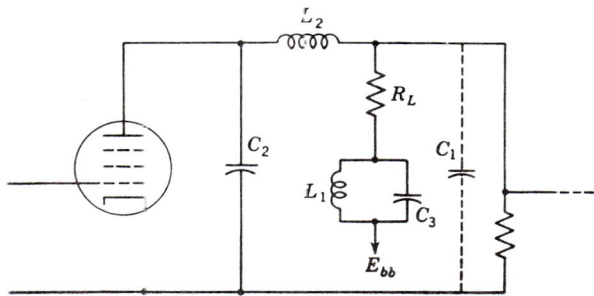

FIG. 35. Reversed modified series-shunt-peaked amplifier for reversed capacitance ratio. (From G. M. Glasford, "Fundamentals of Television Engineering," McGraw-Hill Book Company, New York, 1955. Used with permission.)

The circuits previously described can fulfill many of the needs of wideband or fast-rise-time amplifiers. A study has given optimum values for wideband and most rapid rise time with restricted overshoot.[20] A one-port (Fig. 36) and a two-port (Fig. 37) with mutual coupling are described. The two-port had greater gain bandwidths than any previous circuit. These circuits have been studied from the standpoint of broadest frequency response with the maximally flat magnitude response and

[20] Frank A. Muller, High Frequency Compensation of RC Amplifiers, *Proc. IRE*, vol. 42, pp. 1271–1276, August, 1954

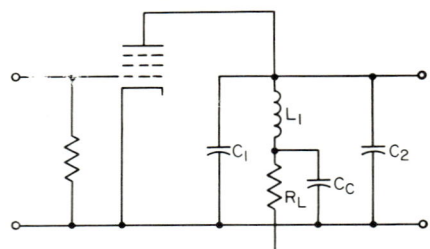

Fig. 36. Circuit of a second modified shunt-peaked amplifier. (*From F. A. Muller, High Frequency Compensation of RC Amplifiers, Proc. IRE, August, 1954.*)

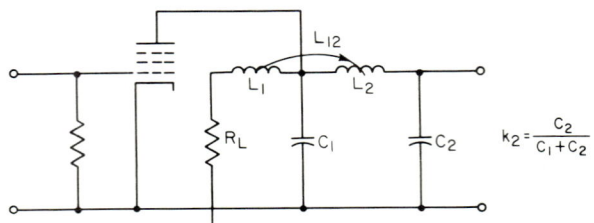

Fig. 37. Circuit of a shunt-series-peaked compensation circuit using mutual coupling. (*From F. A. Muller, High Frequency Compensation of RC Amplifiers, Proc. IRE, August, 1954.*)

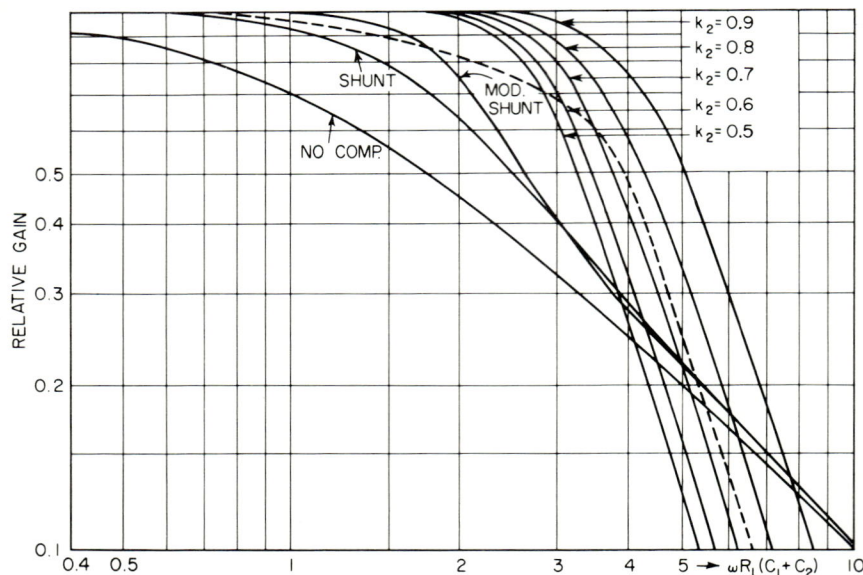

Fig. 38. Frequency response of a second modified shunt-peaked amplifier and the shunt-series-peaking amplifier using mutual inductance, compared with a simple RC amplifier frequency response. The dotted line gives the gain-frequency characteristics of the peaking network with mutual coupling for 1 per cent tolerance pulse compensation. (*From F. A. Muller, High Frequency Compensation of RC Amplifiers, Proc. IRE, August, 1954.*)

the maximally flat time delay. Consideration was also given to compensation based only on the transient response. The performance of several versions of the networks of Figs. 36 and 37 is illustrated in Fig. 38. The compensation is for the most constant amplitude except for the dashed line, which is the amplitude response of a 1 per cent tolerance compensation network, and the curve marked "mod. shunt" is the response for the circuit of Fig. 36. The response to a unit step of the last two circuits designed to several criteria is illustrated in Fig. 39.

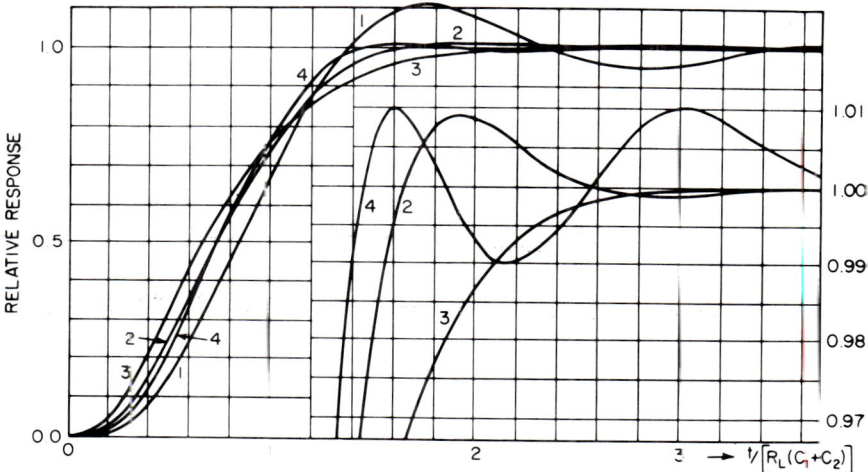

Fig. 39. Response to a unit step of (1) constant gain compensation, (2) linear phase compensation, (3) critical compensation, (4) 1 per cent tolerance compensation. Details of the curve in the region where overshoot takes place are shown in the insert time-response figure. (*From F. A. Muller, High Frequency Compensation of RC Amplifiers, Proc. IRE, August, 1954.*)

The parameters of the circuits previously described are categorized in Table 1 for a number of significant types of compensation, specifically the approximate maximally flat amplitude, and maximally flat delay responses. Note that the maximally flat delay will give about 1 per cent overshoot per stage while the maximally flat amplitude response will give an overshoot of about 9 to 12 per cent per stage.

Rise time is ordinarily defined as the time required for the response to a unit step to go from its 10 per cent to its 90 per cent value. When considering design based on transient response, a tolerance must be specified. For this purpose, the rise time will be understood to be the time elapsing between the departure of the transient response from the tolerance region around the zero value and the entrance into the tolerance region around the final value. For a given tolerance, the transient response with the shortest rise time will, in general, be tangent to the two tolerance limits around the final value as many times as possible; that is, the oscillations at the end of the transient response have equal amplitudes and increasing oscillation period. Responses obtained in this way will be called "tolerance compensation." A series of tables follows which describes some parameters and performance for compensation networks designed in this manner. Table 2 lists this information for the shunt-peaked compensation, Table 3 for the circuit of Fig. 36, and Table 4 is for the circuit of Fig. 37.

In addition to the two families of coupling networks described, a large amount of work has been done with more complicated networks. Increased speed or bandwidth is, however, realized slowly with increasing complexity. Among the factors that significantly affect the behavior of the four-terminal networks is their sensitivity to the

Table 1

Circuit type	Parameters						Response characteristic
	m_1 $\dfrac{L_1}{(C_1+C_2)R_L^2}$	m_2 $\dfrac{L_2}{(C_1+C_2)R_L^2}$	k_1 $\dfrac{C_c}{C_1+C_2}$ $\dfrac{L_{12}}{\sqrt{L_1L_2}}$	k_2 $\dfrac{C_2}{C_1+C_2}$	$\dfrac{f_{3\,db}}{f_2}$	$\dfrac{C_2}{C_1}$	
No compensation................	0.0	.....	.....	.....	1	.....	
Single-inductor compensation (Fig. 19)......	0.25	.....	.....	.....	1.53	.....	Critical compensation
	0.32	.....	.....	.....	1.7	.....	Max flat delay
	0.414	.....	.....	.....	1.8	.....	Approx max flat amplitude
Modified-inductor compensation (Fig. 22)...	0.35	.....	0.22	.....	1.8	.....	Approx max flat delay
	0.414	.....	0.353	.....	1.82	.....	Approx max flat amplitude
Series-inductor compensation (Fig. 25)........	.....	0.353	.....	0.75	2.0	3.0	
	.....	0.700	.....	0.67	2.2	2.0	
Series-shunt compensation (Fig. 29)..........	0.147	0.604	.....	0.585	2.7	1.4	Approx max flat amplitude
	0.1334	0.467	.....	0.667	2.4	2.0	Approx max flat delay
Modified series-shunt compensation (Fig. 32)	0.173	0.547	0.125	0.558	3.2	1.25	Approx max flat amplitude
	0.1334	0.467	0.1467	0.667	2.3	2.0	Approx max flat delay
Shunt-series compensation using negative mutual coupling	0.147	0.529	−0.237	0.5	2.8	1	Constant-gain compensation
	0.180	0.525	−0.550	0.7	3.25	2.33	Constant-gain compensation
	0.219	0.804	−0.736	0.9	4.25	9.0	Constant-gain compensation
	0.129	0.357	−0.513	0.5	.....	1	Linear-phase compensation
	0.176	0.373	−0.710	0.7	.....	2.33	Linear-phase compensation
	0.219	0.618	−0.822	0.9	.....	9.0	Linear-phase compensation

See figures for components in parameters.
f_2 is 3-db frequency for uncompensated case.

25–28

Table 2. Shunt-peaked Rise Time Δt in Units $R_L(C_1 + C_2)$*

Design tolerance, %	Parameter m_1	Rise time for a tolerance of			
		0.3%	1%	3%	10%
0	0	5.81	4.00	3.48	2.21
0	0.250	3.67	2.99	2.32	1.54
0.3	0.315	2.54	2.31	1.95	1.38
1	0.350	....	2.05	1.80	1.30
3	0.412	....	....	1.53	1.21
10	0.56	....	....	....	1.08

*From F. A. Muller, High Frequency Compensation of RC Amplifiers, *Proc. IRE*, August, 1954.

Table 3. Revised Shunt-peaked Rise Time Δt in Units $R_L(C_1 + C_2)$*

Design tolerance, %	Parameters		Rise time for a tolerance of			
	m_1	k_1	0.3%	1%	3%	10%
0	0.422	0.125	3.20	2.63	2.09	1.40
0.3	0.593	0.225	2.02	1.88	1.65	1.20
1	0.661	0.280	....	1.66	1.50	1.17
3	0.765	1.380	....	....	1.42	1.12

*From F. A. Muller, High Frequency Compensation of RC Amplifiers, *Proc. IRE*, August, 1954.

Table 4. Shunt-series (with Mutual Compensation) Rise Time Δt in Units $R_L(C_1 + C_2)$*

Design tolerance, %	Parameters				Rise time for			
	k_2	m_1	m_2	L_{12}	0.3%	1%	3%	10%
0	0.5	0.125	0.250	−0.707	2.37	2.00	1.50	1.05
1	0.5	0.170	0.373	−0.314	....	1.24	1.08	0.80
0	0.6	0.150	0.257	−0.767	2.21	1.86	1.43	0.98
1	0.6	0.185	0.364	−0.459	....	1.16	1.01	0.75
0	0.7	0.175	0.282	−0.812	2.04	1.72	1.32	0.91
1	0.7	0.203	0.386	−0.565	....	1.08	0.95	0.70
0	0.8	0.199	0.339	−0.848	1.73	1.55	1.19	0.81
1	0.8	0.222	0.449	−0.653	....	0.98	0.86	0.64
0	0.9	0.221	0.498	−0.882	1.54	1.30	1.00	0.68
1	0.9	0.237	0.635	−0.731	....	0.82	0.73	0.54

*From F. A. Muller, High Frequency Compensation of RC Amplifiers, *Proc. IRE*, August, 1954.

ratio C_2/C_1. This sensitivity resulted in either a considerably increased rise time or a tendency to severe overshoot when C_2/C_1 deviated from its optimum value. Two networks that have a reduced sensitivity to the C_2/C_1 ratio and their response to a unit step are shown in Figs. 40 to 43.

3.3. Stagger Peaking. The principle of stagger tuning has been well known and applied in the design of bandpass amplifiers since the early 1940s.[21] Modern network

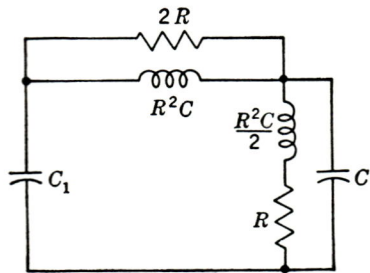

Fig. 40. Series-shunt-peaked circuit which reduces sensitivity to capacitance-ratio variations. (*From G. E. Valley, Jr., and H. Wallman, "Vacuum Tube Amplifiers," McGraw-Hill Book Company, New York, 1948. Used with permission.*)

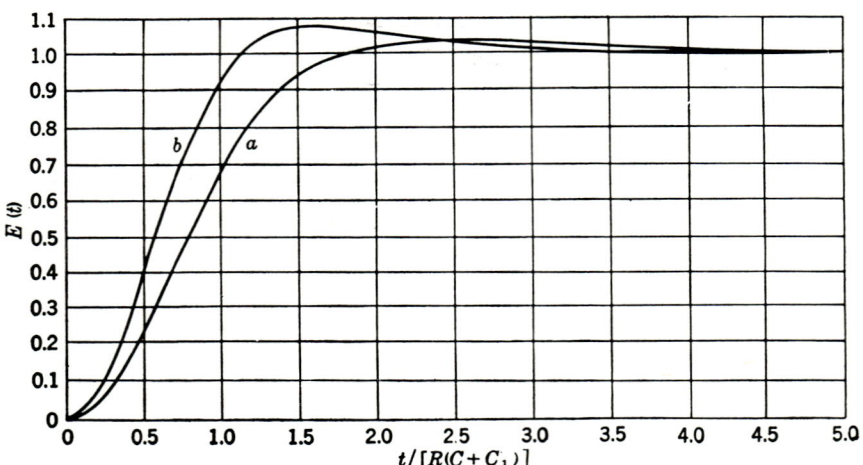

Fig. 41. Response to current step function of series-shunt-peaked circuit shown in Fig. 40. (a) $C_1 = C$. (b) $C_1 = C/2$. (*From G. E. Valley, Jr., and H. Wallman, "Vacuum Tube Amplifiers," McGraw-Hill Book Company, New York, 1948. Used with permission.*)

theory indicated the way to the application of the technique to wideband amplifiers in the late 1940s.[22,23] Since these amplifiers were designed on the basis of the amplitude-frequency characteristic, their response to transients had an excessive amount of overshoot. The advent of the design techniques to realize constant time delay made possible the design of stagger-peaked wideband amplifiers with reduced overshoot in their transient response. The technique has been applied to pulse amplifiers using

[21] H. Wallman, Stagger Tuned I-F Amplifiers, *MIT Rad. Lab. Rept.* 524, February, 1944.
[22] W. H. Huggins, The Natural Behavior of Broad Band Circuits, *Electron. Res. Lab. Rept.* E5013A, Air Materiel Command, November, 1948.
[23] A. Easton, Stagger Peaked Video Amplifier, *Electronics*, vol. 22, p. 118, February, 1949.

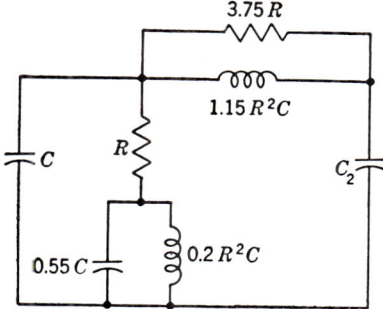

Fig. 42. Four-terminal network designed for 1:1 capacitance ratio. (*From G. E. Valley, Jr., and H. Wallman, "Vacuum Tube Amplifiers," McGraw-Hill Book Company, New York, 1948. Used with permission.*)

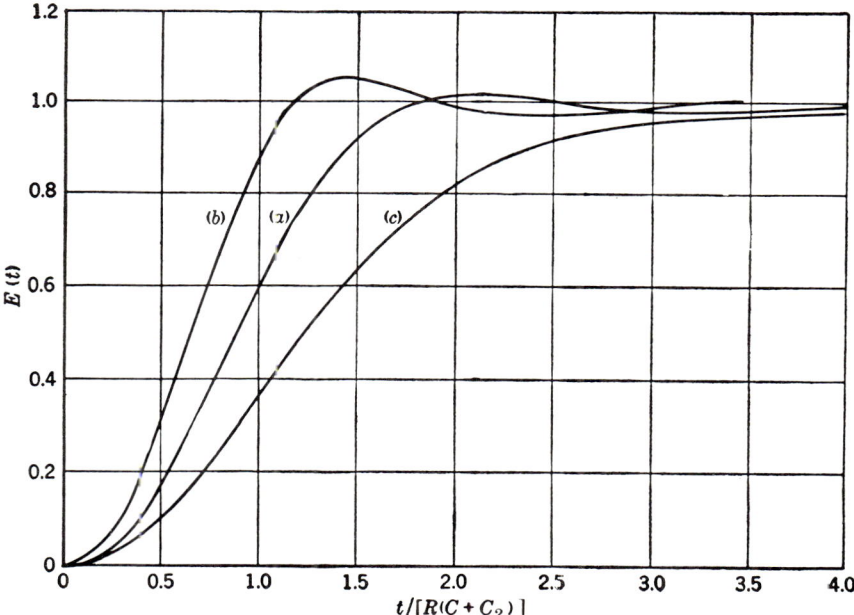

Fig. 43. Response of circuit of Fig. 42 for several capacitance ratios. (*a*) $C_2 = C$. (*b*) $C_2 = C/2$. (*c*) $C_2 = 2C$. (*From G. E. Valley, Jr., and H. Wallman, "Vacuum Tube Amplifiers," McGraw-Hill Book Company, New York, 1948. Used with permission.*)

the simple inductance-peaked stage circuit of Fig. 19.[20] For such an amplifier we can define two parameters m_i and r_i of the ith stage:

$$m_i = L_i/R_{Li}{}^2 C_i{}^* \quad \text{and} \quad r_i = R_{Li} C_i{}^*/\tau_m \qquad (17)$$

where R_{Li} and L_i are the load resistance and peaking coil inductance, respectively, of the ith stage, $C_i{}^* = C_1 - C_2$ of this stage, and τ_m is an overall mean time constant

$$\tau_m = (R_{L1} C_1{}^* R_{L2} C_2{}^* \cdots R_{Ln} C_n{}^*)^{1/n} \qquad (18)$$

The values of m_i and r_i are dictated by (1) the total number of uncompensated stages, (2) the total number of compensated stages, (3) the desired overall rise time, and (4) the permissible tolerance around the unit value (overshoot, ripple).

Table 5 gives values of these parameters for two to five stages of the circuit of Fig. 19, with up to three compensated. In this table the overall rise time is given in terms of τ_m, the mean time constant, defined above. For example, a three-stage amplifier with all stages compensated would have a rise time of $2.76\tau_m$ if designed to have ± 1 per cent tolerance (overshoot). If a 3 per cent tolerance can be accepted this design will give a shorter rise time $2.42\tau_m$, illustrating the interrelationship between rise time and per cent overshoot. The table also gives the appropriate values of r_1, r_2, r_3, m_1, m_2, and m_3 for this specific design. From these parameters, and knowing the total interstage capacitance C_i*, the appropriate values of load resistance and coil inductance can be obtained for each stage from Eqs. (17).

Table 5. Rise Times for Multistage Amplifiers*

Type of amplifier	Design tolerance, %	r_1	m_1	r_2	m_2	r_3	m_3	r_4	0.1%	0.3%	1%	3%	10%
2 stages, 1 coil	0	1.414	0.250	0.707	0				7.44	6.49	5.40	4.23	2.84
	0.1	1.047	0.466	0.950	0				4.26	4.06	3.68	3.11	2.21
	0.3	0.962	0.559	1.040	0					3.62	3.34	2.90	2.10
	1	0.838	0.753	1.190	0						2.93	2.59	1.91
2 stages, 2 coils	0	1.000	0.250	1.000	0.250				5.85	5.11	4.24	3.39	2.32
	0.1	0.624	0.741	1.596	0.292				3.31	3.15	2.87	2.46	1.78
	0.3	0.564	0.953	1.772	0.305					2.88	2.68	2.32	1.69
	1	0.482	1.400	2.073	0.325						2.40	2.13	1.58
3 stages, 1 coil	0	1.588	0.250	0.794	0	0.794	0				7.04	5.64	3.75
	1	0.864	0.932	1.076	0	1.076	0				3.86	3.40	2.50
3 stages, 2 coils	0	1.260	0.250	1.260	0.250	0.630	0				6.14	4.88	3.31
	1	0.335	3.087	1.580	0.565	1.889	0				2.90	2.62	1.92
3 stages, 3 coils	0	1.000	0.250	1.000	0.250	1.000	0.250				5.25	4.21	2.83
	1	0.252	4.434	1.058	0.953	3.754	0.318				2.76	2.42	1.78
4 stages, 2 coils	1	0.296	4.562	1.451	0.612	1.525	0	1.525			3.57	3.09	2.25
4 stages, 3 coils	1	0.1635	10.29	0.724	1.625	2.623	0.541	3.184			3.16	2.74	2.02
5 stages, 3 coils	1	0.135	16.5	0.63	1.90	2.23	0.51	2.30†			3.55	3.11	2.28

* From F. A. Muller, High Frequency Compensation of RC Amplifiers, *Proc. IRE*, August, 1954.

† $r_5 = r_4$.

Table 6 shows the relationship between overall voltage amplification, ith-stage load resistor, and overall rise time, in terms of the above parameters and the g_m of the individual stages.

The order of the circuits is of little importance if the amplifier is to handle only low-level signals, except that amplifiers with higher r_i and consequently higher gain offer the opportunity of obtaining a higher signal-to-noise ratio. In a high-level output stage a larger r_i will provide a larger output voltage. It may not be possible, however, to take full advantage of this, as high-r_i amplifiers generally have a large overshoot in their input which may drive them into grid current.

3.4. Phase Compensation. When a large number of amplifier stages must be cascaded, as in a television system, the rise time can increase considerably because of bandwidth shrinkage if the circuits are minimum-phase, designed for flat delay-frequency characteristic. On the other hand, if the circuits are minimum-phase and

Table 6

Parameter	Nonidentical g_m and C per stage	Identical g_m and C per stage
Voltage amplification	$A_n = \left(\dfrac{\tau_r}{\tau_m}\right)^n \prod_1^n \dfrac{g_{mi}}{C_i^*}$	$A_n = \left(\dfrac{\tau_r g_m}{\tau_m C^*}\right)^n$
ith-stage load resistor for given A_n	$R_L = \dfrac{r_i}{C_i^*}\left[\dfrac{A_n}{\prod_1^n (g_{mi}/C_i^*)}\right]^{1/n}$	$R_{Li} = \dfrac{r_i(A_n)^{1/n}}{g_m}$
Actual rise time for given A_n	$\tau_r = \tau_m\left[\dfrac{A_n}{\prod_1^n (g_{mi}/C_i^*)}\right]^{1/n}$	$\tau_r = \tau_m \dfrac{(A_n)^{1/n}}{g_m/C^*}$

designed for flat frequency response, there will be severe overshoot caused by a combination of somewhat rectangular high-frequency and accumulated phase or delay distortion in the high-frequency portion of the frequency range. If networks are included to compensate for the delay distortion but have no effect on the amplitude characteristic, then overshoot will be obtained but it will not be so severe. The system response can then be altered to reduce this overshoot without severely reducing the overall rise time.

The response desired is properly the subject of system engineering and will therefore not be pursued here. The technique of phase compensation will, however, be briefly considered.

Two networks that are useful for phase compensation are the bridged-T[24,25,26] and the lattice. Specific forms of these networks can be designed to have a complex frequency response of the form

$$f(s) = H\frac{(s - s_1^*)(s - s_1)}{(s + s_1)(s + s_1^*)} \qquad (19)$$

where s_1^* is the complex conjugate of s_1 and $s_1 = \sigma_1 + j\omega_1$. Consideration of this function leads to the conclusion that the amplitude is not affected when the frequency changes, as a consequence of which the network is called an all-pass network. The phase characteristic, however, changes twice as rapidly as with a single-pole pair network; therefore, if a flatter delay frequency characteristic is obtained by adding a network designed to the above complex frequency characteristic, the price paid is increased time delay in the response of such networks.

The schematic diagrams of two all-pass networks are shown in Fig. 44. The formulas with the figures give the values to determine the pole and zero locations. The poles are located at $-\sigma_1 \pm j\omega_1$ and the zeros are at $\sigma_1 \pm j\omega_1$. On the basis of the location of these poles and zeros, the time delay of a system can be equalized so that a constant time delay will be obtained at the same time as a uniform amplitude response is obtained.

A simple lattice that can be used for time-delay equalization is shown in Fig. 45.

[24] Ref. 11, p. 40; Ref. 22, p. 12.
[25] T. C. Nuttal, Some Aspects of Television Circuit Technique; Phase Correction and Gamma Correction, *Bull. Assoc. Suisse Elec.*, vol. 30, Audium, 1949.
[26] A. E. Brain, The Compensation for Phase Errors in Wideband Video Amplifiers, *Proc. IRE*, vol. 97, pp. 243-251, July, 1950.

25-34 AMPLIFIER CIRCUITS

This circuit has a time delay of

$$t_d = (1/\pi f) \tan^{-1} (\omega L/R_o) \tag{20}$$

where $R_o = \sqrt{L/C}$.

As an example of the result that can be obtained using this simple equalizer to compensate for the variation of delay obtained with simple shunt inductive peaking,

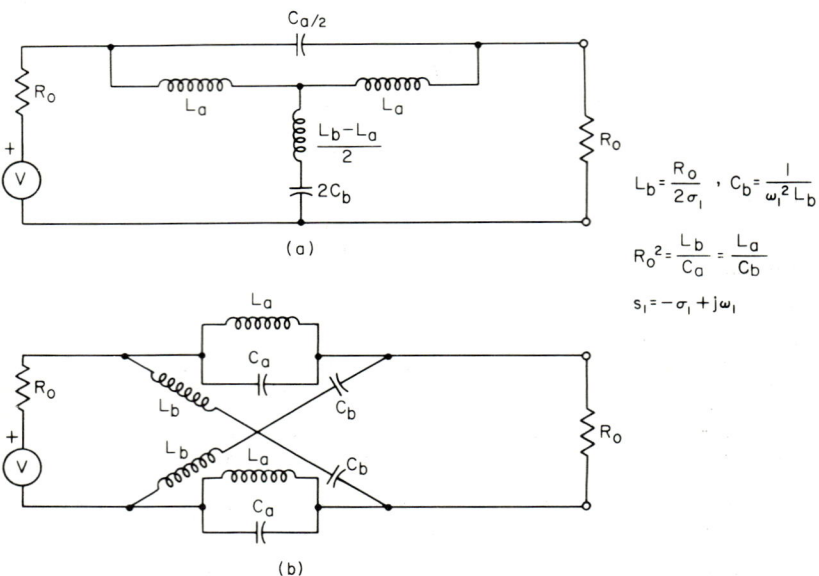

Fig. 44. Two-phase compensation all-pass networks. (a) Bridged T. (b) The lattice.

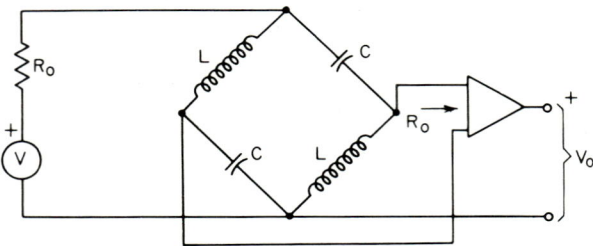

Fig. 45. A simple lattice all-pass-connected for phase compensation. (*From G. M. Glasford, "Fundamentals of Television Engineering," McGraw-Hill Book Company, New York, 1955. Used with permission.*)

consider Fig. 46. In this example, $m_1 = 0.5$ and f_2 is the 3-db frequency with no peaking. The time-delay variation has been decreased by a factor of 4.

3.5. Electron-tube Distributed Amplifiers. The demands on an amplifier in bandwidth and gain are sometimes so severe that the conventional amplifier stages have a voltage amplification of less than 1. Cascading amplifiers of this type can result only in a decrease of amplification. Another amplification technique that adds

BROADBAND AMPLIFIERS 25-35

the contribution of each stage can be used to surmount this limitation.[27,28] This can be understood easily from Fig. 47. The grid and plate each have a lumped transmission line into which they are connected at appropriate node points. Several conditions are necessary for this amplifier to function properly; namely, each transmission

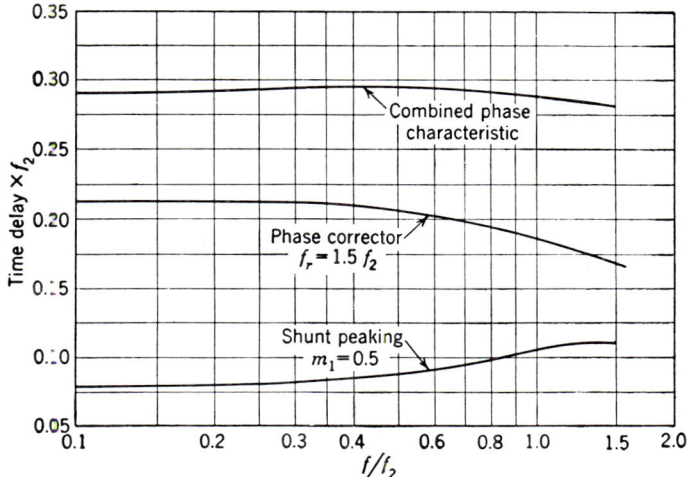

FIG. 46. Example of phase correction obtained with circuit diagram of Fig. 45. (*From G. M. Glasford, "Fundamentals of Television Engineering," McGraw-Hill Book Company, New York, 1955. Used with permission.*)

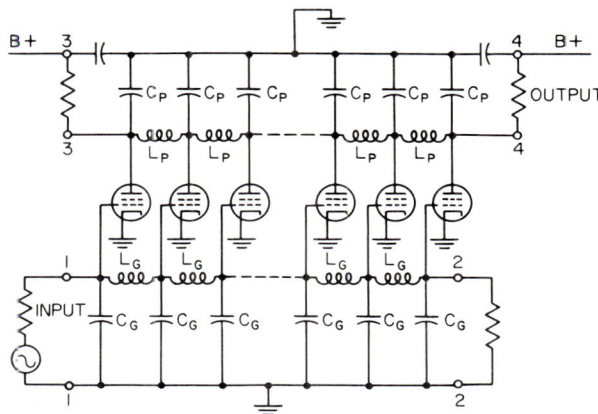

FIG. 47. Circuit diagram of a distributed amplifier. (*From E. L. Ginzton, W. R. Hewlett, J. H. Jasberg, and J. D. Noe, Distributed Amplification, Proc. IRE, August, 1948.*)

line must be properly terminated so that reflections will not be produced, and the delay between grids of particular tubes should be identical with the delay between the plates of the same amplifier tubes.

[27] E. L. Ginzton, W. R. Hewlett, J. H. Jasberg, and J. D. Noe, Distributed Amplification, *Proc. IRE*, vol. 36, pp. 956–969, August, 1948.
[28] W. H. Horton, J. H. Jasberg, and J. D. Noe, Distributed Amplifiers, Practical Considerations and Experimental Results, *Proc. IRE*, vol. 38, pp. 748–753, July, 1950.

If, for the moment, the effects of input grid loading of the tubes are neglected, then the frequency behavior of the constant-K networks used for isolating the tube capacitances from each other results in voltage amplification.

$$A_v = [(ng_m/2)R_o]^m[1 - (f/f_o)^2]^{-m/2} \qquad (21)$$

where $R_o = 1/\pi f_c C$ (the characteristic impedance), n is the number of sections in a stage, m is the number of stages, and f_c is the cutoff frequency of the line. The $\frac{1}{2}$ factor is due to use of only $\frac{1}{2}$ of the traveling-wave set up in the line because each end must be terminated to absorb reflections, only one end supplying an output signal. The minimum number of tubes is needed to achieve a desired total voltage amplification if each stage is designed to produce a voltage amplification of ϵ (2.718). From Eq. (21), it follows that the amplification increases proportionally to the sum of the number of tubes. This is so even if the voltage amplification of each tube is less than 1. This type of amplification is therefore sometimes called additive amplification.

The constant-K lumped transmission line has several shortcomings for use as a coupling network in a distributed amplifier: the termination is not a constant resist-

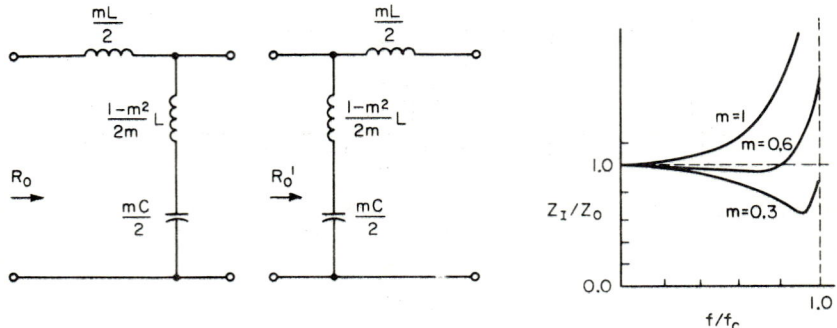

Fig. 48. Terminating half section and image impedance characteristic as a function of frequency.

ance or conductance, the impedance increases with frequency, resulting in a high-frequency peak, and the time delay also changes considerably with frequency. If the termination of either or both the plate and grid line was a constant resistance, changing reflections with changing frequency would result in a highly variable frequency response. The nonconstant phase delay with changing frequency would result in an inferior transient response. All three problems can be met with m-derived filters.

Terminating m-derived half sections and their image impedance characteristics are shown in Fig. 48. A property of such a half section with $m = 0.6$ is that a constant resistance of R_o terminating it results in an impedance match to the adjoining constant-K section up to nearly the cutoff frequency. The time delay per section can be made nearly constant by negative mutual inductance between the sections. Such a section and its time delay–frequency behavior for several values of m are shown in Fig. 49. The value of $m = 1.27$ results in a time delay that is nearly constant over a large portion of the frequency range. This type of m-derived network has no effect on the characteristic impedance of the lumped transmission line, making an $m = 0.6$ terminating half section necessary to match to a constant value of resistance. The combination of m-derived terminating half sections and phase or time-delay equalization sections is illustrated in Fig. 50. Although the approximations used in designing the terminating half sections are rough and become worse as the cutoff is approached, the frequency-amplitude response is quite flat at very high frequencies. A designable transient response, however, is more difficult to obtain, and amplifiers have to be adjusted experimentally. The amplitude characteristic can be compensated by other methods. One of these parallels either plates or grids of adjacent tubes in pairs. This technique is referred to as the paired-plate or paired-grid connection. Since the output capacitance of tubes is smaller than the input capacitance, the paralleling of adjacent plates in pairs will result in no loss of high-frequency response in the plate line as it is generally necessary to add capacitance when the grid lines and plate lines have the same Z_0. Still another technique of compensation is the use of a bridged-T

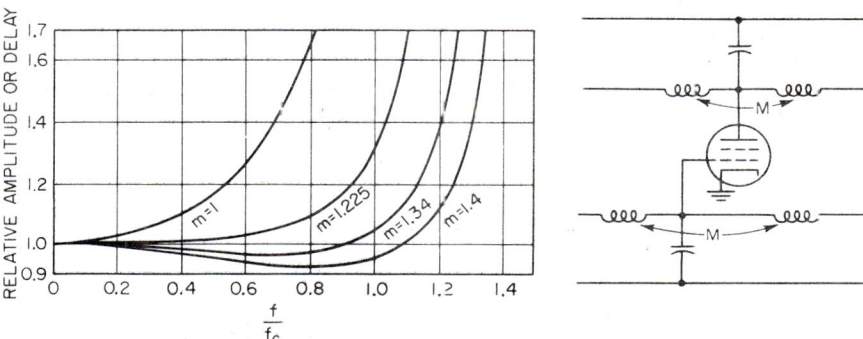

Fig. 49. m-derived section used for time-delay compensation. (*From E. L. Ginzton, W. R. Hewlett, J. H. Jasberg, and J. D. Noe, Distributed Amplification, Proc. IRE, August, 1948.*)

with the bridging element a capacitor. Neither of these techniques realizes the kind or amount of control obtained with the negative mutual inductance between sections, particularly with regard to constant time delay. Consequently, they are used only in special cases because of transit time and cathode lead inductance.

The input loading of the tube introduces a shunt conductance across each capacitor. Other causes of losses are the coil resistance, grid or plate lead inductance, and distributed capacitance in the coils. The input loading of the tube is the most impor-

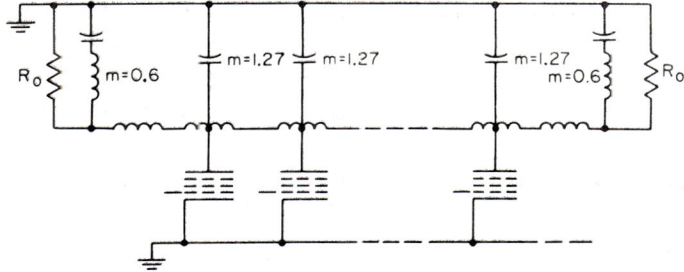

Fig. 50. Circuit using m-derived terminating section and m-derived time-delay compensating sections.

tant cause of attenuation. This cause of attenuation can be used to advantage to design the transmission lines so that the increasing impedance of the constant-K network is compensated. The flattest frequency response is obtained if the attenuation constant α of the lumped transmission line is described by $n\alpha = 1$.

The input loading could be compensated by taking advantage of the Miller-effect feedback from screen to control grid. If the screen has a small amount of inductance in it, the feedback through control-grid to screen-grid capacitance would be phased so as to produce a negative conductance in parallel with the input of the tube as long as the screen inductance is below resonance. Figure 51 is an example of the application of this technique to an experimental amplifier. Notice that with screen lead regeneration the high-frequency fall-off is much more abrupt and could be explained as due to additional positive conductive grid loading on the high-frequency side of the screen-circuit inductances resonant frequency.

Two examples of commercial distributed amplifiers are shown in Figs. 52 and 54 with their respective frequency responses in Figs. 53 and 55, respectively. The second shown is designed to have a peak pulse output of as much as 125 volts. A large number of tubes are required in one stage of the amplifier to produce the large output swing.

The previous material is based on design around constant-K networks and compensation for their shortcomings. Efforts have been made to use[29,30] modern network-synthesis techniques to treat the distributed amplifier, but the problem was found to be too unwieldly of solution with more than two tubes. Some work was done with

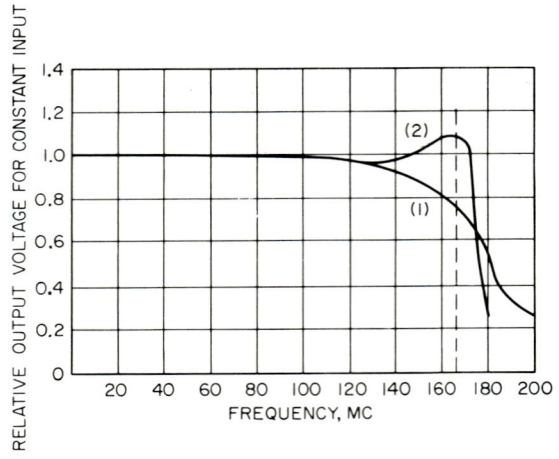

FIG. 51. Example of use of Miller-effect regeneration to compensate for input loading at high frequencies. (*From W. H. Horton, J. H. Jasberg, and J. D. Noe, Distributed Amplifiers: Practical Considerations and Experimental Results, Proc. IRE, July, 1960.*)

distributed pairs and results were realized that were of some interest. For example, one such synthesis technique gave a gain bandwidth 2.5 times that obtainable using the constant-K technique.

3.6. Electron-tube Terminating Techniques. When a wideband amplifier is to drive a transmission line some special provisions must be made. This can be accomplished in a number of ways, the chosen technique depending on the bandwidth and the requirements of the load. If a transmission line is to be driven from a source of the same impedance as Z_o, a cathode follower properly adjusted, a pentode amplifier with the desired terminating resistance as a load, a transformer, or a properly designed distributed amplifier are all possibilities. It should be realized that if the desired source resistance is obtained by use of an added resistor, power output will be consumed in the source impedance-determining resistor. This cost for eliminating reflections at the amplifier output is because the combination of the resistor and transmission line presents the amplifier with a load impedance of $R_o/2$ assuming the line is low-loss and properly terminated at the receiving end.

[29] D. O. Pederson, The Distributed Pair, *IRE PGCT*-1, December, 1952.
[30] A. D. Moore, Synthesis of Distributed Amplifiers for Prescribed Amplitude Response, *Stanford Univ. Electron Res. Lab. Tech. Rept.* 53, August, 1952.

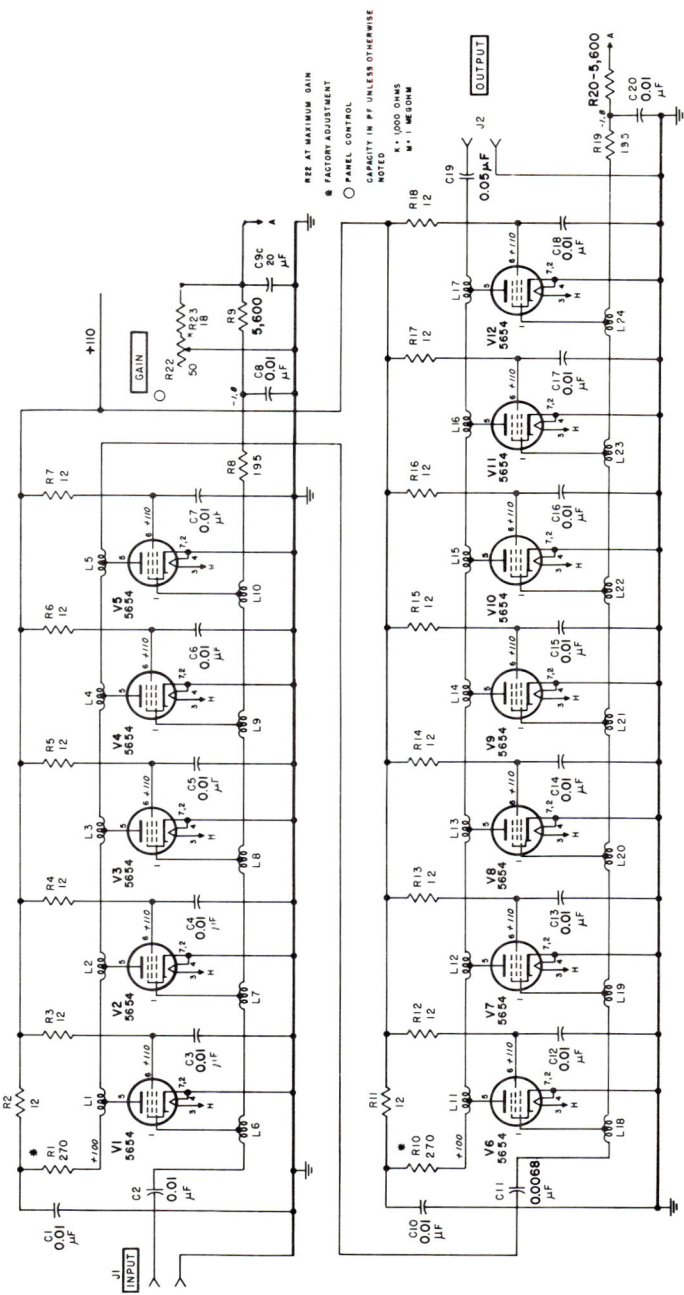

Fig. 52. Circuit diagram of a commercially distributed amplifier. (*From the Operating and Servicing Manual of the Hewlett Packard 460 AR Wide Band Amplifier, reproduced with permission.*)

The cathode follower shown in Fig. 56 has an output impedance very nearly $1/g_m$ if $R_k \gg 1/g_m$. If it is to match an impedance R_o, then $1/g_m = R_o$. For an impedance of 75 ohms, the required g_m would be 13,300 micromhos. If the tubes available do not have the required g_m, then the source impedance of the cathode follower can be adjusted by the addition of a shunt or series resistor of the proper value. The cathode follower has a voltage amplification

$$A_v = \frac{Z_L}{1/g_m + Z_L} \tag{22}$$

If $Z_L = 1/g_m$, then $A_v = \frac{1}{2}$. Under these operating conditions, little advantage would be realized from the inverse feedback in such an amplifier as the above con-

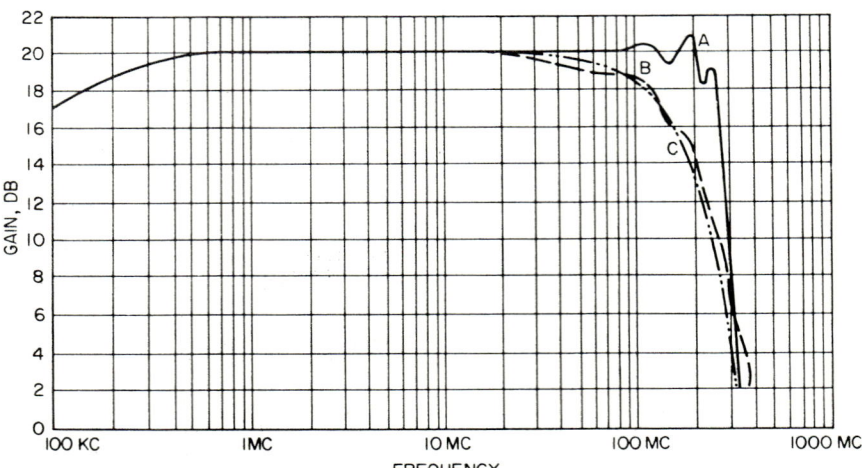

Fig. 53. Amplitude vs. frequency response of the circuit of Fig. 52. *(From the Operation and Servicing Manual of the Hewlett Packard 460 AR Wide Band Amplifier, reproduced with permission.)*

dition would have given an open-loop voltage amplification of 1. The potential peak-to-peak output voltage would be very nearly $2I_{dc}R_o$.

The cathode follower shown in Fig. 57,[31] a two-stage amplifier with 100 per cent feedback, offers another dimension of control. Its voltage amplification is

$$A_v = \frac{g_{m_1}[R_{k_1}R_o/(R_{k_1} + R_o)](1 + g_{m_2}R_{L_1})}{1 + g_{m_1}[R_{k_1}R_o/(R_{k_1} + R_o)](1 + g_{m_2}R_{L_1})}$$

$$\approx \frac{g_{m_1}R_o(1 + g_{m_2}R_{L_1})}{1 + g_{m_1}R_o(1 + g_{m_2}R_{L_1})} \tag{23}$$

where R_{k_1} is any resistor in the cathode of the first tube necessary for determining the d-c operating conditions. The source impedance of this amplifier is

$$R_s = \frac{1}{1/R_{k_1} + g_{m_1}(1 + g_{m_2}R_{L_1})} \tag{24}$$

If R_s is made equal to R_o and R_k is high enough to be neglected, then $A_v = \frac{1}{2}$ as in the case of the ordinary cathode follower. The output impedance of this amplifier is $1/g_m{'}$ where $g_m{'} = g_{m_1} + g_{m_1}g_{m_2}R_{L_1}$. The same output impedance can be obtained

[31] Ref. 1, p. 313.

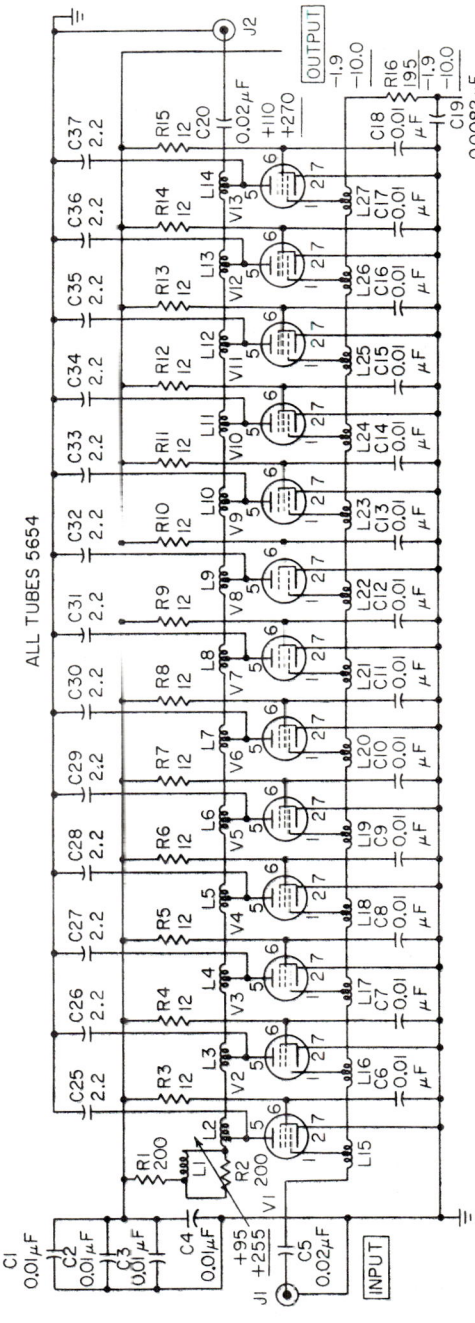

Fig. 54. Circuit diagram of distributed power output amplifier. (From the *Operation and Servicing Manual of the Hewlett Packard 460 BR Wide Band Amplifier*, reproduced with permission.)

with this amplifier as with the regular cathode follower, by using tubes of lower g_m and taking advantage of the $(1 + g_{m_2}R_{L_1})$ factor to multiply g_{m_1} to the desired value. A further advantage of this circuit is that the output current comes from the two tubes in a push-pull–like action; that is, as the current from one tube increases that from the other tube decreases. This, in addition to the linearization due to the inverse feedback, will give more linear operation. The peak-to-peak output voltage will be very nearly $2I_{dc}Z_o$.

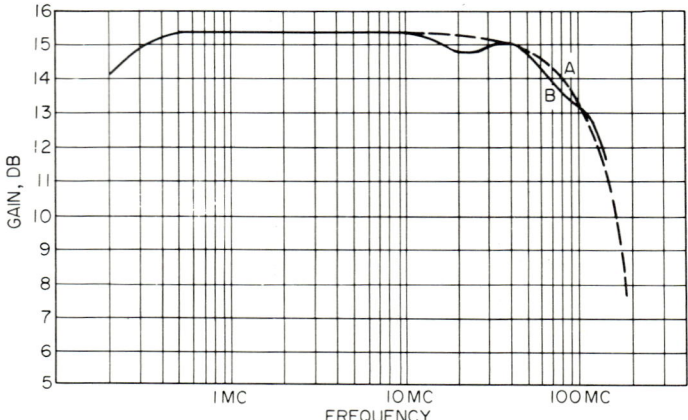

Fig. 55. Frequency response of the circuit of Fig. 54. (*From the Operation and Servicing Manual of the Hewlett Parkard 460 BR Wide Band Amplifier, reproduced with permission.*)

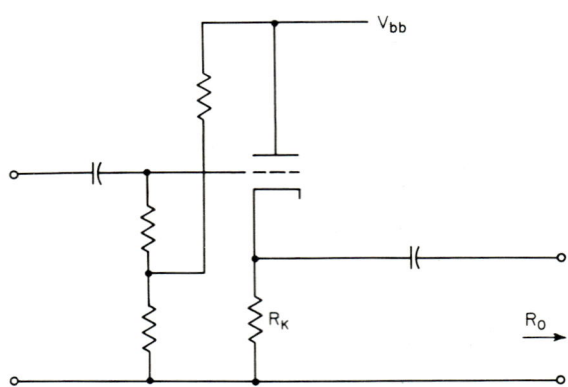

Fig. 56. Cathode-follower circuit.

The circuit of Fig. 58 shows one version of a pentode transmission-line driver. The plate resistor, being equal in value to Z_o, would absorb reflections. The voltage amplification of such an amplifier would be

$$A_v = -g_m Z_o/2 \tag{25}$$

and the output peak-to-peak voltage swing would be $I_{dc}Z_o$.

An amplifier with parallel-parallel feedback accomplished with a resistor connected from output to input as shown in Fig. 59 is still another way to obtain a low output

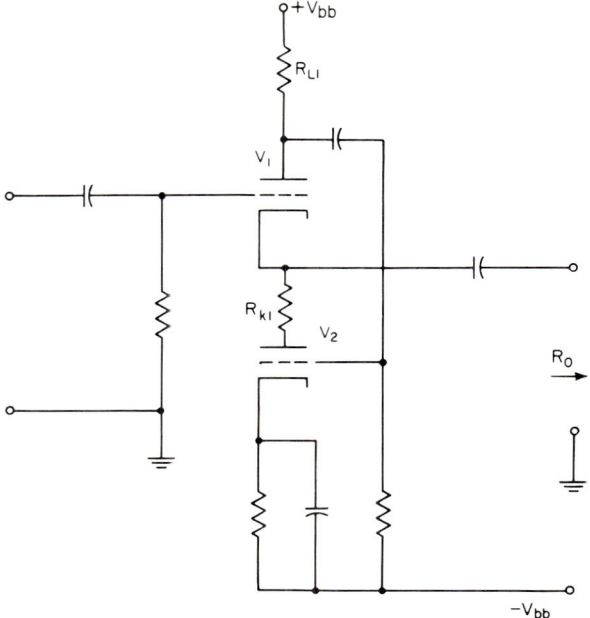

Fig. 57. Two-stage cathode-follower circuit.

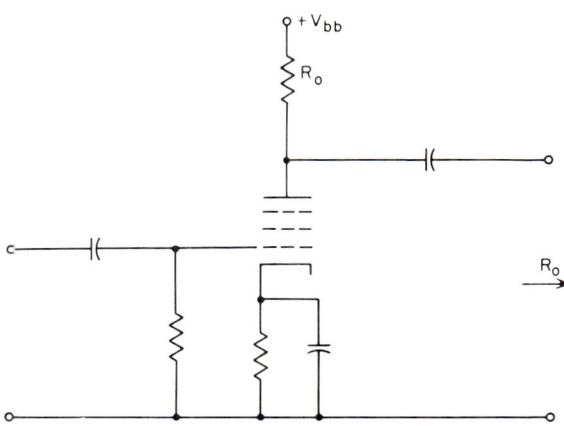

Fig. 58. Simple pentode transmission-line driver.

impedance. In the circuit shown, the feedback is around the output coupling capacitor in the interest of reducing the required value. Such an amplifier would have an output impedance of

$$Z_s = \frac{R_L + (1/j\omega C)}{1 + g_m R_L [R_1/(R_1 + R_f)]} \tag{26}$$

The lower limit that Z_s could take on is $1/g_m$, an unlikely value. Z_s could, however, be reduced by including additional amplification inside the feedback loop. In this case, if the conditions for stable amplification are met,

$$Z_s = \frac{R_L + (1/j\omega C)}{1 + g_m R_L A[R_1/(R_1 + R_f)]} \qquad (27)$$

where A is the additional voltage amplification inside the feedback loop. The peak-to-peak output voltage would be nearly $2I_{dc}Z_o$ where I_{dc} is the plate current of the output stage.

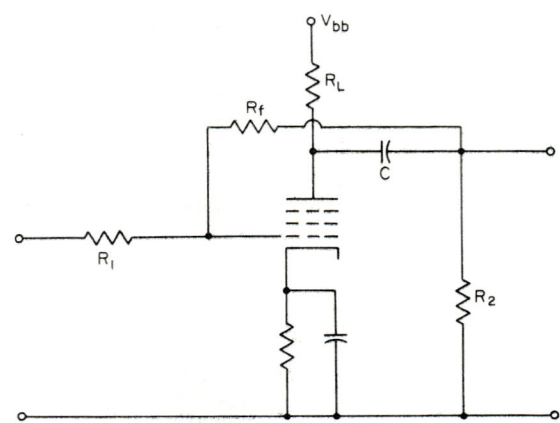

Fig. 59. Transmission-line driver circuit using parallel feedback.

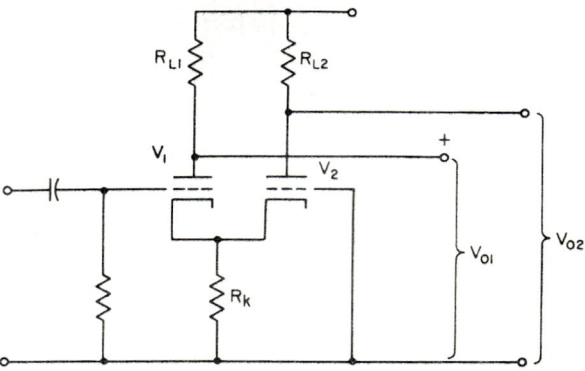

Fig. 60. Wideband phase-inverter circuit.

3.7. Electron-tube Phase Inversion. It is sometimes necessary to have a wideband balanced-output amplifier, one application of which is to drive the plates of a cathode-ray tube. A number of phase inverters are described in the section on audio amplifiers. Most of these are inadequate for wideband application for one reason or another. One circuit that has been used is shown in Fig. 60. In this circuit, the effective g_m to V_{o_1} is $g_m/2$ as is the effective g_m to output V_{o_2}. The difference in outputs is due to the effect of reduced degeneration in the first tube due to R_k increasing its effective g_m and reduced signal to the second tube due to the signal shunting effect of R_k. If

$$R_k \gg \frac{r_{p2} + R_{L_2}}{\mu_2 + 1}$$

then the outputs will be very nearly equal and out of phase. In those cases where the above inequality does not hold the outputs can be made equal in amplitude by increasing the value of R_{L_2}. If the two tubes are operated under the same d-c conditions and are identical, then the ratio of voltage amplifications will be

$$\frac{A_{v2}}{A_{v1}} = \frac{R_{L_2}}{R_{L_1}} \frac{R_k}{R_k + (r_p + R_{L_2})/(\mu + 1)} \qquad (28)$$

and if the voltage amplifications are to be the same then

$$R_{L_2} = R_{L_1} \frac{R_k(\mu + 1) + r_p}{R_k(\mu + 1) - R_{L_1}} \qquad (29)$$

3.8. Electron-tube Power Amplifiers. Power amplification over large frequency ranges can be obtained in several ways. One is the extension of low-frequency power-amplifier techniques such as transformer-coupled amplifiers, but using power tubes with a large gain-bandwidth product. Amplifiers of this type become difficult to design with frequency responses that extend beyond several megacycles. This is because of the increasing importance of the leakage reactance[32] of the transformers and the distributed capacitance of the tubes. Using the best techniques and present materials in the design of transformers, this cutoff frequency can be extended to about 15 Mc.

An example of a transformer-coupled power amplifier extending this technique is shown in Fig. 61.[33] The frequency response of this amplifier is shown in Fig. 62. The output transformer for this amplifier was wound on a ferrite toroid using a technique appropriate for wideband application. The above performance, amplification to 30 Mc with substantial power output, was possible through application of network synthesis considering the transformer parameters as part of an overall impedance-matching network. This technique is similar to video broadbanding, by separating interstage shunt capacitance with an inductor of necessary inductance.

Power amplifiers of wider bandwidth and larger power outputs are possible by application of distributed amplifier techniques. The larger tubes necessary for the larger power outputs cannot fully exploit the potential of distributed amplifiers because of parasitic capacitance and inductance. The introduction of additional compensation circuits can reduce these effects.[34] Amplifiers of 500-Mc bandwidth and several hundred watts output have been developed using this technique.

4. TRANSISTOR CHARACTERIZATION AND AMPLIFIERS

In calculating the performance of transistor amplifiers, the choice of transistor representation is very dependent on the desired end result and the technique for achieving it. For example, the availability of a digital computer and computer program makes practical the handling of a large number of detailed calculations. If the transistor can be described in terms of its two-port characteristics as a function of frequency (h and y parameters are most often used), then only the two-port is needed as an equivalent circuit. On the other hand, many applications do not require the power of a digital computer, or such a computer may not be available or economically feasible. For such amplifiers, an equivalent circuit that is a "reasonable" approxi-

[32] T. R. O'Meara, Wide-band Transformers and Associated Coupling Networks, *Electro-Technol.*, September, 1962.
[33] T. R. O'Meara, Wide-band Power Amplifiers and Transmitters, *Proc. Natl. Electron Conf.*, vol. II, pp. 251–267, 1955.
[34] B. F. Barton, L. A. Beattie, and P. H. Rogers, Some Useful Techniques for Overcoming the Frequency Limitations of Conventional Distributed Amplifiers, *1957 IRE Natl. Conv. Record*, pt. 2, Circuit Theory, pp. 97–106.

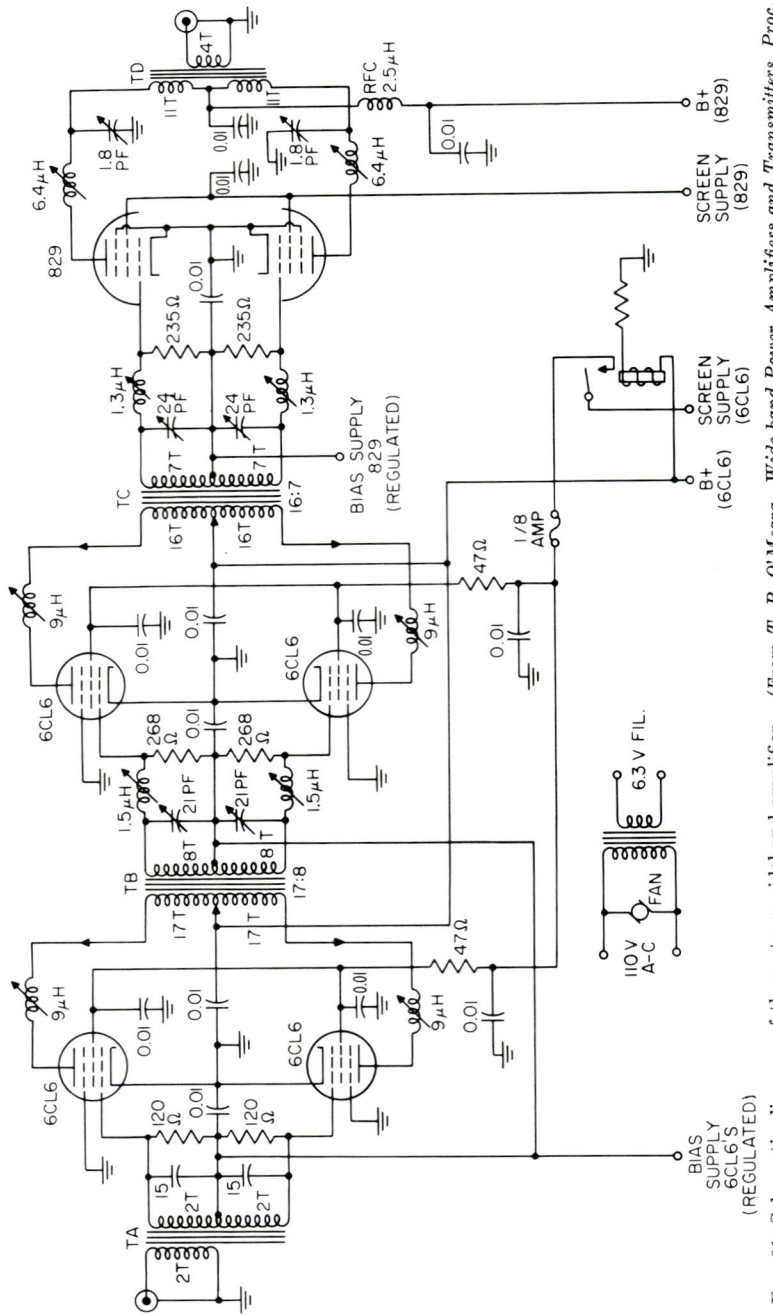

FIG. 61. Schematic diagram of three-stage wideband amplifier. (*From T. R. O'Meara, Wide-band Power Amplifiers and Transmitters, Proc. Natl. Electron. Conf., 1955.*)

mation makes possible the description of amplifier performance in closed-form expressions with some limitations on the unimportant detail.

The most popular equivalent circuit in use in the past years is the hybrid π described in detail in Chap. 10. A useful version of this equivalent circuit for wideband use is shown in Fig. 63. This equivalent circuit is useful to approximately $f_t/4$.

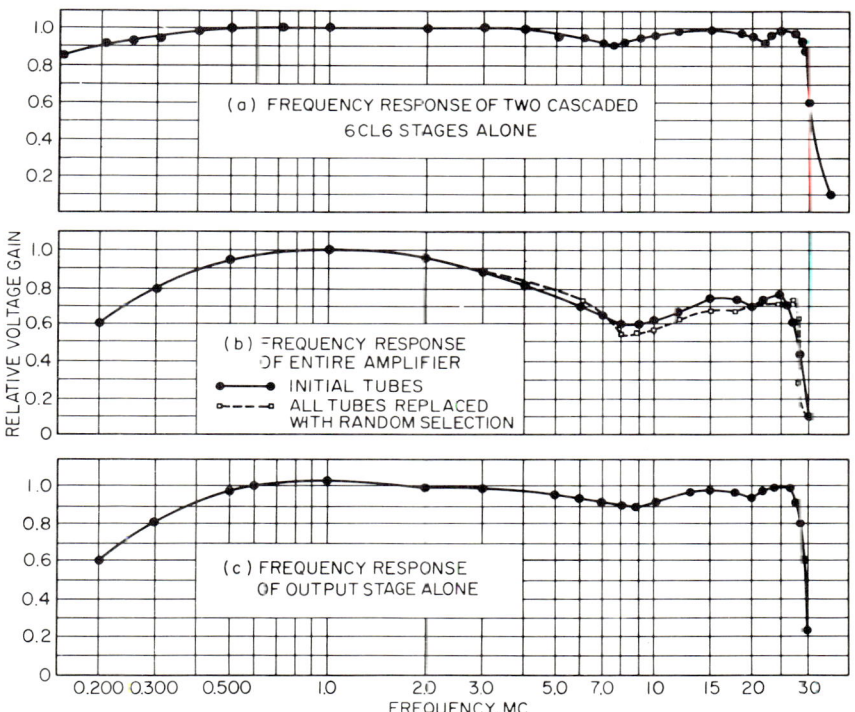

FIG. 62. Frequency response of three-stage wideband power amplifier. (a) Frequency response of two cascaded 6CL6 stages alone. (b) Frequency response of entire amplifier. (c) Frequency response of output stage alone. (*From T. R. O'Meara, Wide-band Power Amplifiers and Transmitters, Proc. Natl. Electron. Conf. 1955.*)

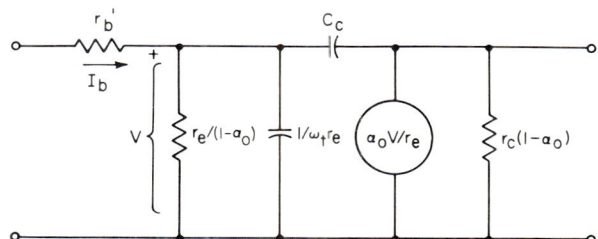

FIG. 63. Approximate hybrid π equivalent circuit of transistor.

In the above circuit, r_b' is the base spreading resistance, r_e is the incremental resistance of the emitter-base diode, α_o is the low-frequency common-base current gain, ω_t is the radian frequency for which the common-emitter current gain is unity, C_c is the collector-to-base capacitance, and r_c is the common-base collector-to-base incremental resistance. The incremental resistance r_e is approximately kT/qI_E, where k is Boltzmann's constant, T is the junction temperature in degrees Kelvin, q is the charge of the electron, and I_E is the d-c emitter current. At room temperature $r_e \approx 25/I_E$, where I_E is in milliamperes.

It should be noted that ω_t is somewhat dependent on the operating bias of the transistor, reaching a maximum value for many combinations of V_{CE} and I_E.

4.1. Low-frequency Response for Transistors. The low-frequency response of a transistor amplifier is limited by two factors, the coupling capacitor and the emitter bypass capacitor when a bypassed emitted resistor is used, as shown in the circuit of Fig. 64. These two effects are interrelated. This follows from the fact that, when the bypass capacitor is small, the incremental input impedance is increased by the resultant emitter degeneration, with the consequence that the cutoff frequency due to the coupling capacitor decreases. On the other hand, if the coupling circuit time constant is too short, it reduces the cutoff frequency due to the emitter bypass capacitor by raising the incremental emitter resistance which is being bypassed.

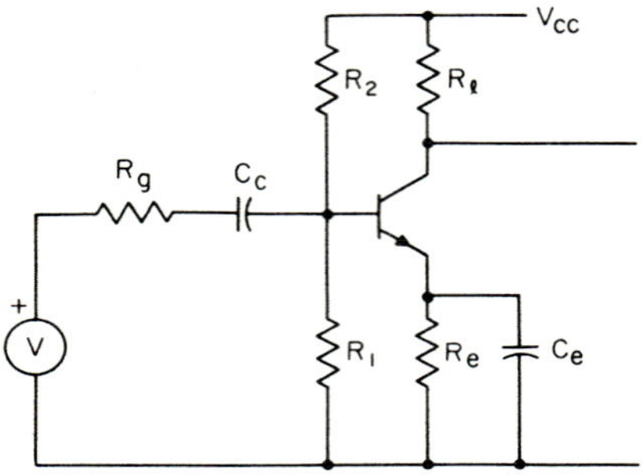

If C_e is large enough so that the time constant of it and the a-c impedance between emitter and ground is much longer than that associated with C_c, then the low-frequency half-power point is at

$$f = 1/2\pi R_T C_c \qquad (30)$$

where $R_T = R_g + \dfrac{\beta_o r_e R_1 R_2/(R_1 + R_2)}{\beta_o r_e + R_1 R_2/(R_1 + R_2)}$

FIG. 64. Circuit diagram of transistor amplifier showing components affecting low-frequency response.

Note that $\beta_o r_e$ could include any additional impedance in the emitter circuit such as that due to a value of C_e, the effect of which is not negligible. In this case, the effect of the reactance would also have to be considered. For this case, the effective value of r_e would be larger with a consequent increase of R_T and, therefore, a decreased cutoff frequency.

On the other hand, if C_c is assumed infinite, the low-frequency cutoff due to C_e would take place at

$$f = 1/2\pi R_p C_e \qquad (31)$$

where $R_p = \dfrac{R_e[r_e + (r_b' + R_b)/\beta_o]}{R_e + r_e + (r_b' + R_b)/\beta_o}$

$R_b = \dfrac{R_1 R_2 R_g}{R_1 R_2 + R_1 R_g + R_2 R_g}$

At those frequencies for which C_c's reactance increases, R_b would increase with a consequent increase in R_p and, therefore, a decrease in the cutoff frequency.

In the above cases, β figures prominently in the resultant value of R_T and R_p. Since β is quite a variable parameter the resultant cutoff frequencies will be quite variable. Thus, while compensation for the low-frequency response could theoretically be obtained in a manner similar to that used with electron tubes, the influence of β on the relationship makes it impractical. Consideration of R_T and R_p does, however, indicate that whereas in the case of the coupling capacitor, a decrease of β increases the low-frequency cutoff frequency, the effect on the bypass capacitor is to decrease the low-frequency cutoff frequency. Because of this interaction, if

$$R_p C_e = R_T C_c = RC$$

the gain would be less than 6 db down at $f = 1/(2\pi RC)$.

BROADBAND AMPLIFIERS

The design of coupling circuits and bypass circuits with very low frequency response requires that very large capacitance capacitors (compared with those needed with tubes) need be used. This is offset somewhat by the fact that considerably lower voltages are needed in transistor amplifiers, which allows the use of lower-voltage capacitors.

An improvement over the previous techniques of designing the low-frequency[35] response of transistor amplifiers can be achieved by going to a different scheme of biasing. This biasing scheme is illustrated in Fig. 65.

It has been shown that the bias stability of this circuit is the same as that of the circuit of Fig. 64 if $R_t = R_1 R_2/(R_1 + R_2)$. The capacitor C_d in this circuit must bypass

$$\frac{R_e[R_b + R_g \beta_o r_e/(R_g + \beta_o r_e)]}{R_e + R_t + R_g \beta_o r_e/(R_g + \beta_o r_e)} \quad (32)$$

while in the previous circuit, it must bypass the impedance looking into the emitter, that is, R_p, which is likely to be $\frac{1}{10}$ or less than is required for this biasing scheme. The emitter bypassing circuit would therefore require a capacitance C_e that would be about ten or more times greater than C_d. The following variation of the above circuit gives even better frequency response. With this circuit (Fig. 66),

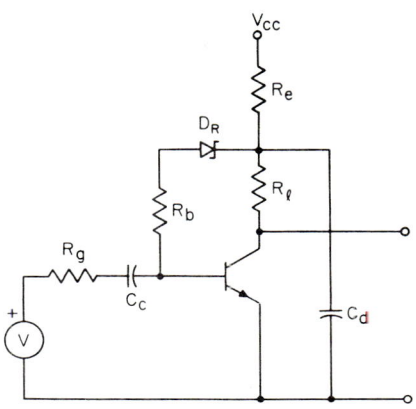

Fig. 65. Transistor biasing circuit requiring smaller capacitors than that of Fig. 64. (From P. J. Bénéteau and J. A. MacIntosh, Low-frequency Compensation of Transistor Amplifiers, Fairchild Application Data, APP-40, 1961.)

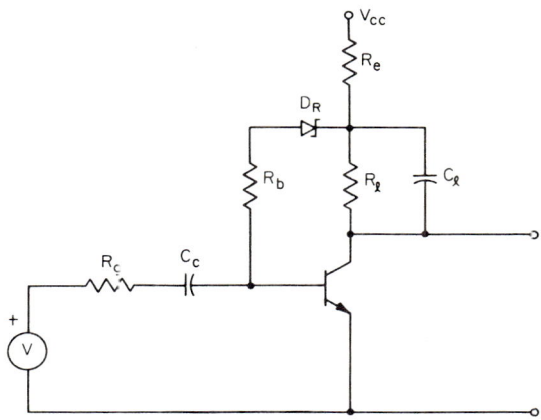

Fig. 66. Transistor biasing circuit adapted to give low-frequency compensation. (From P. J. Bénéteau and J. A. MacIntosh, Low-frequency Compensation of Transistor Amplifiers, Fairchild Application Data, APP-40, 1961.)

define the following time constants $\tau_l = R_l C_l$ and $\tau_g = (R_g + h_{ie}) C_c$ for the conditions $h_{re} = 0$, $h_{fe} \gg 1$, and $R_g \gg h_{ie}$.

[35] P. J. Bénéteau and J. A. MacIntosh, Low-frequency Compensation of Transistor Amplifiers, Fairchild Application Data APP-40, 1961.

A constant can be described

$$k = \frac{1 + R_l/R_e}{\tau_l/\tau_g}$$

For a value of $k = 1$, it can be shown that the low-frequency half-power point takes place at a frequency

$$f = 1/2\pi R_l C_l \qquad (33)$$

If $R_l/R_e > 1$, then for $k = 1$

$$R_l/R_e \approx \tau_l/\tau_g$$

from which it follows that for increasing values of R_l, τ_l must increase with a corresponding reduction in the cutoff frequency.

Values of $k < 1$ result in a higher value of low cutoff frequency while $k > 1$ results in peaking at the low frequencies with some depression of the low cutoff frequency. It should be noted that the overcompression of $k > 1$ could result in a rising or rounded top of a square wave.

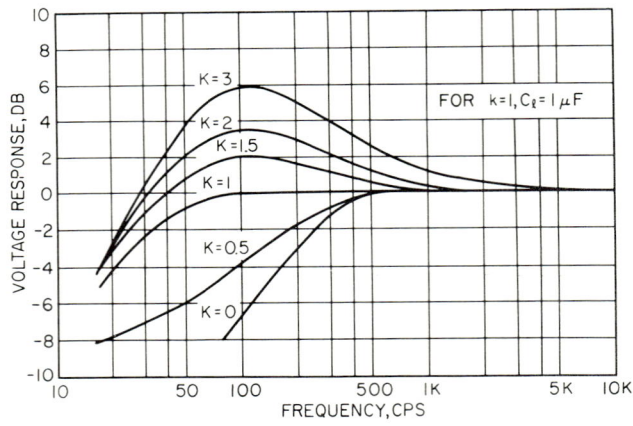

Fig. 67. Low-frequency response obtained using circuit of Fig. 66. (*From P. J. Bénéteau and J. A. MacIntosh, Low-frequency Compensation of Transistor Amplifiers, Fairchild Application Data, APP-40, 1961.*)

The improvement possible is illustrated by Fig. 67 for a circuit in which $R_l = 5.1$ kilohms, $R_e = 1$ kilohm, $R_b = 82$ kilohms, $R_g = 1$ kilohm, $C_c = 0.5$ μf, and $C_l = 1.0$ μf.

4.2. High-frequency Response for Transistors. The simplest transistor video circuit is shown in Fig. 68a. It will be[36] considered one of a cascade of similar stages. Since attention is being focused on the high-frequency response of wideband amplifiers, the circuits that affect the low-frequency response will be ignored, as this results in a simpler circuit. The equivalent circuit is shown in Fig. 68b.

It can be shown that the above amplifier has a low-frequency voltage gain per stage of

$$A_v(0) = \frac{-R_L/r_e}{1 + (r_b' + R_L)/(\beta_o r_e)} \qquad (34)$$

The 3-db frequency of the above amplifier can be shown to be

$$f_{3\,db} = \frac{R_L + r_b' + R_i}{2\pi R_i (R_L + r_b') C_i} \qquad (35)$$

[36] G. Brunn, Common-emitter Transistor Video Amplifiers, *Proc. IRE*, vol. 44, pp. 1561–1572, November, 1956.

From these, the gain bandwidth can be shown to be

$$GB = \frac{1/r_e}{[2\pi(R_L + r_b')/R_L][1/\omega_t r_e + C_c(1 + R_L/r_e)]} \tag{36}$$

As was the case with the triode electron-tube amplifier, the input-to-output capacitance relates the voltage amplification in the approximation (R_L/r_e) to the gain bandwidth. Thus the GB is similar in form to that of the triode electron tube except for

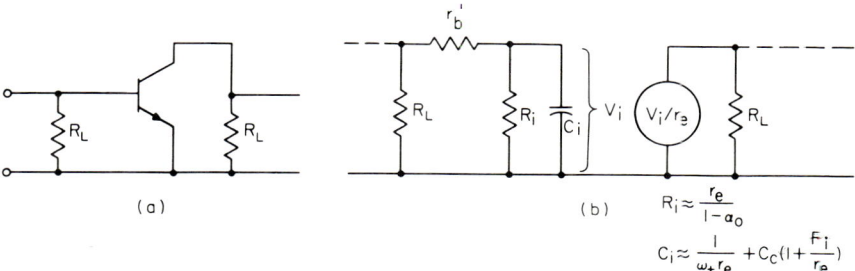

Fig. 68. Circuit diagram and equivalent circuit diagram of transistor amplifier showing device parameters that influence high-frequency response. (*From G. Brunn, Common-emitter Transistor Video Amplifiers, Proc. IRE, November, 1956.*)

the added term $(R_L + r_b')/R_L$. It follows from the equation that for any particular value of r_e and ω_t, a greater GB is obtained when the device has a smaller C_c. Increased bandwidth is obtained by decreasing the value of R_L. This reduces the effect of the C_c term. However, R_L also appears in the $R_L/(R_L + r_b')$ in a way that GB has a maximum for a value of R_L and decreases as R_L becomes smaller than this value.

Several circuit techniques can be used to increase GB.[37] These are shown in Figs. 69 to 71. The first of these, the shunt-peaking circuit, has the most constant GB of

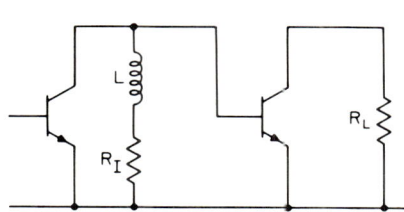

Fig. 69. Shunt-peak transistor circuit. (*From D. O. Pederson and R. S. Pepper, An Evaluation of Transistor Lowpass Broadbanding Techniques, 1959 IRE Wescon Conv. Record.*)

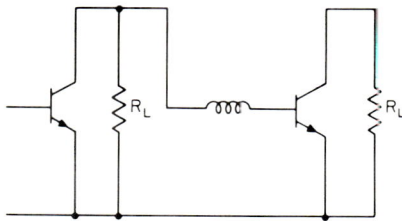

Fig. 70. Series-peaking transistor circuit.

any of the above circuits. This can be seen from Fig. 72. While the curves are plotted for specific device parameters, the quality of amplifier behavior should be consistent with these results with devices having other parameters.

The curve of GB for the shunt-inductance-peaked coupling circuit was calculated for the maximally flat frequency response. This particular response was chosen because it gave the maximum GB.

[37] D. O. Pederson and R. S. Pepper, An Evaluation of Transistor Lowpass Broadbanding Techniques, *1959 IRE Wescon Conv. Record*, pt. 2, pp. 111–127.

In some applications where overshoot is particularly critical, the shunt-inductance compensation can be designed by a zero-pole cancellation technique. The resultant response is then determined by a single pole on the negative real axis. While at first sight this could appear the same as the simple resistive broadbanding, it gives only slightly less than the maximally flat amplitude GB for the same device parameters.

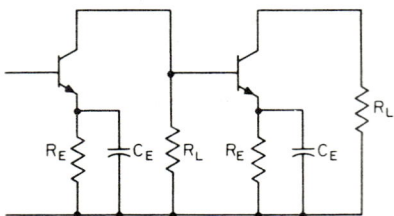

Fig. 71. Emitter bypass peaking transistor circuit.

GB in the example reaches a value of $0.9\omega_t$ at about $20\omega_\beta$ bandwidth and remains at this value to beyond $80\omega_\beta$, only slightly less than one obtains for maximally flat compensation.

The performance of the series-inductance-compensated circuit of Fig. 70 is somewhat different. Not only is the GB less than for shunt inductive peaking, but the maximum GB is not obtained with a maximally flat response.

The techniques of compensation by the emitter time constant illustrated in Fig. 71 are inferior only to the inductive-shunt-compensated case. It can be shown that

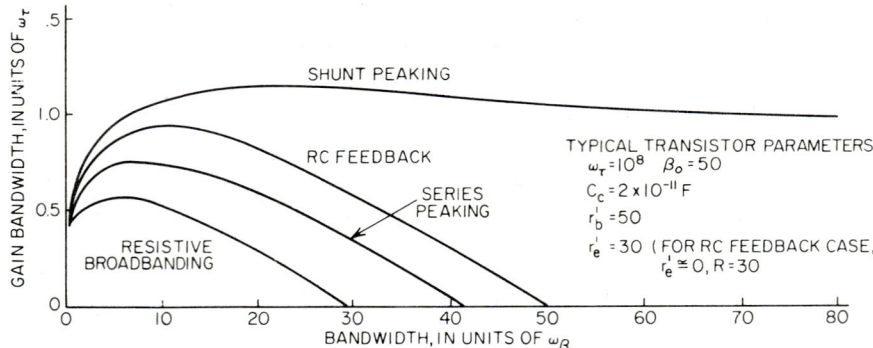

Fig. 72. Example of gain bandwidth vs. bandwidth obtained for shunt peaking, emitter bypass, series peaking, and simple resistive broadbanding with a typical transistor. (*From D. O. Pederson and R. S. Pepper, An Evaluation of Transistor Lowpass Broadbanding Techniques, 1959 IRE Wescon Conv. Record.*)

as r_b' decreases toward zero the GB approaches the performance obtained with the shunt-inductive-peaked case. The general performance of GB is bandwidth peaks at about $10\omega_\beta$ going to zero at $50\omega_\beta$.

In light of the fact that shunt-inductive compensation is superior to the other techniques of compensation, attention will be centered on two procedures of designing this type of circuit. The voltage amplification of one stage of an iterative amplifier using inductive-shunt compensation is described by

$$A_v(s) = \frac{\beta_o \omega_\beta (s + R_I/L)}{D[s^2 + s(R_I + r_b')/L + \omega_\beta/D) + \omega_\beta(R_I + r_b' + \beta_o r_e)/LD]} \qquad (37)$$

where the circuit components are illustrated by Fig. 69, and $D = 1 + \omega_t C_c(r_e + R_L)$. The above equation is of the form

$$A_v(s) = A \frac{s(s - z_1)}{(s + s_1)(s + s_2)} \tag{38}$$

Pole-zero cancellation yields a one-pole response which not only is simpler than the maximally flat response but also has no overshoot in its response to the unit-step function.

The design of the amplifier requires a knowledge of the parameters of the transistor equivalent circuit shown in Fig. 63. The design can start with an assumed value of R_L from which the quantity D, as given above, represents the degradation due to the feedback through C_c. It should be noted that, in the case of cascaded amplifier stages, R_L is not the R_L of Fig. 68 but the parallel combination of R_I and the incremental input impedance of the following stage $r_b' + \beta_o r_e$. This value of D can be used to calculate the gain-bandwidth from the equation

$$A_v(0)\text{BW} = \omega_t/D \tag{39}$$

which with the result of

$$A_v(0) = \frac{-R_L/r_e}{1 + (r_b' + R_I)/\beta_o r_e} \tag{40}$$

yields the bandwidth. If the bandwidth is too small, the above calculations can be repeated using a lower value of R_L, etc. One or two trials should give a value close enough for practical purposes. Using the last value of D, the required value of L can be calculated from

$$L = r_b'/(\text{BW} - \omega_\beta/D) \tag{41}$$

This type of amplifier design is not a direct synthesis process because it concerns a number of performance parameters that are interdependent. The synthesis process consists of repeated analysis with changes being made to accomplish the desired result.

The design of the shunt-peaked case for the maximally flat response is somewhat more complicated, requiring that the following equations be satisfied:

$$\mu^2(1 + \epsilon) + 2\mu(1 + \epsilon - 1/A_o) + 1 - 1/A_o^2 = 0 \tag{42}$$

where
$$\epsilon = r_b'/R_I \tag{43}$$
$$A_o = |A_v(0)|/\beta_o \tag{44}$$

and
$$L = (R_I/u)D/\omega_\beta \tag{45}$$

The 3-db bandwidth is

$$\text{BW} = \frac{\omega_\beta[1 + (1 + 4\mu^2 A_o)^{\frac{1}{2}}]^{\frac{1}{2}}}{D\sqrt{2} A_o} \tag{46}$$

One starts with a required bandwidth and assumes a GBW from which an initial gain is calculated. This value of gain, the transistor parameters and Eq. (40) yields a value of R_I. Equation (43) gives ϵ and Eq. (44) gives A_o. Substituting these into Eq. (42), the value of μ can be calculated. All the parameters necessary to calculate BW with Eq. (46) and thus GBW are now available. If the GBW is significantly different, a new GBW is assumed using the calculated value as a guide, and these calculations will give a reasonably close design for most applications. The above assumes r_e as a given transistor parameter. The choice of r_e, iterated several times for the desired bandwidth, would lead to a maximum GBW.

There are many methods of compensation by use of inverse feedback. One technique based on single-stage compensation is shown in the schematic diagram of Fig. 71. It can be shown that when r_e can be neglected compared with the parallel combination of R_e and C_e the gain is

$$A_v(s) = \frac{R_L \beta_o \omega_\beta (s + 1/R_e C_e)}{D(R_I + r_b)\{s^2 + s(1/R_e C_e + \omega_\beta/D) + \omega_\beta(R_e \beta_o + R_I + r_b')/[(DR_e C_e)(R_I + r_b')]\}} \quad (47)$$

This type of compensation gives results exceeded only by shunt peaking and for some conditions common-collector–common-emitter combinations. One distinction of this amplifier is that it was the first transistor amplifier applied to stagger peaking.

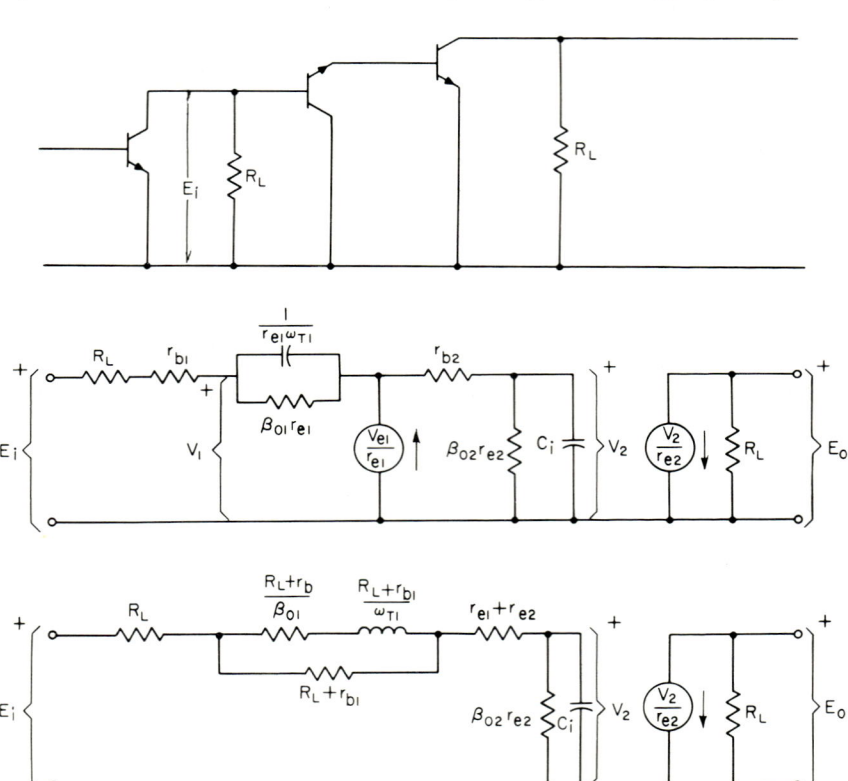

FIG. 73. Circuit diagram and two equivalent circuits of common-emitter common-collector cascade amplifier. (*From D. O. Pederson and R. S. Pepper, An Evaluation of Transistor Lowpass Broadbanding Techniques, 1959 IRE Wescon Conv. Record.*)

The compensation of amplifiers can be carried out in still another manner, that is, connection of a common-collector–connected transistor followed by a common-emitter–connected transistor. This amplifier connection is shown in Fig. 73, which also shows two ways of describing the equivalent circuit. Its desirable properties follow from the fact that the low output impedance of the common-collector transistor shunts the input capacitance of the common-emitter transistor. The input impedance of the emitter follower is quite high and of reduced capacitance. It therefore does not load preceding amplifiers. It has been shown that the output impedance of the common-collector stage can be considered to have an inductive term (see Fig. 73) which can be combined with the input capacitance of the common-emitter stage to yield a compensated frequency response. The frequency behavior of this type of amplifier has been calculated to be

$$A_v(s) = \frac{R_L}{R_L + 2r_b'D} \frac{s/\omega_t + 1}{\left(\frac{s}{\omega_t}\right)^2 + \frac{s}{\omega_t}\frac{R_L + \beta_o(r_b' + r_{e1})}{\beta_o(R_L + 2r_b')} + \frac{R_L + \beta_o r_b'}{\beta_o(R_L + 2r_b')D} + \frac{r_{e2}}{R_L + 2r_b'D}} \quad (48)$$

where $D = 1 + C_c\omega_t(R_L + r_{e2})$. The low-frequency gain can be shown to be

$$A_v(0) = R_L/r_{e2} \quad (49)$$

and the 3-db bandwidth

$$\omega_{3\text{ db}} = [r_{e2}/(R_L + 2r_b')D]^{\frac{1}{2}} \quad (50)$$

The performance of this amplifier can be compared with the previous types of compensation by defining a gain bandwidth normalized to one stage.

This figure of merit, mean gain-bandwidth factor, is written

$$F = [A_v(0)]^{\frac{1}{2}}\omega_{3\text{ db}}/\omega_t \quad (51)$$

which for the common-collector–common-emitter amplifier is

$$F = [R_L/(R_L + r_b')D]^{\frac{1}{2}} \quad (52)$$

Applying standard techniques, the value of R_L for maximum F can be shown to be

$$R_{L\text{ opt}} = [(2r_b')/(C_c\omega_t)]^{\frac{1}{2}} \quad (53)$$

The figure of merit of this amplifier is compared with the other compensation techniques in Fig. 74. At large gain bandwidths, the performance of the common-collector–common-emitter amplifier is exceeded only by the shunt-peaked amplifier. Note that this is the value of R_L that maximizes F, not $A_v(0)$ or $\omega_{3\text{ db}}$. High values of r_b', undesirable in most other applications, are still undesirable because the higher values of R_L restrict the high-frequency response.

4.3. Stagger Peaking with Transistor Circuits. Stagger peaking of transistor wideband amplifiers has been suggested to counteract the reduction of bandwidth of cascaded amplifiers. One writer suggested the use of RC degeneration in the emitter,[38] another the use of shunt-inductance peaking.[39] Both have one thing in common, namely, that a single stage can be described by an equation of the form

$$A_v(s) = H\frac{(s + s_2)}{(s + s_1)(s + s_3)} \quad (54)$$

A two-pole response can be obtained by locating the poles of one stage so that the zeros of the two stages will be canceled. The two poles of the remaining stage can then be located to give the desired response. For example, consider two stagger-tuned stages of shunt-inductance-peaked stages. Referring to Eq. (37) describing the shunt-peaked amplifier, the zeros will be canceled if

$$R_{L1}/L_1 = (\omega_{\beta 1}/D_1)(\beta_o r_{e1} + r_{b1}') \quad (55)$$
and
$$R_{l2}/L_2 = (r_{b1}'/L_1)(\omega_{\beta 1}/D_1) \quad (56)$$

where the subscript 1 or 2 indicates the stage of the amplifier. The performance of this amplifier is then described by

$$A_v(s) = \frac{(\omega_{t1}/D_1)(\omega_{t2}/D_2)}{s^2 + [(R_{l2} + r_{b2}')/L_2 + \omega_{\beta 2}/D_2]s + \omega_{\beta 2}/D_2[(R_{l2} + r_{b2}' + \beta_{o2}r_{e2})/L_2]} \quad (57)$$

---

[38] V. C. Grinick, Stagger-tuned Transistor Amplifiers, *IRE Trans. Broadcast Television Receivers*, vol. BTR-2, no. 3, pp. 53–56, October, 1956.

[39] L. Scott, An Integral Design Technique for Wideband Multistage Transistor Amplifiers, *UCRL* no 100009, Apr. 27, 1962.

where the frequency characteristic is determined by the coefficients of s in the denominator. The condition for a maximally flat amplitude response is

$$(R_{I2} + r_{b2}')/L_2 + \omega_{\beta 2}/D_2 = 2\omega_{\beta 2}/D_2[(R_{I2} + \beta_o r_{e2} + r_{b2}')/L_2]^{\frac{1}{2}} \quad (58)$$

The 3-db frequency is

$$\omega_{3\text{ db}} = [(\omega_{\beta 2}/D_2)(R_{I2} + \beta_{o2} r_{e2} + r_{b2}')/L_2]^{\frac{1}{2}} \quad (59)$$

from which

$$A_r(0) = (\omega_{t1}\omega_{t2})/(D_1 D_2 \omega^2_{3\text{ db}}) \quad (60)$$

The design procedure can start from a choice of bandwidth or of gain the choice of which is substituted into Eq. (60). From Eqs. (58) and (59), R_{I2} and L_2 are deter-

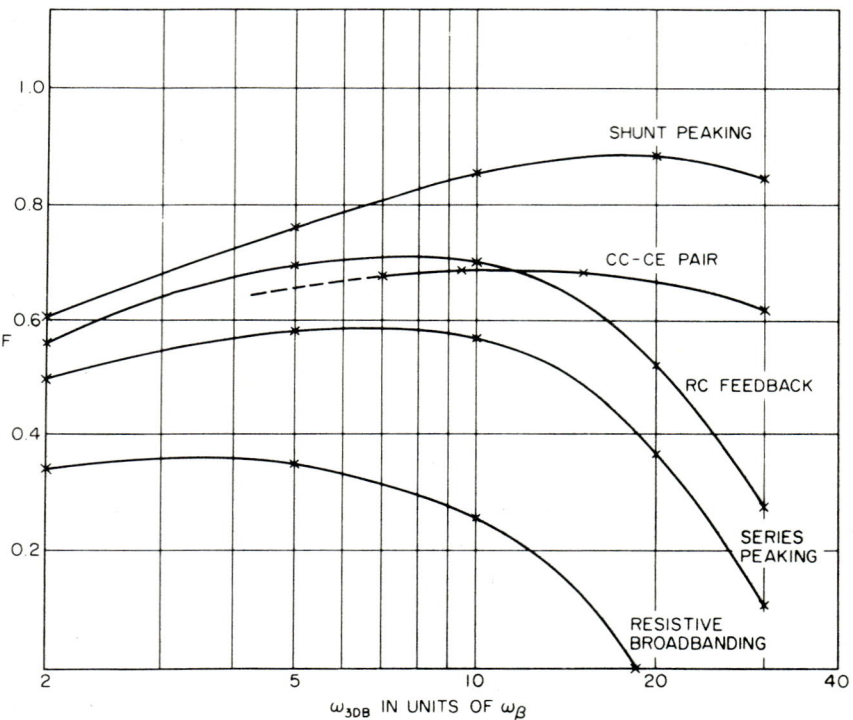

FIG. 74. Gain-bandwidth factor vs. bandwidth for several compensation amplifier pairs. (*From D. O. Pederson and R. S. Pepper, An Evaluation of Transistor Lowpass Broadbanding Techniques, 1959 IRE Wescon Conv. Record.*)

mined. These results and Eqs. (55) and (56) are used to obtain R_{L1} and L_1. Similar techniques could be applied to higher-order approximations; for example, the third-order polynomial. In this case, three zeros would have to be canceled, leaving one stage to supply two complex conjugate poles and two stages to supply three zero-canceling poles in addition to the needed pole on the negative real axis.

Many other applications of stagger peaking are possible; for example, the feedback-pair amplifiers have three poles for each feedback pairs. The pole on the negative real axis can be located so its effect is negligible, leaving a pole pair to intersperse with the three poles of a feedback pair. Details of such an amplifier have not as yet been worked out, but the techniques are indicated by knowledge of necessary pole locations and the theory of the amplifiers.

An example of the application of this technique is shown in Fig. 75. This amplifier was designed to be maximally flat to a frequency of 48 Mc and had a current amplification of 66.

4.4. Wideband Feedback Amplifiers Using Transistors. The bandwidth of amplifiers can be increased by the application of inverse feedback. With the transistors presently available, some amplifiers have been designed with bandwidths exceeding 200 Mc.

The most general approach to feedback amplifiers can be implemented by writing the equations for the circuit being considered. Bode and Blackman have developed

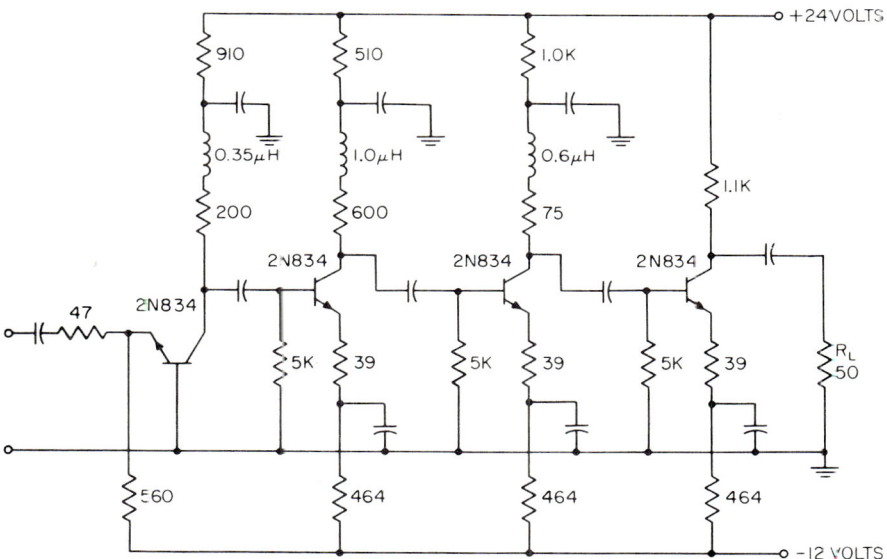

Fig. 75. Circuit diagram of amplifier designed using stagger peaking on three stages. Cutoff frequency 48 Mc, current amplification 66. (*From An Integral Design Technique for Wideband Multistage Transistor Amplifiers, UCRL-10009, Apr. 27, 1962.*)

techniques for finding the return difference and the input and output immittances. The power of this approach is that it requires that no simplifying assumptions be made through consideration of the significant and negligible terms. This technique is described in the chapter on feedback and will therefore not be considered here. The effects of feedback on the terminal impedance or admittance are, however, often important and will therefore be commented on.

Simple[40] techniques are available for making these approximations. From these it can be shown that

$$K_c = K/(1 - T) \qquad (61)$$

where K = voltage or current amplification before feedback
T = open-loop amplification of the feedback loop
K_c = voltage amplification after feedback

[40] S. S. Hakim, "Junction Transistor Circuit Analysis," John Wiley & Sons, Inc., New York, 1962.

25-58　AMPLIFIER CIRCUITS

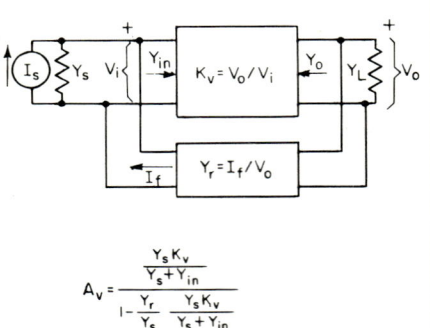

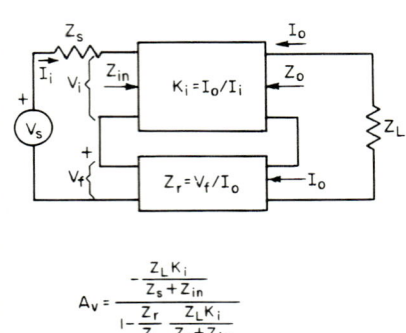

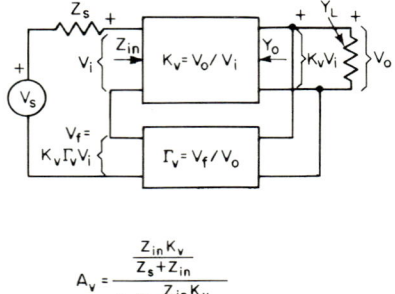

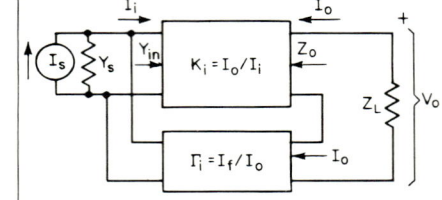

Fig. 76. Four ways of applying feedback and the influence on the amplification, input immittance, and output immittance. (*Adapted from S. S. Hakim, "Junction Transistor Circuit Analysis, John Wiley & Sons, Inc. Used with permission.*)

The terminal impedances can be described in the following terms:
 I_{tf} = terminal immittance after the application of feedback
 I_{sl} = source or load immittance
 I_t = terminal immittance before feedback
A general formula can be written describing the effect on terminal immittance,

$$(I_{tf} + I_{sl})/(I_t + I_{sl}) = 1 - T \tag{62}$$

or
$$I_{tf} = (1 - T)I_t - I_{sl}T \tag{63}$$

When the feedback to a source or load is in parallel the immittance being considered takes the convenient form of an admittance. When the feedback is in series with the source or load, the immittance being considered takes the convenient form of an impedance. In the above context, negative feedback results in an increase of the immittance while positive feedback results in a decrease in the resultant immittance. A combination of positive series feedback and shunt negative feedback can yield terminal impedance that is very nearly zero.[41] Figure 76 shows four feedback arrangements and formulas describing their performance.

Conversely, a combination of parallel positive feedback and series negative feedback could be used to obtain an extremely high input impedance. In both the above techniques of terminal impedance control, the contribution of the two feedback paths must be carefully proportioned, in particular at the frequency extremes, to ensure stability.

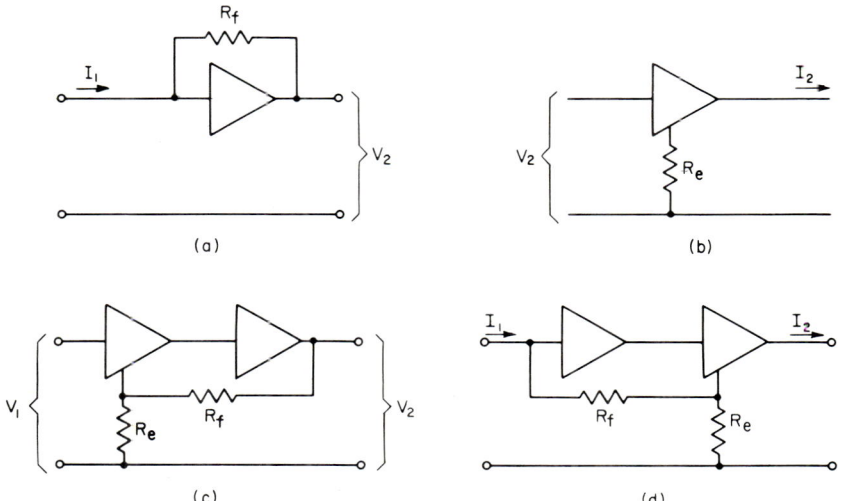

Fig. 77. Four wideband amplifier feedback arrangements. (a) Transimpedance amplifier. (b) Transadmittance amplifier. (c) Series-input parallel-output feedback amplifier. (d) Parallel-input series-output feedback amplifier. (From F. D. Waldhauer, Wide Band Feedback Amplifiers, IRE Trans. Circuit Theory, vol. CT-4, September, 1957.)

Another approach starts by considering the configuration in Fig. 77.[42,43] In this figure, each of the amplifier symbols represents a phase-reversing amplifier, the signal moving from left to right, the output to R_e having a voltage amplification of 1 with no phase reversal. It can be shown that for large amplifications the performance of each configuration at low frequencies will be dictated by the resistors and relationships between them.

The circuit of Fig. 77a behaves like a transfer resistance of value $-R_f$. This follows from the fact that if the gain is very high, then the amplifier will require very little input current to produce V_2. Therefore, most of I_1 will flow into R_f. This type of

[41] H. F. Mayer, Control of the Effective Internal Impedance of Amplifiers by Means of Feedback, Proc. IRE, vol. 27, p. 213, March, 1939.
[42] F. D. Waldhauer, Wide Band Feedback Amplifiers, IRE Trans. Circuit Theory, vol. CT-4, pp. 178–190, September, 1957.
[43] An excellent source of detailed derivations on many configurations can be found in S. S. Hakim, "Junction Transistor Circuit Analysis," John Wiley & Sons, Inc., New York, 1962.

feedback, that is, shunt at the input and the output, results in both impedances being reduced.

The circuit of Fig. 77b on the other hand, behaves like a transconductance of value $-1/R_e$. This follows from the fact that the current in R_e is very nearly I_2 and the voltage across R_e is, to a good approximation, V_1. Since series feedback is applied to the input and output circuit of the amplifier, both are decreased in admittance.

The circuit of Fig. 77c can be shown to have a voltage amplification of approximately

$$A_v = 1 + R_f/R_e \tag{64a}$$

by virtue of the fact that the voltage V_2 is proportional to $R_f + R_e$ and the voltage across R_e is proportional to V_1. Having series feedback on the input and parallel feedback on the output the input conductance is reduced as is the output impedance. In this circuit, the feedback pair takes advantage of the open-loop voltage amplification of two stages to obtain greater benefit from the inverse feedback.

Figure 77d shows the schematic diagram of another transistor feedback pair. It has dual terminal-impedance properties compared with the previous two-stage amplifier, that is, a reduced input impedance and a reduced output conductance. The

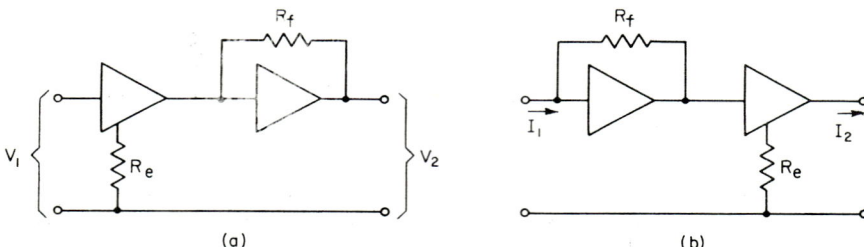

FIG. 78. (a) Voltage amplifier. (b) Current amplifier.

transistor pair, again, takes advantage of the open-loop amplification of both amplifiers' stages. By virtue of the fact that the signal voltages across R_f and R_e are equal and that I_2 equals the sum of currents through R_f and R_e, it can be shown that

$$A_i = 1 + R_f/R_e \tag{64b}$$

A cascade of a transconductance and a transresistance amplifier (Fig. 78a) has a voltage gain of approximately $A_v = R_f/R_e$. This is because the output of the first stage supplies a current and the input of the second realizes the maximum advantage from feedback if supplied with a current. It has low input admittance and low output impedance as the combination has series feedback on the input and parallel feedback on the output.

A cascade of a transresistance and a transconductance amplifier (Fig. 78b) would have a current gain of approximately $A_i = R_f/R_e$, with a low input impedance and a low output conductance. Here again, as was the case with the previous cascade, the two stages are "natural" to cascade; that is, a transistance amplifier supplies a voltage to the input of the transconductance amplifier. Further, the impedances that the two amplifier stages present each other allow each to realize maximum advantage of the individual stage inverse feedback.

The circuits of Fig. 77a and b can be extended to three, four, and more stage configurations provided stability conditions are met. Examples of the three-stage amplifier forms are shown in Fig. 79. The transfer impedance of the circuit of Fig. 79a can be shown to be

$$V_o/I_i = -R_1 R_2 (1/R_1 + 1/R_2 + 1/R_3) \tag{65}$$

Significant properties of this amplifier are low input and output impedances. The dual of the circuit of Fig. 79a is that in Fig. 79b. It can be shown to have a transfer admittance of

$$I_o/V_i = -G_1G_2(1/G_1 + 1/G_2 + 1/G_3) \tag{66}$$

This amplifier has a reduced input and output admittance. In passing, note that the T network of the circuit of Fig. 79a could be replaced with a single impedance of value

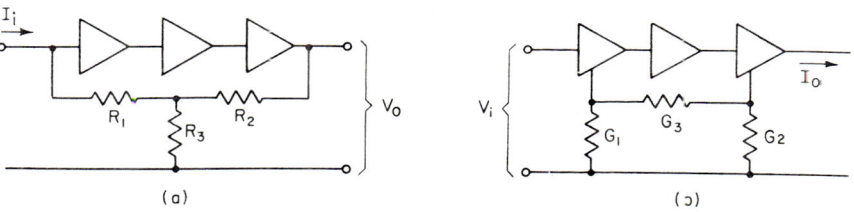

FIG. 79. (a) Three-stage transimpedance amplifier. (b) Three-stage transadmittance amplifier.

$R_1R_2(1/R_1 + 1/R_2 + 1/R_3)$ connected between input and output. Further, the network of the circuit of Fig. 79b could be replaced with a single common admittance having a value of $G_1G_2(1/G_1 + 1/G_2 + 1/G_3)$. These configurations, however, lend themselves more conveniently to the two functions of frequency equalization and stable biasing.

The general technique described in Fig. 77 has been developed analytically.[44,45] The single series-series feedback stage and the approximate equivalent circuit used are shown in Fig. 80. Assuming the collector-to-base capacitance is negligible, the

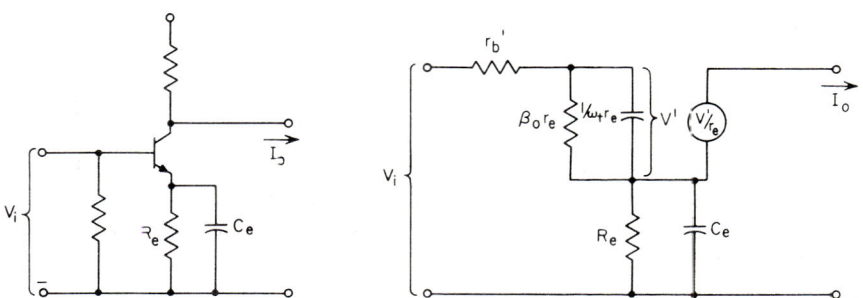

FIG. 80. (a) Circuit of series-series-feedback stage (transadmittance amplifier). (b) Equivalent circuit used in analysis. Note collector-to-base capacitance is neglected. (From E. M. Cherry and D. E. Hooper, The Design of Wide-band Transistor Feedback Amplifiers, Proc. IEE, vol. 110, no. 2, February, 1963.)

transadmittance is

$$\frac{i_o(s)}{v_i(s)} = \frac{-\alpha_o}{r_b'\beta_o + r_e + R_e} \frac{1 + sR_eC_e}{1 + s\left[R_eC_e + \dfrac{r_b'/\omega_t + R_e(1/\omega_t - R_eC_e)}{r_b'/\beta_o + r_e + R_e}\right] + s^2 \dfrac{r_b'R_eC_e/\omega_t}{r_b'/\beta_o + r_e + R_e}} \tag{67}$$

[44] E. M. Cherry and D. C. Hooper, The Design of Wideband Transistor Feedback Amplifiers, Proc. IEE, vol. 110, no. 2, February, 1963.
[45] V. G. K. Reddi, Transistor Pulse Amplifiers, Semicond. Prod., August, 1961, p. 23, also Fairchild Application Data APP 32/2.

and
$$\frac{1}{y_i} = \frac{v_i(s)}{i_i(s)} = r_b' + \beta_o \frac{r_e}{1+s/\omega_t} \frac{R_e}{1+sR_eC_e} \frac{1+s/\omega_t}{1+s\beta_o/\omega_t} \quad (68)$$

A simplifying assumption
$$R_eC_e = 1/\omega_t$$

allows the above two equations to simplify to

$$\frac{i_o(s)}{v_i(s)} = \frac{-\alpha_o}{r_b'/\beta_o + r_e + R_e} \frac{1/R_e}{1+s\dfrac{r_b'/\omega_t}{r_b'/\beta_o + r_e + R_e}} \quad (69)$$

and
$$\frac{1}{y_i} = \frac{v_i(s)}{i_i(s)} = r_b' + \frac{\beta_o(r_e + R_e)}{1+s\beta_o/\omega_t} \quad (70)$$

Significant amounts of feedback are described by the condition $R_e \gg r_b'/\beta_o + r_e$ for which the above simplify further to

$$\frac{i_o(s)}{v_i(s)} = \frac{-1/R_e}{1+sr_b'/\omega_t R_e} \quad (71)$$

and
$$\frac{1}{y_i(s)} = \frac{\beta_o R_e}{1+s\beta_o/\omega_t} \quad (72)$$

The single parallel-parallel feedback stage and the approximate equivalent circuit

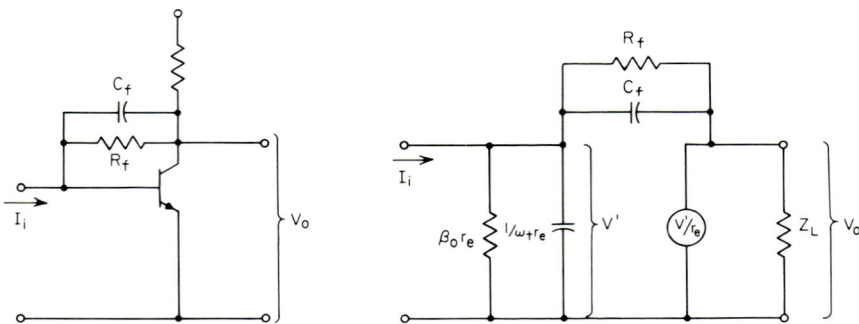

FIG. 81. (a) Schematic of parallel–parallel-feedback amplifier (transimpedance amplifier). (b) Approximate equivalent circuit used to obtain equations. (From E. M. Cherry and D. E. Hooper, The Design of Wide-band Transistor Feedback Amplifiers, Proc. IEE, vol. 110, no. 2, February, 1963.)

used are shown in Fig. 81. For the case of the general load Z_l,

$$\frac{v_o(s)}{i_i(s)} = \frac{-Z_f(s)}{1+\dfrac{Z_f(s)}{\beta(s)Z_p'(s)}} \quad (73)$$

provided $|Z_f(s)| \gg |\beta(s)r_e|$ where Z_f is the feedback impedance and
$$Z_p' = Z_l Z_f/(Z_l + Z_f)$$

For the case where Z_l consists of R_l and C_l in parallel, Z_f consists of R_f and C_f in parallel (where $C_f = C_f' + C_c$), and $\beta_o R_f R_l/(R_f + R_l) \gg R_f$, the above becomes

$$\frac{v_o(s)}{i_i(s)} = \frac{-R_f}{1+\dfrac{R_f R_l}{\beta_o R_l}} \frac{1}{1+s\left[R_fC_f + \dfrac{R_f + R_l}{R_l\omega_t} + \dfrac{R_f(C_f+C_l)}{\beta_o}\right] + s^2\dfrac{R_f(C_f+C_l)}{\omega_t}} \quad (74)$$

BROADBAND AMPLIFIERS

A rough approximation to the above, useful in the early stages of design, is

$$\frac{v_o(s)}{i_i(s)} = \frac{-R_f}{1 + sR_fC_f + s^2[R_f(C_f + C_l)/\omega_t]} \tag{75}$$

from which it follows that the two poles lie on a circle of

$$\text{Radius } |s| = \frac{1}{[R_f(C_f + C_l)/\omega_t]^{\frac{1}{2}}} \tag{76}$$

the real part of which is

$$\sigma = \frac{C_f \omega_t}{2(C_f + C_l)} \tag{77}$$

A more exact value for the above obtained from Eq. (74) is

$$\sigma = \frac{-R_fC_f + (R_f + R_l)/(R_l\omega_t) + R_f(C_f + C_l)/\beta_o}{2R_f(C_f + C_l)/\omega_t} \tag{78}$$

When a transimpedance amplifier is followed by a transadmittance amplifier, then the R_l of Eq. (74) becomes

$$R_l \approx \frac{\beta_o R_e R_c}{\beta_o R_e + R_c} \tag{79}$$

where R_c is the collector resistor of the transresistance amplifier, and Eq. (74) becomes

$$\frac{v_o(s)}{i_i(s)} = \frac{-R_f}{1 + sK + s^2R_f/\omega_t^2 R_e} \tag{80}$$

where $K = R_fC_f + (R_f + R_c)/R_c\omega_t + 2R_f/R_c\beta_o\omega_t$. A cascade of a transimpedance stage followed by a transadmittance stage is then described by

$$\frac{i_o(s)}{i_i(s)} = \frac{R_f/R_e}{1 + sr_b'/(R_e\omega_t)} \cdot \frac{1}{1 + sK + s^2R_f/R_e\omega_t^2} \tag{81}$$

If R_e is chosen sufficiently large the pole of the transadmittance amplifier will be on the negative real axis and of negligible influence. The frequency characteristic is then determined principally by the components of the transimpedance amplifier. For critical damping, that is, the two poles coincident on the negative real axis it is necessary that

$$K = (4R_f/R_e\omega_t^2)^{\frac{1}{2}} \tag{82}$$

and for the maximally flat response

$$K = (2R_f/R_e\omega_t^2)^{\frac{1}{2}} \tag{83}$$

A treatment of the transimpedance amplifier using inductive compensation has been given assuming Z_l is a resistance of low value (say 50 to 100 ohms) and the collector-to-base capacitance of the transistor is negligible. The schematic diagram and equivalent circuit used are shown in Fig. 82.

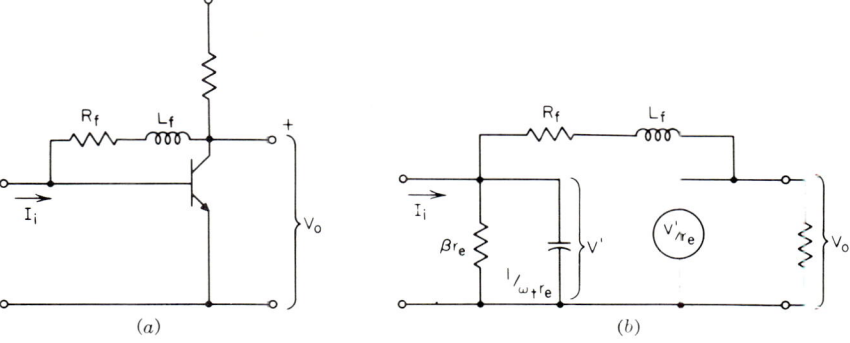

Fig. 82. (a) Circuit of inductively compensated transimpedance amplifier. (b) Equivalent circuit used in obtaining equations. (From E. M. Cherry and D. E. Hooper, The Design of Wide-band Transistor Feedback Amplifiers, Proc. IEE, vol. 110, no. 2, February, 1963.)

Fulfilling the above two conditions,

$$\frac{v_o(s)}{i_i(s)} = \frac{-R_f}{1 + \dfrac{R_f}{\beta_o R_l}} \frac{1 + s(L_f/R_f)}{1 + s\left(\dfrac{L_f}{R_f} + \dfrac{\beta_o}{\omega_t}\right)\dfrac{R_f}{R_f + \beta_o R_l} + s^2 \dfrac{\beta_o L_f}{\omega_t R_f} \dfrac{R_f}{R_f + \beta_o R_l}} \tag{84}$$

if $\beta_o R_l \gg R_f$ and if β_o is sufficiently high to make its terms negligible

$$\frac{v_o(s)}{i_i(s)} = \frac{-R_f(1 + sL_f/R_f)}{1 + sR_f/R_l\omega_t + s^2(L_f/R_f\omega_t)(R_f/R_l)} \tag{85}$$

The poles lie on a circle of diameter

$$|s| = \frac{1}{(L_f/R_l\omega_t)^{\frac{1}{2}}} \tag{86}$$

and have a real part of

$$\sigma = -R_f/(2L_f) - \omega_t/(2\beta_o) \approx -R_f/2L_f \tag{87}$$

There is also a zero at

$$\sigma' = -R_f/L_f \tag{88}$$

It is generally found that the amplification of a two-stage cascade as either a voltage or current amplifier will have a low-frequency amplification of 10 to 20 per cent

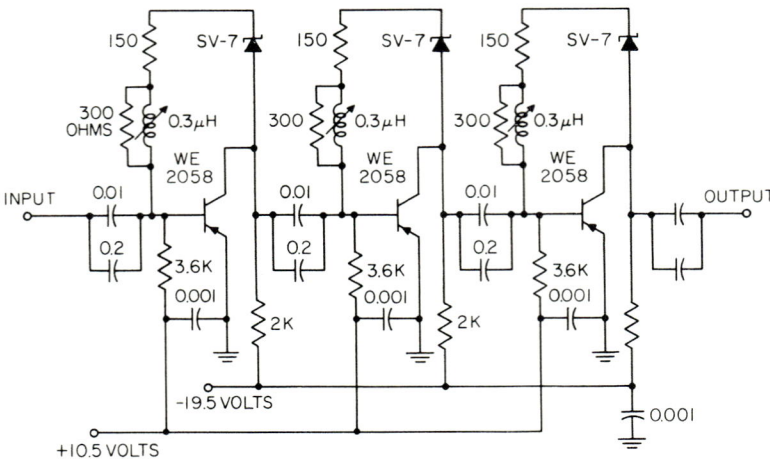

FIG. 83. Three-stage cascade of transimpedance amplifier. (*From W. E. Ballentine and F. H. Blecher, Broadband Transistor Video Amplifiers, Trans. 1959 Solid State Circuits Conf.*)

less than the ratio R_f/R_e. This can be accounted for by the r_e term as well as emitter body and contact resistance that were neglected in the theoretical development. Another factor neglected is the Miller-effect capacitance, which reduces the effective β cutoff frequency ω_β' as well as ω_t. This effect can be described by

$$\omega_\beta' = \omega_\beta/(1 + K\omega_t R_l C) \tag{89}$$

where K depends on the type of transistor, being approximately 0.5 to 0.7 for the transimpedance amplifier and 0.7 to 1.0 for the transadmittance amplifier, ω_t is the radian frequency for which $\beta = 1.0$, R_l is the stage load resistor, and C is the total capacitance from base to collector. An effective ω_t can also be described by

$$\omega_t' = \omega_t/(1 + K\omega_t R_l C) \qquad (90)$$

A number of other amplifier designs have appeared that depend on feedback around a single stage. One such amplifier uses cascaded transimpedance amplifiers using inductive compensation.[46] The schematic diagram is shown in Fig. 83 and the performance in Fig. 84. It is of some interest to note that one study revealed that the

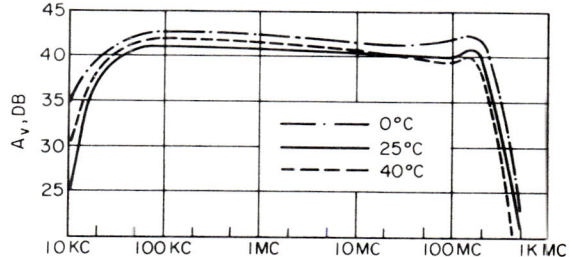

Fig. 84. Frequency-response performance of amplifier of Fig. 83. (*From W. E. Ballentine and F. H. Blecher, Broadband Transistor Video Amplifiers, Trans. 1959 Solid State Circuits Conf.*)

circuit of Fig. 85 has superior gain bandwidth[47] to the circuit of Fig. 83. Note that this circuit had an input series impedance to determine the signal current applied to the transistor base. This resistor in a sense is taking the place of a driving transadmittance amplifier as in Fig. 78b. The current-determining impedance has, however, a relatively low value, with the consequence that it would be a substantial load to previous stages. The use of alternate transadmittance and transimpedance ampli-

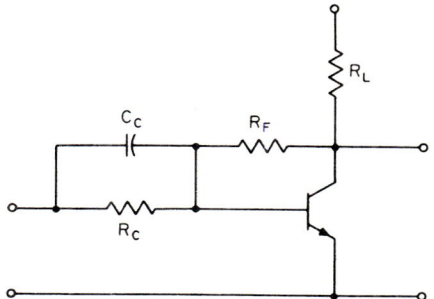

Fig. 85. Capacitance-compensated transimpedance amplifier. (*From M. S. Ghausi and D. O. Pederson, A New Feedback Broadbanding Technique for Transistor Amplifiers, Proc. Natl. Electron. Conf., 1962.*)

fiers offers the advantage of reduced interstage interaction and, therefore, simpler design.

Figure 86 is an example of a transadmittance-transimpedance cascade amplifier and Fig. 87 is the frequency response obtained. Figure 88 is the schematic diagram of a feedback pair. While comparison is not definitive since the devices are not the

---

[46] W. E. Ballentine and F. H. Blecher, "Broadband Transistor Video Amplifier," Digest of Technical Papers, 1959, Solid State Circuits Conference, pp. 42–43.

[47] M. S. Ghausi and D. O. Pederson, A New Feedback Broadbanding Technique for Transistor Amplifiers, *Proc. Natl. Electron. Conf.*, vol. 18, pp. 127–140, Oct. 8, 1962.

same, the devices were of the same class. The GB of the transconductance-transresistance amplifier was superior. The voltage amplification of the feedback pair was more definitely determined by resistor ratios than in the case of the transadmittance-transimpedance amplifier.

In very wideband amplifiers, feedback around more than one or two stages gives no advantage in gain bandwidth. This is because the needs of stability require that open-loop amplification be controlled at the high frequency in such a way that the open-loop amplification can be used to advantage only at the lower frequencies. Even

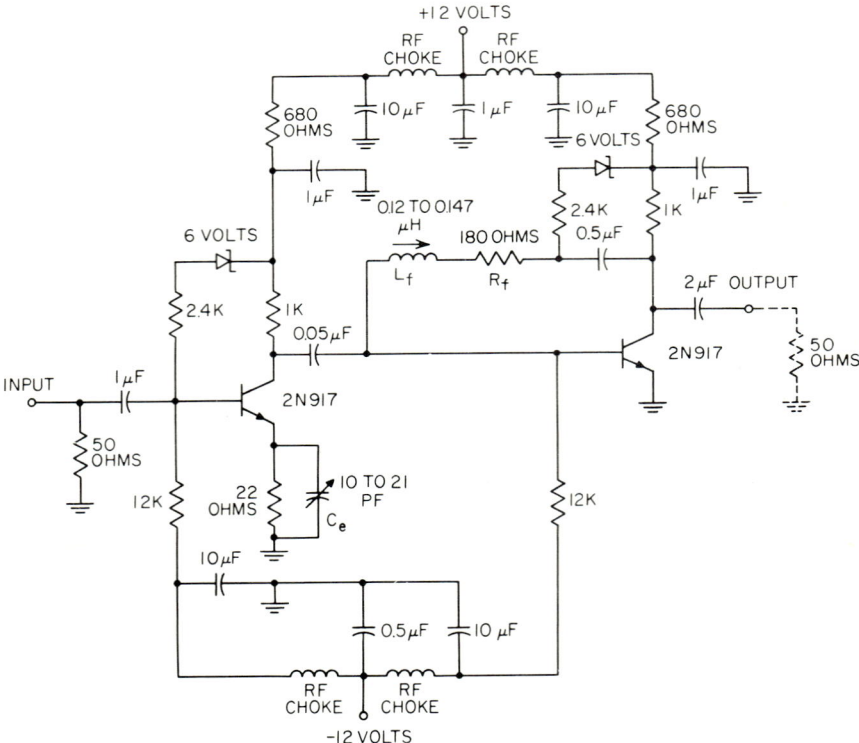

Fig. 86. Schematic of the transadmittance—transimpedance cascaded amplifier. (*From P. J. Bénéteau and J. A. MacIntosh, A 2 Nanosecond Video Amplifier, Fairchild Application Data APP-38.*)

if feedback is used around only two stages of amplification, the amplifier becomes more sensitive to small changes in capacitance. This sensitivity is illustrated by the circuit of Fig. 88. In Fig. 89, curves D and E show that a change in capacitance across the feedback resistor of only 5 pf causes a change in the regenerative peak of about 12 db. Such capacitance changes in a transconductance-transresistance amplifier would produce considerably less change in peaking at the high-frequency end of the amplifier's response.

Before going on to the root-locus technique of feedback amplifier design, several examples will be given of the results of the Bode-Nyquist technique.[48] The first of these (Fig. 90) is a typical three-stage parallel-parallel feedback amplifier fed from a

[48] W. E. Ballentine, V. R. Saari, and F. J. Witt, The Solid-state Receiver in the TL Radio System, *Bell System Tech. J.*, vol. 41, no. 6, pp. 1831–1863, November, 1962.

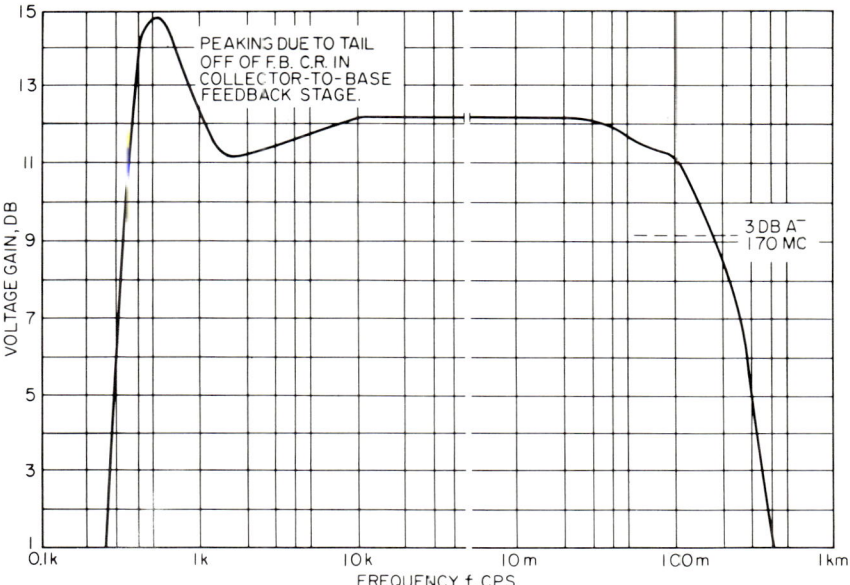

Fig. 87. Frequency response of transadmittance-transimpedance amplifier. (*From P. J. Bénéteau and J. A. MacIntosh, A 2 Nanosecond Video Amplifier, Fairchild Application Data APP-38.*)

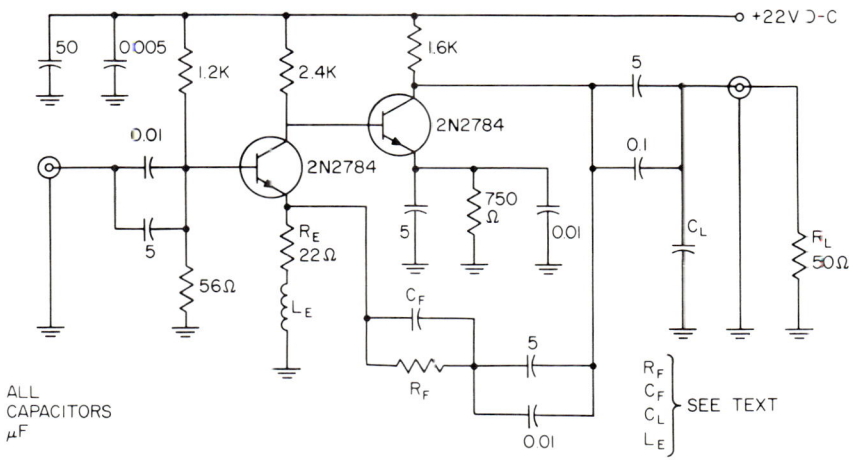

Fig. 88. Feedback pair using series-input parallel-output feedback. (*From A Wideband Amplifier Using Sylvania Micropower Planar Transistors, Circuit Loops 13, Sylvania Electric Products, Inc., Semiconductor Division.*)

current source and terminated in a 75-ohm load. Of particular interest in this circuit is the output stage of Q_3, a p-n-p germanium transistor of 15-Mc common-emitter current-gain cutoff, and Q_4, an n-p-n silicon transistor of 3-Mc common-emitter current-gain cutoff. Both transistors contribute substantially to the load current out to 6 Mc. The resistor R_1 adjusts the relative direct currents of Q_3 and Q_4.

Resistors R_6 and R_7 ensure proper load sharing by controlling their input impedances. The second stage uses local parallel-parallel feedback, the output impedance reduction of which reduces distortion caused by variation of the input impedance of the output stage under large-signal conditions. It also prevents instability of the amplifier caused by high-gain transistors in the output stage. Resistor R_5 can be varied to control the gain over a 5-db range and, R_2 is used to adjust the output voltage of Q_3 and Q_4. The variation of R_5 changes the transfer impedance of the feedback network from 13 to 25 kilohms. The output impedance is approximately 5 ohms because of the large amount of feedback. To terminate a 75-ohm line properly, a 70-ohm resistor is connected in series with the output. The current gain

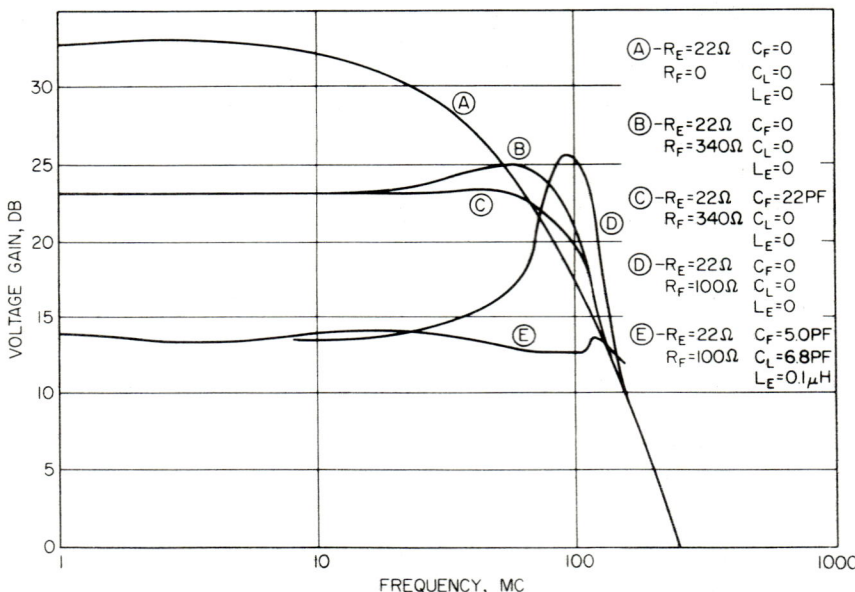

Fig. 89. Frequency response of feedback-pair amplifier. (*From A Wideband Amplifier Using Sylvania Micropower Planar Transistors, Circuit Loops 13, Sylvania Electric Products, Inc., Semiconductor Division.*)

ranges from 13,000/145 to 25,000/145. The open-loop gain and phase characteristics as well as the closed-loop response at a gain setting of 42 db for several temperatures are shown in Fig. 91.

Another three-stage transistor feedback amplifier is shown in Fig. 92. Of particular interest in this amplifier is a stability-aiding cutoff frequency (due to C_1) that goes down if the gain of transistors Q_2 and Q_3 rise. A 75-ohm resistor is added to the input because of the very low input impedance of the common-base transistor and the parallel feedback. The capacitive load of 35 pf is isolated from the feedback network by resistor R_1 with a consequent improved phase margin. The open-loop and closed-loop performance of this amplifier is shown in Fig. 93.

The design of feedback amplifiers must deal with two problems, stable operation and realization of the desired response. The determination of stability can be made by the Routh-Hurwitz, the Nyquist-Bode, or the root-locus method.

The Routh-Hurwitz criterion is a system criterion which indicates whether any poles of the amplifier, with feedback loop closed, are in the right half plane. It gives no information other than the existence or nonexistence of positive real parts of the amplifier's closed-loop poles. This information is obtained by considering the coefficients of the system's differential equations. The Nyquist-Bode techniques obtain the same information from consideration of the amplitude and phase vs. frequency response of the open-loop transfer function of the amplifier and feedback networks. The closed-loop response of the feedback amplifier can be obtained from careful consideration of the amplifier and feedback transfer functions.

The root-locus technique of feedback circuit design yields methods of directly designing for stable operation and a desired transfer-response function. This follows

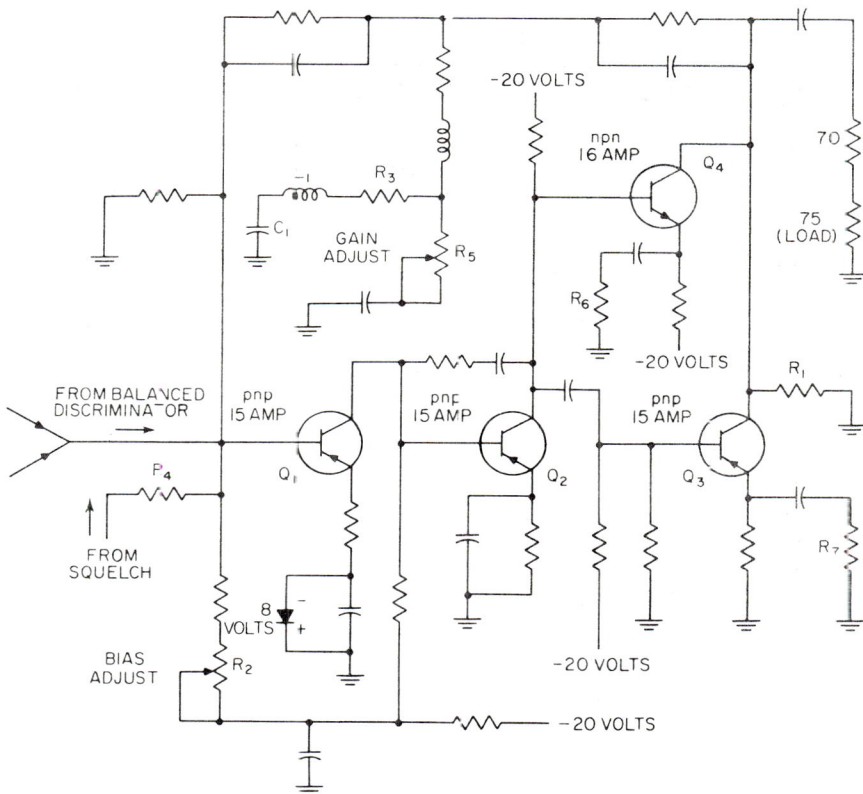

FIG. 90. Three-stage parallel-input parallel-output feedback amplifier. The output transistors in this amplifier take advantage of complementary symmetry. (*From W. E. Ballentine, V. R. Saari, and F. J. Witt, The Solid-state Receiver in the TL Radio System, Bell System Tech. J., November, 1962, reprinted with permission.*)

from the fact that it is concerned with the paths of the poles and zeros of the amplifier's closed-loop frequency response as the amount of feedback is changed. The root-locus technique will therefore be given consideration in direct application to video amplifiers.[49]

The performance of a feedback amplifier can be described by the formula

$$A(s) = \frac{\mu(s)}{1 - B(s)\mu(s)} = \frac{\mu(s)}{1 + T(s)} \qquad (91)$$

[49] M. S. Ghausi and D. O. Pederson, A New Design Approach for Feedback Amplifiers, *IRE Trans. Circuit Theory*, vol. CT-9, no. 3, pp. 274–284, September, 1961.

$\mu(s)$ is assumed to be the unilateral voltage or current amplification and a function of complex frequency s. $B(s)$ is also a function of complex frequency and is the proportion of the output voltage or current that is fed back. It can be bilateral or unilateral. Equation (91) does not indicate anything about impedance-transforming properties of the feedback. When $B(s) = 0$, the poles and zeros of A are the poles and zeros of $\mu(s)$. As the scaling portion of $B(s)$ increases, the zeros and poles move in the s plane.

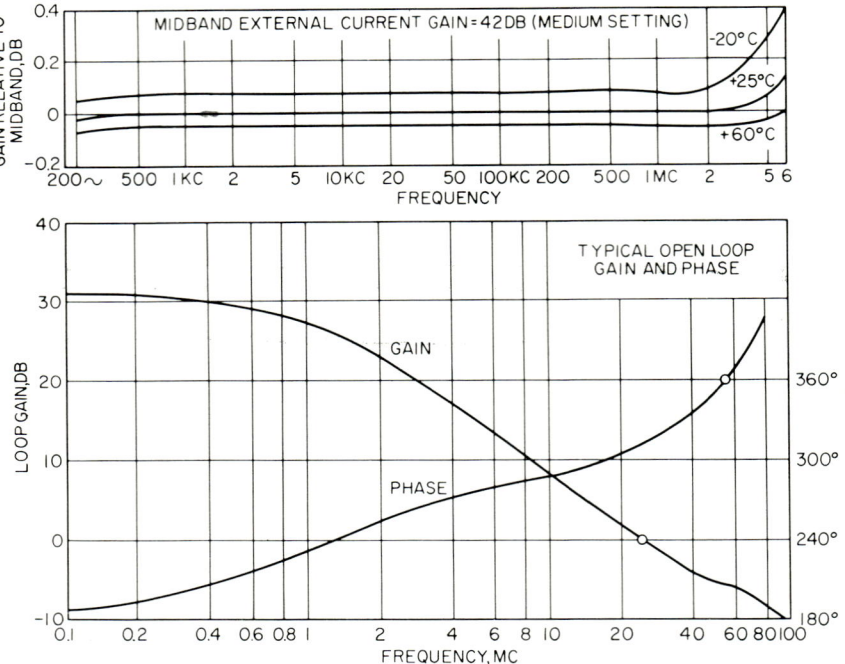

FIG. 91. Frequency-response performance of amplifier of Fig. 90. (*From W. E. Ballentine, V. R. Saari, and F. J. Witt, The Solid-state Receiver in the TL Radio System, Bell System Tech. J., November, 1962, reprinted with permission.*)

The paths of these poles and zeros can be described by the following root-locus construction rules:

1. The loci start from the poles of $T(s)$.
2. The loci are symmetrical about the real axis.
3. The loci on the real axis lie to the left of an odd number of poles or zeros of $T(s)$.
4. The loci which terminate on the zeros of $T(s)$ at infinity approach infinity on lines having an angle of $n\pi/\gamma$, where $n = 1, 3, 5$ and γ is the number of finite poles minus the number of finite zeros. The asymptotes intersect the real axis at σ_o where

$$\sigma_o = \frac{\Sigma \text{poles} - \Sigma \text{zeros}}{\gamma}$$

5. The breakaway point σ_b from the real axis is found by the use of argument $[B(s)\mu(s)] = -180°$ and the use of small-angle approximations.
6. For the arrival angle at the zero z_x,

$$a = \sum_{i=1}^{m} \arg(z_x - p_i) - \sum_{\substack{j=1 \\ j \neq x}}^{n} \arg(z_x - z_j) \pm 180°$$

BROADBAND AMPLIFIERS

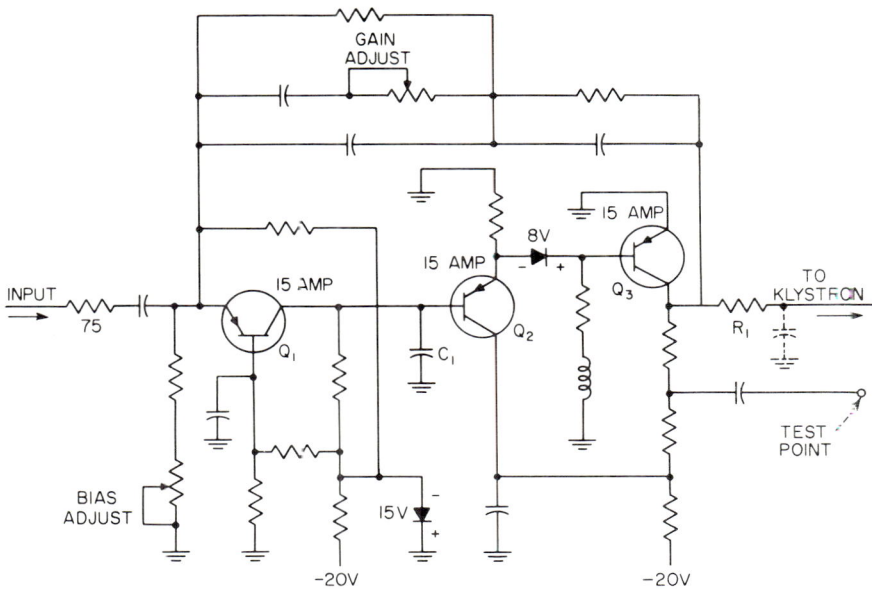

Fig. 92. Three-stage amplifier using parallel-feedback input and parallel feedback at output. Capacitor C_1 produces self-adjusting high-frequency cutoff in open-loop response. (*From W. E. Ballentine, V. R. Saari, and F. J. Witt, The Solid-state Receiver in the TL Radio System, Bell System Tech. J., November, 1962, reprinted with permission.*)

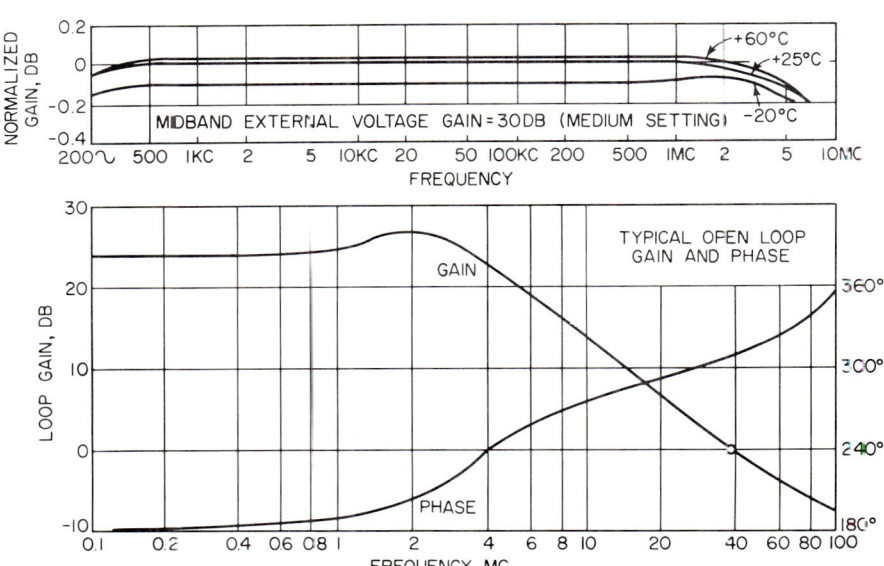

Fig. 93. Open- and closed-loop frequency behavior of circuit in Fig. 92. (*From W. E. Ballentine, V. R. Saari, and F. J. Witt, The Solid-state Receiver in the TL Radio System, Bell System Tech. J., November, 1962, reprinted with permission.*)

The angle of departure from a pole p_x is given by

$$d = \sum_{j=1}^{n} \arg (p_x - z_j) - \sum_{\substack{i=1 \\ i \neq x}}^{m} \arg (p_x - p_i) \pm 180°$$

where p_i are the finite poles of $B(s)\mu(s)$, and z_j are the finite zeros of $B(s)\mu(s)$; n and m are the number of finite zeros and poles, respectively.

The application of these rules is best illustrated by an example. Consider for this purpose the three-stage amplifier of Fig. 94. The load impedance of the first and second stage can be designed using shunt inductive peaking to increase the gain bandwidth. This compensation technique offers a very large range of compensation, of which certain choices are more desirable in specific situations. The examples to be described are neither exhaustive nor optimum but rather examples that are indicative.

The simplest feedback network in the example under consideration is shown in Fig. 94. Increased gain and/or bandwidth can be obtained by shunt-inductive peak-

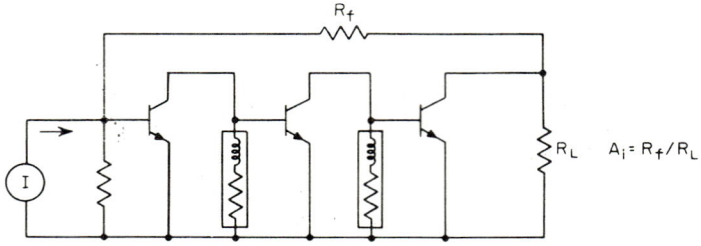

Fig. 94. Three-stage amplifier with parallel-parallel feedback. (*From M. S. Ghausi and D. O. Pederson, A New Design Approach for Feedback Amplifiers, IRE Trans. Circuit Theory, vol. CT-9, no. 3, September, 1961.*)

ing of two stages. Pole-zero cancellation is used. The remaining stage is narrowed with a shunt capacitor. The return difference for such an amplifier can be described by

$$T(s) = \frac{T(0)(-s_1)(-s_2)^2}{(s - s_1)(s - s_2)^2} \qquad (92)$$

Since $T(s)$ has no finite zero, the poles of $T(s)$ are not restricted in their movement in the complex plane. A consequence of this is that a relatively small amount of T results in the poles of A reaching their desired positions. If R_F is shunted with a capacitor, say C_F, it can be shown that

$$T(s) = \frac{T(0)(-s_1)(-s_2)^2(s - z_1)}{(-z_1)(s - s_1)(s - s_2)^2} \qquad (93)$$

For this case by the conditions resulting in the root-locus behavior, one of the poles of $T(s)$ will terminate at z_1 where $z_1 = -1/(R_f C_f)$. This restriction on the root locus of one of the poles of $T(s)$ allows the application of larger amounts of feedback before instability results. The zero of $T(s)$ does not appear in $A(s)$. This zero is therefore called a *phantom zero*. It can be shown that design of an amplifier with complex phantom zeros results in desensitivity of both low-frequency and band-edge frequency response as well as a larger level of feedback without instability. The band-edge ferquency-response desensitivity is a consequence of termination of certain loci very nearly at the complex-phantom-zero frequencies. It is necessary that the poles and zeros determining the frequency response be dominant; that is, the loci of other roots

BROADBAND AMPLIFIERS

are far from the phantom zero. Such a return difference could be described by

$$T(s) = T(0) \frac{(-s_1)(-s_2)^2}{(s-s_1)(s-s_2)^2} \frac{(-s_f)(s-z_1)(s-\bar{z}_1)}{|z_1|^2(s-s_f)} \quad (94)$$

where z_1 and $\bar{z}_1$ are the complex conjugate phantom zero and s_f is the pole of the $B(s)$ network. The resultant gain characteristic would then be written as

$$A_I = H \frac{(s-s_f)}{(s-s_a)(s-\bar{s}_a)}$$

these being the dominant roots.

A three-pole maximally flat response can be obtained in the case of the real phantom zero if the zero at z_1 cancels one of the poles at s_2. A pole of $A_I(s)$ is nevertheless obtained at z_1. The other two poles of $A_I(s)$, s_a, and $\bar{s}_a$ should lie at $z_1/2 \pm j(\sqrt{\frac{3}{4}} z_1)$. The absolute value of the radius from the origin to all the poles is the same and equal to the radian cutoff frequency

$$\omega_{3\,db} = |z_1|$$

The loci of s_1 and s_2 where $s_1 \ll s_2$ move toward each other on the real axis meeting at the breakaway point,

$$\sigma_b = (s_1 + s_2)/2 \quad (95)$$

from which they move to the positions s_a and s_b. $T(0)$ can be found from Eq. (95) by evaluating the equation at s_a or s_b on the locus $|T(s)| = 1$, taking into consideration the conditions $|s_1| \ll |s_2|$, $\sigma_b \approx s_2/2$, and $s_2 = z_1$. The value obtained is

$$T(0) = \left|\frac{z_1}{s_1}\right| \left[\left(1 - \left|\frac{\sigma_b}{z_1}\right|\right)^2 + 3\left|\frac{\sigma_b}{z_1}\right|^2\right]^{\frac{1}{2}} \left[\left(\frac{|s_1| - |\sigma_b|}{|z_1|}\right)^2 + 3\left|\frac{\sigma_b}{z_1}\right|^2\right]^{\frac{1}{2}} \approx \frac{\omega_{3\,db}}{|s_1|} \quad (96)$$

It should be noted that the effects of the input and output admittance of the feedback network have been neglected.

A bridged T such as Fig. 95 can be designed to give the needed complex zeros from the following equations:

$$y_{12} = -H \frac{s^2 + \delta s + 1}{s + \gamma} \quad (97)$$

where the resistance R_1 is normalized to unity and the frequency is normalized such that the zeros have unit magnitude. Then $\delta = 2 \cos \psi$ and $\gamma = s_f/z_1$ where ψ is the angle of the zero radial to the negative real axis and γ is the negative real pole. Physical realizability requires $\delta > \gamma$ and $1/\gamma - (\delta - \gamma) > 0$. For these conditions and if $R_1 = 1$, $C_1 = \gamma/2$, $C_2 = 2/\gamma - 2(\delta - \gamma)$, $R_2 = (\delta - \gamma)/C_2$. The effects of the input and output admittance of the feedback network will be neglected for the present. A useful set of values are in the following table.

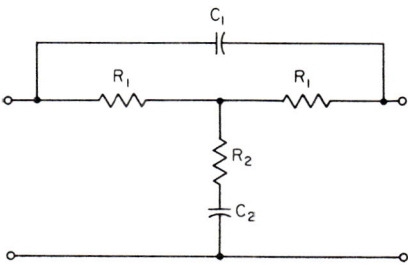

FIG. 95. Bridged-T feedback network to yield complex phantom zero. (From M. S. Ghausi and D. O. Pederson, A New Design Approach for Feedback Amplifiers, IRE Trans. Circuit Theory, vol. CT-8, no. 3, September, 1961.)

ψ	γ	$\omega_{3\,db}$
45°	0°	1.0
35	1.2	1.2
30	1.0	1.27
25	0.88	1.35

25-74 AMPLIFIER CIRCUITS

For general purposes, the reverse short-circuit transfer admittance is

$$y_{12} = -C_1 \frac{s^2 + s\dfrac{2R_1C_1 + R_2C_2}{R_1{}^2C_1C_2 + 2R_1R_2C_1C_2} + \dfrac{1}{R_1{}^2C_1C_2 + 2R_1R_2C_1C_2}}{s + 2/(2R_2C_2 + R_1C_1)} \tag{98}$$

In designing the above network for the maximally flat magnitude response, the zeros of the above network z_1 and $\bar{z}_1$ must be predistorted in consideration of the fact that the loci do not terminate on the above zeros for a finite amount of gain. Further, the root s_f should be chosen slightly less than the normalized value $\gamma = s_f/|z_1|$ as a part of this predistortion.

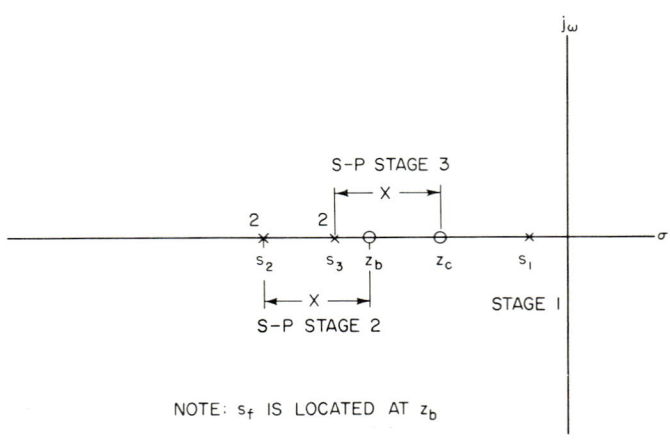

FIG. 96. Open-loop two-zero locations for example being considered. (*From M. S. Ghausi and D. O. Pederson, A New Design Approach for Feedback Amplifiers, IRE Trans. Circuit Theory, vol. CT-9, no. 3, September, 1961.*)

As an example, predistortion would require a 10 per cent increase in $|z_1|$ and a 10 per cent decrease in s_f. For instance, if $\psi = 35°$ is chosen, then the predistorted values of $|z_1|$ and s_f are $|z_1| = 1.1(\omega_{3\,\text{db}}/1.2) \approx 0.9\omega_{3\,\text{db}}$ and $s_f = 0.9(1.2|z_1|) \approx \omega_{3\,\text{db}}$.

Continuing with this example in determining the poles and zeros of the amplifiers, let the first stage be an uncompensated common-emitter amplifier with $s_1 = \omega_\beta$. Next, the pole-zero separation of the shunt-peaked stages can be found from the arrival angle of approach to z_1, $\bar{z}_1$ combined with a cut-and-try solution, thus yielding the maximum basic amplifier gain $A_i(0)$. For simplicity z_b, the second-stage pole, is chosen equal to s_f and it is separated from the second-stage zero by $x = 0.395\omega_{3\,\text{db}}$. For $\psi = 35°$ $z_c = -0.855\omega_{3\,\text{db}}$ with the same zero separation factor x as the second stage. The pole-zero locations for the three stages are tabulated below, and their locations and paths are illustrated in Figs. 96 and 97.

Stage	Poles	Zeros
1	$-\omega/D$	
2	$-1.39\omega_{3\,\text{db}}$ (double)	$-\omega_{3\,\text{db}}$
3	$-1.25\omega_{3\,\text{db}}$ (double)	$-0.855\omega_{3\,\text{db}}$

A number of second-order effects must now be considered. Consider the loci in the illustration of Fig. 97. The poles pair at -1.39ω is shown moving almost vertically off the figure. For very large values of $T(0)$ these nondominant poles will move to the right and if $T(0)$ is sufficiently large will cross the $\pm j\omega$ axis, resulting in instability. This then limits the upper value of $T(0)$. Another nondominant pole due to

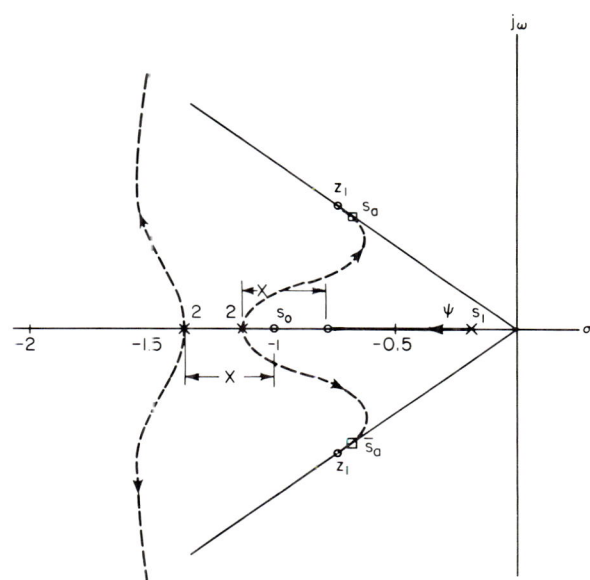

$$T(s) = T(0) \frac{(4.2\omega_B)(s+1)(s+0.855)[(s+0.75)^2+(0.525)^2]}{(s+\omega_B)(s+1.39)^2(s+1.25)^2(s+1)}$$

$$A(s)\bigg|_{s \leq |z_1|} \approx H \frac{(s-s_0)}{(s-s_0)(s-\bar{s}_0)}$$

FIG. 97. Locus of poles in the example being considered. (*From M. S. Chausi and D. O. Pederson, A New Design Approach for Feedback Amplifiers, IRE Trans. Circuit Theory, vol. CT-9, no. 3, September, 1961.*)

the effect of C_c and the effect of excess phase has been ignored. It can be shown that this equivalent pole lies at

$$-D\omega_t/(m + D - 1) \tag{99}$$

where m is the radians of excess phase of $\beta(j\omega)$ at ω_t. This will result in a pole for each stage far out on the negative real axis. In one example considered these poles were represented as a triple pole at $-17\omega_{3 \text{ db}}$.

There are additional nondominant poles and zeros in $T(s)$ due to y_{11} and y_{22} of the feedback network for the complex-phantom-zero design. These can be approximated by $2R_1$ in parallel with C_1. At the amplifier output for $B(0) \ll 1$, the effect of $2R_1$ on R_L is usually negligible but C_1 produces a nondominant pole at $-1/R_L C_1$. At the input of the amplifier there are two poles at $-1/[C_1(r_b' + R_{in}) + C_{in}R_{in}]$, $-1/(C_s \bar{r}_b)$, where $C_s = C_1 C_{in}/(C_1 + C_{in})$, C_{in} is the base-emitter capacitance of the input tran-

25-76 AMPLIFIER CIRCUITS

sistor, and $R_{in} = \beta_o r_e$ of the input transistor. In the above, C_1 could include additional capacitance at the transistor input. Also $2R_1$ is much larger than R_{in} or r_b' where $r_b' < R_{in}$. If $C_1 \ll C_{in}$ then

$$-1/[C_1(r_b' + R_{in}) + C_{in}R_{in}] \approx -\omega_\beta/D \tag{100}$$

and is highly dominant compared with the other natural frequencies of the input circuit. It can, nevertheless, be neglected in most cases.

Examples of the application of several of the root-locus techniques are tabulated below.

| | Resistive feedback | Real phantom zero | Complex phantom zero ||
			Pole-zero cancellation	Double pole, no cancellation
Closed-loop gain $A_I(0)$	100	100	100	100
3-db BW of $A_I(j\omega)$, radians/sec	10^7	10^7	10^7	10^7
Amount of feedback $1 + T(0)$	17.6 (24.9 db)	100 (40 db)	23.3 (27.3 db)	37.5 (31.5 db)
Feedback bandwidth ω_f, radians/sec	3.6×10^5	10^5	10^6	10^6
$T(0)\omega_f/\omega_t$	0.061	0.1	0.233	0.375

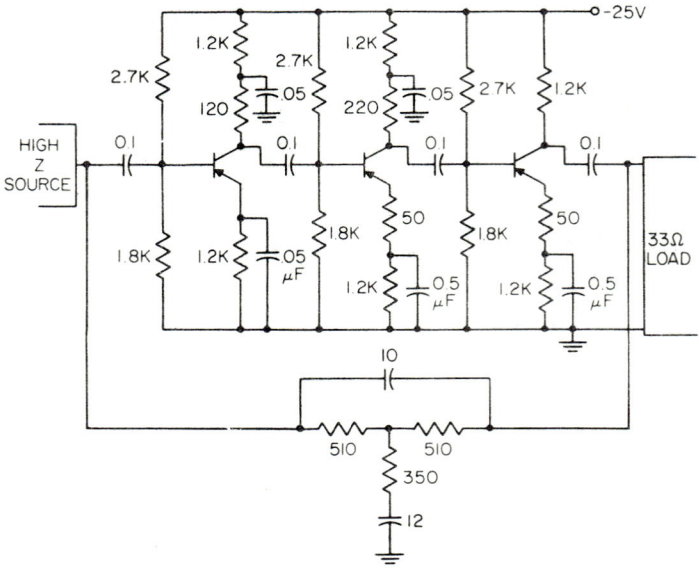

FIG. 98. Circuit diagram of wideband phantom-zero feedback amplifier. (*From M. S. Ghausi and D. O. Pederson, A New Design Approach for Feedback Amplifiers, IRE Trans. Circuit Theory, vol. CT-9, no. 3, September, 1961.*)

This table shows the greater $T(0)\omega_f/\omega_t$ obtained using double-pole shunt peaking in two stages with no zero or pole cancellation. Another illustration follows in which high-frequency transistors were used.

The averaged parameters of these transistors, 2N1142 (germanium), were $f_t \approx 400$ Mc, $C_{cb} \approx 2.4$ pf, $\beta_o \approx 50$, $r_b' \approx 75$ ohms, at $V_{CE} = 7$ volts and $I_C \approx 5$ ma. The schematic diagram and frequency response of this amplifier are shown in Figs.

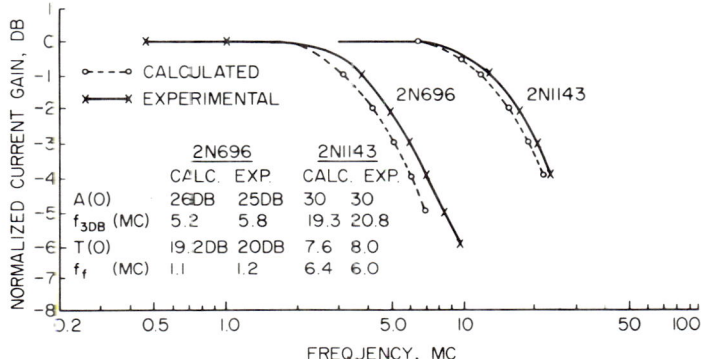

Fig. 99. Amplitude-frequency performance of circuit of Fig. 98. (*From M. S. Ghausi and D. O. Pederson, A New Design Approach for Feedback Amplifiers, IRE Trans. Circuit Theory, vol. CT-9, no. 3, September, 1961.*)

98 and 99. Also shown is the performance of a similar amplifier using 2N696 transistors characterized as $f_t \approx 50$ Mc, $C_{cb} \approx 20$ pf, $\beta_o \approx 30$, at $V_{CE} \approx 10$ and $I_C \approx 5$ ma.

Another technique for designing feedback amplifiers in a direct manner to realize a desired closed-loop response starts by assuming a unit output current or voltage with a chosen type of feedback amplifier.[50] As an example consider the circuit of Fig. 100. Considering first a single stage, assuming a unit output current and computing the

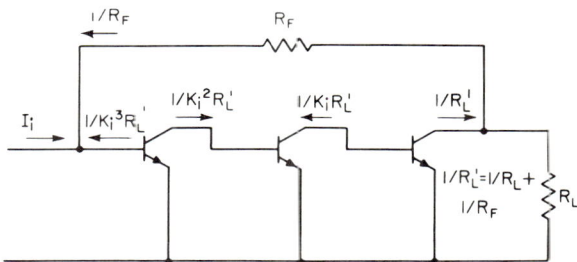

Fig. 100. Three-stage feedback amplifier with parallel-parallel feedback. (*From F. D. Waldhauer, Transistor Circuit Design—A Status Report, NEREM Record 1963, reprinted with permission.*)

input current one obtains

$$i_{in} = 1/\beta_o + s/\omega_1 \tag{101}$$

where $1/\omega_1 = 1/\omega_t - R_L C_c$. The performance of the three stages before feedback can be described by

$$1/K_i{}^3 = 1/\beta_o{}^3 + 3s/\beta_o{}^2\omega_1 + 3s^2/\beta_o\omega_1{}^2 + s^3/\omega_1{}^3 \tag{102}$$

[50] F. D. Waldhauer, Transistor Feedback Design, paper TAM 7.2, *NEREM Record*, 1963.

The performance of the amplifier after feedback can be described by

$$-Z_{12} = K_1^3 \frac{R_L R_F}{R_L + R_F} = K_1^3 R_L' = \frac{1}{a_o + a_1 s + a_2 s^2 + a_3 s^3} \qquad (103)$$

where the constants depend on the desired performance, that is, maximally flat amplitude, maximally flat delay, etc. The constants of the polynomial can be adjusted by circuit elements in the feedback network as shown in Fig. 101.

The a_o term in the polynomial is altered by $1/R_F$, the a_1 term by C_F, the a_2 term by

$$K_s^2 = C_2^2 R_1 s^2 / (1 + 2R_1 C_2 s) \approx C_2^2 R_1 s^2 \qquad (104)$$

and the a_3 term is unaltered by the feedback elements being still determined by the transistor parameters and load resistor.

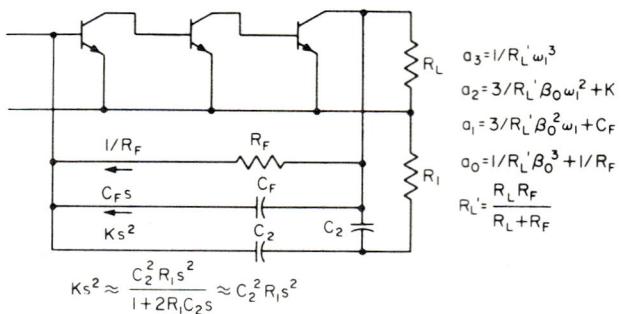

Fig. 101. Three-stage feedback amplifier with components added to give desired frequency response. (*From F. D. Waldhauer, Transistor Circuit Design—A Status Report, NEREM Record 1963, reprinted with permission.*)

In terms of the circuit elements, the transfer impedance of the above circuit can be written

$$Z_{12}(s) = \frac{-R_F}{\left(\dfrac{R_L + R_F}{R_L \beta_o^3} + 1\right) + \left[\dfrac{3(R_L + R_F)}{R_L \beta_o^2 \omega_1} + C_1 R_F\right] s + \left[\dfrac{3(R_L + R_F)}{R_L \beta_o \omega_1^2} + C_2^2 R_1 R_F\right] s^2 + \dfrac{R_L + R_F}{R_L \omega_1^3} s^3} \qquad (105)$$

Each of the above polynomial coefficients can be equated to the value of the coefficient required to obtain the desired response. The four equations can then be solved to determine necessary values of R_F, $C_1 C_2$, and R_1.

4.5. Transistor Distributed Amplifiers. The design of transistor distributed amplifiers is more difficult than it is for electron tubes as the transistor input impedance is quite low. It often has a value approximately that of the constant-K transmission line at a frequency at which amplification is desired. Further, the amplification parameter is also a function of frequency.

Several approaches have been used to treat the above problem. The simplest[51] approach is to connect an RC network in series with the base of each transistor with a time constant equal to $\beta_o/2\pi f_t$. This reduces the effective β of each transistor, making it approximately independent of frequency up to about $\frac{1}{2} f_t$. The voltage amplification of such an amplifier is

$$A_v = (n\beta_o R_o / 2R)^m \qquad (106)$$

[51] P. J. Bénéteau and L. Blaser, A 175 MC Distributed Amplifier Using Silicon Mesa Transistors, *Fairchild Application Data* APP-14.

where m is the number of stages, n is the number of transistors per stage, R is the resistance in series with the base, and R_o is the characteristic impedance of the lumped transmission at zero frequency. Figure 102 shows the circuit of one stage of such an amplifier using 2N706 transistors, and Fig. 103 is the frequency response. The rising high-frequency impedance of the constant-K networks helps to keep the frequency

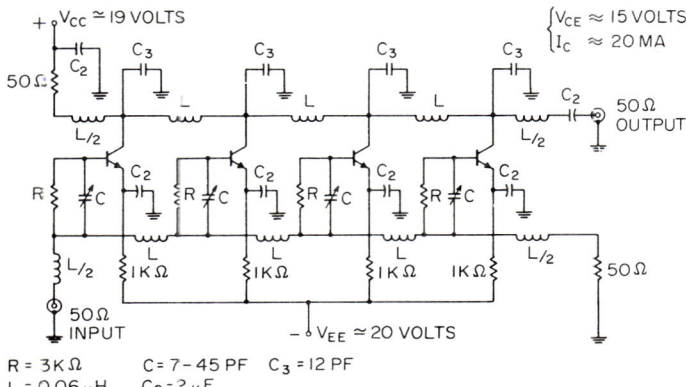

Fig. 102. One stage of distributed amplifier. (*From P. J. Bénéteau and L. Blaser, A 175 MC Distributed Amplifier Using Silicon Mesa Transistors, Fairchild Application Data APP-14.*)

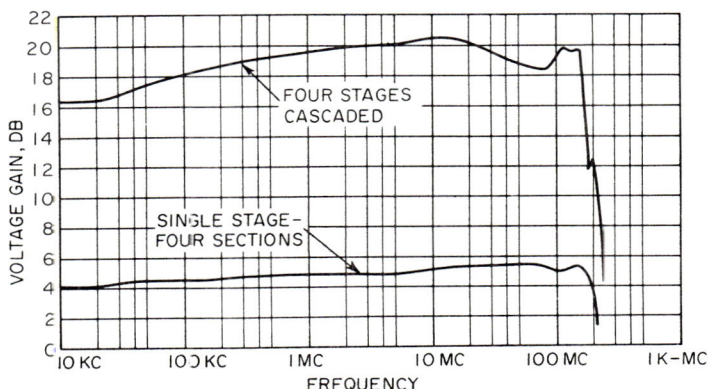

Fig. 103. Frequency response of three cascaded stages of previous circuit. (*From P. J. Bénéteau and L. Blaser, A 175 MC Distributed Amplifier Using Silicon Mesa Transistors, Fairchild Application Data APP-14.*)

response flat; otherwise the reduced impedance of the transistor base RC would lower the high-frequency response as it loads the lumped transmission line. The transient response was not so good as would be anticipated from the frequency response. Negative mutual coupling as described in the section on tube distributed amplifiers would have yielded improved pulse response.

Another approach starts with the hybrid π equivalent circuit,[52] a useful version of

[52] L. F. Roeshot, UHF Broadband Transistor Amplifiers, *Elec. Design News*, January, 1963, pp. 50–60; February 1963, pp. 24–29; March, 1963, pp. 84–89.

which is shown in Fig. 104. From this circuit a G_{in} and B_{in} (ωC_{in}) are used to describe the transistor input parameters. Both of these change considerably with frequency. The conductance and capacitance are then added to conductance G_k and capacitance C_k, respectively, the shunt conductance and capacitance of the constant-K line.

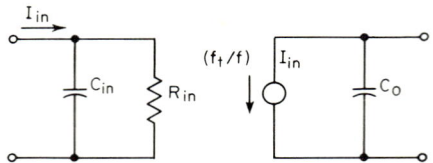

FIG. 104. Equivalent circuit used in distributed amplifier. (*From L. F. Roeshot, UHF Broadband Transistor Amplifiers, Elec. Design News, January, 1963, to March, 1963, reproduced with permission.*)

These are used to describe an input Q for each transistor.

$$Q_{in} = \omega(C_{in} + C_k)/(G_{in} + G_k) \tag{107}$$

A frequency f_Q is then defined as the frequency of minimum Q. When using transistors with an $f_t > 500$ Mc, $f_Q \approx f_t/3$. Figure 105 shows a plot of Q_{in} as a function of

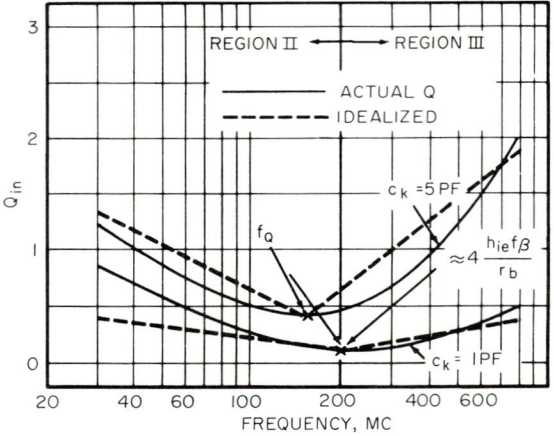

FIG. 105. Q of transistor input circuit vs. frequency. (*From L. F. Roeshot, UHF Broadband Transistor Amplifiers, Elec. Design News, January, 1963, to March, 1963, reproduced with permission.*)

frequency. Approximating the actual curves with straight lines, Q_{in} may be described below f_Q by

$$Q_{in} = (f_Q/f)Q_Q \tag{108}$$

where $Q_Q = Q_{in}(\omega_Q) \approx \omega_Q r_b'(C_{in} + C_k)$, and above f_Q by

$$Q_{in} = (f/f_Q)Q_Q \tag{109}$$

From these the attenuation per transmission-line section, the coil losses being negligible,

$$a = [(f/f_c)/Q_{in}][1 - (f/f_c)^2]^{-\frac{1}{2}} \tag{110}$$

When $f_\beta < f < f_t$, the voltage amplification is

$$A_v = (f_t/f)Z_o \qquad (111)$$

Since for a low-pass constant-K filter

$$Z_o = R_o/[1 - (f/f_c)^2]^{\frac{1}{2}} \qquad (112)$$

and half the signal is consumed in a termination at one end of the transmission line, the lossless voltage amplification

$$A_L = f_t R_o/\{2f[1 - (f/f_c)^2]^{\frac{1}{2}}\} \qquad (113)$$

which does not consider the losses in the transmission line. The losses are dealt with by considering three frequency regions:

I $\quad f_\beta > f_c$
II $\quad f_\beta < f_c < f_Q$
III $\quad f_c > f_Q$

In region I, the transistor input impedance is high and constant. The amplifier

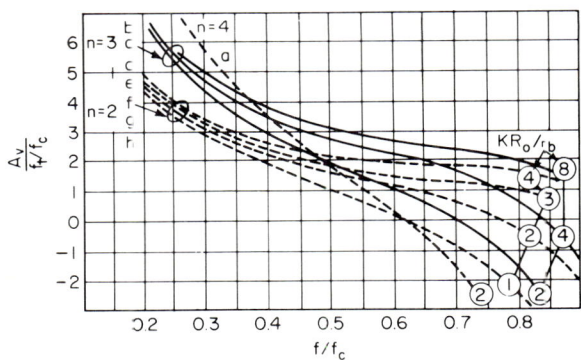

Fig. 106. Normalized gain vs. normalized frequency in region II. (*From L. F. Roeshot, UHF Broadband Transistor Amplifiers, Elec. Design News, January, 1963, to March, 1963, reproduced with permission.*)

behaves like a tube amplifier and supplies high-voltage amplification. In regions II and III, the voltage amplifications vary and can be described by

$$A_{vII} = A_L n(1 - nK_{II}R_o(f/f_c)^2/\{4r_b'[1 - (f/f_c)^2]^{\frac{1}{2}}\}) \qquad (114)$$
and $\quad A_{vIII} = A_L n(1 - nK_{III}R_o/\{4r_b'[1 - (f/f_c)^2]^{\frac{1}{2}}\}) \qquad (115)$

where $K_{II} = (f_c/f_Q)^2$ and $K_{III} = 1$. Figures 106 and 107 are curves for several values of n and KR_o/r_b'.

The following example illustrates the design procedure. Starting with the high-frequency cutoff frequency f_c choose the number of transistors per stage $n = 3$ using transistors with $f_t = 500$ Mc and $f_Q = 150$ Mc. Calculate K_{II}

$$K_{II} = (f_c/f_Q)^2 = 4$$

For $n = 3$, select a normalized frequency-response characteristic in region II, in this example curve d. From $f_Q = 150$ Mc and $f/f_c = 0.5$ (Fig. 106) the normalized voltage amplification $A_v/(f_t/f_c)$ is found to be 2, which gives a voltage amplification of

$$A_v = A_{f_t}/f_c = 2 \times 500/300 = 3.33$$

The above curve was for $K_{II}R_o/r_b' = 2$, and since $K_{II} = (f_c/f_Q)^2 = (\frac{300}{150})^2 = 4$, $R_o/r_b' = 0.5$.

Consider now the conditions in region III. The curve to choose in this region is the one for $R_o/r_b' = 0.5$ and $n = 3$ in Fig. 107. The normalized amplification at f/f_c

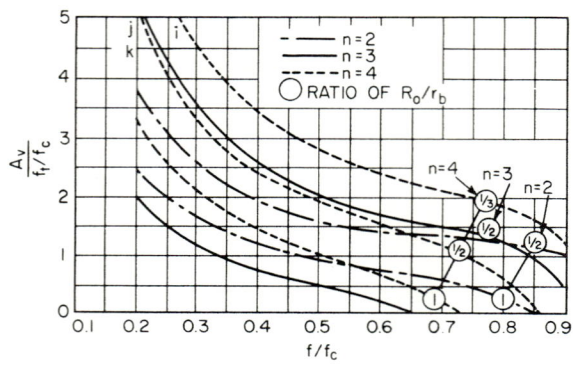

Fig. 107. Normalized gain vs. normalized frequency in region III. (*From L. F. Roeshot, UHF Broadband Transistor Amplifiers, Elec. Design News, January, 1963, to March, 1963, reproduced with permission.*)

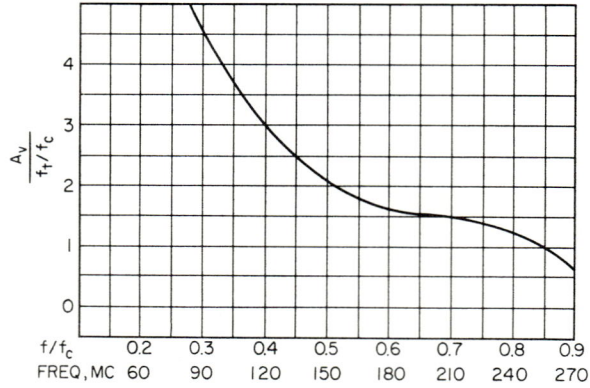

Fig. 108. Composite normalized gain vs. normalized frequency in regions II and III. (*From L. F Roeshot, UHF Broadband Transistor Amplifiers, Elec. Design News, January, 1963, to March, 1963, reproduced with permission.*)

should be the same as in region II, that is, 2, but the shape is that of the curve of region II for $f/f_c < 0.5$ and the region III curve for $f/f_c > 0.5$, yielding the composite normalized curve of Fig. 108. This frequency response is not flat, but a considerable improvement can be obtained by adjusting the terminating networks. The schematic diagram of the amplifier designed by the above technique is shown in Fig. 109, and the frequency response is shown in Fig. 110. The adjustment of these amplifiers involves a good deal of art. Other amplifiers of this type have yielded much flatter frequency responses, one example of which is shown in Fig. 111.

The theory as demonstrated in Figs. 106 and 107 indicates that the use of a large number of transistors in a distributed amplifier stage does not obtain full advantage of the technique at the high-frequency end of the operating range. This is because

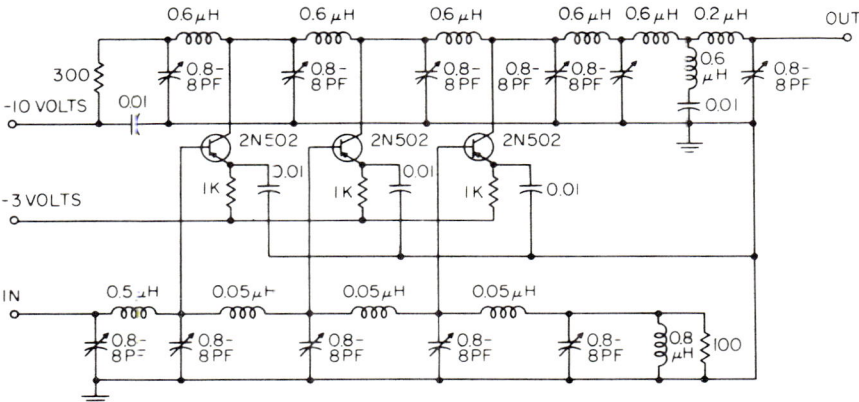

Fig. 109. Circuit of 30- to 300-Mc distributed amplifier. (*From L. F. Roeshot, UHF Broadband Transistor Amplifiers, Elec. Design News, January, 1963, to March, 1963, reproduced with permission.*)

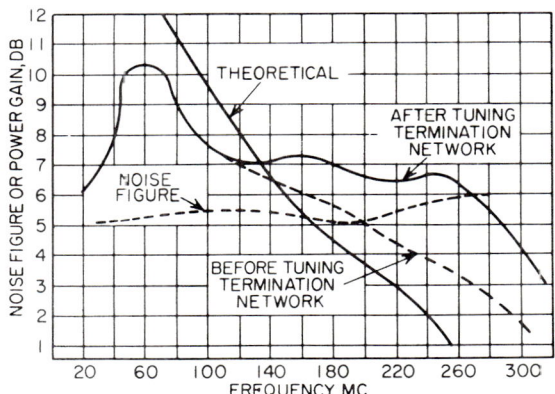

Fig. 110. Frequency response of previous amplifier. (*From L. F. Roeshot, UHF Broadband Transistor Amplifiers, Elec. Design News, January, 1963, to March, 1963, reproduced with permission.*)

of high-frequency losses caused by the increased input loading of the transistors. If only two transistors are used per stage and no termination is used to absorb the reflected wave, a condition can be found that makes possible the design of a useful amplifier. An illustration of one such amplifier is the schematic diagram shown in Fig. 112, and its frequency response is shown in Fig. 113.

One other technique is available to obtain very efficient wideband amplification. It consists of cascaded wideband bandpass amplifiers with a combination of positive and negative feedback. The theory involves too much detail to consider here, but an example of one such amplifier and its performance is shown in Figs. 114 and 115.

4.6. Transistor Driving Stages. Driving a transmission line or any other low-impedance load can be accomplished with a transistor connected in either the common-emitter or the common-collector connection. Achievement of this objective requires that this stage be a power amplifier as the realization of significant voltage across a low impedance requires delivering substantial currents to the load. Operation in a manner where the peak-to-peak swing is a small portion of the available supply voltage will result in inefficient operation of this amplifier unless an impedance-matching transformer can be used. If the impedance that the amplifier presents to the load must be restricted to some value such as that required to absorb transmission-line reflections, but the use of a matching transformer is not possible, then the potential

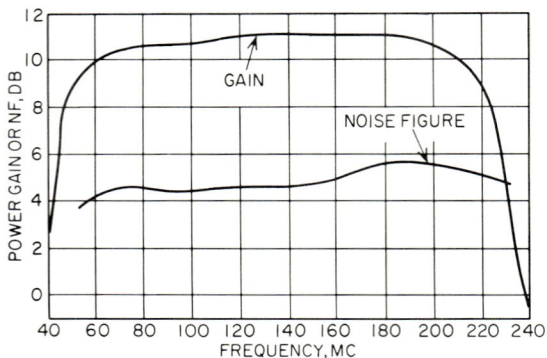

Fig. 111. Performance obtainable from distributed amplifier. (*From L. F. Roeshot, UHF Broadband Transistor Amplifiers, Elec. Design News, January, 1963, to March, 1963, reproduced with permission.*)

output swing due to d-c operating conditions will be restricted. In this case, the dissipation of the output amplifier stage could be reduced by restricting the d-c supply voltage if convenient for particular currents. Thus, while the output will not be increased at least useless power will not be dissipated in the transistor.

One other possibility does exist. This is the use of combinations of feedback so that the internal impedance can be designed to match the operating point of the transistor to the load impedance.

The simplest circuit that will have a source impedance equal to the load R_o is shown in Fig. 116. The total load to the amplifier is $R_o/2$; therefore, the voltage amplification in the low-frequency region will be $R_o/2R_e$ and the peak-to-peak output voltage swing will be $I_{dc}R_o$. Since the voltage amplification is low ($R_o/2R_e$), if $R_e \approx R_o$, the feedback will result in little Miller-effect capacitance being a part of the input impedance. This reduced Miller-effect capacitance makes possible a very wide frequency response, 3 db down at a frequency of nearly f_t.

The common-collector connection is often used as either an input or an output amplifier. It is useful in these applications because it has a relatively high input impedance and a relatively low output impedance. Referring to the circuit of Fig. 117, the low-frequency input impedance is approximately

$$1/R_{in} = 1/r_e + 1/[\beta_o(r_e + R_e)] \tag{116}$$

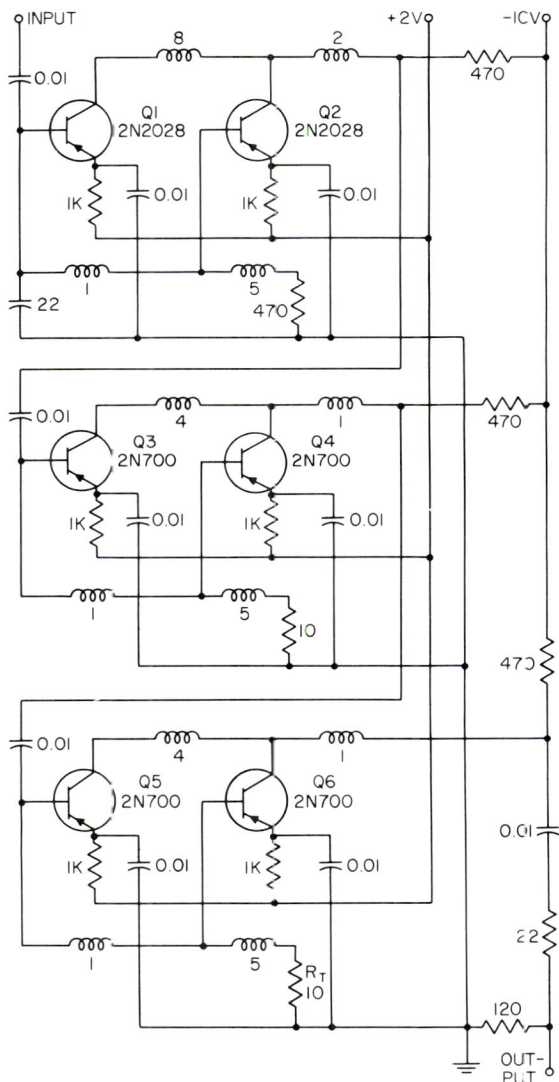

FIG. 112. Three-stage distributed-pair transistor amplifier. (*From L. F. Roeshot, UHF Broadband Transistor Amplifiers, Elec. Design News, January, 1963, to March, 1963 reproduced with permission.*)

and the low-frequency output impedance is

$$1/R_{oc} = \beta_o/(\beta_o r_e + R_s + r_b') + 1/R_e \qquad (117)$$

These values change substantially at higher frequencies, because of the effect of parasitic capacitances. While all these capacitances become significant at the higher frequencies, the emitter-to-collector capacitance in league with emitter-to-base feedback can result in a negative component of input conductance. A stability factor

$K = 1/R_e C_e$ can be defined such that the common-collector amplifier will be stable if $K_2 < K < K_1$, where

$$K_1, K_2 = [1 + 2(\beta_o R_e/R_s)] \pm (2\beta_o R_e/R_s)(1 + R_s/\beta_o R_e)^{\frac{1}{2}} \tag{118}$$

and ω_β is the β 3-db radian frequency. If the emitter follower is operated so that

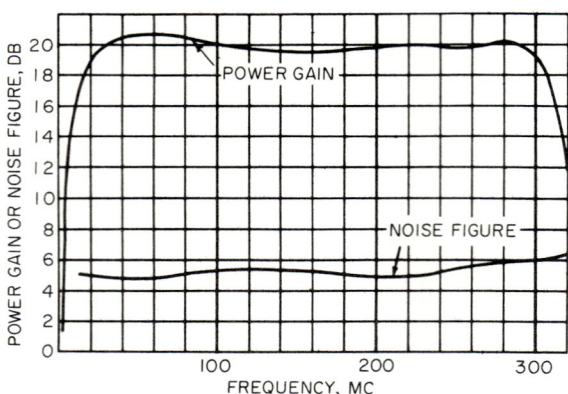

FIG. 113. Frequency response of three-stage distributed-pair transistor amplifier. (*From L. F. Roeshot, UHF Broadband Transistor Amplifiers, Elec. Design News, January, 1963, to March, 1963, reproduced with permission.*)

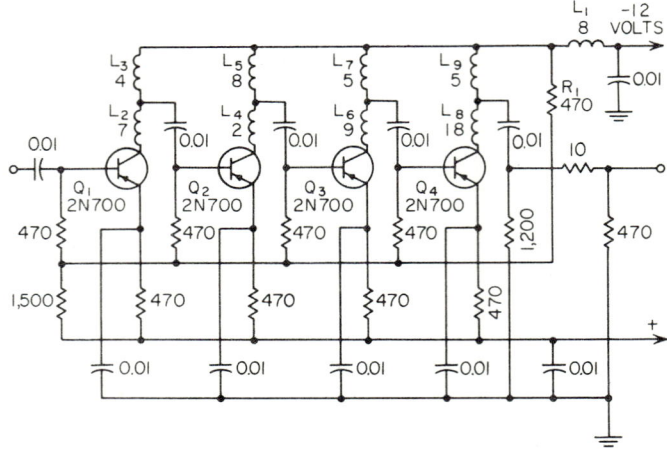

FIG. 114. Circuit diagram of multiple-feedback amplifier. (*From L. F. Roeshot, UHF Broadband Transistor Amplifiers, Elec. Design News, January, 1963, to March, 1963, reproduced with permission.*)

$R_s \ll \beta_o R_e$, the above simplify to

$$K_1, K_2 = 1, 1 + (1 + 4\beta_o R_e/R_s) \tag{119}$$

The input impedance of an emitter follower will vary considerably with frequency since β changes with frequency and $z_{in} \approx \beta R_e$.

BROADBAND AMPLIFIERS 25–87

Compensation for this can be obtained by connecting an inductance of approximately

$$L_e = R_e/\omega_\beta$$

in series with R_e. An increase of this inductance by a factor of 15 to 50 per cent will be necessary because the C_c was neglected in arriving at L_e as was the effect of

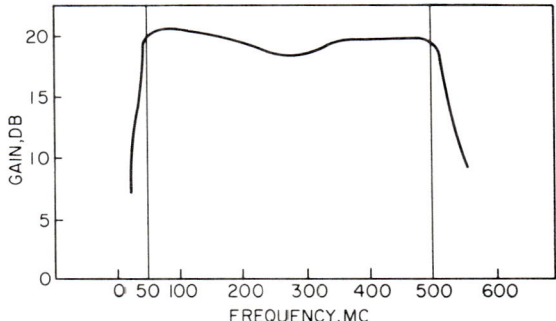

Fig. 115. Frequency characteristic of multiple-feedback amplifier. (*From L. F. Roeshot, UHF Broadband Transistor Amplifiers, Elec. Design News, January, 1963, to March, 1963, reproduced with permission.*)

distributed capacitance of L_e. The above compensation holds for the condition that $R_L \gg R_e$.

The peak-to-peak low-frequency undistorted output of such an amplifier is $2I_{dc}R_e/(R_o + R_e)$. If the output impedance R_{oc} of the emitter follower is equal to R_o and $R_e \gg R_o$ then the peak-to-peak output swing will be approximately $2I_{dc}R_o$, but the

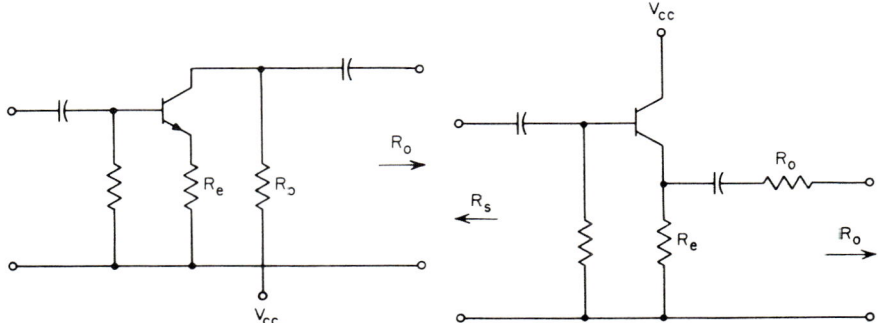

Fig. 116. Simple transmission-line driving amplifier.

Fig. 117. Common-collector amplifier.

voltage amplification will be ½ and the low-frequency input impedance will be only $\beta_o(r_e + R_o)$.

The above limitation in output swing can be reduced by two techniques, both of which depend on driving the load from different devices on the two halves of the input cycle. Figure 118 shows a circuit analogous to the two-tube cathode follower in which the output current is supplied by the upper transistor Q_1 on the positive half-cycle and by the lower transistor on the negative half-cycle. It should be noted that this circuit will not give particularly useful performance if designed to have an output impedance of R_o. It is therefore more useful to match it to the load R_o by designing it to have a very low output impedance, taking full advantage of the two stages of inverse feedback and inserting a resistor of value R_o between it and the load.

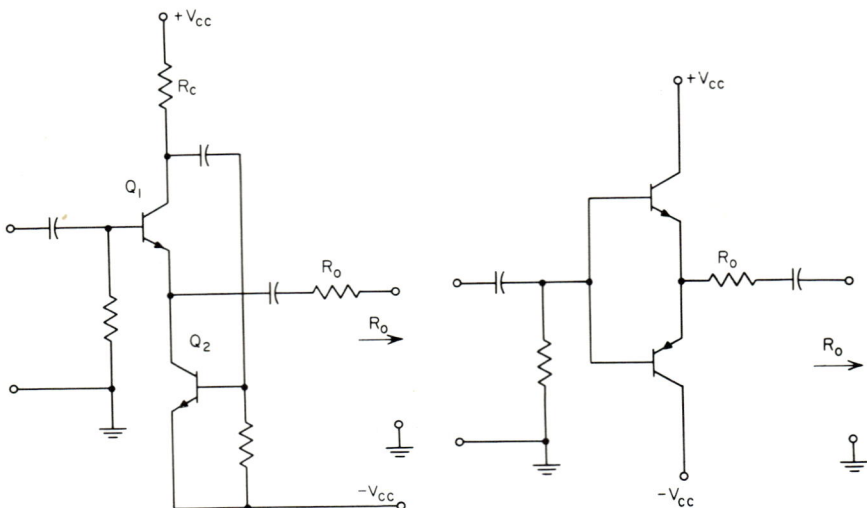

FIG. 118. Two-stage emitter-follower amplifier.

FIG. 119. Complementary-symmetry emitter follower.

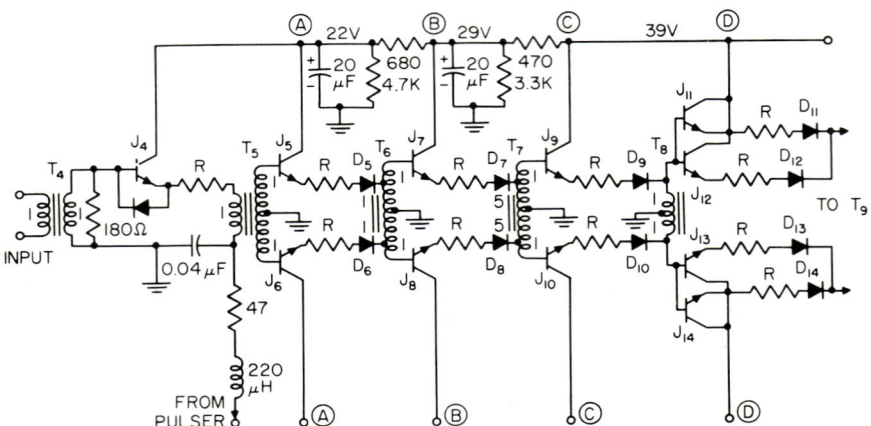

FIG. 120. Driver stage for wideband high-power amplifier. (*From C. A. Franklin, A 100 Watt Wideband Amplifier for the 1–12 Mc Frequency Range, Dig. Tech. Papers, 1962 Intern. Solid-State Circuits Conf., used with permission.*)

The resistor R_c should have a value such that it, in combination with $+V_{CC}$, will allow the desired load current before limiting. The other circuit (Fig. 119) is a complementary-symmetry emitter follower which can operate class B to increase output swing. In this circuit, it is also desirable, when matching to a load R_o, to operate

BROADBAND AMPLIFIERS

the two devices so as to realize a very low output impedance and then determine the amplifier-stage source impedance with a series resistor R_o.

Still another way of building a high-output amplifier for wide bandwidths is to use the distributed-amplifier technique. This is useful because it enables parallel connection of devices without paralleling their parasitic capacitances. Three or four

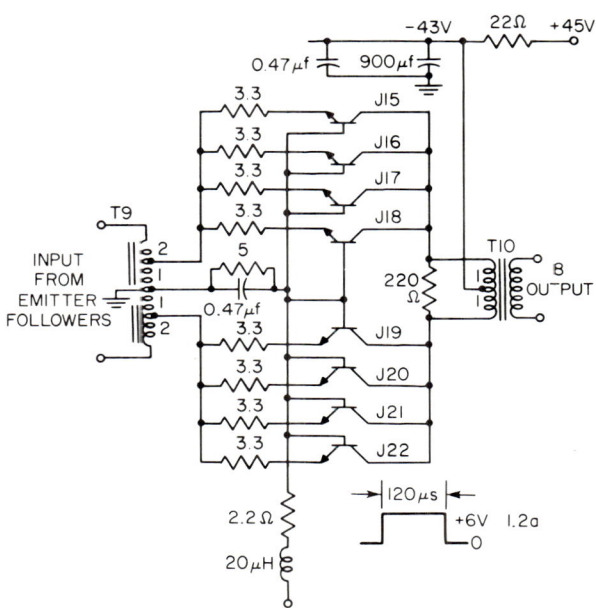

FIG. 121. Output stage for wideband 100-watt amplifier. (*From C. A. Franklin, A 100 Watt Wideband Amplifier for the 1–12 Mc Frequency Range, Dig. Tech. Papers, 1962 Intern. Solid-State Circuits Conf., used with permission.*)

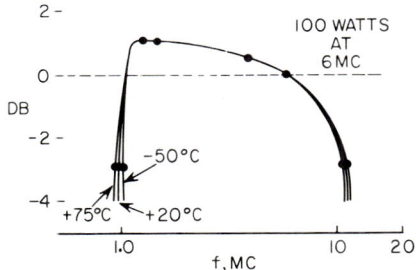

FIG. 122. Frequency response of 100-watt wideband amplifier into a 400-ohm resistive load. (*From C. A. Franklin, A 100 Watt Wideband Amplifier for the 1–12 Mc Frequency Range, Dig. Tech. Papers, 1962 Intern. Solid-State Circuits Conf., used with permission.*)

transistors in a wideband stage, 100 Mc or more bandwidth, is the practical limit because the rapid increase of input loading at frequencies greater than 100 Mc results in attenuation of the input signal so that additional transistors do not get sufficient drive to supply significant power outputs.

4.7. Transistor Power Amplifiers. High-power amplifiers can be made with transistors having very wide bandwidths. Amplifiers of this type have been developed to produce a power output of 300 watts, 3 db down at 180 kc, and 8 db down at 15 Mc.[53] The basic technique is illustrated by the 100-watt amplifier described below,[54] which is designed to operate with a two-thirds of 1 per cent duty cycle amplifying a 100-μsec burst of sine waves.

The schematic diagrams for the driver and power output stages of the 100-watt amplifier are shown in Figs. 120 and 121. Figure 122 is the frequency response of the amplifier into a 400-ohm resistive load.

A number of factors are particularly important in achievement of this performance. The power-output stage used the transistor in the common-base connection because of the greater breakdown voltage and lower output capacitance. The transistors also have no serious protection problem when the load impedance falls to zero. Since the amplifier operated with a very small duty cycle, the thermal time constant of the transistors was particularly important, being approximately 600 μsec.

The common-base connection requires more care because of potential instability, which becomes more severe at higher power levels. This stability problem can be controlled by driving the power-output stage from a sufficiently low impedance. This low impedance was obtained by using transistors in the common-collector connection to drive the output-stage emitters. A further reduction of driving impedance was obtained by use of a step-down autotransformer.

[53] Personal communication from C. A. Franklin.
[54] C. A. Franklin, A 100 Watt Wideband Amplifier for the 1–12 Mc Frequency Range, *Proc. 1962 Intern. Solid-State Circuits Conf.*, paper FM 10.2.

Chapter 26

NONLINEAR AMPLIFIERS

WILLIAM M. TRENHOLME[*]

CONTENTS

1. Introduction.. 26-2
2. Use of Nonlinear Device Characteristics..................... 26-2
   - 2.1. Thermionic Diodes..................................... 26-2
   - 2.2. Thermionic Triodes, Tetrodes, and Pentodes............ 26-3
   - 2.3. Special Thermionic Devices............................ 26-11
   - 2.4. Semiconductor Diodes.................................. 26-14
   - 2.5. Semiconductor Triodes and Tetrodes.................... 26-17
   - 2.6. Varistors and Thermistors............................. 26-25
   - 2.7. Vacuum Thermoelements................................. 26-27
   - 2.8. Opto-electronic Devices............................... 26-27
   - 2.9. Galvanomagnetic Devices............................... 26-29
3. Use of Piecewise-linear-approximation Techniques........... 26-35
   - 3.1. Device Considerations................................. 26-35
   - 3.2. Logarithmic Function Generators....................... 26-36
   - 3.3. Square-law Function Generators........................ 26-40
   - 3.4. Variable Function Generators.......................... 26-42
4. Use of Servo Techniques.................................... 26-44
   - 4.1. Servosystem Components................................ 26-44
   - 4.2. Nonlinear Function Generation by Servo Techniques..... 26-49
   - 4.3. Servomultipliers...................................... 26-51
5. Circuits with Amplitude Limiting........................... 26-53
   - 5.1. Simple Diode-limiter Circuits......................... 26-55
   - 5.2. Precision Limiter Circuits............................ 26-55
6. Circuits with Amplitude Compression and/or Expansion....... 26-58
   - 6.1. General-purpose AGC Circuits with Compression Only.... 26-58
   - 6.2. AGC Circuits for Entertainment Radio Receivers........ 26-63
   - 6.3. High-frequency AGC Circuits 26-65
   - 6.4. Compressors and Expanders for Audio Systems........... 26-73
7. Circuits Having a Logarithmic Amplitude Response........... 26-76
   - 7.1. D-C Logarithmic Amplifiers............................ 26-76
   - 7.2. A-C Logarithmic Amplifiers............................ 26-81
8. Circuits Having a Logarithmic Frequency or Pulse-rate Response 26-85

[*] Massachusetts Institute of Technology, Department of Nuclear Engineering, Cambridge, Mass. (Present affiliation: Arizona Atomic Energy Commission Phoenix, Ariz.

9. Amplitude Squaring and Square-rooting Circuits	26-87
10. Circuits for Analog Multiplication and Division	26-92
10.1. Multipliers and Dividers Using Logarithmic Elements	26-92
10.2. Quarter-square Multipliers and Dividers	26-94
10.3. Other Multipliers and Dividers	26-94

1. INTRODUCTION

Before going into the details of nonlinear amplifier design it is desirable to define clearly the types of circuits and techniques that will be treated.

Six types of amplifier circuits will be covered in this chapter. They are: (1) circuits with amplitude limiting; (2) circuits with amplitude compression and/or expansion; (3) circuits having a logarithmic amplitude response; (4) amplifiers having a logarithmic frequency or pulse-rate response; (5) circuits having a square-law response; (6) circuits useful in analog multiplication and division. Although the same techniques may be found useful in the design of several of these types of amplifier circuits, each type listed requires somewhat different design considerations and hence will be treated separately.

There is no distinction in the above listing as to the range of frequencies to be handled or the types of devices that are used in the circuits. These considerations will be introduced in the text, and examples will be discussed of amplifiers covering various frequency ranges and using various types of devices.

As a preliminary to the discussion of specific circuits, the following three sections of this chapter will deal with the major techniques that are used to achieve nonlinear transfer characteristics. These techniques are: (1) use of nonlinear device characteristics; (2) use of piecewise-linear-approximation techniques; (3) use of servo techniques. Other techniques that are sometimes useful in nonlinear amplifier design will be discussed in the sections dealing with the various types of circuits.

2. USE OF NONLINEAR DEVICE CHARACTERISTICS

There are two general types of nonlinearities exhibited by electronic devices which are useful in the design of nonlinear amplifier circuits. In devices having nonlinearities of the first type, there is a direct nonlinearity in the amplitude transfer characteristic between current and voltage or between input current or voltage and output current or voltage. As an example, thermionic diodes and triodes can exhibit this type of nonlinearity. In devices having nonlinearities of the second type, the electrical resistance or the transfer characteristics of the device are caused to be altered by a control signal in order to achieve nonlinear amplification. As an example, photoconductive cell–light source combinations and tetrode transistors can exhibit this type of nonlinearity.

In the following paragraphs the various devices which exhibit useful nonlinear characteristics will be discussed, with distinctions made as to which of the above types of nonlinearities it possesses. The range of amplitudes and frequencies over which the devices may be used will also be brought out.

2.1. Thermionic Diodes. A high-vacuum thermionic diode is one of the most useful nonlinear devices available. In the transition region between the reverse-bias and forward-bias regions of its current-voltage characteristic, the plate voltage V is related to the diode current I by the following equation:

$$I = I_0 \exp (q/kT)(V - V_0) \tag{1}$$

where I_0 = diode current for zero diode voltage
q = electron charge = 4.77×10^{-10} esu
k = Boltzmann's constant = 1.37×10^{-16}
T = cathode absolute temperature
V_0 = plate-to-cathode contact potential

Taking the logarithm of Eq. (1), rearranging the terms, and simplifying,

$$V = K_0 T(\log_{10} I - \log_{10} I_0) + V_0 \qquad (2)$$

where K_0 is a constant, showing that the change of plate voltage is a logarithmic function of the diode current. For a cathode temperature of 1000°K, the plate voltage increases approximately 0.2 volt/decade increase in current. This direct nonlinearity applies over a range of up to nine decades of diode current for some diodes, making thermionic diodes very useful for the measurement of wide ranges of currents in d-c logarithmic amplifiers.

A set of characteristic curves for the acorn-type 9004 diode, which has been widely used in logarithmic amplifiers, is shown in Fig. 1. The logarithmic characteristic is seen to extend from approximately 10^{-13} to 10^{-4} amp for a heater voltage of 3.0 volts d-c. These curves also indicate the basic problem in using a thermionic diode as a

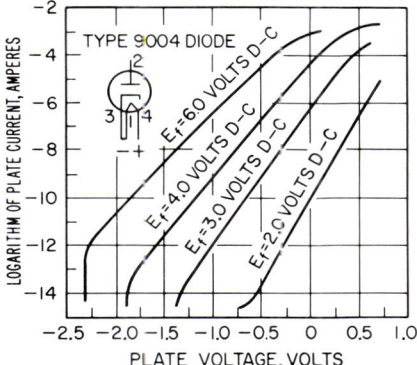

FIG. 1. Characteristic curves for a type 9004 diode. [From H. D. Huskey and G. A. Korn (eds.), "Computer Handbook," p. 3-42, Fig. 3.2.3, McGraw-Hill Book Company, New York, 1962.]

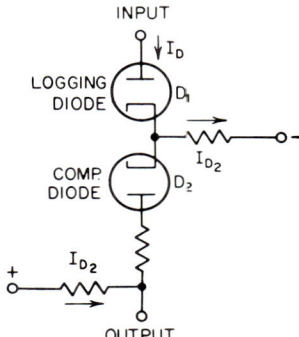

FIG. 2. Compensation scheme for thermionic logarithmic diodes. (From IRE Natl. Conv. Record, pt. 9, p. 79, 1954.)

logarithmic element; it is quite sensitive to cathode temperature variations. The temperature variations cause the term I_0 in Eq. (2) to have variations, and either a very stable d-c heater supply is required or some type of compensation scheme must be employed. A simple technique for reducing the temperature sensitivity is to insert a relatively large resistance in series with the heater supply so that the filament is fed from a current source rather than a voltage source. By use of this technique the amount of change in the logarithmic characteristics of the diode for a given percentage change in the heater supply can be made lower than it would be if the heater supply were connected directly to the diode filament leads.

Figure 2 shows a practical compensation scheme which employs two similar diodes of the same type. The diodes are connected back to back, with one serving as the logarithmic element and the other as a compensating element. A large constant current I_{D_2} is fed through the compensating diodes, so that the current to be measured I_{D_1}, while flowing through both diodes, causes a change of plate voltage only in the logarithmic element. Since the filaments of both diodes are fed from the same heater supply, changes on the plate voltages of the two diodes caused by a change of heater voltage will tend to cancel each other. This method of diode-filament compensation has been used with type 5647 diodes to give approximately a factor of 10 improvement in the stability of the logarithmic characteristic.

2.2. Thermionic Triodes, Tetrodes, and Pentodes. Several useful nonlinearities are exhibited by thermionic triodes, tetrodes, and pentodes. First, they may be connected as diodes and operated as direct logarithmic devices in the manner discussed in Sec. 2.1. The tubes that have been most widely used in this manner are the type 5886 tetrode electrometer tube and its close relative, the 5889. These tubes may be diode-connected in either of two ways: (1) the plate, screen grid, and control grid may all be connected together to form the diode anode; (2) the control grid may be used as the diode anode, with the plate and the screen connected together and maintained at the same potential as the filament or slightly higher. Use of the latter method of connection has been shown to give an extension of the logarithmic range and a lowering of the effective diode capacitance.

The 5886 and 5889 may be used as diode-connected logarithmic elements down to currents of 10^{-13} amp, but they lose their logarithmic characteristic at currents greater than about 10^{-6} amp. The type 6923/EA52 has been successfully used to extend the logarithmic characteristic to currents of greater than 10^{-4} amp, but it is not useful for currents less than 10^{-12} amp. Recently several nuvistor triodes and tetrodes have been evaluated as diode-connected logarithmic devices. They do not match the capabilities of the 5886 or 5889 for low input currents, but they have excellent logarithmic characteristics over the range from 10^{-11} to 10^{-4} amp and do not need so much selection as some of the older types.

A second useful direct nonlinear relationship exhibited by thermionic triodes and triode-connected tetrodes and pentodes is that between the plate current and the grid current of certain tubes for small plate currents. To establish this relation, the experimentally verified equation relating the grid and plate voltages to the plate current I_P is

$$I_P = K_1(V_G + V_P/\mu)^{\frac{3}{2}} \tag{3}$$

where K_1 = a constant whose value depends on the tube geometry
V_G = grid-to-cathode voltage
V_P = plate-to-cathode voltage
μ = amplification factor of the tube

and the grid current I_G is related to the grid voltage V_G by an equation similar to the diode relation of Eq. (2),

$$V_G = K_2 T(\log_{10} I_G - \log_{10} I_{G0}) + V_{G0} \tag{4}$$

where K_2 = a constant whose value depends on the tube geometry
T = cathode absolute temperature
I_{G0} = grid current for zero grid-to-cathode voltage
V_{G0} = grid-to-cathode contact potential

Substituting Eq. (4) into Eq. (3) and simplifying,

$$\begin{aligned} I_P &= K_1(K_2 T \log_{10} I_G - K_2 T \log_{10} I_{G0} + V_{G0} + V_P/\mu)^{\frac{3}{2}} \\ &= (M \log_{10} I_G + N)^{\frac{3}{2}} \end{aligned} \tag{5}$$

where M = a quantity whose value depends on the tube geometry and T
N = a quantity whose value depends on the tube geometry, T, I_{G0}, V_{G0}, V_P, and μ

On a plot of I_P versus $\log_{10} I_G$, M is the slope of the curve and N is the equivalent quiescent current. Since M and N are both functions of the cathode temperature, the logarithmic characteristic might be expected to be quite sensitive to variations of the heater supply voltage. However, there is an inherent canceling effect of the changes due to cathode temperature variations that is not present in diodes or diode-connected triodes, tetrodes, and pentodes. Although the plate current tends to increase when the cathode temperature increases because of increased emission, the grid voltage tends to decrease because of the arrival of more electrons at the grid, which in turn tends to decrease the plate current. This canceling effect is not perfect, but it helps considerably in the application of tubes utilizing the nonlinear relationship of Eq. (5).

Figure 3 shows a set of characteristic curves for a triode-connected 5889 tube. These curves indicate the effect of plate supply voltage on the logarithmic characteristic and also show that improved performance may be obtained when there is a relatively large load resistance in the plate circuit. The effect of increasing the load resistance is to decrease the value of V_P as I_P increases, which makes Eq. (5) become a better approximation to a true logarithmic relationship. However, the accuracy is still not so good as that obtainable with the same tube in a diode connection.

The principal advantage of the logarithmic triode connection over the diode connections, in addition to its somewhat lower sensitivity to cathode temperature changes, is that it can serve as both a logarithmic element and an amplifier. This has led to its use in such applications as ionization-chamber survey instruments where simplicity and low cost are of paramount importance.

A third useful direct nonlinear relationship exhibited by thermionic tetrodes and pentodes is that between the screen-grid voltage and the control-grid current when

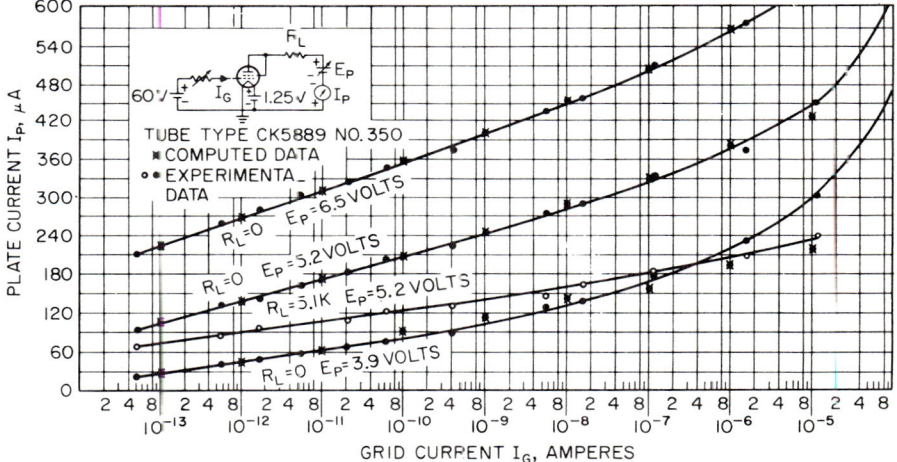

FIG. 3. Characteristic curves for a triode-connected CK5889 electrometer tube. (*From Rev. Sci. Instr., vol. 30, p. 1088, 1959.*)

the plate current is held constant. To establish this relation, consider the ratio of the d-c transconductance from control grid to plate G_G, to the d-c transconductance from screen grid to plate G_S:

$$\frac{G_G}{G_S} = \frac{\Delta I_P/\Delta V_G}{\Delta I_P/\Delta V_S} \qquad (6)$$

If the tube is operated such that I_P is held constant, Eq. (6) reduces to

$$\Delta V_S = -(G_{PG}/G_{PS})\, \Delta V_G \qquad (7)$$

Since the control-grid current I_G is related to the control-grid voltage V_G in the same manner as in a triode, an equation similar to Eq. (4) may be written for ΔV_G. Thus

$$\Delta V_G = K_2 T (\log_{10} I_{G2} - \log_{10} I_{G1}) \qquad (8)$$

where K_2 = a constant whose value depends on the tube geometry
T = cathode absolute temperature

Substituting Eq. (8) into (7) and rearranging,

$$V_{S2} = -K_2 T (G_G/G_S)(\log_{10} I_{G2} - \log_{10} I_{G1}) + V_{S1} \qquad (9)$$

The change in screen-grid voltage is therefore a logarithmic function of the change in control-grid current. Logarithmic amplifiers using this relationship exhibit ten to twenty times lower sensitivity to cathode temperature changes than do uncompensated amplifiers using the diode relationship. This improved stability results because there is no I_0 term in Eq. (9) such as there is in Eq. (2). Although G_G and G_S in Eq. (9) are both quite sensitive to cathode temperature changes, they tend to change together and their ratio thus remains essentially constant.

Figure 4 shows a comparison of the logarithmic characteristics obtained with several types of pentodes. The screen voltage change per decade change of control-grid current is seen to vary from about 6.0 volts for the WE-404A to about 2.5 volts for the 5A/162D, and the logarithmic range is seen to be over eight decades for the latter type. The type 7587 nuvistor tetrode has been shown to give a reasonably good logarithmic relation between screen-grid voltage and control-grid current over an input current range from 10^{-10} to 10^{-4} amp without the need for selection.

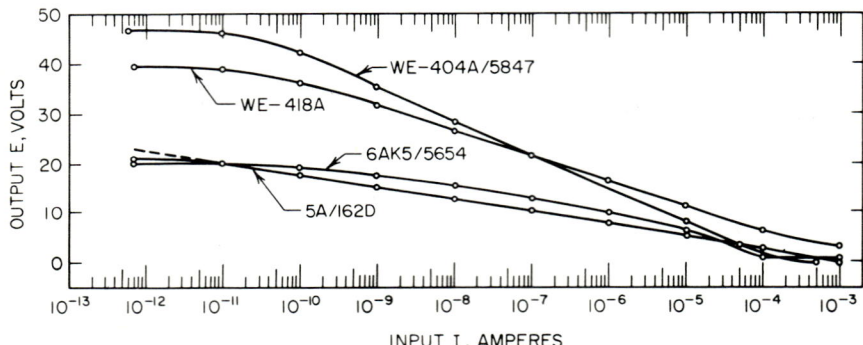

Fig. 4. Characteristics of various pentode logarithmic converters. (*From Naval Res. Lab. Rept. 5025, p. 9, 1957.*)

Although the use of the logarithmic relationship of Eq. (9) gives better stability against heater supply variations than either the diode or the triode relationships, the range of input currents that can be covered by the tetrode or pentode connection is more limited, and the conversion accuracy is generally poorer than that exhibited by the diode connection.

There are other useful nonlinearities exhibited by thermionic triodes, tetrodes, and pentodes which are of the second type and apply only for relatively small variations of the input signal about the d-c operating point of the tube. Foremost among these is the variation of plate resistance and/or transconductance when the d-c control-grid voltage V_G is varied by a control signal. In this case V_G is related to the small-signal a-c plate current i_p by the relation

$$i_p = k_1 V_G{}^s \qquad (10)$$

where k_1 = a constant whose value depends on the tube geometry
s = a quantity whose value depends on both the tube geometry and the d-c operating point of the tube

From the relationship of Eq. (10) the following expression may be developed for the small-signal plate resistance r_p in terms of V_G:

$$r_p = dv_p/di_p = \mu(dV_G/di_p) = \mu/k_2 V_G{}^{s-1} \qquad (11)$$

where v_p = small-signal plate voltage
μ = amplification factor of the tube

Since the transconductance g_m is defined as μ/r_p, Eq. (11) may be rewritten as

$$g_m = k_2 V_G{}^{s-1} \qquad (12)$$

These relationships are often used in the design of volume compression and expansion circuits for communications equipment.

For remote-cutoff variable-mu tubes the general expression of Eq. (10) may be simplified to

$$i_p = A \exp (bV_G) \qquad (13)$$

where b is a constant, and the expression for g_m then becomes

$$g_m = di_p/dV_G = Ab \exp (bV_G) \qquad (14)$$

Figure 5 shows the plate current vs. grid voltage characteristics of a typical remote-cutoff pentode. Amplitudes up to several volts and frequencies up to several hundred megacycles may be handled by tubes exhibiting the characteristics described by Eqs. (13) and (14), making them very useful in the design of a-c logarithmic amplifiers. A typical design will be discussed in Sec. 7.

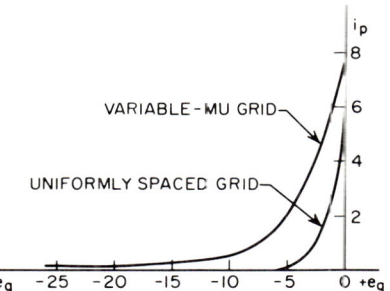

FIG. 5. Plate current vs. grid voltage characteristics of normal and remote-cutoff vacuum tubes. (*From Bureau of Naval Personnel Publication NAVPERS 10087-A, 1962.*)

The direct nonlinearity in a remote-cutoff tetrode or pentode between a-c grid voltage and a-c plate current with a fixed d-c grid bias may also be used to achieve logarithmic amplification. A simple circuit using this type of nonlinearity is shown in Fig. 6. The diode rectifies only that portion of the input signal which has been compressed by the curvature in the transfer characteristic of the tube. A choke is used as an a-c plate load in order to present a low-resistance path for the diode current. This prevents building up across the coupling capacitor of a rectified potential which could allow the introduction of transients in the output circuit when the input

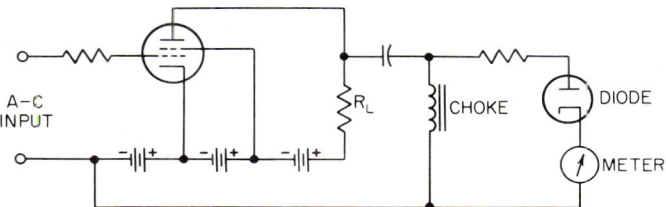

FIG. 6. Circuit for direct logarithmic amplification using a remote-cutoff tetrode. (*From Rev. Sci. Instr., vol. 4, p. 672, 1933.*)

signal changes suddenly. A range of approximately a decade can be covered by a stage of this type, and several stages may be cascaded to obtain more range.

The general relationship between grid voltage and plate current in thermionic triodes, tetrodes, and pentodes that have uniformly spaced control grids may be expressed mathematically as follows:

$$I_P = \alpha + \beta(V_G - V_O) + \gamma(V_G - V_O)^2 + \delta(V_G - V_O)^3 + \cdots \qquad (15)$$

where $\alpha, \beta, \gamma, \delta, \ldots$ are constants of the particular tube and V_O is a reference voltage from which grid voltage changes are measured. The linear term $(V_G - V_O)$ may always be eliminated from Eq. (15) by proper choice of V_O, and if all tube constants

26-8 AMPLIFIER CIRCUITS

beyond γ are negligible, the following equation applies for changes in the grid voltage:

$$\Delta I_P = \gamma(\Delta V_G)^2 \qquad (16)$$

Tubes satisfying Eq. (16) therefore provide an output plate-current change which is proportional to the square of the input grid voltage change. Figure 7 shows a

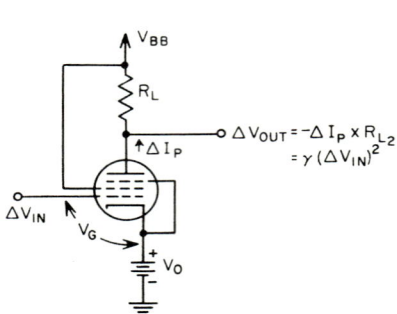

Fig. 7. Simple vacuum-tube amplitude-squaring circuit. (*From B. Chance et al.,* "*Waveforms,*" *vol. 19, p. 680, Fig. 19.15, MIT Radiation Laboratory Series, McGraw-Hill Book Company, New York, 1949.*)

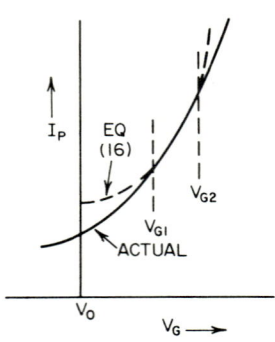

Fig. 8. Approximation to ideal square-law characteristic obtained with circuit of Fig. 7. (*From B. Chance et al.,* "*Waveforms,*" *vol. 19, p. 680, Fig. 19.16, MIT Radiation Laboratory Series, McGraw-Hill Book Company, New York, 1949.*)

simple circuit that may be used to implement this relationship; an output voltage change proportional to ΔI_P and hence to the square of ΔV_{in} is obtained across the plate resistor of the tube. This relationship can be satisfied only over a range V_{G1} to V_{G2} which is not large relative to the value of $(V_{G1} - V_O)$, and it becomes invalid when $(V_G - V_O)$ approaches zero. Figure 8 illustrates these limitations. However, feedback techniques may be used which will allow operation of a simple vacuum-tube squaring circuit near $V_G = V_O$. If the proper amplitude of input signal is subtracted from the plate signal, the resultant voltage will be, within a constant value, proportional to the square of the input signal. Figure 9 shows a circuit for accomplishing this. Both R_1 and R_2 must be much larger than R_L in order for the circuit to function properly.

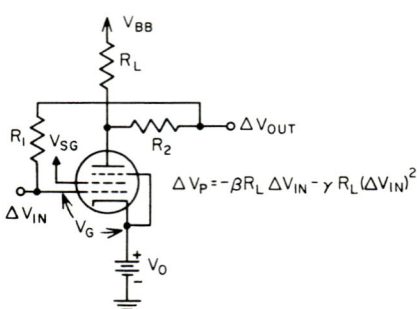

Fig. 9. Feedback compensation circuit for improving square-law characteristics of vacuum tubes. (*From B. Chance et al.,* "*Waveforms,*" *vol. 19, p. 681, Fig. 19.17, MIT Radiation Laboratory Series, McGraw-Hill Book Company, New York, 1949.*)

Table 1 lists several tube types that have proved suitable for use in amplitude-squaring circuits. The I_P versus V_G characteristic of these tubes closely approximates a parabolic shape. Since $g_m = \Delta I_P / \Delta V_G$, it follows that g_m of a suitable tube will be an approximately linear function of V_G. These points are illustrated by the curves of Fig. 10.

Tubes satisfying the relationship of Eq. (16) may also be used to obtain an output voltage change which is roughly proportional to the square root of an input voltage change. To accomplish this the plate current of the tube is made the independent variable and the grid-to-cathode voltage is made the output or dependent variable.

Table 1. Tubes with Useful Square-law Characteristics

Type	Characteristic of interest	Remarks
6B8	I_P vs. V_G	Small linear contribution; 7-volt useful grid swing
1S5	I_P vs. V_G	Small linear contribution; only 2- or 3-volt useful grid swing
6SN7	I_P vs. V_G	Fair with 10-kilohm plate resistor
6D6, 6U7G	I_P vs. V_{SG}	40-volt swing usable

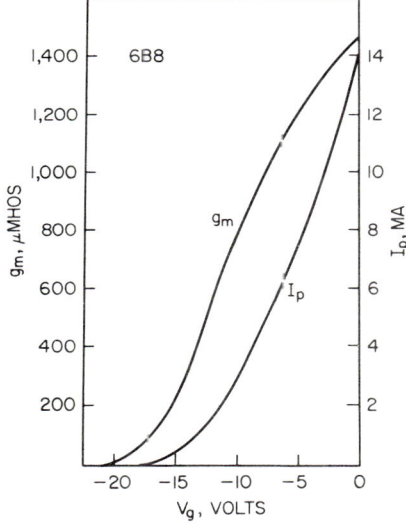

Fig. 10. Characteristics of a typical vacuum tube suitable for amplitude squaring. (From B. Chance et al., "Waveforms," vol. 19, p. 681, Fig. 19.18, MIT Radiation Laboratory Series, McGraw-Hill Book Company, New York, 1949.)

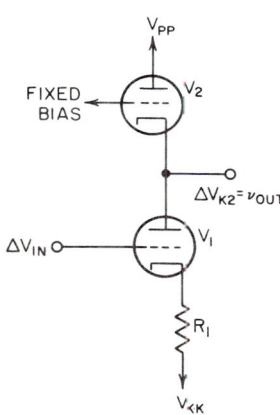

Fig. 11. Simple vacuum-tube square-root-extracting circuit. (From B. Chance et al., "Waveforms," vol. 19, p. 686, Fig. 19.24, MIT Radiation Laboratory Series, McGraw-Hill Book Company, New York, 1949.)

Figure 11 shows a simple square-root-extracting circuit of this type. Tube V_2 has a parabolic transfer characteristic between ΔI_P and ΔV_{GK}, and its plate current is controlled by a constant-current tube V_1, so that ΔI_P is proportional to ΔV_{in}. The grid-to-cathode voltage of V_2 assumes a value consistent with the value of ΔI_P. Since the grid of V_2 is held at a fixed voltage level, ΔV_{K2} is the desired output signal. An extension of this circuit which gives greater precision and more freedom in the choice of bias voltages is discussed in Sec. 9.

Certain thermionic triodes give an output plate current which is roughly proportional to the product of the grid voltage and the plate voltage. Hence they may be used in simple analog multiplier circuits over a narrow range of inputs. Figure 12 shows an ideal multiplier characteristic of this type in comparison with the characteristics of a triode-connected 6K6. Figure 13 shows two simple circuits for utilizing the above relationship to achieve an analog multiplication circuit. In the circuit shown as a, input v_2 must be able to supply the tube current, and the transformer

26–10 AMPLIFIER CIRCUITS

must have a response suitable for passing the output waveform without undue distortion. The transformer must also have a low-impedance primary winding so that the output voltage will not be superimposed on the plate voltage. The circuit shown as *b* makes use of a cathode follower V_2, with the multiplying tube V_1 as the cathode impedance. The cathode of V_2 assumes a voltage approximately equal to v_2, and the

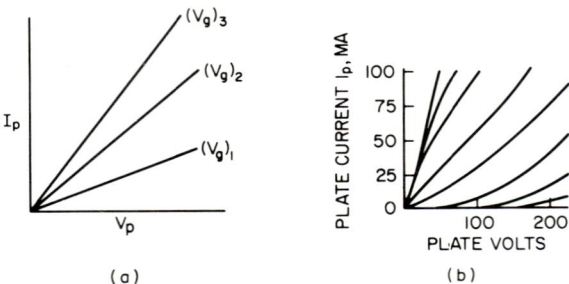

Fig. 12. Ideal and actual characteristics of a vacuum tube useful in amplitude-multiplication circuits. (*a*) Ideal characteristic. (*b*) 6K6 characteristic. (*From B. Chance et al., "Waveforms," vol. 19, p. 669, Fig. 19.2, MIT Radiation Laboratory Series, McGraw-Hill Book Company, New York, 1949.*)

plate current through V_1 and V_2 is thus proportional to $v_1 v_2$. This circuit has the advantage over that of *a* in that it is possible to measure the output at a different point from the v_2 input. It also has the advantage that it may be used for d-c as well as a-c inputs.

A triode-connected 6K6 may be used as a multiplying tube in the range where

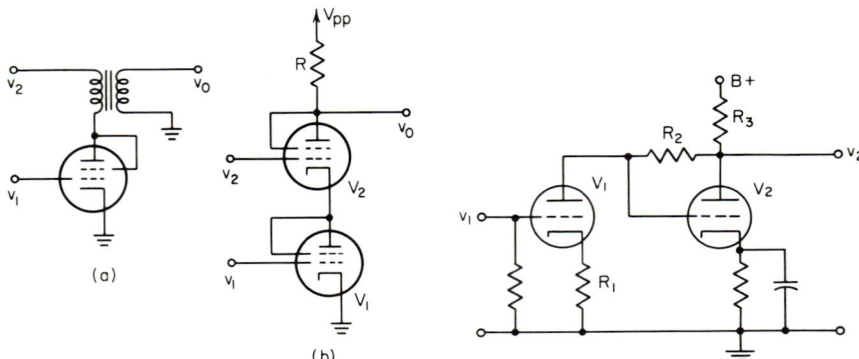

Fig. 13. Simple vacuum-tube multiplying circuits. (*From B. Chance et al., "Waveforms," vol. 19, p. 670, Fig. 19.3, MIT Radiation Laboratory Series, McGraw-Hill Book Company, New York, 1949.*)

Fig. 14. Direct-coupled triode pair configuration for nonlinear pulse amplication. (*From Mound Laboratory Paper MLM-851, 1953.*)

V_P is less than 25 volts and ΔV_G is less than 20 volts. Other tubes such as the 6L6 and the 6V6 roughly satisfy the multiplier relationship, but none is very good over a wide range.

The final nonlinear relation that will be presented in this section is that between the input and output pulse voltages of a direct-coupled triode pair when the input signal is negative. The basic configuration to be considered is shown in Fig. 14, and the equivalent circuits and equations used for computing the voltage amplification of the configuration are shown in Fig. 15. The calculated gain-compression data for this configuration, using a 12AT7 dual triode and with $R_1 = 200$ ohms, $R_2 = 10{,}000$ ohms, and $R_3 = 5{,}000$ ohms, are shown in Table 2, and a comparison of the calculated voltage-amplification curve with two experimentally determined curves is shown in Fig. 16.

$$\frac{E_2}{E_1} = A_1' = \frac{\mu_1 \left(R_1 + R_2 + \left(\frac{R_3 r_{p_2}}{R_3 + r_{p_2}}\right)\right)}{R_1(\mu_1+1) + R_2 + \left(\frac{R_3 r_{p_2}}{R_3 + r_{p_2}}\right) + r_{p_1}}$$

(a)

$$\frac{E_3}{E_2} = A_2' = \frac{\mu_2 R_3 (R_1 + R_2 + r_{p_1})}{(R_1 + R_2 + r_{p_1})(R_3 + r_{p_2}) - \mu_2 R_3 (r_{p_1} + R_1)}$$

(b)

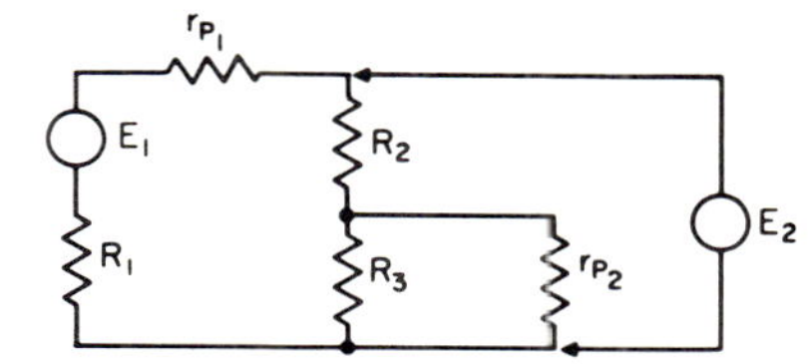

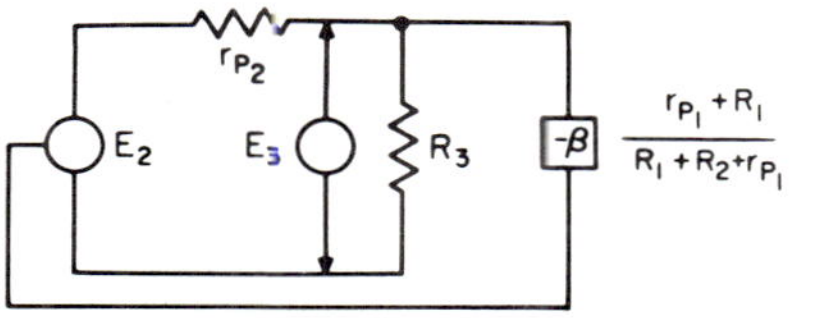

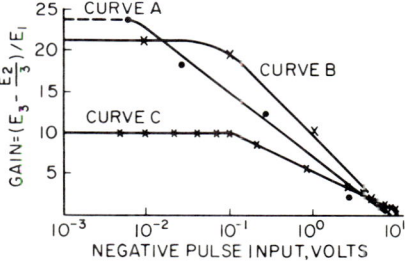

FIG. 15. Equivalent circuit of direct-coupled triode pair. *(From Mound Laboratory Paper MLM-851, 1953.)*

2.3. Special Thermionic Devices. There are several special thermionic devices which are useful in the design of nonlinear amplifiers. Multigrid tubes such as the 6SA7 and the 6AS6 are constructed such that the relationship

$$\Delta I_P = a\, \Delta V_{G_1}\, \Delta V_{G_3} \tag{17}$$

is approximately satisfied, where a is a constant, ΔI_P is the small-signal plate current,

FIG. 16. Comparison of the calculated values of voltage amplication as a function of pulse amplitude for a direct-coupled triode pair with those obtained experimentally. Curve *A*: Actual measurement 68-ohm cathode. Curve *B*: Theoretical 200-ohm cathode. Curve *C*: Actual measurement 200-ohm cathode. *(From Mound Laboratory Paper MLM-851, 1953.)*

and ΔV_{G_1} and ΔV_{G_3} are the small-signal voltages applied to the two control grids of the tube. Although these tubes can be used as analog multipliers, their operation as squarers places less stringent restrictions on the characteristic.

A simple multigrid squaring circuit is shown in Fig. 17. ΔV_{G_1} is made to be pro-

Table 2. Gain-compression Data of Direct-coupled Pair (Calculated)

E_1, mv	μ_1	r_{p_1}	i_{p_1}, ma	μ_2	r_{p_2}	i_{p_2}, ma	E_2, mv	E_3, mv	E_3/E_2	Gain $\frac{E_3 - (E_2/3)}{E_1}$
10	42	26,500	3	44	50,000	0.7	125	253	2.02	21.2
100	42	26,500	3	35	25,000	1.2	1,260	2,380	1.89	19.6
1,000	42	40,000	1.5	34	16,000	2.0	9,430	13,400	1.42	10.2
5,000	34	2×10^5	...	34	16,000	2.0	10,700	13,000	1.21	1.89

$R_1 = 200$ ohms, $R_2 = 10{,}000$ ohms, $R_3 = 5{,}000$ ohms.

portional to ΔV_{G_3}, and the g_m of the tube is varied in proportion to ΔV_{G_1}. Since the change in plate current ΔI_P is proportional to ΔV_{G_3} times g_m,

$$\Delta I_P = k(\Delta V_{G_3})^2 \tag{18}$$

where k is a constant. The curves of Fig. 18 show the performance of such a circuit. The derivative of the output, which should be a linear function, is compared with a straight line.

A special beam-deflection tube, the QK-329, is available which can produce an output current which is proportional to the square of the input voltage at frequencies from direct current up to as high as 40 Mc. This tube achieves its square-law output characteristic by sweeping a beam across a parabolic aperture. Figure 19 shows a

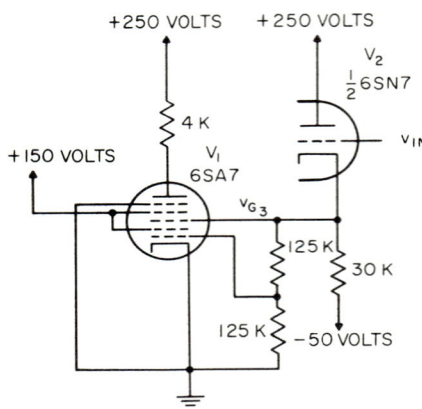

Fig. 17. Simple multigrid vacuum-tube squaring circuit. [From B. Chance et al., "Waveforms," vol. 19, p. 682, Fig. 19.19(a), MIT Radiation Laboratory Series, McGraw-Hill Book Company, New York, 1949.]

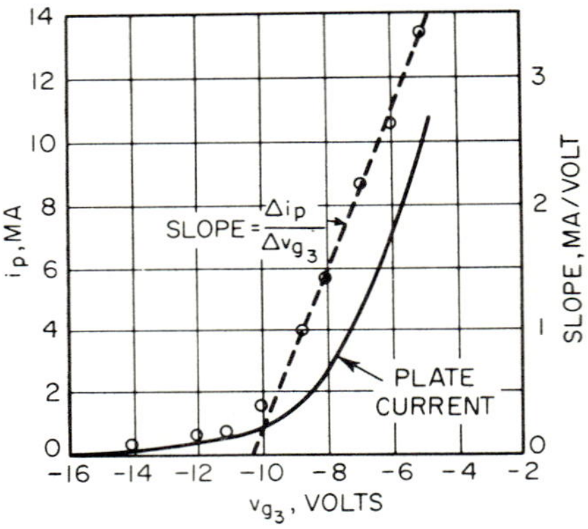

Fig. 18. Performance of the circuit of Fig. 17. [From B. Chance et al., "Waveforms," vol. 19, p. 682, Fig. 19.19(b), MIT Radiation Laboratory Series, McGraw-Hill Book Company, New York, 1949.]

cutaway view of the QK-329. It operates with normal receiving-tube voltages; so its physical size is relatively small as compared with most beam-deflection tubes. The basic equation satisfied by the tube is

$$I_{out} \cong I_0 + K(V_0 + V_{in})^2 \tag{19}$$

where I_{out} = output current from the tube
  I_0 = output current when beam is centered with respect to the target
  K = a constant of the tube
  V_0 = voltage necessary to center beam with respect to the target
  V_{in} = input voltage applied to the tube

Figure 20 shows typical static characteristics of the QK-329, and Fig. 21 gives a comparison of the curved portions of Fig. 20 with a parabola. An analog multiplier circuit which uses the characteristics of the QK-329 is described in Sec. 10.

Multistage electron-multiplier tubes exhibit a useful nonlinear relationship between the applied voltage and the current amplification or gain. The gain G is given approximately by the expression

$$G = (kV)^n \tag{20}$$

where k = a constant of the tube
  V = voltage applied to the tube
  n = number of multiplying stages

NONLINEAR AMPLIFIERS 26-13

The relation between gain and applied voltage for the 931A tube, which has nine stages, is shown in Fig. 22. The gain is seen to be an almost exponential function of voltage over several decades. This characteristic has been used in the design of

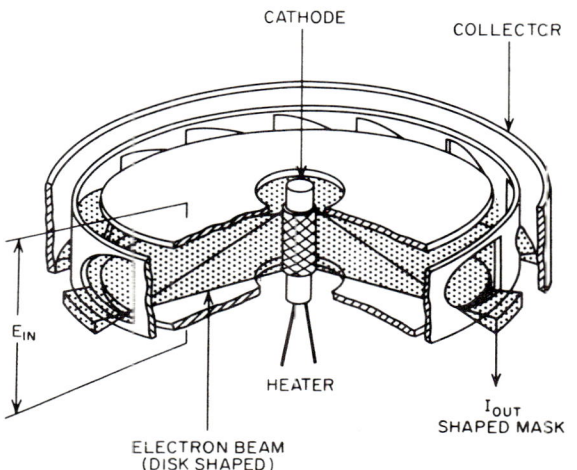

Fig. 19. Cutaway view of the QK-329 beam-deflection tube. (*Courtesy of Raytheon Manufacturing Co.*)

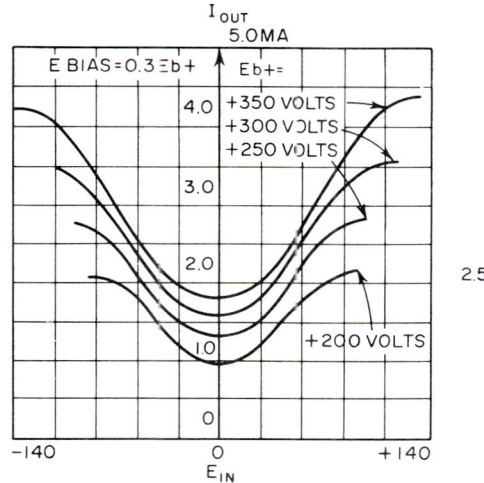

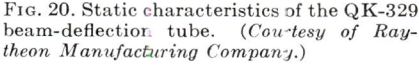

Fig. 20. Static characteristics of the QK-329 beam-deflection tube. (*Courtesy of Raytheon Manufacturing Company.*)

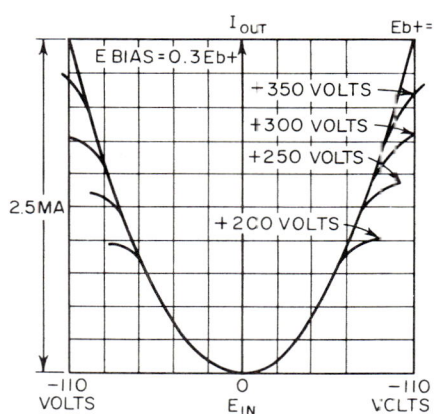

Fig. 21. Comparison of the curved portions of Fig. 20 with a parabola. (*Courtesy of Raytheon Manufacturing Company.*)

wide-range logarithmic radiation meters for nuclear applications, in which case the electron multiplier has a photosensitive cathode and is used with a scintillation crystal.

The final thermionic device exhibiting useful nonlinearities to be considered is the magnetron. A simple magnetron, consisting of a long straight wire emitter surrounded by a coaxial cylindrical plate, both of which are placed in a uniform magnetic field directed along the axis, exhibits the following relationship:

$$V_{co} = KH^2 \qquad (21)$$

where V_{co} is the plate potential at the point of current cutoff, K is a constant, and H is the applied magnetic field strength. Figure 23 shows the current-voltage characteristics of such a device for two values of H. Since the value of H is proportional to the current in a solenoidal coil producing the field, a magnetron together with a solenoid can form the nucleus of a squaring circuit.

2.4. Semiconductor Diodes. Semiconductor diodes of various types have quite useful nonlinear characteristics. The basic relation between voltage and current for a semiconductor p-n junction diode is given by the following equation:

$$I = I_0\{\exp[(q/mkT)(V - IR_s)] - 1\} \qquad (22)$$

where I_0 = reverse saturation current of diode
q = electron charge = 4.77×10^{-10} esu
k = Boltzmann's constant = 1.37×10^{-16}
T = absolute temperature of diode junction
R_s = equivalent series internal resistance of diode

The factor m in this equation represents the effects of diffusion-current flow and carrier generation-recombination effects. It usually has a value between 1.0

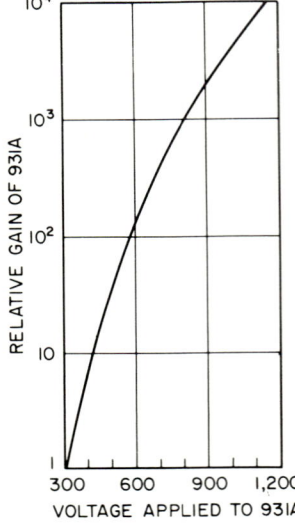

FIG. 22. Curve showing gain as a function of applied voltage for a typical electron multiplier tube. (*From Rev. Sci. Instr., vol. 23, p. 301, 1952.*)

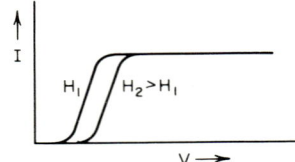

FIG. 23. Current-voltage characteristics of a magnetron for different values of magnetic field strength. (*From B. Chance et al., "Waveforms," vol. 19, p. 692, Fig. 29.32, MIT Radiation Laboratory Series, McGraw-Hill Book Company, New York, 1949.*)

and 2.0 and can vary slightly as the current changes. For silicon diodes Eq. (22) may be simplified to the following expression over a considerable range of forward current:

$$I = I_0 \exp(q/mkT)V \qquad (23)$$

Equation (23) may then be rewritten as

$$V = mK_0 T(\log_{10} I - \log_{10} I_0) \qquad (24)$$

where K_0 is a constant. Thus the change of diode voltage is seen to be a direct logarithmic function of the change of diode current. This relationship holds over a range of up to seven decades of diode current for some silicon diodes, as shown by the curves of Fig. 24.

Figure 24 also shows some of the limitations of using the forward characteristics of semiconductor diodes for logarithmic-function generation. The most obvious of these is the change of diode voltage with temperature for a constant diode current, which for silicon diodes is about -2.4 mv/°C temperature rise. The most satisfactory method of overcoming this limitation is to maintain the diodes at a constant temperature by the use of an oven or a cooler. Use of a cooler is preferable, since the diodes have a greater logarithmic range at lower temperatures. When it is not feasible to maintain the diodes at a constant temperature, various types of compensation networks may be used to reduce the amount of the voltage shift. Figure 25 shows a simple compensation technique which may be used in some applications. The compensating diodes and the load resistor form a temperature-sensitive voltage divider that tends to compensate for the voltage change across the logging diodes as the temperature changes.

The deviations from a true logarithmic characteristic at low and high diode currents in the curves of Fig. 24 may be better understood by examining Eq. (22) in more detail. As V approaches zero it is seen that I also approaches zero, thereby explaining the downward slope of the curves at low diode currents. At high values of diode current the term containing R_s becomes important and causes the voltage to increase at a higher rate. The deviations in the center parts of the curves result from changes in the value of the factor m as the current is changed. The use of several diodes in series to form a composite logarithmic element reduces the effect of these deviations but cannot completely eliminate them.

Semiconductor diodes may also be used to give nonlinearities of the second type.

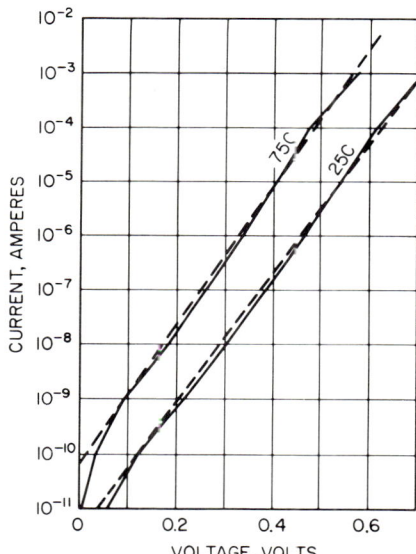

Fig. 24. Current-voltage characteristics of a typical silicon junction diode. (*From Electronics, Aug. 1, 1957, p. 198.*)

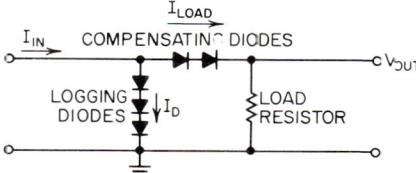

Fig. 25. Simple circuit for temperature compensation of silicon logarithmic diodes.

If a d-c control current I is applied to a silicon diode exhibiting the logarithmic relationship of Eq. (23) the dynamic resistance R_{AC} will be related to I as follows:

$$R_{AC} = dV/dI = mK_0T/I \qquad (25)$$

Thus diodes of this type may be used to provide amplitude compression or expansion in an a-c amplifier circuit by making the current I a function of the output or the input signal. Figure 26 illustrates the application of this technique in an automatic gain control circuit. The input a-c current divides between the a-c diode resistance R_D and the a-c input resistance of the transistor stage. As the input signal increases, the control voltage V_{AGC} is caused to increase, and hence R_D decreases, which lowers the gain of the stage. In order for this circuit to perform satisfactorily, the signal source must have a high resistance and the input resistance of the transistor must be relatively high. A design example of a complete low-noise low-level amplifier using this technique is given in Sec. 6.

The inverse relationship between the d-c current through a silicon diode and its dynamic resistance may also be used to advantage in the design of analog divider circuits. If a diode is driven by both d-c and a-c input current sources, as shown in the diagram of Fig. 27, the a-c voltage output across the diode $V_{out_{ac}}$ will be approximately proportional to the quotient of $V_{in_{ac}}$ and $V_{in_{dc}}$. An example of a complete analog divider using this principle is given in Sec. 10.

The reverse characteristics of silicon breakdown diodes have been used to obtain nonlinearities of both the first and second types. Figure 28 shows the overall current-voltage characteristic of a typical breakdown or "zener" diode. In region 3 of the plot, which is the transition region between the forward and the zener regions, the

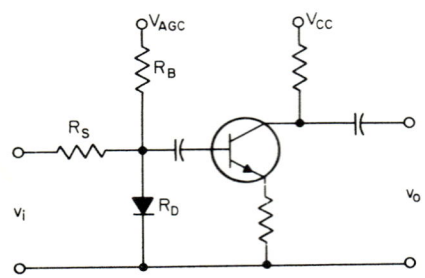

Fig. 26. Automatic gain control circuit using a silicon diode for amplitude compression. [*From Texas Instruments, Inc., "Transistor Circuit Design," p. 177, Fig. 11.4(d), McGraw-Hill Book Company, New York, 1963.*]

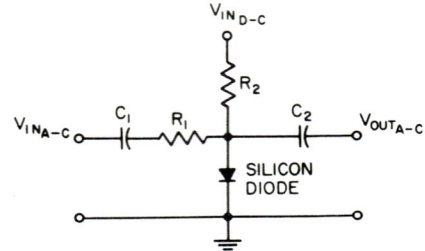

Fig. 27. Basic analog divider circuit using a silicon diode.

voltage is nearly a logarithmic function of current over as many as five decades of current. Representative plots for two 5-volt zener diodes are shown in Fig. 29. Note that the voltage change per decade change of current is approximately 0.8 to 0.9 volt compared to less than 0.1 volt for silicon diodes operated in the forward direction. However, the mechanism responsible for the logarithmic characteristics of these reverse-biased zener diodes is not clearly understood, and the logarithmic characteristics have not proved to be so reproducible in manufacture as the forward logarithmic characteristics.

Germanium point-contact diodes, although obsoleted for many applications by the development of high-quality silicon junction diodes, are still used in many high-frequency gain-control circuits because of their low shunt capacitance. A 70-Mc solid-state i-f amplifier that uses diodes of this type in a feedback control circuit to maintain constant output level is described in Sec. 6.

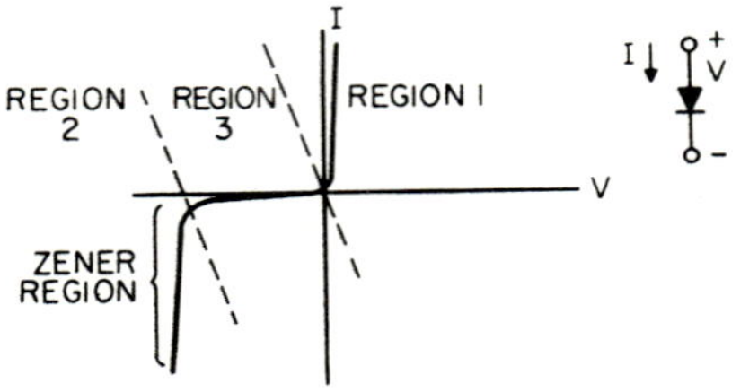

Fig. 28. Overall current-voltage characteristic of a silicon breakdown diode.

The *p-i-n* diode has nearly ideal characteristics for use as a gain-control device at high frequencies. In the frequency band from 50 to 250 Mc this device looks like a resistor whose resistance is controlled by the d-c control current through conductivity modulation. The junction is essentially bypassed by the junction capacitance at these frequencies. The control characteristics of a typical *p-i-n* diode are illustrated by the curves of Fig. 30.

Certain types of germanium backward diodes have a forward current-voltage characteristic that closely approximates the expression $I = kV^2$. Figure 31 shows the current-voltage characteristics of a typical backward diode plotted on a loglog scale. It is seen that this diode has a square-law characteristic over a voltage range from about 50 to 500 mv, with the constant k having a value of about 135 ma/volt². Thus devices of this type could prove useful in low-level squaring and square-root-extracting circuits.

2.5. Semiconductor Triodes and Tetrodes. There are several useful nonlinear relations exhibited by semiconductor triodes and tetrodes. Some of these are similar

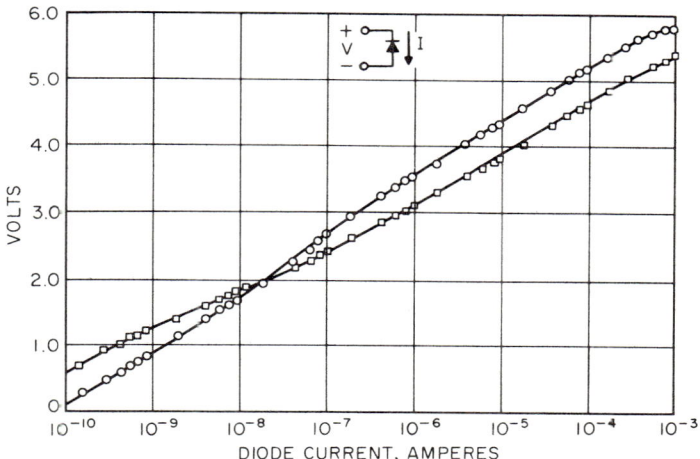

FIG. 29. Logarithmic characteristics of two 5-volt silicon breakdown diodes. (*Courtesy of MIT.*)

to relations that have been discussed for semiconductor diodes, while others are of a type that is possible only with devices having more than two terminals.

As the emitter current of a silicon unijunction transistor (double-base diode) is increased, the interbase resistance of the device will decrease because of conductivity modulation of the silicon in the interbase region. This property of the unijunction transistor makes it quite useful as a wide-band variable-resistance element in amplitude compression or expansion circuits. The interbase resistance can be varied from approximately 8,000 ohms with no emitter current to less than 100 ohms when the emitter current is on the order of 10 ma, and it is linearly related to the emitter current in the range of 1 to 5 ma.

The unijunction transistor is particularly useful as a gain-control element at frequencies higher than 1 Mc, as the interbase resistance then becomes independent of the base-to-base voltage. Pulses having 0.5-nsec rise time and 50-nsec width have been controlled in amplitude over a 10 to 1 range with no measurable degradation of rise time or pulse shape. The circuit shown in Fig. 32 illustrates the application of the unijunction transistor to high-frequency circuits.

When the collector voltage of a silicon planar transistor is held constant, the base-emitter junction exhibits a voltage-current relationship similar to that exhibited by a silicon junction diode. If the transistor is connected as a feedback element in an

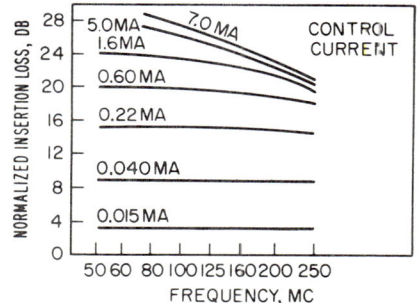

FIG. 30. Control characteristics of a typical *p-i-n* diode as a function of frequency. (*From NEREM Record, 1963, p. 61.*)

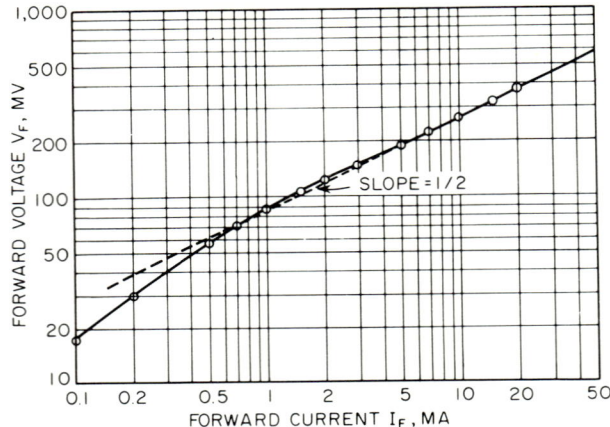

Fig. 31. Forward square-law characteristics of a germanium backward diode. (GE type number BD-1.) $I_p = 1$ ma, $C_v = 2.1$ pf. (*Courtesy of General Electric Co.*)

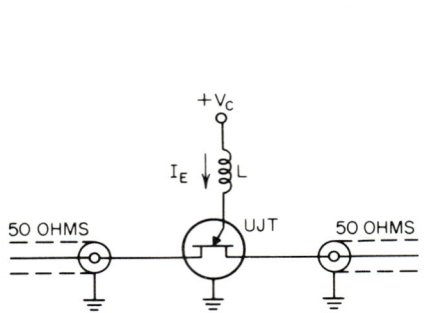

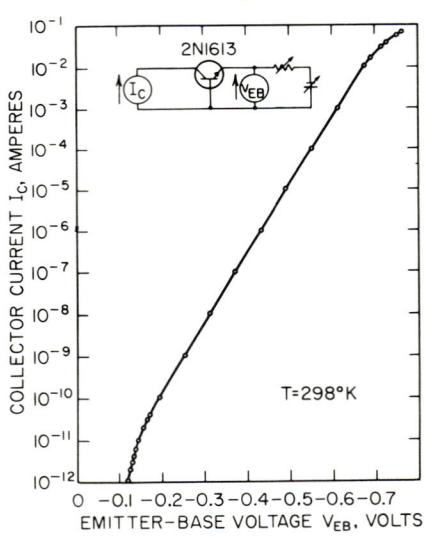

Fig. 32. Application of a unijunction transistor as a gain-control element for high-frequency circuits. (*Courtesy of General Electric Co.*)

Fig. 33. Logarithmic current-voltage characteristic for a common-base silicon junction transistor. (*From SSCC Digest of Technical Papers, 1963, p. 39.*)

operational amplifier, with the output of the amplifier fed to the base of the transistor, an improved performance for large input currents may be achieved over that obtainable with diodes. However, the variable factor m of Eq. (23) still limits the accuracy obtainable with this connection. The application of this transistor relationship to logarithmic amplifiers is discussed in Sec. 7, and extensions to square-law circuits and multiplier/divider circuits are given in Secs. 9 and 10.

If the collector-base junction of a silicon junction transistor is held at zero voltage, the collector current I_C is approximately related to the emitter-base voltage V_E in the following manner:

$$I_C = I_S \exp{(q/kT)V_E} \qquad (26)$$

where I_S is a constant. It will be noted that this equation differs from that of Eq. (23) for diodes in that the dependence on the factor m is eliminated. Thus an improved direct logarithmic characteristic may be obtained, as shown by the plot of Fig. 33. An example of the use of a silicon transistor in a logarithmic amplifier cir-

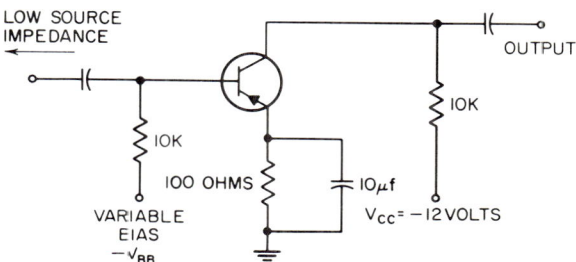

Fig. 34. Variable-gain transistor stage. (*From Hunter, "Handbook of Semiconductor Electronics," 2d ed., p. 11–93, Fig. 11.128, McGraw-Hill Book Company. New York, 1962.*)

cuit is given in Sec. 7, and the application to multiplication circuits is shown in Sec. 10.

The a-c transfer characteristics of a semiconductor triode or tetrode tend to vary with the d-c voltages or currents in the circuit containing the device. If one of these d-c voltages or currents is controlled by the amplitude of the a-c signal input to or output from the circuit containing the device, the a-c transfer characteristics of the device will be nonlinear. Several relationships of this type will be discussed in the following paragraphs.

At low and medium frequencies the value of the transistor parameter h_{ib} increases significantly as the d-c base current is decreased. Since the gain of an amplifier stage can be made to decrease as the value of h_{ib} increases, this relationship is sometimes used to achieve amplitude compression or expansion in an amplifier stage of the type shown in Fig. 34. The gain may be controlled over a range of about 20 db by this technique, as shown by the plot of Fig. 35. Complete amplifier circuits using this and similar techniques are described in Sec. 6.

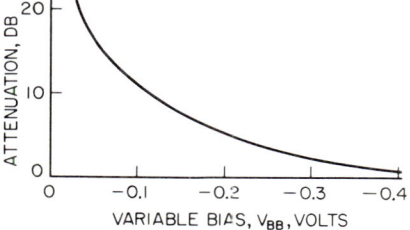

Fig. 35. Control characteristic for transistor stage of Fig. 34. (*From Hunter, "Handbook of Semiconductor Electronics," 2d ed., p. 11–93, Fig. 11.129, McGraw-Hill Book Company, New York, 1962.*)

Another method of using transistors to achieve amplitude compression or expansion at low and medium frequencies is to cause their dynamic resistance to vary in response to a control signal. Figure 36 shows a simple circuit of this type, in which transistor Q_2 serves as a variable shunt resistor to control the amplification of the a-c signal applied to transistor Q_1. Most of the control action of Q_2 is in the region of its I_C-V_{CE} characteristics near saturation, where the slope of the characteristic is changing most rapidly.

At high frequencies the value of the parameter h_{fe} in germanium mesa transistors varies considerably as the d-c collector current and/or the d-c collector-to-emitter voltage changes. This variation in h_{fe} may be used in either of two ways to provide automatic gain control in high-frequency amplifiers. First, the gain of a transistor stage may be controlled by forcing the collector-to-emitter voltage to remain relatively constant and causing the d-c collector current to decrease as the a-c input signal increases. This technique is sometimes called "reverse AGC." Second, the gain may be controlled by forcing the d-c collector-to-emitter voltage to decrease and the d-c collector current to increase as the a-c input signal increases. This technique is sometimes called "forward AGC." The curves and waveforms of Fig. 37 illustrate how these two techniques differ for a mesa transistor operating at a frequency of 100 Mc.

Consider first the case of typical "reverse AGC" conditions. Assume that the d-c collector-to-emitter voltage is held at approximately 10 volts and a "starting" d-c collector current of 2 ma is flowing. At this starting point, labeled A on the plot

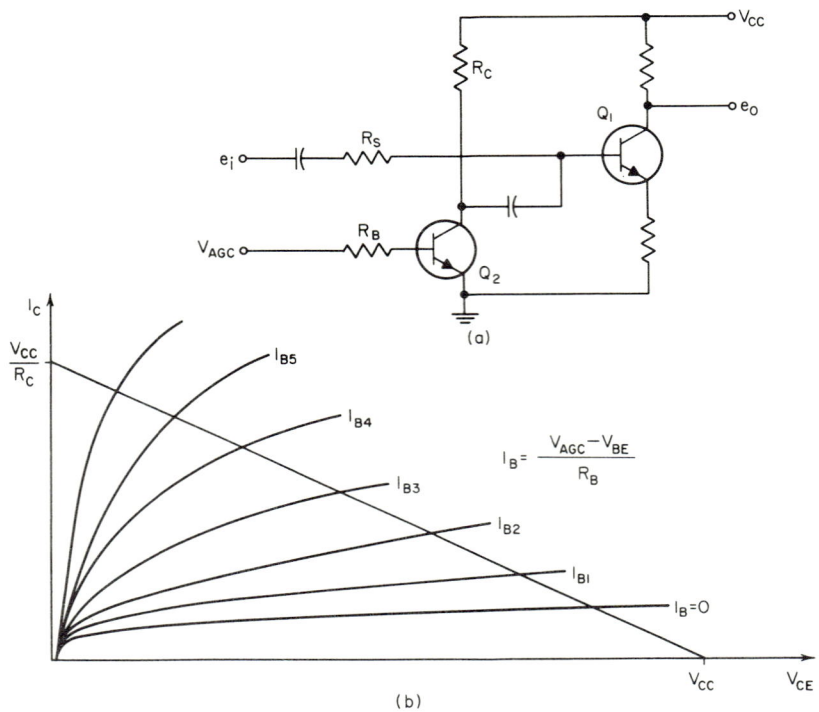

Fig. 36. Gain-control stage using a transistor as a variable shunt resistor. (*From Texas Instruments, Inc., "Transistor Circuit Design," p. 178, Fig. 11.5, McGraw-Hill Book Company, New York, 1963.*)

of Fig. 37, a 100-Mc input signal voltage e_{s1} is shown to produce a small-signal collector current of i_{c1}. If the d-c collector current is caused to decrease to a "stopping" value of 0.25 ma, labeled B on the plot, the 100-Mc gain is seen to be reduced. Under these conditions the maximum input signal which can be handled without excessive distortion is about 60 mv rms, as shown by e_{s2} and i_{c2}.

Now consider the case of typical "forward AGC" conditions. Assume that a d-c supply voltage of 12 volts is used and a total d-c load resistance of 1,200 ohms is in the circuit. This gives a "starting" d-c collector-to-emitter voltage of 6 volts at a collector current of 5 ma. At this starting point, labeled C on the plot of Fig. 37, a 100-Mc input signal voltage e_{s3} is shown to produce a small-signal collector current of i_{c3}. If the d-c collector current is caused to increase to a "stopping" value of 9 ma, labeled D on the plot, the d-c collector-to-emitter voltage will have dropped to 1.25

NONLINEAR AMPLIFIERS 26-21

volts, and the 100-Mc gain is seen to be reduced from that at point C. Under these conditions input signals up to 200 mv rms may be handled without excessive distortion, as shown by e_{s4} and i_{c4}. Thus the "forward AGC" technique is seen to be more desirable than the "reverse AGC" method from the standpoint of signal-handling capability. For low input signal levels either technique is satisfactory.

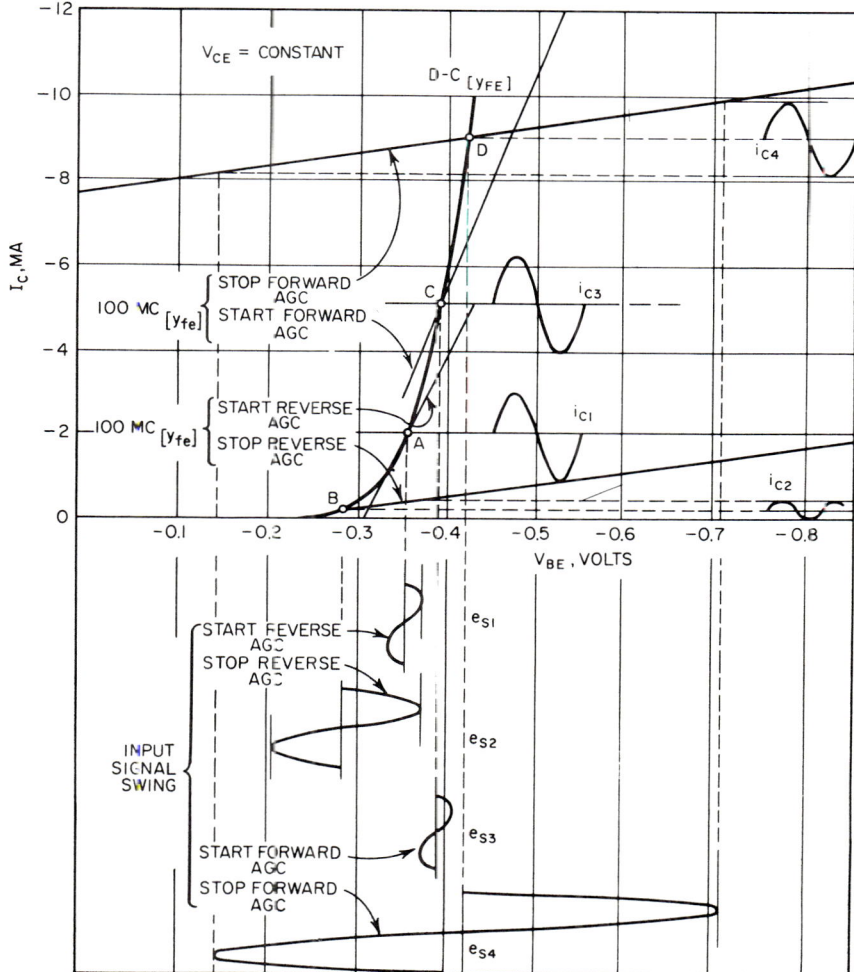

FIG. 37. Curves and waveforms illustrating "forward" and "reverse" AGC characteristics of germanium mesa transistors at high frequencies. (*From Texas Instruments, Inc., "Transistor Circuit Design," p. 330, Fig. 24.1, McGraw-Hill Book Company, New York, 1963.*)

High-frequency transistors are now available which have a gradually decreasing value of 100-Mc h_{fe} as their emitter current is increased. A comparison of this characteristic with that exhibited by most high-frequency transistors is shown in Fig. 38. Use of transistors having the characteristic shown as type 1 in this plot will allow the a-c amplification in an AGC circuit to be easily reduced while the emitter current is increased, thus providing a larger dynamic range to be handled without excessive signal distortion. It is possible to fabricate both germanium and silicon transistors that exhibit this type of h_{fe} vs. emitter current characteristic.

Tetrode transistors may be used to achieve a variable-gain characteristic which is superior in some ways to that obtainable with semiconductor triodes. The two base leads of a tetrode may be viewed as a two-terminal resistor through which a control may be passed to produce a transverse voltage drop. If a positive control current I_{B2} is applied to base 2, the common-emitter current gain of the device h_{fe} will increase with increasing I_{B2} up to a certain point and then will start decreasing, giving an overall curve of h_{fe} versus I_{B2} as shown in Fig. 39. The effect of increasing I_{B2} causes the common-emitter input resistance h_{ie} to increase in the manner shown by Fig. 40.

The voltage-amplification factor A_v of a tetrode transistor amplifier stage at audio frequencies is given approximately by the expression

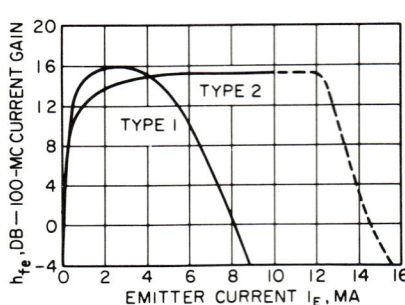

FIG. 38. Current-amplification factor as a function of emitter current for two types of high-frequency transistors. (*Courtesy of Texas Instruments, Inc.*)

$$A_v = \frac{h_{fe}R_L}{h_{ie} + R_g} \tag{27}$$

where R_L = effective collector load resistance
R_g = generator series resistance

If the amplifier is driven by a voltage source, $R_g \cong 0$, and the expression for A_v reduces to

$$A_v \cong y_{fe}R_L \tag{28}$$

where $y_{fe} = h_{fe}/h_{ie}$.

The curve of Fig. 41, obtained from those of Figs. 39 and 40, gives a good indication of the expected behavior of a silicon tetrode in an audio AGC stage.

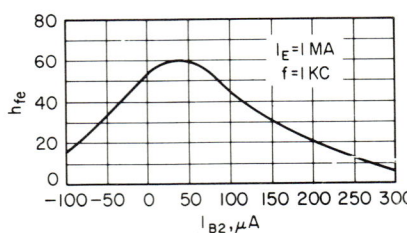

FIG. 39. Current-amplification factor as a function of base-2 current for a silicon tetrode transistor. (*Courtesy of Texas Instruments, Inc.*)

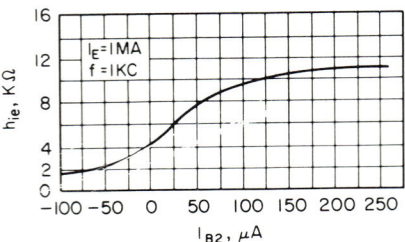

FIG. 40. Common-emitter input resistance as a function of base-2 current for a silicon tetrode transistor. (*Courtesy of Texas Instruments, Inc.*)

The control range of a single tetrode stage is limited by the low BV_{EBO} of some devices. Since the base-to-base voltage drop of a tetrode is approximately I_{B2} times the base-to-base resistance $R_{B\text{-}B}$, the maximum control current allowable before base-emitter breakdown might occur in this device is

$$I_{B2(max)} = \frac{BV_{EBO} + V_{BE(on)}}{R_{B\text{-}B}} \tag{29}$$

For a typical device such as the 3N34, which has a BV_{EBO} of 1 volt, a $V_{BE(o1)}$ of about 0.75 volt, and an R_{B-B} of about 15 kilohms, the value of $I_{B2(max)}$ is about 115 μa. The available control range with this device is slightly less than 20 db. A device with a higher BV_{EBO}, such as the TI874, can give a substantially greater range of con-

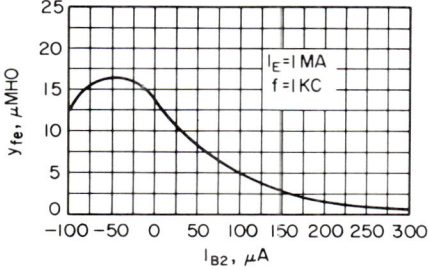
FIG. 41. y_{fe} as a function of base-2 current for a silicon tetrode transistor. (*Courtesy of Texas Instruments, Inc.*)

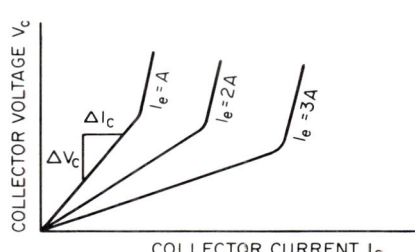
FIG. 42. Collector voltage–collector current characteristics of a point-contact transistor. (*From Electronic Equipment Eng., August, 1960, p. 66.*)

trol. An amplifier circuit using this type of variable-gain characteristic is described in Sec. 6.

In high-frequency amplifier applications, tetrode transistors such as the 2N34 are particularly useful because increases of I_{B2} and I_E both produce decreases in the h_{fe} of the device. Furthermore, by varying both I_{B2} and I_E, the input and output

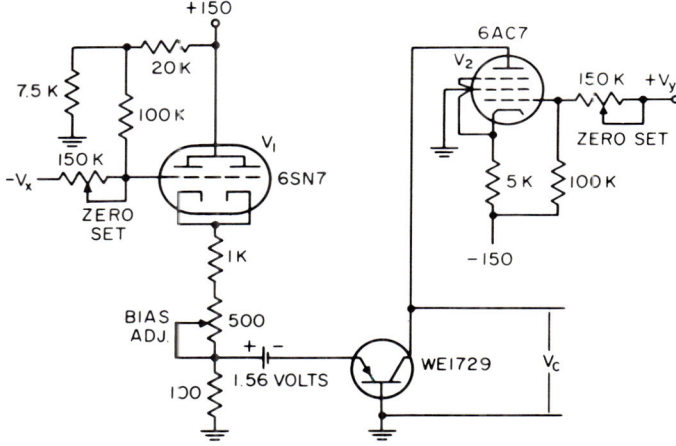
FIG. 43. Amplitude-multiplier circuit using a point-contact transistor. (*From Electronic Equipment Eng., August, 1960, p. 66.*)

impedance variations can be made to compensate for each other, and thus a flat frequency response can more easily be maintained. Examples of high-frequency amplifiers using this technique are given in Sec. 6.

The collector characteristics of a point-contact transistor may be used in the construction of a simple analog multiplier. As shown by the curves of Fig. 42, the resistance of the collector-base diode is a linear function of the emitter current-collector voltage characteristics over a relatively wide range. The collector current I_C flowing through this collector-base resistance R_C creates a linearly increasing collector-to-base voltage drop V_C; so

$$V_C = I_C \times R_C = kI_E I_C \tag{30}$$

where k is a constant.

A circuit which may be used to implement this relationship is shown in Fig. 43. The emitter current is controlled by a voltage V_x fed to the grid of V_1. The collector current is supplied from tube V_2 serving as a constant-current generator. The frequency response of this multiplier circuit is quite good, as it is limited only by the f_{co} characteristics of the transistor and by the wiring and tube interelectrode capacitances. The accuracy is about 5 per cent of full-scale output for a range of V_y from 0 to 25 volts and of V_x from -2 to -15 volts. The maximum output is -10 volts.

The field-effect transistor has several useful nonlinear characteristics that may be applied to nonlinear amplifier circuits. First, there is an inherent quasi-logarithmic relationship between the a-c voltage amplification A_v in a simple amplifier stage and the d-c gate bias V_{GS}. The nature of this relationship is shown by the curve of Fig. 44.

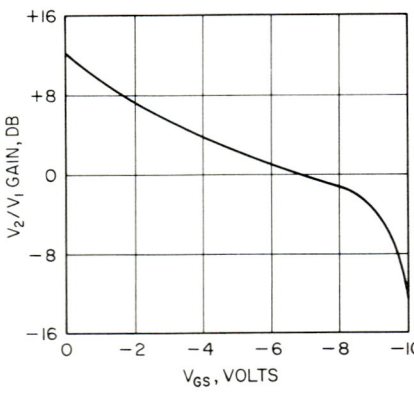

FIG. 44. Gain of a field-effect transistor stage as a function of d-c gate bias. (*Courtesy of Crystalonics, Inc.*)

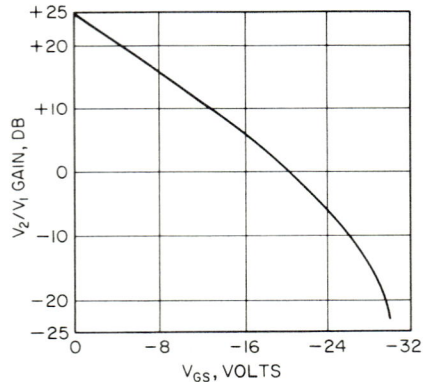

FIG. 45. Control characteristics of a transistor stage employing a field-effect transistor as a voltage-variable resistor. (*Courtesy of Crystalonics. Inc.*)

Another characteristic is the action of the field-effect transistor below pinch-off as a voltage-variable resistance. In this region, a variation of the gate bias produces a corresponding change in the dynamic resistance from drain to source. By using a field-effect transistor as a variable emitter resistor in an *n-p-n* transistor amplifier stage, the stage gain may be controlled over a relatively wide range. Figure 45 shows the relationship between the stage gain and the field-effect transistor gate bias for a typical stage of this type.

The final use of field-effect transistors to be discussed is in amplitude-multiplying circuits for very short pulse signals. The multiplication effect is based on the proportionality of the channel conductance of a junction gate field-effect transistor to the gate voltage. The simplest schemes use two field-effect transistors in a bridge configuration. One variable, a voltage, is applied with opposite polarities to two appropriately biased gates, increasing the conductance of one channel and decreasing the conductance of the other one. Another variable is applied as a voltage or a current to the channels. The difference in channel currents is then proportional to the product of the two variables. A four-quadrant pulse amplitude multiplier based on this principle is described in Sec. 10.

NONLINEAR AMPLIFIERS   26-25

2.6. Varistors and Thermistors. Varistors and thermistors are nonlinear devices which are sometimes useful in nonlinear amplifier circuits. Varistor is the general name given to voltage-sensitive resistors, while thermistors, as the name implies, are resistive elements primarily sensitive to temperature.

The most common types of varistors are made by pressing together silicon carbide granules with a suitable binder under high pressure and firing the mixture at high temperatures. The resultant devices exhibit the following type of nonlinear relationship:

$$I = kV^n \tag{31}$$

where I = instantaneous a-c or d-c current through the device
$\quad k$ = a constant which depends on the resistivity and dimensions of the varistor under question
$\quad V$ = instantaneous a-c or d-c voltage applied to the device
$\quad n$ = an exponent which usually varies between 2 and 5

The exponent n is approximately constant for a given varistor type but is usually not independent of k. Higher values of n are associated with low values of k. Fig-

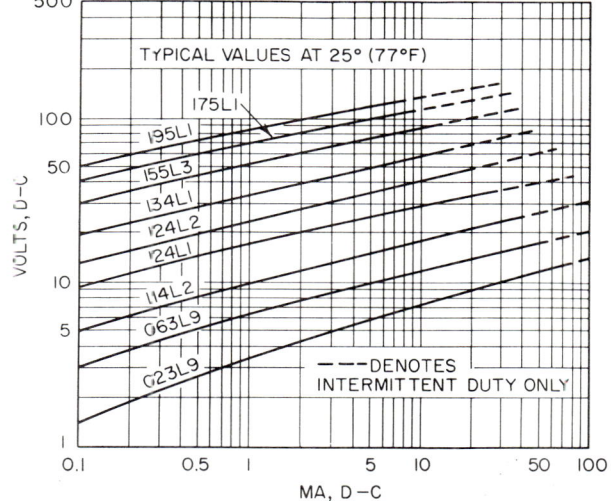

Fig. 46. Voltage-current characteristics of several commercial silicon carbide varistors. (*Courtesy of Victory Engineering Corporation.*)

ure 46 shows the voltage-current characteristics of several commercial silicon carbide varistors.

Varistors may be used as the control elements in simple amplitude compression or expansion circuits. Figure 47 shows a circuit which uses a varistor to limit the dynamic range of any negative input signal without allowing it to reach a threshold or saturation level. The varistor used in this circuit satisfies the relationship $V = KI^{\frac{1}{3}}$. Since transistor Q_1 delivers a current to the varistor which is proportional to the input voltage, and since the varistor voltage determines the current into the base of the second transistor, the output voltage of the circuit is approximately proportional to the cube root of the input voltage. This circuit will handle negative input signals up to 200 mv and furnish outputs up to 3 volts. The frequency response of the circuit as shown is limited to the audio range, but frequencies up to several megacycles can be handled if a suitable inductance is placed in series with the varis-

26-26   AMPLIFIER CIRCUITS

tor and if high-frequency transistors are used. Another circuit using varistors as a voltage-controlled variable attenuator is described in Sec. 6.

By adding a fixed resistance in series with a varistor, the voltage-current relationship of Eq. (31) may be modified to make the exponent n take on a value of exactly 2. Thus varistors may be used as squaring and square-rooting elements in low-speed

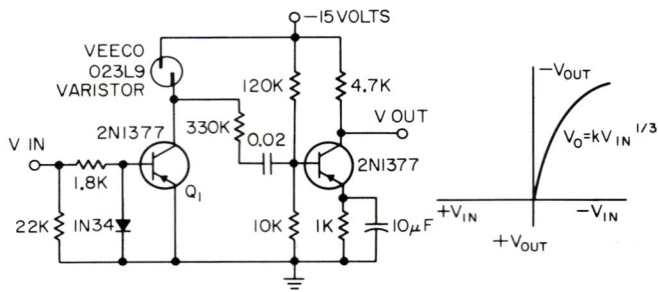

FIG. 47. Circuit using a silicon carbide varistor as a dynamic range compressor. (*From Electronic Equipment Eng., February, 1963, p. 25.*)

analog computing circuits. The chief disadvantage of the use of varistors for these applications is their relatively poor long-term stability as compared with p-n junction semiconductor devices.

Thermistors are usually constructed from the oxides of manganese, nickel, and cobalt. By blending these a wide variety of specific resistances and negative temperature coefficients may be obtained. The resistance as a function of temperature is given approximately by the equation

$$R = R_0 \exp[\beta(1/T - 1/T_0)] \quad (32)$$

where R_0 = resistance of the device at the reference temperature
β = a constant of the material
T = temperature of the device
T_0 = reference temperature

In Fig. 48 the logarithm of the resistance is plotted vs. $1/T$ for several thermistors. The curves are seen to approximate straight lines of slope β. β decreases slightly at very low temperatures and increases at very high temperatures.

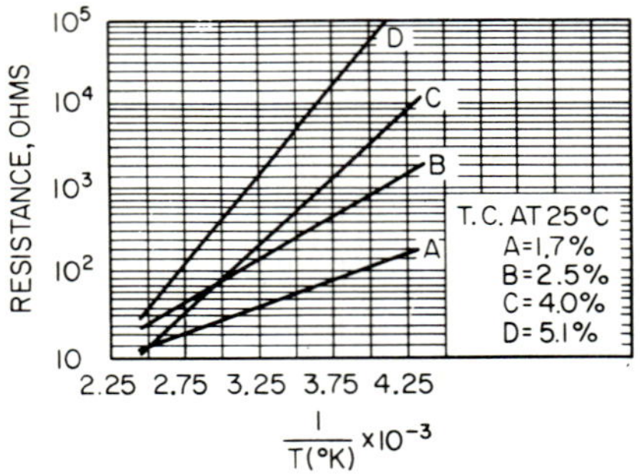

FIG. 48. Resistance vs. temperature plots of several commercial thermistors. (*Courtesy of Thermistor Corp. of America.*)

Both directly heated and indirectly heated thermistors are available, and each may be used in gain-control applications. A directly heated unit may be connected directly into either an attenuating circuit or a negative-feedback circuit of an amplifier in such a manner that its temperature, and hence its resistance, will decrease as the input signal amplitude increases, thus giving an automatic gain control action. An indirectly heated unit may also be used for automatic gain control, as shown in Fig. 49. The negative feedback in this arrangement is accomplished by thermal-energy transfer between the heater and the temperature-sensitive element.

Varistors and thermistors may be used in combination to obtain a temperature-compensated square-law element. Figure 50 shows the simplified circuit of a function-generator element that is offered commercially. It has a maximum rating of 100 volts, a maximum input resistance of 20 kilohms, and a maximum output impedance of 4 kilohms. The rms noise is on the order of 50 mv.

2.7. Vacuum Thermoelements. Vacuum thermoelements are useful nonlinear devices, particularly in instrument applications. These devices produce an output voltage proportional to the heating effect of an applied input signal, thus giving a square-law relationship. The basic characteristics of a line of vacuum thermoelements are listed in Table 3. It will be noted that these devices are useful from direct current up to quite high frequencies, making them ideal for wideband applications.

A thermal converter circuit employing vacuum thermocouples is shown in Fig 51. It is self-balancing, with the output voltage of thermocouple TC-2 always maintained at a constant level. With no input signal the output of this thermocouple is maintained by a d-c bias current applied to its heater. When an input signal is applied, its heating effect unbalances the circuit instantaneously. Balance is restored by causing a reduction of the d-c bias current, with the change in bias current being proportional to the rms value of the input signal. A second thermocouple, TC-1,

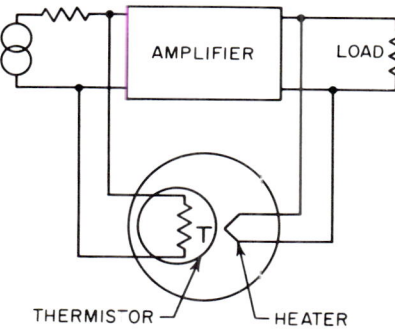

FIG. 49. One method of using an indirectly heated thermistor to provide automatic gain control for an amplifier. (*From Electronic Equipment, June, 1956.*)

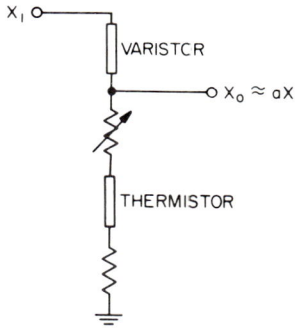

FIG. 50. Simplified circuit of a commercial function-generator element using a varistor and a thermistor. [*From H. D. Huskey and G. A. Korn (eds.), "Computer Handbook," p. 3-75, Fig. 3.3.17, McGraw-Hill Book Company, New York, 1962.*]

serves as an ambient-temperature compensator. This circuit has a frequency response from 15 cps to 50 Mc and is used in a highly accurate true rms voltmeter.

2.8. Opto-electronic Devices. An opto-electronic device is one in which a photoconductive cell is combined with a controlled light source, with the combination mounted in a light-tight case. Devices of this type provide ideal gain-control characteristics for many applications. The cell resistance may be controlled over quite a wide range of values, and the devices may be used as control elements over a frequency range from direct current up to several hundred kilocycles. The response time may be made as short as a few milliseconds by the use of ionized-gas light sources. Table 4 lists the characteristics of a line of commercial opto-electronic devices.

Two methods of applying opto-electronic devices as amplitude-compression or -expansion elements are shown in Fig. 52. In the circuit shown as (*a*), the photoconductive cell is used to give shunt control of the input current to the transistor. As shown in the output vs. control voltage plot, a control range of 10 db may be achieved by this connection. In the circuit shown as (*b*), the photocell shunts the collector load resistor to provide gain control. As shown in the output vs. control voltage plot, this method of application allows control ranges up to 30 db. A transistorized audio AGC amplifier employing two opto-electronic devices as control elements is described in Sec. 3.

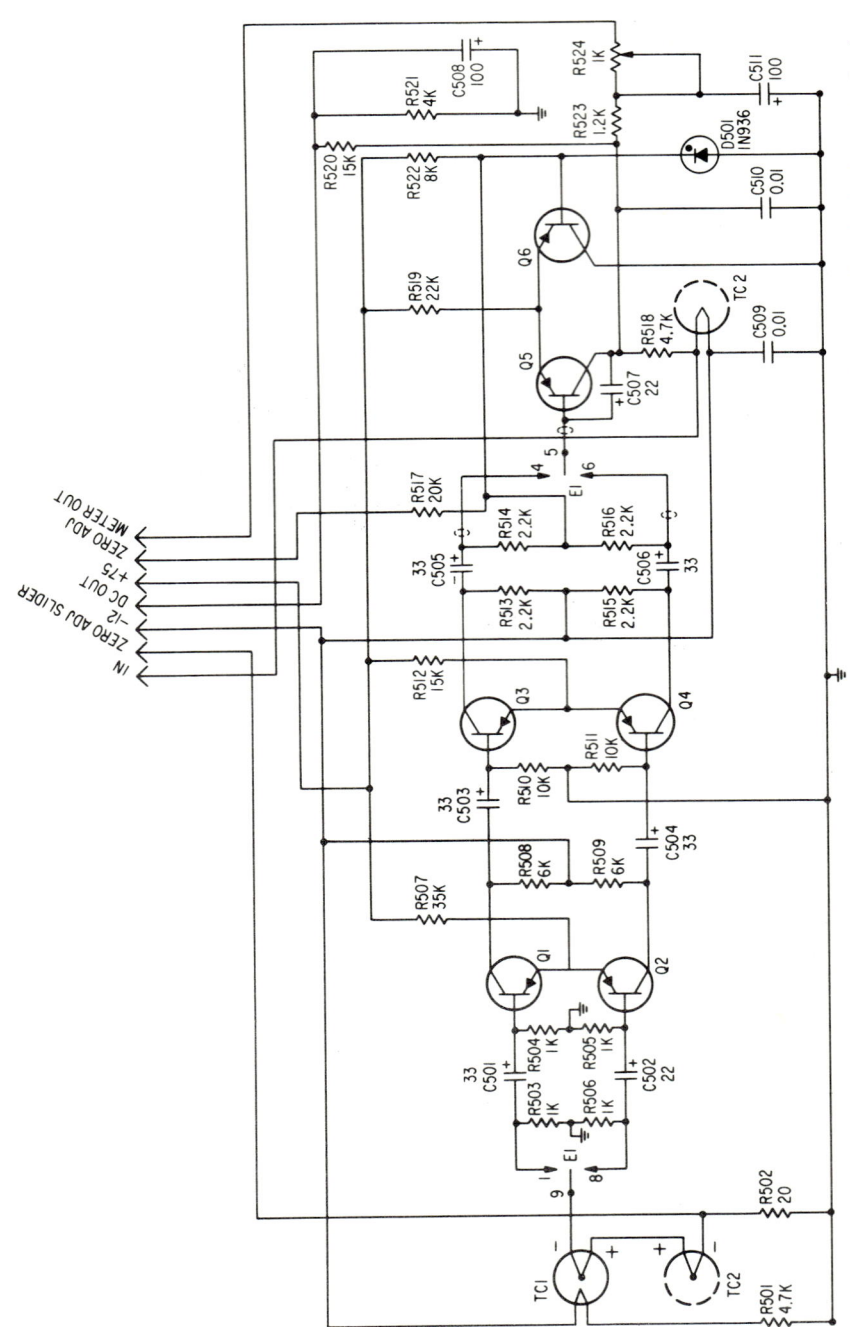

Fig. 51. Circuit diagram of a square-law thermal converter module using vacuum thermocouples. (*Courtesy of Keithley Instruments, Inc.*)

Table 3. Characteristics of Vacuum Thermoelements*

Range rated, ma†	Heater ohms ±10%	Approx heater drop at rated current, mv	Couple ohms ±10% Contact type	Couple ohms ±10% Insulated type	Max safe heater current, ma	Frequency for 2% error, Mc§
1.5	1,365	2,000	7.5	10	3.2	300
2	750	1,500	6	10	5.0	300
2.5	475	1,200	6	10	6.2	300
4	95	380	9	10	9.0	300
5	90	450	6	10	10.0	300
7.5	40‡	300	6	10	16	300
10	27‡	270	6	10	25	300
15	13	195	6	10	40	300
20	8.4	170	6	10	50	300
25	7.0	175	6	10	62	300
30	5.8	175	3	5	75	300
37.5	4.6	170	3	5	85	300
50	3.3	165	3	5	115	300
75	1.36	100	3	5	170	200
100	1.03	100	3	5	220	200
150	0.66	100	3	5	320	150
200	0.44	90	3	5	420	75
250	0.39	100	3	5	510	75
300	0.33	100	3	5	610	50
400	0.25	100	3	5	800	35
500	0.20	100	3	5	1,000	35

* From K. Henney (ed.), "Radio Engineering Handbook," 5th ed., p. 13-15, Table 5, McGraw-Hill Book Company, New York, 1959.
† Rated heater current will produce 5 to 6 mv direct current at couple terminals.
‡ Subject to ±15%.
§ Error due to skin effect only.

2.9. Galvanomagnetic Devices. There are two types of galvanomagnetic effects in intermetallic semiconductors such as indium arsenide or indium antimonide that provide useful nonlinear relationships. The first is the Hall effect, in which an output voltage is developed which is proportional to the product of the magnetic field density normal to the Hall-element surface and the current through the element. This relationship is described mathematically by the following equation:

$$\bar{V}_h = \frac{R_h(\bar{I}_c \times \bar{B})}{d} = \frac{R_h}{d} (I_c B) \sin \theta \qquad (33)$$

where V_h = Hall voltage
R_h = Hall coefficient
I_c = d-c or a-c current through the element
B = magnetic field density normal to the element
d = thickness of the material in the direction of the magnetic field
θ = angle between $\bar{I}_c$ and $\bar{B}$

Figure 53 illustrates a typical physical configuration for a Hall-effect device. These devices provide direct analog multiplication of two input variables and hence are useful in analog computational circuits as either multipliers or squarers.

Figure 54 shows a schematic diagram of a squaring circuit utilizing the Hall-effect

Table 4. Characteristics of Opto-electronic Devices*

Control lamp		Signal—photocell							General		
Voltage	Current, ma	Resistance, ohms — On Typical	On Max	Off 50 volts Typical	Off 50 volts Min	Voltage max	Max power, mw	Typical switching time, sec — On	Typical switching time, sec — Off	Shunt capacity, pf	Coupling capacity, pf	Weight, oz	Light source type†
120	1–3	600	1,000	5×10^7	10^7	60	75	0.0012	0.060	4	0.001	0.2	N
0–1	0–50	550	700	10^7	10^6	60	50	0.020	0.300	4	0.003	0.2	I
0–5	0–175	55	150	10^7	10^6	60	50	0.020	0.800	4	0.005	0.2	I
0–25	0–37	55	150	10^7	10^6	60	50	0.010	0.450	4	0.01	0.2	I
120	1–3	14 kilohms	30 kilohms	7×10^8	2.5×10^7	300	75	0.002	0.105	0.8	0.002	0.2	N
0–10	0–17	340	700	10^7	10^6	60	50	0.030	0.400	4	0.04	0.2	I
0–1	0–15	450	850	3×10^8	10^6	60	50	0.018	0.060	1.5	1.0	0.04	I
0–4	0–53	100	200	10^8	10^7	200	100	0.020	0.300	2	1.0	0.04	I
0–4	0–12	200	350	10^8	10^7	200	100	0.010	0.300	2	1.0	0.04	I
0–5	0–55	100	150	10^8	10^7	200	100	0.004	0.250	2	0.1	0.1	I
0–10	0–17	650	1,000	10^8	10^7	200	100	0.030	0.225	2	0.1	0.1	I
0–25	0–36	100	150	10^8	10^7	200	100	0.010	0.300	2	0.1	0.1	I
150	0.1–1.5	2,500	3,000	10^8	10^7	200	100	0.005	0.100	2	0.1	0.1	N

* Courtesy of Raytheon Manufacturing Company.
† I, incandescent; N, neon.

NONLINEAR AMPLIFIERS 26-31

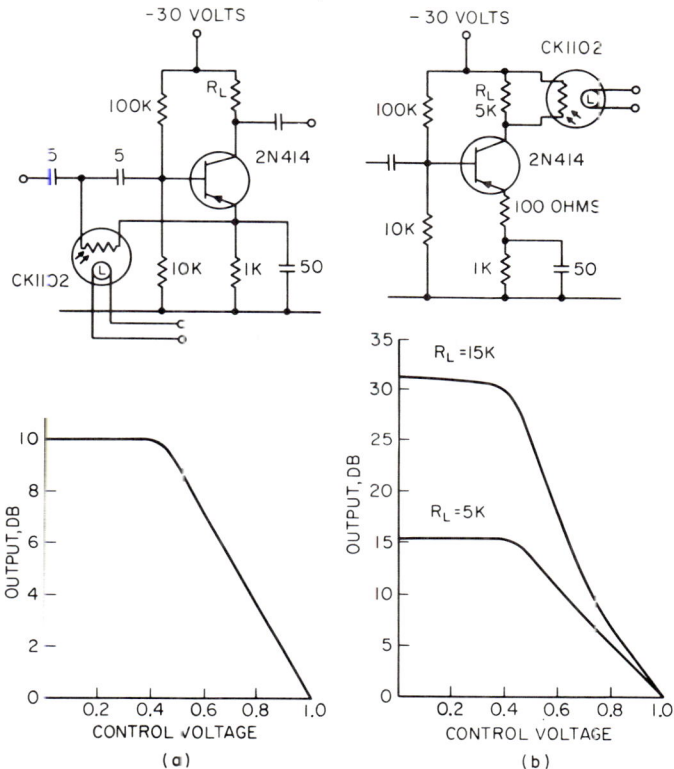

FIG. 52. Circuit diagrams and characteristics for two methods of applying opto-electronic devices as control elements. (*Courtesy of Raytheon Manufacturing Company.*)

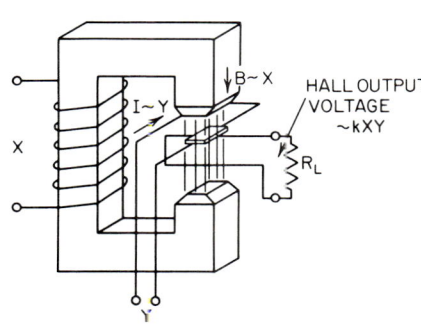

FIG. 53. Typical physical configuration for a Hall-effect device. [*From H. D. Huskey and G. A. Kern (eds.), "Computer Handbook," p. 7-19, Fig. 7.1.12a, McGraw-Hill Book Company, New York, 1962.*]

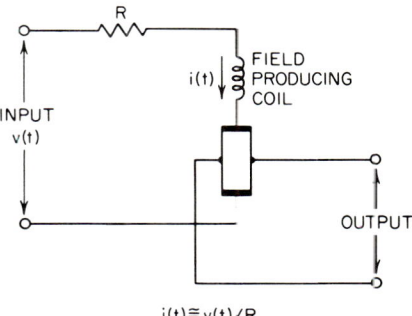

FIG. 54. Schematic diagram of a Hall-effect squaring circuit. (*From IRE Intern. Conv. Record, pt. 9, 1962.*)

Table 5. Input-Output Characteristics of a Hall-effect Squarer

1,000 cps input current, ma	D-C output voltage, mv
0.316	0.003
1.00	0.03
3.16	0.3
10.0	3.0
31.6	30.0
100.0	300.0

relationship in an InSb device. The input-output characteristics of this circuit are listed in Table 5. The lower limit of the input current is in general determined by the sensitivity of the readout device. The practical upper limit of current is established by considerations of Hall-element heating and Hall-coefficient variations. The frequency response of a typical Hall-effect squaring circuit is constant up to about 200 kc/sec, and is limited primarily by core losses and capacitance associated with the field-producing coil.

Figure 55 shows a simple Hall-effect multiplier circuit using a transistor push-pull amplifier for driving the Hall element. The use of current feedback in the magnetic

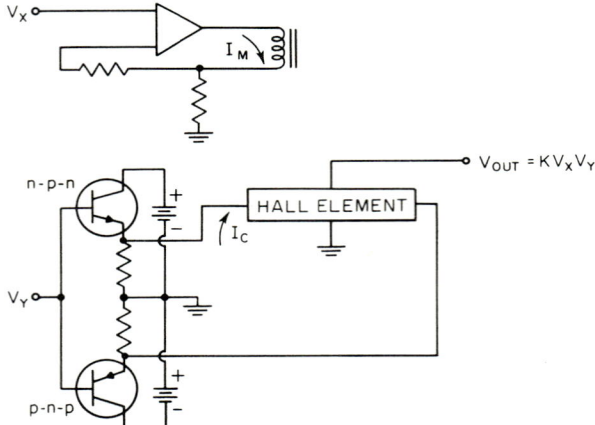

FIG. 55. Simple Hall-effect amplitude-multiplication circuit. (*From Electronic Equipment Eng., February, 1960, p. 39.*)

circuit allows the bandwidth of the multiplier to be extended up to several hundred kilocycles. The current-voltage characteristics of this multiplier are shown in Fig. 56.

There are several intrinsic and extrinsic errors associated with a Hall-effect multiplier. The magnetoresistance effect, in which the resistance of the element changes in proportion to the square of the magnetic field density, is one intrinsic source of error. Since the magnetoresistance is proportional to the square of the electron mobility in the semiconductor material, this effect may be minimized by using material of low mobility. However, the power-transfer efficiency also depends on mobility; so it is not practical to use materials of very low mobility. Thin-film InSb offers a good compromise between low magnetoresistance and high power-transfer efficiency. Figure 57 shows plots of input and output resistance vs. flux density for this material.

The resistive null voltage causes another intrinsic error in Hall-effect multipliers. Part of this error is due to misalignment of the output leads and part is due to variations in the thickness and composition of the material. It is possible to compensate for most of the resistive-null-voltage error, as shown by the plots of Fig. 58.

NONLINEAR AMPLIFIERS 26–33

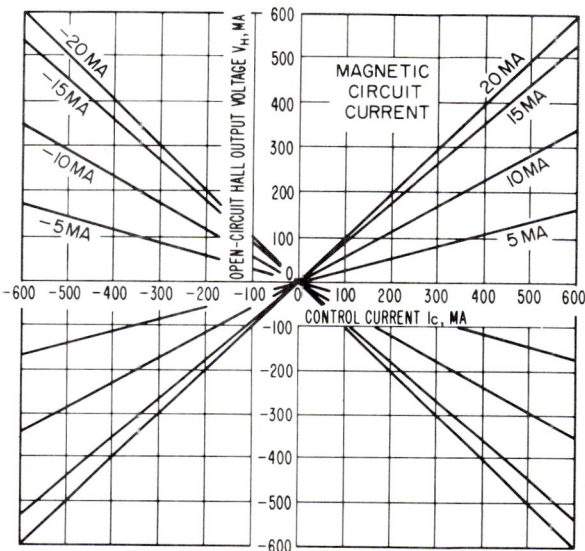

Fig. 56. Output voltage-control current characteristics of Hall-effect element used in circuit of Fig. 55 for various values of magnetic-circuit current. (*From Electronic Equipment Eng.*, February, 1960, p. 39.)

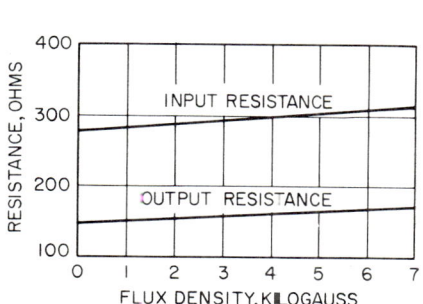

Fig. 57. Input and output resistances of a thin-film InSb Hall-effect element as a function of magnetic flux density. (*From Solid State Design*, October, 1963, p. 18.)

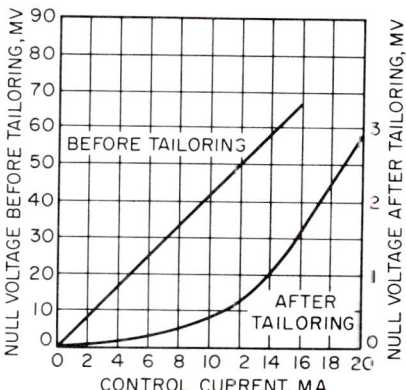

Fig. 58. Resistive null voltage as a function of control current for a thin-film InSb Hall-effect element before and after tailoring. (*From Solid State Design*, October, 1963, p. 18.)

The temperature dependence of a Hall element for two conditions of drive is illustrated by the plots of Fig. 59. The constant voltage drive is seen to be preferable from this point of view.

Some of the extrinsic errors other than frequency limitations in a Hall-effect multiplier are caused by hysteresis and retentivity in the magnetic materials, plus temperature effects in the magnetic materials and circuits. These errors may be reduced to less than 0.2 per cent by proper choice of the magnetic material, appropriate use of heat sinks, and use of constant voltage drive to the magnetic circuit.

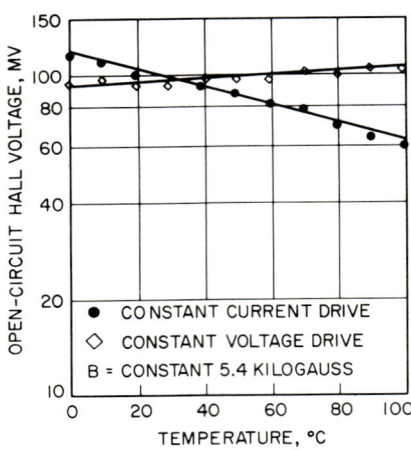

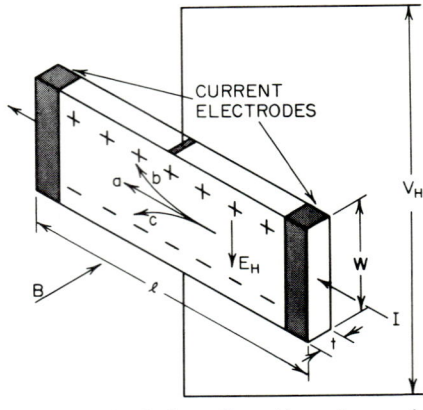

FIG. 59. Comparison of the temperature characteristics of a Hall-effect element for constant-current and constant-voltage drives. (*From Solid State Design, October, 1963, p. 20.*)

FIG. 60. Physical configuration of a semiconductor element exhibiting physical magnetoresistance. (*From IRE Intern. Conv. Record, pt. 9, 1962.*)

The second galvanomagnetic effect useful in nonlinear amplifier circuits is that of magnetoresistance, which was seen to be a source of error in Hall-effect devices.

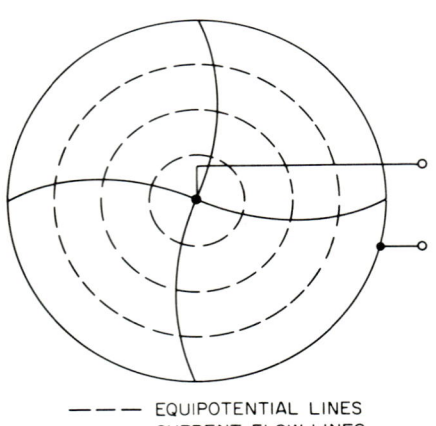

FIG. 61. Equipotential and current-flow lines in a Corbino disk. (*From IRE Intern. Conv. Record, pt. 9, 1962.*)

There are two types of magnetoresistance, both of which provide useful nonlinear relationships. If a high-mobility semiconductor device is of the shape shown in Fig. 60, the resistance changes in proportion to the square of the applied magnetic field for small fields, but for high fields the change of resistance becomes linear with increasing field intensity. This effect is called *physical magnetoresistance*. Resistive bridge devices that can perform analog multiplication and division have been constructed using this type of magnetoresistance.

If the device is made in the form of a Corbino disk, which has equipotential and current-flow lines as shown in Fig. 61, there is an increase of resistivity in the radial direction as the applied magnetic field transverse to the disk increases. The change of resistivity is proportional to the square of the applied magnetic field over a wide range of field intensities. This effect is called *geometrical magnetoresistance*.

Figure 62 shows a schematic diagram of a squaring circuit utilizing the geometrical magnetoresistance-effect relationship in an InSb device. The input-output characteristics of this circuit are listed in Table 6. The lower and upper current limits and the frequency-response limitations are established by considerations similar to those which apply to Hall-effect devices.

Table 6. Input-Output Characteristics of a Geometrical Magnetoresistance-effect Squarer

1,000-cps input current, ma	D-C output voltage, mv
1.0	0.0026
3.16	0.026
10.0	0.260
31.6	2.60
100.0	26.0

3. USE OF PIECEWISE-LINEAR-APPROXIMATION TECHNIQUES

When a high degree of accuracy and good long-term stability are primary requirements in a nonlinear amplifier, it is often desirable to use piecewise-linear-approximation techniques for the generation of the required nonlinear function. In amplifiers using these techniques, electronic devices having a sharp break in their current-voltage characteristics, such as thermionic diodes and semiconductor diodes, are employed as

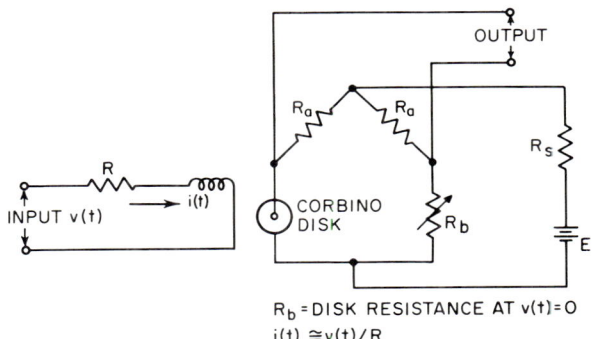

R_b = DISK RESISTANCE AT v(t)=0
$i(t) \cong v(t)/R$

FIG. 62. Schematic diagram of a magnetoresistance-effect squaring circuit. (*From IRE Intern. Conv. Record, pt. 9, 1962.*)

controlled linear elements which are either conducting or nonconducting, depending on the voltage or current level of the input signal. The degree of accuracy obtainable is generally proportional to the number of such devices used.

In the following paragraphs devices which are useful in the application of piecewise-linear-approximation techniques will first be discussed, and then several types of function-generator circuits will be described.

3.1. Device Considerations. An ideal electronic device for use in piecewise-linear-approximation applications would have zero resistance when forward-biased and infinite resistance when reverse-biased. It would also have zero shunt capacitance so that it could be used in high-frequency circuits without introducing errors. Although no such device exists, certain types of thermionic and semiconductor diodes closely approximate these ideal characteristics, as shown by the data in Table 7.

Neither thermionic diodes nor semiconductor diodes have their effective break points exactly at zero bias; the break-point bias is slightly negative for thermionic diodes and slightly positive (0.1 to 0.6 volt) for semiconductor diodes. In critical applications break points must be stabilized through filament-voltage regulation (for thermionic diodes) or through the use of temperature-sensitive resistors in the bias networks (for semiconductor diodes). Constant-temperature ovens or coolers are sometimes used to stabilize semiconductor diodes further.

Low-cost semiconductor diodes are now available which have higher ratios of

Table 7. Characteristics of Ideal and Actual Devices Suitable for Use in Piecewise-linear-approximation Applications

Type of device	Forward resistance, ohms	Reverse resistance, megohms	Capacitance at 10 volts reverse, pf
Ideal....................................	0	∞	0
Thermionic diodes (e.g., 6AL5).................	200–1,500	50–700	2–10
Silicon junction diodes (e.g., 1N302A)...........	5–50	50–200	10–50
High-speed silicon junction diodes (e.g., 1N643)...	100–300	0.1–100	1–8
Germanium point-contact diodes (e.g., 1N39B)...	10–100	0.1–0.5	1–5

forward-to-reverse resistance, lower capacitance, and greater reliability than thermionic diodes. These facts, coupled with the obvious advantage of requiring no filament power, have led to the almost exclusive use of semiconductor diodes as breakpoint devices in nonlinear amplifier circuits using piecewise-linear-approximation techniques.

Circuit resistances associated with the diodes used in piecewise-linear-approximation applications should be designed so as to "swamp out" variations in the forward and/or reverse resistances of the diodes. The use of feedback techniques often helps in the reduction of errors due to these variations and sometimes permits the operation of linear-approximation circuits at higher frequencies.

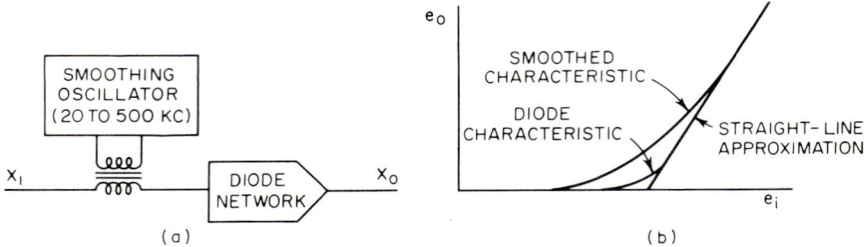

FIG. 63. Use of a smoothing oscillator to smooth diode-network transfer characteristics. [From H. D. Huskey and G. A. Korn (eds.), "Computer Handbook," p. 3–73, Fig. 3.3.13, McGraw-Hill Book Company, New York, 1962.]

The transfer characteristics of a diode network may be smoothed out by inserting a high-frequency oscillator signal in series with a d-c or low-frequency input signal. Figure 63 shows how this may be done and the resultant effect on the input-output characteristic. Note that a certain amount of smoothing results from the nonideal device characteristics even without the use of a smoothing oscillator.

3.2. Logarithmic Function Generators. There are several types of piecewise-linear-approximation function-generator circuits that may be used to give an output signal which is proportional to the logarithm of the input signal. Some of these are simple logarithmic attenuators, while others are used in combination with linear amplifiers of various types. Most logarithmic function generators may also be used to obtain an antilog function, and they may also be used in certain types of analog multiplication and division circuits.

A logarithmic attenuator that covers five decades of input current is shown in Fig. 64. The operation of the circuit is as follows: With very small input currents, semiconductor diodes D_1 through D_7 are all biased in the reverse direction by the action of voltage source V_B and biasing resistors R_8 through R_{14}. Thus these paths are all practically closed to the input current. If the value of resistor R_C is large compared with that of resistor R_A, essentially all the input current will flow through R_A to a low-input-impedance load. As the input current increases, the voltage increases, and a point will be reached where diode D_1 becomes biased in the forward direction. Thus some of the input current will be diverted through this path, the

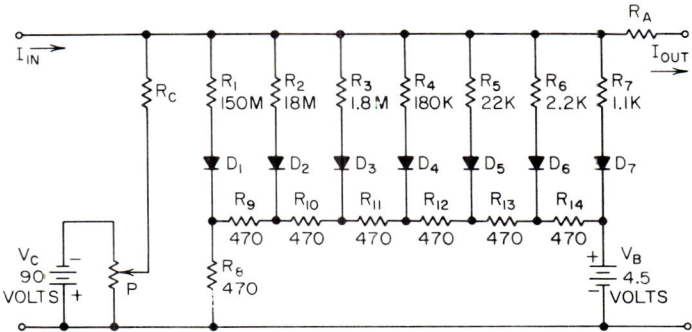

FIG. 64. Wide-range logarithmic attenuator for current inputs. (*Courtesy of General Electric Co.*)

exact amount being controlled by the value of resistor R_1 with respect to that of R_A. This process is repeated with the other paths as the input current increases further, until at maximum input all the diodes are biased in their forward direction.

The resultant input-output current characteristic of the logarithmic attenuator of Fig. 64 is shown in Fig. 65. A closer approximation to a true logarithmic characteristic could be obtained by adding more parallel diode paths. The path in Fig. 64 containing resistor R_C, potentiometer P, and voltage source V_C is required for low-current applications in order to compensate for the finite reverse leakage currents through the semiconductor diodes. Without this compensating path there would be an output current through R_A with zero input current. By proper adjustment of potentiometer P, the output current may be made zero when the input current is zero. A logarithmic preamplifier that uses an attenuator similar to the one of Fig. 64 is described in Sec. 7.

A commercial logarithmic function generator in which the diodes are all forward-biased with zero input voltage is shown in Fig. 66. For low input voltages, the effective resistance loading the input is the combination of R_A and R_1 through R_6, all in parallel, and the gain of the operational amplifier system is at a maximum. As the input voltage increases, the diodes become sequentially reverse-biased, causing the gain to become lower and lower, until at maximum input the gain is determined strictly by resistors R_A and R_B. The output vs. input characteristic of the system is a seven-step approximation to a logarithmic curve, as shown by Fig. 67. This

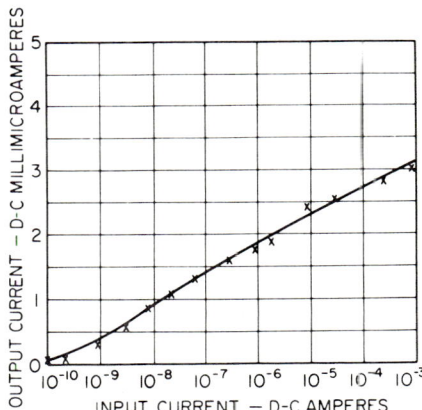

FIG. 65. Output-input characteristic of logarithmic attenuator shown in Fig. 64 (*Courtesy of General Electric Co.*)

function generator has a typical static error of only ±0.5 per cent of full scale over an input range of 1 to 10 volts and an output range of 0 to 10 volts. It is available for either positive or negative inputs and may be used in the various ways shown in Fig. 68.

A temperature-compensated logarithmic function generator that allows a static error of only ±0.2 per cent of the input voltage value when housed in a temperature-stabilized oven is also available commercially. Figure 69 shows the technique used in this function generator to generate a typical segment. For low input voltage diodes $CR1$ and $CR2$ are both forward-biased and the network containing R_D, $CR1$, $CR2$, R_{S1}, and R_{S2} is in parallel with resistor R_{in}, making the gain of the operational

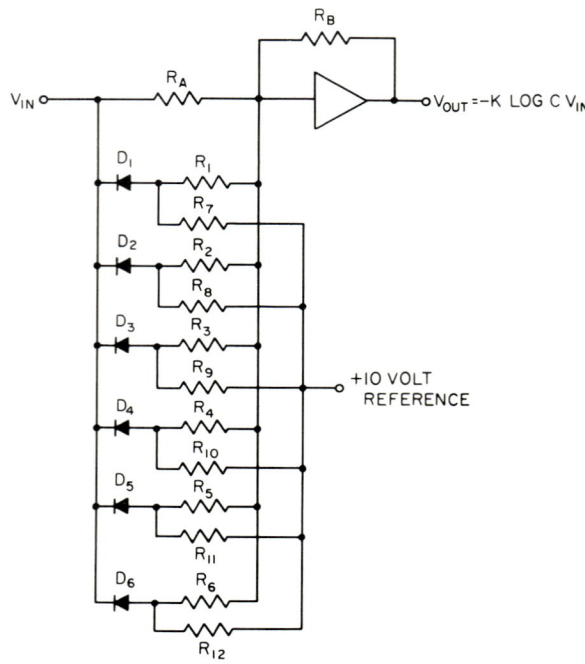

FIG. 66. Logarithmic function-generator circuit of moderate accuracy. (*Courtesy of Electronic Associates, Inc.*)

amplifier system high. As the positive input voltage increases, diode $CR2$ is eventually switched off by the voltage-divider action of R_D, $CR1$, R_T, R_{B1}, and R_{B2}. This decreases the resistance in parallel with R_{in} and hence decreases the gain of the system. The slope of the segment is trimmed by adjusting R_{S2} and the break point is adjusted by R_{B2}. Temperature compensation for diode $CR2$ is provided by the action of thermistor R_T in combination with the compensating effect of $CR1$. There are 14 such segments in the function generator to give a 15-segment approximation to a logarithmic curve. Full-scale calibration is accomplished by adjusting the parallax potentiometer R_P. The oven in which the function generator is mounted controls the temperature at 140 ± 5°F.

The final two logarithmic function generators to be described use amplifiers with controlled gain in conjunction with biased diodes to generate the output-input relationship. A block diagram of a system which gives an output current which is a quasi-logarithmic function of the input d-c or pulse voltage over a range of six decades

NONLINEAR AMPLIFIERS 26-39

is shown in Fig. 70. Its operation is as follows: The output current is the sum of the currents through resistors R_1 through R_6. Each of the amplifier stages has a voltage amplification of 10 and has a diode clamping circuit at its output to limit the output voltage change to a maximum of 10 volts. For input voltages of less than 100 μv, all five stages are active, but above this level one or more of the stages becomes clamped, causing a compression of the output current vs. input current characteristic. The overall current-voltage curve is of the shape shown in Fig. 71. A closer approximation to a true logarithmic relationship could be achieved by using more amplifier stages and summing resistors.

A logarithmic function-generator system which gives an output d-c current proportional to the logarithm of the average rectified value of an a-c input signal is shown in Fig. 72. The operation of the system is as follows: The a-c input voltage is amplified by several cascaded stages of accurately determined gain which allow the output to reach a certain maximum

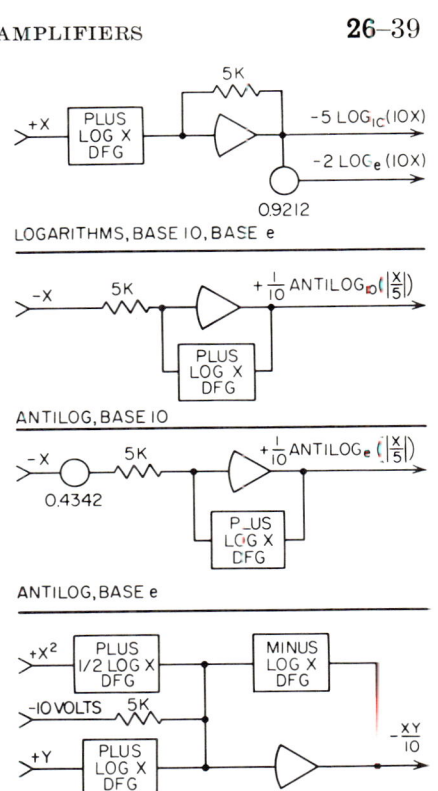

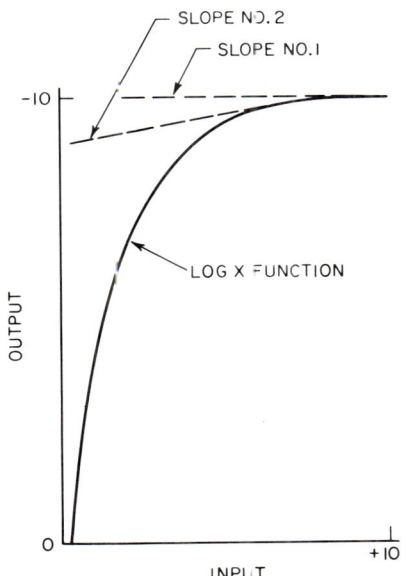

Fig. 67. Output-input characteristic of function generator shown in Fig. 66. (*Courtesy of Electronic Associates, Inc.*)

Fig. 68. Typical uses of a logarithmic function generator of the type shown in Fig. 66. (*Courtesy of Electronic Associates, Inc.*)

value and then hold it at that point. Connected to the output of each a-c amplifier stage are one or more rectifier and filter circuits, each followed by a biased-diode limiter circuit. The d-c outputs of all the biased-diode limiters are summed into a d-c operational amplifier. By proper choice of the amplification factors of the a-c amplifier stages and the break points of the biased-diode circuits the d-c output can be a close approximation to a logarithmic function of the a-c input voltage. When several decades of input amplitude are to be handled, this approach to a-c logarithmic amplification is superior to the more conventional one in which the a-c voltage is merely amplified, rectified, and filtered, and then fed to a d-c logarithmic attenuator or amplifier, since it does not require the rectifiers to work over so wide a dynamic range.

3.3. Square-law Function Generators. The same basic techniques used to generate logarithmic functions apply for piecewise linear approximation to square-law functions. Most of the logarithmic function generators described could be modified to give an output signal which is proportional to the square root of the input signal. However, the more common method of achieving a square-root relationship is to use a diode network in the feedback path of an operational amplifier. For obtaining an output which is proportional to the square of the input, the same diode network can be inserted as the input resistance to an operational amplifier.

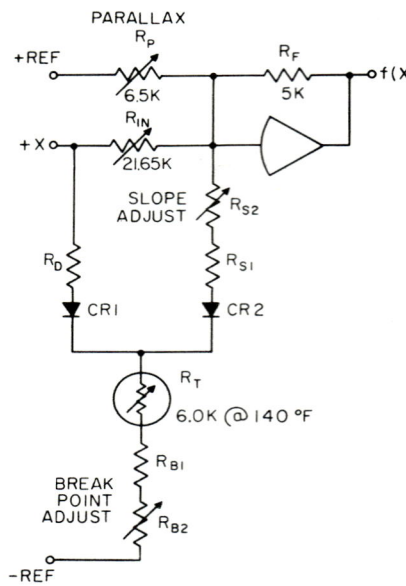

FIG. 69. Typical segment of a high-accuracy logarithmic function-generator circuit using compensating diodes and thermistors. (*Courtesy of Electronic Associates, Inc.*)

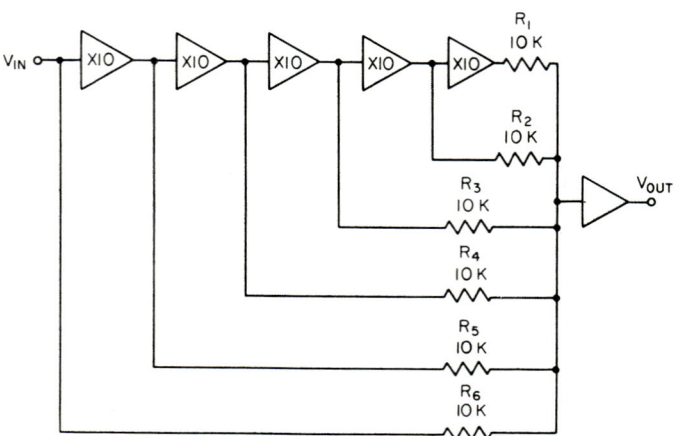

FIG. 70. Simplified schematic of a logarithmic function generator using clamped amplifier stages and summing resistors.

NONLINEAR AMPLIFIERS

Figure 73 shows a simplified diagram of a typical commercial diode network as used in amplitude-squaring circuits. For low input voltages, all the diodes, D_1 through D_6, are reverse-biased, and the gain of the operational-amplifier system is at a minimum, being determined strictly by R_A and R_B. As the input voltage increases, the diodes become sequentially forward-biased, causing the gain to become higher and higher, until at maximum input all the diodes are conducting and the gain is at a maximum. The output vs. input characteristic is a seven-step approximation to a square-law curve, as shown by Fig. 74. This function generator has a typical static error of only ±0.2 per cent of full scale over an input range of 0 to 10 volts. It is available for either positive or negative inputs, the only difference being in the polarity of the reference voltage and the directions in which the diodes are inserted. The application of a diode network of this type to a commercial square-root converter will be discussed in Sec. 9.

Another type of commercial diode squaring network is shown in Fig. 75. For low input voltages, diode D_1 is forward-biased and diodes D_2 through D_{10} are all reverse-biased by the voltage-divider network containing resistors R_4 through R_{14} from the negative reference supply. The output current under these conditions is limited to that flowing through resistor R_2. As the input voltage increases, diode D_1 becomes reverse-biased and diodes D_3 through D_{10} become sequentially forward-biased. As each diode starts conducting, an addi-

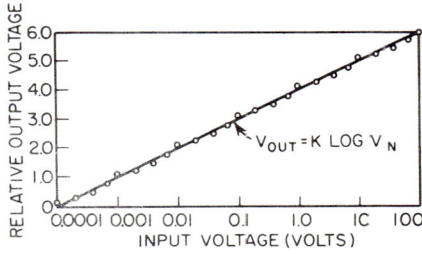

FIG. 71. Output-input characteristic obtained from circuit of Fig. 70.

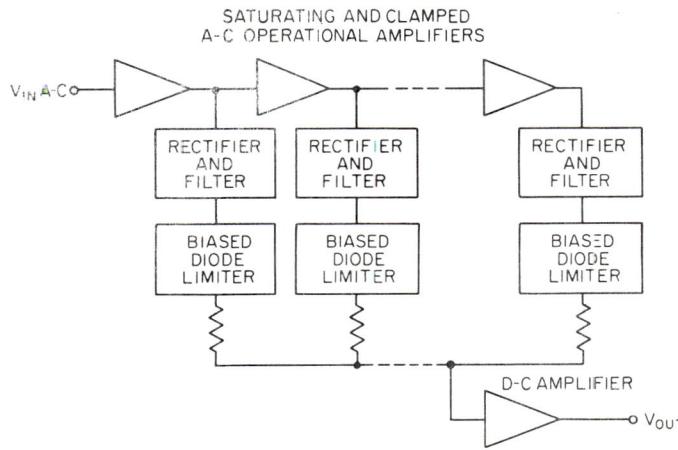

FIG. 72. Logarithmic function generator system for wide-range a-c inputs.

tional path for current flow in parallel with that through R_2 becomes active, causing the output current to increase as the square of the input voltage. This same type of circuit may be used for negative inputs by reversing the polarities of the reference supplies and reversing the directions in which the diodes are inserted. A four-quadrant analog multiplier which uses one positive input squaring network and one negative input network, plus three operational amplifiers, is described in Sec. 10.

A function-generator network that uses silicon break-down diodes to provide an output d-c current approximately proportional to the square of the input d-c voltage over an input range of about 30 to 100 volts is shown in Fig. 76. For input voltages of less than 30 volts there is a very low output current. However, when the input reaches a level of approximately 30 volts, diodes D_1, D_2, D_3 break down and allow some current to flow through resistors R_4, R_5, R_6, and R_7 in parallel with R_T to the load. When the input reaches a level of approximately 50 volts, diode D_4 breaks down, and diodes D_5 and D_6 break down at successively higher input voltage levels. Thus a four-step approximation to the desired curve is generated. The parallel thermistor-resistor combination provides a correction for the small positive temperature coefficient of the breakdown diodes.

The function generator described above is used in an a-c and d-c power transducer manufactured by the General Electric Company. The a-c input voltage is rectified in a center-tap transformer circuit and an a-c voltage derived from the a-c input is alternately added to and subtracted from the half-wave rectified input voltage signal. The resultant signal is

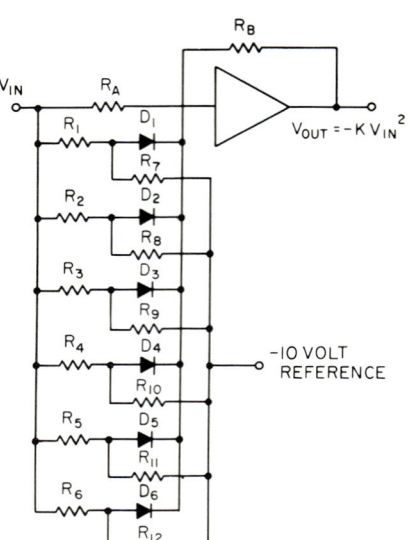

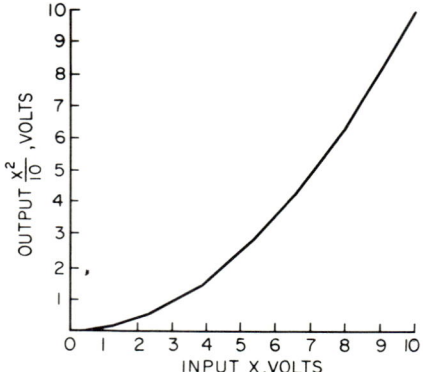

Fig. 73. Typical function-generator system for obtaining an output proportional to the square of the input. (*Courtesy of Electronic Associates, Inc.*)

Fig. 74. Output-input characteristic obtained from system of Fig. 73. (*Courtesy of Electronic Associates, Inc.*)

applied to the square-law function generator and a d-c output proportional to the product of input a-c voltage and input a-c current is obtained across a balancing resistor. A simplified circuit of the overall power transducer is shown in Fig. 77.

3.4. Variable Function Generators. When a nonlinear amplifier circuit requires an accurately controlled nonlinear characteristic of a type other than logarithmic or square-law, it is desirable to use a variable diode function generator. These are basically similar to the fixed function generators described already, but they usually have more break-point devices and more adjusting resistors.

Figure 78 shows a simplified schematic diagram of a typical commercial variable diode function-generator system, and the waveforms resulting from an input to one segment are also shown. The ratio of the resistors between input and reference determines the break point of the diode, or starting point along the X axis of the segment. When the diode conducts, the signal is applied to the wiper of the SLOPE potentiometer. The signals which occur at the ends of the potentiometer are applied through a direct channel and an inverting channel to the input of the summing

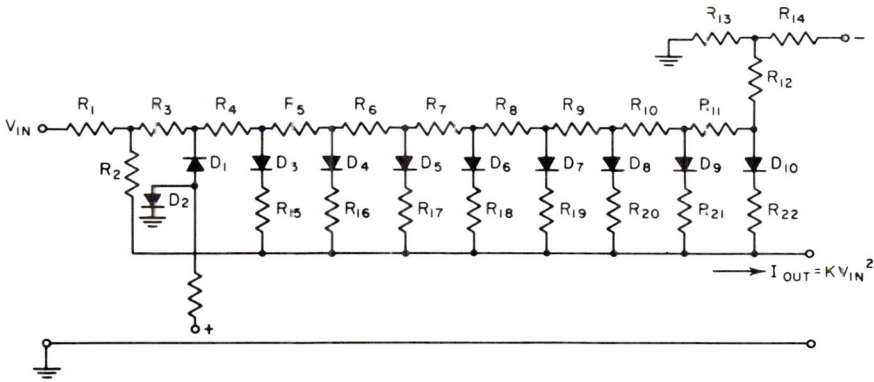

FIG. 75. Biased-diode network which gives an output proportional to the square of the input. (*Courtesy of Systron-Donner Corporation.*)

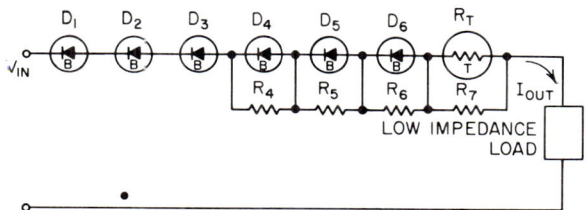

FIG. 76. A function-generator network which uses silicon breakdown diodes to give an output proportional to the square of the input. (*Courtesy of General Electric Co.*)

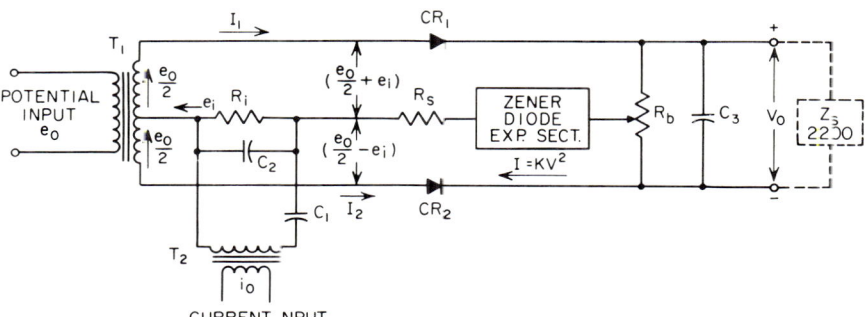

FIG. 77. Simplified schematic of an a-c to d-c power transducer using the function generator of Fig. 76. (*Courtesy of General Electric Co.*)

amplifier. The setting of the SLOPE potentiometer determines the slope of each segment by controlling the ratio of signal levels of the direct and inverting channels. For example, if the wiper of the SLOPE potentiometer is set near the top, a larger signal is applied to the direct channel than to the inverting channel (amplifier 1). Therefore, the input to the summing amplifier (amplifier 2) is positive, or its output is negative. Additional segments are generated in a similar manner and summed to produce the desired function.

4. USE OF SERVO TECHNIQUES

The third major method for achieving nonlinear transfer characteristics is to use servomechanisms. These may be used to establish various nonlinear input-output relationships and may also be used in the design of amplitude multipliers.

In the following paragraphs the various components of servosystems will be briefly discussed, several types of nonlinear function-generator systems will be described, and the application of servo techniques to analog multipliers will be discussed.

4.1. Servosystem Components. The principal components of nonlinear servosystems are potentiometers, servo amplifiers, servomotors, and mechanical components such as gears and limit stops. Several types of potentiometers are used in nonlinear servosystems. For nonlinear function generation the potentiometers usually have taps welded to the turns of wire on a linear resistance element or else have the wire wound so as to give a nonlinear relationship between shaft rotation and resistance. In analog multiplier applications untapped linear potentiometers of good line-

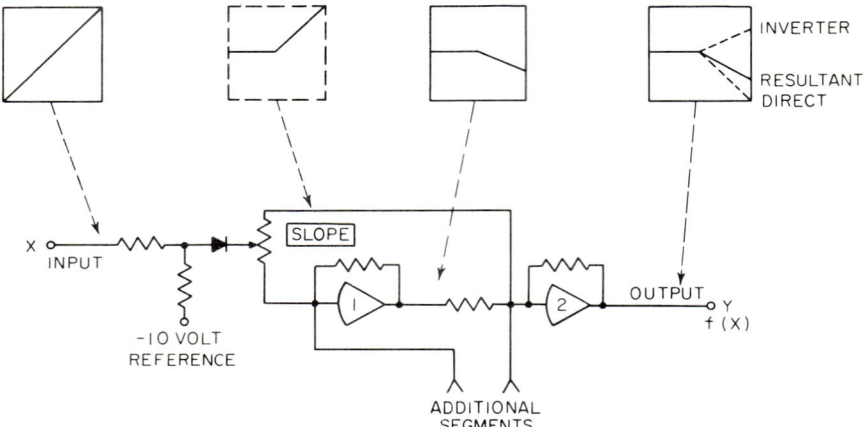

FIG. 78. Simplified schematic illustrating the operation of a typical segment of a variable diode function generator. (*Courtesy of Electronic Associates, Inc.*)

arity and low noise are used. Table 8 gives a comparison of the performance factors of the three types of linear potentiometers most commonly used in servomultipliers.

When tapped linear potentiometers are used, loading (padding) resistors are connected to the taps in such a manner that the output voltage from the potentiometer is a piecewise linear approximation to a desired nonlinear function. Techniques and circuits applying to the use of potentiometers are described in more detail in Sec. 4.2.

A typical vacuum-tube a-c servo amplifier is shown schematically in Fig. 79. It has a gain in the operating frequency range of about 66 db and the output power is 10 watts. A push-pull circuit is used to eliminate the d-c component in the output, and the output stage is tuned to increase efficiency. Transistorized servo amplifiers are now widely used, since they give good efficiency coupled with design simplicity. A simple amplifier for use with d-c inputs and a-c servomotors is shown in the schematic diagram of Fig. 80. It consists of a silicon diode ring modulator followed by a three-stage silicon transistor a-c amplifier. The circuit is simplified by having a single-ended class AB output stage directly driving the control winding of the servomotor. An adequate driving waveform is obtained by shunting the control winding with a tuning capacitor. Resistor R_1 and capacitor C_1 provide stabilization for the system.

NONLINEAR AMPLIFIERS

Table 8. Comparison of Servo-potentiometer Characteristics

Performance factors	Wirewound 10-turn	Wirewound One-turn	Film One-turn
Diameter, in	1.75	2	1.25
Resistance, kilohms	20	50	50
Resistance tolerance, %	±2	±2	±10
Tap resistance, ohms	5	5	300
Conformity, %	±0.05	±0.2	±0.2
Resolution, wire turns	12,000	2,100	Essentially infinite, less than 0.005?
Noise, peak-to-peak mv at 1.5 rpm	50	133	133
Max velocity, volts/sec	150	1,500	1,500
Power dissipation, watts	5 at 40°C	1 at 40°C	1 at 20°C
Breakaway torque, oz-in./section	2.0	1.7	0.5
Moment of inertia, g-cm²/section	14	5	10

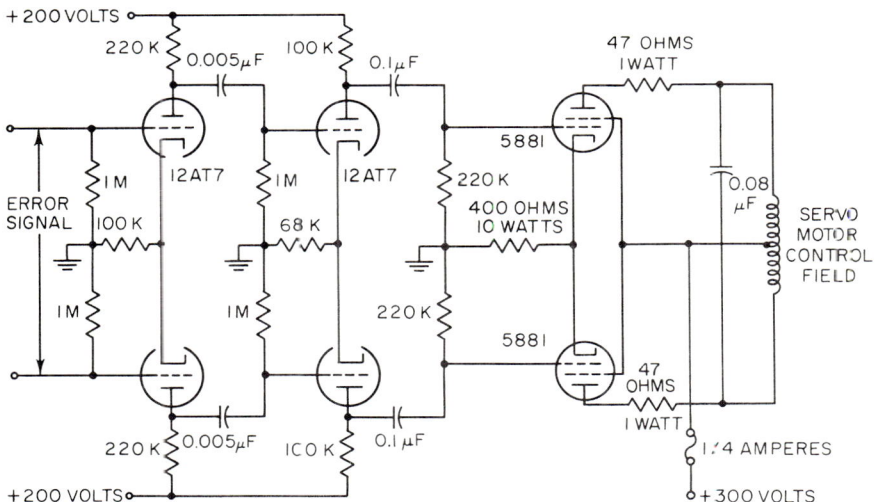

FIG. 79. Typical vacuum-tube servo amplifier for an all a-c system. [*From H. D. Huskey and G. A. Korn (eds.), "Computer Handbook," p. 3-10, Fig. 3.1.8, McGraw-Hill Book Company, New York, 1961.*]

Figure 81 shows a servo amplifier for an all d-c system. It contains a drift-stabilized input amplifier followed by a transistor output stage driving a d-c motor. A series resistor or amplitude limiter must be used to protect the transistors from an excessive output of the input amplifier.

The method by which a servo amplifier such as the ones shown in Figs. 80 and 81 is used in an operational-amplifier system is shown in Fig. 82. The output E_B of such a system is approximately $(R_R/R_A)E_A$ and is limited to a magnitude of E_{T2}. Nonlinear amplification may be achieved either by using nonlinear resistances for R_A or R_B, or by varying R_A and/or R_B through servo drives.

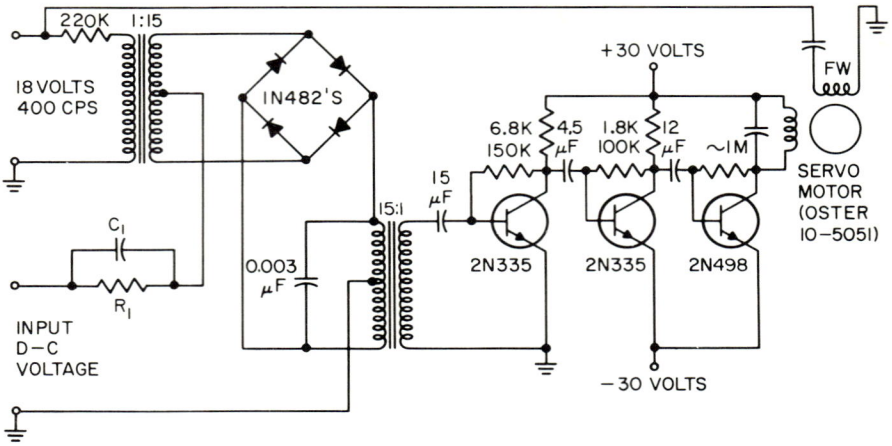

Fig. 80. Transistorized servo amplifier for use with d-c inputs and a-c servomotors. (*Courtesy of General Electric Co.*)

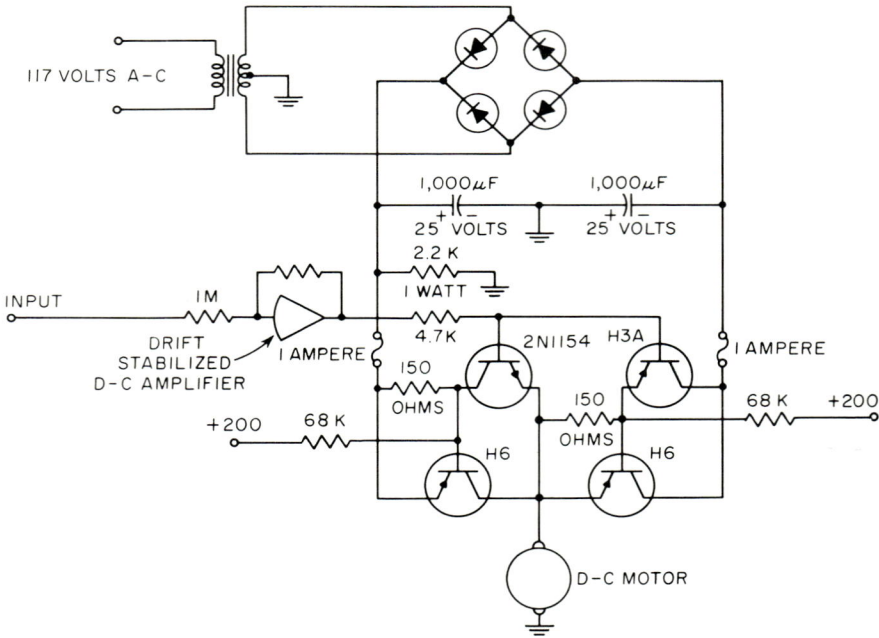

Fig. 81. Servo amplifier with a transistor output stage for an all d-c system. [*From H. D. Huskey and G. A. Korn (eds.), "Computer Handbook," p. 3-11, Fig. 3.1.9, McGraw-Hill Book Company, New York, 1962.*]

NONLINEAR AMPLIFIERS

An a-c servosystem is usually driven by a two-phase induction motor, with one winding excited by a fixed current and the second, which is the control winding, connected to the output of the servo amplifier. Figure 80 illustrates the electrical connections. The motor develops a torque proportional to the current flowing through the control winding when the two currents are 90° out of phase. In-phase currents

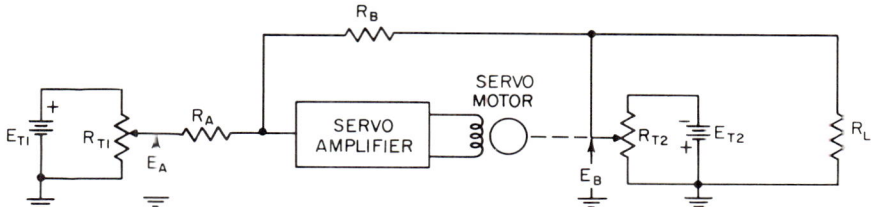

FIG. 82. Simplified schematic of a typical servo operational amplifier system.

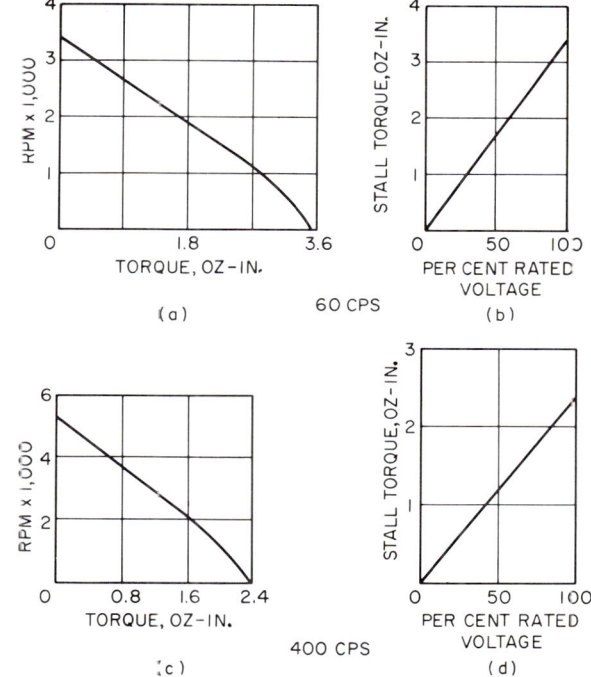

FIG. 83. Performance curves of typical 60-cps and 400-cps servomotors. (a,b) 60 cps. (c,d) 400 cps. [From H. D. Huskey and G. A. Korn (eds.), "Computer Handbook," p. 3–12, Fig. 3.1.10, McGraw-Hill Book Company, New York, 1962.]

are undesirable because they produce no torque and contribute to the saturation of the magnetic field.

Characteristic curves for typical 60- and 400-cps servomotors are shown in Fig. 83. For curves a and c, rated voltage is applied to each winding. For curves b and d, rated voltage is applied to the fixed winding and the voltage of the control winding is varied. Other important performance factors are: (1) maximum motor velocity; (2)

maximum motor acceleration; (3) breakaway current, which is the minimum signal below which the motor will not turn the load. At the breakaway current the torque becomes equal to the static friction.

An all d-c servosystem is usually driven by a d-c permanent-magnet motor, with the output of the servo amplifier connected to the armature. Figure 81 illustrates the method of connection. The torque-speed relationship of a typical d-c servomotor

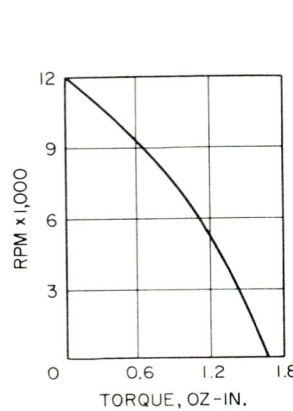

FIG. 84. Speed-torque curve of a typical d-c servomotor. [From H. D. Huskey and G. A. Korn (eds.), "Computer Handbook," p. 3–13, Fig. 3.1.11, McGraw-Hill Book Company, New York, 1962.]

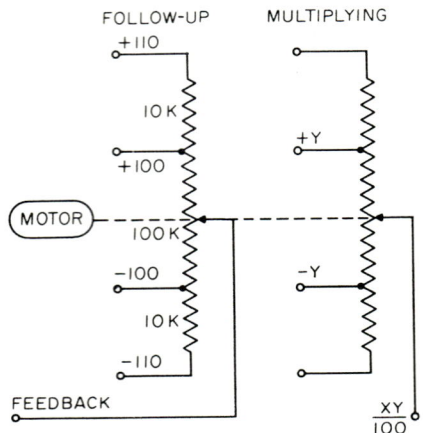

FIG. 85. Scheme for providing electrical limit stops in a multiplying servosystem. [From H. D. Huskey and G. A. Korn (eds.), "Computer Handbook," p. 3–14, Fig. 3.1.12, McGraw-Hill Book Company, New York, 1962.]

is shown in Fig. 84. Table 9 gives a comparison of several characteristics of the servomotors whose performance curves are shown in Figs. 83 and 84.

Table 9. Comparison of Characteristics of Representative Servomotors
(See also Figs. 83 and 84)

Characteristics	Fig. 84	Fig. 83a,b	Fig. 83c,d
Frequency, cps	d-c	60	400
Max control power, watts	20	8	9
Rotor inertia, g-cm^2	4.5	4.0	4.0
Torque at stall, oz-in	1.7	3.5	2.5
No-load speed, rpm	12,000	3,400	4,800
Time constant, msec	46	6	12
Theoretical acceleration at stall, radians/sec^2	27,000	62,000	40,000
Weight, oz	4	12	12

Precision mechanical components are required for accurate nonlinear amplifier systems using servo techniques. Backlash in gears must be kept to a minimum. If wirewound potentiometers are used, the maximum backlash of the gear train measured at the follow-up potentiometer shaft should be less than the angular resolution of one-half wire turn. For example, a servosystem using a 10-turn follow-up potentiometer with 10,000 wire turns satisfies this requirement if it has less than 11 min of backlash. Attempts have been made to reduce backlash by means of antibacklash gears, but these gears also increase the moment of inertia of the system.

NONLINEAR AMPLIFIERS

Limit stops are required in single-turn potentiometers because the resistance element covers slightly less than 360°, leaving a gap between the ends. If the arm of the follow-up travels into this gap the servo will lose its feedback signal and run away. Electrical limit-stop action may be achieved in a servosystem by providing taps near each end of the potentiometers. Figure 85 illustrates how this technique is applied to a multiplying servosystem. The advantage of this scheme is that the full scale of the master variable (servo position) is ±100 volts.

4.2. Nonlinear Function Generation by Servo Techniques. The ability of a servosystem to generate nonlinear output-input relationships is determined primarily by the types of potentiometers used and the loading applied to the potentiometers. Simple relationships may be approximated merely by loading a linear untapped potentiometer, as shown by the diagrams and curves of Fig. 86. However, tapped linear potentiometers or nonlinear resistance elements are required for accurate function generation of logarithmic and square-law relationships.

Two basic circuits for padding a tapped potentiometer are shown in Fig. 87. In a, the loading resistors are connected in series from tap to tap to reduce the voltage

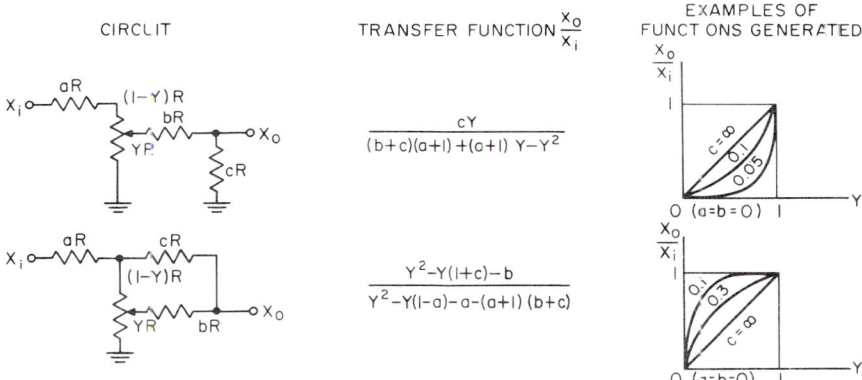

FIG. 86. Use of loaded linear potentiometers for approximating simple nonlinear functions. [*From H. D. Huskey and G. A. Korn* (eds.), "*Computer Handbook,*" *p. 3-32, Fig. 3.120a, McGraw-Hill Book Company, New York, 1962.*]

gradient across each segment to the value required by the slope of the function. In b, the loading resistors are connected from each tap to one end of the potentiometer. Method a has the advantage of minimum current drain from the voltage source and also minimizes the danger of burning out the potentiometer. However, it may be used only for monotonic functions. Method b may be used to generate nonmonotonic as well as monotonic functions, but it requires each tap to be fused to protect the potentiometer. Combinations of these two loading methods are sometimes used as shown by the circuit and curve of Fig. 88.

There are two simple techniques for determining the padding resistors to furnish a desired output-input relationship. The first is a low-cost practical method in which the resistance values of the padders are computed; then the resistors are adjusted to the calculated values using, for example, a Wheatstone bridge. The calculations are a simple and direct application of Kirchhoff's laws at each tap position. The function is divided into monotonic sections. The calculations always start with the segment that has the steepest slope; this segment is unpadded and its current is determined immediately from the given slope. The calculations then proceed in turn to succeeding segments.

In the second method, the servo is driven in turn to consecutive tap locations and at each tap the neighboring padding resistor is adjusted until the output becomes

26-50   AMPLIFIER CIRCUITS

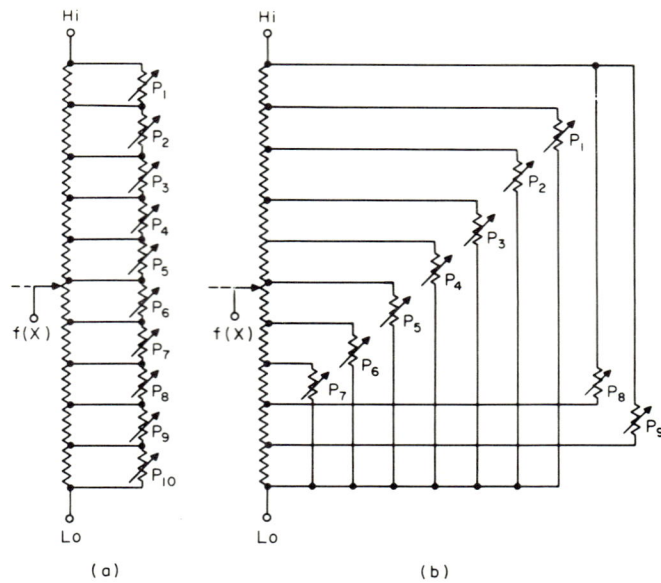

FIG. 87. Two types of loading circuits for tapped linear potentiometers. [*From H. D. Huskey and G. A. Korn (eds.), "Computer Handbook," p. 3-37, Fig. 3.1.33, McGraw-Hill Book Company, New York, 1962.*]

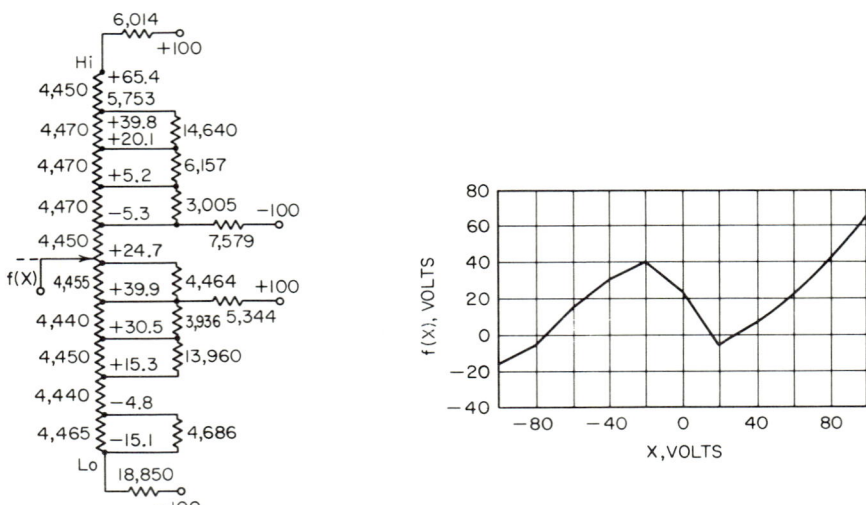

FIG. 88. Circuit and curve illustrating use of both tap-to-tap and tap-to-end loading of a linear potentiometer to generate a desired function. [*From H. D. Huskey and G. A. Korn (eds.), "Computer Handbook," p. 3-37, Fig. 3.1.34, McGraw-Hill Book Company, New York, 1962.*]

equal to the desired value. The inherent shortcoming of this method is that in general there will be interaction between the padding resistors. If, for example, the servo is returned to tap 1 after potentiometers $P1$ and $P2$ (Fig. 87) have been adjusted, the output at this tap will no longer be correct. This means that a long and tedious iteration procedure is required unless a constant-current source is used to drive the potentiometer network. Figure 89 shows a practical circuit using this technique. A positive-feedback circuit is used to generate a current $I = V/R_1$ through the tapped potentiometer. If r is the resistance of the padded potentiometer, the equation

$$r < \frac{R_1 R_2}{R_1 + R_2}\left(\frac{100}{|V|} - 1\right) \quad (34)$$

must be satisfied to prevent the amplifier output from exceeding 100 volts.

With a properly loaded tapped-potentiometer system, the overall accuracy can be made as good as ±0.05 per cent of the reference voltage. A logarithmic voltmeter-converter using servo techniques and a tapped potentiometer is described in Sec. 7.

A servo instrumentation system using specially wound exponential potentiometers to achieve an output signal proportional to the logarithm of the input d-c current over a range of four decades is shown in Fig. 90. Input currents down to 10^{-8} amp may be handled by this system without significant effect on instrument accuracy or response. The maximum error is 0.3 per cent of full scale, and the repeatability is to within 0.05 per cent. The frequency response is better than 5 cps, with a 30-msec time constant.

As illustrated in Fig. 90, the input signal is balanced out by a servo-driven feedback potentiometer. The error signal at the chopper (input signal less feedback voltage) is amplified, mixed with a tachometer feedback signal, further amplified, and then fed to a servomotor in such a phase as to decrease the error between the feedback voltage and the input signal.

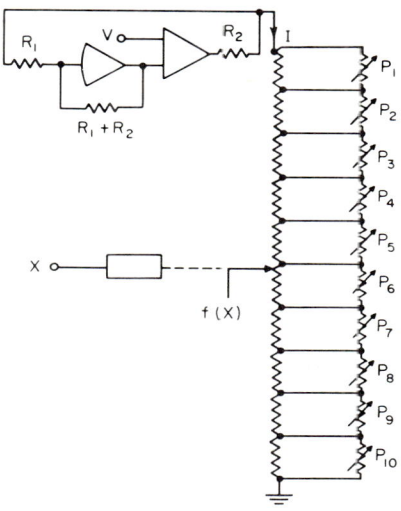

Fig. 89. Circuit for converting a voltage input to a current source driving a tapped linear potentiometer network. [From H. D. Huskey and G. A. Korn (eds.), Computer Handbook," p. 3–38, Fig. 3.1.35, McGraw-Hill Book Company, New York, 1962.]

The feedback potentiometer R_1 is a high-accuracy four-decade exponential device, and it is mounted on a common shaft with the output potentiometer. Attenuators R_2 and R_3 are two-decade exponential potentiometers and are also mounted on the common shaft. Since the feedback potentiometer is exponential, the rotation of the shaft is proportional to the logarithm of the input signal.

As measured at point A, a 1 per cent proportional input error at the top of the instrument range is 10,000 times larger than a 1 per cent proportional error at the bottom of the instrument range. In order to compensate for this, the attenuator potentiometers decrease the gain between points B and C by a factor of 10,000 as the instrument moves from the bottom to the top of its range. This keeps the sensitivity and response time of the instrument constant as the input current level changes.

The "electronic end stop" shown in Fig. 90 attenuates the error signal by a factor of 1 or 550, depending upon a set of preselected conditions. It prevents the application of full output torque against the mechanical end stop but permits maximum acceleration away from the end stop.

4.3. Servomultipliers. Figure 91 shows a block diagram of a typical four-quadrant servomultiplier. The difference between the input signal E_i and the follow-up voltage E_f is modulated, amplified, and fed to the control winding of a servomotor. The motor drives the follow-up potentiometer and the two computing potentiometers. A and B are voltages varying with time; so output voltages proportional to AE_i and BE_i are obtained.

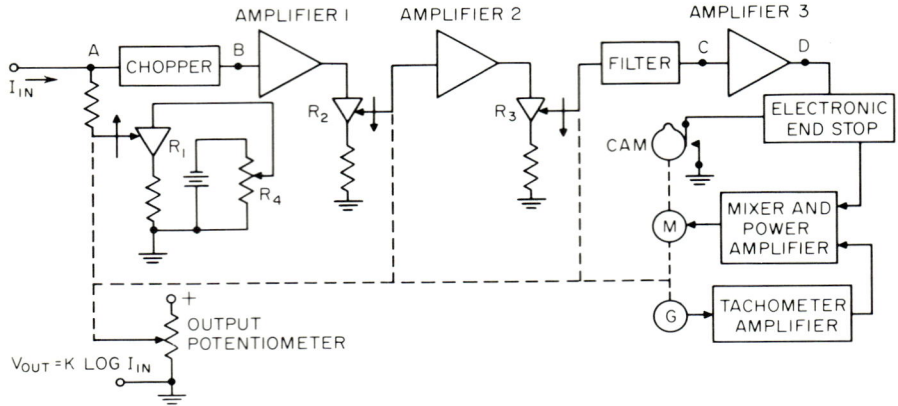

Fig. 90. Simplified schematic of a servo logarithmic amplifier for wide-range current inputs. (*From IRE Trans. Nuclear Sci., vol. NS-6, No. 1 pp. 11–14, March, 1959.*)

The choice of the carrier frequency determines the theoretical upper limit on the bandwidth of a servomultiplier. The loop gain at a frequency of one-half the carrier must be attenuated. In actual practice the bandwidth is usually limited to less than one-tenth the frequency of the carrier. A carrier frequency of 400 cps is commonly used in servomultipliers requiring fairly good frequency response.

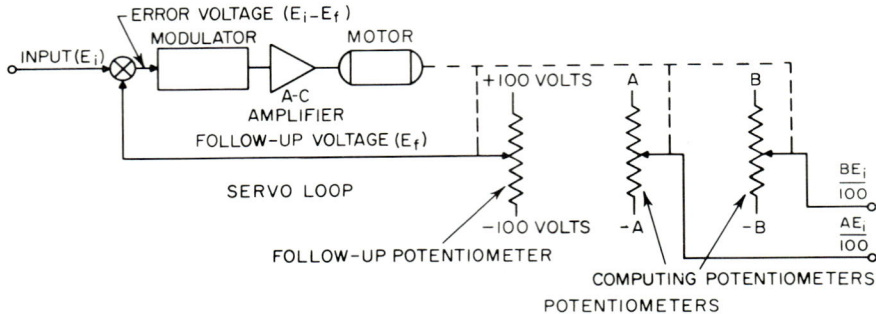

Fig. 91. Simplified schematic of a typical servomultiplier with two computing potentiometers. (*Courtesy of Beckman Instruments, Inc.*)

The principal advantages stemming from the use of servo techniques in analog multipliers are: (1) low cost per product; (2) no error due to the slave variables; and (3) normally no adjustments required. Their chief disadvantage is the dynamic limitations imposed on the master variable, which include velocity and acceleration limits as well as that of bandwidth.

5. CIRCUITS WITH AMPLITUDE LIMITING

There are several methods by which amplitude limiting may be accomplished. For example, automatic gain control circuits achieve limiting action by reducing the gain of an amplifier stage as the input amplitude increases. However, in this section the only types of limiting circuits described are those which use the on-off characteristics of diodes and transistors to achieve the limiting action. Other types of limiting circuits are described in Secs. 6, 7, and 8.

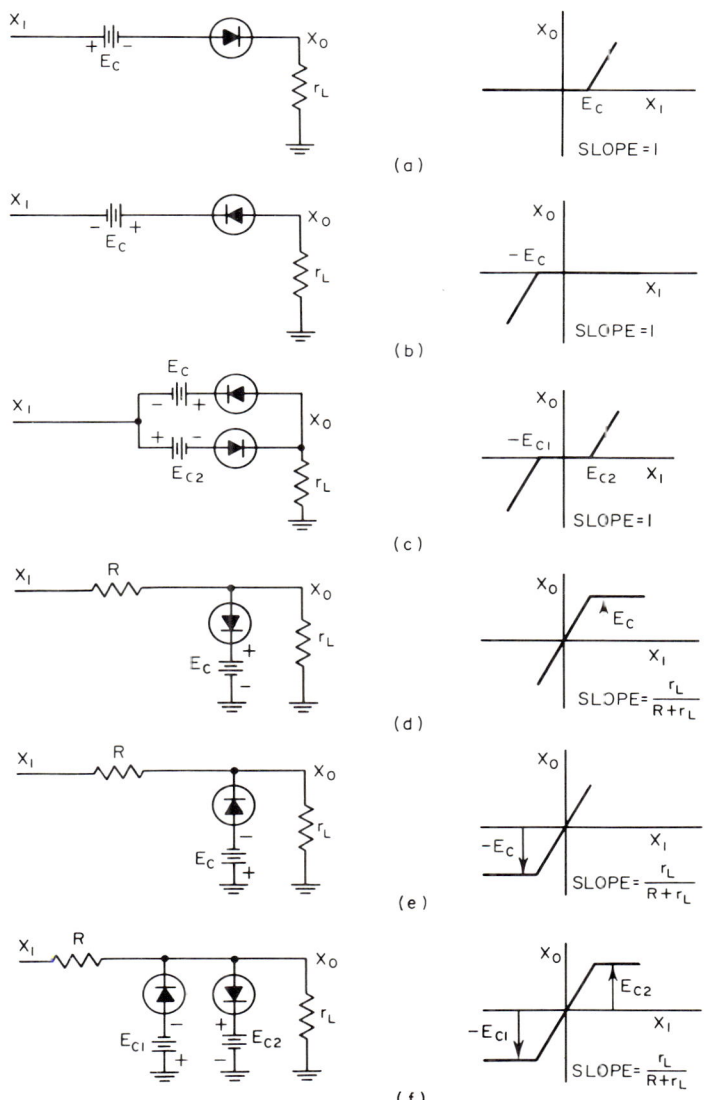

FIG. 92. Basic diode-limiter circuits. [From H. D. Huskey and G. A. Korn (eds.), "Computer Handbook," p. 3-64, Fig. 3.3.1a-f, McGraw-Hill Book Company, New York, 1962.]

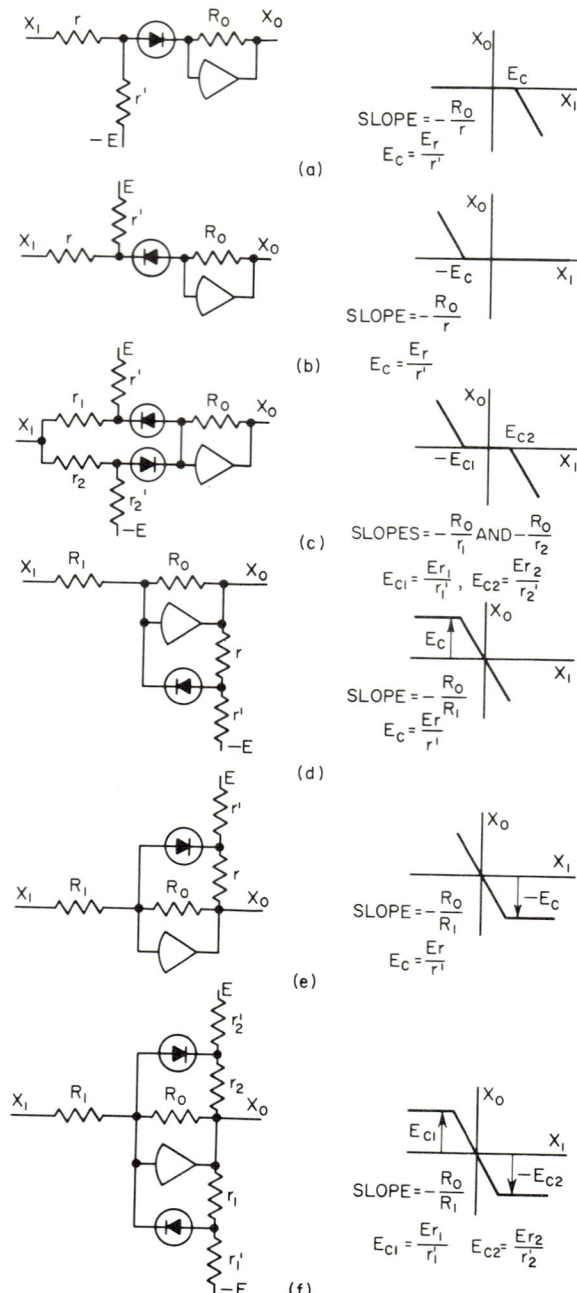

Fig. 93. Simple operational-amplifier circuits using diode limiters. [*From H. D. Huskey and G. A. Korn* (eds.), *"Computer Handbook,"* pp. 3–64 and 3–65, Fig. 3.3.1 g–m, McGraw-Hill Book Company, New York, 1962.*]

5.1. Simple Diode-limiter Circuits. The basic diode-limiter circuits of Fig. 92 utilize the nonlinear current-voltage characteristics of thermionic or semiconductor diodes. The transfer characteristics shown assume ideal diodes and zero-impedance voltage sources. The series limiters (Fig. 92a, b, c) generally provide more accurate limiting action than the shunt limiters (Fig. 92d, e, f).

Figure 93 shows some simple limiter circuits in which diodes are used in conjunction with operational amplifiers. In the feedback limiters, d to f, the limiting por-

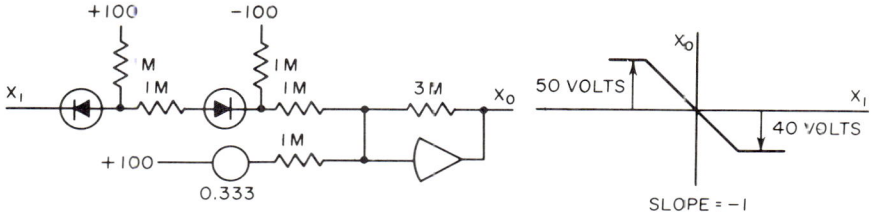

FIG. 94. Dual-series limiter. [From H. D. Huskey and G. A. Korn (eds.), "Computer Handbook," p. 3-66, Fig. 3.3.2, McGraw-Hill Book Company, New York, 1962.]

tions of each characteristic are not ideally flat but have a finite slope given by $-\rho R_0/R_1(R_0 + \rho)$, where ρ is the sum of the diode forward resistance R_D and r, r_1, or r_2. For good limiting action, the value of E should be large, and r and r_1 should be as small as possible without undue loading of the amplifier.

Two simple diode-limiter circuits which are especially useful are the dual-series limiter, shown in Fig. 94, and the bridge limiter, shown in Fig. 95. These are both series limiters whose transfer characteristics are similar to those of dual shunt or feedback limiters.

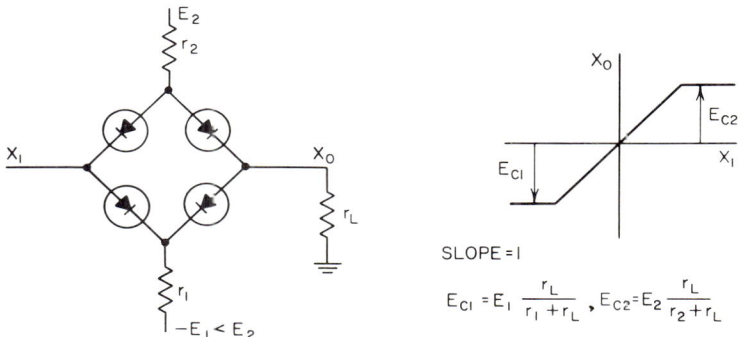

FIG. 95. Bridge limiter. [From H. D. Huskey and G. A. Korn (eds.), "Computer Handbook," p. 3-66, Fig. 3.3.3, McGraw-Hill Book Company, New York, 1962.]

5.2. Precision Limiter Circuits. While the limiter circuits of Figs. 92 to 95 may involve static errors of between 0.2 and 1.0 volt, depending on the particular application, it is possible to achieve much better accuracies through the use of special feedback techniques. The circuits shown in Fig. 96 produce static errors of less than 0.05 volt with the aid of chopper-stabilized high-gain d-c amplifiers. When each diode D_1 conducts, it is inside a high-gain degenerative feedback loop, so that the diode forward resistance and any "built-in" bias voltages are divided by the loop gain. As diodes D_1 cease to conduct in the circuits of Fig. 96a and b, the open-loop

26-56   AMPLIFIER CIRCUITS

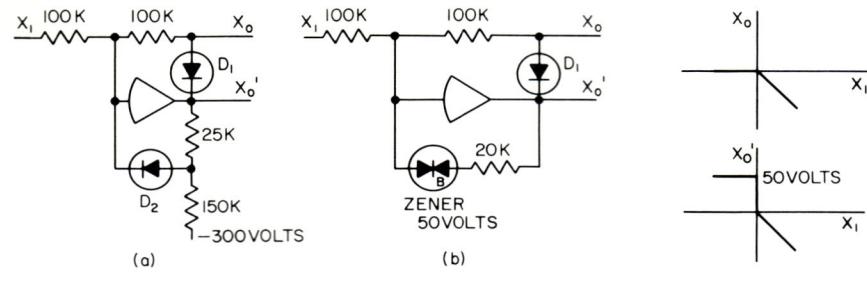

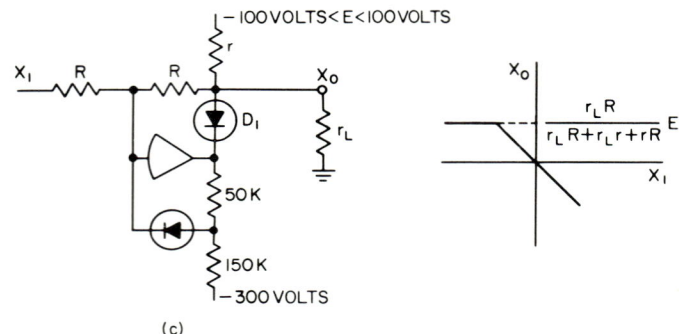

Fig. 96. Precision limiter circuits. [*From H. D. Huskey and G. A. Korn (eds.), "Computer Handbook," p. 3–67, Fig. 3.3.4a,c, McGraw-Hill Book Company, New York, 1962.*]

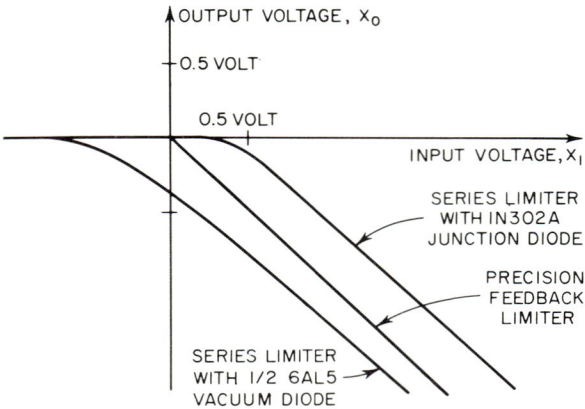

Fig. 97. Comparison of the transfer characteristics of simple and precision diode-limiter circuits. [*From H. D. Huskey and G. A. Korn (eds.), "Computer Handbook," p. 3–68, Fig. 3.3.5, McGraw-Hill Book Company, New York, 1962.*]

gain of the amplifiers produces a sharp step at X and causes diodes D_1 to cut off decisively.

Figure 97 shows the actual d-c transfer characteristics of the precision-limiter circuit of Fig. 96a in comparison with those obtained from a simple series-limiter circuit

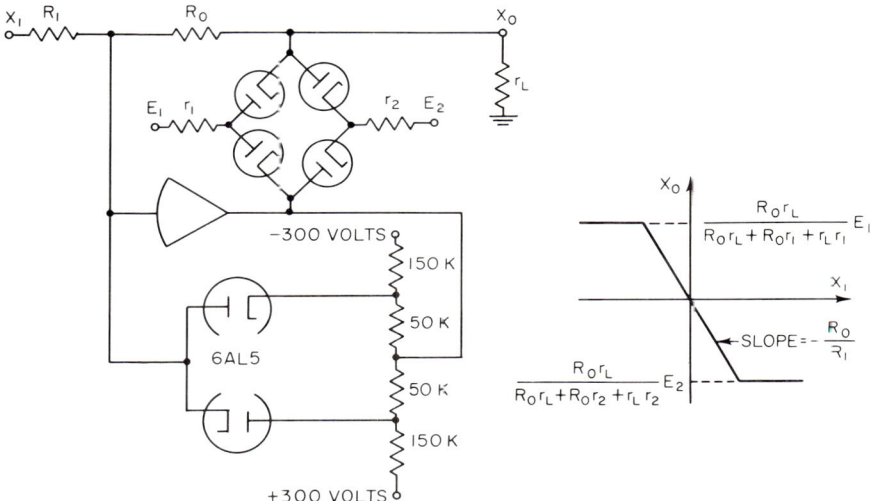

FIG. 98. Precision dual limiter circuit using a single operational amplifier. [*From H. D. Huskey and G. A. Korn* (eds.), "*Computer Handbook*," *p. 3–69, Fig. 3.3.6a, McGraw-Hill Book Company, New York, 1962.*]

of the type shown in Fig. 93a (with $E_c = 0$). These curves also illustrate the differences between the use of thermionic and semiconductor diodes in simple limiters and function generators.

A dual-limiter circuit using a single amplifier to produce precision limiter action is shown in Fig. 98. With suitable amplifiers and diodes, circuits such as this may be used up to high audio frequencies as components of triangle-integration multipliers.

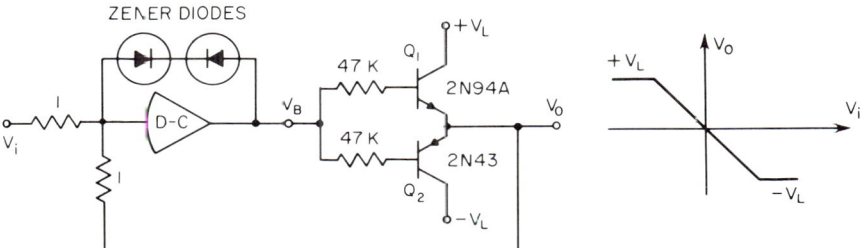

FIG. 99. High-accuracy bidirectional limiter using transistor switches. [*From H. D. Huskey and G. A. Korn* (eds.), "*Computer Handbook*," *p. 7–21, Fig. 7.1.14, McGraw-Hill Book Company, New York, 1962.*]

Transistor switches may be profitably used in the design of high-accuracy bidirectional limiters, in the manner shown by the circuit of Fig. 99. A complementary-transistor voltage switch is connected into the forward path of a low-drift d-c amplifier. The output voltage V_o follows faithfully (with inverted sign) the input voltage

V_i up to the point where $V_i = V_L$. When the magnitude of V_i exceeds the magnitude of V_L, one of the two transistors will saturate, giving the desired limiting action. The breakdown diodes are used to prevent the d-c amplifier from saturating on large input signals. This action is necessary to prevent voltage V_B from rising to the point where the reverse-voltage ratings of the transistors are exceeded, and it also allows the d-c amplifier to respond immediately to a drop in the magnitude of V_i.

The slope of the V_o versus V_i curve may be adjusted simply by changing the ratio between the input and feedback resistances. The break points in the transfer characteristic are defined precisely because the circuit becomes regenerative at the point when the magnitude of V_i reaches and exceeds the magnitude of V_L. Since the maximum value of V_L may be as high as 50 to 60 volts, the dynamic range of this limiter is on the order of 5×10^4.

6. CIRCUITS WITH AMPLITUDE COMPRESSION AND/OR EXPANSION

In this section several types of automatic gain control (AGC) circuits will be described first, followed by a description of some compressor and expander circuits for audio systems. Other sections of this chapter contain material on circuits having a logarithmic amplitude response and those useful in squaring and square-root extracting applications.

Nonlinear devices used in amplitude-compression and -expansion circuits are described in Sec. 2. Some of the devices useful in circuits of this type are not mentioned in the material below, as the purpose of this section is to give only a few examples of practical circuits having amplitude compression and/or expansion.

6.1. General-purpose AGC Circuits with Compression Only. The block diagrams of Fig. 100 show two methods of designing AGC systems. In a, the input

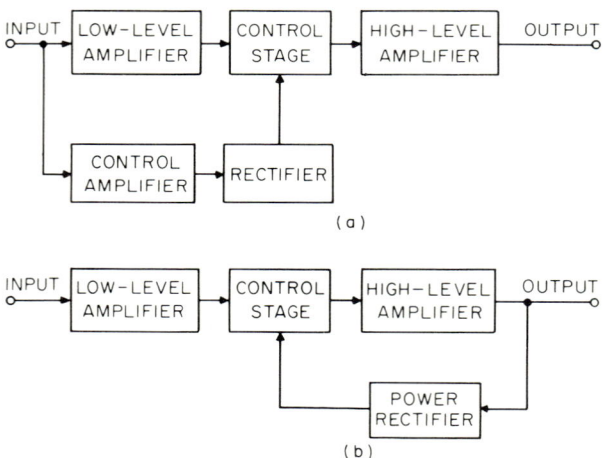

Fig. 100. Block diagrams showing two methods of designing AGC systems.

signal is separately amplified, rectified, and applied to a control stage. The control stage serves to provide a gain reduction in the main signal path as the input signal increases. In b, the control signal is obtained from the output of the main amplifier, rectified, and fed back to the control stage to provide the gain reduction in the main signal path. The latter system usually saves components and is more widely used. In many applications the low-level amplifier stages shown in the diagrams of Fig. 100 are not required, and the input signal is fed directly to the control stage. Examples

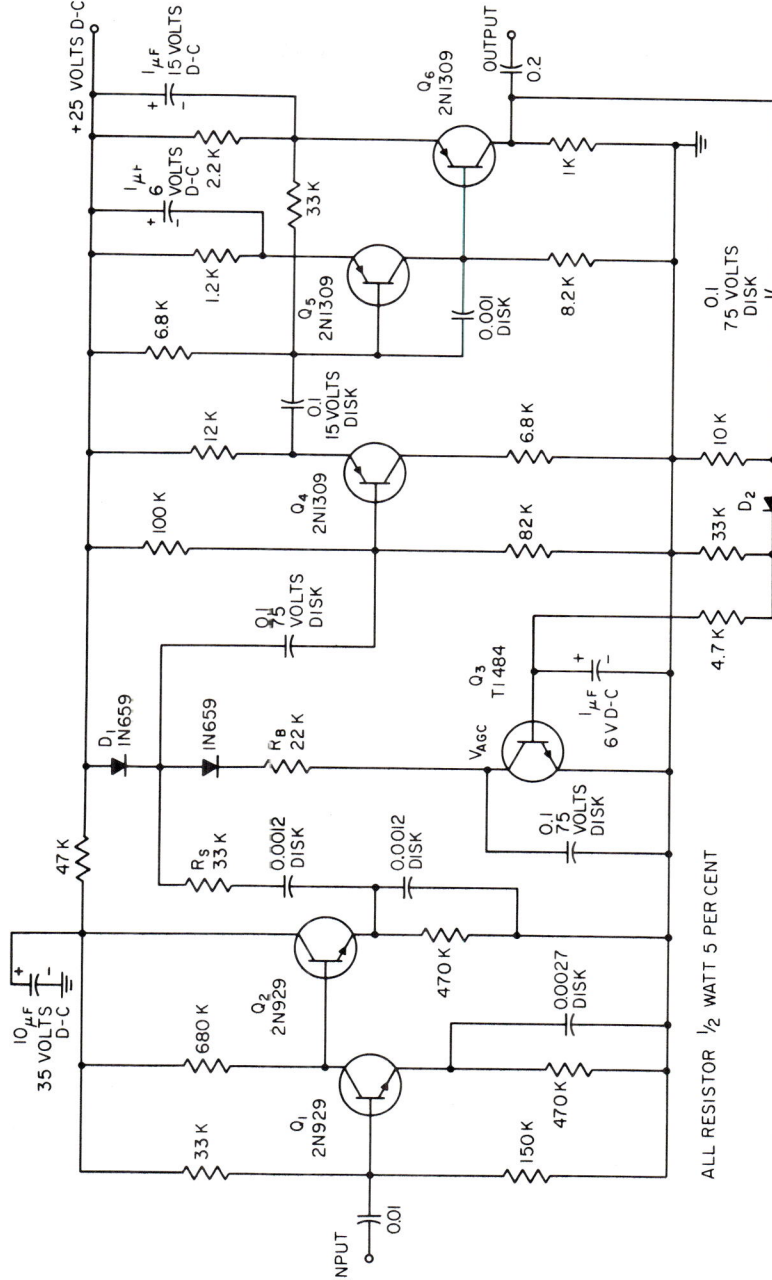

FIG. 101 Low-noise low-level amplifier using a silicon diode as a gain-control device. (*From Texas Instruments, Inc., "Transistor Circuit Design," p. 179, Fig. 11.6, McGraw-Hill Book Company, New York, 1963.*)

of AGC circuits using several different nonlinear devices in the control stage are given in the following paragraphs.

Figure 101 shows a low-noise low-level amplifier circuit using a silicon diode as the control device. Transistors Q_1 and Q_2 are the active elements in a low-noise low-level stage, designed to operate with a source resistance of from 500 to 50,000 ohms. D_1 is the control diode, with its bias current controlled by transistor Q_3. The base of Q_3 is driven from the rectifier circuit containing diode D_2 and its associated low-pass filter. Transistors Q_5 and Q_6 form a high-level amplifier for furnishing output power plus AGC drive, while transistor Q_4 is an emitter follower whose function is to present a high resistance to the AGC network.

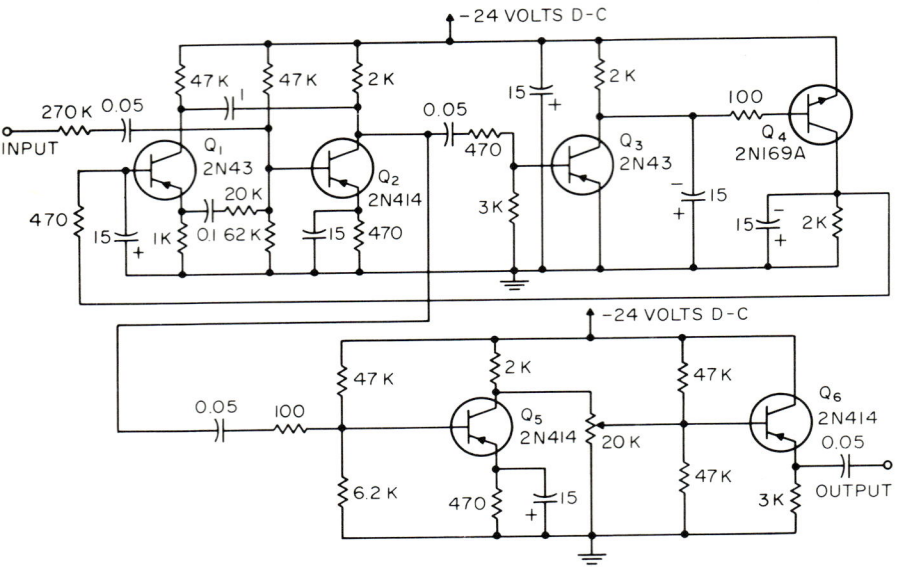

Fig. 102. AGC amplifier using a germanium transistor in a feedback loop as the control device. (*From Electronic Equipment Eng., January, 1962, p. 35.*)

The specifications of this amplifier when driven from a 500-ohm source are given below:

Open-loop gain, db.................... 97
AGC range, db........................ 60
Noise figure, db...................... ≤ 6
Maximum input signal, mv............. 2
Maximum output signal, volts......... 1

An AGC amplifier using a germanium-alloy transistor as the control device is shown in Fig. 102. Rather than having the control transistor in the main circuit of the amplifier, which could introduce distortion for high input signal levels, this amplifier has the control transistor in a feedback path around the controlled stage. Transistor Q_2 is the active element in the controlled stage, and the control transistor Q_1 is connected between the collector and the base of Q_2. A variation in the base current of Q_1 will cause a change in the effective resistance from collector to base and hence will vary the feedback. Control current for Q_1 is derived from the output of Q_2 by rectification and d-c amplification through transistors Q_3 and Q_4. The a-c amplifier stage containing transistors Q_5 and Q_6 brings the output voltage to the desired level.

NONLINEAR AMPLIFIERS

This amplifier accommodates an input signal of from 2 to 15 volts peak-to-peak with a corresponding output variation of from 8 to 10 volts peak-to-peak. The output variation could be further reduced by using a higher-gain d-c amplifier in the control portion of the circuit. The operating frequency range of this circuit is from approximately 5 to 100 kc/sec, and this range could be easily extended with suitable component changes.

As discussed in Sec. 2.5, tetrode transistors are quite useful variable-gain devices. The circuit shown in Fig. 103 uses a 3N45 tetrode transistor with a transverse bias between the two bases as the control device in a constant-output a-c amplifier. The tetrode transistor Q_1 is operated in the common-base configuration for direct current

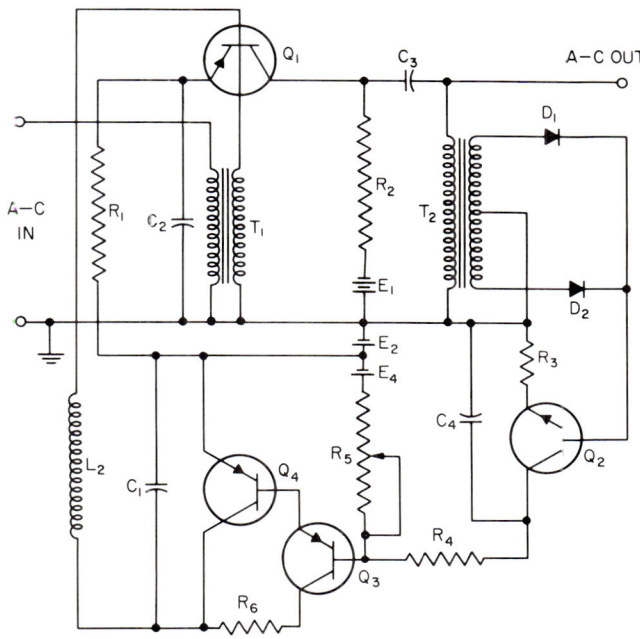

FIG. 103. Constant-output a-c amplifier using a tetrode transistor for gain and as the control device. (*From Electronic Equipment Eng., August, 1960, p. 52.*)

to keep the variation in d-c operating point as small as possible, and it is operated in the common-emitter configuration for alternating current to take advantage of the maximum available gain. This method of circuit design allows h_{fe} to change by a ratio of 15:1 while h_{FB} changes only 25 per cent (4:3).

Diodes D_1 and D_2, transistor Q_2, resistors R_3 and R_4, and capacitor C_4 comprise the rectification, control amplification, and filtering parts of the circuit. The maximum output current of Q_2 is limited by the values of R_3 and R_4. Voltage source E_4 and resistor R_5 determine the reference level for the output. Since the collector current of Q_2 is proportional to the a-c output, when

$$[I_C(Q_2) + I_{CBO}(Q_3)]R_5 > E_4 \tag{35}$$

transistor Q_3 will be turned on. If it is less than that, Q_3 will be biased off. Transistors Q_3 and Q_4 are the active elements in a d-c amplifier stage for the difference signal between the a-c output and the reference level. The output of this stage provides the transverse bias for tetrode Q_1. Use of inductor L_2 and capacitor C_1 ensures that the d-c amplifier stage is isolated from the a-c signal applied to base 2 of Q_1.

This circuit will maintain the output voltage constant to within ±5 per cent over a 10:1 range of input amplitudes. Still closer control could be obtained by using addi-

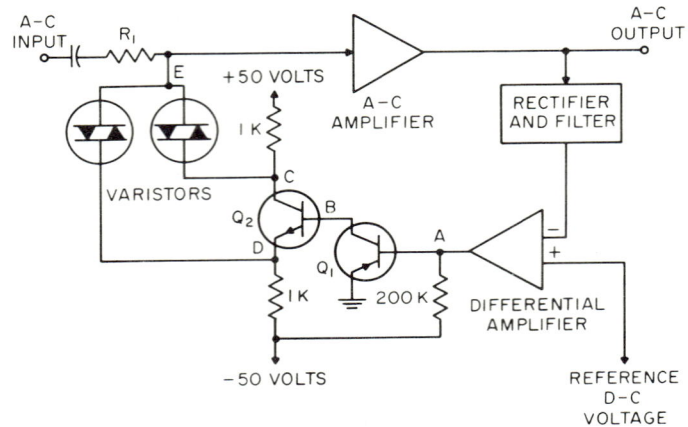

Fig. 104. AGC amplifier using a varistor network to give AGC action. (*From Electronic Equipment Eng., March, 1962, p. 29.*)

tional amplification between the controlled tetrode and the output or by adding another stage to the control circuit, as there is no phase-shift problem in this type of feedback control. In contrast to some AGC systems, the control system in this amplifier tends to decrease rather than increase the distortion.

An AGC system using a network containing two varistors as a d-c controlled variable attenuator is shown in Fig. 104. For low a-c input voltages, the output voltage of the differential amplifier is close to zero and transistors Q_1 and Q_2 are saturated. This causes points A, B, C, and D to all be near ground potential, and the effective shunting resistance of the varistor network is approximately 500 kilohms. As the input voltage increases, Q_1 and Q_2 are driven toward cutoff by the positive output of the differential amplifier, causing the potential at points B and D to go negative and the potential at point C to go positive. The effective resistance of the varistor network as the output of the differential amplifier varies is shown in the plot of Fig. 105. Since the collector and emitter resistors of Q_2 are equal in value, the voltages at points C and D are always equal but opposite in polarity, therefore main-

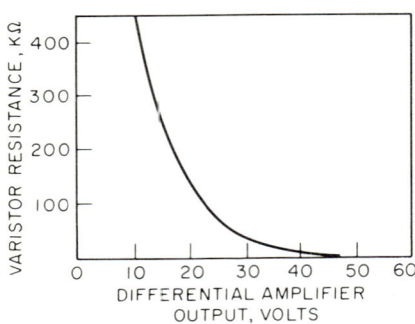

Fig. 105. Effective resistance of the varistor network of Fig. 104 as a function of the output of the differential amplifier. (*From Electronic Equipment Eng., March, 1962, p. 29.*)

taining point E at a zero d-c potential throughout the control range.

At the maximum input level, both Q_1 and Q_2 are cut off, causing points B and D to be at a potential of −50 volts and point C to be at a potential of +50 volts. This makes the effective shunting resistance of the varistor network approximately 2 kilohms. An attenuation range of 50 to 1 is therefore achieved when resistor R_I has a value of 120 kilohms. A greater range could be achieved with higher-voltage transistors by increasing the magnitude of the power-supply voltages and increasing the value of R_I. The signal frequency in this type of system is limited by the frequency response of the varistors to a few kilocycles if large ratios of gain attenuation are desired.

6.2. AGC Circuits for Entertainment Radio Receivers. There are three functions that an AGC circuit for entertainment radio receivers should perform: (1) It should have sufficient delay in its action so that the sensitivity of the receiver is not affected. (2) It should maintain a fairly constant signal strength at the input to the detector so that a constant audio output is maintained under all conditions. (3) It should have enough control to limit the signal strength on strong signals to a degree where overloading does not occur in the circuits preceding the manual volume control.

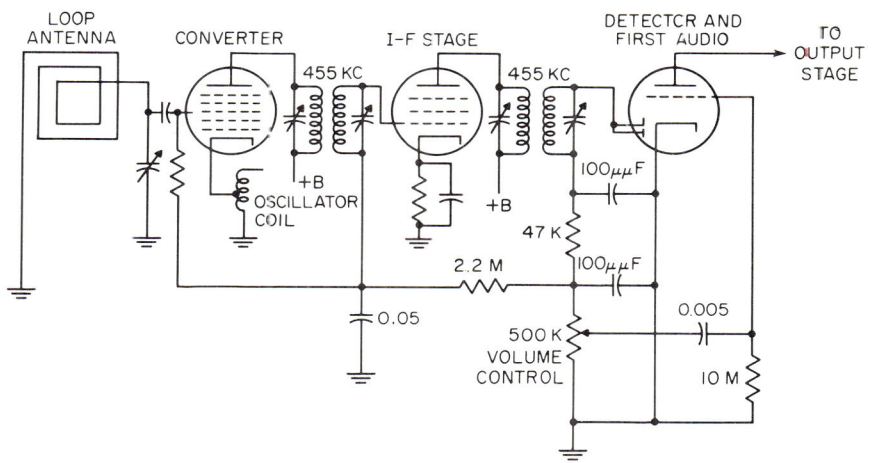

Fig. 106. Typical vacuum-tube AGC circuit for a five-tube radio receiver. [*From K. Henney (ed.), "Radio Engineering Handbook," 5th ed., p. 19–52, Fig. 19, McGraw-Hill Book Company, New York, 1959.*]

Generally, the more stages that can be controlled by the AGC system, the better it can perform the above three functions. However, economic considerations in small receivers often limit the use of AGC to that of controlling only one or two stages. A typical vacuum-tube AGC circuit for a five-tube receiver is shown in Fig. 106. A single diode circuit and a single load resistor are used to develop both the AGC bias and the audio output. In operation, electron current flows from the cathode to the diode plates of the detector tube during positive peaks of the i-f signal. The current then flows through the secondary winding of the i-f transformer, through the 47-kilohm filter resistor, and through the 500-kilohm volume control to the cathode and ground. When modulation is present, this current has both d-c and audio-frequency components. The desired amount of audio-frequency voltage is picked off by the tap on the volume-control potentiometer and applied to the grid of the first audio amplifier tube (actually this is the same tube as used for detection).

The rectified d-c voltage across the load resistor acts through the 2.2-megohm filter resistor in the AGC bus to determine the grid potential of the two tubes which are controlled. As the signal strength at the antenna increases, more signal is rectified by the diode detector and a greater negative bias is applied to the grids of the controlled tubes. This change of bias reduces the gain of these tubes, with the result that the proportionate change in the output of the receiver is much smaller than that of the input.

A block diagram of a typical six-transistor portable radio receiver is shown in Fig. 107, and a schematic diagram of a single-action AGC circuit often used in such a receiver is shown in Fig. 108. The AGC voltage is developed at the diode detector and fed back to the base of transistor Q_1 in Fig. 108. This type of AGC circuit is simple and inexpensive, but it has two basic disadvantages. First, the initial gain of the stage is a direct function of the h_{FE} variations of the transistor used; therefore,

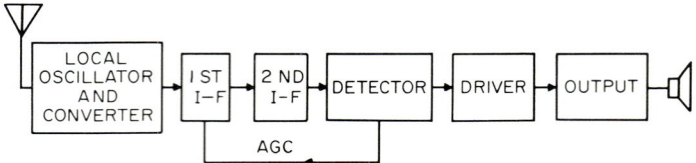

FIG. 107. Block diagram of a typical six-transistor radio receiver.

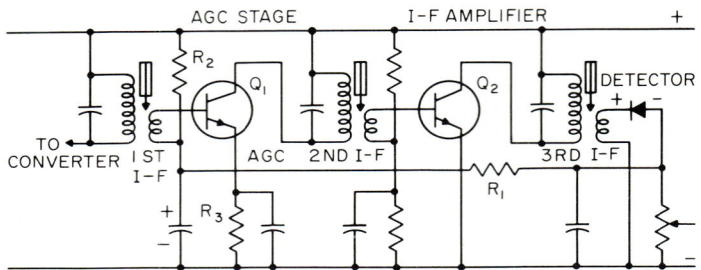

FIG. 108. Schematic diagram of a single-action AGC circuit for a transistor radio receiver. (*Courtesy of General Electric Co.*)

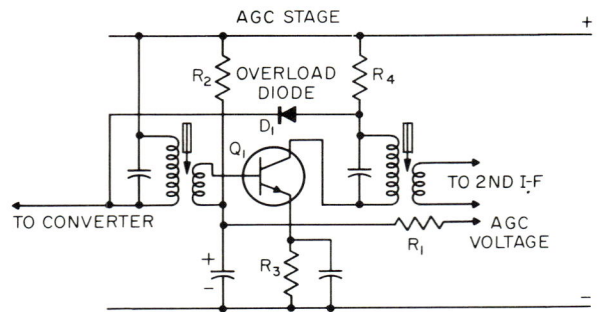

FIG. 109. Primary-plus-delayed AGC circuit for a transistor radio receiver. (*Courtesy of General Electric Co.*)

rather heavy stabilization must be used with an accompanying i-f bypass capacitor in order to eliminate signal degeneration. The second and more serious disadvantage of this circuit is the inability of the transistor to cut off the signal effectively; therefore, it gives rather poor AGC action.

A more adequate AGC action may be achieved by adding a delayed AGC circuit to supplement the primary AGC circuit. Referring to Fig. 109, it is seen that under weak signal conditions the voltage drop across the resistor R_4 causes diode D_1 to be reverse-biased, and it therefore presents a high impedance to the signal. As the signal increases, primary AGC action causes the emitter current of the transistor to decrease, thus decreasing the voltage drop across resistor R_4. At some predetermined point the diode will become forward-biased, causing some of the signal to be shunted through it. This combined circuit, while giving good AGC action when properly adjusted, still has disadvantages in that the point at which the diode starts conducting is directly dependent on the h_{FE} of the AGC transistor. Thus resistors R_2 and/or R_4 in the circuit of Fig. 109 must be adjusted or selected to be compatible with the h_{FE} characteristics of Q_1. The extra diode and resistor also add cost to the circuit.

Figure 110 shows an AGC stage which gives better performance than the stage of Fig. 109, with no more components than the stage shown in Fig. 108. Transistor Q_1 is in the common-base configuration for i-f signals, but it is in the common-emitter configuration for the d-c control voltage. Two types of AGC action occur in this stage as the input signal increases, both caused by the decrease in emitter current of Q_1. First,

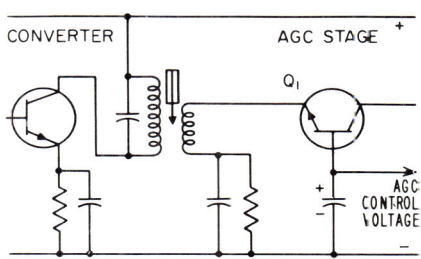

FIG. 110. Common-base AGC circuit for a transistor radio receiver that gives superior performance. (*Courtesy of General Electric Co.*)

the a-c gain of the stage itself is decreased, as in the other circuits. Second, the input impedance of the transistor increases enough to cause an impedance mismatch between the output of the converter stage and the input to AGC stage, which results in further gain reduction. The curves of Fig. 111 show the AGC characteristics of the three control stages under discussion. When using the common-base stage, the cir-

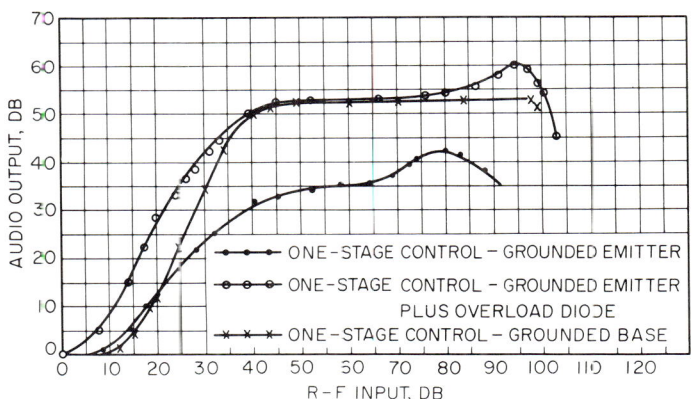

FIG. 111. AGC characteristics of the control stages shown in Figs. 108 to 110. (*Courtesy of General Electric Co.*)

cuit is observed to provide a good delay characteristic and a quite flat response for medium and high input signal strengths.

6.3. High-frequency AGC Circuits. In this section three AGC circuits which are useful for high-frequency applications will be described, and test results will be given for each. The first circuit is a high-frequency i-f amplifier using silicon tetrode transistors as control devices. Referring to Fig. 112, the i-f amplifier is seen to be a three-stage, double-tuned, transitionally-coupled circuit, with the emitter current and

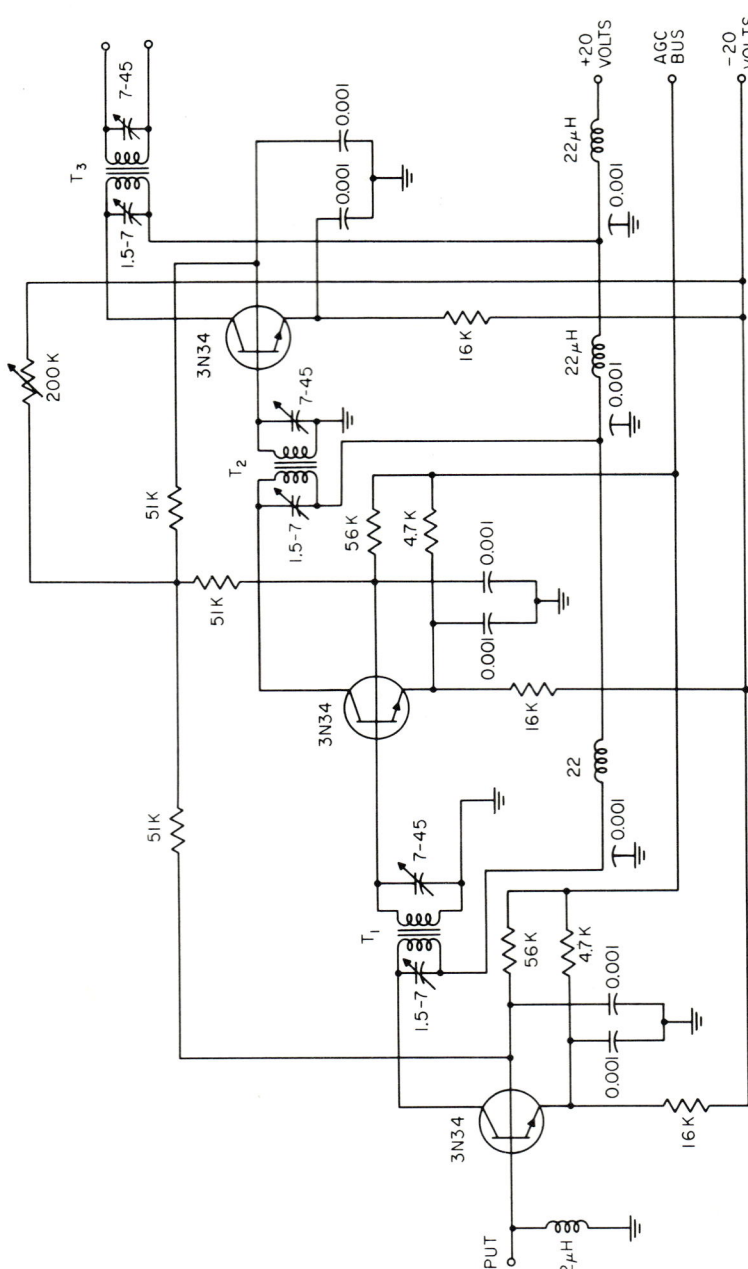

FIG. 112. 20-Mc i-f amplifier using silicon tetrode transistors for gain and as control devices. (*Courtesy of Texas Instruments, Inc.*)

the base-2 current of each tetrode controlled by the AGC bus potential through resistance paths. Figure 113 shows the variation of amplifier gain with frequency for various values of I_E and I_{B2} of each tetrode. The appropriate values of resistance needed in the various AGC control paths may be determined from these curves. Figure 114 shows the center-frequency gain reduction as a function of the AGC bus potential, where the center frequency is 23 Mc.

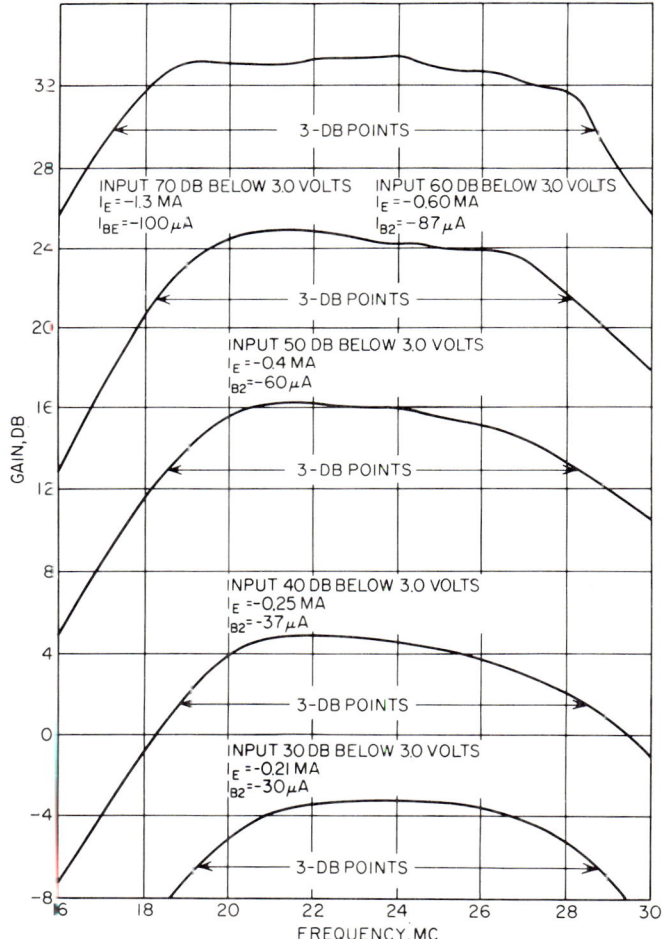

FIG. 113. Curves showing variation of gain with frequency for various AGC conditions in amplifier of Fig. 112. (*Courtesy of Texas Instruments, Inc.*)

A practical power detector circuit for use with the circuit of Fig. 112 is shown in Fig. 115. To avoid complicated AGC circuitry, the base-to-emitter diode of a 2N715 or 2N716 transistor is used as both a signal detector and an AGC detector. The high gain realizable with this power detector gives good AGC action in closed-loop performance without further d-c amplification. Table 10 lists the optimum open-loop and closed-loop performance of the circuits of Figs. 112 and 115 when used together. These optimum results are obtained with $R1 = 11$ kilohms, $R2 = 5$ kilohms, and $R3 = 3.9$ kilohms (Fig. 115).

A wideband FM radio receiver circuit containing a 70-Mc i-f amplifier is shown in block-diagram form in Fig. 116. The i-f preamplifier includes two variolosser networks containing germanium point-contact diodes which maintain a constant level out of the main i-f amplifier. Figure 117 shows the gain-control circuitry as a and the detector and AGC amplifier which drives the variolossers as b.

The impedance of each variolosser diode to signal frequencies is controlled by the direct current passing through it. Direct current is caused to flow in such a manner that as the current increases in the shunt diodes, it decreases in the series diodes (and vice versa). Since the diodes appear in pi networks for the i-f, variable loss is obtained without producing much variation of the input and output impedances of the variolossers. The currents are controlled by the output of the AGC amplifier, which responds to amplitude-modulation frequencies ranging from direct current up to about 50 cps. Diodes D_7, D_8, and D_9 are used as a -2.0-volt d-c supply and are forward-biased through resistor R_3. (These diodes become starved of current under fade conditions, and the reduced voltage ensures a low minimum i-f loss in the variolossers.) The series and shunt diode sets are connected in parallel for incremental currents supplied from the K lead; but the -2.0-volt source is placed within a loop passing through all six attenuator diodes, thus providing a condition wherein their dynamic resistances can simultaneously equal about 130 ohms. This condition occurs when the input d-c lead from the AGC amplifier carries no current, and it corresponds to a medium-loss condition in the variolossers. The two variolossers are so interconnected that they conduct the same direct currents, thereby forcing their loss values to track together. Inductors L_1 and L_2 carry control current and also counteract the effect

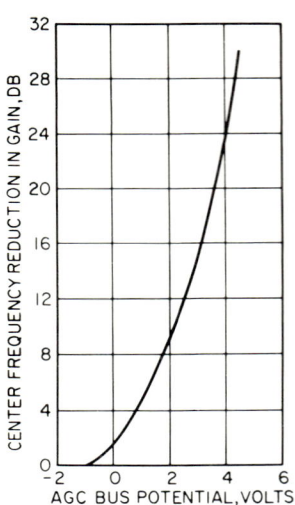

FIG. 114. Center-frequency gain reduction vs. AGC bus voltage for circuit of Fig. 112. (*Courtesy of Texas Instruments, Inc.*)

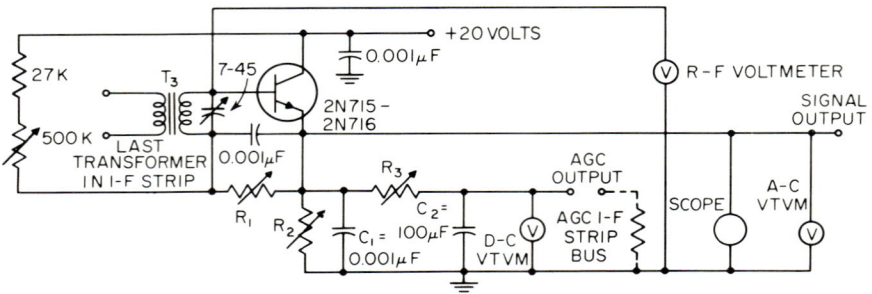

FIG. 115. Practical power detector circuit for use with amplifier of Fig. 112. (*Courtesy of Texas Instruments, Inc.*)

of the series-diode capacitance, which is important when the diode resistances reach their highest level. Resistors R_1 and R_2 provide an upper limit for the series path impedance, thereby forcing the shunt diodes to carry more of the loss burden at higher i-f input signal levels.

NONLINEAR AMPLIFIERS

Table 10. Test Results on Circuits of Figs. 112 and 115 with Optimum Values of Circuit Resistances

Lowest r-f input before signal distorts, volts	Highest r-f input before signal distorts, volts	Modulation, %	Modulation frequency, cps	Signal output, volts
\multicolumn{5}{c}{AGC Loop Open}				

Lowest r-f input before signal distorts, volts	Highest r-f input before signal distorts, volts	Modulation, %	Modulation frequency, cps	Signal output, volts
0.0090		40 max	400	0.0035
0.0090		60 max	1,000	0.0042
	1.25	40	400	0.90
	1.40	60	1,000	1.30
	1.30	70 max	400	1.40
	1.70	90 max	1,000	1.90

AGC Loop Closed (input increased 19 db before signal distorted on both 400- and 1,000-cps modulation)

Lowest r-f input before signal distorts, volts	Highest r-f input before signal distorts, volts	Modulation, %	Modulation frequency, cps	Signal output, volts
0.0090		40	400	0.0037
0.0090		60	1,000	0.0047
	1.45	40	400	1.20
	1.80	60	1,000	1.70
	1.75	70 max	400	2.00
	2.30	90 max	1,000	2.60

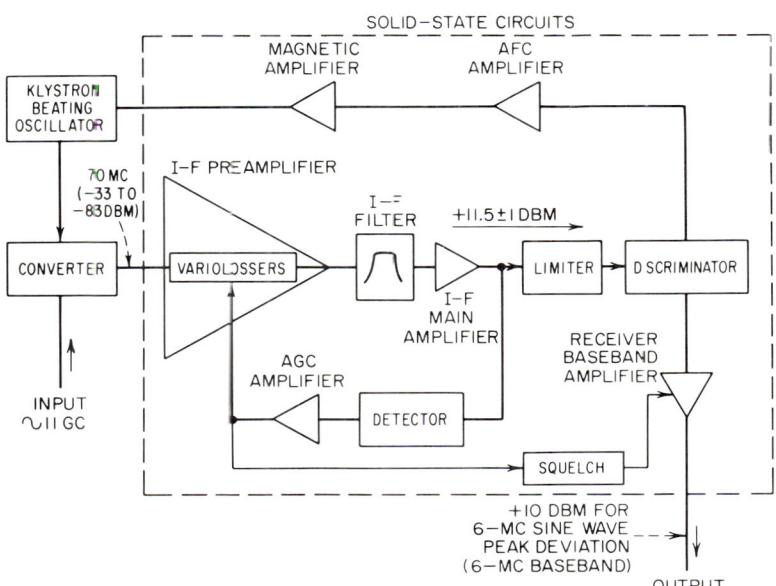

FIG. 116. Block diagram of a wideband FM radio receiver using a gain-controlled 70-Mc i-f amplifier. (*From Bell System Technical J., vol. 41, p. 1832, 1962.*)

A network containing a thermistor applies a temperature-dependent bias voltage in series with the shunt diodes of the variolossers. The effect is to equalize the drift which tends to occur in the AGC amplifier output current, and thus to prevent drift

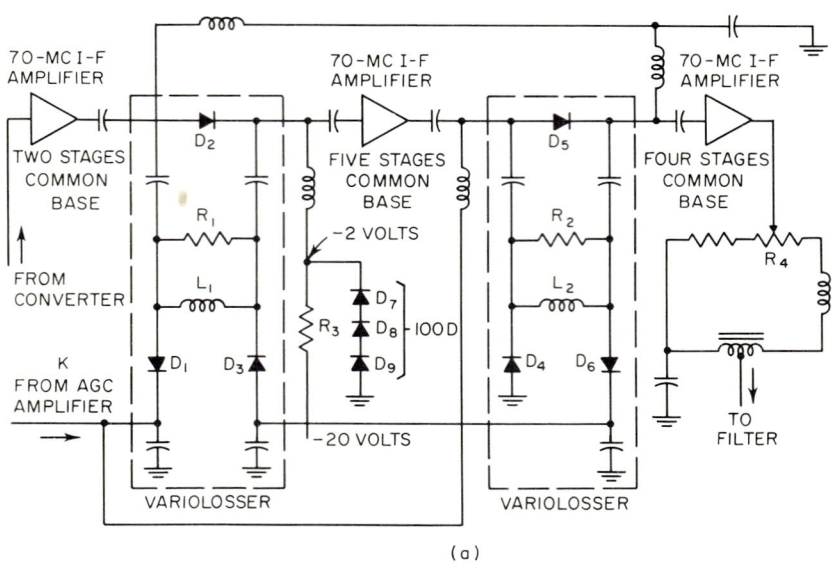

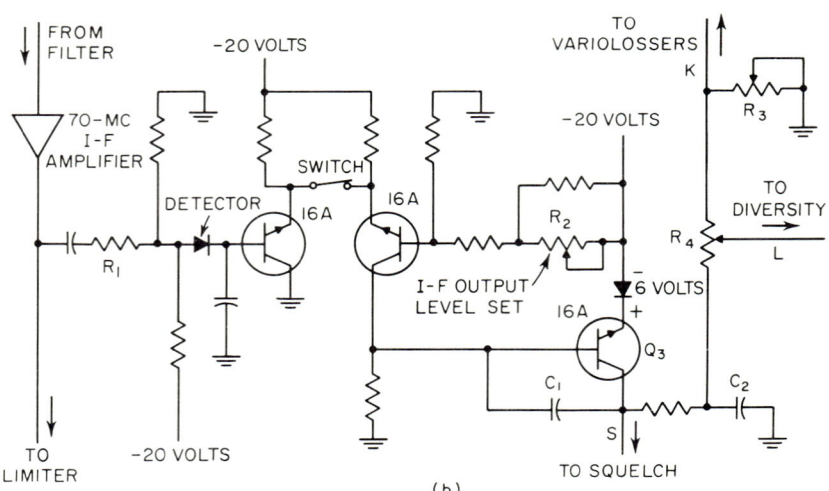

Fig. 117. Gain-control circuit, detector, and AGC amplifier of radio receiver shown in Fig. 116. (*From Bell System Technical J., vol. 41, pp. 1841, 1843, 1962.*)

in the squelch firing level. The thermistor network, not shown in Fig. 117a, is inserted between diode D_4 and ground.

The AGC amplifier has sufficient gain to multiply the small current increment drawn from the detector to the relatively large current needed to drive the variolossers. These circuits hold the output of the main i-f amplifier to $+11.5$ dbm $\pm$ 1 db

NONLINEAR AMPLIFIERS 26-71

for i-f input levels ranging from -83 to -33 dbm over a temperature range from -20 to 60°C. Figure 118 demonstrates graphically the tightness of this AGC system.

Figure 119 shows a block and gain-level diagram of an i-f amplifier using two variolosser networks to maintain constant output level over a frequency band of

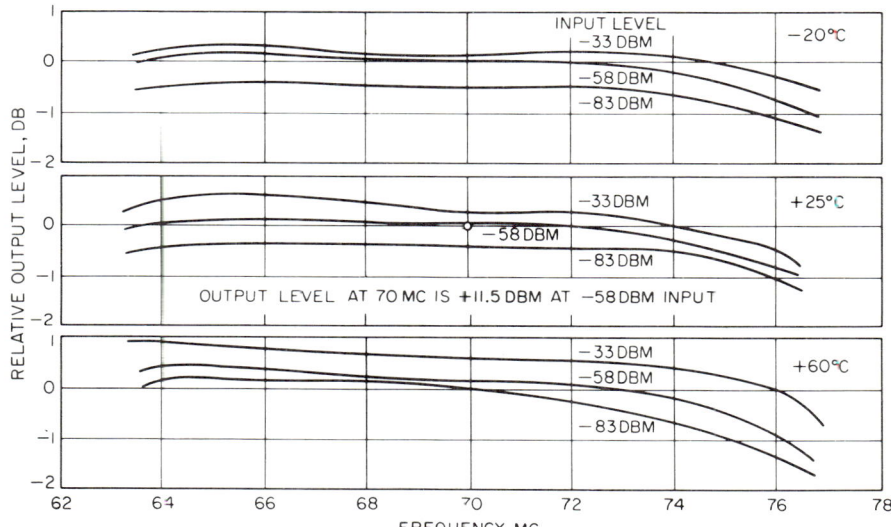

Fig. 118. I-F output level vs. frequency for AGC system of Fig. 117. (*From Bell System Tech. J., vol. 41, p. 1853, 1962.*)

65 to 115 Mc. The basic design of the i-f amplifier is shown schematically in Fig. 120. Diffused-base germanium p-n-p transistors are used for amplification in all the stages shown. The variolosser networks each contain a low-capacitance germanium-alloy diode plus two series resistors. The resistors are large enough in value essentially to

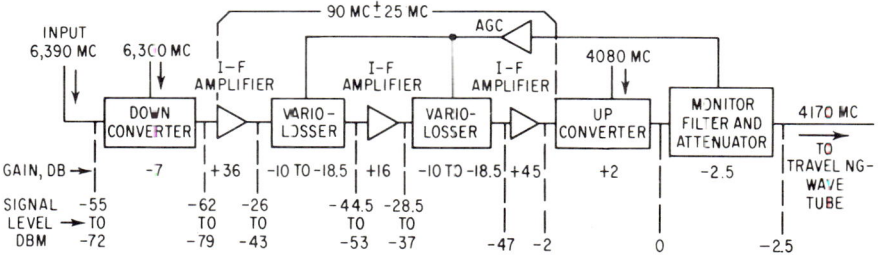

Fig. 119. Block and level diagram of an i-f amplifier system that maintains constant output level over a frequency range of 65 to 115 Mc. (*From Bell System Tech. J., vol. 42, p. 840, 1963.*)

mask out the reactive input and output impedances of the adjacent common-emitter shunt-feedback stages. Their values are small enough, however, to assure a low minimum loss for the variolosser networks.

The loss of the variolosser networks is controlled by having the direct-coupled AGC amplifier, shown schematically in Fig. 121, provide control currents to the shunt diodes which are proportional to the difference between the d-c output from the waveguide monitor and a d-c reference voltage. The AGC amplifier uses diffused-base silicon n-p-n transistors. The differential input stage is followed by a common-emitter stage and an output stage, with one variolosser diode being driven from the emitter

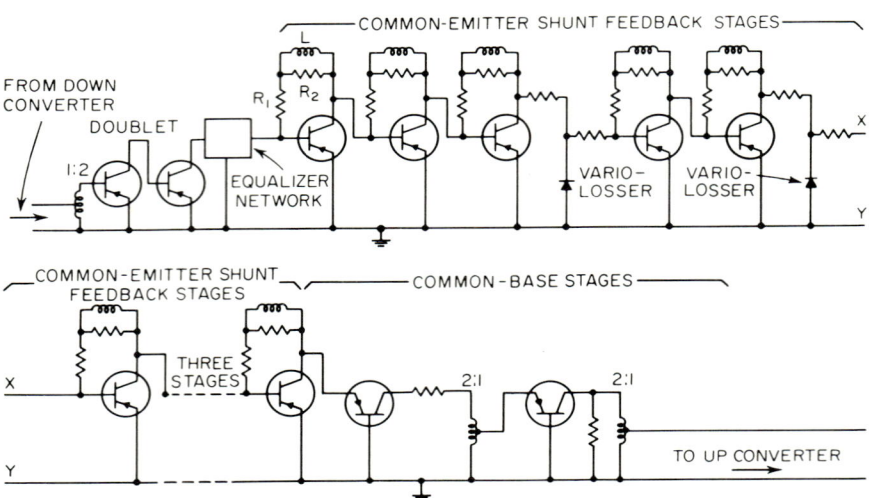

Fig. 120. Basic design of i-f amplifiers used in system of Fig. 119. (*From Bell System Tech. J.*, vol. 42, p. 841, 1963.)

and the other from the collector of the same stage. Thus approximately equal currents flow through each diode. This diode-driving technique and the use of high-frequency transistors in the AGC amplifier result in the frequency response of the AGC loop being controlled almost exclusively by capacitor C (Fig. 121).

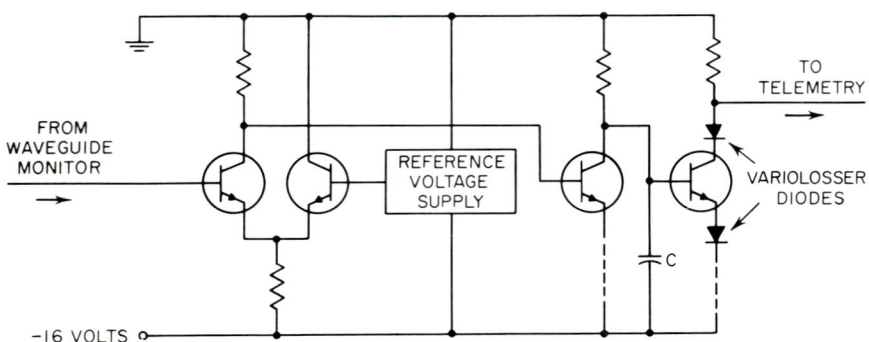

Fig. 121. AGC amplifier used in system of Fig. 119. (*From Bell System Tech. J.*, vol. 42, p. 843, 1963.)

The tightness of this AGC system is indicated by the curves of Fig. 122. For received signals ranging between -55 and -72 dbm over a temperature range of 0 to 60°C, the output level of the i-f amplifier system, including the down-up converters, varies less than ± 0.8 db.

6.4. Compressors and Expanders for Audio Systems. Two examples of compression and expansion circuits for audio systems will be given in this section. The first is a transistorized audio AGC amplifier suitable for use in broadcast studios to control the output level of radio and television audio transmission. Figure 123 shows a simplified schematic diagram of the amplifier, which uses two opto-electronic devices (photoconductive cell–light source combinations) as control elements. When switch S_1 is closed, a greater than normal strength signal applied to the input of the amplifier will cause a positive d-c voltage to be developed at the base of transistor Q_{12}, which will bring both Q_{12} and Q_{13} out of cutoff and cause current to flow through the lamp in R_{1A}. This causes the resistance of the photocell in R_{1A} to decrease, which reduces the a-c gain of the amplifier. The greater the magnitude of the input signal, the greater the amount of gain compression. The typical attack time of this circuit is 20 msec, and the recovery time is approximately 1 sec. The level at which gain reduction occurs and the desired output level of the amplifier may be controlled by adjustment of potentiometer R_2. The output may be held approximately constant at any level from +4 to +30 dbm.

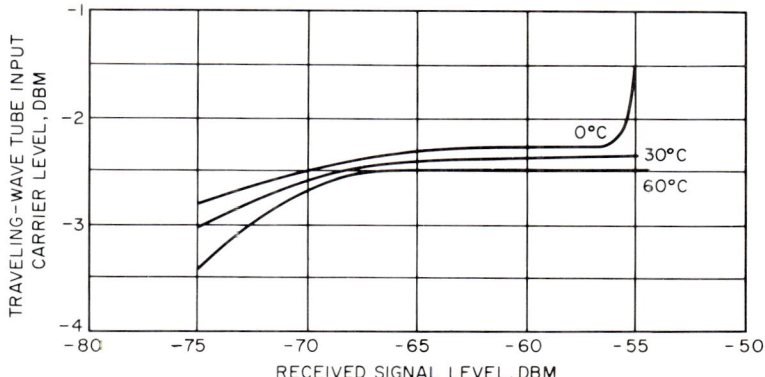

FIG. 122. Output level vs. signal input level for system of Fig. 119. (*From Bell System Tech. J.*, vol. 42, p. 844, 1963.)

The amplifier has gain-expansion capabilities in order to provide noise reduction at low signal levels. When switch S_2 is closed and when no signal is applied to the amplifier, the d-c amplifier stages containing transistors Q_{10} and Q_{11} provide maximum current to the lamp in R_{1B}, thus causing the resistance of the photocell in R_{1B} to be at a minimum and the gain of the amplifier to be reduced. As the signal input increases the photocell resistance increases, causing the amplifier gain to increase toward its normal-value. The threshold for expansion, which is controlled by potentiometer R_{10}, is usually chosen so that there is a 20-db range of linear operation before the threshold for compression is reached.

The gain of this amplifier is 75 db when it is operating in its linear range. Its frequency response is rated at ±1 db over a range of 30 to 15,000 cps under any condition of gain reduction up to 30 db. The distortion rating is 1½ per cent or less over this frequency range, again under any condition of gain reduction up to 30 db, and the signal-to-noise ratio is 65 db or better.

The second example of compression and expansion circuits for audio systems is a unit used in short-haul telephone carrier systems. This unit is called a compandor, as it contains both a compressor and an expandor, with the former used in transmission and the latter in reception. The use of compandors in systems of this type provides a significant reduction in noise and crosstalk over noncompanded systems.

26-74 AMPLIFIER CIRCUITS

Figure 124 is a schematic diagram of the compandor unit, with the compressor shown as *a* and the expandor as *b*. The compressor consists of a three-stage amplifier, a variolosser, and a signal-level detector. This unit reduces the dynamic range of the output signal (voice) to one-half of the range of the input signal. This allows

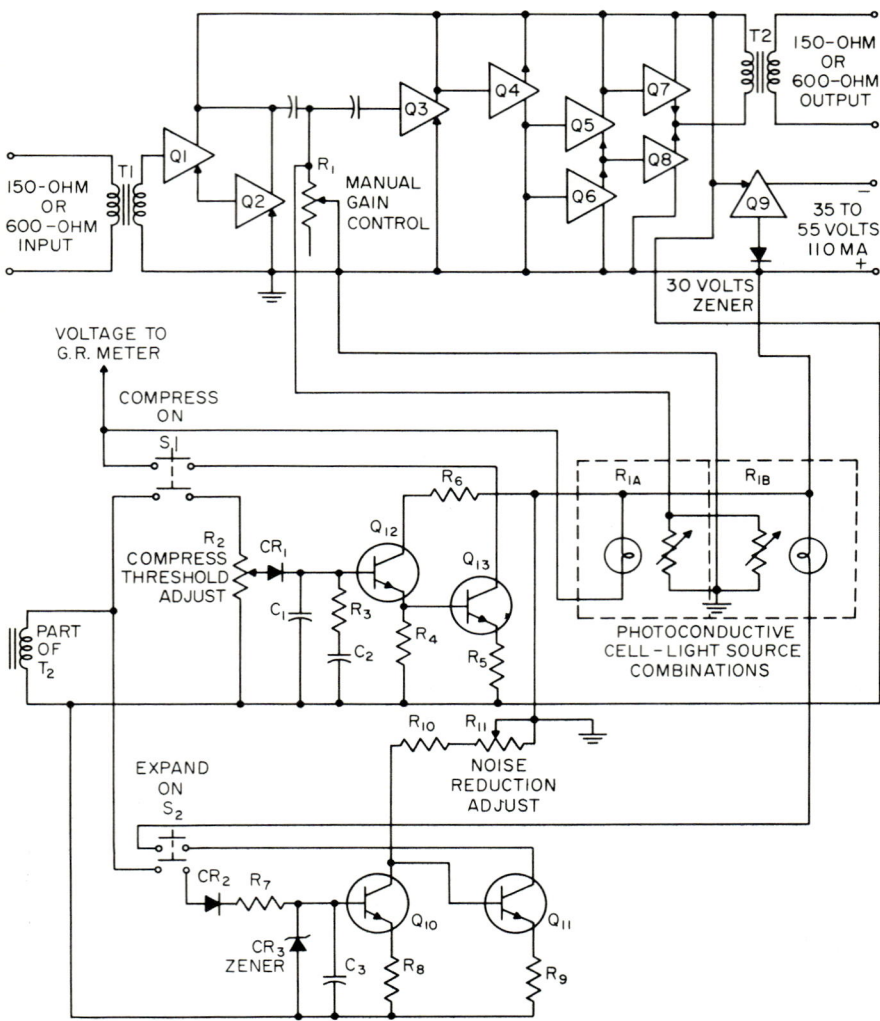

FIG. 123. Simplified schematic diagram of a transistorized audio amplifier with the capability for both compression and expansion of gain. (*Courtesy of General Electric Co.*)

weak signals to be amplified a greater amount than strong signals and consequently greatly improves signal-to-noise ratio. The variolosser operates by means of a pair of diodes shunting a high-impedance circuit and providing loss proportional to a control current in the diodes. The desired 2:1 compression of input signal is obtained by controlling the diode impedance with a signal derived from the output of the

NONLINEAR AMPLIFIERS   26-75

voice amplifier. The diodes in the variolosser are a diffused-junction silicon type having extremely uniform impedance vs. current characteristics and very low values of body resistance. These diodes are the key to the nearly perfect compression over a 60-db input signal range. The three-stage amplifier has approximately 40 db of feedback, which provides the necessary stability and required transmission charac-

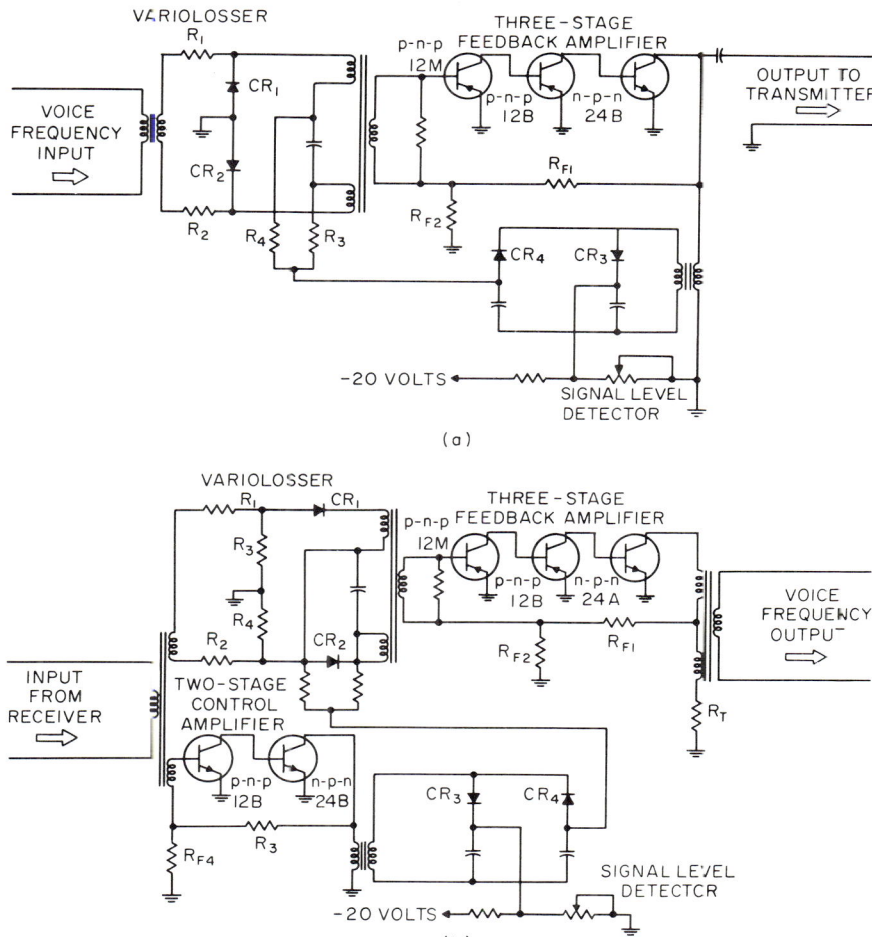

Fig. 124. Schematic diagram of the compression and expansion circuits used in a telephone transmission and reception system. (*From Bell Lab. Record, vol. 42, p. 129, 1964.*)

teristics. The signal-level detector, which provides the control current to the variolosser, uses a step-up transformer and a rectifier circuit. This is a very high impedance circuit which approaches a linear current source and contributes negligible error to the near-perfect compression.

The expandor circuit does the inverse of the compressor by restoring the compressed signal to its original range. It consists of a variolosser, a three-stage amplifier, a two-stage control amplifier, and a signal-level detector. The variolosser is composed of the same type of silicon diodes used in the compressor, but the associated circuitry places the diodes in series in a low-impedance transmission path. The loss introduced by a variolosser is, in this case, inversely proportional to the control current in the diodes. The control current is obtained from a signal-level detector similar to that of the compressor. This detector is driven from a separate two-stage control amplifier which receives its signal from the input to the expandor. In this manner, the expandor variolosser decreases the loss as the input signal increases, effecting a net dynamic expansion of the input signal 2:1. The output of the variolosser is connected to a three-stage feedback amplifier, similar to the compressor amplifier, except that the output is arranged to provide a 600-ohm source impedance to the voice frequency line. By using the same variolosser diodes and signal-level detector operating at the same current levels, both circuits will track over an input signal range of 60 db with an accuracy of ± 0.4 db.

7. CIRCUITS HAVING A LOGARITHMIC AMPLITUDE RESPONSE

As brought out in Secs. 2, 3, and 4, logarithmic amplitude conversion may be achieved through the use of nonlinear-device characteristics, by piecewise linear approximation, and by means of servo techniques. In this section methods of incorporating logarithmic converters of several types into complete amplifier circuits will be described. Circuits which are designed to perform logarithmic amplification of d-c signal inputs are covered first, followed by examples of circuits designed primarily for a-c inputs.

7.1. D-C Logarithmic Amplifiers. In the design of d-c logarithmic amplifiers, the choice often must be made whether to use the logarithmic device or network as a direct attenuator or whether to apply it as a feedback element around an operational amplifier. When devices such as thermionic and semiconductor diodes are used as logarithmic elements, the latter approach is usually desirable. It allows either voltage or current inputs to be easily handled and generally improves the response time of the system. Thus most of the circuits that are described in the following paragraphs use the logarithmic device or network as a feedback element around a d-c amplifier.

An amplifier circuit which furnishes an output d-c voltage proportional to the logarithm of the input d-c current over a range of 10^{-12} to 10^{-6} amp is shown in Fig. 125. Referring to a, the circuit uses a thermionic diode in the feedback loop of a low-level d-c amplifier to obtain its logarithmic response, with a second diode of the same type serving as a compensation element. (See Fig. 2 in Sec. 2.1.) The d-c amplifier (Fig. 125b) is a combined electrometer tube–transistor design which has a very high input impedance. A stable operating level is obtained by connecting the electrometer tubes as tetrodes and supplying the screen currents from the common emitters of the type 2N328 transistors, with the plates of the tubes connected directly to the bases of the transistors. This circuit was designed primarily for the measurement of low-d-c currents from neutron ionization chambers but could be used in various low-level measurement applications.

Figure 126 shows the basic and detailed circuit diagrams of a radiation-tolerant logarithmic amplifier that was designed to measure neutron-sensitive ionization-chamber currents from 10^{-8} to 10^{-4} amp. This amplifier design makes use of the logarithmic relation between the screen-grid voltage and the control-grid current of a thermionic pentode when the plate current is held constant, which is discussed in Sec. 2.2. Referring to Fig. 126a, the d-c amplifier controls the screen-grid voltage so as to maintain a constant plate voltage, and the amplifier output is thus proportional to the logarithm of the pentode input current. The complete circuit of the amplifier is shown in Fig. 126b. High-quality tubes are used, and the resistors are all wire-wound types because of their good radiation resistance. The curves of Fig. 127 show radiation-test results on the type ME-1400 pentode used in the amplifier. Although there was a temporary deviation from a true logarithmic response at low input currents when the tube was subjected to a neutron flux of 5×10^{11} nv, this does not represent a serious problem in the application of the amplifier, since low input currents would not need to be measured under high-flux conditions.

A d-c logarithmic amplifier which covers an input-current range of 5×10^{-11} to 5×10^{-4} amp using solid-state components throughout is shown in Fig. 128. The log diode network contains a combination of silicon and germanium diodes to achieve an improved logarithmic characteristic at higher input currents. The amplifier con-

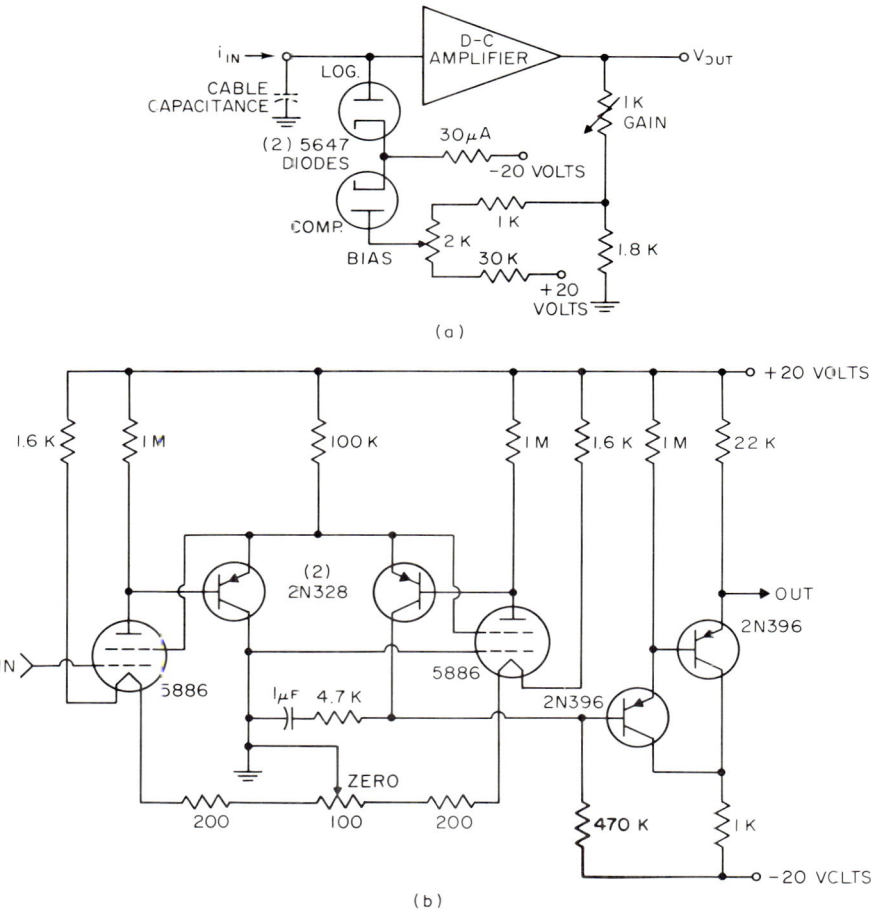

FIG. 125. Low-level logarithmic amplifiers using thermionic diodes and electrometer tubes. (*From IRE Trans. Nucl. Sci., vol. NS-6, no. 2, p. 54, June, 1959.*)

tains a reverse-biased silicon-diode dielectric modulator driven by a 10-kc square-wave generator as an input d-c to a-c converter. The modulator is followed by a simple a-c amplifier, a diode demodulator, and two d-c amplifier stages. With help from the temperature-compensating diode network, the output drift of this amplifier is less than ±0.005 decade/°C from 0 to 50°C over the entire seven decades of operation. This type of approach to d-c amplification is at present required when an all-solid-state system is required for d-c amplification of very small input currents.

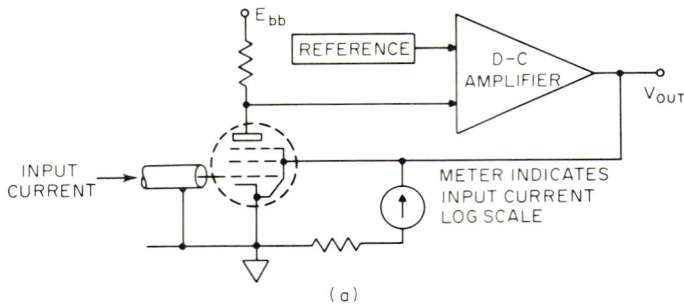

(a)

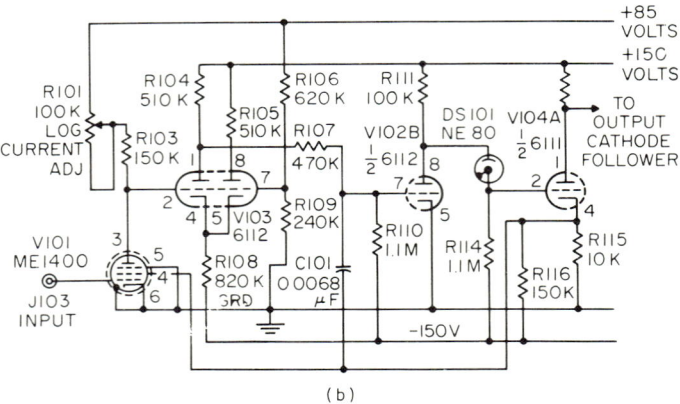

(b)

Fig. 126. Radiation-tolerant amplifier using a thermionic pentode as the logarithmic element. (*From IRE Trans. Nucl. Sci.*, vol. NS-9, no. 1, p. 171, January, 1962.)

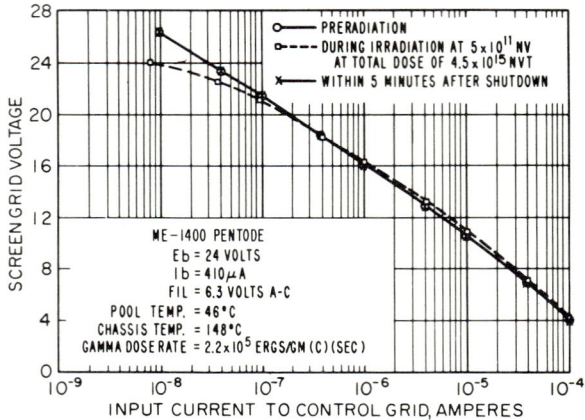

Fig. 127. Curves showing effects of neutron irradiation on pentode used in amplifier of Fig. 126. (*From IRE Trans. Nucl. Sci.*, vol. NS-9, no. 1, p. 172, January, 1962.)

NONLINEAR AMPLIFIERS

The development of high-quality field-effect transistors may allow straight d-c amplifiers to be designed which can equal or exceed this performance.

The circuit shown in Fig. 129a provides an output voltage proportional to the logarithm of the input current (or ratio of two input currents) over an input range from 10^{-8} to 5×10^{-3} amp. Silicon planar transistors are used as feedback logarithmic elements around transistor operational amplifiers. It is possible to handle higher input currents by this circuit than by similar circuits using silicon diodes as feedback elements. This is due to the fact that when transistors are used in the connection shown, the impedance loading the operational-amplifier output is increased by approximately a factor of h_{FE} over that when diodes are used. Inclusion of the second logarithmic network provides temperature compensation as well as allowing a second input if desired. Figure 129b shows test results on the circuit at room temperature. The logarithmic conformity is to within 1.67 per cent of point over the entire range, and the temperature drift is only 0.003 decade/°C. The low-current limit of this

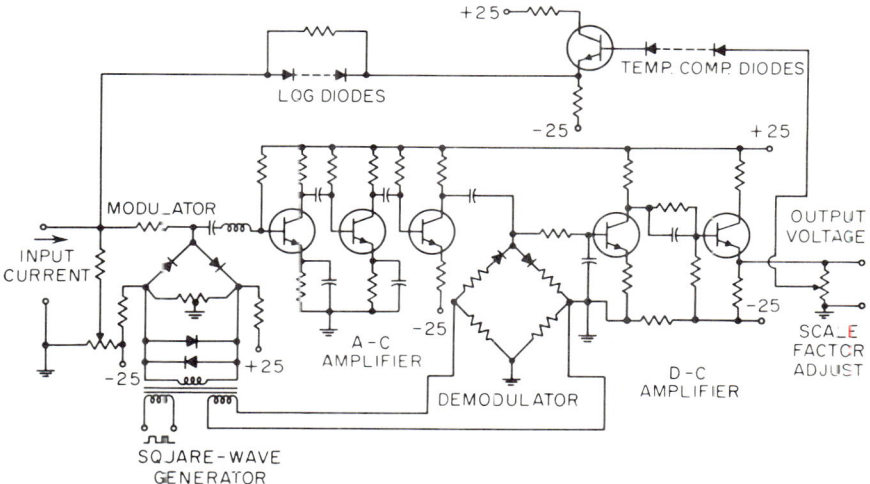

FIG. 128. Modulator-type logarithmic amplifier using solid-state devices throughout. (*From IEEE Trans. Nucl. Sci. vol. NS-10, no. 3, p. 87, July, 1963.*)

circuit is set by the capabilities of the transistor d-c amplifier and could be extended by use of a lower-drift amplifier.

Figure 130 shows three circuits in which silicon transistors are connected in the feedback loops of operational amplifiers in such a manner that the collector-base junctions are held at zero voltage. With this connection, Eq. (26) of Sec. 2.5 applies and the logarithmic characteristics of the devices are similar to those shown in Fig. 33. The basic circuit of Fig. 130a may be used only for positive input signals, and the transistor might be damaged if the input were suddenly reversed. The double-diode circuit of Fig. 130b corrects this deficiency and has a much lower leakage-current path than a protective circuit containing only a single diode. The complementary-transistor circuit of Fig. 130c is useful when the input is an a-c signal. With the use of low-offset and low-drift operational amplifiers input currents down to 10^{-11} amp or less can be amplified by these circuits.

The final d-c logarithmic amplifier to be described is a combined vacuum tube and semiconductor design and uses two biased-diode networks separated by a d-c amplifier to achieve logarithmic compression by piecewise linear approximation. Figure 131 shows a block diagram of the amplifier, which is designed to accommodate a-c as well as d-c inputs. A-c inputs are amplified, rectified, and filtered before application to the d-c part of the amplifier. Rectification of the a-c inputs presents a design problem, since a 50-db input amplitude range is handled. A compromise must be made between small-signal curvature at one limit of the conduction curve of the rectifier diodes and large-signal power-handling capacity at the other limit. A small

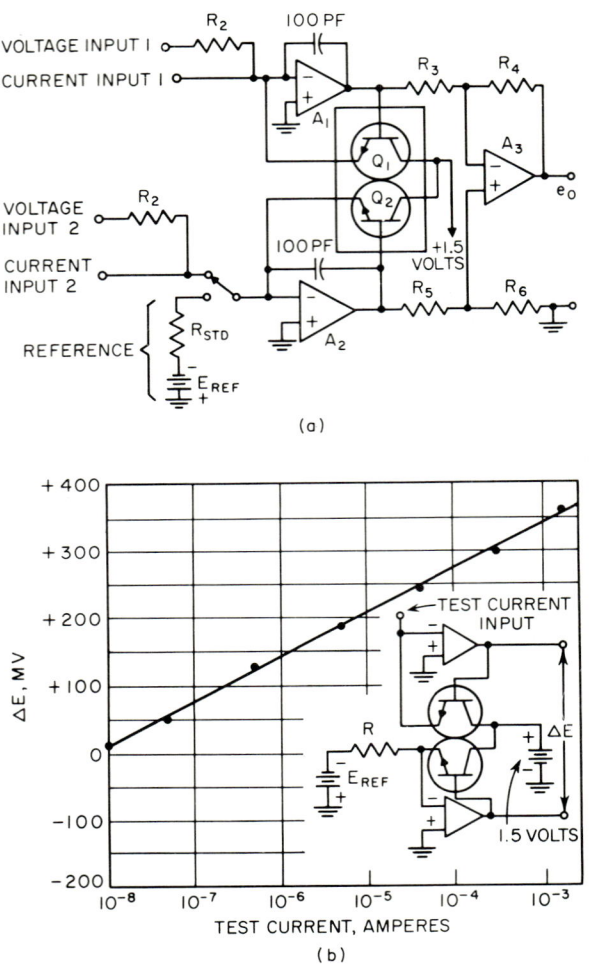

FIG. 129. Circuit and performance curve of a solid-state amplifier using the base-emitter junctions of silicon transistors as feedback logarithmic elements. (*From Electronic Design, Aug. 2, 1963, p. 64.*)

d-c forward bias is used to bring the silicon diodes in the rectifier circuit to the onset of conduction with no signal input. An active filter is used to obtain a fast response and to achieve linearity for high-level inputs. The d-c input range of the amplifier is from -10 to $+40$ db, with the reference of 0 db equal to 0.6 volt. The logarithmic shaper networks are similar in design to the logarithmic attenuator of Fig. 64 but operate at somewhat higher current levels. This amplifier was designed for use in the measurement of complex audio frequency and noise signals, where a readout in decibels is desirable. It is also useful for measuring decay rates of exponential phenomena. Linear operation of the amplifier is also possible by merely switching the logarithmic shaper networks out of the circuit.

7.2. A-C Logarithmic Amplifiers. The following paragraphs describe the circuits of two logarithmic amplifiers designed primarily for a-c input signals, although the second amplifier has provisions for also accepting d-c inputs.

Figure 132 is a schematic diagram of a commercial r-f amplifier which uses the logarithmic relationship between a-c plate current and the d-c control-grid voltage of variable-mu pentodes to obtain its transfer characteristics. This amplifier was designed for use in making response measurements of attenuation of networks, filters, amplifiers, etc., in the frequency range from 500 kc to 100 Mc. It has selectable logarithmic ranges of 0 to 40, 0 to 60, and 0 to 80 db, plus a variable linear range of 0 to 20 db. The overall dynamic range with 1 volt rms input is 90 db, and the log presentation accuracy is $\pm 1\frac{1}{2}$ db.

The circuit consists of nine stages of wideband r-f amplification, a diode detector doubler, and a two-stage d-c amplifier. The r-f amplifier design utilizes four tube pairs, each pair incorporating series shunt and series peaking, with cathode degeneration, both a-c and d-c, in each amplifier.

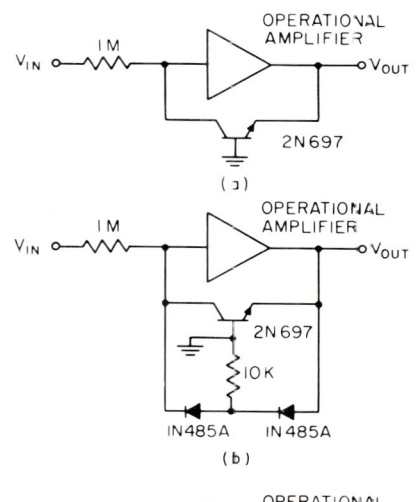

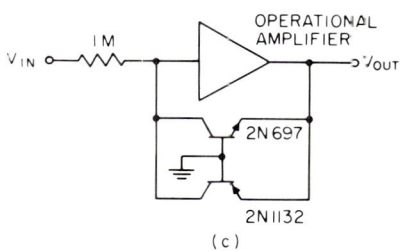

Fig. 130. Three circuits for using silicon transistors operated with zero collector-to-base voltage as feedback logarithmic elements. (*From Rev. Sci. Instr., vol. 34, p. 1313, 1963.*)

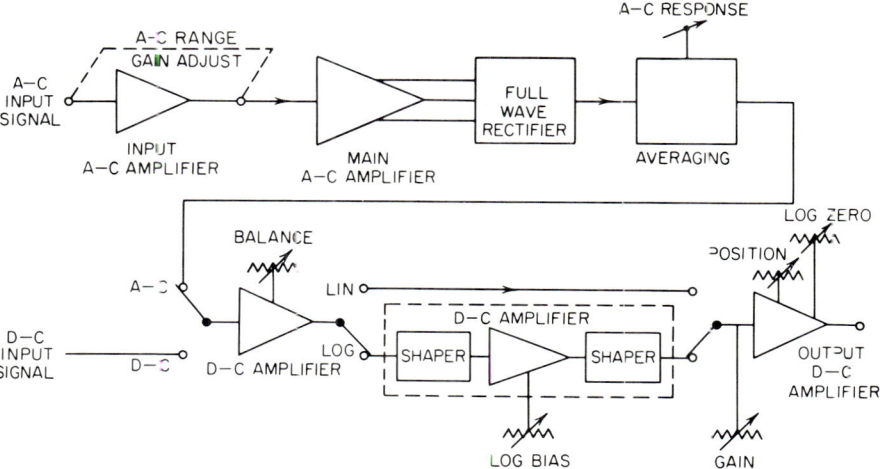

Fig. 131. Block diagram of a logarithmic amplifier using piecewise-linear-approximation techniques. (*Courtesy of Sanborn Company.*)

This type of degeneration provides gain stability with tube aging and filament temperature fluctuation and ensures r-f gain similarity between units. To provide linearity over the full input range, the first r-f amplifier is a remote-cutoff type 6EH7. The last stage in the string is operated at full gain with fixed grid bias in order to maintain a sufficient r-f voltage level for linear diode detection. The voltage developed across the cathode resistor of this stage, because of its low impedance characteristic, is utilized to supply bias for the curve-shaping diodes in the AGC amplifier section.

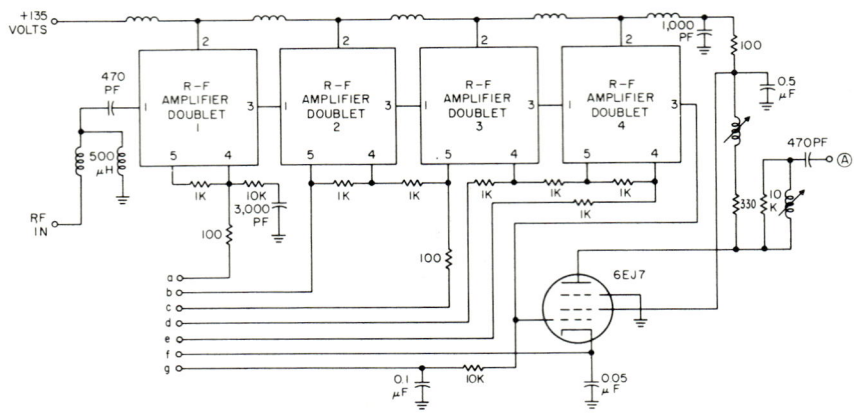

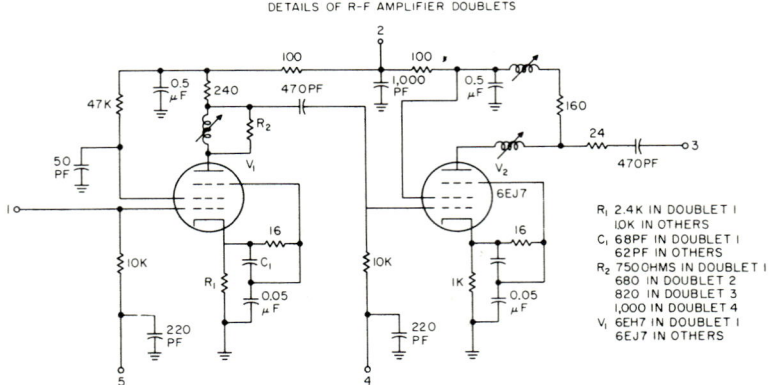

FIG. 132. Schematic diagram of an r-f amplifier using variable-mu pentodes to achieve logarithmic amplitude response. (*Courtesy of Jerrold Electronics Corp.*)

The output of the negative-going detector is fed to the first section of the d-c amplifier which provides AGC gain to the second d-c amplifier stage, along with a low output impedance and phase inversion for the required positive-going meter and output circuits. The second section of the d-c amplifier was designed as an operational amplifier in order to satisfy the following stringent requirements: stability, phase inversion for the required AGC signal polarity, a low and precise output impedance for driving the AGC line, a high-level input signal-handling capability, a convenient and accurate gain adjustment, and permissible gain settings above and below unity.

NONLINEAR AMPLIFIERS

A simplified schematic diagram of a logarithmic voltmeter converter using servo techniques that is manufactured commercially is shown in Fig. 133a. The input signal is attenuated and applied directly to a buffer amplifier if it is a-c, while d-c inputs are applied to a 60-cps chopper ahead of the buffer amplifier. The output of the buffer is applied to a variable-gain preamplifier which uses feedback to accommodate differences in the average value of square waves as compared with sine waves. The output of the preamplifier is applied to the log conversion unit, which consists of a

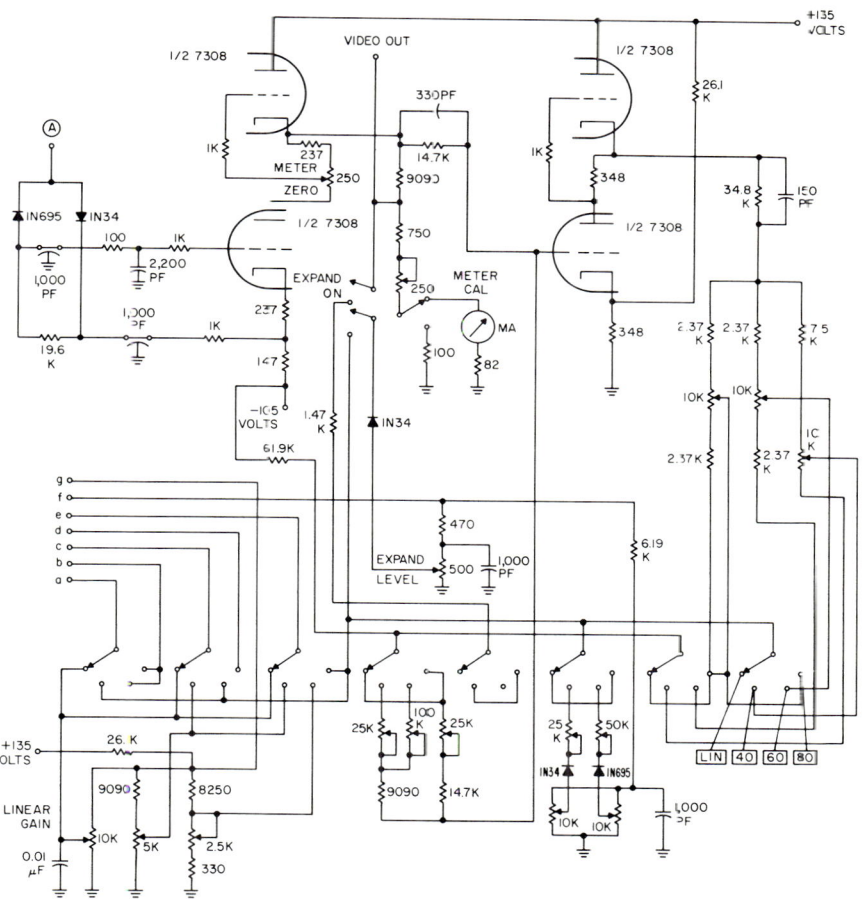

Fig. 132. (*Continued*.)

linear potentiometer having taps equally spaced to correspond to various decibel levels. The potentiometer wiper arm is driven by a servo amplifier-motor combination which senses the error signal derived from the preamplifier output after comparison with an 80-mv reference signal. An electrical output signal is obtained from a readout potentiometer driven by the balancing motor. The accuracy of the logarithmic approximation is greatly improved through the proper choice of resistance loading the wiper arm of the tapped potentiometer, as shown by the curve in Fig. 133b. An absolute accuracy of 0.2 db over a 70-db range is specified for this instrument.

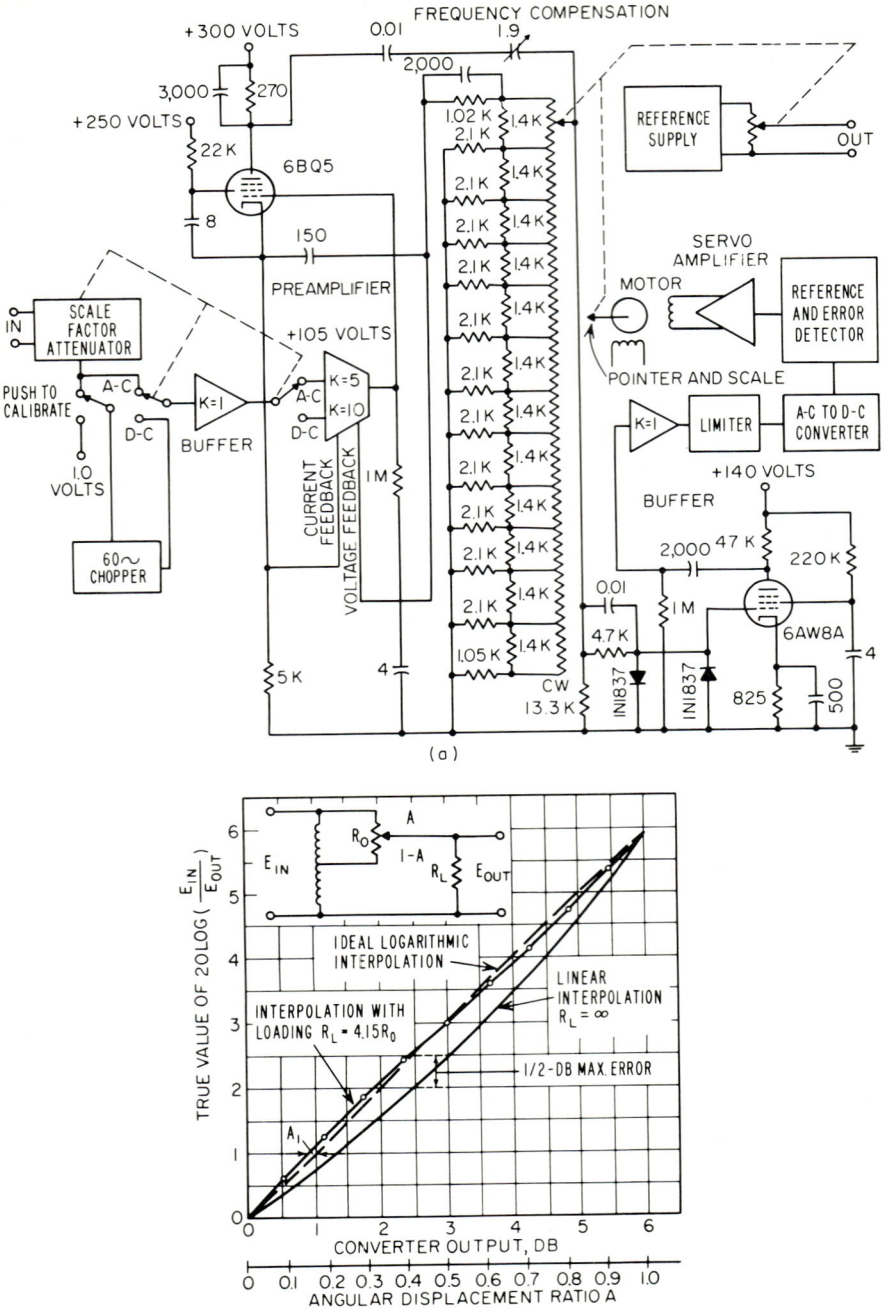

Fig. 133. Simplified circuit and performance curve of a servo-type logarithmic voltmeter-converter. (*Courtesy of Houston Instrument Corp.*)

8. CIRCUITS HAVING A LOGARITHMIC FREQUENCY OR PULSE-RATE RESPONSE

This section deals with circuits that have an output d-c voltage or current proportional to the logarithm of the input frequency or pulse rate. One method of designing logarithmic circuits of this type is to use a circuit such as a diode pump, which gives an output d-c signal linearly proportional to the input frequency or pulse rate, followed by a d-c logarithmic amplifier. However, all three circuits described in this section use piecewise-linear-approximation techniques to obtain the desired results.

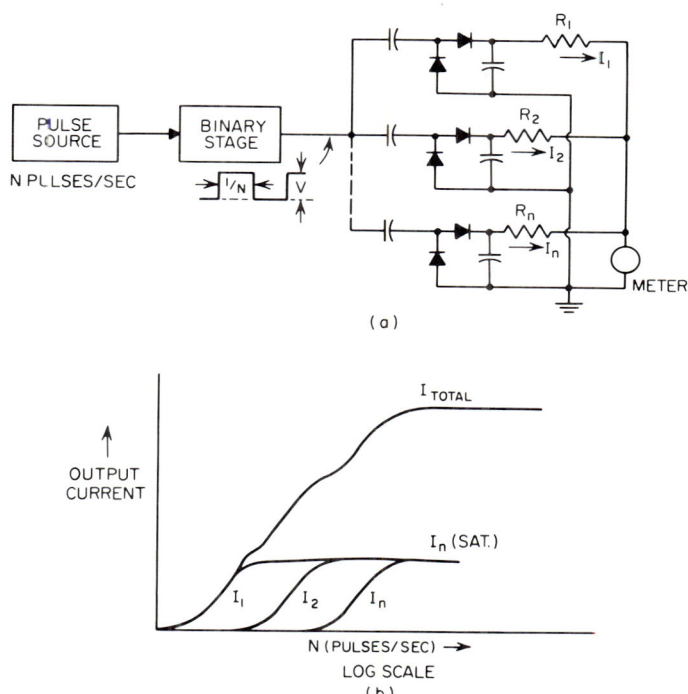

FIG. 134. Pulse integrator using multiple diode-pump circuits to achieve logarithmic conversion. (*From IRE Trans. Nucl. Sci., vol. NS-6, no. 2, p. 74, June, 1959.*)

Figure 134a shows the basic circuit of a pulse integrator frequently used in logarithmic count-rate amplifiers for nuclear measurements. The logarithmic response is obtained by using several diode-pump circuits connected in parallel, with the individual output currents summed into a common meter or readout circuit. The saturated current contributions from each diode-pump circuit are equal, but the time constants are spaced at uniform logarithmic increments, giving the individual and composite curves shown in Fig. 134b. A binary stage is used to shape the driving waveform in order to adjust the pulse length automatically to the value appropriate for the pulse rate of the input. The range of operation of this type of circuit may be extended easily by using additional diode-pump circuits. The degree of accuracy for a given range may be improved by making a smaller logarithmic increment between time constants, thus requiring more diode-pump circuits.

Figure 135 shows another pulse integrator that has been used in logarithmic count-rate amplifiers. The circuit of a shows negative input pulses fed to a vacuum-tube bistable multivibrator circuit, with outputs taken from the plates through networks $Y(s)$ to single diode-pump circuits having a common output. (A simpler version could be designed with only one output taken from the multivibrator.) Networks of the type shown in b are used for $Y(s)$ to control the manner in which the output d-c current I builds up. If the saturated current contributions from each RC combination are made equal, and if the time constants are spaced at uniform logarithmic increments, the buildup of output current will occur in the same manner as shown in

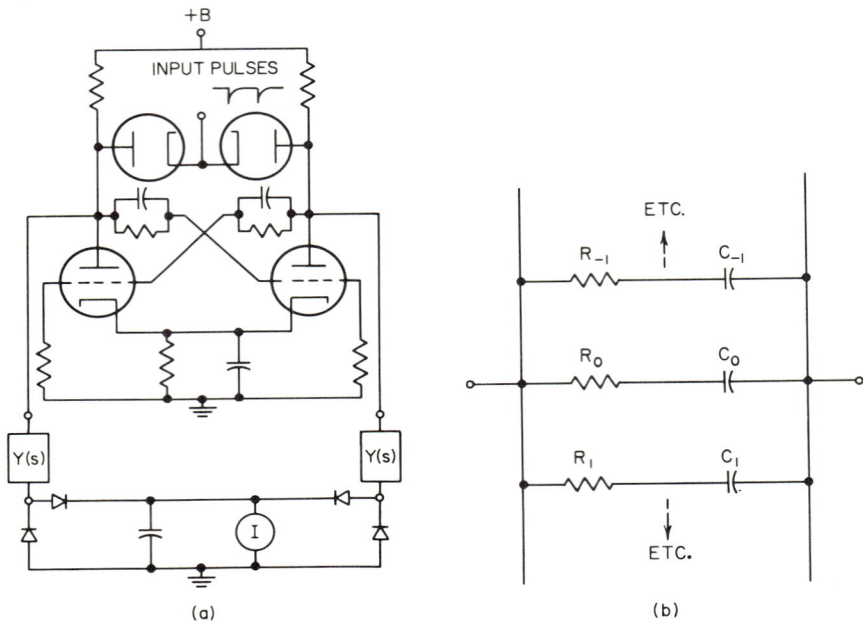

Fig. 135. Pulse integrator using multiple resistance-capacitance combinations with single diode-pump circuits to achieve logarithmic conversion. (*From J. Appl. Phys., vol. 28, p. 988, 1957.*)

Fig. 134b. The advantage of this circuit over that of Fig. 134a is that it requires fewer diodes and capacitors to achieve the same range and accuracy.

A simplified schematic of a logarithmic frequency converter designed to cover a frequency range of 20 cps to 20 kc for input signals having amplitudes anywhere between 0.5 and 150 volts is shown in Fig. 136a. The logarithmic conversion is achieved simply through combinations of series resistors and shunt capacitors. The circuits preceding the logarithmic conversion network use a series of Schmitt triggers and wave-shaping circuits to ensure that the amplitude of the input to the network is a predetermined value despite irregularities in waveshape. The output of the network is an a-c signal inversely proportional to the logarithm of the input frequency, as shown by the curve of Fig. 136b. This signal is amplified, rectified, and filtered, and then a positive bucking voltage is applied to make the output d-c voltage vary directly as the logarithm of the input frequency. This system has an accuracy of better than 2 per cent at all points within the frequency range covered.

NONLINEAR AMPLIFIERS

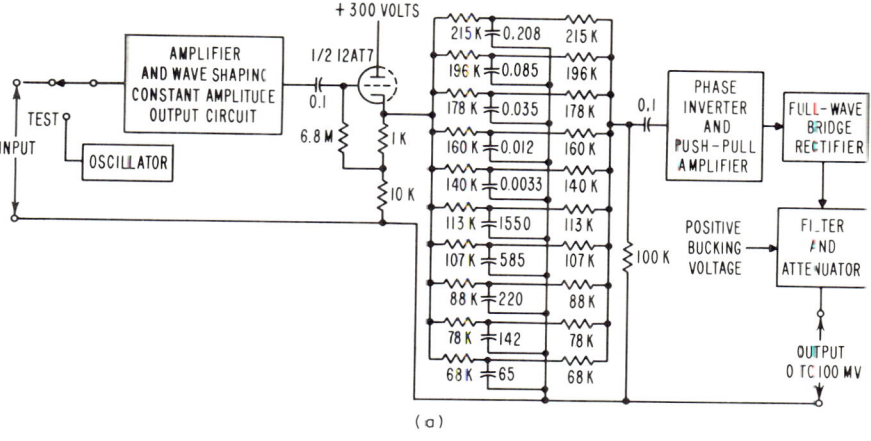

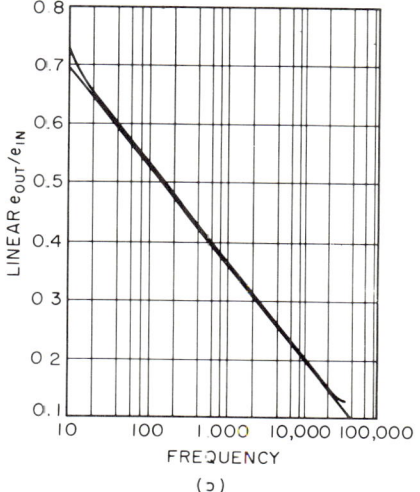

Fig. 136. Simplified circuit and performance curve of a logarithmic frequency converter using only resistance-capacitance combinations for conversion. (*Courtesy of Houston Instrument Corp.*)

9. AMPLITUDE SQUARING AND SQUARE-ROOTING CIRCUITS

Several basic circuits and networks for obtaining a square-law amplitude response have been described in Secs. 2 and 3. In this section a few examples will be given of complete square-root-extracting circuits, some of which use a squaring circuit in a feedback loop of an amplifier.

Figure 137a shows in block-diagram form one method of designing a feedback square-root-extracting circuit. The output of the squaring element is forced to be equal to the signal whose square root is sought; the input to the squaring circuit is then the desired output signal. An adding circuit is used to develop a signal at the

26–88 AMPLIFIER CIRCUITS

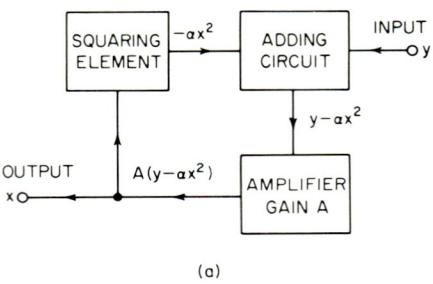

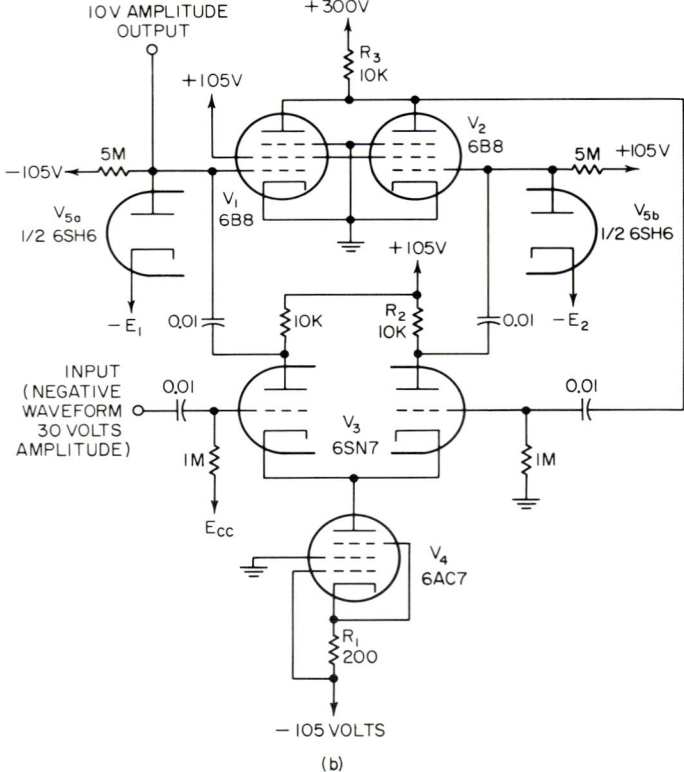

Fig. 137. Feedback square-root-extracting circuit using two thermionic pentodes as the quadratic elements. (*From B. Chance et al., "Waveforms," MIT Radiation Laboratory Series, vol. 19, pp. 689, 691, Figs. 19.28, 19.30, McGraw-Hill Book Company, New York, 1949.*)

amplifier input equal to the difference between the magnitude of the input y and the output ax^2 of the squaring element. If the amplifier gain A is large enough, the output is approximately equal to $\sqrt{y/a}$, as desired.

A practical circuit of this type is shown in Fig. 137b. Tubes V_1 and V_2 constitute the quadratic element, which is driven in push-pull from the plates of a differential

NONLINEAR AMPLIFIERS 26–89

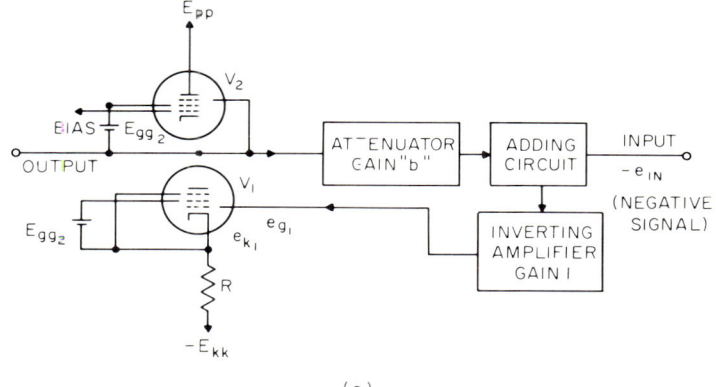

(c)

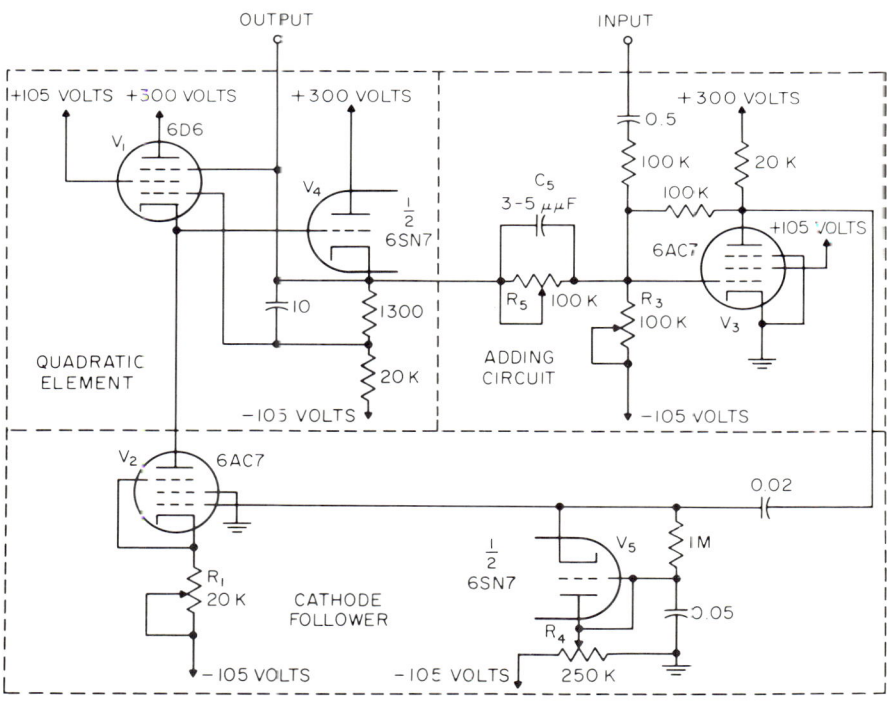

(b)

Fig. 138. Square-root circuit employing as the quadratic element a tube for which $\triangle I_p = \beta e_{in} + \gamma e_{in}^2$. (From P. Chance et al., "Waveforms," MIT Radiation Laboratory Series, vol. 19, pp. 688–689, Figs. 19.26, 19.27, McGraw-Hill Book Company, New York, 1949.)

amplifier (V_3 and V_4). This circuit was designed for negative pulse inputs of 100 μsec duration and 30 volts peak amplitude.

An extension of the simple square-root-extracting circuit of Fig. 11 is shown in Fig. 138a. The tube used for V_2 does not have to give a strictly parabolic transfer characteristic between ΔI_P and ΔV_{GK}; it merely has to satisfy the relation

$$\Delta I_P = \beta(-\Delta V_{GK}) + \gamma(-\Delta V_{GK})^2 \tag{36}$$

By use of an adding circuit, an inverting amplifier, and an attenuator of gain b, in addition to tubes V_1 and V_2, an output voltage can be obtained which is equal to $\sqrt{e_{in}/R\gamma}$, provided the factor b is made equal to $R\beta$.

Figure 138b shows a practical circuit embodying this principle. It is designed to accept waveforms of about 100 μsec duration and 40 volts peak amplitude. If the

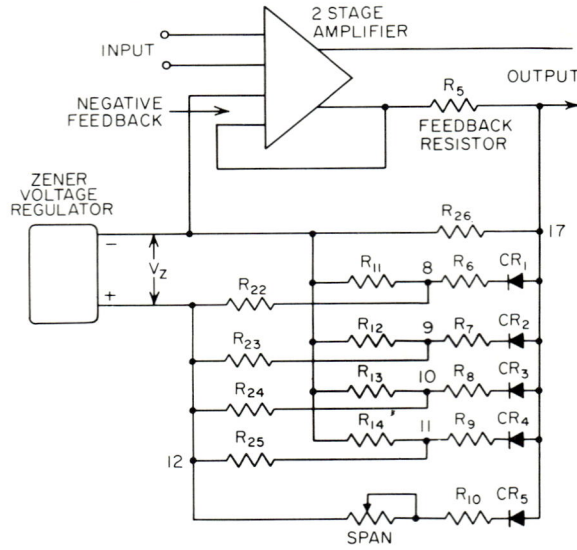

Fig. 139. Square-root converter using a diode squaring network in a feedback circuit around an operational amplifier. (*Courtesy of Foxboro Company.*)

input is a parabolic waveform ($e_{in} = k_1 t^2$) rising to a peak of 40 volts after 100 μsec, the output will be a sawtooth waveform ($e_2 = k_2 t$) of the same peak amplitude and duration.

The application of a diode squaring network similar to the one described in Sec. 3.3 (Fig. 73) to a commercial square-root converter is shown in Fig. 139. The diode network is in the feedback circuit of a two-stage magnetic amplifier. Voltage V_Z from the zener voltage regulator establishes different voltages at points 8, 9, 10, 11, and 12, depending on the values of the resistor dividers; and these voltages in relation to the voltage at point 17 determine when diodes CR_1, CR_2, CR_3, CR_4, and CR_5 will conduct. The net resistance of the feedback circuit decreases with increasing input, giving the desired output-input relationship.

In the practical application of square-root converters of this type, it is desirable to have a low-signal cutoff circuit to eliminate instabilities that would occur for very low inputs. A cutoff circuit to go with the converter of Fig. 139 is shown in the right-hand portion of Fig. 140a, and the effect it gives on the output-input characteristic is shown by the curve of Fig. 140b. The circuit uses a silicon controlled rectifier driving a reed-type relay to achieve the cutoff action.

The final example is a generalized power/rooter circuit using feedback logarithmic elements. The block diagram of Fig. 141 illustrates how an output proportional to a desired power or root of the input may be obtained. The square blocks of the

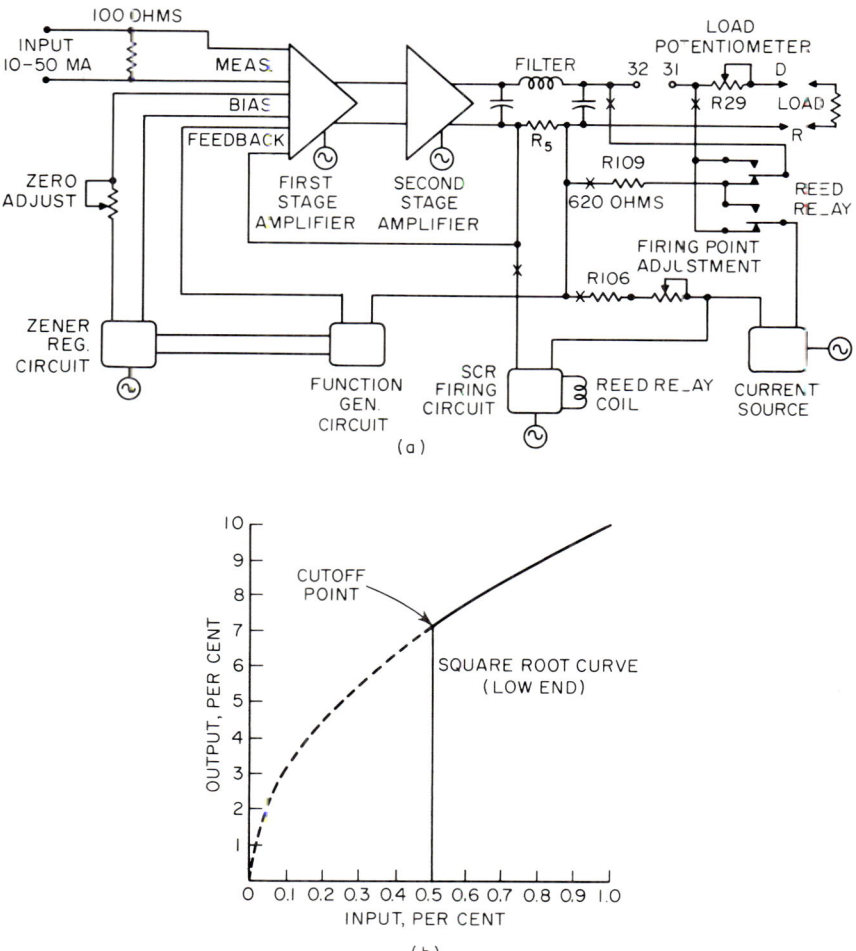

FIG. 140. Low-signal cutoff circuit used with square-root converter of Fig. 139 and curve showing its effect on the output-input characteristic of the converter. (*Courtesy of Foxboro Company.*)

diagram represent the dual-transistor elements shown in Fig. 129, and the triangular blocks are differential-input operational amplifiers. An output proportional to the square of the input could be obtained by making $m = 4$ and $n = \frac{1}{2}$ in the circuit of Fig. 141, while a square-root conversion could be obtained by making $m = 2$ and $n = \frac{1}{4}$.

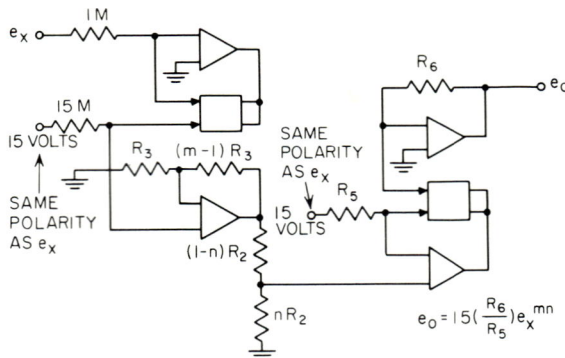

Fig. 141. Block diagram of a power/rooter circuit using feedback logarithmic elements of the type shown in Fig. 129. (*Courtesy of Nexus Research Laboratory, Inc.*)

10. CIRCUITS FOR ANALOG MULTIPLICATION AND DIVISION

There are a great many techniques for obtaining products and quotients of two input variables. In this section examples are given of the application of a few of these techniques to complete circuits. Other techniques have been briefly described in Secs. 2 and 4, while many techniques, particularly those involving time-modulation or time-division principles, are not covered at all in this chapter.

10.1. Multipliers and Dividers Using Logarithmic Elements. Figure 142 shows a block diagram of an implicit-solution multiplier/divider using feedback loga-

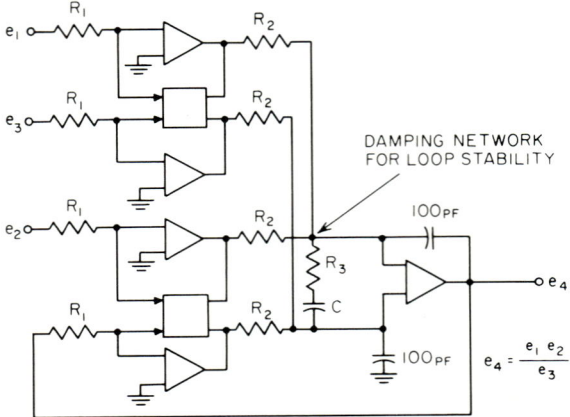

Fig. 142. Block diagram of a multiplier/divider using feedback logarithmic elements of the type shown in Fig. 129. (*Courtesy of Nexus Research Laboratory, Inc.*)

rithmic elements. Two silicon transistors connected as shown in the circuit of Fig. 129 are in each square block, and the triangular blocks represent transistor d-c amplifiers. The outputs from the amplifiers associated with inputs e_1 and e_2 are added together to form one input to the output amplifier. The output of this amplifier is connected back to the input of the lower left amplifier, where it is logarithmically amplified and added to the output of the amplifier associated with input e_3 to form the other input to the output amplifier. This feedback connection provides the desired relationship between e_4 and the various inputs.

A generalized multiplier using transistor logarithmic elements in conjunction with d-c operational amplifiers is shown in Fig. 143. Note that the transistors in this circuit are all operated with the bases grounded and at zero collector-to-base voltage,

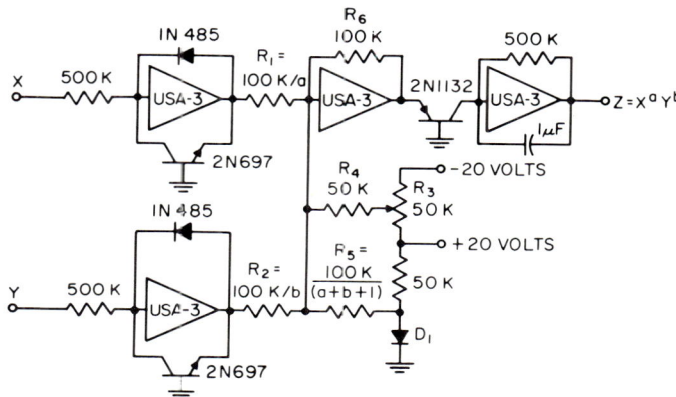

FIG. 143. Generalized multiplier circuit using transistors with zero collector-to-base voltage as logarithmic elements. (*From Rev. Sci. Instr., vol. 34, p. 1315, 1963.*)

and thus the relationship of Eq. (26) applies to their logarithmic behavior. The constants a and b in the output expression for this multiplier are determined by the values of R_1 and R_2. Variation of the current through R_4 by adjustment of R_3 determines the constant factor (i.e., the units of multiplication). Resistor R_5 and diode D_1 provide temperature compensation. Table 11 indicates the order of accuracy achievable with this type of multiplier. The results in a are for $Z = XY/10$ and those in b are for $Z = X^2Y^{0.5}/100$.

Table 11. Test Results on Multiplier of Fig. 143
a. Output of Multiplier, $Z = XY/10$

Y	X = 100	30	10	3	1	0.3	0.1	0.03	0.01	0.003
100	.....	.....	99.27	30.01	10.03	3.015	1.003	0.298	0.096	0.026
30	.....	89.33	29.88	8.984	2.998	0.900	0.299	0.089	0.029	0.008
10	100 4	30.03	10.00	3.000	1.000	0.300	0.100	0.030	0.010	0.003
3	30 39	9.040	3.005	0.901	0.300	0.090	0.030	0.009	0.003	0.001

b. Output of Generalized Multiplier, $Z = X^2Y^{0.5}/100$

Y	X = 70.71	50	20	10	5	2	1	0.5	0.2	0.1
100	.....	.....	40.30	10.10	2.528	0.404	0.101	0.025	0.004	
64	.....	.....	32.14	8.052	2.013	0.322	0.081	0.020	0.003	
36	.....	.....	24.04	6.021	1.506	0.241	0.060	0.015	0.003	
25	.....	124.1	20.02	5.010	1.252	0.200	0.050	0.012	0.002	0.001
16	.....	99.46	16.01	4.006	1.001	0.160	0.040	0.010	0.002	0.001
9	.....	74.78								
4	99.60	49.99								

AMPLIFIER CIRCUITS

10.2. Quarter-square Multipliers and Dividers. A common method of designing analog multiplication and division circuits is to simulate the following algebraic equation:

$$Z = XY = \frac{(X+Y)^2}{4} - \frac{(X-Y)^2}{4} \tag{37}$$

Through the use of squaring networks of the type shown in Fig. 75, plus standard operational amplifiers, this relationship can be directly implemented, as shown by the solid lines of Fig. 144. An output voltage is obtained that is proportional to the product of two input voltages, regardless of the polarity of the input signals. In order to obtain the quotient of two inputs with these components, the multiplier is effectively placed in the feedback loop of the output operational amplifier. The dashed lines of Fig. 144 show the circuit connections to obtain an output proportional to X/Y. For stable operation, the voltage fed back to the summing junction of the

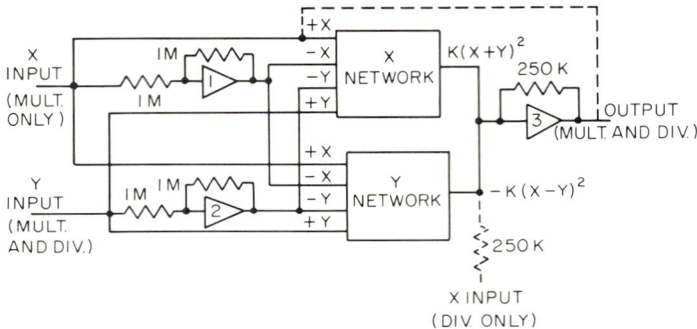

FIG. 144. Block diagram of a quarter-square multiplier-divider using diode squaring networks. For division add connections shown in dotted lines; omit 250-kilohm resistor across amplifier 3. (*Courtesy of Systron-Donner Corp.*)

output operational amplifier through the diode squaring networks must be of the same polarity as the amplifier output voltage. This condition requires that input Y be positive when this circuit is used for division. The accuracy of this circuit as a multiplier is 0.25 per cent of full scale for 0- to 100-volt inputs and 0.5 per cent for 0- to 10-volt inputs. As a divider the accuracy is 0.25 per cent of full scale for $X = Y = 100$ volts, becoming poorer for lower input levels.

A quarter-square multiplier circuit using the square-law characteristics of the QK-329 beam-deflection tube, described in Sec. 2.3, is shown in Fig. 145a. The two inputs are applied in both normal and inverted polarity to a resistive sum and difference network, with two outputs taken to each of the two QK-329 tubes. The final result is obtained by subtracting the outputs of the square-law stages in an operational-amplifier circuit. Figure 145b shows plots of the output-input characteristics. This circuit gave an instantaneous output accuracy within ± 0.5 per cent of maximum product over the input operating range of 25 volts. The frequency response was flat from direct current to 90 kc, with a gradual rolloff at higher frequencies. Long-term drift was within 1 per cent after an initial settling period of about 3 hr, using conventional regulated power supplies fed by a 2 per cent a-c line regulator.

10.3. Other Multipliers and Dividers. A solid-state multiplier-divider which is basically equivalent in principle to one that is servo-driven is available commercially. The diagrams of Fig. 146 illustrate this equivalence. A magnetic control circuit in the solid-state version (Fig. 146b) is seen to replace the servo drive of Fig. 146a. The bridges contain semiconductor elements exhibiting physical magnetoresistance (see Sec. 2.9), which change resistance as the magnetic field is varied. Thus the bridges may be balanced as the inputs vary by changing the control current in the magnetic

NONLINEAR AMPLIFIERS

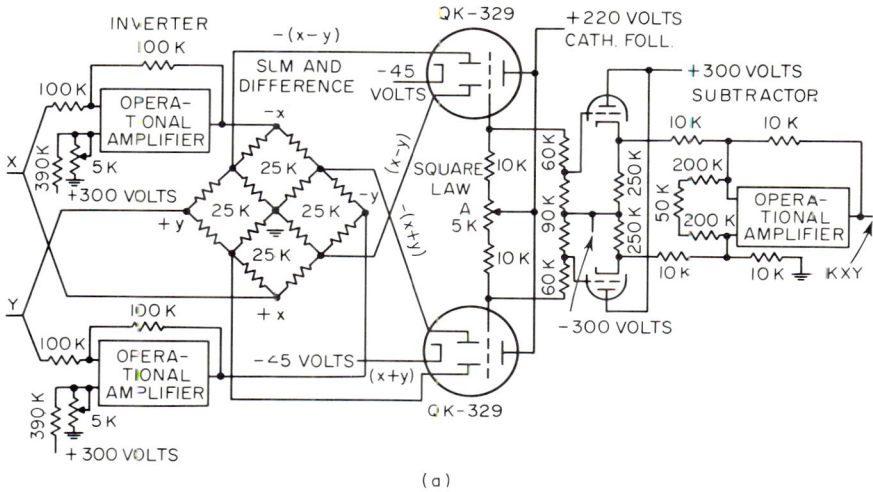

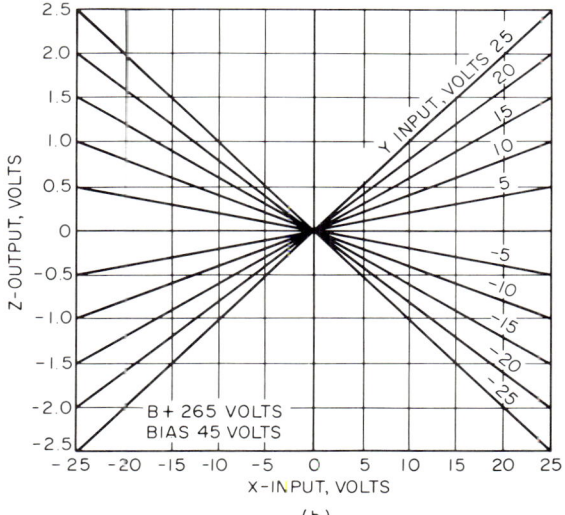

Fig. 145. Circuit diagram and output-input characteristics of a quarter-square multiplier using QK-329 beam-deflection tubes. (*From Electronics, vol. 28, no. 2, p. 163.*)

circuit. Figure 146c shows a block diagram of the entire multiplier-divider, which contains a high-gain driving amplifier to sense the error signal and provide the balancing action necessary to give the desired output-input relationship. Inputs A and C must be a-c signals of less than 100 cps, while input B may be of any frequency from direct current up to several kilocycles per second. The output D will have the same frequency as input B. The accuracy of this multiplier-divider depends almost entirely on how well the two bridges are matched in their characteristics. A maximum error of 0.2 per cent of full scale over a temperature range of 15 to 60°C is claimed by the manufacturer.

26-96 AMPLIFIER CIRCUITS

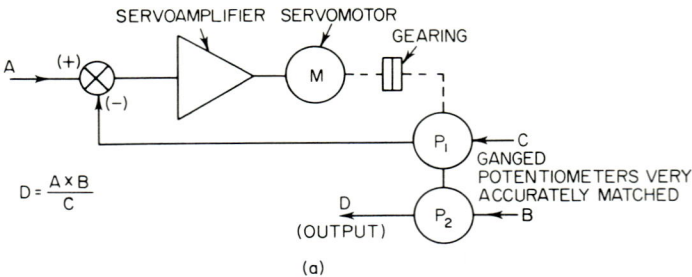

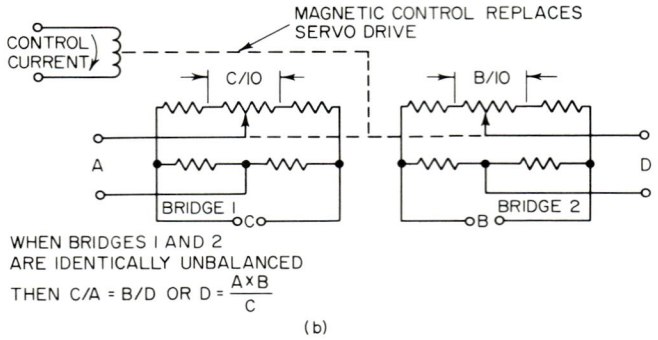

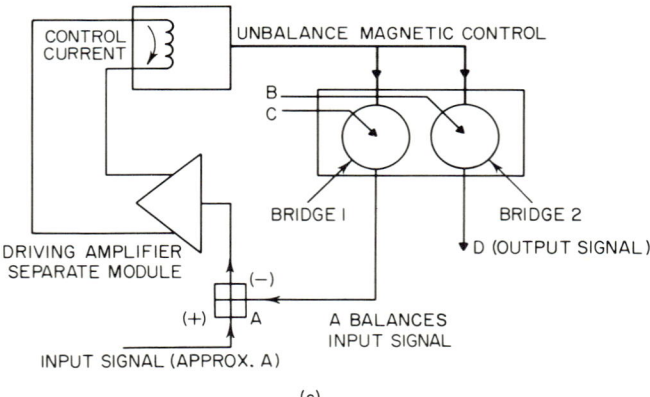

Fig. 146. Diagrams showing the operation of a solid-state multiplier-divider using magneto-resistive elements. (*Courtesy of Elasco, Inc.*)

A complete diode divider circuit which is an extension of the basic diode divider circuit of Fig. 27 is shown in Fig. 147. The output of the basic circuit is fed through an a-c amplifier and rectified by a diode-ring demodulator. This circuit accepts a-c inputs from 0 to 2 volts and d-c inputs from 0.5 to 5 volts. An accuracy of ± 2 per cent of the full-scale output is possible over the input ranges stated above. Figure 148 shows the extension of the basic divider circuit to a circuit which provides an

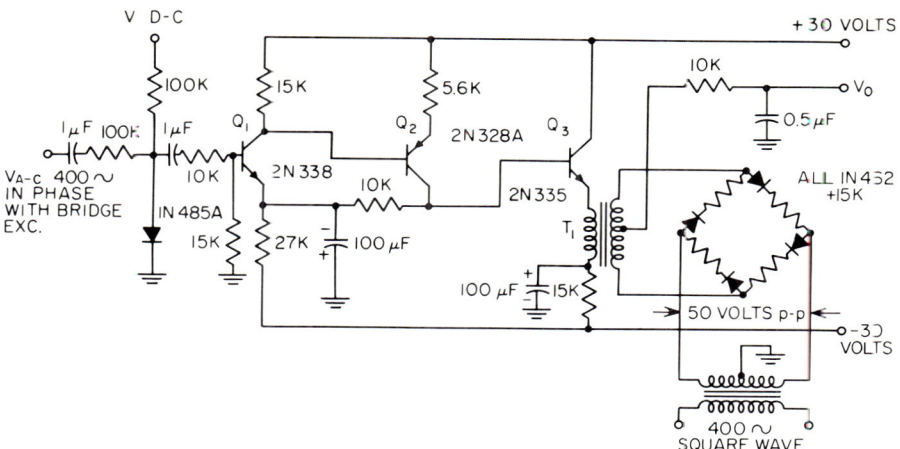

Fig. 147. Complete analog divider circuit using a silicon diode. (*Courtesy of General Electric Co.*)

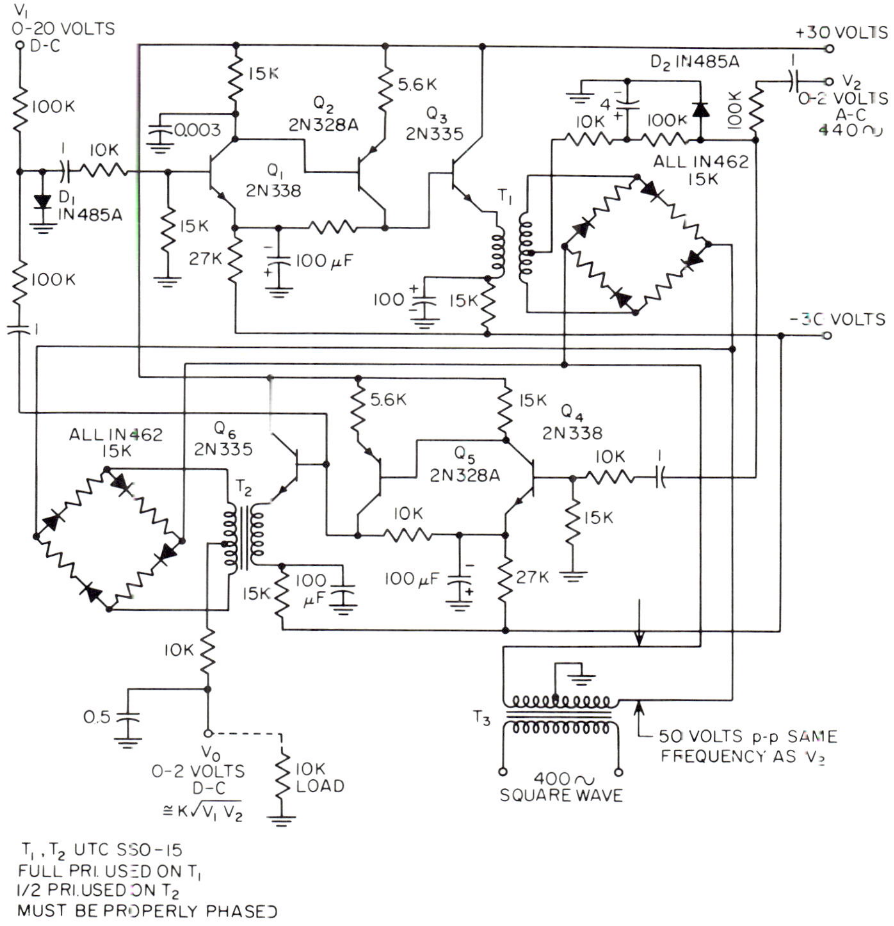

Fig. 148. Circuit using diode dividers that provides an output proportional to the square root of the product of two inputs. (*Courtesy of General Electric Co.*)

output voltage proportional to the square root of the product of two input variables. This operation is possible with only two divider circuits and two amplifier-demodulator combinations. An accuracy of ±2 per cent of full scale is possible over a range of 1 to 10 volts for V_1 and a range of 0 to 2 volts for V_2.

Figure 149a shows a simplified block diagram of a multiplier-divider which uses a complicated diode bridge in conjunction with a sawtooth wave generator and an output operational amplifier to obtain the relation $e_o = k e_1 e_2 / e_3$. The bridge arrange-

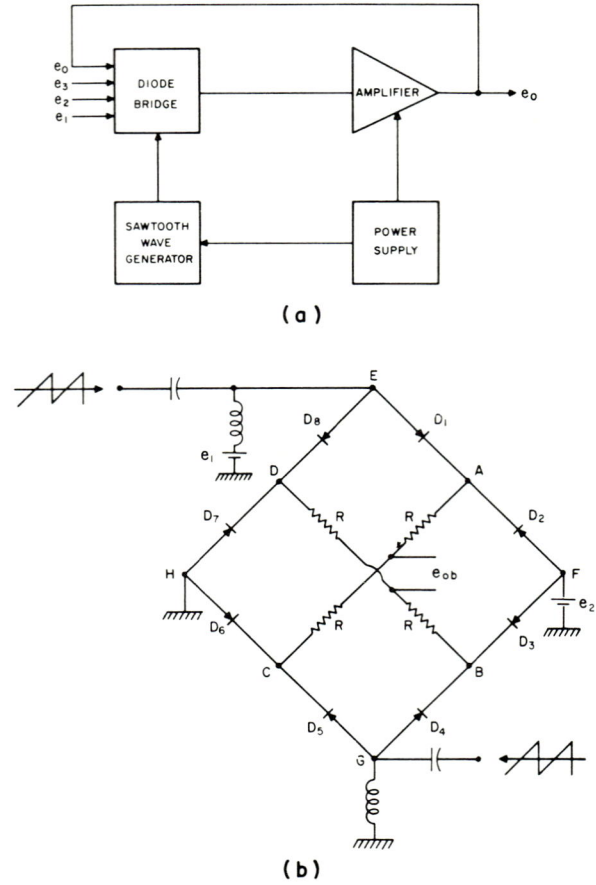

Fig. 149. Block diagram of a multiplier-divider using a diode bridge driven by a sawtooth generator and simplified diagram of the bridge circuit. (*Courtesy of Foxboro Company.*)

ment for obtaining multiplication is shown in simplified form in Fig. 149b. The output signal of this bridge e_{ob} is a trapezoidal-shaped waveform having an integrated value proportional to the product of e_1 and e_2. Thus when the bridge output is amplified and filtered the desired d-c output-input relationship exists. To obtain division as well as multiplication, an additional bridge of this type is needed, with the outputs of the two bridges combined at the input to the amplifier. This multiplier-divider was designed for process-instrumentation applications and accepts inputs of 10 to 50 ma direct current into 100 ohms, with an output of 10 to 50 ma into 600 ohms.

A four-quadrant pulse-amplitude multiplier using field-effect transistors is shown in

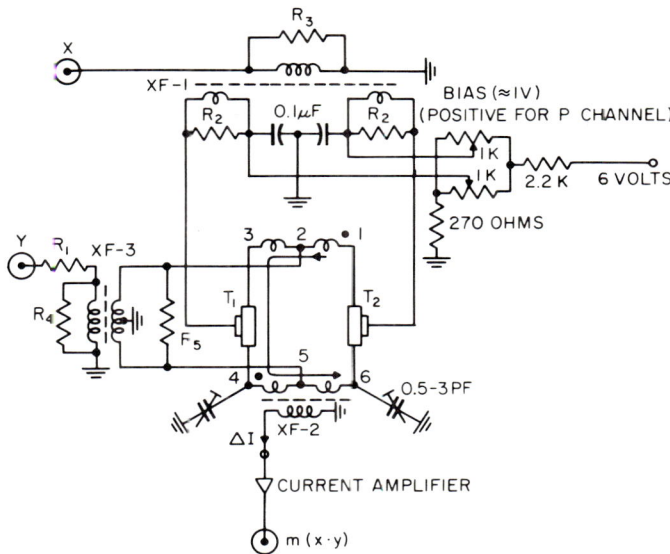

Fig. 150. Schematic diagram of a four-quadrant pulse-amplitude multiplier using field-effect transistors. (*From IEEE Trans. Nucl. Sci., vol. NS-11, no. 1, p. 365, January, 1964.*)

Table 12. Performance Data and Component Values for Two Multipliers of Type Shown in Fig. 150

Multiplier	A	B
Solution time	0.1 μsec	20 nsec
Pulse-length range	0.1–2 μsec	20–100 nsec
x input max	1.5 volts	1.5 volts
y input max	i_y = 0.6 mA, or 5 volts through R_1	1.5 volts
Output ΔI max to current amplifier	≈ 125 μa	250 μa
Amplifier output max	250 mv	125 mv
Residual signal for $x = x_{max}$ $y = 0$	<0.3%	<3%
Transformers XF-1, 2, 3		
Core	Pot S 35/23-3E Ferroxcube	Toroid 208F250-3C Ferroxcube
Electrostatic shield between primary and secondary	Yes	No
XF-1, XF-3		
Windings (number of turns)		
Primary	30 } No. 30	4 } No. 28
Secondary	2 × 15	2 × 2
XF-2		
Primary	4 × 2.5 } No. 28	4 × 1 } No. 28
Secondary	10	2
Resistors		
R_1	8.2 kilohms	30 ohms
R_2	1 kilohm	51 ohms
R_3	1 kilohm	100 ohms
R_4	∞	20 ohms
R_5	∞	1 kilohm
Field-effect transistors used	2N2608, Siliconix	

Fig. 150. The multiplication effect is based on the proportionality of the channel conductance of a junction gate field-effect transistor to the gate voltage, as discussed in Sec. 2.5. Pertinent data on the components and performance of two multiplier circuits, both of the type shown in Fig. 150 but having different ranges of pulse length to solution time ratio, are given in Table 12. It is seen that the use of field-effect transistors makes possible analog multiplication of reasonable accuracy at very high speeds.

BIBLIOGRAPHY

Ballantine, S.: Variable-mu Tetrodes in Logarithmic Recording, *Electronics*, vol. 2, pp. 472–474, 1931.
Bowers, J. C.: A Wide-range AGC Circuit, *Electron. Equipment Eng.*, November, 1963, pp. 58–59.
Chance, B., et al.: "Waveforms," vol. 19, pp. 668–692, MIT Radiation Laboratory Series, McGraw-Hill Book Company, New York, 1949.
Chao, S. K.: Logarithmic Characteristics of Triode Electrometer Circuits, *Rev. Sci. Instr.*, vol. 30, pp. 1087–1092, 1959.
Cooke-Yarborough, E. H., and E. W. Pulsford: An Accurate Logarithmic Counting-rate Meter Covering a Wide Range, *Proc. Inst. Elec. Engrs. (London)*, vol. 98, pt. 2, pp. 196–203, 1951.
DeBolt, H. E.: A Simplified Logarithmic Integrator Circuit, *IRE Trans. Nucl. Sci.*, vol. NS-6, no. 2, pp. 74–77, June, 1959.
Epstein, M., and L. J. Greenstein: True RMS Measurements Utilizing the Galvanomagnetic Effects in Semiconductors, *IRE Intern. Conv. Record*, pt. 9, pp. 186–191, 1962.
Geigel, A. A., H. A. Hageman, and F. J. Witt: "Ultra Wideband Variolossers," IEEE conference paper 63-429, January, 1963.
Glinski, G. S., and J. P. Landolt: Theory and Practice of Hall Effect Multipliers, *IRE Conv. Record*, vol. 9, pt. 2, pp. 143–163, 1961.
Gray, T. S., and H. B. Frey: Acorn Diode has Logarithmic Range of 10^9, *Rev. Sci. Instr.*, vol. 22, no. 2, p. 117, 1951.
Hartin, W. J.: Servologarithmic Amplifier for Reactor Instrumentation, *IRE Trans. Nucl. Sci.*, vol. NS-6, no. 1, pp. 11–14, March, 1959.
Heid, K., and D. Silverman: Stabilization of the Hall Effect Multiplier, *Solid State Design*, vol. 4, no. 10, pp. 17–22, October, 1963.
Henney, K. (ed.): "Radio Engineering Handbook," 5th ed., chaps. 13, 15, 19, McGraw-Hill Book Company, New York, 1959.
Hoge, R. R.: Radiation-tolerant Logarithmic Amplifier, *IRE Trans. Nucl. Sci.*, vol. NS-9, no. 1, pp. 167–173, January, 1962.
Huskey, H. D., and G. A. Korn (eds.): "Computer Handbook," Secs. 3, 7, McGraw-Hill Book Company, New York, 1962.
Izumi, I., and M. Okano: An Improved Solid-state Logarithmic Amplifier, *IEEE Trans. Nucl. Sci.*, vol. NS-10, no. 3, pp. 82–90, July, 1963.
Kuebler, W. P.: N2 Carrier Solid State Circuitry, *Bell Lab. Record*, vol. 42, pp. 127–131, April, 1964.
Lichtenstein, R. M.: Random Interchange of Circuits with Applications to Counting Rate Meters and Function Generators, *J. Appl. Phys.*, vol. 28, no. 9, pp. 984–989, 1957.
Lubkin, Y. J.: Gain Controlled Log Amplifier, *Electron. Equipment Eng.*, September, 1962, p. 91.
Miller, J. A., A. S. Soltes, and R. E. Scott: Wide-band Analog Function Multiplier, *Electronics*, vol. 28, pp. 160–163, February, 1955.
Mollinga, T.: Amplifier with DC Controlled Gain, *Electronic Equipment Eng.*, pp. 92–93, May, 1963.
Paterson, W. L.: Multiplication and Logarithmic Conversion by Operational Amplifier-Transistor Circuits, *Rev. Sci. Instr.*, vol. 34, no. 12, pp. 1311–1316, 1963.
Radeka, V.: Fast Analogue Multipliers with Field-effect Transistors, *IEEE Trans. Nucl. Sci.*, vol. NS-11, no. 1, pp. 302–307, January, 1964.
Sikorsky, E.: An Evaluation of Vacuum-tubes for Log N Amplifiers, *IEEE Trans. Nucl. Sci.*, vol. NS-10, no. 2, pp. 21–33, April, 1963.
Texas Instruments, Inc.: "Transistor Circuit Design," pp. 176–179, 329–344, McGraw-Hill Book Company, New York, 1963.
Ullman, E. M.: How to Use Thermistors, *Electron. Equipment Eng.*, June, 1956, pp. 38–41.
Vice, D.: Designing Variable-gain Amplifiers, *Electron. Equipment Eng.*, January, 1960, pp. 50–54.
Wade, E. J., and D. S. Davidson: Transistorized Log-period Amplifier, *IRE Trans. Nucl. Sci.*, vol. NS-6, no. 2, pp. 53–56, June, 1959.

Chapter 27

MICROWAVE AMPLIFIERS

CARL J. EICHENAUER, JR.*

CONTENTS

1. Introduction... 27-2
   1.1. Microwave Amplifiers..................................... 27-3
   1.2. Microwave Oscillators.................................... 27-4
   1.3. Microwave Frequency Multipliers......................... 27-5
   1.4. Microwave Frequency Translators......................... 27-5
   1.5. The Microwave Amplifier Complex......................... 27-7
2. Functions of Microwave Amplifier Complexes.................. 27-10
   2.1. CW Amplifier Complexes.................................. 27-10
   2.2. Communication Amplifier Complexes....................... 27-10
   2.3. Pulse Amplifier Complexes............................... 27-17
3. Characteristics of Microwave Amplifier Complexes............ 27-19
   3.1. R-F Output Amplitude vs. Frequency Characteristic....... 27-19
   3.2. R-F Output Phase Delay vs. Frequency Characteristic..... 27-21
   3.3. R-F Output Amplitude vs. Voltage(s) Characteristic...... 27-21
   3.4. R-F Output Phase vs. Voltage(s) Characteristic.......... 27-22
   3.5. R-F Input/Output Linearity Characteristics.............. 27-23
   3.6. R-F Input/Output Phase Characteristic................... 27-24
   3.7. R-F Output Amplitude vs. Load Characteristic............ 27-24
   3.8. R-F Output Phase vs. Load Characteristic................ 27-24
   3.9. R-F Harmonic Output Characteristic...................... 27-25
   3.10. Noise Output Characteristic............................ 27-25
   3.11. R-F Peak-power-output Characteristic................... 27-25
   3.12. R-F Average-power-output Characteristic................ 27-26
   3.13. Pulse-duration and Duty-cycle Characteristics.......... 27-26
   3.14. Change of Characteristics with Time.................... 27-26
   3.15. Reproducibility Characteristic......................... 27-26
   3.16. Summary.. 27-27
4. Typical Microwave Amplifier Complexes....................... 27-27
   4.1. A Typical Single-output-phase Power Amplifier........... 27-27
   4.2. A Typical Phased-array Complex.......................... 27-31
   4.3. A Typical Multipurpose Power Amplifier Complex.......... 27-34
5. Power-amplifier Tubes and Their Assemblies.................. 27-36
   5.1. The Negative-grid Power Amplifier....................... 27-37

* General Electric Co., Heavy Military Electronics Department, Syracuse, N.Y.

AMPLIFIER CIRCUITS

  5.2. The Linear-beam Power Amplifier........................ 27–40
  5.3. The Crossed-field Amplifier............................. 27–43
 6. Oscillators... 27–45
  6.1. Magnetron Oscillators.................................. 27–45
  6.2. Negative-grid Oscillators.............................. 27–47
  6.3. Velocity-modulated Oscillators......................... 27–49
  6.4. High-stability Oscillators............................. 27–50
 7. Frequency Translators....................................... 27–50
  7.1. Solid-state Frequency Translators...................... 27–51
  7.2. Tube-type Frequency Translators........................ 27–52
 8. Frequency Multipliers....................................... 27–54
  8.1. Tube-type Frequency Multipliers........................ 27–54
  8.2. Solid-state Frequency Multipliers...................... 27–55
 9. Transmission-line Elements.................................. 27–55
  9.1. Coaxial Transmission Lines............................. 27–55
  9.2. Waveguide Transmission Lines........................... 27–57
  9.3. Directional Couplers................................... 27–60
  9.4. Attenuators and Isolators.............................. 27–62
  9.5. Dummy Loads.. 27–62
  9.6. Power-combining and -splitting Devices................. 27–63
  9.7. Duplexers.. 27–64
  9.8. Filters.. 27–65
  9.9. Rotary Joints.. 27–65
  9.10. Horns... 27–66
 10. Power Supplies... 27–67
  10.1. Unregulated D-C Power Supplies........................ 27–70
  10.2. Regulated D-C Power Supplies.......................... 27–72
  10.3. High-voltage Power Supplies........................... 27–73
 11. Modulators and Pulsers..................................... 27–74
  11.1. Communications Modulators............................. 27–74
  11.2. Line-type Pulsers..................................... 27–77
  11.3. Hard-tube Pulsers..................................... 27–83
 12. Protective Equipment....................................... 27–87
  12.1. Power-amplifier-tube Protection....................... 27–87
  12.2. Transmission-line Protection.......................... 27–88
  12.3. Pulser and Modulator Protection....................... 27–89
  12.4. Personnel Protection.................................. 27–90
 13. Monitoring Equipment....................................... 27–90
  13.1. R-F Monitoring.. 27–90
  13.2. D-C and Low-frequency A-C Monitoring.................. 27–91
  13.3. Pulse Monitoring...................................... 27–91
 14. Control Equipment.. 27–92
  14.1. Manual Control.. 27–92
  14.2. Automatic Control..................................... 27–93

1. INTRODUCTION

 The effective amplification of microwave power involves the application of a wide range of diverse components and techniques. Since the term microwave is subject to a broad range of interpretations, the following definition[1] is presented to express its current range of meanings. "Microwave—a very short electromagnetic wave, especially in the frequency range of 1000 to 30,000 megacycles, or in the wavelength range of 30 centimeters to 1 centimeter. . . . A different approach to the problem of

[1] R. I. Sarbacher, "Encyclopedic Dictionary of Electronic and Nuclear Engineering," Prentice-Hall, Inc., Englewood Cliffs, N.J., 1959.

defining the microwave region is to accept that there are no fixed frequency limits for microwaves and to refer to microwaves in any applications in which distributed constant circuits enclosed by conducting boundaries are used as opposed to conventional lumped-constant elements (i.e., capacitors, inductors and resistors)." From the distributed-constant-circuit definition standpoint, the frequency range of existing microwave amplifiers actually extends from well below 1,000 to above 30,000 Mc.

1.1. Microwave Amplifiers. The components used in the fabrication of lower-frequency r-f amplifiers are often inappropriate for use in microwave amplifiers. Conventional negative-grid amplifier tubes exemplify this situation. The electron transit time of a conventional tube becomes appreciable compared with the period of a microwave r-f cycle, and as a result, severely reduced performance may ensue. Some negative-grid tubes, employing advanced techniques, allow operation over a portion of the microwave range before the transit-time limitations become serious. Despite improved technology, the power-handling capabilities of this class of tube fall off rapidly as a function of frequency.

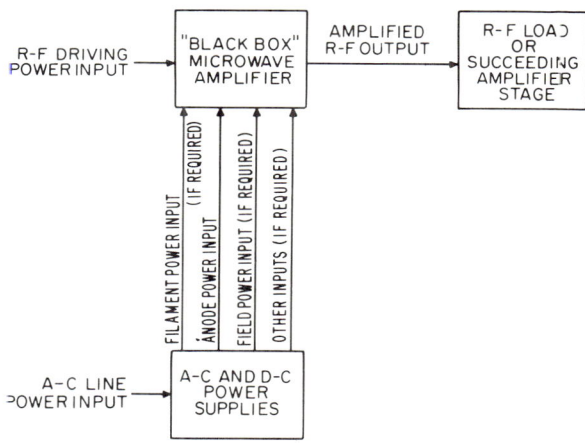

Fig. 1. Typical connections to "black-box" electron beam interaction tubes.

In an effort to overcome the limitations of conventional negative-grid tubes at microwave frequencies, the development of the electron-beam-interaction class of tubes came about. These tubes utilize the electron-transit-time phenomenon to provide progressive interaction with several distributed-constant circuits strategically placed along the path of the electron beam. Usually a circuit near the point of electron emission is excited by the signal power to be amplified, and the amplified output is extracted from a subsequent circuit. In addition, a high-level external magnetic field is usually employed for electron-beam control. Such tubes, by the combined nature of the transit-time effect, the distributed-constant electromagnetic-field aspects, and the external magnetic-field-control considerations that enter into their configuration, are more difficult to design, to fabricate, and to comprehend than are low-frequency tubes. From an application standpoint, however, they offer a degree of simplicity since they are generally packaged with all the r-f circuits inside the vacuum envelope, hence producing an effective "black box." Their operation entails only the connection of drive (or signal) power, the connection of an appropriate output load, and the connection of the necessary power supplies. This is shown in block-diagram form in Fig 1. This simplicity can be deceiving, since more careful control of the components external to the tube is generally required than is the case with negative-grid tubes. This class of tubes is typified by the multicavity klystron, the traveling-wave tube, and the crossed-field amplifier. Beam-interaction tubes enjoy a wide range of application in the microwave region, from the amplification of microvolt signal levels from a receiving antenna, to the generation of multimegawatt r-f carriers to be radiated by a transmitting antenna.

An effect similar to vacuum-tube transit time is present in semiconductor amplifiers, because of the time required for the injected carriers to diffuse across the barrier junction. The requirement for extending semiconductor amplifiers into the microwave region resulted in, among other devices, the development of the parametric amplifier. The semiconductor employed in these amplifiers operates on a principle considerably different from that of the transistors generally employed at lower r-f frequencies. A p-n junction diffused silicon diode, called a varactor, is the heart of the parametric amplifier. Its action is such that the capacitance of the junction varies with the amplitude of the voltage applied across the terminals of the device, hence producing in effect a variable reactance. The nonlinear characteristic created thereby is utilized in a manner similar to the nonlinear characteristic used in conventional diode mixers; however, the outstanding difference obtained by use of the varactor is that

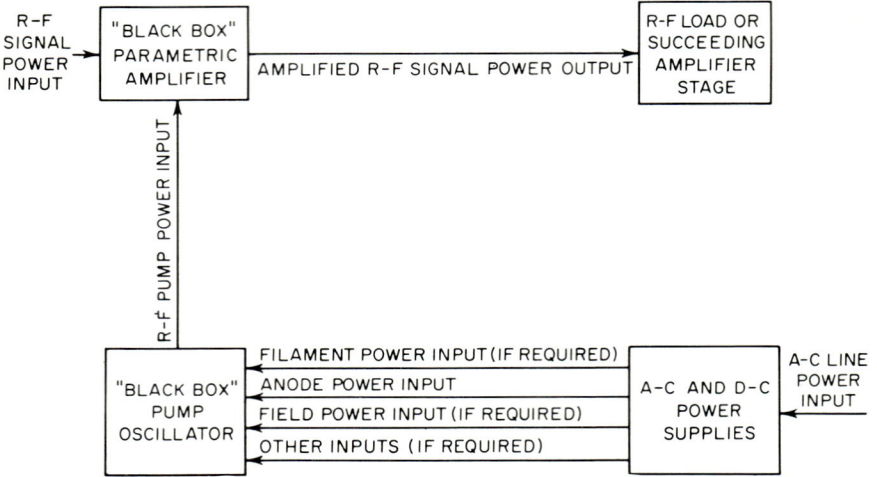

FIG. 2. Typical connection to "black-box" parametric amplifier.

mixing with resultant gain occurs. In the same sense that the electron-beam interaction tube can be considered a "black-box" device, a common form of the parametric amplifier may also be given that connotation (but to a lesser degree). For the benefit of this preliminary discussion, it can be visualized in terms of the block diagram as shown in Fig. 2. Instead of conventional power supplies (as in Fig. 1) external power is supplied by means of an independent r-f source (or pump). The parametric amplifier is currently enjoying wide usage, as a small-signal amplifier, over a significant portion of the microwave region.

1.2. Microwave Oscillators. Microwave oscillators are so intrinsically a part of most microwave-amplifier arrangements that their consideration is of major importance. Historically, microwave oscillators of the negative-grid feedback type, the magnetron type, and the reflex-klystron type were the trend setters for the later development of most microwave-amplifier tubes as we know them today. When the oscillator is considered in terms of an amplifier circuit in which some of the output-circuit energy is returned to the input circuit in such a way as to sustain oscillations, and it is further considered that feedback in most electron tubes is difficult to control (especially at microwave frequencies), it is evident why this sequence of events took place.

MICROWAVE AMPLIFIERS

There are a number of reasons why microwave oscillators still hold a position of prominence. When a microwave power source is required for many typical functions (e.g., a local oscillator to provide mixing action in a microwave receiver, a simple high-powered microwave radar transmitter, a pump for a microwave parametric amplifier) it can usually be shown that a microwave oscillator can produce this power with greater economy, greater overall efficiency, less overall bulk, and less overall weight than any other technique. If a high degree of frequency stability is required, more sophisticated techniques must usually be employed. These techniques customarily involve the use of more stable low-frequency oscillators (often crystal-controlled) in conjunction with frequency-multiplier or frequency-translator circuits, to provide output at the required microwave frequency.

1.3. Microwave Frequency Multipliers. The microwave frequency multiplier is another close relative of the microwave amplifier, and it functions to provide, at its output terminals, power at one or more of the harmonics of the frequency applied at its input terminals. This relationship is so inherent that identical tubes may often be used as either amplifiers or frequency multipliers. This fact is not too surprising when one investigates the high harmonic content in the output of many microwave-amplifier tubes. Low-power traveling-wave tubes are often prime examples of this situation. Such tubes typically possess an electronic bandwidth of 100 per cent (for instance, 1,000 to 2,000 Mc with no tuning adjustment). When the typical tube used in this instance is excited with sufficient 1,000-Mc drive power, output will be observed at both 1,000 and 2,000 Mc, and in fact, enough 2,000-Mc output will often appear as to still indicate the tube is producing an overall power gain as a frequency doubler. Similarly, had the tube been excited with sufficient 500-Mc drive power, appreciable output would be likely to appear at 1,000, 1,500, 2,000, and higher harmonics of the excitation frequency. With some emphasis placed on tubes specifically designed for multiplier service, multiplication ratios of 10:1 are readily available. The use of such tubes will allow extremely stable low-frequency oscillators to have their frequencies multiplied into the microwave region and still deliver output with a high degree of frequency stability. Some form of resonant distributed-constant circuit is often employed at the output of broadband frequency multipliers to attenuate output at undesired harmonics.

Solid-state frequency multipliers also receive wide application. Since the nonlinear characteristic of any sufficiently excited diode will result in harmonic components of the driving frequency being present in the diode's anode current, this characteristic is frequently utilized to provide frequency multiplication without the use of external power supplies. While this form of frequency multiplication bears no resemblance to amplification, the net loss of gain encountered by using several such diode multiplier stages in cascade is easily overcome by a single stage of high-gain tube amplification at the final output frequency. Figure 3 illustrates both tube-type and solid-state-type frequency multipliers in block form.

1.4. Microwave Frequency Translators. The frequency translator is another fundamental element of many microwave-amplifier arrangements. It functions to translate (or move) the signal frequency applied to the input terminals of the device to either a higher or lower (or both) frequency(s) at the output terminals, by an amount related to a third frequency usually injected by a local power source. This process may or may not be accomplished with resultant gain. The common diode mixer used in microwave receivers (in which the signal from an antenna is mixed with a local oscillator to produce output at some lower intermediate frequency) is an example of a typical solid-state frequency translator. A form of parametric amplifier may be used in similar fashion to provide translation with resultant gain, rather than with loss of gain as does conventional diode mixing.

Microwave tubes are also used as translators. Conventional klystron and traveling-wave tubes may be employed as synchrodyne translators. This form of translator functions by injecting the offset frequency in series with one of the beam-control voltages of the tube to produce velocity modulation of the drive signal (which is introduced at the normal input terminals of the tube), thereby producing translated frequencies at the output terminals of the tube. Frequency translators normally

AMPLIFIER CIRCUITS

TWT HARMONIC OUTPUT CHARACTERISTICS

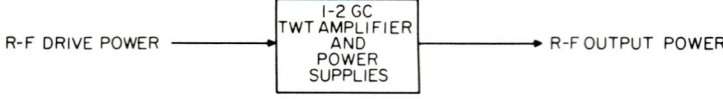

DRIVE FREQUENCY	OUTPUT FREQUENCIES
1 GC	1 AND 2 GC
0.5 GC	1, 1.5 AND 2 GC

TWT MULTIPLIER CHARACTERISTIC

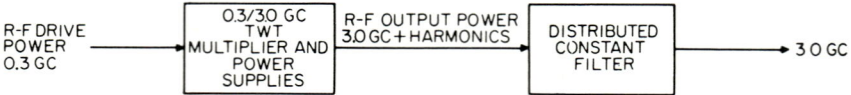

SOLID-STATE MULTIPLIER COUNTERPART OF TWT MULTIPLIER

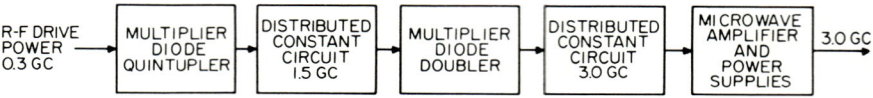

FIG. 3. Typical frequency-multiplier arrangements.

SOLID STATE FREQUENCY TRANSLATOR

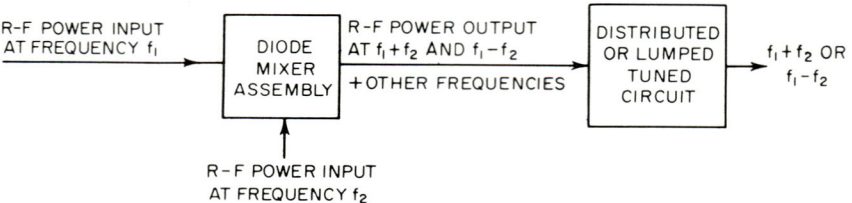

MICROWAVE TUBE TRANSLATOR

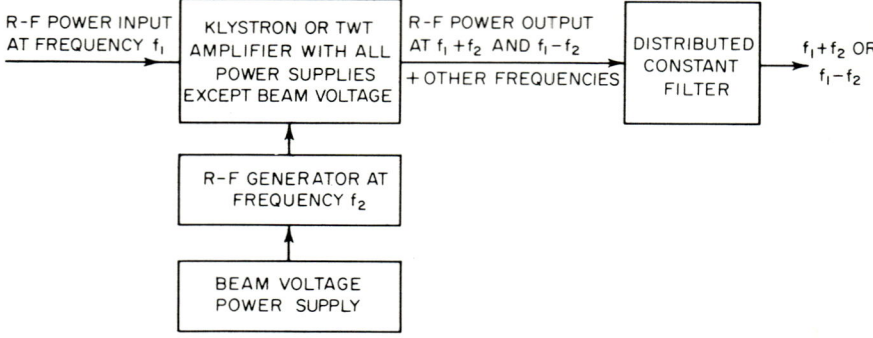

FIG. 4. Typical frequency-translator arrangements.

have their output power passed through some form of resonant distributed-constant circuit since frequencies other than the desired product of translation are usually present. Figure 4 illustrates block diagrams of several forms of frequency translators typically used in conjunction with microwave amplifiers.

1.5. The Microwave Amplifier Complex. The preceding brief discussions of the elementary microwave amplifiers, oscillators, multipliers, and translators illustrate a number of the major active components encountered in typical microwave amplifier complexes. The term complex is used here to denote a group of components used to perform some useful microwave function, such as a microwave transmitter complex, a microwave receiver complex, and a combined microwave transmitter and receiver complex. It is often quite difficult to comprehend the nature of the characteristics required of the individual components employed in a given complex unless the function the complex is to perform is clearly understood. For this reason overall consideration should be given to both the active and passive microwave components as well as the active and passive nonmicrowave components of the complex if optimum functional performance is to be obtained.

To illustrate the interdependence of the individual components on the function being performed, consider a hypothetical radar complex whose function is the detection of moving aircraft. Assume that after a careful analysis of the functional requirements of this complex, it was determined that a highly stable 2,790-Mc transmitter having an output of 1 megawatt of peak power, a rectangular pulse duration of 1 μsec, and a pulse repetition frequency of 1,000 pulses/sec was required. Further, assume that a receiver having a very low noise figure and incorporating the moving-target indicator (MTI) concept was also required. For the conditions stated, a complex resembling Fig. 5 could reasonably result. The following description will attempt to indicate only a few of the ways in which component characteristics are related to the function being performed.

The master oscillator should in this case operate at as high a frequency as possible (consistent with the pulse-to-pulse and the long-term frequency-stability requirements of the function) in order to minimize the frequency multiplication necessary to achieve microwave drive power for the transmitter. It could reasonably consist of a temperature-controlled crystal oscillator operating at about 50 Mc. Since the frequency multiplier would then have to provide a step-up of sixty times, this could be accomplished by two negative-grid multipliers in cascade acting as a frequency tripler and a frequency doubler, respectively, followed by a traveling-wave multiplier tube capable of ten times multiplication. Such an arrangement would undoubtedly have frequencies other than 3,000 Mc at its output terminals; hence a distributed-constant filter for adequately attenuating the undesired output frequencies would be required. The 3,000-Mc filtered output of the multiplier would then pass through a power splitter, one path for the subsequent transmitter stages and the other for receiver translation purposes.

The next component of the transmitter chain could be a small traveling-wave or klystron tube used as a frequency translator of the synchrodyne type. A 30-Mc offset oscillator and amplifier arrangement could translate the 3,000-Mc drive power to provide the three outputs indicated in Fig. 5. A distributed-constant filter would then be employed at the output of the translator to remove the undesired translation by products and leave only the desired 2,970-Mc content.

Up to this point, all the stages of the transmitter chain would have operated at low power levels (under 1 watt) and in a CW mode, with the exception of the offset oscillator's amplifier stage. This stage would be pulsed with a pulse length considerably longer than the required output pulse duration (and also bracketing the output pulse). Since synchrodyne action ceases when the pulse is removed, this eliminates the possibility of any stray 2,970-Mc output leaking into the receiver channel during the major portion of the interpulse period. As a result of the essentially CW operation of all stages, the bandwidth of all the components involved could (and if possible should) have been very narrow.

The final two stages of the transmitter are high-power amplifiers and will be subjected to 1-μsec pulses. If the pulses applied to these stages were truly rectangular

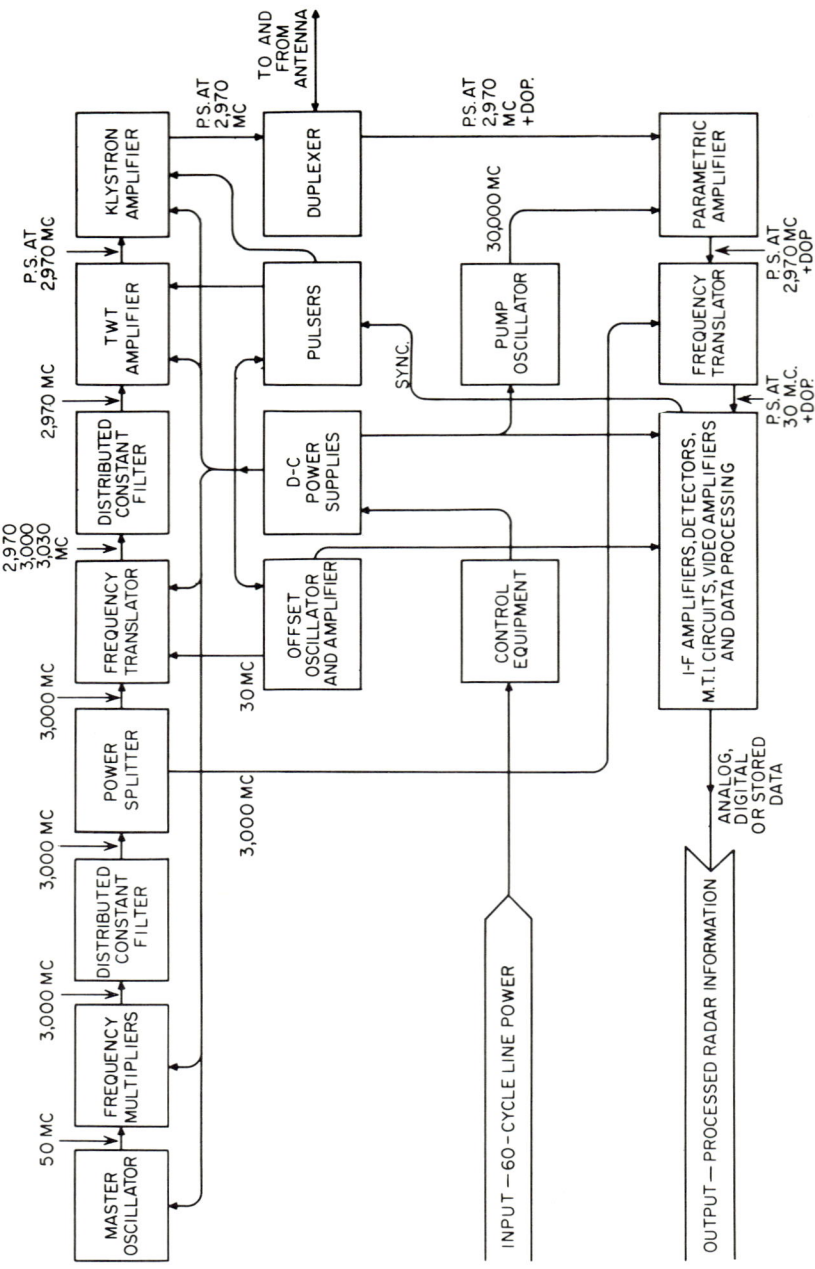

Fig. 5. Block diagram of hypothetical radar complex.

and if each amplifier stage had infinite bandwidth, the output microwave energy would appear in the form of a frequency spectrum whose envelope would be in the form of a sin x/x curve. The first point of zero output would appear 1 Mc either side of the 2,790-Mc carrier frequency, and subsequent crossovers would occur at 1-Mc intervals on either side thereafter. This is the significance of the pulse-spectrum notation at the output of these two tubes in Fig. 5. In actual practice, the output of the pulsers (which in this case are assumed to be pulsing the beam-accelerating voltages of both amplifiers for 1 μsec) will have finite rise and fall times, and the amplifier tubes will have finite bandwidth, hence resulting in a somewhat less severe spectrum distribution than indicated. It is important, however, that the tubes have sufficient bandwidth to maintain an adequate amount of this spectrum width or a severe departure from a truly rectangular r-f pulse will result. In addition, the voltage pulses from the pulsers must resemble quite closely the intended pure rectangular shape or velocity-modulation effects (similar to synchrodyne action) will produce undesired frequency components in the transmitter output. A similar consideration holds true for ripple and other voltage variations present in the d-c power supplies delivering power to the preceding transmitter stages.

At this point of the block diagram of Fig. 5, the 1 megawatt peak power, 1 μsec duration, microwave pulses are directed into the duplexer one thousand times per second. The duplexer functions to pass this pulse power directly to the antenna for radiation. The effectiveness of a duplexer is measured in part by how well it does this without letting appreciable leakage power enter the sensitive receiver input components. The transmitter energy of a given pulse is then radiated into space where it may strike objects and have some minuscule amount returned from each object to the antenna and then further returned from there to the duplexer. The duplexer's second function is then one of directing this energy to the receiver input without losing an appreciable amount in the transmitter output circuits. In both its transmitting and receiving roles, the duplexer must possess adequate bandwidth to ensure pulse transmission without distortion.

At this point Fig. 5 shows the received energy to have a pulse spectrum centered at 2,970 Mc plus some doppler frequency. The doppler frequency would not be present had the transmitted pulse impinged upon some stationary object; however, had it impinged upon a moving target (such as an aircraft whose detection was the function of the complex), the returned energy would have its center frequency (and all the other spectrum components therein) shifted by an amount

$$f_d \approx 89.4v/\lambda \qquad (1)$$

where v is the target radial speed in miles per hour, λ is the transmitted wavelength in centimeters, and f_d is the doppler frequency in cycles per second.

Since the requirement was established for a low-noise-figure receiver, the input stage is of the utmost importance. The probable choice then would be a parametric amplifier because of its excellence in this area as compared with the crystal mixer customarily used for receiver inputs in this frequency range in the past. The pump frequency should ideally be as high as possible since this will provide maximum amplifier gain and minimum amplifier noise figure. The parametric amplifier would reasonably be followed by a conventional crystal mixer, where the offset frequency of 3,000 Mc has been obtained from the second power-splitter output of the low-power transmitter stages and combined with the parametric amplifier output to provide a 30-Mc intermediate frequency to the receiver for amplification. The remainder of the receiver may take many forms, but typically some form of delay line, whose delay is equal to the interval between pulses (in this case 1 msec), is used to compare a returned echo with the previous echo from the same target. Fixed targets, since they return an echo with no doppler-frequency component present, are canceled out by the MTI when compared with the effective transmitted frequency. Moving targets only are then processed and passed on for subsequent amplification and data processing. Although this portion of the overall complex function is not performed at microwave frequencies, the effectiveness of the MTI performance can be seen to depend directly on the phase and frequency stability obtained from the microwave components of the overall complex.

2. FUNCTIONS OF MICROWAVE AMPLIFIER COMPLEXES

It was indicated previously that unless the function a microwave amplifier complex must perform is clearly established, the characteristics required of the active and passive components comprising the complex may be improperly specified. This can result in either inadequate performance of the complex on the one hand, or adequate performance of the complex at the expense of undue complication (and often excessive cost) on the other hand.

The functions of microwave amplifier complexes can be classified in many ways. Perhaps the most straightforward classification relates to the character of the wave to be amplified. Three basic amplifier classes can then be established. They are, first, *the CW amplifier*, in which a continuous wave of constant amplitude and constant frequency is amplified; second, *the communications amplifier*, in which a wave upon which intelligence has been imposed (usually by means of either wave-amplitude variations, wave-frequency variations, or both these types of variations) is amplified; and third, *the pulse amplifier*, in which a wave that is periodically turned on and off (usually with fixed on and off time intervals) is amplified. While it may be reasoned that the pulse amplifier is a form of communications amplifier, in practice the pulsed wave is not customarily used to convey communications intelligence directly.

The above three classes of amplifiers may further be described in terms of their basic types, their modes of operation, their uses, and the nature of the frequency spectrum they must be capable of amplifying. These considerations determine, in large part, the characteristics the active and passive components of the amplifier complex must possess.

2.1. CW Amplifier Complexes. This type of amplifier is commonly found in receiver, transmitter, and test-equipment complexes. Typical uses for such complexes in the microwave region are signal generators for calibrating microwave equipment, high-power amplifiers used for testing the thermal and power-handling capabilities of microwave transmission-line components, microwave heating and cooking equipment, jamming equipment for electronic-countermeasures applications, and local oscillator-amplifier arrangements for use in transmitter and receiver functions, in addition to many others.

The r-f spectrum of a CW amplifier is represented, in theory, by a single frequency. In practice, perfect amplitude and frequency stability is never completely achieved, because of modulation resulting from such external effects as thermal noise, vibration, and power-supply ripple imposed on amplifying elements, and from such internal causes as emission shot and flicker effects. In many of the uses to which CW amplifier complexes are put, these effects can be minimized to a sufficient degree without great effort. In the case of microwave amplifiers requiring a very high degree of output purity, extreme care in component selection, application, and fabrication is necessary.

It may also be noted that in theory, a CW complex is in effect a device that may operate with essentially zero-bandwidth components. This would imply that in practice infinite-Q tuned circuits could be employed without affecting the performance of the complex. One precaution that should be noted in this regard is that the basic stability of oscillator components is often an order of magnitude or more worse (compared with low-frequency performance in absolute terms) at microwave frequencies. This can easily result in an oscillator drifting out of the bandwidth of a succeeding amplifier stage possessing very high-Q tuned circuits or equipped with narrow-bandpass filters. Conversely, assuming the oscillator had absolute stability, it is conceivable that an amplifier having extremely high-Q tuned circuits might have thermal characteristics which could cause the circuits to detune with time, thus resulting in an effect similar to that of the drifting oscillator.

2.2. Communication Amplifier Complexes. This type of amplifier is similar in many respects to the CW amplifier just described. However, since intelligence must be transmitted by a communications amplifier, some sort of modulation must be applied to the carrier wave. All forms of carrier modulation place the requirement on the amplifier for a bandwidth sufficient to pass the sidebands generated by the modulation process. Typical modulation modes employed at microwave frequencies are conventional on-off keyed-code transmission of a constant-frequency carrier, frequency-shift keyed-code transmission of a carrier, conventional double-sideband amplitude modulation, double-sideband suppressed-carrier amplitude modulation, single-sideband amplitude modulation, single-sideband suppressed-carrier modulation, conventional CW frequency modulation, digital data modulation (which in itself involves many submodes of operation such as pulse time modulation, pulse amplitude modulation, and pulse code modulation), and spread-spectrum modulation (which may also take on a number of submodes). The bandwidth necessary for the transmission of any of the above modes of modulation is determined both by the

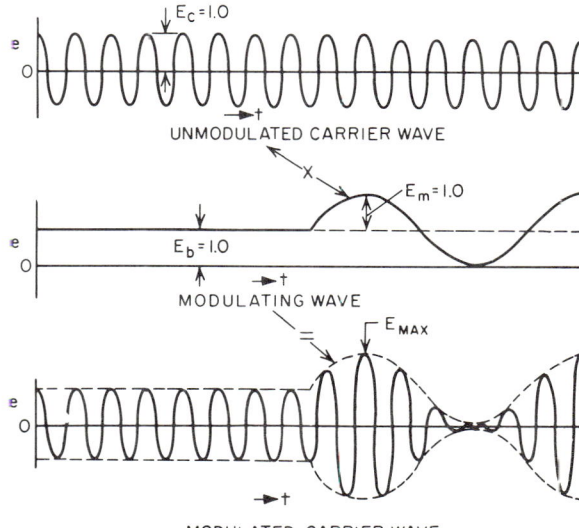

Fig. 6. Graphical representation of amplitude modulation in the time domain.

nature of the intelligence to be transmitted and by the mode of modulation selected. In effect, all the modes may be visualized as various applications of amplitude and angular modulation principles, and the spectrum characteristics associated with these basic forms allow the bandwidth required for a particular modulation mode to be ascertained. A brief summary of the general characteristics of amplitude and angular modulation is presented below.

Amplitude Modulation. Conventional double-sideband modulation of a carrier wave is represented in the time domain by Fig. 6. Intelligence consisting of a single sine wave is the modulating function in this case. It is often useful to visualize amplitude modulation as merely a process whereby an unmodulated carrier wave is multiplied by a modulating wave. Figure 6 illustrates this concept in graphical form in that at each instant of time the carrier wave is multiplied by the modulating wave to produce the modulated output wave shown at the bottom of the figure. This output wave depicts the classical time-domain representation of a 100 per cent amplitude-modulated carrier. It is significant to note that in this waveform the individual half-cycles of the modulated wave are not of a pure sinusoidal form factor. As a result, frequencies other than the carrier frequency are generated by the modulation process. This can be shown mathematically by the same multiplication process as used in the graphical approach.

Figure 7 illustrates the process of multiplication as applied to the frequency domain. The upper figure represents a carrier of unit amplitude and oscillating at an angular frequency of ω_c radians/sec. The center figure represents a modulating function of unit amplitude, oscillating at an angular frequency of ω_m radians/sec, and offset from the zero amplitude level by a unity bias voltage level. If the mathematical expressions for these two waves are multiplied together, three separate frequencies can be seen to result as shown in the lower portion of Fig. 7. The multiplication is noted below the figure. In its simplest form, this procedure illustrates the need for bandwidth in an AM amplifier.

In actual practice two conditions depart from the simple 100 per cent modulated case described in the figures. The first condition is that the departure of the modu-

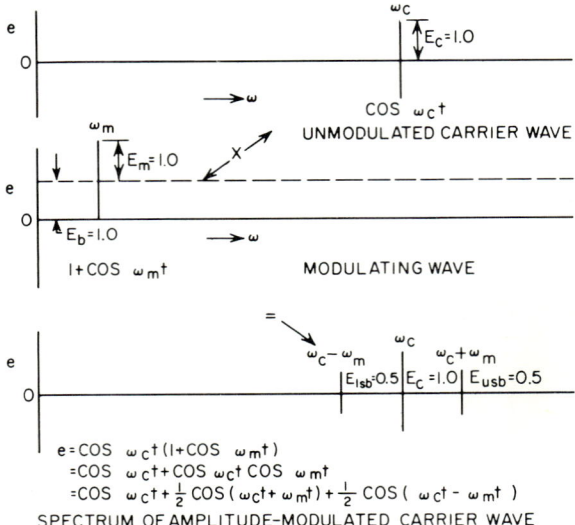

Fig. 7. Mathematical representation of amplitude modulation in the frequency domain.

lating wave from the bias level E_b is not usually so extreme as noted in the figures. This gives rise to a modulation factor which expresses the degree of modulation. It is expressed in terms of the time representation shown in the lower portion of Fig. 6 as

$$m = \text{degree of modulation} = \frac{E_{max} - E_c}{E_c} \tag{2}$$

and the expression for the modulated waveform then becomes

$$e = E_c \cos \omega_c t + (mE_c/2) \cos (\omega_c + \omega_m)t + (mE_c/2) \cos (\omega_c - \omega_m)t \tag{3}$$

The introduction of the degree-of-modulation factor can be seen to affect only the amplitude of the upper and lower sidebands and not their positions in the spectrum; hence the bandwidth required of the amplifier remains a function of the modulating frequency ω_m.

The second condition of departure from the elementary case shown in the figures is concerned with the fact that the modulating waveform is usually of some form other than a single-frequency sine wave. Since it is possible to represent any periodic complex modulating waveform by means of the Fourier series (a summation of the harmonically related components of the basic wave period) the same process of multiplication of the carrier wave by the components of the modulating waveform applies. A complex waveform by its nature then will produce a number of additional pairs of sidebands on either side of the carrier, and if this waveform is applied to an amplifier, the amplifier must have sufficient bandwidth to pass all the significant sidebands. If insufficient bandwidth is employed, distortion of the intelligence will result. An example of a broad-bandwidth amplifier requirement occurs when television picture transmission is considered. In this case the maximum sideband frequency that must be transmitted is given by the approximate expression

$$\tfrac{1}{2}(w/h)n^2 N \tag{4}$$

where w and h represent the ratio of width to height of the picture, n represents the number of lines, and N equals the number of pictures per second.[2] Another case requiring broad-bandwidth amplifiers will be illustrated in the later discussion of pulse amplifiers.

Single-sideband modulation modes operate on the principle that intelligence can be transmitted effectively by eliminating one set of the sidebands, hence reducing the amount of power and bandwidth handled by an amplifier stage. These techniques employ somewhat more complex concepts in the modulation process than double-sideband modulators; however, for a given modulating signal, only one half of the amplifier bandwidth is required, and the spectrum would appear identical to one half of a double-sideband modulated carrier. Additional techniques may be employed to transmit single-sideband with or without the carrier; however, carrier removal does not appreciably affect the bandwidth required of the amplifier although it does reduce its power-handling requirements.

Communication amplifiers used for the direct power amplification of modulated waveforms must be linear amplifiers if distortion of the modulating intelligence is to be avoided. (A linear amplifier is some form of tuned amplifier and operates so that the amplified output voltage that is developed is proportional to the exciting voltage applied to the amplifier.) Not all the common microwave amplifying elements fall into this classification. Where negative-grid tubes may be employed, it is possible to obtain quite linear amplifying properties up to a significant percentage of their maximum power-handling capabilities. Other typical microwave amplifiers, notably the klystron and traveling-wave-tube types, may be operated as linear amplifiers so long as the saturation drive level of the tubes is not too closely approached.

Angular Modulation. Angular modulation modes impose a different set of characteristics on the amplifying devices they work in conjunction with. The two basic forms of angular modulation are frequency modulation and phase modulation. In frequency modulation the modulating intelligence as a function of time is made to control the instantaneous frequency of the carrier in accordance with the modulating signal amplitude, while in phase modulation the modulating intelligence as a function of time is made to control the instantaneous phase of the carrier in accordance with the modulating signal amplitude. In both cases the modulated carrier has a constant-voltage amplitude of the form

$$e = E_c \cos f(t) \tag{5}$$

where the function of time (the total phase angle of the carrier wave) is described in terms of the form of modulation.

In the case of frequency modulation, the instantaneous frequency of the carrier is given by

$$\omega(t) = \omega_c - \Delta\omega m(t) \tag{6}$$

where ω_c is the carrier frequency in radians per second, $\Delta\omega$ is the maximum instantaneous frequency excursion in radians per second from ω_c, and $m(t)$ is the modulating intelligence signal as a function of time. It may be noted that there is a similarity to

[2] F. E. Terman, "Radio Engineers' Handbook," McGraw-Hill Book Company, New York, 1943.

amplitude modulation in that modulation is again expressed as a product of two terms, and this is shown graphically in Fig. 8. For the case where a single sine wave is the modulating intelligence, the form of $m(t)$ is $\cos \omega_m t$ where ω_m is the modulating frequency in radians per second. Since $f(t)$ is the total phase angle of the carrier as a

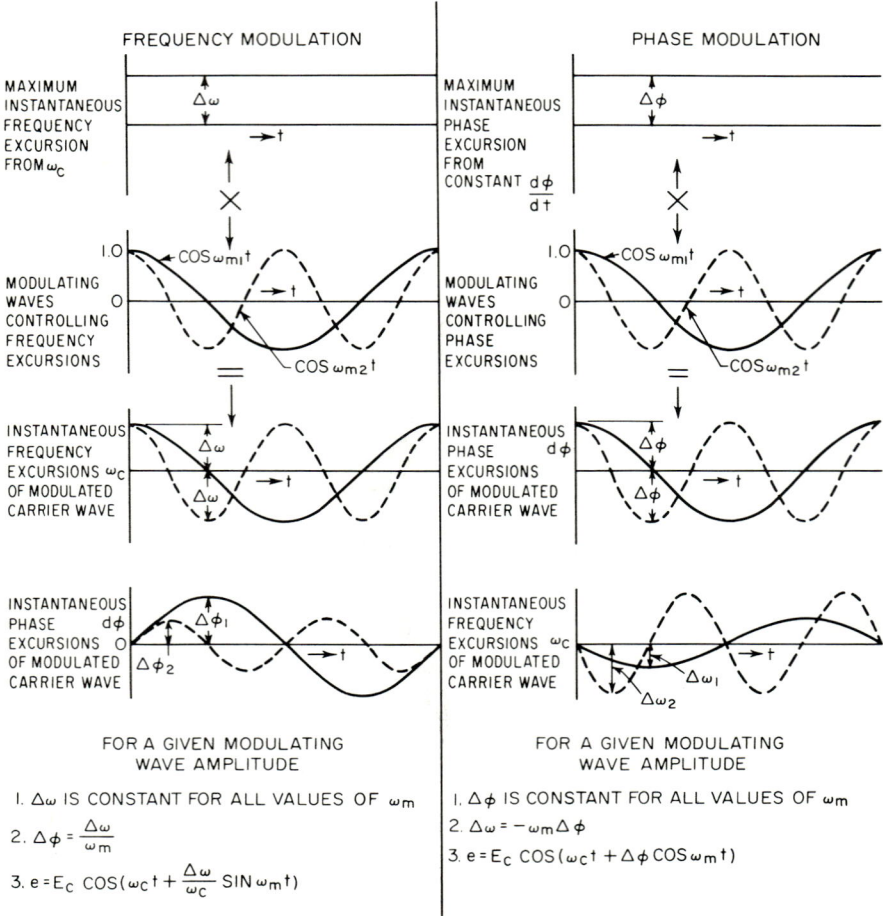

FIG. 8. Comparison of the angular modulation modes of frequency and phase modulation.

function of time, and since phase as a function of time is the integral of frequency as a function of time,

$$f(t) = \int \omega(t)\, dt = \int (\omega_c + \Delta\omega \cos \omega_m t)\, dt = \omega_c t + (\Delta\omega/\omega_m) \sin \omega_m t \qquad (7)$$

Translating this expression to the basic angular-modulation expression gives

$$e = E_c \cos [\omega_c t + (\Delta\omega/\omega_m) \sin \omega_m t] \qquad (8)$$

as the expression for a frequency-modulated wave. The expression $\Delta\omega/\omega_m$ is called the modulation index and has a significance somewhat similar to the degree of modulation in an amplitude-modulated wave. In addition, it expresses the maximum phase excursion of the wave in radians.

A similar approach is employed in describing a phase-modulated wave. In the case of phase modulation the modulating signal intelligence is made to control the instantaneous phase excursions of the carrier by the product of $\Delta\phi$ (the maximum phase excursion) and $m(t)$. This is also shown in Fig. 8. As before, $m(t)$ is expressed as $\cos \omega_m t$. Then $f(t) = (\omega_c t + \Delta\phi \cos \omega_m t)$ and the expression for a phase-modulated wave becomes

$$e = E_c \cos (\omega_c t + \Delta\phi \cos \omega_m t) \tag{9}$$

Since the instantaneous frequency of a wave is given by the derivative of its total phase,

$$\omega(t) = df(t)/dt = d(\omega_c t + \Delta\phi \cos \omega_m t)/dt = \omega_c - \omega_m \Delta\phi \sin \omega_m t \tag{10}$$

and since

$$\Delta\omega = \omega(t) - \omega_c = \omega_c - \omega_m \Delta\phi \sin \omega_m t - \omega_c = -\omega_m \Delta\phi \sin \omega_m t \tag{11}$$

the maximum frequency excursion in phase modulation is given by $\Delta\omega = -\omega_m \Delta\phi$. The term $\Delta\phi$ is termed the modulation index for a phase-modulated wave.

The basic distinction between phase and frequency modulation is that in the former the modulation index is independent of modulating frequency, while in the latter it is inversely proportional to the modulating frequency. Hence, with a fixed modulating voltage the band required to accommodate a phase-modulated signal is proportional to the modulating frequency, while with frequency modulation the band occupied is independent of modulating frequency, except when the modulating index is small. This is indicated in part by Fig. 8. What is not evident from this figure is the nature of the spectrum of angular-modulated waves.

The expansion of the expressions for angular-modulated waves to illustrate the sideband components is considerably more complicated than in the case of amplitude-modulated waves. This results from the fact that they contain a variable of the form $\cos (\Delta\phi \cos \omega_m t)$, and this in turn makes necessary the use of Bessel's functions to complete the expansion. More than a single set of sidebands (separated by $\pm \omega_m$ from the carrier) results from single frequency modulation, and hence the spectrum is considerably broader than is the case in amplitude modulation. The expansion of the waveform developed for the phase-modulated wave becomes

$$\begin{aligned} e = E_c \{ & J_0(\Delta\phi) \cos \omega_c t - J_1(\Delta\phi)[\sin (\omega_c t + \omega_m t) + \sin (\omega_c t - \omega_m t)] \\ & - J_2(\Delta\phi)[\cos (\omega_c t + 2\omega_m t) + \cos (\omega_c t - 2\omega_m t)] \\ & + J_3(\Delta\phi)[\sin (\omega_c t + 3\omega_m t) + \sin (\omega_c t - 3\omega_m t)] \\ & + \cdots \} \end{aligned} \tag{12}$$

where $J_n(\Delta\phi)$ is the Bessel function of the first kind and nth order with the argument $\Delta\phi$, and the values of these functions express the relative amplitudes of the carrier and sidebands. Figure 9 presents the values of $J_n(\Delta\phi)$ as the vertical axis on the charts [i.e., $J_0(\Delta\phi)$ is the carrier amplitude, $J_1(\Delta\phi)$ is the amplitude of the first pair of sidebands, etc.], and of $\Delta\phi$, the modulation index, as the horizontal axis

The expansion of the waveform developed for a frequency-modulated wave will produce an expression very similar to that of the phase-modulated wave except for the signs of some terms. The frequency components are the same as for a phase-modulated wave having the same modulation index. If the modulation index is less than 0.5 (phase deviation of less than $\frac{1}{2}$ radian) only one set of sidebands of significant amplitude will result. For each additional radian in the modulation index, an additional set of sidebands having appreciable magnitude will result.

As contrasted to the amplifiers used for the amplification of amplitude-modulated carriers, the amplifiers for frequency-modulated waves need not be linear amplifiers. This makes most of the typical microwave-amplifier tubes particularly attractive for use with angular modulation. However, the much greater bandwidth often encountered with angular-modulated intelligence may impose limitations with some forms of microwave tubes.

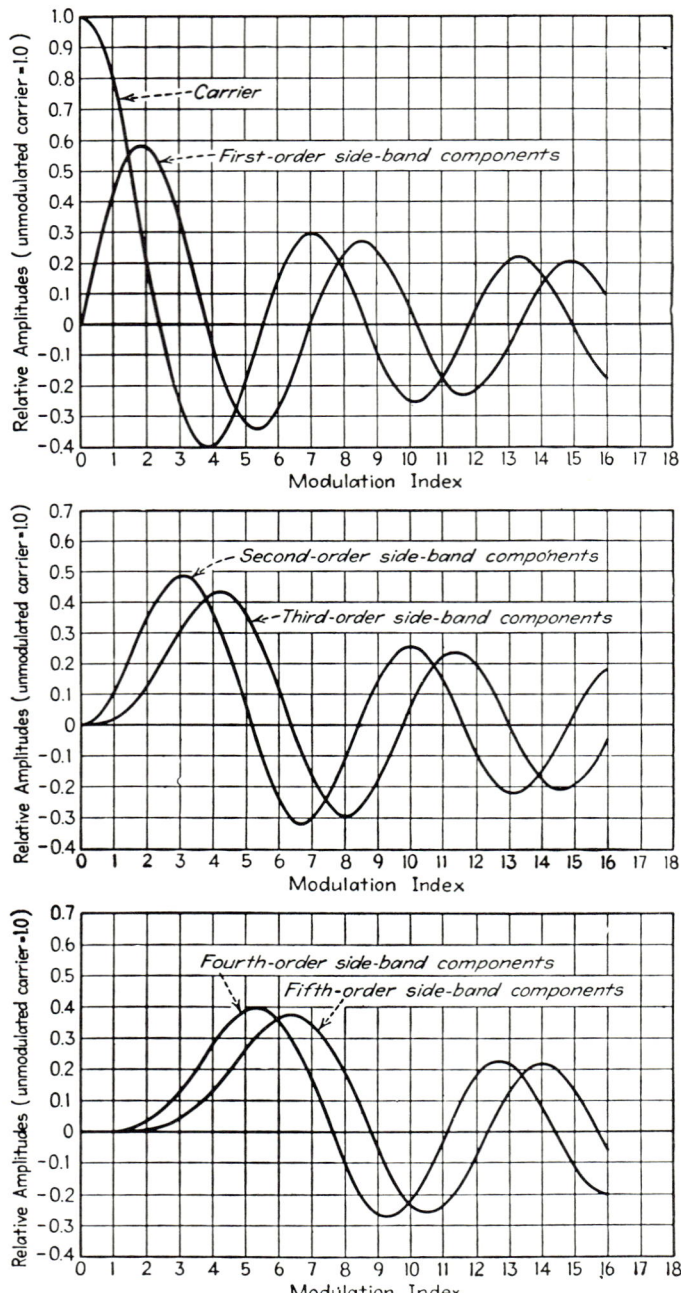

FIG. 9. Amplitudes of frequency components of a frequency- or phase-modulated wave. In the case of the side-bands the amplitude shown is the amplitude of the individual side-band component and not of the pair of companion side-bands taken together.

2.3. Pulse Amplifier Complexes. A pulse amplifier can be viewed from several standpoints. It may be used for direct amplification of a pulsed r-f waveform generated at low power levels, in which case the amplifier must possess bandwidth and linear amplification capabilities similar to those described for the amplification of AM communication inputs. Alternatively, the amplifier may be used to form the pulse within itself, by the application of a pulse of voltage to the anode or beam-accelerating structure, while the input of the amplifier is being driven by a constant-frequency CW carrier wave. In the latter case, the amplifier need not possess linear characteristics. In either case, the pulse spectrum characteristics are determined by the pulse duration, the pulse repetition rate, and the shape factor of the pulse involved.

Perhaps the most common application of pulsed amplifiers in the microwave region is in the field of radar pulse transmission. Pulses ranging from fractions of microseconds to many milliseconds in duration are commonly encountered. When negative-grid amplifiers are utilized, the lower-powered stages of an amplifier complex frequently employ the pulse r-f drive technique noted earlier. These stages are normally

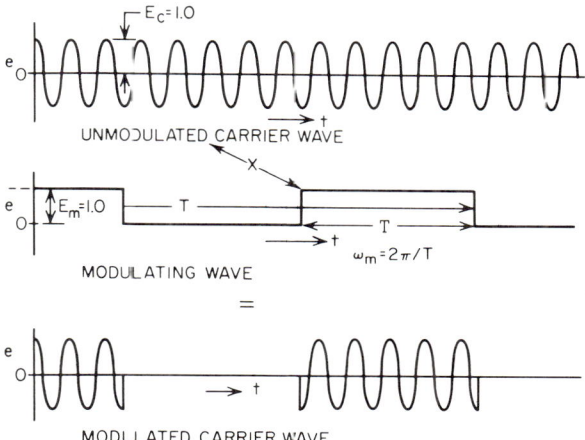

Fig. 10. Graphical representation of pulse modulation in the time domain.

grid-biased below cutoff and hence no anode current flows during the period between drive pulses. High-power-level negative-grid tubes normally employ anode pulsing techniques since higher power output and lower interpulse noise levels are usually achieved from a given tube with this method of operation. Beam-interaction tubes are normally either beam-pulsed or pulsed on some form of control electrode (to cause the cessation of beam current during the interpulse period) since in general, their overall efficiency would be quite poor otherwise.

Traditional radar transmitters have utilized rectangular pulses almost exclusively. In addition to being difficult to generate, this pulse waveshape requires a greater spectrum bandwidth than any other type. Because of its wide usage, a brief review of the characteristics of this form of modulation follows.

The same process of multiplication of the carrier by the modulating wave (as described under amplitude modulation) can be applied to pulse modulation. Figure 10 illustrates this multiplication in the time domain. In this case the carrier is assumed to be a constant-frequency wave, and the modulating wave is assumed to be a rectangular wave with equal on and off times. Such a high duty cycle (50 per cent) is not typical of the usual applications in radar equipment (duty cycles typically run much less than 10 per cent); however, this selection serves to simplify the example greatly. Furthermore, the carrier wave indicated is of much lower frequency than would ever be encountered in normal practice, but this too has been used to simplify the illustration. From Fig. 10, the time-domain multiplication is quite evident, and the formation of the modulated carrier wave is shown in the lower illustration.

Figure 11 illustrates the same multiplication (in a mathematical sense) for development of the components present in the frequency domain. A unity-amplitude unmodulated carrier wave is in this case multiplied by the Fourier series of the unity-ampli-

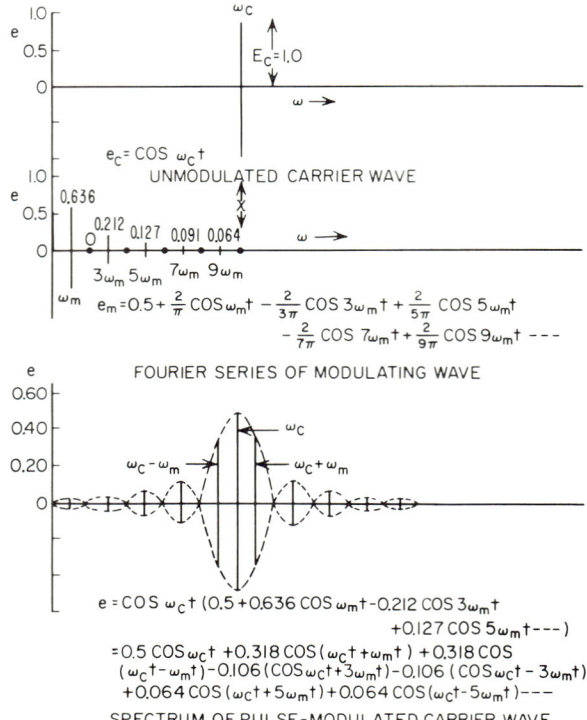

FIG. 11. Mathematical representation of pulse modulation in the frequency domain.

tude rectangular modulating wave of the previous figure. The individual coefficients of the Fourier series are obtained from the expression

$$C_n = 2A_{av} \frac{\sin n\pi\tau/T}{n\pi\tau/T} \qquad (13)$$

where A_{av} is the average value of the modulating wave, τ is the pulse duration in seconds, T is the time between pulses in seconds, and n is the harmonic number of the fundamental component of the modulating wave ($\omega_m = 2\pi/T$). The series thus developed is shown beneath the second illustration of Fig. 11. If these waves were added together about the zero time axis of the center illustration of Fig. 10 they would be seen to form the modulating wave train. The multiplication of the top two expressions of Fig. 11 then produces the spectrum of the pulse-modulated carrier wave shown in the lower illustration. The multiplication is shown below that illustration.

Several interesting observations may be made about the spectrum thus obtained. First, the bandwidth required to pass the pulse energy between the first two spectrum zero-amplitude points (noted by the dotted-line crossovers) on either side of the carrier is given by the expression $\omega = 2/\tau$. It can be shown that over 90 per cent of the energy is concentrated within this bandwidth; however, if only this amount of bandwidth were present in an amplifier passing such a pulse, the detected r-f waveform in the time domain would depart radically from a rectangular shape. Second, if the duty cycle were reduced but the pulse duration remained the same (i.e., the pulse repetition rate made higher), the shape of the spectrum would remain the same; however, the number of sidelobes under each loop would increase, the amplitude of the loops would decrease, and the frequency-axis intercepts of the envelope crossovers would remain the same. Third, if the pulse duration were shortened, the bandwidth required to pass the pulse would become greater, since even though the same-shaped spectrum would exist, its envelope intercepts on the frequency axis would spread out on either side of the carrier.

Other types of pulse shapes are sometimes employed in an effort to reduce the bandwidth required for pulse-transmission complexes. Such modulating waveforms as half-cosine, cosine-squared, cosine-cubed, and gaussian shapes have been effectively utilized. While a significant reduction in bandwidth is achieved by the use of such waveforms, great care must be exercised in the selection of the amplifiers employed. This is particularly true when anode or beam pulsing is used in conjunction with beam-interaction tubes. Such tubes usually possess a quite nonlinear applied-voltage vs. r-f output-voltage characteristic. This is not to be confused with the linear amplifier characteristics of the tubes but rather is analogous to the type of linearity that must be present when plate modulation of a class C negative-grid amplifier is attempted.

Another mode of operation sometimes encountered in pulse operation of microwave amplifiers is the so-called "chirped" pulse mode. In this case the frequency of the carrier is varied during the pulse, often in a linear fashion. This gives rise to a much more complex spectrum. Amplitude- and frequency-modulation considerations must enter into the analysis of both the resulting spectrum and the amplifier complex components.

3. CHARACTERISTICS OF MICROWAVE AMPLIFIER COMPLEXES

The previous discussion of the functions of microwave amplifier complexes was based on the assumption that the microwave amplifying components (as well as their associated components) were ideal elements. This might imply that any characteristic required in a given amplifier could be obtained if desired. To some extent this is true; however, the state of the art of a number of microwave components limits their capabilities to a level considerably below the classification of ideal. It is quite important that these limitations be established in the early stages of any investigation which may ultimately result in the design and fabrication of a microwave amplifier complex. At best, if a typical design is established based on "off-the-shelf" components, it will probably result in a time-consuming and costly endeavor. If the characteristics of these "off-the-shelf" components are not carefully investigated in the design phase, performance of an inferior nature can readily result. This in turn can produce such undesirable conditions as late delivery of the required equipment, additional expenditures for higher-quality "off-the-shelf" components, or the funding of research and development programs to produce components of adequate quality.

As a general rule the amplifier tubes (or solid-state amplifying devices) themselves offer the greatest source of possible difficulty. This is not intended as a criticism of the manufacturers of these devices; rather, it is the inevitable result of the extremely rapid rate at which the state of the art has advanced in this field. Accurate and meaningful test and evaluation data on microwave amplifiers are difficult to achieve and expensive to acquire. It is important that all such available data, related to the areas of potential importance, be obtained at the outset of a microwave-amplifier-complex design program. Several of the more important areas that might reasonably be investigated will be discussed briefly.

3.1. R-F Output Amplitude vs. Frequency Characteristic. There are finite bandwidth limitations with all practical microwave-amplifier tubes. Although some amplifiers have extremely broad bandwidths, their power output amplitude (when the tube is excited with constant-amplitude drive power across the frequency band of interest) will usually show marked variations. This situation comes about because the input and output assemblies of these devices almost always take the form of distributed-constant circuits. In the simplest cases these may consist of nothing more than single-tuned resonant circuits. In beam-interaction tubes there are, in addition to the input and output circuits, a number of intermediate resonant circuits. Such multiple-tuned-circuit types generally produce a bandpass effect similar to that observed in a low-frequency stagger-tuned multiple-stage amplifier arrangement. Figure 12 illustrates ideal and typical examples of this characteristic.

Negative-grid-tube microwave amplifiers appear in two fundamental forms, amplifiers with externally attached distributed-constant circuits and amplifiers with integrally attached distributed-constant circuits. In the former case the equipment designer is usually responsible for the characteristics of the circuits and hence for the amplitude vs. frequency characteristics. In the latter case the circuits are usually

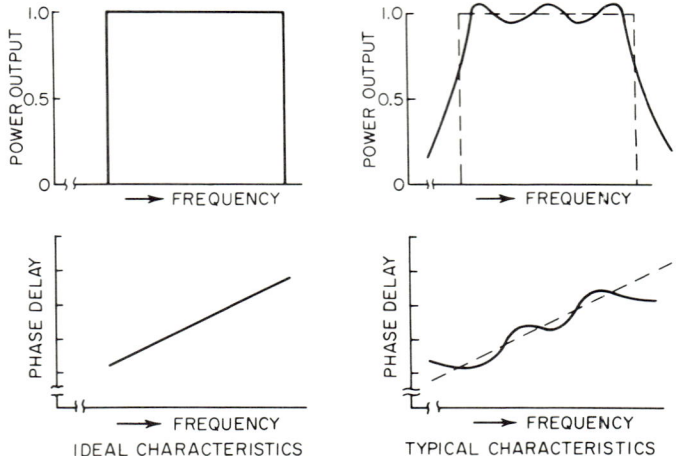

FIG. 12. Ideal and typical amplitude and phase vs. frequency characteristics.

pretuned at the tube manufacturer's plant, and as a result, detailed information is available on the bandpass characteristics of the amplifier. With a fixed value of output loading, negative-grid amplifiers do not have their bandpass characteristics materially affected by changes in anode voltage.

Klystron microwave amplifiers appear in three fundamental forms, tubes with externally attached distributed-constant circuits, tubes with integrally attached tunable distributed-constant circuits, and tubes with integrally attached fixed-tuned distributed-constant circuits. In the former two cases some degree of freedom is available to the equipment designer in adjusting the bandpass characteristics. In the latter case the bandpass characteristics are usually well defined by the manufacturer, for specific operating conditions. This class of tubes has its bandpass characteristic rather seriously affected when the tube is operated at other than its rated value of beam current, since the loading of the distributed-constant tuned circuits is a function of the tube's beam current.

Traveling-wave tubes and crossed-field amplifier tubes are practically always of the integrally attached distributed-constant-circuit type. Both types are therefore supplied with quite detailed bandpass characteristic information by the manufacturer. In general, the traveling-wave tube must operate within a critical range of beam voltages (to maintain proper interaction-circuit loading) while the crossed-field amplifier

MICROWAVE AMPLIFIERS

must operate within a critical range of beam currents (to provide proper electric- and magnetic-field interaction) if their bandpass characteristics are to be maintained. Both types are somewhat more critical than the klystron in this regard.

3.2. R-F Output Phase Delay vs. Frequency Characteristic. From a very elementary standpoint, all microwave amplifiers may be regarded as short lengths of transmission line into which a drive signal is introduced at one end and from which an amplified replica of the drive signal is extracted at the other end. Had the characteristics of this transmission line been ideal, that is, a characteristic such as one would find with a length of ideal coaxial cable, the phase-delay characteristic vs. frequency would be a straight line whose slope was proportional to the length of the line. With practical microwave-amplifier tubes this general characteristic is clearly noted; however, variations are usually superimposed on the otherwise linear characteristic. These variations are in large part the result of the distributed-constant tuned circuits which form a part of the transmission path through the amplifier. It is generally the case that for a given amplifier tube arrangement, the phase variations (from linearity) vs. frequency are coincident with the output amplitude variations vs. frequency. This characteristic is illustrated for ideal and typical cases in Fig. 12.

The negative-grid tube possesses the shortest phase delay of any microwave amplifier. Its phase vs. frequency characteristic is largely a function of the input and output resonant circuits. The crossed-field amplifier has phase delays of intermediate value. Its phase vs. frequency characteristic is usually quite linear because of the large number of interaction segments present in its structure. The klystron is probably the next tube typically in order of phase delay. Since it usually possesses a relatively small number of interaction structures, it may exhibit some appreciable departures from linearity; however, this is largely a matter of how the cavities are tuned. The traveling-wave tube typically possesses the longest phase delay, but generally it will have a fairly linear phase vs. frequency characteristic because of the large number of interaction elements.

The preceding statements are all quite general since each specific tube type noted may take on a number of forms. The manufacturer of a specific tube of interest should be consulted for data. It is usually much more difficult to obtain accurate phase vs. frequency data than it is to obtain accurate amplitude vs. frequency data because of the problems associated with the measurement techniques. Unfortunately, many current applications of amplifier tubes demand explicit information as to the phase characteristics of these tubes. As a result much more emphasis is now being placed on the measurement techniques associated with this characteristic.

3.3. R-F Output Amplitude vs. Voltage(s) Characteristic. All microwave amplifiers have their output amplitude affected in a critical manner with regard to externally applied power supply or pulse voltages. The manner in which these voltages affect the output is largely a function of the tube type. Typical characteristics are shown in Fig. 13. The negative-grid tube behaves much the same at microwaves as it does at lower frequencies in this regard. In general, $P_o = kE_b^2$ where k is a constant related to tube characteristics and efficiency, and E_b is the d-c anode voltage

Klystron and traveling-wave-tube amplifiers in theory would follow a power output characteristic in which $P_o = \eta k E_a^{3/2}$, where η is the tube efficiency, k is the microperveance of the electron gun, and E_a is the accelerating voltage in volts. In practice, this expression really represents the power input to the tube times the efficiency, and since the efficiency generally decreases rapidly when the beam voltage is reduced below its optimum value, the expression is useful only over a very limited range. It does serve to illustrate that this class of tubes is more critical of applied voltage than is the negative-grid class.

The crossed-field amplifier tube acts as a biased diode; that is, it draws very little current from its anode power supply until a critical value of voltage is reached. Once this knee is reached, the increase of current in response to an increase in anode voltage is very rapid. The power output is approximately equal to $P_o = \eta E I^2$ where η is the efficiency of the tube, E is the biased diode level in volts, and I is the d-c anode current in amperes.

27-22 AMPLIFIER CIRCUITS

The above generalizations cover the main voltage variations contributing to changes in power output amplitude. Other voltages can also have a pronounced effect. Various forms of control electrodes are often introduced in microwave amplifiers to control their output. Their control characteristics vary widely and only through contact with the manufacturer of a particular tube can these characteristics be accurately established. Two potential sources of undesired output variation are also present. They are variations in filament voltage and variations in externally applied magnetic field. Careful control of these two sources must be established if unwanted modulation of the output amplitude is to be avoided.

3.4. R-F Output Phase vs. Voltage(s) Characteristic. All microwave amplifiers are subject to the laws of motion of electrons in electric fields. In practical

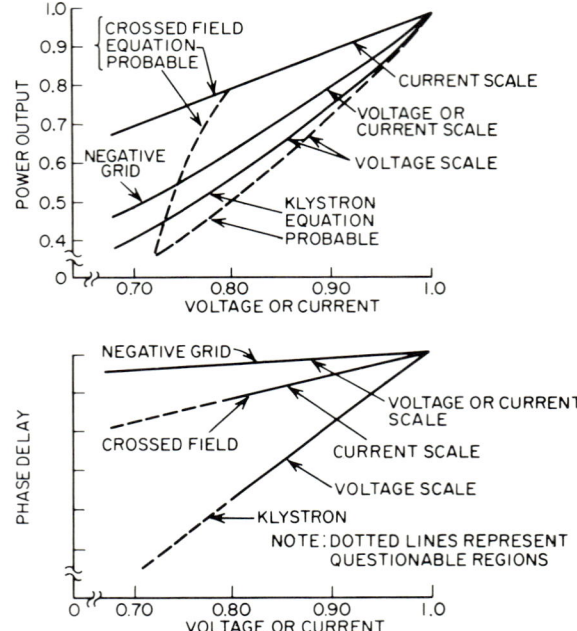

FIG. 13. Approximate amplitude and phase vs. beam voltage or current characteristic.

units, the velocity acquired by an electron in falling through a given potential is $V = 5.97 \times 10^{-7} \times E$ where the potential E is given in volts. Since the phase delay through an amplifier stage is a function of the velocity of the electron beam, it is readily apparent that the external voltages applied to such amplifiers need special attention if undesired phase changes are to be avoided. Examples of this characteristic are shown in Fig. 13.

Because the phase length of microwave negative-grid tubes is quite short, the change in phase with change in anode voltage is quite small (on the order of 0.5° for a 1 per cent change in voltage). The negative-grid tube typically exhibits the lowest $d\phi/dV$ characteristic of all microwave amplifiers.

The crossed-field amplifier is generally the type exhibiting the next lowest phase vs. voltage characteristic. These tubes show a phase change on the order of 0.5° for a 1 per cent change in anode current. It should be noted here that since these tubes operate essentially as biased diodes, current rather than voltage is taken as the variable. In the case of both the negative-grid tube and the crossed-field amplifier the gain per amplifier stage is generally low (on the order of 5 to 20); hence a greater number of stages with a resultant greater overall phase departure is possible. This fact should be kept in mind when comparing this characteristic with higher-gain tubes having higher $d\phi/dV$ characteristics.

Klystron and traveling-wave tubes are subject to much higher $d\phi/dV$ characteristics than are the two preceding types. This results in part from the fact that there are often a number of r-f cycles of delay between the driving circuits and the output circuits. At the same time the gain of these tubes may be much higher, often on the order of 30 to 60 db; hence fewer stages are required to contribute to overall phase variations. A figure of perhaps 10° for a 1 per cent change in accelerating voltage would be typical for a medium-gain klystron. A somewhat higher figure might be representative of traveling-wave tubes having similar gain.

High-power klystrons and traveling-wave tubes often operate at quite high accelerating potentials. The equation noted in the first paragraph is valid until the velocity of a charged particle is no longer negligible compared with the velocity of light. Then a change must be made for the relativistic change of mass of the electron.[3] Tubes operating at accelerating potentials in excess of about 2,500 volts are subject to this effect and hence are not subject to so great a $d\phi/dV$ effect as are lower-voltage tubes.

As noted in the amplitude vs. voltage discussion, other control electrodes can also have a pronounced effect on the phase characteristics of microwave tubes. These voltages, as well as filament and magnetic-field control voltages, deserve careful consideration if unwanted phase modulation of the amplifier output is to be avoided.

3.5. R-F Input/Output Linearity Characteristics. The need for linear-amplifier capabilities in microwave tubes was described briefly in connection with a previous discussion on amplitude modulation. It should be noted that a deficiency in this characteristic may result in the production of intermodulation products and hence may cause undesired spectrum spreading at the output of the amplifier. In addition, in the case of multiplex operation of an amplifier, cross modulation of intelligence can result.

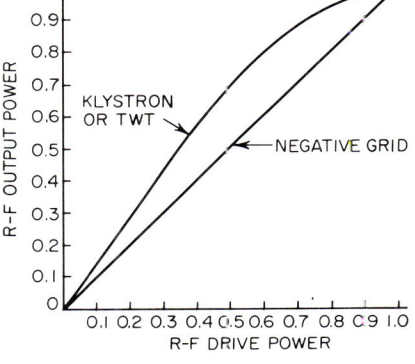
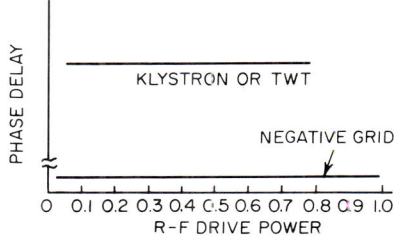

Fig. 14. Approximate amplitude and phase linearity characteristics.

The negative-grid tube can, with proper operation, provide excellent linear-amplifier characteristics while still operating at a high percentage of its maximum power-handling capabilities. Klystrons and traveling-wave tubes are also used as linear amplifiers, but since their saturation curves (input-output) characteristics approximate a Bessel function of the first kind and first order, linear operation limits the tube power-handling capabilities to a level consistent with the allowable level of intermodulation by-products. Approximate characteristics are shown in Fig. 14. In many cases this causes the overall efficiency of these tubes to be quite low. Some forms of crossed-field amplifiers are capable, at least in theory, of linear-amplifier operation. This type of amplifier is typified by the distributed-emission magnetic

[3] Terman, op. cit.

amplifier (Dematron),[4] which in addition is characterized as possessing anode current vs. r-f drive operation similar to a class B negative-grid amplifier.

3.6. R-F Input/Output Phase Characteristic. In an ideal amplifier the phase delay between the input and output should remain constant as a fixed-frequency drive signal is changed in amplitude. Practical microwave amplifiers depart from this characteristic to some degree, apparently most seriously at small-signal and at large-signal drive levels. This area is one in which very little data seem to be available, however, with some of the modes of modulation used in conjunction with linear amplifiers, degradation of performance can result. As with the phase data described previously, measuring techniques are difficult in this area, but presently more emphasis is being placed on improved instrumentation.

3.7. R-F Output Amplitude vs. Load Characteristic. All microwave power amplifiers should be operated in a manner which will allow them to "look into" their optimum r-f load impedance. It is generally assumed that a given tube will deliver its power into a coaxial or waveguide transmission line whose characteristic impedance is mated to the output characteristics of the tube. It is further assumed that the load at the receiving end of this line has a value equal to the characteristic impedance of the transmission line and that the load is nonreactive. In actual practice this exact load-matching situation is quite difficult to achieve, and as a result some maximum transmission-line voltage standing-wave ratio (VSWR) is usually specified by the microwave-amplifier manufacturer. In addition to the fact that the power delivered to the load will be lower when the load is mismatched, it is important that the consequences of the power reflected back down the transmission line be taken into account. In some cases this may lead to amplifier tube instabilities, or even permanent damage to the tube structure.

At least three conditions can arise from excessive VSWR. The first is simply that excessive voltage may appear across the transmission line at certain points along its length, and this excess may cause actual arc-over on the line. This can cause damage to the line, but more important, line arc-over will cause a total reflection of the forward power back down the line and into the amplifier. The second condition concerns the "window," or the isolating medium between the vacuum of the tube and the transmission-line dielectric. This window barrier is a critical element in all microwave tubes, and particularly so in high-power tubes. Excessive VSWR, particularly that due to line or load arcing, can puncture the window, rendering the tube useless. The third condition concerns the effect of reflected power on the tube's distributed-constant circuits. Some tubes, notably the TWT and crossed-field amplifier types, will allow the reflected power to pass directly down the slow wave structure and back to the drive connections to the tube. To prevent this problem from reaching serious proportions, internal isolators are inserted in the structure to attenuate this level sufficiently to prevent both instability and damage to the tube, provided the VSWR is held below some specified maximum. For these and other reasons it is important that careful consideration be given to the load-matching conditions associated with microwave amplifiers.

3.8. R-F Output Phase vs. Load Characteristic. From the above discussion of the effects of reflected power on the operation of many microwave amplifiers, it is apparent that the phase characteristics of such amplifiers are also affected by changes in the amplifier's load. As with other discussions of phase characteristics, this case is no exception in that meaningful data are difficult to come by. This is due to both the highly complex nature of the analysis of the internal functions of most microwave amplifiers (using a profusion of distributed-constant circuits) and the difficulty of achieving reliable measuring techniques in evaluating tubes from an experimental standpoint.

In general, the negative-grid amplifier and the klystron amplifier are less subject to adverse phase changes since both types have very little coupling between the input and output circuits. For this reason the phase-vs.-load characteristic is largely determined by the reaction of the reflected power on the output distributed-constant

---

[4] Trade name of Litton Industries, San Carlos, Calif.

circuits; hence these tubes yield to analytical treatment in some cases. In the case of traveling-wave tubes and crossed-field amplifiers, analysis becomes quite complex. One effective way in which the consequences of reflected power are minimized is by the use of an external load isolator. These devices take many forms, one conventional form being the ferrite isolator. They function to present a very small amount of attenuation to the forward passage of power down a transmission line and a very high attenuation to reflected power.

3.9. R-F Harmonic Output Characteristic. All microwave amplifiers are subject to the production of power output at harmonics of their fundamental frequency. In most cases it is important to be aware of this effect and have knowledge of the relative magnitude of the amplitudes of prominent harmonics in relation to the amplitude at the fundamental frequency of operation. In many cases it is imperative to select tubes having low harmonic output, and then further reduce this output by means of transmission-line line filters, if interference to other services and functions is to be avoided. This is becoming a consideration of increasing importance at microwave frequencies, because of the greatly increased use of this region for varied applications, many of them operating at extremely high power levels.

In general, broadband power amplifiers produce the highest level of harmonic output content. As noted previously, the second-harmonic content of many traveling-wave tubes may be down less than 20 db from the fundamental. In fact, it is probably safe to say that most current microwave amplifiers have second-harmonic output content less than 40 db down from the fundamental. As a result it is most important to investigate any interference requirements which may exist and correlate them with the harmonic characteristics of the amplifier being selected for a given application.

3.10. Noise Output Characteristic. The molecules, atoms, and free charges comprising a resistor are in continual random motion, the energy of this motion being the thermal energy of the resistor. The rms value, in volts, of the frequency components of this noise lying in any given bandwidth Δf is given by the formula $E_n = 4kTR \Delta f$, where k is 1.37×10^{-23}, T is the temperature of the resistor in degrees Kelvin, R is the resistance in ohms, and Δf is the bandwidth in cycles per second. Since all amplifiers have, in effect, some sort of resistor at their input, even an ideal amplifier will produce amplified noise at its output terminals. In addition, all practical amplifiers produce noise within themselves. For instance, with a negative-grid tube, shot noise is encountered because electrons are not emitted in a perfectly uniform manner. This results in irregularities in the electron stream, which, when it impinges on the anode, gives rise to an effective noise current. The components of this shot noise are distributed uniformly in frequency in the same manner as the thermal-agitation noise occurring in a resistor. When an additional grid (such as a screen grid) is inserted in a negative-grid tube, it is subject to shot effect due to current interception, and because of the resulting irregularity in the division of current between the two electrodes this in turn causes an apparent increase in the shot noise observed at the plate. This results in tetrodes usually being noisier than triodes. A third effect noted, induced grid noise, is caused by the flow of noise currents induced in the grid circuits of the tube.

Most microwave amplifiers are subject to the above internally generated noise effects in addition to others. Since cascaded amplifiers are usually the case in microwave amplifier complexes, and since the output signal-to-noise ratio is determined largely by the noise introduced by the first amplifier stage, it is therefore important that this stage have the smallest possible noise and the largest possible gain.

Another source of noise output, especially in radar transmitters, results from incomplete cutoff of electron emission by the control elements within the amplifier tube. With very high powered transmitting tubes, a leakage current of only a fraction of a microampere can give rise to appreciable noise output during the interpulse period of operation. With the advent of extremely low noise figure receivers, this noise source must be reduced to negligible proportions if the advantages of such receivers are to be fully utilized.

3.11. R-F Peak-power-output Characteristic. In pulse operation of r-f amplifiers, the peak power output is defined as the average power during a pulse. The pulses are assumed to be repeated at a regular pulse repetition rate, and the average power output is defined as the average power over a pulse repetition period.

Microwave tubes designed specifically for high peak-power-output capabilities (usually for use in radar and similar applications) are often characterized by cathode emission which is sufficient only for the pulse duration of a single pulse without a rest period (interpulse time) before the next emission period, circuit insulation and placement of elements consistent with the higher voltages associated with high peak-power operation, and (in the case of high-power tubes) the generation of harmful X rays which must be shielded for personnel protection. From these and other considerations which often affect the design of such tubes, pulse-rated microwave tubes are often ineffective when operated at other than their rated conditions, even though the relatively high average power output ratings of some tubes may indicate they have CW capabilities.

3.12. R-F Average-power-output Characteristic. Pulse-rated microwave amplifiers have both a peak and average power output rating, the latter not necessarily indicative of CW capabilities in the tube. CW rated microwave tubes, in general, have only average power ratings, although they are occasionally rated for intermittent peak-power capabilities up to several times their average power for short periods. This increased rating is usually much lower than the ratio usually associated with peak and average power ratings in pulse tubes. In some cases, particularly with negative-grid tubes, a fairly high peak-to-average power ratio is established for primarily CW tubes of special design. Usually this is associated with designs having increased anode insulation paths.

3.13. Pulse-duration and Duty-cycle Characteristics. The duty cycle of a pulse tube is defined as the on time divided by the on-plus-off time, or the average power divided by the peak power rating of the tube. It is easy to be misled in this regard since often in condensed listings of tube characteristics, the pulse-duration limitations are not indicated. Since many microwave tubes are designed for short pulse durations at very high pulse repetition rates, it is sometimes assumed that the same tubes can be operated at long pulse durations and low pulse rates. This is usually not the case, since the cathode-emission capabilities of such tubes are often inadequate. The manufacturer can usually provide information on the maximum pulse-duration capabilities in such cases.

3.14. Change of Characteristics with Time. Perhaps the greatest contributor to the change of characteristics observed in microwave tubes as a function of time is change of cathode emission. This characteristic of aging is quite important when tubes are operated in parallel and in similar applications where two or more tubes should retain identical characteristics. It appears hard to predict accurate depletion curves since such data can be obtained only with large sampling lot tests and closely controlled production techniques. Currently much emphasis is being placed on this and related characteristics, and it is anticipated much better information will be available in the future.

Another characteristic which varies with life is the presence of gas in tubes. Again this is often a function of manufacturing techniques and types of getting materials introduced. On the larger microwave tubes, ion pumps are sometimes incorporated. These devices, installed as a part of the tube wall, act on the vacuum within the tube and absorb internally released gas to ensure a high degree of vacuum. They are usually operated from a small power supply external to the tube.

A third change of characteristics possible as a function of time is detuning of the distributed-constant circuits of the tube. This can generally be avoided if the thermal environment of the amplifier is carefully controlled.

3.15. Reproducibility Characteristic. This characteristic pertains to the degree to which the first tube of a given type can be made to duplicate all subsequent tubes of the same type with regard to all its electrical and mechanical characteristics. This is primarily a quality-control problem. It is currently becoming a problem of the utmost importance because of the use of large numbers of presumably identical tubes in phased-array systems.

In a phased-array transmitting system for instance, a large number (often several thousand) of relatively small tubes are spaced on approximately half-wavelength center lines, each tube delivering its output to some form of individual radiator. By proper phasing of the drive signals to this wall of tubes, an antenna beam can be formed and positioned in a specific direction. When the individual phases are changed to the tubes in an appropriate manner, the beam can be directed to any of a very large number of positions. Identical amplifier complexes are required for most (if not all) of the individual positions. This type of operation emphasizes the need for tubes with highly reproducible characteristics. Furthermore it emphasizes the need for tubes with the least possible changes in characteristics with time. As a result of the work now going on in this field of endeavor it is anticipated that significant advances in the state of the art will come about in the near future, and that more meaningful reproducibility data will be available on all microwave tubes.

3.16. Summary. The preceding 15 areas relating to the characteristics of microwave tube are by no means complete. Each application may dictate additional areas where information of both an electrical and a mechanical nature must be obtained. Such characteristics as gain, efficiency, size, weight, type of cooling required, and many others may be of major importance in one application and of little significance in others. As pointed out earlier, it is important that all available data, related to the areas of potential importance, be investigated at the outset of a microwave-amplifier design program, if the most effective end result is to be achieved.

4. TYPICAL MICROWAVE AMPLIFIER COMPLEXES[5]

The three preceding sections have briefly described the nature of typical active microwave-amplifier elements, the functions these elements are commonly called upon to perform, and the characteristics these elements must possess as a result of the functions they are called upon to perform. When the overall complex of which the amplifier forms a part is considered, a number of additional active and passive components assume prominent degrees of importance with regard to the overall performance of the complex. This was brought out in some detail by the complex described in Sec. 1.5 and by references to other components in the two preceding sections.

The purpose of this section is to describe several hypothetical (although typical) power amplifier complexes as a means of illustrating the basic components which form a part of such arrangements. The active microwave components such as the amplifiers, the oscillators, the frequency translators, and the frequency multipliers form the heart of any such complex. A second class of microwave components, the transmission-line elements, such as coaxial and wave-guide transmission lines, directional couplers, attenuators, isolators, dummy loads, power splitting and combining elements, duplexers, filters, rotary joints, and horns provide the means of transporting, monitoring, dissipating, and directing the power generated by the active elements in such a way as to allow the required useful function of the complex to be accomplished. A third set of components (usually nonmicrowave in nature) vitally linked to the performance of microwave amplifier complexes, is composed of the power class, namely, the power supplies, the modulators, and the pulsers, required as the external energy source for the amplifier elements. A fourth class of components (also usually nonmicrowave in nature) those relating to the protection, monitoring, and control of the three previous classes of components, also form an inseparable part of every microwave amplifier complex. This section will illustrate how the above classes of components are often employed. Later sections will describe in greater detail the characteristics and applications of the various components.

4.1. A Typical Single-output-phase Power Amplifier. The term single output phase is used to describe an amplifier complex in which a single effective phase delay exists between the exciting (or driving) function and the output function of the complex, when operating at a given frequency. This is in contrast to phase-array arrangements where multiple phases exist at the outputs of the complex when compared with the common driving function.

[5] Work for this material supported by Rome Air Development Center under Contract AF30(602)-2654.

Figure 15 illustrates a single-output-phase complex which frequently arises in practice. As is often the case in the microwave field, the demand required at the output of a microwave transmitter is often in excess of that available from a single tube. For this reason several tubes are required to supply the necessary power output. In the example shown, it is assumed that the requirements were such that four klystron amplifier tubes were necessary. If the nature of the individual transmission-line components is passed over for the moment, it can be seen that the r-f drive enters at the left of the figure (r-f is shown by the heavy lines), and this drive is split two ways. The two-way split is then followed by a second two-way split which results in four "in-phase" drive sources for exciting the four klystron amplifiers. Following amplification in the four tubes, the output power of the upper two and the lower two tubes is independently combined, and then the outputs of these pairs are again combined to form the single-output power phase. This power in turn passes through several more transmission-line components to eventually provide r-f radiation at the antenna.

By taking a second trip through this r-f path, consider the functions of the transmission-line components. Starting with the r-f drive, the first component encountered would be some form of transmission line, usually of the flexible coaxial type if the power is low and the runs are short. With klystrons, it is important not to underdrive or overdrive the tubes; hence an attenuator is the first element introduced into the line, so that optimum drive power can be supplied to the tubes. The second element is an isolator, a device which introduces low losses to the passage of forward power to the klystrons but which introduces a high attenuation to the flow of reflected power. This device is often necessary since the input cavities of klystrons do not always present a good match to the transmission-line elements; hence an appreciable amount of drive power may be reflected, often with adverse effects on the stage providing the r-f drive. The next element indicated is a three-port transmission-line element (power splitter 2) used to split the drive power. One of the two output legs of this splitter has in it a phase shifter and attenuator element. This element is employed to compensate for small differences in the splitting ratio of the three-port device, and small differences in the characteristics of the lines running to power splitters 1 and 3. These two power splitters, in turn, have one output leg equipped with phase-shifter and attenuator combinations to ensure that exact phase and amplitude relations may be accomplished at the drive terminals of the four klystrons. Assuming the four klystrons have identical amplification properties, their outputs are now connected to four directional couplers (usually by means of waveguide transmission line because of its lower loss and greater power-handling capabilities), which function to allow a small calibrated sample of unidirectional energy flow to be taken out of the guide for monitoring purposes. Often two directional couplers are employed at this spot to allow monitoring of both the forward and the reflected power to determine the transmission-line VSWR. From the output of the directional coupler elements, waveguide runs are indicated to power combiners 1 and 3. These are four-port devices and function in such a way that if the power output from the pair of amplifiers is equal and of proper phase, the combined output will appear along the heavy-line output path. If this condition of equality does not exist some output will be delivered to the dummy loads shown attached to each combiner. Power combiner 2 functions in like fashion with the two outputs from combiners 1 and 3 to deliver output from all four klystrons to an additional directional coupler (or pair of same) to monitor the combined single-phase output of the complex. Up until this time the concepts pointed out would be essentially the same had the complex been for a communications transmitter or a radar transmitter. The diagram of Fig. 15 is indicative of a radar transmitter; hence the next transmission-line element is a duplexer which functions to deliver the transmitted power upward to the antenna with a minimum of leakage downward to the receiver input circuits. Since most microwave power amplifiers have rather high harmonic content, a transmission-line filter is customarily employed to attenuate these radiations. The next element, a transmission-line rotary joint, functions to transform from conventional rigid waveguide to some model allowing rotation, since practically all radar and many communications antennas are required to direct their

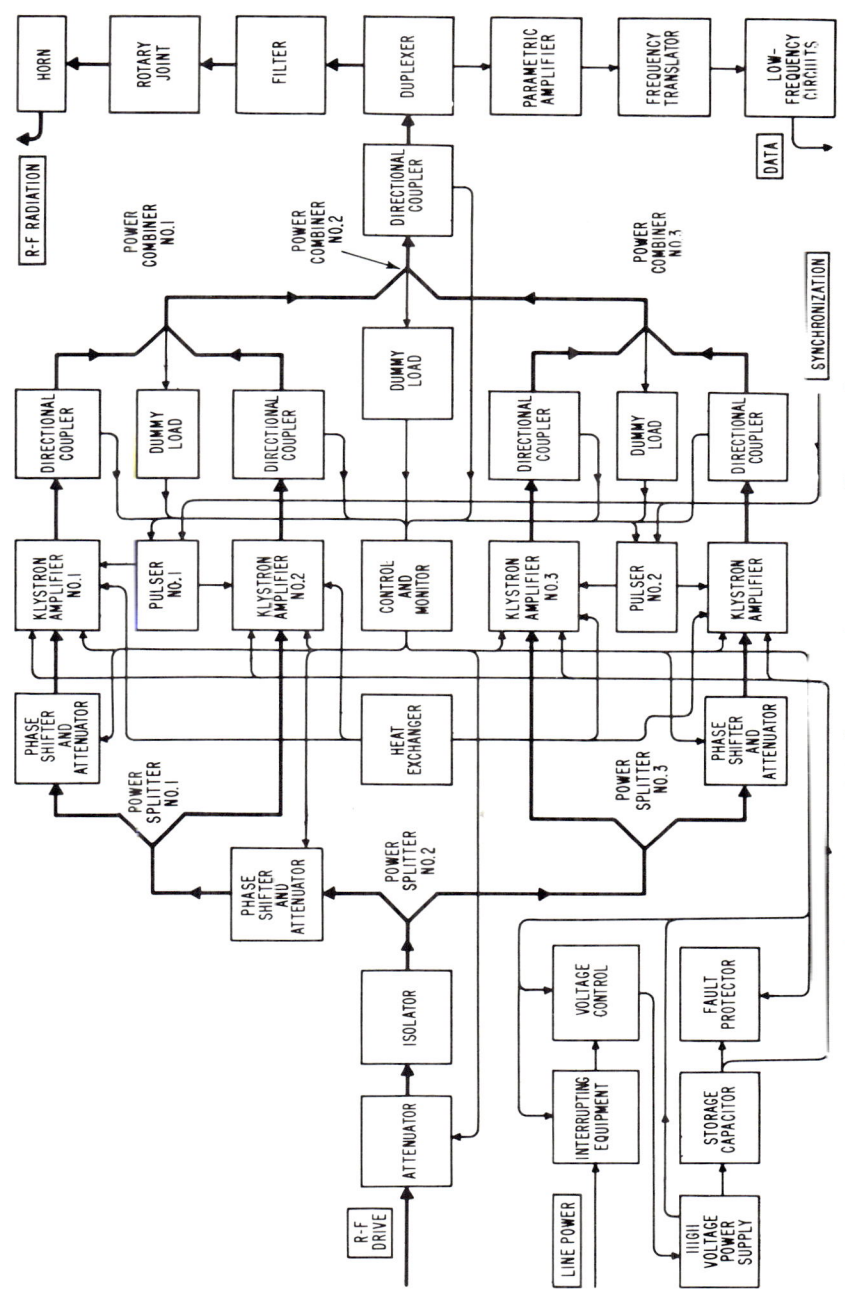

FIG. 15. A typical single-output-phase power-amplifier complex.

narrow-beam radiation in many specific directions. The final transmission-line element is a horn, a device which functions to match the final sections of the transmission line to the atmosphere. Generally this horn directs the radiated energy against some form of parabolic reflector to form a narrow beam of transmitted microwave energy. In the radar mode described here, the pulse thus transmitted will, after reflection against distant objects, have minute portions returned (after varying time delays) to the antenna. These echoes will return through the final transmission-line elements to the duplexer, which will now direct them into the receiver elements for amplification and data processing to provide the necessary data output.

The next class of components to be considered is the power-supply and pulser elements. It will be noted that between each set of klystron amplifier tubes a pulser block is indicated. It is assumed in this case that the klystrons are of the type incorporating modulating anodes. The modulating anode is an element which allows operation similar to positive control-grid action; that is, when the modulating anode is at cathode potential the beam current of the klystron is cut off, and when the modulating anode is pulsed to a predetermined positive voltage level with respect to the cathode the beam will be turned on fully. The pulsers perform this function in response to the synchronizing triggers shown entering each block. In addition a control signal is shown entering each block for the purpose of adjusting the level to which the modulating anode is pulsed. The high-voltage-beam power supply is shown in the lower portion of the figure, and this supply delivers the necessary beam voltage to all four klystrons. The supply is usually a three-phase affair which minimizes the amount of ripple filtering required across its d-c output terminals. In pulse service a rather large filter capacitor is often employed for a somewhat different reason. This capacitor, called the storage capacitor on the block diagram, usually has a value of approximately $I_b\tau/\Delta V_b$ µf, where I_b is the total beam current (in amperes) drawn during a pulse, τ is the pulse duration (in microseconds), and ΔV_b is the allowable droop in beam voltage (in volts) during a pulse. The high peak pulse current is thus drawn from the capacitor during the pulse period, and the power supply then recharges the capacitor during the interpulse period. Were it not for this type of operation severe loading of the power lines would result during the pulse period with the probable result of adverse regulation effects being imposed on other equipment supplies from the same power lines.

The fourth class of components indicated is those associated with the functions of control, protection, and monitoring. A block is indicated at approximately the center of the diagram to indicate the control and monitoring position for a power amplifier complex of this type. Considering first the control function, it will be noted that actions are indicated for each of the klystron amplifiers. If tunable klystrons are employed these control functions would consist of the tuning-motor drives for adjusting the individual klystron cavities for proper operation. Lines are also shown going to each phase shifter and attenuator to allow remote adjustment of each of these functions from the central control and monitoring position. Control lines also lead to the voltage control (usually a variable transformer or inductrol for adjusting the primary voltage to the high-voltage rectifier transformer of the high-voltage power supply) and line-interrupting equipment (usually a large line contactor or circuit breaker) to allow remote control of these power elements at their remote positions. As far as monitoring functions are concerned, each of the directional couplers would provide signals in calibrated relation to the r-f power in each portion of the waveguide circuit being checked. These readings are of the utmost importance both in the initial adjustment of the klystrons and for checking subsequent drift from the initial tuneup conditions. In addition each of the combiner dummy loads provides an indication to tell how well the combining action is taking place. Although not specifically indicated, several other monitoring points are usually provided on each klystron to indicate tuning adjustments and temperature variations and to check the "optics" of the electron beam through the tubes. The high-voltage power supply will of course provide monitoring indications of its voltages and currents. In the area of protection, most of the monitoring functions will assume a protective role when the function being monitored exceeds some normal limits. This may be simply the actuation of the main circuit breaker, to remove all power in the most severe cases. In less severe cases where a gradual change of function being monitored is taking place, manual or automatic control functions may take over; for instance, if the output of klystron 1 started to drop off with time as indicated by the directional coupler at the output of that tube, adjustment of the phase shifter and attenuator at the input of that tube might manually or automatically take place to correct the condition. One power-protective feature should also be noted. That is the fault protector shown connected across the storage capacitor of the high-voltage power supply. This device often consists of a large gas thyratron or ignitron which shorts out the storage capacitor in the event one of the klystron amplifiers should arc over internally during operation. Such an arc could be highly destructive to the tube (because of the large amount of energy customarily stored in the capacitor) unless the capacitor energy is diverted by an alternate path. The necessary monitoring circuits detect such arcs and fire the gas-tube short circuit in such cases. This of course makes it necessary to open the line-interrupting equipment to clear the overall fault condition.

While the above description of the components associated with a high-power four-tube amplifier complex is by no means complete in all details, it serves to illustrate the major considerations involved. Several others are of importance, however. A heat exchanger for circulating coolant to the tubes is usually an important and complicated consideration in such complexes. From the protective standpoint the considerations associated with shielding the klystrons to prevent X-ray radiation can be of vital importance when high-voltage tubes are employed. As noted earlier, additional control and monitoring functions would undoubtedly be incorporated to ensure reliable and "fail-safe" operation of the complex. These considerations and others can and must be investigated thoroughly at the outset of any proposed design of this type in light of the considerations involved with the specific application at hand.

4.2. A Typical Phased-array Complex. The term phased array is used to describe an amplifier complex in which a number of phase delays exist between the exciting (or driving) function and the outputs of the complex. Transmitter complexes of the phased-array type are employed to form an effective antenna beam without the aid of the usual forms of parabolic reflectors normally employed at microwave frequencies. This is accomplished by positioning a large number of independent power-amplifier outputs across a plane area with the horizontal and vertical center lines of these amplifiers spaced at approximately one-half wavelength intervals. The principle involved is much the same as that used in spacing the elements of phased-array antennas at lower frequencies, except that in this case the elements usually become individual transmitter modules operating into individual radiators, and the phase delay between elements is controlled by means of variations in the relative drive phases applied to the individual modules.

Figure 16 illustrates in block-diagram form a simplified multiple-output-phase (phased-array) transmitter complex. Typical phased arrays may require from several tens to several thousands of individual elements, depending on the application. Quite often, for purposes of simplicity, all the amplifiers operate at identical power outputs; however, they may operate at tapered outputs across the face of the array as a means of controlling the beam shape.

The main r-f paths of Fig. 16 are shown by the heavy lines. It will be noted that the common r-f drive signal for the array enters and passes through a phase-delay steering element. This device introduces a predetermined phase delay in the individual drive transmission lines going to each module, for a given antenna pointing direction. It is important to note that this delay must account for the variations in length of the lines running to the various tubes as well as variations in the characteristics of these lines with frequency and with time. Assuming that each module is supplied with drive power of the proper phase, and passing over the transmission-line components of each module for the moment, it can be seen that each module consists of two active microwave-amplifier elements. These are a first-stage amplifier using a traveling-wave tube and a second-stage amplifier using a crossed-field device. Following amplification in each module, the individually phased r-f outputs are fed to individual horns for radiation in the total form of an antenna beam.

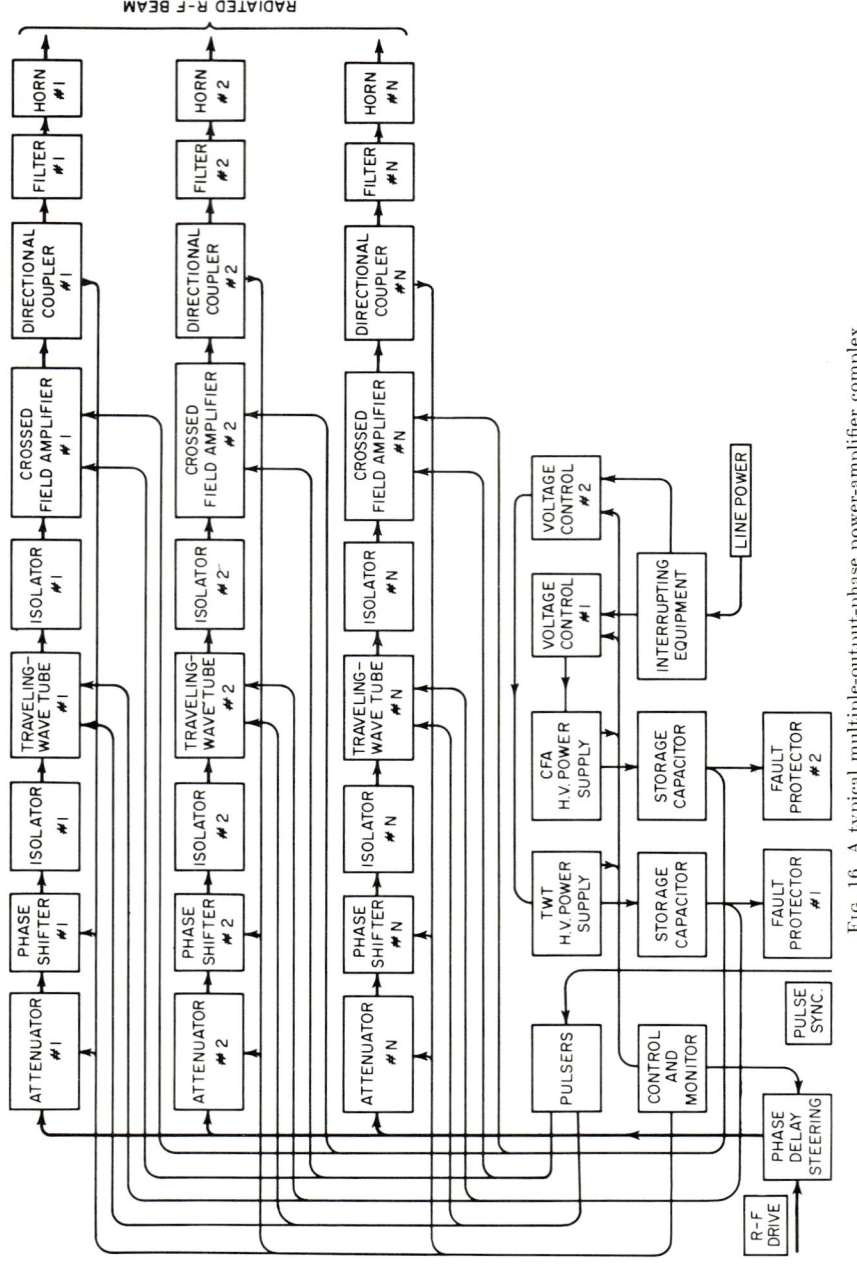

Fig. 16. A typical multiple-output-phase power-amplifier complex.

By taking a second trip through the r-f paths, consider the functions of the transmission-line components. It will be noted that the first element is an attenuator. Since the first amplifier stage is a traveling-wave tube, an attenuator will be required to set the drive level of this stage to its proper point on its saturated drive curve. This is necessary since it is likely that the drive transmission lines would be of unequal lengths, and even if the TWT's had identical drive characteristics, the attenuations introduced by the drive lines would be unequal. The next element noted is an individual phase shifter for each module. In theory, if ideal and identical elements were used to fabricate each module, this device would not be necessary. However, since this situation has not yet been realized in practice, each module would be preadjusted to a definite and identical phase length, before installation in the array, by means of this phase shifter. The next element, an isolator, serves the function of inserting a very low attenuation in the line delivering drive power to the TWT, but a very high attenuation to any reflected power coming back down the line as a result of a mismatch at the input of the TWT. Following the TWT amplifier a second isolator will be noted for the same purpose. In both cases the isolators function to reduce the reflected power to their respective driving stages to ensure the most stable form of operation to the overall amplifier chains. Following the second amplifier stage directional couplers are indicated in the line. These perform the dual roles of monitoring the output power level and the reflected power level. This allows measurement of power outputs of each module and also the VSWR into which the second amplifier operates. In some cases, if this VSWR is too high for the crossed-field amplifier employed, a third isolator would be employed in the output line. The next element noted is a transmission-line filter. Since both traveling-wave tubes and crossed-field amplifiers are generally broad-bandwidth tubes, and since broad bandwidth is customarily associated with rather high harmonic output, this filter would perform the function of attenuating these harmonics to prevent interference with other microwave services. Following the filter the amplified r-f output would go to a horn whose purpose is to match the transmission line to the atmosphere and radiate the power generated by the module.

The next class of components to be considered is the power-supply and pulser elements. It will be noted that two high-voltage power supply and storage capacitor assemblies are indicated, one to supply the TWT driver-stage beam-accelerating voltages, and one to supply the crossed-field amplifier anode voltages. If a very large number of modules were employed it is likely that more than one set of supplies would be used since the redundancy would ensure higher reliability in the event of a failure of one or both of the supplies. Since the modules of the phased array are generally fabricated as small "plug-in" assemblies (for easy maintenance) it is highly desirable to employ amplifier tubes utilizing the lowest possible electrode voltages (to simplify the quick-disconnect features of this approach). The considerations associated with the storage capacitor are similar to those noted in Sec. 4.1. Both amplifier tubes indicated for the individual modules employ control electrode pulsing. Low-power TWT stages are often provided with grids; hence a pulsing scheme similar to that noted in Sec. 4.1 would be appropriate for this application. A common pulser could be utilized to supply the necessary pulse duration to a large number of modules since the power required for grid control is quite small. The second-stage amplifier could employ a crossed-field tube possessing two different forms of control-electrode pulsing. One would consist of an electrode similar to a modulating anode, and the second would employ an electrode which would orient the tube into an amplification mode by means of applying r-f drive power. In the second case the tube would return to its nonamplifying state following the application of a single spike of voltage to the control electrode, whereas in the first case this would occur when the grid pulse amplitude was removed. With properly rated amplifier tubes it would of course be possible to operate a phased-array arrangement for some communications modes without recourse to the pulsers just noted.

The fourth class of components indicated is those associated with the functions of control, protection, and monitoring. Most of these functions are identical to those noted in Sec. 4.1 except for the difference in power levels involved between the two complexes. The differences are largely associated with the control of phase on the individual modules. As an example, it may be desired to provide servo control of the phase delay on an individual module basis by means of sampling the input and output of a given module and correcting for changes by means of the module phase-shifter element. While this may be possible, it can be seen to involve an extremely complex situation fraught with cable lengths, thermal characteristics, servo drive speeds, and many other considerations.

The above considerations include only a few of the major features involved with an actual phased-array system design. The nature of the phased-array receiver considerations is similar although not identical. A separate receiving array is often employed and a rather complex scheme of synchronizing these two functions is thereby involved. In any event the emphasis in phased-array complexes is placed heavily upon the area of obtaining identical microwave components. From these efforts much more accurate and useful information will undoubtedly result, with time, to advance the state of the art on virtually all the components associated with microwave amplifier complexes.

4.3. A Typical Multipurpose Power Amplifier Complex.[6] The term multipurpose is used to denote a power amplifier complex which is capable of both communications and pulse-transmission modes. In general, communications and pulse transmitters are viewed as separate and distinct devices; however, it has been shown to be quite feasible to combine these two functions into a single power amplifier complex. This sort of combination can provide a distinct advantage, for many applications, in terms of equipment simplification. For instance, it may be possible to combine the modes of radar tracking of, and communication with, a distant moving object by means of a single installation rather than employing two separate installations.

A number of possible configurations are realizable by using various types of microwave-amplifier tubes, provided that too wide a range of communication modes is not attempted. Perhaps one of the most versatile forms of this concept is shown in Fig. 17. Observing first the main path of r-f amplification as noted by the heavy lines, a configuration similar to that described in Sec. 1.5 will be observed. A notable difference with this arrangement, however, is that negative-grid power-amplifier tubes have been employed for the two final power-amplifier stages. If very high power output is required, this choice may limit the maximum frequency of the complex to the lower microwave ranges, but the use of negative-grid tubes allows the advantages of both linear-amplifier operation and high efficiency per stage.

Of particular interest is the low-frequency driving function of the frequency translator. It can be seen that three separate video inputs must be handled by a multipurpose transmitter, namely, radar synchronization, communications video, and some sort of time-sharing signal to determine which will take precedence. This function is performed by the time-sharing circuits block of Fig. 17. The output of this block will provide either the radar synchronization or communications video to the modulator circuits block. In addition, in the case of radar synchronization taking precedence, synchronizing triggers will be sent to the high-level pulsers. In either case, the modulator circuits will provide the necessary modulation waveforms to modulate properly the 30-Mc amplifier, which provides the low-frequency drive to the frequency translator. The particular value of 30 Mc has no particular significance other than this being a typical intermediate frequency for radar receivers, and further, it is an adequate and convenient frequency at which to perform the modulation functions associated with most communications modes. In any event, the filtered output of the frequency translator will be either a communication or pulse spectrum at the required output frequency of the complex. This spectrum will then be amplified by the two subsequent linear power amplifiers to the required output power level. All other r-f functions are similar to those noted previously.

The power and pulse components perform in a considerably different manner from those described in previous complexes. Consider the first power-amplifier stage

[6] Work for this material supported by Rome Air Development Center under Contract AF30(602)-2654.

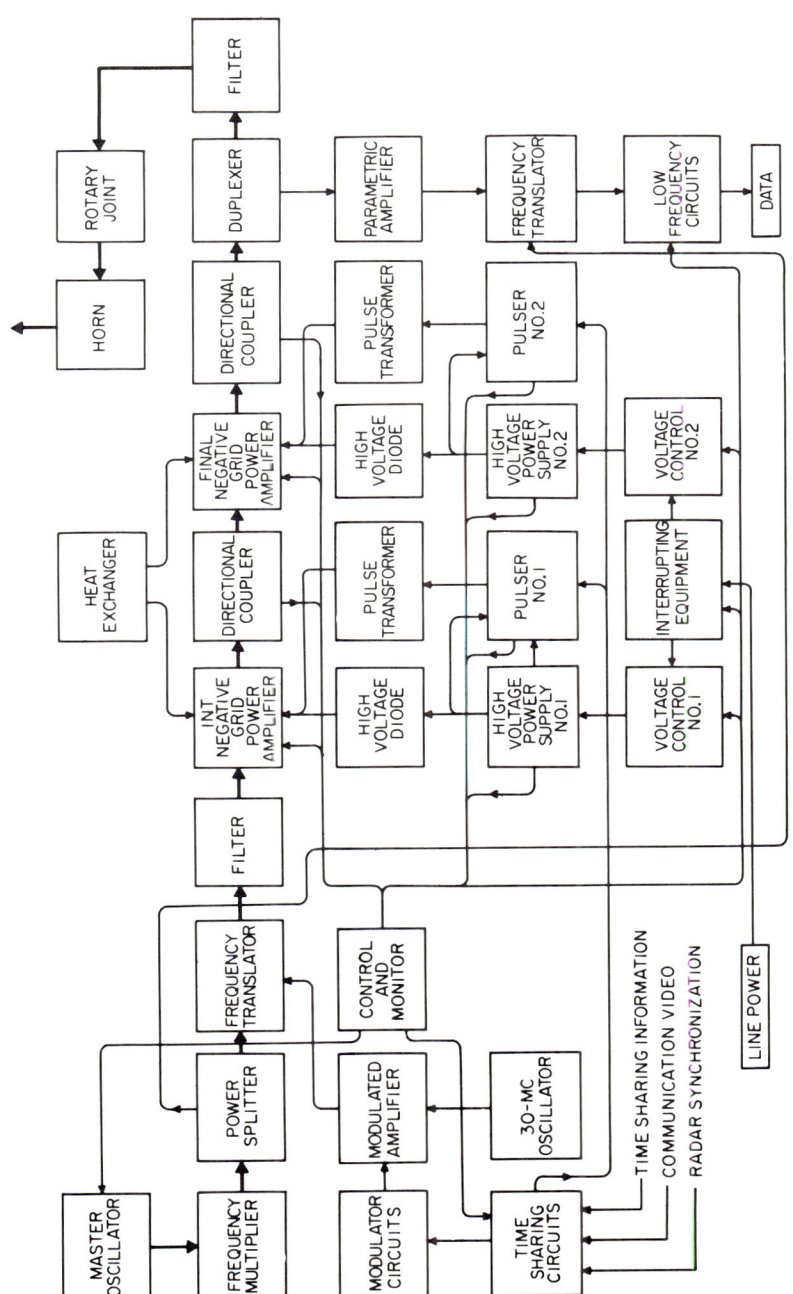

Fig. 17. A typical multipurpose power-amplifier complex.

27–35

(intermediate negative-grid power-amplifier block). When the time-sharing information selects the communications mode, high-voltage power supply 1 will supply anode voltage to this stage through a high-voltage blocking diode. Similar action will take place with the final negative-grid amplifier stage and high-voltage power supply 2. It is assumed that these two power supplies are set to provide the proper anode potentials to these tubes for the power outputs required of the communications modes. It will be recalled from an earlier discussion that some negative-grid tubes can operate at much higher anode voltages for pulse conditions than for CW conditions. Advantage is taken of that fact when the time-sharing information selects the pulse mode. It will be noted that for this mode the same high-voltage power supply as used on the communications modes now becomes the power supply for the pulser. Furthermore, the pulse thus generated (in response to the trigger from the time-sharing circuits) will be stepped up by means of a pulse transformer to provide a higher-voltage pulse to the negative-grid amplifier stage and hence provide a much greater power output level. This higher level is almost totally limited by the capabilities of the amplifier tubes involved. Similar action takes place in both stages. At the conclusion of the pulse output, the amplifiers are almost instantaneously returned to the condition receptive to communications transmissions.

The components associated with control, protection, and monitoring are in general quite similar to those described in connection with the previous two complexes. One difference exists in the receiver area. It will be noted that an interconnection is shown to the low-frequency circuits. It is assumed that the receiver can be used for both the reception of communications and radar information; hence, the most effective bandwidth of the intermediate-frequency amplifiers would be selected as a function of the spectrum being handled. A second function involves radar reception during communications transmission. If this multipurpose mode is employed, it will be necessary to blank the communications video during the known radar return period of operation. It is assumed that only a single radar target is being tracked during this mode; otherwise excessive perforation of the communications video might result.

The field of multipurpose operation has not received a great deal of consideration until recently, and as a result, a number of limitations may be noted from the above description. It is quite likely that all dual systems cannot be replaced by multipurpose complexes; however, due consideration of the possibilities of this basic technique may reveal considerable advantages in terms of complexes of reduced size, weight, and cost for many applications.

5. POWER-AMPLIFIER TUBES AND THEIR ASSEMBLIES

There are a number of basic similarities between all microwave power-amplifier tubes and their associated structures. Considering the three main categories of these components, negative-grid tubes, linear-beam tubes, and crossed-field amplifier tubes, all are similar in the sense that an external source of anode voltage is required, although this is often referred to as beam voltage or accelerating voltage. Most of the tubes are similar in that an external source of filament voltage is required, although certain types of crossed-field amplifiers have cathodes whose initial operating temperature is that of the room ambient temperature and, under some conditions, may actually require cooling of the cathode structure after continued operation. Unlike low-frequency amplifiers, many of the tubes require a magnetic field (generated external to the vacuum envelope) to control the path taken by the electron beam through the tube structure. As noted in the introduction to this chapter, by definition, all microwave tubes have distributed-constant resonant circuits; however, the forms these circuits take vary widely from type to type (and between versions of the same type), and depending on circumstances, the circuits may either be an integral part of the tube or a demountable (and reusable) part of the tube. Finally, many microwave tubes possess additional elements, such as grids, modulating anodes, and helix structures, which are brought out of the vacuum to allow some special control feature to be accomplished. It is the function of this section to point out some of the more important consideration associated with each major tube type. Of necessity, this discussion will be fairly general because of the wide variations encountered between tubes of the same basic type.

5.1. The Negative-grid Power Amplifier. In the microwave range, negative-grid power amplifiers enjoy a wide range of application up to frequencies of several thousand megacycles. In general, the achievable power output drops off appreciably as frequency increases. While peak power outputs of many megawatts and average power outputs of hundreds of kilowatts may be obtained from a single tube at frequencies up to 500 Mc, these values generally drop down to power levels of hundreds of watts and tens of watts, respectively, as the frequency of interest approaches 2,000 or 3,000 Mc. These figures are based on currently available tubes, and continuing development work on tubes for higher frequencies will undoubtedly result in devices with significantly greater power-generating capability. It is questionable if the tubes will ever become competitive, powerwise, with linear-beam and crossed-field amplifiers at the higher microwave ranges.

There are a number of characteristics which highly recommend the negative-grid tube for microwave power-amplifier service. Such characteristics as the capability of producing a high r-f peak power output with modest d-c anode voltage, an excellent r-f phase stability despite anode voltage variation, and excellent linear-amplifier properties are examples of such areas. Intrinsic interelectrode capacitance, present in all electronic devices, must be recognized as a limiting factor in applying this tube type to microwave-amplifier circuits.

In a multiple-stage amplifier chain, the overall bandwidth is a function of that of the individual stages. Cascading several stages shrinks the bandwidth from that of an individual stage by the expression

$$BW_n/BW_1 = (2^{1/n} - 1)^{1/2k} \qquad (14)$$

where n = number of cascaded stages
k = number of coupled tuning elements per stage (multipole filters)

For example, to design a two-stage chain employing single-tuned circuits and possessing an overall bandwidth of 50 Mc, each of the stages must be designed with a bandwidth of 500/0.643 or 78 Mc. Basically then, achieving broadband chains involves maximizing the bandwidth of each individual stage.

For a given tube, two figures of merit may be used to describe the tube. These are:
1. Gain-bandwidth product, which is

$$\frac{\text{Transconductance}}{2\pi(C_{input} + C_{output})} \qquad (15)$$

2. Power-bandwidth product, which is

$$\frac{I_{ac}^2}{2\pi(C_{input} + C_{output})} \qquad (16)$$

where I_{ac} is the fundamental a-c component of the allowable plate current. These two figures are relatively fixed numbers which are basic in the tube design. The bandwidth noted in each figure of merit is that of a single-tuned circuit attached to the tube.

UHF amplifier stages are generally run in grounded-grid circuits. As such, the input cathode circuit is usually wider in bandwidth than the plate circuit by a factor of 5 or 10. Thus the effects of the input circuit are ignored in all but the cases requiring extremely wide bandwidth stages, and only the effect of the plate circuit is normally considered.

Further examining a typical tetrode tube, for instance, the GL7399 and its figures of merit, we find:
1. C_{out} = 9.3 pf
2. C_{in} = 21.5 pf
3. Transconductance = 100,000 μmhos (pulse service)
4. I_{ac} = 13.5 amp rms maximum for 100-μsec pulse

Gain-bandwidth product = 515 Mc
Power-bandwidth product = 0.94 × 10⁶ (watt)(Mc)

Hence, in the theory, if a 10-Mc bandwidth is required, a gain of 51.5 and a power output of 94 kw could be achieved. These figures are theoretically independent of the operating frequency. When applying the power-bandwidth product to this specific tube, the peak allowable element voltages will limit the minimum bandwidths at which operation may be achieved, while driving the tube to its maximum 13.5 amp current level. These figures of merit may be approached in lower-frequency amplifiers where lumped constants are used and the stray capacitances are minimized. However, when the tube is used in microwave service, practical amplifiers require cavities. These cavities store r-f energy which raises the Q and decreases the bandwidth.

Thus the first approach to maximizing bandwidth in an amplifier with a given tube is to design the cavity to have a minimum of stored energy. The graph of Fig. 18 shows numerically how the stored energy within the cavity decreases the available bandwidth. For instance, to design a tube to operate at 900 Mc, where physically a ¾ wave cavity is required (because of the internal lead lengths within the tube), a tube having an output capacitance of 10 pf would have a capacitive plate reactance of 18 ohms. If the characteristic surge impedance of the plate were 40 ohms, $Z_o/X_c = 2.2$ and the theoretical bandwidth is decreased by a factor of 5. If cavity impedances are chosen

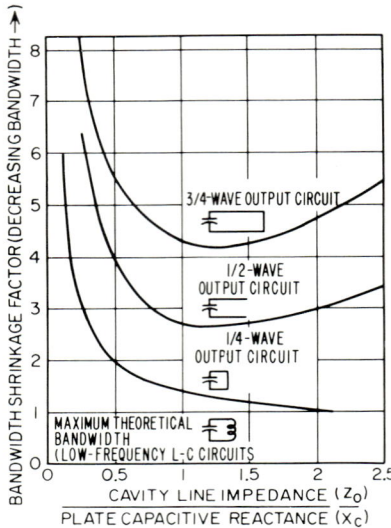

FIG. 18. Bandwidth shrinkage effects of a cavity vs. theoretical available bandwidth.

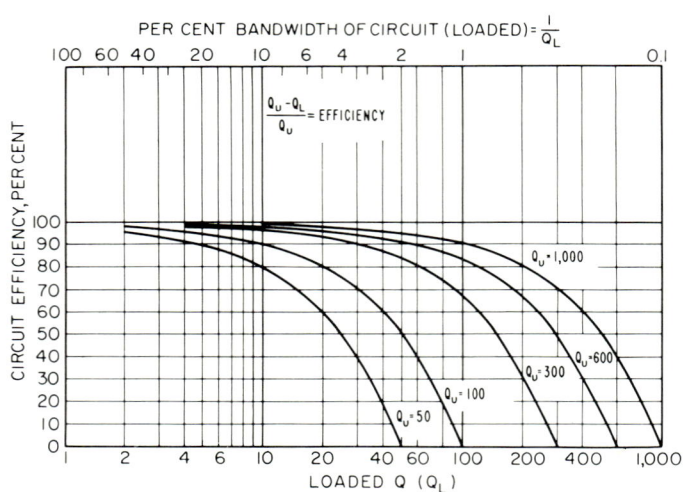

FIG. 19. Circuit efficiencies vs. circuit Q's.

properly during the design, the stored energy may be minimized and bandwidth maximized.

Another approach to bandwidth improvement is the use of multipole filter networks or multiple-coupled cavities for the resonant-circuit elements. For example, in a given amplifier which has a 1-db bandwidth of 20 Mc, simply changing from a single-tuned cavity to a double-tuned coupled cavity increases the 1-db bandwidth to 38 Mc.

The efficiency of an amplifier chain (excluding filament power) is a function of the following items:
1. Output plate-cavity efficiency
2. Operating parameters and gain desired
3. Gain of the final stage

The output plate-cavity efficiency is defined as the ratio of the power delivered by the tube to the plate cavity to the power delivered to the load. The graph of Fig. 19 illustrates how this circuit efficiency is affected by the bandwidths required and the unloaded Q's of the cavities As may be noted, in broadband circuits, the unloaded $Q(Q_u)$ has relatively little effect on the circuit efficiency. As the bandwidths approach 5 per cent or narrower, Q_u has a significant effect. Here is where cavity design may be optimized to maintain a high Q_u and thus maximum circuit efficiency.

To show how the operating parameters and gain affect the stage efficiency, recently measured values on a tetrode amplifier stage operating at 1,200 Mc illustrate these changes.

	Mode 1	Mode 2
Plate voltage, kv	8	8
Plate current, amp	9	9
Power to load, kw	39	28
R-F drive power, kw	5.3	2.2
Grid bias, volts	−80	0
Gain	7.3	13
Overall plate efficiency, %	54	39

These operating parameters readily illustrate the tradeoff between gain and efficiency. In mode 1, with a high grid bias, the angle of plate current flow is narrower and the efficiency is higher; however more r-f drive power must be applied to drive the tube into conduction and thus the stage gain is decreased. Of these various parameters, each must be examined in turn and a compromise value chosen which best suits the situation. If efficiency is a prime consideration, perhaps an extra amplifier stage is warranted to compensate for the loss in gain.

Finally, the gain of the final stage has an overall effect on the efficiency of a chain. Should the gain be 8 or 9, the efficiency of the driver stage has much more of an overall effect on the efficiency than if the final stage has a gain of 20.

Broadband amplifier performance can be optimized by the design of specialized tubes in which complete double-tuned cavity resonators and associated grid-controlled electronics systems are contained within a single vacuum envelope. RCA has pioneered this modular approach to superpower r-f amplification with its family of Coaxitrons, triode or tetrode devices comprising a completely engineered package capable of full-power operation over design bandwidths on the order of 10 per cent without tuning adjustments, either mechanical or electrical. Figure 20 shows a simplified cross-sectional sketch of a Coaxitron triode provided with coaxial input terminals for r-f driving-power, vacuum-contained double-tuned cavity resonators (with associated electronics systems), and a waveguide delivering the r-f output power.

Coaxitron stages can be cascaded. It appears inevitable, however, that the next step in a logical sequence of development will involve the incorporation of multistage Coaxitrons in a common vacuum vessel. Thus, by means of a "totem-pole" configuration it will be possible to cascade stages to optimize compactness and broadband performance. A "totem-pole" arrangement permits the transfer of energy from the output of a driving stage to the input circuit of a driven stage without the complexities of interconnecting link lines

27-40 AMPLIFIER CIRCUITS

5.2. The Linear-beam Power Amplifier. The linear-beam tube is a microwave amplifier which operates on the principle that the electrons in an electron beam may be velocity-modulated, by means of an externally injected r-f drive signal, to produce an amplified replica of the drive signal at the output of the tube. Klystron and traveling-wave-tube amplifiers are the two common forms of this tube type. These tubes are widely used in the microwave region as both low- and high-power amplifiers. Since the length of the tube becomes almost a direct function of the wavelength of operation, the tubes are not generally employed below about 200 Mc because of the excessive size proportions reached. As with the negative-grid tube, these amplifiers are subject to power output vs. frequency limitations; however, because of the principles involved

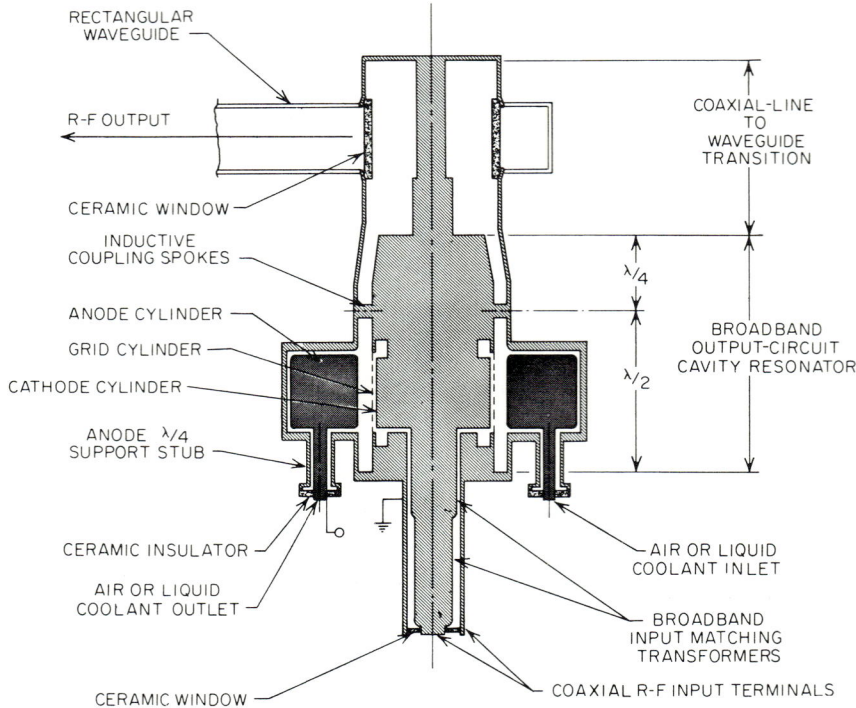

Fig. 20. Simplified cross section of Coaxitron. (*By permission of Radio Corp. of America.*)

in their structure, the limitations are not so severe at the higher frequency ranges. For instance, a factor is often expressed which describes the capabilities of these tubes in terms of PF^2, where P is the average power output in megawatts, and F is the frequency in gigacycles. A figure of 10 is probably representative of the 1964 maximum extension of the state of the art for a single klystron tube. This figure of merit by no means indicates that tubes up to this capability are available at any microwave frequency but rather that such tubes could probably be developed in a relatively short time period. It is somewhat more difficult to express a similar figure of merit for peak-power-output capabilities because of the effect of pulse duration on the determination of this characteristic. Values of about 50 megawatts have been achieved at around 3 gc and at relatively long pulse durations. Presumably, higher and lower values of this base number could be achieved as the frequency of operation is considered at lower and higher values, respectively.

MICROWAVE AMPLIFIERS   27-41

The high r-f power-handling and high-gain capabilities of linear-beam tubes are their main points of excellence. As compared with negative-grid tubes, these characteristics are obtained at the expense of a relatively poor ratio of r-f power output to d-c beam voltage, a relatively poor ratio of phase-delay variation to beam-voltage variation, and relatively poor linear-amplifier characteristics. The first area, that of the power output to beam voltage ratio, is at present receiving considerable attention, and as a result of hollow-beam and multiple-beam tube developments, it now appears likely that this will not be a significant problem in future tubes.

Figure 21 shows a typical representation of a linear-beam tube in cross-sectional view. Typically, a filament heats some form of concave cathode surface to provide for a powerful source of electron emission. Because of the basic shape of the cathode surface and the field created between the cathode and the anode of the tube by the beam-voltage power supply, a quite narrow beam of electrons is formed and injected into the beam-interaction structure. The nature of this interaction structure varies widely between the klystron and the TWT. In the klystron, the interaction structure consists of a series of cavities connected by drift tubes. The drift tubes form gaps within the cavities to allow interaction of the linear electron beam passing

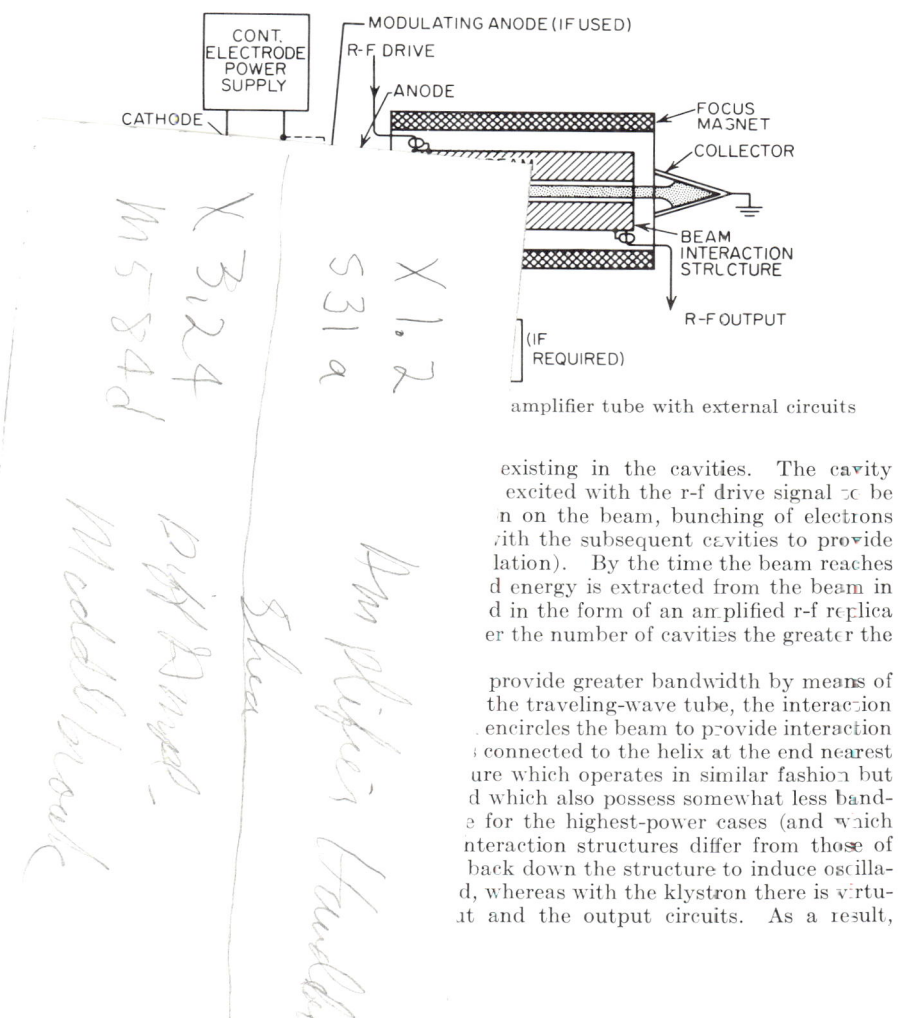

amplifier tube with external circuits

existing in the cavities. The cavity excited with the r-f drive signal to be n on the beam, bunching of electrons ith the subsequent cavities to provide lation). By the time the beam reaches d energy is extracted from the beam in d in the form of an amplified r-f replica er the number of cavities the greater the

provide greater bandwidth by means of the traveling-wave tube, the interaction encircles the beam to provide interaction connected to the helix at the end nearest ure which operates in similar fashion but d which also possess somewhat less band- e for the highest-power cases (and which nteraction structures differ from those of back down the structure to induce oscilla- d, whereas with the klystron there is virtu- it and the output circuits. As a result,

TWT's tend to be somewhat lower in efficiency than klystrons; however, this shortcoming is somewhat compensated for by the fact that considerably greater bandwidth is achievable from the TWT slow-wave structure. In the case of either klystrons or TWT'S a significant portion of the energy of the beam is not removed by the output circuit; hence this energy passes from the interaction structure and is dissipated in the collector as heat.

It is rare that the efficiency of a linear-beam tube reaches 50 per cent. Some operate in the region between 40 and 50 per cent, typically those operating at fairly high power levels and fairly narrow bandwidths. Most large klystrons operate with between 30 and 40 per cent efficiency as do fairly large TWT's where the bandwidth is not too great (less than 10 per cent). Many smaller klystrons and TWT's operate between 20 and 30 per cent, with some even falling as low as 10 to 20 per cent. It must be recognized that linear-beam tubes are not, on the whole, particularly efficient even when neglecting the other circuit losses associated with them, and considering only beam efficiency.

Figure 21 also indicates the typical mounting configuration and auxiliary circuits associated with linear-beam tubes. It is first of all necessary to provide some form of focus-magnet structure. This magnet is strictly for beam control and functions to confine the small-diameter beam, initially established between the cathode and anode structure of the tube, to essentially the same cross-sectional area as it passes through the interaction structure of the tube. This is an important function, for if the beam were allowed to spread out (as it would without focusing) it would tend to impinge partially on the beam-interaction structure with resultant structure heating and loss of bunching efficiency in the tube. Both electromagnets and permanent magnets are used for this purpose. While electromagnets are quite lossy (they often consume as much power as the tube delivers) they usually result in a lower-priced tube package since they may be reused when the tube is replaced. Permanent magnets, when employed, are usually furnished as an integral part of the tube.

A second feature of importance in the application of linear-beam tubes is to note that the interaction structure is practically always operated at ground potential. Usually the collector is connected to the interaction structure; however, in some cases a small amount of isolation is provided between the collector and the body of the tube to allow independent indications of body-interception current and collector current to be obtained. Occasionally the collector is insulated for quite high voltages when depressed collector operation is contemplated. This form of operation may result in somewhat greater efficiency being obtained from the tube. All other electrodes, the cathode, the filament, the control grid, and/or the modulating anode (if used) are at high potential with respect to ground; hence adequately insulated circuits must be used for their control and operation.

Regarding control electrodes, both control grids and modulating anodes are employed on certain tubes requiring their capabilities. The control grid is usually of the intercepting type; that is, some grid current must be supplied (perhaps 10 to 20 per cent of beam current) by the controlling means. This characteristic limits the use of control grids generally to the lower power ranges since grid cooling is quite difficult. Typically, a negative bias is applied to cut off beam current (usually a few per cent of beam voltage is required) and the beam is turned fully on with a few per cent positive voltage applied to the tube (both control voltages being referred to cathode potential). The modulating anode functions in a manner similar to the control grid; however, when it is used instead of a grid, the interception current is quite low (1 or 2 per cent) but the positive potential required to turn the tube on fully usually amounts to 50 to 100 per cent of beam voltage. The negative bias required to turn the beam off fully by means of the modulating anode is usually 1 or 2 per cent of beam voltage. Again these values are with respect to cathode.

Linear-beam tubes leave few complex design considerations to the user except for the proper application and control of the necessary potentials. All basic r-f design characteristics are defined and manufactured into the tube by the producer, as contrasted to negative-grid tubes where the distributed-constant tuned circuits are often designed and fabricated by the user. It is important, however, that the external user-supplied circuits of linear-beam tubes be properly fabricated. Such considerations as supplying adequate cooling, adequate X-ray shielding, and adequate protective circuits are of the utmost importance if the tube is to perform in an effective and long-lived manner. Some of these factors are discussed in later sections.

5.3. The Crossed-field Amplifier. The crossed-field amplifier operates with the same fundamental elements involved as does the linear-beam tube; however, the way in which these elements act to provide amplification is often considerably different. Specifically, a cathode supplies electron emission, and this emission reacts on a beam-interaction structure in accordance with the applied d-c potential, the r-f drive signal, and the external magnetic field. In the case of the linear-beam tube these elements act in such a way that the d-c energy is first converted into kinetic energy by imparting velocity to an electron beam which subsequently delivers a portion of this energy when it interacts with the excited beam-interaction structure. In the case of typical crossed-field amplifier tubes, the first step, that of conversion from direct current to kinetic energy, is not present; hence a direct conversion of power-supply energy to radio frequency results. This is largely due to the fact that the magnetic field forms a fundamental part of this conversion process as contrasted to its role in linear-beam tubes where it merely controls beam dimensions. Virtues of the direct conversion process utilized by crossed-field amplifiers show up in terms of high power-handling capabilities, broad bandwidth, and high efficiency. Drawbacks encountered as a result of the simpler conversion process are low gain and relatively poor stability. The PF^2 factor noted previously is probably as appropriate in describing crossed-field amplifier performance as it is in the case of linear-beam tubes. At present it is difficult to set a value for this figure of merit since considerably less factual data exist on which to base it quantitatively. This is due to the fact that the principles leading to the crossed-field amplifier have not been so fully developed in terms of hardware as in the case of the linear-beam tube. It is evident that the factor should be considerably higher than that ascribed to nega-

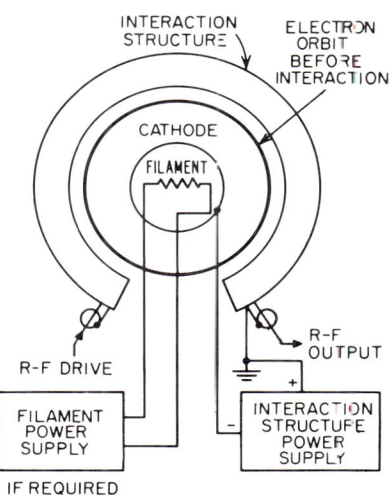

FIG. 22. Cross section of a crossed-field Amplitron (trade name of Raytheon Co., Watham, Mass.) Note: Magnetic field exists perpendicular to cross section.

tive-grid tubes but somewhat less than the factor of 10 quoted for linear-beam tubes. Powers on the order of 10 megawatts peak and tens of kilowatts average have been realized in the 3-Gc region. Considerable effort is currently taking place on tubes of improved characteristics which may greatly exceed these power capabilities.

Figure 22 is representative of the elements comprising a form of crossed-field amplifier known as the Amplitron. It operates in such a way that when r-f drive is applied to the tube, the degree of r-f interaction is a function of the interaction-structure power-supply level. Typically, if this voltage is too low the electrons emitted by the centrally located cathode will orbit concentrically around the cathode without interaction. As the level of the supply is raised the diameter of the orbits will become greater as will the speed of the electrons until a voltage level is reached which is coincident with the speed of the wave applied to the interaction structure (an orbit just prior to reaching the interaction level is shown in the figure). When the interaction level is reached the electrons begin to deliver their energy to the field of the interaction structure. As this happens the electrons come closer to the structure, and after delivering a large percentage of their energy they are collected on the structure. The energy delivered by the interaction process is supplied as r-f output to a load, and the remainder of the energy is dissipated as heat in the interaction structure.

Other crossed-field amplifier types take on variations of the above configuration, some considerably different. For instance, the injected-beam crossed-field amplifier is similar to that shown in Fig. 22, except that, instead of being circular in nature, it is often made in planar form, with a flat cathode surface (called the sole in this case) and a parallel interaction structure placed above it. A beam of electrons is injected between these two electrodes (from a second cathode arranged much as in a linear-beam tube). With the proper values of interaction supply voltage and cross magnetic field applied, interaction and energy transfer takes place in a manner similar to that just described. Instead of collecting the unconverted energy on the interaction structure, a collector is utilized near the r-f output connection.

It was noted in the discussion of linear-beam tubes that the klystron may be considered unconditionally stable since there is no feedback circuit associated with the interaction structure. The traveling-wave tube does not possess this degree of isolation, and as a result some form of attenuation must be provided to ensure that energy reflected back down the interaction structure will not cause unwanted oscillations. Crossed-field amplifiers tend to exhibit this tendency toward oscillation to a much greater extent than do linear-beam tubes. This comes about because many forms of these tubes are reentrant; that is, some of the output energy may be coupled back into the input circuit, particularly when the circular mode of construction is employed. This effect can be minimized to varying degrees by the insertion of internal attenuator sections and similar techniques; however, it is usually the case that a great amount of care must be employed in applying these tubes if instability is to be avoided. For instance, it is usually necessary to apply r-f drive before interaction-structure d-c voltage is applied, and further, it is necessary to remove the d-c voltage before r-f drive is removed. If this precaution is not observed, and if the level of r-f drive applied to the tube is not carefully established within a particular range, it is likely that the tube will not be properly locked to the drive frequency, and spurious r-f outputs will occur rather than an output which is an amplified replica of the r-f drive.

It is important to realize that crossed-field amplifiers are essentially constant-current devices rather than constant-voltage devices. They exhibit many of the properties of a biased diode; that is, until the applied voltage reaches the approximate level at which operation can take place, the tubes draw very little current. As the voltage is increased very slightly above this level large variations in power-supply current (and r-f power output) will occur. If pulsed operation is contemplated, the characteristics obtainable from the line-type modulator are quite beneficial in providing the characteristics required, since this form of pulser acts as a constant-power-output device over a significant portion of its output loading range. If CW operation of a crossed-field amplifier is contemplated, a power supply having adequate regulating characteristics must be provided to prevent excessive current variations from influencing the amplifier's performance.

There are a number of characteristics which highly recommend the crossed-field amplifier for many applications. First, these tubes exhibit a very low value of phase-delay change for a change in power-supply current. This may be as low as 0.2° for a 1 per cent change in current, a value an order of magnitude less than obtainable with linear-beam tubes (but one comparable with that obtained with negative-grid tubes). Second, the tubes generally operate at relatively low power-supply-voltage levels for a given power output level. This not only assures reduced size and weight in the power-supply area, but it also reduces the amount of shielding required for X rays. Third, this class of tubes is generally capable of very high overall efficiency because of the internal energy-conversion processes involved and the fact that the magnetic field is generally of the permanent-magnet type (hence avoiding the use of an external power supply for this purpose). Fourth, a number of these tubes operate as cold-cathode devices; that, is they require no external source of filament heating power. Finally, many of these tubes can be used as feedthrough devices. This means that with no high voltage applied, the input drive power may be fed directly through the interaction structure to appear at the output terminals in virtually an unattenuated form. Similarly, power fed into the output terminal will appear unattenuated at the input terminals of the device. This feedthrough feature can be of considerable benefit for many applications. With all these potential advantages available, it is important that the crossed-field amplifier be carefully considered for any given application. It is also important that the potential instability considerations associated with this class of tubes be carefully weighed against the advantages.

6. OSCILLATORS[7]

Oscillators form a fundamental element in almost every microwave amplifier complex. In the early days of microwave activity virtually all power generation was by means of oscillators alone, and even at present a significant percentage of microwave power-generating equipment relies solely upon some form of microwave oscillator device. The requirement for higher degrees of frequency stability than are achievable with the simple microwave oscillator in some modern amplifier complexes has resulted in the wider use of oscillator/amplifier complexes; however, the need for

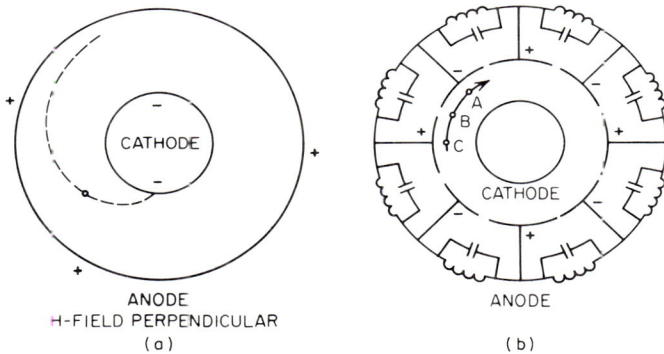

Fig. 23. Magnetron oscillator action. (a) Electron orbit. (b) Electron bunching. (By permission of General Electric Co.)

the basic oscillator element has not been eliminated. As a result many new and improved forms of microwave oscillators are currently available and still further improved forms will undoubtedly appear in the future. The classes of oscillators closely parallel the classes of microwave amplifiers, namely, the magnetron (or crossed-field) type, the negative-grid type, and the velocity-modulated type. The following discussion will cover a few of the more important characteristics associated with these fundamental types.

6.1. Magnetron Oscillators. The conventional magnetron oscillator is a diode in which the electron current from the cathode is influenced by a magnetic field parallel to and coaxial with the cathode and acting at right angles to the applied electric field. When electrons travel in a direction perpendicular to the magnetic field, that field imposes a force at right angles to the direction in which the electrons are moving. This causes the electrons to spiral into circular orbit. The velocity of the electrons is proportionate to the electric field applied between cathode and anode. A simplified schematic illustrating this effect is shown in Fig. 23a.

Random noise present in the tube induces some radio-frequency voltage on the anode segments. This, in turn, builds up the intensity of the radio-frequency power on the anode structure. Figure 23b depicts three electrons rotating in the interaction space at an instant when adjacent anode segments are negatively and positively

---

[7] Parts of this section have been reproduced with permission of the General Electric Company.

charged. In the example, electron A is in a reduced electric field region caused by the radial component of the radio-frequency electric field acting in opposition to the d-c electric field. Electron B is passing through an area of unmodified radial electric field and therefore maintains its initial velocity. Electron C is advanced since it is in a higher electric field region where the radial radio-frequency electric field augments the applied d-c field. The electrons form into a number of bunches equal to one-half the number of segments in the interaction space. The continuous passage of these electron bunches past the anode segments induces radio-frequency power into the anode structure. This induced power is in turn coupled out to deliver r-f energy to a load.

Magnetrons are credited with being the first microwave tube developed. Currently, tubes capable of quite high power outputs are available, typically up to 5 megawatts peak power in the 0.5- to 3-Gc range and dropping off to about 1 megawatt peak power at 10 Gc. Over the same frequency ranges average power outputs of around 5 and 0.5 kw, respectively, are within the state of the art. These tubes are characterized by a fairly desirable mechanical form factor (which allows for convenient mounting) and a relatively low cost per watt (as compared with most other microwave tubes).

Typical characteristics of the conventional magnetron include efficiencies of 40 to 55 per cent, mechanical tunability of up to 12.5 per cent, a temperature coefficient vs. frequency of around 0.003 per cent per °C change in ambient, relatively long tube life (with careful application), and quite low operating voltage for a given power output level. A characteristic of importance is that of pulling, which is defined as the variation in frequency occurring when a mismatch of 1.5/1.0 VSWR is varied through 360° of phase in the load circuit. Typical values of this figure are approximately 5 Mc at 1 Gc, 10 Mc at 3 Gc, and 15 Mc at 10 Gc. Another characteristic of importance is that of pushing, which is defined as the variation of frequency with variation in operating current during a pulse and/or from pulse to pulse. This value typically runs from 0.5 to 2 Mc/amp, depending largely on the structure of the tube involved.

Conventional magnetrons are available in either the pulsed versions or in CW versions. When pulsed magnetrons are employed it is important that the pulser provide characteristics compatible with the tube's starting and running characteristics. In particular, since magnetrons exhibit a biased-diode characteristic, their semiconstant current characteristic is effectively accommodated by the use of a line-type pulser. The characteristics of this type of pulser are described in Sec. 11.2. Another characteristic of considerable importance is the pulse rise time. If the pulse rises too rapidly a situation known as "moding" can occur wherein the operation of the tube is in a mode having a different frequency than that for which the tube was designed. Poor efficiency, arcing, or even no output at all may result from improper rise-time control.

Another class of magnetron oscillator consists of the electronically tunable types. These are typified by the voltage-tunable magnetron and the injected-beam backward-wave oscillator (M-type carcinatron). The voltage-tunable magnetron, as its name implies, can be tuned without mechanical means simply by changing its anode voltage. The CW power output of these devices currently runs from several hundred milliwatts to several hundred watts, values of 100 watts at 3,000 Mc being readily obtainable. The interactions in a voltage-tunable magnetron are similar to those in a conventional magnetron in that there is an emitter producing electrons which enter into an interaction space where they are bunched and induce radio-frequency currents in an anode structure. Voltage-tuned operation of a magnetron, however, requires a heavily loaded r-f anode circuit to reduce the loaded Q. Moreover, the number of electrons injected into the interaction space must be limited. As a result, an external electron injection cathode and injection electrode, not present on a conventional magnetron, are required in the VTM.

Pulling in the VTM will affect both the frequency and power output, and at a single frequency the effect of pulling is much the same as that on a conventional magnetron. If adverse effects on the tuning curve are to be avoided it is advisable to use an isolator in the output circuit. Pushing is present also as in the conventional magnetron and the prime factors causing this effect are temperature of the hot cathode, injection electrode voltage, anode voltage, and variations of the cross magnetic field. Of these, injection electrode voltage and cathode temperature are the most important. For this reason well-filtered d-c power supplies are required for both elements. In general, with care in the application of the VTM, the tuning curve of anode voltage vs. frequency is essentially linear. Wideband units have frequency-range ratios of as great as 3:1. With good commercial power supplies, frequency variations in VTM's can be held to ±0.03 per cent. Where frequency must be held to tighter limits, feedback circuits must be used as frequency-control devices.

The M-type injected-beam backward-wave oscillator has characteristics quite similar to the VTM. The structure of the crossed-field BWO is quite similar to that of the injected-beam crossed-field amplifier. Its operation differs in that in the amplifier version both the electron beam and the energy of the wave are propagated in the forward direction, whereas in the BWO energy propagation is in the reverse direction to electron-beam flow. State-of-the-art BWO's of this type are capable of average power outputs of several hundred watts in the 10,000-Mc region.

6.2. Negative-grid Oscillators. In the early days of radar virtually all the transmitters operating below 1,000 Mc employed negative-grid tube oscillators as their r-f power-generating elements. Power outputs in excess of 1 megawatt peak and tens of kilowatts average were obtainable from these oscillators. Even greater amounts of power output can be readily obtained from present-day triodes and tetrodes; however, in most applications a higher degree of frequency stability is required than is obtainable from such high-power devices. As a result, most current applications employ a stable oscillator which in turn drives some form of amplifier chain to achieve the necessary power output.

The triode oscillator circuit may be considered as a power amplifier with its r-f drive supplied by feeding back a portion of its output in proper phase. Proper design therefore consists of making the best possible power amplifier and then adding a feedback which delivers the correct fraction of the output back to the input with proper phase. The power output from an oscillator would be expected to be less than that from a similarly operating power amplifier by the amount of the driving power. Since the feedback inherently present in the tube interelectrode capacitances is usually inadequate to make a good power oscillator, loops, probes, or holes between the input and output circuits are the usual means of obtaining additional feedback.

Low-power negative-grid oscillators are still widely employed for numerous applications. The tube types normally employed in these oscillators are of the planar type where emphasis is placed on close-spaced electrodes having the minimum of lead length with which to make connection to the external distributed-constant circuits Figure 24a is representative of a microwave oscillator using such a tube. The particular tube type shown is a GL 6442, which measures approximately $\frac{1}{2}$ in. in diameter and $2\frac{1}{2}$ in. in length. The oscillator configuration shown is commonly referred to as a reentrant oscillator and is particularly useful above 2,000 Mc. Figure 24b shows a set of curves of the essential dimensions of the coaxial circuit elements vs. frequency

In the reentrant circuit, the radio-frequency fields from the grid-plate space are coupled directly back to the grid-cathode space. The r-f field between the grid and anode is propagated along the grid-plate line to a point at which the grid cylinder is terminated. At this point, the presence of the choke, or plunger, causes this energy to propagate along the grid-cathode line in the direction of the base of the tube. A certain amount of power propagated in this direction is coupled to the output by the output probe, and enough additional power to supply the tube loss is fed back into the grid-cathode circuit, thus sustaining oscillation. The length of the grid cylinder L_g is the most important factor in determining frequency. The position of the choke with respect to the anode of the tube has some small effect upon frequency, but for the most part, the choke is primarily important in its ability to change the phase of the feedback to the cathode. In this respect, its position affects quite materially the efficiency of the oscillator. The length of the choke L_c is made one quarter wavelength at the center of the band to be tuned. This provides essentially a r-f short circuit on the tube side of the choke, yet provides for isolation of d-c voltages.

To establish relative magnitudes, a small oscillator of the type described can deliver

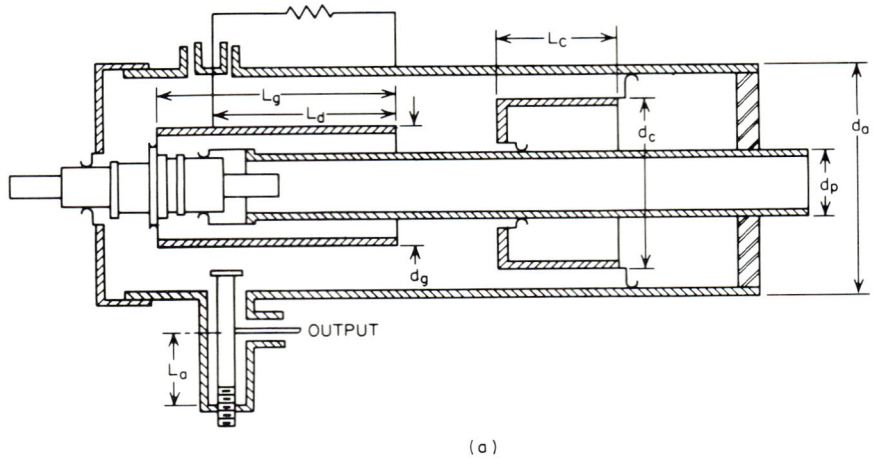

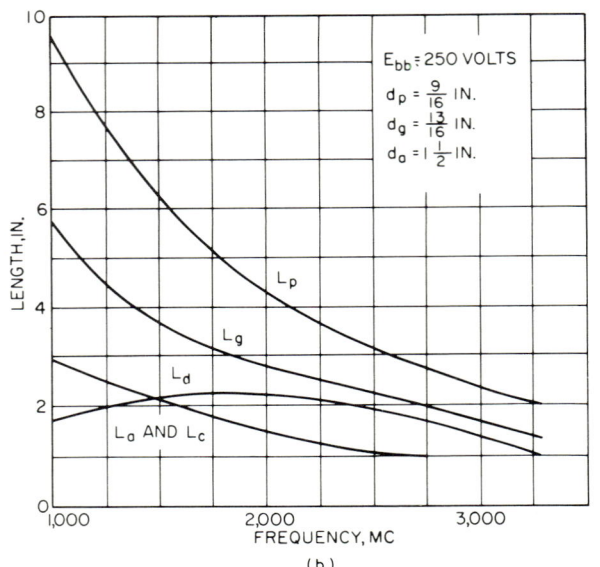

Fig. 24. A typical low-power negative-grid oscillator arrangement. (a) Reentrant oscillator circuit. (b) Dimensions vs. frequency for circuit. (*By permission of General Electric Co.*)

pulse output powers on the order of several kilowatts at 3,000 Mc, and average powers of several watts. The long-term stability obtainable is largely a matter of temperature variations within the circuit. With care this variation can be held to a few hundred kilocycles for a 50°C change in ambient, and with controlled temperature (and an initial warmup time) it may be held to tens of kilocycles. Short-term pulse-to-pulse stabilities of less than 10 cps are achievable with well-regulated commercially available power supplies.

6.3. Velocity-modulated Oscillators.

Two oscillator variations of the linear-beam amplifier types of klystron and TWT are the reflex klystron and the O-type backward-wave oscillator, respectively. These two tube types are particularly useful when power output is required at frequencies in excess of those at which the negative-grid tube can produce appreciable power output. The reflex klystron was first developed in the early days of radar as a low-power receiver local oscillator for use in equipment where high-frequency magnetrons were used as transmitting elements. These initial-reflex klystrons have undergone extensive modification with time and application until now they are perhaps one of the most widely diversified types of microwave tubes available. They are available in the form of oscillators capable of providing power output at almost any desired frequency in the microwave range from around 1 up to around 100 Gc. Generally the power output available is on the order of several watts or less. The O-type backward-wave oscillator is a relatively

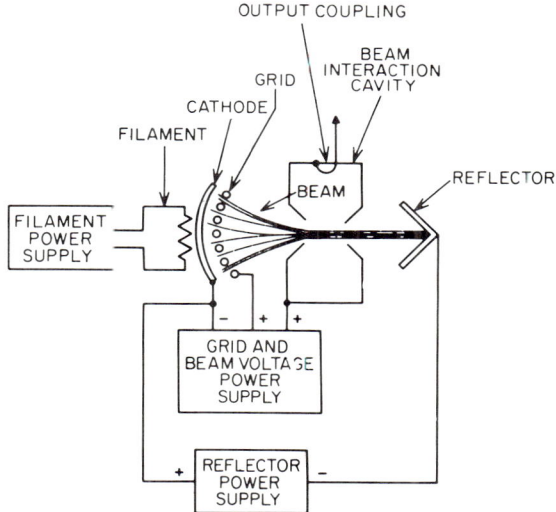

FIG. 25. Cross section of basic reflex klystron.

recent innovation by comparison. Since it is a derivation of the traveling-wave tube in its usual form it is capable of providing power output over essentially the portions of the microwave spectrum where TWT's are practical devices. Both types of tubes can be used as voltage-tunable devices.

A basic form of the reflex klystron is shown in Fig. 25. The electron-gun portion of the tube is similar to that of the conventional klystron amplifier discussed previously. The grid and beam interaction cavities are also similar in their initial function and voltage requirements. The action differs when the klystron is used in an oscillator role in that usually only one cavity is employed, and also, the collector element is operated at a negative potential with respect to cathode. The collector is then referred to as a reflector since it functions to reverse the electron beam coming through the cavity in the forward direction and redirect it back into the cavity, hence providing a feedback action. When the bunched beam is returned to the cavity, a positive field extracts energy from the beam since the direction of electron motion has now been reversed. A reflex klystron will oscillate only within certain ranges of reflector voltage for which the transit time from the center of the gap to the reflection point has the proper values to sustain the oscillation. There are usually a number of these ranges, and the output over each range is referred to as an operating mode. It is possible to vary the reflector voltage around the center value within a given mode and hence control the frequency of oscillation over a small range (usually less than a few per cent). Free-running klystron oscillators have short-term frequency instabilities as a result of inherent random noise in the electron beam as well as from external disturbances. If the oscillator is to be used as a single-frequency device, it is often advisable to apply an external stabilizing cavity. Such cavities are commercially available for many frequency ranges, and their use can provide a stability improvement of 100 or more. Some insertion loss is incurred as a result of the use of these cavities, usually on the order of 10 db.

The O-type backward-wave oscillator (which notation differentiates it from the M-type employing crossed-field principles) has the same basic structure as the helix-type traveling-wave tube (the basic form of which is shown in Fig. 21) but employs a different type of propagation, referred to as backward-wave, because the direction in which phase is increasing is opposite to the direction of energy flow. To obtain interaction with a beam moving from cathode to collector, r-f energy must be fed in at the collector end and propagate toward the cathode end. (Power is taken out of the input connector shown.) Unlike the conventional TWT forward wave, this backward wave has a velocity which varies sharply with frequency. As a result, the frequency of interaction can be varied by as much as two octaves by adjusting the accelerating voltage. These tubes are applied in two ways. First, oscillations which are voltage-tunable may be obtained if the beam current is increased somewhat above the value appropriate for amplification. A second mode of operation, primarily for fixed frequency, is obtained when a small amount of feedback is provided between the output and input connection of the tube. This feedback element is often a delay line, and with proper application of this technique a stability improvement of about 10:1 can be obtained over the first method of operation. As with the stabilized-reflex klystron, some sacrifice of power output results from the second form of operation.

Since velocity-modulated oscillators vary so widely in their manufacture, it is difficult to make any general statements as to their absolute frequency stability. Information of this type can best be obtained by making specific reference to the literature furnished by the manufacturer of the particular device of interest.

6.4. High-stability Oscillators. When higher degrees of frequency stability are necessary than can be obtained from the previously described self-controlled oscillators, crystal oscillators are usually employed. Quartz crystal plates are piezoelectric resonators, and in their realizable forms may have natural frequencies of oscillation ranging from a few thousand cycles to 10 or 15 Mc. The resonant frequency depends on the kind of crystal, the way it is cut from the basic mass of crystalline structure as found in nature, and the dimensions of the plate itself. Compared with the best LC resonant circuits, crystals can provide a Q an order of magnitude higher. In addition to the usual resonant mode of operation which limits the frequency of operation of these devices to 10 to 15 Mc, overtone crystals are also available. These devices can provide output at frequencies up to around 100 Mc.

Crystal oscillators can readily provide short-term stabilities of 1 part in 10^8 and long-term stabilities of better than 1 part in 10^5. With the utmost care taken to minimize all possible causes of frequency variation, a long-term stability of 1 part in 10^8 can be achieved. This, incidentally, is the guaranteed stability of the transmissions of WWV of the National Bureau of Standards.

It can be seen that to utilize the greater stability inherent in crystal oscillators, some form of frequency translator or multiplying circuit must be employed to raise the output frequency to the microwave region. The succeeding sections will describe such devices.

7. FREQUENCY TRANSLATORS

It is common to encounter the terms mixer, converter, modulator, and translator in connection with devices performing the identical microwave function. This function is the combining of two sources of microwave power, each at separate frequencies, to provide a source of microwave power output at a third frequency. This process should ideally be accomplished without the generation of output power at other than the desired sum or difference frequency required at the output of the translator. In actual practice, the elimination of undesired output signals is never completely achieved; however, the use of single-sideband techniques can reduce these undesired signals by appreciable amounts, and with the use of adequate filters they can be further reduced to almost any desired degree.

Frequency translators are commonly employed for both "up conversion" (taking the sum of the two input frequencies as the output signal) and for "down conversion" (taking the difference of the two input frequencies as the output signal). When the sum-frequency mode is employed, an advantage of the process is that if one of the input frequencies has relatively poor frequency stability, this instability is not multi-

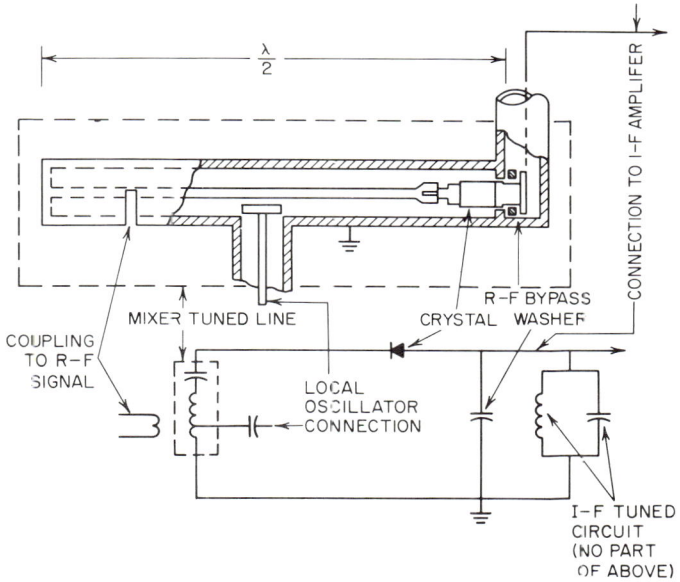

Fig. 26. 3,000-Mc mixer assembly and equivalent circuit.

plied. This is in contrast to frequency-multiplier operation where the instability would of course be increased by the multiplication factor if the same final frequency were to be obtained with either method.

7.1. Solid-state Frequency Translators. Microwave mixers of the unbalanced variety are exemplified by the type shown in Fig. 26. This form of mixer was first used as a means of providing receiver frequency translation for 3,000-Mc radar signals to typically a 30-Mc intermediate frequency by heterodyning with a local oscillator operating at a frequency displaced 30 Mc from the radar signal. The solid-state mixer in the traditional case was a silicon crystal with a point contact consisting of a tungsten wire. The mixer shown is a tuned section of transmission line one half wavelength long and functions to match the crystal to the signal and local oscillator inputs to provide the required i-f output. The manner in which the half-wavelength transmission line forms a series-resonant circuit is evident in the equivalent circuit. Local oscillator injection is by means of a probe, while signal injection is by means of a slot in the coaxial assembly. This slot in turn would normally be inserted in the duplexer waveguide assembly with proper orientation to provide coupling of the returned radar signal. In this application, the unwanted signals at the output of the translator (the carrier, the local oscillator signal, and the sum of these two signals) are effectively eliminated by the resonant lumped-constant circuit tuned to the intermediate frequency. This assembly could also be used in a reverse sense, where a local oscillator at 30 Mc could be injected into the low-frequency leg, a high-frequency signal injected into the slot or the probe, and a high-frequency offset signal removed from the unused high-frequency connection. In this case it becomes apparent that the high-frequency injected signal and the sum and difference frequencies are all quite close together; hence a very high Q circuit would be required to select the desired product of translation while eliminating the other two.

A balanced mixer (or modulator) is one in which the carrier can be eliminated in a mixing process such as that just described. It is shown in one of its forms in Fig. 27. It can be seen by tracing the f_c path around the loops that if the diodes have identical characteristics and if the tap on $T1$ is properly balanced, then the f_c component induced in $T3$ will be zero. However, the conditions for series mixing of f_c and f_m still are present and therefore the sum and difference frequencies are present in $T3$. With this arrangement the problem of eliminating the undesired products of translation are not so severe as in the case of the unbalanced mixer since a filter to remove

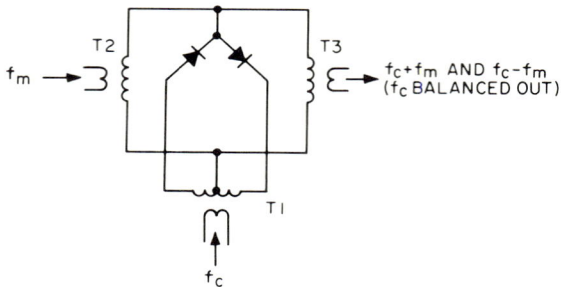

Fig. 27. Schematic diagram of two-diode balanced modulator.

the undesired sideband would not have to have so great an effective Q. An additional advantage of this arrangement is that externally introduced noise on the f_c input would be canceled out.

The problem of obtaining a sufficiently narrow bandpass filter at microwave frequencies is often quite difficult. As a result, a single-sideband translator without filters is often desirable, and with adequate precaution exercised in its construction, it can provide quite satisfactory performance. Figure 28 shows a block diagram of such a device. It employs two balanced modulators arranged in such a way that the carrier and modulating signals have properly introduced 90° phase shifts. While the mechanism by which the cancellation is obtained is not obvious, the phasing of the output of each balanced modulator is indicated. It can be seen that if the outputs of the individual modulators are applied to a summing network, the indicated single-sideband output would result.

In an actual application of these techniques at microwave frequencies,[8] 40-db attenuation of the undesired sideband can be readily achieved if a fixed carrier frequency and a narrow band of modulation frequencies are employed. For operation over a broad band of carrier frequencies, attenuation of the undesired sideband in excess of 20 db over a 17.5 per cent bandwidth has been achieved.

7.2. Tube-type Frequency Translators. The unbalanced-, the balanced-, and the phased-type single-sideband translators previously described in solid-state form are all applicable to negative-grid-tube-type applications. The limitations encountered when applying tubes are largely those encountered when using tubes in ampli-

[8] H. G. Pascalar, A Microwave Single Sideband Modulator, *MIT Lincoln Lab. Group Rept.* 48.6.

MICROWAVE AMPLIFIERS　　27-53

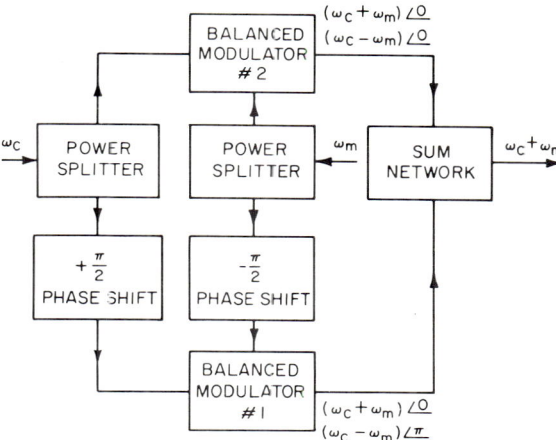

FIG. 28. Phase-shift single-sideband modulator. Note $\omega_c - \omega_m$ can be obtained by changing the sign of either phase-shift block.

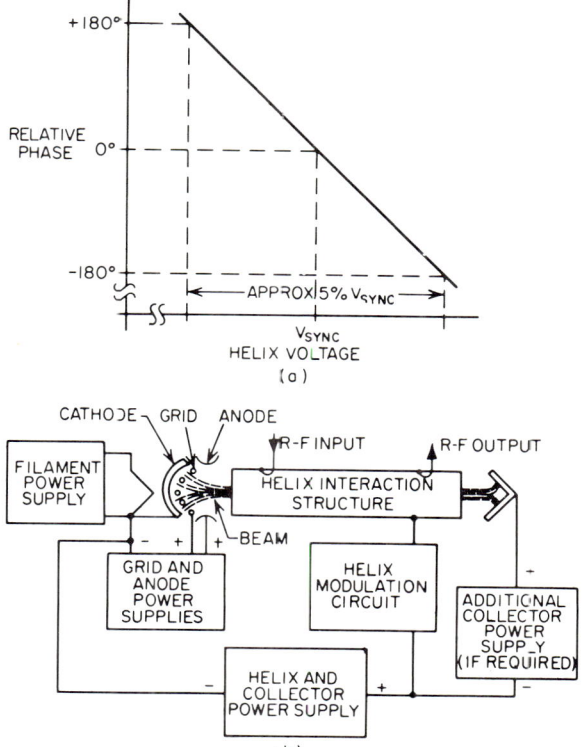

FIG. 29. TWT frequency translator characteristic and block diagram.

fier applications at microwave frequencies. Where their use is appropriate, some conversion gain is usually possible with tube circuits.

Linear-beam tubes, particularly traveling-wave tubes, allow a number of interesting frequency-translation effects to be achieved. Figure 29a illustrates the phase vs. voltage characteristic obtainable from many TWT's. It will be noted that a relatively small swing in helix voltage (about some reference level) will provide a relatively large advancement or retardation of the phase of the r-f output signal relative to the constant-frequency r-f input signal. The nature of the waveform of the modulating voltage has an appreciable effect on the sidebands produced in the r-f output. As is typically the case with phase modulation, if the waveform applied by the helix modulation circuit of Fig. 29b is sinusoidal, there will be modulation sidebands set up at multiples of the modulation frequency on either side of the carrier in the r-f output of the tube. This form of modulation is usually referred to as synchrodyne modulation. The desired product of translation must then be selected from the others present by means of a high-Q filter circuit. Another waveform often applied (in theory) is that of a linear sawtooth with infinitely fast flyback time. If such a waveform were applied at the modulation frequency, it is possible to show that single-sideband modulation will result. The r-f output of the tube is translated either higher or lower than the r-f input frequency by the value of the repetition rate of the sawtooth voltage. A negative slope on the modulating voltage results in a decrease in the frequency output as compared with the input frequency, while a positive slope results in the reverse effect. This form of operation is referred to as serrodyne translation.[9] With adequate precautions applied to the linearity of a practical sawtooth wave, the position of operation on the TWT characteristics, and the greatest possible minimization of the sawtooth flyback time, it is claimed that unwanted sideband attenuation of 30 to 40 db can be obtained. It is possible to obtain frequency shifts on the order of tens of megacycles, the basic limitation being with the characteristics obtainable from a sawtooth voltage generator.

8. FREQUENCY MULTIPLIERS

Frequency multipliers are a necessity if the high degrees of stability obtainable from low-frequency crystal oscillators are to be utilized for the frequency control of microwave amplifiers. It will be recalled from the previous discussion of high-stability crystal oscillators that the upper limit of their capabilities is around 100 Mc; hence, a multiplication of 1,000 or so would be necessary to reach the upper limits of the microwave region by this means. Multiplication of frequency by such factors is possible; however, it involves the use of a large number of stages, and it requires that careful attention be paid to the problems of spurious outputs generated in the process of multiplication. Both solid-state and vacuum-tube multipliers may be employed in the multiplication process. Currently much emphasis is being placed on solid-state multipliers because of their relatively small size and relatively high efficiency. It is possible that this technique may ultimately replace the stabilized klystron local oscillator for most of the applications so widely occupied by that device currently. A careful analysis of the stability, complexity, reliability, and cost considerations of a given application should be undertaken before such a judgment is made.

8.1. Tube-type Frequency Multipliers. Negative-grid tubes have capabilities as frequency multipliers closely related to their capabilities as power amplifiers. The circuit-design principles closely follow those of amplifiers, with the exception that the output-cavity resonant frequency must be the desired multiple of the drive frequency. Whenever possible, the output cavity should be working in the $\lambda/4$ mode. Since this is not always possible, at output frequencies above 2,500 Mc, the $3\lambda/4$ mode is an alternative; however, if the stage is to be used as a frequency tripler, there is the danger that the stage may jump into the mode where the fundamental frequency alone may be amplified. Generalizing, in high-harmonic multiplier applications, the proposed cavity dimensions should be examined with a view to avoiding the possibility of multiplying by a factor lower than intended; this can be done by encouraging

[9] R. C. Cumming, The Serrodyne Frequency Translator, *Proc. IRE*, February, 1957.

the formation of voltage modes in the output cavity. In summation, negative-grid tubes can be made to provide effective multipliers for the frequencies at which they can be made to provide effective amplification, but at the expense of reduced efficiency and more complicated resonant-circuit design. They are customarily not employed at frequencies in excess of several thousand megacycles, and at powers of more than a few watts.

Linear-beam tubes likewise can be used as effective frequency-multiplier devices. As was noted in the introduction to this chapter, the TWT is usually a fairly potent frequency multiplier without effort being made to increase its capabilities. However, if high multiplication factors are necessary, tubes having two or more individual helix structures may be fabricated in such a way that a common beam passes successively through them to provide multiplications of 10 or more, hence providing large orders of multiplication from a common envelope. It is important with such tubes that an adequate filter be employed in the output circuit so that only the harmonic of interest will be passed to the microwave circuit employing this energy. Klystrons are also sometimes used as frequency multipliers. In this case the output cavity is often tuned to the harmonic frequency rather than the drive frequency. In summation, linear-beam tubes can provide effective multiplication of stable low-frequency power generators, and their output can be obtained effectively throughout the microwave region. They are customarily used at low power levels, typically at under 1 watt.

8.2. Solid-state Frequency Multipliers. Solid-state frequency multipliers rely at present on the use of transistors and varactor diodes as frequency-multiplier elements. At present, transistors possessing high-frequency characteristics are available to deliver outputs of up to several watts at up to several hundred megacycles. Above this region, multiplication is generally accomplished by varactor diodes with appropriate distributed-constant circuits. The improvements in varactor diodes are presently appearing at such a rapid rate that it is almost meaningless to state limitations on these devices based on current practice. In general, multiplication ratios of 5 or more can be obtained in the hundreds of megacycles region at conversion efficiencies of 10 to 20 per cent and with power-output levels in the range of watts to tens of watts. As the frequency of operation increases to the thousands of megacycles region, the devices are usually employed as doublers. At frequencies in the 10,000-Mc region, current doublers are capable of delivering tens of milliwatts output at conversion efficiencies in the 1 to 10 per cent region. While it seems unlikely that these diodes will ever become capable of really high power outputs at the high microwave frequencies (up to 100 Gc) they offer the advantage of relatively small size and relatively high reliability as compared with tubes. Even with relatively low output levels, it is possible to use such multipliers to provide stable fixed-frequency locked-oscillator operation to VTM's and BWO's capable of high output in the regions where the capabilities of varactor diodes are presently limited.

9. TRANSMISSION-LINE ELEMENTS

This classification of components serves the primary function of transporting microwave power between various active amplifier and oscillator elements. In addition, these elements serve such functions as monitoring the behavior of the waves along the lines, attenuating or isolating the flow of power along the lines, loading the lines, splitting or combining the flow of power in separate lines, filtering the harmonics from the power flow in a line, or matching the power flowing from a line to free space. While many other instances of transmission-line-element applications could be cited, these fundamental areas are typical of most of the power amplifier complexes described in this chapter. The following sections will describe the general role of these components, their general characteristics, and the general manner in which they are applied.

9.1. Coaxial Transmission Lines. Three forms of transmission lines are commonly employed for microwave purposes. These are open-wire lines, coaxial lines, and waveguide lines. The open two-wire transmission line is not widely used where the distance of transmission is very great since it has high induction losses and high radiation losses; however, it is often used for such purposes as joining the adjacent

27-56 AMPLIFIER CIRCUITS

elements of antenna arrays or for distributed-constant circuits in oscillator and amplifier resonant tanks. When so employed it is usually physically arranged so that the line conductors are balanced with respect to a common ground plane.

Coaxial transmission lines possess the advantage that the electromagnetic field of the line is confined to the space between the inner and outer conductors, hence eliminating the losses due to induction and radiation. While somewhat more difficult to fabricate than an open-wire line, the electrical performance at microwave frequencies far offsets this disadvantage. Two characteristics are of importance in all lines, losses and power-handling capabilities. The losses in turn may be broken down into copper

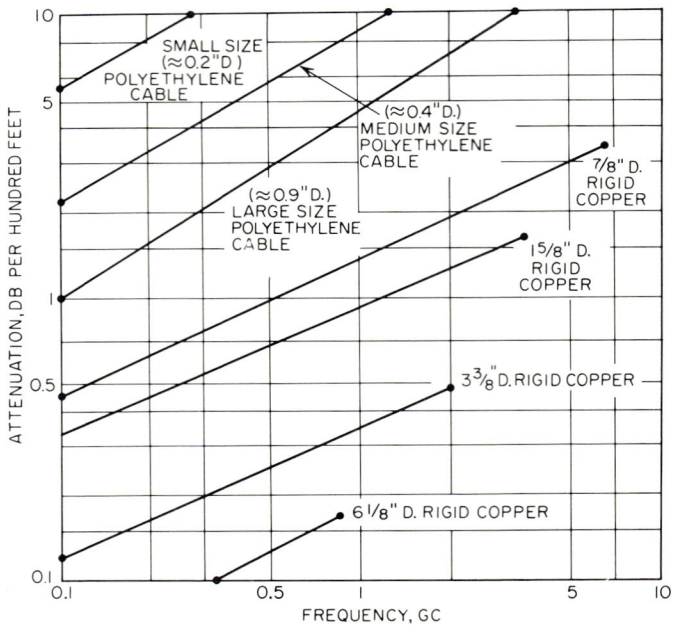

Fig. 30. Approximate attenuation of various coaxial lines.

losses and dielectric losses. Copper (or metallic) losses can simply be stated as being a function of I^2R where R is the d-c resistance of the line as given by the expression

$$R = \rho l / A \tag{17}$$

where ρ is the resistivity of the conductor material in meter-ohms, l is the conductor length in meters, and A is the cross-sectional area in square meters. In the microwave region practically all this current flows on a very shallow skin layer of the conductors involved, and the current depth of penetration is given by the expression

$$\delta = \sqrt{\rho/\pi\mu f} \quad \text{meters} \tag{18}$$

where ρ is the resistivity of the conductor in meter-ohms, μ is its permeability in henrys per meter, and f is the frequency in cycles per second. As an example, copper conductors have a depth of penetration of about 50 μin. at 3,000 Mc/sec.

Copper, brass, and aluminum are typical conductors employed for coaxial lines. Occasionally these lines are silver-plated to reduce the losses further, plating being quite effective since the depth of penetration of current is so small. This method of loss reduction, while significant, can be seen from the above equations to be related to both the square root of frequency and resistivity. For example, changing from copper to brass (a 4:1 increase in resistivity) results in only a 2:1 increase in losses, because of the greater depth of penetration of current in the brass line.

Since it is necessary to support the inner conductor of a coaxial line mechanically, some form of dielectric losses is usually present. In the larger rigid forms of coaxial lines, the dielectric is air with occasional insulators (or beads) for mechanical support. These insulators form the major loss element of such lines, and in addition, since the dielectric of the insulators is invariably different from that of air, line breakdowns usually occur in the vicinity of these elements. Solid dielectric transmission lines, typically employing polyethylene dielectric and flexible inner and outer copper conductors, are widely used because of the alignment problems avoided thereby. This flexibility is provided at the expense of higher line losses than are present in rigid lines. In some cases stub-supported transmission lines are employed. These are air dielectric lines in which the center conductor is supported at intervals by means of quarterwave short-circuited coaxial line lengths of the same material as the line. These stubs present a very high support impedance at their resonant frequency; hence the dielectric losses are virtually negligible. Such lines are of course frequency-sensitive, and this, in conjunction with their mechanical complexity, limits their application.

The power-handling capabilities of coaxial lines are largely a function of their diameter, since larger diameter allows a given characteristic impedance (usually either 50 or 73 ohms) to be obtained with greater cross-sectional conductor areas (for low losses) and greater conductor spacing (for voltage holdoff). Either heating or voltage breakdown may be the limiting factor, depending on the type of line used. Figure 30 is indicative of the line-loss characteristics typical with current materials. Figure 31 is indicative of the average power-handling capabilities of the same lines. Where the peak power is much higher than the average power (particularly in pulse work), the voltage capabilities of the line must be investigated if arc-over is to be avoided.

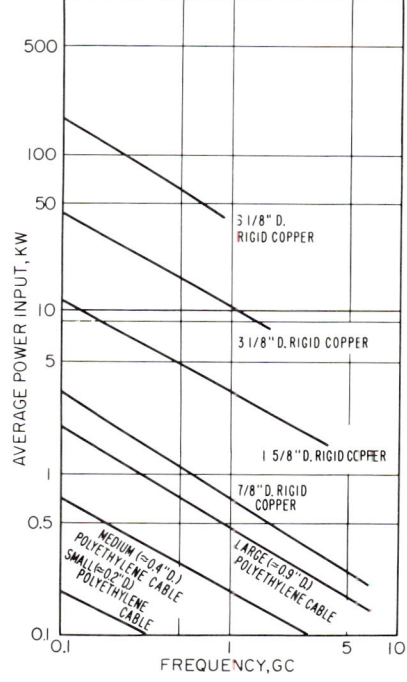

FIG. 31. Approximate average power limitations (at 1.0 VSWR) of various coaxial lines.

9.2. Waveguide Transmission Lines.

Waveguides, unlike open or coaxial transmission lines, have only one conductor and this is usually arranged in the form of a hollow rectangular or circular pipe. At first glance the propagation of power on ordinary two-conductor lines and via a single conductor pipe may seem two separate and distinct situations; however, they are actually both examples of the same situation, namely, the propagation of electromagnetic waves in a dielectric. Technically then, open-wire lines and coaxial cables are both versions of waveguides, but actually this nomenclature is usually reserved for the hollow pipe in its role of guiding electromagnetic waves. Currently the most widely used form of waveguide is the form possessing a rectangular cross section.

The advantages of waveguide over two-wire lines are numerous and include complete shielding, essentially no dielectric loss, copper loss less than that of coaxial cable of the same size when operated at the same frequency, greater power-handling capability than coaxial cable of the same size, simpler construction than coaxial cable, and reasonably small size over most of the microwave region. Against this impressive list of advantages, several disadvantages must be listed such as excessive size in the lower ranges of the microwave region, relatively narrow frequency ranges over which the line is effective, and greater complexity in coupling power into and out of the line.

A number of operating modes are possible using waveguides. These define the electromagnetic field conditions within the energized pipe. If all the components of the electric field within the guide lie within the plane transverse to the direction of propagation of power down the pipe, the mode of operation is called transverse electric or TE mode. This is the most common mode of operation with rectangular-cross-section guide. If all components of the magnetic field lie in the transverse plane, this mode of operation is called transverse magnetic or TM mode. This is the most common mode of operation for circular guide. These designations usually are amended by two subscripts, such as $TE_{1,0}$, where the first subscript designates the number of half-period variations in the electric field across the longer dimension of the guide's cross section and the second subscript designates the number of half-period variations in electric field across the shorter cross-sectional dimension of a rectangular guide. It so happens that the $TE_{1,0}$ mode is the most common mode used in rectangular guide. To permit propagation of this mode the larger cross sectional of the guide must exceed $\lambda_a/2$ where λ_a is the wavelength of the propagated signal in air; however, if the $TE_{2,0}$ mode is to be excluded this dimension must be less than λ_a. Similarly, the smaller cross-sectional dimension may be as small as desired without barring the $TE_{1,0}$ mode, but it must not exceed $\lambda_a/2$ if the $TE_{0,1}$ mode is to be excluded. Restating the above,

$$0 < a < \lambda_a/2$$
$$\lambda_a/2 < b < \lambda_a \qquad (19)$$

where b is the larger dimension of the guide and a is the smaller dimension of the guide. These inequalities can be satisfied by many different values, but in general the practical values are approximately

$$b \approx 0.7\lambda_a$$
$$a \approx 0.5b \approx 0.35\lambda_a \qquad (20)$$

These values represent a good compromise between the attenuation of undesired modes and minimization of loss in the guide.

A number of configurations are available to ensure the desired mode is properly introduced into the guide for propagation. In rectangular waveguide, the $TE_{0,1}$ mode can be effectively introduced by means of a coaxial-to-waveguide transition. In its simplest form this consists of connecting the coaxial line to the center of the longer side wall of the guide in such a way that the center conductor of the coax projects into the interior in the form of an antenna. A short circuit is then placed across the open end of the guide a small distance from the introduction of the probe. Propagation is then down the guide away from the direction of the short circuit. The length of the probe antenna and the distance of the short circuit from the antenna can then be adjusted for most effective matching and power transfer down the guide. A similar arrangement may be used for taking power out at the far end of the waveguide run.

Table 1 shows the characteristics of several commercially available rectangular waveguides. Several terms require definition. λ_g is the wavelength of the waves propagated within the guide, and for rectangular guide operating in the $TE_{1,0}$ mode its value is defined by

$$\lambda_g = \frac{\lambda_a}{\sqrt{1 - (\lambda_g/2b)^2}} \qquad (21)$$

It will be noted that when λ_a is much smaller than $2b$, λ_g is approximately equal to λ_a. As λ_a approaches $2b$, λ_g increases without limit. In practice when λ_a exceeds $2b$ propagation ceases and this value is called the cutoff wavelength or λ_c. A waveguide acts as a filter for frequencies corresponding to those lower than this wavelength value. This effect is often used as a means of preventing radiation from cer-

Table 1. Typical Waveguide Characteristics

	1.12–1.7	1.7–2.6	2.6–3.95	3.95–5.85	5.85–8.2	7.05–10	8.2–12.4
1. Frequency $TE_{1,0}$ mode, Gc							
2. JAN designation							
Brass	RG-69/U	RG-104/U	RG-48/U	RG-49/U	RG-50/U	RG-51/U	RG 52/U
Aluminum	RG-103/U	RG-105/U	RG-75/U	RG-95/U	RG-106/U	RG-68/U	RG-67/U
3. Dimensions, in.							
ID	$6.5 \times 3.25 \pm 0.005$	$4.3 \times 2.15 \pm 0.005$	$2.84 \times 1.34 \pm 0.005$	$1.372 \times 0.872 \pm 0.005$	$1.372 \times 0.622 \pm 0.004$	$1.122 \times 0.497 \pm 0.004$	$0.9 \times 0.4 \pm 0.003$
OD	$6.66 \times 3.41 \pm 0.005$	$4.46 \times 2.31 \pm 0.005$	$3.00 \times 1.50 \pm 0.005$	$2 \times 1 \pm 0.005$	$1.50 \times 0.75 \pm 0.004$	$1.25 \times 0.625 \pm 0.004$	$1.0 \times 0.5 \pm 0.003$
4. Flanges							
Cover							
Brass	UG-417 A/U	UG-435 A/U	UG-53/U	UG-149 A/U	UG-344/U	UG-51/U	UG-39/U
Aluminum	UG 118 A/U	UG-437 A/U	UG-584/U	UG-407/U	UG-441/U	UG-138/U	UG-135/U
Choke							
Brass			UG-54 A/U	UG-118 B/U	UG-343 A/U	UG-52/U	UG-40 A/U
Aluminum			UG-585/U	UG-406 A/U	UG-440 A/U	UG-137 A/U	UG-136 A/U
5. Cutoff frequency λ_c, Gc	0.908	1.375	2.080	3.155	4.285	5.26	6.55
6. λ_g (over range of 1), in	17.93–8.593	11.81–5.361	7.54–3.51	4.956–2.395	2.975–1.691	2.514–1.388	2.397–1.121
7. Attenuation (over range of 1), theoretical, db/100 ft							
Brass	0.424–0.284	0.788–0.516	1.478–1.008	2.79–1.93	3.85–3.08	5.51–4.31	8.64–6.02
Aluminum	0.269–0.178	0.501–0.33	0.940–0.641	1.77–1.22	2.45–1.94	3.50–2.74	5.49–3.83
8. CW power rating (over range of 1), theoretical, megawatts, brass	11.9–17.2	5.2–7.5	2.2–3.2	0.94–1.32	0.56–0.71	0.35–0.46	0.2–0.29

tain types of microwave enclosures while still allowing ventilating holes for cooling equipment within such enclosures.

If a long run of waveguide is required, several commercially available lengths must frequently be coupled together. This is commonly accomplished by means of choke joints. These joints consist of two flanges fixed to the waveguide ends and facing each other. One flange is usually machined flat and the other is slotted at a distance away from the guide. Coupling between the two flanges is then possible without mechanical contact, and for properly designed joints the mismatch introduced and the power lost are very small. In general it may be said that these joints offer the advantages of good electrical connection, isolation of vibration from one part of the system from other more sensitive parts, and easy removal of a section for repair or replacement. Typical joints are also indicated in Table 1.

Other common rectangular waveguide configurations are bends and twists. Bends are accomplished by either a slow gradual bend of the guide (in order to minimize reflections), right-angle E-plane bends, or right-angle H-plane bends. When right-

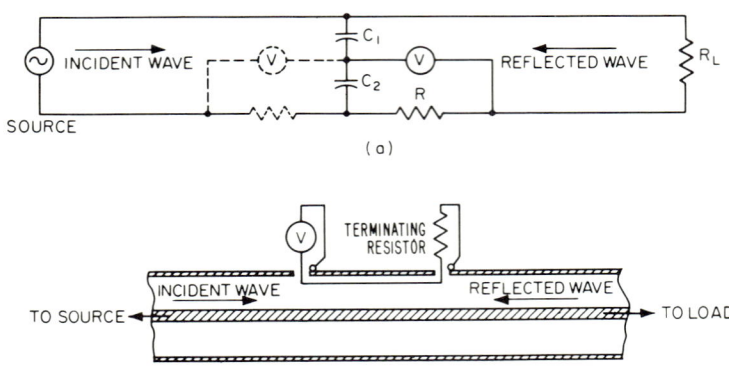

FIG. 32. Two forms of directional couplers. (a) Parallel-line type. (b) Coaxial-line loop type (cross section).

angle bends are employed, reflections are reduced by angling the outer corner of the configuration by an appropriate amount. Flexible waveguide sections are also available for such applications. Waveguides are sometimes twisted to change the direction of polarization (which remains parallel to the shorter side of the cross section). If the twist is gradual it produces a negligible amount of reflection.

9.3. Directional Couplers. A directional coupler is a device employed for monitoring either the incident or the reflected waves appearing on transmission lines. Typical uses of these devices include monitoring of the voltage standing-wave ratio (VSWR) and monitoring of the incident power delivered to the load attached to the line. The following expressions describe the voltage and current at any point along a transmission line in terms of the incident and reflected waves:

$$E = k_1 E_i + k_2 E_r \qquad (22)$$
$$I = (1/Z_o)(k_1 E_i - k_2 E_r) \qquad (23)$$

where E_i is the rms value of the incident wave proceeding toward the load, E_r is the wave reflected from the load because of line mismatching, Z_o is the characteristic impedance of the line, and k_1 and k_2 are parameters that express the phases of the incident and reflected waves at any point on the line (these constants have an rms value of 1.0).

Figure 32a illustrates a fundamental form of the directional-coupler principle. This form of coupler is often applied to open-wire transmission lines where the frequency of operation is not very high. With this arrangement the value of R is chosen to be quite low compared with the characteristic impedance of the line and the capacitor divider ratio is selected so that $R/Z_o = C_1/C_2$. From the circuit configuration it is apparent that the r-f voltmeter will read the sum of the voltage across the bottom leg of the capacitor voltage divider (consisting of C_1 and C_2) and the voltage across R. This sum can be obtained by substituting the appropriate values in the original transmission-line equations to obtain

$$V = k_1 E_i (C_1/C_2 - R/Z_o) + k_2 E_r (C_1/C_2 + R/Z_o) \qquad (24)$$

Since $C_1/C_2 = R/Z_o$, the first term becomes zero, and it can be seen that the voltmeter reading is directly proportional to the reflected wave. The value of the voltage will be

$$V = (2R/Z_o) E_r \qquad (25)$$

If the same process is performed with the position of the resistor R placed in the alternate position shown by the dotted lines of Fig. 32a, it will be seen that the directional coupler will give a reading on the voltmeter directly proportional to the incident wave and that this value will be

$$V = (2R/Z_o) E_i \qquad (26)$$

Since the voltage standing-wave ratio is given by the expression

$$\text{VSWR} = \frac{E_i + E_r}{E_i - E_r} \qquad (27)$$

the usefulness of the directional coupler in determining this important factor is quite obvious.

The particular form of directional coupler just discussed is not particularly applicable to coaxial lines for reasons of mechanical complexity. The type shown in Fig. 32b is often used with such lines. It consists of a loop of wire running through two insulated holes in the outer wall. If arranged as shown in the figure, and if the proper value of terminating resistor is connected to the device, the loop coupling due to the electric and magnetic fields within the coax will be such that the device is responsive only to the incident wave. The r-f voltmeter reading will indicate a value proportional to this wave. If the positions of the voltmeter and the terminating resistor are reversed, the r-f voltmeter will indicate according to the same constant of proportionality, but with respect to the reflected wave.

Other forms of directional couplers operate on the principle of hole coupling alone, without the use of loops. A coaxial form of such a device can be fabricated by placing a short section of coaxial line (often of the same type as the main transmission line) parallel and adjacent to the line carrying power. The two lines are coupled by means of a common hole pierced through the outer walls of both. Electromagnetic coupling then exists to the short section of line, and this coupling is a function of the hole size, hole shape, and orientation of sections. If one end of the short line is terminated in the characteristic impedance of the line and if the electric and magnetic fields are adjusted to be equal (by means of hole size, shape, and skewing of one coax with respect to the other), an r-f voltmeter connected to the opposite end of the short line will read either incident or reflected power in accordance with the relative positions of the voltmeter and load. Other forms of hole-type couplers employ more than one hole (usually spaced at odd multiples of a quarter wavelength apart) to achieve similar results. When directional couplers are applied to waveguides, the hole-coupling principle is usually employed.

Directional couplers are often employed as power-measuring devices in addition to their VSWR role. Since the power delivered to the load can be expressed as

$$P = \frac{E_i^2 - E_r^2}{R_L} \qquad (28)$$

(where R_L = the load resistance) the readings taken from the r-f voltmeter can be converted to load power by a simple calculation. If the VSWR is small (less than 1.5

or so) the contribution of reflected power is small and is often neglected. A direct-reading power scale can then be added to the incident-wave reading of the r-f voltmeter. The coupling coefficient of the directional coupler as well as the line losses between the coupler and the load must of course be properly incorporated in such a direct-reading scale calibration.

9.4. Attenuators and Isolators. Attenuators find a wide range of application in microwave transmission-line configurations. They are used as standards of attenuation in power and attenuation measurement, as a means of extending the range of existing measuring equipment, as a means of reducing the power level of existing sources, as buffer isolating pads between microwave elements, and for many other purposes.

Coaxial line attenuators have two basic forms. The first uses one or more T sections, usually made with lumped resistive film sections. These sections perform in a manner analogous to audio-frequency attenuators. T-section attenuators are frequency-limited because the element lengths become an appreciable portion of a wavelength long in the higher microwave region. For this reason a second type called the distributed attenuator is often used for higher-frequency applications. It consists of a coaxial line section in which the inner conductor is an electrically long resistive film. This film is often broken into three sections, the center section to provide the main bulk of attenuation, and the sections at either end to provide matching to the impedance of the line itself. A well-designed attenuator should always have its input impedance matched to the impedance of the line itself to avoid reflections.

Waveguide attenuators typically employ a thin sheet of resistive film deposited on a dielectric plate. The film is inserted into the guide in such a way as to parallel the electric field lines of the $TE_{1,0}$ propagation down the line. Attenuation can be controlled by either varying the distance of the film from the short wall or by varying the degree of insertion through the long wall. Matching of such an attenuator to the line impedance is of equal importance to that requirement for coaxial lines. This is accomplished either by tapering the leading and trailing edges of the film, or by changing the resistive properties of the leading or trailing edge by means of varying the properties of the surface coating.

The requirements for isolators are in many ways similar to those of attenuators. The isolator differs from the attenuator in that it inserts a negligible amount of attenuation into the line for the incident wave but inserts a large amount of attenuation in the line for the reflected wave; hence, if isolation is the prime purpose of placing a pad in a transmission line, the isolator is usually preferable to the resistive attenuator. Most isolators make use of ferrite material. One class of this material is referred to as nonreciprocal, which means that the phase and amplitude of a signal applied to the material are affected differently for opposite directions of application. An external magnetic field applied to the ferrite material is required to provide nonreciprocal action.

As applied to isolators the ferrite will produce practically no attenuation in one direction and will produce as much as 30 db of attenuation in the reverse direction. Attenuators are usually low power-handling devices (up to a few watts dissipation), whereas ferrites are capable of handling quite high power levels. A common application of ferrites is in the output line of high-power transmitting tubes. Should the r-f load fault, the high reflected power that would result is prevented from returning to the tube and is instead absorbed in the ferrite material. Ferrites are in general quite temperature-sensitive.

9.5. Dummy Loads. Resistive terminations are required for numerous microwave applications. These include loading of r-f generators for test purposes, resonant-circuit loading for bandwidth control, loading mismatch energy in multiple-port junctions, along with many other applications. In general these loads employ techniques similar to those noted for attenuators. Since the resistive elements employed are normally at the end of a transmission-line section, cooling is much more easily accomplished than in the case of attenuators. This is usually accomplished by means of finned structures for air-cooled loads or by means of circulating liquids in higher-powered cases. The configurations employed with the film-resistor strips typically employed are quite diverse; however, all have as their objective low-temperature rise of the surface of the film and the achievement of as low a line VSWR as possible. Some high-power loads use water or treated ceramic to form a lossy dielectric employed for such loads. Quite high power-handling capabilities are possible (particularly so in the liquid case where the water can carry away its own dielectric losses) and quite low VSWR's can be achieved.

9.6. Power-combining and -splitting Devices. Many microwave applications require that power from two separate sources be combined to form a single output, while other applications require that power from a single source be split to provide two separate sources. Still other applications require power flow through one path when it is delivered to a load, but that the power will flow through an alternate path when it is reflected from the load. These and other applications make use of the hybrid-balance principle. Shown in a simple low-frequency form employing iron-core transformers, Fig. 33a depicts such a device. For the power-combining case, consider that two independent power sources, each producing equal amounts of power and having the same frequency and phase relationships, were connected to inputs 2 and 4

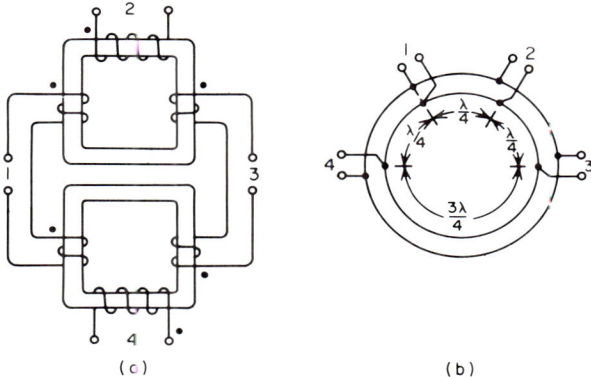

Fig. 33. Two forms of hybrid balance. (a) Iron-core version. (b) Open-wire version.

of the hybrid balance. It is then apparent by transformer action that the power of these two sources would be combined and would be delivered to an appropriate load connected to terminals 1, while no power would be delivered to terminals 3, even if a load identical to that connected to terminals 1 were present. In practice such a load would be connected to terminals 3, since should either of the generators connected to 2 or 4 cease delivering power, the remaining generator would still be able to provide half of its power to the load at 1. Conversely, if a single source of power were introduced at terminals 1, one half of this power would be delivered to each of equal loads attached to terminals 2 and 4 while no power would be delivered to the load attached to terminals 3. To illustrate a radar application, consider that an antenna is connected to terminals 2, a transmitter to terminals 1, a receiver to terminals 3, and a load to terminals 4. If the impedances of these devices were properly matched, it can be shown that when the transmitter is actuated, its power divides equally between the antenna at 2 and the load at 4 while no power enters the receiver at 3. When the antenna returns a signal, this signal would then divide equally between the transmitter at 1 and the receiver at 3 with no power entering the load at 4. This can be regarded as one form of duplexer.

Microwave hybrids can be visualized by reference to Fig. 33b, which shows an open-wire line equivalent of the iron-core version. Practically, open-wire hybrids are seldom used, but coaxial and waveguide forms of these four-port devices, exhibiting the characteristics of the low-frequency versions discussed, are widely employed.

9.7. Duplexers. In many microwave applications it is often desirable from both an economic as well as an operational point of view to use the same antenna system for both transmitting and receiving on or near the same frequency. In most of these applications it is desirable to be able to change from the transmitting mode to the receiving mode (and vice versa) practically instantaneously. This requirement generally rules out the use of a mechanical switching arrangement for making the antenna connection changeover; hence a class of devices known as duplexers has come into being to perform this function by electrical means. These devices typically make use of spark gaps (which function upon the application of transmitter power to the transmission line feeding the antenna) to bypass the transmitter power past the receiver connection to the transmission line. The effectiveness of a duplexer is in large part determined by how effectively this bypassing action is accomplished. The

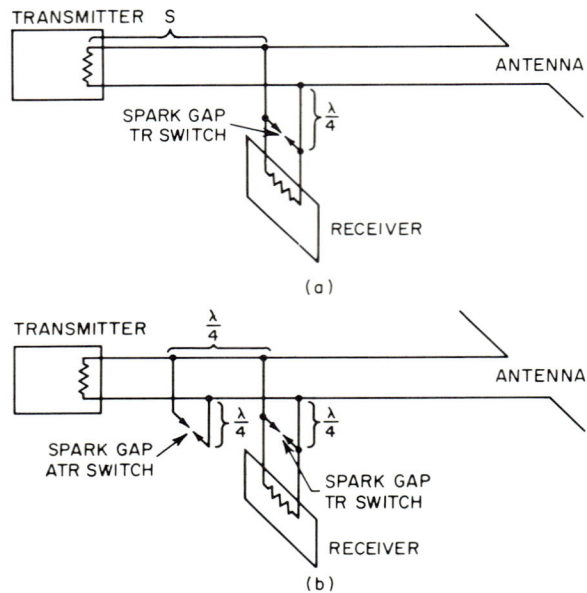

FIG. 34. Two basic duplexing techniques.

second function of a duplexer is to ensure that the received signal is channeled back to the receiver input with the least possible portion being diverted into the transmitter output circuits.

Figure 34a shows in simplified form how a balanced two-wire transmission-line duplexer could be implemented. It will be noted that a spark gap (often referred to as TR or transmit-receive switch or box) is placed at the end of a quarter-wavelength stub attached to some point along the main transmission line feeding power from the transmitter to the antenna. When the transmitter is actuated, this gap is so selected that it will break down almost instantaneously, thus providing two functions, first, connecting an effective open circuit across the line to ensure that virtually all the transmitter power will be diverted from the stub path and directed toward the antenna, and second, placing a virtual short circuit across the line from the gap to the receiver input to minimize the chances of incident power destroying the sensitive input circuits associated with the receiver. Practically as soon as the transmitter is turned off the gap will extinguish itself and signals from the antenna can then return to the receiver. With this arrangement it is important that the transmission-line lengths be selected so that the impedance, looking down the line from the quarter-wavelength stub toward the transmitter, will be quite high compared with the receiver input impedance as viewed from the same junction.

In order to overcome this shortcoming of TR switch placement, the arrangement of Fig. 34b is often used. This makes use of the so-called ATR or anti-transmit-receive switch. This arrangement functions in such a way that when the transmitter is actuated both the TR and ATR spark gaps break down almost instantaneously. Since both switches are at the end of quarter-wavelength stubs, these stubs reflect back virtually open circuits to the main line; hence essentially all the power of the transmitter is delivered to the antenna. As before, when the transmitter is turned off the gaps will extinguish quite rapidly to allow signals from the antenna to return to the receiver. In this case, however, the ATR line reflects back a short circuit at its main line junction, and one quarter wavelength away at the TR stub junction this in turn looks like an open circuit, thus ensuring the returned signal from the antenna a direct path to the receiver input. While the above descriptions covered only open-wire lines (which are seldom used in normal practice) the same general principles apply to both coaxial and waveguide duplexers.

All the above comments have assumed that the transmission lines are matched to both the antenna and receiver and that the spark gaps look like total short circuits when they break down. The extent to which these assumptions can be achieved in actual operation is often quite close, but additional complications are often necessary to the simple circuits shown. Careful control of the nature and pressure of the gas in the vicinity of the electrodes of the gaps is important. In addition, rather than placing the gap directly at the end of the quarter-wavelength stubs, they are sometimes reflected to this position by means of a line transformer section to further reduce the gap's effective impedance. Multiple gaps are also employed in considerably more complicated duplexing arrangements. Duplexers are currently in operation which allow transmitters operating at tens of megawatts of peak power to operate on the same antenna system as today's most sensitive receivers without appreciable degradation of receiver performance.

9.8. Filters. Transmission-line filters are widely employed in both bandpass and low-pass forms. Because of the rapidly increasing utilization of the microwave spectrum by both military and civilian systems, great emphasis has been placed on the use of low-pass (harmonic-suppression) filters, and it is inevitable that this emphasis will become greater with the passage of time. Commercially available filters capable of handling megawatts of peak power and tens of kilowatts of average power are currently available for transmitter applications. The principles involved in microwave filters are extremely diverse, ranging from the use of a simple high-Q resonator, to the use of multiple-section distributed-constant filters. In a very general sense it can be stated that such filters employ the many characteristics attainable with transmission-line sections used as distributed-constant tuned elements, the cutoff characteristic inherent in waveguide structures, and the absorbing characteristics inherent in certain media when placed in the electromagnetic field of the transmission line.

9.9. Rotary Joints. The requirement for connecting the output power of a rigid transmitting line to a rotating load circuit frequently exists in microwave equipment, particularly in the case of mechanically steerable antenna systems. In the case of open-wire lines this requirement can frequently be accomplished by means of inductively coupled circuits, thus avoiding the necessity of slip rings and similar arrangements. For the cases of coaxial lines and waveguide lines, the problem is somewhat more difficult, particularly if continuous rotation of the antenna is to be accomplished.

Figure 35a indicates an elementary form of coaxial rotary joint often called the capacitive type. An axial hole is drilled in the center conductor of the rotary portion of the line and a pin is inserted axially from the fixed portion of the line, in such a way as to represent a quarter-wavelength transmission-line section. In its ideal form, the open-circuited quarter-wavelength stub will then present a short-circuited impedance at point A, thus making the center conductor appear as a continuous run. By similar action the outer conductor of the fixed line is flared out around the outer conductor of the rotary line, as a quarter-wave line section, to present a short-circuited impedance at point *B*, thus making the outer conductor appear as a continuous run. Disadvantages of this form of joint are that some radiation can take place through the outer flanged assembly and pressurization of the line is impractical.

An elementary form of waveguide rotary joint is shown in Fig. 35*b*. In this case the usual form of rectangular guide, propagating the $TE_{1,0}$ mode, must be transformed to some form of circular waveguide mode which will allow continuous rotation. This is accomplished by means of an antenna matching section projecting through a hole

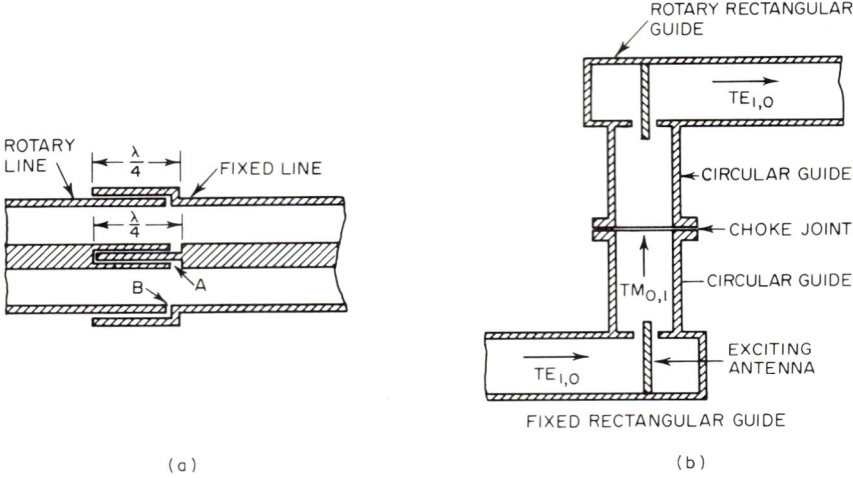

Fig. 35. Two elementary rotary joints. (*a*) Cross-section view of a coaxial rotary joint. (*b*) Cross-section view of waveguide rotary joint.

in the top wall of the fixed guide to set up the $TM_{0,1}$ mode in the circular guide. The two portions of the circular guide are then separated by means of a choke joint which allows continuous rotation. From the upper section of the circular guide the propagation is again converted to the $TE_{1,0}$ mode in the upper rectangular guide by means of a second antenna matching section.

Many more advanced rotary joint configurations exist than the two illustrated. As a result of these improved configurations the shielding and power-handling capabilities of the joint are often quite comparable with the capabilities of the transmission line itself.

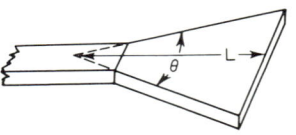

Fig. 36. Sectoral horn.

9.10. Horns. In the cases of open-wire and coaxial transmission lines, radiation of power into space by a transmitter is readily accomplished by matching the output impedance of the line to the impedance of space by means of an appropriate antenna. In the case of waveguides, some form of matching device is usually required. Horns of various configurations and lengths are often employed to accomplish the matching function as well as to provide an effective power gain.

Figure 36 illustrates a form of horn for use with rectangular waveguide which is propagating transverse electric modes. It will be noted that this form of horn retains the smaller dimension of the waveguide through its flare. With such a horn, for a ratio of L/λ of 6 and an angle θ of about 40°, the radiated beam width would be 67° and the power gain about 23 times. With a longer horn, for L/λ of 50 and an angle θ of about 15°, the radiated beamwidth would be about 22° and the power gain about 72. The directivity in a plane perpendicular to the flare is less pronounced than in the plane of the flare. The overall directivity may be improved with other horn configurations. In addition to direct radiation by means of horns, these devices are also used to illuminate the reflecting surfaces of large parabolic reflectors for the achievement of very high gain microwave radiating devices.

10. POWER SUPPLIES

Both a-c and d-c power supplies are employed to deliver the necessary external power required for successful generation of radio frequency by means of microwave amplifier complexes. Line-frequency a-c power supplies are frequently used as filament supplies, and the only precautions necessary in their application are concerned with proper regulation, limiting inrush currents, and similar considerations common to those encountered with low-frequency amplifier complexes. Practically all the other power supplies employed in a microwave amplifier complex are d-c power supplies, and in fact it is often necessary to utilize d-c power supplies for filament circuits if extremely good phase characteristics are to be obtained from the amplifiers. A block diagram illustrating the elements of a complete d-c power supply is shown in Fig. 37. Not all supplies will contain every element shown, but the necessity of the

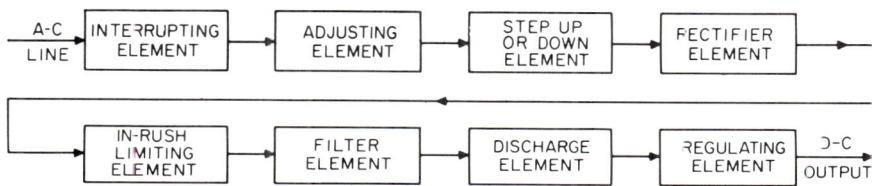

Fig. 37 Elements of a complete d-c power supply.

elements should be considered before their elimination is considered. These elements will be described briefly.

An interrupting element of some sort should be available on all d-c supplies. In the simplest case this may be only a switch and fuse. In more elaborate supplies a circuit breaker may be employed, and on supplies requiring automatic control, both a circuit breaker and a line contactor (or relay) may be employed. In any event, an interrupting element capable of providing the required degree of protection and operating flexibility is a practical necessity.

An adjusting element may or may not be an absolute necessity. It is generally wise to provide at least the simplest type of adjusting element (in the form of a tapped primary or secondary winding on the rectifier transformer) to allow a limited degree of adjustment in the output d-c voltage to compensate for line voltage and component variations. If a wide degree of adjustment is required in the d-c output, some form of adjustable autotransformer is advisable. Such devices customarily take the form of adjustable brush-type units or non-brush-type units. The later is commonly referred to as an induction voltage regulator even though it is in effect simply a brushless autotransformer. Because of its greater fault current-handling capabilities and relative ease of maintenance it is often employed as the adjusting element in power supplies operating at quite high power levels.

A step-down or step-up transformer is practically always required in d-c power-supply arrangements. In addition to its voltage-changing role, an important advantage afforded by this element is that of isolation of circuit grounding from power-line grounding. Without such isolation it is often difficult to achieve accurate sensing, monitoring, and control circuits. It is important that the characteristics of this transformer have the proper impedance characteristics with regard to the other power-supply elements.

The rectifier element of the power supply can take on many configurations in terms

Table 2. Common Rectifier-circuit Characteristics

TYPE OF CIRCUIT	SINGLE-PHASE CENTER TAP (2-1-1-C)	SINGLE-PHASE BRIDGE (4-1-1-B)	THREE-PHASE STAR (WYE) (3-1-1-Y)	THREE-PHASE BRIDGE (6-1-1-B)	SIX-PHASE STAR (THREE PHASE DIAMETRIC) (6-1-1-S)	THREE-PHASE DOUBLE WYE WITH INTERPHASE TRANSFORMER (6-1-1-Y)
PRIMARY						
SECONDARY						
WAVEFORM OF OUTPUT VOLTAGE TO CIRCUIT						
WAVEFORM OF CURRENT THROUGH RECTIFIER LEGS						
NUMBER OF RECTIFIER ELEMENTS IN CIRCUIT	2	4	3	6	6	6

		Single-phase center tap	Single-phase bridge	Three-phase star (wye)	Three-phase bridge	Six-phase star (three-phase diametric)	Three-phase double wye with interphase transformer
Avg d-c volts output	= avg d-c voltage output	1.00	1.00	1.00	1.00	1.00	1.00
RMS d-c volts output	= avg d-c voltage output	1.11	1.11	1.02	1.00	1.00	1.00
Peak d-c volts output	= avg d-c voltage output	1.57	1.57	1.21	1.05	1.05	1.05
Peak reverse volts per rectifier element	= avg d-c voltage output	3.14	1.57	2.09	1.05	2.09	2.42
	= rms secondary volts per transformer leg	2.82	1.41	2.45	2.45	2.83	2.83
	= rms secondary volts line-to-line	1.41	1.41	1.41	1.41	1.41	1.41
Avg d-c output current	= avg d-c output current	1.00	1.00	1.00	1.00	1.00	1.00 (diametric)
Avg d-c output current per rectifier element	= avg d-c output current	0.500	0.500	0.333	0.333	0.167	0.167
RMS current per rectifier element							
Resistive load	= avg d-c output current	0.785	0.785	0.587	0.579	0.409	0.293
Inductive load	= avg d-c output current	0.707	0.707	0.578	0.578	0.408	0.289
Peak current per rectifier element							
Resistive load	= avg d-c output current	1.57	1.57	1.21	1.05	1.05	0.525
Inductive load	= avg d-c output current	1.00	1.00	1.00	1.00	1.00	0.500
Ratio peak to avg current per element							
Resistive load		3.14	3.14	3.63	3.15	6.30	3.15
Inductive load		2.00	2.00	3.00	3.00	6.00	3.00
% ripple $\left(\dfrac{\text{rms of ripple}}{\text{avg output voltage}}\right) \times 100$		48	48	18.3	4.3	4.3	4.3

Inductive Load or Large Choke Input Filter

		Single-phase center tap	Single-phase bridge	Three-phase star (wye)	Three-phase bridge	Six-phase star (three-phase diametric)	Three-phase double wye with interphase transformer
Transformer secondary rms volts per leg	= avg d-c voltage output	1.11 (to center tap)	1.11 (total)	0.855 (to neutral)	0.428 (to neutral)	0.740 (to neutral)	0.855 (to neutral)
Transformer secondary rms volts line-to-line	= avg d-c voltage output	2.22	1.11	1.48	0.740	1.48 (max)	1.71 (max, no load)
Secondary line current	= avg d-c output current	0.707	1.00	0.578	0.816	0.408	0.289
Transformer secondary volt-amperes per leg	= d-c watts output	1.57	1.11	1.48	1.05	1.81	1.48
Transformer primary rms amperes per leg	= avg d-c output current	1.00	1.00	0.471	0.816	0.577	0.408
Transformer primary volt-amperes per leg	= d-c watts output	1.11	1.11	1.21	1.05	1.28	1.05
Avg of primary and secondary volt-amperes	= $\dfrac{\text{avg load current} \times \text{secondary leg voltage}}{\text{primary line voltage}}$	1.34	1.11	1.35	1.05	1.55	1.26
Primary line current		1.00	1.00	0.817	1.41	0.817	0.707
Line power factor		0.900	0.900	0.826	0.955	0.955	0.955

of single-phase and multiple-phase arrangements. In addition, the devices used for rectification are diverse, such as vacuum tubes, gas tubes, and solid-state diodes. Multiple-phase arrangements of course have the advantage of much lower ripple output; however, it must be borne in mind that line-to-line unbalances in the incoming a-c power source can appreciably increase the low-frequency ripple output of such supplies. In microwave applications it is often necessary to analyze such line unbalances carefully to ensure that the correct degree of filtering is incorporated, since many microwave tubes reflect this power-supply ripple in the form of severe phase modulation.

In many applications it is necessary to start a d-c power supply across the line at an appreciable percentage of its full power output level. Unless adequate precautions are taken, such operation may result in excessive inrush currents, especially through the rectifiers. This can be particularly hard on some solid-state rectifiers. As a result, some sort of inrush-limiting device (or devices) is usually provided. The impedance of the adjusting element, the step-down or step-up transformer, the rectifiers, as well as the power lines themselves act to limit these inrush currents. Often a step-start resistor is inserted in series with the output of the rectifiers to limit inrush current further. (This resistor is then automatically shorted out once the inrush has subsided.) The filter reactor, if incorporated, also can provide a degree of effective inrush-current limiting. It is important that the effects of all the above elements be properly taken into account if maximum component life is to be obtained.

Some form of filter element forms a part of practically all d-c power supplies. The proper combination of both inductance and capacitance to provide an LC product, adequate for the amount of rectifier output ripple reduction, is the customary method by which this accomplished. As noted earlier, it is important to account for line-to-line unbalances in the selection of this LC product. For some applications a series regulator can be employed for ripple reduction, provided its speed of response is sufficiently high.

The discharge element is an important element in all d-c power supplies. Without such a device the energy stored in the filter capacitor may remain for a long period of time before draining off because of circuit leakage. In the case of power supplies with large energy-storage elements, this may be an extreme personnel hazard. In the simplest case, the discharge element may be a bleeder resistor permanently connected across the filter capacitor. In higher-power applications a relay and resistor arrangement is usually provided whereby, when line power is removed, the discharge relay will connect a resistor across the capacitor.

The final element shown is a regulator element. This is typically a tube or transistor either in series or in parallel with the filtered output of the supply. It is controlled in accordance with the constant to be held, for instance, regulation of output voltage in the face of changing output current. Since the regulator is in effect either a variable series or shunt resistance, it is usually called upon to operate over the minimum range possible since such operation will result in the minimum-sized regulating element and the highest power-supply efficiency.

The following sections will describe several of the important considerations encountered in the design of typical d-c power supplies for microwave tube applications.

10.1. Unregulated D-C Power Supplies. Table 2 illustrates the conventional configurations of single- and multiple-phase power supplies encountered in microwave amplifier complexes. The choice of which arrangement best fits a given application is largely determined by the characteristics of available rectifying elements, the amount of ripple required at the output of the supply, and the characteristics of the available a-c lines. In general, single-phase power supplies are employed for low-power applications and multiple-phase power supplies are employed for high power levels. It will be noted that the table of characteristics does not account for rectifier voltage drop or line and transformer impedance voltage drops. In addition the effects of turn-on and turn-off transients must be considered. These considerations are of the greatest importance when dealing with solid-state rectifiers.

The charts of Fig. 38 will serve as a useful guide to determine the effects of LC filters on ripple reduction. While the charts describe the ripple-reduction factor for a single-section LC filter, the total reduction in ripple, when the filter consists of more than one section, is very nearly equal to the product of the factors of the individual sections. If the reduction factor is taken from the decibel chart the total reduction in decibels is the sum of the attenuations of the individual sections.

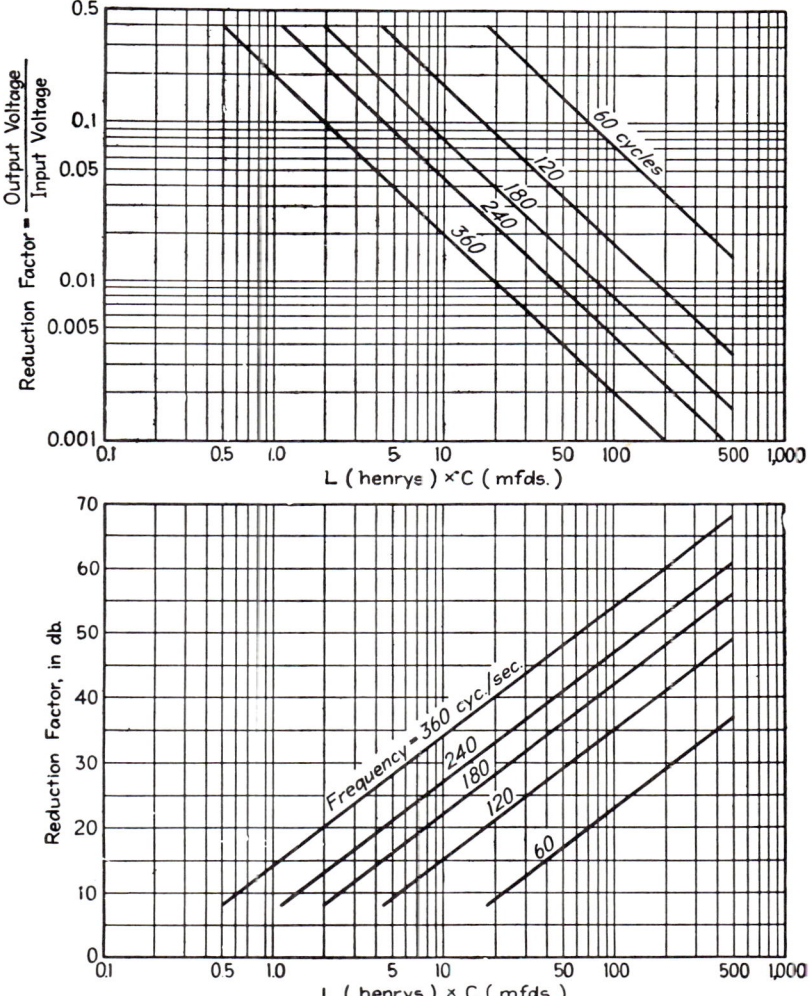

Fig. 38. Reduction factor in ripple voltage produced by a single-section inductance-capacitance filter having inductance L and capacitance C. (*From F. E. Terman, "Radio Engineers' Handbook," McGraw-Hill Book Company, New York, 1943.*)

Figure 39 will serve as a guide to determine the effects of line-to-line unbalances in incoming three-phase power lines. It can be seen that these unbalances require filtering at a much lower basic frequency and hence produce a drastic increase in the amount of filtering required. In applying this consideration to determine the total amount of filtering required for a given application it is important to include the unbalances introduced by transformer secondary voltage-ratio errors, variations in rectifier element drops, and similar effects, if a total appraisal of the situation is to be accomplished.

10.2. Regulated D-C Power Supplies. Regulation of the output of a d-c power supply can be accomplished either by adjustment of the input a-c line voltage to the supply, by phase adjustment of the rectifier elements, or by series or shunt control of the rectified d-c output voltage. These are indicated in block-diagram form in Fig. 40. The usual forms of regulation have as their purpose the maintenance of either constant voltage or constant current across or through the load connected to the output terminals of the supply.

Regulation of the input a-c voltage to the supply can be accomplished by either mechanical or electronic means. Mechanically, a drive motor can be automatically controlled to position an autotransformer line-voltage-adjusting element and hence provide a correcting input a-c voltage in accordance with variations in the output d-c voltage. This method is widely employed where the speed of response of the regulator does not have to be very rapid. Electronically, a pair of "back-to-back" tubes can be inserted in series with the a-c lines, one tube to conduct on the positive half-cycle and the other to conduct during the negative half-cycle. In the mode of operation where each tube conducts over 180° the supply will deliver its maximum voltage. If the tubes are phased to conduct for some number of degrees less than 180, it can be seen that the output of the supply will be controllable either upward or downward in voltage by means of phase variations. Regulation by means of this method can be quite effective; however, it does introduce a distorted line waveform to the rectifier transformer which must be taken into account in the design of the transformer. In addition, the output of the rectifier may be somewhat more difficult to filter because of the chopped waveform produced. An advantage provided by the "back-to-back" switches is that they can also be used for extremely rapid line interruption since it is only necessary to phase both switches off in the event of a fault in the load. As such, they are sometimes used for this function alone since speeds of interruption greater than available with contactors or circuit breakers can be obtained by this electronic means.

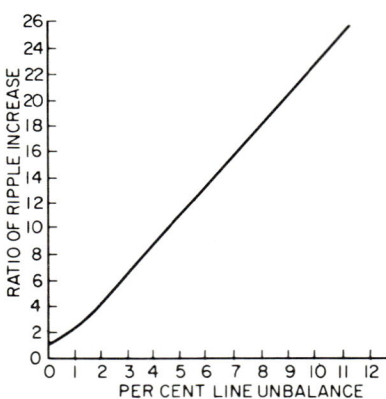

Fig. 39. Line voltage unbalance effect on ripple from three-phase full-wave rectifier. This relation is valid for filters which reduce the ripple voltage to 0.05 per cent or less at zero line unbalance. (*From C. K. Hooper and C. H. McAdie, Effects of Supply Line Unbalance on Filtered Output Ripple of Polyphase Rectifiers, AIEE Technical Paper 50–125.*)

If grid-controlled rectifiers are employed for the power supply, phase control of these elements can provide effective control of the d-c output of the supply. As with the "back-to-back" line control tubes, attention must be given to waveform distortion in the design of such a supply. The elements used in either case are usually low-drop devices such as gas thyratrons, ignitrons, or silicon controlled rectifiers.

Perhaps the most common method of controlling the output of a d-c supply is by means of a series or shunt tube regulator. The use of the series hard-tube regulator has been firmly established over many years of usage in both small and relatively large power supplies. The series tube regulator is not too effective if load is completely removed from the supply at periodic intervals. In this case the shunt tube regulator is recommended. With its use the supply can be loaded to any required degree to maintain the necessary output voltage. In the case of complete loss of load on the supply, a tube of considerable plate-dissipation rating will of course be required. As noted earlier, if the servosystem has sufficient speed of response this form of regulator can be very effectively employed as a ripple clipper or filter.

10.3. High-voltage Power Supplies. While the design problems associated with high-voltage power supplies (generally taken as those delivering greater than several kilovolts of d-c output voltage) are essentially identical to those involved in low-voltage power supplies, greater consideration is due since the components involved in such supplies comprise a major portion of the cost of a given microwave amplifier

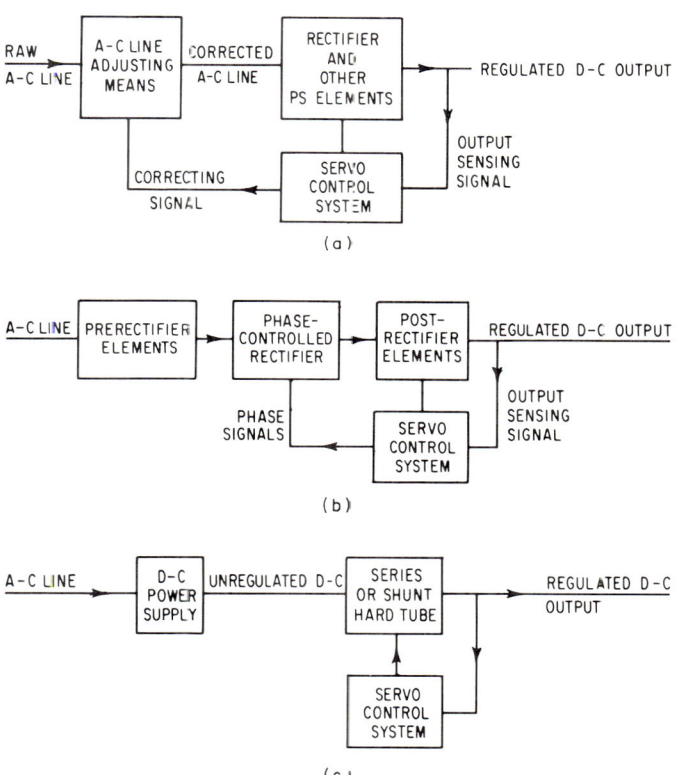

Fig. 40. Three forms of regulated power supplies. (a) A-C line regulator. (b) Phased rectifier regulator. (c) D-C output regulator.

complex. Two areas of importance are those associated with insulation and energy storage.

At voltages exceeding 5 or 10 kv the use of both oil insulation and air insulation is frequently encountered. Since high voltage is usually synonymous with high power, it is common that iron-cored units (rectifier transformers, filter reactors, variable autotransformers, etc.) are designed as oil-immersed assemblies. This approach results in both superior cooling and insulating characteristics in these assemblies. In some cases, where the voltage is quite high (in excess of 100 kv), the rectifier assembly is also oil-immersed. Where the assemblies are partially in oil and partially in air, a good rule of thumb is to allow at least 1 in. of clearance (in air) for each 10 kv involved in a given high-voltage bus run. In addition it is important to use conductors of large diameter if corona is to be avoided.

The second consideration, energy storage, comes about in most part as a function of the filter elements involved in the power supply. In the case of pulse-modulator power supplies an exceptionally large amount of filter capacitance is necessary when long pulse durations requiring low pulse droop are encountered. Since microwave amplifiers often have relatively fragile structures internally, it is necessary to provide means whereby the large energy storage of the filter capacitor does not discharge through the tube in the event of an internal tube arc. To accomplish this an electronic "crowbar" is often employed with high-voltage power supplies. This consists of a short circuit (in the form of a triggered gas tube or triggered spark gap) which is applied across the capacitor bank in the event of an arc in the microwave amplifier. Since a small resistance is usually inserted between the amplifier and the crowbar, practically all the fault current is diverted through the crowbar path. It can be seen that such operation stresses all the power-supply components to fault conditions, and it is of course necessary to interrupt the a-c power lines as rapidly as possible when such faults occur. It is quite important that due consideration be given to voltage and current transients in the case of all high-voltage power supplies, especially so when crowbar circuits are to be incorporated.

11. MODULATORS AND PULSERS

A device used to effect the process of modulation of a microwave carrier is customarily referred to as a modulator. Often, especially in the case of radar pulse modulation, the device is also referred to as a pulser. In either case, the device frequently has many of the characteristics of an amplifier or in fact is an amplifier.

Modulators perform their basic function over a wide range of power levels. In the case of applying amplitude modulation to the r-f carrier generated by a negative-grid tube amplifier chain, either high- or low-level modulation may be employed. It is possible to use low-level modulation because these tube types, in general, have the capabilities of both good linear amplification and good efficiency when they are operated as class B amplifier stages. High-level modulation of negative-grid tubes usually results in higher overall efficiency from the chain, but this is obtained at the expense of a more complex and powerful modulator. As was pointed out in an earlier section, most other microwave amplifier tubes do not possess class B amplifier characteristics; hence their use in low-level modulation arrangements usually results in quite poor overall efficiency from a chain comprised of such tubes. This poor efficiency limitation is inherent and unavoidable when employing many communication modulation modes; however, most forms of pulse modulation will allow the tubes to be high-level-modulated with a decided improvement in efficiency. When angular-modulation modes are employed for communication purposes, low-level modulation is practically always employed. Several of the considerations associated with the characteristics and applications of typical modulators and pulsers will be presented in the following sections.

11.1. Communications Modulators. Most of the modulator considerations associated with the typical forms of communications intelligence transmitted at the lower frequencies are quite similar in their application at microwave frequencies. Figure 41a illustrates a block diagram of a high-level double-sideband amplitude-modulated negative-grid amplifier chain. A stable oscillator and several driver stages are indicated in a typical amplifier chain which ultimately drives a high-power microwave amplifier. The high-level anode modulator in this case would consist of an audio (or video) amplifier arrangement whose final stage must be capable of producing a modulation intelligence power output of approximately one-half the d-c power input (for sinusoidal modulating waveforms) to the final amplifier stage (if 100 per cent modulation is to be achieved). Typically, this modulator output would be series-connected between the power supply and the anode of the negative-grid amplifier tube. Should the final stage be a tetrode, simultaneous modulation of the anode and the screen of the power amplifier is usually required. This technique is identical to high-level modulation as used on low-frequency transmitters; however, several limitations may be encountered at microwaves. At microwave frequencies, because of the greater r-f bandwidth available, much more intelligence may be required to be transmitted, hence greatly increasing the bandwidth required of the high-level modulator. Since the means of coupling the output of the modulator to the power amplifier is usually an iron-core transformer, it may be difficult to get the required bandpass characteristics from this unit. In addition, tube and circuit shunt capacitances introduced across the secondary of this transformer may also reduce the high-frequency capabilities of such a modulator.

Figure 41b illustrates a low-level modulation approach for the same mode of operation. In this case, the modulation is applied to a low-level stage, hence greatly reducing the problems associated with high modulation frequencies. The following

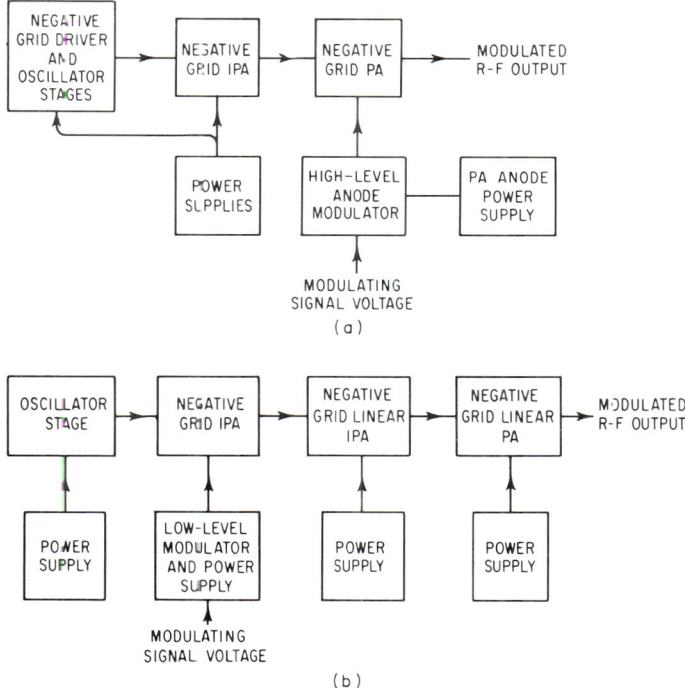

Fig. 41. High-level and low-level double-sideband AM transmitter block diagrams.

two amplifier stages must then be operated as linear amplifier stages. While this mode of operation is more critical of the adjustments of these two stages, most microwave negative-grid tubes have this capability. As is the case at lower frequencies, low-level modulation results in operating a given amplifier tube at somewhat less than the power output possible when used in the high-level modulation mode. In addition the efficiency of an amplifier operated in this manner is somewhat lower.

Control electrode modulation of negative-grid tubes (such as control-grid and screen-grid modulation) is also used at microwave frequencies. A slight variation of this technique is employed with some traveling-wave tubes and klystrons. These tubes are often designed to include a grid or modulating anode element as a part of the electron-beam structure. These are positive-control elements; that is, they control the magnitude of the beam current in such a way that as the control electrode is made more positive with respect to the cathode, the beam current is increased. Figure 42 illustrates how such a tube can be employed for double-sideband amplitude modulation. In this case the control electrode modulator is connected between the cathode and the control electrode to apply voltage excursions in accordance with the modulating signal voltage driving function. Since the beam-interception current on the control electrodes of such tubes is usually quite small, a relatively small amount of power is required to provide the modulation function. A potential difficulty to be anticipated with this technique is that the cathode is practically always operated at full beam voltage with respect to the grounded body structure; hence the modulator

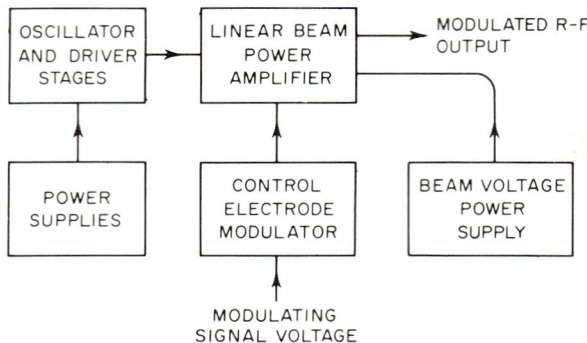

Fig. 42. Control-electrode-modulated linear-beam transmitter block diagram.

must be a floating affair to which the modulating signal voltage must be applied by means of some sort of isolating device.

Linear-beam tubes are also operated in a manner quite similar to that noted in Fig. 41b for negative-grid tubes. In the case shown the IPA and PA stages could be either klystrons or traveling-wave tubes, so adjusted that the linear portion of their r-f drive vs. output characteristic is being utilized. As has been noted previously, this is usually accomplished at the expense of rather low overall efficiency.

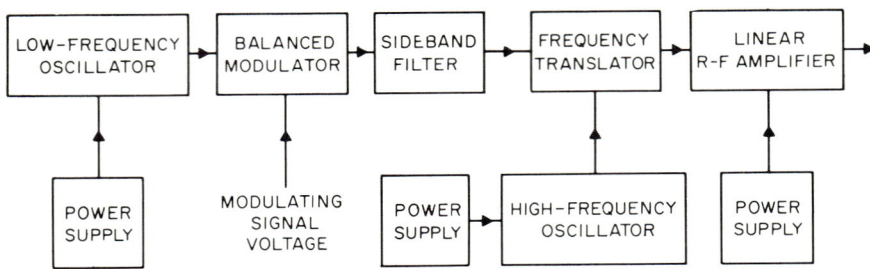

Fig. 43. Single-sideband transmitter block diagram.

The previous examples are typical of double-sideband AM configurations. For a given negative-grid tube used either in the high-level modulated AM mode or in a single-sideband suppressed-carrier mode, it can be shown, in theory, that the use of single-sideband suppressed carrier can provide an effective received signal gain of 9 db over conventional AM. For this reason there is considerable interest in this mode of communication, despite its somewhat greater complexity from both the transmitting and the receiving viewpoint. Figure 43 illustrates one common form of single-sideband generator in block form. This arrangement, called the filter system, functions as follows. A nonmicrowave low-frequency oscillator delivers its CW output to a balanced modulator. The balanced modulator functions to combine the low-frequency carrier and the modulating signal voltage in such a way that at its output only the two modulated sidebands (and no carrier) appear. A number of effective forms of balanced modulator circuits may be found in the literature. The output of the balanced modulator is then passed through a bandpass filter whose characteristics are such that the required sideband is removed. These three block functions are usually performed at relatively low frequencies to simplify the design considerations associated with the balanced modulator and the filter. The single sideband thus obtained at the output of the filter is then passed to a frequency translator where it is effectively mixed with the output of a high-frequency oscillator to provide a translated output at the desired operating frequency. This low-level signal is then amplified up to the required output power level by means of one or more linear amplifier stages. Since no carrier is transmitted, it is necessary to reinsert the carrier in its proper frequency relationship at the receiving position. This is often quite difficult even at relatively low frequencies and is therefore increasingly difficult as the microwave frequency of interest is increased. For this reason the oscillators noted must be of the highest order of stability as must also be the stability of the receiving circuits.

Angular-modulation modes (frequency and phase modulation) are accomplished in a manner identical with low-frequency practice. The problem of obtaining sufficient frequency deviation is usually not difficult if a sufficiently low frequency oscillator is employed as the master oscillator, since its permissible degree of phase modulation can be multiplied extensively to reach the microwave region. As noted previously, practically all the forms of microwave amplifiers are quite effective for use as angular-modulation amplifiers, because of their superior behavior with constant-amplitude driving signals.

11.2. Line-type Pulsers. Two fundamental types of pulse modulators are frequently encountered in microwave pulsing service. They are the line-type pulser and the hard-tube pulser. The line-type pulser is most frequently employed in applications where pulses of constant time duration, repeated at a relatively constant pulse repetition rate, are required. The hard-tube pulser is most appropriate for applications requiring pulses whose durations and interpulse periods are variable as a function of time.

The line-type pulser is a device which stores only the energy required for a single pulse during each interpulse period. This energy is stored in a pulse-forming network (an artificial transmission line from which the pulser derives its name) usually consisting of lumped-constant inductances and capacitors. When this line is discharged once each interpulse period, virtually all the stored energy is delivered to the amplifier being pulsed.

Figure 44 shows a typical line-type pulser in simplified schematic-diagram form. In brief, it functions as follows: the high-voltage d-c power supply charges up the capacitance of the pulse-forming network (via a path including the charging reactor, a charging diode, and the pulse transformer) to a value of approximately twice the potential of the power supply, by means of d-c resonant charging circuit, and the switch tube is ready to be triggered in accordance with an amplified synchronizing trigger; the switch tube (when triggered) discharges the network into the load impedance reflected via the pulse transformer (as a result of the reflected impedance of the amplifier being pulsed); at the conclusion of the pulse duration (which is determined by the relative values of L and C in the network) the network will be almost completely discharged, the switch tube will soon cease conducting, and the network will start to recharge; the cycle of events just described will then repeat itself. This description, while somewhat idealized, correctly describes the events as they occur in a properly performing line-type pulser. A more detailed description of the component parts will now be presented.

Pulse-forming Network. This element forms the heart of any line-type modulator. Its role as the storage element can most clearly be visualized by comparing the characteristics inherent in an actual transmission line with those simulated by the artificial transmission line used as the pulse-forming network. Consider an actual two-conductor air-insulated line having uniformly spaced conductors, an inherent characteristic impedance Z_o, no resistance in its conductors, and no connections at either end. If this line is first charged by means of a d-c source to a voltage E_{ch}, second has the charging source disconnected, and third has a pure resistive termination equal to Z_o instantaneously connected across one end, the following pulse phenomenon will occur at the resistive termination. The voltage will rise instantaneously to a value $E_{ch}/2$; the voltage will then remain at this value for a period of time τ; and following this finite period, the voltage will drop to zero, at which time the line will have completely discharged all its stored energy into the resistive termination. The length of the pulse thus generated is a direct function of the line length involved, and for the case of the ideal transmission line described, the pulse length would be approximately equal to 0.00204 μsec/ft of length. It can be seen that for only a 1-μsec pulse a line length of some 492 ft would be required for its generation. For this reason,

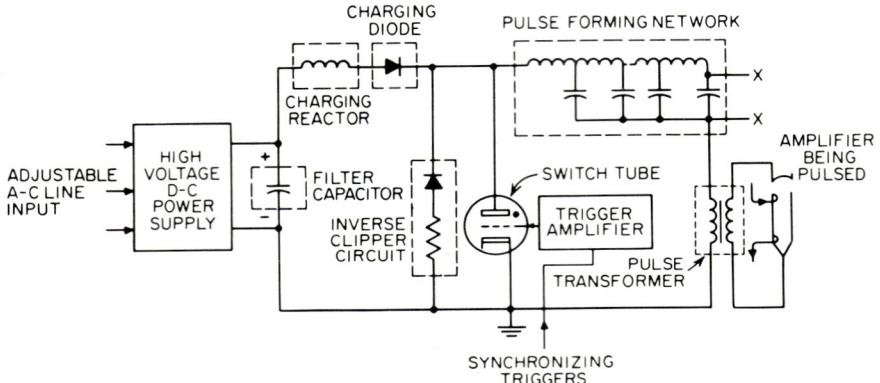

Fig. 44. Simplified schematic diagram of a line-type pulser.

actual lines are very rarely used in practice; rather, lumped constant lines are used to perform the function of the pulse-forming network.

The expression for the total amount of capacitance required for an artificial line is

$$C_n = \tau/2Z_n \tag{29}$$

where τ is the required pulse duration in microseconds, Z_n is the required characteristic impedance in ohms, and C_n is expressed in microfarads. Similarly, the total value of inductance required for such a line is

$$L_n = \tau Z_n/2 \tag{30}$$

where L_n is expressed in microhenrys. As contrasted to the ideal line, an artificial line will have finite rise and fall times which are approximately expressed by $\tau_r = \tau/2n$, where n is the number of sections into which the artificial line is divided. A common form of pulse-forming network is the type E line in which the total capacitance C_n is divided into n equal sections, the coil is wound in the form of a continuous solenoid with n equally spaced taps (except for the end sections), and the length/diameter ratio of the individual coil sections is approximately 1.35:1. In practical networks, slight adjustment of the taps from these approximate initial positions is usually made to provide a pulse with minimum overshoot and minimum pulse top ripple. With care, these imperfections can be reduced to values of well below 1 per cent of the peak value of the pulse. If still lower values of ripple are required they can be achieved by biased-diode ripple-clipping circuits.

In a practical line-type pulser, the load reflected at the pulse transformer primary terminals is usually made to match almost exactly the characteristic impedance of the pulse-forming network (in order to obtain maximum transfer of energy). Assuming the maximum peak power required by the tube being pulsed is known, the value of network impedance is given by the expression

$$Z_n \approx \frac{E_c^2 \eta_n \eta_t}{P_L(2 + 1/k + k)} \qquad (31)$$

where E_c is the voltage to which the pulse-forming network is charged, η_n is the per cent discharge efficiency (expressed decimally) of the network, η_t is the per cent efficiency of the pulse transformer (expressed decimally), P_L is the peak power delivered to the load, and $k = Z_n/Z_p$ (where Z_p is the load impedance reflected at the primary of the pulse transformer). In well-designed high-power circuits, the two efficiency factors are nearly equal to 1.0 and the load is nearly matched to the pulse-forming network; hence the value of Z_n closely approximates

$$Z_n \approx E_c^2/4P_L \qquad (32)$$

The above relationship clearly establishes the close interdependence between the network impedance and the charging voltage for a given application. More correctly stated, it establishes a fundamental relationship between the network and the switching element of the line-type pulser. This will be the next element considered.

Switch Tube. Gas-tube switches are the most commonly employed switching elements in line-type modulators. Typically, these may be either hydrogen thyratrons, xenon thyratrons, mercury-vapor thyratrons, or ignitrons. Each of these basic tube types differs appreciably from the others; however, they all possess certain maximum ratings which define their effective capabilities when used in line-type modulator circuits. The major ratings involved are the maximum forward holdoff voltage the maximum peak current, the maximum average current, the maximum rms current, and the maximum pulse repetition rate. These ratings are quite interrelated although they vary with the type tube being employed. In the case of hydrogen thyratrons, a figure of merit is usually expressed in terms of the product of the peak forward holdoff voltage, the peak current, and the pulse repetition rate. This product must not exceed a stipulated maximum value even though all the individual factors may be less than their maximum specified values for a given application. In addition, the rms current and average current ratings must also not be exceeded.

It is generally the case that switch tubes are operated quite close to their maximum forward holdoff rating. This is done since such operation results in the lowest pulse transformer step-up ratio, the lowest average current, and the lowest rms current, all factors which usually simplify the overall design considerations encountered in the modulator. In addition, from the expression for the network impedance presented in the previous discussion, the highest permissible value of forward holdoff voltage (which is the same as the network charging voltage) will result in the highest value of network impedance for a given peak modulator power. This consideration often makes possible a more effective pulse-forming network design when high-peak-power and short-pulse-length applications are involved.

Spark-gap switches are sometimes employed as modulator switching elements. A rotary spark gap, driven by a motor which rotates the arc contacts at the proper rate, can be employed in systems where external synchronization of the modulator firing is not required. A triggered spark gap (usually consisting of two or more ball gaps with some form of needle gap triggering electrode between them) may be triggered from an external source in much the same manner as gas switch tubes. Spark-gap switches are characterized by much higher pulse-to-pulse time jitter and in addition require considerable maintenance as compared with switch tubes.

Pulse Transformer. Most line-type modulators employ pulse transformers, although occasionally direct connection (via the switch tube) from the network to the load is employed. Two prime attributes of the pulse transformer are that it can be used to reverse the polarity of the pulse delivered by the network and that it can be used as a voltage step-up device. Since any line-type modulator design quite clearly begins with the amplifier being pulsed (its peak voltage times peak current product determines the impedance of the pulse-forming network as noted earlier), the transformer must have a secondary winding voltage, current, and pulse-duration capability adequate for the application at hand. The required turns ratio is given by

$$N = (E_s/I_s Z_p)^{1/2} \tag{33}$$

where N is the step-up ratio between the primary and the secondary, E_s is the secondary pulse voltage, and I_s is the secondary pulse current. The respective primary voltage and current will be given by

$$E_p = E_s/N \sqrt{\eta_t} \quad \text{and} \quad I_p = I_s N/\sqrt{\eta_t} \tag{34}$$

Typical pulse transformers have efficiencies in excess of 90 per cent, the lower 90's being associated with short-pulse and low-power transformers, and the upper 90's range being consistent with long-pulse and high-power transformers.

Pulse transformer characteristics of prime importance are associated with the rise time and overshoot characteristics of a given configuration. To a first-order approximation these characteristics are a function of the leakage inductance and the shunt capacitance inherent in a given unit and its associated circuitry. If it is assumed that the pulse-forming network delivers a step-function pulse to the primary of the pulse transformer, it can be shown that a critical compromise of these two factors will be achieved (the rise time is sufficiently slow as to just prevent any overshoot) when

$$Z_n = 0.707 \sqrt{L_l/C_s} \tag{35}$$

where L_l is the leakage inductance of the transformer as viewed from the primary and C_s is the total shunt capacitance viewed from the primary (including the shunt capacitance of the transformer as well as the load circuit). In practice, a primary step waveform of the type described is not achievable, and in addition, the pulse transformer contains a number of second-order circuit paths which can produce pulse variations (if an extremely flat top characteristic is desired). As a result, the above expression is a guide, and precise waveform adjustment is usually achieved by controlling circuit configurations (primarily the pulse-forming network) after the modulator is assembled.

A final characteristic of importance is the droop inherent in all pulse transformers. As in any transformer, the exciting inductance of a pulse transformer appears essentially as a shunt inductance across the primary driving source. This will result in a shunting effect on the load, with the result that a droop will appear across the load unless proper precautions are taken. The most common way in which this effect is overcome is to incorporate a rising characteristic into the pulse-forming network. This is done by designing the network so that its characteristic impedance is not constant, but rather that the impedance decreases as the length of the network is traversed. An alternative method by which droop is overcome is to employ a biased-diode clamping circuit across one of the transformer windings.

Inverse Clipper Circuit. The role of this circuit is to remove the energy remaining on the pulse-forming network following the main pulse output period of the modulator's cycle of events. Two conditions are of importance in this regard, first the network energy conditions following a normal pulse and second the energy conditions following a faulted load pulse. With normal pulse conditions, it was noted earlier that the network is usually quite closely matched to the load impedance reflected back by the pulse transformer's turns ratio. In practice this may be either a positive or a negative match; that is, the reflected load impedance may be either slightly higher or slightly lower than the network impedance. The negative-match condition is frequently employed since this sort of mismatch will result in the network's developing a reverse polarity across its terminals immediately after the main pulse conduction period, and this polarity is of such a nature as to cause the switch tube to cease conduction and begin to deionize. It can also be shown that in most cases a positive match will also cause the network voltage to reverse (although not so rapidly since in this case a resonance effect between the exciting inductance of the pulse transformer and the capacitance of the pulse-forming network is the reversal medium) to ensure switch-tube deionization. In either case the switch tube will deionize (in a properly designed line-type modulator) after a relatively short portion of the interpulse period and inverse voltage will be left on the network. The function of the inverse diode circuit in this case is to remove the energy left in the network as rapidly as possible without interfering with switch-tube deionization. If this were the only function to be performed, the requirement for the clipper circuit would be that the RC time constant (of the effective clipper resistance and network capacitance circuit) be no more than approximately 10 or 20 per cent of the interpulse time.

Turning now from the normal pulse to the fault pulse condition, somewhat more stringent conditions are placed upon the clipper. The fault condition referred to in this case is that of the amplifier being pulsed, arcing, and hence reflecting back a short circuit at the pulse transformer primary. This condition will result in a network reversal in which the voltage level is approximately equal to the initial forward charge voltage. The inverse clipper will function as before; however, the current and voltage involved are far greater. Also, if the amplitude of the next charging cycle is to be limited to a value not appreciably in excess of its normal level, a considerably lower value of inverse circuit resistance should be employed. The value sometimes selected is approximately equal to the characteristic impedance of the pulse-forming network. When such low values of clipping impedance are used, backswing clipping diodes are often placed across the pulse transformer primary to prevent reverse-direction arcing of the amplifier being pulsed.

An alternate clipping position is to place the diode and resistor combination across the last section of the pulse-forming network in position x-x. In this case the clipper is referred to as an "end-of-the-line clipper," and in some cases it offers advantages not possible otherwise. One of the advantages claimed for this location is the complete dissipation of fault energy before the high inverse voltage can be applied across the switch tube. In practice this condition is quite difficult to achieve, although a close approach is possible with careful design.

The diodes used in inverse clipping circuits cover the range of vacuum types, gas-filled types, and solid-state types. The normal clipper circuit employs noninductive resistors, although numerous variations are possible and frequently nonlinear resistance and even inductance is inserted in series with the resistors to achieve special inverse voltage-control effects.

Charging Circuit. The function of the charging circuit is to transfer energy from the high-voltage power supply to the pulse-forming network during the interpulse period, and to effect this transfer in the most efficient and stable fashion. The technique generally employed utilizes some form of resonant charging, usually d-c resonant charging since the high-voltage power supply is most often a d-c supply. Figure 45 illustrates several forms this charging may take. For the form of resonant charging shown in Fig. 45c the charging action occupies the complete interpulse period. It is frequently used in line-type modulators operating at a fixed pulse repetition rate with hydrogen thyratron switch tubes. The charging diode indicated in Fig. 44 is not required in this case since the switch tube is synchronized to fire at the instant the wave reaches its maximum value. The value of the charging reactor in henrys for this case will be

$$L_c = 1/\pi^2 (\text{PRR})^2 C_n \qquad (36)$$

where PRR is the pulse repetition rate in pulses per second and C_n is the network capacitance in farads. The charging current associated with this type of operation is shown in Fig. 45d.

Figure 45b illustrates d-c resonant charging with a holdoff (or charging diode) in the circuit. In this case the charging inductance will of necessity be smaller than that of the previous type (assuming the same time scale for both figures), and were it not for the action of the diode the resonant charging action would continue to charge and discharge the network at the basic frequency of resonance of the elements. With this mode of operation it is possible to operate a line-type modulator at a variable pulse repetition rate, so long as the switch-tube firing period always occurs during the period

27-82 AMPLIFIER CIRCUITS

indicated as the holdoff time. The value of the charging reactor in henrys in this case will be

$$L_c = \tau_c^2/\pi^2 C_n \qquad (37)$$

where τ_c is the desired charging period in seconds. Since the period of charging is shorter than before, the charging current will be proportionately higher, as noted in Fig. 45d.

Figure 45c illustrates delayed resonant charging. For this type of operation some form of unidirectional switching device is required in place of the charging diode. Both vacuum tubes and gas switch tubes have been employed for this purpose. The

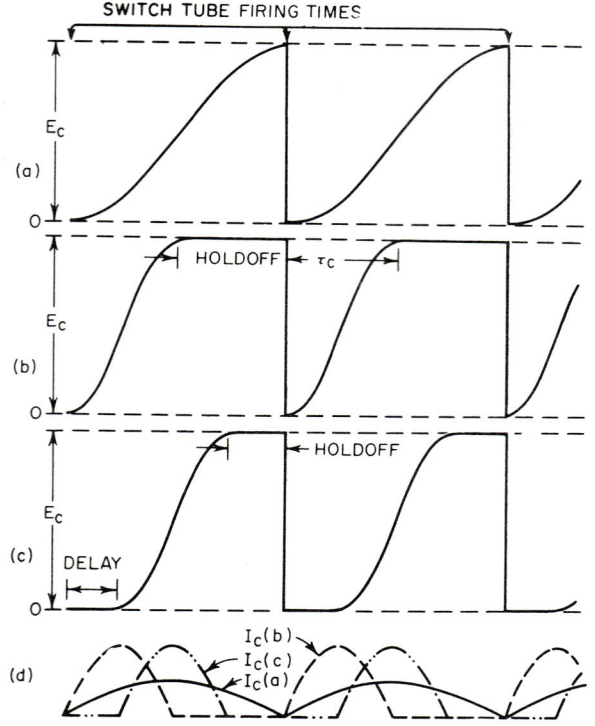

FIG. 45. Typical charging-voltage and charging-current waveforms.

advantage of this type of operation is that a long period of time during which no voltage is applied to the main switch tube can be obtained, thus allowing switch tubes with poor deionization characteristics to be employed in modulators where they would otherwise be ineffective. Another manner in which this action can be achieved is to insert a nonlinear (saturable) reactor in series with the linear charging reactor. The saturable reactor must of course have a very high unsaturated inductance (compared with the charging reactor) if the duplication of action is to compare closely with that of the tube delay method.

In the figures shown, it is assumed that all inverse voltage is removed from the pulse-forming network before charging commences. This is virtually an impossible condition to achieve under all conditions of operation of a line-type modulator. If this energy is removed within a few per cent of the time taken by the charging period, the voltage to which the network will charge can be shown to be

$$E_c = 2E_p(1 - \pi/4Q) \qquad (38)$$

where E_p is the d-c power-supply voltage and Q is the effective quality factor of the charging reactor. The expression $(1 - \pi/4Q)$ is actually the efficiency of the charging reactor. Typical values for the linear iron-core or air-core reactors used for this purpose run in excess of 90 per cent.

Filter Capacitor. While the filter capacitor shown in Fig. 44 is usually made a part of the high-voltage power supply, its role is somewhat different from that of a filter capacitor supplying a constant-resistance load. As was noted earlier, the loading on the power supply occurs in the form of half sinusoids of charging current. If too small a value of filter capacitance is employed, surges will be pulled from the a-c lines supplying the power supply. The actual selection of this capacitor is of course a compromise between line surges and cost of capacitance. It is also a quite complicated analysis to perform since it involves all the characteristics of the elements comprising the power supply. A typical compromise value is to make the filter capacitance equal to approximately ten times the network capacitance.

High-voltage D-C Power Supply. The characteristics of d-c power supplies of the type needed for line-type modulators are similar to those of conventional high-voltage supplies with the exception of the filter capacitor selection just noted. The d-c output voltage level is given by

$$E_p = E_c/2(1 - \pi/4Q) \qquad (39)$$

as noted earlier, and the d-c output current rating is given by the expression

$$I_c \approx E_c C_n(\mathrm{PRR}) \qquad (40)$$

The current expression does not take into account the charging of any stray circuit capacitance. In addition the expression for the network capacitance given earlier did not take into account the additional energy always required to supply rise-time requirements (assuming the pulse length is measured across the flat top of the pulse). These two effects may in a practical situation cause this value of current to be a few per cent too low.

11.3. Hard-tube Pulsers. The hard-tube (vacuum tube) pulser is so called to differentiate it from the types employing gas tubes (soft tubes) as the switching elements. In essence the hard-tube pulser represents a fundamental approach to any switching problem, namely, the use of a unidirectional series switch. The switching elements often employed for this purpose are conventional negative-grid power-amplifier tubes. These tubes are normally grid-biased well beyond cutoff during the "off time" of the pulser and gated into conduction during the "on time" by heavily driving their grids into the positive-conduction region. Both triodes and tetrodes are employed extensively. In addition to conventional power-amplifier tubes (which are usually rated for considerably higher current and voltage operation than when they are used in CW service) some specially designed hard-tube switches are also available. These vary from tubes which depart only slightly from the principles employed in conventional tubes, to types which relate more closely to the principles involved in high-power klystron amplifiers.

The hard-tube pulser is characterized by much greater apparent simplicity than the line-type pulser, and by considerably greater flexibility than is possessed by the line-type pulser. The first characterization is sometimes misleading, especially if very high peak power operation at very high voltage levels is contemplated. The second characterization is usually unqualified (provided the grid gating circuits have the necessary flexibility) since essentially any pulse duration at any pulse repetition rate is readily possible (so long as the hard tube's ratings are not exceeded) simply by changing the character of the grid gates. When such flexibility is necessary, as it sometimes is for pulse chains consisting of a large number of pulses of varying durations and at close spacings in a burst mode, the line-type pulser is completely inappropriate unless great complexity is resorted to. A basic limitation must be considered when employing hard-tube pulsers, however, and that is that the switch tube always draws its pulse energy from a reservoir; hence, equal-amplitude pulses can be obtained only if the reservoir is replenished to its original level between pulses. This is not possible during burst modes, with the result that either a droop appears on the chain of pulses or elaborate variable-level gating circuits must be employed to counteract this droop. The droop problem is of course present in any hard-tube pulsing mode,

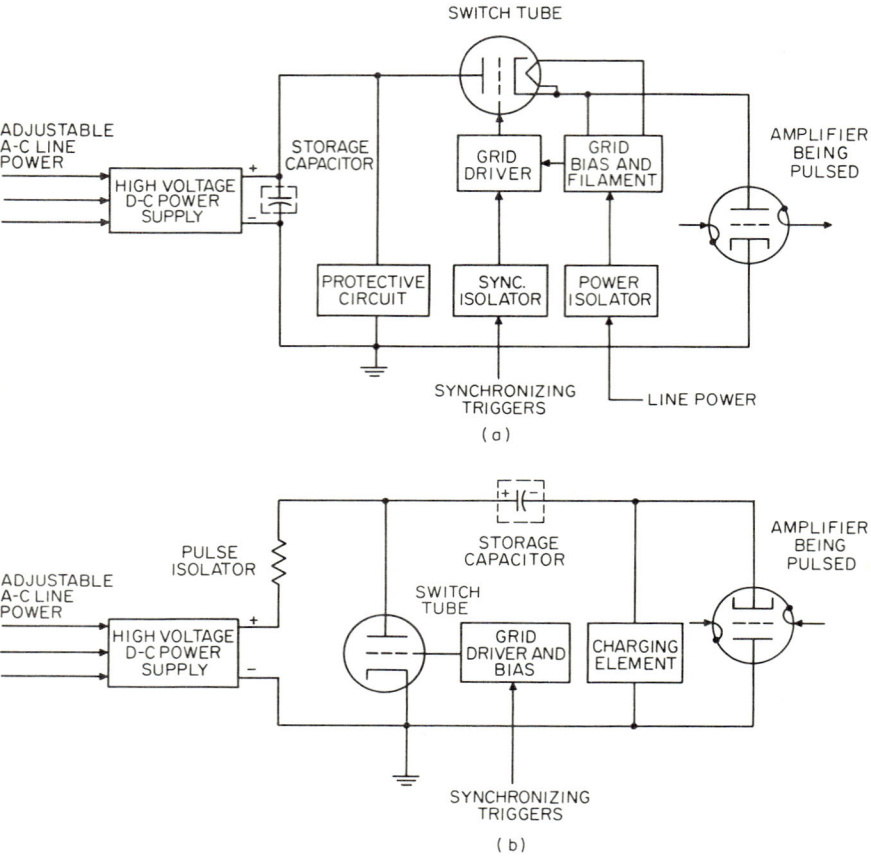

FIG. 46. Two basic forms of hard-tube pulsers.

but it may be more easily counteracted in the cases where a pulse of constant duration, repeated at a constant pulse repetition rate, is involved.

Figure 46a represents a commonly encountered form of the hard-tube pulser employing the series switching mode of operation. In this arrangement it can be seen that the series switch is simply gated on by the grid driver and that the storage capacitor reservoir will then deliver pulse energy to the tube being pulsed for the period of time during which the gate exists. If the droop on the pulse is to be limited to some relatively low level during the switch tube on time, the storage capacitor must have sufficient energy storage according to the approximate relationship

$$\text{Stored energy} = \text{peak power} \times \text{pulse duration}/\text{droop} \times 2 \qquad (41)$$

MICROWAVE AMPLIFIERS   27-85

where the stored energy required is the energy level at the beginning of the pulse, in joules; the peak power is the power to be delivered to the tube being pulsed, in watts; the pulse duration is in seconds; and the droop is expressed as a decimal (1 per cent droop as 0.01).

If long effective pulse durations are involved in a given application, the energy storage involved can be quite significant, particularly when very low droop requirements are coincident with this condition. As a result a protective circuit is often required. This comes about because it is possible that the hard-tube switch may arc, thus causing the total stored energy of the capacitor bank to discharge into the amplifier being pulsed. To prevent the damage which might easily result to the amplifier being pulsed, a short-circuiting device (often called a crowbar) is actuated to discharge the capacitor bank and provide the necessary measure of protection. This of course necessitates disconnection of the high-voltage power supply from the power lines until the shorting circuit is removed. It is also possible (and probably more likely) that the amplifier being pulsed may arc, in which case extremely high fault currents (limited only by the hard tube itself) will tend to flow into the fault. Normally it should be possible to eliminate this condition quickly by some form of quick gate-killing circuit on the hard tube. If for any reason this is not effective, the crowbar may also be fired to prevent damage to both tubes.

In the circuit of Fig. 46a it should be noted that the switch tube is operated with none of its elements at ground potential. This necessitates the use of an isolated platform for the hard tube's grid and cathode circuits; hence some form of isolation must be employed to bring up synchronization and power from ground level. Synchronizing signals are often delivered by such means as properly insulated pulse transformers, light beams, or r-f beams. A low-capacitance isolation transformer is employed to supply the necessary a-c power. Regarding stray capacitance in general, the stray capacitance associated with the switch-tube circuits should have the minimum possible values to ground, since this capacitance essentially shunts the tube being pulsed and can seriously affect the pulse rise time if present in excessive amounts.

It should be noted that the circuit arrangement shown can also be employed in the reverse direction; that is, a negative power supply can be employed and the direction of the tube reversed, thus making the arrangement a negative pulser. This connection considerably reduces the stray capacitance considerations.

Figure 46b illustrates the traditional negative-polarity hard-tube pulser configuration, originated in the early days of radar, and still frequently employed. It uses a floating storage capacitor rather than a floating tube. Its advantage lies in the fact that the hard tube has all its critical gating circuits as well as its filament circuits at ground potential. It is important in this case that the storage capacitor have as low stray capacitance to ground as possible, if good rise time is to be achieved. The pulse isolator (whose function it is to prevent power supply short circuiting during the on period) is typically either a noninductive resistor or a critically inductive resistor. The charging element may be a diode, a resistor, an inductor, or a combination of elements. Proper combination of these elements will provide fast fall time on the pulse by causing a resonant discharge action of the stray pulse circuit capacitance during the postpulse period. This is particularly important if the tube being pulsed is a magnetron or similar biased-diode-action-type tube.

Figure 47 shows another manner in which hard-tube switches are commonly employed. It is often referred to as the floating-deck pulser. This arrangement finds its main application when tubes employing beam-control electrodes are employed for r-f generation. In the example a modulating-anode-type klystron is shown as the tube being pulsed. The action is such that if the klystron's modulating anode is connected to its cathode, the beam current is effectively turned off. This is accomplished by causing the "off" switch tube to conduct. Similarly, the klystron will have its beam current turned on by turning this switch tube off and simultaneously causing the other switch tube to conduct, thus raising the modulating anode to the required operating voltage level. This voltage level is set by the divider resistors and typically lies somewhere between one-half and full beam voltage. The modulating anode interception current is usually quite low, on the order of a per cent or so of beam current. This might imply that relatively small switch tubes could be used to control a quite large amount of beam current. To some extent this is true; however, the effect of importance in this regard is the stray capacitance C_s. Each time the modulating anode is switched either up or down in voltage, this capacitance must be either charged or discharged. It is quite readily apparent, therefore, that each switch tube will have to dissipate approximately

$$P_a = \tfrac{1}{2} C_s V_m{}^2 (\text{PRR}) \tag{42}$$

(where V_m is the required swing in modulating anode voltage in volts, C_s is in farads, and PRR is in pulses per second) from the capacitance dissipation standpoint alone. To this must be added the dissipation due to beam-current interception to arrive at the total anode dissipation of each tube. At beam voltages of several hundred kilovolts (typical for multimegawatt klystrons) this dissipation can be quite significant and often becomes a critical factor in the choice of the switch tubes.

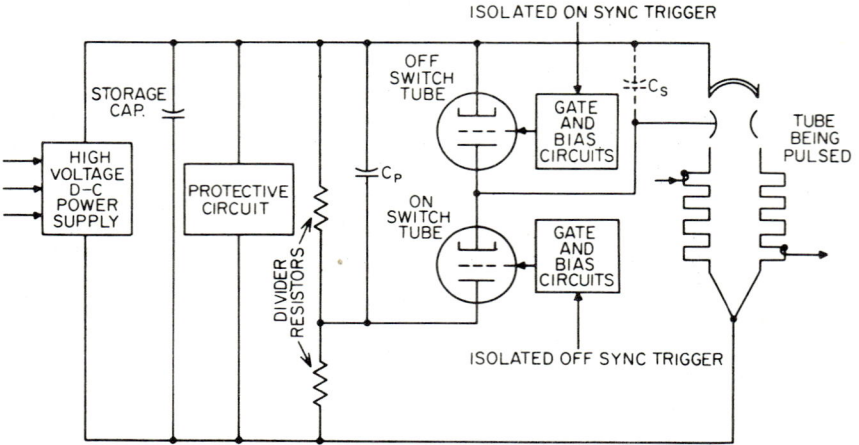

FIG. 47. Floating-deck hard-tube pulser.

The rise time and fall times obtainable from this method of modulation are also a function of the value of C_s. If the maximum pulse current capability of a given switch tube is known, the achievable rise time is given approximately by

$$\tau_r = C_s V_m / I_p \tag{43}$$

where I_p is the peak value of current in amperes (and the current is assumed to be constant during the rise and fall times).

The type tube frequently employed for this switching role at the higher voltage levels (over 100 kv) is the so-called beam-switch tube. This tube type has characteristics quite similar to those of conventional pentode negative-grid tubes, although their construction is similar to that of the klystron, and their connections are similar to those of triodes. The other characteristics of the floating-deck modulator are quite similar to those of conventional hard-tube modulators. The storage capacitor and the fault protective circuit in particular follow the requirements noted previously. Capacitor C_p is a secondary storage capacitor which functions to limit the modulating anode droop to a lesser value than that of the main storage capacitor. The considerations associated with the gating, filament, bias, and trigger isolation circuits are quite similar to those in the series hard-tube modulator. The importance of keeping the stray capacitance of these circuits low (in the case of the "on" switch tube) is quite significant since this value usually forms the major component of the total value of C_s.

MICROWAVE AMPLIFIERS   27-87

12. PROTECTIVE EQUIPMENT

A great deal of costly equipment is included in the typical microwave power amplifier complex, and for this reason it is essential that consideration be given to the problem of affording this equipment adequate protection. Perhaps the most expensive single group of components in most complexes is the microwave amplifier tubes themselves. Protection of these amplifiers from both faults generated within themselves as well as faults and malfunctions occurring in their associated equipment is the logical starting point for any protective-circuitry analysis. This analysis will then include the power supplies and pulsers supplying energy to the amplifiers as well as the transmission-line components which handle the generated radio frequency delivered by the amplifiers. Finally, since most amplifier complexes require operating personnel to control their functions, it is vital that adequate protective circuits exist to safeguard them from both the high voltages as well as the potentially dangerous radiations associated with the amplifier complexes. The following discussion will point out some of the major areas of concern in typical high-power microwave amplifier complexes.

12.1. Power-amplifier-tube Protection. Circuit interlocking, usually provided by means of relay contacts connected in series with the main high-voltage power-supply on-off circuit, is a method by which protection is provided to the power amplifier. In some cases, this interlocking appears in other critical circuits which will be noted later. As a means of approaching the protective functions in a logical fashion a sequence of starting events will be described. Some of the usual functions of importance in a protective sequence are shown in Fig. 48.

In microwave power amplifiers of any appreciable power level some form of cooling is usually required. This may be accomplished by means of forced air in the smaller tubes or by circulated liquids in the case of larger tubes. Two sensors are usually required, monitoring the absolute flow of coolant and absolute temperature of the surface of the tube being cooled. These two sensors should provide two of the interlocks in the high-voltage power-supply starting circuit either to prevent starting the amplifier initially or to remove power from the amplifier in the event the cooling fails during operation. An additional point where such interlocking may be required often appears with tubes having very high filament power requirements. It is then sometimes required that coolant flow be properly established even before filament power is applied to the tube if resultant thermal damage is to be avoided.

Two other considerations concerning the application of filament power are of importance. Most microwave tube filaments have certain fairly close tolerances to which filament voltage or current must be held. Sensors should be provided to remove high voltage in the event these limits are exceeded. In addition, some tube types require the reduction (or even removal) of external filament heating power once the amplifier is operating at its specified power level. These circuits must also be interlocked with the high-voltage circuits by means of the proper sensor elements.

Many microwave amplifiers achieve their electron-beam-control characteristics by means of external-focusing solenoids. It is important that these magnetic elements be properly functioning if the tube is to be afforded adequate protection. This is often done by means of current sensors in the focusing-coil circuits. These sensors then actuate permissive interlocks in the high-voltage control circuits if the coil current values are within proper limits. In addition, on some tubes, body-interception current is monitored, and if the beam control is not proper, excessive body-interception current is used to actuate a sensing element to remove the applied high voltage.

In order to prevent possible damage to certain microwave tubes (in particular crossed-field amplifiers) it is important that r-f drive power be applied before the high voltage is applied. In this case appropriate sensor and interlocking functions must be provided. It is necessary that these circuits be quite fast in action, particularly in the case of pulsed amplifiers.

Some, but not all, tunable amplifiers require protection for improper adjustments of the tuning elements. In general, tunable klystrons cannot be damaged (by having their cavities detuned) since they have adequate collector dissipation to withstand

27-88   AMPLIFIER CIRCUITS

full power input (from the power supply) without excessive heating. It may be necessary for other reasons, however, to prevent improper tuning, particularly when several klystrons are operated in parallel. With negative-grid tubes this internal dissipation capability is not always present; hence some form of sensor of the r-f output must be provided which can be compared with the d-c input to assure the tube is remaining within its plate dissipation ratings. Other means, such as thermal detection, may also be employed. In any event, where necessary, these sensors should be provided with interlock elements in the high-voltage power supply.

    The high-voltage power supply itself (as well as auxiliary bias supplies if required) are usually equipped with their own overvoltage and overcurrent sensors which will remove the high voltage in a matter of a few tens of milliseconds (relay operating time). Fast protection functions are sometimes incorporated to prevent damage to the microwave amplifier in the event it should arc internally. These fast protective functions are usually by means of the so-called crowbar protective circuit which shorts out the power supply within a matter of few microseconds after the detection of the fault.

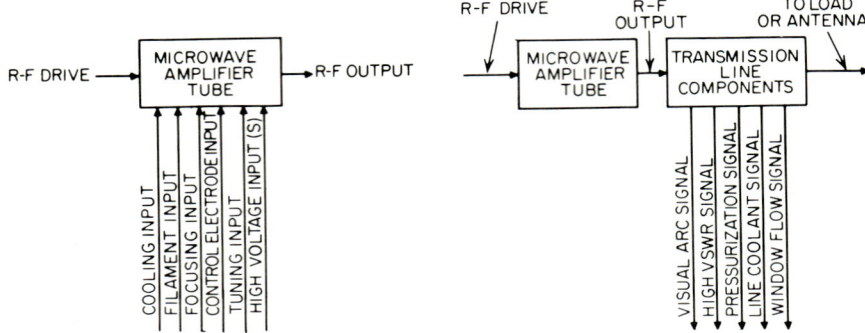

Fig. 48. Inputs requiring integrated protective circuits for a typical microwave amplifier.

Fig. 49. Typical protective sensing signals from transmission-line components.

12.2. Transmission-line Protection. Transmission-line protective circuits are provided for two reasons, first to protect the power-amplifier tube in the event of line arcs, and second to protect the line from the destructive effects of internal arcing. Some of the required sensors are shown in Fig. 49. As was mentioned in a previous section, a transmission-line arc can subject many high-power microwave amplifiers to destructive forces due to the total reflection of the line power back into the tube and its associated internal circuits. Two basic forms of line sensors are common to prevent this occurrence. The first is to use some form of photoelectric device to "look" up the otherwise dark transmission line and detect the visible light generated by the arc. This photoelectric device can then be used to either turn off high voltage by means of interruption of its interlock circuit or, if the interruption must be very rapid, to fire the crowbar protective circuit. An alternative protective means sometimes employed is to have the sensor rapidly remove r-f drive. The second sensing method is to employ a VSWR detection device to note the occurrence of a high amount of reflected power. Directional couplers in the line can be used to detect such a line arc, but it is important that such couplers be located at a line position prior to where the arc occurs. The high VSWR signal so detected would then be used for protection in the same manner as the photoelectric signal noted previously.

    Transmission lines themselves are protected in accordance with the requirements at hand. For instance, very high power waveguide or coaxial lines are often pressurized to enhance their power-handling capabilities. It is necessary in such cases to provide a pressure interlock to prevent the application of high voltage if this pressure is not of the proper quality. Cooling of high-power lines is also applied at high power levels. This may take the form of liquid or air cooling. In either case the necessary sensors and interlocks should be provided for the high-voltage control circuit. A similar situation often arises at the window of high-power klystrons It is frequently necessary to provide an air blast internal to the waveguide to afford additional cooling at the window surface. This protective feature should be handled in similar fashion to those noted above.

12.3. Pulser and Modulator Protection. The previous two sections have assumed that the high-voltage source for the amplifier was a d-c power supply. If instead the source is a pulser, several other protective measures are of consequence. These considerations were touched on in part in the previous section on pulsers and modulators.

It will be recalled that a line-type pulser stores only the energy required for a single pulse during each pulsing period. In a very real sense this feature, plus the fact that the energy is stored in a transmission line, makes this form of pulser virtually self-protecting insofar as arcs in the microwave amplifier are concerned. This comes

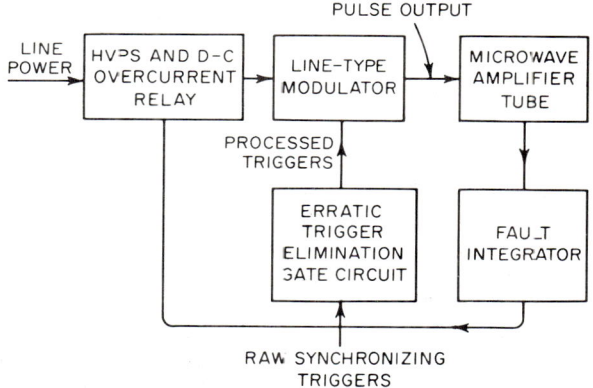

Fig. 50 Typical line-type pulser protective circuits.

about because, when a shorted load is presented to the pulser, the current in the fault can only approximately double under the worst conditions of operation. In effect, the bulk of the energy stored in the network is then reflected and dissipated in the inverse load circuit. Even with this limited energy being delivered to the faulted tube, it is generally the case that the pulser is turned off after a prescribed number of fault pulses. This is done by means of an integrator circuit which uses as its sensing means the double-amplitude load-current pulses so detected, and it ordinarily turns off the high-voltage power supply (Fig. 50).

Two basic deficiencies in the synchronizing triggers controlling the line-type pulser are of importance. The first of these concerns triggers of the improper pulse repetition rate. If the modulator has a holdoff diode circuit, and if the triggers are of too low a PRR, no particular harm can result to the pulser from such operation. If the triggers are of an excessive PRR, protection is generally afforded by the d-c overcurrent relay in the d-c power supply. This assumes that the relay is properly set for tripping action at the particular level at which the pulser is operated. If precise protection from this cause is to be afforded at all power levels, it may be necessary to include a variable calibration feature as a function of the power level at which the pulser is operated.

The second basic deficiency of synchronization is that of erratic triggers. This condition is considerably harder to combat, and in fact is a major source of the down time encountered in line-type modulators. This comes about since an erratic trigger often comes in and fires the switch tube during the time network charging is taking place, and in so doing it may cause the switch tube to fault. Since the triggers are quite difficult to monitor it is often assumed that such a modulator malfunction is a result of the modulator rather than the triggers themselves. It is advisable, if possible, to provide a carefully synchronized gate circuit for the modulator to prevent this condition from occurring, as shown in Fig. 50.

The usual protective functions for hard-tube pulsers, detecting excessive pulse duration, and high fault currents, require special gating circuits and/or crowbar circuits. These considerations were touched upon in the previous discussion of this type of modulator.

12.4. Personnel Protection. It is imperative that adequate personnel protection be afforded on any high-power microwave equipment. Three potential areas of danger exist in most equipments. These are concerned with high voltage, radio frequency, and X rays.

High-voltage protection is afforded in most cases by electrical interlocking of the cabinets and enclosures in which high-voltage equipment is housed. These interlocks are so connected that the source of high voltage will have its primary power removed when the cabinet door or access covering is opened or removed. In addition, these measures are sometimes backed up by mechanical interlocks which directly short out the high-voltage source when the enclosure is opened. A final protective measure in this regard is to provide shorting rods to allow manual discharge of all high-voltage points to ground when the enclosure is opened. There is some difference of opinion as to the level of voltage above which interlocking should be provided. This varies from as low as 40 volts to as high as several hundred volts. In no case should the upper limit of several hundred volts be exceeded without adequate protective circuits being incorporated.

Serious and irreparable damage can be inflected to the human body by excessive exposure to microwave r-f radiation and the radiation of X rays. Microwave amplifiers and their associated equipment can be the source of both types of radiation when operated at high power levels. The manufacturers of microwave and other high-voltages tube should be consulted as to the required shielding means necessary for adequate personnel protection. Once the required shields have been established they should be adequately interlocked so that the equipment cannot be operated if these barriers are removed. Finally, before the equipment is subjected to full-power operation, an extensive survey should be performed by properly qualified radiation-measurement personnel to assure that the measures taken are completely adequate.

13. MONITORING EQUIPMENT

The sensors employed to detect operating conditions in a microwave amplifier complex are items of monitoring equipment. The sensing of such conditions as voltage, current, temperature, pressure, or light may be utilized in an open-loop manner (where visual observation of the particular condition is merely displayed) or in a closed-loop manner (where the signal detected performs some corrective or protective function). Three broad areas of monitoring are encountered in microwave amplifier complexes, r-f monitoring, d-c and low-frequency a-c monitoring, and pulse monitoring. The means whereby these monitoring functions are typically accomplished will be discussed in the following sections.

13.1. R-F Monitoring. Monitoring of r-f transmission-line conditions at the input and output of each r-f amplifier stage is the typical manner in which this area of monitoring is accomplished. The sensing element may be a probe, a directional coupler, or a calibrated r-f load. The functions being monitored are typically VSWR, power output, and spectrum characteristics.

Voltage standing-wave ratio (VSWR) may be detected either by means of a slotted line or by the use of directional couplers. In the former case a line section equipped with a movable probe is used to detect successive maximum and minimum r-f field amplitudes within the line. The r-f signal detected by the probe is usually rectified (and in some cases amplified) to provide a d-c voltage which may be viewed on either a meter or an oscilloscope. The latter case generally employs two directional couplers, one to detect the incident wave on the line and the other to detect the reflected wave on the line. Again the two outputs are rectified and compared by means of calibrated meter or oscilloscope readings. In either case the ratio determines the VSWR. In some cases the presence of an excessive VSWR is used to actuate protective circuits as noted in the previous section.

R-F power output can be measured by means of the indirect bridge method or the thermal method. In the bridge method calibrated directional couplers are utilized, and the rectified output from the couplers is used as the input to some form of calibrated bridge arrangement. The bridge is usually equipped with a direct-reading power meter. A method which usually provides somewhat more accurate results is possible with a calibrated load of either the air-cooled or water-cooled variety. In either case the temperature rise with the application of power is indicated in the form of a calibrated meter reading. The directional-coupler method of course has the advantage that it can provide continuous monitoring while the equipment is in operation.

Spectrum monitoring is accomplished by the use of appropriate receiver equipment. It is possible to detect the components of some spectrum signatures by means of sufficiently selective receivers attached to an attenuated directional-coupler output. This detection may be by either audible or visual means. The spectrum analyzer is a special form of receiver in which a visual display is presented on an oscilloscope. The usual mode of operation with these devices is to adjust the controls in such a way that the receiver automatically tunes the required radiated spectrum bandwidth, and then adjust to make this tuning repeat periodically at a rate of a large number of sweeps per second. The detected output of the receiver is then displayed on the scope in the form of the radiated spectrum envelope. Spectrum analyzers are particularly valuable for viewing pulse spectrum envelopes.

13.2. D-C and Low-frequency A-C Monitoring. The monitoring functions required for microwave amplifier complexes in this area are much the same as those encountered in low-frequency equipments. Line voltage, current, and power are usually monitored, particularly in the case of large power supplies. D-C output voltage and current are monitored in the case of practically all levels of power supplies since in general the r-f tubes being used at microwaves are considerably more costly than their low-frequency counterparts; hence it is important to know the exact conditions of all elements in all stages. Since many microwave tubes operate with their filaments at high voltage rather than at ground level, filament voltage and current are sometimes measured at the primary of the high-voltage insulated filament transformer rather than at the secondary. While this provides convenience in the mounting of the monitoring meters, it is important that the meters be properly calibrated to reflect the transformer's loading characteristics.

Automatic regulators are frequently employed to provide constant power-supply output when circuit components heat up. This is a consideration in the case of focus-coil power supplies. A current-sensing signal is usually detected by means of current shunt or a current transformer, and this actuates a small servosystem to regulate the primary voltage of the supply appropriately to hold constant D-C output current. Similar requirements often occur in other parts of amplifier-complex circuitry.

13.3. Pulse Monitoring. Pulse monitoring is accomplished by means of voltage dividers, current transformers, and current shunts. These elements must be given considerably more attention than the similar elements employed for low-frequency a-c monitoring if adequate pulse response is to be obtained.

Voltage dividers for observing pulse waveforms may be of the resistive, the capacitive, or the compensated RC types. Generally, resistive dividers (comprised of non-inductive resistive elements) are used for viewing long pulse durations (milliseconds) or charging waves. Capacitance dividers are most often used where the best definition is required. They are usually limited to short pulse durations (less than 100 μsec) since their inherent droop characteristic becomes excessive at long pulse lengths. RC dividers are used in intermediate areas.

Pulse currents are viewed by means of either a properly designed current transformer or a resistive shunt. The current transformer has the advantage that it may be placed in high-voltage circuits because of its isolating feature. It is important to know the droop characteristics of the unit employed, as well as its reset characteristics, if error is to be avoided in the results as viewed on an oscilloscope. The noninductive current shunt is recommended where precise measurements are required. Its use is often impractical when current monitoring is required at high voltage and grounded visual detection is employed.

R-F pulse monitoring is accomplished by rectifying the output of a directional coupler. It is important to take into account the linearity and time constant of the rectifier circuit when viewing the waveform monitored by this means.

14. CONTROL EQUIPMENT

By its usual connotation, control equipment functions to provide the capability of making a microwave amplifier complex perform in the desired manner. This capability may result from three possible input information sources, either in conjunction with each other or separately. These sources are operator inputs, protective inputs, or monitoring inputs. When operator inputs are employed the control is usually of the manual type. When control results from either protective or monitoring inputs the usual type of control device employed is of the automatic variety. Some typical control functions will be described in each classification.

14.1. Manual Control. Three main categories of manual control are those involved with the starting, the tuning, and the output power-level adjustment of an amplifier complex. In the most automated case, it may be possible simply to actuate a single switch manually and have all other control functions occur thereafter automatically; however, this degree of refinement is seldom employed except where totally unmanned sites are required. In the usual case a manual on-off circuit is provided to apply high voltage to a given amplifier stage. Before this application of power is possible it is necessary that certain permissive functions be accomplished (such as first having filament voltage applied, drive power applied, focusing power applied, etc.). In many cases starting control is implemented by means of a pushbutton start circuit similar to that shown in the top line of the simple control circuit of Fig. 51. It will be noted that when the start button is pressed contactor $K1$ will energize and seal in by means of one of its own contacts (provided that all permissive functions are proper, as indicated by their interlocks providing a complete circuit path). At the same time contacts of $K1$ will apply line voltage to the autotransformer connected to the primary of the power supply (as shown by the one-line diagram at the bottom of the figure).

A simple tuning function is indicated by the next pair of "across the line" circuits shown. It is readily apparent how the d-c tuning motor (indicated as $M1$) is controlled in response to the lever switch. It is important to note the use of limit switches in series with relay coils $K2$ and $K3$. These switch contacts would be mechanically actuated by the tuning mechanism to prevent cavity damage (often with resultant loss of tube vacuum) due to overtravel. In a sense this is a combination of automatic and manual control functions.

The third circuit shown is similar to that just described and is used to control the power output of the amplifier. The variable autotransformer is in this case driven by a second d-c motor to raise or lower the primary voltage applied to the high-voltage power supply. Use of limit switches is again necessary to prevent motor overtravel. An additional feature indicated is automatic runback control. When the stop button is actuated it will be noted that relay $K5$ (the voltage lower relay) is actuated by a normally closed contact of $K1$ to cause the autotransformer to return to its minimum output position. This is sometimes a requirement with certain types of power amplifiers.

In this elementary control diagram d-c control power has been indicated. This provides certain advantages such as quieter relay operation (no a-c buzz) plus the often superior torque, starting, and stopping characteristics of d-c drive motors. A-C control power is also widely employed with reversible a-c drive motors employed where required.

14.2. Automatic Control.

Manual control relies on operator judgment first to observe the potential need for equipment adjustment (by means of monitor equipment observation), and second to accomplish the desired adjustment (by means of a manual control circuit). It is readily apparent that lack of sufficiently rapid operator response time can cause appreciable loss in terms of both quality of performance and equipment life. For this reason it is often advisable to provide automatic control to provide corrective adjustments. A simple example of this function might be a preset voltage run-up control which automatically runs an operating voltage to a predeter-

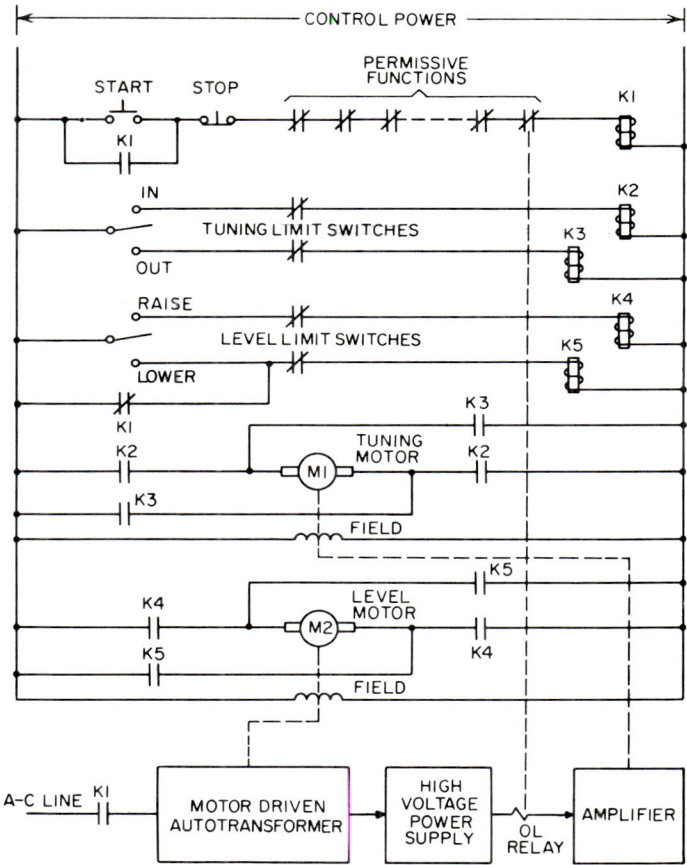

FIG. 51. Elementary control circuit.

mined value once the start button is actuated, thus eliminating inaccuracies caused by an operator's improperly reading a monitoring device. A more complex role might be the automatic tuning for a transmitter circuit. By detection of a change in r-f power output of the amplifier (by means of a directional coupler) a servosystem could be arranged to receive the input signal necessary to actuate the amplifier tuning adjustment required to return it to its proper operating conditions. It is not difficult to visualize how a number of these and similar functions could be combined in the control circuits to provide virtually complete automatic control. In this regard it is important to balance carefully the advantages of simplified and more consistent operation against the disadvantages of greater complexity and higher initial cost.

Chapter 28

DIODE AND PARAMETRIC AMPLIFIERS

CHANG S. KIM,* HARRY J. PEPPIATT,† and HSIUNG HSU‡|

CONTENTS

1. Tunnel-diode Amplifiers.................................... 28–1
   1.1. Introduction... 28–1
   1.2. Stability Criteria.................................... 28–2
   1.3. Low-frequency Models................................. 28–4
   1.4. Various Amplifier Configurations..................... 28–5
   1.5. Amplifier Design..................................... 28–9
   1.6. Gain, Bandwidth, Sensitivity, and Noise Figure....... 28–18
2. Capacitance-diode Amplifiers............................. 28–27
   2.1. General Formulas for Capacitance-diode Amplifiers and Converters...................................... 28–28
   2.2. Bandwidths and Gain Bandwidths of Capacitance-diode Amplifiers... 28–32
   2.3. The Four-pole Parameters............................. 28–33
   2.4. Typical Amplifier Design............................. 28–36
3. Traveling-wave Parametric Amplifiers..................... 28–41
   3.1. General Considerations............................... 28–41
   3.2. Fundamental Mechanism of Traveling-wave Interactions... 28–41
   3.3. The Wideband Traveling-wave Parametric Amplifier..... 28–43
   3.4. The Electronically Tunable Traveling-wave Parametric Amplifier.. 28–45
   3.5. Forward and Backward Traveling-wave Interactions..... 28–47
   3.6. Three-dimensional Parametric Interactions............ 28–49
   3.7. Comments on Selection Rules.......................... 28–50

1. TUNNEL-DIODE AMPLIFIERS

1.1. Introduction. Since a tunnel-diode amplifier is a one-port negative-conductance device, the problems associated with this type of amplifier are quite

---

* General Electric Company Electronics Laboratory, Syracuse, N.Y.
† General Electric Company, Communications Products Department, Lynchburg, Va.
‡ Antenna Laboratory and Department of Electrical Engineering, The Ohio State University, Columbus, Ohio.
|| Section 1 was written by C. S. Kim, Section 2 by H. J. Peppiatt, and Section 3 by H. Hsu.

AMPLIFIER CIRCUITS

different from those with ordinary tube and transistor amplifiers. As mentioned in Chap. 12, a tunnel diode provides a negative-conductance characteristic over a very wide frequency range (from direct current to microwaves), and as a result, stabilization of a tunnel diode as an amplifier becomes a primary concern. It is a fact that a tunnel-diode device gives the largest gain-bandwidth product of any device at present.

A tunnel-diode amplifier does not provide the isolation between input and output terminals which is generally given by an ordinary amplifier. There is no distinction between input and output termination in a tunnel-diode amplifier, since all terminations exist across the device terminals. This lack of isolation could provide amplified power loss at the input termination or reradiated back to the antenna in the case of a receiver. Thus overall efficiency and power gain would be reduced. Furthermore, noise from the output termination can be amplified in the amplifier. Thus special attention is needed to resolve this problem, associated with one-port negative-conductance devices, as mentioned in Chap. 7, Sec. 4.

The isolation can be obtained if special techniques are used, such as the use of a circulator; this will be discussed in more detail later in this chapter.

1.2. Stability Criteria. Several works on the stability conditions of tunnel-diode circuits have been reported.[1,2] In the case where the expression for the external circuit across the diode becomes rather complicated, it is convenient to determine the stability condition in the following way: Representing a tunnel diode by the equivalent circuit of Fig. 1 and connecting a physically realizable external circuit of impedance

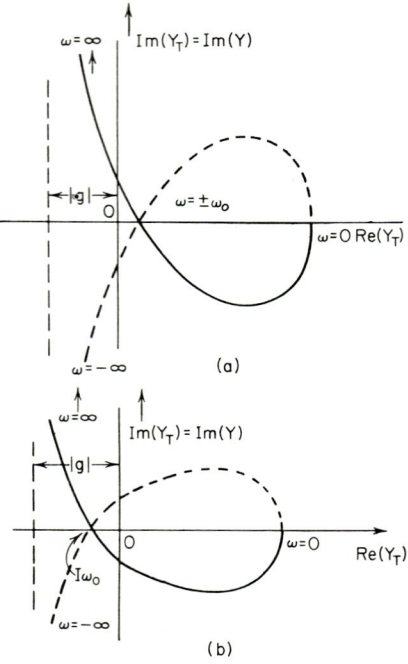

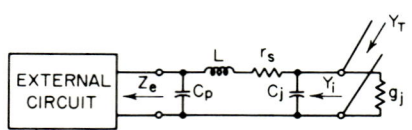

Fig. 1. Tunnel diode equivalent circuit.

Fig. 2. Plot of Y_T.

Z_e across the diode, the total admittance Y_T across the negative conductance g_j can be expressed by

$$Y_T(s) = Y_i(s) - |g_j| = p(s)/q(s) \qquad (1)$$

where $Y_i(s)$, the admittance facing g_j, is a positive real function.[3,*]

As Bode has pointed out,[4,5] Y_T is short-circuit-stable since Y_i is connected in parallel with a short-circuit-stable device g_j. Since $q(s)$ is always a Hurwitz polynomial,[5] the stability condition is that $p(s)$ must be a Hurwitz polynomial.

The graphical interpretation of the stability condition is as follows: The plot† of $Y_T(\omega)$ can be obtained from the plot of $Y_i(\omega)$ by shifting the imaginary axis of $Y(\omega)$ by $|g_j|$ along the positive real axis of $Y(\omega)$ as shown in Fig. 2a and b. Since $q(s)$ has no roots in the right half of the s plane, any encirclement of the origin by the plot of

* A parallel inductance across g_j is not allowable.
† Here the plot of $Y_i(\omega)$ represents the case where the external impedance $Z_e(\omega)$ is shorted. However, similar plots can be obtained for more general cases.

DIODE AND PARAMETRIC AMPLIFIERS

$Y_T(\omega)$ must come from the right-half-plane roots of $p(s)$ only. Therefore, the circuit will be stable if, and only if, the plot of $Y_T(\omega)$ does not encircle the origin. An example of a stable case is shown in Fig. 2a for a small $|g_j|$ value. On the other hand, if the value of $|g_j|$ increases so that the origin is encircled by the $Y_T(\omega)$ plot as in Fig. 2b, the circuit will be unstable.

The plot of $Y_T(\omega)$ becomes very convenient as the rational function $Y_T(s)$ becomes of higher order where the analytical solution is difficult to obtain, especially if the external circuit consists of distributed elements.

It may be possible to find the stability condition at the diode terminal, looking into the total impedance[6] at this point. However, in this case, the short-circuit-stable device is connected in series with a passive network, and the total impedance at 2-2' could have zeros and poles in the right half plane of s. In this case, the determination of the stability condition from the plot of the total impedance becomes cumbersome, since an encirclement of the origin due to the right-half-plane zeros and poles might have been canceled.

The conductance g_j mentioned above is a small-signal dynamic conductance ($g_j = dI/dV$) and can be considered a constant for a given bias, provided that no spurious oscillations are being generated. The maximum value $|g_{jI}|$ which occurs at the inflection point of the tunnel-diode IV characteristic will provide the most critical case for stability. This results from the fact that stability is a direct function of $|g_j|$ and that the larger the value of $|g_j|$ the less the stability. This can be seen from examining Eq (1) and Fig. 1. If the stability criterion is met with the junction conductance $|g_{jI}|$ at the inflection point, any disturbance which changes the value of $|g_{jI}|$ will not upset the stability criteria since any other $|g_j|$ will be smaller than $|g_{jI}|$.

In the unstable case where oscillations occur, the small-signal representation of g_j is no longer valid. This is due to the fact that oscillation is a nonlinear process. Actually, the dynamic conductance for the large-signal case is different from the small-signal dynamic conductance g_j and is a function of the oscillation voltage.

If the plot $Y_T(\omega)$ does not indicate conditional stability, similar to Bode's[4] case applied to a transfer function, the stability condition for the generalized tunnel-diode circuit can be stated as follows:

1. Re $\{Y_i\} > |g_j|$ over the entire frequency range or
2. In the frequency where Re $\{Y_i\} \leq |g_j|$ Im $\{Y_i\} \neq 0$ where g_j is negative.

Conversely, oscillatory conditions* are given by

1. Re $\{Y_i\} \leq |g_j|$ at the oscillation frequency, where g_j is negative.
2. Im $\{Y_i\} = 0$.

In the case where $Z_e = R_e + sL_e$, Y_T of Eq. (1) becomes

$$Y_T(s) = \frac{s^2 L C_j + s(CR - L|g_j|) + 1 - R|g_j|}{sL + R} \tag{2}$$

where $L = L_i + L_e$
$R = r_s + R_e$

The roots (s_1 s_2) of the polynomial of the numerator of $Y_T(s)$ can be expressed in complex quantities as

$$\begin{aligned} s_1 &= \delta_1 + j\omega_1 \\ s_2 &= \delta_1 - j\omega_1 \\ \delta_1 &= \tfrac{1}{2}(|g_j|/C_j - R/L) \\ \omega_1 &= \sqrt{(1/LC_j)(1 - |g_j|R) - \tfrac{1}{4}[(|g_j|/C_j) - R/L]^2} \end{aligned} \tag{3}$$

where

The conditions for stability are given by

1. $\delta_1 < 0$.
2. $j\omega_1 < \delta_1$ if ω_1 is imaginary or any ω_1 if any ω_1 is real. Then

$$|g_j|/C_j < R/L \quad \text{and} \quad 1 > |g_j|R$$

---

* For steady-state sinusoidal oscillations, Im $\{Y_i\}/\partial\omega$ must be positive in addition to the oscillatory condition (see Ref. 5).

28-4 AMPLIFIER CIRCUITS

1.3. Low-frequency Models. In order to gain a basic understanding of a one-port negative-conductance device and of the effect of an external circuit, a low-frequency model will be considered. Assume the tunnel diode is a pure nonlinear-conductance device whose conductance values are determined by the slope of the diode IV curve.

If a conductance G_p is placed across a diode, the overall IV characteristic will become that of Fig. 3. These curves result from adding currents in G_p and the diode for common voltages. It should be noted that the overall IV characteristic is always short-circuit-stable if $G_p \leq |g_{jI}|$ (curves 1 and 2). However, if $G_p > |g_{jI}|$, the overall characteristic becomes a nonlinear device without a region of negative conductance (curve 3).

On the other hand, if a resistance R_s is placed in series with a diode, the overall IV curves of Fig. 4 are obtained. Here voltages across the diode and R_s are added for common currents. If $R_s \leq |r_{jI}| = 1/|g_{jI}|$ (minimum value of $1/|g_j|$) (curves 1 and 2) the overall IV characteristic is still short-circuit-stable. However, if $R_s > |r_{jI}| = 1/|g_{jI}|$ (curve 3) the small-signal conductance of the overall IV characteristic has two zeros as well as two poles and thus becomes unstable. As the voltage on the diode is slowly varied, current jumps represented by a and b on curve 3 will take place.

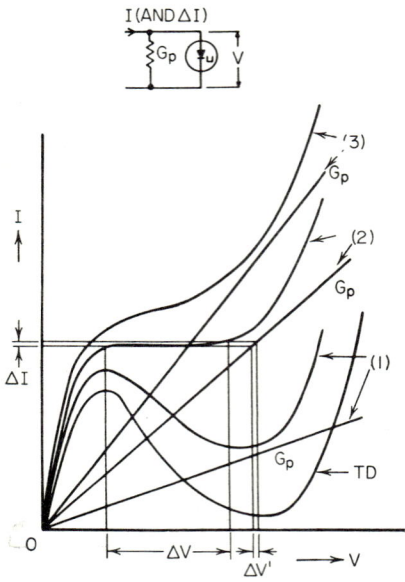

Fig. 3. IV characteristics of parallel connection.

If an a-c current source ΔI is introduced, an a-c voltage ΔV across the parallel G_p and the diode will be produced, as shown in Fig. 3. With the same ΔI an a-c voltage $\Delta V'$ would be obtained across G_p without the tunnel diode. In this case the a-c

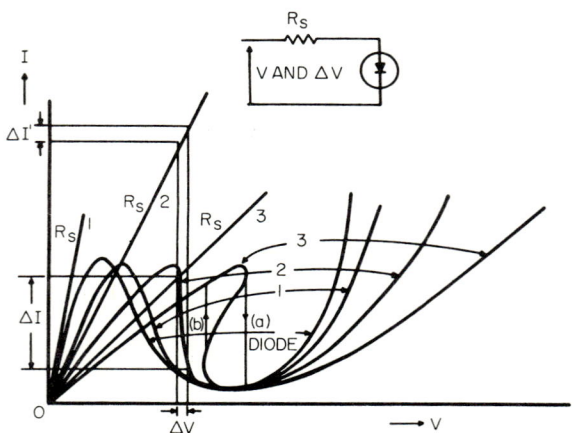

Fig. 4. IV characteristics of series connection.

insertion power gain P_{IG} due to the diode is given by

$$P_{IG} = |\Delta V / \Delta V'|^2$$

In the dual case, referring to Fig. 4, a-c power insertion gain P_{IG} for the series connection of R_s becomes

$$P_{IG} = |\Delta I / \Delta I'|^2$$

1.4. Various Amplifier Configurations. The external circuit Z_e shown in Fig. 1 can include a source impedance Z_s and a load impedance Z_L to form a generalized

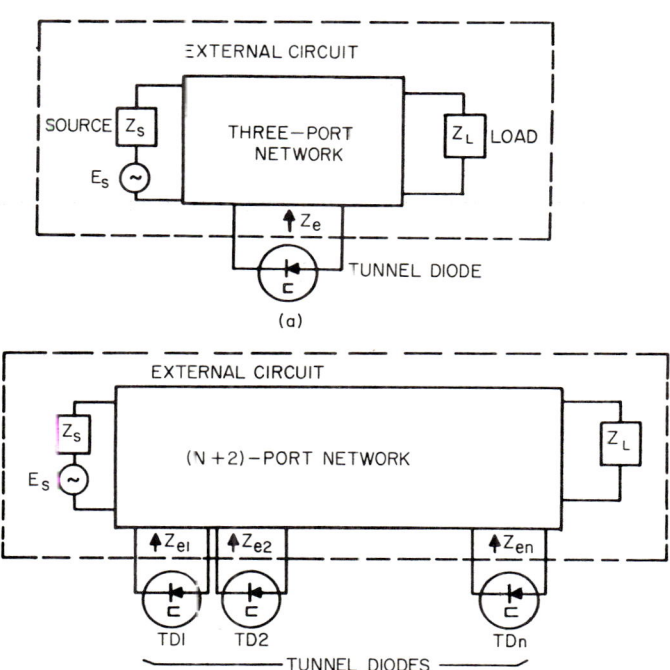

Fig. 5. Generalized tunnel-diode amplifier.

amplifier circuit as shown in Fig. 5. The stability condition of having no encirclements of the new origin in the $Y_T(s)$ plot is met in the generalized amplifier circuit as long as $Y_T(s)$ is formed of the ratio of two Hurwitz polynomials. Three-port networks can include nonreciprocal circuit elements such as circulators and isolators.

The amplifier of Fig. 5a, employing a single tunnel diode, can be extended to that of Fig. 5b where multiple tunnel diodes are used. The stability condition for this case can be found by considering individual $Y_T(s)$ plots; for example, the $Y_{T1}(s)$ plot represents the $Y_T(s)$ plot of the tunnel diode TD 1. If all the individual $Y_T(s)$ plots do not encircle the new origin, the overall network will then be stable.

Bilateral Tunnel-diode Amplifier. If the three-port tunnel-diode amplifier of Fig. 5a is cascaded with another amplifier without going through a nonreciprocal circuit such as a circulator and isolator, the combination can be represented by Fig. 6. This is a typical representation of a tunnel-diode amplifier cascade at low frequencies where a circulator is generally not used. Y_c can be a parallel-tuned circuit and the d-c bias can be applied through the inductive element of this tuned circuit. Special care must be taken in choosing a low resistive-cutoff-frequency diode so that the circuit can be stabilized with lumped circuit elements. If a high-cutoff-frequency diode is used, the stray capacitance produced by lumped circuit elements may produce a short across the tunnel diode at high frequencies which may result in the generation of high-frequency oscillations.   Various configurations resulting from combinations of series and parallel connections of source $Y_s = 1/Z_s$, load $Y_L = 1/Z_L$, and tuned circuit as well as of sources, current, or voltage can be realized.

In general, regardless of the configurations used, a one-port amplifier without unilateral elements is inferior with respect to gain and noise figure to the corresponding amplifier with unilateral elements.

The amplified signal from a one-port tunnel-diode amplifier goes not only to the second-stage amplifier (or load) but also to the source termination; a portion of the amplified signal is thus wasted.   Although a more detailed discussion will be given later on the noise figure of the cascade amplifier combination of Fig. 6, it is appropriate at this time to mention briefly a few points on this noise figure.

In Chap. 7, it was stated that noise figure and exchangeable noise figure are independent of the output termination; thus the noise figure of a tunnel-diode amplifier can be considered as independent of the second stage.   Also, it was mentioned in Chap. 7 that the cascade noise formula is always valid for a terminal impedance of the cascaded network having either a positive or negative real part.

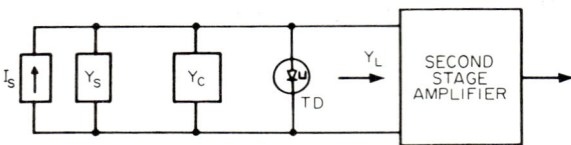

Fig. 6. Bilateral tunnel-diode amplifier cascaded with other ordinary amplifier without nonreciprocal circuit element.

In the configuration of Fig. 6, the first term of the cascaded noise formula is the noise figure F_1 of the tunnel-diode amplifier and is independent of the second stage. However, the second term of the formula is related to the negative conductance of the first stage. If the input termination of the second stage has a negative conductance, noise contributed by this stage is amplified and in the second term of the noise formula becomes larger compared with that of a cascaded network having the same first-stage gain and a positive real part of the second-stage input termination. Consequently, the overall noise figure is not very low.

These disadvantages made the tunnel-diode amplifier of Fig. 6 and similar nondirectional amplifier configurations unattractive for low-frequency applications where other types of unilateral amplifiers having low noise are available.

Circulator-coupled Tunnel-diode Amplifier. A tunnel-diode amplifier coupled to an ordinary amplifier through a circulator is shown in Fig. 7. This is an example of Fig. 5b for nonreciprocal circuit elements. Ideally, the input signal is sent unidirectionally from the source at terminal 1 to the tunnel-diode amplifier at terminal 2 and then to the second-stage amplifier (or mixer) at terminal 3.   If an impedance match exists at the second-stage amplifier at terminal 3, reflections will not take place and the matched termination at terminal 4 will not be needed. A three-port circulator could then replace the four-port circulator shown in Fig. 7.

As will be discussed later in more detail, the circulator-coupled tunnel-diode amplifier surpasses the bilateral tunnel-diode amplifier in performance and has certain advantages over other types of tunnel-diode amplifiers. At present, circulator-coupled tunnel-diode amplifiers are used mainly in microwave applications where small-sized circulators are readily obtained and where good performance from other types of amplifiers is difficult to achieve.

The advantages of a circulator-coupled tunnel-diode amplifier result from the fact that, ideally, it is possible to convert a bilateral one-port amplifier into an ideal unilateral two-port amplifier.

Some degree of isolation between the source and the load (second-stage amplifier) can be achieved by using an isolator-coupled tunnel-diode amplifier as shown in Fig. 8. Here isolators I and II provide unilateral transmission of signal power. However, the amplified signal from a tunnel-diode amplifier will not only be fed to the second stage through isolator II with little loss but will also be fed to the output port of isolator I. The latter part of the power transmission will be absorbed by isolator I and ideally will not reach the source. This absorbed power is a loss and thus the

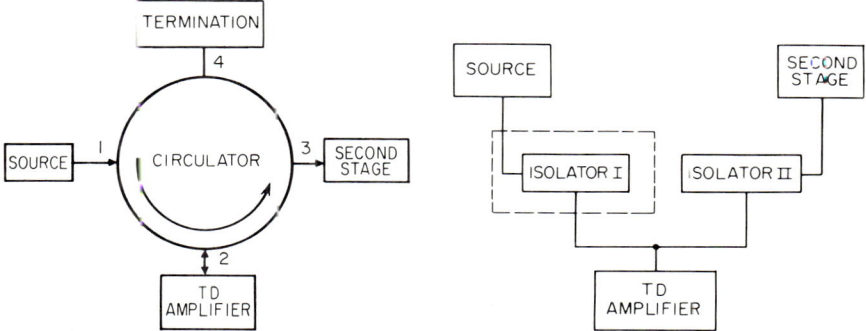

FIG. 7. Circulator-coupled tunnel-diode amplifier.

FIG. 8. Isolator-coupled tunnel-diode amplifier.

amplifier does not operate so efficiently as would a circulator-coupled tunnel-diode amplifier. In a circulator-coupled tunnel-diode amplifier, the entire amplified signal is, ideally, fed to the second stage without feeding back to the source side of the circulator. A more detailed discussion of the actual design procedure for a circulator-coupled tunnel-diode amplifier and an analysis of the gain, bandwidth, and noise figure of this type of amplifier will be given later.

Hybrid-coupled Tunnel-diode Amplifier. Unilateral amplification is possible by using a matched pair of tunnel diodes and a 3-db directional coupler in a hybrid-

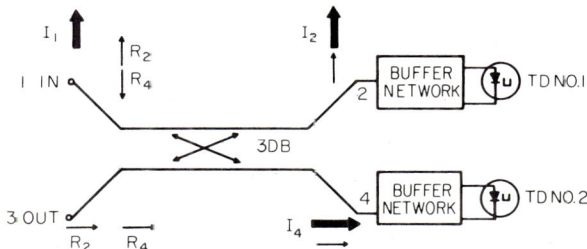

FIG. 9. Hybrid-coupled tunnel-diode amplifier. I_1 = incident signal at input terminal of 3-db coupler. $I_{2,4}$ = incident signals at diodes 2 and 4. $R_{2,4}$ = reflected signals from diodes 2 and 4.

coupled amplifier as illustrated in Fig. 9. This amplifier can be represented by Fig. 5b. As shown in the vector representation of the signals, the input incident signal power splits into two components (3 db) with no phase shift at port 2 and with 90° phase shift at port 4. The incident signals at ports 2 and 4 are amplified and reflected back to the coupler. Considering these reflected signals as new incident signals at ports 2 and 4, and applying the same conditions of power splitting and phase shifting mentioned above, the signals reflected by the diodes add at the output port 3 while they cancel at the input. Buffer networks as shown in Fig. 9 are usually constructed of bias circuits, matching sections, and filters.

This amplifier can be designed to have an extremely large bandwidth. However, the operation of the hybrid amplifier requires a well-matched pair of tunnel diodes, a good 3-db directional coupler which can satisfy the power-splitting and phase-shifting conditions mentioned above, and a matched source and load (next stage) whose impedance should be the characteristic impedance of the 3-db coupler. Because of these requirements, this type of amplifier becomes unattractive in practice.

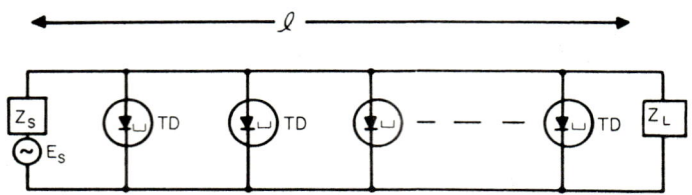

FIG. 10. Distributed tunnel-diode amplifier.

Distributed Tunnel-diode Amplifier. A typical distributed tunnel-diode amplifier is shown in Fig. 10. The tunnel diodes may be lumped diodes spaced a distance less than $\lambda/4$, or distributed tunnel diodes.* If the series resistance r_s and inductance L of the tunnel diode are small, and the loss of the transmission line is assumed negligible, the equivalent circuit of Fig. 10 can be presented schematically by Fig. 11. Here L, $-|g_j|$, and C are, respectively, series inductance, a negative conductance, and parallel capacitance of the diode per unit length of transmission line. C includes both the diode junction capacitance and the inherent parallel capacitance of the transmission line, per unit length of transmission line.

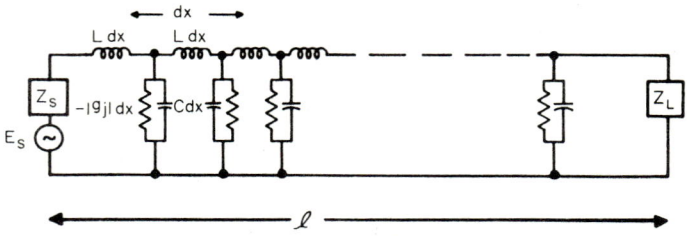

FIG. 11. Equivalent circuit of Fig. 10.

If the operating frequency f is larger than $|g_j|/2\pi C$, the propagation constant γ and the characteristic impedance Z_o of a distributed tunnel-diode amplifier can be given by

$$\gamma = \sqrt{ZY} = \sqrt{j\omega L}\sqrt{-|g_j| + j\omega C} = (j\omega\sqrt{LC})(1 - |g_j|/2j\omega C + \cdots)$$
$$\simeq j\omega\sqrt{LC} - (|g_j|/2)\sqrt{L/C} \quad (4)$$
$$Z_o = \sqrt{j\omega L}/\sqrt{-|g_j| + j\omega C} = (\sqrt{L/C})(1 + |g_j|/2j\omega C + \cdots) \approx \sqrt{L/C} \quad (5)$$

From Eq. (4) for a negative real part of γ, the forward and backward waves have a gain of

$$A = \exp[(|g_j|/2)\sqrt{L/C}\,l] = \exp[(|g_j|/2)Z_o l] \quad (6)$$

* Distributed tunnel diodes are in the experimental stage.

where l is the length of the line. This is essentially a bilateral network. If the load impedance Z_L is equal to Z_o, there will be no reflection from the load, and voltage and current gain will be given by Eq. (6). Furthermore, if the input source impedance Z_s is equal to Z_o, the reflection from the source will be also eliminated. Ideally, then, it is possible to construct a very wideband amplifier under these conditions. However, in practice, matching of the load and source over a wide frequency range is usually not possible and instabilities resulting from mismatches can easily result. This can be seen from Fig. 5b, which represents a general case of Fig. 10. As mentioned previously, the stability condition of each individual tunnel diode must be satisfied. Theoretically, by employing nonreciprocal circuit elements, it is possible to provide the unilateral gain and to improve stability.

Cascaded Circulator-coupled Tunnel-diode Amplifier. Several stages of tunnel-diode amplifiers can be connected in cascade through circulators as shown in Fig. 12. This configuration can be obtained by successively replacing each second stage of the basic amplifier configuration of Fig. 7 with circulator-coupled tunnel-diode amplifiers. Each circulator-coupled tunnel-diode amplifier section can be considered an amplifier having unilateral gain and with positive real parts of input and

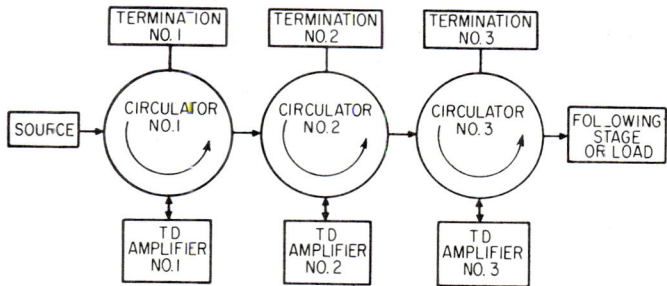

FIG. 12. Cascaded circulator-coupled tunnel-diode amplifiers.

output admittances. Thus it is possible to cascade these amplifiers without generating instabilities provided the stability conditions of Fig. 5 are satisfied.

This type of amplifier is especially attractive as a receiver front end (e.g., such as for use with phased-array radar antennas) where low noise and high dynamic range* are required. Operating with moderate gains (less than 10 db), both noise figure and dynamic range can be improved, as will be seen later. Furthermore, the gain and phase stabilities can be improved with a cascaded amplifier compared with a single high-gain circulator-coupled tunnel-diode amplifier.

Extending the concept of Fig. 12 to the distributed networks discussed previously, a distributed tunnel-diode amplifier with unilateral gain can be constructed as shown in Fig. 13. The ring-shaped strip line has multiple T sections which contain the tunnel diodes. Ferrite cores are placed at each T-section joint or a continuous ferrite core, linked together in a common magnetic circuit, is placed in the ring, effectively transforming each T section into a three-port circulator. With the electrical length of the strip-line ring made equal to one wavelength at the operating frequency, the amplification at the center frequency may be increased if the amplified signal is not completely fed to the load through an external circulator and a portion of the signal is fed back to the ring. Diode bias may be supplied through the termination of the external circulator without disturbing high-frequency performance.

1.5. Amplifier Design. As mentioned above, a circulator-coupled tunnel-diode amplifier is most desirable. Therefore, instead of presenting a brief description of the design procedures for various tunnel-diode amplifiers, more detailed design information

---

* The dynamic range is defined as the range of the input signal from the terminal noise level to a signal level which gives a 1-db gain compression of the output signal. This is related to the intermodulation term in an amplifier.

will be given for the circulator-coupled tunnel-diode amplifier. As will be seen later, the material presented in Sec. 1.2 regarding stability criteria will be used extensively in the design of this type of amplifier. The design goal is to find an external circuit and a tunnel diode such that the plot of $Y_T(\omega)$ of Fig. 1 will not encircle the new origin (the point $|g_j|$ of the original coordinate system), and the portion of the $Y_T(\omega)$ curve over the operating frequency range of the amplifier will come sufficiently close to the new origin to provide suitable gain. More detailed discussions on this point will be given later.

The selection of a circulator to be used in the amplifier is very important. In general, it is desirable to choose a circulator with a wide bandwidth since it is easier to stabilize an amplifier having a wideband circulator. Of course, losses encountered in a circulator should be minimized. The transmission-line circuit shown in Fig. 14 will be investigated as a possible amplifier. The design problem lies in choosing transmission-line lengths l_1, l_2, and l_3, as well as their characteristic admittances, and in selecting a tunnel diode for a given circulator.

A normalized ($Y_o = \frac{1}{50}$ mho) circulator admittance Y_1 as a function of frequency at port 2 of Fig. 14, with matched terminations at all other ports, is given in Fig. 15. Here it is assumed that line length l_1 is attached to the circulator when the admittance characteristic is measured, so that its effects are included in the admittance plot. At

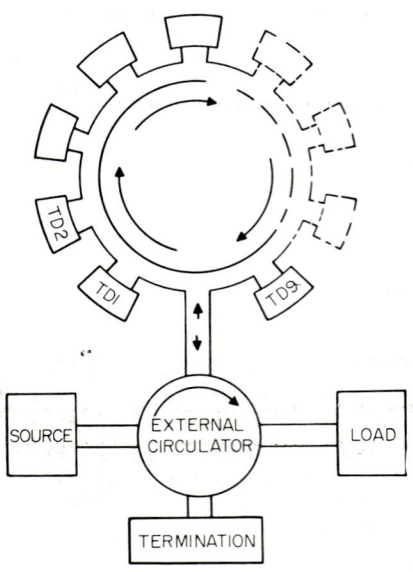

FIG. 13. Unilateral circular ring-shaped distributed tunnel-diode amplifier.

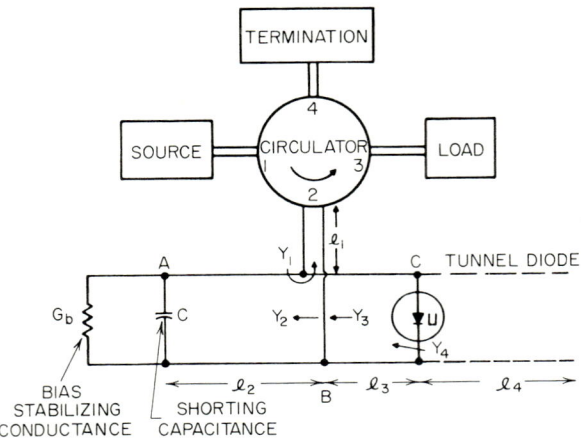

FIG. 14. Circulator-coupled tunnel-diode amplifier using simple transmission-line circuit.

L band and below, the effects of the stray capacitance C_s, the series inductance L, and the series resistance r_s on the overall admittance Y_i across the junction conductance g_j (see Fig. 1) are relatively small and therefore will be neglected in this

DIODE AND PARAMETRIC AMPLIFIERS

preliminary analysis. If these effects become appreciable, as in higher-frequency applications, values of C_s, L, and r_s must be known for this analysis. However, the parallel combination of C_j (junction capacitance) and g_j will, in general, characterize the tunnel diode and will therefore be used for this design procedure.

Under the above assumption, the imaginary part of Y_i at the center of the operating frequency band should tune out the effect of C_j and the real part of Y_i should

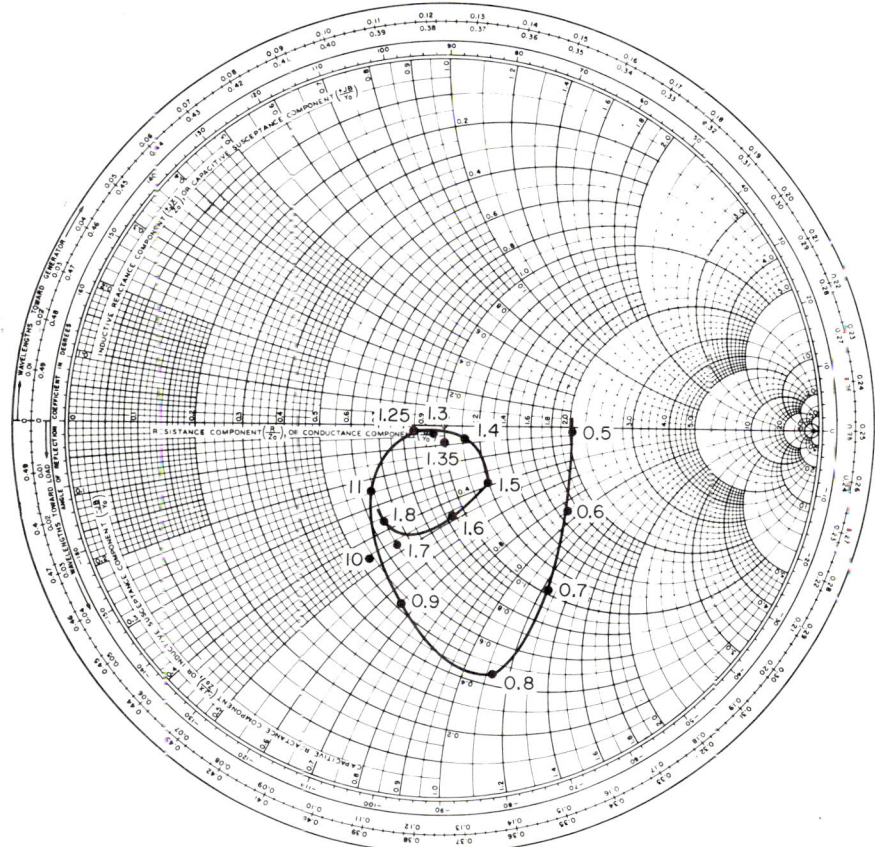

Fig. 15. Terminal admittance of an L-band circulator at port 2 with matched terminations.

be slightly greater than $|g_j|$. Furthermore, the plot of Y_i should not encircle the point $|g_j|$ on the admittance plot.

The following diode parameters will be assumed:

$$C_j = 4 \text{ pf}$$
$$|g_j|/Y_o = 0.5$$

The center frequency of 1.3 gc will be used.

The normalized diode admittance becomes $0.5 + j1.64$. Then the desired normalized external admittance $Y_4/Y_o = Y_e/Y_o$ should be $0.6 - j1.64$ at $f = 1.3$ gc for a gain of 121, which will be explained later.

28-12 AMPLIFIER CIRCUITS

A capacitance C having very little inductance is placed at terminal A in Fig. 14 to produce a short at high frequency (above 500 Mc). At the same time, a low-inductance bias-stabilizing conductance G_b is placed across C.

The value of G_b is larger than $|g_j|$ in order to bias the diode in the negative-conductance region. The admittance of the parallel combination of G_b and C at terminal A will be nearly equal to G_b at low frequencies and the lengths of l_2 and l_3 are

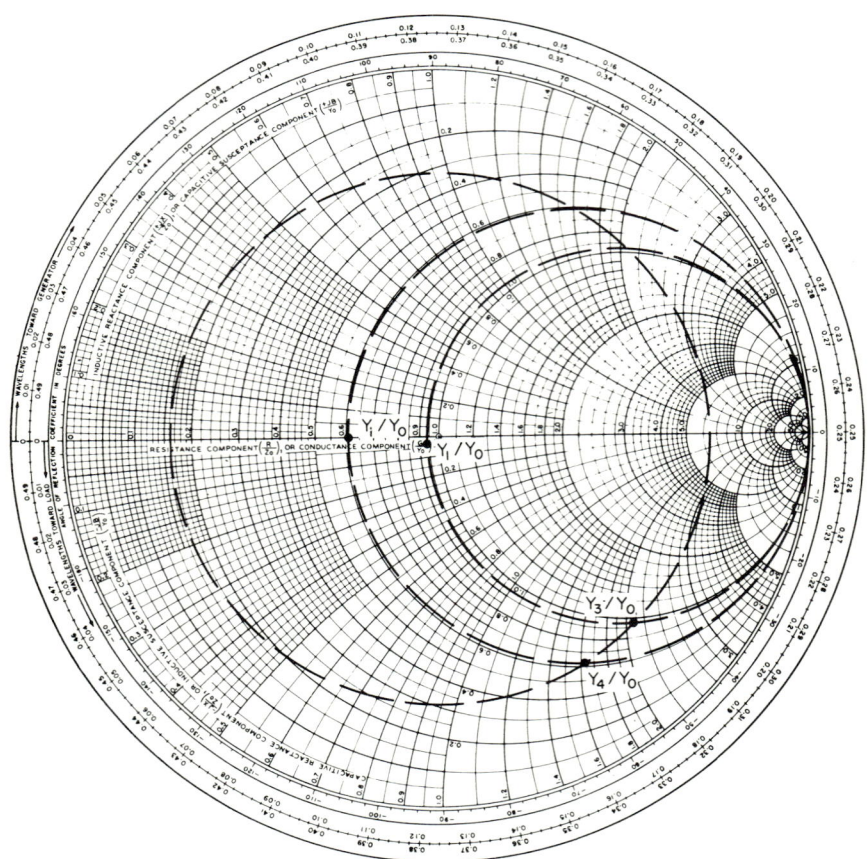

Fig. 16. Preliminary amplifier design technique.

made short enough so that potential low-frequency instabilities produced by the circulator can be avoided. At high frequencies, the short-circuited transmission line, shorted at terminal A, provides an admittance $Y_2 = -jY_o \cot \beta l_2$ at terminal B. Adding Y_2 to the circulator admittance Y_1 at terminal B, this admittance will be transformed for various frequencies on constant-conductance circles on the Smith chart to obtain the admittance Y_3 as a function of frequency at terminal B. Connecting Y_3 to another transmission line (characteristic admittance Y_o), Y_3 will be transformed on constant standing-wave ratio (SWR) circles in the clockwise direction on the Smith chart to form Y_4 and thus can provide some transformation of the real part of Y_4 to match the junction conductance $|g_j|$ for a high-gain condition.

DIODE AND PARAMETRIC AMPLIFIERS

There is no general procedure available utilizing the above transformation steps, which will provide an optimum design for the wideband case. However, a preliminary design procedure which satisfies desirable conditions at a center frequency of 1.3 gc can easily be formulated as follows:

1. Construct a constant circle through the circulator admittance Y_1/Y_o, at the center frequency of 1.3 gc as shown in Fig. 16.

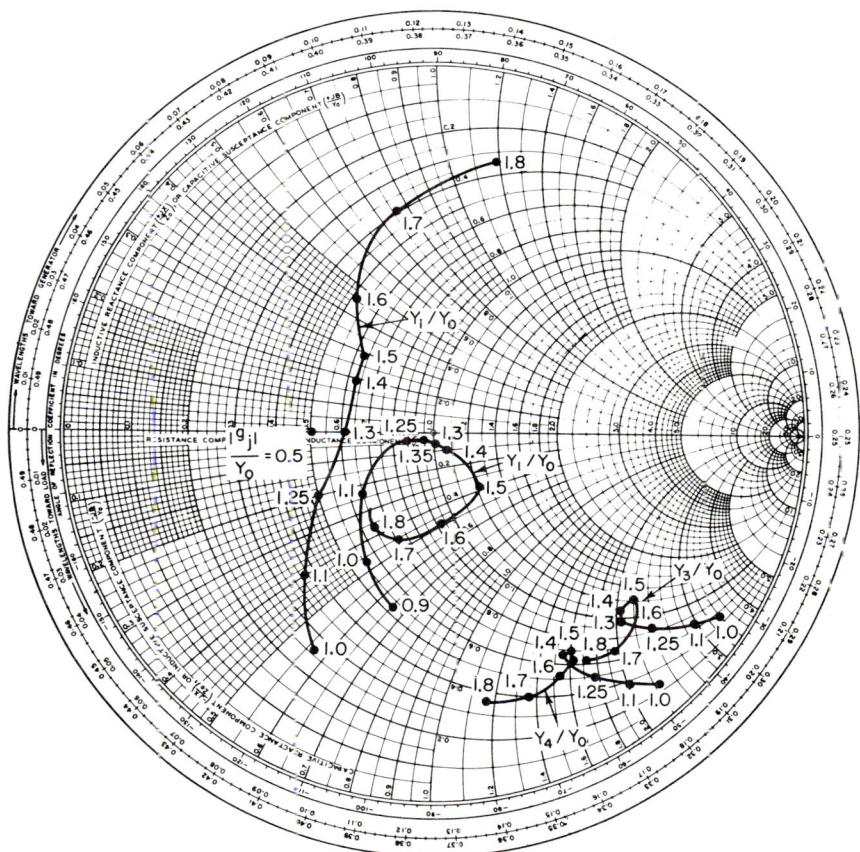

Fig. 17. Plots of various admittance transformations in an L-band tunnel-diode amplifier.

2. Construct a constant VSWR circle on the Smith chart through the desired external admittance $Y_4/Y_o = 0.6 - j1.64$ as shown in Fig. 16.

3. The intersection of the two circles represents Y_3/Y_o at 1.3 gc. This intersection provides information for calculating l_2 and l_3. According to Fig. 16, Y_3/Y_o is given by $0.92 - j2.17$. Then $Y_2/Y_o = -j \cot \beta l_2 = (Y_3 - Y_1)/Y_o$ becomes $-j2.13$ at 1.3 gc where the value of Y_1 is obtained from Fig. 15. From the Smith chart the value of l_2 becomes 0.070λ or 1.51 cm. Then the transmission-line length l_3 to transfer Y_3 into Y_4 is found from Fig. 16 to be 0.021λ or 0.58 cm.

Using values of l_2 and l_3 obtained from the above design procedure, all other admittance plots Y_2, Y_3, and Y_4 over the wide frequency range can be obtained from

28-14 AMPLIFIER CIRCUITS

Fig. 17. Finally adding $j\omega C_j/Y_o$ to Y_4, Y_i is obtained. As seen from Fig. 17, the plot* of Y_i does not encircle $|g_j|/Y_o = 0.5$ and Y_i/Y_o at 1.3 gc is 0.6, which is very close to $|g_j|/Y_o = 0.5$ for large gain.

This example turned out to be quite close to the optimum design for this particular circuit configuration. However, in general, it requires a trial-and-error process to select l_2 and l_3; at times it may be required to determine the length of the l_1 and l_4 sections of the transmission line to make a stable amplifier.

With careful adjustment of the lengths of the transmission lines, and the choice of a tunnel diode, it is possible to build a tunnel-diode amplifier using the above configuration. However, the circuit shown in Fig. 14 has rather limited adjustment

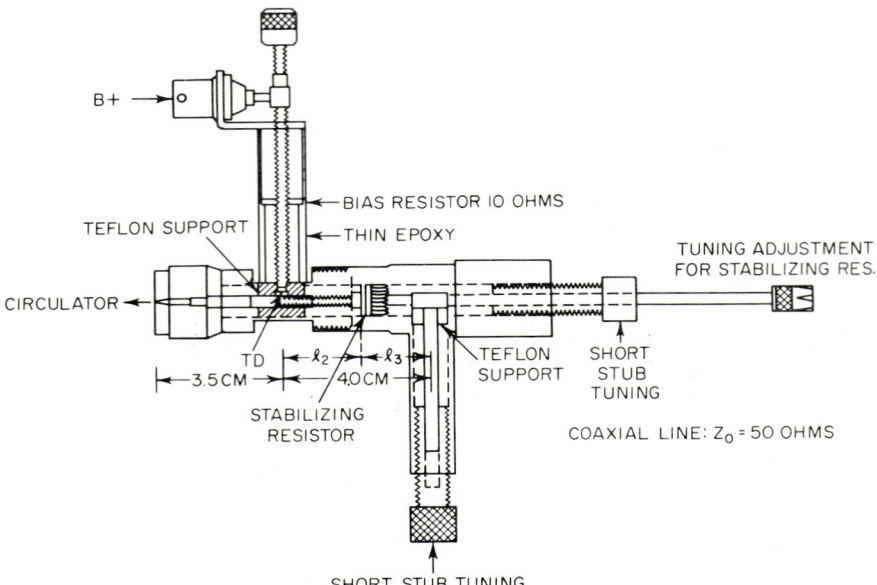

FIG. 18. Circulator-coupled S-band tunnel-diode amplifier.

capabilities. Consequently the replacement of tunnel diodes in the amplifier may be rather critical. Furthermore, if a tunnel diode is used with a very high resistive cutoff frequency f_r and very high self-oscillating frequency f_x compared with the operating frequency f, the circuit will have a tendency to oscillate at very high frequencies. This can be seen from the fact that, if the plot of Y_i/Y_o in Fig. 17 is extended much into the high-frequency region, it is possible that the very high frequency portion of the plot may encircle the $|g_j|/Y_o$ point.

The above amplifier design is an example of a simple design procedure; many different circulator-coupled amplifiers can be designed to improve certain desired amplifier characteristics. Another type of circulator-coupled S-band amplifier is shown in Fig. 18.

The positions of the stabilizing resistor and two shorting stubs can be varied externally. The resistance is shorted at the operating frequency by proper adjustment of two stubs to avoid power losses in the resistance and noise contributions from the resistance in the bandpass region of the amplifier. At the same time, by properly adjusting the two shorting stubs, the resistance acts as a damping resistance at higher frequencies where the unstabilized amplifier otherwise tends to oscillate.

* The extreme high and low ends of the admittance plots are not shown.

DIODE AND PARAMETRIC AMPLIFIERS

The measured circulator admittance characteristic Y_1 with matched circulator terminations is shown in Fig. 19. With the circulator admittance approximately matched to the coaxial line at the operating frequency, the transmission-line length l_2 of the amplifier in Fig. 18 is adjusted to cancel out the reactance of the junction capacitance C_j. A stabilizing resistance of 50 ohms is selected and at an operating frequency $f_o = 3$ gc, cot $\beta l_2 = \omega_o C_j / Y_o$, and l_2 is 1.5 cm for the line having a characteristic impedance of 50 ohms and $C_j = 0.85$ pf. Again the series resistance r_s, series

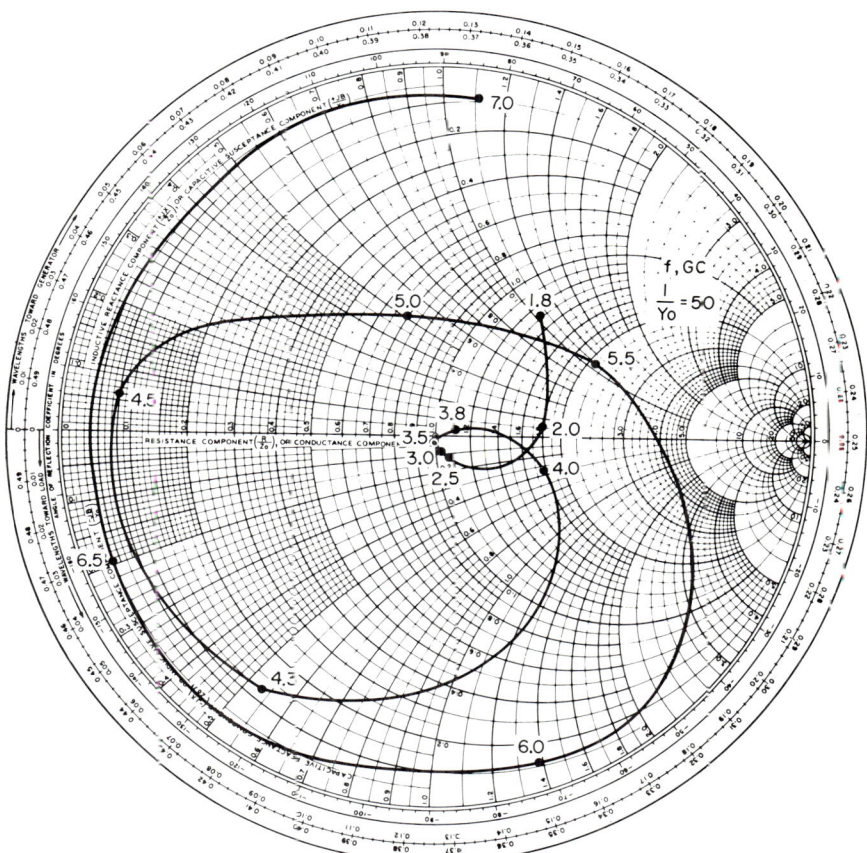

Fig. 19. S-band circulator admittance.

inductance L_s, and package capacitance C_s are neglected. Placing the junction point of the two stubs at a distance $l_3 = 2.5$ cm from the stabilizing resistance, and selecting the lengths* of the two stubs at 2.64 and 2.36 cm from the junction point, the plot of the total normalized admittance Y_2/Y_o at the stabilizing resistance looking toward the stubs and including the stabilizing resistance is given in Fig. 20a. The length l_2 transforms Y_2/Y_o into Y_3/Y_o across the diode; the plot of Y_3/Y_o is shown in Fig. 20a.

* As was stated previously, these stub lengths are adjusted to produce a short at the stabilizing resistance, at the operating frequency, and at the same time to stop possible oscillations at higher frequencies. There is no unique way to choose these lengths except by a trial-and-error process.

Assuming the circulator admittance Y_1 of Fig. 18 is presented across the tunnel diode, the total normalized admittance $Y_4/Y_o = (Y_1 + Y_3)/Y_o$ across the tunnel diode, excluding tunnel-diode admittance itself, is plotted in Fig. 20b. Finally, by adding the susceptance of the junction capacitance C_j to Y_4, the normalized admittance $Y_i/Y_o = Y_4/Y_o + j\omega C_j/Y_o$ across the diode conductance can be obtained as shown in Fig. 21. If a junction conductance $|g_j| = 0.016$ mho is assumed, the $Y_i(\omega)/Y_o$ plot will not encircle the $|g_j|/Y_o$ point on the Smith chart, and a gain of about 20 db at the center frequency will result.

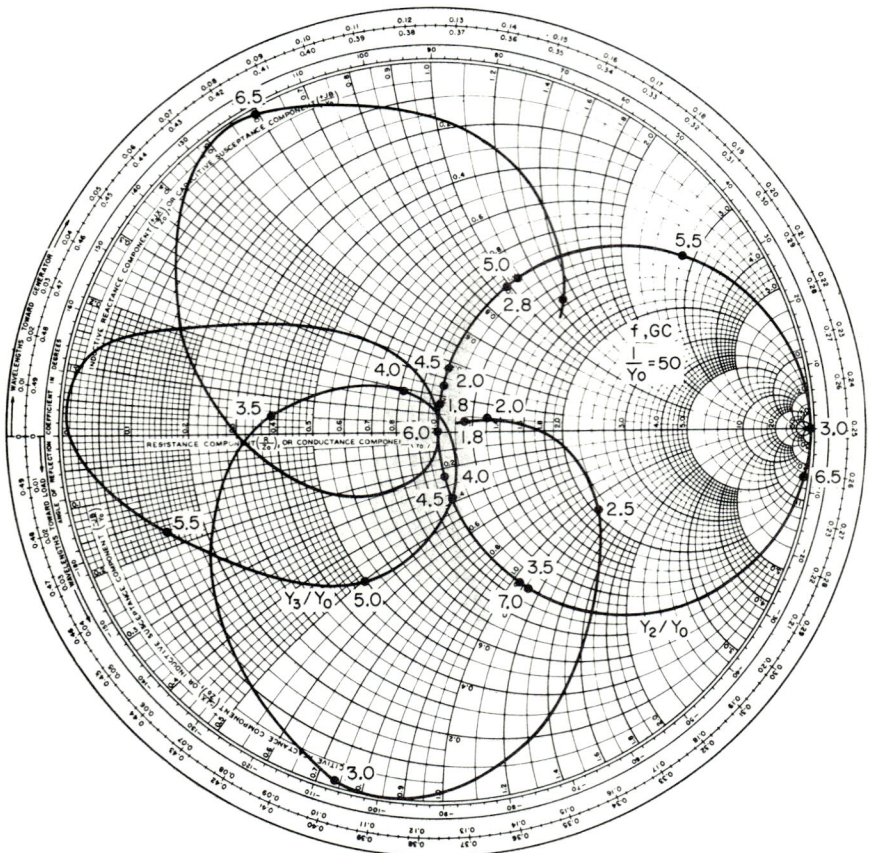

FIG. 20a. Admittance plots (Y_2/Y_o and Y_3/Y_o) of S-band tunnel-diode amplifier.

A photograph of an amplifier of the type just described is shown in Fig. 22. This amplifier had the following performance characteristics:

Insertion gain, db.......... 20
Bandwidth (3 db), Mc...... 70
Tuning range, gc.......... 3.0–3.4
Noise figure, db........... 4 ± 0.5

The nominal parameter values of the tunnel diode (General Electric) used in the amplifier are as follows:

DIODE AND PARAMETRIC AMPLIFIERS

I_p, ma	I_v, ma	V_p, mv	V_{fp}, mv	C_v, pf	$|r_j|$, ohms	r_s, ohms
1.89	0.21	76	566	0.82	0.68	2.37

This diode was biased at around 160 mv.

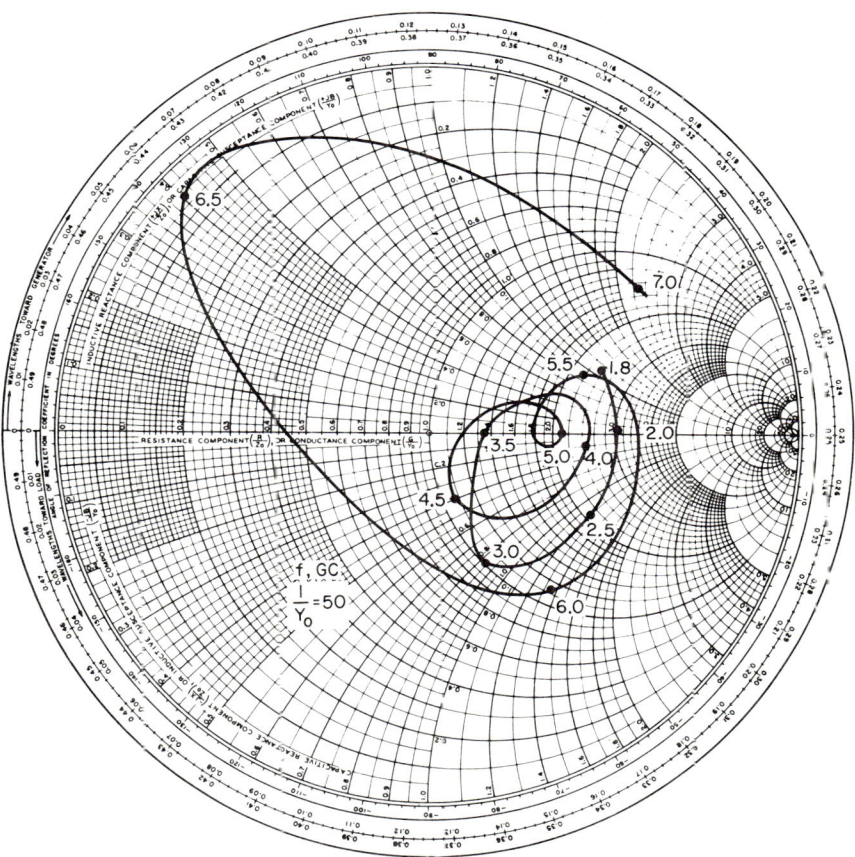

Fig. 20b. Admittance plots Y_4/Y_o of S-band tunnel-diode amplifier.

Bandwidth and linearity plots are shown in Figs. 23 and 24.

As was mentioned above, this amplifier was designed to compensate for changes in the amplifier and diode parameters by adjusting the position of the stabilizing resistance, and by varying the two shorting stubs.

By means of these adjustments, it was possible to interchange tunnel diodes having parameter values within the following ranges:

$$55 < |r_j| > 70 \text{ ohms}$$
$$0.5 < C_j > 1.07 \text{ pf}$$

28-18 AMPLIFIER CIRCUITS

This amplifier was not designed to provide a wide bandwidth but rather to provide a capability for flexible adjustment. The design of a wideband amplifier will be discussed later.

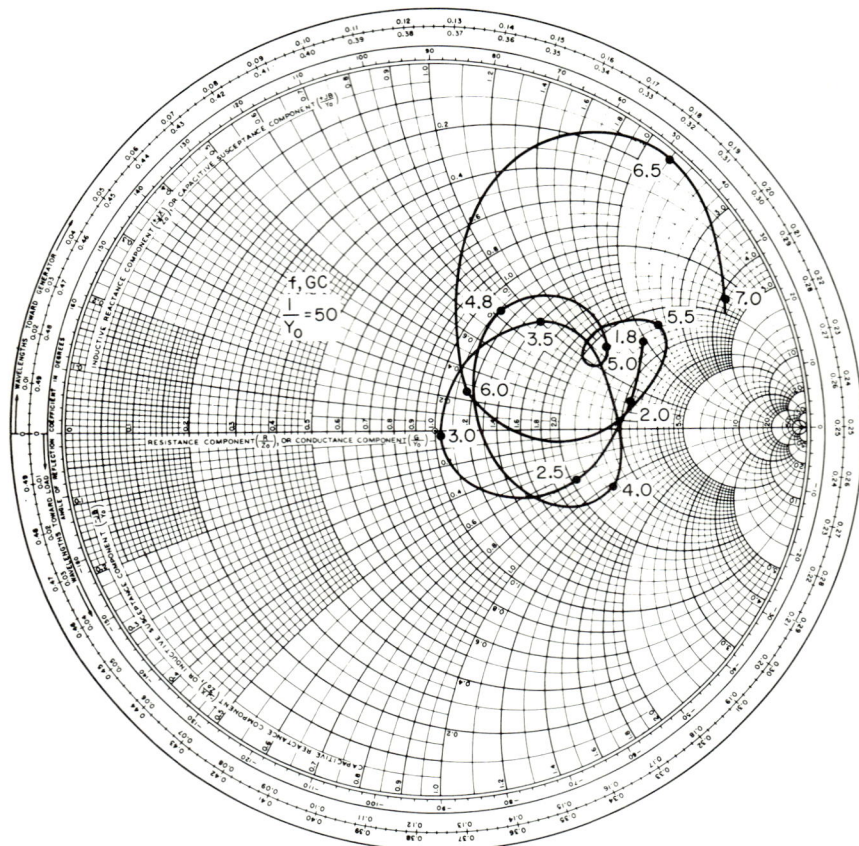

Fig. 21. Admittance plot (Y_i/Y_o) of S-band tunnel-diode amplifier.

1.6. Gain, Bandwidth, Sensitivity, and Noise Figure. *Gain.* The transducer power gain P_T of the bilateral tunnel-diode amplifier shown in Fig. 6 is given by

$$P_T = \frac{\left|\dfrac{I_S}{Y_S + Y_C - |g_j| + Y_L} G_L\right|^2}{|I_S|^2/4G_S} = \frac{4G_S G_L}{|G_T + jB_T|^2} \tag{7}$$

where $Y_S = G_S + jB_S$ = source admittance
 $Y_C = G_C + jB_C$ = tuning-circuit admittance
 $|g_j|$ = junction conductance (series resistance r_s, series inductance L, and the stray capacitance C_s are assumed negligible. The effect of the junction capacitance C_j can be included in Y_C)
 $Y_L = G_L + jB_L$ = load admittance (or second-stage input admittance)
 $G_T = G_S + G_C - |g_j| + G_L$
 $B_T = B_S + B_C + B_L$

DIODE AND PARAMETRIC AMPLIFIERS

By tuning out the total susceptance across the diode conductance, i.e., making $B_T = 0$, the maximum gain becomes

$$P_{TO} = 4G_S G_L / G_T^2 \qquad (8)$$

As G_T approaches zero, P_{TO} becomes infinite.

It should be pointed out that the amplified power P_S is dissipated at the source and can be considered a loss. The expression for P_S is given by

$$P_S = |I_S/Y_T|^2 G_S$$

In circulator-coupled and hybrid-coupled tunnel-diode amplifiers, where the circulator and hybrid coupler can be considered as lossless circuit elements, it is convenient to use the scattering-matrix technique[7,8,9,10,11] in determining the transmission and reflection of power between ports. It can be shown that the power transmission from the ith terminal to the jth terminal of an n-terminal network can be expressed by the network scattering-matrix element $|S_{ij}|^2$; $|S_{ij}|^2$ is identical with insertion gain. If the circulator and hybrid are assumed ideal, the insertion power gain P_I of circulator-coupled and hybrid-coupled TD amplifiers can then be expressed[8] by (refer to Figs. 7 and 9)*

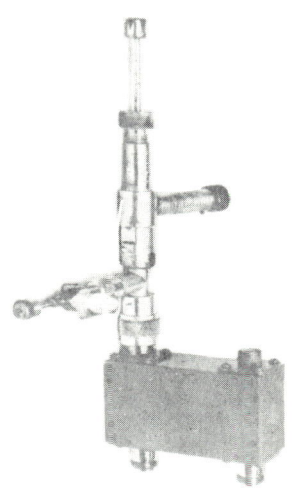

Fig. 22. S-band amplifier.

$$P_I = |\Gamma_2|^2 = |S_{13}|^2 \qquad (9)$$

where $\Gamma_2 = \dfrac{Y_o - Y_d}{Y_o + Y_d}$ = reflection coefficient at terminal 2 where TD amplifier is connected

Y_d = input admittance of TD amplifier
Y_o = characteristic admittance of circulator

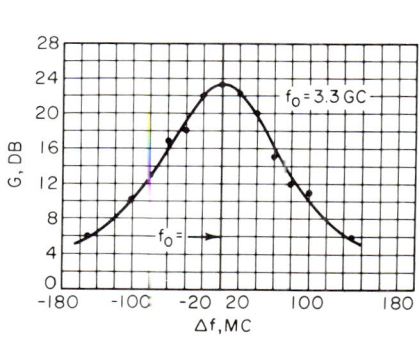

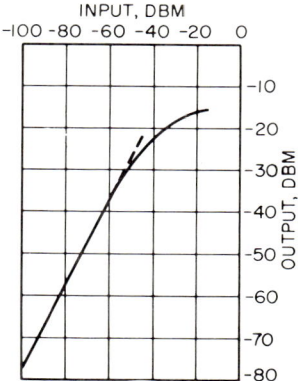

Fig. 23. Bandwidth plot of S-band tunnel-diode amplifier.

Fig. 24. Linearity plot of S-band tunnel-diode amplifier.

An ideal circulator is a lossless network. Thus if the amplifier circuit elements are lossless and the series resistance of the diode is neglected as an approximation, the

* In a hybrid-coupled TD amplifier, $|\Gamma_3|^2$, the reflection coefficient of terminal 3, can be used for P_I instead of $|\Gamma_2|^2$, since $|\Gamma_2|^2 = |\Gamma_3|^2$.

power gain P_I of Eq. (7) should be

$$P_I = \left| \frac{Y_i + |g_j|}{Y_i - |g_j|} \right|^2 = |\Gamma|^2 \qquad (10)$$

where Y_i = total admittance across the junction conductance g_j, excluding g_j (refer to Fig. 1).

The reflection coefficient Γ defined by Eq. (10) is always greater than 1 and can become infinite.

Defining Γ' as the normalized reflection coefficient in a passive circuit,

$$\Gamma' = \frac{Y_i - g_o}{Y_i + g_o} \qquad (11)$$

or where $g_o = |g_j|$

it can be shown that

$$\Gamma' = 1/\Gamma \qquad (12)$$

Then there exists the following relationship between reflection coefficient and voltage standing-wave ratio (VSWR):

$$\Gamma = \frac{\rho + 1}{\rho - 1} = \frac{1}{\Gamma'} \qquad \text{or} \qquad \rho = \frac{1 + \Gamma'}{1 - \Gamma'} = \frac{\Gamma + 1}{\Gamma - 1} = \frac{Y_i}{g_o} \qquad (13)$$

It should be noted that as g_o decreases, Γ' and ρ increase; as g_o goes to zero, Γ' and ρ are confined in the right half portion of the admittance plane. On the other hand, if $g_o = g_j$ becomes negative, the plots of Γ and ρ should be in the left portion of the admittance plane. ρ decreases from infinity to 1 and Γ increases from 1 to infinity, respectively, as $|g_j|$ increases from zero to $|g_j| = Y_i$.

As an example, from the design of the amplifier described in Fig. 14, for $|g_j|/Y_o = 0.5$ and $Y_i/Y_o = 0.6$, the power gain of the amplifier is given by

$$P_I = \left(\frac{0.6 + 0.5}{0.6 - 0.5} \right)^2 = 121 \qquad (14)$$

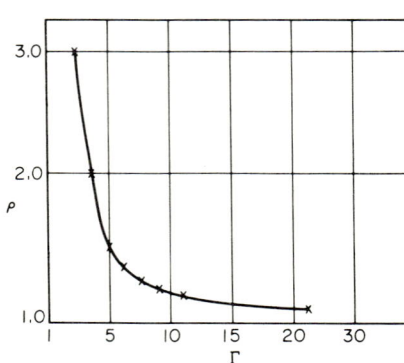

Fig. 25. Relationship between VSWR (ρ) and reflection coefficient Γ.

Transforming the plots of Γ and ρ from the left half to the right half of the admittance place, the same ρ from the right half plane can be used to represent ρ in the left half plane. Here the corresponding reflection coefficient must be Γ instead of Γ'.

This transformation makes it possible to use the Smith chart to represent Γ and ρ. Normalizing the Y_i plot with respect to $|g_j|$, the Y_i plot on the constant-ρ circle in the Smith chart represents the constant-gain requirement for the amplifier.

The relationship between ρ and $|\Gamma|$ is given in Fig. 25.

Bandwidth. The bandwidth of the bilateral tunnel-diode amplifier of Fig. 6 can be derived from the ratio of Eqs. (7) to (8) as follows:

$$\frac{P_T}{P_{TO}} = \frac{1}{1 + (B_T/G_T)^2} = \frac{1}{2} \qquad (15)$$

The bandwidth B is defined as the frequency range $f_2 - f_1$ where Eq. (15) is satisfied. If $Y_T = G_T + jB_T$ consists of a single-tuned circuit so that

$$Y_T = G_T + j(\omega C_T - 1/\omega L_T)$$

Eq. (15) can be expressed by

$$P_T/P_{TO} = 1/[1 + (2\delta Q_1)^2] = \frac{1}{2} \quad (16)$$

where $\delta = \Delta\omega/\omega_o \quad (\delta^2 \ll 1)$
$Q = \omega_o C_T/G_T$
$f_o = \omega_o/2\pi \quad$ (center frequency)

Then the normalized bandwidth 2δ and the bandwidth B become

$$2\delta = 1/Q \quad (17)$$
$$B = f_o/Q$$

The gain-bandwidth product is given by

$$\sqrt{P_{TO}}\, B = \sqrt{G_S G_L}/\pi C_T \quad (18)$$

If C_T consists of only the junction capacitance C_j, the gain-bandwidth product becomes very large.[8,*]

To obtain a large bandwidth, multituned circuits are required. Analytically, this can be realized by making the gain $|\Gamma|^2$ of Eq. (10) a Butterworth or Chebyshev[8,10,12] response.

For a low-pass Butterworth response, the gain $|\Gamma|^2$ can be represented by[3]

$$|\Gamma|^2 = \frac{\Gamma_o^2 + k'^2 \omega^{2n}}{1 + \omega^{2n}} \quad (19)$$

where Γ_o^2 = gain at $\omega = 0$
ω = normalized angular frequency
k' = constant
n = positive integer

Rewriting Eq. (19) in the complex domain,

$$|\Gamma|^2 = \frac{1}{|\Gamma'|^2} = \frac{(\Gamma_o + jk'\omega^n)(\Gamma_o - jk'\omega^n)}{(1 + j\omega^n)(1 - j\omega^n)}$$
$$= \frac{[\Gamma_o + jk'(s/j)^n][\Gamma_o - jk'(s/j)^n]}{[1 + j(s/j)^n][1 - j(s/j)^n]} \quad (20)$$

Roots of polynomial $\Gamma_o + jk'(s/j)^n = A_1(s)$ and $\Gamma_o - jk'(s/j)^n = A_2(s)$.

and
$$s_k = [j(k'/\Gamma_o)^{-1/n}]e^{j(4k-3)(\pi/2n)} \quad \text{for } A_1(s)$$
$$s_r = [j(k'/\Gamma_o)^{-1/n}]e^{-j(4r-3)(\pi/2n)} \quad \text{for } A_2(s) \quad (21)$$

where $1 \leq k \leq n$
$1 \leq r \leq n$
$\sqrt{-1} = j$

As seen from Eq. (21), the roots s_k and s_r are in the left and right half of the s plane, respectively. Since the roots of the denominator of Γ', the reflection coefficient of the passive network, must lie in the left half of the s plane to be physically realizable,[13] the roots of the numerator of Γ must be in the left half of the s plane [i.e., $A_1(s)$]. On the other hand, the roots of the numerator of Γ' or the roots of the denominator

* For example, for $C_j = \pi \times 10^{-1}$ pf, $G_S = 0.01$, and $G_L = 0.01$ mho, $\sqrt{P_{TO}}\, B$ becomes 10^9.

28–22 AMPLIFIER CIRCUITS

of Γ may be in either the left or right half of the s plane.

$$Y_i = \frac{g_o(\Gamma+1)}{\Gamma-1}$$

$$= \frac{g_o\left(\dfrac{\Gamma_o + jk'\omega^n}{1+j\omega^n}+1\right)}{\dfrac{\Gamma_o+jk'\omega^n}{1+j\omega^n}-1} = \frac{g_o\rho_o\left(1+j\dfrac{k'+1}{\Gamma_o+1}\omega^n\right)}{1+j\dfrac{k'-1}{\Gamma_o-1}\omega^n} \quad \text{for case 1}$$

$$= \frac{g_o\left(\dfrac{\Gamma_o + jk'\omega^n}{1-j\omega^n}+1\right)}{\dfrac{\Gamma_o+jk'\omega^n}{1-j\omega^n}-1} = \frac{g_o\rho_o\left(1+j\dfrac{k'-1}{\Gamma_o+1}\omega^n\right)}{1+j\dfrac{k'+1}{\Gamma_o-1}\omega^n} \quad \text{for case 2} \quad (22)$$

where $$\rho_o = \frac{\Gamma_o+1}{\Gamma_o-1}$$

Here cases 1 and 2 represent situations where the zeros of Γ' are in the left and right half planes of s, respectively.

For a Butterworth response, the admittance Y_i from Eq. (22), normalized with respect to $g_o = |g_j|$, should be on the constant VSWR circle ρ_o at $\omega = 0$ for a maximally flat response of Eq. (22) which will match the constant ρ_o circle.

In order for Y_i to be physically realizable, it must be a positive real function, and the numerator and denominator of Y_i should be Hurwitz polynomials where the order of polynomials can be different from 1.[3] Thus k' must be a constant which is larger than 1.

Similarly, the gain $|\Gamma|^2$ having a Chebyshev response can be expressed by[3]

$$|\Gamma|^2 = \frac{\Gamma_o{}^2 + a^2 T_n{}^2(\omega)}{1 + b^2 T_n{}^2(\omega)} \quad (23)$$

where $T_n(\omega) = \cos(n \cos^{-1}\omega)$
$a =$ constant
$b =$ constant

again, $|\Gamma|^2$ can be rewritten as

$$|\Gamma|^2 = \frac{1}{|\Gamma'|^2} = \frac{[\Gamma_o + jaT_n(s/j)][\Gamma_o - jaT_n(s/j)]}{[1+jbT_n(s/j)][1-jbT_n(s/j)]}$$

Roots of $\Gamma_o + jaT_n(s/j) = A_1(s)$ and $\Gamma_o - jaT_n(s/j) = A_2(s)$ are in the left and right half s plane, respectively, and can be represented by

$$\begin{aligned}s_k =\ & -\sin(2k-1)(\pi/2n)\sinh[(1/n)\sinh^{-1}(1/a)] \\ & +j\cos(2k-1)(\pi/2n)\cosh[(1/n)\sinh^{-1}(1/a)] \quad \text{for } A_1(s) \\ s_r =\ & \sin(2r-1)(\pi/2n)\sinh[(1/n)\sinh^{-1}(1/a)] \\ & +j\cos(2r-1)(\pi/2n)\cosh[(1/n)\sinh^{-1}(1/a)] \quad \text{for } A_2(s) \end{aligned} \quad (24)$$

where $1 \le k \le n$
$1 \le r \le n$

Y_i for a Chebyshev response can then be given as

$$Y_i = \frac{g_o(\Gamma+1)}{\Gamma-1}$$

$$= \frac{g_o\rho_o\left[1+j\left(\dfrac{a+b}{\Gamma_o+1}\right)T_n(\omega)\right]}{1+j\left(\dfrac{a-b}{\Gamma_o-1}\right)T_n(\omega)} \quad \text{for case 1}$$

$$= \frac{g_o\rho_o\left[1+j\left(\dfrac{a-b}{\Gamma_o+1}\right)T_n(\omega)\right]}{1+j\left(\dfrac{a+b}{\Gamma_o-1}\right)T_n(\omega)} \quad \text{for case 2} \quad (25)$$

Again, Y_i for a Chebyshev response should be on the ρ_o circle with equiripples. For a physically realizable condition, a must be greater than b, as in the Butterworth case. The above analysis provides information on low-pass responses. For a bandpass response applying the standard low-pass to bandpass frequency transformations, the corresponding bandpass Butterworth and Chebyshev responses can be obtained. Here again the normalized admittance $Y_i/|g_j|$ should be located close to the constant VSWR ρ_o circle in the passband region.

Of course, the plot of $Y_i(\omega)$ should not encircle $|g_j| = g_o$, to ensure the stability conditions mentioned previously.

A wideband amplifier (bandwidth greater than 20 per cent of the center frequency) can be designed by placing a bandpass filter having a wide bandwidth between the circulator and the tunnel diode, and by placing a one-port attenuator across the diode as shown in Fig. 26. The attenuator provides an open circuit in the passband region but provides a large conductance at the outside of the bandwidth across the diode.

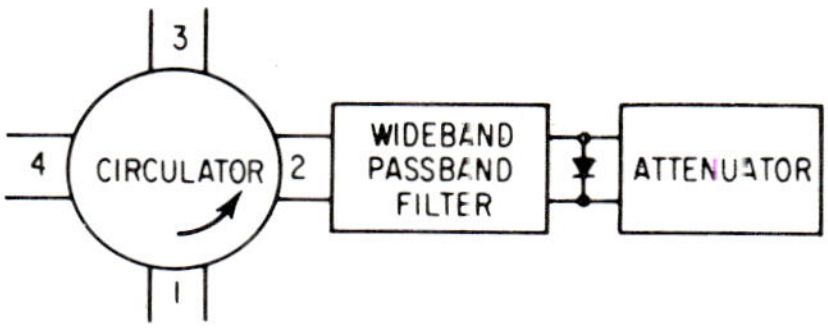

The wideband admittance Y_i across the junction conductance $|g_j|$ includes not only the wideband passband filter but also the circulator and the diode parameters, except $|g_j|$. The attenuator is placed to prevent oscillations which might occur for frequencies outside the range of the passband.

FIG. 26. A possible wideband tunnel-diode amplifier.

Sensitivity. An important measure of performance of a negative-conductance amplifier is the sensitivity[15] S of the amplifier with respect to the negative-conductance variation of the tunnel diode. This is defined by the ratio of the per-unit change in gain to the per-unit change in negative conductance, and can be expressed by

$$S = \frac{\Delta P/P}{\Delta|g_j|/|g_j|} \qquad (26)$$

where P = power gain. Sensitivity S for a bilateral tunnel-diode amplifier at the center frequency can be derived from Eq. (7) and is given by

$$S = 2|g_j|/G_T \qquad (27)$$

where $G_T = G_S + G_C - |g_j| + G_L$

Sensitivity S becomes more of a problem for high gain and becomes infinite at infinite gain. This implies that, for a high-gain amplifier, the fluctuation of $|g_j|$ must be kept to a minimum in order to reduce the amplitude and phase variations in the amplifier response.

In the case of circulator-coupled and hybrid-coupled tunnel-diode amplifiers, S can be derived from Eq. (10) and is given as

$$S = \frac{4|Y_i|\,|g_j|}{|(Y_i - |g_j|)^2|\,|\Gamma|} \qquad (28)$$

Again, for a large gain, $Y_i \to |g_j|$ and S becomes infinite.

Other expressions for S defined by ratios of the per-unit change in gain to the per-unit change of other parameters such as input termination admittance, and output termination admittance have been derived. The sensitivity of a bilateral tunnel-diode amplifier, defined for variations of terminal admittance, is similar to the expression given in Eq. (27).

On the other hand, in a four-port circulator or a three-port circulator with an isolator on the outside coupled to a tunnel-diode amplifier, the variation of terminal admittance has less effect on the variation of Y_i. Consequently, the sensitivity, defined for variation of terminal admittance, in this case is less critical than the sensitivity defined for a variation of $|g_j|$. This can be shown by expressing Y_i in terms of terminal admittances, in $|\Gamma|^2 = P$ and ρ by solving S for the variation of terminal admittance.

This is another advantage of using a circulator-coupled tunnel-diode amplifier.

Noise Figure. Detailed discussions on noise produced by a tunnel diode were given in Chap. 12. Referring to Fig. 15 of Chap. 12, a noise equivalent circuit for the bilateral tunnel-diode amplifier of Fig. 6 can be represented by Fig. 27a. Expressing the noise contributions of the second stage as a pair of uncorrelated voltage and current noise generators at the input side with an appropriate correlation admittance

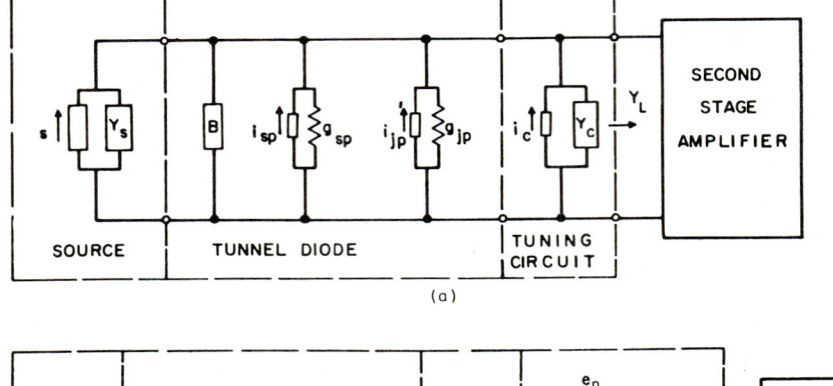

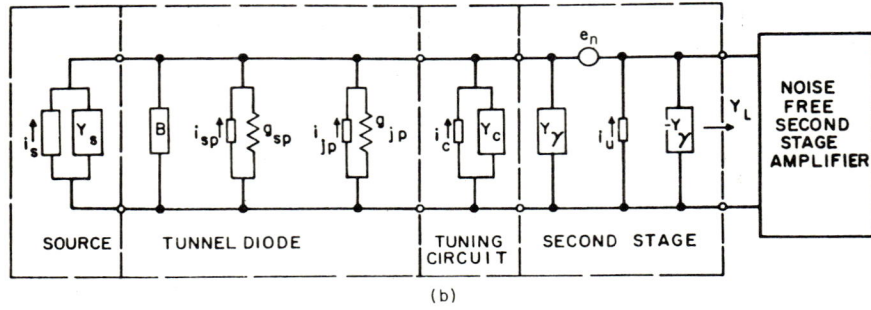

Fig. 27. (a) Noise equivalent circuit of a bilateral tunnel-diode amplifier. (b) Noise equivalent circuit of a bilateral tunnel-diode amplifier with second stage.

Y_γ, the noise equivalent circuit of Fig. 27 can be modified to that of Fig. 27b (see Chap. 7).[15] Then the overall noise figure F_{12} of Fig. 27b can be expressed as

$$F_{12} = 1 + \frac{\overline{i_{sp}^2} + \overline{i_{jp}^2} + \overline{i_c^2} + \overline{i_u^2} + |Y_t|^2 \overline{e_n^2}}{\overline{i_s^2}}$$

$$= F_1 + \frac{F_{2e} - 1}{P_{ei}} = F_1 + \frac{\overline{i_u^2} + |Y_t|^2 \overline{e_n^2}}{\overline{i_s^2}} \qquad (29)$$

where $\overline{i_s^2} = 4kT_o G_S \Delta f$
$\quad\quad\quad Y_S = G_S + jB_S$ = source admittance
$\quad\quad\quad \overline{i_{jp}^2} = 4kT(G_{eq}/|g_{jp}|)|g_j|\Delta f$
$\quad\quad\quad G_{eq} = eI_{eq}/2kT$
$\quad\quad\quad \overline{i_{sp}^2} = 4kT g_{sp} \Delta f$

$\overline{i_c^2} = 4kTG_c \Delta f$
$Y_c = G_c + jB_c$ = tuning-circuit admittance
$\overline{e_n^2} = 4kTR_n \Delta f$
R_n = equivalent noise resistance of noise voltage generator e_n
$\overline{i_u^2} = 4kTG_u \Delta f$
G_u = equivalent noise conductance of noise current generator i_u
$Y_t = Y_s + B + g_{sp} - |g_{jp}| + Y_c + Y_\gamma$
$Y_\gamma = G_\gamma + jB_\gamma$ = correlation admittance
F_1 = noise figure of bilateral tunnel-diode amplifier
F_{e2} = exchangeable noise figure of second stage
P_{ei} = exchangeable gain of first-stage tunnel-diode amplifier
T = circuits and devices temperature
T_o = source temperature

F_1 can be easily identified from Eq. (16) of Chap. 7, and Eq. (29) as

$$F_1 = 1 + \frac{\overline{i^2_{sp}} + \overline{i_{jp}^2} + \overline{i_c^2}}{\overline{i_s^2}} = 1 + \frac{(G_{SP} + G_{eq}|g_{jp}|/|g_j| + G_c)T}{G_S T_o} \quad * \tag{30}$$

where T_o = source temperature
Then the following relation must hold:

$$\frac{F_{e2} - 1}{P_{ei}} = \frac{\overline{i_u^2} + |Y_t|^2\overline{e_n^2}}{\overline{i_s^2}} \tag{31}$$

Referring to Eq. (17) of Chap. 7, P_{ei} for a tuned circuit is given by

$$P_{ei} = \frac{S_{eo}}{S_{ei}} = \frac{4kTG_S \Delta f/G_{S2}}{4kT \Delta f} = \frac{G_S}{G_{S2}} \tag{32}$$

where S_{eo} = exchangeable signal power at the output of the first stage
S_{ei} = exchangeable signal power at the source
$G_{S2} = G_S - |g_{jp}| + G_c + |G_{SP}|$ = source conductance of the second stage
Referring to Eq. (18) of Chap. 7, the exchangeable noise figure of the second stage can be expressed as

$$F_{e2} = 1 + \frac{N_{e2}}{N_i} = 1 + \frac{\overline{i_u^2} + |Y_t|^2\overline{e_n^2}}{4kTG_{S2} \Delta f} \tag{33}$$

where N_{e2} = exchangeable noise power contributed by the second stage, referred to the input of the second stage with no noise contributions from the input circuit of the second stage (this corresponds to N_{eN}/G_z)
$N_i = kT \Delta f$
Then, from Eqs. (32) and (33), the following relation is obtained:

$$\frac{F_{e2} - 1}{P_{ei}} = \frac{\overline{i_u^2} + |Y_t|^2\overline{e_n^2}}{4kTG_S \Delta f} \tag{34}$$

Equation (34) is identical with Eq. (31).

It should be pointed out that if $G_{S2} < 0$, P_{ei} and $F_{e2} - 1$ are negative, but $(F_{e2} - 1)/P_{ei}$ is always positive. Furthermore, it is interesting to note that the magnitude of $F_{e2} - 1$ from Eq. (33) is the same for both negative and positive values of source conductance G_{S2}. Similarly for Eq. (28) in Chap. 7, the minimum value of $|F_{e2} - 1|$ is obtained for an optimum value of source conductance $G_{S2} = G_o$ as follows:

$$F_{e2o} - 1 = 2R_n(G_\gamma + G_o) \tag{35}$$

where
$$G_{S2} = \text{Re }\{Y_{S2}\} = G_o$$
$$= \pm(G_\gamma^2 + G_u/R_n)^{\frac{1}{2}} \tag{36}$$

* The thermal-noise contribution from the tuning circuit is included in the noise figure of a bilateral tunnel-diode amplifier.

A convenient form for the exchangeable noise figure is given by

$$F_{e2} = F_{e2o} + (R_n/G_{S2})[(G_{S2} - G_o)^2 + (B_{S2} - B_o)^2] \tag{37}$$

where $B_{S2} = \text{Im}\{Y_{S2}\}$
$B_o = -B_\gamma$

It should be noted that the magnitude of Eq. (35) can be different for positive or negative G_o depending upon the sign of the correlation conductance G_γ. Therefore, the magnitude of F_{e2o} can be made the same for both positive and negative values of source admittance.

The overall noise figure F_{12} of a bilateral tunnel-diode amplifier will now be discussed in the light of the above discussion. It should be noted that the second term of Eq. (29) or (34) is really not a function of P_{ei}, since G_{S2}, which provides the power gain for P_{ei} in Eq. (32), cancels the G_{S2} in Eq. (33). Thus this term can contribute to F_{12} if $\overline{i_u^2}$ and $|Y_t|^2\overline{e_n^2}$ are comparable with $\overline{i_s^2}$. A simple physical explanation of this result can be given as follows: If the exchangeable gain P_{ei} is large, the noise from the source noise generator i_S of the first stage is amplified. The same amplification occurs for the noise from noise generators i_u and e_n since the first-stage tunnel-diode amplifier is bilateral. Thus the relative noise-power contribution due to $\overline{i_s^2}$, $\overline{i_u^2}$, and $|Y_t|^2\overline{e_n^2}$ remains unchanged.

Making $|Y_t| = 0$, F_{12} of Eq. (29) can be reduced to

$$F_{12} = F_1 + \overline{i_u^2}/\overline{i_s^2} \tag{38}$$

For a circulator-coupled tunnel-diode amplifier, the situation is quite different. The noise equivalent circuit for this amplifier followed by a second stage can be represented in Fig. 28 following Fig. 27 presentation and an ideal circulator.

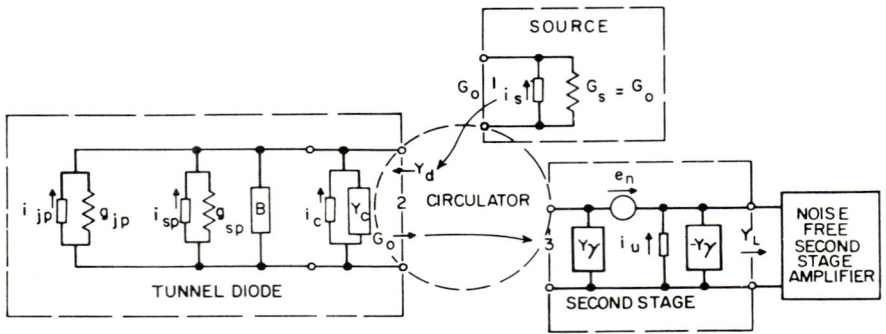

FIG. 28. Noise equivalent circuit of a circulator-coupled tunnel-diode amplifier with second stage.

Referring to Eq. (7) and Fig. 28, the noise figure F_1 of the first stage computed at port 3 can be given for an ideal circulator as

$$F_1 = \frac{\text{total mean-square noise currents appearing at port 3}}{\text{mean-square noise current appearing at port 3 contributed by the source}}$$

$$= \frac{\overline{i_s^2}|\Gamma|^2 + \overline{i_2^2}|\Gamma - 1|^2}{\overline{i_s^2}|\Gamma|^2} = 1 + \frac{\overline{i_2^2}|\Gamma - 1|^2}{\overline{i_s^2}|\Gamma|^2} \tag{39}$$

where
$$\Gamma = \frac{Y_o - Y_d}{Y_o + Y_d} = \frac{G_o - Y_d}{G_o + Y_d}$$

Y_d = input admittance of tunnel-diode amplifier $\tag{40}$

$$\overline{i_s^2} = 4kT_oG_o\,\Delta f$$
$$\overline{i_2^2} = \overline{i_{sp}^2} + \overline{i_{jp}^2} + \overline{i_c^2}$$

Then the overall noise figure F_{12} computed at port 3 becomes

$$F_{12} = F_1 + (F_2 - 1)/P_1 \tag{41}$$

where the noise figure F_2 of the second stage is

$$F_2 = \frac{\overline{i_u^2} + |Y_t|^2 \overline{e_n^2}}{4kTG_o \Delta f} \tag{42}$$

and the available power gain P_1 of the tuned first stage is given by

$$P_1 = \frac{4kTG_o \Delta f/(G_o + G_d)}{4kT \Delta f} = \frac{G_o}{G_o + G_d} \tag{43}$$

where $G_o + G_d \geq 0$ for stability

$$\text{Re}[Y_d] = G_d = |g_{sp}| - |g_{jp}| + G_c$$

In this case, making $G_o + G_d$ very small or making P_1 large, the second term of Eq. (41) (the noise contribution from the second stage) can be made much smaller than the first term.

This difference results from the fact that the source impedance of the second stage in a circulator-coupled tunnel-diode amplifier is always G_o and is independent of P_1.

Making Γ approach 1 (i.e., no-gain case) as seen from Eq. (39), F_1 will become unity as the limit. Therefore, a minimum overall noise figure F_{12} can be obtained at the gain for which a best compromise of the noise contribution from the first and second stages is obtained.

As an example, in the cascaded circulator-coupled tunnel-diode amplifier of Fig. 12 the gain of the first stage would be adjusted for a low value in order to minimize the overall noise figure.

As seen from Eq. (39), if $|\Gamma|^2$ approaches infinity, the noise figure of a tunnel-diode amplifier will become equal to F_o of Eq. (18)[17] in Chap. 12.

2. CAPACITANCE-DIODE AMPLIFIERS

In this section the important formulas useful in the synthesis and design of parametric amplifiers will be derived in a brief and general manner. Then an example of a typical design will be treated.

A list of symbols is presented to facilitate the use of the following design equations.

- MAG = maximum available gain
- ω_1 = input or signal frequency
- ω_p = pump frequency
- $\omega_2 = f_p - f_1$ = output frequency
- ω_Q = ratio of one-fourth of the net elastance change in the diode to the diode series resistance
- R_s = series resistance of the diode
- R_g' = generator resistance for maximum available gain
- R_L' = load resistance for maximum available gain
- C' = a factor defined in Eq. (97)
- C'' = a factor defined in Eq. (100)
- F' = noise figure corresponding to the generator resistance R_g', i.e., the noise figure corresponding to maximum available gain
- F'' = minimum noise figure
- R_g'' = generator impedance which must be used to give the minimum noise figure F''
- m_0 = ratio of the average S_0 to the available elastance swing ($S_{max} - S_{min}$)
- m_1 = ratio of one-half the sinusoidal peak elastance at the pump frequency to the available elastance swing ($S_{max} - S_{min}$)
- $w'_{3\,db}$ = fractional bandwidth when the available gain is a maximum assuming independent lumped inductance tuning at ω_1 and ω_2
- $w''_{3\,db}$ = the same as $w'_{3\,db}$ except that it applies to the minimum-noise-figure operation of the up-converter

2.1. General Formulas for Capacitance-diode Amplifiers and Converters.
In the treatment to follow, it is assumed that
  1. The diode is pumped by a circuit which presents an open circuit at all harmonics of the pump.
  2. The series resistance R_s is a constant, independent of level and frequency.
  3. The pump power level is much greater than the input and output signal levels.
  4. The impedance presented to all undesirable mixing products is open-circuit.

Under the above assumptions it can be shown that any parametric amplifier or converter can be represented by a linear two-port impedance matrix with the following elements:

$$\begin{aligned} z_{11} &= r_{11} + jx_{11} = R_s + jx_{11} \\ z_{12} &= jx_{12} \\ z_{21} &= jx_{21} \\ z_{22} &= r_{22} + jx_{22} = R_s + jx_{22} \end{aligned} \qquad (44)$$

where $r_{11} = r_{22} = R_s$.

The actual values of the x_{ij}'s for the various types of amplifiers will be derived in a later section. Note that the transfer impedance does not contain any real terms.

In deriving the matrix equation for certain parametric devices some of the input and output variables must be written in complex conjugate form. In the formulas to follow this will be ignored, but in the later sections where these devices are treated, the proper formulas will be quoted (their derivation is similar in every detail).

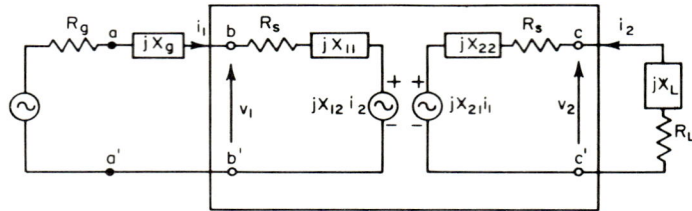

Fig. 29. General representation of a parametric amplifier or converter.

A schematic representation of a parametric amplifier or converter is shown in Fig. 29. If the device is unconditionally stable, the maximum available gain (MAG) is given by[18]

$$\text{MAG} = \frac{x_{21}}{x_{12}} \frac{\sqrt{1 + x_{12}x_{21}/r_{11}r_{22}} - 1}{\sqrt{1 + x_{12}x_{21}/r_{11}r_{22}} + 1} \qquad (45)$$

and the corresponding optimum source impedance Z_g' is

$$Z_g' = r_{11}\sqrt{1 + x_{12}x_{21}/r_{11}r_{22}} - jx_{11} \qquad (46)$$

This gain can be realized if the load impedance Z_L' is

$$Z_L' = r_{22}\sqrt{1 + x_{12}x_{21}/r_{11}r_{22}} - jx_{22} \qquad (47)$$

The transducer gain G_T of the amplifier is given by

$$G_T = \frac{4R_g R_L |x_{21}|^2}{|(R_s + Z_g + jx_{11})(R_s + Z_L + jx_{11}) + x_{12}x_{21}|^2} \qquad (48)$$

This equation applies to both stable and potentially unstable devices. This can be written more compactly as follows:

$$G_T = \frac{4R_g R_L |x_{21}|^2}{|D|^2} \qquad (49)$$

by defining
$$Z_1 = Z_g + jx_{11}$$
$$Z_2 = Z_L + jx_{22}$$
$$D = (R_s + Z_1)(R_s + Z_2) - x_{12}x_{21}$$

Any four-terminal network with a negative input resistance can also be used as a one-port reflection-type amplifier. The transducer gain in this case can be defined as the ratio of the power delivered to the source due to reflections from the one-port to the power available from the source.

Consider the case of a four-terminal network having an input impedance Z_{in} with a negative real part, connected to a source e, through characteristic impedance Z_0. The input current is i, the input voltage to the network v. Consider now the expression

$$\begin{aligned}|a|^2 &= |(v + Z_0 i)/2\sqrt{Z_0}|^2 \\ &= (1/4Z_0)|(e - Z_0 i + Z_0 i)|^2 \\ &= (e^2/4Z_0)\end{aligned} \tag{50}$$

It can be seen from the above that $|a|^2$ is the available power from the source. Next, consider the expression

$$\begin{aligned}|b|^2 &= |(v - Z_0 i)/2\sqrt{Z_0}|^2 \\ &= (1/4Z_0)|(e - 2Z_0 i)|^2 \\ &= (e^2/4Z_0)|1 - 2Z_0 i/e|^2\end{aligned} \tag{51}$$

from which it is evident that $|b|^2$ represents the power delivered to the source because of reflection only. Hence the transducer gain is

$$\begin{aligned}G_T = |b|^2/|a|^2 &= |1 - 2Z_0 i/e|^2 \\ &= \left|1 - \frac{2Z_0}{Z_0 + Z_{in}}\right|^2 \\ &= \left|\frac{Z_0 - Z_{in}}{Z_0 + Z_{in}}\right|^2\end{aligned} \tag{52}$$

It will be noted that this is also the expression for the magnitude squared of the reflection coefficient Γ on a line of characteristic impedance Z_0 and input impedance Z_{in}. Also it can be shown that a is the incident voltage on such a line and b the reflected voltage. It has become common practice to use $|\Gamma|^2$ as the gain, even though the amplifier is made up entirely of lumped components. Of course, to realize this gain a device which can separate the "incident" and "reflected" waves must be inserted between the source and the input terminal. One such device is a circulator.

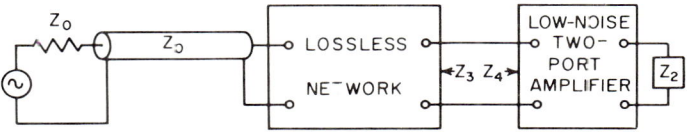

Fig. 30. General method of connecting a low-noise amplifier to a transmission line.

If (for matching or tuning purposes) a lossless network is inserted between the source impedance Z_0 and the input to the two-port (see Fig. 30) the gain is given by

$$\begin{aligned}|\Gamma|^2 &= \left|\frac{Z_{in} - Z_0}{Z_{in} + Z_0}\right|^2 \\ &= \left|\frac{Z_3 - Z_4^*}{Z_3 + Z_4}\right|^2\end{aligned} \tag{53}$$

If Eq. (53) is applied to the general two-port amplifier of Fig. 29 at aa', then

$$|\Gamma|^2 = \left| 1 - \frac{2R_g(R_s + Z_2)}{D} \right|^2 \tag{54}$$

since the input impedance at aa' is

$$Z_{aa'} = D/(R_s + Z_2) - R_g$$

If the gain is large $|\Gamma|^2$ can be approximated by

$$|\Gamma|^2 \approx 4R_g{}^2|(R_s + Z_2)/D|^2 \tag{55}$$

The noise figure F of the two-port can be derived by considering the noise currents that flow through the load R_L. The noise figure can be defined as the mean-square noise current that flows through R_L due to all noise sources divided by that due to R_g only. In Fig. 31, the two-port is redrawn with the noise sources shown. The

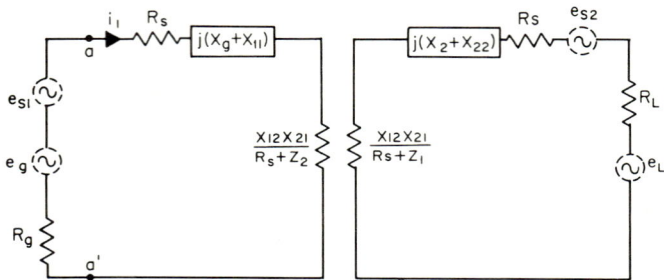

Fig. 31. Parametric amplifier circuit showing noise sources.

voltage sources $jx_{12}i_2$ and $jx_{21}i_1$ are replaced by impedances, a change which is obvious from the general expressions for the input and output impedances ($r_{12} = r_{21} = 0$),

$$z_{11} + x_{12}x_{21}/(z_{22} + Z_L) \qquad z_{22} + x_{12}x_{21}/(z_{11} + Z_g)$$

The mean-square noise current at the output due to R_g is

$$\overline{i_g{}^2} = G_T \overline{e_g{}^2}/4R_L R_g \tag{56}$$

from the definition of G_T. Using Eq. (49), this becomes

$$\overline{i_g{}^2} = |x_{21}|^2 \overline{e_g{}^2}/|D|^2 = 4kT_0 \, \delta f (x_{21}{}^2/|D|^2) R_g \tag{57}$$

Similarly the mean-square noise current due to R_s in the input circuit is

$$\overline{i_{s1}{}^2} = 4kT_0 \, \delta f (x_{21}{}^2/|D|^2) R_s \tag{58}$$

From Fig. 31, the mean-square current due to R_s in the output is

$$\overline{i_{s2}{}^2} = \frac{4kT_0 \, \delta f \, R_s}{|R_s + R_L + j(x_L + x_{22}) + [x_{12}x_{21}/(R_s + Z_1)]|^2}$$

$$= \frac{4kT_0 \, \delta f \, R_s |R_s + Z_1|^2}{|D|^2} \tag{59}$$

Similarly the noise due to R_L is

$$\overline{i_L{}^2} = (4kT_0 \, \delta f \, R_L/|D|^2)|R_s + Z_1|^2 \tag{60}$$

From the above three equations, the noise-figure expression is

$$F = \frac{\overline{i_g^2} + \overline{i_{s1}^2} + \overline{i_{s2}^2} + \overline{i_L^2}}{\overline{i_g^2}}$$

$$= 1 + \frac{R_s}{R_g} + \frac{R_s + R_L}{R_g} \frac{|R_s + Z_1|^2}{x_{21}^2} \tag{61}$$

This expression applies to both stable and potentially unstable parametric devices. In the case of stable devices, the noise from R_L is attributed to the succeeding stage and

$$F = 1 + \frac{R_s}{R_g}\left[1 + \frac{|R_s + R_g + j(x_g + x_{11})|^2}{x_{21}^2}\right] \tag{62}$$

This expression is minimized if $x_g = -x_{11}$ and if

$$R_g = R_g'' = R_s\sqrt{1 + x_{21}^2/R_s^2} \tag{63}$$

The expression for F_{min} is then

$$F_{min} = 1 + \frac{2}{\sqrt{1 + x_{21}^2/R_s^2} - 1} \tag{64}$$

The noise figure of the device when operated as a negative resistance can be defined as the mean-square noise voltage reflected from the device due to all noise sources divided by that due to R_g only. The mean-square voltage reflected from the terminals aa' is

$$|b|^2 = |\Gamma|^2|a|^2 = (1/4R_g)|v_{aa'} - R_g i_{aa'}|^2 \tag{65}$$

where $v_{aa'}$ is the voltage at the terminals aa', $i_{aa'}$ is the current at the terminals aa', and a is the incident voltage. The incident and reflected voltages in Eq. (65) are normalized so that their mean-square values also represent the incident and reflected powers. Hence the mean-square reflected voltage due to R_g is

$$\overline{b_g^2} = kT_0\,\delta f|\Gamma|^2 \tag{66}$$

and that due to R_s at the input is

$$\overline{b_{s1}^2} = kT_0\,\delta f|\Gamma|^2(R_s/R_g) \tag{67}$$

For noise voltages at the output

$$v_{aa'} - R_g i_{aa'} = -2R_g i_{aa'}$$

and from Fig. 29

$$i_{aa'} = i_1 = -[jx_{12}i_2/(R_s + Z_1)]$$

For Fig. 31, the output current due to R_s is

$$\frac{e_{s2}}{R_s + Z_2 + x_{12}x_{21}/(R_s + Z_1)}$$

and $i_{aa'} = -(jx_{12}/D)$. The corresponding mean-square reflected voltage is

$$\overline{b_{s2}^2} = \frac{1}{4R_g} \frac{4kT_0\,\delta f\,R_s 4R_g^2 x_{12}^2}{|D|^2}$$
$$= 4kT_0\,\delta f\,R_s R_g(x_{12}^2/|D|^2) \tag{68}$$

Similarly, the reflected mean-square noise voltage due to R_L is

$$\overline{b_L^2} = 4kT_0\,\delta f\,R_L R_g(x_{12}^2/|D|^2) \tag{69}$$

From the above expressions, then

$$F = \frac{\overline{b_g{}^2} + \overline{b_{s1}{}^2} + \overline{b_{s2}{}^2} + \overline{b_L{}^2}}{\overline{b_g{}^2}}$$

$$= 1 + \frac{R_s}{R_g} + \frac{4(R_s + R_L)R_g x_{12}{}^2}{|D|^2|\Gamma|^2} \tag{70}$$

This expression is exact for all gain and input impedance conditions. If the gain is large, an approximate expression for F is obtained by substituting Eq. (55) into Eq. (70):

$$F \approx 1 + \frac{R_s}{R_g} + \frac{(R_s + R_L)x_{12}{}^2}{R_g|R_s + Z_2|^2} \tag{71}$$

The high-gain condition requires that

$$D = (R_s + Z_1)(R_s + Z_2) + x_{12}x_{21} \approx 0 \tag{72}$$

or

$$-x_{12}/(R_s + Z_2) \approx (R_s + Z_1)/x_{21} \tag{73}$$

in which case it is seen that the noise figure of the reflection-type amplifier and the converter [see Eq. (61)] are approximately equal. The expression for F in Eq. (71) can be further simplified if it is noted that from Eq. (72)

$$|R_s + Z_2|^2 \approx -x_{12}x_{21}\frac{R_s + R_L}{R_s + R_g} \tag{74}$$

in which case

$$F = 1 + \frac{R_s}{R_g} - \frac{x_{12}}{x_{21}}\frac{R_s + R_g}{R_g} \tag{75}$$

It would appear from this expression that to obtain a good noise figure R_g must be made as large as possible, but this would reduce the gain drastically. The value of R_g is determined by the large-gain condition $D \approx 0$. If the reactive parts of Z_1 and Z_2 are zero and if R_L approaches zero, from Eq. (54)

$$R_g \approx -(R_s{}^2 + x_{12}x_{21})/R_s \tag{76}$$

which is the largest possible R_g consistent with the large-gain condition. If (76) is substituted into (75)

$$F = F_m = 1 + [R_s{}^2/(|x_{12}x_{21}| - R_s{}^2)](1 + x_{12}{}^2/R_s{}^2) \tag{77}$$

2.2. Bandwidths and Gain Bandwidths of Capacitance-diode Amplifiers. A rigorous treatment of the optimum bandwidths of capacitance-diode amplifiers is very difficult. The discussion here will be limited to the cases where the input and output are single-tuned by lumped inductances. In many cases, this is sufficient bandwidth, and also this discussion forms a basis for evaluation of practical devices which use more complicated tuning methods.

The transducer gain of the converter-type diode amplifier can be written [see Eq. (48)] as follows:

$$G_T = \frac{4R_L R_g x_{21}{}^2}{|(R_s + R_g + jX_1)(R_s + R_L + jX_2) + x_{12}x_{21}|^2}$$

$$= \frac{4R_L R_g x_{21}{}^2}{|(R_s + R_g)(R_s + R_L) + x_{12}x_{21} - X_1 X_2 + j[(R_s + R_g)X_1 + (R_s + R_L)X_2]|^2} \tag{78}$$

If we use the narrowband approximation with single lumped inductances to tune out the input and output capacitances of the diode then X_1 and X_2 are proportional to w, the fractional bandwidth. The $X_1 X_2$ term in the real part of the denominator of Eq. (78) is a negligible part of the whole real term since it involves w^2. (This fact will be verified when specific converters are discussed in a later section.) This approximation is good for even moderately large fractional bandwidths, and G_T becomes

$$G_T = \frac{4R_g R_L x_{21}^2}{|(R_s + R_g)(R_s + R_L) + x_{12}x_{21} + j[(R_s + R_g)X_2 + (R_s + R_L)X_1]|^2} \quad (79)$$

The gain of the negative-resistance amplifier can be written [see Eq. (55)] as

$$|\Gamma|^2 = \frac{4R_g^2}{\left|(R_s + R_g + jX_1) + \dfrac{x_{12}x_{21}}{R_s + R_L + jX_2}\right|^2}$$

$$= \frac{4R_g^2}{\left|(R_s + R_g) + \dfrac{x_{12}x_{21}(R_s + R_L)}{(R_s + R_L)^2 + X_2^2} + j\left[X_1 - \dfrac{x_{12}x_2 X_2}{(R_s + R_L)^2 + X_2^2}\right]\right|^2} \quad (80)$$

Under the approximations mentioned above, X_1 and X_2 are again proportional to the fractional bandwidth w. Also, it will be shown (when the four-pole parameters z_{11}, z_{22}, x_{12}, x_{21} are derived) that the term X_2^2 (which is proportional to w^2) is negligible with respect to $(R_s + R_L)^2$. The gain then becomes

$$|\Gamma|^2 = \frac{4R_g^2}{\left|(R_s + R_g) + \dfrac{x_{12}x_{21}(R_s + R_L)}{(R_s + R_L)^2} + j\left[X_1 - \dfrac{x_{12}x_{21}X_2}{(R_s + R_L)^2 + X_2^2}\right]\right|^2} \quad (81)$$

Equations (79) and (81) can be represented in the form

$$\text{Power gain} = \frac{\gamma^2}{|\alpha + j\beta w|^2} = \frac{\gamma^2}{\alpha^2} \frac{1}{|1 + (\beta/\alpha)w|^2} \quad (82)$$

where α, β, and γ are assumed to be independent of frequency; i.e., the frequency dependence of x_{12} and x_{21} is neglected.

Here γ^2/α^2 represents the gain at band center ($w = 0$), and the fractional bandwidth at the 3-db points is

$$w_{3\text{ db}} = \alpha/\beta \quad (83)$$

The voltage gain bandwidth is then

$$\sqrt{\text{Gain}}\, w_{3\text{ db}} = \gamma/\beta \quad (84)$$

The values of α, β, and γ can be determined by the inspection of Eq. (79) or (81).

2.3. The Four-pole Parameters. The linear four-pole equation for a capacitance diode can be derived by expressing the small-signal voltages and currents in a Taylor-series expansion about the pump signal.[19] The fundamental relation

$$V = IR_s + \int SI\, dt$$

between the voltage V across the diode (here the diode is represented as resistance R_s in series with a variable capacitance or elastance S) and the current I through the diode becomes

$$v_{ss} = R_s i_{ss} + S \int i_{ss}\, dt \quad (85)$$

where v_{ss}, i_{ss} are the small-signal voltage across and current through the diode, respectively. Because of the small-signal assumption, the elastance S is determined entirely by the high-level pump current. Here we assume the diode is pumped with a sinusoidal current and S can be represented by

$$S(t) = S_{max} \sum_{n=-\infty}^{\infty} m_n e^{jn\omega_p t} \quad (86)$$

In the case of an abrupt junction $m_0 = 0.5$, $m_1 = m_{-1} = 0.25$ and all other m_i's are zero. For the graded junction[19] $m_0 = 0.637$, $m_1 = m_{-1} = 0.212$ and $m_k = 1/\pi(4k^2 - 1)$. As stated in a previous section all other types of pumping will not be treated because of the practical difficulties in realizing the pump circuit. Also only three-frequency devices will be treated. In this case, the small-signal current and voltage can be represented by

$$i_{ss} = I_1 e^{j\omega_1 t} + I_1^* e^{-j\omega_1 t} + I_2 e^{j\omega_2 t} + I_2^* e^{-j\omega_2 t} \tag{87}$$
$$v_{ss} = V_1 e^{j\omega_1 t} + V_1^* e^{-j\omega_1 t} + V_2 e^{j\omega_2 t} + V_2^* e^{-j\omega_2 t} \tag{88}$$

It is not possible to neglect the complex conjugate terms as is customary in linear circuit analysis since they are needed to give the proper phase relationships between the currents and voltages at different frequencies. If Eqs. (86), (87), and (88) are substituted into Eq. (85), the following four-pole equations result:

$$V_1 = (R_s + S_0/j\omega_1)I_1 + (S_1/j\omega_2)I_2 \tag{89a}$$
$$V_2 = (S_1/j\omega_1)I_1 + (R_s + S_0/j\omega_2)I_2 \quad \text{if } \omega_p = \omega_2 - \omega_1 \tag{89b}$$

and
$$V_1 = (R_s + S_0/j\omega_1)I_1 + (S_1/j\omega_2)I_2^* \tag{90a}$$
$$V_2 = (S_1/j\omega_1)I_1^* + (R_s + S_0/j\omega_2)I_2 \quad \text{if } \omega_p = \omega_1 + \omega_2 \tag{90b}$$

where $S_0 = m_0 S_{max}$
$S_1 = m_1 S_{max}$

Since $S_0 = (m_0/m_1)S_1$ and $\omega_Q = S_{max}/4R_s = S_1/4R_s m_1$

the above equations can be written

$$V_1 = (R_s + 4m_0\omega_Q R_s/j\omega_1)I_1 + (4R_s m_1\omega_Q/j\omega_2)I_2 \tag{91a}$$
$$V_2 = (4R_s m_1\omega_Q/\omega_1)I_1 + (R_s + 4m_0\omega_Q R_s/j\omega_2)I_2 \tag{91b}$$

and
$$V_1 = (R_s + 4m_0\omega_Q R_s/j\omega_1)I_1 + (4R_s m_1\omega_Q/j\omega_2)I_2^* \tag{92a}$$
$$V_2 = (4R_s m_1\omega_Q/j\omega_1)I_1^* + (R_s + 4m_0\omega_Q R_s/j\omega_2)I_2 \tag{92b}$$

Equations (91) apply to the positive-resistance up-converter (if $\omega_2 > \omega_1$) and the positive-resistance down-converter (if $\omega_2 < \omega_1$). Equations (92) apply to the negative-resistance amplifier and to the negative-resistance up-converter (if $\omega_2 > \omega_1$) and to the negative-resistance down-converter (if $\omega_2 < \omega_1$).

Recall that Z_1 and Z_2 are defined as

$$Z_1 = R_g + jX_g + jx_{11} = R_1 + jX_1$$
$$Z_2 = R_L + jX_L + jx_{22} = R_2 + jX_2$$

If single series inductances L_1 and L_2 are used to tune out x_{11} and x_{22}, respectively, then X_1 can be represented by

$$X_1 = (4m_0\omega_Q R_s/j\omega_1) + j\omega_1 L_1$$
$$= -j(4m_0\omega_Q R_s/\omega_{10})(\omega_{10}/\omega_1 - \omega_1/\omega_{10})$$
$$\simeq -j(4m_0\omega_Q R_s/\omega_{10})w \tag{93}$$

where w is the percentage bandwidth $\Delta\omega_1/\omega_{10}$ and ω_{10} is the center frequency. Similarly

$$X_2 = -j(4m_0\omega_Q R_s/\omega_{20})(\Delta\omega_2/\omega_{20})$$
$$= j(4m_0\omega_Q\omega_{10}/\omega_{20}^2)w \tag{94}$$

for the negative-resistance converter or amplifier and

$$X_2 = -j(4m_0\omega_Q\omega_{10}R_s/\omega_{20}^2)w \tag{95}$$

for the positive-resistance up-converter. With Eqs. (91) to (95) and the results of Sec. 2.1, the pertinent formulas for the various amplifiers can be written as follows:

DIODE AND PARAMETRIC AMPLIFIERS

Positive-resistance Up-converter. In this case $\omega_p = \omega_2 - \omega_1$. The four-pole equations are

$$V_1 = (R_s + 4m_0\omega_Q R_s/j\omega_1)I_1 + (4m_1\omega_Q R_s/j\omega_2)I_2 \quad (96a)$$
$$V_2 = (4m_1\omega_Q R_s/j\omega_1)I_1 + (R_s + 4m_0\omega_Q R_s/j\omega_2)I_2 \quad (96b)$$

and from Eqs (45) to (84), then

$$\text{MAG} = \frac{f_2}{f_1} \frac{\sqrt{1 + 16m_1{}^2 f_Q{}^2/f_1 f_2} - 1}{\sqrt{1 + 16m_1{}^2 f_Q{}^2/f_1 f_2} + 1} = \frac{f_2}{f_1}\left(\frac{C'-1}{C'+1}\right) \quad (97)$$

$$R_g' = R_L' = R_s C' \quad (98)$$

$$F' = 1 + \frac{1}{C'}\left(1 + \frac{f_1}{f_2}\frac{C'+1}{C'-1}\right) \quad (99)$$

$$R_g'' = R_s\sqrt{1 + 16m_1{}^2 f_Q{}^2/f_1{}^2} = R_s C'' \quad (100)$$

$$F'' = F_{min} = 1 + 2/(C'' - 1) \quad (101)$$

$$w'_{3\,db} \simeq \frac{C'}{2m_0(f_Q/f_1 + f_Q f_1/f_2{}^2)} \quad (102)$$

$$w''_{3\,db} \simeq \frac{C''(1 + f_1/f_2) + 1 - f_1/f_2}{4m_0(f_Q/f_1 + f_Q f_1/f_2{}^2)} \quad (103)$$

The prime quantities refer to the maximum-available-gain operation of the up-converter and the double-primed refer to minimum-noise-figure operation.

Negative-resistance Amplifier. In this case, the four-pole equations are

$$V_1 = (R_s + 4m_0\omega_Q R_s/j\omega_1)I_1 - (4m_1\omega_Q/j\omega_2)R_s I_2^* \quad (104a)$$
$$V_2 = -(4m_1\omega_Q/j\omega_1)R_s I_1^* + (R_s + 4m_0\omega_Q R_s/j\omega_1)I_2 \quad (104b)$$

The gain is given by

$$|\Gamma|^2 = |1 - 2R_g(R_s + Z_2^*)/D|^2 \quad (105)$$

where
$$D = (R_s + Z_1)(R_s + Z_2^*) - 16m_1{}^2 f_Q{}^2 R_s{}^2/f_1 f_2 \quad (106)$$

and under high-gain conditions

$$|\Gamma|^2 \simeq \frac{4R_g{}^2}{\left| R_s + Z_1 - \dfrac{16m_1{}^2 f_Q{}^2 R_s{}^2}{f_1 f_2(R_s + Z_2^*)} \right|^2} \quad (107)$$

From these relations and formulas developed in the general discussion, it can be shown that

$$F_m = 1 + \frac{1}{16m_1{}^2\omega_Q{}^2/\omega_1\omega_2 - 1}(1 + 16m_1{}^2 f_Q{}^2/f_2{}^2) \quad (108)$$

$$F_{min} = 1 + 2/(C'' - 1) \quad (109)$$
$$\omega_{2\,opt} = \omega_1(C'' - 1) \quad (110)$$

$$R_{in} = R_s - RP\left[\frac{16m_1{}^2\omega_Q{}^2 R_s{}^2}{\omega_1\omega_2(R_s + Z_2^*)}\right] \quad (111)$$

where RP signifies "real part of."

$$\Gamma_0 = \frac{2R_g}{R_s + R_g - [16m_1{}^2\omega_Q{}^2 R_s{}^2/\omega_1\omega_2(R_s + R_L)]} \quad (112)$$

$$\Gamma_0 w_{3\,db} = \frac{\Gamma_0[16m_1{}^2(\omega_Q{}^2/\omega_1\omega_2) - 1]}{2m_0(\Gamma_0 - 2)(\omega_Q/\omega_1 + 16m_1{}^2\omega_Q{}^3/\omega_2{}^3)} \quad (113)$$

where ω_1 = signal frequency
ω_2 = idler frequency
F_m = minimum noise figure under high-gain conditions for a given ω_1 and ω_2
F_{min} = minimum noise figure under high-gain conditions for a given $\omega_1(\omega_2 = \omega_{2\ opt})$
$\omega_{2\ opt}$ = optimum idler frequency
R_{in} = real part of the impedance looking into the amplifier at the signal frequency
Γ_0 = signal voltage gain at the center frequency
$w_{3\ db}$ = fractional bandwidth at the 3-db points, assuming lumped inductive tuning
C'' = a factor defined in Eq. (100)
m_0 = ratio of the average elastance S_0 to the available elastance swing $(S_{max} - S_{min})$
m_1 = ratio of one-half the sinusoidal peak elastance at the pump frequency to the available elastance swing $(S_{max} - S_{min})$

In the derivation of the gain bandwidth, it is assumed that the idler load $R_L = 0$. The voltage gain bandwidth can be increased by using a finite R_L at the expense of

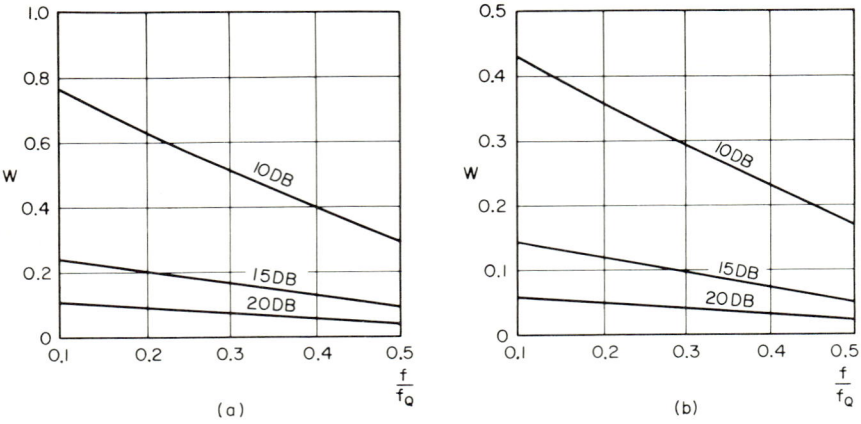

FIG. 32. Fractional bandwidth of a sinusoidal current pumped parametric amplifier with lumped inductance tuning. (a) Abrupt junction. (b) Graded junction.

an increase in noise figure. If $f_{2\ opt}$ is substituted for f_2 into Eq. (113) the voltage gain bandwidth under the minimum-noise-figure condition can be obtained. In Fig. 32 the fractional bandwidth is plotted (for gains of 10, 15, and 20 db) as a function of the center frequency normalized to f_Q.

2.4. Typical Amplifier Design. There is no unique circuit configuration for a given parametric amplifier requirement. However, many of the design problems are common to all circuits. Hence rather than discuss several different designs briefly we shall concentrate on a thorough treatment of one particular design.

The amplifier to be designed is a 450-Mc nondegenerate negative-resistance amplifier using a circulator. Also, assume that a graded-junction diode is to be used. The noise figure should be close to 1 db to be useful in this frequency range.

First, the diode must be chosen. The theoretical design equations do not specify the choice of the average elastance S_0. From the practical point of view, a consideration of signal impedance levels does place limits on S_0. For example, coaxial-line systems are usually 50-ohm and circuits which require transformations to impedance levels greatly different from this may be necessarily lossy or unrealizable in practice.

From Eq. (107), the generator impedance which gives reasonably large gain R_g''' is

$$R_g''' = (16 m_1^2 f_Q^2 / f_1 f_2) R_s$$

and if the amplifier is operated at the optimum idler frequency

$$f_{2\,opt} = f_1\sqrt{1 + 16m_1^2 f_Q^2/f_1^2} - f_1$$
$$\simeq f_1(4m_1 f_Q/f_1)$$
$$\simeq 4m_1 f_Q$$

and hence
$$R_g''' \simeq (4m_1 f_Q/f_1)R_s = m_1 S_{max}/\omega_1$$

but $S_0 = m_0 S_{max}$; therefore
$$R_g''' \simeq m_1 S_0/m_0\omega_1$$

Hence
$$R_g''' \approx \tfrac{1}{2}(S_0/\omega_1) \approx \tfrac{1}{4}(S_{max}/\omega_1) \qquad (114a)$$

for an abrupt junction and

$$R_g''' \approx \tfrac{1}{3}(S_0/\omega_1) \approx 0.212(S_{max}/\omega_1) \qquad (114b)$$

for a graded junction, assuming sinusoidal current pumping. If S_{max} (or S_0) is chosen so that R_g''' as given by Eq. (114) lies somewhere in the range

$$Z_0/4 < R_g''' < 4Z_0$$

where Z_0 is the characteristic impedance of the transmission line, then difficulties due to the impedance level should not be too great. Note that if an idler frequency f_2 less than the optimum idler frequency is used, R_g''' increases by the factor $f_{2\,opt}/f_2$.

Now, assume that a graded capacitance diode with $S_{max} = 4.45 \times 10^{11}$ darafs, $R_s = 5.15$ ohms, and a series inductance $L_p = 4.5 \times 10^{-9}$ henry is available. The average elastance S_0 for full sinusoidal current pumping is 2.84×10^{11} darafs. The series resonance of the diode is then 1.26 gc. The optimum idler frequency for this diode is 2.5 gc. However, amplifier complexity is usually less and the bandwidth is usually greater if the idler frequency is near the series-resonant frequency. Hence we choose an idler frequency of 1.45 gc. Using Eqs. (104) to (113), the performance and impedance levels to be expected are

$$F_m = 1.26 \text{ db}$$
$$R_{in} = -62.1 \text{ ohms}$$
$$w_{3\,db} = 0.127 \text{ (for 16-db gain)}$$
$$R_g = 85.5 \text{ ohms (for 16-db gain)}$$

where R_{in} is the real part of the input impedance and $w_{3\,db}$ is the fractional bandwidth for independent lumped inductive tuning of the signal and idler circuits.

The lumped circuit equivalent of the signal and idler portion of the amplifier to be described is shown in Fig. 33. The parallel combination $C_5 L_6$ along with S_0 and L_p

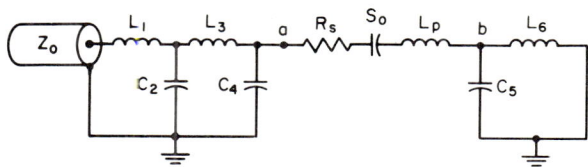

Fig. 33. Lumped-circuit equivalent of uhf parametric amplifier (pump circuit not shown).

allows for tuning to zero reactance at both the signal and idler center frequencies. The circuit to the left of R_s provides the signal impedance transformation.

The plot of the reactance vs. frequency as seen looking toward the right at point a is shown in Fig. 34. The pole frequency ω_5 is determined by the $C_5 L_6$ tank circuit. It can be shown that $\omega_{res} = \sqrt{S_0/L_p}$ must lie between ω_1 and ω_2. This signal-idler

AMPLIFIER CIRCUITS

tuning circuit can be easily synthesized from the following equations:

$$\omega_5 = \omega_1\omega_2/\omega_{res} \tag{115a}$$
$$\omega'^2 = 1/L_pC_5 = \omega_1^2 + \omega_2^2 - \omega_{res}^2 - \omega_5^2 \tag{115b}$$
$$L_6 = 1/\omega_5^2 C_5 \tag{115c}$$

The synthesis of the four-section impedance-matching circuit (L_1, C_2, L_3, C_4) can be accomplished by examining the impedance Z_g looking into C_4 at a point a, expressed in terms of the $ABCD$ matrix,

$$Z_g = \frac{AZ_0 + B}{CZ_0 + D}$$

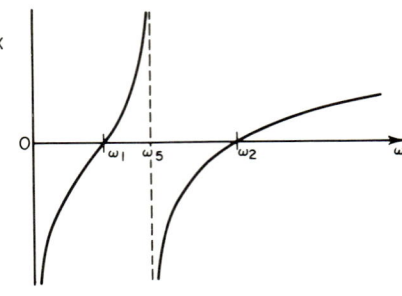

Fig. 34. Reactance plot of signal idler tuning circuit.

where B, C are imaginary and A, D are real for a lossless network. If C is made zero, then

$$Z_g = (A/D)Z_0 + B/D$$

and A/D represents the impedance transformation ratio.

If the above is applied to the pi network C_4, L_3, C_2, it can be shown by examining its $ABCD$ matrix that

$$A/D = C_2^2/C_4^2 \tag{116}$$
$$L_3C_4 = (1 + \sqrt{D/A})/\omega_{10}^2 \tag{117}$$

where ω_{10} is the center frequency of the design (the frequency at which $C = 0$). If L_1 is added in series the overall $ABCD$ matrix is

$$\begin{bmatrix} A & B \\ 0 & D \end{bmatrix} \begin{bmatrix} 1 & j\omega L_1 \\ 0 & 1 \end{bmatrix} = \begin{bmatrix} A & B + j\omega_1 L_1 A \\ 0 & D \end{bmatrix}$$

and the reactive term in Z_g can be made zero at ω_{10} if

$$L_1 = \sqrt{A/D}\, L_3 \tag{118}$$

Equations (116) to (118) can now be used to obtain the proper values for a given impedance match. Note that one of L_1, C_2, L_3, C_4 must be chosen from other considerations such as bandwidth, for example. However, in this case, if C_4 is chosen sufficiently large the idler tuning is independent of the transformer; i.e., idler tuning will be dependent only on S_0, L_p, C_5, and L_5. If the above equations are applied to the example under consideration the component values of Fig. 33 are (assuming $C_4 = 25$ pf)

$$L_1 = 11.51 \text{ nh}$$
$$L_3 = 8.82 \text{ nh}$$
$$L_5 = 8.4 \text{ nh}$$
$$C_2 = 32.7 \text{ pf}$$
$$C_4 = 25 \text{ pf}$$
$$C_5 = 11.2 \text{ pf}$$

The transmission-line equivalent of the circuit can be realized approximately if the series inductances are represented by short lengths of high-impedance lines and the shunt capacitances are represented by short lengths of low-impedance lines according to the following equations:

$$l_i = v_i L_i / Z_{0i} \quad \text{(for series inductance)}$$
$$= v_i C_i / Y_{0i} \quad \text{(for shunt capacitance)}$$

where l_i, v_i, and Z_{0i} ($= 1/Y_{0i}$), respectively, are the length, phase velocity, and characteristic impedance of the ith length of line. The distributed line equivalent of the amplifier is shown in Fig. 35. The low- and high-impedance lines have 5 ohms and 100 ohms, respectively. The low-impedance lines contain a dielectric of dielectric constant 2. The performance of a matching section quite similar to the one in Fig. 35 was computed and measured, and the comparison is shown in Fig. 36. The return power (expressed relative to the input power) was measured by replacing the diode and signal idler circuit by an 82-ohm resistor. A portion of the small discrepancy can

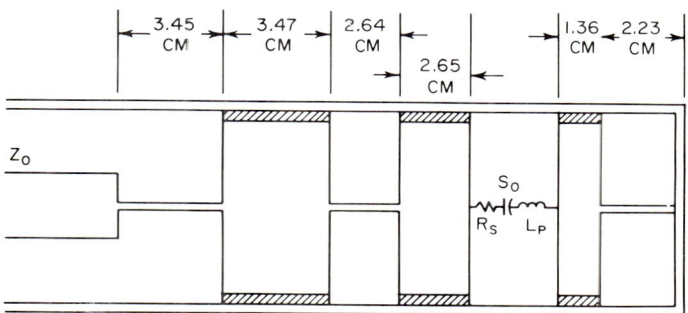

Fig. 35. Distributed line equivalent of uhf parametric amplifier (pump circuit not shown).

be attributed to the loss of the circulator which was used to measure the return loss. The signal-idler tuning circuit can be tested by the method outlined by Kurokawa.

If it is assumed that the diode is pumped by some means with a sinusoidal current source (without influencing the signal or idler circuits) the response of the amplifier can be computed, and the result is shown in Fig. 37.

Because of the low-frequency approximations involved in the transformation to the distributed lines the expected center frequency is approximately 443 Mc. The peaks

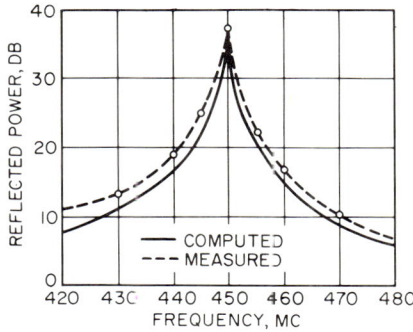

Fig. 36. Reflected power (decibels below incident) for distributed line-matching circuit.

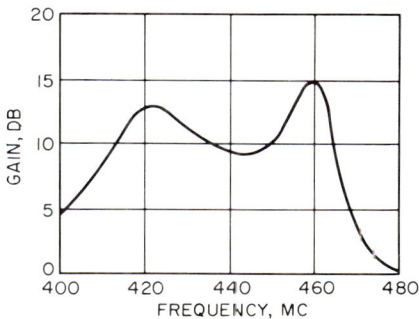

Fig. 37. Computed response of uhf parametric amplifier.

in gain at approximately 420 and 460 Mc are due to the matching section rather than the signal-idler circuits. This is evident from the response shown in Fig. 38 for which the signal impedance transformer was replaced by a flat 85.5-ohm signal line with a zero impedance at the idler frequency. The gain in this case is quite close to the expected 16 db. The bandwidth is only 3.8 per cent compared with the 12.7 per cent calculated, assuming the signal and idler circuits are tuned independently with a lumped inductance (a difficult design to realize). The bandwidth with the transformer is 12 per cent, although it is difficult to estimate because of the peaks in gain.

28-40 AMPLIFIER CIRCUITS

Of course, to realize a practical amplifier in the form shown in Fig. 35, a means must be devised for pumping the varactor without altering the signal or idler circuits. This can be done approximately by placing the diode in shunt across a low-impedance waveguide which passes the pump signal and is cut off to the signal and idler frequencies (see Fig. 39). However, the impedance as seen by the diode below the cutoff

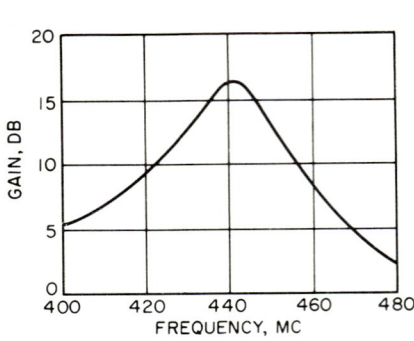

FIG. 38. Computed response of uhf parametric amplifier with flat input matching.

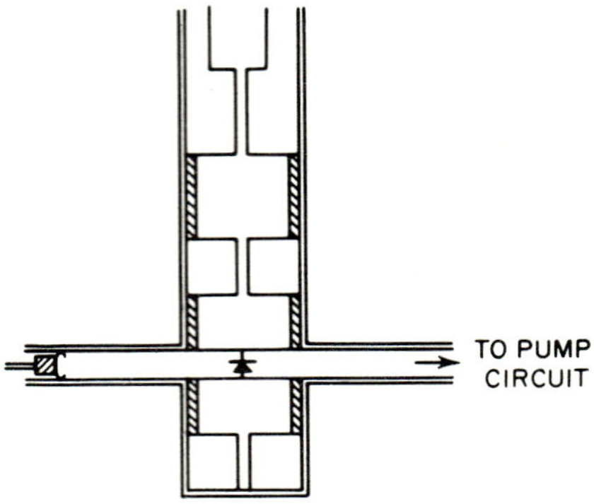

FIG. 39. Parametric amplifier showing waveguide pump circuit.

frequency is appreciable and therefore cannot be neglected. A good approximation to the impedance at the center of a low-impedance guide below cutoff is

$$Z \simeq j \frac{2b}{a} \frac{377}{(f_c^2/f^2 - 1)^{\frac{1}{2}}} \tag{119}$$

where b is the narrow dimension and a the wide dimension and f_c is the cutoff frequency of the guide. A comparison of this expression with the measured reactance is given in Fig. 40.

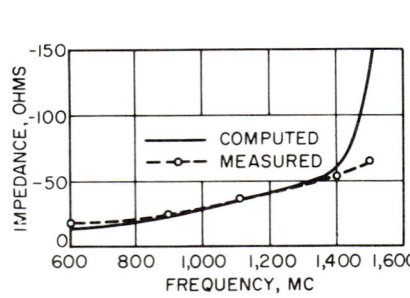

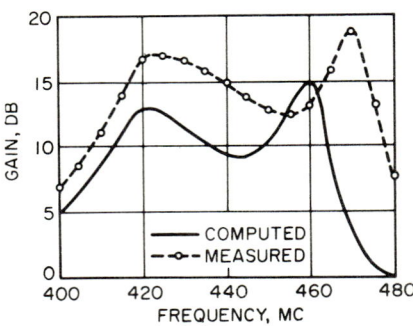

FIG. 40. Impedance at the center of a long waveguide at frequencies below f_c (= 1,600 Mc).

FIG. 41. Comparison of computed and measured response of parametric amplifier.

Assuming for the example under consideration ($f_p = 1.9$ gc, $f_2 = 1.45$ gc), a cutoff frequency of 1.7 gc is chosen, then

$$Z_2 = -j616(b/a) \quad \text{ohms}$$
$$Z_1 = -j106(b/a) \quad \text{ohms}$$

where Z_2 is the impedance at the idler frequency f_2 and Z_1 is the impedance at the signal frequency. These impedances are substantially different from the desirable short-circuit value even for a b/a ratio as small as 0.1. Actually, if this impedance is included in the computation of the signal-idler circuit, the signal center frequency increases about 1 per cent but no idler resonance appears below the cutoff frequency (1.7 gc). If the ratio b/a is reduced to $\frac{1}{100}$, about the practical limit, the idler resonance is still above 1.45 gc.

To circumvent this problem, the coaxial line can be moved from the center of the waveguide to the side so that a portion of the coaxial outer conductor coincides with the side of the waveguide. In this way the signal-idler circuit tuning is practically independent of the waveguide circuit tuning required to give proper match of the pump signal to the diode. The design as outlined above was fabricated and approached the computed performance. A comparison of the computed and measured response of this design is shown in Fig. 41. The gain is greater than 12.5 db over a bandwidth of 14 per cent. The noise figure, measured by conventional noise diode and sensitivity measurements, was below 2 db. Special noise-measurement[21] techniques are needed to make a more accurate determination.

3. TRAVELING-WAVE PARAMETRIC AMPLIFIERS

3.1. General Considerations. The traveling-wave parametric amplifier is an outgrowth of the cavity-type parametric amplifier. As far as its fundamental mechanism is concerned, the standing waves in the cavity amplifier can always be resolved into traveling waves. Thus one might consider the parametric amplification as the interaction of traveling waves, irrespective of whether the actual device is a cavity or a traveling-wave type. The practical difference is that the interaction occurs in a cavity at the same one or two active elements a great number of times, whereas in the traveling-wave type the interaction occurs at a large number of active elements and usually each element is perturbed only once by the traveling waves. The practical advantages of a traveling-wave amplifier over the cavity type are the large bandwidth or electronic tunability and the unidirectional amplification.

3.2. Fundamental Mechanism of Traveling-wave Interactions. The interaction of traveling waves in a traveling-wave parametric amplifier can be constructed from the interactions in a cavity-type parametric amplifier. In a cavity amplifier, a condition for obtaining amplification is that the pump frequency be equal to the sum of the signal and idling frequencies. That is,

$$\omega_p = \omega_i + \omega_s \tag{120}$$

The active element, for instance, a variable-reactance diode, sees the standing waves of these three frequencies. Now let us move the cavity at a fixed-velocity v. Then, to an outside stationary observer, the standing waves become traveling waves moving at velocity v. Correspondingly, we can designate the phase constants of the traveling waves as β_p, β_i, and β_s for the pump, idler, and signal frequencies. In this way, we constructed a model of traveling-wave parametric interactions. We observe that

$$\omega_p/\beta_p = \omega_i/\beta_i = \omega_s/\beta_s = v \tag{121}$$

and, as a result,

$$\beta_p = \beta_i + \beta_s \tag{122}$$

Equation (122) is Tien's assumption. Equation (121) states that the phase velocities of the three traveling waves should be equal. The limitation implied by Eq. (121) is removed in subsequent discussion.

It is obvious that a cavity amplifier will operate irrespective of whether the cavity is stationary or moving. We can also assume that, for the three traveling waves having the same phase velocity, the traveling waves will perform parametric amplification. Figure 42a bears out that conclusion. Here, ω_p', ω_s', and ω_i' are the apparent frequencies as seen by observers who are no longer stationary but moving at a certain velocity not equal to the phase-velocity v_ϕ. These apparent frequencies are in fact the Doppler frequencies. Based upon the doppler relations, or the simple geometry of Fig. 42a, we conclude that

$$\omega_p' = \omega_i' + \omega_s'$$

That is, Eq. (120) is automatically satisfied in any moving system. Thus the interaction and the parametric amplification of the traveling waves are assured.

Figure 42b is a similar diagram for three traveling waves with different phase velocities. Here, to an observer, the Doppler frequencies no longer satisfy Eq. (120). Thus there can be no parametric interaction in this case. In particular, for signal waves traveling in opposite directions, there is no amplification. Therefore, any amplification that occurs can be unidirectional.

Now, let us consider Eqs. (121) and (122) and show that the limitation of Eq. (121) can be considerably relaxed.

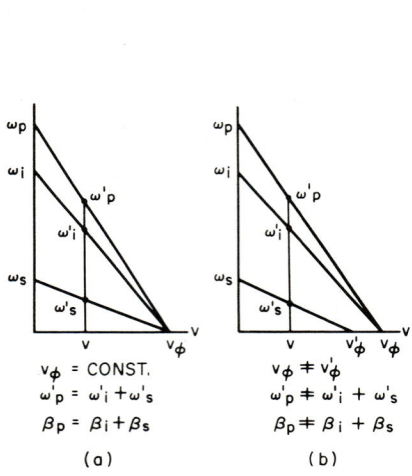

Fig. 42. Doppler effect of traveling waves.

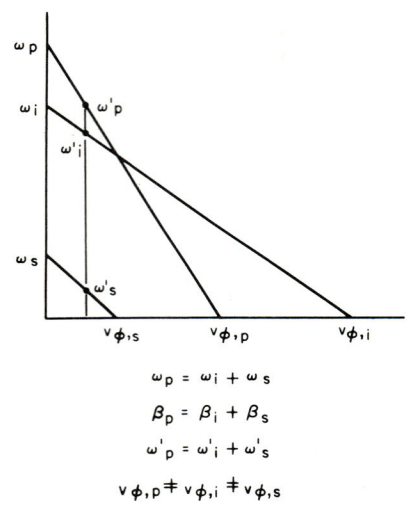

Fig. 43. Doppler effect of interacting traveling waves with unequal phase velocities.

From Fig. 42a, using Eqs. (120) and (122), we have

$$\begin{aligned} \omega_p' &= \omega_p - \beta_p v \\ \omega_i' &= \omega_i - \beta_i v \\ \omega_s' &= \omega_s - \beta_s v \end{aligned} \tag{123}$$

Or, as expected,

$$\omega_p' = \omega_i' + \omega_s' \tag{124}$$

Now, if Eqs. (123) are modified as follows:

$$\begin{aligned} \omega_p' &= \omega_p - \beta_p v \\ \omega_i' &= \omega_i - (\beta_i + \Delta\beta)v \\ \omega_s' &= \omega_s - (\beta_s - \Delta\beta)v \end{aligned} \tag{125}$$

we again have

$$\omega_p' = \omega_i' + \omega_s'$$

but observe that

$$\omega_p/\beta_p \neq \omega_i/(\beta_i + \Delta\beta) \neq \omega_s/(\beta_s - \Delta\beta) \tag{126}$$

Thus the phase velocities corresponding to Eq. (126) are no longer equal and Eq. (121) is violated. Nevertheless, Eq. (124) is satisfied and we expect parametric amplification to result. Since Eq. (122) still holds for the modified phase constants in Eq. (125), we conclude that the requirement of equal phase velocities of Eq. (121) is no longer necessary as long as Eqs. (120) and (122) are satisfied. Figure 43 illustrates the Doppler relationship of interacting traveling waves with different phase velocities. By comparing Fig. 43 with Fig. 42a and remembering Eqs. (123) and (125), we can see that Eqs. (120) and (122) are indeed the necessary conditions for coupling of traveling waves. Equations (120) and (122) are often referred to as the selection rules for traveling-wave parametric interactions. For parametric converters and for parametric interactions involving four or more frequencies, the corresponding selection rules can also be derived from the above Doppler considerations. In the design of parametric devices, these selection rules dictate the desired propagation characteristics of traveling-wave structures.

It should be pointed out that traveling waves of different frequencies cannot be coupled together in a linear system even though they satisfy the selection rules. They are coupled through parametric interaction only when the propagating medium has the proper nonlinearity. Whereas in a cavity parametric amplifier the nonlinearity appears as the variable reactance, the corresponding nonlinearity in a traveling-wave parametric amplifier may occur as the variable propagation constant of the traveling waves because of the nonlinear distributed capacitance or inductance of the transmission line or the nonlinear dielectric constant or permeability of the propagating medium.

The criterion of the design of traveling-wave parametric devices is to provide the proper propagation characteristics for a nonlinear medium as determined by the selection rules. The significance of the selection rules can be demonstrated by the large bandwidth or electronic tunability of traveling-wave parametric amplifiers.

3.3. The Wideband Traveling-wave Parametric Amplifier. When the selection rules of Eqs. (120) and (122) are satisfied over a very wide band at a fixed pump frequency, a large bandwidth can be obtained for the traveling-wave parametric amplifier with no tuning in the circuit. The actual bandwidth can be made comparable to the bandwidth of a traveling-wave tube, being much wider than the bandwidth of a cavity-type amplifier. Furthermore, there is no limitation with respect to the gain-bandwidth product as there usually is for cavity-type amplifiers. The large bandwidth is not obtained at the expense of sacrificing gain.

The condition for wideband operation can be obtained from the selection rules. Let $\Delta\beta_i$ and $\Delta\beta_s$ be the changes in the phase constants corresponding to the difference between Eqs. (121) and (126). Then the modification of Eqs. (123) to (125) can also be expressed as

$$\Delta\beta_i = \Delta\beta$$
$$\Delta\beta_s = -\Delta\beta$$

and
$$\Delta\beta_i + \Delta\beta_s = 0 \tag{127}$$

Similarly, it is possible to change the signal and idling frequencies in Eq. (125) without disturbing Eq. (124), provided

$$\Delta\omega_i + \Delta\omega_s = 0 \tag{128}$$

If successive changes in the phase constants and frequencies are made in Eq. (125), Eqs. (127) and (128) are the conditions required for each change in order to maintain parametric amplification. Equations (127) and (128) may be combined as

$$\Delta\omega_i/\Delta\beta_i = \Delta\omega_s/\Delta\beta_s \tag{129}$$

Equation (129) holds even if the system is dispersive. If the perturbations are small, or if the system is not dispersive, Eq. (129) reduces to

$$\delta\omega_i/\delta\beta_i = \delta\omega_s/\delta\beta_s$$
or
$$(v_g)_i = (v_g)_s \tag{130}$$

where v_g is the group velocity and Eq. (130) indicates the condition of identical group velocity for the signal and idler waves. Equations (127) to (130) can also be obtained from the selection rules of Eqs. (120) and (122) by direct differentiation. In this way, we may regard the variations in frequency and phase constants as the differences between any two operating conditions. For wideband operations, Eq. (129) or (130) should be satisfied over the whole band. These equations are the conditions for wide bandwidth of traveling-wave parametric amplifiers.

The propagation characteristics of traveling waves are usually expressed as the ω-β or Brillouin diagram of the traveling-wave structures. The ω-β diagram is extremely important because it provides all the information for selection-rule considerations. Figure 44 illustrates the propagation characteristics of an ideal unloaded TEM-mode transmission line, a hypothetical waveguide, and a moving cavity. For the TEM-mode transmission line, the condition of equal phase velocities of Eq. (121) is satisfied. In the case of waveguides, Eq. (121) can be satisfied for the set of conditions corresponding to the dotted line. Nevertheless, $\Delta\omega$ and $\Delta\beta$ satisfy Eqs. (127) through (129), and the selection rules are satisfied over a wide band. Thus wideband operation becomes possible Now, returning to the conventional cavity, we see

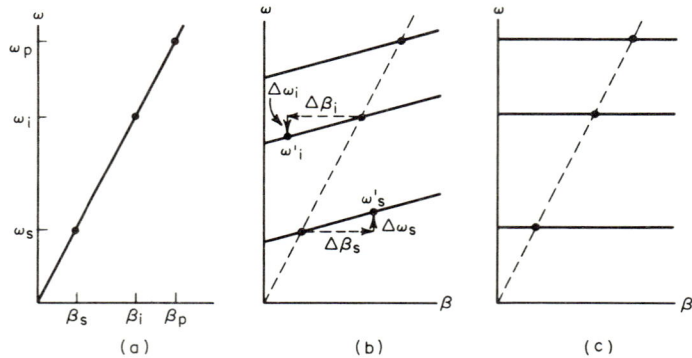

Fig. 44. The ω-β diagram of interacting waves in (a) TEM mode, (b) waveguide mode, (c) cavity.

that a moving cavity is in fact very similar to the hypothetical waveguide. The limited bandwidth is due to the limitation in the cavity resonances at ω_i and ω_s.

From Fig. 44, it can be seen that the characteristics of the ω-β diagram serve as the basis for the design and synthesis of the active circuits. Various practical considerations are involved. For example, Fig. 44a holds only for structures using a continuous parametric medium. If individual parametric diodes are used to form a periodically loaded transmission line, as is often the case in the diode-type parametric amplifiers, the propagation characteristic will become highly dispersive. The criterion in the design is to match the resulting characteristics of the idler and signal waves following the demonstration of Fig. 44b. All unwanted modes should be eliminated or avoided in the design of the structure. In the TEM mode of Fig. 44a, the upper sideband is also present in the traveling-wave structure and can be coupled to the parametric amplifier resulting in frequency conversion to the upper sideband frequencies. Such additional parametric interactions are usually detrimental to the performance of the amplifier unless special purposes are served.

Traveling-wave parametric amplifiers are usually designed with balanced structures to isolate the pump and signal circuits. Figure 45 shows the equivalent circuit of a balanced parametric amplifier using parametric diodes. Possible microwave structures for this circuit are shown in Fig. 46a and b. In Fig. 46a, a split coaxial structure is used. In Fig. 46b, a stripline structure is inserted in the equipotential plane of a waveguide. In these structures, the idler wave may propagate in the mode of transmission of either the signal or pump depending upon the frequency range. It is often easier to design a traveling-wave parametric amplifier near the degenerate mode of operation with the idler and signal propagating along the same path.

The structure of Fig. 46b also allows variation in the design of the signal circuit. For example, it is possible to propagate the signal and idler waves as backward waves for which the group velocities become negative while the phase velocities remain positive. Then, the directions of the idler and signal waves in Fig. 45 are reversed. The situation corresponds to a change in the signs of $\Delta\omega_i$ and $\Delta\omega_s$ in Fig. 44b. Nevertheless, Eqs. (127) and (128) remain valid and wideband parametric amplification should be still possible.

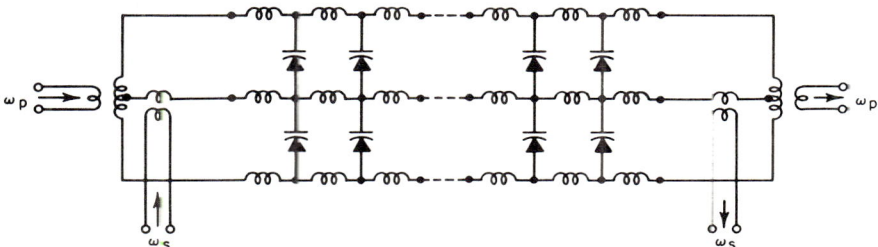

FIG. 45. The equivalent circuit of a balanced parametric amplifier.

In the design of traveling-wave parametric amplifiers, it is always important to provide proper impedance matching at the input and output terminals over the complete band. Any reflections at the terminals would cause gain fluctuations and even instabilities of the amplifier. In a properly designed and terminated amplifier, one obtains unidirectional amplification. The electrical characteristics of these wideband traveling-wave parametric amplifiers are analyzed in Sec. 3.5. The characteristics depend on the pump factor ξ as defined by the equations for the pump. The length of the amplifier is z. The fact that the gain is an exponential function of z and ξ is expected because the interaction of parametric amplification increases linearly with z.

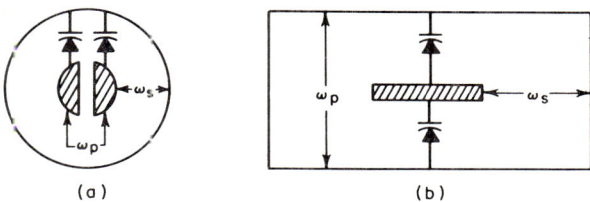

(a)   (b)

FIG. 46. Microwave structures for balanced parametric amplifier. [From R. S. Engelbrecht, Nonlinear-reactance (Parametric) Traveling-wave Amplifiers for UHF, Digest of Technical Papers, 1959 Solid State Circuits Conference, Philadelphia, Pa.]

As typical electrical characteristics, a balanced structure consisting of 16 pairs of parametric diodes had about 10 db of gain over a 200-Mc bandwidth centered at about 700 Mc. The reverse gain was between 1 and −2 db depending on frequency. The single-sideband noise figure was 3.5 db.

3.4. The Electronically Tunable Traveling-wave Parametric Amplifier. In the Doppler relationship of Fig. 43, if the phase velocities of the pump, idler, and signal waves are constant, the relationship will remain valid when all the frequencies are varied proportionally. Thus it is possible to tune the signal by varying the pump frequency. In practice, the arrangement of Fig. 43 would be complicated because of the requirement of three separate transmission lines with different propagation characteristics. A much preferred scheme is one requiring only two transmission lines such that the balanced structure of Fig. 45 can be applied. A typical electronically tunable parametric amplifier of this type is the backward traveling-wave parametric amplifier. This amplifier has a narrow bandwidth, but the signal frequency can be tuned over a wide range solely by varying the pump frequency. There is no change in the circuit structure while tuning. The tuning can be accomplished at a rapid rate by means of an electronically tunable pump source.

Figure 47 shows the Doppler relationship for the backward traveling-wave parametric amplifier. The changes from Fig. 42a through Fig. 43 to Fig. 47 correspond to the gradual increase of $\Delta\beta$ in Eq. (125). For Fig. 47, the value of $\Delta\beta$ is so large that the resulting phase constant of the signal becomes negative. The signal is now a backward traveling wave for which both the phase velocity and group velocity are negative. It is often desirable to propagate the signal and idler in the same transmission line so that the balanced structure of Figs. 45 and 46 can be applied. In this arrangement, the direction of the signal in Fig. 45 is actually reversed. Remembering

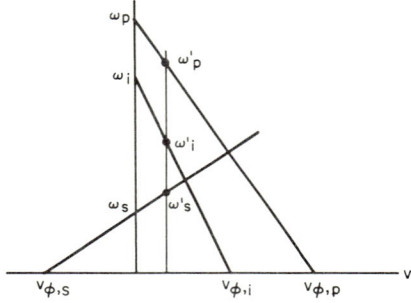
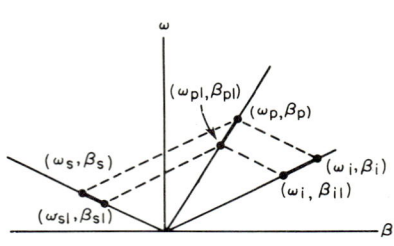

FIG. 47. Doppler effect of traveling waves for backward traveling-wave parametric amplification

$\omega_p = \omega_i + \omega_s$
$\beta_p = \beta_i + \beta_s \qquad (\beta_s < 0)$
$\omega_p' = \omega_i' + \omega_s'$
$v_{\phi p} \neq v_{\phi i} \neq v_{\phi s} \qquad (v_s < 0)$

FIG. 48. Characteristics of backward traveling-wave parametric amplifier.

that the idler wave propagates along the opposite direction from the signal, the amplifier offers the possible practical advantage of terminating the idler to the antenna and thus effectively cooling the idler to the sky temperature.

The tuning characteristics can be illustrated further with the ω-β diagram as shown in Fig. 48. The phase velocities of the three traveling waves are assumed to be constant, corresponding to TEM mode of propagations. It is seen that the selection rules of Eqs. (120) and (122) are satisfied for the set of points corresponding to ω_p, ω_i, ω_s, as well as for the set corresponding to ω_{p1}, ω_{i1}, ω_{s1}. Thus the signal frequency can be tuned solely by varying the pump frequency. Figure 48 also shows that the amplifier is basically a narrowband device.

The arrangement of Fig. 48 requires a proportional change in the signal and pump frequencies. The tuning range of this scheme is often limited by the available tuning of the pump source as well as the desired separation of the frequency bands of the pump, idler, and signal. It is possible to extend the tuning range in many different ways. The criterion in the design is to keep the idler frequency essentially constant.

Figure 49 shows a typical ω-β diagram of the loaded transmission lines of a backward traveling-wave parametric amplifier. The PQ branch belongs to the pump. The P_sQ_s and P_iQ_i branches belong to the signal and idler which propagate in the same transmission line. It is assumed that the periodic loading of parametric diodes occurs at intervals of a.

In Fig. 49, the positions of P_iQ_i are chosen such that the idler frequency is very close to the cutoff frequency, and thus the idler frequency is essentially constant over the whole tuning range. As is shown in Fig. 49, the signal-frequency tuning range (P_s,Q_s) is larger than $2:1$. The tuning range can also be extended further by properly shaping the propagation characteristics of the pump (PQ), for example, by increasing the periodic loading of the transmission line of the pump with dummy diodes.

With an essentially constant idler frequency, the fractional change in pump frequency becomes smaller. This results in simpler and possibly faster electronic tuning. The noise figure is also improved because of the possible choice of higher ω_i/ω_s ratio.

Parametric interactions with an essentially constant idler frequency can be achieved also in other schemes. For example, it is possible to provide separate idler circuits with a constant idler frequency. It is also possible to operate the idler near the cutoff frequency of the pump circuit when a loaded waveguide is used.

From Figs. 48 and 49, it can be seen that the tuning characteristics of a backward traveling-wave parametric amplifier are a function of the relative phase velocities of the pump, idler, and signal waves. The enhanced tuning range for Fig. 49 as-compared with Fig. 48, for example, is a result of varying the idler phase velocity while keeping the phase velocities of the signal and pump nearly constant. The effect of the variations of the phase velocities on the tuning range can be calculated from the graphs. With the balanced scheme of Fig. 45, it is possible to vary the relative phase velocities of these traveling waves, for example, by introducing various types of meandering striplines for the structure of Fig. 46b.

The backward traveling-wave parametric amplifiers, like the conventional backward-wave tubes, have inherent regeneration and may become unstable because of the backward-wave interaction between the opposite-traveling signal and idler waves. In the design of these amplifiers, it is important to achieve proper impedance matching over the tuning range for all traveling-wave circuits as well as at the input and output terminals. The matching requirement is often the practical limitation on the tuning range of the amplifier.

In a properly matched and terminated backward traveling-wave parametric amplifier, one obtains unidirectional amplification.

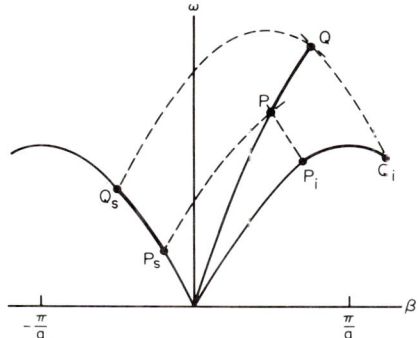

FIG. 49. Backward traveling-wave parametric amplifier with essentially constant idler frequency for enhanced tuning range. The dotted curves which are parallel to the ω-β curve show that the sets of points (P,P_s,P_i) and (Q,Q_s,Q_i) satisfy the selection rules. [*From H. Hsu and S. Wanuga, The Wide Tuning Range of Backward Traveling-wave Parametric Amplifiers, Proc. IRE, (correspondence), vol. 49, p. 1339, 1961.*]

3.5. Forward and Backward Traveling-wave Interactions. The basic concept of forward and backward traveling-wave parametric interactions can be explained as follows with electromagnetic waves as the typical example. The parametric medium is assumed to be homogeneous and isotropic, although the analysis can be extended to anisotropic materials. The wave equation of the electric field E can be obtained from Maxwell's equation as

$$\nabla^2 E - \mu_0(\partial^2/\partial t^2)(\epsilon E) = 0 \qquad (131)$$

where
$$\epsilon = \epsilon_0[1 + \xi \cos(\omega_p t - \beta_p z)]$$

i.e.,
$$\nabla^2 E - (\partial^2/\partial t^2)[(1/v)^2 E] = 0 \qquad (132)$$

with
$$(1/v)^2 = \mu_0\epsilon_0[1 + \xi \cos(\omega_p t - \beta_p z)] \qquad (133)$$

Let
$$E = E_0 e^{-j(\omega_s t - \beta z)} \sum_{-\infty}^{\infty} a_n e^{-jn(\omega_p t - \beta_p z)} \qquad (134)$$

where β is now the perturbed phase constant of the signal (the unperturbed phase constant is β_s, as defined previously). Then

$$a_n \left[1 - \frac{(n\beta_p + \beta)^2}{\mu_0 \epsilon_0 (\omega_s + n\omega_p)^2} \right] + \tfrac{1}{2}\xi(a_{n-1} + a_{n+1}) = 0 \tag{135}$$

Assume there are only pump, signal, and idler waves; therefore, all terms except $n = 0$ and $n = -1$ are zero. With $n = 0$, we have

$$a_0(1 - \beta^2/\beta_s^2) + \tfrac{1}{2}\xi a_{-1} = 0 \tag{136}$$

With $n = -1$, we have

$$\tfrac{1}{2}\xi a_0 + a_{-1}\left[1 - \left(\frac{\beta_p - \beta}{\beta_p - \beta_s}\right)^2 \right] = 0 \tag{137}$$

where
$$\beta_s^2 \equiv \mu_0 \epsilon_0 \omega_s^2$$
and
$$(\beta_p - \beta_s)^2 \equiv \mu_0 \epsilon_0 (\omega_p - \omega_s)^2$$

By eliminating a_0 and a_{-1}, we get

$$\beta = \beta_s \pm \tfrac{1}{4}\xi \sqrt{\beta_s(\beta_s - \beta_p)} \quad \text{and} \quad a_{-1}/a_0 = \pm \sqrt{(\beta_s - \beta_p)/\beta_s} \tag{138}$$

For forward traveling-wave amplifiers, $\beta_p > \beta_s > 0$,

$$\beta = \beta_s \pm \tfrac{1}{4}j\xi \sqrt{\beta_s(\beta_p - \beta_s)} = \beta_s \pm \tfrac{1}{4}j\xi \sqrt{\beta_s \beta_i} \tag{139}$$

and
$$a_{-1}/a_0 = \pm j \sqrt{(\omega_p - \omega_s)/\omega_s} = \pm j \sqrt{\omega_i/\omega_s} \tag{140}$$

For backward traveling-wave amplifiers, $\beta_s < 0$; therefore,

$$\beta = \beta_s \pm \tfrac{1}{4}\xi \sqrt{|\beta_s|(\beta_p + |\beta_s|)} = \beta_s \pm \tfrac{1}{4}\xi \sqrt{|\beta_s|\beta_i} \tag{141}$$

and
$$a_{-1}/a_0 = \pm \sqrt{(\omega_p - \omega_s)/\omega_s} = \pm \sqrt{\omega_i/\omega_s} \tag{142}$$

Assume that, at $t = 0$ and $z = 0$,

$$E = E_0 \text{ for signal wave} \quad \text{and} \quad E = 0 \text{ for idler wave}$$

Then, for the forward traveling-wave amplifier

$$\omega_p = \omega_i + \omega_s \qquad \beta_p = \beta_i + \beta_s$$

$$E = E_0 \cosh\left(\tfrac{1}{4}\xi \sqrt{\beta_s \beta_i}\, z\right) \cos(\omega_s t - \beta_s z)$$
$$+ \sqrt{\omega_i/\omega_s}\, E_0 \sinh\left(\tfrac{1}{4}\xi \sqrt{\beta_s \beta_i}\, z\right) \sin(\omega_i t - \beta_i z) \tag{143}$$

For backward traveling-wave amplifiers,

$$\omega_p = \omega_i + \omega_s \qquad \beta_p = \beta_i - \beta_s$$

$$E = E_0 \cos\left(\tfrac{1}{4}\xi \sqrt{|\beta_s|\beta_i}\, z\right) \cos(\omega_s t + |\beta_s|z)$$
$$+ (-\sqrt{\omega_i/\omega_s})E_0 \sin\left(\tfrac{1}{4}\xi \sqrt{|\beta_s|\beta_i}\, z\right) \sin(\omega_i t - \beta_i z) \tag{144}$$

Whereas the forward traveling-wave amplifier has exponentially growing waves, the power gain for the backward traveling-wave amplifiers is

$$\text{Gain} \cong [\cos(\tfrac{1}{4}\xi \sqrt{|\beta_s|\beta_i}\,)z]^{-2} \tag{145}$$

For both types of amplifiers, assuming lossless circuits,

$$\text{Noise figure} \cong 1 + (\omega_s/\omega_i)(T_i/T_0) \tag{146}$$

The present analysis can be applied to converters by choosing terms of $n = 0$ and $n = +1$, or to four-frequency devices with appropriate selection of terms of the series expansion.

3.6. Three-dimensional Parametric Interactions.

The concept of three-dimensional parametric interactions is important to the understanding and interpretation of parametric interactions. This concept is also very important in the development of solid-state parametric devices, for example, optical parametric devices.

For three-dimensional parametric interactions, $\vec{\beta}$ becomes a vector quantity. We limit our discussion to the simple case of a parametric amplifier with the pump, signal, and idler at ω_p, ω_s, ω_i, and $\vec{\beta}_p$, $\vec{\beta}_s$, $\vec{\beta}_i$, respectively. The selection rules become

and
$$\omega_p = \omega_i + \omega_s$$
$$\vec{\beta}_p = \vec{\beta}_i + \vec{\beta}_s$$

For optical frequencies,
$$\vec{\beta} = \omega \vec{n}/c$$

where $\vec{n}$ is the index of refraction and c is the velocity of light. The wave equation of the previous section becomes

$$\nabla^2 \sum_{p,s,i}(E + E^*) - (\partial^2/\partial t^2)\left[\sum_{p,s,i} \mu\epsilon(E + E^*)\right] = 0 \qquad (147)$$

Since
$$\epsilon = \epsilon_0[1 + \tfrac{1}{2}\xi(\omega_p, \vec{\beta}_p)]$$
$$\mu = \mu_0$$

we have

$$\nabla^2 \sum_{p,s,i}(E + E^*) - (\partial^2/\partial t^2)\left\{\mu_0\epsilon_0[1 + \tfrac{1}{2}\xi(\omega_p, \vec{\beta}_p)] \sum_{p,s,i}(E + E^*)\right\} = 0 \qquad (148)$$

i.e.,
$$\nabla^2 E_s - \mu_0\epsilon_0 \ddot{E}_s = \mu_0\epsilon_0(\partial^2/\partial t^2)[\tfrac{1}{2}\xi(\omega_p, \vec{\beta}_p)E_i^*(-\omega_i, -\vec{\beta}_i)] \qquad (149)$$

and
$$\nabla^2 E_i^* - \mu_0\epsilon_0 \ddot{E}_i^* = \mu_0\epsilon_0(\partial^2/\partial t^2)[\tfrac{1}{2}\xi(\omega_p, \vec{\beta}_p)E_s(+\omega_s, +\vec{\beta}_s)] \qquad (150)$$

The above equations are the coupled-mode equations of a three-dimensional parametric amplifier. The terms on the left-hand side indicate the normal-mode traveling waves at $(\omega_s, \vec{\beta}_s)$ and $(\omega_i, \vec{\beta}_i)$. The right-hand terms correspond to the parametric excitation which produces the coupling between the normal-mode waves.

Let $\vec{\beta}_s'$ and $\vec{\beta}_i'$ be the propagation constants of the signal and idler, perturbed because of parametric excitation. Then, remembering the selection rules, we may define the perturbation of the propagation constants $\Delta\vec{\beta}$ as

and
$$\vec{\beta}_s' = \vec{\beta}_s + \Delta\vec{\beta}$$
$$\vec{\beta}_i' = \vec{\beta}_i - \Delta\vec{\beta}$$

The value of $\Delta\vec{\beta}$ can be solved from the coupled-mode equations as

$$\Delta\vec{\beta} = |\Delta\vec{\beta}|\hat{r} = \pm \frac{i}{4} \frac{\xi|\vec{\beta}_i| \, |\vec{\beta}_s|}{\sqrt{(\vec{\beta}_i \cdot \hat{r})(\vec{\beta}_s \cdot \hat{r})}} \hat{r} \qquad (151)$$

where $\hat{r}$ is the unit vector of the perturbation of the propagation constant. Since the perturbation is caused by the pump, the direction of $\hat{r}$ is usually along the direction of the pump traveling wave or very close to it.

According to the $\Delta\vec{\beta}$ equation, the performance of the parametric amplification can be classified as either forward or backward traveling-wave interactions, depending upon the relative directions of the three traveling waves. In the case of forward interactions, all three waves travel along the same general direction and the denominator of the $\vec{\Delta\beta}$ equation becomes positive. Thus the perturbed $\vec{\beta}'$ become complex quantities and the waves are amplified with exponential gain. The backward interaction corresponds to the case when one of the traveling waves, say the signal wave, travels nearly along the opposite direction. Then the denominator of the $\vec{\Delta\beta}$ equation becomes imaginary and the value of $\vec{\Delta\beta}$ is a real quantity. The effect of the perturbation produces an inherent regeneration among these traveling waves. The operation has high gain and may become unstable.

It should be pointed out that the above analysis serves only to demonstrate the simple three-dimensional parametric interactions. In actual cases, it is often necessary to take into consideration the anisotropic and dispersive characteristics of the propagation medium. The selection rules, and the basic characteristics of the forward and backward traveling-wave interactions, are nevertheless usually not affected by these complications. For periodic media, such as crystals, the selection rule for $\vec{\beta}$ has an additional factor because of the reciprocal lattice.

3.7. Comments on Selection Rules. In a complicated resonating system, there are often more than three resonant frequencies. It is possible that a combination of parametric coupling of modes exists simultaneously. Some of these modes may be coupled as frequency converters, while the other modes are coupled as amplifiers. Consider a system of n modes. The coupled-mode equations become[29]

$$da_i/dz = \sum_{j=-n}^{+n} C_{ij}(\omega,\vec{\beta})a_j \qquad (i = 0 \pm 1 \cdots \pm n) \tag{152}$$

Then a_i and a_j are coupled together, if

$$\omega = \omega_i \pm \omega_j$$

and

$$\vec{\beta} = \vec{\beta}_i \pm \vec{\beta}_j$$

From the Manley-Rowe relationship, we have

	Forward traveling wave	Backward traveling wave
As converter......	$C_{ij} = -(\omega_i/\omega_j)C_{ji}^*$	$C_{ij} = +(\omega_i/\omega_j)C_{ji}^*$
As amplifier........	$C_{ij} = +(\omega_i/\omega_j)C_{ji}^*$	$C_{ij} = -(\omega_i/\omega_j)C_{ji}^*$

If one mode is coupled to several other modes, then there are a number of components of the corresponding C_{ij} or C_{ji}. Furthermore, if a higher-order nonlinearity is present, the selection rules may have to be modified correspondingly.

In general, the selection rules can be expressed as[30]

$$\hbar\omega = \hbar\omega_i \pm \hbar\omega_j \tag{153}$$

and

$$\hbar\vec{\beta} = \hbar\vec{\beta}_i \pm \hbar\vec{\beta}_j \tag{154}$$

or, for periodic structures, such as crystals,

$$\hbar\vec{\beta} = \hbar\vec{\beta}_i \pm \hbar\vec{\beta}_j + \hbar\vec{g} \tag{155}$$

where $\hbar$ is Planck's constant divided by 2π. When g is nonzero, the interaction corresponds to the so-called "Umklapp" process in solids. These equations indicate that the selection rules can be regarded as the conservation laws of energy and momentum (or crystal momentum) as in all scattering processes.

Thus, when many frequency components of waves are involved in the interaction, the selection rules can be obtained readily from the conservation laws as

$$\sum_i \hbar\omega_i = \sum_s \hbar\omega_s \qquad (156)$$

and

$$\sum_i \hbar\vec{\beta}_i = \sum_s \hbar\vec{\beta}_s \qquad (157)$$

or

$$\sum_i \hbar\vec{\beta}_i = \sum_s \hbar\vec{\beta}_s + \hbar\vec{g} \qquad (158)$$

where the index i stands for all incoming waves and s stands for all scattered waves.

REFERENCES

1. M. E. Hines, High-frequency Negative-resistance Circuit Principles for Esaki Diode Application, *Bell System Tech. J.*, vol. 39, pp. 477–513, May, 1960.
2. L. I. Sniden and D. C. Yonla, Stability Criteria for Tunnel Diode, *Proc. IRE* (correspondence), vol. 49, pp. 1206–1207, July, 1961.
3. E. A. Guillemin, "Synthesis of Passive Network," John Wiley & Sons, Inc., New York, 1957.
4. H. W. Bode, "Network Analysis and Feedback Amplifier Design," D. Van Nostrand Company, Inc., Princeton, N.J., 1945.
5. C. S. Kim and A. Brandli, High Frequency High-power Operation of Tunnel Diodes, *IRE Trans. Circuit Theory*, December, 1962, pp. 416–426.
6. U. S. Davidson, Y. C. Hwang, and G. B. Ober, Designing with Tunnel Diodes, Part I, *Elec. Design*, February, 1960, pp. 50–55.
7. T. T. Sie, Absolutely Stable Hybrid Coupler Tunnel Diode Amplifiers, *Proc. IRE*, vol. 48, no. 7, p. 1321, July, 1960.
8. E. S. Kuh and J. D. Patterson, Design Theory of Optimum Negative Resistance Amplifiers, Electronics Research Laboratory, University of California, Dec. 6, 1960.
9. H. J. Carlin and R. La Rosa, Broadband Reflectionless Matching with Minimum Insertion Loss, *Proc. Symposium on Modern Network Synthesis*, Polytechnic Institute of Brooklyn, 1952, p. 161.
10. L. I. Sniden and D. C. Yonla, Exact Theory and Synthesis of a Class of Diode Amplifiers, *Proc. Natl. Electron. Conf.*, vol. 16, p. 376, 1960.
11. C. G. Montgomery, R. H. Dicke, and E. M. Purcell, "Principles of Microwave Circuits," MIT Radiation Laboratory Series, McGraw-Hill Book Company, New York, 1948.
12. B. T. Henoch and Y. Konerna, Broadband Tunnel Diode Amplifier, Stanford Electronics Laboratory, *Tech. Rept.* 213-2, SEL-62-099, 1962.
13. R. M. Fano, Theoretical Limitations on the Broadband Matching of an Arbitrary Impedance, *J. Franklin Inst.*, vol. 269, nos. 1–2, pp. 57–83, 139–154, January, February, 1950.
14. J. G. Truxal, "Automatic Feedback Control System Synthesis," p. 120, McGraw-Hill Book Company, New York, 1955.
15. H. Rothe and W. Dahlke, Theory of Noisy Four-poles, *Proc. IRE*, vol. 44, pp. 811–818, June, 1956.
16. H. A. Haus, Noise Figure of Negative Source Resistance, *Proc. IRE*, vol. 50, pp. 2135–2136, October, 1962.
17. E. G. Nielsen, Noise Performance of Tunnel Diodes, *Proc. IRE*, vol. 48, no. 11, p. 1903, November, 1960.
18. A. W. Lo et al., "Transistor Electronics," Prentice-Hall, Inc., Englewood Cliffs, N.J., 1955.
19. P. Penfield and R. P. Rafuse, "Varactor Application," The M.I.T. Press, Cambridge, Mass., 1962.
20. K. Kurokawa, On the Use of Passive Circuit Measurements for the Adjustment of Variable Capacitance Amplifiers, *Bell System Tech. J.*, vol. 41, January, 1962, pp. 361–381.
21. IRE Standards on Methods of Measuring Noise in Linear Twoports, 1959 (59IRE20 S1), *Proc. IRE*, January, 1960, pp. 61–68.

28-52　　AMPLIFIER CIRCUITS

Additional References on Parametric Amplifiers

As pointed out previously in this chapter, there is no unique synthesis of a parametric amplifier, and space does not allow us to consider every design which has appeared in the literature. However, the following references are included as a starting point in the investigation of other special types of amplifiers described in the literature.

22. K. Kurokawa and M. Uenohara, Minimum Noise Figure of the Variable-capacitance Amplifier, *Bell System Tech. J.*, vol. 40, no. 3, p. 695, May, 1961.

This is a very complete analysis of the most practical types of parametric amplifiers. Included is a discussion of the performance to be expected upon cooling various parts of the amplifier. The noise figure of several experimental amplifiers is compared with theoretical results. The figure of merit used in this article is $\tilde{Q}$, the dynamic quality factor, which can be equated to $4m_1 f_q/f$ for comparison with the results given in this chapter.

23. Paul Penfield, Jr. and Robert P. Rafuse, "Varactor Applications," The M.I.T. Press, Cambridge, Mass., 1962.

This text gives a thorough theoretical treatment of the optimization of parametric amplifier performance. In Chap. 10 the efficiency of the high-level up-converter is thoroughly treated. The results of this text can be compared with those in this chapter if f_c is substituted for $4f_q$. The optimum pump frequency for a given diode is within 30 per cent of f_q for all amplifiers with noise figures less than 3 db. This is the justification for the definition of f_q.

24. L. A. Blackwell and K. L. Kotzebue, "Semiconductor Parametric Amplifiers," Prentice-Hall, Inc., Englewood Cliffs, N.J., 1961.

A brief discussion of broadbanding techniques is given. Some practical designs are described briefly. A pre-1961 bibliography is included.

25. W. J. Getsinger and G. L. Matthaei, Some Aspects of the Design of Wide-band Up-converters and Nondegenerate Parametric Amplifiers, *IRE Trans. Microwave Theory Tech.* vol. MTT-12, no. 1, pp. 77–87, January, 1964.

A nondegenerate amplifier in which the diode series resistance is the only dissipative load is designed and measured results are given. This is a somewhat more sophisticated example (similar to the one given in this chapter) of how passive network synthesis and computer analysis can be used to produce a broadband amplifier. This article also contains references to previous broadband synthesis work.

26. S. Hayasi and T. Kurokawa, A Balanced-type Parametric Amplifier, *IRE Trans. Microwave Theory Tech.*, vol. MTT-10, no. 3, pp. 185–190, May, 1962.

This paper illustrates the simplification in circuit complexity which can result in using a balanced structure. Also included are references to previous work in this area, as well as a reference to a bandwidth analysis equivalent to that given in this chapter.

27. D. C. Hanson, H. J. Fink, and M. Uenohara, "Varactor Diode Amplifier at Liquid Helium Temperature," Digest of Technical Papers, ISSCC, February, 1963.

This paper gives measurements of the change in the small-signal impedance of GaAs varactors at 298°K, 77°K, and 4.2°K. From this the f_q and the average elastance can be estimated at various temperatures for the design of cooled parametric amplifiers.

28. P. K. Tien, Parametric Amplification and Frequency Mixing in Propagating Circuits, *J. Appl. Phys.*, vol. 29, p. 1345, September, 1958.
29. W. H. Louisell, "Coupled Mode and Parametric Electronics," John Wiley & Sons, Inc., New York, 1960.
30. H. Hsu, Three-dimensional Parametric Interactions of Waves and Quasi-particles, *Proc. IRE* (correspondence), vol. 50, p. 1977, September, 1962.
31. N. M. Kroll, Parametric Amplification in Spatially Extended Media and Application to the Design of Tunable Oscillations at Optical Frequencies, *Phys. Rev.*, vol. 127, p. 1204, 1963; or *Proc. IEEE*, vol. 51, p. 110, 1963.
32. N. Bloembergen, "Nonlinear Optics," W. A. Benjamin, Inc., New York, 1965.
33. C. W. Barnes, Conservative Coupling Between Modes of Propagation, a Tabular Summary, *Proc. IEEE*, vol. 52, pp. 64, 295, January, 1964.

Chapter 29

INDUCED-EMISSION AMPLIFIERS AND OSCILLATORS (MASERS AND LASERS)

EDMUND B. TUCKER[*]

CONTENTS

1. Introduction.. 29–2
2. The Beam Maser... 29–2
   2.1. The Effuser or Beam Collimator...................... 29–3
   2.2. Beam Focusers....................................... 29–5
   2.3. Beam Maser Cavities................................. 29–6
   2.4. The Ammonia-beam Maser............................. 29–8
   2.5. Beam Maser Using Separated Oscillatory Fields........ 29–9
   2.6. The Atomic-hydrogen Maser........................... 29–10
   2.7. Gas-beam Maser Amplifiers........................... 29–11
3. Three-level Solid-state Cavity Masers..................... 29–11
   3.1. Reflection-cavity Masers for 400-Mc Operation........ 29–11
   3.2. Reflection-cavity Masers for L Band................ 29–12
   3.3. S-band Reflection Masers.......................... 29–16
   3.4. A 6-Gc Maser.. 29–16
   3.5. X-band Reflection-cavity Masers................... 29–19
   3.6. CW Millimeter-wave Masers........................... 29–21
   3.7. Masers Using an Optical Pump........................ 29–21
   3.8. High-temperature Operation of Masers................. 29–21
4. Traveling-wave Masers.................................... 29–22
   4.1. The Comb Structure.................................. 29–24
   4.2. A Traveling-wave Maser for 3.0 Gc................... 29–25
   4.3. A Traveling-wave Maser for 5.8 Gc................... 29–27
   4.4. Dielectric Slowing.................................. 29–29
5. Masers with Intermittent Operating Characteristics........ 29–29
   5.1. Adiabatic Rapid Passage............................. 29–29
   5.2. Adiabatic Rapid Passage with Pulsed Fields.......... 29–29
   5.3. Pulsed-field Maser—Three-level Inversion............ 29–31
6. The CW Gas Laser... 29–32
   6.1. Radio-frequency Discharges.......................... 29–33
   6.2. D-C Discharges...................................... 29–34
   6.3. External-mirror Geometries.......................... 29–34
   6.4. Mode Selection...................................... 29–35

[*] General Electric Company, Research and Development Center, Schenectady, N.Y

      6.5. Mechanical Considerations.............................. 29–35
      6.6. Pulsed Operation of Gas Lasers......................... 29–36
      6.7. Typical Gas Systems................................... 29–36
  7. Metallic-vapor Lasers.. 29–41
      7.1. The Cesium Amplifier.................................. 29–41
      7.2. The Cesium Oscillator................................. 29–42
  8. Solid-state Lasers—Ruby..................................... 29–43
      8.1. Pulsed Ruby Lasers.................................... 29–43
      8.2. The Ruby-laser Output................................. 29–44
      8.3. Pump Energy Density in a Dielectric Cylinder........... 29–46
      8.4. CW Ruby Laser.. 29–47
      8.5. Solar-pumped Laser.................................... 29–48
      8.6. The Giant-pulse Laser................................. 29–48
      8.7. A Ruby-laser Amplifier................................ 29–50
      8.8. Other Iron-group Laser Materials...................... 29–52
  9. Rare-earth Lasers... 29–52
      9.1. Rare-earth Laser Characteristics....................... 29–52
 10. Semiconductor or Injection Lasers........................... 29–58
     10.1. Construction of GaAs Injection Lasers................. 29–58
     10.2. Radiation Properties of GaAs Injection Lasers......... 29–59
     10.3. Harmonic Content of GaAs Laser Output................ 29–63
     10.4. Efficiency of GaAs Lasers............................. 29–63
     10.5. Other Injection Lasers................................ 29–63
     10.6. Magnetic Field Effects in Injection Lasers............ 29–63
 11. The Raman Laser... 29–65
 12. The Organic Laser... 29–67

1. INTRODUCTION

In a new field it seems best to use, as examples of configurations which may be useful for design purposes, those used by the early explorers of the area. The circuits, and this is perhaps not the best terminology, presented in this chapter are representative of those used in experimental models and of a few actual operational devices. The *maser* field is mature enough that representative configurations from the early work are used to supplement the operational units—primarily for radio astronomy. The *laser* area is, however, expanding rapidly, and an effort has been made to give credit to the first work of a given type. This is forced, in many instances, by the fact that the field is just developing, as, for example, in such potentially important areas as millimeter-wave generation and Raman interactions.

2. THE BEAM MASER

In addition to a suitable gas, the beam maser requires four main components. They are:

1. A beam-forming device or diffuser consisting of an array of long parallel tubes through which the neutral gas molecules flow because of the pressure differential between input and output.

2. The beam focuser which retains or focuses the upper-level molecules while defocusing and discarding the molecules in the lower level.

3. The resonant cavity whose dimensions are governed by the frequency and mode chosen according to well-known cavity-design formulas. The mode must be one which will provide sufficient (usually maximum) electric or magnetic field along the path traveled by the molecules as they pass through the cavity.

4. The vacuum pumping system to maintain as low a background pressure as possible.

INDUCED-EMISSION AMPLIFIERS AND OSCILLATORS   **29-3**

Items 3 and 4 will be mentioned only incidentally, and reference to items in the general bibliography at the end of this chapter is recommended to those persons unfamiliar with these aspects of the problem. The levels for use in the gas-beam maser must possess either electric- or magnetic-dipole moments. Most gas masers have used the former but one example of the latter, the hydrogen maser, is considered in Sec. 2.6. The electric-dipole moment must exist in the states, such as one of the pyramidal configurations of NH_3 (Chap. 14, Fig. 2), from which the symmetric and antisymmetric states are formed and gives rise to the necessary transition probabilities even though the symmetric and antisymmetric states themselves possess no permanent electric-dipole moment.

2.1. The Effuser or Beam Collimator. The beam collimator has generally taken the form of an aligned collection of small-bore tubes through which the gas is forced by the pressure difference between the source and the background pressure of the apparatus. Giordmaine and Wang[1] have made a systematic study of the properties of such a system. The discussion which follows draws heavily from their work.

Consider the gas flow in an individual tube of small inside diameter. The optimum collimation will be obtained if the mean free path of the gas molecules is larger than the length L of the tube (and also greater than the radius a, assumed $\ll L$). In this case the peak intensity in a solid angle $d\omega$ will be

$$I(0)\, d\omega = (n_0 \bar{v} a^2/4)\, d\omega \tag{1}$$

where n_0 = molecular density at the high-pressure end of tube
$\bar{v}$ = average molecular velocity

The flow rate for this system will be

$$n_t = (2\pi/3)(n_0 \bar{v} a^3/L) \qquad \text{molecules/sec} \tag{2}$$

The half width, or the angle at which the intensity drops to one-half, is

$$\theta_{\frac{1}{2}} = 1.68 a/L \tag{3}$$

Because of the high peak intensities ($\simeq 10^{19}/\text{sec}^{-1}\,\text{cm}^{-2}\,\text{steradian}^{-1}$) required for beam masers, source pressures of a few tenths of a millimeter are required [see Eq. (1)], and at this pressure mean free paths are about 10^{-2} mm. Tubes with inside diameters of 10^{-3} mm diameter ($\simeq 40 \times 10^{-6}$ in.) are difficult to obtain (see latter part of this section), and the usual mode of operation is one of the following cases.

The usual relationship between the molecular mean free path and the individual tube dimensions may be separated into two categories: (1) $a \ll \lambda \leq L$ throughout the whole tube, or (2) $\lambda \leq a \ll L$ at the high-pressure or input end with condition 1 holding near the output. That is, Knudsen flow ($\lambda \gg a$) is obtained in the output section of each individual tube in either case. Giordmaine and Wang[1] find the peak intensity for case 2 to be

$$I(0) = \left(\frac{3 n_t a \bar{v}}{64 \sqrt{2}\, \pi \sigma^2} \right)^{\frac{1}{2}} \tag{4}$$

where σ = molecular diameter. This formula holds for

$$L > 2.5 \left(\frac{2^{\frac{3}{2}} \bar{v} a^3}{3 n_t \sigma^2} \right)^{\frac{1}{2}} \tag{5}$$

Increasing L beyond the equality condition has little effect on the properties of the source. Note that the source pressure has no effect on the peak intensity $I(0)$.

Backtracking to condition 1, intermediate between the assumed conditions for Eqs. (1) and (4), it is found that

$$I(0) = \frac{\pi a^2 \bar{v}}{8 \pi \sigma} \left(\frac{2^{\frac{1}{2}} n_0}{L} \right)^{\frac{1}{2}} \tag{6}$$

The source pressure is a factor and generally will be, if $\lambda \gg a$ for the whole tube length. The beam half width for condition 1 is

$$\theta_{\frac{1}{2}} \simeq \frac{1}{1.78}\left(\frac{2^{\frac{7}{3}}3\sigma^2 n_t}{a\bar{v}}\right)^{\frac{1}{2}} \tag{7}$$

Experimental results for CO_2 agree reasonably well with this formula, which has been empirically adjusted and is double the theoretically half width. The experimental results obtained for CO_2 with the configurations of Table 1 are given in Fig. 1. Further information on the same multiple tube bundles at a variety of pressures is contained in Table 2.

Table 1

Source	Tube length, cm	Radius of single tube, cm	No. of tubes	Overall source diam, cm	Shape of total cross section
A	0.66	1.65×10^{-2}	224	0.51	Hexagonal
B	0.31	2.35×10^{-3}	1.28×10^4	1.3	Circular
C	0.95	2.69×10^{-3}	1.80×10^4	1.1	Triangular

Table 2

Source	Pressure at input	Peak intensity, molecules/sec /steradian	Total flow, molecules/sec	Half width at half intensity, deg
A	0.024	1.42×10^{17}	8.91×10^{16}	5.0
	0.060	3.19×10^{17}	1.87×10^{17}	8.3
	0.110	5.70×10^{17}	5.26×10^{17}	15.0
	0.190	9.66×10^{17}	1.21×10^{18}	21.1
B	0.03	9.16×10^{17}	2.69×10^{17}	3.5
	0.075	1.49×10^{18}	4.56×10^{17}	4.3
	0.15	2.80×10^{18}	1.25×10^{18}	6.0
	0.25	3.73×10^{18}	2.28×10^{18}	9.1
	0.44	5.01×10^{18}	4.61×10^{18}	15.7
C	0.035	3.05×10^{17}	5.42×10^{16}	2.5
	0.13	7.87×10^{17}	1.81×10^{17}	3.3
	0.265	1.30×10^{18}	3.88×10^{17}	4.5
	0.61	2.29×10^{18}	1.03×10^{18}	5.7
	1.28	3.74×10^{18}	2.18×10^{18}	7.8
	1.90	4.79×10^{18}	3.58×10^{18}	11.3

Besides the above configurations, effusers actually used have taken several additional forms. Gordon, Zeiger, and Townes[2] formed a thin foil in roughly sinusoidal fashion by deforming it around small-diameter wires. This was placed next to a plane foil and the combination, flat on one side, corrugated on the other, was wound into a tight spiral. Helmer, Jacobus, and Sturrock[3] have used a technique which may be useful down to inside tube diameters of 0.001 in. Aluminum wire is copper-plated, cut in lengths, and stacked in a copper tube. The tube is swaged until the spaces between wires are filled in because of deformation of the wire cross section. The aluminum is etched out, leaving the copper diffuser. Transparency of 85 per cent with 0.005-in. wires has been obtained. It is worth noting that these same authors suggest that

$$I_{max} \simeq (t/2Da)^{\frac{1}{2}} \tag{8}$$

where t = transparency
  D = overall diameter of diffuser
  a = individual tube diameter

INDUCED-EMISSION AMPLIFIERS AND OSCILLATORS 29–5

2.2. Beam Focusers. The coupling of the beam from effuser to cavity is accomplished by means of the focuser. It, in effect, has two purposes; first to provide an inverted population to the cavity; second to deliver to the cavity as large a fraction of the upper-state molecules emitted from the effuser as possible. In considering the second of these it is evident that, using the definitions of Fig. 2, the number of molecules N entering the focuser and subsequently the cavity is

$$n_t = I_1\Omega_1 A_1 = I_2\Omega_2 A_2 \tag{9}$$

The solid angle into which the effuser emits is generally larger than Ω_1, and only a reduction of A_1 while maintaining Ω_2, A_2, etc., will reduce the gas inflow to the system.

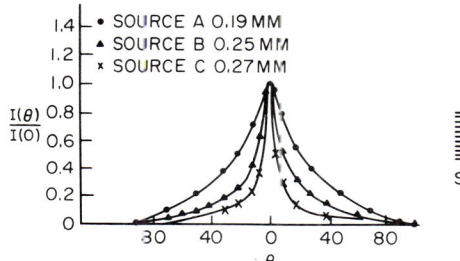

Fig. 1. Experimental results of beam-collimation properties for the three sources of Table 1 at the pressures noted. (*After Giordmaine and Wang*,[1] *published by permission of the editor, J. Appl. Phys.*)

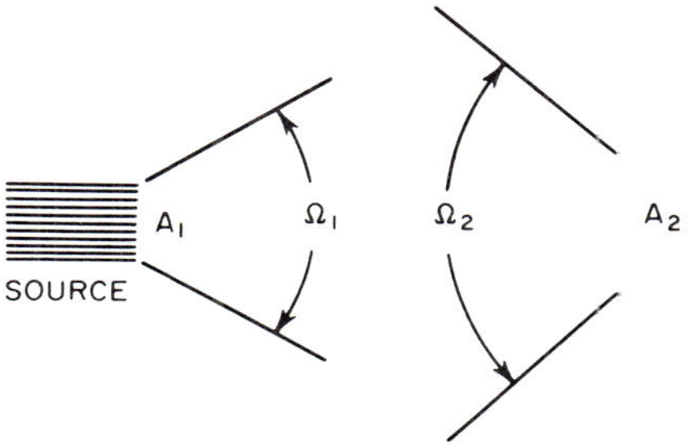

Fig. 2. Input and output aperture areas and solid angles.

The constant-cross-section focusers such as the four-pole one, biased alternately plus and minus, used in the first maser[2] (Fig. 3) have $A_1 = A_2$. Flexibility may be added by shaping the inside of the focusing bars. A paraboloid of revolution has been suggested[3] as a solution to the problem. If the input and output apertures of

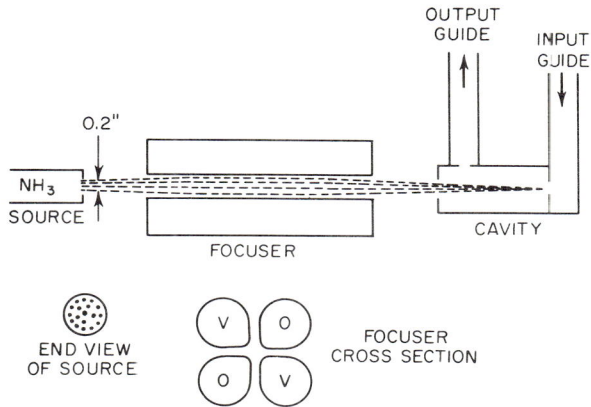

Fig. 3. Schematic of NH_3-beam maser using linear focuser, with hyperbolic cross section. (*After Gordon, Zeiger, and Townes*,[2] *published by permission of the editor, Phys. Rev.*)

the focuser are circular and of radius R_1 and R_2, respectively, and the effuser is assumed to occupy the entire input aperture, the beam radius as a function of distance z from the input is given by

$$r = [(z/L)(R_2{}^2 - R_1{}^2) + R_1{}^2]^{\frac{1}{2}} \tag{10}$$

where L is the length of the focuser. The focusing electrodes are made with circular cross section and radius to match the beam radius at that point. The inner radius of the focus electrode lies on the parabolic envelope which encompasses the beam. A comparison of a linear focuser and a parabolic focuser can be obtained by analyzing the data of Figs. 4 and 5. For a given power the parabolic system requires about one-eighth the gas flow needed by the linear system.

For maximum beam transmission the trajectories must change slowly along the beam path. Empirically this condition will hold[3] for arbitrarily high potentials so long as, in a uniform focuser of radius R_2, the length L is at least a quarter of the

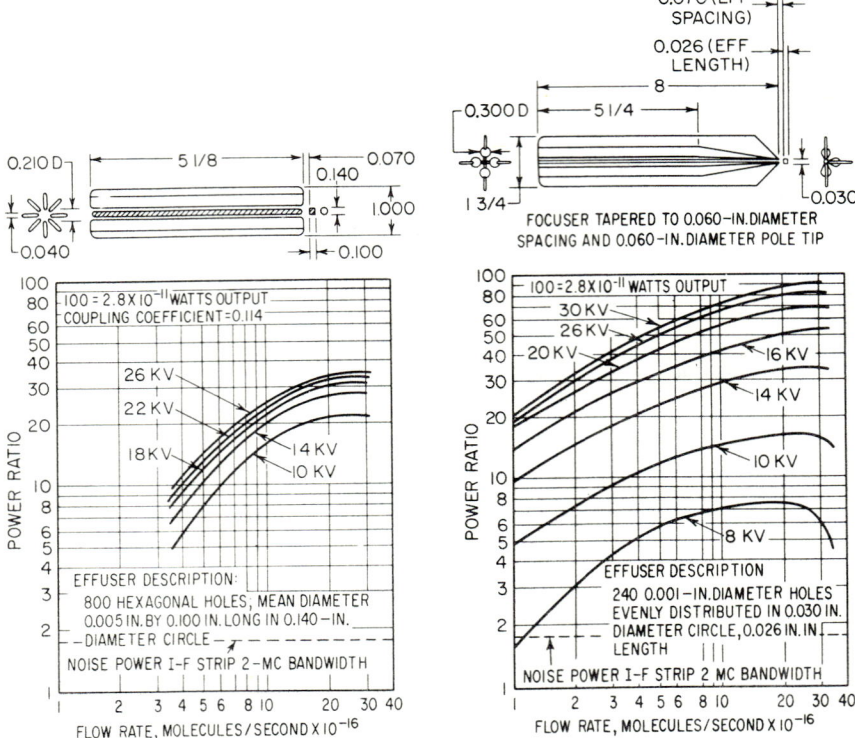

FIG. 4. Operating characteristics of an ammonia-beam maser using a linear focuser. (*After Helmer, Jacobus, and Sturrock*,[3] *published by permission of the editor, J. Appl. Phys.*)

FIG. 5. Operating characteristics of an ammonia-beam maser using a parabolic focuser. (*After Helmer, Jacobus, and Sturrock*,[3] *published by permission of the editor, J. Appl. Phys.*)

trajectory period. The length then is the same for the two types of focusing, linear and parabolic, but the gas utilization is better in the parabolic because of the smaller input area. Dimensions and operating voltages are given Figs. 4 and 5.

A schematic of the original ammonia maser[2] is given in Fig. 3. The focuser cross section is hyperbolic to match the equipotential lines. The conditions under which focusers of this type were operated are given in Table 3.

2.3. Beam Maser Cavities. The power emitted by a beam of n molecules/sec in its passage through the cavity is given by[4,5]

$$\Delta P = \frac{x^2 n L^2}{4 \langle v \rangle^2} \frac{\sin^2 (\theta^2 + \delta^2)^{\frac{1}{2}} h \nu_0}{\theta^2 + \delta^2} \tag{11}$$

where $\theta^2 = E^2\bar{\mu}^2L^2/4\langle v\rangle^2\hbar^2 = \mathcal{E}/\mathcal{E}_c$, $\mathcal{E}_c = h^2Av^2/8\pi^3(\bar{\mu})^2L$
$\bar{\mu}$ = the electric-dipole matrix element
$\delta = (\omega - \omega_0)(L/2\langle v\rangle)$
$\langle v \rangle$ = average velocity
$x = E\bar{\mu}/\hbar$

It should be noted that $\mathcal{E} = E^2LA/8\pi$. At the oscillation threshold this power must equal that dissipated in the cavity, i.e.,

$$\Delta P = \omega\mathcal{E}/Q = x^2\hbar^2\nu_0LA/16\pi^2Q(\bar{\mu})^2$$

and the threshold molecular flow for oscillation is

$$n_t = h\nu_0{}^2A/4\pi^2 2QL \quad (12)$$

taking $\sin\theta = \theta$, i.e., low powers, and using $\langle v_0 \rangle$ as the effective average velocity. Hence a figure of merit for beam maser cavities may be defined including both the length and the unloaded cavity Q.

$$M = (LQ_0/A)(8/\pi^2)^m \quad (13)$$

for $m = 0$ or 1. m refers to the mode number designating the number of field nodes along the cavity length. The transition efficiency decreases as the field becomes nonuniform. Values of m larger than 1 are normally undesirable since the line splits into

Table 3

Radius of beam, in.	Nearest electrode separation, in.	Length, in.	Voltages, kv
0.4	0.08	22	<15
0.4	0.16	8	<30

two for $m = 2$. In fact, the lowest modes of resonance give the best ratio of Q_0/A because of the radial dependence of Q_0 being proportional to the radius and A being proportional to (radius)^2. Table 4 compares several cylindrical and rectangular cavities using as criteria the values of $M\epsilon$ when ϵ is the skin depth multiplied by the specific permeability of the walls μ

$$\epsilon = (c/2\pi)(\mu\rho/\nu)^{\frac{1}{2}}$$

where ρ is the electrical resistivity.

Table 4

Type		Radius, cm	Narrow beam A_e/area	Narrow beam $M\epsilon$	Broad beam A_e/area	Broad beam $M\epsilon$	Q_0
Cylindrical	TE_{111}	0.37	0.48	12.2	1.00	5.9	6,100
Cylindrical	TM_{010}	0.48	0.27	28.4	1.00	7.7	10,800
Cylindrical	TM_{011}	0.48	0.27	22.2	1.00	6.0	10,400
Cylindrical	TE_{211}	0.61	....	....	1.00	2.9	8,100
Cylindrical	TE_{011}	0.76	....	....	1.00	4.1	17,800
		Cross section					
Rectangular	TE_{011}	0.63 × 0.31	0.50	15.5	....	7.8	3,700
Rectangular	TE_{011}	0.63 × 0.63	0.50	10.3	....	5.2	4,900
Rectangular	TM_{110}	0.89 × 0.89	0.25	26.0	....	6.5	10,100

29–8 AMPLIFIER CIRCUITS

For copper at 24×10^{10} cps, $\epsilon = 4.27 \times 10^{-5}$ cm. The ratio A_e/area represents the fraction of the cross section filled by the beam. In practice the operating condition falls between the two cases (narrow and broad beam) considered. For the cylindrical cavities the TM_{010} mode with its uniform axial electric field is most efficient for electric-dipole transitions. If magnetic-dipole transitions were to be used one of the TE modes such as TE_{011} would be preferred.

2.4. The Ammonia-beam Maser. Operating conditions of an ammonia-beam maser are represented reasonably well by those of the first maser of Gordon, Zeiger, and Townes.[2] A pressure of a few millimeters of mercury forces ammonia molecules out of the effuser, described in Sec. 2.1, into the vacuum of about 10^{-5} mm Hg at a rate of 10^{15} per second. Of these about 6 per cent are in the desired upper level; most are focused, with 4×10^{13} entering the cavity. Output powers of about 5×10^{-10} watts were observed when used as an oscillator. The frequency of oscillation, 23.870 Gc, was stable to a few parts in 10^{12} over times on the order of a second while a stability of $1/10^{10}$ was obtained for times on the order of an hour. Two such oscillators will beat together to give an audio note.

The frequency is affected by cavity tuning as predicted in Chap 14, Sec. 16.1. Source pressure and focuser voltage also change the frequency because of the change in numbers of molecules in the cavity. A further cause of frequency shift is the so-called traveling-wave effect. Since the major virtue of the ammonia-beam maser is its high-frequency stability there is reason to discuss some of the reasons for frequency variations due to causes other than cavity tuning.

For typical ammonia 3-3 maser cavities $\mathcal{E}_c$ [Eq. (11)] is on the order of 5×10^{-11} ergs. It can be shown[5] that the frequency deviation from the molecular resonance frequency ω_0 is related to the cavity frequency by the equation

$$\frac{\omega - \omega_0}{\omega_0} = \frac{Q_L}{Q_b} \frac{1}{2.8} \left[\frac{1 - \cos 2\theta}{1 - (\sin 2\theta)/2\theta} \right] \frac{\omega_c - \omega_0}{\omega_0} \tag{14}$$

where Q_L is the cavity loaded Q without the beam

$$Q_b = \nu_0/2\Delta\nu = Q \text{ of the line}$$

For the case of no saturation or $\theta \ll 1$, (14) reduces to

$$\frac{\omega - \omega_0}{\omega_0} = 1.07 \frac{Q_L}{Q_b} \frac{\omega_c - \omega_0}{\omega_0} \tag{14a}$$

which is equivalent to Eq. (91) in Chap. 14. However, Eq. (14) includes effects due to saturation through the value of θ. The saturation has an effect only if the cavity and molecular frequencies differ, and furthermore, since the bracketed quantity decreases as θ increases, i.e., as the energy in the cavity becomes larger, the saturation effect decreases at higher powers and larger beam intensities. Frequency fluctuations due to source pressure and focuser fields may be attributed to this cause,[5] although on occasion the penetration of focusing fields into the cavity can cause a shift of resonance frequency.

Another source of frequency shift is due to the cavity standing wave being set up by two waves, one traveling in the direction of the beam velocity and the other in the opposite direction. A net flow of power in one direction is due to one traveling wave being of larger amplitude. The interaction of the molecular beam with this energy excess in one direction leads to a shift in frequency. The fractional frequency shift due to this traveling-wave effect when expressed in the nomenclature of Eq. (11) becomes

$$\frac{\omega - \omega_0}{\omega_0} = \frac{E_1}{E_0} \frac{2kv}{\omega_0} \left(1 - \frac{\theta \sin^2 \theta}{2\theta - \sin 2\theta} \right) \tag{15}$$

$E_1 \cos(\omega t - kz)$ represents the flow of power in one direction while $E_0 \cos \omega t$ represents the standing-wave amplitude. The beam velocity v is positive in the direction of net energy flow—the positive z direction. A reversal of v changes the sign of the frequency shift. Frequency shifts due to this cause are best determined from knowledge of the Q's involved. If we define the output coupling Q to be

$$Q_{out} = \omega_1 W/S = (2\pi^2 L/\lambda^2 k)(E_0/E_1)$$

where S is the power flow $(2Kkc^2 E_0 E_1/\omega_0)$ and the stored energy W is KLx^2, Eq. (15) may be rewritten as

$$\frac{\omega - \omega_0}{\omega_0} = \frac{2\pi}{0.89} \frac{1}{Q_t Q_b} \frac{L^2}{\lambda^2} \left(1 - \frac{\theta \sin^2 \theta}{2\theta - \sin 2\theta}\right) \qquad (16)$$

For the values $Q_t = 3 \times 10^4$, $Q_b = 4 \times 10^6$, $L = 10$ cm, $\lambda = 1.25$ cm the fractional frequency shift is 2×10^{-9}, and since this is directly proportional to the output coupling, a 5 per cent change of output coupling will cause a frequency shift of $10^{-9}\omega_0$. This type of shift may be minimized by placing the coupling hole in the middle of the cavity. Even then the effect is not completely eliminated because of nonuniform emission from the beam during its travel through the cavity. At low powers the emission tends to occur toward the end of the trajectory in the cavity while at high powers it occurs in the first part of the path. Hence the direction of power flow will differ, depending on power level in the symmetric coupling case.

Shimoda[6] utilized the 3-2N^{14}H$_3$ line, which is not troubled by unresolved hyperfine structure as is the 3-3 line. The focuser of steel rods 3 mm in diameter and 40 cm in length having an aperture of 1 cm is fed by a beam effuser consisting of a cluster of 1-mm-diameter holes. The TM_{010} cavity is 12 cm long and is made tunable by lengthwise slotting so that it may be deformed. Beam inlet and outlet holes are 8 mm in diameter and 10 mm long. The loaded Q is 9,000 with a coupling iris 2.1 mm in diameter situated in the center of the cavity. The line center may be found to about 1 part in 10^9 or an order of magnitude better than for the 3-3 line. The same frequency could be reset to within ± 2 to 3 cps with constant focuser voltage. If Eq. (15) is taken as a measure of the deviation from the line center it is possible to determine the line center to within 1 or 2 parts in 10^{10}.

2.5. Beam Maser Using Separated Oscillatory Fields. Using Ramsey's separated-oscillatory-field technique, Holuj, Daams, and Kalra[8] have constructed a maser shown schematically in Fig. 6. The use of the separated cavities provides a

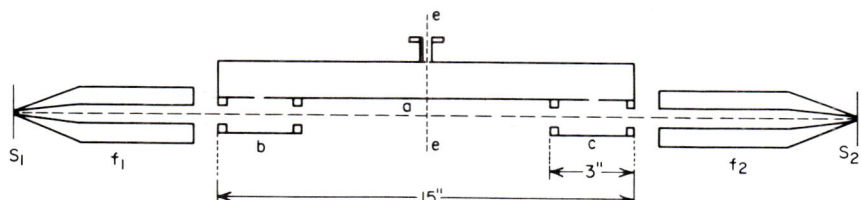

FIG. 6. Schematic of separated oscillatory field N^{14}H$_3$ maser. (*After Holuj, Daams, and Kalra,[8] published by permission of the editor, J. Appl. Phys.*)

sharper line than the usual single cavity, and the oppositely directed beams, in principle, remove the doppler shift. In fact, a single hyperfine component of the 3-3 transition can be isolated. Care is taken to maintain symmetry with the identical cavities b and c locked to the master cavity a. Operation of cavity a in the TM_{015} mode with b and c in the TM_{011} mode provides good coupling and assures uniform frequency shifts with temperature.

In this case the cavity pulling is given by[8]

$$\omega - \omega_0 = \frac{2Q_L}{2\tau_2 \omega_0} \frac{1 - \cos 2\theta}{1 - (\sin 2\theta)/2\theta\tau_2 + (\tau_1/2\tau_2)\theta \sin \theta} (\omega_c - \omega_0) \qquad (17)$$

where τ_1 is the molecular time of flight between cavities b and c, and $2\tau_2$ is the combined time of flight through b and c. This equation holds for $\theta \gg (\omega - \omega_0)\tau_2/2$ and $\tau_1(\omega - \omega_0)/2 \to 0$. For low powers $\theta \to 0$ and Eq. (17) reduces to the normal cavity pulling expression given in Eq. (91) of Chap. 14, but because of the lower Q's used the pulling is less drastic than in the single-cavity case. The low Q's do require a higher beam intensity for oscillation. To within 10 cps the maser exhibits no dependence on beam flux.

2.6. The Atomic-hydrogen Maser. The use of a gas possessing a magnetic-dipole transition in a beam maser requires compensation for the size of the magnetic-dipole moment. The ratio of the magnetic-dipole to electric-dipole moment is 10^{-2} and the ratio of transition probabilities is then 10^{-4}. The transition probability is directly proportional to the length of time spent in the field, and a considerable increase in this interaction time, over the single-cavity transit of the ammonia maser, has been achieved by Ramsey[9,10,11] and his coworkers. By thus buying transition probability with time they have succeeded in making an oscillator utilizing the $\mathbf{F} = 1$, $m = 0$ to the $\mathbf{F} = 0$, $m = 0$ transition of atomic hydrogen at approximately 1,420.405 Mc. As compared with the ammonia maser, the atomic-hydrogen maser possesses some very definite advantages: The line is not complicated by structure as is the ammonia inversion line; the time of observation is very long, resulting in a narrow line; and the storage method is such that the first-order doppler shift is practically eliminated.

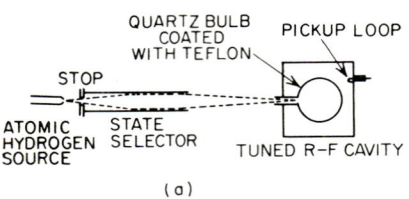

The general features of the apparatus are illustrated in Fig. 7a. The six-pole focuser uses magnetic rather than electric fields but is otherwise identical to those described in Sec. 2.2. It results in a magnetic field proportional to radius with focusing and defocusing properties for appropriate levels. The $\mathbf{F} = 1$, $m = 0$ and $\mathbf{F} = 1$, $m = 1$ levels are focused into the cavity tuned to the $\mathbf{F} = 1$, $m = 0 \cdots \mathbf{F} = 0$, $m = 0$ transition (see Fig. 7b). The real point of interest is the lined quartz bulb which serves as a storage box, prolonging the interaction time between the atoms and the magnetic fields of the cavity.

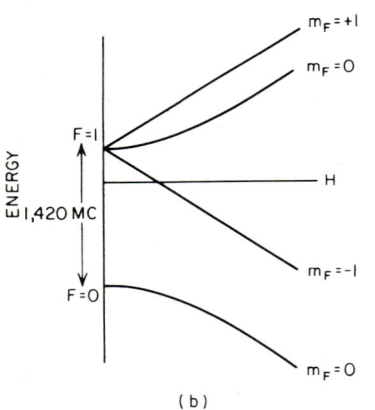

Fig. 7. (a) Hydrogen maser schematic. (*After Kleppner, Goldenberg and Ramsey,*[10] *published by permission of the editor, Phys. Rev.*) (b) Energy levels of hydrogen ground state ($^2S_{\frac{1}{2}}$).

The storage bulb is effective with several linings, the criterion of usefulness being that a collision with the wall should perturb the energy levels negligibly. Linings which have been successful in this respect are Paraflint,* dimethyldichlorosilane,† and Teflon. With these the storage time is limited, not by the wall collisions, but by the reemergence of the atom from the entrance aperture, and interaction times on the order of a second can be achieved. This time is given by $T_b = 4V/A_e \bar{v}$, where V is the volume of the bulb, A_e the entrance aperture, and $\bar{v}$ the mean velocity. For a bulb 16 cm in diameter and an exit iris (assumed to be thin) 2 mm in diameter the storage time is 1 sec. As a result of this long storage 4×10^{12} atoms/sec will generate 10^{-12} watts of power in a cavity with loaded Q of 60,000.

A measure of the effectiveness of the walls is the phase shift per collision

$$\delta\phi = \int (\delta\varepsilon/h)\, dt$$

* Trade name Paraflint by Moore and Munger, New York, N.Y.
† General Electric Dri-Film, SC-02.

The frequency shift is given by $\delta\nu \simeq \delta\phi/2\pi t_0$ where t_0 is the time between collisions. These collisions are random, giving rise to a broadening of the line by a factor $1 + \frac{1}{2}n(\delta\phi)^2$, where n represents the mean number of collisions made by an atom. Estimates of $\delta\phi$ are near 6×10^{-8} radians/collision, and for a 15-cm bulb the resonant frequency shift $\delta\nu/\nu \simeq 3 \times 10^{-12}$. Experimentally the value of $\delta\phi$ is less than 10^{-4} radians/collision. The probability of magnetic interaction during a wall collision producing a transition to a neighboring Zeeman state is estimated to be 6×10^{-13}, and therefore the lifetime for this process is $t_0/(6 \times 10^{-13}) = 5 \times 10^7$ sec for a 15-cm-diameter bulb, making the process important from the point of view of frequency shift but not of lifetime.

A spectral purity of 10^{-5} over a 10-sec period is predicted with resetability limited by residual magnetic fields, aging of the walls, and cavity pulling. The first-order doppler effect is essentially zero because the average velocity of the atom in the cavity is very nearly zero. The second-order doppler effect given by $v^2/2c^2 = 3kT/2mc^2$ is ever present, giving a fractional shift of about $1.4 \times 10^{-13}T$. The experimental precision is such that the frequency of the hydrogen line can be assigned to within ± 4 cps, with the line widths much sharper than that.

2.7. Gas-beam Maser Amplifiers. Because of their narrow bandwidth and lack of tuning range beam masers have been used little as amplifiers. They have, however, been used[12] as low-noise preamplifiers in such applications as paramagnetic-resonance spectrometers.

3. THREE-LEVEL SOLID-STATE CAVITY MASERS

The largest class of maser amplifier is that utilizing three paramagnetic energy levels and based on Bloembergen's continuous-inversion scheme.[13] The frequency range is limited by availability of suitable energy-level systems, relaxation times, and pump sources. The differences from one frequency to the next tend to be those of detail, and many of the common features are not discussed below.

Radio astronomy has benefited greatly by the advent of the maser. It is one of the few fields where the low noise figure is usable, and where the problems associated with the use of liquid helium are considered manageable. A number of the masers described have been designed for such use.

Masers at frequencies higher than X band have been of most interest as oscillators, i.e., as millimeter-wave generators.

3.1. Reflection-cavity Masers for 400-Mc Operation. Such low-frequency operation leads to use of small magnetic fields, relatively large cavities, and large samples. The latter is necessary as well as possible since the power output of an inverted spin system is given by Eq. (35) (Chap. 14)

$$P_M = (n_2 - n_1)h\nu W_{12}$$

The output then is proportional to frequency, and in order to obtain reasonable gain at low frequency the value of $n_2 - n_1$, the population difference, must be correspondingly increased.

Figure 8 illustrates a type of cavity used by both R. H. Kingston[14] and G. K. Wessel[15] in the vicinity of 400 Mc. The details of operation are given in Table 5. The operation is not very dependent on the orientation of the crystal with respect to the magnetic field, and in addition, Wessel found that the signal could be tuned from 380 to 450 Mc without adjustment of magnetic field or pump. The use of the lumped-constant circuit for the signal results in a very good filling factor. The high Q for this circuit is due to the use of a lead plating on the coils. At the temperatures and fields used lead provides a superconducting, very low loss surface. Use of this as a device to lower loss is more difficult the higher the frequency and the larger the magnetic field required, although success with high-field superconductors indicates that the latter restriction may be solvable. The fields required at this frequency can easily be provided by small permanent magnets with variable-gap spacing used to adjust the field magnitude.

29-12   AMPLIFIER CIRCUITS

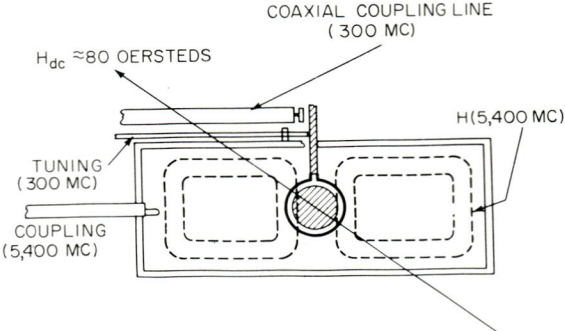

FIG. 8. Cavity arrangement for a maser operating in the 300- to 500-Mc region. The signal circuit is of lumped-constant design with a superconducting coating. (*After Kingston*,[14b] *published by permission of the editor, IEEE*.)

Table 5

Host crystal................	$K_3Co(CN)_6$	Al_2O_3
Paramagnetic impurity........	0.5% Cr^{3+}	0.05% Cr^{3+}
Signal frequency, Mc.........	300	380–450
Pump frequency, Mc.........	5,120	11,800
Magnetic field, oersteds......	50–100	80
Pump mode.................	TE_{112}	TE_{011} (dielectrically loaded with Teflon)
Pump power, watts..........	0.010	
Signal circuit Q.............	10,000	
Signal filling factor, %.......	50	
Crystal volume, cu cm........	....	4
Gain, db....................	10	15
Bandwidth, kc...............	100	100
Operating temperature, °K....	1.6	1.7

3.2. Reflection-cavity Masers for L band. The maser has been extremely useful for radio-astronomy purposes, and a number of circuits have been used to investigate the 1,420-Mc hydrogen line. Since the cavity arrangements have been different, several of these devices are illustrated.

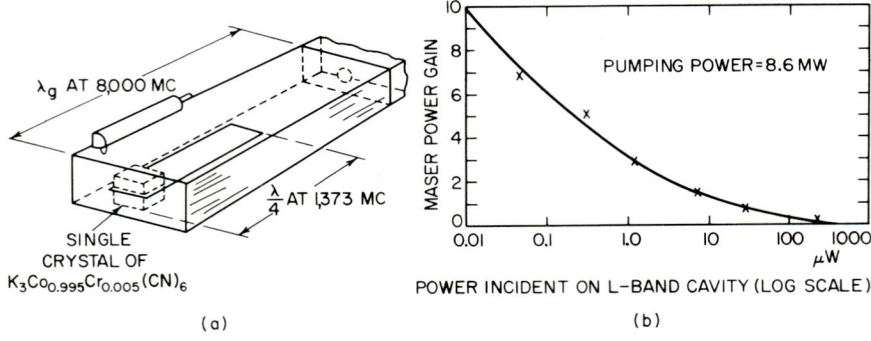

FIG. 9. (*a*) Schematic of 1,373-Mc maser. (*After Artman, Bloembergen, and Shapiro*,[16] *published by permission of the editor, Phys. Rev.*) (*b*) Saturation characteristics of Artman, Bloembergen, and Shapiro[16] maser. (*Published by permission of the editor, Phys. Rev.*)

Figure 9a illustrates the earliest of these amplifiers. Built by Artman, Bloembergen, and Shapiro,[16] it uses 0.5 per cent Cr^{3-} in $K_3Co(CN)_6$ with a strip-line resonator for the signal. Oscillation occurred at pump powers of 9 mw, and the saturation properties at 8.6 mw of pump power are shown in Fig. 9b. The decrease of gain with increasing signal level, as illustrated in this figure, is due to the variation of the population difference giving rise to the amplification. Such behavior is typical of masers and is a result of the nonfulfillment of the condition for linear amplification; i.e., the inverted population must remain essentially constant [Chap. 14, Eq. (35)]. Comparison with other L-band masers is given in Table 6.

Autler and McAvoy[17] used a completely different type of cavity (Fig. 10) to obtain the characteristics given in Table 6.

The most recent, and also the first operational, design is that of Jelley and Cooper[18] illustrated in Fig. 11. The active material is ruby operated at $\theta = 90°$. The pump cavity is actually a piece of Ku-band waveguide which, because of the dielectric loading, is quite satisfactory at the pump frequency of 11.28 Gc. The pump resonator frequency is adjusted by varying the position of the movable iris. The signal is coupled to the $\lambda/4$ resonant strip line by the loop at the end of the signal coax. This strip-line resonator is bounded by ruby (on the broad surfaces) for about three quarters of its length. Signal tuning is accomplished by adjusting the position of the L-band (Teflon) tuning slug. In practice a magnetic Q_M of -180 is obtained, giving a gain of 20 db at a bandwidth of 2 Mc. Magnetic field orientation is critical in that a variation of 0.5° in θ causes a gain change of 3 db. Rigidity in mounting is a necessity. The temperature used is 4.2°K because of the trouble caused by having to pump the liquid helium in order to reduce the operating temperature.

This maser has been operated with an automatic gain control (Fig. 12), in which

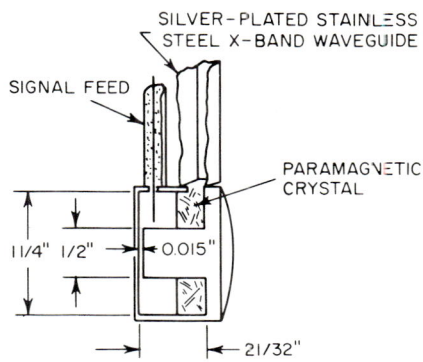

Fig. 10. Cavity schematic of 1,382-Mc maser. (*After Autler and McAvoy,*[17] *published by permission of the editor, Phys. Rev.*)

Table 6

	Artman, Bloembergen, Shapiro[16]	Autler, McAvoy[17]	Jelley, Cooper[8]
Host crystal	$K_3Co(CN)_6$	$K_3Co(CN)_6$	Al_2O_3
Paramagnetic impurity	0.5% Cr^{3+}	0.5% Cr^{3+}	0.05% Cr^{3+}
Signal frequency, Mc	1,373	1,382	1,420
Pump frequency, Mc	8,000	9,070	11,270
Magnetic field, gauss	....	1,200	2,000
Pump mode	TE_{102}	....	TE_{012}
Pump Q	5,000		
Pump power, mw	9	28	
Signal mode	$\lambda/4$ strip resonator	....	$\lambda/4$ strip resonator
Signal Q	2,000		
Crystal volume cu cm	1		
Filling factor	....	0.5	
Gain, db	....	....	20
Bandwidth, Mc	....	....	2
$G\frac{1}{2}B$	....	1.85×10^6 Mc	20×10^6 Mc
Operating temp, °K	<2	1.25	4.2

29-14 AMPLIFIER CIRCUITS

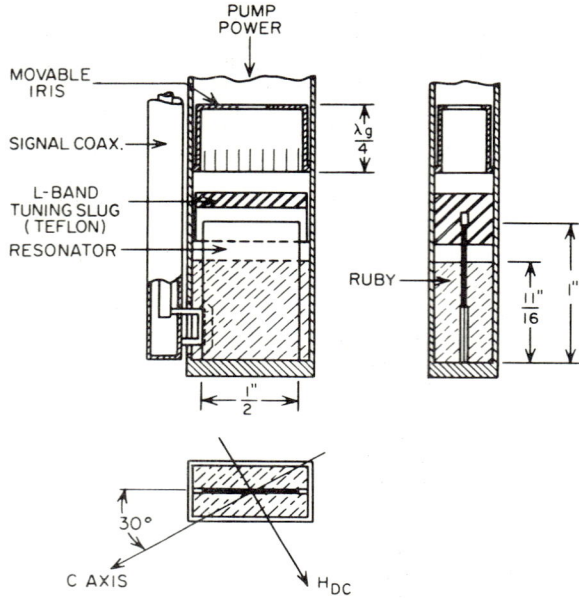

Fig. 11. Cavity for L-band radio-astronomy maser. (*After Jelley and Cooper,*[18] *published by permission of the editor, Rev. Sci. Instr.*)

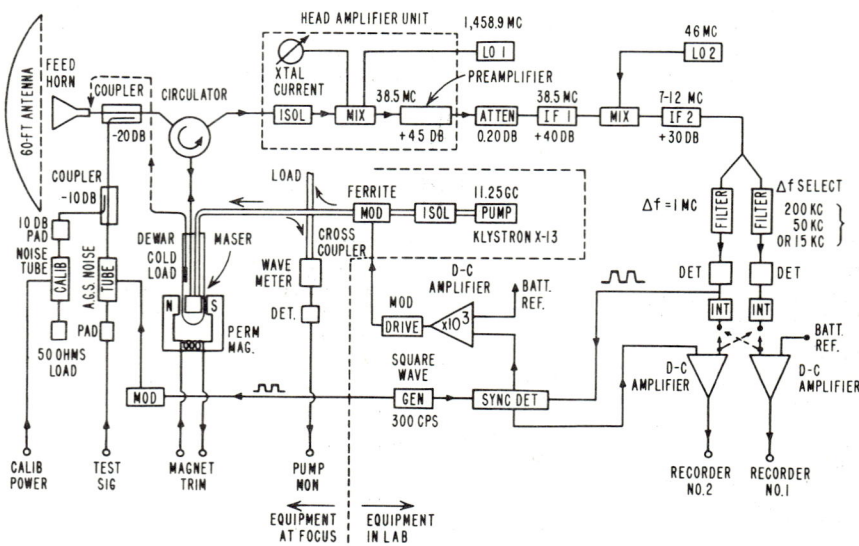

Fig. 12. Block diagram of radiometer using maser with automatic gain control.[18] (*Published by permission of the editor, Rev. Sci. Instr.*)

INDUCED-EMISSION AMPLIFIERS AND OSCILLATORS 29-15

case the maser is operated about 2 db from saturation and the pump power is controlled by the ferrite modulator. A modulated noise tube at the input gives rise to a pulsed noise output which when synchronously detected, results in the control signal for the ferrite modulator. The noise sources for this mode of operation are compared with the more recent circuit in Table 7. The automatic gain control adds 63°K to the noise.

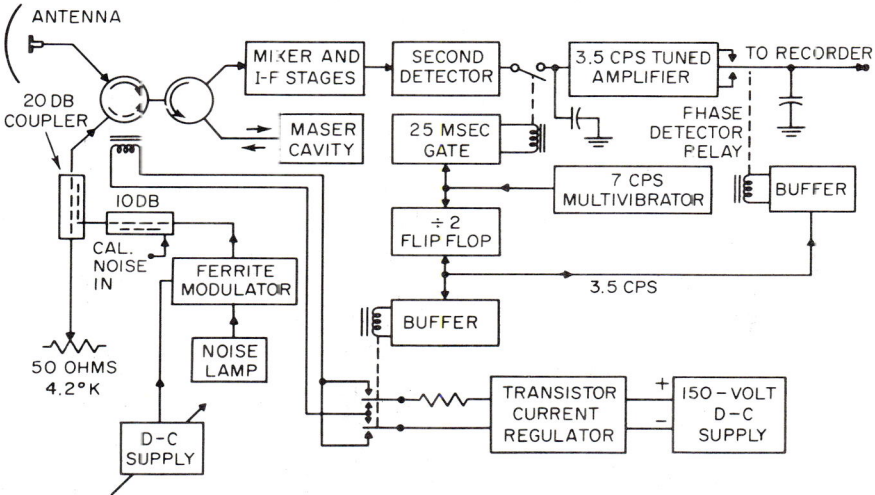

FIG. 13. Block diagram of maser used with Dicke radiometer. (*After Cooper,*[19] *published by permission of the editor, Rev. Sci. Instr.*)

The essential elements of the more recent Dicke radiometer comparison system[19] are given in block-diagram form in Fig. 13, and its noise characteristics are in Table 7. Without the maser the system noise was 1000°K; so the maser results in an improvement of a factor of 10.

The magnetic field is provided by a permanent magnet with trimming coils to give ±100 gauss variation. The weight of the system is of interest in this case and is

Table 7

	Automatic-gain-control system	Dicke radiometer system
Antenna spillover, °K............................	20	20
Input coax, °K..................................	<4	
Input coupler for AGC system (0.07 db), °K......	5	
Switched circulator, °K..........................	...	25
Maser circulator, °K.............................	25	25
Maser input coax (0.1 db), °K...................	7	
Spontaneous emission, °K........................	2	2
Noise from AGC system, °K.....................	63	
Second-stage noise (1000°K with 20-db maser gain), °K...	10	10
	136	97

broken down into

	Lb
Cryostat	35
Magnet	60
Circulator	9
Maser	2
Pump circuit and waveguides	10
Motors for tuning and coupling	10
Other	75

The stainless-steel cryostat has nitrogen capacity of 6.2 liters and helium capacity of 3.3 liters, which last for 30 and 16 hr, respectively.

Another maser,[20] designed for radio astronomy and space communications, is illustrated in Fig. 14. Ruby is used as the active material, with the magnetic field oriented at 90° to the c axis. Amplification was obtained over the frequency range of 900 to 3,000 Mc with pump frequencies of from 10.5 to 13 Gc. The dual-purpose input coax is unusual. The signal is used in the TEM coaxial mode while the pump is transmitted in the TE_{11} waveguide mode in the same coaxial line. The use of different modes makes separation of the pump power at the circulator possible. The center conductor of the coax is movable in order to make the coupling adjustable.

The $G^{\frac{1}{2}}B$ product is 25 Mc, i.e., 2.5 Mc at 20 db gain. The noise temperature at the circulator input is 25 to 30°K. An interesting feature of this system is that a parametric amplifier is used immediately following the maser.

3.3. S-band Reflection Masers. The reflection-cavity maser constructed for use at 2,800 Mc by McWhorter and Meyer[21] is illustrated in Fig. 15. The cavity is resonant to the pump frequency of 9,400 Mc in the TE_{113} mode while it is one-half wavelength long in the TEM coaxial mode at the signal frequency. The loop coupling from coaxial line to the cavity at 2,800 Mc is adjustable by rotation or change of depth of immersion. Operation was at 1.25°K with 0.5 per cent Cr^{3+} in $K_3Co(CN)_6$. The energy levels are illustrated in Fig. 16.

Amplifier characteristics are shown in Fig. 17 for various saturating powers at two different values of coupling or external Q at the signal frequency. The higher external Q is represented by the G-1, B-1 curves indicating higher gain and smaller bandwidth. With 10 per cent filling factor, the voltage gain bandwidth $G^{\frac{1}{2}}B$ was found to be 1.8×10^6 sec^{-1} as against the theoretical value of 2.6×10^6 sec^{-1}.

The properties as an oscillator are illustrated by Fig. 18. The maximum efficiency $P(2,800 \text{ Mc})/P(9,400 \text{ Mc})$ is 0.14 per cent at about 1 mw of pump power.

The microwave instrumentation is given in block-diagram form in Fig. 19.

3.4. A 6-Gc Maser. Figure 20 illustrates the configuration used by Scovil, Feher, and Seidel[22,23,24] operating with a pump frequency of 11.5 Gc and a signal transition

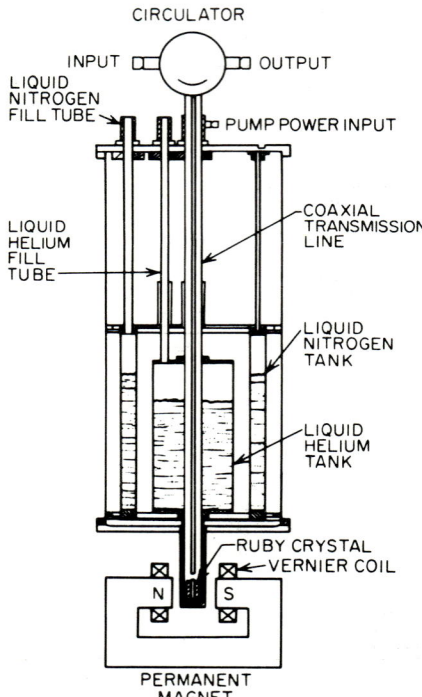

Fig. 14. Cross-section view of maser and cryostat. Double-duty coaxial input is noteworthy. (*After Higa*,[20] *published by permission of the editor, Solid-State Electron.*)

at 6 Gc. The $-\frac{1}{2}$ to $-\frac{5}{2}$ transition of Gd^{3+} in lanthanum ethyl sulfate with the magnetic field oriented at 90° to the crystal symmetry axis is used for the pump, while the $-\frac{3}{2}$ to $-\frac{5}{2}$ transition serves for the signal. This is the system mentioned in Chap. 14, Sec. 6.5, in which Ce^{3+} is used to modify the relaxation time of the $-\frac{3}{2}$ to $-\frac{1}{2}$ transition.

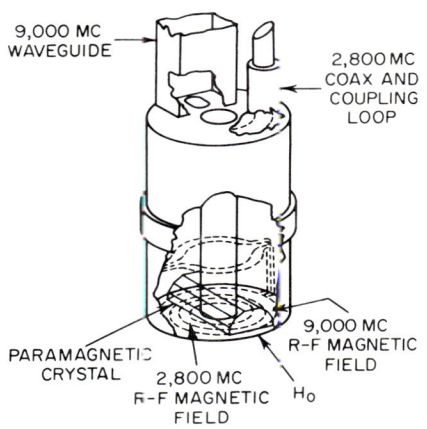

Fig. 15. S-band maser. (After McWhorter and Meyer,[21] published by permission of the editor, Phys. Rev.)

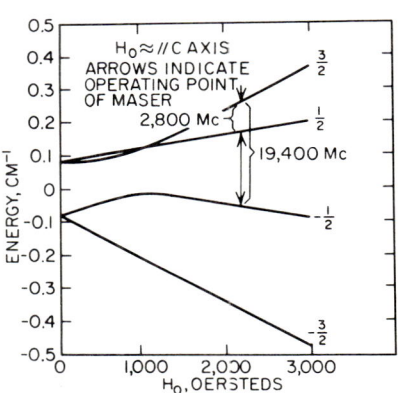

Fig. 16. Energy levels of $K_3Co(CN)_6$ showing operating conditions[21] for maser of Fig. 15. (Published by permission of the editor, Phys. Rev.)

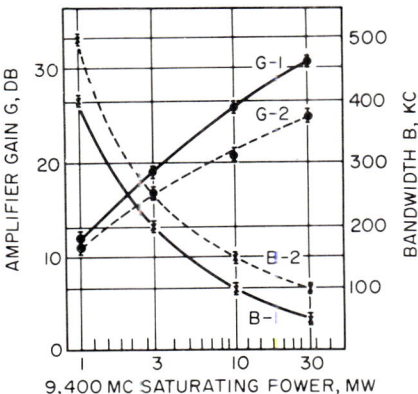

Fig. 17. Gain-bandwidth characteristics of McWhorter and Meyer maser[21] as a function of pump power at two values of coupling. G-1 and B-1 represent the behavior at the higher value of external Q (smaller coupling coefficient). (Published by permission of the editor, Phys. Rev.)

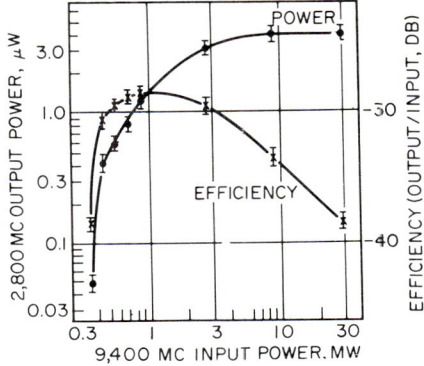

Fig. 18. Oscillator characteristics of McWhorter and Meyer maser.[21] (Published by permission of the editor, Phys. Rev.)

Operation at a signal frequency of 9 Gc with the same type of cavity has also been reported[22,23] with a pump frequency of 17.5 Gc.

The pump power is transmitted through the waveguide to the $\frac{3}{4}\lambda$ cavity. The signal energy travels on the strip line to the $\frac{1}{2}\lambda$ resonant length situated in the pump

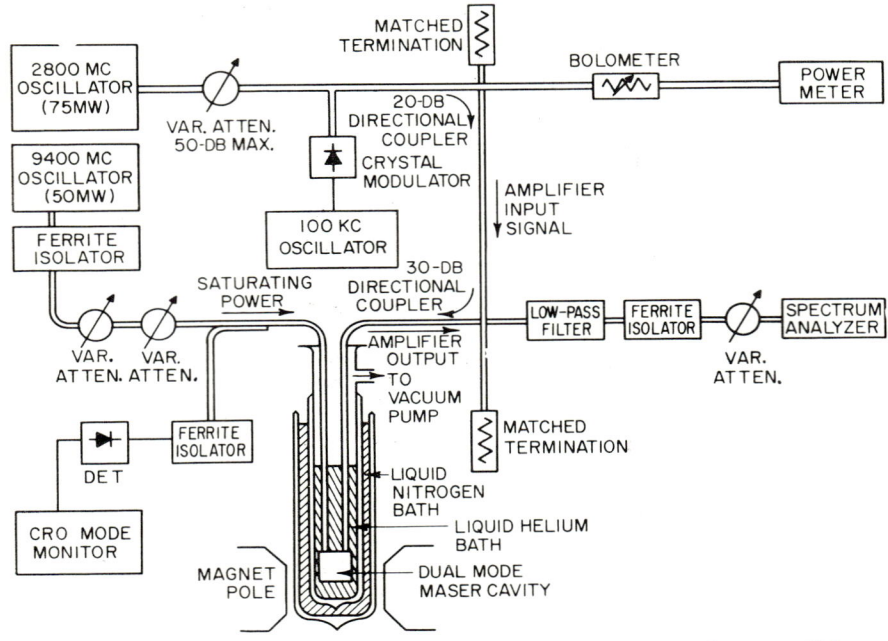

Fig. 19. Block diagram of microwave instrumentation used with the McWhorter and Meyer maser.[21] For test purposes the directional coupler and the isolator replace the circulator; to get comparable sensitivity the coupler must be 3 db. (*Published by permission of the editor, Phys. Rev.*)

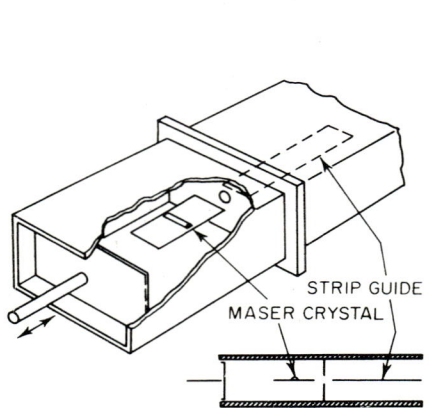

Fig. 20. A maser cavity arrangement used at 6 Gc and at 9 Gc. The signal resonant circuit and transmission line consist of the strip-line sections while the pump utilizes the tunable cavity. (*After Scovil,*[24] *published by permission of the editor, IEEE.*)

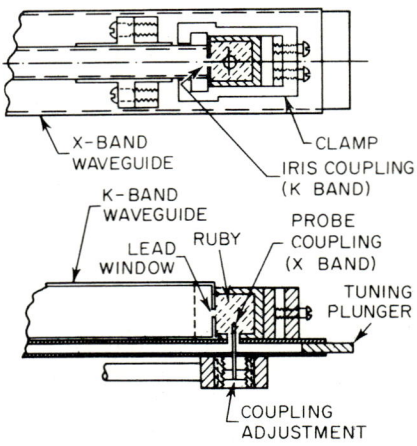

Fig. 21. X-band maser cavity. (*After Giordmaine, Alsop, Mayer, and Townes,*[25] *published by permission of the editor, IEEE.*)

cavity. The crystal is situated in the high-field regions of both signal and pump fields. The high-frequency version had an oscillator output at 9 Gc of 15 µw at a pump power of 200 mw. The low-frequency device operated as an amplifier with 20-db gain at 100-kc bandwidth with a pump power input of 38 mw. This was not optimized, and more efficient operation is certainly possible. The system noise temperature was 150°K with 35°K contributed by the maser itself (operating at a spin temperature of −4°), the remainder coming from the circulator and other components at room temperature.

3.5. X-band Reflection-cavity Masers. Ruby is the most common of the materials for maser use, and both the X-band systems outlined below follow the pattern.

The cavity arrangement of Fig. 21 is one of the versions used on the Naval Research Laboratory radio telescope in Washington, D.C., designed and built by Giordmaine, Alsop, Mayer, and Townes.[25] The characteristics are given in Table 8. The density of modes at K band in the ruby-filled cavity is high, and no attempt has been made to identify the pump modes. The coupling and tuning of the signal mode are accomplished as indicated in the sketch with the added feature of being able to move the ruby with respect to an unshown dielectric spacer in the cavity.

Table 8

	Giordmaine et al.[25]	Cook et al.[26]
$G^{\frac{1}{2}}B$, Mc	50	≃300
Magnetic Q, Q_M	−380	−200
Signal frequency, Gc	9,500	8,700
Pump frequency, Gc	23.5	22.3
Orientation, deg	55	55
H_{DC}	3,740	3,850
Q_0	7,000	≃4,000
Temp, °K	1.4	4.2
Q_{pump}	4,000	
Pump power, mw	30	
Signal mode	TE_{101}	TE_{111}
Cavity volume, cu cm	0.48	1.55
Crystal volume, cu cm	0.35	1.55
$N(Cr^{3+}$ ions)	8×10^{18}	3.5×10^{19}
Filling factor	0.8	1.0

Two types of problems have been encountered. The gain is affected by the output impedance to the extent that for a 20-db maser gain a voltage-reflection coefficient change of 1 per cent causes a gain variation of ±0.8 db. This is the limiting factor in gain stability, although obviously variation of magnetic field or certain mechanical motions such as the movement of the crystal with respect to the magnetic field can cause trouble. A ferrite circulator has been used to insulate the maser from this problem. The signal waveguide is filled with polystyrene foam, displacing the liquid helium, in order to avoid reflections from the surface of the refrigerant liquid, whose level changes with time can lead to gain variations.

The pumping arrangement is push-pull at the symmetric $\theta = 55°$ angle for the pink (0.05 per cent Cr) ruby used.

At higher frequencies the use of dielectric cavities has become common practice. Figure 22 shows the ruby cavity used by Cook, Cross, Bair, and Terhune[26] on the University of Michigan radio telescope. The cavity walls consist of three thin coats of silver paint (Hanovia No. 32A), each coat being baked for 30 min at 700°C before the next is applied. The cavity Q so obtained is about 4,000. Operation is in the push-pull pump mode with $\theta = 54.7°$.

29-20 AMPLIFIER CIRCUITS

The use of a coupling cavity between the waveguide and the ruby signal cavity gives this maser the gain vs. frequency response shown in Fig. 23. The width of the coupling cavity is typically 250 Mc, and the separation $\Delta\nu_k$ between peaks of the gain curve is given by $\Delta\nu_k = k\nu_0$ where k is the coupling coefficient between the two cavities. In practice this is adjustable. The cavity operates in a semicoaxial mode, and the two silver pins in the Teflon disk change the frequency of the coupling cavity. The

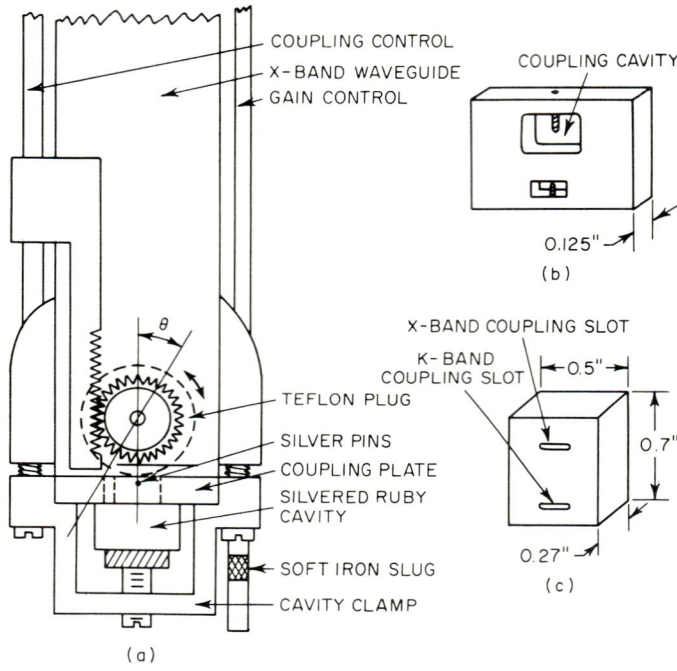

FIG. 22. X-band maser. (a) Assembly. (b) Coupling plate including cavities. (c) Silver-plated ruby cavity. (*After Cook, Cross, Bair, and Terhune,*[26] *published by permission of the editor, IEEE.*)

voltage-gain–bandwidth product for a coupled-cavity arrangement such as this is given by

$$(G^{\frac{1}{2}} - 1)B = 2B_b G^{\frac{1}{4}} \qquad (18)$$

Using gains of 20 to 40 db, voltage-gain–bandwidth products of 200 to 550 Mc have been obtained. The high gains lead to gain instabilities given approximately by

$$\delta G/G = G^{\frac{1}{2}}(\delta x/x)$$

(for high gains), where x may be magnetic bandwidth, coupling bandwidth, or VSWR. Since at 30-db gain this gives a multiplying factor of 32 on any of these variations, high-gain operation is avoided. The gain of the system is adjustable during operation by varying the position of the (0.125 in. diameter by 0.150 in. long) soft-iron slug. This perturbs the magnetic field, changes its homogeneity, broadens the resonance line, and decreases the effective number of spins. Gain can be varied from 10 to 30 db in this manner. This method is an alternative to changing the cavity coupling and is said to have some stability advantages. Both signal and pump waveguides are filled with polystyrene foam to eliminate problems due to changing liquid-helium level. Some of the properties of this maser are listed in Table 8.

3.6. CW Millimeter-wave Masers. The large zero-field splitting of both Cr^{3+} and Fe^{3+} in TiO_2 (rutile) makes possible millimeter-wave masers with moderate magnetic fields. Foner and Momo[27] and Carter[28] have used Fe^{3+} in rutile while Gerritsen and Lewis[29] make use of Cr^{3+} doped rutile for millimeter-wave amplification. The highest-frequency operation was obtained by Carter, and therefore his system will be outlined.

Rutile has a dielectric constant of about 100; so that even a small piece is equivalent to a reasonably large cavity. In all cases a dielectric cavity as discussed by Okaya[30] is used in a high-order mode. Crystals, of a size such that the mode separation is 5 to 20 Mc, are supported in polyfoam in a copper box 0.6 in. on a side. The pump and signal leads are 0.074×0.148 (M-band) waveguides. The coupling is varied by changing the size of the box, whose bottom is in the form of a movable plunger. Pump power at 78.2 Gc was supplied by a Raytheon QKK 866 reflex klystron with an output of 50 mw. Push-pull pumping was used as illustrated in the energy-level diagram of Fig. 24. The resulting signal frequency is 49 to 57 Gc. The signal frequency could be varied by changing magnetic field magnitude since the pump levels

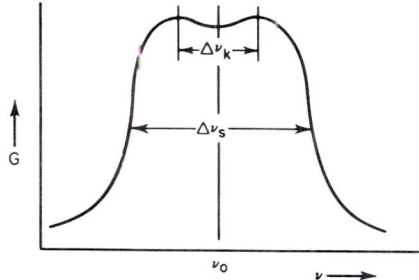
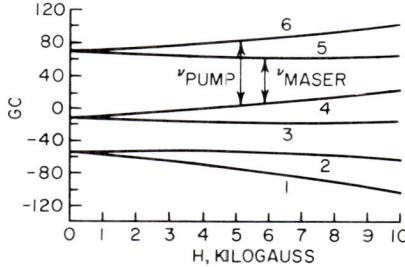

Fig. 23. Gain vs. frequency curve[26] for maser of Fig. 21. (*Published by permission of the editor, IEEE.*)

Fig. 24. Energy levels of Fe^{3+} in TiO_2, as a function of magnetic field for the field pointing 73° from the [110] direction in the (001) plane. Pump and signal frequencies and transitions are indicated.

maintain the same separation to within ½ per cent from 5,500 to 7,200 gauss. The available pump power was insufficient for crystals of 0.15 cu cm at 4.2°K but was large enough to operate at 1.6°K. Oscillations were observed throughout the signal frequency range.

Cross-relaxation effects in the 10-energy-level system were found to be unimportant at the concentration of 0.12 per cent Fe^{3+} used.

Foner and Momo[27] measured root-gain bandwidth for the same type of system at a signal frequency of 26 to 39 Gc, pump powers of 2 to 10 mw at about 70 Gc, and crystal size of 0.26 by 0.15 by 0.35 in. $G^{\frac{1}{2}}B$ was found to be between 10 and 40×10^6 cps.

3.7. Masers Using an Optical Pump.[31] As discussed in Chap. 14, Sec. 7.2, it is possible to pump the ground state of ruby by using a ruby laser as the pump. The energy levels at the operating point are given in Fig. 11 of Chap. 14. The experimental arrangement is illustrated in Fig. 25. The cavity is a section of 0.050- by 0.130-in. waveguide loaded with a piece of ruby 0.78 in. long. The dielectrically loaded cavity operates in the TE_{011} mode with a Q_0 of about 1,600 and filling factor of 0.96. Both amplification and oscillation were obtained. The high pump frequency would be expected to make high-temperature operation possible.

3.8. High-temperature Operation of Masers. Only the problem of maintaining sufficient inverted population needs to be solved to enable masers to operate at temperatures above the typical helium range. Szabo[32] has operated ruby at 77°K by using a laser pump in a fashion similar to that described above. The achieved inverted population corresponded to a spin temperature of −13°K.

29–22   AMPLIFIER CIRCUITS

Operation of ruby masers in the region of liquid-nitrogen temperature has also been achieved by using push-pull pumping.[33,34] The performance characteristics are not so good as those obtained at lower temperatures. Using 0.2 per cent concentration, Maiman[34] obtained a gain-bandwidth product of 14 Mc and an effective noise temperature of 100°K at an operating temperature of 77°K. Maser action up to 195°K was reported at the same time.

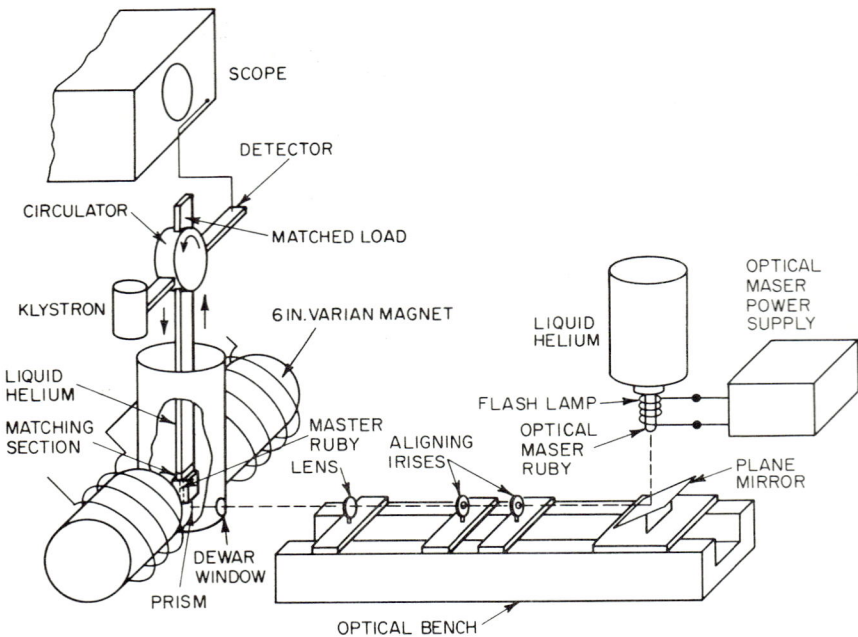

Fig. 25. Apparatus used for laser pumping of ruby maser. (*After Devor, D'Haenens, and Asawa,*[31] *published by permission of the editor, Phys. Rev. Letters.*)

4. TRAVELING-WAVE MASERS

The properties of structures to transmit electromagnetic energy at a velocity less than the free-space velocity, i.e., slow-wave structures, have been studied extensively,[35] particularly in connection with their use in traveling-wave tubes and accelerators. These analyses are frequently primarily concerned with the phase velocity, and it must be borne in mind that the concern in traveling-wave masers is the group velocity and its variation as a function of frequency. A second factor of major importance for the traveling-wave maser is the region of circular polarization. Spin systems respond to only one circularly polarized component; the other oppositely rotating one has negligible effect. It is essential to the efficient operation of a traveling-wave maser that areas of quite pure circularly polarized magnetic fields exist in a plane perpendicular to the applied d-c magnetic field (assuming the signal to operate in a $\Delta m = \pm 1$ transition). Because of the reciprocal nature of these fields—the direction of polarization reverses with the direction of propagation—it is possible to locate the amplifying material so that the amplification is essentially unidirectional (forward) and at the same time place magnetic materials in the other regions of polarization to give almost unidirectional backward loss. The forward-directed gain and oppositely directed loss are most helpful.

INDUCED-EMISSION AMPLIFIERS AND OSCILLATORS **29-23**

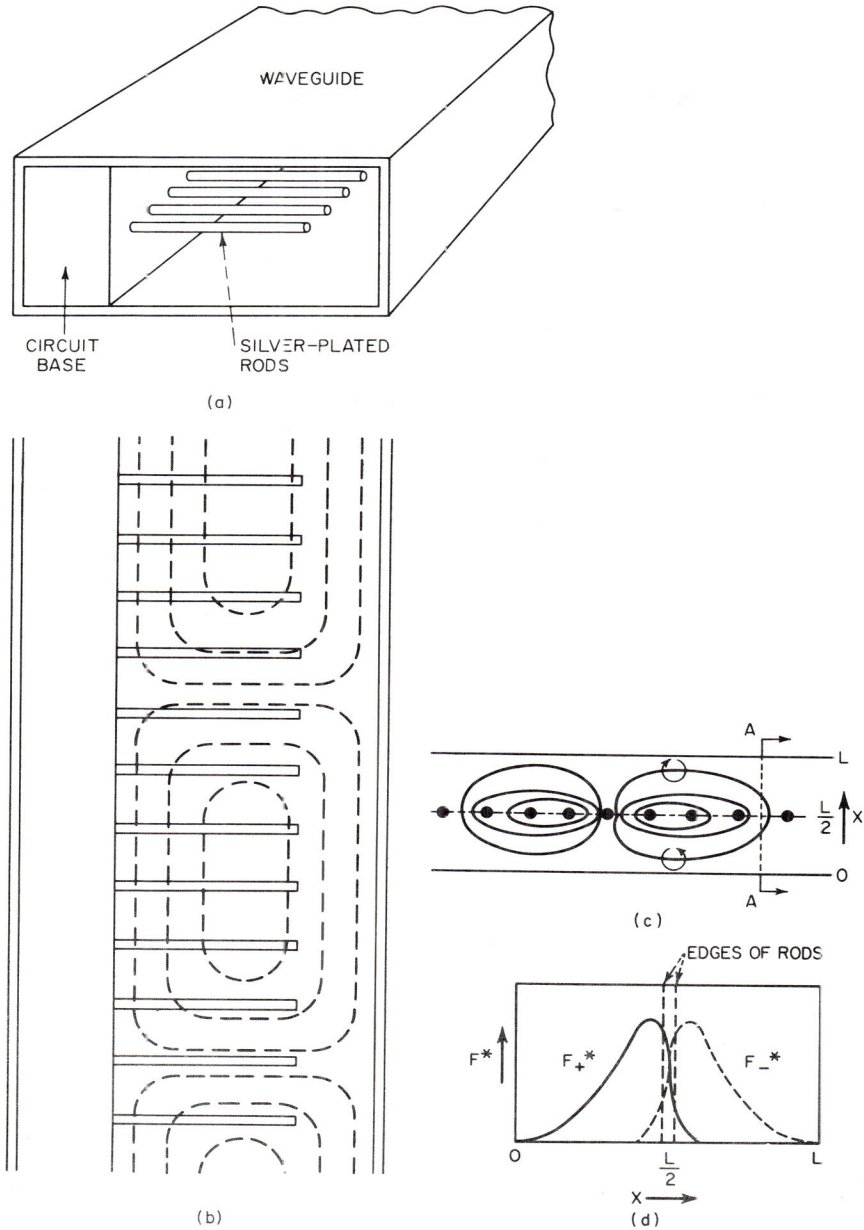

FIG. 26. The comb structure. (a) General view. (b) View from the top showing the pump magnetic field. (c) Side view showing the signal magnetic field lines and the positions of the circularly polarized components. (d) The filling factors or selective amplitudes as a function of position in the guide. (a, c, and d reprinted by permission of the copyright owner American Telephone and Telegraph Company, and the authors. This article originally appeared[36] in Bell System Tech. J., vol. 38, March, 1959.)

One further requirement must be recognized. The structure must allow the pump power to propagate with a field distribution conducive to saturating the pump transition. The propagation need only be marginal, as it is in some accelerator cavities, if operation is in a resonant mode.

4.1. The Comb Structure. This structure, illustrated in Fig. 26a, has relatively ideal properties for this purpose. It consists of cylindrical rods projecting from the narrow side of the waveguide. The pump is transmitted in the conventional TE_{01} mode, and the rods, being perpendicular to the electric field, perturb the flow of pump energy very little so that either a one-pass (continuous-flow) pump or a resonant pump is possible. The field configurations are illustrated in Fig. 26b, c, and d. The signal fields as shown in c have areas of circularly polarized magnetic fields situated at the base and to either side of the rods as indicated. The intensity of the circular polarizations is illustrated by the data of Fig. 26d from De Grasse, Schulz-DuBois, and Scovil,[36] whose treatment of traveling-wave masers we shall generally follow. Consideration of the Poynting vector shows that propagation down the structure at signal frequency depends on the fringing fields at the free ends of the posts.

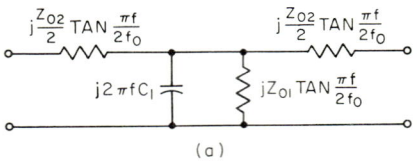

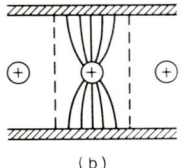

(b)

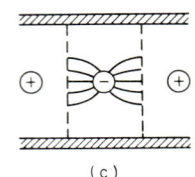

(c)

FIG. 27. The comb structure. (a) Equivalent circuit. C_1 represents the fringe capacitance at the ends of the fingers. (b) Electric field configuration at low-frequency cutoff. (c) At high-frequency cutoff. (Reprinted by permission of the copyright owner, American Telephone and Telegraph Company, and the authors. This article originally appeared[36] in Bell System Tech. J., vol. 38, March, 1959.)

An equivalent circuit for the comb structure is shown in Fig. 27 along with the electric field configurations for the low- and high-frequency cutoffs, with impedances designated as Z_{++} and Z_{+-}, respectively. These cutoff impedances, which may be experimentally determined from tank tests or analytically, are related to the Z_{01} and Z_{02} of the equivalent circuit as follows:

$$Z_{01} = Z_{++}$$
$$Z_{02} = \frac{4Z_{+-}}{1 - Z_{+-}/Z_{++}} \tag{19}$$

The properties of the circuit are determined from the equation

$$\frac{1}{j2\pi\nu C_1}\left(\frac{1}{Z_{01}} + \frac{2}{Z_{02}}\sin^2\frac{\phi}{2}\right) = -\left(\tan\frac{\pi\nu}{2\nu_0}\right)^{-1} \tag{20}$$

giving the relationship between ϕ, the phase shift per section, and the frequency ν. ν_0 is the frequency at which the posts are $\lambda/4$ long. Equation (20) may be rewritten as

$$\frac{1}{2\pi\nu_0 C_1}\left(\frac{1}{Z_{01}} + \frac{2}{Z_{02}}\sin\frac{2\phi}{2}\right) = -\frac{j\nu}{\nu_0}\left(\tan\frac{\pi\nu}{2\nu_0}\right)^{-1} \tag{21}$$

or
$$X_{c0}Y(\phi) = f(\nu/\nu_0) \tag{22}$$

in which form Fig. 28 is plotted. Solution of (20) then is accomplished by calculating the left-hand side of Eq. (21) and determining the value of ν/ν_0 from Fig. 28. Coupling between nonadjacent posts has been neglected, but nevertheless reasonable agreement between the calculated and experimental ϕ versus ν curve is obtained.

The factors ZL^* needed to evaluate the traveling-wave maser can be obtained from

* See Chap. 14, Sec. 13. Z, the slowing factor, is defined as the velocity of light divided by the group velocity (c/v_g), and L is the length of the structure measured in free-space wavelengths ($l\nu/c$).

the measured curve of ϕ vs. frequency. Using the definition of group velocity

$$v_g = \frac{\partial \omega}{\partial k} = \frac{\partial \nu}{\partial(1/\lambda)}$$

the phase shift per section is

$$\phi = 2\pi\nu(l_s/v_{ph})$$
$$\partial\phi/\partial\nu = 2\pi l_s(\partial/\partial\nu)(\nu/v_{ph}) \quad (23)$$

and
$$ZL = l\nu/v_g = l\nu(\partial/\partial\nu)(1/\lambda) = l\nu(\partial/\partial\nu)(\nu/v_{ph})$$
Therefore
$$ZL = (l/l_s)(\nu/2\pi)(\partial\phi/\partial\nu) = N_S(\nu/2\pi)(\partial\phi/\partial\nu)$$

where l_s = length of a section
N_S = number of sections

Thus the properties of the structure may be derived from ϕ versus ν plots such as Fig. 30.

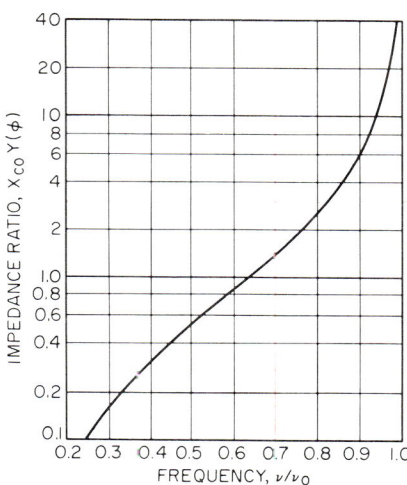

FIG. 28. The curve of $X_{co}Y(\varphi)$ versus ν/ν_0 used to solve Eq. (22). (Reprinted by permission of the copyright owner, American Telephone and Telegraph Company, and the authors. This article originally appeared[36] in Bell System. Tech. J., vol. 38, March, 1959.)

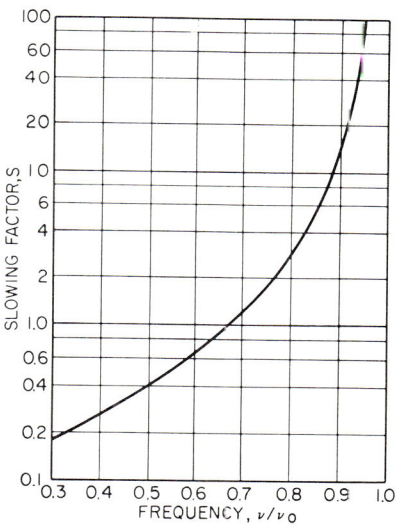

FIG. 29. The slowing factor $s = ZLX_{co} \sin \varphi / N_s Z_{02}$ as a function of ν/ν_0. Reprinted by permission of the copyright owner, American Telephone and Telegraph Company, and the authors. This article originally appeared[36] in Bell System Tech. J., vol. 38, March, 1959.)

From the theoretical point of view the curve of Fig. 29 may be used to determine ZL from the definition of the slowing factor

$$S = \frac{ZLX_{e0} \sin \phi}{N_s Z_{02}}$$

plotted there against frequency and the values of ϕ, X_{c0}, and Z_{02} used to determine ν/ν_0 from Fig. 27.

The curves of Fig. 30 illustrate the comparison of theoretical phase-shift curves with those determined experimentally. Similar results achieved with the ruby traveling-wave maser are illustrated in Sec. 4.3.

4.2. A Traveling-wave Maser for 3.0 Gc.[37] A maser using ruby as the amplifying material and yttrium iron garnet as the isolator with the comb structure is illustrated in Fig. 31. The ruby and YIG are situated in areas of opposite circular polarization on either side of the posts. The overall length is 12 cm with a slowing factor of 150 and a passband of 150 Mc. The YIG disks with aspect ratio (thickness/diameter ratio) of 0.1 introduce a 4-db forward loss along with the 60-db backward attenuation.

The pink ruby (0.036 atom per cent Cr^{3+} to Al^{3+} ratio) is used at a 90° orientation. The gain, bandwidth, and pumping conditions are given in Table 9, in

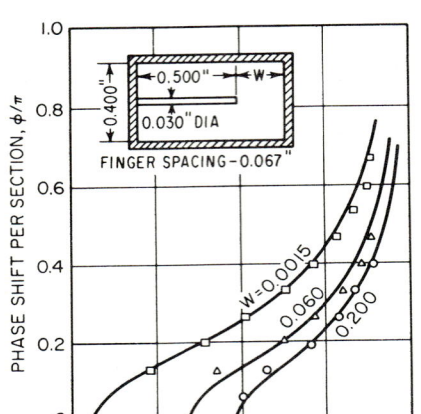

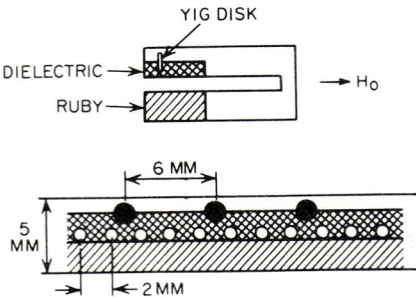

FIG. 30. Experimental points and theoretical curves for φ versus ν in comb structure shown. (Reprinted by permission of the copyright owner, American Telephone and Telegraph Company, and the authors. This article originally appeared[36] in Bell System Tech. J., vol. 38, March, 1959.)

FIG. 31. Schematic of traveling-wave maser for 3,000 Mc. (After Walling.[37] Figure reproduced by courtesy of Mullard Research Laboratories and the editor, Solid-State Electron.)

which the advantage of the higher pump frequency, in providing better inversion ratios and gain, is very evident. The operating temperature is 1.4°K with pump powers of tens of milliwatts, operating in a nonresonant pump mode.

Table 9

Pump	Pump frequency, Mc	Gain, db	Bandwidth, Mc	Inversion
1–3	13,900	29	17	2.7
1–3, 3–4	13,900, 12,520	31	16	2.9
1–4	26,420	47	12	4.4

The assumption that pure circularly polarized radio frequency is the optimum for both the forward amplification and backward attenuation is based on an idealized spin system responding only to $\Delta m = \pm 1$ transitions. In ruby, with a crystalline axis of trigonal symmetry as well as the magnetic field, only in the special situation of H_0 aligned along the c axis will pure $\Delta m = \pm 1$ transitions occur. At other orientations mixtures of $\Delta m = \pm 1$ and $\Delta m = 0$ transitions will be induced in ratios determined by both the orientation and value of the d-c magnetic field H_0. For the case of ruby, if one quantizes with respect to the crystalline c axis, as done by Chang and Siegman,[38] a linear r-f field along the c axis induces $\Delta m = 0$ transitions and one perpendicular to the c axis produces $\Delta m = \pm 1$ transitions. Using the orientation of fields indicated in Fig. 32, the r-f fields will produce both $\Delta m = 0$ and $\Delta m = \pm 1$ transitions with the transition probability

$$[(-2.0957 - j1.3247)H_{rf}]^2 \tag{24}$$

for $H_0 = 3,000$ gauss in the transition (1–2) used. The optimum transition probability is not induced by using a circularly polarized H_{rf} but rather an elliptically polarized field with major and minor axes matching the matrix elements. Both circular polarizations induce transitions, the right-hand circularly polarized field having a transition probability proportional to $(3.4)^2 H_{rf}^2$ and the left circular polarization one proportional to $(0.77)^2 H_{rf}^2$. Depending on the system used the net result is that one may gain by putting the ruby on both sides of the rods, producing gain in both directions, and hence requiring that the isolators, YIG in this case, have sufficient backward attenuation to overcome the round-trip gain. This scheme has been used in traveling-wave masers, and for further information the literature should be consulted.

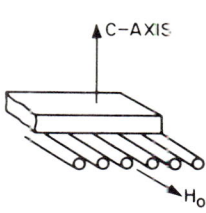

FIG. 32. Details of ruby alignment in TWM of Fig. 31.[37] (*Figure reproduced by courtesy of Mullard Research Laboratories and the editor, Solid-State Electron.*)

4.3. A Traveling-wave Maser for 5.8 Gc.[36] The paramagnetic material used is ruby, 0.05 per cent Cr^{3+} for the amplifying material and 1 per cent Cr^{3+} concentration for the isolators. A cutaway drawing of the structure is reproduced in Fig. 33. The overall length is 5 in., with 62 brass rods 0.4 in. long forming the comb structure. Signal input and output are coaxial, with Fig. 34 illustrating a possible matching arrangement which is quite satisfactory over the passband where ZL is constant.

The phase characteristics of the structure with and without the ruby loading are illustrated in Fig. 35. The slowing was improved by not completely filling the post-to-wall gap with ruby. The slope $d\phi/d\nu$ is constant over about 0.6 Gc, thus giving a constant slowing [Eq. (23)].

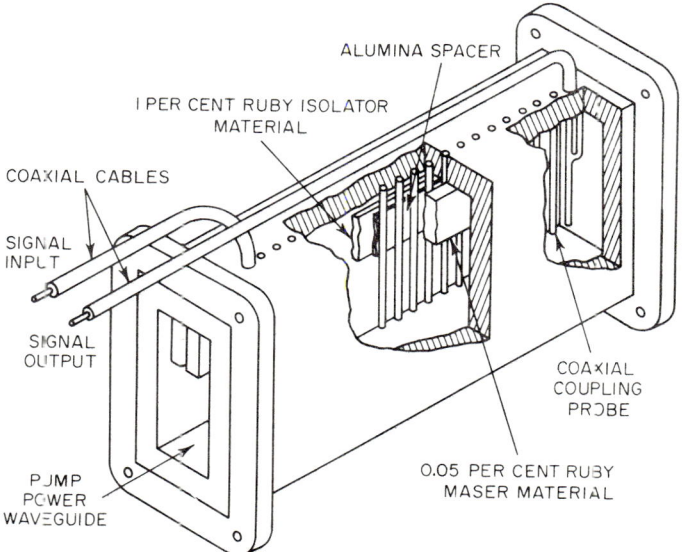

FIG. 33. Cutaway view of De Grasse, Schulz-DuBois, and Scovil traveling-wave maser using ruby. (*Reprinted by permission of the copyright owner, American Telephone and Telegraph Company, and the authors. The article originally appeared*[36] *in Bell System Tech. J., vol. 38, March, 1959.*)

The pump power is used in a resonant mode; so that the flanges of Fig. 33 are attached on one end to a coupling iris and on the other to a sliding short. Pump powers are typically less than 10 mw. The 1 per cent ruby, serving as an isolator because the high concentration makes it impossible to invert the spin system with the pump powers used, is spaced from the rods by a dielectric spacer to put it in a more optimum field. This accounts for the isolator filling factor of 8.5 as compared with 3.5 for the amplifying material.

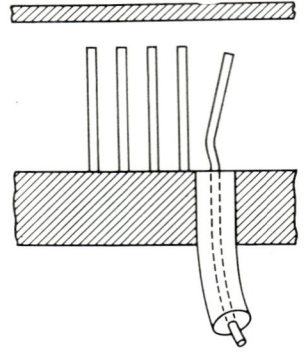

FIG. 34. Coax to comb-structure impedance-matching section. (*Reprinted by permission of the copyright owner, American Telephone and Telegraph Company, and the authors. The article originally appeared[36] in Bell System Tech. J., vol. 38, March, 1959.*)

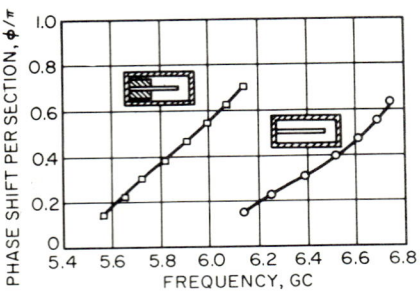

FIG. 35. Phase characteristic of TWM of Fig. 33 with and without the ruby loading. (*Reprinted by permission of the copyright owner, American Telephone and Telegraph Company, and the authors. The article originally appeared[36] in Bell System Tech. J., vol. 38, March, 1959.*)

The electronic gain of the 0.05 per cent material was 30 db while the reverse loss due to the isolator was 35 db. The net result was a 23-db observed forward gain with 29-db reverse loss. The loss of the structure in zero magnetic field was 3 db. Bandwidths of 25 Mc may be tuned over the passband of the structure (350 Mc) by changing magnetic field and pump frequency. Table 10 summarizes the characteristics.

Table 10

Pump	Pump frequency, Gc	Signal frequency, Gc	Gain, db	B, Mc
1–3	18.9–19.5	5.75–6.1	23 13	25 67 (stagger-tuned)

The results of stagger tuning, achieved by varying the orientation of the three ruby pieces used with respect to the magnetic field, are of interest. The values of $G^{\frac{1}{2}}B$ are then 350 Mc for the narrowband case and 300 Mc for the stagger-tuned case. These are similar to the results obtained by Cook et al., given in Table 8, obtained with coupled cavities.

One of the great advantages of the traveling-wave maser is its high power output without saturation or decrease of gain such as is illustrated in Fig. 9b for a cavity maser. For the traveling-wave maser the gain is down 0.5 db for a CW signal power output of −22 dbm. For pulse operation with 10-μsec pulses at 100 cps the maximum power output is +8 dbm for the same gain reduction. This is directly attributable to the long spin-lattice relaxation time (0.1 sec.) of ruby, which allows large inversions to be built up between pulses.

4.4. Dielectric Slowing.[40] Even near cutoff in a normal waveguide the velocity of propagation slows appreciably, but its frequency variation is rapid and hence the bandwidth possibilities are limited. The group velocity in a waveguide or transmission line is given by[39]

$$v_g = \frac{\sqrt{1 - (k_c^2/\omega^2\epsilon\mu)}}{\sqrt{\epsilon\mu}}$$

where ϵ = dielectric constant
μ = magnetic permeability
k_c = cutoff propagation constant or wave vector = $2\pi/\lambda_c$

and thus the group velocity is slowed appreciably by a high dielectric constant.

A traveling-wave maser utilizing dielectric slowing is illustrated in Fig. 36. The active material is Cr^{3+} in rutile. The maser length of 1.9 cm provided 3.4 db of gain at 4.2°K, 6 db at 1.4°K with bandwidths of 20 Mc tunable over 15 Gc. No reverse-loss mechanism is provided and stability depends on elimination of reflections and use of external isolators.

5. MASERS WITH INTERMITTENT OPERATING CHARACTERISTICS

Masers falling into this classification tend to be experimental. At lower frequencies they are useful for making certain types of measurements, such as relaxation rates as a function of spin temperature. At higher frequencies the use of pulsed fields makes possible generation of coherent millimeter waves even if only on a short-term basis.

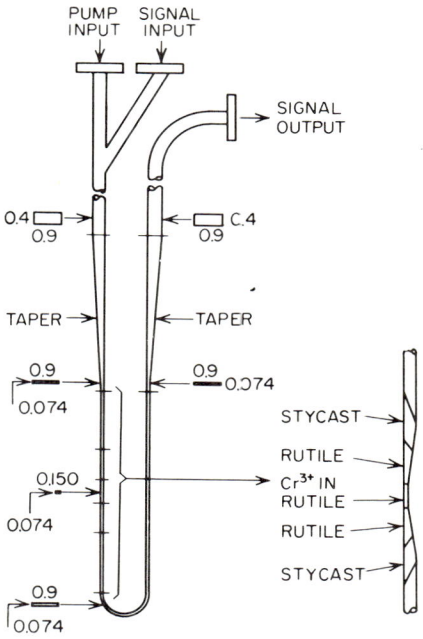

FIG. 36. Traveling-wave maser with dielectric slowing. (*After Gerritsen,*[40] *published by permission of the editor, Appl. Opt.*)

5.1. Adiabatic Rapid Passage. Chester, Wagner, and Castle[41] have done adiabatic-rapid-passage experiments on several spin $\frac{1}{2}$ systems. The magnetic field is sinusoidally modulated to sweep through the resonance corresponding to the cavity frequency periodically. Microwave power of about $\frac{1}{2}$ watt, of duration 50 to 100 μsec, is applied during one of these passages through resonance. This power in the X-band cavity, of loaded $Q \simeq 6,000$, together with the magnetic field sweep inverts the spin-system populations. During succeeding passages through resonance low probing powers are used to determine the properties of the system. The rate at which the magnetic field is modulated is determined by the relaxation time of the spin system since it is necessary to return to resonance in less than the spin-lattice relaxation time.

Because the inverted spin system returns to equilibrium with the lattice, the reflections from the cavity are of the form of Fig. 37. If the probing power is of low enough intensity to cause minor population changes, in a time comparable with the spin-lattice relaxation time, the system changes from an amplifying to an absorbing one. If then the system were inverted and the field returned to the resonance value there would be the possibility of oscillation, amplification, and certainly attenuation as a function of time. The system must be allowed to come to thermal equilibrium before another inversion begins because the populations are merely exchanged.

5.2. Adiabatic Rapid Passage with Pulsed Fields. An obvious adjunct for use with adiabatic rapid passage is that of pulsed magnetic fields. The procedure[42,43] is illustrated in Fig. 38. At time zero the field is pulsed down and an inverting pulse of 0.5 watt of radio frequency at 9,000 Mc successively inverts (staircase inversion) levels 2–3 and 3–4. The circuit used is illustrated in Fig. 39. The cavity is made 75 per cent of sapphire (pure Al_2O_3) and 25 per cent of ruby with a silver coating.

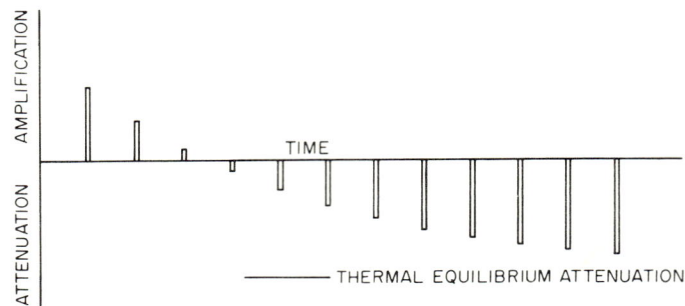

FIG. 37. The temporal behavior of population-inversion amplification and attenuation as a function of time for a pulsed maser.

Coupling to the waveguide is best accomplished by means of an iris. This type of cavity allows the magnetic field to change rapidly without major problems due to eddy currents in the cavity walls. The pulsed field is produced by a condenser discharge, resulting in a sinusoidal field variation with the coil and condenser forming a

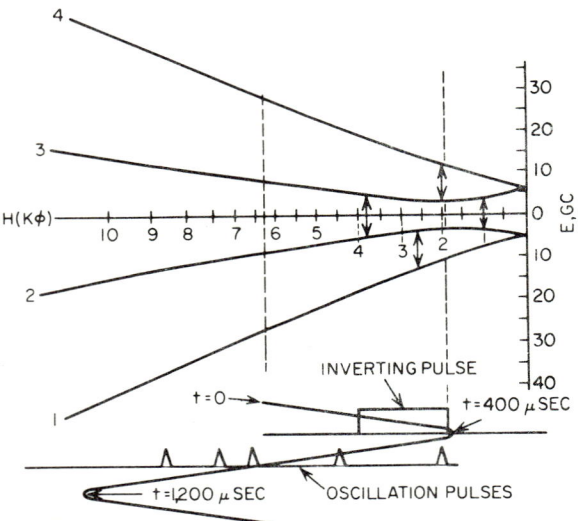

FIG. 38. Ruby energy levels near $\theta = 60°$ and magnetic field variation pulse.[43] (*Published by permission of the editor, Phys. Rev. Letters.*)

resonant circuit with a 0.0016-sec period. Most of the energy returns to the capacitor at the end of the cycle and a condenser bank of 30 μf at 2 kv provides a peak field of 18,000 gauss in the 3-cm-diameter Helmholtz pair.

The properties of such a system, if the field is returned to that corresponding to the cavity resonance at the inversion frequency, are as discussed above. The observation of the output over the cycle illustrated in Fig. 38 results in oscillation at a number of points where the frequency of the inverted transitions corresponds to various cavity modes.

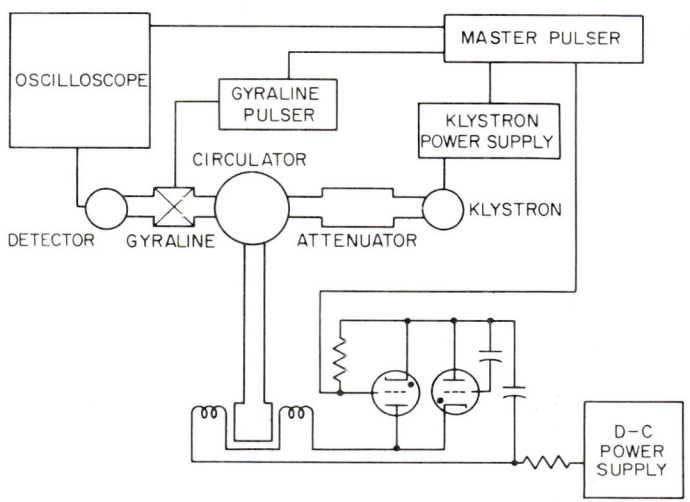

Fig. 39. Adiabatic-rapid-passage and pulsed-field apparatus. The second thyratron provides the second half-cycle if desired.

This system has been used up to 28 Gc, but no effort has been made to stabilize the magnetic field at the amplifying point. Hoskins and Birnbaum[42] have suggested using adiabatic fast passage to invert levels with a zero-field splitting and to then reduce the magnetic field to zero, depending on the zero-field transition for the amplifying segment of the cycle.

A point which is of importance here is the effect of cross relaxation (see Chap. 14, Sec. 6.5). If after inversion the magnetic field passes through a point where the inverted frequency corresponds to one of the noninverted transitions, spin-spin relaxation will tend to equilibrate the two levels, resulting in at least a lower spin population inversion or even a positive spin temperature. This is a hazard characteristic of multilevel systems, which do have some real advantages in that for a given frequency a lower magnetic field may be required because of the zero-field splitting.

5.3. Pulsed-field Maser—Three-level Inversion. Masers which operate with a fixed magnetic field require the pump frequency to be higher than the signal. Schemes to bypass this limitation have been proposed,

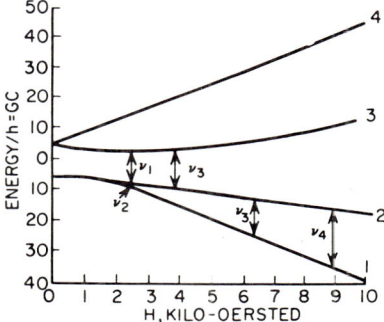

Fig. 40. Energy levels of Cr^{3+} in ruby $\theta = 90°$ marked with frequencies appropriate to three-level inversion followed by pulsed fields. Pump frequency $\nu_3 = \nu_1 + \nu_2$ oscillating output at ν_3 and ν_4.

but none has been successful, because of noncompatibility of demands and energy levels. A variable magnetic field does enable one to obtain transient amplification or oscillation at frequencies higher than the pump.

A method utilizing standard three-level inversion techniques has been used by Foner et al.[44] Using pink ruby with the magnetic field perpendicular to the c axis a pump frequency $\nu_{13} = \nu_3$ (Fig. 40) is used to invert levels 1 and 2. The magnetic field is then pulsed to values corresponding to ν_3, ν_4, ν_5, etc., in a time which is short compared with the spin-lattice relaxation time T_1. The bottom two levels of the spin system are still in an amplifying condition at this point and will amplify or oscillate at the resonant frequency ν_3, ν_4, or ν_5, etc.

The apparatus is sketched in Fig. 41. The reflection-type maser cavity is silver-plated ruby with resonant modes TE_{101} at 12.7 Gc (ν_3) and TE_{102} at 17.2 Gc (ν_4) and higher modes at which output was observed but which were not identified. The iris-coupled cavity (undercoupled) had Q's of 2,000 and 4,800 at ν_3 and ν_4, respectively.

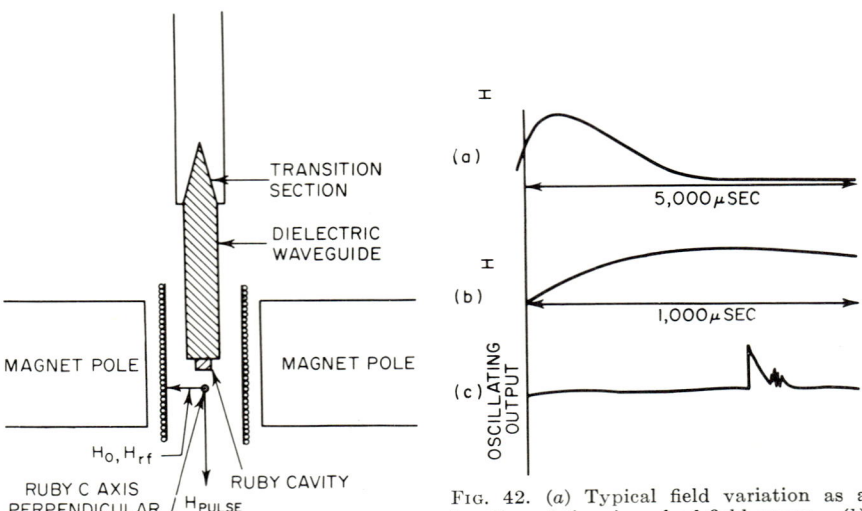

Fig. 41. Sketch of pulsed-field maser assembly.

Fig. 42. (a) Typical field variation as a function of time in pulsed-field maser. (b) Same as in a but different time base. (c) 19.5-Gc oscillations using envelope detector. Time base as in b. Oscillations occur during both rise and fall.

The dielectric transition section (also silver-plated) served as pump waveguide during the inversion and as signal waveguide during the amplifying or oscillating portion of the cycle.

The pulsed field is oriented perpendicular to both the crystal c axis and the fixed magnetic field used during the pump cycle. The solenoid (13 cm long by 7 cm ID) is energized by discharging a condenser bank. A 2,000-μf capacitor charged to 1,000 volts produced 9,400 oersteds with a time variation shown in Fig. 42. If the field passes through a resonance such as ν_3 fast enough no oscillations will build up and operation may then be possible to ν_4.

In this case only oscillation output was sought and achieved over a range of 12 to 70 Gc. In principle amplification is then possible at the various cavity resonances if the magnetic field can be held to the proper value. This is difficult and has not been achieved. In any case so long as pulsed fields of this type are used only very transient operation can be obtained. This method also of course suffers from the same spin-lattice relaxation time limitations as the adiabatic-rapid-passage work.

6. THE CW GAS LASER

The gas laser powered by a gas discharge is usually a low-power device. The attributes of a CW output, an extremely well defined frequency, and a well-collimated output make the gas laser a precise instrument. The active medium is more easily varied and controlled, and the output is better understood than in the case of the solid-state laser.

To a large extent the various component functions may be separated from one another. The discharge mechanism, the mirror geometry, the active materials and methods of mode selection are considered more or less independently of one another in the following sections. The diagrams used generally illustrate a complete system with unusual features commented on either in the text or in figure captions.

6.1. Radio-frequency Discharges. The first gas laser, utilizing a 10:1 ratio of He to Ne, with a total pressure of 1 mm of mercury, made use of an r-f discharge to provide the fast electrons necessary for inversion. Figure 43 illustrates the type of apparatus used. The r-f power is applied to the three external electrodes as shown, generating a discharge whose extent depends on the power level. Normal excitation requires about 50 watts, in this case at 28 Mc. The mirrors (fused silica, flat to within $\frac{1}{100}\lambda$) are coated with 13-layer dielectric films giving 98.9 per cent reflectance and 0.3 per cent transmission. Commercial gases (reagent grade) are generally acceptable, and the main source of contamination is outgassing over a period of time. This is minimized by bakeout, but the dielectric mirrors restrict the temperature which may

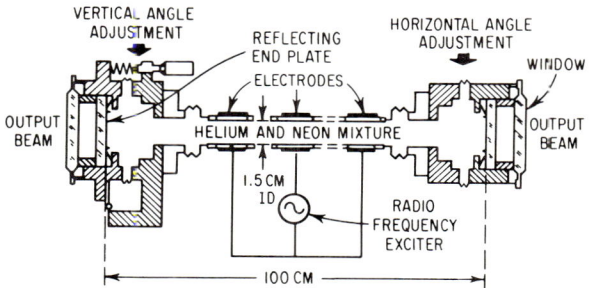

Fig. 43. Schematic of He, Ne optical maser. The electrons are external to the quartz discharge tube. The degree of freedom for one mirror is at right angles to the other. (*After Herriott*,[46] *published by permission of the editor, J. Opt. Soc. Am.*)

be used to $\simeq 150°C$. Since the dielectric coatings are easily damaged, the mirrors may be made accessible via metallic O-ring seals. The alignment of the Fabry-Perot plates is varied by the two micrometer screw adjustments. Initial alignment, to within 6 sec of arc, is accomplished by using an autocollimator. Typical gains for this geometry are near 6 per cent per pass, and since the loss increases by 1 per cent per second of mirror misalignment, parallelism of a few seconds must be obtained. A major source of noise and instability is the variation of both distance between mirrors, controlling the frequency, and alignment, affecting the cavity Q.

The tube diameter is important since gain is inversely proportional to diameter because of the necessity of lower-level ions diffusing to the wall for quenching. This has been discussed by Gordon and White,[47a,b] who derive for the optimum gain conditions

$$G_{max} = (pD)^2 \left(1 + \frac{pD}{D}\right)^{-1} H(pD)/D^2 \qquad (25)$$

H is a monotonically decreasing function of pD, the normalized pressure times tube diameter. For pressures below 1 Torr (1 mm Hg) for pD = constant, (electron density)/(optimum pressure) is constant and G_{max} is proportional to D^{-2}. For pressures above 1 Torr, G_{max} increases less rapidly, becoming proportional to D^{-1}. For G_{max} optimized with respect to pD, it has been shown that the value of pD at optimum G_{max} must increase slowly and monotonically with D.[47] Experimentally optimum gain for diameters of 1 to 15 mm follows the general rule

$$pD \simeq 2.9 \text{ to } 3.6 \text{ Torr mm}$$
$$p_{He} \simeq 5 p_{Ne} \tag{26}$$

The lower numerical value is applicable to smaller diameters.

All the gas-laser work requiring high-frequency stability has utilized r-f discharge excitation. It tends to be less noisy than the d-c discharge. The internal mirror geometry is more suitable for lasers intended to have high-frequency stability since fluctuations in the air path between Brewster-angle windows and mirrors are a source of noise.

6.2. D-C Discharges. Considerable power-supply simplification may be possible by use of d-c discharges. Hot-cathode d-c discharges have been used by a number of people, and a sketch of a possible arrangement is shown in Fig. 44. The discharge extends the full length of the tube. There are two problems with which to be concerned in d-c discharges. The first is the tendency for cataphoresis, or gas separation, to take place. This is a relatively high pressure phenomenon[48] but can be troublesome. The second problem is that of pumping by the discharge. The hot cathode circumvents this, but in cold-cathode discharges the ion bombardment of the cathode sputters material which tends to cover and entrap ions on neighboring surfaces. In order to obtain long life in cold-cathode discharges some means such as suitably biasing areas near the cathode must be employed to prevent loss of active material. As will be mentioned later, cold-cathode discharges are used in some cases, especially in pulsed operation.

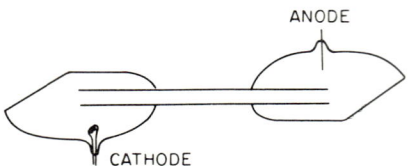

Fig. 44. Sketch of hot-cathode d-c discharge gas laser with Brewster-angle windows and external mirrors (not shown). Discharge will extend whole length of the tubulation.

6.3. External-mirror Geometries.[49,50] The internal mirror of Fig. 43 has a number of disadvantages, the most obvious and most troublesome of which are the probability of damage to the mirror surfaces and the lack of flexibility. An important step toward improving the latter problem was the confinement of the discharge to a tube terminating in Brewster-angle windows and placement of the mirrors outside the envelope. Figure 45 illustrates the main features of this arrangement.

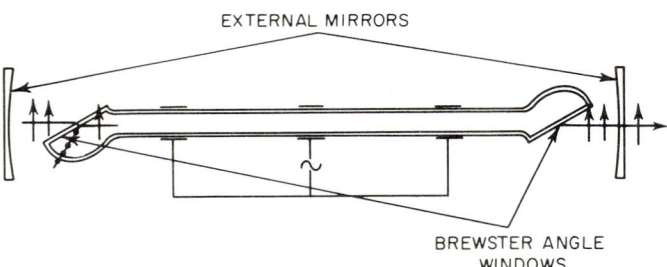

Fig. 45. Laser tube with Brewster-angle end windows and external mirrors. The output in this case will be vertically polarized, the horizontal polarization having been reflected out of the resonator. R-F excitation using external electrodes is illustrated.

The Brewster-angle windows pass the radiation which is plane-polarized perpendicular to the surface while reflecting that polarized parallel to the face. This serves, then, to remove the degeneracy between these polarized outputs by removing one of them. Use of plane-parallel windows leads to unwanted reflections, resulting in losses and the presence of undesirable fixed cavities. There are some problems associated with distortions of the mode patterns by the Brewster-angle windows, but the flexibility they introduce into the device makes them very desirable in many instances. For example, bakeout is easily accomplished in the discharge tube by itself. The concave mirrors, as pointed out in Chap. 14, have a number of advantages, including lower diffraction loss and simpler mode spectrum, when compared with the plane-parallel mirrors.

As mentioned in Sec. 6.1 the fluctuations in the air path between the Brewster-angle windows and the mirrors introduce noise or frequency fluctuations into the laser cavity. For high stability the air path should be kept short or the region should be evacuated, eliminating the air path.

The choice between a partially transmitting mirror and a completely opaque one with a hole in the middle can be made on the basis of loss. In the infrared region, for example, a much higher Q and higher power output result from using the hole—in some cases all the way through the mirror. This decision must be based on the particular frequency to be used.

6.4. Mode Selection. Except in the vicinity of threshold the laser output tends to consist of a number of modes, representing both radial and longitudinal field distribution patterns, separated in frequency by amounts which depend on the geometry used. Diffraction losses tend to limit oscillation to the even symmetric modes with the TEM_{00} being dominant. In the case of the plane-parallel mirrors a longitudinal mode separation of $c/2d$ (d = mirror separation) is expected, with near neighbors of higher-order modes, while for the confocal arrangement the longitudinal mode separation should be $c/4d$, each mode having a high degree of degeneracy which may be lifted by slight misadjustment. For most purposes these longitudinal modes are the ones desired. The ideal of a single frequency independent of power level is difficult to achieve and requires introduction of an element which is extremely frequency-sensitive. Near threshold the exciting power can be limited to a value such that only the mode nearest the line center is above threshold.

On a somewhat coarser scale the frequency of oscillation may be selected by using a prism[51] to deflect a dominant wavelength out of the apparatus leaving the desired frequency to oscillate. This, for example, has been useful in removing the dominant $3S_2-3P_4$ Ne line at 33,913 angstroms and allowing the $3S_2-2P_4$ visible line at 6,328 angstroms to take over. In fact, lines differing from one another by 30 angstroms can be separated in this way. The use of mirrors with high reflectivity over a given frequency range limits the cavity resonances to this same frequency range because of the necessity of a reasonably high Q to ensure overall gain. When external-mirror geometry is used, as it is also for the method mentioned in the previous paragraph, this provides an easy frequency-selective mechanism.

Of a somewhat different nature is the limitation to higher-order modes achieved for both "rectangular" and axisymmetric concave-mirror modes.[51,52,53] The first takes advantage of the astigmatism of the Brewster-angle windows for the large-angular-spread beam of the concentric-concave-mirror arrangement. By adjusting gain and mirror positions it is possible to have only one high-order mode in oscillation. Again only the spatial distribution is limited and frequencies corresponding to different longitudinal modes may exist. The better the quality of the Brewster-angle window the less the distortion. In the second example, use of a concave-plane-mirror arrangement (5.2 m radius of curvature 1.75 m from a plane mirror) separates the axisymmetric modes by using a circular iris. Experimentally, the highest-order mode, which the laser gain and iris diffraction loss allow, will oscillate. It is not completely clear why this occurs, but the outer ring of a confocal mode is of greatest intensity, and as soon as the iris is made large enough to reduce the diffraction losses of this mode enough, the lower-order modes disappear.

6.5. Mechanical Considerations. The stability of the output frequency of the laser depends on the cavity frequency and hence dimensions through Eq. (92) of Chap. 14

$$\nu \simeq \nu_c + (\nu_l - \nu_c)(Q_b/Q_c)$$

The mechanical mounting of the mirrors must achieve the fractional length stability which is expected of the laser frequency. Theoretically this is on the order of $\nu/\Delta\nu = 1.6 \times 10^{16}$ or a length stability of 6×10^{-15} cm for a 1-m resonator.

Both stability and flexibility of adjustment must be provided. A fine adjustment of 10 sec of arc from the perfectly aligned position is generally sufficient, with orthogonal independent adjustments most desirable (see Fig. 43). The main structure is best constructed of some low-thermal-expansion material such as Invar or quartz. Figure 46 illustrates a support used by Bennett and Kindlemann[54] of Nilvar, a low-thermal-expansion alloy, and which makes use of its magnetostrictive properties in order to adjust the alignment. This design has the potentialities of serving as part of an automatic-frequency-control system using electrical negative feedback. It is possible in this design, if working in the linear magnetostrictive range, to have the length adjustment necessary for frequency control independent of the angular adjustments. For details of coils and coil connections the original paper should be consulted.

Isolation from vibration requires considerable care, but its necessity for high precision is illustrated by Javan's results.[55] Some of his work has been carried out in an old wine cellar, where both temperature changes (using an Invar support) and vibration are minimized. Frequency stabilities of tens of cycles per second over periods of several seconds were obtained, and audio beats between two lasers could be observed for several minutes at a time. The stability, however, is affected by vibration caused by surf on the shores some distance away.

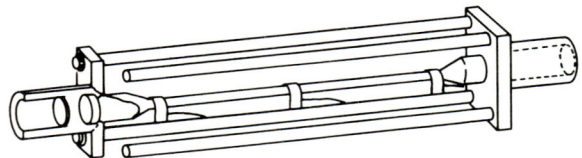

FIG. 46. Magnetostrictively tuned mirror support. Constructed of low-thermal-expansion Nilvar with coils around the four longitudinal stringers to provide the magnetic field, this type of support is suitable for lasers utilizing either internal or external mirrors. (*After Bennett and Kindlemann,*[54] *published by permission of the editor, Rev. Sci. Instr.*)

6.6. Pulsed Operation of Gas Lasers. Gas lasers are usually operated in CW fashion and derive therefrom much of the stability for which they have become noted. There have been some attempts at pulsing gas-discharge laser systems, and notable success has been achieved. Pulsed powers of as much as 84 watts have been obtained[56,57] from the helium-neon system using 5 Torr Ne and 240 Torr He pressures. One-microsecond d-c pulses of 35 kv at 35 amp are applied either by tungsten electrodes inserted through the side of the tube or by use of external ring electrodes (similar to the r-f electrodes shown in Fig. 45). The oscillations at four frequencies near 1.1 microns occur about 1 μsec after the d-c pulse and last a few microseconds. They are apparently due to an inversion produced by recombination. Work at lower pressures results in a lower power output and a longer delay between pulse and output.

Two other types of pulsed-gas-system results should be noted. W. E. Bell[58] has obtained a number of oscillating frequencies, including 6,149 and 5677.2 angstroms in the visible, from a mercury-helium pulsed cold-cathode discharge. The output, however, occurs during the pulse and the average output power increases with pulse length, leading to the expectation that CW operation may be possible. Still another type of pulse behavior is that obtained in nitrogen. Some of the results[59] are similar to those described for the pulsed He-Ne system with infrared radiation observed following the pulse. Pulses of ultraviolet radiation which occur during the pulse have also been observed in nitrogen.[60] This mode of operation, applicable to molecular as well as to atomic systems, should have wide application with frequencies well down into the infrared as well as up into the ultraviolet accessible.

6.7. Typical Gas Systems. A wide variety of gas masers using pure gases and gas mixtures has been operated. In most cases a number of transitions can be made to oscillate. The wavelengths range from the ultraviolet through the visible to the infrared. Table 11 lists a number of representative examples. It should be construed only as representative since both new systems and new frequencies are reported with regularity.

The best-known system is that of the helium-neon mixture, and neon itself In either case the levels of neon are the ones of interest. Neon has an electronic configuration of $(2s)^2(2p)^6$ in its ground state with excited levels formed by promoting one or more electrons to another quantum level. Figure 47 illustrates the levels

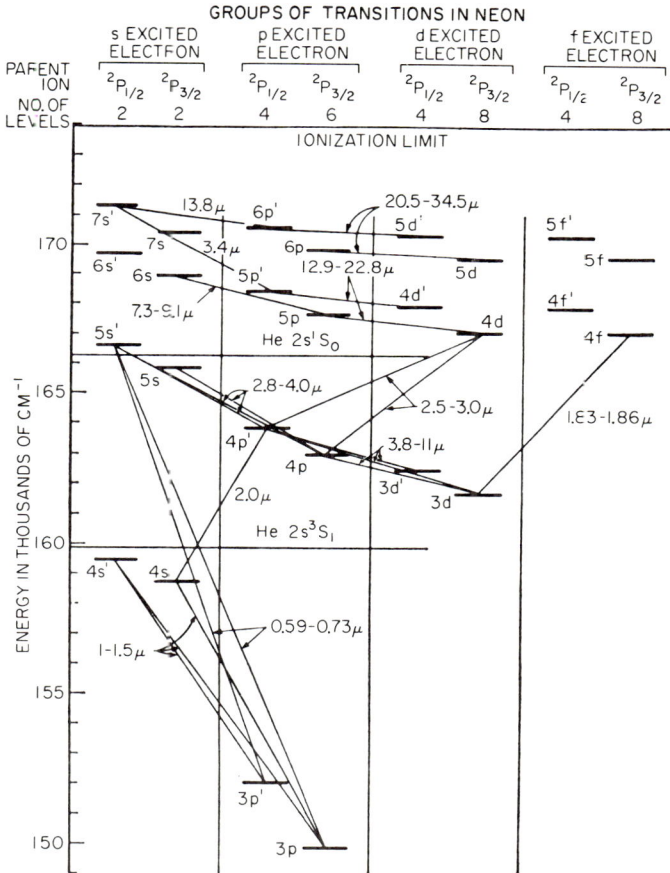

Fig. 47. Energy levels of neon laser transitions with wavelength ranges indicated. (After Faust, McFarlane, Patel, and Garrett,[61] published by permission of the editor, Phys. Rev.)

formed for the configurations $(2s)^2(2p)^5nl$. For some of the major transitions observed in the laser output a number of frequencies are given, e.g., $7p$–$6d$ and $4s$–$3p$. These should be taken as representative of the possible transition frequencies between the major levels of Fig. 47. The same variety is observed for most of the other gas systems such as xenon, nitrogen, and CO, and only wave-number ranges are listed in these latter cases.

The selection rules allow transitions for which ΔJ or $\Delta l = \pm 1$, and it has been shown that the highest transition probability exists for $\Delta k = \Delta l = \Delta J$ for neon, where Racah notation is used. Racah coupling was developed for the case when the orbital angu-

Table 11

Gas	Wavelength, microns	Transition, Racah or Paschen	Pressures Active gas	Pressures Other	Remarks	Ref.
Neon	0.6118	$5s'[\frac{1}{2}]_1^o - 3p[\frac{3}{2}]_2$	0.1	He 0.4		51
	0.6293	$5s'[\frac{1}{2}]_1^o - 3p[\frac{3}{2}]_1$	0.1	He 0.4		62
	0.6328	$5s'[\frac{1}{2}]_1^o - 3p[\frac{3}{2}]_2$	0.1	He 0.5		51
	0.6401	$5s'[\frac{1}{2}]_1^o - 3p[\frac{1}{2}]_1$	0.1	He 0.4		63
	1.0798	$4s'[\frac{1}{2}]_0^o - 3p[\frac{3}{2}]_1$	0.1–0.2	He 1.0–2.0		63
	1.0845	$4s'[\frac{1}{2}]_0^o - 3p[\frac{3}{2}]_1$			In afterglow only	64
	1.1409	$4s'[\frac{1}{2}]_1^o - 3p[\frac{3}{2}]_2$				65
	1.1523	$4s'[\frac{1}{2}]_1^o - 3p[\frac{3}{2}]_2$	0.1	He 1.0	Gain 12% per meter, 20 mw power output	65
	1.1523	$4s'[\frac{1}{2}]_1^o - 3p[\frac{3}{2}]_2$	0.08		Gain 4% per meter, 1 mw power output	66
	1.1601	$4s'[\frac{1}{2}]_1^o - 3p[\frac{3}{2}]_0$				64
	1.1614	$4s'[\frac{1}{2}]_0^o - 3p[\frac{3}{2}]_1$	0.1–0.2	He 1.0–2.0	Gain 2% per meter, 1 mw power output	65
	1.1767	$4s'[\frac{1}{2}]_1^o - 3p[\frac{1}{2}]_1$	0.1–0.2	He 1.0–2.0		63
	1.1985	$4s'[\frac{1}{2}]_1^o - 3p[\frac{1}{2}]_1$	0.1–0.2	He 1.0–2.0		65
	1.5231	$4s'[\frac{1}{2}]_0^o - 3p[\frac{1}{2}]_0$	0.1–0.2	He 1.0–2.0	Gain 6% per meter, 3 mw power output	63
	1.1143	$4s[\frac{3}{2}]_1^o - 3p[\frac{5}{2}]_2$	0.1–0.2	He 1.0–2.0		63
	1.1177	$4s[\frac{3}{2}]_2^o - 3p[\frac{5}{2}]_3$	0.1–0.2	He 1.0–2.0		65
	1.1390	$4s[\frac{3}{2}]_2^o - 3p[\frac{3}{2}]_2$	0.1–0.2	He 1.0–2.0		63
	1.2066	$4s[\frac{3}{2}]_2^o - 3p[\frac{3}{2}]_2$	0.1–0.2	He 1.0–2.0		65
	2.3958	$4p'[\frac{3}{2}]_2 - 4s'[\frac{1}{2}]_1^o$	0.1	He 0.5	Observed only when 3.39 and 1.15 transitions oscillate	67
	3.3912	$5s'[\frac{1}{2}]_1^o - 4p'[\frac{1}{2}]_1$	0.005–0.05	He 0.5	Gain 1.6×10^4 per meter, 10 mw power output	68
	1.8281–1.8602	$4f - 3d$				69
	2.500–3.0276	$4d - 4p$				61
	2.7826–3.9817	$5s - 4p$				61
	7.3228–9.0896	$6s - 5p$				61
	3.3813–3.3849	$7s - 5p$				61
	3.7746–10.981	$4p - 3d$				61
	2.0356–2.0359	$4p - 4s$				61
	11.861	$5p - 5s$				61
	13.759	$7s - 6p$				61

Element	Wavelength	Transition	(col)	(col)	Notes	Ref
Neon	13.759	$4d-4f$				61
	7.4237–22.836	$5p-4d$				61
	20.480–41.741	$6p-5d$				70, 61
	35.602–85.047	$7p-6d$				70, 71, 61
Oxygen	0.8446	3^3P_2–3^3S_1	0.014	Ne 0.35	Gain 3% per meter, 1 mw power output	72
	0.8446	3^3P_2–3^3S_1	0.036	Ar 1.3	Gain 3% per meter, 1 mw power output	72
Helium	2.0603	7^3D–4^3P	8		Gain 5% per meter, 3 mw power output	71
Argon	1.6941	$3d[\frac{3}{2}]^o_2$–$4p[\frac{3}{2}]_2$	0.035		Gain 3% per meter, 0.5 mw power output	71
	2.0616	$3d[\frac{3}{2}]^o_2$–$4p'[\frac{3}{2}]_2$	0.035		Gain 3% per meter, 1 mw power output	71
	1.6941–2.3973	$3d-4p$				61
	1.6180	$5s-4p$				61
	2.5014	$6d-6p$	0.05			61
	2.5512–3.1333	$5p-5s$				61
	2.5494–3.0996	$5p-3d$				61
	4.9213–5.4694	$5d-4f$				61
	4.9160–5.1216	$6p-4d$				61
	5.8477–7.2166	$6p-6s$				61
	6.0531–6.9429	$4d-5p$				61
	7.8003–7.8023	$4f-4d$				61
	12.141–26.944	$4d-4f$				61
	15.037 15.012	$5d-5f$				61
Krypton	1.6900–4.8773	$4d-5p$				71, 73
	1.9211	$8s-6p$	0.035			71
	2.6288	$7p-4d$				61
	2.8618–3.3409	$6p-6s$				61
	2.9845–3.4895	$6p-5d$				61
	3.4883	$6p-7s$				61
	4.3748–5.5700	$5d-6p$				61
	5.5863–5.6306	$6d-4f$				61
	7.0581	$4f-5d$	0.02	He 5		61
Xenon	2.0261	$5d[\frac{3}{2}]_1^o$–$6p[\frac{3}{2}]_1$			Gain 120% per meter, 10 mw power output	71
	2.0261	$5d[\frac{3}{2}]_1^o$–$6p[\frac{3}{2}]_1$	0.02		Gain 10% per meter, 5 mw power output	71
	2.0261–12.913	$5d-6p$				71, 73
	3.4345–3.6518	$7p-7s$				61
	11.299–18.506	$5d-4f$				61
	3.5070	$5d[\frac{7}{2}]_3^o$–$6p[\frac{5}{2}]_2$	0.01	He 1.0	Gain 60 db/meter	73

Table II. (Continued)

Gas	Wavelength, microns	Transition, Racah or Paschen	Pressures Active gas	Pressures Other	Remarks	Ref.
Carbon	1.069	$3p^3D_3$–$3s^3P_2^o$			He in afterglow	56
		$3p^1P_1$–$3s^1P_1^o$			He in afterglow	56
Carbon monoxide	0.6614–0.6596	Band spectrum	2.0			74
	0.6074–0.6063	Band spectrum	2.0			74
	0.5604–0.5591	Band spectrum	2.0			74
	0.5186–0.5198	Band spectrum	2.0			74
Nitrogen	0.3–0.4	Band spectrum			Gain 60 db/meter, pulsed power 10^4 watts	60
	0.86835–0.87099	Band spectrum				59
	0.8844–0.8879	Band spectrum				59
	0.88865–0.89093	Band spectrum				59
	1.04493–1.05052	Band spectrum				59
	1.2303–1.2347	Band spectrum				59
Mercury	0.6150	$7p^2P_{\frac{3}{2}}$–$7s^2S_{\frac{1}{2}}$				58
	0.5677	$5f^2P_{\frac{7}{2}}^o$–$6d^2D_{\frac{5}{2}}$				58
	0.7346	$7d^2D_{\frac{5}{2}}^o$–$7p^2P_{\frac{3}{2}}$				58
	1.0583	$8s^2S_{\frac{1}{2}}$–$7p^2P_{\frac{3}{2}}$				58
	1.5295	$6p'^3P_2^o$–$7s^3S_1$				75
Sulfur	1.813					75a
	1.0455					61
Bromine	1.0628					61
Iodine	0.8446					61
Chlorine	3.236–3.431					75a
	1.975–2.020					75b

29–40

lar momentum of the electron is added to the orbital momentum of the kernel to form **k**, which is then added to the electron spin to form **J**. The numbers in the square brackets give the k value and the following subscript the J value. The preceding number and letter give the nl quantum numbers of the electron involved. The superscript c indicates odd parity.

With the exception of the pulsed operation described in Sec. 6.6 these lasers are low-power, continuously operating devices. The power outputs are typically in the vicinity of a few milliwatts. The maximum amplification is represented by a figure of 60 db/m for He-Ne r-f discharges at 3.5070 microns, and from Table 11 it is evident that a few per cent per meter is more typical. The continuous operation allows the temperature to stabilize, and the very narrow bandwidth outputs such as those mentioned in Chap. 14 have been achieved using this type of laser.

7. METALLIC-VAPOR LASERS

This type of active material was suggested in Schawlow and Townes's original article on infrared and optical masers, but the problems associated with the use of the alkali metals (potassium was originally suggested) are numerous and difficult. Cesium is the only metallic vapor which will be considered here, although mercury, iodine, bromine, etc., have been used in gas-discharge systems (see Table 11).

7.1. The Cesium Amplifier.[76] In contrast to the gas-discharge pumped gas lasers of Sec. 6 the cesium laser is optically pumped. This allows the laser to be operated with short pumping pulses and with high peak power outputs. On the other hand the optical pumping adds the complication of having to provide efficient energy transfer to the cesium.

Amplification has been demonstrated in the system illustrated in Fig. 48. The intense 3.338-angstrom helium line from the He lamps P_1 and P_2 excites the $6S_{\frac{1}{2}}-8P_{\frac{1}{2}}$

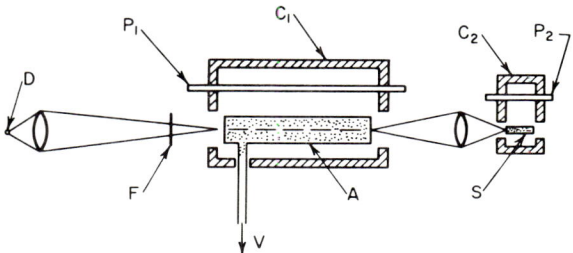

FIG. 48. Schematic of amplification arrangement using cesium. P_1 and P_2 are helium lamps, C_1 and C_2 reflecting cavities, S cesium source cell, A cesium amplifier cell, F bandpass filter, D detector, V to vacuum pump. (*After Jacobs, Gould, and Rabinowitz*,[76] *published by permission of the editor, Phys. Rev. Letters.*)

transition illustrated in Fig. 49. The resulting 3.2- and 7.18-micron radiations from the 9-cm-long source are amplified in the 90-cm-long amplifying section A. The source and amplifier diameters are 4 and 10 mm, respectively. Optimum amplifier temperature is 165°C and continuous pumping is utilized to diminish impurity-poisoning problems.

The inverted populations achieved are given in Table 12. Both the $8P_{\frac{1}{2}}-8S_{\frac{1}{2}}$ and

Table 12

Cs level	Population density (10⁶ atoms/cu cm)
$8P_{\frac{1}{2}}$	100
$8P_{\frac{3}{2}}$	30
$8S_{\frac{1}{2}}$	10
$6D_{\frac{3}{2}}$	3
$5D_{\frac{5}{2}}$	200

the $8P_{1/2}-6D_{3/2}$ transitions are well inverted. The 3.2-micron line (investigated because of the sensitivity of liquid-nitrogen-cooled PbS detectors) showed an amplification of 3 per cent.

7.2. The Cesium Oscillator.[77]

Construction of a cesium oscillator is illustrated in Fig. 50. The helium lamp (r-f excited) and the 90-cm cesium tube are at the foci of the elliptical reflector. BaF_2 windows, fitted to the glass tube with optically polished surfaces so that no seal was necessary, are set, one at the Brewster angle and the other at 45°. The latter causes about ½ per cent of the coherent, plane-polarized beam to be deflected to the output. Both the cesium cell and the surroundings are evacuated, as indicated. The external mirrors, flat and concave with 396 ± 1 cm radius of curvature, are separated by 193 ± 1 cm, giving an axial-mode separation of 75 Mc and an 18-Mc separation between adjacent angular modes. The BaF_2 windows and the concave mirror are flat to within $\lambda/10$ and the flat mirror to within $\lambda/30$ (both measured in the visible). The mirrors are silvered so as to be opaque.

The helium lamp used is 4 mm ID by 150 mm long, filled with 4 mm He, and produces about 8 mw/sq cm steradian at 80 watts r-f excitation. The gain of the cesium-filled tube is 1 per cent per centimeter at a temperature of 175°C. The beam diameters (there are two because of reflections from both surfaces of the 45° BaF_2 window) are 3.2 ± 0.1 mm with angular divergence of 10^{-3} radians. Beam power is about 25 μw in each beam. Only the lowest-order TE_{00m} modes are observed, and the zero-field hyperfine splitting limits the number to 3 for a power of about 8 μw per

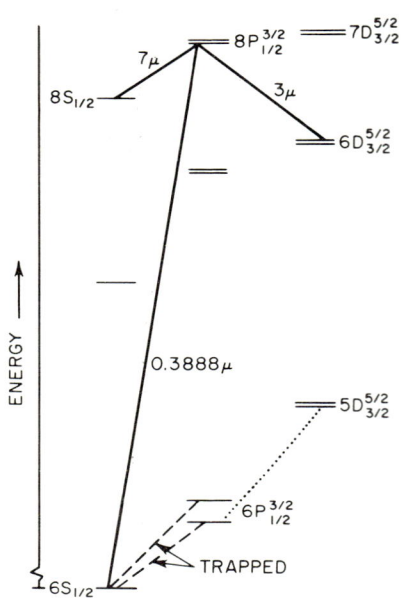

Fig. 49. Energy levels of cesium. (*After Jacobs, Gould, and Rabinowitz,[76] published by permission of the editor, Phys. Rev. Letters.*)

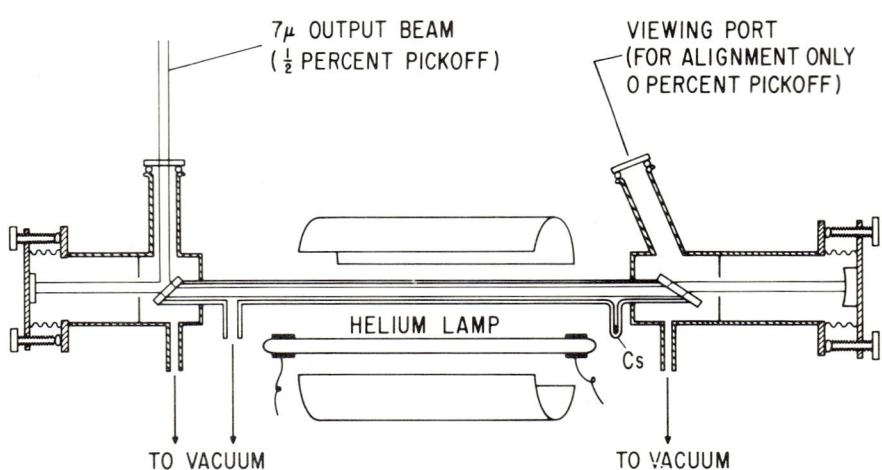

Fig. 50. Schematic of optically pumped cesium laser. (*After Rabinowitz, Jacobs, and Gould,[77] published by permission of the editor, Appl. Opt.*)

mode. It is estimated that seven-eighths of the power emitted is lost by absorption; then the total emitted power per mode is 64 µw, which should correspond to a frequency width of 0.003 cps. This design is intended for 7.18-micron operation, but use of sapphire windows should allow operation at 3.20 microns since the loss at 7.18 microns will then more than make up the $6\frac{1}{2}$ times higher threshold for the 3.20 µ line (at equal cavity Q's). The output power, albeit in more modes, will be increased by using plane mirrors and higher output coupling to perhaps 10^{-3} watts. Pulsed He discharge tubes should further improve this output to 1 watt peak.

8. SOLID-STATE LASERS—RUBY

Cavity lasers or optical masers using solid active media are generally divisible into two categories—pulsed and continuous. The pulsed operation is usually forced either by the lack of sufficient pump power on a continuous basis or by heating of the crystal, although in certain applications the pulsed output is desirable.

8.1. Pulsed Ruby Lasers. Historically the first laser was that using the three-level ruby[78] scheme, and it is one of the most commonly used types. The energy levels of Cr^{3+} in Al_2O_3 are given in Fig. 51. Optical pumping corresponding to the $^4A_2 - {}^4T_2$ and $^4A_2 - {}^4T_1$ transitions and a fast nonradiative decay to the $\bar{E}$ states serve to populate the $2\bar{A}$ and $\bar{E}$ levels. These rapidly thermalize, leaving the $\bar{E}$ the most populous and the R_1 line at 6,943 angstroms as the laser output. Alternatively, using frequency-selective methods the R_1 loss may be made higher than that for the R_2 line, resulting in the 6,929-angstrom output of the $2\bar{A} \to {}^4A_2$ transition. Both the R_1 and R_2 lines are doublets because of the 11.7-Gc splitting of the ground state by the crystalline electric field.

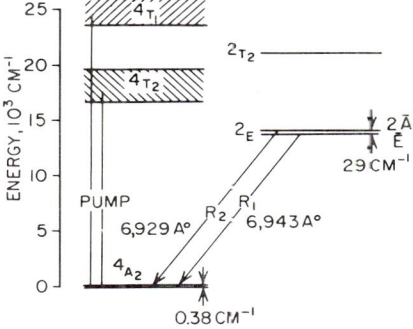

FIG. 51. Energy levels of Cr^{3+} in Al_2O_3.

A sketch of the relevant components used in a ruby laser is given in Fig. 52. A helical xenon flash lamp such as the FT-524 has been shown, although the types illustrated in later figures are sometimes more satisfactory. The flash lamp is pulsed from a condenser bank and will dissipate an energy of several thousand joules in the few milliseconds during which it operates. The reflecting walls of the pump chamber cause a large fraction of the light eventually to pass through the ruby. Energy

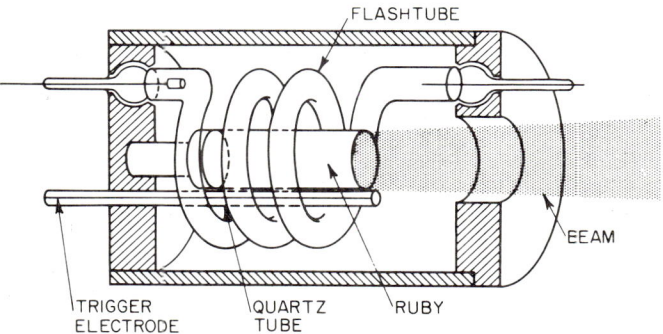

FIG. 52. Essential components of a solid-state laser employing a helical flash tube. The reflection from the surface can be either specular or diffuse, the latter being preferable. (*After Maiman*,[79] *published by permission of the editor, Solid-State Electron.*)

corresponding to the pumping-band absorption $^4A_2 \rightarrow {}^4T_1$ and $^4T_2$ will contribute to inversion while other absorbed energy leads primarily to thermal heating. In some cases filters can usefully be employed to prevent the undesirable frequencies from reaching the crystal. In most cases it is helpful to cool the crystal.

Reflection from the ends of the ruby is aided by silvering or use of dielectric coatings.* One end must be partially transparent in order to get an output. This may be accomplished by allowing partial transmission over the whole end surface or by leaving a small area of the end unsilvered. It will be seen in later sections that there are devices which operate without adding reflectivity to the ends of laser crystals, and in fact, amplifier operation requires that one minimize the reflection due to the crystal interface.

The flash lamps are energized by means of capacitor banks charged to voltages necessary to provide the energy of up to several joules necessary for pumping. The voltage used depends on the modulator. In a system such as that illustrated in Fig. 52 the voltage is placed across the flash tube, a trigger pulse from a pulse transformer or other source ignites the lamp, and the energy of the condenser bank supplies the discharge. In this case the peak voltage is limited by the maximum voltage the flash tube will hold off prior to triggering. Pulse modulators or systems using a holdoff switch in series with the voltage avoid this limitation. Reflecting walls are made of aluminum or aluminum foil, forming a specular reflector or a diffuse reflector such as magnesium oxide. The latter is preferred for the helical flash-tube reflector. The use of a linear flash lamp and crystal at the foci of an elliptical specularly reflecting chamber is the alternative arrangement. The efficiencies of the two pumping systems are comparable if they are correctly designed.

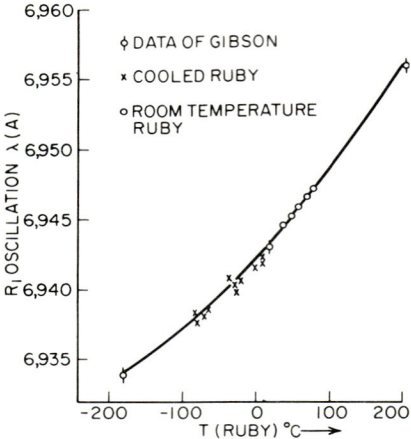

FIG. 53. R_1 ruby laser output as a function of temperature. (After Abella and Cummins,[80] published by permission of the editor, J. Appl. Phys.)

The temperature of operation influences both the output frequency and the threshold power required. The oscillation frequency of the R_1 line as a function of temperature is given in Fig. 53 from the work of Abella and Cummins[80] and Gibson.[81] The ruby line width is also a function of temperature,[82] varying from 2×10^{11} cps at room temperature to about one-fifth this value at 77°K (nitrogen temperature).

The spectrum of Cr^{3+} in Al_2O_3 with concentrations of 0.5 per cent, and above, contains emission lines which have been interpreted to be due to exchange-coupled pairs. The main transition frequencies are 7,009 and 7,040 angstroms. Laser action has been observed at both frequencies (simultaneously).[83,84,85] The simultaneity gives some information concerning the two lines since for a common upper level or close-spaced upper levels, between which thermal equilibrium is rapidly established (such as between $2\bar{A}$ and $\bar{E}$), the dominant frequency would establish an oscillation and the populations would then not become favorable for the second frequency.

8.2. The Ruby-laser Output. The typical pulsed ruby-laser output consists of a series of short (≈ 0.5 μsec) pulses at intervals of a few microseconds from the time

* The parallel reflecting surfaces are the most common, but the "chisel-tip" or roof-prism configuration has been used. The roof-prism end is cut and polished with each half at 45° to the rod axis and perpendicular to the other half. The action is then similar to the radar corner reflector, with reflections from each side of the roof being totally internally reflected, and energy which travels down one side is reflected by the two sloped surfaces to travel back on the other side of the crystal.

the oscillation conditions are reached until the inverted population decreases to the point where the gain is insufficient to overcome cavity losses. These pulses have been investigated by a number of groups[86-92] spectroscopically, photographically, and with photomixing tubes. The general properties so determined are as follows:

1. The near-field pattern (using no lens) when investigated using a Fabry-Perot interferometer shows a concentric-ring pattern.
2. The far-field pattern obtained by focusing the beam onto a photographic plate, also shows a concentric-ring pattern through a Fabry-Perot interferometer.
3. The first output pulses originate near the center of the rod, as is to be expected from the focusing properties of dielectric cylinders.[89]
4. The frequency of the emitted light tends to decrease in successive pulses (see Fig. 54).

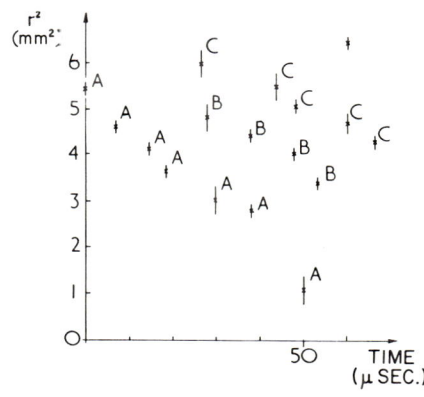

FIG. 54. Output of a ruby laser as a function of time. The abscissa is the (radius)^2 in a Fabry-Perot etalon. The difference from mode to mode corresponds to a (radius)^2 difference of 0.50 mm^2. Successive spikes, indicated by the crosses, are, within the experimental error, successive modes with the trend being toward longer wavelengths as time progresses. A, B, and C are different sequences, with B and C appearing at nearly the same time. (After Hughes and Young,[89] published by permission of the editor, Nature.)

5. Beats corresponding to mode differences are observed and confirm the Fox and Li predictions for plane and confocal end mirrors.
6. Effects which can be attributed to frequency pulling by the two R_1 components, and hence deviations from the cavity-mode frequencies have been observed at low temperatures.
7. The radiation patterns or areas of oscillation tend to differ from one 0.5-μsec pulse to the next; i.e., different regions of the rod are independent of one another.
8. With pump powers high above threshold the oscillation patterns are more complicated and more frequencies are observed.
9. There is a correlation between good optical quality of the ruby resonator and simple behavior.
10. Polarization—for ruby resonators with the c axis along the resonator-rod axis (0° orientation) the output is not polarized. For orientations of 60 or 90° the R_1 output is plane-polarized with the E vector perpendicular to the plane defined by the crystal and rod axes. This agrees with the polarization expected on the basis of both the fluorescent and absorption spectra of the R_1 line and is therefore attributable at the selection rules for the transition.

The picture which results from the above facts is not completely clear,[86,87] but the general outline seems to be as follows: After the application of pump power the populations of the upper states build up and the first frequency to oscillate is that cavity resonance nearest the peak of the R_1 line. The inverted population at that frequency lasts only the 0.5 μsec or so and emission stops.* A few microseconds later the frequency which corresponds to the next lower cavity mode is sufficiently inverted to oscillate. The process continues with regions farther from the line center becoming active in sequence until at some point the line center again oscillates and the sequence starts over. Thus the frequency changes from pulse to pulse in a way which is somewhat regular near the beginning of the pump pulse—later the behavior is more complicated and seemingly more random. The sequence for each pump pulse generally follows this pattern, but there are often differences in detail (Fig. 54). There is an added frequency shift due to the change of refractive index with increase of temperature during the pump pulse, and it is probable that this determines the direction of the frequency shift with time. The frequency width of each spike is a few megacycles and appears to be determined by the length of the pulse, e.g., 2 Mc for 0.5-μsec pulses.†

Observation of ring patterns in both near- and far-field experiments indicates that light is scattered from the lowest-order symmetric mode in essentially all directions. It is known that 20 to 30 per cent scattering loss per pass is characteristic of some

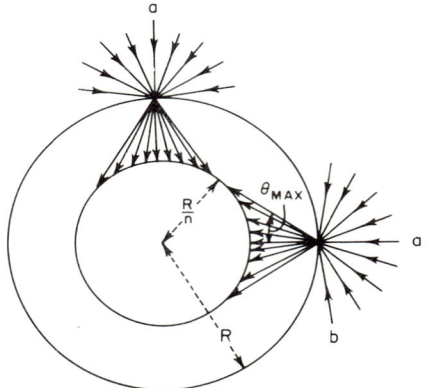

FIG. 55. Light paths in a dielectric crystal. (After Devlin, McKenna, May, and Schawlow,[94] published by permission of the editor, Appl. Opt.)

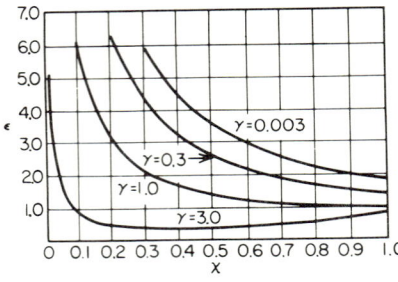

FIG. 56. Normalized energy density vs. normalized radius for a polished dielectric cylinder with pump light incident radially. This is a family of curves calculated from Eq. (27) for various values of $\gamma = aR =$ attenuation x radius of rod. (After Sooy and Stitch,[95] published by permission of the editor, J. Appl. Phys.)

ruby rods. This light, at the frequency of the uniform mode and usually a small fraction of its intensity, coming from different angles on the crystal face is responsible for the ring formation.

The observation that there is a tendency for emission to occur first at the rod center is further confirmation of pumping-efficiency predictions covered in Sec. 8.3.

Because of the puisating nature of the output the measurement of output power is really not meaningful. The total emitted energy per flash-lamp pulse can and has been measured[93] for several lasers. The radiated energy increases as the flash-tube energy is increased. The measured ratios varied from 0.00005 to 0.0005 (joules output)/(joule input). The value will vary with the spectral characteristic of the lamp and the efficiency of the pump housing. This measure of efficiency is more useful when applied to the giant pulse laser as in Sec. 8.6.

8.3. Pump Energy Density in a Dielectric Cylinder. The path of radiation in a dielectric cylinder, such as is commonly used for a laser crystal, is such that the

* The theory of such pulsation of fluorescence has been treated by R. W. Hellwarth, *Phys. Rev. Letters*, vol. 6, p. 9, 1961.

† An explanation of the pulsating output based on spatial inhomogeneity of the modes and lateral diffusion of energy has been proposed by H. Statz and C. L. Tang, *J. Appl. Phys.*, vol. 35, p. 1377, 1964.

energy density is much higher near the cylinder axis. Consider, for example, radiation traveling in a plane perpendicular to the axis of the cylinder as illustrated in Fig. 55. Rays such as those marked a travel undeflected. All others are refracted so as to approach the axis more closely. The ray which passes farthest from the axis is initially traveling tangentially to the cylinder. Its angle of refraction is $\theta = \sin^{-1}(1/n)$ where n is the refractive index of the dielectric, and hence its distance of closest approach to the axis is R/n. It is apparent that there is then a concentration of energy density in a cylinder of radius R/n where R is the outer diameter of the crystal.

This effect has been used by Devlin et al.,[94] who, by coating a ruby cylinder with sapphire to focus the pumping energy more effectively into the ruby, were able to reduce the energy required for the flash lamp by almost a factor of 2. In this case the surface area available for cooling is also increased considerably.

In applying this principle to a laser crystal the attenuation of the pump energy must be included. This case has been considered by Sooy and Stitch[95] for two cases. The first assumes radiation incident radially, a fairly good approximation for elliptic geometries with small eccentricity using a linear flash lamp. The energy density ratio may then be written as

$$\epsilon = \frac{\mathcal{E}}{\mathcal{E}_s} = \frac{4n^2}{(n+1)^2} \frac{\cosh \gamma \chi}{2 \sinh \chi + [4n/(n+1)^2] e^{-\gamma}} \frac{1}{\chi} \quad (27)$$

where $\mathcal{E}_s$ = energy density at R in the absence of the rod
$\chi = r/R$
$\gamma = \alpha R$
α = attenuation, defined by $p/p_0 = e^{-\alpha x}$
$\mathcal{E}_s = P_s/\pi R c$, P_s = incident power/unit length of rod

This has been calculated for ruby including the internal reflection (8 per cent for ruby). Figure 56 illustrates the behavior of (27) for a number of $\gamma = \alpha R$ values. The concentration of energy density near the axis is appreciable.

Of interest is the more general form of (27) giving the energy density in terms of the internal-reflection coefficient $\delta = [(n-1)/(n+1)]^2$

$$\mathcal{E} = \frac{P_s}{2\pi R} \frac{n}{c} \frac{2t \cosh \alpha r}{2 \sinh \alpha R + \delta e^{-\alpha R}} \quad (28)$$

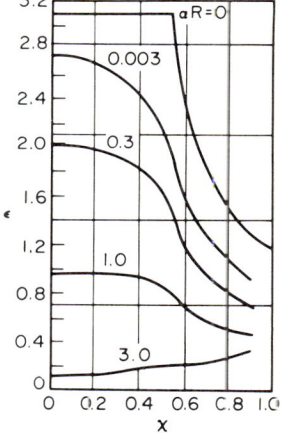

FIG. 57. Normalized energy density vs. normalized radius for a polished dielectric cylinder with pump light in the plane perpendicular to the axis and including internal reflections for $R = 0$ curve, with isotropic pump light but not including internal reflections for $\alpha R \neq 0$ curves. Data derived from Refs. 94 and 95.

The second approximation assumes isotropic pump radiation but neglects internal reflections and is illustrated by the lower four curves of Fig. 57. The upper curve for $\alpha = 0$ is a plot of the results assuming the radiation in the plane perpendicular to the axis but not necessarily radial. The $\alpha R = 0.003$ should compare quite well with it and the difference between the two, approximately $7\frac{1}{2}$ per cent, is due to the neglect of internal reflections in the lower curve and $4\frac{1}{2}$ per cent due to the differing assumptions concerning the incident radiation. The lower curves of Fig. 57 will be quite accurate because at the high attenuations the internal reflection has a negligible effect.

8.4. CW Ruby Laser.[95] The pump power limitations of ruby have been overcome by the scheme illustrated in Fig. 58. The high-pressure mercury-xenon lamp (Hanovia 941B) emitting about 900 watts/sq cm/steradian in the 3,500- to 6,000-angstrom band is located so that the pump light is focused into the sapphire trumpet. The system collects in a solid angle of approximately 0.73 steradian, thus providing more than the estimated 400 watts/sq cm/steradian needed to pump the ruby. The trumpet angle along with the polished walls ensures internal reflection so that pump light makes a double pass through the ruby. The ruby shank (diameter 0.061 cm by 1.15 cm long) and the sapphire cone (0.150 cm at the larger end by 1.05 cm long) are immersed in liquid nitrogen.

Continuous oscillating output is obtained but the output contains the same type of spiking observed in the pulsed ruby laser. This is not explained, and details as to frequency-stability and time-variation information, such as discussed in Sec. 8.2 for the pulsed ruby, are not available.

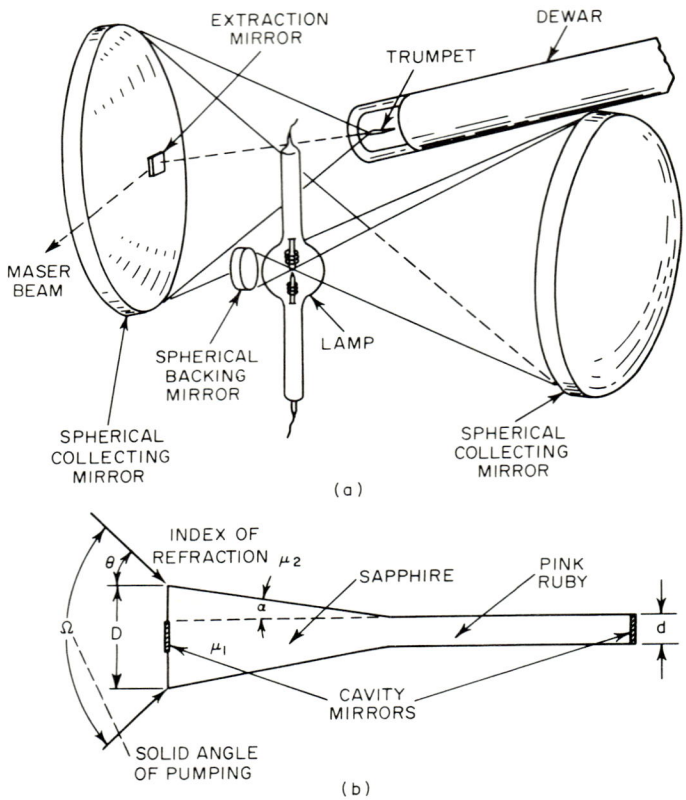

FIG. 58. (a) Schematic of system used to obtain CW operation in a ruby laser. (b) The ruby sapphire trumpet used in a. (*After Nelson and Boyle,*[96] *published by permission of the editor, Appl. Opt.*)

8.5. Solar-pumped Laser. The same sort of optical system and pumping arrangement as illustrated above is quite well suited to solar pumping of lasers. A similar type of cone geometry feeding a spherical pump chamber containing the crystal has been described by Keck et al.[97]

8.6. The Giant-pulse Laser. A method of circumventing the pulsating output of the ruby laser and compressing the power into a single pulse of very high energy is that of Q switching. The Q of the laser resonator is, by some device, kept too low for oscillation to take place until the pump has created a maximum population inversion. The Q is suddenly increased, and the laser produces a burst of energy much higher than under normal operation because of the greater inversion at the start of the pulse. The Q switch is cut off before the population can rebuild to the oscillation point, resulting in a single very high powered pulse of short duration. The total energy output during a single shot of a ruby laser is, in fact, decreased in this mode of operation, but since the emission is all contained in a single pulse the power is considerably increased.

Two of the most important characteristics of the giant-pulse laser, total radiated energy and pulse rise time, may be estimated easily.[98,99] The ruby is quite efficient in that essentially all ions excited to level 3 drop to level 2 so that each absorbed pump photon gives rise to either a spontaneous or an induced output photon of frequency ν_{R_1} (see Fig. 51). If n_i and n_f represent the initial and final population inversion, at beginning and end of pulse, the total output energy during the pulse is

$$\mathcal{E} = \tfrac{1}{2}h\nu(n_i - n_f) \tag{29}$$

The one-half is required since the transfer of one ion from level 2 to level 1 changes Δn by 2. The value of n_f can be expressed in terms of the loss factor per pass γ and the gain per unit length α, due to a population difference of Δn so that $\alpha/\Delta n = \alpha_0/n_0$. This latter quantity α_0 is the gain for complete inversion or the attenuation in normal thermal equilibrium when only the ground state is populated. Since near threshold gain equals loss, for a length l

and then
$$\alpha l = \gamma$$
$$n_f = n\gamma/\alpha_0 l \tag{30}$$

Numerically for pink ruby $n \simeq 1.6 \times 10^{19}$/cu cm, $\alpha_0 \simeq 10$ cm^{-1} at 77°K, and $\gamma \simeq 0.1$ (for some ruby samples it is higher than this because of scattering losses). For a 5-cm rod

$$n_f = n(0.1l/10l) = 0.010n$$

or is negligible compared with the $0.25n$ which can be achieved in initial inversion. The value of $\mathcal{E}$ is then

$$\mathcal{E} = \tfrac{1}{2}h\nu n_i = \tfrac{1}{8}h\nu n \simeq 0.5\mathcal{E} \text{ joule/cu cm}$$

a value which is not achieved in practice and which for a 0.1-μsec pulse would correspond to 5.6 megawatts/cu cm for a square pulse. Wagner and Lengyel[100] give as the peak radiated power

$$P = \tfrac{1}{2}[n_p \log (n_p/n_i) + n_i - n_p]V\aleph_0(h\nu/\tau_l) \tag{31}$$

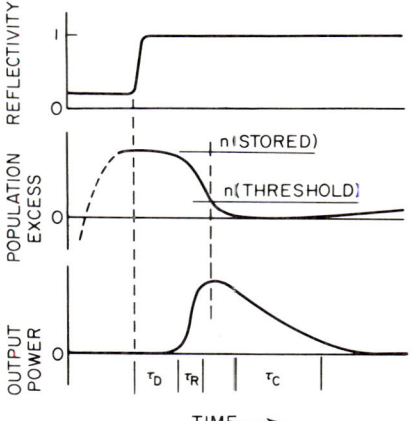

FIG. 59. Time sequence of events in giant-pulse laser. The reflectivity must drop again in order to prevent the system from building up to a second, smaller pulse. (*After McLung and Hellwarth,*[99] *published by permission of the editor, Proc. IEEE.*)

where $\tau_l = l/c\gamma$ = photon lifetime in Fabry-Perot interferometer, and n_p is the population inversion at threshold. For pink ruby this yields about 10^7 watts/cu cm.

The time sequence of events is illustrated in Fig. 59. The times τ_D, τ_R, and τ_c are of interest and have been estimated to be[93]

$$\begin{aligned}\tau_D &\simeq \tau \ln\left(1 + \frac{1}{1 - n_i/n}\frac{\Delta\alpha}{\alpha_i\omega\tau}\right) \\ \tau_R &\simeq t_1/\alpha_i l \\ \tau_c &\simeq t_1/\gamma \end{aligned} \quad \text{for } \alpha_i l \gg \gamma \tag{32}$$
$$\tau_R \simeq \tau_c \simeq t_1/(\alpha_i l - \gamma) \quad \text{for } \alpha_i l \simeq \gamma$$

where $\tau = t_1/(\alpha_i l - \gamma)$ = time constant for photon density rise
t_1 = time for a single pass
$\Delta\alpha/\alpha_i$ = change of gain at which pulse rise begins $\simeq 0.1$
w = probability per second for a pump phonon to excite level 3
α_i = gain before switching

In the case of populations being just above threshold, the pulse is symmetric, while if pumped well above threshold, before pulsing the shutter, the rise time will be shorter than the decay time. In fact both rise and fall will be shorter than for the less inverted situation, and the decay is effectively that governed by the cavity Q.

A diagram of the essential components of the giant-pulse laser is given in Fig. 60. The ruby output is polarized perpendicular to the plane containing both the c and the rod axes. The Kerr cell is operated with the electric field at 45° to the plane of polarization of the ruby output and the parallel polarizer and with the electric field at its 45° rotation value. With the field on, the cavity Q is considerably reduced since light passing through the Kerr cell to the mirror and back is rotated 90° into the nonpass orientation of the polarizer. Momentary removal of the field from the Kerr cell increases the resonator Q to the point where the losses are those due to reflectivities, absorption, and scattering losses. The effectiveness of the action is determined by the change of Q produced by the Kerr cell or other shutter arrangement. Utilizing the polarization of the ruby radiation with no polarizer, McLung

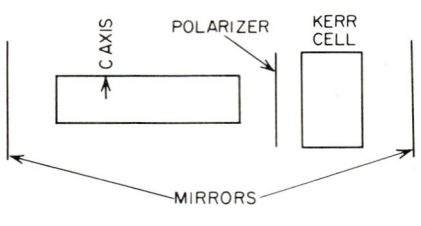

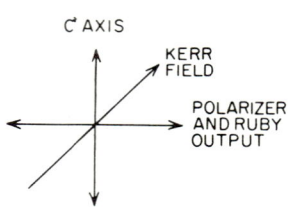

Fig. 60. Block diagram of giant-pulse laser. When used with ruby, polarizer should be lined up perpendicular to the c axis so as to agree with the normal polarization of the ruby output. Kerr cell field is oriented at 45° to polarizer as shown below.

Fig. 61. Block diagram of traveling-wave laser amplifier. One of the two 45° polarizers can be omitted with little effect. The necessity of a Faraday rotator-polarizer combination in input and output depends on the reflections encountered. In general they should not be necessary.

and Hellwarth[98] obtained giant-pulse operation even though the threshold differential was not extremely large.

The use of a Glan-Thomson prism or any of the other devices for rotating the plane of polarization is quite satisfactory as long as the high powers do not destroy the effectiveness. For example, dielectric layers on the mirrors are far superior to silvered ones for this purpose because the high loss in the silver results in their destruction by a single high-powered pulse. Other schemes for Q switching involving rotating mirrors, mechanical shutters, ultrasonic diffraction shutters, etc., are all effective methods of giant-pulse generation if the necessary synchronization is provided. Dye films[101] have also been used in a mode in which the film because of its high loss is destroyed at a low level of oscillation corresponding to a very high inverted population with low Q. After loss of the film the Q is large and the pulse forms as above.

8.7. A Ruby-laser Amplifier. Laser amplifiers such as the cesium amplifier illustrated in Fig. 48 must be designed with precautions to prevent oscillation. As an example, the solution to the problem using a ruby laser will be sketched.

For a reciprocal net power gain of G, which the maser and laser exhibit without the addition of nonreciprocal elements, such as the ferrite of the traveling-wave maser (see Chap. 14, Sec. 13.1), and a reflection coefficient r the amplifier will be stable, i.e., not oscillate if $Gr < 1$. For a ruby rod with an air-ruby reflection of $r = 0.07$ the gain is limited to 14. This may be improved by inserting nonreciprocal loss,

INDUCED-EMISSION AMPLIFIERS AND OSCILLATORS 29-51

such as a Faraday isolator, rotating the angle of polarization so as to allow passage in one direction but not in the other, and using antireflection coatings on all surfaces.

The block diagram of such a system[102] is given in Fig. 61. With uncoated ends each amplifying section can have just less than 14 db gain, and the whole system gain is determined by the number of sections used.

The isolator is illustrated in Fig. 62a. The angle of rotation of the plane of polarization is given by

$$\theta = VHl \tag{33}$$

where H = magnetic field
  l = length
  V = the Verdet constant = $\pi\nu_0(n_+ - n_-)/Hc$
  n_+, n_- = refractive indices for right- and left-handed circularly polarized light

ZnS and PbO glass are both possible materials for the rotator. The Verdet constants are 0.09 min/cm/gauss and 0.22 min/cm/gauss for high-density PbO glass

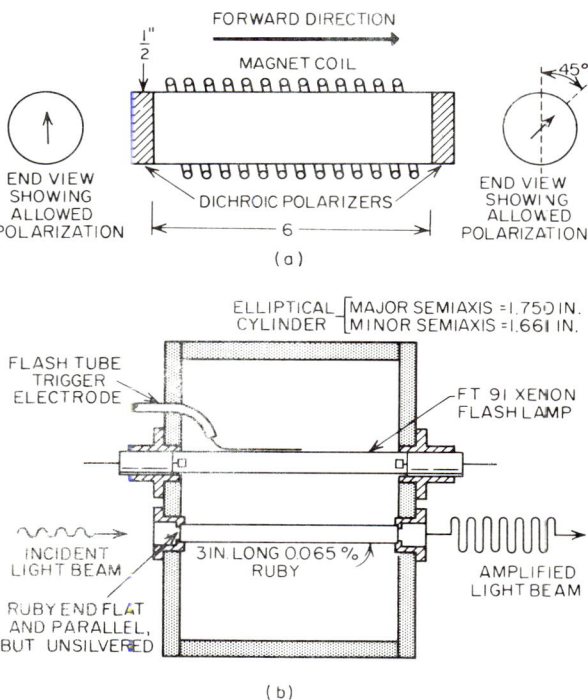

Fig. 62. (a) Faraday rotator for traveling-wave laser amplifier. The material is PbO glass. (b) Ruby amplifier section. (*After Geusic and Scovil,*[102] *reprinted by permission of the copyright owner, American Telephone and Telegraph Company, and the authors. This article originally appeared in Bell System Tech. J., vol. 41, July, 1962.*)

and ZnS(β), respectively. The attenuation of PbO glass is 0.08 db/cm at 14,400 cm^{-1} (R_1 line) while for ZnS the attenuation is negligible at this frequency. With Polaroid polarizers (type HN-38) the forward loss of 3 db is to be compared with the reverse loss of 16 db. This can be improved, primarily in the forward direction, by use of antireflection coatings.

The amplifier section is illustrated in Fig. 62b. Using a 250-joule input to selected FT-91 flash tubes a net gain of 10 db for the 3-in. unit ($\simeq$130 db/m) was obtained without the use of cooling. This, of course, is higher than typical gains of gas systems, and the problems associated with preventing oscillation are therefore greater.

8.8. Other Iron-group Laser Materials. Although Cr^{3+} is the most common of the iron-group impurities used for laser operation, others are possibilities, and it is worth pointing out the merits of one which has been used. The ion is Ni^{++} in MgF_2.[103] In this case the transition involved (Fig. 63) requires the emission of a phonon with each laser quantum. The normal fluorescent spectrum of Ni^{++} in MgF_2 consists of a number of transitions in the region of 6,500 cm^{-1} with a broad smear at slightly lower energies due to interaction with the lattice. From the laser point of view the energy levels of Fig. 63 are those of a four-level system if the laser transition terminates on the phonon level at 340 cm^{-1}. Since the pumping requirements for four-level systems are so much lower than for three-level systems, such as ruby, it is not surprising that the 6,164 cm^{-1} laser transition dominates.

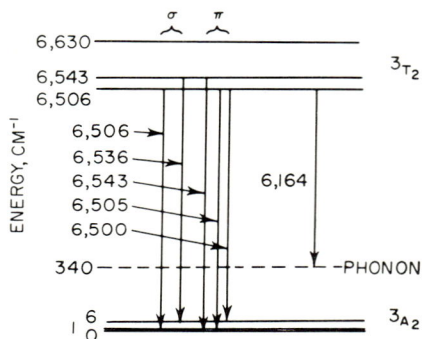

Fig. 63. Energy levels of Ni^{++} in MgF_2.

Utilizing this type of transition it is possible to consider lasers in which the frequency may be varied over a considerable range by using frequency-selective devices to choose the region of the phonon band in which operation is desired. In this particular case the band extends from 6.5×10^3 to 5.5×10^3 cm^{-1} or from approximately 1.53 to 1.82 microns wavelength.

9. RARE-EARTH LASERS

Except for Cr^{3+} (in Al_2O_3) the ions generally involved in laser action are of the $4f$ rare-earth and the $5f$ rare-earth or actinide series. These were considered as possible laser materials very early in the development of the field because of the sharp spectral lines known to be characteristic of their presence in crystalline solids. The $4f$ rare-earth ions have the $4f$ valence electrons ($n = 4$, $l = 3$) shielded from the crystal field effects by the outer electrons (two $5s$ and six $5p$), the $5d$ and $6s$ electrons of the neutral atom having been removed in the ionization process. In a similar fashion the $5f$ valence electrons of the actinide series are shielded by the outer $6s$ and $6p$ electrons. Transitions involving only $4d$ electrons (considering the $4f$ rare-earth series) are sharp while those requiring $5d \rightarrow 4f$ transitions are broadened by the crystal field effects on the $5d$ levels. The former are then ideal laser oscillating or amplifying transitions while the latter are ideal for pumping. For the doubly ionized rare earths these bands are at a convenient frequency for pumping, but in the triply ionized state they are in the ultraviolet and have not been utilized. The shielding is not so complete in the actinide series with its two stable elements uranium and thorium, but the principles are the same.

9.1. Rare-earth Laser Characteristics. From the point of view of laser operation the important difference between this group and the Cr^{3+} in Al_2O_3 is that four-

INDUCED-EMISSION AMPLIFIERS AND OSCILLATORS 29–53

level operation is quite common in the rare earths. As pointed out in Chap. 14, this reduces the pump power requirement by a considerable factor, making continuous operation more readily attainable. The majority of the oscillating transitions are in the infrared.

One achievement with rare-earth ions should be pointed out. Certain of them have been successfully used for laser work as impurities in glass. This has a tremendous effect on the ease of obtaining large active volumes. Not that manufacture of large pieces of doped glass to be uniform and strain-free is easy, but it seems to be easier than growing large perfect strain-free crystals.

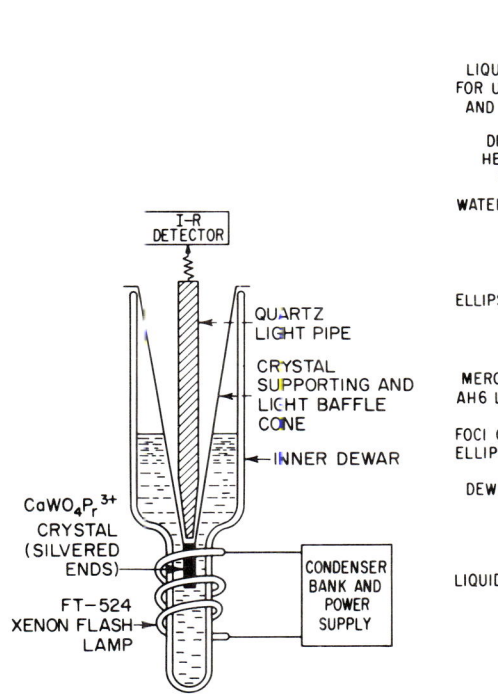

Fig. 64. Configuration for pulse laser operation. (*After Yariv and Gordon*,[113] published by permission of the editor, *IEEE*.)

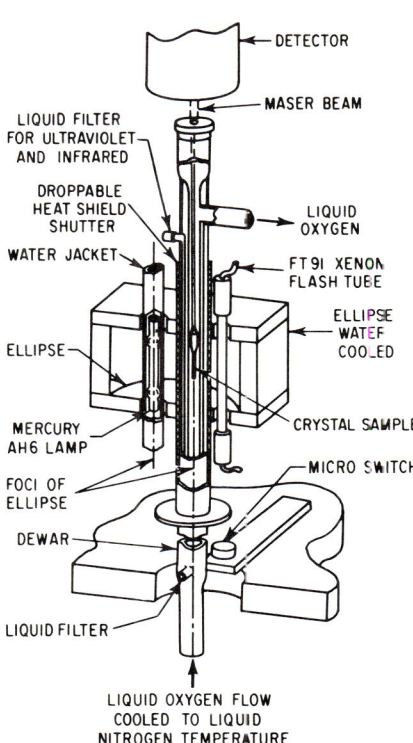

Fig. 65. Experimental arrangement for CW operation. (*After Johnson, Boyd, Nassau, and Sodden*,[115] published by permission of the editor, *Phys. Rev.*)

Table 13 gives the properties of a number of laser transitions. The pulse thresholds quoted are for an FT524 flash lamp (Fig. 64), and comparison from one transition to another should be valid at least over a limited range. Actually the threshold depends on the ratio of pump pulse length to fluorescence lifetime; obviously it appears high if the exciting pulse length is long compared with the fluorescent lifetime. For wide variations of threshold power the relative lengths are unlikely to be comparable, and hence the threshold figures must be used with care.

For CW operation a system such as that of Fig. 65 with a linear mercury lamp or one of the compact xenon arc lamps and specular elliptical reflector is satisfactory and in fact was used for many of the CW measurements noted in Table 13. The mercury lamp has a higher proportion of its energy output in the visible while the

Table 13

Host	Ion	λ, microns	τ_{spont}, msec	Threshold, joules	Pump wavelength, microns	Terminal level, cm^{-1}	Laser transition	Temp, °K	Remarks	References
Ca(NbO$_3$)$_2$	Pr^{3+}	1.04	...	20–25	...	...	...	77	Pulsed. All thresholds determined in same apparatus. CaWO$_4$:Nd^{3+} in same apparatus. Has threshold of 1.0 joule	104
	Nd^{3+}	1.060	0.12	2.0	...	...	...	77		104
	Ho^{3+}	2.047	2.2	90	...	...	...	77		104
	Er^{3+}	1.61	...	800	...	...	...	77		104
	Tm^{3+}	1.91	...	125	...	...	...	77		104
CaF$_2$	Nd^{3+}	1.0457	...	60	0.70–0.80 0.56–0.58	2,000	$^4F_{\frac{3}{2}}-^4I_{\frac{13}{2}}$	77	Pulsed	105
	Sm^{2+}	0.708	0.002	0.01	0.425–0.5 0.59–0.65	263	$^5D_0-^7F_1$	20	Pulsed, no spiking	106
	Dy^{2+}	2.36	10 (77°)	20 (77°)	0.8–1.0	35	$^5I_7-^5I_8$	90	CW—no spiking; spiking in pulsed operation	107, 108
	Ho^{3+}	2.092	...	260	0.4–0.66	0	$^5I_7-^5I_8$	77	Pulsed	109
	Tm^{2+}	1.116	4	50 (4°)	0.28–0.34	...	$^2F_{\frac{5}{2}}-^2F_{\frac{7}{2}}$	4		110
		1.115	...	450	...	...	$^2F_{\frac{5}{2}}-^2F_{\frac{7}{2}}$	20		109
				800			$^2F_{\frac{5}{2}}-^2F_{\frac{7}{2}}$	77		109
	U^{3+}	2.613	0.13 (77°)	1 (77°)	0.9	609	$^4I_{\frac{11}{2}}-^4I_{\frac{9}{2}}$	300	Pulsed (spiking)	111
								100	CW	111
		2.438	...	...	...	...	...	77	Pulsed	112
		2.511	...	6	...	...	...	77	Pulsed	113
		2.223	...	2,000	...	0	...	77	Pulsed	113
CaWO$_4$	Pr^{3+}	1.047	0.05 (77°)	15	0.45–0.5	377	$^1G_4-^3H_4$	77	Pulsed	114
	Nd^{3+}	1.065	...	1.5 (77°)	0.57–0.6	2,000	$^4F_{\frac{3}{2}}-^4I_{\frac{11}{2}}$	300	CW—no spiking; spiking in pulsed operation	115
		1.063	14 (77°)	14 (77°)	0.57–0.6	2,000	$^4F_{\frac{3}{2}}-^4I_{\frac{11}{2}}$	...	Pulsed operation	116
		1.066	0.1 (77°)	6 (77°)	0.57–0.6	2,000	$^4F_{\frac{3}{2}}-^4I_{\frac{11}{2}}$	...	Pulsed	116
		1.058	...	80 (77°)	0.57–0.6	2,000	$^4F_{\frac{3}{2}}-^4I_{\frac{11}{2}}$	...	Pulsed	116
		1.064	...	7 (77°)	0.57–0.6	2,000	$^4F_{\frac{3}{2}}-^4I_{\frac{11}{2}}$	...	Pulsed	116
	Ho^{3+}	2.046	...	80 (77°)	0.44–0.46	230	$^5I_7-^5I_8$	77	Pulsed	109

	Ho³⁺	2.059		250 (77°)			77	Pulsed	117
	Er³⁺	1.612		800 (77°)	0.38 0.52	375	$4I_{13/2}{-}4I_{15/2}$	77	Pulsed	118
	Tm³⁺	1.911		60 (77°)	0.46–0.48 1.7–1.8	325	$3H_4{-}3H_6$	77	Pulsed	119
		1.916		73 (77°)	0.46–0.48 1.7–1.8	325	$3H_4{-}3H_6$	77	Pulsed	119
		1.918			0.46–0.48 1.7–1.8	325	$3H_4{-}3H_6$	77	Pulsed	119
SrWO₄	Nd³⁺	1.057		4.7	0.57–0.60	2,000	$4F_{3/2}{-}4I_{11/2}$	77		120
		1.063		5.1	0.57–0.60	2,000	$4F_{3/2}{-}4I_{11/2}$	77		120
		1.061		7.6	0.57–0.60	2,000	$4F_{3/2}{-}4I_{11/2}$	77		120
		1.063		180	0.57–0.60	2,000	$4F_{3/2}{-}4I_{11/2}$	77		120
		1.067		45	0.57–0.59	2,000	$4F_{3/2}{-}4I_{11/2}$	295		109
CaMoO₄	Nd³⁺	1.0673		100	0.57–0.59	2,000	$4F_{3/2}{-}4I_{11/2}$	77		109
PbMoO₄	Nd³⁺	1.0586		60	0.57–0.59	2,000	$4F_{3/2}{-}4I_{11/2}$	295		120
SrMoO₄	Nd³⁺	1.064		17	0.57–0.60	2,000	$4F_{3/2}{-}4I_{11/2}$	295		115
SrMoO₄	Nd³⁺	1.0652		70	0.57–0.60	2,000	$4F_{3/2}{-}4I_{11/2}$	77		115
		1.059		150	0.57–0.60	2,000	$4F_{3/2}{-}4I_{11/2}$			115
		1.0627		170	0.57–0.60	2,000	$4F_{3/2}{-}4I_{11/2}$			115
		1.0611		500	0.57–0.60	2,000	$4F_{3/2}{-}4I_{11/2}$			115
		1.0643		125	0.57–0.60	2,000	$4F_{3/2}{-}4I_{11/2}$	295		115
		1.0576		45	0.57–0.60	2,000	$4F_{3/2}{-}4I_{11/2}$	77		109
SrF₂	Nd³⁺	1.0437		150	0.72–0.75	2,000	$4F_{3/2}{-}4I_{11/2}$	295		109
		1.0370		480	0.78–0.81	2,000	$4F_{3/2}{-}4I_{11/2}$	77		109
BaF₂	Tm³⁺	1.972		1,600		325	$3H_4{-}3H_6$	77		109
	Nd³⁺	1.060		1,600	0.57–0.60	2,000	$4F_{3/2}{-}4I_{11/2}$	77		109
LaF₂	Nd³⁺	1.0631		93	0.5–0.6	2,000	$4F_{3/2}{-}4I_{11/2}$	77		109
		1.0394		75	0.5–0.6	2,000	$4F_{3/2}{-}4I_{11/2}$	295		109
		1.0633		150	0.5–0.6	2,000	$4F_{3/2}{-}4I_{11/2}$	300		109
Glass	Nd³⁺	1.06		50	0.5–0.6	2,000	$4F_{3/2}{-}4I_{11/2}$	77		121
	Gd³⁺	0.3125	4 (300°)		0.274 0.277		$6P_{7/2}{-}8S_{7/2}$		Pulsed	122
Ho³⁺		1.95	0.07 (77°)	3,600 (77°)	0.44–0.46		$5I_7{-}5I_8$	77	Pulsed	123
Yb³⁺		1.015	1.5	1,300	0.91 0.95 0.96		$2F_{5/2}{-}2F_{7/2}$	77	Pulsed	124

29-55

29-56   AMPLIFIER CIRCUITS

xenon has more infrared. The choice between the two is determined by the pumping frequency required. Cooling is accomplished by flowing a liquid coolant past the crystal. It is essential that the liquid be precooled (e.g., liquid oxygen cooled to liquid nitrogen temperature of 77°K) so that transfer of energy will not result in a layer of gas surrounding the crystal, insulating it from the coolant. An outer jacket is provided to allow use of a liquid filter to eliminate unwanted frequencies in the pump spectrum. For example, the ultraviolet has been found to be harmful for Nd^{3+} in $CaWO_4$ operation because of formation of a radiation-damage center in addition to

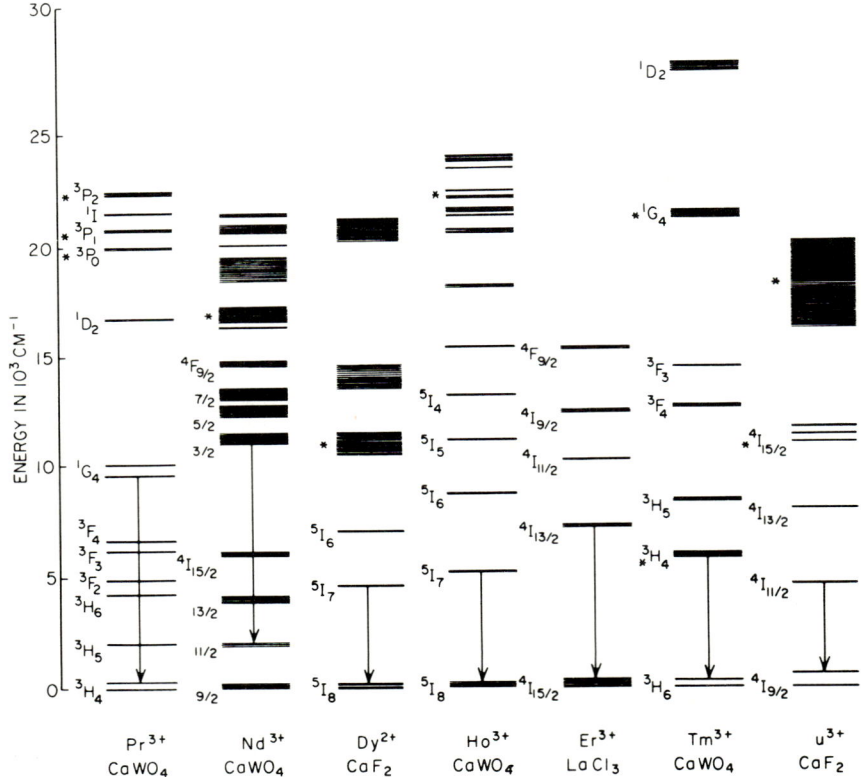

FIG. 66. Some of the rare-earth levels. There is little variation from one lattice to another. Arrows indicate laser transitions, and the pumping levels are marked with an asterisk.

the useless heating which results from noneffective parts of the pump spectrum (the major source of heating in many lasers).

As an example of the details in the operation of one of the rare-earth lasers, Fig. 68 illustrates the energy levels $4F_{3/2}$ and $4I_{3/2}$ involved in the Nd^{3+} laser in $CaWO_4$. The two heavy transitions are dominant, the higher-energy one at room temperature where the 11,469 cm^{-1} level is partially populated and the 9,390 cm^{-1} (1.065-micron) transition at 77°K, at which temperature the thermal equilibrium in the $4F_{3/2}$ level leaves only the 11,406 cm^{-1} level significantly populated. These energy levels are appropriate to Na^+ compensated Nd^{3+} in the divalent Ca^{2+} site of $CaWO_4$.[109] Other charge compensation will change frequencies slightly and account for the variety of energies quoted for the Nd^{3+} laser. If the crystal is grown without specific steps to

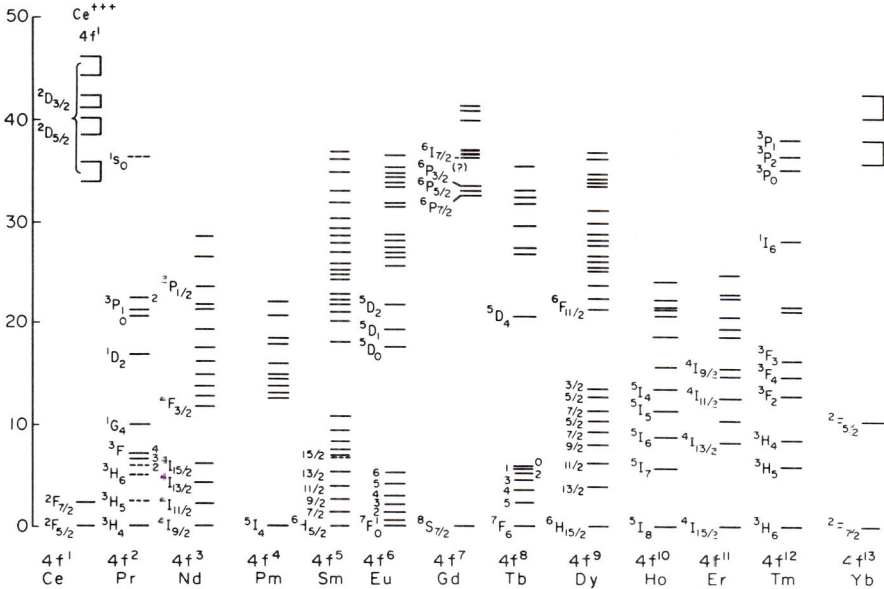

FIG. 67. Triply ionized rare-earth levels. Much of the data is derived from McClure's review.[125]

control the compensation a wide variety of local defects, e.g., calcium vacancies, interstitial oxygen, pairing of neodymium ions, interstitial alkali metals, or alkali metals substituted for Ca^{2+} in neighboring sites, will be contained in the crystal

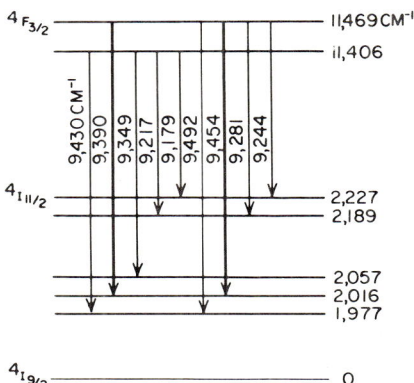

FIG. 68. Energy levels and laser transitions of Nd^{3+} in $CaWO_4$ with Na^+ compensation. (After Johnson,[109] published by permission of the editor, J. Appl. Phys.)

near Nd^{3+} sites, and the resulting lines will then have a number of components. In fact, even for the Na^+ compensated crystals three frequencies in the vicinity of 1.065 microns (9,390 cm^{-1}) are observed to oscillate, indicating either other compensation mechanisms or more likely different sites for the Na^+ relative to the Nd^{3+}.

10. SEMICONDUCTOR OR INJECTION LASERS

Characterized by its small size and high efficiency the injection laser falls short of the frequency-stability characteristic of the gas and solid-state lasers. The power output is limited by the necessity of dissipating the excess injection power in the extremely small junction volume and by the problem of conducting this heat to a sink. The beam is usually fan-shaped with an angular range of approximately 10° in the plane perpendicular to the junction and less than 1° in the plane of the junction itself. Efficiencies on the order of 50 per cent for high-power injection lasers and 75 to 80 per cent for lower-power ones along with the modulation capability make semiconductor lasers of considerable interest.

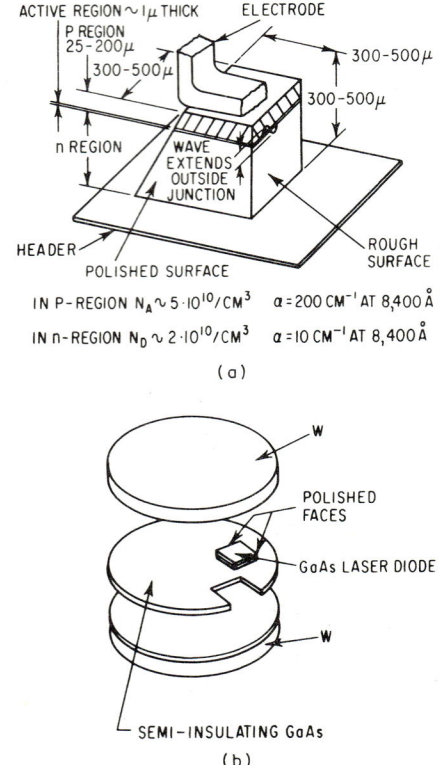

FIG. 69. (a) Construction of typical GaAs p-n junction. (*After Fenner and Kingsley*,[127] *published by permission of the editor, J. Appl. Phys.*) (b) High-powered GaAs p-n junction. (*After Engeler and Garfinkel*,[128] *published by permission of the editor, J. Appl. Phys.*)

10.1. Construction of GaAs Injection Lasers. Two junction lasers are illustrated in Fig. 69. One is illustrative of a typical Fabry-Perot resonator construction suitable for pulse operation. The other sketch is of a laser designed for continuous operation at 20°K.

Typical manufacture involves diffusing an acceptor such as Zn into Te or Si-doped (n-type) GaAs with concentrations of 3×10^{17} to 3×10^{18} per cu cm. Diffusion at 850°C for 16 hr forms a junction a few mils below the surface. The diodes are then cut and polished or cleaved to provide the two parallel Fabry-Perot faces. All four faces are sometimes finished, resulting in a lower threshold but a more complicated radiation pattern due to the numerous possible resonant modes. The surfaces could be coated to increase reflectivity, but the gain is found to be so high that normally the GaAs air or liquid interfaces with reflectivity of approximately 30 per cent make the resonator end plates.

For high-power operation the main problem is heat removal from the junction volume. In the device sketched this is accomplished by silver-soldering the thin GaAs crystal (they can be constructed with thicknesses down to 120 microns) to the tungsten disks, of ultra-high-purity material to provide good heat conduction at low temperatures. The assembly is then soldered to a copper finger which may be attached to a cryostat. The bulk GaAs of the diode provides the main thermal impedance at hydrogen temperature and below. The choice of tungsten next to the

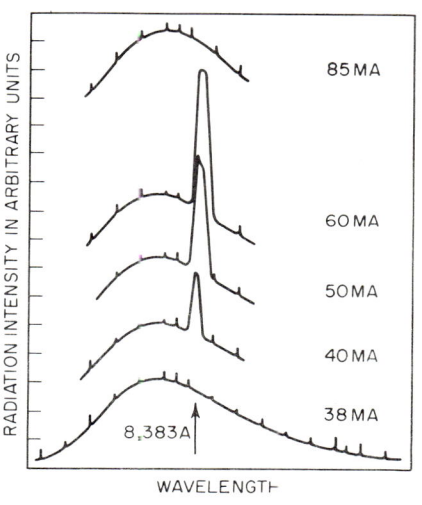

FIG. 70. Relative CW output of a GaAs p-n junction at different currents. Each successive curve is moved upward by two divisions. The lower and upper curves correspond to spontaneous emission, the latter because of heating in the junction. (After Howard, Fang, Dill, and Nathan,[129] published by permission of the editor, IBM J.)

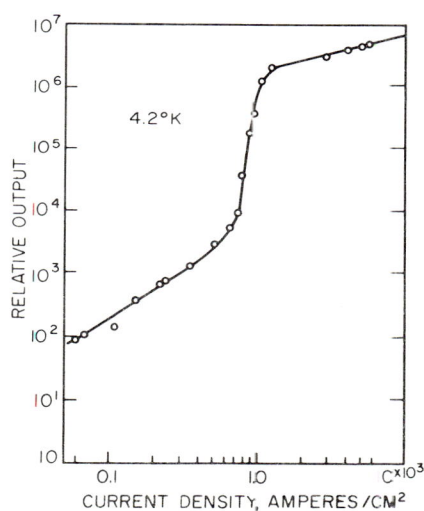

FIG. 71. Variation of peak amplitude of GaAs output with current. Threshold was 800 amp/sq cm for this diode. (After Engeler and Garfinkel,[130] published by permission of the editor, J. Appl. Phys.)

diode is the result of the necessity of matching thermal-expansion coefficients, and for this purpose molybdenum and tungsten are most suitable.

10.2. Radiation Properties of GaAs Injection Lasers. The radiation from the GaAs junction is a function of the current density passing through the forward-biased junction. The sequence of changes is illustrated in Fig. 70. Below threshold the broad spontaneous-emission line is observed. Just below threshold the radiation traveling along the plane of the junction is amplified and the intensity begins to peak in this direction while the frequency begins to narrow because of the selective amplification—highest near the maximum of the spontaneous line. At threshold the gain is sufficient to overcome all losses, the intensity peaks in the forward direction and is far more intense than the isotropically radiated spontaneous emission, and the oscillating line is much narrower in width than the line below threshold.

Figure 71 provides a plot of the peak output intensity as a function of current density in the junction. The slightly faster than linear rise below threshold, at lower

29-60 AMPLIFIER CIRCUITS

current densities, is the result of the spectral narrowing, and the very precipitous increase in intensity just at threshold is due primarily to spatial narrowing of the beam but is partly accounted for by the increased efficiency of the diode above threshold.

The absorption of the n and p regions at 8,400 angstroms is high (10 cm^{-1} and 200 cm^{-1}, respectively). This combined with the limited diffusion of carriers restricts

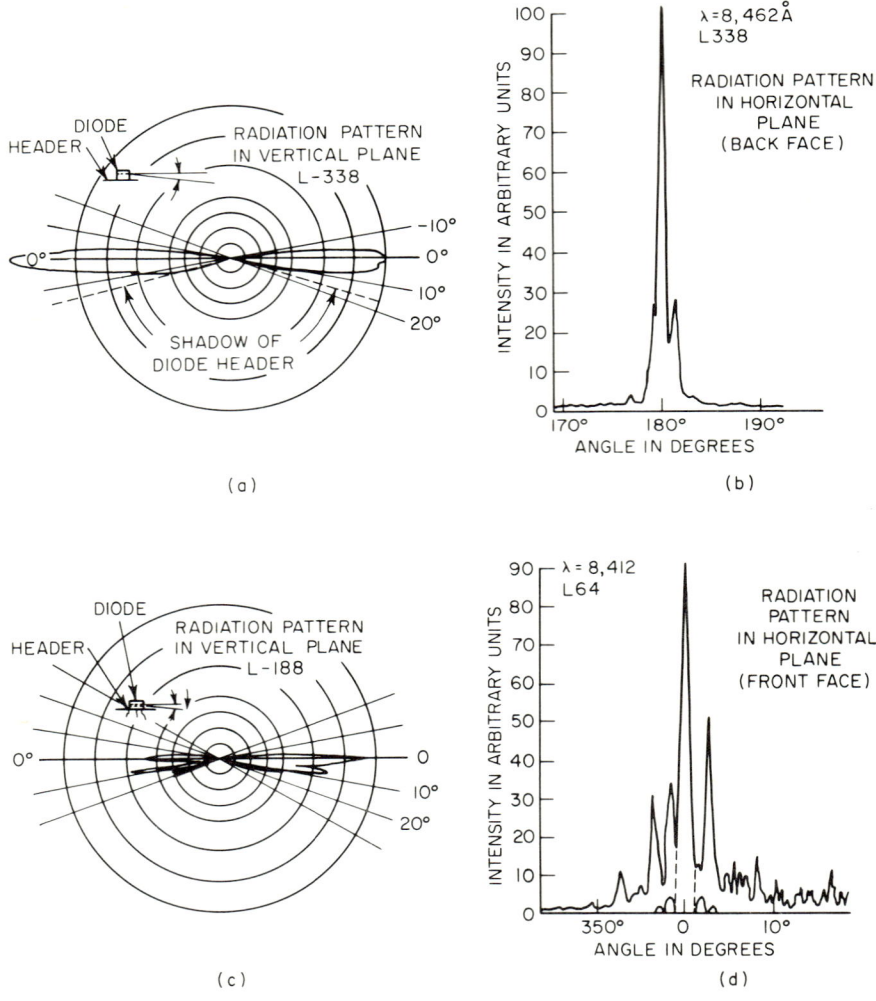

Fig. 72. Radiation patterns from GaAs diodes. a and c represent relatively ideal patterns for vertical and horizontal directions as defined in Fig. 69a. (*After Fenner and Kingsley*,[127] *published by permission of the editor, J. Appl. Phys.*)

the oscillating fields to a narrow region, a few microns in thickness. The majority of the light is emitted on the p side of the junction.[131] The external radiation then appears to originate at the junction edges, with patterns as illustrated in Fig. 72. a and b of this figure show radiation patterns in the vertical (perpendicular to junction) plane where the total radiating width is limited by the junction thickness.

c and d illustrate radiation patterns in the plane of the junction. From the width of the latter it can be estimated that typical junctions are oscillating coherently over lengths considerably less than the maximum length along the mirror face, i.e., perhaps 50 to 100λ along the junction rather than uniformly along the whole length.

The output frequency of the GaAs laser is just less than that corresponding to the band gap of the material. At room temperature the output is 9,000 angstroms, decreasing to about 8,400 angstroms at liquid helium temperatures. This variation is plotted in Fig. 73, where a comparison is made with the value of the energy gap,

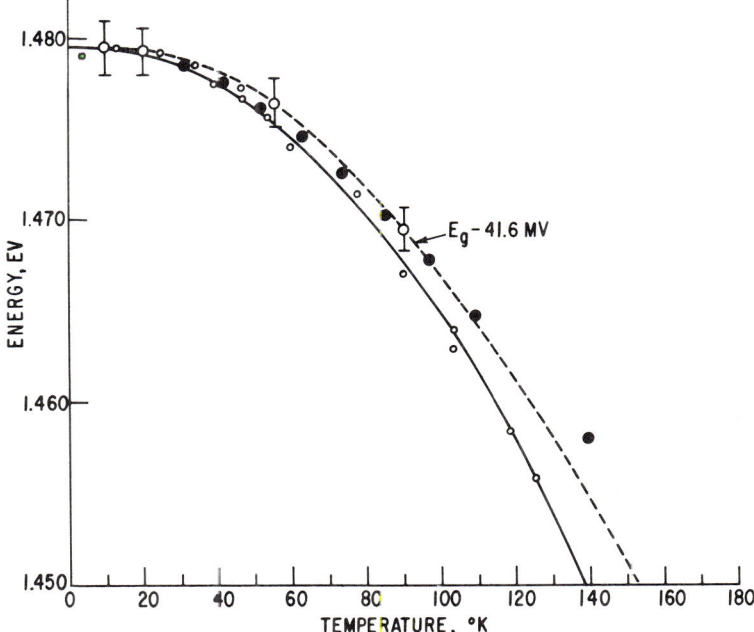

FIG. 73. Frequency or energy of output just above threshold at various temperatures. ° experimental data, • data corrected for temperature rise during pulse, Φ energy gap less 41.6 ev. (After Engeler and Garfinkel,[130] published by permission of the editor, J. Appl. Phys.)

as it is known to vary with temperature, minus 41.6 mv. The agreement is rather good, indicating a relationship between output frequency and the band gap—but the exact initial and final levels are not clearly delineated.

Actually, as the temperature is changed two phenomena occur: the energy changes as described above and the properties of the resonator change primarily because of the change in refractive index with temperature as the band gap changes. Engeler and Garfinkel[130] give the value of

$$\frac{d(h\nu)}{dT} = \frac{n_0 - n}{n_0}\frac{dE_g}{dT} \tag{34}$$

where $n_0 = n + \nu(\partial n/\partial \nu)$, with n being the refractive index at frequency ν and n_0, being defined as $c/(2l\,\Delta\nu)$ when $\Delta\nu$ is the mode separation of the resonator, is the refractive index observed in the diode, as measured by mode separation. Numerically this result is

$$d(h\nu)/dT = 0.35(dE_g/dT) \tag{35}$$

and dE_g/dT is given by Fig. 73. Thus the shift of emission peak is about 2.9 times as fast as the mode-frequency shift. The result is that a change of temperature will cause the output frequency to shift, which shift is in fact accomplished by mode jumping. This type of behavior is illustrated in Fig. 74, in which the outputs of a number of modes are plotted with current as a parameter. The shift to lower energy is attributed to an increase in temperature while the increased number of modes is attributable to the higher pumping level.

Figure 74 is also an excellent representation of the type of output obtained in semiconduction injection lasers. Normally, above threshold a number of modes will oscillate. The separation between the modes depends on cavity dimensions in the usual way while the width should be determined by the limitation [Chap. 14, Eq. (95)] set by spontaneous emission into the mode. Since cavity dimensions are typically

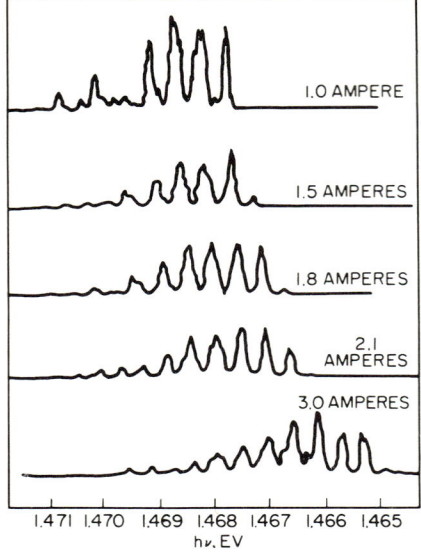

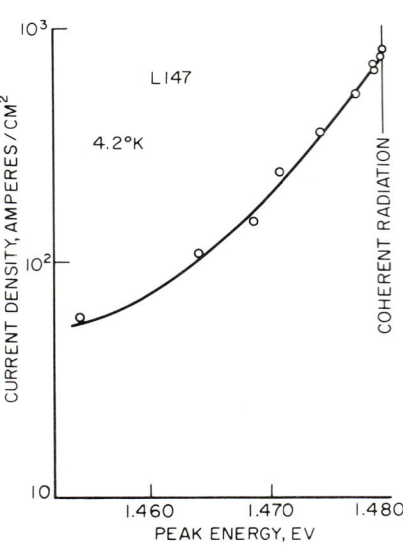

FIG. 74. Spectral output of a GaAs diode for a number of currents well above the threshold of 0.26 amp. The shift of mode frequency due to heating as well as the overall shift of oscillating frequency is evident. (*After Engeler and Garfinkel*,[128] *published by permission of the editor, J. Appl. Phys.*)

FIG. 75. The peak of the spontaneous-emission line as it varies with current density. At the point of oscillation the output frequency is not a function of current density. (*After Engeler and Garfinkel*,[130] *published by permission of the editor, J. Appl. Phys.*)

0.3 mm, mode separations are in the vicinity of 0.0005 ev, while the mode widths observed are near 300 Mc or $\simeq 10^{-6}$ ev.

Because of temperature changes during the pulse, the output of the pulsed injection laser drifts with time during the pulse. This shift takes the form of mode jumping in going to longer wavelengths as the temperature rises during the pulse. This is analogous to the frequency changes occurring in other pulsed solid-state lasers.

There is, in addition, some variation of emitted frequency from diode to diode. This can be attributed to a variation of the threshold current density since one can physically consider that, at higher currents, higher energy levels of the conduction band are occupied and are responsible for the emission. In effect the quasi-Fermi level of the injected electrons rises with injected current density. The result is a shift of frequency with current density below threshold, as illustrated in Fig. 75. The output is constant above threshold because of the equilibrium between the number of injected carriers and output power. If, however, for some reason, such as scattering or reflection losses, the thresholds differ from diode to diode, the output frequencies will also differ.

10.3. Harmonic Content of GaAs Laser Output. Because of anharmonic forces in the diode itself the output, in fact, consists of all possible sum and difference frequencies caused by the main transition beating with itself. There is noticeable blue output which increases rapidly with fundamental power output (proportional to P^2) as shown in Fig. 76.

10.4. Efficiency of GaAs Lasers. Efficiencies at threshold are about 5 per cent and for the diode of Fig. 69b with a threshold of 0.26 amp they rise to 36.5 per cent at 1 amp and to 46.5 per cent at 4 amp for a diode base temperature of 20°K. This measurement made by a thermodynamic comparison method includes all radiant

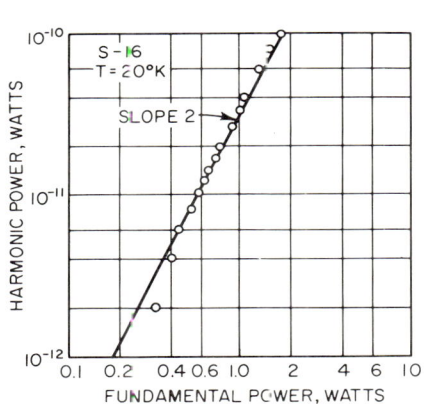

FIG. 76. Harmonic power output ≃0.42 micron as a function of the fundamental 0.84 micron power. (After Engeler and Garfinkel,[128] published by permission of the editor, J. Appl. Phys.)

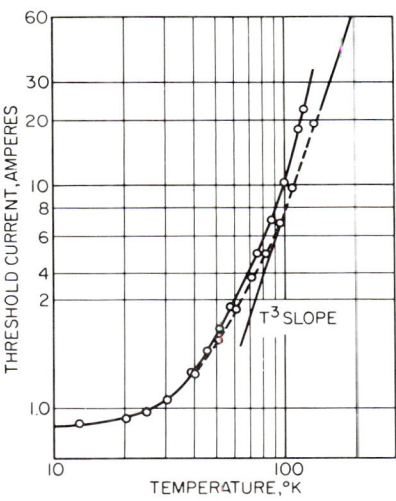

FIG. 77. Typical variation of threshold current vs. temperature for GaAs diodes. (After Engeler and Garfinkel,[130] published by permission of the editor, J. Appl. Phys.)

power, which was found to be 0.57 watt at 1 amp and 3.2 watts at 4 amp. Using the 5 per cent threshold figure it can be estimated that 20 mw of incoherent light is emitted.

The threshold current, however, is also a function of temperature as illustrated in Fig. 77. At temperatures above about 100°K the variation is proportional to T^3. For a given base temperature and thermal-conductivity path from the junction to that base temperature only a certain maximum power can be dissipated in the junction if the temperature is to be kept low enough for oscillation to take place at all. For instance, many diodes made in the form of Fig. 69b will not oscillate with the base temperature of 77° because the threshold power causes the junction temperature to rise so far that oscillation is impossible.

10.5. Other Injection Lasers. A number of different semiconducting n-p junctions are useful as injection lasers. Some, as is true of GaAs, are relatively fixed in frequency, but others, alloys such as $GaAs_{1-x}P_x$, have an output which depends on composition since the band gap varies with the value of x.

10.6. Magnetic Field Effects in Injection Lasers. Application of a magnetic field to an injection laser has two effects. The first, the reduction of threshold current, is mentioned in Table 14 in connection with InSb and $InGa_{1-x}As_x$ thresholds

Table 14

Material	Temp, °K	λ, microns	Energy, ev	Remarks	Reference
GaAs	4	0.84	1.479		132, 133, 134
	77	0.843	1.470		132, 133, 134
	300	0.90	1.377		132, 133, 134
InAs	4	3.1	0.401	CW	135
	77				135
InP	4	0.906	1.37		136
	77				136
InSb	1.7	5.2	0.239	Uses magnetic field to reduce threshold	137
$Ga(As_{1-x}P_x)$	77	0.62–0.84	2.0–1.48	λ depends on the value of x	138
$InP_{1-x}As_x$	77	1.6	0.775	For $x = 0.51$; could vary from 3.1 to 0.906 microns depending on x	139
$InGa_{1-x}As_x$	1.9	2.09	0.58	$x = 0.75$; magnetic field used to reduce threshold	140
		1.77	0.7	$x = 0.65$; magnetic field used to reduce threshold	140

and has been observed in other materials. This effect takes two forms, depending on the direction of the magnetic field relative to the current flow. For the two specific cases mentioned above the field is parallel to the direction of current flow, giving rise to the variation of threshold with field shown in Fig. 78. It is probable that this effect is due to a change in density of states produced by the magnetic field.

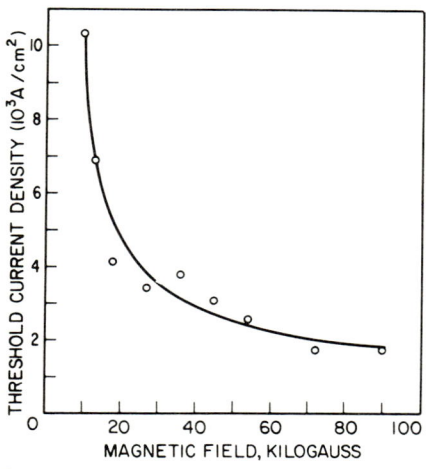

FIG. 78. Threshold current density for laser action as a function of a magnetic field parallel to the direction of current flow in an InSb diode. Temperature 2°K. (*After Phelan, Calawa, Rediker, Keyes, and Lax*,[137] published by permission of the editor, *Appl. Phys. Letters*.)

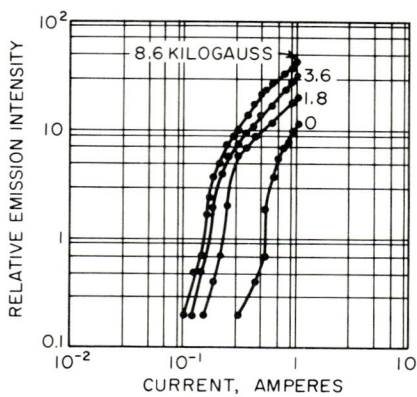

FIG. 79. Threshold current as a function of magnetic field perpendicular to the direction of current flow for pulse operation (0.4 μsec) of an InAs diode. Temperature 4.2°. (*After Melngailis and Rediker*,[141] published by permission of the editor, *Appl. Phys. Letters*.)

A magnetic field in the plane of the junction also reduces the threshold,[141] as illustrated in Fig. 79 for InAs. In this case the effect is attributed to a decrease of the diffusion constant because of the circular carrier motion produced by the magnetic field. This tends to concentrate the induced emission. Too high a magnetic field could be detrimental, i.e., cause cutoff, in this case.

Magnetic fields also tend to cause energy shifts of the output frequency. In the case of GaAs the effect is small but is quadratically dependent on magnetic field. For materials such as InSb the tuning is linear, is quite large, and is interpreted as a shift of the energy gap with magnetic field. This latter situation is illustrated in Fig. 80 from the work of Phelan et al.[137]

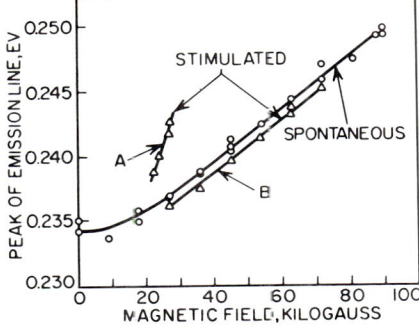

Fig. 80. Energy of both spontaneous and coherent emission peaks as a function of magnetic field parallel to current flow for InSb diode at 1.7°K. (*After Phelan, Calawa, Rediker, Keyes, and Lax,*[137] *published by permission of the editor, Appl. Phys. Letters.*)

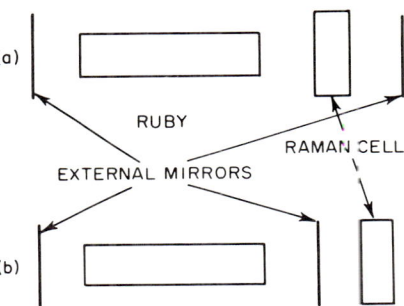

Fig. 81. (a) Raman laser schematic using an internal Raman cell. The mirrors serve to define the resonant cavity for both the ruby laser and the Raman frequencies. (b) The Raman cell used externally so that a single-pass traveling-wave maser-type interaction takes place in the Raman cell.

11. THE RAMAN LASER

The inclusion of a Raman active material within the laser cavity of giant-pulse ruby lasers leads to a number of coherent frequencies, other than the usual ruby 6,934-angstrom output, because of induced Raman scattering from totally symmetric vibrations. In contrast to other lasers the population is not inverted and the two-quantum Raman process leads to an induced, coherent output if the density of photons in the inducing transition is made high enough to provide more induced than spontaneous emission. This may be accomplished by placing the cell containing the Raman material within the external mirrors of the laser configuration as in Fig. 81a. If the laser output is sufficiently high, normal pulse operation or CW operation should result in laser action. Some of the Raman materials and frequencies observed along with the original laser frequency are listed in Table 15. This promises to be a convenient method of shifting frequencies by a predetermined amount.

Although the most convenient method of observation using a giant-pulse laser with the Raman material within the cavity is to use the Raman material in the Kerr cell, this will obviously work only with certain materials. For substances not suitable for use in the Kerr cell a Raman cell and a shutter arrangement (Kerr cell, Pockels cell, rotating shutter, etc.) must be provided. In this case the shutter arrangement should be non-Raman-active since some materials such as nitrobenzene are so strongly Raman-active that other Raman laser transitions are suppressed.[142]

Conversion efficiencies as high as 30 per cent, i.e., ruby laser to Raman frequency, using a giant-pulse laser have been observed in benzene (25 per cent to 7,455 angstroms and 5 per cent to 8,052 angstroms from the ruby 6,943 angstroms).

With high enough pulse energies the Raman laser effect can be observed in materials

Table 15

Raman-active material	Shift from 6,943-angstrom ruby line, cm^{-1}	Reference
Ethyl benzene, C_8H_{10}	1,002	142
Acetone, C_3H_6O	2,921	142
Cyclohexanone, $C_6H_{10}O$	2,863	142
	2,945	142
Piperidine, $C_5H_{11}N$	2,933	142
	2,936	142
	2,940	142
	2,943	142
Paraxylene, C_8H_{10}	2,988	142
Metaxylene, C_8H_{10}	2,933	142
Orthoxylene, C_8H_{10}	2,913	142
	2,922	142
	2,933	142
1,1,2,2 Tetrachloroethane, $C_2H_2Cl_4$	2,984	142
Nitrobenzene, $C_6H_5NO_2$	1,344	143
Benzene, C_6H_6	3,064	143
	990	143
Toluene, $C_6H_5CH_3$	1,004	143
1 Bromonaphthalene, $C_{10}H_7Br$	1,368	143
Pyridine, N:CHCH:CHCH:CH	992	143
Cyclohexane, C_6H_{12}	2,852	143
Deuterated benzene, C_6D_6	944	143

placed external to the cavity, and then it manifests itself as a quantum frequency transfer both up and down. Results obtained[144] in this fashion (Fig. 81b) are given in Fig. 82 for benzene. The remarkable aspects of this are the existence of the high harmonics of both Stokes ($\omega < \omega$ incident) and anti-Stokes ($\omega > \omega$ incident) lines,

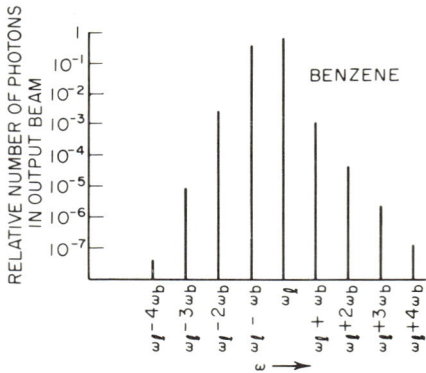

FIG. 82. Stokes ($\omega_l - n\omega_b$) and anti-Stokes ($\omega_l + n\omega_b$) lines observed in Raman laser action from benzene. (*After Terhune,*[144] *published by permission of the editor, Solid State Design.*)

and the high intensity of the anti-Stokes lines. Stoicheff[145] has observed similar results in liquid oxygen, hydrogen, and nitrogen with the added observation that no transitions corresponding to excitation of the second vibrational level through Stokes radiation are observed, indicating that the population of the first vibrational level is small. A classical theory which accounts for the results has been given by Garmire.

Pandarese, and Townes.[143] By utilizing the nonlinear polarizability of the Raman-active molecules to mix the frequencies of the incoming laser beam, the coherent induced Stokes emission, and the energy in the anti-Stokes line, the conditions under which multiple harmonics of each of the Raman-shifted lines will appear can be derived. There is every indication that the observations are completely compatible with this type of mixing and that the coherence of the laser beam, although not a necessary condition for the stimulated Raman Stokes emission, is required to obtain results similar to those illustrated in Fig. 82.

12. THE ORGANIC LASER

The use of organic molecules, specifically aromatic hydrocarbons, to provide the necessary energy levels for laser action is attractive both because of the variety of levels which are available and because of the glasslike host. The inclusion of a rare-earth ion in organic complexes such as the chelates has resulted in a certain amount of success,[147,148] although these are not true organic molecules. By and large the success with aromatic molecules is disappointing. Morantz[149,150] has reported on experiments with naphthalene, benzophenone, and a number of other organic molecules. In all cases the potential laser transition is from the metastable triplet to the ground singlet, and although a change in phosphorescent lifetime is claimed no bona fide net stimulated emission seems to have been obtained.

REFERENCES

1. J. A. Giordmaine and T. C. Wang, *J. Appl. Phys.*, vol. 31, p. 463, 1960.
2. J. P. Gordon, H. J. Zeiger, and C. H. Townes, *Phys. Rev.*, vol. 99, p. 1264, 1955.
3. J. C. Helmer, F. B. Jacobus, and P. A. Sturrock, *J. Appl. Phys.*, vol. 31, p. 458, 1960.
4. K. Shimoda, *J. Phys. Soc. Japan*, vol. 12, p. 1006, 1957.
5. K. Shimoda, T. C. Wang, and C. H. Townes, *Phys. Rev.*, vol. 102, p. 1308, 1956.
6. K. Shimoda, in C. H. Townes (ed.), "Quantum Electronics," p. 25, Columbia University Press, New York, 1960.
7. N. F. Ramsey, *Phys. Rev.*, vol. 78, p. 695, 1950.
8. F. Holtj, H. Daams, and S. N. Kalra, *J. Appl. Phys.*, vol. 33, p. 2370, 1962.
9. H. M. Goldenberg, D Kleppner, and N. F. Ramsey, *Phys. Rev. Letters*, vol. 5, p. 361 1960.
10. Daniel Kleppner, H. M. Goldenberg, and Norman F. Ramsey, *Phys. Rev.*, vol. 126, p. 603, 1962.
11. Daniel Kleppner, H. Mark Goldenberg, and Norman F. Ramsey, *Appl. Opt.*, vol. 1 p. 55, 1962.
12. W. A. Gambling and T. H. Wilmshurst, Third International Conference on Quantum Electronics, Paris, 1963; in P. Grivet and N. Bloembergen (ed.), 'Quantum Electronics III,' p. 401, Columbia University Press, N. Y., 1964.
13. N. Bloembergen, *Phys. Rev.*, vol. 104, p. 324, 1956.
14a. R. H. Kingston, *Proc. IRE*, vol. 46, p. 916, 1958.
14b. R. H. Kingston, *IRE Trans. PGMTT*, vol. 7, p. 92, 1959.
15. G. K. Wessel, *Proc. IRE*, vol. 47, p. 590, 1959.
16. J. O. Artman, N. Bloembergen, and S. Shapiro, *Phys. Rev.*, vol. 109, p. 1392, 1958.
17. S. H. Autler and Nelson McAvoy, *Phys. Rev.*, vol. 110, p. 280, 1958.
18. J. V. Jelley and B. F. C. Cooper, *Rev. Sci. Instr.*, vol. 32, p. 166, 1961.
19. B. F. C. Cooper, *Rev. Sci. Instr.*, vol. 32, p. 202, 1961.
20. Walter H. Higa, *Solid-State Electron.*, vol. 4, p. 296, 1962.
21. A. I. McWhorter and J. W. Meyer, *Phys. Rev.*, vol. 109, p. 312, 1958.
22. H. E. D. Scovil, G. Feher, and H. Seidel, *Phys. Rev.*, vol. 105, p. 762, 1957.
23. G. Feher and H. E. D. Scovil, *Phys. Rev.*, vol. 105, p. 760, 1957.
24. H. E. D. Scovil, *IRE Trans. PGMTT*, vol. 6, p. 29, 1958.
25. J. A. Giordmaine, L. E. Alsop, C. H. Mayer, and C. H. Townes, *Proc. IRE*, vol. 47, p. 1062, 1959.
26. J. J. Cook, L. G. Cross, M. E. Bair, and R. W. Terhune, *Proc. IRE*, vol. 49, p. 768, 1961.
27. S. Foner and L. R. Momo, *J. Appl. Phys.*, vol. 31, p. 742, 1960.

28. David L. Carter, *J. Appl. Phys.*, vol. 32, p. 2541, 1961.
29. H. J. Gerritsen and H. R. Lewis, *J. Appl. Phys.*, vol. 31, p. 608, 1960.
30. Akira Okaya, *Proc. IRE*, vol. 48, p. 1921, 1960.
31. D. P. Devor, I. J. D'Haenens, and C. K. Asawa, *Phys. Rev. Letters*, vol. 8, p. 432, 1962.
32. A. Szabo, *Proc. IEEE*, vol. 51, p. 1037, 1963.
33. C. R. Ditchfield and P. A. Forrester, *Phys. Rev. Letters*, vol. 1, p. 448, 1958.
34. T. H. Maiman, *J. Appl. Phys.*, vol. 31, p. 222, 1960.
35. See, for example, J. R. Pierce, *IRE Trans. PGED*, vol. 2, January, 1955, and "Traveling Wave Tubes," D. Van Nostrand Company, Inc., Princeton, N.J. 1950; E. L. Chu and W. W. Hansen, *J. Appl. Phys.*, vol. 18, p. 996, 1947; L. Brillouin, *J. Appl. Phys.*, vol. 19, p. 1023, 1948.
36. R. W. De Grasse, E. O. Schulz-DuBois, and H. E. D. Scovil, *Bell System Tech. J.*, vol. 38, p. 305, 1959.
37. J. C. Walling, *Solid-State Electron.*, vol. 4, p. 225, 1962.
38. W. S. Chang and A. E. Siegman, *Stanford Electron Lab. Tech. Rept.* 156-2, Sept. 30, 1958. The matrix elements from this paper have been republished by J. Weber, *Rev. Mod. Phys.*, vol. 31, p. 681, 1959.
39. J. C. Slater, "Microwave Electronics," D. Van Nostrand Company, Inc., Princeton, N.J., 1950.
40. H. J. Gerritsen, *Appl. Opt.*, vol. 1, p. 37, 1962.
41. P. F. Chester, P. E. Wagner, and J. G. Castle, Jr., *Phys. Rev.*, vol. 110, p. 281, 1948.
42. R. H. Hoskins and G. Birnbaum, in C. H. Townes (ed.), "Quantum Electronics," p. 499, Columbia University Press, New York, 1960.
43. R. H. Hoskins, *Phys. Rev. Letters*, vol. 3, p. 174, 1959.
44. S. Foner, L. R. Momo, and A. Mayer, *Phys. Rev. Letters*, vol. 3, p. 36, 1959.
45. S. Foner, L. R. Momo, A. Mayer, and R. A. Myers, "Quantum Electronics," p. 487, Columbia University Press, New York, 1960.
46. D. R. Herriott, *J. Opt. Soc. Am.*, vol. 52, p. 31, 1962.
47a. A. D. White and E. I. Gordon, *Appl. Phys. Letters*, vol. 3, p. 197, 1963.
47b. E. I. Gordon and A. D. White, *Appl. Phys. Letters*, vol. 3, p. 199, 1963.
48. H. J. Oskam and V. R. Mittelstadt, *Phys. Rev.*, vol. 132, p. 1435, 1963.
49. W. W. Rigrod, H. Kogelnik, D. J. Brangaccio, and D. R. Herriott, *J. Appl. Phys.*, vol. 33, p. 743, 1962.
50. D. J. Brangaccio, *Rev. Sci. Instr.*, vol. 33, p. 921, 1962.
51. Arnold L. Bloom, *Appl. Phys. Letters*, vol. 2, p. 101, 1963.
52. H. Kogelnik and W. W. Rigrod, *Proc. IRE* (c), vol. 50, p. 220, 1962.
53. W. W. Rigrod, *Appl. Phys. Letters*, vol. 2, p. 51, 1963.
54. W. R. Bennett, Jr., and P. J. Kindlemann, *Rev. Sci. Instr.*, vol. 33, p. 601, 1962.
55. A. Javan, Solid State Design, vol. 4, no. 11, p. 22, November, 1963; T. S. Jaseja, A. Javan, and C. H. Townes, *Phys. Rev. Letters*, vol. 10, p. 165, 1964.
56. H. A. H. Boot and D. M. Clunie, *Nature*, vol. 197, p. 173, 1963.
57. H. A. H. Boot, D. M. Clunie, and R. S. A. Thorn, *Nature*, vol. 198, p. 773, 1963.
58. W. E. Bell, *Appl. Phys. Letters*, vol. 4, p. 34, 1964.
59. L. E. S. Mathias and J. T. Parker, *Appl. Phys. Letters*, vol. 3, p. 16, 1963.
60. Harry G. Heard, *Bull. Am. Phys. Soc.*, ser. 2, vol. 9, p. 65, 1964.
61. W. L. Faust, R. A. McFarlane, C. K. N. Patel, and C. G. B. Garrett, *Phys. Rev.*, vol. 133, p. A1476, 1964.
62. A. D. White and J. D. Ridgen, *Proc. IRE*, vol. 50, p. 1697, 1962.
63. R. A. McFarlane, C. K. N. Patel, W. R. Bennett, Jr., and W. L. Faust, *Proc. IRE*, vol. 50, p. 2111, 1962.
64. A. D. White and J. D. Ridgen, *Proc. IRE*, vol. 50, p. 2366, 1962.
65. A. Javan, W. R. Bennett, Jr., and D. R. Herriott, *Phys. Rev. Letters*, vol. 6, p. 106, 1961.
66. C. K. N. Patel, *J. Appl. Phys.*, vol. 33, p. 3194, 1962.
67. H. J. Gerritsen and P. V. Goedertier, *Appl. Phys. Letters*, vol. 4, p. 20, 1964.
68. A. L. Bloom, W. E. Bell, and R. E. Rempel, *Appl. Opt.*, vol. 2, p. 317, 1963.
69. R. A. McFarlane, W. L. Faust, and C. K. N. Patel, *Proc. IEEE*, vol. 51, p. 468, 1963.
70. C. K. N. Patel, W. L. Faust, R. A. McFarlane, and C. G. B. Garrett, *Appl. Phys. Letters*, vol. 4, p. 18, 1964.
71. C. K. N. Patel, W. R. Bennett, Jr., W. L. Faust, and R. A. McFarlane, *Phys. Rev. Letters*, vol. 9, p. 102, 1962.
72. W. R. Bennett, Jr., W. L. Faust, R. A. McFarlane, and C. K. N. Patel, *Phys. Rev. Letters*, vol. 8, p. 470, 1962.
73. W. L. Faust, R. A. McFarlane, C. K. N. Patel, and C. G. B. Garrett, *Appl. Phys. Letters*, vol. 1, p. 85, 1962.

74. L. E. S. Mathias and J. J. Parker, *Phys. Letters*, vol. 7, p. 194, 1963.
75. R. A. Paananen, C. L. Tang, F. A. Harrigan, and H. Statz, *J. Appl. Phys.*, vol. 34, p. 3148, 1963.
75a. J. D. Ridgen and A. D. White, *Nature*, vol. 198, p. 774, 1963.
75b. R. A. Paananen, C. L. Tang, and F. A. Harrigan, *Appl. Phys. Letters*, vol. 3, p. 154 1963.
76. S. Jacobs, G. Gould, and P. Rabinowitz, *Phys. Rev. Letters*, vol. 7, p. 415, 1961.
77. P. Rabinowitz, S. Jacobs, and G. Gould, *Appl. Opt.*, vol. 1, p. 513, 1962.
78. T. H. Maiman, *Nature*, vol. 187, p. 493, 1960.
79. T. H. Maiman, *Solid-State Electron.*, vol. 4, p. 236, 1962.
80. I. D. Abella and H. Z. Cummins, *J. Appl. Phys.*, vol. 32, p. 1177, 1961.
81. K. S. Gibson, *Phys. Rev.*, vol. 8, p. 38, 1916.
82. R. J. Collins, D. F. Nelson, A. L. Schawlow, W. Bond, C. G. B. Garrett, and W. Kaiser, *Phys. Rev. Letters*, vol. 5, p. 303, 1960.
83. Irwin Weider and Lynn R. Sarles, *Phys. Rev. Letters*, vol. 6, p. 95, 1961.
84. A. L. Schawlow and G. E. Devlin, *Phys. Rev. Letters*, vol. 6, p. 96, 1961.
85. F. J. McLung, S. E. Schwarz, and F. J. Meyers, *J. Appl. Phys.*, vol. 33, p. 3139, 1962.
86. R. G. Hanes and B. P. Stoicheff, *Nature*, vol. 195, p. 587, 1962.
87. I. D. Abella and C. H. Townes, *Nature*, vol. 192, p. 957, 1961.
88. B. P. Stoicheff and A. Szabo, *Appl. Opt.*, vol. 2, p. 811, 1963.
89. T. P. Hughes and K. M. Young, *Nature*, vol. 196, p. 332, 1962.
90. J. McMurtry, *Appl. Opt.*, vol. 2, p. 767, 1963.
91. C. Martin Stickley, *Appl. Opt.*, vol. 2, p. 855, 1963.
92. R. L. Aagard, D. L. Hardwick, and J. F. Ready, *Appl. Opt.*, vol. 1, p. 537, 1962.
93. S. Koozekanani, P. P. Debye, A. Krutchkoff, and M. Ciftan, *Proc. IRE*, vol. 50, p. 207, 1962.
94. G. E. Devlin, J. McKenna, A. D. May, and A. L. Schawlow, *Appl. Opt.*, vol. 1, p. 11, 1962.
95. W. R. Sooy and M. L. Stitch, *J. Appl. Phys.*, vol. 34, p. 1719, 1963.
96. D. F. Nelson and W. S. Boyle, *Appl. Opt.*, vol. 1, p. 181, 1962.
97. P. H. Keck, J. J. Redmann, C. E. White, and R. E. DeKinder, Jr., *Appl. Opt.*, vol. 2, p. 827, 1963.
98. F. J. McLung and R. W. Hellwarth, *J. Appl. Phys.*, vol. 33, p. 828, 1962.
99. F. J. McLung and R. W. Hellwarth, *Proc. IRE*, vol. 51, p. 46, 1963.
100. William G. Wagner and Bela A. Lengyel, *J. Appl. Phys.*, vol. 34 p. 2040, 1963.
101. J. I. Masters, J. Ward, and E. Hartouni, *Rev. Sci. Instr.*, vol. 34, p. 365, 1963.
102. J. E. Geusic and H. E. D. Scovil, *Bell System Tech. J.*, vol. 41, p. 1371, 1962.
103. L. F. Johnson, R. E. Dietz, and H. J. Guggenheim, *Phys. Rev. Letters*, vol. 11, p. 318, 1963.
104. A. A. Ballman, S. P. S. Porto, and A. Yariv, *J. Appl. Phys.*, vol. 34, p. 3155, 1963.
105. L. F. Johnson, *J. Appl. Phys.*, vol. 33, p. 756, 1962.
106. P. P. Sorokin and M. J. Stevenson, in J. R. Singer (ed.), "Advances in Quantum Electronics," p. 65, Columbia University Press, New York, 1961.
107. L. F. Johnson, *Proc. IRE*, vol. 50, p. 1691, 1962.
108. Z. J. Kiss and R. C. Duncan, *Proc. IRE*, vol. 50, p. 1531, 1962.
109. L. F. Johnson, *J. Appl. Phys.*, vol. 34, p. 897, 1963.
110. Z. J. Kiss and R. C. Duncan, *Proc. IRE*, vol. 50, p. 1532, 1962.
111. G. D. Boyd, R. J. Collins, S. P. S. Porto, A. Yariv, and W. A. Hargreaves, *Phys. Rev Letters*, vol. 8, p. 269, 1962.
112. P. P. Sorokin and M. J. Stevenson, *Phys. Rev. Letters*, vol. 5, p. 557, 1960.
113. A. Yariv and J. P. Gordon, *Proc. IEEE*, vol. 51, p. 4, 1963.
114. A. Yariv, S. P. S. Porto, and K. Nassau, *J. Appl. Phys.*, vol. 33, p. 2519, 1962.
115. L. F. Johnson, G. D. Boyd, K. Nassau, and R. R. Soden, *Phys. Rev.*, vol. 126, p. 1406, 1962.
116. L. F. Johnson and K. Nassau, *Proc. IRE*, vol. 49, p. 1704, 1961.
117. L. F. Johnson, G. D. Boyd, and K. Nassau, *Proc. IRE*, vol. 50, p. 87, 1962.
118. Z. J. Kiss and R. C. Duncan, *Proc. IEEE*, vol. 50, p. 1531, 1962.
119. L. F. Johnson, G. D. Boyd, and K. Nassau, *Proc. IRE*, vol. 50, p. 86, 1962.
120. L. F. Johnson and R. R. Soden, *J. Appl. Phys.*, vol. 33, p. 757, 1962.
121. E. Snitzer, *Phys. Rev. Letters*, vol. 7, p. 444, 1961.
122. H. W. Gandy and R. J. Ginther, *Appl. Phys. Letters*, vol. 1, p. 25, 1962.
123. H. W. Gandy and R. J. Ginther, *Proc. IRE*, vol. 50, p. 2113, 1962.
124. H. W. Etzel, H. W. Gandy, and R. J. Ginther, *NRL Progr. Repts.* 27, 1962.
125. Donald S. McClure, "Solid State Physics," vol. 9, Academic Press Inc., New York, 1959.

126. Donald S. McClure, "Solid State Physics," vol. 12, Academic Press Inc., New York, 1959.
127. G. E. Fenner and J. D. Kingsley, *J. Appl. Phys.*, vol. 34, p. 3204, 1963.
128. W. E. Engeler and M. Garfinkel, *J. Appl. Phys.*, vol. 35, p. 1734, 1964.
129. W. E. Howard, F. F. Fang, F. H. Dill, Jr., and M. I. Nathan, *IBM J. Res. Develop.*, vol. 7, p. 74, 1963.
130. W. E. Engeler and M. Garfinkel, *J. Appl. Phys.*, vol. 34, p. 2746, 1963.
131. A. E. Michel, E. J. Walker, and M. I. Nathan, *IBM J.*, vol. 7, p. 70, 1963.
132. R. N. Hall, G. E. Fenner, J. D. Kingsley, T. J. Soltys, and R. O. Carlson, *Phys. Rev. Letters*, vol. 9, p. 366, 1962.
133. M. I. Nathan, W. P. Dumke, G. Burns, F. H. Dill, and G. J. Lasher, *Appl. Phys. Letters*, vol. 1, p. 62, 1962.
134. T. M. Quist, R. H. Rediker, R. J. Keyes, W. E. Krag, B. Lax, A. L. McWhorter, and H. J. Zeiger, *Appl. Phys. Letters*, vol. 1, p. 91, 1962.
135. I. Melngailis, *Appl. Phys. Letters*, vol. 2, p. 176, 1963.
136. K. Weiser, *Appl. Phys. Letters*, vol. 2, p. 178, 1963.
137. R. J. Phelan, A. R. Calawa, R. H. Rediker, R. J. Keyes, and B. Lax, *Appl. Phys. Letters*, vol. 3, p. 143, 1963.
138. Nick Holonyak, Jr., and S. F. Bevacqua, *Appl. Phys. Letters*, vol. 1, p. 82, 1962.
139. F. B. Alexander, V. R. Bird, D. R. Carpenter, G. W. Manley, P. S. McDermott, J. R. Peloke, H. F. Quinn, R. J. Riley, and L. R. Yetter, *Appl. Phys. Letters*, vol. 4, p. 13, 1964.
140. I. Melngailis, A. J. Strauss, and R. H. Rediker, *Proc. IEEE*, vol. 51, p. 1154, 1963.
141. I. Melngailis and R. H. Rediker, *Appl. Phys. Letters*, vol. 2, p. 202, 1963.
142. M. Geller, D. P. Bortfield, and W. R. Sooy, *Appl. Phys. Letters*, vol. 3, p. 36, 1963.
143. Gisela Eckhardt, R. W. Hellwarth, F. J. McLung, S. E. Schwarz, D. Weiner, and E. J. Woodbury, *Phys. Rev. Letters*, vol. 9, p. 455, 1962.
144. R. W. Terhune, *Solid State Design*, vol. 4, p. 38, November, 1963.
145. B. P. Stoicheff, *Phys. Letters*, vol. 7, p. 186, 1963.
146. E. Garmire, F. Pandarese, and C. H. Townes, *Phys. Rev. Letters*, vol. 11, p. 160, 1962.
147. E. H. Huffman, *Phys. Letters*, vol. 7, p. 237, 1963.
148. A. Lempicki and H. Samelson, *Phys. Letters*, vol. 4, p. 133, 1963.
149. D. J. Morantz, B. G. White, and A. J. C. Wright, *Phys. Rev. Letters*, vol. 8, p. 23, 1962.
150. D. J. Morantz, B. G. White, and A. J. C. Wright, *J. Phys. Chem.*, vol. 37, p. 2041, 1962.

BIBLIOGRAPHY

Dushman, Saul: J. M. Lafferty (ed.), "Scientific Foundations of Vacuum Technique," 2d ed., John Wiley & Sons, Inc., New York, 1962.

Roberts, Richard W., and Thomas A. Vanderslide: "Ultrahigh Vacuum and Its Applications," Prentice-Hall, Inc., Englewood Cliffs, N. J., 1963.

Condon, E. U., and G. H. Shortley: "The Theory of Atomic Spectra," Cambridge University Press, New York, 1951.

See also Bibliography of Chap. 14.

Chapter 30

ACOUSTIC-WAVE AMPLIFIERS

STEPHEN W. TEHON*

CONTENTS

1. Introduction... 30-1
2. Principles of Operation....................................... 30-2
3. Linearized Analysis.. 30-3
4. Interpretation of the Linear Theory........................... 30-5
5. Design Procedures.. 30-7

1. INTRODUCTION

In 1961, Hutson, McFee, and White[1] reported the discovery of a traveling-wave interaction in single crystals of CdS between a bias current of negative-charge carriers and a signal in the form of an ultrasonic wave. The CdS crystals, which are n-type semiconductors, were doped to provide desirable ranges of conductivity and illuminated to provide conduction electrons through photon excitation. In the interaction, power was transferred from the stress wave to the source of bias field if the drift velocity was less than the speed of sound, or with bias field increased to produce drift velocity greater than the speed of sound, power was transferred into the ultrasonic signal. A power gain of 54 db/cm was reported for pulsed 45-Mc ultrasonic waves.

White[2] and Blotekjaer and Quate[3] have published theoretical studies of the amplification effect, and are in agreement in predicting large theoretical gains for high-frequency signals with large bandwidths. In CdS, which has received the greatest attention among a number of possible materials, a maximum gain of as much as 6 per cent of amplitude per wavelength is possible.

The basic amplifier structure, shown in Fig. 1, consists of an input transducer for converting input electrical signals to stress waves, a buffer for isolating bias voltage from the signal circuit, the CdS crystal with ohmic contacts and bias supply, a buffer, and an output transducer for converting the amplified stress wave into an electrical output signal. Variations of this structure have been built, chiefly through incorporation of isolating transformers to eliminate one or both buffers. An external light source, not shown here, is required for adjustment of conductivity.

In spite of the large amounts of signal gain which can be achieved, the acoustic-wave amplifier has two basic problems which have been sufficient to limit its present

* General Electric Co., Electronics Laboratory, Syracuse, N.Y.

status to that of a laboratory device. In CdS, the low value of electron mobility requires fields on the order of 1,000 volts/cm to achieve drift velocities greater than the speed of sound, and the resultant power dissipation produces excessive heating. The second problem, which applies to any material which might be used, is that ultrasonic transducers with appreciable bandwidths tend to have large conversion loss, so that the ultrasonic gain is offset by a considerable margin.

At frequencies below about 100 Mc, it is feasible to operate a CdS crystal with continuous bias through careful design of the amplifier for heat transfer into a coolant. Wanuga[4] has described an experimental 60-Mc CW amplifier, operating with a CdS crystal 10 mm long and 3 mm by 2 mm in cross section, immersed in liquid nitrogen. The continuously applied drift field, 810 volts/cm, produced approximately 1 watt of dissipated power.

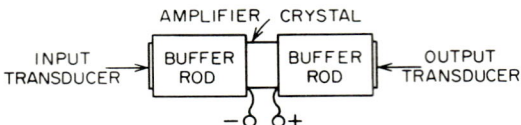

Fig. 1. Typical structure for acoustic-amplification measurements. The buffer rods are fused silica, and the transducers are quartz crystal plates.

A conventional quartz-crystal transducer produces a conversion loss on the order of 20 db. The effective gain of an acoustic amplifier is thereby reduced by about 40 db through the input and output transducers. In laboratory amplifiers, with short amplifying crystals or inefficient transducers, the net gain from electrical input to electrical output has frequently been an actual loss. Development of structures with positive net gain has been achieved, however, as described by Hickernell and Sakiotis,[5] and by Wanuga.[4]

Considerable research activity is being devoted to improvements in amplifier crystals and in high-frequency transducers. Success with new types of transducers is providing much more efficient conversion for frequencies extending well above 1,000 Mc.

2. PRINCIPLES OF OPERATION

The interaction between stress waves and space charge takes place through piezoelectric coupling. The amplifying medium must simultaneously be an elastic medium capable of supporting stress-wave propagation, a semiconductor, and a piezoelectric material oriented to provide coupling between the wave and the electric field along the propagation axis. It is convenient to use incident light as a control of the density of charge carriers.

The basic properties of importance in the amplifying crystal are a high value of electromechanical coupling coefficient and a high value of mobility. A number of crystal types are piezoelectric semiconductors. Cubic crystals with the zinc blende structure, such as GaAs, InSb, GaP, InAs, AlP, and the cubic forms of CdS, CdSe, ZnS, ZnSe, and ZnTe, meet this requirement but have not been used because of their very low coupling coefficients. More promising crystals, with higher coupling coefficients but rather low mobilities, are members of the wurtzite family, which includes ZnO, CdS, ZnS, CdSe, CdTe, and ZnTe. The wurtzite crystals have hexagonal symmetry about the c axis, which is a direction of relatively strong piezoelectric coupling.

In cubic crystals of the zinc blende symmetry, the necessary parallel coupling occurs for longitudinal propagation and electric field in the (111) direction; shear-wave coupling occurs for propagation in the (110) direction with particle motion along one of the cubic axes, (001), (010), or (100).

Hexagonal crystals with the wurtzite structure are highly oriented. The c direction is an axis of polar symmetry for determination of the elastic, dielectric, and piezoelectric properties, and the matrices describing these properties are identical in form with those described in Chap. 16 for polarized ceramics, with the same arrangement of zero terms and equal-valued terms. The necessary conditions for amplification are met either by an electric field along the c axis, coupled to a longitudinal wave traveling along the c axis, or by an electric field perpendicular to the c axis, coupled to a shear wave traveling in the same direction and with particle motion along the c axis.

Figure 2 shows the variation of stress and induced field intensity produced by an ultrasonic wave traveling in the x direction in a piezoelectric medium. Force on an electron is along the axis of propagation at points where the field intensity is negative and toward $-x$ where E is positive. Electrons drifting through the crystal therefore have a tendency to bunch in regions of stable equilibrium, producing a component

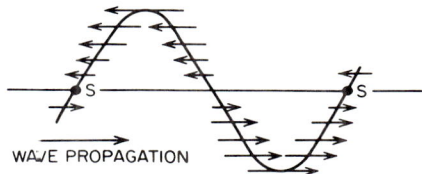

Fig. 2. Traveling wave of electric field intensity, induced by the traveling stress wave. Arrows indicate the directions and relative magnitudes of force on a drifting electron. Electrons tend to bunch at stable equilibrium postions (S).

of field directed toward each bunch. Electrons which accelerate in bunching derive kinetic energy from the alternating field-and-stress wave; electrons which decelerate give up energy to the wave. The bunching may be observed as progressive attenuation or amplification of the wave. With attenuation, the increase in electron velocity is observed[7] as an increase in drift current; with gain, the drift current decreases. This change in drift current, termed the "acoustoelectric effect" by Parmenter,[8] is produced primarily in the large-signal portion of the crystal, i.e., near the stress-wave input in a highly attenuating crystal or near the output when gain is large.

The piezoelectric interaction can be controlled by a bias field applied by means of electrodes on the crystal end surfaces. No energy transfer takes place when the applied field produces an average drift velocity just equal to the speed of stress-wave propagation. With less than this critical field, acoustic attenuation is observed, and more than critical field produces acoustic power gain.

Since the probability that an electron will bunch is greater for average electron velocity equal to the speed of sound, maximum gain and attenuation are observed for bias fields near the critical value. A stress wave traveling against the electron motion is attenuated but not as much as a forward wave can be. For some values of bias field, forward gain can exceed backward attenuation, so that a stress wave reflecting from the crystal ends can build up under the net round-trip gain. This gives rise to acoustic oscillations, generally of a random nature, termed "acoustic flux." With buildup of acoustic flux, acoustic gain for applied coherent signals and bias current are reduced. McFee[9] has described the gain reduction observed in experiments employing pulsed-bias field, and Wanuga[4] has noted significant gain reduction in going from pulsed to continuous bias. Current saturation with the acoustic-flux buildup appears as a kink in the voltage-current characteristic.

3. LINEARIZED ANALYSIS

The most generally applicable theory of acoustic-wave amplification, due to White,[2] treats the semiconducting crystal as a piezoelectric continuum, described by the

30–4 AMPLIFIER CIRCUITS

classical (one-dimensional) piezoelectric equations*

$$T = c^E S - eE \\ D = eS + \epsilon^S E \qquad (1)$$

where T is stress, S is strain, E is electric field intensity, D is dielectric displacement, c^E is the elastic modulus under constant field, ϵ^S is the dielectric permittivity under constant strain, and e is the piezoelectric constant. With this notation, the electromechanical coupling factor k is given by the expression

$$\frac{k^2}{1-k^2} = \frac{e^2}{c^E \epsilon^S} \qquad (2)$$

The average velocity of a charge carrier (electron) is μE, where μ is the drift mobility. Critical field intensity, for which drift velocity equals the speed of sound v_s, is equal to v_s/μ. In CdS, μ has a value of approximately 250 cm²/volt-sec.

Charge distribution is calculated by (one-dimensional) application of Poisson's equation:

$$\text{div } D = q = \frac{\partial D}{\partial x} \qquad (3)$$

Charge continuity:

$$\frac{\partial J}{\partial x} = -\frac{\partial q}{\partial t} \qquad (4)$$

and the equation for current density in an n-type semiconductor when the frequency of collisions is much larger than the acoustic frequency:

$$J = q_e \mu n_c E + q_e D_n \frac{\partial n_c}{\partial x} \qquad (5)$$

where J is current density, q_e is the magnitude of electron charge, n_c is the density of electrons in the conduction band, and D_n is the electron diffusion constant. The density of space charge is

$$q = -q_e n_s \qquad (6)$$

and the equilibrium density of charge in the crystal, which maintains electrical neutrality when no stress wave is present, is $-q_e n_0$. Trapping is assumed to remove the fraction f of space charge from the conduction band, so

$$n_c = n_0 + f n_s \qquad (7)$$

For one dimensional longitudinal or shear waves, the equation of motion for particle displacement $u(x,t)$ is

$$\rho \frac{\partial^2 u}{\partial t^2} = \frac{\partial T}{\partial x} \qquad (8)$$

By simultaneous solution of Eqs. (1) to (8), White derives the third-order equation

$$-\frac{\partial^2 D}{\partial x \partial t} = \mu \frac{\partial}{\partial x}\left[\left(q n_0 - f \frac{\partial D}{\partial x}\right) E\right] - f D_n \frac{\partial^3 D}{\partial x^3} \qquad (9)$$

which can be solved explicitly under the small-signal assumptions that

$$u = A e^{\left(-\alpha + j\frac{\omega}{v_s}\right) x} e^{-j\omega t} \qquad (10)$$

and

$$E = E_0 + E_1 e^{\left(-\alpha + j\frac{\omega}{v_s}\right) x} e^{-j\omega t} \qquad (11)$$

Here A is a constant, α is wave attenuation in nepers/cm, v_s is phase velocity, E_0 is the steady-bias field intensity, and E_1 is the intensity of the alternating field produced by the wave of stress and bunched charge.

* See Chap. 16 for a detailed discussion of the piezoelectric relations and the stress-wave equation.

D can be eliminated from Eq. (9) by substitution of Eq. (1). Then Eqs. (10) and (11), substituted in (9) yield approximate expressions for attenuation and phase velocity:

$$\alpha = \frac{1}{2}\left(\frac{k^2}{1-k^2}\right)\frac{\omega_c}{v_s\gamma}\frac{1}{1+\frac{\omega_c^2}{\gamma^2\omega^2}\left(1+\frac{\omega^2}{\omega_C\omega_D}\right)^2} \quad (12)$$

$$v_s = \sqrt{\frac{c^E}{\rho}}\left[1+\frac{1}{2}\frac{k^2}{1-k^2}\frac{1+\frac{\omega_c}{\omega_D\gamma^2}+\frac{\omega^2}{\omega_D^2\gamma^2}}{1+\frac{\omega_c^2}{\gamma^2\omega^2}\left(1+\frac{\omega^2}{\omega_C\omega_D}\right)^2}\right] \quad (13)$$

where ω_D is a diffusion frequency, defined for convenience as

$$\omega_D = \frac{v_s^2}{fD_n} = \frac{v_s^2 q_e}{f\mu KT} \quad (14)$$

and

$$\omega_C = \frac{\sigma}{\epsilon^S} = \frac{n_0 q_e \mu}{\epsilon^S} \quad (15)$$

is the dielectric relaxation frequency, or conductivity frequency.

The ratio of electron-drift velocity μE_0 to phase velocity v_s, which determines the degree of attenuation or gain, is represented by the variable

$$\gamma = 1 - \frac{f\mu E_0}{v_s} \quad (16)$$

Equation (12) indicates that attenuation is zero when γ is zero. Since the fraction f is equal to 1 when there is no trapping and would approach zero in the limit of complete trapping, the solution predicts that a drift velocity greater than v_s must be produced by the bias field to offset attenuation when trapping occurs.

4. INTERPRETATION OF THE LINEAR THEORY

The linearized theory describes attenuation and phase velocity in terms of the parameters ω_c, ω_D, and γ. Reference (3) has pointed out that the analysis should be accurate to frequencies on the order of 10^{10} cps, and experimental results from a number of investigators are in excellent agreement with theory. Others have reported experimental deviations in the range of γ for which maximum gain is predicted.

The dielectric relaxation frequency, proportional to crystal conductivity σ, can be controlled by means of the equilibrium charge density $n_0 q_e$. In photoconductive materials such as CdS, σ can range upward from the value of dark conductivity, depending on the level of incident light intensity, and dark conductivity can be influenced by doping the crystal, as by diffusing sulfur into CdS.

The diffusion frequency ω_D, determined by the sound velocity v_s, electron charge q_e, trapping factor f, drift mobility μ, Boltzmann constant K, and absolute temperature T, is not an adjustable parameter. May[10] points out that for most materials $\omega_D/2\pi > 10^9$. Gain can therefore be treated as a function of ω, ω_c, and γ.

The variation of attenuation with bias field can be employed as a gain control; when α is negative, the wave is amplified. Equation (12) can be rewritten in the convenient form

$$\alpha = \frac{1}{2mv_s}\left(\frac{k^2}{1-k^2}\right)\frac{1}{\xi + 1/\xi}$$

where

$$m = \frac{1}{\omega} + \frac{\omega}{\omega_C\omega_D} = \frac{1}{\sqrt{\omega_C\omega_D}}\left(\frac{\sqrt{\omega_C\omega_D}}{\omega}+\frac{\omega}{\sqrt{\omega_C\omega_D}}\right)$$

and

$$\xi = \frac{\gamma}{\frac{\omega_c}{\omega}+\frac{\omega}{\omega_D}} \quad (17)$$

In Eq. (17), variation of α with γ is described by the factor $1/(\xi + 1/\xi)$, plotted in Fig. 3. This is symmetrical about the value $\xi = 0$ for which γ and α are zero, and has maximum attenuation at $\xi = 1$, maximum gain at $\xi = -1$. In decibels per centimeter, gain is $-8.68\,\alpha$, and maximum gain (or attenuation) available through variation of bias field is equal to

$$\frac{8.68}{4mv_s}\left(\frac{k^2}{1-k^2}\right) \quad \text{db/cm}$$

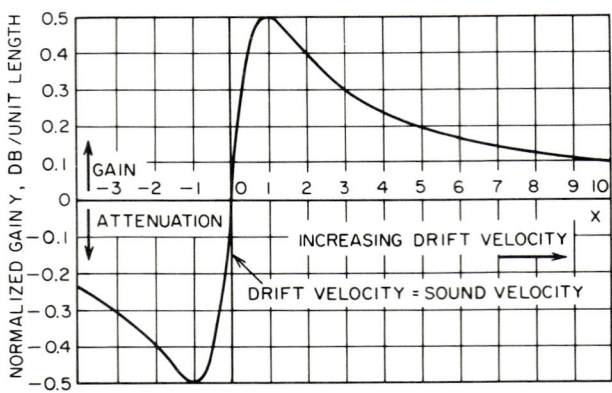

Fig. 3. Normalized plot of gain as a function of bias field; the normalized variables X and Y are interpreted as

$$\text{Gain in db/cm} = Y\left[\frac{4.34\omega_D k^2/(1-k^2)}{\omega_D/\omega - \omega/\omega_C}\right]$$

$$X = \frac{\gamma}{\omega_C/\omega - \omega/\omega_D}$$

At zero field, $\gamma = 1$. At zero gain, $\gamma = -1$.

Maximum gain, a function of m and hence of the frequency for which gain is maximized, is greatest when m is a minimum at $\omega = \sqrt{\omega_c \omega_D}$, and equal to

$$\frac{8.68}{8v_s}\left(\frac{k^2}{1-k^2}\right)\sqrt{\omega_c \omega_D} \quad \text{db/cm}$$

This can be a sizable amount of gain. For example, with shear-wave propagation in CdS,

$$\begin{aligned} k &= 0.188 \\ v_s &= 1.75 \times 10^5 \text{cm/sec} \end{aligned} \tag{18}$$

At a frequency of 1 Gc, assuming this equals $\sqrt{\omega_c \omega_D}/2\pi$ as a result of conductivity adjustment, the predicted gain is 1,430 db/cm.

For the case of fixed ω_c, variation of gain with frequency can be examined by rewriting Eq. (12) in the form

$$\alpha = \frac{1}{2}\left(\frac{k^2}{1-k^2}\right)\left(\frac{\omega_D \gamma}{v_s}\right)\frac{1}{\gamma^2 \frac{\omega_D}{\omega_c} + \left(\zeta + \frac{1}{\zeta}\right)^2}$$

where

$$\zeta = \frac{\omega}{\sqrt{\omega_c \omega_D}} \tag{19}$$

Maximum gain and maximum attenuation then occur when $(\zeta + 1/\zeta)$ is a minimum, at $\omega = \sqrt{\omega_c \omega_D}$. The frequency of maximum gain can therefore be selected by conductivity control, and is independent of γ and bias field. Large values of $\gamma^2 \omega_D/\omega_c$

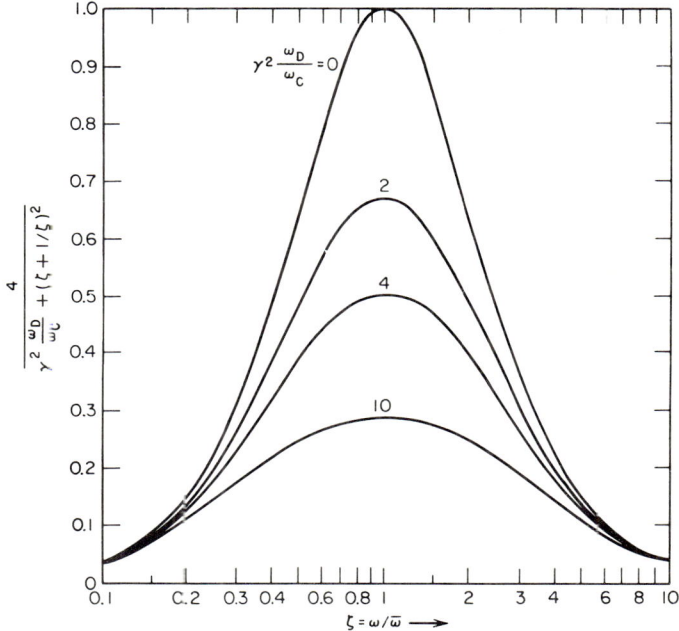

FIG. 4. Normalized plots of gain vs. frequency, showing the geometric symmetry of frequency response, and the effects of bias and the critical-frequency values ω_C, ω_D on bandwidth.

produce large fractional bandwidths. Figure 4 is a normalized gain plot, computed from Eq. (19).

5. DESIGN PROCEDURES

In the early work on acoustic amplifiers, a number of investigators have found discrepancies between the predictions of White's linearized theory and the measured performance of CdS crystals. The discrepancies have been largely explained as due to crystal imperfections or the results of trapping. The simplified theory adopts the trapping factor f which appears in the expression for γ, as in Eq. (16). More recent studies are showing[11] that trapping, when an effective mechanism, is field-dependent, is most effective in highly resistive crystals with low values of σ, and produces most pronounced modifications of gain for applied bias voltages intended to give maximum gain. The net result is flattening of the high-gain portion of the bias response curve, at levels less than expected gain, as in Fig. 5. It has been suggested that, for minimum trapping effects in CdS crystals, conductivity should be fairly high, and the crystals should be of maximum purity, with sulfur infusion as required for compensation and control of dark conductivity. There is recent experimental evidence that in crystals under nonuniform illumination the applied bias voltage is distributed spatially in a highly nonuniform fashion along the crystal. In regions of intense field, for example, trapping effects and acoustic-flux generation, as well as current saturation, produce marked nonlinear effects.

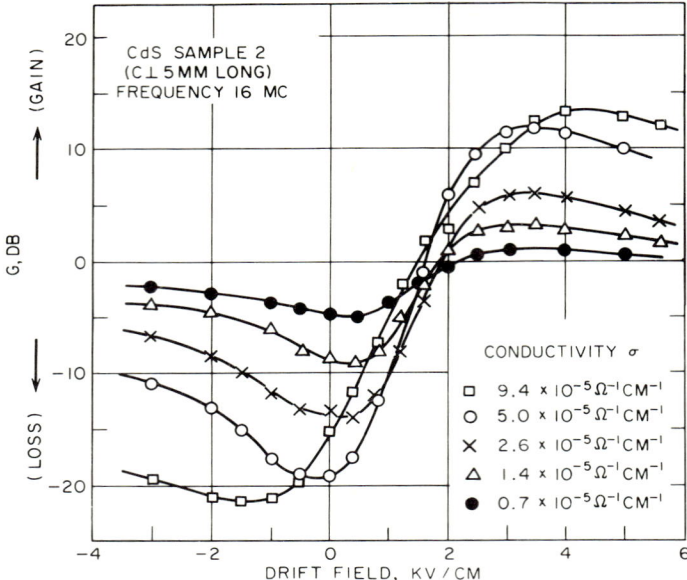

Fig. 5. Observed gain and loss vs. drift field at 16 Mc under various intensities of illumination. After Ishiguro et al.[11]

Nonlinear operation tends to become apparent whenever current saturation appears, which in turn is dependent on light intensity. At bias field values below those for maximum gain, operation generally follows the White theory, and in some cases[12] well-compensated crystals have shown close agreement with the theory over the field range extending even well beyond maximum-gain values. Since inadequate gain per unit length requires additional crystal length and higher bias voltage, as well as dissipated power, successful development of acoustic amplifiers will depend primarily on production of crystals essentially free from the effects of trapping, current saturation in nonuniform field distributions, and acoustic-flux generation.

If it is assumed that no trapping takes place, and that nonlinear effects are negligible, then performance can be analyzed in considerable detail with White's theory to provide design information. Table 1 summarizes the constants, in some cases with some suspected inaccuracies, as noted, for materials regarded as most suitable for acoustic-amplifier construction. Cadmium sulfide has received the greatest experi-

Table 1. Room-temperature (300°K) Constants

Material (mode)	ϵ/ϵ_0 Dielectric constant*	Sound velocity v_s, cm/sec*	Coupling factor k*	Approx. diffusion frequency $\omega_D/2\pi$	Approx. drift mobility μ, cm²/volt-sec	Approx. gain factor $\dfrac{2.7\omega_D}{v_s}\dfrac{k^2}{1-k^2}$	Acoustic impedance ρc, kg/m²-sec
CdS(L)....	9.53	4.50 × 10⁵	0.154	4.96 × 10⁹	250	3,680	21.7 × 10⁶
CdS(S)....	9.02	1.80 × 10⁵	0.188	0.796 × 10⁹	250	2,210	8.68 × 10⁶
CdSe(L)...	10.2	3.86 × 10⁵	0.124	1.83 × 10⁹	500	1,011	21.9 × 10⁶
CdSe(S)...	9.53	1.54 × 10⁵	0.130	0.292 × 10⁹	500	444	8.75 × 10⁶
ZnO(L)....	8.84	6.40 × 10⁵	0.282	12.6 × 10⁹	200	23,200	36.3 × 10⁶
ZnO(S)....	8.33	2.945 × 10⁵	0.316	2.66 × 10⁹	200	13,690	16.73 × 10⁶
GaAs(S)...	13.5	3.35 × 10⁵	0.0354	0.11 × 10⁹	5800	53.7	17.85 × 10⁶

* Jaffe and Berlincourt,[19] who summarize values published by a number of investigators.

mental attention, but its drift mobility and the consequently calculated values for diffusion frequency are still not precisely determined. Generally, it has been noted that drift mobility tends to be somewhat less than Hall mobility. Cadmium selenide is not generally so attractive, except for low-frequency operation, because of low coupling coefficient and low ω_D. Zinc oxide crystals, with exceptionally high values of coupling coefficient and diffusion frequency, exhibit relatively high sound velocities which require large bias fields but are the most promising for most frequencies. However, ZnO crystals have proved difficult to grow, and a supply of crystals with sufficient perfection has not been available for experimental investigation in detail. Gallium arsenide, with high sound velocity, low coupling factor, and low diffusion frequency has only its high mobility to offer and is not a likely material for amplifier design.

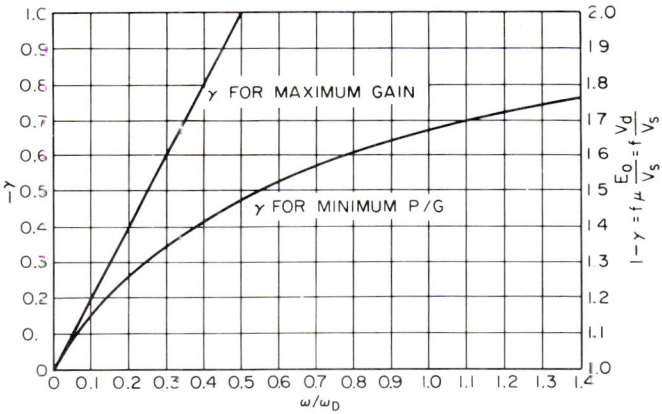

Fig. 6. Values of γ which produce maximum gain, and minimum dissipated bias power per db of gain. These curves apply at the maximum gain frequency, where $\omega = \sqrt{\omega_C \omega_D}$ for any crystal.

Calculations of power delivered by the bias field for any of these materials indicate that dissipation of the resultant heat will be a serious problem, while gain tends to be very large. It is therefore almost mandatory to operate the crystal at the conditions of conductivity and bias field which provide minimum dissipation per decibel of gain. White[2] showed that this most efficient operation occurs when two conditions are met simultaneously. Conductivity, adjusted by doping the crystal and subsequently adjusting the intensity of incident light, should produce a value of ω_c such that the angular frequency for maximum gain, $\bar{\omega} = \sqrt{\omega_c \omega_D}$, falls at the frequency of operation. Also, the bias field should be adjusted so that γ satisfies the condition

$$\left(\frac{\bar{\omega}}{\omega_D}\right)^2 = \frac{\gamma^2 - 3\gamma^3}{4(\gamma + 1)} \qquad (20)$$

In decibels, the gain is then given by the equation

$$G = -8.68\alpha$$
$$= -\frac{2.17\omega_D}{v_s} \frac{k^2}{1-k^2} \left(\frac{\bar{\omega}}{\omega_D} \frac{1}{\gamma\omega_D/2\bar{\omega} + 2\bar{\omega}/\gamma\omega_D}\right) \qquad \text{db/cm} \qquad (21)$$

where γ and $\bar{\omega}/\omega_D$ are related by Eq. (20).

Figure 6, plotted from Eq. (20) for most efficient operation and from the relation

$$-\gamma = \frac{\omega_c}{\bar{\omega}} + \frac{\bar{\omega}}{\omega_D} = 2\frac{\bar{\omega}}{\omega_D} \qquad (22)$$

AMPLIFIER CIRCUITS

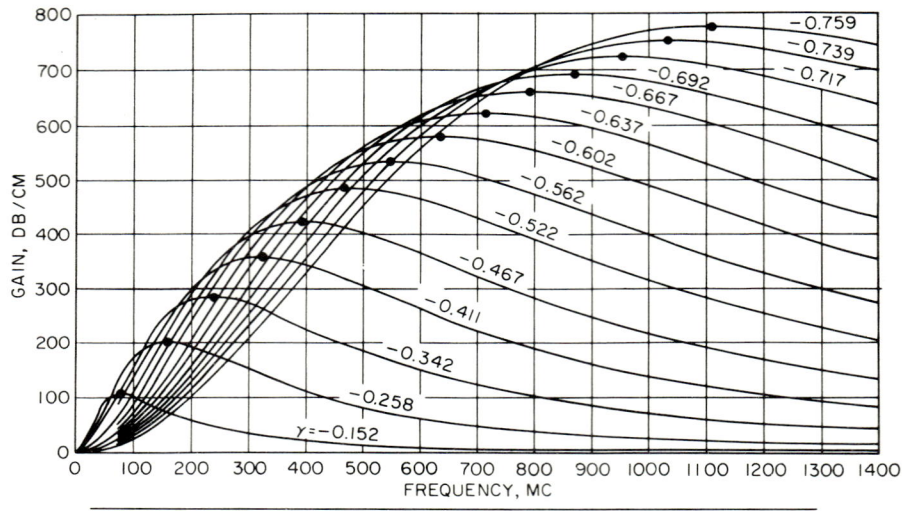

$-\gamma$	f_c	Watts/mm³	$\bar{\omega}/\omega_D$	Volts/cm	$\bar{f}$
0.152	7.9	0.027	0.1	829	79.2
0.258	31.7	0.13	0.2	906	159
0.342	71.6	0.33	0.3	966	239
0.411	127	0.66	0.4	1,016	318
0.467	195	1.09	0.5	1,056	394
0.522	291	1.75	0.6	1,096	481
0.562	385	2.45	0.7	1,125	554
0.602	508	3.39	0.8	1,153	636
0.637	647	4.51	0.9	1,179	718
0.667	798	5.77	1.0	1,200	797
0.692	951	7.09	1.1	1,218	870
0.717	1,139	8.73	1.2	1,236	952
0.739	1,339	10.53	1.3	1,252	1,032
0.759	1,558	12.54	1.4	1,267	1,114

FIG. 7. Frequency-response characteristics for CdS shear-wave amplification, calculated for $f_D = 796$ Mc. Along each curve, γ and ω_C are fixed, corresponding to most efficient operation at a maximum-gain frequency chosen here as an integral multiple of $0.1\,f_D$. At the circled maximum-gain point on each curve, gain per watt is a maximum with frequency and with γ. For any given frequency, values of γ and ω_C corresponding to other curves can produce more or less gain, but with lower efficiency.

for bias which produces maximum possible gain, indicates that efficient operation calls for less than maximum-gain bias by an amount which is more pronounced at higher frequencies.

Figure 7, which is computed with Eq. (12) for cadmium sulfide amplifying shear waves under most efficient conditions, shows that for frequencies greater than 100 Mc considerable gain and bandwidth can be achieved, at least in theory. Here, each curve has been calculated for fixed values of ω_c which corresponds to most efficient operation at a multiple of $0.1\omega_D$; the corresponding value of γ was taken from the curve of Fig. 6. Power dissipation, indicated for each curve, was computed as

$$P_{DC} = ALE_0^2 \sigma = AL\left[\frac{v_s}{\mu}(1-\gamma)\right]^2 \epsilon \frac{\bar{\omega}^2}{\omega_D} \qquad (23)$$

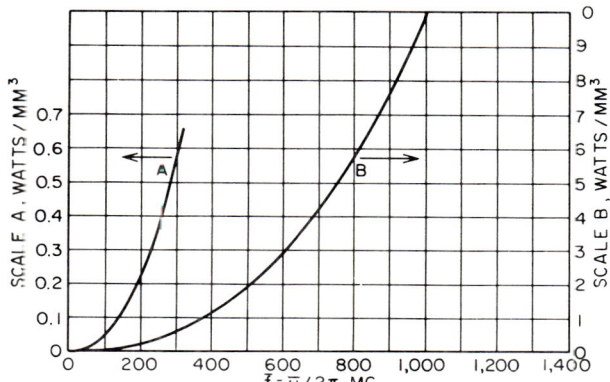

Fig. 8. Bias power dissipation in shear mode CdS crystals, based on the data in Table 1. At each value of frequency, it is assumed that ω_C and γ are adjusted for most efficient operation.

Figure 8, plotted from this equation, indicates the importance of power dissipation as a design consideration. Most experiments have utilized pulsed-bias field to avoid difficulties with dissipation and trapping effects.

Figure 9 gives an example of gain and dissipation computed for 100-db gain at a center frequency of 500 Mc. Since input and output transducers will contribute insertion loss and when tuned for minimum insertion loss will have relatively narrow passbands, the net gain of the amplifier will be less than 100 db, and the frequency response will be approximately that of the transducers.

Coincident with the development of acoustic amplification techniques, there has been intensive effort devoted to improvements in transducers. Foster[13,14] has developed two new types of transducers, which are much more suitable for high-frequency operation than the conventional piezoelectric plate bonded to the acoustic medium. One approach involves the diffusion of acceptors, such as sulfur into CdS, in a very thin region at the end face of a conductive semiconductor crystal with

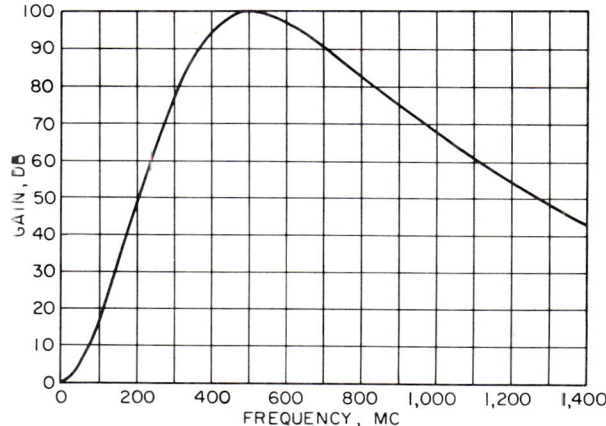

Fig. 9. Gain for a shear mode CdS crystal with 100-db acoustic gain at 500 Mc. Calculations are based on Table 1. Length of crystal = 2.01 mm. Bias voltage = 222 volts. Crystal resistivity = 6.34 ohm-cm. Power dissipation = 0.387 watts per mm^2 cross section.

piezoelectric coupling. Signals are then applied between the conductive body of the crystal and an external metal-electrode film deposited on the end face. Signal field is then concentrated within the high-resistivity doped region, which acts as a

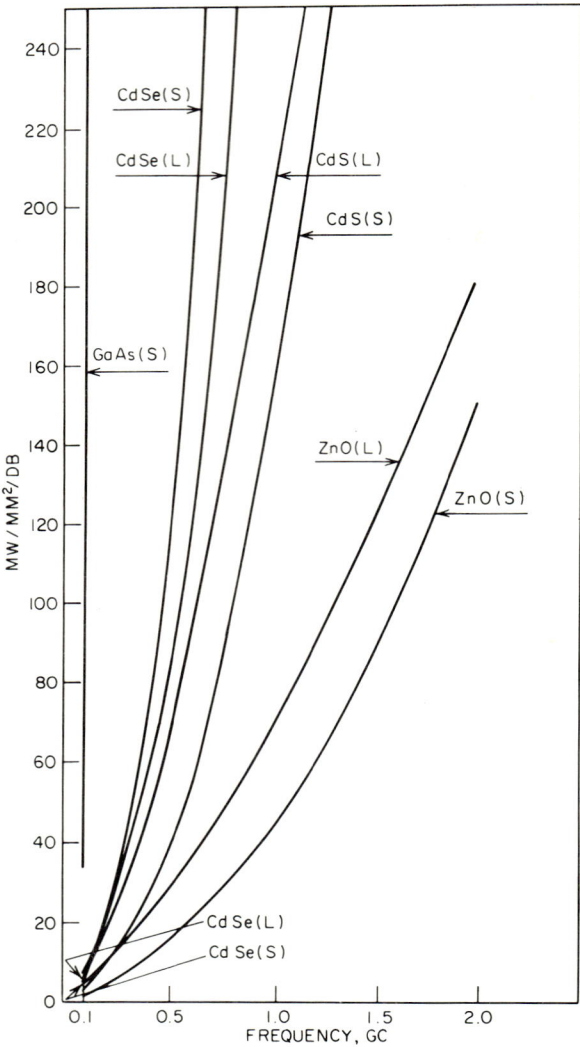

Fig. 10. Comparison of minimum values of P/G ratio vs. frequency for various materials. Conditions: f_C chosen so that maximum gain occurs at frequencies plotted, effect of acoustic attenuation included in all curves. After May.[10]

piezoelectric transducer with center frequency at half-wave for the layer thickness. The second technique utilizes a thin film of evaporated piezoelectric CdS, ZnS, or CdSe, deposited over a thin metal film on the amplifier end face. A second metal film is then deposited over the piezoelectric layer. Signal field, applied by means of these metal electrodes excites the piezoelectric layer as a half-wave transducer.

Since resistance of the crystal reduces the field in the compensated layer type of transducer, the performance of the deposited-film transducer is generally superior, except at low frequencies, and a number of workers have studied film-transducer deposition. Shaw and Winslow[15] report insertion loss as low as 7 db, each way, with evaporated CdS films in thin film sandwiches which utilize the acoustic impedance transformations provided in passage through quarter-wave films of evaporated SiO, or Ti. Wanuga, Midford, and Dietz[16] have recently reported successful sputtered deposition of ZnO film transducers, which may be expected to provide low insertion loss through the higher coupling factor of this material. To date, none of the transducer techniques can be regarded as fully developed, since improvements are being achieved rapidly. Given a film transducer with known thickness and coupling factor, it is possible to compute insertion loss by the methods described in Chap. 16, treating the film as a plate with parallel faces, and perfect bonding to the adjoining materials.

May[10] has extended the basic analysis of acoustic amplifiers to account for the effects of nonelectronic attenuation in the crystal. A few measurements have reported attenuation increasing with frequency approximately as $\omega^{1.3}$. This shifts the frequency of maximum gain slightly downward from $\bar{\omega}$, thereby increasing the required values of $-\gamma$ and bias field. Power dissipation is also increased as a result. Figures 10 and 11 give his results for power at the resultant conditions for most efficient operation.

Power dissipation in the amplifying crystal is directly proportional to cross-sectional area A, although gain is independent of A. Therefore, it is desirable to design the amplifier with minimum cross section, subject to power-handling limitations. May has assumed that the maximum power level occurs at saturation, when all electrons in the conduction band are involved in bunching. This maximum, a function of frequency and cross-sectional area for each crystal type and mode, corresponds to the saturation level of acoustic flux. At the conditions for most efficient operation, Tables 2 and 3 indicate the results. Since maximum power appears only at the output end of the crystal in normal operation, a crystal utilizes its bias power most effectively only at that end, and the most efficient amplifier therefore will have very low gain. Table 4 gives computed efficiency for cadmium sulfide shear amplifiers at 1,000 Mc, one with net gain of 40 db and one for a power amplifier with net gain of 8.7 db. Performance for the high-gain amplifier without transducers, including the effects of acoustic attenuation and the backward wave attenuation, are shown in Fig 12. Backward attenuation is computed with the gain expression, using a negative value for the bias field. In this case, the bias voltage is 66 volts, A is 0.56 mm^2, dissipated power is 2.04 watts, and the 3-db bandwidth is 700 Mc.

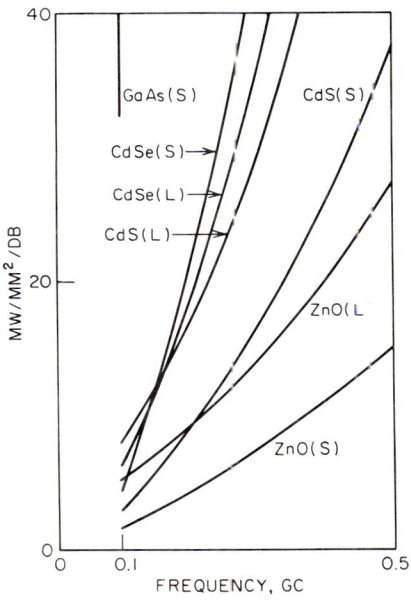

Fig. 11. Low-frequency portion of Fig. 10, expanded scale. After May.[10]

In some cases, backward wave attenuation may not be sufficiently high to effectively eliminate round-trip reflected signals, and under some operating conditions there may be net loop gain, accompanied by spontaneous buildup of acoustic flux to the saturation power level. Under these circumstances, May includes a length of crystal in the bias path which has a gain of 0 db in the forward direction, and high backward

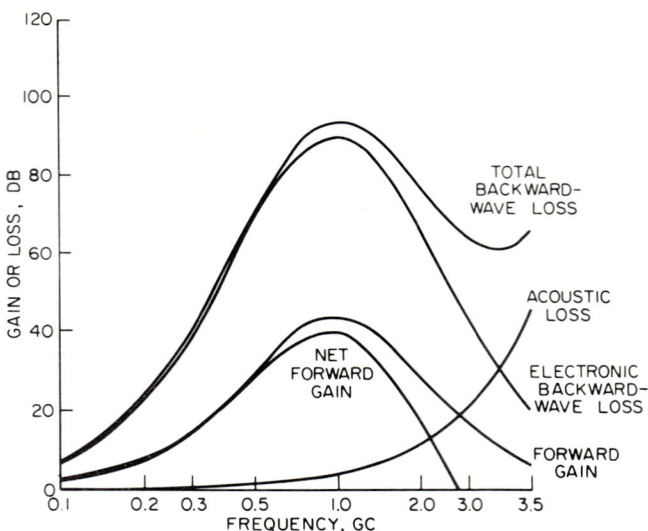

Fig. 12. Calculated gain and loss curves, 1-Gc CdS(S) amplifier, length = 0.68 mm. After May.[10]

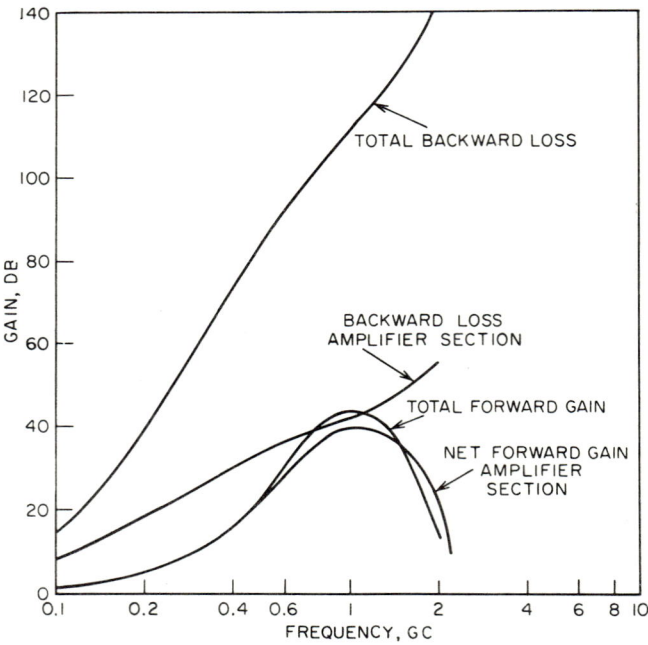

Fig. 13. Calculated gain and loss curves. 1-Gc CdS(L) amplifier with isolator section (1.58 mm). After May.[10]

wave attenuation. Figure 13 summarizes the design of a 40-db amplifier for such a case.

May has also computed the noise figure to be expected, based on theories by Hanlon and White,[17] and by Quate.[8] Since the noise arises effectively at the input end of the amplifying crystal and since input-transducer loss must be offset with additional acoustic amplification of that noise, the net amplifier noise figure is degraded by input-transducer loss. Table 5 gives noise-figure estimates for the shear-mode cadmium sulfide amplifier.

Table 2. Estimated Saturation Acoustic Flux*

Operating condition $f_c = f^2/f_D$, $f = 100$ Mc

	Acoustic flux, mw/mm²	D-C power/gain, mw/mm²/db
CdS (L)	0.85	7.8
CdS (S)	1.0	2.9
CdSe (L)	1.0	6.3
CdSe (S)	2.0	4.3
ZnO (L)	0.19	4.5
ZnO (S)	0.23	1.8
GaAs (L)	0.037	10.6†
GaAs (S)	0.87	33

* After May.[10]
† Effect of acoustic attenuation not included.

Table 3. Estimated Saturation Acoustic Flux*

Operating condition $f_c = f^2/f_D$, $f = 1,000$ Mc

	Acoustic flux, mw/mm²	D-C power/gain, mw/mm²/db
CdS (L)	76	207
CdS (S)	72	142
CdSe (L)	159	410
CdSe (S)	281	756
ZnO (L)	20	71
ZnO (S)	19	43

* After May.[10]

Table 4. Estimated Efficiency for 1,000-Mc CdS(S) Amplifier*

Transducer loss, db (one transducer)	Gain, db	β	Efficiency, %
0	40	1.9	2.7
2	42	1.9	1.9
6	46	1.9	0.6
15	55	1.9	0.06
0	8.7	2.0	13.0
2	10.7	2.0	6.5
6	14.7	2.0	1.9
15	23.7	2.0	0.6

* After May.[10]

Table 5. Calculated Noise Figure for 1,000-Mc CdS(S) Amplifier*

Insertion loss, db (two transducers)	Acoustic gain, db	Noise figure, db	Equivalent noise temperature, °K
0	40	3	296
4	40	5	635
30	40	18.1	18,400

* After May.[10]

REFERENCES

1. A. R. Hutson, J. H. McFee, and D. L. White, Ultrasonic Amplification in CdS, *Phys. Rev. Letters*, vol. 7, no. 6, pp. 237–239, September 15, 1961. See also A. R. Hutson and D. L. White, Propagation of Elastic Waves in Piezoelectric Semiconductors, *J. Appl. Phys.*, vol. 33, no. 1, pp. 40–47, January, 1962.
2. D. L. White, Amplification of Ultrasonic Waves in Piezoelectric Semiconductors, *J. Appl. Phys.*, vol. 33, no. 8, pp. 2547–2554, August, 1962.
3. K. Blotekjaer and C. F. Quate, The Coupled Modes of Acoustic Waves and Drifting Carriers in Piezoelectric Crystals, *Proc. IEEE*, vol. 52, no. 4, pp. 360–377, April, 1965.
4. S. Wanuga, CW Acoustic Amplifier, *Proc. IEEE* (Corres.), vol. 53, no. 5, p. 555, May, 1965.
5. F. S. Hickernell and N. G. Sakiotis, An Electroacoustic Amplifier with Net Electrical Gain, *Proc. IEEE* (Corres.), vol. 52, no. 2, pp. 194–195, February, 1964.
6. S. Wanuga, Theoretical and Experimental Evaluation of Electroacoustic Amplifiers, *Solid State Research and Development Conf.*, Boulder, Colo., July, 1964.
7. W. C. Wang, Strong Electroacoustic Effect in CdS, *Phys. Rev. Letters*, vol. 9, no. 11, pp. 443–445, December 1, 1962.
8. R. H. Parmenter, The Acousto-electric Effect, *Phy. Rev.*, vol. 89, no. 5, pp. 990–998, March 1, 1953.
9. J. H. McFee, Ultrasonic Amplification and Non-ohmic Behavior in CdS and ZnO, *J. Appl. Phys.*, vol. 34, no. 5, pp. 1548–1553, May, 1963.
10. John E. May, Jr., Electronic Signal Amplification in the UHF Range with the Ultrasonic Traveling Wave Amplifier, *Proc. IEEE*, vol. 53, no. 10, pp. 1465–1485, October, 1965.
11. T. Ishiguro, I. Uchida, and T. Suzuki, Ultrasonic Amplification Characteristics and Nonlinearity in CdS, *IEEE Intern. Conv. Record*, pt. 2, vol. 12, pp. 93–100, March, 1964.
12. D. L. White and E. T. Handelman, Ultrasonic Amplification in Sulfur Doped CdS, paper D34, *5th Intern. Congr. on Acoustics*, Liege, Belgium, September, 1965.
13. N. F. Foster, Diffusion Layer Ultrasonic Transducer, *J. Appl. Phys.*, vol. 34, pt. 1, no. 4, pp. 990–991, April, 1963.
14. N. F. Foster, Ultra-high Frequency Cadmium-sulphide Transducers, *IEEE Trans. on Sonics and Ultrasonics*, vol. SU-11, no. 2, pp. 63–68, November, 1964.
15. H. J. Shaw and D. K. Winslow, "Multiple Film Transducers and Brillouin Scattering at Microwave Frequencies," presented at AFCRL Symposium on Microwave Acoustics, Bedford, Mass., October, 1965.
16. S. Wanuga, T. A. Midford, and J. P. Dietz, Zinc Oxide Film Transducers, paper I-6, *1965 Ultrasonics Symp.*, Boston, Mass., December, 1965.
17. J. T. Hanlon and D. L. White, private communication, quoted by May.[10]
18. C. F. Quate, Coupled Mode Theory of Acoustic Wave Amplifiers, Stanford University, Stanford, Calif., M. L. Rept. 889, February, 1962.
19. H. Jaffe and D. A. Berlincourt, Piezoelectric Transducer Materials, *Proc. IEEE*, vol. 53, no. 10, pp. 1372–1386, October, 1965.

Chapter 31

INTEGRATED CIRCUITS

RICHARD J. PATCH*

CONTENTS

1. Introduction..... 31-2
2. Manufacturing Techniques..... 31-2
   - 2.1. Planar Process..... 31-2
   - 2.2. Epitaxial Process..... 31-2
   - 2.3. Isolation Techniques..... 31-2
   - 2.4. Interconnection Techniques..... 31-4
   - 2.5. Packaging Techniques..... 31-4
   - 2.6. Test Procedure..... 31-5
3. Integrated Components..... 31-6
   - 3.1. n-p-n Transistors..... 31-7
   - 3.2. p-n-p Transistors..... 31-8
   - 3.3. Field-effect Transistors..... 31-9
   - 3.4. Four-layer Devices..... 31-10
   - 3.5. Diodes..... 31-11
   - 3.6. Zener Diodes..... 31-12
   - 3.7. Resistors..... 31-12
   - 3.8. Capacitors..... 31-14
4. Integrated-circuit Design Considerations..... 31-15
   - 4.1. Yield Considerations..... 31-16
   - 4.2. Component Parameter Study..... 31-17
   - 4.3. Configuration Selection..... 31-18
   - 4.4. Breadboarding Techniques..... 31-20
5. Multipurpose Circuits..... 31-20
   - 5.1. Differential Amplifier..... 31-21
   - 5.2. Dual Common-emitter Darlington Amplifier..... 31-22
   - 5.3. Dual Darlington Emitter-follower Amplifier..... 31-23
   - 5.4. Darlington Common-base Amplifier..... 31-23
   - 5.5. Dual Zener Reference..... 31-23
   - 5.6. Buffered Zener Regulator..... 31-24
   - 5.7. Full-wave Transformerless Demodulator..... 31-25
   - 5.8. Transformerless Modulator..... 31-26
   - 5.9. Transformerless Quadrature Rejector..... 31-26
   - 5.10. Precision High-input-impedance A-C Amplifier..... 31-28
   - 5.11. Miscellaneous Nonlinear-amplifier Applications..... 31-31

* General Electric Co., Semiconductor Products Dept., Syracuse, N.Y.

31–2 AMPLIFIER CIRCUITS

    6. Direct-coupled Amplifiers................................. 31–32
       6.1. Differential Amplifiers............................... 31–33
    7. Audio Amplifiers.. 31–39
    8. Servoamplifiers... 31–42
    9. Precision A-C Summing Amplifiers........................ 31–42
 10. I-F Amplifiers.. 31–44
 11. Video Amplifiers.. 31–45

1. INTRODUCTION

In the rapidly expanding field of integrated circuits, technology is expected to render obsolete many circuits that are presently using conventional components. Reduced costs, increased reliability, smaller size, lighter weight, and in some cases, higher performance are influencing almost all semiconductor-component manufacturers to prepare for high-volume production of integrated circuits, with primary emphasis on digital types. While linear integrated circuits are no more difficult to manufacture, standard high-usage linear designs are presently lacking. A great deal of effort has to be expended toward developing a family of such designs, capable of being integrated, before the user can economically afford to use linear integrated circuits. It is the purpose of this chapter to describe how integrated circuits can be applied to the design of linear amplifiers, and it is hoped that this work will help develop the needed family of standard linear circuits.

2. MANUFACTURING TECHNIQUES

The planar[1] and epitaxial[2] processes have made possible the manufacture of large numbers of complete circuit functions in a single silicon crystal, commonly referred to as integrated circuits. The types of components that can be integrated include n-p-n and p-n-p transistors; p- and n-channel field-effect transistors;[3] MOS (metal-oxide silicon) field-effect transistors;[4] two-, three-, and four-terminal four-layer devices;[5] diodes and zener diodes;[6] resistors; and capacitors. This section will explain the techniques involved in the integration of the above components in a single silicon crystal.

2.1. Planar Process. The planar process consists of first selectively exposing a photosensitized oxide-coated silicon wafer, using photographic pattern masks. The unexposed silicon oxide is then dissolved and washed away. Thereafter the silicon wafer is placed in a diffusion furnace containing gaseous dopants which diffuse[7] into those areas from which the original silicon-oxide covering was removed. The diffusion depth is proportional to time and temperature and can be controlled with extreme accuracy. The open silicon areas are then reoxidized by supplying oxygen, directly or in the form of steam. If the wafer is again photosensitized, the above process can be repeated as many times as necessary to form the desired devices.

The advantage of the planar process lies in the fact that the dopants diffuse laterally under the oxide, as well as inward. The junctions formed intersect the oxide in a region that has never been a surface, since the oxide was formed by converting the silicon into silicon oxide thermally, from the surface inward.

2.2. Epitaxial Process. The epitaxial process consists of vapor deposition of a very thin layer of high-resistivity silicon on top of a polished surface of very low-resistivity silicon. Using the planar process described in Sec. 2.1, transistors with exceptional characteristics can be formed in the very thin epitaxial layer. The planar epitaxial process can be utilized to produce high-voltage transistors with low saturation resistance and low capacitance, thus resulting in an excellent high-frequency amplification device.

2.3. Isolation Techniques. In an integrated circuit, all components are diffused into a single silicon crystal. Isolation between the various components can be achieved using back-biased diodes, resulting in a relatively high degree of isolation. Diode a-c resistance values of 10 to 100 megohms at low frequencies are common, with the capacitance limiting the degree of isolation at high frequencies. In some cases, the diode leakage current can become troublesome, especially at high temperatures. Typical leakage currents at 25°C are in the range of 10^{-9} amp with the current doubling every 8 to 10°C. An isolation technique used in the early development of integrated circuits relied on the high resistance of the silicon bulk material. Components were spaced at relatively long intervals and resulted in a much lower degree of isolation than with back-biased diodes. The back-biased diode method of isolation was proved to be far superior and hence was adopted by almost all semiconductor-component manufacturers.

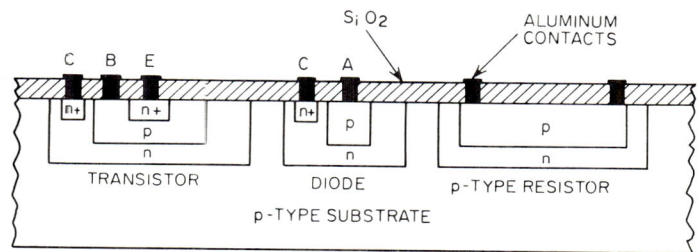

FIG. 1. Example of integrated structure.

Figure 1 shows an example of an integrated structure using the back-biased diode isolation technique. The structure contains an n-p-n transistor, a diode, and a resistor. The p isolation region completely encloses all components within the block of silicon and acts as an insulator between the components. Figure 2 shows the equivalent circuit of the integrated structure. From any one component to any other component in the structure, two diodes back to back provide the needed isolation. For example, the collector, base, or emitter of the n-p-n transistor can be more positive or negative than the cathode or anode of the diode with no appreciable parasitic cur-

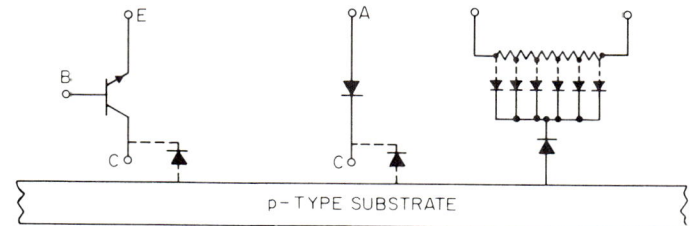

FIG. 2. Equivalent circuit of integrated structure.

rent flow between the transistor and diode. Since the resistor is diffused into the n-type structure a set of back-to-back diodes exists between the p isolation region and the resistor. Many p-type resistors can be diffused into the same n-type structure, resulting in back-to-back diodes between resistors and isolation between the resistors. Note that a distributed set of diodes exists between the p-type resistor and its adjoining n-type region, yielding a distributed RC network.

If the p-type substrate is floating with respect to a-c ground, some extraneous[8] signals will be transmitted between components, resulting in noise or unwanted feedback paths. This effect is much more noticeable at high frequencies, because of the isolation diode capacitances. It is most desirable to connect the p-type substrate to the most negative potential in the system to obtain the highest degree of isolation.

2.4. Interconnection Techniques. Two types of interconnection exist in manufacturing of integrated circuits. The first is the interconnection between the various diffused components[9] within the single-crystal circuit; second is the connection between the single-crystal circuit and the external leads connected to the header of the package. Interconnections have been proved to be the greatest cause of electronic-system failures, resulting in the expenditure of a great deal of time and money to improve the techniques of their manufacture.

Some of the first integrated circuits utilized thermocompression bonds with flying leads to interconnect the various diffused components within the single-crystal circuits. This method resulted in poor yield and reliability and has since been replaced by a deposited-aluminum interconnection technique. The deposited-aluminum pattern is formed by first photosensitizing the oxide-covered wafer containing the integrated components. The wafer is then exposed, using a photographic mask that covers the component contact areas. The unexposed silicon oxide is then dissolved and washed away, leaving nonoxidized contact areas. A thin film of metal, usually aluminum, is deposited over the entire wafer, making contacts to the nonoxidized areas. The surface is again photosensitized and exposed, using a mask appropriate for the particular interconnection desired. The unexposed metal is then dissolved and washed away, leaving the desired interconnection pattern with relatively large pads clustered on the outside periphery of the individual integrated circuits for external connections. The silicon oxide, which is an excellent insulator, covers all components except the etched-away contact areas. Crossover interconnection problems which exist in some of the more complex circuits are eliminated by using an underground-tunneling technique. The technique consists of using a low-value resistor especially designed for underground tunneling for one conductor, with the other conductor passing over the resistor.

When the metal-interconnection pattern is completed, the wafer is scribed into the individual integrated circuits and mounted into the desired package. The circuit is then ready for the second type of interconnection between the integrated chip and the external leads on the package header. This type of interconnection is accomplished by connecting a very thin wire, approximately 1/1,000 in. in diameter, usually gold, between the pads on the integrated chip and the header leads. The ends of the wire are secured, using thermocompression-bonding techniques. The gold-to-aluminum bond forms a brittle structure which sometimes fractures under severe mechanical stress, resulting in an open circuit. At present, this is the predominant failure mechanism in integrated circuits, and an effort is being made to eliminate the gold-wire connection. A method under development completely eliminates the very thin wire by making a direct contact between the integrated chip pads and header leads. This method, if automated, could result in lower cost and higher reliability.

2.5. Packaging Techniques. Since the initial concept of the transistor, many packages capable of containing semiconductor devices have been developed. Some of the more popular types used to package integrated circuits are described here. The type of package used depends on such factors as cost, reliability, manufacturability, size, weight, heat transfer, number of leads, and system adaptability. Varying emphasis on the above factors has resulted in a wide variety of packages that are presently available. Without a doubt more will be developed in the future.

One of the earlier packages is the well-known TO-5 that has been used for many years to contain transistor elements reliably. The TO-5 package is a cylinder approximately 0.36 in. in diameter by 0.24 in. high having a maximum of 12 external leads mounted on the header. The package cost is very low, and at the same time it is a readily adaptable container for many simple integrated circuits having chip sizes up to 0.1 by 0.1 in. The TO-5 package is readily adaptable to systems using conventional printed-wiring boards and has relatively good heat-transfer capabilities but is not optimum from the standpoint of size and weight. Other popular cylindrical packages which are smaller in size and weight include the TO-18, TO-46, and TO-47. These packages are restricted to smaller chip sizes and have fewer leads available, with eight leads being a practical upper limit.

Since the integrated circuit is a very thin rectangular structure, it appears that the most desirable package form would also be thin and rectangular. A large variety of such packages have been developed in ceramic, glass, or metal having good thermal properties. They are capable of a larger number of external leads and at the same time result in a major reduction in size and weight. One of the smaller packages measures $\frac{1}{8}$ by $\frac{1}{4}$ by 0.035 in. thick, having 14 external leads with the capability of containing a 0.06- by 0.12-in. integrated chip. Other packages measure $\frac{1}{4}$ by $\frac{1}{4}$, $\frac{1}{4}$ by $\frac{3}{8}$, and $\frac{3}{8}$ by $\frac{3}{8}$, all approximately 0.035 in. thick. The $\frac{3}{8}$ by $\frac{3}{8}$ package has up to 22 leads and is capable of containing a 0.2- by 0.2-in. integrated chip.

It is difficult to state at this time which packages will become standards in industry, although the flat rectangular type has certain size advantages. The major factors that will determine the standard are cost, reliability, and systems adaptability.

2.6. Test Procedure. When manufacturing high-quality electronic equipment utilizing discrete components, the manufacturer usually tests the individual components prior to assembly. After assembly, the circuits are given a functional test to

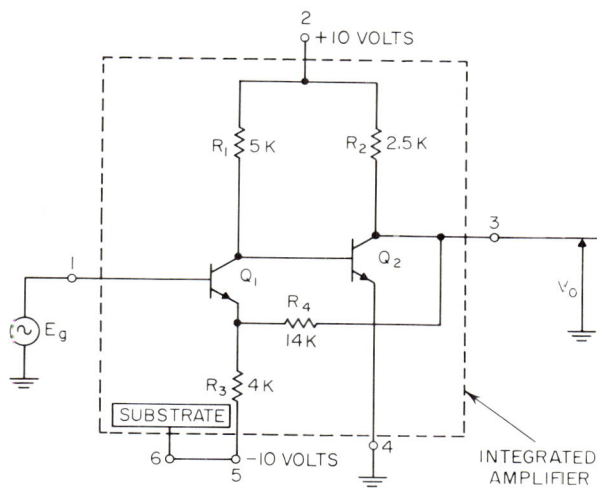

FIG. 3. Example of test procedure.

assure proper operation. In integrated circuits, only the functional test is possible. If the manufacturing process is under control, the functional testing of the integrated circuits is essentially the same as that for circuits made up of discrete components. When the manufacturing process is not under control, the test procedure differs significantly. It is sometimes very difficult or impossible to determine which component or components are outside the design limits. Unfortunately, there is no universal test procedure for determining faulty components in integrated circuits. Each circuit type must be treated as a special case.

An example of an integrated-circuit test procedure will be worked out to illustrate the difficulty involved. Figure 3 shows a schematic diagram of an integrated direct-coupled amplifier designed to have a voltage amplification of approximately 4.5. The object is to measure the resistors and determine if the transistors have transistor action. All external leads except those used for testing will be disconnected while measuring.

Resistor R_2 can be determined by connecting a d-c voltage between pin 3 and pin 2 with pin 3 positive, then measuring the current that flows. Resistor R_2 will equal the voltage divided by the current. The value of applied voltage should be less than the base-to-emitter breakdown voltage of transistor Q_1, to avoid the stray leakage current through R_4 and R_1. If R_2 is known then resistor R_1 can be determined by connecting a d-c voltage between pin 3 and pin 2 with pin 2 positive. Resistor R_1 can be expressed as

$$R_1 = \frac{(V_{23} - V_{BC2})R_2}{I_2 R_2 - V_{23}} \qquad (1)$$

where V_{23} = applied d-c voltage pin 2 to pin 3
I_2 = applied d-c current pin 2
V_{BC2} = forward base-to-collector drop of Q_2, approximately 0.65 volt

Resistor R_3 can be determined by connecting a d-c voltage between pin 1 and pin 5 with pin 1 positive and measuring the current that flows. Resistor R_3 can be expressed as

$$R_3 = \frac{V_{15} - V_{BE1}}{I_1} \qquad (2)$$

where V_{15} = applied d-c voltage pin 1 to pin 5
I_1 = applied d-c current pin 1
V_{BE1} = forward base-to-emitter drop of Q_1, approximately 0.65 volt

If R_3 is known, then resistor R_4 can be determined by connecting a d-c voltage between pin 3 and pin 5, with pin 3 positive. Resistor R_4 can be expressed as

$$R_4 = V_{35}/I_3 - R_3 \qquad (3)$$

where V_{35} = applied d-c voltage pin 3 to pin 5
I_3 = applied d-c current pin 3

If 5 per cent measurement accuracies are adequate, a transistor curve tracer can be used to measure the VI characteristics directly on the scope. Integrated resistors are voltage-dependent, and the scope display shows the range where the resistor departs from the linear region.

The quality of transistors Q_1 and Q_2 can best be determined by connecting the circuit as shown in Fig. 3. If the a-c output signal measured at pin 3 is approximately 4.5 times greater than the a-c input voltage E_g, then the transistors are operating units. The d-c common-emitter current amplification h_{FE1} of transistor Q_1 can be easily measured by replacing E_g with a microammeter. If the base current of Q_2 is much less than the collector current of Q_1, then h_{FE1} can be approximately expressed as

$$h_{FE1} = \frac{10 - V_{BE2}}{R_1 I_B} \qquad (4)$$

where V_{BE2} = forward base-to-emitter drop of Q_2, approximately 0.65 volt
I_B = base input current pin 1

The current amplification of transistor Q_2 cannot be measured. Statistical data show that two transistors in close proximity have approximately the same amplifications; thus it can be assumed that the current amplification of transistor Q_2 is approximately equal to that of Q_1.

If a nonfunctioning circuit is much more complex than the preceding example, it may not be possible to determine which component, or components, are out of specifications. In this case, the corrective action to be taken may be difficult to define.

3. INTEGRATED COMPONENTS

As the state of the art of integrated circuits advances, it will be possible, but not always practical, to have all forms of diffused components contained on a single silicon wafer. The types of components that can be integrated are n-p-n and p-n-p transistors, n- and p-channel field-effect transistors, n- and p-channel metal-oxide silicon field-effect transistors, four- and five-layer devices, diodes, zener diodes, resistors, and capacitors. The semiconductor manufacturer can vary the bogey value and the range of component parameters for each device by changing geometry, resistivity, doping materials, or doping levels. Since all components are made during a single process, all device parameters are interrelated in a rather complex fashion.

The range of component parameters available to the circuit designer decreases drastically as more different types of integrated components are added to the single-crystal structure. The circuit designer must develop new techniques to continue to design high-quality circuits with the lower range of component parameters.

The limitations of the power-handling capability of integrated circuits are essentially the same as for conventional semiconductor devices. The maximum power dissipation is related to the maximum allowable chip temperature, the heat-sink ambient temperature, and the thermal resistance between the chip and the ambient heat sink. Integrated circuits capable of dissipating hundreds of watts should be readily attainable in the future. The maximum current-handling capability is related to the maximum current density and the cross-sectional area of the device. Integrated structures containing large-area devices capable of passing currents in the hundreds of amperes will be available as time progresses. The epitaxial process described in Sec. 2.2 can be used to produce breakdown voltages in the range of hundreds of volts. Thus it appears that integrated circuits will be readily capable of controlling power of tens or even hundreds of kilowatts.

When using conventional components in high-frequency amplifiers, the maximum frequency response is limited by the component cutoff frequencies, stray capacitance, and lead inductance. Integrated circuits have the same limitations. Because integrated components are very small and are in close proximity to each other, the stray capacitance and leakage inductance can be very low in value, thus requiring less power. Small-geometry active devices have high gains at low current levels and very high cutoff frequencies, with present-day devices approaching the 1-gc region. The future upper frequency of operation will be limited by the practical minimum component size that can be achieved. Improved masking registration techniques should result in integrated circuits capable of operating well beyond 1 gc.

3.1. n-p-n Transistors. Integrated n-p-n transistors have essentially the same characteristics as discrete n-p-n planar transistors. In general, the integrated n-p-n transistor requires three diffusion steps, compared with only two for the discrete n-p-n planar transistor. Figure 4 shows a cross-sectional view of two n-p-n transistors integrated within a p-type isolation region. The three diffusion steps are the n-type collectors, the p-type bases, and finally the n-type emitters. The fabrication technique is described in Sec. 2.

The breakdown voltages[10] of the various p-n junctions are generally fixed for a given manufacturing process and are primarily a function of the resistivity of the silicon material. The higher-resistivity materials result in higher breakdown voltages and also higher saturation resistance. A very thin, high-resistivity epitaxial layer grown on top of the p-type substrate can produce high breakdown voltages and reasonably low saturation resistance. As impurities are diffused into the epitaxial layer forming the various p-type and n-type elements, the resistivity of the material is continually decreased. Thus the collector-to-base breakdown is much greater than the emitter-to-base because of the higher impurity concentration in the collector region. The breakdown voltage from collector to substrate or between the collectors of two transistors is very near that of the collector-to-base breakdown.

As shown in Fig. 4, the n-p-n transistor and the p-type substrate form a four-layer device. If the collector region is made relatively wide, compared with the base width, then the gain of the p-n-p transistor can be made very low, resulting in negligible four-layer action. If four-layer action does exist, and a sufficiently large, low-impedance voltage is applied between the emitter of the n-p-n transistor and the p-type substrate, the circuit can become permanently damaged.

The degree of isolation between the n-p-n transistors is highly related to the p-type substrate termination. If the p-type substrate is floating, it will assume a potential approximately equal to the most negative collector potential. For example, if the collector potential of a low-level stage is +2 volts and all other collectors are at a higher potential, then the p-type substrate will be virtually clamped to the collector of the low-level stage. This condition can produce heavy capacitive loading of the low-level stage or oscillation in high-gain amplifiers. In addition, the circuit may be very sensitive to noise pickup between components within the integrated circuit or from external sources. This undesirable situation can be easily corrected by connecting the p-type substrate to the most negative circuit potential, also being a-c ground. Doing this will assure that all p-n junctions will be back-biased, thus minimizing stray capacitance and stray pickup between components within the integrated circuit or from external sources.

3.2. p-n-p Transistors. p-n-p transistors can be included in an integrated circuit containing n-p-n transistors, resistors, and capacitors, but at the expense of altering the characteristics of the n-p-n transistors. Figure 5[11] shows a cross-sectional view of a p-n-p and an n-p-n transistor contained within a single silicon substrate. The

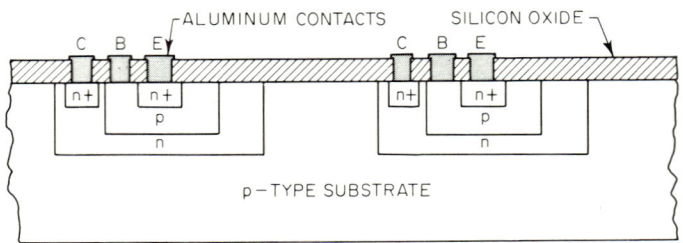

Fig. 4. Two integrated n-p-n transistors.

p-type wafer thickness is approximately 80 microns. During a 24-hr diffusion period the n-type collector region of the n-p-n transistor is deeply diffused into the p-type substrate. The n-type base region of the p-n-p transistor has a much shallower diffusion depth. The p-type diffusion is that of the n-type emitter of the n-p-n transistor. Thus the entire process requires only three diffusion cycles. By proper control of diffusion depths, the gains of the two transistors can be closely matched. The collector-to-base breakdown voltage of the p-n-p transistor will be slightly greater than that of the n-p-n because of the higher resistivity of the p-type substrate. Fig-

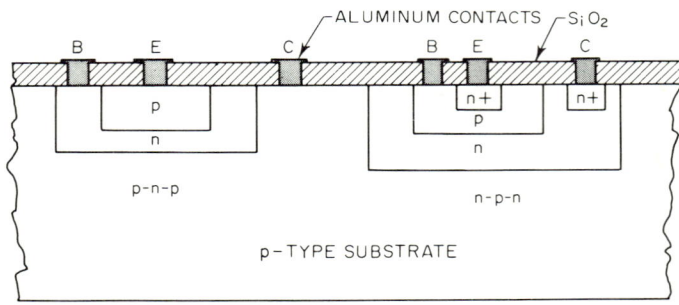

Fig. 5. Cross-sectional view of p-n-p and n-p-n. (*Courtesy of Signetics Corporation.*)

ure 6 shows an equivalent circuit of the p-n-p and n-p-n transistors. Since the p-type substrate is the collector of the p-n-p transistor, the two devices are electrically isolated only if the collector of the n-p-n is more positive than the collector of the p-n-p. Fortunately this is valid in most of the popular complementary n-p-n, p-n-p connections.

A quadruple diffusion process can also be used to produce p-n-p and n-p-n transistors on the same substrate as shown in Fig. 7. In this case, an n-type substrate with a grown epitaxial layer is used in place of the previously described p-type substrate. Four diffusion steps are required to complete the structure. No polarity restrictions exist between the two transistor types. The gains of the p-n-p and n-p-n transistors can be made relatively high and closely matched. The collector-to-base breakdown voltage of the p-n-p will be slightly higher than that of the n-p-n, because of the lower impurity concentrations in the collector of the p-n-p. For maximum isolation between the two devices, the n-type substrate should be connected to the most positive circuit potential. The quadruple diffusion process can be used to produce either n-type or p-type resistors, each having different resistivities. The temperature coefficient of the n-type resistors corresponding to the emitter diffusion of the n-p-n transistor will be lower than that of the p-type resistors corresponding to the emitter diffusion of the p-n-p transistor. Since the resistivity of the p type is higher, larger-valued resistors can be obtained in a smaller area. The particular application will dictate the optimum choice.

3.3. Field-effect Transistors. Unlike the conventional bipolar transistor, the unipolar field-effect transistor (FET) exhibits relatively high input impedance similar to that of a vacuum tube, with the additional feature of having a very low internally generated noise figure. Many amplifier configurations have been developed using the FET as the input stage, followed

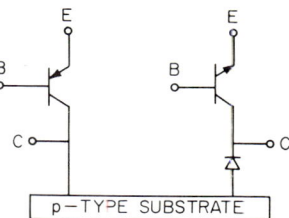

Fig. 6. Equivalent circuit of p-n-p and n-p-n.

by one or more bipolar transistor stages. Amplifiers of this type with feedback connections have resulted in an input impedance exceeding 100 megohms. Circuits requiring FET and bipolar transistors with resistors, diodes, and zener diodes can be readily integrated.

Figure 8 shows a top view of one of many possible integrated p-channel FET structures, and Fig. 9 shows the cross-sectional view. The FET is a device having a semiconductor current path whose resistance R is modulated by the application of a

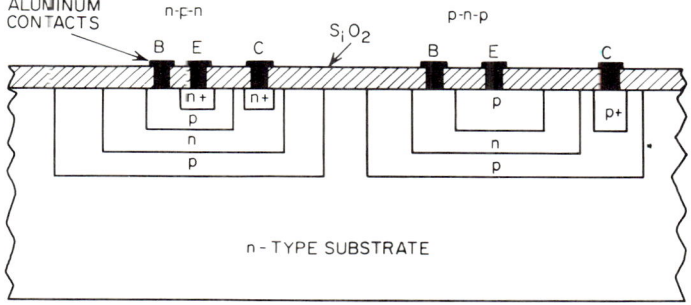

Fig. 7. Quadruple diffused n-p-n and p-n-p. (*Courtesy of Texas Instruments, Inc.*)

transverse electric field. To a first-order approximation, the resistance from drain to source R_{DS} can be approximated as

$$R_{DS} = \rho L/WT \tag{5}$$

where L = channel length
W = channel width
T = channel thickness
ρ = channel-material resistivity

For a given structure, the channel length, width, and resistivity can be considered as constant. If the gate of the p channel is made more positive than the source and drain the effective channel thickness T will decrease, thus raising the drain-to-source resistance R_{DS}. Under the above conditions, if a d-c potential is applied between the drain and source and a time-varying voltage is applied between the gate and the source, the drain-to-source current will also be time-varying. The operation is very similar to that of a vacuum tube. The d-c and a-c input gate current is very low, essentially that of a reverse-biased silicon planar diode; hence the device is capable of very large power gains. Since the leakage current of a silicon diode doubles every 8 to 10°C increase in temperature, it follows that the input gate current of the FET will also increase with temperature.

If an integrated circuit contains n-p-n transistors, p-channel FET's, zener diodes, and resistors, the parameters of all the devices will be interrelated. For example, the base width of the n-p-n transistor will equal the channel thickness of the FET. The optimum base width of the transistor which affects gain, high-frequency response,

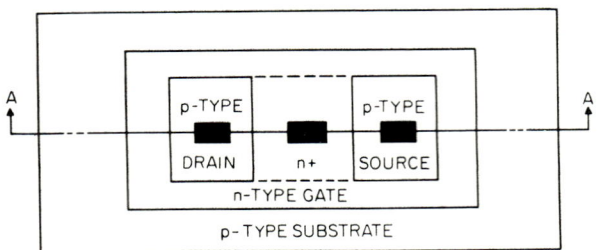

Fig. 8. Top view, integrated FET.

and other parameters does not necessarily correspond to the optimum channel thickness of the FET. Also, the resistivity in the base region will determine the n-p-n transistor and FET parameters, the breakdown voltage of the zener diode, and the value of p-type resistors, as well as the temperature coefficient of the resistors. Thus it is seen that the parameters of these devices are closely interrelated and one device cannot be optimized without considering what is happening to the other device parameters.

The p-type substrate must be connected to a d-c potential more negative than the gate to prevent shorting of the gate input signal. In addition, this will reduce stray coupling between the FET and other components in the circuit.

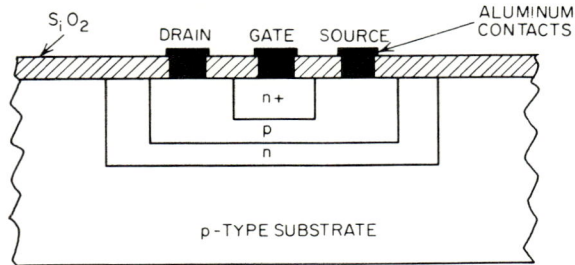

Fig. 9. Cross section AA of FET.

3.4. Four-layer Devices. Four-layer devices have found wide usage from the high-power control field to low-level switching. Amplifiers operating in the switching mode can deliver very high output power at efficiencies greater than 90 per cent. Transistors or four-layer devices can be used in the output stages. Four-layer devices can be readily manufactured, using the quadruple-diffusion process described in Sec. 3.2. Figure 10 shows a cross-sectional view of an integrated four-leaded four-layer device. The device will operate as a four-layer diode if connections are made between aluminum contacts 1 and 4. The peak-point firing voltage will be determined primarily by the resistivity of the p-type region corresponding to the second diffusion region. If a high-resistivity epitaxial layer is first grown on the p-type substrate, followed by lightly doped impurities in the first and second diffusion steps, then high-voltage four-layer diodes can be manufactured. The peak-point firing current, on the other hand, will be heavily dependent on the effective diffusion depths of the p and $n+$ layers.

If a three-leaded four-layer device, commonly referred to as a silicon-controlled rectifier, is desired, contacts 1, 3, and 4 may be used. A large number of useful circuits[12] have been developed which use the three-leaded device. If the four leads

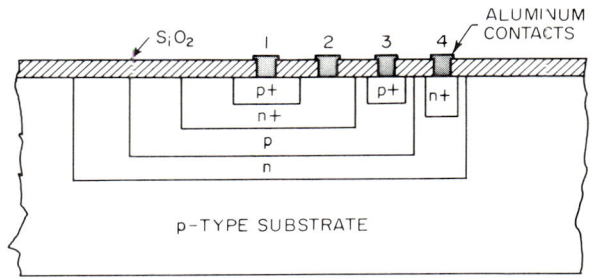

Fig. 10. Four-layer device.

are available to the circuit designer, an even larger number of applications can be generated. To minimize cross-coupling between the four-layer device and other components in the circuit, the p-type substrate should be connected to the most negative d-c potential in the circuit.

3.5. Diodes. Either the collector-base or emitter-base junctions can be used as diodes. A cross-sectional view of the diode structure is shown in Fig. 11. Because of the high resistivity of the collector junction, the collector-base diode will have a higher reverse breakdown voltage. Unfortunately, the forward voltage drop at a given current will also be slightly higher, because of the relatively long high-resistance path between the collector and the metal contact on top of the integrated circuit. The base-emitter diode has a lower breakdown voltage, but also a lower forward drop when compared with the collector-base diode. If the diode inverse voltage requirement is less than the base-to-emitter breakdown voltage, then the base-emitter junction is the preferred choice.

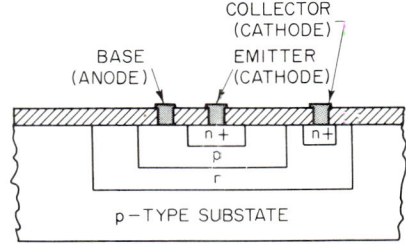

Fig. 11. Available diodes.

Another type of diode can be fabricated by shorting the collector-base terminals of the transistor and using this as the diode anode. The remaining emitter lead forms the cathode. In this connection the transistor is operating in the active region, with most of the anode current flowing through the collector. The base current is simply the collector current divided by the common-emitter current amplification. The inverse breakdown voltage of this connection is very near that of the base-emitter diode. Figure 12 shows an example of the forward diode characteristics for the three connections.

Many diodes can be formed with a common cathode connection by diffusing several p-type areas into the n-type region corresponding to the collector region of the transistor. These diodes will have high breakdown voltage and slightly higher forward voltage drops. If a common anode connection is desired, several n-type areas can be diffused into the p-type region corresponding to the base region of the transistor. The diodes in this connection will have low breakdown voltage and relatively low forward voltage drops. In addition the forward diode voltage drops will have good matching characteristics. If high-voltage common-anode diodes are desired, several n-type regions can be diffused into the p-type substrate followed by a p-type diffusion into each of the n-type regions. All the p-type regions are interconnected by aluminum deposition forming the common-anode terminal. Each n-type region forms the cathode of each of the diodes.

3.6. Zener Diodes. The reverse VI characteristics of the transistor base-to-emitter junction with the collector floating exhibits extremely sharp voltage breakdown. Figure 13 shows the breakdown characteristics of a commercially available unit. Most of the types currently available (mid-1963) have breakdown voltages in the range of 5 to 10 volts. The voltage temperature coefficient (TC) of these devices is approximately the same as that of a discrete zener diode. A 5- to 6-volt zener diode has approximately zero TC, with voltages greater than 6 having a positive TC. One

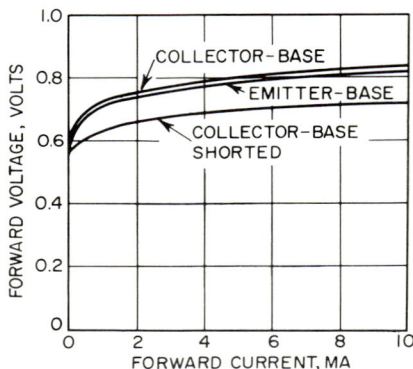
FIG. 12. Typical forward diode characteristics.

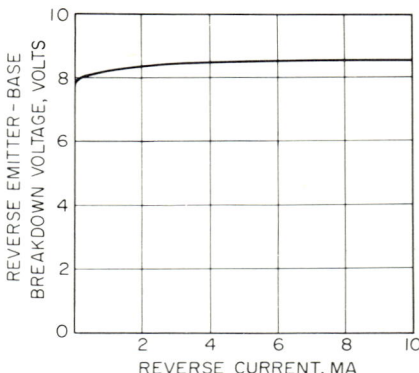
FIG. 13. Emitter-base breakdown voltage.

or more forward-biased diodes can be placed in series with the higher-voltage zener diodes to reduce the TC value. If one forward-biased diode is placed in series with a 7-volt zener, the resultant TC will be very near zero.

Arrays of common-cathode or common-anode zener diodes can be fabricated in the same way that diode arrays are fabricated, as explained in Sec. 3.5. The breakdown voltages of the zener-diode arrays will have very good matching characteristics, with typically less than 1 per cent mismatch. For a given process, the major portion of the units will have breakdown voltages falling within a ±5 per cent band. If high-voltage zeners are required, the lower-voltage units may be connected in series. The maximum attainable value will be limited by the breakdown voltage of the isolation region.

3.7. Resistors. Integrated resistors are fabricated at the same time that the other components are being made in the integrated circuit. The integrated resistors may be either p-type or n-type, as desired. Figure 14 shows a cross-sectional view of a p-type resistor. The resistor value can be expressed as

$$R = \rho_s(L/W) \qquad (6)$$

where ρ_s = sheet resistivity, ohms per square
L = resistor length, mils (0.001 in.)
W = resistor width, mils

The sheet resistivity is directly proportional to the resistivity and inversely proportional to the diffusion depth. In general, the sheet resistivity can be controlled to an absolute accuracy of better than 5 per cent. Two or more resistors in close proximity in the same isolation region will have very nearly the same sheet resistivity. These resistors, therefore, will track well with temperature, typically in the range of 0.005 per cent per degree centigrade. Also the ratios of the resistors will be almost entirely a function of the length and width of the resistors. The errors in length and width

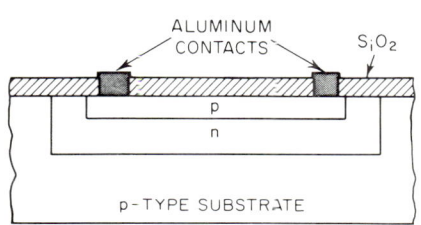

Fig 14. p-type resistor.

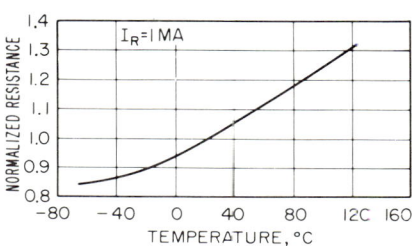

Fig. 15. Silicon resistor normalized temperature dependence.

are caused by errors in mask dimensions, misalignment of masks, photographic resolution, and improper etching. Resistors having very wide and long dimensions are more accurate than short, narrow resistors, but they also require more area. In Sec. 4.1 it is shown that yield is inversely proportional to circuit area; hence large-geometry resistors will have poorer yield, resulting in higher circuit cost. Absolute accuracies of 20 per cent and ratios of 5 per cent appear to be a good compromise of accuracy vs. yield.

Since the resistivity of silicon is temperature-dependent, resistors will also be temperature-dependent. Figure 15 shows the normalized temperature variation of a typical integrated silicon resistor. The temperature coefficient is dependent on the manufacturer's process. It can be changed, although at the expense of altering the transistor characteristics.

At high frequencies, the impedance of integrated resistors can change in a manner similar to a distributed RC network. Figure 16 shows an equivalent circuit of a p-type resistor. It should be noted that a distributed set of diodes exists between the p-type resistor and the n-type isolation region. Depending on the potential difference between n-type isolation region and all points on the p-type resistor, the distributed diodes may be forward- or reverse-biased. If all the distributed diodes

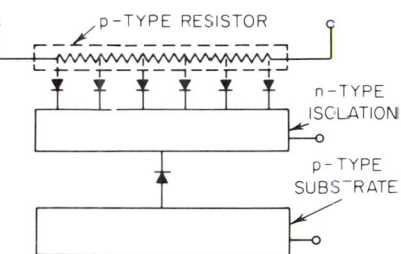

Fig. 16. Equivalent circuit of p-type resistor.

are reverse-biased, they form a distributed RC network. At high frequencies this distributed effect will decrease the impedance of the resistor. Since diode capacitance is voltage-dependent, a large reverse-bias voltage between the resistor and n-type isolation will result in the minimum capacitance and hence maximum resistor frequency cutoff. If the n-type isolation region is floating and the p-type resistor is more positive than the p-type substrate, the capacitance between the isolation region and the substrate becomes tied to the most positive terminal of the resistor. If the most positive terminal of the resistor is connected to the B+ supply, the frequency characteristics of the resistor will not be altered. On the other hand, if the positive terminal of the resistor is being driven by an a-c source, this added capacitance will shunt the a-c source and the resistor. This effect is difficult to avoid if several resistors existing in different n-type isolation regions are connected in series.

Since the leakage currents of diodes are voltage- and temperature-dependent, it is relatively easy to understand why the parasitic leakage currents associated with integrated resistors are also voltage- and temperature-dependent. A qualitative understanding can be realized by drawing an equivalent circuit similar to that shown in Fig. 16 with indicated potentials at all points in the circuit. By noting which diodes are forward- and reverse-biased, the nature of the parasitic leakage currents can be readily determined.

The maximum power-handling capability of integrated resistors depends on many factors, such as reliability, ambient temperature, thermal resistance between the resistor and the ambient, and allowed performance degradation. The performance degradation can be caused by resistor value change by the elevated temperature or by the

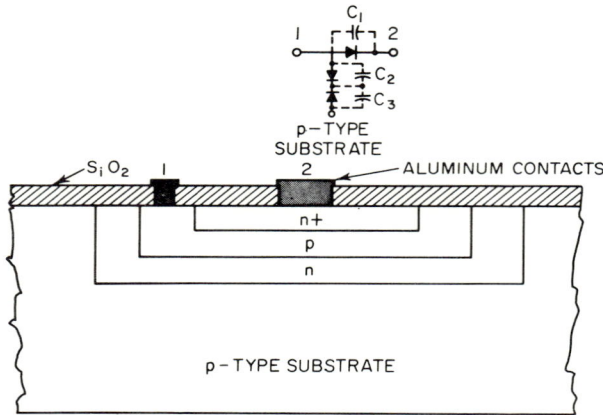

Fig. 17. Cross section of diode junction capacitor.

heating of adjacent components. If resistors are operated at extremely high temperatures, their characteristics may become permanently altered. To date, very little work has been published on this subject.

3.8. Capacitors. In integrated circuits, large-valued capacitors cannot presently be fabricated, although newly developed thin-film techniques should help to eliminate this deficiency. Integrated capacitors can be formed using diode junctions[13] or metal-oxide silicon (MOS) layers. Figure 17 shows a cross-section view of a diode-junction-type capacitor. In this configuration, terminal 2 must always be more positive than terminal 1; thus the capacitor is polarized. The diode capacitor is voltage-dependent, with the maximum capacitance occurring near zero bias and the minimum capacitance at maximum reverse bias. To achieve maximum capacitance for C_1, the p-type substrate should be connected to a relatively high negative potential, thus minimizing the stray C_2 and C_3 effect. At the same time a low reverse bias between terminals 1 and 2 should be used. In general, terminal 1 should be connected to a-c ground or the low-driving-source-impedance side of the circuit to reduce the loading effects of the stray capacitors C_2 and C_3. Diode capacitors such as C_1 have a capacitance in the range of 0.5 to 1.0 pf/sq mil at a reverse bias of 1 volt. If a 1,000-pf capacitor is needed, it may consume an area of from 1,000 to 2,000 sq mils, which is very costly when compared with either transistors or resistors. Large-value diode-junction capacitors should be avoided whenever possible.

INTEGRATED CIRCUITS 31-15

The MOS-type capacitor has essentially no voltage dependence. Figure 18 shows a cross-sectional view of a MOS-type capacitor. One electrode is a heavily doped N+ region, and the other is a thin metal film of deposited aluminum. The two electrodes are separated by a thin SiO_2 layer which forms the dielectric. The capacitance and breakdown voltage of the capacitor are determined by the thickness of the SiO_2

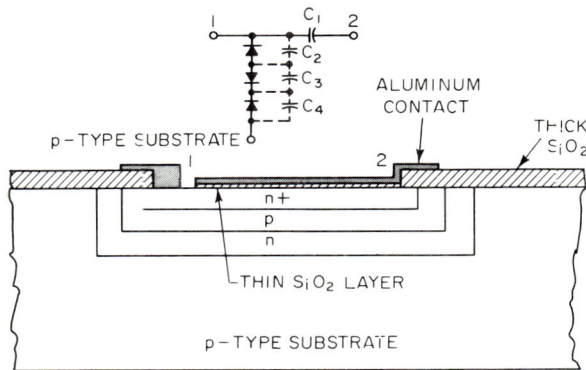

FIG. 18. Cross section of MOS capacitor.

layer. Figure 19 shows the capacitance and breakdown voltage vs. oxide thickness for the MOS capacitor. The capacitance per unit area for the MOS-type capacitor is slightly less than that of the diode-junction type, although if the oxide thickness is reduced below 500 angstroms, the capacitance can be made larger, but at the expense of accuracy. It should be noted that the stray capacitance for the MOS structure is considerably less than for the junction type, because of the series effect of C_2, C_3, and C_4. To minimize C_2, C_3, and C_4, the p-type substrate should be connected to a relatively high negative potential. Also terminal 1 should be connected to a-c ground or the low-driving-source-impedance side of the circuit to reduce the loading effects of the stray capacitors.

4. INTEGRATED-CIRCUIT DESIGN CONSIDERATIONS

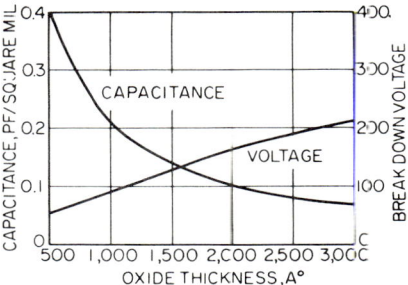

FIG. 19. Capacitance and breakdown voltage vs. oxide thickness for MOS capacitor. (*Courtesy of Fairchild Semiconductor Division, "Custom Microcircuit Design Handbook," 1963.*)

During the past decade, a new breed of circuit designers has evolved concurrently with the transistor itself. Prior to this era, all electronic circuits used high-impedance vacuum tubes. When low-impedance transistors were first introduced to industry, the designer found that new circuit-design techniques had to be developed. For example, the d-c properties of transistors varied considerably with temperature, which caused large shifts in the operating points. In RC-coupled amplifiers, low-valued input-biasing resistors were used to reduce the operating-point variations with temperature. This class of amplifier used relatively large capacitors and resulted in rather poor gain efficiencies. Later direct-coupled amplifiers with d-c and a-c feedback were developed and found to be far superior when compared with the RC-coupled types. The direct-coupled amplifiers had better bias stability, higher gain efficiency and gain stability, used few resistors and capacitors, and most important, were less expensive. Many of the techniques, such as the one illustrated, can be applied when designing integrated circuits.

When designing transistor circuits using discrete components, the circuit designer has unusual freedom. He may choose any combination from the many thousands of semiconductor components available. The circuit designed can use highly stable and accurate resistors, inductors, and capacitors with very low temperature coefficients and good high-frequency characteristics. Stray capacitance and leakage between the various discrete components are almost immeasurable. If the circuit designer finds that his finished product does not function properly, he can quite often correct the situation by changing a resistor value or a semiconductor component at small cost and with little time wasted. Unfortunately, the integrated-circuit designer will not have the same degree of freedom as the transistor-circuit designer who uses discrete components.

The designer of integrated circuits has relatively few types of active devices in the circuit. Very high-valued or very low-valued resistors are expensive and should be avoided. The resistor values vary appreciably with temperature, voltage, and frequency. At present, no inductors are available, and only low-valued capacitors can be obtained. In general, the range of all parameters available is low, and to make things worse, the absolute value of the parameters is quite uncertain. Stray capacitance and leakage between components is very high when compared with discrete component circuits. Integrated-circuit tooling charges are high, and if the circuit designer finds that his finished product does not function properly, he cannot correct the situation at low cost. In general, poor circuit designs can be costly and may result in considerable manufacturing-time delays. After considering the above disadvantages of integrated circuits one might ask "Why bother?" The answer is that because many integrated circuits can be made simultaneously in a batch process, the unit cost should be appreciably less than that of the discrete-component-type circuits. In addition, reliability data indicate that integrated circuits will be one to three orders of magnitude more reliable than discrete-type circuits. Thus cost and reliability are causing industry to shift from discrete to integrated circuits, although many years are likely to pass before the transition becomes stabilized.

4.1. Yield Considerations. It can be shown that the cost of integrated circuits is directly related to the number of useful circuits obtained from the silicon wafer. Yield, therefore, is one of the most important factors to be considered by the designer. Factors which can cause poor yield include opens in the metallization pattern, surface scratches, dislocation in the crystalline structure, and pinholes in the oxide. These failure modes can be considered to be distributed at random over the silicon wafer. The probability of obtaining one of these failure modes on a large area circuit can be very high. As the circuit area approaches zero, the probability of obtaining a non-functioning circuit is the ratio of the area containing failure modes to the total area of the wafer. Yield for very small circuits can be expressed as

$$\text{Yield} = \frac{\text{wafer area}}{\text{circuit area}} \times \frac{\text{area not containing failure modes}}{\text{wafer area}}$$

More simply expressed, the yield becomes equal to the ratio of the area not containing failure modes to the total circuit area. For example, if a 0.700-sq-in. wafer has an area not containing failure modes of 0.500 sq in. and the integrated circuit area is 0.001 sq in., then the yield will be 500 circuits. If the circuit area on the same wafer were 0.700 in. no yield would result.

Another factor affecting yield is the performance specification placed on the circuit. If the distribution of all component parameters is accurately known, the final yield can be readily computed for a given set of performance specifications. The circuit designer should examine his system needs carefully before he writes the final circuit performance specification. Tight performance specifications can result in poor yield and correspondingly increased cost. If the component manufacturer provides the circuit designer with component-parameter-distribution information, the circuit designer can then statistically compute his probable yield according to the end-performance specification.

The components used in the integrated circuits can be expressed in terms of area, and as explained above, yield can be expressed as a function of area. From this it appears that yield can be related to the number and types of integrated components. Several manufacturers have developed component-weighting functions for each of their integrated components. Cost is then directly related to the weighting function. One such weighting system is the component point-count method illustrated in Table 1. The maximum practical number of points is defined as 100. One point is equal to approximately 25 sq mils in area. If the number of points exceeds 100 the price increases sharply. Note that a 3-kilohm resistor or a 40-ohm resistor is approximately the same price as a type 1 small-geometry transistor. It should be emphasized that each manufacturing process has a fixed point-count system. For example, if sheet resistance is increased more resistance is obtained per unit area. Hence the high-valued resistor weighting function decreases in magnitude. At the same time, the weighting function for low-value resistors increases.

Table 1. Component Point Count*

μET-1 type transistor	$2(N - 1) + 5$
μET-2 type transistor	$5(N - 1) + 9$
Common-cathode collector-base diodes or common-anode emitter-base diodes	$(N - 1) + 4$
Resistor >300 ohms†	$K + 2$
Resistor <300 ohms†	$120/R + 2$
Supply voltage	$(V - 5)$

N = number of transistors or diodes in a single isolation region (i.e., common collector). Applies for $N < 5$
K = number of kilohms per each resistor greater than 300 ohms
R = number of ohms per each resistor less than 300 ohms
V = number of volts of maximum impressed power supply

* Courtesy of Fairchild Semiconductor Division from "Custom Microcircuit Design Handbook," 1963.

† These points apply to the case where all resistors are located in a common isolation area.

4.2. Component Parameter Study. The integrated-circuit design begins with the selection of a suitable family of components, defined as those components which evolve from a fixed manufacturing process. Varying the combinations of wafer resistivities, doping materials, doping concentrations, diffusion depths, and the number of diffusion steps will produce different families of components. In terms of both time and money, developing a repeatable manufacturing process is expensive. In small-volume jobs, cost will force the circuit designer to choose one of the few available manufacturing processes. If, however, the circuit has high-volume potential it may be more economical to pay the process-developing charge, thereby obtaining the optimum family of components. Semiconductor manufacturers assuredly will develop many new processes resulting in large numbers of different families of components available to the circuit designer.

When the most satisfactory process has been selected, the circuit designer can define the nature of each of the components at his disposal. At moderately priced masking fees, the component geometries can be varied at will. Resistors can be made wide or narrow, short or long, to best suit the application. In general, the transistor will be made as small as practical to minimize the total circuit area. The transistor shape also can be modified to produce the most optimum parameters. Both bipolar and unipolar transistors can be made by changing geometries. Diode junction or MOS (metal, oxide, silicon) capacitors can be made any shape, with area being proportional to capacitance. The size and shape of diodes and zener diodes can vary for low- and high-current applications.

Since the size and shape of the semiconductor components determine many of its electrical parameters, the circuit designer will be forced to tabulate a list of components with parameter range and tolerance. The list should include the voltage, frequency, and temperature dependence of each component. When the list is compiled, weighting functions as described in Sec. 4.1 can be assigned to each component. This list will allow the circuit designer to determine quickly the circuit size and the resulting relative cost compared with other designs.

4.3. Configuration Selection. Once the list of available components is established as described in Sec. 4.2, the circuit designer can proceed to choose the optimum configuration to suit the desired circuit function. The performance of the circuit configuration should not be sensitive to absolute values of the integrated resistor. As explained previously, integrated resistors have a large temperature coefficient which can result in performance degradation with temperature. On the other hand, tolerance and temperature tracking of resistor ratios are reasonably good, ratio tolerances being good to 5 per cent, with temperature tracking better than 0.005 per cent per degree centigrade. Hence, whenever possible, resistor ratios should be relied on in place of absolute resistor values.

In discrete-component-type circuits, it is often found to be economically desirable to have an external adjustment to trim the circuit operation. Typical instances are gain adjustments, hum-balancing adjustments, and operating-point balance adjustment for push-pull output stages. Similar adjustments can be made in integrated circuits, but certain cautions must be exercised. For example, if a near-zero temperature coefficient external resistor is used to adjust an internal-integrated resistor, the circuit performance is likely to change considerably with temperature. In the case of gain adjustments in a feedback amplifier it may be far more desirable not to integrate the two gain-determining resistors but rather to connect them externally to the circuit.

A trial-and-error method can be used when seeking the optimum configuration to satisfy a particular design requirement. A configuration is selected which is capable of satisfying the circuit function. A quick analytical design using rule-of-thumb techniques follows. Each component is tabulated and the weighting functions are applied to determine the overall yield and cost. The list is then carefully reviewed to determine which components are the most expensive. Quite often it will be found that very high- or very low-value resistors or capacitors are many times more expensive than a transistor. When the analytical evaluation is complete on the initially chosen configuration, the designer will have a better perspective on the relative densities of components that should be used. A new and, hopefully, better configuration is selected, followed by another quick analytical design. The weighting functions are again applied to determine the overall yield and the components are again reviewed to determine which are most costly. The above procedure is repeated as many times as practical for the particular design. In the end, all designs are compared to see which best satisfies the desired performance and, at the same time, gives the highest yield. The time spent in choosing the best configuration can range from a few hours to a few months, depending to a great extent on the complexity of the circuit and the expected quantity that will be manufactured.

As previously mentioned, high-value and low-value resistors are sometimes many times more expensive than a transistor. In direct-coupled amplifiers, it is desirable to have the collector biasing resistor much greater than the input impedance of the following stage to assure maximum current transfer. It may be much less expensive to use low-valued collector resistors and add more stages to make up for the gain lost.

Figure 20 shows a schematic diagram of a direct-coupled feedback amplifier designed to have a gain of 100,000 volts/amp. The amplifier must be capable of operating from a $+28$-volt and a -10-volt supply and have a gain stability of better than 0.1 per cent. For the sake of simplicity, the high-frequency gain-shaping networks to assure stable crossover are not shown. Because of accuracy restraints, resistor R_f will not be integrated. Also capacitor C_1 will not be integrated. The weighting functions for the family of components available are tabulated in Table 2. Applying the appropriate weighting functions yields 436 points. Resistor R_2 costs more than 40 times that of transistor Q_1. At this point, it appears that more transistors could be used in place of high-valued resistors, at much lower cost.

Figure 21 shows another configuration which will also satisfy the amplifier require-

INTEGRATED CIRCUITS

Table 2. Component Weighting Functions

Component	Weighting function
Small geometry:	
Transistors (Q_1, Q_2)................	$5N$
Large geometry:	
Transistor (Q_3)......................	$10N$
Diodes...............................	$5N$
Zener diodes, 7 volts at 1 ma..........	$5N$
Resistors >2 kilohms.................	$2K + 4$
Resistors <2 kilohms.................	$500/R + 4$

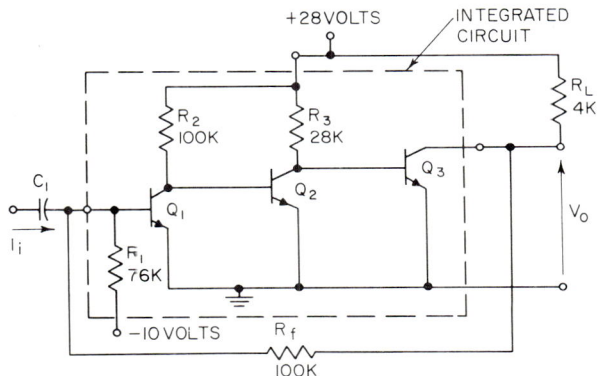

FIG. 20. High-cost direct-coupled feedback amplifier.

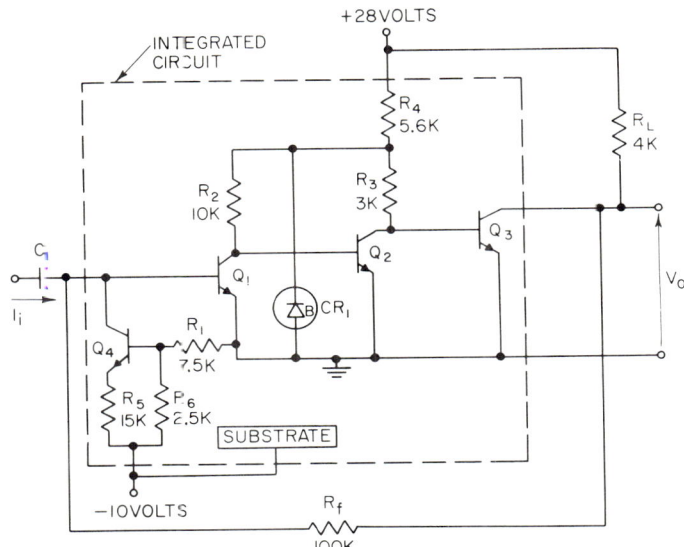

FIG. 21. Lower-cost direct-coupled feedback amplifier.

ment. At first glance, it appears that the second trial is much more complex and costly than the first. If discrete components were used, this would definitely be true. Applying the appropriate weighting functions to the second example yields 121.2 points. Thus it is seen that the second trial results in a 3-to-1 reduction in the number of points. This would probably represent more than a 3-to-1 reduction in cost. The low-cost amplifier will have approximately 20 per cent less loop gain than the high-cost version, primarily because R_2 and R_3 were reduced in value, but both amplifiers are capable of having gain stabilities of better than 0.1 per cent. The resistors in Fig. 21 can be further reduced in value, but the module power dissipation will begin to increase sharply.

4.4. Breadboarding Techniques. When the optimum circuit configuration is selected and a thorough analytical evaluation is completed, the circuit can be breadboarded and tested in the laboratory. The most desirable laboratory model would consist of a circuit having exactly the same electrical, mechanical, and thermal properties as the final manufactured circuit. Unfortunately, models such as this cannot be readily fabricated. It is not too difficult to construct a model that closely simulates the electrical properties of the final circuits. Most semiconductor-component manufacturers have available integrated component-part kits that closely resemble a given manufacturing process. When properly interconnected, the circuit will exhibit electrical characteristics similar to the final manufactured circuit, except that the frequency response of the model will be lower, primarily because of added wiring capacitance and inductance. In building the model, the p-type isolation region should be terminated to the most negative potential in the circuit. Likewise, all n-type isolation regions should be terminated to the most positive potential in the circuit. This will result in minimum crosstalk between components and minimum capacitance between the components and the substrate. Unfortunately, this termination of the isolation regions does impose higher breakdown voltage requirement on the n- and p-type substrates.

It should be emphasized that the above-described model does not simulate the thermal properties of the circuits. Some final manufactured circuits, especially integrated-power amplifiers, may oscillate, because of varying thermal gradients between components. Generally a careful analytical study will eliminate most of these problems before the final production circuit is built. In low-level circuits, no large thermal gradients exist and all components are at approximately the same temperature. To simulate this situation with the kit components, all components can be thermally tied to a metal plate.

5. MULTIPURPOSE CIRCUITS

A universal multipurpose circuit which can be externally pin-programmed to perform a large variety of circuit functions can greatly reduce the cost of large-scale electronic systems. For example, if one or two linear amplifiers are required in a microelectronic system, it may not be economical to pay the tooling charges for a special set of masks to mechanize the low-usage circuit. Much more desirable would be a multipurpose circuit which could be utilized to satisfy other low-usage applications in the same system. Figure 22 shows the schematic diagram of a multipurpose circuit consisting of four n-p-n transistors and four 9-kilohm resistors contained in a single silicon chip.

A flat container measuring $\frac{1}{4}$ by $\frac{3}{8}$ by 0.035 in. with 15 external leads was chosen to package the circuit. Note that pin 15 is connected to the p-type isolation region. To reduce stray pickup between the various components within the chip, pin 15 should always be terminated at the most negative potential in the circuit.

This section will illustrate some of the more popular applications using only one multipurpose circuit chip. If more than one chip is used, an infinite combination of circuit functions can be developed. It is beyond the scope of this book to show more than the single-chip examples. Many of the applications require the use of discrete components such as resistors or capacitors. The integrated resistors have a large temperature coefficient, of from $+0.15$ to $+0.30$ per cent per degree centigrade.

Circuit performance may change considerably with varying ambient temperature if integrated resistors are used with conventional nonsilicon discrete resistors.

5.1. Differential Amplifier. Figure 23 shows the schematic diagram of a connection which will yield a differential amplifier. The circuit requires one additional

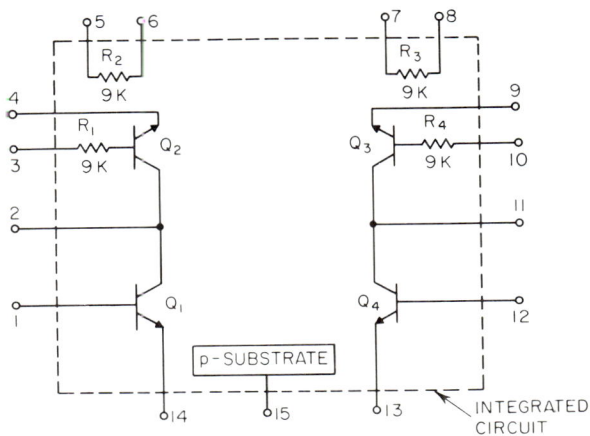

Fig. 22. Universal multipurpose circuit. (*Courtesy of General Electric Co., 4JP904.*)

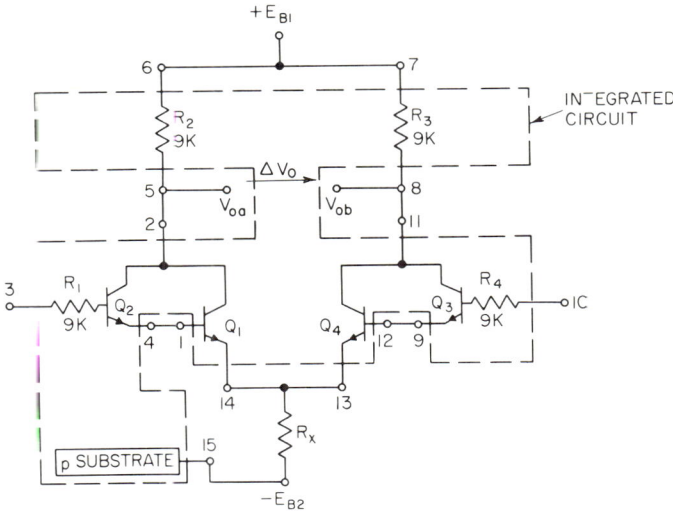

Fig. 23. D-C differential amplifier.

resistor R_x, which can be derived from another integrated chip; or a conventional resistor may be used. Caution should be exercised when using a conventional resistor with integrated resistors. Because of the different temperature coefficients, the operating points can shift considerably if the circuit is exposed to wide temperature variations.

AMPLIFIER CIRCUITS

If the circuit is balanced and both inputs grounded the collector voltage V_o with respect to ground of the output stages can be approximated as

$$V_o = E_{B1} - \frac{(E_{B2} - 1.0)9K}{2R_x} \qquad (7)$$

If $E_{B1} = 10$ volts, $E_{B2} = 10$ volts, and $R_x = 8.1$ kilohms, then the collector voltages of the output stage will equal approximately 5 volts and will yield the maximum undistorted output. The differential voltage gain in this case will be approximately 85. Figure 24 shows input vs. output data taken on an experimental circuit with $E_{B1} = E_{B2} = 10$ volts and $R_x = 8.1$ kilohms. The differential temperature drift referred to the input from 25 to 100°C measured 10 $\mu v/°C$.

The differential gain is heavily dependent on the transistor emitter currents in the output stages, which, to a first-order approximation, are controlled by the negative supply voltage E_{B2} and the resistor R_x. Since the gain can be controlled by adjusting the voltage E_{B2}, a simple AGC function can be mechanized if E_{B2} is made to be a function of the a-c output of the final stage in a cascaded amplifier.

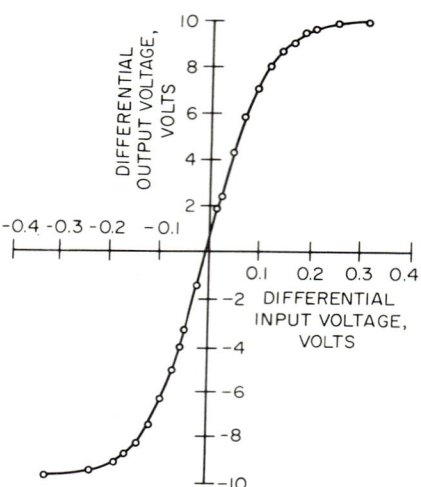

Fig. 24. Linearity of experimental differential amplifier.

5.2. Dual Common-emitter Darlington Amplifier.
Darlington common-emitter amplifiers exhibit moderately high input resistance, high current amplification and voltage amplification, and have been used in a large variety of linear and nonlinear amplifier applications. To a first-order approximation, the current amplification is the product of the current amplifications of the individual stages.

Figure 25 shows the connection for the dual Darlington common-emitter ampli-

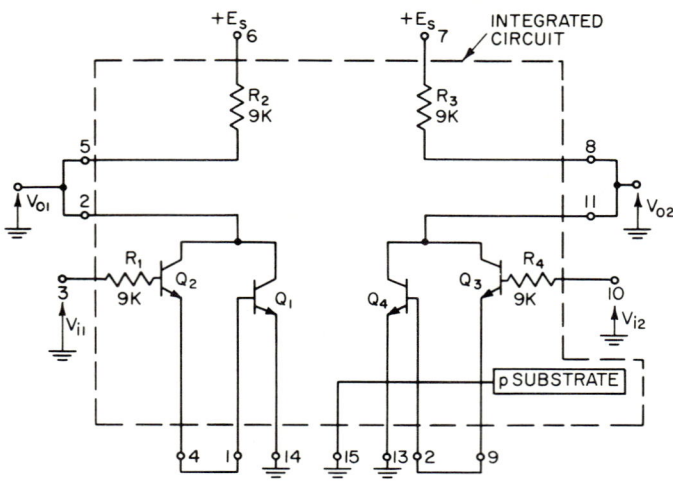

Fig. 25. Dual Darlington common-emitter amplifier.

fiers, which can be effectively used as elements in a direct-coupled cascaded amplifier with overall d-c feedback for bias stabilization. With current amplifications of 50 and 100 for the input and output stages, respectively, and 1 ma of emitter current in the output stage the Darlington amplifier would have an approximate current amplification of 5,000, a voltage amplification of 200, and an input resistance in the range of 200 kilohms.

5.3. Dual Darlington Emitter-Follower Amplifier. Darlington emitter-follower amplifiers exhibit very high input resistance, very low output resistance, high current amplification and approximately unity voltage amplification. The circuit is widely utilized either to drive relatively low values of load resistance or to be driven by relatively high values of source resistance. It is commonly referred to as an impedance converter.

Figure 26 shows the connection for the dual Darlington emitter-follower amplifier using typical integrated transistors and $E_{S1} = 10$ volts, $E_{S2} = 10$ volts, and the input

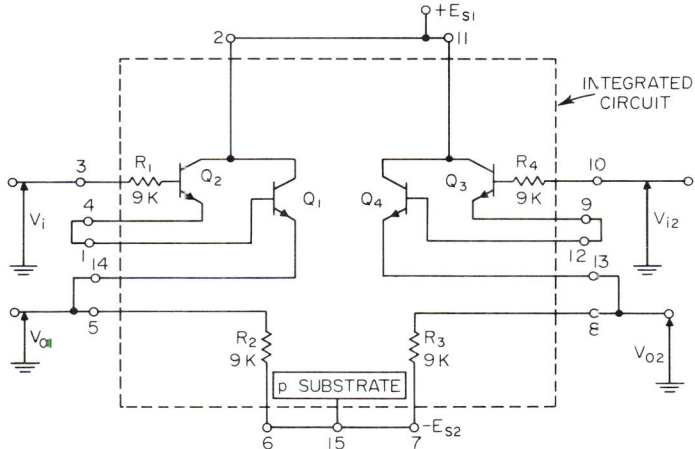

Fig. 26. Dual Darlington emitter-follower amplifier.

signal referenced to d-c ground. The input resistance will be in the range of megohms, the output resistance tens of ohms, the current amplification 5,000, the voltage amplification approximately unity, and the d-c input current will be approximately 0.2 μa.

5.4. Darlington Common-base Amplifier. Darlington common-base amplifiers have very low input resistance, very high output resistance, high voltage amplification, and very nearly unity current amplification. They are frequently used in precise constant-current source applications.

Figure 27 shows the connection for the Darlington common-base amplifier. Using typical integrated transistors with $E_B = 20$ volts and the d-c input current equal to 1 ma, the input resistance will be 50 to 75 ohms, the output resistance 10 to 100 megohms, the voltage amplification 120 to 180, and the current amplification 0.9998. If the driving-source impedance is much greater than 75 ohms, then the d-c output voltage will change at the rate of approximately +0.2 per cent per degree centigrade increase in temperature because of the temperature coefficient of the 9-kilohm resistor R_2. The load resistor at the output (see Fig. 22) can be changed to either 4.5 or 18 kilohms by the parallel or series connection of R_2 and R_3, respectively, thus changing the voltage amplification by a factor of 1 to 4 as desired.

5.5. Dual Zener Reference. Zener diodes are frequently used in d-c amplifiers or in direct-coupled a-c amplifiers for stabilizing unregulated power or for voltage-level shifting. The breakdown voltage of the base-emitter junction can be accurately controlled by the semiconductor component manufacturer to a tolerance of better than 10 per cent with the bulk of the yield within a 5 per cent band.

Figure 28 shows the connection for the dual zener reference circuit. Unregulated supply voltages $+E_{B1}$ and $-E_{B2}$ represent the inputs and $+V_{O1}$ and $-V_{O2}$ the positive and negative regulated outputs. Transistor Q_1 can be used as a negative supply regulator by interchanging pins 1 and 14, or Q_4 can be used as a positive supply regu-

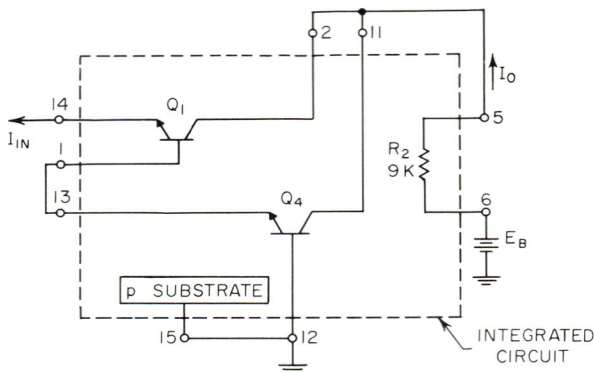

FIG. 27. Darlington common-base amplifier.

lator by interchanging pins 12 and 13. Thus the circuit can be pin-programmed to yield either a positive and negative regulator, or two independent negative regulators, or two independent positive regulators. If the breakdown voltage of the base-to-emitter junction is 7 volts, then an unregulated supply voltage of from 15 to 45 volts will yield a regulated output of approximately 7 volts. In applications where it is

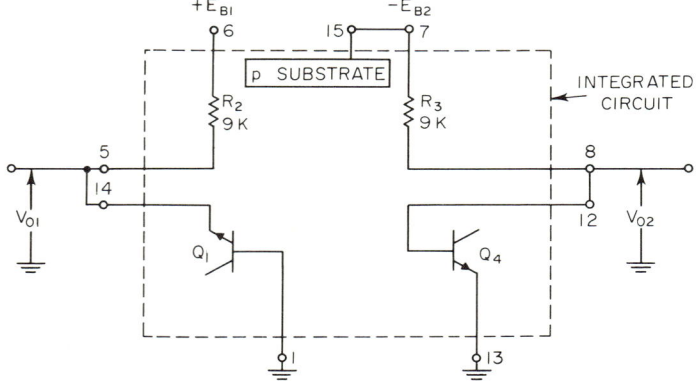

FIG. 28. Dual zener references.

desired to use zener diodes as biasing elements, then Q_1 and Q_4 may be used directly by making connections to pins 14 and 1, and pins 13 and 12, respectively.

5.6. Buffered Zener Regulator. In some amplifier applications it may be desired to supply accurately more regulated output current than can be supplied from the zener references shown in Fig. 28. The multipurpose circuit can be pin-programmed to form a buffered zener regulator as shown in Fig. 29, which is capable of supplying

INTEGRATED CIRCUITS   31-25

relatively large variations of output current without appreciably changing the current through the reference element Q_1.

5.7. Full-wave Transformerless Demodulator. In many servo-amplifier applications, it is necessary to demodulate an amplitude-modulated signal accurately. In

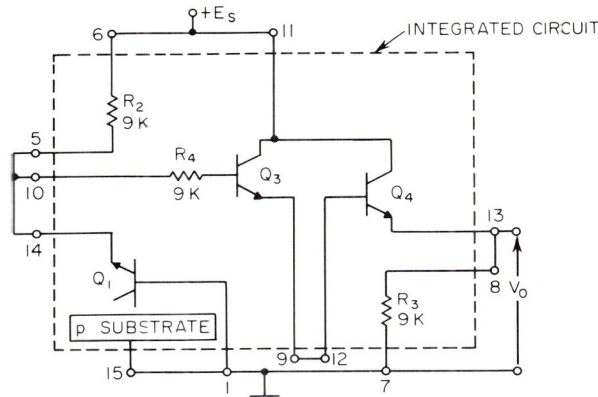

FIG. 29. Buffered zener regulator.

the past, demodulators have required either one or two well-shielded transformers. The transformer-type demodulators are costly, bulky, and often require special factory adjustments to reduce the initial offsets. In some cases special temperature compensation is required.

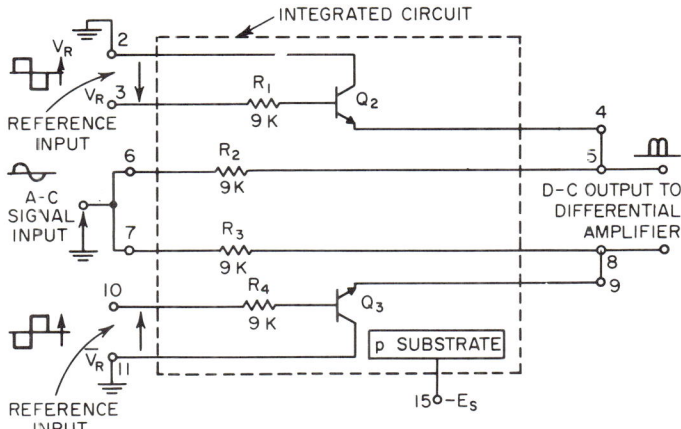

FIG. 30. Full-wave transformerless demodulator.

Figure 30 shows the connection for a full-wave transformerless demodulator. The circuit requires a two-phase reference input V_R and $\bar{V}_R$ which can be easily derived from the two outputs of a flip-flop that is driven by a single-phase reference. Reference V_R is a positive- and negative-going square wave with respect to ground, which drives transistor Q_2 into saturation and cutoff, respectively. In a like manner transistor Q_3 is being driven into cutoff and saturation by reference $\bar{V}_R$. When Q_2 is saturated, Q_3 is cut off. The amplitude-modulated a-c input signal is applied to pins 6 and 7 referenced to ground. The d-c output is connected to a differential amplifier which converts the difference signal to a full-wave d-c output proportional to the amplitude-modulated a-c input signal. The saturated offset voltages of transistors Q_2 and Q_3 tend to cancel and produce a very small output offset which is proportional to the difference of the offset voltages.

Figure 31 shows the input vs. output transfer function of an experimental model of the transformerless full-wave demodulator. The square-wave reference input was a 5-volt positive- and 5-volt negative-going voltage, referenced to ground. The measured offset voltage was 41 μv and the gain 0.9 volt d-c per volt a-c rms. The nonlinearity at large input signals is a function of the amplitude of the square-wave reference drive and the breakdown voltage of the base-emitter junction of transistors Q_2 and Q_3. A capacitor may be connected between pin 5 and pin 8 to filter the output, producing a smooth direct current to the input terminals of the differential amplifier.

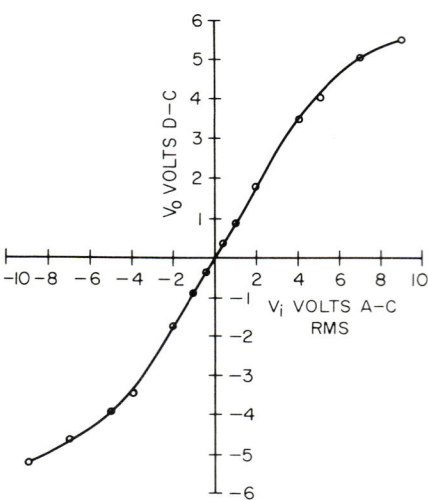

Fig. 31. Transformerless full-wave demodulator transfer function.

5.8. Transformerless Modulator. Transistor-type modulators have found wide usage in servo applications when it is necessary to convert a d-c signal to an amplitude-modulated a-c signal. Also, choppers are used in some d-c amplifiers requiring low drift and offset. Modulators having a differential input can be more universally adapted to system-wide usage in that they are not appreciably sensitive to ground noise between the source ground and the modulator ground. Single-ended modulators modulate ground noise and appreciably deteriorate the true signal that is being modulated.

Figure 32 shows the connection for a transformerless modulator which has a differential input. The circuit requires a single-phase positive- and negative-going reference drive voltage with respect to ground. The output is connected to a differential amplifier which converts the difference a-c signal to a single-ended a-c output proportional to the d-c differential input of the modulator.

The circuit operation can be explained as follows: When the reference voltage is positive, transistors Q_2 and Q_3 are saturated and the output voltage is the difference of the saturation voltages of transistors Q_2 and Q_3 which is very near zero volts. When the reference voltage is negative, transistors Q_2 and Q_3 are cut off and the output voltage is equal to the input voltage. Thus to a first-order approximation, the peak-to-peak output voltage is the input d-c signal plus or minus the difference of the collector-to-emitter saturation voltages of transistors Q_2 and Q_3. In integrated form, the transistors Q_2 and Q_3 will have very similar characteristics, resulting in exceptionally low offset voltage.

The results of an experimental model are shown in Fig. 33. The reference input was a +5-volt and −5-volt square wave with respect to ground. The differential output voltage was read with a phase-sensitive voltmeter. The offset output voltage with both inputs shorted to ground measured 10 μv rms. The gain of the modulator measured approximately 0.45 volt rms output per volt d-c input. The modulation frequency was chosen for convenience to be 800 cps.

5.9. Transformerless Quadrature Rejector. Many servoamplifiers have input signals that are derived from electromagnetic transducers such as resolvers or synchros. In general, the transducers have a relatively large quadrature output voltage at the mechanical null position, which can cause saturation problems in very high-gain servoamplifiers. One solution to this problem is to connect a quadrature rejector between the output of the transducer and the input of the servo amplifier. The

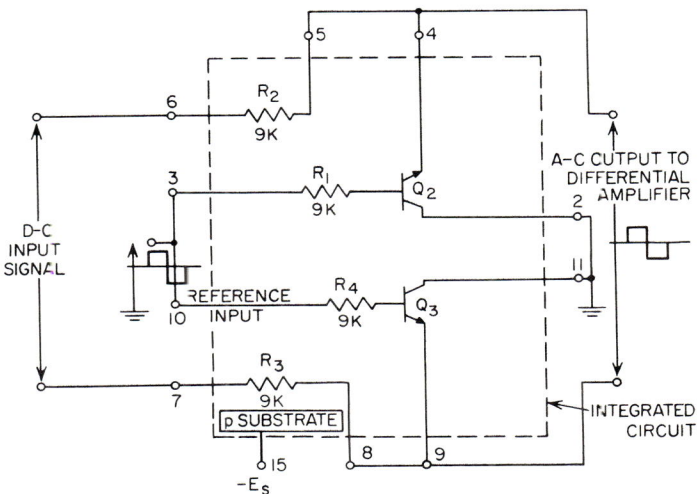

FIG. 32. Transformerless modulator.

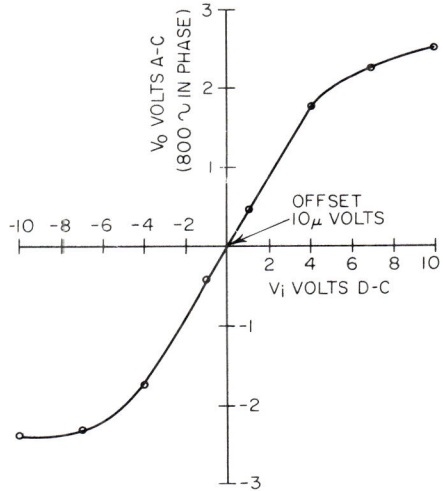

FIG. 33. Transformerless modulator transfer characteristics.

quadrature rejector will attenuate the quadrature component, random noise, and even-order harmonics while passing the in-phase component with negligible attenuation. Quadrature rejectors with 100 db attenuation to quadrature voltages have been built.

Figure 34 shows the connection for a transformerless quadrature rejector. The circuit requires two external capacitors C_x and C_y and a two-phase positive- and negative-going reference drive voltage with respect to ground. The two-phase reference voltage can be derived from the two outputs of a flip-flop. The quadrature rejector has two outputs, a voltage output when driving into a high-input-impedance series-feedback amplifier and/or a current output when driving into a low-input-impedance shunt-feedback amplifier.

The circuit operation can be explained as follows: When V_R is positive, transistor Q_2 is saturated and Q_3 is cut off. If at this time the a-c input signal connected to pin 6 is positive, the voltage across capacitor C_x will be a positive value proportional to the average value of a half wave of the in-phase component of the signal. During the other half-cycle Q_3 will be saturated and Q_2 cut off, and the voltage across C_y will be negative and proportional to the average value of the negative half of the in-phase

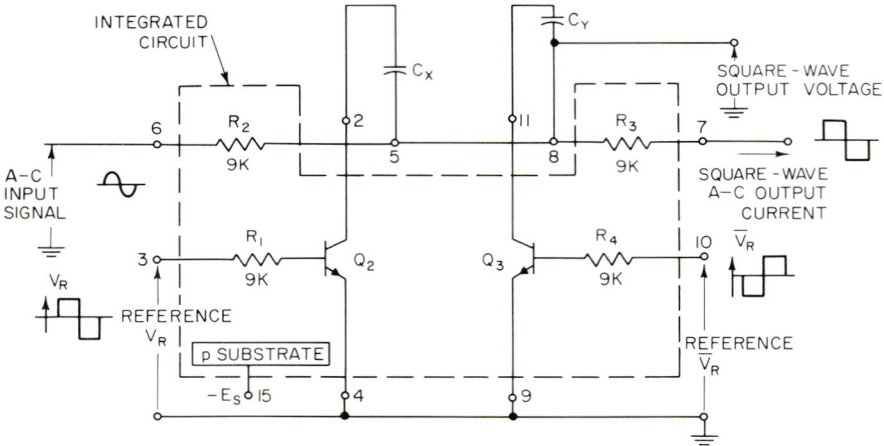

Fig. 34. Transformerless quadrature rejector.

component of the signal. If the voltage output is used and pin 7 is floating, then the square-wave output vs. sine-wave input can be approximated as

$$\frac{V_o}{V_i} = \frac{1.8 \text{ volts peak-to-peak}}{\text{volts rms}} \quad (8)$$

The output voltage with respect to quadrature input voltage can be approximated as

$$V_o = \frac{V_Q}{2\pi f C R_2} \quad \text{volts rms} \quad (9)$$

where f = frequency, cps
V_Q = quadrature input voltage rms
$C = C_x = C_y$

If pin 7 is connected to a low-impedance current summing point then the square-wave output current vs. sine-wave input voltage can be approximated as

$$I_o = \frac{(R_2 + R_3)V_Q}{4\pi f C R_2 R_3^2} \quad \text{amp rms} \quad (10)$$

5.10. Precision High-input-impedance A-C Amplifier. Two-stage direct-coupled series-voltage-feedback amplifiers have found wide usage in applications requiring accurate amplification of voltages originating from a high-impedance source

INTEGRATED CIRCUITS 31-29

with the additional requirement of driving a low value of load impedance. The circuit connection shown in Fig. 35 is an extension of the two-stage circuit using a Darlington common-emitter pair in place of each single transistor, resulting in considerably more gain and hence in more loop gain and higher closed-loop gain accuracy. The circuit requires three external resistors and one capacitor, which can be varied to yield an infinite number of different operating points and gain values to suit a variety of applications.

If the input terminal pin 3 is terminated to a signal source having zero d-c potential referenced to ground, then the output d-c voltage at pin 11 referenced to ground

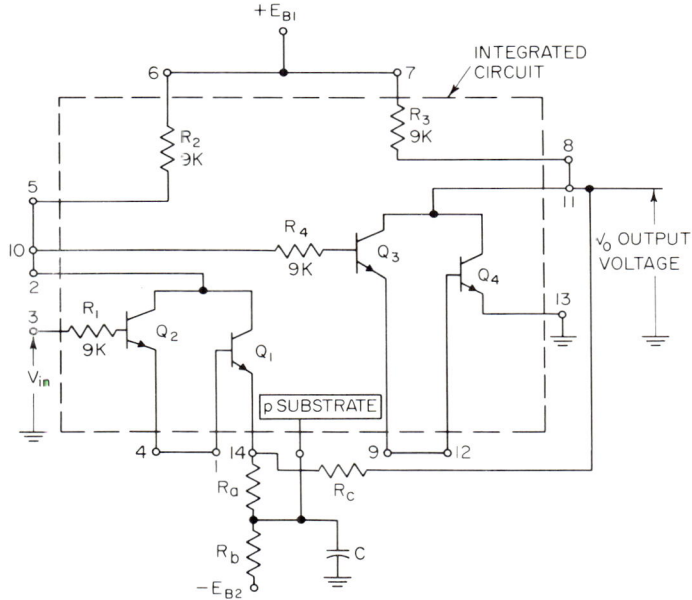

FIG. 35. Precision high-input-impedance a-c amplifier.

can be approximated as

$$V_o(\text{d-c}) = \frac{(E_{B2} - 1)R_c}{R_a + R_b} - \frac{(E_{B1} - 1)R_c}{R_2} - 1 \tag{11}$$

The second term has a ratio of two resistors R_c/R_2. If R_c is a conventional component with a temperature coefficient considerably different from that of the silicon resistor R_2 (+0.2 per cent per degree centigrade), the operating point can shift considerably with temperature variations. To correct this condition R_a, R_b, and R_c should be silicon resistors, or the first term should be made much larger than the second term by making $E_{B2} \gg E_{B1}$ or $R_a + R_b \ll R_2$. If the loop gain is large, the closed-loop voltage amplification A_V can be approximated as

$$A_V = (R_a + R_c)/R_a \tag{12}$$

Thus, to a first-order approximation, the voltage amplification is independent of the characteristics of the integrated circuits. Gain accuracies of better than 1 per cent can be readily designed.

The value of capacitor C determines the amplifier low-frequency response. If

31-30   AMPLIFIER CIRCUITS

$R_b \gg R_a$, the 3-db low-frequency break cutoff f_l can be expressed as

$$f_l = 1/2\pi R_a C \tag{13}$$

The amplification A_V at very low frequencies where $X_C \gg R_b$ can be approximated as

$$A_{V_l} \cong \frac{R_a + R_b + R_c}{R_a + R_b} \tag{14}$$

To avoid large shifts in the operating point of the output stage with temperature, it is desirable to maintain a low value of very-low-frequency amplification A_{V_l}. The

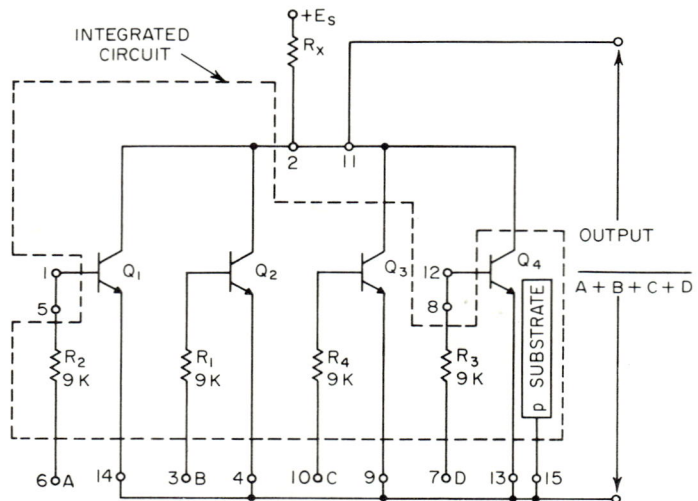

Fig. 36. Four-input NOR gate.

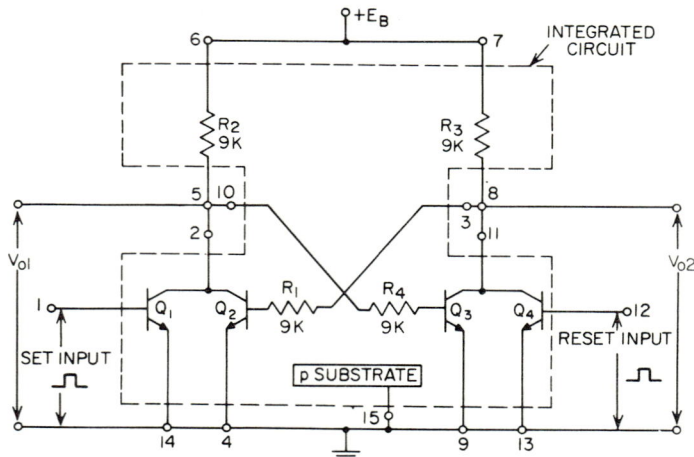

Fig. 37. High-sensitivity flip-flop.

variations of the base-to-emitter forward-voltage drops of transistors Q_1 and Q_2 will be amplified by the low-frequency amplification A_{V_l} and result in a shift in the operating point of the output stage. In general A_{V_l} should be kept below 10.

5.11. Miscellaneous Nonlinear-amplifier Applications. In future systems it will not be practical to isolate the linear analog circuit designs from the nonlinear

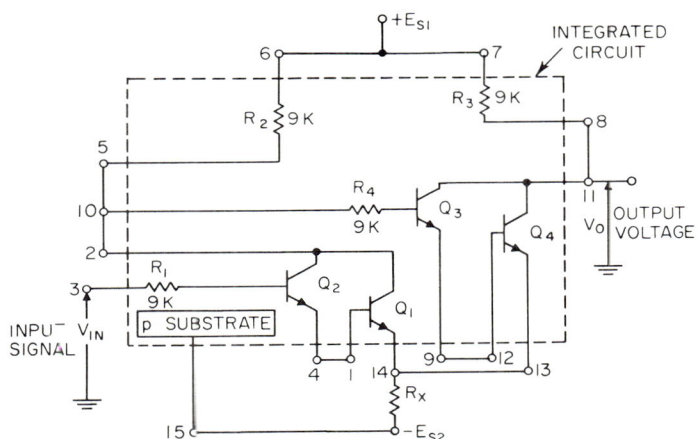

FIG. 38. Darlington Schmitt trigger (high sensitivity, low speed).

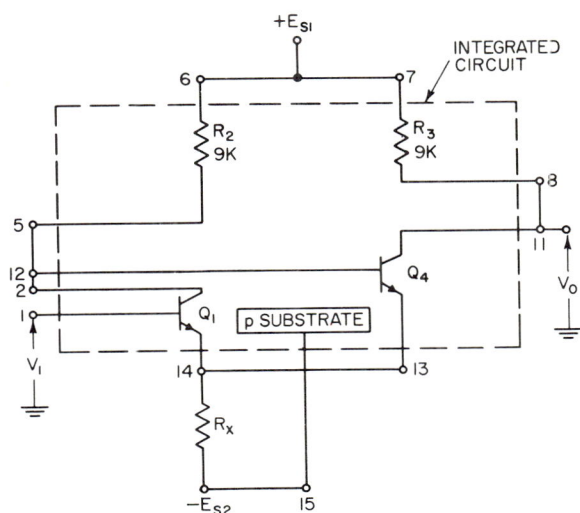

FIG. 39. High-speed Schmitt trigger.

digital designs, when quite often one circuit can be used in both applications. Figures 36 through 45 illustrate 10 additional connections of the multipurpose circuit which do not fall in the area of linear amplifier applications but nevertheless should be mentioned to realize fully the usefulness of the single-chip configuration. The connections include a four-input NOR gate, a high-sensitivity flip-flop, a Darlington

Schmitt trigger, a high-speed Schmitt trigger, a triggered one-shot, a free-running multivibrator, a voltage-controlled oscillator, a dual high-sensitivity relay driver, a dual lamp driver, and a bistable indicator lamp driver.

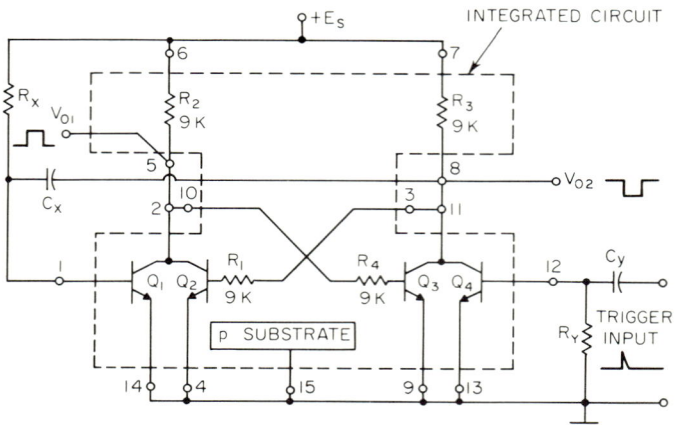

Fig. 40. Triggered one-shot.

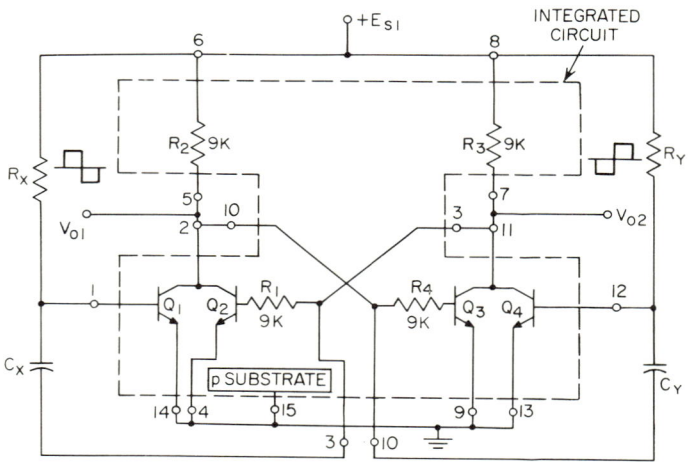

Fig. 41. Free-running multivibrator.

6. DIRECT-COUPLED AMPLIFIERS

In low-frequency integrated amplifiers, RC-coupled amplifiers are extremely impractical, since high-value capacitors require large areas. In general, RC-coupled amplifiers are used when the d-c output drift reaches a condition where the wanted output swing saturates or cuts off, causing distortion. In other words, the a-c gain is made greater than the d-c gain by breaking the d-c path with the capacitor. If a d-c amplifier with no drift, or very small drift, were available, there would be no need for RC-coupled amplifiers. This section will discuss some integrated low-drift direct-coupled amplifiers that could be used as a-c amplifiers or as d-c amplifiers.

INTEGRATED CIRCUITS

6.1. Differential Amplifiers. In the past, differential amplifiers were used in d-c amplifier applications requiring low drift. Differential amplifiers can also be used in integrated a-c amplifier applications to eliminate the need for RC-coupled amplifiers.

Figure 46 shows the schematic diagram of one type of integrated differential amplifier that uses 10 resistors totaling 135 kilohms, one zener diode, and five n-p-n tran-

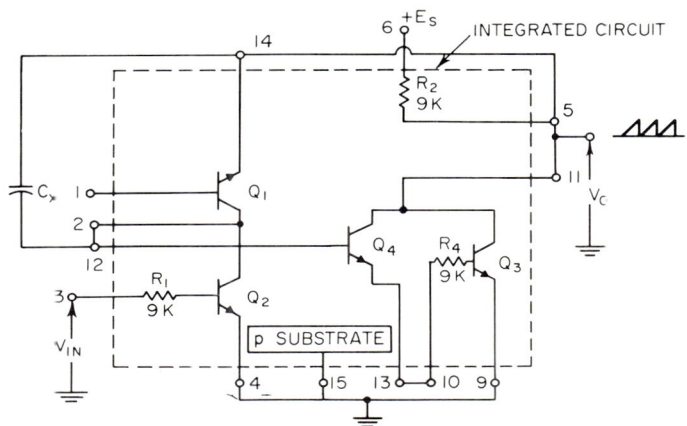

Fig. 42. Voltage-controlled oscillator.

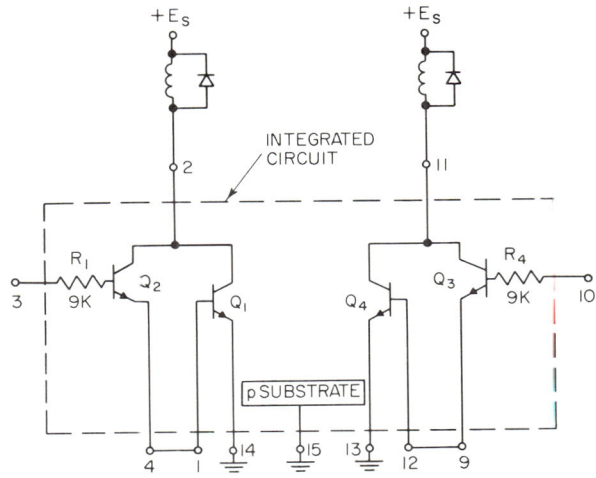

Fig. 43. Dual high-sensitivity relay driver.

sistors. The integrated chip area measures 0.11 by 0.09 in., which is approximately 0.01 sq in., and is mounted in a 12-pin TO-5 package. The circuit is made up of dual two-stage feedback pairs connected differentially and driven by a constant-current generator Q_5. The circuit was designed to operate with two 9-volt supplies E_{B1} and E_{B2} and consume a total power of approximately 50 mw.

To a first-order approximation, assuming a balanced circuit with both inputs

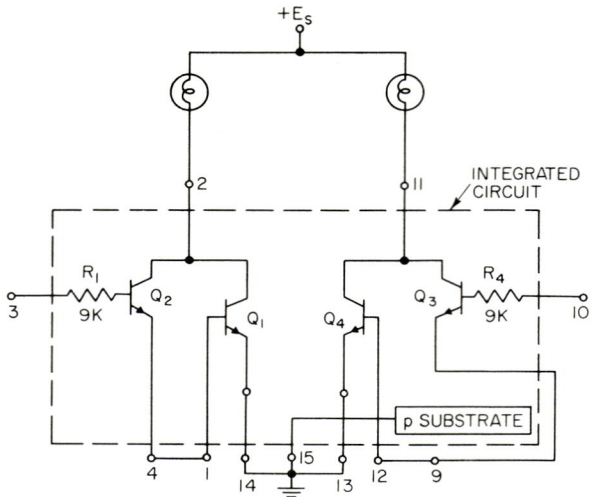

Fig. 44. Dual lamp driver.

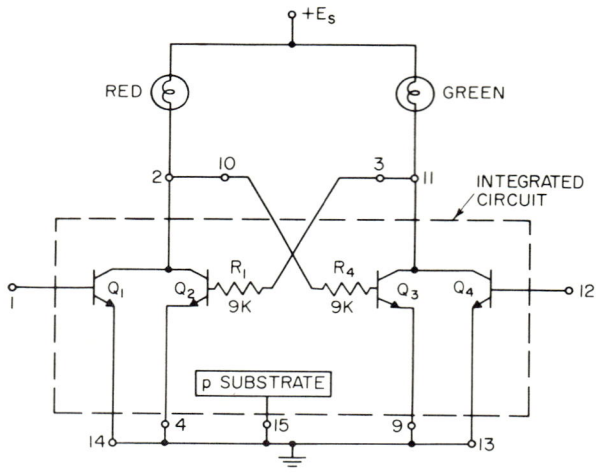

Fig. 45. Bistable indicator lamp driver.

shorted to ground, the d-c collector voltage of transistor Q_3 (V_{C3} referenced to ground) can be expressed as

$$V_{C3} = R_{F1}\left[\frac{V_{CR1} - V_{BE5}}{2(R_S + R_X)} - \frac{E_{B1} - V_{BE3}}{R_{C1}}\right] - V_{BE1} \quad (15)$$

and the collector voltage of transistor Q_5 referenced to ground is approximately

$$V_{C4} = R_{F2}\left[\frac{V_{CR1} - V_{BE5}}{2(R_S + R_X)} - \frac{E_{B1} - V_{BE4}}{R_{C2}}\right] - V_{BE2} \quad (16)$$

where V_{CR1} = breakdown voltage of CR_1 = 7 volts
V_{BEN} = forward base-to-emitter voltage drop of transistor number N
E_{B1} = plus supply voltage
R_x = external trim resistor

Examination of the above equations will show that V_{C3} and V_{C4} can be raised or lowered by varying the external trim resistor R_x, which is connected between pins 7 and 8. For any desired value of V_{C3} and V_{C4}, R_x could be reduced to zero if the ratios of R_{F1}/R_S, R_{F2}/R_S, R_{F1}/R_{C1}, and R_{F2}/R_{C2} were under complete control along with the supply voltage E_{B1}, zener voltage of CR_1, and the transistor base-to-emitter diode drops. As process controls improve, it is expected that high yields of resistors having ratios of better than 1 per cent will be feasible along with zener diodes with absolute tolerances of better than 5 per cent. When these controls are realized, external trim resistors such as R_x can be eliminated.

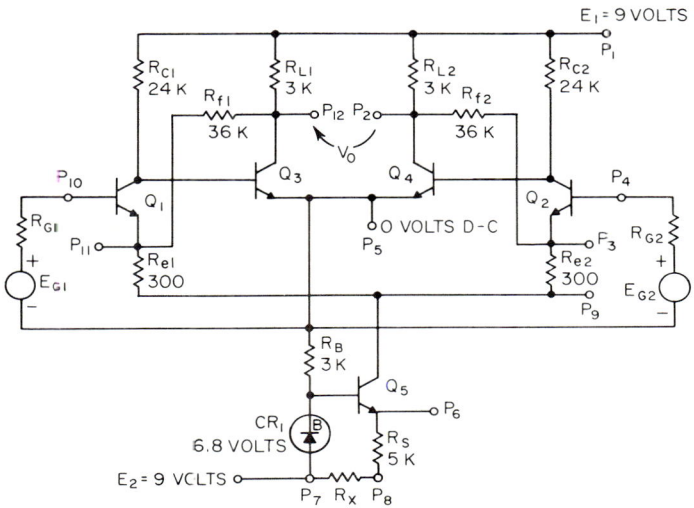

Fig. 46. Integrated differential amplifier. (*Courtesy of General Electric Co.*)

The incremental gain of the differential amplifier is highly dependent on the feedback elements R_{F1}, R_{F2}, R_{E1}, and R_{E2} and the loop gain of the amplifier. If the loop gain is high, and $R_{F1} = R_{F2} = R_F$ and $R_{E1} = R_{E2} = R_E$, the differential gain, defined as the differential output voltage divided by the differential input voltage, can be approximated as the ratio of R_F to R_E.

Data recorded on one of the early samples of the integrated differential amplifier are shown in Figs. 47 and 48. Figure 47 shows the input vs. output voltage-transfer characteristics. The amplification is seen to be approximately 73, with an output differential offset of approximately 12 mv. Figure 48 shows a plot of the input drift variation with temperature from 25 to 100°C with both inputs terminated to ground. The data indicate a variation of approximately 130 $\mu v/°C$. The bulk of this drift is attributed to the change of the integrated resistors vs. temperature. Later units were reported to have variations of 40 $\mu v/°C$. A breadboarded model was built using conventional components, and a temperature variation of 4 $\mu v/°C$ was measured. It is expected that integrated amplifiers of this type, with drifts of better than 10 $\mu v/°C$, will be readily available as the integrated-resistor technology is improved.

The 12-pin package was chosen to make the circuit as flexible as possible, and as a result the circuit has many applications. The additional applications are outlined below.

High-gain A-C Amplifier. Figure 49 shows the connection for pin-programming the differential amplifier into a high-gain a-c amplifier configuration. Capacitor C_1 isolates the first half of the differential amplifier from the second half. Resistor R_1 provides a d-c ground bias for transistor Q_2. Capacitor C_2 couples the output of transistor Q_3 to the input of transistor Q_2.

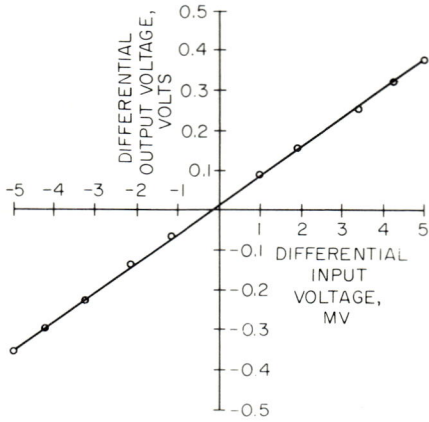

FIG. 47. Integrated differential amplifier transfer characteristics.

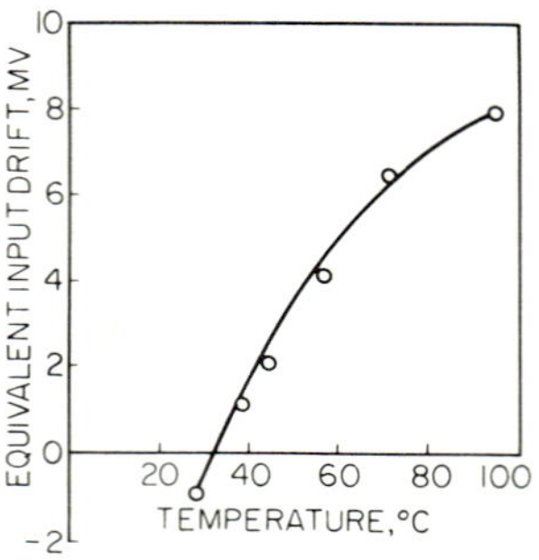

FIG. 48. Integrated differential amplifier temperature characteristics.

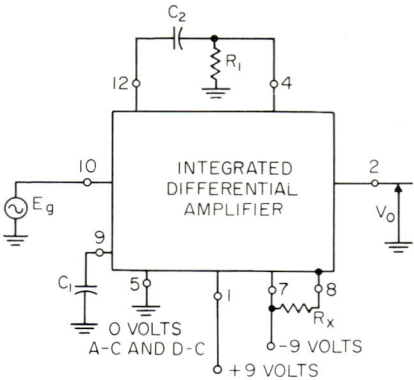

FIG. 49. High-gain a-c amplifier connection.

The signal flow is as follows: The input signal is connected to the base of transistor Q_1, pin 10; is amplified by transistors Q_1 and Q_3 and capacitance-coupled to the base of transistor Q_2, pin 4; and again is amplified by transistors Q_2 and Q_4. The final output is taken from the collector of transistor Q_4, pin 2. The overall gain of the amplifier is approximately the square of the differential gain; thus

$$A_v = (R_F/R_E)^2 \qquad (17)$$

Buffered Relaxation Oscillator. Figure 50 shows the connection for pin-programming the differential amplifier into a relaxation oscillator with an isolated buffered output. The output V_{o1}, pin 12, will have a sawtooth waveform and the output V_{o2}, pin 2, will be a rectangular pulse with fast rise and fall times. The duration of the pulse output at pin 2 can be voltage-controlled by applying a d-c voltage to pin 4.

Dual-relaxation Oscillator. Figure 51 shows the connection for pin-programming the differential amplifier into a dual-relaxation-oscillator configuration, which

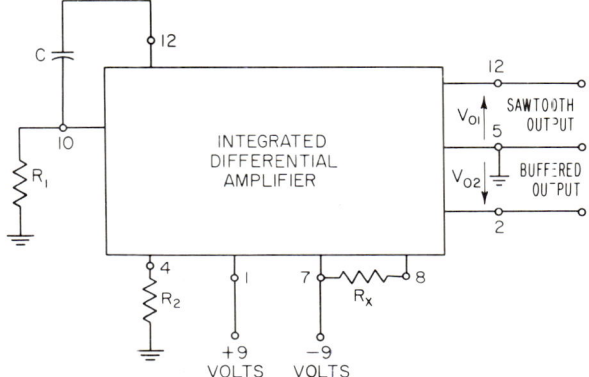

FIG. 50. Buffered relaxation oscillator.

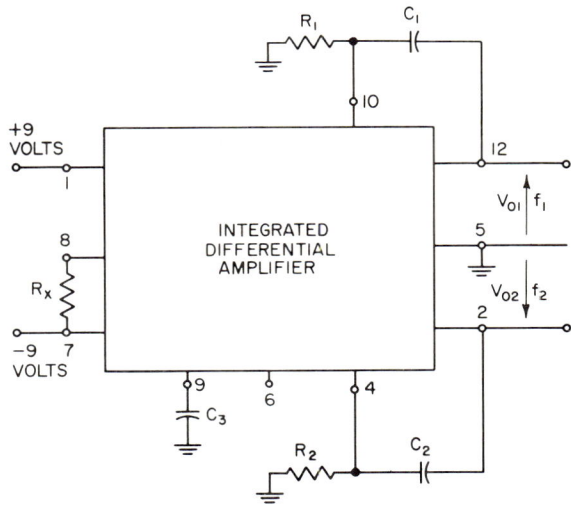

FIG. 51. Dual-relaxation oscillators.

has two output voltages of two independently different frequencies. Components R_1C_1 and R_2C_2 determine the frequencies f_1 and f_2, respectively. The outputs will be sawtooth waveforms. The frequencies can be time-modulated by applying a current source to pin 6.

Dual Flip-flops. Figure 52 shows the connection for dual bistable flip-flops which are completely independent. Each flip-flop has a single output and a single input. If a positive trigger is applied at the input, the output goes positive and remains in the positive state until a negative trigger is applied to the input, at which time the output saturates and remains in the saturated state until the next positive pulse is applied to the input.

Schmitt Trigger. Figure 53 shows the connection to pin-program the differential amplifier into a Schmitt-trigger configuration. This configuration requires no additional components and uses only resistors R_{G1}, R_{L1}, R_{E1}, R_{F1}, and transistors Q_1 and Q_3. Resistor R_{F1} produces negative feedback, but not enough to affect the cir-

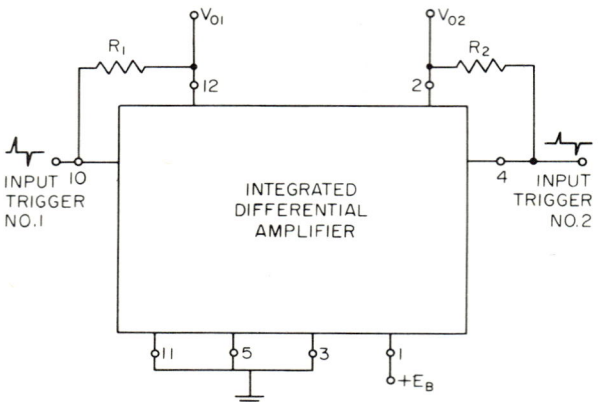

Fig. 52. Dual flip-flops.

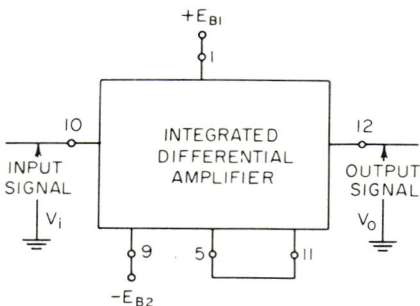

Fig. 53. Schmitt trigger.

cuit operation appreciably. The peak- and valley-point voltages of the input characteristic are heavily related to the negative supply voltage E_{B2}.

Applications of Schmitt triggers include wave shaping, triggered-pulse generators, free-running pulse generators, threshold detectors, and many other too numerous to mention.

Operational Amplifier. Operational amplifiers have found wide usage in many electronic systems of the past. In analog computers, the operational amplifier is the key element and can be programmed to add, subtract, integrate, and differentiate, using resistors and capacitors as input-summing or feedback networks. In addition, operational amplifiers are commonly used in analog-to-digital converters and as high-accuracy feedback amplifiers.

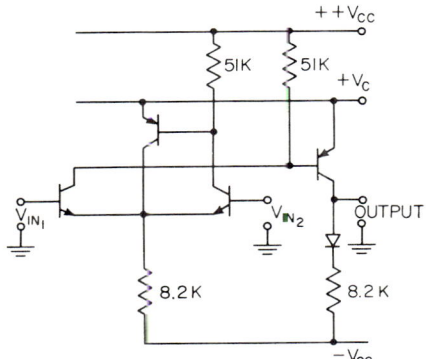
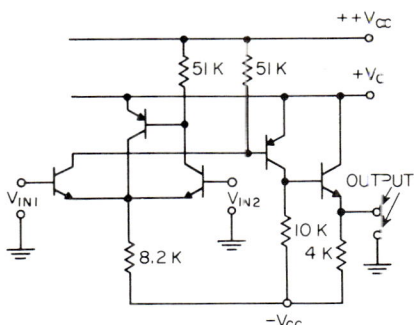

FIG. 54. Basic operational amplifier SN521. (*Courtesy of Texas Instruments, Inc.*)

FIG. 55. Basic operational amplifier with emitter-follower output SN522. (*Courtesy of Texas Instruments, Inc.*)

Figure 54 shows the schematic diagram of an integrated operational amplifier TI type SN521 which uses two p-n-p and two n-p-n transistors, one diode, and four resistors totaling 118.4 kilohms.

Figure 55 shows the schematic diagram of the SN522, which is essentially the same amplifier with an emitter-follower output for driving higher output power.

Each circuit type is packaged in a 10-pin hermetically sealed container measuring $\frac{1}{4}$ by $\frac{1}{8}$ by $\frac{3}{32}$ in. and weighs less than 0.1 gram.

The typical characteristics of some early units are tabulated below.

	SN521	SN522
Open-loop voltage gain, db	62	62
Common-mode rejection, db	58	58
D-C offset referred to input, mv	2	2
Half-power frequency point, kc	60	60
Input-impedance, differential, kilohms	18	18
Input impedance to ground, kilohms	10	10
Output impedance, kilohms	8	150
D-C drift referred to input, $\mu v/°C$	10	10

7. AUDIO AMPLIFIERS

The amplifiers described in this section are those designed to amplify audio-type signals, such as voice or music, and hence are probably the most widely used type of electronic equipment in existence. Audio amplifiers are used in all radio, television, and telephone communications, and almost all phonograph equipment. Amplifiers with a frequency response flat from 20 cps to 20 kc, and with harmonic distortion of less than 5 per cent, can be considered high-quality audio amplifiers.

Figure 56 shows a schematic diagram of an amplifier containing two unipolar and two bipolar transistors[14] interconnected to satisfy a wide variety of audio-amplifier applications. The gate-to-drain connection of unipolar Q_1 results in constant-current drive which supplies current to the drain of unipolar Q_2 and base drive to bipolar Q_3. Unipolar Q_2 is generally connected to operate in the common-source connection, which yields a reasonably high voltage gain and has high input impedance in the range of

31-40 AMPLIFIER CIRCUITS

3 megohms. Bipolar transistors Q_3 and Q_4 are connected in a Darlington common-emitter connection which also yields a high voltage and current amplification with high input impedance.

Figure 57 shows the collector characteristics, and Fig. 58 the transfer characteristics, of a typical unit measured on a transistor curve tracer with +9 volts connected to pin 1, −3 volts connected to pin 3, and pin 5 grounded. A d-c voltage V_{GE} was applied to pin 2 to make the output current at pin 4 denoted as I_C equal to zero. Step input voltages ΔV_{GE} were applied to pin 2 and the output varied as shown in the two figures. The characteristics indicate that g_m is exceptionally linear, with a value of approximately 23,000 micromhos.

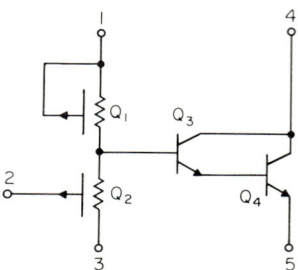

FIG. 56. Unibi audio amplifier. (*Courtesy of Westinghouse Electric Corporation.*)

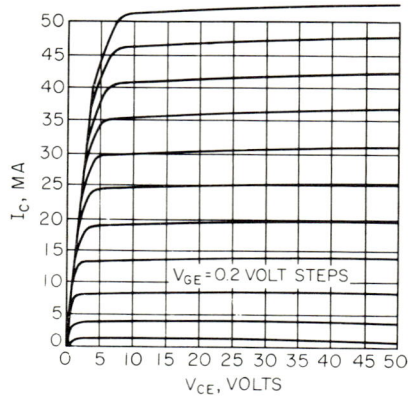

FIG. 57. Collector characteristics.

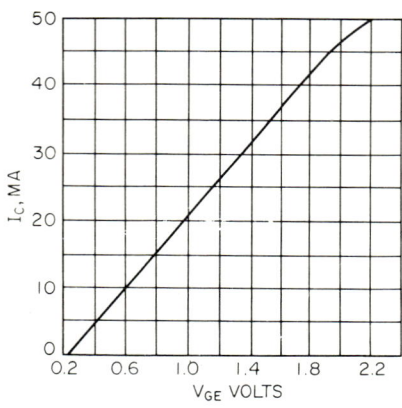

FIG. 58. Transfer characteristics.

The typical performance characteristics driven from a 600-ohm source and the circuit biased as explained above are tabulated below.

Input resistance, megohms	3
Input capacitance, pf	Less than 100
Transconductance, micromhos	20,000
Frequency response (−6 db), cps	0–100,000
Power sensitivity, mw/volt^2	75
Harmonic distortion, %	2

The maximum d-c ratings at 25°C of the device are $V_{13} = 18$ volts, $V_{23} = 9$ volts, $V_{43} = 30$ volts, $V_{45} = 30$ volts, $V_{52} = 7$ volts, and $I_4 = 100$ ma. The circuit is designed to operate over a temperature range of −55 to +125°C with a maximum power dissipation of 0.5 watt at 25°C.

Figure 59 shows the circuit connected as a typical audio amplifier. As shown, the circuit could be used as the audio amplifier for a small broadcast receiver with a detector input and a speaker load with approximately 20 mw of undistorted output

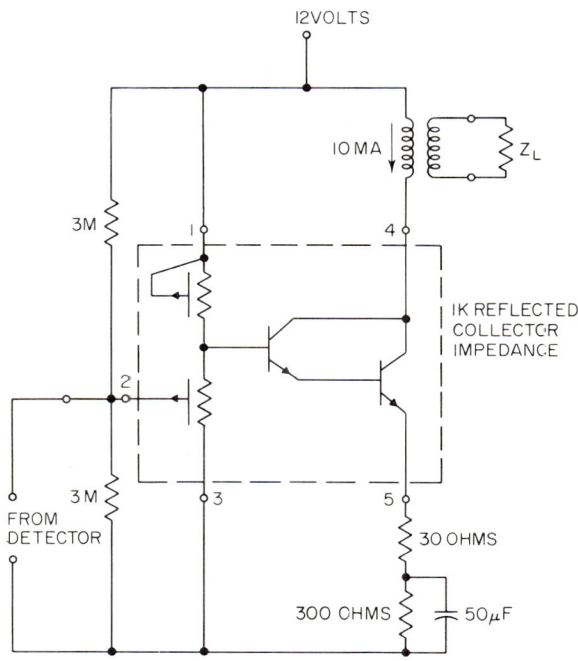

Fig. 59. Typical circuit application. (*Courtesy of Westinghouse Electric Corporation.*)

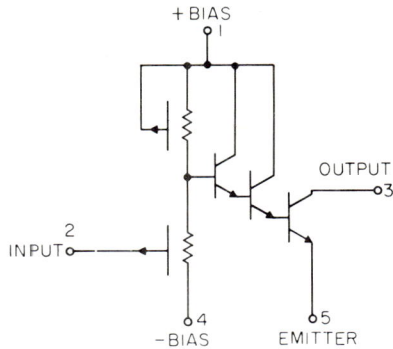

Fig. 30. Audio amplifier. (*Courtesy of Westinghouse Electric Corporation.*)

power. If more output power is required the audio-amplifier circuit shown in Fig. 60 can be used. The circuit contains two unipolar and three bipolar transistors. With the amplifier connected to a ceramic phonograph pickup, the circuit can deliver 3 watts of output power with a total harmonic distortion of less than 5 per cent. The circuit is packaged in a 5-pin 0.9-in.-diameter stud-mounted package.

8. SERVOAMPLIFIERS

Servo amplifiers are used extensively in analog computers and control systems for both industrial and military applications. Servo amplifiers may be required to amplify either time-varying d-c or amplitude-modulated single-frequency a-c carrier-type signals. In both cases, the amplifier usually drives a servomotor-type load. In general, the a-c servoamplifier is simpler and does not suffer from d-c drift. For a more detailed explanation of servo amplifiers refer to Chap. 21.

Figure 61 shows the schematic diagram of an integrated 1.5-watt direct-coupled differential servo amplifier[15] designed to drive 36 volts rms into a 400-cps size 8 or size 11 a-c servomotor. The complete single-silicon-chip amplifier measures 0.170 by 17.00 in. and contains 10 transistors, 5 zener diodes, and 27 resistors. The circuit requires a single 28-volt supply. To assure stable performance, the circuit utilizes

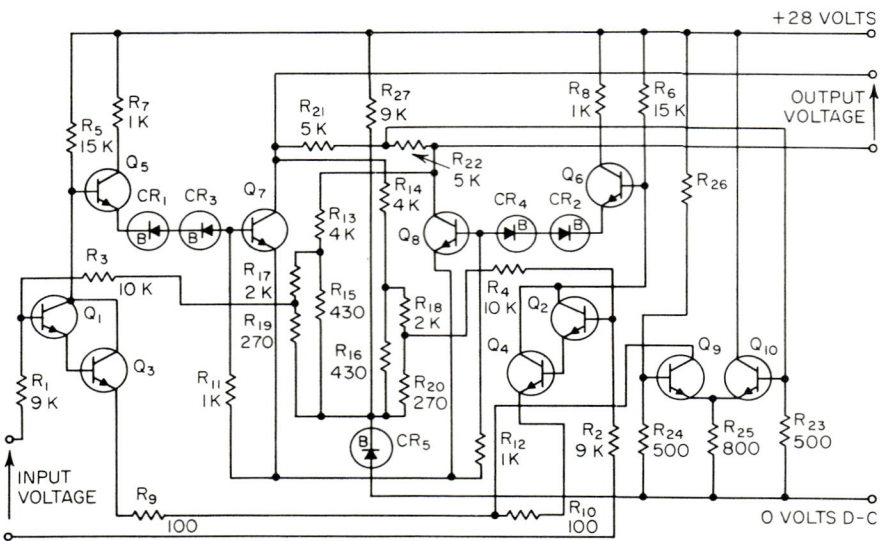

FIG. 61. Integrated servo amplifier. (*Courtesy of Norden Division, United Aircraft Corporation.*)

negative differential a-c feedback, common-mode d-c feedback, thermal feedback, and thermoelectric generation of stabilizing potentials. Since the amplifier must be capable of delivering 36 volts rms, the collector-to-collector breakdown voltage was designed to be greater than 120 volts using planar epitaxial techniques. Figure 62 shows an experimental graph of the transfer characteristics of the completely integrated servo amplifier driving a tuned servomotor whose primary is open to simulate a stalled condition.

9. PRECISION A-C SUMMING AMPLIFIERS

Precision a-c summing amplifiers are widely used in high-accuracy analog computers and in control systems for industrial and military applications. The summing accuracy of a summing amplifier is independent of the amplifier characteristic and dependent only on the accuracy of the input-summing resistors. On the other hand, the overall gain of the amplifier is a function of all summing resistors, source impedances, and amplifier characteristics. If shunt-voltage-feedback amplifiers have large loop gains, the overall voltage-gain accuracy is primarily a function of the ratio of the feedback resistor to the input-summing resistor. Since integrated-resistor tolerances are rather poorly controlled and vary considerably with temperature, they should not be used as feedback or summing resistors for amplifiers requiring absolute accuracies better than 5 per cent. In general, the feedback and summing resistors should be of the high-quality discrete type, such as wirewound or metal film. The remaining resistors and transistors that make up the amplifier can be readily integrated into a single silicon substrate. Since large loop gains are required in high-accuracy feedback amplifiers, resistor-capacitor networks are generally needed to control the gain-phase characteristics, thus preventing unwanted oscillations. In most cases, the value of the capacitors is too large for a practical integrated component. Either discrete or thin-film capacitors would be a more practical choice. Thus it appears that a precision a-c summing amplifier might consist of a network of high-quality discrete resistors, plus a few discrete or thin-film capacitors and an integrated

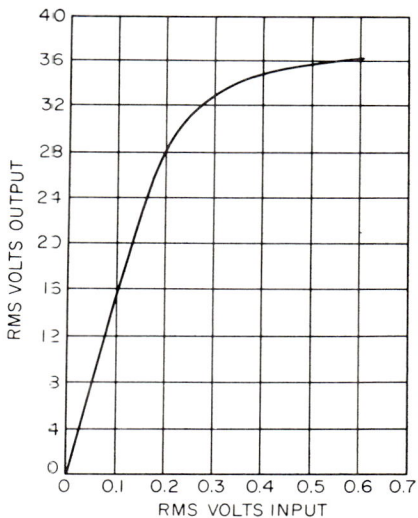

Fig. 62. Servo-amplifier transfer characteristic. (*Courtesy of Norden Division, United Aircraft Corporation.*)

circuit containing resistors, transistors, and possibly zener diodes. In some cases, large-valued tantalum capacitors may also be needed, depending to a great extent on the characteristics of the source and load.

Figure 63 shows a schematic diagram of a three-stage direct-coupled precision a-c summing amplifier that can be partially integrated. Figure 64 shows the loop gain vs. frequency plot with a feedback resistor equal to 100 kilohms and a load resistor of 5 kilohms. The circuit is seen to have a loop gain greater than 80 db from d-c to approximately 2 kc. Using high-accuracy discrete feedback and summing resistors, the amplifier could maintain a gain accuracy of better than 0.01 per cent at frequencies up to 2 kc. All components in the circuit can be readily integrated except the feedback resistor R_f, the input-summing resistors, and the four capacitors C_1 through C_4. To maintain the high-gain accuracy of 0.01 per cent, the capacitive reactance of C_1 must be very small compared with the total shunt resistance of the parallel combination of all input-summing resistors. If all inputs are at zero volts d-c and the parallel combination of all input-summing resistors is large, the capacitor C_1 can be eliminated from the circuit. Capacitors C_2, C_3, and C_4 are used to provide the

31-44 AMPLIFIER CIRCUITS

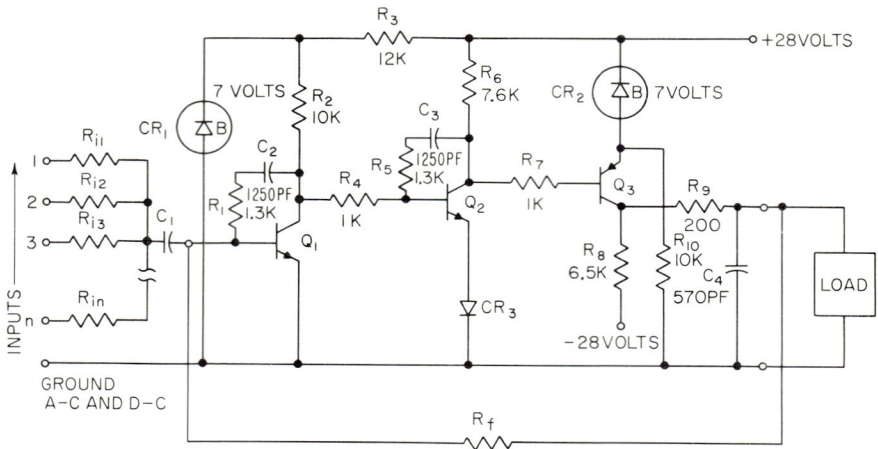

Fig. 63. Precision a-c summing amplifier.

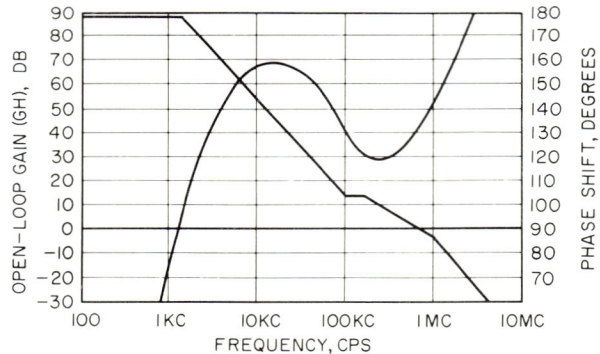

Fig. 64. Precision a-c summing amplifier loop gain vs. frequency.

proper frequency shaping of the loop-gain characteristics to assure stability from oscillations.

10. I-F AMPLIFIERS

Intermediate-frequency (i-f) amplifiers are tuned to provide a selective narrow-frequency band amplification and large attenuation to frequencies below and above the desired band. I-F amplifiers are used in almost all types of wireless communications receivers, such as radio, television, and telephone. Without a doubt, the i-f amplifier is one of the most widely used in the world. Considering the potential market, integrated i-f amplifiers could result in millions of dollars of savings to the users.

Figure 65 shows a schematic diagram of an integrated i-f amplifier consisting of a single-chip silicon planar-epitaxial amplifier with two direct-coupled stages with degenerative feedback used to provide a stable gain, and a lead-zirconate-titanate (PZT) ceramic filter to provide the selective attenuation. The amplifier has an input impedance between 100 and 200 ohms and an output impedance between 1,000 and

INTEGRATED CIRCUITS

2,000 ohms. It is designed to drive 1.5 volts rms into a PZT ceramic filter with a gain of 30 to 40 db at frequencies up to 3 mc. The single-chip circuit is packaged in an eight-lead package measuring $\frac{1}{4}$ by $\frac{1}{4}$ by 0.035 in.

Figure 66 shows the frequency response of an early model of the amplifier–ceramic filter combination. The amplifier can be operated with supply-voltage variations of 8 to 16 volts, with a total block-current drain of from 3 to 7 ma. The circuit will operate over a temperature range of -55 to $+125°C$ and can be stored at temperatures up to $+200°C$.

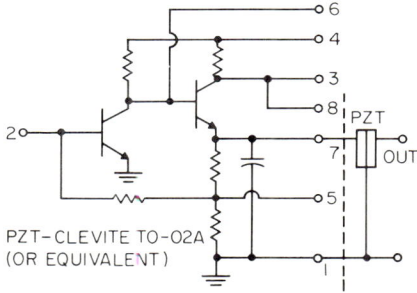

Fig. 65. I-F amplifier. (*Courtesy of Westinghouse Electric Corporation.*)

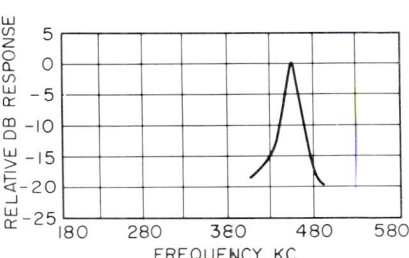

Fig. 66. Typical frequency response of the amplifier-filter combination.

11. VIDEO AMPLIFIERS

Video amplifiers provide relatively constant gain over a wide band, ranging from a few cycles per second up to several megacycles per second, with low value of time delay. One of the most popular uses of a video amplifier is the amplification of video signals in a television receiver, which requires reasonably flat gain from 60 cps to about 4 Mc. Video amplifiers also find wide usage in modern radar equipment to amplify linearly pulses with fast rise and fall times.

Figure 67 shows the schematic diagram of an integrated video amplifier which is of the same configuration as the i-f amplifier shown in Fig. 65. The video amplifier has an input impedance between 100 and 200 ohms and an output impedance between 1,500 and 2,700 ohms with an insertion gain greater than 20 db measured at 455 kc and an undistorted output voltage greater than 1 volt rms. The single-chip circuit is contained in an eight-lead package measuring $\frac{1}{4}$ by $\frac{1}{4}$ by 0.035 in.

Figure 68 shows a typical frequency response of the video amplifier, which is seen to be down 3 db at approximately 12 Mc. The amplifier can be operated with

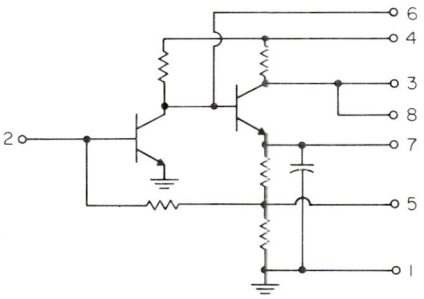

Fig. 67. Video amplifier WM-1106. (*Courtesy of Westinghouse Electric Corporation.*)

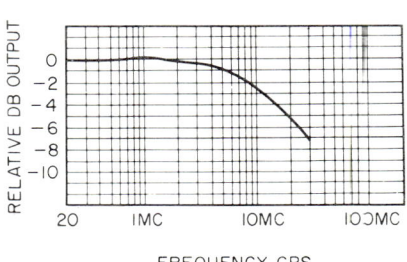

Fig. 68. Frequency response of video amplifier.

supply-voltage variations of 8 to 16 volts with a total block-current drain of from 3 to 7 ma. The circuit will operate over a temperature range of from -55 to $+125°C$ and can be stored at temperatures up to 200°C.

REFERENCES

1. J. A. Hoerni, "Planar Silicon Diodes and Transistors," Professional Group on Electron Devices, Washington, D.C., October, 1960.
2. L. P. Hunter (ed.), "Handbook of Semiconductor Electronics," 2d ed., p. 7–27, McGraw-Hill Book Company, New York, 1962.
3. W. Shockley, A Unipolar "Field Effect" Transistor, *Proc. IRE*, vol. 40, pp. 1365–1376, November, 1952.
4. F. M. Wanlass and C. T. Sah, "Nanowatt Logic Using Field-effect Metal-oxide Semiconductor Triodes," p. 32, International Solid-State Circuits Conference, February, 1963.
5. R. W. Aldrich and N. Holonyak, Jr., Multiterminal P-N-P-N Switches, *Proc. IRE*, vol. 46, pp. 1236–1239, June, 1958.
6. C. N. Wulfsberg, Zener-voltage Breakdown Uses in Silicon Diodes, *Electronics*, vol. 28, pp. 182–192, December, 1955.
7. E. Keonjian, "Microelectronics," p. 268, McGraw-Hill Book Company, New York, 1963.
8. H. Dicken, "Parasitic Effects in Integrated Circuits," p. 98, International Solid-State Circuits Conference, February, 1963.
9. E. Keonjian, "Microelectronics," p. 295, McGraw-Hill Book Company, New York, 1963.
10. A. B. Phillips, "Transistor Engineering," pp. 133–139, McGraw-Hill Book Company, New York, 1962.
11. *Electronic News*, Apr. 22, 1963, p. 4.
12. "Silicon Controlled Rectifier Manual," General Electric Company, 1961.
13. A. B. Phillips, "Transistor Engineering," pp. 112–124, McGraw-Hill Book Company, New York, 1962.
14. H. C. Lin, M. J. Geisler, and K. K. Yu, "A Unipolar-Bipolar Transistor Configuration for Integrated Audio Amplifiers," p. 100, International Solid-State Circuits Conference, February, 1963.
15. M. W. Aarons, "A 1.5 Watt Molecular Servo Amplifier," Norden Division, United Aircraft Corporation, Norwalk, Conn., March, 1963.

INDEX

Absorption, **14**-4
Accelerator, **9**-53
Acceptor, **10**-3
Acoustic flux, **30**-3
Acoustic response, **17**-81
Acoustoelectric effect, **30**-3
Adaptive circuits, solions in, **22**-14
Adiabatic constants, **16**-8
Adiabatic fast passage, **14**-11 **29**-29
Adler tube, **9**-81, **9**-96
Admittance, **17**-5
 driving-point, **4**-33
 input, **16**-34, **16**-49, **17**-5
 motional, **16**-49
 output, **17**-5
 transfer, **4**-34, **19**-5
Admittance circle, **16**-38
Admittance loci, for ceramic transducers, **16**-36
AGC (automatic gain control), **9**-19
 with compression, **26**-58
 delayed, **26**-64
 forward, **26**-20
 high-frequency, **26**-65
 reverse, **26**-20
 using tetrode transistors, **26**-22
Aging, **16**-4
AlP (cubic crystal), **30**-2
Ammonia-beam maser, **29**-5
Amplification, **17**-5
 additive, **25**-36
 current, **17**-5
 light, stimulated emission, **14**-7
 of microwaves, **27**-2
  stimulated emission, **14**-7
 voltage, **17**-5
Amplification factor, **2**-18, **2**-19, **9**-48
 common-base short-circuit, **10**-6
 common-emitter short-circuit, **10**-6
Amplifier(s), acoustic-wave, **30**-1 to **30**-16
 bilateral, **1**-16
 broadband, **25**-1
 capacitance-diode, **28**-27 to **28**-41
  bandwidth, **28**-32
  gain-bandwidth, **28**-32
 cascade, **25**-54, **25**-66
 chopper, **18**-25, **19**-20, **20**-33
 class B tuned, **24**-15
 class C tuned, **24**-15
 complementary-pair feedback, **23**-7 to **23**-9
 conditionally stable, **6**-24, **6**-31
 Continental 300-kw H-F, **20**-57

Amplifier(s), crossed-field, **9**-80, **9**-81, **9**-91 to **9**-96, **27**-3
 current, **25**-60
 d-c field-effect transistor, **18**-69
 d-c wideband, **19**-11
 differential, **19**-8
  cathode-follower input, **18**-15
  drift considerations, **18**-15
  electron tube, **18**-14
  long-tailed pair, **18**-15, **18**-16, **18**-22
  n-p-n-p-n-p pairs, **18**-19
  p-n-p-n-p-n pairs, **18**-18
  transistor, **18**-17
 direct-coupled, **18**-2
  transistor, **18**-18
 distributed, electron-tube, **25**-34
  transistor, **25**-78
 electron tube, equivalent circuit of, **25**-10
  frequency response, **25**-10
 feedback, **6**-1 to **6**-56, **23**-7 to **23**-9, **25**-57 to **25**-78
  return ratio, **6**-9
  series-series connection, **6**-6
  series-shunt connection, **6**-7
  shunt-series connection, **6**-8
  shunt-shunt connection, **6**-7
  specification of performance of, **6**-31
  two-port analysis, **6**-5
  wideband, **25**-57
 field-effect transistor, **17**-33, **17**-34
 filter, **5**-1
  realization of, **5**-28
  for FM and TV transmitters, **20**-59
 gas-beam maser, **29**-11
 Goldberg, **18**-66
 high-impedance-input, **17**-28, **17**-33
 high-power, audio 200-watt, **20**-26
  class C 100-watt 3-Mc, **20**-30
  class C 100-watt 10-Mc, **20**-26
  circuit configurations, **20**-3
  Continental 300-kw, **20**-57
  sonar 10-kw, **20**-24
  switching-type, **20**-31
  transistor, **20**-2
  utilizing SCR's, **20**-32, **20**-36, **20**-38
 ideal, **1**-13
 i-f, integrated-circuit, **31**-44
 induced-emission, **14**-3. **14**-7, **14**-8
 integrated-circuit, audio, **31**-39
  Darlington common-base, **31**-23
  d-c differential, **31**-21
  differential, **31**-33, **31**-35
  direct-coupled feedback, **31**-19

1

INDEX

Amplifier(s), integrated-circuit, dual common-emitter Darlington, **31**-22
  dual Darlington emitter-follower, **31**-23
  high-impedance-input, **31**-28
  i-f, **31**-44
  Unibi, **31**-40
  universal multipurpose, **31**-21
  video, **31**-45
intermediate-power, **20**-49
inverter, **20**-33
linear, **27**-13
logarithmic, a-c, **26**-81
  d-c, **26**-76
    using operational amplifier, **26**-79
    using thermionic diode, **26**-76
    using thermionic pentode, **26**-76
  r-f, **26**-81
M-type, **9**-80, **9**-81, **9**-91 to **9**-96
magnetic, **8**-3
  self-balancing potentiometer, **21**-29
microwave, **27**-3
  typical complexes, **27**-27 to **27**-36
microwave power, multipurpose, **27**-34
  single-output-phase, **27**-27
Miller-type integrator, **8**-10
modulated, **20**-49
modulated-carrier, **18**-25
multistage, **17**-23
negative-resistance, **28**-35
operational, **19**-3
  electron-tube, **19**-8
  field-effect transistor, **19**-23
  frequency response, **19**-16
  integrated-circuit, **31**-38
  as oscillator, **19**-31
  pulse response, **19**-17
  transistor, **19**-14
overstaggered, **5**-3
parallel-input series-output feedback, **25**-59
parametric, **1**-6, **3**-1, **27**-4, **28**-28
  backward traveling-wave, **28**-46
  balanced, **28**-45
  cavity-type, **28**-41
  degenerate mode of operation, **28**-45
  electronically tunable, **28**-45
  large-signal analysis, **13**-10
  maximum available gain, **28**-28
  with noise sources, **28**-30
  pump circuit, **13**-2
  transducer gain, **28**-28, **28**-32
  traveling-wave, **28**-41 to **28**-51
  tuning, **28**-46
  typical design, **28**-36
  uhf, **28**-39
phase-controlled, **20**-33
power, **17**-40
  class A transistor, **17**-43, **17**-44
  class A tube, **17**-41
  class AB transistor, **17**-60
  class AB tube, **17**-57
  class B transistor, **17**-60
  crossed-field, **27**-43
  design charts, **17**-67
  electron-tube, **25**-45
  linear-beam, **27**-40

Amplifier(s), power, negative-grid, **27**-37
  resistance-coupled tube, **17**-41
  tetrode transistor, **17**-73
  transformer-coupled tube, **17**-41
  typical, **17**-71
precision a-c summing, **31**-42
public address, **17**-70
pulse, **11**-11
quadrupole, **9**-81, **9**-96
RC-coupled, **17**-18, **17**-42
resistance-coupled, **9**-64
ruby-laser, **29**-50 to **29**-52
series-input parallel-output feedback, **25**-59
servo, alternating-current, **23**-6 to **23**-14
  with chopper input, **23**-4
  direct-coupled, **23**-2
  electron-tube, **23**-2, **26**-44
  nonlinear, **23**-16
  3-watt, **23**-10
  6-watt, **23**-10
  transistor, **23**-3, **26**-44
  using SCR's in switching mode, **23**-24
  using silicon controlled rectifiers, **23**-20
  using thyratrons, **23**-18
servomotor, 3-watt, **23**-10
  6-watt, **23**-10
single-stage electron-tube, **17**-8
single-stage transformer-coupled, **17**-15
single-stage transistor, **17**-9
solion-stabilized wideband, **22**-9
stereo, electron-tube, **17**-79
  transistor, **17**-90
synchronously tuned, **5**-3
transadmittance, **5**-1, **25**-59, **25**-61
transformerless, **17**-73
transimpedance, **25**-59, **25**-61
transistor power, **17**-73, **25**-89
  10-watt, **17**-75
  12-watt, **17**-77
  125-watt, **17**-78
transistor video, **25**-50
for transmitters, **20**-40, **20**-59
traveling-wave, gain, **28**-48
  noise figure, **28**-48
  phase constant, **28**-48
tuned, **24**-1 to **24**-21
  class A, **24**-2
  class B, **24**-15
  class C, **24**-15
    high-efficiency, **24**-18
  class S, **24**-18
  design procedure, **24**-12
tunnel-diode, **28**-1 to **28**-27
  bandwidth, **28**-20
  bilateral, **28**-5
  circulator-coupled, **28**-6, **28**-9
  design of, **28**-9 to **28**-18
  distributed, **28**-8
  gain, **28**-18
  gain-bandwidth product, **28**-21
  hybrid-coupled, **28**-7
  isolator-coupled, **28**-7
  L-band, **28**-13
  noise equivalent circuit, **28**-26
  noise figure, **28**-24

INDEX

Amplifier(s), tunnel-diode, S-band, **28**-14
    sensitivity, **28**-23
    two-stage, **17**-23
    unconditionally stable, **6**-30
    vacuum-tube, as a two-port network, **2**-17
    variable-pulse-width, **20**-33
    video, integrated-circuit, **31**-45
    voltage, **25**-60
    wideband, **19**-19
    wideband feedback, **25**-57
    wideband transistor, **25**-48
    wideband traveling-wave parametric, **28**-43
Amplifier complexes, communication, **27**-10
    CW, **27**-10
    pulse, **27**-17
Amplistat, **8**-3, **8**-4, **21**-5
    control characteristics, **21**-15
    design considerations, **21**-30
    feedback, **21**-29
    full-wave bridge, **21**-26
    half-wave, **21**-6
    operation into inductive load, **21**-25
    push-pull, **21**-26
    with slave action, **21**-22, **21**-24
    voltage-controlled, **21**-17
Amplitron, **9**-80, **27**-43
Analog division, **26**-2
    memory, solion, **22**-15
    multiplication, **26**-2
    recorder, **19**-24
Annealing, effect on hysteresis curve, **8**-14
Anode, of four-layer diode, **11**-2
    modulating **27**-30
Antenna array, characteristics of, **20**-50
Antiresonance, **5**-25
Approximation, piecewise linear, **26**-35
Asymptote, high-frequency, of feedback amplifier, **6**-35
Attenuation, **16**-27 to **16**-31
    acoustic, **30**-3, **30**-5, **30**-15
    backward, **30**-13
    piezoelectric, **16**-30
Attenuator, logarithmic, **26**-37
    microwave, **27**-33, **27**-62
Automatic gain control (*see* AGC)
Autotune, **20**-61
Avalanche, current, **13**-2
    high-frequency, **13**-2
    multiplication, **13**-2
Avalanche multiplication, **10**-5, **10**-6, **13**-2
Averager, signal, **22**-12

Backing, absorptive, **16**-57
Backward-wave circuits, **9**-95
Backward waves, in parametric amplifiers, **28**-45
Bandwidth, **24**-4
    acoustic-amplifier, **30**-10
    of cascaded stages, **24**-11
    of ceramic transducer, **16**-37, **16**-40
    of constant-D resonators, **16**-45
    delay line transducer, **16**-57
    fractional, piezoelectric, **16**-31
    laser, **14**-34

Bandwidth, maser, **14**-26, **14**-29, **14**-31
    of radial-mode disk resonators, **16**-48
    resonance-antiresonance, **16**-39
Bar, constant-E, **16**-32
    end-plated ceramic, **16**-46
    resonant free, **16**-29
    side-plated ceramic, **16**-32
Barium titanate, **16**-3
    single-crystal, **16**-3
Bartlett's bisection theorem, **16**-50
Base, of a transistor, **10**-3
Beam collimator, **29**-3
Beam deflection, **9**-54
Beam pentode, **9**-51
    shadow-grid, **9**-52
Beam splitter, **9**-54
Bessel polynomials, **5**-3, **5**-6 to **5**-8
Bessel's equation, **16**-47
Bias, transistor, **17**-11
    class B, **17**-64
    equations for, **17**-11, **17**-12
    graphical determination of, **17**-12
    power amplifier, **17**-46
Bias resistors, effect of, **17**-29
Bimatron, **9**-80
Binistor, **11**-19
Bitermitron, **9**-80
Blackout, **9**-12
Blower, **20**-51
Bode, H. W., **6**-11, **6**-35
Bode plot, **6**-25, **25**-11
Boltzmann ratio, **14**-8
Bonds, ceramic, **16**-57
Bootstrapping, **17**-29
    application of, **17**-30, **17**-32, **17**-33
Breakdown, **20**-8
    second, **20**-9
Brewster-angle windows, in lasers, **29**-34, **29**-35
Bridge balance, **19**-24
Brillouin diagram, **28**-44
Brillouin focusing, **9**-82
Buffer, **20**-49
Bunching, electron, **9**-83, **9**-89, **30**-3
Butterfly curve, **16**-5
Butterworth characteristic, **5**-3
Butterworth filters, **5**-4
Butterworth function, **5**-4
Butterworth polynomials, **5**-3, **5**-4, **5**-7, **5**-8
Butterworth response, **23**-21
Butterworth-Thomson characteristic, **5**-11
Butterworth-Thomson transitional filter, **5**-13
BV_{CBO}, **10**-6
BV_{CEO}, **10**-6
BV_{CES}, **10**-6

Campbell, G. A., network representation due to, **4**-44
Capacitance, bridging, **16**-52
    clamped, **16**-48
    collector, **10**-7, **10**-9
    collector diffusion, **10**-9, **10**-11
    depletion-layer, **13**-2
    diffusion, **13**-1

INDEX

Capacitance, emitter, **10**-7
  junction, **10**-13, **11**-6
  negative, **16**-41
  nonlinear, **13**-1
  small-signal junction, **13**-2
Capacitance ratio, of a ceramic transducer, **16**-39
Carriers, minority, **10**-4
Cascaded stages, bandwidth of, **24**-11
Cathode, dispenser, **9**-3
  of four-layer diode, **11**-2
  matrix, **9**-3
Cathode emission, **20**-41
Cathode follower, **17**-7, **25**-42
  equivalent circuit, **17**-8
Cathode interface, **9**-46
Cathode poison, **9**-3
Cathode resistance, effect of, **17**-7
Cauchy-Riemann conditions, **25**-5
Cavities, **14**-39
  laser, **14**-39
  maser, **14**-39
Cavity, beam maser, **29**-6
  microwave maser, **14**-5
  penultimate, **9**-83
    action, **9**-86
Cavity tuning, **9**-82
C_c, **10**-7
CdS (Wurtzite crystal), **30**-1, **30**-2
CdSe (Wurtzite crystal), **30**-2
CdTe (Wurtzite crystal), **30**-2
C_e, **10**-7
Ceramic, poled, **16**-4
Ceramic filter, **16**-32
Ceramic pickup, **17**-34
Ceramic resonator, **16**-32
Cesium energy levels, **14**-17
Cesium oscillator, **29**-42
Characteristic(s), amplistat control, **21**-15
  delay time, **11**-6
  dynamic, **20**-11
  of electron tubes, **9**-16, **9**-26
  equiripple, **6**-30
  forward saturation, **11**-4
  frequency, **17**-7
  gate, of a four-layer diode, **11**-13
  ideal cutoff, **6**-35
  input, measurement of, **20**-11
  logarithmic, **26**-2 to **26**-7, **26**-19
  maximally flat amplitude, **5**-7
  maximally flat delay, **5**-11, **6**-30
  maximally flat magnitude, **6**-30
  monotonic, **5**-4
  nonlinear, of tunnel diodes, **12**-12
  nonlinear device, **26**-2 to **26**-35
  optimum magnitude, **5**-4
  optimum phase, **5**-4
  output, measurement of, **20**-11
  plate families, **9**-30
  power transistor, measurement of, **20**-9
  pulse, **20**-12
  reverse-leakage-current, **11**-4
  SCR, **11**-13
  screen families, **9**-30
  steady-state thermal resistance, **11**-5
  transfer, **10**-5

Characteristic(s), transfer curves, **9**-30
  transient thermal resistance, **11**-5
  of transistors, **10**-5
Characteristic equation, **1**-18
Characteristic polynomial, **1**-18
Charge indicator, **22**-12
Chebyshev inverse characteristic, **5**-6
Chebyshev polynomials, **5**-3, **5**-5, **5**-13, **5**-14
Chebyshev rational functions, **5**-6, **5**-20 to **5**-22
Chebyshev response, **28**-22
Choke joint, **27**-60
Choppers, **18**-25, **18**-26
  balanced-transistor, **18**-54
  bipolar-transistor, **18**-49
  capacitor-coupled, **18**-27
  diode, **18**-58
  electromechanical, **18**-47
  field-effect-transistor, **18**-55
  photoconductor, **18**-65
  transformer-coupled, **18**-33
  transistor, **18**-49, **23**-4
Circuit(s), audio-frequency, for transmitters, **20**-51
  charging, **27**-81
  double-tuned, **24**-8
  equivalent, **2**-17, **16**-27 to **16**-31
  single-tuned, **24**-3
  transmission, **16**-37
Circuit breakers, high-speed, **20**-47
Circuit configurations, high-power amplifiers, **20**-3
Circuit efficiency, high-power amplifiers, **20**-3
Circular polarization, in masers, **29**-22
Circulator, **28**-5
Class A operating point, determination of, **17**-43, **17**-44
Class C power amplifier, **20**-26, **20**-30
Class S amplifier, **24**-18
  load line of, **24**-19
Clipper, inverse, **27**-80
Clipping, peak, **17**-46
Coaxitron, **27**-39
Coefficient, coupling, **16**-12, **16**-38, **16**-45, **16**-46
  elastic-compliance, **16**-6
  electromechanical coupling, **16**-13
  forward transmission, **4**-6
  input reflection, **4**-20
  permittivity, **16**-6
  reflection, **3**-16, **4**-3, **4**-6, **4**-15, **4**-17
  reverse transmission, **4**-6
  scattering, **4**-3, **4**-4, **4**-6, **4**-7
  transmission, **3**-16, **4**-15
Coefficients, poled ceramic, **16**-10
Coercive force, **8**-4
Coil configurations, of a saturable reactor, **8**-25
Collector, of a transistor, **10**-3
Collector impedance, effect of, **17**-30
Collector multiplication factor, **10**-6
Collision, inelastic, **14**-19
Comb structure, maser, **29**-24
Common-mode rejection, **18**-16, **18**-24
Common-mode suppression, **19**-8, **19**-11

INDEX

Compactron, **9**-55, **9**-57
Compandor, **26**-73
Compensation, **30**-7
  low-frequency, **25**-13, **25**-49
Compensation theorem, **2**-3
Complementary symmetry, **25**-88
Complex frequency, **1**-5
Complex variable, **1**-4
Compression, **26**-2
  in audio systems, **26**-73
  logarithmic, **26**-79
  using transistors, **26**-19
  using varistors, **26**-25
Condensation, pivotal, **4**-39
Conductance collector, **10**-9
  collector diffusion, **10**-9
  mutual, **2**-18, **17**-8
  negative, **12**-12
Conduction, in electrolytic solutions, **15**-2
  thermal, **17**-53
Connection, of parallel networks, **4**-39
Conservation law, **28**-50
Constant-D conditions, **16**-25
Constant-E conditions, **16**-25
Constants, of acoustic-amplifier materials, **30**-8
Control, bass, **17**-35, **17**-40, **17**-79
  gain, **17**-81, **18**-5
  incandescent light, **20**-39
  level, **18**-5
  loudness, **17**-81
  shunt-wound d-c motor, **20**-40
  treble, **17**-85
  tone, **17**-35, **17**-40, **17**-79
  universal motor, **20**-39
  volume, **17**-40
Control equipment, automatic, **27**-93
  manual, **27**-92
Control winding, of saturable reactor, **8**-17, **21**-3
Controls, transmitter, **20**-51
Convection, **17**-53
Conversion factors, electron-tube parameters, **9**-35, **9**-37
Converter, **11**-18
  capacitance-diode, **28**-28
  logarithmic voltmeter, **26**-83
  negative-immittance, **19**-26
Cooling, **20**-42
  air, **20**-42
  convection, **20**-42
  evaporation, **20**-43
  radiation, **20**-42
  transistor, **17**-53
  water, **20**-43
Core configurations, of a saturable reactor, **8**-25
Correlation impedance, **7**-11
Coulombmeter, mercury, **15**-14, **22**-12
Counter, pulse, **22**-12
Coupler, directional, **27**-28, **27**-60
Coupling, critical, **24**-8
  piezoelectric, **16**-6
  transitional, **24**-8
  transverse longitudinal, **16**-25
  two-port, **5**-30

Coupling, two-terminal, **5**-28
Coupling coefficient, electromechanical, **16**-12, **30**-2
Coupling factor, electromechanical, **16**-12, **30**-4
  piezoelectric, **16**-14
Critical coupling, **24**-8
Critical frequencies, **1**-8
Critical values, **1**-8
Crossover distortion, **17**-61
Crowbar, **20**-54
Crystal(s), amplifying, **30**-2
  quartz, **16**-18
  Wurtzite, **30**-2
Crystal imperfections, **30**-7
Curie-Weiss law, **16**-3, **16**-4
Current(s), contact-potential grid, **9**-5
  control-grid, **26**-6
  equivalent shot-noise, **12**-10
  excess, tunnel diode, **12**-5
  forward, **10**-5, **11**-4
  forward surge, **11**-4
  free even-harmonic, **8**-17, **8**-19
  gate trigger, **11**-13
  grid emission, **9**-4
  grid leakage, **9**-4
  holding, **11**-4
  inflection-point, tunnel diode, **12**-6
  leakage, **11**-4
  peak, tunnel diode, **12**-5
  reverse, **10**-5, **11**-4
  saturation, **10**-5
  suppressed even-harmonic, **8**-17, **8**-19
  valley, tunnel diode, **12**-5
Current saturation, **30**-3
Curve(s), constant-current, **24**-16
  frequency-response, **24**-12
  selectivity, **24**-12

Damping, critical, **25**-63
  stagger, **24**-10
Damping ratio, **6**-27, **6**-33
Darlington configuration, **17**-30
Darlington pairs, **18**-19, **18**-22
Deflector, **9**-53
Delay, **20**-35
  envelope, **2**-11
  group, **2**-11
  phase, **2**-11
  time, **2**-11
Delay line, ultrasonic, **16**-41, **16**-57
Delta configuration, **4**-34
Deltamax, **8**-7, **8**-14
Dematron, **9**-80
Demodulator, **18**-26, **18**-39, **18**-43, **18**-44
  integrated-circuit transformerless, **31**-25
Density, critical inversion, **14**-38
Density modulation, **9**-80
Depletion, **10**-23
Detection, linear time-variant, **19**-35
  phase, **19**-35
Determinant(s), network, **6**-11
  of network matrices, **4**-31

Device(s), electrochemical, **22**-12
  electromechanical, **22**-12
  galvanomagnetic, **26**-29
  Hall-effect, **26**-29
  ionic, **22**-1, **22**-12
  opto-electronic, **26**-27 to **26**-29
Dicke radiometer, **29**-15
Dielectric constant, relative, **16**-2, **16**-3
Dielectric cylinder, pump energy density in, **29**-46
Dielectric slowing, in traveling-wave masers, **29**-29
Differentiator, **19**-31, **19**-33
Diffuser, maser, **29**-2
Diffusion, of carriers, **10**-4, **10**-10
  of impurities, **10**-13
Diffusion depth, **10**-13
Diffusion length, **10**-10
Diode, abrupt-junction, **13**-3
  acorn-type, **26**-3
  backward, **12**-2, **12**-5
  as square-law device, **26**-16
  capacitor, **13**-1
    parameters of, **28**-33
    pill-type, **13**-9
  Esaki, **12**-1
  four-layer, **11**-1
  germanium point-contact, for gain control, **26**-16
  impedance of, **13**-7
  logarithmic, **26**-2
  p-i-n, as gain-control device, **26**-16
  semiconductor, as amplitude compressor, **26**-15
    as amplitude expander, **26**-15
    as break-point device, **26**-36
    as nonlinear device, **26**-14
  Shockley, **11**-1, **11**-19
  silicon, in analog circuits, **26**-15
    as analog divider, **26**-15
  silicon breakdown, as function generator, **26**-42
    as logarithmic device, **26**-16
    as nonlinear device, **26**-16
  thermionic, **26**-2, **26**-76
  tunnel, **12**-1
  variable-reactance, **28**-41
  zener, **11**-19
Diode limiter, **26**-55
Diode pump, **26**-85
Dipole, **1**-6
  electric, **14**-11
  magnetic, **14**-11
Dipole moment, **14**-8
Dipole transition, **14**-7
Discrimination ratio of operational amplifier, **19**-10, **19**-12
Dispersion, **16**-24, **16**-25
Dissipation, tube, **20**-41
Dissociative excitation transfer, **14**-20
Distortion, crossover, **17**-61
  in class A amplifiers, **17**-45
  effect of feedback on, **6**-4
  harmonic, versus power output, **17**-81
  intermodulation, **17**-81
  in tunnel diodes, **12**-16, **12**-20

Distributed tunnel-diode amplifier, **28**-8
Divider, diode, **26**-96
  quarter-square, **26**-94
  solid-state, **26**-94
Division, analog, **26**-2, **26**-15
  using logarithmic feedback elements, **26**-94
Domain, **16**-2
Domain walls, **16**-3
Dominant pole pair, **6**-31
Donor, **10**-3
Doppler effect of traveling waves, **28**-42
Doppler frequency(ies), **27**-9, **28**-42
Doppler shift, **14**-6
Dosimeter, **22**-12
Doubler, **21**-6
Doubler circuit, **21**-8, **21**-19
Doublet, **14**-8
Drift, **18**-6
  in d-c amplifiers, **19**-3
  due to filament-voltage variations, **18**-6
  due to grid current, **18**-7
  due to power-supply variation, **18**-7
  parameters which affect, **18**-22
  of single stage, **18**-10
Drift compensation, **18**-21
Drift field, **30**-2
Drift mobility, **30**-9
Drift reduction, **19**-14
Drift stabilization, **19**-17
Drift velocity, **30**-2
Driver, transistor wideband, **25**-84
Driving-point function of reciprocal n-terminal network, **4**-41
Dual formulation, **4**-10
Dummy load, **27**-62
Duplexer, **27**-28, **27**-64
Dynamic range with tunnel diodes, **12**-21

Efficiency, of acoustic amplifiers, **30**-13
  of class B amplifier, **17**-66
  high, tuned amplifiers, **24**-17
  of high-power amplifiers, **20**-3
  idealized, of class B tuned amplifier, **24**-15
  maximum, of class B tuned amplifier, **24**-15
Effuser, **29**-3
EIA (Electronic Industries Association), Joint Electron Device Engineering Council of the, **9**-14
Einstein coefficients, **14**-7
Elapsed time indicator, **22**-12
Elastance, junction, **13**-2, **13**-10
  small-signal, **13**-3, **13**-5
Electric field, **10**-4
Electrode, end-free, **16**-41
  polarized, **15**-2
  ring-and-dot, **16**-49
Electromechanical analogy, **16**-27
Electrometer, tetrode, **26**-4
  triode-connected, **26**-4
Electron, **10**-3
Electron beam, **9**-81
Electron emission, **9**-2

INDEX

Electron gun, 9-81
Electron impact, 14-18
Electronic Industries Association (EIA),
    Joint Electron Device Engineering
    Council of the, 9-14
Electrostriction, 16-5
Elements, n-terminal, 4-2
Emission, coherent, 14-6
  electron, 9-2
  grid, 9-4
  induced, 14-3, 14-4
  primary, 9-3
  secondary, 9-3, 9-54
  spontaneous, 14-4
Emitter, of a transistor, 10-3
Emitter efficiency, 10-6, 10-10, 10-11
Emitter impedance, effect of, 17-21, 17-29
Enclosures, power transistor, 20-5
Energy, stored, 16-12, 16-16
Energy diagram, 10-9
Energy extraction, electron, 9-84
Energy levels 14-8
Energy per mode, 14-3
Engineering Council, 9-14
Enhancement, 10-23
Epitaxial technique, 10-20
Epoch, signal 19-4
Equalization, time delay, 25-33
Equations, of motion, 16-19, 16-46
  of state, 16-46
Equilibrium, thermodynamic, 14-7
Equiripple in pass- and stopbands, 5-20
Equiripple magnitude characteristic, 5-3,
  5-5, 5-13
Equivalent circuit, of capacitance diode,
  13-4
  cathode-follower, 17-8
  ceramic filter, 16-61
  constant-D resonator, 16-44
  constant-D transducers, 16-41, 16-43
  constant-E transducers, 16-34, 16-40
  delay-line transducer, 16-57, 16-59
  electron-tube, 2-17, 15-9, 17-6
  hybrid π transistor, 6-14
  piezoelectric element, 16-33
  RC-coupled stage, 17-19
  symmetrical ceramic transformer, 16-50
  transformer-coupled stage, 17-15 to 17-17
  transistor, 2-22, 25-47
  tunnel diode, 12-6
    noise, 12-10
Error polynomial, 19-6
Error response, 19-4
Esaki diode, 12-1
Excited state, 14-20
Exciting interval, of saturable reactor, 8-18
Expansion, 26-2
  in audio systems, 26-73
  using transistors, 26-19
  using varistors, 26-25
Exposure meter, 22-13

Fabry-Perot interferometer, 14-3, 14-39
Fabry-Perot plates, 29-33
Fabry-Perot resonator, 29-58

Factor, gain-bandwidth, 24-5
  planar coupling, 16-46
Faraday rotator, 29-51
Faraday's law, 15-2
Feedback, amount of, 6-31
  in amplistats, 21-29
  current, 6-8
  effect, on distortion, 6-4
    on gain, 6-3
    on noise, 6-4
  magnetic, 21-29
  maximum amount of, 6-24, 6-36
  multiple, 25-86
  negative, 6-3
  unwanted, 6-23
  voltage, 6-8
  in wideband transistor amplifiers, 25-57
Feedback chains, 6-41 to 6-55
  alternating transistor, 6-49
  transistor, 6-44
  transistor feedback-pair, 6-52
  vacuum-tube, 6-42
Feedback factor, 6-3
Feedback pairs, 25-67
Feedback theory, 6-2 to 6-9
FET (field-effect transistors) as pulse-
    amplitude multiplier, 26-98 to 26-100
Field-effect transistors (see FET)
Field intensity, induced, 30-3
Figure of merit, ceramic resonator, 16-40
  parametric-amplifier diode, 13-8
Filament, thoriated tungsten, 20-41
Filling factor, 14-25
Filter(s), 19-26, 20-50
  band-rejection, 19-27
  bandpass, 19-27
  ceramic, 16-32, 16-41, 16-61 to 16-69
  cutoff, 16-65
  equivalent circuit, 16-61
  image-impedance analysis, 16-62
  mid-series image impedance, 16-65
  mid-shunt image admittance, 16-25
  notch, 16-53
  passband, 16-61
  resonant ladder, 16-67
  ripple, 16-64
  series-C, shunt-resonator ladder, 16-64
  series-resonator, shunt-C ladder, 16-66
  stability, 16-69
  stopband attenuation, 16-65
  elliptic-function, 5-6
  high-pass, 5-6, 19-27
  low-pass, 5-6, 5-14, 5-15, 19-27
  operational, 19-27
  power-supply, 20-47
  transmission-line, 27-28, 27-65
Firing, saturable reactor, 21-6
Firing angle, 21-6, 21-9
Flip-flop, 11-11, 19-35
  dual, integrated, 31-37
  high-sensitivity, integrated, 31-30
Flowmeter, 22-13
Flux, acoustic, 30-13
Flux density, 8-4
  intrinsic, 8-5
  residual, 8-4

Flux reset, **21**-5
Focuser, linear, **29**-6
  parabolic, **29**-6
Form factor, amplistat, **21**-32
Four-legged construction, of saturable reactor, **8**-26
Four-pole, **1**-6
Fraction, continued, **19**-5
Frame grid, **9**-50
Frequency, alpha cutoff, **10**-7
  antiresonant, **16**-29, **16**-48
  conductivity, **30**-5
  corner, **19**-27
  cutoff, **16**-24, **16**-26, **20**-12
  dielectric relaxation, **30**-5
  diffusion, **30**-5
  natural, **1**-8, **1**-18
  resistive cutoff, **12**-6
  resonant, **16**-29, **16**-48
  self-resonance, **12**-6
Frequency response, of acoustic amplifiers, **30**-10, **30**-11
Frequency shift, maser, **29**-8
Fresnel number, **14**-41
f_T, **10**-7
Function, all-pass, **2**-13
  conjugate, **2**-12
  consinusoidal, **19**-4
  damped-wave, **19**-4
  driving-point, **1**-5, **1**-7
  excitation, **19**-4
  gain, **2**-10
  image loss, **3**-12
  image propagation, **3**-12
  magnitude, **2**-13
  minimum-phase, **2**-13, **2**-14
  positive real, **5**-22
  propagation, **2**-10
  ramp, **19**-4
  reactance, **2**-14
  real rational, **1**-8, **5**-23
  reciprocal system, **1**-7, **1**-8, **3**-8
  sinusoidal, **19**-4
  step, **19**-4
  system, **1**-7, **2**-8, **2**-10
  transfer, **1**-5, **1**-7
  two-terminal, **1**-7
Function generator, logarithmic, **26**-36
  square-law, **26**-40
  using servo techniques, **26**-49
  variable, **26**-42

GaAs (cubic crystal), **30**-2
Gain, **3**-15, **17**-5
  acoustic-amplifier, **30**-2, **30**-6, **30**-10
  available, **3**-25, **7**-13
  common-emitter current, **6**-38
  effect of feedback on, **6**-3
  exponential, in parametric amplifiers, **28**-50
  insertion, **3**-25
  loop, **6**-3
  of maser, **14**-25, **14**-29, **14**-30
  maximum of tuned amplifier, **24**-2

Gain, maximum available, **3**-25, **3**-26
  of parametric amplifier, **28**-28
  midband, **24**-9
  optimum, **24**-3
  power, **3**-25
    of amplistat, **21**-32
    dynamic, **21**-32
  of power amplifier, **17**-66
  transducer, **3**-15, **3**-24, **7**-21, **17**-5
  of parametric amplifier, **28**-28, **28**-32
  transducer power, **3**-15, **4**-16, **4**-21
  voltage, **24**-4
Gain-bandwidth factor, **24**-5, **24**-10
Gain-bandwidth product, **24**-4, **24**-10, **25**-14, **25**-51, **25**-52, **25**-56
  of parametric amplifier, **28**-43
Gain control, automatic, **26**-20, **26**-63
  delayed, **26**-64
  high-frequency, **26**-65
  in radio receivers, **26**-63
  using alloy transistor, **26**-60
  using silicon diode, **26**-60
  using tetrode transistor, **26**-22, **26**-61
  using varistors, **26**-62
  high-frequency, **26**-16
Gain margin, **6**-32
Gain modulation of a maser, **14**-27 to **14**-29
Gain stability, in d-c amplifiers, **18**-23
  of transistor stage, **18**-12
Galvanomagnetic devices, **26**-29
GaP (cubic crystal), **30**-2
Gap, ball, **20**-54
  horn, **20**-54
Gas systems, in lasers, **29**-36
Gate, NOR, four-input, integrated, **31**-30
Gate angle, **21**-6
Gate-turn-off (GTO) device, **11**-12, **11**-16
Gate winding, of saturable reactor, **8**-17, **21**-2
Gating, in saturable reactors, **21**-6
Gaussian line, **14**-6
Gaussian noise, **7**-46
General Electric stereo amplifier, **17**-81
Generator, function, **22**-12
  pulse, **11**-11
Germanium, **10**-3
Glan-Thomson prism, **29**-50
Goldberg circuit, **18**-66, **19**-11, **19**-17
Grid film, **9**-11
Ground state, **14**-16
GTO (gate-turn-off) device, **11**-12, **11**-16
Gyrator, **1**-6, **1**-11
Gyromagnetic ratio, **14**-10

Hall effect, **26**-29
  in analog multiplication, **26**-29
  squaring circuit, **26**-29
Hall mobility, **30**-9
Harmonic analysis, **24**-16
Hearing aids, **17**-24 to **17**-28
  Otarion, **17**-24 to **17**-28
Heat dissipator, **17**-55
Heat sink, **17**-55, **20**-6, **20**-21
  selection of, **20**-16
Helium energy levels, **14**-18

INDEX

Hermitian matrix, 4-17
h_f, 17-10
h_{fb}, 10-6
h_{fe}, 10-6
h_i, 17-10
Hilbert transform, 2-12, 2-16
h_o, 17-10
Hole, 10-3
Horn, microwave, 27-30, 27-66
h_r, 17-10
Hybrid, microwave, 27-63
Hymu, 8-7
Hypernik, 8-7
Hypersil, 8-7
Hysteresigraph, 8-7
Hysteresis, dielectric, 16-3
  mechanical, 16-30
Hysteresis loop, dynamic, 8-8, 21-12
  measurement, of 8-8
  minor, 21-12
  piezoelectric, 16-4
  rectangular, 21-6
  of selected materials, 8-12
  static, 8-4

I_{CBO}, 10-6
I_{CEO}, 10-6
I_{CER}, 10-6
I_{CES}, 10-6
Idler, 28-46
Ignitron, 27-31
Impedance, 17-5
  in active networks, 6-17
  characteristic, 4-3
  input, 17-5
    components of, 17-28
    piezoelectric resonator, 16-44
  mechanical, 16-30, 16-34
  negative, 1-17
  open-circuit, 1-20, 3-3
  output, 17-5, 17-8
  reference, 4-4
  source, class S amplifier, 24-20
  transfer, 4-37
Impedance matching, acoustic transducers, 16-55
Impurity distribution, 10-12
InAs (cubic crystal), 30-2
Induced emission, 14-3, 14-6
Inflection point, tunnel diode, operation at, 12-15, 12-18
InSb (cubic crystal), 30-2
Instability, potential, 3-22
Instantuner, 20-61
Integrated circuit components, 31-6 to 31-15
  capacitors 31-14
  diodes, 31-11
  field-effect transistors, 31-9
  four-layer devices, 31-10
  n-p-n transistors, 31-7
  p-n-p transistors, 31-7
  resistors, 31-11
  Zener diodes, 31-11

Integrated circuit design considerations, 31-15
  breadboarding, 31-20
  component parameter study, 31-17
  component point count, 31-17
  configuration selection, 31-17
  yield, 31-16
Integrated circuit housings, cylindrical packages, 31-4
  flat pack, 31-5, 31-20
Integrated circuits, 31-1 to 31-46
  epitaxial process, 31-2
  fabrication, 31-2
  interconnection techniques, 31-4
  isolation techniques, 31-2
  multipurpose, 31-20
  packaging techniques, 31-4
  planar process, 31-2
  test procedure, 31-5
Integrated structures, 31-3
Integrator, 19-31
  ionic, 15-6
    two-terminal, 15-10
Interaction(s), backward traveling-wave, 28-47
  electric, 14-6
  forward traveling-wave, 28-47
  magnetic, 14-6
  traveling-wave, 30-1
Intermodulation, with tunnel diodes, 12-23
Interstage coupling, 18-3
Inversion, gas discharge, 14-18
  semiconducting junction, 14-20
  staircase, 14-14
Inversion efficiency, 14-14
Inverter, 11-11, 11-18
  phase, 17-81
Ion, triiodide, 15-3
Iraser, 14-7
Isolation, between stages, 24-3
Isolator, 27-28, 27-33, 27-62, 28-5
Isothermal constants, 16-8
Isotropy, transverse, 16-4, 16-8, 16-18

JEDEC (Joint Electron Device Engineering Council) of the Electric Industries Association, 9-14
JEDEC registration format, 20-5
Joint, rotary, 27-28, 27-65
Joint Electron Device Engineering Council (JEDEC) of the Electronic Industries Association, 9-14
Junction, abrupt, 10-13
  linear-graded, 13-4
  p-n, 10-5
  sharp, 10-13

Kerr cell, 29-50
Klystron, 9-80, 9-81
  multicavity, 9-81, 27-3
  multiple-beam, 9-87
  reflex, 27-4, 27-49
Kramers-Kronig relation, 14-26
kvar, per-unit, 20-45

INDEX

Ladder network, **1**-7
Lamé's constants, **16**-18
Lamp driver, integrated, **31**-34
Laplace transform, **1**-14, **19**-4
Laser, cesium, **29**-41
  CW gas, **29**-32
    d-c discharges, **29**-34
  gas, pulsed operation of, **29**-36
  giant-pulse, **29**-48
  injection, **29**-58 to **29**-65
    GaAs, **29**-58 to **29**-63
    magnetic field effects, **29**-63
  metallic-vapor, **29**-41 to **29**-43
    cesium, **29**-41
  neon-oxygen, **14**-20
  optically pumped, **29**-41
  organic, **29**-67
  Raman, **14**-21, **29**-65 to **29**-67
  rare-earth, **29**-52 to **29**-57
  ruby, CW, **29**-47, **29**-48
  semiconductor, **29**-58 to **29**-65
    GaAs, **29**-58 to **29**-63
    junction inversion, **14**-7
  solar-pumped, **29**-48
  solid-state, **14**-7, **29**-43 to **29**-52
    ruby, **29**-43 to **29**-52
    pulsed, **29**-43, **29**-44
Laser characteristics, **14**-34
Laser crystals, **14**-34
Lattice, crystalline, **14**-8
  equivalent, **16**-50
Lattice network, **1**-7
Lattice vibration, **14**-6
Laurent series, **19**-5, **19**-6
Layer, space-charge, **10**-13
Leakage, grid current, **9**-4
  heater-cathode, **9**-6
Leakage current, **20**-43
Learning device, **22**-14
Level, of a function, **1**-8
Life, electron-tube, **9**-42
    effect, of envelope temperature, **9**-46
    of power dissipation, **9**-44
    end points, **9**-42
Lifetime, semiconductor, **10**-3
Limiter, bidirectional, **26**-57
  precision, **26**-55
Limiting, amplitude, **26**-2, **26**-53
  with diodes, **26**-53, **26**-55
  with transistors, **26**-53
Line shape, effect of, **14**-6
Linearity, **20**-37
Load, dummy, **27**-62
Loading, dynamic, **20**-18
  periodic, **28**-46
Logarithmic amplitude response, **26**-2
Logarithmic characteristic, **26**-19
Logarithmic diode, **26**-2
  compensation scheme, **26**-3
Logarithmic frequency response, **26**-2, **26**-85
Logarithmic function generator, **26**-36
Logarithmic pulse-rate response, **26**-2, **26**-85
Logarithmic servo system, **26**-51
Long-tailed pair, **18**-15, **23**-3
Lorentz line, **14**-6
Loser, **14**-7

Loss, insertion, **16**-55
  return, **4**-17
  transducer, **16**-55

MagAmp (*see* Magnetic amplifier)
Magnesil, **8**-7
Magnestat, **8**-4
Magnetic alloys, **8**-7
Magnetic amplifier (MagAmp), **8**-3, **8**-4
  self-saturating, **8**-4
  simple, **8**-4
  static, **8**-4
Magnetic amplifier core materials, **8**-6
Magnetic moment, **14**-8
Magnetic pickup, **17**-34
Magnetization, transverse, **14**-23
Magnetization curve, d-c, **8**-4, **8**-5
Magnetizing force, **8**-4
Magnetoresistance, **26**-34
  geometrical, **26**-34
  as multiplier-divider, **26**-94
  physical, **26**-34
  use in squaring circuit, **26**-34
Magnetostrictive alignment of lasers, **29**-36
Magnetron, **27**-4, **27**-45
  as nonlinear device, **26**-13
Manley-Rowe relationship, **28**-50
Margin, gain, **6**-32
  phase, **6**-32
Maser, ammonia, **14**-8, **14**-35
  atomic-hydrogen, **29**-10
  beam, **29**-2
    ammonia, **29**-5, **29**-8
    using separated oscillatory fields, **29**-9
  cavity, solid-state, **29**-11
  double-quantum, **14**-21
  high-temperature operation of, **29**-21
  with intermittent operating characteristics, **29**-29 to **29**-32
  millimeter-wave, **29**-21
  optical, **14**-8, **29**-33
  phonon, **14**-7
  pulsed-field, **29**-29 to **29**-32
  radio-astronomy, **29**-14
  reflection-cavity, **14**-25, **14**-29
    for 400 Mc, **29**-11
    L-band, **29**-12, **29**-13
    S-band, **29**-16
    X-band, **29**-19
  ruby, **29**-25
  transmission-cavity, **14**-25
  traveling-wave, **14**-7, **14**-25, **14**-30, **29**-22 to **29**-29
    3.0 Gc, **29**-25
    5.8 Gc, **29**-27
  two-level gas, **14**-8
  using optical pump, **29**-21
  X-ray, **14**-7
Maser types, comparison of, **14**-32
Match, image, **3**-10, **4**-5
  reflectionless, **3**-10
Matching, **3**-22, **4**-10
  impedance, **25**-38
Materials, photoconductive, **30**-5

INDEX

Matrix (matrices), chain, 3-8, 3-30
   elastic, 16-7
   g, 3-7
   h, 3-7
   Hermitian, 4-17
   identity, 4-11
   image parameter, 3-9
   indefinite admittance, 4-2, 4-35, 4-40
   for n-port network, 3-3
   node-to-datum, 4-35
   Pauli spin, 4-25
   permittivity, 16-8
   piezoelectric, 16-8
   reference-impedance, 4-7
   scattering, 4-1, 4-11
   second-order, 4-27
   transfer scattering, 4-23
   for two-port network, 3-3
   unit, 4-11, 4-17
   zero-sum, 4-2, 4-35
Matrix representation, electron-tube, 17-6
Maxwell's equations, 28-47
Memistor, 15-12, 22-12
Mercury coulombmeter, 15-14
Microbarometer, solion, 22-10
Microphone, ceramic, 16-41
Microphonics, 9-7
Microsil, 8-7
Microwave amplifiers, characteristics of, 27-19 to 27-27
   harmonic output, 27-25
   linearity, 27-23
   noise output, 27-25
   output, vs. frequency, 27-19
      vs. load, 27-24
      vs. voltage, 27-19
   phase vs. voltage, 27-22
Microwave power amplification, 27-2
Midband values, 6-3
Miller capacitance, 19-20
Miller effect, 9-72, 19-10, 25-37
Mirror absorption, 14-37
Mirror geometry, in lasers, 29-34
Mirrors, confocal, 14-41
   dielectric-coatings, 14-42
   metal-film, 14-42
   plane-parallel, 14-41
Mixer(s), 19-35
   microwave 27-51
   tunnel-diode, 12-21
Mixing, at a tunnel-diode inflection point, 12-21
Mobility, drift, 30-9
   electron, 30-2
   Hall, 30-9
Mode, lowest longitudinal, 16-24
   natural, 1-8
   propagation, 16-24, 16-26
Mode density, 14-3
Mode selection, laser, 29-35
Modulation, amplitude, 27-11
   angular, 27-13
   frequency, 27-13
   phase, 27-15
   pulse, 27-17

Modulator, communications, 27-74
   dielectric, silicon diode, 26-77
   diode, 18-58, 23-4, 27-52
   electromechanical chopper, 18-47
   field-effect transistor, 18-55
   Hall-effect, 18-68
   integrated-circuit transformerless, 31-26
   ring, 18-59
   second-harmonic, 21-2
   transistor, 18-49, 23-5
   varactor, 18-61
   variable-capacitance diode, 18-61
   vibrating-reed capacitance, 18-64
Molecular dissociation, 14-20
Molypermalloy, 8-7
Monitoring equipment, 27-90
   d-c, 27-91
   low-frequency, 27-91
   pulse, 27-91
   r-f, 27-90
Mo-Permalloy, 8-7
Motorola stereo amplifier, 17-92
Moving-target indicator, 27-7
Multiplication, analog, 26-2, 26-9
   using Hall effect, 26-29
   using logarithmic feedback elements, 26-92
   using point-contact transistor, 26-23
   using transistor logarithmic elements, 26-93
   frequency, 27-12
Multiplier(s), constant, 1-8
   diode bridge, 26-98
   frequency, 27-54, 27-55
      solid-state, 27-55
      tube, 27-54
   microwave frequency, 27-5
   pulse-amplitude, 26-98
   quarter-square, 26-94
      using beam-deflection tube, 26-94
      solid-state, 26-94
Multivibrator, 19-35
   free-running, integrated-circuit, 31-32
Mumetal, 8-7

NAB frequency compensation, 17-79
NAB tape characteristic, 17-34
Neon energy levels, 14-18
Neonoval, 9-55, 9-57
Network(s), all-pass, 24-34
   augmented, 4-8
   bridged-T, 25-73
   cascaded, 25-5
   complementary, 1-23
   constant-K, 25-36, 25-81
   constant-resistance, 1-23
   coupling, calculation of, 24-14
   dipole, 5-22
   distributed parameter, 1-6, 1-9
   dual, 1-24
   equalization, 23-14
   equivalent, 1-23
   inverse, 1-24
   ladder, 1-7, 19-5
   lattice, 1-7

Network(s), *LC*, **1**-6
  linear passive constant, **1**-9
  lossless, **28**-29
  lumped constant, **1**-6
  lumped parameter, **1**-6, **1**-9
  m-derived, **25**-36
  multiterminal, **4**-32
  n-terminal, **1**-6
  n-terminal pair, **1**-6
  nonlinear passive constant, **1**-9
  nonreciprocal, **1**-11
  normalized augmented, **4**-8
  notch, **23**-15
  one-port, **1**-5
  planar, **1**-25
  polar-zero, **19**-7
  pulse-forming, **27**-77
  RC, **1**-6
  realizable, **1**-4
  reciprocal, **1**-11, **1**-24, **4**-10
  RLC, **1**-6
  RLCM, **1**-6
  single-tuned coupling, **24**-5
  tandem, **17**-22
  terminated, equations for, **3**-13
  three-terminal, **1**-6
  two-element-kind, **1**-6
  two-port, **1**-5
  variable phase-shift, **23**-18
Network analysis, **1**-4
Network equations, **17**-5
Network matrices, interrelationships among, **4**-37
Network synthesis, **1**-4, **5**-22
Neutralization, **24**-2
Nichols chart, **6**-27
Noise, **9**-7
  effect of feedback on, **6**-4
  flicker, **9**-11
  gaussian, **7**-46
  of induced-emission amplifier, **14**-23, **14**-28, **14**-29, **14**-31
  induced grid, **9**-7
  in microwave amplifiers, **27**-25
  partition, **9**-7, **9**-11
  Rayleigh, **7**-46
  shot, **7**-7, **9**-7, **9**-11
  thermal, **7**-6, **9**-7
  in transistors, **10**-12
  tube, **9**-7
  in tunnel diode, **12**-9
  in two-ports, **7**-8
Noise bandwidth, **7**-22, **7**-31
Noise conditions, limiting, **7**-52
Noise conductance, equivalent, **7**-7, **7**-11
Noise currents, **7**-7, **7**-8
Noise diode, **7**-36
Noise factor, **7**-12, **9**-8
  average, **7**-21
  for cascaded two-ports, **7**-14
  excess, **7**-14
  extended, **7**-15
  multiple-response, **7**-13
  operating, **7**-48
  single-response, **7**-13
  spot, **7**-21

Noise factor meters, **7**-40
Noise figure, **7**-12, **9**-8
  acoustic-amplifier, **30**-15
  of parametric amplifier, **28**-30
  of tunnel-diode amplifier, **28**-24
Noise measurements, on amplifiers, **7**-32
  instrumentation, **7**-35
Noise meter, low-frequency, **22**-13
Noise parameters, measurement of, **7**-29, **7**-32
  two-port, **7**-19
Noise power, per unit bandwidth, **14**-23, **14**-28
Noise resistance, equivalent, **7**-7, **7**-11
Noise source, **7**-9
  diode, **7**-36
  equivalent circuit, **7**-9, **7**-11
  gas discharge, **7**-39
  thermal, **7**-39
Noise temperature, **7**-8, **7**-21, **7**-23
Noise voltages, **7**-7
Nonlinear effects, of tunnel diodes, **12**-12
Nonlinear operation, in acoustic amplifiers, **30**-8
Nonlinearity, direct, **26**-2
NOR gate, four-input, integrated, **31**-30
Normalizing impedance, **4**-6
Normalizing number, **4**-6
Norton theorem, **2**-3
Novar, **9**-55, **9**-57
n-tuple, flat-staggered, **5**-3, **5**-4
Null detector, saturable reactor, **21**-5
Nuvistor, **9**-55, **9**-56
  double-ended, **9**-80
Nuvistor tetrode, **26**-4
Nuvistor triode, **26**-4
Nyquist plot, **6**-25

Offset compensation, **18**-53
One-shot, integrated, **31**-32
Optoelectronic devices, **26**-17
  for compression, **26**-27, **26**-73
  for expansion, **26**-27, **26**-73
  for gain control, **26**-27
Order, of a function, **1**-8
Orthonol, **8**-7
Oscillation conditions, maser, lasers, **14**-37
Oscillations, spurious, **24**-8
Oscillator(s), **19**-26, **20**-48
  buffered relaxation, integrated, **31**-36
  induced-emission, **14**-35
  magnetron, **27**-45
  microwave, **27**-4
  monostable, integrated, **31**-32
  multivibrator, integrated, **31**-32
  negative-grid, **27**-47
  sawtooth, **11**-9
  using operational amplifiers, **19**-31
  velocity-modulated, **27**-49
  voltage-controlled, **19**-17
    integrated, **31**-36
Output, power, of a tuned amplifier, **24**-17
Output stage, class A, **17**-40
  class AB, **17**-40
  class B, **17**-40

INDEX

Output stage, class C, **17**-40
  power, **17**-40
Overlap, **18**-29
Overshoot, **6**-23
  pulse, **19**-17, **25**-5
Overvoltage, **15**-2

Paraconjugate matching, **4**-21
Parallel operation of transistors, **20**-33
Paramagnetic spin, **14**-6
Paramagnetic system, two-level, **14**-9
Paramagnetic three-level system, **14**-12
Parameters, backward diode, **12**-5
  chain-matrix, **3**-8
  differential-coefficient, **2**-17
  equivalent noise, **9**-8
  general circuit, **3**-8
  h, **10**-7, **17**-10
  hybrid, **10**-8
  incremental, **2**-17
  tunnel diode, **12**-5
  variational, **2**-17
Parametric amplifiers [see Amplifier(s), parametric]
Parametric interactions, three-dimensional, **28**-49
Part, even, **2**-9
  imaginary, **2**-8
  odd, **2**-9
  real, **2**-8
  of system function, **2**-8
Passband, **2**-6, **5**-20 to **5**-22
Passivated surfaces, **10**-19
Passivation, **10**-19
Pauli spin matrix, **4**-25
Peak clipping, **17**-46
Peaking, emitter-bypass, **25**-52
  series, **25**-23, **25**-51
  shunt, **25**-15, **25**-51
  shunt-series, **25**-26
  stagger, **25**-30, **25**-55
Pentode, **2**-19
  power output, **9**-37
  thermionic, **26**-3, **26**-76
    connected as logarithmic diode, **26**-4
    logarithmic characteristics, **26**-6
Permalloy, **8**-14
Permeability, differential, **8**-5
  normal, **8**-5
Permendur, **8**-7
Permittivity, free space, **16**-2
Phase, excess, **6**-37
Phase angle, **2**-10
Phase correction, **25**-35
Phase detector, **19**-35
Phase function, **2**-10
Phase inversion, **25**-44
Phase margin, **6**-32, **19**-15
Phase shift, **2**-10, **20**-35
Phase velocity, acoustic, **30**-5
Phased-array complex, **27**-31
Philbrick, G. A., **19**-8
Phonograph pickup, **17**-34
Pickup, ceramic, **16**-41, **17**-34
  magnetic, **17**-34

Pickup, phonograph, **17**-34
Piezoelectric ceramics, properties of, **16**-68
Piezoelectric conversion, **16**-16
Piezoelectric coupling, **16**-6
Piezoelectric equations of state, **16**-7
Piezoelectric interaction, **16**-20
Piezoelectric matrices, **16**-8
Piezoelectric matrix relationships, **16**-7
p-i-n diode, **26**-16
Planck's constant, **14**-4
Planck's distribution law, **14**-4
Planck's expression, **14**-23
Planimeter, **22**-13
Plate, thickness-shear, **16**-41
p-n-p-n structure, **11**-3
Poisson's ratio, **16**-11, **16**-46
Polarization, **15**-2, **16**-2, **16**-3
  circular, **14**-31
  spontaneous, **16**-3
Polarization catastrophe, **16**-3
Pole(s), **1**-6, **1**-8, **1**-18, **25**-72
Pole-zero network, **19**-7
Polynomial(s), **1**-8
  Bessel, **5**-3, **5**-6 to **5**-11
  Butterworth, **5**-3, **5**-4, **5**-7, **5**-8
  characteristic, **1**-18, **1**-20, **5**-3
  Chebyshev, **5**-3, **5**-5, **5**-13 to **5**-20
  surplus, **5**-28
Population, **14**-3, **14**-6
Population difference, **14**-15
Population inversion, **14**-3, **14**-8
  ground state, **14**-16
Port, **1**-5
Position error, **19**-5
Potential, concentration, **15**-2, **22**-2
  contact, **9**-4, **9**-5
  bias, **9**-5, **9**-21
Potentiometer, capacitive, **22**-13
  exponential, **26**-51
  tapped linear, **26**-44
Power, absorbed, **4**-17
  available, **3**-15, **7**-6, **17**-5
  average gate, **11**-13
  dissipated bias in acoustic amplifiers, **30**-9
  exchangeable, **7**-15
  incident, **4**-17
  maximum available, **3**-24
  output, **17**-65 to **17**-67
  peak gate, **11**-13
  reflected, **4**-17
Power circuit, **26**-91
Power handling capability, **20**-2
Power level, maximum, acoustic-amplifier, **30**-13
Power output, laser oscillator, **14**-39
Power splitter, **27**-28, **27**-63
Power supply (supplies), **27**-67
  high-voltage, **27**-73
  regulated, **19**-24
  regulated d-c, **27**-72
    using GTO device, **20**-39
  three-phase, **20**-44
  unregulated d-c, **27**-70
Power supply considerations, **20**-44
Power transfer, **3**-22

Preamplifier, electron-tube, **17**-34, **17**-35
 servo, **23**-11
 transistor, **17**-38, **17**-91
 tube stereo, **17**-35, **17**-85
Presetting, **21**-6
Probability, absorption, **14**-5
 induced-emission, **14**-5
 spontaneous-emission, **14**-5
 transition, **14**-5
Product, gain-bandwidth, **24**-4
Product circuit, **19**-35
Propagation, parallel to poling axis, **16**-21
 perpendicular to poling axis, **16**-19, **16**-22
 stress-wave, **16**-18
 torsional-wave, **16**-26
Proportional operating range, of saturable reactor, **8**-21
Protection, modulator, **27**-89
 personnel, **27**-90
 power tube, **27**-87
 pulser, **27**-89
 transmission-line, **27**-88
Protective equipment, **27**-87 to **27**-90
Pulse, repetitive, on power transistor, **20**-20
 single, on power transistor, **20**-18
Pulse conditions, **20**-21
Pulse counter, **22**-12
Pulse firing, **11**-15
Pulse overshoot, **19**-17
Pulser, **27**-30, **27**-33
 hard-tube, **27**-83
 line-type, **27**-27
Pump, parametric amplifier, **13**-2
Pump factor, in parametric amplifiers, **28**-45
Pumping, laser, **14**-16
 optical, **14**-7, **14**-15, **14**-17
 push-push, **14**-14
 push-pull, **14**-14
Punch-through, **10**-6
Pylistor, **11**-12
Pyroelectric effect, primary, **16**-10
 secondary, **16**-10
Pyroelectricity, **16**-8

Q multiplication, **19**-28
Quadrature rejector, integrated-circuit transformerless, **31**-26
Quadripole(s), **1**-6
 interconnections of, **3**-26
 terminated, **3**-14
Quadrupole electric field, **14**-9
Quality factor, mechanical, **16**-30
Quantum of radiation, **14**-5

Radiation, spectral, **14**-6
 thermal, **17**-53
Radio-frequency discharge, in lasers, **29**-33
Radiometer, maser, **29**-14
Ratio, transformation, **24**-6
Rating, average gate power-dissipation, **11**-13
 di/dt, **11**-14
 dv/dt, **11**-13
 gate trigger current, **11**-13

Rating, gate trigger voltage, **11**-13
 maximum operating junction temperature, **11**-6
 maximum storage temperature, **11**-6
 peak forward voltage, **11**-13
 peak gate power dissipation, **11**-13
 peak reverse gate voltage, **11**-13
Ratings, absolute maximum, **9**-17
 continuous commercial service (CCS), **9**-18
 design-center, **9**-17
 design-maximum, **9**-17
 of electron tubes, **9**-16
 envelope temperature, **9**-22
 grid current, **9**-21
 heater-cathode voltage, **9**-20
 intermittent commercial and amateur service (ICAS), **9**-18
 intermittent mobile service (IMS), **9**-18
 plate dissipation, **9**-19
 of power transistors, **20**-8
 methods, **20**-12
 screen dissipation, **9**-19
Rayleigh approximation, **16**-25
Rayleigh noise, **7**-46
r_b, **10**-7
r_b', **10**-7
RC-coupled stages, **17**-18 to **17**-22
Reactance, current-limiting, **20**-44
 leakage, **8**-26, **8**-27
 nonlinear, **13**-1
 per-unit, **20**-45
 time-varying, **1**-6
Reactor, saturable, **8**-3, **8**-4, **8**-17, **21**-1
 with blocked intrinsic feedback, **8**-4
 generalized characteristics of, **8**-22
 low-power circuits, **21**-4
 parallel-connected, **8**-4, **8**-19
 series-connected, **8**-17, **8**-23, **21**-3
Real frequency, **1**-4
Realizability condition, **4**-17
Reciprocal n-port, **4**-6
Reciprocity, **16**-6, **16**-57
Reciprocity theorem, **2**-5, **4**-33
Recombination time, **10**-3
Recorder, analog, **19**-24
Rectifier, selenium, **21**-17
 silicon controlled (*see* SCR)
Redox system, **15**-3
Reference condition, of feedback amplifier, **6**-15, **6**-19
Reflection coefficient, **3**-16, **4**-3
 of tunnel-diode amplifier, **28**-19
Reflector, parabolic, **27**-30
Reflex klystron, **27**-4, **27**-49
Regulation, **20**-45
 screen supply, **17**-43
Regulator, alternator, **20**-40
Relaxation, **14**-6
Relay driver, high-sensitivity, integrated, **31**-33
Repeater, **1**-16
Reset, degree of, **21**-6
Reset flux level, **21**-6
Resetting, **21**-6
Resetting half-cycle, **21**-6

INDEX

Resistance, base, **10**-7
  base spreading, **10**-7
  effective transfer, **2**-28
  emitter diffusion, **10**-9
  input, of subminiature pentode, **9**-13
  negative, **1**-15, **1**-17, **4**-21
  net mutual, **2**-28
  nonlinear, **13**-2
  plate, **2**-18 **9**-48
  thermal, **17**-48, **17**-53, **17**-54, **17**-55
Resistance-limited region, of saturable reactor, **8**-22
Resonance, **5**-25
  constant-L transducer, **16**-43
  constant-E transducer, **16**-43
  mechanical, **16**-18
Resonator, ceramic, **16**-32
  ceramic-disk, **16**-46 to **16**-49
Resonator admittance, **16**-37
Response, Chebyshev amplitude, **25**-4
  closed-loop, **6**-32
  equiripple, **25**-4
  gaussian, **25**-2
  high-frequency, **25**-2, **25**-13, **25**-50
  to impulse function, **25**-2, **25**-3
  low-frequency, **25**-7, **25**-48
  low-pass, **25**-2
  maximally flat amplitude, **25**-4
  maximally flat delay, **25**-4
  to step function, **25**-2, **25**-3, **25**-12, **25**-18, **25**-23, **25**-27
  transient, **5**-32
Return difference, **6**-11
Return loss, **4**-17
Return ratio **6**-9
  for general reference, **6**-14
  for transistor, **6**-12
  for vacuum tube, **6**-11
RIAA frequency compensation, **17**-79
RIAA recording characteristic, **17**-34
Root-locus design, **25**-69
Root-locus method, **6**-29
  for unconditionally stable amplifier, **6**-30
Rooter, **26**-91
Rotating coordinate system, **14**-10
Rumble filter, **17**-35, **17**-38, **17**-81

Saturation in tunnel diodes, **12**-16, **12**-23
Saturation interval of saturable reactor, **8**-18
Saturation region of magnetization curve, **8**-5
Scattering matrix, **4**-1, **4**-11
Schmitt trigger, Darlington, integrated, **31**-31, **31**-32
  high-speed, integrated, **31**-31, **31**-32
  using integrated differential amplifier, **31**-38
SCR (silicon controlled rectifier), **11**-12
  to control servo amplifier, **23**-20
  di/dt rating, **11**-14
  dv/dt rating, **11**-13, **11**-14
  gate characteristics, **11**-13
  holding current, **11**-13
  as phase-control device, **11**-18

SCR (silicon controlled rectifier), trigger characteristics, **11**-13, **11**-14
  turn-off characteristics, **11**-13
  turn-off techniques, **11**-17
  typical circuits, **11**-17
  voltage ratings, **11**-13
SCR amplifier, **20**-32
  maximum operating frequency, **20**-32
  phase-controlled, **20**-32
SCR triggering, **23**-22
Scratch filter, **17**-35, **17**-38, **17**-85
SCS (silicon controlled switch), **11**-19
Second breakdown, in power transistors, **10**-21
Selectivity, of tuned amplifier, **24**-12
  variable, **19**-28
Selenium rectifier, **21**-17
Self-saturation, in magnetic amplifier, **8**-4
Semiconductor, **10**-3
  gallium arsenide, **10**-3, **10**-12
  germanium, **10**-3, **10**-12
  impurity, **10**-3
  indium antimonide, **10**-12
  n region, **10**-3
  n-type, **10**-3
  p region, **10**-3
  p-type, **10**-3
  piezoelectric, **16**-20, **30**-2
  p-n junction, **10**-3
  silicon, **10**-3
  silicon carbide, **10**-12
Sensitivity, in feedback amplifiers, **6**-15 to **6**-17
  limiting-temperature, **14**-24
Serrasoid modulator, **20**-58
Servo amplifiers, integrated-circuit, **31**-42
Servomechanism, **19**-4
Servomotor, **23**-10, **26**-47
  a-c, **23**-4
Servomultiplier, **26**-51
Servosystems, **26**-44
  linear, **26**-44
  logarithmic, **26**-51
Servo tester, **22**-13
Shear, **16**-26
Sherwood amplifier, **17**-79
Shockley diode, **11**-1, **11**-19
Signal-to-noise ratio, **7**-12
Silectron, **8**-7
Silicon, **10**-3
Silicon controlled rectifier (*see* SCR)
Silicon controlled switch (SCS), **11**-19
Silicon dioxide, **10**-18
Silicon steel, oriented, **8**-17
Sleeping sickness, **9**-46
Solid, isotropic, **16**-18
Solion, **15**-3, **22**-1
  in adaptive circuits, **22**-14
  in d-c amplifiers, **22**-3
Solion analog memory, **22**-15
Solion instrument amplifier, **22**-7
Solion integrator, **15**-6, **15**-10, **22**-2
Solion microbarometer, **22**-10
Solion tetrode, **15**-5, **22**-2
Solion transducers, **15**-11, **22**-7
Sonar amplifier, **20**-24

INDEX

Source, controlled, **1**-12, **1**-14
  current, **1**-12
  dependent, **1**-12
  independent, **1**-12
  voltage, **1**-12
Sources, laser pump, **14**-44
    exploding wires, **14**-44
    lamps, **14**-44
    xenon tube, **14**-44
Space-charge layer, **10**-4
Space harmonics in microwave tubes, **9**-96
Speed control system, **23**-23
Spontaneous emission, **14**-4
Square-law characteristics of tubes, **26**-9
Square-law response, **26**-2
Square mu, **8**-7
Square root of product, **26**-98
Square-root converter, **26**-90
Square-root extraction, **26**-9, **26**-87
Squaring circuit, **26**-8
  backward diode, **26**-16
  Hall-effect, **26**-29
  magnetoresistance, **26**-34
  multigrid, **26**-11
  using breakdown diodes, **26**-42
  using piece-wise linear approximation, **26**-40
Stability, closed-loop, **20**-38
  conditional, **6**-23 to **6**-25
  of maser or laser, **14**-37
  of multiloop feedback amplifier, **6**-22
  open-circuit, **1**-21
  open-loop, **20**-37
  short-circuit, **1**-21
  of single-loop feedback amplifier, **6**-21
  unconditional, **6**-23 to **6**-25
Stability factor, **3**-26, **17**-11, **17**-47, **24**-2
  significance of, **17**-14
  voltage, **17**-48
Stabilization of d-c amplifiers, **19**-3
Stages, cascaded, **17**-22
  double-tuned, **24**-8
  transitionally coupled, **24**-10
Stagger damping, **24**-10
Stagger tuning, **24**-10
  in masers, **29**-28
Standing waves, acoustic, **16**-18
Step function, **19**-4
Stereo amplifiers, General Electric, **17**-81
  Motorola, **17**-92
  Sherwood, **17**-79
  transistor, **17**-90
  tube, **17**-79
Stokes lines in laser action, **29**-66
Stopband, **5**-6, **5**-20 to **5**-22
Storage bulb, maser, **29**-10
Strain, **16**-4, **16**-5
  longitudinal, **16**-18
  shear, **16**-18
Stress(es), acoustic, **16**-4, **16**-5, **30**-3
  environmental, **9**-41
Structures, parametric amplifier, balanced, **28**-44
    microwave, **28**-45
    split coaxial, **28**-44
    stripline, **28**-44

Sublimation of metallic elements, **9**-5
Supermalloy, **8**-7, **8**-12
Supermendur, **8**-7
Superposition, principle of, **2**-2
Suppressor, **9**-54
Surface recombination, **10**-11
Switch, light-activated, **11**-12
  silicon controlled (SCS), **11**-19
  symmetrical diode, **11**-3
  transistor, as limiter, **26**-57
Switching speed, **20**-11
Switching time, **20**-11
Symmetric and nonsymmetric matrices, **4**-6
Symmetry, transverse, **16**-7
Synthesis, direct, **5**-23
  network, **5**-22
System(s), feedback, multiloop, **6**-24
  passive constant nonlinear, **1**-11
  second-order, **6**-33
System function, **1**-7, **2**-8, **2**-10

Tangent, hyperbolic, **16**-34
Tecnetron, **10**-22
Temperature, collector junction, **20**-8
  determination of, **20**-15
  effective noise, **14**-24
  junction, **17**-48
  negative, **14**-23
  noise, **7**-8, **7**-21, **7**-23, **14**-25
    average, **7**-25
    combinations of, **7**-49
    effective input, **7**-23
    operating, **7**-23
  spin, **14**-23
  standard noise, **7**-6, **7**-12
Temperature stabilization, with diodes, **17**-49
  with thermistors, **17**-49
Terminal pair, **1**-5
Terminal properties, effects of feedback on, **6**-5
Terminations, optimum, for two-port, **3**-26
Tetrode, semiconductor, as nonlinear device, **26**-17
  solion, **15**-5
  thermionic, **26**-3
  connected as log diodes, **26**-4
Theorem, compensation, **2**-3
  Norton, **2**-3
  reciprocity, **2**-5, **4**-33
  Thévenin, **2**-3
Thermal considerations, high-power transistors, **20**-14
Thermal expansion, **16**-9
Thermal pressure, **16**-9
Thermal resistance, **17**-48, **17**-53 to **17**-57, **20**-14, **20**-17, **20**-18
Thermal runaway, **17**-51, **17**-52
Thermionic integrated micro module (TIMM), **9**-2, **9**-55, **9**-58
Thermistors, control, **17**-50
  directly heated, **26**-26
  indirectly heated, **26**-26
  as nonlinear devices, **26**-25 to **26**-27
  as square-law elements, **26**-26

INDEX

Thermistors, stabilization with, **17**-49
Thermoelements, vacuum, **26**-27
    as nonlinear devices, **26**-27
Thévenin theorem, **2**-3
Three-phase power, control of, **21**-3
Three-pole, analysis of, **4**-33
Tien's assumption, **28**-41
Time, delay, **20**-11
    in four-layer diodes, **11**-6
  fall, **20**-11
    in four-layer diodes, **11**-6
  relaxation, **14**-15
  rise, **20**-11
    for multistage amplifiers, **25**-32
    relation to bandwidth, **25**-4
  storage, **20**-11
  turn-off, in four-layer diodes, **11**-6
  turn-on, in four-layer diodes, **11**-6
Time constant of Amplistat, **21**-32
Timer, ionic, **22**-12
  solion, **22**-13
TIMM (thermionic integrated micro module), **9**-2, **9**-55, **9**-58
Tone control, **17**-35
Torsion, **16**-26
  simple, **16**-26
Totem pole arrangement, **17**-91
Trancor, **8**-7
Transconductance, **2**-18, **9**-48
Transducer(s), acoustic, **30**-2, **30**-11
  constant-D, **16**-41 to **16**-46, **16**-60
  constant-E, **16**-59
  delay line, **16**-57 to **16**-61
    frequency response, **16**-60
    insertion loss, **16**-60
    resonances, **16**-60
  high-frequency, **30**-11
  n-terminal, **4**-2
  solion, **15**-11, **22**-9
    with negative feedback, **22**-12
  sonar, **16**-41
  ultrasonic constant-E, **16**-32
Transfer admittance, **1**-7
Transfer characteristics, **20**-33
Transfer current ratio, **1**-7
Transfer function, **1**-5
  of reciprocal n-terminal network, **4**-41
Transfer impedance, **1**-7
Transfer ratio, large-signal, measurement of, **20**-11
  low-frequency reverse, **10**-9
Transfer voltage ratio, **1**-7
Transformer(s), ceramic, **16**-49 to **16**-57
    bandwidth, **16**-54
    cascaded, **16**-63
    effective circuit Q, **16**-54
    equivalent circuit, **16**-50
    frequency response of, **16**-51
    geometries, **16**-49
    transducer loss, **16**-56
    transmission loci for, **16**-52
    unsymmetrical, **16**-55
  current, **8**-17
  direct-current, **21**-4
  output, design considerations, **17**-43
  potential, **8**-17

Transformer(s), pulse, **27**-79
Transformer coupling, **17**-15 to **17**-18, **17**-41
Transient response, **6**-32
Transients, switching, **18**-52
Transistor bias, **17**-11
Transistor configurations, **2**-26
  common-base, **2**-26
  common-collector, **2**-26, **2**-30
  common-emitter, **2**-26, **2**-30
Transistor equivalent circuits, **2**-22, **10**-7
  high-frequency, **2**-31
  hybrid-pi, **10**-9
  T-equivalent, **10**-7
Transistor performance, alloy, **10**-14
Transistor structure, alloy, **10**-12
Transistors, alloy-fused, **10**-2, **10**-12
  avalanche mode, **10**-7, **10**-23
  bipolar, **10**-3
  complementary, **17**-38
  design theory, **10**-9
  diffused, **10**-16
    mesa, **10**-2, **10**-17
    mixed alloy-diffused, **10**-16
    planar, **10**-18
    p-n-i-p, **10**-17
    post-alloy diffused, **10**-17
  field-effect, **10**-3, **10**-21
    as amplitude multiplier, **26**-24
    characteristics, **10**-22
    depletion-mode MOS (Metal-Oxide-Semiconductor), **10**-23
    drain, **10**-22
    enhancement-mode MOS, **10**-23
    gate, **10**-22
    MOS, **10**-22, **10**-23
    as nonlinear device, **26**-24
    pinch-off, **10**-22
    as pulse amplitude multiplier, **26**-98
    source, **10**-22
    thin-film MOS, **10**-25
    as variable resistor, **26**-24
  functional blocks, **10**-25
  grown-junction, **10**-2, **10**-24
  high-frequency, for AGC, **26**-21
  high-frequency effects, **10**-11
  high-power, **20**-4
    characteristics of, **20**-4
    ratings, **20**-8
    typical, **20**-14
  integrated, **10**-25
  junction, **10**-3
  melt-back, **10**-23
  metal-base, **10**-25
  micrologic, **10**-25
  multiple, **10**-24
  as network elements, **17**-10
  planar, **10**-2
  point-contact, **10**-2
  power, **10**-20
    n-p-i-n, **10**-20
    power vs. frequency, **10**-21
    second breakdown, **10**-21
  rate-grown, **10**-23, **10**-24
  silicon planar, as nonlinear device, **26**-17

18 INDEX

Transistors, silicon unijunction, as variable-resistance device, **26**-17
  surface-barrier, **10**-14
    ECDC (ElectroChemical-Diffused Collector), **10**-16
    MADT (MicroAlloy-Diffused Transistor), **10**-14
    temperature effects, **10**-11
  tetrodes, **10**-23
  thin-film, **10**-25
  unipolar, **10**-3, **10**-21
Transit-time effects, in electron tubes, **9**-13, **9**-80
Transition equalizer, **19**-7
Transition frequency, **10**-7
Transitional coupling, **24**-8
Transitions, polarization-dependent, **14**-16
Translator(s), frequency, **27**-50
  solid-state, **27**-51
  tube, **27**-52
  microwave frequency, **17**-5
  synchrodyne, **27**-5
Transmission coefficient, **3**-16
Transmission line, **1**-9, **27**-28, **27**-55
  coaxial, **27**-55
  waveguide, **27**-57
Transmission-line driver, **25**-44
Transmission zero, **19**-6
Transmitter(s), amplifiers for, **20**-40
  FM, **20**-58, **20**-63
  phased-array, **27**-31
  short-wave, **20**-61
  standard broadcast, **20**-47
  superpower, **20**-54
  typical 50-kw BC, **20**-52
  TV, **20**-59, **20**-62
  250-kw short-wave, **20**-55
Transport factor, **10**-6, **10**-10
Transwitch, **11**-19
Trapping, **30**-4
Trapping factor, **30**-7
Traveling-wave interactions, **28**-41
Traveling-wave solutions, **16**-27
Traveling-wave tubes, **9**-80, **9**-88
  low-noise, **9**-90
  phase velocity, **9**-90
  ω-β diagram, **9**-90
Traveling waves, **16**-21, **16**-24
Triggering, by four-layer diode, **11**-17
  of four-layer diode, **11**-17
  SCR, **11**-14
  by unijunction firing circuit, **11**-15
Trigistor, **11**-19
Triode, semiconductor, as nonlinear device, **26**-17
  thermionic, **26**-3
    connected as log diode, **26**-4
    as logarithmic device, **26**-5
Tube(s), aligned-grid, **9**-51
  amplitron, **9**-80
  audio-frequency amplifier, **9**-58, **9**-62, **9**-70
  audio power output, **9**-51, **9**-70
  backward-wave circuits, **9**-95
  beam deflection, **26**-12
    as multiplier, **26**-94

Tube(s), beam deflection, square-law characteristic, **26**-12
  beam power, **17**-42
  bimatron, **9**-80
  bitermitron, **9**-80
  compactrons, **9**-55, **9**-57
  crossed-field amplifier, **27**-36, **27**-43
  d-c, **9**-58
  dematron, **9**-80
  electrometer, **9**-58, **9**-60
  electron, as network element, **9**-48
  electron-beam-interaction, **27**-3
  electron-multiplier, **26**-12
    nonlinear characteristic, **26**-12
  gated-beam, **9**-53
  i-f, **9**-61, **9**-74
  klystron, **9**-80, **9**-81, **9**-87
  lighthouse, **9**-55
  linear-beam, **27**-36, **27**-40
  M-type, **9**-80, **9**-81, **9**-91
  metal-ceramic, **9**-55, **9**-56
  microwave, **9**-78, **9**-80
    power amplifiers, **27**-36 to **27**-45
  multielectrode, **2**-20
  negative-grid, **27**-36, **27**-37
  neonovals, **9**-55, **9**-57
  novars, **9**-55, **9**-57
  nuvistors, **9**-55, **9**-56
  O-type, **9**-80
  pencil, **9**-55, **9**-56
  pentode, **17**-42
  planar, **9**-80
  power, **17**-42
  pulse ratings of, **20**-41
  receiving, **9**-47, **9**-49
  r-f, **9**-61, **9**-74
    power-amplifier, **9**-77
  rocket, **9**-55
  sheet-beam, **9**-53
  switch, **27**-79
  TIMM, **9**-2, **9**-55, **9**-58
  transit time, **9**-80
  transmitting, **9**-97
    construction, **9**-98
    cooling, **9**-102
    pulse ratings, **9**-101
    ratings, **9**-100
  traveling-wave, **9**-80, **9**-88, **27**-3, **27**-5
  uhf, **9**-78, **9**-80
  vlf amplifier, **9**-58
Tuned amplifier [*see* Amplifier(s), tuned]
Tuning, stagger, **24**-10
Tunnel diode, **12**-1
  coaxial ring, **12**-9
  in combination, with parallel conductance, **28**-4
    with series resistance, **28**-4
  equivalent circuit, **28**-2
  impedance, **28**-2
  stability criteria, **28**-2
Tunneling mechanism, **12**-1
Tunneling probability, **12**-1
Turn-off, of four-layer diode, **11**-7
  of SCR, **11**-17
Two-port, **1**-6

INDEX

Ultrasonic cleaning, **16**-41
Ulug, M. E., **9**-52
Ulug tube, **9**-52
"Umklapp" process, **28**-50
Underlap, **18**-29
Unibi amplifier, **31**-40
Unijunction oscillator, **23**-20
Unijunction transistor, as trigger, **11**-15
　as variable-resistance device, **26**-17
Unit impulse, **1**-18
Upconverter, positive-resistance, **28**-35

Vapotron, **20**-55
Varactor, **27**-4
Variable, extensive, **16**-13
　intensive, **16**-13
Variable-mu tubes, remote cut-off, **26**-7
Variable-resistance element, **26**-17
Variolosser, **26**-68, **26**-71, **26**-74
Varistors, for AGC, **26**-62
　as compressors, **26**-25
　as expanders, **26**-25
　as nonlinear devices, **26**-25
　as square-law elements, **26**-26
Vector potential, **14**-3
Velocity, dilatational, **16**-18
　distortional wave, **16**-18
　drift, **30**-2
　group, **16**-24, **16**-26
　longitudinal, **16**-19
　phase, **16**-24, **16**-26
　shear stress wave, **16**-19
Velocity modulation, **9**-83, **27**-5
Voltage, breakdown, **10**-5, **20**-8
　forward, **11**-5
　forward breakover, **11**-5
　forward point, tunnel diode, **12**-5
　gate reverse, **11**-13
　gate trigger, **11**-13
　holding, **11**-5
　incident, **4**-5
　inflection-point, tunnel diode, **12**-5
　peak, tunnel diode, **12**-5
　peak reverse, **11**-5
　reflected, **4**-5
　screen-grid, as logarithmic function, **26**-6

Voltage, valley, tunnel diode, **12**-5
Voltage ratio, **4**-37
Voltage standing-wave ratio, relation to reflection coefficient, **28**-20
　of tunnel-diode amplifier, **28**-20
Voltmeter, logarithmic, **26**-83
V_{PC}, **10**-6

Wave(s), extensional, **16**-23, **16**-25
　incident, **4**-2
　plane logitudinal, **16**-19, **16**-21
　plane-polarized, plane shear, **16**-22, **16**-23
　reflected, **4**-2
　and resonance, **16**-17 to **16**-26
　standing, **16**-27 to **16**-31
　torsional, **16**-25
　ultrasonic, **30**-1
Wave equations, **16**-19 to **16**-26
Wave vector, **14**-3
Wavefront, **16**-18
Waveguide, **27**-28, **27**-57
Wiener-Paley criterion, **19**-5, **19**-8, **19**-31
Williamson amplifier, **17**-11

Xenon tube, **14**-44

Young's modulus, **16**-11
Yttrium iron garnet as isolator, **29**-25

Zener diode, **11**-19
Zener-diode coupling, **18**-8
Zener reference, buffered integrated-circuit, **31**-24
　dual integrated-circuit, **31**-23
Zero, **1**-8, **1**-18, **5**-16, **25**-72
　phantom, **25**-72, **25**-73
Zinc blende, **30**-2
ZnO (Wurtzite crystal), **30**-2
ZnS (Wurtzite crystal), **30**-2
ZnSe (crystal), cubic form of, **30**-2
ZnTe (Wurtzite crystal), **30**-2